$194.95 per copy (in United States)
Price is subject to change without prior notice.

0214

The 2014 *Plumbing Cost Data,* 37th annual edition, and all of the RSMeans annual cost data books for 2014, are dedicated to the memory of our respected colleague and friend, Paula Reale-Camelio.

RSMeans®
Plumbing
Cost Data

37TH ANNUAL EDITION

RSMeans
A division of Reed Construction Data, LLC
Construction Publishers & Consultants
700 Longwater Drive
Norwell, MA 02061
USA
1-800-334-3509
www.rsmeans.com

Copyright 2013 by RSMeans
A division of Reed Construction Data, LLC
All rights reserved.

Printed in the United States of America
ISSN 1537-8411
ISBN 978-1-940238-15-9

The authors, editors, and engineers of RSMeans apply diligence and judgment in locating and using reliable sources for the information published. However, RSMeans makes no express or implied warranty or guarantee in connection with the content of the information contained herein, including the accuracy, correctness, value, sufficiency, or completeness of the data, methods, and other information contained herein. RSMeans makes no express or implied warranty of merchantability or fitness for a particular purpose. RSMeans shall have no liability to any customer or third party for any loss, expense, or damage, including consequential, incidental, special, or punitive damage, including lost profits or lost revenue, caused directly or indirectly by any error or omission, or arising out of, or in connection with, the information contained herein.

No part of this publication may be reproduced, stored in a retrieval system, or transmitted in any form or by any means without prior written permission of RSMeans. The cost data contained in this publication is valuable and proprietary information of RSMeans and others and there is no transfer to you of any ownership rights in the cost data or any license granted to you to create derivative works based on or which utilize the cost data.

Senior Editor
Melville J. Mossman, PE *(21, 22, 23)**

Contributing Editors
Christopher Babbitt
Adrian C. Charest, PE *(26, 27, 28, 48)*
Gary W. Christensen
David G. Drain, PE *(33)*
Cheryl Elsmore
David Fiske
Robert Fortier, PE *(2, 31, 32, 34, 35, 46)*
Robert J. Kuchta *(8)*
Thomas Lane
Robert C. McNichols
Genevieve Medeiros
Jeannene D. Murphy
Marilyn Phelan, AIA *(9, 10, 11, 12)*
Stephen C. Plotner *(3, 5)*
Siobhan Soucie
Phillip R. Waier, PE *(1,13)*

Cover Design
Wendy McBay
Paul Robertson
Wayne Engebretson

Chief Operating Officer
Richard Remington

**Vice President,
Business Development**
Steven Ritchie

**Vice President,
Finance & Operations**
Renée Pinczes

Senior Director, Inside Sales
Derek F. Dean

Sales Manager
Reid Clanton

Director, Engineering
Bob Mewis, CCC *(4, 6, 7, 14, 41, 44)*

**Senior Director, Product
Management**
Jennifer Johnson

**Senior Product
Marketing Manager**
Andrea Sillah

**Senior Director,
Operations**
Rodney L. Sharples

Production Manager
Michael Kokernak

Production
Jill Goodman
Jonathan Forgit
Debbie Panarelli
Sara Rutan
Mary Lou Geary

Technical Support
Christopher Anderson
Gary L. Hoitt
Kathryn S. Rodriguez

*Numbers in italics are the divisional responsibilities for each editor.
Please contact the designated editor directly with any questions.*

This book is printed on 90% recycled paper using soy-based ink and is entirely recyclable.

Related RSMeans Products and Services

The engineers at RSMeans suggest the following products and services as companion information resources to *RSMeans Plumbing Cost Data*:

Construction Cost Data Books
Assemblies Cost Data 2014
Square Foot Costs 2014

Reference Books
Estimating Building Costs
RSMeans Estimating Handbook
Green Building: Project Planning & Estimating
How to Estimate with RSMeans Data
Plan Reading & Material Takeoff
Project Scheduling & Management for Construction
Universal Design Ideas for Style, Comfort & Safety

Seminars and In-House Training
Unit Price Estimating
RSMeans Online® Training
RSMeans Data for Job Order Contracting (JOC)
Practical Project Management for Construction Professionals
Scheduling with MSProject for Construction Professionals
Mechanical & Electrical Estimating

RSMeans Online Bookstore
Visit RSMeans at **www.rsmeans.com** for a one-stop portal for the most reliable and current resources available on the market. Here you can learn about our more than 20 annual cost data books, also available on CD and online versions, as well as our library of reference books and RSMeans seminars.

RSMeans Electronic Data
Get the information found in RSMeans cost books electronically at **RSMeansOnline.com**.

RSMeans Business Solutions
Cost engineers and analysts conduct facility life cycle and benchmark research studies and predictive cost modeling, as well as offer consultation services for conceptual estimating, real property management, and job order contracting. Research studies are designed with the application of proprietary cost and project data from Reed Construction Data and RSMeans' extensive North American databases. Analysts offer building product manufacturers qualitative and quantitative market research, as well as time/motion studies for new products, and Web-based dashboards for market opportunity analysis. Clients are from both the public and private sectors, including federal, state, and municipal agencies; corporations; institutions; construction management firms; hospitals; and associations.

RSMeans for Job Order Contracting (JOC)
Best practice job order contracting (JOC) services support the owner in the development of their program. RSMeans develops all required contracting documents, works with stakeholders, and develops qualified and experienced contractor lists. RSMeans JOCWorks® software is the best of class in the functional tools that meet JOC specific requirements. RSMeans data is organized to meet JOC cost estimating requirements.

Construction Costs for Software Applications
More than 25 unit price and assemblies cost databases are available through a number of leading estimating and facilities management software providers (listed below). For more information see the "Other RSMeans Products" pages at the back of this publication.

RSMeansData™ is also available to federal, state, and local government agencies as multi-year, multi-seat licenses.

- 4Clicks-Solutions, LLC
- Assetworks
- Beck Technology
- BSD – Building Systems Design, Inc.
- CMS – Construction Management Software
- Corecon Technologies, Inc.
- CorVet Systems
- Hard Dollar Corporation
- Maxwell Systems
- MC^2
- MOCA Systems
- Parsons Corporation
- ProEst
- Refined Data
- Sage Timberline Office
- UDA Technologies, Inc.
- US Cost, Inc.
- VFA – Vanderweil Facility Advisers
- WinEstimator, Inc.
- R & K Solutions
- Tririga

Table of Contents

Foreword	iv
How the Book Is Built: An Overview	v
Absolute Essentials for a Quick Start	vi
Estimating with RSMeans Unit Price Cost Data	vii
How to Use the Book: The Details	ix
Unit Price Section	1
Assemblies Section	431
Reference Section	543
Construction Equipment Rental Costs	545
Crew Listings	557
Historical Cost Indexes	593
City Cost Indexes	594
Location Factors	637
Reference Tables	643
Change Orders	680
Square Foot Costs	685
Abbreviations	695
Index	699
Other RSMeans Products and Services	724
Labor Trade Rates including	
Overhead & Profit	Inside Back Cover

Foreword

Our Mission

Since 1942, RSMeans has been actively engaged in construction cost publishing and consulting throughout North America.

Today, more than 70 years after RSMeans began, our primary objective remains the same: to provide you, the construction and facilities professional, with the most current and comprehensive construction cost data possible.

Whether you are a contractor, owner, architect, engineer, facilities manager, or anyone else who needs a reliable construction cost estimate, you'll find this publication to be a highly useful and necessary tool.

With the constant flow of new construction methods and materials today, it's difficult to find the time to look at and evaluate all the different construction cost possibilities. In addition, because labor and material costs keep changing, last year's cost information is not a reliable basis for today's estimate or budget.

That's why so many construction professionals turn to RSMeans. We keep track of the costs for you, along with a wide range of other key information, from city cost indexes . . . to productivity rates . . . to crew composition . . . to contractor's overhead and profit rates.

RSMeans performs these functions by collecting data from all facets of the industry and organizing it in a format that is instantly accessible to you. From the preliminary budget to the detailed unit price estimate, you'll find the data in this book useful for all phases of construction cost determination.

The Staff, the Organization, and Our Services

When you purchase one of RSMeans' publications, you are, in effect, hiring the services of a full-time staff of construction and engineering professionals.

Our thoroughly experienced and highly qualified staff works daily at collecting, analyzing, and disseminating comprehensive cost information for your needs. These staff members have years of practical construction experience and engineering training prior to joining the firm. As a result, you can count on them not only for accurate cost figures, but also for additional background reference information that will help you create a realistic estimate.

The RSMeans organization is always prepared to help you solve construction problems through its variety of data solutions, including online, CD, and print book formats, as well as cost estimating expertise available via our business solutions, training, and seminars.

Besides a full array of construction cost estimating books, RSMeans also publishes a number of other reference works for the construction industry. Subjects include construction estimating and project and business management, special topics such as green building and job order contracting, and a library of facility management references.

In addition, you can access all of our construction cost data electronically in convenient CD format or on the web. Visit RSMeansOnline.com for more information on our 24/7 online cost data.

What's more, you can increase your knowledge and improve your construction estimating and management performance with an RSMeans construction seminar or in-house training program. These two-day seminar programs offer unparalleled opportunities for everyone in your organization to become updated on a wide variety of construction-related issues.

RSMeans is also a worldwide provider of construction cost management and analysis services for commercial and government owners.

In short, RSMeans can provide you with the tools and expertise for constructing accurate and dependable construction estimates and budgets in a variety of ways.

Robert Snow Means Established a Tradition of Quality That Continues Today

Robert Snow Means spent years building RSMeans, making certain he always delivered a quality product.

Today, at RSMeans, we do more than talk about the quality of our data and the usefulness of our books. We stand behind all of our data, from historical cost indexes to construction materials and techniques to current costs.

If you have any questions about our products or services, please call us toll-free at 1-800-334-3509. Our customer service representatives will be happy to assist you. You can also visit our website at **www.rsmeans.com**

How the Book is Built: An Overview

The Construction Specifications Institute (CSI) and Construction Specifications Canada (CSC) have produced the 2012 edition of MasterFormat®, a system of titles and numbers used extensively to organize construction information.

All unit price data in the RSMeans cost data books is now arranged in the 50-division MasterFormat® 2012 system.

A Powerful Construction Tool

You have in your hands one of the most powerful construction tools available today. A successful project is built on the foundation of an accurate and dependable estimate. This book will enable you to construct just such an estimate.

For the casual user the book is designed to be:

- quickly and easily understood so you can get right to your estimate.
- filled with valuable information so you can understand the necessary factors that go into the cost estimate.

For the regular user, the book is designed to be:

- a handy desk reference that can be quickly referred to for key costs.
- a comprehensive, fully reliable source of current construction costs and productivity rates so you'll be prepared to estimate any project.
- a source book for preliminary project cost, product selections, and alternate materials and methods.

To meet all of these requirements, we have organized the book into the following clearly defined sections.

Quick Start

See our "Quick Start" instructions on the following page to get started right away.

Estimating with RSMeans Unit Price Cost Data

Please refer to these steps for guidance on completing an estimate using RSMeans unit price cost data.

How to Use the Book: The Details

This section contains an in-depth explanation of how the book is arranged... and how you can use it to determine a reliable construction cost estimate. It includes information about how we develop our cost figures and how to completely prepare your estimate.

Unit Price Section

All cost data has been divided into the 50 divisions according to the MasterFormat system of classification and numbering. For a listing of these divisions and an outline of their subdivisions, see the Unit Price Section Table of Contents.

Estimating tips are included at the beginning of each division.

Assemblies Section

The cost data in this section has been organized in an "Assemblies" format. These assemblies are the functional elements of a building and are arranged according to the 7 elements of the UNIFORMAT II classification system. For a complete explanation of a typical "Assemblies" page, see "How RSMeans Assemblies Data Works."

Reference Section

This section includes information on Equipment Rental Costs, Crew Listings, Historical Cost Indexes, City Cost Indexes, Location Factors, Reference Tables, Change Orders, Square Foot Costs, and a listing of Abbreviations.

Equipment Rental Costs: This section contains the average costs to rent and operate hundreds of pieces of construction equipment.

Crew Listings: This section lists all of the crews referenced in the book. For the purposes of this book, a crew is composed of more than one trade classification and/or the addition of power equipment to any trade classification. Power equipment is included in the cost of the crew. Costs are shown both with bare labor rates and with the installing contractor's overhead and profit added. For each, the total crew cost per eight-hour day and the composite cost per labor-hour are listed.

Historical Cost Indexes: These indexes provide you with data to adjust construction costs over time.

City Cost Indexes: All costs in this book are U.S. national averages. Costs vary because of the regional economy. You can adjust costs by CSI Division to over 700 locations throughout the U.S. and Canada by using the data in this section.

Location Factors: You can adjust total project costs to over 900 locations throughout the U.S. and Canada by using the data in this section.

Reference Tables: At the beginning of selected major classifications in the Unit Price section are reference numbers shown in a shaded box. These numbers refer you to related information in the Reference Section. In this section, you'll find reference tables, explanations, estimating information that support how we develop the unit price data, technical data, and estimating procedures.

Change Orders: This section includes information on the factors that influence the pricing of change orders.

Square Foot Costs: This section contains costs for 59 different building types that allow you to make a rough estimate for the overall cost of a project or its major components.

Abbreviations: A listing of abbreviations used throughout this book, along with the terms they represent, is included in this section.

Index

A comprehensive listing of all terms and subjects in this book will help you quickly find what you need when you are not sure where it occurs in MasterFormat.

The Scope of This Book

This book is designed to be as comprehensive and as easy to use as possible. To that end we have made certain assumptions and limited its scope in two key ways:

1. We have established material prices based on a national average.
2. We have computed labor costs based on a 30-city national average of union wage rates.

For a more detailed explanation of how the cost data is developed, see "How To Use the Book: The Details."

Project Size/Type

The material prices in RSMeans data cost books are "contractor's prices." They are the prices that contractors can expect to pay at the lumberyards, suppliers/distributers warehouses, etc. Small orders of speciality items would be higher than the costs shown, while very large orders, such as truckload lots, would be less. The variation would depend on the size, timing, and negotiating power of the contractor. The labor costs are primarily for new construction or major renovation rather than repairs or minor alterations.

With reasonable exercise of judgment, the figures can be used for any building work.

Absolute Essentials for a Quick Start

If you feel you are ready to use this book and don't think you will need the detailed instructions that begin on the following page, this Absolute Essentials for a Quick Start page is for you. These steps will allow you to get started estimating in a matter of minutes.

1 Scope
Think through the project that you will be estimating, and identify the many individual work tasks that will need to be covered in your estimate.

2 Quantify
Determine the number of units that will be required for each work task that you identified.

3 Pricing
Locate individual Unit Price line items that match the work tasks you identified. The Unit Price Section Table of Contents that begins on page 1 and the Index in the back of the book will help you find these line items.

4 Multiply
Multiply the Total Incl O&P cost for a Unit Price line item in the book by your quantity for that item. The price you calculate will be an estimate for a completed item of work performed by a subcontractor. Keep adding line items in this manner to build your estimate.

5 Project Overhead
Include project overhead items in your estimate. These items are needed to make the job run and are typically, but not always, provided by the General Contractor. They can be found in Division 1. An alternate method of estimating project overhead costs is to apply a percentage of the total project cost.

Include rented tools not included in crews, waste, rubbish handling, and cleanup.

6 Estimate Summary
Include General Contractor's markup on subcontractors, General Contractor's office overhead and profit, and sales tax on materials and equipment.

Adjust your estimate to the project's location by using the City Cost Indexes or Location Factors found in the Reference Section.

Editors' Note: We urge you to spend time reading and understanding the supporting material in the front of this book. An accurate estimate requires experience, knowledge, and careful calculation. The more you know about how we at RSMeans developed the data, the more accurate your estimate will be. In addition, it is important to take into consideration the reference material in the back of the book such as Equipment Listings, Crew Listings, City Cost Indexes, Location Factors, and Reference Tables.

Estimating with RSMeans Unit Price Cost Data

Following these steps will allow you to complete an accurate estimate using RSMeans Unit Price cost data.

1 Scope Out the Project

- Identify the individual work tasks that will need to be covered in your estimate.
- The Unit Price data in this book has been divided into 50 Divisions according to the CSI MasterFormat 2012—their titles are listed on the back cover of your book.
- Think through the project and identify those CSI Divisions needed in your estimate.
- The Unit Price Section Table of Contents on page 1 may also be helpful when scoping out your project.
- Experienced estimators find it helpful to begin with Division 2 and continue through completion. Division 1 can be estimated after the full project scope is known.

2 Quantify

- Determine the number of units required for each work task that you identified.
- Experienced estimators include an allowance for waste in their quantities. (Waste is not included in RSMeans Unit Price line items unless so stated.)

3 Price the Quantities

- Use the Unit Price Table of Contents, and the Index, to locate individual Unit Price line items for your estimate.
- Reference Numbers indicated within a Unit Price section refer to additional information that you may find useful.
- The crew indicates who is performing the work for that task. Crew codes are expanded in the Crew Listings in the Reference Section to include all trades and equipment that comprise the crew.
- The Daily Output is the amount of work the crew is expected to do in one day.
- The Labor-Hours value is the amount of time it will take for the crew to install one unit of work.
- The abbreviated Unit designation indicates the unit of measure upon which the crew, productivity, and prices are based.
- Bare Costs are shown for materials, labor, and equipment needed to complete the Unit Price line item. Bare costs do not include waste, project overhead, payroll insurance, payroll taxes, main office overhead, or profit.
- The Total Incl O&P cost is the billing rate or invoice amount of the installing contractor or subcontractor who performs the work for the Unit Price line item.

4 Multiply

- Multiply the total number of units needed for your project by the Total Incl O&P cost for each Unit Price line item.
- Be careful that your take off unit of measure matches the unit of measure in the Unit column.
- The price you calculate is an estimate for a completed item of work.
- Keep scoping individual tasks, determining the number of units required for those tasks, matching each task with individual Unit Price line items in the book, and multiply quantities by Total Incl O&P costs.
- An estimate completed in this manner is priced as if a subcontractor, or set of subcontractors, is performing the work. The estimate does not yet include Project Overhead or Estimate Summary components such as General Contractor markups on subcontracted work, General Contractor office overhead and profit, contingency, and location factor.

5 Project Overhead

- Include project overhead items from Division 1 – General Requirements.
- These items are needed to make the job run. They are typically, but not always, provided by the General Contractor. Items include, but are not limited to, field personnel, insurance, performance bond, permits, testing, temporary utilities, field office and storage facilities, temporary scaffolding and platforms, equipment mobilization and demobilization, temporary roads and sidewalks, winter protection, temporary barricades and fencing, temporary security, temporary signs, field engineering and layout, final cleaning and commissioning.
- Each item should be quantified, and matched to individual Unit Price line items in Division 1, and then priced and added to your estimate.
- An alternate method of estimating project overhead costs is to apply a percentage of the total project cost, usually 5% to 15% with an average of 10% (see General Conditions, page ix).
- Include other project related expenses in your estimate such as:
  - Rented equipment not itemized in the Crew Listings
  - Rubbish handling throughout the project (see 02 41 19.19)

6 Estimate Summary

- Include sales tax as required by laws of your state or county.
- Include the General Contractor's markup on self-performed work, usually 5% to 15% with an average of 10%.
- Include the General Contractor's markup on subcontracted work, usually 5% to 15% with an average of 10%.
- Include General Contractor's main office overhead and profit:
  - RSMeans gives general guidelines on the General Contractor's main office overhead (see section 01 31 13.50 and Reference Number R013113-50).
  - RSMeans gives no guidance on the General Contractor's profit.
  - Markups will depend on the size of the General Contractor's operations, his projected annual revenue, the level of risk he is taking on, and on the level of competition in the local area and for this project in particular.
- Include a contingency, usually 3% to 5%, if appropriate.
- Adjust your estimate to the project's location by using the City Cost Indexes or the Location Factors in the Reference Section:
  - Look at the rules on the pages for How to Use the City Cost Indexes to see how to apply the Indexes for your location.
  - When the proper Index or Factor has been identified for the project's location, convert it to a multiplier by dividing it by 100, and then multiply that multiplier by your estimated total cost. The original estimated total cost will now be adjusted up or down from the national average to a total that is appropriate for your location.

Editors' Notes:

1) *We urge you to spend time reading and understanding the supporting material in the front of this book. An accurate estimate requires experience, knowledge, and careful calculation. The more you know about how we at RSMeans developed the data, the more accurate your estimate will be. In addition, it is important to take into consideration the reference material in the back of the book such as Equipment Listings, Crew Listings, City Cost Indexes, Location Factors, and Reference Tables.*

2) *Contractors who are bidding or are involved in JOC, DOC, SABER, or IDIQ type contracts are cautioned that workers' compensation insurance, federal and state payroll taxes, waste, project supervision, project overhead, main office overhead, and profit are not included in bare costs. The coefficient or multiplier must cover these costs.*

How to Use the Book: The Details

What's Behind the Numbers? The Development of Cost Data

The staff at RSMeans continually monitors developments in the construction industry in order to ensure reliable, thorough, and up-to-date cost information. While overall construction costs may vary relative to general economic conditions, price fluctuations within the industry are dependent upon many factors. Individual price variations may, in fact, be opposite to overall economic trends. Therefore, costs are constantly tracked and complete updates are published yearly. Also, new items are frequently added in response to changes in materials and methods.

Costs—$ (U.S.)

All costs represent U.S. national averages and are given in U.S. dollars. The RSMeans City Cost Indexes can be used to adjust costs to a particular location. The City Cost Indexes for Canada can be used to adjust U.S. national averages to local costs in Canadian dollars. No exchange rate conversion is necessary.

G The processes or products identified by the green symbol in our publications have been determined to be environmentally responsible and/or resource-efficient solely by the RSMeans engineering staff. The inclusion of the green symbol does not represent compliance with any specific industry association or standard.

Material Costs

The RSMeans staff contacts manufacturers, dealers, distributors, and contractors all across the U.S. and Canada to determine national average material costs. If you have access to current material costs for your specific location, you may wish to make adjustments to reflect differences from the national average. Included within material costs are fasteners for a normal installation. RSMeans engineers use manufacturers' recommendations, written specifications, and/or standard construction practice for size and spacing of fasteners. Adjustments to material costs may be required for your specific application or location. The manufacturer's warranty is assumed. Extended warranties are not included in the material costs. Material costs do not include sales tax..

Labor Costs

Labor costs are based on the average of wage rates from 30 major U.S. cities. Rates are determined from labor union agreements or prevailing wages for construction trades for the current year. Rates, along with overhead and profit markups, are listed on the inside back cover of this book.

- If wage rates in your area vary from those used in this book, or if rate increases are expected within a given year, labor costs should be adjusted accordingly.

Labor costs reflect productivity based on actual working conditions. In addition to actual installation, these figures include time spent during a normal weekday on tasks such as, material receiving and handling, mobilization at site, site movement, breaks, and cleanup.

Productivity data is developed over an extended period so as not to be influenced by abnormal variations and reflects a typical average.

Equipment Costs

Equipment costs include not only rental, but also operating costs for equipment under normal use. The operating costs include parts and labor for routine servicing such as repair and replacement of pumps, filters, and worn lines. Normal operating expendables, such as fuel, lubricants, tires, and electricity (where applicable), are also included. Extraordinary operating expendables with highly variable wear patterns, such as diamond bits and blades, are excluded. These costs are included under materials. Equipment rental rates are obtained from industry sources throughout North America—contractors, suppliers, dealers, manufacturers, and distributors.

Rental rates can also be treated as reimbursement costs for contractor-owned equipment. Owned equipment costs include depreciation, loan payments, interest, taxes, insurance, storage and major repairs.

Equipment costs do not include operators' wages; nor do they include the cost to move equipment to a job site (mobilization) or from a job site (demobilization).

Equipment Cost/Day——The cost of power equipment required for each crew is included in the Crew Listings in the Reference Section (small tools that are considered as essential everyday tools are not listed out separately). The Crew Listings itemize specialized tools and heavy equipment along with labor trades. The daily cost of itemized equipment included in a crew is based on dividing the weekly bare rental rate by 5 (number of working days per week) and then adding the hourly operating cost times 8 (the number of hours per day). This Equipment Cost/Day is shown in the last column of the Equipment Rental Cost pages in the Reference Section.

Mobilization/Demobilization——The cost to move construction equipment from an equipment yard or rental company to the job site and back again is not included in equipment costs. Mobilization (to the site) and demobilization (from the site) costs can be found in the Unit Price Section. If a piece of equipment is already at the job site, it is not appropriate to utilize mob./demob. costs again in an estimate.

Overhead and Profit

Total Cost including O&P for the *Installing Contractor* is shown in the last column on the Unit Price and/or the Assemblies pages of this book. This figure is the sum of the bare material cost plus 10% for profit, the bare labor cost plus total overhead and profit, and the bare equipment cost plus 10% for profit. Details for the calculation of Overhead and Profit on labor are shown on the inside back cover and in the Reference Section of this book. (See "How RSMeans Data Works" for an example of this calculation.)

General Conditions

Cost data in this book is presented in two ways: Bare Costs and Total Cost including O&P (Overhead and Profit). General Conditions, or General Requirements, of the contract should also be added to the Total Cost including O&P when applicable. Costs for General Conditions are listed in Division 1 of the Unit Price Section and the Reference

Section of this book. General Conditions for the *Installing Contractor* may range from 0% to 10% of the Total Cost including O&P. For the *General* or *Prime Contractor*, costs for General Conditions may range from 5% to 15% of the Total Cost including O&P, with a figure of 10% as the most typical allowance. If applicable, the Assemblies and Models sections of this book use costs that include the installing contractor's overhead and profit (O&P).

Factors Affecting Costs

Costs can vary depending upon a number of variables. Here's how we have handled the main factors affecting costs.

Quality——The prices for materials and the workmanship upon which productivity is based represent sound construction work. They are also in line with U.S. government specifications.

Overtime——We have made no allowance for overtime. If you anticipate premium time or work beyond normal working hours, be sure to make an appropriate adjustment to your labor costs.

Productivity——The productivity, daily output, and labor-hour figures for each line item are based on working an eight-hour day in daylight hours in moderate temperatures. For work that extends beyond normal work hours or is performed under adverse conditions, productivity may decrease. (See "How RSMeans Data Works" for more on productivity.)

Size of Project——The size, scope of work, and type of construction project will have a significant impact on cost. Economies of scale can reduce costs for large projects. Unit costs can often run higher for small projects.

Location——Material prices in this book are for metropolitan areas. However, in dense urban areas, traffic and site storage limitations may increase costs. Beyond a 20-mile radius of large cities, extra trucking or transportation charges may also increase the material costs slightly. On the other hand, lower wage rates may be in effect. Be sure to consider both of these factors when preparing an estimate, particularly if the job site is located in a central city or remote rural location. In addition, highly specialized subcontract items may require travel and per-diem expenses for mechanics.

Other Factors——
- season of year
- contractor management
- weather conditions
- local union restrictions
- building code requirements
- availability of:
    - adequate energy
    - skilled labor
    - building materials
- owner's special requirements/restrictions
- safety requirements
- environmental considerations

Unpredictable Factors——General business conditions influence "in-place" costs of all items. Substitute materials and construction methods may have to be employed. These may affect the installed cost and/or life cycle costs. Such factors may be difficult to evaluate and cannot necessarily be predicted on the basis of the job's location in a particular section of the country. Thus, where these factors apply, you may find significant but unavoidable cost variations for which you will have to apply a measure of judgment to your estimate.

Rounding of Costs

In general, all unit prices in excess of $5.00 have been rounded to make them easier to use and still maintain adequate precision of the results. The rounding rules we have chosen are in the following table.

Prices from . . .	Rounded to the nearest . . .
$.01 to $5.00	$.01
$5.01 to $20.00	$.05
$20.01 to $100.00	$.50
$100.01 to $300.00	$1.00
$300.01 to $1,000.00	$5.00
$1,000.01 to $10,000.00	$25.00
$10,000.01 to $50,000.00	$100.00
$50,000.01 and above	$500.00

How Subcontracted Items Affect Costs

A considerable portion of all large construction jobs is usually subcontracted. In fact, the percentage done by subcontractors is constantly increasing and may run over 90%. Since the workers employed by these companies do nothing else but install their particular product, they soon become expert in that line. The result is, installation by these firms is accomplished so efficiently that the total in-place cost, even adding the general contractor's overhead and profit, is no more, and often less, than if the principal contractor had handled the installation himself/herself. Companies that deal with construction specialties are anxious to have their product perform well and, consequently, the installation will be the best possible.

Contingencies

The allowance for contingencies generally provides for unforeseen construction difficulties. On alterations or repair jobs, 20% is not too much. If drawings are final and only field contingencies are being considered, 2% or 3% is probably sufficient, and often nothing need be added. Contractually, changes in plans will be covered by extras. The contractor should consider inflationary price trends and possible material shortages during the course of the job. These escalation factors are dependent upon both economic conditions and the anticipated time between the estimate and actual construction. If drawings are not complete or approved, or a budget cost is wanted, it is wise to add 5% to 10%. Contingencies, then, are a matter of judgment. Additional allowances for contingencies are shown in Division 1.

Important Estimating Considerations

The "productivity," or daily output, of each craftsman includes mobilization and cleanup time, break time and plan layout time, as well as an allowance to carry stock from the storage trailer or location on the job site up to 200' into the building and to the first or second floor. If material has to be transported over greater distances or to higher floors, an additional allowance should be considered by the estimator. An allowance has also been included in the piping and fittings installation time for leak checking and minor tightening. Equipment installation time includes the following applicable items: positioning, leveling and securing the unit in place, connecting all associated piping, ducts, vents, etc., which shall have been estimated separately, connecting to an adjacent power source, filling/bleeding, startup, adjusting the controls up and down to ensure proper response, setting the integral controls/valves/regulators/thermostats for proper operation (does not include external building type control systems, DDC systems, etc.), explaining/training owner's operator, and warranty. A reasonable breakdown of the labor costs is as follows:

1. Movement into building, installation/setting of equipment — 35%
2. Connection to piping/duct/power, etc. — 25%
3. Filling/flushing/cleaning/touchup, etc. — 15%
4. Startup/running adjustments — 5%
5. Training owner's representative — 5%
6. Warranty/call back/service — 15%

Note that cranes or other lifting equipment are not included on any lines in the Mechanical divisions. For example, if a crane is required to lift a heavy piece of pipe into place high above a gym floor, or to put a rooftop unit on the roof of a four-story building, etc., it must be added. Due to the potential for extreme variation—from nothing additional required to a major crane or helicopter—we feel that including a nominal amount for "lifting contingency" would be useless and detract from the accuracy of the estimate. When using equipment rental from RSMeans, remember to include the cost of the operator(s).

Estimating Labor-Hours

The labor-hours expressed in this publication are based on Average Installation time, using an efficiency level of approximately 60%–65% (see item 7), which has been found reasonable and acceptable by many contractors. The book uses this national efficiency average to establish a consistent benchmark. For bid situations, adjustments to this efficiency level should be the responsibility of the contractor bidding the project. The unit labor-hour is divided in the following manner. A typical day for a journeyman might be:

1.	Study Plans	3%	14.4 min.
2.	Material Procurement	3%	14.4 min.
3.	Receiving and Storing	3%	14.4 min.
4.	Mobilization	5%	24.0 min.
5.	Site Movement	5%	24.0 min.
6.	Layout and Marking	8%	38.4 min.
7.	Actual Installation	64%	307.2 min.
8.	Cleanup	3%	14.4 min.
9.	Breaks, Nonproductive	6%	28.8 min.
		100%	480.0 min.

If any of the percentages expressed in this breakdown do not apply to the particular work or project situation, then that percentage or a portion of it may be deducted from or added to labor-hours.

Final Checklist

Estimating can be a straightforward process provided you remember the basics. Here's a checklist of some of the steps you should remember to complete before finalizing your estimate.

Did you remember to . . .

- factor in the City Cost Index for your locale?
- take into consideration which items have been marked up and by how much?
- mark up the entire estimate sufficiently for your purposes?
- read the background information on techniques and technical matters that could impact your project time span and cost?
- include all components of your project in the final estimate?
- double check your figures for accuracy?
- call RSMeans if you have any questions about your estimate or the data you've found in our publications? Remember, RSMeans stands behind its publications. If you have any questions about your estimate . . . about the costs you've used from our books . . . or even about the technical aspects of the job that may affect your estimate, feel free to call the RSMeans editors at 1-800-334-3509.

Free Quarterly Updates

Stay up-to-date throughout 2014 with RSMeans' free cost data updates four times a year. Sign up online to make sure you have access to the newest data. Every quarter we provide the city cost adjustment factors for hundreds of cities and key materials. Register at: **www.rsmeans2014updates.com**

Unit Price Section

Table of Contents

Sect. No.		Page
	General Requirements	
01 11	Summary of Work	8
01 21	Allowances	8
01 31	Project Management and Coordination	10
01 32	Construction Progress Documentation	11
01 41	Regulatory Requirements	11
01 51	Temporary Utilities	11
01 52	Construction Facilities	12
01 54	Construction Aids	12
01 55	Vehicular Access and Parking	15
01 56	Temporary Barriers and Enclosures	15
01 58	Project Identification	17
01 74	Cleaning and Waste Management	17
01 91	Commissioning	17
01 93	Facility Maintenance	17
	Existing Conditions	
02 21	Surveys	22
02 41	Demolition	22
02 65	Underground Storage Tank Removal	24
02 81	Transportation and Disposal of Hazardous Materials	24
02 82	Asbestos Remediation	25
02 85	Mold Remediation	28
	Concrete	
03 11	Concrete Forming	32
03 15	Concrete Accessories	32
03 21	Reinforcement Bars	35
03 22	Fabric and Grid Reinforcing	36
03 30	Cast-In-Place Concrete	36
03 31	Structural Concrete	37
03 35	Concrete Finishing	38
03 63	Epoxy Grouting	38
03 82	Concrete Boring	39
	Metals	
05 05	Common Work Results for Metals	42
05 12	Structural Steel Framing	46
05 54	Metal Floor Plates	46
	Wood, Plastics & Composites	
06 11	Wood Framing	48
06 16	Sheathing	48
	Thermal & Moisture Protection	
07 65	Flexible Flashing	50
07 72	Roof Accessories	51
07 76	Roof Pavers	51
07 84	Firestopping	52
07 91	Preformed Joint Seals	53
07 92	Joint Sealants	54
	Finishes	
09 22	Supports for Plaster and Gypsum Board	58
09 91	Painting	58
	Specialties	
10 21	Compartments and Cubicles	62
10 28	Toilet, Bath, and Laundry Accessories	65
10 44	Fire Protection Specialties	66
	Equipment	
11 11	Vehicle Service Equipment	70
11 19	Detention Equipment	70
11 21	Mercantile and Service Equipment	70
11 23	Commercial Laundry and Dry Cleaning Equipment	70
11 24	Maintenance Equipment	71
11 26	Unit Kitchens	71
11 27	Photographic Processing Equipment	72
11 31	Residential Appliances	72

Sect. No.		Page
11 41	Foodservice Storage Equipment	73
11 44	Food Cooking Equipment	74
11 47	Ice Machines	74
11 48	Cleaning and Disposal Equipment	74
11 53	Laboratory Equipment	74
11 71	Medical Sterilizing Equipment	75
11 73	Patient Care Equipment	75
11 74	Dental Equipment	76
11 76	Operating Room Equipment	76
11 78	Mortuary Equipment	76
11 91	Religious Equipment	76
	Furnishings	
12 32	Manufactured Wood Casework	78
	Special Construction	
13 11	Swimming Pools	80
13 17	Tubs and Pools	80
13 24	Special Activity Rooms	81
13 34	Fabricated Engineered Structures	82
13 42	Building Modules	82
	Conveying Equipment	
14 45	Vehicle Lifts	84
	Fire Suppression	
21 05	Common Work Results for Fire Suppression	86
21 11	Facility Fire-Suppression Water-Service Piping	86
21 12	Fire-Suppression Standpipes	89
21 13	Fire-Suppression Sprinkler Systems	93
21 21	Carbon-Dioxide Fire-Extinguishing Systems	95
21 22	Clean-Agent Fire-Extinguishing Systems	96
21 31	Centrifugal Fire Pumps	96
	Plumbing	
22 01	Operation and Maintenance of Plumbing	100
22 05	Common Work Results for Plumbing	101
22 07	Plumbing Insulation	124
22 11	Facility Water Distribution	144
22 12	Facility Potable-Water Storage Tanks	248
22 13	Facility Sanitary Sewerage	249
22 14	Facility Storm Drainage	270
22 15	General Service Compressed-Air Systems	273
22 31	Domestic Water Softeners	274
22 32	Domestic Water Filtration Equipment	274
22 33	Electric Domestic Water Heaters	275
22 34	Fuel-Fired Domestic Water Heaters	277
22 41	Residential Plumbing Fixtures	279
22 42	Commercial Plumbing Fixtures	286
22 43	Healthcare Plumbing Fixtures	291
22 45	Emergency Plumbing Fixtures	292
22 46	Security Plumbing Fixtures	293
22 47	Drinking Fountains and Water Coolers	294
22 51	Swimming Pool Plumbing Systems	296
22 52	Fountain Plumbing Systems	296
22 62	Vacuum Systems for Laboratory and Healthcare Facilities	297
22 63	Gas Systems for Laboratory and Healthcare Facilities	297
22 66	Chemical-Waste Systems for Lab. and Healthcare Facilities	298
	Heating Ventilation Air Conditioning	
23 05	Common Work Results for HVAC	306
23 07	HVAC Insulation	313
23 09	Instrumentation and Control for HVAC	315

Sect. No.		Page
23 11	Facility Fuel Piping	316
23 12	Facility Fuel Pumps	317
23 13	Facility Fuel-Storage Tanks	317
23 21	Hydronic Piping and Pumps	320
23 22	Steam and Condensate Piping and Pumps	329
23 23	Refrigerant Piping	329
23 34	HVAC Fans	329
23 35	Special Exhaust Systems	330
23 38	Ventilation Hoods	332
23 51	Breechings, Chimneys, and Stacks	332
23 52	Heating Boilers	339
23 56	Solar Energy Heating Equipment	341
23 57	Heat Exchangers for HVAC	344
23 72	Air-to-Air Energy Recovery Equipment	344
23 81	Decentralized Unitary HVAC Equipment	344
23 82	Convection Heating and Cooling Units	346
23 83	Radiant Heating Units	347
23 91	Prefabricated Equipment Supports	352
	Electrical	
26 05	Common Work Results for Electrical	354
26 24	Switchboards and Panelboards	362
26 29	Low-Voltage Controllers	363
26 42	Cathodic Protection	363
	Electronic Safety & Security	
28 16	Intrusion Detection	366
28 31	Fire Detection and Alarm	366
28 33	Gas Detection and Alarm	367
	Earthwork	
31 23	Excavation and Fill	370
	Exterior Improvements	
32 12	Flexible Paving	386
32 84	Planting Irrigation	387
	Utilities	
33 01	Operation and Maintenance of Utilities	392
33 05	Common Work Results for Utilities	395
33 11	Water Utility Distribution Piping	396
33 12	Water Utility Distribution Equipment	402
33 16	Water Utility Storage Tanks	406
33 21	Water Supply Wells	406
33 31	Sanitary Utility Sewerage Piping	407
33 36	Utility Septic Tanks	408
33 41	Storm Utility Drainage Piping	409
33 42	Culverts	412
33 44	Storm Utility Water Drains	412
33 46	Subdrainage	413
33 49	Storm Drainage Structures	414
33 51	Natural-Gas Distribution	414
33 52	Liquid Fuel Distribution	415
33 61	Hydronic Energy Distribution	416
33 63	Steam Energy Distribution	417
	Material Processing & Handling Equipment	
41 22	Cranes and Hoists	422
	Pollution Control Equipment	
44 11	Particulate Control Equipment	424
	Water and Wastewater Equipment	
46 07	Packaged Water and Wastewater Treatment Equipment	428
46 23	Grit Removal And Handling Equipment	428
46 25	Oil and Grease Separation and Removal Equipment	429

How RSMeans Unit Price Data Works

All RSMeans unit price data is organized in the same way. It is important to understand the structure, so that you can find information easily and use it correctly.

RSMeans **Line Numbers** consist of 12 characters, which identify a unique location in the database for each task. The first 6 or 8 digits conform to the Construction Specifications Institute MasterFormat® 2012. The remainder of the digits are a further breakdown by RSMeans in order to arrange items in understandable groups of similar tasks. Line numbers are consistent across all RSMeans publications, so a line number in any RSMeans product will always refer to the same unit of work.

RSMeans engineers have created **reference** information to assist you in your estimate. If there is information that applies to a section, it will be indicated at the start of the section. In this case, R033105-10 provides information on the proportionate quantities of formwork, reinforcing, and concrete used in cast-in-place concrete items such as footings, slabs, beams, and columns. The Reference Section is located in the back of the book on the pages with a gray edge.

RSMeans **Descriptions** are shown in a hierarchical structure to make them readable. In order to read a complete description, read up through the indents to the top of the section. Include everything that is above and to the left that is not contradicted by information below. For instance, the complete description for line 03 30 53.40 3550 is "Concrete in place, including forms (4 uses), Grade 60 rebar, concrete (Portland cement Type 1), placement and finishing unless otherwise indicated; Equipment pad (3000 psi), 4' x 4' x 6" thick."

When using **RSMeans data**, it is important to read through an entire section to ensure that you use the data that most closely matches your work. Note that sometimes there is additional information shown in the section that may improve your price. There are frequently lines that further describe, add to, or adjust data for specific situations.

03 30 Cast-In-Place Concrete

03 30 53 – Miscellaneous Cast-In-Place Concrete

03 30 53.40 Concrete In Place

0010	**CONCRETE IN PLACE**	R033105-10
0020	Including forms (4 uses), Grade 60 rebar, concrete (Portland cement	R033105-20
0050	Type I), placement and finishing unless otherwise indicated	R033105-50
0300	Beams (3500 psi), 5 kip per L.F., 10' span	R033105-65
0350	25' span	R033105-70
0500	Chimney foundations (5000 psi), over 5 C.Y.	R033105-85
0510	(3500 psi), under 5 C.Y.	
0700	Columns, square (4000 psi), 12" x 12", less than 2% reinforcing	
3450	Over 10,000 S.F.	
3500	Add per floor for 3 to 6 stories high	
3520	For 7 to 20 stories high	
3540	Equipment pad (3000 psi), 3' x 3' x 6" thick	
3550	4' x 4' x 6" thick	
3560	5' x 5' x 8" thick	
3570	6' x 6' x 8" thick	

The data published in RSMeans print books represents a "national average" cost. This data should be modified to the project location using the **City Cost Indexes** or **Location Factors** tables found in the Reference Section (see pages 594–642). Use the location factors to adjust estimate totals if the project covers multiple trades. Use the city cost indexes (CCI) for single trade projects or projects where a more detailed analysis is required. All figures in the two tables are derived from the same research. The last row of data in the CCI, the weighted average, is the same as the numbers reported for each location in the location factor table.

Crews include labor or labor and equipment necessary to accomplish each task. In this case, Crew C-14H is used. RSMeans selects a crew to represent the workers and equipment that are typically used for that task. In this case, Crew C-14H consists of one carpenter foreman (outside), two carpenters, one rodman, one laborer, one cement finisher, and one gas engine vibrator. Details of all crews can be found in the reference section.

Crews

Crew No.	Bare Costs		Incl. Subs O&P		Cost Per Labor-Hour	
Crew C-14H	Hr.	Daily	Hr.	Daily	Bare Costs	Incl. O&P
1 Carpenter Foreman (outside)	$47.85	$382.80	$73.85	$590.80	$45.15	$69.42
2 Carpenters	45.85	733.60	70.75	1132.00		
1 Rodman (reinf.)	50.65	405.20	79.55	636.40		
1 Laborer	36.65	293.20	56.55	452.40		
1 Cement Finisher	44.05	352.40	65.10	520.80		
1 Gas Engine Vibrator		33.00		36.30	.69	.76
48 L.H., Daily Totals		$2200.20		$3368.70	$45.84	$70.18

The **Daily Output** is the amount of work that the crew can do in a normal 8-hour workday, including mobilization, layout, movement of materials, and cleanup. In this case, crew C-14H can install thirty 4' x 4' x 6" thick concrete pads in a day. Daily output is variable, based on many factors, including the size of the job, location, and environmental conditions. RSMeans data represents work done in daylight (or adequate lighting) and temperate conditions.

Bare Costs are the costs of materials, labor, and equipment that the installing contractor pays. They represent the cost, in U.S. dollars, for one unit of work. They do not include any markups for profit or labor burden.

Crew	Daily Output	Labor-Hours	Unit	Material	2014 Bare Costs Labor	Equipment	Total	Total Incl O&P
C-14A	15.62	12.804	C.Y.	315	590	47.50	952.50	1,325
"	18.55	10.782		335	495	40	870	1,175
C-14C	32.22	3.476		147	153	1.01	301.01	400
"	23.71	4.724		173	207	1.37	381.37	515
C-14A	11.96	16.722		355	770	62	1,187	1,675
	2200	.025		.80	1.04	.32	2.16	2.83
	31800	.002			.07	.02	.09	.13
	21200	.003			.11	.03	.14	.20
C-14H	45	1.067	Ea.	44.50	48	.74	93.24	124
	30	1.600		66.50	72	1.10	139.60	185
	18	2.667		119	120	1.84	240.84	320
	14	3.429		160	155	2.37	317.37	415

The **Total Incl O&P column** is the total cost, including overhead and profit, that the installing contractor will charge the customer. This represents the cost of materials plus 10% profit, the cost of labor plus labor burden and 10% profit, and the cost of equipment plus 10% profit. It does not include the general contractor's overhead and profit. Note: See the inside back cover for details of how RSMeans calculates labor burden.

The **Total column** represents the total bare cost for the installing contractor, in U.S. dollars. In this case, the sum of $66.50 for material + $72.00 for labor + $1.10 for equipment is $139.60.

The figure in the **Labor Hours** column is the amount of labor required to perform one unit of work—in this case the amount of labor required to construct one 4' x 4' equipment pad. This figure is calculated by dividing the number of hours of labor in the crew by the daily output (48 labor hours divided by 30 pads = 1.6 hours of labor per pad). Multiply 1.600 times 60 to see the value in minutes: 60 x 1.6 = 96 minutes. Note: the labor hour figure is not dependent on the crew size. A change in crew size will result in a corresponding change in daily output, but the labor hours per unit of work will not change.

All RSMeans unit cost data includes the typical **Unit of Measure** used for estimating that item. For concrete-in-place the typical unit is cubic yards (C.Y.) or each (Ea.). For installing broadloom carpet it is square yard, and for gypsum board it is square foot. The estimator needs to take special care that the unit in the data matches the unit in the take-off. Unit conversions may be found in the Reference Section.

How RSMeans Unit Price Data Works (Continued)

Sample Estimate

This sample demonstrates the elements of an estimate, including a tally of the RSMeans data lines, and a summary of the markups on a contractor's work to arrive at a total cost to the owner. The RSMeans Location Factor is added at the bottom of the estimate to adjust the cost of the work to a specific location.

Work Performed: The body of the estimate shows the RSMeans data selected, including line number, a brief description of each item, its take-off unit and quantity, and the bare costs of materials, labor, and equipment. This estimate also includes a column titled "SubContract." This data is taken from the RSMeans column "Total Incl O&P," and represents the total that a subcontractor would charge a general contractor for the work, including the sub's markup for overhead and profit.

Division 1, General Requirements: This is the first division numerically, but the last division estimated. Division 1 includes project-wide needs provided by the general contractor. These requirements vary by project, but may include temporary facilities and utilities, security, testing, project cleanup, etc. For small projects a percentage can be used, typically between 5% and 15% of project cost. For large projects the costs may be itemized and priced individually.

Bonds: Bond costs should be added to the estimate. The figures here represent a typical performance bond, ensuring the owner that if the general contractor does not complete the obligations in the construction contract the bonding company will pay the cost for completion of the work.

Location Adjustment: RSMeans published data is based on national average costs. If necessary, adjust the total cost of the project using a location factor from the "Location Factor" table or the "City Cost Index" table. Use location factors if the work is general, covering multiple trades. If the work is by a single trade (e.g., masonry) use the more specific data found in the "City Cost Indexes."

This estimate is based on an interactive spreadsheet. A copy of this spreadsheet is located on the RSMeans website at
http://www.reedconstructiondata.com/rsmeans/extras/546011.
You are free to download it and adjust it to your methodology.

Project Name: Pre-Engineered Steel Building			Architect: As Shown	
Location:	Anywhere, USA			
Line Number	Description	Qty	Unit	Material
03 30 53.40 3940	Strip footing, 12" x 24", reinforced	34	C.Y.	$4,624.00
03 30 53.40 3950	Strip footing, 12" x 36", reinforced	15	C.Y.	$1,950.00
03 11 13.65 3000	Concrete slab edge forms	500	L.F.	$155.00
03 22 11.10 0200	Welded wire fabric reinforcing	150	C.S.F.	$2,580.00
03 31 13.35 0300	Ready mix concrete, 4000 psi for slab on grade	278	C.Y.	$28,912.00
03 31 13.70 4300	Place, strike off & consolidate concrete slab	278	C.Y.	$0.00
03 35 13.30 0250	Machine float & trowel concrete slab	15,000	S.F.	$0.00
03 15 16.20 0140	Cut control joints in concrete slab	950	L.F.	$57.00
03 39 23.13 0300	Sprayed concrete curing membrane	150	C.S.F.	$1,642.50
Division 03	**Subtotal**			**$39,920.50**
08 36 13.10 2650	Manual 10' x 10' steel sectional overhead door	8	Ea.	$8,800.00
08 36 13.10 2860	Insulation and steel back panel for OH door	800	S.F.	$3,760.00
Division 08	**Subtotal**			**$12,560.00**
13 34 19.50 1100	Pre-Engineered Steel Building, 100' x 150' x 24'	15,000	SF Flr.	$0.00
13 34 19.50 6050	Framing for PESB door opening, 3' x 7'	4	Opng.	$0.00
13 34 19.50 6100	Framing for PESB door opening, 10' x 10'	8	Opng.	$0.00
13 34 19.50 6200	Framing for PESB window opening, 4' x 3'	6	Opng.	$0.00
13 34 19.50 5750	PESB door, 3' x 7', single leaf	4	Opng.	$2,340.00
13 34 19.50 7750	PESB sliding window, 4' x 3' with screen	6	Opng.	$2,250.00
13 34 19.50 6550	PESB gutter, eave type, 26 ga., painted	300	L.F.	$2,010.00
13 34 19.50 8650	PESB roof vent, 12" wide x 10' long	15	Ea.	$540.00
13 34 19.50 6900	PESB insulation, vinyl faced, 4" thick	27,400	S.F.	$10,960.00
Division 13	**Subtotal**			**$18,100.00**
	Subtotal			$70,580.50
Division 01	**General Requirements @ 7%**			**4,940.64**
	Estimate Subtotal			$75,521.14
	Sales Tax @ 5%			3,776.06
	Subtotal A			79,297.19
	GC O & P			7,929.72
	Subtotal B			87,226.91
	Contingency @ 5%			
	Subtotal C			
	Bond @ $12/1000 +10% O&P			
	Subtotal D			
	Location Adjustment Factor			
	Grand Total			

This example shows the cost to construct a pre-engineered steel building. The foundation, doors, windows, and insulation will be installed by the general contractor. A subcontractor will install the structural steel, roofing, and siding.

Labor	Equipment	SubContract	01/01/14 Estimate Total	STD
$3,468.00	$23.12	$0.00		
$1,230.00	$8.10	$0.00		
$1,160.00	$0.00	$0.00		
$3,900.00	$0.00	$0.00		
$0.00	$0.00	$0.00		
$4,642.60	$166.80	$0.00		
$8,700.00	$450.00	$0.00		
$370.50	$95.00	$0.00		
$922.50	$0.00	$0.00		
$24,393.60	**$743.02**	**$0.00**	**$65,057.12**	**Division 03**
$3,280.00	$0.00	$0.00		
$0.00	$0.00	$0.00		
$3,280.00	**$0.00**	**$0.00**	**$15,840.00**	**Division 08**
$0.00	$0.00	$345,000.00		
$0.00	$0.00	$2,260.00		
$0.00	$0.00	$9,200.00		
$0.00	$0.00	$3,300.00		
$656.00	$0.00	$0.00		
$558.00	$65.70	$0.00		
$768.00	$0.00	$0.00		
$3,060.00	$0.00	$0.00		
$8,768.00	$0.00	$0.00		
$13,810.00	**$65.70**	**$359,760.00**	**$391,735.70**	**Division 13**
$41,483.60	$808.72	$359,760.00	$472,632.32	Subtotal
2,903.85	56.61	25,183.20		Gen. Requirements
$44,387.45	$865.33	$384,943.20	$472,632.82	Estimate Subtotal
	43.27	9,623.58		Sales tax
44,387.45	908.60	394,566.78		Subtotal
24,368.71	90.86	39,456.68		GC O&P
68,756.16	999.46	434,023.46	$591,005.99	Subtotal
			29,550.30	Contingency
			$620,556.29	Subtotal
			8,191.34	Bond
			$628,747.63	Subtotal
102.30			14,461.20	Location Adjustment
			$643,208.83	**Grand Total**

Sales Tax: If the work is subject to state or local sales taxes, the amount must be added to the estimate. Sales tax may be added to material costs, equipment costs, and subcontracted work. In this case, sales tax was added in all three categories. It was assumed that approximately half the subcontracted work would be material cost, so the tax was applied to 50% of the subcontract total.

GC O&P: This entry represents the general contractor's markup on material, labor, equipment, and subcontractor costs. RSMeans' standard markup on materials, equipment, and subcontracted work is 10%. In this estimate, the markup on the labor performed by the GC's workers uses "Skilled Workers Average" shown in Column F on the table "Installing Contractor's Overhead & Profit," which can be found on the inside-back cover of the book.

Contingency: A factor for contingency may be added to any estimate to represent the cost of unknowns that may occur between the time that the estimate is performed and the time the project is constructed. The amount of the allowance will depend on the stage of design at which the estimate is done, and the contractor's assessment of the risk involved. Refer to section 01 21 16.50 for contigency allowances.

Estimating Tips

01 20 00 Price and Payment Procedures

- Allowances that should be added to estimates to cover contingencies and job conditions that are not included in the national average material and labor costs are shown in section 01 21.
- When estimating historic preservation projects (depending on the condition of the existing structure and the owner's requirements), a 15%–20% contingency or allowance is recommended, regardless of the stage of the drawings.

01 30 00 Administrative Requirements

- Before determining a final cost estimate, it is a good practice to review all the items listed in Subdivisions 01 31 and 01 32 to make final adjustments for items that may need customizing to specific job conditions.
- Requirements for initial and periodic submittals can represent a significant cost to the General Requirements of a job. Thoroughly check the submittal specifications when estimating a project to determine any costs that should be included.

01 40 00 Quality Requirements

- All projects will require some degree of Quality Control. This cost is not included in the unit cost of construction listed in each division. Depending upon the terms of the contract, the various costs of inspection and testing can be the responsibility of either the owner or the contractor. Be sure to include the required costs in your estimate.

01 50 00 Temporary Facilities and Controls

- Barricades, access roads, safety nets, scaffolding, security, and many more requirements for the execution of a safe project are elements of direct cost. These costs can easily be overlooked when preparing an estimate. When looking through the major classifications of this subdivision, determine which items apply to each division in your estimate.
- Construction Equipment Rental Costs can be found in the Reference Section in section 01 54 33. Operators' wages are not included in equipment rental costs.
- Equipment mobilization and demobilization costs are not included in equipment rental costs and must be considered separately in section 01 54 36.50.
- The cost of small tools provided by the installing contractor for his workers is covered in the "Overhead" column on the "Installing Contractor's Overhead and Profit" table that lists labor trades, base rates and markups and, therefore, is included in the "Total Incl. O&P" cost of any Unit Price line item. For those users who are constrained by contract terms to use only bare costs, there are two line items in section 01 54 39.70 for small tools as a percentage of bare labor cost. If some of those users are further constrained by contract terms to refrain from using line items from Division 1, you are advised to cover the cost of small tools within your coefficient or multiplier.

01 70 00 Execution and Closeout Requirements

- When preparing an estimate, thoroughly read the specifications to determine the requirements for Contract Closeout. Final cleaning, record documentation, operation and maintenance data, warranties and bonds, and spare parts and maintenance materials can all be elements of cost for the completion of a contract. Do not overlook these in your estimate.

Reference Numbers

Reference numbers are shown in shaded boxes at the beginning of some major classifications. These numbers refer to related items in the Reference Section. The reference information may be an estimating procedure, an alternate pricing method, or technical information.

Note: Not all subdivisions listed here necessarily appear in this publication.

No part of this publication may be reproduced, stored in a retrieval system, or transmitted in any form or by any means without prior written permission of RSMeans, a division of Reed Construction Data.

01 11 Summary of Work

01 11 31 – Professional Consultants

01 11 31.10 Architectural Fees

		Crew	Daily Output	Labor-Hours	Unit	Material	2014 Bare Costs Labor	Equipment	Total	Total Incl O&P
0010	**ARCHITECTURAL FEES**									
0020	For new construction									
0060	Minimum				Project				4.90%	4.90%
0090	Maximum								16%	16%
0100	For alteration work, to $500,000, add to new construction fee								50%	50%
0150	Over $500,000, add to new construction fee								25%	25%

01 11 31.20 Construction Management Fees

0010	**CONSTRUCTION MANAGEMENT FEES**									
0020	$1,000,000 job, minimum				Project				4.50%	4.50%
0050	Maximum								7.50%	7.50%
0300	$50,000,000 job, minimum								2.50%	2.50%
0350	Maximum								4%	4%

01 11 31.30 Engineering Fees

0010	**ENGINEERING FEES**	R011110-30								
0020	Educational planning consultant, minimum				Project				.50%	.50%
0100	Maximum				"				2.50%	2.50%
0200	Electrical, minimum				Contrct				4.10%	4.10%
0300	Maximum								10.10%	10.10%
0400	Elevator & conveying systems, minimum								2.50%	2.50%
0500	Maximum								5%	5%
0600	Food service & kitchen equipment, minimum								8%	8%
0700	Maximum								12%	12%
0800	Landscaping & site development, minimum								2.50%	2.50%
0900	Maximum								6%	6%
1000	Mechanical (plumbing & HVAC), minimum								4.10%	4.10%
1100	Maximum								10.10%	10.10%
1200	Structural, minimum				Project				1%	1%
1300	Maximum				"				2.50%	2.50%

01 21 Allowances

01 21 16 – Contingency Allowances

01 21 16.50 Contingencies

0010	**CONTINGENCIES**, Add to estimate									
0020	Conceptual stage				Project				20%	20%
0050	Schematic stage								15%	15%
0100	Preliminary working drawing stage (Design Dev.)								10%	10%
0150	Final working drawing stage								3%	3%

01 21 53 – Factors Allowance

01 21 53.50 Factors

0010	**FACTORS** Cost adjustments	R012153-10								
0100	Add to construction costs for particular job requirements									
0500	Cut & patch to match existing construction, add, minimum				Costs	2%	3%			
0550	Maximum					5%	9%			
0800	Dust protection, add, minimum					1%	2%			
0850	Maximum					4%	11%			
1100	Equipment usage curtailment, add, minimum					1%	1%			
1150	Maximum					3%	10%			
1400	Material handling & storage limitation, add, minimum					1%	1%			
1450	Maximum					6%	7%			
1700	Protection of existing work, add, minimum					2%	2%			

01 21 Allowances

01 21 53 – Factors Allowance

01 21 53.50 Factors

		Crew	Daily Output	Labor-Hours	Unit	Material	2014 Bare Costs Labor	Equipment	Total	Total Incl O&P
1750	Maximum				Costs	5%	7%			
2000	Shift work requirements, add, minimum						5%			
2050	Maximum						30%			
2300	Temporary shoring and bracing, add, minimum					2%	5%			
2350	Maximum					5%	12%			
2400	Work inside prisons and high security areas, add, minimum						30%			
2450	Maximum						50%			

01 21 55 – Job Conditions Allowance

01 21 55.50 Job Conditions

0010	**JOB CONDITIONS** Modifications to applicable									
0020	cost summaries									
0100	Economic conditions, favorable, deduct				Project				2%	2%
0200	Unfavorable, add								5%	5%
0300	Hoisting conditions, favorable, deduct								2%	2%
0400	Unfavorable, add								5%	5%
0700	Labor availability, surplus, deduct								1%	1%
0800	Shortage, add								10%	10%
0900	Material storage area, available, deduct								1%	1%
1000	Not available, add								2%	2%
1100	Subcontractor availability, surplus, deduct								5%	5%
1200	Shortage, add								12%	12%
1300	Work space, available, deduct								2%	2%
1400	Not available, add								5%	5%

01 21 57 – Overtime Allowance

01 21 57.50 Overtime

0010	**OVERTIME** for early completion of projects or where	R012909-90								
0020	labor shortages exist, add to usual labor, up to				Costs		100%			

01 21 61 – Cost Indexes

01 21 61.10 Construction Cost Index

0010	**CONSTRUCTION COST INDEX** (Reference) over 930 zip code locations in									
0020	the U.S. and Canada, total bldg. cost, min. (Longview, TX)				%					73.50%
0050	Average									100%
0100	Maximum (New York, NY)									131.10%

01 21 61.20 Historical Cost Indexes

0010	**HISTORICAL COST INDEXES** (See Reference Section)									

01 21 61.30 Labor Index

0010	**LABOR INDEX** (Reference) For over 930 zip code locations in									
0020	the U.S. and Canada, minimum (Longview, TX)				%		40.50%			
0050	Average						100%			
0100	Maximum (New York, NY)						166.40%			

01 21 61.50 Material Index

0010	**MATERIAL INDEX** (Reference) For over 930 zip code locations in									
0020	the U.S. and Canada, minimum (Vale, OR)				%	91.30%				
0040	Average					100%				
0060	Maximum (Ketchikan, AK)					132%				

01 21 63 – Taxes

01 21 63.10 Taxes

0010	**TAXES**	R012909-80								
0020	Sales tax, State, average				%	5.06%				
0050	Maximum	R012909-85				7.25%				

01 21 Allowances

01 21 63 – Taxes

01 21 63.10 Taxes		Crew	Daily Output	Labor-Hours	Unit	Material	2014 Bare Costs Labor	Equipment	Total	Total Incl O&P
0200	Social Security, on first $115,800 of wages				%		7.65%			
0300	Unemployment, combined Federal and State, minimum						.60%			
0350	Average						7.80%			
0400	Maximum						13.50%			

01 31 Project Management and Coordination

01 31 13 – Project Coordination

01 31 13.20 Field Personnel

		Crew	Daily Output	Labor-Hours	Unit	Material	Labor	Equipment	Total	Total Incl O&P
0010	**FIELD PERSONNEL**									
0020	Clerk, average				Week		435		435	675
0100	Field engineer, minimum						1,025		1,025	1,600
0120	Average						1,350		1,350	2,100
0140	Maximum						1,525		1,525	2,375
0160	General purpose laborer, average						1,475		1,475	2,250
0180	Project manager, minimum						1,925		1,925	2,975
0200	Average						2,225		2,225	3,425
0220	Maximum						2,525		2,525	3,925
0240	Superintendent, minimum						1,875		1,875	2,900
0260	Average						2,050		2,050	3,200
0280	Maximum						2,350		2,350	3,625
0290	Timekeeper, average						1,200		1,200	1,850

01 31 13.30 Insurance

					Unit				Total	Total Incl O&P
0010	**INSURANCE**	R013113-40								
0020	Builders risk, standard, minimum				Job				.24%	.24%
0050	Maximum	R013113-60							.64%	.64%
0200	All-risk type, minimum								.25%	.25%
0250	Maximum								.62%	.62%
0400	Contractor's equipment floater, minimum				Value				.50%	.50%
0450	Maximum				"				1.50%	1.50%
0600	Public liability, average				Job				2.02%	2.02%
0800	Workers' compensation & employer's liability, average									
0850	by trade, carpentry, general				Payroll		15.38%			
1000	Electrical						5.83%			
1150	Insulation						11.67%			
1450	Plumbing						7.08%			
1550	Sheet metal work (HVAC)						8.79%			

01 31 13.50 General Contractor's Mark-Up

					Unit				Total	Total Incl O&P
0010	**GENERAL CONTRACTOR'S MARK-UP** on Change Orders									
0200	Extra work, by subcontractors, add				%				10%	10%
0250	By General Contractor, add								15%	15%
0400	Omitted work, by subcontractors, deduct all but								5%	5%
0450	By General Contractor, deduct all but								7.50%	7.50%
0600	Overtime work, by subcontractors, add								15%	15%
0650	By General Contractor, add								10%	10%

01 31 13.60 Installing Contractor's Main Office Overhead

					Unit				Total	Total Incl O&P
0010	**INSTALLING CONTRACTOR'S MAIN OFFICE OVERHEAD**									
0020	As percent of direct costs, minimum				%					5%
0050	Average									13%
0100	Maximum									30%

01 31 Project Management and Coordination

01 31 13 – Project Coordination

01 31 13.80 Overhead and Profit

		Crew	Daily Output	Labor-Hours	Unit	Material	2014 Bare Costs Labor	Equipment	Total	Total Incl O&P
0010	**OVERHEAD & PROFIT** Allowance to add to items in this									
0020	book that do not include Subs O&P, average				%				25%	
0100	Allowance to add to items in this book that									
0110	do include Subs O&P, minimum				%				5%	5%
0150	Average								10%	10%
0200	Maximum								15%	15%
0300	Typical, by size of project, under $100,000								30%	
0350	$500,000 project								25%	
0400	$2,000,000 project								20%	
0450	Over $10,000,000 project								15%	

01 31 13.90 Performance Bond

0010	**PERFORMANCE BOND**	R013113-80								
0020	For buildings, minimum				Job				.60%	.60%
0100	Maximum				"				2.50%	2.50%

01 32 Construction Progress Documentation

01 32 33 – Photographic Documentation

01 32 33.50 Photographs

		Crew	Daily Output	Labor-Hours	Unit	Material	Labor	Equipment	Total	Total Incl O&P
0010	**PHOTOGRAPHS**									
0020	8" x 10", 4 shots, 2 prints ea., std. mounting				Set	475			475	525
0100	Hinged linen mounts					530			530	580
0200	8" x 10", 4 shots, 2 prints each, in color					425			425	470
0300	For I.D. slugs, add to all above					5.30			5.30	5.85
1500	Time lapse equipment, camera and projector, buy				Ea.	2,650			2,650	2,925
1550	Rent per month				"	1,600			1,600	1,750
1700	Cameraman and film, including processing, B.&W.				Day	1,225			1,225	1,350
1720	Color				"	1,425			1,425	1,550

01 41 Regulatory Requirements

01 41 26 – Permit Requirements

01 41 26.50 Permits

0010	**PERMITS**									
0020	Rule of thumb, most cities, minimum				Job				.50%	.50%
0100	Maximum				"				2%	2%

01 51 Temporary Utilities

01 51 13 – Temporary Electricity

01 51 13.80 Temporary Utilities

		Crew	Daily Output	Labor-Hours	Unit	Material	Labor	Equipment	Total	Total Incl O&P
0010	**TEMPORARY UTILITIES**									
0100	Heat, incl. fuel and operation, per week, 12 hrs. per day	1 Skwk	100	.080	CSF Flr	28	3.78		31.78	36.50
0200	24 hrs. per day	"	60	.133		53.50	6.30		59.80	68.50
0350	Lighting, incl. service lamps, wiring & outlets, minimum	1 Elec	34	.235		2.80	12.55		15.35	22
0360	Maximum	"	17	.471		5.85	25		30.85	44
0400	Power for temp lighting only, 6.6 KWH, per month								.92	1.01
0430	11.8 KWH, per month								1.65	1.82
0450	23.6 KWH, per month								3.30	3.63
0600	Power for job duration incl. elevator, etc., minimum								47	51.50

01 51 Temporary Utilities

01 51 13 – Temporary Electricity

01 51 13.80 Temporary Utilities

		Crew	Daily Output	Labor-Hours	Unit	Material	2014 Bare Costs Labor	Equipment	Total	Total Incl O&P
0650	Maximum				CSF Flr				110	121
1000	Toilet, portable, see Equip. Rental 01 54 33 in Reference Section									

01 52 Construction Facilities

01 52 13 – Field Offices and Sheds

01 52 13.20 Office and Storage Space

		Crew	Daily Output	Labor-Hours	Unit	Material	Labor	Equipment	Total	Total Incl O&P
0010	**OFFICE AND STORAGE SPACE**									
0020	Office trailer, furnished, no hookups, 20' x 8', buy	2 Skwk	1	16	Ea.	9,150	755		9,905	11,300
0250	Rent per month					169			169	185
0300	32' x 8', buy	2 Skwk	.70	22.857		14,000	1,075		15,075	17,100
0350	Rent per month					203			203	223
0400	50' x 10', buy	2 Skwk	.60	26.667		22,400	1,250		23,650	26,600
0450	Rent per month					305			305	335
0500	50' x 12', buy	2 Skwk	.50	32		27,000	1,525		28,525	32,100
0550	Rent per month					350			350	385
0700	For air conditioning, rent per month, add					48.50			48.50	53.50
0800	For delivery, add per mile				Mile	11			11	12.10
0890	Delivery each way				Ea.	200			200	220
0900	Bunk house trailer, 8' x 40' duplex dorm with kitchen, no hookups, buy	2 Carp	1	16		37,300	735		38,035	42,100
0910	9 man with kitchen and bath, no hookups, buy		1	16		38,000	735		38,735	42,900
0920	18 man sleeper with bath, no hookups, buy		1	16		49,000	735		49,735	55,000
1000	Portable buildings, prefab, on skids, economy, 8' x 8'		265	.060	S.F.	29.50	2.77		32.27	37
1100	Deluxe, 8' x 12'		150	.107	"	24	4.89		28.89	34
1200	Storage boxes, 20' x 8', buy	2 Skwk	1.80	8.889	Ea.	2,775	420		3,195	3,700
1250	Rent per month					78.50			78.50	86
1300	40' x 8', buy	2 Skwk	1.40	11.429		3,700	540		4,240	4,900
1350	Rent per month					99			99	109

01 54 Construction Aids

01 54 16 – Temporary Hoists

01 54 16.50 Weekly Forklift Crew

		Crew	Daily Output	Labor-Hours	Unit	Material	Labor	Equipment	Total	Total Incl O&P
0010	**WEEKLY FORKLIFT CREW**									
0100	All-terrain forklift, 45' lift, 35' reach, 9000 lb. capacity	A-3P	.20	40	Week		1,875	2,650	4,525	5,775

01 54 19 – Temporary Cranes

01 54 19.50 Daily Crane Crews

		Crew	Daily Output	Labor-Hours	Unit	Material	Labor	Equipment	Total	Total Incl O&P
0010	**DAILY CRANE CREWS** for small jobs, portal to portal									
0100	12-ton truck-mounted hydraulic crane	A-3H	1	8	Day		400	875	1,275	1,575
0200	25-ton	A-3I	1	8			400	1,000	1,400	1,700
0300	40-ton	A-3J	1	8			400	1,250	1,650	1,975
0400	55-ton	A-3K	1	16			750	1,650	2,400	2,975
0500	80-ton	A-3L	1	16			750	2,400	3,150	3,775
0600	100-ton	A-3M	1	16			750	2,375	3,125	3,775
0900	If crane is needed on a Saturday, Sunday or Holiday									
0910	At time-and-a-half, add				Day		50%			
0920	At double time, add				"		100%			

01 54 19.60 Monthly Tower Crane Crew

		Crew	Daily Output	Labor-Hours	Unit	Material	Labor	Equipment	Total	Total Incl O&P
0010	**MONTHLY TOWER CRANE CREW**, excludes concrete footing									
0100	Static tower crane, 130' high, 106' jib, 6200 lb. capacity	A-3N	.05	176	Month		8,850	24,200	33,050	40,000

01 54 Construction Aids

01 54 23 – Temporary Scaffolding and Platforms

01 54 23.70 Scaffolding

		Crew	Daily Output	Labor-Hours	Unit	Material	2014 Bare Costs Labor	Equipment	Total	Total Incl O&P
0010	**SCAFFOLDING** R015423-10									
0015	Steel tube, regular, no plank, labor only to erect & dismantle									
0090	Building exterior, wall face, 1 to 5 stories, 6'-4" x 5' frames	3 Carp	8	3	C.S.F.		138		138	212
0200	6 to 12 stories	4 Carp	8	4			183		183	283
0301	13 to 20 stories	5 Clab	8	5			183		183	283
0460	Building interior, wall face area, up to 16' high	3 Carp	12	2			91.50		91.50	142
0560	16' to 40' high		10	2.400			110		110	170
0800	Building interior floor area, up to 30' high		150	.160	C.C.F.		7.35		7.35	11.30
0900	Over 30' high	4 Carp	160	.200	"		9.15		9.15	14.15
0906	Complete system for face of walls, no plank, material only rent/mo				C.S.F.	35			35	38.50
0908	Interior spaces, no plank, material only rent/mo				C.C.F.	3.35			3.35	3.68
0910	Steel tubular, heavy duty shoring, buy									
0920	Frames 5' high 2' wide				Ea.	81.50			81.50	90
0925	5' high 4' wide					93			93	102
0930	6' high 2' wide					93.50			93.50	103
0935	6' high 4' wide					109			109	120
0940	Accessories									
0945	Cross braces				Ea.	15.50			15.50	17.05
0950	U-head, 8" x 8"					19.10			19.10	21
0955	J-head, 4" x 8"					13.90			13.90	15.30
0960	Base plate, 8" x 8"					15.50			15.50	17.05
0965	Leveling jack					33.50			33.50	36.50
1000	Steel tubular, regular, buy									
1100	Frames 3' high 5' wide				Ea.	72.50			72.50	80
1150	5' high 5' wide					86.50			86.50	95
1200	6'-4" high 5' wide					118			118	129
1350	7'-6" high 6' wide					158			158	173
1500	Accessories cross braces					15.50			15.50	17.05
1550	Guardrail post					16.40			16.40	18.05
1600	Guardrail 7' section					6.30			6.30	6.95
1650	Screw jacks & plates					21.50			21.50	24
1700	Sidearm brackets					25.50			25.50	28.50
1750	8" casters					31			31	34
1800	Plank 2" x 10" x 16'-0"					52			52	57.50
1900	Stairway section					276			276	305
1910	Stairway starter bar					32			32	35
1920	Stairway inside handrail					53			53	58
1930	Stairway outside handrail					81.50			81.50	90
1940	Walk-thru frame guardrail					41.50			41.50	46
2000	Steel tubular, regular, rent/mo.									
2100	Frames 3' high 5' wide				Ea.	5			5	5.50
2150	5' high 5' wide					5			5	5.50
2200	6'-4" high 5' wide					5.15			5.15	5.65
2250	7'-6" high 6' wide					7			7	7.70
2500	Accessories, cross braces					1			1	1.10
2550	Guardrail post					1			1	1.10
2600	Guardrail 7' section					1			1	1.10
2650	Screw jacks & plates					2			2	2.20
2700	Sidearm brackets					2			2	2.20
2750	8" casters					8			8	8.80
2800	Outrigger for rolling tower					3			3	3.30
2850	Plank 2" x 10" x 16'-0"					6			6	6.60

01 54 Construction Aids

01 54 23 – Temporary Scaffolding and Platforms

01 54 23.70 Scaffolding

		Crew	Daily Output	Labor-Hours	Unit	Material	2014 Bare Costs Labor	Equipment	Total	Total Incl O&P
2900	Stairway section				Ea.	35.50			35.50	39
2940	Walk-thru frame guardrail				↓	2.50			2.50	2.75
3000	Steel tubular, heavy duty shoring, rent/mo.									
3250	5' high 2' & 4' wide				Ea.	5			5	5.50
3300	6' high 2' & 4' wide					5			5	5.50
3500	Accessories, cross braces					1			1	1.10
3600	U - head, 8" x 8"					1			1	1.10
3650	J - head, 4" x 8"					1			1	1.10
3700	Base plate, 8" x 8"					1			1	1.10
3750	Leveling jack					2			2	2.20
5700	Planks, 2" x 10" x 16'-0", labor only to erect & remove to 50' H	3 Carp	72	.333			15.30		15.30	23.50
5800	Over 50' high	4 Carp	80	.400	↓		18.35		18.35	28.50
6000	Heavy duty shoring for elevated slab forms to 8'-2" high, floor area									
6100	Labor only to erect & dismantle	4 Carp	16	2	C.S.F.		91.50		91.50	142
6110	Materials only, rent.mo				"	29.50			29.50	32.50
6500	To 14'-8" high									
6600	Labor only to erect & dismantle	4 Carp	10	3.200	C.S.F.		147		147	226
6610	Materials only, rent/mo				"	43			43	47.50

01 54 23.75 Scaffolding Specialties

		Crew	Daily Output	Labor-Hours	Unit	Material	Labor	Equipment	Total	Total Incl O&P
0010	**SCAFFOLDING SPECIALTIES**									
1200	Sidewalk bridge, heavy duty steel posts & beams, including									
1210	parapet protection & waterproofing (material cost is rent/month)									
1220	8' to 10' wide, 2 posts	3 Carp	15	1.600	L.F.	37	73.50		110.50	154
1230	3 posts	"	10	2.400	"	56.50	110		166.50	232
1500	Sidewalk bridge using tubular steel scaffold frames including									
1510	planking (material cost is rent/month)	3 Carp	45	.533	L.F.	5.60	24.50		30.10	43.50
1600	For 2 uses per month, deduct from all above					50%				
1700	For 1 use every 2 months, add to all above					100%				
1900	Catwalks, 20" wide, no guardrails, 7' span, buy				Ea.	133			133	146
2000	10' span, buy					186			186	205
3720	Putlog, standard, 8' span, with hangers, buy					66.50			66.50	73
3730	Rent per month					10			10	11
3750	12' span, buy					99			99	109
3755	Rent per month					15			15	16.50
3760	Trussed type, 16' span, buy					229			229	252
3770	Rent per month					20			20	22
3790	22' span, buy					274			274	300
3795	Rent per month					30			30	33
3800	Rolling ladders with handrails, 30" wide, buy, 2 step					231			231	254
4000	7 step					760			760	835
4050	10 step					1,050			1,050	1,150
4100	Rolling towers, buy, 5' wide, 7' long, 10' high					1,150			1,150	1,275
4200	For 5' high added sections, to buy, add				↓	204			204	224
4300	Complete incl. wheels, railings, outriggers,									
4350	21' high, to buy				Ea.	1,975			1,975	2,175
4400	Rent/month = 5% of purchase cost				"	170			170	187

01 54 36 – Equipment Mobilization

01 54 36.50 Mobilization

			Crew	Daily Output	Labor-Hours	Unit	Material	Labor	Equipment	Total	Total Incl O&P
0010	**MOBILIZATION** (Use line item again for demobilization)	R015433-10									
0015	Up to 25 mi. haul dist. (50 mi. RT for mob/demob crew)										
0020	Dozer, loader, backhoe, excav., grader, paver, roller, 70 to 150 H.P.		B-34N	4	2	Ea.		75	142	217	272
0100	Above 150 HP		B-34K	3	2.667	↓		100	320	420	510

01 54 Construction Aids

01 54 36 – Equipment Mobilization

01 54 36.50 Mobilization

		Crew	Daily Output	Labor-Hours	Unit	Material	2014 Bare Costs Labor	Equipment	Total	Total Incl O&P
0900	Shovel or dragline, 3/4 C.Y.	B-34K	3.60	2.222	Ea.		83.50	267	350.50	420
1100	Small equipment, placed in rear of, or towed by pickup truck	A-3A	8	1			36.50	21	57.50	78.50
1150	Equip up to 70 HP, on flatbed trailer behind pickup truck	A-3D	4	2			73	70.50	143.50	189
2000	Crane, truck-mounted, up to 75 ton, (driver only, one-way)	1 Eqhv	7.20	1.111			56		56	84.50
2100	Crane, truck-mounted, over 75 ton	A-3E	2.50	6.400			281	67.50	348.50	500
2200	Crawler-mounted, up to 75 ton	A-3F	2	8			350	500	850	1,075
2300	Over 75 ton	A-3G	1.50	10.667			470	755	1,225	1,550
2500	For each additional 5 miles haul distance, add						10%	10%		
3000	For large pieces of equipment, allow for assembly/knockdown									
3100	For mob/demob of micro-tunneling equip, see Section 33 05 23.19									

01 54 39 – Construction Equipment

01 54 39.70 Small Tools

0010	**SMALL TOOLS** R013113-50									
0020	As % of contractor's bare labor cost for project, minimum					Total	.50%			
0100	Maximum					"	2%			

01 55 Vehicular Access and Parking

01 55 23 – Temporary Roads

01 55 23.50 Roads and Sidewalks

		Crew	Daily Output	Labor-Hours	Unit	Material	Labor	Equipment	Total	Total Incl O&P
0010	**ROADS AND SIDEWALKS** Temporary									
0050	Roads, gravel fill, no surfacing, 4" gravel depth	B-14	715	.067	S.Y.	3.98	2.60	.51	7.09	8.95
0100	8" gravel depth	"	615	.078	"	7.95	3.02	.59	11.56	14.05
1000	Ramp, 3/4" plywood on 2" x 6" joists, 16" O.C.	2 Carp	300	.053	S.F.	1.44	2.45		3.89	5.35
1100	On 2" x 10" joists, 16" O.C.	"	275	.058	"	1.99	2.67		4.66	6.30

01 56 Temporary Barriers and Enclosures

01 56 13 – Temporary Air Barriers

01 56 13.60 Tarpaulins

		Crew	Daily Output	Labor-Hours	Unit	Material	Labor	Equipment	Total	Total Incl O&P
0010	**TARPAULINS**									
0020	Cotton duck, 10 oz. to 13.13 oz. per S.Y., 6'x8'				S.F.	.75			.75	.83
0050	30'x30'					.43			.43	.47
0100	Polyvinyl coated nylon, 14 oz. to 18 oz., minimum					1.30			1.30	1.43
0150	Maximum					1.19			1.19	1.31
0200	Reinforced polyethylene 3 mils thick, white					.03			.03	.03
0300	4 mils thick, white, clear or black					.09			.09	.10
0400	5.5 mils thick, clear					.17			.17	.19
0500	White, fire retardant					.41			.41	.45
0600	12 mils, oil resistant, fire retardant					.27			.27	.30
0700	8.5 mils, black					.57			.57	.63
0710	Woven polyethylene, 6 mils thick					.17			.17	.19
0730	Polyester reinforced w/integral fastening system 11 mils thick					.24			.24	.26
0740	Mylar polyester, non-reinforced, 7 mils thick					1.17			1.17	1.29

01 56 13.90 Winter Protection

0010	**WINTER PROTECTION**									
0100	Framing to close openings	2 Clab	500	.032	S.F.	.41	1.17		1.58	2.26
0200	Tarpaulins hung over scaffolding, 8 uses, not incl. scaffolding		1500	.011		.25	.39		.64	.88
0250	Tarpaulin polyester reinf. w/integral fastening system 11 mils thick		1600	.010		.24	.37		.61	.83
0300	Prefab fiberglass panels, steel frame, 8 uses		1200	.013		2.20	.49		2.69	3.17

01 56 Temporary Barriers and Enclosures

01 56 16 – Temporary Dust Barriers

01 56 16.10 Dust Barriers, Temporary

		Crew	Daily Output	Labor-Hours	Unit	Material	2014 Bare Costs Labor	Equipment	Total	Total Incl O&P
0010	**DUST BARRIERS, TEMPORARY**									
0020	Spring loaded telescoping pole & head, to 12', erect and dismantle	1 Clab	240	.033	Ea.		1.22		1.22	1.88
0025	Cost per day (based upon 250 days)				Day	.24			.24	.26
0030	To 21', erect and dismantle	1 Clab	240	.033	Ea.		1.22		1.22	1.88
0035	Cost per day (based upon 250 days)				Day	.38			.38	.42
0040	Accessories, caution tape reel, erect and dismantle	1 Clab	480	.017	Ea.		.61		.61	.94
0045	Cost per day (based upon 250 days)				Day	.36			.36	.40
0060	Foam rail and connector, erect and dismantle	1 Clab	240	.033	Ea.		1.22		1.22	1.88
0065	Cost per day (based upon 250 days)				Day	.10			.10	.11
0070	Caution tape	1 Clab	384	.021	C.L.F.	2.70	.76		3.46	4.15
0080	Zipper, standard duty		60	.133	Ea.	15.95	4.89		20.84	25
0090	Heavy duty		48	.167	"	18.90	6.10		25	30.50
0100	Polyethylene sheet, 4 mil		37	.216	Sq.	2.91	7.90		10.81	15.45
0110	6 mil		37	.216	"	3.98	7.90		11.88	16.65
1000	Dust partition, 6 mil polyethylene, 1" x 3" frame	2 Carp	2000	.008	S.F.	.29	.37		.66	.89
1080	2" x 4" frame	"	2000	.008	"	.32	.37		.69	.92

01 56 23 – Temporary Barricades

01 56 23.10 Barricades

		Crew	Daily Output	Labor-Hours	Unit	Material	Labor	Equipment	Total	Total Incl O&P
0010	**BARRICADES**									
0020	5' high, 3 rail @ 2" x 8", fixed	2 Carp	20	.800	L.F.	5.75	36.50		42.25	63
0150	Movable		30	.533		4.77	24.50		29.27	43
1000	Guardrail, wooden, 3' high, 1" x 6", on 2" x 4" posts		200	.080		1.29	3.67		4.96	7.05
1100	2" x 6", on 4" x 4" posts		165	.097		2.35	4.45		6.80	9.45
1200	Portable metal with base pads, buy					15.55			15.55	17.10
1250	Typical installation, assume 10 reuses	2 Carp	600	.027		2.55	1.22		3.77	4.70
1300	Barricade tape, polyethylene, 7 mil, 3" wide x 500' long roll				Ea.	25			25	27.50
3000	Detour signs, set up and remove									
3010	Reflective aluminum, MUTCD, 24" x 24", post mounted	1 Clab	20	.400	Ea.	2.35	14.65		17	25

01 56 26 – Temporary Fencing

01 56 26.50 Temporary Fencing

		Crew	Daily Output	Labor-Hours	Unit	Material	Labor	Equipment	Total	Total Incl O&P
0010	**TEMPORARY FENCING**									
0020	Chain link, 11 ga., 4' high	2 Clab	400	.040	L.F.	2.95	1.47		4.42	5.50
0100	6' high		300	.053		2.95	1.95		4.90	6.25
0200	Rented chain link, 6' high, to 1000' (up to 12 mo.)		400	.040		4.39	1.47		5.86	7.10
0250	Over 1000' (up to 12 mo.)		300	.053		4.29	1.95		6.24	7.75
0350	Plywood, painted, 2" x 4" frame, 4' high	A-4	135	.178		6.25	7.80		14.05	18.75
0400	4" x 4" frame, 8' high	"	110	.218		11.45	9.55		21	27
0500	Wire mesh on 4" x 4" posts, 4' high	2 Carp	100	.160		9.75	7.35		17.10	22
0550	8' high	"	80	.200		14.70	9.15		23.85	30.50

01 56 29 – Temporary Protective Walkways

01 56 29.50 Protection

		Crew	Daily Output	Labor-Hours	Unit	Material	Labor	Equipment	Total	Total Incl O&P
0010	**PROTECTION**									
0020	Stair tread, 2" x 12" planks, 1 use	1 Carp	75	.107	Tread	4.57	4.89		9.46	12.60
0100	Exterior plywood, 1/2" thick, 1 use		65	.123		1.79	5.65		7.44	10.65
0200	3/4" thick, 1 use		60	.133		2.64	6.10		8.74	12.35
2200	Sidewalks, 2" x 12" planks, 2 uses		350	.023	S.F.	.76	1.05		1.81	2.46
2300	Exterior plywood, 2 uses, 1/2" thick		750	.011		.30	.49		.79	1.08
2400	5/8" thick		650	.012		.37	.56		.93	1.28
2500	3/4" thick		600	.013		.44	.61		1.05	1.42

01 56 Temporary Barriers and Enclosures

01 56 32 – Temporary Security

01 56 32.50 Watchman

		Crew	Daily Output	Labor-Hours	Unit	Material	2014 Bare Costs Labor	Equipment	Total	Total Incl O&P
0010	**WATCHMAN**									
0020	Service, monthly basis, uniformed person, minimum				Hr.				25	27.50
0100	Maximum								45.50	50
0200	Person and command dog, minimum								31	34
0300	Maximum				▼				54.50	60
0500	Sentry dog, leased, with job patrol (yard dog), 1 dog				Week				290	320
0600	2 dogs				"				390	430
0800	Purchase, trained sentry dog, minimum				Ea.				1,375	1,500
0900	Maximum				"				2,725	3,000

01 58 Project Identification

01 58 13 – Temporary Project Signage

01 58 13.50 Signs

		Crew	Daily Output	Labor-Hours	Unit	Material	Labor	Equipment	Total	Total Incl O&P
0010	**SIGNS**									
0020	High intensity reflectorized, no posts, buy				S.F.	31.50			31.50	34.50

01 74 Cleaning and Waste Management

01 74 13 – Progress Cleaning

01 74 13.20 Cleaning Up

		Crew	Daily Output	Labor-Hours	Unit	Material	Labor	Equipment	Total	Total Incl O&P
0010	**CLEANING UP**									
0020	After job completion, allow, minimum				Job				.30%	.30%
0040	Maximum				"				1%	1%
0050	Cleanup of floor area, continuous, per day, during const.	A-5	24	.750	M.S.F.	.81	27.50	2.81	31.12	46.50
0100	Final by GC at end of job	"	11.50	1.565	"	.86	57.50	5.85	64.21	96
1000	Mechanical demolition, see Section 02 41 19									

01 91 Commissioning

01 91 13 – General Commissioning Requirements

01 91 13.50 Building Commissioning

		Crew	Daily Output	Labor-Hours	Unit	Material	Labor	Equipment	Total	Total Incl O&P
0010	**BUILDING COMMISSIONING**									
0100	Basic building commissioning, minimum				%				.25%	.25%
0150	Maximum								.50%	.50%
0200	Enhanced building commissioning, minimum								.50%	.50%
0250	Maximum				▼				1%	1%

01 93 Facility Maintenance

01 93 13 – Facility Maintenance Procedures

01 93 13.15 Mechanical Facilities Maintenance

		Crew	Daily Output	Labor-Hours	Unit	Material	Labor	Equipment	Total	Total Incl O&P
0010	**MECHANICAL FACILITIES MAINTENANCE**									
0100	Air conditioning system maintenance									
0130	Belt, replace	1 Stpi	15	.533	Ea.		31		31	47
0170	Fan, clean		16	.500			29.50		29.50	44
0180	Filter, remove, clean, replace		12	.667			39		39	59
0190	Flexible coupling alignment, inspect		40	.200			11.70		11.70	17.65
0200	Gas leak locate and repair	▼	4	2	▼		117		117	177

01 93 Facility Maintenance

01 93 13 – Facility Maintenance Procedures

01 93 13.15 Mechanical Facilities Maintenance

		Crew	Daily Output	Labor-Hours	Unit	Material	2014 Bare Costs Labor	Equipment	Total	Total Incl O&P
0250	Pump packing gland, remove and replace	1 Stpi	11	.727	Ea.		42.50		42.50	64.50
0270	Tighten		32	.250			14.65		14.65	22
0290	Pump disassemble and assemble		4	2			117		117	177
0300	Air pressure regulator, disassemble, clean, assemble	1 Skwk	4	2			94.50		94.50	147
0310	Repair or replace part	"	6	1.333			63		63	97.50
0320	Purging system	1 Stpi	16	.500			29.50		29.50	44
0400	Compressor, air, remove or install fan wheel	1 Skwk	20	.400			18.90		18.90	29.50
0410	Disassemble or assemble 2 cylinder, 2 stage		4	2			94.50		94.50	147
0420	4 cylinder, 4 stage		1	8			380		380	585
0430	Repair or replace part		2	4			189		189	293
0700	Demolition, for mech. demolition see Section 23 05 05.10 or 22 05 05.10									
0800	Ductwork, clean									
0810	Rectangular									
0820	6" [G]	1 Shee	187.50	.043	L.F.		2.33		2.33	3.57
0830	8" [G]		140.63	.057			3.11		3.11	4.75
0840	10" [G]		112.50	.071			3.89		3.89	5.95
0850	12" [G]		93.75	.085			4.67		4.67	7.15
0860	14" [G]		80.36	.100			5.45		5.45	8.30
0870	16" [G]		70.31	.114			6.20		6.20	9.50
0900	Round									
0910	4" [G]	1 Shee	358.10	.022	L.F.		1.22		1.22	1.87
0920	6" [G]		238.73	.034			1.83		1.83	2.80
0930	8" [G]		179.05	.045			2.44		2.44	3.73
0940	10" [G]		143.24	.056			3.06		3.06	4.67
0950	12" [G]		119.37	.067			3.67		3.67	5.60
0960	16" [G]		89.52	.089			4.89		4.89	7.45
1000	Expansion joint, not screwed, install or remove	1 Stpi	3	2.667	Ea.		156		156	236
1010	Repack	"	6	1.333	"		78		78	118
1200	Fire protection equipment									
1220	Fire hydrant, replace	Q-1	3	5.333	Ea.		276		276	415
1230	Service, lubricate, inspect, flush, clean	1 Plum	7	1.143			66		66	99.50
1240	Test	"	11	.727			42		42	63
1310	Inspect valves, pressure, nozzle,	1 Spri	4	2	System		111		111	167
1500	Instrumentation preventive maintenance, rule of thumb									
1510	Annual cost per instrument for calibration & trouble shooting									
1520	Includes flow meters, analytical transmitters, flow control									
1530	chemical feed systems, recorders and distrib. control sys.				Ea.					325
1800	Plumbing fixtures, for installation see Section 22 41 00									
1801	Plumbing fixtures									
1840	Open drain with snake	1 Plum	13	.615	Ea.		35.50		35.50	53.50
1850	Open drain with toilet auger		16	.500			29		29	43.50
1860	Grease trap, clean		2	4			230		230	350
1870	Plaster trap, clean		6	1.333			76.50		76.50	116
1881	Clean commode	1 Clab	20	.400		.37	14.65		15.02	23
1882	Clean commode seat		48	.167		.18	6.10		6.28	9.65
1883	Clean single sink		28	.286		.24	10.45		10.69	16.40
1884	Clean double sink		20	.400		.37	14.65		15.02	23
1885	Clean lavatory		28	.286		.24	10.45		10.69	16.40
1886	Clean bathtub		12	.667		.73	24.50		25.23	38.50
1887	Clean fiberglass shower		12	.667		.73	24.50		25.23	38.50
1888	Clean fiberglass tub/shower		8	1		.92	36.50		37.42	57.50
1889	Clean faucet set		32	.250		.15	9.15		9.30	14.30
1891	Clean shower head		80	.100		.02	3.67		3.69	5.65

01 93 Facility Maintenance

01 93 13 – Facility Maintenance Procedures

01 93 13.15 Mechanical Facilities Maintenance		Crew	Daily Output	Labor-Hours	Unit	Material	2014 Bare Costs Labor	Equipment	Total	Total Incl O&P
1893	Clean water heater	1 Clab	16	.500	Ea.	1.28	18.35		19.63	30
1900	Relief valve, test and adjust	1 Stpi	20	.400			23.50		23.50	35.50
1910	Clean pump, heater, or motor for whirlpool	1 Clab	16	.500		.18	18.35		18.53	28.50
1920	Clean thermal cover for whirlpool	"	16	.500		1.83	18.35		20.18	30.50
2000	Repair or replace-steam trap	1 Stpi	8	1			58.50		58.50	88.50
2020	Y-type or bell strainer		6	1.333			78		78	118
2040	Water trap or vacuum breaker, screwed joints		13	.615			36		36	54.50
2100	Steam specialties, clean									
2120	Air separator with automatic trap, 1" fittings	1 Stpi	12	.667	Ea.		39		39	59
2130	Bucket trap, 2" pipe		7	1.143			67		67	101
2150	Drip leg, 2" fitting		45	.178			10.40		10.40	15.70
2200	Thermodynamic trap, 1" fittings		50	.160			9.35		9.35	14.15
2210	Thermostatic		65	.123			7.20		7.20	10.85
2240	Screen and seat in y-type strainer, plug type.		25	.320			18.70		18.70	28.50
2242	Screen and seat in y-type strainer, flange type.		12	.667			39		39	59
2500	Valve, replace broken handwheel		24	.333			19.50		19.50	29.50
3000	Valve, overhaul, regulator, relief, flushometer, mixing									
3040	Cold water, gas	1 Stpi	5	1.600	Ea.		93.50		93.50	141
3050	Hot water, steam		3	2.667			156		156	236
3080	Globe, gate, check up to 4" cold water, gas		10	.800			47		47	70.50
3090	Hot water, steam		5	1.600			93.50		93.50	141
3100	Over 4" ID hot or cold line		1.40	5.714			335		335	505
3120	Remove and replace, gate, globe or check up to 4"	Q-5	6	2.667			140		140	212
3130	Over 4"	"	2	8			420		420	635
3150	Repack up to 4"	1 Stpi	13	.615			36		36	54.50
3160	Over 4"	"	4	2			117		117	177

Division Notes

		CREW	DAILY OUTPUT	LABOR-HOURS	UNIT	BARE COSTS MAT.	LABOR	EQUIP.	TOTAL	TOTAL INCL O&P

Estimating Tips

02 30 00 Subsurface Investigation

In preparing estimates on structures involving earthwork or foundations, all information concerning soil characteristics should be obtained. Look particularly for hazardous waste, evidence of prior dumping of debris, and previous stream beds.

02 40 00 Demolition and Structure Moving

The costs shown for selective demolition do not include rubbish handling or disposal. These items should be estimated separately using RSMeans data or other sources.

- Historic preservation often requires that the contractor remove materials from the existing structure, rehab them, and replace them. The estimator must be aware of any related measures and precautions that must be taken when doing selective demolition and cutting and patching. Requirements may include special handling and storage, as well as security.
- In addition to Subdivision 02 41 00, you can find selective demolition items in each division. Example: Roofing demolition is in Division 7.

02 40 00 Building Deconstruction

This section provides costs for the careful dismantling and recycling of most of low-rise building materials.

02 50 00 Containment of Hazardous Waste

This section addresses on-site hazardous waste disposal costs.

02 80 00 Hazardous Material Disposal/ Remediation

This subdivision includes information on hazardous waste handling, asbestos remediation, lead remediation, and mold remediation. See reference R028213-20 and R028319-60 for further guidance in using these unit price lines.

02 90 00 Monitoring Chemical Sampling, Testing Analysis

This section provides costs for on-site sampling and testing hazardous waste.

Reference Numbers

Reference numbers are shown in shaded boxes at the beginning of some major classifications. These numbers refer to related items in the Reference Section. The reference information may be an estimating procedure, an alternate pricing method, or technical information.

Note: Not all subdivisions listed here necessarily appear in this publication.

No part of this publication may be reproduced, stored in a retrieval system, or transmitted in any form or by any means without prior written permission of RSMeans, a division of Reed Construction Data.

02 21 Surveys

02 21 13 – Site Surveys

02 21 13.09 Topographical Surveys

		Crew	Daily Output	Labor-Hours	Unit	Material	2014 Bare Costs Labor	Equipment	Total	Total Incl O&P
0010	**TOPOGRAPHICAL SURVEYS**									
0020	Topographical surveying, conventional, minimum	A-7	3.30	7.273	Acre	19.20	365	21.50	405.70	610
0100	Maximum	A-8	.60	53.333	"	58.50	2,600	119	2,777.50	4,200

02 21 13.13 Boundary and Survey Markers

		Crew	Daily Output	Labor-Hours	Unit	Material	Labor	Equipment	Total	Total Incl O&P
0010	**BOUNDARY AND SURVEY MARKERS**									
0300	Lot location and lines, large quantities, minimum	A-7	2	12	Acre	34	605	35.50	674.50	1,000
0320	Average	"	1.25	19.200		54	970	57	1,081	1,600
0400	Small quantities, maximum	A-8	1	32	↓	71.50	1,575	71.50	1,718	2,550
0600	Monuments, 3' long	A-7	10	2.400	Ea.	34	121	7.15	162.15	231
0800	Property lines, perimeter, cleared land	"	1000	.024	L.F.	.04	1.21	.07	1.32	1.98
0900	Wooded land	A-8	875	.037	"	.06	1.79	.08	1.93	2.90

02 21 13.16 Aerial Surveys

		Crew	Daily Output	Labor-Hours	Unit	Material	Labor	Equipment	Total	Total Incl O&P
0010	**AERIAL SURVEYS**									
1500	Aerial surveying, including ground control, minimum fee, 10 acres				Total					4,700
1510	100 acres									9,400
1550	From existing photography, deduct									1,625
1600	2' contours, 10 acres				Acre					470
1850	100 acres									94
2000	1000 acres									90
2050	10,000 acres				↓					85

02 41 Demolition

02 41 13 – Selective Site Demolition

02 41 13.17 Demolish, Remove Pavement and Curb

			Crew	Daily Output	Labor-Hours	Unit	Material	Labor	Equipment	Total	Total Incl O&P
0010	**DEMOLISH, REMOVE PAVEMENT AND CURB**	R024119-10									
5010	Pavement removal, bituminous roads, up to 3" thick		B-38	690	.058	S.Y.		2.41	1.86	4.27	5.75
5050	4" to 6" thick		"	420	.095	"		3.96	3.05	7.01	9.40

02 41 13.33 Minor Site Demolition

			Crew	Daily Output	Labor-Hours	Unit	Material	Labor	Equipment	Total	Total Incl O&P
0010	**MINOR SITE DEMOLITION**	R024119-10									
0015	No hauling, abandon catch basin or manhole		B-6	7	3.429	Ea.		138	52	190	269
0020	Remove existing catch basin or manhole, masonry			4	6			241	91.50	332.50	470
0030	Catch basin or manhole frames and covers, stored			13	1.846			74	28	102	145
0040	Remove and reset		↓	7	3.429			138	52	190	269
0900	Hydrants, fire, remove only		B-21A	5	8			365	96	461	665
0950	Remove and reset		"	2	20	↓		915	241	1,156	1,675
2900	Pipe removal, sewer/water, no excavation, 12" diameter		B-6	175	.137	L.F.		5.50	2.09	7.59	10.75
2930	15"-18" diameter		B-12Z	150	.160			6.60	10.90	17.50	22
2960	21"-24" diameter			120	.200			8.25	13.65	21.90	27.50
3000	27"-36" diameter		↓	90	.267			11	18.20	29.20	37
3200	Steel, welded connections, 4" diameter		B-6	160	.150			6	2.28	8.28	11.70
3300	10" diameter			80	.300	↓		12.05	4.57	16.62	23.50
4000	Sidewalk removal, bituminous, 2" thick			350	.069	S.Y.		2.75	1.04	3.79	5.35
4010	2-1/2" thick			325	.074			2.96	1.12	4.08	5.80
4100	Concrete, plain, 4"			160	.150			6	2.28	8.28	11.70
4110	Plain, 5"			140	.171			6.90	2.61	9.51	13.40
4120	Plain, 6"			120	.200	↓		8	3.04	11.04	15.65

02 41 Demolition

02 41 19 – Selective Demolition

02 41 19.19 Selective Facility Services Demolition

		Crew	Daily Output	Labor-Hours	Unit	Material	2014 Bare Costs Labor	Equipment	Total	Total Incl O&P
0010	**SELECTIVE FACILITY SERVICES DEMOLITION**, Rubbish Handling R024119-10									
0020	The following are to be added to the demolition prices									
0600	Dumpster, weekly rental, 1 dump/week, 6 C.Y. capacity (2 Tons)				Week	415			415	455
0700	10 C.Y. capacity (3 Tons)					480			480	530
0725	20 C.Y. capacity (5 Tons) R024119-20					565			565	625
0800	30 C.Y. capacity (7 Tons)					730			730	800
0840	40 C.Y. capacity (10 Tons)					775			775	850
2000	Load, haul, dump and return, 50' haul, hand carried	2 Clab	24	.667	C.Y.		24.50		24.50	37.50
2005	Wheeled		37	.432			15.85		15.85	24.50
2040	51' to 100' haul, hand carried		16.50	.970			35.50		35.50	55
2045	Wheeled		25	.640			23.50		23.50	36
2080	Over 100' haul, add per 100 L.F., hand carried		35.50	.451			16.50		16.50	25.50
2085	Wheeled		54	.296			10.85		10.85	16.75
2120	In elevators, per 10 floors, add		140	.114			4.19		4.19	6.45
2130	Load, haul, dump and return, up to 50' haul, incl. up to 5 rsr stair, hand		23	.696			25.50		25.50	39.50
2135	Wheeled		35	.457			16.75		16.75	26
2140	6 – 10 riser stairs, hand carried		22	.727			26.50		26.50	41
2145	Wheeled		34	.471			17.25		17.25	26.50
2150	11 – 20 riser stairs, hand carried		20	.800			29.50		29.50	45
2155	Wheeled		31	.516			18.90		18.90	29
2160	21 – 40 riser stairs, hand carried		16	1			36.50		36.50	56.50
2165	Wheeled		24	.667			24.50		24.50	37.50
2170	100' haul, incl. 5 riser stair, hand carried		15	1.067			39		39	60.50
2175	Wheeled		23	.696			25.50		25.50	39.50
2180	6 – 10 riser stair, hand carried		14	1.143			42		42	64.50
2185	Wheeled		21	.762			28		28	43
2190	11 – 20 riser stair, hand carried		12	1.333			49		49	75.50
2195	Wheeled		18	.889			32.50		32.50	50.50
2200	21 – 40 riser stair, hand carried		8	2			73.50		73.50	113
2205	Wheeled		12	1.333			49		49	75.50
2210	Over 100' haul, add per 100 L.F., hand carried		35.50	.451			16.50		16.50	25.50
2215	Wheeled		54	.296			10.85		10.85	16.75
2220	For each additional flight of stairs, up to 5 risers, add		550	.029	Flight		1.07		1.07	1.65
2225	6 – 10 risers, add		275	.058			2.13		2.13	3.29
2230	11 – 20 risers, add		138	.116			4.25		4.25	6.55
2235	21 – 40 risers, add		69	.232			8.50		8.50	13.10
3000	Loading & trucking, including 2 mile haul, chute loaded	B-16	45	.711	C.Y.		26.50	15.40	41.90	58
3040	Hand loading truck, 50' haul	"	48	.667			25	14.40	39.40	54.50
3080	Machine loading truck	B-17	120	.267			10.55	6.50	17.05	23.50
5000	Haul, per mile, up to 8 C.Y. truck	B-34B	1165	.007			.26	.59	.85	1.04
5100	Over 8 C.Y. truck	"	1550	.005			.19	.45	.64	.79

02 41 19.20 Selective Demolition, Dump Charges

		Crew	Daily Output	Labor-Hours	Unit	Material	Labor	Equipment	Total	Total Incl O&P
0010	**SELECTIVE DEMOLITION, DUMP CHARGES** R024119-10									
0100	Building construction materials				Ton	74			74	81
0300	Rubbish only					63			63	69.50
0500	Reclamation station, usual charge					74			74	81

02 41 19.27 Selective Demolition, Torch Cutting

		Crew	Daily Output	Labor-Hours	Unit	Material	Labor	Equipment	Total	Total Incl O&P
0010	**SELECTIVE DEMOLITION, TORCH CUTTING** R024119-10									
0020	Steel, 1" thick plate	E-25	333	.024	L.F.	.76	1.28	.03	2.07	3.12
0040	1" diameter bar	"	600	.013	Ea.	.13	.71	.02	.86	1.41
1000	Oxygen lance cutting, reinforced concrete walls									
1040	12" to 16" thick walls	1 Clab	10	.800	L.F.		29.50		29.50	45

02 41 Demolition

02 41 19 – Selective Demolition

02 41 19.27 Selective Demolition, Torch Cutting		Crew	Daily Output	Labor-Hours	Unit	Material	2014 Bare Costs Labor	Equipment	Total	Total Incl O&P
1080	24" thick walls	1 Clab	6	1.333	L.F.		49		49	75.50

02 65 Underground Storage Tank Removal

02 65 10 – Underground Tank and Contaminated Soil Removal

02 65 10.30 Removal of Underground Storage Tanks

			Crew	Daily Output	Labor-Hours	Unit	Material	Labor	Equipment	Total	Total Incl O&P
0010	**REMOVAL OF UNDERGROUND STORAGE TANKS**	R026510-20									
0011	Petroleum storage tanks, non-leaking										
0100	Excavate & load onto trailer										
0110	3000 gal. to 5000 gal. tank	G	B-14	4	12	Ea.		465	91.50	556.50	815
0120	6000 gal. to 8000 gal. tank	G	B-3A	3	13.333			520	345	865	1,175
0130	9000 gal. to 12000 gal. tank	G	"	2	20	↓		780	515	1,295	1,775
0190	Known leaking tank, add					%				100%	100%
0200	Remove sludge, water and remaining product from tank bottom										
0201	of tank with vacuum truck										
0300	3000 gal. to 5000 gal. tank	G	A-13	5	1.600	Ea.		75.50	154	229.50	284
0310	6000 gal. to 8000 gal. tank	G	↓	4	2			94	193	287	355
0320	9000 gal. to 12000 gal. tank	G	↓	3	2.667	↓		125	257	382	475
0390	Dispose of sludge off-site, average					Gal.				6.25	6.80
0400	Insert inert solid CO_2 "dry ice" into tank										
0401	For cleaning/transporting tanks (1.5 lb./100 gal. cap)	G	1 Clab	500	.016	Lb.	1.08	.59		1.67	2.09
1020	Haul tank to certified salvage dump, 100 miles round trip										
1023	3000 gal. to 5000 gal. tank					Ea.				760	830
1026	6000 gal. to 8000 gal. tank									880	960
1029	9,000 gal. to 12,000 gal. tank									1,050	1,150
1100	Disposal of contaminated soil to landfill										
1110	Minimum					C.Y.				145	160
1111	Maximum					"				400	440
1120	Disposal of contaminated soil to										
1121	bituminous concrete batch plant										
1130	Minimum					C.Y.				80	88
1131	Maximum					"				115	125
2010	Decontamination of soil on site incl poly tarp on top/bottom										
2011	Soil containment berm, and chemical treatment										
2020	Minimum	G	B-11C	100	.160	C.Y.	7.80	6.85	3.65	18.30	23
2021	Maximum	G	"	100	.160	↓	10.10	6.85	3.65	20.60	25.50
2050	Disposal of decontaminated soil, minimum									135	150
2055	Maximum					↓				400	440

02 81 Transportation and Disposal of Hazardous Materials

02 81 20 – Hazardous Waste Handling

02 81 20.10 Hazardous Waste Cleanup/Pickup/Disposal

						Unit	Material	Labor	Equipment	Total	Total Incl O&P
0010	**HAZARDOUS WASTE CLEANUP/PICKUP/DISPOSAL**										
0100	For contractor rental equipment, i.e., Dozer,										
0110	Front end loader, Dump truck, etc., see 01 54 33 Reference Section										
1000	Solid pickup										
1100	55 gal. drums					Ea.				240	265
1120	Bulk material, minimum					Ton				190	210
1130	Maximum					"				595	655
1200	Transportation to disposal site										

02 81 Transportation and Disposal of Hazardous Materials

02 81 20 – Hazardous Waste Handling

02 81 20.10 Hazardous Waste Cleanup/Pickup/Disposal

		Crew	Daily Output	Labor-Hours	Unit	Material	2014 Bare Costs Labor	Equipment	Total	Total Incl O&P
1220	Truckload = 80 drums or 25 C.Y. or 18 tons									
1260	Minimum				Mile				3.95	4.45
1270	Maximum				"				7.25	7.35
3000	Liquid pickup, vacuum truck, stainless steel tank									
3100	Minimum charge, 4 hours									
3110	1 compartment, 2200 gallon				Hr.				140	155
3120	2 compartment, 5000 gallon				"				200	225
3400	Transportation in 6900 gallon bulk truck				Mile				7.95	8.75
3410	In teflon lined truck				"				10.20	11.25
5000	Heavy sludge or dry vacuumable material				Hr.				140	160
6000	Dumpsite disposal charge, minimum				Ton				140	155
6020	Maximum				"				415	455

02 82 Asbestos Remediation

02 82 13 – Asbestos Abatement

02 82 13.39 Asbestos Remediation Plans and Methods

		Crew	Daily Output	Labor-Hours	Unit	Material	Labor	Equipment	Total	Total Incl O&P
0010	**ASBESTOS REMEDIATION PLANS AND METHODS**									
0100	Building Survey-Commercial Building				Ea.				2,200	2,400
0200	Asbestos Abatement Remediation Plan				"				1,350	1,475

02 82 13.41 Asbestos Abatement Equip.

		Crew	Daily Output	Labor-Hours	Unit	Material	Labor	Equipment	Total	Total Incl O&P
0010	**ASBESTOS ABATEMENT EQUIP.** R028213-20									
0011	Equipment and supplies, buy									
0200	Air filtration device, 2000 CFM				Ea.	940			940	1,025
0250	Large volume air sampling pump, minimum					340			340	375
0260	Maximum					330			330	360
0300	Airless sprayer unit, 2 gun					4,425			4,425	4,850
0350	Light stand, 500 watt					48			48	53
0400	Personal respirators									
0410	Negative pressure, 1/2 face, dual operation, min.				Ea.	26			26	28.50
0420	Maximum					29			29	31.50
0450	P.A.P.R., full face, minimum					122			122	134
0460	Maximum					162			162	178
0470	Supplied air, full face, incl. air line, minimum					165			165	182
0480	Maximum					395			395	435
0500	Personnel sampling pump					222			222	244
1500	Power panel, 20 unit, incl. GFI					590			590	650
1600	Shower unit, including pump and filters					1,250			1,250	1,375
1700	Supplied air system (type C)					3,350			3,350	3,675
1750	Vacuum cleaner, HEPA, 16 gal., stainless steel, wet/dry					1,175			1,175	1,300
1760	55 gallon					1,225			1,225	1,350
1800	Vacuum loader, 9 – 18 ton/hr.					92,500			92,500	102,000
1900	Water atomizer unit, including 55 gal. drum					275			275	305
2000	Worker protection, whole body, foot, head cover & gloves, plastic					14.50			14.50	15.95
2500	Respirator, single use					24.50			24.50	27
2550	Cartridge for respirator					7.15			7.15	7.85
2570	Glove bag, 7 mil, 50" x 64"					16.25			16.25	17.90
2580	10 mil, 44" x 60"					5.65			5.65	6.20
2590	6 mil, 44" x 60"					5.65			5.65	6.20
3000	HEPA vacuum for work area, minimum					1,400			1,400	1,550
3050	Maximum					1,450			1,450	1,600
6000	Disposable polyethylene bags, 6 mil, 3 C.F.					1.65			1.65	1.82

02 82 Asbestos Remediation

02 82 13 – Asbestos Abatement

02 82 13.41 Asbestos Abatement Equip.

		Crew	Daily Output	Labor-Hours	Unit	Material	2014 Bare Costs Labor	Equipment	Total	Total Incl O&P
6300	Disposable fiber drums, 3 C.F.				Ea.	17.85			17.85	19.65
6400	Pressure sensitive caution labels, 3" x 5"					3.40			3.40	3.74
6450	11" x 17"					7.20			7.20	7.90
6500	Negative air machine, 1800 CFM					830			830	915

02 82 13.42 Preparation of Asbestos Containment Area

		Crew	Daily Output	Labor-Hours	Unit	Material	Labor	Equipment	Total	Total Incl O&P
0010	**PREPARATION OF ASBESTOS CONTAINMENT AREA** R028213-20									
0100	Pre-cleaning, HEPA vacuum and wet wipe, flat surfaces	A-9	12000	.005	S.F.	.02	.27		.29	.45
0200	Protect carpeted area, 2 layers 6 mil poly on 3/4" plywood	"	1000	.064		2	3.28		5.28	7.30
0300	Separation barrier, 2" x 4" @ 16", 1/2" plywood ea. side, 8' high	2 Carp	400	.040		3.25	1.83		5.08	6.40
0310	12' high		320	.050		3.25	2.29		5.54	7.10
0320	16' high		200	.080		2.25	3.67		5.92	8.15
0400	Personnel decontam. chamber, 2" x 4" @ 16", 3/4" ply ea. side		280	.057		4.25	2.62		6.87	8.70
0450	Waste decontam. chamber, 2" x 4" studs @ 16", 3/4" ply ea. side		360	.044		4.25	2.04		6.29	7.80
0500	Cover surfaces with polyethylene sheeting									
0501	Including glue and tape									
0550	Floors, each layer, 6 mil	A-9	8000	.008	S.F.	.04	.41		.45	.69
0551	4 mil		9000	.007		.03	.36		.39	.60
0560	Walls, each layer, 6 mil		6000	.011		.04	.55		.59	.90
0561	4 mil		7000	.009		.03	.47		.50	.76
0570	For heights above 12', add						20%			
0575	For heights above 20', add						30%			
0580	For fire retardant poly, add					100%				
0590	For large open areas, deduct					10%	20%			
0600	Seal floor penetrations with foam firestop to 36 sq. in.	2 Carp	200	.080	Ea.	9.25	3.67		12.92	15.85
0610	36 sq. in. to 72 sq. in.		125	.128		18.50	5.85		24.35	29.50
0615	72 sq. in. to 144 sq. in.		80	.200		37	9.15		46.15	55
0620	Wall penetrations, to 36 square inches		180	.089		9.25	4.08		13.33	16.50
0630	36 sq. in. to 72 sq. in.		100	.160		18.50	7.35		25.85	32
0640	72 sq. in. to 144 sq. in.		60	.267		37	12.25		49.25	60
0800	Caulk seams with latex	1 Carp	230	.035	L.F.	.17	1.59		1.76	2.65
0900	Set up neg. air machine, 1-2k CFM/25 M.C.F. volume	1 Asbe	4.30	1.860	Ea.		95		95	148
0950	Set up and remove portable shower unit	2 Asbe	4	4	"		205		205	320

02 82 13.43 Bulk Asbestos Removal

		Crew	Daily Output	Labor-Hours	Unit	Material	Labor	Equipment	Total	Total Incl O&P
0010	**BULK ASBESTOS REMOVAL**									
0020	Includes disposable tools and 2 suits and 1 respirator filter/day/worker									
0200	Boiler insulation	A-9	480	.133	S.F.	.67	6.85		7.52	11.40
0210	With metal lath, add				%				50%	50%
0300	Boiler breeching or flue insulation	A-9	520	.123	S.F.	.56	6.30		6.86	10.40
0310	For active boiler, add				%				100%	100%
0400	Duct or AHU insulation	A-10B	440	.073	S.F.	.33	3.73		4.06	6.15
0500	Duct vibration isolation joints, up to 24 sq. in. duct	A-9	56	1.143	Ea.	5.15	58.50		63.65	96.50
0520	25 sq. in. to 48 sq. in. duct		48	1.333		6	68.50		74.50	113
0530	49 sq. in. to 76 sq. in. duct		40	1.600		7.25	82		89.25	136
0600	Pipe insulation, air cell type, up to 4" diameter pipe		900	.071	L.F.	.32	3.64		3.96	6
0610	4" to 8" diameter pipe		800	.080		.36	4.10		4.46	6.80
0620	10" to 12" diameter pipe		700	.091		.41	4.68		5.09	7.75
0630	14" to 16" diameter pipe		550	.116		.53	5.95		6.48	9.85
0650	Over 16" diameter pipe		650	.098	S.F.	.44	5.05		5.49	8.35
0700	With glove bag up to 3" diameter pipe		200	.320	L.F.	8.95	16.40		25.35	35.50
1000	Pipe fitting insulation up to 4" diameter pipe		320	.200	Ea.	.90	10.25		11.15	16.95
1100	6" to 8" diameter pipe		304	.211		.95	10.80		11.75	17.85
1110	10" to 12" diameter pipe		192	.333		1.51	17.05		18.56	28

02 82 Asbestos Remediation

02 82 13 – Asbestos Abatement

02 82 13.43 Bulk Asbestos Removal

		Crew	Daily Output	Labor-Hours	Unit	Material	2014 Bare Costs Labor	Equipment	Total	Total Incl O&P
1120	14" to 16" diameter pipe	A-9	128	.500	Ea.	2.26	25.50		27.76	42.50
1130	Over 16" diameter pipe		176	.364	S.F.	1.64	18.60		20.24	31
1200	With glove bag, up to 8" diameter pipe		75	.853	L.F.	6.15	43.50		49.65	75
2000	Scrape foam fireproofing from flat surface		2400	.027	S.F.	.12	1.37		1.49	2.26
2100	Irregular surfaces		1200	.053		.24	2.73		2.97	4.51
3000	Remove cementitious material from flat surface		1800	.036		.16	1.82		1.98	3.01
3100	Irregular surface		1000	.064		.21	3.28		3.49	5.35
6000	Remove contaminated soil from crawl space by hand		400	.160	C.F.	.72	8.20		8.92	13.55
6100	With large production vacuum loader	A-12	700	.091	"	.41	4.68	1.10	6.19	8.95
7000	Radiator backing, not including radiator removal	A-9	1200	.053	S.F.	.24	2.73		2.97	4.51
9000	For type B (supplied air) respirator equipment, add				%				10%	10%

02 82 13.44 Demolition In Asbestos Contaminated Area

		Crew	Daily Output	Labor-Hours	Unit	Material	Labor	Equipment	Total	Total Incl O&P
0010	**DEMOLITION IN ASBESTOS CONTAMINATED AREA**									
0200	Ceiling, including suspension system, plaster and lath	A-9	2100	.030	S.F.	.14	1.56		1.70	2.58
0210	Finished plaster, leaving wire lath		585	.109		.49	5.60		6.09	9.25
0220	Suspended acoustical tile		3500	.018		.08	.94		1.02	1.55
0230	Concealed tile grid system		3000	.021		.10	1.09		1.19	1.81
0240	Metal pan grid system		1500	.043		.19	2.19		2.38	3.61
0250	Gypsum board		2500	.026		.12	1.31		1.43	2.17
0260	Lighting fixtures up to 2' x 4'		72	.889	Ea.	4.01	45.50		49.51	75.50
0400	Partitions, non load bearing									
0410	Plaster, lath, and studs	A-9	690	.093	S.F.	1.02	4.75		5.77	8.50
0450	Gypsum board and studs	"	1390	.046	"	.21	2.36		2.57	3.90
9000	For type B (supplied air) respirator equipment, add				%				10%	10%

02 82 13.45 OSHA Testing

		Crew	Daily Output	Labor-Hours	Unit	Material	Labor	Equipment	Total	Total Incl O&P
0010	**OSHA TESTING**									
0100	Certified technician, minimum				Day				200	220
0110	Maximum								300	330
0120	Industrial hygienist, minimum								250	250
0130	Maximum								400	440
0200	Asbestos sampling and PCM analysis, NIOSH 7400, minimum	1 Asbe	8	1	Ea.	2.90	51		53.90	82.50
0210	Maximum		4	2		3.20	102		105.20	163
1000	Cleaned area samples		8	1		2.75	51		53.75	82.50
1100	PCM air sample analysis, NIOSH 7400, minimum		8	1		31	51		82	114
1110	Maximum		4	2		3.35	102		105.35	163
1200	TEM air sample analysis, NIOSH 7402, minimum								80	106
1210	Maximum								360	450

02 82 13.46 Decontamination of Asbestos Containment Area

		Crew	Daily Output	Labor-Hours	Unit	Material	Labor	Equipment	Total	Total Incl O&P
0010	**DECONTAMINATION OF ASBESTOS CONTAINMENT AREA**									
0100	Spray exposed substrate with surfactant (bridging)									
0200	Flat surfaces	A-9	6000	.011	S.F.	.35	.55		.90	1.24
0250	Irregular surfaces		4000	.016	"	.30	.82		1.12	1.61
0300	Pipes, beams, and columns		2000	.032	L.F.	.55	1.64		2.19	3.16
1000	Spray encapsulate polyethylene sheeting		8000	.008	S.F.	.33	.41		.74	1
1100	Roll down polyethylene sheeting		8000	.008	"		.41		.41	.64
1500	Bag polyethylene sheeting		400	.160	Ea.	.79	8.20		8.99	13.60
2000	Fine clean exposed substrate, with nylon brush		2400	.027	S.F.		1.37		1.37	2.13
2500	Wet wipe substrate		4800	.013			.68		.68	1.06
2600	Vacuum surfaces, fine brush		6400	.010			.51		.51	.80
3000	Structural demolition									
3100	Wood stud walls	A-9	2800	.023	S.F.		1.17		1.17	1.82
3500	Window manifolds, not incl. window replacement		4200	.015			.78		.78	1.21

02 82 Asbestos Remediation

02 82 13 – Asbestos Abatement

02 82 13.46 Decontamination of Asbestos Containment Area	Crew	Daily Output	Labor-Hours	Unit	Material	2014 Bare Costs Labor	Equipment	Total	Total Incl O&P
3600 Plywood carpet protection	A-9	2000	.032	S.F.		1.64		1.64	2.55
4000 Remove custom decontamination facility	A-10A	8	3	Ea.	15.10	154		169.10	257
4100 Remove portable decontamination facility	3 Asbe	12	2	"	12.80	102		114.80	173
5000 HEPA vacuum, shampoo carpeting	A-9	4800	.013	S.F.	.07	.68		.75	1.14
9000 Final cleaning of protected surfaces	A-10A	8000	.003	"		.15		.15	.24

02 82 13.47 Asbestos Waste Pkg., Handling, and Disp.

	Crew	Daily Output	Labor-Hours	Unit	Material	Labor	Equipment	Total	Total Incl O&P
0010 **ASBESTOS WASTE PACKAGING, HANDLING, AND DISPOSAL**									
0100 Collect and bag bulk material, 3 C.F. bags, by hand	A-9	400	.160	Ea.	1.65	8.20		9.85	14.55
0200 Large production vacuum loader	A-12	880	.073		.84	3.72	.88	5.44	7.70
1000 Double bag and decontaminate	A-9	960	.067		1.65	3.41		5.06	7.10
2000 Containerize bagged material in drums, per 3 C.F. drum	"	800	.080		17.85	4.10		21.95	26
3000 Cart bags 50' to dumpster	2 Asbe	400	.040			2.05		2.05	3.18
5000 Disposal charges, not including haul, minimum				C.Y.				61	67
5020 Maximum				"				355	395
9000 For type B (supplied air) respirator equipment, add				%				10%	10%

02 82 13.48 Asbestos Encapsulation With Sealants

	Crew	Daily Output	Labor-Hours	Unit	Material	Labor	Equipment	Total	Total Incl O&P
0010 **ASBESTOS ENCAPSULATION WITH SEALANTS**									
0100 Ceilings and walls, minimum	A-9	21000	.003	S.F.	.30	.16		.46	.57
0110 Maximum		10600	.006	"	.44	.31		.75	.96
0300 Pipes to 12" diameter including minor repairs, minimum		800	.080	L.F.	.42	4.10		4.52	6.85
0310 Maximum		400	.160	"	1.10	8.20		9.30	13.95

02 85 Mold Remediation

02 85 16 – Mold Remediation Preparation and Containment

02 85 16.50 Preparation of Mold Containment Area

	Crew	Daily Output	Labor-Hours	Unit	Material	Labor	Equipment	Total	Total Incl O&P
0010 **PREPARATION OF MOLD CONTAINMENT AREA**									
0100 Pre-cleaning, HEPA vacuum and wet wipe, flat surfaces	A-9	12000	.005	S.F.	.02	.27		.29	.45
0300 Separation barrier, 2" x 4" @ 16", 1/2" plywood ea. side, 8' high	2 Carp	400	.040		3.25	1.83		5.08	6.40
0310 12' high		320	.050		3.25	2.29		5.54	7.10
0320 16' high		200	.080		2.25	3.67		5.92	8.15
0400 Personnel decontam. chamber, 2" x 4" @ 16", 3/4" ply ea. side		280	.057		4.25	2.62		6.87	8.70
0450 Waste decontam. chamber, 2" x 4" studs @ 16", 3/4" ply each side		360	.044		4.25	2.04		6.29	7.80
0500 Cover surfaces with polyethylene sheeting									
0501 Including glue and tape									
0550 Floors, each layer, 6 mil	A-9	8000	.008	S.F.	.04	.41		.45	.69
0551 4 mil		9000	.007		.03	.36		.39	.60
0560 Walls, each layer, 6 mil		6000	.011		.04	.55		.59	.90
0561 4 mil		7000	.009		.03	.47		.50	.76
0570 For heights above 12', add						20%			
0575 For heights above 20', add						30%			
0580 For fire retardant poly, add					100%				
0590 For large open areas, deduct					10%	20%			
0600 Seal floor penetrations with foam firestop to 36 sq. in.	2 Carp	200	.080	Ea.	9.25	3.67		12.92	15.85
0610 36 sq. in. to 72 sq. in.		125	.128		18.50	5.85		24.35	29.50
0615 72 sq. in. to 144 sq. in.		80	.200		37	9.15		46.15	55
0620 Wall penetrations, to 36 square inches		180	.089		9.25	4.08		13.33	16.50
0630 36 sq. in. to 72 sq. in.		100	.160		18.50	7.35		25.85	32
0640 72 sq. in. to 144 sq. in.		60	.267		37	12.25		49.25	60
0800 Caulk seams with latex caulk	1 Carp	230	.035	L.F.	.17	1.59		1.76	2.65
0900 Set up neg. air machine, 1-2k CFM/25 M.C.F. volume	1 Asbe	4.30	1.860	Ea.		95		95	148

02 85 Mold Remediation

02 85 33 – Removal and Disposal of Materials with Mold

02 85 33.50 Demolition in Mold Contaminated Area	Crew	Daily Output	Labor-Hours	Unit	Material	2014 Bare Costs Labor	Equipment	Total	Total Incl O&P
0010 **DEMOLITION IN MOLD CONTAMINATED AREA**									
0200 Ceiling, including suspension system, plaster and lath	A-9	2100	.030	S.F.	.14	1.56		1.70	2.58
0210 Finished plaster, leaving wire lath		585	.109		.49	5.60		6.09	9.25
0220 Suspended acoustical tile		3500	.018		.08	.94		1.02	1.55
0230 Concealed tile grid system		3000	.021		.10	1.09		1.19	1.81
0240 Metal pan grid system		1500	.043		.19	2.19		2.38	3.61
0250 Gypsum board		2500	.026		.12	1.31		1.43	2.17
0255 Plywood		2500	.026		.12	1.31		1.43	2.17
0260 Lighting fixtures up to 2' x 4'		72	.889	Ea.	4.01	45.50		49.51	75.50
0400 Partitions, non load bearing									
0410 Plaster, lath, and studs	A-9	690	.093	S.F.	1.02	4.75		5.77	8.50
0450 Gypsum board and studs		1390	.046		.21	2.36		2.57	3.90
0465 Carpet & pad		1390	.046		.21	2.36		2.57	3.90
0600 Pipe insulation, air cell type, up to 4" diameter pipe		900	.071	L.F.	.32	3.64		3.96	6
0610 4" to 8" diameter pipe		800	.080		.36	4.10		4.46	6.80
0620 10" to 12" diameter pipe		700	.091		.41	4.68		5.09	7.75
0630 14" to 16" diameter pipe		550	.116		.53	5.95		6.48	9.85
0650 Over 16" diameter pipe		650	.098	S.F.	.44	5.05		5.49	8.35
9000 For type B (supplied air) respirator equipment, add				%				10%	10%

Division Notes

	CREW	DAILY OUTPUT	LABOR-HOURS	UNIT	MAT.	LABOR	EQUIP.	TOTAL	TOTAL INCL O&P

Estimating Tips

General

- Carefully check all the plans and specifications. Concrete often appears on drawings other than structural drawings, including mechanical and electrical drawings for equipment pads. The cost of cutting and patching is often difficult to estimate. See Subdivision 03 81 for Concrete Cutting, Subdivision 02 41 19.16 for Cutout Demolition, Subdivision 03 05 05.10 for Concrete Demolition, and Subdivision 02 41 19.19 for Rubbish Handling (handling, loading and hauling of debris).
- Always obtain concrete prices from suppliers near the job site. A volume discount can often be negotiated, depending upon competition in the area. Remember to add for waste, particularly for slabs and footings on grade.

03 10 00 Concrete Forming and Accessories

- A primary cost for concrete construction is forming. Most jobs today are constructed with prefabricated forms. The selection of the forms best suited for the job and the total square feet of forms required for efficient concrete forming and placing are key elements in estimating concrete construction. Enough forms must be available for erection to make efficient use of the concrete placing equipment and crew.
- Concrete accessories for forming and placing depend upon the systems used. Study the plans and specifications to ensure that all special accessory requirements have been included in the cost estimate, such as anchor bolts, inserts, and hangers.
- Included within costs for forms-in-place are all necessary bracing and shoring.

03 20 00 Concrete Reinforcing

- Ascertain that the reinforcing steel supplier has included all accessories, cutting, bending, and an allowance for lapping, splicing, and waste. A good rule of thumb is 10% for lapping, splicing, and waste. Also, 10% waste should be allowed for welded wire fabric.
- The unit price items in the subdivisions for Reinforcing In Place, Glass Fiber Reinforcing, and Welded Wire Fabric include the labor to install accessories such as beam and slab bolsters, high chairs, and bar ties and tie wire. The material cost for these accessories is not included; they may be obtained from the Accessories Division.

03 30 00 Cast-In-Place Concrete

- When estimating structural concrete, pay particular attention to requirements for concrete additives, curing methods, and surface treatments. Special consideration for climate, hot or cold, must be included in your estimate. Be sure to include requirements for concrete placing equipment, and concrete finishing.
- For accurate concrete estimating, the estimator must consider each of the following major components individually: forms, reinforcing steel, ready-mix concrete, placement of the concrete, and finishing of the top surface. For faster estimating, Subdivision 03 30 53.40 for Concrete-In-Place can be used; here, various items of concrete work are presented that include the costs of all five major components (unless specifically stated otherwise).

03 40 00 Precast Concrete
03 50 00 Cast Decks and Underlayment

- The cost of hauling precast concrete structural members is often an important factor. For this reason, it is important to get a quote from the nearest supplier. It may become economically feasible to set up precasting beds on the site if the hauling costs are prohibitive.

Reference Numbers

Reference numbers are shown in shaded boxes at the beginning of some major classifications. These numbers refer to related items in the Reference Section. The reference information may be an estimating procedure, an alternate pricing method, or technical information.

Note: Not all subdivisions listed here necessarily appear in this publication.

No part of this publication may be reproduced, stored in a retrieval system, or transmitted in any form or by any means without prior written permission of RSMeans, a division of Reed Construction Data.

03 11 Concrete Forming

03 11 13 – Structural Cast-In-Place Concrete Forming

03 11 13.40 Forms In Place, Equipment Foundations

		Crew	Daily Output	Labor-Hours	Unit	Material	2014 Bare Costs Labor	Equipment	Total	Total Incl O&P
0010	**FORMS IN PLACE, EQUIPMENT FOUNDATIONS**									
0020	1 use	C-2	160	.300	SFCA	3.09	13.40		16.49	24
0050	2 use		190	.253		1.70	11.30		13	19.25
0100	3 use		200	.240		1.24	10.70		11.94	17.90
0150	4 use		205	.234		1.01	10.45		11.46	17.25

03 11 13.45 Forms In Place, Footings

		Crew	Daily Output	Labor-Hours	Unit	Material	Labor	Equipment	Total	Total Incl O&P
0010	**FORMS IN PLACE, FOOTINGS**									
0020	Continuous wall, plywood, 1 use	C-1	375	.085	SFCA	6.25	3.72		9.97	12.60
0050	2 use		440	.073		3.42	3.17		6.59	8.65
0100	3 use		470	.068		2.49	2.97		5.46	7.30
0150	4 use		485	.066		2.03	2.87		4.90	6.65
5000	Spread footings, job-built lumber, 1 use		305	.105		2	4.57		6.57	9.25
5050	2 use		371	.086		1.11	3.76		4.87	7
5100	3 use		401	.080		.80	3.48		4.28	6.25
5150	4 use		414	.077		.65	3.37		4.02	5.90

03 11 13.65 Forms In Place, Slab On Grade

		Crew	Daily Output	Labor-Hours	Unit	Material	Labor	Equipment	Total	Total Incl O&P
0010	**FORMS IN PLACE, SLAB ON GRADE**									
3000	Edge forms, wood, 4 use, on grade, to 6" high	C-1	600	.053	L.F.	.31	2.32		2.63	3.92
6000	Trench forms in floor, wood, 1 use		160	.200	SFCA	1.68	8.70		10.38	15.30
6050	2 use		175	.183		.92	7.95		8.87	13.30
6100	3 use		180	.178		.67	7.75		8.42	12.70
6150	4 use		185	.173		.55	7.55		8.10	12.20

03 15 Concrete Accessories

03 15 05 – Concrete Forming Accessories

03 15 05.75 Sleeves and Chases

		Crew	Daily Output	Labor-Hours	Unit	Material	Labor	Equipment	Total	Total Incl O&P
0010	**SLEEVES AND CHASES**									
0100	Plastic, 1 use, 12" long, 2" diameter	1 Carp	100	.080	Ea.	2.44	3.67		6.11	8.35
0150	4" diameter		90	.089		6.65	4.08		10.73	13.60
0200	6" diameter		75	.107		14.25	4.89		19.14	23.50
0250	12" diameter		60	.133		46.50	6.10		52.60	60.50
5000	Sheet metal, 2" diameter [G]		100	.080		1.36	3.67		5.03	7.15
5100	4" diameter [G]		90	.089		1.70	4.08		5.78	8.15
5150	6" diameter [G]		75	.107		1.90	4.89		6.79	9.65
5200	12" diameter [G]		60	.133		3.63	6.10		9.73	13.45
6000	Steel pipe, 2" diameter [G]		100	.080		4.65	3.67		8.32	10.75
6100	4" diameter [G]		90	.089		16.70	4.08		20.78	24.50
6150	6" diameter [G]		75	.107		36.50	4.89		41.39	47.50
6200	12" diameter [G]		60	.133		86.50	6.10		92.60	104

03 15 16 – Concrete Construction Joints

03 15 16.20 Control Joints, Saw Cut

		Crew	Daily Output	Labor-Hours	Unit	Material	Labor	Equipment	Total	Total Incl O&P
0010	**CONTROL JOINTS, SAW CUT**									
0100	Sawcut control joints in green concrete									
0120	1" depth	C-27	2000	.008	L.F.	.04	.35	.09	.48	.66
0140	1-1/2" depth		1800	.009		.06	.39	.10	.55	.75
0160	2" depth		1600	.010		.07	.44	.11	.62	.85
0180	Sawcut joint reservoir in cured concrete									
0182	3/8" wide x 3/4" deep, with single saw blade	C-27	1000	.016	L.F.	.06	.70	.18	.94	1.29
0184	1/2" wide x 1" deep, with double saw blades		900	.018		.11	.78	.19	1.08	1.49
0186	3/4" wide x 1-1/2" deep, with double saw blades		800	.020		.22	.88	.22	1.32	1.79

03 15 Concrete Accessories

03 15 16 – Concrete Construction Joints

03 15 16.20 Control Joints, Saw Cut		Crew	Daily Output	Labor-Hours	Unit	Material	2014 Bare Costs Labor	Equipment	Total	Total Incl O&P
0190	Water blast joint to wash away laitance, 2 passes	C-29	2500	.003	L.F.		.12	.03	.15	.21
0200	Air blast joint to blow out debris and air dry, 2 passes	C-28	2000	.004	↓		.18	.01	.19	.27
0300	For backer rod, see Section 07 91 23.10									
0342	For joint sealant, see Section 07 92 13.20									

03 15 19 – Cast-In Concrete Anchors

03 15 19.10 Anchor Bolts

			Crew	Daily Output	Labor-Hours	Unit	Material	Labor	Equipment	Total	Incl O&P
0010	**ANCHOR BOLTS**										
0015	Made from recycled materials										
0025	Single bolts installed in fresh concrete, no templates										
0030	Hooked w/nut and washer, 1/2" diameter, 8" long	G	1 Carp	132	.061	Ea.	1.49	2.78		4.27	5.95
0040	12" long	G		131	.061		1.66	2.80		4.46	6.15
0050	5/8" diameter, 8" long	G		129	.062		3.25	2.84		6.09	7.95
0060	12" long	G		127	.063		4	2.89		6.89	8.85
0070	3/4" diameter, 8" long	G		127	.063		4	2.89		6.89	8.85
0080	12" long	G	↓	125	.064	↓	5	2.93		7.93	10.05
0090	2-bolt pattern, including job-built 2-hole template, per set										
0100	J-type, incl. hex nut & washer, 1/2" diameter x 6" long	G	1 Carp	21	.381	Set	5.60	17.45		23.05	33
0110	12" long	G		21	.381		6.30	17.45		23.75	34
0120	18" long	G		21	.381		7.30	17.45		24.75	35
0130	3/4" diameter x 8" long	G		20	.400		10.95	18.35		29.30	40.50
0140	12" long	G		20	.400		12.95	18.35		31.30	43
0150	18" long	G		20	.400		15.95	18.35		34.30	46
0160	1" diameter x 12" long	G		19	.421		22	19.30		41.30	54.50
0170	18" long	G		19	.421		26	19.30		45.30	58.50
0180	24" long	G		19	.421		31	19.30		50.30	64
0190	36" long	G		18	.444		41.50	20.50		62	77
0200	1-1/2" diameter x 18" long	G		17	.471		65.50	21.50		87	106
0210	24" long	G		16	.500		77.50	23		100.50	121
0300	L-type, incl. hex nut & washer, 3/4" diameter x 12" long	G		20	.400		12.20	18.35		30.55	42
0310	18" long	G		20	.400		14.80	18.35		33.15	45
0320	24" long	G		20	.400		17.40	18.35		35.75	47.50
0330	30" long	G		20	.400		21.50	18.35		39.85	52
0340	36" long	G		20	.400		24	18.35		42.35	55
0350	1" diameter x 12" long	G		19	.421		18.90	19.30		38.20	51
0360	18" long	G		19	.421		23	19.30		42.30	55
0370	24" long	G		19	.421		27.50	19.30		46.80	60
0380	30" long	G		19	.421		32	19.30		51.30	65
0390	36" long	G		18	.444		36	20.50		56.50	71
0400	42" long	G		18	.444		43	20.50		63.50	79
0410	48" long	G		18	.444		48	20.50		68.50	84.50
0420	1-1/4" diameter x 18" long	G		18	.444		33	20.50		53.50	68
0430	24" long	G		18	.444		39	20.50		59.50	74
0440	30" long	G		17	.471		44.50	21.50		66	82.50
0450	36" long	G	↓	17	.471		50	21.50		71.50	88.50
0460	42" long	G	2 Carp	32	.500		56.50	23		79.50	97.50
0470	48" long	G		32	.500		64	23		87	106
0480	54" long	G		31	.516		74.50	23.50		98	119
0490	60" long	G		31	.516		82	23.50		105.50	127
0500	1-1/2" diameter x 18" long	G		33	.485		48	22		70	87
0510	24" long	G		32	.500		55.50	23		78.50	96.50
0520	30" long	G		31	.516		62	23.50		85.50	105
0530	36" long	G	↓	30	.533		71	24.50		95.50	116

03 15 Concrete Accessories

03 15 19 – Cast-In Concrete Anchors

03 15 19.10 Anchor Bolts		Crew	Daily Output	Labor-Hours	Unit	Material	2014 Bare Costs Labor	Equipment	Total	Total Incl O&P
0540	42" long	G 2 Carp	30	.533	Set	80.50	24.50		105	126
0550	48" long	G	29	.552		90	25.50		115.50	138
0560	54" long	G	28	.571		109	26		135	161
0570	60" long	G	28	.571		119	26		145	172
0580	1-3/4" diameter x 18" long	G	31	.516		67.50	23.50		91	111
0590	24" long	G	30	.533		78.50	24.50		103	124
0600	30" long	G	29	.552		91	25.50		116.50	139
0610	36" long	G	28	.571		103	26		129	155
0620	42" long	G	27	.593		116	27		143	169
0630	48" long	G	26	.615		127	28		155	183
0640	54" long	G	26	.615		156	28		184	216
0650	60" long	G	25	.640		169	29.50		198.50	232
0660	2" diameter x 24" long	G	27	.593		105	27		132	157
0670	30" long	G	27	.593		117	27		144	171
0680	36" long	G	26	.615		129	28		157	186
0690	42" long	G	25	.640		143	29.50		172.50	204
0700	48" long	G	24	.667		164	30.50		194.50	228
0710	54" long	G	23	.696		195	32		227	263
0720	60" long	G	23	.696		209	32		241	279
0730	66" long	G	22	.727		224	33.50		257.50	298
0740	72" long	G	21	.762		245	35		280	325
1000	4-bolt pattern, including job-built 4-hole template, per set									
1100	J-type, incl. hex nut & washer, 1/2" diameter x 6" long	G 1 Carp	19	.421	Set	8.25	19.30		27.55	39
1110	12" long	G	19	.421		9.60	19.30		28.90	40.50
1120	18" long	G	18	.444		11.60	20.50		32.10	44.50
1130	3/4" diameter x 8" long	G	17	.471		18.95	21.50		40.45	54.50
1140	12" long	G	17	.471		23	21.50		44.50	59
1150	18" long	G	17	.471		29	21.50		50.50	65.50
1160	1" diameter x 12" long	G	16	.500		41	23		64	80.50
1170	18" long	G	15	.533		49	24.50		73.50	91.50
1180	24" long	G	15	.533		59	24.50		83.50	103
1190	36" long	G	15	.533		80	24.50		104.50	126
1200	1-1/2" diameter x 18" long	G	13	.615		128	28		156	185
1210	24" long	G	12	.667		152	30.50		182.50	215
1300	L-type, incl. hex nut & washer, 3/4" diameter x 12" long	G	17	.471		21.50	21.50		43	57
1310	18" long	G	17	.471		26.50	21.50		48	63
1320	24" long	G	17	.471		32	21.50		53.50	68.50
1330	30" long	G	16	.500		39.50	23		62.50	79
1340	36" long	G	16	.500		45	23		68	85
1350	1" diameter x 12" long	G	16	.500		35	23		58	74
1360	18" long	G	15	.533		42.50	24.50		67	84.50
1370	24" long	G	15	.533		52	24.50		76.50	94.50
1380	30" long	G	15	.533		61	24.50		85.50	105
1390	36" long	G	15	.533		69	24.50		93.50	114
1400	42" long	G	14	.571		83.50	26		109.50	132
1410	48" long	G	14	.571		93	26		119	143
1420	1-1/4" diameter x 18" long	G	14	.571		63.50	26		89.50	110
1430	24" long	G	14	.571		74.50	26		100.50	123
1440	30" long	G	13	.615		86	28		114	138
1450	36" long	G	13	.615		97.50	28		125.50	151
1460	42" long	G 2 Carp	25	.640		110	29.50		139.50	167
1470	48" long	G	24	.667		125	30.50		155.50	184
1480	54" long	G	23	.696		146	32		178	210

03 15 Concrete Accessories

03 15 19 – Cast-In Concrete Anchors

03 15 19.10 Anchor Bolts

			Crew	Daily Output	Labor-Hours	Unit	Material	2014 Bare Costs Labor	Equipment	Total	Total Incl O&P
1490	60" long	G	2 Carp	23	.696	Set	161	32		193	226
1500	1-1/2" diameter x 18" long	G		25	.640		92.50	29.50		122	148
1510	24" long	G		24	.667		108	30.50		138.50	165
1520	30" long	G		23	.696		121	32		153	182
1530	36" long	G		22	.727		139	33.50		172.50	205
1540	42" long	G		22	.727		158	33.50		191.50	226
1550	48" long	G		21	.762		177	35		212	249
1560	54" long	G		20	.800		215	36.50		251.50	294
1570	60" long	G		20	.800		236	36.50		272.50	315
1580	1-3/4" diameter x 18" long	G		22	.727		132	33.50		165.50	197
1590	24" long	G		21	.762		154	35		189	223
1600	30" long	G		21	.762		179	35		214	251
1610	36" long	G		20	.800		204	36.50		240.50	281
1620	42" long	G		19	.842		228	38.50		266.50	310
1630	48" long	G		18	.889		251	41		292	340
1640	54" long	G		18	.889		310	41		351	405
1650	60" long	G		17	.941		335	43		378	435
1660	2" diameter x 24" long	G		19	.842		206	38.50		244.50	287
1670	30" long	G		18	.889		232	41		273	320
1680	36" long	G		18	.889		254	41		295	345
1690	42" long	G		17	.941		283	43		326	375
1700	48" long	G		16	1		325	46		371	430
1710	54" long	G		15	1.067		385	49		434	500
1720	60" long	G		15	1.067		415	49		464	530
1730	66" long	G		14	1.143		445	52.50		497.50	570
1740	72" long	G		14	1.143		485	52.50		537.50	615
1990	For galvanized, add					Ea.	75%				

03 15 19.45 Machinery Anchors

			Crew	Daily Output	Labor-Hours	Unit	Material	2014 Bare Costs Labor	Equipment	Total	Total Incl O&P
0010	**MACHINERY ANCHORS**, heavy duty, incl. sleeve, floating base nut,										
0020	lower stud & coupling nut, fiber plug, connecting stud, washer & nut.										
0030	For flush mounted embedment in poured concrete heavy equip. pads.										
0200	Stud & bolt, 1/2" diameter	G	E-16	40	.400	Ea.	72.50	21	3.55	97.05	121
0300	5/8" diameter	G		35	.457		80.50	24	4.06	108.56	135
0500	3/4" diameter	G		30	.533		93	28	4.74	125.74	156
0600	7/8" diameter	G		25	.640		101	33.50	5.70	140.20	176
0800	1" diameter	G		20	.800		117	41.50	7.10	165.60	210
0900	1-1/4" diameter	G		15	1.067		141	55.50	9.45	205.95	263

03 21 Reinforcement Bars

03 21 11 – Plain Steel Reinforcement Bars

03 21 11.60 Reinforcing In Place

			Crew	Daily Output	Labor-Hours	Unit	Material	2014 Bare Costs Labor	Equipment	Total	Total Incl O&P
0010	**REINFORCING IN PLACE**, 50-60 ton lots, A615 Grade 60										
0020	Includes labor, but not material cost, to install accessories										
0030	Made from recycled materials										
0502	Footings, #4 to #7	G	4 Rodm	4200	.008	Lb.	.50	.39		.89	1.16
0552	#8 to #18	G		7200	.004		.50	.22		.72	.90
0602	Slab on grade, #3 to #7	G		4200	.008		.50	.39		.89	1.16
0900	For other than 50 – 60 ton lots										
1000	Under 10 ton job, #3 to #7, add						25%	10%			
1010	#8 to #18, add						20%	10%			
1050	10 – 50 ton job, #3 to #7, add						10%				

35

03 21 Reinforcement Bars

03 21 11 – Plain Steel Reinforcement Bars

03 21 11.60 Reinforcing In Place	Crew	Daily Output	Labor-Hours	Unit	Material	2014 Bare Costs Labor	Equipment	Total	Total Incl O&P
1060 #8 to #18, add					5%				
1100 60 – 100 ton job, #3 to #7, deduct					5%				
1110 #8 to #18, deduct					10%				
1150 Over 100 ton job, #3 to #7, deduct					10%				
1160 #8 to #18, deduct					15%				

03 22 Fabric and Grid Reinforcing

03 22 11 – Plain Welded Wire Fabric Reinforcing

03 22 11.10 Plain Welded Wire Fabric

0010 PLAIN WELDED WIRE FABRIC ASTM A185									
0020 Includes labor, but not material cost, to install accessories									
0030 Made from recycled materials									
0050 Sheets									
0100 6 x 6 - W1.4 x W1.4 (10 x 10) 21 lb. per C.S.F. G	2 Rodm	35	.457	C.S.F.	14.50	23		37.50	52.50

03 22 13 – Galvanized Welded Wire Fabric Reinforcing

03 22 13.10 Galvanized Welded Wire Fabric

0010 GALVANIZED WELDED WIRE FABRIC									
0100 Add to plain welded wire pricing for galvanized welded wire				Lb.	.23			.23	.25

03 22 16 – Epoxy-Coated Welded Wire Fabric Reinforcing

03 22 16.10 Epoxy-Coated Welded Wire Fabric

0010 EPOXY-COATED WELDED WIRE FABRIC									
0100 Add to plain welded wire pricing for epoxy-coated welded wire				Lb.	.21			.21	.23

03 30 Cast-In-Place Concrete

03 30 53 – Miscellaneous Cast-In-Place Concrete

03 30 53.40 Concrete In Place

	Crew	Daily Output	Labor-Hours	Unit	Material	Labor	Equipment	Total	Total Incl O&P
0010 CONCRETE IN PLACE									
0020 Including forms (4 uses), Grade 60 rebar, concrete (Portland cement									
0050 Type I), placement and finishing unless otherwise indicated									
3540 Equipment pad (3000 psi), 3' x 3' x 6" thick	C-14H	45	1.067	Ea.	44.50	48	.74	93.24	124
3550 4' x 4' x 6" thick		30	1.600		66.50	72	1.10	139.60	185
3560 5' x 5' x 8" thick		18	2.667		119	120	1.84	240.84	320
3570 6' x 6' x 8" thick		14	3.429		160	155	2.37	317.37	415
3580 8' x 8' x 10" thick		8	6		335	271	4.14	610.14	790
3590 10' x 10' x 12" thick		5	9.600		575	435	6.60	1,016.60	1,300
3800 Footings (3000 psi), spread under 1 C.Y.	C-14C	28	4	C.Y.	164	176	1.16	341.16	455
3825 1 C.Y. to 5 C.Y.		43	2.605		198	114	.76	312.76	395
3850 Over 5 C.Y. R033053-60		75	1.493		180	65.50	.43	245.93	299
3900 Footings, strip (3000 psi), 18" x 9", unreinforced	C-14L	40	2.400		120	103	.82	223.82	291
3920 18" x 9", reinforced	C-14C	35	3.200		144	141	.93	285.93	375
3925 20" x 10", unreinforced	C-14L	45	2.133		117	91.50	.73	209.23	269
3930 20" x 10", reinforced	C-14C	40	2.800		136	123	.81	259.81	340
3935 24" x 12", unreinforced	C-14L	55	1.745		116	74.50	.59	191.09	243
3940 24" x 12", reinforced	C-14C	48	2.333		136	102	.68	238.68	310
3945 36" x 12", unreinforced	C-14L	70	1.371		112	58.50	.47	170.97	214
3950 36" x 12", reinforced	C-14C	60	1.867		130	82	.54	212.54	271
4000 Foundation mat (3000 psi), under 10 C.Y.		38.67	2.896		204	127	.84	331.84	420
4050 Over 20 C.Y.		56.40	1.986		178	87	.58	265.58	330

03 30 Cast-In-Place Concrete

03 30 53 – Miscellaneous Cast-In-Place Concrete

03 30 53.40 Concrete In Place

		Crew	Daily Output	Labor-Hours	Unit	Material	2014 Bare Costs Labor	Equipment	Total	Total Incl O&P
4650	Slab on grade (3500 psi), not including finish, 4" thick	C-14E	60.75	1.449	C.Y.	120	65.50	.55	186.05	234
4700	6" thick	"	92	.957		115	43	.36	158.36	194
4701	Thickened slab edge (3500 psi), for slab on grade poured									
4702	monolithically with slab; depth is in addition to slab thickness;									
4703	formed vertical outside edge, earthen bottom and inside slope									
4705	8" deep x 8" wide bottom, unreinforced	C-14L	2190	.044	L.F.	3.33	1.88	.01	5.22	6.55
4710	8" x 8", reinforced	C-14C	1670	.067		5.70	2.95	.02	8.67	10.80
4715	12" deep x 12" wide bottom, unreinforced	C-14L	1800	.053		6.75	2.28	.02	9.05	11
4720	12" x 12", reinforced	C-14C	1310	.086		11.05	3.76	.02	14.83	18
4725	16" deep x 16" wide bottom, unreinforced	C-14L	1440	.067		11.40	2.85	.02	14.27	16.95
4730	16" x 16", reinforced	C-14C	1120	.100		16.50	4.39	.03	20.92	25
4735	20" deep x 20" wide bottom, unreinforced	C-14L	1150	.083		17.25	3.57	.03	20.85	24.50
4740	20" x 20", reinforced	C-14C	920	.122		24	5.35	.04	29.39	34.50
4745	24" deep x 24" wide bottom, unreinforced	C-14L	930	.103		24.50	4.42	.04	28.96	34
4750	24" x 24", reinforced	C-14C	740	.151		33	6.65	.04	39.69	46.50

03 31 Structural Concrete

03 31 13 – Heavyweight Structural Concrete

03 31 13.35 Heavyweight Concrete, Ready Mix

		Crew	Daily Output	Labor-Hours	Unit	Material	Labor	Equipment	Total	Total Incl O&P
0010	**HEAVYWEIGHT CONCRETE, READY MIX,** delivered									
0012	Includes local aggregate, sand, Portland cement (Type I) and water									
0015	Excludes all additives and treatments									
0020	2000 psi				C.Y.	93			93	102
0100	2500 psi					95.50			95.50	105
0150	3000 psi					99			99	109
0200	3500 psi					101			101	111
0300	4000 psi					104			104	114
1000	For high early strength (Portland cement Type III), add					10%				
1300	For winter concrete (hot water), add					4.50			4.50	4.95
1400	For hot weather concrete (ice), add					7.65			7.65	8.40
1410	For mid-range water reducer, add					3.69			3.69	4.06
1420	For high-range water reducer/superplasticizer, add					5.70			5.70	6.30
1430	For retarder, add					3.06			3.06	3.37
1440	For non-Chloride accelerator, add					5.65			5.65	6.20
1450	For Chloride accelerator, per 1%, add					3.32			3.32	3.65
1460	For fiber reinforcing, synthetic (1 lb./C.Y.), add					6.95			6.95	7.60
1500	For Saturday delivery, add					4.93			4.93	5.40
1510	For truck holding/waiting time past 1st hour per load, add				Hr.	97			97	107
1520	For short load (less than 4 C.Y.), add per load				Ea.	77.50			77.50	85
2000	For all lightweight aggregate, add				C.Y.	45%				

03 31 13.70 Placing Concrete

		Crew	Daily Output	Labor-Hours	Unit	Material	Labor	Equipment	Total	Total Incl O&P
0010	**PLACING CONCRETE**									
0020	Includes labor and equipment to place, level (strike off) and consolidate									
1900	Footings, continuous, shallow, direct chute	C-6	120	.400	C.Y.		15.30	.55	15.85	24
1950	Pumped	C-20	150	.427			16.80	5.20	22	31
2000	With crane and bucket	C-7	90	.800			32	13.60	45.60	63.50
2100	Footings, continuous, deep, direct chute	C-6	140	.343			13.10	.47	13.57	20.50
2150	Pumped	C-20	160	.400			15.75	4.85	20.60	29.50
2200	With crane and bucket	C-7	110	.655			26	11.15	37.15	52.50
2400	Footings, spread, under 1 C.Y., direct chute	C-6	55	.873			33.50	1.20	34.70	52.50
2450	Pumped	C-20	65	.985			39	11.95	50.95	72.50

03 31 Structural Concrete

03 31 13 – Heavyweight Structural Concrete

03 31 13.70 Placing Concrete

		Crew	Daily Output	Labor-Hours	Unit	Material	2014 Bare Costs Labor	Equipment	Total	Total Incl O&P
2500	With crane and bucket	C-7	45	1.600	C.Y.		63.50	27.50	91	128
2600	Over 5 C.Y., direct chute	C-6	120	.400			15.30	.55	15.85	24
2650	Pumped	C-20	150	.427			16.80	5.20	22	31
2700	With crane and bucket	C-7	100	.720			28.50	12.25	40.75	57.50
2900	Foundation mats, over 20 C.Y., direct chute	C-6	350	.137			5.25	.19	5.44	8.20
2950	Pumped	C-20	400	.160			6.30	1.94	8.24	11.80
3000	With crane and bucket	C-7	300	.240			9.55	4.09	13.64	19.10

03 35 Concrete Finishing

03 35 13 – High-Tolerance Concrete Floor Finishing

03 35 13.30 Finishing Floors, High Tolerance

		Crew	Daily Output	Labor-Hours	Unit	Material	Labor	Equipment	Total	Total Incl O&P
0010	**FINISHING FLOORS, HIGH TOLERANCE**									
0012	Finishing requires that concrete first be placed, struck off & consolidated									
0015	Basic finishing for various unspecified flatwork									
0100	Bull float only	C-10	4000	.006	S.F.		.25		.25	.37
0125	Bull float & manual float		2000	.012			.50		.50	.75
0150	Bull float, manual float, & broom finish, w/edging & joints		1850	.013			.54		.54	.81
0200	Bull float, manual float & manual steel trowel		1265	.019			.79		.79	1.18
0210	For specified Random Access Floors in ACI Classes 1, 2, 3 and 4 to achieve									
0215	Composite Overall Floor Flatness and Levelness values up to FF35/FL25									
0250	Bull float, machine float & machine trowel (walk-behind)	C-10C	1715	.014	S.F.		.58	.03	.61	.90
0300	Power screed, bull float, machine float & trowel (walk-behind)	C-10D	2400	.010			.42	.05	.47	.67
0350	Power screed, bull float, machine float & trowel (ride-on)	C-10E	4000	.006			.25	.07	.32	.44
0352	For specified Random Access Floors in ACI Classes 5, 6, 7 and 8 to achieve									
0354	Composite Overall Floor Flatness and Levelness values up to FF50/FL50									
0356	Add for two-dimensional restraightening after power float	C-10	6000	.004	S.F.		.17		.17	.25
0358	For specified Random or Defined Access Floors in ACI Class 9 to achieve									
0360	Composite Overall Floor Flatness and Levelness values up to FF100/FL100									
0362	Add for two-dimensional restraightening after bull float & power float	C-10	3000	.008	S.F.		.33		.33	.50
0364	For specified Superflat Defined Access Floors in ACI Class 9 to achieve									
0366	Minimum Floor Flatness and Levelness values of FF100/FL100									
0368	Add for 2-dim'l restraightening after bull float, power float, power trowel	C-10	2000	.012	S.F.		.50		.50	.75

03 63 Epoxy Grouting

03 63 05 – Grouting of Dowels and Fasteners

03 63 05.10 Epoxy Only

		Crew	Daily Output	Labor-Hours	Unit	Material	Labor	Equipment	Total	Total Incl O&P
0010	**EPOXY ONLY**									
1500	Chemical anchoring, epoxy cartridge, excludes layout, drilling, fastener									
1530	For fastener 3/4" diam. x 6" embedment	2 Skwk	72	.222	Ea.	5.55	10.50		16.05	22.50
1535	1" diam. x 8" embedment		66	.242		8.30	11.45		19.75	27
1540	1-1/4" diam. x 10" embedment		60	.267		16.60	12.60		29.20	38
1545	1-3/4" diam. x 12" embedment		54	.296		27.50	14		41.50	52
1550	14" embedment		48	.333		33	15.75		48.75	61
1555	2" diam. x 12" embedment		42	.381		44.50	18		62.50	76.50
1560	18" embedment		32	.500		55.50	23.50		79	97.50

03 82 Concrete Boring

03 82 13 – Concrete Core Drilling

03 82 13.10 Core Drilling

		Crew	Daily Output	Labor-Hours	Unit	Material	2014 Bare Costs Labor	Equipment	Total	Total Incl O&P
0010	**CORE DRILLING**									
0015	Includes bit cost, layout and set-up time									
0020	Reinforced concrete slab, up to 6" thick									
0100	1" diameter core	B-89A	17	.941	Ea.	.18	39.50	6.85	46.53	68.50
0150	For each additional inch of slab thickness in same hole, add		1440	.011		.03	.47	.08	.58	.84
0200	2" diameter core		16.50	.970		.29	40.50	7.05	47.84	71
0250	For each additional inch of slab thickness in same hole, add		1080	.015		.05	.62	.11	.78	1.13
0300	3" diameter core		16	1		.39	42	7.25	49.64	73.50
0350	For each additional inch of slab thickness in same hole, add		720	.022		.07	.93	.16	1.16	1.69
0500	4" diameter core		15	1.067		.50	45	7.75	53.25	78
0550	For each additional inch of slab thickness in same hole, add		480	.033		.08	1.40	.24	1.72	2.52
0700	6" diameter core		14	1.143		.78	48	8.30	57.08	84
0750	For each additional inch of slab thickness in same hole, add		360	.044		.13	1.87	.32	2.32	3.38
0900	8" diameter core		13	1.231		1.08	51.50	8.95	61.53	91
0950	For each additional inch of slab thickness in same hole, add		288	.056		.18	2.33	.40	2.91	4.25
1100	10" diameter core		12	1.333		1.51	56	9.70	67.21	99
1150	For each additional inch of slab thickness in same hole, add		240	.067		.25	2.80	.48	3.53	5.15
1300	12" diameter core		11	1.455		1.79	61	10.55	73.34	108
1350	For each additional inch of slab thickness in same hole, add		206	.078		.30	3.26	.56	4.12	6
1500	14" diameter core		10	1.600		2.11	67	11.60	80.71	119
1550	For each additional inch of slab thickness in same hole, add		180	.089		.35	3.73	.65	4.73	6.85
1700	18" diameter core		9	1.778		2.84	74.50	12.90	90.24	132
1750	For each additional inch of slab thickness in same hole, add		144	.111		.47	4.66	.81	5.94	8.60
1754	24" diameter core		8	2		3.99	84	14.50	102.49	150
1756	For each additional inch of slab thickness in same hole, add		120	.133		.67	5.60	.97	7.24	10.45
1760	For horizontal holes, add to above						20%	20%		
1770	Prestressed hollow core plank, 8" thick									
1780	1" diameter core	B-89A	17.50	.914	Ea.	.24	38.50	6.65	45.39	67
1790	For each additional inch of plank thickness in same hole, add		3840	.004		.03	.18	.03	.24	.33
1794	2" diameter core		17.25	.928		.38	39	6.75	46.13	68
1796	For each additional inch of plank thickness in same hole, add		2880	.006		.05	.23	.04	.32	.45
1800	3" diameter core		17	.941		.52	39.50	6.85	46.87	69
1810	For each additional inch of plank thickness in same hole, add		1920	.008		.07	.35	.06	.48	.68
1820	4" diameter core		16.50	.970		.66	40.50	7.05	48.21	71.50
1830	For each additional inch of plank thickness in same hole, add		1280	.013		.08	.52	.09	.69	1
1840	6" diameter core		15.50	1.032		1.04	43.50	7.50	52.04	76.50
1850	For each additional inch of plank thickness in same hole, add		960	.017		.13	.70	.12	.95	1.35
1860	8" diameter core		15	1.067		1.44	45	7.75	54.19	79
1870	For each additional inch of plank thickness in same hole, add		768	.021		.18	.87	.15	1.20	1.72
1880	10" diameter core		14	1.143		2.01	48	8.30	58.31	85.50
1890	For each additional inch of plank thickness in same hole, add		640	.025		.25	1.05	.18	1.48	2.10
1900	12" diameter core		13.50	1.185		2.39	50	8.60	60.99	89
1910	For each additional inch of plank thickness in same hole, add		548	.029		.30	1.23	.21	1.74	2.46

03 82 16 – Concrete Drilling

03 82 16.10 Concrete Impact Drilling

		Crew	Daily Output	Labor-Hours	Unit	Material	Labor	Equipment	Total	Total Incl O&P
0010	**CONCRETE IMPACT DRILLING**									
0020	Includes bit cost, layout and set-up time, no anchors									
0050	Up to 4" deep in concrete/brick floors/walls									
0100	Holes, 1/4" diameter	1 Carp	75	.107	Ea.	.06	4.89		4.95	7.60
0150	For each additional inch of depth in same hole, add		430	.019		.02	.85		.87	1.34
0200	3/8" diameter		63	.127		.05	5.80		5.85	9.05
0250	For each additional inch of depth in same hole, add		340	.024		.01	1.08		1.09	1.67

03 82 Concrete Boring

03 82 16 – Concrete Drilling

03 82 16.10 Concrete Impact Drilling		Crew	Daily Output	Labor-Hours	Unit	Material	2014 Bare Costs Labor	Equipment	Total	Total Incl O&P
0300	1/2" diameter	1 Carp	50	.160	Ea.	.05	7.35		7.40	11.35
0350	For each additional inch of depth in same hole, add		250	.032		.01	1.47		1.48	2.27
0400	5/8" diameter		48	.167		.08	7.65		7.73	11.90
0450	For each additional inch of depth in same hole, add		240	.033		.02	1.53		1.55	2.38
0500	3/4" diameter		45	.178		.11	8.15		8.26	12.70
0550	For each additional inch of depth in same hole, add		220	.036		.03	1.67		1.70	2.60
0600	7/8" diameter		43	.186		.15	8.55		8.70	13.30
0650	For each additional inch of depth in same hole, add		210	.038		.04	1.75		1.79	2.74
0700	1" diameter		40	.200		.16	9.15		9.31	14.35
0750	For each additional inch of depth in same hole, add		190	.042		.04	1.93		1.97	3.02
0800	1-1/4" diameter		38	.211		.23	9.65		9.88	15.15
0850	For each additional inch of depth in same hole, add		180	.044		.06	2.04		2.10	3.20
0900	1-1/2" diameter		35	.229		.32	10.50		10.82	16.50
0950	For each additional inch of depth in same hole, add	▼	165	.048	▼	.08	2.22		2.30	3.52
1000	For ceiling installations, add						40%			

Estimating Tips

05 05 00 Common Work Results for Metals

- Nuts, bolts, washers, connection angles, and plates can add a significant amount to both the tonnage of a structural steel job and the estimated cost. As a rule of thumb, add 10% to the total weight to account for these accessories.
- Type 2 steel construction, commonly referred to as "simple construction," consists generally of field-bolted connections with lateral bracing supplied by other elements of the building, such as masonry walls or x-bracing. The estimator should be aware, however, that shop connections may be accomplished by welding or bolting. The method may be particular to the fabrication shop and may have an impact on the estimated cost.

05 10 00 Structural Steel

- Steel items can be obtained from two sources: a fabrication shop or a metals service center. Fabrication shops can fabricate items under more controlled conditions than can crews in the field. They are also more efficient and can produce items more economically. Metal service centers serve as a source of long mill shapes to both fabrication shops and contractors.
- Most line items in this structural steel subdivision, and most items in 05 50 00 Metal Fabrications, are indicated as being shop fabricated. The bare material cost for these shop fabricated items is the "Invoice Cost" from the shop and includes the mill base price of steel plus mill extras, transportation to the shop, shop drawings and detailing where warranted, shop fabrication and handling, sandblasting and a shop coat of primer paint, all necessary structural bolts, and delivery to the job site. The bare labor cost and bare equipment cost for these shop fabricated items is for field installation or erection.
- Line items in Subdivision 05 12 23.40 Lightweight Framing, and other items scattered in Division 5, are indicated as being field fabricated. The bare material cost for these field fabricated items is the "Invoice Cost" from the metals service center and includes the mill base price of steel plus mill extras, transportation to the metals service center, material handling, and delivery of long lengths of mill shapes to the job site. Material costs for structural bolts and welding rods should be added to the estimate. The bare labor cost and bare equipment cost for these items is for both field fabrication and field installation or erection, and include time for cutting, welding and drilling in the fabricated metal items. Drilling into concrete and fasteners to fasten field fabricated items to other work are not included and should be added to the estimate.

05 20 00 Steel Joist Framing

- In any given project the total weight of open web steel joists is determined by the loads to be supported and the design. However, economies can be realized in minimizing the amount of labor used to place the joists. This is done by maximizing the joist spacing, and therefore minimizing the number of joists required to be installed on the job. Certain spacings and locations may be required by the design, but in other cases maximizing the spacing and keeping it as uniform as possible will keep the costs down.

05 30 00 Steel Decking

- The takeoff and estimating of metal deck involves more than simply the area of the floor or roof and the type of deck specified or shown on the drawings. Many different sizes and types of openings may exist. Small openings for individual pipes or conduits may be drilled after the floor/roof is installed, but larger openings may require special deck lengths as well as reinforcing or structural support. The estimator should determine who will be supplying this reinforcing. Additionally, some deck terminations are part of the deck package, such as screed angles and pour stops, and others will be part of the steel contract, such as angles attached to structural members and cast-in-place angles and plates. The estimator must ensure that all pieces are accounted for in the complete estimate.

05 50 00 Metal Fabrications

- The most economical steel stairs are those that use common materials, standard details, and most importantly, a uniform and relatively simple method of field assembly. Commonly available A36 channels and plates are very good choices for the main stringers of the stairs, as are angles and tees for the carrier members. Risers and treads are usually made by specialty shops, and it is most economical to use a typical detail in as many places as possible. The stairs should be pre-assembled and shipped directly to the site. The field connections should be simple and straightforward to be accomplished efficiently, and with minimum equipment and labor.

Reference Numbers

Reference numbers are shown in shaded boxes at the beginning of some major classifications. These numbers refer to related items in the Reference Section. The reference information may be an estimating procedure, an alternate pricing method, or technical information. *Note:* Not all subdivisions listed here necessarily appear in this publication.

No part of this publication may be reproduced, stored in a retrieval system, or transmitted in any form or by any means without prior written permission of RSMeans, a division of Reed Construction Data.

05 05 Common Work Results for Metals

05 05 19 – Post-Installed Concrete Anchors

05 05 19.10 Chemical Anchors

		Crew	Daily Output	Labor-Hours	Unit	Material	2014 Bare Costs Labor	Equipment	Total	Total Incl O&P
0010	**CHEMICAL ANCHORS**									
0020	Includes layout & drilling									
1430	Chemical anchor, w/rod & epoxy cartridge, 3/4" diam. x 9-1/2" long	B-89A	27	.593	Ea.	9.75	25	4.30	39.05	54
1435	1" diameter x 11-3/4" long		24	.667		16.70	28	4.84	49.54	67
1440	1-1/4" diameter x 14" long		21	.762		36	32	5.55	73.55	95.50
1445	1-3/4" diameter x 15" long		20	.800		64.50	33.50	5.80	103.80	129
1450	18" long		17	.941		77.50	39.50	6.85	123.85	154
1455	2" diameter x 18" long		16	1		102	42	7.25	151.25	185
1460	24" long		15	1.067		132	45	7.75	184.75	224

05 05 19.20 Expansion Anchors

			Crew	Daily Output	Labor-Hours	Unit	Material	Labor	Equipment	Total	Total Incl O&P
0010	**EXPANSION ANCHORS**										
0100	Anchors for concrete, brick or stone, no layout and drilling										
0200	Expansion shields, zinc, 1/4" diameter, 1-5/16" long, single	G	1 Carp	90	.089	Ea.	.48	4.08		4.56	6.85
0300	1-3/8" long, double	G		85	.094		.59	4.32		4.91	7.30
0400	3/8" diameter, 1-1/2" long, single	G		85	.094		.81	4.32		5.13	7.55
0500	2" long, double	G		80	.100		1.32	4.59		5.91	8.55
0600	1/2" diameter, 2-1/16" long, single	G		80	.100		1.62	4.59		6.21	8.90
0700	2-1/2" long, double	G		75	.107		2.10	4.89		6.99	9.85
0800	5/8" diameter, 2-5/8" long, single	G		75	.107		2.85	4.89		7.74	10.70
0900	2-3/4" long, double	G		70	.114		2.99	5.25		8.24	11.40
1000	3/4" diameter, 2-3/4" long, single	G		70	.114		3.25	5.25		8.50	11.70
1100	3-15/16" long, double	G		65	.123		5.80	5.65		11.45	15.10
2100	Hollow wall anchors for gypsum wall board, plaster or tile										
2300	1/8" diameter, short	G	1 Carp	160	.050	Ea.	.23	2.29		2.52	3.79
2400	Long	G		150	.053		.23	2.45		2.68	4.02
2500	3/16" diameter, short	G		150	.053		.40	2.45		2.85	4.21
2600	Long	G		140	.057		.58	2.62		3.20	4.68
2700	1/4" diameter, short	G		140	.057		.62	2.62		3.24	4.72
2800	Long	G		130	.062		.61	2.82		3.43	5
3000	Toggle bolts, bright steel, 1/8" diameter, 2" long	G		85	.094		.19	4.32		4.51	6.85
3100	4" long	G		80	.100		.25	4.59		4.84	7.40
3200	3/16" diameter, 3" long	G		80	.100		.26	4.59		4.85	7.40
3300	6" long	G		75	.107		.35	4.89		5.24	7.95
3400	1/4" diameter, 3" long	G		75	.107		.34	4.89		5.23	7.90
3500	6" long	G		70	.114		.50	5.25		5.75	8.65
3600	3/8" diameter, 3" long	G		70	.114		.82	5.25		6.07	9
3700	6" long	G		60	.133		1.24	6.10		7.34	10.80
3800	1/2" diameter, 4" long	G		60	.133		1.84	6.10		7.94	11.45
3900	6" long	G		50	.160		2.40	7.35		9.75	13.95
4000	Nailing anchors										
4100	Nylon nailing anchor, 1/4" diameter, 1" long		1 Carp	3.20	2.500	C	16.35	115		131.35	195
4200	1-1/2" long			2.80	2.857		21	131		152	226
4300	2" long			2.40	3.333		25.50	153		178.50	264
4400	Metal nailing anchor, 1/4" diameter, 1" long	G		3.20	2.500		17.60	115		132.60	196
4500	1-1/2" long	G		2.80	2.857		22.50	131		153.50	227
4600	2" long	G		2.40	3.333		28	153		181	267
5000	Screw anchors for concrete, masonry,										
5100	stone & tile, no layout or drilling included										
5700	Lag screw shields, 1/4" diameter, short	G	1 Carp	90	.089	Ea.	.37	4.08		4.45	6.70
5800	Long	G		85	.094		.41	4.32		4.73	7.10
5900	3/8" diameter, short	G		85	.094		.65	4.32		4.97	7.35
6000	Long	G		80	.100		.78	4.59		5.37	7.95

05 05 Common Work Results for Metals

05 05 19 – Post-Installed Concrete Anchors

05 05 19.20 Expansion Anchors

		Crew	Daily Output	Labor-Hours	Unit	Material	2014 Bare Costs Labor	Equipment	Total	Total Incl O&P
6100	1/2" diameter, short	G 1 Carp	80	.100	Ea.	1.05	4.59		5.64	8.25
6200	Long	G	75	.107		1.18	4.89		6.07	8.85
6300	5/8" diameter, short	G	70	.114		1.81	5.25		7.06	10.10
6400	Long	G	65	.123		2	5.65		7.65	10.90
6600	Lead, #6 & #8, 3/4" long	G	260	.031		.18	1.41		1.59	2.38
6700	#10 - #14, 1-1/2" long	G	200	.040		.29	1.83		2.12	3.15
6800	#16 & #18, 1-1/2" long	G	160	.050		.39	2.29		2.68	3.97
6900	Plastic, #6 & #8, 3/4" long	G	260	.031		.04	1.41		1.45	2.22
7000	#8 & #10, 7/8" long	G	240	.033		.04	1.53		1.57	2.40
7100	#10 & #12, 1" long	G	220	.036		.05	1.67		1.72	2.63
7200	#14 & #16, 1-1/2" long		160	.050		.07	2.29		2.36	3.62
8000	Wedge anchors, not including layout or drilling									
8050	Carbon steel, 1/4" diameter, 1-3/4" long	G 1 Carp	150	.053	Ea.	.40	2.45		2.85	4.21
8100	3-1/4" long	G	140	.057		.53	2.62		3.15	4.62
8150	3/8" diameter, 2-1/4" long	G	145	.055		.49	2.53		3.02	4.44
8200	5" long	G	140	.057		.86	2.62		3.48	4.99
8250	1/2" diameter, 2-3/4" long	G	140	.057		.97	2.62		3.59	5.10
8300	7" long	G	125	.064		1.66	2.93		4.59	6.35
8350	5/8" diameter, 3-1/2" long	G	130	.062		1.69	2.82		4.51	6.20
8400	8-1/2" long	G	115	.070		3.60	3.19		6.79	8.90
8450	3/4" diameter, 4-1/4" long	G	115	.070		2.81	3.19		6	8
8500	10" long	G	95	.084		6.40	3.86		10.26	13
8550	1" diameter, 6" long	G	100	.080		9.05	3.67		12.72	15.60
8575	9" long	G	85	.094		11.75	4.32		16.07	19.60
8600	12" long	G	75	.107		12.70	4.89		17.59	21.50
8650	1-1/4" diameter, 9" long	G	70	.114		24	5.25		29.25	34
8700	12" long	G	60	.133		30.50	6.10		36.60	43
8750	For type 303 stainless steel, add					350%				
8800	For type 316 stainless steel, add					450%				
8950	Self-drilling concrete screw, hex washer head, 3/16" diam. x 1-3/4" long	G 1 Carp	300	.027	Ea.	.20	1.22		1.42	2.11
8960	2-1/4" long	G	250	.032		.23	1.47		1.70	2.51
8970	Phillips flat head, 3/16" diam. x 1-3/4" long	G	300	.027		.20	1.22		1.42	2.11
8980	2-1/4" long	G	250	.032		.23	1.47		1.70	2.51

05 05 21 – Fastening Methods for Metal

05 05 21.15 Drilling Steel

		Crew	Daily Output	Labor-Hours	Unit	Material	Labor	Equipment	Total	Total Incl O&P
0010	**DRILLING STEEL**									
1910	Drilling & layout for steel, up to 1/4" deep, no anchor									
1920	Holes, 1/4" diameter	1 Sswk	112	.071	Ea.	.08	3.65		3.73	6.55
1925	For each additional 1/4" depth, add		336	.024		.08	1.22		1.30	2.24
1930	3/8" diameter		104	.077		.09	3.93		4.02	7.05
1935	For each additional 1/4" depth, add		312	.026		.09	1.31		1.40	2.41
1940	1/2" diameter		96	.083		.09	4.26		4.35	7.60
1945	For each additional 1/4" depth, add		288	.028		.09	1.42		1.51	2.61
1950	5/8" diameter		88	.091		.13	4.65		4.78	8.35
1955	For each additional 1/4" depth, add		264	.030		.13	1.55		1.68	2.88
1960	3/4" diameter		80	.100		.17	5.10		5.27	9.20
1965	For each additional 1/4" depth, add		240	.033		.17	1.70		1.87	3.19
1970	7/8" diameter		72	.111		.22	5.70		5.92	10.25
1975	For each additional 1/4" depth, add		216	.037		.22	1.89		2.11	3.58
1980	1" diameter		64	.125		.23	6.40		6.63	11.55
1985	For each additional 1/4" depth, add		192	.042		.23	2.13		2.36	4.01
1990	For drilling up, add					40%				

05 05 Common Work Results for Metals

05 05 21 – Fastening Methods for Metal

05 05 21.90 Welding Steel

		Crew	Daily Output	Labor-Hours	Unit	Material	2014 Bare Costs Labor	Equipment	Total	Total Incl O&P
0010	**WELDING STEEL**, Structural R050521-20									
0020	Field welding, 1/8" E6011, cost per welder, no operating engineer	E-14	8	1	Hr.	4.33	53	17.75	75.08	118
0200	With 1/2 operating engineer	E-13	8	1.500		4.33	76.50	17.75	98.58	153
0300	With 1 operating engineer	E-12	8	2	↓	4.33	100	17.75	122.08	189
0500	With no operating engineer, 2# weld rod per ton	E-14	8	1	Ton	4.33	53	17.75	75.08	118
0600	8# E6011 per ton	"	2	4		17.30	212	71	300.30	470
0800	With one operating engineer per welder, 2# E6011 per ton	E-12	8	2		4.33	100	17.75	122.08	189
0900	8# E6011 per ton	"	2	8	↓	17.30	400	71	488.30	755
1200	Continuous fillet, down welding									
1300	Single pass, 1/8" thick, 0.1#/L.F.	E-14	150	.053	L.F.	.22	2.83	.95	4	6.30
1400	3/16" thick, 0.2#/L.F.		75	.107		.43	5.65	1.89	7.97	12.55
1500	1/4" thick, 0.3#/L.F.		50	.160		.65	8.50	2.84	11.99	18.85
1610	5/16" thick, 0.4#/L.F.		38	.211		.87	11.20	3.74	15.81	25
1800	3 passes, 3/8" thick, 0.5#/L.F.		30	.267		1.08	14.15	4.73	19.96	31.50
2010	4 passes, 1/2" thick, 0.7#/L.F.		22	.364		1.52	19.30	6.45	27.27	43
2200	5 to 6 passes, 3/4" thick, 1.3#/L.F.		12	.667		2.81	35.50	11.85	50.16	78.50
2400	8 to 11 passes, 1" thick, 2.4#/L.F.	↓	6	1.333		5.20	71	23.50	99.70	157
2600	For vertical joint welding, add						20%			
2700	Overhead joint welding, add						300%			
2900	For semi-automatic welding, obstructed joints, deduct						5%			
3000	Exposed joints, deduct				↓		15%			
4000	Cleaning and welding plates, bars, or rods									
4010	to existing beams, columns, or trusses	E-14	12	.667	L.F.	1.08	35.50	11.85	48.43	76.50

05 05 23 – Metal Fastenings

05 05 23.30 Lag Screws

			Crew	Daily Output	Labor-Hours	Unit	Material	Labor	Equipment	Total	Total Incl O&P
0010	**LAG SCREWS**										
0020	Steel, 1/4" diameter, 2" long	G	1 Carp	200	.040	Ea.	.09	1.83		1.92	2.93
0100	3/8" diameter, 3" long	G		150	.053		.29	2.45		2.74	4.09
0200	1/2" diameter, 3" long	G		130	.062		.63	2.82		3.45	5.05
0300	5/8" diameter, 3" long	G	↓	120	.067	↓	1.13	3.06		4.19	5.95

05 05 23.35 Machine Screws

			Crew	Daily Output	Labor-Hours	Unit	Material	Labor		Total	Total Incl O&P
0010	**MACHINE SCREWS**										
0020	Steel, round head, #8 x 1" long	G	1 Carp	4.80	1.667	C	3.79	76.50		80.29	122
0110	#8 x 2" long	G		2.40	3.333		6.10	153		159.10	243
0200	#10 x 1" long	G		4	2		6	91.50		97.50	149
0300	#10 x 2" long	G	↓	2	4	↓	8.50	183		191.50	292

05 05 23.50 Powder Actuated Tools and Fasteners

			Crew	Daily Output	Labor-Hours	Unit	Material	Labor		Total	Total Incl O&P
0010	**POWDER ACTUATED TOOLS & FASTENERS**										
0020	Stud driver, .22 caliber, single shot					Ea.	147			147	161
0100	.27 caliber, semi automatic, strip					"	435			435	475
0300	Powder load, single shot, .22 cal, power level 2, brown					C	5.30			5.30	5.80
0400	Strip, .27 cal, power level 4, red						7.60			7.60	8.35
0600	Drive pin, .300 x 3/4" long	G	1 Carp	4.80	1.667		4.86	76.50		81.36	123
0700	.300 x 3" long with washer	G	"	4	2	↓	12.50	91.50		104	156

05 05 23.55 Rivets

			Crew	Daily Output	Labor-Hours	Unit	Material	Labor		Total	Total Incl O&P
0010	**RIVETS**										
0100	Aluminum rivet & mandrel, 1/2" grip length x 1/8" diameter	G	1 Carp	4.80	1.667	C	7.65	76.50		84.15	126
0200	3/16" diameter	G		4	2		11.40	91.50		102.90	155
0300	Aluminum rivet, steel mandrel, 1/8" diameter	G		4.80	1.667		10.25	76.50		86.75	129
0400	3/16" diameter	G		4	2		16.45	91.50		107.95	160
0500	Copper rivet, steel mandrel, 1/8" diameter	G	↓	4.80	1.667	↓	9.10	76.50		85.60	128

05 05 Common Work Results for Metals

05 05 23 – Metal Fastenings

05 05 23.55 Rivets

		Crew	Daily Output	Labor-Hours	Unit	Material	2014 Bare Costs Labor	Equipment	Total	Total Incl O&P
0800	Stainless rivet & mandrel, 1/8" diameter [G]	1 Carp	4.80	1.667	C	26	76.50		102.50	147
0900	3/16" diameter [G]		4	2		36	91.50		127.50	182
1000	Stainless rivet, steel mandrel, 1/8" diameter [G]		4.80	1.667		15.65	76.50		92.15	135
1100	3/16" diameter [G]		4	2		25.50	91.50		117	170
1200	Steel rivet and mandrel, 1/8" diameter [G]		4.80	1.667		7.40	76.50		83.90	126
1300	3/16" diameter [G]		4	2		12	91.50		103.50	155
1400	Hand riveting tool, standard				Ea.	71.50			71.50	78.50
1500	Deluxe					380			380	420
1600	Power riveting tool, standard					570			570	625
1700	Deluxe					3,600			3,600	3,950

05 05 23.70 Structural Blind Bolts

		Crew	Daily Output	Labor-Hours	Unit	Material	Labor	Equipment	Total	Total Incl O&P
0010	**STRUCTURAL BLIND BOLTS**									
0100	1/4" diameter x 1/4" grip [G]	1 Sswk	240	.033	Ea.	1.24	1.70		2.94	4.37
0150	1/2" grip [G]		216	.037		1.33	1.89		3.22	4.80
0200	3/8" diameter x 1/2" grip [G]		232	.034		1.75	1.76		3.51	5.05
0250	3/4" grip [G]		208	.038		1.84	1.97		3.81	5.50
0300	1/2" diameter x 1/2" grip [G]		224	.036		3.99	1.82		5.81	7.60
0350	3/4" grip [G]		200	.040		5.60	2.04		7.64	9.75
0400	5/8" diameter x 3/4" grip [G]		216	.037		8.25	1.89		10.14	12.45
0450	1" grip [G]		192	.042		9.50	2.13		11.63	14.20

05 05 23.90 Welding Rod

		Crew	Daily Output	Labor-Hours	Unit	Material	Labor	Equipment	Total	Total Incl O&P
0010	**WELDING ROD**									
0020	Steel, type 6011, 1/8" diam., less than 500#				Lb.	2.16			2.16	2.38
0100	500# to 2,000#					1.95			1.95	2.15
0200	2,000# to 5,000#					1.83			1.83	2.02
0300	5/32" diameter, less than 500#					2.20			2.20	2.42
0310	500# to 2,000#					1.98			1.98	2.18
0320	2,000# to 5,000#					1.86			1.86	2.05
0400	3/16" diam., less than 500#					2.46			2.46	2.71
0500	500# to 2,000#					2.22			2.22	2.44
0600	2,000# to 5,000#					2.09			2.09	2.30
0620	Steel, type 6010, 1/8" diam., less than 500#					2.22			2.22	2.44
0630	500# to 2,000#					2			2	2.20
0640	2,000# to 5,000#					1.88			1.88	2.07
0650	Steel, type 7018 Low Hydrogen, 1/8" diam., less than 500#					2.15			2.15	2.37
0660	500# to 2,000#					1.94			1.94	2.13
0670	2,000# to 5,000#					1.82			1.82	2.01
0700	Steel, type 7024 Jet Weld, 1/8" diam., less than 500#					2.43			2.43	2.67
0710	500# to 2,000#					2.19			2.19	2.41
0720	2,000# to 5,000#					2.06			2.06	2.26
1550	Aluminum, type 4043 TIG, 1/8" diam., less than 10#					5.10			5.10	5.65
1560	10# to 60#					4.61			4.61	5.05
1570	Over 60#					4.33			4.33	4.77
1600	Aluminum, type 5356 TIG, 1/8" diam., less than 10#					5.45			5.45	6
1610	10# to 60#					4.90			4.90	5.40
1620	Over 60#					4.61			4.61	5.05
1900	Cast iron, type 8 Nickel, 1/8" diam., less than 500#					22			22	24
1910	500# to 1,000#					19.75			19.75	21.50
1920	Over 1,000#					18.55			18.55	20.50
2000	Stainless steel, type 316/316L, 1/8" diam., less than 500#					7.10			7.10	7.85
2100	500# to 1000#					6.40			6.40	7.05
2220	Over 1000#					6.05			6.05	6.65

05 12 Structural Steel Framing

05 12 23 – Structural Steel for Buildings

05 12 23.40 Lightweight Framing

			Crew	Daily Output	Labor-Hours	Unit	Material	2014 Bare Costs Labor	Equipment	Total	Total Incl O&P
0010	**LIGHTWEIGHT FRAMING**										
0015	Made from recycled materials										
0400	Angle framing, field fabricated, 4" and larger	G	E-3	440	.055	Lb.	.77	2.82	.32	3.91	6.20
0450	Less than 4" angles	G		265	.091		.80	4.69	.54	6.03	9.75
0600	Channel framing, field fabricated, 8" and larger	G		500	.048		.80	2.49	.28	3.57	5.55
0650	Less than 8" channels	G		335	.072		.80	3.71	.42	4.93	7.90
1000	Continuous slotted channel framing system, shop fab, simple framing	G	2 Sswk	2400	.007		4.11	.34		4.45	5.10
1200	Complex framing	G	"	1600	.010		4.64	.51		5.15	6
1250	Plate & bar stock for reinforcing beams and trusses	G					1.46			1.46	1.60
1300	Cross bracing, rods, shop fabricated, 3/4" diameter	G	E-3	700	.034		1.59	1.78	.20	3.57	5.10
1310	7/8" diameter	G		850	.028		1.59	1.46	.17	3.22	4.51
1320	1" diameter	G		1000	.024		1.59	1.24	.14	2.97	4.10
1330	Angle, 5" x 5" x 3/8"	G		2800	.009		1.59	.44	.05	2.08	2.59
1350	Hanging lintels, shop fabricated	G		850	.028		1.59	1.46	.17	3.22	4.51
1380	Roof frames, shop fabricated, 3'-0" square, 5' span	G	E-2	4200	.013		1.59	.67	.36	2.62	3.29
1400	Tie rod, not upset, 1-1/2" to 4" diameter, with turnbuckle	G	2 Sswk	800	.020		1.72	1.02		2.74	3.69
1420	No turnbuckle	G		700	.023		1.66	1.17		2.83	3.88
1500	Upset, 1-3/4" to 4" diameter, with turnbuckle	G		800	.020		1.72	1.02		2.74	3.69
1520	No turnbuckle	G		700	.023		1.66	1.17		2.83	3.88

05 12 23.60 Pipe Support Framing

			Crew	Daily Output	Labor-Hours	Unit	Material	Labor	Equipment	Total	Total Incl O&P
0010	**PIPE SUPPORT FRAMING**										
0020	Under 10#/L.F., shop fabricated	G	E-4	3900	.008	Lb.	1.78	.42	.04	2.24	2.74
0200	10.1 to 15#/L.F.	G		4300	.007		1.75	.38	.03	2.16	2.64
0400	15.1 to 20#/L.F.	G		4800	.007		1.72	.34	.03	2.09	2.53
0600	Over 20#/L.F.	G		5400	.006		1.70	.31	.03	2.04	2.44

05 54 Metal Floor Plates

05 54 13 – Floor Plates

05 54 13.20 Checkered Plates

			Crew	Daily Output	Labor-Hours	Unit	Material	Labor	Equipment	Total	Total Incl O&P
0010	**CHECKERED PLATES**, steel, field fabricated										
0015	Made from recycled materials										
0020	1/4" & 3/8", 2000 to 5000 S.F., bolted	G	E-4	2900	.011	Lb.	.87	.57	.05	1.49	2.01
0100	Welded	G		4400	.007	"	.83	.38	.03	1.24	1.61
0300	Pit or trench cover and frame, 1/4" plate, 2' to 3' wide	G		100	.320	S.F.	11.05	16.50	1.42	28.97	42.50
0400	For galvanizing, add	G				Lb.	.29			.29	.32
0500	Platforms, 1/4" plate, no handrails included, rectangular	G	E-4	4200	.008		3.21	.39	.03	3.63	4.26
0600	Circular	G	"	2500	.013		4.01	.66	.06	4.73	5.65

Estimating Tips

06 05 00 Common Work Results for Wood, Plastics, and Composites

- Common to any wood-framed structure are the accessory connector items such as screws, nails, adhesives, hangers, connector plates, straps, angles, and hold-downs. For typical wood-framed buildings, such as residential projects, the aggregate total for these items can be significant, especially in areas where seismic loading is a concern. For floor and wall framing, the material cost is based on 10 to 25 lbs. per MBF. Hold-downs, hangers, and other connectors should be taken off by the piece.

  Included with material costs are fasteners for a normal installation. RSMeans engineers use manufacturer's recommendations, written specifications, and/or standard construction practice for size and spacing of fasteners. Prices for various fasteners are shown for informational purposes only. Adjustments should be made if unusual fastening conditions exist.

06 10 00 Carpentry

- Lumber is a traded commodity and therefore sensitive to supply and demand in the marketplace. Even in "budgetary" estimating of wood-framed projects, it is advisable to call local suppliers for the latest market pricing.

- Common quantity units for wood-framed projects are "thousand board feet" (MBF). A board foot is a volume of wood, 1" x 1' x 1', or 144 cubic inches. Board-foot quantities are generally calculated using nominal material dimensions—dressed sizes are ignored. Board foot per lineal foot of any stick of lumber can be calculated by dividing the nominal cross-sectional area by 12. As an example, 2,000 lineal feet of 2 x 12 equates to 4 MBF by dividing the nominal area, 2 x 12, by 12, which equals 2, and multiplying by 2,000 to give 4,000 board feet. This simple rule applies to all nominal dimensioned lumber.

- Waste is an issue of concern at the quantity takeoff for any area of construction. Framing lumber is sold in even foot lengths, i.e., 10', 12', 14', 16', and depending on spans, wall heights, and the grade of lumber, waste is inevitable. A rule of thumb for lumber waste is 5%–10% depending on material quality and the complexity of the framing.

- Wood in various forms and shapes is used in many projects, even where the main structural framing is steel, concrete, or masonry. Plywood as a back-up partition material and 2x boards used as blocking and cant strips around roof edges are two common examples. The estimator should ensure that the costs of all wood materials are included in the final estimate.

06 20 00 Finish Carpentry

- It is necessary to consider the grade of workmanship when estimating labor costs for erecting millwork and interior finish. In practice, there are three grades: premium, custom, and economy. The RSMeans daily output for base and case moldings is in the range of 200 to 250 L.F. per carpenter per day. This is appropriate for most average custom-grade projects. For premium projects, an adjustment to productivity of 25%–50% should be made, depending on the complexity of the job.

Reference Numbers

Reference numbers are shown in shaded boxes at the beginning of some major classifications. These numbers refer to related items in the Reference Section. The reference information may be an estimating procedure, an alternate pricing method, or technical information.

Note: Not all subdivisions listed here necessarily appear in this publication.

No part of this publication may be reproduced, stored in a retrieval system, or transmitted in any form or by any means without prior written permission of RSMeans, a division of Reed Construction Data.

06 11 Wood Framing

06 11 10 – Framing with Dimensional, Engineered or Composite Lumber

06 11 10.24 Miscellaneous Framing

		Crew	Daily Output	Labor-Hours	Unit	Material	2014 Bare Costs Labor	Equipment	Total	Total Incl O&P
0010	**MISCELLANEOUS FRAMING**									
8500	Firestops, 2" x 4"	2 Carp	.51	31.373	M.B.F.	570	1,450		2,020	2,850
8505	Pneumatic nailed		.62	25.806		575	1,175		1,750	2,450
8520	2" x 6"		.60	26.667		570	1,225		1,795	2,500
8525	Pneumatic nailed		.73	21.858		575	1,000		1,575	2,175
8540	2" x 8"		.60	26.667		645	1,225		1,870	2,575
8560	2" x 12"		.70	22.857		770	1,050		1,820	2,475
8600	Nailers, treated, wood construction, 2" x 4"		.53	30.189		735	1,375		2,110	2,925
8605	Pneumatic nailed		.64	25.157		745	1,150		1,895	2,600
8620	2" x 6"		.75	21.333		760	980		1,740	2,325
8625	Pneumatic nailed		.90	17.778		765	815		1,580	2,100

06 16 Sheathing

06 16 36 – Wood Panel Product Sheathing

06 16 36.10 Sheathing

		Crew	Daily Output	Labor-Hours	Unit	Material	Labor	Equipment	Total	Total Incl O&P
0010	**SHEATHING**									
0012	Plywood on roofs, CDX									
0050	3/8" thick	2 Carp	1525	.010	S.F.	.50	.48		.98	1.29
0055	Pneumatic nailed		1860	.009		.50	.39		.89	1.16
0200	5/8" thick		1300	.012		.74	.56		1.30	1.69
0205	Pneumatic nailed		1586	.010		.74	.46		1.20	1.53
0300	3/4" thick		1200	.013		.88	.61		1.49	1.91
0305	Pneumatic nailed		1464	.011		.88	.50		1.38	1.74
0500	Plywood on walls, with exterior CDX, 3/8" thick		1200	.013		.50	.61		1.11	1.49
0505	Pneumatic nailed		1488	.011		.50	.49		.99	1.31
0700	5/8" thick		1050	.015		.74	.70		1.44	1.90
0705	Pneumatic nailed		1302	.012		.74	.56		1.30	1.69
0800	3/4" thick		975	.016		.88	.75		1.63	2.13
0805	Pneumatic nailed		1209	.013		.88	.61		1.49	1.91

Estimating Tips

07 10 00 Dampproofing and Waterproofing

- Be sure of the job specifications before pricing this subdivision. The difference in cost between waterproofing and dampproofing can be great. Waterproofing will hold back standing water. Dampproofing prevents the transmission of water vapor. Also included in this section are vapor retarding membranes.

07 20 00 Thermal Protection

- Insulation and fireproofing products are measured by area, thickness, volume or R-value. Specifications may give only what the specific R-value should be in a certain situation. The estimator may need to choose the type of insulation to meet that R-value.

07 30 00 Steep Slope Roofing
07 40 00 Roofing and Siding Panels

- Many roofing and siding products are bought and sold by the square. One square is equal to an area that measures 100 square feet. This simple change in unit of measure could create a large error if the estimator is not observant. Accessories necessary for a complete installation must be figured into any calculations for both material and labor.

07 50 00 Membrane Roofing
07 60 00 Flashing and Sheet Metal
07 70 00 Roofing and Wall Specialties and Accessories

- The items in these subdivisions compose a roofing system. No one component completes the installation, and all must be estimated. Built-up or single-ply membrane roofing systems are made up of many products and installation trades. Wood blocking at roof perimeters or penetrations, parapet coverings, reglets, roof drains, gutters, downspouts, sheet metal flashing, skylights, smoke vents, and roof hatches all need to be considered along with the roofing material. Several different installation trades will need to work together on the roofing system. Inherent difficulties in the scheduling and coordination of various trades must be accounted for when estimating labor costs.

07 90 00 Joint Protection

- To complete the weather-tight shell, the sealants and caulkings must be estimated. Where different materials meet—at expansion joints, at flashing penetrations, and at hundreds of other locations throughout a construction project—they provide another line of defense against water penetration. Often, an entire system is based on the proper location and placement of caulking or sealants. The detailed drawings that are included as part of a set of architectural plans show typical locations for these materials. When caulking or sealants are shown at typical locations, this means the estimator must include them for all the locations where this detail is applicable. Be careful to keep different types of sealants separate, and remember to consider backer rods and primers if necessary.

Reference Numbers

Reference numbers are shown in shaded boxes at the beginning of some major classifications. These numbers refer to related items in the Reference Section. The reference information may be an estimating procedure, an alternate pricing method, or technical information.

Note: Not all subdivisions listed here necessarily appear in this publication.

No part of this publication may be reproduced, stored in a retrieval system, or transmitted in any form or by any means without prior written permission of RSMeans, a division of Reed Construction Data.

07 65 Flexible Flashing

07 65 10 – Sheet Metal Flashing

07 65 10.10 Sheet Metal Flashing and Counter Flashing

		Crew	Daily Output	Labor-Hours	Unit	Material	2014 Bare Costs Labor	Equipment	Total	Total Incl O&P
0010	**SHEET METAL FLASHING AND COUNTER FLASHING**									
0011	Including up to 4 bends									
0020	Aluminum, mill finish, .013" thick	1 Rofc	145	.055	S.F.	.79	2.16		2.95	4.55
0030	.016" thick		145	.055		.91	2.16		3.07	4.68
0060	.019" thick		145	.055		1.15	2.16		3.31	4.95
0100	.032" thick		145	.055		1.35	2.16		3.51	5.15
0200	.040" thick		145	.055		2.31	2.16		4.47	6.20
0300	.050" thick		145	.055		2.43	2.16		4.59	6.35
0325	Mill finish 5" x 7" step flashing, .016" thick		1920	.004	Ea.	.15	.16		.31	.45
0350	Mill finish 12" x 12" step flashing, .016" thick		1600	.005	"	.50	.20		.70	.88
0400	Painted finish, add				S.F.	.29			.29	.32
1000	Mastic-coated 2 sides, .005" thick	1 Rofc	330	.024		1.82	.95		2.77	3.62
1100	.016" thick		330	.024		2.01	.95		2.96	3.83
1600	Copper, 16 oz., sheets, under 1000 lb.		115	.070		6.95	2.72		9.67	12.30
1700	Over 4000 lb.		155	.052		6.95	2.02		8.97	11.10
1900	20 oz. sheets, under 1000 lb.		110	.073		9.85	2.85		12.70	15.70
2000	Over 4000 lb.		145	.055		9.35	2.16		11.51	14
2200	24 oz. sheets, under 1000 lb.		105	.076		12.10	2.98		15.08	18.45
2300	Over 4000 lb.		135	.059		11.50	2.32		13.82	16.60
2500	32 oz. sheets, under 1000 lb.		100	.080		16.15	3.13		19.28	23
2600	Over 4000 lb.		130	.062		15.35	2.41		17.76	21
5800	Lead, 2.5 lb. per S.F., up to 12" wide		135	.059		4.99	2.32		7.31	9.45
5900	Over 12" wide		135	.059		3.99	2.32		6.31	8.35
8900	Stainless steel sheets, 32 ga., .010" thick		155	.052		3	2.02		5.02	6.75
9000	28 ga., .015" thick		155	.052		3.99	2.02		6.01	7.85
9100	26 ga., .018" thick		155	.052		4.25	2.02		6.27	8.10
9200	24 ga., .025" thick		155	.052		4.75	2.02		6.77	8.70
9400	Terne coated stainless steel, .015" thick, 28 ga		155	.052		7.35	2.02		9.37	11.55
9500	.018" thick, 26 ga		155	.052		8.30	2.02		10.32	12.55
9600	Zinc and copper alloy (brass), .020" thick		155	.052		8.55	2.02		10.57	12.85
9700	.027" thick		155	.052		10.70	2.02		12.72	15.20
9800	.032" thick		155	.052		13.65	2.02		15.67	18.50
9900	.040" thick		155	.052		17.10	2.02		19.12	22

07 65 13 – Laminated Sheet Flashing

07 65 13.10 Laminated Sheet Flashing

		Crew	Daily Output	Labor-Hours	Unit	Material	Labor	Equipment	Total	Total Incl O&P
0010	**LAMINATED SHEET FLASHING**, Including up to 4 bends									
0500	Aluminum, fabric-backed 2 sides, .004" thick	1 Rofc	330	.024	S.F.	1.54	.95		2.49	3.31
0700	.005" thick		330	.024		1.82	.95		2.77	3.62
0750	Mastic-backed, self adhesive		460	.017		3.51	.68		4.19	5
0800	Mastic-coated 2 sides, .004" thick		330	.024		1.54	.95		2.49	3.31
2800	Copper, paperbacked 1 side, 2 oz.		330	.024		1.67	.95		2.62	3.46
2900	3 oz.		330	.024		1.97	.95		2.92	3.79
3100	Paperbacked 2 sides, 2 oz.		330	.024		1.72	.95		2.67	3.51
3150	3 oz.		330	.024		2.07	.95		3.02	3.90
3200	5 oz.		330	.024		2.69	.95		3.64	4.58
3250	7 oz.		330	.024		4.81	.95		5.76	6.90
3400	Mastic-backed 2 sides, copper, 2 oz.		330	.024		1.65	.95		2.60	3.44
3500	3 oz.		330	.024		2	.95		2.95	3.82
3700	5 oz.		330	.024		3	.95		3.95	4.92
3800	Fabric-backed 2 sides, copper, 2 oz.		330	.024		1.93	.95		2.88	3.74
4000	3 oz.		330	.024		2.19	.95		3.14	4.03
4100	5 oz.		330	.024		3.13	.95		4.08	5.05

07 65 Flexible Flashing

07 65 13 – Laminated Sheet Flashing

07 65 13.10 Laminated Sheet Flashing		Crew	Daily Output	Labor-Hours	Unit	Material	2014 Bare Costs Labor	Equipment	Total	Total Incl O&P
4300	Copper-clad stainless steel, .015" thick, under 500 lb.	1 Rofc	115	.070	S.F.	5.80	2.72		8.52	11.05
4400	Over 2000 lb.		155	.052		5.55	2.02		7.57	9.55
4600	.018" thick, under 500 lb.		100	.080		6.55	3.13		9.68	12.55
4700	Over 2000 lb.		145	.055		6.20	2.16		8.36	10.55
8550	Shower pan, 3 ply copper and fabric, 3 oz.		155	.052		3.13	2.02		5.15	6.90
8600	7 oz.		155	.052		3.99	2.02		6.01	7.85
9300	Stainless steel, paperbacked 2 sides, .005" thick		330	.024		3.62	.95		4.57	5.60

07 65 19 – Plastic Sheet Flashing

07 65 19.10 Plastic Sheet Flashing and Counter Flashing

		Crew	Daily Output	Labor-Hours	Unit	Material	Labor	Equipment	Total	Total Incl O&P
0010	**PLASTIC SHEET FLASHING AND COUNTER FLASHING**									
7300	Polyvinyl chloride, black, .010" thick	1 Rofc	285	.028	S.F.	.21	1.10		1.31	2.10
7400	.020" thick		285	.028		.28	1.10		1.38	2.18
7600	.030" thick		285	.028		.35	1.10		1.45	2.26
7700	.056" thick		285	.028		.85	1.10		1.95	2.81
7900	Black or white for exposed roofs, .060" thick		285	.028		1.80	1.10		2.90	3.85
8060	PVC tape, 5" x 45 mils, for joint covers, 100 L.F./roll				Ea.	170			170	187
8850	Polyvinyl chloride, .030" thick	1 Rofc	160	.050	S.F.	1.30	1.96		3.26	4.76

07 72 Roof Accessories

07 72 53 – Snow Guards

07 72 53.10 Snow Guard Options

		Crew	Daily Output	Labor-Hours	Unit	Material	Labor	Equipment	Total	Total Incl O&P
0010	**SNOW GUARD OPTIONS**									
0100	Slate & asphalt shingle roofs, fastened with nails	1 Rofc	160	.050	Ea.	12.55	1.96		14.51	17.15
0200	Standing seam metal roofs, fastened with set screws		48	.167		16.95	6.55		23.50	30
0300	Surface mount for metal roofs, fastened with solder		48	.167		6.05	6.55		12.60	17.75
0400	Double rail pipe type, including pipe		130	.062	L.F.	32	2.41		34.41	39

07 72 73 – Pitch Pockets

07 72 73.10 Pitch Pockets, Variable Sizes

		Crew	Daily Output	Labor-Hours	Unit	Material	Labor	Equipment	Total	Total Incl O&P
0010	**PITCH POCKETS, VARIABLE SIZES**									
0100	Adjustable, 4" to 7", welded corners, 4" deep	1 Rofc	48	.167	Ea.	15.75	6.55		22.30	28.50
0200	Side extenders, 6"	"	240	.033	"	2.15	1.30		3.45	4.59

07 76 Roof Pavers

07 76 16 – Roof Decking Pavers

07 76 16.10 Roof Pavers and Supports

		Crew	Daily Output	Labor-Hours	Unit	Material	Labor	Equipment	Total	Total Incl O&P
0010	**ROOF PAVERS AND SUPPORTS**									
1000	Roof decking pavers, concrete blocks, 2" thick, natural	1 Clab	115	.070	S.F.	3.37	2.55		5.92	7.65
1100	Colors		115	.070	"	3.73	2.55		6.28	8.05
1200	Support pedestal, bottom cap		960	.008	Ea.	2.99	.31		3.30	3.76
1300	Top cap		960	.008		4.79	.31		5.10	5.70
1400	Leveling shims, 1/16"		1920	.004		1.19	.15		1.34	1.55
1500	1/8"		1920	.004		1.19	.15		1.34	1.55
1600	Buffer pad		960	.008		2.49	.31		2.80	3.21
1700	PVC legs (4" SDR 35)		2880	.003	Inch	.12	.10		.22	.29
2000	Alternate pricing method, system in place		101	.079	S.F.	7	2.90		9.90	12.20

07 84 Firestopping

07 84 13 – Penetration Firestopping

07 84 13.10 Firestopping		Crew	Daily Output	Labor-Hours	Unit	Material	2014 Bare Costs Labor	Equipment	Total	Total Incl O&P
0010	**FIRESTOPPING**									
0100	Metallic piping, non insulated									
0110	Through walls, 2" diameter	1 Carp	16	.500	Ea.	15.90	23		38.90	53
0120	4" diameter		14	.571		24.50	26		50.50	67
0130	6" diameter		12	.667		32.50	30.50		63	83
0140	12" diameter		10	.800		57.50	36.50		94	120
0150	Through floors, 2" diameter		32	.250		9.65	11.45		21.10	28.50
0160	4" diameter		28	.286		13.85	13.10		26.95	35.50
0170	6" diameter		24	.333		18.20	15.30		33.50	43.50
0180	12" diameter		20	.400		31	18.35		49.35	62.50
0190	Metallic piping, insulated									
0200	Through walls, 2" diameter	1 Carp	16	.500	Ea.	22.50	23		45.50	60.50
0210	4" diameter		14	.571		31	26		57	74.50
0220	6" diameter		12	.667		39	30.50		69.50	90
0230	12" diameter		10	.800		64	36.50		100.50	127
0240	Through floors, 2" diameter		32	.250		16.30	11.45		27.75	35.50
0250	4" diameter		28	.286		20.50	13.10		33.60	42.50
0260	6" diameter		24	.333		25	15.30		40.30	51
0270	12" diameter		20	.400		31	18.35		49.35	62.50
0280	Non metallic piping, non insulated									
0290	Through walls, 2" diameter	1 Carp	12	.667	Ea.	65.50	30.50		96	119
0300	4" diameter		10	.800		82	36.50		118.50	147
0310	6" diameter		8	1		114	46		160	197
0330	Through floors, 2" diameter		16	.500		51.50	23		74.50	92
0340	4" diameter		6	1.333		63.50	61		124.50	165
0350	6" diameter		6	1.333		76.50	61		137.50	179
0370	Ductwork, insulated & non insulated, round									
0380	Through walls, 6" diameter	1 Carp	12	.667	Ea.	33.50	30.50		64	83.50
0390	12" diameter		10	.800		66	36.50		102.50	129
0400	18" diameter		8	1		108	46		154	190
0410	Through floors, 6" diameter		16	.500		18.20	23		41.20	55.50
0420	12" diameter		14	.571		33.50	26		59.50	77
0430	18" diameter		12	.667		57.50	30.50		88	111
0440	Ductwork, insulated & non insulated, rectangular									
0450	With stiffener/closure angle, through walls, 6" x 12"	1 Carp	8	1	Ea.	27.50	46		73.50	102
0460	12" x 24"		6	1.333		36.50	61		97.50	135
0470	24" x 48"		4	2		105	91.50		196.50	257
0480	With stiffener/closure angle, through floors, 6" x 12"		10	.800		14.95	36.50		51.45	73
0490	12" x 24"		8	1		27	46		73	101
0500	24" x 48"		6	1.333		52.50	61		113.50	153
0510	Multi trade openings									
0520	Through walls, 6" x 12"	1 Carp	2	4	Ea.	57.50	183		240.50	345
0530	12" x 24"	"	1	8		234	365		599	820
0540	24" x 48"	2 Carp	1	16		935	735		1,670	2,150
0550	48" x 96"	"	.75	21.333		3,750	980		4,730	5,625
0560	Through floors, 6" x 12"	1 Carp	2	4		57.50	183		240.50	345
0570	12" x 24"	"	1	8		234	365		599	820
0580	24" x 48"	2 Carp	.75	21.333		935	980		1,915	2,525
0590	48" x 96"	"	.50	32		3,750	1,475		5,225	6,400
0600	Structural penetrations, through walls									
0610	Steel beams, W8 x 10	1 Carp	8	1	Ea.	36.50	46		82.50	111
0620	W12 x 14		6	1.333		57.50	61		118.50	158
0630	W21 x 44		5	1.600		116	73.50		189.50	241

07 84 Firestopping

07 84 13 – Penetration Firestopping

07 84 13.10 Firestopping

		Crew	Daily Output	Labor-Hours	Unit	Material	2014 Bare Costs Labor	Equipment	Total	Total Incl O&P
0640	W36 x 135	1 Carp	3	2.667	Ea.	282	122		404	500
0650	Bar joists, 18" deep		6	1.333		53.50	61		114.50	153
0660	24" deep		6	1.333		66	61		127	167
0670	36" deep		5	1.600		99.50	73.50		173	222
0680	48" deep	↓	4	2	↓	116	91.50		207.50	270
0690	Construction joints, floor slab at exterior wall									
0700	Precast, brick, block or drywall exterior									
0710	2" wide joint	1 Carp	125	.064	L.F.	8.30	2.93		11.23	13.65
0720	4" wide joint	"	75	.107	"	16.55	4.89		21.44	26
0730	Metal panel, glass or curtain wall exterior									
0740	2" wide joint	1 Carp	40	.200	L.F.	19.60	9.15		28.75	35.50
0750	4" wide joint	"	25	.320	"	27	14.65		41.65	52
0760	Floor slab to drywall partition									
0770	Flat joint	1 Carp	100	.080	L.F.	8.15	3.67		11.82	14.60
0780	Fluted joint		50	.160		16.55	7.35		23.90	29.50
0790	Etched fluted joint	↓	75	.107	↓	10.75	4.89		15.64	19.40
0800	Floor slab to concrete/masonry partition									
0810	Flat joint	1 Carp	75	.107	L.F.	18.20	4.89		23.09	27.50
0820	Fluted joint	"	50	.160	"	21.50	7.35		28.85	35
0830	Concrete/CMU wall joints									
0840	1" wide	1 Carp	100	.080	L.F.	9.95	3.67		13.62	16.55
0850	2" wide		75	.107		18.20	4.89		23.09	27.50
0860	4" wide	↓	50	.160	↓	34.50	7.35		41.85	49.50
0870	Concrete/CMU floor joints									
0880	1" wide	1 Carp	200	.040	L.F.	4.99	1.83		6.82	8.35
0890	2" wide		150	.053		9.10	2.45		11.55	13.75
0900	4" wide	↓	100	.080	↓	17.40	3.67		21.07	25

07 91 Preformed Joint Seals

07 91 13 – Compression Seals

07 91 13.10 Compression Seals

		Crew	Daily Output	Labor-Hours	Unit	Material	2014 Bare Costs Labor	Equipment	Total	Total Incl O&P
0010	**COMPRESSION SEALS**									
4900	O-ring type cord, 1/4"	1 Bric	472	.017	L.F.	.48	.77		1.25	1.71
4910	1/2"		440	.018		1.79	.83		2.62	3.23
4920	3/4"		424	.019		3.62	.86		4.48	5.30
4930	1"		408	.020		6.95	.89		7.84	9
4940	1-1/4"		384	.021		13.75	.95		14.70	16.55
4950	1-1/2"		368	.022		17.40	.99		18.39	20.50
4960	1-3/4"		352	.023		36.50	1.04		37.54	41.50
4970	2"	↓	344	.023	↓	64.50	1.06		65.56	72

07 91 16 – Joint Gaskets

07 91 16.10 Joint Gaskets

		Crew	Daily Output	Labor-Hours	Unit	Material	2014 Bare Costs Labor	Equipment	Total	Total Incl O&P
0010	**JOINT GASKETS**									
4400	Joint gaskets, neoprene, closed cell w/adh, 1/8" x 3/8"	1 Bric	240	.033	L.F.	.28	1.52		1.80	2.63
4500	1/4" x 3/4"		215	.037		.59	1.70		2.29	3.24
4700	1/2" x 1"		200	.040		1.30	1.82		3.12	4.21
4800	3/4" x 1-1/2"	↓	165	.048	↓	1.45	2.21		3.66	4.97

07 91 Preformed Joint Seals

07 91 23 – Backer Rods

07 91 23.10 Backer Rods

		Crew	Daily Output	Labor-Hours	Unit	Material	2014 Bare Costs Labor	Equipment	Total	Total Incl O&P
0010	**BACKER RODS**									
0030	Backer rod, polyethylene, 1/4" diameter	1 Bric	4.60	1.739	C.L.F.	2.43	79.50		81.93	124
0050	1/2" diameter		4.60	1.739		3.66	79.50		83.16	125
0070	3/4" diameter		4.60	1.739		5.55	79.50		85.05	127
0090	1" diameter	▼	4.60	1.739	▼	8.80	79.50		88.30	131

07 91 26 – Joint Fillers

07 91 26.10 Joint Fillers

		Crew	Daily Output	Labor-Hours	Unit	Material	Labor	Equipment	Total	Incl O&P
0010	**JOINT FILLERS**									
4360	Butyl rubber filler, 1/4" x 1/4"	1 Bric	290	.028	L.F.	.23	1.26		1.49	2.17
4365	1/2" x 1/2"		250	.032		.92	1.46		2.38	3.23
4370	1/2" x 3/4"		210	.038		1.38	1.74		3.12	4.17
4375	3/4" x 3/4"		230	.035		2.08	1.59		3.67	4.70
4380	1" x 1"	▼	180	.044	▼	2.77	2.03		4.80	6.15
4390	For coloring, add					12%				
4980	Polyethylene joint backing, 1/4" x 2"	1 Bric	2.08	3.846	C.L.F.	12	175		187	280
4990	1/4" x 6"		1.28	6.250	"	28	285		313	465
5600	Silicone, room temp vulcanizing foam seal, 1/4" x 1/2"		1312	.006	L.F.	.33	.28		.61	.79
5610	1/2" x 1/2"		656	.012		.67	.56		1.23	1.59
5620	1/2" x 3/4"		442	.018		1	.83		1.83	2.36
5630	3/4" x 3/4"		328	.024		1.50	1.11		2.61	3.35
5640	1/8" x 1"		1312	.006		.33	.28		.61	.79
5650	1/8" x 3"		442	.018		1	.83		1.83	2.36
5670	1/4" x 3"		295	.027		2.01	1.24		3.25	4.09
5680	1/4" x 6"		148	.054		4.01	2.46		6.47	8.15
5690	1/2" x 6"		82	.098		8	4.45		12.45	15.60
5700	1/2" x 9"		52.50	.152		12.05	6.95		19	24
5710	1/2" x 12"	▼	33	.242	▼	16.05	11.05		27.10	34.50

07 92 Joint Sealants

07 92 13 – Elastomeric Joint Sealants

07 92 13.20 Caulking and Sealant Options

		Crew	Daily Output	Labor-Hours	Unit	Material	Labor	Equipment	Total	Incl O&P
0010	**CAULKING AND SEALANT OPTIONS**									
0050	Latex acrylic based, bulk				Gal.	29.50			29.50	32.50
0055	Bulk in place 1/4" x 1/4" bead	1 Bric	300	.027	L.F.	.09	1.22		1.31	1.95
0060	1/4" x 3/8"		294	.027		.16	1.24		1.40	2.06
0065	1/4" x 1/2"		288	.028		.21	1.27		1.48	2.16
0075	3/8" x 3/8"		284	.028		.23	1.28		1.51	2.21
0080	3/8" x 1/2"		280	.029		.31	1.30		1.61	2.33
0085	3/8" x 5/8"		276	.029		.39	1.32		1.71	2.44
0095	3/8" x 3/4"		272	.029		.46	1.34		1.80	2.55
0100	1/2" x 1/2"		275	.029		.41	1.33		1.74	2.48
0105	1/2" x 5/8"		269	.030		.52	1.36		1.88	2.64
0110	1/2" x 3/4"		263	.030		.62	1.39		2.01	2.79
0115	1/2" x 7/8"		256	.031		.72	1.43		2.15	2.97
0120	1/2" x 1"		250	.032		.83	1.46		2.29	3.13
0125	3/4" x 3/4"		244	.033		.93	1.50		2.43	3.31
0130	3/4" x 1"		225	.036		1.24	1.62		2.86	3.84
0135	1" x 1"	▼	200	.040		1.66	1.82		3.48	4.60
0190	Cartridges				Gal.	26.50			26.50	29
0200	11 fl. oz. cartridge				Ea.	2.28			2.28	2.51

07 92 Joint Sealants

07 92 13 – Elastomeric Joint Sealants

07 92 13.20 Caulking and Sealant Options

		Crew	Daily Output	Labor-Hours	Unit	Material	2014 Bare Costs Labor	Equipment	Total	Total Incl O&P
0500	1/4" x 1/2"	1 Bric	288	.028	L.F.	.19	1.27		1.46	2.13
0600	1/2" x 1/2"		275	.029		.37	1.33		1.70	2.43
0800	3/4" x 3/4"		244	.033		.84	1.50		2.34	3.20
0900	3/4" x 1"		225	.036		1.12	1.62		2.74	3.70
1000	1" x 1"		200	.040		1.40	1.82		3.22	4.32
1400	Butyl based, bulk				Gal.	34			34	37.50
1500	Cartridges				"	38.50			38.50	42.50
1700	1/4" x 1/2", 154 L.F./gal.	1 Bric	288	.028	L.F.	.22	1.27		1.49	2.17
1800	1/2" x 1/2", 77 L.F./gal.	"	275	.029	"	.44	1.33		1.77	2.51
2300	Polysulfide compounds, 1 component, bulk				Gal.	61			61	67
2600	1 or 2 component, in place, 1/4" x 1/4", 308 L.F./gal.	1 Bric	300	.027	L.F.	.20	1.22		1.42	2.07
2700	1/2" x 1/4", 154 L.F./gal.		288	.028		.40	1.27		1.67	2.37
2900	3/4" x 3/8", 68 L.F./gal.		272	.029		.90	1.34		2.24	3.03
3000	1" x 1/2", 38 L.F./gal.		250	.032		1.61	1.46		3.07	3.99
3200	Polyurethane, 1 or 2 component				Gal.	51			51	56
3300	Cartridges				"	64			64	70.50
3500	Bulk, in place, 1/4" x 1/4"	1 Bric	300	.027	L.F.	.17	1.22		1.39	2.03
3655	1/2" x 1/4"		288	.028		.33	1.27		1.60	2.29
3800	3/4" x 3/8"		272	.029		.75	1.34		2.09	2.87
3900	1" x 1/2"		250	.032		1.33	1.46		2.79	3.68
4100	Silicone rubber, bulk				Gal.	45.50			45.50	50
4200	Cartridges				"	45			45	49.50

07 92 16 – Rigid Joint Sealants

07 92 16.10 Rigid Joint Sealants

		Crew	Daily Output	Labor-Hours	Unit	Material	Labor	Equipment	Total	Total Incl O&P
0010	**RIGID JOINT SEALANTS**									
5800	Tapes, sealant, PVC foam adhesive, 1/16" x 1/4"				C.L.F.	6			6	6.60
5900	1/16" x 1/2"					8			8	8.80
5950	1/16" x 1"					15			15	16.50
6000	1/8" x 1/2"					9			9	9.90

07 92 19 – Acoustical Joint Sealants

07 92 19.10 Acoustical Sealant

		Crew	Daily Output	Labor-Hours	Unit	Material	Labor	Equipment	Total	Total Incl O&P
0010	**ACOUSTICAL SEALANT**									
0020	Acoustical sealant, elastomeric, cartridges				Ea.	8.50			8.50	9.35
0025	In place, 1/4" x 1/4"	1 Bric	300	.027	L.F.	.35	1.22		1.57	2.23
0030	1/4" x 1/2"		288	.028		.69	1.27		1.96	2.69
0035	1/2" x 1/2"		275	.029		1.39	1.33		2.72	3.54
0040	1/2" x 3/4"		263	.030		2.08	1.39		3.47	4.40
0045	3/4" x 3/4"		244	.033		3.12	1.50		4.62	5.70
0050	1" x 1"		200	.040		5.55	1.82		7.37	8.90

Division Notes

	CREW	DAILY OUTPUT	LABOR-HOURS	UNIT	BARE COSTS MAT.	LABOR	EQUIP.	TOTAL	TOTAL INCL O&P

Estimating Tips

General
- Room Finish Schedule: A complete set of plans should contain a room finish schedule. If one is not available, it would be well worth the time and effort to obtain one.

09 20 00 Plaster and Gypsum Board
- Lath is estimated by the square yard plus a 5% allowance for waste. Furring, channels, and accessories are measured by the linear foot. An extra foot should be allowed for each accessory miter or stop.
- Plaster is also estimated by the square yard. Deductions for openings vary by preference, from zero deduction to 50% of all openings over 2 feet in width. The estimator should allow one extra square foot for each linear foot of horizontal interior or exterior angle located below the ceiling level. Also, double the areas of small radius work.
- Drywall accessories, studs, track, and acoustical caulking are all measured by the linear foot. Drywall taping is figured by the square foot. Gypsum wallboard is estimated by the square foot. No material deductions should be made for door or window openings under 32 S.F.

09 60 00 Flooring
- Tile and terrazzo areas are taken off on a square foot basis. Trim and base materials are measured by the linear foot. Accent tiles are listed per each. Two basic methods of installation are used. Mud set is approximately 30% more expensive than thin set. In terrazzo work, be sure to include the linear footage of embedded decorative strips, grounds, machine rubbing, and power cleanup.
- Wood flooring is available in strip, parquet, or block configuration. The latter two types are set in adhesives with quantities estimated by the square foot. The laying pattern will influence labor costs and material waste. In addition to the material and labor for laying wood floors, the estimator must make allowances for sanding and finishing these areas unless the flooring is prefinished.
- Sheet flooring is measured by the square yard. Roll widths vary, so consideration should be given to use the most economical width, as waste must be figured into the total quantity. Consider also the installation methods available, direct glue down or stretched.

09 70 00 Wall Finishes
- Wall coverings are estimated by the square foot. The area to be covered is measured, length by height of wall above baseboards, to calculate the square footage of each wall. This figure is divided by the number of square feet in the single roll which is being used. Deduct, in full, the areas of openings such as doors and windows. Where a pattern match is required allow 25%–30% waste.

09 80 00 Acoustic Treatment
- Acoustical systems fall into several categories. The takeoff of these materials should be by the square foot of area with a 5% allowance for waste. Do not forget about scaffolding, if applicable, when estimating these systems.

09 90 00 Painting and Coating
- A major portion of the work in painting involves surface preparation. Be sure to include cleaning, sanding, filling, and masking costs in the estimate.
- Protection of adjacent surfaces is not included in painting costs. When considering the method of paint application, an important factor is the amount of protection and masking required. These must be estimated separately and may be the determining factor in choosing the method of application.

Reference Numbers
Reference numbers are shown in shaded boxes at the beginning of some major classifications. These numbers refer to related items in the Reference Section. The reference information may be an estimating procedure, an alternate pricing method, or technical information.

Note: Not all subdivisions listed here necessarily appear in this publication.

No part of this publication may be reproduced, stored in a retrieval system, or transmitted in any form or by any means without prior written permission of RSMeans, a division of Reed Construction Data.

09 22 Supports for Plaster and Gypsum Board

09 22 03 – Fastening Methods for Finishes

09 22 03.20 Drilling Plaster/Drywall

		Crew	Daily Output	Labor-Hours	Unit	Material	2014 Bare Costs Labor	Equipment	Total	Total Incl O&P
0010	**DRILLING PLASTER/DRYWALL**									
1100	Drilling & layout for drywall/plaster walls, up to 1" deep, no anchor									
1200	Holes, 1/4" diameter	1 Carp	150	.053	Ea.	.01	2.45		2.46	3.78
1300	3/8" diameter		140	.057		.01	2.62		2.63	4.05
1400	1/2" diameter		130	.062		.01	2.82		2.83	4.36
1500	3/4" diameter		120	.067		.01	3.06		3.07	4.74
1600	1" diameter		110	.073		.02	3.33		3.35	5.15
1700	1-1/4" diameter		100	.080		.03	3.67		3.70	5.70
1800	1-1/2" diameter	↓	90	.089		.04	4.08		4.12	6.35
1900	For ceiling installations, add				↓		40%			

09 91 Painting

09 91 23 – Interior Painting

09 91 23.52 Miscellaneous, Interior

		Crew	Daily Output	Labor-Hours	Unit	Material	2014 Bare Costs Labor	Equipment	Total	Total Incl O&P
0010	**MISCELLANEOUS, INTERIOR**									
3800	Grilles, per side, oil base, primer coat, brushwork	1 Pord	520	.015	S.F.	.13	.61		.74	1.06
3850	Spray		1140	.007		.13	.28		.41	.57
3880	Paint 1 coat, brushwork		520	.015		.22	.61		.83	1.16
3900	Spray		1140	.007		.24	.28		.52	.68
3920	Paint 2 coats, brushwork		325	.025		.42	.97		1.39	1.92
3940	Spray		650	.012		.48	.49		.97	1.26
3950	Prime & paint 1 coat		325	.025		.34	.97		1.31	1.84
3960	Prime & paint 2 coats		270	.030		.34	1.17		1.51	2.13
4500	Louvers, one side, primer, brushwork		524	.015		.08	.60		.68	1
4520	Paint one coat, brushwork		520	.015		.10	.61		.71	1.02
4530	Spray		1140	.007		.11	.28		.39	.54
4540	Paint two coats, brushwork		325	.025		.19	.97		1.16	1.66
4550	Spray		650	.012		.21	.49		.70	.96
4560	Paint three coats, brushwork		270	.030		.27	1.17		1.44	2.06
4570	Spray	↓	500	.016		.31	.63		.94	1.29
5000	Pipe, 1" - 4" diameter, primer or sealer coat, oil base, brushwork	2 Pord	1250	.013	L.F.	.09	.51		.60	.86
5100	Spray		2165	.007		.08	.29		.37	.53
5200	Paint 1 coat, brushwork		1250	.013		.11	.51		.62	.88
5300	Spray		2165	.007		.10	.29		.39	.55
5350	Paint 2 coats, brushwork		775	.021		.19	.82		1.01	1.44
5400	Spray		1240	.013		.22	.51		.73	1.01
5420	Paint 3 coats, brushwork		775	.021		.29	.82		1.11	1.55
5450	5" - 8" diameter, primer or sealer coat, brushwork		620	.026		.18	1.02		1.20	1.73
5500	Spray		1085	.015		.29	.58		.87	1.20
5550	Paint 1 coat, brushwork		620	.026		.30	1.02		1.32	1.86
5600	Spray		1085	.015		.33	.58		.91	1.24
5650	Paint 2 coats, brushwork		385	.042		.39	1.64		2.03	2.90
5700	Spray		620	.026		.43	1.02		1.45	2.01
5720	Paint 3 coats, brushwork		385	.042		.58	1.64		2.22	3.10
5750	9" - 12" diameter, primer or sealer coat, brushwork		415	.039		.26	1.52		1.78	2.58
5800	Spray		725	.022		.35	.87		1.22	1.70
5850	Paint 1 coat, brushwork		415	.039		.30	1.52		1.82	2.62
6000	Spray		725	.022		.33	.87		1.20	1.68
6200	Paint 2 coats, brushwork		260	.062		.58	2.43		3.01	4.30
6250	Spray		415	.039		.65	1.52		2.17	3
6270	Paint 3 coats, brushwork	↓	260	.062	↓	.86	2.43		3.29	4.61

09 91 Painting

09 91 23 – Interior Painting

09 91 23.52 Miscellaneous, Interior		Crew	Daily Output	Labor-Hours	Unit	Material	2014 Bare Costs Labor	Equipment	Total	Total Incl O&P
6300	13" - 16" diameter, primer or sealer coat, brushwork	2 Pord	310	.052	L.F.	.35	2.04		2.39	3.46
6350	Spray		540	.030		.39	1.17		1.56	2.19
6400	Paint 1 coat, brushwork		310	.052		.40	2.04		2.44	3.51
6450	Spray		540	.030		.44	1.17		1.61	2.25
6500	Paint 2 coats, brushwork		195	.082		.78	3.25		4.03	5.75
6550	Spray	↓	310	.052	↓	.86	2.04		2.90	4.02
6600	Radiators, per side, primer, brushwork	1 Pord	520	.015	S.F.	.08	.61		.69	1.01
6620	Paint, one coat		520	.015		.08	.61		.69	1.01
6640	Two coats		340	.024		.19	.93		1.12	1.60
6660	Three coats	↓	283	.028	↓	.27	1.12		1.39	1.98

Division Notes

	CREW	DAILY OUTPUT	LABOR-HOURS	UNIT	BARE COSTS MAT.	LABOR	EQUIP.	TOTAL	TOTAL INCL O&P

Estimating Tips

General

- The items in this division are usually priced per square foot or each.
- Many items in Division 10 require some type of support system or special anchors that are not usually furnished with the item. The required anchors must be added to the estimate in the appropriate division.
- Some items in Division 10, such as lockers, may require assembly before installation. Verify the amount of assembly required. Assembly can often exceed installation time.

10 20 00 Interior Specialties

- Support angles and blocking are not included in the installation of toilet compartments, shower/dressing compartments, or cubicles. Appropriate line items from Divisions 5 or 6 may need to be added to support the installations.
- Toilet partitions are priced by the stall. A stall consists of a side wall, pilaster, and door with hardware. Toilet tissue holders and grab bars are extra.
- The required acoustical rating of a folding partition can have a significant impact on costs. Verify the sound transmission coefficient rating of the panel priced to the specification requirements.
- Grab bar installation does not include supplemental blocking or backing to support the required load. When grab bars are installed at an existing facility, provisions must be made to attach the grab bars to solid structure.

Reference Numbers

Reference numbers are shown in shaded boxes at the beginning of some major classifications. These numbers refer to related items in the Reference Section. The reference information may be an estimating procedure, an alternate pricing method, or technical information.

Note: Not all subdivisions listed here necessarily appear in this publication.

No part of this publication may be reproduced, stored in a retrieval system, or transmitted in any form or by any means without prior written permission of RSMeans, a division of Reed Construction Data.

10 21 Compartments and Cubicles

10 21 13 – Toilet Compartments

10 21 13.13 Metal Toilet Compartments

		Crew	Daily Output	Labor-Hours	Unit	Material	2014 Bare Costs Labor	Equipment	Total	Total Incl O&P
0010	**METAL TOILET COMPARTMENTS**									
0110	Cubicles, ceiling hung									
0200	Powder coated steel	2 Carp	4	4	Ea.	515	183		698	850
0500	Stainless steel	"	4	4		1,075	183		1,258	1,450
0600	For handicap units, incl. 52" grab bars, add					430			430	475
0900	Floor and ceiling anchored									
1000	Powder coated steel	2 Carp	5	3.200	Ea.	590	147		737	870
1300	Stainless steel	"	5	3.200		1,150	147		1,297	1,500
1400	For handicap units, incl. 52" grab bars, add					296			296	325
1610	Floor anchored									
1700	Powder coated steel	2 Carp	7	2.286	Ea.	585	105		690	800
2000	Stainless steel	"	7	2.286		1,400	105		1,505	1,700
2100	For handicap units, incl. 52" grab bars, add					310			310	345
2200	For juvenile units, deduct					41.50			41.50	45.50
2450	Floor anchored, headrail braced									
2500	Powder coated steel	2 Carp	6	2.667	Ea.	380	122		502	605
2804	Stainless steel	"	4.60	3.478		1,050	159		1,209	1,400
2900	For handicap units, incl. 52" grab bars, add					370			370	410
3000	Wall hung partitions, powder coated steel	2 Carp	7	2.286		635	105		740	860
3300	Stainless steel	"	7	2.286		1,650	105		1,755	1,975
3400	For handicap units, incl. 52" grab bars, add					370			370	410
4000	Screens, entrance, floor mounted, 58" high, 48" wide									
4200	Powder coated steel	2 Carp	15	1.067	Ea.	242	49		291	340
4500	Stainless steel	"	15	1.067	"	910	49		959	1,075
4650	Urinal screen, 18" wide									
4704	Powder coated steel	2 Carp	6.15	2.602	Ea.	206	119		325	410
5004	Stainless steel	"	6.15	2.602	"	580	119		699	825
5100	Floor mounted, head rail braced									
5300	Powder coated steel	2 Carp	8	2	Ea.	230	91.50		321.50	395
5600	Stainless steel	"	8	2	"	595	91.50		686.50	795
5750	Pilaster, flush									
5800	Powder coated steel	2 Carp	10	1.600	Ea.	278	73.50		351.50	420
6100	Stainless steel		10	1.600		615	73.50		688.50	790
6300	Post braced, powder coated steel		10	1.600		153	73.50		226.50	281
6600	Stainless steel		10	1.600		450	73.50		523.50	610
6700	Wall hung, bracket supported									
6800	Powder coated steel	2 Carp	10	1.600	Ea.	153	73.50		226.50	281
7100	Stainless steel		10	1.600		227	73.50		300.50	365
7400	Flange supported, powder coated steel		10	1.600		106	73.50		179.50	229
7700	Stainless steel		10	1.600		310	73.50		383.50	455
7800	Wedge type, powder coated steel		10	1.600		134	73.50		207.50	261
8100	Stainless steel		10	1.600		575	73.50		648.50	750

10 21 13.16 Plastic-Laminate-Clad Toilet Compartments

		Crew	Daily Output	Labor-Hours	Unit	Material	Labor	Equipment	Total	Total Incl O&P
0010	**PLASTIC-LAMINATE-CLAD TOILET COMPARTMENTS**									
0110	Cubicles, ceiling hung									
0300	Plastic laminate on particle board	2 Carp	4	4	Ea.	520	183		703	855
0600	For handicap units, incl. 52" grab bars, add				"	430			430	475
0900	Floor and ceiling anchored									
1100	Plastic laminate on particle board	2 Carp	5	3.200	Ea.	785	147		932	1,100
1400	For handicap units, incl. 52" grab bars, add				"	296			296	325
1610	Floor mounted									
1800	Plastic laminate on particle board	2 Carp	7	2.286	Ea.	490	105		595	700

10 21 Compartments and Cubicles

10 21 13 – Toilet Compartments

10 21 13.16 Plastic-Laminate-Clad Toilet Compartments

	10 21 13.16 Plastic-Laminate-Clad Toilet Compartments	Crew	Daily Output	Labor-Hours	Unit	Material	2014 Bare Costs Labor	Equipment	Total	Total Incl O&P
2450	Floor mounted, headrail braced									
2600	Plastic laminate on particle board	2 Carp	6	2.667	Ea.	750	122		872	1,025
3400	For handicap units, incl. 52" grab bars, add					370			370	410
4300	Entrance screen, floor mtd., plas. lam., 58" high, 48" wide	2 Carp	15	1.067		610	49		659	745
4800	Urinal screen, 18" wide, ceiling braced, plastic laminate		8	2		184	91.50		275.50	345
5400	Floor mounted, headrail braced		8	2		190	91.50		281.50	350
5900	Pilaster, flush, plastic laminate		10	1.600		505	73.50		578.50	675
6400	Post braced, plastic laminate		10	1.600		300	73.50		373.50	445
6700	Wall hung, bracket supported									
6900	Plastic laminate on particle board	2 Carp	10	1.600	Ea.	92.50	73.50		166	215
7450	Flange supported									
7500	Plastic laminate on particle board	2 Carp	10	1.600	Ea.	230	73.50		303.50	365

10 21 13.19 Plastic Toilet Compartments

		Crew	Daily Output	Labor-Hours	Unit	Material	Labor	Equipment	Total	Total Incl O&P
0010	**PLASTIC TOILET COMPARTMENTS**									
0110	Cubicles, ceiling hung									
0250	Phenolic	2 Carp	4	4	Ea.	865	183		1,048	1,225
0600	For handicap units, incl. 52" grab bars, add				"	430			430	475
0900	Floor and ceiling anchored									
1050	Phenolic	2 Carp	5	3.200	Ea.	825	147		972	1,125
1400	For handicap units, incl. 52" grab bars, add				"	296			296	325
1610	Floor mounted									
1750	Phenolic	2 Carp	7	2.286	Ea.	795	105		900	1,025
2100	For handicap units, incl. 52" grab bars, add					310			310	345
2200	For juvenile units, deduct					41.50			41.50	45.50
2450	Floor mounted, headrail braced									
2550	Phenolic	2 Carp	6	2.667	Ea.	795	122		917	1,075

10 21 13.40 Stone Toilet Compartments

		Crew	Daily Output	Labor-Hours	Unit	Material	Labor	Equipment	Total	Total Incl O&P
0010	**STONE TOILET COMPARTMENTS**									
0100	Cubicles, ceiling hung, marble	2 Marb	2	8	Ea.	1,625	340		1,965	2,300
0600	For handicap units, incl. 52" grab bars, add					430			430	475
0800	Floor & ceiling anchored, marble	2 Marb	2.50	6.400		2,025	270		2,295	2,650
1400	For handicap units, incl. 52" grab bars, add					296			296	325
1600	Floor mounted, marble	2 Marb	3	5.333		1,200	225		1,425	1,675
2400	Floor mounted, headrail braced, marble	"	3	5.333		1,150	225		1,375	1,625
2900	For handicap units, incl. 52" grab bars, add					370			370	410
4100	Entrance screen, floor mounted marble, 58" high, 48" wide	2 Marb	9	1.778		755	75		830	945
4600	Urinal screen, 18" wide, ceiling braced, marble	D-1	6	2.667		755	110		865	1,000
5100	Floor mounted, head rail braced									
5200	Marble	D-1	6	2.667	Ea.	645	110		755	880
5700	Pilaster, flush, marble		9	1.778		840	73.50		913.50	1,025
6200	Post braced, marble		9	1.778		825	73.50		898.50	1,025

10 21 16 – Shower and Dressing Compartments

10 21 16.10 Partitions, Shower

		Crew	Daily Output	Labor-Hours	Unit	Material	Labor	Equipment	Total	Total Incl O&P
0010	**PARTITIONS, SHOWER** floor mounted, no plumbing									
0400	Cabinet, one piece, fiberglass, 32" x 32"	2 Carp	5	3.200	Ea.	510	147		657	785
0420	36" x 36"		5	3.200		610	147		757	895
0440	36" x 48"		5	3.200		1,300	147		1,447	1,650
0460	Acrylic, 32" x 32"		5	3.200		299	147		446	555
0480	36" x 36"		5	3.200		1,000	147		1,147	1,325
0500	36" x 48"		5	3.200		1,500	147		1,647	1,875
0520	Shower door for above, clear plastic, 24" wide	1 Carp	8	1		177	46		223	266
0540	28" wide		8	1		201	46		247	293

63

10 21 Compartments and Cubicles

10 21 16 – Shower and Dressing Compartments

10 21 16.10 Partitions, Shower	Crew	Daily Output	Labor-Hours	Unit	Material	2014 Bare Costs Labor	Equipment	Total	Total Incl O&P
0560	Tempered glass, 24" wide	1 Carp	8	1	Ea.	191	46		237	281
0580	28" wide	▼	8	1		216	46		262	310
2400	Glass stalls, with doors, no receptors, chrome on brass	2 Shee	3	5.333		1,650	292		1,942	2,275
2700	Anodized aluminum	"	4	4		1,150	219		1,369	1,600
2900	Marble shower stall, stock design, with shower door	2 Marb	1.20	13.333		2,475	565		3,040	3,575
3000	With curtain		1.30	12.308		2,200	520		2,720	3,200
3200	Receptors, precast terrazzo, 32" x 32"		14	1.143		330	48.50		378.50	440
3300	48" x 34"		9.50	1.684		440	71		511	595
3500	Plastic, simulated terrazzo receptor, 32" x 32"		14	1.143		148	48.50		196.50	237
3600	32" x 48"		12	1.333		215	56.50		271.50	325
3800	Precast concrete, colors, 32" x 32"		14	1.143		186	48.50		234.50	279
3900	48" x 48"	▼	8	2		251	84.50		335.50	405
4100	Shower doors, economy plastic, 24" wide	1 Shee	9	.889		144	48.50		192.50	233
4200	Tempered glass door, economy		8	1		260	54.50		314.50	370
4400	Folding, tempered glass, aluminum frame		6	1.333		390	73		463	540
4500	Sliding, tempered glass, 48" opening		6	1.333		540	73		613	700
4700	Deluxe, tempered glass, chrome on brass frame, minimum		8	1		385	54.50		439.50	505
4800	Maximum		1	8		625	440		1,065	1,350
4850	On anodized aluminum frame, minimum		2	4		540	219		759	925
4900	Maximum	▼	1	8	▼	625	440		1,065	1,350
5100	Shower enclosure, tempered glass, anodized alum. frame									
5120	2 panel & door, corner unit, 32" x 32"	1 Shee	2	4	Ea.	1,025	219		1,244	1,475
5140	Neo-angle corner unit, 16" x 24" x 16"	"	2	4		1,075	219		1,294	1,525
5200	Shower surround, 3 wall, polypropylene, 32" x 32"	1 Carp	4	2		455	91.50		546.50	640
5220	PVC, 32" x 32"		4	2		380	91.50		471.50	560
5240	Fiberglass		4	2		420	91.50		511.50	600
5250	2 wall, polypropylene, 32" x 32"		4	2		315	91.50		406.50	490
5270	PVC		4	2		395	91.50		486.50	575
5290	Fiberglass	▼	4	2		400	91.50		491.50	580
5300	Tub doors, tempered glass & frame, minimum	1 Shee	8	1		229	54.50		283.50	335
5400	Maximum		6	1.333		535	73		608	700
5600	Chrome plated, brass frame, minimum		8	1		305	54.50		359.50	420
5700	Maximum		6	1.333		745	73		818	925
5900	Tub/shower enclosure, temp. glass, alum. frame, minimum		2	4		410	219		629	785
6200	Maximum		1.50	5.333		855	292		1,147	1,375
6500	On chrome-plated brass frame, minimum		2	4		565	219		784	960
6600	Maximum	▼	1.50	5.333		1,200	292		1,492	1,775
6800	Tub surround, 3 wall, polypropylene	1 Carp	4	2		256	91.50		347.50	425
6900	PVC		4	2		390	91.50		481.50	570
7000	Fiberglass, minimum		4	2		400	91.50		491.50	580
7100	Maximum	▼	3	2.667	▼	680	122		802	940

10 28 Toilet, Bath, and Laundry Accessories

10 28 13 – Toilet Accessories

10 28 13.13 Commercial Toilet Accessories		Crew	Daily Output	Labor-Hours	Unit	Material	2014 Bare Costs Labor	Equipment	Total	Total Incl O&P
0010	**COMMERCIAL TOILET ACCESSORIES**									
0200	Curtain rod, stainless steel, 5' long, 1" diameter	1 Carp	13	.615	Ea.	26.50	28		54.50	72.50
0300	1-1/4" diameter	"	13	.615	"	30.50	28		58.50	77
0500	Dispenser units, combined soap & towel dispensers,									
0510	mirror and shelf, flush mounted	1 Carp	10	.800	Ea.	310	36.50		346.50	395
0600	Towel dispenser and waste receptacle,									
0610	18 gallon capacity	1 Carp	10	.800	Ea.	286	36.50		322.50	370
0800	Grab bar, straight, 1-1/4" diameter, stainless steel, 18" long		24	.333		29	15.30		44.30	55.50
0900	24" long		23	.348		29.50	15.95		45.45	57
1000	30" long		22	.364		31.50	16.65		48.15	60
1100	36" long		20	.400		38.50	18.35		56.85	71
1105	42" long		20	.400		46	18.35		64.35	79
1120	Corner, 36" long		20	.400		85.50	18.35		103.85	123
1200	1-1/2" diameter, 24" long		23	.348		31	15.95		46.95	58.50
1300	36" long		20	.400		33.50	18.35		51.85	65.50
1310	42" long		18	.444		43	20.50		63.50	79
1500	Tub bar, 1-1/4" diameter, 24" x 36"		14	.571		92.50	26		118.50	143
1600	Plus vertical arm		12	.667		97.50	30.50		128	154
1900	End tub bar, 1" diameter, 90° angle, 16" x 32"		12	.667		109	30.50		139.50	167
2300	Hand dryer, surface mounted, electric, 115 volt, 20 amp		4	2		430	91.50		521.50	615
2400	230 volt, 10 amp		4	2		745	91.50		836.50	960
2450	Hand dryer, touch free, 1400 watt, 81,000 rpm		4	2		1,025	91.50		1,116.50	1,275
2600	Hat and coat strip, stainless steel, 4 hook, 36" long		24	.333		75	15.30		90.30	106
2700	6 hook, 60" long		20	.400		131	18.35		149.35	173
3000	Mirror, with stainless steel 3/4" square frame, 18" x 24"		20	.400		47	18.35		65.35	80
3100	36" x 24"		15	.533		103	24.50		127.50	151
3200	48" x 24"		10	.800		139	36.50		175.50	210
3300	72" x 24"		6	1.333		256	61		317	375
3500	With 5" stainless steel shelf, 18" x 24"		20	.400		186	18.35		204.35	234
3600	36" x 24"		15	.533		217	24.50		241.50	277
3700	48" x 24"		10	.800		245	36.50		281.50	325
3800	72" x 24"		6	1.333		300	61		361	425
4100	Mop holder strip, stainless steel, 5 holders, 48" long		20	.400		89	18.35		107.35	127
4200	Napkin/tampon dispenser, recessed		15	.533		525	24.50		549.50	615
4220	Semi-recessed		6.50	1.231		310	56.50		366.50	425
4250	Napkin receptacle, recessed		6.50	1.231		165	56.50		221.50	269
4300	Robe hook, single, regular		36	.222		18.10	10.20		28.30	35.50
4400	Heavy duty, concealed mounting		36	.222		21	10.20		31.20	38.50
4600	Soap dispenser, chrome, surface mounted, liquid		20	.400		46.50	18.35		64.85	79.50
4700	Powder		20	.400		55	18.35		73.35	89
5000	Recessed stainless steel, liquid		10	.800		137	36.50		173.50	208
5600	Shelf, stainless steel, 5" wide, 18 ga., 24" long		24	.333		68.50	15.30		83.80	99
5700	48" long		16	.500		135	23		158	185
5800	8" wide shelf, 18 ga., 24" long		22	.364		69	16.65		85.65	102
5900	48" long		14	.571		121	26		147	174
6000	Toilet seat cover dispenser, stainless steel, recessed		20	.400		167	18.35		185.35	212
6050	Surface mounted		15	.533		34.50	24.50		59	75
6100	Toilet tissue dispenser, surface mounted, SS, single roll		30	.267		17.80	12.25		30.05	38.50
6200	Double roll		24	.333		23.50	15.30		38.80	49.50
6240	Plastic, twin/jumbo dbl. roll		24	.333		26	15.30		41.30	52
6400	Towel bar, stainless steel, 18" long		23	.348		41.50	15.95		57.45	70.50
6500	30" long		21	.381		111	17.45		128.45	150
6700	Towel dispenser, stainless steel, surface mounted		16	.500		44.50	23		67.50	84.50

10 28 Toilet, Bath, and Laundry Accessories

10 28 13 – Toilet Accessories

10 28 13.13 Commercial Toilet Accessories

		Crew	Daily Output	Labor-Hours	Unit	Material	2014 Bare Costs Labor	Equipment	Total	Total Incl O&P
6800	Flush mounted, recessed	1 Carp	10	.800	Ea.	257	36.50		293.50	340
6900	Plastic, touchless, battery operated		16	.500		80.50	23		103.50	124
7000	Towel holder, hotel type, 2 guest size		20	.400		51.50	18.35		69.85	85
7200	Towel shelf, stainless steel, 24" long, 8" wide		20	.400		59.50	18.35		77.85	94
7400	Tumbler holder, tumbler only		30	.267		15.90	12.25		28.15	36.50
7410	Tumbler holder		20	.400		9.10	18.35		27.45	38.50
7500	Soap, tumbler & toothbrush		30	.267		19.60	12.25		31.85	40.50
7510	Tumbler & toothbrush holder		20	.400		12.60	18.35		30.95	42.50
7700	Wall urn ash receiver, surface mount, 11" long		12	.667		95	30.50		125.50	151
7800	7-1/2", long		18	.444		128	20.50		148.50	172
8000	Waste receptacles, stainless steel, with top, 13 gallon		10	.800		295	36.50		331.50	380
8100	36 gallon		8	1		395	46		441	505

10 28 16 – Bath Accessories

10 28 16.20 Medicine Cabinets

		Crew	Daily Output	Labor-Hours	Unit	Material	Labor	Equipment	Total	Total Incl O&P
0010	**MEDICINE CABINETS**									
0020	With mirror, sst frame, 16" x 22", unlighted	1 Carp	14	.571	Ea.	95	26		121	146
0100	Wood frame		14	.571		136	26		162	191
0300	Sliding mirror doors, 20" x 16" x 4-3/4", unlighted		7	1.143		125	52.50		177.50	218
0400	24" x 19" x 8-1/2", lighted		5	1.600		179	73.50		252.50	310
0600	Triple door, 30" x 32", unlighted, plywood body		7	1.143		320	52.50		372.50	430
0700	Steel body		7	1.143		350	52.50		402.50	465
0900	Oak door, wood body, beveled mirror, single door		7	1.143		195	52.50		247.50	295
1000	Double door		6	1.333		390	61		451	525
1200	Hotel cabinets, stainless, with lower shelf, unlighted		10	.800		200	36.50		236.50	276
1300	Lighted		5	1.600		305	73.50		378.50	450

10 44 Fire Protection Specialties

10 44 13 – Fire Protection Cabinets

10 44 13.53 Fire Equipment Cabinets

			Crew	Daily Output	Labor-Hours	Unit	Material	Labor	Equipment	Total	Total Incl O&P
0010	**FIRE EQUIPMENT CABINETS**, not equipped, 20 ga. steel box	D4020-310									
0040	recessed, D.S. glass in door, box size given										
1000	Portable extinguisher, single, 8" x 12" x 27", alum. door & frame		Q-12	8	2	Ea.	151	99.50		250.50	320
1100	Steel door and frame			8	2		105	99.50		204.50	266
1200	Stainless steel door and frame			8	2		196	99.50		295.50	365
2000	Portable extinguisher, large, 8" x 12" x 36", alum. door & frame			8	2		257	99.50		356.50	435
2100	Steel door and frame	D4020-330		8	2		155	99.50		254.50	320
2200	Stainless steel door and frame			8	2		257	99.50		356.50	435
2500	8" x 16" x 38", aluminum door & frame			8	2		218	99.50		317.50	390
2600	Steel door and frame			8	2		223	99.50		322.50	395
2700	Fire blanket & extinguisher cab, inc blanket, rec stl., 14" x 40" x 8"			7	2.286		179	114		293	370
2800	Fire blanket cab, inc blanket, surf mtd, stl, 15"x10"x5", w/pwdr coat fin			8	2		90	99.50		189.50	250
3000	Hose rack assy., 1-1/2" valve & 100' hose, 24" x 40" x 5-1/2"										
3100	Aluminum door and frame		Q-12	6	2.667	Ea.	350	133		483	585
3200	Steel door and frame	D4020-410		6	2.667		234	133		367	460
3300	Stainless steel door and frame			6	2.667		475	133		608	720
4000	Hose rack assy., 2-1/2" x 1-1/2" valve, 100' hose, 24" x 40" x 8"										
4100	Aluminum door and frame		Q-12	6	2.667	Ea.	355	133		488	590
4200	Steel door and frame	R211226-10		6	2.667		241	133		374	465
4300	Stainless steel door and frame			6	2.667		475	133		608	720
5000	Hose rack assy., 2-1/2" x 1-1/2" valve, 100' hose										

10 44 Fire Protection Specialties

10 44 13 – Fire Protection Cabinets

10 44 13.53 Fire Equipment Cabinets

		Crew	Daily Output	Labor-Hours	Unit	Material	2014 Bare Costs Labor	Equipment	Total	Total Incl O&P
5010	and extinguisher, 30" x 40" x 8"									
5100	Aluminum door and frame	Q-12	5	3.200	Ea.	455	160		615	740
5200	Steel door and frame		5	3.200		300	160		460	570
5300	Stainless steel door and frame	↓	5	3.200	↓	550	160		710	845
6000	Hose rack assy., 1-1/2" valve, 100' hose									
6010	and 2-1/2" FD valve, 24" x 44" x 8"									
6100	Aluminum door and frame	Q-12	5	3.200	Ea.	385	160		545	660
6200	Steel door and frame		5	3.200		246	160		406	510
6300	Stainless steel door and frame	↓	5	3.200	↓	535	160		695	830
7000	Hose rack assy., 1-1/2" valve & 100' hose, 2-1/2" FD valve	R211226-20								
7010	and extinguisher, 30" x 44" x 8"									
7100	Aluminum door and frame	Q-12	5	3.200	Ea.	475	160		635	765
7200	Steel door and frame		5	3.200		315	160		475	585
7300	Stainless steel door and frame	↓	5	3.200	↓	635	160		795	940
8000	Valve cabinet for 2-1/2" FD angle valve, 18" x 18" x 8"									
8100	Aluminum door and frame	Q-12	12	1.333	Ea.	156	66.50		222.50	272
8200	Steel door and frame		12	1.333		129	66.50		195.50	242
8300	Stainless steel door and frame	↓	12	1.333	↓	210	66.50		276.50	330

10 44 16 – Fire Extinguishers

10 44 16.13 Portable Fire Extinguishers

		Crew	Daily Output	Labor-Hours	Unit	Material	Labor	Equipment	Total	Total Incl O&P
0010	**PORTABLE FIRE EXTINGUISHERS**									
0140	CO_2, with hose and "H" horn, 10 lb.				Ea.	268			268	295
0160	15 lb.					345			345	380
0180	20 lb.				↓	380			380	420
1000	Dry chemical, pressurized									
1040	Standard type, portable, painted, 2-1/2 lb.				Ea.	36.50			36.50	40
1060	5 lb.					50			50	55
1080	10 lb.					79.50			79.50	87.50
1100	20 lb.					134			134	148
1120	30 lb.					395			395	435
1300	Standard type, wheeled, 150 lb.					2,350			2,350	2,600
2000	ABC all purpose type, portable, 2-1/2 lb.					21			21	23.50
2060	5 lb.					26.50			26.50	29
2080	9-1/2 lb.					44			44	48
2100	20 lb.					79.50			79.50	87.50
5000	Pressurized water, 2-1/2 gallon, stainless steel					102			102	112
5060	With anti-freeze					106			106	116
9400	Installation of extinguishers, 12 or more, on nailable surface	1 Carp	30	.267			12.25		12.25	18.85
9420	On masonry or concrete	"	15	.533	↓		24.50		24.50	37.50

10 44 16.16 Wheeled Fire Extinguisher Units

		Crew	Daily Output	Labor-Hours	Unit	Material	Labor	Equipment	Total	Total Incl O&P
0010	**WHEELED FIRE EXTINGUISHER UNITS**									
0350	CO_2, portable, with swivel horn									
0360	Wheeled type, cart mounted, 50 lb.				Ea.	1,100			1,100	1,200
0400	100 lb.				"	4,125			4,125	4,550
2200	ABC all purpose type									
2300	Wheeled, 45 lb.				Ea.	735			735	810
2360	150 lb.				"	1,850			1,850	2,025

Division Notes

		CREW	DAILY OUTPUT	LABOR-HOURS	UNIT	BARE COSTS MAT.	LABOR	EQUIP.	TOTAL	TOTAL INCL O&P

Division 11 - Equipment

Estimating Tips

General

- The items in this division are usually priced per square foot or each. Many of these items are purchased by the owner for installation by the contractor. Check the specifications for responsibilities and include time for receiving, storage, installation, and mechanical and electrical hookups in the appropriate divisions.

- Many items in Division 11 require some type of support system that is not usually furnished with the item. Examples of these systems include blocking for the attachment of casework and support angles for ceiling-hung projection screens. The required blocking or supports must be added to the estimate in the appropriate division.

- Some items in Division 11 may require assembly or electrical hookups. Verify the amount of assembly required or the need for a hard electrical connection and add the appropriate costs.

Reference Numbers

Reference numbers are shown in shaded boxes at the beginning of some major classifications. These numbers refer to related items in the Reference Section. The reference information may be an estimating procedure, an alternate pricing method, or technical information.

Note: Not all subdivisions listed here necessarily appear in this publication.

No part of this publication may be reproduced, stored in a retrieval system, or transmitted in any form or by any means without prior written permission of RSMeans, a division of Reed Construction Data.

11 11 Vehicle Service Equipment

11 11 13 – Compressed-Air Vehicle Service Equipment

11 11 13.10 Compressed Air Equipment

		Crew	Daily Output	Labor-Hours	Unit	Material	2014 Bare Costs Labor	Equipment	Total	Total Incl O&P
0010	**COMPRESSED AIR EQUIPMENT**									
0030	Compressors, electric, 1-1/2 H.P., standard controls	L-4	1.50	16	Ea.	450	690		1,140	1,575
0550	Dual controls		1.50	16		800	690		1,490	1,950
0600	5 H.P., 115/230 volt, standard controls		1	24		2,550	1,025		3,575	4,400
0650	Dual controls		1	24		3,100	1,025		4,125	5,000

11 11 19 – Vehicle Lubrication Equipment

11 11 19.10 Lubrication Equipment

0010	**LUBRICATION EQUIPMENT**									
3000	Lube equipment, 3 reel type, with pumps, not including piping	L-4	.50	48	Set	8,700	2,075		10,775	12,800
3700	Pump lubrication, pneumatic, not incl. air compressor									
3710	Oil/gear lube	Q-1	9.60	1.667	Ea.	850	86.50		936.50	1,075
3720	Grease	"	9.60	1.667	"	1,075	86.50		1,161.50	1,300

11 11 33 – Vehicle Spray Painting Equipment

11 11 33.10 Spray Painting Equipment

0010	**SPRAY PAINTING EQUIPMENT**									
4000	Spray painting booth, 26' long, complete	L-4	.40	60	Ea.	16,200	2,575		18,775	21,800

11 19 Detention Equipment

11 19 30 – Detention Cell Equipment

11 19 30.10 Cell Equipment

0010	**CELL EQUIPMENT**									
3000	Toilet apparatus including wash basin, average	L-8	1.50	13.333	Ea.	3,350	645		3,995	4,675

11 21 Mercantile and Service Equipment

11 21 53 – Barber and Beauty Shop Equipment

11 21 53.10 Barber Equipment

0010	**BARBER EQUIPMENT**									
0020	Chair, hydraulic, movable, minimum	1 Carp	24	.333	Ea.	555	15.30		570.30	635
0050	Maximum	"	16	.500		3,425	23		3,448	3,800
0500	Sink, hair washing basin, rough plumbing not incl.	1 Plum	8	1		490	57.50		547.50	625
1000	Sterilizer, liquid solution for tools					159			159	175

11 23 Commercial Laundry and Dry Cleaning Equipment

11 23 13 – Dry Cleaning Equipment

11 23 13.13 Dry Cleaners

0010	**DRY CLEANERS** Not incl. rough-in									

11 23 16 – Drying and Conditioning Equipment

11 23 16.13 Dryers

0010	**DRYERS**, Not including rough-in									
1500	Industrial, 30 lb. capacity	1 Plum	2	4	Ea.	3,125	230		3,355	3,800
1600	50 lb. capacity	"	1.70	4.706	"	3,350	271		3,621	4,100

11 23 Commercial Laundry and Dry Cleaning Equipment

11 23 26 – Commercial Washers and Extractors

11 23 26.13 Washers and Extractors	Crew	Daily Output	Labor-Hours	Unit	Material	2014 Bare Costs Labor	Equipment	Total	Total Incl O&P
0010 **WASHERS AND EXTRACTORS**, not including rough-in									
6000 Combination washer/extractor, 20 lb. capacity	L-6	1.50	8	Ea.	5,875	450		6,325	7,125
6100 30 lb. capacity		.80	15		9,150	840		9,990	11,400
6200 50 lb. capacity		.68	17.647		10,700	990		11,690	13,300
6300 75 lb. capacity		.30	40		20,200	2,250		22,450	25,600
6350 125 lb. capacity		.16	75		27,300	4,200		31,500	36,500

11 23 33 – Coin-Operated Laundry Equipment

11 23 33.13 Coin Operated Washers and Dryers	Crew	Daily Output	Labor-Hours	Unit	Material	Labor	Equipment	Total	Total Incl O&P
0010 **COIN OPERATED WASHERS AND DRYERS**									
0990 Dryer, gas fired									
1000 Commercial, 30 lb. capacity, coin operated, single	1 Plum	3	2.667	Ea.	3,250	153		3,403	3,800
1100 Double stacked	"	2	4	"	7,200	230		7,430	8,275
5290 Clothes washer									
5300 Commercial, coin operated, average	1 Plum	3	2.667	Ea.	1,250	153		1,403	1,600

11 24 Maintenance Equipment

11 24 19 – Vacuum Cleaning Systems

11 24 19.10 Vacuum Cleaning	Crew	Daily Output	Labor-Hours	Unit	Material	Labor	Equipment	Total	Total Incl O&P
0010 **VACUUM CLEANING**									
0020 Central, 3 inlet, residential	1 Skwk	.90	8.889	Total	1,050	420		1,470	1,825
0200 Commercial		.70	11.429		1,225	540		1,765	2,150
0400 5 inlet system, residential		.50	16		1,475	755		2,230	2,800
0600 7 inlet system, commercial		.40	20		1,675	945		2,620	3,325
0800 9 inlet system, residential		.30	26.667		3,700	1,250		4,950	6,025
4010 Rule of thumb: First 1200 S.F., installed								1,425	1,575
4020 For each additional S.F., add				S.F.					.26

11 26 Unit Kitchens

11 26 13 – Metal Unit Kitchens

11 26 13.10 Commercial Unit Kitchens	Crew	Daily Output	Labor-Hours	Unit	Material	Labor	Equipment	Total	Total Incl O&P
0010 **COMMERCIAL UNIT KITCHENS**									
1500 Combination range, refrigerator and sink, 30" wide, minimum	L-1	2	8	Ea.	1,100	445		1,545	1,875
1550 Maximum		1	16		1,450	885		2,335	2,925
1570 60" wide, average		1.40	11.429		1,525	635		2,160	2,625
1590 72" wide, average		1.20	13.333		1,600	740		2,340	2,850
1600 Office model, 48" wide		2	8		2,000	445		2,445	2,875
1620 Refrigerator and sink only		2.40	6.667		2,500	370		2,870	3,300
1640 Combination range, refrigerator, sink, microwave									
1660 Oven and ice maker	L-1	.80	20	Ea.	4,500	1,100		5,600	6,625

11 27 Photographic Processing Equipment

11 27 13 – Darkroom Processing Equipment

11 27 13.10 Darkroom Equipment		Crew	Daily Output	Labor-Hours	Unit	Material	2014 Bare Costs Labor	Equipment	Total	Total Incl O&P
0010	**DARKROOM EQUIPMENT**									
0020	Developing sink, 5" deep, 24" x 48"	Q-1	2	8	Ea.	4,100	415		4,515	5,150
0050	48" x 52"		1.70	9.412		4,375	490		4,865	5,550
0200	10" deep, 24" x 48"		1.70	9.412		1,525	490		2,015	2,400
0250	24" x 108"		1.50	10.667		3,625	555		4,180	4,800
3500	Washers, round, minimum sheet 11" x 14"		2	8		3,225	415		3,640	4,175
3550	Maximum sheet 20" x 24"		1	16		3,575	830		4,405	5,175
3800	Square, minimum sheet 20" x 24"		1	16		3,125	830		3,955	4,675
3900	Maximum sheet 50" x 56"		.80	20		4,850	1,025		5,875	6,900
4500	Combination tank sink, tray sink, washers, with									
4510	Dry side tables, average	Q-1	.45	35.556	Ea.	10,600	1,850		12,450	14,500

11 31 Residential Appliances

11 31 13 – Residential Kitchen Appliances

11 31 13.13 Cooking Equipment

		Crew	Daily Output	Labor-Hours	Unit	Material	Labor	Equipment	Total	Total Incl O&P
0010	**COOKING EQUIPMENT**									
0020	Cooking range, 30" free standing, 1 oven, minimum	2 Clab	10	1.600	Ea.	435	58.50		493.50	570
0050	Maximum		4	4		2,000	147		2,147	2,425
0150	2 oven, minimum		10	1.600		1,200	58.50		1,258.50	1,425
0200	Maximum		10	1.600		2,525	58.50		2,583.50	2,875

11 31 13.23 Refrigeration Equipment

0010	**REFRIGERATION EQUIPMENT**									
5200	Icemaker, automatic, 20 lb. per day	1 Plum	7	1.143	Ea.	885	66		951	1,075
5350	51 lb. per day	"	2	4	"	1,575	230		1,805	2,075

11 31 13.33 Kitchen Cleaning Equipment

0010	**KITCHEN CLEANING EQUIPMENT**									
2750	Dishwasher, built-in, 2 cycles, minimum	L-1	4	4	Ea.	235	222		457	595
2800	Maximum		2	8		430	445		875	1,125
2950	4 or more cycles, minimum		4	4		370	222		592	745
2960	Average		4	4		495	222		717	880
3000	Maximum		2	8		1,075	445		1,520	1,875

11 31 13.43 Waste Disposal Equipment

0010	**WASTE DISPOSAL EQUIPMENT**									
3300	Garbage disposal, sink type, minimum	L-1	10	1.600	Ea.	86.50	88.50		175	228
3350	Maximum	"	10	1.600	"	217	88.50		305.50	370

11 31 13.53 Kitchen Ventilation Equipment

0010	**KITCHEN VENTILATION EQUIPMENT**									
4150	Hood for range, 2 speed, vented, 30" wide, minimum	L-3	5	3.200	Ea.	69.50	160		229.50	320
4200	Maximum		3	5.333		770	266		1,036	1,250
4300	42" wide, minimum		5	3.200		166	160		326	425
4330	Custom		5	3.200		1,650	160		1,810	2,050
4350	Maximum		3	5.333		2,000	266		2,266	2,600
4500	For ventless hood, 2 speed, add					18.45			18.45	20.50
4650	For vented 1 speed, deduct from maximum					49.50			49.50	54.50

11 31 23 – Residential Laundry Appliances

11 31 23.13 Washers

0010	**WASHERS**									
5000	Residential, 4 cycle, average	1 Plum	3	2.667	Ea.	865	153		1,018	1,175
6650	Washing machine, automatic, minimum		3	2.667		480	153		633	755

11 31 Residential Appliances

11 31 23 – Residential Laundry Appliances

11 31 23.13 Washers

		Crew	Daily Output	Labor-Hours	Unit	Material	2014 Bare Costs Labor	Equipment	Total	Total Incl O&P
6700	Maximum	1 Plum	1	8	Ea.	1,500	460		1,960	2,350

11 31 23.23 Dryers

		Crew	Daily Output	Labor-Hours	Unit	Material	Labor	Equipment	Total	Incl O&P
0010	**DRYERS**									
0500	Gas fired residential, 16 lb. capacity, average	1 Plum	3	2.667	Ea.	665	153		818	965
7450	Vent kits for dryers	1 Carp	10	.800	"	38.50	36.50		75	98.50

11 31 33 – Miscellaneous Residential Appliances

11 31 33.13 Sump Pumps

		Crew	Daily Output	Labor-Hours	Unit	Material	Labor	Equipment	Total	Incl O&P
0010	**SUMP PUMPS**									
6400	Cellar drainer, pedestal, 1/3 H.P., molded PVC base	1 Plum	3	2.667	Ea.	135	153		288	380
6450	Solid brass	"	2	4	"	289	230		519	670
6460	Sump pump, see also Section 22 14 29.16									

11 31 33.23 Water Heaters

		Crew	Daily Output	Labor-Hours	Unit	Material	Labor	Equipment	Total	Incl O&P
0010	**WATER HEATERS**									
6900	Electric, glass lined, 30 gallon, minimum	L-1	5	3.200	Ea.	430	177		607	735
6950	Maximum		3	5.333		595	296		891	1,100
7100	80 gallon, minimum		2	8		1,225	445		1,670	2,025
7150	Maximum		1	16		1,700	885		2,585	3,175
7180	Gas, glass lined, 30 gallon, minimum	2 Plum	5	3.200		805	184		989	1,175
7220	Maximum		3	5.333		1,125	305		1,430	1,700
7260	50 gallon, minimum		2.50	6.400		845	370		1,215	1,475
7300	Maximum		1.50	10.667		1,175	615		1,790	2,200
7310	Water heater, see also Section 22 33 30.13									

11 31 33.43 Air Quality

		Crew	Daily Output	Labor-Hours	Unit	Material	Labor	Equipment	Total	Incl O&P
0010	**AIR QUALITY**									
2450	Dehumidifier, portable, automatic, 15 pint	1 Elec	4	2	Ea.	150	107		257	325
2550	40 pint					207			207	228
3550	Heater, electric, built-in, 1250 watt, ceiling type, minimum	1 Elec	4	2		106	107		213	276
3600	Maximum		3	2.667		172	142		314	405
3700	Wall type, minimum		4	2		170	107		277	345
3750	Maximum		3	2.667		183	142		325	415
3900	1500 watt wall type, with blower		4	2		170	107		277	345
3950	3000 watt		3	2.667		350	142		492	600
4850	Humidifier, portable, 8 gallons per day					131			131	144
5000	15 gallons per day					208			208	229

11 41 Foodservice Storage Equipment

11 41 13 – Refrigerated Food Storage Cases

11 41 13.20 Refrigerated Food Storage Equipment

		Crew	Daily Output	Labor-Hours	Unit	Material	Labor	Equipment	Total	Incl O&P
0010	**REFRIGERATED FOOD STORAGE EQUIPMENT**									
2350	Cooler, reach-in, beverage, 6' long	Q-1	6	2.667	Ea.	3,550	138		3,688	4,100
4300	Freezers, reach-in, 44 C.F.		4	4		4,350	207		4,557	5,125
4500	68 C.F.		3	5.333		4,775	276		5,051	5,675

11 44 Food Cooking Equipment

11 44 13 – Commercial Ranges

11 44 13.10 Cooking Equipment

		Crew	Daily Output	Labor-Hours	Unit	Material	2014 Bare Costs Labor	Equipment	Total	Total Incl O&P
0010	**COOKING EQUIPMENT**									
0020	Bake oven, gas, one section	Q-1	8	2	Ea.	5,550	104		5,654	6,250
0300	Two sections		7	2.286		8,975	118		9,093	10,100
0600	Three sections		6	2.667		12,400	138		12,538	13,800
0900	Electric convection, single deck	L-7	4	7		6,150	310		6,460	7,225
6350	Kettle, w/steam jacket, tilting, w/positive lock, SS, 20 gallons		7	4		8,125	177		8,302	9,225
6600	60 gallons		6	4.667		10,800	207		11,007	12,200

11 47 Ice Machines

11 47 10 – Commercial Ice Machines

11 47 10.10 Commercial Ice Equipment

		Crew	Daily Output	Labor-Hours	Unit	Material	Labor	Equipment	Total	Total Incl O&P
0010	**COMMERCIAL ICE EQUIPMENT**									
5800	Ice cube maker, 50 pounds per day	Q-1	6	2.667	Ea.	1,600	138		1,738	1,950
6050	500 pounds per day	"	4	4	"	2,750	207		2,957	3,350

11 48 Cleaning and Disposal Equipment

11 48 13 – Commercial Dishwashers

11 48 13.10 Dishwashers

		Crew	Daily Output	Labor-Hours	Unit	Material	Labor	Equipment	Total	Total Incl O&P
0010	**DISHWASHERS**									
2700	Dishwasher, commercial, rack type									
2720	10 to 12 racks per hour	Q-1	3.20	5	Ea.	3,475	259		3,734	4,225
2730	Energy star rated, 35 to 40 racks/hour [G]		1.30	12.308		4,650	640		5,290	6,075
2740	50 to 60 racks/hour [G]		1.30	12.308		10,100	640		10,740	12,100
2800	Automatic, 190 to 230 racks per hour	L-6	.35	34.286		13,800	1,925		15,725	18,000
2820	235 to 275 racks per hour		.25	48		31,100	2,700		33,800	38,300
2840	8,750 to 12,500 dishes per hour		.10	120		54,500	6,750		61,250	70,000

11 53 Laboratory Equipment

11 53 13 – Laboratory Fume Hoods

11 53 13.13 Recirculating Laboratory Fume Hoods

		Crew	Daily Output	Labor-Hours	Unit	Material	Labor	Equipment	Total	Total Incl O&P
0010	**RECIRCULATING LABORATORY FUME HOODS**									
0670	Service fixtures, average				Ea.	237			237	261
0680	For sink assembly with hot and cold water, add	1 Plum	1.40	5.714	"	725	330		1,055	1,300

11 53 19 – Laboratory Sterilizers

11 53 19.13 Sterilizers

		Crew	Daily Output	Labor-Hours	Unit	Material	Labor	Equipment	Total	Total Incl O&P
0010	**STERILIZERS**									
0700	Glassware washer, undercounter, minimum	L-1	1.80	8.889	Ea.	6,250	495		6,745	7,625
0710	Maximum	"	1	16		13,100	885		13,985	15,700
1850	Utensil washer-sanitizer	1 Plum	2	4		11,100	230		11,330	12,600

11 53 23 – Laboratory Refrigerators

11 53 23.13 Refrigerators

		Crew	Daily Output	Labor-Hours	Unit	Material	Labor	Equipment	Total	Total Incl O&P
0010	**REFRIGERATORS**									
1200	Blood bank, 28.6 C.F. emergency signal				Ea.	9,550			9,550	10,500
1210	Reach-in, 16.9 C.F.				"	8,425			8,425	9,275

11 53 Laboratory Equipment

11 53 33 – Emergency Safety Appliances

11 53 33.13 Emergency Equipment

		Crew	Daily Output	Labor-Hours	Unit	Material	2014 Bare Costs Labor	Equipment	Total	Total Incl O&P
0010	**EMERGENCY EQUIPMENT**									
1400	Safety equipment, eye wash, hand held				Ea.	405			405	445
1450	Deluge shower				"	760			760	840

11 53 43 – Service Fittings and Accessories

11 53 43.13 Fittings

		Crew	Daily Output	Labor-Hours	Unit	Material	Labor	Equipment	Total	Total Incl O&P
0010	**FITTINGS**									
1600	Sink, one piece plastic, flask wash, hose, free standing	1 Plum	1.60	5	Ea.	1,925	288		2,213	2,550
1610	Epoxy resin sink, 25" x 16" x 10"	"	2	4	"	218	230		448	590
8000	Alternate pricing method: as percent of lab furniture									
8050	Installation, not incl. plumbing & duct work				% Furn.				22%	22%
8100	Plumbing, final connections, simple system								10%	10%
8110	Moderately complex system								15%	15%
8120	Complex system								20%	20%
8150	Electrical, simple system								10%	10%
8160	Moderately complex system								20%	20%
8170	Complex system								35%	35%

11 71 Medical Sterilizing Equipment

11 71 10 – Medical Sterilizers & Distillers

11 71 10.10 Sterilizers and Distillers

		Crew	Daily Output	Labor-Hours	Unit	Material	Labor	Equipment	Total	Total Incl O&P
0010	**STERILIZERS AND DISTILLERS**									
0700	Distiller, water, steam heated, 50 gal. capacity	1 Plum	1.40	5.714	Ea.	19,000	330		19,330	21,400
5600	Sterilizers, floor loading, 26" x 62" x 42", single door, steam					121,000			121,000	133,000
5650	Double door, steam					204,000			204,000	224,000
5800	General purpose, 20" x 20" x 38", single door					12,900			12,900	14,200
6000	Portable, counter top, steam, minimum					3,500			3,500	3,850
6020	Maximum					4,275			4,275	4,700
6050	Portable, counter top, gas, 17" x 15" x 32-1/2"					39,300			39,300	43,200
6150	Manual washer/sterilizer, 16" x 16" x 26"	1 Plum	2	4		54,000	230		54,230	59,500
6200	Steam generators, electric 10 kW to 180 kW, freestanding									
6250	Minimum	1 Elec	3	2.667	Ea.	8,675	142		8,817	9,775
6300	Maximum	"	.70	11.429		28,800	610		29,410	32,600
8200	Bed pan washer-sanitizer	1 Plum	2	4		7,425	230		7,655	8,500

11 73 Patient Care Equipment

11 73 10 – Patient Treatment Equipment

11 73 10.10 Treatment Equipment

		Crew	Daily Output	Labor-Hours	Unit	Material	Labor	Equipment	Total	Total Incl O&P
0010	**TREATMENT EQUIPMENT**									
1800	Heat therapy unit, humidified, 26" x 78" x 28"				Ea.	3,550			3,550	3,925
2100	Hubbard tank with accessories, stainless steel,									
2110	125 GPM at 45 psi water pressure				Ea.	26,800			26,800	29,500
2150	For electric overhead hoist, add					2,925			2,925	3,225
3600	Paraffin bath, 126°F, auto controlled					1,100			1,100	1,200
4600	Station, dietary, medium, with ice					16,500			16,500	18,200
8400	Whirlpool bath, mobile, sst, 18" x 24" x 60"					4,850			4,850	5,325
8450	Fixed, incl. mixing valves	1 Plum	2	4		9,575	230		9,805	10,900

75

11 74 Dental Equipment

11 74 10 – Dental Office Equipment

11 74 10.10 Diagnostic and Treatment Equipment

		Crew	Daily Output	Labor-Hours	Unit	Material	2014 Bare Costs Labor	Equipment	Total	Total Incl O&P
0010	**DIAGNOSTIC AND TREATMENT EQUIPMENT**									
0020	Central suction system, minimum	1 Plum	1.20	6.667	Ea.	1,550	385		1,935	2,275
0100	Maximum	"	.90	8.889		4,400	510		4,910	5,625
0300	Air compressor, minimum	1 Skwk	.80	10		3,000	475		3,475	4,025
0400	Maximum		.50	16		8,850	755		9,605	10,900
0600	Chair, electric or hydraulic, minimum		.50	16		2,225	755		2,980	3,625
0700	Maximum		.25	32		7,725	1,525		9,250	10,900
2000	Light, ceiling mounted, minimum		8	1		1,150	47.50		1,197.50	1,350
2100	Maximum		8	1		2,000	47.50		2,047.50	2,275
2200	Unit light, minimum	2 Skwk	5.33	3.002		730	142		872	1,025
2210	Maximum		5.33	3.002		1,550	142		1,692	1,950
2220	Track light, minimum		3.20	5		1,550	237		1,787	2,100
2230	Maximum		3.20	5		2,625	237		2,862	3,275
2300	Sterilizers, steam portable, minimum					1,300			1,300	1,425
2350	Maximum					10,400			10,400	11,400
2600	Steam, institutional					3,250			3,250	3,550
2650	Dry heat, electric, portable, 3 trays					1,225			1,225	1,325

11 76 Operating Room Equipment

11 76 10 – Operating Room Equipment

11 76 10.10 Surgical Equipment

		Crew	Daily Output	Labor-Hours	Unit	Material	Labor	Equipment	Total	Total Incl O&P
0010	**SURGICAL EQUIPMENT**									
5000	Scrub, surgical, stainless steel, single station, minimum	1 Plum	3	2.667	Ea.	4,250	153		4,403	4,900
5100	Maximum				"	6,900			6,900	7,600

11 78 Mortuary Equipment

11 78 13 – Mortuary Refrigerators

11 78 13.10 Mortuary and Autopsy Equipment

		Crew	Daily Output	Labor-Hours	Unit	Material	Labor	Equipment	Total	Total Incl O&P
0010	**MORTUARY AND AUTOPSY EQUIPMENT**									
0015	Autopsy table, standard	1 Plum	1	8	Ea.	9,625	460		10,085	11,300
0020	Deluxe	"	.60	13.333	"	14,900	765		15,665	17,600

11 91 Religious Equipment

11 91 13 – Baptisteries

11 91 13.10 Baptistry

		Crew	Daily Output	Labor-Hours	Unit	Material	Labor	Equipment	Total	Total Incl O&P
0010	**BAPTISTRY**									
0150	Fiberglass, 3'-6" deep, x 13'-7" long,									
0160	steps at both ends, incl. plumbing, minimum	L-8	1	20	Ea.	5,650	965		6,615	7,700
0200	Maximum	"	.70	28.571		9,225	1,375		10,600	12,200
0250	Add for filter, heater and lights					1,825			1,825	2,025

11 91 23 – Sanctuary Equipment

11 91 23.10 Sanctuary Furnishings

		Crew	Daily Output	Labor-Hours	Unit	Material	Labor	Equipment	Total	Total Incl O&P
0010	**SANCTUARY FURNISHINGS**									
0020	Altar, wood, custom design, plain	1 Carp	1.40	5.714	Ea.	2,525	262		2,787	3,175
0050	Deluxe	"	.20	40	"	12,200	1,825		14,025	16,200

Estimating Tips
General
- The items in this division are usually priced per square foot or each. Most of these items are purchased by the owner and installed by the contractor. Do not assume the items in Division 12 will be purchased and installed by the contractor. Check the specifications for responsibilities and include receiving, storage, installation, and mechanical and electrical hookups in the appropriate divisions.
- Some items in this division require some type of support system that is not usually furnished with the item. Examples of these systems include blocking for the attachment of casework and heavy drapery rods. The required blocking must be added to the estimate in the appropriate division.

Reference Numbers
Reference numbers are shown in shaded boxes at the beginning of some major classifications. These numbers refer to related items in the Reference Section. The reference information may be an estimating procedure, an alternate pricing method, or technical information.

Note: Not all subdivisions listed here necessarily appear in this publication.

No part of this publication may be reproduced, stored in a retrieval system, or transmitted in any form or by any means without prior written permission of RSMeans, a division of Reed Construction Data.

12 32 Manufactured Wood Casework

12 32 23 – Hardwood Casework

12 32 23.10 Manufactured Wood Casework, Stock Units

		Crew	Daily Output	Labor-Hours	Unit	Material	2014 Bare Costs Labor	Equipment	Total	Total Incl O&P
0010	**MANUFACTURED WOOD CASEWORK, STOCK UNITS**									
0700	Kitchen base cabinets, hardwood, not incl. counter tops,									
0710	24" deep, 35" high, prefinished									
0800	One top drawer, one door below, 12" wide	2 Carp	24.80	.645	Ea.	254	29.50		283.50	325
0840	18" wide		23.30	.687		287	31.50		318.50	365
0880	24" wide		22.30	.717		350	33		383	435
1000	Four drawers, 12" wide		24.80	.645		267	29.50		296.50	340
1040	18" wide		23.30	.687		298	31.50		329.50	380
1060	24" wide		22.30	.717		330	33		363	410
1200	Two top drawers, two doors below, 27" wide		22	.727		375	33.50		408.50	465
1260	36" wide		20.30	.788		435	36		471	535
1300	48" wide		18.90	.847		495	39		534	605
1500	Range or sink base, two doors below, 30" wide		21.40	.748		335	34.50		369.50	425
1540	36" wide		20.30	.788		380	36		416	470
1580	48" wide	↓	18.90	.847		415	39		454	520
1800	For sink front units, deduct				↓	154			154	169
9000	For deluxe models of all cabinets, add					40%				
9500	For custom built in place, add					25%	10%			
9558	Rule of thumb, kitchen cabinets not including									
9560	appliances & counter top, minimum	2 Carp	30	.533	L.F.	168	24.50		192.50	223
9600	Maximum	"	25	.640	"	380	29.50		409.50	460

12 32 23.30 Manufactured Wood Casework Vanities

		Crew	Daily Output	Labor-Hours	Unit	Material	Labor	Equipment	Total	Total Incl O&P
0010	**MANUFACTURED WOOD CASEWORK VANITIES**									
8000	Vanity bases, 2 doors, 30" high, 21" deep, 24" wide	2 Carp	20	.800	Ea.	297	36.50		333.50	380
8050	30" wide		16	1		355	46		401	460
8100	36" wide		13.33	1.200		345	55		400	465
8150	48" wide	↓	11.43	1.400	↓	450	64		514	595
9000	For deluxe models of all vanities, add to above					40%				
9500	For custom built in place, add to above					25%	10%			

Estimating Tips

General

- The items and systems in this division are usually estimated, purchased, supplied, and installed as a unit by one or more subcontractors. The estimator must ensure that all parties are operating from the same set of specifications and assumptions, and that all necessary items are estimated and will be provided. Many times the complex items and systems are covered, but the more common ones, such as excavation or a crane, are overlooked for the very reason that everyone assumes nobody could miss them. The estimator should be the central focus and be able to ensure that all systems are complete.
- Another area where problems can develop in this division is at the interface between systems. The estimator must ensure, for instance, that anchor bolts, nuts, and washers are estimated and included for the air-supported structures and pre-engineered buildings to be bolted to their foundations. Utility supply is a common area where essential items or pieces of equipment can be missed or overlooked due to the fact that each subcontractor may feel it is another's responsibility. The estimator should also be aware of certain items which may be supplied as part of a package but installed by others, and ensure that the installing contractor's estimate includes the cost of installation. Conversely, the estimator must also ensure that items are not costed by two different subcontractors, resulting in an inflated overall estimate.

13 30 00 Special Structures

- The foundations and floor slab, as well as rough mechanical and electrical, should be estimated, as this work is required for the assembly and erection of the structure. Generally, as noted in the book, the pre-engineered building comes as a shell. Pricing is based on the size and structural design parameters stated in the reference section. Additional features, such as windows and doors with their related structural framing, must also be included by the estimator. Here again, the estimator must have a clear understanding of the scope of each portion of the work and all the necessary interfaces.

Reference Numbers

Reference numbers are shown in shaded boxes at the beginning of some major classifications. These numbers refer to related items in the Reference Section. The reference information may be an estimating procedure, an alternate pricing method, or technical information.

Note: Not all subdivisions listed here necessarily appear in this publication.

13 11 Swimming Pools

13 11 13 – Below-Grade Swimming Pools

13 11 13.50 Swimming Pools		Crew	Daily Output	Labor-Hours	Unit	Material	2014 Bare Costs Labor	Equipment	Total	Total Incl O&P
0010	**SWIMMING POOLS** Residential in-ground, vinyl lined, concrete									
0020	Sides including equipment, sand bottom	B-52	300	.187	SF Surf	21.50	7.95	1.98	31.43	38
0100	Metal or polystyrene sides	R131113-20	B-14	410	.117		17.75	4.53	.89	23.17	27.50
0200	Add for vermiculite bottom						1.35			1.35	1.49
0500	Gunite bottom and sides, white plaster finish										
0600	12' x 30' pool	B-52	145	.386	SF Surf	39.50	16.40	4.11	60.01	73
0720	16' x 32' pool		155	.361		35.50	15.35	3.84	54.69	66.50
0750	20' x 40' pool		250	.224		32	9.50	2.38	43.88	52
0810	Concrete bottom and sides, tile finish										
0820	12' x 30' pool	B-52	80	.700	SF Surf	40	30	7.45	77.45	97.50
0830	16' x 32' pool		95	.589		33	25	6.25	64.25	82
0840	20' x 40' pool		130	.431		26.50	18.30	4.58	49.38	62
1100	Motel, gunite with plaster finish, incl. medium										
1150	capacity filtration & chlorination	B-52	115	.487	SF Surf	48.50	20.50	5.20	74.20	90.50
1200	Municipal, gunite with plaster finish, incl. high										
1250	capacity filtration & chlorination	B-52	100	.560	SF Surf	63	24	5.95	92.95	113
1350	Add for formed gutters				L.F.	92.50			92.50	102
1360	Add for stainless steel gutters				"	274			274	300
1600	For water heating system, see Section 23 52 28.10									
1700	Filtration and deck equipment only, as % of total				Total				20%	20%
1800	Deck equipment, rule of thumb, 20' x 40' pool				SF Pool				1.18	1.30
1900	5000 S.F. pool				"				1.73	1.90
3000	Painting pools, preparation + 3 coats, 20' x 40' pool, epoxy	2 Pord	.33	48.485	Total	1,775	1,925		3,700	4,825
3100	Rubber base paint, 18 gallons	"	.33	48.485		1,325	1,925		3,250	4,325
3500	42' x 82' pool, 75 gallons, epoxy paint	3 Pord	.14	171		7,500	6,775		14,275	18,500
3600	Rubber base paint	"	.14	171		5,450	6,775		12,225	16,200

13 11 46 – Swimming Pool Accessories

13 11 46.50 Swimming Pool Equipment		Crew	Daily Output	Labor-Hours	Unit	Material	Labor	Equipment	Total	Total Incl O&P
0010	**SWIMMING POOL EQUIPMENT**									
0020	Diving stand, stainless steel, 3 meter	2 Carp	.40	40	Ea.	14,500	1,825		16,325	18,700
0300	1 meter	"	2.70	5.926		8,825	272		9,097	10,100
2100	Lights, underwater, 12 volt, with transformer, 300 watt	1 Elec	1	8		335	425		760	1,000
2200	110 volt, 500 watt, standard		1	8		293	425		718	960
2400	Low water cutoff type		1	8		300	425		725	970
2800	Heaters, see Section 23 52 28.10									

13 17 Tubs and Pools

13 17 13 – Hot Tubs

13 17 13.10 Redwood Hot Tub System

		Crew	Daily Output	Labor-Hours	Unit	Material	Labor	Equipment	Total	Total Incl O&P
0010	**REDWOOD HOT TUB SYSTEM**									
7050	4' diameter x 4' deep	Q-1	1	16	Ea.	3,225	830		4,055	4,800
7100	5' diameter x 4' deep		1	16		4,100	830		4,930	5,750
7150	6' diameter x 4' deep		.80	20		4,950	1,025		5,975	7,000
7200	8' diameter x 4' deep		.80	20		7,250	1,025		8,275	9,550

13 17 33 – Whirlpool Tubs

13 17 33.10 Whirlpool Bath

		Crew	Daily Output	Labor-Hours	Unit	Material	Labor	Equipment	Total	Total Incl O&P
0010	**WHIRLPOOL BATH**									
6000	Whirlpool, bath with vented overflow, molded fiberglass									
6100	66" x 36" x 24"	Q-1	1	16	Ea.	3,325	830		4,155	4,900
6400	72" x 36" x 21"		1	16		1,800	830		2,630	3,250

13 17 Tubs and Pools

13 17 33 – Whirlpool Tubs

13 17 33.10 Whirlpool Bath		Crew	Daily Output	Labor-Hours	Unit	Material	2014 Bare Costs Labor	Equipment	Total	Total Incl O&P
6500	60" x 34" x 21"	Q-1	1	16	Ea.	1,700	830		2,530	3,125
6600	72" x 42" x 23"	↓	1	16		2,175	830		3,005	3,650
6710	For color add					10%				
6711	For designer colors and trim add					25%				

13 24 Special Activity Rooms

13 24 16 – Saunas

13 24 16.50 Saunas and Heaters		Crew	Daily Output	Labor-Hours	Unit	Material	Labor	Equipment	Total	Total Incl O&P
0010	**SAUNAS AND HEATERS**									
0020	Prefabricated, incl. heater & controls, 7' high, 6' x 4', C/C	L-7	2.20	12.727	Ea.	5,375	565		5,940	6,800
0050	6' x 4', C/P		2	14		4,875	620		5,495	6,325
0400	6' x 5', C/C		2	14		5,675	620		6,295	7,200
0450	6' x 5', C/P		2	14		5,150	620		5,770	6,600
0600	6' x 6', C/C		1.80	15.556		6,275	690		6,965	7,950
0650	6' x 6', C/P		1.80	15.556		5,725	690		6,415	7,350
0800	6' x 9', C/C		1.60	17.500		7,750	775		8,525	9,725
0850	6' x 9', C/P		1.60	17.500		7,000	775		7,775	8,900
1000	8' x 12', C/C		1.10	25.455		11,600	1,125		12,725	14,500
1050	8' x 12', C/P		1.10	25.455		10,500	1,125		11,625	13,200
1200	8' x 8', C/C		1.40	20		9,150	885		10,035	11,500
1250	8' x 8', C/P		1.40	20		8,450	885		9,335	10,600
1400	8' x 10', C/C		1.20	23.333		10,100	1,025		11,125	12,800
1450	8' x 10', C/P		1.20	23.333		9,225	1,025		10,250	11,800
1600	10' x 12', C/C		1	28		12,300	1,250		13,550	15,400
1650	10' x 12', C/P	↓	1	28		10,900	1,250		12,150	13,900
2500	Heaters only (incl. above), wall mounted, to 200 C.F.					675			675	745
2750	To 300 C.F.					910			910	1,000
3000	Floor standing, to 720 C.F., 10,000 watts, w/controls	1 Elec	3	2.667		2,900	142		3,042	3,425
3250	To 1,000 C.F., 16,000 watts	"	3	2.667		3,700	142		3,842	4,300

13 24 26 – Steam Baths

13 24 26.50 Steam Baths and Components		Crew	Daily Output	Labor-Hours	Unit	Material	Labor	Equipment	Total	Total Incl O&P
0010	**STEAM BATHS AND COMPONENTS**									
0020	Heater, timer & head, single, to 140 C.F.	1 Plum	1.20	6.667	Ea.	1,450	385		1,835	2,175
0500	To 300 C.F.		1.10	7.273		1,550	420		1,970	2,325
1000	Commercial size, with blow-down assembly, to 800 C.F.		.90	8.889		7,350	510		7,860	8,850
1500	To 2500 C.F.	↓	.80	10		9,125	575		9,700	10,900
2000	Multiple, motels, apts., 2 baths, w/blow-down assm., 500 C.F.	Q-1	1.30	12.308		7,950	640		8,590	9,725
2500	4 baths	"	.70	22.857		10,700	1,175		11,875	13,600
2700	Conversion unit for residential tub, including door					3,600			3,600	3,950

13 34 Fabricated Engineered Structures

13 34 23 – Fabricated Structures

13 34 23.10 Comfort Stations

		Crew	Daily Output	Labor-Hours	Unit	Material	2014 Bare Costs Labor	Equipment	Total	Total Incl O&P
0010	**COMFORT STATIONS** Prefab., stock, w/doors, windows & fixt.									
0100	Not incl. interior finish or electrical									
0300	Mobile, on steel frame, 2 unit				S.F.	196			196	215
0350	7 unit					249			249	274
0400	Permanent, including concrete slab, 2 unit	B-12J	50	.320		249	13.90	17.65	280.55	315
0500	6 unit	"	43	.372		191	16.15	20.50	227.65	258
0600	Alternate pricing method, mobile, 2 fixture				Fixture	5,575			5,575	6,150
0650	7 fixture					8,975			8,975	9,875
0700	Permanent, 2 unit	B-12J	.70	22.857		20,500	995	1,250	22,745	25,500
0750	6 unit	"	.50	32		17,400	1,400	1,775	20,575	23,300

13 42 Building Modules

13 42 63 – Detention Cell Modules

13 42 63.16 Steel Detention Cell Modules

		Crew	Daily Output	Labor-Hours	Unit	Material	2014 Bare Costs Labor	Equipment	Total	Total Incl O&P
0010	**STEEL DETENTION CELL MODULES**									
2000	Cells, prefab., 5' to 6' wide, 7' to 8' high, 7' to 8' deep,									
2010	bar front, cot, not incl. plumbing	E-4	1.50	21.333	Ea.	9,650	1,100	94.50	10,844.50	12,700

Estimating Tips

General
- Many products in Division 14 will require some type of support or blocking for installation not included with the item itself. Examples are supports for conveyors or tube systems, attachment points for lifts, and footings for hoists or cranes. Add these supports in the appropriate division.

14 10 00 Dumbwaiters
14 20 00 Elevators
- Dumbwaiters and elevators are estimated and purchased in a method similar to buying a car. The manufacturer has a base unit with standard features. Added to this base unit price will be whatever options the owner or specifications require. Increased load capacity, additional vertical travel, additional stops, higher speed, and cab finish options are items to be considered. When developing an estimate for dumbwaiters and elevators, remember that some items needed by the installers may have to be included as part of the general contract. Examples are:
  — shaftway
  — rail support brackets
  — machine room
  — electrical supply
  — sill angles
  — electrical connections
  — pits
  — roof penthouses
  — pit ladders

  Check the job specifications and drawings before pricing.
- Installation of elevators and handicapped lifts in historic structures can require significant additional costs. The associated structural requirements may involve cutting into and repairing finishes, mouldings, flooring, etc. The estimator must account for these special conditions.

14 30 00 Escalators and Moving Walks
- Escalators and moving walks are specialty items installed by specialty contractors. There are numerous options associated with these items. For specific options, contact a manufacturer or contractor. In a method similar to estimating dumbwaiters and elevators, you should verify the extent of general contract work and add items as necessary.

14 40 00 Lifts
14 90 00 Other Conveying Equipment
- Products such as correspondence lifts, chutes, and pneumatic tube systems, as well as other items specified in this subdivision, may require trained installers. The general contractor might not have any choice as to who will perform the installation or when it will be performed. Long lead times are often required for these products, making early decisions in scheduling necessary.

Reference Numbers
Reference numbers are shown in shaded boxes at the beginning of some major classifications. These numbers refer to related items in the Reference Section. The reference information may be an estimating procedure, an alternate pricing method, or technical information.

Note: Not all subdivisions listed here necessarily appear in this publication.

No part of this publication may be reproduced, stored in a retrieval system, or transmitted in any form or by any means without prior written permission of RSMeans, a division of Reed Construction Data.

14 45 Vehicle Lifts

14 45 10 – Hydraulic Vehicle Lifts

14 45 10.10 Hydraulic Lifts		Crew	Daily Output	Labor-Hours	Unit	Material	2014 Bare Costs Labor	Equipment	Total	Total Incl O&P
0010	**HYDRAULIC LIFTS**									
2200	Single post, 8000 lb. capacity	L-4	.40	60	Ea.	5,500	2,575		8,075	10,100
2810	Double post, 6000 lb. capacity		2.67	8.989		7,750	385		8,135	9,125
2815	9000 lb. capacity		2.29	10.480		18,400	450		18,850	20,900
2820	15,000 lb. capacity		2	12		20,800	515		21,315	23,600
2822	26,000 lb. capacity		1.80	13.333		13,700	575		14,275	16,000
2825	30,000 lb. capacity		1.60	15		45,900	645		46,545	51,500
2830	Ramp style, 4 post, 25,000 lb. capacity		2	12		18,000	515		18,515	20,600
2835	35,000 lb. capacity		1	24		83,500	1,025		84,525	93,500
2840	50,000 lb. capacity		1	24		93,500	1,025		94,525	104,500
2845	75,000 lb. capacity		1	24		108,500	1,025		109,525	121,000
2850	For drive thru tracks, add, minimum					1,150			1,150	1,275
2855	Maximum					1,975			1,975	2,175
2860	Ramp extensions, 3′ (set of 2)					945			945	1,050
2865	Rolling jack platform					3,300			3,300	3,625
2870	Electric/hydraulic jacking beam					8,825			8,825	9,700
2880	Scissor lift, portable, 6000 lb. capacity					8,650			8,650	9,525

Division 21 - Fire Suppression

Estimating Tips

Pipe for fire protection and all uses is located in Subdivisions 21 11 13 and 22 11 13.

The labor adjustment factors listed in Subdivision 22 01 02.20 also apply to Division 21.

Many, but not all, areas in the U.S. require backflow protection in the fire system. It is advisable to check local building codes for specific requirements.

For your reference, the following is a list of the most applicable Fire Codes and Standards which may be purchased from the NFPA, 1 Batterymarch Park, Quincy, MA 02169-7471.

NFPA 1: Uniform Fire Code
NFPA 10: Portable Fire Extinguishers
NFPA 11: Low-, Medium-, and High-Expansion Foam
NFPA 12: Carbon Dioxide Extinguishing Systems (Also companion 12A)
NFPA 13: Installation of Sprinkler Systems (Also companion 13D, 13E, and 13R)
NFPA 14: Installation of Standpipe and Hose Systems
NFPA 15: Water Spray Fixed Systems for Fire Protection
NFPA 16: Installation of Foam-Water Sprinkler and Foam-Water Spray Systems
NFPA 17: Dry Chemical Extinguishing Systems (Also companion 17A)
NFPA 18: Wetting Agents
NFPA 20: Installation of Stationary Pumps for Fire Protection
NFPA 22: Water Tanks for Private Fire Protection
NFPA 24: Installation of Private Fire Service Mains and their Appurtenances
NFPA 25: Inspection, Testing and Maintenance of Water-Based Fire Protection

Reference Numbers

Reference numbers are shown in shaded boxes at the beginning of some major classifications. These numbers refer to related items in the Reference Section. The reference information may be an estimating procedure, an alternate pricing method, or technical information.

Note: Not all subdivisions listed here necessarily appear in this publication.

Note: **Trade Service,** *in part, has been used as a reference source for some of the material prices used in Division 21.*

21 05 Common Work Results for Fire Suppression

21 05 23 – General-Duty Valves for Water-Based Fire-Suppression Piping

21 05 23.50 General-Duty Valves

		Crew	Daily Output	Labor-Hours	Unit	Material	2014 Bare Costs Labor	Equipment	Total	Total Incl O&P
0010	**GENERAL-DUTY VALVES**, for water-based fire suppression									
6200	Valves and components									
6210	Alarm, includes									
6220	retard chamber, trim, gauges, alarm line strainer									
6260	3" size	Q-12	3	5.333	Ea.	1,550	266		1,816	2,100
6280	4" size	"	2	8		1,625	400		2,025	2,375
6300	6" size	Q-13	4	8		1,800	420		2,220	2,650
6320	8" size	"	3	10.667	↓	2,175	565		2,740	3,250
6500	Check, swing, C.I. body, brass fittings, auto. ball drip									
6520	4" size	Q-12	3	5.333	Ea.	335	266		601	765
6540	6" size	Q-13	4	8		615	420		1,035	1,325
6580	8" size	"	3	10.667		1,175	565		1,740	2,125
6800	Check, wafer, butterfly type, C.I. body, bronze fittings									
6820	4" size	Q-12	4	4	Ea.	460	199		659	805
6840	6" size	Q-13	5.50	5.818		785	305		1,090	1,325
6860	8" size		5	6.400		1,400	340		1,740	2,050
6880	10" size	↓	4.50	7.111		2,150	375		2,525	2,950
8800	Flow control valve, includes trim and gauges, 2" size	Q-12	2	8		4,475	400		4,875	5,525
8820	3" size	"	1.50	10.667		4,925	530		5,455	6,225
8840	4" size	Q-13	2.80	11.429		5,450	605		6,055	6,900
8860	6" size	"	2	16		6,300	845		7,145	8,225
9200	Pressure operated relief valve, brass body	1 Spri	18	.444	↓	550	24.50		574.50	640
9600	Waterflow indicator, with recycling retard and									
9610	two single pole retard switches, 2" thru 6" pipe size	1 Spri	8	1	Ea.	129	55.50		184.50	226

21 11 Facility Fire-Suppression Water-Service Piping

21 11 13 – Facility Water Distribution Piping

21 11 13.16 Pipe, Plastic

		Crew	Daily Output	Labor-Hours	Unit	Material	Labor	Equipment	Total	Total Incl O&P
0010	**PIPE, PLASTIC**									
0020	CPVC, fire suppression, (C-UL-S. FM, NFPA 13, 13D & 13R)									
0030	Socket joint, no couplings or hangers									
0100	SDR 13.5, (ASTM F442)									
0120	3/4" diameter	Q-12	420	.038	L.F.	1.62	1.90		3.52	4.65
0130	1" diameter		340	.047		2.50	2.35		4.85	6.30
0140	1-1/4" diameter		260	.062		3.90	3.07		6.97	8.95
0150	1-1/2" diameter		190	.084		5.40	4.20		9.60	12.25
0160	2" diameter		140	.114		8.60	5.70		14.30	18.05
0170	2-1/2" diameter		130	.123		15.85	6.15		22	26.50
0180	3" diameter	↓	120	.133		24	6.65		30.65	36.50

21 11 13.18 Pipe Fittings, Plastic

		Crew	Daily Output	Labor-Hours	Unit	Material	Labor	Equipment	Total	Total Incl O&P
0010	**PIPE FITTINGS, PLASTIC**									
0020	CPVC, fire suppression, (C-UL-S. FM, NFPA 13, 13D & 13R)									
0030	Socket joint									
0100	90° Elbow									
0120	3/4"	1 Plum	26	.308	Ea.	2.30	17.70		20	29
0130	1"		22.70	.352		5.05	20.50		25.55	36
0140	1-1/4"		20.20	.396		6.40	23		29.40	41.50
0150	1-1/2"	↓	18.20	.440		9.10	25.50		34.60	48
0160	2"	Q-1	33.10	.483		11.30	25		36.30	50.50
0170	2-1/2"		24.20	.661		21.50	34.50		56	75.50
0180	3"	↓	20.80	.769		29.50	40		69.50	92.50

21 11 Facility Fire-Suppression Water-Service Piping

21 11 13 – Facility Water Distribution Piping

21 11 13.18 Pipe Fittings, Plastic		Crew	Daily Output	Labor-Hours	Unit	Material	2014 Bare Costs Labor	Equipment	Total	Total Incl O&P
0200	45° Elbow									
0210	3/4"	1 Plum	26	.308	Ea.	3.17	17.70		20.87	30
0220	1"		22.70	.352		3.71	20.50		24.21	34.50
0230	1-1/4"		20.20	.396		5.35	23		28.35	40.50
0240	1-1/2"		18.20	.440		7.50	25.50		33	46.50
0250	2"	Q-1	33.10	.483		9.30	25		34.30	48.50
0260	2-1/2"		24.20	.661		16.75	34.50		51.25	70
0270	3"		20.80	.769		24	40		64	86.50
0300	Tee									
0310	3/4"	1 Plum	17.30	.462	Ea.	3.17	26.50		29.67	43.50
0320	1"		15.20	.526		6.25	30.50		36.75	52.50
0330	1-1/4"		13.50	.593		9.40	34		43.40	62
0340	1-1/2"		12.10	.661		13.80	38		51.80	72.50
0350	2"	Q-1	20	.800		20.50	41.50		62	85
0360	2-1/2"		16.20	.988		33	51		84	114
0370	3"		13.90	1.151		52	59.50		111.50	147
0400	Tee, reducing x any size									
0420	1"	1 Plum	15.20	.526	Ea.	5.25	30.50		35.75	51.50
0430	1-1/4"		13.50	.593		9.70	34		43.70	62
0440	1-1/2"		12.10	.661		11.75	38		49.75	70.50
0450	2"	Q-1	20	.800		18.80	41.50		60.30	83
0460	2-1/2"		16.20	.988		26	51		77	106
0470	3"		13.90	1.151		30	59.50		89.50	123
0500	Coupling									
0510	3/4"	1 Plum	26	.308	Ea.	2.21	17.70		19.91	29
0520	1"		22.70	.352		2.93	20.50		23.43	33.50
0530	1-1/4"		20.20	.396		4.27	23		27.27	39
0540	1-1/2"		18.20	.440		6.10	25.50		31.60	44.50
0550	2"	Q-1	33.10	.483		8.20	25		33.20	47
0560	2-1/2"		24.20	.661		12.55	34.50		47.05	65.50
0570	3"		20.80	.769		16.35	40		56.35	78
0600	Coupling, reducing									
0610	1" x 3/4"	1 Plum	22.70	.352	Ea.	2.93	20.50		23.43	33.50
0620	1-1/4" x 1"		20.20	.396		4.43	23		27.43	39.50
0630	1-1/2" x 3/4"		18.20	.440		6.65	25.50		32.15	45.50
0640	1-1/2" x 1"		18.20	.440		6.40	25.50		31.90	45
0650	1-1/2" x 1-1/4"		18.20	.440		6.10	25.50		31.60	44.50
0660	2" x 1"	Q-1	33.10	.483		8.55	25		33.55	47.50
0670	2" x 1-1/2"	"	33.10	.483		8.20	25		33.20	47
0700	Cross									
0720	3/4"	1 Plum	13	.615	Ea.	4.98	35.50		40.48	59
0730	1"		11.30	.708		6.25	40.50		46.75	68.50
0740	1-1/4"		10.10	.792		8.60	45.50		54.10	78.50
0750	1-1/2"		9.10	.879		11.35	50.50		61.85	89
0760	2"	Q-1	16.60	.964		19.45	50		69.45	97
0770	2-1/2"	"	12.10	1.322		42.50	68.50		111	150
0800	Cap									
0820	3/4"	1 Plum	52	.154	Ea.	1.28	8.85		10.13	14.75
0830	1"		45	.178		1.90	10.25		12.15	17.55
0840	1-1/4"		40	.200		3.30	11.50		14.80	21
0850	1-1/2"		36.40	.220		4.27	12.65		16.92	24
0860	2"	Q-1	66	.242		6.40	12.55		18.95	26
0870	2-1/2"		48.40	.331		9.25	17.10		26.35	36

21 11 Facility Fire-Suppression Water-Service Piping

21 11 13 – Facility Water Distribution Piping

21 11 13.18 Pipe Fittings, Plastic

		Crew	Daily Output	Labor-Hours	Unit	Material	2014 Bare Costs Labor	Equipment	Total	Total Incl O&P
0880	3"	Q-1	41.60	.385	Ea.	14.95	19.90		34.85	46.50
0900	Adapter, sprinkler head, female w/metal thd. insert, (s x FNPT)									
0920	3/4" x 1/2"	1 Plum	52	.154	Ea.	5.85	8.85		14.70	19.75
0930	1" x 1/2"		45	.178		6.15	10.25		16.40	22.50
0940	1" x 3/4"	↓	45	.178	↓	9.70	10.25		19.95	26

21 11 16 – Facility Fire Hydrants

21 11 16.50 Fire Hydrants for Buildings

		Crew	Daily Output	Labor-Hours	Unit	Material	Labor	Equipment	Total	Total Incl O&P
0010	**FIRE HYDRANTS FOR BUILDINGS**									
3750	Hydrants, wall, w/caps, single, flush, polished brass									
3800	2-1/2" x 2-1/2"	Q-12	5	3.200	Ea.	193	160		353	455
3840	2-1/2" x 3"		5	3.200		410	160		570	690
3860	3" x 3"	↓	4.80	3.333		330	166		496	615
3900	For polished chrome, add				↓	20%				
3950	Double, flush, polished brass									
4000	2-1/2" x 2-1/2" x 4"	Q-12	5	3.200	Ea.	545	160		705	840
4040	2-1/2" x 2-1/2" x 6"		4.60	3.478		785	173		958	1,125
4080	3" x 3" x 4"		4.90	3.265		1,175	163		1,338	1,525
4120	3" x 3" x 6"	↓	4.50	3.556		1,175	177		1,352	1,550
4200	For polished chrome, add				↓	10%				
4350	Double, projecting, polished brass									
4400	2-1/2" x 2-1/2" x 4"	Q-12	5	3.200	Ea.	242	160		402	505
4450	2-1/2" x 2-1/2" x 6"	"	4.60	3.478	"	495	173		668	805
4460	Valve control, dbl. flush/projecting hydrant, cap &									
4470	chain, extension rod & cplg., escutcheon, polished brass	Q-12	8	2	Ea.	288	99.50		387.50	465
4480	Four-way square, flush, polished brass									
4540	2-1/2"(4) x 6"	Q-12	3.60	4.444	Ea.	2,925	222		3,147	3,550

21 11 19 – Fire-Department Connections

21 11 19.50 Connections for the Fire-Department

		Crew	Daily Output	Labor-Hours	Unit	Material	Labor	Equipment	Total	Total Incl O&P
0010	**CONNECTIONS FOR THE FIRE-DEPARTMENT**									
6000	Roof manifold, horiz., brass, without valves & caps									
6040	2-1/2" x 2-1/2" x 4"	Q-12	4.80	3.333	Ea.	160	166		326	425
6060	2-1/2" x 2-1/2" x 6"		4.60	3.478		173	173		346	455
6080	2-1/2" x 2-1/2" x 2-1/2" x 4"		4.60	3.478		256	173		429	545
6090	2-1/2" x 2-1/2" x 2-1/2" x 6"	↓	4.60	3.478		263	173		436	550
7000	Sprinkler line tester, cast brass				↓	23.50			23.50	25.50
7140	Standpipe connections, wall, w/plugs & chains									
7160	Single, flush, brass, 2-1/2" x 2-1/2"	Q-12	5	3.200	Ea.	151	160		311	410
7180	2-1/2" x 3"	"	5	3.200	"	156	160		316	415
7240	For polished chrome, add					15%				
7280	Double, flush, polished brass									
7300	2-1/2" x 2-1/2" x 4"	Q-12	5	3.200	Ea.	495	160		655	785
7330	2-1/2" x 2-1/2" x 6"		4.60	3.478		690	173		863	1,025
7340	3" x 3" x 4"		4.90	3.265		900	163		1,063	1,225
7370	3" x 3" x 6"	↓	4.50	3.556	↓	1,075	177		1,252	1,450
7400	For polished chrome, add					15%				
7440	For sill cock combination, add				Ea.	86			86	94.50
7580	Double projecting, polished brass									
7600	2-1/2" x 2-1/2" x 4"	Q-12	5	3.200	Ea.	450	160		610	735
7630	2-1/2" x 2-1/2" x 6"	"	4.60	3.478	"	765	173		938	1,100
7680	For polished chrome, add					15%				
7900	Three way, flush, polished brass									
7920	2-1/2" (3) x 4"	Q-12	4.80	3.333	Ea.	1,575	166		1,741	2,000

21 11 Facility Fire-Suppression Water-Service Piping

21 11 19 – Fire-Department Connections

21 11 19.50 Connections for the Fire-Department		Crew	Daily Output	Labor-Hours	Unit	Material	2014 Bare Costs Labor	Equipment	Total	Total Incl O&P
7930	2-1/2" (3) x 6"	Q-12	4.80	3.333	Ea.	1,575	166		1,741	2,000
8000	For polished chrome, add				↓	9%				
8020	Three way, projecting, polished brass									
8040	2-1/2"(3) x 4"	Q-12	4.80	3.333	Ea.	1,400	166		1,566	1,800
8070	2-1/2" (3) x 6"	"	4.60	3.478		1,400	173		1,573	1,800
8100	For polished chrome, add				↓	12%				
8200	Four way, square, flush, polished brass,									
8240	2-1/2"(4) x 6"	Q-12	3.60	4.444	Ea.	1,250	222		1,472	1,700
8300	For polished chrome, add				"	10%				
8550	Wall, vertical, flush, cast brass									
8600	Two way, 2-1/2" x 2-1/2" x 4"	Q-12	5	3.200	Ea.	395	160		555	675
8660	Four way, 2-1/2"(4) x 6"		3.80	4.211		1,300	210		1,510	1,750
8680	Six way, 2-1/2"(6) x 6"	↓	3.40	4.706		1,550	235		1,785	2,075
8700	For polished chrome, add				↓	10%				
8800	Sidewalk siamese unit, polished brass, two way									
8820	2-1/2" x 2-1/2" x 4"	Q-12	2.50	6.400	Ea.	605	320		925	1,150
8850	2-1/2" x 2-1/2" x 6"		2	8		740	400		1,140	1,425
8860	3" x 3" x 4"		2.50	6.400		860	320		1,180	1,425
8890	3" x 3" x 6"	↓	2	8	↓	1,150	400		1,550	1,875
8940	For polished chrome, add					12%				
9100	Sidewalk siamese unit, polished brass, three way									
9120	2-1/2" x 2-1/2" x 2-1/2" x 6"	Q-12	2	8	Ea.	960	400		1,360	1,650
9160	For polished chrome, add				"	15%				

21 12 Fire-Suppression Standpipes

21 12 13 – Fire-Suppression Hoses and Nozzles

21 12 13.50 Fire Hoses and Nozzles			Crew	Daily Output	Labor-Hours	Unit	Material	Labor	Equipment	Total	Total Incl O&P
0010	**FIRE HOSES AND NOZZLES**										
0200	Adapters, rough brass, straight hose threads	R211226-10									
0220	One piece, female to male, rocker lugs										
0240	1" x 1"	R211226-20				Ea.	46			46	50.50
0260	1-1/2" x 1"						45.50			45.50	50
0280	1-1/2" x 1-1/2"						12.90			12.90	14.20
0300	2" x 1-1/2"						62.50			62.50	68.50
0320	2" x 2"						42			42	46
0340	2-1/2" x 1-1/2"						13.95			13.95	15.35
0360	2-1/2" x 2"						36			36	39.50
0380	2-1/2" x 2-1/2"						19			19	21
0400	3" x 2-1/2"						119			119	131
0420	3" x 3"						83.50			83.50	92
0500	For polished brass, add						50%				
0520	For polished chrome, add					↓	75%				
0700	One piece, female to male, hexagon										
0740	1-1/2" x 3/4"					Ea.	38.50			38.50	42.50
0760	2" x 1-1/2"						95			95	104
0780	2-1/2" x 1"						154			154	169
0800	2-1/2" x 1-1/2"						53			53	58
0820	2-1/2" x 2"						36			36	39.50
0840	3" x 2-1/2"						67			67	73.50
0900	For polished chrome, add					↓	75%				
1100	Swivel, female to female, pin lugs										

21 12 Fire-Suppression Standpipes

21 12 13 – Fire-Suppression Hoses and Nozzles

21 12 13.50 Fire Hoses and Nozzles		Crew	Daily Output	Labor-Hours	Unit	Material	2014 Bare Costs Labor	Equipment	Total	Total Incl O&P
1120	1-1/2" x 1-1/2"				Ea.	58.50			58.50	64.50
1200	2-1/2" x 2-1/2"				"	114			114	125
1260	For polished brass, add					50%				
1280	For polished chrome, add					75%				
1400	Couplings, sngl. & dbl. jacket, pin lug or rocker lug, cast brass									
1410	1-1/2"				Ea.	50.50			50.50	56
1420	2-1/2"				"	66.50			66.50	73
1500	For polished brass, add					20%				
1520	For polished chrome, add					40%				
1580	Reducing, F x M, interior installation, cast brass									
1590	2" x 1-1/2"				Ea.	62.50			62.50	68.50
1600	2-1/2" x 1-1/2"					13.95			13.95	15.35
1680	For polished brass, add					50%				
1720	For polished chrome, add					75%				
2200	Hose, less couplings									
2260	Synthetic jacket, lined, 300 lb. test, 1-1/2" diameter	Q-12	2600	.006	L.F.	3.09	.31		3.40	3.86
2280	2-1/2" diameter		2200	.007		5.30	.36		5.66	6.40
2360	High strength, 500 lb. test, 1-1/2" diameter		2600	.006		3.18	.31		3.49	3.96
2380	2-1/2" diameter		2200	.007		5.65	.36		6.01	6.75
5000	Nipples, straight hose to tapered iron pipe, brass									
5060	Female to female, 1-1/2" x 1-1/2"				Ea.	18.75			18.75	20.50
5100	2-1/2" x 2-1/2"					35			35	38.50
5190	For polished chrome, add					75%				
5200	Double male or male to female, 1" x 1"					40			40	44
5220	1-1/2" x 1"					59			59	64.50
5230	1-1/2" x 1-1/2"					13.60			13.60	14.95
5260	2" x 1-1/2"					64			64	70.50
5270	2" x 2"					80			80	88
5280	2-1/2" x 1-1/2"					53			53	58
5300	2-1/2" x 2"					43.50			43.50	47.50
5310	2-1/2" x 2-1/2"					24.50			24.50	27
5340	For polished chrome, add					75%				
5600	Nozzles, brass									
5620	Adjustable fog, 3/4" booster line				Ea.	106			106	116
5630	1" booster line					125			125	137
5640	1-1/2" leader line					75.50			75.50	83
5660	2-1/2" direct connection					143			143	158
5680	2-1/2" playpipe nozzle					209			209	230
5780	For chrome plated, add					8%				
5850	Electrical fire, adjustable fog, no shock									
5900	1-1/2"				Ea.	390			390	430
5920	2-1/2"					525			525	575
5980	For polished chrome, add					6%				
6200	Heavy duty, comb. adj. fog and str. stream, with handle									
6210	1" booster line				Ea.	375			375	410
6240	1-1/2"					420			420	465
6260	2-1/2", for playpipe					750			750	825
6280	2-1/2" direct connection					525			525	575
6300	2-1/2" playpipe combination					525			525	575
6480	For polished chrome, add					7%				
6500	Plain fog, polished brass, 1-1/2"					109			109	120
6540	Chrome plated, 1-1/2"					111			111	122
6700	Plain stream, polished brass, 1-1/2" x 10"					45.50			45.50	50

21 12 Fire-Suppression Standpipes

21 12 13 – Fire-Suppression Hoses and Nozzles

21 12 13.50 Fire Hoses and Nozzles

		Crew	Daily Output	Labor-Hours	Unit	Material	2014 Bare Costs Labor	Equipment	Total	Total Incl O&P
6760	2-1/2" x 15" x 7/8" or 1-1/2"				Ea.	95.50			95.50	105
6860	For polished chrome, add					20%				
7000	Underwriters playpipe, 2-1/2" x 30" with 1-1/8" tip				Ea.	233			233	256
7040	Less tip					170			170	187
9200	Storage house, hose only, primed steel					815			815	895
9220	Aluminum					1,775			1,775	1,950
9280	Hose and hydrant house, primed steel					1,100			1,100	1,200
9300	Aluminum					1,600			1,600	1,750
9340	Tools, crowbar and brackets	1 Carp	12	.667		66.50	30.50		97	120
9360	Combination hydrant wrench and spanner					30			30	33
9380	Fire axe and brackets									
9400	6 lb.	1 Carp	12	.667	Ea.	110	30.50		140.50	168

21 12 16 – Fire-Suppression Hose Reels

21 12 16.50 Fire-Suppression Hose Reels

		Crew	Daily Output	Labor-Hours	Unit	Material	Labor	Equipment	Total	Total Incl O&P
0010	**FIRE-SUPPRESSION HOSE REELS**									
2990	Hose reel, swinging, for 1-1/2" polyester neoprene lined hose									
3000	50' long	Q-12	14	1.143	Ea.	125	57		182	223
3020	100' long		14	1.143		175	57		232	278
3060	For 2-1/2" cotton rubber hose, 75' long		14	1.143		202	57		259	310
3100	150' long		14	1.143		236	57		293	345

21 12 19 – Fire-Suppression Hose Racks

21 12 19.50 Fire Hose Racks

		Crew	Daily Output	Labor-Hours	Unit	Material	Labor	Equipment	Total	Total Incl O&P
0010	**FIRE HOSE RACKS**									
2600	Hose rack, swinging, for 1-1/2" diameter hose,									
2620	Enameled steel, 50' & 75' lengths of hose	Q-12	20	.800	Ea.	58	40		98	125
2640	100' and 125' lengths of hose		20	.800		58	40		98	125
2680	Chrome plated, 50' and 75' lengths of hose		20	.800		93.50	40		133.50	164
2700	100' and 125' lengths of hose		20	.800		99.50	40		139.50	170
2780	For hose rack nipple, 1-1/2" polished brass, add					26.50			26.50	29
2820	2-1/2" polished brass, add					49			49	54
2840	1-1/2" polished chrome, add					37			37	41
2860	2-1/2" polished chrome, add					52			52	57.50

21 12 23 – Fire-Suppression Hose Valves

21 12 23.70 Fire Hose Valves

			Crew	Daily Output	Labor-Hours	Unit	Material	Labor	Equipment	Total	Total Incl O&P
0010	**FIRE HOSE VALVES**										
0020	Angle, combination pressure adjust/restricting, rough brass	R211226-10									
0030	1-1/2"		1 Spri	12	.667	Ea.	81	37		118	146
0040	2-1/2"	R211226-20	"	7	1.143	"	167	63.50		230.50	279
0042	Nonpressure adjustable/restricting, rough brass										
0044	1-1/2"		1 Spri	12	.667	Ea.	50.50	37		87.50	112
0046	2-1/2"		"	7	1.143	"	85.50	63.50		149	190
0050	For polished brass, add						30%				
0060	For polished chrome, add						40%				
0080	Wheel handle, 300 lb., 1-1/2"		1 Spri	12	.667	Ea.	86	37		123	151
0090	2-1/2"		"	7	1.143	"	167	63.50		230.50	279
0100	For polished brass, add						35%				
0110	For polished chrome, add						50%				
1000	Ball drip, automatic, rough brass, 1/2"		1 Spri	20	.400	Ea.	15.45	22		37.45	50.50
1010	3/4"		"	20	.400	"	17.30	22		39.30	52.50
1100	Ball, 175 lb., sprinkler system, FM/UL, threaded, bronze										
1120	Slow close										

21 12 Fire-Suppression Standpipes

21 12 23 – Fire-Suppression Hose Valves

21 12 23.70 Fire Hose Valves		Crew	Daily Output	Labor-Hours	Unit	Material	2014 Bare Costs Labor	Equipment	Total	Total Incl O&P
1150	1" size	1 Spri	19	.421	Ea.	234	23.50		257.50	293
1160	1-1/4" size		15	.533		253	29.50		282.50	325
1170	1-1/2" size		13	.615		335	34		369	415
1180	2" size		11	.727		405	40.50		445.50	505
1190	2-1/2" size	Q-12	15	1.067		545	53		598	680
1230	For supervisory switch kit, all sizes									
1240	One circuit, add	1 Spri	48	.167	Ea.	177	9.25		186.25	209
1280	Quarter turn for trim									
1300	1/2" size	1 Spri	22	.364	Ea.	31	20		51	64.50
1310	3/4" size		20	.400		33.50	22		55.50	70
1320	1" size		19	.421		37	23.50		60.50	75.50
1330	1-1/4" size		15	.533		60.50	29.50		90	112
1340	1-1/2" size		13	.615		76	34		110	136
1350	2" size		11	.727		90.50	40.50		131	161
1400	Caps, polished brass with chain, 3/4"					40			40	44
1420	1"					51			51	56
1440	1-1/2"					13.60			13.60	14.95
1460	2-1/2"					20.50			20.50	22.50
1480	3"					32			32	35
1900	Escutcheon plate, for angle valves, polished brass, 1-1/2"					15.20			15.20	16.70
1920	2-1/2"					24.50			24.50	27
1940	3"					31			31	34
1980	For polished chrome, add					15%				
2000	Foam, control valve, 3"	1 Spri	6	1.333		1,825	74		1,899	2,125
2020	Supply valve, 2-1/2"		7	1.143		147	63.50		210.50	257
2040	Proportioner, 8"		2	4		2,850	222		3,072	3,475
2060	Oscillating foam monitor with electric remote control	Q-12	5.33	3.002		10,800	150		10,950	12,000
3000	Gate, hose, wheel handle, N.R.S., rough brass, 1-1/2"	1 Spri	12	.667		128	37		165	197
3040	2-1/2", 300 lb.	"	7	1.143		180	63.50		243.50	294
3080	For polished brass, add					40%				
3090	For polished chrome, add					50%				
3800	Hydrant, screw type, crank handle, brass									
3840	2-1/2" size	Q-12	11	1.455	Ea.	335	72.50		407.50	480
3880	For chrome, same price									
4200	Hydrolator, vent and draining, rough brass, 1-1/2"	1 Spri	12	.667	Ea.	68	37		105	131
4280	For polished brass, add					50%				
4290	For polished chrome, add					90%				
5000	Pressure restricting, adjustable rough brass, 1-1/2"	1 Spri	12	.667		114	37		151	181
5020	2-1/2"	"	7	1.143		160	63.50		223.50	272
5080	For polished brass, add					30%				
5090	For polished chrome, add					45%				
8000	Wye, leader line, ball type, swivel female x male x male									
8040	2-1/2" x 1-1/2" x 1-1/2" polished brass				Ea.	245			245	270
8060	2-1/2" x 1-1/2" x 1-1/2" polished chrome				"	245			245	270

21 13 Fire-Suppression Sprinkler Systems

21 13 13 – Wet-Pipe Sprinkler Systems

21 13 13.50 Wet-Pipe Sprinkler System Components		Crew	Daily Output	Labor-Hours	Unit	Material	2014 Bare Costs Labor	Equipment	Total	Total Incl O&P
0010	**WET-PIPE SPRINKLER SYSTEM COMPONENTS**									
1100	Alarm, electric pressure switch (circuit closer)	R211313-10	1 Spri	26	.308	Ea.	84	17.05		101.05	119
1140	For explosion proof, max 20 PSI, contacts close or open			26	.308		505	17.05		522.05	580
1220	Water motor, complete with gong	R211313-30		4	2		390	111		501	595
1800	Firecycle system, controls, includes panel,										
1820	batteries, solenoid valves and pressure switches		Q-13	1	32	Ea.	19,800	1,700		21,500	24,300
1860	Detector		1 Spri	16	.500	"	680	27.50		707.50	790
1900	Flexible sprinkler head connectors										
1910	Braided stainless steel hose with mounting bracket										
1920	1/2" and 3/4" outlet size										
1940	36" length		1 Spri	30	.267	Ea.	70.50	14.75		85.25	100
1960	60" length		"	22	.364	"	81	20		101	120
1982	May replace hard-pipe armovers										
1984	For wet, pre-action, deluge or dry pipe systems										
2000	Release, emergency, manual, for hydraulic or pneumatic system		1 Spri	12	.667	Ea.	205	37		242	282
2060	Release, thermostatic, for hydraulic or pneumatic release line			20	.400		635	22		657	735
2200	Sprinkler cabinets, 6 head capacity			16	.500		74.50	27.50		102	124
2260	12 head capacity			16	.500		79	27.50		106.50	129
2340	Sprinkler head escutcheons, standard, brass tone, 1" size			40	.200		2.69	11.10		13.79	19.70
2360	Chrome, 1" size			40	.200		2.87	11.10		13.97	19.90
2400	Recessed type, brass tone			40	.200		3.16	11.10		14.26	20
2440	Chrome or white enamel			40	.200		3.52	11.10		14.62	20.50
2600	Sprinkler heads, not including supply piping										
3700	Standard spray, pendent or upright, brass, 135°F to 286°F										
3720	1/2" NPT, 3/8" orifice		1 Spri	16	.500	Ea.	13.85	27.50		41.35	57
3730	1/2" NPT, 7/16" orifice			16	.500		14.35	27.50		41.85	58
3740	1/2" NPT, 1/2" orifice			16	.500		9.35	27.50		36.85	52.50
3760	1/2" NPT, 17/32" orifice			16	.500		12.20	27.50		39.70	55.50
3780	3/4" NPT, 17/32" orifice			16	.500		11.20	27.50		38.70	54.50
3800	For open sprinklers, deduct						15%				
3840	For chrome, add					Ea.	3.13			3.13	3.44
3860	For wax and lead coating, add						33			33	36
3880	For wax coating, add						19.75			19.75	21.50
3900	For lead coating, add						21.50			21.50	23.50
3920	For 360°F, same cost										
3930	For 400°F					Ea.	87.50			87.50	96.50
3940	For 500°F					"	87.50			87.50	96.50
4200	Sidewall, vertical brass, 135°F to 286°F										
4240	1/2" NPT, 1/2" orifice		1 Spri	16	.500	Ea.	23.50	27.50		51	68
4280	3/4" NPT, 17/32" orifice		"	16	.500		66	27.50		93.50	115
4360	For satin chrome, add						2.63			2.63	2.89
4400	For 360°F, same cost										
4500	Sidewall, horizontal, brass, 135°F to 286°F										
4520	1/2" NPT, 1/2" orifice		1 Spri	16	.500	Ea.	23.50	27.50		51	68
4540	For 360°F, same cost										
4800	Recessed pendent, brass, 135°F to 286°F										
4820	1/2" NPT, 3/8" orifice		1 Spri	10	.800	Ea.	45	44.50		89.50	117
4830	1/2" NPT, 7/16" orifice			10	.800		22	44.50		66.50	91
4840	1/2" NPT, 1/2" orifice			10	.800		18.15	44.50		62.65	87
4860	1/2" NPT, 17/32" orifice			10	.800		45	44.50		89.50	117
4900	For satin chrome, add						3.63			3.63	3.99
5000	Recessed-vertical sidewall, brass, 135°F to 286°F										
5020	1/2" NPT, 3/8" orifice		1 Spri	10	.800	Ea.	29.50	44.50		74	99

93

21 13 Fire-Suppression Sprinkler Systems

21 13 13 – Wet-Pipe Sprinkler Systems

21 13 13.50 Wet-Pipe Sprinkler System Components		Crew	Daily Output	Labor-Hours	Unit	Material	2014 Bare Costs Labor	Equipment	Total	Total Incl O&P
5030	1/2" NPT, 7/16" orifice	1 Spri	10	.800	Ea.	29.50	44.50		74	99
5040	1/2" NPT, 1/2" orifice	↓	10	.800	↓	29.50	44.50		74	99
5100	For bright nickel, same cost									
5600	Concealed, complete with cover plate									
5620	1/2" NPT, 1/2" orifice, 135°F to 212°F	1 Spri	9	.889	Ea.	22.50	49		71.50	99
5800	Window, brass, 1/2" NPT, 1/4" orifice		16	.500		32	27.50		59.50	77
5810	1/2" NPT, 5/16" orifice		16	.500		32	27.50		59.50	77
5820	1/2" NPT, 3/8" orifice		16	.500		32	27.50		59.50	77
5830	1/2" NPT, 7/16" orifice		16	.500		32	27.50		59.50	77
5840	1/2" NPT, 1/2" orifice	↓	16	.500		34	27.50		61.50	79
5860	For polished chrome, add					4.71			4.71	5.20
5880	3/4" NPT, 5/8" orifice	1 Spri	16	.500		40	27.50		67.50	86
5890	3/4 NPT, 3/4" orifice	"	16	.500		40	27.50		67.50	86
6000	Sprinkler head guards, bright zinc, 1/2" NPT					5.70			5.70	6.25
6020	Bright zinc, 3/4" NPT				↓	5.70			5.70	6.25
6025	Residential sprinkler components, (one and two family)									
6026	Water motor alarm, with strainer	1 Spri	4	2	Ea.	390	111		501	595
6027	Fast response, glass bulb, 135°F to 155°F									
6028	1/2" NPT, pendent, brass	1 Spri	16	.500	Ea.	25.50	27.50		53	70
6029	1/2" NPT, sidewall, brass		16	.500		25.50	27.50		53	70
6030	1/2" NPT, pendent, brass, extended coverage		16	.500		22.50	27.50		50	66.50
6031	1/2" NPT, sidewall, brass, extended coverage		16	.500		20.50	27.50		48	64.50
6032	3/4" NPT sidewall, brass, extended coverage	↓	16	.500	↓	23	27.50		50.50	67
6033	For chrome, add					15%				
6034	For polyester/teflon coating add					20%				
6100	Sprinkler head wrenches, standard head				Ea.	23			23	25
6120	Recessed head					35.50			35.50	39
6160	Tamper switch, (valve supervisory switch)	1 Spri	16	.500	↓	99.50	27.50		127	151

21 13 16 – Dry-Pipe Sprinkler Systems

21 13 16.50 Dry-Pipe Sprinkler System Components

0010	**DRY-PIPE SPRINKLER SYSTEM COMPONENTS**									
0600	Accelerator	1 Spri	8	1	Ea.	715	55.50		770.50	870
0800	Air compressor for dry pipe system, automatic, complete									
0820	30 gal. system capacity, 3/4 HP	1 Spri	1.30	6.154	Ea.	735	340		1,075	1,325
0860	30 gal. system capacity, 1 HP		1.30	6.154		760	340		1,100	1,350
0910	30 gal. system capacity, 1-1/2 HP		1.30	6.154		795	340		1,135	1,400
0920	30 gal. system capacity, 2 HP		1.30	6.154		840	340		1,180	1,450
0960	Air pressure maintenance control		24	.333		325	18.45		343.45	390
1600	Dehydrator package, incl. valves and nipples	R211313-20	12	.667	↓	670	37		707	795
2600	Sprinkler heads, not including supply piping									
2640	Dry, pendent, 1/2" orifice, 3/4" or 1" NPT									
2660	1/2" to 6" length	1 Spri	14	.571	Ea.	117	31.50		148.50	176
2670	6-1/4" to 8" length		14	.571		122	31.50		153.50	182
2680	8-1/4" to 12" length		14	.571		127	31.50		158.50	187
2690	12-1/4" to 15" length		14	.571		132	31.50		163.50	193
2700	15-1/4" to 18" length		14	.571		137	31.50		168.50	199
2710	18-1/4" to 21" length		13	.615		142	34		176	208
2720	21-1/4" to 24" length		13	.615		147	34		181	214
2730	24-1/4" to 27" length		13	.615		153	34		187	220
2740	27-1/4" to 30" length		13	.615		158	34		192	226
2750	30-1/4" to 33" length		13	.615		163	34		197	232
2760	33-1/4" to 36" length	↓	13	.615	↓	169	34		203	238

21 13 Fire-Suppression Sprinkler Systems

21 13 16 – Dry-Pipe Sprinkler Systems

21 13 16.50 Dry-Pipe Sprinkler System Components	Crew	Daily Output	Labor-Hours	Unit	Material	2014 Bare Costs Labor	Equipment	Total	Total Incl O&P
2780 36-1/4" to 39" length	1 Spri	12	.667	Ea.	174	37		211	247
2790 39-1/4" to 42" length	↓	12	.667		179	37		216	253
2800 For each inch or fraction, add				↓	2.84			2.84	3.12
6330 Valves and components									
6340 Alarm test/shut off valve, 1/2"	1 Spri	20	.400	Ea.	20.50	22		42.50	56
8000 Dry pipe air check valve, 3" size	Q-12	2	8		1,700	400		2,100	2,450
8200 Dry pipe valve, incl. trim and gauges, 3" size		2	8		2,500	400		2,900	3,350
8220 4" size	↓	1	16		2,725	800		3,525	4,200
8240 6" size	Q-13	2	16		3,200	845		4,045	4,775
8280 For accelerator trim with gauges, add	1 Spri	8	1	↓	228	55.50		283.50	335

21 13 20 – Fire-Cycle Sprinkler Systems

21 13 20.50 Firecycle Fire-Suppression Sprinkler Systems

	Crew	Daily Output	Labor-Hours	Unit	Material	Labor	Equipment	Total	Total Incl O&P
0010 **FIRECYCLE FIRE-SUPPRESSION SPRINKLER SYSTEMS**									
8400 Firecycle package, includes swing check									
8420 and flow control valves with required trim									
8440 2" size	Q-12	2	8	Ea.	4,400	400		4,800	5,450
8460 3" size		1.50	10.667		4,825	530		5,355	6,125
8480 4" size	↓	1	16		5,400	800		6,200	7,125
8500 6" size	Q-13	1.40	22.857	↓	6,200	1,200		7,400	8,650

21 13 26 – Deluge Fire-Suppression Sprinkler Systems

21 13 26.50 Deluge Fire-Suppression Sprinkler Sys. Comp.

	Crew	Daily Output	Labor-Hours	Unit	Material	Labor	Equipment	Total	Total Incl O&P
0010 **DELUGE FIRE-SUPPRESSION SPRINKLER SYSTEM COMPONENTS**									
1400 Deluge system, monitoring panel w/deluge valve & trim	1 Spri	18	.444	Ea.	10,300	24.50		10,324.50	11,300
6200 Valves and components									
7000 Deluge, assembly, incl. trim, pressure									
7020 operated relief, emergency release, gauges									
7040 2" size	Q-12	2	8	Ea.	3,500	400		3,900	4,450
7060 3" size		1.50	10.667		3,925	530		4,455	5,100
7080 4" size	↓	1	16		4,475	800		5,275	6,125
7100 6" size	Q-13	1.80	17.778	↓	5,250	940		6,190	7,200
7800 Pneumatic actuator, bronze, required on all									
7820 pneumatic release systems, any size deluge	1 Spri	18	.444	Ea.	390	24.50		414.50	465

21 13 39 – Foam-Water Systems

21 13 39.50 Foam-Water System Components

	Crew	Daily Output	Labor-Hours	Unit	Material	Labor	Equipment	Total	Total Incl O&P
0010 **FOAM-WATER SYSTEM COMPONENTS**									
2600 Sprinkler heads, not including supply piping									
3600 Foam-water, pendent or upright, 1/2" NPT	1 Spri	12	.667	Ea.	198	37		235	273

21 21 Carbon-Dioxide Fire-Extinguishing Systems

21 21 16 – Carbon-Dioxide Fire-Extinguishing Equipment

21 21 16.50 CO2 Fire Extinguishing System

	Crew	Daily Output	Labor-Hours	Unit	Material	Labor	Equipment	Total	Total Incl O&P
0010 **CO_2 FIRE EXTINGUISHING SYSTEM**									
0042 For detectors and control stations, see Section 28 31 23.50									
0100 Control panel, single zone with batteries (2 zones det., 1 suppr.)	1 Elec	1	8	Ea.	1,675	425		2,100	2,500
0150 Multizone (4) with batteries (8 zones det., 4 suppr.)	"	.50	16		3,200	855		4,055	4,775
1000 Dispersion nozzle, CO_2, 3" x 5"	1 Plum	18	.444		65	25.50		90.50	110
2000 Extinguisher, CO_2 system, high pressure, 75 lb. cylinder	Q-1	6	2.667		1,250	138		1,388	1,550
2100 100 lb. cylinder	"	5	3.200		1,275	166		1,441	1,650
3000 Electro/mechanical release	L-1	4	4	↓	163	222		385	515

21 21 Carbon-Dioxide Fire-Extinguishing Systems

21 21 16 – Carbon-Dioxide Fire-Extinguishing Equipment

21 21 16.50 CO2 Fire Extinguishing System

		Crew	Daily Output	Labor-Hours	Unit	Material	2014 Bare Costs Labor	Equipment	Total	Total Incl O&P
3400	Manual pull station	1 Plum	6	1.333	Ea.	58.50	76.50		135	181
4000	Pneumatic damper release	"	8	1	↓	217	57.50		274.50	325

21 22 Clean-Agent Fire-Extinguishing Systems

21 22 16 – Clean-Agent Fire-Extinguishing Equipment

21 22 16.50 FM200 Fire Extinguishing System

		Crew	Daily Output	Labor-Hours	Unit	Material	Labor	Equipment	Total	Total Incl O&P
0010	**FM200 FIRE EXTINGUISHING SYSTEM**									
1100	Dispersion nozzle FM200, 1-1/2"	1 Plum	14	.571	Ea.	65	33		98	121
2400	Extinguisher, FM200 system, filled, with mounting bracket									
2460	26 lb. container	Q-1	8	2	Ea.	2,225	104		2,329	2,600
2480	44 lb. container		7	2.286		2,975	118		3,093	3,450
2500	63 lb. container		6	2.667		3,475	138		3,613	4,025
2520	101 lb. container		5	3.200		4,650	166		4,816	5,350
2540	196 lb. container	↓	4	4	↓	7,550	207		7,757	8,625
6000	Average FM200 system, minimum				C.F.	1.71			1.71	1.88
6020	Maximum				"	3.40			3.40	3.74

21 31 Centrifugal Fire Pumps

21 31 13 – Electric-Drive, Centrifugal Fire Pumps

21 31 13.50 Electric-Drive Fire Pumps

		Crew	Daily Output	Labor-Hours	Unit	Material	Labor	Equipment	Total	Total Incl O&P
0010	**ELECTRIC-DRIVE FIRE PUMPS** Including controller, fittings and relief valve									
3100	250 GPM, 55 psi, 15 HP, 3550 RPM, 2" pump	Q-13	.70	45.714	Ea.	15,200	2,400		17,600	20,500
3200	500 GPM, 50 psi, 27 HP, 1770 RPM, 4" pump		.68	47.059		15,500	2,475		17,975	20,900
3250	500 GPM, 100 psi, 47 HP, 3550 RPM, 3" pump		.66	48.485		16,700	2,550		19,250	22,200
3300	500 GPM, 125 psi, 64 HP, 3550 RPM, 3" pump		.62	51.613		18,500	2,725		21,225	24,400
3350	750 GPM, 50 psi, 44 HP, 1770 RPM, 5" pump		.64	50		16,200	2,650		18,850	21,800
3400	750 GPM, 100 psi, 66 HP, 3550 RPM, 4" pump		.58	55.172		18,700	2,900		21,600	24,900
3450	750 GPM, 165 psi, 120 HP, 3550 RPM, 4" pump		.56	57.143		25,800	3,025		28,825	33,000
3500	1000 GPM, 50 psi, 48 HP 1770 RPM, 5" pump		.60	53.333		17,100	2,825		19,925	23,100
3550	1000 GPM, 100 psi, 86 HP, 3550 RPM, 5" pump		.54	59.259		23,500	3,125		26,625	30,600
3600	1000 GPM, 150 psi, 142 HP, 3550 RPM, 5" pump		.50	64		27,800	3,375		31,175	35,700
3650	1000 GPM, 200 psi, 245 HP, 1770 RPM, 6" pump		.36	88.889		40,400	4,700		45,100	51,500
3660	1250 GPM, 75 psi, 75 HP, 1770 RPM, 5" pump		.55	58.182		22,000	3,075		25,075	28,800
3700	1500 GPM, 50 psi, 66 HP, 1770 RPM, 6" pump		.50	64		20,900	3,375		24,275	28,100
3750	1500 GPM, 100 psi, 139 HP, 1770 RPM, 6" pump		.46	69.565		27,000	3,675		30,675	35,300
3800	1500 GPM, 150 psi, 200 HP, 1770 RPM, 6" pump		.36	88.889		43,800	4,700		48,500	55,500
3850	1500 GPM, 200 psi, 279 HP, 1770 RPM, 6" pump		.32	100		46,600	5,275		51,875	59,000
3900	2000 GPM, 100 psi, 167 HP, 1770 RPM, 6" pump		.34	94.118		29,500	4,975		34,475	40,000
3950	2000 GPM, 150 psi, 292 HP, 1770 RPM, 6" pump		.28	114		44,300	6,025		50,325	58,000
4000	2500 GPM, 100 psi, 213 HP, 1770 RPM, 8" pump		.30	106		36,800	5,625		42,425	49,000
4040	2500 GPM, 135 psi, 339 HP, 1770 RPM, 8" pump		.26	123		48,700	6,500		55,200	63,500
4100	3000 GPM, 100 psi, 250 HP, 1770 RPM, 8" pump		.28	114		52,500	6,025		58,525	67,000
4150	3000 GPM, 140 psi, 428 HP, 1770 RPM, 10" pump		.24	133		57,000	7,025		64,025	73,000
4200	3500 GPM, 100 psi, 300 HP, 1770 RPM, 10" pump		.26	123		53,000	6,500		59,500	68,000
4250	3500 GPM, 140 psi, 450 HP, 1770 RPM, 10" pump	↓	.24	133		64,500	7,025		71,525	81,000
5000	For jockey pump 1", 3 HP, with control, add	Q-12	2	8	↓	2,525	400		2,925	3,375

21 31 Centrifugal Fire Pumps

21 31 16 – Diesel-Drive, Centrifugal Fire Pumps

21 31 16.50 Diesel-Drive Fire Pumps		Crew	Daily Output	Labor-Hours	Unit	Material	2014 Bare Costs Labor	Equipment	Total	Total Incl O&P
0010	**DIESEL-DRIVE FIRE PUMPS** Including controller, fittings and relief valve									
0050	500 GPM, 50 psi, 27 HP, 4" pump	Q-13	.64	50	Ea.	35,300	2,650		37,950	42,800
0100	500 GPM, 100 psi, 62 HP, 4" pump		.60	53.333		39,900	2,825		42,725	48,200
0150	500 GPM, 125 psi, 78 HP, 4" pump		.56	57.143		39,900	3,025		42,925	48,500
0200	750 GPM, 50 psi, 44 HP, 5" pump		.60	53.333		36,300	2,825		39,125	44,200
0250	750 GPM, 100 psi, 80 HP, 4" pump		.56	57.143		39,500	3,025		42,525	48,000
0300	750 GPM, 165 psi, 203 HP, 5" pump		.52	61.538		53,000	3,250		56,250	63,500
0350	1000 GPM, 50 psi, 48 HP, 5" pump		.58	55.172		39,800	2,900		42,700	48,100
0400	1000 GPM, 100 psi, 89 HP, 4" pump		.56	57.143		39,800	3,025		42,825	48,300
0450	1000 GPM, 150 psi, 148 HP, 4" pump		.48	66.667		51,000	3,525		54,525	62,000
0470	1000 GPM, 200 psi, 280 HP, 5" pump		.40	80		62,500	4,225		66,725	75,500
0480	1250 GPM, 75 psi, 75 HP, 5" pump		.54	59.259		40,300	3,125		43,425	49,000
0500	1500 GPM, 50 psi, 66 HP, 6" pump		.50	64		38,600	3,375		41,975	47,600
0550	1500 GPM, 100 psi, 140 HP, 6" pump		.46	69.565		50,500	3,675		54,175	61,000
0600	1500 GPM, 150 psi, 228 HP, 6" pump		.42	76.190		57,000	4,025		61,025	68,500
0650	1500 GPM, 200 psi, 279 HP, 6" pump		.38	84.211		70,500	4,450		74,950	84,000
0700	2000 GPM, 100 psi, 167 HP, 6" pump		.34	94.118		50,500	4,975		55,475	63,000
0750	2000 GPM, 150 psi, 284 HP, 6" pump		.30	106		64,500	5,625		70,125	79,500
0800	2500 GPM, 100 psi, 213 HP, 8" pump		.32	100		54,000	5,275		59,275	67,500
0820	2500 GPM, 150 psi, 365 HP, 8" pump		.26	123		69,500	6,500		76,000	86,500
0850	3000 GPM, 100 psi, 250 HP, 8" pump		.28	114		81,000	6,025		87,025	98,500
0900	3000 GPM, 150 psi, 384 HP, 10" pump		.20	160		86,000	8,450		94,450	107,500
0950	3500 GPM, 100 psi, 300 HP, 10" pump		.24	133		70,500	7,025		77,525	88,000
1000	3500 GPM, 150 psi, 518 HP, 10" pump	↓	.20	160	↓	99,000	8,450		107,450	122,000

Division Notes

	CREW	DAILY OUTPUT	LABOR-HOURS	UNIT	BARE COSTS MAT.	LABOR	EQUIP.	TOTAL	TOTAL INCL O&P

Estimating Tips

22 10 00 Plumbing Piping and Pumps

This subdivision is primarily basic pipe and related materials. The pipe may be used by any of the mechanical disciplines, i.e., plumbing, fire protection, heating, and air conditioning.

Note: CPVC plastic piping approved for fire protection is located in 21 11 13.

- The labor adjustment factors listed in Subdivision 22 01 02.20 apply throughout Divisions 21, 22, and 23. CAUTION: the correct percentage may vary for the same items. For example, the percentage add for the basic pipe installation should be based on the maximum height that the craftsman must install for that particular section. If the pipe is to be located 14' above the floor but it is suspended on threaded rod from beams, the bottom flange of which is 18' high (4' rods), then the height is actually 18' and the add is 20%. The pipe coverer, however, does not have to go above the 14', and so his or her add should be 10%.
- Most pipe is priced first as straight pipe with a joint (coupling, weld, etc.) every 10' and a hanger usually every 10'. There are exceptions with hanger spacing such as for cast iron pipe (5') and plastic pipe (3 per 10'). Following each type of pipe there are several lines listing sizes and the amount to be subtracted to delete couplings and hangers. This is for pipe that is to be buried or supported together on trapeze hangers. The reason that the couplings are deleted is that these runs are usually long, and frequently longer lengths of pipe are used. By deleting the couplings, the estimator is expected to look up and add back the correct reduced number of couplings.
- When preparing an estimate, it may be necessary to approximate the fittings. Fittings usually run between 25% and 50% of the cost of the pipe. The lower percentage is for simpler runs, and the higher number is for complex areas, such as mechanical rooms.
- For historic restoration projects, the systems must be as invisible as possible, and pathways must be sought for pipes, conduit, and ductwork. While installations in accessible spaces (such as basements and attics) are relatively straightforward to estimate, labor costs may be more difficult to determine when delivery systems must be concealed.

22 40 00 Plumbing Fixtures

- Plumbing fixture costs usually require two lines: the fixture itself and its "rough-in, supply, and waste."
- In the Assemblies Section (Plumbing D2010) for the desired fixture, the System Components Group at the center of the page shows the fixture on the first line. The rest of the list (fittings, pipe, tubing, etc.) will total up to what we refer to in the Unit Price section as "Rough-in, supply, waste, and vent." Note that for most fixtures we allow a nominal 5' of tubing to reach from the fixture to a main or riser.
- Remember that gas- and oil-fired units need venting.

Reference Numbers

Reference numbers are shown in shaded boxes at the beginning of some major classifications. These numbers refer to related items in the Reference Section. The reference information may be an estimating procedure, an alternate pricing method, or technical information.

Note: Not all subdivisions listed here necessarily appear in this publication.

Note: **Trade Service,** *in part, has been used as a reference source for some of the material prices used in Division 22.*

22 01 Operation and Maintenance of Plumbing

22 01 02 – Labor Adjustments

22 01 02.10 Boilers, General

		Crew	Daily Output	Labor-Hours	Unit	Material	2014 Bare Costs Labor	Equipment	Total	Total Incl O&P
0010	**BOILERS, GENERAL**, Prices do not include flue piping, elec. wiring,									
0020	gas or oil piping, boiler base, pad, or tankless unless noted									
0100	Boiler H.P.: 10 KW = 34 lb./steam/hr. = 33,475 BTU/hr.									
0150	To convert SFR to BTU rating: Hot water, 150 x SFR;									
0160	Forced hot water, 180 x SFR; steam, 240 x SFR									

22 01 02.20 Labor Adjustment Factors

0010	**LABOR ADJUSTMENT FACTORS**, (For Div. 21, 22 and 23) R220102-20									
0100	Labor factors, The below are reasonable suggestions, however									
0110	each project must be evaluated for its own peculiarities, and									
0120	the adjustments be increased or decreased depending on the									
0130	severity of the special conditions.									
1000	Add to labor for elevated installation (Above floor level)									
1080	10' to 14.5' high R221113-70						10%			
1100	15' to 19.5' high						20%			
1120	20' to 24.5' high						25%			
1140	25' to 29.5' high						35%			
1160	30' to 34.5' high						40%			
1180	35' to 39.5' high						50%			
1200	40' and higher						55%			
2000	Add to labor for crawl space									
2100	3' high						40%			
2140	4' high						30%			
3000	Add to labor for multi-story building									
3100	Add per floor for floors 3 thru 19						2%			
3140	Add per floor for floors 20 and up						4%			
4000	Add to labor for working in existing occupied buildings									
4100	Hospital						35%			
4140	Office building						25%			
4180	School						20%			
4220	Factory or warehouse						15%			
4260	Multi dwelling						15%			
5000	Add to labor, miscellaneous									
5100	Cramped shaft						35%			
5140	Congested area						15%			
5180	Excessive heat or cold						30%			
9000	Labor factors, The above are reasonable suggestions, however									
9010	each project should be evaluated for its own peculiarities.									
9100	Other factors to be considered are:									
9140	Movement of material and equipment through finished areas									
9180	Equipment room									
9220	Attic space									
9260	No service road									
9300	Poor unloading/storage area									
9340	Congested site area/heavy traffic									

22 05 Common Work Results for Plumbing

22 05 05 – Selective Plumbing Demolition

22 05 05.10 Plumbing Demolition

		Crew	Daily Output	Labor-Hours	Unit	Material	2014 Bare Costs Labor	2014 Bare Costs Equipment	Total	Total Incl O&P
0010	**PLUMBING DEMOLITION**	R220105-10								
0400	Air compressor, up thru 2 H.P.	Q-1	10	1.600	Ea.		83		83	125
0410	3 H.P. thru 7-1/2 H.P.		5.60	2.857			148		148	224
0420	10 H.P. thru 15 H.P.		1.40	11.429			590		590	895
0430	20 H.P. thru 30 H.P.	Q-2	1.30	18.462			990		990	1,500
0500	Backflow preventer, up thru 2" diameter	1 Plum	17	.471			27		27	41
0510	2-1/2" thru 3" diameter	Q-1	10	1.600			83		83	125
0520	4" thru 6" diameter	"	5	3.200			166		166	250
0530	8" thru 10" diameter	Q-2	3	8			430		430	650
0700	Carriers and supports									
0710	Fountains, sinks, lavatories and urinals	1 Plum	14	.571	Ea.		33		33	49.50
0720	Water closets	"	12	.667	"		38.50		38.50	58
0730	Grinder pump or sewage ejector system									
0732	Simplex	Q-1	7	2.286	Ea.		118		118	179
0734	Duplex	"	2.80	5.714			296		296	445
0738	Hot water dispenser	1 Plum	36	.222			12.80		12.80	19.30
0740	Hydrant, wall		26	.308			17.70		17.70	26.50
0744	Ground		12	.667			38.50		38.50	58
0760	Cleanouts and drains, up thru 4" pipe diameter		10	.800			46		46	69.50
0764	5" thru 8" pipe diameter	Q-1	10	1.600			83		83	125
0780	Industrial safety fixtures	1 Plum	8	1			57.50		57.50	87
1020	Fixtures, including 10' piping									
1100	Bathtubs, cast iron	1 Plum	4	2	Ea.		115		115	174
1120	Fiberglass		6	1.333			76.50		76.50	116
1140	Steel		5	1.600			92		92	139
1150	Bidet	Q-1	7	2.286			118		118	179
1200	Lavatory, wall hung	1 Plum	10	.800			46		46	69.50
1220	Counter top		8	1			57.50		57.50	87
1300	Sink, single compartment		8	1			57.50		57.50	87
1320	Double compartment		7	1.143			66		66	99.50
1340	Shower, stall and receptor	Q-1	6	2.667			138		138	209
1350	Group	"	7	2.286			118		118	179
1400	Water closet, floor mounted	1 Plum	8	1			57.50		57.50	87
1420	Wall mounted	"	7	1.143			66		66	99.50
1440	Wash fountain, 36" diameter	Q-2	8	3			161		161	243
1442	54" diameter	"	7	3.429			184		184	278
1500	Urinal, floor mounted	1 Plum	4	2			115		115	174
1520	Wall mounted	"	7	1.143			66		66	99.50
1590	Whirl pool or hot tub	Q-1	2.60	6.154			320		320	480
1600	Water fountains, free standing	1 Plum	8	1			57.50		57.50	87
1620	Wall or deck mounted		6	1.333			76.50		76.50	116
1800	Medical gas specialties		8	1			57.50		57.50	87
1810	Plumbing demo, Floor drain, Remove		12	.667			38.50		38.50	58
1820	Plumbing demo, Roof drain, Remove		9	.889			51		51	77
1900	Piping fittings, single connection, up thru 1-1/2" diameter		30	.267			15.35		15.35	23
1910	2" thru 4" diameter		14	.571			33		33	49.50
1980	Pipe hanger/support removal		80	.100			5.75		5.75	8.70
1990	Glass pipe with fittings, 1" thru 3" diameter		200	.040	L.F.		2.30		2.30	3.48
1992	4" thru 6" diameter		150	.053			3.07		3.07	4.63
2000	Piping, metal, up thru 1-1/2" diameter		200	.040			2.30		2.30	3.48
2050	2" thru 3-1/2" diameter		150	.053			3.07		3.07	4.63
2100	4" thru 6" diameter	2 Plum	100	.160			9.20		9.20	13.90
2150	8" thru 14" diameter	"	60	.267			15.35		15.35	23

22 05 Common Work Results for Plumbing

22 05 05 – Selective Plumbing Demolition

22 05 05.10 Plumbing Demolition

		Crew	Daily Output	Labor-Hours	Unit	Material	2014 Bare Costs Labor	Equipment	Total	Total Incl O&P
2153	16" thru 20" diameter	Q-18	70	.343	L.F.		18.70	.79	19.49	29.50
2155	24" thru 26" diameter		55	.436			24	1	25	37
2156	30" thru 36" diameter		40	.600			33	1.38	34.38	51
2160	Plastic pipe with fittings, up thru 1-1/2" diameter	1 Plum	250	.032			1.84		1.84	2.78
2162	2" thru 3" diameter	"	200	.040			2.30		2.30	3.48
2164	4" thru 6" diameter	Q-1	200	.080			4.14		4.14	6.25
2166	8" thru 14" diameter		150	.107			5.55		5.55	8.35
2168	16" diameter		100	.160			8.30		8.30	12.50
2170	Prison fixtures, lavatory or sink		18	.889	Ea.		46		46	69.50
2172	Shower		5.60	2.857			148		148	224
2174	Urinal or water closet		13	1.231			64		64	96.50
2180	Pumps, all fractional horse-power		12	1.333			69		69	104
2184	1 H.P. thru 5 H.P.		6	2.667			138		138	209
2186	7-1/2 H.P. thru 15 H.P.		2.50	6.400			330		330	500
2188	20 H.P. thru 25 H.P.	Q-2	4	6			320		320	485
2190	30 H.P. thru 60 H.P.		.80	30			1,600		1,600	2,425
2192	75 H.P. thru 100 H.P.		.60	40			2,150		2,150	3,250
2194	150 H.P.		.50	48			2,575		2,575	3,900
2198	Pump, sump or submersible	1 Plum	12	.667			38.50		38.50	58
2200	Receptors and interceptors, up thru 20 GPM	"	8	1			57.50		57.50	87
2204	25 thru 100 GPM	Q-1	6	2.667			138		138	209
2208	125 thru 300 GPM	"	2.40	6.667			345		345	520
2211	325 thru 500 GPM	Q-2	2.60	9.231			495		495	750
2212	Deduct for salvage, aluminum scrap				Ton				700	770
2214	Brass scrap								2,450	2,675
2216	Copper scrap								3,200	3,525
2218	Lead scrap								520	570
2220	Steel scrap								180	200
2230	Temperature maintenance cable	1 Plum	1200	.007	L.F.		.38		.38	.58
2240	Toilet partitions, see Section 10 21 13									
2250	Water heater, 40 gal.	1 Plum	6	1.333	Ea.		76.50		76.50	116
3100	Tanks, water heaters and liquid containers									
3110	Up thru 45 gallons	Q-1	22	.727	Ea.		37.50		37.50	57
3120	50 thru 120 gallons		14	1.143			59		59	89.50
3130	130 thru 240 gallons		7.60	2.105			109		109	165
3140	250 thru 500 gallons		5.40	2.963			153		153	232
3150	600 thru 1000 gallons	Q-2	1.60	15			805		805	1,225
3160	1100 thru 2000 gallons		.70	34.286			1,850		1,850	2,775
3170	2100 thru 4000 gallons		.50	48			2,575		2,575	3,900
6000	Remove and reset fixtures, minimum	1 Plum	6	1.333			76.50		76.50	116
6100	Maximum	"	4	2			115		115	174
6500	Solar heating system									
6510	Solar panel, hot water system	1 Plum	304	.026	S.F.		1.51		1.51	2.29
7910	Solar panel for pool	"	300	.027	"		1.53		1.53	2.32
8000	Sprinkler system									
8100	Exposed wet/dry	1 Plum	2500	.003	SF Flr.		.18		.18	.28
8200	Concealed wet/dry	"	1250	.006	"		.37		.37	.56
9100	Valve, metal valves, strainers and similar, up thru 1-1/2" diameter	1 Stpi	28	.286	Ea.		16.70		16.70	25
9110	2" thru 3" diameter	Q-1	11	1.455			75.50		75.50	114
9120	4" thru 6" diameter	"	8	2			104		104	156
9130	8" thru 14" diameter	Q-2	8	3			161		161	243
9140	16" thru 20" diameter		2	12			645		645	975
9150	24" diameter		1.20	20			1,075		1,075	1,625

22 05 Common Work Results for Plumbing

22 05 05 – Selective Plumbing Demolition

22 05 05.10 Plumbing Demolition

		Crew	Daily Output	Labor-Hours	Unit	Material	2014 Bare Costs Labor	Equipment	Total	Total Incl O&P
9200	Valve, plastic, up thru 1-1/2" diameter	1 Plum	42	.190	Ea.		10.95		10.95	16.55
9210	2" thru 3" diameter		15	.533			30.50		30.50	46.50
9220	4" thru 6" diameter		12	.667			38.50		38.50	58
9300	Vent flashing and caps		55	.145			8.35		8.35	12.65
9350	Water filter, commercial, 1" thru 1-1/2"	Q-1	2	8			415		415	625
9360	2" thru 2-1/2"	"	1.60	10			520		520	780
9400	Water heaters									
9410	Up thru 245 GPH	Q-1	2.40	6.667	Ea.		345		345	520
9420	250 thru 756 GPH		1.60	10			520		520	780
9430	775 thru 1640 GPH		.80	20			1,025		1,025	1,575
9440	1650 thru 4000 GPH	Q-2	.50	48			2,575		2,575	3,900
9470	Water softener	Q-1	2	8			415		415	625

22 05 23 – General-Duty Valves for Plumbing Piping

22 05 23.10 Valves, Brass

			Crew	Daily Output	Labor-Hours	Unit	Material	Labor	Equipment	Total	Total Incl O&P
0010	**VALVES, BRASS**										
0032	For motorized valves, see Section 23 09 53.10										
0500	Gas cocks, threaded										
0510	1/4"		1 Plum	26	.308	Ea.	14.70	17.70		32.40	42.50
0520	3/8"			24	.333		14.70	19.20		33.90	45
0530	1/2"			24	.333		13.15	19.20		32.35	43.50
0540	3/4"			22	.364		15.50	21		36.50	48.50
0550	1"			19	.421		31	24		55	70.50
0560	1-1/4"			15	.533		44.50	30.50		75	95.50
0570	1-1/2"			13	.615		61	35.50		96.50	121
0580	2"			11	.727		103	42		145	177
0672	For larger sizes use lubricated plug valve, Section 23 05 23.70										

22 05 23.20 Valves, Bronze

			Crew	Daily Output	Labor-Hours	Unit	Material	Labor	Equipment	Total	Total Incl O&P
0010	**VALVES, BRONZE**	R220523-80									
1020	Angle, 150 lb., rising stem, threaded										
1030	1/8"	R220523-90	1 Plum	24	.333	Ea.	135	19.20		154.20	177
1040	1/4"			24	.333		135	19.20		154.20	177
1050	3/8"			24	.333		135	19.20		154.20	177
1060	1/2"			22	.364		135	21		156	180
1070	3/4"			20	.400		184	23		207	237
1080	1"			19	.421		265	24		289	330
1090	1-1/4"			15	.533		340	30.50		370.50	420
1100	1-1/2"			13	.615		445	35.50		480.50	545
1110	2"			11	.727		715	42		757	850
1300	Ball										
1304	Soldered										
1312	3/8"		1 Plum	21	.381	Ea.	14.30	22		36.30	48.50
1316	1/2"			18	.444		14.30	25.50		39.80	54
1320	3/4"			17	.471		23.50	27		50.50	67
1324	1"			15	.533		29.50	30.50		60	79
1328	1-1/4"			13	.615		49.50	35.50		85	108
1332	1-1/2"			11	.727		63.50	42		105.50	133
1336	2"			9	.889		79.50	51		130.50	165
1340	2-1/2"			7	1.143		375	66		441	515
1344	3"			5	1.600		435	92		527	615
1350	Single union end										
1358	3/8"		1 Plum	21	.381	Ea.	20.50	22		42.50	55.50
1362	1/2"			18	.444		22	25.50		47.50	63

22 05 Common Work Results for Plumbing
22 05 23 – General-Duty Valves for Plumbing Piping

22 05 23.20 Valves, Bronze		Crew	Daily Output	Labor-Hours	Unit	Material	2014 Bare Costs Labor	Equipment	Total	Total Incl O&P
1366	3/4"	1 Plum	17	.471	Ea.	36.50	27		63.50	81.50
1370	1"		15	.533		49.50	30.50		80	101
1374	1-1/4"		13	.615		84	35.50		119.50	146
1378	1-1/2"		11	.727		105	42		147	179
1382	2"		9	.889		165	51		216	258
1398	Threaded, 150 psi									
1400	1/4"	1 Plum	24	.333	Ea.	13.55	19.20		32.75	44
1430	3/8"		24	.333		13.55	19.20		32.75	44
1450	1/2"		22	.364		13.55	21		34.55	46.50
1460	3/4"		20	.400		22.50	23		45.50	59.50
1470	1"		19	.421		32	24		56	71.50
1480	1-1/4"		15	.533		56	30.50		86.50	108
1490	1-1/2"		13	.615		69.50	35.50		105	130
1500	2"		11	.727		88	42		130	160
1510	2-1/2"		9	.889		295	51		346	400
1520	3"		8	1		450	57.50		507.50	575
1600	Butterfly, 175 psi, full port, solder or threaded ends									
1610	Stainless steel disc and stem									
1620	1/4"	1 Plum	24	.333	Ea.	16.50	19.20		35.70	47
1630	3/8"		24	.333		16.50	19.20		35.70	47
1640	1/2"		22	.364		17.35	21		38.35	50.50
1650	3/4"		20	.400		28	23		51	66
1660	1"		19	.421		34.50	24		58.50	74
1670	1-1/4"		15	.533		56	30.50		86.50	108
1680	1-1/2"		13	.615		71.50	35.50		107	132
1690	2"		11	.727		89.50	42		131.50	161
1750	Check, swing, class 150, regrinding disc, threaded									
1800	1/8"	1 Plum	24	.333	Ea.	63	19.20		82.20	98.50
1830	1/4"		24	.333		63	19.20		82.20	98.50
1840	3/8"		24	.333		67	19.20		86.20	103
1850	1/2"		24	.333		72	19.20		91.20	108
1860	3/4"		20	.400		95	23		118	139
1870	1"		19	.421		137	24		161	187
1880	1-1/4"		15	.533		198	30.50		228.50	265
1890	1-1/2"		13	.615		230	35.50		265.50	305
1900	2"		11	.727		340	42		382	435
1910	2-1/2"	Q-1	15	1.067		760	55.50		815.50	920
1920	3"	"	13	1.231		1,025	64		1,089	1,225
2000	For 200 lb., add					5%	10%			
2040	For 300 lb., add					15%	15%			
2060	Check swing, 300#, sweat, 3/8" size	1 Plum	24	.333	Ea.	63	19.20		82.20	98
2070	1/2"		24	.333		63	19.20		82.20	98
2080	3/4"		20	.400		79.50	23		102.50	123
2090	1"		19	.421		114	24		138	162
2100	1-1/4"		15	.533		159	30.50		189.50	222
2110	1-1/2"		13	.615		194	35.50		229.50	267
2120	2"		11	.727		296	42		338	390
2130	2-1/2"	Q-1	15	1.067		705	55.50		760.50	860
2140	3"	"	13	1.231		940	64		1,004	1,125
2850	Gate, N.R.S., soldered, 125 psi									
2900	3/8"	1 Plum	24	.333	Ea.	54.50	19.20		73.70	89
2920	1/2"		24	.333		46.50	19.20		65.70	80.50
2940	3/4"		20	.400		53	23		76	93

22 05 Common Work Results for Plumbing

22 05 23 – General-Duty Valves for Plumbing Piping

22 05 23.20 Valves, Bronze		Crew	Daily Output	Labor-Hours	Unit	Material	2014 Bare Costs Labor	Equipment	Total	Total Incl O&P
2950	1"	1 Plum	19	.421	Ea.	75	24		99	119
2960	1-1/4"		15	.533		115	30.50		145.50	173
2970	1-1/2"		13	.615		129	35.50		164.50	196
2980	2"	↓	11	.727		182	42		224	264
2990	2-1/2"	Q-1	15	1.067		455	55.50		510.50	590
3000	3"	"	13	1.231	↓	595	64		659	750
3350	Threaded, class 150									
3410	1/4"	1 Plum	24	.333	Ea.	78	19.20		97.20	115
3420	3/8"		24	.333		78	19.20		97.20	115
3430	1/2"		24	.333		73.50	19.20		92.70	110
3440	3/4"		20	.400		84.50	23		107.50	128
3450	1"		19	.421		106	24		130	153
3460	1-1/4" size		15	.533		143	30.50		173.50	205
3470	1-1/2"		13	.615		205	35.50		240.50	280
3480	2"	↓	11	.727		245	42		287	335
3490	2-1/2"	Q-1	15	1.067		575	55.50		630.50	715
3500	3" size	"	13	1.231	↓	815	64		879	990
3600	Gate, flanged, 150 lb.									
3610	1"	1 Plum	7	1.143	Ea.	1,300	66		1,366	1,525
3620	1-1/2"		6	1.333		1,625	76.50		1,701.50	1,900
3630	2"	↓	5	1.600		2,400	92		2,492	2,775
3634	2-1/2"	Q-1	5	3.200		3,700	166		3,866	4,325
3640	3"	"	4.50	3.556	↓	4,300	184		4,484	5,025
3850	Rising stem, soldered, 300 psi									
3900	3/8"	1 Plum	24	.333	Ea.	149	19.20		168.20	193
3920	1/2"		24	.333		122	19.20		141.20	163
3940	3/4"		20	.400		139	23		162	188
3950	1"		19	.421		189	24		213	245
3960	1-1/4"		15	.533		261	30.50		291.50	335
3970	1-1/2"		13	.615		315	35.50		350.50	405
3980	2"	↓	11	.727		505	42		547	625
3990	2-1/2"	Q-1	15	1.067		1,075	55.50		1,130.50	1,275
4000	3"	"	13	1.231	↓	1,675	64		1,739	1,950
4250	Threaded, class 150									
4310	1/4"	1 Plum	24	.333	Ea.	71	19.20		90.20	108
4320	3/8"		24	.333		71	19.20		90.20	108
4330	1/2"		24	.333		65	19.20		84.20	101
4340	3/4"		20	.400		76	23		99	119
4350	1"		19	.421		102	24		126	149
4360	1-1/4"		15	.533		139	30.50		169.50	200
4370	1-1/2"		13	.615		175	35.50		210.50	246
4380	2"	↓	11	.727		235	42		277	320
4390	2-1/2"	Q-1	15	1.067		550	55.50		605.50	690
4400	3"	"	13	1.231		765	64		829	935
4500	For 300 psi, threaded, add					100%	15%			
4540	For chain operated type, add				↓	15%				
4850	Globe, class 150, rising stem, threaded									
4920	1/4"	1 Plum	24	.333	Ea.	102	19.20		121.20	141
4940	3/8" size		24	.333		101	19.20		120.20	140
4950	1/2"		24	.333		101	19.20		120.20	140
4960	3/4"		20	.400		135	23		158	183
4970	1"		19	.421		212	24		236	270
4980	1-1/4"	↓	15	.533		335	30.50		365.50	415

22 05 Common Work Results for Plumbing

22 05 23 – General-Duty Valves for Plumbing Piping

22 05 23.20 Valves, Bronze		Crew	Daily Output	Labor-Hours	Unit	Material	2014 Bare Costs Labor	Equipment	Total	Total Incl O&P
4990	1-1/2"	1 Plum	13	.615	Ea.	410	35.50		445.50	505
5000	2"	▼	11	.727		615	42		657	740
5010	2-1/2"	Q-1	15	1.067		1,225	55.50		1,280.50	1,425
5020	3"	"	13	1.231	▼	1,750	64		1,814	2,025
5120	For 300 lb. threaded, add					50%	15%			
5600	Relief, pressure & temperature, self-closing, ASME, threaded									
5640	3/4"	1 Plum	28	.286	Ea.	253	16.45		269.45	305
5650	1"		24	.333		405	19.20		424.20	475
5660	1-1/4"		20	.400		895	23		918	1,025
5670	1-1/2"		18	.444		1,250	25.50		1,275.50	1,425
5680	2"	▼	16	.500	▼	1,350	29		1,379	1,550
5950	Pressure, poppet type, threaded									
6000	1/2"	1 Plum	30	.267	Ea.	68.50	15.35		83.85	98.50
6040	3/4"	"	28	.286	"	73.50	16.45		89.95	106
6400	Pressure, water, ASME, threaded									
6440	3/4"	1 Plum	28	.286	Ea.	116	16.45		132.45	152
6450	1"		24	.333		260	19.20		279.20	315
6460	1-1/4"		20	.400		390	23		413	465
6470	1-1/2"		18	.444		570	25.50		595.50	665
6480	2"		16	.500		820	29		849	950
6490	2-1/2"	▼	15	.533	▼	3,150	30.50		3,180.50	3,525
6900	Reducing, water pressure									
6920	300 psi to 25-75 psi, threaded or sweat									
6940	1/2"	1 Plum	24	.333	Ea.	395	19.20		414.20	465
6950	3/4"		20	.400		395	23		418	470
6960	1"		19	.421		615	24		639	710
6970	1-1/4"		15	.533		1,100	30.50		1,130.50	1,250
6980	1-1/2"		13	.615		1,650	35.50		1,685.50	1,875
6990	2"	▼	11	.727	▼	2,400	42		2,442	2,700
7100	For built-in by-pass or 10-35 psi, add					43			43	47.50
7700	High capacity, 250 psi to 25-75 psi, threaded									
7740	1/2"	1 Plum	24	.333	Ea.	540	19.20		559.20	625
7780	3/4"		20	.400		540	23		563	630
7790	1"		19	.421		775	24		799	890
7800	1-1/4"		15	.533		1,350	30.50		1,380.50	1,525
7810	1-1/2"		13	.615		1,975	35.50		2,010.50	2,225
7820	2"		11	.727		2,875	42		2,917	3,225
7830	2-1/2"		9	.889		4,100	51		4,151	4,600
7840	3"	▼	8	1		4,925	57.50		4,982.50	5,475
7850	3" flanged (iron body)	Q-1	10	1.600		4,850	83		4,933	5,450
7860	4" flanged (iron body)	"	8	2	▼	6,300	104		6,404	7,075
7920	For higher pressure, add					25%				
8000	Silent check, bronze trim									
8010	Compact wafer type, for 125 or 150 lb. flanges									
8020	1-1/2"	1 Plum	11	.727	Ea.	335	42		377	430
8021	2"	"	9	.889		355	51		406	465
8022	2-1/2"	Q-1	9	1.778		370	92		462	550
8023	3"		8	2		415	104		519	610
8024	4"	▼	5	3.200		710	166		876	1,025
8025	5"	Q-2	6	4		940	215		1,155	1,350
8026	6"	"	5	4.800	▼	1,100	258		1,358	1,625
8050	For 250 or 300 lb. flanges, thru 6" no change									
8060	Full flange wafer type, 150 lb.									

22 05 Common Work Results for Plumbing

22 05 23 – General-Duty Valves for Plumbing Piping

22 05 23.20 Valves, Bronze

		Crew	Daily Output	Labor-Hours	Unit	Material	2014 Bare Costs Labor	Equipment	Total	Total Incl O&P
8063	1-1/2"	1 Plum	11	.727	Ea.	440	42		482	550
8064	2"	"	9	.889		595	51		646	725
8065	2-1/2"	Q-1	9	1.778		670	92		762	875
8066	3"		8	2		740	104		844	970
8067	4"		5	3.200		1,100	166		1,266	1,450
8068	5"	Q-2	6	4		1,425	215		1,640	1,875
8069	6"	"	5	4.800		1,850	258		2,108	2,450
8080	For 300 lb., add					40%	10%			
8100	Globe type, 150 lb.									
8110	2"	1 Plum	9	.889	Ea.	690	51		741	835
8111	2-1/2"	Q-1	9	1.778		855	92		947	1,075
8112	3"		8	2		1,025	104		1,129	1,275
8113	4"		5	3.200		1,425	166		1,591	1,800
8114	5"	Q-2	6	4		1,725	215		1,940	2,225
8115	6"	"	5	4.800		2,400	258		2,658	3,050
8130	For 300 lb., add					20%	10%			
8140	Screwed end type, 250 lb.									
8141	1/2"	1 Plum	24	.333	Ea.	62.50	19.20		81.70	97.50
8142	3/4"		20	.400		62.50	23		85.50	104
8143	1"		19	.421		71	24		95	115
8144	1-1/4"		15	.533		98.50	30.50		129	155
8145	1-1/2"		13	.615		110	35.50		145.50	175
8146	2"		11	.727		149	42		191	227
8350	Tempering, water, sweat connections									
8400	1/2"	1 Plum	24	.333	Ea.	109	19.20		128.20	149
8440	3/4"	"	20	.400	"	126	23		149	174
8650	Threaded connections									
8700	1/2"	1 Plum	24	.333	Ea.	126	19.20		145.20	168
8740	3/4"		20	.400		770	23		793	880
8750	1"		19	.421		865	24		889	990
8760	1-1/4"		15	.533		1,350	30.50		1,380.50	1,525
8770	1-1/2"		13	.615		1,475	35.50		1,510.50	1,650
8780	2"		11	.727		2,200	42		2,242	2,500

22 05 23.40 Valves, Lined, Corrosion Resistant/High Purity

		Crew	Daily Output	Labor-Hours	Unit	Material	2014 Bare Costs Labor	Equipment	Total	Total Incl O&P
0010	**VALVES, LINED, CORROSION RESISTANT/HIGH PURITY**									
3500	Check lift, 125 lb., cast iron flanged									
3510	Horizontal PPL or SL lined									
3530	1"	1 Plum	14	.571	Ea.	635	33		668	750
3540	1-1/2"		11	.727		770	42		812	910
3550	2"		8	1		895	57.50		952.50	1,075
3560	2-1/2"	Q-1	5	3.200		1,175	166		1,341	1,550
3570	3"		4.50	3.556		1,450	184		1,634	1,875
3590	4"		3	5.333		1,925	276		2,201	2,550
3610	6"	Q-2	3	8		3,275	430		3,705	4,250
3620	8"	"	2.50	9.600		7,225	515		7,740	8,725
4250	Vertical PPL or SL lined									
4270	1"	1 Plum	14	.571	Ea.	570	33		603	680
4290	1-1/2"		11	.727		705	42		747	840
4300	2"		8	1		820	57.50		877.50	990
4310	2-1/2"	Q-1	5	3.200		1,250	166		1,416	1,625
4320	3"		4.50	3.556		1,250	184		1,434	1,625
4340	4"		3	5.333		1,625	276		1,901	2,225

22 05 Common Work Results for Plumbing

22 05 23 – General-Duty Valves for Plumbing Piping

22 05 23.40 Valves, Lined, Corrosion Resistant/High Purity

		Crew	Daily Output	Labor-Hours	Unit	Material	2014 Bare Costs Labor	Equipment	Total	Total Incl O&P
4360	6"	Q-2	3	8	Ea.	2,750	430		3,180	3,675
4370	8"	"	2.50	9.600		4,975	515		5,490	6,225
5000	Clamp type, ductile iron, 150 lb. flanged									
5010	TFE lined									
5030	1" size, lever handle	1 Plum	9	.889	Ea.	1,350	51		1,401	1,550
5050	1-1/2" size, lever handle		6	1.333		1,775	76.50		1,851.50	2,075
5060	2" size, lever handle		5	1.600		2,150	92		2,242	2,525
5080	3" size, lever handle	Q-1	4.50	3.556		2,725	184		2,909	3,275
5100	4" size, gear operated	"	3	5.333		3,900	276		4,176	4,725
5120	6" size, gear operated	Q-2	3	8		13,500	430		13,930	15,600
5130	8" size, gear operated	"	2.50	9.600		14,600	515		15,115	16,800
6000	Diaphragm type, cast iron, 125 lb. flanged									
6010	PTFE or VITON, lined									
6030	1" size, handwheel operated	1 Plum	9	.889	Ea.	315	51		366	425
6050	1-1/2" size, handwheel operated		6	1.333		385	76.50		461.50	535
6060	2" size, handwheel operated		5	1.600		460	92		552	645
6080	3" size, handwheel operated	Q-1	4.50	3.556		760	184		944	1,125
6100	4" size, handwheel operated	"	3	5.333		1,375	276		1,651	1,950
6120	6" size, handwheel operated	Q-2	3	8		2,350	430		2,780	3,250
6130	8" size, handwheel operated	"	2.50	9.600		4,750	515		5,265	6,000

22 05 23.60 Valves, Plastic

		Crew	Daily Output	Labor-Hours	Unit	Material	Labor	Equipment	Total	Total Incl O&P
0010	**VALVES, PLASTIC** R220523-90									
1100	Angle, PVC, threaded									
1110	1/4"	1 Plum	26	.308	Ea.	74	17.70		91.70	108
1120	1/2"		26	.308		74	17.70		91.70	108
1130	3/4"		25	.320		87.50	18.40		105.90	125
1140	1"		23	.348		106	20		126	147
1150	Ball, PVC, socket or threaded, single union									
1230	1/2"	1 Plum	26	.308	Ea.	28	17.70		45.70	57.50
1240	3/4"		25	.320		32	18.40		50.40	63.50
1250	1"		23	.348		40	20		60	74.50
1260	1-1/4"		21	.381		54	22		76	92.50
1270	1-1/2"		20	.400		64.50	23		87.50	106
1280	2"		17	.471		93.50	27		120.50	144
1290	2-1/2"	Q-1	26	.615		395	32		427	485
1300	3"		24	.667		480	34.50		514.50	575
1310	4"		20	.800		810	41.50		851.50	955
1360	For PVC, flanged, add					100%	15%			
1450	Double union 1/2"	1 Plum	26	.308		36	17.70		53.70	66
1460	3/4"		25	.320		42.50	18.40		60.90	74.50
1470	1"		23	.348		50.50	20		70.50	85.50
1480	1-1/4"		21	.381		69.50	22		91.50	110
1490	1-1/2"		20	.400		84	23		107	128
1500	2"		17	.471		111	27		138	163
1650	CPVC, socket or threaded, single union									
1700	1/2"	1 Plum	26	.308	Ea.	56	17.70		73.70	88
1720	3/4"		25	.320		70	18.40		88.40	105
1730	1"		23	.348		84	20		104	123
1750	1-1/4"		21	.381		134	22		156	181
1760	1-1/2"		20	.400		134	23		157	183
1770	2"		17	.471		185	27		212	244
1780	3"	Q-1	24	.667		760	34.50		794.50	885

22 05 Common Work Results for Plumbing

22 05 23 – General-Duty Valves for Plumbing Piping

22 05 23.60 Valves, Plastic		Crew	Daily Output	Labor-Hours	Unit	Material	2014 Bare Costs Labor	Equipment	Total	Total Incl O&P
1840	For CPVC, flanged, add				Ea.	65%	15%			
1880	For true union, socket or threaded, add				↓	50%	5%			
2050	Polypropylene, threaded									
2100	1/4"	1 Plum	26	.308	Ea.	45	17.70		62.70	76
2120	3/8"		26	.308		45	17.70		62.70	76
2130	1/2"		26	.308		45	17.70		62.70	76
2140	3/4"		25	.320		54.50	18.40		72.90	88
2150	1"		23	.348		62.50	20		82.50	98.50
2160	1-1/4"		21	.381		84	22		106	126
2170	1-1/2"		20	.400		103	23		126	149
2180	2"	↓	17	.471		138	27		165	193
2190	3"	Q-1	24	.667		315	34.50		349.50	395
2200	4"	"	20	.800	↓	550	41.50		591.50	670
2550	PVC, three way, socket or threaded									
2600	1/2"	1 Plum	26	.308	Ea.	62.50	17.70		80.20	95
2640	3/4"		25	.320		71	18.40		89.40	106
2650	1"		23	.348		76.50	20		96.50	114
2660	1-1/2"		20	.400		181	23		204	234
2670	2"	↓	17	.471		206	27		233	268
2680	3"	Q-1	24	.667		430	34.50		464.50	525
2740	For flanged, add				↓	60%	15%			
3150	Ball check, PVC, socket or threaded									
3200	1/4"	1 Plum	26	.308	Ea.	44	17.70		61.70	74.50
3220	3/8"		26	.308		44	17.70		61.70	74.50
3240	1/2"		26	.308		44	17.70		61.70	74.50
3250	3/4"		25	.320		49	18.40		67.40	82
3260	1"		23	.348		61.50	20		81.50	97.50
3270	1-1/4"		21	.381		103	22		125	147
3280	1-1/2"		20	.400		103	23		126	149
3290	2"	↓	17	.471		140	27		167	195
3310	3"	Q-1	24	.667		390	34.50		424.50	480
3320	4"	"	20	.800		550	41.50		591.50	670
3360	For PVC, flanged, add				↓	50%	15%			
3750	CPVC, socket or threaded									
3800	1/2"	1 Plum	26	.308	Ea.	67.50	17.70		85.20	101
3840	3/4"		25	.320		80.50	18.40		98.90	117
3850	1"		23	.348		95.50	20		115.50	135
3860	1-1/2"		20	.400		164	23		187	215
3870	2"	↓	17	.471		164	27		191	221
3880	3"	Q-1	24	.667		595	34.50		629.50	705
3920	4"	"	20	.800		800	41.50		841.50	945
3930	For CPVC, flanged, add					40%	15%			
4340	Polypropylene, threaded									
4360	1/2"	1 Plum	26	.308	Ea.	44	17.70		61.70	75
4400	3/4"		25	.320		61	18.40		79.40	95
4440	1"		23	.348		65	20		85	102
4450	1-1/2"		20	.400		125	23		148	172
4460	2"	↓	17	.471		158	27		185	215
4500	For polypropylene flanged, add				↓	200%	15%			
4850	Foot valve, PVC, socket or threaded									
4900	1/2"	1 Plum	34	.235	Ea.	73	13.55		86.55	101
4930	3/4"		32	.250		83	14.40		97.40	113
4940	1"	↓	28	.286		107	16.45		123.45	143

22 05 Common Work Results for Plumbing

22 05 23 – General-Duty Valves for Plumbing Piping

22 05 23.60 Valves, Plastic		Crew	Daily Output	Labor-Hours	Unit	Material	2014 Bare Costs Labor	Equipment	Total	Total Incl O&P
4950	1-1/4"	1 Plum	27	.296	Ea.	206	17.05		223.05	253
4960	1-1/2"		26	.308		206	17.70		223.70	254
4970	2"		24	.333		365	19.20		384.20	430
4980	3"		20	.400		835	23		858	955
4990	4"	▼	18	.444		1,275	25.50		1,300.50	1,475
5000	For flanged, add				▼	25%	10%			
5050	CPVC, socket or threaded									
5060	1/2"	1 Plum	34	.235	Ea.	72	13.55		85.55	100
5070	3/4"		32	.250		83	14.40		97.40	113
5080	1"		28	.286		102	16.45		118.45	138
5090	1-1/4"		27	.296		164	17.05		181.05	206
5100	1-1/2"		26	.308		164	17.70		181.70	207
5110	2"		24	.333		211	19.20		230.20	261
5120	3"		20	.400		430	23		453	505
5130	4"	▼	18	.444		780	25.50		805.50	895
5140	For flanged, add				▼	25%	10%			
5280	Needle valve, PVC, threaded									
5300	1/4"	1 Plum	26	.308	Ea.	51	17.70		68.70	82.50
5340	3/8"		26	.308		59	17.70		76.70	91.50
5360	1/2"	▼	26	.308		59	17.70		76.70	91.50
5380	For polypropylene, add				▼	10%				
5800	Y check, PVC, socket or threaded									
5820	1/2"	1 Plum	26	.308	Ea.	55	17.70		72.70	87
5840	3/4"		25	.320		89	18.40		107.40	126
5850	1"		23	.348		96.50	20		116.50	136
5860	1-1/4"		21	.381		151	22		173	200
5870	1-1/2"		20	.400		165	23		188	217
5880	2"		17	.471		205	27		232	267
5890	2-1/2"	▼	15	.533		435	30.50		465.50	520
5900	3"	Q-1	24	.667		435	34.50		469.50	525
5910	4"	"	20	.800		710	41.50		751.50	845
5960	For PVC flanged, add				▼	45%	15%			
6350	Y sediment strainer, PVC, socket or threaded									
6400	1/2"	1 Plum	26	.308	Ea.	55	17.70		72.70	87
6440	3/4"		24	.333		58.50	19.20		77.70	93
6450	1"		23	.348		69.50	20		89.50	107
6460	1-1/4"		21	.381		117	22		139	161
6470	1-1/2"		20	.400		117	23		140	163
6480	2"		17	.471		144	27		171	199
6490	2-1/2"	▼	15	.533		350	30.50		380.50	430
6500	3"	Q-1	24	.667		350	34.50		384.50	435
6510	4"	"	20	.800		585	41.50		626.50	705
6560	For PVC, flanged, add				▼	55%	15%			

22 05 29 – Hangers and Supports for Plumbing Piping and Equipment

22 05 29.10 Hangers & Supp. for Plumb'g/HVAC Pipe/Equip.		Crew	Daily Output	Labor-Hours	Unit	Material	Labor	Equipment	Total	Total Incl O&P
0010	**HANGERS AND SUPPORTS FOR PLUMB'G/HVAC PIPE/EQUIP.**									
0011	TYPE numbers per MSS-SP58									
0050	Brackets									
0060	Beam side or wall, malleable iron, TYPE 34									
0070	3/8" threaded rod size	1 Plum	48	.167	Ea.	2.96	9.60		12.56	17.75
0080	1/2" threaded rod size		48	.167		4.18	9.60		13.78	19.10
0090	5/8" threaded rod size	▼	48	.167		7.85	9.60		17.45	23

22 05 Common Work Results for Plumbing

22 05 29 – Hangers and Supports for Plumbing Piping and Equipment

22 05 29.10 Hangers & Supp. for Plumb'g/HVAC Pipe/Equip.		Crew	Daily Output	Labor-Hours	Unit	Material	2014 Bare Costs Labor	Equipment	Total	Total Incl O&P
0100	3/4" threaded rod size	1 Plum	48	.167	Ea.	10.75	9.60		20.35	26.50
0110	7/8" threaded rod size		48	.167		11.10	9.60		20.70	27
0120	For concrete installation, add						30%			
0150	Wall, welded steel, medium, TYPE 32									
0160	0 size, 12" wide, 18" deep	1 Plum	34	.235	Ea.	164	13.55		177.55	201
0170	1 size, 18" wide, 24" deep		34	.235		201	13.55		214.55	242
0180	2 size, 24" wide, 30" deep		34	.235		276	13.55		289.55	325
0200	Beam attachment, welded, TYPE 22									
0202	3/8"	Q-15	80	.200	Ea.	6.80	10.35	.69	17.84	24
0203	1/2"		76	.211		6.90	10.90	.73	18.53	25
0204	5/8"		72	.222		8.05	11.50	.77	20.32	27
0205	3/4"		68	.235		11.70	12.20	.81	24.71	32
0206	7/8"		64	.250		15	12.95	.87	28.82	37
0207	1"		56	.286		30	14.80	.99	45.79	56.50
0300	Clamps									
0310	C-clamp, for mounting on steel beam flange, w/locknut, TYPE 23									
0320	3/8" threaded rod size	1 Plum	160	.050	Ea.	1.95	2.88		4.83	6.50
0330	1/2" threaded rod size		160	.050		2.09	2.88		4.97	6.65
0340	5/8" threaded rod size		160	.050		3.52	2.88		6.40	8.20
0350	3/4" threaded rod size		160	.050		4.34	2.88		7.22	9.10
0352	7/8" threaded rod size		140	.057		14.30	3.29		17.59	20.50
0400	High temperature to 1050°F, alloy steel									
0410	4" pipe size	Q-1	106	.151	Ea.	25.50	7.80		33.30	40
0420	6" pipe size		106	.151		43.50	7.80		51.30	59.50
0430	8" pipe size		97	.165		48	8.55		56.55	66
0440	10" pipe size		84	.190		70.50	9.85		80.35	92.50
0450	12" pipe size		72	.222		79	11.50		90.50	104
0460	14" pipe size		64	.250		199	12.95		211.95	239
0470	16" pipe size		56	.286		219	14.80		233.80	263
0480	Beam clamp, flange type, TYPE 25									
0482	For 3/8" bolt	1 Plum	48	.167	Ea.	3.26	9.60		12.86	18.10
0483	For 1/2" bolt		44	.182		4.18	10.45		14.63	20.50
0484	For 5/8" bolt		40	.200		4.75	11.50		16.25	22.50
0485	For 3/4" bolt		36	.222		4.75	12.80		17.55	24.50
0486	For 1" bolt		32	.250		4.75	14.40		19.15	27
0500	I-beam, for mounting on bottom flange, strap iron, TYPE 21									
0510	2" flange size	1 Plum	96	.083	Ea.	10.40	4.80		15.20	18.70
0520	3" flange size		95	.084		11.80	4.85		16.65	20.50
0530	4" flange size		93	.086		5.45	4.95		10.40	13.50
0540	5" flange size		92	.087		6	5		11	14.15
0550	6" flange size		90	.089		6.80	5.10		11.90	15.20
0560	7" flange size		88	.091		7.55	5.25		12.80	16.25
0570	8" flange size		86	.093		8.15	5.35		13.50	17.05
0600	One hole, vertical mounting, malleable iron									
0610	1/2" pipe size	1 Plum	160	.050	Ea.	1.03	2.88		3.91	5.50
0620	3/4" pipe size		145	.055		1.12	3.18		4.30	6
0630	1" pipe size		136	.059		1.19	3.39		4.58	6.40
0640	1-1/4" pipe size		128	.063		2.09	3.60		5.69	7.75
0650	1-1/2" pipe size		120	.067		2.40	3.84		6.24	8.45
0660	2" pipe size		112	.071		3.47	4.11		7.58	10
0670	2-1/2" pipe size		104	.077		6.75	4.43		11.18	14.10
0680	3" pipe size		96	.083		8.15	4.80		12.95	16.20
0690	3-1/2" pipe size		90	.089		9.95	5.10		15.05	18.65

22 05 Common Work Results for Plumbing

22 05 29 – Hangers and Supports for Plumbing Piping and Equipment

22 05 29.10 Hangers & Supp. for Plumb'g/HVAC Pipe/Equip.		Crew	Daily Output	Labor-Hours	Unit	Material	2014 Bare Costs Labor	Equipment	Total	Total Incl O&P
0700	4" pipe size	1 Plum	84	.095	Ea.	11.35	5.50		16.85	21
0750	Riser or extension pipe, carbon steel, TYPE 8									
0756	1/2" pipe size	1 Plum	52	.154	Ea.	2.40	8.85		11.25	16
0760	3/4" pipe size		48	.167		2.40	9.60		12	17.15
0770	1" pipe size		47	.170		2.52	9.80		12.32	17.55
0780	1-1/4" pipe size		46	.174		3.06	10		13.06	18.45
0790	1-1/2" pipe size		45	.178		3.26	10.25		13.51	19.05
0800	2" pipe size		43	.186		3.38	10.70		14.08	19.85
0810	2-1/2" pipe size		41	.195		3.51	11.25		14.76	21
0820	3" pipe size		40	.200		3.81	11.50		15.31	21.50
0830	3-1/2" pipe size		39	.205		4.28	11.80		16.08	22.50
0840	4" pipe size		38	.211		5.20	12.10		17.30	24
0850	5" pipe size		37	.216		6.50	12.45		18.95	26
0860	6" pipe size		36	.222		7.95	12.80		20.75	28
0870	8" pipe size		34	.235		13.90	13.55		27.45	36
0880	10" pipe size		32	.250		21.50	14.40		35.90	45.50
0890	12" pipe size	▼	28	.286	▼	28.50	16.45		44.95	56.50
0900	For plastic coating 3/4" to 4", add					190%				
0910	For copper plating 3/4" to 4", add					58%				
0950	Two piece, complete, carbon steel, medium weight, TYPE 4									
0960	1/2" pipe size	Q-1	137	.117	Ea.	2.18	6.05		8.23	11.55
0970	3/4" pipe size		134	.119		2.18	6.20		8.38	11.75
0980	1" pipe size		132	.121		2.46	6.30		8.76	12.20
0990	1-1/4" pipe size		130	.123		2.67	6.40		9.07	12.60
1000	1-1/2" pipe size		126	.127		3.46	6.60		10.06	13.75
1010	2" pipe size		124	.129		4.26	6.70		10.96	14.80
1020	2-1/2" pipe size		120	.133		4.61	6.90		11.51	15.50
1030	3" pipe size		117	.137		5.10	7.10		12.20	16.30
1040	3-1/2" pipe size		114	.140		9.55	7.25		16.80	21.50
1050	4" pipe size		110	.145		6.75	7.55		14.30	18.85
1060	5" pipe size		106	.151		16.55	7.80		24.35	30
1070	6" pipe size		104	.154		18.30	7.95		26.25	32
1080	8" pipe size		100	.160		22	8.30		30.30	37
1090	10" pipe size		96	.167		44	8.65		52.65	61
1100	12" pipe size		89	.180		53	9.30		62.30	72.50
1110	14" pipe size		82	.195		70	10.10		80.10	92.50
1120	16" pipe size	▼	68	.235		90.50	12.20		102.70	118
1130	For galvanized, add				▼	45%				
1150	Insert, concrete									
1160	Wedge type, carbon steel body, malleable iron nut, galvanized									
1170	1/4" threaded rod size	1 Plum	96	.083	Ea.	2.53	4.80		7.33	10.05
1180	3/8" threaded rod size		96	.083		2.53	4.80		7.33	10.05
1190	1/2" threaded rod size		96	.083		2.85	4.80		7.65	10.40
1200	5/8" threaded rod size		96	.083		3.38	4.80		8.18	10.95
1210	3/4" threaded rod size		96	.083		7.05	4.80		11.85	15.05
1220	7/8" threaded rod size	▼	96	.083	▼	7.85	4.80		12.65	15.90
1250	Pipe guide sized for insulation									
1260	No. 1, 1" pipe size, 1" thick insulation	1 Stpi	26	.308	Ea.	118	18		136	157
1270	No. 2, 1-1/4"-2" pipe size, 1" thick insulation		23	.348		144	20.50		164.50	189
1280	No. 3, 1-1/4"-2" pipe size, 1-1/2" thick insulation		21	.381		144	22.50		166.50	192
1290	No. 4, 2-1/2"-3-1/2" pipe size, 1-1/2" thick insulation		18	.444		144	26		170	198
1300	No. 5, 4"-5" pipe size, 1-1/2" thick insulation	▼	16	.500		163	29.50		192.50	224
1310	No. 6, 5"-6" pipe size, 2" thick insulation	Q-5	21	.762	▼	189	40		229	269

22 05 Common Work Results for Plumbing

22 05 29 – Hangers and Supports for Plumbing Piping and Equipment

	22 05 29.10 Hangers & Supp. for Plumb'g/HVAC Pipe/Equip.	Crew	Daily Output	Labor-Hours	Unit	Material	2014 Bare Costs Labor	Equipment	Total	Total Incl O&P
1320	No. 7, 8" pipe size, 2" thick insulation	Q-5	16	1	Ea.	239	52.50		291.50	345
1330	No. 8, 10" pipe size, 2" thick insulation		12	1.333		355	70		425	495
1340	No. 9, 12" pipe size, 2" thick insulation	Q-6	17	1.412		355	77		432	505
1350	No. 10, 12"-14" pipe size, 2-1/2" thick insulation		16	1.500		415	82		497	585
1360	No. 11, 16" pipe size, 2-1/2" thick insulation		10.50	2.286		415	125		540	650
1370	No. 12, 16"-18" pipe size, 3" thick insulation		9	2.667		610	146		756	890
1380	No. 13, 20" pipe size, 3" thick insulation		7.50	3.200		610	175		785	935
1390	No. 14, 24" pipe size, 3" thick insulation		7	3.429		805	187		992	1,175
1400	Bands									
1410	Adjustable band, carbon steel, for non-insulated pipe, TYPE 7									
1420	1/2" pipe size	Q-1	142	.113	Ea.	.29	5.85		6.14	9.10
1430	3/4" pipe size		140	.114		.29	5.90		6.19	9.25
1440	1" pipe size		137	.117		.29	6.05		6.34	9.45
1450	1-1/4" pipe size		134	.119		.32	6.20		6.52	9.70
1460	1-1/2" pipe size		131	.122		.32	6.35		6.67	9.90
1470	2" pipe size		129	.124		.32	6.40		6.72	10.05
1480	2-1/2" pipe size		125	.128		.57	6.65		7.22	10.65
1490	3" pipe size		122	.131		.61	6.80		7.41	10.90
1500	3-1/2" pipe size		119	.134		.99	6.95		7.94	11.60
1510	4" pipe size		114	.140		.99	7.25		8.24	12.10
1520	5" pipe size		110	.145		1.58	7.55		9.13	13.15
1530	6" pipe size		108	.148		1.85	7.65		9.50	13.65
1540	8" pipe size		104	.154		2.79	7.95		10.74	15.10
1550	For copper plated, add					50%				
1560	For galvanized, add					30%				
1570	For plastic coating, add					30%				
1600	Adjusting nut malleable iron, steel band, TYPE 9									
1610	1/2" pipe size, galvanized band	Q-1	137	.117	Ea.	3.21	6.05		9.26	12.70
1620	3/4" pipe size, galvanized band		135	.119		3.21	6.15		9.36	12.80
1630	1" pipe size, galvanized band		132	.121		3.32	6.30		9.62	13.15
1640	1-1/4" pipe size, galvanized band		129	.124		3.33	6.40		9.73	13.35
1650	1-1/2" pipe size, galvanized band		126	.127		3.44	6.60		10.04	13.75
1660	2" pipe size, galvanized band		124	.129		3.59	6.70		10.29	14.05
1670	2-1/2" pipe size, galvanized band		120	.133		5.70	6.90		12.60	16.70
1680	3" pipe size, galvanized band		117	.137		5.95	7.10		13.05	17.20
1690	3-1/2" pipe size, galvanized band		114	.140		9.45	7.25		16.70	21.50
1700	4" pipe size, cadmium plated band		110	.145		10.60	7.55		18.15	23
1740	For plastic coated band, add					35%				
1750	For completely copper coated, add					45%				
1800	Clevis, adjustable, carbon steel, for non-insulated pipe, TYPE 1									
1810	1/2" pipe size	Q-1	137	.117	Ea.	.89	6.05		6.94	10.15
1820	3/4" pipe size		135	.119		.89	6.15		7.04	10.25
1830	1" pipe size		132	.121		.93	6.30		7.23	10.50
1840	1-1/4" pipe size		129	.124		1.05	6.40		7.45	10.85
1850	1-1/2" pipe size		126	.127		1.16	6.60		7.76	11.25
1860	2" pipe size		124	.129		1.37	6.70		8.07	11.60
1870	2-1/2" pipe size		120	.133		2.27	6.90		9.17	12.95
1880	3" pipe size		117	.137		2.46	7.10		9.56	13.40
1890	3-1/2" pipe size		114	.140		2.86	7.25		10.11	14.15
1900	4" pipe size		110	.145		3.09	7.55		10.64	14.80
1910	5" pipe size		106	.151		4.27	7.80		12.07	16.50
1920	6" pipe size		104	.154		5.15	7.95		13.10	17.75
1930	8" pipe size		100	.160		7.55	8.30		15.85	21

22 05 Common Work Results for Plumbing

22 05 29 – Hangers and Supports for Plumbing Piping and Equipment

22 05 29.10 Hangers & Supp. for Plumb'g/HVAC Pipe/Equip.		Crew	Daily Output	Labor-Hours	Unit	Material	2014 Bare Costs Labor	Equipment	Total	Total Incl O&P
1940	10" pipe size	Q-1	96	.167	Ea.	13.55	8.65		22.20	28
1950	12" pipe size		89	.180		17.50	9.30		26.80	33.50
1960	14" pipe size		82	.195		29.50	10.10		39.60	48
1970	16" pipe size		68	.235		47	12.20		59.20	70.50
1971	18" pipe size		54	.296		57	15.35		72.35	85.50
1972	20" pipe size		38	.421		117	22		139	162
1980	For galvanized, add					66%				
1990	For copper plated 1/2" to 4", add					77%				
2000	For light weight 1/2" to 4", deduct					13%				
2010	Insulated pipe type, 3/4" to 12" pipe, add					180%				
2020	Insulated pipe type, chrome-moly U-strap, add					530%				
2250	Split ring, malleable iron, for non-insulated pipe, TYPE 11									
2260	1/2" pipe size	Q-1	137	.117	Ea.	4.18	6.05		10.23	13.75
2270	3/4" pipe size		135	.119		4.22	6.15		10.37	13.90
2280	1" pipe size		132	.121		4.47	6.30		10.77	14.40
2290	1-1/4" pipe size		129	.124		5.55	6.40		11.95	15.80
2300	1-1/2" pipe size		126	.127		6.75	6.60		13.35	17.35
2310	2" pipe size		124	.129		7.60	6.70		14.30	18.50
2320	2-1/2" pipe size		120	.133		11	6.90		17.90	22.50
2330	3" pipe size		117	.137		13.75	7.10		20.85	26
2340	3-1/2" pipe size		114	.140		14.35	7.25		21.60	27
2350	4" pipe size		110	.145		14.35	7.55		21.90	27
2360	5" pipe size		106	.151		19.05	7.80		26.85	33
2370	6" pipe size		104	.154		42.50	7.95		50.45	59
2380	8" pipe size		100	.160		62.50	8.30		70.80	81
2390	For copper plated, add					8%				
2400	Channels, steel, 3/4" x 1-1/2"	1 Plum	80	.100	L.F.	4.83	5.75		10.58	14
2404	1-1/2" x 1-1/2"		70	.114		6.35	6.60		12.95	16.95
2408	1-7/8" x 1-1/2"		60	.133		22.50	7.65		30.15	36
2412	3" x 1-1/2"		50	.160		38	9.20		47.20	55.50
2416	Hangers, trapeze channel support, 12" wide, steel		8.80	.909	Ea.	34	52.50		86.50	116
2418	18" wide, steel		8.30	.964		37	55.50		92.50	125
2430	Spring nuts, long, 1/4"		120	.067		1.37	3.84		5.21	7.30
2432	3/8"		100	.080		1.44	4.60		6.04	8.55
2434	1/2"		80	.100		1.56	5.75		7.31	10.40
2436	5/8"		80	.100		3.62	5.75		9.37	12.70
2438	3/4"		75	.107		6.30	6.15		12.45	16.20
2440	Spring nuts, short, 1/4"		120	.067		1.36	3.84		5.20	7.30
2442	3/8"		100	.080		1.40	4.60		6	8.50
2444	1/2"		80	.100		1.67	5.75		7.42	10.55
2500	Washer, flat steel									
2502	3/8"	1 Plum	240	.033	Ea.	.09	1.92		2.01	3
2503	1/2"		220	.036		.19	2.09		2.28	3.37
2504	5/8"		200	.040		.19	2.30		2.49	3.69
2505	3/4"		180	.044		.14	2.56		2.70	4.01
2506	7/8"		160	.050		.26	2.88		3.14	4.64
2507	1"		140	.057		.23	3.29		3.52	5.20
2508	1-1/4"		120	.067		.46	3.84		4.30	6.30
2520	Nut, steel, hex									
2522	3/8"	1 Plum	200	.040	Ea.	1.02	2.30		3.32	4.60
2523	1/2"		180	.044		1.97	2.56		4.53	6.05
2524	5/8"		160	.050		3.33	2.88		6.21	8
2525	3/4"		140	.057		5.05	3.29		8.34	10.50

22 05 Common Work Results for Plumbing

22 05 29 – Hangers and Supports for Plumbing Piping and Equipment

	22 05 29.10 Hangers & Supp. for Plumb'g/HVAC Pipe/Equip.	Crew	Daily Output	Labor-Hours	Unit	Material	2014 Bare Costs Labor	Equipment	Total	Total Incl O&P
2526	7/8"	1 Plum	120	.067	Ea.	8	3.84		11.84	14.60
2527	1"		100	.080		11.60	4.60		16.20	19.70
2528	1-1/4"		80	.100		22	5.75		27.75	33
2532	Turnbuckle, TYPE 13									
2534	3/8"	1 Plum	80	.100	Ea.	5.45	5.75		11.20	14.70
2535	1/2"		72	.111		6.35	6.40		12.75	16.60
2536	5/8"		64	.125		10.20	7.20		17.40	22
2537	3/4"		56	.143		15.60	8.20		23.80	29.50
2538	7/8"		48	.167		39.50	9.60		49.10	58
2539	1"		40	.200		54	11.50		65.50	77
2540	1-1/4"		32	.250		98	14.40		112.40	130
2650	Rods, carbon steel									
2660	Continuous thread									
2670	1/4" thread size	1 Plum	144	.056	L.F.	1.62	3.20		4.82	6.60
2680	3/8" thread size		144	.056		1.72	3.20		4.92	6.70
2690	1/2" thread size		144	.056		2.72	3.20		5.92	7.80
2700	5/8" thread size		144	.056		3.86	3.20		7.06	9.10
2710	3/4" thread size		144	.056		6.80	3.20		10	12.35
2720	7/8" thread size		144	.056		8.55	3.20		11.75	14.25
2721	1" thread size	Q-1	160	.100		17.80	5.20		23	27.50
2722	1-1/8" thread size	"	120	.133		24	6.90		30.90	37
2725	1/4" thread size, bright finish	1 Plum	144	.056		2.13	3.20		5.33	7.15
2726	1/2" thread size, bright finish	"	144	.056		3.83	3.20		7.03	9.05
2730	For galvanized, add					40%				
2750	Both ends machine threaded 18" length									
2760	3/8" thread size	1 Plum	240	.033	Ea.	4.83	1.92		6.75	8.20
2770	1/2" thread size		240	.033		7.15	1.92		9.07	10.80
2780	5/8" thread size		240	.033		6.40	1.92		8.32	9.95
2790	3/4" thread size		240	.033		11.45	1.92		13.37	15.50
2800	7/8" thread size		240	.033		11.85	1.92		13.77	15.95
2810	1" thread size		240	.033		15.10	1.92		17.02	19.50
2820	Rod couplings									
2821	3/8"	1 Plum	60	.133	Ea.	1.36	7.65		9.01	13.10
2822	1/2"		54	.148		1.90	8.55		10.45	14.95
2823	5/8"		48	.167		2.38	9.60		11.98	17.10
2824	3/4"		44	.182		3.35	10.45		13.80	19.50
2825	7/8"		40	.200		5.65	11.50		17.15	23.50
2826	1"		34	.235		7.50	13.55		21.05	29
2827	1-1/8"		30	.267		15.55	15.35		30.90	40
2860	Pipe hanger assy, adj. clevis, saddle, rod, clamp, insul. allowance									
2864	1/2" pipe size	Q-5	35	.457	Ea.	22.50	24		46.50	61.50
2866	3/4" pipe size		34.80	.460		23.50	24		47.50	62.50
2868	1" pipe size		34.60	.462		17.90	24.50		42.40	56.50
2869	1-1/4" pipe size		34.30	.466		18.30	24.50		42.80	57
2870	1-1/2" pipe size		33.90	.472		18.60	25		43.60	58
2872	2" pipe size		33.30	.480		20.50	25.50		46	60.50
2874	2-1/2" pipe size		32.30	.495		24	26		50	66
2876	3" pipe size		31.20	.513		29	27		56	73
2880	4" pipe size		30.70	.521		31.50	27.50		59	76
2884	6" pipe size		29.80	.537		40	28.50		68.50	86.50
2888	8" pipe size		28	.571		50.50	30		80.50	101
2892	10" pipe size		25.20	.635		57.50	33.50		91	114
2896	12" pipe size		23.20	.690		84.50	36.50		121	148

22 05 Common Work Results for Plumbing
22 05 29 – Hangers and Supports for Plumbing Piping and Equipment

22 05 29.10 Hangers & Supp. for Plumb'g/HVAC Pipe/Equip.		Crew	Daily Output	Labor-Hours	Unit	Material	2014 Bare Costs Labor	Equipment	Total	Total Incl O&P
2900	Rolls									
2910	Adjustable yoke, carbon steel with CI roll, TYPE 43									
2918	2" pipe size	Q-1	140	.114	Ea.	10.10	5.90		16	20
2920	2-1/2" pipe size		137	.117		10.10	6.05		16.15	20.50
2930	3" pipe size		131	.122		10.70	6.35		17.05	21.50
2940	3-1/2" pipe size		124	.129		15.05	6.70		21.75	26.50
2950	4" pipe size		117	.137		15.05	7.10		22.15	27.50
2960	5" pipe size		110	.145		21.50	7.55		29.05	35.50
2970	6" pipe size		104	.154		29	7.95		36.95	44
2980	8" pipe size		96	.167		41.50	8.65		50.15	58.50
2990	10" pipe size		80	.200		51	10.35		61.35	71.50
3000	12" pipe size		68	.235		81.50	12.20		93.70	108
3010	14" pipe size		56	.286		154	14.80		168.80	193
3020	16" pipe size		48	.333		185	17.25		202.25	230
3050	Chair, carbon steel with CI roll									
3060	2" pipe size	1 Plum	68	.118	Ea.	11.25	6.75		18	22.50
3070	2-1/2" pipe size		65	.123		12.20	7.10		19.30	24
3080	3" pipe size		62	.129		13.10	7.45		20.55	25.50
3090	3-1/2" pipe size		60	.133		16.95	7.65		24.60	30
3100	4" pipe size		58	.138		15.45	7.95		23.40	29
3110	5" pipe size		56	.143		18.15	8.20		26.35	32.50
3120	6" pipe size		53	.151		24	8.70		32.70	39.50
3130	8" pipe size		50	.160		33	9.20		42.20	50
3140	10" pipe size		48	.167		45	9.60		54.60	64
3150	12" pipe size		46	.174		68.50	10		78.50	90.50
3170	Single pipe roll, (see line 2650 for rods), TYPE 41, 1" pipe size	Q-1	137	.117		7.50	6.05		13.55	17.40
3180	1-1/4" pipe size		131	.122		7.75	6.35		14.10	18.10
3190	1-1/2" pipe size		129	.124		8.05	6.40		14.45	18.55
3200	2" pipe size		124	.129		8.30	6.70		15	19.25
3210	2-1/2" pipe size		118	.136		8.65	7		15.65	20
3220	3" pipe size		115	.139		8.95	7.20		16.15	21
3230	3-1/2" pipe size		113	.142		9.80	7.35		17.15	22
3240	4" pipe size		112	.143		10.10	7.40		17.50	22.50
3250	5" pipe size		110	.145		11.55	7.55		19.10	24
3260	6" pipe size		101	.158		24.50	8.20		32.70	39.50
3270	8" pipe size		90	.178		25.50	9.20		34.70	42
3280	10" pipe size		80	.200		33	10.35		43.35	52
3290	12" pipe size		68	.235		55	12.20		67.20	78.50
3300	Saddles (add vertical pipe riser, usually 3" diameter)									
3310	Pipe support, complete, adjust., CI saddle, TYPE 36									
3320	2-1/2" pipe size	1 Plum	96	.083	Ea.	88.50	4.80		93.30	104
3330	3" pipe size		88	.091		89.50	5.25		94.75	106
3340	3-1/2" pipe size		79	.101		91	5.85		96.85	109
3350	4" pipe size		68	.118		132	6.75		138.75	155
3360	5" pipe size		64	.125		135	7.20		142.20	159
3370	6" pipe size		59	.136		135	7.80		142.80	160
3380	8" pipe size		53	.151		140	8.70		148.70	167
3390	10" pipe size		50	.160		163	9.20		172.20	193
3400	12" pipe size		48	.167		172	9.60		181.60	204
3450	For standard pipe support, one piece, CI deduct					34%				
3460	For stanchion support, CI with steel yoke, deduct					60%				
3550	Insulation shield 1" thick, 1/2" pipe size, TYPE 40	1 Asbe	100	.080	Ea.	2.60	4.09		6.69	9.20
3560	3/4" pipe size		100	.080		2.91	4.09		7	9.55

22 05 Common Work Results for Plumbing

22 05 29 – Hangers and Supports for Plumbing Piping and Equipment

22 05 29.10 Hangers & Supp. for Plumb'g/HVAC Pipe/Equip.

		Crew	Daily Output	Labor-Hours	Unit	Material	2014 Bare Costs Labor	Equipment	Total	Total Incl O&P
3570	1" pipe size	1 Asbe	98	.082	Ea.	3.14	4.18		7.32	9.95
3580	1-1/4" pipe size		98	.082		3.35	4.18		7.53	10.20
3590	1-1/2" pipe size		96	.083		3.28	4.26		7.54	10.25
3600	2" pipe size		96	.083		3.48	4.26		7.74	10.50
3610	2-1/2" pipe size		94	.085		3.59	4.35		7.94	10.70
3620	3" pipe size		94	.085		4.32	4.35		8.67	11.50
3630	2" thick, 3-1/2" pipe size		92	.087		4.61	4.45		9.06	11.95
3640	4" pipe size		92	.087		6.45	4.45		10.90	13.95
3650	5" pipe size		90	.089		8.70	4.55		13.25	16.70
3660	6" pipe size		90	.089		9.95	4.55		14.50	18.05
3670	8" pipe size		88	.091		16.15	4.65		20.80	25
3680	10" pipe size		88	.091		20.50	4.65		25.15	30
3690	12" pipe size		86	.093		23.50	4.76		28.26	33
3700	14" pipe size		86	.093		38.50	4.76		43.26	50
3710	16" pipe size		84	.095		41	4.87		45.87	53
3720	18" pipe size		84	.095		46	4.87		50.87	58
3730	20" pipe size		82	.098		50.50	4.99		55.49	63.50
3732	24" pipe size		80	.100		59	5.10		64.10	73
3733	30" pipe size		78	.103		505	5.25		510.25	565
3735	36" pipe size		75	.107		770	5.45		775.45	860
3750	Covering protection saddle, TYPE 39									
3760	1" covering size									
3770	3/4" pipe size	1 Plum	68	.118	Ea.	4.39	6.75		11.14	15.05
3780	1" pipe size		68	.118		4.39	6.75		11.14	15.05
3790	1-1/4" pipe size		68	.118		4.39	6.75		11.14	15.05
3800	1-1/2" pipe size		66	.121		4.69	7		11.69	15.70
3810	2" pipe size		66	.121		4.69	7		11.69	15.70
3820	2-1/2" pipe size		64	.125		4.69	7.20		11.89	16
3830	3" pipe size		64	.125		6.30	7.20		13.50	17.75
3840	3-1/2" pipe size		62	.129		6.95	7.45		14.40	18.85
3850	4" pipe size		62	.129		6.95	7.45		14.40	18.85
3860	5" pipe size		60	.133		6.95	7.65		14.60	19.25
3870	6" pipe size		60	.133		8.35	7.65		16	21
3900	1-1/2" covering size									
3910	3/4" pipe size	1 Plum	68	.118	Ea.	5.25	6.75		12	15.95
3920	1" pipe size		68	.118		5.25	6.75		12	15.95
3930	1-1/4" pipe size		68	.118		5.25	6.75		12	15.95
3940	1-1/2" pipe size		66	.121		5.25	7		12.25	16.30
3950	2" pipe size		66	.121		5.70	7		12.70	16.80
3960	2-1/2" pipe size		64	.125		6.85	7.20		14.05	18.40
3970	3" pipe size		64	.125		6.85	7.20		14.05	18.40
3980	3-1/2" pipe size		62	.129		7.20	7.45		14.65	19.15
3990	4" pipe size		62	.129		7.20	7.45		14.65	19.15
4000	5" pipe size		60	.133		7.20	7.65		14.85	19.55
4010	6" pipe size		60	.133		9.75	7.65		17.40	22.50
4020	8" pipe size		58	.138		12.25	7.95		20.20	25.50
4022	10" pipe size		56	.143		12.25	8.20		20.45	26
4024	12" pipe size		54	.148		30	8.55		38.55	46
4028	2" covering size									
4029	2-1/2" pipe size	1 Plum	62	.129	Ea.	6.90	7.45		14.35	18.80
4032	3" pipe size		60	.133		8.15	7.65		15.80	20.50
4033	4" pipe size		58	.138		8.15	7.95		16.10	21
4034	6" pipe size		56	.143		11.65	8.20		19.85	25

117

22 05 Common Work Results for Plumbing

22 05 29 – Hangers and Supports for Plumbing Piping and Equipment

22 05 29.10 Hangers & Supp. for Plumb'g/HVAC Pipe/Equip.		Crew	Daily Output	Labor-Hours	Unit	Material	2014 Bare Costs Labor	Equipment	Total	Total Incl O&P
4035	8" pipe size	1 Plum	54	.148	Ea.	13.75	8.55		22.30	28
4080	10" pipe size		58	.138		14.60	7.95		22.55	28
4090	12" pipe size		56	.143		33.50	8.20		41.70	49.50
4100	14" pipe size		56	.143		33.50	8.20		41.70	49.50
4110	16" pipe size		54	.148		43.50	8.55		52.05	61
4120	18" pipe size		54	.148		43.50	8.55		52.05	61
4130	20" pipe size		52	.154		48.50	8.85		57.35	66.50
4150	24" pipe size		50	.160		56.50	9.20		65.70	76
4160	30" pipe size		48	.167		62	9.60		71.60	82.50
4180	36" pipe size		45	.178		72	10.25		82.25	94.50
4186	2-1/2" covering size									
4187	3" pipe size	1 Plum	58	.138	Ea.	8.25	7.95		16.20	21
4188	4" pipe size		56	.143		8.95	8.20		17.15	22.50
4189	6" pipe size		52	.154		12.20	8.85		21.05	27
4190	8" pipe size		48	.167		15.10	9.60		24.70	31
4191	10" pipe size		44	.182		16.60	10.45		27.05	34
4192	12" pipe size		40	.200		35.50	11.50		47	56.50
4193	14" pipe size		36	.222		35.50	12.80		48.30	58.50
4194	16" pipe size		32	.250		45.50	14.40		59.90	71.50
4195	18" pipe size		28	.286		51	16.45		67.45	81
4200	Sockets									
4210	Rod end, malleable iron, TYPE 16									
4220	1/4" thread size	1 Plum	240	.033	Ea.	1.35	1.92		3.27	4.39
4230	3/8" thread size		240	.033		1.35	1.92		3.27	4.39
4240	1/2" thread size		230	.035		1.78	2		3.78	4.98
4250	5/8" thread size		225	.036		3.33	2.05		5.38	6.75
4260	3/4" thread size		220	.036		4.68	2.09		6.77	8.30
4270	7/8" thread size		210	.038		6.55	2.19		8.74	10.50
4290	Strap, 1/2" pipe size, TYPE 26	Q-1	142	.113		1.47	5.85		7.32	10.40
4300	3/4" pipe size		140	.114		1.54	5.90		7.44	10.65
4310	1" pipe size		137	.117		2.26	6.05		8.31	11.65
4320	1-1/4" pipe size		134	.119		2.30	6.20		8.50	11.90
4330	1-1/2" pipe size		131	.122		2.44	6.35		8.79	12.25
4340	2" pipe size		129	.124		2.47	6.40		8.87	12.40
4350	2-1/2" pipe size		125	.128		3.78	6.65		10.43	14.15
4360	3" pipe size		122	.131		4.98	6.80		11.78	15.75
4370	3-1/2" pipe size		119	.134		5.15	6.95		12.10	16.20
4380	4" pipe size		114	.140		5.40	7.25		12.65	16.95
4400	U-bolt, carbon steel									
4410	Standard, with nuts, TYPE 42									
4420	1/2" pipe size	1 Plum	160	.050	Ea.	.93	2.88		3.81	5.35
4430	3/4" pipe size		158	.051		.98	2.91		3.89	5.50
4450	1" pipe size		152	.053		1.02	3.03		4.05	5.70
4460	1-1/4" pipe size		148	.054		1.29	3.11		4.40	6.10
4470	1-1/2" pipe size		143	.056		1.37	3.22		4.59	6.35
4480	2" pipe size		139	.058		1.48	3.31		4.79	6.65
4490	2-1/2" pipe size		134	.060		2.41	3.44		5.85	7.85
4500	3" pipe size		128	.063		2.72	3.60		6.32	8.45
4510	3-1/2" pipe size		122	.066		2.93	3.77		6.70	8.90
4520	4" pipe size		117	.068		2.97	3.94		6.91	9.20
4530	5" pipe size		114	.070		3.49	4.04		7.53	9.95
4540	6" pipe size		111	.072		6.65	4.15		10.80	13.55
4550	8" pipe size		109	.073		7.25	4.22		11.47	14.40

22 05 Common Work Results for Plumbing

22 05 29 – Hangers and Supports for Plumbing Piping and Equipment

22 05 29.10 Hangers & Supp. for Plumb'g/HVAC Pipe/Equip.		Crew	Daily Output	Labor-Hours	Unit	Material	2014 Bare Costs Labor	Equipment	Total	Total Incl O&P
4560	10" pipe size	1 Plum	107	.075	Ea.	12.55	4.30		16.85	20.50
4570	12" pipe size	↓	104	.077	↓	18.25	4.43		22.68	26.50
4580	For plastic coating on 1/2" thru 6" size, add					150%				
4700	U-hook, carbon steel, requires mounting screws or bolts									
4710	3/4" thru 2" pipe size									
4720	6" long	1 Plum	96	.083	Ea.	.73	4.80		5.53	8.05
4730	8" long		96	.083		1.06	4.80		5.86	8.40
4740	10" long		96	.083		1.64	4.80		6.44	9.05
4750	12" long	↓	96	.083		2.24	4.80		7.04	9.70
4760	For copper plated, add				↓	50%				
7000	Roof supports									
7006	Duct									
7010	Rectangular, open, 12" off roof									
7020	To 18" wide	Q-9	26	.615	Ea.	150	30.50		180.50	212
7030	To 24" wide	"	22	.727		175	36		211	248
7040	To 36" wide	Q-10	30	.800		200	41		241	283
7050	To 48" wide		28	.857		225	44		269	315
7060	To 60" wide	↓	24	1	↓	250	51		301	355
7100	Equipment									
7120	Equipment support	Q-5	20	.800	Ea.	59	42		101	129
7300	Pipe									
7310	Roller type									
7320	Up to 2-1/2" diam. pipe									
7324	3-1/2" off roof	Q-5	24	.667	Ea.	14	35		49	68.50
7326	Up to 10" off roof	"	20	.800	"	19	42		61	84.50
7340	2-1/2" to 3-1/2" diam. pipe									
7342	Up to 16" off roof	Q-5	18	.889	Ea.	45.50	47		92.50	121
7360	4" to 5" diam. pipe									
7362	Up to 12" off roof	Q-5	16	1	Ea.	65	52.50		117.50	151
7400	Strut/channel type									
7410	Up to 2-1/2" diam. pipe									
7424	3-1/2" off roof	Q-5	24	.667	Ea.	12	35		47	66
7426	Up to 10" off roof	"	20	.800	"	17	42		59	82
7440	Strut and roller type									
7452	2-1/2" to 3-1/2" diam. pipe									
7454	Up to 16" off roof	Q-5	18	.889	Ea.	65.50	47		112.50	143
7460	Strut and hanger type									
7470	Up to 3" diam. pipe									
7474	Up to 8" off roof	Q-5	19	.842	Ea.	41.50	44.50		86	113
8000	Pipe clamp, plastic, 1/2" CTS	1 Plum	80	.100		.23	5.75		5.98	8.95
8010	3/4" CTS		73	.110		.24	6.30		6.54	9.75
8020	1" CTS		68	.118		.54	6.75		7.29	10.80
8080	Economy clamp, 1/4" CTS		175	.046		.05	2.63		2.68	4.03
8090	3/8" CTS		168	.048		.07	2.74		2.81	4.22
8100	1/2" CTS		160	.050		.07	2.88		2.95	4.43
8110	3/4" CTS		145	.055		.14	3.18		3.32	4.94
8200	Half clamp, 1/2" CTS		80	.100		.08	5.75		5.83	8.80
8210	3/4" CTS		73	.110		.12	6.30		6.42	9.65
8300	Suspension clamp, 1/2" CTS		80	.100		.23	5.75		5.98	8.95
8310	3/4" CTS		73	.110		.24	6.30		6.54	9.75
8320	1" CTS		68	.118		.54	6.75		7.29	10.80
8400	Insulator, 1/2" CTS		80	.100		.37	5.75		6.12	9.10
8410	3/4" CTS	↓	73	.110	↓	.39	6.30		6.69	9.95

22 05 Common Work Results for Plumbing

22 05 29 – Hangers and Supports for Plumbing Piping and Equipment

22 05 29.10 Hangers & Supp. for Plumb'g/HVAC Pipe/Equip.		Crew	Daily Output	Labor-Hours	Unit	Material	2014 Bare Costs Labor	Equipment	Total	Total Incl O&P
8420	1" CTS	1 Plum	68	.118	Ea.	.40	6.75		7.15	10.65
8500	J hook clamp with nail 1/2" CTS		240	.033		.13	1.92		2.05	3.04
8501	3/4" CTS	↓	240	.033	↓	.13	1.92		2.05	3.04
8800	Wire cable support system									
8810	Cable with hook terminal and locking device									
8830	2 mm, (.079") dia cable, (100 lb. cap.)									
8840	1 m, (3.3') length, with hook	1 Shee	96	.083	Ea.	3.03	4.56		7.59	10.30
8850	2 m, (6.6') length, with hook		84	.095		3.50	5.20		8.70	11.80
8860	3 m, (9.9') length, with hook	↓	72	.111		3.97	6.10		10.07	13.65
8870	5 m, (16.4') length, with hook	Q-9	60	.267		4.97	13.15		18.12	25.50
8880	10 m, (32.8') length, with hook	"	30	.533	↓	7.20	26.50		33.70	48
8900	3mm, (.118") dia cable, (200 lb. cap.)									
8910	1 m, (3.3') length, with hook	1 Shee	96	.083	Ea.	3.88	4.56		8.44	11.20
8920	2 m, (6.6') length, with hook		84	.095		4.36	5.20		9.56	12.75
8930	3 m, (9.9') length, with hook	↓	72	.111		4.81	6.10		10.91	14.60
8940	5 m, (16.4') length, with hook	Q-9	60	.267		5.95	13.15		19.10	26.50
8950	10 m, (32.8') length, with hook	"	30	.533	↓	8.55	26.50		35.05	49.50
9000	Cable system accessories									
9010	Anchor bolt, 3/8", with nut	1 Shee	140	.057	Ea.	1.16	3.13		4.29	6.05
9020	Air duct corner protector		160	.050		.52	2.74		3.26	4.75
9030	Air duct support attachment	↓	140	.057	↓	.99	3.13		4.12	5.85
9040	Flange clip, hammer-on style									
9044	For flange thickness 3/32" - 9/64", 160 lb. cap.	1 Shee	180	.044	Ea.	.29	2.43		2.72	4.03
9048	For flange thickness 1/8" - 1/4", 200 lb. cap.		160	.050		.29	2.74		3.03	4.50
9052	For flange thickness 5/16" - 1/2", 200 lb. cap.		150	.053		.56	2.92		3.48	5.10
9056	For flange thickness 9/16" - 3/4", 200 lb. cap.		140	.057	↓	.75	3.13		3.88	5.60
9060	Wire insulation protection tube	↓	180	.044	L.F.	.34	2.43		2.77	4.08
9070	Wire cutter				Ea.	41.50			41.50	45.50

22 05 33 – Heat Tracing for Plumbing Piping

22 05 33.20 Temperature Maintenance Cable

0010	**TEMPERATURE MAINTENANCE CABLE**									
0040	Components									
0080	Heating cable									
0100	208 V									
0150	140°F	Q-1	1060.80	.015	L.F.	9.20	.78		9.98	11.35
0200	120 V									
0220	125°F	Q-1	1060.80	.015	L.F.	8.10	.78		8.88	10.10
0300	Power kit w/1 end seal	1 Elec	48.80	.164	Ea.	76.50	8.75		85.25	97
0310	Splice kit		35.50	.225		90.50	12		102.50	118
0320	End seal		160	.050		8.80	2.67		11.47	13.70
0330	Tee kit w/1 end seal		26.80	.299		96	15.95		111.95	129
0340	Powered splice w/2 end seals		20	.400		102	21.50		123.50	144
0350	Powered tee kit w/3 end seals		18	.444		131	23.50		154.50	180
0360	Cross kit w/2 end seals	↓	18.60	.430	↓	135	23		158	184
0500	Recommended thickness of fiberglass insulation									
0510	Pipe size									
0520	1/2" - 1" use 1" insulation									
0530	1-1/4" - 2" use 1-1/2" insulation									
0540	2-1/2" - 6" use 2" insulation									
0560	NOTE: For pipe sizes 1-1/4" and smaller use 1/4" larger diameter									
0570	insulation to allow room for installation over cable.									

22 05 Common Work Results for Plumbing

22 05 48 – Vibration and Seismic Controls for Plumbing Piping and Equipment

22 05 48.40 Vibration Absorbers		Crew	Daily Output	Labor-Hours	Unit	Material	2014 Bare Costs Labor	Equipment	Total	Total Incl O&P
0010	**VIBRATION ABSORBERS**									
0100	Hangers, neoprene flex									
0200	10 – 120 lb. capacity				Ea.	25.50			25.50	28.50
0220	75 – 550 lb. capacity					36			36	39.50
0240	250 – 1100 lb. capacity					66.50			66.50	73
0260	1000 – 4000 lb. capacity					124			124	137
0500	Spring flex, 60 lb. capacity					34.50			34.50	38
0520	450 lb. capacity					74.50			74.50	82
0540	900 lb. capacity					67			67	73.50
0560	1100 – 1300 lb. capacity					66.50			66.50	73
0600	Rubber in shear									
0610	45 – 340 lb., up to 1/2" rod size	1 Stpi	22	.364	Ea.	14.30	21.50		35.80	48
0620	130 – 700 lb., up to 3/4" rod size		20	.400		31	23.50		54.50	69.50
0630	50 – 1000 lb., up to 3/4" rod size		18	.444		36.50	26		62.50	79.50
1000	Mounts, neoprene, 45 – 380 lb. capacity					16.70			16.70	18.40
1020	250 – 1100 lb. capacity					47.50			47.50	52.50
1040	1000 – 4000 lb. capacity					107			107	117
1100	Spring flex, 60 lb. capacity					69			69	76
1120	165 lb. capacity					70			70	77
1140	260 lb. capacity					72			72	79
1160	450 lb. capacity					95.50			95.50	105
1180	600 lb. capacity					98			98	108
1200	750 lb. capacity					104			104	114
1220	900 lb. capacity					105			105	116
1240	1100 lb. capacity					108			108	119
1260	1300 lb. capacity					108			108	119
1280	1500 lb. capacity					171			171	188
1300	1800 lb. capacity					184			184	202
1320	2200 lb. capacity					184			184	202
1340	2600 lb. capacity					184			184	202
1399	Spring type									
1400	2 Piece									
1410	50 – 1000 lb.	1 Stpi	12	.667	Ea.	130	39		169	202
1420	1100 – 1600 lb.	"	12	.667	"	141	39		180	214
1500	Double spring open									
1510	150 – 450 lb.	1 Stpi	24	.333	Ea.	89	19.50		108.50	128
1520	500 – 1000 lb.		24	.333		89	19.50		108.50	128
1530	1100 – 1600 lb.		24	.333		89	19.50		108.50	128
1540	1700 – 2400 lb.		24	.333		97	19.50		116.50	136
1550	2500 – 3400 lb.		24	.333		165	19.50		184.50	212
2000	Pads, cork rib, 18" x 18" x 1", 10-50 psi					173			173	190
2020	18" x 36" x 1", 10-50 psi					345			345	380
2100	Shear flexible pads, 18" x 18" x 3/8", 20-70 psi					75.50			75.50	83
2120	18" x 36" x 3/8", 20-70 psi					173			173	190
2150	Laminated neoprene and cork									
2160	1" thick	1 Stpi	16	.500	S.F.	86	29.50		115.50	139
2200	Neoprene elastomer isolation bearing pad									
2230	464 PSI, 5/8" x 19-11/16" x 39-3/8"	1 Stpi	16	.500	Ea.	285	29.50		314.50	360
3000	Note overlap in capacities due to deflections									

22 05 Common Work Results for Plumbing

22 05 53 – Identification for Plumbing Piping and Equipment

22 05 53.10 Piping System Identification Labels

		Crew	Daily Output	Labor-Hours	Unit	Material	2014 Bare Costs Labor	Equipment	Total	Total Incl O&P
0010	**PIPING SYSTEM IDENTIFICATION LABELS**									
0100	Indicate contents and flow direction									
0106	Pipe markers									
0110	Plastic snap around									
0114	1/2" pipe	1 Plum	80	.100	Ea.	6.50	5.75		12.25	15.85
0116	3/4" pipe		80	.100		6.50	5.75		12.25	15.85
0118	1" pipe		80	.100		6.50	5.75		12.25	15.85
0120	2" pipe		75	.107		8.10	6.15		14.25	18.15
0122	3" pipe		70	.114		12.40	6.60		19	23.50
0124	4" pipe		60	.133		12.60	7.65		20.25	25.50
0126	6" pipe		60	.133		13.20	7.65		20.85	26
0128	8" pipe		56	.143		17.40	8.20		25.60	31.50
0130	10" pipe		56	.143		17.40	8.20		25.60	31.50
0200	Over 10" pipe size		50	.160		21	9.20		30.20	37
1110	Self adhesive									
1114	1" pipe	1 Plum	80	.100	Ea.	3.20	5.75		8.95	12.20
1116	2" pipe		75	.107		3.20	6.15		9.35	12.75
1118	3" pipe		70	.114		5.10	6.60		11.70	15.55
1120	4" pipe		60	.133		5	7.65		12.65	17.10
1122	6" pipe		60	.133		5	7.65		12.65	17.10
1124	8" pipe		56	.143		7.20	8.20		15.40	20.50
1126	10" pipe		56	.143		7.20	8.20		15.40	20.50
1200	Over 10" pipe size		50	.160		10.20	9.20		19.40	25
2000	Valve tags									
2010	Numbered plus identifying legend									
2100	Brass, 2" diameter	1 Plum	40	.200	Ea.	3	11.50		14.50	20.50
2200	Plastic, 1-1/2" diam.	"	40	.200	"	2.70	11.50		14.20	20.50

22 05 76 – Facility Drainage Piping Cleanouts

22 05 76.10 Cleanouts

		Crew	Daily Output	Labor-Hours	Unit	Material	Labor	Equipment	Total	Total Incl O&P
0010	**CLEANOUTS**									
0060	Floor type									
0080	Round or square, scoriated nickel bronze top									
0100	2" pipe size	1 Plum	10	.800	Ea.	183	46		229	271
0120	3" pipe size		8	1		274	57.50		331.50	385
0140	4" pipe size		6	1.333		274	76.50		350.50	415
0160	5" pipe size		4	2		345	115		460	555
0180	6" pipe size	Q-1	6	2.667		345	138		483	590
0200	8" pipe size	"	4	4		615	207		822	990
0340	Recessed for tile, same price									
0980	Round top, recessed for terrazzo									
1000	2" pipe size	1 Plum	9	.889	Ea.	183	51		234	278
1080	3" pipe size		6	1.333		274	76.50		350.50	415
1100	4" pipe size		4	2		274	115		389	475
1120	5" pipe size	Q-1	6	2.667		345	138		483	590
1140	6" pipe size		5	3.200		345	166		511	630
1160	8" pipe size		4	4		615	207		822	990
2000	Round scoriated nickel bronze top, extra heavy duty									
2060	2" pipe size	1 Plum	9	.889	Ea.	249	51		300	350
2080	3" pipe size		6	1.333		340	76.50		416.50	490
2100	4" pipe size		4	2		340	115		455	550
2120	5" pipe size	Q-1	6	2.667		415	138		553	665
2140	6" pipe size		5	3.200		415	166		581	705

22 05 Common Work Results for Plumbing

22 05 76 – Facility Drainage Piping Cleanouts

22 05 76.10 Cleanouts

		Crew	Daily Output	Labor-Hours	Unit	Material	2014 Bare Costs Labor	Equipment	Total	Total Incl O&P
2160	8" pipe size	Q-1	4	4	Ea.	680	207		887	1,075
4000	Wall type, square smooth cover, over wall frame									
4060	2" pipe size	1 Plum	14	.571	Ea.	298	33		331	380
4080	3" pipe size		12	.667		320	38.50		358.50	415
4100	4" pipe size		10	.800		345	46		391	450
4120	5" pipe size		9	.889		500	51		551	625
4140	6" pipe size		8	1		555	57.50		612.50	695
4160	8" pipe size	Q-1	11	1.455		740	75.50		815.50	925
5000	Extension, C.I.; bronze countersunk plug, 8" long									
5040	2" pipe size	1 Plum	16	.500	Ea.	157	29		186	216
5060	3" pipe size		14	.571		182	33		215	250
5080	4" pipe size		13	.615		185	35.50		220.50	257
5100	5" pipe size		12	.667		289	38.50		327.50	380
5120	6" pipe size		11	.727		350	42		392	450

22 05 76.20 Cleanout Tees

		Crew	Daily Output	Labor-Hours	Unit	Material	Labor	Equipment	Total	Total Incl O&P
0010	**CLEANOUT TEES**									
0100	Cast iron, B&S, with countersunk plug									
0200	2" pipe size	1 Plum	4	2	Ea.	248	115		363	445
0220	3" pipe size		3.60	2.222		270	128		398	490
0240	4" pipe size		3.30	2.424		335	140		475	580
0260	5" pipe size	Q-1	5.50	2.909		730	151		881	1,025
0280	6" pipe size	"	5	3.200		910	166		1,076	1,250
0300	8" pipe size	Q-3	5	6.400		1,050	350		1,400	1,675
0500	For round smooth access cover, same price									
0600	For round scoriated access cover, same price									
0700	For square smooth access cover, add				Ea.	60%				
2000	Cast iron, no hub									
2010	Cleanout tee, with 2 couplings									
2012	2"	Q-1	22	.727	Ea.	43	37.50		80.50	105
2014	3"		19	.842		55	43.50		98.50	127
2016	4"		16.50	.970		76.50	50		126.50	161
2018	6"	Q-2	20	1.200		173	64.50		237.50	288
2040	Cleanout plug, no hub, with 1 coupling									
2042	2"	Q-1	44	.364	Ea.	11.70	18.85		30.55	41.50
2046	3"		38	.421		18.15	22		40.15	53
2048	4"		33	.485		30	25		55	71
2050	6"	Q-2	40	.600		108	32		140	168
2052	8"	"	33	.727		182	39		221	259
4000	Plastic, tees and adapters. Add plugs									
4010	ABS, DWV									
4020	Cleanout tee, 1-1/2" pipe size	1 Plum	15	.533	Ea.	20	30.50		50.50	68.50
4030	2" pipe size	Q-1	27	.593		22.50	30.50		53	71.50
4040	3" pipe size		21	.762		43.50	39.50		83	108
4050	4" pipe size		16	1		87	52		139	174
4100	Cleanout plug, 1-1/2" pipe size	1 Plum	32	.250		2.97	14.40		17.37	25
4110	2" pipe size	Q-1	56	.286		3.90	14.80		18.70	27
4120	3" pipe size		36	.444		6.30	23		29.30	42
4130	4" pipe size		30	.533		11	27.50		38.50	53.50
4180	Cleanout adapter fitting, 1-1/2" pipe size	1 Plum	32	.250		4.70	14.40		19.10	26.50
4190	2" pipe size	Q-1	56	.286		7.20	14.80		22	30.50
4200	3" pipe size		36	.444		18.45	23		41.45	55.50
4210	4" pipe size		30	.533		34	27.50		61.50	79

22 05 Common Work Results for Plumbing

22 05 76 – Facility Drainage Piping Cleanouts

22 05 76.20 Cleanout Tees

		Crew	Daily Output	Labor-Hours	Unit	Material	2014 Bare Costs Labor	Equipment	Total	Total Incl O&P
5000	PVC, DWV									
5010	Cleanout tee, 1-1/2" pipe size	1 Plum	15	.533	Ea.	12.75	30.50		43.25	60.50
5020	2" pipe size	Q-1	27	.593		14.85	30.50		45.35	63
5030	3" pipe size		21	.762		29	39.50		68.50	91
5040	4" pipe size	↓	16	1		51.50	52		103.50	135
5090	Cleanout plug, 1-1/2" pipe size	1 Plum	32	.250		3.04	14.40		17.44	25
5100	2" pipe size	Q-1	56	.286		3.30	14.80		18.10	26
5110	3" pipe size		36	.444		6.05	23		29.05	41.50
5120	4" pipe size		30	.533		8.90	27.50		36.40	51.50
5130	6" pipe size	↓	24	.667		29.50	34.50		64	84.50
5170	Cleanout adapter fitting, 1-1/2" pipe size	1 Plum	32	.250		4.06	14.40		18.46	26
5180	2" pipe size	Q-1	56	.286		5.20	14.80		20	28.50
5190	3" pipe size		36	.444		14.70	23		37.70	51
5200	4" pipe size		30	.533		23.50	27.50		51	67.50
5210	6" pipe size	↓	24	.667	↓	70.50	34.50		105	130
5300	Cleanout tee, with plug									
5310	4"	Q-1	14	1.143	Ea.	74	59		133	171
5320	6"	"	8	2		63	104		167	225
5330	8"	Q-2	11	2.182		132	117		249	325
5340	12"	"	10	2.400	↓	320	129		449	550
5400	Polypropylene, Schedule 40									
5410	Cleanout tee with plug									
5420	1-1/2"	1 Plum	10	.800	Ea.	35.50	46		81.50	109
5430	2"	Q-1	17	.941		38	49		87	116
5440	3"		11	1.455		80.50	75.50		156	203
5450	4"	↓	9	1.778	↓	109	92		201	259

22 07 Plumbing Insulation

22 07 16 – Plumbing Equipment Insulation

22 07 16.10 Insulation for Plumbing Equipment

			Crew	Daily Output	Labor-Hours	Unit	Material	Labor	Equipment	Total	Total Incl O&P
0010	**INSULATION FOR PLUMBING EQUIPMENT**										
2900	Domestic water heater wrap kit										
2920	1-1/2" with vinyl jacket, 20-60 gal.	G	1 Plum	8	1	Ea.	18.60	57.50		76.10	108
2925	50 to 80 gallons	G	"	8	1	"	23.50	57.50		81	113

22 07 19 – Plumbing Piping Insulation

22 07 19.10 Piping Insulation

		Crew	Daily Output	Labor-Hours	Unit	Material	Labor	Equipment	Total	Total Incl O&P
0010	**PIPING INSULATION**									
0100	Rule of thumb, as a percentage of total mechanical costs				Job				10%	10%
0110	Insulation req'd. is based on the surface size/area to be covered									
0230	Insulated protectors, (ADA)									
0235	For exposed piping under sinks or lavatories									
0240	Vinyl coated foam, velcro tabs									
0245	P Trap, 1-1/4" or 1-1/2"	1 Plum	32	.250	Ea.	44.50	14.40		58.90	70.50
0260	Valve and supply cover									
0265	1/2", 3/8", and 7/16" pipe size	1 Plum	32	.250	Ea.	22	14.40		36.40	45.50
0280	Tailpiece offset (wheelchair)									
0285	1-1/4" pipe size	1 Plum	32	.250	Ea.	18.20	14.40		32.60	41.50
0600	Pipe covering (price copper tube one size less than IPS)									
1000	Mineral wool									
1010	Preformed, 1200°F, plain									

22 07 Plumbing Insulation

22 07 19 – Plumbing Piping Insulation

22 07 19.10 Piping Insulation		Crew	Daily Output	Labor-Hours	Unit	Material	2014 Bare Costs Labor	Equipment	Total	Total Incl O&P
1014	1" wall									
1016	1/2" iron pipe size G	Q-14	230	.070	L.F.	1.65	3.20		4.85	6.80
1018	3/4" iron pipe size G		220	.073		1.72	3.35		5.07	7.10
1022	1" iron pipe size G		210	.076		1.77	3.51		5.28	7.40
1024	1-1/4" iron pipe size G		205	.078		1.82	3.59		5.41	7.60
1026	1-1/2" iron pipe size G		205	.078		1.90	3.59		5.49	7.70
1028	2" iron pipe size G		200	.080		2.32	3.68		6	8.30
1030	2-1/2" iron pipe size G		190	.084		2.46	3.88		6.34	8.75
1032	3" iron pipe size G		180	.089		2.55	4.09		6.64	9.15
1034	4" iron pipe size G		150	.107		3.14	4.91		8.05	11.10
1036	5" iron pipe size G		140	.114		3.55	5.25		8.80	12.10
1038	6" iron pipe size G		120	.133		3.79	6.15		9.94	13.70
1040	7" iron pipe size G		110	.145		4.43	6.70		11.13	15.25
1042	8" iron pipe size G		100	.160		6.10	7.35		13.45	18.15
1044	9" iron pipe size G		90	.178		6.65	8.20		14.85	20
1046	10" iron pipe size G		90	.178		7.50	8.20		15.70	21
1050	1-1/2" wall									
1052	1/2" iron pipe size G	Q-14	225	.071	L.F.	2.82	3.27		6.09	8.20
1054	3/4" iron pipe size G		215	.074		2.89	3.43		6.32	8.55
1056	1" iron pipe size G		205	.078		2.93	3.59		6.52	8.80
1058	1-1/4" iron pipe size G		200	.080		2.99	3.68		6.67	9.05
1060	1-1/2" iron pipe size G		200	.080		3.14	3.68		6.82	9.20
1062	2" iron pipe size G		190	.084		3.32	3.88		7.20	9.70
1064	2-1/2" iron pipe size G		180	.089		3.62	4.09		7.71	10.35
1066	3" iron pipe size G		165	.097		3.77	4.46		8.23	11.10
1068	4" iron pipe size G		140	.114		4.53	5.25		9.78	13.20
1070	5" iron pipe size G		130	.123		4.84	5.65		10.49	14.10
1072	6" iron pipe size G		110	.145		4.99	6.70		11.69	15.90
1074	7" iron pipe size G		100	.160		5.90	7.35		13.25	17.95
1076	8" iron pipe size G		90	.178		6.55	8.20		14.75	19.95
1078	9" iron pipe size G		85	.188		7.55	8.65		16.20	22
1080	10" iron pipe size G		80	.200		8.30	9.20		17.50	23.50
1082	12" iron pipe size G		75	.213		9.45	9.80		19.25	25.50
1084	14" iron pipe size G		70	.229		10.80	10.50		21.30	28.50
1086	16" iron pipe size G		65	.246		12.30	11.35		23.65	31
1088	18" iron pipe size G		60	.267		13.85	12.25		26.10	34.50
1090	20" iron pipe size G		55	.291		15.25	13.40		28.65	38
1092	22" iron pipe size G		50	.320		17.45	14.75		32.20	42
1094	24" iron pipe size G		45	.356		18.65	16.35		35	46
1100	2" wall									
1102	1/2" iron pipe size G	Q-14	220	.073	L.F.	3.65	3.35		7	9.20
1104	3/4" iron pipe size G		210	.076		3.78	3.51		7.29	9.60
1106	1" iron pipe size G		200	.080		4.01	3.68		7.69	10.15
1108	1-1/4" iron pipe size G		190	.084		4.30	3.88		8.18	10.80
1110	1-1/2" iron pipe size G		190	.084		4.48	3.88		8.36	11
1112	2" iron pipe size G		180	.089		4.70	4.09		8.79	11.50
1114	2-1/2" iron pipe size G		170	.094		5.30	4.33		9.63	12.60
1116	3" iron pipe size G		160	.100		5.70	4.60		10.30	13.40
1118	4" iron pipe size G		130	.123		6.60	5.65		12.25	16.10
1120	5" iron pipe size G		120	.133		7.35	6.15		13.50	17.65
1122	6" iron pipe size G		100	.160		7.65	7.35		15	19.85
1124	8" iron pipe size G		80	.200		9.25	9.20		18.45	24.50
1126	10" iron pipe size G		70	.229		11.35	10.50		21.85	29

22 07 Plumbing Insulation

22 07 19 – Plumbing Piping Insulation

22 07 19.10 Piping Insulation		Crew	Daily Output	Labor-Hours	Unit	Material	2014 Bare Costs Labor	Equipment	Total	Total Incl O&P
1128	12" iron pipe size	G Q-14	65	.246	L.F.	12.55	11.35		23.90	31.50
1130	14" iron pipe size	G	60	.267		14.05	12.25		26.30	34.50
1132	16" iron pipe size	G	55	.291		15.90	13.40		29.30	38.50
1134	18" iron pipe size	G	50	.320		17.20	14.75		31.95	42
1136	20" iron pipe size	G	45	.356		19.80	16.35		36.15	47.50
1138	22" iron pipe size	G	45	.356		21.50	16.35		37.85	49
1140	24" iron pipe size	G	40	.400		22.50	18.40		40.90	53.50
1150	4" wall									
1152	1-1/4" iron pipe size	G Q-14	170	.094	L.F.	11.65	4.33		15.98	19.55
1154	1-1/2" iron pipe size	G	165	.097		12.10	4.46		16.56	20.50
1156	2" iron pipe size	G	155	.103		12.35	4.75		17.10	21
1158	4" iron pipe size	G	105	.152		15.60	7		22.60	28
1160	6" iron pipe size	G	75	.213		18.40	9.80		28.20	35.50
1162	8" iron pipe size	G	60	.267		21	12.25		33.25	42
1164	10" iron pipe size	G	50	.320		25	14.75		39.75	50.50
1166	12" iron pipe size	G	45	.356		27.50	16.35		43.85	56
1168	14" iron pipe size	G	40	.400		29.50	18.40		47.90	61
1170	16" iron pipe size	G	35	.457		33.50	21		54.50	69
1172	18" iron pipe size	G	32	.500		36	23		59	75.50
1174	20" iron pipe size	G	30	.533		38.50	24.50		63	80.50
1176	22" iron pipe size	G	28	.571		43.50	26.50		70	88.50
1178	24" iron pipe size	G	26	.615		46	28.50		74.50	94.50
4280	Cellular glass, closed cell foam, all service jacket, sealant,									
4281	working temp. (-450°F to +900°F), 0 water vapor transmission									
4284	1" wall									
4286	1/2" iron pipe size	G Q-14	120	.133	L.F.	5	6.15		11.15	15.05
4300	1-1/2" wall,									
4301	1" iron pipe size	G Q-14	105	.152	L.F.	8.70	7		15.70	20.50
4304	2-1/2" iron pipe size	G	90	.178		15.40	8.20		23.60	29.50
4306	3" iron pipe size	G	85	.188		15.40	8.65		24.05	30.50
4308	4" iron pipe size	G	70	.229		16.30	10.50		26.80	34.50
4310	5" iron pipe size	G	65	.246		19	11.35		30.35	38.50
4320	2" wall,									
4322	1" iron pipe size	G Q-14	100	.160	L.F.	11.60	7.35		18.95	24
4324	2-1/2" iron pipe size	G	85	.188		18.70	8.65		27.35	34
4326	3" iron pipe size	G	80	.200		18.70	9.20		27.90	35
4328	4" iron pipe size	G	65	.246		20.50	11.35		31.85	40
4330	5" iron pipe size	G	60	.267		23.50	12.25		35.75	44.50
4332	6" iron pipe size	G	50	.320		27.50	14.75		42.25	53.50
4336	8" iron pipe size	G	40	.400		32	18.40		50.40	63.50
4338	10" iron pipe size	G	35	.457		39	21		60	75.50
4350	2-1/2" wall,									
4360	12" iron pipe size	G Q-14	32	.500	L.F.	53	23		76	94.50
4362	14" iron pipe size	G	"	28	.571	"	57	26.50		83.50	104
4370	3" wall,									
4378	6" iron pipe size	G Q-14	48	.333	L.F.	36.50	15.35		51.85	64
4380	8" iron pipe size	G	38	.421		45.50	19.40		64.90	80
4382	10" iron pipe size	G	33	.485		54	22.50		76.50	93.50
4384	16" iron pipe size	G	25	.640		72.50	29.50		102	126
4386	18" iron pipe size	G	22	.727		82.50	33.50		116	143
4388	20" iron pipe size	G	20	.800		89.50	37		126.50	156
4400	3-1/2" wall,									
4412	12" iron pipe size	G Q-14	27	.593	L.F.	69	27.50		96.50	118

22 07 Plumbing Insulation
22 07 19 – Plumbing Piping Insulation

22 07 19.10 Piping Insulation		Crew	Daily Output	Labor-Hours	Unit	Material	2014 Bare Costs Labor	Equipment	Total	Total Incl O&P
4414	14" iron pipe size	G Q-14	25	.640	L.F.	71	29.50		100.50	124
4430	4" wall,									
4446	16" iron pipe size	G Q-14	22	.727	L.F.	92.50	33.50		126	154
4448	18" iron pipe size	G	20	.800		101	37		138	169
4450	20" iron pipe size	G	18	.889		114	41		155	189
4480	Fittings, average with fabric and mastic									
4484	1" wall,									
4486	1/2" iron pipe size	G 1 Asbe	40	.200	Ea.	5.65	10.25		15.90	22
4500	1-1/2" wall,									
4502	1" iron pipe size	G 1 Asbe	38	.211	Ea.	7.50	10.75		18.25	25
4504	2-1/2" iron pipe size	G	32	.250		13.60	12.80		26.40	35
4506	3" iron pipe size	G	30	.267		13.60	13.65		27.25	36
4508	4" iron pipe size	G	28	.286		17.15	14.60		31.75	41.50
4510	5" iron pipe size	G	24	.333		21	17.05		38.05	49.50
4520	2" wall,									
4522	1" iron pipe size	G 1 Asbe	36	.222	Ea.	10.90	11.35		22.25	29.50
4524	2-1/2" iron pipe size	G	30	.267		17.15	13.65		30.80	40
4526	3" iron pipe size	G	28	.286		17.15	14.60		31.75	41.50
4528	4" iron pipe size	G	24	.333		21	17.05		38.05	49.50
4530	5" iron pipe size	G	22	.364		31	18.60		49.60	63
4532	6" iron pipe size	G	20	.400		44	20.50		64.50	80
4536	8" iron pipe size	G	12	.667		73.50	34		107.50	134
4538	10" iron pipe size	G	8	1		85	51		136	173
4550	2-1/2" wall,									
4560	12" iron pipe size	G 1 Asbe	6	1.333	Ea.	156	68		224	278
4562	14" iron pipe size	G "	4	2	"	206	102		308	385
4570	3" wall,									
4578	6" iron pipe size	G 1 Asbe	16	.500	Ea.	63	25.50		88.50	110
4580	8" iron pipe size	G	10	.800		94.50	41		135.50	168
4582	10" iron pipe size	G	6	1.333		128	68		196	246
4900	Calcium silicate, with 8 oz. canvas cover									
5100	1" wall, 1/2" iron pipe size	G Q-14	170	.094	L.F.	3.30	4.33		7.63	10.40
5130	3/4" iron pipe size	G	170	.094		3.33	4.33		7.66	10.40
5140	1" iron pipe size	G	170	.094		3.23	4.33		7.56	10.30
5150	1-1/4" iron pipe size	G	165	.097		3.30	4.46		7.76	10.60
5160	1-1/2" iron pipe size	G	165	.097		3.35	4.46		7.81	10.65
5170	2" iron pipe size	G	160	.100		3.84	4.60		8.44	11.35
5180	2-1/2" iron pipe size	G	160	.100		4.09	4.60		8.69	11.65
5190	3" iron pipe size	G	150	.107		4.42	4.91		9.33	12.50
5200	4" iron pipe size	G	140	.114		5.50	5.25		10.75	14.25
5210	5" iron pipe size	G	135	.119		5.80	5.45		11.25	14.85
5220	6" iron pipe size	G	130	.123		6.15	5.65		11.80	15.55
5280	1-1/2" wall, 1/2" iron pipe size	G	150	.107		3.57	4.91		8.48	11.60
5310	3/4" iron pipe size	G	150	.107		3.64	4.91		8.55	11.65
5320	1" iron pipe size	G	150	.107		3.96	4.91		8.87	12
5330	1-1/4" iron pipe size	G	145	.110		4.26	5.10		9.36	12.60
5340	1-1/2" iron pipe size	G	145	.110		4.57	5.10		9.67	12.95
5350	2" iron pipe size	G	140	.114		5.05	5.25		10.30	13.75
5360	2-1/2" iron pipe size	G	140	.114		5.50	5.25		10.75	14.25
5370	3" iron pipe size	G	135	.119		5.75	5.45		11.20	14.85
5380	4" iron pipe size	G	125	.128		6.65	5.90		12.55	16.50
5390	5" iron pipe size	G	120	.133		7.50	6.15		13.65	17.80
5400	6" iron pipe size	G	110	.145		7.75	6.70		14.45	18.95

22 07 Plumbing Insulation

22 07 19 – Plumbing Piping Insulation

	22 07 19.10 Piping Insulation		Crew	Daily Output	Labor-Hours	Unit	Material	2014 Bare Costs Labor	Equipment	Total	Total Incl O&P
5460	2" wall, 1/2" iron pipe size	G	Q-14	135	.119	L.F.	5.50	5.45		10.95	14.55
5490	3/4" iron pipe size	G		135	.119		5.80	5.45		11.25	14.85
5500	1" iron pipe size	G		135	.119		6.10	5.45		11.55	15.20
5510	1-1/4" iron pipe size	G		130	.123		6.55	5.65		12.20	16
5520	1-1/2" iron pipe size	G		130	.123		6.80	5.65		12.45	16.30
5530	2" iron pipe size	G		125	.128		7.20	5.90		13.10	17.10
5540	2-1/2" iron pipe size	G		125	.128		8.55	5.90		14.45	18.55
5550	3" iron pipe size	G		120	.133		8.65	6.15		14.80	19.05
5560	4" iron pipe size	G		115	.139		9.95	6.40		16.35	21
5570	5" iron pipe size	G		110	.145		11.35	6.70		18.05	23
5580	6" iron pipe size	G		105	.152		12.40	7		19.40	24.50
5600	Calcium silicate, no cover										
5720	1" wall, 1/2" iron pipe size	G	Q-14	180	.089	L.F.	3.01	4.09		7.10	9.65
5740	3/4" iron pipe size	G		180	.089		3.01	4.09		7.10	9.65
5750	1" iron pipe size	G		180	.089		2.88	4.09		6.97	9.50
5760	1-1/4" iron pipe size	G		175	.091		2.93	4.21		7.14	9.75
5770	1-1/2" iron pipe size	G		175	.091		2.95	4.21		7.16	9.80
5780	2" iron pipe size	G		170	.094		3.41	4.33		7.74	10.50
5790	2-1/2" iron pipe size	G		170	.094		3.60	4.33		7.93	10.70
5800	3" iron pipe size	G		160	.100		3.87	4.60		8.47	11.40
5810	4" iron pipe size	G		150	.107		4.84	4.91		9.75	12.95
5820	5" iron pipe size	G		145	.110		5.05	5.10		10.15	13.45
5830	6" iron pipe size	G		140	.114		5.30	5.25		10.55	14
5900	1-1/2" wall, 1/2" iron pipe size	G		160	.100		3.17	4.60		7.77	10.65
5920	3/4" iron pipe size	G		160	.100		3.23	4.60		7.83	10.70
5930	1" iron pipe size	G		160	.100		3.52	4.60		8.12	11
5940	1-1/4" iron pipe size	G		155	.103		3.79	4.75		8.54	11.55
5950	1-1/2" iron pipe size	G		155	.103		4.08	4.75		8.83	11.90
5960	2" iron pipe size	G		150	.107		4.49	4.91		9.40	12.60
5970	2-1/2" iron pipe size	G		150	.107		4.90	4.91		9.81	13.05
5980	3" iron pipe size	G		145	.110		5.10	5.10		10.20	13.55
5990	4" iron pipe size	G		135	.119		5.95	5.45		11.40	15
6000	5" iron pipe size	G		130	.123		6.70	5.65		12.35	16.15
6010	6" iron pipe size	G		120	.133		6.80	6.15		12.95	17.05
6020	7" iron pipe size	G		115	.139		8	6.40		14.40	18.75
6030	8" iron pipe size	G		105	.152		9.05	7		16.05	21
6040	9" iron pipe size	G		100	.160		10.85	7.35		18.20	23.50
6050	10" iron pipe size	G		95	.168		11.95	7.75		19.70	25
6060	12" iron pipe size	G		90	.178		14.20	8.20		22.40	28.50
6070	14" iron pipe size	G		85	.188		16.05	8.65		24.70	31
6080	16" iron pipe size	G		80	.200		17.95	9.20		27.15	34
6090	18" iron pipe size	G		75	.213		19.85	9.80		29.65	37.50
6120	2" wall, 1/2" iron pipe size	G		145	.110		5	5.10		10.10	13.40
6140	3/4" iron pipe size	G		145	.110		5.30	5.10		10.40	13.70
6150	1" iron pipe size	G		145	.110		5.55	5.10		10.65	14.05
6160	1-1/4" iron pipe size	G		140	.114		5.95	5.25		11.20	14.75
6170	1-1/2" iron pipe size	G		140	.114		6.25	5.25		11.50	15.05
6180	2" iron pipe size	G		135	.119		6.55	5.45		12	15.70
6190	2-1/2" iron pipe size	G		135	.119		7.85	5.45		13.30	17.15
6200	3" iron pipe size	G		130	.123		7.90	5.65		13.55	17.50
6210	4" iron pipe size	G		125	.128		9.15	5.90		15.05	19.20
6220	5" iron pipe size	G		120	.133		10.40	6.15		16.55	21
6230	6" iron pipe size	G		115	.139		11.40	6.40		17.80	22.50

22 07 Plumbing Insulation

22 07 19 – Plumbing Piping Insulation

22 07 19.10 Piping Insulation		Crew	Daily Output	Labor-Hours	Unit	Material	2014 Bare Costs Labor	Equipment	Total	Total Incl O&P
6240	7" iron pipe size	Q-14	110	.145	L.F.	12.35	6.70		19.05	24
6250	8" iron pipe size		105	.152		13.55	7		20.55	26
6260	9" iron pipe size		100	.160		15.35	7.35		22.70	28.50
6270	10" iron pipe size		95	.168		17.05	7.75		24.80	31
6280	12" iron pipe size		90	.178		18.95	8.20		27.15	34
6290	14" iron pipe size		85	.188		21	8.65		29.65	36.50
6300	16" iron pipe size		80	.200		23	9.20		32.20	40
6310	18" iron pipe size		75	.213		25	9.80		34.80	43
6320	20" iron pipe size		65	.246		31	11.35		42.35	51.50
6330	22" iron pipe size		60	.267		34	12.25		46.25	56.50
6340	24" iron pipe size		55	.291		35	13.40		48.40	59.50
6360	3" wall, 1/2" iron pipe size		115	.139		9.10	6.40		15.50	19.95
6380	3/4" iron pipe size		115	.139		9.15	6.40		15.55	20
6390	1" iron pipe size		115	.139		9.20	6.40		15.60	20
6400	1-1/4" iron pipe size		110	.145		9.35	6.70		16.05	20.50
6410	1-1/2" iron pipe size		110	.145		9.45	6.70		16.15	21
6420	2" iron pipe size		105	.152		9.75	7		16.75	21.50
6430	2-1/2" iron pipe size		105	.152		11.45	7		18.45	23.50
6440	3" iron pipe size		100	.160		11.55	7.35		18.90	24
6450	4" iron pipe size		95	.168		14.75	7.75		22.50	28.50
6460	5" iron pipe size		90	.178		15.95	8.20		24.15	30.50
6470	6" iron pipe size		90	.178		17.95	8.20		26.15	32.50
6480	7" iron pipe size		85	.188		20	8.65		28.65	35.50
6490	8" iron pipe size		85	.188		21.50	8.65		30.15	37
6500	9" iron pipe size		80	.200		24.50	9.20		33.70	41
6510	10" iron pipe size		75	.213		26	9.80		35.80	44
6520	12" iron pipe size		70	.229		28.50	10.50		39	48
6530	14" iron pipe size		65	.246		32	11.35		43.35	53
6540	16" iron pipe size		60	.267		35.50	12.25		47.75	58
6550	18" iron pipe size		55	.291		39	13.40		52.40	64
6560	20" iron pipe size		50	.320		46.50	14.75		61.25	74
6570	22" iron pipe size		45	.356		50	16.35		66.35	80.50
6580	24" iron pipe size		40	.400		54	18.40		72.40	88
6600	Fiberglass, with all service jacket									
6640	1/2" wall, 1/2" iron pipe size	Q-14	250	.064	L.F.	.74	2.95		3.69	5.40
6660	3/4" iron pipe size		240	.067		.83	3.07		3.90	5.70
6670	1" iron pipe size		230	.070		.86	3.20		4.06	5.95
6680	1-1/4" iron pipe size		220	.073		.92	3.35		4.27	6.20
6690	1-1/2" iron pipe size		220	.073		1.06	3.35		4.41	6.35
6700	2" iron pipe size		210	.076		1.13	3.51		4.64	6.70
6710	2-1/2" iron pipe size		200	.080		1.18	3.68		4.86	7.05
6840	1" wall, 1/2" iron pipe size		240	.067		.89	3.07		3.96	5.75
6860	3/4" iron pipe size		230	.070		.97	3.20		4.17	6.05
6870	1" iron pipe size		220	.073		1.04	3.35		4.39	6.35
6880	1-1/4" iron pipe size		210	.076		1.13	3.51		4.64	6.70
6890	1-1/2" iron pipe size		210	.076		1.22	3.51		4.73	6.80
6900	2" iron pipe size		200	.080		1.32	3.68		5	7.20
6910	2-1/2" iron pipe size		190	.084		1.50	3.88		5.38	7.70
6920	3" iron pipe size		180	.089		1.60	4.09		5.69	8.10
6930	3-1/2" iron pipe size		170	.094		1.74	4.33		6.07	8.65
6940	4" iron pipe size		150	.107		2.12	4.91		7.03	10
6950	5" iron pipe size		140	.114		2.39	5.25		7.64	10.85
6960	6" iron pipe size		120	.133		2.54	6.15		8.69	12.35

129

22 07 Plumbing Insulation

22 07 19 – Plumbing Piping Insulation

22 07 19.10 Piping Insulation		Crew	Daily Output	Labor-Hours	Unit	Material	2014 Bare Costs Labor	Equipment	Total	Total Incl O&P
6970	7" iron pipe size	Q-14 [G]	110	.145	L.F.	3.02	6.70		9.72	13.70
6980	8" iron pipe size	[G]	100	.160		4.15	7.35		11.50	16
6990	9" iron pipe size	[G]	90	.178		4.38	8.20		12.58	17.55
7000	10" iron pipe size	[G]	90	.178		4.40	8.20		12.60	17.60
7010	12" iron pipe size	[G]	80	.200		4.80	9.20		14	19.65
7020	14" iron pipe size	[G]	80	.200		5.75	9.20		14.95	20.50
7030	16" iron pipe size	[G]	70	.229		7.30	10.50		17.80	24.50
7040	18" iron pipe size	[G]	70	.229		8.10	10.50		18.60	25.50
7050	20" iron pipe size	[G]	60	.267		9.05	12.25		21.30	29
7060	24" iron pipe size	[G]	60	.267		11.05	12.25		23.30	31.50
7080	1-1/2" wall, 1/2" iron pipe size	[G]	230	.070		1.69	3.20		4.89	6.85
7100	3/4" iron pipe size	[G]	220	.073		1.69	3.35		5.04	7.05
7110	1" iron pipe size	[G]	210	.076		1.81	3.51		5.32	7.45
7120	1-1/4" iron pipe size	[G]	200	.080		1.97	3.68		5.65	7.90
7130	1-1/2" iron pipe size	[G]	200	.080		2.07	3.68		5.75	8.05
7140	2" iron pipe size	[G]	190	.084		2.28	3.88		6.16	8.55
7150	2-1/2" iron pipe size	[G]	180	.089		2.44	4.09		6.53	9.05
7160	3" iron pipe size	[G]	170	.094		2.56	4.33		6.89	9.55
7170	3-1/2" iron pipe size	[G]	160	.100		2.80	4.60		7.40	10.25
7180	4" iron pipe size	[G]	140	.114		2.91	5.25		8.16	11.40
7190	5" iron pipe size	[G]	130	.123		3.26	5.65		8.91	12.40
7200	6" iron pipe size	[G]	110	.145		3.44	6.70		10.14	14.20
7210	7" iron pipe size	[G]	100	.160		3.82	7.35		11.17	15.65
7220	8" iron pipe size	[G]	90	.178		4.82	8.20		13.02	18.05
7230	9" iron pipe size	[G]	85	.188		5	8.65		13.65	19
7240	10" iron pipe size	[G]	80	.200		5.20	9.20		14.40	20
7250	12" iron pipe size	[G]	75	.213		5.90	9.80		15.70	22
7260	14" iron pipe size	[G]	70	.229		7.05	10.50		17.55	24
7270	16" iron pipe size	[G]	65	.246		9.15	11.35		20.50	28
7280	18" iron pipe size	[G]	60	.267		10.30	12.25		22.55	30.50
7290	20" iron pipe size	[G]	55	.291		10.50	13.40		23.90	32.50
7300	24" iron pipe size	[G]	50	.320		12.85	14.75		27.60	37
7320	2" wall, 1/2" iron pipe size	[G]	220	.073		2.62	3.35		5.97	8.10
7340	3/4" iron pipe size	[G]	210	.076		2.70	3.51		6.21	8.40
7350	1" iron pipe size	[G]	200	.080		2.88	3.68		6.56	8.90
7360	1-1/4" iron pipe size	[G]	190	.084		3.04	3.88		6.92	9.40
7370	1-1/2" iron pipe size	[G]	190	.084		3.17	3.88		7.05	9.55
7380	2" iron pipe size	[G]	180	.089		3.34	4.09		7.43	10
7390	2-1/2" iron pipe size	[G]	170	.094		3.60	4.33		7.93	10.70
7400	3" iron pipe size	[G]	160	.100		3.82	4.60		8.42	11.35
7410	3-1/2" iron pipe size	[G]	150	.107		4.14	4.91		9.05	12.20
7420	4" iron pipe size	[G]	130	.123		4.45	5.65		10.10	13.70
7430	5" iron pipe size	[G]	120	.133		5.10	6.15		11.25	15.15
7440	6" iron pipe size	[G]	100	.160		5.25	7.35		12.60	17.25
7450	7" iron pipe size	[G]	90	.178		6	8.20		14.20	19.35
7460	8" iron pipe size	[G]	80	.200		6.40	9.20		15.60	21.50
7470	9" iron pipe size	[G]	75	.213		7	9.80		16.80	23
7480	10" iron pipe size	[G]	70	.229		7.65	10.50		18.15	25
7490	12" iron pipe size	[G]	65	.246		8.55	11.35		19.90	27
7500	14" iron pipe size	[G]	60	.267		11.05	12.25		23.30	31.50
7510	16" iron pipe size	[G]	55	.291		12.10	13.40		25.50	34.50
7520	18" iron pipe size	[G]	50	.320		13.50	14.75		28.25	38
7530	20" iron pipe size	[G]	45	.356		15.35	16.35		31.70	42.50

22 07 Plumbing Insulation

22 07 19 – Plumbing Piping Insulation

22 07 19.10 Piping Insulation		Crew	Daily Output	Labor-Hours	Unit	Material	2014 Bare Costs Labor	Equipment	Total	Total Incl O&P
7540	24" iron pipe size [G]	Q-14	40	.400	L.F.	16.45	18.40		34.85	46.50
7560	2-1/2" wall, 1/2" iron pipe size [G]		210	.076		3.10	3.51		6.61	8.85
7562	3/4" iron pipe size [G]		200	.080		3.24	3.68		6.92	9.30
7564	1" iron pipe size [G]		190	.084		3.38	3.88		7.26	9.75
7566	1-1/4" iron pipe size [G]		185	.086		3.50	3.98		7.48	10.05
7568	1-1/2" iron pipe size [G]		180	.089		3.68	4.09		7.77	10.40
7570	2" iron pipe size [G]		170	.094		3.86	4.33		8.19	11
7572	2-1/2" iron pipe size [G]		160	.100		4.45	4.60		9.05	12.05
7574	3" iron pipe size [G]		150	.107		4.68	4.91		9.59	12.80
7576	3-1/2" iron pipe size [G]		140	.114		5.10	5.25		10.35	13.80
7578	4" iron pipe size [G]		120	.133		5.35	6.15		11.50	15.45
7580	5" iron pipe size [G]		110	.145		6.40	6.70		13.10	17.45
7582	6" iron pipe size [G]		90	.178		7.70	8.20		15.90	21.50
7584	7" iron pipe size [G]		80	.200		7.75	9.20		16.95	23
7586	8" iron pipe size [G]		70	.229		8.25	10.50		18.75	25.50
7588	9" iron pipe size [G]		65	.246		9	11.35		20.35	27.50
7590	10" iron pipe size [G]		60	.267		9.80	12.25		22.05	30
7592	12" iron pipe size [G]		55	.291		11.90	13.40		25.30	34
7594	14" iron pipe size [G]		50	.320		13.90	14.75		28.65	38.50
7596	16" iron pipe size [G]		45	.356		15.95	16.35		32.30	43
7598	18" iron pipe size [G]		40	.400		17.30	18.40		35.70	47.50
7602	24" iron pipe size [G]		30	.533		23	24.50		47.50	63
7620	3" wall, 1/2" iron pipe size [G]		200	.080		3.97	3.68		7.65	10.10
7622	3/4" iron pipe size [G]		190	.084		4.20	3.88		8.08	10.65
7624	1" iron pipe size [G]		180	.089		4.45	4.09		8.54	11.25
7626	1-1/4" iron pipe size [G]		175	.091		4.54	4.21		8.75	11.55
7628	1-1/2" iron pipe size [G]		170	.094		4.74	4.33		9.07	11.95
7630	2" iron pipe size [G]		160	.100		5.10	4.60		9.70	12.80
7632	2-1/2" iron pipe size [G]		150	.107		5.35	4.91		10.26	13.50
7634	3" iron pipe size [G]		140	.114		5.65	5.25		10.90	14.45
7636	3-1/2" iron pipe size [G]		130	.123		6.25	5.65		11.90	15.70
7638	4" iron pipe size [G]		110	.145		6.75	6.70		13.45	17.80
7640	5" iron pipe size [G]		100	.160		7.60	7.35		14.95	19.85
7642	6" iron pipe size [G]		80	.200		8.15	9.20		17.35	23.50
7644	7" iron pipe size [G]		70	.229		9.30	10.50		19.80	26.50
7646	8" iron pipe size [G]		60	.267		10.15	12.25		22.40	30.50
7648	9" iron pipe size [G]		55	.291		10.95	13.40		24.35	33
7650	10" iron pipe size [G]		50	.320		11.75	14.75		26.50	36
7652	12" iron pipe size [G]		45	.356		14.60	16.35		30.95	41.50
7654	14" iron pipe size [G]		40	.400		16.85	18.40		35.25	47
7656	16" iron pipe size [G]		35	.457		19.10	21		40.10	53.50
7658	18" iron pipe size [G]		32	.500		20.50	23		43.50	58.50
7660	20" iron pipe size [G]		30	.533		22	24.50		46.50	62.50
7662	24" iron pipe size [G]		28	.571		28	26.50		54.50	72
7664	26" iron pipe size [G]		26	.615		31	28.50		59.50	78
7666	30" iron pipe size [G]		24	.667		37	30.50		67.50	88.50
7800	For fiberglass with standard canvas jacket, deduct					5%				
7802	For fittings, add 3 L.F. for each fitting									
7804	plus 4 L.F. for each flange of the fitting									
7810	Finishes									
7814	For single layer of felt, add					10%	10%			
7816	For roofing paper, 45 lb. to 55 lb., add					25%	10%			
7820	Polyethylene tubing flexible closed cell foam, UV resistant									

22 07 Plumbing Insulation

22 07 19 – Plumbing Piping Insulation

22 07 19.10 Piping Insulation		Crew	Daily Output	Labor-Hours	Unit	Material	2014 Bare Costs Labor	Equipment	Total	Total Incl O&P
7828	Standard temperature (-90°F to +212°F)									
7830	3/8" wall, 1/8" iron pipe size G	1 Asbe	130	.062	L.F.	.22	3.15		3.37	5.15
7831	1/4" iron pipe size G		130	.062		.23	3.15		3.38	5.15
7832	3/8" iron pipe size G		130	.062		.26	3.15		3.41	5.20
7833	1/2" iron pipe size G		126	.063		.29	3.25		3.54	5.35
7834	3/4" iron pipe size G		122	.066		.34	3.35		3.69	5.55
7835	1" iron pipe size G		120	.067		.39	3.41		3.80	5.75
7836	1-1/4" iron pipe size G		118	.068		.47	3.47		3.94	5.90
7837	1-1/2" iron pipe size G		118	.068		.60	3.47		4.07	6.05
7838	2" iron pipe size G		116	.069		.79	3.53		4.32	6.35
7839	2-1/2" iron pipe size G		114	.070		1.01	3.59		4.60	6.70
7840	3" iron pipe size G		112	.071		1.56	3.65		5.21	7.40
7842	1/2" wall, 1/8" iron pipe size G		120	.067		.33	3.41		3.74	5.65
7843	1/4" iron pipe size G		120	.067		.35	3.41		3.76	5.70
7844	3/8" iron pipe size G		120	.067		.39	3.41		3.80	5.75
7845	1/2" iron pipe size G		118	.068		.43	3.47		3.90	5.85
7846	3/4" iron pipe size G		116	.069		.49	3.53		4.02	6.05
7847	1" iron pipe size G		114	.070		.55	3.59		4.14	6.20
7848	1-1/4" iron pipe size G		112	.071		.65	3.65		4.30	6.40
7849	1-1/2" iron pipe size G		110	.073		.77	3.72		4.49	6.65
7850	2" iron pipe size G		108	.074		1.11	3.79		4.90	7.10
7851	2-1/2" iron pipe size G		106	.075		1.37	3.86		5.23	7.50
7852	3" iron pipe size G		104	.077		1.89	3.93		5.82	8.20
7853	3-1/2" iron pipe size G		102	.078		2.14	4.01		6.15	8.60
7854	4" iron pipe size G		100	.080		2.44	4.09		6.53	9.05
7855	3/4" wall, 1/8" iron pipe size G		110	.073		.50	3.72		4.22	6.35
7856	1/4" iron pipe size G		110	.073		.53	3.72		4.25	6.40
7857	3/8" iron pipe size G		108	.074		.61	3.79		4.40	6.55
7858	1/2" iron pipe size G		106	.075		.73	3.86		4.59	6.80
7859	3/4" iron pipe size G		104	.077		.88	3.93		4.81	7.05
7860	1" iron pipe size G		102	.078		1.01	4.01		5.02	7.35
7861	1-1/4" iron pipe size G		100	.080		1.38	4.09		5.47	7.85
7862	1-1/2" iron pipe size G		100	.080		1.56	4.09		5.65	8.05
7863	2" iron pipe size G		98	.082		1.98	4.18		6.16	8.70
7864	2-1/2" iron pipe size G		96	.083		2.39	4.26		6.65	9.30
7865	3" iron pipe size G		94	.085		3.15	4.35		7.50	10.20
7866	3-1/2" iron pipe size G		92	.087		3.51	4.45		7.96	10.75
7867	4" iron pipe size G		90	.089		3.92	4.55		8.47	11.40
7868	1" wall, 1/4" iron pipe size G		100	.080		1.06	4.09		5.15	7.50
7869	3/8" iron pipe size G		98	.082		1.08	4.18		5.26	7.70
7870	1/2" iron pipe size G		96	.083		1.35	4.26		5.61	8.15
7871	3/4" iron pipe size G		94	.085		1.57	4.35		5.92	8.50
7872	1" iron pipe size G		92	.087		1.91	4.45		6.36	9
7873	1-1/4" iron pipe size G		90	.089		2.17	4.55		6.72	9.50
7874	1-1/2" iron pipe size G		90	.089		2.48	4.55		7.03	9.85
7875	2" iron pipe size G		88	.091		3.23	4.65		7.88	10.80
7876	2-1/2" iron pipe size G		86	.093		4.22	4.76		8.98	12.05
7877	3" iron pipe size G		84	.095		4.74	4.87		9.61	12.80
7878	Contact cement, quart can G				Ea.	10.50			10.50	11.55
7879	Rubber tubing, flexible closed cell foam									
7880	3/8" wall, 1/4" iron pipe size G	1 Asbe	120	.067	L.F.	.33	3.41		3.74	5.65
7900	3/8" iron pipe size G		120	.067		.36	3.41		3.77	5.70
7910	1/2" iron pipe size G		115	.070		.41	3.56		3.97	6

22 07 Plumbing Insulation

22 07 19 – Plumbing Piping Insulation

22 07 19.10 Piping Insulation

		Crew	Daily Output	Labor-Hours	Unit	Material	2014 Bare Costs Labor	Equipment	Total	Total Incl O&P
7920	3/4" iron pipe size	1 Asbe G	115	.070	L.F.	.46	3.56		4.02	6.05
7930	1" iron pipe size	G	110	.073		.52	3.72		4.24	6.35
7940	1-1/4" iron pipe size	G	110	.073		.59	3.72		4.31	6.45
7950	1-1/2" iron pipe size	G	110	.073		.73	3.72		4.45	6.60
8100	1/2" wall, 1/4" iron pipe size	G	90	.089		.54	4.55		5.09	7.70
8120	3/8" iron pipe size	G	90	.089		.60	4.55		5.15	7.75
8130	1/2" iron pipe size	G	89	.090		.67	4.60		5.27	7.90
8140	3/4" iron pipe size	G	89	.090		.75	4.60		5.35	8
8150	1" iron pipe size	G	88	.091		.82	4.65		5.47	8.15
8160	1-1/4" iron pipe size	G	87	.092		.95	4.70		5.65	8.35
8170	1-1/2" iron pipe size	G	87	.092		1.15	4.70		5.85	8.55
8180	2" iron pipe size	G	86	.093		1.47	4.76		6.23	9
8190	2-1/2" iron pipe size	G	86	.093		1.85	4.76		6.61	9.45
8200	3" iron pipe size	G	85	.094		2.06	4.81		6.87	9.75
8210	3-1/2" iron pipe size	G	85	.094		2.86	4.81		7.67	10.65
8220	4" iron pipe size	G	80	.100		3.06	5.10		8.16	11.30
8230	5" iron pipe size	G	80	.100		4.17	5.10		9.27	12.55
8240	6" iron pipe size	G	75	.107		4.17	5.45		9.62	13.10
8300	3/4" wall, 1/4" iron pipe size	G	90	.089		.85	4.55		5.40	8.05
8320	3/8" iron pipe size	G	90	.089		.92	4.55		5.47	8.10
8330	1/2" iron pipe size	G	89	.090		1.10	4.60		5.70	8.35
8340	3/4" iron pipe size	G	89	.090		1.35	4.60		5.95	8.65
8350	1" iron pipe size	G	88	.091		1.54	4.65		6.19	8.95
8360	1-1/4" iron pipe size	G	87	.092		2.05	4.70		6.75	9.55
8370	1-1/2" iron pipe size	G	87	.092		2.32	4.70		7.02	9.85
8380	2" iron pipe size	G	86	.093		2.68	4.76		7.44	10.35
8390	2-1/2" iron pipe size	G	86	.093		3.58	4.76		8.34	11.35
8400	3" iron pipe size	G	85	.094		4.08	4.81		8.89	12
8410	3-1/2" iron pipe size	G	85	.094		4.81	4.81		9.62	12.80
8420	4" iron pipe size	G	80	.100		5.10	5.10		10.20	13.60
8430	5" iron pipe size	G	80	.100		6.15	5.10		11.25	14.75
8440	6" iron pipe size	G	80	.100		7.45	5.10		12.55	16.15
8444	1" wall, 1/2" iron pipe size	G	86	.093		2.05	4.76		6.81	9.65
8445	3/4" iron pipe size	G	84	.095		2.48	4.87		7.35	10.35
8446	1" iron pipe size	G	84	.095		2.89	4.87		7.76	10.80
8447	1-1/4" iron pipe size	G	82	.098		3.23	4.99		8.22	11.30
8448	1-1/2" iron pipe size	G	82	.098		3.76	4.99		8.75	11.90
8449	2" iron pipe size	G	80	.100		4.95	5.10		10.05	13.40
8450	2-1/2" iron pipe size	G	80	.100		6.45	5.10		11.55	15.05
8456	Rubber insulation tape, 1/8" x 2" x 30'	G			Ea.	12.05			12.05	13.25
8460	Polyolefin tubing, flexible closed cell foam, UV stabilized, work									
8462	temp.-165°F to +210°F, 0 water vapor transmission									
8464	3/8" wall, 1/8" iron pipe size	1 Asbe G	140	.057	L.F.	.47	2.92		3.39	5.05
8466	1/4" iron pipe size	G	140	.057		.49	2.92		3.41	5.10
8468	3/8" iron pipe size	G	140	.057		.54	2.92		3.46	5.15
8470	1/2" iron pipe size	G	136	.059		.60	3.01		3.61	5.35
8472	3/4" iron pipe size	G	132	.061		.71	3.10		3.81	5.60
8474	1" iron pipe size	G	130	.062		.81	3.15		3.96	5.80
8476	1-1/4" iron pipe size	G	128	.063		.99	3.20		4.19	6.05
8478	1-1/2" iron pipe size	G	128	.063		1.21	3.20		4.41	6.30
8480	2" iron pipe size	G	126	.063		1.35	3.25		4.60	6.55
8482	2-1/2" iron pipe size	G	123	.065		1.96	3.33		5.29	7.35
8484	3" iron pipe size	G	121	.066		2.17	3.38		5.55	7.65

22 07 Plumbing Insulation

22 07 19 – Plumbing Piping Insulation

22 07 19.10 Piping Insulation

		Crew	Daily Output	Labor-Hours	Unit	Material	Labor	Equipment	Total	Total Incl O&P
8486	4" iron pipe size G	1 Asbe	118	.068	L.F.	4.22	3.47		7.69	10.05
8500	1/2" wall, 1/8" iron pipe size G		130	.062		.68	3.15		3.83	5.65
8502	1/4" iron pipe size G		130	.062		.73	3.15		3.88	5.70
8504	3/8" iron pipe size G		130	.062		.79	3.15		3.94	5.75
8506	1/2" iron pipe size G		128	.063		.86	3.20		4.06	5.95
8508	3/4" iron pipe size G		126	.063		.99	3.25		4.24	6.15
8510	1" iron pipe size G		123	.065		1.14	3.33		4.47	6.45
8512	1-1/4" iron pipe size G		121	.066		1.35	3.38		4.73	6.75
8514	1-1/2" iron pipe size G		119	.067		1.62	3.44		5.06	7.15
8516	2" iron pipe size G		117	.068		2.05	3.50		5.55	7.70
8518	2-1/2" iron pipe size G		114	.070		2.77	3.59		6.36	8.65
8520	3" iron pipe size G		112	.071		3.60	3.65		7.25	9.65
8522	4" iron pipe size G		110	.073		4.98	3.72		8.70	11.30
8534	3/4" wall, 1/8" iron pipe size G		120	.067		1.04	3.41		4.45	6.45
8536	1/4" iron pipe size G		120	.067		1.10	3.41		4.51	6.50
8538	3/8" iron pipe size G		117	.068		1.27	3.50		4.77	6.85
8540	1/2" iron pipe size G		114	.070		1.40	3.59		4.99	7.15
8542	3/4" iron pipe size G		112	.071		1.52	3.65		5.17	7.35
8544	1" iron pipe size G		110	.073		2.13	3.72		5.85	8.15
8546	1-1/4" iron pipe size G		108	.074		2.85	3.79		6.64	9.05
8548	1-1/2" iron pipe size G		108	.074		3.33	3.79		7.12	9.55
8550	2" iron pipe size G		106	.075		4.22	3.86		8.08	10.65
8552	2-1/2" iron pipe size G		104	.077		5.30	3.93		9.23	11.95
8554	3" iron pipe size G		102	.078		6.60	4.01		10.61	13.55
8556	4" iron pipe size G		100	.080		8.75	4.09		12.84	15.95
8570	1" wall, 1/8" iron pipe size G		110	.073		1.89	3.72		5.61	7.90
8572	1/4" iron pipe size G		108	.074		1.94	3.79		5.73	8.05
8574	3/8" iron pipe size G		106	.075		1.98	3.86		5.84	8.20
8576	1/2" iron pipe size G		104	.077		2.10	3.93		6.03	8.40
8578	3/4" iron pipe size G		102	.078		2.74	4.01		6.75	9.25
8580	1" iron pipe size G		100	.080		3.18	4.09		7.27	9.85
8582	1-1/4" iron pipe size G		97	.082		3.61	4.22		7.83	10.50
8584	1-1/2" iron pipe size G		97	.082		4.13	4.22		8.35	11.10
8586	2" iron pipe size G		95	.084		5.85	4.31		10.16	13.10
8588	2-1/2" iron pipe size G		93	.086		7.60	4.40		12	15.20
8590	3" iron pipe size G		91	.088		9.25	4.50		13.75	17.20
8606	Contact adhesive (R-320) G				Qt.	26.50			26.50	29.50
8608	Contact adhesive (R-320) G				Gal.	109			109	120
8610	NOTE: Preslit/preglued vs unslit, same price									

22 07 19.30 Piping Insulation Protective Jacketing, PVC

0010	**PIPING INSULATION PROTECTIVE JACKETING, PVC**									
0100	PVC, white, 48" lengths cut from roll goods									
0120	20 mil thick									
0140	Size based on OD of insulation									
0150	1-1/2" ID	Q-14	270	.059	L.F.	.20	2.73		2.93	4.46
0152	2" ID		260	.062		.27	2.83		3.10	4.71
0154	2-1/2" ID		250	.064		.32	2.95		3.27	4.93
0156	3" ID		240	.067		.39	3.07		3.46	5.20
0158	3-1/2" ID		230	.070		.44	3.20		3.64	5.45
0160	4" ID		220	.073		.50	3.35		3.85	5.75
0162	4-1/2" ID		210	.076		.56	3.51		4.07	6.05
0164	5" ID		200	.080		.61	3.68		4.29	6.40

22 07 Plumbing Insulation

22 07 19 – Plumbing Piping Insulation

22 07 19.30 Piping Insulation Protective Jacketing, PVC		Crew	Daily Output	Labor-Hours	Unit	Material	2014 Bare Costs Labor	Equipment	Total	Total Incl O&P
0166	5-1/2" ID	Q-14	190	.084	L.F.	.67	3.88		4.55	6.80
0168	6" ID		180	.089		.73	4.09		4.82	7.15
0170	6-1/2" ID		175	.091		.79	4.21		5	7.40
0172	7" ID		170	.094		.84	4.33		5.17	7.65
0174	7-1/2" ID		164	.098		.91	4.49		5.40	8
0176	8" ID		161	.099		.96	4.57		5.53	8.15
0178	8-1/2" ID		158	.101		1.02	4.66		5.68	8.35
0180	9" ID		155	.103		1.08	4.75		5.83	8.60
0182	9-1/2" ID		152	.105		1.14	4.85		5.99	8.80
0184	10" ID		149	.107		1.19	4.94		6.13	9
0186	10-1/2" ID		146	.110		1.25	5.05		6.30	9.25
0188	11" ID		143	.112		1.31	5.15		6.46	9.45
0190	11-1/2" ID		140	.114		1.37	5.25		6.62	9.70
0192	12" ID		137	.117		1.42	5.40		6.82	9.90
0194	12-1/2" ID		134	.119		1.49	5.50		6.99	10.20
0195	13" ID		132	.121		1.54	5.60		7.14	10.40
0196	13-1/2" ID		132	.121		1.60	5.60		7.20	10.45
0198	14" ID		130	.123		1.66	5.65		7.31	10.65
0200	15" ID		128	.125		1.78	5.75		7.53	10.90
0202	16" ID		126	.127		1.89	5.85		7.74	11.20
0204	17" ID		124	.129		2.01	5.95		7.96	11.45
0206	18" ID		122	.131		2.13	6.05		8.18	11.75
0208	19" ID		120	.133		2.24	6.15		8.39	12
0210	20" ID		118	.136		2.35	6.25		8.60	12.30
0212	21" ID		116	.138		2.47	6.35		8.82	12.60
0214	22" ID		114	.140		2.58	6.45		9.03	12.90
0216	23" ID		112	.143		2.69	6.60		9.29	13.20
0218	24" ID		110	.145		2.81	6.70		9.51	13.50
0220	25" ID		108	.148		2.93	6.80		9.73	13.80
0222	26" ID		106	.151		3.04	6.95		9.99	14.15
0224	27" ID		104	.154		3.16	7.10		10.26	14.50
0226	28" ID		102	.157		3.27	7.20		10.47	14.85
0228	29" ID		100	.160		3.39	7.35		10.74	15.20
0230	30" ID		98	.163		3.51	7.50		11.01	15.55
0300	For colors, add				Ea.	10%				
1000	30 mil thick									
1010	Size based on OD of insulation									
1020	2" ID	Q-14	260	.062	L.F.	.40	2.83		3.23	4.85
1022	2-1/2" ID		250	.064		.48	2.95		3.43	5.10
1024	3" ID		240	.067		.58	3.07		3.65	5.40
1026	3-1/2" ID		230	.070		.65	3.20		3.85	5.70
1028	4" ID		220	.073		.74	3.35		4.09	6
1030	4-1/2" ID		210	.076		.82	3.51		4.33	6.35
1032	5" ID		200	.080		.91	3.68		4.59	6.75
1034	5-1/2" ID		190	.084		.99	3.88		4.87	7.15
1036	6" ID		180	.089		1.08	4.09		5.17	7.55
1038	6-1/2" ID		175	.091		1.16	4.21		5.37	7.85
1040	7" ID		170	.094		1.25	4.33		5.58	8.15
1042	7-1/2" ID		164	.098		1.34	4.49		5.83	8.45
1044	8" ID		161	.099		1.43	4.57		6	8.65
1046	8-1/2" ID		158	.101		1.51	4.66		6.17	8.90
1048	9" ID		155	.103		1.60	4.75		6.35	9.15
1050	9-1/2" ID		152	.105		1.68	4.85		6.53	9.40

22 07 Plumbing Insulation

22 07 19 – Plumbing Piping Insulation

22 07 19.30 Piping Insulation Protective Jacketing, PVC		Crew	Daily Output	Labor-Hours	Unit	Material	2014 Bare Costs Labor	Equipment	Total	Total Incl O&P
1052	10" ID	Q-14	149	.107	L.F.	1.77	4.94		6.71	9.65
1054	10-1/2" ID		146	.110		1.85	5.05		6.90	9.90
1056	11" ID		143	.112		1.94	5.15		7.09	10.15
1058	11-1/2" ID		140	.114		2.02	5.25		7.27	10.40
1060	12" ID		137	.117		2.11	5.40		7.51	10.65
1062	12-1/2" ID		134	.119		2.20	5.50		7.70	10.95
1063	13" ID		132	.121		2.29	5.60		7.89	11.20
1064	13-1/2" ID		132	.121		2.37	5.60		7.97	11.30
1066	14" ID		130	.123		2.46	5.65		8.11	11.50
1068	15" ID		128	.125		2.63	5.75		8.38	11.85
1070	16" ID		126	.127		2.80	5.85		8.65	12.20
1072	17" ID		124	.129		2.97	5.95		8.92	12.50
1074	18" ID		122	.131		3.15	6.05		9.20	12.85
1076	19" ID		120	.133		3.32	6.15		9.47	13.20
1078	20" ID		118	.136		3.49	6.25		9.74	13.55
1080	21" ID		116	.138		3.66	6.35		10.01	13.95
1082	22" ID		114	.140		3.83	6.45		10.28	14.25
1084	23" ID		112	.143		4	6.60		10.60	14.65
1086	24" ID		110	.145		4.17	6.70		10.87	15
1088	25" ID		108	.148		4.34	6.80		11.14	15.35
1090	26" ID		106	.151		4.51	6.95		11.46	15.75
1092	27" ID		104	.154		4.68	7.10		11.78	16.15
1094	28" ID		102	.157		4.85	7.20		12.05	16.60
1096	29" ID		100	.160		5	7.35		12.35	16.95
1098	30" ID		98	.163		5.20	7.50		12.70	17.40
1300	For colors, add				Ea.	10%				
2000	PVC, white, fitting covers									
2020	Fiberglass insulation inserts included with sizes 1-3/4" thru 9-3/4"									
2030	Size is based on OD of insulation									
2040	90° Elbow fitting									
2060	1-3/4"	Q-14	135	.119	Ea.	.47	5.45		5.92	9
2062	2"		130	.123		.58	5.65		6.23	9.45
2064	2-1/4"		128	.125		.65	5.75		6.40	9.65
2068	2-1/2"		126	.127		.71	5.85		6.56	9.90
2070	2-3/4"		123	.130		.77	6		6.77	10.15
2072	3"		120	.133		.82	6.15		6.97	10.45
2074	3-3/8"		116	.138		.93	6.35		7.28	10.90
2076	3-3/4"		113	.142		1.04	6.50		7.54	11.30
2078	4-1/8"		110	.145		1.31	6.70		8.01	11.85
2080	4-3/4"		105	.152		1.58	7		8.58	12.65
2082	5-1/4"		100	.160		1.84	7.35		9.19	13.45
2084	5-3/4"		95	.168		2.43	7.75		10.18	14.70
2086	6-1/4"		90	.178		4.05	8.20		12.25	17.20
2088	6-3/4"		87	.184		4.27	8.45		12.72	17.85
2090	7-1/4"		85	.188		5.35	8.65		14	19.40
2092	7-3/4"		83	.193		5.60	8.85		14.45	19.95
2094	8-3/4"		80	.200		7.20	9.20		16.40	22.50
2096	9-3/4"		77	.208		9.65	9.55		19.20	25.50
2098	10-7/8"		74	.216		10.75	9.95		20.70	27.50
2100	11-7/8"		71	.225		12.35	10.35		22.70	29.50
2102	12-7/8"		68	.235		15.55	10.85		26.40	34
2104	14-1/8"		66	.242		17.80	11.15		28.95	37
2106	15-1/8"		64	.250		19.15	11.50		30.65	39

22 07 Plumbing Insulation

22 07 19 – Plumbing Piping Insulation

22 07 19.30 Piping Insulation Protective Jacketing, PVC

		Crew	Daily Output	Labor-Hours	Unit	Material	Labor	Equipment	Total	Total Incl O&P
2108	16-1/8"	Q-14	63	.254	Ea.	21	11.70		32.70	41
2110	17-1/8"		62	.258		23	11.90		34.90	44
2112	18-1/8"		61	.262		30	12.05		42.05	52
2114	19-1/8"		60	.267		40.50	12.25		52.75	63.50
2116	20-1/8"		59	.271		52	12.50		64.50	77
2200	45° Elbow fitting									
2220	1-3/4" thru 9-3/4" same price as 90° Elbow fitting									
2320	10-7/8"	Q-14	74	.216	Ea.	10.75	9.95		20.70	27.50
2322	11-7/8"		71	.225		11.85	10.35		22.20	29
2324	12-7/8"		68	.235		13.25	10.85		24.10	31.50
2326	14-1/8"		66	.242		15.10	11.15		26.25	34
2328	15-1/8"		64	.250		16.25	11.50		27.75	36
2330	16-1/8"		63	.254		18.55	11.70		30.25	38.50
2332	17-1/8"		62	.258		21	11.90		32.90	41.50
2334	18-1/8"		61	.262		25.50	12.05		37.55	47
2336	19-1/8"		60	.267		35	12.25		47.25	57.50
2338	20-1/8"		59	.271		39.50	12.50		52	63
2400	Tee fitting									
2410	1-3/4"	Q-14	96	.167	Ea.	.88	7.65		8.53	12.90
2412	2"		94	.170		1	7.85		8.85	13.30
2414	2-1/4"		91	.176		1.09	8.10		9.19	13.80
2416	2-1/2"		88	.182		1.19	8.35		9.54	14.30
2418	2-3/4"		85	.188		1.31	8.65		9.96	14.95
2420	3"		82	.195		1.42	9		10.42	15.55
2422	3-3/8"		79	.203		1.63	9.30		10.93	16.30
2424	3-3/4"		76	.211		1.85	9.70		11.55	17.15
2426	4-1/8"		73	.219		2.19	10.10		12.29	18.10
2428	4-3/4"		70	.229		2.73	10.50		13.23	19.35
2430	5-1/4"		67	.239		3.28	11		14.28	20.50
2432	5-3/4"		63	.254		4.36	11.70		16.06	23
2434	6-1/4"		60	.267		5.75	12.25		18	25.50
2436	6-3/4"		59	.271		7.10	12.50		19.60	27.50
2438	7-1/4"		57	.281		11.45	12.90		24.35	32.50
2440	7-3/4"		54	.296		12.55	13.65		26.20	35
2442	8-3/4"		52	.308		15.30	14.15		29.45	39
2444	9-3/4"		50	.320		18	14.75		32.75	43
2446	10-7/8"		48	.333		18.25	15.35		33.60	44
2448	11-7/8"		47	.340		20.50	15.65		36.15	47
2450	12-7/8"		46	.348		22.50	16		38.50	49.50
2452	14-1/8"		45	.356		24.50	16.35		40.85	52.50
2454	15-1/8"		44	.364		26.50	16.75		43.25	55
2456	16-1/8"		43	.372		28.50	17.15		45.65	57.50
2458	17-1/8"		42	.381		30.50	17.55		48.05	61
2460	18-1/8"		41	.390		33.50	17.95		51.45	65
2462	19-1/8"		40	.400		36.50	18.40		54.90	68.50
2464	20-1/8"		39	.410		40	18.90		58.90	73.50
4000	Mechanical grooved fitting cover, including insert									
4020	90° Elbow fitting									
4030	3/4" & 1"	Q-14	140	.114	Ea.	3.70	5.25		8.95	12.25
4040	1-1/4" & 1-1/2"		135	.119		4.68	5.45		10.13	13.65
4042	2"		130	.123		7.40	5.65		13.05	16.95
4044	2-1/2"		125	.128		8.20	5.90		14.10	18.15
4046	3"		120	.133		9.20	6.15		15.35	19.65

22 07 Plumbing Insulation

22 07 19 – Plumbing Piping Insulation

22 07 19.30 Piping Insulation Protective Jacketing, PVC

		Crew	Daily Output	Labor-Hours	Unit	Material	2014 Bare Costs Labor	Equipment	Total	Total Incl O&P
4048	3-1/2"	Q-14	115	.139	Ea.	10.65	6.40		17.05	21.50
4050	4"		110	.145		11.85	6.70		18.55	23.50
4052	5"		100	.160		14.75	7.35		22.10	27.50
4054	6"		90	.178		22	8.20		30.20	37
4056	8"		80	.200		23.50	9.20		32.70	40.50
4058	10"		75	.213		30.50	9.80		40.30	49
4060	12"		68	.235		43.50	10.85		54.35	65
4062	14"		65	.246		43.50	11.35		54.85	65.50
4064	16"		63	.254		59.50	11.70		71.20	83.50
4066	18"	▼	61	.262	▼	81.50	12.05		93.55	109
4100	45° Elbow fitting									
4120	3/4" & 1"	Q-14	140	.114	Ea.	3.29	5.25		8.54	11.80
4130	1-1/4" & 1-1/2"		135	.119		4.19	5.45		9.64	13.10
4140	2"		130	.123		6.60	5.65		12.25	16.05
4142	2-1/2"		125	.128		7.30	5.90		13.20	17.20
4144	3"		120	.133		8.20	6.15		14.35	18.55
4146	3-1/2"		115	.139		8.65	6.40		15.05	19.45
4148	4"		110	.145		10.55	6.70		17.25	22
4150	5"		100	.160		13.10	7.35		20.45	26
4152	6"		90	.178		19.35	8.20		27.55	34.50
4154	8"		80	.200		21	9.20		30.20	37.50
4156	10"		75	.213		27	9.80		36.80	45
4158	12"		68	.235		39	10.85		49.85	59.50
4160	14"		65	.246		50.50	11.35		61.85	73
4162	16"		63	.254		61	11.70		72.70	85.50
4164	18"	▼	61	.262	▼	72	12.05		84.05	98.50
4200	Tee fitting									
4220	3/4" & 1"	Q-14	93	.172	Ea.	4.82	7.90		12.72	17.60
4230	1-1/4" & 1-1/2"		90	.178		6.10	8.20		14.30	19.45
4240	2"		87	.184		9.60	8.45		18.05	23.50
4242	2-1/2"		84	.190		10.70	8.75		19.45	25.50
4244	3"		80	.200		11.95	9.20		21.15	27.50
4246	3-1/2"		77	.208		12.65	9.55		22.20	29
4248	4"		73	.219		15.45	10.10		25.55	32.50
4250	5"		67	.239		17.05	11		28.05	36
4252	6"		60	.267		25	12.25		37.25	46.50
4254	8"		54	.296		27.50	13.65		41.15	51
4256	10"		50	.320		39.50	14.75		54.25	66.50
4258	12"		46	.348		56.50	16		72.50	87.50
4260	14"		43	.372		68.50	17.15		85.65	102
4262	16"		42	.381		82	17.55		99.55	118
4264	18"	▼	41	.390	▼	90.50	17.95		108.45	128

22 07 19.40 Pipe Insulation Protective Jacketing, Aluminum

0010	**PIPE INSULATION PROTECTIVE JACKETING, ALUMINUM**									
0100	Metal roll jacketing									
0120	Aluminum with polykraft moisture barrier									
0140	Smooth, based on OD of insulation, .016" thick									
0180	1/2" ID	Q-14	220	.073	L.F.	.20	3.35		3.55	5.40
0190	3/4" ID		215	.074		.26	3.43		3.69	5.65
0200	1" ID		210	.076		.32	3.51		3.83	5.80
0210	1-1/4" ID		205	.078		.38	3.59		3.97	6
0220	1-1/2" ID	▼	202	.079	▼	.44	3.65		4.09	6.15

22 07 Plumbing Insulation

22 07 19 – Plumbing Piping Insulation

	22 07 19.40 Pipe Insulation Protective Jacketing, Aluminum	Crew	Daily Output	Labor-Hours	Unit	Material	2014 Bare Costs Labor	Equipment	Total	Total Incl O&P
0230	1-3/4" ID	Q-14	199	.080	L.F.	.50	3.70		4.20	6.30
0240	2" ID		195	.082		.56	3.78		4.34	6.50
0250	2-1/4" ID		191	.084		.62	3.86		4.48	6.70
0260	2-1/2" ID		187	.086		.69	3.94		4.63	6.90
0270	2-3/4" ID		184	.087		.75	4		4.75	7.10
0280	3" ID		180	.089		.82	4.09		4.91	7.25
0290	3-1/4" ID		176	.091		.88	4.18		5.06	7.45
0300	3-1/2" ID		172	.093		.94	4.28		5.22	7.70
0310	3-3/4" ID		169	.095		1	4.36		5.36	7.90
0320	4" ID		165	.097		1.06	4.46		5.52	8.10
0330	4-1/4" ID		161	.099		1.12	4.57		5.69	8.35
0340	4-1/2" ID		157	.102		1.19	4.69		5.88	8.60
0350	4-3/4" ID		154	.104		1.25	4.78		6.03	8.85
0360	5" ID		150	.107		1.31	4.91		6.22	9.10
0370	5-1/4" ID		146	.110		1.37	5.05		6.42	9.35
0380	5-1/2" ID		143	.112		1.43	5.15		6.58	9.55
0390	5-3/4" ID		139	.115		1.49	5.30		6.79	9.90
0400	6" ID		135	.119		1.55	5.45		7	10.20
0410	6-1/4" ID		133	.120		1.61	5.55		7.16	10.35
0420	6-1/2" ID		131	.122		1.67	5.60		7.27	10.60
0430	7" ID		128	.125		1.79	5.75		7.54	10.90
0440	7-1/4" ID		125	.128		1.85	5.90		7.75	11.20
0450	7-1/2" ID		123	.130		1.91	6		7.91	11.40
0460	8" ID		121	.132		2.03	6.10		8.13	11.70
0470	8-1/2" ID		119	.134		2.15	6.20		8.35	12
0480	9" ID		116	.138		2.27	6.35		8.62	12.40
0490	9-1/2" ID		114	.140		2.39	6.45		8.84	12.70
0500	10" ID		112	.143		2.51	6.60		9.11	13
0510	10-1/2" ID		110	.145		2.63	6.70		9.33	13.30
0520	11" ID		107	.150		2.75	6.90		9.65	13.75
0530	11-1/2" ID		105	.152		2.88	7		9.88	14.05
0540	12" ID		103	.155		3	7.15		10.15	14.45
0550	12-1/2" ID		100	.160		3.12	7.35		10.47	14.90
0560	13" ID		99	.162		3.24	7.45		10.69	15.15
0570	14" ID		98	.163		3.49	7.50		10.99	15.55
0580	15" ID		96	.167		3.73	7.65		11.38	16.05
0590	16" ID		95	.168		3.98	7.75		11.73	16.45
0600	17" ID		93	.172		4.22	7.90		12.12	16.95
0610	18" ID		92	.174		4.38	8		12.38	17.25
0620	19" ID		90	.178		4.70	8.20		12.90	17.90
0630	20" ID		89	.180		4.95	8.30		13.25	18.35
0640	21" ID		87	.184		5.20	8.45		13.65	18.85
0650	22" ID		86	.186		5.45	8.55		14	19.30
0660	23" ID		84	.190		5.70	8.75		14.45	19.90
0670	24" ID		83	.193		5.90	8.85		14.75	20.50
0710	For smooth .020" thick, add					27%	10%			
0720	For smooth .024" thick, add					52%	20%			
0730	For smooth .032" thick, add					104%	33%			
0800	For stucco embossed, add					1%				
0820	For corrugated, add					2.50%				
0900	White aluminum with polysurlyn moisture barrier									
0910	Smooth, % is an add to polykraft lines of same thickness									
0940	For smooth .016" thick, add				L.F.	35%				

22 07 Plumbing Insulation

22 07 19 – Plumbing Piping Insulation

22 07 19.40 Pipe Insulation Protective Jacketing, Aluminum	Crew	Daily Output	Labor-Hours	Unit	Material	2014 Bare Costs Labor	Equipment	Total	Total Incl O&P
0960 For smooth .024" thick, add				L.F.	22%				
1000 Aluminum fitting covers									
1010 Size is based on OD of insulation									
1020 90° LR elbow, 2 piece									
1100 1-1/2"	Q-14	140	.114	Ea.	3.45	5.25		8.70	12
1110 1-3/4"		135	.119		3.45	5.45		8.90	12.30
1120 2"		130	.123		4.34	5.65		9.99	13.55
1130 2-1/4"		128	.125		4.34	5.75		10.09	13.70
1140 2-1/2"		126	.127		4.34	5.85		10.19	13.85
1150 2-3/4"		123	.130		4.34	6		10.34	14.05
1160 3"		120	.133		4.71	6.15		10.86	14.75
1170 3-1/4"		117	.137		4.71	6.30		11.01	15
1180 3-1/2"		115	.139		5.55	6.40		11.95	16.10
1190 3-3/4"		113	.142		5.70	6.50		12.20	16.45
1200 4"		110	.145		6	6.70		12.70	17
1210 4-1/4"		108	.148		6.15	6.80		12.95	17.35
1220 4-1/2"		106	.151		6.15	6.95		13.10	17.55
1230 4-3/4"		104	.154		6.15	7.10		13.25	17.75
1240 5"		102	.157		6.95	7.20		14.15	18.85
1250 5-1/4"		100	.160		8	7.35		15.35	20.50
1260 5-1/2"		97	.165		9.95	7.60		17.55	23
1270 5-3/4"		95	.168		9.95	7.75		17.70	23
1280 6"		92	.174		8.50	8		16.50	22
1290 6-1/4"		90	.178		8.50	8.20		16.70	22
1300 6-1/2"		87	.184		10.25	8.45		18.70	24.50
1310 7"		85	.188		12.40	8.65		21.05	27
1320 7-1/4"		84	.190		12.40	8.75		21.15	27.50
1330 7-1/2"		83	.193		18.25	8.85		27.10	34
1340 8"		82	.195		13.60	9		22.60	29
1350 8-1/2"		80	.200		25	9.20		34.20	42
1360 9"		78	.205		25	9.45		34.45	42
1370 9-1/2"		77	.208		18.50	9.55		28.05	35.50
1380 10"		76	.211		18.50	9.70		28.20	35.50
1390 10-1/2"		75	.213		18.90	9.80		28.70	36.50
1400 11"		74	.216		18.90	9.95		28.85	36.50
1410 11-1/2"		72	.222		21.50	10.25		31.75	39.50
1420 12"		71	.225		21.50	10.35		31.85	39.50
1430 12-1/2"		69	.232		38	10.65		48.65	58.50
1440 13"		68	.235		38	10.85		48.85	59
1450 14"		66	.242		51.50	11.15		62.65	74
1460 15"		64	.250		53.50	11.50		65	77
1470 16"		63	.254		58.50	11.70		70.20	82.50
2000 45° Elbow, 2 piece									
2010 2-1/2"	Q-14	126	.127	Ea.	3.59	5.85		9.44	13.05
2020 2-3/4"		123	.130		3.59	6		9.59	13.25
2030 3"		120	.133		4.05	6.15		10.20	14
2040 3-1/4"		117	.137		4.05	6.30		10.35	14.25
2050 3-1/2"		115	.139		4.62	6.40		11.02	15.05
2060 3-3/4"		113	.142		4.62	6.50		11.12	15.25
2070 4"		110	.145		4.70	6.70		11.40	15.55
2080 4-1/4"		108	.148		5.35	6.80		12.15	16.50
2090 4-1/2"		106	.151		5.35	6.95		12.30	16.70
2100 4-3/4"		104	.154		5.35	7.10		12.45	16.90

22 07 Plumbing Insulation

22 07 19 – Plumbing Piping Insulation

22 07 19.40 Pipe Insulation Protective Jacketing, Aluminum		Crew	Daily Output	Labor-Hours	Unit	Material	2014 Bare Costs Labor	Equipment	Total	Total Incl O&P
2110	5"	Q-14	102	.157	Ea.	6.30	7.20		13.50	18.20
2120	5-1/4"		100	.160		6.30	7.35		13.65	18.40
2130	5-1/2"		97	.165		6.60	7.60		14.20	19.05
2140	6"		92	.174		6.60	8		14.60	19.70
2150	6-1/2"		87	.184		9.10	8.45		17.55	23
2160	7"		85	.188		9.10	8.65		17.75	23.50
2170	7-1/2"		83	.193		9.20	8.85		18.05	24
2180	8"		82	.195		9.20	9		18.20	24
2190	8-1/2"		80	.200		11.45	9.20		20.65	27
2200	9"		78	.205		11.45	9.45		20.90	27.50
2210	9-1/2"		77	.208		15.90	9.55		25.45	32.50
2220	10"		76	.211		15.90	9.70		25.60	32.50
2230	10-1/2"		75	.213		14.85	9.80		24.65	31.50
2240	11"		74	.216		14.85	9.95		24.80	32
2250	11-1/2"		72	.222		17.85	10.25		28.10	35.50
2260	12"		71	.225		17.85	10.35		28.20	36
2270	13"		68	.235		21	10.85		31.85	40
2280	14"		66	.242		25.50	11.15		36.65	46
2290	15"		64	.250		39.50	11.50		51	61.50
2300	16"		63	.254		43	11.70		54.70	65.50
2310	17"		62	.258		42	11.90		53.90	64.50
2320	18"		61	.262		48	12.05		60.05	71.50
2330	19"		60	.267		58.50	12.25		70.75	83.50
2340	20"		59	.271		56.50	12.50		69	81.50
2350	21"	↓	58	.276	↓	61	12.70		73.70	87
3000	Tee, 4 piece									
3010	2-1/2"	Q-14	88	.182	Ea.	29	8.35		37.35	45
3020	2-3/4"		86	.186		29	8.55		37.55	45.50
3030	3"		84	.190		32.50	8.75		41.25	49.50
3040	3-1/4"		82	.195		32.50	9		41.50	50
3050	3-1/2"		80	.200		34.50	9.20		43.70	52
3060	4"		78	.205		34.50	9.45		43.95	52.50
3070	4-1/4"		76	.211		36.50	9.70		46.20	55
3080	4-1/2"		74	.216		36.50	9.95		46.45	55.50
3090	4-3/4"		72	.222		36.50	10.25		46.75	56
3100	5"		70	.229		37.50	10.50		48	58
3110	5-1/4"		68	.235		37.50	10.85		48.35	58.50
3120	5-1/2"		66	.242		39.50	11.15		50.65	61
3130	6"		64	.250		39.50	11.50		51	61.50
3140	6-1/2"		60	.267		43.50	12.25		55.75	66.50
3150	7"		58	.276		43.50	12.70		56.20	67.50
3160	7-1/2"		56	.286		48	13.15		61.15	73.50
3170	8"		54	.296		48	13.65		61.65	74
3180	8-1/2"		52	.308		49.50	14.15		63.65	76.50
3190	9"		50	.320		49.50	14.75		64.25	77.50
3200	9-1/2"		49	.327		36.50	15.05		51.55	64
3210	10"		48	.333		36.50	15.35		51.85	64.50
3220	10-1/2"		47	.340		39	15.65		54.65	67.50
3230	11"		46	.348		39	16		55	68
3240	11-1/2"		45	.356		41	16.35		57.35	71
3250	12"		44	.364		41	16.75		57.75	71.50
3260	13"		43	.372		43	17.15		60.15	74
3270	14"	↓	42	.381	↓	45.50	17.55		63.05	77.50

22 07 Plumbing Insulation

22 07 19 – Plumbing Piping Insulation

22 07 19.40 Pipe Insulation Protective Jacketing, Aluminum

		Crew	Daily Output	Labor-Hours	Unit	Material	2014 Bare Costs Labor	Equipment	Total	Total Incl O&P
3280	15"	Q-14	41	.390	Ea.	49	17.95		66.95	82
3290	16"		40	.400		50.50	18.40		68.90	84
3300	17"		39	.410		58	18.90		76.90	93
3310	18"		38	.421		60	19.40		79.40	96
3320	19"		37	.432		66.50	19.90		86.40	104
3330	20"		36	.444		68	20.50		88.50	107
3340	22"		35	.457		86.50	21		107.50	128
3350	23"		34	.471		89	21.50		110.50	131
3360	24"		31	.516		91	24		115	137

22 07 19.50 Pipe Insulation Protective Jacketing, St. Stl.

		Crew	Daily Output	Labor-Hours	Unit	Material	2014 Bare Costs Labor	Equipment	Total	Total Incl O&P
0010	**PIPE INSULATION PROTECTIVE JACKETING, STAINLESS STEEL**									
0100	Metal roll jacketing									
0120	Type 304 with moisture barrier									
0140	Smooth, based on OD of insulation, .010" thick									
0260	2-1/2" ID	Q-14	250	.064	L.F.	2.12	2.95		5.07	6.90
0270	2-3/4" ID		245	.065		2.31	3.01		5.32	7.20
0280	3" ID		240	.067		2.50	3.07		5.57	7.55
0290	3-1/4" ID		235	.068		2.69	3.13		5.82	7.85
0300	3-1/2" ID		230	.070		2.88	3.20		6.08	8.15
0310	3-3/4" ID		225	.071		3.07	3.27		6.34	8.50
0320	4" ID		220	.073		3.26	3.35		6.61	8.80
0330	4-1/4" ID		215	.074		3.44	3.43		6.87	9.15
0340	4-1/2" ID		210	.076		3.63	3.51		7.14	9.45
0350	5" ID		200	.080		4.01	3.68		7.69	10.15
0360	5-1/2" ID		190	.084		4.39	3.88		8.27	10.90
0370	6" ID		180	.089		4.76	4.09		8.85	11.60
0380	6-1/2" ID		175	.091		5.15	4.21		9.36	12.20
0390	7" ID		170	.094		5.50	4.33		9.83	12.80
0400	7-1/2" ID		164	.098		5.90	4.49		10.39	13.50
0410	8" ID		161	.099		6.25	4.57		10.82	14
0420	8-1/2" ID		158	.101		6.65	4.66		11.31	14.55
0430	9" ID		155	.103		7.05	4.75		11.80	15.15
0440	9-1/2" ID		152	.105		7.40	4.85		12.25	15.70
0450	10" ID		149	.107		7.80	4.94		12.74	16.25
0460	10-1/2" ID		146	.110		8.15	5.05		13.20	16.85
0470	11" ID		143	.112		8.55	5.15		13.70	17.40
0480	12" ID		137	.117		9.30	5.40		14.70	18.55
0490	13" ID		132	.121		10.05	5.60		15.65	19.75
0500	14" ID		130	.123		10.80	5.65		16.45	20.50
0700	For smooth .016" thick, add					45%	33%			
1000	Stainless steel, Type 316, fitting covers									
1010	Size is based on OD of insulation									
1020	90° LR Elbow, 2 piece									
1100	1-1/2"	Q-14	126	.127	Ea.	12.50	5.85		18.35	23
1110	2-3/4"		123	.130		12.50	6		18.50	23
1120	3"		120	.133		13.05	6.15		19.20	24
1130	3-1/4"		117	.137		13.05	6.30		19.35	24
1140	3-1/2"		115	.139		13.75	6.40		20.15	25
1150	3-3/4"		113	.142		14.50	6.50		21	26
1160	4"		110	.145		15.95	6.70		22.65	28
1170	4-1/4"		108	.148		21	6.80		27.80	34
1180	4-1/2"		106	.151		21	6.95		27.95	34.50

22 07 Plumbing Insulation

22 07 19 – Plumbing Piping Insulation

22 07 19.50 Pipe Insulation Protective Jacketing, St. Stl.		Crew	Daily Output	Labor-Hours	Unit	Material	2014 Bare Costs Labor	Equipment	Total	Total Incl O&P
1190	5"	Q-14	102	.157	Ea.	21.50	7.20		28.70	35.50
1200	5-1/2"		97	.165		32.50	7.60		40.10	47.50
1210	6"		92	.174		35.50	8		43.50	51.50
1220	6-1/2"		87	.184		49	8.45		57.45	66.50
1230	7"		85	.188		49	8.65		57.65	67
1240	7-1/2"		83	.193		57	8.85		65.85	76.50
1250	8"		80	.200		57	9.20		66.20	77
1260	8-1/2"		80	.200		59	9.20		68.20	79.50
1270	9"		78	.205		89.50	9.45		98.95	113
1280	9-1/2"		77	.208		88	9.55		97.55	112
1290	10"		76	.211		88	9.70		97.70	112
1300	10-1/2"		75	.213		105	9.80		114.80	131
1310	11"		74	.216		101	9.95		110.95	127
1320	12"		71	.225		114	10.35		124.35	142
1330	13"		68	.235		158	10.85		168.85	191
1340	14"	▼	66	.242	▼	159	11.15		170.15	192
2000	45° Elbow, 2 piece									
2010	2-1/2"	Q-14	126	.127	Ea.	10.40	5.85		16.25	20.50
2020	2-3/4"		123	.130		10.40	6		16.40	21
2030	3"		120	.133		11.20	6.15		17.35	22
2040	3-1/4"		117	.137		11.20	6.30		17.50	22
2050	3-1/2"		115	.139		11.35	6.40		17.75	22.50
2060	3-3/4"		113	.142		11.35	6.50		17.85	22.50
2070	4"		110	.145		14.60	6.70		21.30	26.50
2080	4-1/4"		108	.148		20.50	6.80		27.30	33
2090	4-1/2"		106	.151		20.50	6.95		27.45	33.50
2100	4-3/4"		104	.154		20.50	7.10		27.60	33.50
2110	5"		102	.157		21	7.20		28.20	34.50
2120	5-1/2"		97	.165		21	7.60		28.60	35
2130	6"		92	.174		24.50	8		32.50	39.50
2140	6-1/2"		87	.184		24.50	8.45		32.95	40
2150	7"		85	.188		42	8.65		50.65	59.50
2160	7-1/2"		83	.193		42.50	8.85		51.35	61
2170	8"		82	.195		42.50	9		51.50	61
2180	8-1/2"		80	.200		50	9.20		59.20	69.50
2190	9"		78	.205		50	9.45		59.45	69.50
2200	9-1/2"		77	.208		58.50	9.55		68.05	79.50
2210	10"		76	.211		58.50	9.70		68.20	79.50
2220	10-1/2"		75	.213		70.50	9.80		80.30	93
2230	11"		74	.216		70.50	9.95		80.45	93
2240	12"		71	.225		76.50	10.35		86.85	100
2250	13"	▼	68	.235	▼	90	10.85		100.85	116

22 11 Facility Water Distribution

22 11 13 – Facility Water Distribution Piping

22 11 13.14 Pipe, Brass

		Crew	Daily Output	Labor-Hours	Unit	Material	2014 Bare Costs Labor	Equipment	Total	Total Incl O&P
0010	**PIPE, BRASS**, Plain end									
0900	Field threaded, coupling & clevis hanger assembly 10' O.C.									
0920	Regular weight									
1120	1/2" diameter	1 Plum	48	.167	L.F.	6.40	9.60		16	21.50
1140	3/4" diameter		46	.174		8.60	10		18.60	24.50
1160	1" diameter		43	.186		12.50	10.70		23.20	30
1180	1-1/4" diameter	Q-1	72	.222		19.10	11.50		30.60	38.50
1200	1-1/2" diameter		65	.246		23	12.75		35.75	44.50
1220	2" diameter		53	.302		32	15.65		47.65	59
1240	2-1/2" diameter		41	.390		50.50	20		70.50	86
1260	3" diameter		31	.516		70.50	26.50		97	118
1300	4" diameter	Q-2	37	.649		125	35		160	190
1930	To delete coupling & hanger, subtract									
1940	1/2" diam.					14%	46%			
1950	3/4" diam. to 1-1/2" diam.					8%	47%			
1960	2" diam. to 4" diam.					10%	37%			

22 11 13.16 Pipe Fittings, Brass

		Crew	Daily Output	Labor-Hours	Unit	Material	2014 Bare Costs Labor	Equipment	Total	Total Incl O&P
0010	**PIPE FITTINGS, BRASS**, Rough bronze, threaded.									
1000	Standard wt., 90° Elbow									
1040	1/8"	1 Plum	13	.615	Ea.	18	35.50		53.50	73.50
1060	1/4"		13	.615		18	35.50		53.50	73.50
1080	3/8"		13	.615		18	35.50		53.50	73.50
1100	1/2"		12	.667		18	38.50		56.50	78
1120	3/4"		11	.727		24	42		66	89.50
1140	1"		10	.800		39	46		85	113
1160	1-1/4"	Q-1	17	.941		63.50	49		112.50	143
1180	1-1/2"		16	1		78.50	52		130.50	164
1200	2"		14	1.143		126	59		185	229
1220	2-1/2"		11	1.455		305	75.50		380.50	450
1240	3"		8	2		465	104		569	670
1260	4"	Q-2	11	2.182		945	117		1,062	1,225
1280	5"		8	3		2,625	161		2,786	3,125
1300	6"		7	3.429		3,850	184		4,034	4,525
1500	45° Elbow, 1/8"	1 Plum	13	.615		22	35.50		57.50	78
1540	1/4"		13	.615		22	35.50		57.50	78
1560	3/8"		13	.615		22	35.50		57.50	78
1580	1/2"		12	.667		22	38.50		60.50	82.50
1600	3/4"		11	.727		31.50	42		73.50	97.50
1620	1"		10	.800		53.50	46		99.50	129
1640	1-1/4"	Q-1	17	.941		85	49		134	167
1660	1-1/2"		16	1		107	52		159	196
1680	2"		14	1.143		173	59		232	281
1700	2-1/2"		11	1.455		330	75.50		405.50	480
1720	3"		8	2		510	104		614	715
1740	4"	Q-2	11	2.182		1,125	117		1,242	1,425
1760	5"		8	3		2,125	161		2,286	2,600
1780	6"		7	3.429		2,950	184		3,134	3,525
2000	Tee, 1/8"	1 Plum	9	.889		21	51		72	101
2040	1/4"		9	.889		21	51		72	101
2060	3/8"		9	.889		21	51		72	101
2080	1/2"		8	1		21	57.50		78.50	111
2100	3/4"		7	1.143		30	66		96	133

22 11 Facility Water Distribution

22 11 13 – Facility Water Distribution Piping

22 11 13.16 Pipe Fittings, Brass		Crew	Daily Output	Labor-Hours	Unit	Material	2014 Bare Costs Labor	Equipment	Total	Total Incl O&P
2120	1"	1 Plum	6	1.333	Ea.	54.50	76.50		131	176
2140	1-1/4"	Q-1	10	1.600		93.50	83		176.50	228
2160	1-1/2"		9	1.778		105	92		197	255
2180	2"		8	2		175	104		279	350
2200	2-1/2"		7	2.286		415	118		533	640
2220	3"		5	3.200		635	166		801	950
2240	4"	Q-2	7	3.429		1,575	184		1,759	2,000
2260	5"		5	4.800		3,275	258		3,533	4,000
2280	6"		4	6		5,150	320		5,470	6,125
2500	Coupling, 1/8"	1 Plum	26	.308		15.05	17.70		32.75	43
2540	1/4"		22	.364		15.05	21		36.05	48
2560	3/8"		18	.444		15.05	25.50		40.55	55
2580	1/2"		15	.533		15.05	30.50		45.55	63
2600	3/4"		14	.571		21	33		54	73
2620	1"		13	.615		36	35.50		71.50	93
2640	1-1/4"	Q-1	22	.727		60	37.50		97.50	124
2660	1-1/2"		20	.800		78.50	41.50		120	149
2680	2"		18	.889		130	46		176	212
2700	2-1/2"		14	1.143		221	59		280	335
2720	3"		10	1.600		305	83		388	460
2740	4"	Q-2	12	2		630	107		737	855
2760	5"		10	2.400		1,175	129		1,304	1,500
2780	6"		9	2.667		1,675	143		1,818	2,075
3000	Union, 125 lb.									
3020	1/8"	1 Plum	12	.667	Ea.	51.50	38.50		90	115
3040	1/4"		12	.667		51.50	38.50		90	115
3060	3/8"		12	.667		51.50	38.50		90	115
3080	1/2"		11	.727		51.50	42		93.50	120
3100	3/4"		10	.800		70.50	46		116.50	147
3120	1"		9	.889		106	51		157	194
3140	1-1/4"	Q-1	16	1		154	52		206	247
3160	1-1/2"		15	1.067		183	55.50		238.50	285
3180	2"		13	1.231		247	64		311	370
3200	2-1/2"		10	1.600		675	83		758	865
3220	3"		7	2.286		1,050	118		1,168	1,325
3240	4"	Q-2	10	2.400		3,250	129		3,379	3,775

22 11 13.23 Pipe/Tube, Copper										
0010	**PIPE/TUBE, COPPER**, Solder joints R221113-50									
0100	Solder									
0120	Solder, lead free, roll				Lb.	10.10			10.10	11.15
1000	Type K tubing, couplings & clevis hanger assemblies 10' O.C.									
1100	1/4" diameter R221113-70	1 Plum	84	.095	L.F.	4.62	5.50		10.12	13.40
1120	3/8" diameter		82	.098		3.99	5.60		9.59	12.90
1140	1/2" diameter		78	.103		4.47	5.90		10.37	13.80
1160	5/8" diameter		77	.104		5.95	6		11.95	15.55
1180	3/4" diameter		74	.108		7.80	6.20		14	18
1200	1" diameter		66	.121		10.60	7		17.60	22
1220	1-1/4" diameter		56	.143		13.25	8.20		21.45	27
1240	1-1/2" diameter		50	.160		17.15	9.20		26.35	33
1260	2" diameter		40	.200		26.50	11.50		38	46.50
1280	2-1/2" diameter	Q-1	60	.267		41	13.80		54.80	66.50
1300	3" diameter		54	.296		58	15.35		73.35	86.50

145

22 11 Facility Water Distribution

22 11 13 – Facility Water Distribution Piping

22 11 13.23 Pipe/Tube, Copper

		Crew	Daily Output	Labor-Hours	Unit	Material	2014 Bare Costs Labor	Equipment	Total	Total Incl O&P
1320	3-1/2" diameter	Q-1	42	.381	L.F.	82	19.75		101.75	120
1330	4" diameter		38	.421		101	22		123	144
1340	5" diameter		32	.500		203	26		229	262
1360	6" diameter	Q-2	38	.632		305	34		339	385
1380	8" diameter	"	34	.706		480	38		518	590
1390	For other than full hard temper, add					13%				
1440	For silver solder, add						15%			
1800	For medical clean, (oxygen class), add					12%				
1950	To delete cplgs. & hngrs., 1/4"-1" pipe, subtract					27%	60%			
1960	1-1/4"-3" pipe, subtract					14%	52%			
1970	3-1/2"-5" pipe, subtract					10%	60%			
1980	6"-8" pipe, subtract					19%	53%			
2000	Type L tubing, couplings & clevis hanger assemblies 10' O.C.									
2100	1/4" diameter	1 Plum	88	.091	L.F.	2.30	5.25		7.55	10.45
2120	3/8" diameter		84	.095		3.13	5.50		8.63	11.75
2140	1/2" diameter		81	.099		3.37	5.70		9.07	12.30
2160	5/8" diameter		79	.101		5.35	5.85		11.20	14.65
2180	3/4" diameter		76	.105		5.15	6.05		11.20	14.80
2200	1" diameter		68	.118		7.85	6.75		14.60	18.85
2220	1-1/4" diameter		58	.138		11.25	7.95		19.20	24.50
2240	1-1/2" diameter		52	.154		14.35	8.85		23.20	29
2260	2" diameter		42	.190		22	10.95		32.95	40.50
2280	2-1/2" diameter	Q-1	62	.258		35.50	13.35		48.85	59
2300	3" diameter		56	.286		48.50	14.80		63.30	76
2320	3-1/2" diameter		43	.372		70.50	19.25		89.75	107
2340	4" diameter		39	.410		85.50	21.50		107	126
2360	5" diameter		34	.471		168	24.50		192.50	222
2380	6" diameter	Q-2	40	.600		244	32		276	315
2400	8" diameter	"	36	.667		370	36		406	465
2410	For other than full hard temper, add					21%				
2590	For silver solder, add						15%			
2900	For medical clean, (oxygen class), add					12%				
2940	To delete cplgs. & hngrs., 1/4"-1" pipe, subtract					37%	63%			
2960	1-1/4"-3" pipe, subtract					12%	53%			
2970	3-1/2"-5" pipe, subtract					12%	63%			
2980	6"-8" pipe, subtract					24%	55%			
3000	Type M tubing, couplings & clevis hanger assemblies 10' O.C.									
3100	1/4" diameter	1 Plum	90	.089	L.F.	2.58	5.10		7.68	10.55
3120	3/8" diameter		87	.092		2.65	5.30		7.95	10.90
3140	1/2" diameter		84	.095		2.65	5.50		8.15	11.20
3160	5/8" diameter		81	.099		4.30	5.70		10	13.35
3180	3/4" diameter		78	.103		3.99	5.90		9.89	13.30
3200	1" diameter		70	.114		6.50	6.60		13.10	17.10
3220	1-1/4" diameter		60	.133		9.55	7.65		17.20	22
3240	1-1/2" diameter		54	.148		12.85	8.55		21.40	27
3260	2" diameter		44	.182		20	10.45		30.45	38
3280	2-1/2" diameter	Q-1	64	.250		31.50	12.95		44.45	54
3300	3" diameter		58	.276		43	14.30		57.30	69
3320	3-1/2" diameter		45	.356		64.50	18.40		82.90	98.50
3340	4" diameter		40	.400		79	20.50		99.50	119
3360	5" diameter		36	.444		168	23		191	220
3370	6" diameter	Q-2	42	.571		244	30.50		274.50	315
3380	8" diameter	"	38	.632		370	34		404	460

22 11 Facility Water Distribution

22 11 13 – Facility Water Distribution Piping

22 11 13.23 Pipe/Tube, Copper

		Crew	Daily Output	Labor-Hours	Unit	Material	Labor	Equipment	Total	Total Incl O&P
3440	For silver solder, add						15%			
3960	To delete cplgs. & hngrs., 1/4"-1" pipe, subtract					35%	65%			
3970	1-1/4"-3" pipe, subtract					19%	56%			
3980	3-1/2"-5" pipe, subtract					13%	65%			
3990	6"-8" pipe, subtract					28%	58%			
4000	Type DWV tubing, couplings & clevis hanger assemblies 10' O.C.									
4100	1-1/4" diameter	1 Plum	60	.133	L.F.	9.95	7.65		17.60	22.50
4120	1-1/2" diameter		54	.148		12.45	8.55		21	26.50
4140	2" diameter		44	.182		17.05	10.45		27.50	34.50
4160	3" diameter	Q-1	58	.276		35	14.30		49.30	60
4180	4" diameter		40	.400		62	20.50		82.50	99.50
4200	5" diameter		36	.444		154	23		177	205
4220	6" diameter	Q-2	42	.571		230	30.50		260.50	300
4240	8" diameter	"	38	.632		540	34		574	640
4730	To delete cplgs. & hngrs., 1-1/4"-2" pipe, subtract					16%	53%			
4740	3"-4" pipe, subtract					13%	60%			
4750	5"-8" pipe, subtract					23%	58%			
5200	ACR tubing, type L, hard temper, cleaned and									
5220	capped, no couplings or hangers									
5240	3/8" OD				L.F.	2.47			2.47	2.72
5250	1/2" OD					3.02			3.02	3.32
5260	5/8" OD					4.26			4.26	4.69
5270	3/4" OD					4.73			4.73	5.20
5280	7/8" OD					4.73			4.73	5.20
5290	1-1/8" OD					6.85			6.85	7.55
5300	1-3/8" OD					9.25			9.25	10.15
5310	1-5/8" OD					11.80			11.80	12.95
5320	2-1/8" OD					27.50			27.50	30.50
5330	2-5/8" OD					27.50			27.50	30.50
5340	3-1/8" OD					37			37	40.50
5350	3-5/8" OD					48.50			48.50	53
5360	4-1/8" OD					61.50			61.50	67.50
5380	ACR tubing, type L, hard, cleaned and capped									
5381	No couplings or hangers									
5384	3/8"	1 Stpi	160	.050	L.F.	2.47	2.93		5.40	7.15
5385	1/2"		160	.050		3.02	2.93		5.95	7.75
5386	5/8"		160	.050		4.26	2.93		7.19	9.10
5387	3/4"		130	.062		4.73	3.60		8.33	10.65
5388	7/8"		130	.062		4.73	3.60		8.33	10.65
5389	1-1/8"		115	.070		6.85	4.07		10.92	13.70
5390	1-3/8"		100	.080		9.25	4.68		13.93	17.20
5391	1-5/8"		90	.089		11.80	5.20		17	21
5392	2-1/8"		80	.100		27.50	5.85		33.35	39.50
5393	2-5/8"	Q-5	125	.128		27.50	6.75		34.25	40.50
5394	3-1/8"		105	.152		37	8		45	52.50
5395	4-1/8"		95	.168		61.50	8.85		70.35	81
5800	Refrigeration tubing, dryseal, 50' coils									
5840	1/8" OD				Coil	34.50			34.50	38
5850	3/16" OD					40.50			40.50	44.50
5860	1/4" OD					46.50			46.50	51
5870	5/16" OD					62.50			62.50	68.50
5880	3/8" OD					67.50			67.50	74.50
5890	1/2" OD					93.50			93.50	103

22 11 Facility Water Distribution

22 11 13 – Facility Water Distribution Piping

22 11 13.23 Pipe/Tube, Copper

		Crew	Daily Output	Labor-Hours	Unit	Material	2014 Bare Costs Labor	Equipment	Total	Total Incl O&P
5900	5/8" OD				Coil	125			125	138
5910	3/4" OD					150			150	165
5920	7/8" OD					305			305	335
5930	1-1/8" OD					410			410	450
5940	1-3/8" OD					555			555	610
5950	1-5/8" OD				▼	710			710	780

22 11 13.25 Pipe/Tube Fittings, Copper

		Crew	Daily Output	Labor-Hours	Unit	Material	Labor	Equipment	Total	Total Incl O&P
0010	**PIPE/TUBE FITTINGS, COPPER**, Wrought unless otherwise noted									
0040	Solder joints, copper x copper									
0070	90° elbow, 1/4"	1 Plum	22	.364	Ea.	9.70	21		30.70	42
0090	3/8"		22	.364		9.25	21		30.25	41.50
0100	1/2"		20	.400		3.08	23		26.08	38.50
0110	5/8"		19	.421		10.65	24		34.65	48
0120	3/4"		19	.421		6.95	24		30.95	44
0130	1"		16	.500		17	29		46	62
0140	1-1/4"		15	.533		25.50	30.50		56	74.50
0150	1-1/2"		13	.615		40.50	35.50		76	98
0160	2"	▼	11	.727		72.50	42		114.50	143
0170	2-1/2"	Q-1	13	1.231		146	64		210	258
0180	3"		11	1.455		195	75.50		270.50	330
0190	3-1/2"		10	1.600		685	83		768	875
0200	4"		9	1.778		500	92		592	690
0210	5"	▼	6	2.667		2,100	138		2,238	2,500
0220	6"	Q-2	9	2.667		2,800	143		2,943	3,300
0230	8"	"	8	3		10,300	161		10,461	11,600
0250	45° elbow, 1/4"	1 Plum	22	.364		17.30	21		38.30	50.50
0270	3/8"		22	.364		14.80	21		35.80	48
0280	1/2"		20	.400		5.65	23		28.65	41.50
0290	5/8"		19	.421		27	24		51	66
0300	3/4"		19	.421		9.90	24		33.90	47.50
0310	1"		16	.500		25	29		54	71
0320	1-1/4"		15	.533		33.50	30.50		64	83.50
0330	1-1/2"		13	.615		40.50	35.50		76	98
0340	2"	▼	11	.727		67.50	42		109.50	138
0350	2-1/2"	Q-1	13	1.231		144	64		208	256
0360	3"		13	1.231		214	64		278	335
0370	3-1/2"		10	1.600		600	83		683	785
0380	4"		9	1.778		455	92		547	640
0390	5"	▼	6	2.667		1,775	138		1,913	2,150
0400	6"	Q-2	9	2.667		2,775	143		2,918	3,300
0410	8"	"	8	3		9,450	161		9,611	10,600
0450	Tee, 1/4"	1 Plum	14	.571		19.40	33		52.40	71
0470	3/8"		14	.571		15.60	33		48.60	66.50
0480	1/2"		13	.615		5.25	35.50		40.75	59.50
0490	5/8"		12	.667		34	38.50		72.50	95.50
0500	3/4"		12	.667		12.70	38.50		51.20	72
0510	1"		10	.800		39.50	46		85.50	113
0520	1-1/4"		9	.889		53.50	51		104.50	136
0530	1-1/2"		8	1		82.50	57.50		140	178
0540	2"	▼	7	1.143		129	66		195	242
0550	2-1/2"	Q-1	8	2		260	104		364	440
0560	3"	▼	7	2.286	▼	395	118		513	615

22 11 Facility Water Distribution

22 11 13 – Facility Water Distribution Piping

22 11 13.25 Pipe/Tube Fittings, Copper		Crew	Daily Output	Labor-Hours	Unit	Material	2014 Bare Costs Labor	Equipment	Total	Total Incl O&P
0570	3-1/2"	Q-1	6	2.667	Ea.	1,150	138		1,288	1,475
0580	4"		5	3.200		960	166		1,126	1,300
0590	5"	↓	4	4		3,150	207		3,357	3,775
0600	6"	Q-2	6	4		4,300	215		4,515	5,050
0610	8"	"	5	4.800		16,600	258		16,858	18,700
0612	Tee, reducing on the outlet, 1/4"	1 Plum	15	.533		33	30.50		63.50	82.50
0613	3/8"		15	.533		31	30.50		61.50	80.50
0614	1/2"		14	.571		27	33		60	79.50
0615	5/8"		13	.615		54.50	35.50		90	114
0616	3/4"		12	.667		15	38.50		53.50	74.50
0617	1"		11	.727		39.50	42		81.50	107
0618	1-1/4"		10	.800		58	46		104	133
0619	1-1/2"		9	.889		61.50	51		112.50	145
0620	2"	↓	8	1		97.50	57.50		155	194
0621	2-1/2"	Q-1	9	1.778		305	92		397	475
0622	3"		8	2		365	104		469	560
0623	4"		6	2.667		685	138		823	960
0624	5"	↓	5	3.200		3,150	166		3,316	3,700
0625	6"	Q-2	7	3.429		4,075	184		4,259	4,775
0626	8"	"	6	4		16,600	215		16,815	18,500
0630	Tee, reducing on the run, 1/4"	1 Plum	15	.533		41	30.50		71.50	92
0631	3/8"		15	.533		47	30.50		77.50	98
0632	1/2"		14	.571		37.50	33		70.50	90.50
0633	5/8"		13	.615		55	35.50		90.50	114
0634	3/4"		12	.667		15	38.50		53.50	74.50
0635	1"		11	.727		47	42		89	115
0636	1-1/4"		10	.800		74.50	46		120.50	152
0637	1-1/2"		9	.889		123	51		174	212
0638	2"	↓	8	1		168	57.50		225.50	271
0639	2-1/2"	Q-1	9	1.778		415	92		507	600
0640	3"		8	2		560	104		664	770
0641	4"		6	2.667		1,275	138		1,413	1,600
0642	5"	↓	5	3.200		2,975	166		3,141	3,525
0643	6"	Q-2	7	3.429		4,550	184		4,734	5,275
0644	8"	"	6	4		16,100	215		16,315	18,000
0650	Coupling, 1/4"	1 Plum	24	.333		2.25	19.20		21.45	31.50
0670	3/8"		24	.333		2.98	19.20		22.18	32.50
0680	1/2"		22	.364		2.34	21		23.34	34
0690	5/8"		21	.381		7.05	22		29.05	41
0700	3/4"		21	.381		4.70	22		26.70	38
0710	1"		18	.444		9.60	25.50		35.10	49
0715	1-1/4"		17	.471		16.60	27		43.60	59.50
0716	1-1/2"		15	.533		22	30.50		52.50	70.50
0718	2"	↓	13	.615		36.50	35.50		72	94
0721	2-1/2"	Q-1	15	1.067		78	55.50		133.50	169
0722	3"		13	1.231		117	64		181	226
0724	3-1/2"		8	2		224	104		328	405
0726	4"		7	2.286		246	118		364	450
0728	5"	↓	6	2.667		605	138		743	875
0731	6"	Q-2	8	3		1,000	161		1,161	1,350
0732	8"	"	7	3.429	↓	3,250	184		3,434	3,850
0741	Coupling, reducing, concentric									
0743	1/2"	1 Plum	23	.348	Ea.	6.10	20		26.10	36.50

149

22 11 Facility Water Distribution

22 11 13 – Facility Water Distribution Piping

22 11 13.25 Pipe/Tube Fittings, Copper		Crew	Daily Output	Labor-Hours	Unit	Material	2014 Bare Costs Labor	Equipment	Total	Total Incl O&P
0745	3/4"	1 Plum	21.50	.372	Ea.	11.75	21.50		33.25	45.50
0747	1"		19.50	.410		15.80	23.50		39.30	53
0748	1-1/4"		18	.444		23.50	25.50		49	64.50
0749	1-1/2"		16	.500		31	29		60	77.50
0751	2"		14	.571		65	33		98	121
0752	2-1/2"	▼	13	.615		134	35.50		169.50	202
0753	3"	Q-1	14	1.143		162	59		221	268
0755	4"	"	8	2		340	104		444	530
0757	5"	Q-2	7.50	3.200		1,800	172		1,972	2,250
0759	6"		7	3.429		2,875	184		3,059	3,450
0761	8"	▼	6.50	3.692	▼	5,975	198		6,173	6,850
0771	Cap, sweat									
0773	1/2"	1 Plum	40	.200	Ea.	2.25	11.50		13.75	19.90
0775	3/4"		38	.211		4.18	12.10		16.28	23
0777	1"		32	.250		9.90	14.40		24.30	32.50
0778	1-1/4"		29	.276		13.25	15.90		29.15	38.50
0779	1-1/2"		26	.308		19.40	17.70		37.10	48
0781	2"	▼	22	.364	▼	36	21		57	71
0791	Flange, sweat									
0793	3"	Q-1	22	.727	Ea.	495	37.50		532.50	600
0795	4"		18	.889		690	46		736	830
0797	5"	▼	12	1.333		1,275	69		1,344	1,525
0799	6"	Q-2	18	1.333		1,350	71.50		1,421.50	1,575
0801	8"	"	16	1.500		3,250	80.50		3,330.50	3,700
0850	Unions, 1/4"	1 Plum	21	.381		70.50	22		92.50	111
0870	3/8"		21	.381		71	22		93	111
0880	1/2"		19	.421		38	24		62	78
0890	5/8"		18	.444		161	25.50		186.50	216
0900	3/4"		18	.444		47.50	25.50		73	91
0910	1"		15	.533		81.50	30.50		112	137
0920	1-1/4"		14	.571		142	33		175	206
0930	1-1/2"		12	.667		154	38.50		192.50	228
0940	2"	▼	10	.800		320	46		366	420
0950	2-1/2"	Q-1	12	1.333		695	69		764	870
0960	3"	"	10	1.600		1,800	83		1,883	2,100
0980	Adapter, copper x male IPS, 1/4"	1 Plum	20	.400		32	23		55	70.50
0990	3/8"		20	.400		15.95	23		38.95	52.50
1000	1/2"		18	.444		6.45	25.50		31.95	45.50
1010	3/4"		17	.471		10.85	27		37.85	53
1020	1"		15	.533		28	30.50		58.50	77.50
1030	1-1/4"		13	.615		41.50	35.50		77	99
1040	1-1/2"		12	.667		47.50	38.50		86	111
1050	2"	▼	11	.727		80.50	42		122.50	152
1060	2-1/2"	Q-1	10.50	1.524		310	79		389	460
1070	3"		10	1.600		410	83		493	575
1080	3-1/2"		9	1.778		485	92		577	675
1090	4"		8	2		555	104		659	765
1200	5"	▼	6	2.667		3,275	138		3,413	3,800
1210	6"	Q-2	8.50	2.824	▼	3,650	152		3,802	4,225
1214	Adapter, copper x female IPS									
1216	1/2"	1 Plum	18	.444	Ea.	10.20	25.50		35.70	49.50
1218	3/4"		17	.471		14.05	27		41.05	56.50
1220	1"	▼	15	.533		32.50	30.50		63	82.50

22 11 Facility Water Distribution

22 11 13 – Facility Water Distribution Piping

22 11 13.25	Pipe/Tube Fittings, Copper	Crew	Daily Output	Labor-Hours	Unit	Material	2014 Bare Costs Labor	Equipment	Total	Total Incl O&P
1221	1-1/4"	1 Plum	13	.615	Ea.	47.50	35.50		83	106
1222	1-1/2"		12	.667		74.50	38.50		113	140
1224	2"		11	.727		101	42		143	174
1250	Cross, 1/2"		10	.800		51.50	46		97.50	126
1260	3/4"		9.50	.842		99.50	48.50		148	183
1270	1"		8	1		169	57.50		226.50	273
1280	1-1/4"		7.50	1.067		243	61.50		304.50	360
1290	1-1/2"		6.50	1.231		345	71		416	485
1300	2"	↓	5.50	1.455		655	83.50		738.50	845
1310	2-1/2"	Q-1	6.50	2.462		1,500	128		1,628	1,850
1320	3"	"	5.50	2.909	↓	1,775	151		1,926	2,175
1500	Tee fitting, mechanically formed, (Type 1, 'branch sizes up to 2 in.')									
1520	1/2" run size, 3/8" to 1/2" branch size	1 Plum	80	.100	Ea.		5.75		5.75	8.70
1530	3/4" run size, 3/8" to 3/4" branch size		60	.133			7.65		7.65	11.60
1540	1" run size, 3/8" to 1" branch size		54	.148			8.55		8.55	12.85
1550	1-1/4" run size, 3/8" to 1-1/4" branch size		48	.167			9.60		9.60	14.50
1560	1-1/2" run size, 3/8" to 1-1/2" branch size		40	.200			11.50		11.50	17.40
1570	2" run size, 3/8" to 2" branch size		35	.229			13.15		13.15	19.85
1580	2-1/2" run size, 1/2" to 2" branch size		32	.250			14.40		14.40	21.50
1590	3" run size, 1" to 2" branch size		26	.308			17.70		17.70	26.50
1600	4" run size, 1" to 2" branch size	↓	24	.333	↓		19.20		19.20	29
1640	Tee fitting, mechanically formed, (Type 2, branches 2-1/2" thru 4")									
1650	2-1/2" run size, 2-1/2" branch size	1 Plum	12.50	.640	Ea.		37		37	55.50
1660	3" run size, 2-1/2" to 3" branch size		12	.667			38.50		38.50	58
1670	3-1/2" run size, 2-1/2" to 3-1/2" branch size		11	.727			42		42	63
1680	4" run size, 2-1/2" to 4" branch size		10.50	.762			44		44	66
1698	5" run size, 2" to 4" branch size		9.50	.842			48.50		48.50	73
1700	6" run size, 2" to 4" branch size		8.50	.941			54		54	82
1710	8" run size, 2" to 4" branch size	↓	7	1.143	↓		66		66	99.50
1800	ACR fittings, OD size									
1802	Tee, straight									
1808	5/8"	1 Stpi	12	.667	Ea.	5.25	39		44.25	65
1810	3/4"		12	.667		34	39		73	96.50
1812	7/8"		10	.800		12.70	47		59.70	84.50
1813	1"		10	.800		92	47		139	172
1814	1-1/8"		10	.800		39.50	47		86.50	114
1816	1-3/8"		9	.889		53.50	52		105.50	138
1818	1-5/8"		8	1		82.50	58.50		141	180
1820	2-1/8"	↓	7	1.143		129	67		196	243
1822	2-5/8"	Q-5	8	2		260	105		365	445
1824	3-1/8"		7	2.286		395	120		515	615
1826	4-1/8"	↓	5	3.200	↓	960	168		1,128	1,300
1830	90° elbow									
1836	5/8"	1 Stpi	19	.421	Ea.	10.35	24.50		34.85	48.50
1838	3/4"		19	.421		18.75	24.50		43.25	57.50
1840	7/8"		16	.500		18.65	29.50		48.15	64.50
1842	1-1/8"		16	.500		25	29.50		54.50	71.50
1844	1-3/8"		15	.533		25.50	31		56.50	75
1846	1-5/8"		13	.615		40	36		76	98.50
1848	2-1/8"	↓	11	.727		73	42.50		115.50	145
1850	2-5/8"	Q-5	13	1.231		146	65		211	259
1852	3-1/8"		11	1.455		195	76.50		271.50	330
1854	4-1/8"	↓	9	1.778	↓	500	93.50		593.50	690

22 11 Facility Water Distribution

22 11 13 – Facility Water Distribution Piping

22 11 13.25 Pipe/Tube Fittings, Copper		Crew	Daily Output	Labor-Hours	Unit	Material	2014 Bare Costs Labor	Equipment	Total	Total Incl O&P
1860	Coupling									
1866	5/8"	1 Stpi	21	.381	Ea.	2.34	22.50		24.84	36
1868	3/4"		21	.381		6.70	22.50		29.20	41
1870	7/8"		18	.444		4.70	26		30.70	44.50
1871	1"		18	.444		15.95	26		41.95	57
1872	1-1/8"		18	.444		9.45	26		35.45	50
1874	1-3/8"		17	.471		16.60	27.50		44.10	60
1876	1-5/8"		15	.533		22	31		53	71
1878	2-1/8"		13	.615		36.50	36		72.50	95
1880	2-5/8"	Q-5	15	1.067		82.50	56		138.50	176
1882	3-1/8"		13	1.231		120	65		185	230
1884	4-1/8"		7	2.286		276	120		396	485
2000	DWV, solder joints, copper x copper									
2030	90° Elbow, 1-1/4"	1 Plum	13	.615	Ea.	29.50	35.50		65	86
2050	1-1/2"		12	.667		44.50	38.50		83	107
2070	2"		10	.800		58.50	46		104.50	134
2090	3"	Q-1	10	1.600		144	83		227	284
2100	4"	"	9	1.778		1,000	92		1,092	1,250
2150	45° Elbow, 1-1/4"	1 Plum	13	.615		27.50	35.50		63	83.50
2170	1-1/2"		12	.667		22.50	38.50		61	83
2180	2"		10	.800		52	46		98	127
2190	3"	Q-1	10	1.600		111	83		194	247
2200	4"	"	9	1.778		515	92		607	705
2250	Tee, Sanitary, 1-1/4"	1 Plum	9	.889		59	51		110	142
2270	1-1/2"		8	1		73.50	57.50		131	168
2290	2"		7	1.143		85.50	66		151.50	194
2310	3"	Q-1	7	2.286		370	118		488	590
2330	4"	"	6	2.667		945	138		1,083	1,250
2400	Coupling, 1-1/4"	1 Plum	14	.571		13.95	33		46.95	65
2420	1-1/2"		13	.615		17.35	35.50		52.85	72.50
2440	2"		11	.727		24	42		66	89.50
2460	3"	Q-1	11	1.455		49	75.50		124.50	168
2480	4"	"	10	1.600		149	83		232	288
2602	Traps, see Section 22 13 16.60									
3500	Compression joint fittings									
3510	As used for plumbing and oil burner work									
3520	Fitting price includes nuts and sleeves									
3540	Sleeve, 1/8"				Ea.	.22			.22	.24
3550	3/16"					.22			.22	.24
3560	1/4"					.08			.08	.09
3570	5/16"					.23			.23	.25
3580	3/8"					.46			.46	.51
3600	1/2"					.61			.61	.67
3620	Nut, 1/8"					.32			.32	.35
3630	3/16"					.34			.34	.37
3640	1/4"					.33			.33	.36
3650	5/16"					.49			.49	.54
3660	3/8"					.54			.54	.59
3670	1/2"					.95			.95	1.05
3710	Union, 1/8"	1 Plum	26	.308		2	17.70		19.70	28.50
3720	3/16"		24	.333		1.92	19.20		21.12	31
3730	1/4"		24	.333		2.01	19.20		21.21	31
3740	5/16"		23	.348		2.58	20		22.58	33

22 11 Facility Water Distribution

22 11 13 – Facility Water Distribution Piping

22 11 13.25	Pipe/Tube Fittings, Copper	Crew	Daily Output	Labor-Hours	Unit	Material	2014 Bare Costs Labor	Equipment	Total	Total Incl O&P
3750	3/8"	1 Plum	22	.364	Ea.	2.60	21		23.60	34.50
3760	1/2"		22	.364		4	21		25	36
3780	5/8"		21	.381		4.43	22		26.43	38
3820	Union tee, 1/8"		17	.471		7.60	27		34.60	49.50
3830	3/16"		16	.500		4.42	29		33.42	48.50
3840	1/4"		15	.533		4	30.50		34.50	51
3850	5/16"		15	.533		5.65	30.50		36.15	52.50
3860	3/8"		15	.533		5.70	30.50		36.20	53
3870	1/2"		15	.533		9.65	30.50		40.15	57
3910	Union elbow, 1/4"		24	.333		2.47	19.20		21.67	31.50
3920	5/16"		23	.348		4.34	20		24.34	35
3930	3/8"		22	.364		4.19	21		25.19	36
3940	1/2"		22	.364		6.85	21		27.85	39
3980	Female connector, 1/8"		26	.308		1.55	17.70		19.25	28
4000	3/16" x 1/8"		24	.333		1.98	19.20		21.18	31
4010	1/4" x 1/8"		24	.333		1.75	19.20		20.95	31
4020	1/4"		24	.333		2.35	19.20		21.55	31.50
4030	3/8" x 1/4"		22	.364		2.35	21		23.35	34
4040	1/2" x 3/8"		22	.364		3.75	21		24.75	35.50
4050	5/8" x 1/2"		21	.381		5.30	22		27.30	39
4090	Male connector, 1/8"		26	.308		1.25	17.70		18.95	28
4100	3/16" x 1/8"		24	.333		1.29	19.20		20.49	30.50
4110	1/4" x 1/8"		24	.333		1.34	19.20		20.54	30.50
4120	1/4"		24	.333		1.34	19.20		20.54	30.50
4130	5/16" x 1/8"		23	.348		1.58	20		21.58	31.50
4140	5/16" x 1/4"		23	.348		1.79	20		21.79	32
4150	3/8" x 1/8"		22	.364		1.87	21		22.87	33.50
4160	3/8" x 1/4"		22	.364		2.15	21		23.15	34
4170	3/8"		22	.364		2.15	21		23.15	34
4180	3/8" x 1/2"		22	.364		2.95	21		23.95	35
4190	1/2" x 3/8"		22	.364		2.86	21		23.86	34.50
4200	1/2"		22	.364		3.21	21		24.21	35
4210	5/8" x 1/2"		21	.381		3.71	22		25.71	37
4240	Male elbow, 1/8"		26	.308		2.58	17.70		20.28	29.50
4250	3/16" x 1/8"		24	.333		2.85	19.20		22.05	32
4260	1/4" x 1/8"		24	.333		2.15	19.20		21.35	31.50
4270	1/4"		24	.333		2.45	19.20		21.65	31.50
4280	3/8" x 1/4"		22	.364		3.11	21		24.11	35
4290	3/8"		22	.364		5.35	21		26.35	37.50
4300	1/2" x 1/4"		22	.364		5.15	21		26.15	37
4310	1/2" x 3/8"		22	.364		8.10	21		29.10	40.50
4340	Female elbow, 1/8"		26	.308		5.55	17.70		23.25	32.50
4350	1/4" x 1/8"		24	.333		3.15	19.20		22.35	32.50
4360	1/4"		24	.333		4.13	19.20		23.33	33.50
4370	3/8" x 1/4"		22	.364		4.07	21		25.07	36
4380	1/2" x 3/8"		22	.364		6.95	21		27.95	39
4390	1/2"		22	.364		9.35	21		30.35	42
4420	Male run tee, 1/4" x 1/8"		15	.533		3.86	30.50		34.36	51
4430	5/16" x 1/8"		15	.533		6.65	30.50		37.15	54
4440	3/8" x 1/4"		15	.533		5.95	30.50		36.45	53
4480	Male branch tee, 1/4" x 1/8"		15	.533		4.47	30.50		34.97	51.50
4490	1/4"		15	.533		4.81	30.50		35.31	52
4500	3/8" x 1/4"		15	.533		5.60	30.50		36.10	52.50

153

22 11 Facility Water Distribution

22 11 13 – Facility Water Distribution Piping

22 11 13.25	Pipe/Tube Fittings, Copper	Crew	Daily Output	Labor-Hours	Unit	Material	2014 Bare Costs Labor	Equipment	Total	Total Incl O&P
4510	1/2" x 3/8"	1 Plum	15	.533	Ea.	8.70	30.50		39.20	56
4520	1/2"	↓	15	.533	↓	13.95	30.50		44.45	62
4800	Flare joint fittings									
4810	Refrigeration fittings									
4820	Flare joint nuts and labor not incl. in price. Add 1 nut per jnt.									
4830	90° Elbow, 1/4"				Ea.	2.32			2.32	2.55
4840	3/8"					3.32			3.32	3.65
4850	1/2"					6.20			6.20	6.85
4860	5/8"					9.10			9.10	10
4870	3/4"					13.80			13.80	15.15
5030	Tee, 1/4"					2.99			2.99	3.29
5040	5/16"					3.58			3.58	3.94
5050	3/8"					3.89			3.89	4.28
5060	1/2"					5.85			5.85	6.45
5070	5/8"					7.45			7.45	8.20
5080	3/4"					8.50			8.50	9.35
5140	Union, 3/16"					1.37			1.37	1.51
5150	1/4"					.94			.94	1.03
5160	5/16"					1.70			1.70	1.87
5170	3/8"					1.58			1.58	1.74
5180	1/2"					2.35			2.35	2.59
5190	5/8"					3.78			3.78	4.16
5200	3/4"					10.55			10.55	11.60
5260	Long flare nut, 3/16"	1 Stpi	42	.190		1.74	11.15		12.89	18.75
5270	1/4"		41	.195		.96	11.40		12.36	18.30
5280	5/16"		40	.200		1.74	11.70		13.44	19.55
5290	3/8"		39	.205		2.09	12		14.09	20.50
5300	1/2"		38	.211		3.34	12.30		15.64	22.50
5310	5/8"		37	.216		5.85	12.65		18.50	25.50
5320	3/4"		34	.235		11.65	13.75		25.40	34
5380	Short flare nut, 3/16"		42	.190		2.88	11.15		14.03	20
5390	1/4"		41	.195		.68	11.40		12.08	18
5400	5/16"		40	.200		.81	11.70		12.51	18.55
5410	3/8"		39	.205		1.06	12		13.06	19.25
5420	1/2"		38	.211		1.58	12.30		13.88	20.50
5430	5/8"		36	.222		2.20	13		15.20	22
5440	3/4"	↓	34	.235		7.35	13.75		21.10	29
5500	90° Elbow flare by MIPS, 1/4"					2.59			2.59	2.85
5510	3/8"					2.82			2.82	3.10
5520	1/2"					4.97			4.97	5.45
5530	5/8"					8.20			8.20	9
5540	3/4"					13.35			13.35	14.70
5600	Flare by FIPS, 1/4"					4.73			4.73	5.20
5610	3/8"					5.15			5.15	5.65
5620	1/2"					7.35			7.35	8.10
5670	Flare by sweat, 1/4"					2.54			2.54	2.79
5680	3/8"					3.32			3.32	3.65
5690	1/2"					7.25			7.25	7.95
5700	5/8"					22			22	24
5760	Tee flare by IPS, 1/4"					4.73			4.73	5.20
5770	3/8"					6.65			6.65	7.30
5780	1/2"					11.10			11.10	12.25
5790	5/8"				↓	7.55			7.55	8.30

22 11 Facility Water Distribution

22 11 13 – Facility Water Distribution Piping

22 11 13.25 Pipe/Tube Fittings, Copper		Crew	Daily Output	Labor-Hours	Unit	Material	2014 Bare Costs Labor	Equipment	Total	Total Incl O&P
5850	Connector, 1/4"				Ea.	1.02			1.02	1.12
5860	3/8"					1.53			1.53	1.68
5870	1/2"					2.49			2.49	2.74
5880	5/8"					3.85			3.85	4.24
5890	3/4"					7.05			7.05	7.75
5950	Seal cap, 1/4"					.84			.84	.92
5960	3/8"					1.01			1.01	1.11
5970	1/2"					1.26			1.26	1.39
5980	5/8"					2.62			2.62	2.88
5990	3/4"				↓	7.65			7.65	8.40
6000	Water service fittings									
6010	Flare joints nut and labor are included in the fitting price.									
6020	90° Elbow, C x C, 3/8"	1 Plum	19	.421	Ea.	125	24		149	174
6030	1/2"		18	.444		125	25.50		150.50	176
6040	3/4"		16	.500		162	29		191	222
6050	1"		15	.533		305	30.50		335.50	380
6080	2"		10	.800		1,050	46		1,096	1,225
6090	90° Elbow, C x MPT, 3/8"		19	.421		91.50	24		115.50	138
6100	1/2"		18	.444		91.50	25.50		117	140
6110	3/4"		16	.500		106	29		135	161
6120	1"		15	.533		289	30.50		319.50	365
6130	1-1/4"		13	.615		595	35.50		630.50	710
6140	1-1/2"		12	.667		595	38.50		633.50	715
6150	2"		10	.800		720	46		766	865
6160	90° Elbow, C x FPT, 3/8"		19	.421		98	24		122	145
6170	1/2"		18	.444		98	25.50		123.50	147
6180	3/4"		16	.500		122	29		151	178
6190	1"		15	.533		280	30.50		310.50	355
6200	1-1/4"		13	.615		292	35.50		327.50	375
6210	1-1/2"		12	.667		1,125	38.50		1,163.50	1,275
6220	2"		10	.800		1,625	46		1,671	1,850
6230	Tee, C x C x C, 3/8"		13	.615		185	35.50		220.50	258
6240	1/2"		12	.667		185	38.50		223.50	262
6250	3/4"		11	.727		217	42		259	300
6260	1"		10	.800		385	46		431	495
6330	Tube nut, C x nut seat, 3/8"		40	.200		37.50	11.50		49	58.50
6340	1/2"		38	.211		37.50	12.10		49.60	59.50
6350	3/4"		34	.235		37.50	13.55		51.05	61.50
6360	1"		32	.250		68	14.40		82.40	96.50
6380	Coupling, C x C, 3/8"		19	.421		114	24		138	162
6390	1/2"		18	.444		114	25.50		139.50	164
6400	3/4"		16	.500		144	29		173	203
6410	1"		15	.533		266	30.50		296.50	340
6420	1-1/4"		12	.667		430	38.50		468.50	535
6430	1-1/2"		12	.667		650	38.50		688.50	780
6440	2"		10	.800		980	46		1,026	1,150
6450	Adapter, C x FPT, 3/8"		19	.421		86	24		110	131
6460	1/2"		18	.444		86	25.50		111.50	133
6470	3/4"		16	.500		104	29		133	158
6480	1"		15	.533		222	30.50		252.50	291
6490	1-1/4"		13	.615		470	35.50		505.50	575
6500	1-1/2"		12	.667		470	38.50		508.50	580
6510	2"		10	.800	↓	615	46		661	745

22 11 Facility Water Distribution

22 11 13 – Facility Water Distribution Piping

22 11 13.25 Pipe/Tube Fittings, Copper

		Crew	Daily Output	Labor-Hours	Unit	Material	2014 Bare Costs Labor	Equipment	Total	Total Incl O&P
6520	Adapter, C x MPT, 3/8"	1 Plum	19	.421	Ea.	77	24		101	121
6530	1/2"		18	.444		77	25.50		102.50	123
6540	3/4"		16	.500		106	29		135	161
6550	1"		15	.533		187	30.50		217.50	253
6560	1-1/4"		13	.615		470	35.50		505.50	570
6570	1-1/2"		12	.667		470	38.50		508.50	575
6580	2"		10	.800		590	46		636	720
6992	Tube connector fittings, See Section 22 11 13.76 for plastic ftng.									
7000	Insert type Brass/copper, 100 psi @ 180°F, CTS									
7010	Adapter MPT 3/8" x 3/8" CTS	1 Plum	29	.276	Ea.	2.93	15.90		18.83	27
7020	1/2" x 1/2"		26	.308		2.95	17.70		20.65	30
7030	3/4" x 1/2"		26	.308		3.77	17.70		21.47	30.50
7040	3/4" x 3/4"		25	.320		4.35	18.40		22.75	33
7050	Adapter CTS 1/2" x 1/2" sweat		24	.333		3.96	19.20		23.16	33.50
7060	3/4" x 3/4" sweat		22	.364		1.39	21		22.39	33
7070	Coupler center set 3/8" CTS		25	.320		1.44	18.40		19.84	29.50
7080	1/2" CTS		23	.348		3.96	20		23.96	34.50
7090	3/4" CTS		22	.364		1.39	21		22.39	33
7100	Elbow 90°, copper 3/8"		25	.320		3.16	18.40		21.56	31.50
7110	1/2" CTS		23	.348		2.02	20		22.02	32
7120	3/4" CTS		22	.364		2.46	21		23.46	34
7130	Tee copper 3/8" CTS		17	.471		3.75	27		30.75	45
7140	1/2" CTS		15	.533		2.58	30.50		33.08	49.50
7150	3/4" CTS		14	.571		3.96	33		36.96	54
7160	3/8" x 3/8" x 1/2"		16	.500		2.98	29		31.98	47
7170	1/2" x 3/8" x 1/2"		15	.533		1.93	30.50		32.43	48.50
7180	3/4" x 1/2" x 3/4"		14	.571		2.98	33		35.98	53

22 11 13.27 Pipe/Tube, Grooved Joint for Copper

		Crew	Daily Output	Labor-Hours	Unit	Material	Labor	Equipment	Total	Total Incl O&P
0010	**PIPE/TUBE, GROOVED JOINT FOR COPPER**									
4000	Fittings: coupling material required at joints not incl. in fitting price.									
4001	Add 1 selected coupling, material only, per joint for installed price.									
4010	Coupling, rigid style									
4018	2" diameter	1 Plum	50	.160	Ea.	19.80	9.20		29	36
4020	2-1/2" diameter	Q-1	80	.200		22.50	10.35		32.85	40
4022	3" diameter		67	.239		25	12.35		37.35	46
4024	4" diameter		50	.320		37.50	16.60		54.10	66.50
4026	5" diameter		40	.400		63.50	20.50		84	102
4028	6" diameter	Q-2	50	.480		84	26		110	131
4100	Elbow, 90° or 45°									
4108	2" diameter	1 Plum	25	.320	Ea.	35	18.40		53.40	66.50
4110	2-1/2" diameter	Q-1	40	.400		38	20.50		58.50	73
4112	3" diameter		33	.485		53	25		78	96.50
4114	4" diameter		25	.640		122	33		155	185
4116	5" diameter		20	.800		350	41.50		391.50	445
4118	6" diameter	Q-2	25	.960		555	51.50		606.50	690
4200	Tee									
4208	2" diameter	1 Plum	17	.471	Ea.	57.50	27		84.50	105
4210	2-1/2" diameter	Q-1	27	.593		61	30.50		91.50	114
4212	3" diameter		22	.727		91	37.50		128.50	157
4214	4" diameter		17	.941		200	49		249	294
4216	5" diameter		13	1.231		560	64		624	710
4218	6" diameter	Q-2	17	1.412		690	76		766	875

22 11 Facility Water Distribution

22 11 13 – Facility Water Distribution Piping

22 11 13.27 Pipe/Tube, Grooved Joint for Copper

		Crew	Daily Output	Labor-Hours	Unit	Material	2014 Bare Costs Labor	Equipment	Total	Total Incl O&P
4300	Reducer, concentric									
4310	3" x 2-1/2" diameter	Q-1	35	.457	Ea.	50.50	23.50		74	91.50
4312	4" x 2-1/2" diameter		32	.500		103	26		129	153
4314	4" x 3" diameter		29	.552		103	28.50		131.50	157
4316	5" x 3" diameter		25	.640		291	33		324	370
4318	5" x 4" diameter		22	.727		291	37.50		328.50	375
4320	6" x 3" diameter	Q-2	28	.857		315	46		361	415
4322	6" x 4" diameter		26	.923		315	49.50		364.50	420
4324	6" x 5" diameter		24	1		315	53.50		368.50	425
4350	Flange, w/groove gasket									
4351	ANSI class 125 and 150									
4355	2" diameter	1 Plum	23	.348	Ea.	152	20		172	197
4356	2-1/2" diameter	Q-1	37	.432		158	22.50		180.50	208
4358	3" diameter		31	.516		166	26.50		192.50	223
4360	4" diameter		23	.696		182	36		218	255
4362	5" diameter		19	.842		244	43.50		287.50	335
4364	6" diameter	Q-2	23	1.043		266	56		322	380

22 11 13.44 Pipe, Steel

		Crew	Daily Output	Labor-Hours	Unit	Material	Labor	Equipment	Total	Total Incl O&P
0010	**PIPE, STEEL** R221113-50									
0020	All pipe sizes are to Spec. A-53 unless noted otherwise R221113-70									
0032	Schedule 10, see Line 22 11 13.48 0500									
0050	Schedule 40, threaded, with couplings, and clevis hanger									
0060	assemblies sized for covering, 10' O.C.									
0540	Black, 1/4" diameter	1 Plum	66	.121	L.F.	4.51	7		11.51	15.50
0550	3/8" diameter		65	.123		5.10	7.10		12.20	16.30
0560	1/2" diameter		63	.127		2.73	7.30		10.03	14.05
0570	3/4" diameter		61	.131		3.17	7.55		10.72	14.90
0580	1" diameter		53	.151		4.64	8.70		13.34	18.20
0590	1-1/4" diameter	Q-1	89	.180		5.75	9.30		15.05	20.50
0600	1-1/2" diameter		80	.200		6.70	10.35		17.05	23
0610	2" diameter		64	.250		8.95	12.95		21.90	29.50
0620	2-1/2" diameter		50	.320		14.05	16.60		30.65	40.50
0630	3" diameter		43	.372		18	19.25		37.25	49
0640	3-1/2" diameter		40	.400		24.50	20.50		45	58.50
0650	4" diameter		36	.444		27	23		50	65
0809	A-106, gr. A/B, seamless w/cplgs. & clevis hanger assemblies									
0811	1/4" diameter	1 Plum	66	.121	L.F.	7.75	7		14.75	19.05
0812	3/8" diameter		65	.123		7.70	7.10		14.80	19.15
0813	1/2" diameter		63	.127		8.90	7.30		16.20	21
0814	3/4" diameter		61	.131		10.80	7.55		18.35	23.50
0815	1" diameter		53	.151		11.90	8.70		20.60	26
0816	1-1/4" diameter	Q-1	89	.180		14.45	9.30		23.75	30
0817	1-1/2" diameter		80	.200		22	10.35		32.35	40
0819	2" diameter		64	.250		24	12.95		36.95	46
0821	2-1/2" diameter		50	.320		29	16.60		45.60	57
0822	3" diameter		43	.372		32	19.25		51.25	64
0823	4" diameter		36	.444		49	23		72	89
1220	To delete coupling & hanger, subtract									
1230	1/4" diam. to 3/4" diam.					31%	56%			
1240	1" diam. to 1-1/2" diam.					23%	51%			
1250	2" diam. to 4" diam.					23%	41%			
1280	All pipe sizes are to Spec. A-53 unless noted otherwise									

22 11 Facility Water Distribution

22 11 13 – Facility Water Distribution Piping

22 11 13.44 Pipe, Steel		Crew	Daily Output	Labor-Hours	Unit	Material	2014 Bare Costs Labor	Equipment	Total	Total Incl O&P
1281	Schedule 40, threaded, with couplings and clevis hanger									
1282	assemblies sized for covering, 10' O.C.									
1290	Galvanized, 1/4" diameter	1 Plum	66	.121	L.F.	6.40	7		13.40	17.60
1300	3/8" diameter		65	.123		7.05	7.10		14.15	18.45
1310	1/2" diameter		63	.127		3.76	7.30		11.06	15.20
1320	3/4" diameter		61	.131		4.45	7.55		12	16.30
1330	1" diameter		53	.151		6.10	8.70		14.80	19.80
1340	1-1/4" diameter	Q-1	89	.180		7.70	9.30		17	22.50
1350	1-1/2" diameter		80	.200		9.05	10.35		19.40	25.50
1360	2" diameter		64	.250		12.15	12.95		25.10	33
1370	2-1/2" diameter		50	.320		20	16.60		36.60	47
1380	3" diameter		43	.372		25.50	19.25		44.75	57.50
1390	3-1/2" diameter		40	.400		33	20.50		53.50	68
1400	4" diameter		36	.444		38	23		61	76.50
1750	To delete coupling & hanger, subtract									
1760	1/4" diam. to 3/4" diam.					31%	56%			
1770	1" diam. to 1-1/2" diam.					23%	51%			
1780	2" diam. to 4" diam.					23%	41%			
1900	Pipe nipple std black 2"lg 1/2"	1 Plum	19	.421	Ea.	.97	24		24.97	37.50
1910	Pipe nipple std black 2"lg 1" diameter	"	15	.533		1.78	30.50		32.28	48.50
1920	3"lg 2-1/2"	Q-1	16	1		11.10	52		63.10	90.50
2000	Welded, sch. 40, on yoke & roll hanger assy's, sized for covering, 10' O.C.									
2040	Black, 1" diameter	Q-15	93	.172	L.F.	3.91	8.90	.60	13.41	18.40
2050	1-1/4" diameter		84	.190		5.10	9.85	.66	15.61	21
2060	1-1/2" diameter		76	.211		5.70	10.90	.73	17.33	23.50
2070	2" diameter		61	.262		7.05	13.60	.91	21.56	29.50
2080	2-1/2" diameter		47	.340		12.65	17.65	1.18	31.48	41.50
2090	3" diameter		43	.372		15.40	19.25	1.29	35.94	47.50
2100	3-1/2" diameter		39	.410		19.60	21.50	1.42	42.52	55
2110	4" diameter		37	.432		21.50	22.50	1.50	45.50	59
2120	5" diameter		32	.500		34	26	1.73	61.73	78.50
2130	6" diameter	Q-16	36	.667		42.50	36	1.53	80.03	102
2140	8" diameter		29	.828		68	44.50	1.90	114.40	144
2150	10" diameter		24	1		88.50	53.50	2.30	144.30	181
2160	12" diameter		19	1.263		104	68	2.91	174.91	219
2170	14" diameter, (two rod roll type hanger for 14" diam. and up)		15	1.600		96.50	86	3.68	186.18	240
2180	16" diameter, (two rod roll type hanger)		13	1.846		147	99	4.25	250.25	315
2190	18" diameter, (two rod roll type hanger)		11	2.182		139	117	5	261	335
2200	20" diameter, (two rod roll type hanger)		9	2.667		140	143	6.15	289.15	375
2220	24" diameter, (two rod roll type hanger)		8	3		180	161	6.90	347.90	450
2560	To delete hanger, subtract									
2570	1" diam. to 1-1/2" diam.					15%	34%			
2580	2" diam. to 3-1/2" diam.					9%	21%			
2590	4" diam. to 12" diam.					5%	12%			
2596	14" diam. to 36" diam.					3%	10%			
3250	Flanged, 150 lb. weld neck, on yoke & roll hangers									
3260	sized for covering, 10' O.C.									
3290	Black, 1" diameter	Q-15	70	.229	L.F.	9.10	11.85	.79	21.74	29
3300	1-1/4" diameter		64	.250		10.25	12.95	.87	24.07	32
3310	1-1/2" diameter		58	.276		10.85	14.30	.95	26.10	34.50
3320	2" diameter		45	.356		13	18.40	1.23	32.63	43.50
3330	2-1/2" diameter		36	.444		19.35	23	1.54	43.89	58
3340	3" diameter		32	.500		22.50	26	1.73	50.23	66

22 11 Facility Water Distribution

22 11 13 – Facility Water Distribution Piping

22 11 13.44 Pipe, Steel

		Crew	Daily Output	Labor-Hours	Unit	Material	2014 Bare Costs Labor	Equipment	Total	Total Incl O&P
3350	3-1/2" diameter	Q-15	29	.552	L.F.	29	28.50	1.91	59.41	76.50
3360	4" diameter		26	.615		30	32	2.13	64.13	83.50
3370	5" diameter		21	.762		48	39.50	2.64	90.14	115
3380	6" diameter	Q-16	25	.960		58	51.50	2.21	111.71	144
3390	8" diameter		19	1.263		95	68	2.91	165.91	210
3400	10" diameter		16	1.500		135	80.50	3.45	218.95	275
3410	12" diameter		14	1.714		163	92	3.94	258.94	325
3470	For 300 lb. flanges, add					63%				
3480	For 600 lb. flanges, add					310%				
3960	To delete flanges & hanger, subtract									
3970	1" diam. to 2" diam.					76%	65%			
3980	2-1/2" diam. to 4" diam.					62%	59%			
3990	5" diam. to 12" diam.					60%	46%			
4750	Schedule 80, threaded, with couplings, and clevis hanger assemblies									
4760	sized for covering, 10' O.C.									
4790	Black, 1/4" diameter	1 Plum	54	.148	L.F.	12.65	8.55		21.20	27
4800	3/8" diameter		53	.151		16.30	8.70		25	31
4810	1/2" diameter		52	.154		4.11	8.85		12.96	17.85
4820	3/4" diameter		50	.160		5.25	9.20		14.45	19.65
4830	1" diameter		45	.178		6.60	10.25		16.85	22.50
4840	1-1/4" diameter	Q-1	75	.213		8.80	11.05		19.85	26.50
4850	1-1/2" diameter		69	.232		10.15	12		22.15	29.50
4860	2" diameter		56	.286		13.75	14.80		28.55	37.50
4870	2-1/2" diameter		44	.364		21	18.85		39.85	52
4880	3" diameter		38	.421		27.50	22		49.50	63.50
4890	3-1/2" diameter		35	.457		34	23.50		57.50	73.50
4900	4" diameter		32	.500		33	26		59	75
5430	To delete coupling & hanger, subtract									
5440	1/4" diam. to 1/2" diam.					31%	54%			
5450	3/4" diam. to 1-1/2" diam.					28%	49%			
5460	2" diam. to 4" diam.					21%	40%			
5510	Galvanized, 1/4" diameter	1 Plum	54	.148	L.F.	5.80	8.55		14.35	19.25
5520	3/8" diameter		53	.151		6.90	8.70		15.60	20.50
5530	1/2" diameter		52	.154		4.08	8.85		12.93	17.85
5540	3/4" diameter		50	.160		6.30	9.20		15.50	21
5550	1" diameter		45	.178		8.60	10.25		18.85	25
5560	1-1/4" diameter	Q-1	75	.213		11.55	11.05		22.60	29.50
5570	1-1/2" diameter		69	.232		13.50	12		25.50	33
5580	2" diameter		56	.286		18.55	14.80		33.35	43
5590	2-1/2" diameter		44	.364		28.50	18.85		47.35	60
5600	3" diameter		38	.421		37.50	22		59.50	74.50
5610	3-1/2" diameter		35	.457		45	23.50		68.50	85.50
5620	4" diameter		32	.500		53.50	26		79.50	98
5930	To delete coupling & hanger, subtract									
5940	1/4" diam. to 1/2" diam.					31%	54%			
5950	3/4" diam. to 1-1/2" diam.					28%	49%			
5960	2" diam. to 4" diam.					21%	40%			
6000	Welded, on yoke & roller hangers									
6010	sized for covering, 10' O.C.									
6040	Black, 1" diameter	Q-15	85	.188	L.F.	6.65	9.75	.65	17.05	23
6050	1-1/4" diameter		79	.203		8.80	10.50	.70	20	26.50
6060	1-1/2" diameter		72	.222		10.15	11.50	.77	22.42	29.50
6070	2" diameter		57	.281		13.10	14.55	.97	28.62	37.50

22 11 Facility Water Distribution

22 11 13 – Facility Water Distribution Piping

22 11 13.44 Pipe, Steel

		Crew	Daily Output	Labor-Hours	Unit	Material	2014 Bare Costs Labor	Equipment	Total	Total Incl O&P
6080	2-1/2" diameter	Q-15	44	.364	L.F.	20	18.85	1.26	40.11	52
6090	3" diameter		40	.400		26	20.50	1.38	47.88	62
6100	3-1/2" diameter		34	.471		32	24.50	1.63	58.13	74
6110	4" diameter		33	.485		30	25	1.68	56.68	73
6120	5" diameter, A-106B		26	.615		47.50	32	2.13	81.63	103
6130	6" diameter, A-106B	Q-16	30	.800		66	43	1.84	110.84	140
6140	8" diameter, A-106B		25	.960		99.50	51.50	2.21	153.21	189
6150	10" diameter, A-106B		20	1.200		182	64.50	2.76	249.26	300
6160	12" diameter, A-106B		15	1.600		360	86	3.68	449.68	535
6540	To delete hanger, subtract									
6550	1" diam. to 1-1/2" diam.					30%	14%			
6560	2" diam. to 3" diam.					23%	9%			
6570	3-1/2" diam. to 5" diam.					12%	6%			
6580	6" diam. to 12" diam.					10%	4%			
7250	Flanged, 300 lb. weld neck, on yoke & roll hangers									
7260	sized for covering, 10' O.C.									
7290	Black, 1" diameter	Q-15	66	.242	L.F.	13.10	12.55	.84	26.49	34.50
7300	1-1/4" diameter		61	.262		15.20	13.60	.91	29.71	38.50
7310	1-1/2" diameter		54	.296		16.60	15.35	1.03	32.98	42.50
7320	2" diameter		42	.381		21.50	19.75	1.32	42.57	55
7330	2-1/2" diameter		33	.485		29.50	25	1.68	56.18	72.50
7340	3" diameter		29	.552		35.50	28.50	1.91	65.91	84
7350	3-1/2" diameter		24	.667		47.50	34.50	2.31	84.31	107
7360	4" diameter		23	.696		45.50	36	2.41	83.91	108
7370	5" diameter		19	.842		71.50	43.50	2.91	117.91	148
7380	6" diameter	Q-16	23	1.043		90	56	2.40	148.40	186
7390	8" diameter		17	1.412		140	76	3.25	219.25	273
7400	10" diameter		14	1.714		256	92	3.94	351.94	425
7410	12" diameter		12	2		455	107	4.60	566.60	670
7470	For 600 lb. flanges, add					100%				
7940	To delete flanges & hanger, subtract									
7950	1" diam. to 1-1/2" diam.					75%	66%			
7960	2" diam. to 3" diam.					62%	60%			
7970	3-1/2" diam. to 5" diam.					54%	66%			
7980	6" diam. to 12" diam.					55%	62%			
9000	Threading pipe labor, one end, all schedules through 80									
9010	1/4" through 3/4" pipe size	1 Plum	80	.100	Ea.		5.75		5.75	8.70
9020	1" through 2" pipe size		73	.110			6.30		6.30	9.50
9030	2-1/2" pipe size		53	.151			8.70		8.70	13.10
9040	3" pipe size		50	.160			9.20		9.20	13.90
9050	3-1/2" pipe size	Q-1	89	.180			9.30		9.30	14.05
9060	4" pipe size		73	.219			11.35		11.35	17.15
9070	5" pipe size		53	.302			15.65		15.65	23.50
9080	6" pipe size		46	.348			18		18	27
9090	8" pipe size		29	.552			28.50		28.50	43
9100	10" pipe size		21	.762			39.50		39.50	59.50
9110	12" pipe size		13	1.231			64		64	96.50
9200	Welding labor per joint									
9210	Schedule 40,									
9230	1/2" pipe size	Q-15	32	.500	Ea.		26	1.73	27.73	41
9240	3/4" pipe size		27	.593			30.50	2.05	32.55	49
9250	1" pipe size		23	.696			36	2.41	38.41	57
9260	1-1/4" pipe size		20	.800			41.50	2.77	44.27	65.50

22 11 Facility Water Distribution

22 11 13 – Facility Water Distribution Piping

22 11 13.44 Pipe, Steel

		Crew	Daily Output	Labor-Hours	Unit	Material	2014 Bare Costs Labor	Equipment	Total	Total Incl O&P
9270	1-1/2" pipe size	Q-15	19	.842	Ea.		43.50	2.91	46.41	69
9280	2" pipe size		16	1			52	3.46	55.46	82
9290	2-1/2" pipe size		13	1.231			64	4.26	68.26	101
9300	3" pipe size		12	1.333			69	4.61	73.61	109
9310	4" pipe size		10	1.600			83	5.55	88.55	131
9320	5" pipe size		9	1.778			92	6.15	98.15	146
9330	6" pipe size		8	2			104	6.90	110.90	164
9340	8" pipe size		5	3.200			166	11.05	177.05	262
9350	10" pipe size		4	4			207	13.85	220.85	330
9360	12" pipe size		3	5.333			276	18.45	294.45	435
9370	14" pipe size		2.60	6.154			320	21.50	341.50	505
9380	16" pipe size		2.20	7.273			375	25	400	600
9390	18" pipe size		2	8			415	27.50	442.50	655
9400	20" pipe size		1.80	8.889			460	31	491	730
9410	22" pipe size		1.70	9.412			490	32.50	522.50	770
9420	24" pipe size		1.50	10.667			555	37	592	875
9450	Schedule 80,									
9460	1/2" pipe size	Q-15	27	.593	Ea.		30.50	2.05	32.55	49
9470	3/4" pipe size		23	.696			36	2.41	38.41	57
9480	1" pipe size		20	.800			41.50	2.77	44.27	65.50
9490	1-1/4" pipe size		19	.842			43.50	2.91	46.41	69
9500	1-1/2" pipe size		18	.889			46	3.08	49.08	73
9510	2" pipe size		15	1.067			55.50	3.69	59.19	87.50
9520	2-1/2" pipe size		12	1.333			69	4.61	73.61	109
9530	3" pipe size		11	1.455			75.50	5.05	80.55	120
9540	4" pipe size		8	2			104	6.90	110.90	164
9550	5" pipe size		6	2.667			138	9.25	147.25	219
9560	6" pipe size		5	3.200			166	11.05	177.05	262
9570	8" pipe size		4	4			207	13.85	220.85	330
9580	10" pipe size		3	5.333			276	18.45	294.45	435
9590	12" pipe size		2	8			415	27.50	442.50	655
9600	14" pipe size	Q-16	2.60	9.231			495	21	516	775
9610	16" pipe size		2.30	10.435			560	24	584	870
9620	18" pipe size		2	12			645	27.50	672.50	1,000
9630	20" pipe size		1.80	13.333			715	30.50	745.50	1,100
9640	22" pipe size		1.60	15			805	34.50	839.50	1,275
9650	24" pipe size		1.50	16			860	37	897	1,350

22 11 13.45 Pipe Fittings, Steel, Threaded

		Crew	Daily Output	Labor-Hours	Unit	Material	Labor	Equipment	Total	Total Incl O&P
0010	**PIPE FITTINGS, STEEL, THREADED**									
0020	Cast Iron									
0040	Standard weight, black									
0060	90° Elbow, straight									
0070	1/4"	1 Plum	16	.500	Ea.	9	29		38	53.50
0080	3/8"		16	.500		13	29		42	58
0090	1/2"		15	.533		5.70	30.50		36.20	53
0100	3/4"		14	.571		5.95	33		38.95	56
0110	1"		13	.615		7.05	35.50		42.55	61.50
0120	1-1/4"	Q-1	22	.727		10	37.50		47.50	68
0130	1-1/2"		20	.800		13.80	41.50		55.30	77.50
0140	2"		18	.889		21.50	46		67.50	93
0150	2-1/2"		14	1.143		51.50	59		110.50	147
0160	3"		10	1.600		84.50	83		167.50	218

161

22 11 Facility Water Distribution

22 11 13 – Facility Water Distribution Piping

22 11 13.45 Pipe Fittings, Steel, Threaded

		Crew	Daily Output	Labor-Hours	Unit	Material	2014 Bare Costs Labor	Equipment	Total	Total Incl O&P
0170	3-1/2"	Q-1	8	2	Ea.	230	104		334	410
0180	4"	↓	6	2.667	↓	157	138		295	380
0250	45° Elbow, straight									
0260	1/4"	1 Plum	16	.500	Ea.	11.10	29		40.10	55.50
0270	3/8"		16	.500		11.90	29		40.90	56.50
0280	1/2"		15	.533		8.65	30.50		39.15	56
0300	3/4"		14	.571		8.75	33		41.75	59
0320	1"	↓	13	.615		10.30	35.50		45.80	65
0330	1-1/4"	Q-1	22	.727		13.85	37.50		51.35	72
0340	1-1/2"		20	.800		23	41.50		64.50	87.50
0350	2"		18	.889		26.50	46		72.50	98.50
0360	2-1/2"		14	1.143		68.50	59		127.50	165
0370	3"		10	1.600		109	83		192	245
0380	3-1/2"		8	2		251	104		355	430
0400	4"	↓	6	2.667	↓	226	138		364	460
0500	Tee, straight									
0510	1/4"	1 Plum	10	.800	Ea.	14.05	46		60.05	85
0520	3/8"		10	.800		13.65	46		59.65	84.50
0530	1/2"		9	.889		8.90	51		59.90	87
0540	3/4"		9	.889		10.35	51		61.35	88.50
0550	1"	↓	8	1		9.20	57.50		66.70	97
0560	1-1/4"	Q-1	14	1.143		16.80	59		75.80	108
0570	1-1/2"		13	1.231		22	64		86	121
0580	2"		11	1.455		30.50	75.50		106	148
0590	2-1/2"		9	1.778		79	92		171	226
0600	3"		6	2.667		121	138		259	340
0610	3-1/2"		5	3.200		245	166		411	520
0620	4"	↓	4	4	↓	245	207		452	585
0660	Tee, reducing, run or outlet									
0661	1/2"	1 Plum	9	.889	Ea.	22	51		73	101
0662	3/4"		9	.889		19	51		70	98
0663	1"	↓	8	1		21.50	57.50		79	111
0664	1-1/4"	Q-1	14	1.143		30	59		89	123
0665	1-1/2"		13	1.231		35.50	64		99.50	136
0666	2"		11	1.455		55.50	75.50		131	175
0667	2-1/2"		9	1.778		104	92		196	253
0668	3"		6	2.667		156	138		294	380
0669	3-1/2"		5	3.200		370	166		536	655
0670	4"	↓	4	4	↓	410	207		617	765
0674	Reducer, concentric									
0675	3/4"	1 Plum	18	.444	Ea.	15.90	25.50		41.40	56
0676	1"	"	15	.533		10.80	30.50		41.30	58.50
0677	1-1/4"	Q-1	26	.615		32	32		64	83.50
0678	1-1/2"		24	.667		50.50	34.50		85	108
0679	2"		21	.762		57.50	39.50		97	123
0680	2-1/2"		18	.889		82	46		128	160
0681	3"		14	1.143		129	59		188	231
0682	3-1/2"		12	1.333		220	69		289	345
0683	4"	↓	10	1.600	↓	239	83		322	390
0687	Reducer, eccentric									
0688	3/4"	1 Plum	16	.500	Ea.	33.50	29		62.50	80.50
0689	1"	"	14	.571		36	33		69	89.50
0690	1-1/4"	Q-1	25	.640	↓	58	33		91	114

22 11 Facility Water Distribution

22 11 13 – Facility Water Distribution Piping

22 11 13.45 Pipe Fittings, Steel, Threaded

		Crew	Daily Output	Labor-Hours	Unit	Material	2014 Bare Costs Labor	Equipment	Total	Total Incl O&P
0691	1-1/2"	Q-1	22	.727	Ea.	76.50	37.50		114	141
0692	2"		20	.800		110	41.50		151.50	184
0693	2-1/2"		16	1		151	52		203	245
0694	3"		12	1.333		239	69		308	365
0695	3-1/2"		10	1.600		335	83		418	495
0696	4"	↓	9	1.778	↓	430	92		522	615
0700	Standard weight, galvanized cast iron									
0720	90° Elbow, straight									
0730	1/4"	1 Plum	16	.500	Ea.	15.05	29		44.05	60
0740	3/8"		16	.500		15.05	29		44.05	60
0750	1/2"		15	.533		17.15	30.50		47.65	65.50
0760	3/4"		14	.571		16.85	33		49.85	68
0770	1"	↓	13	.615		19.50	35.50		55	75
0780	1-1/4"	Q-1	22	.727		30.50	37.50		68	90.50
0790	1-1/2"		20	.800		41.50	41.50		83	109
0800	2"		18	.889		61.50	46		107.50	137
0810	2-1/2"		14	1.143		126	59		185	229
0820	3"		10	1.600		192	83		275	335
0830	3-1/2"		8	2		305	104		409	490
0840	4"	↓	6	2.667	↓	350	138		488	595
0900	45° Elbow, straight									
0910	1/4"	1 Plum	16	.500	Ea.	17.15	29		46.15	62.50
0920	3/8"		16	.500		17.15	29		46.15	62.50
0930	1/2"		15	.533		17.15	30.50		47.65	65.50
0940	3/4"		14	.571		19.65	33		52.65	71
0950	1"	↓	13	.615		24	35.50		59.50	80
0960	1-1/4"	Q-1	22	.727		36	37.50		73.50	96.50
0970	1-1/2"		20	.800		51	41.50		92.50	119
0980	2"		18	.889		71.50	46		117.50	148
0990	2-1/2"		14	1.143		115	59		174	217
1000	3"		10	1.600		225	83		308	370
1010	3-1/2"		8	2		415	104		519	610
1020	4"	↓	6	2.667	↓	415	138		553	665
1100	Tee, straight									
1110	1/4"	1 Plum	10	.800	Ea.	18.15	46		64.15	89.50
1120	3/8"		10	.800		18.15	46		64.15	89.50
1130	1/2"		9	.889		18.15	51		69.15	97
1140	3/4"		9	.889		24	51		75	104
1150	1"	↓	8	1		26.50	57.50		84	116
1160	1-1/4"	Q-1	14	1.143		46	59		105	140
1170	1-1/2"		13	1.231		61	64		125	164
1180	2"		11	1.455		76	75.50		151.50	198
1190	2-1/2"		9	1.778		154	92		246	310
1200	3"		6	2.667		345	138		483	590
1210	3-1/2"		5	3.200		435	166		601	730
1220	4"	↓	4	4	↓	485	207		692	850
1300	Extra heavy weight, black									
1310	Couplings, steel straight									
1320	1/4"	1 Plum	19	.421	Ea.	3.77	24		27.77	40.50
1330	3/8"		19	.421		4.10	24		28.10	41
1340	1/2"		19	.421		5.55	24		29.55	42.50
1350	3/4"		18	.444		5.90	25.50		31.40	45
1360	1"	↓	15	.533		7.55	30.50		38.05	55

22 11 Facility Water Distribution

22 11 13 – Facility Water Distribution Piping

22 11 13.45 Pipe Fittings, Steel, Threaded

		Crew	Daily Output	Labor-Hours	Unit	Material	2014 Bare Costs Labor	Equipment	Total	Total Incl O&P
1370	1-1/4"	Q-1	26	.615	Ea.	12.10	32		44.10	61.50
1380	1-1/2"		24	.667		12.10	34.50		46.60	65.50
1390	2"		21	.762		18.45	39.50		57.95	80
1400	2-1/2"		18	.889		27.50	46		73.50	99.50
1410	3"		14	1.143		32.50	59		91.50	126
1420	3-1/2"		12	1.333		43.50	69		112.50	152
1430	4"		10	1.600		51.50	83		134.50	182
1510	90° Elbow, straight									
1520	1/2"	1 Plum	15	.533	Ea.	29.50	30.50		60	79
1530	3/4"		14	.571		30	33		63	82.50
1540	1"		13	.615		36	35.50		71.50	93
1550	1-1/4"	Q-1	22	.727		54	37.50		91.50	116
1560	1-1/2"		20	.800		66.50	41.50		108	136
1580	2"		18	.889		82.50	46		128.50	160
1590	2-1/2"		14	1.143		200	59		259	310
1600	3"		10	1.600		264	83		347	415
1610	4"		6	2.667		540	138		678	805
1650	45° Elbow, straight									
1660	1/2"	1 Plum	15	.533	Ea.	42	30.50		72.50	92.50
1670	3/4"		14	.571		40.50	33		73.50	94
1680	1"		13	.615		48.50	35.50		84	107
1690	1-1/4"	Q-1	22	.727		80	37.50		117.50	145
1700	1-1/2"		20	.800		88	41.50		129.50	159
1710	2"		18	.889		125	46		171	207
1720	2-1/2"		14	1.143		216	59		275	325
1800	Tee, straight									
1810	1/2"	1 Plum	9	.889	Ea.	46	51		97	128
1820	3/4"		9	.889		46	51		97	128
1830	1"		8	1		55.50	57.50		113	148
1840	1-1/4"	Q-1	14	1.143		82.50	59		141.50	180
1850	1-1/2"		13	1.231		106	64		170	214
1860	2"		11	1.455		132	75.50		207.50	259
1870	2-1/2"		9	1.778		281	92		373	450
1880	3"		6	2.667		385	138		523	635
1890	4"		4	4		750	207		957	1,150
4000	Standard weight, black									
4010	Couplings, steel straight, merchants									
4030	1/4"	1 Plum	19	.421	Ea.	1.01	24		25.01	37.50
4040	3/8"		19	.421		1.22	24		25.22	38
4050	1/2"		19	.421		1.30	24		25.30	38
4060	3/4"		18	.444		1.64	25.50		27.14	40.50
4070	1"		15	.533		2.31	30.50		32.81	49
4080	1-1/4"	Q-1	26	.615		2.95	32		34.95	51.50
4090	1-1/2"		24	.667		3.73	34.50		38.23	56
4100	2"		21	.762		5.35	39.50		44.85	65.50
4110	2-1/2"		18	.889		16.75	46		62.75	88
4120	3"		14	1.143		23.50	59		82.50	116
4130	3-1/2"		12	1.333		41.50	69		110.50	150
4140	4"		10	1.600		41.50	83		124.50	171
4166	Plug, 1/4"	1 Plum	38	.211		2.53	12.10		14.63	21
4167	3/8"		38	.211		2.53	12.10		14.63	21
4168	1/2"		38	.211		2.53	12.10		14.63	21
4169	3/4"		32	.250		7.65	14.40		22.05	30

22 11 Facility Water Distribution

22 11 13 – Facility Water Distribution Piping

22 11 13.45 Pipe Fittings, Steel, Threaded		Crew	Daily Output	Labor-Hours	Unit	Material	2014 Bare Costs Labor	Equipment	Total	Total Incl O&P
4170	1"	1 Plum	30	.267	Ea.	8.15	15.35		23.50	32
4171	1-1/4"	Q-1	52	.308		9.40	15.95		25.35	34.50
4172	1-1/2"		48	.333		13.30	17.25		30.55	40.50
4173	2"		42	.381		17.15	19.75		36.90	49
4176	2-1/2"		36	.444		25.50	23		48.50	63
4180	4"		20	.800		44	41.50		85.50	111
4200	Standard weight, galvanized									
4210	Couplings, steel straight, merchants									
4230	1/4"	1 Plum	19	.421	Ea.	1.16	24		25.16	38
4240	3/8"		19	.421		1.49	24		25.49	38
4250	1/2"		19	.421		1.58	24		25.58	38
4260	3/4"		18	.444		1.99	25.50		27.49	40.50
4270	1"		15	.533		2.78	30.50		33.28	49.50
4280	1-1/4"	Q-1	26	.615		3.55	32		35.55	52
4290	1-1/2"		24	.667		4.41	34.50		38.91	57
4300	2"		21	.762		6.60	39.50		46.10	67
4310	2-1/2"		18	.889		20.50	46		66.50	92
4320	3"		14	1.143		27.50	59		86.50	120
4330	3-1/2"		12	1.333		48	69		117	157
4340	4"		10	1.600		48	83		131	178
4370	Plug, galvanized, square head									
4374	1/2"	1 Plum	38	.211	Ea.	6.10	12.10		18.20	25
4375	3/4"		32	.250		6.10	14.40		20.50	28.50
4376	1"		30	.267		6.10	15.35		21.45	30
4377	1-1/4"	Q-1	52	.308		10.20	15.95		26.15	35
4378	1-1/2"		48	.333		13.80	17.25		31.05	41
4379	2"		42	.381		17.10	19.75		36.85	49
4380	2-1/2"		36	.444		36	23		59	75
4381	3"		28	.571		46.50	29.50		76	95.50
4382	4"		20	.800		127	41.50		168.50	203
4700	Nipple, black									
4710	1/2" x 4" long	1 Plum	19	.421	Ea.	2.21	24		26.21	39
4712	3/4" x 4" long		18	.444		2.67	25.50		28.17	41.50
4714	1" x 4" long		15	.533		3.70	30.50		34.20	50.50
4716	1-1/4" x 4" long	Q-1	26	.615		4.63	32		36.63	53
4718	1-1/2" x 4" long		24	.667		5.45	34.50		39.95	58
4720	2" x 4" long		21	.762		7.60	39.50		47.10	68
4722	2-1/2" x 4" long		18	.889		20.50	46		66.50	92.50
4724	3" x 4" long		14	1.143		26	59		85	119
4726	4" x 4" long		10	1.600		35.50	83		118.50	164
4800	Nipple, galvanized									
4810	1/2" x 4" long	1 Plum	19	.421	Ea.	2.71	24		26.71	39.50
4812	3/4" x 4" long		18	.444		3.35	25.50		28.85	42
4814	1" x 4" long		15	.533		4.49	30.50		34.99	51.50
4816	1-1/4" x 4" long	Q-1	26	.615		5.50	32		37.50	54
4818	1-1/2" x 4" long		24	.667		7	34.50		41.50	59.50
4820	2" x 4" long		21	.762		8.85	39.50		48.35	69.50
4822	2-1/2" x 4" long		18	.889		23.50	46		69.50	95
4824	3" x 4" long		14	1.143		31	59		90	124
4826	4" x 4" long		10	1.600		41.50	83		124.50	171
5000	Malleable iron, 150 lb.									
5020	Black									
5040	90° elbow, straight									

22 11 Facility Water Distribution

22 11 13 – Facility Water Distribution Piping

22 11 13.45 Pipe Fittings, Steel, Threaded

		Crew	Daily Output	Labor-Hours	Unit	Material	2014 Bare Costs Labor	Equipment	Total	Total Incl O&P
5060	1/4"	1 Plum	16	.500	Ea.	4.42	29		33.42	48.50
5070	3/8"		16	.500		4.42	29		33.42	48.50
5080	1/2"		15	.533		3.06	30.50		33.56	50
5090	3/4"		14	.571		3.70	33		36.70	53.50
5100	1"		13	.615		6.40	35.50		41.90	60.50
5110	1-1/4"	Q-1	22	.727		10.60	37.50		48.10	68.50
5120	1-1/2"		20	.800		13.95	41.50		55.45	78
5130	2"		18	.889		24	46		70	96
5140	2-1/2"		14	1.143		53.50	59		112.50	149
5150	3"		10	1.600		78.50	83		161.50	211
5160	3-1/2"		8	2		215	104		319	395
5170	4"		6	2.667		168	138		306	395
5250	45° elbow, straight									
5270	1/4"	1 Plum	16	.500	Ea.	6.65	29		35.65	51
5280	3/8"		16	.500		6.65	29		35.65	51
5290	1/2"		15	.533		5.05	30.50		35.55	52
5300	3/4"		14	.571		6.25	33		39.25	56.50
5310	1"		13	.615		7.90	35.50		43.40	62
5320	1-1/4"	Q-1	22	.727		13.95	37.50		51.45	72.50
5330	1-1/2"		20	.800		17.25	41.50		58.75	81.50
5340	2"		18	.889		26	46		72	98
5350	2-1/2"		14	1.143		75.50	59		134.50	173
5360	3"		10	1.600		174	83		257	315
5370	3-1/2"		8	2		192	104		296	365
5380	4"		6	2.667		192	138		330	420
5450	Tee, straight									
5470	1/4"	1 Plum	10	.800	Ea.	6.40	46		52.40	76.50
5480	3/8"		10	.800		6.40	46		52.40	76.50
5490	1/2"		9	.889		4.09	51		55.09	81.50
5500	3/4"		9	.889		5.85	51		56.85	83.50
5510	1"		8	1		10.05	57.50		67.55	98
5520	1-1/4"	Q-1	14	1.143		16.30	59		75.30	107
5530	1-1/2"		13	1.231		20.50	64		84.50	119
5540	2"		11	1.455		34.50	75.50		110	152
5550	2-1/2"		9	1.778		74.50	92		166.50	221
5560	3"		6	2.667		110	138		248	330
5570	3-1/2"		5	3.200		254	166		420	530
5580	4"		4	4		265	207		472	605
5601	Tee, reducing, on outlet									
5602	1/2"	1 Plum	9	.889	Ea.	8.95	51		59.95	87
5603	3/4"		9	.889		10.10	51		61.10	88
5604	1"		8	1		16.15	57.50		73.65	105
5605	1-1/4"	Q-1	14	1.143		28.50	59		87.50	121
5606	1-1/2"		13	1.231		32	64		96	132
5607	2"		11	1.455		44.50	75.50		120	163
5608	2-1/2"		9	1.778		112	92		204	262
5609	3"		6	2.667		154	138		292	380
5610	3-1/2"		5	3.200		360	166		526	645
5611	4"		4	4		305	207		512	650
5650	Coupling									
5670	1/4"	1 Plum	19	.421	Ea.	5.50	24		29.50	42.50
5680	3/8"		19	.421		5.50	24		29.50	42.50
5690	1/2"		19	.421		4.24	24		28.24	41

22 11 Facility Water Distribution
22 11 13 – Facility Water Distribution Piping

22 11 13.45 Pipe Fittings, Steel, Threaded		Crew	Daily Output	Labor-Hours	Unit	Material	2014 Bare Costs Labor	Equipment	Total	Total Incl O&P
5700	3/4"	1 Plum	18	.444	Ea.	4.97	25.50		30.47	44
5710	1"	▼	15	.533		7.45	30.50		37.95	54.50
5720	1-1/4"	Q-1	26	.615		9.65	32		41.65	58.50
5730	1-1/2"		24	.667		13	34.50		47.50	66.50
5740	2"		21	.762		19.25	39.50		58.75	80.50
5750	2-1/2"		18	.889		53	46		99	128
5760	3"		14	1.143		72	59		131	169
5770	3-1/2"		12	1.333		145	69		214	263
5780	4"	▼	10	1.600		145	83		228	284
5840	Reducer, concentric, 1/4"	1 Plum	19	.421		5.80	24		29.80	43
5850	3/8"		19	.421		7.40	24		31.40	44.50
5860	1/2"		19	.421		5.85	24		29.85	43
5870	3/4"		16	.500		6.90	29		35.90	51
5880	1"	▼	15	.533		11.70	30.50		42.20	59.50
5890	1-1/4"	Q-1	26	.615		13.05	32		45.05	62.50
5900	1-1/2"		24	.667		18.75	34.50		53.25	72.50
5910	2"		21	.762		27	39.50		66.50	89.50
5911	2-1/2"		18	.889		60.50	46		106.50	136
5912	3"		14	1.143		73.50	59		132.50	170
5913	3-1/2"		12	1.333		234	69		303	360
5914	4"	▼	10	1.600		176	83		259	320
5981	Bushing, 1/4"	1 Plum	19	.421		1.15	24		25.15	38
5982	3/8"		19	.421		5.75	24		29.75	43
5983	1/2"		19	.421		5.75	24		29.75	43
5984	3/4"		16	.500		5.65	29		34.65	50
5985	1"	▼	15	.533		8.05	30.50		38.55	55.50
5986	1-1/4"	Q-1	26	.615		10	32		42	59
5987	1-1/2"		24	.667		8.65	34.50		43.15	61.50
5988	2"		21	.762		10.75	39.50		50.25	71.50
5989	Cap, 1/4"	1 Plum	38	.211		4.44	12.10		16.54	23
5991	3/8"		38	.211		5.45	12.10		17.55	24.50
5992	1/2"		38	.211		3.03	12.10		15.13	21.50
5993	3/4"		32	.250		4.19	14.40		18.59	26
5994	1"	▼	30	.267		5.05	15.35		20.40	28.50
5995	1-1/4"	Q-1	52	.308		6.65	15.95		22.60	31.50
5996	1-1/2"		48	.333		9.20	17.25		26.45	36
5997	2"	▼	42	.381		13.40	19.75		33.15	45
6000	For galvanized elbows, tees, and couplings add				▼	20%				
6058	For galvanized reducers, caps and bushings add					20%				
6100	90° Elbow, galvanized, 150 lb., reducing									
6110	3/4" x 1/2"	1 Plum	15.40	.519	Ea.	8.80	30		38.80	54.50
6112	1" x 3/4"		14	.571		11.30	33		44.30	62
6114	1" x 1/2"	▼	14.50	.552		12	32		44	61
6116	1-1/4" x 1"	Q-1	24.20	.661		18.70	34.50		53.20	72
6118	1-1/4" x 3/4"		25.40	.630		23	32.50		55.50	74.50
6120	1-1/4" x 1/2"		26.20	.611		24	31.50		55.50	74.50
6122	1-1/2" x 1-1/4"		21.60	.741		30.50	38.50		69	91.50
6124	1-1/2" x 1"		23.50	.681		30.50	35.50		66	87
6126	1-1/2" x 3/4"		24.60	.650		30.50	33.50		64	84.50
6128	2" x 1-1/2"		20.50	.780		35.50	40.50		76	100
6130	2" x 1-1/4"		21	.762		40.50	39.50		80	104
6132	2" x 1"		22.80	.702		42	36.50		78.50	102
6134	2" x 3/4"	▼	23.90	.669		43	34.50		77.50	99.50

22 11 Facility Water Distribution

22 11 13 – Facility Water Distribution Piping

22 11 13.45 Pipe Fittings, Steel, Threaded		Crew	Daily Output	Labor-Hours	Unit	Material	2014 Bare Costs Labor	Equipment	Total	Total Incl O&P
6136	2-1/2" x 2"	Q-1	12.30	1.301	Ea.	120	67.50		187.50	234
6138	2-1/2" x 1-1/2"		12.50	1.280		134	66.50		200.50	247
6140	3" x 2-1/2"		8.60	1.860		218	96.50		314.50	385
6142	3" x 2"		11.80	1.356		189	70		259	315
6144	4" x 3"	▼	8.20	1.951	▼	495	101		596	700
6160	90° Elbow, black, 150 lb., reducing									
6170	1" x 3/4"	1 Plum	14	.571	Ea.	7.70	33		40.70	58
6174	1-1/2" x 1"	Q-1	23.50	.681		18.30	35.50		53.80	73.50
6178	1-1/2" x 3/4"		24.60	.650		21	33.50		54.50	74
6182	2" x 1-1/2"		20.50	.780		26.50	40.50		67	90
6186	2" x 1"		22.80	.702		30	36.50		66.50	88
6190	2" x 3/4"		23.90	.669		32	34.50		66.50	87.50
6194	2-1/2" x 2"	▼	12.30	1.301		72.50	67.50		140	182
7000	Union, with brass seat									
7010	1/4"	1 Plum	15	.533	Ea.	23.50	30.50		54	72.50
7020	3/8"		15	.533		14.95	30.50		45.45	63
7030	1/2"		14	.571		13.55	33		46.55	64.50
7040	3/4"		13	.615		15.55	35.50		51.05	70.50
7050	1"	▼	12	.667		20.50	38.50		59	80.50
7060	1-1/4"	Q-1	21	.762		29	39.50		68.50	91.50
7070	1-1/2"		19	.842		36.50	43.50		80	106
7080	2"		17	.941		42	49		91	120
7090	2-1/2"		13	1.231		126	64		190	235
7100	3"	▼	9	1.778	▼	151	92		243	305
7120	Union, galvanized									
7124	1/2"	1 Plum	14	.571	Ea.	17.25	33		50.25	68.50
7125	3/4"		13	.615		20.50	35.50		56	76
7126	1"	▼	12	.667		27	38.50		65.50	87.50
7127	1-1/4"	Q-1	21	.762		39	39.50		78.50	103
7128	1-1/2"		19	.842		47.50	43.50		91	118
7129	2"		17	.941		54.50	49		103.50	134
7130	2-1/2"		13	1.231		192	64		256	310
7131	3"	▼	9	1.778		265	92		357	430
7500	Malleable iron, 300 lb.									
7520	Black									
7540	90° Elbow, straight, 1/4"	1 Plum	16	.500	Ea.	15.50	29		44.50	60.50
7560	3/8"		16	.500		13.80	29		42.80	58.50
7570	1/2"		15	.533		17.85	30.50		48.35	66
7580	3/4"		14	.571		20	33		53	71.50
7590	1"	▼	13	.615		26	35.50		61.50	82
7600	1-1/4"	Q-1	22	.727		37.50	37.50		75	98.50
7610	1-1/2"		20	.800		44.50	41.50		86	112
7620	2"		18	.889		64	46		110	140
7630	2-1/2"		14	1.143		165	59		224	272
7640	3"		10	1.600		190	83		273	335
7650	4"	▼	6	2.667		445	138		583	700
7700	45° Elbow, straight, 1/4"	1 Plum	16	.500		23	29		52	69
7720	3/8"		16	.500		25.50	29		54.50	72
7730	1/2"		15	.533		25.50	30.50		56	75
7740	3/4"		14	.571		29	33		62	81.50
7750	1"	▼	13	.615		31.50	35.50		67	88.50
7760	1-1/4"	Q-1	22	.727		50.50	37.50		88	113
7770	1-1/2"		20	.800		66	41.50		107.50	135

22 11 Facility Water Distribution

22 11 13 – Facility Water Distribution Piping

22 11 13.45 Pipe Fittings, Steel, Threaded		Crew	Daily Output	Labor-Hours	Unit	Material	2014 Bare Costs Labor	Equipment	Total	Total Incl O&P
7780	2"	Q-1	18	.889	Ea.	100	46		146	180
7790	2-1/2"		14	1.143		223	59		282	335
7800	3"		10	1.600		293	83		376	450
7810	4"	▼	6	2.667		665	138		803	945
7850	Tee, straight, 1/4"	1 Plum	10	.800		19.55	46		65.55	91
7870	3/8"		10	.800		20.50	46		66.50	92.50
7880	1/2"		9	.889		26.50	51		77.50	106
7890	3/4"		9	.889		28.50	51		79.50	109
7900	1"	▼	8	1		34	57.50		91.50	125
7910	1-1/4"	Q-1	14	1.143		49	59		108	144
7920	1-1/2"		13	1.231		57	64		121	160
7930	2"		11	1.455		84.50	75.50		160	207
7940	2-1/2"		9	1.778		205	92		297	365
7950	3"		6	2.667		292	138		430	530
7960	4"	▼	4	4		835	207		1,042	1,225
8050	Couplings, straight, 1/4"	1 Plum	19	.421		14	24		38	52
8070	3/8"		19	.421		14	24		38	52
8080	1/2"		19	.421		15.60	24		39.60	53.50
8090	3/4"		18	.444		17.95	25.50		43.45	58.50
8100	1"	▼	15	.533		20.50	30.50		51	69
8110	1-1/4"	Q-1	26	.615		24.50	32		56.50	75
8120	1-1/2"		24	.667		36.50	34.50		71	92
8130	2"		21	.762		52.50	39.50		92	118
8140	2-1/2"		18	.889		97	46		143	177
8150	3"		14	1.143		140	59		199	244
8160	4"	▼	10	1.600	▼	254	83		337	405
8200	Galvanized									
8220	90° Elbow, straight, 1/4"	1 Plum	16	.500	Ea.	28	29		57	74
8222	3/8"		16	.500		28.50	29		57.50	74.50
8224	1/2"		15	.533		33.50	30.50		64	83.50
8226	3/4"		14	.571		38	33		71	91
8228	1"	▼	13	.615		48.50	35.50		84	107
8230	1-1/4"	Q-1	22	.727		76.50	37.50		114	141
8232	1-1/2"		20	.800		83	41.50		124.50	154
8234	2"		18	.889		139	46		185	223
8236	2-1/2"		14	1.143		310	59		369	430
8238	3"		10	1.600		365	83		448	530
8240	4"	▼	6	2.667	▼	1,100	138		1,238	1,400
8280	45° Elbow, straight									
8282	1/2"	1 Plum	15	.533	Ea.	50	30.50		80.50	102
8284	3/4"		14	.571		58	33		91	114
8286	1"	▼	13	.615		63.50	35.50		99	123
8288	1-1/4"	Q-1	22	.727		98.50	37.50		136	165
8290	1-1/2"		20	.800		127	41.50		168.50	203
8292	2"	▼	18	.889		176	46		222	264
8310	Tee, straight, 1/4"	1 Plum	10	.800		39.50	46		85.50	113
8312	3/8"		10	.800		40.50	46		86.50	114
8314	1/2"		9	.889		50.50	51		101.50	133
8316	3/4"		9	.889		55.50	51		106.50	138
8318	1"	▼	8	1		68.50	57.50		126	163
8320	1-1/4"	Q-1	14	1.143		98	59		157	198
8322	1-1/2"		13	1.231		102	64		166	209
8324	2"	▼	11	1.455		162	75.50		237.50	293

169

22 11 Facility Water Distribution

22 11 13 – Facility Water Distribution Piping

22 11 13.45 Pipe Fittings, Steel, Threaded

		Crew	Daily Output	Labor-Hours	Unit	Material	2014 Bare Costs Labor	Equipment	Total	Total Incl O&P
8326	2-1/2"	Q-1	9	1.778	Ea.	515	92		607	705
8328	3"		6	2.667		585	138		723	850
8330	4"		4	4		1,500	207		1,707	1,975
8380	Couplings, straight, 1/4"	1 Plum	19	.421		18.20	24		42.20	56.50
8382	3/8"		19	.421		26	24		50	65.50
8384	1/2"		19	.421		27	24		51	66
8386	3/4"		18	.444		30.50	25.50		56	72
8388	1"		15	.533		40	30.50		70.50	90.50
8390	1-1/4"	Q-1	26	.615		52.50	32		84.50	106
8392	1-1/2"		24	.667		74	34.50		108.50	134
8394	2"		21	.762		89.50	39.50		129	158
8396	2-1/2"		18	.889		235	46		281	330
8398	3"		14	1.143		280	59		339	400
8399	4"		10	1.600		430	83		513	600
8529	Black									
8530	Reducer, concentric, 1/4"	1 Plum	19	.421	Ea.	18.50	24		42.50	57
8531	3/8"		19	.421		18.50	24		42.50	57
8532	1/2"		17	.471		22.50	27		49.50	65.50
8533	3/4"		16	.500		32	29		61	78.50
8534	1"		15	.533		38	30.50		68.50	88.50
8535	1-1/4"	Q-1	26	.615		57	32		89	111
8536	1-1/2"		24	.667		59	34.50		93.50	117
8537	2"		21	.762		85	39.50		124.50	153
8550	Cap, 1/4"	1 Plum	38	.211		14.35	12.10		26.45	34
8551	3/8"		38	.211		14.35	12.10		26.45	34
8552	1/2"		34	.235		14.25	13.55		27.80	36
8553	3/4"		32	.250		17.95	14.40		32.35	41.50
8554	1"		30	.267		23.50	15.35		38.85	49
8555	1-1/4"	Q-1	52	.308		26	15.95		41.95	52.50
8556	1-1/2"		48	.333		38	17.25		55.25	68
8557	2"		42	.381		53	19.75		72.75	88.50
8570	Plug, 1/4"	1 Plum	38	.211		3.28	12.10		15.38	22
8571	3/8"		38	.211		3.44	12.10		15.54	22
8572	1/2"		34	.235		3.60	13.55		17.15	24.50
8573	3/4"		32	.250		4.49	14.40		18.89	26.50
8574	1"		30	.267		6.85	15.35		22.20	30.50
8575	1-1/4"	Q-1	52	.308		14.15	15.95		30.10	39.50
8576	1-1/2"		48	.333		16.20	17.25		33.45	44
8577	2"		42	.381		25.50	19.75		45.25	58
9500	Union with brass seat, 1/4"	1 Plum	15	.533		31.50	30.50		62	81.50
9530	3/8"		15	.533		32	30.50		62.50	82
9540	1/2"		14	.571		26	33		59	78
9550	3/4"		13	.615		28.50	35.50		64	85
9560	1"		12	.667		37.50	38.50		76	99.50
9570	1-1/4"	Q-1	21	.762		61	39.50		100.50	127
9580	1-1/2"		19	.842		64	43.50		107.50	137
9590	2"		17	.941		79.50	49		128.50	161
9600	2-1/2"		13	1.231		256	64		320	380
9610	3"		9	1.778		330	92		422	505
9620	4"		5	3.200		1,050	166		1,216	1,400
9630	Union, all iron, 1/4"	1 Plum	15	.533		49	30.50		79.50	101
9650	3/8"		15	.533		49	30.50		79.50	101
9660	1/2"		14	.571		54.50	33		87.50	109

22 11 Facility Water Distribution

22 11 13 – Facility Water Distribution Piping

22 11 13.45 Pipe Fittings, Steel, Threaded		Crew	Daily Output	Labor-Hours	Unit	Material	2014 Bare Costs Labor	Equipment	Total	Total Incl O&P
9670	3/4"	1 Plum	13	.615	Ea.	55.50	35.50		91	115
9680	1"	▼	12	.667		54	38.50		92.50	117
9690	1-1/4"	Q-1	21	.762		104	39.50		143.50	175
9700	1-1/2"		19	.842		128	43.50		171.50	206
9710	2"		17	.941		162	49		211	252
9720	2-1/2"		13	1.231		286	64		350	410
9730	3"	▼	9	1.778		545	92		637	740
9750	For galvanized unions, add				▼	15%				
9757	Forged steel, 3000 lb.									
9758	Black									
9760	90° Elbow, 1/4"	1 Plum	16	.500	Ea.	19.35	29		48.35	65
9761	3/8"		16	.500		19.35	29		48.35	65
9762	1/2"		15	.533		14.80	30.50		45.30	63
9763	3/4"		14	.571		18.60	33		51.60	70
9764	1"	▼	13	.615		27.50	35.50		63	84
9765	1-1/4"	Q-1	22	.727		52.50	37.50		90	115
9766	1-1/2"		20	.800		67.50	41.50		109	137
9767	2"	▼	18	.889		82	46		128	160
9780	45° Elbow 1/4"	1 Plum	16	.500		24.50	29		53.50	70.50
9781	3/8"		16	.500		24.50	29		53.50	70.50
9782	1/2"		15	.533		24	30.50		54.50	73
9783	3/4"		14	.571		28	33		61	80.50
9784	1"	▼	13	.615		38	35.50		73.50	95.50
9785	1-1/4"	Q-1	22	.727		52.50	37.50		90	115
9786	1-1/2"		20	.800		75.50	41.50		117	146
9787	2"	▼	18	.889		104	46		150	185
9800	Tee, 1/4"	1 Plum	10	.800		23.50	46		69.50	95.50
9801	3/8"		10	.800		23.50	46		69.50	95.50
9802	1/2"		9	.889		21	51		72	101
9803	3/4"		9	.889		28.50	51		79.50	109
9804	1"	▼	8	1		37.50	57.50		95	129
9805	1-1/4"	Q-1	14	1.143		73	59		132	170
9806	1-1/2"		13	1.231		85	64		149	190
9807	2"	▼	11	1.455		110	75.50		185.50	234
9820	Reducer, concentric, 1/4"	1 Plum	19	.421		11.80	24		35.80	49.50
9821	3/8"		19	.421		12.30	24		36.30	50
9822	1/2"		17	.471		12.30	27		39.30	54.50
9823	3/4"		16	.500		14.60	29		43.60	59.50
9824	1"	▼	15	.533		19.05	30.50		49.55	67.50
9825	1-1/4"	Q-1	26	.615		33.50	32		65.50	84.50
9826	1-1/2"		24	.667		36	34.50		70.50	92
9827	2"	▼	21	.762		52.50	39.50		92	118
9840	Cap, 1/4"	1 Plum	38	.211		6.90	12.10		19	26
9841	3/8"		38	.211		7.25	12.10		19.35	26.50
9842	1/2"		34	.235		7.05	13.55		20.60	28.50
9843	3/4"		32	.250		9.85	14.40		24.25	32.50
9844	1"	▼	30	.267		15.10	15.35		30.45	39.50
9845	1-1/4"	Q-1	52	.308		24.50	15.95		40.45	51
9846	1-1/2"		48	.333		29	17.25		46.25	58
9847	2"	▼	42	.381		42	19.75		61.75	76
9860	Plug, 1/4"	1 Plum	38	.211		3.28	12.10		15.38	22
9861	3/8"		38	.211		3.44	12.10		15.54	22
9862	1/2"	▼	34	.235		3.60	13.55		17.15	24.50

22 11 Facility Water Distribution

22 11 13 – Facility Water Distribution Piping

22 11 13.45 Pipe Fittings, Steel, Threaded

		Crew	Daily Output	Labor-Hours	Unit	Material	2014 Bare Costs Labor	Equipment	Total	Total Incl O&P
9863	3/4"	1 Plum	32	.250	Ea.	4.49	14.40		18.89	26.50
9864	1"	▼	30	.267		6.85	15.35		22.20	30.50
9865	1-1/4"	Q-1	52	.308		14.15	15.95		30.10	39.50
9866	1-1/2"		48	.333		16.20	17.25		33.45	44
9867	2"	▼	42	.381		25.50	19.75		45.25	58
9880	Union, bronze seat, 1/4"	1 Plum	15	.533		53.50	30.50		84	106
9881	3/8"		15	.533		53.50	30.50		84	106
9882	1/2"		14	.571		51.50	33		84.50	106
9883	3/4"		13	.615		69	35.50		104.50	129
9884	1"	▼	12	.667		77.50	38.50		116	144
9885	1-1/4"	Q-1	21	.762		151	39.50		190.50	226
9886	1-1/2"		19	.842		155	43.50		198.50	237
9887	2"	▼	17	.941		184	49		233	277
9900	Coupling, 1/4"	1 Plum	19	.421		7.55	24		31.55	45
9901	3/8"		19	.421		7.55	24		31.55	45
9902	1/2"		17	.471		6.20	27		33.20	48
9903	3/4"		16	.500		8.05	29		37.05	52.50
9904	1"	▼	15	.533		14	30.50		44.50	62
9905	1-1/4"	Q-1	26	.615		23.50	32		55.50	73.50
9906	1-1/2"		24	.667		30	34.50		64.50	85
9907	2"	▼	21	.762	▼	37.50	39.50		77	101

22 11 13.47 Pipe Fittings, Steel

		Crew	Daily Output	Labor-Hours	Unit	Material	2014 Bare Costs Labor	Equipment	Total	Total Incl O&P
0010	**PIPE FITTINGS, STEEL**, Flanged, Welded & Special									
0020	Flanged joints, C.I., standard weight, black. One gasket & bolt									
0040	set, mat'l only, required at each joint, not included (see line 0620)									
0060	90° Elbow, straight, 1-1/2" pipe size	Q-1	14	1.143	Ea.	510	59		569	650
0080	2" pipe size		13	1.231		300	64		364	425
0090	2-1/2" pipe size		12	1.333		325	69		394	460
0100	3" pipe size		11	1.455		270	75.50		345.50	410
0110	4" pipe size		8	2		330	104		434	515
0120	5" pipe size	▼	7	2.286		790	118		908	1,050
0130	6" pipe size	Q-2	9	2.667		525	143		668	790
0140	8" pipe size		8	3		905	161		1,066	1,250
0150	10" pipe size		7	3.429		1,975	184		2,159	2,450
0160	12" pipe size	▼	6	4	▼	4,000	215		4,215	4,725
0171	90° Elbow, reducing									
0172	2-1/2" by 2" pipe size	Q-1	12	1.333	Ea.	990	69		1,059	1,200
0173	3" by 2-1/2" pipe size		11	1.455		920	75.50		995.50	1,125
0174	4" by 3" pipe size		8	2		740	104		844	970
0175	5" by 3" pipe size	▼	7	2.286		1,600	118		1,718	1,925
0176	6" by 4" pipe size	Q-2	9	2.667		990	143		1,133	1,325
0177	8" by 6" pipe size		8	3		1,350	161		1,511	1,725
0178	10" by 8" pipe size		7	3.429		2,550	184		2,734	3,075
0179	12" by 10" pipe size	▼	6	4		4,900	215		5,115	5,725
0200	45° Elbow, straight, 1-1/2" pipe size	Q-1	14	1.143		615	59		674	765
0220	2" pipe size		13	1.231		435	64		499	570
0230	2-1/2" pipe size		12	1.333		465	69		534	615
0240	3" pipe size		11	1.455		450	75.50		525.50	610
0250	4" pipe size		8	2		505	104		609	710
0260	5" pipe size	▼	7	2.286		1,225	118		1,343	1,500
0270	6" pipe size	Q-2	9	2.667		830	143		973	1,125
0280	8" pipe size	▼	8	3	▼	1,200	161		1,361	1,575

22 11 Facility Water Distribution

22 11 13 – Facility Water Distribution Piping

22 11 13.47 Pipe Fittings, Steel

		Crew	Daily Output	Labor-Hours	Unit	Material	2014 Bare Costs Labor	Equipment	Total	Total Incl O&P
0290	10" pipe size	Q-2	7	3.429	Ea.	2,550	184		2,734	3,100
0300	12" pipe size	↓	6	4	↓	3,900	215		4,115	4,625
0310	Cross, straight									
0311	2-1/2" pipe size	Q-1	6	2.667	Ea.	950	138		1,088	1,250
0312	3" pipe size		5	3.200		1,000	166		1,166	1,350
0313	4" pipe size		4	4		1,300	207		1,507	1,750
0314	5" pipe size	↓	3	5.333		2,725	276		3,001	3,425
0315	6" pipe size	Q-2	5	4.800		2,725	258		2,983	3,400
0316	8" pipe size		4	6		4,400	320		4,720	5,325
0317	10" pipe size		3	8		5,225	430		5,655	6,400
0318	12" pipe size	↓	2	12		7,975	645		8,620	9,750
0350	Tee, straight, 1-1/2" pipe size	Q-1	10	1.600		585	83		668	770
0370	2" pipe size		9	1.778		330	92		422	500
0380	2-1/2" pipe size		8	2		480	104		584	680
0390	3" pipe size		7	2.286		335	118		453	550
0400	4" pipe size		5	3.200		510	166		676	810
0410	5" pipe size	↓	4	4		1,375	207		1,582	1,825
0420	6" pipe size	Q-2	6	4		740	215		955	1,150
0430	8" pipe size		5	4.800		1,250	258		1,508	1,800
0440	10" pipe size		4	6		3,400	320		3,720	4,200
0450	12" pipe size	↓	3	8	↓	5,325	430		5,755	6,525
0459	Tee, reducing on outlet									
0460	2-1/2" by 2" pipe size	Q-1	8	2	Ea.	995	104		1,099	1,250
0461	3" by 2-1/2" pipe size		7	2.286		995	118		1,113	1,275
0462	4" by 3" pipe size		5	3.200		1,050	166		1,216	1,400
0463	5" by 4" pipe size	↓	4	4		2,275	207		2,482	2,850
0464	6" by 4" pipe size	Q-2	6	4		1,000	215		1,215	1,425
0465	8" by 6" pipe size		5	4.800		1,575	258		1,833	2,125
0466	10" by 8" pipe size		4	6		4,250	320		4,570	5,150
0467	12" by 10" pipe size	↓	3	8	↓	6,225	430		6,655	7,500
0476	Reducer, concentric									
0477	3" by 2-1/2"	Q-1	12	1.333	Ea.	600	69		669	765
0478	4" by 3"		9	1.778		675	92		767	880
0479	5" by 4"	↓	8	2		1,050	104		1,154	1,300
0480	6" by 4"	Q-2	10	2.400		905	129		1,034	1,200
0481	8" by 6"		9	2.667		1,125	143		1,268	1,475
0482	10" by 8"		8	3		2,050	161		2,211	2,525
0483	12" by 10"	↓	7	3.429	↓	4,050	184		4,234	4,725
0492	Reducer, eccentric									
0493	4" by 3"	Q-1	8	2	Ea.	1,125	104		1,229	1,375
0494	5" by 4"	"	7	2.286		1,675	118		1,793	2,025
0495	6" by 4"	Q-2	9	2.667		1,025	143		1,168	1,350
0496	8" by 6"		8	3		1,350	161		1,511	1,725
0497	10" by 8"		7	3.429		3,325	184		3,509	3,950
0498	12" by 10"	↓	6	4		4,525	215		4,740	5,300
0500	For galvanized elbows and tees, add					100%				
0520	For extra heavy weight elbows and tees, add					140%				
0620	Gasket and bolt set, 150#, 1/2" pipe size	1 Plum	20	.400		2.50	23		25.50	38
0622	3/4" pipe size		19	.421		2.64	24		26.64	39.50
0624	1" pipe size		18	.444		2.67	25.50		28.17	41.50
0626	1-1/4" pipe size		17	.471		2.80	27		29.80	44
0628	1-1/2" pipe size		15	.533		2.98	30.50		33.48	50
0630	2" pipe size	↓	13	.615	↓	4.85	35.50		40.35	59

22 11 Facility Water Distribution

22 11 13 – Facility Water Distribution Piping

22 11 13.47 Pipe Fittings, Steel		Crew	Daily Output	Labor-Hours	Unit	Material	2014 Bare Costs Labor	Equipment	Total	Total Incl O&P
0640	2-1/2" pipe size	1 Plum	12	.667	Ea.	5.60	38.50		44.10	64
0650	3" pipe size		11	.727		5.80	42		47.80	69.50
0660	3-1/2" pipe size		9	.889		9.40	51		60.40	87.50
0670	4" pipe size		8	1		10.40	57.50		67.90	98.50
0680	5" pipe size		7	1.143		17.10	66		83.10	118
0690	6" pipe size		6	1.333		18.10	76.50		94.60	136
0700	8" pipe size		5	1.600		19.55	92		111.55	161
0710	10" pipe size		4.50	1.778		35.50	102		137.50	193
0720	12" pipe size		4.20	1.905		37.50	110		147.50	208
0730	14" pipe size		4	2		39.50	115		154.50	218
0740	16" pipe size		3	2.667		41.50	153		194.50	278
0750	18" pipe size		2.70	2.963		80.50	171		251.50	345
0760	20" pipe size		2.30	3.478		131	200		331	445
0780	24" pipe size		1.90	4.211		164	242		406	545
0790	26" pipe size		1.60	5		222	288		510	680
0810	30" pipe size		1.40	5.714		430	330		760	970
0830	36" pipe size		1.10	7.273		810	420		1,230	1,525
0850	For 300 lb. gasket set, add					40%				
2000	Flanged unions, 125 lb., black, 1/2" pipe size	1 Plum	17	.471	Ea.	76.50	27		103.50	125
2040	3/4" pipe size		17	.471		101	27		128	152
2050	1" pipe size		16	.500		97.50	29		126.50	151
2060	1-1/4" pipe size	Q-1	28	.571		115	29.50		144.50	172
2070	1-1/2" pipe size		27	.593		107	30.50		137.50	164
2080	2" pipe size		26	.615		125	32		157	186
2090	2-1/2" pipe size		24	.667		170	34.50		204.50	239
2100	3" pipe size		22	.727		192	37.50		229.50	268
2110	3-1/2" pipe size		18	.889		325	46		371	430
2120	4" pipe size		16	1		261	52		313	365
2130	5" pipe size		14	1.143		610	59		669	760
2140	6" pipe size	Q-2	19	1.263		570	68		638	730
2150	8" pipe size	"	16	1.500		1,325	80.50		1,405.50	1,575
2200	For galvanized unions, add					150%				
2290	Threaded flange									
2300	Cast iron									
2310	Black, 125 lb., per flange									
2320	1" pipe size	1 Plum	27	.296	Ea.	43.50	17.05		60.55	74
2330	1-1/4" pipe size	Q-1	44	.364		52.50	18.85		71.35	86
2340	1-1/2" pipe size		40	.400		48.50	20.50		69	84.50
2350	2" pipe size		36	.444		48	23		71	88
2360	2-1/2" pipe size		28	.571		56.50	29.50		86	107
2370	3" pipe size		20	.800		72.50	41.50		114	143
2380	3-1/2" pipe size		16	1		102	52		154	190
2390	4" pipe size		12	1.333		98	69		167	212
2400	5" pipe size		10	1.600		138	83		221	277
2410	6" pipe size	Q-2	14	1.714		157	92		249	310
2420	8" pipe size		12	2		247	107		354	435
2430	10" pipe size		10	2.400		440	129		569	675
2440	12" pipe size		8	3		975	161		1,136	1,325
2460	For galvanized flanges, add					95%				
2490	Blind flange									
2492	Cast iron									
2494	Black, 125 lb., per flange									
2496	1" pipe size	1 Plum	27	.296	Ea.	69.50	17.05		86.55	103

22 11 Facility Water Distribution

22 11 13 – Facility Water Distribution Piping

22 11 13.47 Pipe Fittings, Steel

		Crew	Daily Output	Labor-Hours	Unit	Material	2014 Bare Costs Labor	Equipment	Total	Total Incl O&P
2500	1-1/2" pipe size	Q-1	40	.400	Ea.	78	20.50		98.50	118
2502	2" pipe size		36	.444		88	23		111	132
2504	2-1/2" pipe size		28	.571		94	29.50		123.50	148
2506	3" pipe size		20	.800		116	41.50		157.50	191
2508	4" pipe size		12	1.333		149	69		218	268
2510	5" pipe size		10	1.600		242	83		325	390
2512	6" pipe size	Q-2	14	1.714		263	92		355	430
2514	8" pipe size		12	2		410	107		517	615
2516	10" pipe size		10	2.400		610	129		739	865
2518	12" pipe size		8	3		1,150	161		1,311	1,525
2520	For galvanized flanges, add					80%				
2570	Threaded flange									
2580	Forged steel,									
2590	Black 150 lb., per flange									
2600	1/2" pipe size	1 Plum	30	.267	Ea.	26.50	15.35		41.85	52
2610	3/4" pipe size		28	.286		26.50	16.45		42.95	54
2620	1" pipe size		27	.296		26.50	17.05		43.55	55
2630	1-1/4" pipe size	Q-1	44	.364		26.50	18.85		45.35	57.50
2640	1-1/2" pipe size		40	.400		26.50	20.50		47	60.50
2650	2" pipe size		36	.444		29.50	23		52.50	67.50
2660	2-1/2" pipe size		28	.571		36	29.50		65.50	84.50
2670	3" pipe size		20	.800		36.50	41.50		78	103
2690	4" pipe size		12	1.333		42.50	69		111.50	151
2700	5" pipe size		10	1.600		66.50	83		149.50	199
2710	6" pipe size	Q-2	14	1.714		73	92		165	220
2720	8" pipe size		12	2		126	107		233	300
2730	10" pipe size		10	2.400		228	129		357	445
2860	Black 300 lb., per flange									
2870	1/2" pipe size	1 Plum	30	.267	Ea.	29	15.35		44.35	55
2880	3/4" pipe size		28	.286		29	16.45		45.45	57
2890	1" pipe size		27	.296		29	17.05		46.05	58
2900	1-1/4" pipe size	Q-1	44	.364		29	18.85		47.85	60.50
2910	1-1/2" pipe size		40	.400		29	20.50		49.50	63.50
2920	2" pipe size		36	.444		34	23		57	72.50
2930	2-1/2" pipe size		28	.571		47.50	29.50		77	96.50
2940	3" pipe size		20	.800		50.50	41.50		92	118
2960	4" pipe size		12	1.333		73	69		142	184
2970	6" pipe size	Q-2	14	1.714		136	92		228	289
3000	Weld joint, butt, carbon steel, standard weight									
3040	90° elbow, long radius									
3050	1/2" pipe size	Q-15	16	1	Ea.	45	52	3.46	100.46	131
3060	3/4" pipe size		16	1		45	52	3.46	100.46	131
3070	1" pipe size		16	1		21	52	3.46	76.46	105
3080	1-1/4" pipe size		14	1.143		21	59	3.95	83.95	117
3090	1-1/2" pipe size		13	1.231		21	64	4.26	89.26	124
3100	2" pipe size		10	1.600		22.50	83	5.55	111.05	156
3110	2-1/2" pipe size		8	2		27.50	104	6.90	138.40	194
3120	3" pipe size		7	2.286		29	118	7.90	154.90	220
3130	4" pipe size		5	3.200		48	166	11.05	225.05	315
3136	5" pipe size		4	4		103	207	13.85	323.85	445
3140	6" pipe size	Q-16	5	4.800		106	258	11.05	375.05	520
3150	8" pipe size		3.75	6.400		200	345	14.70	559.70	755
3160	10" pipe size		3	8		400	430	18.40	848.40	1,100

22 11 Facility Water Distribution
22 11 13 – Facility Water Distribution Piping

22 11 13.47 Pipe Fittings, Steel		Crew	Daily Output	Labor-Hours	Unit	Material	2014 Bare Costs Labor	Equipment	Total	Total Incl O&P
3170	12" pipe size	Q-16	2.50	9.600	Ea.	590	515	22	1,127	1,450
3180	14" pipe size		2	12		970	645	27.50	1,642.50	2,075
3190	16" pipe size		1.50	16		1,325	860	37	2,222	2,825
3191	18" pipe size		1.25	19.200		1,525	1,025	44	2,594	3,275
3192	20" pipe size		1.15	20.870		2,450	1,125	48	3,623	4,450
3194	24" pipe size	↓	1.02	23.529	↓	3,450	1,275	54	4,779	5,750
3200	45° Elbow, long									
3210	1/2" pipe size	Q-15	16	1	Ea.	62.50	52	3.46	117.96	150
3220	3/4" pipe size		16	1		62.50	52	3.46	117.96	150
3230	1" pipe size		16	1		22	52	3.46	77.46	106
3240	1-1/4" pipe size		14	1.143		22	59	3.95	84.95	118
3250	1-1/2" pipe size		13	1.231		22	64	4.26	90.26	126
3260	2" pipe size		10	1.600		22	83	5.55	110.55	156
3270	2-1/2" pipe size		8	2		26.50	104	6.90	137.40	193
3280	3" pipe size		7	2.286		27.50	118	7.90	153.40	218
3290	4" pipe size		5	3.200		49.50	166	11.05	226.55	315
3296	5" pipe size	↓	4	4		75	207	13.85	295.85	415
3300	6" pipe size	Q-16	5	4.800		97.50	258	11.05	366.55	510
3310	8" pipe size		3.75	6.400		163	345	14.70	522.70	715
3320	10" pipe size		3	8		325	430	18.40	773.40	1,025
3330	12" pipe size		2.50	9.600		460	515	22	997	1,300
3340	14" pipe size		2	12		615	645	27.50	1,287.50	1,675
3341	16" pipe size		1.50	16		1,100	860	37	1,997	2,575
3342	18" pipe size		1.25	19.200		1,575	1,025	44	2,644	3,325
3343	20" pipe size		1.15	20.870		1,625	1,125	48	2,798	3,525
3345	24" pipe size		1.05	22.857		2,450	1,225	52.50	3,727.50	4,600
3346	26" pipe size		.85	28.235		2,700	1,525	65	4,290	5,350
3347	30" pipe size		.45	53.333		2,975	2,875	123	5,973	7,725
3349	36" pipe size	↓	.38	63.158	↓	3,275	3,400	145	6,820	8,900
3350	Tee, straight									
3352	For reducing tees and concentrics see starting line 4600									
3360	1/2" pipe size	Q-15	10	1.600	Ea.	110	83	5.55	198.55	252
3370	3/4" pipe size		10	1.600		110	83	5.55	198.55	252
3380	1" pipe size		10	1.600		54.50	83	5.55	143.05	191
3390	1-1/4" pipe size		9	1.778		68.50	92	6.15	166.65	221
3400	1-1/2" pipe size		8	2		68.50	104	6.90	179.40	239
3410	2" pipe size		6	2.667		54.50	138	9.25	201.75	279
3420	2-1/2" pipe size		5	3.200		75.50	166	11.05	252.55	345
3430	3" pipe size		4	4		84	207	13.85	304.85	425
3440	4" pipe size		3	5.333		118	276	18.45	412.45	565
3446	5" pipe size	↓	2.50	6.400		195	330	22	547	740
3450	6" pipe size	Q-16	3	8		203	430	18.40	651.40	895
3460	8" pipe size		2.50	9.600		355	515	22	892	1,200
3470	10" pipe size		2	12		695	645	27.50	1,367.50	1,775
3480	12" pipe size		1.60	15		975	805	34.50	1,814.50	2,350
3481	14" pipe size		1.30	18.462		1,700	990	42.50	2,732.50	3,425
3482	16" pipe size		1	24		1,900	1,300	55	3,255	4,100
3483	18" pipe size		.80	30		3,025	1,600	69	4,694	5,825
3484	20" pipe size		.75	32		4,750	1,725	73.50	6,548.50	7,900
3486	24" pipe size		.70	34.286		6,150	1,850	79	8,079	9,600
3487	26" pipe size		.55	43.636		6,775	2,350	100	9,225	11,100
3488	30" pipe size		.30	80		7,425	4,300	184	11,909	14,900
3490	36" pipe size	↓	.25	96	↓	8,200	5,150	221	13,571	17,000

22 11 Facility Water Distribution
22 11 13 – Facility Water Distribution Piping

22 11 13.47 Pipe Fittings, Steel

		Crew	Daily Output	Labor-Hours	Unit	Material	2014 Bare Costs Labor	Equipment	Total	Total Incl O&P
3491	Eccentric reducer, 1-1/2" pipe size	Q-15	14	1.143	Ea.	63.50	59	3.95	126.45	164
3492	2" pipe size		11	1.455		43	75.50	5.05	123.55	167
3493	2-1/2" pipe size		9	1.778		50	92	6.15	148.15	201
3494	3" pipe size		8	2		59	104	6.90	169.90	229
3495	4" pipe size		6	2.667		75.50	138	9.25	222.75	300
3496	6" pipe size	Q-16	5	4.800		232	258	11.05	501.05	655
3497	8" pipe size		4	6		345	320	13.80	678.80	880
3498	10" pipe size		3	8		435	430	18.40	883.40	1,150
3499	12" pipe size		2.50	9.600		605	515	22	1,142	1,475
3501	Cap, 1-1/2" pipe size	Q-15	28	.571		24.50	29.50	1.98	55.98	73.50
3502	2" pipe size		22	.727		22	37.50	2.52	62.02	84
3503	2-1/2" pipe size		18	.889		30	46	3.08	79.08	105
3504	3" pipe size		16	1		33.50	52	3.46	88.96	118
3505	4" pipe size		12	1.333		45	69	4.61	118.61	159
3506	6" pipe size	Q-16	10	2.400		92	129	5.50	226.50	300
3507	8" pipe size		8	3		138	161	6.90	305.90	405
3508	10" pipe size		6	4		212	215	9.20	436.20	570
3509	12" pipe size		5	4.800		283	258	11.05	552.05	710
3511	14" pipe size		4	6		293	320	13.80	626.80	825
3512	16" pipe size		4	6		455	320	13.80	788.80	1,000
3513	18" pipe size		3	8		665	430	18.40	1,113.40	1,400
3517	Weld joint, butt, carbon steel, extra strong									
3519	90° elbow, long									
3520	1/2" pipe size	Q-15	13	1.231	Ea.	55.50	64	4.26	123.76	163
3530	3/4" pipe size		12	1.333		55.50	69	4.61	129.11	171
3540	1" pipe size		11	1.455		27	75.50	5.05	107.55	150
3550	1-1/4" pipe size		10	1.600		27	83	5.55	115.55	161
3560	1-1/2" pipe size		9	1.778		27	92	6.15	125.15	176
3570	2" pipe size		8	2		27.50	104	6.90	138.40	194
3580	2-1/2" pipe size		7	2.286		38.50	118	7.90	164.40	230
3590	3" pipe size		6	2.667		49.50	138	9.25	196.75	274
3600	4" pipe size		4	4		81.50	207	13.85	302.35	420
3606	5" pipe size		3.50	4.571		195	237	15.80	447.80	590
3610	6" pipe size	Q-16	4.50	5.333		206	287	12.25	505.25	675
3620	8" pipe size		3.50	6.857		390	370	15.75	775.75	1,000
3630	10" pipe size		2.50	9.600		830	515	22	1,367	1,725
3640	12" pipe size		2.25	10.667		1,025	575	24.50	1,624.50	2,025
3650	45° Elbow, long									
3660	1/2" pipe size	Q-15	13	1.231	Ea.	62	64	4.26	130.26	169
3670	3/4" pipe size		12	1.333		62	69	4.61	135.61	177
3680	1" pipe size		11	1.455		29	75.50	5.05	109.55	151
3690	1-1/4" pipe size		10	1.600		29	83	5.55	117.55	163
3700	1-1/2" pipe size		9	1.778		29	92	6.15	127.15	177
3710	2" pipe size		8	2		29	104	6.90	139.90	195
3720	2-1/2" pipe size		7	2.286		64	118	7.90	189.90	258
3730	3" pipe size		6	2.667		37	138	9.25	184.25	260
3740	4" pipe size		4	4		58.50	207	13.85	279.35	395
3746	5" pipe size		3.50	4.571		139	237	15.80	391.80	530
3750	6" pipe size	Q-16	4.50	5.333		160	287	12.25	459.25	625
3760	8" pipe size		3.50	6.857		277	370	15.75	662.75	875
3770	10" pipe size		2.50	9.600		535	515	22	1,072	1,400
3780	12" pipe size		2.25	10.667		790	575	24.50	1,389.50	1,750
3800	Tee, straight									

22 11 Facility Water Distribution
22 11 13 – Facility Water Distribution Piping

22 11 13.47 Pipe Fittings, Steel		Crew	Daily Output	Labor-Hours	Unit	Material	2014 Bare Costs Labor	Equipment	Total	Total Incl O&P
3810	1/2" pipe size	Q-15	9	1.778	Ea.	136	92	6.15	234.15	295
3820	3/4" pipe size		8.50	1.882		131	97.50	6.50	235	298
3830	1" pipe size		8	2		52	104	6.90	162.90	221
3840	1-1/4" pipe size		7	2.286		52	118	7.90	177.90	245
3850	1-1/2" pipe size		6	2.667		52	138	9.25	199.25	277
3860	2" pipe size		5	3.200		59.50	166	11.05	236.55	330
3870	2-1/2" pipe size		4	4		99	207	13.85	319.85	440
3880	3" pipe size		3.50	4.571		124	237	15.80	376.80	515
3890	4" pipe size		2.50	6.400		149	330	22	501	690
3896	5" pipe size	↓	2.25	7.111		375	370	24.50	769.50	990
3900	6" pipe size	Q-16	2.25	10.667		286	575	24.50	885.50	1,200
3910	8" pipe size		2	12		545	645	27.50	1,217.50	1,600
3920	10" pipe size		1.75	13.714		845	735	31.50	1,611.50	2,100
3930	12" pipe size	↓	1.50	16		1,225	860	37	2,122	2,700
4000	Eccentric reducer, 1-1/2" pipe size	Q-15	10	1.600		17.30	83	5.55	105.85	150
4010	2" pipe size		9	1.778		47.50	92	6.15	145.65	198
4020	2-1/2" pipe size		8	2		71.50	104	6.90	182.40	242
4030	3" pipe size		7	2.286		59	118	7.90	184.90	253
4040	4" pipe size		5	3.200		95.50	166	11.05	272.55	365
4046	5" pipe size	↓	4.70	3.404		266	176	11.80	453.80	570
4050	6" pipe size	Q-16	4.50	5.333		259	287	12.25	558.25	735
4060	8" pipe size		3.50	6.857		390	370	15.75	775.75	1,000
4070	10" pipe size		2.50	9.600		665	515	22	1,202	1,525
4080	12" pipe size		2.25	10.667		845	575	24.50	1,444.50	1,825
4090	14" pipe size		2.10	11.429		1,575	615	26.50	2,216.50	2,675
4100	16" pipe size	↓	1.90	12.632		2,025	680	29	2,734	3,275
4151	Cap, 1-1/2" pipe size	Q-15	24	.667		24.50	34.50	2.31	61.31	81.50
4152	2" pipe size		18	.889		22	46	3.08	71.08	97
4153	2-1/2" pipe size		16	1		30	52	3.46	85.46	114
4154	3" pipe size		14	1.143		33.50	59	3.95	96.45	130
4155	4" pipe size	↓	10	1.600		45	83	5.55	133.55	181
4156	6" pipe size	Q-16	9	2.667		92	143	6.15	241.15	325
4157	8" pipe size		7	3.429		138	184	7.90	329.90	440
4158	10" pipe size		5	4.800		212	258	11.05	481.05	635
4159	12" pipe size	↓	4	6	↓	283	320	13.80	616.80	810
4190	Weld fittings, reducing, standard weight									
4200	Welding ring w/spacer pins, 2" pipe size				Ea.	2.40			2.40	2.64
4210	2-1/2" pipe size					2.45			2.45	2.70
4220	3" pipe size					2.50			2.50	2.75
4230	4" pipe size					2.65			2.65	2.92
4236	5" pipe size					3.35			3.35	3.69
4240	6" pipe size					3.60			3.60	3.96
4250	8" pipe size					4.20			4.20	4.62
4260	10" pipe size					4.70			4.70	5.15
4270	12" pipe size					5.40			5.40	5.95
4280	14" pipe size					6.45			6.45	7.10
4290	16" pipe size					7.60			7.60	8.35
4300	18" pipe size					8.60			8.60	9.45
4310	20" pipe size					8.95			8.95	9.85
4330	24" pipe size					11.10			11.10	12.20
4340	26" pipe size					13.70			13.70	15.05
4350	30" pipe size					16.75			16.75	18.45
4370	36" pipe size				↓	20.50			20.50	22.50

22 11 Facility Water Distribution
22 11 13 – Facility Water Distribution Piping

22 11 13.47 Pipe Fittings, Steel

		Crew	Daily Output	Labor-Hours	Unit	Material	2014 Bare Costs Labor	Equipment	Total	Total Incl O&P
4600	Tee, reducing on outlet									
4601	2-1/2" x 2" pipe size	Q-15	5	3.200	Ea.	100	166	11.05	277.05	375
4602	3" x 2-1/2" pipe size		4	4		118	207	13.85	338.85	460
4604	4" x 3" pipe size		3	5.333		124	276	18.45	418.45	575
4605	5" x 4" pipe size		2.50	6.400		268	330	22	620	820
4606	6" x 5" pipe size	Q-16	3	8		370	430	18.40	818.40	1,075
4607	8" x 6" pipe size		2.50	9.600		460	515	22	997	1,300
4608	10" x 8" pipe size		2	12		850	645	27.50	1,522.50	1,950
4609	12" x 10" pipe size		1.60	15		1,250	805	34.50	2,089.50	2,650
4610	16" x 12" pipe size		1.50	16		2,000	860	37	2,897	3,550
4611	14" x 12" pipe size		1.52	15.789		1,900	850	36.50	2,786.50	3,400
4618	Reducer, concentric									
4619	2-1/2" by 2" pipe size	Q-15	10	1.600	Ea.	42.50	83	5.55	131.05	178
4620	3" by 2-1/2" pipe size		9	1.778		35	92	6.15	133.15	184
4621	3-1/2" by 3" pipe size		8	2		89.50	104	6.90	200.40	262
4622	4" by 2-1/2" pipe size		7	2.286		54.50	118	7.90	180.40	248
4623	5" by 3" pipe size		7	2.286		114	118	7.90	239.90	315
4624	6" by 4" pipe size	Q-16	6	4		94	215	9.20	318.20	440
4625	8" by 6" pipe size		5	4.800		124	258	11.05	393.05	540
4626	10" by 8" pipe size		4	6		240	320	13.80	573.80	765
4627	12" by 10" pipe size		3	8		296	430	18.40	744.40	995
4660	Reducer, eccentric									
4662	3" x 2" pipe size	Q-15	8	2	Ea.	59	104	6.90	169.90	229
4664	4" x 3" pipe size		6	2.667		71.50	138	9.25	218.75	298
4666	4" x 2" pipe size		6	2.667		87.50	138	9.25	234.75	315
4670	6" x 4" pipe size	Q-16	5	4.800		148	258	11.05	417.05	565
4672	6" x 3" pipe size		5	4.800		232	258	11.05	501.05	655
4676	8" x 6" pipe size		4	6		188	320	13.80	521.80	705
4678	8" x 4" pipe size		4	6		345	320	13.80	678.80	880
4682	10" x 8" pipe size		3	8		247	430	18.40	695.40	940
4684	10" x 6" pipe size		3	8		435	430	18.40	883.40	1,150
4688	12" x 10" pipe size		2.50	9.600		385	515	22	922	1,225
4690	12" x 8" pipe size		2.50	9.600		605	515	22	1,142	1,475
4691	14" x 12" pipe size		2.20	10.909		675	585	25	1,285	1,650
4693	16" x 14" pipe size		1.80	13.333		1,025	715	30.50	1,770.50	2,225
4694	16" x 12" pipe size		2	12		1,375	645	27.50	2,047.50	2,500
4696	18" x 16" pipe size		1.60	15		1,100	805	34.50	1,939.50	2,500
5000	Weld joint, socket, forged steel, 3000 lb., schedule 40 pipe									
5010	90° elbow, straight									
5020	1/4" pipe size	Q-15	22	.727	Ea.	23.50	37.50	2.52	63.52	86
5030	3/8" pipe size		22	.727		23.50	37.50	2.52	63.52	86
5040	1/2" pipe size		20	.800		13.20	41.50	2.77	57.47	80
5050	3/4" pipe size		20	.800		13.70	41.50	2.77	57.97	80.50
5060	1" pipe size		20	.800		17.60	41.50	2.77	61.87	85
5070	1-1/4" pipe size		18	.889		34	46	3.08	83.08	110
5080	1-1/2" pipe size		16	1		39.50	52	3.46	94.96	125
5090	2" pipe size		12	1.333		58	69	4.61	131.61	173
5100	2-1/2" pipe size		10	1.600		165	83	5.55	253.55	310
5110	3" pipe size		8	2		284	104	6.90	394.90	480
5120	4" pipe size		6	2.667		750	138	9.25	897.25	1,050
5130	45° Elbow, straight									
5134	1/4" pipe size	Q-15	22	.727	Ea.	24	37.50	2.52	64.02	86
5135	3/8" pipe size		22	.727		24	37.50	2.52	64.02	86

22 11 Facility Water Distribution

22 11 13 – Facility Water Distribution Piping

22 11 13.47 Pipe Fittings, Steel

		Crew	Daily Output	Labor-Hours	Unit	Material	2014 Bare Costs Labor	Equipment	Total	Total Incl O&P
5136	1/2" pipe size	Q-15	20	.800	Ea.	17.55	41.50	2.77	61.82	85
5137	3/4" pipe size		20	.800		20.50	41.50	2.77	64.77	88
5140	1" pipe size		20	.800		26.50	41.50	2.77	70.77	95
5150	1-1/4" pipe size		18	.889		36.50	46	3.08	85.58	113
5160	1-1/2" pipe size		16	1		45	52	3.46	100.46	131
5170	2" pipe size		12	1.333		72	69	4.61	145.61	189
5180	2-1/2" pipe size		10	1.600		190	83	5.55	278.55	340
5190	3" pipe size		8	2		315	104	6.90	425.90	510
5200	4" pipe size		6	2.667		620	138	9.25	767.25	905
5250	Tee, straight									
5254	1/4" pipe size	Q-15	15	1.067	Ea.	26	55.50	3.69	85.19	116
5255	3/8" pipe size		15	1.067		26	55.50	3.69	85.19	116
5256	1/2" pipe size		13	1.231		16.35	64	4.26	84.61	119
5257	3/4" pipe size		13	1.231		19.95	64	4.26	88.21	123
5260	1" pipe size		13	1.231		27	64	4.26	95.26	131
5270	1-1/4" pipe size		12	1.333		41.50	69	4.61	115.11	155
5280	1-1/2" pipe size		11	1.455		55	75.50	5.05	135.55	181
5290	2" pipe size		8	2		80	104	6.90	190.90	251
5300	2-1/2" pipe size		6	2.667		229	138	9.25	376.25	470
5310	3" pipe size		5	3.200		550	166	11.05	727.05	860
5320	4" pipe size		4	4		880	207	13.85	1,100.85	1,300
5350	For reducing sizes, add					60%				
5450	Couplings									
5451	1/4" pipe size	Q-15	23	.696	Ea.	15	36	2.41	53.41	73.50
5452	3/8" pipe size		23	.696		15	36	2.41	53.41	73.50
5453	1/2" pipe size		21	.762		6.95	39.50	2.64	49.09	70
5454	3/4" pipe size		21	.762		8.95	39.50	2.64	51.09	72.50
5460	1" pipe size		20	.800		9.90	41.50	2.77	54.17	76.50
5470	1-1/4" pipe size		20	.800		17.55	41.50	2.77	61.82	85
5480	1-1/2" pipe size		18	.889		19.40	46	3.08	68.48	94.50
5490	2" pipe size		14	1.143		31	59	3.95	93.95	128
5500	2-1/2" pipe size		12	1.333		69.50	69	4.61	143.11	186
5510	3" pipe size		9	1.778		156	92	6.15	254.15	315
5520	4" pipe size		7	2.286		234	118	7.90	359.90	445
5570	Union, 1/4" pipe size		21	.762		32	39.50	2.64	74.14	98
5571	3/8" pipe size		21	.762		32	39.50	2.64	74.14	98
5572	1/2" pipe size		19	.842		27	43.50	2.91	73.41	98.50
5573	3/4" pipe size		19	.842		31.50	43.50	2.91	77.91	104
5574	1" pipe size		19	.842		40.50	43.50	2.91	86.91	114
5575	1-1/4" pipe size		17	.941		65.50	49	3.26	117.76	149
5576	1-1/2" pipe size		15	1.067		71	55.50	3.69	130.19	166
5577	2" pipe size		11	1.455		102	75.50	5.05	182.55	232
5600	Reducer, 1/4" pipe size		23	.696		38.50	36	2.41	76.91	99
5601	3/8" pipe size		23	.696		40.50	36	2.41	78.91	102
5602	1/2" pipe size		21	.762		24.50	39.50	2.64	66.64	89.50
5603	3/4" pipe size		21	.762		24.50	39.50	2.64	66.64	89.50
5604	1" pipe size		21	.762		34.50	39.50	2.64	76.64	100
5605	1-1/4" pipe size		19	.842		44	43.50	2.91	90.41	118
5607	1-1/2" pipe size		17	.941		47	49	3.26	99.26	129
5608	2" pipe size		13	1.231		52	64	4.26	120.26	159
5612	Cap, 1/4" pipe size		46	.348		14.30	18	1.20	33.50	44
5613	3/8" pipe size		46	.348		14.30	18	1.20	33.50	44
5614	1/2" pipe size		42	.381		8.75	19.75	1.32	29.82	41

22 11 Facility Water Distribution

22 11 13 – Facility Water Distribution Piping

22 11 13.47 Pipe Fittings, Steel

		Crew	Daily Output	Labor-Hours	Unit	Material	2014 Bare Costs Labor	Equipment	Total	Total Incl O&P
5615	3/4" pipe size	Q-15	42	.381	Ea.	10.20	19.75	1.32	31.27	42.50
5616	1" pipe size		42	.381		15.45	19.75	1.32	36.52	48.50
5617	1-1/4" pipe size		38	.421		18.20	22	1.46	41.66	54.50
5618	1-1/2" pipe size		34	.471		25.50	24.50	1.63	51.63	67
5619	2" pipe size		26	.615		39	32	2.13	73.13	93.50
5630	T-O-L, 1/4" pipe size, nozzle		23	.696		7.10	36	2.41	45.51	65
5631	3/8" pipe size, nozzle		23	.696		7.20	36	2.41	45.61	65
5632	1/2" pipe size, nozzle		22	.727		7.10	37.50	2.52	47.12	67.50
5633	3/4" pipe size, nozzle		21	.762		8.20	39.50	2.64	50.34	71.50
5634	1" pipe size, nozzle		20	.800		9.55	41.50	2.77	53.82	76
5635	1-1/4" pipe size, nozzle		18	.889		15.10	46	3.08	64.18	89.50
5636	1-1/2" pipe size, nozzle		16	1		15.10	52	3.46	70.56	98.50
5637	2" pipe size, nozzle		12	1.333		17.20	69	4.61	90.81	128
5638	2-1/2" pipe size, nozzle		10	1.600		57.50	83	5.55	146.05	195
5639	4" pipe size, nozzle		6	2.667		131	138	9.25	278.25	365
5640	W-O-L, 1/4" pipe size, nozzle		23	.696		16.80	36	2.41	55.21	75.50
5641	3/8" pipe size, nozzle		23	.696		15.85	36	2.41	54.26	74.50
5642	1/2" pipe size, nozzle		22	.727		14.70	37.50	2.52	54.72	76
5643	3/4" pipe size, nozzle		21	.762		15.50	39.50	2.64	57.64	79.50
5644	1" pipe size, nozzle		20	.800		16.20	41.50	2.77	60.47	83.50
5645	1-1/4" pipe size, nozzle		18	.889		19.25	46	3.08	68.33	94
5646	1-1/2" pipe size, nozzle		16	1		19.25	52	3.46	74.71	103
5647	2" pipe size, nozzle		12	1.333		19.45	69	4.61	93.06	131
5648	2-1/2" pipe size, nozzle		10	1.600		44.50	83	5.55	133.05	180
5649	3" pipe size, nozzle		8	2		48.50	104	6.90	159.40	217
5650	4" pipe size, nozzle		6	2.667		61.50	138	9.25	208.75	287
5651	5" pipe size, nozzle		5	3.200		152	166	11.05	329.05	430
5652	6" pipe size, nozzle		4	4		177	207	13.85	397.85	525
5653	8" pipe size, nozzle		3	5.333		340	276	18.45	634.45	810
5654	10" pipe size, nozzle		2.60	6.154		490	320	21.50	831.50	1,050
5655	12" pipe size, nozzle		2.20	7.273		930	375	25	1,330	1,625
5674	S-O-L, 1/4" pipe size, outlet		23	.696		9.45	36	2.41	47.86	67.50
5675	3/8" pipe size, outlet		23	.696		9.45	36	2.41	47.86	67.50
5676	1/2" pipe size, outlet		22	.727		8.80	37.50	2.52	48.82	69.50
5677	3/4" pipe size, outlet		21	.762		8.95	39.50	2.64	51.09	72.50
5678	1" pipe size, outlet		20	.800		9.90	41.50	2.77	54.17	76.50
5679	1-1/4" pipe size, outlet		18	.889		16.55	46	3.08	65.63	91
5680	1-1/2" pipe size, outlet		16	1		16.55	52	3.46	72.01	100
5681	2" pipe size, outlet	↓	12	1.333	↓	18.80	69	4.61	92.41	130
6000	Weld-on flange, forged steel									
6020	Slip-on, 150 lb. flange (welded front and back)									
6050	1/2" pipe size	Q-15	18	.889	Ea.	17.50	46	3.08	66.58	92
6060	3/4" pipe size		18	.889		17.50	46	3.08	66.58	92
6070	1" pipe size		17	.941		17.50	49	3.26	69.76	96.50
6080	1-1/4" pipe size		16	1		17.50	52	3.46	72.96	101
6090	1-1/2" pipe size		15	1.067		17.50	55.50	3.69	76.69	107
6100	2" pipe size		12	1.333		19.45	69	4.61	93.06	131
6110	2-1/2" pipe size		10	1.600		25	83	5.55	113.55	159
6120	3" pipe size		9	1.778		26.50	92	6.15	124.65	175
6130	3-1/2" pipe size		7	2.286		33	118	7.90	158.90	224
6140	4" pipe size		6	2.667		33	138	9.25	180.25	256
6150	5" pipe size	↓	5	3.200	↓	57.50	166	11.05	234.55	325
6160	6" pipe size	Q-16	6	4		54.50	215	9.20	278.70	395

22 11 Facility Water Distribution

22 11 13 – Facility Water Distribution Piping

22 11 13.47 Pipe Fittings, Steel

		Crew	Daily Output	Labor-Hours	Unit	Material	2014 Bare Costs Labor	Equipment	Total	Total Incl O&P
6170	8" pipe size	Q-16	5	4.800	Ea.	82.50	258	11.05	351.55	495
6180	10" pipe size		4	6		143	320	13.80	476.80	655
6190	12" pipe size		3	8		212	430	18.40	660.40	905
6191	14" pipe size		2.50	9.600		281	515	22	818	1,125
6192	16" pipe size	↓	1.80	13.333	↓	440	715	30.50	1,185.50	1,600
6200	300 lb. flange									
6210	1/2" pipe size	Q-15	17	.941	Ea.	23.50	49	3.26	75.76	103
6220	3/4" pipe size		17	.941		23.50	49	3.26	75.76	103
6230	1" pipe size		16	1		23.50	52	3.46	78.96	107
6240	1-1/4" pipe size		13	1.231		23.50	64	4.26	91.76	127
6250	1-1/2" pipe size		12	1.333		23.50	69	4.61	97.11	135
6260	2" pipe size		11	1.455		31	75.50	5.05	111.55	154
6270	2-1/2" pipe size		9	1.778		35	92	6.15	133.15	184
6280	3" pipe size		7	2.286		38	118	7.90	163.90	230
6290	4" pipe size		6	2.667		55.50	138	9.25	202.75	281
6300	5" pipe size	↓	4	4		95	207	13.85	315.85	435
6310	6" pipe size	Q-16	5	4.800		95.50	258	11.05	364.55	505
6320	8" pipe size		4	6		163	320	13.80	496.80	680
6330	10" pipe size		3.40	7.059		278	380	16.25	674.25	900
6340	12" pipe size	↓	2.80	8.571		340	460	19.70	819.70	1,100
6400	Welding neck, 150 lb. flange									
6410	1/2" pipe size	Q-15	40	.400	Ea.	25.50	20.50	1.38	47.38	61
6420	3/4" pipe size		36	.444		25.50	23	1.54	50.04	64.50
6430	1" pipe size		32	.500		25.50	26	1.73	53.23	69
6440	1-1/4" pipe size		29	.552		25.50	28.50	1.91	55.91	73
6450	1-1/2" pipe size		26	.615		25.50	32	2.13	59.63	78.50
6460	2" pipe size		20	.800		29.50	41.50	2.77	73.77	97.50
6470	2-1/2" pipe size		16	1		32	52	3.46	87.46	117
6480	3" pipe size		14	1.143		35	59	3.95	97.95	132
6500	4" pipe size		10	1.600		42.50	83	5.55	131.05	178
6510	5" pipe size	↓	8	2		66.50	104	6.90	177.40	237
6520	6" pipe size	Q-16	10	2.400		64	129	5.50	198.50	271
6530	8" pipe size		7	3.429		113	184	7.90	304.90	410
6540	10" pipe size		6	4		180	215	9.20	404.20	535
6550	12" pipe size		5	4.800		262	258	11.05	531.05	690
6551	14" pipe size		4.50	5.333		380	287	12.25	679.25	865
6552	16" pipe size		3	8		570	430	18.40	1,018.40	1,300
6553	18" pipe size		2.50	9.600		800	515	22	1,337	1,675
6554	20" pipe size		2.30	10.435		975	560	24	1,559	1,950
6556	24" pipe size		2	12		1,300	645	27.50	1,972.50	2,450
6557	26" pipe size		1.70	14.118		1,425	760	32.50	2,217.50	2,750
6558	30" pipe size		.90	26.667		1,650	1,425	61.50	3,136.50	4,050
6559	36" pipe size	↓	.75	32	↓	1,900	1,725	73.50	3,698.50	4,750
6560	300 lb. flange									
6570	1/2" pipe size	Q-15	36	.444	Ea.	31	23	1.54	55.54	70.50
6580	3/4" pipe size		34	.471		31	24.50	1.63	57.13	73
6590	1" pipe size		30	.533		31	27.50	1.85	60.35	77.50
6600	1-1/4" pipe size		28	.571		31	29.50	1.98	62.48	80.50
6610	1-1/2" pipe size		24	.667		31	34.50	2.31	67.81	88.50
6620	2" pipe size		18	.889		38.50	46	3.08	87.58	115
6630	2-1/2" pipe size		14	1.143		44	59	3.95	106.95	142
6640	3" pipe size		12	1.333		44.50	69	4.61	118.11	158
6650	4" pipe size	↓	8	2		73	104	6.90	183.90	244

22 11 Facility Water Distribution

22 11 13 – Facility Water Distribution Piping

22 11 13.47 Pipe Fittings, Steel		Crew	Daily Output	Labor-Hours	Unit	Material	2014 Bare Costs Labor	Equipment	Total	Total Incl O&P
6660	5" pipe size	Q-15	7	2.286	Ea.	111	118	7.90	236.90	310
6670	6" pipe size	Q-16	9	2.667		113	143	6.15	262.15	350
6680	8" pipe size		6	4		195	215	9.20	419.20	550
6690	10" pipe size		5	4.800		355	258	11.05	624.05	790
6700	12" pipe size	↓	4	6	↓	455	320	13.80	788.80	1,000
7740	Plain ends for plain end pipe, mechanically coupled									
7750	Cplg. & labor required at joints not included, add 1 per									
7760	joint for installed price, see line 9180									
7770	Malleable iron, painted, unless noted otherwise									
7800	90° Elbow 1"				Ea.	105			105	116
7810	1-1/2"					125			125	137
7820	2"					186			186	204
7830	2-1/2"					217			217	239
7840	3"					226			226	248
7860	4"					256			256	281
7870	5" welded steel					294			294	325
7880	6"					370			370	410
7890	8" welded steel					700			700	765
7900	10" welded steel					890			890	975
7910	12" welded steel					990			990	1,075
7970	45° Elbow 1"					63			63	69
7980	1-1/2"					94			94	103
7990	2"					197			197	217
8000	2-1/2"					197			197	217
8010	3"					226			226	248
8030	4"					235			235	259
8040	5" welded steel					294			294	325
8050	6"					335			335	370
8060	8"					395			395	435
8070	10" welded steel					410			410	450
8080	12" welded steel					735			735	810
8140	Tee, straight 1"					118			118	130
8150	1-1/2"					152			152	167
8160	2"					152			152	167
8170	2-1/2"					198			198	217
8180	3"					305			305	335
8200	4"					435			435	480
8210	5" welded steel					595			595	655
8220	6"					520			520	570
8230	8" welded steel					745			745	820
8240	10" welded steel					1,175			1,175	1,275
8250	12" welded steel				↓	1,375			1,375	1,500
8340	Segmentally welded steel, painted									
8390	Wye 2"				Ea.	189			189	207
8400	2-1/2"					189			189	207
8410	3"					211			211	232
8430	4"					315			315	345
8440	5"					385			385	420
8450	6"					530			530	580
8460	8"					685			685	750
8470	10"					1,000			1,000	1,100
8480	12"					1,575			1,575	1,725
8540	Wye, lateral 2"				↓	205			205	225

22 11 Facility Water Distribution

22 11 13 – Facility Water Distribution Piping

22 11 13.47 Pipe Fittings, Steel		Crew	Daily Output	Labor-Hours	Unit	Material	2014 Bare Costs Labor	Equipment	Total	Total Incl O&P
8550	2-1/2"				Ea.	236			236	259
8560	3"					280			280	310
8580	4"					385			385	425
8590	5"					655			655	720
8600	6"					675			675	740
8610	8"					1,125			1,125	1,250
8620	10"					1,125			1,125	1,250
8630	12"					2,150			2,150	2,375
8690	Cross, 2"					199			199	219
8700	2-1/2"					199			199	219
8710	3"					238			238	262
8730	4"					325			325	360
8740	5"					460			460	510
8750	6"					610			610	670
8760	8"					780			780	860
8770	10"					1,150			1,150	1,250
8780	12"					1,650			1,650	1,800
8800	Tees, reducing 2" x 1"					129			129	142
8810	2" x 1-1/2"					129			129	142
8820	3" x 1"					129			129	142
8830	3" x 1-1/2"					130			130	143
8840	3" x 2"					131			131	144
8850	4" x 1"					190			190	209
8860	4" x 1-1/2"					190			190	209
8870	4" x 2"					188			188	207
8880	4" x 2-1/2"					193			193	213
8890	4" x 3"					193			193	213
8900	6" x 2"					281			281	310
8910	6" x 3"					305			305	335
8920	6" x 4"					330			330	360
8930	8" x 2"					345			345	380
8940	8" x 3"					350			350	385
8950	8" x 4"					375			375	410
8960	8" x 5"					405			405	445
8970	8" x 6"					440			440	480
8980	10" x 4"					580			580	640
8990	10" x 6"					580			580	640
9000	10" x 8"					615			615	680
9010	12" x 6"					895			895	985
9020	12" x 8"					900			900	990
9030	12" x 10"				▼	930			930	1,025
9080	Adapter nipples 3" long									
9090	1"				Ea.	18.20			18.20	20
9100	1-1/2"					18.20			18.20	20
9110	2"					18.20			18.20	20
9120	2-1/2"					21			21	23
9130	3"					25.50			25.50	28
9140	4"					41.50			41.50	46
9150	6"				▼	110			110	121
9180	Coupling, mechanical, plain end pipe to plain end pipe or fitting									
9190	1"	Q-1	29	.552	Ea.	63.50	28.50		92	113
9200	1-1/2"		28	.571		63.50	29.50		93	115
9210	2"	▼	27	.593	▼	63.50	30.50		94	117

22 11 Facility Water Distribution

22 11 13 – Facility Water Distribution Piping

22 11 13.47 Pipe Fittings, Steel

		Crew	Daily Output	Labor-Hours	Unit	Material	2014 Bare Costs Labor	Equipment	Total	Total Incl O&P
9220	2-1/2"	Q-1	26	.615	Ea.	63.50	32		95.50	118
9230	3"		25	.640		93	33		126	152
9240	3-1/2"		24	.667		107	34.50		141.50	170
9250	4"		22	.727		107	37.50		144.50	175
9260	5"	Q-2	28	.857		153	46		199	238
9270	6"		24	1		187	53.50		240.50	286
9280	8"		19	1.263		330	68		398	460
9290	10"		16	1.500		425	80.50		505.50	590
9300	12"		12	2		535	107		642	750
9310	Outlets for precut holes through pipe wall									
9331	Strapless type, with gasket									
9332	4" to 8" pipe x 1/2"	1 Plum	13	.615	Ea.	69.50	35.50		105	130
9333	4" to 8" pipe x 3/4"		13	.615		74	35.50		109.50	135
9334	10" pipe and larger x 1/2"		11	.727		69.50	42		111.50	140
9335	10" pipe and larger x 3/4"		11	.727		74	42		116	145
9341	Thermometer wells with gasket									
9342	4" to 8" pipe, 6" stem	1 Plum	14	.571	Ea.	105	33		138	165
9343	8" pipe and larger, 6" stem	"	13	.615	"	105	35.50		140.50	169
9940	For galvanized fittings for plain end pipe, add					20%				

22 11 13.48 Pipe, Fittings and Valves, Steel, Grooved-Joint

		Crew	Daily Output	Labor-Hours	Unit	Material	2014 Bare Costs Labor	Equipment	Total	Total Incl O&P
0010	**PIPE, FITTINGS AND VALVES, STEEL, GROOVED-JOINT** R221113-70									
0012	Fittings are ductile iron. Steel fittings noted.									
0020	Pipe includes coupling & clevis type hanger assemblies, 10' O.C.									
0500	Schedule 10, black									
0550	2" diameter	1 Plum	43	.186	L.F.	6.15	10.70		16.85	23
0560	2-1/2" diameter	Q-1	61	.262		7.85	13.60		21.45	29
0570	3" diameter		55	.291		9.10	15.05		24.15	33
0580	3-1/2" diameter		53	.302		11.65	15.65		27.30	36.50
0590	4" diameter		49	.327		12.25	16.90		29.15	39
0600	5" diameter		40	.400		14.80	20.50		35.30	48
0610	6" diameter	Q-2	46	.522		19.45	28		47.45	64
0620	8" diameter	"	41	.585		26.50	31.50		58	77
0700	To delete couplings & hangers, subtract									
0710	2" diam. to 5" diam.					25%	20%			
0720	6" diam. to 8" diam.					27%	15%			
1000	Schedule 40, black									
1040	3/4" diameter	1 Plum	71	.113	L.F.	4.62	6.50		11.12	14.90
1050	1" diameter		63	.127		4.41	7.30		11.71	15.90
1060	1-1/4" diameter		58	.138		5.70	7.95		13.65	18.25
1070	1-1/2" diameter		51	.157		6.50	9.05		15.55	21
1080	2" diameter		40	.200		7.85	11.50		19.35	26
1090	2-1/2" diameter	Q-1	57	.281		13.45	14.55		28	37
1100	3" diameter		50	.320		16.55	16.60		33.15	43
1110	4" diameter		45	.356		22.50	18.40		40.90	53
1120	5" diameter		37	.432		36.50	22.50		59	74.50
1130	6" diameter	Q-2	42	.571		47	30.50		77.50	98.50
1140	8" diameter		37	.649		77	35		112	137
1150	10" diameter		31	.774		104	41.50		145.50	177
1160	12" diameter		27	.889		116	48		164	200
1170	14" diameter		20	1.200		119	64.50		183.50	229
1180	16" diameter		17	1.412		179	76		255	310
1190	18" diameter		14	1.714		178	92		270	335

22 11 Facility Water Distribution

22 11 13 – Facility Water Distribution Piping

22 11 13.48 Pipe, Fittings and Valves, Steel, Grooved-Joint		Crew	Daily Output	Labor-Hours	Unit	Material	2014 Bare Costs Labor	Equipment	Total	Total Incl O&P
1200	20" diameter	Q-2	12	2	L.F.	208	107		315	390
1210	24" diameter	↓	10	2.400	↓	234	129		363	450
1740	To delete coupling & hanger, subtract									
1750	3/4" diam. to 2" diam.					65%	27%			
1760	2-1/2" diam. to 5" diam.					41%	18%			
1770	6" diam. to 12" diam.					31%	13%			
1780	14" diam. to 24" diam.					35%	10%			
1800	Galvanized									
1840	3/4" diameter	1 Plum	71	.113	L.F.	5.85	6.50		12.35	16.25
1850	1" diameter		63	.127		5.55	7.30		12.85	17.15
1860	1-1/4" diameter		58	.138		7.20	7.95		15.15	19.95
1870	1-1/2" diameter		51	.157		8.40	9.05		17.45	23
1880	2" diameter	↓	40	.200		12	11.50		23.50	30.50
1890	2-1/2" diameter	Q-1	57	.281		18.20	14.55		32.75	42
1900	3" diameter		50	.320		23	16.60		39.60	50.50
1910	4" diameter		45	.356		32	18.40		50.40	63.50
1920	5" diameter	↓	37	.432		58.50	22.50		81	98.50
1930	6" diameter	Q-2	42	.571		63.50	30.50		94	117
1940	8" diameter		37	.649		74.50	35		109.50	135
1950	10" diameter		31	.774		120	41.50		161.50	194
1960	12" diameter	↓	27	.889	↓	146	48		194	233
2540	To delete coupling & hanger, subtract									
2550	3/4" diam. to 2" diam.					36%	27%			
2560	2-1/2" diam. to 5" diam.					19%	18%			
2570	6" diam. to 12" diam.					14%	13%			
2600	Schedule 80, black									
2610	3/4" diameter	1 Plum	65	.123	L.F.	6.25	7.10		13.35	17.60
2650	1" diameter		61	.131		7.35	7.55		14.90	19.50
2660	1-1/4" diameter		55	.145		9.60	8.35		17.95	23
2670	1-1/2" diameter		49	.163		11.15	9.40		20.55	26.50
2680	2" diameter	↓	38	.211		14.30	12.10		26.40	34
2690	2-1/2" diameter	Q-1	54	.296		21	15.35		36.35	46.50
2700	3" diameter		48	.333		27.50	17.25		44.75	56.50
2710	4" diameter		44	.364		32	18.85		50.85	64
2720	5" diameter	↓	35	.457		51	23.50		74.50	92
2730	6" diameter	Q-2	40	.600		69	32		101	125
2740	8" diameter		35	.686		105	37		142	172
2750	10" diameter		29	.828		190	44.50		234.50	276
2760	12" diameter	↓	24	1	↓	370	53.50		423.50	490
3240	To delete coupling & hanger, subtract									
3250	3/4" diam. to 2" diam.					30%	25%			
3260	2-1/2" diam. to 5" diam.					14%	17%			
3270	6" diam. to 12" diam.					12%	12%			
3300	Galvanized									
3310	3/4" diameter	1 Plum	65	.123	L.F.	7.30	7.10		14.40	18.75
3350	1" diameter		61	.131		9.35	7.55		16.90	21.50
3360	1-1/4" diameter		55	.145		12.35	8.35		20.70	26
3370	1-1/2" diameter		46	.174		14.50	10		24.50	31
3380	2" diameter	↓	38	.211		19.50	12.10		31.60	40
3390	2-1/2" diameter	Q-1	54	.296		29	15.35		44.35	55
3400	3" diameter		48	.333		38	17.25		55.25	68
3410	4" diameter		44	.364		53.50	18.85		72.35	87.50
3420	5" diameter	↓	35	.457	↓	69.50	23.50		93	113

22 11 Facility Water Distribution

22 11 13 – Facility Water Distribution Piping

22 11 13.48 Pipe, Fittings and Valves, Steel, Grooved-Joint

		Crew	Daily Output	Labor-Hours	Unit	Material	Labor	Equipment	Total	Total Incl O&P
3430	6" diameter	Q-2	40	.600	L.F.	95.50	32		127.50	154
3440	8" diameter		35	.686		145	37		182	216
3450	10" diameter		29	.828		262	44.50		306.50	355
3460	12" diameter	▼	24	1	▼	520	53.50		573.50	650
3920	To delete coupling & hanger, subtract									
3930	3/4" diam. to 2" diam.					30%	25%			
3940	2-1/2" diam. to 5" diam.					15%	17%			
3950	6" diam. to 12" diam.					11%	12%			
3990	Fittings: coupling material required at joints not incl. in fitting price.									
3994	Add 1 selected coupling, material only, per joint for installed price.									
4000	Elbow, 90° or 45°, painted									
4030	3/4" diameter	1 Plum	50	.160	Ea.	56	9.20		65.20	76
4040	1" diameter		50	.160		30	9.20		39.20	47
4050	1-1/4" diameter		40	.200		30	11.50		41.50	50.50
4060	1-1/2" diameter		33	.242		30	13.95		43.95	54
4070	2" diameter	▼	25	.320		30	18.40		48.40	61
4080	2-1/2" diameter	Q-1	40	.400		30	20.50		50.50	64.50
4090	3" diameter		33	.485		53	25		78	96.50
4100	4" diameter		25	.640		57.50	33		90.50	114
4110	5" diameter	▼	20	.800		137	41.50		178.50	214
4120	6" diameter	Q-2	25	.960		161	51.50		212.50	256
4130	8" diameter		21	1.143		335	61.50		396.50	465
4140	10" diameter		18	1.333		590	71.50		661.50	755
4150	12" diameter		15	1.600		940	86		1,026	1,150
4170	14" diameter		12	2		860	107		967	1,100
4180	16" diameter	▼	11	2.182		1,125	117		1,242	1,400
4190	18" diameter	Q-3	15	2.133		1,425	117		1,542	1,725
4200	20" diameter		13	2.462		1,875	135		2,010	2,250
4210	24" diameter	▼	11	2.909	▼	2,700	159		2,859	3,225
4250	For galvanized elbows, add					26%				
4690	Tee, painted									
4700	3/4" diameter	1 Plum	38	.211	Ea.	60.50	12.10		72.60	85
4740	1" diameter		33	.242		46.50	13.95		60.45	72.50
4750	1-1/4" diameter		27	.296		46.50	17.05		63.55	77.50
4760	1-1/2" diameter		22	.364		46.50	21		67.50	83
4770	2" diameter	▼	17	.471		46.50	27		73.50	92.50
4780	2-1/2" diameter	Q-1	27	.593		46.50	30.50		77	98
4790	3" diameter		22	.727		64	37.50		101.50	127
4800	4" diameter		17	.941		97	49		146	181
4810	5" diameter	▼	13	1.231		226	64		290	345
4820	6" diameter	Q-2	17	1.412		261	76		337	400
4830	8" diameter		14	1.714		570	92		662	770
4840	10" diameter		12	2		1,150	107		1,257	1,400
4850	12" diameter		10	2.400		1,600	129		1,729	1,950
4851	14" diameter		9	2.667		1,500	143		1,643	1,900
4852	16" diameter	▼	8	3		1,550	161		1,711	1,950
4853	18" diameter	Q-3	11	2.909		1,625	159		1,784	2,050
4854	20" diameter		10	3.200		2,325	175		2,500	2,850
4855	24" diameter	▼	8	4	▼	3,550	219		3,769	4,225
4900	For galvanized tees, add					24%				
4906	Couplings, rigid style, painted									
4908	1" diameter	1 Plum	100	.080	Ea.	23	4.60		27.60	32.50
4909	1-1/4" diameter	↓	100	.080	↓	23	4.60		27.60	32.50

187

22 11 Facility Water Distribution

22 11 13 – Facility Water Distribution Piping

22 11 13.48 Pipe, Fittings and Valves, Steel, Grooved-Joint		Crew	Daily Output	Labor-Hours	Unit	Material	2014 Bare Costs Labor	Equipment	Total	Total Incl O&P
4910	1-1/2" diameter	1 Plum	67	.119	Ea.	23	6.85		29.85	36
4912	2" diameter		50	.160		29.50	9.20		38.70	46.50
4914	2-1/2" diameter	Q-1	80	.200		33.50	10.35		43.85	52.50
4916	3" diameter		67	.239		39	12.35		51.35	61.50
4918	4" diameter		50	.320		54.50	16.60		71.10	85
4920	5" diameter		40	.400		70	20.50		90.50	109
4922	6" diameter	Q-2	50	.480		93	26		119	141
4924	8" diameter		42	.571		146	30.50		176.50	208
4926	10" diameter		35	.686		211	37		248	289
4928	12" diameter		32	.750		236	40.50		276.50	320
4930	14" diameter		24	1		257	53.50		310.50	365
4931	16" diameter		20	1.200		335	64.50		399.50	470
4932	18" diameter		18	1.333		385	71.50		456.50	535
4933	20" diameter		16	1.500		530	80.50		610.50	700
4934	24" diameter		13	1.846		675	99		774	895
4940	Flexible, standard, painted									
4950	3/4" diameter	1 Plum	100	.080	Ea.	16.60	4.60		21.20	25
4960	1" diameter		100	.080		16.60	4.60		21.20	25
4970	1-1/4" diameter		80	.100		22	5.75		27.75	32.50
4980	1-1/2" diameter		67	.119		24	6.85		30.85	36.50
4990	2" diameter		50	.160		25.50	9.20		34.70	42
5000	2-1/2" diameter	Q-1	80	.200		29.50	10.35		39.85	48
5010	3" diameter		67	.239		33	12.35		45.35	55
5020	3-1/2" diameter		57	.281		47	14.55		61.55	73.50
5030	4" diameter		50	.320		47.50	16.60		64.10	77
5040	5" diameter		40	.400		72	20.50		92.50	111
5050	6" diameter	Q-2	50	.480		85	26		111	133
5070	8" diameter		42	.571		138	30.50		168.50	199
5090	10" diameter		35	.686		225	37		262	305
5110	12" diameter		32	.750		256	40.50		296.50	340
5120	14" diameter		24	1		365	53.50		418.50	480
5130	16" diameter		20	1.200		480	64.50		544.50	625
5140	18" diameter		18	1.333		560	71.50		631.50	725
5150	20" diameter		16	1.500		880	80.50		960.50	1,100
5160	24" diameter		13	1.846		965	99		1,064	1,225
5176	Lightweight style, painted									
5178	1-1/2" diameter	1 Plum	67	.119	Ea.	21	6.85		27.85	33.50
5180	2" diameter	"	50	.160		25.50	9.20		34.70	42
5182	2-1/2" diameter	Q-1	80	.200		29.50	10.35		39.85	48
5184	3" diameter		67	.239		33	12.35		45.35	55
5186	3-1/2" diameter		57	.281		40	14.55		54.55	66
5188	4" diameter		50	.320		47.50	16.60		64.10	77
5190	5" diameter		40	.400		72	20.50		92.50	111
5192	6" diameter	Q-2	50	.480		85	26		111	133
5194	8" diameter		42	.571		138	30.50		168.50	199
5196	10" diameter		35	.686		279	37		316	360
5198	12" diameter		32	.750		315	40.50		355.50	405
5200	For galvanized couplings, add					33%				
5220	Tee, reducing, painted									
5225	2" x 1-1/2" diameter	Q-1	38	.421	Ea.	98	22		120	141
5226	2-1/2" x 2" diameter		28	.571		98	29.50		127.50	153
5227	3" x 2-1/2" diameter		23	.696		86.50	36		122.50	150
5228	4" x 3" diameter		18	.889		117	46		163	198

22 11 Facility Water Distribution

22 11 13 – Facility Water Distribution Piping

22 11 13.48 Pipe, Fittings and Valves, Steel, Grooved-Joint		Crew	Daily Output	Labor-Hours	Unit	Material	2014 Bare Costs Labor	Equipment	Total	Total Incl O&P
5229	5" x 4" diameter	Q-1	15	1.067	Ea.	249	55.50		304.50	360
5230	6" x 4" diameter	Q-2	18	1.333		275	71.50		346.50	410
5231	8" x 6" diameter		15	1.600		570	86		656	760
5232	10" x 8" diameter		13	1.846		720	99		819	945
5233	12" x 10" diameter		11	2.182		1,100	117		1,217	1,375
5234	14" x 12" diameter		10	2.400		840	129		969	1,125
5235	16" x 12" diameter		9	2.667		1,050	143		1,193	1,375
5236	18" x 12" diameter	Q-3	12	2.667		1,250	146		1,396	1,600
5237	18" x 16" diameter		11	2.909		1,575	159		1,734	1,975
5238	20" x 16" diameter		10	3.200		2,050	175		2,225	2,550
5239	24" x 20" diameter		9	3.556		3,250	195		3,445	3,875
5240	Reducer, concentric, painted									
5241	2-1/2" x 2" diameter	Q-1	43	.372	Ea.	35	19.25		54.25	67.50
5242	3" x 2-1/2" diameter		35	.457		42	23.50		65.50	82.50
5243	4" x 3" diameter		29	.552		51	28.50		79.50	99
5244	5" x 4" diameter		22	.727		70.50	37.50		108	135
5245	6" x 4" diameter	Q-2	26	.923		81.50	49.50		131	165
5246	8" x 6" diameter		23	1.043		211	56		267	315
5247	10" x 8" diameter		20	1.200		415	64.50		479.50	555
5248	12" x 10" diameter		16	1.500		740	80.50		820.50	935
5255	Eccentric, painted									
5256	2-1/2" x 2" diameter	Q-1	42	.381	Ea.	73	19.75		92.75	111
5257	3" x 2-1/2" diameter		34	.471		83.50	24.50		108	129
5258	4" x 3" diameter		28	.571		102	29.50		131.50	157
5259	5" x 4" diameter		21	.762		138	39.50		177.50	212
5260	6" x 4" diameter	Q-2	25	.960		161	51.50		212.50	255
5261	8" x 6" diameter		22	1.091		325	58.50		383.50	445
5262	10" x 8" diameter		19	1.263		855	68		923	1,050
5263	12" x 10" diameter		15	1.600		1,175	86		1,261	1,425
5270	Coupling, reducing, painted									
5272	2" x 1-1/2" diameter	1 Plum	52	.154	Ea.	34	8.85		42.85	50.50
5274	2-1/2" x 2" diameter	Q-1	82	.195		44	10.10		54.10	63.50
5276	3" x 2" diameter		69	.232		50	12		62	73
5278	4" x 2" diameter		52	.308		78.50	15.95		94.45	111
5280	5" x 4" diameter		42	.381		89	19.75		108.75	128
5282	6" x 4" diameter	Q-2	52	.462		134	25		159	185
5284	8" x 6" diameter	"	44	.545		200	29.50		229.50	265
5290	Outlet coupling, painted									
5294	1-1/2" x 1" pipe size	1 Plum	65	.123	Ea.	44.50	7.10		51.60	59.50
5296	2" x 1" pipe size	"	48	.167		45.50	9.60		55.10	64.50
5298	2-1/2" x 1" pipe size	Q-1	78	.205		70	10.65		80.65	93
5300	2-1/2" x 1-1/4" pipe size		70	.229		79	11.85		90.85	105
5302	3" x 1" pipe size		65	.246		89.50	12.75		102.25	118
5304	4" x 3/4" pipe size		48	.333		99	17.25		116.25	135
5306	4" x 1-1/2" pipe size		46	.348		141	18		159	182
5308	6" x 1-1/2" pipe size	Q-2	44	.545		198	29.50		227.50	263
5320	Outlet, strap-on T, painted									
5324	Threaded female branch									
5330	2" pipe x 1-1/2" branch size	1 Plum	50	.160	Ea.	32.50	9.20		41.70	49.50
5334	2-1/2" pipe x 1-1/2" branch size	Q-1	80	.200		39.50	10.35		49.85	59
5338	3" pipe x 1-1/2" branch size		67	.239		41	12.35		53.35	63.50
5342	3" pipe x 2" branch size		59	.271		47	14.05		61.05	73
5346	4" pipe x 2" branch size		50	.320		50	16.60		66.60	80

22 11 Facility Water Distribution

22 11 13 – Facility Water Distribution Piping

22 11 13.48 Pipe, Fittings and Valves, Steel, Grooved-Joint

		Crew	Daily Output	Labor-Hours	Unit	Material	2014 Bare Costs Labor	Equipment	Total	Total Incl O&P
5350	4" pipe x 2-1/2" branch size	Q-1	47	.340	Ea.	52	17.65		69.65	83.50
5354	4" pipe x 3" branch size		43	.372		55.50	19.25		74.75	90
5358	5" pipe x 3" branch size	↓	40	.400		68	20.50		88.50	107
5362	6" pipe x 2" branch size	Q-2	50	.480		62.50	26		88.50	108
5366	6" pipe x 3" branch size		47	.511		71	27.50		98.50	120
5370	6" pipe x 4" branch size		44	.545		79	29.50		108.50	132
5374	8" pipe x 2-1/2" branch size		42	.571		109	30.50		139.50	167
5378	8" pipe x 4" branch size	↓	38	.632	↓	118	34		152	181
5390	Grooved branch									
5394	2" pipe x 1-1/4" branch size	1 Plum	78	.103	Ea.	32.50	5.90		38.40	44.50
5398	2" pipe x 1-1/2" branch size	"	50	.160		32.50	9.20		41.70	49.50
5402	2-1/2" pipe x 1-1/2" branch size	Q-1	80	.200		39.50	10.35		49.85	59
5406	3" pipe x 1-1/2" branch size		67	.239		41	12.35		53.35	63.50
5410	3" pipe x 2" branch size		59	.271		47	14.05		61.05	73
5414	4" pipe x 2" branch size		50	.320		50	16.60		66.60	80
5418	4" pipe x 2-1/2" branch size		47	.340		52	17.65		69.65	83.50
5422	4" pipe x 3" branch size		43	.372		55.50	19.25		74.75	90
5426	5" pipe x 3" branch size	↓	40	.400		68	20.50		88.50	107
5430	6" pipe x 2" branch size	Q-2	50	.480		62.50	26		88.50	108
5434	6" pipe x 3" branch size	Q-1	78	.205		71	10.65		81.65	94.50
5438	6" pipe x 4" branch size	Q-2	44	.545		79	29.50		108.50	132
5442	8" pipe x 2-1/2" branch size		42	.571		109	30.50		139.50	167
5446	8" pipe x 4" branch size	↓	38	.632	↓	118	34		152	181
5750	Flange, w/groove gasket, black steel									
5754	See Line 22 11 13.47 0620 for gasket & bolt set									
5760	ANSI class 125 and 150, painted									
5780	2" pipe size	1 Plum	23	.348	Ea.	99.50	20		119.50	139
5790	2-1/2" pipe size	Q-1	37	.432		123	22.50		145.50	170
5800	3" pipe size		31	.516		133	26.50		159.50	188
5820	4" pipe size		23	.696		178	36		214	251
5830	5" pipe size	↓	19	.842		206	43.50		249.50	293
5840	6" pipe size	Q-2	23	1.043		225	56		281	335
5850	8" pipe size		17	1.412		254	76		330	395
5860	10" pipe size		14	1.714		400	92		492	580
5870	12" pipe size		12	2		525	107		632	740
5880	14" pipe size		10	2.400		1,025	129		1,154	1,325
5890	16" pipe size		9	2.667		1,200	143		1,343	1,525
5900	18" pipe size		6	4		1,475	215		1,690	1,950
5910	20" pipe size		5	4.800		1,775	258		2,033	2,350
5920	24" pipe size	↓	4.50	5.333	↓	2,250	287		2,537	2,900
5940	ANSI class 350, painted									
5946	2" pipe size	1 Plum	23	.348	Ea.	124	20		144	167
5948	2-1/2" pipe size	Q-1	37	.432		144	22.50		166.50	192
5950	3" pipe size		31	.516		197	26.50		223.50	257
5952	4" pipe size		23	.696		261	36		297	345
5954	5" pipe size	↓	19	.842		297	43.50		340.50	390
5956	6" pipe size	Q-2	23	1.043		345	56		401	465
5958	8" pipe size		17	1.412		395	76		471	550
5960	10" pipe size	↓	14	1.714		630	92		722	835
5962	12" pipe size	1 Plum	12	.667	↓	675	38.50		713.50	800
6100	Cross, painted									
6110	2" diameter	1 Plum	12.50	.640	Ea.	83	37		120	147
6112	2-1/2" diameter	Q-1	20	.800		83	41.50		124.50	154

22 11 Facility Water Distribution

22 11 13 – Facility Water Distribution Piping

22 11 13.48 Pipe, Fittings and Valves, Steel, Grooved-Joint		Crew	Daily Output	Labor-Hours	Unit	Material	2014 Bare Costs Labor	Equipment	Total	Total Incl O&P
6114	3" diameter	Q-1	15.50	1.032	Ea.	148	53.50		201.50	243
6116	4" diameter	↓	12.50	1.280		245	66.50		311.50	370
6118	6" diameter	Q-2	12.50	1.920		640	103		743	860
6120	8" diameter		10.50	2.286		845	123		968	1,125
6122	10" diameter		9	2.667		1,375	143		1,518	1,750
6124	12" diameter	↓	7.50	3.200	↓	2,000	172		2,172	2,450
7400	Suction diffuser									
7402	Grooved end inlet x flanged outlet									
7410	3" x 3"	Q-1	50	.320	Ea.	805	16.60		821.60	910
7412	4" x 4"		38	.421		1,100	22		1,122	1,225
7414	5" x 5"	↓	30	.533		1,275	27.50		1,302.50	1,450
7416	6" x 6"	Q-2	38	.632		1,600	34		1,634	1,825
7418	8" x 8"		27	.889		3,000	48		3,048	3,375
7420	10" x 10"		20	1.200		4,075	64.50		4,139.50	4,575
7422	12" x 12"		16	1.500		6,700	80.50		6,780.50	7,500
7424	14" x 14"		15	1.600		7,575	86		7,661	8,475
7426	16" x 14"	↓	14	1.714	↓	7,800	92		7,892	8,725
7500	Strainer, tee type, painted									
7506	2" pipe size	1 Plum	38	.211	Ea.	540	12.10		552.10	610
7508	2-1/2" pipe size	Q-1	62	.258		565	13.35		578.35	640
7510	3" pipe size		50	.320		635	16.60		651.60	720
7512	4" pipe size		38	.421		715	22		737	825
7514	5" pipe size	↓	30	.533		1,050	27.50		1,077.50	1,200
7516	6" pipe size	Q-2	38	.632		1,125	34		1,159	1,300
7518	8" pipe size		27	.889		1,725	48		1,773	1,975
7520	10" pipe size		20	1.200		2,550	64.50		2,614.50	2,900
7522	12" pipe size		16	1.500		3,275	80.50		3,355.50	3,725
7524	14" pipe size		15	1.600		10,200	86		10,286	11,400
7526	16" pipe size	↓	14	1.714	↓	12,700	92		12,792	14,100
7570	Expansion joint, max. 3" travel									
7572	2" diameter	1 Plum	38	.211	Ea.	680	12.10		692.10	765
7574	3" diameter	Q-1	50	.320		780	16.60		796.60	880
7576	4" diameter	"	38	.421		1,025	22		1,047	1,150
7578	6" diameter	Q-2	38	.632	↓	1,600	34		1,634	1,800
7800	Ball valve w/handle, carbon steel trim									
7810	1-1/2" pipe size	1 Plum	50	.160	Ea.	135	9.20		144.20	162
7812	2" pipe size	"	38	.211		149	12.10		161.10	182
7814	2-1/2" pipe size	Q-1	62	.258		325	13.35		338.35	380
7816	3" pipe size		50	.320		525	16.60		541.60	605
7818	4" pipe size	↓	38	.421		815	22		837	930
7820	6" pipe size	Q-2	30	.800	↓	2,500	43		2,543	2,825
7830	With gear operator									
7834	2-1/2" pipe size	Q-1	62	.258	Ea.	660	13.35		673.35	745
7836	3" pipe size		50	.320		935	16.60		951.60	1,050
7838	4" pipe size	↓	38	.421		1,200	22		1,222	1,325
7840	6" pipe size	Q-2	30	.800	↓	2,800	43		2,843	3,150
7870	Check valve									
7874	2-1/2" pipe size	Q-1	62	.258	Ea.	257	13.35		270.35	305
7876	3" pipe size		50	.320		305	16.60		321.60	360
7878	4" pipe size		38	.421		320	22		342	390
7880	5" pipe size	↓	30	.533		535	27.50		562.50	625
7882	6" pipe size	Q-2	38	.632		630	34		664	745
7884	8" pipe size	↓	27	.889	↓	865	48		913	1,025

22 11 Facility Water Distribution

22 11 13 – Facility Water Distribution Piping

22 11 13.48 Pipe, Fittings and Valves, Steel, Grooved-Joint		Crew	Daily Output	Labor-Hours	Unit	Material	2014 Bare Costs Labor	Equipment	Total	Total Incl O&P
7886	10" pipe size	Q-2	20	1.200	Ea.	2,300	64.50		2,364.50	2,650
7888	12" pipe size	↓	16	1.500	↓	2,725	80.50		2,805.50	3,125
7900	Plug valve, balancing, w/lever operator									
7906	3" pipe size	Q-1	50	.320	Ea.	635	16.60		651.60	725
7908	4" pipe size	"	38	.421		785	22		807	895
7909	6" pipe size	Q-2	30	.800	↓	1,200	43		1,243	1,400
7916	With gear operator									
7920	3" pipe size	Q-1	50	.320	Ea.	1,200	16.60		1,216.60	1,350
7922	4" pipe size	"	38	.421		1,250	22		1,272	1,400
7924	6" pipe size	Q-2	38	.632		1,700	34		1,734	1,900
7926	8" pipe size		27	.889		2,600	48		2,648	2,925
7928	10" pipe size		20	1.200		3,275	64.50		3,339.50	3,700
7930	12" pipe size	↓	16	1.500	↓	5,000	80.50		5,080.50	5,625
8000	Butterfly valve, 2 position handle, with standard trim									
8010	1-1/2" pipe size	1 Plum	50	.160	Ea.	242	9.20		251.20	280
8020	2" pipe size	"	38	.211		242	12.10		254.10	284
8030	3" pipe size	Q-1	50	.320		345	16.60		361.60	405
8050	4" pipe size	"	38	.421		380	22		402	455
8070	6" pipe size	Q-2	38	.632		770	34		804	895
8080	8" pipe size		27	.889		1,025	48		1,073	1,225
8090	10" pipe size	↓	20	1.200	↓	1,700	64.50		1,764.50	1,975
8200	With stainless steel trim									
8240	1-1/2" pipe size	1 Plum	50	.160	Ea.	305	9.20		314.20	350
8250	2" pipe size	"	38	.211		305	12.10		317.10	355
8270	3" pipe size	Q-1	50	.320		410	16.60		426.60	480
8280	4" pipe size	"	38	.421		445	22		467	525
8300	6" pipe size	Q-2	38	.632		835	34		869	970
8310	8" pipe size		27	.889		2,075	48		2,123	2,350
8320	10" pipe size		20	1.200		3,450	64.50		3,514.50	3,900
8322	12" pipe size		16	1.500		3,950	80.50		4,030.50	4,475
8324	14" pipe size		15	1.600		5,075	86		5,161	5,700
8326	16" pipe size	↓	14	1.714		7,325	92		7,417	8,200
8328	18" pipe size	Q-3	12	2.667		9,000	146		9,146	10,100
8330	20" pipe size		11	2.909		11,100	159		11,259	12,400
8332	24" pipe size	↓	10	3.200	↓	14,800	175		14,975	16,600
8336	Note: sizes 8" up w/manual gear operator									
9000	Cut one groove, labor									
9010	3/4" pipe size	Q-1	152	.105	Ea.		5.45		5.45	8.25
9020	1" pipe size		140	.114			5.90		5.90	8.95
9030	1-1/4" pipe size		124	.129			6.70		6.70	10.10
9040	1-1/2" pipe size		114	.140			7.25		7.25	11
9050	2" pipe size		104	.154			7.95		7.95	12.05
9060	2-1/2" pipe size		96	.167			8.65		8.65	13.05
9070	3" pipe size		88	.182			9.40		9.40	14.20
9080	3-1/2" pipe size		83	.193			10		10	15.10
9090	4" pipe size		78	.205			10.65		10.65	16.05
9100	5" pipe size		72	.222			11.50		11.50	17.40
9110	6" pipe size		70	.229			11.85		11.85	17.90
9120	8" pipe size		54	.296			15.35		15.35	23
9130	10" pipe size		38	.421			22		22	33
9140	12" pipe size		30	.533			27.50		27.50	41.50
9150	14" pipe size		20	.800			41.50		41.50	62.50
9160	16" pipe size	↓	19	.842	↓		43.50		43.50	66

22 11 Facility Water Distribution

22 11 13 – Facility Water Distribution Piping

22 11 13.48 Pipe, Fittings and Valves, Steel, Grooved-Joint

		Crew	Daily Output	Labor-Hours	Unit	Material	2014 Bare Costs Labor	Equipment	Total	Total Incl O&P
9170	18" pipe size	Q-1	18	.889	Ea.		46		46	69.50
9180	20" pipe size		17	.941			49		49	73.50
9190	24" pipe size	↓	15	1.067	↓		55.50		55.50	83.50
9210	Roll one groove									
9220	3/4" pipe size	Q-1	266	.060	Ea.		3.12		3.12	4.71
9230	1" pipe size		228	.070			3.64		3.64	5.50
9240	1-1/4" pipe size		200	.080			4.14		4.14	6.25
9250	1-1/2" pipe size		178	.090			4.66		4.66	7.05
9260	2" pipe size		116	.138			7.15		7.15	10.80
9270	2-1/2" pipe size		110	.145			7.55		7.55	11.40
9280	3" pipe size		100	.160			8.30		8.30	12.50
9290	3-1/2" pipe size		94	.170			8.80		8.80	13.30
9300	4" pipe size		86	.186			9.65		9.65	14.55
9310	5" pipe size		84	.190			9.85		9.85	14.90
9320	6" pipe size		80	.200			10.35		10.35	15.65
9330	8" pipe size		66	.242			12.55		12.55	18.95
9340	10" pipe size		58	.276			14.30		14.30	21.50
9350	12" pipe size		46	.348			18		18	27
9360	14" pipe size		30	.533			27.50		27.50	41.50
9370	16" pipe size		28	.571			29.50		29.50	44.50
9380	18" pipe size		27	.593			30.50		30.50	46.50
9390	20" pipe size		25	.640			33		33	50
9400	24" pipe size	↓	23	.696	↓		36		36	54.50

22 11 13.60 Tubing, Stainless Steel

		Crew	Daily Output	Labor-Hours	Unit	Material	Labor	Equipment	Total	Total Incl O&P
0010	**TUBING, STAINLESS STEEL**									
5000	Tubing									
5010	Type 304, no joints, no hangers									
5020	.035 wall									
5021	1/4"	1 Plum	160	.050	L.F.	3.07	2.88		5.95	7.75
5022	3/8"		160	.050		3.87	2.88		6.75	8.60
5023	1/2"		160	.050		4.99	2.88		7.87	9.85
5024	5/8"		160	.050		6.10	2.88		8.98	11.05
5025	3/4"		133	.060		7.35	3.46		10.81	13.35
5026	7/8"		133	.060		7.75	3.46		11.21	13.80
5027	1"	↓	114	.070	↓	8.55	4.04		12.59	15.50
5040	.049 wall									
5041	1/4"	1 Plum	160	.050	L.F.	3.60	2.88		6.48	8.30
5042	3/8"		160	.050		4.99	2.88		7.87	9.85
5043	1/2"		160	.050		5.50	2.88		8.38	10.40
5044	5/8"		160	.050		7.45	2.88		10.33	12.55
5045	3/4"		133	.060		7.95	3.46		11.41	14
5046	7/8"		133	.060		8.30	3.46		11.76	14.35
5047	1"	↓	114	.070		8.80	4.04		12.84	15.80
5060	.065 wall									
5061	1/4"	1 Plum	160	.050	L.F.	4.04	2.88		6.92	8.80
5062	3/8"		160	.050		5.55	2.88		8.43	10.50
5063	1/2"		160	.050		6.75	2.88		9.63	11.75
5064	5/8"		160	.050		8.20	2.88		11.08	13.35
5065	3/4"		133	.060		8.70	3.46		12.16	14.80
5066	7/8"		133	.060		10	3.46		13.46	16.25
5067	1"	↓	114	.070	↓	9.75	4.04		13.79	16.85
5210	Type 316									

22 11 Facility Water Distribution

22 11 13 – Facility Water Distribution Piping

22 11 13.60 Tubing, Stainless Steel

		Crew	Daily Output	Labor-Hours	Unit	Material	Labor	Equipment	Total	Total Incl O&P
5220	.035 wall									
5221	1/4"	1 Plum	160	.050	L.F.	3.96	2.88		6.84	8.70
5222	3/8"		160	.050		5	2.88		7.88	9.85
5223	1/2"		160	.050		7	2.88		9.88	12.05
5224	5/8"		160	.050		9.10	2.88		11.98	14.35
5225	3/4"		133	.060		10.50	3.46		13.96	16.80
5226	7/8"		133	.060		15	3.46		18.46	22
5227	1"		114	.070		11.90	4.04		15.94	19.20
5240	.049 wall									
5241	1/4"	1 Plum	160	.050	L.F.	5.45	2.88		8.33	10.35
5242	3/8"		160	.050		7.15	2.88		10.03	12.20
5243	1/2"		160	.050		7.85	2.88		10.73	13
5244	5/8"		160	.050		10.90	2.88		13.78	16.35
5245	3/4"		133	.060		10.50	3.46		13.96	16.80
5246	7/8"		133	.060		13.50	3.46		16.96	20
5247	1"		114	.070		12.30	4.04		16.34	19.65
5260	.065 wall									
5261	1/4"	1 Plum	160	.050	L.F.	5.20	2.88		8.08	10.05
5262	3/8"		160	.050		8.30	2.88		11.18	13.45
5263	1/2"		160	.050		8.90	2.88		11.78	14.15
5264	5/8"		160	.050		12.40	2.88		15.28	18
5265	3/4"		133	.060		13.35	3.46		16.81	19.95
5266	7/8"		133	.060		15.80	3.46		19.26	22.50
5267	1"		114	.070		15.05	4.04		19.09	22.50

22 11 13.61 Tubing Fittings, Stainless Steel

		Crew	Daily Output	Labor-Hours	Unit	Material	Labor	Equipment	Total	Total Incl O&P
0010	**TUBING FITTINGS, STAINLESS STEEL**									
8200	Tube fittings, compression type									
8202	Type 316									
8204	90° elbow									
8206	1/4"	1 Plum	24	.333	Ea.	16.30	19.20		35.50	47
8207	3/8"		22	.364		20	21		41	53.50
8208	1/2"		22	.364		33	21		54	68
8209	5/8"		21	.381		37	22		59	73.50
8210	3/4"		21	.381		58.50	22		80.50	97
8211	7/8"		20	.400		89.50	23		112.50	134
8212	1"		20	.400		112	23		135	158
8220	Union tee									
8222	1/4"	1 Plum	15	.533	Ea.	23	30.50		53.50	72
8224	3/8"		15	.533		30	30.50		60.50	79.50
8225	1/2"		15	.533		46	30.50		76.50	97
8226	5/8"		14	.571		51.50	33		84.50	106
8227	3/4"		14	.571		69	33		102	125
8228	7/8"		13	.615		139	35.50		174.50	207
8229	1"		13	.615		147	35.50		182.50	216
8234	Union									
8236	1/4"	1 Plum	24	.333	Ea.	11.30	19.20		30.50	41.50
8237	3/8"		22	.364		16.10	21		37.10	49.50
8238	1/2"		22	.364		24	21		45	58
8239	5/8"		21	.381		31.50	22		53.50	67.50
8240	3/4"		21	.381		39	22		61	76
8241	7/8"		20	.400		63.50	23		86.50	105
8242	1"		20	.400		67.50	23		90.50	109

22 11 Facility Water Distribution

22 11 13 – Facility Water Distribution Piping

22 11 13.61 Tubing Fittings, Stainless Steel		Crew	Daily Output	Labor-Hours	Unit	Material	2014 Bare Costs Labor	Equipment	Total	Total Incl O&P
8250	Male connector									
8252	1/4" x 1/4"	1 Plum	24	.333	Ea.	7.25	19.20		26.45	37
8253	3/8" x 3/8"		22	.364		11.30	21		32.30	44
8254	1/2" x 1/2"		22	.364		16.75	21		37.75	50
8256	3/4" x 3/4"		21	.381		25.50	22		47.50	61.50
8258	1" x 1"	▼	20	.400	▼	45	23		68	84.50

22 11 13.64 Pipe, Stainless Steel

		Crew	Daily Output	Labor-Hours	Unit	Material	Labor	Equipment	Total	Total Incl O&P
0010	**PIPE, STAINLESS STEEL** R221113-70									
0020	Welded, with clevis type hanger assemblies, 10' O.C.									
0500	Schedule 5, type 304									
0540	1/2" diameter	Q-15	128	.125	L.F.	7.35	6.50	.43	14.28	18.35
0550	3/4" diameter		116	.138		9.30	7.15	.48	16.93	21.50
0560	1" diameter		103	.155		10.35	8.05	.54	18.94	24
0570	1-1/4" diameter		93	.172		14.40	8.90	.60	23.90	30
0580	1-1/2" diameter		85	.188		17.80	9.75	.65	28.20	35
0590	2" diameter		69	.232		30	12	.80	42.80	52
0600	2-1/2" diameter		53	.302		32.50	15.65	1.04	49.19	60.50
0610	3" diameter		48	.333		39.50	17.25	1.15	57.90	71
0620	4" diameter		44	.364		50	18.85	1.26	70.11	85
0630	5" diameter	▼	36	.444		102	23	1.54	126.54	150
0640	6" diameter	Q-16	42	.571		95	30.50	1.31	126.81	152
0650	8" diameter		34	.706		146	38	1.62	185.62	219
0660	10" diameter		26	.923		201	49.50	2.12	252.62	298
0670	12" diameter	▼	21	1.143	▼	265	61.50	2.63	329.13	385
0700	To delete hangers, subtract									
0710	1/2" diam. to 1-1/2" diam.					8%	19%			
0720	2" diam. to 5" diam.					4%	9%			
0730	6" diam. to 12" diam.					3%	4%			
0750	For small quantities, add				L.F.	10%				
1250	Schedule 5, type 316									
1290	1/2" diameter	Q-15	128	.125	L.F.	9.50	6.50	.43	16.43	20.50
1300	3/4" diameter		116	.138		13.95	7.15	.48	21.58	26.50
1310	1" diameter		103	.155		15.65	8.05	.54	24.24	30
1320	1-1/4" diameter		93	.172		19.45	8.90	.60	28.95	35.50
1330	1-1/2" diameter		85	.188		27	9.75	.65	37.40	45
1340	2" diameter		69	.232		41.50	12	.80	54.30	64.50
1350	2-1/2" diameter		53	.302		53.50	15.65	1.04	70.19	83.50
1360	3" diameter		48	.333		68.50	17.25	1.15	86.90	102
1370	4" diameter		44	.364		85	18.85	1.26	105.11	123
1380	5" diameter	▼	36	.444		162	23	1.54	186.54	215
1390	6" diameter	Q-16	42	.571		158	30.50	1.31	189.81	222
1400	8" diameter		34	.706		238	38	1.62	277.62	320
1410	10" diameter		26	.923		340	49.50	2.12	391.62	450
1420	12" diameter	▼	21	1.143	▼	440	61.50	2.63	504.13	575
1490	For small quantities, add					10%				
1940	To delete hanger, subtract									
1950	1/2" diam. to 1-1/2" diam.					5%	19%			
1960	2" diam. to 5" diam.					3%	9%			
1970	6" diam. to 12" diam.					2%	4%			
2000	Schedule 10, type 304									
2040	1/4" diameter	Q-15	131	.122	L.F.	4.63	6.35	.42	11.40	15.10
2050	3/8" diameter		128	.125	▼	5.15	6.50	.43	12.08	15.95

22 11 Facility Water Distribution

22 11 13 – Facility Water Distribution Piping

22 11 13.64 Pipe, Stainless Steel

		Crew	Daily Output	Labor-Hours	Unit	Material	2014 Bare Costs Labor	Equipment	Total	Total Incl O&P
2060	1/2" diameter	Q-15	125	.128	L.F.	5.95	6.65	.44	13.04	17.05
2070	3/4" diameter		113	.142		7	7.35	.49	14.84	19.35
2080	1" diameter		100	.160		10.35	8.30	.55	19.20	24.50
2090	1-1/4" diameter		91	.176		11.90	9.10	.61	21.61	27.50
2100	1-1/2" diameter		83	.193		13.30	10	.67	23.97	30.50
2110	2" diameter		67	.239		16.30	12.35	.83	29.48	37.50
2120	2-1/2" diameter		51	.314		21	16.25	1.09	38.34	48.50
2130	3" diameter		46	.348		25.50	18	1.20	44.70	56.50
2140	4" diameter		42	.381		31.50	19.75	1.32	52.57	66
2150	5" diameter		35	.457		42.50	23.50	1.58	67.58	84
2160	6" diameter	Q-16	40	.600		50.50	32	1.38	83.88	106
2170	8" diameter		33	.727		88	39	1.67	128.67	157
2180	10" diameter		25	.960		121	51.50	2.21	174.71	213
2190	12" diameter		21	1.143		151	61.50	2.63	215.13	261
2250	For small quantities, add					10%				
2650	To delete hanger, subtract									
2660	1/4" diam. to 3/4" diam.					9%	22%			
2670	1" diam. to 2" diam.					4%	15%			
2680	2-1/2" diam. to 5" diam.					3%	8%			
2690	6" diam. to 12" diam.					3%	4%			
2750	Schedule 10, type 316									
2790	1/4" diameter	Q-15	131	.122	L.F.	5.75	6.35	.42	12.52	16.35
2800	3/8" diameter		128	.125		6.90	6.50	.43	13.83	17.85
2810	1/2" diameter		125	.128		7.65	6.65	.44	14.74	18.95
2820	3/4" diameter		113	.142		9.15	7.35	.49	16.99	21.50
2830	1" diameter		100	.160		13.30	8.30	.55	22.15	28
2840	1-1/4" diameter		91	.176		17.10	9.10	.61	26.81	33
2850	1-1/2" diameter		83	.193		19	10	.67	29.67	37
2860	2" diameter		67	.239		23	12.35	.83	36.18	45
2870	2-1/2" diameter		51	.314		30	16.25	1.09	47.34	58.50
2880	3" diameter		46	.348		36	18	1.20	55.20	68
2890	4" diameter		42	.381		45.50	19.75	1.32	66.57	81.50
2900	5" diameter		35	.457		55.50	23.50	1.58	80.58	98.50
2910	6" diameter	Q-16	40	.600		68	32	1.38	101.38	125
2920	8" diameter		33	.727		129	39	1.67	169.67	203
2930	10" diameter		25	.960		170	51.50	2.21	223.71	267
2940	12" diameter		21	1.143		213	61.50	2.63	277.13	330
2990	For small quantities, add					10%				
3430	To delete hanger, subtract									
3440	1/4" diam. to 3/4" diam.					6%	22%			
3450	1" diam. to 2" diam.					3%	15%			
3460	2-1/2" diam. to 5" diam.					2%	8%			
3470	6" diam. to 12" diam.					2%	4%			
3500	Threaded, couplings and clevis hanger assemblies, 10' O.C.									
3520	Schedule 40, type 304									
3540	1/4" diameter	1 Plum	54	.148	L.F.	10.10	8.55		18.65	24
3550	3/8" diameter		53	.151		10.35	8.70		19.05	24.50
3560	1/2" diameter		52	.154		12.05	8.85		20.90	26.50
3570	3/4" diameter		51	.157		12.55	9.05		21.60	27.50
3580	1" diameter		45	.178		18.50	10.25		28.75	36
3590	1-1/4" diameter	Q-1	76	.211		26	10.90		36.90	45.50
3600	1-1/2" diameter		69	.232		28.50	12		40.50	49.50
3610	2" diameter		57	.281		44	14.55		58.55	70.50

22 11 Facility Water Distribution

22 11 13 – Facility Water Distribution Piping

22 11 13.64 Pipe, Stainless Steel		Crew	Daily Output	Labor-Hours	Unit	Material	2014 Bare Costs Labor	Equipment	Total	Total Incl O&P
3620	2-1/2" diameter	Q-1	44	.364	L.F.	76	18.85		94.85	113
3630	3" diameter	↓	38	.421		99.50	22		121.50	142
3640	4" diameter	Q-2	51	.471		137	25.50		162.50	188
3740	For small quantities, add				↓	10%				
4200	To delete couplings & hangers, subtract									
4210	1/4" diam. to 3/4" diam.					15%	56%			
4220	1" diam. to 2" diam.					18%	49%			
4230	2-1/2" diam. to 4" diam.					34%	40%			
4250	Schedule 40, type 316									
4290	1/4" diameter	1 Plum	54	.148	L.F.	10.90	8.55		19.45	25
4300	3/8" diameter		53	.151		12	8.70		20.70	26.50
4310	1/2" diameter		52	.154		14.70	8.85		23.55	29.50
4320	3/4" diameter		51	.157		17.15	9.05		26.20	32.50
4330	1" diameter	↓	45	.178		24.50	10.25		34.75	42.50
4340	1-1/4" diameter	Q-1	76	.211		33	10.90		43.90	53
4350	1-1/2" diameter		69	.232		38	12		50	59.50
4360	2" diameter		57	.281		52.50	14.55		67.05	79.50
4370	2-1/2" diameter		44	.364		90	18.85		108.85	127
4380	3" diameter	↓	38	.421		115	22		137	160
4390	4" diameter	Q-2	51	.471		152	25.50		177.50	205
4490	For small quantities, add				↓	10%				
4900	To delete couplings & hangers, subtract									
4910	1/4" diam. to 3/4" diam.					12%	56%			
4920	1" diam. to 2" diam.					14%	49%			
4930	2-1/2" diam. to 4" diam.					27%	40%			
5000	Schedule 80, type 304									
5040	1/4" diameter	1 Plum	53	.151	L.F.	17.40	8.70		26.10	32.50
5050	3/8" diameter		52	.154		19.95	8.85		28.80	35.50
5060	1/2" diameter		51	.157		24.50	9.05		33.55	40
5070	3/4" diameter		48	.167		25.50	9.60		35.10	43
5080	1" diameter	↓	43	.186		32.50	10.70		43.20	51.50
5090	1-1/4" diameter	Q-1	73	.219		43.50	11.35		54.85	65
5100	1-1/2" diameter		67	.239		55.50	12.35		67.85	79.50
5110	2" diameter	↓	54	.296		69	15.35		84.35	99
5190	For small quantities, add				↓	10%				
5700	To delete couplings & hangers, subtract									
5710	1/4" diam. to 3/4" diam.					10%	53%			
5720	1" diam. to 2" diam.					14%	47%			
5750	Schedule 80, type 316									
5790	1/4" diameter	1 Plum	53	.151	L.F.	21	8.70		29.70	36
5800	3/8" diameter		52	.154		25.50	8.85		34.35	41.50
5810	1/2" diameter		51	.157		30.50	9.05		39.55	47
5820	3/4" diameter		48	.167		33.50	9.60		43.10	51.50
5830	1" diameter	↓	43	.186		44	10.70		54.70	64.50
5840	1-1/4" diameter	Q-1	73	.219		63.50	11.35		74.85	87
5850	1-1/2" diameter		67	.239		68	12.35		80.35	93
5860	2" diameter	↓	54	.296		85.50	15.35		100.85	117
5950	For small quantities, add				↓	10%				
7000	To delete couplings & hangers, subtract									
7010	1/4" diam. to 3/4" diam.					9%	53%			
7020	1" diam. to 2" diam.					14%	47%			
8000	Weld joints with clevis type hanger assemblies, 10' O.C.									
8010	Schedule 40, type 304									

22 11 Facility Water Distribution

22 11 13 – Facility Water Distribution Piping

22 11 13.64 Pipe, Stainless Steel		Crew	Daily Output	Labor-Hours	Unit	Material	2014 Bare Costs Labor	Equipment	Total	Total Incl O&P
8050	1/8" pipe size	Q-15	126	.127	L.F.	8	6.60	.44	15.04	19.25
8060	1/4" pipe size		125	.128		8.90	6.65	.44	15.99	20.50
8070	3/8" pipe size		122	.131		8.90	6.80	.45	16.15	20.50
8080	1/2" pipe size		118	.136		10.15	7	.47	17.62	22.50
8090	3/4" pipe size		109	.147		10	7.60	.51	18.11	23
8100	1" pipe size		95	.168		14.35	8.70	.58	23.63	29.50
8110	1-1/4" pipe size		86	.186		19.55	9.65	.64	29.84	37
8120	1-1/2" pipe size		78	.205		21	10.65	.71	32.36	40
8130	2" pipe size		62	.258		32	13.35	.89	46.24	56
8140	2-1/2" pipe size		49	.327		47.50	16.90	1.13	65.53	78.50
8150	3" pipe size		44	.364		60.50	18.85	1.26	80.61	96.50
8160	3-1/2" pipe size		44	.364		71.50	18.85	1.26	91.61	109
8170	4" pipe size		39	.410		82	21.50	1.42	104.92	124
8180	5" pipe size		32	.500		109	26	1.73	136.73	161
8190	6" pipe size	Q-16	37	.649		127	35	1.49	163.49	194
8200	8" pipe size		29	.828		179	44.50	1.90	225.40	266
8210	10" pipe size		24	1		300	53.50	2.30	355.80	420
8220	12" pipe size		20	1.200		490	64.50	2.76	557.26	640
8300	Schedule 40, type 316									
8310	1/8" pipe size	Q-15	126	.127	L.F.	8.50	6.60	.44	15.54	19.80
8320	1/4" pipe size		125	.128		9.45	6.65	.44	16.54	21
8330	3/8" pipe size		122	.131		10.25	6.80	.45	17.50	22
8340	1/2" pipe size		118	.136		12.35	7	.47	19.82	24.50
8350	3/4" pipe size		109	.147		14.05	7.60	.51	22.16	27.50
8360	1" pipe size		95	.168		19.45	8.70	.58	28.73	35.50
8370	1-1/4" pipe size		86	.186		25	9.65	.64	35.29	43
8380	1-1/2" pipe size		78	.205		29	10.65	.71	40.36	48.50
8390	2" pipe size		62	.258		37.50	13.35	.89	51.74	62.50
8400	2-1/2" pipe size		49	.327		55.50	16.90	1.13	73.53	87.50
8410	3" pipe size		44	.364		68	18.85	1.26	88.11	105
8420	3-1/2" pipe size		44	.364		79.50	18.85	1.26	99.61	117
8430	4" pipe size		39	.410		86	21.50	1.42	108.92	129
8440	5" pipe size		32	.500		119	26	1.73	146.73	172
8450	6" pipe size	Q-16	37	.649		138	35	1.49	174.49	205
8460	8" pipe size		29	.828		183	44.50	1.90	229.40	271
8470	10" pipe size		24	1		245	53.50	2.30	300.80	355
8480	12" pipe size		20	1.200		330	64.50	2.76	397.26	465
8500	Schedule 80, type 304									
8510	1/4" pipe size	Q-15	110	.145	L.F.	16	7.55	.50	24.05	29.50
8520	3/8" pipe size		109	.147		18.35	7.60	.51	26.46	32
8530	1/2" pipe size		106	.151		22.50	7.80	.52	30.82	37
8540	3/4" pipe size		96	.167		23	8.65	.58	32.23	39
8550	1" pipe size		87	.184		28	9.55	.64	38.19	45.50
8560	1-1/4" pipe size		81	.198		33	10.25	.68	43.93	52.50
8570	1-1/2" pipe size		74	.216		43	11.20	.75	54.95	64.50
8580	2" pipe size		58	.276		52.50	14.30	.95	67.75	80
8590	2-1/2" pipe size		46	.348		67.50	18	1.20	86.70	102
8600	3" pipe size		41	.390		78	20	1.35	99.35	117
8610	4" pipe size		33	.485		95.50	25	1.68	122.18	145
8630	6" pipe size	Q-16	30	.800		170	43	1.84	214.84	254
8640	Schedule 80, type 316									
8650	1/4" pipe size	Q-15	110	.145	L.F.	19.20	7.55	.50	27.25	33
8660	3/8" pipe size		109	.147		24	7.60	.51	32.11	38

22 11 Facility Water Distribution

22 11 13 – Facility Water Distribution Piping

22 11 13.64 Pipe, Stainless Steel

		Crew	Daily Output	Labor-Hours	Unit	Material	Labor	Equipment	Total	Total Incl O&P
8670	1/2" pipe size	Q-15	106	.151	L.F.	28	7.80	.52	36.32	43.50
8680	3/4" pipe size		96	.167		30	8.65	.58	39.23	46.50
8690	1" pipe size		87	.184		38	9.55	.64	48.19	56.50
8700	1-1/4" pipe size		81	.198		50	10.25	.68	60.93	71
8710	1-1/2" pipe size		74	.216		52	11.20	.75	63.95	74.50
8720	2" pipe size		58	.276		63.50	14.30	.95	78.75	92.50
8730	2-1/2" pipe size		46	.348		79	18	1.20	98.20	115
8740	3" pipe size		41	.390		91.50	20	1.35	112.85	133
8760	4" pipe size		33	.485		121	25	1.68	147.68	173
8770	6" pipe size	Q-16	30	.800		219	43	1.84	263.84	305
9100	Threading pipe labor, sst, one end, schedules 40 & 80									
9110	1/4" through 3/4" pipe size	1 Plum	61.50	.130	Ea.		7.50		7.50	11.30
9120	1" through 2" pipe size		55.90	.143			8.25		8.25	12.45
9130	2-1/2" pipe size		41.50	.193			11.10		11.10	16.75
9140	3" pipe size		38.50	.208			11.95		11.95	18.05
9150	3-1/2" pipe size	Q-1	68.40	.234			12.10		12.10	18.30
9160	4" pipe size		73	.219			11.35		11.35	17.15
9170	5" pipe size		40.70	.393			20.50		20.50	31
9180	6" pipe size		35.40	.452			23.50		23.50	35.50
9190	8" pipe size		22.30	.717			37		37	56
9200	10" pipe size		16.10	.994			51.50		51.50	77.50
9210	12" pipe size		12.30	1.301			67.50		67.50	102
9250	Welding labor per joint for stainless steel									
9260	Schedule 5 and 10									
9270	1/4" pipe size	Q-15	36	.444	Ea.		23	1.54	24.54	36.50
9280	3/8" pipe size		35	.457			23.50	1.58	25.08	37.50
9290	1/2" pipe size		35	.457			23.50	1.58	25.08	37.50
9300	3/4" pipe size		28	.571			29.50	1.98	31.48	46.50
9310	1" pipe size		25	.640			33	2.21	35.21	52.50
9320	1-1/4" pipe size		22	.727			37.50	2.52	40.02	60
9330	1-1/2" pipe size		21	.762			39.50	2.64	42.14	62.50
9340	2" pipe size		18	.889			46	3.08	49.08	73
9350	2-1/2" pipe size		12	1.333			69	4.61	73.61	109
9360	3" pipe size		9.73	1.644			85	5.70	90.70	135
9370	4" pipe size		7.37	2.171			112	7.50	119.50	178
9380	5" pipe size		6.15	2.602			135	9	144	214
9390	6" pipe size		5.71	2.802			145	9.70	154.70	230
9400	8" pipe size		3.69	4.336			225	15	240	355
9410	10" pipe size		2.91	5.498			285	19	304	450
9420	12" pipe size		2.31	6.926			360	24	384	565
9500	Schedule 40									
9510	1/4" pipe size	Q-15	28	.571	Ea.		29.50	1.98	31.48	46.50
9520	3/8" pipe size		27	.593			30.50	2.05	32.55	49
9530	1/2" pipe size		25.40	.630			32.50	2.18	34.68	52
9540	3/4" pipe size		22.22	.720			37.50	2.49	39.99	59
9550	1" pipe size		20.25	.790			41	2.73	43.73	65
9560	1-1/4" pipe size		18.82	.850			44	2.94	46.94	69.50
9570	1-1/2" pipe size		17.78	.900			46.50	3.11	49.61	74
9580	2" pipe size		15.09	1.060			55	3.67	58.67	87
9590	2-1/2" pipe size		7.96	2.010			104	6.95	110.95	165
9600	3" pipe size		6.43	2.488			129	8.60	137.60	204
9610	4" pipe size		4.88	3.279			170	11.35	181.35	268
9620	5" pipe size		4.26	3.756			195	13	208	310

22 11 Facility Water Distribution

22 11 13 – Facility Water Distribution Piping

22 11 13.64 Pipe, Stainless Steel

		Crew	Daily Output	Labor-Hours	Unit	Material	2014 Bare Costs Labor	Equipment	Total	Total Incl O&P
9630	6" pipe size	Q-15	3.77	4.244	Ea.		220	14.70	234.70	345
9640	8" pipe size		2.44	6.557			340	22.50	362.50	540
9650	10" pipe size		1.92	8.333			430	29	459	680
9660	12" pipe size		1.52	10.526			545	36.50	581.50	865
9750	Schedule 80									
9760	1/4" pipe size	Q-15	21.55	.742	Ea.	38.50	2.57		41.07	61
9770	3/8" pipe size		20.75	.771		40	2.67		42.67	63.50
9780	1/2" pipe size		19.54	.819		42.50	2.83		45.33	67
9790	3/4" pipe size		17.09	.936		48.50	3.24		51.74	76.50
9800	1" pipe size		15.58	1.027		53	3.55		56.55	84.50
9810	1-1/4" pipe size		14.48	1.105		57	3.82		60.82	90.50
9820	1-1/2" pipe size		13.68	1.170		60.50	4.05		64.55	96
9830	2" pipe size		11.61	1.378		71.50	4.77		76.27	113
9840	2-1/2" pipe size		6.12	2.614		135	9.05		144.05	215
9850	3" pipe size		4.94	3.239		168	11.20		179.20	265
9860	4" pipe size		3.75	4.267		221	14.75		235.75	350
9870	5" pipe size		3.27	4.893		253	16.95		269.95	405
9880	6" pipe size		2.90	5.517		286	19.10		305.10	450
9890	8" pipe size		1.87	8.556		445	29.50		474.50	705
9900	10" pipe size		1.48	10.811		560	37.50		597.50	885
9910	12" pipe size		1.17	13.675		710	47.50		757.50	1,125
9920	Schedule 160, 1/2" pipe size		17	.941		49	3.26		52.26	77
9930	3/4" pipe size		14.81	1.080		56	3.74		59.74	88.50
9940	1" pipe size		13.50	1.185		61.50	4.10		65.60	97
9950	1-1/4" pipe size		12.55	1.275		66	4.41		70.41	104
9960	1-1/2" pipe size		11.85	1.350		70	4.67		74.67	111
9970	2" pipe size		10	1.600		83	5.55		88.55	131
9980	3" pipe size		4.28	3.738		194	12.95		206.95	305
9990	4" pipe size		3.25	4.923		255	17.05		272.05	405

22 11 13.66 Pipe Fittings, Stainless Steel

		Crew	Daily Output	Labor-Hours	Unit	Material	Labor	Equipment	Total	Total Incl O&P
0010	**PIPE FITTINGS, STAINLESS STEEL**									
0100	Butt weld joint, schedule 5, type 304									
0120	90° Elbow, long									
0140	1/2"	Q-15	17.50	.914	Ea.	19.80	47.50	3.16	70.46	97
0150	3/4"		14	1.143		19.80	59	3.95	82.75	116
0160	1"		12.50	1.280		21	66.50	4.43	91.93	128
0170	1-1/4"		11	1.455		28.50	75.50	5.05	109.05	151
0180	1-1/2"		10.50	1.524		24	79	5.25	108.25	151
0190	2"		9	1.778		28.50	92	6.15	126.65	177
0200	2-1/2"		6	2.667		66	138	9.25	213.25	292
0210	3"		4.86	3.292		72.50	171	11.40	254.90	350
0220	3-1/2"		4.27	3.747		182	194	12.95	388.95	505
0230	4"		3.69	4.336		99	225	15	339	465
0240	5"		3.08	5.195		385	269	17.95	671.95	850
0250	6"	Q-16	4.29	5.594		297	300	12.85	609.85	795
0260	8"		2.76	8.696		625	465	20	1,110	1,425
0270	10"		2.18	11.009		970	590	25.50	1,585.50	2,000
0280	12"		1.73	13.873		1,375	745	32	2,152	2,675
0320	For schedule 5, type 316, add					30%				
0600	45° Elbow, long									
0620	1/2"	Q-15	17.50	.914	Ea.	19.80	47.50	3.16	70.46	97
0630	3/4"		14	1.143		19.80	59	3.95	82.75	116

22 11 Facility Water Distribution

22 11 13 – Facility Water Distribution Piping

22 11 13.66 Pipe Fittings, Stainless Steel		Crew	Daily Output	Labor-Hours	Unit	Material	2014 Bare Costs Labor	Equipment	Total	Total Incl O&P
0640	1"	Q-15	12.50	1.280	Ea.	21	66.50	4.43	91.93	128
0650	1-1/4"		11	1.455		28.50	75.50	5.05	109.05	151
0660	1-1/2"		10.50	1.524		24	79	5.25	108.25	151
0670	2"		9	1.778		28.50	92	6.15	126.65	177
0680	2-1/2"		6	2.667		66	138	9.25	213.25	292
0690	3"		4.86	3.292		72.50	171	11.40	254.90	350
0700	3-1/2"		4.27	3.747		182	194	12.95	388.95	505
0710	4"		3.69	4.336		83.50	225	15	323.50	450
0720	5"		3.08	5.195		310	269	17.95	596.95	765
0730	6"	Q-16	4.29	5.594		209	300	12.85	521.85	700
0740	8"		2.76	8.696		440	465	20	925	1,200
0750	10"		2.18	11.009		770	590	25.50	1,385.50	1,775
0760	12"		1.73	13.873		965	745	32	1,742	2,200
0800	For schedule 5, type 316, add					25%				
1100	Tee, straight									
1130	1/2"	Q-15	11.66	1.372	Ea.	59.50	71	4.75	135.25	178
1140	3/4"		9.33	1.715		59.50	89	5.95	154.45	206
1150	1"		8.33	1.921		62.50	99.50	6.65	168.65	226
1160	1-1/4"		7.33	2.183		50.50	113	7.55	171.05	235
1170	1-1/2"		7	2.286		49.50	118	7.90	175.40	242
1180	2"		6	2.667		51.50	138	9.25	198.75	276
1190	2-1/2"		4	4		124	207	13.85	344.85	465
1200	3"		3.24	4.938		198	256	17.10	471.10	620
1210	3-1/2"		2.85	5.614		260	291	19.40	570.40	750
1220	4"		2.46	6.504		141	335	22.50	498.50	690
1230	5"		2	8		445	415	27.50	887.50	1,150
1240	6"	Q-16	2.85	8.421		350	450	19.35	819.35	1,100
1250	8"		1.84	13.043		750	700	30	1,480	1,900
1260	10"		1.45	16.552		1,200	890	38	2,128	2,725
1270	12"		1.15	20.870		1,675	1,125	48	2,848	3,600
1320	For schedule 5, type 316, add					25%				
2000	Butt weld joint, schedule 10, type 304									
2020	90° elbow, long									
2040	1/2"	Q-15	17	.941	Ea.	13.50	49	3.26	65.76	92
2050	3/4"		14	1.143		13.50	59	3.95	76.45	109
2060	1"		12.50	1.280		14.25	66.50	4.43	85.18	121
2070	1-1/4"		11	1.455		19.50	75.50	5.05	100.05	141
2080	1-1/2"		10.50	1.524		16.50	79	5.25	100.75	143
2090	2"		9	1.778		19.50	92	6.15	117.65	167
2100	2-1/2"		6	2.667		45	138	9.25	192.25	269
2110	3"		4.86	3.292		42	171	11.40	224.40	315
2120	3-1/2"		4.27	3.747		124	194	12.95	330.95	445
2130	4"		3.69	4.336		70.50	225	15	310.50	435
2140	5"		3.08	5.195		263	269	17.95	549.95	715
2150	6"	Q-16	4.29	5.594		203	300	12.85	515.85	690
2160	8"		2.76	8.696		430	465	20	915	1,200
2170	10"		2.18	11.009		660	590	25.50	1,275.50	1,650
2180	12"		1.73	13.873		940	745	32	1,717	2,175
2500	45° elbow, long									
2520	1/2"	Q-15	17.50	.914	Ea.	13.50	47.50	3.16	64.16	90
2530	3/4"		14	1.143		13.50	59	3.95	76.45	109
2540	1"		12.50	1.280		14.25	66.50	4.43	85.18	121
2550	1-1/4"		11	1.455		19.50	75.50	5.05	100.05	141

22 11 Facility Water Distribution

22 11 13 – Facility Water Distribution Piping

22 11 13.66 Pipe Fittings, Stainless Steel		Crew	Daily Output	Labor-Hours	Unit	Material	2014 Bare Costs Labor	Equipment	Total	Total Incl O&P
2560	1-1/2"	Q-15	10.50	1.524	Ea.	16.50	79	5.25	100.75	143
2570	2"		9	1.778		19.50	92	6.15	117.65	167
2580	2-1/2"		6	2.667		45	138	9.25	192.25	269
2590	3"		4.86	3.292		34	171	11.40	216.40	310
2600	3-1/2"		4.27	3.747		124	194	12.95	330.95	445
2610	4"		3.69	4.336		57	225	15	297	420
2620	5"		3.08	5.195		210	269	17.95	496.95	655
2630	6"	Q-16	4.29	5.594		143	300	12.85	455.85	625
2640	8"		2.76	8.696		300	465	20	785	1,050
2650	10"		2.18	11.009		525	590	25.50	1,140.50	1,500
2660	12"		1.73	13.873		655	745	32	1,432	1,875
2670	Reducer, concentric									
2674	1" x 3/4"	Q-15	13.25	1.208	Ea.	29.50	62.50	4.18	96.18	131
2676	2" x 1-1/2"	"	9.75	1.641		24	85	5.70	114.70	161
2678	6" x 4"	Q-16	4.91	4.888		72	263	11.25	346.25	485
2680	Caps									
2682	1"	Q-15	25	.640	Ea.	30	33	2.21	65.21	85.50
2684	1-1/2"		21	.762		36	39.50	2.64	78.14	102
2685	2"		18	.889		34	46	3.08	83.08	110
2686	4"		7.38	2.168		54	112	7.50	173.50	238
2687	6"	Q-16	8.58	2.797		86.50	150	6.45	242.95	330
3000	Tee, straight									
3030	1/2"	Q-15	11.66	1.372	Ea.	40.50	71	4.75	116.25	157
3040	3/4"		9.33	1.715		40.50	89	5.95	135.45	185
3050	1"		8.33	1.921		43	99.50	6.65	149.15	204
3060	1-1/4"		7.33	2.183		49	113	7.55	169.55	233
3070	1-1/2"		7	2.286		49	118	7.90	174.90	241
3080	2"		6	2.667		51	138	9.25	198.25	275
3090	2-1/2"		4	4		35.50	207	13.85	256.35	370
3100	3"		3.24	4.938		67	256	17.10	340.10	475
3110	3-1/2"		2.85	5.614		98.50	291	19.40	408.90	570
3120	4"		2.46	6.504		96	335	22.50	453.50	640
3130	5"		2	8		305	415	27.50	747.50	990
3140	6"	Q-16	2.85	8.421		240	450	19.35	709.35	970
3150	8"		1.84	13.043		510	700	30	1,240	1,650
3151	10"		1.45	16.552		825	890	38	1,753	2,300
3152	12"		1.15	20.870		1,150	1,125	48	2,323	3,000
3154	For schedule 10, type 316, add					25%				
3281	Butt weld joint, schedule 40, type 304									
3284	90° Elbow, long, 1/2"	Q-15	12.70	1.260	Ea.	15.75	65.50	4.36	85.61	121
3288	3/4"		11.10	1.441		15.75	74.50	4.99	95.24	136
3289	1"		10.13	1.579		16.50	82	5.45	103.95	148
3290	1-1/4"		9.40	1.702		22	88	5.90	115.90	163
3300	1-1/2"		8.89	1.800		17.25	93	6.25	116.50	167
3310	2"		7.55	2.119		25	110	7.35	142.35	201
3320	2-1/2"		3.98	4.020		47.50	208	13.90	269.40	380
3330	3"		3.21	4.984		61	258	17.25	336.25	475
3340	3-1/2"		2.83	5.654		225	293	19.55	537.55	710
3350	4"		2.44	6.557		105	340	22.50	467.50	655
3360	5"		2.13	7.512		355	390	26	771	1,000
3370	6"	Q-16	2.83	8.481		310	455	19.50	784.50	1,050
3380	8"		1.83	13.115		615	705	30	1,350	1,775
3390	10"		1.44	16.667		1,300	895	38.50	2,233.50	2,825

22 11 Facility Water Distribution

22 11 13 – Facility Water Distribution Piping

22 11 13.66 Pipe Fittings, Stainless Steel		Crew	Daily Output	Labor-Hours	Unit	Material	2014 Bare Costs Labor	Equipment	Total	Total Incl O&P
3400	12"	Q-16	1.14	21.053	Ea.	1,650	1,125	48.50	2,823.50	3,575
3410	For schedule 40, type 316, add					25%				
3460	45° Elbow, long, 1/2"	Q-15	12.70	1.260	Ea.	15.75	65.50	4.36	85.61	121
3470	3/4"		11.10	1.441		15.75	74.50	4.99	95.24	136
3480	1"		10.13	1.579		16.50	82	5.45	103.95	148
3490	1-1/4"		9.40	1.702		22	88	5.90	115.90	163
3500	1-1/2"		8.89	1.800		17.25	93	6.25	116.50	167
3510	2"		7.55	2.119		25	110	7.35	142.35	201
3520	2-1/2"		3.98	4.020		47.50	208	13.90	269.40	380
3530	3"		3.21	4.984		48	258	17.25	323.25	460
3540	3-1/2"		2.83	5.654		225	293	19.55	537.55	710
3550	4"		2.44	6.557		75	340	22.50	437.50	625
3560	5"	↓	2.13	7.512		248	390	26	664	890
3570	6"	Q-16	2.83	8.481		218	455	19.50	692.50	950
3580	8"		1.83	13.115		430	705	30	1,165	1,575
3590	10"		1.44	16.667		910	895	38.50	1,843.50	2,400
3600	12"	↓	1.14	21.053	↓	1,150	1,125	48.50	2,323.50	3,025
3610	For schedule 40, type 316, add					25%				
3660	Tee, straight 1/2"	Q-15	8.46	1.891	Ea.	40.50	98	6.55	145.05	200
3670	3/4"		7.40	2.162		40.50	112	7.50	160	222
3680	1"		6.74	2.374		43	123	8.20	174.20	242
3690	1-1/4"		6.27	2.552		100	132	8.85	240.85	320
3700	1-1/2"		5.92	2.703		46	140	9.35	195.35	272
3710	2"		5.03	3.181		68.50	165	11	244.50	335
3720	2-1/2"		2.65	6.038		86.50	315	21	422.50	590
3730	3"		2.14	7.477		87	385	26	498	710
3740	3-1/2"		1.88	8.511		164	440	29.50	633.50	880
3750	4"		1.62	9.877		164	510	34	708	995
3760	5"	↓	1.42	11.268		380	585	39	1,004	1,350
3770	6"	Q-16	1.88	12.766		365	685	29.50	1,079.50	1,450
3780	8"		1.22	19.672		735	1,050	45.50	1,830.50	2,450
3790	10"		.96	25		1,425	1,350	57.50	2,832.50	3,675
3800	12"	↓	.76	31.579	↓	1,900	1,700	72.50	3,672.50	4,725
3810	For schedule 40, type 316, add					25%				
3820	Tee, reducing on outlet, 3/4" x 1/2"	Q-15	7.73	2.070	Ea.	59.50	107	7.15	173.65	235
3822	1" x 1/2"		7.24	2.210		58.50	114	7.65	180.15	246
3824	1" x 3/4"		6.96	2.299		54	119	7.95	180.95	248
3826	1-1/4" x 1"		6.43	2.488		132	129	8.60	269.60	350
3828	1-1/2" x 1/2"		6.58	2.432		79.50	126	8.40	213.90	287
3830	1-1/2" x 3/4"		6.35	2.520		74.50	131	8.70	214.20	288
3832	1-1/2" x 1"		6.18	2.589		57	134	8.95	199.95	275
3834	2" x 1"		5.50	2.909		102	151	10.05	263.05	350
3836	2" x 1-1/2"		5.30	3.019		85.50	156	10.45	251.95	340
3838	2-1/2" x 2"		3.15	5.079		129	263	17.55	409.55	555
3840	3" x 1-1/2"		2.72	5.882		131	305	20.50	456.50	625
3842	3" x 2"		2.65	6.038		109	315	21	445	615
3844	4" x 2"		2.10	7.619		272	395	26.50	693.50	925
3846	4" x 3"		1.77	9.040		197	470	31.50	698.50	955
3848	5" x 4"	↓	1.48	10.811		455	560	37.50	1,052.50	1,375
3850	6" x 3"	Q-16	2.19	10.959		505	590	25	1,120	1,475
3852	6" x 4"		2.04	11.765		435	630	27	1,092	1,475
3854	8" x 4"		1.46	16.438		1,025	885	38	1,948	2,500
3856	10" x 8"	↓	.69	34.783	↓	1,725	1,875	80	3,680	4,800

22 11 Facility Water Distribution

22 11 13 – Facility Water Distribution Piping

22 11 13.66 Pipe Fittings, Stainless Steel

		Crew	Daily Output	Labor-Hours	Unit	Material	2014 Bare Costs Labor	Equipment	Total	Total Incl O&P
3858	12" x 10"	Q-16	.55	43.636	Ea.	2,275	2,350	100	4,725	6,175
3950	Reducer, concentric, 3/4" x 1/2"	Q-15	11.85	1.350		30	70	4.67	104.67	144
3952	1" x 3/4"		10.60	1.509		31.50	78	5.20	114.70	158
3954	1-1/4" x 3/4"		10.19	1.570		77.50	81.50	5.45	164.45	214
3956	1-1/4" x 1"		9.76	1.639		39	85	5.65	129.65	177
3958	1-1/2" x 3/4"		9.88	1.619		65.50	84	5.60	155.10	205
3960	1-1/2" x 1"		9.47	1.690		52.50	87.50	5.85	145.85	196
3962	2" x 1"		8.65	1.850		30	96	6.40	132.40	185
3964	2" x 1-1/2"		8.16	1.961		26.50	102	6.80	135.30	189
3966	2-1/2" x 1"		5.71	2.802		113	145	9.70	267.70	355
3968	2-1/2" x 2"		5.21	3.071		58	159	10.65	227.65	315
3970	3" x 1"		4.88	3.279		78	170	11.35	259.35	355
3972	3" x 1-1/2"		4.72	3.390		40	176	11.75	227.75	320
3974	3" x 2"		4.51	3.548		37.50	184	12.25	233.75	335
3976	4" x 2"		3.69	4.336		53.50	225	15	293.50	415
3978	4" x 3"		2.77	5.776		40	299	20	359	515
3980	5" x 3"		2.56	6.250		237	325	21.50	583.50	775
3982	5" x 4"		2.27	7.048		191	365	24.50	580.50	785
3984	6" x 3"	Q-16	3.57	6.723		122	360	15.45	497.45	695
3986	6" x 4"		3.19	7.524		101	405	17.30	523.30	740
3988	8" x 4"		2.44	9.836		315	530	22.50	867.50	1,175
3990	8" x 6"		2.22	10.811		236	580	25	841	1,150
3992	10" x 6"		1.91	12.565		440	675	29	1,144	1,550
3994	10" x 8"		1.61	14.907		365	800	34.50	1,199.50	1,650
3995	12" x 6"		1.63	14.724		675	790	34	1,499	1,975
3996	12" x 8"		1.41	17.021		620	915	39	1,574	2,100
3997	12" x 10"		1.27	18.898		435	1,025	43.50	1,503.50	2,050
4000	Socket weld joint, 3000 lb., type 304									
4100	90° Elbow									
4140	1/4"	Q-15	13.47	1.188	Ea.	44	61.50	4.11	109.61	146
4150	3/8"		12.97	1.234		57	64	4.27	125.27	164
4160	1/2"		12.21	1.310		62.50	68	4.53	135.03	177
4170	3/4"		10.68	1.498		72	77.50	5.20	154.70	202
4180	1"		9.74	1.643		108	85	5.70	198.70	254
4190	1-1/4"		9.05	1.768		189	91.50	6.10	286.60	355
4200	1-1/2"		8.55	1.871		229	97	6.45	332.45	405
4210	2"		7.26	2.204		370	114	7.65	491.65	585
4300	45° Elbow									
4340	1/4"	Q-15	13.47	1.188	Ea.	83	61.50	4.11	148.61	189
4350	3/8"		12.97	1.234		83	64	4.27	151.27	192
4360	1/2"		12.21	1.310		83	68	4.53	155.53	199
4370	3/4"		10.68	1.498		94	77.50	5.20	176.70	226
4380	1"		9.74	1.643		136	85	5.70	226.70	285
4390	1-1/4"		9.05	1.768		224	91.50	6.10	321.60	390
4400	1-1/2"		8.55	1.871		225	97	6.45	328.45	400
4410	2"		7.26	2.204		410	114	7.65	531.65	630
4500	Tee									
4540	1/4"	Q-15	8.97	1.784	Ea.	58	92.50	6.15	156.65	210
4550	3/8"		8.64	1.852		69	96	6.40	171.40	228
4560	1/2"		8.13	1.968		85	102	6.80	193.80	255
4570	3/4"		7.12	2.247		98	116	7.80	221.80	293
4580	1"		6.48	2.469		132	128	8.55	268.55	345
4590	1-1/4"		6.03	2.653		234	137	9.20	380.20	475

22 11 Facility Water Distribution

22 11 13 – Facility Water Distribution Piping

22 11 13.66 Pipe Fittings, Stainless Steel		Crew	Daily Output	Labor-Hours	Unit	Material	2014 Bare Costs Labor	Equipment	Total	Total Incl O&P
4600	1-1/2"	Q-15	5.69	2.812	Ea.	335	146	9.75	490.75	600
4610	2"	↓	4.83	3.313	↓	510	172	11.45	693.45	830
5000	Socket weld joint, 3000 lb., type 316									
5100	90° Elbow									
5140	1/4"	Q-15	13.47	1.188	Ea.	54	61.50	4.11	119.61	157
5150	3/8"		12.97	1.234		63	64	4.27	131.27	170
5160	1/2"		12.21	1.310		76	68	4.53	148.53	191
5170	3/4"		10.68	1.498		100	77.50	5.20	182.70	233
5180	1"		9.74	1.643		142	85	5.70	232.70	291
5190	1-1/4"		9.05	1.768		253	91.50	6.10	350.60	425
5200	1-1/2"		8.55	1.871		286	97	6.45	389.45	470
5210	2"	↓	7.26	2.204	↓	485	114	7.65	606.65	715
5300	45° Elbow									
5340	1/4"	Q-15	13.47	1.188	Ea.	104	61.50	4.11	169.61	212
5350	3/8"		12.97	1.234		104	64	4.27	172.27	215
5360	1/2"		12.21	1.310		104	68	4.53	176.53	222
5370	3/4"		10.68	1.498		121	77.50	5.20	203.70	256
5380	1"		9.74	1.643		182	85	5.70	272.70	335
5390	1-1/4"		9.05	1.768		258	91.50	6.10	355.60	430
5400	1-1/2"		8.55	1.871		292	97	6.45	395.45	475
5410	2"	↓	7.26	2.204	↓	435	114	7.65	556.65	660
5500	Tee									
5540	1/4"	Q-15	8.97	1.784	Ea.	73	92.50	6.15	171.65	227
5550	3/8"		8.64	1.852		89.50	96	6.40	191.90	250
5560	1/2"		8.13	1.968		99.50	102	6.80	208.30	271
5570	3/4"		7.12	2.247		123	116	7.80	246.80	320
5580	1"		6.48	2.469		188	128	8.55	324.55	410
5590	1-1/4"		6.03	2.653		299	137	9.20	445.20	550
5600	1-1/2"		5.69	2.812		420	146	9.75	575.75	695
5610	2"	↓	4.83	3.313		670	172	11.45	853.45	1,000
5700	For socket weld joint, 6000 lb., type 304 and 316, add				↓	100%				
6000	Threaded companion flange									
6010	Stainless steel, 150 lb., type 304									
6020	1/2" diam.	1 Plum	30	.267	Ea.	45	15.35		60.35	72.50
6030	3/4" diam.		28	.286		50	16.45		66.45	80
6040	1" diam.	↓	27	.296		55	17.05		72.05	86.50
6050	1-1/4" diam.	Q-1	44	.364		71	18.85		89.85	107
6060	1-1/2" diam.		40	.400		71	20.50		91.50	110
6070	2" diam.		36	.444		93	23		116	137
6080	2-1/2" diam.		28	.571		130	29.50		159.50	188
6090	3" diam.		20	.800		136	41.50		177.50	213
6110	4" diam.	↓	12	1.333		186	69		255	310
6130	6" diam.	Q-2	14	1.714		335	92		427	505
6140	8" diam.	"	12	2	↓	645	107		752	870
6150	For type 316 add					40%				
6260	Weld flanges, stainless steel, type 304									
6270	Slip on, 150 lb. (welded, front and back)									
6280	1/2" diam.	Q-15	12.70	1.260	Ea.	40	65.50	4.36	109.86	147
6290	3/4" diam.		11.11	1.440		41	74.50	4.98	120.48	163
6300	1" diam.		10.13	1.579		45	82	5.45	132.45	180
6310	1-1/4" diam.		9.41	1.700		61	88	5.90	154.90	206
6320	1-1/2" diam.		8.89	1.800		61	93	6.25	160.25	215
6330	2" diam.	↓	7.55	2.119		79	110	7.35	196.35	261

22 11 Facility Water Distribution

22 11 13 – Facility Water Distribution Piping

22 11 13.66 Pipe Fittings, Stainless Steel

		Crew	Daily Output	Labor-Hours	Unit	Material	2014 Bare Costs Labor	Equipment	Total	Total Incl O&P
6340	2-1/2" diam.	Q-15	3.98	4.020	Ea.	111	208	13.90	332.90	450
6350	3" diam.		3.21	4.984		119	258	17.25	394.25	540
6370	4" diam.	▼	2.44	6.557		162	340	22.50	524.50	720
6390	6" diam.	Q-16	1.89	12.698		247	680	29	956	1,325
6400	8" diam.	"	1.22	19.672	▼	465	1,050	45.50	1,560.50	2,175
6410	For type 316, add					40%				
6530	Weld neck 150 lb.									
6540	1/2" diam.	Q-15	25.40	.630	Ea.	20.50	32.50	2.18	55.18	74.50
6550	3/4" diam.		22.22	.720		25	37.50	2.49	64.99	86
6560	1" diam.		20.25	.790		27.50	41	2.73	71.23	95.50
6570	1-1/4" diam.		18.82	.850		43.50	44	2.94	90.44	117
6580	1-1/2" diam.		17.78	.900		39	46.50	3.11	88.61	117
6590	2" diam.		15.09	1.060		41.50	55	3.67	100.17	133
6600	2-1/2" diam.		7.96	2.010		68	104	6.95	178.95	240
6610	3" diam.		6.43	2.488		68.50	129	8.60	206.10	280
6630	4" diam.		4.88	3.279		104	170	11.35	285.35	380
6640	5" diam.	▼	4.26	3.756		134	195	13	342	455
6650	6" diam.	Q-16	5.66	4.240		150	228	9.75	387.75	520
6652	8" diam.		3.65	6.575		284	355	15.10	654.10	860
6654	10" diam.		2.88	8.333		400	450	19.15	869.15	1,125
6656	12" diam.	▼	2.28	10.526	▼	765	565	24	1,354	1,725
6670	For type 316 add					23%				
7000	Threaded joint, 150 lb., type 304									
7030	90° elbow									
7040	1/8"	1 Plum	13	.615	Ea.	25.50	35.50		61	81.50
7050	1/4"		13	.615		25.50	35.50		61	81.50
7070	3/8"		13	.615		29.50	35.50		65	86
7080	1/2"		12	.667		26.50	38.50		65	87
7090	3/4"		11	.727		31.50	42		73.50	97.50
7100	1"	▼	10	.800		43	46		89	117
7110	1-1/4"	Q-1	17	.941		66.50	49		115.50	147
7120	1-1/2"		16	1		77	52		129	163
7130	2"		14	1.143		111	59		170	213
7140	2-1/2"		11	1.455		271	75.50		346.50	410
7150	3"	▼	8	2		395	104		499	590
7160	4"	Q-2	11	2.182	▼	670	117		787	910
7180	45° elbow									
7190	1/8"	1 Plum	13	.615	Ea.	37	35.50		72.50	94.50
7200	1/4"		13	.615		37	35.50		72.50	94.50
7210	3/8"		13	.615		37.50	35.50		73	94.50
7220	1/2"		12	.667		38	38.50		76.50	99.50
7230	3/4"		11	.727		41	42		83	108
7240	1"	▼	10	.800		47.50	46		93.50	122
7250	1-1/4"	Q-1	17	.941		66	49		115	146
7260	1-1/2"		16	1		85.50	52		137.50	172
7270	2"		14	1.143		121	59		180	223
7280	2-1/2"		11	1.455		380	75.50		455.50	535
7290	3"	▼	8	2		560	104		664	770
7300	4"	Q-2	11	2.182		1,000	117		1,117	1,275
7320	Tee, straight									
7330	1/8"	1 Plum	9	.889	Ea.	39	51		90	120
7340	1/4"		9	.889		39	51		90	120
7350	3/8"		9	.889	▼	41.50	51		92.50	123

22 11 Facility Water Distribution

22 11 13 – Facility Water Distribution Piping

22 11 13.66 Pipe Fittings, Stainless Steel		Crew	Daily Output	Labor-Hours	Unit	Material	2014 Bare Costs Labor	Equipment	Total	Total Incl O&P
7360	1/2"	1 Plum	8	1	Ea.	39.50	57.50		97	131
7370	3/4"		7	1.143		42	66		108	146
7380	1"		6.50	1.231		54	71		125	167
7390	1-1/4"	Q-1	11	1.455		91	75.50		166.50	214
7400	1-1/2"		10	1.600		116	83		199	253
7410	2"		9	1.778		145	92		237	298
7420	2-1/2"		7	2.286		390	118		508	610
7430	3"		5	3.200		590	166		756	900
7440	4"	Q-2	7	3.429		1,475	184		1,659	1,875
7460	Coupling, straight									
7470	1/8"	1 Plum	19	.421	Ea.	10.25	24		34.25	48
7480	1/4"		19	.421		12.10	24		36.10	50
7490	3/8"		19	.421		14.50	24		38.50	52.50
7500	1/2"		19	.421		19.30	24		43.30	57.50
7510	3/4"		18	.444		25.50	25.50		51	67
7520	1"		15	.533		41.50	30.50		72	92
7530	1-1/4"	Q-1	26	.615		66.50	32		98.50	122
7540	1-1/2"		24	.667		74.50	34.50		109	134
7550	2"		21	.762		123	39.50		162.50	196
7560	2-1/2"		18	.889		287	46		333	385
7570	3"		14	1.143		390	59		449	520
7580	4"	Q-2	16	1.500		550	80.50		630.50	725
7600	Reducer, concentric, 1/2"	1 Plum	12	.667		20.50	38.50		59	81
7610	3/4"		11	.727		26.50	42		68.50	92.50
7612	1"		10	.800		44.50	46		90.50	119
7614	1-1/4"	Q-1	17	.941		94.50	49		143.50	178
7616	1-1/2"		16	1		103	52		155	191
7618	2"		14	1.143		161	59		220	267
7620	2-1/2"		11	1.455		400	75.50		475.50	555
7622	3"		8	2		445	104		549	645
7624	4"	Q-2	11	2.182		735	117		852	985
7710	Union									
7720	1/8"	1 Plum	12	.667	Ea.	47.50	38.50		86	111
7730	1/4"		12	.667		47.50	38.50		86	111
7740	3/8"		12	.667		55.50	38.50		94	119
7750	1/2"		11	.727		69.50	42		111.50	140
7760	3/4"		10	.800		96	46		142	175
7770	1"		9	.889		138	51		189	229
7780	1-1/4"	Q-1	16	1		340	52		392	455
7790	1-1/2"		15	1.067		370	55.50		425.50	490
7800	2"		13	1.231		465	64		529	610
7810	2-1/2"		10	1.600		880	83		963	1,100
7820	3"		7	2.286		1,175	118		1,293	1,475
7830	4"	Q-2	10	2.400		1,625	129		1,754	1,975
7838	Caps									
7840	1/2"	1 Plum	24	.333	Ea.	12.90	19.20		32.10	43
7841	3/4"		22	.364		18.80	21		39.80	52
7842	1"		20	.400		30	23		53	68
7843	1-1/2"	Q-1	32	.500		76	26		102	123
7844	2"	"	28	.571		96	29.50		125.50	151
7845	4"	Q-2	22	1.091		380	58.50		438.50	505
7850	For 150 lb., type 316, add					25%				

207

22 11 Facility Water Distribution

22 11 13 – Facility Water Distribution Piping

22 11 13.74 Pipe, Plastic

		Crew	Daily Output	Labor-Hours	Unit	Material	2014 Bare Costs Labor	Equipment	Total	Total Incl O&P
0010	**PIPE, PLASTIC** R221113-70									
0020	Fiberglass reinforced, couplings 10' O.C., clevis hanger assy's, 3 per 10'									
0080	General service									
0120	2" diameter	Q-1	59	.271	L.F.	10.20	14.05		24.25	32
0140	3" diameter	↓	52	.308		13.70	15.95		29.65	39
0150	4" diameter		48	.333		18.50	17.25		35.75	46.50
0160	6" diameter	↓	39	.410		34	21.50		55.50	69.50
0170	8" diameter	Q-2	49	.490		53.50	26.50		80	98.50
0180	10" diameter		41	.585		78	31.50		109.50	134
0190	12" diameter	↓	36	.667	↓	101	36		137	165
0600	PVC, high impact/pressure, cplgs. 10' O.C., clevis hanger assy's, 3 per 10'									
1020	Schedule 80									
1070	1/2" diameter	1 Plum	50	.160	L.F.	4.62	9.20		13.82	19
1080	3/4" diameter		47	.170		5.65	9.80		15.45	21
1090	1" diameter		43	.186		7.25	10.70		17.95	24
1100	1-1/4" diameter		39	.205		9.45	11.80		21.25	28.50
1110	1-1/2" diameter	↓	34	.235		11	13.55		24.55	32.50
1120	2" diameter	Q-1	55	.291		14.25	15.05		29.30	38.50
1140	3" diameter		50	.320		29	16.60		45.60	56.50
1150	4" diameter		46	.348		41	18		59	72
1170	6" diameter	↓	38	.421	↓	92.50	22		114.50	135
1730	To delete coupling & hangers, subtract									
1740	1/4" diam. to 1/2" diam.					62%	80%			
1750	3/4" diam. to 1-1/4" diam.					58%	73%			
1760	1-1/2" diam. to 6" diam.					40%	57%			
1770	8" diam. to 12" diam.					34%	50%			
1800	PVC, couplings 10' O.C., clevis hanger assemblies, 3 per 10'									
1820	Schedule 40									
1860	1/2" diameter	1 Plum	54	.148	L.F.	3.02	8.55		11.57	16.20
1870	3/4" diameter		51	.157		3.33	9.05		12.38	17.30
1880	1" diameter		46	.174		3.96	10		13.96	19.45
1890	1-1/4" diameter		42	.190		4.94	10.95		15.89	22
1900	1-1/2" diameter	↓	36	.222		4.76	12.80		17.56	24.50
1910	2" diameter	Q-1	59	.271		5.95	14.05		20	27.50
1920	2-1/2" diameter		56	.286		9.15	14.80		23.95	32.50
1930	3" diameter		53	.302		11.05	15.65		26.70	35.50
1940	4" diameter		48	.333		14.20	17.25		31.45	41.50
1950	5" diameter		43	.372		28	19.25		47.25	60
1960	6" diameter	↓	39	.410		29.50	21.50		51	64.50
1970	8" diameter	Q-2	48	.500		42	27		69	87
1980	10" diameter		43	.558		82	30		112	136
1990	12" diameter		42	.571		103	30.50		133.50	160
2000	14" diameter		31	.774		161	41.50		202.50	240
2010	16" diameter	↓	23	1.043	↓	228	56		284	335
2340	To delete coupling & hangers, subtract									
2360	1/2" diam. to 1-1/4" diam.					65%	74%			
2370	1-1/2" diam. to 6" diam.					44%	57%			
2380	8" diam. to 12" diam.					41%	53%			
2390	14" diam. to 16" diam.					48%	45%			
2420	Schedule 80									
2440	1/4" diameter	1 Plum	58	.138	L.F.	2.20	7.95		10.15	14.40
2450	3/8" diameter	↓	55	.145	↓	2.20	8.35		10.55	15.05

22 11 Facility Water Distribution

22 11 13 – Facility Water Distribution Piping

22 11 13.74 Pipe, Plastic		Crew	Daily Output	Labor-Hours	Unit	Material	2014 Bare Costs Labor	Equipment	Total	Total Incl O&P
2460	1/2" diameter	1 Plum	50	.160	L.F.	3.08	9.20		12.28	17.30
2470	3/4" diameter		47	.170		3.53	9.80		13.33	18.70
2480	1" diameter		43	.186		4.20	10.70		14.90	21
2490	1-1/4" diameter		39	.205		5.25	11.80		17.05	23.50
2500	1-1/2" diameter	↓	34	.235		5.65	13.55		19.20	26.50
2510	2" diameter	Q-1	55	.291		6.85	15.05		21.90	30.50
2520	2-1/2" diameter		52	.308		9.55	15.95		25.50	34.50
2530	3" diameter		50	.320		13.75	16.60		30.35	40
2540	4" diameter		46	.348		18.85	18		36.85	47.50
2550	5" diameter		42	.381		27.50	19.75		47.25	60.50
2560	6" diameter	↓	38	.421		43.50	22		65.50	80.50
2570	8" diameter	Q-2	47	.511		48	27.50		75.50	94.50
2580	10" diameter		42	.571		98.50	30.50		129	156
2590	12" diameter	↓	38	.632	↓	123	34		157	186
2830	To delete coupling & hangers, subtract									
2840	1/4" diam. to 1/2" diam.					66%	80%			
2850	3/4" diam. to 1-1/4" diam.					61%	73%			
2860	1-1/2" diam. to 6" diam.					41%	57%			
2870	8" diam. to 12" diam.					31%	50%			
2900	Schedule 120									
2910	1/2" diameter	1 Plum	50	.160	L.F.	3.28	9.20		12.48	17.50
2950	3/4" diameter		47	.170		3.79	9.80		13.59	18.95
2960	1" diameter		43	.186		4.55	10.70		15.25	21
2970	1-1/4" diameter		39	.205		5.35	11.80		17.15	23.50
2980	1-1/2" diameter	↓	33	.242		6.45	13.95		20.40	28
2990	2" diameter	Q-1	54	.296		8.05	15.35		23.40	32
3000	2-1/2" diameter		52	.308		13	15.95		28.95	38.50
3010	3" diameter		49	.327		16.10	16.90		33	43
3020	4" diameter		45	.356		24	18.40		42.40	54.50
3030	6" diameter	↓	37	.432	↓	46.50	22.50		69	85
3240	To delete coupling & hangers, subtract									
3250	1/2" diam. to 1-1/4" diam.					52%	74%			
3260	1-1/2" diam. to 4" diam.					30%	57%			
3270	6" diam.					17%	50%			
3300	PVC, pressure, couplings 10' O.C., clevis hanger assy's, 3 per 10'									
3310	SDR 26, 160 psi									
3350	1-1/4" diameter	1 Plum	42	.190	L.F.	3.77	10.95		14.72	20.50
3360	1-1/2" diameter	"	36	.222		4.07	12.80		16.87	24
3370	2" diameter	Q-1	59	.271		4.80	14.05		18.85	26.50
3380	2-1/2" diameter		56	.286		7.20	14.80		22	30.50
3390	3" diameter		53	.302		9.65	15.65		25.30	34
3400	4" diameter		48	.333		13.65	17.25		30.90	41
3420	6" diameter	↓	39	.410		28	21.50		49.50	63
3430	8" diameter	Q-2	48	.500	↓	42	27		69	87
3660	To delete coupling & clevis hanger assy's, subtract									
3670	1-1/4" diam.					63%	68%			
3680	1-1/2" diam. to 4" diam.					48%	57%			
3690	6" diam. to 8" diam.					60%	54%			
3720	SDR 21, 200 psi, 1/2" diameter	1 Plum	54	.148	L.F.	2.60	8.55		11.15	15.70
3740	3/4" diameter		51	.157		2.92	9.05		11.97	16.85
3750	1" diameter		46	.174		3.18	10		13.18	18.60
3760	1-1/4" diameter		42	.190		3.95	10.95		14.90	21
3770	1-1/2" diameter	↓	36	.222	↓	4.29	12.80		17.09	24

22 11 Facility Water Distribution

22 11 13 – Facility Water Distribution Piping

22 11 13.74 Pipe, Plastic

		Crew	Daily Output	Labor-Hours	Unit	Material	2014 Bare Costs Labor	Equipment	Total	Total Incl O&P
3780	2" diameter	Q-1	59	.271	L.F.	5.15	14.05		19.20	26.50
3790	2-1/2" diameter		56	.286		9.25	14.80		24.05	32.50
3800	3" diameter		53	.302		10.45	15.65		26.10	35
3810	4" diameter		48	.333		17.20	17.25		34.45	45
3830	6" diameter		39	.410		31	21.50		52.50	66.50
3840	8" diameter	Q-2	48	.500		47	27		74	92
4000	To delete coupling & hangers, subtract									
4010	1/2" diam. to 3/4" diam.					71%	77%			
4020	1" diam. to 1-1/4" diam.					63%	70%			
4030	1-1/2" diam. to 6" diam.					44%	57%			
4040	8" diam.					46%	54%			
4100	DWV type, schedule 40, couplings 10' O.C., clevis hanger assy's, 3 per 10'									
4210	ABS, schedule 40, foam core type									
4212	Plain end black									
4214	1-1/2" diameter	1 Plum	39	.205	L.F.	2.96	11.80		14.76	21
4216	2" diameter	Q-1	62	.258		3.40	13.35		16.75	23.50
4218	3" diameter		56	.286		6.10	14.80		20.90	29
4220	4" diameter		51	.314		8.70	16.25		24.95	34
4222	6" diameter		42	.381		20.50	19.75		40.25	52.50
4240	To delete coupling & hangers, subtract									
4244	1-1/2" diam. to 6" diam.					43%	48%			
4400	PVC									
4410	1-1/4" diameter	1 Plum	42	.190	L.F.	3.41	10.95		14.36	20.50
4420	1-1/2" diameter	"	36	.222		2.98	12.80		15.78	22.50
4460	2" diameter	Q-1	59	.271		3.39	14.05		17.44	24.50
4470	3" diameter		53	.302		6.05	15.65		21.70	30
4480	4" diameter		48	.333		8.35	17.25		25.60	35
4490	6" diameter		39	.410		17.25	21.50		38.75	51
4500	8" diameter	Q-2	48	.500		27.50	27		54.50	71
4510	To delete coupling & hangers, subtract									
4520	1-1/4" diam. to 1-1/2" diam.					48%	60%			
4530	2" diam. to 8" diam.					42%	54%			
4550	PVC, schedule 40, foam core type									
4552	Plain end, white									
4554	1-1/2" diameter	1 Plum	39	.205	L.F.	2.72	11.80		14.52	21
4556	2" diameter	Q-1	62	.258		3.07	13.35		16.42	23.50
4558	3" diameter		56	.286		5.45	14.80		20.25	28.50
4560	4" diameter		51	.314		7.40	16.25		23.65	32.50
4562	6" diameter		42	.381		15.40	19.75		35.15	47
4564	8" diameter	Q-2	51	.471		24.50	25.50		50	65
4568	10" diameter		48	.500		29.50	27		56.50	73
4570	12" diameter		46	.522		34	28		62	80
4580	To delete coupling & hangers, subtract									
4582	1-1/2" dia to 2" dia					58%	54%			
4584	3" dia to 12" dia					46%	42%			
4800	PVC, clear pipe, cplgs. 10' O.C., clevis hanger assy's 3 per 10', Sched. 40									
4840	1/4" diameter	1 Plum	59	.136	L.F.	2.76	7.80		10.56	14.85
4850	3/8" diameter		56	.143		3.10	8.20		11.30	15.80
4860	1/2" diameter		54	.148		3.75	8.55		12.30	16.95
4870	3/4" diameter		51	.157		4.43	9.05		13.48	18.50
4880	1" diameter		46	.174		5.80	10		15.80	21.50
4890	1-1/4" diameter		42	.190		7.20	10.95		18.15	24.50
4900	1-1/2" diameter		36	.222		8.30	12.80		21.10	28.50

22 11 Facility Water Distribution

22 11 13 − Facility Water Distribution Piping

22 11 13.74 Pipe, Plastic		Crew	Daily Output	Labor-Hours	Unit	Material	2014 Bare Costs Labor	Equipment	Total	Total Incl O&P
4910	2" diameter	Q-1	59	.271	L.F.	10.60	14.05		24.65	32.50
4920	2-1/2" diameter		56	.286		16.55	14.80		31.35	41
4930	3" diameter		53	.302		21	15.65		36.65	46.50
4940	3-1/2" diameter		50	.320		30	16.60		46.60	58
4950	4" diameter		48	.333		31	17.25		48.25	60
5250	To delete coupling & hangers, subtract									
5260	1/4" diam. to 3/8" diam.					60%	81%			
5270	1/2" diam. to 3/4" diam.					41%	77%			
5280	1" diam. to 1-1/2" diam.					26%	67%			
5290	2" diam. to 4" diam.					16%	58%			
5300	CPVC, socket joint, couplings 10' O.C., clevis hanger assemblies, 3 per 10'									
5302	Schedule 40									
5304	1/2" diameter	1 Plum	54	.148	L.F.	3.65	8.55		12.20	16.85
5305	3/4" diameter		51	.157		4.30	9.05		13.35	18.40
5306	1" diameter		46	.174		5.50	10		15.50	21
5307	1-1/4" diameter		42	.190		6.90	10.95		17.85	24
5308	1-1/2" diameter		36	.222		8.05	12.80		20.85	28
5309	2" diameter	Q-1	59	.271		9.90	14.05		23.95	32
5310	2-1/2" diameter		56	.286		16.60	14.80		31.40	41
5311	3" diameter		53	.302		20	15.65		35.65	45.50
5312	4" diameter		48	.333		27.50	17.25		44.75	56.50
5314	6" diameter		43	.372		53	19.25		72.25	87
5318	To delete coupling & hangers, subtract									
5319	1/2" diam. to 3/4" diam.					37%	77%			
5320	1" diam. to 1-1/4" diam.					27%	70%			
5321	1-1/2" diam. to 3" diam.					21%	57%			
5322	4" diam. to 6" diam.					16%	57%			
5324	Schedule 80									
5325	1/2" diameter	1 Plum	50	.160	L.F.	3.88	9.20		13.08	18.15
5326	3/4" diameter		47	.170		4.43	9.80		14.23	19.70
5327	1" diameter		43	.186		5.65	10.70		16.35	22.50
5328	1-1/4" diameter		39	.205		7.80	11.80		19.60	26.50
5329	1-1/2" diameter		34	.235		9.15	13.55		22.70	30.50
5330	2" diameter	Q-1	55	.291		11.65	15.05		26.70	36
5331	2-1/2" diameter		52	.308		18.85	15.95		34.80	45
5332	3" diameter		50	.320		22	16.60		38.60	49.50
5333	4" diameter		46	.348		31	18		49	61
5334	6" diameter		38	.421		62	22		84	102
5335	8" diameter	Q-2	47	.511		118	27.50		145.50	171
5339	To delete couplings & hangers, subtract									
5340	1/2" diam. to 3/4" diam.					44%	77%			
5341	1" diam. to 1-1/4" diam.					32%	71%			
5342	1-1/2" diam. to 4" diam.					25%	58%			
5343	6" diam. to 8" diam.					20%	53%			
5360	CPVC, threaded, couplings 10' O.C., clevis hanger assemblies, 3 per 10'									
5380	Schedule 40									
5460	1/2" diameter	1 Plum	54	.148	L.F.	4.45	8.55		13	17.75
5470	3/4" diameter		51	.157		5.70	9.05		14.75	19.90
5480	1" diameter		46	.174		6.95	10		16.95	22.50
5490	1-1/4" diameter		42	.190		8	10.95		18.95	25.50
5500	1-1/2" diameter		36	.222		9	12.80		21.80	29
5510	2" diameter	Q-1	59	.271		11.10	14.05		25.15	33
5520	2-1/2" diameter		56	.286		17.85	14.80		32.65	42

22 11 Facility Water Distribution

22 11 13 – Facility Water Distribution Piping

22 11 13.74 Pipe, Plastic		Crew	Daily Output	Labor-Hours	Unit	Material	2014 Bare Costs Labor	Equipment	Total	Total Incl O&P
5530	3" diameter	Q-1	53	.302	L.F.	22	15.65		37.65	47.50
5540	4" diameter		48	.333		34.50	17.25		51.75	64
5550	6" diameter	↓	43	.372	↓	56.50	19.25		75.75	91
5730	To delete coupling & hangers, subtract									
5740	1/2" diam. to 3/4" diam.					37%	77%			
5750	1" diam. to 1-1/4" diam.					27%	70%			
5760	1-1/2" diam. to 3" diam.					21%	57%			
5770	4" diam. to 6" diam.					16%	57%			
5800	Schedule 80									
5860	1/2" diameter	1 Plum	50	.160	L.F.	4.68	9.20		13.88	19.05
5870	3/4" diameter		47	.170		5.80	9.80		15.60	21
5880	1" diameter		43	.186		7.10	10.70		17.80	24
5890	1-1/4" diameter		39	.205		8.90	11.80		20.70	27.50
5900	1-1/2" diameter	↓	34	.235		10.10	13.55		23.65	31.50
5910	2" diameter	Q-1	55	.291		12.80	15.05		27.85	37
5920	2-1/2" diameter		52	.308		20	15.95		35.95	46
5930	3" diameter		50	.320		24	16.60		40.60	51.50
5940	4" diameter		46	.348		38	18		56	69
5950	6" diameter	↓	38	.421		65.50	22		87.50	105
5960	8" diameter	Q-2	47	.511	↓	117	27.50		144.50	170
6060	To delete couplings & hangers, subtract									
6070	1/2" diam. to 3/4" diam.					44%	77%			
6080	1" diam. to 1-1/4" diam.					32%	71%			
6090	1-1/2" diam. to 4" diam.					25%	58%			
6100	6" diam. to 8" diam.					20%	53%			
6240	CTS, 1/2" diameter	1 Plum	54	.148	L.F.	2.20	8.55		10.75	15.25
6250	3/4" diameter		51	.157		2.79	9.05		11.84	16.70
6260	1" diameter		46	.174		4.21	10		14.21	19.75
6270	1 1/4"		42	.190		5.50	10.95		16.45	22.50
6280	1 1/2" diameter	↓	36	.222		6.80	12.80		19.60	27
6290	2" diameter	Q-1	59	.271	↓	10.55	14.05		24.60	32.50
6370	To delete coupling & hangers, subtract									
6380	1/2" diam.					51%	79%			
6390	3/4" diam.					40%	76%			
6392	1" thru 2" diam.					72%	68%			
6500	Residential installation, plastic pipe									
6510	Couplings 10' O.C., strap hangers 3 per 10'									
6520	PVC, Schedule 40									
6530	1/2" diameter	1 Plum	138	.058	L.F.	1.29	3.34		4.63	6.45
6540	3/4" diameter		128	.063		1.48	3.60		5.08	7.05
6550	1" diameter		119	.067		2.28	3.87		6.15	8.35
6560	1-1/4" diameter		111	.072		2.94	4.15		7.09	9.50
6570	1-1/2" diameter	↓	104	.077		2.70	4.43		7.13	9.65
6580	2" diameter	Q-1	197	.081		3.35	4.21		7.56	10.05
6590	2-1/2" diameter		162	.099		6.25	5.10		11.35	14.60
6600	4" diameter	↓	123	.130	↓	9.10	6.75		15.85	20.50
6700	PVC, DWV, Schedule 40									
6720	1-1/4" diameter	1 Plum	100	.080	L.F.	2.25	4.60		6.85	9.45
6730	1-1/2" diameter	"	94	.085		1.83	4.90		6.73	9.40
6740	2" diameter	Q-1	178	.090		2.19	4.66		6.85	9.45
6760	4" diameter	"	110	.145	↓	6.45	7.55		14	18.50
7280	PEX, flexible, no couplings or hangers									
7282	Note: For labor costs add 25% to the couplings and fittings labor total.									

22 11 Facility Water Distribution

22 11 13 – Facility Water Distribution Piping

22 11 13.74 Pipe, Plastic

		Crew	Daily Output	Labor-Hours	Unit	Material	2014 Bare Costs Labor	Equipment	Total	Total Incl O&P
7285	For fittings see section 23 83 16.10 7000									
7300	Non-barrier type, hot/cold tubing rolls									
7310	1/4" diameter x 100'				L.F.	.47			.47	.52
7350	3/8" diameter x 100'					.52			.52	.57
7360	1/2" diameter x 100'					.58			.58	.64
7370	1/2" diameter x 500'					.58			.58	.64
7380	1/2" diameter x 1000'					.58			.58	.64
7400	3/4" diameter x 100'					1.05			1.05	1.16
7410	3/4" diameter x 500'					1.05			1.05	1.16
7420	3/4" diameter x 1000'					1.05			1.05	1.16
7460	1" diameter x 100'					1.82			1.82	2
7470	1" diameter x 300'					1.82			1.82	2
7480	1" diameter x 500'					1.82			1.82	2
7500	1-1/4" diameter x 100'					3.08			3.08	3.39
7510	1-1/4" diameter x 300'					3.08			3.08	3.39
7540	1-1/2" diameter x 100'					4.19			4.19	4.61
7550	1-1/2" diameter x 300'					4.19			4.19	4.61
7596	Most sizes available in red or blue									
7700	Non-barrier type, hot/cold tubing straight lengths									
7710	1/2" diameter x 20'				L.F.	.58			.58	.64
7750	3/4" diameter x 20'					1.06			1.06	1.17
7760	1" diameter x 20'					1.82			1.82	2
7770	1-1/4" diameter x 20'					3.08			3.08	3.39
7780	1-1/2" diameter x 20'					4.19			4.19	4.61
7790	2" diameter					8.20			8.20	9
7796	Most sizes available in red or blue									
9000	Polypropylene pipe									
9002	For fusion weld fittings and accessories see line 22 11 13.76 9400									
9004	Note: sizes 1/2" thru 4" use socket fusion									
9005	Sizes 6" thru 10" use butt fusion									
9010	SDR 7.4, (domestic hot water piping)									
9011	Enhanced to minimize thermal expansion and high temperature life									
9016	13' lengths, size is ID, includes joints 13' O.C. and hangers 3 per 10'									
9020	3/8" diameter	1 Plum	53	.151	L.F.	2.27	8.70		10.97	15.60
9022	1/2" diameter		52	.154		2.72	8.85		11.57	16.35
9024	3/4" diameter		50	.160		3.23	9.20		12.43	17.45
9026	1" diameter		45	.178		4.25	10.25		14.50	20
9028	1-1/4" diameter		40	.200		6.05	11.50		17.55	24
9030	1-1/2" diameter		35	.229		8.65	13.15		21.80	29.50
9032	2" diameter	Q-1	58	.276		12.25	14.30		26.55	35
9034	2-1/2" diameter		55	.291		16.90	15.05		31.95	41.50
9036	3" diameter		52	.308		23	15.95		38.95	49.50
9038	3-1/2" diameter		49	.327		32.50	16.90		49.40	61
9040	4" diameter		46	.348		35.50	18		53.50	66
9042	6" diameter		39	.410		46	21.50		67.50	82.50
9044	8" diameter	Q-2	48	.500		69.50	27		96.50	117
9046	10" diameter	"	43	.558		109	30		139	166
9050	To delete joint & hangers, subtract									
9052	3/8" diam. to 1" diam.					45%	65%			
9054	1-1/4" diam. to 4" diam.					15%	45%			
9056	6" diam. to 10" diam.					5%	24%			
9060	SDR 11, (domestic cold water piping)									
9062	13' lengths, size is ID, includes joints 13' O.C. and hangers 3 per 10'									

213

22 11 Facility Water Distribution

22 11 13 – Facility Water Distribution Piping

22 11 13.74 Pipe, Plastic

		Crew	Daily Output	Labor-Hours	Unit	Material	2014 Bare Costs Labor	Equipment	Total	Total Incl O&P
9064	1/2" diameter	1 Plum	57	.140	L.F.	2.40	8.10		10.50	14.85
9066	3/4" diameter		54	.148		2.72	8.55		11.27	15.85
9068	1" diameter		49	.163		3.32	9.40		12.72	17.85
9070	1-1/4" diameter		45	.178		4.72	10.25		14.97	20.50
9072	1-1/2" diameter		40	.200		6.30	11.50		17.80	24.50
9074	2" diameter	Q-1	62	.258		8.90	13.35		22.25	30
9076	2-1/2" diameter		59	.271		11.70	14.05		25.75	34
9078	3" diameter		56	.286		16.40	14.80		31.20	40.50
9080	3-1/2" diameter		53	.302		24	15.65		39.65	49.50
9082	4" diameter		50	.320		27	16.60		43.60	54.50
9084	6" diameter		47	.340		31	17.65		48.65	60.50
9086	8" diameter	Q-2	51	.471		47.50	25.50		73	90.50
9088	10" diameter	"	46	.522		72	28		100	122
9090	To delete joint & hangers, subtract									
9092	3/8" diam. to 1" diam.					45%	65%			
9094	1-1/4" diam. to 4" diam.					15%	45%			
9096	6" diam. to 10" diam.					5%	24%			

22 11 13.76 Pipe Fittings, Plastic

		Crew	Daily Output	Labor-Hours	Unit	Material	Labor	Equipment	Total	Total Incl O&P
0010	**PIPE FITTINGS, PLASTIC**									
0030	Epoxy resin, fiberglass reinforced, general service									
0090	Elbow, 90°, 2"	Q-1	33.10	.483	Ea.	102	25		127	150
0100	3"		20.80	.769		113	40		153	184
0110	4"		16.50	.970		115	50		165	202
0120	6"		10.10	1.584		225	82		307	370
0130	8"	Q-2	9.30	2.581		415	139		554	665
0140	10"		8.50	2.824		520	152		672	805
0150	12"		7.60	3.158		750	170		920	1,075
0160	45° Elbow, same as 90°									
0170	Elbow, 90°, flanged									
0172	2"	Q-1	23	.696	Ea.	197	36		233	272
0173	3"		16	1		226	52		278	325
0174	4"		13	1.231		294	64		358	420
0176	6"		8	2		535	104		639	740
0177	8"	Q-2	9	2.667		965	143		1,108	1,300
0178	10"		7	3.429		1,325	184		1,509	1,725
0179	12"		5	4.800		1,775	258		2,033	2,350
0186	Elbow, 45°, flanged									
0188	2"	Q-1	23	.696	Ea.	197	36		233	272
0189	3"		16	1		226	52		278	325
0190	4"		13	1.231		294	64		358	420
0192	6"		8	2		535	104		639	740
0193	8"	Q-2	9	2.667		915	143		1,058	1,225
0194	10"		7	3.429		1,150	184		1,334	1,550
0195	12"		5	4.800		1,500	258		1,758	2,050
0290	Tee, 2"	Q-1	20	.800		54	41.50		95.50	122
0300	3"		13.90	1.151		64.50	59.50		124	161
0310	4"		11	1.455		89.50	75.50		165	212
0320	6"		6.70	2.388		225	124		349	435
0330	8"	Q-2	6.20	3.871		1,600	208		1,808	2,075
0340	10"		5.70	4.211		1,800	226		2,026	2,350
0350	12"		5.10	4.706		2,200	253		2,453	2,800
0352	Tee, flanged									

22 11 Facility Water Distribution

22 11 13 – Facility Water Distribution Piping

22 11 13.76 Pipe Fittings, Plastic		Crew	Daily Output	Labor-Hours	Unit	Material	2014 Bare Costs Labor	Equipment	Total	Total Incl O&P
0354	2"	Q-1	17	.941	Ea.	267	49		316	370
0355	3"		10	1.600		355	83		438	515
0356	4"		8	2		395	104		499	590
0358	6"		5	3.200		680	166		846	1,000
0359	8"	Q-2	6	4		1,300	215		1,515	1,750
0360	10"		5	4.800		1,875	258		2,133	2,450
0361	12"		4	6		2,575	320		2,895	3,325
0365	Wye, flanged									
0367	2"	Q-1	17	.941	Ea.	530	49		579	655
0368	3"		10	1.600		725	83		808	925
0369	4"		8	2		970	104		1,074	1,225
0371	6"		5	3.200		1,375	166		1,541	1,750
0372	8"	Q-2	6	4		2,250	215		2,465	2,800
0373	10"		5	4.800		3,475	258		3,733	4,225
0374	12"		4	6		5,075	320		5,395	6,050
0380	Couplings									
0410	2"	Q-1	33.10	.483	Ea.	22.50	25		47.50	63
0420	3"		20.80	.769		24.50	40		64.50	87
0430	4"		16.50	.970		33	50		83	113
0440	6"		10.10	1.584		78.50	82		160.50	210
0450	8"	Q-2	9.30	2.581		133	139		272	355
0460	10"		8.50	2.824		195	152		347	445
0470	12"		7.60	3.158		262	170		432	545
0473	High corrosion resistant couplings, add					30%				
0474	Reducer, concentric, flanged									
0475	2" x 1-1/2"	Q-1	30	.533	Ea.	510	27.50		537.50	605
0476	3" x 2"		24	.667		560	34.50		594.50	665
0477	4" x 3"		19	.842		625	43.50		668.50	755
0479	6" x 4"		15	1.067		625	55.50		680.50	775
0480	8" x 6"	Q-2	16	1.500		1,025	80.50		1,105.50	1,250
0481	10" x 8"		13	1.846		955	99		1,054	1,200
0482	12" x 10"		11	2.182		1,475	117		1,592	1,800
0486	Adapter, bell x male or female									
0488	2"	Q-1	28	.571	Ea.	25.50	29.50		55	72.50
0489	3"		20	.800		38	41.50		79.50	105
0491	4"		17	.941		49.50	49		98.50	128
0492	6"		12	1.333		107	69		176	222
0493	8"	Q-2	15	1.600		154	86		240	300
0494	10"	"	11	2.182		221	117		338	420
0528	Flange									
0532	2"	Q-1	46	.348	Ea.	36.50	18		54.50	67
0533	3"		32	.500		46.50	26		72.50	90
0534	4"		26	.615		60.50	32		92.50	115
0536	6"		16	1		114	52		166	204
0537	8"	Q-2	18	1.333		175	71.50		246.50	300
0538	10"		14	1.714		248	92		340	410
0539	12"		10	2.400		310	129		439	535
2100	PVC schedule 80, socket joint									
2110	90° elbow, 1/2"	1 Plum	30.30	.264	Ea.	2.48	15.20		17.68	25.50
2130	3/4"		26	.308		3.17	17.70		20.87	30
2140	1"		22.70	.352		4.95	20.50		25.45	36
2150	1-1/4"		20.20	.396		6.85	23		29.85	42
2160	1-1/2"		18.20	.440		7.30	25.50		32.80	46

22 11 Facility Water Distribution
22 11 13 − Facility Water Distribution Piping

22 11 13.76 Pipe Fittings, Plastic

		Crew	Daily Output	Labor-Hours	Unit	Material	2014 Bare Costs Labor	Equipment	Total	Total Incl O&P
2170	2"	Q-1	33.10	.483	Ea.	8.85	25		33.85	47.50
2180	3"		20.80	.769		23.50	40		63.50	85.50
2190	4"		16.50	.970		35.50	50		85.50	115
2200	6"		10.10	1.584		101	82		183	235
2210	8"	Q-2	9.30	2.581		277	139		416	515
2250	45° elbow, 1/2"	1 Plum	30.30	.264		4.67	15.20		19.87	28
2270	3/4"		26	.308		7.15	17.70		24.85	34.50
2280	1"		22.70	.352		10.70	20.50		31.20	42.50
2290	1-1/4"		20.20	.396		13.60	23		36.60	49.50
2300	1-1/2"		18.20	.440		16.10	25.50		41.60	55.50
2310	2"	Q-1	33.10	.483		21	25		46	61
2320	3"		20.80	.769		53.50	40		93.50	119
2330	4"		16.50	.970		96	50		146	182
2340	6"		10.10	1.584		121	82		203	257
2350	8"	Q-2	9.30	2.581		262	139		401	495
2400	Tee, 1/2"	1 Plum	20.20	.396		7	23		30	42
2420	3/4"		17.30	.462		7.35	26.50		33.85	48
2430	1"		15.20	.526		9.15	30.50		39.65	55.50
2440	1-1/4"		13.50	.593		25.50	34		59.50	79.50
2450	1-1/2"		12.10	.661		25.50	38		63.50	85.50
2460	2"	Q-1	20	.800		31.50	41.50		73	97
2470	3"		13.90	1.151		43	59.50		102.50	137
2480	4"		11	1.455		49.50	75.50		125	169
2490	6"		6.70	2.388		169	124		293	375
2500	8"	Q-2	6.20	3.871		395	208		603	745
2510	Flange, socket, 150 lb., 1/2"	1 Plum	55.60	.144		13.60	8.30		21.90	27.50
2514	3/4"		47.60	.168		14.50	9.65		24.15	30.50
2518	1"		41.70	.192		16.15	11.05		27.20	34.50
2522	1-1/2"		33.30	.240		16.65	13.85		30.50	39.50
2526	2"	Q-1	60.60	.264		22.50	13.70		36.20	45.50
2530	4"		30.30	.528		49	27.50		76.50	95.50
2534	6"		18.50	.865		77	45		122	152
2538	8"	Q-2	17.10	1.404		138	75.50		213.50	265
2550	Coupling, 1/2"	1 Plum	30.30	.264		4.49	15.20		19.69	28
2570	3/4"		26	.308		6.05	17.70		23.75	33
2580	1"		22.70	.352		6.25	20.50		26.75	37.50
2590	1-1/4"		20.20	.396		9.50	23		32.50	45
2600	1-1/2"		18.20	.440		10.25	25.50		35.75	49.50
2610	2"	Q-1	33.10	.483		11	25		36	50
2620	3"		20.80	.769		31	40		71	94
2630	4"		16.50	.970		39	50		89	119
2640	6"		10.10	1.584		84	82		166	216
2650	8"	Q-2	9.30	2.581		114	139		253	335
2660	10"		8.50	2.824		400	152		552	670
2670	12"		7.60	3.158		450	170		620	750
2700	PVC (white), schedule 40, socket joints									
2760	90° elbow, 1/2"	1 Plum	33.30	.240	Ea.	.48	13.85		14.33	21.50
2770	3/4"		28.60	.280		.54	16.10		16.64	25
2780	1"		25	.320		.97	18.40		19.37	29
2790	1-1/4"		22.20	.360		1.71	20.50		22.21	33.50
2800	1-1/2"		20	.400		1.85	23		24.85	37
2810	2"	Q-1	36.40	.440		2.89	23		25.89	37.50
2820	2-1/2"		26.70	.599		8.80	31		39.80	56.50

22 11 Facility Water Distribution

22 11 13 – Facility Water Distribution Piping

22 11 13.76 Pipe Fittings, Plastic		Crew	Daily Output	Labor-Hours	Unit	Material	2014 Bare Costs Labor	Equipment	Total	Total Incl O&P
2830	3"	Q-1	22.90	.699	Ea.	10.50	36		46.50	66
2840	4"		18.20	.879		18.80	45.50		64.30	89.50
2850	5"		12.10	1.322		48.50	68.50		117	157
2860	6"	↓	11.10	1.441		60	74.50		134.50	179
2870	8"	Q-2	10.30	2.330		154	125		279	360
2980	45° elbow, 1/2"	1 Plum	33.30	.240		.80	13.85		14.65	22
2990	3/4"		28.60	.280		1.23	16.10		17.33	26
3000	1"		25	.320		1.47	18.40		19.87	29.50
3010	1-1/4"		22.20	.360		2.03	20.50		22.53	33.50
3020	1-1/2"	↓	20	.400		2.60	23		25.60	38
3030	2"	Q-1	36.40	.440		3.38	23		26.38	38
3040	2-1/2"		26.70	.599		8.80	31		39.80	56.50
3050	3"		22.90	.699		13.65	36		49.65	69.50
3060	4"		18.20	.879		24.50	45.50		70	96
3070	5"		12.10	1.322		48.50	68.50		117	157
3080	6"	↓	11.10	1.441		60.50	74.50		135	180
3090	8"	Q-2	10.30	2.330		146	125		271	350
3180	Tee, 1/2"	1 Plum	22.20	.360		.60	20.50		21.10	32
3190	3/4"		19	.421		.69	24		24.69	37.50
3200	1"		16.70	.479		1.29	27.50		28.79	43
3210	1-1/4"		14.80	.541		2.01	31		33.01	49
3220	1-1/2"	↓	13.30	.602		2.45	34.50		36.95	55
3230	2"	Q-1	24.20	.661		3.56	34.50		38.06	55.50
3240	2-1/2"		17.80	.899		11.75	46.50		58.25	83.50
3250	3"		15.20	1.053		15.45	54.50		69.95	99.50
3260	4"		12.10	1.322		28	68.50		96.50	134
3270	5"		8.10	1.975		67.50	102		169.50	229
3280	6"	↓	7.40	2.162		94	112		206	272
3290	8"	Q-2	6.80	3.529		218	190		408	525
3380	Coupling, 1/2"	1 Plum	33.30	.240		.32	13.85		14.17	21.50
3390	3/4"		28.60	.280		.44	16.10		16.54	25
3400	1"		25	.320		.77	18.40		19.17	29
3410	1-1/4"		22.20	.360		1.06	20.50		21.56	32.50
3420	1-1/2"	↓	20	.400		1.13	23		24.13	36
3430	2"	Q-1	36.40	.440		1.72	23		24.72	36.50
3440	2-1/2"		26.70	.599		3.81	31		34.81	51
3450	3"		22.90	.699		5.95	36		41.95	61
3460	4"		18.20	.879		8.65	45.50		54.15	78.50
3470	5"		12.10	1.322		15.80	68.50		84.30	120
3480	6"	↓	11.10	1.441		27.50	74.50		102	143
3490	8"	Q-2	10.30	2.330		51	125		176	245
3600	Cap, schedule 40, PVC socket 1/2"	1 Plum	60.60	.132		.44	7.60		8.04	11.95
3610	3/4"		51.90	.154		.50	8.85		9.35	13.95
3620	1"		45.50	.176		.80	10.10		10.90	16.20
3630	1-1/4"		40.40	.198		1.13	11.40		12.53	18.45
3640	1-1/2"	↓	36.40	.220		1.23	12.65		13.88	20.50
3650	2"	Q-1	66.10	.242		1.49	12.55		14.04	20.50
3660	2-1/2"		48.50	.330		4.73	17.10		21.83	31
3670	3"		41.60	.385		5.10	19.90		25	35.50
3680	4"		33.10	.483		11.75	25		36.75	51
3690	6"	↓	20.20	.792		28	41		69	93
3700	8"	Q-2	18.60	1.290	↓	71	69.50		140.50	183
3710	Reducing insert, schedule 40, socket weld									

22 11 Facility Water Distribution

22 11 13 – Facility Water Distribution Piping

22 11 13.76 Pipe Fittings, Plastic		Crew	Daily Output	Labor-Hours	Unit	Material	2014 Bare Costs Labor	Equipment	Total	Total Incl O&P
3712	3/4"	1 Plum	31.50	.254	Ea.	.50	14.60		15.10	22.50
3713	1"		27.50	.291		.92	16.75		17.67	26.50
3715	1-1/2"	↓	22	.364		1.31	21		22.31	33
3716	2"	Q-1	40	.400		2.17	20.50		22.67	34
3717	4"		20	.800		11.45	41.50		52.95	75
3718	6"	↓	12.20	1.311		28.50	68		96.50	134
3719	8"	Q-2	11.30	2.124	↓	112	114		226	295
3730	Reducing insert, socket weld x female/male thread									
3732	1/2"	1 Plum	38.30	.209	Ea.	2.28	12		14.28	20.50
3733	3/4"		32.90	.243		1.41	14		15.41	22.50
3734	1"		28.80	.278		1.98	16		17.98	26
3736	1-1/2"	↓	23	.348		3.56	20		23.56	34
3737	2"	Q-1	41.90	.382		3.81	19.80		23.61	34
3738	4"	"	20.90	.766	↓	36.50	39.50		76	100
3742	Male adapter, socket weld x male thread									
3744	1/2"	1 Plum	38.30	.209	Ea.	.44	12		12.44	18.65
3745	3/4"		32.90	.243		.48	14		14.48	21.50
3746	1"		28.80	.278		.87	16		16.87	25
3748	1-1/2"	↓	23	.348		1.41	20		21.41	31.50
3749	2"	Q-1	41.90	.382		1.85	19.80		21.65	32
3750	4"	"	20.90	.766	↓	10.20	39.50		49.70	71
3754	Female adapter, socket weld x female thread									
3756	1/2"	1 Plum	38.30	.209	Ea.	.54	12		12.54	18.75
3757	3/4"		32.90	.243		.69	14		14.69	22
3758	1"		28.80	.278		.80	16		16.80	25
3760	1-1/2"	↓	23	.348		1.41	20		21.41	31.50
3761	2"	Q-1	41.90	.382		1.90	19.80		21.70	32
3762	4"	"	20.90	.766	↓	10.70	39.50		50.20	72
3800	PVC, schedule 80, socket joints									
3810	Reducing insert									
3812	3/4"	1 Plum	28.60	.280	Ea.	1.45	16.10		17.55	26
3813	1"		25	.320		4.14	18.40		22.54	32.50
3815	1-1/2"	↓	20	.400		9	23		32	45
3816	2"	Q-1	36.40	.440		12.60	23		35.60	48.50
3817	4"		18.20	.879		48	45.50		93.50	122
3818	6"	↓	11.10	1.441		67	74.50		141.50	187
3819	8"	Q-2	10.20	2.353	↓	385	126		511	615
3830	Reducing insert, socket weld x female/male thread									
3832	1/2"	1 Plum	34.80	.230	Ea.	9.05	13.25		22.30	30
3833	3/4"		29.90	.268		5.55	15.40		20.95	29.50
3834	1"		26.10	.307		8.70	17.65		26.35	36
3836	1-1/2"	↓	20.90	.383		10.95	22		32.95	45.50
3837	2"	Q-1	38	.421		16	22		38	50.50
3838	4"	"	19	.842	↓	77.50	43.50		121	152
3844	Adapter, male socket x male thread									
3846	1/2"	1 Plum	34.80	.230	Ea.	3.50	13.25		16.75	24
3847	3/4"		29.90	.268		3.86	15.40		19.26	28
3848	1"		26.10	.307		6.65	17.65		24.30	34
3850	1-1/2"	↓	20.90	.383		11.20	22		33.20	46
3851	2"	Q-1	38	.421		16.20	22		38.20	51
3852	4"	"	19	.842	↓	36.50	43.50		80	106
3860	Adapter, female socket x female thread									
3862	1/2"	1 Plum	34.80	.230	Ea.	4.20	13.25		17.45	24.50

22 11 Facility Water Distribution

22 11 13 – Facility Water Distribution Piping

22 11 13.76 Pipe Fittings, Plastic

		Crew	Daily Output	Labor-Hours	Unit	Material	2014 Bare Costs Labor	Equipment	Total	Total Incl O&P
3863	3/4"	1 Plum	29.90	.268	Ea.	6.25	15.40		21.65	30.50
3864	1"		26.10	.307		9.20	17.65		26.85	36.50
3866	1-1/2"	↓	20.90	.383		18.25	22		40.25	53.50
3867	2"	Q-1	38	.421		32	22		54	68
3868	4"	"	19	.842	↓	97.50	43.50		141	173
3872	Union, socket joints									
3874	1/2"	1 Plum	25.80	.310	Ea.	9.20	17.85		27.05	37
3875	3/4"		22.10	.362		11.70	21		32.70	44.50
3876	1"		19.30	.415		13.35	24		37.35	50.50
3878	1-1/2"	↓	15.50	.516		30	29.50		59.50	78
3879	2"	Q-1	28.10	.569	↓	40.50	29.50		70	89
3888	Cap									
3890	1/2"	1 Plum	54.50	.147	Ea.	4.39	8.45		12.84	17.60
3891	3/4"		46.70	.171		4.63	9.85		14.48	20
3892	1"		41	.195		8.20	11.25		19.45	26
3894	1-1/2"	↓	32.80	.244		9.90	14.05		23.95	32
3895	2"	Q-1	59.50	.269		26	13.95		39.95	49.50
3896	4"		30	.533		79	27.50		106.50	129
3897	6"	↓	18.20	.879		196	45.50		241.50	285
3898	8"	Q-2	16.70	1.437	↓	252	77		329	395
4500	DWV, ABS, non pressure, socket joints									
4540	1/4 Bend, 1-1/4"	1 Plum	20.20	.396	Ea.	6.15	23		29.15	41.50
4560	1-1/2"	"	18.20	.440		4.69	25.50		30.19	43
4570	2"	Q-1	33.10	.483		7.45	25		32.45	46
4580	3"		20.80	.769		18.30	40		58.30	80
4590	4"		16.50	.970		37.50	50		87.50	118
4600	6"	↓	10.10	1.584	↓	164	82		246	305
4650	1/8 Bend, same as 1/4 Bend									
4800	Tee, sanitary									
4820	1-1/4"	1 Plum	13.50	.593	Ea.	8.15	34		42.15	60.50
4830	1-1/2"	"	12.10	.661		7.15	38		45.15	65.50
4840	2"	Q-1	20	.800		11	41.50		52.50	74.50
4850	3"		13.90	1.151		30.50	59.50		90	124
4860	4"		11	1.455		53.50	75.50		129	173
4862	Tee, sanitary, reducing, 2" x 1-1/2"		22	.727		11.15	37.50		48.65	69.50
4864	3" x 2"		15.30	1.046		22	54		76	107
4868	4" x 3"	↓	12.10	1.322	↓	52.50	68.50		121	161
4870	Combination Y and 1/8 bend									
4872	1-1/2"	1 Plum	12.10	.661	Ea.	17.80	38		55.80	77
4874	2"	Q-1	20	.800		22	41.50		63.50	87
4876	3"		13.90	1.151		41	59.50		100.50	136
4878	4"		11	1.455		87	75.50		162.50	210
4880	3" x 1-1/2"		15.50	1.032		53.50	53.50		107	140
4882	4" x 3"	↓	12.10	1.322		82.50	68.50		151	194
4900	Wye, 1-1/4"	1 Plum	13.50	.593		9.40	34		43.40	62
4902	1-1/2"	"	12.10	.661		11.60	38		49.60	70.50
4904	2"	Q-1	20	.800		14.30	41.50		55.80	78
4906	3"		13.90	1.151		33.50	59.50		93	127
4908	4"		11	1.455		68.50	75.50		144	190
4910	6"		6.70	2.388		204	124		328	410
4918	3" x 1-1/2"		15.50	1.032		27	53.50		80.50	111
4920	4" x 3"		12.10	1.322		54	68.50		122.50	163
4922	6" x 4"	↓	6.90	2.319	↓	168	120		288	365

22 11 Facility Water Distribution

22 11 13 – Facility Water Distribution Piping

22 11 13.76 Pipe Fittings, Plastic

		Crew	Daily Output	Labor-Hours	Unit	Material	2014 Bare Costs Labor	Equipment	Total	Total Incl O&P
4930	Double Wye, 1-1/2"	1 Plum	9.10	.879	Ea.	33	50.50		83.50	113
4932	2"	Q-1	16.60	.964		39.50	50		89.50	119
4934	3"		10.40	1.538		89.50	79.50		169	219
4936	4"		8.25	1.939		175	100		275	345
4940	2" x 1-1/2"		16.80	.952		34.50	49.50		84	112
4942	3" x 2"		10.60	1.509		66	78		144	191
4944	4" x 3"		8.45	1.893		139	98		237	300
4946	6" x 4"		7.25	2.207		214	114		328	410
4950	Reducer bushing, 2" x 1-1/2"		36.40	.440		3.84	23		26.84	38.50
4952	3" x 1-1/2"		27.30	.586		16.20	30.50		46.70	64
4954	4" x 2"		18.20	.879		31	45.50		76.50	103
4956	6" x 4"		11.10	1.441		87	74.50		161.50	209
4960	Couplings, 1-1/2"	1 Plum	18.20	.440		2.33	25.50		27.83	40.50
4962	2"	Q-1	33.10	.483		3.14	25		28.14	41.50
4963	3"		20.80	.769		9.05	40		49.05	70
4964	4"		16.50	.970		16.35	50		66.35	94
4966	6"		10.10	1.584		68	82		150	199
4970	2" x 1-1/2"		33.30	.480		6.70	25		31.70	45
4972	3" x 1-1/2"		21	.762		18.55	39.50		58.05	80
4974	4" x 3"		16.70	.958		29.50	49.50		79	108
4978	Closet flange, 4"	1 Plum	32	.250		16.20	14.40		30.60	39.50
4980	4" x 3"	"	34	.235		18.50	13.55		32.05	41
5000	DWV, PVC, schedule 40, socket joints									
5040	1/4 bend, 1-1/4"	1 Plum	20.20	.396	Ea.	12.55	23		35.55	48.50
5060	1-1/2"	"	18.20	.440		3.45	25.50		28.95	42
5070	2"	Q-1	33.10	.483		5.45	25		30.45	44
5080	3"		20.80	.769		16	40		56	77.50
5090	4"		16.50	.970		31.50	50		81.50	111
5100	6"		10.10	1.584		111	82		193	246
5105	8"	Q-2	9.30	2.581		224	139		363	455
5106	10"	"	8.50	2.824		290	152		442	550
5110	1/4 bend, long sweep, 1-1/2"	1 Plum	18.20	.440		8.30	25.50		33.80	47
5112	2"	Q-1	33.10	.483		8.95	25		33.95	48
5114	3"		20.80	.769		21	40		61	83.50
5116	4"		16.50	.970		40	50		90	121
5150	1/8 bend, 1-1/4"	1 Plum	20.20	.396		8.20	23		31.20	43.50
5170	1-1/2"	"	18.20	.440		3.38	25.50		28.88	41.50
5180	2"	Q-1	33.10	.483		4.87	25		29.87	43.50
5190	3"		20.80	.769		14.35	40		54.35	76
5200	4"		16.50	.970		25	50		75	104
5210	6"		10.10	1.584		102	82		184	236
5215	8"	Q-2	9.30	2.581		158	139		297	380
5216	10"		8.50	2.824		206	152		358	455
5217	12"		7.60	3.158		284	170		454	565
5250	Tee, sanitary 1-1/4"	1 Plum	13.50	.593		12.30	34		46.30	65
5254	1-1/2"	"	12.10	.661		6.25	38		44.25	64.50
5255	2"	Q-1	20	.800		9.20	41.50		50.70	72.50
5256	3"		13.90	1.151		23	59.50		82.50	115
5257	4"		11	1.455		43	75.50		118.50	161
5259	6"		6.70	2.388		179	124		303	385
5261	8"	Q-2	6.20	3.871		535	208		743	905
5264	2" x 1-1/2"	Q-1	22	.727		8.10	37.50		45.60	66
5266	3" x 1-1/2"		15.50	1.032		17.05	53.50		70.55	100

22 11 Facility Water Distribution

22 11 13 – Facility Water Distribution Piping

22 11 13.76 Pipe Fittings, Plastic

		Crew	Daily Output	Labor-Hours	Unit	Material	2014 Bare Costs Labor	Equipment	Total	Total Incl O&P
5268	4" x 3"	Q-1	12.10	1.322	Ea.	52	68.50		120.50	161
5271	6" x 4"	↓	6.90	2.319	↓	173	120		293	370
5276	Tee, sanitary, reducing									
5281	2" x 1-1/2" x 1-1/2"	Q-1	23	.696	Ea.	8.10	36		44.10	63.50
5282	2" x 1-1/2" x 2"		22	.727		9.80	37.50		47.30	68
5283	2" x 2" x 1-1/2"		22	.727		8.10	37.50		45.60	66
5284	3" x 3" x 1-1/2"		15.50	1.032		17.05	53.50		70.55	100
5285	3" x 3" x 2"		15.30	1.046		18.05	54		72.05	102
5286	4" x 4" x 1-1/2"		12.30	1.301		46	67.50		113.50	153
5287	4" x 4" x 2"		12.20	1.311		38.50	68		106.50	145
5288	4" x 4" x 3"		12.10	1.322		48	68.50		116.50	156
5291	6" x 6" x 4"	↓	6.90	2.319		173	120		293	370
5294	Tee, double sanitary									
5295	1-1/2"	1 Plum	9.10	.879	Ea.	13.90	50.50		64.40	92
5296	2"	Q-1	16.60	.964		18.75	50		68.75	96
5297	3"		10.40	1.538		52.50	79.50		132	178
5298	4"	↓	8.25	1.939	↓	84.50	100		184.50	245
5303	Wye, reducing									
5304	2" x 1-1/2" x 1-1/2"	Q-1	23	.696	Ea.	15.75	36		51.75	72
5305	2" x 2" x 1-1/2"		22	.727		13.75	37.50		51.25	72
5306	3" x 3" x 2"		15.30	1.046		44.50	54		98.50	131
5307	4" x 4" x 2"		12.20	1.311		32.50	68		100.50	139
5309	4" x 4" x 3"	↓	12.10	1.322		44.50	68.50		113	152
5314	Combination Y & 1/8 bend, 1-1/2"	1 Plum	12.10	.661		15.10	38		53.10	74
5315	2"	Q-1	20	.800		16.15	41.50		57.65	80.50
5317	3"		13.90	1.151		40	59.50		99.50	134
5318	4"		11	1.455		79.50	75.50		155	202
5319	6"	↓	6.70	2.388		294	124		418	510
5320	8"	Q-2	6.20	3.871		575	208		783	950
5321	10"		5.70	4.211		705	226		931	1,125
5322	12"	↓	5.10	4.706	↓	820	253		1,073	1,275
5324	Combination Y & 1/8 bend, reducing									
5325	2" x 2" x 1-1/2"	Q-1	22	.727	Ea.	21	37.50		58.50	80.50
5327	3" x 3" x 1-1/2"		15.50	1.032		38	53.50		91.50	123
5328	3" x 3" x 2"		15.30	1.046		26.50	54		80.50	111
5329	4" x 4" x 2"	↓	12.20	1.311		43	68		111	151
5331	Wye, 1-1/4"	1 Plum	13.50	.593		16.10	34		50.10	69
5332	1-1/2"	"	12.10	.661		11.80	38		49.80	70.50
5333	2"	Q-1	20	.800		11.20	41.50		52.70	75
5334	3"		13.90	1.151		30	59.50		89.50	123
5335	4"		11	1.455		55	75.50		130.50	175
5336	6"	↓	6.70	2.388		165	124		289	370
5337	8"	Q-2	6.20	3.871		385	208		593	740
5338	10"		5.70	4.211		830	226		1,056	1,250
5339	12"	↓	5.10	4.706		1,350	253		1,603	1,875
5341	2" x 1-1/2"	Q-1	22	.727		13.85	37.50		51.35	72.50
5342	3" x 1-1/2"		15.50	1.032		19.85	53.50		73.35	103
5343	4" x 3"		12.10	1.322		44.50	68.50		113	152
5344	6" x 4"	↓	6.90	2.319		122	120		242	315
5345	8" x 6"	Q-2	6.40	3.750		291	201		492	625
5347	Double wye, 1-1/2"	1 Plum	9.10	.879		25.50	50.50		76	105
5348	2"	Q-1	16.60	.964		29.50	50		79.50	108
5349	3"	↓	10.40	1.538	↓	60	79.50		139.50	186

221

22 11 Facility Water Distribution

22 11 13 – Facility Water Distribution Piping

22 11 13.76 Pipe Fittings, Plastic

		Crew	Daily Output	Labor-Hours	Unit	Material	2014 Bare Costs Labor	Equipment	Total	Total Incl O&P
5350	4"	Q-1	8.25	1.939	Ea.	122	100		222	286
5353	Double wye, reducing									
5354	2" x 2" x 1-1/2" x 1-1/2"	Q-1	16.80	.952	Ea.	26	49.50		75.50	103
5355	3" x 3" x 2" x 2"		10.60	1.509		44.50	78		122.50	167
5356	4" x 4" x 3" x 3"		8.45	1.893		96.50	98		194.50	254
5357	6" x 6" x 4" x 4"	▼	7.25	2.207		325	114		439	530
5374	Coupling, 1-1/4"	1 Plum	20.20	.396		7.60	23		30.60	43
5376	1-1/2"	"	18.20	.440		1.72	25.50		27.22	40
5378	2"	Q-1	33.10	.483		2.24	25		27.24	40.50
5380	3"		20.80	.769		7.70	40		47.70	68.50
5390	4"		16.50	.970		13.65	50		63.65	91
5400	6"	▼	10.10	1.584		44	82		126	173
5402	8"	Q-2	9.30	2.581		109	139		248	330
5404	2" x 1-1/2"	Q-1	33.30	.480		4.99	25		29.99	43
5406	3" x 1-1/2"		21	.762		15.15	39.50		54.65	76
5408	4" x 3"		16.70	.958		24.50	49.50		74	102
5410	Reducer bushing, 2" x 1-1/4"		36.50	.438		7.05	22.50		29.55	42.50
5411	2" x 1-1/2"		36.40	.440		2.95	23		25.95	38
5412	3" x 1-1/2"		27.30	.586		14.15	30.50		44.65	61.50
5413	3" x 2"		27.10	.590		7.40	30.50		37.90	54
5414	4" x 2"		18.20	.879		24.50	45.50		70	96
5415	4" x 3"		16.70	.958		12.55	49.50		62.05	89
5416	6" x 4"	▼	11.10	1.441		65.50	74.50		140	185
5418	8" x 6"	Q-2	10.20	2.353		142	126		268	345
5425	Closet flange 4"	Q-1	32	.500		17.35	26		43.35	58
5426	4" x 3"	"	34	.471	▼	17.50	24.50		42	56.50
5450	Solvent cement for PVC, industrial grade, per quart				Qt.	35.50			35.50	39
5500	CPVC, Schedule 80, threaded joints									
5540	90° Elbow, 1/4"	1 Plum	32	.250	Ea.	12.15	14.40		26.55	35
5560	1/2"		30.30	.264		7.05	15.20		22.25	31
5570	3/4"		26	.308		10.55	17.70		28.25	38
5580	1"		22.70	.352		14.80	20.50		35.30	47
5590	1-1/4"		20.20	.396		28.50	23		51.50	66
5600	1-1/2"	▼	18.20	.440		30.50	25.50		56	72
5610	2"	Q-1	33.10	.483		41	25		66	83
5620	2-1/2"		24.20	.661		128	34.50		162.50	192
5630	3"		20.80	.769		137	40		177	210
5640	4"		16.50	.970		214	50		264	310
5650	6"	▼	10.10	1.584	▼	252	82		334	400
5660	45° Elbow same as 90° Elbow									
5700	Tee, 1/4"	1 Plum	22	.364	Ea.	23.50	21		44.50	57.50
5702	1/2"		20.20	.396		23.50	23		46.50	60.50
5704	3/4"		17.30	.462		34	26.50		60.50	77.50
5706	1"		15.20	.526		36.50	30.50		67	86
5708	1-1/4"		13.50	.593		37	34		71	92
5710	1-1/2"	▼	12.10	.661		38.50	38		76.50	100
5712	2"	Q-1	20	.800		43	41.50		84.50	110
5714	2-1/2"		16.20	.988		211	51		262	310
5716	3"		13.90	1.151		245	59.50		304.50	360
5718	4"		11	1.455		580	75.50		655.50	750
5720	6"	▼	6.70	2.388		670	124		794	920
5730	Coupling, 1/4"	1 Plum	32	.250		15.45	14.40		29.85	38.50
5732	1/2"		30.30	.264	▼	12.75	15.20		27.95	37

22 11 Facility Water Distribution

22 11 13 – Facility Water Distribution Piping

22 11 13.76 Pipe Fittings, Plastic

		Crew	Daily Output	Labor-Hours	Unit	Material	2014 Bare Costs Labor	Equipment	Total	Total Incl O&P
5734	3/4"	1 Plum	26	.308	Ea.	20.50	17.70		38.20	49
5736	1"		22.70	.352		23.50	20.50		44	56.50
5738	1-1/4"		20.20	.396		25	23		48	61.50
5740	1-1/2"		18.20	.440		26.50	25.50		52	67.50
5742	2"	Q-1	33.10	.483		31.50	25		56.50	72.50
5744	2-1/2"		24.20	.661		56.50	34.50		91	114
5746	3"		20.80	.769		65.50	40		105.50	132
5748	4"		16.50	.970		133	50		183	223
5750	6"		10.10	1.584		182	82		264	325
5752	8"	Q-2	9.30	2.581		390	139		529	635
5900	CPVC, Schedule 80, socket joints									
5904	90° Elbow, 1/4"	1 Plum	32	.250	Ea.	11.55	14.40		25.95	34
5906	1/2"		30.30	.264		4.53	15.20		19.73	28
5908	3/4"		26	.308		5.80	17.70		23.50	33
5910	1"		22.70	.352		9.20	20.50		29.70	40.50
5912	1-1/4"		20.20	.396		19.85	23		42.85	56.50
5914	1-1/2"		18.20	.440		22	25.50		47.50	62.50
5916	2"	Q-1	33.10	.483		26.50	25		51.50	67.50
5918	2-1/2"		24.20	.661		61.50	34.50		96	119
5920	3"		20.80	.769		69.50	40		109.50	137
5922	4"		16.50	.970		126	50		176	214
5924	6"		10.10	1.584		252	82		334	400
5926	8"		9.30	1.720		620	89		709	815
5930	45° Elbow, 1/4"	1 Plum	32	.250		17.20	14.40		31.60	40.50
5932	1/2"		30.30	.264		5.55	15.20		20.75	29
5934	3/4"		26	.308		7.95	17.70		25.65	35.50
5936	1"		22.70	.352		12.75	20.50		33.25	44.50
5938	1-1/4"		20.20	.396		25	23		48	62
5940	1-1/2"		18.20	.440		25.50	25.50		51	66
5942	2"	Q-1	33.10	.483		29	25		54	70
5944	2-1/2"		24.20	.661		59	34.50		93.50	116
5946	3"		20.80	.769		75.50	40		115.50	143
5948	4"		16.50	.970		104	50		154	190
5950	6"		10.10	1.584		305	82		387	460
5952	8"		9.30	1.720		665	89		754	865
5960	Tee, 1/4"	1 Plum	22	.364		10.60	21		31.60	43
5962	1/2"		20.20	.396		10.60	23		33.60	46
5964	3/4"		17.30	.462		10.80	26.50		37.30	52
5966	1"		15.20	.526		13.25	30.50		43.75	60
5968	1-1/4"		13.50	.593		28	34		62	82
5970	1-1/2"		12.10	.661		32	38		70	92.50
5972	2"	Q-1	20	.800		36	41.50		77.50	102
5974	2-1/2"		16.20	.988		90.50	51		141.50	178
5976	3"		13.90	1.151		90.50	59.50		150	190
5978	4"		11	1.455		121	75.50		196.50	247
5980	6"		6.70	2.388		315	124		439	530
5982	8"	Q-2	6.20	3.871		900	208		1,108	1,300
5990	Coupling, 1/4"	1 Plum	32	.250		12.30	14.40		26.70	35
5992	1/2"		30.30	.264		4.79	15.20		19.99	28.50
5994	3/4"		26	.308		6.70	17.70		24.40	34
5996	1"		22.70	.352		9	20.50		29.50	40.50
5998	1-1/4"		20.20	.396		13.50	23		36.50	49.50
6000	1-1/2"		18.20	.440		17	25.50		42.50	56.50

22 11 Facility Water Distribution

22 11 13 – Facility Water Distribution Piping

22 11 13.76 Pipe Fittings, Plastic

		Crew	Daily Output	Labor-Hours	Unit	Material	2014 Bare Costs Labor	Equipment	Total	Total Incl O&P
6002	2"	Q-1	33.10	.483	Ea.	19.75	25		44.75	59.50
6004	2-1/2"		24.20	.661		44	34.50		78.50	100
6006	3"		20.80	.769		47.50	40		87.50	113
6008	4"		16.50	.970		62.50	50		112.50	145
6010	6"		10.10	1.584		147	82		229	286
6012	8"	Q-2	9.30	2.581		400	139		539	650
6200	CTS, 100 psi at 180°F, hot and cold water									
6230	90° Elbow, 1/2"	1 Plum	20	.400	Ea.	.48	23		23.48	35.50
6250	3/4"		19	.421		.80	24		24.80	37.50
6251	1"		16	.500		2.19	29		31.19	46
6252	1-1/4"		15	.533		4.65	30.50		35.15	51.50
6253	1-1/2"		14	.571		8.40	33		41.40	59
6254	2"	Q-1	23	.696		16.10	36		52.10	72
6260	45° Elbow, 1/2"	1 Plum	20	.400		.62	23		23.62	35.50
6280	3/4"		19	.421		1.06	24		25.06	37.50
6281	1"		16	.500		2.71	29		31.71	46.50
6282	1-1/4"		15	.533		5.75	30.50		36.25	53
6283	1-1/2"		14	.571		8.30	33		41.30	58.50
6284	2"	Q-1	23	.696		17.40	36		53.40	73.50
6290	Tee, 1/2"	1 Plum	13	.615		.62	35.50		36.12	54
6310	3/4"		12	.667		1.17	38.50		39.67	59.50
6311	1"		11	.727		6.10	42		48.10	70
6312	1-1/4"		10	.800		9.40	46		55.40	80
6313	1-1/2"		10	.800		12.25	46		58.25	83
6314	2"	Q-1	17	.941		19.80	49		68.80	95.50
6320	Coupling, 1/2"	1 Plum	22	.364		.40	21		21.40	32
6340	3/4"		21	.381		.53	22		22.53	33.50
6341	1"		18	.444		2.38	25.50		27.88	41
6342	1-1/4"		17	.471		3.07	27		30.07	44.50
6343	1-1/2"		16	.500		4.34	29		33.34	48.50
6344	2"	Q-1	28	.571		8.25	29.50		37.75	53.50
6360	Solvent cement for CPVC, commercial grade, per quart				Qt.	44.50			44.50	49
7340	PVC flange, slip-on, Sch 80 std., 1/2"	1 Plum	22	.364	Ea.	13.60	21		34.60	46.50
7350	3/4"		21	.381		14.50	22		36.50	49
7360	1"		18	.444		16.15	25.50		41.65	56.50
7370	1-1/4"		17	.471		16.65	27		43.65	59.50
7380	1-1/2"		16	.500		17	29		46	62
7390	2"	Q-1	26	.615		22.50	32		54.50	73
7400	2-1/2"		24	.667		35	34.50		69.50	90.50
7410	3"		18	.889		38.50	46		84.50	112
7420	4"		15	1.067		49	55.50		104.50	138
7430	6"		10	1.600		77	83		160	210
7440	8"	Q-2	11	2.182		138	117		255	330
7550	Union, schedule 40, socket joints, 1/2"	1 Plum	19	.421		5.20	24		29.20	42.50
7560	3/4"		18	.444		5.35	25.50		30.85	44.50
7570	1"		15	.533		5.55	30.50		36.05	52.50
7580	1-1/4"		14	.571		16.65	33		49.65	68
7590	1-1/2"		13	.615		18.60	35.50		54.10	74
7600	2"	Q-1	20	.800		25	41.50		66.50	90
7992	Polybutyl/polyethyl pipe, for copper fittings see Line 22 11 13.25 7000									
8000	Compression type, PVC, 160 psi cold water									
8010	Coupling, 3/4" CTS	1 Plum	21	.381	Ea.	3.48	22		25.48	37
8020	1" CTS		18	.444		4.32	25.50		29.82	43.50

22 11 Facility Water Distribution

22 11 13 – Facility Water Distribution Piping

22 11 13.76 Pipe Fittings, Plastic

		Crew	Daily Output	Labor-Hours	Unit	Material	2014 Bare Costs Labor	Equipment	Total	Total Incl O&P
8030	1-1/4" CTS	1 Plum	17	.471	Ea.	6.10	27		33.10	47.50
8040	1-1/2" CTS		16	.500		8.30	29		37.30	52.50
8050	2" CTS		15	.533		11.65	30.50		42.15	59.50
8060	Female adapter, 3/4" FPT x 3/4" CTS		23	.348		5.70	20		25.70	36.50
8070	3/4" FPT x 1" CTS		21	.381		6.60	22		28.60	40.50
8080	1" FPT x 1" CTS		20	.400		6.60	23		29.60	42.50
8090	1-1/4" FPT x 1-1/4" CTS		18	.444		8.65	25.50		34.15	48
8100	1-1/2" FPT x 1-1/2" CTS		16	.500		9.90	29		38.90	54.50
8110	2" FPT x 2" CTS		13	.615		14.35	35.50		49.85	69.50
8130	Male adapter, 3/4" MPT x 3/4" CTS		23	.348		4.74	20		24.74	35
8140	3/4" MPT x 1" CTS		21	.381		5.65	22		27.65	39
8150	1" MPT x 1" CTS		20	.400		5.65	23		28.65	41
8160	1-1/4" MPT x 1-1/4" CTS		18	.444		7.60	25.50		33.10	47
8170	1-1/2" MPT x 1-1/2" CTS		16	.500		9.15	29		38.15	53.50
8180	2" MPT x 2" CTS		13	.615		11.85	35.50		47.35	66.50
8200	Spigot adapter, 3/4" IPS x 3/4" CTS		23	.348		2.70	20		22.70	33
8210	3/4" IPS x 1" CTS		21	.381		3.30	22		25.30	36.50
8220	1" IPS x 1" CTS		20	.400		3.30	23		26.30	38.50
8230	1-1/4" IPS x 1-1/4" CTS		18	.444		4.99	25.50		30.49	44
8240	1-1/2" IPS x 1-1/2" CTS		16	.500		5.25	29		34.25	49.50
8250	2" IPS x 2" CTS		13	.615		6.45	35.50		41.95	60.50
8270	Price includes insert stiffeners									
8280	250 psi is same price as 160 psi									
8300	Insert type, nylon, 160 & 250 psi, cold water									
8310	Clamp ring stainless steel, 3/4" IPS	1 Plum	115	.070	Ea.	2.38	4		6.38	8.65
8320	1" IPS		107	.075		2.43	4.30		6.73	9.15
8330	1-1/4" IPS		101	.079		2.45	4.56		7.01	9.60
8340	1-1/2" IPS		95	.084		3.28	4.85		8.13	10.90
8350	2" IPS		85	.094		3.77	5.40		9.17	12.35
8370	Coupling, 3/4" IPS		22	.364		.89	21		21.89	32.50
8380	1" IPS		19	.421		.92	24		24.92	37.50
8390	1-1/4" IPS		18	.444		1.37	25.50		26.87	40
8400	1-1/2" IPS		17	.471		1.62	27		28.62	43
8410	2" IPS		16	.500		3.60	29		32.60	47.50
8430	Elbow, 90°, 3/4" IPS		22	.364		1.77	21		22.77	33.50
8440	1" IPS		19	.421		1.96	24		25.96	38.50
8450	1-1/4" IPS		18	.444		2.19	25.50		27.69	41
8460	1-1/2" IPS		17	.471		2.58	27		29.58	44
8470	2" IPS		16	.500		3.60	29		32.60	47.50
8490	Male adapter, 3/4" IPS x 3/4" MPT		25	.320		.89	18.40		19.29	29
8500	1" IPS x 1" MPT		21	.381		.92	22		22.92	34
8510	1-1/4" IPS x 1-1/4" MPT		20	.400		1.45	23		24.45	36.50
8520	1-1/2" IPS x 1-1/2" MPT		18	.444		1.62	25.50		27.12	40.50
8530	2" IPS x 2" MPT		15	.533		3.11	30.50		33.61	50
8550	Tee, 3/4" IPS		14	.571		1.71	33		34.71	51.50
8560	1" IPS		13	.615		2.23	35.50		37.73	56
8570	1-1/4" IPS		12	.667		3.48	38.50		41.98	62
8580	1-1/2" IPS		11	.727		3.95	42		45.95	67.50
8590	2" IPS		10	.800		7.80	46		53.80	78
8610	Insert type, PVC, 100 psi @ 180°F, hot & cold water									
8620	Coupler, male, 3/8" CTS x 3/8" MPT	1 Plum	29	.276	Ea.	.72	15.90		16.62	25
8630	3/8" CTS x 1/2" MPT		28	.286		.72	16.45		17.17	26
8640	1/2" CTS x 1/2" MPT		27	.296		.72	17.05		17.77	27

225

22 11 Facility Water Distribution

22 11 13 – Facility Water Distribution Piping

22 11 13.76 Pipe Fittings, Plastic

		Crew	Daily Output	Labor-Hours	Unit	Material	2014 Bare Costs Labor	Equipment	Total	Total Incl O&P
8650	1/2" CTS x 3/4" MPT	1 Plum	26	.308	Ea.	2.29	17.70		19.99	29
8660	3/4" CTS x 1/2" MPT		25	.320		2.07	18.40		20.47	30.50
8670	3/4" CTS x 3/4" MPT		25	.320		2.46	18.40		20.86	30.50
8700	Coupling, 3/8" CTS x 1/2" CTS		25	.320		4.58	18.40		22.98	33
8710	1/2" CTS		23	.348		5.50	20		25.50	36
8730	3/4" CTS		22	.364		11.60	21		32.60	44.50
8750	Elbow 90°, 3/8" CTS		25	.320		4	18.40		22.40	32.50
8760	1/2" CTS		23	.348		4.73	20		24.73	35
8770	3/4" CTS		22	.364		7.25	21		28.25	39.50
8800	Rings, crimp, copper, 3/8" CTS		120	.067		.22	3.84		4.06	6.05
8810	1/2" CTS		117	.068		.22	3.94		4.16	6.20
8820	3/4" CTS		115	.070		.29	4		4.29	6.35
8850	Reducer tee, bronze, 3/8" x 3/8" x 1/2" CTS		17	.471		4.35	27		31.35	46
8860	1/2" x 1/2" x 3/4" CTS		15	.533		5.55	30.50		36.05	52.50
8870	3/4" x 1/2" x 1/2" CTS		14	.571		5.55	33		38.55	55.50
8890	3/4" x 3/4" x 1/2" CTS		14	.571		5.65	33		38.65	56
8900	1" x 1/2" x 1/2" CTS		14	.571		9.35	33		42.35	60
8930	Tee, 3/8" CTS		17	.471		3.31	27		30.31	44.50
8940	1/2" CTS		15	.533		2.95	30.50		33.45	50
8950	3/4" CTS	▼	14	.571	▼	3.50	33		36.50	53.50
8960	Copper rings included in fitting price									
9000	Flare type, assembled, acetal, hot & cold water									
9010	Coupling, 1/4" & 3/8" CTS	1 Plum	24	.333	Ea.	3.36	19.20		22.56	32.50
9020	1/2" CTS		22	.364		3.84	21		24.84	35.50
9030	3/4" CTS		21	.381		5.70	22		27.70	39.50
9040	1" CTS		18	.444		7.25	25.50		32.75	46.50
9050	Elbow 90°, 1/4" CTS		26	.308		3.73	17.70		21.43	30.50
9060	3/8" CTS		24	.333		4	19.20		23.20	33.50
9070	1/2" CTS		22	.364		4.73	21		25.73	36.50
9080	3/4" CTS		21	.381		7.25	22		29.25	41
9090	1" CTS		18	.444		9.15	25.50		34.65	48.50
9110	Tee, 1/4" CTS		16	.500		4.09	29		33.09	48
9114	3/8" CTS		15	.533		4.13	30.50		34.63	51
9120	1/2" CTS		14	.571		5.25	33		38.25	55.50
9130	3/4" CTS		13	.615		8.25	35.50		43.75	62.50
9140	1" CTS	▼	12	.667	▼	11.05	38.50		49.55	70
9400	Polypropylene, fittings and accessories									
9404	Fittings fusion welded, sizes are I.D.									
9408	Note: sizes 1/2" thru 4" use socket fusion									
9410	Sizes 6" thru 10" use butt fusion									
9416	Coupling									
9420	3/8"	1 Plum	39	.205	Ea.	.88	11.80		12.68	18.80
9422	1/2"		37.40	.214		1.15	12.30		13.45	19.85
9424	3/4"		35.40	.226		1.28	13		14.28	21
9426	1"		29.70	.269		1.69	15.50		17.19	25.50
9428	1-1/4"		27.60	.290		2.02	16.70		18.72	27
9430	1-1/2"	▼	24.80	.323		4.25	18.55		22.80	32.50
9432	2"	Q-1	43	.372		8.50	19.25		27.75	38.50
9434	2-1/2"		35.60	.449		9.50	23.50		33	45.50
9436	3"		30.90	.518		21	27		48	63.50
9438	3-1/2"		27.80	.576		34	30		64	82.50
9440	4"	▼	25	.640	▼	44.50	33		77.50	99
9442	Reducing coupling, female to female									

226

22 11 Facility Water Distribution

22 11 13 – Facility Water Distribution Piping

22 11 13.76 Pipe Fittings, Plastic

		Crew	Daily Output	Labor-Hours	Unit	Material	2014 Bare Costs Labor	Equipment	Total	Total Incl O&P
9446	2" to 1-1/2"	Q-1	49	.327	Ea.	12.25	16.90		29.15	39
9448	2-1/2" to 2"		41.20	.388		13.45	20		33.45	45.50
9450	3" to 2-1/2"		33.10	.483		18.15	25		43.15	58
9470	Reducing bushing, female to female									
9472	1/2" to 3/8"	1 Plum	38.20	.209	Ea.	1.15	12.05		13.20	19.45
9474	3/4" to 3/8" or 1/2"		36.50	.219		1.28	12.60		13.88	20.50
9476	1" to 3/4" or 1/2"		33.10	.242		1.71	13.90		15.61	23
9478	1-1/4" to 3/4" or 1"		28.70	.279		2.63	16.05		18.68	27
9480	1-1/2" to 1/2" thru 1-1/4"		26.20	.305		4.36	17.55		21.91	31.50
9482	2" to 1/2" thru 1-1/2"	Q-1	43	.372		8.70	19.25		27.95	38.50
9484	2-1/2" to 1/2" thru 2"		41.20	.388		9.75	20		29.75	41
9486	3" to 1-1/2" thru 2-1/2"		33.10	.483		21.50	25		46.50	62
9488	3-1/2" to 2" thru 3"		29.20	.548		34.50	28.50		63	81
9490	4" to 2-1/2" thru 3-1/2"		26.20	.611		54	31.50		85.50	108
9491	6" to 4" SDR 7.4		16.50	.970		65	50		115	148
9492	6" to 4" SDR 11		16.50	.970		65	50		115	148
9493	8" to 6" SDR 7.4		10.10	1.584		95.50	82		177.50	229
9494	8" to 6" SDR 11		10.10	1.584		87.50	82		169.50	221
9495	10" to 8" SDR 7.4		7.80	2.051		131	106		237	305
9496	10" to 8" SDR 11		7.80	2.051		114	106		220	286
9500	90° Elbow									
9504	3/8"	1 Plum	39	.205	Ea.	.97	11.80		12.77	18.90
9506	1/2"		37.40	.214		1.23	12.30		13.53	19.95
9508	3/4"		35.40	.226		1.58	13		14.58	21.50
9510	1"		29.70	.269		2.28	15.50		17.78	26
9512	1-1/4"		27.60	.290		3.51	16.70		20.21	29
9514	1-1/2"		24.80	.323		7.55	18.55		26.10	36.50
9516	2"	Q-1	43	.372		11.60	19.25		30.85	42
9518	2-1/2"		35.60	.449		26	23.50		49.50	63.50
9520	3"		30.90	.518		43	27		70	87.50
9522	3-1/2"		27.80	.576		61	30		91	112
9524	4"		25	.640		93.50	33		126.50	153
9526	6" SDR 7.4		5.55	2.883		107	149		256	345
9528	6" SDR 11		5.55	2.883		83.50	149		232.50	320
9530	8" SDR 7.4	Q-2	8.10	2.963		415	159		574	700
9532	8" SDR 11		8.10	2.963		355	159		514	630
9534	10" SDR 7.4		7.50	3.200		635	172		807	960
9536	10" SDR 11		7.50	3.200		575	172		747	895
9551	45° Elbow									
9554	3/8"	1 Plum	39	.205	Ea.	.94	11.80		12.74	18.90
9556	1/2"		37.40	.214		1.23	12.30		13.53	19.95
9558	3/4"		35.40	.226		1.58	13		14.58	21.50
9564	1"		29.70	.269		2.28	15.50		17.78	26
9566	1-1/4"		27.60	.290		3.51	16.70		20.21	29
9568	1-1/2"		24.80	.323		7.55	18.55		26.10	36.50
9570	2"	Q-1	43	.372		11.45	19.25		30.70	41.50
9572	2-1/2"		35.60	.449		25.50	23.50		49	63
9574	3"		30.90	.518		47	27		74	92
9576	3-1/2"		27.80	.576		67	30		97	119
9578	4"		25	.640		103	33		136	163
9580	6" SDR 7.4		5.55	2.883		119	149		268	355
9582	6" SDR 11		5.55	2.883		98	149		247	335
9584	8" SDR 7.4	Q-2	8.10	2.963		325	159		484	600

22 11 Facility Water Distribution

22 11 13 – Facility Water Distribution Piping

22 11 13.76 Pipe Fittings, Plastic

		Crew	Daily Output	Labor-Hours	Unit	Material	2014 Bare Costs Labor	Equipment	Total	Total Incl O&P
9586	8" SDR 11	Q-2	8.10	2.963	Ea.	284	159		443	550
9588	10" SDR 7.4		7.50	3.200		525	172		697	835
9590	10" SDR 11		7.50	3.200		435	172		607	735
9600	Tee									
9604	3/8"	1 Plum	26	.308	Ea.	1.22	17.70		18.92	28
9606	1/2"		24.90	.321		1.66	18.50		20.16	30
9608	3/4"		23.70	.338		2.28	19.45		21.73	32
9610	1"		19.90	.402		2.90	23		25.90	38
9612	1-1/4"		18.50	.432		4.43	25		29.43	42.50
9614	1-1/2"		16.60	.482		12.65	27.50		40.15	56
9616	2"	Q-1	26.80	.597		18.10	31		49.10	66.50
9618	2-1/2"		23.80	.672		30.50	35		65.50	86
9620	3"		20.60	.777		50	40		90	116
9622	3-1/2"		18.40	.870		78	45		123	154
9624	4"		16.70	.958		104	49.50		153.50	189
9626	6" SDR 7.4		3.70	4.324		129	224		353	480
9628	6" SDR 11		3.70	4.324		98	224		322	450
9630	8" SDR 7.4	Q-2	5.40	4.444		355	239		594	750
9632	8" SDR 11		5.40	4.444		310	239		549	705
9634	10" SDR 7.4		5	4.800		610	258		868	1,050
9636	10" SDR 11		5	4.800		530	258		788	970
9638	For reducing tee use same tee price									
9660	End cap									
9662	3/8"	1 Plum	78	.103	Ea.	1.22	5.90		7.12	10.25
9664	1/2"		74.60	.107		1.80	6.15		7.95	11.30
9666	3/4"		71.40	.112		2.28	6.45		8.73	12.25
9668	1"		59.50	.134		2.76	7.75		10.51	14.75
9670	1-1/4"		55.60	.144		4.36	8.30		12.66	17.30
9672	1-1/2"		49.50	.162		6	9.30		15.30	20.50
9674	2"	Q-1	80	.200		10.05	10.35		20.40	26.50
9676	2-1/2"		71.40	.224		14.55	11.60		26.15	33.50
9678	3"		61.70	.259		33	13.45		46.45	56.50
9680	3-1/2"		55.20	.290		39.50	15		54.50	66
9682	4"		50	.320		60.50	16.60		77.10	91.50
9684	6" SDR 7.4		26.50	.604		83	31.50		114.50	138
9686	6" SDR 11		26.50	.604		83	31.50		114.50	138
9688	8" SDR 7.4	Q-2	16.30	1.472		83	79		162	211
9690	8" SDR 11		16.30	1.472		91.50	79		170.50	219
9692	10" SDR 7.4		14.90	1.611		124	86.50		210.50	268
9694	10" SDR 11		14.90	1.611		106	86.50		192.50	247
9800	Accessories and tools									
9802	Pipe clamps for suspension, not including rod or beam clamp									
9804	3/8"	1 Plum	74	.108	Ea.	2.13	6.20		8.33	11.75
9805	1/2"		70	.114		2.67	6.60		9.27	12.90
9806	3/4"		68	.118		3.07	6.75		9.82	13.60
9807	1"		66	.121		3.35	7		10.35	14.25
9808	1-1/4"		64	.125		3.44	7.20		10.64	14.65
9809	1-1/2"		62	.129		3.76	7.45		11.21	15.35
9810	2"	Q-1	110	.145		4.66	7.55		12.21	16.55
9811	2-1/2"		104	.154		5.95	7.95		13.90	18.60
9812	3"		98	.163		6.45	8.45		14.90	19.80
9813	3-1/2"		92	.174		7.05	9		16.05	21.50
9814	4"		86	.186		7.90	9.65		17.55	23.50

22 11 Facility Water Distribution

22 11 13 – Facility Water Distribution Piping

22 11 13.76 Pipe Fittings, Plastic

		Crew	Daily Output	Labor-Hours	Unit	Material	2014 Bare Costs Labor	Equipment	Total	Total Incl O&P
9815	6"	Q-1	70	.229	Ea.	9.65	11.85		21.50	28.50
9816	8"	Q-2	100	.240		34.50	12.90		47.40	57.50
9817	10"	"	94	.255	↓	39.50	13.70		53.20	64
9820	Pipe cutter									
9822	For 3/8" thru 1-1/4"				Ea.	118			118	129
9824	For 1-1/2" thru 4"				"	310			310	340
9826	Note: Pipes may be cut with standard									
9827	iron saw with blades for plastic.									
9830	Manual welding device, 800 W, 110 V									
9831	For 3/8" thru 2"				Ea.	375			375	410
9834	Manual welding device, 1400 W, 110 V									
9835	For 1-1/2" thru 4"				Ea.	730			730	805
9838	Butt welding machine									
9839	For 6" thru 10"				Ea.	25,400			25,400	27,900
9982	For plastic hangers see Line 22 05 29.10 8000									
9986	For copper/brass fittings see Line 22 11 13.25 7000									

22 11 13.78 Pipe, High Density Polyethylene Plastic (HDPE)

		Crew	Daily Output	Labor-Hours	Unit	Material	Labor	Equipment	Total	Total Incl O&P
0010	**PIPE, HIGH DENSITY POLYETHYLENE PLASTIC (HDPE)**									
0020	Not incl. hangers, trenching, backfill, hoisting or digging equipment.									
0030	Standard length is 40', add a weld for each joint									
0040	Single wall									
0050	Straight									
0054	1" diameter DR 11				L.F.	.79			.79	.87
0058	1-1/2" diameter DR 11					1			1	1.10
0062	2" diameter DR 11					1.67			1.67	1.84
0066	3" diameter DR 11					2.01			2.01	2.21
0070	3" diameter DR 17					1.61			1.61	1.77
0074	4" diameter DR 11					3.36			3.36	3.70
0078	4" diameter DR 17					3.36			3.36	3.70
0082	6" diameter DR 11					8.35			8.35	9.20
0086	6" diameter DR 17					5.55			5.55	6.10
0090	8" diameter DR 11					13.90			13.90	15.30
0094	8" diameter DR 26					6.50			6.50	7.15
0098	10" diameter DR 11					22			22	24
0102	10" diameter DR 26					10.05			10.05	11.05
0106	12" diameter DR 11					32			32	35
0110	12" diameter DR 26					15.05			15.05	16.60
0114	16" diameter DR 11					48.50			48.50	53.50
0118	16" diameter DR 26					22			22	24
0122	18" diameter DR 11					62			62	68
0126	18" diameter DR 26					28.50			28.50	31.50
0130	20" diameter DR 11					75.50			75.50	83
0134	20" diameter DR 26					33.50			33.50	37
0138	22" diameter DR 11					92			92	101
0142	22" diameter DR 26					42			42	46
0146	24" diameter DR 11					109			109	120
0150	24" diameter DR 26					48.50			48.50	53.50
0154	28" diameter DR 17					100			100	111
0158	28" diameter DR 26					67			67	73.50
0162	30" diameter DR 21					94			94	103
0166	30" diameter DR 26					77			77	84.50
0170	36" diameter DR 26				↓	77			77	84.50

22 11 Facility Water Distribution

22 11 13 – Facility Water Distribution Piping

22 11 13.78 Pipe, High Density Polyethylene Plastic (HDPE)

		Crew	Daily Output	Labor-Hours	Unit	Material	2014 Bare Costs Labor	Equipment	Total	Total Incl O&P
0174	42" diameter DR 26				L.F.	149			149	164
0178	48" diameter DR 26					196			196	216
0182	54" diameter DR 26				↓	246			246	271
0300	90° Elbow									
0304	1" diameter DR 11				Ea.	5.60			5.60	6.15
0308	1-1/2" diameter DR 11					7			7	7.70
0312	2" diameter DR 11					7			7	7.70
0316	3" diameter DR 11					14			14	15.40
0320	3" diameter DR 17					14			14	15.40
0324	4" diameter DR 11					19.60			19.60	21.50
0328	4" diameter DR 17					19.60			19.60	21.50
0332	6" diameter DR 11					45			45	49.50
0336	6" diameter DR 17					45			45	49.50
0340	8" diameter DR 11					111			111	122
0344	8" diameter DR 26					98			98	108
0348	10" diameter DR 11					415			415	455
0352	10" diameter DR 26					380			380	420
0356	12" diameter DR 11					435			435	480
0360	12" diameter DR 26					395			395	435
0364	16" diameter DR 11					515			515	570
0368	16" diameter DR 26					510			510	560
0372	18" diameter DR 11					650			650	715
0376	18" diameter DR 26					610			610	670
0380	20" diameter DR 11					765			765	840
0384	20" diameter DR 26					740			740	810
0388	22" diameter DR 11					790			790	870
0392	22" diameter DR 26					765			765	840
0396	24" diameter DR 11					890			890	980
0400	24" diameter DR 26					865			865	950
0404	28" diameter DR 17					1,075			1,075	1,200
0408	28" diameter DR 26					1,025			1,025	1,125
0412	30" diameter DR 17					1,525			1,525	1,675
0416	30" diameter DR 26					1,400			1,400	1,550
0420	36" diameter DR 26					1,775			1,775	1,950
0424	42" diameter DR 26					2,300			2,300	2,525
0428	48" diameter DR 26					2,675			2,675	2,950
0432	54" diameter DR 26				↓	6,375			6,375	7,000
0500	45° Elbow									
0512	2" diameter DR 11				Ea.	5.60			5.60	6.15
0516	3" diameter DR 11					14			14	15.40
0520	3" diameter DR 17					14			14	15.40
0524	4" diameter DR 11					19.60			19.60	21.50
0528	4" diameter DR 17					19.60			19.60	21.50
0532	6" diameter DR 11					45			45	49.50
0536	6" diameter DR 17					45			45	49.50
0540	8" diameter DR 11					111			111	122
0544	8" diameter DR 26					64.50			64.50	71
0548	10" diameter DR 11					415			415	455
0552	10" diameter DR 26					380			380	420
0556	12" diameter DR 11					435			435	480
0560	12" diameter DR 26					395			395	435
0564	16" diameter DR 11					229			229	252
0568	16" diameter DR 26				↓	216			216	238

22 11 Facility Water Distribution

22 11 13 – Facility Water Distribution Piping

22 11 13.78 Pipe, High Density Polyethylene Plastic (HDPE)

		Crew	Daily Output	Labor-Hours	Unit	Material	2014 Bare Costs Labor	Equipment	Total	Total Incl O&P
0572	18" diameter DR 11				Ea.	243			243	268
0576	18" diameter DR 26					223			223	245
0580	20" diameter DR 11					370			370	405
0584	20" diameter DR 26					350			350	385
0588	22" diameter DR 11					510			510	560
0592	22" diameter DR 26					485			485	530
0596	24" diameter DR 11					625			625	685
0600	24" diameter DR 26					600			600	660
0604	28" diameter DR 17					710			710	780
0608	28" diameter DR 26					690			690	755
0612	30" diameter DR 17					880			880	965
0616	30" diameter DR 26					855			855	940
0620	36" diameter DR 26					1,075			1,075	1,200
0624	42" diameter DR 26					1,400			1,400	1,525
0628	48" diameter DR 26					1,525			1,525	1,675
0632	54" diameter DR 26					2,025			2,025	2,225
0700	Tee									
0704	1" diameter DR 11				Ea.	7.30			7.30	8.05
0708	1-1/2" diameter DR 11					10.25			10.25	11.30
0712	2" diameter DR 11					8.80			8.80	9.65
0716	3" diameter DR 11					16.10			16.10	17.70
0720	3" diameter DR 17					16.10			16.10	17.70
0724	4" diameter DR 11					23.50			23.50	26
0728	4" diameter DR 17					23.50			23.50	26
0732	6" diameter DR 11					58.50			58.50	64.50
0736	6" diameter DR 17					58.50			58.50	64.50
0740	8" diameter DR 11					145			145	159
0744	8" diameter DR 17					145			145	159
0748	10" diameter DR 11					430			430	475
0752	10" diameter DR 17					430			430	475
0756	12" diameter DR 11					575			575	635
0760	12" diameter DR 17					575			575	635
0764	16" diameter DR 11					305			305	335
0768	16" diameter DR 17					251			251	276
0772	18" diameter DR 11					430			430	470
0776	18" diameter DR 17					355			355	390
0780	20" diameter DR 11					525			525	575
0784	20" diameter DR 17					430			430	470
0788	22" diameter DR 11					670			670	740
0792	22" diameter DR 17					530			530	585
0796	24" diameter DR 11					835			835	915
0800	24" diameter DR 17					705			705	775
0804	28" diameter DR 17					1,350			1,350	1,500
0812	30" diameter DR 17					1,550			1,550	1,725
0820	36" diameter DR 17					2,575			2,575	2,825
0824	42" diameter DR 26					2,875			2,875	3,150
0828	48" diameter DR 26					3,075			3,075	3,375
1000	Flange adptr, w/back-up ring and 1/2 cost of plated bolt set									
1004	1" diameter DR 11				Ea.	28			28	30.50
1008	1-1/2" diameter DR 11					28			28	30.50
1012	2" diameter DR 11					17.55			17.55	19.35
1016	3" diameter DR 11					20.50			20.50	22.50
1020	3" diameter DR 17					20.50			20.50	22.50

22 11 Facility Water Distribution

22 11 13 – Facility Water Distribution Piping

22 11 13.78 Pipe, High Density Polyethylene Plastic (HDPE)

		Crew	Daily Output	Labor-Hours	Unit	Material	2014 Bare Costs Labor	Equipment	Total	Total Incl O&P
1024	4" diameter DR 11				Ea.	28			28	30.50
1028	4" diameter DR 17					28			28	30.50
1032	6" diameter DR 11					39.50			39.50	43.50
1036	6" diameter DR 17					39.50			39.50	43.50
1040	8" diameter DR 11					57			57	63
1044	8" diameter DR 26					57			57	63
1048	10" diameter DR 11					91			91	100
1052	10" diameter DR 26					91			91	100
1056	12" diameter DR 11					133			133	147
1060	12" diameter DR 26					133			133	147
1064	16" diameter DR 11					286			286	315
1068	16" diameter DR 26					286			286	315
1072	18" diameter DR 11					370			370	410
1076	18" diameter DR 26					370			370	410
1080	20" diameter DR 11					515			515	565
1084	20" diameter DR 26					515			515	565
1088	22" diameter DR 11					560			560	615
1092	22" diameter DR 26					555			555	610
1096	24" diameter DR 17					605			605	665
1100	24" diameter DR 32.5					605			605	665
1104	28" diameter DR 15.5					820			820	900
1108	28" diameter DR 32.5					820			820	900
1112	30" diameter DR 11					950			950	1,050
1116	30" diameter DR 21					950			950	1,050
1120	36" diameter DR 26					1,050			1,050	1,150
1124	42" diameter DR 26					1,175			1,175	1,300
1128	48" diameter DR 26					1,450			1,450	1,600
1132	54" diameter DR 26				▼	1,750			1,750	1,925
1200	Reducer									
1208	2" x 1-1/2" diameter DR 11				Ea.	8.40			8.40	9.25
1212	3" x 2" diameter DR 11					8.40			8.40	9.25
1216	4" x 2" diameter DR 11					9.80			9.80	10.80
1220	4" x 3" diameter DR 11					12.60			12.60	13.85
1224	6" x 4" diameter DR 11					29.50			29.50	32.50
1228	8" x 6" diameter DR 11					45			45	49.50
1232	10" x 8" diameter DR 11					77			77	85
1236	12" x 8" diameter DR 11					126			126	139
1240	12" x 10" diameter DR 11					101			101	111
1244	14" x 12" diameter DR 11					112			112	123
1248	16" x 14" diameter DR 11					143			143	157
1252	18" x 16" diameter DR 11					174			174	191
1256	20" x 18" diameter DR 11					345			345	380
1260	22" x 20" diameter DR 11					425			425	465
1264	24" x 22" diameter DR 11					475			475	525
1268	26" x 24" diameter DR 11					560			560	615
1272	28" x 24" diameter DR 11					715			715	785
1276	32" x 28" diameter DR 17					925			925	1,025
1280	36" x 32" diameter DR 17				▼	1,250			1,250	1,375
4000	Welding labor per joint, not including welding machine									
4010	Pipe joint size (cost based on thickest wall for each diam.)									
4030	1" pipe size	4 Skwk	273	.117	Ea.		5.55		5.55	8.60
4040	1-1/2" pipe size		175	.183			8.65		8.65	13.40
4050	2" pipe size	▼	128	.250	▼		11.85		11.85	18.30

22 11 Facility Water Distribution

22 11 13 – Facility Water Distribution Piping

22 11 13.78 Pipe, High Density Polyethylene Plastic (HDPE)

		Crew	Daily Output	Labor-Hours	Unit	Material	2014 Bare Costs Labor	Equipment	Total	Total Incl O&P
4060	3" pipe size	4 Skwk	100	.320	Ea.		15.15		15.15	23.50
4070	4" pipe size		77	.416			19.65		19.65	30.50
4080	6" pipe size	5 Skwk	63	.635			30		30	46.50
4090	8" pipe size		48	.833			39.50		39.50	61
4100	10" pipe size		40	1			47.50		47.50	73.50
4110	12" pipe size	6 Skwk	41	1.171			55.50		55.50	86
4120	16" pipe size		34	1.412			67		67	103
4130	18" pipe size		32	1.500			71		71	110
4140	20" pipe size	8 Skwk	37	1.730			82		82	127
4150	22" pipe size		35	1.829			86.50		86.50	134
4160	24" pipe size		34	1.882			89		89	138
4170	28" pipe size		33	1.939			91.50		91.50	142
4180	30" pipe size		32	2			94.50		94.50	147
4190	36" pipe size		31	2.065			97.50		97.50	151
4200	42" pipe size		30	2.133			101		101	156
4210	48" pipe size	9 Skwk	33	2.182			103		103	160
4220	54" pipe size	"	31	2.323			110		110	170
4300	Note: Cost for set up each time welder is moved.									
4301	Add 50% of a weld cost									
4310	Welder usually remains stationary with pipe moved through it.									
4340	Weld machine, rental per day based on diam. capacity									
4350	1" thru 2" diameter				Ea.			40.50	40.50	44.50
4360	3" thru 4" diameter							46	46	50.50
4370	6" thru 8" diameter							103	103	113
4380	10" thru 12" diameter							178	178	196
4390	16" thru 18" diameter							259	259	285
4400	20" thru 24" diameter							500	500	550
4410	28" thru 32" diameter							545	545	600
4420	36" diameter							570	570	625
4430	42" thru 54" diameter							890	890	980
5000	Dual wall contained pipe									
5040	Straight									
5054	1" DR 11 x 3" DR 11				L.F.	6.75			6.75	7.40
5058	1" DR 11 x 4" DR 11					7.20			7.20	7.90
5062	1-1/2" DR 11 x 4" DR 17					7.40			7.40	8.15
5066	2" DR 11 x 4" DR 17					7.95			7.95	8.75
5070	2" DR 11 x 6" DR 17					11.90			11.90	13.10
5074	3" DR 11 x 6" DR 17					13.30			13.30	14.60
5078	3" DR 11 x 6" DR 26					11.40			11.40	12.55
5086	3" DR 17 x 8" DR 17					16.85			16.85	18.55
5090	4" DR 11 x 8" DR 17					19.75			19.75	21.50
5094	4" DR 17 x 8" DR 26					15.60			15.60	17.15
5098	6" DR 11 x 10" DR 17					30.50			30.50	33.50
5102	6" DR 17 x 10" DR 26					23			23	25
5106	6" DR 26 x 10" DR 26					21.50			21.50	24
5110	8" DR 17 x 12" DR 26					32.50			32.50	36
5114	8" DR 26 x 12" DR 32.5					27.50			27.50	30.50
5118	10" DR 17 x 14" DR 26					45			45	50
5122	10" DR 17 x 16" DR 26					49.50			49.50	54.50
5126	10" DR 26 x 16" DR 26					45			45	49.50
5130	12" DR 26 x 16" DR 26					51			51	56.50
5134	12" DR 17 x 18" DR 26					65.50			65.50	72
5138	12" DR 26 x 18" DR 26					58.50			58.50	64.50

22 11 Facility Water Distribution

22 11 13 – Facility Water Distribution Piping

22 11 13.78 Pipe, High Density Polyethylene Plastic (HDPE)

		Crew	Daily Output	Labor-Hours	Unit	Material	2014 Bare Costs Labor	Equipment	Total	Total Incl O&P
5142	14" DR 26 x 20" DR 32.5				L.F.	64			64	70.50
5146	16" DR 26 x 22" DR 32.5					97			97	107
5150	18" DR 26 x 24" DR 32.5					89			89	98
5154	20" DR 32.5 x 28" DR 32.5					115			115	127
5158	22" DR 32.5 x 30" DR 32.5					115			115	127
5162	24" DR 32.5 x 32" DR 32.5					147			147	162
5166	36" DR 32.5 x 42" DR 32.5				▼	249			249	274
5300	Force transfer coupling									
5354	1" DR 11 x 3" DR 11				Ea.	243			243	267
5358	1" DR 11 x 4" DR 17					255			255	280
5362	1-1/2" DR 11 x 4" DR 17					267			267	293
5366	2" DR 11 x 4" DR 17					279			279	305
5370	2" DR 11 x 6" DR 17					405			405	445
5374	3" DR 11 x 6" DR 17					405			405	445
5378	3" DR 11 x 6" DR 26					405			405	445
5382	3" DR 11 x 8" DR 11					430			430	470
5386	3" DR 11 x 8" DR 17					430			430	470
5390	4" DR 11 x 8" DR 17					430			430	470
5394	4" DR 17 x 8" DR 26					375			375	415
5398	6" DR 11 x 10" DR 17					545			545	595
5402	6" DR 17 x 10" DR 26					445			445	490
5406	6" DR 26 x 10" DR 26					445			445	490
5410	8" DR 17 x 12" DR 26					595			595	655
5414	8" DR 26 x 12" DR 32.5					480			480	530
5418	10" DR 17 x 14" DR 26					780			780	855
5422	10" DR 17 x 16" DR 26					780			780	855
5426	10" DR 26 x 16" DR 26					780			780	855
5430	12" DR 26 x 16" DR 26					975			975	1,075
5434	12" DR 17 x 18" DR 26					1,000			1,000	1,100
5438	12" DR 26 x 18" DR 26					1,000			1,000	1,100
5442	14" DR 26 x 20" DR 32.5					1,050			1,050	1,150
5446	16" DR 26 x 22" DR 32.5					1,100			1,100	1,200
5450	18" DR 26 x 24" DR 32.5					1,175			1,175	1,300
5454	20" DR 32.5 x 28" DR 32.5					1,425			1,425	1,575
5458	22" DR 32.5 x 30" DR 32.5					1,575			1,575	1,750
5462	24" DR 32.5 x 32" DR 32.5					1,725			1,725	1,900
5466	36" DR 32.5 x 42" DR 32.5				▼	2,850			2,850	3,125
5600	90° Elbow									
5654	1" DR 11 x 3" DR 11				Ea.	200			200	220
5658	1" DR 11 x 4" DR 17					192			192	211
5662	1-1/2" DR 11 x 4" DR 17					210			210	231
5666	2" DR 11 x 4" DR 17					222			222	244
5670	2" DR 11 x 6" DR 17					283			283	310
5674	3" DR 11 x 6" DR 17					325			325	355
5678	3" DR 17 x 6" DR 26					258			258	284
5682	3" DR 17 x 8" DR 11					520			520	575
5686	3" DR 17 x 8" DR 17					405			405	445
5690	4" DR 11 x 8" DR 17					480			480	525
5694	4" DR 17 x 8" DR 26					380			380	415
5698	6" DR 11 x 10" DR 17					635			635	700
5702	6" DR 17 x 10" DR 26					485			485	535
5706	6" DR 26 x 10" DR 26					455			455	500
5710	8" DR 17 x 12" DR 26					725			725	795

22 11 Facility Water Distribution

22 11 13 – Facility Water Distribution Piping

22 11 13.78 Pipe, High Density Polyethylene Plastic (HDPE)

		Crew	Daily Output	Labor-Hours	Unit	Material	2014 Bare Costs Labor	Equipment	Total	Total Incl O&P
5714	8" DR 26 x 12" DR 32.5				Ea.	670			670	735
5718	10" DR 17 x 14" DR 26					1,150			1,150	1,250
5722	10" DR 17 x 16" DR 26					1,100			1,100	1,200
5726	10" DR 26 x 16" DR 26					1,000			1,000	1,100
5730	12" DR 26 x 16" DR 26					1,125			1,125	1,225
5734	12" DR 17 x 18" DR 26					1,325			1,325	1,475
5738	12" DR 26 x 18" DR 26					1,225			1,225	1,350
5742	14" DR 26 x 20" DR 32.5					1,625			1,625	1,775
5746	16" DR 26 x 22" DR 32.5					1,625			1,625	1,800
5750	18" DR 26 x 24" DR 32.5					2,000			2,000	2,200
5754	20" DR 32.5 x 28" DR 32.5					2,300			2,300	2,525
5758	22" DR 32.5 x 30" DR 32.5					3,150			3,150	3,450
5762	24" DR 32.5 x 32" DR 32.5					3,050			3,050	3,350
5766	36" DR 32.5 x 42" DR 32.5					4,950			4,950	5,450
5800	45° Elbow									
5804	1" DR 11 x 3" DR 11				Ea.	120			120	132
5808	1" DR 11 x 4" DR 17					119			119	131
5812	1-1/2" DR 11 x 4" DR 17					128			128	141
5816	2" DR 11 x 4" DR 17					140			140	154
5820	2" DR 11 x 6" DR 17					175			175	193
5824	3" DR 11 x 6" DR 17					195			195	214
5828	3" DR 17 x 6" DR 26					161			161	177
5832	3" DR 17 x 8" DR 11					298			298	330
5836	3" DR 17 x 8" DR 17					238			238	262
5840	4" DR 11 x 8" DR 17					278			278	305
5844	4" DR 17 x 8" DR 26					227			227	250
5848	6" DR 11 x 10" DR 17					365			365	405
5852	6" DR 17 x 10" DR 26					288			288	315
5856	6" DR 26 x 10" DR 26					271			271	299
5860	8" DR 17 x 12" DR 26					440			440	480
5864	8" DR 26 x 12" DR 32.5					410			410	450
5868	10" DR 17 x 14" DR 26					685			685	755
5872	10" DR 17 x 16" DR 26					670			670	740
5876	10" DR 26 x 16" DR 26					630			630	690
5880	12" DR 26 x 16" DR 26					710			710	780
5884	12" DR 17 x 18" DR 26					835			835	920
5888	12" DR 26 x 18" DR 26					780			780	855
5892	14" DR 26 x 20" DR 32.5					1,000			1,000	1,100
5896	16" DR 26 x 22" DR 32.5					1,050			1,050	1,150
5900	18" DR 26 x 24" DR 32.5					1,250			1,250	1,375
5904	20" DR 32.5 x 28" DR 32.5					1,525			1,525	1,675
5908	22" DR 32.5 x 30" DR 32.5					1,925			1,925	2,125
5912	24" DR 32.5 x 32" DR 32.5					1,925			1,925	2,125
5916	36" DR 32.5 x 42" DR 32.5					3,000			3,000	3,275
6000	Access port with 4" riser									
6050	1" DR 11 x 4" DR 17				Ea.	262			262	288
6054	1-1/2" DR 11 x 4" DR 17					267			267	294
6058	2" DR 11 x 6" DR 17					325			325	360
6062	3" DR 11 x 6" DR 17					330			330	360
6066	3" DR 17 x 6" DR 26					320			320	355
6070	3" DR 17 x 8" DR 11					370			370	405
6074	3" DR 17 x 8" DR 17					350			350	385
6078	4" DR 11 x 8" DR 17					365			365	400

22 11 Facility Water Distribution

22 11 13 – Facility Water Distribution Piping

22 11 13.78 Pipe, High Density Polyethylene Plastic (HDPE)

		Crew	Daily Output	Labor-Hours	Unit	Material	2014 Bare Costs Labor	Equipment	Total	Total Incl O&P
6082	4" DR 17 x 8" DR 26				Ea.	350			350	385
6086	6" DR 11 x 10" DR 17					455			455	500
6090	6" DR 17 x 10" DR 26					420			420	460
6094	6" DR 26 x 10" DR 26					410			410	450
6098	8" DR 17 x 12" DR 26					440			440	485
6102	8" DR 26 x 12" DR 32.5				▼	420			420	460
6200	End termination with vent plug									
6204	1" DR 11 x 3" DR 11				Ea.	217			217	238
6208	1" DR 11 x 4" DR 17					232			232	255
6212	1-1/2" DR 11 x 4" DR 17					232			232	255
6216	2" DR 11 x 4" DR 17					252			252	277
6220	2" DR 11 x 6" DR 17					287			287	315
6224	3" DR 11 x 6" DR 17					287			287	315
6228	3" DR 17 x 6" DR 26					287			287	315
6232	3" DR 17 x 8" DR 11					400			400	435
6236	3" DR 17 x 8" DR 17					400			400	435
6240	4" DR 11 x 8" DR 17					400			400	435
6244	4" DR 17 x 8" DR 26					345			345	380
6248	6" DR 11 x 10" DR 17					470			470	520
6252	6" DR 17 x 10" DR 26					390			390	425
6256	6" DR 26 x 10" DR 26					390			390	425
6260	8" DR 17 x 12" DR 26					575			575	630
6264	8" DR 26 x 12" DR 32.5					460			460	505
6268	10" DR 17 x 14" DR 26					595			595	655
6272	10" DR 17 x 16" DR 26					595			595	655
6276	10" DR 26 x 16" DR 26					595			595	655
6280	12" DR 26 x 16" DR 26					595			595	655
6284	12" DR 17 x 18" DR 26					785			785	865
6288	12" DR 26 x 18" DR 26					785			785	865
6292	14" DR 26 x 20" DR 32.5					800			800	880
6296	16" DR 26 x 22" DR 32.5					930			930	1,025
6300	18" DR 26 x 24" DR 32.5					960			960	1,050
6304	20" DR 32.5 x 28" DR 32.5					1,225			1,225	1,350
6308	22" DR 32.5 x 30" DR 32.5					1,350			1,350	1,475
6312	24" DR 32.5 x 32" DR 32.5					1,500			1,500	1,650
6316	36" DR 32.5 x 42" DR 32.5				▼	2,175			2,175	2,400
6600	Tee									
6604	1" DR 11 x 3" DR 11				Ea.	221			221	243
6608	1" DR 11 x 4" DR 17					274			274	300
6612	1-1/2" DR 11 x 4" DR 17					305			305	335
6616	2" DR 11 x 4" DR 17					335			335	365
6620	2" DR 11 x 6" DR 17					405			405	445
6624	3" DR 11 x 6" DR 17					465			465	510
6628	3" DR 17 x 6" DR 26					455			455	500
6632	3" DR 17 x 8" DR 11					550			550	605
6636	3" DR 17 x 8" DR 17					520			520	575
6640	4" DR 11 x 8" DR 17					570			570	625
6644	4" DR 17 x 8" DR 26					545			545	600
6648	6" DR 11 x 10" DR 17					710			710	780
6652	6" DR 17 x 10" DR 26					675			675	740
6656	6" DR 26 x 10" DR 26					665			665	735
6660	8" DR 17 x 12" DR 26					665			665	735
6664	8" DR 26 x 12" DR 32.5				▼	870			870	955

22 11 Facility Water Distribution

22 11 13 − Facility Water Distribution Piping

22 11 13.78 Pipe, High Density Polyethylene Plastic (HDPE)		Crew	Daily Output	Labor-Hours	Unit	Material	2014 Bare Costs Labor	Equipment	Total	Total Incl O&P
6668	10" DR 17 x 14" DR 26				Ea.	1,000			1,000	1,100
6672	10" DR 17 x 16" DR 26					1,050			1,050	1,150
6676	10" DR 26 x 16" DR 26					1,050			1,050	1,150
6680	12" DR 26 x 16" DR 26					1,150			1,150	1,250
6684	12" DR 17 x 18" DR 26					1,150			1,150	1,250
6688	12" DR 26 x 18" DR 26					1,250			1,250	1,400
6692	14" DR 26 x 20" DR 32.5					1,675			1,675	1,825
6696	16" DR 26 x 22" DR 32.5					2,050			2,050	2,250
6700	18" DR 26 x 24" DR 32.5					2,225			2,225	2,450
6704	20" DR 32.5 x 28" DR 32.5					2,800			2,800	3,100
6708	22" DR 32.5 x 30" DR 32.5					3,325			3,325	3,675
6712	24" DR 32.5 x 32" DR 32.5					4,000			4,000	4,375
6716	36" DR 32.5 x 42" DR 32.5					8,125			8,125	8,950
6800	Wye									
6816	2" DR 11 x 4" DR 17				Ea.	420			420	460
6820	2" DR 11 x 6" DR 17					475			475	525
6824	3" DR 11 x 6" DR 17					480			480	530
6828	3" DR 17 x 6" DR 26					465			465	515
6832	3" DR 17 x 8" DR 11					585			585	645
6836	3" DR 17 x 8" DR 17					545			545	600
6840	4" DR 11 x 8" DR 17					595			595	655
6844	4" DR 17 x 8" DR 26					565			565	625
6848	6" DR 11 x 10" DR 17					740			740	815
6852	6" DR 17 x 10" DR 26					685			685	755
6856	6" DR 26 x 10" DR 26					685			685	750
6860	8" DR 17 x 12" DR 26					980			980	1,075
6864	8" DR 26 x 12" DR 32.5					945			945	1,050
6868	10" DR 17 x 14" DR 26					1,175			1,175	1,300
6872	10" DR 17 x 16" DR 26					1,250			1,250	1,375
6876	10" DR 26 x 16" DR 26					1,225			1,225	1,350
6880	12" DR 26 x 16" DR 26					1,375			1,375	1,525
6884	12" DR 17 x 18" DR 26					1,550			1,550	1,725
6888	12" DR 26 x 18" DR 26					1,575			1,575	1,750
6892	14" DR 26 x 20" DR 32.5					1,925			1,925	2,125
6896	16" DR 26 x 22" DR 32.5					2,775			2,775	3,050
6900	18" DR 26 x 24" DR 32.5					3,250			3,250	3,575
6904	20" DR 32.5 x 28" DR 32.5					4,075			4,075	4,500
6908	22" DR 32.5 x 30" DR 32.5					4,650			4,650	5,125
6912	24" DR 32.5 x 32" DR 32.5					5,300			5,300	5,825
9000	Welding labor per joint, not including welding machine									
9010	Pipe joint size, outer pipe (cost based on the thickest walls)									
9020	Straight pipe									
9050	3" pipe size	4 Skwk	96	.333	Ea.		15.75		15.75	24.50
9060	4" pipe size	"	77	.416			19.65		19.65	30.50
9070	6" pipe size	5 Skwk	60	.667			31.50		31.50	49
9080	8" pipe size	"	40	1			47.50		47.50	73.50
9090	10" pipe size	6 Skwk	41	1.171			55.50		55.50	86
9100	12" pipe size		39	1.231			58		58	90
9110	14" pipe size		38	1.263			60		60	92.50
9120	16" pipe size		35	1.371			65		65	100
9130	18" pipe size	8 Skwk	45	1.422			67.50		67.50	104
9140	20" pipe size		42	1.524			72		72	112
9150	22" pipe size		40	1.600			75.50		75.50	117

237

22 11 Facility Water Distribution

22 11 13 – Facility Water Distribution Piping

22 11 13.78 Pipe, High Density Polyethylene Plastic (HDPE)

		Crew	Daily Output	Labor-Hours	Unit	Material	2014 Bare Costs Labor	Equipment	Total	Total Incl O&P
9160	24" pipe size	8 Skwk	38	1.684	Ea.		79.50		79.50	123
9170	28" pipe size		37	1.730			82		82	127
9180	30" pipe size		36	1.778			84		84	130
9190	32" pipe size		35	1.829			86.50		86.50	134
9200	42" pipe size		32	2			94.50		94.50	147
9300	Note: Cost for set up each time welder is moved.									
9301	Add 100% of weld labor cost									
9310	For handling between fitting welds add 50% of weld labor cost									
9320	Welder usually remains stationary with pipe moved through it									
9360	Weld machine, rental per day based on diam. capacity									
9380	3" thru 4" diameter				Ea.			63.50	63.50	70
9390	6" thru 8" diameter							207	207	228
9400	10" thru 12" diameter							310	310	340
9410	14" thru 18" diameter							420	420	460
9420	20" thru 24" diameter							750	750	825
9430	28" thru 32" diameter							880	880	970
9440	42" thru 58" diameter							910	910	1,000

22 11 19 – Domestic Water Piping Specialties

22 11 19.10 Flexible Connectors

		Crew	Daily Output	Labor-Hours	Unit	Material	Labor	Equipment	Total	Total Incl O&P
0010	**FLEXIBLE CONNECTORS**, Corrugated, 7/8" O.D., 1/2" I.D.									
0050	Gas, seamless brass, steel fittings									
0200	12" long	1 Plum	36	.222	Ea.	17.40	12.80		30.20	38.50
0220	18" long		36	.222		21.50	12.80		34.30	43.50
0240	24" long		34	.235		25.50	13.55		39.05	48.50
0260	30" long		34	.235		27.50	13.55		41.05	51
0280	36" long		32	.250		30.50	14.40		44.90	55
0320	48" long		30	.267		38.50	15.35		53.85	65.50
0340	60" long		30	.267		46	15.35		61.35	73.50
0360	72" long		30	.267		53	15.35		68.35	81.50
2000	Water, copper tubing, dielectric separators									
2100	12" long	1 Plum	36	.222	Ea.	17.35	12.80		30.15	38.50
2220	15" long		36	.222		19.30	12.80		32.10	40.50
2240	18" long		36	.222		21	12.80		33.80	42.50
2260	24" long		34	.235		26	13.55		39.55	49

22 11 19.14 Flexible Metal Hose

		Crew	Daily Output	Labor-Hours	Unit	Material	Labor	Equipment	Total	Total Incl O&P
0010	**FLEXIBLE METAL HOSE**, Connectors, standard lengths									
0100	Bronze braided, bronze ends									
0120	3/8" diameter x 12"	1 Stpi	26	.308	Ea.	19.60	18		37.60	48.50
0140	1/2" diameter x 12"		24	.333		19.80	19.50		39.30	51.50
0160	3/4" diameter x 12"		20	.400		29.50	23.50		53	68
0180	1" diameter x 18"		19	.421		35.50	24.50		60	76
0200	1-1/2" diameter x 18"		13	.615		48	36		84	107
0220	2" diameter x 18"		11	.727		63	42.50		105.50	134
1000	Carbon steel ends									
1020	1/4" diameter x 12"	1 Stpi	28	.286	Ea.	15.50	16.70		32.20	42
1040	3/8" diameter x 12"		26	.308		15.90	18		33.90	44.50
1060	1/2" diameter x 12"		24	.333		17.55	19.50		37.05	49
1080	1/2" diameter x 24"		24	.333		46	19.50		65.50	80
1120	3/4" diameter x 12"		20	.400		28.50	23.50		52	67
1140	3/4" diameter x 24"		20	.400		57	23.50		80.50	98
1160	3/4" diameter x 36"		20	.400		67.50	23.50		91	110
1180	1" diameter x 18"		19	.421		35.50	24.50		60	76

22 11 Facility Water Distribution

22 11 19 – Domestic Water Piping Specialties

22 11 19.14 Flexible Metal Hose		Crew	Daily Output	Labor-Hours	Unit	Material	2014 Bare Costs Labor	Equipment	Total	Total Incl O&P
1200	1" diameter x 30"	1 Stpi	19	.421	Ea.	55.50	24.50		80	98
1220	1" diameter x 36"		19	.421		79	24.50		103.50	124
1240	1-1/4" diameter x 18"		15	.533		49.50	31		80.50	101
1260	1-1/4" diameter x 36"		15	.533		106	31		137	163
1280	1-1/2" diameter x 18"		13	.615		69.50	36		105.50	131
1300	1-1/2" diameter x 36"		13	.615		115	36		151	181
1320	2" diameter x 24"		11	.727		96.50	42.50		139	171
1340	2" diameter x 36"		11	.727		142	42.50		184.50	221
1360	2-1/2" diameter x 24"		9	.889		221	52		273	320
1380	2-1/2" diameter x 36"		9	.889		525	52		577	655
1400	3" diameter x 24"		7	1.143		305	67		372	435
1420	3" diameter x 36"	▼	7	1.143	▼	62.50	67		129.50	170
2000	Carbon steel braid, carbon steel solid ends									
2100	1/2" diameter x 12"	1 Stpi	24	.333	Ea.	41.50	19.50		61	75
2120	3/4" diameter x 12"		20	.400		62.50	23.50		86	105
2140	1" diameter x 12"		19	.421		89.50	24.50		114	136
2160	1-1/4" diameter x 12"		15	.533		26	31		57	75.50
2180	1-1/2" diameter x 12"	▼	13	.615	▼	28	36		64	85
3000	Stainless steel braid, welded on carbon steel ends									
3100	1/2" diameter x 12"	1 Stpi	24	.333	Ea.	56.50	19.50		76	91.50
3120	3/4" diameter x 12"		20	.400		69	23.50		92.50	112
3140	3/4" diameter x 24"		20	.400		79.50	23.50		103	123
3160	3/4" diameter x 36"		20	.400		92	23.50		115.50	137
3180	1" diameter x 12"		19	.421		82	24.50		106.50	128
3200	1" diameter x 24"		19	.421		99.50	24.50		124	146
3220	1" diameter x 36"		19	.421		114	24.50		138.50	162
3240	1-1/4" diameter x 12"		15	.533		124	31		155	183
3260	1-1/4" diameter x 24"		15	.533		135	31		166	196
3280	1-1/4" diameter x 36"		15	.533		153	31		184	215
3300	1-1/2" diameter x 12"		13	.615		41	36		77	99.50
3320	1-1/2" diameter x 24"		13	.615		143	36		179	212
3340	1-1/2" diameter x 36"	▼	13	.615	▼	173	36		209	245
3400	Metal stainless steel braid, over corrugated stainless steel, flanged ends									
3410	150 PSI									
3420	1/2" diameter x 12"	1 Stpi	24	.333	Ea.	120	19.50		139.50	162
3430	1" diameter x 12"		20	.400		180	23.50		203.50	234
3440	1-1/2" diameter x 12"		15	.533		204	31		235	271
3450	2-1/2" diameter x 9"		12	.667		108	39		147	178
3460	3" diameter x 9"		9	.889		93	52		145	181
3470	4" diameter x 9"		7	1.143		113	67		180	225
3480	4" diameter x 30"		5	1.600		560	93.50		653.50	755
3490	4" diameter x 36"		4.80	1.667		580	97.50		677.50	785
3500	6" diameter x 11"		5	1.600		188	93.50		281.50	350
3510	6" diameter x 36"		3.80	2.105		685	123		808	935
3520	8" diameter x 12"		4	2		410	117		527	625
3530	10" diameter x 13"	▼	3	2.667		625	156		781	925
3540	12" diameter x 14"	Q-5	4	4	▼	950	211		1,161	1,375
6000	Molded rubber with helical wire reinforcement									
6010	150 PSI									
6020	1-1/2" diameter x 12"	1 Stpi	15	.533	Ea.	88	31		119	144
6030	2" diameter x 12"		12	.667		240	39		279	325
6040	3" diameter x 12"		8	1		251	58.50		309.50	365
6050	4" diameter x 12"	▼	6	1.333	▼	320	78		398	470

239

22 11 Facility Water Distribution

22 11 19 – Domestic Water Piping Specialties

22 11 19.14 Flexible Metal Hose

		Crew	Daily Output	Labor-Hours	Unit	Material	2014 Bare Costs Labor	Equipment	Total	Total Incl O&P
6060	6" diameter x 18"	1 Stpi	4	2	Ea.	475	117		592	700
6070	8" diameter x 24"		3	2.667		665	156		821	970
6080	10" diameter x 24"		2	4		810	234		1,044	1,250
6090	12" diameter x 24"	Q-5	3	5.333		925	281		1,206	1,450
7000	Molded teflon with stainless steel flanges									
7010	150 PSI									
7020	2-1/2" diameter x 3-3/16"	Q-1	7.80	2.051	Ea.	2,325	106		2,431	2,700
7030	3" diameter x 3-5/8"		6.50	2.462		1,625	128		1,753	1,975
7040	4" diameter x 3-5/8"		5	3.200		2,100	166		2,266	2,575
7050	6" diameter x 4"		4.30	3.721		2,950	193		3,143	3,550
7060	8" diameter x 6"		3.80	4.211		4,700	218		4,918	5,500

22 11 19.18 Mixing Valve

		Crew	Daily Output	Labor-Hours	Unit	Material	Labor	Equipment	Total	Total Incl O&P
0010	**MIXING VALVE**, Automatic, water tempering.									
0040	1/2" size	1 Stpi	19	.421	Ea.	555	24.50		579.50	645
0050	3/4" size		18	.444		555	26		581	650
0100	1" size		16	.500		845	29.50		874.50	975
0120	1-1/4" size		13	.615		1,150	36		1,186	1,325
0140	1-1/2" size		10	.800		1,400	47		1,447	1,600
0160	2" size		8	1		1,750	58.50		1,808.50	2,025
0170	2-1/2" size		6	1.333		1,750	78		1,828	2,050
0180	3" size		4	2		4,125	117		4,242	4,725
0190	4" size		3	2.667		4,125	156		4,281	4,775

22 11 19.22 Pressure Reducing Valve

		Crew	Daily Output	Labor-Hours	Unit	Material	Labor	Equipment	Total	Total Incl O&P
0010	**PRESSURE REDUCING VALVE**, Steam, pilot operated.									
0100	Threaded, iron body									
0200	1-1/2" size	1 Stpi	8	1	Ea.	1,775	58.50		1,833.50	2,050
0220	2" size	"	5	1.600	"	2,025	93.50		2,118.50	2,375
1000	Flanged, iron body, 125 lb. flanges									
1020	2" size	1 Stpi	8	1	Ea.	2,050	58.50		2,108.50	2,350
1040	2-1/2" size	"	4	2		2,475	117		2,592	2,875
1060	3" size	Q-5	4.50	3.556		3,000	187		3,187	3,550
1080	4" size	"	3	5.333		4,300	281		4,581	5,150
1500	For 250 lb. flanges, add					5%				

22 11 19.26 Pressure Regulators

		Crew	Daily Output	Labor-Hours	Unit	Material	Labor	Equipment	Total	Total Incl O&P
0010	**PRESSURE REGULATORS**									
0100	Gas appliance regulators									
0106	Main burner and pilot applications									
0108	Rubber seat poppet type									
0109	1/8" pipe size	1 Stpi	24	.333	Ea.	15.50	19.50		35	46.50
0110	1/4" pipe size		24	.333		17.80	19.50		37.30	49
0112	3/8" pipe size		24	.333		18.90	19.50		38.40	50.50
0113	1/2" pipe size		24	.333		21	19.50		40.50	52.50
0114	3/4" pipe size		20	.400		24.50	23.50		48	62.50
0122	Lever action type									
0123	3/8" pipe size	1 Stpi	24	.333	Ea.	23	19.50		42.50	54.50
0124	1/2" pipe size		24	.333		23	19.50		42.50	54.50
0125	3/4" pipe size		20	.400		45	23.50		68.50	85
0126	1" pipe size		19	.421		45	24.50		69.50	86.50
0132	Double diaphragm type									
0133	3/8" pipe size	1 Stpi	24	.333	Ea.	34.50	19.50		54	67.50
0134	1/2" pipe size		24	.333		47.50	19.50		67	82
0135	3/4" pipe size		20	.400		79.50	23.50		103	123

22 11 Facility Water Distribution

22 11 19 – Domestic Water Piping Specialties

22 11 19.26 Pressure Regulators

		Crew	Daily Output	Labor-Hours	Unit	Material	2014 Bare Costs Labor	Equipment	Total	Total Incl O&P
0136	1" pipe size	1 Stpi	19	.421	Ea.	97.50	24.50		122	144
0137	1-1/4" pipe size		15	.533		340	31		371	415
0138	1-1/2" pipe size		13	.615		620	36		656	740
0139	2" pipe size		11	.727		620	42.50		662.50	750
0140	2-1/2" pipe size	Q-5	15	1.067		1,200	56		1,256	1,400
0141	3" pipe size		13	1.231		1,200	65		1,265	1,425
0142	4" pipe size (flanged)		8	2		2,150	105		2,255	2,525
0160	Main burner only									
0162	Straight-thru-flow design									
0163	1/2" pipe size	1 Stpi	24	.333	Ea.	39	19.50		58.50	72
0164	3/4" pipe size		20	.400		53	23.50		76.50	94
0165	1" pipe size		19	.421		75.50	24.50		100	120
0166	1-1/4" pipe size		15	.533		75.50	31		106.50	130
0200	Oil, light, hot water, ordinary steam, threaded									
0220	Bronze body, 1/4" size	1 Stpi	24	.333	Ea.	149	19.50		168.50	194
0230	3/8" size		24	.333		161	19.50		180.50	207
0240	1/2" size		24	.333		200	19.50		219.50	250
0250	3/4" size		20	.400		239	23.50		262.50	298
0260	1" size		19	.421		365	24.50		389.50	435
0270	1-1/4" size		15	.533		490	31		521	585
0320	Iron body, 1/4" size		24	.333		117	19.50		136.50	158
0330	3/8" size		24	.333		137	19.50		156.50	181
0340	1/2" size		24	.333		145	19.50		164.50	190
0350	3/4" size		20	.400		177	23.50		200.50	230
0360	1" size		19	.421		230	24.50		254.50	290
0370	1-1/4" size		15	.533		320	31		351	395
9002	For water pressure regulators, see Section 22 05 23.20									

22 11 19.34 Sleeves and Escutcheons

		Crew	Daily Output	Labor-Hours	Unit	Material	Labor	Equipment	Total	Total Incl O&P
0010	**SLEEVES & ESCUTCHEONS**									
0100	Pipe sleeve									
0110	Steel, w/water stop, 12" long, with link seal									
0120	2" diam. for 1/2" carrier pipe	1 Plum	8.40	.952	Ea.	58.50	55		113.50	148
0130	2-1/2" diam. for 3/4" carrier pipe		8	1		65.50	57.50		123	159
0140	2-1/2" diam. for 1" carrier pipe		8	1		62.50	57.50		120	156
0150	3" diam. for 1-1/4" carrier pipe		7.20	1.111		79.50	64		143.50	184
0160	3-1/2" diam. for 1-1/2" carrier pipe		6.80	1.176		80	67.50		147.50	190
0170	4" diam. for 2" carrier pipe		6	1.333		86.50	76.50		163	211
0180	4" diam. for 2-1/2" carrier pipe		6	1.333		87.50	76.50		164	213
0190	5" diam. for 3" carrier pipe		5.40	1.481		99	85.50		184.50	238
0200	6" diam. for 4" carrier pipe		4.80	1.667		119	96		215	276
0210	10" diam. for 6" carrier pipe	Q-1	8	2		238	104		342	420
0220	12" diam. for 8" carrier pipe		7.20	2.222		320	115		435	525
0230	14" diam. for 10" carrier pipe		6.40	2.500		335	130		465	560
0240	16" diam. for 12" carrier pipe		5.80	2.759		370	143		513	620
0250	18" diam. for 14" carrier pipe		5.20	3.077		805	159		964	1,125
0260	24" diam. for 18" carrier pipe		4	4		1,100	207		1,307	1,525
0270	24" diam. for 20" carrier pipe		4	4		1,000	207		1,207	1,425
0280	30" diam. for 24" carrier pipe		3.20	5		1,450	259		1,709	2,000
0500	Wall sleeve									
0510	Ductile iron with rubber gasket seal									
0520	3"	1 Plum	8.40	.952	Ea.	700	55		755	855
0530	4"		7.20	1.111		750	64		814	920

22 11 Facility Water Distribution

22 11 19 – Domestic Water Piping Specialties

22 11 19.34 Sleeves and Escutcheons

		Crew	Daily Output	Labor-Hours	Unit	Material	2014 Bare Costs Labor	Equipment	Total	Total Incl O&P
0540	6"	1 Plum	6	1.333	Ea.	920	76.50		996.50	1,150
0550	8"		4	2		1,125	115		1,240	1,425
0560	10"		3	2.667		1,375	153		1,528	1,750
0570	12"	↓	2.40	3.333	↓	1,625	192		1,817	2,100
5000	Escutcheon									
5100	Split ring, pipe									
5110	Chrome plated									
5120	1/2"	1 Plum	160	.050	Ea.	1.09	2.88		3.97	5.55
5130	3/4"		160	.050		1.21	2.88		4.09	5.70
5140	1"		135	.059		1.30	3.41		4.71	6.60
5150	1-1/2"		115	.070		1.83	4		5.83	8.05
5160	2"		100	.080		2.03	4.60		6.63	9.20
5170	4"		80	.100		4.50	5.75		10.25	13.65
5180	6"	↓	68	.118	↓	7.50	6.75		14.25	18.45
5400	Shallow flange type									
5410	Chrome plated steel									
5420	1/2" CTS	1 Plum	180	.044	Ea.	.32	2.56		2.88	4.21
5430	3/4" CTS		180	.044		.34	2.56		2.90	4.23
5440	1/2" IPS		180	.044		.36	2.56		2.92	4.26
5450	3/4" IPS		180	.044		.40	2.56		2.96	4.30
5460	1" IPS		175	.046		.46	2.63		3.09	4.48
5470	1-1/2" IPS		170	.047		.76	2.71		3.47	4.93
5480	2" IPS	↓	160	.050	↓	.91	2.88		3.79	5.35

22 11 19.38 Water Supply Meters

		Crew	Daily Output	Labor-Hours	Unit	Material	Labor	Equipment	Total	Total Incl O&P
0010	**WATER SUPPLY METERS**									
1000	Detector, serves dual systems such as fire and domestic or									
1020	process water, wide range cap., UL and FM approved									
1100	3" mainline x 2" by-pass, 400 GPM	Q-1	3.60	4.444	Ea.	7,225	230		7,455	8,300
1140	4" mainline x 2" by-pass, 700 GPM	"	2.50	6.400		7,225	330		7,555	8,450
1180	6" mainline x 3" by-pass, 1600 GPM	Q-2	2.60	9.231		11,100	495		11,595	13,000
1220	8" mainline x 4" by-pass, 2800 GPM		2.10	11.429		16,400	615		17,015	18,900
1260	10" mainline x 6" by-pass, 4400 GPM		2	12		23,400	645		24,045	26,800
1300	10" x 12" mainlines x 6" by-pass, 5400 GPM	↓	1.70	14.118	↓	31,800	760		32,560	36,200
2000	Domestic/commercial, bronze									
2020	Threaded									
2060	5/8" diameter, to 20 GPM	1 Plum	16	.500	Ea.	48.50	29		77.50	97
2080	3/4" diameter, to 30 GPM		14	.571		88.50	33		121.50	147
2100	1" diameter, to 50 GPM	↓	12	.667	↓	134	38.50		172.50	206
2300	Threaded/flanged									
2340	1-1/2" diameter, to 100 GPM	1 Plum	8	1	Ea.	330	57.50		387.50	445
2360	2" diameter, to 160 GPM	"	6	1.333	"	445	76.50		521.50	605
2600	Flanged, compound									
2640	3" diameter, 320 GPM	Q-1	3	5.333	Ea.	3,025	276		3,301	3,750
2660	4" diameter, to 500 GPM		1.50	10.667		4,850	555		5,405	6,150
2680	6" diameter, to 1,000 GPM		1	16		7,725	830		8,555	9,750
2700	8" diameter, to 1,800 GPM	↓	.80	20	↓	12,100	1,025		13,125	14,900
7000	Turbine									
7260	Flanged									
7300	2" diameter, to 160 GPM	1 Plum	7	1.143	Ea.	590	66		656	750
7320	3" diameter, to 450 GPM	Q-1	3.60	4.444		1,225	230		1,455	1,700
7340	4" diameter, to 650 GPM	"	2.50	6.400		2,050	330		2,380	2,750
7360	6" diameter, to 1800 GPM	Q-2	2.60	9.231		3,650	495		4,145	4,775

22 11 Facility Water Distribution

22 11 19 – Domestic Water Piping Specialties

22 11 19.38 Water Supply Meters

		Crew	Daily Output	Labor-Hours	Unit	Material	2014 Bare Costs Labor	Equipment	Total	Total Incl O&P
7380	8" diameter, to 2500 GPM	Q-2	2.10	11.429	Ea.	5,800	615		6,415	7,300
7400	10" diameter, to 5500 GPM	↓	1.70	14.118	↓	7,825	760		8,585	9,750

22 11 19.42 Backflow Preventers

		Crew	Daily Output	Labor-Hours	Unit	Material	Labor	Equipment	Total	Total Incl O&P
0010	**BACKFLOW PREVENTERS**, Includes valves									
0020	and four test cocks, corrosion resistant, automatic operation									
1000	Double check principle									
1010	Threaded, with ball valves									
1020	3/4" pipe size	1 Plum	16	.500	Ea.	231	29		260	298
1030	1" pipe size		14	.571		264	33		297	340
1040	1-1/2" pipe size		10	.800		560	46		606	690
1050	2" pipe size	↓	7	1.143	↓	655	66		721	820
1080	Threaded, with gate valves									
1100	3/4" pipe size	1 Plum	16	.500	Ea.	1,075	29		1,104	1,250
1120	1" pipe size		14	.571		1,100	33		1,133	1,250
1140	1-1/2" pipe size		10	.800		1,400	46		1,446	1,625
1160	2" pipe size	↓	7	1.143	↓	1,725	66		1,791	2,000
1200	Flanged, valves are gate									
1210	3" pipe size	Q-1	4.50	3.556	Ea.	2,800	184		2,984	3,350
1220	4" pipe size	"	3	5.333		2,900	276		3,176	3,625
1230	6" pipe size	Q-2	3	8		4,275	430		4,705	5,350
1240	8" pipe size		2	12		7,775	645		8,420	9,525
1250	10" pipe size	↓	1	24	↓	11,500	1,300		12,800	14,700
1300	Flanged, valves are OS&Y									
1370	1" pipe size	1 Plum	5	1.600	Ea.	1,175	92		1,267	1,450
1374	1-1/2" pipe size		5	1.600		1,500	92		1,592	1,825
1378	2" pipe size	↓	4.80	1.667		1,825	96		1,921	2,175
1380	3" pipe size	Q-1	4.50	3.556		2,875	184		3,059	3,425
1400	4" pipe size	"	3	5.333		3,475	276		3,751	4,250
1420	6" pipe size	Q-2	3	8		5,450	430		5,880	6,650
1430	8" pipe size	"	2	12	↓	10,800	645		11,445	12,800
4000	Reduced pressure principle									
4100	Threaded, bronze, valves are ball									
4120	3/4" pipe size	1 Plum	16	.500	Ea.	445	29		474	535
4140	1" pipe size		14	.571		480	33		513	575
4150	1-1/4" pipe size		12	.667		845	38.50		883.50	990
4160	1-1/2" pipe size		10	.800		960	46		1,006	1,125
4180	2" pipe size	↓	7	1.143		1,075	66		1,141	1,300
5000	Flanged, bronze, valves are OS&Y									
5060	2-1/2" pipe size	Q-1	5	3.200	Ea.	4,025	166		4,191	4,675
5080	3" pipe size		4.50	3.556		4,650	184		4,834	5,375
5100	4" pipe size	↓	3	5.333		5,500	276		5,776	6,475
5120	6" pipe size	Q-2	3	8	↓	8,750	430		9,180	10,300
5200	Flanged, iron, valves are gate									
5210	2-1/2" pipe size	Q-1	5	3.200	Ea.	2,650	166		2,816	3,175
5220	3" pipe size		4.50	3.556		2,775	184		2,959	3,325
5230	4" pipe size	↓	3	5.333		3,750	276		4,026	4,550
5240	6" pipe size	Q-2	3	8		5,275	430		5,705	6,475
5250	8" pipe size		2	12		9,475	645		10,120	11,400
5260	10" pipe size	↓	1	24	↓	13,400	1,300		14,700	16,700
5600	Flanged, iron, valves are OS&Y									
5660	2-1/2" pipe size	Q-1	5	3.200	Ea.	3,025	166		3,191	3,575
5680	3" pipe size	↓	4.50	3.556	↓	3,175	184		3,359	3,775

22 11 Facility Water Distribution

22 11 19 – Domestic Water Piping Specialties

22 11 19.42 Backflow Preventers

		Crew	Daily Output	Labor-Hours	Unit	Material	2014 Bare Costs Labor	Equipment	Total	Total Incl O&P
5700	4" pipe size	Q-1	3	5.333	Ea.	3,975	276		4,251	4,800
5720	6" pipe size	Q-2	3	8		5,775	430		6,205	7,000
5740	8" pipe size	↓	2	12		10,100	645		10,745	12,100
5760	10" pipe size		1	24		13,600	1,300		14,900	16,900

22 11 19.50 Vacuum Breakers

		Crew	Daily Output	Labor-Hours	Unit	Material	Labor	Equipment	Total	Total Incl O&P
0010	**VACUUM BREAKERS** R221113-40									
0013	See also backflow preventers Section 22 11 19.42									
1000	Anti-siphon continuous pressure type									
1010	Max. 150 PSI - 210°F									
1020	Bronze body									
1030	1/2" size	1 Stpi	24	.333	Ea.	187	19.50		206.50	236
1040	3/4" size		20	.400		187	23.50		210.50	242
1050	1" size		19	.421		194	24.50		218.50	250
1060	1-1/4" size		15	.533		380	31		411	465
1070	1-1/2" size		13	.615		470	36		506	570
1080	2" size	↓	11	.727		485	42.50		527.50	595
1200	Max. 125 PSI with atmospheric vent									
1210	Brass, in-line construction									
1220	1/4" size	1 Stpi	24	.333	Ea.	117	19.50		136.50	159
1230	3/8" size	"	24	.333		117	19.50		136.50	159
1260	For polished chrome finish, add				↓	13%				
2000	Anti-siphon, non-continuous pressure type									
2010	Hot or cold water 125 PSI - 210°F									
2020	Bronze body									
2030	1/4" size	1 Stpi	24	.333	Ea.	64	19.50		83.50	100
2040	3/8" size		24	.333		64	19.50		83.50	100
2050	1/2" size		24	.333		72.50	19.50		92	110
2060	3/4" size		20	.400		86.50	23.50		110	131
2070	1" size		19	.421		134	24.50		158.50	184
2080	1-1/4" size		15	.533		235	31		266	305
2090	1-1/2" size		13	.615		276	36		312	360
2100	2" size		11	.727		430	42.50		472.50	535
2110	2-1/2" size		8	1		1,225	58.50		1,283.50	1,450
2120	3" size	↓	6	1.333	↓	1,625	78		1,703	1,925
2150	For polished chrome finish, add					50%				

22 11 19.54 Water Hammer Arresters/Shock Absorbers

		Crew	Daily Output	Labor-Hours	Unit	Material	Labor	Equipment	Total	Total Incl O&P
0010	**WATER HAMMER ARRESTERS/SHOCK ABSORBERS**									
0490	Copper									
0500	3/4" male I.P.S. For 1 to 11 fixtures	1 Plum	12	.667	Ea.	28	38.50		66.50	89
0600	1" male I.P.S. For 12 to 32 fixtures		8	1		45.50	57.50		103	137
0700	1-1/4" male I.P.S. For 33 to 60 fixtures		8	1		47	57.50		104.50	139
0800	1-1/2" male I.P.S. For 61 to 113 fixtures		8	1		67.50	57.50		125	161
0900	2" male I.P.S. For 114 to 154 fixtures		8	1		98.50	57.50		156	195
1000	2-1/2" male I.P.S. For 155 to 330 fixtures	↓	4	2	↓	305	115		420	510
4000	Bellows type									
4010	3/4" FNPT, to 11 fixture units	1 Plum	10	.800	Ea.	241	46		287	335
4020	1" FNPT, to 32 fixture units		8.80	.909		485	52.50		537.50	615
4030	1" FNPT, to 60 fixture units		8.80	.909		730	52.50		782.50	885
4040	1" FNPT, to 113 fixture units		8.80	.909		1,825	52.50		1,877.50	2,100
4050	1" FNPT, to 154 fixture units		6.60	1.212		2,075	70		2,145	2,375
4060	1-1/2" FNPT, to 300 fixture units	↓	5	1.600		2,550	92		2,642	2,950

22 11 Facility Water Distribution

22 11 19 – Domestic Water Piping Specialties

22 11 19.64 Hydrants

		Crew	Daily Output	Labor-Hours	Unit	Material	2014 Bare Costs Labor	Equipment	Total	Total Incl O&P
0010	**HYDRANTS**									
0050	Wall type, moderate climate, bronze, encased									
0200	3/4" IPS connection	1 Plum	16	.500	Ea.	705	29		734	825
0300	1" IPS connection		14	.571		810	33		843	940
0500	Anti-siphon type, 3/4" connection	↓	16	.500	↓	610	29		639	715
1000	Non-freeze, bronze, exposed									
1100	3/4" IPS connection, 4" to 9" thick wall	1 Plum	14	.571	Ea.	480	33		513	575
1120	10" to 14" thick wall		12	.667		515	38.50		553.50	630
1140	15" to 19" thick wall		12	.667		580	38.50		618.50	695
1160	20" to 24" thick wall	↓	10	.800	↓	625	46		671	760
1200	For 1" IPS connection, add					15%	10%			
1240	For 3/4" adapter type vacuum breaker, add				Ea.	60			60	66
1280	For anti-siphon type, add				"	126			126	139
2000	Non-freeze bronze, encased, anti-siphon type									
2100	3/4" IPS connection, 5" to 9" thick wall	1 Plum	14	.571	Ea.	1,225	33		1,258	1,375
2120	10" to 14" thick wall		12	.667		1,250	38.50		1,288.50	1,425
2140	15" to 19" thick wall		12	.667		1,300	38.50		1,338.50	1,500
2160	20" to 24" thick wall	↓	10	.800	↓	1,350	46		1,396	1,575
2200	For 1" IPS connection, add					10%	10%			
3000	Ground box type, bronze frame, 3/4" IPS connection									
3080	Non-freeze, all bronze, polished face, set flush									
3100	2 feet depth of bury	1 Plum	8	1	Ea.	905	57.50		962.50	1,075
3120	3 feet depth of bury		8	1		970	57.50		1,027.50	1,150
3140	4 feet depth of bury		8	1		1,050	57.50		1,107.50	1,225
3160	5 feet depth of bury		7	1.143		1,125	66		1,191	1,325
3180	6 feet depth of bury		7	1.143		1,175	66		1,241	1,400
3200	7 feet depth of bury		6	1.333		1,250	76.50		1,326.50	1,500
3220	8 feet depth of bury		5	1.600		1,325	92		1,417	1,600
3240	9 feet depth of bury		4	2		1,400	115		1,515	1,700
3260	10 feet depth of bury	↓	4	2		1,450	115		1,565	1,775
3400	For 1" IPS connection, add					15%	10%			
3450	For 1-1/4" IPS connection, add					325%	14%			
3500	For 1-1/2" connection, add					370%	18%			
3550	For 2" connection, add					445%	24%			
3600	For tapped drain port in box, add				↓	82			82	90
4000	Non-freeze, CI body, bronze frame & scoriated cover									
4010	with hose storage									
4100	2 feet depth of bury	1 Plum	7	1.143	Ea.	1,675	66		1,741	1,925
4120	3 feet depth of bury		7	1.143		1,750	66		1,816	2,000
4140	4 feet depth of bury		7	1.143		1,800	66		1,866	2,075
4160	5 feet depth of bury		6.50	1.231		1,825	71		1,896	2,100
4180	6 feet depth of bury		6	1.333		1,850	76.50		1,926.50	2,175
4200	7 feet depth of bury		5.50	1.455		1,925	83.50		2,008.50	2,250
4220	8 feet depth of bury		5	1.600		2,000	92		2,092	2,350
4240	9 feet depth of bury		4.50	1.778		2,050	102		2,152	2,400
4260	10 feet depth of bury	↓	4	2		2,125	115		2,240	2,500
4280	For 1" IPS connection, add					365			365	405
4300	For tapped drain port in box, add				↓	82			82	90
5000	Moderate climate, all bronze, polished face									
5020	and scoriated cover, set flush									
5100	3/4" IPS connection	1 Plum	16	.500	Ea.	625	29		654	735
5120	1" IPS connection	"	14	.571	↓	770	33		803	895

22 11 Facility Water Distribution

22 11 19 – Domestic Water Piping Specialties

22 11 19.64 Hydrants

		Crew	Daily Output	Labor-Hours	Unit	Material	2014 Bare Costs Labor	Equipment	Total	Total Incl O&P
5200	For tapped drain port in box, add				Ea.	82			82	90
6000	Ground post type, all non-freeze, all bronze, aluminum casing									
6010	guard, exposed head, 3/4" IPS connection									
6100	2 feet depth of bury	1 Plum	8	1	Ea.	885	57.50		942.50	1,050
6120	3 feet depth of bury		8	1		955	57.50		1,012.50	1,125
6140	4 feet depth of bury		8	1		1,025	57.50		1,082.50	1,200
6160	5 feet depth of bury		7	1.143		1,100	66		1,166	1,300
6180	6 feet depth of bury		7	1.143		1,175	66		1,241	1,400
6200	7 feet depth of bury		6	1.333		1,250	76.50		1,326.50	1,500
6220	8 feet depth of bury		5	1.600		1,350	92		1,442	1,625
6240	9 feet depth of bury		4	2		1,425	115		1,540	1,725
6260	10 feet depth of bury		4	2		1,500	115		1,615	1,825
6300	For 1" IPS connection, add					40%	10%			
6350	For 1-1/4" IPS connection, add					140%	14%			
6400	For 1-1/2" IPS connection, add					225%	18%			
6450	For 2" IPS connection, add					315%	24%			

22 11 23 – Domestic Water Pumps

22 11 23.10 General Utility Pumps

		Crew	Daily Output	Labor-Hours	Unit	Material	Labor	Equipment	Total	Total Incl O&P
0010	**GENERAL UTILITY PUMPS**									
2000	Single stage									
3000	Double suction,									
3140	5"D. x 6"S., 50 HP	Q-2	.33	72.727	Ea.	9,175	3,900		13,075	16,000
3180	6"D. x 8"S., 60 HP	Q-3	.30	106		13,400	5,850		19,250	23,500
3190	75 HP, to 2500 GPM		.28	114		20,100	6,275		26,375	31,600
3220	100 HP, to 3000 GPM		.26	123		25,500	6,750		32,250	38,300
3240	150 HP, to 4000 GPM		.24	133		36,000	7,300		43,300	50,500
4000	Centrifugal, end suction, mounted on base									
4010	Horizontal mounted, with drip proof motor, rated @ 100' head									
4020	Vertical split case, single stage									
4040	100 GPM, 5 HP, 1-1/2" discharge	Q-1	1.70	9.412	Ea.	3,675	490		4,165	4,750
4050	200 GPM, 10 HP, 2" discharge		1.30	12.308		4,000	640		4,640	5,375
4060	250 GPM, 10 HP, 3" discharge		1.28	12.500		4,450	650		5,100	5,875
4070	300 GPM, 15 HP, 2" discharge	Q-2	1.56	15.385		5,225	825		6,050	7,000
4080	500 GPM, 20 HP, 4" discharge		1.44	16.667		6,025	895		6,920	7,975
4090	750 GPM, 30 HP, 4" discharge		1.20	20		6,350	1,075		7,425	8,600
4100	1050 GPM, 40 HP, 5" discharge		1	24		9,175	1,300		10,475	12,100
4110	1500 GPM, 60 HP, 6" discharge		.60	40		13,400	2,150		15,550	18,000
4120	2000 GPM, 75 HP, 6" discharge		.50	48		14,600	2,575		17,175	19,900
4130	3000 GPM, 100 HP, 8" discharge		.40	60		18,900	3,225		22,125	25,700
4200	Horizontal split case, single stage									
4210	100 GPM, 7.5 HP, 1-1/2" discharge	Q-1	1.70	9.412	Ea.	3,725	490		4,215	4,825
4220	250 GPM, 15 HP, 2-1/2" discharge	"	1.30	12.308		7,900	640		8,540	9,650
4230	500 GPM, 20 HP, 4" discharge	Q-2	1.60	15		9,625	805		10,430	11,800
4240	750 GPM, 25 HP, 5" discharge		1.54	15.584		10,100	835		10,935	12,400
4250	1000 GPM, 40 HP, 5" discharge		1.20	20		14,000	1,075		15,075	16,900
4260	1500 GPM, 50 HP, 6" discharge	Q-3	1.42	22.535		16,400	1,225		17,625	20,000
4270	2000 GPM, 75 HP, 8" discharge		1.14	28.070		22,300	1,550		23,850	26,900
4280	3000 GPM, 100 HP, 10" discharge		.96	33.333		25,000	1,825		26,825	30,300
4290	3500 GPM, 150 HP, 10" discharge		.86	37.209		32,900	2,050		34,950	39,300
4300	4000 GPM, 200 HP, 10" discharge		.66	48.485		39,300	2,650		41,950	47,200
4330	Horizontal split case, two stage, 500' head									
4340	100 GPM, 40 HP, 1-1/2" discharge	Q-2	1.70	14.118	Ea.	16,100	760		16,860	19,000

22 11 Facility Water Distribution

22 11 23 – Domestic Water Pumps

22 11 23.10 General Utility Pumps

		Crew	Daily Output	Labor-Hours	Unit	Material	2014 Bare Costs Labor	Equipment	Total	Total Incl O&P
4350	200 GPM, 50 HP, 1-1/2" discharge	Q-2	1.44	16.667	Ea.	16,900	895		17,795	20,000
4360	300 GPM, 75 HP, 2" discharge	Q-3	1.57	20.382		23,200	1,125		24,325	27,300
4370	400 GPM, 100 HP, 3" discharge		1.14	28.070		27,500	1,550		29,050	32,500
4380	800 GPM, 200 HP, 4" discharge	↓	.86	37.209	↓	36,400	2,050		38,450	43,100
5000	Centrifugal, in-line									
5006	Vertical mount, iron body, 125 lb. flgd, 1800 RPM TEFC mtr									
5010	Single stage									
5012	.5 HP, 1-1/2" suction & discharge	Q-1	3.20	5	Ea.	1,400	259		1,659	1,925
5014	.75 HP, 2" suction & discharge		2.80	5.714		2,575	296		2,871	3,275
5015	1 HP, 3" suction & discharge		2.50	6.400		2,975	330		3,305	3,775
5016	1.5 HP, 3" suction & discharge		2.30	6.957		3,025	360		3,385	3,875
5018	2 HP, 4" suction & discharge		2.10	7.619		3,775	395		4,170	4,775
5020	3 HP, 5" suction & discharge		1.90	8.421		4,425	435		4,860	5,525
5030	5 HP, 6" suction & discharge		1.60	10		5,050	520		5,570	6,325
5040	7.5 HP, 6" suction & discharge	↓	1.30	12.308		5,800	640		6,440	7,350
5050	10 HP, 8" suction & discharge	Q-2	1.80	13.333		7,725	715		8,440	9,575
5060	15 HP, 8" suction & discharge		1.70	14.118		9,025	760		9,785	11,100
5064	20 HP, 8" suction & discharge		1.60	15		10,500	805		11,305	12,700
5070	30 HP, 8" suction & discharge		1.50	16		13,300	860		14,160	16,000
5080	40 HP, 8" suction & discharge		1.40	17.143		16,000	920		16,920	19,000
5090	50 HP, 8" suction & discharge		1.30	18.462		16,200	990		17,190	19,400
5094	60 HP, 8" suction & discharge	↓	.90	26.667		17,100	1,425		18,525	21,000
5100	75 HP, 8" suction & discharge	Q-3	.60	53.333		19,500	2,925		22,425	25,900
5110	100 HP, 8" suction & discharge	"	.50	64	↓	22,200	3,500		25,700	29,800

22 11 23.11 Miscellaneous Pumps

		Crew	Daily Output	Labor-Hours	Unit	Material	Labor	Equipment	Total	Total Incl O&P
0010	**MISCELLANEOUS PUMPS**									
0020	Water pump, portable, gasoline powered									
0100	170 GPH, 2" discharge	Q-1	11	1.455	Ea.	600	75.50		675.50	775
0110	343 GPH, 3" discharge		10.50	1.524		735	79		814	925
0120	608 GPH, 4" discharge	↓	10	1.600	↓	1,875	83		1,958	2,175
0500	Pump, propylene body, housing and impeller									
0510	22 GPM, 1/3 HP, 40' HD	1 Plum	5	1.600	Ea.	825	92		917	1,050
0520	33 GPM, 1/2 HP, 40' HD	Q-1	5	3.200		965	166		1,131	1,300
0530	53 GPM, 3/4 HP, 40' HD	"	4	4	↓	1,200	207		1,407	1,625
0600	Rotary pump, CI									
0614	282 GPH, 3/4 HP, 1" discharge	Q-1	4.50	3.556	Ea.	1,225	184		1,409	1,625
0618	277 GPH, 1 HP, 1" discharge		4	4		1,225	207		1,432	1,675
0624	1100 GPH, 1.5 HP, 1-1/4" discharge		3.60	4.444		3,225	230		3,455	3,900
0628	1900 GPH, 2 HP, 1-1/4" discharge	↓	3.20	5		4,550	259		4,809	5,425
1000	Turbine pump, CI									
1010	50 GPM, 2 HP, 3" discharge	Q-1	.80	20	Ea.	3,500	1,025		4,525	5,425
1020	100 GPM, 3 HP, 4" discharge	Q-2	.96	25		5,750	1,350		7,100	8,350
1030	250 GPM, 15 HP, 6" discharge	"	.94	25.532		7,125	1,375		8,500	9,925
1040	500 GPM, 25 HP, 6" discharge	Q-3	1.22	26.230		7,200	1,425		8,625	10,100
1050	1000 GPM, 50 HP, 8" discharge		1.14	28.070		9,925	1,550		11,475	13,200
1060	2000 GPM, 100 HP, 10" discharge		1	32		11,800	1,750		13,550	15,600
1070	3000 GPM, 150 HP, 10" discharge		.80	40		18,900	2,200		21,100	24,100
1080	4000 GPM, 200 HP, 12" discharge		.70	45.714		20,700	2,500		23,200	26,600
1090	6000 GPM, 300 HP, 14" discharge		.60	53.333		31,600	2,925		34,525	39,100
1100	10,000 GPM, 300 HP, 18" discharge		.58	55.172	↓	36,500	3,025		39,525	44,700
2000	Centrifugal stainless steel pumps									
2100	100 GPM, 100' TDH, 3 HP	Q-1	1.80	8.889	Ea.	10,400	460		10,860	12,200

22 1.1 Facility Water Distribution

22 11 23 – Domestic Water Pumps

22 11 23.11 Miscellaneous Pumps

		Crew	Daily Output	Labor-Hours	Unit	Material	2014 Bare Costs Labor	Equipment	Total	Total Incl O&P
2130	250 GPM, 100' TDH, 10 HP	Q-1	1.28	12.500	Ea.	14,600	650		15,250	17,000
2160	500 GPM, 100' TDH, 20 HP	Q-2	1.44	16.667	↓	16,700	895		17,595	19,700
2200	Vertical turbine stainless steel pumps									
2220	100 GPM, 100' TDH, 7.5 HP	Q-2	.95	25.263	Ea.	58,500	1,350		59,850	66,000
2240	250 GPM, 100' TDH, 15 HP		.94	25.532		64,500	1,375		65,875	73,000
2260	500 GPM, 100' TDH, 20 HP	↓	.93	25.806		71,000	1,375		72,375	80,000
2280	750 GPM, 100' TDH, 30 HP	Q-3	1.19	26.891		77,000	1,475		78,475	87,000
2300	1000 GPM, 100' TDH, 40 HP	"	1.16	27.586		83,500	1,500		85,000	94,000
2320	100 GPM, 200' TDH, 15 HP	Q-2	.94	25.532		89,500	1,375		90,875	100,500
2340	250 GPM, 200' TDH, 25 HP	Q-3	1.22	26.230		96,000	1,425		97,425	107,500
2360	500 GPM, 200' TDH, 40 HP		1.16	27.586		102,000	1,500		103,500	115,000
2380	750 GPM, 200' TDH, 60 HP		1.13	28.319		103,000	1,550		104,550	115,500
2400	1000 GPM, 200' TDH, 75 HP	↓	1.12	28.571	↓	106,000	1,575		107,575	119,000

22 11 23.13 Domestic-Water Packaged Booster Pumps

		Crew	Daily Output	Labor-Hours	Unit	Material	Labor	Equipment	Total	Total Incl O&P
0010	**DOMESTIC-WATER PACKAGED BOOSTER PUMPS**									
0200	Pump system, with diaphragm tank, control, press. switch									
0300	1 HP pump	Q-1	1.30	12.308	Ea.	6,500	640		7,140	8,125
0400	1-1/2 HP pump		1.25	12.800		6,550	665		7,215	8,225
0420	2 HP pump		1.20	13.333		6,725	690		7,415	8,450
0440	3 HP pump	↓	1.10	14.545		6,825	755		7,580	8,650
0460	5 HP pump	Q-2	1.50	16		7,550	860		8,410	9,600
0480	7-1/2 HP pump		1.42	16.901		8,400	910		9,310	10,600
0500	10 HP pump	↓	1.34	17.910	↓	8,800	960		9,760	11,100
2000	Pump system, variable speed, base, controls, starter									
2010	Duplex, 100' head									
2020	400 GPM, 7-1/2HP, 4" discharge	Q-2	.70	34.286	Ea.	38,900	1,850		40,750	45,500
2025	Triplex, 100' head									
2030	1000 GPM, 15HP, 6" discharge	Q-2	.50	48	Ea.	56,500	2,575		59,075	66,500
2040	1700 GPM, 30HP, 6" discharge	"	.30	80	"	68,500	4,300		72,800	81,500

22 12 Facility Potable-Water Storage Tanks

22 12 21 – Facility Underground Potable-Water Storage Tanks

22 12 21.13 Fiberglass, Underground Potable-Water Storage Tanks

		Crew	Daily Output	Labor-Hours	Unit	Material	Labor	Equipment	Total	Total Incl O&P
0010	**FIBERGLASS, UNDERGROUND POTABLE-WATER STORAGE TANKS**									
0020	Excludes excavation, backfill, & piping									
2000	600 gallon capacity	B-21B	3.75	10.667	Ea.	3,150	425	177	3,752	4,300
2010	1,000 gallon capacity		3.50	11.429		4,050	455	189	4,694	5,350
2020	2,000 gallon capacity		3.25	12.308		5,900	490	204	6,594	7,450
2030	4,000 gallon capacity		3	13.333		8,175	530	221	8,926	10,000
2040	6,000 gallon capacity		2.65	15.094		9,175	600	250	10,025	11,300
2050	8,000 gallon capacity		2.30	17.391		11,000	690	288	11,978	13,500
2060	10,000 gallon capacity		2	20		12,600	795	330	13,725	15,500
2070	12,000 gallon capacity		1.50	26.667		17,400	1,050	440	18,890	21,200
2080	15,000 gallon capacity		1	40		20,200	1,600	665	22,465	25,400
2090	20,000 gallon capacity		.75	53.333		26,000	2,125	885	29,010	32,800
2100	25,000 gallon capacity		.50	80		38,800	3,175	1,325	43,300	49,100
2110	30,000 gallon capacity		.35	114		47,100	4,550	1,900	53,550	61,000
2120	40,000 gallon capacity	↓	.30	133	↓	68,000	5,300	2,200	75,500	85,500

22 12 Facility Potable-Water Storage Tanks

22 12 23 – Facility Indoor Potable-Water Storage Tanks

22 12 23.13 Facility Steel, Indoor Pot.-Water Storage Tanks

		Crew	Daily Output	Labor-Hours	Unit	Material	2014 Bare Costs Labor	Equipment	Total	Total Incl O&P
0010	**FACILITY STEEL, INDOOR POT.-WATER STORAGE TANKS**									
2000	Galvanized steel, 15 gal., 14" diam., 26" LOA	1 Plum	12	.667	Ea.	1,425	38.50		1,463.50	1,625
2060	30 gal., 14" diam. x 49" LOA		11	.727		1,700	42		1,742	1,925
2080	80 gal., 20" diam. x 64" LOA		9	.889		2,775	51		2,826	3,125
2100	135 gal., 24" diam. x 75" LOA		6	1.333		3,825	76.50		3,901.50	4,325
2120	240 gal., 30" diam. x 86" LOA		4	2		7,475	115		7,590	8,400
2140	300 gal., 36" diam. x 76" LOA	▼	3	2.667		10,600	153		10,753	11,900
2160	400 gal., 36" diam. x 100" LOA	Q-1	4	4		13,100	207		13,307	14,700
2180	500 gal., 36" diam., x 126" LOA	"	3	5.333		16,200	276		16,476	18,200
3000	Glass lined, P.E., 80 gal., 20" diam. x 60" LOA	1 Plum	9	.889		4,725	51		4,776	5,275
3060	140 gal., 24" diam. x 80" LOA		6	1.333		6,575	76.50		6,651.50	7,350
3080	225 gal., 30" diam. x 78" LOA		4	2		6,775	115		6,890	7,625
3100	325 gal., 36" diam. x 81" LOA	▼	3	2.667		6,975	153		7,128	7,900
3120	460 gal., 42" diam. x 84" LOA	Q-1	4	4		7,350	207		7,557	8,400
3140	605 gal., 48" diam. x 87" LOA		3	5.333		14,000	276		14,276	15,800
3160	740 gal., 54" diam. x 91" LOA		3	5.333		15,800	276		16,076	17,800
3180	940 gal., 60" diam. x 93" LOA		2.50	6.400		18,200	330		18,530	20,600
3200	1330 gal., 66" diam. x 107" LOA		2	8		25,000	415		25,415	28,100
3220	1615 gal., 72" diam. x 110" LOA		1.50	10.667		28,400	555		28,955	32,100
3240	2285 gal., 84" diam. x 128" LOA	▼	1	16		35,200	830		36,030	40,000
3260	3440 gal., 96" diam. x 157" LOA	Q-2	1.50	16	▼	52,000	860		52,860	58,500

22 13 Facility Sanitary Sewerage

22 13 16 – Sanitary Waste and Vent Piping

22 13 16.20 Pipe, Cast Iron

			Crew	Daily Output	Labor-Hours	Unit	Material	Labor	Equipment	Total	Total Incl O&P
0010	**PIPE, CAST IRON**, Soil, on clevis hanger assemblies, 5' O.C.	R221113-70									
0020	Single hub, service wt., lead & oakum joints 10' O.C.										
2120	2" diameter	R221316-10	Q-1	63	.254	L.F.	9.30	13.15		22.45	30
2140	3" diameter			60	.267		12.95	13.80		26.75	35.50
2160	4" diameter	R221316-20	▼	55	.291		16.85	15.05		31.90	41.50
2180	5" diameter		Q-2	76	.316		23	16.95		39.95	51
2200	6" diameter		"	73	.329		29	17.65		46.65	58.50
2220	8" diameter		Q-3	59	.542		44	29.50		73.50	93.50
2240	10" diameter			54	.593		71	32.50		103.50	127
2260	12" diameter			48	.667		101	36.50		137.50	166
2261	15" diameter		▼	40	.800		150	44		194	231
2320	For service weight, double hub, add						10%				
2340	For extra heavy, single hub, add						48%	4%			
2360	For extra heavy, double hub, add					▼	71%	4%			
2400	Lead for caulking, (1#/diam. in.)		Q-1	160	.100	Lb.	1.04	5.20		6.24	8.95
2420	Oakum for caulking, (1/8#/diam. in.)		"	40	.400	"	4.80	20.50		25.30	37
2960	To delete hangers, subtract										
2970	2" diam. to 4" diam.						16%	19%			
2980	5" diam. to 8" diam.						14%	14%			
2990	10" diam. to 15" diam.						13%	19%			
3000	Single hub, service wt., push-on gasket joints 10' O.C.										
3010	2" diameter		Q-1	66	.242	L.F.	10.25	12.55		22.80	30
3020	3" diameter			63	.254		14.20	13.15		27.35	35.50
3030	4" diameter		▼	57	.281		18.40	14.55		32.95	42
3040	5" diameter		Q-2	79	.304		25.50	16.30		41.80	52.50

249

22 13 Facility Sanitary Sewerage

22 13 16 – Sanitary Waste and Vent Piping

22 13 16.20 Pipe, Cast Iron

		Crew	Daily Output	Labor-Hours	Unit	Material	2014 Bare Costs Labor	Equipment	Total	Total Incl O&P
3050	6" diameter	Q-2	75	.320	L.F.	31.50	17.20		48.70	61
3060	8" diameter	Q-3	62	.516		49.50	28.50		78	97
3070	10" diameter		56	.571		80.50	31.50		112	136
3080	12" diameter		49	.653		113	36		149	178
3082	15" diameter	↓	40	.800	↓	164	44		208	247
3100	For service weight, double hub, add					65%				
3110	For extra heavy, single hub, add					48%	4%			
3120	For extra heavy, double hub, add					29%	4%			
3130	To delete hangers, subtract									
3140	2" diam. to 4" diam.					12%	21%			
3150	5" diam. to 8" diam.					10%	16%			
3160	10" diam. to 15" diam.					9%	21%			
4000	No hub, couplings 10' O.C.									
4100	1-1/2" diameter	Q-1	71	.225	L.F.	9.15	11.65		20.80	27.50
4120	2" diameter		67	.239		9.40	12.35		21.75	29
4140	3" diameter		64	.250		13.35	12.95		26.30	34
4160	4" diameter	↓	58	.276		16.95	14.30		31.25	40
4180	5" diameter	Q-2	83	.289		23	15.55		38.55	49
4200	6" diameter	"	79	.304		30	16.30		46.30	57.50
4220	8" diameter	Q-3	69	.464		52	25.50		77.50	96
4240	10" diameter		61	.525		85	29		114	137
4244	12" diameter		58	.552		105	30		135	162
4248	15" diameter	↓	52	.615	↓	156	33.50		189.50	222
4280	To delete hangers, subtract									
4290	1-1/2" diam. to 6" diam.					22%	47%			
4300	8" diam. to 10" diam.					21%	44%			
4310	12" diam. to 15" diam.					19%	40%			

22 13 16.30 Pipe Fittings, Cast Iron

		Crew	Daily Output	Labor-Hours	Unit	Material	Labor	Equipment	Total	Total Incl O&P
0010	**PIPE FITTINGS, CAST IRON**, Soil									
0040	Hub and spigot, service weight, lead & oakum joints									
0080	1/4 bend, 2"	Q-1	16	1	Ea.	19.25	52		71.25	99
0120	3"		14	1.143		25.50	59		84.50	118
0140	4"	↓	13	1.231		40	64		104	141
0160	5"	Q-2	18	1.333		56	71.50		127.50	170
0180	6"	"	17	1.412		69.50	76		145.50	192
0200	8"	Q-3	11	2.909		208	159		367	470
0220	10"		10	3.200		305	175		480	600
0224	12"		9	3.556		415	195		610	750
0226	15"	↓	7	4.571		1,425	251		1,676	1,950
0242	Short sweep, CI, 90°, 2"	Q-1	16	1		18.75	52		70.75	98.50
0243	3"		14	1.143		33.50	59		92.50	127
0244	4"	↓	13	1.231		51	64		115	153
0245	6"	Q-2	17	1.412		104	76		180	229
0246	8"	Q-3	11	2.909		225	159		384	490
0247	10"		10	3.200		465	175		640	775
0248	12"	↓	9	3.556	↓	1,075	195		1,270	1,475
0251	Long sweep elbow									
0252	2"	Q-1	16	1	Ea.	29	52		81	110
0253	3"		14	1.143		41.50	59		100.50	135
0254	4"	↓	13	1.231		61	64		125	164
0255	6"	Q-2	17	1.412		122	76		198	250
0256	8"	Q-3	11	2.909		277	159		436	545

22 13 Facility Sanitary Sewerage
22 13 16 – Sanitary Waste and Vent Piping

22 13 16.30 Pipe Fittings, Cast Iron		Crew	Daily Output	Labor-Hours	Unit	Material	2014 Bare Costs Labor	Equipment	Total	Total Incl O&P
0257	10"	Q-3	10	3.200	Ea.	470	175		645	780
0258	12"		9	3.556		760	195		955	1,125
0259	15"	▼	7	4.571		1,975	251		2,226	2,550
0266	Closet bend, 3" diameter with flange 10" x 16"	Q-1	14	1.143		125	59		184	228
0268	16" x 16"		12	1.333		138	69		207	255
0270	Closet bend, 4" diameter, 2-1/2" x 4" ring, 6" x 16"		13	1.231		112	64		176	220
0280	8" x 16"		13	1.231		100	64		164	207
0290	10" x 12"		12	1.333		94.50	69		163.50	208
0300	10" x 18"		11	1.455		130	75.50		205.50	257
0310	12" x 16"		11	1.455		112	75.50		187.50	238
0330	16" x 16"		10	1.600		141	83		224	280
0340	1/8 bend, 2"		16	1		13.65	52		65.65	93
0350	3"		14	1.143		21.50	59		80.50	113
0360	4"	▼	13	1.231		31.50	64		95.50	131
0380	5"	Q-2	18	1.333		43.50	71.50		115	156
0400	6"	"	17	1.412		53	76		129	173
0420	8"	Q-3	11	2.909		158	159		317	415
0440	10"		10	3.200		227	175		402	515
0460	12"		9	3.556		430	195		625	770
0461	15"	▼	7	4.571		1,000	251		1,251	1,475
0500	Sanitary tee, 2"	Q-1	10	1.600		26.50	83		109.50	155
0540	3"		9	1.778		43.50	92		135.50	187
0620	4"	▼	8	2		53	104		157	214
0700	5"	Q-2	12	2		105	107		212	278
0800	6"	"	11	2.182		119	117		236	310
0880	8"	Q-3	7	4.571		315	251		566	730
0881	10"		7	4.571		580	251		831	1,025
0882	12"		6	5.333		1,025	292		1,317	1,575
0883	15"	▼	4	8	▼	2,050	440		2,490	2,925
0900	Sanitary tee, tapped									
0901	2" x 2"	Q-1	9	1.778	Ea.	36	92		128	179
0902	3" x 2"		8	2		40	104		144	200
0903	4" x 2"	▼	7	2.286	▼	79.50	118		197.50	267
0910	Sanitary cross, tapped [double tapped sanitary tee]									
0911	2" x 2"	Q-1	7	2.286	Ea.	57.50	118		175.50	242
0912	3" x 2"		6	2.667		70.50	138		208.50	287
0913	4" x 2"	▼	5	3.200	▼	69	166		235	325
0940	Sanitary Tee, reducing									
0942	3" x 2"	Q-1	10	1.600	Ea.	37	83		120	166
0943	4" x 3"		9	1.778		48.50	92		140.50	193
0944	4" x 2"	▼	9	1.778		48.50	92		140.50	193
0945	5" x 3"	Q-2	12.50	1.920		126	103		229	294
0946	6" x 4"		10	2.400		135	129		264	345
0947	6" x 3"		11	2.182		97	117		214	284
0948	6" x 2"	▼	12.50	1.920		93.50	103		196.50	259
0949	8" x 6"	Q-3	9	3.556		267	195		462	590
0950	8" x 5"		9	3.556		267	195		462	590
0951	8" x 4"		10	3.200		202	175		377	490
0954	10" x 6"		8	4		425	219		644	800
0958	12" x 8"		7	4.571		830	251		1,081	1,300
0962	15" x 10"	▼	5	6.400		1,750	350		2,100	2,450
1000	Tee, 2"	Q-1	10	1.600		38.50	83		121.50	168
1060	3"	▼	9	1.778	▼	57.50	92		149.50	203

251

22 13 Facility Sanitary Sewerage

22 13 16 – Sanitary Waste and Vent Piping

22 13 16.30 Pipe Fittings, Cast Iron		Crew	Daily Output	Labor-Hours	Unit	Material	2014 Bare Costs Labor	Equipment	Total	Total Incl O&P
1120	4"	Q-1	8	2	Ea.	74	104		178	237
1200	5"	Q-2	12	2		157	107		264	335
1300	6"	"	11	2.182		154	117		271	345
1380	8"	Q-3	7	4.571	↓	272	251		523	680
1400	Combination Y and 1/8 bend									
1420	2"	Q-1	10	1.600	Ea.	33.50	83		116.50	162
1460	3"		9	1.778		51	92		143	195
1520	4"	↓	8	2		70	104		174	233
1540	5"	Q-2	12	2		133	107		240	310
1560	6"		11	2.182		168	117		285	360
1580	8"		7	3.429		415	184		599	735
1582	12"	Q-3	6	5.333	↓	850	292		1,142	1,375
1584	Combination Y & 1/8 bend, reducing									
1586	3" x 2"	Q-1	10	1.600	Ea.	38.50	83		121.50	168
1587	4" x 2"		9.50	1.684		52	87		139	189
1588	4" x 3"	↓	9	1.778		61	92		153	206
1589	6" x 2"	Q-2	12.50	1.920		109	103		212	276
1590	6" x 3"		12	2		116	107		223	289
1591	6" x 4"	↓	11	2.182		122	117		239	310
1592	8" x 2"	Q-3	11	2.909		229	159		388	495
1593	8" x 4"		10	3.200		202	175		377	485
1594	8" x 6"	↓	9	3.556		277	195		472	600
1600	Double Y, 2"	Q-1	8	2		58.50	104		162.50	221
1610	3"		7	2.286		74	118		192	261
1620	4"	↓	6.50	2.462		97	128		225	300
1630	5"	Q-2	9	2.667		199	143		342	435
1640	6"	"	8	3		249	161		410	515
1650	8"	Q-3	5.50	5.818		610	320		930	1,150
1660	10"		5	6.400		1,350	350		1,700	2,000
1670	12"	↓	4.50	7.111	↓	1,550	390		1,940	2,300
1676	Combination double Y & 1/8 bend									
1678	2"	Q-1	8	2	Ea.	68.50	104		172.50	232
1680	3"		7	2.286		88	118		206	276
1682	4"	↓	6.50	2.462		144	128		272	350
1684	6"	Q-2	8	3		500	161		661	795
1690	Combination double Y & 1/8 bend, CI, reducing									
1692	3" x 2"	Q-1	8	2	Ea.	75	104		179	239
1694	4" x 2"		7	2.286		89	118		207	277
1696	4" x 3"	↓	7	2.286		101	118		219	290
1698	6" x 4"	Q-2	9	2.667		325	143		468	575
1700	Double Y, CI, reducing									
1702	3" x 2"	Q-1	8	2	Ea.	63	104		167	226
1703	4" x 2"		7	2.286		76	118		194	263
1704	4" x 3"	↓	7	2.286		80	118		198	267
1706	6" x 3"	Q-2	9.50	2.526		194	136		330	420
1707	6" x 4"	"	9	2.667		180	143		323	415
1708	8" x 4"	Q-3	8.50	3.765		425	206		631	775
1709	8" x 6"		8	4		435	219		654	810
1711	10" x 6"		8	4		915	219		1,134	1,325
1712	10" x 8"		6.50	4.923		1,300	270		1,570	1,825
1713	12" x 6"		7.50	4.267		1,650	234		1,884	2,175
1714	12" x 8"	↓	6	5.333		1,500	292		1,792	2,100
1740	Reducer, 3" x 2"	Q-1	15	1.067		18.75	55.50		74.25	104

22 13 Facility Sanitary Sewerage

22 13 16 – Sanitary Waste and Vent Piping

22 13 16.30 Pipe Fittings, Cast Iron		Crew	Daily Output	Labor-Hours	Unit	Material	2014 Bare Costs Labor	Equipment	Total	Total Incl O&P
1750	4" x 2"	Q-1	14.50	1.103	Ea.	21.50	57		78.50	110
1760	4" x 3"		14	1.143		24.50	59		83.50	117
1770	5" x 2"		14	1.143		51	59		110	146
1780	5" x 3"		13.50	1.185		54	61.50		115.50	152
1790	5" x 4"		13	1.231		31.50	64		95.50	131
1800	6" x 2"		13.50	1.185		48.50	61.50		110	146
1810	6" x 3"		13	1.231		50	64		114	152
1830	6" x 4"		12.50	1.280		49	66.50		115.50	154
1840	6" x 5"	↓	11	1.455		53	75.50		128.50	172
1880	8" x 3"	Q-2	13.50	1.778		94	95.50		189.50	247
1900	8" x 4"		13	1.846		82	99		181	240
1920	8" x 5"		12	2		85	107		192	256
1940	8" x 6"		12	2		84.50	107		191.50	255
1942	10" x 4"		12.50	1.920		124	103		227	292
1943	10" x 6"		11.50	2.087		138	112		250	320
1944	10" x 8"		9.50	2.526		138	136		274	355
1945	12" x 4"		11.50	2.087		201	112		313	390
1946	12" x 6"		11	2.182		216	117		333	415
1947	12" x 8"		9	2.667		222	143		365	460
1948	12" x 10"	↓	8.50	2.824		231	152		383	485
1949	15" x 6"	Q-3	12	2.667		420	146		566	685
1950	15" x 8"		10	3.200		455	175		630	770
1951	15" x 10"		9.50	3.368		475	185		660	800
1952	15" x 12"	↓	9	3.556		480	195		675	825
1960	Increaser, 2" x 3"	Q-1	15	1.067		45	55.50		100.50	133
1980	2" x 4"		14	1.143		45	59		104	139
2000	2" x 5"		13	1.231		55	64		119	157
2020	3" x 4"		13	1.231		49.50	64		113.50	151
2040	3" x 5"		13	1.231		55	64		119	157
2060	3" x 6"		12	1.333		66	69		135	177
2070	4" x 5"		13	1.231		58.50	64		122.50	161
2080	4" x 6"	↓	12	1.333		67.50	69		136.50	178
2090	4" x 8"	Q-2	13	1.846		136	99		235	300
2100	5" x 6"	Q-1	11	1.455		99	75.50		174.50	223
2110	5" x 8"	Q-2	12	2		160	107		267	340
2120	6" x 8"		12	2		160	107		267	340
2130	6" x 10"		8	3		288	161		449	560
2140	8" x 10"		6.50	3.692		290	198		488	620
2150	10" x 12"	↓	5.50	4.364		515	234		749	920
2500	Y, 2"	Q-1	10	1.600		24.50	83		107.50	152
2510	3"		9	1.778		45	92		137	189
2520	4"	↓	8	2		60.50	104		164.50	223
2530	5"	Q-2	12	2		107	107		214	280
2540	6"	"	11	2.182		139	117		256	330
2550	8"	Q-3	7	4.571		340	251		591	755
2560	10"		6	5.333		550	292		842	1,050
2570	12"		5	6.400		1,250	350		1,600	1,900
2580	15"	↓	4	8	↓	2,725	440		3,165	3,650
2581	Y, reducing									
2582	3" x 2"	Q-1	10	1.600	Ea.	35	83		118	164
2584	4" x 2"		9.50	1.684		53.50	87		140.50	191
2586	4" x 3"	↓	9	1.778		52	92		144	196
2588	6" x 2"	Q-2	12.50	1.920		93	103		196	258

22 13 Facility Sanitary Sewerage

22 13 16 – Sanitary Waste and Vent Piping

22 13 16.30 Pipe Fittings, Cast Iron

		Crew	Daily Output	Labor-Hours	Unit	Material	2014 Bare Costs Labor	Equipment	Total	Total Incl O&P
2590	6" x 3"	Q-2	12.25	1.959	Ea.	95	105		200	263
2592	6" x 4"	▼	12	2		94	107		201	265
2594	8" x 2"	Q-3	11	2.909		209	159		368	470
2596	8" x 3"		10.50	3.048		208	167		375	480
2598	8" x 4"		10	3.200		179	175		354	460
2600	8" x 6"		9	3.556		222	195		417	540
2602	10" x 3"		10	3.200		310	175		485	610
2604	10" x 4"		9.50	3.368		305	185		490	615
2606	10" x 6"		8	4		335	219		554	695
2608	10" x 8"		8	4		460	219		679	835
2610	12" x 4"		9	3.556		520	195		715	865
2612	12" x 6"		8.50	3.765		535	206		741	895
2614	12" x 8"		8	4		640	219		859	1,025
2616	12" x 10"		7.50	4.267		1,075	234		1,309	1,550
2618	15" x 4"		7	4.571		1,675	251		1,926	2,200
2620	15" x 6"		6.50	4.923		1,750	270		2,020	2,325
2622	15" x 8"		6.50	4.923		1,775	270		2,045	2,350
2624	15" x 10"		5	6.400		1,800	350		2,150	2,525
2626	15" x 12"	▼	4.50	7.111	▼	1,950	390		2,340	2,750
3000	For extra heavy, add					44%	4%			
3600	Hub and spigot, service weight gasket joint									
3605	Note: gaskets and joint labor have									
3606	been included with all listed fittings.									
3610	1/4 bend, 2"	Q-1	20	.800	Ea.	28.50	41.50		70	94
3620	3"		17	.941		38	49		87	115
3630	4"	▼	15	1.067		55.50	55.50		111	145
3640	5"	Q-2	21	1.143		80	61.50		141.50	181
3650	6"	"	19	1.263		94.50	68		162.50	206
3660	8"	Q-3	12	2.667		263	146		409	510
3670	10"		11	2.909		400	159		559	680
3680	12"		10	3.200		535	175		710	855
3690	15"	▼	8	4		1,575	219		1,794	2,050
3692	Short sweep, CI, 90°, 2"	Q-1	20	.800		28	41.50		69.50	93.50
3693	3"		17	.941		45.50	49		94.50	124
3694	4"	▼	15	1.067		66.50	55.50		122	157
3695	6"	Q-2	19	1.263		129	68		197	244
3696	8"	Q-3	12	2.667		280	146		426	530
3697	10"		11	2.909		560	159		719	855
3698	12"		10	3.200		1,200	175		1,375	1,600
3700	Closet bend, 3" diameter with ring 10" x 16"	Q-1	17	.941		138	49		187	225
3710	16" x 16"		15	1.067		150	55.50		205.50	249
3730	Closet bend, 4" diameter, 1" x 4" ring, 6" x 16"		15	1.067		127	55.50		182.50	224
3740	8" x 16"		15	1.067		115	55.50		170.50	211
3750	10" x 12"		14	1.143		110	59		169	211
3760	10" x 18"		13	1.231		145	64		209	257
3770	12" x 16"		13	1.231		128	64		192	238
3780	16" x 16"	▼	12	1.333	▼	156	69		225	276
3786	Long sweep elbow									
3787	2"	Q-1	20	.800	Ea.	38.50	41.50		80	105
3788	3"		17	.941		53.50	49		102.50	133
3789	4"	▼	15	1.067		76	55.50		131.50	167
3790	6"	Q-2	19	1.263		147	68		215	264
3791	8"	Q-3	12	2.667	▼	330	146		476	585

22 13 Facility Sanitary Sewerage

22 13 16 – Sanitary Waste and Vent Piping

22 13 16.30 Pipe Fittings, Cast Iron		Crew	Daily Output	Labor-Hours	Unit	Material	2014 Bare Costs Labor	Equipment	Total	Total Incl O&P
3792	10"	Q-3	11	2.909	Ea.	565	159		724	860
3793	12"		10	3.200		880	175		1,055	1,225
3794	15"	↓	8	4		2,125	219		2,344	2,650
3800	1/8 bend, 2"	Q-1	20	.800		23	41.50		64.50	88
3810	3"		17	.941		33.50	49		82.50	111
3820	4"	↓	15	1.067		46.50	55.50		102	135
3830	5"	Q-2	21	1.143		67.50	61.50		129	167
3840	6"	"	19	1.263		77.50	68		145.50	188
3850	8"	Q-3	12	2.667		213	146		359	455
3860	10"		11	2.909		325	159		484	595
3870	12"		10	3.200		555	175		730	875
3880	15"	↓	8	4		1,150	219		1,369	1,600
3882	Sanitary tee, tapped									
3884	2" x 2"	Q-1	11	1.455	Ea.	54.50	75.50		130	174
3886	3" x 2"		10	1.600		62	83		145	193
3888	4" x 2"	↓	9	1.778	↓	104	92		196	254
3890	Sanitary cross, tapped (double tapped sanitary tee)									
3892	2" x 2"	Q-1	9	1.778	Ea.	76	92		168	223
3894	3" x 2"		8	2		92	104		196	257
3896	4" x 2"		7	2.286		93.50	118		211.50	282
3900	Sanitary Tee, 2"		12	1.333		45.50	69		114.50	154
3910	3"		10	1.600		67.50	83		150.50	200
3920	4"	↓	9	1.778		83.50	92		175.50	231
3930	5"	Q-2	13	1.846		153	99		252	320
3940	6"	"	11	2.182		169	117		286	365
3950	8"	Q-3	8.50	3.765		425	206		631	780
3952	10"		8	4		770	219		989	1,175
3954	12"		7	4.571		1,275	251		1,526	1,775
3956	15"	↓	6	5.333	↓	2,350	292		2,642	3,050
3960	Sanitary Tee, reducing									
3961	3" x 2"	Q-1	10.50	1.524	Ea.	59	79		138	184
3962	4" x 2"		10	1.600		73.50	83		156.50	206
3963	4" x 3"	↓	9.50	1.684		76	87		163	216
3964	5" x 3"	Q-2	13.50	1.778		162	95.50		257.50	320
3965	6" x 2"		13	1.846		128	99		227	291
3966	6" x 3"		12.50	1.920		134	103		237	305
3967	6" x 4"	↓	12	2		175	107		282	355
3968	8" x 4"	Q-3	10.50	3.048		273	167		440	550
3969	8" x 5"		10	3.200		345	175		520	645
3970	8" x 6"		9.50	3.368		345	185		530	660
3971	10" x 6"		9	3.556		545	195		740	895
3972	12" x 8"		8.50	3.765		1,000	206		1,206	1,400
3973	15" x 10"	↓	6	5.333		2,000	292		2,292	2,650
3980	Tee, 2"	Q-1	12	1.333		57.50	69		126.50	167
3990	3"		10	1.600		82	83		165	215
4000	4"	↓	9	1.778		104	92		196	254
4010	5"	Q-2	13	1.846		205	99		304	375
4020	6"	"	11	2.182		204	117		321	400
4030	8"	Q-3	8	4	↓	380	219		599	750
4060	Combination Y and 1/8 bend									
4070	2"	Q-1	12	1.333	Ea.	52.50	69		121.50	162
4080	3"		10	1.600		75.50	83		158.50	208
4090	4"	↓	9	1.778	↓	101	92		193	250

255

22 13 Facility Sanitary Sewerage

22 13 16 – Sanitary Waste and Vent Piping

22 13 16.30 Pipe Fittings, Cast Iron		Crew	Daily Output	Labor-Hours	Unit	Material	2014 Bare Costs Labor	Equipment	Total	Total Incl O&P
4100	5"	Q-2	13	1.846	Ea.	181	99		280	350
4110	6"	"	11	2.182		218	117		335	415
4120	8"	Q-3	8	4		525	219		744	910
4121	12"	"	7	4.571		1,100	251		1,351	1,575
4130	Combination Y & 1/8 bend, reducing									
4132	3" x 2"	Q-1	10.50	1.524	Ea.	60	79		139	185
4134	4" x 2"		10	1.600		76.50	83		159.50	209
4136	4" x 3"		9.50	1.684		88.50	87		175.50	230
4138	6" x 2"	Q-2	13	1.846		143	99		242	305
4140	6" x 3"		12.50	1.920		153	103		256	325
4142	6" x 4"		12	2		162	107		269	340
4144	8" x 2"	Q-3	11	2.909		310	159		469	580
4146	8" x 4"		10.50	3.048		272	167		439	550
4148	8" x 6"		9.50	3.368		355	185		540	670
4160	Double Y, 2"	Q-1	10	1.600		86.50	83		169.50	221
4170	3"		8	2		111	104		215	278
4180	4"		7	2.286		143	118		261	335
4190	5"	Q-2	10	2.400		271	129		400	495
4200	6"	"	9	2.667		325	143		468	570
4210	8"	Q-3	6	5.333		775	292		1,067	1,300
4220	10"		5	6.400		1,625	350		1,975	2,325
4230	12"		4.50	7.111		1,900	390		2,290	2,700
4234	Combination double Y & 1/8 bend									
4235	2"	Q-1	10	1.600	Ea.	96.50	83		179.50	231
4236	3"		8	2		125	104		229	293
4237	4"		7	2.286		190	118		308	390
4238	6"	Q-2	9	2.667		575	143		718	845
4242	Combination double Y & 1/8 bend, CI, reducing									
4243	3" x 2"	Q-1	9	1.778	Ea.	106	92		198	256
4244	4" x 2"		8	2		123	104		227	291
4245	4" x 3"		7.50	2.133		141	111		252	320
4246	6" x 4"	Q-2	10	2.400		380	129		509	615
4248	Double Y, CI, reducing									
4249	3" x 2"	Q-1	9	1.778	Ea.	94	92		186	243
4250	4" x 2"		8	2		110	104		214	277
4251	4" x 3"		7.50	2.133		108	111		219	285
4252	6" x 3"	Q-2	10.50	2.286		243	123		366	450
4253	6" x 4"	"	10	2.400		235	129		364	455
4254	8" x 4"	Q-3	9.50	3.368		510	185		695	840
4255	8" x 6"		9	3.556		540	195		735	890
4256	10" x 6"		8.50	3.765		1,050	206		1,256	1,475
4257	10" x 8"		7.50	4.267		1,500	234		1,734	2,000
4258	12" x 6"		8	4		1,825	219		2,044	2,325
4259	12" x 8"		7	4.571		1,725	251		1,976	2,275
4260	Reducer, 3" x 2"	Q-1	17	.941		40.50	49		89.50	118
4270	4" x 2"		16.50	.970		46	50		96	127
4280	4" x 3"		16	1		52	52		104	135
4290	5" x 2"		16	1		84	52		136	171
4300	5" x 3"		15.50	1.032		90	53.50		143.50	180
4310	5" x 4"		15	1.067		70.50	55.50		126	161
4320	6" x 2"		15.50	1.032		82.50	53.50		136	172
4330	6" x 3"		15	1.067		87	55.50		142.50	179
4336	6" x 4"		14	1.143		89.50	59		148.50	188

22 13 Facility Sanitary Sewerage

22 13 16 – Sanitary Waste and Vent Piping

22 13 16.30 Pipe Fittings, Cast Iron

		Crew	Daily Output	Labor-Hours	Unit	Material	2014 Bare Costs Labor	Equipment	Total	Total Incl O&P
4340	6" x 5"	Q-1	13	1.231	Ea.	102	64		166	209
4360	8" x 3"	Q-2	15	1.600		161	86		247	305
4370	8" x 4"		15	1.600		152	86		238	298
4380	8" x 5"		14	1.714		164	92		256	320
4390	8" x 6"		14	1.714		164	92		256	320
4394	10" x 4"		13.50	1.778		235	95.50		330.50	400
4395	10" x 6"		13	1.846		259	99		358	435
4396	10" x 8"		12.50	1.920		289	103		392	475
4397	12" x 4"		12	2		340	107		447	530
4398	12" x 6"		11.50	2.087		365	112		477	570
4399	12" x 8"		11	2.182		400	117		517	615
4400	12" x 10"		10.50	2.286		450	123		573	680
4401	15" x 6"	Q-3	13	2.462		590	135		725	855
4402	15" x 8"		12	2.667		660	146		806	945
4403	15" x 10"		11	2.909		715	159		874	1,025
4404	15" x 12"		10	3.200		750	175		925	1,075
4430	Increaser, 2" x 3"	Q-1	17	.941		57	49		106	137
4440	2" x 4"		16	1		60	52		112	144
4450	2" x 5"		15	1.067		79	55.50		134.50	171
4460	3" x 4"		15	1.067		64.50	55.50		120	155
4470	3" x 5"		15	1.067		79	55.50		134.50	171
4480	3" x 6"		14	1.143		91	59		150	190
4490	4" x 5"		15	1.067		82.50	55.50		138	175
4500	4" x 6"		14	1.143		92.50	59		151.50	192
4510	4" x 8"	Q-2	15	1.600		191	86		277	340
4520	5" x 6"	Q-1	13	1.231		124	64		188	233
4530	5" x 8"	Q-2	14	1.714		215	92		307	375
4540	6" x 8"		14	1.714		215	92		307	375
4550	6" x 10"		10	2.400		385	129		514	615
4560	8" x 10"		8.50	2.824		385	152		537	655
4570	10" x 12"		7.50	3.200		635	172		807	960
4600	Y, 2"	Q-1	12	1.333		43	69		112	152
4610	3"		10	1.600		69.50	83		152.50	202
4620	4"		9	1.778		91	92		183	239
4630	5"	Q-2	13	1.846		155	99		254	320
4640	6"	"	11	2.182		189	117		306	385
4650	8"	Q-3	8	4		450	219		669	825
4660	10"		7	4.571		740	251		991	1,200
4670	12"		6	5.333		1,500	292		1,792	2,100
4672	15"		5	6.400		3,025	350		3,375	3,850
4680	Y, reducing									
4681	3" x 2"	Q-1	10.50	1.524	Ea.	56.50	79		135.50	182
4682	4" x 2"		10	1.600		78	83		161	211
4683	4" x 3"		9.50	1.684		79.50	87		166.50	220
4684	6" x 2"	Q-2	13	1.846		127	99		226	290
4685	6" x 3"		12.50	1.920		132	103		235	300
4686	6" x 4"		12	2		134	107		241	310
4687	8" x 2"	Q-3	11	2.909		273	159		432	540
4688	8" x 3"		10.75	2.977		275	163		438	545
4689	8" x 4"		10.50	3.048		249	167		416	525
4690	8" x 6"		9.50	3.368		300	185		485	610
4691	10" x 3"		10	3.200		420	175		595	725
4692	10" x 4"		9.50	3.368		420	185		605	740

22 13 Facility Sanitary Sewerage

22 13 16 – Sanitary Waste and Vent Piping

22 13 16.30 Pipe Fittings, Cast Iron		Crew	Daily Output	Labor-Hours	Unit	Material	2014 Bare Costs Labor	Equipment	Total	Total Incl O&P
4693	10" x 6"	Q-3	9	3.556	Ea.	455	195		650	795
4694	10" x 8"		8.80	3.636		610	199		809	970
4695	12" x 4"		9.50	3.368		655	185		840	1,000
4696	12" x 6"		9	3.556		680	195		875	1,050
4697	12" x 8"		8.50	3.765		820	206		1,026	1,200
4698	12" x 10"		8	4		1,300	219		1,519	1,750
4699	15" x 4"		8	4		1,825	219		2,044	2,350
4700	15" x 6"		7.50	4.267		1,900	234		2,134	2,450
4701	15" x 8"		7	4.571		1,975	251		2,226	2,550
4702	15" x 10"		6	5.333		2,050	292		2,342	2,700
4703	15" x 12"	↓	5	6.400	↓	2,225	350		2,575	2,975
4900	For extra heavy, add					44%	4%			
4940	Gasket and making push-on joint									
4950	2"	Q-1	40	.400	Ea.	9.40	20.50		29.90	42
4960	3"		35	.457		12.20	23.50		35.70	49.50
4970	4"	↓	32	.500		15.30	26		41.30	56
4980	5"	Q-2	43	.558		24	30		54	72
4990	6"	"	40	.600		25	32		57	76
5000	8"	Q-3	32	1		55	55		110	144
5010	10"		29	1.103		95.50	60.50		156	197
5020	12"		25	1.280		122	70		192	240
5022	15"	↓	21	1.524	↓	146	83.50		229.50	287
5030	Note: gaskets and joint labor have									
5040	Been included with all listed fittings.									
5990	No hub									
6000	Cplg. & labor required at joints not incl. in fitting									
6010	price. Add 1 coupling per joint for installed price									
6020	1/4 Bend, 1-1/2"				Ea.	9.85			9.85	10.85
6060	2"					10.70			10.70	11.80
6080	3"					15			15	16.50
6120	4"					22			22	24.50
6140	5"					53.50			53.50	58.50
6160	6"					53.50			53.50	59
6180	8"					151			151	166
6181	10"					265			265	292
6182	12"					940			940	1,025
6183	15"					1,225			1,225	1,350
6184	1/4 Bend, long sweep, 1-1/2"					24.50			24.50	27
6186	2"					23.50			23.50	26
6188	3"					28.50			28.50	31.50
6189	4"					45			45	49.50
6190	5"					87.50			87.50	96.50
6191	6"					99.50			99.50	110
6192	8"					269			269	296
6193	10"					550			550	605
6200	1/8 Bend, 1-1/2"					8.35			8.35	9.15
6210	2"					9.25			9.25	10.15
6212	3"					12.40			12.40	13.65
6214	4"					16.25			16.25	17.85
6216	5"					34			34	37
6218	6"					36			36	39.50
6220	8"					104			104	114
6222	10"				↓	197			197	217

22 13 Facility Sanitary Sewerage

22 13 16 – Sanitary Waste and Vent Piping

22 13 16.30 Pipe Fittings, Cast Iron		Crew	Daily Output	Labor-Hours	Unit	Material	2014 Bare Costs Labor	Equipment	Total	Total Incl O&P
6364	Closet flange									
6366	4"				Ea.	15			15	16.50
6370	Closet bend, no hub									
6376	4" x 16"				Ea.	79.50			79.50	87.50
6380	Sanitary Tee, tapped, 1-1/2"					19.60			19.60	21.50
6382	2" x 1-1/2"					17.35			17.35	19.05
6384	2"					18.55			18.55	20.50
6386	3" x 2"					27.50			27.50	30.50
6388	3"					47.50			47.50	52.50
6390	4" x 1-1/2"					24.50			24.50	27
6392	4" x 2"					28			28	30.50
6393	4"					28			28	30.50
6394	6" x 1-1/2"					64			64	70.50
6396	6" x 2"					65			65	71.50
6459	Sanitary Tee, 1-1/2"					13.85			13.85	15.25
6460	2"					14.75			14.75	16.25
6470	3"					18.25			18.25	20
6472	4"					34.50			34.50	38
6474	5"					81			81	89
6476	6"					82.50			82.50	91
6478	8"					335			335	365
6480	10"					410			410	450
6724	Sanitary Tee, reducing									
6725	3" x 2"				Ea.	16.15			16.15	17.80
6726	4" x 3"					26.50			26.50	29.50
6727	5" x 4"					62.50			62.50	68.50
6728	6" x 3"					57			57	63
6729	8" x 4"					176			176	194
6730	Y, 1-1/2"					14			14	15.40
6740	2"					13.70			13.70	15.10
6750	3"					19.95			19.95	22
6760	4"					32			32	35
6762	5"					75.50			75.50	83
6764	6"					85			85	93.50
6768	8"					200			200	220
6769	10"					445			445	490
6770	12"					875			875	960
6771	15"					1,950			1,950	2,150
6791	Y, reducing, 3" x 2"					14.85			14.85	16.30
6792	4" x 2"					21.50			21.50	23.50
6793	5" x 2"					47			47	51.50
6794	6" x 2"					52			52	57.50
6795	6" x 4"					68			68	74.50
6796	8" x 4"					116			116	128
6797	8" x 6"					143			143	157
6798	10" x 6"					320			320	355
6799	10" x 8"					385			385	425
6800	Double Y, 2"					21.50			21.50	24
6920	3"					40			40	44
7000	4"					81.50			81.50	89.50
7100	6"					144			144	158
7120	8"					410			410	450
7200	Combination Y and 1/8 Bend									

22 13 Facility Sanitary Sewerage
22 13 16 — Sanitary Waste and Vent Piping

22 13 16.30 Pipe Fittings, Cast Iron		Crew	Daily Output	Labor-Hours	Unit	Material	2014 Bare Costs Labor	Equipment	Total	Total Incl O&P
7220	1-1/2"				Ea.	14.90			14.90	16.40
7260	2"					15.60			15.60	17.15
7320	3"					24.50			24.50	27
7400	4"					47.50			47.50	52
7480	5"					97			97	107
7500	6"					130			130	143
7520	8"					305			305	335
7800	Reducer, 3" x 2"					7.55			7.55	8.30
7820	4" x 2"					11.70			11.70	12.85
7840	4" x 3"					11.70			11.70	12.85
7842	6" x 3"					31.50			31.50	34.50
7844	6" x 4"					31.50			31.50	34.50
7846	6" x 5"					32			32	35.50
7848	8" x 2"					50			50	55
7850	8" x 3"					46			46	50.50
7852	8" x 4"					48.50			48.50	53.50
7854	8" x 5"					55			55	60.50
7856	8" x 6"					54			54	59.50
7858	10" x 4"					95.50			95.50	105
7860	10" x 6"					101			101	111
7862	10" x 8"					118			118	130
7864	12" x 4"					197			197	217
7866	12" x 6"					211			211	232
7868	12" x 8"					217			217	239
7870	12" x 10"					221			221	243
7872	15" x 4"					410			410	455
7874	15" x 6"					390			390	430
7876	15" x 8"					450			450	495
7878	15" x 10"					465			465	510
7880	15" x 12"				↓	470			470	515
8000	Coupling, standard (by CISPI Mfrs.)									
8020	1-1/2"	Q-1	48	.333	Ea.	13.95	17.25		31.20	41.50
8040	2"		44	.364		15.25	18.85		34.10	45.50
8080	3"		38	.421		17.10	22		39.10	52
8120	4"	↓	33	.485		20	25		45	60
8160	5"	Q-2	44	.545		41	29.50		70.50	89.50
8180	6"	"	40	.600		46.50	32		78.50	100
8200	8"	Q-3	33	.970		97	53		150	187
8220	10"	"	26	1.231	↓	124	67.50		191.50	238
8300	Coupling, cast iron clamp & neoprene gasket (by MG)									
8310	1-1/2"	Q-1	48	.333	Ea.	6.40	17.25		23.65	33
8320	2"		44	.364		8.35	18.85		27.20	37.50
8330	3"		38	.421		12.65	22		34.65	47
8340	4"	↓	33	.485		14.60	25		39.60	54
8350	5"	Q-2	44	.545		20.50	29.50		50	67
8360	6"	"	40	.600		32.50	32		64.50	84.50
8380	8"	Q-3	33	.970		116	53		169	208
8400	10"	"	26	1.231	↓	189	67.50		256.50	310
8410	Reducing, no hub									
8416	2" x 1-1/2"	Q-1	44	.364	Ea.	8.10	18.85		26.95	37.50
8600	Coupling, Stainless steel, heavy duty									
8620	1-1/2"	Q-1	48	.333	Ea.	9.55	17.25		26.80	36.50
8630	2"	↓	44	.364	↓	9.90	18.85		28.75	39.50

22 13 Facility Sanitary Sewerage

22 13 16 – Sanitary Waste and Vent Piping

22 13 16.30 Pipe Fittings, Cast Iron

		Crew	Daily Output	Labor-Hours	Unit	Material	2014 Bare Costs Labor	Equipment	Total	Total Incl O&P
8640	2" x 1-1/2"	Q-1	44	.364	Ea.	12.05	18.85		30.90	42
8650	3"		38	.421		10.75	22		32.75	45
8660	4"		33	.485		12.15	25		37.15	51.50
8670	4" x 3"		33	.485		17.50	25		42.50	57.50
8680	5"	Q-2	44	.545		26	29.50		55.50	73
8690	6"	"	40	.600		29.50	32		61.50	80.50
8700	8"	Q-3	33	.970		49.50	53		102.50	135
8710	10"		26	1.231		62.50	67.50		130	171
8712	12"		22	1.455		97	79.50		176.50	226
8715	15"		18	1.778		114	97.50		211.50	273

22 13 16.40 Pipe Fittings, Cast Iron for Drainage

		Crew	Daily Output	Labor-Hours	Unit	Material	Labor	Equipment	Total	Total Incl O&P
0010	**PIPE FITTINGS, CAST IRON FOR DRAINAGE**, Special									
0020	Cast iron, drainage, threaded, black									
0030	90° Elbow, straight									
0031	1-1/2" pipe size	Q-1	20	.800	Ea.	34	41.50		75.50	100
0032	2" pipe size		18	.889		51.50	46		97.50	126
0033	3" pipe size		10	1.600		194	83		277	340
0034	4" pipe size		6	2.667		305	138		443	545
0040	90° Long turn elbow, straight									
0041	1-1/2" pipe size	Q-1	20	.800	Ea.	48	41.50		89.50	115
0042	2" pipe size		18	.889		71.50	46		117.50	148
0043	3" pipe size		10	1.600		260	83		343	410
0044	4" pipe size		6	2.667		445	138		583	695
0050	90° Street elbow, straight									
0051	1-1/2" pipe size	Q-1	20	.800	Ea.	49	41.50		90.50	116
0052	2" pipe size	"	18	.889	"	64.50	46		110.50	141
0060	45° Elbow									
0061	1-1/2" pipe size	Q-1	20	.800	Ea.	33.50	41.50		75	99.50
0062	2" pipe size		18	.889		48.50	46		94.50	123
0063	3" pipe size		10	1.600		189	83		272	335
0064	4" pipe size		6	2.667		296	138		434	535
0070	45° Street elbow									
0071	1-1/2" pipe size	Q-1	20	.800	Ea.	45.50	41.50		87	113
0072	2" pipe size	"	18	.889	"	73.50	46		119.50	151
0092	Tees, straight									
0093	1-1/2" pipe size	Q-1	13	1.231	Ea.	56	64		120	159
0094	2" pipe size	"	11	1.455	"	93	75.50		168.50	216
0100	TY's, straight									
0101	1-1/2" pipe size	Q-1	13	1.231	Ea.	55	64		119	157
0102	2" pipe size		11	1.455		91	75.50		166.50	214
0103	3" pipe size		6	2.667		340	138		478	580
0104	4" pipe size		4	4		495	207		702	860
0120	45° Y branch, straight									
0121	1-1/2" pipe size	Q-1	13	1.231	Ea.	67.50	64		131.50	171
0122	2" pipe size		11	1.455		79	75.50		154.50	201
0123	3" pipe size		6	2.667		435	138		573	690
0124	4" pipe size		4	4		650	207		857	1,025
0147	Double Y branch, straight									
0148	1-1/2" pipe size	Q-1	10	1.600	Ea.	206	83		289	350
0149	2" pipe size	"	7	2.286	"	191	118		309	390
0160	P trap									
0161	1-1/2" pipe size	Q-1	15	1.067	Ea.	101	55.50		156.50	195

22 13 Facility Sanitary Sewerage
22 13 16 – Sanitary Waste and Vent Piping

22 13 16.40 Pipe Fittings, Cast Iron for Drainage		Crew	Daily Output	Labor-Hours	Unit	Material	2014 Bare Costs Labor	Equipment	Total	Total Incl O&P
0162	2" pipe size	Q-1	13	1.231	Ea.	161	64		225	274
0163	3" pipe size		7	2.286		525	118		643	755
0164	4" pipe size	↓	5	3.200	↓	1,175	166		1,341	1,550
0180	Tucker connection									
0181	1-1/2" pipe size	Q-1	24	.667	Ea.	98	34.50		132.50	160
0182	2" pipe size	"	21	.762	"	120	39.50		159.50	192
0205	Cast iron, drainage, threaded, galvanized									
0206	90° Elbow, straight									
0207	1-1/2" pipe size	Q-1	20	.800	Ea.	50.50	41.50		92	119
0208	2" pipe size		18	.889		73.50	46		119.50	150
0209	3" pipe size		10	1.600		290	83		373	445
0210	4" pipe size	↓	6	2.667	↓	350	138		488	595
0216	90° Long turn elbow, straight									
0217	1-1/2" pipe size	Q-1	20	.800	Ea.	57.50	41.50		99	126
0218	2" pipe size		18	.889		99	46		145	179
0219	3" pipe size		10	1.600		325	83		408	480
0220	4" pipe size	↓	6	2.667	↓	560	138		698	825
0226	90° Street elbow, straight									
0227	1-1/2" pipe size	Q-1	20	.800	Ea.	64	41.50		105.50	133
0228	2" pipe size	"	18	.889	"	101	46		147	181
0236	45° Elbow									
0237	1-1/2" pipe size	Q-1	20	.800	Ea.	50.50	41.50		92	118
0238	2" pipe size		18	.889		71.50	46		117.50	149
0239	3" pipe size		10	1.600		250	83		333	400
0240	4" pipe size	↓	6	2.667	↓	405	138		543	655
0246	45° Street elbow									
0247	1-1/2" pipe size	Q-1	20	.800	Ea.	66.50	41.50		108	136
0248	2" pipe size	"	18	.889	"	100	46		146	180
0268	Tees, straight									
0269	1-1/2" pipe size	Q-1	13	1.231	Ea.	76.50	64		140.50	181
0270	2" pipe size	"	11	1.455	"	128	75.50		203.50	254
0276	TY's, straight									
0277	1-1/2" pipe size	Q-1	13	1.231	Ea.	120	64		184	229
0278	2" pipe size		11	1.455		126	75.50		201.50	253
0279	3" pipe size		6	2.667		505	138		643	770
0280	4" pipe size	↓	4	4	↓	610	207		817	985
0296	45° Y branch, straight									
0297	1-1/2" pipe size	Q-1	13	1.231	Ea.	85.50	64		149.50	191
0298	2" pipe size		11	1.455		172	75.50		247.50	305
0299	3" pipe size		6	2.667		610	138		748	880
0300	4" pipe size	↓	4	4	↓	935	207		1,142	1,350
0323	Double Y branch, straight									
0324	1-1/2" pipe size	Q-1	10	1.600	Ea.	170	83		253	310
0325	2" pipe size	"	7	2.286	"	291	118		409	500
0336	P Trap									
0337	1-1/2" pipe size	Q-1	15	1.067	Ea.	138	55.50		193.50	236
0338	2" pipe size		13	1.231		242	64		306	365
0339	3" pipe size		7	2.286		770	118		888	1,025
0340	4" pipe size	↓	5	3.200	↓	1,700	166		1,866	2,100
0356	Tucker connection									
0357	1-1/2" pipe size	Q-1	24	.667	Ea.	222	34.50		256.50	296
0358	2" pipe size	"	21	.762	"	238	39.50		277.50	320
1000	Drip pan elbow, (safety valve discharge elbow)									

22 13 Facility Sanitary Sewerage

22 13 16 – Sanitary Waste and Vent Piping

22 13 16.40 Pipe Fittings, Cast Iron for Drainage

		Crew	Daily Output	Labor-Hours	Unit	Material	2014 Bare Costs Labor	Equipment	Total	Total Incl O&P
1010	Cast iron, threaded inlet									
1014	2-1/2"	Q-1	8	2	Ea.	1,200	104		1,304	1,475
1015	3"		6.40	2.500		1,300	130		1,430	1,625
1017	4"	↓	4.80	3.333	↓	1,750	173		1,923	2,175
1018	Cast iron, flanged inlet									
1019	6"	Q-2	3.60	6.667	Ea.	1,800	360		2,160	2,550
1020	8"	"	2.60	9.231	"	1,900	495		2,395	2,850

22 13 16.50 Shower Drains

		Crew	Daily Output	Labor-Hours	Unit	Material	Labor	Equipment	Total	Total Incl O&P
0010	**SHOWER DRAINS**									
2780	Shower, with strainer, uniform diam. trap, bronze top									
2800	2" and 3" pipe size	Q-1	8	2	Ea.	395	104		499	590
2820	4" pipe size	"	7	2.286		435	118		553	655
2840	For galvanized body, add				↓	175			175	193
2860	With strainer, backwater valve, drum trap									
2880	1-1/2", 2" & 3" pipe size	Q-1	8	2	Ea.	350	104		454	540
2890	4" pipe size	"	7	2.286		490	118		608	720
2900	For galvanized body, add				↓	175			175	193

22 13 16.60 Traps

		Crew	Daily Output	Labor-Hours	Unit	Material	Labor	Equipment	Total	Total Incl O&P
0010	**TRAPS**									
0030	Cast iron, service weight									
0050	Running P trap, without vent									
1100	2"	Q-1	16	1	Ea.	164	52		216	258
1140	3"		14	1.143		164	59		223	270
1150	4"	↓	13	1.231		164	64		228	277
1160	6"	Q-2	17	1.412	↓	605	76		681	780
1180	Running trap, single hub, with vent									
2080	3" pipe size, 3" vent	Q-1	14	1.143	Ea.	109	59		168	210
2120	4" pipe size, 4" vent	"	13	1.231		149	64		213	261
2140	5" pipe size, 4" vent	Q-2	11	2.182		237	117		354	435
2160	6" pipe size, 4" vent		10	2.400		635	129		764	895
2180	6" pipe size, 6" vent	↓	8	3		690	161		851	1,000
2200	8" pipe size, 4" vent	Q-3	10	3.200		2,900	175		3,075	3,475
2220	8" pipe size, 6" vent	"	8	4		2,275	219		2,494	2,825
2300	For double hub, vent, add				↓	10%	20%			
2800	S trap,									
2850	4" pipe size	Q-1	13	1.231	Ea.	66.50	64		130.50	170
3000	P trap, B&S, 2" pipe size		16	1		32.50	52		84.50	114
3040	3" pipe size		14	1.143		49	59		108	143
3060	4" pipe size	↓	13	1.231		70.50	64		134.50	174
3080	5" pipe size	Q-2	18	1.333		156	71.50		227.50	280
3100	6" pipe size	"	17	1.412		217	76		293	355
3120	8" pipe size	Q-3	11	2.909		655	159		814	960
3130	10" pipe size	"	10	3.200		1,225	175		1,400	1,625
3150	P trap, no hub, 1-1/2" pipe size	Q-1	17	.941		17.70	49		66.70	93
3160	2" pipe size		16	1		16.75	52		68.75	96.50
3170	3" pipe size		14	1.143		36.50	59		95.50	130
3180	4" pipe size	↓	13	1.231		65	64		129	168
3190	6" pipe size	Q-2	17	1.412	↓	158	76		234	289
3350	Deep seal trap, B&S									
3400	1-1/4" pipe size	Q-1	14	1.143	Ea.	51	59		110	146
3410	1-1/2" pipe size		14	1.143		51	59		110	146
3420	2" pipe size	↓	14	1.143		46	59		105	140

22 13 Facility Sanitary Sewerage
22 13 16 – Sanitary Waste and Vent Piping

22 13 16.60 Traps		Crew	Daily Output	Labor-Hours	Unit	Material	2014 Bare Costs Labor	Equipment	Total	Total Incl O&P
3440	3" pipe size	Q-1	12	1.333	Ea.	60	69		129	170
3460	4" pipe size	↓	11	1.455		95.50	75.50		171	219
3500	For trap primer connection, add				↓	125			125	137
3540	For trap with floor cleanout, add					70%	5%			
3580	For trap with adjustable cleanout, add	Q-1	10	1.600	Ea.	284	83		367	435
4700	Copper, drainage, drum trap									
4800	3" x 5" solid, 1-1/2" pipe size	1 Plum	16	.500	Ea.	286	29		315	360
4840	3" x 6" swivel, 1-1/2" pipe size	"	16	.500	"	455	29		484	545
5100	P trap, standard pattern									
5200	1-1/4" pipe size	1 Plum	18	.444	Ea.	211	25.50		236.50	271
5240	1-1/2" pipe size		17	.471		204	27		231	266
5260	2" pipe size		15	.533		315	30.50		345.50	390
5280	3" pipe size	↓	11	.727	↓	760	42		802	900
5340	With cleanout, swivel joint and slip joint									
5360	1-1/4" pipe size	1 Plum	18	.444	Ea.	148	25.50		173.50	202
5400	1-1/2" pipe size		17	.471		310	27		337	380
5420	2" pipe size	↓	15	.533	↓	500	30.50		530.50	600
5750	Chromed brass, tubular, P trap, without cleanout, 20 Ga.									
5800	1-1/4" pipe size	1 Plum	18	.444	Ea.	19.25	25.50		44.75	59.50
5840	1-1/2" pipe size	"	17	.471	"	22	27		49	65
5900	With cleanout, 20 Ga.									
5940	1-1/4" pipe size	1 Plum	18	.444	Ea.	31	25.50		56.50	72.50
6000	1-1/2" pipe size	"	17	.471	"	33.50	27		60.50	78
6350	S trap, without cleanout, 20 Ga.									
6400	1-1/4" pipe size	1 Plum	18	.444	Ea.	50	25.50		75.50	93.50
6440	1-1/2" pipe size	"	17	.471	"	57.50	27		84.50	104
6550	With cleanout, 20 Ga.									
6600	1-1/4" pipe size	1 Plum	18	.444	Ea.	61.50	25.50		87	106
6640	1-1/2" pipe size	"	17	.471		64	27		91	111
6660	Corrosion resistant, glass, P trap, 1-1/2" pipe size	Q-1	17	.941		72	49		121	153
6670	2" pipe size		16	1		94	52		146	181
6680	3" pipe size		14	1.143		193	59		252	300
6690	4" pipe size	↓	13	1.231		287	64		351	410
6700	6" pipe size	Q-2	17	1.412	↓	1,100	76		1,176	1,350
6710	ABS DWV P trap, solvent weld joint									
6720	1-1/2" pipe size	1 Plum	18	.444	Ea.	12.65	25.50		38.15	52.50
6722	2" pipe size		17	.471		19.80	27		46.80	63
6724	3" pipe size		15	.533		78	30.50		108.50	133
6726	4" pipe size	↓	14	.571	↓	161	33		194	227
6732	PVC DWV P trap, solvent weld joint									
6733	1-1/2" pipe size	1 Plum	18	.444	Ea.	11.95	25.50		37.45	51.50
6734	2" pipe size		17	.471		16.15	27		43.15	59
6735	3" pipe size		15	.533		54.50	30.50		85	107
6736	4" pipe size		14	.571		125	33		158	187
6760	PP DWV, dilution trap, 1-1/2" pipe size		16	.500		222	29		251	288
6770	P trap, 1-1/2" pipe size		17	.471		54.50	27		81.50	101
6780	2" pipe size		16	.500		80.50	29		109.50	133
6790	3" pipe size		14	.571		148	33		181	213
6800	4" pipe size		13	.615		249	35.50		284.50	330
6830	S trap, 1-1/2" pipe size		16	.500		45	29		74	93
6840	2" pipe size		15	.533		67.50	30.50		98	121
6850	Universal trap, 1-1/2" pipe size		14	.571		89	33		122	148
6860	PVC DWV hub x hub, basin trap, 1-1/4" pipe size	↓	18	.444		14.60	25.50		40.10	54.50

22 13 Facility Sanitary Sewerage

22 13 16 – Sanitary Waste and Vent Piping

22 13 16.60 Traps

		Crew	Daily Output	Labor-Hours	Unit	Material	2014 Bare Costs Labor	Equipment	Total	Total Incl O&P
6870	Sink P trap, 1-1/2" pipe size	1 Plum	18	.444	Ea.	12.25	25.50		37.75	52
6880	Tubular S trap, 1-1/2" pipe size	↓	17	.471	↓	22.50	27		49.50	66
6890	PVC sch. 40 DWV, drum trap									
6900	1-1/2" pipe size	1 Plum	16	.500	Ea.	49.50	29		78.50	98
6910	P trap, 1-1/2" pipe size		18	.444		12.25	25.50		37.75	52
6920	2" pipe size		17	.471		16.45	27		43.45	59
6930	3" pipe size		15	.533		35.50	30.50		66	85.50
6940	4" pipe size		14	.571		128	33		161	190
6950	P trap w/clean out, 1-1/2" pipe size		18	.444		20	25.50		45.50	60.50
6960	2" pipe size		17	.471		34	27		61	78.50
6970	P trap adjustable, 1-1/2" pipe size		17	.471		22.50	27		49.50	66
6980	P trap adj. w/union & cleanout, 1-1/2" pipe size		16	.500		59.50	29		88.50	109
7000	Trap primer, flow through type, 1/2" diameter		24	.333		45.50	19.20		64.70	79
7100	With sediment strainer	↓	22	.364	↓	49.50	21		70.50	86
7450	Trap primer distribution unit									
7500	2 openings	1 Plum	18	.444	Ea.	23.50	25.50		49	64.50
7540	3 openings		17	.471		30	27		57	74
7560	4 openings	↓	16	.500		35.50	29		64.50	82.50
7850	Trap primer manifold									
7900	2 outlet	1 Plum	18	.444	Ea.	58.50	25.50		84	103
7940	4 outlet		16	.500		94.50	29		123.50	148
7960	6 outlet		15	.533		131	30.50		161.50	191
7980	8 outlet	↓	13	.615	↓	166	35.50		201.50	237

22 13 16.80 Vent Flashing and Caps

		Crew	Daily Output	Labor-Hours	Unit	Material	Labor	Equipment	Total	Total Incl O&P
0010	**VENT FLASHING AND CAPS**									
0120	Vent caps									
0140	Cast iron									
0160	1-1/4" - 1-1/2" pipe	1 Plum	23	.348	Ea.	34	20		54	67.50
0170	2" - 2-1/8" pipe		22	.364		39.50	21		60.50	75
0180	2-1/2" - 3-5/8" pipe		21	.381		45	22		67	82.50
0190	4" - 4-1/8" pipe		19	.421		55	24		79	97
0200	5" - 6" pipe	↓	17	.471	↓	83.50	27		110.50	133
0300	PVC									
0320	1-1/4" - 1-1/2" pipe	1 Plum	24	.333	Ea.	9.20	19.20		28.40	39
0330	2" - 2-1/8" pipe	"	23	.348	"	9.30	20		29.30	40
0900	Vent flashing									
1000	Aluminum with lead ring									
1020	1-1/4" pipe	1 Plum	20	.400	Ea.	11.80	23		34.80	48
1030	1-1/2" pipe		20	.400		14.15	23		37.15	50.50
1040	2" pipe		18	.444		12.70	25.50		38.20	52.50
1050	3" pipe		17	.471		14.05	27		41.05	56.50
1060	4" pipe	↓	16	.500	↓	16.95	29		45.95	62
1350	Copper with neoprene ring									
1400	1-1/4" pipe	1 Plum	20	.400	Ea.	20.50	23		43.50	57.50
1430	1-1/2" pipe		20	.400		20.50	23		43.50	57.50
1440	2" pipe		18	.444		21.50	25.50		47	62
1450	3" pipe		17	.471		25.50	27		52.50	69
1460	4" pipe	↓	16	.500	↓	28	29		57	74
2000	Galvanized with neoprene ring									
2020	1-1/4" pipe	1 Plum	20	.400	Ea.	16.95	23		39.95	53.50
2030	1-1/2" pipe		20	.400		16.95	23		39.95	53.50
2040	2" pipe	↓	18	.444	↓	17.55	25.50		43.05	58

22 13 Facility Sanitary Sewerage

22 13 16 – Sanitary Waste and Vent Piping

22 13 16.80 Vent Flashing and Caps

		Crew	Daily Output	Labor-Hours	Unit	Material	2014 Bare Costs Labor	Equipment	Total	Total Incl O&P
2050	3" pipe	1 Plum	17	.471	Ea.	18.55	27		45.55	61.50
2060	4" pipe	↓	16	.500	↓	21.50	29		50.50	67
2980	Neoprene, one piece									
3000	1-1/4" pipe	1 Plum	24	.333	Ea.	7.45	19.20		26.65	37
3030	1-1/2" pipe		24	.333		7.50	19.20		26.70	37.50
3040	2" pipe		23	.348		7.50	20		27.50	38.50
3050	3" pipe		21	.381		8.65	22		30.65	42.50
3060	4" pipe	↓	20	.400	↓	12.85	23		35.85	49
4000	Lead, 4#, 8" skirt, vent through roof									
4100	2" pipe	1 Plum	18	.444	Ea.	41.50	25.50		67	84.50
4110	3" pipe		17	.471		47.50	27		74.50	93
4120	4" pipe		16	.500		57	29		86	106
4130	6" pipe	↓	14	.571		81	33		114	139

22 13 19 – Sanitary Waste Piping Specialties

22 13 19.13 Sanitary Drains

		Crew	Daily Output	Labor-Hours	Unit	Material	Labor	Equipment	Total	Total Incl O&P
0010	**SANITARY DRAINS**									
0400	Deck, auto park, C.I., 13" top									
0440	3", 4", 5", and 6" pipe size	Q-1	8	2	Ea.	1,325	104		1,429	1,625
0480	For galvanized body, add				"	720			720	790
0800	Promenade, heelproof grate, C.I., 14" top									
0840	2", 3", and 4" pipe size	Q-1	10	1.600	Ea.	510	83		593	685
0860	5" and 6" pipe size		9	1.778		635	92		727	840
0880	8" pipe size	↓	8	2		750	104		854	980
0940	For galvanized body, add					380			380	415
0960	With polished bronze top, 2"-3"-4" diam.				↓	860			860	945
1200	Promenade, heelproof grate, C.I., lateral, 14" top									
1240	2", 3" and 4" pipe size	Q-1	10	1.600	Ea.	665	83		748	860
1260	5" and 6" pipe size		9	1.778		790	92		882	1,000
1280	8" pipe size	↓	8	2		910	104		1,014	1,150
1340	For galvanized body, add					380			380	415
1360	For polished bronze top, add				↓	350			350	385
1500	Promenade, slotted grate, C.I., 11" top									
1540	2", 3", 4", 5", and 6" pipe size	Q-1	12	1.333	Ea.	460	69		529	610
1600	For galvanized body, add					213			213	234
1640	With polished bronze top				↓	730			730	805
2000	Floor, medium duty, C.I., deep flange, 7" diam. top									
2040	2" and 3" pipe size	Q-1	12	1.333	Ea.	186	69		255	310
2080	For galvanized body, add					89.50			89.50	98.50
2120	With polished bronze top				↓	291			291	320
2160	Heavy duty, C.I., 12" dia anti-tilt grate									
2180	2", 3", 4", 5" and 6" pipe size	Q-1	10	1.600	Ea.	540	83		623	720
2220	For galvanized body, add					274			274	300
2240	With polished bronze top				↓	810			810	890
2300	X-Heavy duty, C.I., 15" antitilt grate									
2320	4", 5", 6", and 8" pipe size	Q-1	8	2	Ea.	1,150	104		1,254	1,425
2360	For galvanized body, add				↓	445			445	490
2380	With polished bronze top					1,625			1,625	1,775
2400	Heavy duty, with sediment bucket, C.I., 12" diam. loose grate									
2420	2", 3", 4", 5", and 6" pipe size	Q-1	9	1.778	Ea.	640	92		732	845
2440	For galvanized body, add					405			405	450
2460	With polished bronze top				↓	905			905	995
2500	Heavy duty, cleanout & trap w/bucket, C.I., 15" top									

22 13 Facility Sanitary Sewerage

22 13 19 – Sanitary Waste Piping Specialties

22 13 19.13 Sanitary Drains

		Crew	Daily Output	Labor-Hours	Unit	Material	2014 Bare Costs Labor	Equipment	Total	Total Incl O&P
2540	2", 3", and 4" pipe size	Q-1	6	2.667	Ea.	6,075	138		6,213	6,900
2560	For galvanized body, add					1,550			1,550	1,725
2580	With polished bronze top					6,575			6,575	7,250
2600	Medium duty, with perforated SS basket, C.I., body,									
2610	18" top for refuse container washing area									
2620	2" thru 6" pipe size	Q-1	4	4	Ea.	4,025	207		4,232	4,750
2630	Acid resistant									
2638	PVC									
2640	2", 3" and 4" pipe size	Q-1	16	1	Ea.	310	52		362	420
2644	Cast iron, epoxy coated									
2646	2", 3" and 4" pipe size	Q-1	14	1.143	Ea.	435	59		494	570
2650	PVC or ABS thermoplastic									
2660	3" and 4" pipe size	Q-1	16	1	Ea.	340	52		392	450
2680	Extra heavy duty, oil intercepting, gas seal cone,									
2690	with cleanout, loose grate, C.I., body 16" top									
2700	3" and 4" diameter outlet, 4" slab depth	Q-1	4	4	Ea.	6,575	207		6,782	7,575
2720	4" diameter outlet, 8" slab depth		3	5.333		6,575	276		6,851	7,675
2740	4" diam. outlet, 10"-12" slab depth, 16"top		2	8		7,400	415		7,815	8,775
2910	Prison cell, vandal-proof, 1-1/2", and 2" diam. pipe		12	1.333		410	69		479	560
2920	3" pipe size		10	1.600		460	83		543	630
2930	Trap drain, light duty, backwater valve C.I. top									
2950	8" diameter top, 2" pipe size	Q-1	12	1.333	Ea.	360	69		429	500
2960	10" diameter top, 3" pipe size		10	1.600		495	83		578	670
2970	12" diameter top, 4" pipe size		8	2		690	104		794	915

22 13 19.14 Floor Receptors

		Crew	Daily Output	Labor-Hours	Unit	Material	Labor	Equipment	Total	Total Incl O&P
0010	**FLOOR RECEPTORS**, For connection to 2", 3" & 4" diameter pipe									
0200	12-1/2" square top, 25 sq. in. open area	Q-1	10	1.600	Ea.	900	83		983	1,125
0300	For grate with 4" diam. x 3-3/4" high funnel, add					161			161	177
0400	For grate with 6" diameter x 6" high funnel, add					207			207	227
0500	For full hinged grate with open center, add					75			75	82.50
0600	For aluminum bucket, add					135			135	149
0700	For acid-resisting bucket, add					253			253	279
0900	For stainless steel mesh bucket liner, add					200			200	220
1000	For bronze antisplash dome strainer, add					96.50			96.50	106
1100	For partial solid cover, add					48.50			48.50	53
1200	For trap primer connection, add					82			82	90
2000	12-5/8" diameter top, 40 sq. in. open area	Q-1	10	1.600		715	83		798	910
2100	For options, add same prices as square top									
3000	8" x 4" rectangular top, 7.5 sq. in. open area	Q-1	14	1.143	Ea.	690	59		749	850
3100	For trap primer connections, add					82			82	90
4000	24" x 16" rectangular top, 70 sq. in. open area	Q-1	4	4		4,025	207		4,232	4,775
4100	For trap primer connection, add					161			161	177

22 13 19.15 Sink Waste Treatment

		Crew	Daily Output	Labor-Hours	Unit	Material	Labor	Equipment	Total	Total Incl O&P
0010	**SINK WASTE TREATMENT**, System for commercial kitchens									
0100	includes clock timer, & fittings									
0200	System less chemical, wall mounted cabinet	1 Plum	16	.500	Ea.	550	29		579	650
2000	Chemical, 1 gallon, add					68.50			68.50	75.50
2100	6 gallons, add					310			310	340
2200	15 gallons, add					840			840	925
2300	30 gallons, add					1,575			1,575	1,725
2400	55 gallons, add					2,675			2,675	2,950

22 13 Facility Sanitary Sewerage

22 13 23 – Sanitary Waste Interceptors

22 13 23.10 Interceptors		Crew	Daily Output	Labor-Hours	Unit	Material	2014 Bare Costs Labor	Equipment	Total	Total Incl O&P
0010	**INTERCEPTORS**									
0150	Grease, fabricated steel, 4 GPM, 8 lb. fat capacity	1 Plum	4	2	Ea.	1,100	115		1,215	1,375
0200	7 GPM, 14 lb. fat capacity		4	2		1,525	115		1,640	1,850
1000	10 GPM, 20 lb. fat capacity		4	2		1,800	115		1,915	2,150
1040	15 GPM, 30 lb. fat capacity		4	2		2,650	115		2,765	3,100
1060	20 GPM, 40 lb. fat capacity		3	2.667		3,250	153		3,403	3,800
1080	25 GPM, 50 lb. fat capacity	Q-1	3.50	4.571		3,650	237		3,887	4,375
1100	35 GPM, 70 lb. fat capacity		3	5.333		4,525	276		4,801	5,400
1120	50 GPM, 100 lb. fat capacity		2	8		9,575	415		9,990	11,100
1140	75 GPM, 150 lb. fat capacity		2	8		12,100	415		12,515	13,900
1160	100 GPM, 200 lb. fat capacity		2	8		13,600	415		14,015	15,500
1180	150 GPM, 300 lb. fat capacity		2	8		15,100	415		15,515	17,200
1200	200 GPM, 400 lb. fat capacity		1.50	10.667		21,900	555		22,455	24,900
1220	250 GPM, 500 lb. fat capacity		1.30	12.308		25,500	640		26,140	29,000
1240	300 GPM, 600 lb. fat capacity		1	16		30,000	830		30,830	34,300
1260	400 GPM, 800 lb. fat capacity	Q-2	1.20	20		36,700	1,075		37,775	42,000
1280	500 GPM, 1000 lb. fat capacity	"	1	24		43,500	1,300		44,800	49,800
1580	For seepage pan, add					7%				
3000	Hair, cast iron, 1-1/4" and 1-1/2" pipe connection	1 Plum	8	1	Ea.	415	57.50		472.50	540
3100	For chrome-plated cast iron, add					253			253	279
3200	For polished bronze, add					253			253	279
3400	Lint interceptor, fabricated steel									
3410	Size based on 10 GPM per machine									
3420	30 GPM, 2" pipe size	Q-1	3	5.333	Ea.	4,850	276		5,126	5,750
3430	70 GPM, 3" pipe size		2.50	6.400		5,625	330		5,955	6,700
3440	100 GPM, 4" pipe size		2	8		6,625	415		7,040	7,925
3450	200 GPM, 4" pipe size		1.50	10.667		8,025	555		8,580	9,650
3460	300 GPM, 6" pipe size		1	16		9,325	830		10,155	11,600
3470	400 GPM, 6" pipe size	Q-2	1.20	20		10,600	1,075		11,675	13,300
3480	500 GPM, 6" pipe size	"	1	24		11,900	1,300		13,200	15,100
4000	Oil, fabricated steel, 10 GPM, 2" pipe size	1 Plum	4	2		2,450	115		2,565	2,875
4100	15 GPM, 2" or 3" pipe size		4	2		3,375	115		3,490	3,875
4120	20 GPM, 2" or 3" pipe size		3	2.667		4,075	153		4,228	4,700
4140	25 GPM, 2" or 3" pipe size	Q-1	3.50	4.571		4,425	237		4,662	5,225
4160	35 GPM, 2", 3", or 4" pipe size		3	5.333		5,375	276		5,651	6,350
4180	50 GPM, 2", 3", or 4" pipe size		2	8		7,225	415		7,640	8,575
4200	75 GPM, 3" pipe size		2	8		12,500	415		12,915	14,400
4220	100 GPM, 3" pipe size		2	8		13,600	415		14,015	15,500
4240	150 GPM, 4" pipe size		2	8		16,800	415		17,215	19,100
4260	200 GPM, 4" pipe size		1.50	10.667		23,800	555		24,355	26,900
4280	250 GPM, 5" pipe size		1.30	12.308		27,500	640		28,140	31,300
4300	300 GPM, 5" pipe size		1	16		31,200	830		32,030	35,600
4320	400 GPM, 6" pipe size	Q-2	1.20	20		40,100	1,075		41,175	45,700
4340	500 GPM, 6" pipe size	"	1	24		50,500	1,300		51,800	57,500
5000	Sand interceptor, fabricated steel									
5020	20 GPM, 4" pipe size	Q-1	3	5.333	Ea.	6,050	276		6,326	7,100
5030	50 GPM, 4" pipe size		2.50	6.400		10,300	330		10,630	11,800
5040	150 GPM, 4" pipe size		2	8		31,000	415		31,415	34,800
5050	250 GPM, 6" pipe size		1.30	12.308		32,300	640		32,940	36,500
5060	500 GPM, 6" pipe size	Q-2	1	24		41,600	1,300		42,900	47,700
6000	Solids, precious metals recovery, C.I., 1-1/4" to 2" pipe	1 Plum	4	2		620	115		735	860
6100	Dental Lab., large, C.I., 1-1/2" to 2" pipe	"	3	2.667		2,175	153		2,328	2,625

22 13 Facility Sanitary Sewerage

22 13 26 – Sanitary Waste Separators

22 13 26.10 Separators

		Crew	Daily Output	Labor-Hours	Unit	Material	2014 Bare Costs Labor	Equipment	Total	Total Incl O&P
0010	**SEPARATORS**, Entrainment eliminator, steel body, 150 PSIG									
0100	1/4" size	1 Stpi	24	.333	Ea.	259	19.50		278.50	315
0120	1/2" size		24	.333		268	19.50		287.50	325
0140	3/4" size		20	.400		280	23.50		303.50	345
0160	1" size		19	.421		289	24.50		313.50	355
0180	1-1/4" size		15	.533		305	31		336	385
0200	1-1/2" size		13	.615		340	36		376	425
0220	2" size		11	.727		375	42.50		417.50	480

22 13 29 – Sanitary Sewerage Pumps

22 13 29.13 Wet-Pit-Mounted, Vertical Sewerage Pumps

		Crew	Daily Output	Labor-Hours	Unit	Material	Labor	Equipment	Total	Total Incl O&P
0010	**WET-PIT-MOUNTED, VERTICAL SEWERAGE PUMPS**									
0020	Controls incl. alarm/disconnect panel w/wire. Excavation not included									
0260	Simplex, 9 GPM at 60 PSIG, 91 gal. tank				Ea.	3,325			3,325	3,650
0300	Unit with manway, 26" I.D., 18" high					3,700			3,700	4,050
0340	26" I.D., 36" high					3,750			3,750	4,125
0380	43" I.D., 4' high					3,975			3,975	4,375
0600	Simplex, 9 GPM at 60 PSIG, 150 gal. tank, indoor					3,525			3,525	3,875
0700	Unit with manway, 26" I.D., 36" high					4,225			4,225	4,650
0740	26" I.D., 4' high					4,450			4,450	4,900
2000	Duplex, 18 GPM at 60 PSIG, 150 gal. tank, indoor					7,025			7,025	7,750
2060	Unit with manway, 43" I.D., 4' high					8,100			8,100	8,900
2400	For core only					1,875			1,875	2,050
3000	Indoor residential type installation									
3020	Simplex, 9 GPM at 60 PSIG, 91 gal. HDPE tank				Ea.	3,350			3,350	3,675

22 13 29.14 Sewage Ejector Pumps

		Crew	Daily Output	Labor-Hours	Unit	Material	Labor	Equipment	Total	Total Incl O&P
0010	**SEWAGE EJECTOR PUMPS**, With operating and level controls									
0100	Simplex system incl. tank, cover, pump 15' head									
0500	37 gal. PE tank, 12 GPM, 1/2 HP, 2" discharge	Q-1	3.20	5	Ea.	480	259		739	915
0510	3" discharge		3.10	5.161		520	267		787	975
0530	87 GPM, .7 HP, 2" discharge		3.20	5		735	259		994	1,200
0540	3" discharge		3.10	5.161		795	267		1,062	1,275
0600	45 gal. coated stl. tank, 12 GPM, 1/2 HP, 2" discharge		3	5.333		855	276		1,131	1,350
0610	3" discharge		2.90	5.517		890	286		1,176	1,400
0630	87 GPM, .7 HP, 2" discharge		3	5.333		1,100	276		1,376	1,625
0640	3" discharge		2.90	5.517		1,150	286		1,436	1,700
0660	134 GPM, 1 HP, 2" discharge		2.80	5.714		1,175	296		1,471	1,750
0680	3" discharge		2.70	5.926		1,250	305		1,555	1,850
0700	70 gal. PE tank, 12 GPM, 1/2 HP, 2" discharge		2.60	6.154		920	320		1,240	1,500
0710	3" discharge		2.40	6.667		980	345		1,325	1,600
0730	87 GPM, 0.7 HP, 2" discharge		2.50	6.400		1,200	330		1,530	1,800
0740	3" discharge		2.30	6.957		1,275	360		1,635	1,950
0760	134 GPM, 1 HP, 2" discharge		2.20	7.273		1,300	375		1,675	2,000
0770	3" discharge		2	8		1,375	415		1,790	2,150
0800	75 gal. coated stl. tank, 12 GPM, 1/2 HP, 2" discharge		2.40	6.667		1,025	345		1,370	1,650
0810	3" discharge		2.20	7.273		1,075	375		1,450	1,750
0830	87 GPM, .7 HP, 2" discharge		2.30	6.957		1,300	360		1,660	1,975
0840	3" discharge		2.10	7.619		1,350	395		1,745	2,100
0860	134 GPM, 1 HP, 2" discharge		2	8		1,375	415		1,790	2,150
0880	3" discharge		1.80	8.889		1,450	460		1,910	2,300
1040	Duplex system incl. tank, covers, pumps									
1060	110 gal. fiberglass tank, 24 GPM, 1/2 HP, 2" discharge	Q-1	1.60	10	Ea.	1,850	520		2,370	2,800

22 13 Facility Sanitary Sewerage

22 13 29 – Sanitary Sewerage Pumps

22 13 29.14 Sewage Ejector Pumps

		Crew	Daily Output	Labor-Hours	Unit	Material	2014 Bare Costs Labor	Equipment	Total	Total Incl O&P
1080	3" discharge	Q-1	1.40	11.429	Ea.	1,950	590		2,540	3,025
1100	174 GPM, .7 HP, 2" discharge		1.50	10.667		2,400	555		2,955	3,450
1120	3" discharge		1.30	12.308		2,475	640		3,115	3,700
1140	268 GPM, 1 HP, 2" discharge		1.20	13.333		2,600	690		3,290	3,900
1160	3" discharge		1	16		2,700	830		3,530	4,200
1260	135 gal. coated stl. tank, 24 GPM, 1/2 HP, 2" discharge	Q-2	1.70	14.118		1,900	760		2,660	3,250
2000	3" discharge		1.60	15		2,025	805		2,830	3,450
2640	174 GPM, .7 HP, 2" discharge		1.60	15		2,500	805		3,305	3,975
2660	3" discharge		1.50	16		2,625	860		3,485	4,200
2700	268 GPM, 1 HP, 2" discharge		1.30	18.462		2,700	990		3,690	4,475
3040	3" discharge		1.10	21.818		2,875	1,175		4,050	4,950
3060	275 gal. coated stl. tank, 24 GPM, 1/2 HP, 2" discharge		1.50	16		2,375	860		3,235	3,925
3080	3" discharge		1.40	17.143		2,425	920		3,345	4,050
3100	174 GPM, .7 HP, 2" discharge		1.40	17.143		3,075	920		3,995	4,775
3120	3" discharge		1.30	18.462		3,250	990		4,240	5,075
3140	268 GPM, 1 HP, 2" discharge		1.10	21.818		3,375	1,175		4,550	5,475
3160	3" discharge		.90	26.667		3,550	1,425		4,975	6,075
3260	Pump system accessories, add									
3300	Alarm horn and lights, 115 V mercury switch	Q-1	8	2	Ea.	97	104		201	262
3340	Switch, mag. contactor, alarm bell, light, 3 level control		5	3.200		475	166		641	775
3380	Alternator, mercury switch activated		4	4		855	207		1,062	1,250

22 14 Facility Storm Drainage

22 14 23 – Storm Drainage Piping Specialties

22 14 23.33 Backwater Valves

		Crew	Daily Output	Labor-Hours	Unit	Material	2014 Bare Costs Labor	Equipment	Total	Total Incl O&P
0010	**BACKWATER VALVES**, C.I. Body									
6980	Bronze gate and automatic flapper valves									
7000	3" and 4" pipe size	Q-1	13	1.231	Ea.	1,900	64		1,964	2,175
7100	5" and 6" pipe size	"	13	1.231	"	2,900	64		2,964	3,300
7240	Bronze flapper valve, bolted cover									
7260	2" pipe size	Q-1	16	1	Ea.	555	52		607	690
7280	3" pipe size		14.50	1.103		865	57		922	1,025
7300	4" pipe size		13	1.231		1,075	64		1,139	1,275
7320	5" pipe size	Q-2	18	1.333		1,275	71.50		1,346.50	1,500
7340	6" pipe size	"	17	1.412		1,525	76		1,601	1,800
7360	8" pipe size	Q-3	10	3.200		2,150	175		2,325	2,650
7380	10" pipe size	"	9	3.556		3,525	195		3,720	4,175
7500	For threaded cover, same cost									
7540	Revolving disk type, same cost as flapper type									

22 14 26 – Facility Storm Drains

22 14 26.13 Roof Drains

		Crew	Daily Output	Labor-Hours	Unit	Material	2014 Bare Costs Labor	Equipment	Total	Total Incl O&P
0010	**ROOF DRAINS**									
0140	Cornice, C.I., 45° or 90° outlet									
0200	3" and 4" pipe size	Q-1	12	1.333	Ea.	360	69		429	500
0260	For galvanized body, add					70			70	77
0280	For polished bronze dome, add					87.50			87.50	96.50
3860	Roof, flat metal deck, C.I. body, 12" C.I. dome									
3880	2" pipe size	Q-1	15	1.067	Ea.	266	55.50		321.50	375
3890	3" pipe size		14	1.143		365	59		424	490
3900	4" pipe size		13	1.231		365	64		429	495

22 14 Facility Storm Drainage

22 14 26 – Facility Storm Drains

22 14 26.13 Roof Drains

		Crew	Daily Output	Labor-Hours	Unit	Material	2014 Bare Costs Labor	Equipment	Total	Total Incl O&P
3910	5" pipe size	Q-1	12	1.333	Ea.	495	69		564	650
3920	6" pipe size	↓	10	1.600	↓	635	83		718	825
4280	Integral expansion joint, C.I. body, 12" C.I. dome									
4300	2" pipe size	Q-1	8	2	Ea.	545	104		649	755
4320	3" pipe size		7	2.286		565	118		683	800
4340	4" pipe size		6	2.667		605	138		743	875
4360	5" pipe size		4	4		760	207		967	1,150
4380	6" pipe size		3	5.333		830	276		1,106	1,325
4400	8" pipe size	↓	3	5.333	↓	1,275	276		1,551	1,825
4440	For galvanized body, add					340			340	375
4620	Main, all aluminum, 12" low profile dome									
4640	2", 3" and 4" pipe size	Q-1	14	1.143	Ea.	405	59		464	535
4660	5" and 6" pipe size		13	1.231		535	64		599	680
4680	8" pipe size		10	1.600		665	83		748	855
4690	Main, CI body, 12" poly. dome, 2", 3", & 4" pipe		8	2		300	104		404	485
4710	5" and 6" pipe size		6	2.667		430	138		568	685
4720	8" pipe size		4	4		555	207		762	930
4730	For underdeck clamp, add	↓	22	.727		207	37.50		244.50	285
4740	For vandalproof dome, add					52.50			52.50	58
4750	For galvanized body, add					405			405	450
4760	Main, PVC body and dome, 2" pipe size	Q-1	14	1.143		131	59		190	234
4780	3" pipe size		14	1.143		131	59		190	234
4800	4" pipe size		14	1.143		131	59		190	234
4820	For underdeck clamp, add	↓	24	.667	↓	27	34.50		61.50	82
4900	Terrace planting area, with perforated overflow, C.I.									
4920	2", 3" and 4" pipe size	Q-1	8	2	Ea.	570	104		674	780

22 14 26.16 Facility Area Drains

		Crew	Daily Output	Labor-Hours	Unit	Material	Labor	Equipment	Total	Total Incl O&P
0010	**FACILITY AREA DRAINS**									
4980	Scupper floor, oblique strainer, C.I.									
5000	6" x 7" top, 2", 3" and 4" pipe size	Q-1	16	1	Ea.	262	52		314	365
5100	8" x 12" top, 5" and 6" pipe size	"	14	1.143		510	59		569	650
5160	For galvanized body, add					40%				
5200	For polished bronze strainer, add				↓	85%				

22 14 26.19 Facility Trench Drains

		Crew	Daily Output	Labor-Hours	Unit	Material	Labor	Equipment	Total	Total Incl O&P
0010	**FACILITY TRENCH DRAINS**									
5980	Trench, floor, heavy duty, modular, C.I., 12" x 12" top									
6000	2", 3", 4", 5", & 6" pipe size	Q-1	8	2	Ea.	830	104		934	1,075
6100	For unit with polished bronze top		8	2		1,250	104		1,354	1,500
6200	For 12" extension section, C.I. top		8	2		830	104		934	1,075
6240	For 12" extension section, polished bronze top	↓	8	2	↓	1,300	104		1,404	1,575
6600	Trench, floor, for cement concrete encasement									
6610	Not including trenching or concrete									
6640	Polyester polymer concrete									
6650	4" internal width, with grate									
6660	Light duty steel grate	Q-1	120	.133	L.F.	34	6.90		40.90	48
6670	Medium duty steel grate		115	.139		39	7.20		46.20	54
6680	Heavy duty iron grate	↓	110	.145	↓	60	7.55		67.55	77.50
6700	12" internal width, with grate									
6770	Heavy duty galvanized grate	Q-1	80	.200	L.F.	164	10.35		174.35	196
6800	Fiberglass									
6810	8" internal width, with grate									
6820	Medium duty galvanized grate	Q-1	115	.139	L.F.	96.50	7.20		103.70	117

22 14 Facility Storm Drainage

22 14 26 – Facility Storm Drains

22 14 26.19 Facility Trench Drains		Crew	Daily Output	Labor-Hours	Unit	Material	2014 Bare Costs Labor	Equipment	Total	Total Incl O&P
6830	Heavy duty iron grate	Q-1	110	.145	L.F.	102	7.55		109.55	124

22 14 29 – Sump Pumps

22 14 29.13 Wet-Pit-Mounted, Vertical Sump Pumps

		Crew	Daily Output	Labor-Hours	Unit	Material	Labor	Equipment	Total	Incl O&P
0010	**WET-PIT-MOUNTED, VERTICAL SUMP PUMPS**									
0400	Molded PVC base, 21 GPM at 15' head, 1/3 HP	1 Plum	5	1.600	Ea.	135	92		227	288
0800	Iron base, 21 GPM at 15' head, 1/3 HP		5	1.600		164	92		256	320
1200	Solid brass, 21 GPM at 15' head, 1/3 HP		5	1.600		289	92		381	460
2000	Sump pump, single stage									
2010	25 GPM, 1 HP, 1-1/2" discharge	Q-1	1.80	8.889	Ea.	3,625	460		4,085	4,675
2020	75 GPM, 1-1/2 HP, 2" discharge		1.50	10.667		3,825	555		4,380	5,025
2030	100 GPM, 2 HP, 2-1/2" discharge		1.30	12.308		3,900	640		4,540	5,250
2040	150 GPM, 3 HP, 3" discharge		1.10	14.545		3,900	755		4,655	5,425
2050	200 GPM, 3 HP, 3" discharge		1	16		4,125	830		4,955	5,800
2060	300 GPM, 10 HP, 4" discharge	Q-2	1.20	20		4,450	1,075		5,525	6,525
2070	500 GPM, 15 HP, 5" discharge		1.10	21.818		5,075	1,175		6,250	7,350
2080	800 GPM, 20 HP, 6" discharge		1	24		6,000	1,300		7,300	8,550
2090	1000 GPM, 30 HP, 6" discharge		.85	28.235		6,575	1,525		8,100	9,550
2100	1600 GPM, 50 HP, 8" discharge		.72	33.333		10,300	1,800		12,100	14,000
2110	2000 GPM, 60 HP, 8" discharge	Q-3	.85	37.647		10,500	2,075		12,575	14,700
2202	For general purpose float switch, copper coated float, add	Q-1	5	3.200		108	166		274	370

22 14 29.16 Submersible Sump Pumps

		Crew	Daily Output	Labor-Hours	Unit	Material	Labor	Equipment	Total	Incl O&P
0010	**SUBMERSIBLE SUMP PUMPS**									
7000	Sump pump, automatic									
7100	Plastic, 1-1/4" discharge, 1/4 HP	1 Plum	6	1.333	Ea.	134	76.50		210.50	263
7140	1/3 HP		5	1.600		194	92		286	350
7160	1/2 HP		5	1.600		238	92		330	400
7180	1-1/2" discharge, 1/2 HP		4	2		272	115		387	475
7500	Cast iron, 1-1/4" discharge, 1/4 HP		6	1.333		189	76.50		265.50	325
7540	1/3 HP		6	1.333		222	76.50		298.50	360
7560	1/2 HP		5	1.600		269	92		361	435

22 14 53 – Rainwater Storage Tanks

22 14 53.13 Fiberglass, Rainwater Storage Tank

		Crew	Daily Output	Labor-Hours	Unit	Material	Labor	Equipment	Total	Incl O&P
0010	**FIBERGLASS, RAINWATER STORAGE TANK**									
2000	600 gallon	B-21B	3.75	10.667	Ea.	3,150	425	177	3,752	4,300
2010	1,000 gallon		3.50	11.429		4,050	455	189	4,694	5,350
2020	2,000 gallon		3.25	12.308		5,900	490	204	6,594	7,450
2030	4,000 gallon		3	13.333		8,175	530	221	8,926	10,000
2040	6,000 gallon		2.65	15.094		9,175	600	250	10,025	11,300
2050	8,000 gallon		2.30	17.391		11,000	690	288	11,978	13,500
2060	10,000 gallon		2	20		12,600	795	330	13,725	15,500
2070	12,000 gallon		1.50	26.667		17,400	1,050	440	18,890	21,200
2080	15,000 gallon		1	40		20,200	1,600	665	22,465	25,400
2090	20,000 gallon		.75	53.333		26,000	2,125	885	29,010	32,800
2100	25,000 gallon		.50	80		38,800	3,175	1,325	43,300	49,100
2110	30,000 gallon		.35	114		47,100	4,550	1,900	53,550	61,000
2120	40,000 gallon		.30	133		68,000	5,300	2,200	75,500	85,500

22 15 General Service Compressed-Air Systems

22 15 13 – General Service Compressed-Air Piping

22 15 13.10 Compressor Accessories

		Crew	Daily Output	Labor-Hours	Unit	Material	2014 Bare Costs Labor	Equipment	Total	Total Incl O&P
0010	**COMPRESSOR ACCESSORIES**									
4000	Couplers, air line, sleeve type									
4010	Female, connection size NPT									
4020	1/4"	1 Stpi	38	.211	Ea.	9	12.30		21.30	28.50
4030	3/8"		36	.222		16.60	13		29.60	38
4040	1/2"		35	.229		24	13.35		37.35	46.50
4050	3/4"		34	.235		24	13.75		37.75	47.50
4100	Male									
4110	1/4"	1 Stpi	38	.211	Ea.	9	12.30		21.30	28.50
4120	3/8"		36	.222		11.35	13		24.35	32
4130	1/2"		35	.229		23.50	13.35		36.85	45.50
4140	3/4"		34	.235		24	13.75		37.75	47.50
4150	Coupler, combined male and female halves									
4160	1/2"	1 Stpi	17	.471	Ea.	47.50	27.50		75	94
4170	3/4"	"	15	.533	"	48.50	31		79.50	100

22 15 19 – General Service Packaged Air Compressors and Receivers

22 15 19.10 Air Compressors

		Crew	Daily Output	Labor-Hours	Unit	Material	Labor	Equipment	Total	Total Incl O&P
0010	**AIR COMPRESSORS**									
5250	Air, reciprocating air cooled, splash lubricated, tank mounted									
5300	Single stage, 1 phase, 140 psi									
5303	1/2 HP, 30 gal. tank	1 Stpi	3	2.667	Ea.	1,825	156		1,981	2,250
5305	3/4 HP, 30 gal. tank		2.60	3.077		1,875	180		2,055	2,325
5307	1 HP, 30 gal. tank		2.20	3.636		2,300	213		2,513	2,850
5309	2 HP, 30 gal. tank	Q-5	4	4		2,800	211		3,011	3,400
5310	3 HP, 30 gal. tank		3.60	4.444		3,150	234		3,384	3,800
5314	3 HP, 60 gal. tank		3.50	4.571		3,550	241		3,791	4,275
5320	5 HP, 60 gal. tank		3.20	5		3,725	263		3,988	4,500
5330	5 HP, 80 gal. tank		3	5.333		4,100	281		4,381	4,925
5340	7.5 HP, 80 gal. tank		2.60	6.154		5,300	325		5,625	6,325
5600	2 stage pkg., 3 phase									
5650	6 CFM at 125 psi 1-1/2 HP, 60 gal. tank	Q-5	3	5.333	Ea.	3,375	281		3,656	4,125
5670	10.9 CFM at 125 psi, 3 HP, 80 gal. tank		1.50	10.667		3,800	560		4,360	5,025
5680	38.7 CFM at 125 psi, 10 HP, 120 gal. tank		.60	26.667		6,825	1,400		8,225	9,625
5690	105 CFM at 125 psi, 25 HP, 250 gal. tank	Q-6	.60	40		14,100	2,175		16,275	18,800
5800	With single stage pump									
5850	8.3 CFM at 125 psi, 2 HP, 80 gal. tank	Q-6	3.50	6.857	Ea.	3,775	375		4,150	4,725
5860	38.7 CFM at 125 psi, 10 HP, 120 gal. tank	"	.90	26.667	"	7,525	1,450		8,975	10,500
6000	Reciprocating, 2 stage, tank mtd, 3 Ph., Cap. rated @175 PSIG									
6050	Pressure lubricated, hvy. duty, 9.7 CFM, 3 HP, 120 gal. tank	Q-5	1.30	12.308	Ea.	5,775	650		6,425	7,325
6054	5 CFM, 1-1/2 HP, 80 gal. tank		2.80	5.714		3,575	300		3,875	4,400
6056	6.4 CFM, 2 HP, 80 gal. tank		2	8		3,775	420		4,195	4,775
6058	8.1 CFM, 3 HP, 80 gal. tank		1.70	9.412		3,800	495		4,295	4,925
6059	14.8 CFM, 5 HP, 80 gal. tank		1	16		4,100	840		4,940	5,775
6060	16.5 CFM, 5 HP, 120 gal. tank		1	16		5,900	840		6,740	7,750
6063	13 CFM, 6 HP, 80 gal. tank		.90	17.778		5,875	935		6,810	7,900
6066	19.8 CFM, 7.5 HP, 80 gal. tank		.80	20		5,875	1,050		6,925	8,075
6070	25.8 CFM, 7-1/2 HP, 120 gal. tank		.80	20		8,875	1,050		9,925	11,400
6078	34.8 CFM, 10 HP, 80 gal. tank		.70	22.857		8,875	1,200		10,075	11,600
6080	34.8 CFM, 10 HP, 120 gal. tank		.60	26.667		8,950	1,400		10,350	12,000
6090	53.7 CFM, 15 HP, 120 gal. tank	Q-6	.80	30		10,300	1,650		11,950	13,800
6100	76.7 CFM, 20 HP, 120 gal. tank		.70	34.286		13,500	1,875		15,375	17,600
6104	76.7 CFM, 20 HP, 240 gal. tank		.68	35.294		14,800	1,925		16,725	19,200

22 15 General Service Compressed-Air Systems

22 15 19 – General Service Packaged Air Compressors and Receivers

22 15 19.10 Air Compressors

		Crew	Daily Output	Labor-Hours	Unit	Material	2014 Bare Costs Labor	Equipment	Total	Total Incl O&P
6110	90.1 CFM, 25 HP, 120 gal. tank	Q-6	.63	38.095	Ea.	14,100	2,075		16,175	18,700
6120	101 CFM, 30 HP, 120 gal. tank		.57	42.105		15,200	2,300		17,500	20,200
6130	101 CFM, 30 HP, 250 gal. tank		.52	46.154		16,400	2,525		18,925	21,900
6200	Oil-less, 13.6 CFM, 5 HP, 120 gal. tank	Q-5	.88	18.182		16,900	955		17,855	20,100
6210	13.6 CFM, 5 HP, 250 gal. tank		.80	20		18,200	1,050		19,250	21,600
6220	18.2 CFM, 7.5 HP, 120 gal. tank		.73	21.918		16,900	1,150		18,050	20,400
6230	18.2 CFM, 7.5 HP, 250 gal. tank		.67	23.881		18,200	1,250		19,450	21,900
6250	30.5 CFM, 10 HP, 120 gal. tank		.57	28.070		19,900	1,475		21,375	24,100
6260	30.5 CFM, 10 HP, 250 gal. tank		.53	30.189		21,200	1,600		22,800	25,700
6270	41.3 CFM, 15 HP, 120 gal. tank	Q-6	.70	34.286		21,700	1,875		23,575	26,700
6280	41.3 CFM, 15 HP, 250 gal. tank	"	.67	35.821		22,900	1,950		24,850	28,200

22 31 Domestic Water Softeners

22 31 13 – Residential Domestic Water Softeners

22 31 13.10 Residential Water Softeners

		Crew	Daily Output	Labor-Hours	Unit	Material	Labor	Equipment	Total	Total Incl O&P
0010	**RESIDENTIAL WATER SOFTENERS**									
7350	Water softener, automatic, to 30 grains per gallon	2 Plum	5	3.200	Ea.	400	184		584	720
7400	To 100 grains per gallon	"	4	4	"	650	230		880	1,075

22 31 16 – Commercial Domestic Water Softeners

22 31 16.10 Water Softeners

		Crew	Daily Output	Labor-Hours	Unit	Material	Labor	Equipment	Total	Total Incl O&P
0010	**WATER SOFTENERS**									
5800	Softener systems, automatic, intermediate sizes									
5820	available, may be used in multiples.									
6000	Hardness capacity between regenerations and flow									
6060	40,000 grains, 14 GPM	Q-1	4	4	Ea.	1,625	207		1,832	2,125
6070	50,000 grains, 17 GPM		3.60	4.444		2,525	230		2,755	3,125
6080	90,000 grains, 25 GPM		2.80	5.714		5,300	296		5,596	6,300
6100	150,000 grains, 37 GPM cont., 51 GPM peak		1.20	13.333		5,300	690		5,990	6,900
6200	300,000 grains, 81 GPM cont., 113 GPM peak		1	16		9,375	830		10,205	11,600
6300	750,000 grains, 160 GPM cont., 230 GPM peak		.80	20		12,200	1,025		13,225	15,000
6400	900,000 grains, 185 GPM cont., 270 GPM peak		.70	22.857		19,700	1,175		20,875	23,400
8000	Water treatment, salts, 50 lb. bag									
8020	Salt, water softener, bag, pelletized				Lb.	.35			.35	.39
8030	Salt, water softener, bag, crystal rock salt				"	.31			.31	.34

22 32 Domestic Water Filtration Equipment

22 32 19 – Domestic-Water Off-Floor Cartridge Filters

22 32 19.10 Water Filters

		Crew	Daily Output	Labor-Hours	Unit	Material	Labor	Equipment	Total	Total Incl O&P
0010	**WATER FILTERS**, Purification and treatment.									
1000	Cartridge style, dirt and rust type	1 Plum	12	.667	Ea.	190	38.50		228.50	267
1200	Replacement cartridge		32	.250		16.65	14.40		31.05	40
1600	Taste and odor type		12	.667		225	38.50		263.50	305
1700	Replacement cartridge		32	.250		43	14.40		57.40	68.50
3000	Central unit, dirt/rust/odor/taste/scale		4	2		1,525	115		1,640	1,850
3100	Replacement cartridge, standard		20	.400		252	23		275	310
3600	Replacement cartridge, heavy duty		20	.400		199	23		222	254
8000	Commercial, fully automatic or push button automatic									
8200	Iron removal, 660 GPH, 1" pipe size	Q-1	1.50	10.667	Ea.	2,925	555		3,480	4,050
8240	1500 GPH, 1-1/4" pipe size		1	16		4,925	830		5,755	6,675

22 32 Domestic Water Filtration Equipment

22 32 19 – Domestic-Water Off-Floor Cartridge Filters

22 32 19.10 Water Filters

		Crew	Daily Output	Labor-Hours	Unit	Material	2014 Bare Costs Labor	Equipment	Total	Total Incl O&P
8280	2340 GPH, 1-1/2" pipe size	Q-1	.80	20	Ea.	5,400	1,025		6,425	7,525
8320	3420 GPH, 2" pipe size		.60	26.667		9,975	1,375		11,350	13,100
8360	4620 GPH, 2-1/2" pipe size		.50	32		15,800	1,650		17,450	19,900
8500	Neutralizer for acid water, 780 GPH, 1" pipe size		1.50	10.667		2,825	555		3,380	3,925
8540	1140 GPH, 1-1/4" pipe size		1	16		3,175	830		4,005	4,750
8580	1740 GPH, 1-1/2" pipe size		.80	20		4,600	1,025		5,625	6,650
8620	2520 GPH, 2" pipe size		.60	26.667		6,000	1,375		7,375	8,675
8660	3480 GPH, 2-1/2" pipe size		.50	32		9,975	1,650		11,625	13,500
8800	Sediment removal, 780 GPH, 1" pipe size		1.50	10.667		2,700	555		3,255	3,800
8840	1140 GPH, 1-1/4" pipe size		1	16		3,225	830		4,055	4,800
8880	1740 GPH, 1-1/2" pipe size		.80	20		4,275	1,025		5,300	6,275
8920	2520 GPH, 2" pipe size		.60	26.667		6,175	1,375		7,550	8,875
8960	3480 GPH, 2-1/2" pipe size		.50	32		9,700	1,650		11,350	13,200
9200	Taste and odor removal, 660 GPH, 1" pipe size		1.50	10.667		3,775	555		4,330	4,975
9240	1500 GPH, 1-1/4" pipe size		1	16		6,475	830		7,305	8,375
9280	2340 GPH, 1-1/2" pipe size		.80	20		7,375	1,025		8,400	9,675
9320	3420 GPH, 2" pipe size		.60	26.667		11,300	1,375		12,675	14,600
9360	4620 GPH, 2-1/2" pipe size		.50	32		17,500	1,650		19,150	21,700

22 33 Electric Domestic Water Heaters

22 33 13 – Instantaneous Electric Domestic Water Heaters

22 33 13.10 Hot Water Dispensers

		Crew	Daily Output	Labor-Hours	Unit	Material	Labor	Equipment	Total	Total Incl O&P
0010	**HOT WATER DISPENSERS**									
0160	Commercial, 100 cup, 11.3 amp	1 Plum	14	.571	Ea.	510	33		543	615
3180	Household, 60 cup	"	14	.571	"	269	33		302	345

22 33 13.20 Instantaneous Electric Point-Of-Use Water Heaters

			Crew	Daily Output	Labor-Hours	Unit	Material	Labor	Equipment	Total	Total Incl O&P
0010	**INSTANTANEOUS ELECTRIC POINT-OF-USE WATER HEATERS**										
8965	Point of use, electric, glass lined										
8969	Energy saver										
8970	2.5 gal. single element	G	1 Plum	2.80	2.857	Ea.	245	164		409	520
8971	4 gal. single element	G		2.80	2.857		253	164		417	525
8974	6 gal. single element	G		2.50	3.200		261	184		445	565
8975	10 gal. single element	G		2.50	3.200		330	184		514	640
8976	15 gal. single element	G		2.40	3.333		370	192		562	695
8977	20 gal. single element	G		2.40	3.333		410	192		602	740
8978	30 gal. single element	G		2.30	3.478		480	200		680	830
8979	40 gal. single element	G		2.20	3.636		800	209		1,009	1,200
8988	Commercial (ASHRAE energy std. 90)										
8989	6 gallon	G	1 Plum	2.50	3.200	Ea.	665	184		849	1,000
8990	10 gallon	G		2.50	3.200		710	184		894	1,050
8991	15 gallon	G		2.40	3.333		750	192		942	1,125
8992	20 gallon	G		2.40	3.333		785	192		977	1,150
8993	30 gallon	G		2.30	3.478		1,700	200		1,900	2,175
8995	Under the sink, copper, w/bracket										
8996	2.5 gallon	G	1 Plum	4	2	Ea.	435	115		550	650

22 33 30 – Residential, Electric Domestic Water Heaters

22 33 30.13 Residential, Small-Capacity Elec. Water Heaters

			Crew	Daily Output	Labor-Hours	Unit	Material	Labor	Equipment	Total	Total Incl O&P
0010	**RESIDENTIAL, SMALL-CAPACITY ELECTRIC DOMESTIC WATER HEATERS**										
1000	Residential, electric, glass lined tank, 5 yr., 10 gal., single element	D2020-210	1 Plum	2.30	3.478	Ea.	330	200		530	660
1040	20 gallon, single element			2.20	3.636		410	209		619	765

275

22 33 Electric Domestic Water Heaters

22 33 30 – Residential, Electric Domestic Water Heaters

22 33 30.13 Residential, Small-Capacity Elec. Water Heaters

		Crew	Daily Output	Labor-Hours	Unit	Material	2014 Bare Costs Labor	Equipment	Total	Total Incl O&P
1060	30 gallon, double element	1 Plum	2.20	3.636	Ea.	475	209		684	840
1080	40 gallon, double element		2	4		800	230		1,030	1,225
1100	52 gallon, double element		2	4		895	230		1,125	1,325
1120	66 gallon, double element		1.80	4.444		1,200	256		1,456	1,700
1140	80 gallon, double element		1.60	5		1,350	288		1,638	1,900
1180	120 gallon, double element	↓	1.40	5.714	↓	1,900	330		2,230	2,575

22 33 33 – Light-Commercial Electric Domestic Water Heaters

22 33 33.10 Commercial Electric Water Heaters

		Crew	Daily Output	Labor-Hours	Unit	Material	Labor	Equipment	Total	Total Incl O&P
0010	**COMMERCIAL ELECTRIC WATER HEATERS**									
4000	Commercial, 100° rise. NOTE: for each size tank, a range of									
4010	heaters between the ones shown are available									
4020	Electric									
4100	5 gal., 3 kW, 12 GPH, 208 volt	1 Plum	2	4	Ea.	2,750	230		2,980	3,375
4120	10 gal., 6 kW, 25 GPH, 208 volt		2	4		3,050	230		3,280	3,700
4130	30 gal., 24 kW, 98 GPH, 208 volt		1.92	4.167		4,725	240		4,965	5,550
4136	40 gal., 36 kW, 148 GPH, 208 volt		1.88	4.255		5,650	245		5,895	6,600
4140	50 gal., 9 kW, 37 GPH, 208 volt		1.80	4.444		4,175	256		4,431	4,975
4160	50 gal., 36 kW, 148 GPH, 208 volt		1.80	4.444		6,400	256		6,656	7,400
4180	80 gal., 12 kW, 49 GPH, 208 volt		1.50	5.333		5,150	305		5,455	6,150
4200	80 gal., 36 kW, 148 GPH, 208 volt		1.50	5.333		6,925	305		7,230	8,100
4220	100 gal., 36 kW, 148 GPH, 208 volt		1.20	6.667		7,425	385		7,810	8,750
4240	120 gal., 36 kW, 148 GPH, 208 volt		1.20	6.667		7,750	385		8,135	9,100
4260	150 gal., 15 kW, 61 GPH, 480 volt		1	8		18,800	460		19,260	21,400
4280	150 gal., 120 kW, 490 GPH, 480 volt	↓	1	8		25,500	460		25,960	28,800
4300	200 gal., 15 kW, 61 GPH, 480 volt	Q-1	1.70	9.412		19,700	490		20,190	22,400
4320	200 gal., 120 kW, 490 GPH, 480 volt		1.70	9.412		26,900	490		27,390	30,300
4340	250 gal., 15 kW, 61 GPH, 480 volt		1.50	10.667		20,700	555		21,255	23,600
4360	250 gal., 150 kW, 615 GPH, 480 volt		1.50	10.667		32,500	555		33,055	36,500
4380	300 gal., 30 kW, 123 GPH, 480 volt		1.30	12.308		22,400	640		23,040	25,600
4400	300 gal., 180 kW, 738 GPH, 480 volt		1.30	12.308		37,500	640		38,140	42,300
4420	350 gal., 30 kW, 123 GPH, 480 volt		1.10	14.545		23,600	755		24,355	27,200
4440	350 gal., 180 kW, 738 GPH, 480 volt		1.10	14.545		33,300	755		34,055	37,900
4460	400 gal., 30 kW, 123 GPH, 480 volt		1	16		26,800	830		27,630	30,800
4480	400 gal., 210 kW, 860 GPH, 480 volt		1	16		38,500	830		39,330	43,600
4500	500 gal., 30 kW, 123 GPH, 480 volt		.80	20		31,200	1,025		32,225	36,000
4520	500 gal., 240 kW, 984 GPH, 480 volt		.80	20		46,000	1,025		47,025	52,000
4540	600 gal., 30 kW, 123 GPH, 480 volt	Q-2	1.20	20		23,800	1,075		24,875	27,800
4560	600 gal., 300 kW, 1230 GPH, 480 volt		1.20	20		36,100	1,075		37,175	41,300
4580	700 gal., 30 kW, 123 GPH, 480 volt		1	24		25,000	1,300		26,300	29,500
4600	700 gal., 300 kW, 1230 GPH, 480 volt		1	24		37,600	1,300		38,900	43,400
4620	800 gal., 60 kW, 245 GPH, 480 volt		.90	26.667		27,000	1,425		28,425	32,000
4640	800 gal., 300 kW, 1230 GPH, 480 volt		.90	26.667		38,400	1,425		39,825	44,500
4660	1000 gal., 60 kW, 245 GPH, 480 volt		.70	34.286		29,600	1,850		31,450	35,400
4680	1000 gal., 480 kW, 1970 GPH, 480 volt		.70	34.286		49,200	1,850		51,050	57,000
4700	1200 gal., 60 kW, 245 GPH, 480 volt		.60	40		50,500	2,150		52,650	59,000
4720	1200 gal., 480 kW, 1970 GPH, 480 volt		.60	40		76,500	2,150		78,650	87,500
4740	1500 gal., 60 kW, 245 GPH, 480 volt		.50	48		66,500	2,575		69,075	77,000
4760	1500 gal., 480 kW, 1970 GPH, 480 volt	↓	.50	48	↓	92,000	2,575		94,575	105,500
5400	Modulating step control for under 90 kW, 2-5 steps	1 Elec	5.30	1.509		810	80.50		890.50	1,000
5440	1 through 5 steps beyond standard		3.20	2.500		221	133		354	445
5460	6 through 10 steps beyond standard		2.70	2.963		455	158		613	735
5480	11 through 18 steps beyond standard	↓	1.60	5	↓	680	267		947	1,150

22 34 Fuel-Fired Domestic Water Heaters

22 34 13 – Instantaneous, Tankless, Gas Domestic Water Heaters

22 34 13.10 Instantaneous, Tankless, Gas Water Heaters		Crew	Daily Output	Labor-Hours	Unit	Material	2014 Bare Costs Labor	Equipment	Total	Total Incl O&P
0010	**INSTANTANEOUS, TANKLESS, GAS WATER HEATERS**									
9410	Natural gas/propane, 3.2 GPM [G]	1 Plum	2	4	Ea.	365	230		595	750
9420	6.4 GPM [G]		1.90	4.211		745	242		987	1,175
9430	8.4 GPM [G]		1.80	4.444		890	256		1,146	1,375
9440	9.5 GPM [G]	↓	1.60	5	↓	1,150	288		1,438	1,675

22 34 30 – Residential Gas Domestic Water Heaters

22 34 30.13 Residential, Atmos, Gas Domestic Wtr Heaters

0010	**RESIDENTIAL, ATMOSPHERIC, GAS DOMESTIC WATER HEATERS**									
2000	Gas fired, foam lined tank, 10 yr., vent not incl.									
2040	30 gallon	1 Plum	2	4	Ea.	895	230		1,125	1,325
2060	40 gallon		1.90	4.211		895	242		1,137	1,350
2080	50 gallon		1.80	4.444		935	256		1,191	1,400
2090	60 gallon		1.70	4.706		1,325	271		1,596	1,850
2100	75 gallon		1.50	5.333		1,350	305		1,655	1,975
2120	100 gallon		1.30	6.154		1,525	355		1,880	2,200
2900	Water heater, safety-drain pan, 26" round	↓	20	.400	↓	37	23		60	76

22 34 36 – Commercial Gas Domestic Water Heaters

22 34 36.13 Commercial, Atmos., Gas Domestic Water Htrs.

0010	**COMMERCIAL, ATMOSPHERIC, GAS DOMESTIC WATER HEATERS**									
6000	Gas fired, flush jacket, std. controls, vent not incl.									
6040	75 MBH input, 73 GPH	1 Plum	1.40	5.714	Ea.	3,500	330		3,830	4,350
6060	98 MBH input, 95 GPH		1.40	5.714		5,300	330		5,630	6,325
6080	120 MBH input, 110 GPH		1.20	6.667		5,500	385		5,885	6,600
6100	120 MBH input, 115 GPH		1.10	7.273		6,625	420		7,045	7,925
6120	140 MBH input, 130 GPH		1	8		7,225	460		7,685	8,650
6140	155 MBH input, 150 GPH		.80	10		8,200	575		8,775	9,900
6160	180 MBH input, 170 GPH		.70	11.429		8,675	660		9,335	10,500
6180	200 MBH input, 192 GPH		.60	13.333		8,925	765		9,690	11,000
6200	250 MBH input, 245 GPH	↓	.50	16		9,325	920		10,245	11,700
6220	260 MBH input, 250 GPH	Q-1	.80	20		9,900	1,025		10,925	12,500
6240	360 MBH input, 360 GPH		.80	20		11,700	1,025		12,725	14,500
6260	500 MBH input, 480 GPH		.70	22.857		16,400	1,175		17,575	19,900
6280	725 MBH input, 690 GPH	↓	.60	26.667		19,300	1,375		20,675	23,300
6900	For low water cutoff, add	1 Plum	8	1		340	57.50		397.50	455
6960	For bronze body hot water circulator, add	"	4	2		1,925	115		2,040	2,300

22 34 36.45 Commercial Packaged Water Heater Systems

0010	**COMMERCIAL PACKAGED WATER HEATER SYSTEMS**									
1000	Car wash package, continuous duty, high recovery, gas fired									
1040	100° rise, 180 MBH input, 174 GPH	1 Plum	3	2.667	Ea.	5,900	153		6,053	6,725
1060	280 MBH input, 270 GPH		2.50	3.200		6,300	184		6,484	7,200
1080	400 MBH input, 386 GPH		2.50	3.200		7,250	184		7,434	8,250
1100	480 MBH input, 464 GPH		2	4		7,650	230		7,880	8,775
1120	605 MBH input, 584 GPH		1.50	5.333		8,725	305		9,030	10,100
1140	700 MBH input, 676 GPH	↓	1	8		9,075	460		9,535	10,700
1160	1000 MBH input, 966 GPH	Q-1	1.60	10		10,100	520		10,620	11,900
1180	1200 MBH input, 1159 GPH		1.40	11.429		13,700	590		14,290	16,000
1200	1400 MBH input, 1353 GPH	↓	1.20	13.333	↓	15,000	690		15,690	17,600
3000	Combination dishwasher & general purpose, 2 temp. gas fired									
3040	124 GPH @ 140° rise; 434 GPH @ 40° rise	1 Plum	2	4	Ea.	6,025	230		6,255	6,975
3060	193 GPH @ 140° rise; 677 GPH @ 40° rise		1.60	5		6,875	288		7,163	8,000
3080	276 GPH @ 140° rise; 969 GPH @ 40° rise	↓	1.40	5.714	↓	7,825	330		8,155	9,125

22 34 Fuel-Fired Domestic Water Heaters

22 34 36 – Commercial Gas Domestic Water Heaters

22 34 36.45 Commercial Packaged Water Heater Systems

		Crew	Daily Output	Labor-Hours	Unit	Material	2014 Bare Costs Labor	Equipment	Total	Total Incl O&P
3100	330 GPH @ 140° rise; 1160 GPH @ 40° rise	1 Plum	1.20	6.667	Ea.	8,300	385		8,685	9,700
3120	417 GPH @ 140° rise; 1460 GPH @ 40° rise		1	8		9,375	460		9,835	11,000
3140	481 GPH @ 140° rise; 1685 GPH @ 40° rise	Q-1	1.40	11.429		10,800	590		11,390	12,800
3160	688 GPH @ 140° rise; 2410 GPH @ 40° rise		1.20	13.333		11,900	690		12,590	14,100
3180	828 GPH @ 140° rise; 2790 GPH @ 40° rise		1	16		15,500	830		16,330	18,400
3200	965 GPH @ 140° rise; 3370 GPH @ 40° rise		.80	20		16,800	1,025		17,825	20,000
3960	For unit base, add					465			465	510
5000	Coin laundry units, gas fired, 100° rise									
5020	Single heater,									
5040	280 MBH input, 270 GPH	1 Plum	1.60	5	Ea.	6,875	288		7,163	8,000
5060	400 MBH input, 386 GPH		1.30	6.154		9,075	355		9,430	10,500
5080	480 MBH input, 464 GPH		1	8		9,550	460		10,010	11,200
5100	605 MBH input, 584 GPH	Q-1	1.40	11.429		11,300	590		11,890	13,300
5120	700 MBH input, 676 GPH		1.20	13.333		11,900	690		12,590	14,100
5140	1000 MBH input, 966 GPH		1	16		12,900	830		13,730	15,500
5160	1200 MBH input, 1159 GPH		.90	17.778		16,700	920		17,620	19,800
5180	1400 MBH input, 1353 GPH		.70	22.857		18,600	1,175		19,775	22,200
6000	Multiple heater									
6040	560 MBH input, 540 GPH	1 Plum	1	8	Ea.	10,900	460		11,360	12,700
6060	800 MBH input, 772 GPH	Q-1	1.30	12.308		14,100	640		14,740	16,500
6080	960 MBH input, 928 GPH		1.20	13.333		15,900	690		16,590	18,600
6100	1210 MBH input, 1168 GPH		1	16		17,900	830		18,730	21,000
6120	1400 MBH input, 1352 GPH		1	16		18,700	830		19,530	21,900
6140	1700 MBH input, 1642 GPH		.80	20		21,400	1,025		22,425	25,200
6160	2000 MBH input, 1932 GPH		.70	22.857		25,900	1,175		27,075	30,300
6180	2400 MBH input, 2318 GPH	Q-2	.90	26.667		29,000	1,425		30,425	34,100
6200	2600 MBH input, 2512 GPH		.80	30		32,400	1,600		34,000	38,000
6220	2800 MBH input, 2706 GPH		.70	34.286		36,600	1,850		38,450	43,000
6800	Standard system sizing is based on									
6820	80% of the washers operating at one time.									

22 34 46 – Oil-Fired Domestic Water Heaters

22 34 46.10 Residential Oil-Fired Water Heaters

0010	**RESIDENTIAL OIL-FIRED WATER HEATERS**									
3000	Oil fired, glass lined tank, 5 yr., vent not included, 30 gallon D2020-260	1 Plum	2	4	Ea.	1,175	230		1,405	1,625
3040	50 gallon		1.80	4.444		1,375	256		1,631	1,900
3060	70 gallon		1.50	5.333		1,975	305		2,280	2,650

22 34 46.20 Commercial Oil-Fired Water Heaters

0010	**COMMERCIAL OIL-FIRED WATER HEATERS**									
8000	Oil fired, glass lined, UL listed, std. controls, vent not incl.									
8060	140 gal., 140 MBH input, 134 GPH	Q-1	2.13	7.512	Ea.	19,900	390		20,290	22,500
8080	140 gal., 199 MBH input, 191 GPH		2	8		20,600	415		21,015	23,300
8100	140 gal., 255 MBH input, 247 GPH		1.60	10		21,200	520		21,720	24,100
8120	140 gal., 270 MBH input, 259 GPH		1.20	13.333		26,200	690		26,890	29,900
8140	140 gal., 400 MBH input, 384 GPH		1	16		27,100	830		27,930	31,100
8160	140 gal., 540 MBH input, 519 GPH		.96	16.667		28,100	865		28,965	32,200
8180	140 gal., 720 MBH input, 691 GPH		.92	17.391		28,600	900		29,500	32,900
8200	221 gal., 300 MBH input, 288 GPH		.88	18.182		37,800	940		38,740	43,000
8220	221 gal., 600 MBH input, 576 GPH		.86	18.605		42,200	965		43,165	47,900
8240	221 gal., 800 MBH input, 768 GPH		.82	19.512		42,500	1,000		43,500	48,200
8260	201 gal., 1000 MBH input, 960 GPH	Q-2	1.26	19.048		43,500	1,025		44,525	49,400
8280	201 gal., 1250 MBH input, 1200 GPH		1.22	19.672		43,900	1,050		44,950	49,900
8300	201 gal., 1500 MBH input, 1441 GPH		1.16	20.690		47,800	1,100		48,900	54,000

22 34 Fuel-Fired Domestic Water Heaters

22 34 46 – Oil-Fired Domestic Water Heaters

22 34 46.20 Commercial Oil-Fired Water Heaters		Crew	Daily Output	Labor-Hours	Unit	Material	2014 Bare Costs Labor	Equipment	Total	Total Incl O&P
8320	411 gal., 600 MBH input, 576 GPH	Q-2	1.12	21.429	Ea.	48,200	1,150		49,350	55,000
8340	411 gal., 800 MBH input, 768 GPH		1.08	22.222		48,300	1,200		49,500	55,000
8360	411 gal., 1000 MBH input, 960 GPH		1.04	23.077		51,000	1,250		52,250	58,000
8380	411 gal., 1250 MBH input, 1200 GPH		.98	24.490		52,000	1,325		53,325	59,500
8400	397 gal., 1500 MBH input, 1441 GPH		.92	26.087		52,500	1,400		53,900	60,000
8420	397 gal., 1750 MBH input, 1681 GPH		.86	27.907		57,000	1,500		58,500	65,500
8430	397 gal., 2000 MBH input, 1921 GPH		.82	29.268		62,000	1,575		63,575	70,500
8440	375 gal., 2250 MBH input, 2161 GPH		.76	31.579		63,500	1,700		65,200	72,500
8450	375 gal., 2500 MBH input, 2401 GPH		.82	29.268		66,000	1,575		67,575	75,000
8500	Oil fired, polymer lined									
8510	400 MBH, 125 gallon	Q-2	1	24	Ea.	24,400	1,300		25,700	28,900
8520	400 MBH, 600 gallon		.80	30		45,800	1,600		47,400	53,000
8530	800 MBH, 400 gallon		.67	35.821		41,700	1,925		43,625	48,800
8540	800 MBH, 600 gallon		.60	40		54,000	2,150		56,150	63,000
8550	1000 MBH, 600 gallon		.50	48		56,000	2,575		58,575	65,500
8560	1200 MBH, 900 gallon		.40	60		67,500	3,225		70,725	79,500
8900	For low water cutoff, add	1 Plum	8	1		325	57.50		382.50	445
8960	For bronze body hot water circulator, add	"	4	2		710	115		825	955

22 41 Residential Plumbing Fixtures

22 41 06 – Plumbing Fixtures General

22 41 06.10 Plumbing Fixture Notes

0010	**PLUMBING FIXTURE NOTES**, Incl. trim fittings unless otherwise noted R224000-30									
0080	For rough-in, supply, waste, and vent, see add for each type									
0122	For electric water coolers, see Section 22 47 16.10									
0160	For color, unless otherwise noted, add				Ea.	20%				

22 41 13 – Residential Water Closets, Urinals, and Bidets

22 41 13.10 Bidets

0010	**BIDETS**									
0180	Vitreous china, with trim on fixture	Q-1	5	3.200	Ea.	595	166		761	905
0200	With trim for wall mounting	"	5	3.200		710	166		876	1,025
9590	For color add					40%				
9591	For designer colors and trim add					50%				
9600	For rough-in, supply, waste and vent, add	Q-1	1.78	8.989		380	465		845	1,125

22 41 13.40 Water Closets

0010	**WATER CLOSETS** D2010-110									
0022	For seats, see Section 22 41 13.44									
0032	For automatic flush, see Line 22 42 39.10 0972									
0150	Tank type, vitreous china, incl. seat, supply pipe w/stop, 1.6 gpf or noted									
0200	Wall hung R224000-30									
0400	Two piece, close coupled	Q-1	5.30	3.019	Ea.	630	156		786	925
0960	For rough-in, supply, waste, vent and carrier	"	2.73	5.861	"	810	305		1,115	1,350
0999	Floor mounted									
1020	One piece, low profile	Q-1	5.30	3.019	Ea.	1,100	156		1,256	1,425
1050	One piece		5.30	3.019		1,050	156		1,206	1,375
1100	Two piece, close coupled		5.30	3.019		233	156		389	490
1102	Economy		5.30	3.019		130	156		286	380
1110	Two piece, close coupled, dual flush		5.30	3.019		299	156		455	565
1140	Two piece, close coupled, 1.28 gpf, ADA [G]		5.30	3.019		305	156		461	570
1960	For color, add					30%				

279

22 41 Residential Plumbing Fixtures

22 41 13 – Residential Water Closets, Urinals, and Bidets

22 41 13.40 Water Closets

		Crew	Daily Output	Labor-Hours	Unit	Material	2014 Bare Costs Labor	Equipment	Total	Total Incl O&P
1961	For designer colors and trim, add					55%				
1980	For rough-in, supply, waste and vent	Q-1	3.05	5.246	Ea.	330	272		602	775

22 41 13.44 Toilet Seats

		Crew	Daily Output	Labor-Hours	Unit	Material	Labor	Equipment	Total	Total Incl O&P
0010	**TOILET SEATS**									
0100	Molded composition, white									
0150	Industrial, w/o cover, open front, regular bowl	1 Plum	24	.333	Ea.	21	19.20		40.20	52
0200	With self-sustaining hinge		24	.333		22.50	19.20		41.70	53.50
0220	With self-sustaining check hinge		24	.333		22.50	19.20		41.70	53.50
0240	Extra heavy, with check hinge		24	.333		27.50	19.20		46.70	59.50
0260	Elongated bowl, same price									
0300	Junior size, w/o cover, open front	1 Plum	24	.333	Ea.	40.50	19.20		59.70	73.50
0320	Regular primary bowl, open front		24	.333		40.50	19.20		59.70	73.50
0340	Regular baby bowl, open front, check hinge		24	.333		37	19.20		56.20	69.50
0380	Open back & front, w/o cover, reg. or elongated bowl		24	.333		29	19.20		48.20	61
0400	Residential									
0420	Regular bowl, w/cover, closed front	1 Plum	24	.333	Ea.	29	19.20		48.20	60.50
0440	Open front	"	24	.333	"	26	19.20		45.20	58
0460	Elongated bowl, add					25%				
0500	Self-raising hinge, w/o cover, open front									
0520	Regular bowl	1 Plum	24	.333	Ea.	101	19.20		120.20	140
0540	Elongated bowl	"	24	.333	"	23	19.20		42.20	54.50
0700	Molded wood, white, with cover									
0720	Closed front, regular bowl, square back	1 Plum	24	.333	Ea.	10.90	19.20		30.10	41
0740	Extended back		24	.333		13.80	19.20		33	44
0780	Elongated bowl, square back		24	.333		14.10	19.20		33.30	44.50
0800	Open front		24	.333		15.05	19.20		34.25	45.50
0850	Decorator styles									
0890	Vinyl top, patterned	1 Plum	24	.333	Ea.	22	19.20		41.20	53
0900	Vinyl padded, plain colors, regular bowl		24	.333		22	19.20		41.20	53
0930	Elongated bowl		24	.333		24	19.20		43.20	55.50
1000	Solid plastic, white									
1030	Industrial, w/o cover, open front, regular bowl	1 Plum	24	.333	Ea.	26	19.20		45.20	58
1080	Extra heavy, concealed check hinge		24	.333		18.75	19.20		37.95	49.50
1100	Self-sustaining hinge		24	.333		23	19.20		42.20	54.50
1150	Elongated bowl		24	.333		29	19.20		48.20	61
1170	Concealed check		24	.333		17.25	19.20		36.45	48
1190	Self-sustaining hinge, concealed check		24	.333		51.50	19.20		70.70	85.50
1220	Residential, with cover, closed front, regular bowl		24	.333		45	19.20		64.20	78.50
1240	Elongated bowl		24	.333		55	19.20		74.20	89
1260	Open front, regular bowl		24	.333		40	19.20		59.20	73
1280	Elongated bowl		24	.333		48.50	19.20		67.70	82.50

22 41 16 – Residential Lavatories and Sinks

22 41 16.10 Lavatories

			Crew	Daily Output	Labor-Hours	Unit	Material	Labor	Equipment	Total	Total Incl O&P
0010	**LAVATORIES**, With trim, white unless noted otherwise	D2010-310									
0500	Vanity top, porcelain enamel on cast iron										
0600	20" x 18"	R224000-30	Q-1	6.40	2.500	Ea.	330	130		460	555
0640	33" x 19" oval			6.40	2.500		690	130		820	955
0680	20" x 17" oval			6.40	2.500		172	130		302	385
0720	19" round			6.40	2.500		425	130		555	665
0760	20" x 12" triangular bowl			6.40	2.500		273	130		403	495
0860	For color, add						25%				
0861	For designer colors and trim, add						70%				

22 41 Residential Plumbing Fixtures

22 41 16 - Residential Lavatories and Sinks

22 41 16.10 Lavatories

		Crew	Daily Output	Labor-Hours	Unit	Material	2014 Bare Costs Labor	Equipment	Total	Total Incl O&P
1000	Cultured marble, 19" x 17", single bowl	Q-1	6.40	2.500	Ea.	175	130		305	390
1040	25" x 19", single bowl		6.40	2.500		206	130		336	425
1080	31" x 19", single bowl		6.40	2.500		216	130		346	435
1120	25" x 22", single bowl		6.40	2.500		215	130		345	430
1160	37" x 22", single bowl		6.40	2.500		245	130		375	465
1200	49" x 22", single bowl		6.40	2.500		292	130		422	515
1580	For color, same price									
1900	Stainless steel, self-rimming, 25" x 22", single bowl, ledge	Q-1	6.40	2.500	Ea.	365	130		495	600
1960	17" x 22", single bowl		6.40	2.500		355	130		485	585
2040	18-3/4" round		6.40	2.500		825	130		955	1,100
2600	Steel, enameled, 20" x 17", single bowl		5.80	2.759		197	143		340	435
2660	19" round		5.80	2.759		171	143		314	405
2720	18" round		5.80	2.759		138	143		281	370
2860	For color, add					10%				
2861	For designer colors and trim, add					20%				
2900	Vitreous china, 20" x 16", single bowl	Q-1	5.40	2.963	Ea.	260	153		413	520
2960	20" x 17", single bowl		5.40	2.963		176	153		329	425
3020	19" round, single bowl		5.40	2.963		174	153		327	425
3080	19" x 16", single bowl		5.40	2.963		267	153		420	525
3140	17" x 14", single bowl		5.40	2.963		214	153		367	470
3200	22" x 13", single bowl		5.40	2.963		267	153		420	525
3560	For color, add					50%				
3561	For designer colors and trim, add					100%				
3580	Rough-in, supply, waste and vent for all above lavatories	Q-1	2.30	6.957	Ea.	415	360		775	1,000
4000	Wall hung									
4040	Porcelain enamel on cast iron, 16" x 14", single bowl	Q-1	8	2	Ea.	510	104		614	715
4060	18" x 15" single bowl		8	2		400	104		504	595
4120	19" x 17", single bowl		8	2		475	104		579	680
4180	20" x 18", single bowl		8	2		277	104		381	460
4240	22" x 19", single bowl		8	2		685	104		789	910
4580	For color, add					30%				
4581	For designer colors and trim, add					75%				
6000	Vitreous china, 18" x 15", single bowl with backsplash	Q-1	7	2.286	Ea.	231	118		349	435
6060	19" x 17", single bowl		7	2.286		190	118		308	390
6120	20" x 18", single bowl		7	2.286		273	118		391	480
6210	27" x 20", wheelchair type		7	2.286		815	118		933	1,075
6500	For color, add					30%				
6501	For designer colors and trim, add					50%				
6960	Rough-in, supply, waste and vent for above lavatories	Q-1	1.66	9.639	Ea.	470	500		970	1,275
7000	Pedestal type									
7600	Vitreous china, 27" x 21", white	Q-1	6.60	2.424	Ea.	690	126		816	950
7610	27" x 21", colored		6.60	2.424		865	126		991	1,150
7620	27" x 21", premium color		6.60	2.424		970	126		1,096	1,275
7660	26" x 20", white		6.60	2.424		770	126		896	1,025
7670	26" x 20", colored		6.60	2.424		770	126		896	1,025
7680	26" x 20", premium color		6.60	2.424		970	126		1,096	1,275
7700	24" x 20", white		6.60	2.424		395	126		521	625
7710	24" x 20", colored		6.60	2.424		480	126		606	715
7720	24" x 20", premium color		6.60	2.424		535	126		661	780
7760	21" x 18", white		6.60	2.424		284	126		410	505
7770	21" x 18", colored		6.60	2.424		300	126		426	520
7990	Rough-in, supply, waste and vent for pedestal lavatories		1.66	9.639		470	500		970	1,275

22 41 Residential Plumbing Fixtures

22 41 16 – Residential Lavatories and Sinks

22 41 16.30 Sinks

			Crew	Daily Output	Labor-Hours	Unit	Material	2014 Bare Costs Labor	Equipment	Total	Total Incl O&P
0010	**SINKS**, With faucets and drain	D2010-410									
2000	Kitchen, counter top style, P.E. on C.I., 24" x 21" single bowl		Q-1	5.60	2.857	Ea.	275	148		423	525
2100	31" x 22" single bowl			5.60	2.857		575	148		723	855
2200	32" x 21" double bowl			4.80	3.333		340	173		513	635
2310	For color, add						20%				
2311	For designer colors and trim, add						50%				
3000	Stainless steel, self rimming, 19" x 18" single bowl		Q-1	5.60	2.857	Ea.	590	148		738	875
3100	25" x 22" single bowl			5.60	2.857		660	148		808	950
3200	33" x 22" double bowl			4.80	3.333		960	173		1,133	1,300
3300	43" x 22" double bowl			4.80	3.333		1,125	173		1,298	1,475
3400	22" x 43" triple bowl			4.40	3.636		1,350	188		1,538	1,750
3500	Corner double bowl each 14" x 16"			4.80	3.333		830	173		1,003	1,175
4000	Steel, enameled, with ledge, 24" x 21" single bowl			5.60	2.857		505	148		653	780
4100	32" x 21" double bowl			4.80	3.333		495	173		668	805
4960	For color sinks except stainless steel, add						10%				
4961	For designer colors and trim add						20%				
4980	For rough-in, supply, waste and vent, counter top sinks		Q-1	2.14	7.477		460	385		845	1,100
5000	Kitchen, raised deck, P.E. on C.I.										
5100	32" x 21", dual level, double bowl		Q-1	2.60	6.154	Ea.	410	320		730	930
5200	42" x 21", double bowl & disposer well		"	2.20	7.273		1,175	375		1,550	1,875
5700	For color, add						20%				
5701	For designer colors and trim add						50%				
5790	For rough-in, supply, waste & vent, sinks		Q-1	1.85	8.649		460	450		910	1,175

22 41 19 – Residential Bathtubs

22 41 19.10 Baths

			Crew	Daily Output	Labor-Hours	Unit	Material	Labor	Equipment	Total	Total Incl O&P
0010	**BATHS**	D2010-510									
0100	Tubs, recessed porcelain enamel on cast iron, with trim										
0180	48" x 42"		Q-1	4	4	Ea.	2,625	207		2,832	3,225
0220	72" x 36"		"	3	5.333	"	2,725	276		3,001	3,425
0300	Mat bottom										
0340	4'-6" long		Q-1	5	3.200	Ea.	1,325	166		1,491	1,700
0380	5' long			4.40	3.636		1,100	188		1,288	1,500
0420	5'-6" long			4	4		1,775	207		1,982	2,275
0480	Above floor drain, 5' long			4	4		785	207		992	1,175
0560	Corner 48" x 44"			4.40	3.636		2,625	188		2,813	3,175
0750	For color, add						30%				
0760	For designer colors & trim, add						60%				
2000	Enameled formed steel, 4'-6" long		Q-1	5.80	2.759	Ea.	495	143		638	760
2300	Above floor drain, 5' long		"	5.50	2.909	"	535	151		686	820
2350	For color, add						10%				
4000	Soaking, acrylic, w/pop-up drain 66" x 36" x 20" deep		Q-1	5.50	2.909	Ea.	1,825	151		1,976	2,225
4100	60" x 42" x 20" deep			5	3.200		1,150	166		1,316	1,525
4200	72" x 42" x 23" deep			4.80	3.333		1,750	173		1,923	2,175
4310	For color, add						5%				
4311	For designer colors & trim, add						20%				
4600	Module tub & showerwall surround, molded fiberglass										
4610	5' long x 34" wide x 76" high		Q-1	4	4	Ea.	815	207		1,022	1,200
4620	For color add						10%				
4621	For designer colors and trim add						25%				
4750	Handicap with 1-1/2" OD grab bar, antiskid bottom										
4760	60" x 32-3/4" x 72" high		Q-1	4	4	Ea.	810	207		1,017	1,200
4770	60" x 30" x 71" high with molded seat			3.50	4.571		795	237		1,032	1,225

22 41 Residential Plumbing Fixtures

22 41 19 – Residential Bathtubs

22 41 19.10 Baths		Crew	Daily Output	Labor-Hours	Unit	Material	2014 Bare Costs Labor	Equipment	Total	Total Incl O&P
9600	Rough-in, supply, waste and vent, for all above tubs, add	Q-1	2.07	7.729	Ea.	485	400		885	1,150

22 41 23 – Residential Showers

22 41 23.20 Showers

0010	**SHOWERS** D2010-710									
1500	Stall, with drain only. Add for valve and door/curtain									
1510	Baked enamel, molded stone receptor, 30" square	Q-1	5.20	3.077	Ea.	1,150	159		1,309	1,525
1520	32" square		5	3.200		1,175	166		1,341	1,525
1530	36" square		4.80	3.333		2,925	173		3,098	3,450
1540	Terrazzo receptor, 32" square		5	3.200		1,350	166		1,516	1,725
1560	36" square		4.80	3.333		1,475	173		1,648	1,875
1580	36" corner angle		4.80	3.333		1,725	173		1,898	2,150
1600	For color, add					10%				
1601	For designer colors and trim, add					15%				
1604	For thermostatic valve add				Ea.	585			585	645
3000	Fiberglass, one piece, with 3 walls, 32" x 32" square	Q-1	5.50	2.909		490	151		641	770
3100	36" x 36" square	"	5.50	2.909		505	151		656	785
3200	Handicap, 1-1/2" O.D. grab bars, nonskid floor									
3210	48" x 34-1/2" x 72" corner seat	Q-1	5	3.200	Ea.	690	166		856	1,000
3220	60" x 34-1/2" x 72" corner seat		4	4		720	207		927	1,100
3230	48" x 34-1/2" x 72" fold up seat		5	3.200		2,125	166		2,291	2,575
3250	64" x 65-3/4" x 81-1/2" fold. seat, whlchr.		3.80	4.211		2,200	218		2,418	2,750
3260	For thermostatic valve add					585			585	645
4000	Polypropylene, stall only, w/molded-stone floor, 30" x 30"	Q-1	2	8		635	415		1,050	1,325
4100	32" x 32"	"	2	8		650	415		1,065	1,350
4110	For thermostatic valve add					585			585	645
4200	Rough-in, supply, waste and vent for above showers	Q-1	2.05	7.805		540	405		945	1,200

22 41 23.40 Shower System Components

0010	**SHOWER SYSTEM COMPONENTS**									
4500	Receptor only									
4510	For tile, 36" x 36"	1 Plum	4	2	Ea.	365	115		480	580
4520	Fiberglass receptor only, 32" x 32"		8	1		107	57.50		164.50	205
4530	34" x 34"		7.80	1.026		126	59		185	228
4540	36" x 36"		7.60	1.053		132	60.50		192.50	237
4600	Rectangular									
4620	32" x 48"	1 Plum	7.40	1.081	Ea.	149	62		211	257
4630	32" x 54"		7.20	1.111		196	64		260	310
4640	32" x 60"		7	1.143		208	66		274	330
5000	Built-in, head, arm, 2.5 GPM valve		4	2		80	115		195	262
5200	Head, arm, by-pass, integral stops, handles		3.60	2.222		255	128		383	475
5500	Head, water economizer, 1.6 GPM [G]		24	.333		43	19.20		62.20	76.50
5800	Mixing valve, built-in		6	1.333		153	76.50		229.50	285
5900	Exposed		6	1.333		630	76.50		706.50	810

22 41 36 – Residential Laundry Trays

22 41 36.10 Laundry Sinks

0010	**LAUNDRY SINKS**, With trim D2010-420									
0020	Porcelain enamel on cast iron, black iron frame									
0050	24" x 21", single compartment	Q-1	6	2.667	Ea.	580	138		718	845
0100	26" x 21", single compartment		6	2.667		610	138		748	880
0200	48" x 20", double compartment		5	3.200		830	166		996	1,175
2000	Molded stone, on wall hanger or legs									
2020	22" x 23", single compartment	Q-1	6	2.667	Ea.	134	138		272	355

22 41 Residential Plumbing Fixtures

22 41 36 – Residential Laundry Trays

22 41 36.10 Laundry Sinks

		Crew	Daily Output	Labor-Hours	Unit	Material	2014 Bare Costs Labor	Equipment	Total	Total Incl O&P
2100	45" x 21", double compartment	Q-1	5	3.200	Ea.	330	166		496	615
3000	Plastic, on wall hanger or legs									
3020	18" x 23", single compartment	Q-1	6.50	2.462	Ea.	135	128		263	340
3100	20" x 24", single compartment		6.50	2.462		154	128		282	365
3200	36" x 23", double compartment		5.50	2.909		185	151		336	430
3300	40" x 24", double compartment		5.50	2.909		278	151		429	535
5000	Stainless steel, counter top, 22" x 17" single compartment		6	2.667		64	138		202	280
5200	33" x 22", double compartment		5	3.200		79	166		245	335
9600	Rough-in, supply, waste and vent, for all laundry sinks		2.14	7.477		460	385		845	1,100

22 41 39 – Residential Faucets, Supplies and Trim

22 41 39.10 Faucets and Fittings

		Crew	Daily Output	Labor-Hours	Unit	Material	Labor	Equipment	Total	Total Incl O&P
0010	**FAUCETS AND FITTINGS**									
0150	Bath, faucets, diverter spout combination, sweat	1 Plum	8	1	Ea.	86.50	57.50		144	182
0200	For integral stops, IPS unions, add					109			109	120
0300	Three valve combinations, spout, head, arm, flange, sweat	1 Plum	6	1.333		88	76.50		164.50	213
0400	For integral stops, IPS unions, add				Pr.	64.50			64.50	71
0420	Bath, press-bal mix valve w/diverter, spout, shower head, arm/flange	1 Plum	8	1	Ea.	168	57.50		225.50	272
0500	Drain, central lift, 1-1/2" IPS male		20	.400		71	23		94	114
0600	Trip lever, 1-1/2" IPS male		20	.400		45	23		68	84.50
0700	Pop up, 1-1/2" IPS male		18	.444		53	25.50		78.50	97
0800	Chain and stopper, 1-1/2" IPS male		24	.333		32.50	19.20		51.70	64.50
0810	Bidet									
0812	Fitting, over the rim, swivel spray/pop-up drain	1 Plum	8	1	Ea.	206	57.50		263.50	315
1000	Kitchen sink faucets, top mount, cast spout		10	.800		61.50	46		107.50	137
1100	For spray, add		24	.333		16.15	19.20		35.35	47
1110	For basket strainer w/tail piece, add		24	.333		14.85	19.20		34.05	45.50
1200	Wall type, swing tube spout		10	.800		74.50	46		120.50	151
1240	For soap dish, add					3.60			3.60	3.96
1250	For basket strainer w/tail piece, add					51			51	56
1300	Single control lever handle									
1310	With pull out spray									
1320	Polished chrome	1 Plum	10	.800	Ea.	196	46		242	285
2000	Laundry faucets, shelf type, IPS or copper unions		12	.667		49.50	38.50		88	113
2100	Lavatory faucet, centerset, without drain		10	.800		44.50	46		90.50	119
2120	With pop-up drain		6.66	1.201		62.50	69		131.50	173
2130	For acrylic handles, add					5.15			5.15	5.65
2150	Concealed, 12" centers	1 Plum	10	.800		101	46		147	181
2160	With pop-up drain	"	6.66	1.201		118	69		187	234
2210	Porcelain cross handles and pop-up drain									
2220	Polished chrome	1 Plum	6.66	1.201	Ea.	189	69		258	310
2230	Polished brass	"	6.66	1.201	"	284	69		353	415
2260	Single lever handle and pop-up drain									
2280	Satin nickel	1 Plum	6.66	1.201	Ea.	277	69		346	410
2290	Polished chrome		6.66	1.201		198	69		267	320
2600	Shelfback, 4" to 6" centers, 17 Ga. tailpiece		10	.800		78	46		124	156
2650	With pop-up drain		6.66	1.201		95	69		164	208
2700	Shampoo faucet with supply tube		24	.333		48.50	19.20		67.70	82.50
2800	Self-closing, center set		10	.800		131	46		177	214
2810	Automatic sensor and operator, with faucet head		6.15	1.301		450	75		525	610
4000	Shower by-pass valve with union		18	.444		68.50	25.50		94	114
4100	Shower arm with flange and head		22	.364		19.25	21		40.25	52.50
4140	Shower, hand held, pin mount, massage action, chrome		22	.364		73	21		94	112

22 41 Residential Plumbing Fixtures

22 41 39 – Residential Faucets, Supplies and Trim

22 41 39.10 Faucets and Fittings

		Crew	Daily Output	Labor-Hours	Unit	Material	2014 Bare Costs Labor	Equipment	Total	Total Incl O&P
4142	Polished brass	1 Plum	22	.364	Ea.	136	21		157	182
4144	Shower, hand held, wall mtd, adj. spray, 2 wall mounts, chrome		20	.400		97.50	23		120.50	142
4146	Polished brass		20	.400		185	23		208	239
4148	Shower, hand held head, bar mounted 24", adj. spray, chrome		20	.400		158	23		181	208
4150	Polished brass		20	.400		325	23		348	395
4200	Shower thermostatic mixing valve, concealed, with shower head trim kit		8	1		330	57.50		387.50	445
4220	Shower pressure balancing mixing valve,									
4230	With shower head, arm, flange and diverter tub spout									
4240	Chrome	1 Plum	6.14	1.303	Ea.	350	75		425	500
4250	Satin nickel		6.14	1.303		475	75		550	635
4260	Polished graphite		6.14	1.303		475	75		550	635
5000	Sillcock, compact, brass, IPS or copper to hose		24	.333		9.70	19.20		28.90	39.50
6000	Stop and waste valves, bronze									
6100	Angle, solder end 1/2"	1 Plum	24	.333	Ea.	12.05	19.20		31.25	42.50
6110	3/4"		20	.400		14.15	23		37.15	50.50
6300	Straightway, solder end 3/8"		24	.333		12.35	19.20		31.55	42.50
6310	1/2"		24	.333		12.35	19.20		31.55	42.50
6320	3/4"		20	.400		13.30	23		36.30	49.50
6410	Straightway, threaded 1/2"		24	.333		15.95	19.20		35.15	46.50
6420	3/4"		20	.400		18.30	23		41.30	55
6430	1"		19	.421		13.40	24		37.40	51.50
7800	Water closet, wax gasket		96	.083		1.38	4.80		6.18	8.75
7820	Gasket toilet tank to bowl		32	.250		2.38	14.40		16.78	24
7830	Replacement diaphragm washer assy for ballcock valve		12	.667		2.38	38.50		40.88	60.50
7850	Dual flush valve		12	.667		20	38.50		58.50	80
8000	Water supply stops, polished chrome plate									
8200	Angle, 3/8"	1 Plum	24	.333	Ea.	12.90	19.20		32.10	43
8300	1/2"		22	.364		12.90	21		33.90	45.50
8400	Straight, 3/8"		26	.308		13.55	17.70		31.25	41.50
8500	1/2"		24	.333		13.55	19.20		32.75	44
8600	Water closet, angle, w/flex riser, 3/8"		24	.333		44	19.20		63.20	77.50
9100	Miscellaneous									
9720	Teflon tape, 1/2" x 520" roll				Ea.	1.07			1.07	1.18

22 41 39.70 Washer/Dryer Accessories

		Crew	Daily Output	Labor-Hours	Unit	Material	Labor	Equipment	Total	Total Incl O&P
0010	**WASHER/DRYER ACCESSORIES**									
1020	Valves ball type single lever									
1030	1/2" diam., IPS	1 Plum	21	.381	Ea.	54.50	22		76.50	93
1040	1/2" diam., solder	"	21	.381	"	54.50	22		76.50	93
1050	Recessed box, 16 ga., two hose valves and drain									
1060	1/2" size, 1-1/2" drain	1 Plum	18	.444	Ea.	116	25.50		141.50	167
1070	1/2" size, 2" drain	"	17	.471	"	108	27		135	160
1080	With grounding electric receptacle									
1090	1/2" size, 1-1/2" drain	1 Plum	18	.444	Ea.	127	25.50		152.50	179
1100	1/2" size, 2" drain	"	17	.471	"	138	27		165	192
1110	With grounding and dryer receptacle									
1120	1/2" size, 1-1/2" drain	1 Plum	18	.444	Ea.	157	25.50		182.50	212
1130	1/2" size, 2" drain	"	17	.471	"	159	27		186	216
1140	Recessed box 16 ga., ball valves with single lever and drain									
1150	1/2" size, 1-1/2" drain	1 Plum	19	.421	Ea.	219	24		243	278
1160	1/2" size, 2" drain	"	18	.444	"	197	25.50		222.50	256
1170	With grounding electric receptacle									
1180	1/2" size, 1-1/2" drain	1 Plum	19	.421	Ea.	190	24		214	246

22 41 Residential Plumbing Fixtures

22 41 39 – Residential Faucets, Supplies and Trim

22 41 39.70 Washer/Dryer Accessories

		Crew	Daily Output	Labor-Hours	Unit	Material	2014 Bare Costs Labor	Equipment	Total	Total Incl O&P
1190	1/2" size, 2" drain	1 Plum	18	.444	Ea.	211	25.50		236.50	271
1200	With grounding and dryer receptacles									
1210	1/2" size, 1-1/2" drain	1 Plum	19	.421	Ea.	208	24		232	265
1220	1/2" size, 2" drain	"	18	.444	"	230	25.50		255.50	293
1300	Recessed box, 20 ga., two hose valves and drain (economy type)									
1310	1/2" size, 1-1/2" drain	1 Plum	19	.421	Ea.	90.50	24		114.50	136
1320	1/2" size, 2" drain		18	.444		86	25.50		111.50	133
1330	Box with drain only		24	.333		52.50	19.20		71.70	86.50
1340	1/2" size, 1-1/2" ABS/PVC drain		19	.421		100	24		124	147
1350	1/2" size, 2" ABS/PVC drain		18	.444		108	25.50		133.50	157
1352	Box with drain and 15 A receptacle		24	.333		90.50	19.20		109.70	129
1360	1/2" size, 2" drain ABS/PVC, 15 A receptacle	↓	24	.333	↓	127	19.20		146.20	169
1400	Wall mounted									
1410	1/2" size, 1-1/2" plastic drain	1 Plum	19	.421	Ea.	24.50	24		48.50	63.50
1420	1/2" size, 2" plastic drain	"	18	.444	"	21	25.50		46.50	61.50
1500	Dryer vent kit									
1510	8' flex duct, clamps and outside hood	1 Plum	20	.400	Ea.	15.30	23		38.30	52
1980	Rough-in, supply, waste, and vent for washer boxes	"	3.46	2.310	"	520	133		653	770

22 42 Commercial Plumbing Fixtures

22 42 13 – Commercial Water Closets, Urinals, and Bidets

22 42 13.30 Urinals

			Crew	Daily Output	Labor-Hours	Unit	Material	Labor	Equipment	Total	Total Incl O&P
0010	**URINALS**	D2010-210									
0102	For automatic flush see Line 22 42 39.10 0972										
3000	Wall hung, vitreous china, with hanger & self-closing valve	R224000-30									
3100	Siphon jet type		Q-1	3	5.333	Ea.	315	276		591	765
3120	Blowout type			3	5.333		450	276		726	910
3140	Water saving .5 gpf	G		3	5.333		585	276		861	1,050
3300	Rough-in, supply, waste & vent			2.83	5.654		570	293		863	1,075
5000	Stall type, vitreous china, includes valve			2.50	6.400		715	330		1,045	1,275
6980	Rough-in, supply, waste and vent		↓	1.99	8.040	↓	585	415		1,000	1,275
8000	Waterless (no flush) urinal										
8010	Wall hung										
8014	Fiberglass reinforced polyester										
8020	Standard unit	G	Q-1	21.30	.751	Ea.	385	39		424	480
8030	ADA compliant unit	G	"	21.30	.751		400	39		439	500
8070	For solid color, add	G					48			48	53
8080	For 2" brass flange, (new const.), add	G	Q-1	96	.167	↓	19.20	8.65		27.85	34
8200	Vitreous china										
8220	ADA compliant unit, 14"	G	Q-1	21.30	.751	Ea.	198	39		237	277
8240	ADA compliant unit, 18"	G		21.30	.751		320	39		359	410
8250	ADA compliant unit, 15.5"	G	↓	21.30	.751		272	39		311	360
8270	For solid color, add	G					48			48	53
8290	Rough-in, supply, waste & vent	G	Q-1	2.92	5.479	↓	520	284		804	1,000
8400	Trap liquid										
8410	1 quart	G				Ea.	15.95			15.95	17.55
8420	1 gallon	G				"	58			58	64

22 42 Commercial Plumbing Fixtures

22 42 13 – Commercial Water Closets, Urinals, and Bidets

22 42 13.40 Water Closets

		Crew	Daily Output	Labor-Hours	Unit	Material	2014 Bare Costs Labor	Equipment	Total	Total Incl O&P
0010	**WATER CLOSETS**									
3000	Bowl only, with flush valve, seat, 1.6 gpf unless noted									
3100	Wall hung	Q-1	5.80	2.759	Ea.	755	143		898	1,050
3200	For rough-in, supply, waste and vent, single WC		2.56	6.250		890	325		1,215	1,475
3300	Floor mounted		5.80	2.759		355	143		498	605
3350	With wall outlet		5.80	2.759		530	143		673	795
3360	With floor outlet, 1.28 gpf [G]		5.80	2.759		585	143		728	860
3362	With floor outlet, 1.28 gpf, ADA [G]		5.80	2.759		620	143		763	900
3370	For rough-in, supply, waste and vent, single WC		2.84	5.634		410	292		702	895
3390	Floor mounted children's size, 10-3/4" high									
3392	With automatic flush sensor, 1.6 gpf	Q-1	6.20	2.581	Ea.	610	134		744	870
3396	With automatic flush sensor, 1.28 gpf		6.20	2.581		610	134		744	870
3400	For rough-in, supply, waste and vent, single WC		2.84	5.634		410	292		702	895
3500	Gang side by side carrier system, rough-in, supply, waste & vent									
3510	For single hook-up	Q-1	1.97	8.122	Ea.	1,075	420		1,495	1,825
3520	For each additional hook-up, add	"	2.14	7.477	"	1,025	385		1,410	1,700
3550	Gang back to back carrier system, rough-in, supply, waste & vent									
3560	For pair hook-up	Q-1	1.76	9.091	Pr.	1,700	470		2,170	2,575
3570	For each additional pair hook-up, add	"	1.81	8.840	"	1,650	460		2,110	2,525

22 42 16 – Commercial Lavatories and Sinks

22 42 16.14 Lavatories

		Crew	Daily Output	Labor-Hours	Unit	Material	Labor	Equipment	Total	Total Incl O&P
0010	**LAVATORIES**, With trim, white unless noted otherwise									
0020	Commercial lavatories same as residential. See Section 22 41 16.10									

22 42 16.20 Commercial Sinks

		Crew	Daily Output	Labor-Hours	Unit	Material	Labor	Equipment	Total	Total Incl O&P
0010	**COMMERCIAL SINKS**									
5900	Scullery sink, stainless steel									
5910	1 bowl and drain board, 43" x 22" O.D.	Q-1	5.40	2.963	Ea.	4,050	153		4,203	4,675
5920	2 bowls and drain board, 49" x 22" O.D.		4.60	3.478		6,675	180		6,855	7,600
5930	3 bowls and drain board, 43" x 22" O.D.		4.20	3.810		6,825	197		7,022	7,825
5940	1 bowl and drain board, 50" x 28" O.D., with legs		5.40	2.963		3,025	153		3,178	3,550

22 42 16.30 Classroom Sinks

		Crew	Daily Output	Labor-Hours	Unit	Material	Labor	Equipment	Total	Total Incl O&P
0010	**CLASSROOM SINKS**									
6020	Countertop, stainless steel									
6024	with faucet, bubbler and strainer, ADA compliant									
6036	25" x 17" single bowl	Q-1	5.20	3.077	Ea.	1,150	159		1,309	1,525
6040	28" x 22" single bowl		5.20	3.077		1,200	159		1,359	1,550
6044	31" x 19" single bowl		5.20	3.077		1,300	159		1,459	1,675
6070	37" x 17" double bowl		4.40	3.636		1,700	188		1,888	2,150
6100	For rough-in, supply, waste and vent, counter top classroom sinks		2.14	7.477		460	385		845	1,100

22 42 16.34 Laboratory Countertops and Sinks

		Crew	Daily Output	Labor-Hours	Unit	Material	Labor	Equipment	Total	Total Incl O&P
0010	**LABORATORY COUNTERTOPS AND SINKS**									
0050	Laboratory sinks, corrosion resistant									
1000	Stainless steel sink, bench mounted, with									
1020	plug & waste fitting with 1-1/2" straight threads									
1030	Single bowl, 2 drainboards, backnut & strainer									
1050	18-1/2" x 15-1/2" x 12-1/2" sink, 54" x 24" O.D.	Q-1	3	5.333	Ea.	1,700	276		1,976	2,300
1100	Single bowl, single drainboard, backnut & strainer									
1130	18-1/2" x 15-1/2" x 12-1/2" sink, 47" x 24" O.D.	Q-1	3	5.333	Ea.	1,200	276		1,476	1,725
1146	Double bowl, single drainboard, backnut & strainer									
1150	18-1/2" x 15-1/2" x 12-1/2" sink, 70" x 24" O.D.	Q-1	3	5.333	Ea.	1,875	276		2,151	2,500
1280	Polypropylene									

22 42 Commercial Plumbing Fixtures

22 42 16 – Commercial Lavatories and Sinks

22 42 16.34 Laboratory Countertops and Sinks		Crew	Daily Output	Labor-Hours	Unit	Material	2014 Bare Costs Labor	Equipment	Total	Total Incl O&P
1290	Flanged 1-1/4" wide, rectangular with strainer									
1300	plug & waste fitting, 1-1/2" straight threads									
1320	12" x 12" x 8" sink, 14-1/2" x 14-1/2" O.D.	Q-1	4	4	Ea.	278	207		485	620
1340	16" x 16" x 8" sink, 18-1/2" x 18-1/2" O.D.		4	4		370	207		577	720
1360	21" x 18" x 10" sink, 23-1/2" x 20-1/2" O.D.		4	4		410	207		617	765
1490	For rough-in, supply, waste & vent, add		2.02	7.921		219	410		629	860
1600	Polypropylene									
1620	Cup sink, oval, integral strainers									
1640	6" x 3" I.D., 7" x 4" O.D.	Q-1	6	2.667	Ea.	115	138		253	335
1660	9" x 3" I.D., 10" x 4-1/2" O.D.	"	6	2.667		136	138		274	360
1740	1-1/2" diam. x 11" long					52.50			52.50	58
1980	For rough-in, supply, waste & vent, add	Q-1	1.70	9.412		223	490		713	980

22 42 16.40 Service Sinks

		Crew	Daily Output	Labor-Hours	Unit	Material	Labor	Equipment	Total	Total Incl O&P
0010	**SERVICE SINKS**									
6650	Service, floor, corner, P.E. on C.I., 28" x 28"	Q-1	4.40	3.636	Ea.	990	188		1,178	1,375
6750	Vinyl coated rim guard, add					67.50			67.50	74
6760	Mop sink, molded stone, 24" x 36"	1 Plum	3.33	2.402		273	138		411	510
6770	Mop sink, molded stone, 24" x 36", w/rim 3 sides	"	3.33	2.402		278	138		416	515
6790	For rough-in, supply, waste & vent, floor service sinks	Q-1	1.64	9.756		1,475	505		1,980	2,400
7000	Service, wall, P.E. on C.I., roll rim, 22" x 18"		4	4		750	207		957	1,150
7100	24" x 20"		4	4		835	207		1,042	1,225
7600	For stainless steel rim guard, two sides only, add					88.50			88.50	97.50
7800	For stainless steel rim guard, front only, add					56			56	62
8600	Vitreous china, 22" x 20"	Q-1	4	4		595	207		802	970
8960	For stainless steel rim guard, front or one side, add					56			56	62
8980	For rough-in, supply, waste & vent, wall service sinks	Q-1	1.30	12.308		1,800	640		2,440	2,975

22 42 23 – Commercial Showers

22 42 23.30 Group Showers

		Crew	Daily Output	Labor-Hours	Unit	Material	Labor	Equipment	Total	Total Incl O&P
0010	**GROUP SHOWERS**									
6000	Group, w/pressure balancing valve, rough-in and rigging not included									
6800	Column, 6 heads, no receptors, less partitions	Q-1	3	5.333	Ea.	3,000	276		3,276	3,725
6900	With stainless steel partitions		1	16		9,250	830		10,080	11,500
7600	5 heads, no receptors, less partitions		3	5.333		2,475	276		2,751	3,150
7620	4 heads (1 handicap) no receptors, less partitions		3	5.333		3,975	276		4,251	4,800
7700	With stainless steel partitions		1	16		7,600	830		8,430	9,600
8000	Wall, 2 heads, no receptors, less partitions		4	4		1,375	207		1,582	1,850
8100	With stainless steel partitions		2	8		3,225	415		3,640	4,175

22 42 33 – Wash Fountains

22 42 33.20 Commercial Wash Fountains

			Crew	Daily Output	Labor-Hours	Unit	Material	Labor	Equipment	Total	Total Incl O&P
0010	**COMMERCIAL WASH FOUNTAINS**	D2010-610									
1900	Group, foot control										
2000	Precast terrazzo, circular, 36" diam., 5 or 6 persons		Q-2	3	8	Ea.	5,800	430		6,230	7,025
2100	54" diameter for 8 or 10 persons			2.50	9.600		9,025	515		9,540	10,700
2400	Semi-circular, 36" diam. for 3 persons			3	8		4,775	430		5,205	5,900
2500	54" diam. for 4 or 5 persons			2.50	9.600		6,775	515		7,290	8,225
2700	Quarter circle (corner), 54" for 3 persons			3.50	6.857		6,000	370		6,370	7,150
3000	Stainless steel, circular, 36" diameter			3.50	6.857		4,850	370		5,220	5,875
3100	54" diameter			2.80	8.571		7,050	460		7,510	8,450
3400	Semi-circular, 36" diameter			3.50	6.857		4,800	370		5,170	5,825
3500	54" diameter			2.80	8.571		5,425	460		5,885	6,650
5000	Thermoplastic, pre-assembled, circular, 36" diameter			6	4		3,850	215		4,065	4,575

22 42 Commercial Plumbing Fixtures

22 42 33 – Wash Fountains

22 42 33.20 Commercial Wash Fountains

		Crew	Daily Output	Labor-Hours	Unit	Material	2014 Bare Costs Labor	Equipment	Total	Total Incl O&P
5100	54" diameter	Q-2	4	6	Ea.	4,750	320		5,070	5,700
5400	Semi-circular, 36" diameter		6	4		3,850	215		4,065	4,575
5600	54" diameter		4	6		4,625	320		4,945	5,575
5610	Group, infrared control, barrier free									
5614	Precast terrazzo									
5620	Semi-circular 36" diam. for 3 persons	Q-2	3	8	Ea.	7,150	430		7,580	8,525
5630	46" diam. for 4 persons		2.80	8.571		7,725	460		8,185	9,200
5640	Circular, 54" diam. for 8 persons, button control		2.50	9.600		9,125	515		9,640	10,800
5700	Rough-in, supply, waste and vent for above wash fountains	Q-1	1.82	8.791		660	455		1,115	1,425
6200	Duo for small washrooms, stainless steel		2	8		2,725	415		3,140	3,600
6400	Bowl with backsplash		2	8		2,250	415		2,665	3,100
6500	Rough-in, supply, waste & vent for duo fountains		2.02	7.921		405	410		815	1,075

22 42 39 – Commercial Faucets, Supplies, and Trim

22 42 39.10 Faucets and Fittings

		Crew	Daily Output	Labor-Hours	Unit	Material	Labor	Equipment	Total	Total Incl O&P
0010	**FAUCETS AND FITTINGS**									
0840	Flush valves, with vacuum breaker									
0850	Water closet									
0860	Exposed, rear spud	1 Plum	8	1	Ea.	148	57.50		205.50	250
0870	Top spud		8	1		149	57.50		206.50	250
0880	Concealed, rear spud		8	1		197	57.50		254.50	305
0890	Top spud		8	1		159	57.50		216.50	262
0900	Wall hung		8	1		177	57.50		234.50	282
0910	Dual flush flushometer		12	.667		197	38.50		235.50	274
0912	Flushometer retrofit kit		18	.444		16.30	25.50		41.80	56.50
0920	Urinal									
0930	Exposed, stall	1 Plum	8	1	Ea.	149	57.50		206.50	251
0940	Wall, (washout)		8	1		149	57.50		206.50	250
0950	Pedestal, top spud		8	1		143	57.50		200.50	244
0960	Concealed, stall		8	1		156	57.50		213.50	259
0970	Wall (washout)		8	1		168	57.50		225.50	272
0971	Automatic flush sensor and operator for									
0972	urinals or water closets, standard [G]	1 Plum	8	1	Ea.	450	57.50		507.50	580
0980	High efficiency water saving									
0984	Water closets, 1.28 gpf [G]	1 Plum	8	1	Ea.	415	57.50		472.50	545
0988	Urinals, .5 gpf [G]	"	8	1	"	415	57.50		472.50	545
2790	Faucets for lavatories									
2800	Self-closing, center set	1 Plum	10	.800	Ea.	131	46		177	214
2810	Automatic sensor and operator, with faucet head		6.15	1.301		450	75		525	610
3000	Service sink faucet, cast spout, pail hook, hose end		14	.571		80	33		113	138

22 42 39.30 Carriers and Supports

		Crew	Daily Output	Labor-Hours	Unit	Material	Labor	Equipment	Total	Total Incl O&P
0010	**CARRIERS AND SUPPORTS**, For plumbing fixtures									
0500	Drinking fountain, wall mounted									
0600	Plate type with studs, top back plate	1 Plum	7	1.143	Ea.	83.50	66		149.50	191
0700	Top front and back plate		7	1.143		102	66		168	212
0800	Top & bottom, front & back plates, w/bearing jacks		7	1.143		148	66		214	262
3000	Lavatory, concealed arm									
3050	Floor mounted, single									
3100	High back fixture	1 Plum	6	1.333	Ea.	465	76.50		541.50	625
3200	Flat slab fixture		6	1.333		400	76.50		476.50	555
3220	Paraplegic		6	1.333		515	76.50		591.50	685
3250	Floor mounted, back to back									
3300	High back fixtures	1 Plum	5	1.600	Ea.	660	92		752	865

22 42 Commercial Plumbing Fixtures

22 42 39 – Commercial Faucets, Supplies, and Trim

22 42 39.30 Carriers and Supports		Crew	Daily Output	Labor-Hours	Unit	Material	2014 Bare Costs Labor	Equipment	Total	Total Incl O&P
3400	Flat slab fixtures	1 Plum	5	1.600	Ea.	810	92		902	1,025
3430	Paraplegic	↓	5	1.600	↓	755	92		847	970
3500	Wall mounted, in stud or masonry									
3600	High back fixture	1 Plum	6	1.333	Ea.	274	76.50		350.50	415
3700	Flat slab fixture	"	6	1.333	"	237	76.50		313.50	375
4000	Exposed arm type, floor mounted									
4100	Single high back or flat slab fixture	1 Plum	6	1.333	Ea.	660	76.50		736.50	840
4200	Back to back, high back or flat slab fixtures		5	1.600		1,025	92		1,117	1,300
4300	Wall mounted, High back or flat slab lavatory	↓	6	1.333	↓	455	76.50		531.50	620
4600	Sink, floor mounted									
4650	Exposed arm system									
4700	Single heavy fixture	1 Plum	5	1.600	Ea.	770	92		862	990
4750	Single heavy sink with slab		5	1.600		990	92		1,082	1,250
4800	Back to back, standard fixtures		5	1.600		570	92		662	765
4850	Back to back, heavy fixtures		5	1.600		875	92		967	1,100
4900	Back to back, heavy sink with slab	↓	5	1.600	↓	875	92		967	1,100
4950	Exposed offset arm system									
5000	Single heavy deep fixture	1 Plum	5	1.600	Ea.	735	92		827	950
5100	Plate type system									
5200	With bearing jacks, single fixture	1 Plum	5	1.600	Ea.	845	92		937	1,075
5300	With exposed arms, single heavy fixture		5	1.600		1,075	92		1,167	1,350
5400	Wall mounted, exposed arms, single heavy fixture		5	1.600		395	92		487	575
6000	Urinal, floor mounted, 2" or 3" coupling, blowout type		6	1.333		490	76.50		566.50	655
6100	With fixture or hanger bolts, blowout or washout		6	1.333		350	76.50		426.50	500
6200	With bearing plate		6	1.333		390	76.50		466.50	545
6300	Wall mounted, plate type system	↓	6	1.333	↓	300	76.50		376.50	450
6980	Water closet, siphon jet									
7000	Horizontal, adjustable, caulk									
7040	Single, 4" pipe size	1 Plum	5.33	1.501	Ea.	655	86.50		741.50	850
7050	4" pipe size, paraplegic		5.33	1.501		655	86.50		741.50	850
7060	5" pipe size		5.33	1.501		845	86.50		931.50	1,050
7100	Double, 4" pipe size		5	1.600		1,225	92		1,317	1,500
7110	4" pipe size, paraplegic		5	1.600		1,225	92		1,317	1,500
7120	5" pipe size	↓	5	1.600	↓	1,475	92		1,567	1,775
7160	Horizontal, adjustable, extended, caulk									
7180	Single, 4" pipe size	1 Plum	5.33	1.501	Ea.	885	86.50		971.50	1,100
7200	5" pipe size		5.33	1.501		1,125	86.50		1,211.50	1,350
7240	Double, 4" pipe size		5	1.600		1,550	92		1,642	1,850
7260	5" pipe size	↓	5	1.600	↓	1,875	92		1,967	2,200
7400	Vertical, adjustable, caulk or thread									
7440	Single, 4" pipe size	1 Plum	5.33	1.501	Ea.	905	86.50		991.50	1,125
7460	5" pipe size		5.33	1.501		1,125	86.50		1,211.50	1,350
7480	6" pipe size		5	1.600		1,275	92		1,367	1,550
7520	Double, 4" pipe size		5	1.600		1,350	92		1,442	1,625
7540	5" pipe size		5	1.600		1,550	92		1,642	1,850
7560	6" pipe size	↓	4	2		1,725	115		1,840	2,075
7600	Vertical, adjustable, extended, caulk									
7620	Single, 4" pipe size	1 Plum	5.33	1.501	Ea.	905	86.50		991.50	1,125
7640	5" pipe size		5.33	1.501		1,125	86.50		1,211.50	1,350
7680	6" pipe size		5	1.600		1,275	92		1,367	1,550
7720	Double, 4" pipe size		5	1.600		1,350	92		1,442	1,625
7740	5" pipe size		5	1.600		1,550	92		1,642	1,850
7760	6" pipe size	↓	4	2	↓	1,725	115		1,840	2,075

22 42 Commercial Plumbing Fixtures

22 42 39 – Commercial Faucets, Supplies, and Trim

22 42 39.30 Carriers and Supports

		Crew	Daily Output	Labor-Hours	Unit	Material	2014 Bare Costs Labor	Equipment	Total	Total Incl O&P
7780	Water closet, blow out									
7800	Vertical offset, caulk or thread									
7820	Single, 4" pipe size	1 Plum	5.33	1.501	Ea.	665	86.50		751.50	865
7840	Double, 4" pipe size	"	5	1.600	"	1,150	92		1,242	1,400
7880	Vertical offset, extended, caulk									
7900	Single, 4" pipe size	1 Plum	5.33	1.501	Ea.	835	86.50		921.50	1,050
7920	Double, 4" pipe size	"	5	1.600	"	1,300	92		1,392	1,600
7960	Vertical, for floor mounted back-outlet									
7980	Single, 4" thread, 2" vent	1 Plum	5.33	1.501	Ea.	585	86.50		671.50	775
8000	Double, 4" thread, 2" vent	"	6	1.333	"	1,725	76.50		1,801.50	2,025
8040	Vertical, for floor mounted back-outlet, extended									
8060	Single, 4" caulk, 2" vent	1 Plum	6	1.333	Ea.	585	76.50		661.50	760
8080	Double, 4" caulk, 2" vent	"	6	1.333	"	1,725	76.50		1,801.50	2,025
8200	Water closet, residential									
8220	Vertical centerline, floor mount									
8240	Single, 3" caulk, 2" or 3" vent	1 Plum	6	1.333	Ea.	535	76.50		611.50	700
8260	4" caulk, 2" or 4" vent		6	1.333		690	76.50		766.50	870
8280	3" copper sweat, 3" vent		6	1.333		480	76.50		556.50	640
8300	4" copper sweat, 4" vent		6	1.333		580	76.50		656.50	755
8400	Vertical offset, floor mount									
8420	Single, 3" or 4" caulk, vent	1 Plum	4	2	Ea.	665	115		780	910
8440	3" or 4" copper sweat, vent		5	1.600		665	92		757	875
8460	Double, 3" or 4" caulk, vent		4	2		1,150	115		1,265	1,425
8480	3" or 4" copper sweat, vent		5	1.600		1,150	92		1,242	1,400
9000	Water cooler (electric), floor mounted									
9100	Plate type with bearing plate, single	1 Plum	6	1.333	Ea.	335	76.50		411.50	485
9140	Plate type with bearing plate, back to back	"	4	2	"	470	115		585	695

22 43 Healthcare Plumbing Fixtures

22 43 13 – Healthcare Water Closets

22 43 13.40 Water Closets

			Crew	Daily Output	Labor-Hours	Unit	Material	Labor	Equipment	Total	Total Incl O&P
0010	**WATER CLOSETS**										
1000	Bowl only, 1 piece, w/seat and flush valve, ADA compliant 18" high										
1030	Floor mounted										
1150	With wall outlet	♿	Q-1	5.30	3.019	Ea.	385	156		541	660
1180	For rough-in, supply, waste and vent			2.84	5.634		410	292		702	895
1200	With floor outlet			5.30	3.019		350	156		506	620
1800	For rough-in, supply, waste and vent			3.05	5.246		330	272		602	775
3100	Wall hung	♿		5.80	2.759		755	143		898	1,050
3150	Hospital type, slotted rim for bed pan										
3156	Elongated bowl, top spud		Q-1	5.80	2.759	Ea.	570	143		713	845
3160	Elongated bowl, rear spud			5.80	2.759		420	143		563	675
3200	For rough-in, supply, waste and vent, single WC			2.56	6.250		890	325		1,215	1,475
3300	Floor mounted										
3320	Bariatric, (1,200 lb. capacity), elongated bowl, ADA	♿	Q-1	4.60	3.478	Ea.	2,775	180		2,955	3,325
3360	Hospital type, slotted rim for bed pan										
3370	Elongated bowl, top spud	♿	Q-1	5	3.200	Ea.	370	166		536	660
3380	Elongated bowl, rear spud			5	3.200		405	166		571	695
3500	For rough-in, supply, waste and vent			3.05	5.246		330	272		602	775

22 43 Healthcare Plumbing Fixtures

22 43 16 – Healthcare Sinks

22 43 16.10 Sinks

		Crew	Daily Output	Labor-Hours	Unit	Material	2014 Bare Costs Labor	Equipment	Total	Total Incl O&P
0010	**SINKS**									
0020	Vitreous china									
6702	Hospital type, without trim (see Section 22 41 39.10)									
6710	20" x 18", contoured splash shield	Q-1	8	2	Ea.	93.50	104		197.50	259
6730	28" x 20", surgeon, side decks		8	2		575	104		679	790
6740	28" x 22", surgeon scrub-up, deep bowl		8	2		825	104		929	1,075
6750	20" x 27", patient, wheelchair		7	2.286		475	118		593	700
6760	30" x 22", all purpose		7	2.286		685	118		803	935
6770	30" x 22", plaster work		7	2.286		690	118		808	935
6820	20" x 24" clinic service, liquid/solid waste		6	2.667		875	138		1,013	1,175

22 43 19 – Healthcare Bathtubs

22 43 19.10 Bathtubs

		Crew	Daily Output	Labor-Hours	Unit	Material	Labor	Equipment	Total	Total Incl O&P
0010	**BATHTUBS**									
5002	Hospital type, with trim (see Section 22 41 39.10)									
5050	Bathing pool, porcelain enamel on cast iron, grab bars									
5060	pop-up drain, 72" x 36"	Q-1	3	5.333	Ea.	3,600	276		3,876	4,400
5100	Perineal (sitz), vitreous china		3	5.333		1,250	276		1,526	1,800
5120	For pedestal, vitreous china, add		8	2		241	104		345	420
5300	Whirlpool, porcelain enamel on cast iron, 72" x 36"		1	16		3,450	830		4,280	5,025
5310	For color add					5%				
5311	For designer colors and trim add					15%				

22 43 23 – Healthcare Showers

22 43 23.10 Showers

		Crew	Daily Output	Labor-Hours	Unit	Material	Labor	Equipment	Total	Total Incl O&P
0010	**SHOWERS**									
5950	Module, handicap, SS panel, fixed & hand held head, control									
5960	valves, grab bar, curtain & rod, folding seat	1 Plum	4	2	Ea.	1,900	115		2,015	2,275

22 43 39 – Healthcare Faucets

22 43 39.10 Faucets and Fittings

		Crew	Daily Output	Labor-Hours	Unit	Material	Labor	Equipment	Total	Total Incl O&P
0010	**FAUCETS AND FITTINGS**									
2850	Medical, bedpan cleanser, with pedal valve,	1 Plum	12	.667	Ea.	855	38.50		893.50	1,000
2860	With screwdriver stop valve		12	.667		440	38.50		478.50	545
2870	With self-closing spray valve		12	.667		272	38.50		310.50	355
2900	Faucet, gooseneck spout, wrist handles, grid drain		10	.800		194	46		240	283
2940	Mixing valve, knee action, screwdriver stops		4	2		475	115		590	700

22 45 Emergency Plumbing Fixtures

22 45 13 – Emergency Showers

22 45 13.10 Emergency Showers

		Crew	Daily Output	Labor-Hours	Unit	Material	Labor	Equipment	Total	Total Incl O&P
0010	**EMERGENCY SHOWERS**, Rough-in not included									
5000	Shower, single head, drench, ball valve, pull, freestanding	Q-1	4	4	Ea.	380	207		587	735
5200	Horizontal or vertical supply		4	4		540	207		747	910
6000	Multi-nozzle, eye/face wash combination		4	4		640	207		847	1,025
6400	Multi-nozzle, 12 spray, shower only		4	4		1,950	207		2,157	2,475
6600	For freeze-proof, add		6	2.667		455	138		593	710
8000	Walk-thru decontamination with eye-face wash		2	8		3,275	415		3,690	4,250
8200	For freeze proof, add		4	4		550	207		757	920

22 45 Emergency Plumbing Fixtures

22 45 16 – Eyewash Equipment

22 45 16.10 Eyewash Safety Equipment

		Crew	Daily Output	Labor-Hours	Unit	Material	2014 Bare Costs Labor	Equipment	Total	Total Incl O&P
0010	**EYEWASH SAFETY EQUIPMENT**, Rough-in not included									
1000	Eye wash fountain									
1400	Plastic bowl, pedestal mounted	Q-1	4	4	Ea.	269	207		476	610
1600	Unmounted		4	4		240	207		447	580
1800	Wall mounted		4	4		435	207		642	795
2000	Stainless steel, pedestal mounted		4	4		340	207		547	690
2200	Unmounted		4	4		266	207		473	605
2400	Wall mounted		4	4		282	207		489	625

22 45 19 – Self-Contained Eyewash Equipment

22 45 19.10 Self-Contained Eyewash Safety Equipment

		Crew	Daily Output	Labor-Hours	Unit	Material	Labor	Equipment	Total	Total Incl O&P
0010	**SELF-CONTAINED EYEWASH SAFETY EQUIPMENT**									
3000	Eye wash, portable, self-contained				Ea.	990			990	1,100

22 45 26 – Eye/Face Wash Equipment

22 45 26.10 Eye/Face Wash Safety Equipment

		Crew	Daily Output	Labor-Hours	Unit	Material	Labor	Equipment	Total	Total Incl O&P
0010	**EYE/FACE WASH SAFETY EQUIPMENT**, Rough-in not included									
4000	Eye and face wash, combination fountain									
4200	Stainless steel, pedestal mounted	Q-1	4	4	Ea.	1,050	207		1,257	1,475
4400	Unmounted		4	4		266	207		473	605
4600	Wall mounted		4	4		282	207		489	625

22 46 Security Plumbing Fixtures

22 46 13 – Security Water Closets and Urinals

22 46 13.10 Security Water Closets and Urinals

		Crew	Daily Output	Labor-Hours	Unit	Material	Labor	Equipment	Total	Total Incl O&P
0010	**SECURITY WATER CLOSETS AND URINALS**, Stainless steel									
2000	Urinal, back supply and flush									
2200	Wall hung	Q-1	4	4	Ea.	1,225	207		1,432	1,675
2240	Stall		2.50	6.400		2,100	330		2,430	2,800
2300	For urinal rough-in, supply, waste and vent		1.49	10.738		345	555		900	1,225
3000	Water closet, integral seat, back supply and flush									
3300	Wall hung, wall outlet	Q-1	5.80	2.759	Ea.	985	143		1,128	1,300
3400	Floor mount, wall outlet		5.80	2.759		1,275	143		1,418	1,625
3440	Floor mount, floor outlet		5.80	2.759		1,300	143		1,443	1,650
3480	For recessed tissue holder, add					88			88	96.50
3500	For water closet rough-in, supply, waste and vent	Q-1	1.19	13.445		350	695		1,045	1,425
5000	Water closet and lavatory units, push button filler valves,									
5010	soap & paper holders, seat									
5300	Wall hung	Q-1	5	3.200	Ea.	2,100	166		2,266	2,575
5400	Floor mount		5	3.200		2,000	166		2,166	2,450
6300	For unit rough-in, supply, waste and vent		1	16		405	830		1,235	1,700

22 46 16 – Security Lavatories and Sinks

22 46 16.10 Security Lavatories

		Crew	Daily Output	Labor-Hours	Unit	Material	Labor	Equipment	Total	Total Incl O&P
0010	**SECURITY LAVATORIES**, Stainless steel									
1000	Lavatory, wall hung, push button filler valve									
1100	Rectangular bowl	Q-1	8	2	Ea.	1,075	104		1,179	1,350
1200	Oval bowl		8	2		1,125	104		1,229	1,375
1240	Oval bowl, corner mount		8	2		1,225	104		1,329	1,500
1300	For lavatory rough-in, supply, waste and vent		1.50	10.667		485	555		1,040	1,375

22 46 Security Plumbing Fixtures

22 46 63 – Security Service Sink

22 46 63.10 Security Service Sink		Crew	Daily Output	Labor-Hours	Unit	Material	2014 Bare Costs Labor	Equipment	Total	Total Incl O&P
0010	**SECURITY SERVICE SINK**, Stainless steel									
1700	Service sink, with soap dish									
1740	24" x 19" size	Q-1	3	5.333	Ea.	2,175	276		2,451	2,825
1790	For sink rough-in, supply, waste and vent	"	.89	17.978	"	1,225	930		2,155	2,750

22 46 73 – Security Shower

22 46 73.10 Security Shower		Crew	Daily Output	Labor-Hours	Unit	Material	Labor	Equipment	Total	Total Incl O&P
0010	**SECURITY SHOWER**, Stainless steel									
1800	Shower cabinet, unitized									
1840	36" x 36" x 88"	Q-1	2.20	7.273	Ea.	4,950	375		5,325	6,025
1900	Shower package for built-in									
1940	Hot & cold valves, recessed soap dish	Q-1	6	2.667	Ea.	540	138		678	800

22 47 Drinking Fountains and Water Coolers

22 47 13 – Drinking Fountains

22 47 13.10 Drinking Water Fountains

			Crew	Daily Output	Labor-Hours	Unit	Material	Labor	Equipment	Total	Total Incl O&P
0010	**DRINKING WATER FOUNTAINS**, For connection to cold water supply										
0802	For remote water chiller, see Section 22 47 23.10										
1000	Wall mounted, non-recessed	R224000-30									
1200	Aluminum,										
1280	Dual bubbler type	R224000-50	1 Plum	3.20	2.500	Ea.	2,400	144		2,544	2,875
1400	Bronze, with no back			4	2		990	115		1,105	1,275
1600	Cast iron, enameled, low back, single bubbler	D2010-810		4	2		920	115		1,035	1,175
1640	Dual bubbler type			3.20	2.500		1,475	144		1,619	1,850
1680	Triple bubbler type			3.20	2.500		1,925	144		2,069	2,325
1800	Cast aluminum, enameled, for correctional institutions			4	2		1,625	115		1,740	1,975
2000	Fiberglass, 12" back, single bubbler unit			4	2		1,875	115		1,990	2,225
2040	Dual bubbler			3.20	2.500		2,100	144		2,244	2,550
2080	Triple bubbler			3.20	2.500		2,325	144		2,469	2,775
2200	Polymarble, no back, single bubbler			4	2		705	115		820	950
2240	Dual bubbler			3.20	2.500		1,700	144		1,844	2,100
2280	Triple bubbler			3.20	2.500		2,100	144		2,244	2,525
2400	Precast stone, no back			4	2		950	115		1,065	1,225
2700	Stainless steel, single bubbler, no back			4	2		1,025	115		1,140	1,325
2740	With back			4	2		540	115		655	765
2780	Dual handle & wheelchair projection type			4	2		730	115		845	980
2820	Dual level for handicapped type			3.20	2.500		1,500	144		1,644	1,875
2840	Vandal resistant type			4	2		650	115		765	890
3300	Vitreous china										
3340	7" back		1 Plum	4	2	Ea.	610	115		725	845
3940	For vandal-resistant bottom plate, add						73.50			73.50	81
3960	For freeze-proof valve system, add		1 Plum	2	4		700	230		930	1,125
3980	For rough-in, supply and waste, add		"	2.21	3.620		365	208		573	715
4000	Wall mounted, semi-recessed										
4200	Poly-marble, single bubbler		1 Plum	4	2	Ea.	955	115		1,070	1,225
4600	Stainless steel, satin finish, single bubbler			4	2		1,125	115		1,240	1,425
4900	Vitreous china, single bubbler			4	2		880	115		995	1,150
5980	For rough-in, supply and waste, add			1.83	4.372		365	252		617	780
6000	Wall mounted, fully recessed										
6400	Poly-marble, single bubbler		1 Plum	4	2	Ea.	1,600	115		1,715	1,925
6440	For water glass filler, add						83			83	91.50

22 47 Drinking Fountains and Water Coolers

22 47 13 – Drinking Fountains

22 47 13.10 Drinking Water Fountains

		Crew	Daily Output	Labor-Hours	Unit	Material	2014 Bare Costs Labor	Equipment	Total	Total Incl O&P
6800	Stainless steel, single bubbler	1 Plum	4	2	Ea.	1,525	115		1,640	1,850
6900	Fountain and cuspidor combination		2	4		2,950	230		3,180	3,600
7560	For freeze-proof valve system, add		2	4		795	230		1,025	1,225
7580	For rough-in, supply and waste, add		1.83	4.372		365	252		617	780
7600	Floor mounted, pedestal type									
7700	Aluminum, architectural style, C.I. base	1 Plum	2	4	Ea.	2,250	230		2,480	2,825
7780	Wheelchair handicap unit		2	4		1,625	230		1,855	2,125
8000	Bronze, architectural style		2	4		1,925	230		2,155	2,475
8040	Enameled steel cylindrical column style		2	4		2,125	230		2,355	2,675
8200	Precast stone/concrete, cylindrical column		1	8		1,350	460		1,810	2,200
8240	Wheelchair handicap unit		1	8		2,725	460		3,185	3,700
8400	Stainless steel, architectural style		2	4		1,750	230		1,980	2,275
8600	Enameled iron, heavy duty service, 2 bubblers		2	4		2,575	230		2,805	3,200
8660	4 bubblers		2	4		3,950	230		4,180	4,675
8880	For freeze-proof valve system, add		2	4		665	230		895	1,075
8900	For rough-in, supply and waste, add		1.83	4.372		365	252		617	780
9100	Deck mounted									
9500	Stainless steel, circular receptor	1 Plum	4	2	Ea.	415	115		530	630
9540	14" x 9" receptor		4	2		345	115		460	550
9580	25" x 17" deep receptor, with water glass filler		3	2.667		294	153		447	555
9760	White enameled steel, 14" x 9" receptor		4	2		360	115		475	570
9860	White enameled cast iron, 24" x 16" receptor		3	2.667		445	153		598	720
9980	For rough-in, supply and waste, add		1.83	4.372		365	252		617	780

22 47 16 – Pressure Water Coolers

22 47 16.10 Electric Water Coolers

		Crew	Daily Output	Labor-Hours	Unit	Material	2014 Bare Costs Labor	Equipment	Total	Total Incl O&P
0010	**ELECTRIC WATER COOLERS** D2010-820									
0100	Wall mounted, non-recessed									
0140	4 GPH	Q-1	4	4	Ea.	680	207		887	1,075
0160	8 GPH, barrier free, sensor operated		4	4		1,025	207		1,232	1,475
0180	8.2 GPH		4	4		725	207		932	1,100
0220	14.3 GPH		4	4		830	207		1,037	1,225
0600	8 GPH hot and cold water		4	4		1,050	207		1,257	1,500
0640	For stainless steel cabinet, add R224000-50					84			84	92
1000	Dual height, 8.2 GPH	Q-1	3.80	4.211		2,025	218		2,243	2,550
1040	14.3 GPH	"	3.80	4.211		975	218		1,193	1,400
1240	For stainless steel cabinet, add					171			171	188
2600	Wheelchair type, 8 GPH	Q-1	4	4		915	207		1,122	1,325
3000	Simulated recessed, 8 GPH		4	4		630	207		837	1,000
3040	11.5 GPH		4	4		1,025	207		1,232	1,450
3200	For glass filler, add					99			99	109
3240	For stainless steel cabinet, add					73.50			73.50	81
3300	Semi-recessed, 8.1 GPH	Q-1	4	4		715	207		922	1,100
3320	12 GPH	"	4	4		820	207		1,027	1,225
3340	For glass filler, add					129			129	142
3360	For stainless steel cabinet, add					145			145	160
3400	Full recessed, stainless steel, 8 GPH	Q-1	3.50	4.571		1,925	237		2,162	2,475
3420	11.5 GPH	"	3.50	4.571		1,650	237		1,887	2,150
3460	For glass filler, add					139			139	153
3600	For mounting can only					243			243	268
4600	Floor mounted, flush-to-wall									
4640	4 GPH	1 Plum	3	2.667	Ea.	705	153		858	1,000
4680	8.2 GPH		3	2.667		740	153		893	1,050

22 47 Drinking Fountains and Water Coolers

22 47 16 – Pressure Water Coolers

22 47 16.10 Electric Water Coolers

		Crew	Daily Output	Labor-Hours	Unit	Material	2014 Bare Costs Labor	Equipment	Total	Total Incl O&P
4720	14.3 GPH	1 Plum	3	2.667	Ea.	850	153		1,003	1,175
4960	14 GPH hot and cold water	↓	3	2.667		1,100	153		1,253	1,450
4980	For stainless steel cabinet, add					128			128	141
5000	Dual height, 8.2 GPH	1 Plum	2	4		1,100	230		1,330	1,575
5040	14.3 GPH	"	2	4		1,150	230		1,380	1,625
5120	For stainless steel cabinet, add					186			186	205
5600	Explosion Proof, 16 GPH	1 Plum	3	2.667		1,975	153		2,128	2,400
6000	Refrigerator Compartment Type, 4.5 GPH		3	2.667		1,425	153		1,578	1,775
6600	Bottle Supply Type, 1.0 GPH		4	2		375	115		490	590
6640	Hot and cold, 1.0 GPH		4	2		510	115		625	735
9800	For supply, waste & vent, all coolers	↓	2.21	3.620	↓	365	208		573	715

22 47 23 – Remote Water Coolers

22 47 23.10 Remote Water Coolers

		Crew	Daily Output	Labor-Hours	Unit	Material	Labor	Equipment	Total	Incl O&P
0010	**REMOTE WATER COOLERS**, 80°F inlet									
0100	Air cooled, 50°F outlet, 115 V, 4.1 GPH	1 Plum	6	1.333	Ea.	525	76.50		601.50	695
0200	5.7 GPH		5.50	1.455		835	83.50		918.50	1,050
0300	8.0 GPH		5	1.600		625	92		717	830
0400	10.0 GPH		4.50	1.778		920	102		1,022	1,150
0500	13.4 GPH	↓	4	2		1,425	115		1,540	1,750
0700	29 GPH	Q-1	5	3.200		1,675	166		1,841	2,100
1000	230V, 32 GPH	"	5	3.200	↓	1,750	166		1,916	2,175

22 51 Swimming Pool Plumbing Systems

22 51 19 – Swimming Pool Water Treatment Equipment

22 51 19.50 Swimming Pool Filtration Equipment

		Crew	Daily Output	Labor-Hours	Unit	Material	Labor	Equipment	Total	Incl O&P
0010	**SWIMMING POOL FILTRATION EQUIPMENT**									
0900	Filter system, sand or diatomite type, incl. pump, 6,000 gal./hr.	2 Plum	1.80	8.889	Total	1,925	510		2,435	2,900
1020	Add for chlorination system, 800 S.F. pool		3	5.333	Ea.	177	305		482	660
1040	5,000 S.F. pool	↓	3	5.333	"	1,875	305		2,180	2,550

22 52 Fountain Plumbing Systems

22 52 16 – Fountain Pumps

22 52 16.10 Fountain Water Pumps

		Crew	Daily Output	Labor-Hours	Unit	Material	Labor	Equipment	Total	Incl O&P
0010	**FOUNTAIN WATER PUMPS**									
0100	Pump w/controls									
0200	Single phase, 100' cord, 1/2 H.P. pump	2 Skwk	4.40	3.636	Ea.	1,275	172		1,447	1,675
0300	3/4 H.P. pump		4.30	3.721		2,175	176		2,351	2,675
0400	1 H.P. pump		4.20	3.810		2,350	180		2,530	2,875
0500	1-1/2 H.P. pump		4.10	3.902		2,800	185		2,985	3,350
0600	2 H.P. pump		4	4		3,775	189		3,964	4,475
0700	Three phase, 200' cord, 5 H.P. pump		3.90	4.103		5,175	194		5,369	6,000
0800	7-1/2 H.P. pump		3.80	4.211		9,000	199		9,199	10,200
0900	10 H.P. pump		3.70	4.324		13,300	205		13,505	15,000
1000	15 H.P. pump	↓	3.60	4.444	↓	16,800	210		17,010	18,800
2000	DESIGN NOTE: Use two horsepower per surface acre.									

22 52 Fountain Plumbing Systems

22 52 33 – Fountain Ancillary

22 52 33.10 Fountain Miscellaneous

		Crew	Daily Output	Labor-Hours	Unit	Material	2014 Bare Costs Labor	Equipment	Total	Total Incl O&P
0010	**FOUNTAIN MISCELLANEOUS**									
1100	Nozzles, minimum	2 Skwk	8	2	Ea.	149	94.50		243.50	310
1200	Maximum		8	2		320	94.50		414.50	495
1300	Lights w/mounting kits, 200 watt		18	.889		1,050	42		1,092	1,225
1400	300 watt		18	.889		1,275	42		1,317	1,475
1500	500 watt		18	.889		1,425	42		1,467	1,650
1600	Color blender		12	1.333		555	63		618	710

22 62 Vacuum Systems for Laboratory and Healthcare Facilities

22 62 19 – Vacuum Equipment for Laboratory and Healthcare Facilities

22 62 19.70 Healthcare Vacuum Equipment

		Crew	Daily Output	Labor-Hours	Unit	Material	Labor	Equipment	Total	Total Incl O&P
0010	**HEALTHCARE VACUUM EQUIPMENT**									
0100	Medical, with receiver									
0110	Duplex									
0120	20 SCFM	Q-1	1.14	14.035	Ea.	17,600	725		18,325	20,500
0130	60 SCFM, 10 HP	Q-2	1.20	20	"	32,000	1,075		33,075	36,800
0200	Triplex									
0220	180 SCFM	Q-2	.86	27.907	Ea.	49,600	1,500		51,100	57,000
0300	Dental oral									
0310	Duplex									
0330	165 SCFM with 77 gal. separator	Q-2	1.30	18.462	Ea.	52,000	990		52,990	58,500
1100	Vacuum system									
1110	Vacuum outlet alarm panel	1 Plum	3.20	2.500	Ea.	1,100	144		1,244	1,450

22 63 Gas Systems for Laboratory and Healthcare Facilities

22 63 13 – Gas Piping for Laboratory and Healthcare Facilities

22 63 13.70 Healthcare Gas Piping

		Crew	Daily Output	Labor-Hours	Unit	Material	Labor	Equipment	Total	Total Incl O&P
0010	**HEALTHCARE GAS PIPING**									
1000	Nitrogen or oxygen system									
1010	Cylinder manifold									
1020	5 cylinder	1 Plum	.80	10	Ea.	5,950	575		6,525	7,425
1026	10 cylinder	"	.40	20	"	7,175	1,150		8,325	9,625
3000	Outlets and valves									
3010	Recessed, wall mounted									
3012	Single outlet	1 Plum	3.20	2.500	Ea.	71.50	144		215.50	296
3100	Ceiling outlet									
3190	Zone valve with box									
3210	2" valve size	1 Plum	3.20	2.500	Ea.	420	144		564	675
4000	Alarm panel, medical gases and vacuum									
4010	Alarm panel	1 Plum	3.20	2.500	Ea.	1,100	144		1,244	1,450

22 66 Chemical-Waste Systems for Lab. and Healthcare Facilities

22 66 53 – Laboratory Chemical-Waste and Vent Piping

22 66 53.30 Glass Pipe

		Crew	Daily Output	Labor-Hours	Unit	Material	2014 Bare Costs Labor	Equipment	Total	Total Incl O&P
0010	**GLASS PIPE**, Borosilicate, couplings & clevis hanger assemblies, 10' O.C. R221113-70									
0020	Drainage									
1100	1-1/2" diameter	Q-1	52	.308	L.F.	10.60	15.95		26.55	35.50
1120	2" diameter		44	.364		13.85	18.85		32.70	44
1140	3" diameter		39	.410		18.70	21.50		40.20	52.50
1160	4" diameter		30	.533		33.50	27.50		61	78
1180	6" diameter		26	.615		58	32		90	112
1870	To delete coupling & hanger, subtract									
1880	1-1/2" diam. to 2" diam.					19%	22%			
1890	3" diam. to 6" diam.					20%	17%			
2000	Process supply (pressure), beaded joints									
2040	1/2" diameter	1 Plum	36	.222	L.F.	5.20	12.80		18	25
2060	3/4" diameter		31	.258		5.85	14.85		20.70	29
2080	1" diameter		27	.296		15.75	17.05		32.80	43.50
2100	1-1/2" diameter	Q-1	47	.340		13.90	17.65		31.55	42
2120	2" diameter		39	.410		18.90	21.50		40.40	53
2140	3" diameter		34	.471		25	24.50		49.50	64.50
2160	4" diameter		25	.640		38	33		71	92
2180	6" diameter		21	.762		107	39.50		146.50	178
2860	To delete coupling & hanger, subtract									
2870	1/2" diam. to 1" diam.					25%	33%			
2880	1-1/2" diam. to 3" diam.					22%	21%			
2890	4" diam. to 6" diam.					23%	15%			
3800	Conical joint, transparent									
3980	6" diameter	Q-1	21	.762	L.F.	143	39.50		182.50	217
4500	To delete couplings & hangers, subtract									
4530	6" diam.					22%	26%			

22 66 53.40 Pipe Fittings, Glass

		Crew	Daily Output	Labor-Hours	Unit	Material	Labor	Equipment	Total	Total Incl O&P
0010	**PIPE FITTINGS, GLASS**									
0020	Drainage, beaded ends									
0040	Coupling & labor required at joints not incl. in fitting									
0050	price. Add 1 per joint for installed price									
0070	90° Bend or sweep, 1-1/2"				Ea.	31.50			31.50	35
0090	2"					40			40	44
0100	3"					66			66	72.50
0110	4"					105			105	116
0120	6" (sweep only)					330			330	360
0200	45° Bend or sweep same as 90°									
0350	Tee, single sanitary, 1-1/2"				Ea.	51			51	56.50
0370	2"					51			51	56.50
0380	3"					76.50			76.50	84
0390	4"					137			137	150
0400	6"					365			365	405
0410	Tee, straight, 1-1/2"					63.50			63.50	69.50
0430	2"					63.50			63.50	69.50
0440	3"					91.50			91.50	101
0450	4"					107			107	117
0460	6"					395			395	435
0500	Coupling, stainless steel, TFE seal ring									
0520	1-1/2"	Q-1	32	.500	Ea.	23.50	26		49.50	64.50
0530	2"		30	.533		29.50	27.50		57	73.50
0540	3"		25	.640		39.50	33		72.50	93.50

22 66 Chemical-Waste Systems for Lab. and Healthcare Facilities

22 66 53 – Laboratory Chemical-Waste and Vent Piping

22 66 53.40 Pipe Fittings, Glass

		Crew	Daily Output	Labor-Hours	Unit	Material	Labor	Equipment	Total	Total Incl O&P
0550	4"	Q-1	23	.696	Ea.	67.50	36		103.50	129
0560	6"	↓	20	.800	↓	152	41.50		193.50	230
0600	Coupling, stainless steel, bead to plain end									
0610	1-1/2"	Q-1	36	.444	Ea.	30	23		53	68
0620	2"		34	.471		37	24.50		61.50	78
0630	3"		29	.552		63	28.50		91.50	113
0640	4"		27	.593		94	30.50		124.50	151
0650	6"	↓	24	.667	↓	320	34.50		354.50	400
2350	Coupling, Viton liner, for temperatures to 400°F									
2370	1/2"	Q-1	40	.400	Ea.	46.50	20.50		67	83
2380	3/4"		37	.432		53	22.50		75.50	92.50
2390	1"		35	.457		63	23.50		86.50	106
2400	1-1/2"		32	.500		23	26		49	64.50
2410	2"		30	.533		29.50	27.50		57	74
2420	3"		25	.640		39.50	33		72.50	93.50
2430	4"		23	.696		67.50	36		103.50	129
2440	6"	↓	20	.800	↓	152	41.50		193.50	230
2550	For beaded joint armored fittings, add					200%				
2600	Conical ends. Flange set, gasket & labor not incl. in fitting									
2620	price. Add 1 per joint for installed price.									
2650	90° Sweep elbow, 1"				Ea.	105			105	116
2670	1-1/2"					211			211	232
2680	2"					219			219	241
2690	3"					340			340	375
2700	4"					605			605	665
2710	6"					1,050			1,050	1,175
2750	Cross (straight), add					55%				
2850	Tee, add				↓	20%				

22 66 53.60 Corrosion Resistant Pipe

			Crew	Daily Output	Labor-Hours	Unit	Material	Labor	Equipment	Total	Total Incl O&P
0010	**CORROSION RESISTANT PIPE**, No couplings or hangers	R221113-70									
0020	Iron alloy, drain, mechanical joint										
1000	1-1/2" diameter		Q-1	70	.229	L.F.	46	11.85		57.85	68.50
1100	2" diameter			66	.242		47	12.55		59.55	70.50
1120	3" diameter			60	.267		60.50	13.80		74.30	87.50
1140	4" diameter		↓	52	.308	↓	77.50	15.95		93.45	110
1980	Iron alloy, drain, B&S joint										
2000	2" diameter		Q-1	54	.296	L.F.	55.50	15.35		70.85	84.50
2100	3" diameter			52	.308		71	15.95		86.95	102
2120	4" diameter		↓	48	.333		98	17.25		115.25	134
2140	6" diameter		Q-2	59	.407		158	22		180	206
2160	8" diameter		"	54	.444	↓	297	24		321	360
2980	Plastic, epoxy, fiberglass filament wound, B&S joint										
3000	2" diameter		Q-1	62	.258	L.F.	12	13.35		25.35	33
3100	3" diameter			51	.314		14	16.25		30.25	40
3120	4" diameter			45	.356		20	18.40		38.40	50
3140	6" diameter		↓	32	.500		28	26		54	70
3160	8" diameter		Q-2	38	.632		44	34		78	99.50
3180	10" diameter			32	.750		60	40.50		100.50	127
3200	12" diameter			28	.857		72	46		118	149
3980	Polyester, fiberglass filament wound, B&S joint										
4000	2" diameter		Q-1	62	.258	L.F.	13.05	13.35		26.40	34.50
4100	3" diameter		↓	51	.314	↓	17	16.25		33.25	43

22 66 Chemical-Waste Systems for Lab. and Healthcare Facilities

22 66 53 – Laboratory Chemical-Waste and Vent Piping

22 66 53.60 Corrosion Resistant Pipe		Crew	Daily Output	Labor-Hours	Unit	Material	2014 Bare Costs Labor	Equipment	Total	Total Incl O&P
4120	4" diameter	Q-1	45	.356	L.F.	25	18.40		43.40	55.50
4140	6" diameter	▼	32	.500		36	26		62	78.50
4160	8" diameter	Q-2	38	.632		85.50	34		119.50	145
4180	10" diameter		32	.750		105	40.50		145.50	176
4200	12" diameter	▼	28	.857		125	46		171	208
4980	Polypropylene, acid resistant, fire retardant, schedule 40									
5000	1-1/2" diameter	Q-1	68	.235	L.F.	7.90	12.20		20.10	27
5100	2" diameter		62	.258		12.45	13.35		25.80	33.50
5120	3" diameter		51	.314		22	16.25		38.25	49
5140	4" diameter		45	.356		28	18.40		46.40	59
5160	6" diameter	▼	32	.500	▼	56.50	26		82.50	101
5980	Proxylene, fire retardant, Schedule 40									
6000	1-1/2" diameter	Q-1	68	.235	L.F.	12.40	12.20		24.60	32
6100	2" diameter		62	.258		17	13.35		30.35	38.50
6120	3" diameter		51	.314		30.50	16.25		46.75	58.50
6140	4" diameter		45	.356		43.50	18.40		61.90	75.50
6160	6" diameter	▼	32	.500		73.50	26		99.50	120
6820	For Schedule 80, add				▼	35%	2%			

22 66 53.70 Pipe Fittings, Corrosion Resistant

		Crew	Daily Output	Labor-Hours	Unit	Material	Labor	Equipment	Total	Total Incl O&P
0010	**PIPE FITTINGS, CORROSION RESISTANT**									
0030	Iron alloy									
0050	Mechanical joint									
0060	1/4 Bend, 1-1/2"	Q-1	12	1.333	Ea.	77	69		146	189
0080	2"		10	1.600		126	83		209	264
0090	3"		9	1.778		151	92		243	305
0100	4"		8	2		174	104		278	345
0110	1/8 Bend, 1-1/2"		12	1.333		49	69		118	158
0130	2"		10	1.600		84	83		167	218
0140	3"		9	1.778		112	92		204	262
0150	4"	▼	8	2	▼	150	104		254	320
0160	Tee and Y, sanitary, straight									
0170	1-1/2"	Q-1	8	2	Ea.	84	104		188	249
0180	2"		7	2.286		112	118		230	300
0190	3"		6	2.667		174	138		312	400
0200	4"		5	3.200		320	166		486	600
0360	Coupling, 1-1/2"		14	1.143		47.50	59		106.50	142
0380	2"		12	1.333		54	69		123	164
0390	3"		11	1.455		56.50	75.50		132	177
0400	4"	▼	10	1.600	▼	64.50	83		147.50	196
0500	Bell & Spigot									
0510	1/4 and 1/16 bend, 2"	Q-1	16	1	Ea.	99.50	52		151.50	187
0520	3"		14	1.143		232	59		291	345
0530	4"	▼	13	1.231		238	64		302	360
0540	6"	Q-2	17	1.412		470	76		546	635
0550	8"	"	12	2		1,950	107		2,057	2,275
0620	1/8 bend, 2"	Q-1	16	1		112	52		164	201
0640	3"		14	1.143		207	59		266	320
0650	4"	▼	13	1.231		238	64		302	360
0660	6"	Q-2	17	1.412		395	76		471	550
0680	8"	"	12	2		1,550	107		1,657	1,850
0700	Tee, sanitary, 2"	Q-1	10	1.600		213	83		296	360
0710	3"	▼	9	1.778		690	92		782	900

22 66 Chemical-Waste Systems for Lab. and Healthcare Facilities

22 66 53 – Laboratory Chemical-Waste and Vent Piping

22 66 53.70 Pipe Fittings, Corrosion Resistant		Crew	Daily Output	Labor-Hours	Unit	Material	2014 Bare Costs Labor	Equipment	Total	Total Incl O&P
0720	4"	Q-1	8	2	Ea.	575	104		679	785
0730	6"	Q-2	11	2.182		715	117		832	965
0740	8"	"	8	3		1,950	161		2,111	2,400
1800	Y, sanitary, 2"	Q-1	10	1.600		224	83		307	370
1820	3"		9	1.778		400	92		492	580
1830	4"	↓	8	2		360	104		464	550
1840	6"	Q-2	11	2.182		1,200	117		1,317	1,500
1850	8"	"	8	3	↓	3,300	161		3,461	3,875
3000	Epoxy, filament wound									
3030	Quick-lock joint									
3040	90° Elbow, 2"	Q-1	28	.571	Ea.	97.50	29.50		127	152
3060	3"		16	1		112	52		164	201
3070	4"		13	1.231		153	64		217	265
3080	6"	↓	8	2		223	104		327	400
3090	8"	Q-2	9	2.667		410	143		553	665
3100	10"		7	3.429		520	184		704	850
3110	12"	↓	6	4		740	215		955	1,150
3120	45° Elbow, 2"	Q-1	28	.571		75	29.50		104.50	127
3130	3"		16	1		106	52		158	195
3140	4"		13	1.231		109	64		173	217
3150	6"	↓	8	2		223	104		327	400
3160	8"	Q-2	9	2.667		410	143		553	665
3170	10"		7	3.429		515	184		699	850
3180	12"	↓	6	4		740	215		955	1,150
3190	Tee, 2"	Q-1	19	.842		233	43.50		276.50	320
3200	3"		11	1.455		280	75.50		355.50	425
3210	4"		9	1.778		335	92		427	510
3220	6"	↓	5	3.200		575	166		741	880
3230	8"	Q-2	6	4		645	215		860	1,025
3240	10"		5	4.800		900	258		1,158	1,375
3250	12"	↓	4	6	↓	1,275	320		1,595	1,875
4000	Polypropylene, acid resistant									
4020	Non-pressure, electrofusion joints									
4050	1/4 bend, 1-1/2"	1 Plum	16	.500	Ea.	13.55	29		42.55	58.50
4060	2"	Q-1	28	.571		26	29.50		55.50	73
4080	3"		17	.941		28.50	49		77.50	105
4090	4"		14	1.143		46.50	59		105.50	141
4110	6"	↓	8	2	↓	110	104		214	277
4150	1/4 Bend, long sweep									
4170	1-1/2"	1 Plum	16	.500	Ea.	14.95	29		43.95	60
4180	2"	Q-1	28	.571		26.50	29.50		56	73.50
4200	3"		17	.941		33	49		82	110
4210	4"	↓	14	1.143		47.50	59		106.50	142
4250	1/8 bend, 1-1/2"	1 Plum	16	.500		14	29		43	59
4260	2"	Q-1	28	.571		16.40	29.50		45.90	62.50
4280	3"		17	.941		30	49		79	107
4290	4"		14	1.143		33.50	59		92.50	127
4310	6"	↓	8	2	↓	92.50	104		196.50	258
4400	Tee, sanitary									
4420	1-1/2"	1 Plum	10	.800	Ea.	17.10	46		63.10	88.50
4430	2"	Q-1	17	.941		19.95	49		68.95	95.50
4450	3"		11	1.455		40	75.50		115.50	158
4460	4"	↓	9	1.778		60	92		152	205

22 66 Chemical-Waste Systems for Lab. and Healthcare Facilities

22 66 53 – Laboratory Chemical-Waste and Vent Piping

22 66 53.70 Pipe Fittings, Corrosion Resistant		Crew	Daily Output	Labor-Hours	Unit	Material	2014 Bare Costs Labor	Equipment	Total	Total Incl O&P
4480	6"	Q-1	5	3.200	Ea.	400	166		566	690
4490	Tee, sanitary reducing, 2" x 2" x 1-1/2"		17	.941		19.95	49		68.95	95.50
4491	3" x 3" x 1-1/2"		11	1.455		35.50	75.50		111	153
4492	3" x 3" x 2"		11	1.455		40	75.50		115.50	158
4493	4" x 4" x 2"		10	1.600		53.50	83		136.50	184
4494	4" x 4" x 3"		9	1.778		58.50	92		150.50	203
4496	6" x 6" x 4"	↓	5	3.200		196	166		362	465
4650	Wye 45°, 1-1/2"	1 Plum	10	.800		18.50	46		64.50	90
4652	2"	Q-1	17	.941		25.50	49		74.50	102
4653	3"		11	1.455		42.50	75.50		118	161
4654	4"		9	1.778		62	92		154	207
4656	6"	↓	5	3.200	↓	160	166		326	425
4660	Wye, reducing									
4662	2" x 2" x 1-1/2"	Q-1	17	.941	Ea.	24	49		73	100
4666	3" x 3" x 2"		11	1.455		48.50	75.50		124	168
4668	4" x 4" x 2"		10	1.600		57	83		140	188
4669	4" x 4" x 3"		9	1.778		60.50	92		152.50	206
4671	6" x 6" x 2"		6	2.667		97	138		235	315
4673	6" x 6" x 3"		5.50	2.909		107	151		258	345
4675	6" x 6" x 4"	↓	5	3.200	↓	111	166		277	370
4678	Combination Y & 1/8 bend									
4681	1-1/2"	1 Plum	10	.800	Ea.	27	46		73	99.50
4683	2"	Q-1	17	.941		28.50	49		77.50	105
4684	3"		11	1.455		48.50	75.50		124	168
4685	4"	↓	9	1.778	↓	66	92		158	212
4689	Combination Y & 1/8 bend, reducing									
4692	2" x 2" x 1-1/2"	Q-1	17	.941	Ea.	27	49		76	104
4694	3" x 3" x 1-1/2"		12	1.333		40.50	69		109.50	149
4695	3" x 3" x 2"		11	1.455		42.50	75.50		118	161
4697	4" x 4" x 2"		10	1.600		60.50	83		143.50	192
4699	4" x 4" x 3"	↓	9	1.778	↓	62.50	92		154.50	208
4710	Hub adapter									
4712	1-1/2"	1 Plum	16	.500	Ea.	31.50	29		60.50	78
4713	2"	Q-1	28	.571		35.50	29.50		65	84
4714	3"		17	.941		45	49		94	123
4715	4"	↓	14	1.143	↓	58.50	59		117.50	154
4719	Mechanical joint adapter									
4721	1-1/2"	1 Plum	16	.500	Ea.	19.25	29		48.25	64.50
4722	2"	Q-1	28	.571		19.95	29.50		49.45	66.50
4723	3"		17	.941		28.50	49		77.50	105
4724	4"	↓	14	1.143	↓	42.50	59		101.50	137
4728	Couplings									
4731	1-1/2"	1 Plum	16	.500	Ea.	10.70	29		39.70	55.50
4732	2"	Q-1	28	.571		13.55	29.50		43.05	59.50
4733	3"		17	.941		17.10	49		66.10	92.50
4734	4"		14	1.143		24	59		83	116
4736	6"	↓	8	2	↓	38.50	104		142.50	199

22 66 Chemical-Waste Systems for Lab. and Healthcare Facilities

22 66 83 – Chemical-Waste Tanks

22 66 83.13 Chemical-Waste Dilution Tanks		Crew	Daily Output	Labor-Hours	Unit	Material	Labor	2014 Bare Costs Equipment	Total	Total Incl O&P
0010	**CHEMICAL-WASTE DILUTION TANKS**									
7000	Tanks, covers included									
7800	Polypropylene									
7810	Continuous service to 200°F									
7830	2 gallon, 8" x 8" x 8"	Q-1	20	.800	Ea.	124	41.50		165.50	199
7850	7 gallon, 12" x 12" x 12"		20	.800		189	41.50		230.50	271
7870	16 gallon, 18" x 12" x 18"		17	.941		243	49		292	340
8010	33 gallon, 24" x 18" x 18"		12	1.333		315	69		384	450
8070	44 gallon, 24" x 18" x 24"		10	1.600		380	83		463	545
8080	89 gallon, 36" x 24" x 24"	▼	8	2	▼	400	104		504	600
8150	Polyethylene, heavy duty walls									
8160	Continuous service to 180°F									
8180	5 gallon, 12" X 6" X 18"	Q-1	20	.800	Ea.	48.50	41.50		90	116
8210	15 gallon, 14" I.D. x 27" deep		17	.941		72	49		121	153
8230	55 gallon, 22" I.D. x 36" deep		10	1.600		145	83		228	284
8250	100 gallon 28" I.D. x 42" deep		8	2		370	104		474	565
8270	200 gallon 36" I.D. x 48" deep		6	2.667		545	138		683	810
8290	360 gallon 48" I.D. x 48" deep	▼	5	3.200	▼	670	166		836	990

Division Notes

		CREW	DAILY OUTPUT	LABOR-HOURS	UNIT	BARE COSTS MAT.	LABOR	EQUIP.	TOTAL	TOTAL INCL O&P

Estimating Tips

The labor adjustment factors listed in Subdivision 22 01 02.20 also apply to Division 23.

23 10 00 Facility Fuel Systems

- The prices in this subdivision for above- and below-ground storage tanks do not include foundations or hold-down slabs, unless noted. The estimator should refer to Divisions 3 and 31 for foundation system pricing. In addition to the foundations, required tank accessories, such as tank gauges, leak detection devices, and additional manholes and piping, must be added to the tank prices.

23 50 00 Central Heating Equipment

- When estimating the cost of an HVAC system, check to see who is responsible for providing and installing the temperature control system. It is possible to overlook controls, assuming that they would be included in the electrical estimate.
- When looking up a boiler, be careful on specified capacity. Some manufacturers rate their products on output while others use input.
- Include HVAC insulation for pipe, boiler, and duct (wrap and liner).
- Be careful when looking up mechanical items to get the correct pressure rating and connection type (thread, weld, flange).

23 70 00 Central HVAC Equipment

- Combination heating and cooling units are sized by the air conditioning requirements. (See Reference No. R236000-20 for preliminary sizing guide.)
- A ton of air conditioning is nominally 400 CFM.
- Rectangular duct is taken off by the linear foot for each size, but its cost is usually estimated by the pound. Remember that SMACNA standards now base duct on internal pressure.
- Prefabricated duct is estimated and purchased like pipe: straight sections and fittings.
- Note that cranes or other lifting equipment are not included on any lines in Division 23. For example, if a crane is required to lift a heavy piece of pipe into place high above a gym floor, or to put a rooftop unit on the roof of a four-story building, etc., it must be added. Due to the potential for extreme variation—from nothing additional required to a major crane or helicopter—we feel that including a nominal amount for "lifting contingency" would be useless and detract from the accuracy of the estimate. When using equipment rental cost data from RSMeans, do not forget to include the cost of the operator(s).

Reference Numbers

Reference numbers are shown in shaded boxes at the beginning of some major classifications. These numbers refer to related items in the Reference Section. The reference information may be an estimating procedure, an alternate pricing method, or technical information.

Note: Not all subdivisions listed here necessarily appear in this publication.

Note: **Trade Service,** *in part, has been used as a reference source for some of the material prices used in Division 23.*

23 05 Common Work Results for HVAC

23 05 02 – HVAC General

23 05 02.10 Air Conditioning, General		Crew	Daily Output	Labor-Hours	Unit	Material	2014 Bare Costs Labor	Equipment	Total	Total Incl O&P
0010	**AIR CONDITIONING, GENERAL** Prices are for standard efficiencies (SEER 13)									
0020	for upgrade to SEER 14 add					10%				

23 05 05 – Selective HVAC Demolition

23 05 05.10 HVAC Demolition

		Crew	Daily Output	Labor-Hours	Unit	Material	Labor	Equipment	Total	Total Incl O&P
0010	**HVAC DEMOLITION** R220105-10									
0100	Air conditioner, split unit, 3 ton	Q-5	2	8	Ea.		420		420	635
0150	Package unit, 3 ton	Q-6	3	8			435		435	660
0190	Rooftop, self contained, up to 5 ton	1 Plum	1.20	6.667			385		385	580
0250	Air curtain	Q-9	20	.800	L.F.		39.50		39.50	60
0254	Air filters, up thru 16,000 CFM		20	.800	Ea.		39.50		39.50	60
0256	20,000 thru 60,000 CFM		16	1			49		49	75
0297	Boiler blowdown	Q-5	8	2			105		105	159
0298	Boilers									
0300	Electric, up thru 148 kW	Q-19	2	12	Ea.		635		635	955
0310	150 thru 518 kW	"	1	24			1,275		1,275	1,900
0320	550 thru 2000 kW	Q-21	.40	80			4,350		4,350	6,550
0330	2070 kW and up	"	.30	106			5,800		5,800	8,725
0340	Gas and/or oil, up thru 150 MBH	Q-7	2.20	14.545			810		810	1,225
0350	160 thru 2000 MBH		.80	40			2,225		2,225	3,375
0360	2100 thru 4500 MBH		.50	64			3,575		3,575	5,375
0370	4600 thru 7000 MBH		.30	106			5,950		5,950	8,975
0380	7100 thru 12,000 MBH		.16	200			11,100		11,100	16,800
0390	12,200 thru 25,000 MBH		.12	266			14,900		14,900	22,400
0400	Central station air handler unit, up thru 15 ton	Q-5	1.60	10			525		525	795
0410	17.5 thru 30 ton	"	.80	20			1,050		1,050	1,600
0430	Computer room unit									
0434	Air cooled split, up thru 10 ton	Q-5	.67	23.881	Ea.		1,250		1,250	1,900
0436	12 thru 23 ton		.53	30.189			1,600		1,600	2,400
0440	Chilled water, up thru 10 ton		1.30	12.308			650		650	980
0444	12 thru 23 ton		1	16			840		840	1,275
0450	Glycol system, up thru 10 ton		.53	30.189			1,600		1,600	2,400
0454	12 thru 23 ton		.40	40			2,100		2,100	3,175
0460	Water cooled, not including condenser, up thru 10 ton		.80	20			1,050		1,050	1,600
0464	12 thru 23 ton		.60	26.667			1,400		1,400	2,125
0600	Condenser, up thru 50 ton		1	16			840		840	1,275
0610	51 thru 100 ton	Q-6	.90	26.667			1,450		1,450	2,200
0620	101 thru 1000 ton	"	.70	34.286			1,875		1,875	2,825
0660	Condensing unit, up thru 10 ton	Q-5	1.25	12.800			675		675	1,025
0670	11 thru 50 ton	"	.40	40			2,100		2,100	3,175
0680	60 thru 100 ton	Q-6	.30	80			4,375		4,375	6,600
0700	Cooling tower, up thru 400 ton		.80	30			1,650		1,650	2,475
0710	450 thru 600 ton		.53	45.283			2,475		2,475	3,725
0720	700 thru 1300 ton		.40	60			3,275		3,275	4,950
0780	Dehumidifier, up thru 155 lb./hr.	Q-1	8	2			104		104	156
0790	240 lb./hr. and up	"	2	8			415		415	625
1560	Ductwork									
1570	Metal, steel, sst, fabricated	Q-9	1000	.016	Lb.		.79		.79	1.20
1580	Aluminum, fabricated		485	.033	"		1.62		1.62	2.48
1590	Spiral, prefabricated		400	.040	L.F.		1.97		1.97	3.01
1600	Fiberglass, prefabricated		400	.040			1.97		1.97	3.01
1610	Flex, prefabricated		500	.032			1.58		1.58	2.41
1620	Glass fiber reinforced plastic, prefabricated		280	.057			2.81		2.81	4.30

23 05 Common Work Results for HVAC

23 05 05 – Selective HVAC Demolition

23 05 05.10 HVAC Demolition

		Crew	Daily Output	Labor-Hours	Unit	Material	2014 Bare Costs Labor	2014 Bare Costs Equipment	Total	Total Incl O&P
1630	Diffusers, registers or grills, up thru 20" max dimension	1 Shee	50	.160	Ea.		8.75		8.75	13.35
1640	21 thru 36" max dimension		36	.222			12.15		12.15	18.55
1650	Above 36" max dimension	↓	30	.267			14.60		14.60	22.50
1700	Evaporator, up thru 12,000 BTUH	Q-5	5.30	3.019			159		159	240
1710	12,500 thru 30,000 BTUH	"	2.70	5.926			310		310	470
1720	31,000 BTUH and up	Q-6	1.50	16			875		875	1,325
1730	Evaporative cooler, up thru 5 H.P.	Q-9	2.70	5.926			292		292	445
1740	10 thru 30 H.P.	"	.67	23.881	↓		1,175		1,175	1,800
1750	Exhaust systems									
1760	Exhaust components	1 Shee	8	1	System		54.50		54.50	83.50
1770	Weld fume hoods	"	20	.400	Ea.		22		22	33.50
2120	Fans, up thru 1 H.P. or 2000 CFM	Q-9	8	2			98.50		98.50	150
2124	1-1/2 thru 10 H.P. or 20,000 CFM		5.30	3.019			149		149	227
2128	15 thru 30 H.P. or above 20,000 CFM	↓	4	4			197		197	300
2150	Fan coil air conditioner, chilled water, up thru 7.5 ton	Q-5	14	1.143			60		60	91
2154	Direct expansion, up thru 10 ton		8	2			105		105	159
2158	11 thru 30 ton	↓	2	8			420		420	635
2170	Flue shutter damper	Q-9	8	2			98.50		98.50	150
2200	Furnace, electric	Q-20	2	10			500		500	760
2300	Gas or oil, under 120 MBH	Q-9	4	4			197		197	300
2340	Over 120 MBH	"	3	5.333			263		263	400
2730	Heating and ventilating unit	Q-5	2.70	5.926			310		310	470
2740	Heater, electric, wall, baseboard and quartz	1 Elec	10	.800			42.50		42.50	64
2750	Heater, electric, unit, cabinet, fan and convector	"	8	1			53.50		53.50	80
2760	Heat exchanger, shell and tube type	Q-5	1.60	10			525		525	795
2770	Plate type	Q-6	.60	40	↓		2,175		2,175	3,300
2810	Heat pump									
2820	Air source, split, up thru 10 ton	Q-5	.90	17.778	Ea.		935		935	1,425
2830	15 thru 25 ton	Q-6	.80	30			1,650		1,650	2,475
2850	Single package, up thru 12 ton	Q-5	1	16			840		840	1,275
2860	Water source, up thru 15 ton	"	.90	17.778			935		935	1,425
2870	20 thru 50 ton	Q-6	.80	30			1,650		1,650	2,475
2910	Heat recovery package, up thru 20,000 CFM	Q-5	2	8			420		420	635
2920	25,000 CFM and up		1.20	13.333			700		700	1,050
2930	Heat transfer package, up thru 130 GPM		.80	20			1,050		1,050	1,600
2934	255 thru 800 GPM		.42	38.095			2,000		2,000	3,025
2940	Humidifier		10.60	1.509			79.50		79.50	120
2961	Hydronic unit heaters, up thru 200 MBH		14	1.143			60		60	91
2962	Above 200 MBH		8	2			105		105	159
2964	Valance units	↓	32	.500	↓		26.50		26.50	40
2966	Radiant floor heating									
2967	System valves, controls, manifolds	Q-5	16	1	Ea.		52.50		52.50	79.50
2968	Per room distribution		8	2			105		105	159
2970	Hydronic heating, baseboard radiation		16	1			52.50		52.50	79.50
2976	Convectors and free standing radiators	↓	18	.889			47		47	70.50
2980	Induced draft fan, up thru 1 H.P.	Q-9	4.60	3.478			171		171	262
2984	1-1/2 H.P. thru 7-1/2 H.P.	"	2.20	7.273			360		360	545
2988	Infrared unit	Q-5	16	1	↓		52.50		52.50	79.50
2992	Louvers	1 Shee	46	.174	S.F.		9.50		9.50	14.55
3000	Mechanical equipment, light items. Unit is weight, not cooling.	Q-5	.90	17.778	Ton		935		935	1,425
3600	Heavy items	"	1.10	14.545	"		765		765	1,150
3700	Deduct for salvage (when applicable), minimum				Job				73	80
3710	Maximum				"				455	500

307

23 05 Common Work Results for HVAC

23 05 05 – Selective HVAC Demolition

23 05 05.10 HVAC Demolition

		Crew	Daily Output	Labor-Hours	Unit	Material	2014 Bare Costs Labor	Equipment	Total	Total Incl O&P
3720	Make-up air unit, up thru 6000 CFM	Q-5	3	5.333	Ea.		281		281	425
3730	6500 thru 30,000 CFM	"	1.60	10			525		525	795
3740	35,000 thru 75,000 CFM	Q-6	1	24			1,300		1,300	1,975
3800	Mixing boxes, constant and VAV	Q-9	18	.889			44		44	67
4000	Packaged terminal air conditioner, up thru 18,000 BTUH	Q-5	8	2			105		105	159
4010	24,000 thru 48,000 BTUH	"	2.80	5.714			300		300	455
5000	Refrigerant compressor, reciprocating or scroll									
5010	Up thru 5 ton	1 Stpi	6	1.333	Ea.		78		78	118
5020	5.08 thru 10 ton	Q-5	6	2.667			140		140	212
5030	15 thru 50 ton	"	.40	40			2,100		2,100	3,175
5040	60 thru 130 ton	Q-6	.48	50			2,725		2,725	4,125
5090	Remove refrigerant from system	1 Stpi	40	.200	Lb.		11.70		11.70	17.65
5100	Rooftop air conditioner, up thru 10 ton	Q-5	1.40	11.429	Ea.		600		600	910
5110	12 thru 40 ton	Q-6	1	24			1,300		1,300	1,975
5120	50 thru 140 ton		.50	48			2,625		2,625	3,950
5130	150 thru 300 ton		.30	80			4,375		4,375	6,600
6000	Self contained single package air conditioner, up thru 10 ton	Q-5	1.60	10			525		525	795
6010	15 thru 60 ton	Q-6	1.20	20			1,100		1,100	1,650
6100	Space heaters, up thru 200 MBH	Q-5	10	1.600			84		84	127
6110	Over 200 MBH		5	3.200			168		168	254
6200	Split ductless, both sections		8	2			105		105	159
6300	Steam condensate meter	1 Stpi	11	.727			42.50		42.50	64.50
6600	Thru-the-wall air conditioner	L-2	8	2			80.50		80.50	124
7000	Vent chimney, prefabricated, up thru 12" diameter	Q-9	94	.170	V.L.F.		8.40		8.40	12.80
7010	14" thru 36" diameter		40	.400			19.70		19.70	30
7020	38" thru 48" diameter		32	.500			24.50		24.50	37.50
7030	54" thru 60" diameter	Q-10	14	1.714			87.50		87.50	134
7400	Ventilators, up thru 14" neck diameter	Q-9	58	.276	Ea.		13.60		13.60	20.50
7410	16" thru 50" neck diameter		40	.400			19.70		19.70	30
7450	Relief vent, up thru 24" x 96"		22	.727			36		36	54.50
7460	48" x 60" thru 96" x 144"		10	1.600			79		79	120
8000	Water chiller up thru 10 ton	Q-5	2.50	6.400			335		335	510
8010	15 thru 100 ton	Q-6	.48	50			2,725		2,725	4,125
8020	110 thru 500 ton	Q-7	.29	110			6,150		6,150	9,275
8030	600 thru 1000 ton		.23	139			7,750		7,750	11,700
8040	1100 ton and up		.20	160			8,900		8,900	13,500
8400	Window air conditioner	1 Carp	16	.500			23		23	35.50

23 05 23 – General-Duty Valves for HVAC Piping

23 05 23.30 Valves, Iron Body

		Crew	Daily Output	Labor-Hours	Unit	Material	2014 Bare Costs Labor	Equipment	Total	Total Incl O&P
0010	**VALVES, IRON BODY** R220523-90									
0022	For grooved joint, see Section 22 11 13.48									
0100	Angle, 125 lb.									
0110	Flanged									
0116	2"	1 Plum	5	1.600	Ea.	1,275	92		1,367	1,550
0118	4"	Q-1	3	5.333		2,100	276		2,376	2,725
0120	6"	Q-2	3	8		4,100	430		4,530	5,150
0122	8"	"	2.50	9.600		7,325	515		7,840	8,850
1020	Butterfly, wafer type, gear actuator, 200 lb.									
1030	2"	1 Plum	14	.571	Ea.	96	33		129	155
1040	2-1/2"	Q-1	9	1.778		97.50	92		189.50	246
1050	3"		8	2		101	104		205	267
1060	4"		5	3.200		113	166		279	375

23 05 Common Work Results for HVAC

23 05 23 – General-Duty Valves for HVAC Piping

23 05 23.30 Valves, Iron Body		Crew	Daily Output	Labor-Hours	Unit	Material	2014 Bare Costs Labor	Equipment	Total	Total Incl O&P
1070	5"	Q-2	5	4.800	Ea.	126	258		384	530
1080	6"		5	4.800		143	258		401	550
1090	8"		4.50	5.333		178	287		465	630
1100	10"		4	6		239	320		559	750
1110	12"		3	8		360	430		790	1,050
1200	Wafer type, lever actuator, 200 lb.									
1220	2"	1 Plum	14	.571	Ea.	165	33		198	232
1230	2-1/2"	Q-1	9	1.778		170	92		262	325
1240	3"		8	2		180	104		284	355
1250	4"		5	3.200		219	166		385	490
1260	5"	Q-2	5	4.800		305	258		563	725
1270	6"		5	4.800		365	258		623	795
1280	8"		4.50	5.333		580	287		867	1,075
1290	10"		4	6		865	320		1,185	1,450
1300	12"		3	8		1,575	430		2,005	2,400
1600	Gate, threaded, 125 lb.									
1603	1-1/2"	1 Plum	13	.615	Ea.	245	35.50		280.50	325
1604	2"		11	.727		935	42		977	1,100
1605	2-1/2"		10	.800		685	46		731	825
1606	3"		8	1		970	57.50		1,027.50	1,150
1607	4"		5	1.600		2,225	92		2,317	2,600
1650	Gate, 125 lb., N.R.S.									
2150	Flanged									
2200	2"	1 Plum	5	1.600	Ea.	680	92		772	890
2240	2-1/2"	Q-1	5	3.200		700	166		866	1,025
2260	3"		4.50	3.556		780	184		964	1,150
2280	4"		3	5.333		1,125	276		1,401	1,650
2290	5"	Q-2	3.40	7.059		1,900	380		2,280	2,675
2300	6"		3	8		1,900	430		2,330	2,750
2320	8"		2.50	9.600		3,275	515		3,790	4,375
2340	10"		2.20	10.909		5,775	585		6,360	7,225
2360	12"		1.70	14.118		7,925	760		8,685	9,850
2420	For 250 lb. flanged, add					200%	10%			
3500	OS&Y, 125 lb., (225# noted), threaded									
3504	3/4" (225 lb.)	1 Plum	18	.444	Ea.	440	25.50		465.50	525
3505	1" (225 lb.)		16	.500		500	29		529	595
3506	1-1/2" (225 lb.)		13	.615		730	35.50		765.50	855
3507	2"		11	.727		1,275	42		1,317	1,475
3550	OS&Y, 125 lb., flanged									
3600	2"	1 Plum	5	1.600	Ea.	455	92		547	640
3640	2-1/2"	Q-1	5	3.200		455	166		621	750
3660	3"		4.50	3.556		495	184		679	825
3670	3-1/2"		3	5.333		605	276		881	1,075
3680	4"		3	5.333		730	276		1,006	1,225
3690	5"	Q-2	3.40	7.059		1,175	380		1,555	1,875
3700	6"		3	8		1,175	430		1,605	1,950
3720	8"		2.50	9.600		2,100	515		2,615	3,075
3740	10"		2.20	10.909		3,825	585		4,410	5,100
3760	12"		1.70	14.118		5,225	760		5,985	6,900
3900	For 175 lb., flanged, add					200%	10%			
4350	Globe, OS&Y									
4540	Class 125, flanged									
4550	2"	1 Plum	5	1.600	Ea.	890	92		982	1,125

23 05 Common Work Results for HVAC
23 05 23 – General-Duty Valves for HVAC Piping

23 05 23.30 Valves, Iron Body		Crew	Daily Output	Labor-Hours	Unit	Material	2014 Bare Costs Labor	Equipment	Total	Total Incl O&P
4560	2-1/2"	Q-1	5	3.200	Ea.	895	166		1,061	1,225
4570	3"		4.50	3.556		1,075	184		1,259	1,475
4580	4"	↓	3	5.333		1,550	276		1,826	2,125
4590	5"	Q-2	3.40	7.059		2,825	380		3,205	3,700
4600	6"		3	8		2,825	430		3,255	3,775
4610	8"		2.50	9.600		5,550	515		6,065	6,875
4612	10"		2.20	10.909		8,675	585		9,260	10,400
4614	12"	↓	1.70	14.118	↓	8,450	760		9,210	10,500
5040	Class 250, flanged									
5050	2"	1 Plum	4.50	1.778	Ea.	1,425	102		1,527	1,725
5060	2-1/2"	Q-1	4.50	3.556		1,850	184		2,034	2,300
5070	3"		4	4		1,900	207		2,107	2,425
5080	4"	↓	2.70	5.926		2,800	305		3,105	3,550
5090	5"	Q-2	3	8		5,200	430		5,630	6,375
5100	6"		2.70	8.889		5,025	480		5,505	6,250
5110	8"		2.20	10.909		8,525	585		9,110	10,300
5120	10"		2	12		12,800	645		13,445	15,000
5130	12"	↓	1.60	15		19,200	805		20,005	22,300
5240	Valve sprocket rim w/chain, for 2" valve	1 Stpi	30	.267		127	15.60		142.60	164
5250	2-1/2" valve		27	.296		127	17.35		144.35	166
5260	3-1/2" valve		25	.320		127	18.70		145.70	169
5270	6" valve		20	.400		127	23.50		150.50	176
5280	8" valve		18	.444		156	26		182	211
5290	12" valve		16	.500		156	29.50		185.50	215
5300	16" valve		12	.667		215	39		254	295
5310	20" valve		10	.800		215	47		262	305
5320	36" valve	↓	8	1	↓	380	58.50		438.50	510
5450	Swing check, 125 lb., threaded									
5470	1"	1 Plum	13	.615	Ea.	635	35.50		670.50	755
5500	2"	"	11	.727		410	42		452	515
5540	2-1/2"	Q-1	15	1.067		525	55.50		580.50	665
5550	3"		13	1.231		570	64		634	720
5560	4"	↓	10	1.600		905	83		988	1,125
5950	Flanged									
5994	1"	1 Plum	7	1.143	Ea.	187	66		253	305
5998	1-1/2"		6	1.333		260	76.50		336.50	400
6000	2"	↓	5	1.600		395	92		487	575
6040	2-1/2"	Q-1	5	3.200		365	166		531	650
6050	3"		4.50	3.556		390	184		574	705
6060	4"	↓	3	5.333		610	276		886	1,075
6070	6"	Q-2	3	8		1,050	430		1,480	1,800
6080	8"		2.50	9.600		1,950	515		2,465	2,925
6090	10"		2.20	10.909		3,350	585		3,935	4,550
6100	12"		1.70	14.118		5,200	760		5,960	6,875
6110	18"		1.30	18.462		22,600	990		23,590	26,300
6114	24"	↓	.75	32	↓	50,000	1,725		51,725	57,500
6160	For 250 lb. flanged, add					200%	20%			
6600	Silent check, bronze trim									
6610	Compact wafer type, for 125 or 150 lb. flanges									
6630	1-1/2"	1 Plum	11	.727	Ea.	136	42		178	212
6640	2"	"	9	.889		165	51		216	259
6650	2-1/2"	Q-1	9	1.778		180	92		272	335
6660	3"	↓	8	2	↓	192	104		296	365

23 05 Common Work Results for HVAC

23 05 23 – General-Duty Valves for HVAC Piping

23 05 23.30 Valves, Iron Body

		Crew	Daily Output	Labor-Hours	Unit	Material	Labor	Equipment	Total	Total Incl O&P
6670	4"	Q-1	5	3.200	Ea.	245	166		411	520
6680	5"	Q-2	6	4		325	215		540	680
6690	6"		6	4		430	215		645	800
6700	8"		4.50	5.333		730	287		1,017	1,250
6710	10"		4	6		1,275	320		1,595	1,875
6720	12"		3	8		2,425	430		2,855	3,300
6740	For 250 or 300 lb. flanges, thru 6" no change									
6741	For 8" and 10", add				Ea.	11%	10%			
6750	Twin disc									
6752	2"	1 Plum	9	.889	Ea.	310	51		361	415
6754	4"	Q-1	5	3.200		505	166		671	805
6756	6"	Q-2	5	4.800		760	258		1,018	1,225
6758	8"		4.50	5.333		1,150	287		1,437	1,675
6760	10"		4	6		1,825	320		2,145	2,500
6762	12"		3	8		2,375	430		2,805	3,275
6764	18"		1.50	16		10,300	860		11,160	12,600
6766	24"		.75	32		14,500	1,725		16,225	18,500
6800	Full flange type, 150 lb.									
6900	Globe type, 125 lb.									
6911	2-1/2"	Q-1	9	1.778	Ea.	405	92		497	585
6912	3"		8	2		475	104		579	680
6913	4"		5	3.200		640	166		806	955
6914	5"	Q-2	6	4		890	215		1,105	1,300
6915	6"		5	4.800		1,125	258		1,383	1,625
6916	8"		4.50	5.333		2,050	287		2,337	2,675
6917	10"		4	6		2,575	320		2,895	3,300
6918	12"		3	8		4,275	430		4,705	5,350
6940	For 250 lb., add					40%	10%			
6980	Screwed end type, 125 lb.									
6981	1"	1 Plum	19	.421	Ea.	61.50	24		85.50	104
6982	1-1/4"		15	.533		80	30.50		110.50	135
6983	1-1/2"		13	.615		99	35.50		134.50	163
6984	2"		11	.727		134	42		176	210

23 05 23.70 Valves, Semi-Steel

		Crew	Daily Output	Labor-Hours	Unit	Material	Labor	Equipment	Total	Total Incl O&P
0010	**VALVES, SEMI-STEEL** R220523-90									
1020	Lubricated plug valve, threaded, 200 psi									
1030	1/2"	1 Plum	18	.444	Ea.	90	25.50		115.50	138
1040	3/4"		16	.500		90	29		119	143
1050	1"		14	.571		115	33		148	176
1060	1-1/4"		12	.667		137	38.50		175.50	208
1070	1-1/2"		11	.727		147	42		189	225
1080	2"		8	1		175	57.50		232.50	279
1090	2-1/2"	Q-1	5	3.200		269	166		435	545
1100	3"	"	4.50	3.556		330	184		514	645
6990	Flanged, 200 psi									
7000	2"	1 Plum	8	1	Ea.	217	57.50		274.50	325
7010	2-1/2"	Q-1	5	3.200		310	166		476	590
7020	3"		4.50	3.556		375	184		559	695
7030	4"		3	5.333		470	276		746	935
7036	5"		2.50	6.400		935	330		1,265	1,525
7040	6"	Q-2	3	8		780	430		1,210	1,500
7050	8"		2.50	9.600		1,575	515		2,090	2,500

311

23 05 Common Work Results for HVAC

23 05 23 – General-Duty Valves for HVAC Piping

23 05 23.70 Valves, Semi-Steel

		Crew	Daily Output	Labor-Hours	Unit	Material	2014 Bare Costs Labor	Equipment	Total	Total Incl O&P
7060	10"	Q-2	2.20	10.909	Ea.	2,450	585		3,035	3,550
7070	12"	↓	1.70	14.118	↓	4,700	760		5,460	6,325

23 05 23.90 Valves, Stainless Steel

		Crew	Daily Output	Labor-Hours	Unit	Material	Labor	Equipment	Total	Incl O&P
0010	**VALVES, STAINLESS STEEL** R220523-90									
1700	Check, 200 lb., threaded									
1710	1/4"	1 Plum	24	.333	Ea.	147	19.20		166.20	191
1720	1/2"		22	.364		147	21		168	194
1730	3/4"		20	.400		161	23		184	212
1750	1"		19	.421		205	24		229	262
1760	1-1/2"		13	.615		395	35.50		430.50	490
1770	2"	↓	11	.727	↓	660	42		702	795
1800	150 lb., flanged									
1810	2-1/2"	Q-1	5	3.200	Ea.	1,675	166		1,841	2,100
1820	3"		4.50	3.556		1,675	184		1,859	2,125
1830	4"	↓	3	5.333		2,500	276		2,776	3,175
1840	6"	Q-2	3	8		4,450	430		4,880	5,550
1850	8"	"	2.50	9.600	↓	7,450	515		7,965	8,950
2100	Gate, OS&Y, 150 lb., flanged									
2120	1/2"	1 Plum	18	.444	Ea.	470	25.50		495.50	560
2140	3/4"		16	.500		500	29		529	595
2150	1"		14	.571		620	33		653	730
2160	1-1/2"		11	.727		860	42		902	1,000
2170	2"	↓	8	1		1,000	57.50		1,057.50	1,175
2180	2-1/2"	Q-1	5	3.200		1,600	166		1,766	2,025
2190	3"		4.50	3.556		1,600	184		1,784	2,050
2200	4"		3	5.333		2,450	276		2,726	3,125
2205	5"	↓	2.80	5.714		4,250	296		4,546	5,125
2210	6"	Q-2	3	8		4,250	430		4,680	5,325
2220	8"		2.50	9.600		6,800	515		7,315	8,275
2230	10"		2.30	10.435		10,800	560		11,360	12,700
2240	12"	↓	1.90	12.632		13,800	680		14,480	16,200
2260	For 300 lb., flanged, add				↓	120%	15%			
2600	600 lb., flanged									
2620	1/2"	1 Plum	16	.500	Ea.	213	29		242	278
2640	3/4"		14	.571		230	33		263	305
2650	1"		12	.667		277	38.50		315.50	365
2660	1-1/2"		10	.800		445	46		491	555
2670	2"	↓	7	1.143		610	66		676	770
2680	2-1/2"	Q-1	4	4		5,575	207		5,782	6,475
2690	3"	"	3.60	4.444	↓	5,575	230		5,805	6,500
3100	Globe, OS&Y, 150 lb., flanged									
3120	1/2"	1 Plum	18	.444	Ea.	470	25.50		495.50	560
3140	3/4"		16	.500		525	29		554	620
3150	1"		14	.571		650	33		683	765
3160	1-1/2"		11	.727		970	42		1,012	1,150
3170	2"	↓	8	1		1,175	57.50		1,232.50	1,375
3180	2-1/2"	Q-1	5	3.200		2,550	166		2,716	3,050
3190	3"		4.50	3.556		2,550	184		2,734	3,075
3200	4"	↓	3	5.333		3,125	276		3,401	3,850
3210	6"	Q-2	3	8	↓	5,475	430		5,905	6,675

23 05 Common Work Results for HVAC

23 05 93 – Testing, Adjusting, and Balancing for HVAC

23 05 93.50 Piping, Testing

		Crew	Daily Output	Labor-Hours	Unit	Material	2014 Bare Costs Labor	Equipment	Total	Total Incl O&P
0010	**PIPING, TESTING**									
0100	Nondestructive testing									
0110	Nondestructive hydraulic pressure test, isolate & 1 hr. hold									
0120	1" - 4" pipe									
0140	0 – 250 L.F.	1 Stpi	1.33	6.015	Ea.		350		350	530
0160	250 – 500 L.F.	"	.80	10			585		585	885
0180	500 – 1000 L.F.	Q-5	1.14	14.035			740		740	1,125
0200	1000 – 2000 L.F.	"	.80	20			1,050		1,050	1,600
0300	6" - 10" pipe									
0320	0 – 250 L.F.	Q-5	1	16	Ea.		840		840	1,275
0340	250 – 500 L.F.		.73	21.918			1,150		1,150	1,750
0360	500 – 1000 L.F.		.53	30.189			1,600		1,600	2,400
0380	1000 – 2000 L.F.		.38	42.105			2,225		2,225	3,350
1000	Pneumatic pressure test, includes soaping joints									
1120	1" - 4" pipe									
1140	0 – 250 L.F.	Q-5	2.67	5.993	Ea.	12.20	315		327.20	490
1160	250 – 500 L.F.		1.33	12.030		24.50	635		659.50	980
1180	500 – 1000 L.F.		.80	20		36.50	1,050		1,086.50	1,650
1200	1000 – 2000 L.F.		.50	32		49	1,675		1,724	2,600
1300	6" - 10" pipe									
1320	0 – 250 L.F.	Q-5	1.33	12.030	Ea.	12.20	635		647.20	970
1340	250 – 500 L.F.		.67	23.881		24.50	1,250		1,274.50	1,925
1360	500 – 1000 L.F.		.40	40		49	2,100		2,149	3,225
1380	1000 – 2000 L.F.		.25	64		61	3,375		3,436	5,175
2000	X-Ray of welds									
2110	2" diam.	1 Stpi	8	1	Ea.	12.60	58.50		71.10	102
2120	3" diam.		8	1		12.60	58.50		71.10	102
2130	4" diam.		8	1		18.90	58.50		77.40	110
2140	6" diam.		8	1		18.90	58.50		77.40	110
2150	8" diam.		6.60	1.212		18.90	71		89.90	128
2160	10" diam.		6	1.333		25	78		103	146
3000	Liquid penetration of welds									
3110	2" diam.	1 Stpi	14	.571	Ea.	3.95	33.50		37.45	55
3120	3" diam.		13.60	.588		3.95	34.50		38.45	56.50
3130	4" diam.		13.40	.597		3.95	35		38.95	57.50
3140	6" diam.		13.20	.606		3.95	35.50		39.45	58
3150	8" diam.		13	.615		5.95	36		41.95	61
3160	10" diam.		12.80	.625		5.95	36.50		42.45	61.50

23 07 HVAC Insulation

23 07 13 – Duct Insulation

23 07 13.10 Duct Thermal Insulation

			Crew	Daily Output	Labor-Hours	Unit	Material	Labor	Equipment	Total	Total Incl O&P
0010	**DUCT THERMAL INSULATION**										
0100	Rule of thumb, as a percentage of total mechanical costs					Job				10%	10%
0110	Insulation req'd. is based on the surface size/area to be covered										
3730	Sheet insulation										
3760	Polyethylene foam, closed cell, UV resistant										
3770	Standard temperature (-90°F to +212°F)										
3771	1/4" thick	G	Q-14	450	.036	S.F.	1.74	1.64		3.38	4.46
3772	3/8" thick	G		440	.036		2.48	1.67		4.15	5.35

23 07 HVAC Insulation

23 07 13 – Duct Insulation

23 07 13.10 Duct Thermal Insulation

		Crew	Daily Output	Labor-Hours	Unit	Material	2014 Bare Costs Labor	Equipment	Total	Total Incl O&P
3773	1/2" thick	G Q-14	420	.038	S.F.	3.05	1.75		4.80	6.10
3774	3/4" thick	G	400	.040		4.36	1.84		6.20	7.65
3775	1" thick	G	380	.042		5.90	1.94		7.84	9.45
3776	1-1/2" thick	G	360	.044		9.30	2.05		11.35	13.45
3777	2" thick	G	340	.047		12.30	2.17		14.47	16.90
3778	2-1/2" thick	G	320	.050		15.80	2.30		18.10	21
3779	Adhesive (see line 7878)									
3780	Foam, rubber									
3782	1" thick	G 1 Stpi	50	.160	S.F.	2.89	9.35		12.24	17.35
7000	Board insulation									
7020	Mineral wool, 1200° F									
7022	6 lb. density, plain									
7024	1" thick	G Q-14	370	.043	S.F.	.33	1.99		2.32	3.46
7026	1-1/2" thick	G	350	.046		.50	2.10		2.60	3.82
7028	2" thick	G	330	.048		.67	2.23		2.90	4.21
7030	3" thick	G	300	.053		.83	2.45		3.28	4.73
7032	4" thick	G	280	.057		1.33	2.63		3.96	5.55
7038	8 lb. density, plain									
7040	1" thick	G Q-14	360	.044	S.F.	.40	2.05		2.45	3.62
7042	1-1/2" thick	G	340	.047		.61	2.17		2.78	4.04
7044	2" thick	G	320	.050		.81	2.30		3.11	4.47
7046	3" thick	G	290	.055		1.21	2.54		3.75	5.30
7048	4" thick	G	270	.059		1.61	2.73		4.34	6
7060	10 lb. density, plain									
7062	1" thick	G Q-14	350	.046	S.F.	.55	2.10		2.65	3.88
7064	1-1/2" thick	G	330	.048		.83	2.23		3.06	4.38
7066	2" thick	G	310	.052		1.11	2.38		3.49	4.92
7068	3" thick	G	280	.057		1.66	2.63		4.29	5.90
7070	4" thick	G	260	.062		2.21	2.83		5.04	6.85
7878	Contact cement, quart can				Ea.	10.50			10.50	11.55

23 07 16 – HVAC Equipment Insulation

23 07 16.10 HVAC Equipment Thermal Insulation

		Crew	Daily Output	Labor-Hours	Unit	Material	Labor	Equipment	Total	Total Incl O&P
0010	**HVAC EQUIPMENT THERMAL INSULATION**									
0100	Rule of thumb, as a percentage of total mechanical costs				Job				10%	10%
0110	Insulation req'd. is based on the surface size/area to be covered									
1000	Boiler, 1-1/2" calcium silicate only	G Q-14	110	.145	S.F.	3.77	6.70		10.47	14.55
1020	Plus 2" fiberglass	G "	80	.200	"	4.74	9.20		13.94	19.55
2000	Breeching, 2" calcium silicate									
2020	Rectangular	G Q-14	42	.381	S.F.	7.40	17.55		24.95	35.50
2040	Round	G "	38.70	.413	"	7.70	19.05		26.75	38
2300	Calcium silicate block, + 200°F to +1200°F									
2310	On irregular surfaces, valves and fittings									
2340	1" thick	G Q-14	30	.533	S.F.	3.34	24.50		27.84	41.50
2360	1-1/2" thick	G	25	.640		3.77	29.50		33.27	50
2380	2" thick	G	22	.727		4.94	33.50		38.44	57.50
2400	3" thick	G	18	.889		7.60	41		48.60	72
2410	On plane surfaces									
2420	1" thick	G Q-14	126	.127	S.F.	3.34	5.85		9.19	12.75
2430	1-1/2" thick	G	120	.133		3.77	6.15		9.92	13.70
2440	2" thick	G	100	.160		4.94	7.35		12.29	16.90
2450	3" thick	G	70	.229		7.60	10.50		18.10	24.50

23 09 Instrumentation and Control for HVAC

23 09 13 – Instrumentation and Control Devices for HVAC

23 09 13.60 Water Level Controls

		Crew	Daily Output	Labor-Hours	Unit	Material	2014 Bare Costs Labor	Equipment	Total	Total Incl O&P
0010	**WATER LEVEL CONTROLS**									
3000	Low water cut-off for hot water boiler, 50 psi maximum									
3100	1" top & bottom equalizing pipes, manual reset	1 Stpi	14	.571	Ea.	325	33.50		358.50	410
3200	1" top & bottom equalizing pipes	↓	14	.571		325	33.50		358.50	410
3300	2-1/2" side connection for nipple-to-boiler	↓	14	.571	↓	296	33.50		329.50	375

23 09 53 – Pneumatic and Electric Control System for HVAC

23 09 53.10 Control Components

		Crew	Daily Output	Labor-Hours	Unit	Material	2014 Bare Costs Labor	Equipment	Total	Total Incl O&P
0010	**CONTROL COMPONENTS** R230500-10									
2000	Gauges, pressure or vacuum									
2100	2" diameter dial	1 Stpi	32	.250	Ea.	14.30	14.65		28.95	37.50
2200	2-1/2" diameter dial		32	.250		15.65	14.65		30.30	39.50
2300	3-1/2" diameter dial		32	.250		35.50	14.65		50.15	61.50
2400	4-1/2" diameter dial	↓	32	.250	↓	52.50	14.65		67.15	80
2700	Flanged iron case, black ring									
2800	3-1/2" diameter dial	1 Stpi	32	.250	Ea.	102	14.65		116.65	134
2900	4-1/2" diameter dial		32	.250		105	14.65		119.65	138
3000	6" diameter dial	↓	32	.250	↓	167	14.65		181.65	205
3010	Steel case, 0 – 300 psi									
3012	2" diam. dial	1 Stpi	16	.500	Ea.	9.20	29.50		38.70	54
3014	4" diam. dial	"	16	.500	"	27	29.50		56.50	74
3020	Aluminum case, 0 – 300 psi									
3022	3-1/2" diam. dial	1 Stpi	16	.500	Ea.	10.15	29.50		39.65	55
3024	4-1/2" diam. dial		16	.500		112	29.50		141.50	168
3026	6" diam. dial		16	.500		184	29.50		213.50	246
3028	8-1/2" diam. dial	↓	16	.500		315	29.50		344.50	390
3030	Brass case, 0 – 300 psi									
3032	2" diam. dial	1 Stpi	16	.500	Ea.	44.50	29.50		74	93
3034	4-1/2" diam. dial	"	16	.500	"	110	29.50		139.50	165
3040	Steel case, high pressure, 0 -10,000 psi									
3042	4-1/2" diam. dial	1 Stpi	16	.500	Ea.	291	29.50		320.50	365
3044	6-1/2" diam. dial		16	.500		305	29.50		334.50	380
3046	8-1/2" diam. dial	↓	16	.500	↓	390	29.50		419.50	475
3080	Pressure gauge, differential, magnehelic									
3084	0 – 2" W.C., with air filter kit	1 Stpi	6	1.333	Ea.	105	78		183	234
3300	For compound pressure-vacuum, add					18%				
4000	Thermometers									
4100	Dial type, 3-1/2" diameter, vapor type, union connection	1 Stpi	32	.250	Ea.	298	14.65		312.65	350
4120	Liquid type, union connection		32	.250		580	14.65		594.65	660
4500	Stem type, 6-1/2" case, 2" stem, 1/2" NPT		32	.250		64.50	14.65		79.15	93
4520	4" stem, 1/2" NPT		32	.250		79.50	14.65		94.15	109
4600	9" case, 3-1/2" stem, 3/4" NPT		28	.286		115	16.70		131.70	151
4620	6" stem, 3/4" NPT		28	.286		115	16.70		131.70	151
4640	8" stem, 3/4" NPT		28	.286		206	16.70		222.70	252
4660	12" stem, 1" NPT	↓	26	.308	↓	184	18		202	229
4670	Bi-metal, dial type, steel case brass stem									
4672	2" dial, 4" - 9" stem	1 Stpi	16	.500	Ea.	45	29.50		74.50	93.50
4673	2-1/2" dial, 4" - 9" stem		16	.500		45	29.50		74.50	93.50
4674	3-1/2" dial, 4" - 9" stem	↓	16	.500	↓	49.50	29.50		79	98.50
4680	Mercury filled, industrial, union connection type									
4682	Angle stem, 7" scale	1 Stpi	16	.500	Ea.	67.50	29.50		97	119
4683	9" scale		16	.500		181	29.50		210.50	243
4684	12" scale	↓	16	.500	↓	234	29.50		263.50	300

23 09 Instrumentation and Control for HVAC

23 09 53 – Pneumatic and Electric Control System for HVAC

23 09 53.10 Control Components

		Crew	Daily Output	Labor-Hours	Unit	Material	2014 Bare Costs Labor	Equipment	Total	Total Incl O&P
4686	Straight stem, 7" scale	1 Stpi	16	.500	Ea.	179	29.50		208.50	241
4687	9" scale		16	.500		167	29.50		196.50	228
4688	12" scale		16	.500		206	29.50		235.50	271
4690	Mercury filled, industrial, separable socket type, with well									
4692	Angle stem, with socket, 7" scale	1 Stpi	16	.500	Ea.	69	29.50		98.50	120
4693	9" scale		16	.500		72	29.50		101.50	123
4694	12" scale		16	.500		224	29.50		253.50	291
4696	Straight stem, with socket, 7" scale		16	.500		37.50	29.50		67	85.50
4697	9" scale		16	.500		90.50	29.50		120	144
4698	12" scale		16	.500		98	29.50		127.50	152
6000	Valves, motorized zone									
6100	Sweat connections, 1/2" C x C	1 Stpi	20	.400	Ea.	166	23.50		189.50	218
6110	3/4" C x C		20	.400		166	23.50		189.50	219
6120	1" C x C		19	.421		217	24.50		241.50	275
6140	1/2" C x C, with end switch, 2 wire		20	.400		164	23.50		187.50	216
6150	3/4" C x C, with end switch, 2 wire		20	.400		175	23.50		198.50	229
6160	1" C x C, with end switch, 2 wire		19	.421		200	24.50		224.50	257

23 11 Facility Fuel Piping

23 11 13 – Facility Fuel-Oil Piping

23 11 13.10 Fuel Oil Specialties

		Crew	Daily Output	Labor-Hours	Unit	Material	Labor	Equipment	Total	Total Incl O&P
0010	**FUEL OIL SPECIALTIES**									
0020	Foot valve, single poppet, metal to metal construction									
0040	Bevel seat, 1/2" diameter	1 Stpi	20	.400	Ea.	61.50	23.50		85	104
0060	3/4" diameter		18	.444		71	26		97	118
1000	Oil filters, 3/8" IPT, 20 gal. per hour		20	.400		30	23.50		53.50	68.50
2000	Remote tank gauging system, self contained									
3000	Valve, ball check, globe type, 3/8" diameter	1 Stpi	24	.333	Ea.	14.05	19.50		33.55	45
3500	Fusible, 3/8" diameter		24	.333		13.05	19.50		32.55	44
3600	1/2" diameter		24	.333		32	19.50		51.50	64.50
3610	3/4" diameter		20	.400		70	23.50		93.50	113
3620	1" diameter		19	.421		205	24.50		229.50	263
4000	Nonfusible, 3/8" diameter		24	.333		22.50	19.50		42	54
4500	Shutoff, gate type, lever handle, spring-fusible kit									
4520	1/4" diameter	1 Stpi	14	.571	Ea.	34.50	33.50		68	88.50
4540	3/8" diameter		12	.667		33.50	39		72.50	96
4560	1/2" diameter		10	.800		49.50	47		96.50	125
4570	3/4" diameter		8	1		86	58.50		144.50	183
5000	Vent alarm, whistling signal					26			26	28.50
5500	Vent protector/breather, 1-1/4" diameter	1 Stpi	32	.250		13.60	14.65		28.25	37

23 11 23 – Facility Natural-Gas Piping

23 11 23.10 Gas Meters

		Crew	Daily Output	Labor-Hours	Unit	Material	Labor	Equipment	Total	Total Incl O&P
0010	**GAS METERS**									
4000	Residential									
4010	Gas meter, residential, 3/4" pipe size	1 Plum	14	.571	Ea.	208	33		241	279
4020	Gas meter, residential, 1" pipe size		12	.667		176	38.50		214.50	252
4030	Gas meter, residential, 1-1/4" pipe size		10	.800		177	46		223	265

23 12 Facility Fuel Pumps

23 12 13 – Facility Fuel-Oil Pumps

23 12 13.10 Pump and Motor Sets	Crew	Daily Output	Labor-Hours	Unit	Material	2014 Bare Costs Labor	Equipment	Total	Total Incl O&P
0010 **PUMP AND MOTOR SETS**									
1810 Light fuel and diesel oils									
1820 20 GPH 1/3 HP	Q-5	6	2.667	Ea.	1,050	140		1,190	1,350
1850 145 GPH, 1/2 HP	"	4	4	"	1,100	211		1,311	1,550

23 13 Facility Fuel-Storage Tanks

23 13 13 – Facility Underground Fuel-Oil, Storage Tanks

23 13 13.09 Single-Wall Steel Fuel-Oil Tanks

0010 **SINGLE-WALL STEEL FUEL-OIL TANKS**									
5000 Tanks, steel ugnd., sti-p3, not incl. hold-down bars									
5500 Excavation, pad, pumps and piping not included									
5510 Single wall, 500 gallon capacity, 7 ga. shell	Q-5	2.70	5.926	Ea.	2,050	310		2,360	2,725
5520 1,000 gallon capacity, 7 ga. shell	"	2.50	6.400		3,825	335		4,160	4,700
5530 2,000 gallon capacity, 1/4" thick shell	Q-7	4.60	6.957		6,200	385		6,585	7,400
5535 2,500 gallon capacity, 7 ga. shell	Q-5	3	5.333		6,850	281		7,131	7,950
5540 5,000 gallon capacity, 1/4" thick shell	Q-7	3.20	10		11,500	555		12,055	13,500
5560 10,000 gallon capacity, 1/4" thick shell		2	16		12,300	890		13,190	15,000
5580 15,000 gallon capacity, 5/16" thick shell		1.70	18.824		19,000	1,050		20,050	22,500
5600 20,000 gallon capacity, 5/16" thick shell		1.50	21.333		36,300	1,200		37,500	41,800
5610 25,000 gallon capacity, 3/8" thick shell		1.30	24.615		37,900	1,375		39,275	43,800
5620 30,000 gallon capacity, 3/8" thick shell		1.10	29.091		38,400	1,625		40,025	44,700
5630 40,000 gallon capacity, 3/8" thick shell		.90	35.556		42,000	1,975		43,975	49,200
5640 50,000 gallon capacity, 3/8" thick shell		.80	40		46,600	2,225		48,825	55,000

23 13 13.13 Dbl-Wall Steel, Undrgrnd Fuel-Oil, Stor. Tanks

0010 **DOUBLE-WALL STEEL, UNDERGROUND FUEL-OIL, STORAGE TANKS**									
6200 Steel, underground, 360°, double wall, U.L. listed,									
6210 with sti-P3 corrosion protection,									
6220 (dielectric coating, cathodic protection, electrical									
6230 isolation) 30 year warranty,									
6240 not incl. manholes or hold-downs.									
6250 500 gallon capacity	Q-5	2.40	6.667	Ea.	5,575	350		5,925	6,675
6260 1,000 gallon capactiy	"	2.25	7.111		6,900	375		7,275	8,150
6270 2,000 gallon capacity	Q-7	4.16	7.692		9,050	430		9,480	10,600
6280 3,000 gallon capacity		3.90	8.205		10,900	455		11,355	12,700
6290 4,000 gallon capacity		3.64	8.791		13,900	490		14,390	16,000
6300 5,000 gallon capacity		2.91	10.997		19,900	615		20,515	22,800
6310 6,000 gallon capacity		2.42	13.223		23,600	735		24,335	27,000
6320 8,000 gallon capacity		2.08	15.385		24,600	855		25,455	28,300
6330 10,000 gallon capacity		1.82	17.582		27,700	980		28,680	32,000
6340 12,000 gallon capacity		1.70	18.824		30,600	1,050		31,650	35,300
6350 15,000 gallon capacity		1.33	24.060		31,900	1,350		33,250	37,100
6360 20,000 gallon capacity		1.33	24.060		32,700	1,350		34,050	37,900
6370 25,000 gallon capacity		1.16	27.586		72,500	1,525		74,025	82,500
6380 30,000 gallon capacity		1.03	31.068		87,500	1,725		89,225	98,500
6390 40,000 gallon capacity		.80	40		112,500	2,225		114,725	127,000
6395 50,000 gallon capacity		.73	43.836		136,500	2,450		138,950	154,000
6400 For hold-downs 500-2000 gal., add		16	2	Set	183	111		294	370
6410 For hold-downs 3000-6000 gal., add		12	2.667		370	149		519	635
6420 For hold-downs 8000-12,000 gal., add		11	2.909		445	162		607	735
6430 For hold-downs 15,000 gal., add		9	3.556		640	198		838	1,000
6440 For hold-downs 20,000 gal., add		8	4		735	223		958	1,150

23 13 Facility Fuel-Storage Tanks

23 13 13 – Facility Underground Fuel-Oil, Storage Tanks

23 13 13.13 Dbl-Wall Steel, Undrgrnd Fuel-Oil, Stor. Tanks		Crew	Daily Output	Labor-Hours	Unit	Material	2014 Bare Costs Labor	Equipment	Total	Total Incl O&P
6450	For hold-downs 20,000 gal. plus, add	Q-7	6	5.333	Set	960	297		1,257	1,500
6500	For manways, add				Ea.	1,800			1,800	2,000
6600	In place with hold-downs									
6652	550 gallon capacity	Q-5	1.84	8.696	Ea.	5,775	460		6,235	7,050

23 13 13.23 Glass-Fiber-Reinfcd-Plastic, Fuel-Oil, Storage

		Crew	Daily Output	Labor-Hours	Unit	Material	Labor	Equipment	Total	Incl O&P
0010	**GLASS-FIBER-REINFCD-PLASTIC, UNDERGRND. FUEL-OIL, STORAGE**									
0210	Fiberglass, underground, single wall, U.L. listed, not including									
0220	manway or hold-down strap									
0225	550 gallon capacity	Q-5	2.67	5.993	Ea.	3,025	315		3,340	3,825
0230	1,000 gallon capacity	"	2.46	6.504		3,925	340		4,265	4,825
0240	2,000 gallon capacity	Q-7	4.57	7.002		5,675	390		6,065	6,825
0245	3,000 gallon capacity		3.90	8.205		6,800	455		7,255	8,175
0250	4,000 gallon capacity		3.55	9.014		7,875	500		8,375	9,425
0255	5,000 gallon capacity		3.20	10		8,750	555		9,305	10,500
0260	6,000 gallon capacity		2.67	11.985		8,950	670		9,620	10,900
0270	8,000 gallon capacity		2.29	13.974		10,700	780		11,480	13,000
0280	10,000 gallon capacity		2	16		12,200	890		13,090	14,800
0282	12,000 gallon capacity		1.88	17.021		13,800	950		14,750	16,600
0284	15,000 gallon capacity		1.68	19.048		19,500	1,050		20,550	23,000
0290	20,000 gallon capacity		1.45	22.069		25,100	1,225		26,325	29,500
0300	25,000 gallon capacity		1.28	25		51,500	1,400		52,900	59,000
0320	30,000 gallon capacity		1.14	28.070		73,500	1,575		75,075	83,500
0340	40,000 gallon capacity		.89	35.955		81,500	2,000		83,500	93,000
0360	48,000 gallon capacity		.81	39.506		122,000	2,200		124,200	137,500
0500	For manway, fittings and hold-downs, add					20%	15%			
0600	For manways, add					2,125			2,125	2,350
1000	For helical heating coil, add	Q-5	2.50	6.400		4,550	335		4,885	5,500
1020	Fiberglass, underground, double wall, U.L. listed									
1030	includes manways, not incl. hold-down straps									
1040	600 gallon capacity	Q-5	2.42	6.612	Ea.	6,850	350		7,200	8,050
1050	1,000 gallon capacity	"	2.25	7.111		9,350	375		9,725	10,900
1060	2,500 gallon capacity	Q-7	4.16	7.692		14,100	430		14,530	16,100
1070	3,000 gallon capacity		3.90	8.205		15,300	455		15,755	17,500
1080	4,000 gallon capacity		3.64	8.791		15,500	490		15,990	17,700
1090	6,000 gallon capacity		2.42	13.223		20,100	735		20,835	23,300
1100	8,000 gallon capacity		2.08	15.385		22,400	855		23,255	26,000
1110	10,000 gallon capacity		1.82	17.582		26,000	980		26,980	30,100
1120	12,000 gallon capacity		1.70	18.824		32,400	1,050		33,450	37,300
1122	15,000 gallon capacity		1.52	21.053		43,100	1,175		44,275	49,200
1124	20,000 gallon capacity		1.33	24.060		53,000	1,350		54,350	60,500
1126	25,000 gallon capacity		1.16	27.586		73,000	1,525		74,525	83,000
1128	30,000 gallon capacity		1.03	31.068		87,500	1,725		89,225	99,000
1140	For hold-down straps, add					2%	10%			
1150	For hold-downs 500-4000 gal., add	Q-7	16	2	Set	425	111		536	640
1160	For hold-downs 5000-15000 gal., add		8	4		855	223		1,078	1,275
1170	For hold-downs 20,000 gal., add		5.33	6.004		1,275	335		1,610	1,900
1180	For hold-downs 25,000 gal., add		4	8		1,700	445		2,145	2,550
1190	For hold-downs 30,000 gal., add		2.60	12.308		2,550	685		3,235	3,850
2210	Fiberglass, underground, single wall, U.L. listed, including									
2220	hold-down straps, no manways									
2225	550 gallon capacity	Q-5	2	8	Ea.	3,475	420		3,895	4,425
2230	1,000 gallon capacity	"	1.88	8.511		4,350	450		4,800	5,450

23 13 Facility Fuel-Storage Tanks

23 13 13 – Facility Underground Fuel-Oil, Storage Tanks

23 13 13.23 Glass-Fiber-Reinfcd-Plastic, Fuel-Oil, Storage

		Crew	Daily Output	Labor-Hours	Unit	Material	2014 Bare Costs Labor	Equipment	Total	Total Incl O&P
2240	2,000 gallon capacity	Q-7	3.55	9.014	Ea.	6,100	500		6,600	7,450
2250	4,000 gallon capacity		2.90	11.034		8,300	615		8,915	10,100
2260	6,000 gallon capacity		2	16		9,800	890		10,690	12,200
2270	8,000 gallon capacity		1.78	17.978		11,600	1,000		12,600	14,200
2280	10,000 gallon capacity		1.60	20		13,100	1,125		14,225	16,100
2282	12,000 gallon capacity		1.52	21.053		14,700	1,175		15,875	18,000
2284	15,000 gallon capacity		1.39	23.022		20,300	1,275		21,575	24,300
2290	20,000 gallon capacity		1.14	28.070		26,400	1,575		27,975	31,400
2300	25,000 gallon capacity		.96	33.333		53,500	1,850		55,350	61,500
2320	30,000 gallon capacity		.80	40		76,000	2,225		78,225	87,500
3020	Fiberglass, underground, double wall, U.L. listed									
3030	includes manways and hold-down straps									
3040	600 gallon capacity	Q-5	1.86	8.602	Ea.	7,275	455		7,730	8,675
3050	1,000 gallon capacity	"	1.70	9.412		9,775	495		10,270	11,600
3060	2,500 gallon capacity	Q-7	3.29	9.726		14,500	540		15,040	16,700
3070	3,000 gallon capacity		3.13	10.224		15,700	570		16,270	18,100
3080	4,000 gallon capacity		2.93	10.922		15,900	610		16,510	18,400
3090	6,000 gallon capacity		1.86	17.204		21,000	960		21,960	24,600
3100	8,000 gallon capacity		1.65	19.394		23,300	1,075		24,375	27,200
3110	10,000 gallon capacity		1.48	21.622		26,800	1,200		28,000	31,300
3120	12,000 gallon capacity		1.40	22.857		33,300	1,275		34,575	38,500
3122	15,000 gallon capacity		1.28	25		44,000	1,400		45,400	50,500
3124	20,000 gallon capacity		1.06	30.189		54,500	1,675		56,175	62,500
3126	25,000 gallon capacity		.90	35.556		74,500	1,975		76,475	85,000
3128	30,000 gallon capacity		.74	43.243		90,500	2,400		92,900	103,000

23 13 23 – Facility Aboveground Fuel-Oil, Storage Tanks

23 13 23.16 Horizontal, Stl, Abvgrd Fuel-Oil, Storage Tanks

		Crew	Daily Output	Labor-Hours	Unit	Material	Labor	Equipment	Total	Total Incl O&P
0010	**HORIZONTAL, STEEL, ABOVEGROUND FUEL-OIL, STORAGE TANKS**									
3000	Steel, storage, above ground, including cradles, coating,									
3020	fittings, not including foundation, pumps or piping									
3040	Single wall, 275 gallon	Q-5	5	3.200	Ea.	490	168		658	795
3060	550 gallon	"	2.70	5.926		3,400	310		3,710	4,225
3080	1,000 gallon	Q-7	5	6.400		5,600	355		5,955	6,700
3100	1,500 gallon		4.75	6.737		7,975	375		8,350	9,350
3120	2,000 gallon		4.60	6.957		9,675	385		10,060	11,200
3320	Double wall, 500 gallon capacity	Q-5	2.40	6.667		2,625	350		2,975	3,400
3330	2000 gallon capacity	Q-7	4.15	7.711		9,975	430		10,405	11,700
3340	4000 gallon capacity		3.60	8.889		17,800	495		18,295	20,400
3350	6000 gallon capacity		2.40	13.333		21,000	745		21,745	24,200
3360	8000 gallon capacity		2	16		27,000	890		27,890	31,100
3370	10000 gallon capacity		1.80	17.778		30,200	990		31,190	34,700
3380	15000 gallon capacity		1.50	21.333		45,900	1,200		47,100	52,500
3390	20000 gallon capacity		1.30	24.615		52,500	1,375		53,875	59,500
3400	25000 gallon capacity		1.15	27.826		63,500	1,550		65,050	72,500
3410	30000 gallon capacity		1	32		69,500	1,775		71,275	79,000

23 13 23.26 Horizontal, Conc., Abvgrd Fuel-Oil, Stor. Tanks

		Crew	Daily Output	Labor-Hours	Unit	Material	Labor	Equipment	Total	Total Incl O&P
0010	**HORIZONTAL, CONCRETE, ABOVEGROUND FUEL-OIL, STORAGE TANKS**									
0050	Concrete, storage, above ground, including pad & pump									
0100	500 gallon	F-3	2	20	Ea.	10,000	935	330	11,265	12,800
0200	1,000 gallon	"	2	20		14,000	935	330	15,265	17,200
0300	2,000 gallon	F-4	2	24		18,000	1,100	570	19,670	22,100

23 21 Hydronic Piping and Pumps

23 21 20 – Hydronic HVAC Piping Specialties

23 21 20.10 Air Control

		Crew	Daily Output	Labor-Hours	Unit	Material	2014 Bare Costs Labor	Equipment	Total	Total Incl O&P
0010	**AIR CONTROL**									
0030	Air separator, with strainer									
0040	2" diameter	Q-5	6	2.667	Ea.	1,175	140		1,315	1,500
0080	2-1/2" diameter		5	3.200		1,325	168		1,493	1,700
0100	3" diameter	↓	4	4	↓	2,050	211		2,261	2,575
1000	Micro-bubble separator for total air removal, closed loop system									
1010	Requires bladder type tank in system.									
1020	Water (hot or chilled) or glycol system									
1030	Threaded									
1040	3/4" diameter	1 Stpi	20	.400	Ea.	88.50	23.50		112	133
1050	1" diameter		19	.421		99.50	24.50		124	146
1060	1-1/4" diameter		16	.500		138	29.50		167.50	195
1070	1-1/2" diameter	↓	13	.615		179	36		215	252
1080	2" diameter		11	.727		865	42.50		907.50	1,025
1090	2-1/2" diameter	Q-5	15	1.067		965	56		1,021	1,125
1100	3" diameter		13	1.231		1,375	65		1,440	1,625
1110	4" diameter	↓	10	1.600	↓	1,500	84		1,584	1,775

23 21 20.14 Air Purging Scoop

		Crew	Daily Output	Labor-Hours	Unit	Material	Labor	Equipment	Total	Total Incl O&P
0010	**AIR PURGING SCOOP**, With tappings.									
0020	For air vent and expansion tank connection									
0100	1" pipe size, threaded	1 Stpi	19	.421	Ea.	23.50	24.50		48	63
0110	1-1/4" pipe size, threaded		15	.533		23.50	31		54.50	73
0120	1-1/2" pipe size, threaded		13	.615		52	36		88	112
0130	2" pipe size, threaded	↓	11	.727	↓	59.50	42.50		102	130

23 21 20.18 Automatic Air Vent

		Crew	Daily Output	Labor-Hours	Unit	Material	Labor	Equipment	Total	Total Incl O&P
0010	**AUTOMATIC AIR VENT**									
0020	Cast iron body, stainless steel internals, float type									
0180	1/2" NPT inlet, 250 psi	1 Stpi	10	.800	Ea.	335	47		382	440
0220	3/4" NPT inlet, 250 psi	"	10	.800	"	335	47		382	440
0600	Forged steel body, stainless steel internals, float type									
0640	1/2" NPT inlet, 750 psi	1 Stpi	12	.667	Ea.	1,275	39		1,314	1,450
0680	3/4" NPT inlet, 750 psi	"	12	.667	"	1,275	39		1,314	1,450
1100	Formed steel body, noncorrosive									
1110	1/8" NPT inlet 150 psi	1 Stpi	32	.250	Ea.	12.10	14.65		26.75	35.50
1120	1/4" NPT inlet 150 psi		32	.250		41	14.65		55.65	67
1130	3/4" NPT inlet 150 psi	↓	32	.250	↓	41	14.65		55.65	67
1300	Chrome plated brass, automatic/manual, for radiators									
1310	1/8" NPT inlet, nickel plated brass	1 Stpi	32	.250	Ea.	7.35	14.65		22	30

23 21 20.26 Circuit Setter

		Crew	Daily Output	Labor-Hours	Unit	Material	Labor	Equipment	Total	Total Incl O&P
0010	**CIRCUIT SETTER**, Balance valve									
0018	Threaded									
0019	1/2" pipe size	1 Stpi	22	.364	Ea.	74	21.50		95.50	113
0020	3/4" pipe size	"	20	.400	"	79	23.50		102.50	123

23 21 20.30 Cocks, Drains and Specialties

		Crew	Daily Output	Labor-Hours	Unit	Material	Labor	Equipment	Total	Total Incl O&P
0010	**COCKS, DRAINS AND SPECIALTIES**									
1000	Boiler drain									
1010	Pipe thread to hose									
1020	Bronze									
1030	1/2" size	1 Stpi	36	.222	Ea.	11.90	13		24.90	32.50
1040	3/4" size	"	34	.235	"	12.95	13.75		26.70	35
1100	Solder to hose									
1110	Bronze									

23 21 Hydronic Piping and Pumps

23 21 20 – Hydronic HVAC Piping Specialties

23 21 20.30 Cocks, Drains and Specialties

		Crew	Daily Output	Labor-Hours	Unit	Material	2014 Bare Costs Labor	Equipment	Total	Total Incl O&P
1120	1/2" size	1 Stpi	46	.174	Ea.	11.90	10.15		22.05	28.50
1130	3/4" size	"	44	.182	"	12.95	10.65		23.60	30.50
1600	With built-in vacuum breaker									
1610	1/2" I.P. or solder	1 Stpi	36	.222	Ea.	40	13		53	63.50
1630	With tamper proof vacuum breaker									
1640	1/2" I.P. or solder	1 Stpi	36	.222	Ea.	45.50	13		58.50	69.50
1650	3/4" I.P. or solder	"	34	.235	"	46	13.75		59.75	71.50
3000	Cocks									
3010	Air, lever or tee handle									
3020	Bronze, single thread									
3030	1/8" size	1 Stpi	52	.154	Ea.	10.55	9		19.55	25
3040	1/4" size		46	.174		10.90	10.15		21.05	27.50
3050	3/8" size		40	.200		11.05	11.70		22.75	30
3060	1/2" size		36	.222		13.05	13		26.05	34
3100	Bronze, double thread									
3110	1/8" size	1 Stpi	26	.308	Ea.	13.15	18		31.15	41.50
3120	1/4" size		22	.364		13.55	21.50		35.05	47
3130	3/8" size		18	.444		14.10	26		40.10	55
3140	1/2" size		15	.533		17.25	31		48.25	66
4500	Gauge cock, brass									
4510	1/4" FPT	1 Stpi	24	.333	Ea.	10.90	19.50		30.40	41.50
4512	1/4" MPT	"	24	.333	"	12.95	19.50		32.45	44
4600	Pigtail, steam syphon									
4604	1/4"	1 Stpi	24	.333	Ea.	19.90	19.50		39.40	51.50
4650	Snubber valve									
4654	1/4"	1 Stpi	22	.364	Ea.	12.55	21.50		34.05	46
4660	Nipple, black steel									
4664	1/4" x 3"	1 Stpi	37	.216	Ea.	2.83	12.65		15.48	22

23 21 20.34 Dielectric Unions

		Crew	Daily Output	Labor-Hours	Unit	Material	2014 Bare Costs Labor	Equipment	Total	Total Incl O&P
0010	**DIELECTRIC UNIONS**, Standard gaskets for water and air									
0020	250 psi maximum pressure									
0280	Female IPT to sweat, straight									
0300	1/2" pipe size	1 Plum	24	.333	Ea.	5.60	19.20		24.80	35
0340	3/4" pipe size		20	.400		6.20	23		29.20	42
0360	1" pipe size		19	.421		8.45	24		32.45	46
0380	1-1/4" pipe size		15	.533		13.10	30.50		43.60	61
0400	1-1/2" pipe size		13	.615		19.70	35.50		55.20	75
0420	2" pipe size		11	.727		26.50	42		68.50	92
0580	Female IPT to brass pipe thread, straight									
0600	1/2" pipe size	1 Plum	24	.333	Ea.	12.10	19.20		31.30	42.50
0640	3/4" pipe size		20	.400		13.30	23		36.30	49.50
0660	1" pipe size		19	.421		24.50	24		48.50	63.50
0680	1-1/4" pipe size		15	.533		30	30.50		60.50	79.50
0700	1-1/2" pipe size		13	.615		44	35.50		79.50	102
0720	2" pipe size		11	.727		86	42		128	158
0780	Female IPT to female IPT, straight									
0800	1/2" pipe size	1 Plum	24	.333	Ea.	13.85	19.20		33.05	44
0840	3/4" pipe size		20	.400		15.60	23		38.60	52
0860	1" pipe size		19	.421		21	24		45	59.50
0880	1-1/4" pipe size		15	.533		28	30.50		58.50	77.50
0900	1-1/2" pipe size		13	.615		43	35.50		78.50	101
0920	2" pipe size		11	.727		63	42		105	132

23 21 Hydronic Piping and Pumps

23 21 20 – Hydronic HVAC Piping Specialties

23 21 20.34 Dielectric Unions

		Crew	Daily Output	Labor-Hours	Unit	Material	2014 Bare Costs Labor	Equipment	Total	Total Incl O&P
2000	175 psi maximum pressure									
2180	Female IPT to sweat									
2240	2" pipe size	1 Plum	9	.889	Ea.	142	51		193	233
2260	2-1/2" pipe size	Q-1	15	1.067		155	55.50		210.50	255
2280	3" pipe size		14	1.143		213	59		272	325
2300	4" pipe size	↓	11	1.455	↓	565	75.50		640.50	735
2480	Female IPT to brass pipe									
2500	1-1/2" pipe size	1 Plum	11	.727	Ea.	176	42		218	257
2540	2" pipe size	"	9	.889		207	51		258	305
2560	2-1/2" pipe size	Q-1	15	1.067		305	55.50		360.50	420
2580	3" pipe size		14	1.143		355	59		414	480
2600	4" pipe size	↓	11	1.455	↓	635	75.50		710.50	815

23 21 20.42 Expansion Joints

		Crew	Daily Output	Labor-Hours	Unit	Material	Labor	Equipment	Total	Total Incl O&P
0010	**EXPANSION JOINTS**									
0100	Bellows type, neoprene cover, flanged spool									
0140	6" face to face, 1-1/4" diameter	1 Stpi	11	.727	Ea.	255	42.50		297.50	345
0160	1-1/2" diameter	"	10.60	.755		255	44		299	350
0180	2" diameter	Q-5	13.30	1.203		258	63.50		321.50	380
0190	2-1/2" diameter		12.40	1.290		267	68		335	395
0200	3" diameter		11.40	1.404		299	74		373	440
0210	4" diameter		8.40	1.905		325	100		425	505
0220	5" diameter		7.60	2.105		395	111		506	600
0230	6" diameter		6.80	2.353		405	124		529	635
0240	8" diameter		5.40	2.963		475	156		631	755
0250	10" diameter		5	3.200		650	168		818	970
0260	12" diameter		4.60	3.478		745	183		928	1,100
0480	10" face to face, 2" diameter		13	1.231		370	65		435	510
0500	2-1/2" diameter		12	1.333		390	70		460	535
0520	3" diameter		11	1.455		400	76.50		476.50	555
0540	4" diameter		8	2		455	105		560	660
0560	5" diameter		7	2.286		540	120		660	775
0580	6" diameter		6	2.667		560	140		700	825
0600	8" diameter		5	3.200		670	168		838	990
0620	10" diameter		4.60	3.478		735	183		918	1,075
0640	12" diameter		4	4		910	211		1,121	1,325
0660	14" diameter		3.80	4.211		1,125	222		1,347	1,575
0680	16" diameter		2.90	5.517		1,300	290		1,590	1,875
0700	18" diameter		2.50	6.400		1,475	335		1,810	2,125
0720	20" diameter		2.10	7.619		1,550	400		1,950	2,300
0740	24" diameter		1.80	8.889		1,800	470		2,270	2,675
0760	26" diameter		1.40	11.429		2,025	600		2,625	3,125
0780	30" diameter		1.20	13.333		2,250	700		2,950	3,525
0800	36" diameter	↓	1	16	↓	2,800	840		3,640	4,350

23 21 20.46 Expansion Tanks

		Crew	Daily Output	Labor-Hours	Unit	Material	Labor	Equipment	Total	Total Incl O&P
0010	**EXPANSION TANKS**									
1502	Plastic, corrosion resistant, see Section 22 66 83.13									
1505	Aboveground fuel-oil, storage tanks, see Section 23 13 23									
1507	Underground fuel-oil storage tanks, see Section 23 13 13									
1512	Tank leak detection systems, see Section 28 33 33.50									
2000	Steel, liquid expansion, ASME, painted, 15 gallon capacity	Q-5	17	.941	Ea.	640	49.50		689.50	775
2020	24 gallon capacity		14	1.143		715	60		775	875
2040	30 gallon capacity	↓	12	1.333	↓	715	70		785	890

23 21 Hydronic Piping and Pumps

23 21 20 – Hydronic HVAC Piping Specialties

23 21 20.46 Expansion Tanks

		Crew	Daily Output	Labor-Hours	Unit	Material	2014 Bare Costs Labor	Equipment	Total	Total Incl O&P
2060	40 gallon capacity	Q-5	10	1.600	Ea.	835	84		919	1,050
2360	Galvanized									
2370	15 gallon capacity	Q-5	17	.941	Ea.	1,125	49.50		1,174.50	1,300
2380	24 gallon capacity		14	1.143		1,150	60		1,210	1,375
2390	30 gallon capacity		12	1.333		1,350	70		1,420	1,575
3000	Steel ASME expansion, rubber diaphragm, 19 gal. cap. accept.		12	1.333		2,425	70		2,495	2,775
3020	31 gallon capacity		8	2		2,700	105		2,805	3,125
3040	61 gallon capacity		6	2.667		3,800	140		3,940	4,375

23 21 20.50 Float Valves

		Crew	Daily Output	Labor-Hours	Unit	Material	Labor	Equipment	Total	Total Incl O&P
0010	**FLOAT VALVES**									
0020	With ball and bracket									
0030	Single seat, threaded									
0040	Brass body									
0050	1/2"	1 Stpi	11	.727	Ea.	97	42.50		139.50	172
0060	3/4"		9	.889		108	52		160	198
0070	1"		7	1.143		144	67		211	259
0080	1-1/2"		4.50	1.778		211	104		315	390
0090	2"		3.60	2.222		155	130		285	365
0300	For condensate receivers, CI, in-line mount									
0320	1" inlet	1 Stpi	7	1.143	Ea.	144	67		211	259
0360	For condensate receiver, CI, external float, flanged tank mount									
0370	3/4" inlet	1 Stpi	5	1.600	Ea.	108	93.50		201.50	260

23 21 20.54 Flow Check Control

		Crew	Daily Output	Labor-Hours	Unit	Material	Labor	Equipment	Total	Total Incl O&P
0010	**FLOW CHECK CONTROL**									
0100	Bronze body, soldered									
0110	3/4" size	1 Stpi	20	.400	Ea.	60	23.50		83.50	102
0120	1" size	"	19	.421	"	72.50	24.50		97	117
0200	Cast iron body, threaded									
0210	3/4" size	1 Stpi	20	.400	Ea.	46.50	23.50		70	86.50
0220	1" size		19	.421		52.50	24.50		77	94.50
0230	1-1/4" size		15	.533		63.50	31		94.50	117
0240	1-1/2" size		13	.615		97	36		133	162
0250	2" size		11	.727		140	42.50		182.50	219

23 21 20.58 Hydronic Heating Control Valves

		Crew	Daily Output	Labor-Hours	Unit	Material	Labor	Equipment	Total	Total Incl O&P
0010	**HYDRONIC HEATING CONTROL VALVES**									
0050	Hot water, nonelectric, thermostatic									
0100	Radiator supply, 1/2" diameter	1 Stpi	24	.333	Ea.	68.50	19.50		88	105
0120	3/4" diameter	"	20	.400	"	68.50	23.50		92	111
1000	Manual, radiator supply									
1010	1/2" pipe size, angle union	1 Stpi	24	.333	Ea.	50	19.50		69.50	84.50
1020	3/4" pipe size, angle union	"	20	.400	"	63	23.50		86.50	105
1100	Radiator, balancing, straight, sweat connections									
1110	1/2" pipe size	1 Stpi	24	.333	Ea.	19.05	19.50		38.55	50.50
1120	3/4" pipe size	"	20	.400	"	26.50	23.50		50	65
1200	Steam, radiator, supply									
1210	1/2" pipe size, angle union	1 Stpi	24	.333	Ea.	49	19.50		68.50	83
1220	3/4" pipe size, angle union	"	20	.400	"	53	23.50		76.50	93.50
8000	System balancing and shut-off									
8020	Butterfly, quarter turn, calibrated, threaded or solder									
8040	Bronze, -30°F to +350°F, pressure to 175 psi									
8060	1/2" size	1 Stpi	22	.364	Ea.	17.35	21.50		38.85	51
8070	3/4" size	"	20	.400	"	28	23.50		51.50	66.50

323

23 21 Hydronic Piping and Pumps

23 21 20 – Hydronic HVAC Piping Specialties

23 21 20.66 Monoflow Tee Fitting

		Crew	Daily Output	Labor-Hours	Unit	Material	2014 Bare Costs Labor	Equipment	Total	Total Incl O&P
0010	**MONOFLOW TEE FITTING**									
1100	For one pipe hydronic, supply and return									
1110	Copper, soldered									
1120	3/4" x 1/2" size	1 Stpi	13	.615	Ea.	20.50	36		56.50	77

23 21 20.74 Strainers, Basket Type

		Crew	Daily Output	Labor-Hours	Unit	Material	Labor	Equipment	Total	Total Incl O&P
0010	**STRAINERS, BASKET TYPE**, Perforated stainless steel basket									
0100	Brass or monel available									
2000	Simplex style									
2300	Bronze body									
2320	Screwed, 3/8" pipe size	1 Stpi	22	.364	Ea.	172	21.50		193.50	221
2340	1/2" pipe size		20	.400		284	23.50		307.50	345
2360	3/4" pipe size		17	.471		380	27.50		407.50	455
2380	1" pipe size		15	.533		380	31		411	460
2400	1-1/4" pipe size		13	.615		555	36		591	665
2420	1-1/2" pipe size		12	.667		555	39		594	670
2440	2" pipe size		10	.800		825	47		872	975
2460	2-1/2" pipe size	Q-5	15	1.067		770	56		826	935
2480	3" pipe size	"	14	1.143		1,125	60		1,185	1,325
2600	Flanged, 2" pipe size	1 Stpi	6	1.333		820	78		898	1,025
2620	2-1/2" pipe size	Q-5	4.50	3.556		1,275	187		1,462	1,700
2640	3" pipe size		3.50	4.571		1,450	241		1,691	1,975
2660	4" pipe size		3	5.333		2,450	281		2,731	3,125
2680	5" pipe size	Q-6	3.40	7.059		3,750	385		4,135	4,700
2700	6" pipe size		3	8		4,775	435		5,210	5,900
2710	8" pipe size		2.50	9.600		7,750	525		8,275	9,325
3600	Iron body									
3700	Screwed, 3/8" pipe size	1 Stpi	22	.364	Ea.	98.50	21.50		120	141
3720	1/2" pipe size		20	.400		102	23.50		125.50	148
3740	3/4" pipe size		17	.471		130	27.50		157.50	185
3760	1" pipe size		15	.533		133	31		164	193
3780	1-1/4" pipe size		13	.615		173	36		209	245
3800	1-1/2" pipe size		12	.667		191	39		230	269
3820	2" pipe size		10	.800		228	47		275	320
3840	2-1/2" pipe size	Q-5	15	1.067		305	56		361	420
3860	3" pipe size	"	14	1.143		370	60		430	495
4000	Flanged, 2" pipe size	1 Stpi	6	1.333		355	78		433	510
4020	2-1/2" pipe size	Q-5	4.50	3.556		475	187		662	810
4040	3" pipe size		3.50	4.571		500	241		741	915
4060	4" pipe size		3	5.333		760	281		1,041	1,275
4080	5" pipe size	Q-6	3.40	7.059		1,150	385		1,535	1,850
4100	6" pipe size		3	8		1,475	435		1,910	2,275
4120	8" pipe size		2.50	9.600		2,675	525		3,200	3,750
4140	10" pipe size		2.20	10.909		5,875	595		6,470	7,375
7000	Stainless steel body									
7200	Screwed, 1" pipe size	1 Stpi	15	.533	Ea.	283	31		314	355
7210	1-1/4" pipe size		13	.615		440	36		476	540
7220	1-1/2" pipe size		12	.667		440	39		479	545
7240	2" pipe size		10	.800		655	47		702	790
7260	2-1/2" pipe size	Q-5	15	1.067		925	56		981	1,100
7280	3" pipe size	"	14	1.143		1,275	60		1,335	1,500
7400	Flanged, 2" pipe size	1 Stpi	6	1.333		1,125	78		1,203	1,350
7420	2-1/2" pipe size	Q-5	4.50	3.556		2,025	187		2,212	2,525

23 21 Hydronic Piping and Pumps

23 21 20 – Hydronic HVAC Piping Specialties

23 21 20.74 Strainers, Basket Type

		Crew	Daily Output	Labor-Hours	Unit	Material	2014 Bare Costs Labor	Equipment	Total	Total Incl O&P
7440	3" pipe size	Q-5	3.50	4.571	Ea.	2,075	241		2,316	2,675
7460	4" pipe size	↓	3	5.333		3,275	281		3,556	4,025
7480	6" pipe size	Q-6	3	8		5,700	435		6,135	6,925
7500	8" pipe size	"	2.50	9.600	↓	8,300	525		8,825	9,925
8100	Duplex style									
8200	Bronze body									
8240	Screwed, 3/4" pipe size	1 Stpi	16	.500	Ea.	1,175	29.50		1,204.50	1,325
8260	1" pipe size		14	.571		1,175	33.50		1,208.50	1,325
8280	1-1/4" pipe size		12	.667		2,825	39		2,864	3,150
8300	1-1/2" pipe size		11	.727		2,825	42.50		2,867.50	3,175
8320	2" pipe size	↓	9	.889		3,700	52		3,752	4,125
8340	2-1/2" pipe size	Q-5	14	1.143		4,775	60		4,835	5,350
8420	Flanged, 2" pipe size	1 Stpi	6	1.333		3,975	78		4,053	4,500
8440	2-1/2" pipe size	Q-5	4.50	3.556		5,475	187		5,662	6,300
8460	3" pipe size		3.50	4.571		5,975	241		6,216	6,950
8480	4" pipe size	↓	3	5.333		8,575	281		8,856	9,850
8500	5" pipe size	Q-6	3.40	7.059		18,600	385		18,985	21,000
8520	6" pipe size	"	3	8	↓	19,100	435		19,535	21,700
8700	Iron body									
8740	Screwed, 3/4" pipe size	1 Stpi	16	.500	Ea.	1,050	29.50		1,079.50	1,200
8760	1" pipe size		14	.571		1,050	33.50		1,083.50	1,200
8780	1-1/4" pipe size		12	.667		1,175	39		1,214	1,325
8800	1-1/2" pipe size		11	.727		1,175	42.50		1,217.50	1,350
8820	2" pipe size	↓	9	.889		1,975	52		2,027	2,250
8840	2-1/2" pipe size	Q-5	14	1.143		2,175	60		2,235	2,500
9000	Flanged, 2" pipe size	1 Stpi	6	1.333		2,100	78		2,178	2,450
9020	2-1/2" pipe size	Q-5	4.50	3.556		2,250	187		2,437	2,750
9040	3" pipe size		3.50	4.571		2,450	241		2,691	3,075
9060	4" pipe size	↓	3	5.333		4,150	281		4,431	4,975
9080	5" pipe size	Q-6	3.40	7.059		9,025	385		9,410	10,500
9100	6" pipe size		3	8		9,025	435		9,460	10,600
9120	8" pipe size		2.50	9.600		16,700	525		17,225	19,200
9140	10" pipe size		2.20	10.909		23,900	595		24,495	27,100
9160	12" pipe size		1.70	14.118		26,200	770		26,970	30,100
9170	14" pipe size		1.40	17.143		31,000	935		31,935	35,500
9180	16" pipe size	↓	1	24	↓	38,200	1,300		39,500	44,000
9700	Stainless steel body									
9740	Screwed, 1" pipe size	1 Stpi	14	.571	Ea.	1,900	33.50		1,933.50	2,125
9760	1-1/2" pipe size		11	.727		2,850	42.50		2,892.50	3,225
9780	2" pipe size		9	.889		4,000	52		4,052	4,475
9860	Flanged, 2" pipe size	↓	6	1.333		4,350	78		4,428	4,925
9880	2-1/2" pipe size	Q-5	4.50	3.556		7,300	187		7,487	8,300
9900	3" pipe size		3.50	4.571		7,950	241		8,191	9,125
9920	4" pipe size	↓	3	5.333		10,400	281		10,681	11,800
9940	6" pipe size	Q-6	3	8		15,800	435		16,235	18,100
9960	8" pipe size	"	2.50	9.600	↓	44,200	525		44,725	49,400

23 21 20.76 Strainers, Y Type, Bronze Body

		Crew	Daily Output	Labor-Hours	Unit	Material	Labor	Equipment	Total	Total Incl O&P
0010	**STRAINERS, Y TYPE, BRONZE BODY**									
0050	Screwed, 125 lb., 1/4" pipe size	1 Stpi	24	.333	Ea.	23	19.50		42.50	55
0070	3/8" pipe size		24	.333		27.50	19.50		47	60
0100	1/2" pipe size		20	.400		27.50	23.50		51	66
0120	3/4" pipe size	↓	19	.421	↓	34	24.50		58.50	74.50

23 21 Hydronic Piping and Pumps

23 21 20 – Hydronic HVAC Piping Specialties

23 21 20.76 Strainers, Y Type, Bronze Body

		Crew	Daily Output	Labor-Hours	Unit	Material	2014 Bare Costs Labor	Equipment	Total	Total Incl O&P
0140	1" pipe size	1 Stpi	17	.471	Ea.	40.50	27.50		68	86
0150	1-1/4" pipe size		15	.533		81	31		112	136
0160	1-1/2" pipe size		14	.571		87.50	33.50		121	147
0180	2" pipe size		13	.615		116	36		152	183
0182	3" pipe size	▼	12	.667		845	39		884	990
0200	300 lb., 2-1/2" pipe size	Q-5	17	.941		510	49.50		559.50	635
0220	3" pipe size		16	1		1,000	52.50		1,052.50	1,175
0240	4" pipe size	▼	15	1.067	▼	2,300	56		2,356	2,600
0500	For 300 lb. rating 1/4" thru 2", add					15%				
1000	Flanged, 150 lb., 1-1/2" pipe size	1 Stpi	11	.727	Ea.	475	42.50		517.50	585
1020	2" pipe size	"	8	1		640	58.50		698.50	795
1030	2-1/2" pipe size	Q-5	5	3.200		950	168		1,118	1,300
1040	3" pipe size		4.50	3.556		1,175	187		1,362	1,550
1060	4" pipe size	▼	3	5.333		1,775	281		2,056	2,375
1080	5" pipe size	Q-6	3.40	7.059		1,775	385		2,160	2,525
1100	6" pipe size		3	8		3,400	435		3,835	4,375
1106	8" pipe size	▼	2.60	9.231	▼	3,725	505		4,230	4,850
1500	For 300 lb. rating, add					40%				

23 21 20.78 Strainers, Y Type, Iron Body

		Crew	Daily Output	Labor-Hours	Unit	Material	Labor	Equipment	Total	Total Incl O&P
0010	**STRAINERS, Y TYPE, IRON BODY**									
0050	Screwed, 250 lb., 1/4" pipe size	1 Stpi	20	.400	Ea.	11.05	23.50		34.55	47.50
0070	3/8" pipe size		20	.400		11.05	23.50		34.55	47.50
0100	1/2" pipe size		20	.400		11.05	23.50		34.55	47.50
0120	3/4" pipe size		18	.444		12.95	26		38.95	54
0140	1" pipe size		16	.500		18.25	29.50		47.75	64
0150	1-1/4" pipe size		15	.533		24	31		55	73.50
0160	1-1/2" pipe size		12	.667		30	39		69	91.50
0180	2" pipe size	▼	8	1		44.50	58.50		103	138
0200	2-1/2" pipe size	Q-5	12	1.333		268	70		338	400
0220	3" pipe size		11	1.455		289	76.50		365.50	435
0240	4" pipe size	▼	5	3.200	▼	490	168		658	795
0500	For galvanized body, add					50%				
1000	Flanged, 125 lb., 1-1/2" pipe size	1 Stpi	11	.727	Ea.	155	42.50		197.50	235
1020	2" pipe size	"	8	1		113	58.50		171.50	213
1030	2-1/2" pipe size	Q-5	5	3.200		115	168		283	380
1040	3" pipe size		4.50	3.556		150	187		337	450
1060	4" pipe size	▼	3	5.333		247	281		528	695
1080	5" pipe size	Q-6	3.40	7.059		385	385		770	1,000
1100	6" pipe size		3	8		495	435		930	1,200
1120	8" pipe size		2.50	9.600		830	525		1,355	1,700
1140	10" pipe size		2	12		1,500	655		2,155	2,650
1160	12" pipe size		1.70	14.118		2,275	770		3,045	3,675
1170	14" pipe size		1.30	18.462		4,175	1,000		5,175	6,125
1180	16" pipe size	▼	1	24	▼	5,925	1,300		7,225	8,475
1500	For 250 lb. rating, add					20%				
2000	For galvanized body, add					50%				
2500	For steel body, add					40%				

23 21 20.80 Suction Diffusers

		Crew	Daily Output	Labor-Hours	Unit	Material	Labor	Equipment	Total	Total Incl O&P
0010	**SUCTION DIFFUSERS**									
0100	Cast iron body with integral straightening vanes, strainer									
1000	Flanged									
1010	2" inlet, 1-1/2" pump side	1 Stpi	6	1.333	Ea.	325	78		403	475

23 21 Hydronic Piping and Pumps

23 21 20 – Hydronic HVAC Piping Specialties

23 21 20.80 Suction Diffusers

		Crew	Daily Output	Labor-Hours	Unit	Material	2014 Bare Costs Labor	Equipment	Total	Total Incl O&P
1020	2" pump side	1 Stpi	5	1.600	Ea.	335	93.50		428.50	510
1030	3" inlet, 2" pump side	Q-5	6.50	2.462	↓	405	130		535	640

23 21 20.84 Thermoflo Indicator

0010	**THERMOFLO INDICATOR**, For balancing									
1000	Sweat connections, 1-1/4" pipe size	1 Stpi	12	.667	Ea.	635	39		674	755
1020	1-1/2" pipe size	"	10	.800	"	645	47		692	780

23 21 20.88 Venturi Flow

		Crew	Daily Output	Labor-Hours	Unit	Material	Labor	Equipment	Total	Total Incl O&P
0010	**VENTURI FLOW**, Measuring device									
0050	1/2" diameter	1 Stpi	24	.333	Ea.	281	19.50		300.50	340
0100	3/4" diameter		20	.400		257	23.50		280.50	320
0120	1" diameter		19	.421		276	24.50		300.50	340
0140	1-1/4" diameter		15	.533		340	31		371	420
0160	1-1/2" diameter		13	.615		355	36		391	445
0180	2" diameter	↓	11	.727		365	42.50		407.50	470
0200	2-1/2" diameter	Q-5	16	1		500	52.50		552.50	630
0220	3" diameter		14	1.143		515	60		575	660
0240	4" diameter	↓	11	1.455		775	76.50		851.50	965
0260	5" diameter	Q-6	4	6		1,025	330		1,355	1,625
0280	6" diameter		3.50	6.857		1,125	375		1,500	1,825
0300	8" diameter		3	8		1,450	435		1,885	2,250
0320	10" diameter		2	12		3,425	655		4,080	4,775
0330	12" diameter		1.80	13.333		4,950	730		5,680	6,550
0340	14" diameter		1.60	15		5,600	820		6,420	7,400
0350	16" diameter	↓	1.40	17.143		6,250	935		7,185	8,300
0500	For meter, add				↓	2,125			2,125	2,350

23 21 23 – Hydronic Pumps

23 21 23.13 In-Line Centrifugal Hydronic Pumps

		Crew	Daily Output	Labor-Hours	Unit	Material	Labor	Equipment	Total	Total Incl O&P
0010	**IN-LINE CENTRIFUGAL HYDRONIC PUMPS**									
0600	Bronze, sweat connections, 1/40 HP, in line									
0640	3/4" size	Q-1	16	1	Ea.	212	52		264	310
1000	Flange connection, 3/4" to 1-1/2" size									
1040	1/12 HP	Q-1	6	2.667	Ea.	545	138		683	810
1060	1/8 HP		6	2.667		940	138		1,078	1,225
1100	1/3 HP		6	2.667		1,050	138		1,188	1,350
1140	2" size, 1/6 HP		5	3.200		1,350	166		1,516	1,725
1180	2-1/2" size, 1/4 HP		5	3.200		1,775	166		1,941	2,200
1220	3" size, 1/4 HP		4	4		1,850	207		2,057	2,375
1260	1/3 HP		4	4		2,275	207		2,482	2,825
1300	1/2 HP		4	4		2,325	207		2,532	2,875
1340	3/4 HP		4	4		2,550	207		2,757	3,125
1380	1 HP	↓	4	4	↓	4,100	207		4,307	4,825
2000	Cast iron, flange connection									
2040	3/4" to 1-1/2" size, in line, 1/12 HP	Q-1	6	2.667	Ea.	355	138		493	600
2060	1/8 HP		6	2.667		590	138		728	860
2100	1/3 HP		6	2.667		660	138		798	935
2140	2" size, 1/6 HP		5	3.200		720	166		886	1,050
2180	2-1/2" size, 1/4 HP		5	3.200		960	166		1,126	1,300
2220	3" size, 1/4 HP		4	4		975	207		1,182	1,400
2260	1/3 HP		4	4		1,325	207		1,532	1,775
2300	1/2 HP		4	4		1,375	207		1,582	1,825
2340	3/4 HP		4	4		1,575	207		1,782	2,075
2380	1 HP	↓	4	4	↓	2,275	207		2,482	2,825

23 21 Hydronic Piping and Pumps

23 21 23 - Hydronic Pumps

23 21 23.13 In-Line Centrifugal Hydronic Pumps

		Crew	Daily Output	Labor-Hours	Unit	Material	2014 Bare Costs Labor	Equipment	Total	Total Incl O&P
2600	For nonferrous impeller, add					3%				
3000	High head, bronze impeller									
3030	1-1/2" size 1/2 HP	Q-1	5	3.200	Ea.	1,150	166		1,316	1,525
3040	1-1/2" size 3/4 HP		5	3.200		1,250	166		1,416	1,625
3050	2" size 1 HP		4	4		1,525	207		1,732	2,000
3090	2" size 1-1/2 HP		4	4		1,900	207		2,107	2,400
4000	Close coupled, end suction, bronze impeller									
4040	1-1/2" size, 1-1/2 HP, to 40 GPM	Q-1	3	5.333	Ea.	2,200	276		2,476	2,850
4090	2" size, 2 HP, to 50 GPM		3	5.333		2,600	276		2,876	3,300
4100	2" size, 3 HP, to 90 GPM		2.30	6.957		2,725	360		3,085	3,550
4190	2-1/2" size, 3 HP, to 150 GPM		2	8		2,950	415		3,365	3,875
4300	3" size, 5 HP, to 225 GPM		1.80	8.889		3,375	460		3,835	4,400
4410	3" size, 10 HP, to 350 GPM		1.60	10		4,925	520		5,445	6,200
4420	4" size, 7-1/2 HP, to 350 GPM		1.60	10		4,400	520		4,920	5,600
4520	4" size, 10 HP, to 600 GPM	Q-2	1.70	14.118		5,025	760		5,785	6,675
4530	5" size, 15 HP, to 1000 GPM		1.70	14.118		5,025	760		5,785	6,675
4610	5" size, 20 HP, to 1350 GPM		1.50	16		5,350	860		6,210	7,175
4620	5" size, 25 HP, to 1550 GPM		1.50	16		7,300	860		8,160	9,325
5000	Base mounted, bronze impeller, coupling guard									
5040	1-1/2" size, 1-1/2 HP, to 40 GPM	Q-1	2.30	6.957	Ea.	5,850	360		6,210	6,975
5090	2" size, 2 HP, to 50 GPM		2.30	6.957		6,550	360		6,910	7,750
5100	2" size, 3 HP, to 90 GPM		2	8		7,200	415		7,615	8,525
5190	2-1/2" size, 3 HP, to 150 GPM		1.80	8.889		7,350	460		7,810	8,800
5300	3" size, 5 HP, to 225 GPM		1.60	10		8,875	520		9,395	10,600
5410	4" size, 5 HP, to 350 GPM		1.50	10.667		8,950	555		9,505	10,700
5420	4" size, 7-1/2 HP, to 350 GPM		1.50	10.667		10,900	555		11,455	12,800
5520	5" size, 10 HP, to 600 GPM	Q-2	1.60	15		14,100	805		14,905	16,800
5530	5" size, 15 HP, to 1000 GPM		1.60	15		15,400	805		16,205	18,200
5610	6" size, 20 HP, to 1350 GPM		1.40	17.143		17,100	920		18,020	20,200
5620	6" size, 25 HP, to 1550 GPM		1.40	17.143		19,200	920		20,120	22,500
5800	The above pump capacities are based on 1800 RPM,									
5810	at a 60 foot head. Increasing the RPM									
5820	or decreasing the head will increase the GPM.									

23 21 29 - Automatic Condensate Pump Units

23 21 29.10 Condensate Removal Pump System

			Crew	Daily Output	Labor-Hours	Unit	Material	Labor	Equipment	Total	Total Incl O&P
0010	**CONDENSATE REMOVAL PUMP SYSTEM**										
0020	Pump with 1 gal. ABS tank										
0100	115 V										
0120	1/50 HP, 200 GPH	G	1 Stpi	12	.667	Ea.	191	39		230	269
0140	1/18 HP, 270 GPH	G		10	.800		204	47		251	295
0160	1/5 HP, 450 GPH	G		8	1		455	58.50		513.50	590
0200	230 V										
0240	1/18 HP, 270 GPH	G	1 Stpi	10	.800	Ea.	211	47		258	305
0260	1/5 HP, 450 GPH	G	"	8	1	"	505	58.50		563.50	645

23 22 Steam and Condensate Piping and Pumps

23 22 23 - Steam Condensate Pumps

23 22 23.23 Pumps, Pneumatic Ejector		Crew	Daily Output	Labor-Hours	Unit	Material	2014 Bare Costs Labor	Equipment	Total	Total Incl O&P
0010	**PUMPS, PNEUMATIC EJECTOR**									
0020	With steel receiver, level controls, inlet/outlet gate/check valves									
0030	Cross connect. not incl. compressor, fittings or piping									
0040	Duplex									
0050	30 GPM	Q-2	1.70	14.118	Ea.	33,200	760		33,960	37,700
0060	50 GPM		1.56	15.385		33,900	825		34,725	38,500
0070	100 GPM		1.30	18.462		41,000	990		41,990	46,500
0080	150 GPM		1.10	21.818		79,000	1,175		80,175	89,000
0090	200 GPM		.85	28.235		100,500	1,525		102,025	113,000
0100	250 GPM	Q-3	.91	35.165		129,000	1,925		130,925	145,000
0110	300 GPM	"	.57	56.140		158,000	3,075		161,075	178,000

23 23 Refrigerant Piping

23 23 23 - Refrigerants

23 23 23.10 Anti-Freeze

		Crew	Daily Output	Labor-Hours	Unit	Material	Labor	Equipment	Total	Total Incl O&P
0010	**ANTI-FREEZE**, Inhibited									
0900	Ethylene glycol concentrated									
1000	55 gallon drums, small quantities				Gal.	11.95			11.95	13.10
1200	Large quantities					9.65			9.65	10.60
2000	Propylene glycol, for solar heat, small quantities					15.45			15.45	17
2100	Large quantities					12.40			12.40	13.65

23 34 HVAC Fans

23 34 14 - Blower HVAC Fans

23 34 14.10 Blower Type HVAC Fans

		Crew	Daily Output	Labor-Hours	Unit	Material	Labor	Equipment	Total	Total Incl O&P
0010	**BLOWER TYPE HVAC FANS**									
2500	Ceiling fan, right angle, extra quiet, 0.10" S.P.									
2520	95 CFM	Q-20	20	1	Ea.	287	50		337	390
2540	210 CFM		19	1.053		340	52.50		392.50	455
2560	385 CFM		18	1.111		430	55.50		485.50	560
2640	For wall or roof cap, add	1 Shee	16	.500		287	27.50		314.50	355
2660	For straight thru fan, add					10%				
2680	For speed control switch, add	1 Elec	16	.500		157	26.50		183.50	212

23 34 23 - HVAC Power Ventilators

23 34 23.10 HVAC Power Circulators and Ventilators

		Crew	Daily Output	Labor-Hours	Unit	Material	Labor	Equipment	Total	Total Incl O&P
0010	**HVAC POWER CIRCULATORS AND VENTILATORS**									
6650	Residential, bath exhaust, grille, back draft damper									
6660	50 CFM	Q-20	24	.833	Ea.	61	41.50		102.50	131
6670	110 CFM		22	.909		94.50	45.50		140	173
6680	Light combination, squirrel cage, 100 watt, 70 CFM		24	.833		108	41.50		149.50	182
6700	Light/heater combination, ceiling mounted									
6710	70 CFM, 1450 watt	Q-20	24	.833	Ea.	155	41.50		196.50	235
6800	Heater combination, recessed, 70 CFM		24	.833		64.50	41.50		106	135
6820	With 2 infrared bulbs		23	.870		101	43.50		144.50	177
6940	Residential roof jacks and wall caps									
6944	Wall cap with back draft damper									
6946	3" & 4" diam. round duct	1 Shee	11	.727	Ea.	24.50	40		64.50	88
6948	6" diam. round duct	"	11	.727	"	61.50	40		101.50	129
6958	Roof jack with bird screen and back draft damper									

23 34 HVAC Fans

23 34 23 – HVAC Power Ventilators

23 34 23.10 HVAC Power Circulators and Ventilators

		Crew	Daily Output	Labor-Hours	Unit	Material	2014 Bare Costs Labor	Equipment	Total	Total Incl O&P
6960	3" & 4" diam. round duct	1 Shee	11	.727	Ea.	25	40		65	88.50
6962	3-1/4" x 10" rectangular duct	"	10	.800	"	46.50	44		90.50	118
6980	Transition									
6982	3-1/4" x 10" to 6" diam. round	1 Shee	20	.400	Ea.	31	22		53	67.50

23 35 Special Exhaust Systems

23 35 16 – Engine Exhaust Systems

23 35 16.10 Engine Exhaust Removal Systems

		Crew	Daily Output	Labor-Hours	Unit	Material	Labor	Equipment	Total	Total Incl O&P
0010	**ENGINE EXHAUST REMOVAL SYSTEMS** D3090-320									
0500	Engine exhaust, garage, in-floor system									
0510	Single tube outlet assemblies									
0520	For transite pipe ducting, self-storing tube									
0530	3" tubing adapter plate	1 Shee	16	.500	Ea.	240	27.50		267.50	305
0540	4" tubing adapter plate		16	.500		245	27.50		272.50	310
0550	5" tubing adapter plate	↓	16	.500	↓	245	27.50		272.50	310
0600	For vitrified tile ducting									
0610	3" tubing adapter plate, self-storing tube	1 Shee	16	.500	Ea.	241	27.50		268.50	305
0620	4" tubing adapter plate, self-storing tube		16	.500		245	27.50		272.50	310
0660	5" tubing adapter plate, self-storing tube	↓	16	.500		245	27.50		272.50	310
0800	Two tube outlet assemblies									
0810	For transite pipe ducting, self-storing tube									
0820	3" tubing, dual exhaust adapter plate	1 Shee	16	.500	Ea.	240	27.50		267.50	305
0850	For vitrified tile ducting									
0860	3" tubing, dual exhaust, self-storing tube	1 Shee	16	.500	Ea.	298	27.50		325.50	370
0870	3" tubing, double outlet, non-storing tubes	"	16	.500	"	298	27.50		325.50	370
0900	Accessories for metal tubing, (overhead systems also)									
0910	Adapters, for metal tubing end									
0920	3" tail pipe type				Ea.	49.50			49.50	54.50
0930	4" tail pipe type					53			53	58.50
0940	5" tail pipe type				↓	54			54	59.50
0990	5" diesel stack type					315			315	350
1000	6" diesel stack type				↓	330			330	360
1100	Bullnose (guide) required for in-floor assemblies									
1110	3" tubing size				Ea.	28			28	30.50
1120	4" tubing size					29.50			29.50	32.50
1130	5" tubing size				↓	31.50			31.50	34.50
1150	Plain rings, for tubing end									
1160	3" tubing size				Ea.	24.50			24.50	26.50
1170	4" tubing size				"	41.50			41.50	45.50
1200	Tubing, galvanized, flexible, (for overhead systems also)									
1210	3" ID				L.F.	11.60			11.60	12.75
1220	4" ID					14.20			14.20	15.65
1230	5" ID					16.75			16.75	18.40
1240	6" ID				↓	19.35			19.35	21.50
1250	Stainless steel, flexible, (for overhead system, also)									
1260	3" ID				L.F.	25.50			25.50	28
1270	4" ID					35			35	38.50
1280	5" ID					39.50			39.50	43.50
1290	6" ID				↓	46			46	50.50
1500	Engine exhaust, garage, overhead components, for neoprene tubing									
1510	Alternate metal tubing & accessories see above									

23 35 Special Exhaust Systems

23 35 16 – Engine Exhaust Systems

23 35 16.10 Engine Exhaust Removal Systems	Crew	Daily Output	Labor-Hours	Unit	Material	2014 Bare Costs Labor	Equipment	Total	Total Incl O&P
1550 Adapters, for neoprene tubing end									
1560 3" tail pipe, adjustable, neoprene				Ea.	56			56	61.50
1570 3" tail pipe, heavy wall neoprene					61			61	67.50
1580 4" tail pipe, heavy wall neoprene					100			100	110
1590 5" tail pipe, heavy wall neoprene				↓	105			105	116
1650 Connectors, tubing									
1660 3" interior, aluminum				Ea.	24.50			24.50	26.50
1670 4" interior, aluminum					41.50			41.50	45.50
1710 5" interior, neoprene					61			61	67.50
1750 3" spiralock, neoprene					24.50			24.50	26.50
1760 4" spiralock, neoprene					41.50			41.50	45.50
1780 Y for 3" ID tubing, neoprene, dual exhaust					195			195	215
1790 Y for 4" ID tubing, aluminum, dual exhaust				↓	195			195	215
1850 Elbows, aluminum, splice into tubing for strap									
1860 3" neoprene tubing size				Ea.	48.50			48.50	53.50
1870 4" neoprene tubing size					40.50			40.50	44.50
1900 Flange assemblies, connect tubing to overhead duct				↓	73			73	80
2000 Hardware and accessories									
2020 Cable, galvanized, 1/8" diameter				L.F.	.51			.51	.56
2040 Cleat, tie down cable or rope				Ea.	5.40			5.40	5.90
2060 Pulley					7.40			7.40	8.15
2080 Pulley hook, universal				↓	5.40			5.40	5.90
2100 Rope, nylon, 1/4" diameter				L.F.	.41			.41	.45
2120 Winch, 1" diameter				Ea.	114			114	126
2150 Lifting strap, mounts on neoprene									
2160 3" tubing size				Ea.	27			27	29.50
2170 4" tubing size					27			27	29.50
2180 5" tubing size					27			27	29.50
2190 6" tubing size					27			27	29.50
2200 Tubing, neoprene, 11' lengths									
2210 3" ID				L.F.	10.90			10.90	11.95
2220 4" ID					18.15			18.15	20
2230 5" ID				↓	27			27	29.50
2500 Engine exhaust, thru-door outlet									
2510 3" tube size	1 Carp	16	.500	Ea.	59.50	23		82.50	101
2530 4" tube size	"	16	.500	"	74	23		97	117
3000 Tubing, exhaust, flex hose, with									
3010 coupler, damper and tail pipe adapter									
3020 Neoprene									
3040 3" x 20'	1 Shee	6	1.333	Ea.	330	73		403	470
3050 4" x 15'		5.40	1.481		440	81		521	610
3060 4" x 20'		5	1.600		530	87.50		617.50	720
3070 5" x 15'	↓	4.40	1.818	↓	600	99.50		699.50	810
3100 Galvanized									
3110 3" x 20'	1 Shee	6	1.333	Ea.	330	73		403	475
3120 4" x 17'		5.60	1.429		360	78		438	520
3130 4" x 20'		5	1.600		405	87.50		492.50	580
3140 5" x 17'	↓	4.60	1.739	↓	430	95		525	620

23 35 Special Exhaust Systems

23 35 43 – Welding Fume Elimination Systems

23 35 43.10 Welding Fume Elimination System Components	Crew	Daily Output	Labor-Hours	Unit	Material	2014 Bare Costs Labor	Equipment	Total	Total Incl O&P
0010 **WELDING FUME ELIMINATION SYSTEM COMPONENTS**									
7500 Welding fume elimination accessories for garage exhaust systems									
7600 Cut off (blast gate)									
7610 3" tubing size, 3" x 6" opening	1 Shee	24	.333	Ea.	26	18.25		44.25	56.50
7620 4" tubing size, 4" x 8" opening		24	.333		26	18.25		44.25	56.50
7630 5" tubing size, 5" x 10" opening		24	.333		31.50	18.25		49.75	62.50
7640 6" tubing size		24	.333		32.50	18.25		50.75	63.50
7650 8" tubing size	↓	24	.333	↓	45	18.25		63.25	77.50
7700 Hoods, magnetic, with handle & screen									
7710 3" tubing size, 3" x 6" opening	1 Shee	24	.333	Ea.	107	18.25		125.25	146
7720 4" tubing size, 4" x 8" opening		24	.333		97	18.25		115.25	135
7730 5" tubing size, 5" x 10" opening	↓	24	.333	↓	97	18.25		115.25	135

23 38 Ventilation Hoods

23 38 13 – Commercial-Kitchen Hoods

23 38 13.10 Hood and Ventilation Equipment

	Crew	Daily Output	Labor-Hours	Unit	Material	Labor	Equipment	Total	Total Incl O&P
0010 **HOOD AND VENTILATION EQUIPMENT**									
2970 Exhaust hood, sst, gutter on all sides, 4' x 4' x 2'	1 Carp	1.80	4.444	Ea.	4,675	204		4,879	5,450
2980 4' x 4' x 7'	"	1.60	5	"	7,425	229		7,654	8,525
7800 Vent hood, wall canopy with fire protection	L-3A	9	1.333	L.F.	425	67		492	570
7810 Without fire protection		10	1.200		460	60		520	600
7820 Island canopy with fire protection		7	1.714		790	86		876	1,000
7830 Without fire protection		8	1.500		800	75		875	995
7840 Back shelf with fire protection		11	1.091		400	54.50		454.50	525
7850 Without fire protection	↓	12	1	↓	500	50		550	625
7860 Range hood & CO$_2$ system, minimum	1 Carp	2.50	3.200	Ea.	2,775	147		2,922	3,275
7870 Maximum	"	1	8	"	37,900	365		38,265	42,300

23 51 Breechings, Chimneys, and Stacks

23 51 13 – Draft Control Devices

23 51 13.16 Vent Dampers

	Crew	Daily Output	Labor-Hours	Unit	Material	Labor	Equipment	Total	Total Incl O&P
0010 **VENT DAMPERS**									
5000 Vent damper, bi-metal, gas, 3" diameter	Q-9	24	.667	Ea.	44	33		77	98.50
5010 4" diameter	"	24	.667	"	44	33		77	98.50

23 51 13.19 Barometric Dampers

	Crew	Daily Output	Labor-Hours	Unit	Material	Labor	Equipment	Total	Total Incl O&P
0010 **BAROMETRIC DAMPERS**									
1000 Barometric, gas fired system only, 6" size for 5" and 6" pipes	1 Shee	20	.400	Ea.	67	22		89	107
1020 7" size, for 6" and 7" pipes		19	.421		71.50	23		94.50	114
1040 8" size, for 7" and 8" pipes		18	.444		92.50	24.50		117	139
1060 9" size, for 8" and 9" pipes	↓	16	.500	↓	104	27.50		131.50	156
2000 All fuel, oil, oil/gas, coal									
2020 10" for 9" and 10" pipes	1 Shee	15	.533	Ea.	155	29		184	215
2040 12" for 11" and 12" pipes		15	.533		202	29		231	267
2060 14" for 13" and 14" pipes		14	.571		263	31.50		294.50	335
2080 16" for 15" and 16" pipes		13	.615		370	33.50		403.50	455
2100 18" for 17" and 18" pipes		12	.667		490	36.50		526.50	595
2120 20" for 19" and 21" pipes		10	.800		585	44		629	710
2140 24" for 22" and 25" pipes	Q-9	12	1.333	↓	715	65.50		780.50	885

23 51 Breechings, Chimneys, and Stacks

23 51 13 – Draft Control Devices

23 51 13.19 Barometric Dampers

		Crew	Daily Output	Labor-Hours	Unit	Material	2014 Bare Costs Labor	Equipment	Total	Total Incl O&P
2160	28" for 26" and 30" pipes	Q-9	10	1.600	Ea.	890	79		969	1,100
2180	32" for 31" and 34" pipes	↓	8	2		1,125	98.50		1,223.50	1,400
3260	For thermal switch for above, add	1 Shee	24	.333	↓	64	18.25		82.25	98.50

23 51 23 – Gas Vents

23 51 23.10 Gas Chimney Vents

		Crew	Daily Output	Labor-Hours	Unit	Material	Labor	Equipment	Total	Total Incl O&P
0010	**GAS CHIMNEY VENTS**, Prefab metal, U.L. listed									
0020	Gas, double wall, galvanized steel									
0080	3" diameter	Q-9	72	.222	V.L.F.	5.55	10.95		16.50	23
0100	4" diameter		68	.235		7.20	11.60		18.80	25.50
0120	5" diameter		64	.250		7.80	12.30		20.10	27.50
0140	6" diameter		60	.267		9.45	13.15		22.60	30.50
0160	7" diameter		56	.286		15.10	14.05		29.15	38
0180	8" diameter		52	.308		17.40	15.15		32.55	42
0200	10" diameter		48	.333		36	16.40		52.40	65
0220	12" diameter		44	.364		43	17.90		60.90	75
0240	14" diameter		42	.381		72	18.75		90.75	108
0260	16" diameter		40	.400		103	19.70		122.70	144
0280	18" diameter	↓	38	.421		128	20.50		148.50	172
0300	20" diameter	Q-10	36	.667		150	34		184	217
0320	22" diameter		34	.706		187	36		223	260
0340	24" diameter		32	.750	↓	236	38.50		274.50	320
0600	For 4", 5" and 6" oval, add					50%				
0650	Gas, double wall, galvanized steel, fittings									
0660	Elbow 45°, 3" diameter	Q-9	36	.444	Ea.	13.50	22		35.50	48.50
0670	4" diameter		34	.471		16.35	23		39.35	53.50
0680	5" diameter		32	.500		19.10	24.50		43.60	58.50
0690	6" diameter		30	.533		23.50	26.50		50	66
0700	7" diameter		28	.571		37.50	28		65.50	84.50
0710	8" diameter		26	.615		50	30.50		80.50	102
0720	10" diameter		24	.667		107	33		140	168
0730	12" diameter		22	.727		114	36		150	180
0740	14" diameter		21	.762		179	37.50		216.50	255
0750	16" diameter		20	.800		232	39.50		271.50	315
0760	18" diameter	↓	19	.842		305	41.50		346.50	400
0770	20" diameter	Q-10	18	1.333		340	68		408	480
0780	22" diameter		17	1.412		545	72		617	710
0790	24" diameter	↓	16	1.500	↓	695	76.50		771.50	880
0916	Adjustable length									
0918	3" diameter, to 12"	Q-9	36	.444	Ea.	12.45	22		34.45	47
0920	4" diameter, to 12"		34	.471		14.50	23		37.50	51.50
0924	6" diameter, to 12"		30	.533		18.70	26.50		45.20	60.50
0928	8" diameter, to 12"		26	.615		35	30.50		65.50	85
0930	10" diameter, to 18"		24	.667		107	33		140	168
0932	12" diameter, to 18"		22	.727		110	36		146	176
0936	16" diameter, to 18"		20	.800		222	39.50		261.50	305
0938	18" diameter, to 18"	↓	19	.842		254	41.50		295.50	345
0944	24" diameter, to 18"	Q-10	16	1.500		485	76.50		561.50	650
0950	Elbow 90°, adjustable, 3" diameter	Q-9	36	.444		23	22		45	59
0960	4" diameter		34	.471		27	23		50	65
0970	5" diameter		32	.500		33.50	24.50		58	74.50
0980	6" diameter		30	.533		39	26.50		65.50	83
0990	7" diameter	↓	28	.571		67	28		95	117

23 51 Breechings, Chimneys, and Stacks

23 51 23 – Gas Vents

23 51 23.10 Gas Chimney Vents

		Crew	Daily Output	Labor-Hours	Unit	Material	2014 Bare Costs Labor	Equipment	Total	Total Incl O&P
1010	8" diameter	Q-9	26	.615	Ea.	68	30.50		98.50	122
1020	Wall thimble, 4 to 7" adjustable, 3" diameter		36	.444		14.45	22		36.45	49.50
1022	4" diameter		34	.471		16.25	23		39.25	53.50
1024	5" diameter		32	.500		17.10	24.50		41.60	56.50
1026	6" diameter		30	.533		17.55	26.50		44.05	59.50
1028	7" diameter		28	.571		36.50	28		64.50	83
1030	8" diameter		26	.615		51.50	30.50		82	104
1040	Roof flashing, 3" diameter		36	.444		8.50	22		30.50	43
1050	4" diameter		34	.471		9.90	23		32.90	46.50
1060	5" diameter		32	.500		29.50	24.50		54	70
1070	6" diameter		30	.533		23.50	26.50		50	66
1080	7" diameter		28	.571		31	28		59	77.50
1090	8" diameter		26	.615		33.50	30.50		64	83
1100	10" diameter		24	.667		45	33		78	99.50
1110	12" diameter		22	.727		63	36		99	124
1120	14" diameter		20	.800		143	39.50		182.50	217
1130	16" diameter		18	.889		201	44		245	288
1140	18" diameter		16	1		229	49		278	325
1150	20" diameter	Q-10	18	1.333		305	68		373	440
1160	22" diameter		14	1.714		375	87.50		462.50	550
1170	24" diameter		12	2		440	102		542	635
1200	Tee, 3" diameter	Q-9	27	.593		35	29		64	83.50
1210	4" diameter		26	.615		38.50	30.50		69	89
1220	5" diameter		25	.640		40	31.50		71.50	92
1230	6" diameter		24	.667		45.50	33		78.50	100
1240	7" diameter		23	.696		65.50	34.50		100	125
1250	8" diameter		22	.727		71.50	36		107.50	133
1260	10" diameter		21	.762		191	37.50		228.50	268
1270	12" diameter		20	.800		196	39.50		235.50	275
1280	14" diameter		18	.889		350	44		394	450
1290	16" diameter		16	1		515	49		564	645
1300	18" diameter		14	1.143		625	56.50		681.50	770
1310	20" diameter	Q-10	17	1.412		855	72		927	1,050
1320	22" diameter		13	1.846		1,100	94.50		1,194.50	1,350
1330	24" diameter		12	2		1,250	102		1,352	1,525
1460	Tee cap, 3" diameter	Q-9	45	.356		2.36	17.50		19.86	29
1470	4" diameter		42	.381		2.54	18.75		21.29	31.50
1480	5" diameter		40	.400		3.19	19.70		22.89	33.50
1490	6" diameter		37	.432		4.73	21.50		26.23	37.50
1500	7" diameter		35	.457		7.95	22.50		30.45	43
1510	8" diameter		34	.471		8.70	23		31.70	45
1520	10" diameter		32	.500		70	24.50		94.50	115
1530	12" diameter		30	.533		70	26.50		96.50	117
1540	14" diameter		28	.571		73	28		101	123
1550	16" diameter		25	.640		74.50	31.50		106	130
1560	18" diameter		24	.667		88	33		121	147
1570	20" diameter	Q-10	27	.889		99.50	45.50		145	179
1580	22" diameter		22	1.091		300	55.50		355.50	415
1590	24" diameter		21	1.143		340	58.50		398.50	465
1750	Top, 3" diameter	Q-9	46	.348		13.05	17.10		30.15	40.50
1760	4" diameter		44	.364		13.75	17.90		31.65	42.50
1770	5" diameter		42	.381		17.50	18.75		36.25	48
1780	6" diameter		40	.400		23.50	19.70		43.20	56

23 51 Breechings, Chimneys, and Stacks

23 51 23 – Gas Vents

23 51 23.10 Gas Chimney Vents

		Crew	Daily Output	Labor-Hours	Unit	Material	2014 Bare Costs Labor	Equipment	Total	Total Incl O&P
1790	7" diameter	Q-9	38	.421	Ea.	37.50	20.50		58	73
1800	8" diameter		36	.444		46.50	22		68.50	84.50
1810	10" diameter		34	.471		105	23		128	151
1820	12" diameter		32	.500		154	24.50		178.50	207
1830	14" diameter		30	.533		205	26.50		231.50	265
1840	16" diameter		28	.571		263	28		291	335
1850	18" diameter		26	.615		360	30.50		390.50	445
1860	20" diameter	Q-10	28	.857		550	44		594	670
1870	22" diameter		22	1.091		840	55.50		895.50	1,000
1880	24" diameter		20	1.200		1,050	61.50		1,111.50	1,250
1900	Gas, double wall, galvanized steel, oval									
1904	4" x 1'	Q-9	68	.235	V.L.F.	16.65	11.60		28.25	36
1906	5"		64	.250		39.50	12.30		51.80	62.50
1908	5"/6" x 1'		60	.267		36.50	13.15		49.65	60.50
1910	Oval fittings									
1912	Adjustable length									
1914	4" diameter to 12" lg.	Q-9	34	.471	Ea.	16.10	23		39.10	53
1916	5" diameter to 12" lg.		32	.500		44.50	24.50		69	86.50
1918	5"/6" diameter to 12"		30	.533		40.50	26.50		67	84.50
1920	Elbow 45°									
1922	4"	Q-9	34	.471	Ea.	29.50	23		52.50	68
1924	5"		32	.500		56	24.50		80.50	99
1926	5"/6"		30	.533		53	26.50		79.50	98
1930	Elbow 45°, flat									
1932	4"	Q-9	34	.471	Ea.	29.50	23		52.50	68
1934	5"		32	.500		55.50	24.50		80	99
1936	5"/6"		30	.533		53	26.50		79.50	98
1940	Top									
1942	4"	Q-9	44	.364	Ea.	35	17.90		52.90	66
1944	5"		42	.381		24.50	18.75		43.25	55.50
1946	5"/6"		40	.400		59.50	19.70		79.20	95
1950	Adjustable flashing									
1952	4"	Q-9	34	.471	Ea.	9.15	23		32.15	45.50
1954	5"		32	.500		29	24.50		53.50	69.50
1956	5"/6"		30	.533		30.50	26.50		57	73.50
1960	Tee									
1962	4"	Q-9	26	.615	Ea.	43.50	30.50		74	94.50
1964	5"		25	.640		91.50	31.50		123	149
1966	5"/6"		24	.667		84.50	33		117.50	143
1970	Tee with short snout									
1972	4"	Q-9	26	.615	Ea.	43.50	30.50		74	94.50

23 51 26 – All-Fuel Vent Chimneys

23 51 26.10 All-Fuel Vent Chimneys, Press. Tight, Dbl. Wall

		Crew	Daily Output	Labor-Hours	Unit	Material	Labor	Equipment	Total	Total Incl O&P
0010	**ALL-FUEL VENT CHIMNEYS, PRESSURE TIGHT, DOUBLE WALL**									
3200	All fuel, pressure tight, double wall, 1" insulation, U.L. listed, 1400°F.									
3210	304 stainless steel liner, aluminized steel outer jacket									
3220	6" diameter	Q-9	60	.267	L.F.	56	13.15		69.15	81.50
3221	8" diameter		52	.308		59	15.15		74.15	88
3222	10" diameter		48	.333		66	16.40		82.40	97.50
3223	12" diameter		44	.364		75.50	17.90		93.40	111
3224	14" diameter		42	.381		85	18.75		103.75	122
3225	16" diameter		40	.400		95.50	19.70		115.20	135

23 51 Breechings, Chimneys, and Stacks

23 51 26 – All-Fuel Vent Chimneys

23 51 26.10 All-Fuel Vent Chimneys, Press. Tight, Dbl. Wall		Crew	Daily Output	Labor-Hours	Unit	Material	2014 Bare Costs Labor	Equipment	Total	Total Incl O&P
3226	18" diameter	Q-9	38	.421	L.F.	109	20.50		129.50	152
3227	20" diameter	Q-10	36	.667		123	34		157	187
3228	24" diameter	"	32	.750		158	38.50		196.50	232
3260	For 316 stainless steel liner add					30%				
3280	All fuel, pressure tight, double wall fittings									
3284	304 stainless steel inner, aluminized steel jacket									
3288	Adjustable 20"/29" section									
3292	6" diameter	Q-9	30	.533	Ea.	187	26.50		213.50	246
3293	8" diameter		26	.615		195	30.50		225.50	261
3294	10" diameter		24	.667		220	33		253	292
3295	12" diameter		22	.727		248	36		284	330
3296	14" diameter		21	.762		280	37.50		317.50	370
3297	16" diameter		20	.800		315	39.50		354.50	405
3298	18" diameter		19	.842		355	41.50		396.50	455
3299	20" diameter	Q-10	18	1.333		405	68		473	550
3300	24" diameter	"	16	1.500		520	76.50		596.50	685
3350	Elbow 90° fixed									
3354	6" diameter	Q-9	30	.533	Ea.	390	26.50		416.50	470
3355	8" diameter		26	.615		440	30.50		470.50	530
3356	10" diameter		24	.667		500	33		533	600
3357	12" diameter		22	.727		565	36		601	680
3358	14" diameter		21	.762		640	37.50		677.50	765
3359	16" diameter		20	.800		720	39.50		759.50	855
3360	18" diameter		19	.842		815	41.50		856.50	960
3361	20" diameter	Q-10	18	1.333		925	68		993	1,125
3362	24" diameter	"	16	1.500		1,175	76.50		1,251.50	1,425
3380	For 316 stainless steel liner, add					30%				
3400	Elbow 45°									
3404	6" diameter	Q-9	30	.533	Ea.	195	26.50		221.50	254
3405	8" diameter		26	.615		219	30.50		249.50	288
3406	10" diameter		24	.667		249	33		282	325
3407	12" diameter		22	.727		285	36		321	370
3408	14" diameter		21	.762		320	37.50		357.50	410
3409	16" diameter		20	.800		360	39.50		399.50	455
3410	18" diameter		19	.842		410	41.50		451.50	515
3411	20" diameter	Q-10	18	1.333		460	68		528	615
3412	24" diameter	"	16	1.500		590	76.50		666.50	765
3430	For 316 stainless steel liner, add					30%				
3450	Tee 90°									
3454	6" diameter	Q-9	24	.667	Ea.	229	33		262	300
3455	8" diameter		22	.727		251	36		287	330
3456	10" diameter		21	.762		283	37.50		320.50	370
3457	12" diameter		20	.800		330	39.50		369.50	420
3458	14" diameter		18	.889		375	44		419	475
3459	16" diameter		16	1		410	49		459	525
3460	18" diameter		14	1.143		485	56.50		541.50	615
3461	20" diameter	Q-10	17	1.412		560	72		632	725
3462	24" diameter	"	12	2		700	102		802	925
3480	For Tee Cap, add					35%	20%			
3500	For 316 stainless steel liner, add					30%				
3520	Plate support, galvanized									
3524	6" diameter	Q-9	26	.615	Ea.	115	30.50		145.50	173
3525	8" diameter		22	.727		134	36		170	202

23 51 Breechings, Chimneys, and Stacks

23 51 26 – All-Fuel Vent Chimneys

23 51 26.10 All-Fuel Vent Chimneys, Press. Tight, Dbl. Wall		Crew	Daily Output	Labor-Hours	Unit	Material	2014 Bare Costs Labor	Equipment	Total	Total Incl O&P
3526	10" diameter	Q-9	20	.800	Ea.	146	39.50		185.50	220
3527	12" diameter		18	.889		153	44		197	235
3528	14" diameter		17	.941		181	46.50		227.50	270
3529	16" diameter		16	1		191	49		240	285
3530	18" diameter		15	1.067		201	52.50		253.50	300
3531	20" diameter	Q-10	16	1.500		211	76.50		287.50	350
3532	24" diameter	"	14	1.714		219	87.50		306.50	375
3570	Bellows, lined									
3574	6" diameter	Q-9	30	.533	Ea.	1,250	26.50		1,276.50	1,425
3575	8" diameter		26	.615		1,300	30.50		1,330.50	1,500
3576	10" diameter		24	.667		1,325	33		1,358	1,525
3577	12" diameter		22	.727		1,375	36		1,411	1,550
3578	14" diameter		21	.762		1,425	37.50		1,462.50	1,600
3579	16" diameter		20	.800		1,475	39.50		1,514.50	1,675
3580	18" diameter		19	.842		1,525	41.50		1,566.50	1,750
3581	20" diameter	Q-10	18	1.333		1,575	68		1,643	1,825
3590	For all 316 stainless steel construction, add					55%				
3600	Ventilated roof thimble, 304 stainless steel									
3620	6" diameter	Q-9	26	.615	Ea.	239	30.50		269.50	310
3624	8" diameter		22	.727		248	36		284	330
3625	10" diameter		20	.800		256	39.50		295.50	340
3626	12" diameter		18	.889		263	44		307	355
3627	14" diameter		17	.941		273	46.50		319.50	370
3628	16" diameter		16	1		294	49		343	400
3629	18" diameter		15	1.067		320	52.50		372.50	430
3630	20" diameter	Q-10	16	1.500		335	76.50		411.50	480
3631	24" diameter	"	14	1.714		365	87.50		452.50	535
3650	For 316 stainless steel, add					30%				
3670	Exit cone, 316 stainless steel only									
3674	6" diameter	Q-9	46	.348	Ea.	181	17.10		198.10	225
3675	8" diameter		42	.381		187	18.75		205.75	234
3676	10" diameter		40	.400		197	19.70		216.70	247
3677	12" diameter		38	.421		210	20.50		230.50	263
3678	14" diameter		37	.432		225	21.50		246.50	280
3679	16" diameter		36	.444		282	22		304	345
3680	18" diameter		35	.457		305	22.50		327.50	370
3681	20" diameter	Q-10	28	.857		365	44		409	465
3682	24" diameter	"	26	.923		480	47		527	595
3720	Roof guide, 304 stainless steel									
3724	6" diameter	Q-9	25	.640	Ea.	83	31.50		114.50	139
3725	8" diameter		21	.762		96.50	37.50		134	164
3726	10" diameter		19	.842		106	41.50		147.50	180
3727	12" diameter		17	.941		109	46.50		155.50	191
3728	14" diameter		16	1		127	49		176	215
3729	16" diameter		15	1.067		135	52.50		187.50	228
3730	18" diameter		14	1.143		143	56.50		199.50	243
3731	20" diameter	Q-10	15	1.600		150	81.50		231.50	290
3732	24" diameter	"	13	1.846		157	94.50		251.50	315
3750	For 316 stainless steel, add					30%				
3770	Rain cap with bird screen									
3774	6" diameter	Q-9	46	.348	Ea.	280	17.10		297.10	335
3775	8" diameter		42	.381		325	18.75		343.75	385
3776	10" diameter		40	.400		380	19.70		399.70	445

23 51 Breechings, Chimneys, and Stacks

23 51 26 – All-Fuel Vent Chimneys

	23 51 26.10 All-Fuel Vent Chimneys, Press. Tight, Dbl. Wall	Crew	Daily Output	Labor-Hours	Unit	Material	2014 Bare Costs Labor	Equipment	Total	Total Incl O&P
3777	12" diameter	Q-9	38	.421	Ea.	440	20.50		460.50	510
3778	14" diameter		37	.432		495	21.50		516.50	580
3779	16" diameter		36	.444		565	22		587	655
3780	18" diameter		35	.457		655	22.50		677.50	755
3781	20" diameter	Q-10	28	.857		740	44		784	880
3782	24" diameter	"	26	.923		890	47		937	1,050

23 51 26.30 All-Fuel Vent Chimneys, Double Wall, St. Stl.

		Crew	Daily Output	Labor-Hours	Unit	Material	Labor	Equipment	Total	Total Incl O&P
0010	**ALL-FUEL VENT CHIMNEYS, DOUBLE WALL, STAINLESS STEEL**									
7780	All fuel, pressure tight, double wall, 4" insulation, U.L. listed, 1400°F.									
7790	304 stainless steel liner, aluminized steel outer jacket									
7800	6" diameter	Q-9	60	.267	V.L.F.	62	13.15		75.15	88
7804	8" diameter		52	.308		71	15.15		86.15	101
7806	10" diameter		48	.333		79	16.40		95.40	112
7808	12" diameter		44	.364		90.50	17.90		108.40	127
7810	14" diameter		42	.381		102	18.75		120.75	141
7880	For 316 stainless steel liner add				L.F.	30%				
8000	All fuel, double wall, stainless steel fittings									
8010	Roof support 6" diameter	Q-9	30	.533	Ea.	109	26.50		135.50	160
8030	8" diameter		26	.615		127	30.50		157.50	187
8040	10" diameter		24	.667		135	33		168	198
8050	12" diameter		22	.727		143	36		179	212
8060	14" diameter		21	.762		150	37.50		187.50	223
8100	Elbow 45°, 6" diameter		30	.533		233	26.50		259.50	296
8140	8" diameter		26	.615		262	30.50		292.50	335
8160	10" diameter		24	.667		298	33		331	380
8180	12" diameter		22	.727		340	36		376	430
8200	14" diameter		21	.762		380	37.50		417.50	480
8300	Insulated tee, 6" diameter		30	.533		274	26.50		300.50	340
8360	8" diameter		26	.615		300	30.50		330.50	375
8380	10" diameter		24	.667		335	33		368	420
8400	12" diameter		22	.727		380	36		416	475
8420	14" diameter		20	.800		445	39.50		484.50	550
8500	Boot tee, 6" diameter		28	.571		525	28		553	620
8520	8" diameter		24	.667		570	33		603	675
8530	10" diameter		22	.727		665	36		701	785
8540	12" diameter		20	.800		785	39.50		824.50	925
8550	14" diameter		18	.889		895	44		939	1,050
8600	Rain cap with bird screen, 6" diameter		30	.533		280	26.50		306.50	350
8640	8" diameter		26	.615		325	30.50		355.50	400
8660	10" diameter		24	.667		380	33		413	465
8680	12" diameter		22	.727		440	36		476	535
8700	14" diameter		21	.762		495	37.50		532.50	605
8800	Flat roof flashing, 6" diameter		30	.533		103	26.50		129.50	153
8840	8" diameter		26	.615		112	30.50		142.50	170
8860	10" diameter		24	.667		120	33		153	182
8880	12" diameter		22	.727		132	36		168	200
8900	14" diameter		21	.762		135	37.50		172.50	206

23 52 Heating Boilers

23 52 13 – Electric Boilers

23 52 13.10 Electric Boilers, ASME

		Crew	Daily Output	Labor-Hours	Unit	Material	2014 Bare Costs Labor	Equipment	Total	Total Incl O&P
0010	**ELECTRIC BOILERS, ASME**, Standard controls and trim D3020-102									
1000	Steam, 6 KW, 20.5 MBH	Q-19	1.20	20	Ea.	3,950	1,050		5,000	5,950
1040	9 KW, 30.7 MBH		1.20	20		4,025	1,050		5,075	6,025
1060	18 KW, 61.4 MBH		1.20	20		4,150	1,050		5,200	6,175
1080	24 KW, 81.8 MBH		1.10	21.818		4,800	1,150		5,950	7,025
1120	36 KW, 123 MBH		1.10	21.818		5,325	1,150		6,475	7,575
2000	Hot water, 7.5 KW, 25.6 MBH		1.30	18.462		4,975	975		5,950	6,950
2020	15 KW, 51.2 MBH		1.30	18.462		5,000	975		5,975	6,975
2040	30 KW, 102 MBH		1.20	20		5,325	1,050		6,375	7,450
2060	45 KW, 164 MBH		1.20	20		5,425	1,050		6,475	7,550
2070	60 KW, 205 MBH		1.20	20		5,525	1,050		6,575	7,675
2080	75 KW, 256 MBH		1.10	21.818		5,900	1,150		7,050	8,225

23 52 16 – Condensing Boilers

23 52 16.24 Condensing Boilers

			Crew	Daily Output	Labor-Hours	Unit	Material	Labor	Equipment	Total	Total Incl O&P
0010	**CONDENSING BOILERS**, Cast iron										
0020	Packaged with standard controls, circulator and trim										
0030	Intermittent (spark) pilot, natural or LP gas										
0040	Hot water, DOE MBH output, (AFUE)										
0100	42 MBH, (84.0%)	G	Q-5	1.80	8.889	Ea.	1,500	470		1,970	2,350
0120	57 MBH, (84.3%)	G		1.60	10		1,675	525		2,200	2,650
0140	85 MBH, (84.0%)	G		1.40	11.429		1,850	600		2,450	2,925
0160	112 MBH, (83.7%)	G		1.20	13.333		2,075	700		2,775	3,325
0180	140 MBH, (83.3%)	G	Q-6	1.60	15		2,350	820		3,170	3,800
0200	167 MBH, (83.0%)	G		1.40	17.143		2,625	935		3,560	4,325
0220	194 MBH, (82.7%)	G		1.20	20		2,875	1,100		3,975	4,825

23 52 19 – Pulse Combustion Boilers

23 52 19.20 Pulse Type Combustion Boilers

			Crew	Daily Output	Labor-Hours	Unit	Material	Labor	Equipment	Total	Total Incl O&P
0010	**PULSE TYPE COMBUSTION BOILERS**										
7990	Special feature gas fired boilers										
8000	Pulse combustion, standard controls/trim										
8050	88,000 BTU	G	Q-5	1.40	11.429	Ea.	4,900	600		5,500	6,300
8080	134,000 BTU	G	"	1.20	13.333	"	5,500	700		6,200	7,100

23 52 23 – Cast-Iron Boilers

23 52 23.20 Gas-Fired Boilers

		Crew	Daily Output	Labor-Hours	Unit	Material	Labor	Equipment	Total	Total Incl O&P
0010	**GAS-FIRED BOILERS**, Natural or propane, standard controls, packaged									
1000	Cast iron, with insulated jacket										
2000	Steam, gross output, 81 MBH	Q-7	1.40	22.857	Ea.	2,350	1,275		3,625	4,500
2020	102 MBH		1.30	24.615		2,450	1,375		3,825	4,750
2040	122 MBH		1	32		2,800	1,775		4,575	5,775
2060	163 MBH		.90	35.556		3,175	1,975		5,150	6,500
2080	203 MBH		.90	35.556		3,650	1,975		5,625	7,000
2100	240 MBH		.85	37.647		3,725	2,100		5,825	7,275
2120	280 MBH		.80	40		4,425	2,225		6,650	8,250
2140	320 MBH		.70	45.714		4,700	2,550		7,250	9,025
3000	Hot water, gross output, 80 MBH		1.46	21.918		2,150	1,225		3,375	4,225
3020	100 MBH		1.35	23.704		2,300	1,325		3,625	4,525
3040	122 MBH		1.10	29.091		2,625	1,625		4,250	5,325
3060	163 MBH		1	32		3,000	1,775		4,775	6,000
3080	203 MBH		1	32		3,450	1,775		5,225	6,500
3100	240 MBH		.95	33.684		3,550	1,875		5,425	6,725
3120	280 MBH		.90	35.556		4,175	1,975		6,150	7,575

23 52 Heating Boilers

23 52 23 – Cast-Iron Boilers

23 52 23.20 Gas-Fired Boilers

		Crew	Daily Output	Labor-Hours	Unit	Material	2014 Bare Costs Labor	Equipment	Total	Total Incl O&P
3140	320 MBH	Q-7	.80	40	Ea.	4,500	2,225		6,725	8,325
7000	For tankless water heater, add					10%				
7050	For additional zone valves up to 312 MBH add					188			188	207

23 52 23.40 Oil-Fired Boilers

		Crew	Daily Output	Labor-Hours	Unit	Material	Labor	Equipment	Total	Total Incl O&P
0010	**OIL-FIRED BOILERS**, Standard controls, flame retention burner, packaged									
1000	Cast iron, with insulated flush jacket									
2000	Steam, gross output, 109 MBH	Q-7	1.20	26.667	Ea.	2,175	1,475		3,650	4,650
2020	144 MBH		1.10	29.091		2,475	1,625		4,100	5,175
2040	173 MBH		1	32		2,750	1,775		4,525	5,750
2060	207 MBH		.90	35.556		3,000	1,975		4,975	6,300
3000	Hot water, same price as steam									
4000	For tankless coil in smaller sizes, add				Ea.	15%				

23 52 26 – Steel Boilers

23 52 26.40 Oil-Fired Boilers

		Crew	Daily Output	Labor-Hours	Unit	Material	Labor	Equipment	Total	Total Incl O&P
0010	**OIL-FIRED BOILERS**, Standard controls, flame retention burner									
5000	Steel, with insulated flush jacket									
7000	Hot water, gross output, 103 MBH	Q-6	1.60	15	Ea.	1,775	820		2,595	3,175
7020	122 MBH		1.45	16.506		1,875	900		2,775	3,425
7040	137 MBH		1.36	17.595		2,000	960		2,960	3,650
7060	168 MBH		1.30	18.405		2,100	1,000		3,100	3,850
7080	225 MBH		1.22	19.704		2,575	1,075		3,650	4,450
7340	For tankless coil in steam or hot water, add					7%				

23 52 28 – Swimming Pool Boilers

23 52 28.10 Swimming Pool Heaters

		Crew	Daily Output	Labor-Hours	Unit	Material	Labor	Equipment	Total	Total Incl O&P
0010	**SWIMMING POOL HEATERS**, Not including wiring, external									
0020	piping, base or pad,									
0160	Gas fired, input, 155 MBH	Q-6	1.50	16	Ea.	1,900	875		2,775	3,425
0200	199 MBH		1	24		2,025	1,300		3,325	4,200
0220	250 MBH		.70	34.286		2,200	1,875		4,075	5,250
0240	300 MBH		.60	40		2,300	2,175		4,475	5,850
0260	399 MBH		.50	48		2,600	2,625		5,225	6,800
0280	500 MBH		.40	60		8,475	3,275		11,750	14,300
0300	650 MBH		.35	68.571		9,000	3,750		12,750	15,600
0320	750 MBH		.33	72.727		9,850	3,975		13,825	16,800
0360	990 MBH		.22	109		13,200	5,950		19,150	23,500
0370	1,260 MBH		.21	114		15,500	6,250		21,750	26,400
0380	1,440 MBH		.19	126		16,500	6,900		23,400	28,600
0400	1,800 MBH		.14	171		18,400	9,350		27,750	34,300
0410	2,070 MBH		.13	184		21,700	10,100		31,800	39,100
2000	Electric, 12 KW, 4,800 gallon pool	Q-19	3	8		2,075	425		2,500	2,900
2020	15 KW, 7,200 gallon pool		2.80	8.571		2,100	455		2,555	3,000
2040	24 KW, 9,600 gallon pool		2.40	10		2,425	530		2,955	3,475
2060	30 KW, 12,000 gallon pool		2	12		2,475	635		3,110	3,675
2080	36 KW, 14,400 gallon pool		1.60	15		2,850	795		3,645	4,325
2100	57 KW, 24,000 gallon pool		1.20	20		3,575	1,050		4,625	5,525
9000	To select pool heater: 12 BTUH x S.F. pool area									
9010	X temperature differential = required output									
9050	For electric, KW = gallons x 2.5 divided by 1000									
9100	For family home type pool, double the									
9110	Rated gallon capacity = 1/2°F rise per hour									

23 52 Heating Boilers

23 52 88 – Burners

23 52 88.10 Replacement Type Burners

		Crew	Daily Output	Labor-Hours	Unit	Material	2014 Bare Costs Labor	Equipment	Total	Total Incl O&P
0010	**REPLACEMENT TYPE BURNERS**									
0990	Residential, conversion, gas fired, LP or natural									
1000	Gun type, atmospheric input 50 to 225 MBH	Q-1	2.50	6.400	Ea.	720	330		1,050	1,300
1020	100 to 400 MBH	"	2	8	"	1,200	415		1,615	1,950
3000	Flame retention oil fired assembly, input									
3020	.50 to 2.25 GPH	Q-1	2.40	6.667	Ea.	289	345		634	840
3040	2.0 to 5.0 GPH	"	2	8		345	415		760	1,000
4600	Gas safety, shut off valve, 3/4" threaded	1 Stpi	20	.400		178	23.50		201.50	232
4610	1" threaded		19	.421		172	24.50		196.50	227
4620	1-1/4" threaded		15	.533		194	31		225	260
4630	1-1/2" threaded		13	.615		210	36		246	286
4640	2" threaded		11	.727		235	42.50		277.50	325
4650	2-1/2" threaded	Q-1	15	1.067		269	55.50		324.50	380
4660	3" threaded		13	1.231		370	64		434	500
4670	4" flanged		3	5.333		2,625	276		2,901	3,325
4680	6" flanged	Q-2	3	8		5,875	430		6,305	7,125

23 56 Solar Energy Heating Equipment

23 56 16 – Packaged Solar Heating Equipment

23 56 16.40 Solar Heating Systems

				Crew	Daily Output	Labor-Hours	Unit	Material	Labor	Equipment	Total	Total Incl O&P
0010	**SOLAR HEATING SYSTEMS**	D2020-265										
0020	System/Package prices, not including connecting											
0030	pipe, insulation, or special heating/plumbing fixtures	D2020-270										
0152	For solar ultraviolet pipe insulation see Section 22 07 19.10											
0500	Hot water, standard package, low temperature	D2020-275										
0540	1 collector, circulator, fittings, 65 gal. tank		G	Q-1	.50	32	Ea.	3,575	1,650		5,225	6,425
0580	2 collectors, circulator, fittings, 120 gal. tank	D2020-280	G		.40	40		4,700	2,075		6,775	8,275
0620	3 collectors, circulator, fittings, 120 gal. tank		G		.34	47.059		6,325	2,450		8,775	10,700
0700	Medium temperature package	D2020-285										
0720	1 collector, circulator, fittings, 80 gal. tank		G	Q-1	.50	32	Ea.	4,950	1,650		6,600	7,950
0740	2 collectors, circulator, fittings, 120 gal. tank	D2020-290	G		.40	40		6,400	2,075		8,475	10,200
0780	3 collectors, circulator, fittings, 120 gal. tank		G		.30	53.333		7,225	2,775		10,000	12,100
0980	For each additional 120 gal. tank, add	D2020-295						1,675			1,675	1,850

23 56 19 – Solar Heating Components

23 56 19.50 Solar Heating Ancillary

				Crew	Daily Output	Labor-Hours	Unit	Material	Labor	Equipment	Total	Total Incl O&P
0010	**SOLAR HEATING ANCILLARY**											
2300	Circulators, air	D3010-650										
2310	Blowers											
2330	100-300 S.F. system, 1/10 HP	D3010-660	G	Q-9	16	1	Ea.	249	49		298	350
2340	300-500 S.F. system, 1/5 HP		G		15	1.067		330	52.50		382.50	440
2350	Two speed, 100-300 S.F., 1/10 HP	D3010-675	G		14	1.143		142	56.50		198.50	243
2400	Reversible fan, 20" diameter, 2 speed		G		18	.889		113	44		157	191
2550	Booster fan 6" diameter, 120 CFM		G		16	1		36.50	49		85.50	116
2570	6" diameter, 225 CFM		G		16	1		45	49		94	125
2580	8" diameter, 150 CFM		G		16	1		41	49		90	121
2590	8" diameter, 310 CFM		G		14	1.143		63.50	56.50		120	156
2600	8" diameter, 425 CFM		G		14	1.143		71	56.50		127.50	164
2650	Rheostat		G		32	.500		15.45	24.50		39.95	54.50
2660	Shutter/damper		G		12	1.333		56	65.50		121.50	162
2670	Shutter motor		G		16	1		139	49		188	228

23 56 Solar Energy Heating Equipment

23 56 19 – Solar Heating Components

23 56 19.50 Solar Heating Ancillary		Crew	Daily Output	Labor-Hours	Unit	Material	2014 Bare Costs Labor	Equipment	Total	Total Incl O&P
2800	Circulators, liquid, 1/25 HP, 5.3 GPM	G	Q-1	14	1.143	Ea.	274	59		333	390
2820	1/20 HP, 17 GPM	G		12	1.333		161	69		230	281
2850	1/20 HP, 17 GPM, stainless steel	G		12	1.333		254	69		323	385
2870	1/12 HP, 30 GPM	G	↓	10	1.600	↓	345	83		428	505
3000	Collector panels, air with aluminum absorber plate										
3010	Wall or roof mount										
3040	Flat black, plastic glazing										
3080	4' x 8'	G	Q-9	6	2.667	Ea.	660	131		791	930
3100	4' x 10'	G		5	3.200	"	815	158		973	1,125
3200	Flush roof mount, 10' to 16' x 22" wide	G		96	.167	L.F.	132	8.20		140.20	158
3210	Manifold, by L.F. width of collectors	G	↓	160	.100	"	134	4.92		138.92	156
3300	Collector panels, liquid with copper absorber plate										
3320	Black chrome, tempered glass glazing										
3330	Alum. frame, 4' x 8', 5/32" single glazing	G	Q-1	9.50	1.684	Ea.	995	87		1,082	1,225
3390	Alum. frame, 4' x 10', 5/32" single glazing	G		6	2.667		1,150	138		1,288	1,450
3450	Flat black, alum. frame, 3.5' x 7.5'	G		9	1.778		730	92		822	940
3500	4' x 8'	G		5.50	2.909		915	151		1,066	1,225
3520	4' x 10'	G		10	1.600		1,025	83		1,108	1,250
3540	4' x 12.5'	G	↓	5	3.200	↓	1,225	166		1,391	1,600
3550	Liquid with fin tube absorber plate										
3560	Alum. frame 4' x 8' tempered glass	G	Q-1	10	1.600	Ea.	560	83		643	740
3580	Liquid with vacuum tubes, 4' x 6'-10"	G		9	1.778		895	92		987	1,125
3600	Liquid, full wetted, plastic, alum. frame, 4' x 10'	G		5	3.200		247	166		413	520
3650	Collector panel mounting, flat roof or ground rack	G		7	2.286	↓	244	118		362	445
3670	Roof clamps	G	↓	70	.229	Set	2.80	11.85		14.65	21
3700	Roof strap, teflon	G	1 Plum	205	.039	L.F.	23.50	2.25		25.75	29.50
3900	Differential controller with two sensors										
3930	Thermostat, hard wired	G	1 Plum	8	1	Ea.	101	57.50		158.50	198
3950	Line cord and receptacle	G		12	.667		570	38.50		608.50	685
4050	Pool valve system	G		2.50	3.200		287	184		471	595
4070	With 12 VAC actuator	G		2	4		300	230		530	680
4080	Pool pump system, 2" pipe size	G		6	1.333		187	76.50		263.50	320
4100	Five station with digital read-out	G	↓	3	2.667	↓	236	153		389	490
4150	Sensors										
4200	Brass plug, 1/2" MPT	G	1 Plum	32	.250	Ea.	19.45	14.40		33.85	43
4210	Brass plug, reversed	G		32	.250		26.50	14.40		40.90	50.50
4220	Freeze prevention	G		32	.250		23.50	14.40		37.90	47.50
4240	Screw attached	G		32	.250		9	14.40		23.40	31.50
4250	Brass, immersion	G	↓	32	.250	↓	28.50	14.40		42.90	53
4300	Heat exchanger										
4315	includes coil, blower, circulator										
4316	and controller for DHW and space hot air										
4330	Fluid to air coil, up flow, 45 MBH	G	Q-1	4	4	Ea.	315	207		522	665
4380	70 MBH	G		3.50	4.571		355	237		592	755
4400	80 MBH	G	↓	3	5.333		475	276		751	940
4580	Fluid to fluid package includes two circulating pumps										
4590	expansion tank, check valve, relief valve										
4600	controller, high temperature cutoff and sensors	G	Q-1	2.50	6.400	Ea.	770	330		1,100	1,350
4650	Heat transfer fluid										
4700	Propylene glycol, inhibited anti-freeze	G	1 Plum	28	.286	Gal.	15.45	16.45		31.90	42
4800	Solar storage tanks, knocked down										
4810	Air, galvanized steel clad, double wall, 4" fiberglass insulation										
5120	45 Mil reinforced polypropylene lining,										

23 56 Solar Energy Heating Equipment

23 56 19 – Solar Heating Components

23 56 19.50 Solar Heating Ancillary

		Crew	Daily Output	Labor-Hours	Unit	Material	Labor	Equipment	Total	Total Incl O&P
5140	4' high, 4' x 4' = 64 C.F./450 gallons	Q-9	2	8	Ea.	3,650	395		4,045	4,600
5150	4' x 8' = 128 C.F./900 gallons		1.50	10.667		5,450	525		5,975	6,800
5160	4' x 12' = 190 C.F./1300 gallons		1.30	12.308		7,275	605		7,880	8,925
5170	8' x 8' = 250 C.F./1700 gallons		1	16		7,275	790		8,065	9,200
5190	6'-3" high, 7' x 7' = 306 C.F./2000 gallons	Q-10	1.20	20		14,200	1,025		15,225	17,300
5200	7' x 10'-6" = 459 C.F./3000 gallons		.80	30		17,800	1,525		19,325	22,000
5210	7' x 14' = 613 C.F./4000 gallons		.60	40		21,400	2,050		23,450	26,600
5220	10'-6" x 10'-6" = 689 C.F./4500 gallons		.50	48		21,400	2,450		23,850	27,300
5230	10'-6" x 14' = 919 C.F./6000 gallons	Q-9	.40	60		24,900	3,075		27,975	32,100
5240	14' x 14' = 1225 C.F./8000 gallons	Q-11	.40	80		28,500	4,175		32,675	37,700
5250	14' x 17'-6" = 1531 C.F./10,000 gallons		.30	106		32,100	5,550		37,650	43,800
5260	17'-6" x 17'-6" = 1914 C.F./12,500 gallons		.25	128		35,600	6,675		42,275	49,400
5270	17'-6" x 21' = 2297 C.F./15,000 gallons		.20	160		39,100	8,325		47,425	55,500
5280	21' x 21' = 2756 C.F./18,000 gallons		.18	177		42,700	9,250		51,950	61,000
5290	30 Mil reinforced Hypalon lining, add					.02%				
7000	Solar control valves and vents									
7050	Air purger, 1" pipe size	1 Plum	12	.667	Ea.	46.50	38.50		85	110
7070	Air eliminator, automatic 3/4" size		32	.250		30	14.40		44.40	54.50
7090	Air vent, automatic, 1/8" fitting		32	.250		17.25	14.40		31.65	40.50
7100	Manual, 1/8" NPT		32	.250		25.50	14.40		39.90	49.50
7120	Backflow preventer, 1/2" pipe size		16	.500		75.50	29		104.50	127
7130	3/4" pipe size		16	.500		77	29		106	128
7150	Balancing valve, 3/4" pipe size		20	.400		63	23		86	105
7180	Draindown valve, 1/2" copper tube		9	.889		212	51		263	310
7200	Flow control valve, 1/2" pipe size		22	.364		120	21		141	164
7220	Expansion tank, up to 5 gal.		32	.250		68	14.40		82.40	96.50
7250	Hydronic controller (aquastat)		8	1		155	57.50		212.50	258
7400	Pressure gauge, 2" dial		32	.250		24	14.40		38.40	48
7450	Relief valve, temp. and pressure 3/4" pipe size		30	.267		21	15.35		36.35	46
7500	Solenoid valve, normally closed									
7520	Brass, 3/4" NPT, 24V	1 Plum	9	.889	Ea.	119	51		170	207
7530	1" NPT, 24V		9	.889		186	51		237	282
7750	Vacuum relief valve, 3/4" pipe size		32	.250		29.50	14.40		43.90	54
7800	Thermometers									
7820	Digital temperature monitoring, 4 locations	1 Plum	2.50	3.200	Ea.	137	184		321	430
7900	Upright, 1/2" NPT		8	1		28	57.50		85.50	118
7970	Remote probe, 2" dial		8	1		33	57.50		90.50	124
7990	Stem, 2" dial, 9" stem		16	.500		21.50	29		50.50	67.50
8250	Water storage tank with heat exchanger and electric element									
8270	66 gal. with 2" x 2 lb. density insulation	1 Plum	1.60	5	Ea.	1,600	288		1,888	2,200
8300	80 gal. with 2" x 2 lb. density insulation		1.60	5		1,600	288		1,888	2,200
8380	120 gal. with 2" x 2 lb. density insulation		1.40	5.714		1,825	330		2,155	2,525
8400	120 gal. with 2" x 2 lb. density insul., 40 S.F. heat coil		1.40	5.714		2,325	330		2,655	3,050
8500	Water storage module, plastic									
8600	Tubular, 12" diameter, 4' high	1 Carp	48	.167	Ea.	114	7.65		121.65	137
8610	12" diameter, 8' high		40	.200		175	9.15		184.15	207
8620	18" diameter, 5' high		38	.211		193	9.65		202.65	227
8630	18" diameter, 10' high		32	.250		255	11.45		266.45	299
8640	58" diameter, 5' high	2 Carp	32	.500		495	23		518	580
8650	Cap, 12" diameter					22			22	24
8660	18" diameter					28			28	31

23 57 Heat Exchangers for HVAC

23 57 16 – Steam-to-Water Heat Exchangers

23 57 16.10 Shell/Tube Type Steam-to-Water Heat Exch.	Crew	Daily Output	Labor-Hours	Unit	Material	2014 Bare Costs Labor	Equipment	Total	Total Incl O&P
0010 **SHELL AND TUBE TYPE STEAM-TO-WATER HEAT EXCHANGERS**									
0016 Shell & tube type, 2 or 4 pass, 3/4" O.D. copper tubes,									
0020 C.I. heads, C.I. tube sheet, steel shell									
0100 Hot water 40°F to 180°F, by steam at 10 PSI									
0120 8 GPM	Q-5	6	2.667	Ea.	2,150	140		2,290	2,575
0140 10 GPM	↓	5	3.200	↓	3,250	168		3,418	3,825
0160 40 GPM		4	4		5,025	211		5,236	5,850
0500 For bronze head and tube sheet, add					50%				

23 57 19 – Liquid-to-Liquid Heat Exchangers

23 57 19.13 Plate-Type, Liquid-to-Liquid Heat Exchangers

	Crew	Daily Output	Labor-Hours	Unit	Material	Labor	Equipment	Total	Incl O&P
0010 **PLATE-TYPE, LIQUID-TO-LIQUID HEAT EXCHANGERS**									
3000 Plate type,									
3100 400 GPM	Q-6	.80	30	Ea.	35,600	1,650		37,250	41,700
3120 800 GPM	"	.50	48		61,500	2,625		64,125	71,500
3140 1200 GPM	Q-7	.34	94.118		91,500	5,250		96,750	108,500
3160 1800 GPM	"	.24	133	↓	121,000	7,425		128,425	144,000

23 57 19.16 Shell-Type, Liquid-to-Liquid Heat Exchangers

	Crew	Daily Output	Labor-Hours	Unit	Material	Labor	Equipment	Total	Incl O&P
0010 **SHELL-TYPE, LIQUID-TO-LIQUID HEAT EXCHANGERS**									
1000 Hot water 40°F to 140°F, by water at 200°F									
1020 7 GPM	Q-5	6	2.667	Ea.	2,650	140		2,790	3,100
1040 16 GPM		5	3.200		3,750	168		3,918	4,375
1060 34 GPM		4	4		5,675	211		5,886	6,550
1080 55 GPM		3	5.333		8,225	281		8,506	9,475
1100 74 GPM		1.50	10.667		10,300	560		10,860	12,200
1120 86 GPM	↓	1.40	11.429		13,700	600		14,300	16,000
1140 112 GPM	Q-6	2	12		17,100	655		17,755	19,800
1160 126 GPM		1.80	13.333		21,400	730		22,130	24,600
1180 152 GPM	↓	1	24		27,300	1,300		28,600	32,000

23 72 Air-to-Air Energy Recovery Equipment

23 72 16 – Heat-Pipe Air-To-Air Energy-Recovery Equipment

23 72 16.10 Heat Pipes

	Crew	Daily Output	Labor-Hours	Unit	Material	Labor	Equipment	Total	Incl O&P
0010 **HEAT PIPES**									
8000 Heat pipe type, glycol, 50% efficient									
8010 100 MBH, 1700 CFM	1 Stpi	.80	10	Ea.	4,100	585		4,685	5,400
8020 160 MBH, 2700 CFM		.60	13.333		5,525	780		6,305	7,250
8030 620 MBH, 4000 CFM	↓	.40	20	↓	8,325	1,175		9,500	11,000

23 81 Decentralized Unitary HVAC Equipment

23 81 13 – Packaged Terminal Air-Conditioners

23 81 13.10 Packaged Cabinet Type Air-Conditioners

	Crew	Daily Output	Labor-Hours	Unit	Material	Labor	Equipment	Total	Incl O&P
0010 **PACKAGED CABINET TYPE AIR-CONDITIONERS**, Cabinet, wall sleeve,									
0100 louver, electric heat, thermostat, manual changeover, 208 V									
0200 6,000 BTUH cooling, 8800 BTU heat	Q-5	6	2.667	Ea.	775	140		915	1,075
0220 9,000 BTUH cooling, 13,900 BTU heat		5	3.200		850	168		1,018	1,200
0240 12,000 BTUH cooling, 13,900 BTU heat		4	4		920	211		1,131	1,325
0260 15,000 BTUH cooling, 13,900 BTU heat	↓	3	5.333		1,150	281		1,431	1,700

23 81 Decentralized Unitary HVAC Equipment

23 81 19 – Self-Contained Air-Conditioners

23 81 19.10 Window Unit Air Conditioners

		Crew	Daily Output	Labor-Hours	Unit	Material	2014 Bare Costs Labor	Equipment	Total	Total Incl O&P
0010	**WINDOW UNIT AIR CONDITIONERS**									
4000	Portable/window, 15 amp 125 V grounded receptacle required									
4060	5000 BTUH	1 Carp	8	1	Ea.	284	46		330	380
4340	6000 BTUH		8	1		325	46		371	430
4480	8000 BTUH		6	1.333		445	61		506	585
4500	10,000 BTUH		6	1.333		570	61		631	720
4520	12,000 BTUH	L-2	8	2		695	80.50		775.50	890
4600	Window/thru-the-wall, 15 amp 230 V grounded receptacle required									
4780	18,000 BTUH	L-2	6	2.667	Ea.	930	107		1,037	1,200
4940	25,000 BTUH		4	4		1,175	161		1,336	1,550
4960	29,000 BTUH		4	4		1,300	161		1,461	1,675

23 81 43 – Air-Source Unitary Heat Pumps

23 81 43.10 Air-Source Heat Pumps

		Crew	Daily Output	Labor-Hours	Unit	Material	Labor	Equipment	Total	Total Incl O&P
0010	**AIR-SOURCE HEAT PUMPS**, Not including interconnecting tubing									
1000	Air to air, split system, not including curbs, pads, fan coil and ductwork									
1010	For curbs/pads see Section 23 91 00									
1012	Outside condensing unit only, for fan coil see Section 23 82 19.10									
1015	1.5 ton cooling, 7 MBH heat @ 0°F	Q-5	2.40	6.667	Ea.	2,175	350		2,525	2,925
1020	2 ton cooling, 8.5 MBH heat @ 0°F		2	8		2,300	420		2,720	3,175
1030	2.5 ton cooling, 10 MBH heat @ 0°F		1.60	10		2,550	525		3,075	3,600
1040	3 ton cooling, 13 MBH heat @ 0°F		1.20	13.333		2,800	700		3,500	4,125
1050	3.5 ton cooling, 18 MBH heat @ 0°F		1	16		3,000	840		3,840	4,575
1054	4 ton cooling, 24 MBH heat @ 0°F		.80	20		3,300	1,050		4,350	5,250
1060	5 ton cooling, 27 MBH heat @ 0°F		.50	32		3,625	1,675		5,300	6,525
1500	Single package, not including curbs, pads, or plenums									
1502	1/2 ton cooling, supplementary heat included	Q-5	8	2	Ea.	2,375	105		2,480	2,750
1504	3/4 ton cooling, supplementary heat supplementary heat included		6	2.667		2,500	140		2,640	2,950
1506	1 ton cooling, supplementary heat included		4	4		2,875	211		3,086	3,475
1510	1.5 ton cooling, 5 MBH heat @ 0°F		1.55	10.323		2,750	545		3,295	3,850
1520	2 ton cooling, 6.5 MBH heat @ 0°F		1.50	10.667		2,950	560		3,510	4,100
1540	2.5 ton cooling, 8 MBH heat @ 0°F		1.40	11.429		3,175	600		3,775	4,400
1560	3 ton cooling, 10 MBH heat @ 0°F		1.20	13.333		3,475	700		4,175	4,875
1570	3.5 ton cooling, 11 MBH heat @ 0°F		1	16		3,800	840		4,640	5,450
1580	4 ton cooling, 13 MBH heat @ 0°F		.96	16.667		4,125	880		5,005	5,875
1620	5 ton cooling, 27 MBH heat @ 0°F		.65	24.615		4,375	1,300		5,675	6,775
1640	7.5 ton cooling, 35 MBH heat @ 0°F		.40	40		7,250	2,100		9,350	11,200

23 81 46 – Water-Source Unitary Heat Pumps

23 81 46.10 Water Source Heat Pumps

		Crew	Daily Output	Labor-Hours	Unit	Material	Labor	Equipment	Total	Total Incl O&P
0010	**WATER SOURCE HEAT PUMPS**, Not incl. connecting tubing or water source									
2000	Water source to air, single package									
2100	1 ton cooling, 13 MBH heat @ 75°F	Q-5	2	8	Ea.	1,825	420		2,245	2,625
2120	1.5 ton cooling, 17 MBH heat @ 75°F		1.80	8.889		1,900	470		2,370	2,800
2140	2 ton cooling, 19 MBH heat @ 75°F		1.70	9.412		2,050	495		2,545	3,000
2160	2.5 ton cooling, 25 MBH heat @ 75°F		1.60	10		2,150	525		2,675	3,175
2180	3 ton cooling, 27 MBH heat @ 75°F		1.40	11.429		2,225	600		2,825	3,350
2190	3.5 ton cooling, 29 MBH heat @ 75°F		1.30	12.308		2,525	650		3,175	3,750
2200	4 ton cooling, 31 MBH heat @ 75°F		1.20	13.333		2,700	700		3,400	4,025
2220	5 ton cooling, 29 MBH heat @ 75°F		.90	17.778		3,025	935		3,960	4,750
3960	For supplementary heat coil, add					10%				

23 82 Convection Heating and Cooling Units

23 82 19 – Fan Coil Units

23 82 19.10 Fan Coil Air Conditioning

		Crew	Daily Output	Labor-Hours	Unit	Material	2014 Bare Costs Labor	Equipment	Total	Total Incl O&P
0010	**FAN COIL AIR CONDITIONING**									
0030	Fan coil AC, cabinet mounted, filters and controls									
0940	Direct expansion, for use w/air cooled condensing unit, 1.5 ton cooling	Q-5	5	3.200	Ea.	755	168		923	1,075
0950	2 ton cooling		4.80	3.333		810	176		986	1,150
0960	2.5 ton cooling		4.40	3.636		835	191		1,026	1,200
0970	3 ton cooling		3.80	4.211		1,050	222		1,272	1,475
0980	3.5 ton cooling		3.60	4.444		1,075	234		1,309	1,550
0990	4 ton cooling		3.40	4.706		1,225	248		1,473	1,725
1000	5 ton cooling	↓	3	5.333		1,450	281		1,731	2,025
1500	For hot water coil, add				↓	40%	10%			

23 82 27 – Infrared Units

23 82 27.10 Infrared Type Heating Units

		Crew	Daily Output	Labor-Hours	Unit	Material	Labor	Equipment	Total	Total Incl O&P
0010	**INFRARED TYPE HEATING UNITS**									
0020	Gas fired, unvented, electric ignition, 100% shutoff.									
0030	Piping and wiring not included									
0060	Input, 15 MBH	Q-5	7	2.286	Ea.	555	120		675	790
0100	30 MBH		6	2.667		850	140		990	1,150
0120	45 MBH		5	3.200		1,150	168		1,318	1,525
0140	50 MBH		4.50	3.556		1,175	187		1,362	1,575
0160	60 MBH		4	4		1,200	211		1,411	1,650
0180	75 MBH		3	5.333		1,250	281		1,531	1,800
0200	90 MBH		2.50	6.400		1,375	335		1,710	2,000
0220	105 MBH		2	8		1,475	420		1,895	2,250
0240	120 MBH	↓	2	8	↓	1,500	420		1,920	2,275
1000	Gas fired, vented, electric ignition, tubular									
1020	Piping and wiring not included, 20' to 80' lengths									
1030	Single stage, input, 60 MBH	Q-6	4.50	5.333	Ea.	1,400	291		1,691	2,000
1040	80 MBH		3.90	6.154		1,400	335		1,735	2,050
1050	100 MBH		3.40	7.059		1,400	385		1,785	2,125
1060	125 MBH		2.90	8.276		1,625	450		2,075	2,450
1070	150 MBH		2.70	8.889		1,625	485		2,110	2,500
1080	170 MBH		2.50	9.600		1,625	525		2,150	2,575
1090	200 MBH	↓	2.20	10.909	↓	1,825	595		2,420	2,900
1100	Note: Final pricing may vary due to									
1110	tube length and configuration package selected									
1130	Two stage, input, 60 MBH high, 45 MBH low	Q-6	4.50	5.333	Ea.	1,525	291		1,816	2,125
1140	80 MBH high, 60 MBH low		3.90	6.154		1,525	335		1,860	2,175
1150	100 MBH high, 65 MBH low		3.40	7.059		1,525	385		1,910	2,250
1160	125 MBH high, 95 MBH low		2.90	8.276		1,725	450		2,175	2,575
1170	150 MBH high, 100 MBH low		2.70	8.889		1,725	485		2,210	2,625
1180	170 MBH high, 125 MBH low		2.50	9.600		1,925	525		2,450	2,925
1190	200 MBH high, 150 MBH low	↓	2.20	10.909	↓	1,925	595		2,520	3,025
1220	Note: Final pricing may vary due to									
1230	tube length and configuration package selected									
2000	Electric, single or three phase									
2050	6 kW, 20,478 BTU	1 Elec	2.30	3.478	Ea.	675	186		861	1,025
2100	13.5 KW, 40,956 BTU		2.20	3.636		835	194		1,029	1,200
2150	24 KW, 81,912 BTU	↓	2	4	↓	1,675	213		1,888	2,175
3000	Oil fired, pump, controls, fusible valve, oil supply tank									
3050	91,000 BTU	Q-5	2.50	6.400	Ea.	6,425	335		6,760	7,575
3080	105,000 BTU		2.25	7.111		6,800	375		7,175	8,075
3110	119,000 BTU	↓	2	8	↓	7,450	420		7,870	8,825

23 82 Convection Heating and Cooling Units

23 82 29 – Radiators

23 82 29.10 Hydronic Heating

		Crew	Daily Output	Labor-Hours	Unit	Material	2014 Bare Costs Labor	Equipment	Total	Total Incl O&P
0010	**HYDRONIC HEATING**, Terminal units, not incl. main supply pipe									
1000	Radiation									
1100	Panel, baseboard, C.I., including supports, no covers	Q-5	46	.348	L.F.	36.50	18.30		54.80	67.50
3000	Radiators, cast iron									
3100	Free standing or wall hung, 6 tube, 25" high	Q-5	96	.167	Section	44	8.80		52.80	62
3150	4 tube 25" high		96	.167		33	8.80		41.80	49.50
3200	4 tube, 19" high	↓	96	.167	↓	30.50	8.80		39.30	47
3250	Adj. brackets, 2 per wall radiator up to 30 sections	1 Stpi	32	.250	Ea.	34	14.65		48.65	59.50
3500	Recessed, 20" high x 5" deep, without grille	Q-5	60	.267	Section	32	14.05		46.05	56
3600	For inlet grille, add				"	2.21			2.21	2.43
9500	To convert SFR to BTU rating: Hot water, 150 x SFR									
9510	Forced hot water, 180 x SFR; steam, 240 x SFR									

23 82 33 – Convectors

23 82 33.10 Convector Units

		Crew	Daily Output	Labor-Hours	Unit	Material	Labor	Equipment	Total	Total Incl O&P
0010	**CONVECTOR UNITS**, Terminal units, not incl. main supply pipe									
2204	Convector, multifin, 2 pipe w/cabinet									
2210	17" H x 24" L	Q-5	10	1.600	Ea.	99	84		183	236
2214	17" H x 36" L		8.60	1.860		148	98		246	310
2218	17" H x 48" L		7.40	2.162		198	114		312	390
2222	21" H x 24" L		9	1.778		99	93.50		192.50	250
2228	21" H x 48" L	↓	6.80	2.353	↓	198	124		322	405
2240	For knob operated damper, add					140%				
2241	For metal trim strips, add	Q-5	64	.250	Ea.	19.25	13.15		32.40	41
2243	For snap-on inlet grille, add					10%	10%			
2245	For hinged access door, add	Q-5	64	.250	Ea.	35.50	13.15		48.65	59
2246	For air chamber, auto-venting, add	"	58	.276	"	6.85	14.50		21.35	29.50

23 82 36 – Finned-Tube Radiation Heaters

23 82 36.10 Finned Tube Radiation

		Crew	Daily Output	Labor-Hours	Unit	Material	Labor	Equipment	Total	Total Incl O&P
0010	**FINNED TUBE RADIATION**, Terminal units, not incl. main supply pipe									
1310	Baseboard, pkgd, 1/2" copper tube, alum. fin, 7" high	Q-5	60	.267	L.F.	11.20	14.05		25.25	33.50
1320	3/4" copper tube, alum. fin, 7" high	"	58	.276	"	7.35	14.50		21.85	30
1381	Rough in baseboard panel & fin tube, supply & balance valves	1 Stpi	1.06	7.547	Ea.	325	440		765	1,025
1500	Note: fin tube may also require corners, caps, etc.									

23 83 Radiant Heating Units

23 83 16 – Radiant-Heating Hydronic Piping

23 83 16.10 Radiant Floor Heating

		Crew	Daily Output	Labor-Hours	Unit	Material	Labor	Equipment	Total	Total Incl O&P
0010	**RADIANT FLOOR HEATING**									
0100	Tubing, PEX (cross-linked polyethylene)									
0110	Oxygen barrier type for systems with ferrous materials									
0120	1/2"	Q-5	800	.020	L.F.	1.02	1.05		2.07	2.71
0130	3/4"		535	.030		1.44	1.57		3.01	3.96
0140	1"	↓	400	.040	↓	2.25	2.11		4.36	5.65
0200	Non barrier type for ferrous free systems									
0210	1/2"	Q-5	800	.020	L.F.	.58	1.05		1.63	2.23
0220	3/4"		535	.030		1.05	1.57		2.62	3.54
0230	1"	↓	400	.040	↓	1.46	2.11		3.57	4.79
1000	Manifolds									
1110	Brass									

23 83 Radiant Heating Units

23 83 16 – Radiant-Heating Hydronic Piping

23 83 16.10 Radiant Floor Heating		Crew	Daily Output	Labor-Hours	Unit	Material	2014 Bare Costs Labor	Equipment	Total	Total Incl O&P
1120	With supply and return valves, flow meter, thermometer,									
1122	auto air vent and drain/fill valve.									
1130	1", 2 circuit	Q-5	14	1.143	Ea.	245	60		305	360
1140	1", 3 circuit		13.50	1.185		280	62.50		342.50	405
1150	1", 4 circuit		13	1.231		305	65		370	435
1154	1", 5 circuit		12.50	1.280		365	67.50		432.50	500
1158	1", 6 circuit		12	1.333		395	70		465	540
1162	1", 7 circuit		11.50	1.391		430	73.50		503.50	585
1166	1", 8 circuit		11	1.455		480	76.50		556.50	640
1172	1", 9 circuit		10.50	1.524		515	80		595	685
1174	1", 10 circuit		10	1.600		555	84		639	735
1178	1", 11 circuit		9.50	1.684		575	88.50		663.50	770
1182	1", 12 circuit		9	1.778		635	93.50		728.50	840
1610	Copper manifold header, (cut to size)									
1620	1" header, 12 – 1/2" sweat outlets	Q-5	3.33	4.805	Ea.	91	253		344	480
1630	1-1/4" header, 12 – 1/2" sweat outlets		3.20	5		106	263		369	515
1640	1-1/4" header, 12 – 3/4" sweat outlets		3	5.333		114	281		395	550
1650	1-1/2" header, 12 – 3/4" sweat outlets		3.10	5.161		139	272		411	565
1660	2" header, 12 – 3/4" sweat outlets		2.90	5.517		201	290		491	660
3000	Valves									
3110	Thermostatic zone valve actuator with end switch	Q-5	40	.400	Ea.	38.50	21		59.50	74.50
3114	Thermostatic zone valve actuator	"	36	.444	"	81	23.50		104.50	125
3120	Motorized straight zone valve with operator complete									
3130	3/4"	Q-5	35	.457	Ea.	130	24		154	180
3140	1"		32	.500		141	26.50		167.50	195
3150	1-1/4"		29.60	.541		179	28.50		207.50	240
3500	4 Way mixing valve, manual, brass									
3530	1"	Q-5	13.30	1.203	Ea.	180	63.50		243.50	294
3540	1-1/4"		11.40	1.404		195	74		269	325
3550	1-1/2"		11	1.455		249	76.50		325.50	390
3560	2"		10.60	1.509		350	79.50		429.50	510
3800	Mixing valve motor, 4 way for valves, 1" and 1-1/4"		34	.471		310	25		335	380
3810	Mixing valve motor, 4 way for valves, 1-1/2" and 2"		30	.533		355	28		383	435
5000	Radiant floor heating, zone control panel									
5120	4 Zone actuator valve control, expandable	Q-5	20	.800	Ea.	163	42		205	243
5130	6 Zone actuator valve control, expandable		18	.889		226	47		273	320
6070	Thermal track, straight panel for long continuous runs, 5.333 S.F.		40	.400		26.50	21		47.50	61.50
6080	Thermal track, utility panel, for direction reverse at run end, 5.333 S.F.		40	.400		26.50	21		47.50	61.50
6090	Combination panel, for direction reverse plus straight run, 5.333 S.F.		40	.400		26.50	21		47.50	61.50
7000	PEX tubing fittings									
7100	Compression type									
7116	Coupling									
7120	1/2" x 1/2"	1 Stpi	27	.296	Ea.	6.70	17.35		24.05	33.50
7124	3/4" x 3/4"	"	23	.348	"	10.60	20.50		31.10	42
7130	Adapter									
7132	1/2" x female sweat 1/2"	1 Stpi	27	.296	Ea.	4.36	17.35		21.71	31
7134	1/2" x female sweat 3/4"		26	.308		4.88	18		22.88	32.50
7136	5/8" x female sweat 3/4"		24	.333		7	19.50		26.50	37
7140	Elbow									
7142	1/2" x female sweat 1/2"	1 Stpi	27	.296	Ea.	6.65	17.35		24	33.50
7144	1/2" x female sweat 3/4"		26	.308		7.80	18		25.80	35.50
7146	5/8" x female sweat 3/4"		24	.333		8.75	19.50		28.25	39
7200	Insert type									

23 83 Radiant Heating Units

23 83 16 – Radiant-Heating Hydronic Piping

23 83 16.10 Radiant Floor Heating		Crew	Daily Output	Labor-Hours	Unit	Material	2014 Bare Costs Labor	Equipment	Total	Total Incl O&P
7206	PEX x male NPT									
7210	1/2" x 1/2"	1 Stpi	29	.276	Ea.	2.59	16.15		18.74	27.50
7220	3/4" x 3/4"		27	.296		3.81	17.35		21.16	30
7230	1" x 1"	↓	26	.308	↓	6.45	18		24.45	34
7300	PEX coupling									
7310	1/2" x 1/2"	1 Stpi	30	.267	Ea.	1.76	15.60		17.36	25.50
7320	3/4" x 3/4"		29	.276		2.15	16.15		18.30	27
7330	1" x 1"	↓	28	.286	↓	5.85	16.70		22.55	31.50
7400	PEX stainless crimp ring									
7410	1/2" x 1/2"	1 Stpi	86	.093	Ea.	.34	5.45		5.79	8.55
7420	3/4" x 3/4"		84	.095		.47	5.55		6.02	8.90
7430	1" x 1"	↓	82	.098	↓	.68	5.70		6.38	9.35

23 83 33 – Electric Radiant Heaters

23 83 33.10 Electric Heating

		Crew	Daily Output	Labor-Hours	Unit	Material	Labor	Equipment	Total	Total Incl O&P
0010	**ELECTRIC HEATING**, not incl. conduit or feed wiring									
1100	Rule of thumb: Baseboard units, including control	1 Elec	4.40	1.818	kW	101	97		198	256
1300	Baseboard heaters, 2' long, 350 watt		8	1	Ea.	30	53.50		83.50	113
1400	3' long, 750 watt		8	1		34.50	53.50		88	118
1600	4' long, 1000 watt		6.70	1.194		41	63.50		104.50	141
1800	5' long, 935 watt		5.70	1.404		51	75		126	168
2000	6' long, 1500 watt		5	1.600		56	85.50		141.50	190
2200	7' long, 1310 watt		4.40	1.818		64	97		161	216
2400	8' long, 2000 watt		4	2		69.50	107		176.50	236
2600	9' long, 1680 watt		3.60	2.222		104	119		223	291
2800	10' long, 1875 watt	↓	3.30	2.424	↓	193	129		322	405
2950	Wall heaters with fan, 120 to 277 volt									
3160	Recessed, residential, 750 watt	1 Elec	6	1.333	Ea.	134	71		205	253
3170	1000 watt		6	1.333		134	71		205	253
3180	1250 watt		5	1.600		134	85.50		219.50	275
3190	1500 watt		4	2		134	107		241	305
3600	Thermostats, integral		16	.500		30	26.50		56.50	73
3800	Line voltage, 1 pole		8	1		20.50	53.50		74	103
3810	2 pole	↓	8	1	↓	29.50	53.50		83	113
4000	Heat trace system, 400 degree									
4020	115 V, 2.5 watts per L.F.	1 Elec	530	.015	L.F.	7.90	.81		8.71	9.85
4030	5 watts per L.F.		530	.015		7.90	.81		8.71	9.85
4050	10 watts per L.F.		530	.015		7.90	.81		8.71	9.85
4060	208 V, 5 watts per L.F.		530	.015		7.90	.81		8.71	9.85
4080	480 V, 8 watts per L.F.	↓	530	.015	↓	7.90	.81		8.71	9.85
4200	Heater raceway									
4260	Heat transfer cement									
4280	1 gallon				Ea.	57			57	62.50
4300	5 gallon				"	220			220	242
4320	Snap band, clamp									
4340	3/4" pipe size	1 Elec	470	.017	Ea.		.91		.91	1.36
4360	1" pipe size		444	.018			.96		.96	1.44
4380	1-1/4" pipe size		400	.020			1.07		1.07	1.60
4400	1-1/2" pipe size		355	.023			1.20		1.20	1.80
4420	2" pipe size		320	.025			1.33		1.33	2
4440	3" pipe size		160	.050			2.67		2.67	3.99
4460	4" pipe size		100	.080			4.27		4.27	6.40
4480	Thermostat NEMA 3R, 22 amp, 0-150 Deg, 10' cap.	↓	8	1	↓	241	53.50		294.50	345

23 83 Radiant Heating Units

23 83 33 − Electric Radiant Heaters

23 83 33.10 Electric Heating

		Crew	Daily Output	Labor-Hours	Unit	Material	2014 Bare Costs Labor	Equipment	Total	Total Incl O&P
4500	Thermostat NEMA 4X, 25 amp, 40 Deg, 5-1/2' cap.	1 Elec	7	1.143	Ea.	241	61		302	355
4520	Thermostat NEMA 4X, 22 amp, 25-325 Deg, 10' cap.		7	1.143		650	61		711	805
4540	Thermostat NEMA 4X, 22 amp, 15-140 Deg,		6	1.333		545	71		616	705
4580	Thermostat NEMA 4,7,9, 22 amp, 25-325 Deg, 10' cap.		3.60	2.222		800	119		919	1,050
4600	Thermostat NEMA 4,7,9, 22 amp, 15-140 Deg,		3	2.667		795	142		937	1,075
4720	Fiberglass application tape, 36 yard roll		11	.727		77.50	39		116.50	144
5000	Radiant heating ceiling panels, 2' x 4', 500 watt		16	.500		291	26.50		317.50	360
5050	750 watt		16	.500		310	26.50		336.50	385
5200	For recessed plaster frame, add		32	.250		110	13.35		123.35	141
5300	Infrared quartz heaters, 120 volts, 1000 watts		6.70	1.194		230	63.50		293.50	350
5350	1500 watt		5	1.600		230	85.50		315.50	380
5400	240 volts, 1500 watt		5	1.600		230	85.50		315.50	380
5450	2000 watt		4	2		230	107		337	415
5500	3000 watt		3	2.667		255	142		397	495
5550	4000 watt		2.60	3.077		255	164		419	525
5570	Modulating control		.80	10		275	535		810	1,100
5600	Unit heaters, heavy duty, with fan & mounting bracket									
5650	Single phase, 208-240-277 volt, 3 kW	1 Elec	6	1.333	Ea.	435	71		506	585
5750	5 kW		5.50	1.455		460	77.50		537.50	620
5800	7 kW		5	1.600		710	85.50		795.50	910
5850	10 kW		4	2		810	107		917	1,050
5950	15 kW		3.80	2.105		1,300	112		1,412	1,600
6000	480 volt, 3 kW		6	1.333		480	71		551	635
6020	4 kW		5.80	1.379		490	73.50		563.50	650
6040	5 kW		5.50	1.455		510	77.50		587.50	675
6060	7 kW		5	1.600		755	85.50		840.50	965
6080	10 kW		4	2		845	107		952	1,100
6100	13 kW		3.80	2.105		1,300	112		1,412	1,600
6120	15 kW		3.70	2.162		1,300	115		1,415	1,600
6140	20 kW		3.50	2.286		1,700	122		1,822	2,025
6300	3 phase, 208-240 volt, 5 kW		5.50	1.455		430	77.50		507.50	590
6320	7 kW		5	1.600		670	85.50		755.50	865
6340	10 kW		4	2		750	107		857	985
6360	15 kW		3.70	2.162		1,375	115		1,490	1,675
6380	20 kW		3.50	2.286		1,725	122		1,847	2,075
6400	25 kW		3.30	2.424		2,025	129		2,154	2,425
6500	480 volt, 5 kW		5.50	1.455		605	77.50		682.50	780
6520	7 kW		5	1.600		765	85.50		850.50	970
6540	10 kW		4	2		810	107		917	1,050
6560	13 kW		3.80	2.105		1,475	112		1,587	1,800
6580	15 kW		3.70	2.162		1,475	115		1,590	1,800
6600	20 kW		3.50	2.286		1,925	122		2,047	2,275
6800	Vertical discharge heaters, with fan									
6820	Single phase, 208-240-277 volt, 10 kW	1 Elec	4	2	Ea.	750	107		857	985
6840	15 kW		3.70	2.162		1,250	115		1,365	1,550
6900	3 phase, 208-240 volt, 10 kW		4	2		750	107		857	985
6920	15 kW		3.70	2.162		1,250	115		1,365	1,550
6940	20 kW		3.50	2.286		1,775	122		1,897	2,150
7100	480 volt, 10 kW		4	2		810	107		917	1,050
7120	15 kW		3.70	2.162		1,300	115		1,415	1,600
7140	20 kW		3.50	2.286		1,700	122		1,822	2,025
7160	25 kW		3.30	2.424		2,025	129		2,154	2,425
7900	Cabinet convector heaters, 240 volt									

23 83 Radiant Heating Units

23 83 33 – Electric Radiant Heaters

23 83 33.10 Electric Heating		Crew	Daily Output	Labor-Hours	Unit	Material	2014 Bare Costs Labor	Equipment	Total	Total Incl O&P
7920	3' long, 2000 watt	1 Elec	5.30	1.509	Ea.	1,975	80.50		2,055.50	2,300
7940	3000 watt		5.30	1.509		2,075	80.50		2,155.50	2,400
7960	4000 watt		5.30	1.509		2,125	80.50		2,205.50	2,450
7980	6000 watt		4.60	1.739		2,200	93		2,293	2,575
8000	8000 watt		4.60	1.739		2,300	93		2,393	2,675
8020	4' long, 4000 watt		4.60	1.739		2,150	93		2,243	2,525
8040	6000 watt		4	2		2,225	107		2,332	2,600
8060	8000 watt		4	2		2,325	107		2,432	2,700
8080	10,000 watt		4	2		2,325	107		2,432	2,725
8100	Available also in 208 or 277 volt									
8200	Cabinet unit heaters, 120 to 277 volt, 1 pole,									
8220	wall mounted, 2 kW	1 Elec	4.60	1.739	Ea.	1,925	93		2,018	2,250
8230	3 kW		4.60	1.739		1,950	93		2,043	2,300
8240	4 kW		4.40	1.818		2,000	97		2,097	2,350
8250	5 kW		4.40	1.818		2,075	97		2,172	2,425
8260	6 kW		4.20	1.905		2,100	102		2,202	2,450
8270	8 kW		4	2		2,175	107		2,282	2,550
8280	10 kW		3.80	2.105		2,225	112		2,337	2,600
8290	12 kW		3.50	2.286		2,400	122		2,522	2,825
8300	13.5 kW		2.90	2.759		2,450	147		2,597	2,900
8310	16 kW		2.70	2.963		2,500	158		2,658	2,975
8320	20 kW		2.30	3.478		3,625	186		3,811	4,250
8330	24 kW		1.90	4.211		3,675	225		3,900	4,375
8350	Recessed, 2 kW		4.40	1.818		1,875	97		1,972	2,700
8370	3 kW		4.40	1.818		1,950	97		2,047	2,275
8380	4 kW		4.20	1.905		1,950	102		2,052	2,300
8390	5 kW		4.20	1.905		2,175	102		2,277	2,550
8400	6 kW		4	2		2,275	107		2,382	2,675
8410	8 kW		3.80	2.105		2,275	112		2,387	2,700
8420	10 kW		3.50	2.286		2,800	122		2,922	3,250
8430	12 kW		2.90	2.759		2,800	147		2,947	3,325
8440	13.5 kW		2.70	2.963		2,975	158		3,133	3,500
8450	16 kW		2.30	3.478		3,000	186		3,186	3,575
8460	20 kW		1.90	4.211		3,025	225		3,250	3,650
8470	24 kW		1.60	5		3,600	267		3,867	4,350
8490	Ceiling mounted, 2 kW		3.20	2.500		1,950	133		2,083	2,325
8510	3 kW		3.20	2.500		2,025	133		2,158	2,425
8520	4 kW		3	2.667		2,050	142		2,192	2,475
8530	5 kW		3	2.667		2,150	142		2,292	2,600
8540	6 kW		2.80	2.857		2,175	152		2,327	2,600
8550	8 kW		2.40	3.333		2,250	178		2,428	2,750
8560	10 kW		2.20	3.636		2,225	194		2,419	2,725
8570	12 kW		2	4		2,325	213		2,538	2,900
8580	13.5 kW		1.50	5.333		2,375	285		2,660	3,025
8590	16 kW		1.30	6.154		2,500	330		2,830	3,250
8600	20 kW		.90	8.889		3,500	475		3,975	4,550
8610	24 kW		.60	13.333		3,600	710		4,310	5,025
8630	208 to 480 V, 3 pole									
8650	Wall mounted, 2 kW	1 Elec	4.60	1.739	Ea.	2,075	93		2,168	2,450
8670	3 kW		4.60	1.739		2,175	93		2,268	2,525
8680	4 kW		4.40	1.818		2,225	97		2,322	2,600
8690	5 kW		4.40	1.818		2,300	97		2,397	2,675
8700	6 kW		4.20	1.905		2,300	102		2,402	2,675

23 83 Radiant Heating Units

23 83 33 – Electric Radiant Heaters

23 83 33.10 Electric Heating

		Crew	Daily Output	Labor-Hours	Unit	Material	2014 Bare Costs Labor	Equipment	Total	Total Incl O&P
8710	8 kW	1 Elec	4	2	Ea.	2,400	107		2,507	2,775
8720	10 kW		3.80	2.105		2,425	112		2,537	2,850
8730	12 kW		3.50	2.286		2,475	122		2,597	2,900
8740	13.5 kW		2.90	2.759		2,475	147		2,622	2,950
8750	16 kW		2.70	2.963		2,500	158		2,658	2,975
8760	20 kW		2.30	3.478		3,200	186		3,386	3,775
8770	24 kW		1.90	4.211		3,750	225		3,975	4,450
8790	Recessed, 2 kW		4.40	1.818		2,075	97		2,172	2,450
8810	3 kW		4.40	1.818		2,175	97		2,272	2,525
8820	4 kW		4.20	1.905		2,250	102		2,352	2,625
8830	5 kW		4.20	1.905		2,300	102		2,402	2,675
8840	6 kW		4	2		2,300	107		2,407	2,675
8850	8 kW		3.80	2.105		2,400	112		2,512	2,800
8860	10 kW		3.50	2.286		2,425	122		2,547	2,850
8870	12 kW		2.90	2.759		2,475	147		2,622	2,950
8880	13.5 kW		2.70	2.963		2,525	158		2,683	3,000
8890	16 kW		2.30	3.478		3,650	186		3,836	4,275
8900	20 kW		1.90	4.211		3,750	225		3,975	4,450
8920	24 kW		1.60	5		3,800	267		4,067	4,575
8940	Ceiling mount, 2 kW		3.20	2.500		2,075	133		2,208	2,500
8950	3 kW		3.20	2.500		2,175	133		2,308	2,575
8960	4 kW		3	2.667		2,225	142		2,367	2,675
8970	5 kW		3	2.667		2,300	142		2,442	2,750
8980	6 kW		2.80	2.857		2,300	152		2,452	2,750
8990	8 kW		2.40	3.333		2,400	178		2,578	2,900
9000	10 kW		2.20	3.636		2,425	194		2,619	2,975
9020	13.5 kW		1.50	5.333		2,525	285		2,810	3,200
9030	16 kW		1.30	6.154		3,700	330		4,030	4,575
9040	20 kW		.90	8.889		3,700	475		4,175	4,775
9060	24 kW		.60	13.333		3,725	710		4,435	5,175

23 91 Prefabricated Equipment Supports

23 91 10 – Prefabricated Curbs, Pads and Stands

23 91 10.10 Prefabricated Pads and Stands

		Crew	Daily Output	Labor-Hours	Unit	Material	Labor	Equipment	Total	Total Incl O&P
0010	**PREFABRICATED PADS AND STANDS**									
6000	Pad, fiberglass reinforced concrete with polystyrene foam core									
6050	Condenser, 2" thick, 20" x 38"	1 Shee	8	1	Ea.	26.50	54.50		81	113
6220	30" x 36"	"	8	1		41.50	54.50		96	129
6340	36" x 54"	Q-9	6	2.667		77.50	131		208.50	286

Division 26 - Electrical

Estimating Tips

26 05 00 Common Work Results for Electrical

- Conduit should be taken off in three main categories—power distribution, branch power, and branch lighting—so the estimator can concentrate on systems and components, therefore making it easier to ensure all items have been accounted for.
- For cost modifications for elevated conduit installation, add the percentages to labor according to the height of installation, and only to the quantities exceeding the different height levels, not to the total conduit quantities.
- Remember that aluminum wiring of equal ampacity is larger in diameter than copper and may require larger conduit.
- If more than three wires at a time are being pulled, deduct percentages from the labor hours of that grouping of wires.
- When taking off grounding systems, identify separately the type and size of wire, and list each unique type of ground connection.
- The estimator should take the weights of materials into consideration when completing a takeoff. Topics to consider include: How will the materials be supported? What methods of support are available? How high will the support structure have to reach? Will the final support structure be able to withstand the total burden? Is the support material included or separate from the fixture, equipment, and material specified?
- Do not overlook the costs for equipment used in the installation. If scaffolding or highlifts are available in the field, contractors may use them in lieu of the proposed ladders and rolling staging.

26 20 00 Low-Voltage Electrical Transmission

- Supports and concrete pads may be shown on drawings for the larger equipment, or the support system may be only a piece of plywood for the back of a panelboard. In either case, it must be included in the costs.

26 40 00 Electrical and Cathodic Protection

- When taking off cathodic protections systems, identify the type and size of cable, and list each unique type of anode connection.

26 50 00 Lighting

- Fixtures should be taken off room by room, using the fixture schedule, specifications, and the ceiling plan. For large concentrations of lighting fixtures in the same area, deduct the percentages from labor hours.

Reference Numbers

Reference numbers are shown in shaded boxes at the beginning of some major classifications. These numbers refer to related items in the Reference Section. The reference information may be an estimating procedure, an alternate pricing method, or technical information.

Note: Not all subdivisions listed here necessarily appear in this publication.

Note: **Trade Service,** *in part, has been used as a reference source for some of the material prices used in Division 26.*

26 05 Common Work Results for Electrical

26 05 33 – Raceway and Boxes for Electrical Systems

26 05 33.95 Cutting and Drilling

		Crew	Daily Output	Labor-Hours	Unit	Material	2014 Bare Costs Labor	Equipment	Total	Total Incl O&P
0010	**CUTTING AND DRILLING**									
0100	Hole drilling to 10' high, concrete wall									
0110	8" thick, 1/2" pipe size	R-31	12	.667	Ea.	.24	35.50	3.85	39.59	57.50
0120	3/4" pipe size		12	.667		.24	35.50	3.85	39.59	57.50
0130	1" pipe size		9.50	.842		.38	45	4.87	50.25	73
0140	1-1/4" pipe size		9.50	.842		.38	45	4.87	50.25	73
0150	1-1/2" pipe size		9.50	.842		.38	45	4.87	50.25	73
0160	2" pipe size		4.40	1.818		.52	97	10.50	108.02	157
0170	2-1/2" pipe size		4.40	1.818		.52	97	10.50	108.02	157
0180	3" pipe size		4.40	1.818		.52	97	10.50	108.02	157
0190	3-1/2" pipe size		3.30	2.424		.66	129	14	143.66	210
0200	4" pipe size		3.30	2.424		.66	129	14	143.66	210
0500	12" thick, 1/2" pipe size		9.40	.851		.36	45.50	4.92	50.78	74
0520	3/4" pipe size		9.40	.851		.36	45.50	4.92	50.78	74
0540	1" pipe size		7.30	1.096		.58	58.50	6.35	65.43	95
0560	1-1/4" pipe size		7.30	1.096		.58	58.50	6.35	65.43	95
0570	1-1/2" pipe size		7.30	1.096		.58	58.50	6.35	65.43	95
0580	2" pipe size		3.60	2.222		.78	119	12.85	132.63	192
0590	2-1/2" pipe size		3.60	2.222		.78	119	12.85	132.63	192
0600	3" pipe size		3.60	2.222		.78	119	12.85	132.63	192
0610	3-1/2" pipe size		2.80	2.857		.99	152	16.50	169.49	247
0630	4" pipe size		2.50	3.200		.99	171	18.50	190.49	278
0650	16" thick, 1/2" pipe size		7.60	1.053		.48	56	6.10	62.58	91
0670	3/4" pipe size		7	1.143		.48	61	6.60	68.08	99.50
0690	1" pipe size		6	1.333		.77	71	7.70	79.47	115
0710	1-1/4" pipe size		5.50	1.455		.77	77.50	8.40	86.67	126
0730	1-1/2" pipe size		5.50	1.455		.77	77.50	8.40	86.67	126
0750	2" pipe size		3	2.667		1.05	142	15.40	158.45	231
0770	2-1/2" pipe size		2.70	2.963		1.05	158	17.15	176.20	257
0790	3" pipe size		2.50	3.200		1.05	171	18.50	190.55	278
0810	3-1/2" pipe size		2.30	3.478		1.32	186	20	207.32	300
0830	4" pipe size		2	4		1.32	213	23	237.32	345
0850	20" thick, 1/2" pipe size		6.40	1.250		.60	66.50	7.25	74.35	109
0870	3/4" pipe size		6	1.333		.60	71	7.70	79.30	115
0890	1" pipe size		5	1.600		.96	85.50	9.25	95.71	139
0910	1-1/4" pipe size		4.80	1.667		.96	89	9.65	99.61	145
0930	1-1/2" pipe size		4.60	1.739		.96	93	10.05	104.01	151
0950	2" pipe size		2.70	2.963		1.31	158	17.15	176.46	257
0970	2-1/2" pipe size		2.40	3.333		1.31	178	19.25	198.56	288
0990	3" pipe size		2.20	3.636		1.31	194	21	216.31	315
1010	3-1/2" pipe size		2	4		1.66	213	23	237.66	345
1030	4" pipe size		1.70	4.706		1.66	251	27	279.66	405
1050	24" thick, 1/2" pipe size		5.50	1.455		.72	77.50	8.40	86.62	126
1070	3/4" pipe size		5.10	1.569		.72	83.50	9.05	93.27	136
1090	1" pipe size		4.30	1.860		1.15	99.50	10.75	111.40	162
1110	1-1/4" pipe size		4	2		1.15	107	11.55	119.70	174
1130	1-1/2" pipe size		4	2		1.15	107	11.55	119.70	174
1150	2" pipe size		2.40	3.333		1.57	178	19.25	198.82	289
1170	2-1/2" pipe size		2.20	3.636		1.57	194	21	216.57	315
1190	3" pipe size		2	4		1.57	213	23	237.57	345
1210	3-1/2" pipe size		1.80	4.444		1.99	237	25.50	264.49	385
1230	4" pipe size		1.50	5.333		1.99	285	31	317.99	460
1500	Brick wall, 8" thick, 1/2" pipe size		18	.444		.24	23.50	2.57	26.31	38.50

26 05 Common Work Results for Electrical

26 05 33 – Raceway and Boxes for Electrical Systems

26 05 33.95 Cutting and Drilling		Crew	Daily Output	Labor-Hours	Unit	Material	2014 Bare Costs Labor	Equipment	Total	Total Incl O&P
1520	3/4" pipe size	R-31	18	.444	Ea.	.24	23.50	2.57	26.31	38.50
1540	1" pipe size		13.30	.602		.38	32	3.48	35.86	52
1560	1-1/4" pipe size		13.30	.602		.38	32	3.48	35.86	52
1580	1-1/2" pipe size		13.30	.602		.38	32	3.48	35.86	52
1600	2" pipe size		5.70	1.404		.52	75	8.10	83.62	121
1620	2-1/2" pipe size		5.70	1.404		.52	75	8.10	83.62	121
1640	3" pipe size		5.70	1.404		.52	75	8.10	83.62	121
1660	3-1/2" pipe size		4.40	1.818		.66	97	10.50	108.16	157
1680	4" pipe size		4	2		.66	107	11.55	119.21	173
1700	12" thick, 1/2" pipe size		14.50	.552		.36	29.50	3.19	33.05	48
1720	3/4" pipe size		14.50	.552		.36	29.50	3.19	33.05	48
1740	1" pipe size		11	.727		.58	39	4.20	43.78	63.50
1760	1-1/4" pipe size		11	.727		.58	39	4.20	43.78	63.50
1780	1-1/2" pipe size		11	.727		.58	39	4.20	43.78	63.50
1800	2" pipe size		5	1.600		.78	85.50	9.25	95.53	139
1820	2-1/2" pipe size		5	1.600		.78	85.50	9.25	95.53	139
1840	3" pipe size		5	1.600		.78	85.50	9.25	95.53	139
1860	3-1/2" pipe size		3.80	2.105		.99	112	12.15	125.14	182
1880	4" pipe size		3.30	2.424		.99	129	14	143.99	210
1900	16" thick, 1/2" pipe size		12.30	.650		.48	34.50	3.76	38.74	56.50
1920	3/4" pipe size		12.30	.650		.48	34.50	3.76	38.74	56.50
1940	1" pipe size		9.30	.860		.77	46	4.97	51.74	75
1960	1-1/4" pipe size		9.30	.860		.77	46	4.97	51.74	75
1980	1-1/2" pipe size		9.30	.860		.77	46	4.97	51.74	75
2000	2" pipe size		4.40	1.818		1.05	97	10.50	108.55	158
2010	2-1/2" pipe size		4.40	1.818		1.05	97	10.50	108.55	158
2030	3" pipe size		4.40	1.818		1.05	97	10.50	108.55	158
2050	3-1/2" pipe size		3.30	2.424		1.32	129	14	144.32	211
2070	4" pipe size		3	2.667		1.32	142	15.40	158.72	231
2090	20" thick, 1/2" pipe size		10.70	.748		.60	40	4.32	44.92	65
2110	3/4" pipe size		10.70	.748		.60	40	4.32	44.92	65
2130	1" pipe size		8	1		.96	53.50	5.80	60.26	87.50
2150	1-1/4" pipe size		8	1		.96	53.50	5.80	60.26	87.50
2170	1-1/2" pipe size		8	1		.96	53.50	5.80	60.26	87.50
2190	2" pipe size		4	2		1.31	107	11.55	119.86	174
2210	2-1/2" pipe size		4	2		1.31	107	11.55	119.86	174
2230	3" pipe size		4	2		1.31	107	11.55	119.86	174
2250	3-1/2" pipe size		3	2.667		1.66	142	15.40	159.06	232
2270	4" pipe size		2.70	2.963		1.66	158	17.15	176.81	258
2290	24" thick, 1/2" pipe size		9.40	.851		.72	45.50	4.92	51.14	74
2310	3/4" pipe size		9.40	.851		.72	45.50	4.92	51.14	74
2330	1" pipe size		7.10	1.127		1.15	60	6.50	67.65	98.50
2350	1-1/4" pipe size		7.10	1.127		1.15	60	6.50	67.65	98.50
2370	1-1/2" pipe size		7.10	1.127		1.15	60	6.50	67.65	98.50
2390	2" pipe size		3.60	2.222		1.57	119	12.85	133.42	193
2410	2-1/2" pipe size		3.60	2.222		1.57	119	12.85	133.42	193
2430	3" pipe size		3.60	2.222		1.57	119	12.85	133.42	193
2450	3-1/2" pipe size		2.80	2.857		1.99	152	16.50	170.49	248
2470	4" pipe size		2.50	3.200		1.99	171	18.50	191.49	279
3000	Knockouts to 8' high, metal boxes & enclosures									
3020	With hole saw, 1/2" pipe size	1 Elec	53	.151	Ea.		8.05		8.05	12.05
3040	3/4" pipe size		47	.170			9.10		9.10	13.60
3050	1" pipe size		40	.200			10.65		10.65	15.95

355

26 05 Common Work Results for Electrical

26 05 33 – Raceway and Boxes for Electrical Systems

26 05 33.95 Cutting and Drilling

		Crew	Daily Output	Labor-Hours	Unit	Material	2014 Bare Costs Labor	Equipment	Total	Total Incl O&P
3060	1-1/4" pipe size	1 Elec	36	.222	Ea.		11.85		11.85	17.75
3070	1-1/2" pipe size		32	.250			13.35		13.35	19.95
3080	2" pipe size		27	.296			15.80		15.80	23.50
3090	2-1/2" pipe size		20	.400			21.50		21.50	32
4010	3" pipe size		16	.500			26.50		26.50	40
4030	3-1/2" pipe size		13	.615			33		33	49
4050	4" pipe size		11	.727			39		39	58
4070	With hand punch set, 1/2" pipe size		40	.200			10.65		10.65	15.95
4090	3/4" pipe size		32	.250			13.35		13.35	19.95
4110	1" pipe size		30	.267			14.25		14.25	21.50
4130	1-1/4" pipe size		28	.286			15.25		15.25	23
4150	1-1/2" pipe size		26	.308			16.40		16.40	24.50
4170	2" pipe size		20	.400			21.50		21.50	32
4190	2-1/2" pipe size		17	.471			25		25	37.50
4200	3" pipe size		15	.533			28.50		28.50	42.50
4220	3-1/2" pipe size		12	.667			35.50		35.50	53
4240	4" pipe size		10	.800			42.50		42.50	64
4260	With hydraulic punch, 1/2" pipe size		44	.182			9.70		9.70	14.50
4280	3/4" pipe size		38	.211			11.25		11.25	16.80
4300	1" pipe size		38	.211			11.25		11.25	16.80
4320	1-1/4" pipe size		38	.211			11.25		11.25	16.80
4340	1-1/2" pipe size		38	.211			11.25		11.25	16.80
4360	2" pipe size		32	.250			13.35		13.35	19.95
4380	2-1/2" pipe size		27	.296			15.80		15.80	23.50
4400	3" pipe size		23	.348			18.55		18.55	28
4420	3-1/2" pipe size		20	.400			21.50		21.50	32
4440	4" pipe size		18	.444			23.50		23.50	35.50

26 05 80 – Wiring Connections

26 05 80.10 Motor Connections

		Crew	Daily Output	Labor-Hours	Unit	Material	Labor	Equipment	Total	Total Incl O&P
0010	**MOTOR CONNECTIONS**									
0020	Flexible conduit and fittings, 115 volt, 1 phase, up to 1 HP motor	1 Elec	8	1	Ea.	7.25	53.50		60.75	88
0050	2 HP motor		6.50	1.231		13.75	65.50		79.25	114
0100	3 HP motor		5.50	1.455		12.25	77.50		89.75	129
0110	230 volt, 3 phase, 3 HP motor		6.78	1.180		8.70	63		71.70	104
0112	5 HP motor		5.47	1.463		7.40	78		85.40	125
0114	7-1/2 HP motor		4.61	1.735		10.50	92.50		103	151
0120	10 HP motor		4.20	1.905		23.50	102		125.50	178
0150	15 HP motor		3.30	2.424		23.50	129		152.50	220
0200	25 HP motor		2.70	2.963		33.50	158		191.50	274
0400	50 HP motor		2.20	3.636		67	194		261	365
0600	100 HP motor		1.50	5.333		158	285		443	600

26 05 90 – Residential Applications

26 05 90.10 Residential Wiring

		Crew	Daily Output	Labor-Hours	Unit	Material	Labor	Equipment	Total	Total Incl O&P
0010	**RESIDENTIAL WIRING**									
0020	20' avg. runs and #14/2 wiring incl. unless otherwise noted									
1000	Service & panel, includes 24' SE-AL cable, service eye, meter,									
1010	Socket, panel board, main bkr., ground rod, 15 or 20 amp									
1020	1-pole circuit breakers, and misc. hardware									
1100	100 amp, with 10 branch breakers	1 Elec	1.19	6.723	Ea.	540	360		900	1,125
1110	With PVC conduit and wire		.92	8.696		595	465		1,060	1,350
1120	With RGS conduit and wire		.73	10.959		740	585		1,325	1,700
1150	150 amp, with 14 branch breakers		1.03	7.767		840	415		1,255	1,550

26 05 Common Work Results for Electrical

26 05 90 – Residential Applications

26 05 90.10 Residential Wiring

		Crew	Daily Output	Labor-Hours	Unit	Material	2014 Bare Costs Labor	Equipment	Total	Total Incl O&P
1170	With PVC conduit and wire	1 Elec	.82	9.756	Ea.	955	520		1,475	1,825
1180	With RGS conduit and wire	↓	.67	11.940		1,225	635		1,860	2,300
1200	200 amp, with 18 branch breakers	2 Elec	1.80	8.889		1,125	475		1,600	1,925
1220	With PVC conduit and wire		1.46	10.959		1,225	585		1,810	2,225
1230	With RGS conduit and wire	↓	1.24	12.903		1,600	690		2,290	2,775
1800	Lightning surge suppressor	1 Elec	32	.250	↓	50.50	13.35		63.85	76
2000	Switch devices									
2100	Single pole, 15 amp, Ivory, with a 1-gang box, cover plate,									
2110	Type NM (Romex) cable	1 Elec	17.10	.468	Ea.	12.30	25		37.30	51
2120	Type MC (BX) cable		14.30	.559		23.50	30		53.50	70.50
2130	EMT & wire		5.71	1.401		29.50	75		104.50	145
2150	3-way, #14/3, type NM cable		14.55	.550		15.60	29.50		45.10	61
2170	Type MC cable		12.31	.650		31	34.50		65.50	86
2180	EMT & wire		5	1.600		32.50	85.50		118	164
2200	4-way, #14/3, type NM cable		14.55	.550		22.50	29.50		52	68.50
2220	Type MC cable		12.31	.650		37.50	34.50		72	93.50
2230	EMT & wire		5	1.600		39.50	85.50		125	172
2250	S.P., 20 amp, #12/2, type NM cable		13.33	.600		21	32		53	71.50
2270	Type MC cable		11.43	.700		30	37.50		67.50	89
2280	EMT & wire		4.85	1.649		40	88		128	176
2290	S.P. rotary dimmer, 600W, no wiring		17	.471		27	25		52	67
2300	S.P. rotary dimmer, 600W, type NM cable		14.55	.550		32	29.50		61.50	79
2320	Type MC cable		12.31	.650		43.50	34.50		78	99.50
2330	EMT & wire		5	1.600		51	85.50		136.50	184
2350	3-way rotary dimmer, type NM cable		13.33	.600		31.50	32		63.50	82.50
2370	Type MC cable		11.43	.700		42.50	37.50		80	103
2380	EMT & wire	↓	4.85	1.649	↓	50	88		138	187
2400	Interval timer wall switch, 20 amp, 1-30 min., #12/2									
2410	Type NM cable	1 Elec	14.55	.550	Ea.	54	29.50		83.50	104
2420	Type MC cable		12.31	.650		59	34.50		93.50	117
2430	EMT & wire	↓	5	1.600	↓	73	85.50		158.50	209
2500	Decorator style									
2510	S.P., 15 amp, type NM cable	1 Elec	17.10	.468	Ea.	16.15	25		41.15	55.50
2520	Type MC cable		14.30	.559		27.50	30		57.50	74.50
2530	EMT & wire		5.71	1.401		33.50	75		108.50	149
2550	3-way, #14/3, type NM cable		14.55	.550		19.40	29.50		48.90	65.50
2570	Type MC cable		12.31	.650		34.50	34.50		69	90
2580	EMT & wire		5	1.600		36.50	85.50		122	168
2600	4-way, #14/3, type NM cable		14.55	.550		26.50	29.50		56	73
2620	Type MC cable		12.31	.650		41.50	34.50		76	97.50
2630	EMT & wire		5	1.600		43	85.50		128.50	176
2650	S.P., 20 amp, #12/2, type NM cable		13.33	.600		25	32		57	75.50
2670	Type MC cable		11.43	.700		34	37.50		71.50	93
2680	EMT & wire		4.85	1.649		44	88		132	180
2700	S.P., slide dimmer, type NM cable		17.10	.468		32	25		57	73
2720	Type MC cable		14.30	.559		43.50	30		73.50	92.50
2730	EMT & wire		5.71	1.401		51	75		126	168
2750	S.P., touch dimmer, type NM cable		17.10	.468		29.50	25		54.50	69.50
2770	Type MC cable		14.30	.559		40.50	30		70.50	89
2780	EMT & wire		5.71	1.401		48	75		123	165
2800	3-way touch dimmer, type NM cable		13.33	.600		51.50	32		83.50	105
2820	Type MC cable		11.43	.700		63	37.50		100.50	125
2830	EMT & wire	↓	4.85	1.649	↓	70.50	88		158.50	210

357

26 05 Common Work Results for Electrical

26 05 90 – Residential Applications

26 05 90.10 Residential Wiring

		Crew	Daily Output	Labor-Hours	Unit	Material	2014 Bare Costs Labor	Equipment	Total	Total Incl O&P
3000	Combination devices									
3100	S.P. switch/15 amp recpt., Ivory, 1-gang box, plate									
3110	Type NM cable	1 Elec	11.43	.700	Ea.	24	37.50		61.50	82.50
3120	Type MC cable		10	.800		35.50	42.50		78	103
3130	EMT & wire		4.40	1.818		43	97		140	192
3150	S.P. switch/pilot light, type NM cable		11.43	.700		24	37.50		61.50	82.50
3170	Type MC cable		10	.800		35.50	42.50		78	103
3180	EMT & wire		4.43	1.806		43	96.50		139.50	192
3190	2-S.P. switches, 2-#14/2, no wiring		14	.571		6.95	30.50		37.45	53
3200	2-S.P. switches, 2-#14/2, type NM cables		10	.800		27.50	42.50		70	94.50
3220	Type MC cable		8.89	.900		44.50	48		92.50	121
3230	EMT & wire		4.10	1.951		46.50	104		150.50	207
3250	3-way switch/15 amp recpt., #14/3, type NM cable		10	.800		31.50	42.50		74	98.50
3270	Type MC cable		8.89	.900		46.50	48		94.50	123
3280	EMT & wire		4.10	1.951		48	104		152	209
3300	2-3 way switches, 2-#14/3, type NM cables		8.89	.900		41	48		89	117
3320	Type MC cable		8	1		66	53.50		119.50	153
3330	EMT & wire		4	2		56	107		163	222
3350	S.P. switch/20 amp recpt., #12/2, type NM cable		10	.800		32	42.50		74.50	99.50
3370	Type MC cable		8.89	.900		37	48		85	113
3380	EMT & wire	▼	4.10	1.951	▼	51	104		155	212
3400	Decorator style									
3410	S.P. switch/15 amp recpt., type NM cable	1 Elec	11.43	.700	Ea.	28	37.50		65.50	86.50
3420	Type MC cable		10	.800		39	42.50		81.50	107
3430	EMT & wire		4.40	1.818		46.50	97		143.50	197
3450	S.P. switch/pilot light, type NM cable		11.43	.700		28	37.50		65.50	87
3470	Type MC cable		10	.800		39.50	42.50		82	108
3480	EMT & wire		4.40	1.818		47	97		144	197
3500	2-S.P. switches, 2-#14/2, type NM cables		10	.800		31.50	42.50		74	98.50
3520	Type MC cable		8.89	.900		48.50	48		96.50	125
3530	EMT & wire		4.10	1.951		50	104		154	211
3550	3-way/15 amp recpt., #14/3, type NM cable		10	.800		35	42.50		77.50	103
3570	Type MC cable		8.89	.900		50.50	48		98.50	128
3580	EMT & wire		4.10	1.951		52	104		156	213
3650	2-3 way switches, 2-#14/3, type NM cables		8.89	.900		45	48		93	122
3670	Type MC cable		8	1		69.50	53.50		123	157
3680	EMT & wire		4	2		59.50	107		166.50	226
3700	S.P. switch/20 amp recpt., #12/2, type NM cable		10	.800		36	42.50		78.50	104
3720	Type MC cable		8.89	.900		41	48		89	117
3730	EMT & wire	▼	4.10	1.951	▼	55	104		159	216
4000	Receptacle devices									
4010	Duplex outlet, 15 amp recpt., Ivory, 1-gang box, plate									
4015	Type NM cable	1 Elec	14.55	.550	Ea.	11	29.50		40.50	56
4020	Type MC cable		12.31	.650		22.50	34.50		57	76.50
4030	EMT & wire		5.33	1.501		28	80		108	151
4050	With #12/2, type NM cable		12.31	.650		13.65	34.50		48.15	67
4070	Type MC cable		10.67	.750		22.50	40		62.50	84.50
4080	EMT & wire		4.71	1.699		32.50	90.50		123	172
4100	20 amp recpt., #12/2, type NM cable		12.31	.650		21	34.50		55.50	75
4120	Type MC cable		10.67	.750		29.50	40		69.50	92.50
4130	EMT & wire	▼	4.71	1.699	▼	39.50	90.50		130	180
4140	For GFI see Section 26 05 90.10 line 4300 below									
4150	Decorator style, 15 amp recpt., type NM cable	1 Elec	14.55	.550	Ea.	14.80	29.50		44.30	60.50

26 05 Common Work Results for Electrical

26 05 90 – Residential Applications

26 05 90.10 Residential Wiring

		Crew	Daily Output	Labor-Hours	Unit	Material	2014 Bare Costs Labor	Equipment	Total	Total Incl O&P
4170	Type MC cable	1 Elec	12.31	.650	Ea.	26	34.50		60.50	80.50
4180	EMT & wire		5.33	1.501		32	80		112	155
4200	With #12/2, type NM cable		12.31	.650		17.45	34.50		51.95	71
4220	Type MC cable		10.67	.750		26	40		66	89
4230	EMT & wire		4.71	1.699		36	90.50		126.50	176
4250	20 amp recpt. #12/2, type NM cable		12.31	.650		24.50	34.50		59	79
4270	Type MC cable		10.67	.750		33.50	40		73.50	96.50
4280	EMT & wire		4.71	1.699		43.50	90.50		134	184
4300	GFI, 15 amp recpt., type NM cable		12.31	.650		47.50	34.50		82	104
4320	Type MC cable		10.67	.750		58.50	40		98.50	125
4330	EMT & wire		4.71	1.699		64.50	90.50		155	207
4350	GFI with #12/2, type NM cable		10.67	.750		50	40		90	115
4370	Type MC cable		9.20	.870		58.50	46.50		105	134
4380	EMT & wire		4.21	1.900		68.50	101		169.50	228
4400	20 amp recpt., #12/2 type NM cable		10.67	.750		52	40		92	118
4420	Type MC cable		9.20	.870		61	46.50		107.50	137
4430	EMT & wire		4.21	1.900		71	101		172	230
4500	Weather-proof cover for above receptacles, add	▼	32	.250	▼	4.08	13.35		17.43	24.50
4550	Air conditioner outlet, 20 amp-240 volt recpt.									
4560	30' of #12/2, 2 pole circuit breaker									
4570	Type NM cable	1 Elec	10	.800	Ea.	63	42.50		105.50	133
4580	Type MC cable		9	.889		73	47.50		120.50	152
4590	EMT & wire		4	2		81.50	107		188.50	250
4600	Decorator style, type NM cable		10	.800		67.50	42.50		110	139
4620	Type MC cable		9	.889		78	47.50		125.50	157
4630	EMT & wire	▼	4	2	▼	86.50	107		193.50	255
4650	Dryer outlet, 30 amp-240 volt recpt., 20' of #10/3									
4660	2 pole circuit breaker									
4670	Type NM cable	1 Elec	6.41	1.248	Ea.	64	66.50		130.50	170
4680	Type MC cable		5.71	1.401		69	75		144	188
4690	EMT & wire	▼	3.48	2.299	▼	75.50	123		198.50	267
4700	Range outlet, 50 amp-240 volt recpt., 30' of #8/3									
4710	Type NM cable	1 Elec	4.21	1.900	Ea.	92	101		193	253
4720	Type MC cable		4	2		136	107		243	310
4730	EMT & wire		2.96	2.703		106	144		250	335
4750	Central vacuum outlet, Type NM cable		6.40	1.250		60	66.50		126.50	166
4770	Type MC cable		5.71	1.401		75.50	75		150.50	195
4780	EMT & wire	▼	3.48	2.299	▼	81.50	123		204.50	274
4800	30 amp-110 volt locking recpt., #10/2 circ. bkr.									
4810	Type NM cable	1 Elec	6.20	1.290	Ea.	69	69		138	179
4830	EMT & wire	"	3.20	2.500	"	92	133		225	300
4900	Low voltage outlets									
4910	Telephone recpt., 20' of 4/C phone wire	1 Elec	26	.308	Ea.	10.95	16.40		27.35	36.50
4920	TV recpt., 20' of RG59U coax wire, F type connector	"	16	.500	"	19.20	26.50		45.70	61
4950	Door bell chime, transformer, 2 buttons, 60' of bellwire									
4970	Economy model	1 Elec	11.50	.696	Ea.	58	37		95	120
4980	Custom model		11.50	.696		123	37		160	191
4990	Luxury model, 3 buttons	▼	9.50	.842	▼	305	45		350	400
6000	Lighting outlets									
6050	Wire only (for fixture), type NM cable	1 Elec	32	.250	Ea.	7.20	13.35		20.55	28
6070	Type MC cable		24	.333		13.85	17.80		31.65	41.50
6080	EMT & wire		10	.800		18.65	42.50		61.15	84.50
6100	Box (4"), and wire (for fixture), type NM cable	▼	25	.320	▼	16.10	17.05		33.15	43.50

359

26 05 Common Work Results for Electrical

26 05 90 – Residential Applications

26 05 90.10 Residential Wiring

		Crew	Daily Output	Labor-Hours	Unit	Material	2014 Bare Costs Labor	Equipment	Total	Total Incl O&P
6120	Type MC cable	1 Elec	20	.400	Ea.	23	21.50		44.50	57
6130	EMT & wire	↓	11	.727	↓	27.50	39		66.50	88.50
6200	Fixtures (use with lines 6050 or 6100 above)									
6210	Canopy style, economy grade	1 Elec	40	.200	Ea.	37.50	10.65		48.15	57
6220	Custom grade		40	.200		57.50	10.65		68.15	79
6250	Dining room chandelier, economy grade		19	.421		93.50	22.50		116	137
6260	Custom grade		19	.421		274	22.50		296.50	335
6270	Luxury grade		15	.533		620	28.50		648.50	725
6310	Kitchen fixture (fluorescent), economy grade		30	.267		69.50	14.25		83.75	98
6320	Custom grade		25	.320		190	17.05		207.05	235
6350	Outdoor, wall mounted, economy grade		30	.267		35.50	14.25		49.75	60.50
6360	Custom grade		30	.267		121	14.25		135.25	156
6370	Luxury grade		25	.320		274	17.05		291.05	325
6410	Outdoor PAR floodlights, 1 lamp, 150 watt		20	.400		38	21.50		59.50	74
6420	2 lamp, 150 watt each		20	.400		63	21.50		84.50	101
6430	For infrared security sensor, add		32	.250		127	13.35		140.35	160
6450	Outdoor, quartz-halogen, 300 watt flood		20	.400		46	21.50		67.50	82.50
6600	Recessed downlight, round, pre-wired, 50 or 75 watt trim		30	.267		47	14.25		61.25	73
6610	With shower light trim		30	.267		58	14.25		72.25	85.50
6620	With wall washer trim		28	.286		70	15.25		85.25	100
6630	With eye-ball trim	↓	28	.286		70	15.25		85.25	100
6640	For direct contact with insulation, add					2.49			2.49	2.74
6700	Porcelain lamp holder	1 Elec	40	.200		3.50	10.65		14.15	19.80
6710	With pull switch		40	.200		5.70	10.65		16.35	22.50
6750	Fluorescent strip, 1-20 watt tube, wrap around diffuser, 24"		24	.333		58.50	17.80		76.30	91
6770	2-34 watt tubes, 48"		20	.400		89.50	21.50		111	131
6780	With residential ballast		20	.400		100	21.50		121.50	142
6800	Bathroom heat lamp, 1-250 watt		28	.286		51.50	15.25		66.75	79.50
6810	2-250 watt lamps	↓	28	.286	↓	82	15.25		97.25	114
6820	For timer switch, see Section 26 05 90.10 line 2400									
6900	Outdoor post lamp, incl. post, fixture, 35' of #14/2									
6910	Type NMC cable	1 Elec	3.50	2.286	Ea.	226	122		348	430
6920	Photo-eye, add		27	.296		38	15.80		53.80	65.50
6950	Clock dial time switch, 24 hr., w/enclosure, type NM cable		11.43	.700		70.50	37.50		108	134
6970	Type MC cable		11	.727		82	39		121	148
6980	EMT & wire	↓	4.85	1.649	↓	88	88		176	229
7000	Alarm systems									
7050	Smoke detectors, box, #14/3, type NM cable	1 Elec	14.55	.550	Ea.	35	29.50		64.50	83
7070	Type MC cable		12.31	.650		46.50	34.50		81	103
7080	EMT & wire	↓	5	1.600		48.50	85.50		134	181
7090	For relay output to security system, add				↓	14.10			14.10	15.55
8000	Residential equipment									
8050	Disposal hook-up, incl. switch, outlet box, 3' of flex									
8060	20 amp-1 pole circ. bkr., and 25' of #12/2									
8070	Type NM cable	1 Elec	10	.800	Ea.	32.50	42.50		75	100
8080	Type MC cable	↓	8	1		42	53.50		95.50	127
8090	EMT & wire	↓	5	1.600	↓	53.50	85.50		139	187
8100	Trash compactor or dishwasher hook-up, incl. outlet box,									
8110	3' of flex, 15 amp-1 pole circ. bkr., and 25' of #14/2									
8120	Type NM cable	1 Elec	10	.800	Ea.	24	42.50		66.50	90.50
8130	Type MC cable	↓	8	1		37	53.50		90.50	121
8140	EMT & wire	↓	5	1.600	↓	45	85.50		130.50	178
8150	Hot water sink dispensor hook-up, use line 8100									

26 05 Common Work Results for Electrical

26 05 90 – Residential Applications

26 05 90.10 Residential Wiring

		Crew	Daily Output	Labor-Hours	Unit	Material	2014 Bare Costs Labor	Equipment	Total	Total Incl O&P
8200	Vent/exhaust fan hook-up, type NM cable	1 Elec	32	.250	Ea.	7.20	13.35		20.55	28
8220	Type MC cable		24	.333		13.85	17.80		31.65	41.50
8230	EMT & wire	↓	10	.800	↓	18.65	42.50		61.15	84.50
8250	Bathroom vent fan, 50 CFM (use with above hook-up)									
8260	Economy model	1 Elec	15	.533	Ea.	27	28.50		55.50	72.50
8270	Low noise model		15	.533		37	28.50		65.50	83
8280	Custom model	↓	12	.667	↓	127	35.50		162.50	193
8300	Bathroom or kitchen vent fan, 110 CFM									
8310	Economy model	1 Elec	15	.533	Ea.	75.50	28.50		104	126
8320	Low noise model	"	15	.533	"	93.50	28.50		122	146
8350	Paddle fan, variable speed (w/o lights)									
8360	Economy model (AC motor)	1 Elec	10	.800	Ea.	125	42.50		167.50	201
8362	With light kit		10	.800		166	42.50		208.50	246
8370	Custom model (AC motor)		10	.800		197	42.50		239.50	281
8372	With light kit		10	.800		238	42.50		280.50	325
8380	Luxury model (DC motor)		8	1		390	53.50		443.50	510
8382	With light kit		8	1		430	53.50		483.50	555
8390	Remote speed switch for above, add	↓	12	.667	↓	32.50	35.50		68	88.50
8500	Whole house exhaust fan, ceiling mount, 36", variable speed									
8510	Remote switch, incl. shutters, 20 amp-1 pole circ. bkr.									
8520	30' of #12/2, type NM cable	1 Elec	4	2	Ea.	1,250	107		1,357	1,525
8530	Type MC cable		3.50	2.286		1,275	122		1,397	1,575
8540	EMT & wire	↓	3	2.667	↓	1,275	142		1,417	1,625
8600	Whirlpool tub hook-up, incl. timer switch, outlet box									
8610	3' of flex, 20 amp-1 pole GFI circ. bkr.									
8620	30' of #12/2, type NM cable	1 Elec	5	1.600	Ea.	127	85.50		212.50	267
8630	Type MC cable		4.20	1.905		131	102		233	297
8640	EMT & wire	↓	3.40	2.353	↓	141	126		267	345
8650	Hot water heater hook-up, incl. 1-2 pole circ. bkr., box;									
8660	3' of flex, 20' of #10/2, type NM cable	1 Elec	5	1.600	Ea.	34	85.50		119.50	165
8670	Type MC cable		4.20	1.905		47.50	102		149.50	204
8680	EMT & wire	↓	3.40	2.353	↓	48.50	126		174.50	241
9000	Heating/air conditioning									
9050	Furnace/boiler hook-up, incl. firestat, local on-off switch									
9060	Emergency switch, and 40' of type NM cable	1 Elec	4	2	Ea.	52.50	107		159.50	218
9070	Type MC cable		3.50	2.286		69.50	122		191.50	260
9080	EMT & wire	↓	1.50	5.333	↓	81	285		366	515
9100	Air conditioner hook-up, incl. local 60 amp disc. switch									
9110	3' sealtite, 40 amp, 2 pole circuit breaker									
9130	40' of #8/2, type NM cable	1 Elec	3.50	2.286	Ea.	182	122		304	385
9140	Type MC cable		3	2.667		249	142		391	485
9150	EMT & wire	↓	1.30	6.154	↓	218	330		548	730
9200	Heat pump hook-up, 1-40 & 1-100 amp 2 pole circ. bkr.									
9210	Local disconnect switch, 3' sealtite									
9220	40' of #8/2 & 30' of #3/2									
9230	Type NM cable	1 Elec	1.30	6.154	Ea.	525	330		855	1,075
9240	Type MC cable		1.08	7.407		600	395		995	1,250
9250	EMT & wire	↓	.94	8.511	↓	590	455		1,045	1,325
9500	Thermostat hook-up, using low voltage wire									
9520	Heating only, 25' of #18-3	1 Elec	24	.333	Ea.	9.60	17.80		27.40	37
9530	Heating/cooling, 25' of #18-4	"	20	.400	"	11.75	21.50		33.25	45

26 24 Switchboards and Panelboards

26 24 19 – Motor-Control Centers

26 24 19.40 Motor Starters and Controls		Crew	Daily Output	Labor-Hours	Unit	Material	2014 Bare Costs Labor	Equipment	Total	Total Incl O&P
0010	**MOTOR STARTERS AND CONTROLS**									
0050	Magnetic, FVNR, with enclosure and heaters, 480 volt									
0080	2 HP, size 00	1 Elec	3.50	2.286	Ea.	195	122		317	395
0100	5 HP, size 0		2.30	3.478		236	186		422	535
0200	10 HP, size 1	↓	1.60	5		265	267		532	690
0300	25 HP, size 2	2 Elec	2.20	7.273		500	390		890	1,125
0400	50 HP, size 3		1.80	8.889		810	475		1,285	1,600
0500	100 HP, size 4		1.20	13.333		1,800	710		2,510	3,050
0600	200 HP, size 5		.90	17.778		4,225	950		5,175	6,050
0610	400 HP, size 6		.80	20		18,200	1,075		19,275	21,700
0620	NEMA 7, 5 HP, size 0	1 Elec	1.60	5		1,350	267		1,617	1,900
0630	10 HP, size 1	"	1.10	7.273		1,425	390		1,815	2,125
0640	25 HP, size 2	2 Elec	1.80	8.889		2,300	475		2,775	3,225
0650	50 HP, size 3		1.20	13.333		3,425	710		4,135	4,850
0660	100 HP, size 4		.90	17.778		5,550	950		6,500	7,525
0670	200 HP, size 5	↓	.50	32		13,200	1,700		14,900	17,200
0700	Combination, with motor circuit protectors, 5 HP, size 0	1 Elec	1.80	4.444		765	237		1,002	1,200
0800	10 HP, size 1	"	1.30	6.154		795	330		1,125	1,375
0900	25 HP, size 2	2 Elec	2	8		1,125	425		1,550	1,875
1000	50 HP, size 3		1.32	12.121		1,600	645		2,245	2,750
1200	100 HP, size 4	↓	.80	20		3,500	1,075		4,575	5,450
1220	NEMA 7, 5 HP, size 0	1 Elec	1.30	6.154		2,725	330		3,055	3,475
1230	10 HP, size 1	"	1	8		2,775	425		3,200	3,700
1240	25 HP, size 2	2 Elec	1.32	12.121		3,725	645		4,370	5,050
1250	50 HP, size 3		.80	20		6,125	1,075		7,200	8,350
1260	100 HP, size 4		.60	26.667		9,550	1,425		10,975	12,600
1270	200 HP, size 5	↓	.40	40		20,800	2,125		22,925	26,000
1400	Combination, with fused switch, 5 HP, size 0	1 Elec	1.80	4.444		585	237		822	1,000
1600	10 HP, size 1	"	1.30	6.154		625	330		955	1,175
1800	25 HP, size 2	2 Elec	2	8		1,025	425		1,450	1,775
2000	50 HP, size 3		1.32	12.121		1,725	645		2,370	2,875
2200	100 HP, size 4	↓	.80	20	↓	3,000	1,075		4,075	4,900
3500	Magnetic FVNR with NEMA 12, enclosure & heaters, 480 volt									
3600	5 HP, size 0	1 Elec	2.20	3.636	Ea.	222	194		416	535
3700	10 HP, size 1	"	1.50	5.333		335	285		620	795
3800	25 HP, size 2	2 Elec	2	8		625	425		1,050	1,325
3900	50 HP, size 3		1.60	10		965	535		1,500	1,850
4000	100 HP, size 4		1	16		2,300	855		3,155	3,800
4100	200 HP, size 5	↓	.80	20		5,525	1,075		6,600	7,675
4200	Combination, with motor circuit protectors, 5 HP, size 0	1 Elec	1.70	4.706		735	251		986	1,175
4300	10 HP, size 1	"	1.20	6.667		765	355		1,120	1,375
4400	25 HP, size 2	2 Elec	1.80	8.889		1,150	475		1,625	1,950
4500	50 HP, size 3		1.20	13.333		1,875	710		2,585	3,125
4600	100 HP, size 4	↓	.74	21.622		4,200	1,150		5,350	6,350
4700	Combination, with fused switch, 5 HP, size 0	1 Elec	1.70	4.706		715	251		966	1,150
4800	10 HP, size 1	"	1.20	6.667		740	355		1,095	1,350
4900	25 HP, size 2	2 Elec	1.80	8.889		1,125	475		1,600	1,950
5000	50 HP, size 3		1.20	13.333		1,825	710		2,535	3,075
5100	100 HP, size 4	↓	.74	21.622	↓	3,675	1,150		4,825	5,775
5200	Factory installed controls, adders to size 0 thru 5									
5300	Start-stop push button	1 Elec	32	.250	Ea.	46.50	13.35		59.85	71
5400	Hand-off-auto-selector switch		32	.250		46.50	13.35		59.85	71
5500	Pilot light	↓	32	.250	↓	87.50	13.35		100.85	116

26 24 Switchboards and Panelboards

26 24 19 – Motor-Control Centers

26 24 19.40 Motor Starters and Controls

		Crew	Daily Output	Labor-Hours	Unit	Material	2014 Bare Costs Labor	Equipment	Total	Total Incl O&P
5600	Start-stop-pilot	1 Elec	32	.250	Ea.	134	13.35		147.35	167
5700	Auxiliary contact, NO or NC		32	.250		64	13.35		77.35	90.50
5800	NO-NC	↓	32	.250	↓	128	13.35		141.35	161

26 29 Low-Voltage Controllers

26 29 13 – Enclosed Controllers

26 29 13.20 Control Stations

		Crew	Daily Output	Labor-Hours	Unit	Material	Labor	Equipment	Total	Total Incl O&P
0010	**CONTROL STATIONS**									
0050	NEMA 1, heavy duty, stop/start	1 Elec	8	1	Ea.	134	53.50		187.50	227
0100	Stop/start, pilot light		6.20	1.290		182	69		251	305
0200	Hand/off/automatic		6.20	1.290		99	69		168	212
0400	Stop/start/reverse		5.30	1.509		180	80.50		260.50	320
0500	NEMA 7, heavy duty, stop/start		6	1.333		420	71		491	565
0600	Stop/start, pilot light	↓	4	2	↓	510	107		617	725

26 42 Cathodic Protection

26 42 16 – Passive Cathodic Protection for Underground Storage Tank

26 42 16.50 Cathodic Protection Wiring Methods

		Crew	Daily Output	Labor-Hours	Unit	Material	Labor	Equipment	Total	Total Incl O&P
0010	**CATHODIC PROTECTION WIRING METHODS**									
1000	Anodes, magnesium type, 9 #	R-15	18.50	2.595	Ea.	39	136	17.55	192.55	266
1010	17 #		13	3.692		70.50	193	25	288.50	395
1020	32 #		10	4.800		123	251	32.50	406.50	545
1030	48 #	↓	7.20	6.667		162	350	45	557	755
1100	Graphite type w/epoxy cap, 3" x 60" (32 #)	R-22	8.40	4.438		127	201		328	445
1110	4" x 80" (68 #)		6	6.213		239	282		521	690
1120	6" x 72" (80 #)		5.20	7.169		1,475	325		1,800	2,125
1130	6" x 36" (45 #)	↓	9.60	3.883		750	176		926	1,100
2000	Rectifiers, silicon type, air cooled, 28 V/10 A	R-19	3.50	5.714		2,300	305		2,605	2,975
2010	20 V/20 A		3.50	5.714		2,350	305		2,655	3,025
2100	Oil immersed, 28 V/10 A		3	6.667		3,125	355		3,480	3,975
2110	20 V/20 A	↓	3	6.667	↓	3,150	355		3,505	4,000
3000	Anode backfill, coke breeze	R-22	3850	.010	Lb.	.22	.44		.66	.91
4000	Cable, HMWPE, No. 8		2.40	15.533	M.L.F.	495	705		1,200	1,625
4010	No. 6		2.40	15.533		740	705		1,445	1,875
4020	No. 4		2.40	15.533		1,125	705		1,830	2,300
4030	No. 2		2.40	15.533		1,750	705		2,455	3,000
4040	No. 1		2.20	16.945		2,375	770		3,145	3,800
4050	No. 1/0		2.20	16.945		2,950	770		3,720	4,425
4060	No. 2/0		2.20	16.945		4,575	770		5,345	6,225
4070	No. 4/0	↓	2	18.640	↓	5,950	845		6,795	7,800
5000	Test station, 7 terminal box, flush curb type w/lockable cover	R-19	12	1.667	Ea.	71	89		160	211
5010	Reference cell, 2" dia PVC conduit, cplg., plug, set flush	"	4.80	4.167	"	150	223		373	500

Division Notes

Estimating Tips

- When estimating material costs for electronic safety and security systems, it is always prudent to obtain manufacturers' quotations for equipment prices and special installation requirements that affect the total cost.
- Fire alarm systems consist of control panels, annunciator panels, battery with rack, charger, and fire alarm actuating and indicating devices. Some fire alarm systems include speakers, telephone lines, door closer controls, and other components. Be careful not to overlook the costs related to installation for these items. Also be aware of costs for integrated automation instrumentation and terminal devices, control equipment, control wiring, and programming.
- Security equipment includes items such as CCTV, access control, and other detection and identification systems to perform alert and alarm functions. Be sure to consider the costs related to installation for this security equipment, such as for integrated automation instrumentation and terminal devices, control equipment, control wiring, and programming.

Reference Numbers

Reference numbers are shown in shaded boxes at the beginning of some major classifications. These numbers refer to related items in the Reference Section. The reference information may be an estimating procedure, an alternate pricing method, or technical information.

Note: Not all subdivisions listed here necessarily appear in this publication.

No part of this publication may be reproduced, stored in a retrieval system, or transmitted in any form or by any means without prior written permission of RSMeans, a division of Reed Construction Data.

28 16 Intrusion Detection

28 16 16 – Intrusion Detection Systems Infrastructure

28 16 16.50 Intrusion Detection

		Crew	Daily Output	Labor-Hours	Unit	Material	2014 Bare Costs Labor	Equipment	Total	Total Incl O&P
0010	**INTRUSION DETECTION**, not including wires & conduits									
0100	Burglar alarm, battery operated, mechanical trigger	1 Elec	4	2	Ea.	275	107		382	460
0200	Electrical trigger		4	2		330	107		437	520
0400	For outside key control, add		8	1		83	53.50		136.50	172
0600	For remote signaling circuitry, add		8	1		124	53.50		177.50	216
0800	Card reader, flush type, standard		2.70	2.963		920	158		1,078	1,250
1000	Multi-code		2.70	2.963		1,175	158		1,333	1,525

28 31 Fire Detection and Alarm

28 31 23 – Fire Detection and Alarm Annunciation Panels and Fire Stations

28 31 23.50 Alarm Panels and Devices

		Crew	Daily Output	Labor-Hours	Unit	Material	Labor	Equipment	Total	Total Incl O&P
0010	**ALARM PANELS AND DEVICES**, not including wires & conduits									
3594	Fire, alarm control panel									
3600	4 zone	2 Elec	2	8	Ea.	395	425		820	1,075
3800	8 zone		1	16		775	855		1,630	2,125
4000	12 zone		.67	23.988		2,375	1,275		3,650	4,525
4020	Alarm device	1 Elec	8	1		235	53.50		288.50	340
4050	Actuating device		8	1		335	53.50		388.50	445
4200	Battery and rack		4	2		410	107		517	610
4400	Automatic charger		8	1		580	53.50		633.50	715
4600	Signal bell		8	1		78	53.50		131.50	166
4800	Trouble buzzer or manual station		8	1		82.50	53.50		136	171
5600	Strobe and horn		5.30	1.509		151	80.50		231.50	287
5800	Fire alarm horn		6.70	1.194		60.50	63.50		124	162
6000	Door holder, electro-magnetic		4	2		102	107		209	272
6200	Combination holder and closer		3.20	2.500		122	133		255	335
6600	Drill switch		8	1		365	53.50		418.50	485
6800	Master box		2.70	2.963		6,350	158		6,508	7,200
7000	Break glass station		8	1		55	53.50		108.50	141
7800	Remote annunciator, 8 zone lamp		1.80	4.444		207	237		444	580
8000	12 zone lamp	2 Elec	2.60	6.154		330	330		660	855
8200	16 zone lamp	"	2.20	7.273		415	390		805	1,025

28 31 43 – Fire Detection Sensors

28 31 43.50 Fire and Heat Detectors

		Crew	Daily Output	Labor-Hours	Unit	Material	Labor	Equipment	Total	Total Incl O&P
0010	**FIRE & HEAT DETECTORS**									
5000	Detector, rate of rise	1 Elec	8	1	Ea.	50.50	53.50		104	136
5100	Fixed temperature	"	8	1	"	50.50	53.50		104	136

28 31 46 – Smoke Detection Sensors

28 31 46.50 Smoke Detectors

		Crew	Daily Output	Labor-Hours	Unit	Material	Labor	Equipment	Total	Total Incl O&P
0010	**SMOKE DETECTORS**									
5200	Smoke detector, ceiling type	1 Elec	6.20	1.290	Ea.	109	69		178	223
5400	Duct type	"	3.20	2.500	"	325	133		458	555

28 33 Gas Detection and Alarm

28 33 33 – Gas Detection Sensors

28 33 33.50 Tank Leak Detection Systems	Crew	Daily Output	Labor-Hours	Unit	Material	2014 Bare Costs Labor	Equipment	Total	Total Incl O&P
0010 **TANK LEAK DETECTION SYSTEMS** Liquid and vapor									
0100 For hydrocarbons and hazardous liquids/vapors									
0120 Controller, data acquisition, incl. printer, modem, RS232 port									
0140 24 channel, for use with all probes				Ea.	4,125			4,125	4,525
0160 9 channel, for external monitoring				"	905			905	1,000
0200 Probes									
0210 Well monitoring									
0220 Liquid phase detection				Ea.	470			470	520
0230 Hydrocarbon vapor, fixed position					475			475	525
0240 Hydrocarbon vapor, float mounted					475			475	525
0250 Both liquid and vapor hydrocarbon					475			475	525
0300 Secondary containment, liquid phase									
0310 Pipe trench/manway sump				Ea.	815			815	895
0320 Double wall pipe and manual sump					815			815	895
0330 Double wall fiberglass annular space					320			320	350
0340 Double wall steel tank annular space					335			335	365
0500 Accessories									
0510 Modem, non-dedicated phone line				Ea.	300			300	330
0600 Monitoring, internal									
0610 Automatic tank gauge, incl. overfill				Ea.	1,150			1,150	1,250
0620 Product line				"	1,150			1,150	1,250
0700 Monitoring, special									
0710 Cathodic protection				Ea.	715			715	790
0720 Annular space chemical monitor				"	975			975	1,075

Division Notes

		CREW	DAILY OUTPUT	LABOR-HOURS	UNIT	BARE COSTS MAT.	LABOR	EQUIP.	TOTAL	TOTAL INCL O&P

Estimating Tips

31 05 00 Common Work Results for Earthwork

- Estimating the actual cost of performing earthwork requires careful consideration of the variables involved. This includes items such as type of soil, whether water will be encountered, dewatering, whether banks need bracing, disposal of excavated earth, and length of haul to fill or spoil sites, etc. If the project has large quantities of cut or fill, consider raising or lowering the site to reduce costs, while paying close attention to the effect on site drainage and utilities.
- If the project has large quantities of fill, creating a borrow pit on the site can significantly lower the costs.
- It is very important to consider what time of year the project is scheduled for completion. Bad weather can create large cost overruns from dewatering, site repair, and lost productivity from cold weather.

Reference Numbers

Reference numbers are shown in shaded boxes at the beginning of some major classifications. These numbers refer to related items in the Reference Section. The reference information may be an estimating procedure, an alternate pricing method, or technical information.

Note: Not all subdivisions listed here necessarily appear in this publication.

31 23 Excavation and Fill

31 23 16 – Excavation

31 23 16.13 Excavating, Trench

		Crew	Daily Output	Labor-Hours	Unit	Material	2014 Bare Costs Labor	Equipment	Total	Total Incl O&P
0010	**EXCAVATING, TRENCH**				G1030-805					
0011	Or continuous footing									
0020	Common earth with no sheeting or dewatering included									
0050	1' to 4' deep, 3/8 C.Y. excavator	B-11C	150	.107	B.C.Y.		4.56	2.44	7	9.65
0060	1/2 C.Y. excavator	B-11M	200	.080			3.42	1.99	5.41	7.45
0090	4' to 6' deep, 1/2 C.Y. excavator	"	200	.080			3.42	1.99	5.41	7.45
0100	5/8 C.Y. excavator	B-12Q	250	.064			2.78	2.38	5.16	6.85
0300	1/2 C.Y. excavator, truck mounted	B-12J	200	.080			3.48	4.41	7.89	10.15
0500	6' to 10' deep, 3/4 C.Y. excavator	B-12F	225	.071			3.09	2.94	6.03	7.95
0600	1 C.Y. excavator, truck mounted	B-12K	400	.040			1.74	2.51	4.25	5.40
0900	10' to 14' deep, 3/4 C.Y. excavator	B-12F	200	.080			3.48	3.31	6.79	8.95
1000	1-1/2 C.Y. excavator	B-12B	540	.030			1.29	1.91	3.20	4.07
1300	14' to 20' deep, 1 C.Y. excavator	B-12A	320	.050			2.17	2.55	4.72	6.10
1340	20' to 24' deep, 1 C.Y. excavator	"	288	.056			2.41	2.83	5.24	6.80
1352	4' to 6' deep, 1/2 C.Y. excavator w/trench box	B-13H	188	.085			3.70	5.15	8.85	11.30
1354	5/8 C.Y. excavator	"	235	.068			2.96	4.11	7.07	9.05
1362	6' to 10' deep, 3/4 C.Y. excavator w/trench box	B-13G	212	.075			3.28	3.52	6.80	8.85
1374	10' to 14' deep, 3/4 C.Y. excavator w/trench box	"	188	.085			3.70	3.97	7.67	10
1376	1-1/2 C.Y. excavator	B-13E	508	.032			1.37	2.20	3.57	4.50
1381	14' to 20' deep, 1 C.Y. excavator w/trench box	B-13D	301	.053			2.31	2.99	5.30	6.80
1386	20' to 24' deep, 1 C.Y. excavator w/trench box	"	271	.059			2.57	3.32	5.89	7.55
1400	By hand with pick and shovel 2' to 6' deep, light soil	1 Clab	8	1			36.50		36.50	56.50
1500	Heavy soil	"	4	2			73.50		73.50	113
1700	For tamping backfilled trenches, air tamp, add	A-1G	100	.080	E.C.Y.		2.93	.56	3.49	5.15
1900	Vibrating plate, add	B-18	180	.133	"		4.98	.26	5.24	8
2100	Trim sides and bottom for concrete pours, common earth		1500	.016	S.F.		.60	.03	.63	.95
2300	Hardpan		600	.040	"		1.49	.08	1.57	2.38
5020	Loam & Sandy clay with no sheeting or dewatering included									
5050	1' to 4' deep, 3/8 C.Y. tractor loader/backhoe	B-11C	162	.099	B.C.Y.		4.23	2.25	6.48	8.95
5060	1/2 C.Y. excavator	B-11M	216	.074			3.17	1.84	5.01	6.85
5080	4' to 6' deep, 1/2 C.Y. excavator	"	216	.074			3.17	1.84	5.01	6.85
5090	5/8 C.Y. excavator	B-12Q	276	.058			2.52	2.15	4.67	6.20
5130	1/2 C.Y. excavator, truck mounted	B-12J	216	.074			3.22	4.08	7.30	9.40
5140	6' to 10' deep, 3/4 C.Y. excavator	B-12F	243	.066			2.86	2.72	5.58	7.35
5160	1 C.Y. excavator, truck mounted	B-12K	432	.037			1.61	2.33	3.94	5
5190	10' to 14' deep, 3/4 C.Y. excavator	B-12F	216	.074			3.22	3.07	6.29	8.30
5210	1-1/2 C.Y. excavator	B-12B	583	.027			1.19	1.77	2.96	3.76
5250	14' to 20' deep, 1 C.Y. excavator	B-12A	346	.046			2.01	2.36	4.37	5.65
5300	20' to 24' deep, 1 C.Y. excavator	"	311	.051			2.24	2.62	4.86	6.30
5352	4' to 6' deep, 1/2 C.Y. excavator w/trench box	B-13H	205	.078			3.39	4.71	8.10	10.40
5354	5/8 C.Y. excavator	"	257	.062			2.71	3.76	6.47	8.25
5362	6' to 10' deep, 3/4 C.Y. excavator w/trench box	B-13G	231	.069			3.01	3.23	6.24	8.15
5370	10' to 14' deep, 3/4 C.Y. excavator w/trench box	"	205	.078			3.39	3.64	7.03	9.20
5374	1-1/2 C.Y. excavator	B-13E	554	.029			1.25	2.01	3.26	4.13
5382	14' to 20' deep, 1 C.Y. excavator w/trench box	B-13D	329	.049			2.11	2.73	4.84	6.25
5392	20' to 24' deep, 1 C.Y. excavator w/trench box	"	295	.054			2.36	3.05	5.41	6.95
6020	Sand & gravel with no sheeting or dewatering included									
6050	1' to 4' deep, 3/8 C.Y. excavator	B-11C	165	.097	B.C.Y.		4.15	2.21	6.36	8.80
6060	1/2 C.Y. excavator	B-11M	220	.073			3.11	1.81	4.92	6.75
6080	4' to 6' deep, 1/2 C.Y. excavator	"	220	.073			3.11	1.81	4.92	6.75
6090	5/8 C.Y. excavator	B-12Q	275	.058			2.53	2.16	4.69	6.25
6130	1/2 C.Y. excavator, truck mounted	B-12J	220	.073			3.16	4.01	7.17	9.25
6140	6' to 10' deep, 3/4 C.Y. excavator	B-12F	248	.065			2.80	2.67	5.47	7.20

31 23 Excavation and Fill

31 23 16 – Excavation

31 23 16.13 Excavating, Trench

		Crew	Daily Output	Labor-Hours	Unit	Material	2014 Bare Costs Labor	Equipment	Total	Total Incl O&P
6160	1 C.Y. excavator, truck mounted	B-12K	440	.036	B.C.Y.		1.58	2.28	3.86	4.92
6190	10' to 14' deep, 3/4 C.Y. excavator	B-12F	220	.073			3.16	3.01	6.17	8.15
6210	1-1/2 C.Y. excavator	B-12B	594	.027			1.17	1.74	2.91	3.70
6250	14' to 20' deep, 1 C.Y. excavator	B-12A	352	.045			1.97	2.31	4.28	5.55
6300	20' to 24' deep, 1 C.Y. excavator	"	317	.050			2.19	2.57	4.76	6.20
6352	4' to 6' deep, 1/2 C.Y. excavator w/trench box	B-13H	209	.077			3.33	4.62	7.95	10.20
6354	5/8 C.Y. excavator	"	261	.061			2.66	3.70	6.36	8.15
6362	6' to 10' deep, 3/4 C.Y. excavator w/trench box	B-13G	236	.068			2.95	3.16	6.11	8
6370	10' to 14' deep, 3/4 C.Y. excavator w/trench box	"	209	.077			3.33	3.57	6.90	9.05
6374	1-1/2 C.Y. excavator	B-13E	564	.028			1.23	1.98	3.21	4.05
6382	14' to 20' deep, 1 C.Y. excavator w/trench box	B-13D	334	.048			2.08	2.69	4.77	6.15
6392	20' to 24' deep, 1 C.Y. excavator w/trench box	"	301	.053			2.31	2.99	5.30	6.80
7020	Dense hard clay with no sheeting or dewatering included									
7050	1' to 4' deep, 3/8 C.Y. excavator	B-11C	132	.121	B.C.Y.		5.20	2.77	7.97	10.95
7060	1/2 C.Y. excavator	B-11M	176	.091			3.89	2.26	6.15	8.45
7080	4' to 6' deep, 1/2 C.Y. excavator	"	176	.091			3.89	2.26	6.15	8.45
7090	5/8 C.Y. excavator	B-12Q	220	.073			3.16	2.70	5.86	7.80
7130	1/2 C.Y. excavator, truck mounted	B-12J	176	.091			3.95	5	8.95	11.55
7140	6' to 10' deep, 3/4 C.Y. excavator	B-12F	198	.081			3.51	3.34	6.85	9.05
7160	1 C.Y. excavator, truck mounted	B-12K	352	.045			1.97	2.85	4.82	6.15
7190	10' to 14' deep, 3/4 C.Y. excavator	B-12F	176	.091			3.95	3.76	7.71	10.20
7210	1-1/2 C.Y. excavator	B-12B	475	.034			1.46	2.17	3.63	4.63
7250	14' to 20' deep, 1 C.Y. excavator	B-12A	282	.057			2.47	2.89	5.36	6.95
7300	20' to 24' deep, 1 C.Y. excavator	"	254	.063			2.74	3.21	5.95	7.70

31 23 16.14 Excavating, Utility Trench

		Crew	Daily Output	Labor-Hours	Unit	Material	Labor	Equipment	Total	Total Incl O&P
0010	**EXCAVATING, UTILITY TRENCH** G1030-805									
0011	Common earth									
0050	Trenching with chain trencher, 12 H.P., operator walking									
0100	4" wide trench, 12" deep	B-53	800	.010	L.F.		.47	.09	.56	.81
0150	18" deep		750	.011			.50	.10	.60	.86
0200	24" deep		700	.011			.54	.10	.64	.93
0300	6" wide trench, 12" deep		650	.012			.58	.11	.69	1
0350	18" deep		600	.013			.63	.12	.75	1.08
0400	24" deep		550	.015			.68	.13	.81	1.18
0450	36" deep		450	.018			.84	.16	1	1.44
0600	8" wide trench, 12" deep		475	.017			.79	.15	.94	1.36
0650	18" deep		400	.020			.94	.18	1.12	1.63
0700	24" deep		350	.023			1.08	.20	1.28	1.85
0750	36" deep		300	.027			1.25	.24	1.49	2.16
1000	Backfill by hand including compaction, add									
1050	4" wide trench, 12" deep	A-1G	800	.010	L.F.		.37	.07	.44	.65
1100	18" deep		530	.015			.55	.11	.66	.97
1150	24" deep		400	.020			.73	.14	.87	1.28
1300	6" wide trench, 12" deep		540	.015			.54	.10	.64	.95
1350	18" deep		405	.020			.72	.14	.86	1.27
1400	24" deep		270	.030			1.09	.21	1.30	1.91
1450	36" deep		180	.044			1.63	.31	1.94	2.85
1600	8" wide trench, 12" deep		400	.020			.73	.14	.87	1.28
1650	18" deep		265	.030			1.11	.21	1.32	1.94
1700	24" deep		200	.040			1.47	.28	1.75	2.57
1750	36" deep		135	.059			2.17	.41	2.58	3.81
2000	Chain trencher, 40 H.P. operator riding									

31 23 Excavation and Fill

31 23 16 – Excavation

31 23 16.14 Excavating, Utility Trench

		Crew	Daily Output	Labor-Hours	Unit	Material	Labor	Equipment	Total	Total Incl O&P
2050	6" wide trench and backfill, 12" deep	B-54	1200	.007	L.F.		.31	.28	.59	.79
2100	18" deep		1000	.008			.38	.34	.72	.94
2150	24" deep		975	.008			.39	.35	.74	.97
2200	36" deep		900	.009			.42	.38	.80	1.04
2250	48" deep		750	.011			.50	.45	.95	1.26
2300	60" deep		650	.012			.58	.52	1.10	1.45
2400	8" wide trench and backfill, 12" deep		1000	.008			.38	.34	.72	.94
2450	18" deep		950	.008			.40	.36	.76	.99
2500	24" deep		900	.009			.42	.38	.80	1.04
2550	36" deep		800	.010			.47	.42	.89	1.18
2600	48" deep		650	.012			.58	.52	1.10	1.45
2700	12" wide trench and backfill, 12" deep		975	.008			.39	.35	.74	.97
2750	18" deep		860	.009			.44	.39	.83	1.09
2800	24" deep		800	.010			.47	.42	.89	1.18
2850	36" deep		725	.011			.52	.47	.99	1.30
3000	16" wide trench and backfill, 12" deep		835	.010			.45	.41	.86	1.13
3050	18" deep		750	.011			.50	.45	.95	1.26
3100	24" deep	↓	700	.011	↓		.54	.49	1.03	1.35
3200	Compaction with vibratory plate, add								35%	35%
5100	Hand excavate and trim for pipe bells after trench excavation									
5200	8" pipe	1 Clab	155	.052	L.F.		1.89		1.89	2.92
5300	18" pipe	"	130	.062	"		2.26		2.26	3.48

31 23 19 – Dewatering

31 23 19.20 Dewatering Systems

		Crew	Daily Output	Labor-Hours	Unit	Material	Labor	Equipment	Total	Total Incl O&P
0010	**DEWATERING SYSTEMS**									
0020	Excavate drainage trench, 2' wide, 2' deep	B-11C	90	.178	C.Y.		7.60	4.06	11.66	16.05
0100	2' wide, 3' deep, with backhoe loader	"	135	.119			5.05	2.71	7.76	10.75
0200	Excavate sump pits by hand, light soil	1 Clab	7.10	1.127			41.50		41.50	63.50
0300	Heavy soil	"	3.50	2.286	↓		84		84	129
0500	Pumping 8 hr., attended 2 hrs. per day, including 20 L.F.									
0550	of suction hose & 100 L.F. discharge hose									
0600	2" diaphragm pump used for 8 hours	B-10H	4	3	Day		134	18.95	152.95	226
0650	4" diaphragm pump used for 8 hours	B-10I	4	3			134	31	165	239
0800	8 hrs. attended, 2" diaphragm pump	B-10H	1	12			540	75.50	615.50	905
0900	3" centrifugal pump	B-10J	1	12			540	84.50	624.50	915
1000	4" diaphragm pump	B-10I	1	12			540	124	664	955
1100	6" centrifugal pump	B-10K	1	12	↓		540	365	905	1,225
1300	CMP, incl. excavation 3' deep, 12" diameter	B-6	115	.209	L.F.	8.35	8.35	3.18	19.88	25.50
1400	18" diameter		100	.240	"	12.55	9.65	3.65	25.85	32.50
1600	Sump hole construction, incl. excavation and gravel, pit		1250	.019	C.F.	1.12	.77	.29	2.18	2.73
1700	With 12" gravel collar, 12" pipe, corrugated, 16 ga.		70	.343	L.F.	21.50	13.75	5.20	40.45	51
1800	15" pipe, corrugated, 16 ga.		55	.436		28	17.50	6.65	52.15	65
1900	18" pipe, corrugated, 16 ga.		50	.480		32.50	19.25	7.30	59.05	73.50
2000	24" pipe, corrugated, 14 ga.		40	.600	↓	39	24	9.15	72.15	90
2200	Wood lining, up to 4' x 4', add	↓	300	.080	SFCA	16.25	3.21	1.22	20.68	24
9950	See Section 31 23 19.40 for wellpoints									
9960	See Section 31 23 19.30 for deep well systems									

31 23 19.30 Wells

		Crew	Daily Output	Labor-Hours	Unit	Material	Labor	Equipment	Total	Total Incl O&P
0010	**WELLS**									
0011	For dewatering 10' to 20' deep, 2' diameter									
0020	with steel casing, minimum	B-6	165	.145	V.L.F.	38	5.85	2.21	46.06	53
0050	Average	↓	98	.245	↓	43	9.85	3.73	56.58	66.50

31 23 Excavation and Fill

31 23 19 – Dewatering

31 23 19.30 Wells

		Crew	Daily Output	Labor-Hours	Unit	Material	2014 Bare Costs Labor	Equipment	Total	Total Incl O&P
0100	Maximum	B-6	49	.490	V.L.F.	48.50	19.65	7.45	75.60	91.50
0300	For dewatering pumps see 01 54 33 in Reference Section									
0500	For domestic water wells, see Section 33 21 13.10									

31 23 19.40 Wellpoints

		Crew	Daily Output	Labor-Hours	Unit	Material	Labor	Equipment	Total	Total Incl O&P
0010	**WELLPOINTS** R312319-90									
0011	For equipment rental, see 01 54 33 in Reference Section									
0100	Installation and removal of single stage system									
0110	Labor only, .75 labor-hours per L.F.	1 Clab	10.70	.748	LF Hdr		27.50		27.50	42.50
0200	2.0 labor-hours per L.F.	"	4	2	"		73.50		73.50	113
0400	Pump operation, 4 @ 6 hr. shifts									
0410	Per 24 hour day	4 Eqlt	1.27	25.197	Day			1,175	1,175	1,800
0500	Per 168 hour week, 160 hr. straight, 8 hr. double time		.18	177	Week			8,375	8,375	12,700
0550	Per 4.3 week month		.04	800	Month			37,600	37,600	57,000
0600	Complete installation, operation, equipment rental, fuel &									
0610	removal of system with 2" wellpoints 5' O.C.									
0700	100' long header, 6" diameter, first month	4 Eqlt	3.23	9.907	LF Hdr	159	465		624	880
0800	Thereafter, per month		4.13	7.748		127	365		492	695
1000	200' long header, 8" diameter, first month		6	5.333		144	251		395	540
1100	Thereafter, per month		8.39	3.814		71.50	179		250.50	350
1300	500' long header, 8" diameter, first month		10.63	3.010		55.50	142		197.50	276
1400	Thereafter, per month		20.91	1.530		40	72		112	153
1600	1,000' long header, 10" diameter, first month		11.62	2.754		47.50	130		177.50	249
1700	Thereafter, per month		41.81	.765		24	36		60	80.50
1900	Note: above figures include pumping 168 hrs. per week									
1910	and include the pump operator and one stand-by pump.									

31 23 23 – Fill

31 23 23.15 Borrow, Loading And/Or Spreading

		Crew	Daily Output	Labor-Hours	Unit	Material	Labor	Equipment	Total	Total Incl O&P
0010	**BORROW, LOADING AND/OR SPREADING**									
0020	Material only, bank run gravel				Ton	15.65			15.65	17.20
0500	Haul 2 mi. spread, 200 HP dozer, bank run gravel	B-15	1100	.025			1.03	2.46	3.49	4.29
1000	Hand spread, bank run gravel	A-5	33	.545			20	2.05	22.05	33.50
1800	Delivery charge, minimum 20 tons, 1 hr. round trip, add	B-34B	130	.062			2.31	5.35	7.66	9.40
1820	1-1/2 hr. round trip, add		93	.086			3.23	7.45	10.68	13.15
1840	2 hr. round trip, add		65	.123			4.62	10.65	15.27	18.75

31 23 23.16 Fill By Borrow and Utility Bedding

		Crew	Daily Output	Labor-Hours	Unit	Material	Labor	Equipment	Total	Total Incl O&P
0010	**FILL BY BORROW AND UTILITY BEDDING**									
0049	Utility bedding, for pipe & conduit, not incl. compaction G1030-805									
0050	Crushed or screened bank run gravel	B-6	150	.160	L.C.Y.	25	6.40	2.44	33.84	40
0100	Crushed stone 3/4" to 1/2"		150	.160		28.50	6.40	2.44	37.34	43.50
0200	Sand, dead or bank		150	.160		17.55	6.40	2.44	26.39	32
0500	Compacting bedding in trench	A-1D	90	.089	E.C.Y.		3.26	.40	3.66	5.50
0600	If material source exceeds 2 miles, add for extra mileage.									
0610	See Section 31 23 23.20 for hauling mileage add.									

31 23 23.17 General Fill

		Crew	Daily Output	Labor-Hours	Unit	Material	Labor	Equipment	Total	Total Incl O&P
0010	**GENERAL FILL**									
0011	Spread dumped material, no compaction									
0020	By dozer, no compaction	B-10B	1000	.012	L.C.Y.		.54	1.33	1.87	2.28
0100	By hand	1 Clab	12	.667	"		24.50		24.50	37.50
0500	Gravel fill, compacted, under floor slabs, 4" deep	B-37	10000	.005	S.F.	.42	.19	.02	.63	.77
0600	6" deep		8600	.006		.63	.22	.02	.87	1.05
0700	9" deep		7200	.007		1.05	.26	.02	1.33	1.58

31 23 Excavation and Fill

31 23 23 – Fill

31 23 23.17 General Fill

		Crew	Daily Output	Labor-Hours	Unit	Material	2014 Bare Costs Labor	Equipment	Total	Total Incl O&P
0800	12" deep	B-37	6000	.008	S.F.	1.48	.31	.03	1.82	2.13
1000	Alternate pricing method, 4" deep		120	.400	E.C.Y.	31.50	15.50	1.32	48.32	60.50
1100	6" deep		160	.300		31.50	11.60	.99	44.09	54
1200	9" deep		200	.240		31.50	9.30	.79	41.59	50
1300	12" deep		220	.218		31.50	8.45	.72	40.67	49
1400	Granular fill				L.C.Y.	21			21	23

31 23 23.20 Hauling

		Crew	Daily Output	Labor-Hours	Unit	Material	Labor	Equipment	Total	Total Incl O&P
0010	**HAULING**									
0011	Excavated or borrow, loose cubic yards									
0012	no loading equipment, including hauling, waiting, loading/dumping									
0013	time per cycle (wait, load, travel, unload or dump & return)									
0014	8 C.Y. truck, 15 MPH ave, cycle 0.5 miles, 10 min. wait/Ld./Uld.	B-34A	320	.025	L.C.Y.		.94	1.30	2.24	2.86
0016	cycle 1 mile		272	.029			1.10	1.53	2.63	3.38
0018	cycle 2 miles		208	.038			1.44	2.01	3.45	4.41
0020	cycle 4 miles		144	.056			2.09	2.90	4.99	6.35
0022	cycle 6 miles		112	.071			2.68	3.73	6.41	8.20
0024	cycle 8 miles		88	.091			3.41	4.74	8.15	10.40
0026	20 MPH ave, cycle 0.5 mile		336	.024			.89	1.24	2.13	2.73
0028	cycle 1 mile		296	.027			1.02	1.41	2.43	3.10
0030	cycle 2 miles		240	.033			1.25	1.74	2.99	3.82
0032	cycle 4 miles		176	.045			1.71	2.37	4.08	5.20
0034	cycle 6 miles		136	.059			2.21	3.07	5.28	6.75
0036	cycle 8 miles		112	.071			2.68	3.73	6.41	8.20
0044	25 MPH ave, cycle 4 miles		192	.042			1.56	2.17	3.73	4.78
0046	cycle 6 miles		160	.050			1.88	2.61	4.49	5.75
0048	cycle 8 miles		128	.063			2.35	3.26	5.61	7.15
0050	30 MPH ave, cycle 4 miles		216	.037			1.39	1.93	3.32	4.25
0052	cycle 6 miles		176	.045			1.71	2.37	4.08	5.20
0054	cycle 8 miles		144	.056			2.09	2.90	4.99	6.35
0114	15 MPH ave, cycle 0.5 mile, 15 min. wait/Ld./Uld.		224	.036			1.34	1.86	3.20	4.10
0116	cycle 1 mile		200	.040			1.50	2.09	3.59	4.58
0118	cycle 2 miles		168	.048			1.79	2.48	4.27	5.45
0120	cycle 4 miles		120	.067			2.50	3.48	5.98	7.65
0122	cycle 6 miles		96	.083			3.13	4.35	7.48	9.55
0124	cycle 8 miles		80	.100			3.76	5.20	8.96	11.50
0126	20 MPH ave, cycle 0.5 mile		232	.034			1.29	1.80	3.09	3.96
0128	cycle 1 mile		208	.038			1.44	2.01	3.45	4.41
0130	cycle 2 miles		184	.043			1.63	2.27	3.90	4.98
0132	cycle 4 miles		144	.056			2.09	2.90	4.99	6.35
0134	cycle 6 miles		112	.071			2.68	3.73	6.41	8.20
0136	cycle 8 miles		96	.083			3.13	4.35	7.48	9.55
0144	25 MPH ave, cycle 4 miles		152	.053			1.98	2.74	4.72	6.05
0146	cycle 6 miles		128	.063			2.35	3.26	5.61	7.15
0148	cycle 8 miles		112	.071			2.68	3.73	6.41	8.20
0150	30 MPH ave, cycle 4 miles		168	.048			1.79	2.48	4.27	5.45
0152	cycle 6 miles		144	.056			2.09	2.90	4.99	6.35
0154	cycle 8 miles		120	.067			2.50	3.48	5.98	7.65
0214	15 MPH ave, cycle 0.5 mile, 20 min wait/Ld./Uld.		176	.045			1.71	2.37	4.08	5.20
0216	cycle 1 mile		160	.050			1.88	2.61	4.49	5.75
0218	cycle 2 miles		136	.059			2.21	3.07	5.28	6.75
0220	cycle 4 miles		104	.077			2.89	4.01	6.90	8.80
0222	cycle 6 miles		88	.091			3.41	4.74	8.15	10.40

31 23 Excavation and Fill

31 23 23 – Fill

31 23 23.20 Hauling

		Crew	Daily Output	Labor-Hours	Unit	Material	2014 Bare Costs Labor	Equipment	Total	Total Incl O&P
0224	cycle 8 miles	B-34A	72	.111	L.C.Y.		4.17	5.80	9.97	12.70
0226	20 MPH ave, cycle 0.5 mile		176	.045			1.71	2.37	4.08	5.20
0228	cycle 1 mile		168	.048			1.79	2.48	4.27	5.45
0230	cycle 2 miles		144	.056			2.09	2.90	4.99	6.35
0232	cycle 4 miles		120	.067			2.50	3.48	5.98	7.65
0234	cycle 6 miles		96	.083			3.13	4.35	7.48	9.55
0236	cycle 8 miles		88	.091			3.41	4.74	8.15	10.40
0244	25 MPH ave, cycle 4 miles		128	.063			2.35	3.26	5.61	7.15
0246	cycle 6 miles		112	.071			2.68	3.73	6.41	8.20
0248	cycle 8 miles		96	.083			3.13	4.35	7.48	9.55
0250	30 MPH ave, cycle 4 miles		136	.059			2.21	3.07	5.28	6.75
0252	cycle 6 miles		120	.067			2.50	3.48	5.98	7.65
0254	cycle 8 miles		104	.077			2.89	4.01	6.90	8.80
0314	15 MPH ave, cycle 0.5 mile, 25 min wait/Ld./Uld.		144	.056			2.09	2.90	4.99	6.35
0316	cycle 1 mile		128	.063			2.35	3.26	5.61	7.15
0318	cycle 2 miles		112	.071			2.68	3.73	6.41	8.20
0320	cycle 4 miles		96	.083			3.13	4.35	7.48	9.55
0322	cycle 6 miles		80	.100			3.76	5.20	8.96	11.50
0324	cycle 8 miles		64	.125			4.69	6.50	11.19	14.30
0326	20 MPH ave, cycle 0.5 mile		144	.056			2.09	2.90	4.99	6.35
0328	cycle 1 mile		136	.059			2.21	3.07	5.28	6.75
0330	cycle 2 miles		120	.067			2.50	3.48	5.98	7.65
0332	cycle 4 miles		104	.077			2.89	4.01	6.90	8.80
0334	cycle 6 miles		88	.091			3.41	4.74	8.15	10.40
0336	cycle 8 miles		80	.100			3.76	5.20	8.96	11.50
0344	25 MPH ave, cycle 4 miles		112	.071			2.68	3.73	6.41	8.20
0346	cycle 6 miles		96	.083			3.13	4.35	7.48	9.55
0348	cycle 8 miles		88	.091			3.41	4.74	8.15	10.40
0350	30 MPH ave, cycle 4 miles		112	.071			2.68	3.73	6.41	8.20
0352	cycle 6 miles		104	.077			2.89	4.01	6.90	8.80
0354	cycle 8 miles		96	.083			3.13	4.35	7.48	9.55
0414	15 MPH ave, cycle 0.5 mile, 30 min wait/Ld./Uld.		120	.067			2.50	3.48	5.98	7.65
0416	cycle 1 mile		112	.071			2.68	3.73	6.41	8.20
0418	cycle 2 miles		96	.083			3.13	4.35	7.48	9.55
0420	cycle 4 miles		80	.100			3.76	5.20	8.96	11.50
0422	cycle 6 miles		72	.111			4.17	5.80	9.97	12.70
0424	cycle 8 miles		64	.125			4.69	6.50	11.19	14.30
0426	20 MPH ave, cycle 0.5 mile		120	.067			2.50	3.48	5.98	7.65
0428	cycle 1 mile		112	.071			2.68	3.73	6.41	8.20
0430	cycle 2 miles		104	.077			2.89	4.01	6.90	8.80
0432	cycle 4 miles		88	.091			3.41	4.74	8.15	10.40
0434	cycle 6 miles		80	.100			3.76	5.20	8.96	11.50
0436	cycle 8 miles		72	.111			4.17	5.80	9.97	12.70
0444	25 MPH ave, cycle 4 miles		96	.083			3.13	4.35	7.48	9.55
0446	cycle 6 miles		88	.091			3.41	4.74	8.15	10.40
0448	cycle 8 miles		80	.100			3.76	5.20	8.96	11.50
0450	30 MPH ave, cycle 4 miles		96	.083			3.13	4.35	7.48	9.55
0452	cycle 6 miles		88	.091			3.41	4.74	8.15	10.40
0454	cycle 8 miles		80	.100			3.76	5.20	8.96	11.50
0514	15 MPH ave, cycle 0.5 mile, 35 min wait/Ld./Uld.		104	.077			2.89	4.01	6.90	8.80
0516	cycle 1 mile		96	.083			3.13	4.35	7.48	9.55
0518	cycle 2 miles		88	.091			3.41	4.74	8.15	10.40
0520	cycle 4 miles		72	.111			4.17	5.80	9.97	12.70

31 23 Excavation and Fill

31 23 23 – Fill

31 23 23.20 Hauling

		Crew	Daily Output	Labor-Hours	Unit	Material	2014 Bare Costs Labor	Equipment	Total	Total Incl O&P
0522	cycle 6 miles	B-34A	64	.125	L.C.Y.		4.69	6.50	11.19	14.30
0524	cycle 8 miles		56	.143			5.35	7.45	12.80	16.40
0526	20 MPH ave, cycle 0.5 mile		104	.077			2.89	4.01	6.90	8.80
0528	cycle 1 mile		96	.083			3.13	4.35	7.48	9.55
0530	cycle 2 miles		96	.083			3.13	4.35	7.48	9.55
0532	cycle 4 miles		80	.100			3.76	5.20	8.96	11.50
0534	cycle 6 miles		72	.111			4.17	5.80	9.97	12.70
0536	cycle 8 miles		64	.125			4.69	6.50	11.19	14.30
0544	25 MPH ave, cycle 4 miles		88	.091			3.41	4.74	8.15	10.40
0546	cycle 6 miles		80	.100			3.76	5.20	8.96	11.50
0548	cycle 8 miles		72	.111			4.17	5.80	9.97	12.70
0550	30 MPH ave, cycle 4 miles		88	.091			3.41	4.74	8.15	10.40
0552	cycle 6 miles		80	.100			3.76	5.20	8.96	11.50
0554	cycle 8 miles		72	.111			4.17	5.80	9.97	12.70
1014	12 C.Y. truck, cycle 0.5 mile, 15 MPH ave, 15 min. wait/Ld./Uld.	B-34B	336	.024			.89	2.06	2.95	3.63
1016	cycle 1 mile		300	.027			1	2.31	3.31	4.07
1018	cycle 2 miles		252	.032			1.19	2.75	3.94	4.84
1020	cycle 4 miles		180	.044			1.67	3.85	5.52	6.80
1022	cycle 6 miles		144	.056			2.09	4.81	6.90	8.50
1024	cycle 8 miles		120	.067			2.50	5.75	8.25	10.15
1025	cycle 10 miles		96	.083			3.13	7.20	10.33	12.70
1026	20 MPH ave, cycle 0.5 mile		348	.023			.86	1.99	2.85	3.51
1028	cycle 1 mile		312	.026			.96	2.22	3.18	3.91
1030	cycle 2 miles		276	.029			1.09	2.51	3.60	4.42
1032	cycle 4 miles		216	.037			1.39	3.21	4.60	5.65
1034	cycle 6 miles		168	.048			1.79	4.12	5.91	7.25
1036	cycle 8 miles		144	.056			2.09	4.81	6.90	8.50
1038	cycle 10 miles		120	.067			2.50	5.75	8.25	10.15
1040	25 MPH ave, cycle 4 miles		228	.035			1.32	3.04	4.36	5.35
1042	cycle 6 miles		192	.042			1.56	3.61	5.17	6.35
1044	cycle 8 miles		168	.048			1.79	4.12	5.91	7.25
1046	cycle 10 miles		144	.056			2.09	4.81	6.90	8.50
1050	30 MPH ave, cycle 4 miles		252	.032			1.19	2.75	3.94	4.84
1052	cycle 6 miles		216	.037			1.39	3.21	4.60	5.65
1054	cycle 8 miles		180	.044			1.67	3.85	5.52	6.80
1056	cycle 10 miles		156	.051			1.93	4.44	6.37	7.80
1060	35 MPH ave, cycle 4 miles		264	.030			1.14	2.62	3.76	4.62
1062	cycle 6 miles		228	.035			1.32	3.04	4.36	5.35
1064	cycle 8 miles		204	.039			1.47	3.39	4.86	6
1066	cycle 10 miles		180	.044			1.67	3.85	5.52	6.80
1068	cycle 20 miles		120	.067			2.50	5.75	8.25	10.15
1069	cycle 30 miles		84	.095			3.58	8.25	11.83	14.50
1070	cycle 40 miles		72	.111			4.17	9.60	13.77	16.95
1072	40 MPH ave, cycle 6 miles		240	.033			1.25	2.88	4.13	5.10
1074	cycle 8 miles		216	.037			1.39	3.21	4.60	5.65
1076	cycle 10 miles		192	.042			1.56	3.61	5.17	6.35
1078	cycle 20 miles		120	.067			2.50	5.75	8.25	10.15
1080	cycle 30 miles		96	.083			3.13	7.20	10.33	12.70
1082	cycle 40 miles		72	.111			4.17	9.60	13.77	16.95
1084	cycle 50 miles		60	.133			5	11.55	16.55	20.50
1094	45 MPH ave, cycle 8 miles		216	.037			1.39	3.21	4.60	5.65
1096	cycle 10 miles		204	.039			1.47	3.39	4.86	6
1098	cycle 20 miles		132	.061			2.28	5.25	7.53	9.20

31 23 Excavation and Fill

31 23 23 – Fill

31 23 23.20 Hauling		Crew	Daily Output	Labor-Hours	Unit	Material	2014 Bare Costs Labor	Equipment	Total	Total Incl O&P
1100	cycle 30 miles	B-34B	108	.074	L.C.Y.		2.78	6.40	9.18	11.30
1102	cycle 40 miles		84	.095			3.58	8.25	11.83	14.50
1104	cycle 50 miles		72	.111			4.17	9.60	13.77	16.95
1106	50 MPH ave, cycle 10 miles		216	.037			1.39	3.21	4.60	5.65
1108	cycle 20 miles		144	.056			2.09	4.81	6.90	8.50
1110	cycle 30 miles		108	.074			2.78	6.40	9.18	11.30
1112	cycle 40 miles		84	.095			3.58	8.25	11.83	14.50
1114	cycle 50 miles		72	.111			4.17	9.60	13.77	16.95
1214	15 MPH ave, cycle 0.5 mile, 20 min. wait/Ld./Uld.		264	.030			1.14	2.62	3.76	4.62
1216	cycle 1 mile		240	.033			1.25	2.88	4.13	5.10
1218	cycle 2 miles		204	.039			1.47	3.39	4.86	6
1220	cycle 4 miles		156	.051			1.93	4.44	6.37	7.80
1222	cycle 6 miles		132	.061			2.28	5.25	7.53	9.20
1224	cycle 8 miles		108	.074			2.78	6.40	9.18	11.30
1225	cycle 10 miles		96	.083			3.13	7.20	10.33	12.70
1226	20 MPH ave, cycle 0.5 mile		264	.030			1.14	2.62	3.76	4.62
1228	cycle 1 mile		252	.032			1.19	2.75	3.94	4.84
1230	cycle 2 miles		216	.037			1.39	3.21	4.60	5.65
1232	cycle 4 miles		180	.044			1.67	3.85	5.52	6.80
1234	cycle 6 miles		144	.056			2.09	4.81	6.90	8.50
1236	cycle 8 miles		132	.061			2.28	5.25	7.53	9.20
1238	cycle 10 miles		108	.074			2.78	6.40	9.18	11.30
1240	25 MPH ave, cycle 4 miles		192	.042			1.56	3.61	5.17	6.35
1242	cycle 6 miles		168	.048			1.79	4.12	5.91	7.25
1244	cycle 8 miles		144	.056			2.09	4.81	6.90	8.50
1246	cycle 10 miles		132	.061			2.28	5.25	7.53	9.20
1250	30 MPH ave, cycle 4 miles		204	.039			1.47	3.39	4.86	6
1252	cycle 6 miles		180	.044			1.67	3.85	5.52	6.80
1254	cycle 8 miles		156	.051			1.93	4.44	6.37	7.80
1256	cycle 10 miles		144	.056			2.09	4.81	6.90	8.50
1260	35 MPH ave, cycle 4 miles		216	.037			1.39	3.21	4.60	5.65
1262	cycle 6 miles		192	.042			1.56	3.61	5.17	6.35
1264	cycle 8 miles		168	.048			1.79	4.12	5.91	7.25
1266	cycle 10 miles		156	.051			1.93	4.44	6.37	7.80
1268	cycle 20 miles		108	.074			2.78	6.40	9.18	11.30
1269	cycle 30 miles		72	.111			4.17	9.60	13.77	16.95
1270	cycle 40 miles		60	.133			5	11.55	16.55	20.50
1272	40 MPH ave, cycle 6 miles		192	.042			1.56	3.61	5.17	6.35
1274	cycle 8 miles		180	.044			1.67	3.85	5.52	6.80
1276	cycle 10 miles		156	.051			1.93	4.44	6.37	7.80
1278	cycle 20 miles		108	.074			2.78	6.40	9.18	11.30
1280	cycle 30 miles		84	.095			3.58	8.25	11.83	14.50
1282	cycle 40 miles		72	.111			4.17	9.60	13.77	16.95
1284	cycle 50 miles		60	.133			5	11.55	16.55	20.50
1294	45 MPH ave, cycle 8 miles		180	.044			1.67	3.85	5.52	6.80
1296	cycle 10 miles		168	.048			1.79	4.12	5.91	7.25
1298	cycle 20 miles		120	.067			2.50	5.75	8.25	10.15
1300	cycle 30 miles		96	.083			3.13	7.20	10.33	12.70
1302	cycle 40 miles		72	.111			4.17	9.60	13.77	16.95
1304	cycle 50 miles		60	.133			5	11.55	16.55	20.50
1306	50 MPH ave, cycle 10 miles		180	.044			1.67	3.85	5.52	6.80
1308	cycle 20 miles		132	.061			2.28	5.25	7.53	9.20
1310	cycle 30 miles		96	.083			3.13	7.20	10.33	12.70

31 23 Excavation and Fill

31 23 23 – Fill

31 23 23.20 Hauling		Crew	Daily Output	Labor-Hours	Unit	Material	2014 Bare Costs Labor	Equipment	Total	Total Incl O&P
1312	cycle 40 miles	B-34B	84	.095	L.C.Y.		3.58	8.25	11.83	14.50
1314	cycle 50 miles		72	.111			4.17	9.60	13.77	16.95
1414	15 MPH ave, cycle 0.5 mile, 25 min. wait/Ld./Uld.		204	.039			1.47	3.39	4.86	6
1416	cycle 1 mile		192	.042			1.56	3.61	5.17	6.35
1418	cycle 2 miles		168	.048			1.79	4.12	5.91	7.25
1420	cycle 4 miles		132	.061			2.28	5.25	7.53	9.20
1422	cycle 6 miles		120	.067			2.50	5.75	8.25	10.15
1424	cycle 8 miles		96	.083			3.13	7.20	10.33	12.70
1425	cycle 10 miles		84	.095			3.58	8.25	11.83	14.50
1426	20 MPH ave, cycle 0.5 mile		216	.037			1.39	3.21	4.60	5.65
1428	cycle 1 mile		204	.039			1.47	3.39	4.86	6
1430	cycle 2 miles		180	.044			1.67	3.85	5.52	6.80
1432	cycle 4 miles		156	.051			1.93	4.44	6.37	7.80
1434	cycle 6 miles		132	.061			2.28	5.25	7.53	9.20
1436	cycle 8 miles		120	.067			2.50	5.75	8.25	10.15
1438	cycle 10 miles		96	.083			3.13	7.20	10.33	12.70
1440	25 MPH ave, cycle 4 miles		168	.048			1.79	4.12	5.91	7.25
1442	cycle 6 miles		144	.056			2.09	4.81	6.90	8.50
1444	cycle 8 miles		132	.061			2.28	5.25	7.53	9.20
1446	cycle 10 miles		108	.074			2.78	6.40	9.18	11.30
1450	30 MPH ave, cycle 4 miles		168	.048			1.79	4.12	5.91	7.25
1452	cycle 6 miles		156	.051			1.93	4.44	6.37	7.80
1454	cycle 8 miles		132	.061			2.28	5.25	7.53	9.20
1456	cycle 10 miles		120	.067			2.50	5.75	8.25	10.15
1460	35 MPH ave, cycle 4 miles		180	.044			1.67	3.85	5.52	6.80
1462	cycle 6 miles		156	.051			1.93	4.44	6.37	7.80
1464	cycle 8 miles		144	.056			2.09	4.81	6.90	8.50
1466	cycle 10 miles		132	.061			2.28	5.25	7.53	9.20
1468	cycle 20 miles		96	.083			3.13	7.20	10.33	12.70
1469	cycle 30 miles		72	.111			4.17	9.60	13.77	16.95
1470	cycle 40 miles		60	.133			5	11.55	16.55	20.50
1472	40 MPH ave, cycle 6 miles		168	.048			1.79	4.12	5.91	7.25
1474	cycle 8 miles		156	.051			1.93	4.44	6.37	7.80
1476	cycle 10 miles		144	.056			2.09	4.81	6.90	8.50
1478	cycle 20 miles		96	.083			3.13	7.20	10.33	12.70
1480	cycle 30 miles		84	.095			3.58	8.25	11.83	14.50
1482	cycle 40 miles		60	.133			5	11.55	16.55	20.50
1484	cycle 50 miles		60	.133			5	11.55	16.55	20.50
1494	45 MPH ave, cycle 8 miles		156	.051			1.93	4.44	6.37	7.80
1496	cycle 10 miles		144	.056			2.09	4.81	6.90	8.50
1498	cycle 20 miles		108	.074			2.78	6.40	9.18	11.30
1500	cycle 30 miles		84	.095			3.58	8.25	11.83	14.50
1502	cycle 40 miles		72	.111			4.17	9.60	13.77	16.95
1504	cycle 50 miles		60	.133			5	11.55	16.55	20.50
1506	50 MPH ave, cycle 10 miles		156	.051			1.93	4.44	6.37	7.80
1508	cycle 20 miles		120	.067			2.50	5.75	8.25	10.15
1510	cycle 30 miles		96	.083			3.13	7.20	10.33	12.70
1512	cycle 40 miles		72	.111			4.17	9.60	13.77	16.95
1514	cycle 50 miles		60	.133			5	11.55	16.55	20.50
1614	15 MPH, cycle 0.5 mile, 30 min. wait/Ld./Uld.		180	.044			1.67	3.85	5.52	6.80
1616	cycle 1 mile		168	.048			1.79	4.12	5.91	7.25
1618	cycle 2 miles		144	.056			2.09	4.81	6.90	8.50
1620	cycle 4 miles		120	.067			2.50	5.75	8.25	10.15

31 23 Excavation and Fill

31 23 23 – Fill

31 23 23.20 Hauling

		Crew	Daily Output	Labor-Hours	Unit	Material	2014 Bare Costs Labor	Equipment	Total	Total Incl O&P
1622	cycle 6 miles	B-34B	108	.074	L.C.Y.		2.78	6.40	9.18	11.30
1624	cycle 8 miles		84	.095			3.58	8.25	11.83	14.50
1625	cycle 10 miles		84	.095			3.58	8.25	11.83	14.50
1626	20 MPH ave, cycle 0.5 mile		180	.044			1.67	3.85	5.52	6.80
1628	cycle 1 mile		168	.048			1.79	4.12	5.91	7.25
1630	cycle 2 miles		156	.051			1.93	4.44	6.37	7.80
1632	cycle 4 miles		132	.061			2.28	5.25	7.53	9.20
1634	cycle 6 miles		120	.067			2.50	5.75	8.25	10.15
1636	cycle 8 miles		108	.074			2.78	6.40	9.18	11.30
1638	cycle 10 miles		96	.083			3.13	7.20	10.33	12.70
1640	25 MPH ave, cycle 4 miles		144	.056			2.09	4.81	6.90	8.50
1642	cycle 6 miles		132	.061			2.28	5.25	7.53	9.20
1644	cycle 8 miles		108	.074			2.78	6.40	9.18	11.30
1646	cycle 10 miles		108	.074			2.78	6.40	9.18	11.30
1650	30 MPH ave, cycle 4 miles		144	.056			2.09	4.81	6.90	8.50
1652	cycle 6 miles		132	.061			2.28	5.25	7.53	9.20
1654	cycle 8 miles		120	.067			2.50	5.75	8.25	10.15
1656	cycle 10 miles		108	.074			2.78	6.40	9.18	11.30
1660	35 MPH ave, cycle 4 miles		156	.051			1.93	4.44	6.37	7.80
1662	cycle 6 miles		144	.056			2.09	4.81	6.90	8.50
1664	cycle 8 miles		132	.061			2.28	5.25	7.53	9.20
1666	cycle 10 miles		120	.067			2.50	5.75	8.25	10.15
1668	cycle 20 miles		84	.095			3.58	8.25	11.83	14.50
1669	cycle 30 miles		72	.111			4.17	9.60	13.77	16.95
1670	cycle 40 miles		60	.133			5	11.55	16.55	20.50
1672	40 MPH, cycle 6 miles		144	.056			2.09	4.81	6.90	8.50
1674	cycle 8 miles		132	.061			2.28	5.25	7.53	9.20
1676	cycle 10 miles		120	.067			2.50	5.75	8.25	10.15
1678	cycle 20 miles		96	.083			3.13	7.20	10.33	12.70
1680	cycle 30 miles		72	.111			4.17	9.60	13.77	16.95
1682	cycle 40 miles		60	.133			5	11.55	16.55	20.50
1684	cycle 50 miles		48	.167			6.25	14.40	20.65	25.50
1694	45 MPH ave, cycle 8 miles		144	.056			2.09	4.81	6.90	8.50
1696	cycle 10 miles		132	.061			2.28	5.25	7.53	9.20
1698	cycle 20 miles		96	.083			3.13	7.20	10.33	12.70
1700	cycle 30 miles		84	.095			3.58	8.25	11.83	14.50
1702	cycle 40 miles		60	.133			5	11.55	16.55	20.50
1704	cycle 50 miles		60	.133			5	11.55	16.55	20.50
1706	50 MPH ave, cycle 10 miles		132	.061			2.28	5.25	7.53	9.20
1708	cycle 20 miles		108	.074			2.78	6.40	9.18	11.30
1710	cycle 30 miles		84	.095			3.58	8.25	11.83	14.50
1712	cycle 40 miles		72	.111			4.17	9.60	13.77	16.95
1714	cycle 50 miles		60	.133			5	11.55	16.55	20.50
2000	Hauling, 8 C.Y. truck, small project cost per hour	B-34A	8	1	Hr.		37.50	52	89.50	115
2100	12 C.Y. Truck	B-34B	8	1			37.50	86.50	124	153
2150	16.5 C.Y. Truck	B-34C	8	1			37.50	92.50	130	160
2175	18 C.Y. 8 wheel Truck	B-34I	8	1			37.50	108	145.50	177
2200	20 C.Y. Truck	B-34D	8	1			37.50	94	131.50	162
2300	Grading at dump, or embankment if required, by dozer	B-10B	1000	.012	L.C.Y.		.54	1.33	1.87	2.28
2310	Spotter at fill or cut, if required	1 Clab	8	1	Hr.		36.50		36.50	56.50
9014	18 C.Y. truck, 8 wheels,15 min. wait/Ld./Uld.,15 MPH, cycle 0.5 mi.	B-34I	504	.016	L.C.Y.		.60	1.72	2.32	2.80
9016	cycle 1 mile		450	.018			.67	1.93	2.60	3.14
9018	cycle 2 miles		378	.021			.79	2.29	3.08	3.73

31 23 Excavation and Fill

31 23 23 – Fill

31 23 23.20 Hauling

		Crew	Daily Output	Labor-Hours	Unit	Material	2014 Bare Costs Labor	Equipment	Total	Total Incl O&P
9020	cycle 4 miles	B-34I	270	.030	L.C.Y.		1.11	3.21	4.32	5.25
9022	cycle 6 miles		216	.037			1.39	4.02	5.41	6.55
9024	cycle 8 miles		180	.044			1.67	4.82	6.49	7.85
9025	cycle 10 miles		144	.056			2.09	6.05	8.14	9.85
9026	20 MPH ave, cycle 0.5 mile		522	.015			.58	1.66	2.24	2.71
9028	cycle 1 mile		468	.017			.64	1.85	2.49	3.02
9030	cycle 2 miles		414	.019			.73	2.10	2.83	3.41
9032	cycle 4 miles		324	.025			.93	2.68	3.61	4.36
9034	cycle 6 miles		252	.032			1.19	3.44	4.63	5.60
9036	cycle 8 miles		216	.037			1.39	4.02	5.41	6.55
9038	cycle 10 miles		180	.044			1.67	4.82	6.49	7.85
9040	25 MPH ave, cycle 4 miles		342	.023			.88	2.54	3.42	4.13
9042	cycle 6 miles		288	.028			1.04	3.01	4.05	4.90
9044	cycle 8 miles		252	.032			1.19	3.44	4.63	5.60
9046	cycle 10 miles		216	.037			1.39	4.02	5.41	6.55
9050	30 MPH ave, cycle 4 miles		378	.021			.79	2.29	3.08	3.73
9052	cycle 6 miles		324	.025			.93	2.68	3.61	4.36
9054	cycle 8 miles		270	.030			1.11	3.21	4.32	5.25
9056	cycle 10 miles		234	.034			1.28	3.71	4.99	6.05
9060	35 MPH ave, cycle 4 miles		396	.020			.76	2.19	2.95	3.57
9062	cycle 6 miles		342	.023			.88	2.54	3.42	4.13
9064	cycle 8 miles		288	.028			1.04	3.01	4.05	4.90
9066	cycle 10 miles		270	.030			1.11	3.21	4.32	5.25
9068	cycle 20 miles		162	.049			1.85	5.35	7.20	8.75
9070	cycle 30 miles		126	.063			2.38	6.90	9.28	11.20
9072	cycle 40 miles		90	.089			3.34	9.65	12.99	15.70
9074	40 MPH ave, cycle 6 miles		360	.022			.83	2.41	3.24	3.92
9076	cycle 8 miles		324	.025			.93	2.68	3.61	4.36
9078	cycle 10 miles		288	.028			1.04	3.01	4.05	4.90
9080	cycle 20 miles		180	.044			1.67	4.82	6.49	7.85
9082	cycle 30 miles		144	.056			2.09	6.05	8.14	9.85
9084	cycle 40 miles		108	.074			2.78	8.05	10.83	13.10
9086	cycle 50 miles		90	.089			3.34	9.65	12.99	15.70
9094	45 MPH ave, cycle 8 miles		324	.025			.93	2.68	3.61	4.36
9096	cycle 10 miles		306	.026			.98	2.83	3.81	4.62
9098	cycle 20 miles		198	.040			1.52	4.38	5.90	7.15
9100	cycle 30 miles		144	.056			2.09	6.05	8.14	9.85
9102	cycle 40 miles		126	.063			2.38	6.90	9.28	11.20
9104	cycle 50 miles		108	.074			2.78	8.05	10.83	13.10
9106	50 MPH ave, cycle 10 miles		324	.025			.93	2.68	3.61	4.36
9108	cycle 20 miles		216	.037			1.39	4.02	5.41	6.55
9110	cycle 30 miles		162	.049			1.85	5.35	7.20	8.75
9112	cycle 40 miles		126	.063			2.38	6.90	9.28	11.20
9114	cycle 50 miles		108	.074			2.78	8.05	10.83	13.10
9214	20 min. wait/Ld./Uld.,15 MPH, cycle 0.5 mi.		396	.020			.76	2.19	2.95	3.57
9216	cycle 1 mile		360	.022			.83	2.41	3.24	3.92
9218	cycle 2 miles		306	.026			.98	2.83	3.81	4.62
9220	cycle 4 miles		234	.034			1.28	3.71	4.99	6.05
9222	cycle 6 miles		198	.040			1.52	4.38	5.90	7.15
9224	cycle 8 miles		162	.049			1.85	5.35	7.20	8.75
9225	cycle 10 miles		144	.056			2.09	6.05	8.14	9.85
9226	20 MPH ave, cycle 0.5 mile		396	.020			.76	2.19	2.95	3.57
9228	cycle 1 mile		378	.021			.79	2.29	3.08	3.73

31 23 Excavation and Fill

31 23 23 – Fill

31 23 23.20 Hauling

		Crew	Daily Output	Labor-Hours	Unit	Material	Labor	Equipment	Total	Total Incl O&P
9230	cycle 2 miles	B-34I	324	.025	L.C.Y.		.93	2.68	3.61	4.36
9232	cycle 4 miles		270	.030			1.11	3.21	4.32	5.25
9234	cycle 6 miles		216	.037			1.39	4.02	5.41	6.55
9236	cycle 8 miles		198	.040			1.52	4.38	5.90	7.15
9238	cycle 10 miles		162	.049			1.85	5.35	7.20	8.75
9240	25 MPH ave, cycle 4 miles		288	.028			1.04	3.01	4.05	4.90
9242	cycle 6 miles		252	.032			1.19	3.44	4.63	5.60
9244	cycle 8 miles		216	.037			1.39	4.02	5.41	6.55
9246	cycle 10 miles		198	.040			1.52	4.38	5.90	7.15
9250	30 MPH ave, cycle 4 miles		306	.026			.98	2.83	3.81	4.62
9252	cycle 6 miles		270	.030			1.11	3.21	4.32	5.25
9254	cycle 8 miles		234	.034			1.28	3.71	4.99	6.05
9256	cycle 10 miles		216	.037			1.39	4.02	5.41	6.55
9260	35 MPH ave, cycle 4 miles		324	.025			.93	2.68	3.61	4.36
9262	cycle 6 miles		288	.028			1.04	3.01	4.05	4.90
9264	cycle 8 miles		252	.032			1.19	3.44	4.63	5.60
9266	cycle 10 miles		234	.034			1.28	3.71	4.99	6.05
9268	cycle 20 miles		162	.049			1.85	5.35	7.20	8.75
9270	cycle 30 miles		108	.074			2.78	8.05	10.83	13.10
9272	cycle 40 miles		90	.089			3.34	9.65	12.99	15.70
9274	40 MPH ave, cycle 6 miles		288	.028			1.04	3.01	4.05	4.90
9276	cycle 8 miles		270	.030			1.11	3.21	4.32	5.25
9278	cycle 10 miles		234	.034			1.28	3.71	4.99	6.05
9280	cycle 20 miles		162	.049			1.85	5.35	7.20	8.75
9282	cycle 30 miles		126	.063			2.38	6.90	9.28	11.20
9284	cycle 40 miles		108	.074			2.78	8.05	10.83	13.10
9286	cycle 50 miles		90	.089			3.34	9.65	12.99	15.70
9294	45 MPH ave, cycle 8 miles		270	.030			1.11	3.21	4.32	5.25
9296	cycle 10 miles		252	.032			1.19	3.44	4.63	5.60
9298	cycle 20 miles		180	.044			1.67	4.82	6.49	7.85
9300	cycle 30 miles		144	.056			2.09	6.05	8.14	9.85
9302	cycle 40 miles		108	.074			2.78	8.05	10.83	13.10
9304	cycle 50 miles		90	.089			3.34	9.65	12.99	15.70
9306	50 MPH ave, cycle 10 miles		270	.030			1.11	3.21	4.32	5.25
9308	cycle 20 miles		198	.040			1.52	4.38	5.90	7.15
9310	cycle 30 miles		144	.056			2.09	6.05	8.14	9.85
9312	cycle 40 miles		126	.063			2.38	6.90	9.28	11.20
9314	cycle 50 miles		108	.074			2.78	8.05	10.83	13.10
9414	25 min. wait/Ld./Uld.,15 MPH, cycle 0.5 mi.		306	.026			.98	2.83	3.81	4.62
9416	cycle 1 mile		288	.028			1.04	3.01	4.05	4.90
9418	cycle 2 miles		252	.032			1.19	3.44	4.63	5.60
9420	cycle 4 miles		198	.040			1.52	4.38	5.90	7.15
9422	cycle 6 miles		180	.044			1.67	4.82	6.49	7.85
9424	cycle 8 miles		144	.056			2.09	6.05	8.14	9.85
9425	cycle 10 miles		126	.063			2.38	6.90	9.28	11.20
9426	20 MPH ave, cycle 0.5 mile		324	.025			.93	2.68	3.61	4.36
9428	cycle 1 mile		306	.026			.98	2.83	3.81	4.62
9430	cycle 2 miles		270	.030			1.11	3.21	4.32	5.25
9432	cycle 4 miles		234	.034			1.28	3.71	4.99	6.05
9434	cycle 6 miles		198	.040			1.52	4.38	5.90	7.15
9436	cycle 8 miles		180	.044			1.67	4.82	6.49	7.85
9438	cycle 10 miles		144	.056			2.09	6.05	8.14	9.85
9440	25 MPH ave, cycle 4 miles		252	.032			1.19	3.44	4.63	5.60

31 23 Excavation and Fill

31 23 23 – Fill

31 23 23.20 Hauling

		Crew	Daily Output	Labor-Hours	Unit	Material	2014 Bare Costs Labor	Equipment	Total	Total Incl O&P
9442	cycle 6 miles	B-34I	216	.037	L.C.Y.		1.39	4.02	5.41	6.55
9444	cycle 8 miles		198	.040			1.52	4.38	5.90	7.15
9446	cycle 10 miles		180	.044			1.67	4.82	6.49	7.85
9450	30 MPH ave, cycle 4 miles		252	.032			1.19	3.44	4.63	5.60
9452	cycle 6 miles		234	.034			1.28	3.71	4.99	6.05
9454	cycle 8 miles		198	.040			1.52	4.38	5.90	7.15
9456	cycle 10 miles		180	.044			1.67	4.82	6.49	7.85
9460	35 MPH ave, cycle 4 miles		270	.030			1.11	3.21	4.32	5.25
9462	cycle 6 miles		234	.034			1.28	3.71	4.99	6.05
9464	cycle 8 miles		216	.037			1.39	4.02	5.41	6.55
9466	cycle 10 miles		198	.040			1.52	4.38	5.90	7.15
9468	cycle 20 miles		144	.056			2.09	6.05	8.14	9.85
9470	cycle 30 miles		108	.074			2.78	8.05	10.83	13.10
9472	cycle 40 miles		90	.089			3.34	9.65	12.99	15.70
9474	40 MPH ave, cycle 6 miles		252	.032			1.19	3.44	4.63	5.60
9476	cycle 8 miles		234	.034			1.28	3.71	4.99	6.05
9478	cycle 10 miles		216	.037			1.39	4.02	5.41	6.55
9480	cycle 20 miles		144	.056			2.09	6.05	8.14	9.85
9482	cycle 30 miles		126	.063			2.38	6.90	9.28	11.20
9484	cycle 40 miles		90	.089			3.34	9.65	12.99	15.70
9486	cycle 50 miles		90	.089			3.34	9.65	12.99	15.70
9494	45 MPH ave, cycle 8 miles		234	.034			1.28	3.71	4.99	6.05
9496	cycle 10 miles		216	.037			1.39	4.02	5.41	6.55
9498	cycle 20 miles		162	.049			1.85	5.35	7.20	8.75
9500	cycle 30 miles		126	.063			2.38	6.90	9.28	11.20
9502	cycle 40 miles		108	.074			2.78	8.05	10.83	13.10
9504	cycle 50 miles		90	.089			3.34	9.65	12.99	15.70
9506	50 MPH ave, cycle 10 miles		234	.034			1.28	3.71	4.99	6.05
9508	cycle 20 miles		180	.044			1.67	4.82	6.49	7.85
9510	cycle 30 miles		144	.056			2.09	6.05	8.14	9.85
9512	cycle 40 miles		108	.074			2.78	8.05	10.83	13.10
9514	cycle 50 miles		90	.089			3.34	9.65	12.99	15.70
9614	30 min. wait/Ld./Uld.,15 MPH, cycle 0.5 mi.		270	.030			1.11	3.21	4.32	5.25
9616	cycle 1 mile		252	.032			1.19	3.44	4.63	5.60
9618	cycle 2 miles		216	.037			1.39	4.02	5.41	6.55
9620	cycle 4 miles		180	.044			1.67	4.82	6.49	7.85
9622	cycle 6 miles		162	.049			1.85	5.35	7.20	8.75
9624	cycle 8 miles		126	.063			2.38	6.90	9.28	11.20
9625	cycle 10 miles		126	.063			2.38	6.90	9.28	11.20
9626	20 MPH ave, cycle 0.5 mile		270	.030			1.11	3.21	4.32	5.25
9628	cycle 1 mile		252	.032			1.19	3.44	4.63	5.60
9630	cycle 2 miles		234	.034			1.28	3.71	4.99	6.05
9632	cycle 4 miles		198	.040			1.52	4.38	5.90	7.15
9634	cycle 6 miles		180	.044			1.67	4.82	6.49	7.85
9636	cycle 8 miles		162	.049			1.85	5.35	7.20	8.75
9638	cycle 10 miles		144	.056			2.09	6.05	8.14	9.85
9640	25 MPH ave, cycle 4 miles		216	.037			1.39	4.02	5.41	6.55
9642	cycle 6 miles		198	.040			1.52	4.38	5.90	7.15
9644	cycle 8 miles		180	.044			1.67	4.82	6.49	7.85
9646	cycle 10 miles		162	.049			1.85	5.35	7.20	8.75
9650	30 MPH ave, cycle 4 miles		216	.037			1.39	4.02	5.41	6.55
9652	cycle 6 miles		198	.040			1.52	4.38	5.90	7.15
9654	cycle 8 miles		180	.044			1.67	4.82	6.49	7.85

31 23 Excavation and Fill

31 23 23 – Fill

31 23 23.20 Hauling		Crew	Daily Output	Labor-Hours	Unit	Material	2014 Bare Costs Labor	Equipment	Total	Total Incl O&P
9656	cycle 10 miles	B-34I	162	.049	L.C.Y.		1.85	5.35	7.20	8.75
9660	35 MPH ave, cycle 4 miles		234	.034			1.28	3.71	4.99	6.05
9662	cycle 6 miles		216	.037			1.39	4.02	5.41	6.55
9664	cycle 8 miles		198	.040			1.52	4.38	5.90	7.15
9666	cycle 10 miles		180	.044			1.67	4.82	6.49	7.85
9668	cycle 20 miles		126	.063			2.38	6.90	9.28	11.20
9670	cycle 30 miles		108	.074			2.78	8.05	10.83	13.10
9672	cycle 40 miles		90	.089			3.34	9.65	12.99	15.70
9674	40 MPH ave, cycle 6 miles		216	.037			1.39	4.02	5.41	6.55
9676	cycle 8 miles		198	.040			1.52	4.38	5.90	7.15
9678	cycle 10 miles		180	.044			1.67	4.82	6.49	7.85
9680	cycle 20 miles		144	.056			2.09	6.05	8.14	9.85
9682	cycle 30 miles		108	.074			2.78	8.05	10.83	13.10
9684	cycle 40 miles		90	.089			3.34	9.65	12.99	15.70
9686	cycle 50 miles		72	.111			4.17	12.05	16.22	19.60
9694	45 MPH ave, cycle 8 miles		216	.037			1.39	4.02	5.41	6.55
9696	cycle 10 miles		198	.040			1.52	4.38	5.90	7.15
9698	cycle 20 miles		144	.056			2.09	6.05	8.14	9.85
9700	cycle 30 miles		126	.063			2.38	6.90	9.28	11.20
9702	cycle 40 miles		108	.074			2.78	8.05	10.83	13.10
9704	cycle 50 miles		90	.089			3.34	9.65	12.99	15.70
9706	50 MPH ave, cycle 10 miles		198	.040			1.52	4.38	5.90	7.15
9708	cycle 20 miles		162	.049			1.85	5.35	7.20	8.75
9710	cycle 30 miles		126	.063			2.38	6.90	9.28	11.20
9712	cycle 40 miles		108	.074			2.78	8.05	10.83	13.10
9714	cycle 50 miles		90	.089			3.34	9.65	12.99	15.70

Division Notes

	CREW	DAILY OUTPUT	LABOR-HOURS	UNIT	BARE COSTS MAT.	LABOR	EQUIP.	TOTAL	TOTAL INCL O&P

Estimating Tips

32 01 00 Operations and Maintenance of Exterior Improvements

- Recycling of asphalt pavement is becoming very popular and is an alternative to removal and replacement. It can be a good value engineering proposal if removed pavement can be recycled, either at the project site or at another site that is reasonably close to the project site. Sections on repair of flexible and rigid pavement are included.

32 10 00 Bases, Ballasts, and Paving

- When estimating paving, keep in mind the project schedule. Also note that prices for asphalt and concrete are generally higher in the cold seasons. Lines for pavement markings, including tactile warning systems and fence lines, are included.

32 90 00 Planting

- The timing of planting and guarantee specifications often dictate the costs for establishing tree and shrub growth and a stand of grass or ground cover. Establish the work performance schedule to coincide with the local planting season. Maintenance and growth guarantees can add from 20%–100% to the total landscaping cost and can be contractually cumbersome. The cost to replace trees and shrubs can be as high as 5% of the total cost, depending on the planting zone, soil conditions, and time of year.

Reference Numbers

Reference numbers are shown in shaded boxes at the beginning of some major classifications. These numbers refer to related items in the Reference Section. The reference information may be an estimating procedure, an alternate pricing method, or technical information.

Note: Not all subdivisions listed here necessarily appear in this publication.

No part of this publication may be reproduced, stored in a retrieval system, or transmitted in any form or by any means without prior written permission of RSMeans, a division of Reed Construction Data.

32 12 Flexible Paving

32 12 16 – Asphalt Paving

32 12 16.13 Plant-Mix Asphalt Paving		Crew	Daily Output	Labor-Hours	Unit	Material	2014 Bare Costs Labor	Equipment	Total	Total Incl O&P
0010	**PLANT-MIX ASPHALT PAVING**									
0020	And large paved areas with no hauling included									
0025	See Section 31 23 23.20 for hauling costs									
0080	Binder course, 1-1/2" thick	B-25	7725	.011	S.Y.	5.55	.46	.36	6.37	7.20
0120	2" thick		6345	.014		7.40	.56	.44	8.40	9.50
0130	2-1/2" thick		5620	.016		9.25	.63	.49	10.37	11.65
0160	3" thick		4905	.018		11.10	.72	.56	12.38	13.95
0170	3-1/2" thick		4520	.019		12.95	.78	.61	14.34	16.10
0200	4" thick		4140	.021		14.80	.85	.67	16.32	18.30
0300	Wearing course, 1" thick	B-25B	10575	.009		3.67	.37	.28	4.32	4.92
0340	1-1/2" thick		7725	.012		6.15	.51	.39	7.05	8
0380	2" thick		6345	.015		8.30	.62	.47	9.39	10.55
0420	2-1/2" thick		5480	.018		10.20	.72	.55	11.47	12.95
0460	3" thick		4900	.020		12.20	.80	.61	13.61	15.30
0470	3-1/2" thick		4520	.021		14.30	.87	.66	15.83	17.75
0480	4" thick		4140	.023		16.35	.95	.73	18.03	20
0500	Open graded friction course	B-25C	5000	.010		2.04	.39	.48	2.91	3.38
0800	Alternate method of figuring paving costs									
0810	Binder course, 1-1/2" thick	B-25	630	.140	Ton	68	5.60	4.38	77.98	88.50
0811	2" thick		690	.128		68	5.10	4	77.10	87.50
0812	3" thick		800	.110		68	4.42	3.45	75.87	85.50
0813	4" thick		900	.098		68	3.93	3.07	75	84.50
0850	Wearing course, 1" thick	B-25B	575	.167		68	6.85	5.20	80.05	91
0851	1-1/2" thick		630	.152		68	6.25	4.76	79.01	90
0852	2" thick		690	.139		68	5.70	4.35	78.05	88.50
0853	2-1/2" thick		765	.125		68	5.15	3.92	77.07	87
0854	3" thick		800	.120		68	4.91	3.75	76.66	86.50
1000	Pavement replacement over trench, 2" thick	B-37	90	.533	S.Y.	7.65	20.50	1.77	29.92	42.50
1050	4" thick		70	.686		15.10	26.50	2.27	43.87	60
1080	6" thick		55	.873		24	34	2.89	60.89	81.50

32 12 16.14 Asphaltic Concrete Paving		Crew	Daily Output	Labor-Hours	Unit	Material	Labor	Equipment	Total	Total Incl O&P
0011	**ASPHALTIC CONCRETE PAVING**, parking lots & driveways									
0015	No asphalt hauling included									
0018	Use 6.05 C.Y. per inch per M.S.F. for hauling									
0020	6" stone base, 2" binder course, 1" topping	B-25C	9000	.005	S.F.	1.74	.22	.27	2.23	2.55
0025	2" binder course, 2" topping		9000	.005		2.16	.22	.27	2.65	3
0030	3" binder course, 2" topping		9000	.005		2.57	.22	.27	3.06	3.46
0035	4" binder course, 2" topping		9000	.005		2.98	.22	.27	3.47	3.91
0040	1.5" binder course, 1" topping		9000	.005		1.54	.22	.27	2.03	2.32
0042	3" binder course, 1" topping		9000	.005		2.16	.22	.27	2.65	3
0045	3" binder course, 3" topping		9000	.005		2.98	.22	.27	3.47	3.91
0050	4" binder course, 3" topping		9000	.005		3.39	.22	.27	3.88	4.36
0055	4" binder course, 4" topping		9000	.005		3.80	.22	.27	4.29	4.81
0300	Binder course, 1-1/2" thick		35000	.001		.62	.06	.07	.75	.85
0400	2" thick		25000	.002		.80	.08	.10	.98	1.11
0500	3" thick		15000	.003		1.24	.13	.16	1.53	1.74
0600	4" thick		10800	.004		1.62	.18	.22	2.02	2.31
0800	Sand finish course, 3/4" thick		41000	.001		.32	.05	.06	.43	.48
0900	1" thick		34000	.001		.39	.06	.07	.52	.60
1000	Fill pot holes, hot mix, 2" thick	B-16	4200	.008		.84	.28	.16	1.28	1.54
1100	4" thick		3500	.009		1.23	.34	.20	1.77	2.10
1120	6" thick		3100	.010		1.65	.39	.22	2.26	2.66

32 12 Flexible Paving

32 12 16 – Asphalt Paving

32 12 16.14 Asphaltic Concrete Paving		Crew	Daily Output	Labor-Hours	Unit	Material	2014 Bare Costs Labor	Equipment	Total	Total Incl O&P
1140	Cold patch, 2" thick	B-51	3000	.016	S.F.	.98	.59	.09	1.66	2.08
1160	4" thick		2700	.018		1.86	.66	.10	2.62	3.16
1180	6" thick	↓	1900	.025	↓	2.89	.93	.14	3.96	4.78

32 84 Planting Irrigation

32 84 13 – Drip Irrigation

32 84 13.10 Subsurface Drip Irrigation

			Crew	Daily Output	Labor-Hours	Unit	Material	Labor	Equipment	Total	Total Incl O&P
0010	**SUBSURFACE DRIP IRRIGATION**										
0011	Looped grid, pressure compensating										
0100	Preinserted PE emitter, line, hand bury, irregular area, small	G	3 Skwk	1200	.020	L.F.	.29	.95		1.24	1.79
0150	Medium	G		1800	.013		.29	.63		.92	1.30
0200	Large	G		2520	.010		.29	.45		.74	1.02
0250	Rectangular area, small	G		2040	.012		.29	.56		.85	1.18
0300	Medium	G		2640	.009		.29	.43		.72	.99
0350	Large	G		3600	.007		.29	.32		.61	.81
0400	Install in trench, irregular area, small	G		4050	.006		.29	.28		.57	.75
0450	Medium	G		7488	.003		.29	.15		.44	.56
0500	Large	G		16560	.001		.29	.07		.36	.43
0550	Rectangular area, small	G		8100	.003		.29	.14		.43	.54
0600	Medium	G		21960	.001		.29	.05		.34	.40
0650	Large	G	↓	33264	.001		.29	.03		.32	.37
0700	Trenching and backfill	G	B-53	500	.016			.75	.14	.89	1.30
0750	For vinyl tubing, 1/4", add to above						25%	10%			
0800	Vinyl tubing, 1/4", material only	G					.05			.05	.06
0850	Supply tubing, 1/2", material only, 100' coil	G					.12			.12	.13
0900	500' coil	G				↓	.11			.11	.12
0950	Compression fittings	G	1 Skwk	90	.089	Ea.	.72	4.20		4.92	7.30
1000	Barbed fittings, 1/4"	G		360	.022		.16	1.05		1.21	1.81
1100	Flush risers	G		60	.133		3.40	6.30		9.70	13.50
1150	Flush ends, figure eight	G		180	.044		.52	2.10		2.62	3.83
1200	Ball valve, 4-1/2"	G		20	.400		9.25	18.90		28.15	39.50
1250	4-3/4"	G		20	.400		7.25	18.90		26.15	37.50
1300	Auto flush, spring loaded	G		90	.089		2.20	4.20		6.40	8.90
1350	Volumetric	G		90	.089		7.50	4.20		11.70	14.75
1400	Air relief valve, inline with compensation tee, 1/2"	G		45	.178		16	8.40		24.40	30.50
1450	1"	G		30	.267		17	12.60		29.60	38.50
1500	Round box for flush ends, 6"	G		30	.267		7.75	12.60		20.35	28
1550	Fertilizer injector, non-proportional	G		4	2		12	94.50		106.50	160
1600	Screen filter, 3/4" screen	G		12	.667		10.30	31.50		41.80	60.50
1650	1" disk	G		8	1		16	47.50		63.50	91
1700	1-1/2" disk	G		4	2		88	94.50		182.50	244
1750	2" disk	G	↓	3	2.667	↓	100	126		226	305
1800	Typical installation 18" O.C., small, minimum					S.F.				1.53	2.35
1850	Maximum									1.78	2.75
1900	Large, minimum									1.48	2.28
2000	Maximum					↓				1.71	2.64
2100	For non-pressure compensating systems, deduct									10%	10%

32 84 Planting Irrigation

32 84 23 – Underground Sprinklers

32 84 23.10 Sprinkler Irrigation System

		Crew	Daily Output	Labor-Hours	Unit	Material	2014 Bare Costs Labor	Equipment	Total	Total Incl O&P
0010	**SPRINKLER IRRIGATION SYSTEM** G2050-710									
0011	For lawns									
0100	Golf course with fully automatic system	C-17	.05	1600	9 holes	100,000	76,500		176,500	228,000
0200	24' diam. head at 15' O.C. incl. piping, auto oper., minimum	B-20	70	.343	Head	23.50	14		37.50	47
0300	Maximum		40	.600		56	24.50		80.50	99.50
0500	60' diam. head at 40' O.C. incl. piping, auto oper., minimum		28	.857		70.50	35		105.50	132
0600	Maximum		23	1.043		190	42.50		232.50	275
0800	Residential system, custom, 1" supply		2000	.012	S.F.	.26	.49		.75	1.05
0900	1-1/2" supply		1800	.013	"	.49	.54		1.03	1.38
0990	For renovation work, add to above						50%			
1020	Pop up spray head w/risers, hi-pop, full circle pattern, 4"	2 Skwk	76	.211	Ea.	4.48	9.95		14.43	20.50
1030	1/2 circle pattern, 4"		76	.211		4.48	9.95		14.43	20.50
1040	6", full circle pattern		76	.211		9.85	9.95		19.80	26.50
1050	1/2 circle pattern, 6"		76	.211		8.50	9.95		18.45	25
1060	12", full circle pattern		76	.211		11.60	9.95		21.55	28
1070	1/2 circle pattern, 12"		76	.211		11.50	9.95		21.45	28
1080	Pop up bubbler head w/risers, hi-pop bubbler head, 4"		76	.211		4.35	9.95		14.30	20
1090	6"		76	.211		9.10	9.95		19.05	25.50
1100	12"		76	.211		10.90	9.95		20.85	27.50
1110	Impact full/part circle sprinklers, 28'-54' 25-60 PSI		37	.432		19.60	20.50		40.10	53
1120	Spaced 37'-49' @ 25-50 PSI		37	.432		27.50	20.50		48	62
1130	Spaced 43'-61' @ 30-60 PSI		37	.432		77	20.50		97.50	116
1140	Spaced 54'-78' @ 40-80 PSI		37	.432		133	20.50		153.50	178
1145	Impact rotor pop-up full/part commercial circle sprinklers									
1150	Spaced 42'-65' 35-80 PSI	2 Skwk	25	.640	Ea.	17	30.50		47.50	65.50
1160	Spaced 48'-76' 45-85 PSI	"	25	.640	"	18.70	30.50		49.20	67.50
1165	Impact rotor pop-up part. circle comm., 53'-75', 55-100 PSI, w/accessories									
1170	Plastic case, metal cover	2 Skwk	25	.640	Ea.	82	30.50		112.50	137
1180	Rubber cover		25	.640		63	30.50		93.50	117
1190	Iron case, metal cover		22	.727		135	34.50		169.50	203
1200	Rubber cover		22	.727		143	34.50		177.50	211
1250	Plastic case, 2 nozzle, metal cover		25	.640		128	30.50		158.50	187
1260	Rubber cover		25	.640		131	30.50		161.50	191
1270	Iron case, 2 nozzle, metal cover		22	.727		185	34.50		219.50	258
1280	Rubber cover		22	.727		185	34.50		219.50	258
1282	Impact rotor pop-up full circle commercial, 39'-99', 30-100 PSI									
1284	Plastic case, metal cover	2 Skwk	25	.640	Ea.	138	30.50		168.50	199
1286	Rubber cover		25	.640		158	30.50		188.50	220
1288	Iron case, metal cover		22	.727		203	34.50		237.50	277
1290	Rubber cover		22	.727		210	34.50		244.50	285
1292	Plastic case, 2 nozzle, metal cover		22	.727		150	34.50		184.50	219
1294	Rubber cover		22	.727		150	34.50		184.50	219
1296	Iron case, 2 nozzle, metal cover		20	.800		198	38		236	276
1298	Rubber cover		20	.800		205	38		243	285
1305	Electric remote control valve, plastic, 3/4"		18	.889		17.30	42		59.30	84
1310	1"		18	.889		33	42		75	102
1320	1-1/2"		18	.889		60.50	42		102.50	132
1330	2"		18	.889		86	42		128	160
1335	Quick coupling valves, brass, locking cover									
1340	Inlet coupling valve, 3/4"	2 Skwk	18.75	.853	Ea.	22	40.50		62.50	86.50
1350	1"		18.75	.853		30.50	40.50		71	96.50
1360	Controller valve boxes, 6" round boxes		18.75	.853		7	40.50		47.50	70

32 84 Planting Irrigation

32 84 23 – Underground Sprinklers

32 84 23.10 Sprinkler Irrigation System

		Crew	Daily Output	Labor-Hours	Unit	Material	2014 Bare Costs Labor	Equipment	Total	Total Incl O&P
1370	10" round boxes	2 Skwk	14.25	1.123	Ea.	11.30	53		64.30	95
1380	12" square box	↓	9.75	1.641	↓	15.90	77.50		93.40	137
1388	Electromech. control, 14 day 3-60 min., auto start to 23/day									
1390	4 station	2 Skwk	1.04	15.385	Ea.	185	730		915	1,325
1400	7 station		.64	25		201	1,175		1,376	2,050
1410	12 station		.40	40		220	1,900		2,120	3,175
1420	Dual programs, 18 station		.24	66.667		610	3,150		3,760	5,550
1430	23 station	↓	.16	100	↓	690	4,725		5,415	8,075
1435	Backflow preventer, bronze, 0-175 PSI, w/valves, test cocks									
1440	3/4"	2 Skwk	2	8	Ea.	103	380		483	700
1450	1"		2	8		118	380		498	715
1460	1-1/2"		2	8		280	380		660	895
1470	2"	↓	2	8	↓	345	380		725	965
1475	Pressure vacuum breaker, brass, 15-150 PSI									
1480	3/4"	2 Skwk	2	8	Ea.	60.50	380		440.50	650
1490	1"		2	8		65.50	380		445.50	655
1500	1-1/2"		2	8		270	380		650	880
1510	2"	↓	2	8		325	380		705	945
2000	Pop-up spray, head & nozzle, low/medium volume, plastic	1 Skwk	30	.267		3.25	12.60		15.85	23
2200	Brass, economy		30	.267		11.50	12.60		24.10	32
2400	Heavy duty		30	.267		19.75	12.60		32.35	41
3000	Riser mounted spray, head, low/medium volume, plastic		30	.267		3.25	12.60		15.85	23
3200	Brass		30	.267		10.75	12.60		23.35	31.50
4000	Pop-up impact sprinkler, body & case, plastic low/medium volume		30	.267		22.50	12.60		35.10	44.50
4200	Brass, high/medium volume		30	.267		49.50	12.60		62.10	73.50
4400	High volume		30	.267		132	12.60		144.60	165
5000	Quick coupling valve and key, brass, thread type		25	.320		58	15.15		73.15	87.50
5200	Lug type		25	.320		61	15.15		76.15	90.50
5400	Pop-up gear drive sprinkler, nozzle & case plastic, medium volume	↓	30	.267	↓	26	12.60		38.60	48
6000	Riser mounted gear drive sprinkler, nozzle and case									
6200	Plastic, medium volume	1 Skwk	30	.267	Ea.	14.80	12.60		27.40	36
7000	Riser mounted impact sprinkler, body, part or full circle									
7200	Full circle plastic, low/medium volume	1 Skwk	25	.320	Ea.	14.75	15.15		29.90	40
7400	Brass, low/medium volume		25	.320		39	15.15		54.15	66
7600	Medium volume		25	.320		50	15.15		65.15	78
7800	Female thread	↓	25	.320	↓	92	15.15		107.15	125
8000	Riser mounted impact sprinkler, body, full circle only									
8200	Plastic, low/medium volume	1 Skwk	25	.320	Ea.	8.70	15.15		23.85	33
8400	Brass, low/medium volume		25	.320		37.50	15.15		52.65	65
8600	High volume		25	.320		76.50	15.15		91.65	108
8800	Very high volume	↓	25	.320	↓	150	15.15		165.15	189

Division Notes

		CREW	DAILY OUTPUT	LABOR-HOURS	UNIT	BARE COSTS MAT.	LABOR	EQUIP.	TOTAL	TOTAL INCL O&P

Division 33 - Utilities

Estimating Tips

33 10 00 Water Utilities
33 30 00 Sanitary Sewerage Utilities
33 40 00 Storm Drainage Utilities

- Never assume that the water, sewer, and drainage lines will go in at the early stages of the project. Consider the site access needs before dividing the site in half with open trenches, loose pipe, and machinery obstructions. Always inspect the site to establish that the site drawings are complete. Check off all existing utilities on your drawings as you locate them. Be especially careful with underground utilities because appurtenances are sometimes buried during regrading or repaving operations. If you find any discrepancies, mark up the site plan for further research. Differing site conditions can be very costly if discovered later in the project.

- See also Section 33 01 00 for restoration of pipe where removal/replacement may be undesirable. Use of new types of piping materials can reduce the overall project cost. Owners/design engineers should consider the installing contractor as a valuable source of current information on utility products and local conditions that could lead to significant cost savings.

Reference Numbers

Reference numbers are shown in shaded boxes at the beginning of some major classifications. These numbers refer to related items in the Reference Section. The reference information may be an estimating procedure, an alternate pricing method, or technical information.

Note: Not all subdivisions listed here necessarily appear in this publication.

Note: **Trade Service**, *in part, has been used as a reference source for some of the material prices used in Division 33.*

33 01 Operation and Maintenance of Utilities

33 01 10 – Operation and Maintenance of Water Utilities

33 01 10.10 Corrosion Resistance

		Crew	Daily Output	Labor-Hours	Unit	Material	2014 Bare Costs Labor	Equipment	Total	Total Incl O&P
0010	**CORROSION RESISTANCE**									
0012	Wrap & coat, add to pipe, 4" diameter				L.F.	2.53			2.53	2.78
0020	5" diameter					3.18			3.18	3.50
0040	6" diameter					3.81			3.81	4.19
0060	8" diameter					4.53			4.53	4.98
0080	10" diameter					5.70			5.70	6.25
0100	12" diameter					7.45			7.45	8.20
0120	14" diameter					8.60			8.60	9.45
0140	16" diameter					9.75			9.75	10.70
0160	18" diameter					10.70			10.70	11.80
0180	20" diameter					11.85			11.85	13
0200	24" diameter					14.40			14.40	15.85
0220	Small diameter pipe, 1" diameter, add					1.65			1.65	1.82
0240	2" diameter					1.81			1.81	1.99
0260	2-1/2" diameter					1.99			1.99	2.19
0280	3" diameter					2.39			2.39	2.63
0300	Fittings, field covered, add				S.F.	10.45			10.45	11.45
0500	Coating, bituminous, per diameter inch, 1 coat, add				L.F.	.64			.64	.70
0540	3 coat					1.93			1.93	2.12
0560	Coal tar epoxy, per diameter inch, 1 coat, add					.23			.23	.25
0600	3 coat					.69			.69	.76
1000	Polyethylene H.D. extruded, .025" thk., 1/2" diameter add					.26			.26	.29
1020	3/4" diameter					.29			.29	.32
1040	1" diameter					.33			.33	.36
1060	1-1/4" diameter					.41			.41	.45
1080	1-1/2" diameter					.42			.42	.46
1100	.030" thk., 2" diameter					.46			.46	.51
1120	2-1/2" diameter					.52			.52	.57
1140	.035" thk., 3" diameter					.75			.75	.83
1160	3-1/2" diameter					.80			.80	.88
1180	4" diameter					.85			.85	.94
1200	.040" thk., 5" diameter					1.04			1.04	1.14
1220	6" diameter					1.10			1.10	1.21
1240	8" diameter					1.44			1.44	1.58
1260	10" diameter					1.80			1.80	1.98
1280	12" diameter					2.16			2.16	2.38
1300	.060" thk., 14" diameter					2.99			2.99	3.29
1320	16" diameter					3.56			3.56	3.92
1340	18" diameter					4.12			4.12	4.53
1360	20" diameter					4.68			4.68	5.15
1380	Fittings, field wrapped, add				S.F.	4.80			4.80	5.30

33 01 10.20 Pipe Repair

		Crew	Daily Output	Labor-Hours	Unit	Material	2014 Bare Costs Labor	Equipment	Total	Total Incl O&P
0010	**PIPE REPAIR**									
0020	Not including excavation or backfill									
0100	Clamp, stainless steel, lightweight, for steel pipe									
0110	3" long, 1/2" diameter pipe	1 Plum	34	.235	Ea.	11.90	13.55		25.45	33.50
0120	3/4" diameter pipe		32	.250		12.55	14.40		26.95	35.50
0130	1" diameter pipe		30	.267		13.20	15.35		28.55	37.50
0140	1-1/4" diameter pipe		28	.286		13.95	16.45		30.40	40.50
0150	1-1/2" diameter pipe		26	.308		14.65	17.70		32.35	42.50
0160	2" diameter pipe		24	.333		15.75	19.20		34.95	46.50
0170	2-1/2" diameter pipe		23	.348		21.50	20		41.50	53.50

33 01 Operation and Maintenance of Utilities

33 01 10 – Operation and Maintenance of Water Utilities

33 01 10.20 Pipe Repair		Crew	Daily Output	Labor-Hours	Unit	Material	2014 Bare Costs Labor	Equipment	Total	Total Incl O&P
0180	3" diameter pipe	1 Plum	22	.364	Ea.	19.30	21		40.30	52.50
0190	3-1/2" diameter pipe	↓	21	.381		25	22		47	60.50
0200	4" diameter pipe	B-20	44	.545		25	22.50		47.50	62
0210	5" diameter pipe		42	.571		29.50	23.50		53	68.50
0220	6" diameter pipe		38	.632		34	26		60	77.50
0230	8" diameter pipe		30	.800		40.50	32.50		73	95
0240	10" diameter pipe		28	.857		130	35		165	197
0250	12" diameter pipe		24	1		180	41		221	261
0260	14" diameter pipe		22	1.091		260	44.50		304.50	355
0270	16" diameter pipe		20	1.200		280	49		329	385
0280	18" diameter pipe		18	1.333		310	54.50		364.50	425
0290	20" diameter pipe		16	1.500		330	61.50		391.50	460
0300	24" diameter pipe	↓	14	1.714		350	70		420	495
0360	For 6" long, add					100%	40%			
0370	For 9" long, add					200%	100%			
0380	For 12" long, add					300%	150%			
0390	For 18" long, add					500%	200%			
0400	Pipe Freezing for live repairs of systems 3/8 inch to 6 inch									
0410	Note: Pipe Freezing can also be used to install a valve into a live system									
0420	Pipe Freezing each side 3/8 inch	2 Skwk	8	2	Ea.	530	94.50		624.50	730
0425	Pipe Freezing each side 3/8 inch, second location same kit		8	2		22.50	94.50		117	172
0430	Pipe Freezing each side 3/4 inch		8	2		510	94.50		604.50	705
0435	Pipe Freezing each side 3/4 inch, second location same kit		8	2		22.50	94.50		117	172
0440	Pipe Freezing each side 1-1/2 inch		6	2.667		510	126		636	755
0445	Pipe Freezing each side 1-1/2 inch, second location same kit		6	2.667		22.50	126		148.50	220
0450	Pipe Freezing each side 2 inch		6	2.667		890	126		1,016	1,175
0455	Pipe Freezing each side 2 inch, second location same kit		6	2.667		22.50	126		148.50	220
0460	Pipe Freezing each side 2-1/2 inch-3 inch		6	2.667		925	126		1,051	1,225
0465	Pipe Freeze each side 2-1/2 -3 inch, second location same kit		6	2.667		45.50	126		171.50	245
0470	Pipe Freezing each side 4 inch		4	4		1,475	189		1,664	1,925
0475	Pipe Freezing each side 4 inch, second location same kit		4	4		72	189		261	375
0480	Pipe Freezing each side 5-6 inch		4	4		4,000	189		4,189	4,700
0485	Pipe Freezing each side 5-6 inch, second location same kit	↓	4	4		216	189		405	530
0490	Pipe Freezing extra 20 lb. CO_2 cylinders (3/8" to 2" - 1 ea, 3" -2 ea)					223			223	245
0500	Pipe Freezing extra 50 lb. CO_2 cylinders (4" - 2 ea, 5"-6" -6 ea)				↓	440			440	485
1000	Clamp, stainless steel, with threaded service tap									
1040	Full seal for iron, steel, PVC pipe									
1100	6" long, 2" diameter pipe	1 Plum	17	.471	Ea.	87	27		114	137
1110	2-1/2" diameter pipe		16	.500		90	29		119	143
1120	3" diameter pipe		15.60	.513		152	29.50		181.50	212
1130	3-1/2" diameter pipe	↓	15	.533		157	30.50		187.50	219
1140	4" diameter pipe	B-20	32	.750		164	30.50		194.50	228
1150	6" diameter pipe		28	.857		172	35		207	243
1160	8" diameter pipe		21	1.143		179	46.50		225.50	269
1170	10" diameter pipe		20	1.200		224	49		273	320
1180	12" diameter pipe	↓	17	1.412		261	57.50		318.50	375
1205	For 9" long, add					20%	45%			
1210	For 12" long, add					40%	80%			
1220	For 18" long, add				↓	70%	110%			
1600	Clamp, stainless steel, single section									
1640	Full seal for iron, steel, PVC pipe									
1700	6" long, 2" diameter pipe	1 Plum	17	.471	Ea.	87	27		114	137
1710	2-1/2" diameter pipe	↓	16	.500		90.50	29		119.50	143

33 01 Operation and Maintenance of Utilities

33 01 10 – Operation and Maintenance of Water Utilities

33 01 10.20 Pipe Repair

		Crew	Daily Output	Labor-Hours	Unit	Material	2014 Bare Costs Labor	Equipment	Total	Total Incl O&P
1720	3" diameter pipe	1 Plum	15.60	.513	Ea.	105	29.50		134.50	160
1730	3-1/2" diameter pipe	▼	15	.533		110	30.50		140.50	168
1740	4" diameter pipe	B-20	32	.750		117	30.50		147.50	176
1750	6" diameter pipe		27	.889		145	36.50		181.50	216
1760	8" diameter pipe		21	1.143		171	46.50		217.50	260
1770	10" diameter pipe		20	1.200		224	49		273	320
1780	12" diameter pipe	▼	17	1.412		261	57.50		318.50	375
1800	For 9" long, add					40%	45%			
1810	For 12" long, add					60%	80%			
1820	For 18" long, add				▼	120%	110%			
2000	Clamp, stainless steel, two section									
2040	Full seal, for iron, steel, PVC pipe									
2100	6" long, 4" diameter pipe	B-20	24	1	Ea.	230	41		271	315
2110	6" diameter pipe		20	1.200		264	49		313	365
2120	8" diameter pipe		13	1.846		300	75.50		375.50	450
2130	10" diameter pipe		12	2		305	81.50		386.50	460
2140	12" diameter pipe		10	2.400		390	98		488	580
2200	9" long, 4" diameter pipe		16	1.500		300	61.50		361.50	425
2210	6" diameter pipe		13	1.846		340	75.50		415.50	490
2220	8" diameter pipe		9	2.667		375	109		484	585
2230	10" diameter pipe		8	3		495	123		618	735
2240	12" diameter pipe		7	3.429		565	140		705	840
2250	14" diameter pipe		6.40	3.750		760	153		913	1,075
2260	16" diameter pipe		6	4		710	163		873	1,025
2270	18" diameter pipe		5	4.800		800	196		996	1,175
2280	20" diameter pipe		4.60	5.217		885	213		1,098	1,300
2290	24" diameter pipe	▼	4	6		1,175	245		1,420	1,675
2320	For 12" long, add to 9"					15%	25%			
2330	For 18" long, add to 9"				▼	70%	55%			
8000	For internal cleaning and inspection, see Section 33 01 30.16									
8100	For pipe testing, see Section 23 05 93.50									

33 01 30 – Operation and Maintenance of Sewer Utilities

33 01 30.16 TV Inspection of Sewer Pipelines

		Crew	Daily Output	Labor-Hours	Unit	Material	Labor	Equipment	Total	Total Incl O&P
0010	**TV INSPECTION OF SEWER PIPELINES**									
0100	Pipe internal cleaning & inspection, cleaning, pressure pipe systems									
0120	Pig method, lengths 1000' to 10,000'									
0140	4" diameter thru 24" diameter, minimum				L.F.				3.60	4.14
0160	Maximum				"				18	21
6000	Sewage/sanitary systems									
6100	Power rodder with header & cutters									
6110	Mobilization charge, minimum				Total				695	800
6120	Mobilization charge, maximum				"				9,125	10,600
6140	Cleaning 4"-12" diameter				L.F.				3.39	3.90
6190	14"-24" diameter								3.98	4.59
6240	30" diameter								5.80	6.65
6250	36" diameter								6.75	7.75
6260	48" diameter								7.70	8.85
6270	60" diameter								8.70	9.95
6280	72" diameter				▼				9.65	11.10
9000	Inspection, television camera with video									
9060	up to 500 linear feet				Total				715	820

33 01 Operation and Maintenance of Utilities

33 01 30 – Operation and Maintenance of Sewer Utilities

33 01 30.71 Pipebursting

		Crew	Daily Output	Labor-Hours	Unit	Material	2014 Bare Costs Labor	Equipment	Total	Total Incl O&P
0010	**PIPEBURSTING**									
0011	300' runs, replace with HDPE pipe									
0020	Not including excavation, backfill, shoring, or dewatering									
0100	6" to 15" diameter, minimum				L.F.				104	114
0200	Maximum								214	236
0300	18" to 36" diameter, minimum								208	229
0400	Maximum								435	510
0500	Mobilize and demobilize, minimum				Job				3,000	3,300
0600	Maximum				"				32,100	35,400

33 01 30.72 Relining Sewers

		Crew	Daily Output	Labor-Hours	Unit	Material	Labor	Equipment	Total	Total Incl O&P
0010	**RELINING SEWERS**									
0011	With cement incl. bypass & cleaning									
0020	Less than 10,000 L.F., urban, 6" to 10"	C-17E	130	.615	L.F.	9.55	29.50	.73	39.78	57
0050	10" to 12"		125	.640		11.75	30.50	.76	43.01	61
0070	12" to 16"		115	.696		12.05	33	.82	45.87	65.50
0100	16" to 20"		95	.842		14.15	40	.99	55.14	78.50
0200	24" to 36"		90	.889		15.25	42.50	1.05	58.80	83.50
0300	48" to 72"		80	1		24.50	47.50	1.18	73.18	102

33 01 30.74 HDPE Pipe Lining

		Crew	Daily Output	Labor-Hours	Unit	Material	Labor	Equipment	Total	Total Incl O&P
0010	**HDPE PIPE LINING**, excludes cleaning and video inspection									
0020	Pipe relined with one pipe size smaller than original (4" for 6")									
0100	6" diameter, original size	B-6B	600	.080	L.F.	4.13	2.99	1.55	8.67	10.85
0150	8" diameter, original size		600	.080		8.20	2.99	1.55	12.74	15.30
0200	10" diameter, original size		600	.080		10.65	2.99	1.55	15.19	18
0250	12" diameter, original size		400	.120		20	4.48	2.33	26.81	31.50
0300	14" diameter, original size		400	.120		22.50	4.48	2.33	29.31	34
0350	16" diameter, original size	B-6C	300	.160		38.50	5.95	5.50	49.95	57.50
0400	18" diameter, original size	"	300	.160		44	5.95	5.50	55.45	63
1000	Pipe HDPE lining, make service line taps	B-6	4	6	Ea.	109	241	91.50	441.50	590

33 05 Common Work Results for Utilities

33 05 16 – Utility Structures

33 05 16.13 Precast Concrete Utility Boxes

		Crew	Daily Output	Labor-Hours	Unit	Material	Labor	Equipment	Total	Total Incl O&P
0010	**PRECAST CONCRETE UTILITY BOXES**, 6" thick									
0050	5' x 10' x 6' high, I.D.	B-13	2	28	Ea.	3,675	1,125	375	5,175	6,125
0100	6' x 10' x 6' high, I.D.		2	28		3,800	1,125	375	5,300	6,300
0150	5' x 12' x 6' high, I.D.		2	28		4,025	1,125	375	5,525	6,525
0200	6' x 12' x 6' high, I.D.		1.80	31.111		4,525	1,250	415	6,190	7,325
0250	6' x 13' x 6' high, I.D.		1.50	37.333		5,925	1,500	495	7,920	9,350
0300	8' x 14' x 7' high, I.D.		1	56		6,400	2,225	745	9,370	11,300
0350	Hand hole, precast concrete, 1-1/2" thick									
0400	1'-0" x 2'-0" x 1'-9", I.D., light duty	B-1	4	6	Ea.	410	224		634	795
0450	4'-6" x 3'-2" x 2'-0", O.D., heavy duty	B-6	3	8	"	1,475	320	122	1,917	2,225

33 05 23 – Trenchless Utility Installation

33 05 23.19 Microtunneling

		Crew	Daily Output	Labor-Hours	Unit	Material	Labor	Equipment	Total	Total Incl O&P
0010	**MICROTUNNELING**									
0011	Not including excavation, backfill, shoring,									
0020	or dewatering, average 50'/day, slurry method									
0100	24" to 48" outside diameter, minimum				L.F.				875	965

33 05 Common Work Results for Utilities

33 05 23 – Trenchless Utility Installation

33 05 23.19 Microtunneling

		Crew	Daily Output	Labor-Hours	Unit	Material	2014 Bare Costs Labor	Equipment	Total	Total Incl O&P
0110	Adverse conditions, add				%				50%	50%
1000	Rent microtunneling machine, average monthly lease				Month				97,500	107,000
1010	Operating technician				Day				630	705
1100	Mobilization and demobilization, minimum				Job				41,200	45,900
1110	Maximum				"				445,500	490,500

33 05 23.20 Horizontal Boring

		Crew	Daily Output	Labor-Hours	Unit	Material	Labor	Equipment	Total	Total Incl O&P
0010	**HORIZONTAL BORING**									
0011	Casing only, 100' minimum,									
0020	not incl. jacking pits or dewatering									
0100	Roadwork, 1/2" thick wall, 24" diameter casing	B-42	20	3.200	L.F.	121	132	68.50	321.50	415
0200	36" diameter		16	4		223	165	85.50	473.50	600
0300	48" diameter		15	4.267		310	176	91	577	715
0500	Railroad work, 24" diameter		15	4.267		121	176	91	388	510
0600	36" diameter		14	4.571		223	189	97.50	509.50	650
0700	48" diameter		12	5.333		310	220	114	644	810
0900	For ledge, add								20%	20%
1000	Small diameter boring, 3", sandy soil	B-82	900	.018		22	.74	.09	22.83	25
1040	Rocky soil	"	500	.032		22	1.34	.17	23.51	26
1100	Prepare jacking pits, incl. mobilization & demobilization, minimum				Ea.				3,225	3,700
1101	Maximum				"				22,000	25,500

33 05 26 – Utility Identification

33 05 26.10 Utility Accessories

		Crew	Daily Output	Labor-Hours	Unit	Material	Labor	Equipment	Total	Total Incl O&P
0010	**UTILITY ACCESSORIES**									
0400	Underground tape, detectable, reinforced, alum. foil core, 2"	1 Clab	150	.053	C.L.F.	5.65	1.95		7.60	9.25
0500	6"	"	140	.057	"	27.50	2.09		29.59	33.50

33 11 Water Utility Distribution Piping

33 11 13 – Public Water Utility Distribution Piping

33 11 13.15 Water Supply, Ductile Iron Pipe

		Crew	Daily Output	Labor-Hours	Unit	Material	Labor	Equipment	Total	Total Incl O&P
0010	**WATER SUPPLY, DUCTILE IRON PIPE** R331113-80									
0020	Not including excavation or backfill									
2000	Pipe, class 50 water piping, 18' lengths									
2020	Mechanical joint, 4" diameter	B-21A	200	.200	L.F.	15.60	9.15	2.41	27.16	34
2040	6" diameter		160	.250		17.75	11.45	3.01	32.21	40.50
2060	8" diameter		133.33	.300		24.50	13.75	3.61	41.86	52
2080	10" diameter		114.29	.350		35	16.05	4.21	55.26	67.50
2100	12" diameter		105.26	.380		42.50	17.40	4.57	64.47	78.50
2120	14" diameter		100	.400		60	18.35	4.81	83.16	99.50
2140	16" diameter		72.73	.550		63.50	25	6.60	95.10	116
2160	18" diameter		68.97	.580		74.50	26.50	7	108	130
2170	20" diameter		57.14	.700		85.50	32	8.40	125.90	153
2180	24" diameter		47.06	.850		108	39	10.25	157.25	190
3000	Tyton, push-on joint, 4" diameter		400	.100		17	4.58	1.20	22.78	27
3020	6" diameter		333.33	.120		17.85	5.50	1.44	24.79	29.50
3040	8" diameter		200	.200		23.50	9.15	2.41	35.06	42
3060	10" diameter		181.82	.220		35	10.10	2.65	47.75	57
3080	12" diameter		160	.250		37	11.45	3.01	51.46	62
3100	14" diameter		133.33	.300		41	13.75	3.61	58.36	70
3120	16" diameter		114.29	.350		49.50	16.05	4.21	69.76	83.50
3140	18" diameter		100	.400		55	18.35	4.81	78.16	94

33 11 Water Utility Distribution Piping

33 11 13 – Public Water Utility Distribution Piping

33 11 13.15 Water Supply, Ductile Iron Pipe		Crew	Daily Output	Labor-Hours	Unit	Material	2014 Bare Costs Labor	Equipment	Total	Total Incl O&P
3160	20" diameter	B-21A	88.89	.450	L.F.	57	20.50	5.40	82.90	100
3180	24" diameter	▼	76.92	.520	▼	64	24	6.25	94.25	114
8000	Fittings, mechanical joint									
8006	90° bend, 4" diameter	B-20A	16	2	Ea.	273	89.50		362.50	435
8020	6" diameter		12.80	2.500		405	112		517	615
8040	8" diameter	▼	10.67	2.999		725	134		859	1,000
8060	10" diameter	B-21A	11.43	3.500		980	160	42	1,182	1,375
8080	12" diameter		10.53	3.799		1,375	174	45.50	1,594.50	1,825
8100	14" diameter		10	4		1,850	183	48	2,081	2,350
8120	16" diameter		7.27	5.502		2,400	252	66	2,718	3,075
8140	18" diameter		6.90	5.797		3,225	266	69.50	3,560.50	4,025
8160	20" diameter		5.71	7.005		4,050	320	84.50	4,454.50	5,025
8180	24" diameter	▼	4.70	8.511		6,275	390	102	6,767	7,600
8200	Wye or tee, 4" diameter	B-20A	10.67	2.999		440	134		574	690
8220	6" diameter		8.53	3.751		670	168		838	990
8240	8" diameter	▼	7.11	4.501		1,075	201		1,276	1,475
8260	10" diameter	B-21A	7.62	5.249		1,550	241	63	1,854	2,125
8280	12" diameter		7.02	5.698		2,025	261	68.50	2,354.50	2,700
8300	14" diameter		6.67	5.997		3,250	275	72	3,597	4,075
8320	16" diameter		4.85	8.247		3,600	380	99	4,079	4,650
8340	18" diameter		4.60	8.696		4,875	400	105	5,380	6,100
8360	20" diameter		3.81	10.499		6,825	480	126	7,431	8,400
8380	24" diameter	▼	3.14	12.739		11,600	585	153	12,338	13,800
8398	45° bends, 4" diameter	B-20A	16	2		254	89.50		343.50	415
8400	6" diameter		12.80	2.500		365	112		477	575
8405	8" diameter	▼	10.67	2.999		520	134		654	780
8410	12" diameter	B-21A	10.53	3.799		1,175	174	45.50	1,394.50	1,625
8420	16" diameter		7.27	5.502		2,100	252	66	2,418	2,750
8430	20" diameter		5.71	7.005		3,275	320	84.50	3,679.50	4,200
8440	24" diameter	▼	4.70	8.511		4,575	390	102	5,067	5,750
8450	Decreaser, 6" x 4" diameter	B-20A	14.22	2.250		360	101		461	550
8460	8" x 6" diameter	"	11.64	2.749		525	123		648	765
8470	10" x 6" diameter	B-21A	13.33	3.001		645	138	36	819	960
8480	12" x 6" diameter		12.70	3.150		795	144	38	977	1,125
8490	16" x 6" diameter		10	4		1,425	183	48	1,656	1,900
8500	20" x 6" diameter	▼	8.42	4.751	▼	3,125	218	57	3,400	3,850
8552	For water utility valves see Section 33 12 16									
8700	Joint restraint, ductile iron mechanical joints									
8710	4" diameter	B-20A	32	1	Ea.	36.50	44.50		81	109
8720	6" diameter		25.60	1.250		43.50	56		99.50	133
8730	8" diameter		21.33	1.500		64	67		131	172
8740	10" diameter		18.28	1.751		92	78.50		170.50	220
8750	12" diameter		16.84	1.900		131	85		216	275
8760	14" diameter		16	2		170	89.50		259.50	325
8770	16" diameter		11.64	2.749		227	123		350	435
8780	18" diameter		11.03	2.901		315	130		445	545
8785	20" diameter		9.14	3.501		385	157		542	665
8790	24" diameter	▼	7.53	4.250		505	190		695	845
9600	Steel sleeve with tap, 4" diameter	B-20	3	8		430	325		755	980
9620	6" diameter		2	12		470	490		960	1,275
9630	8" diameter	▼	2	12	▼	515	490		1,005	1,325

33 11 Water Utility Distribution Piping

33 11 13 – Public Water Utility Distribution Piping

33 11 13.20 Water Supply, Polyethylene Pipe, C901

		Crew	Daily Output	Labor-Hours	Unit	Material	2014 Bare Costs Labor	Equipment	Total	Total Incl O&P
0010	**WATER SUPPLY, POLYETHYLENE PIPE, C901**									
0020	Not including excavation or backfill									
1000	Piping, 160 PSI, 3/4" diameter	Q-1A	525	.019	L.F.	.38	1.10		1.48	2.09
1120	1" diameter		485	.021		.52	1.19		1.71	2.37
1140	1-1/2" diameter		450	.022		1.27	1.29		2.56	3.34
1160	2" diameter	↓	365	.027	↓	2.10	1.59		3.69	4.71
2000	Fittings, insert type, nylon, 160 & 250 psi, cold water									
2220	Clamp ring, stainless steel, 3/4" diameter	Q-1A	345	.029	Ea.	1.15	1.68		2.83	3.81
2240	1" diameter		321	.031		1.34	1.81		3.15	4.20
2260	1-1/2" diameter		285	.035		2.40	2.03		4.43	5.70
2280	2" diameter		255	.039		3.15	2.27		5.42	6.90
2300	Coupling, 3/4" diameter		66	.152		1.27	8.80		10.07	14.65
2320	1" diameter		57	.175		1.48	10.15		11.63	17
2340	1-1/2" diameter		51	.196		3	11.35		14.35	20.50
2360	2" diameter		48	.208		4.70	12.05		16.75	23.50
2400	Elbow, 90°, 3/4" diameter		66	.152		2.06	8.80		10.86	15.50
2420	1" diameter		57	.175		2.49	10.15		12.64	18.10
2440	1-1/2" diameter		51	.196		4.10	11.35		15.45	21.50
2460	2" diameter	↓	48	.208	↓	5.75	12.05		17.80	24.50

33 11 13.25 Water Supply, Polyvinyl Chloride Pipe

		Crew	Daily Output	Labor-Hours	Unit	Material	Labor	Equipment	Total	Total Incl O&P
0010	**WATER SUPPLY, POLYVINYL CHLORIDE PIPE** R331113-80									
0020	Not including excavation or backfill, unless specified									
2100	PVC pipe, Class 150, 1-1/2" diameter	Q-1A	750	.013	L.F.	.43	.77		1.20	1.64
2120	2" diameter		686	.015		.67	.84		1.51	2.02
2140	2-1/2" diameter	↓	500	.020	↓	1.28	1.16		2.44	3.16
2160	3" diameter	B-20	430	.056		1.38	2.28		3.66	5.05
3010	AWWA C905, PR 100, DR 25									
3030	14" diameter	B-20A	213	.150	L.F.	13.65	6.70		20.35	25.50
3040	16" diameter		200	.160		19.20	7.15		26.35	32
3050	18" diameter		160	.200		24	8.95		32.95	40
3060	20" diameter		133	.241		30	10.75		40.75	49.50
3070	24" diameter		107	.299		43	13.40		56.40	67.50
3080	30" diameter		80	.400		81	17.90		98.90	117
3090	36" diameter		80	.400		126	17.90		143.90	166
3100	42" diameter		60	.533		168	24		192	222
3200	48" diameter	↓	60	.533		220	24		244	279
3960	Pressure pipe, class 200, ASTM 2241, SDR 21, 3/4" diameter	Q-1A	1000	.010		.21	.58		.79	1.11
3980	1" diameter		900	.011		.30	.64		.94	1.30
4000	1-1/2" diameter		750	.013		.68	.77		1.45	1.92
4010	2" diameter		686	.015		.86	.84		1.70	2.23
4020	2-1/2" diameter	↓	500	.020		1.35	1.16		2.51	3.24
4030	3" diameter	B-20A	430	.074		1.59	3.33		4.92	6.80
4040	4" diameter		375	.085		2.47	3.82		6.29	8.50
4050	6" diameter		316	.101		4.82	4.53		9.35	12.20
4060	8" diameter	↓	260	.123		8.35	5.50		13.85	17.60
4090	Including trenching to 3' deep, 3/4" diameter	Q-1C	300	.080		.21	4.07	11.25	15.53	18.80
4100	1" diameter		280	.086		.30	4.36	12.05	16.71	20
4110	1-1/2" diameter		260	.092		.68	4.69	13	18.37	22
4120	2" diameter		220	.109		.86	5.55	15.35	21.76	26.50
4130	2-1/2" diameter		200	.120		1.35	6.10	16.90	24.35	29
4140	3" diameter		175	.137		1.59	6.95	19.30	27.84	33.50
4150	4" diameter	↓	150	.160		2.47	8.15	22.50	33.12	40

33 11 Water Utility Distribution Piping

33 11 13 − Public Water Utility Distribution Piping

33 11 13.25 Water Supply, Polyvinyl Chloride Pipe		Crew	Daily Output	Labor-Hours	Unit	Material	2014 Bare Costs Labor	Equipment	Total	Total Incl O&P
4160	6" diameter	Q-1C	125	.192	L.F.	4.82	9.75	27	41.57	49.50
4165	Fittings									
4170	Elbow, 90°, 3/4"	Q-1A	114	.088	Ea.	1.57	5.10		6.67	9.45
4180	1"		100	.100		4.57	5.80		10.37	13.80
4190	1-1/2"		80	.125		16.15	7.25		23.40	29
4200	2"		72	.139		25.50	8.05		33.55	40
4210	3"	B-20A	46	.696		31	31		62	81.50
4220	4"		36	.889		38.50	40		78.50	103
4230	6"		24	1.333		70.50	59.50		130	169
4240	8"		14	2.286		145	102		247	315
4250	Elbow, 45°, 3/4"	Q-1A	114	.088		2.64	5.10		7.74	10.60
4260	1"		100	.100		6.35	5.80		12.15	15.70
4270	1-1/2"		80	.125		18.60	7.25		25.85	31.50
4280	2"		72	.139		19.85	8.05		27.90	34
4290	2-1/2"		54	.185		21	10.75		31.75	39
4300	3"	B-20A	46	.696		25	31		56	75
4310	4"		36	.889		37	40		77	101
4320	6"		24	1.333		75.50	59.50		135	174
4330	8"		14	2.286		135	102		237	305
4340	Tee, 3/4"	Q-1A	76	.132		3.20	7.65		10.85	15
4350	1"		66	.152		16.75	8.80		25.55	31.50
4360	1-1/2"		54	.185		24.50	10.75		35.25	42.50
4370	2"		48	.208		29	12.05		41.05	50.50
4380	2-1/2"		36	.278		31.50	16.10		47.60	59.50
4390	3"	B-20A	30	1.067		33.50	47.50		81	109
4400	4"		24	1.333		53	59.50		112.50	150
4410	6"		14.80	2.162		117	96.50		213.50	276
4420	8"		9	3.556		210	159		369	475
4430	Coupling, 3/4"	Q-1A	114	.088		1.47	5.10		6.57	9.30
4440	1"		100	.100		6.60	5.80		12.40	16.05
4450	1-1/2"		80	.125		10.60	7.25		17.85	22.50
4460	2"		72	.139		11.60	8.05		19.65	25
4470	2-1/2"		54	.185		12.80	10.75		23.55	30.50
4480	3"	B-20A	46	.696		16	31		47	65
4490	4"		36	.889		24.50	40		64.50	87.50
4500	6"		24	1.333		38	59.50		97.50	133
4510	8"		14	2.286		86.50	102		188.50	252
4520	Pressure pipe Class 150, SDR 18, AWWA C900, 4" diameter		380	.084	L.F.	2.61	3.77		6.38	8.60
4530	6" diameter		316	.101		5.25	4.53		9.78	12.70
4540	8" diameter		264	.121		8.90	5.40		14.30	18
4550	10" diameter		220	.145		15.45	6.50		21.95	27
4560	12" diameter		186	.172		20.50	7.70		28.20	34.50
8000	Fittings with rubber gasket									
8003	Class 150, DR 18									
8006	90° Bend, 4" diameter	B-20	100	.240	Ea.	42.50	9.80		52.30	61.50
8020	6" diameter		90	.267		74.50	10.90		85.40	99
8040	8" diameter		80	.300		144	12.25		156.25	178
8060	10" diameter		50	.480		330	19.60		349.60	390
8080	12" diameter		30	.800		425	32.50		457.50	515
8100	Tee, 4" diameter		90	.267		58	10.90		68.90	80.50
8120	6" diameter		80	.300		144	12.25		156.25	177
8140	8" diameter		70	.343		184	14		198	224
8160	10" diameter		40	.600		590	24.50		614.50	690

33 11 Water Utility Distribution Piping

33 11 13 – Public Water Utility Distribution Piping

33 11 13.25 Water Supply, Polyvinyl Chloride Pipe

		Crew	Daily Output	Labor-Hours	Unit	Material	2014 Bare Costs Labor	Equipment	Total	Total Incl O&P
8180	12" diameter	B-20	20	1.200	Ea.	765	49		814	915
8200	45° Bend, 4" diameter		100	.240		42	9.80		51.80	61
8220	6" diameter		90	.267		73	10.90		83.90	97
8240	8" diameter		50	.480		144	19.60		163.60	190
8260	10" diameter		50	.480		330	19.60		349.60	390
8280	12" diameter		30	.800		425	32.50		457.50	515
8300	Reducing tee 6" x 4"		100	.240		126	9.80		135.80	153
8320	8" x 6"		90	.267		224	10.90		234.90	263
8330	10" x 6"		90	.267		380	10.90		390.90	430
8340	10" x 8"		90	.267		395	10.90		405.90	445
8350	12" x 6"		90	.267		450	10.90		460.90	505
8360	12" x 8"		90	.267		470	10.90		480.90	530
8400	Tapped service tee (threaded type) 6" x 6" x 3/4"		100	.240		95	9.80		104.80	120
8430	6" x 6" x 1"		90	.267		95	10.90		105.90	122
8440	6" x 6" x 1-1/2"		90	.267		95	10.90		105.90	122
8450	6" x 6" x 2"		90	.267		95	10.90		105.90	122
8460	8" x 8" x 3/4"		90	.267		140	10.90		150.90	171
8470	8" x 8" x 1"		90	.267		140	10.90		150.90	171
8480	8" x 8" x 1-1/2"		90	.267		140	10.90		150.90	171
8490	8" x 8" x 2"		90	.267		140	10.90		150.90	171
8500	Repair coupling 4"		100	.240		22.50	9.80		32.30	40
8520	6" diameter		90	.267		34.50	10.90		45.40	55
8540	8" diameter		50	.480		83	19.60		102.60	122
8560	10" diameter		50	.480		208	19.60		227.60	260
8580	12" diameter		50	.480		233	19.60		252.60	288
8600	Plug end 4"		100	.240		22.50	9.80		32.30	40
8620	6" diameter		90	.267		40.50	10.90		51.40	61.50
8640	8" diameter		50	.480		69	19.60		88.60	107
8660	10" diameter		50	.480		96.50	19.60		116.10	137
8680	12" diameter		50	.480		119	19.60		138.60	162
8700	PVC pipe, joint restraint									
8710	4" diameter	B-20A	32	1	Ea.	46	44.50		90.50	119
8720	6" diameter		25.60	1.250		57	56		113	148
8730	8" diameter		21.33	1.500		79.50	67		146.50	190
8740	10" diameter		18.28	1.751		141	78.50		219.50	275
8750	12" diameter		16.84	1.900		148	85		233	293
8760	14" diameter		16	2		221	89.50		310.50	380
8770	16" diameter		11.64	2.749		274	123		397	485
8780	18" diameter		11.03	2.901		330	130		460	565
8785	20" diameter		9.14	3.501		400	157		557	680
8790	24" diameter		7.53	4.250		465	190		655	800

33 11 13.35 Water Supply, HDPE

		Crew	Daily Output	Labor-Hours	Unit	Material	Labor	Equipment	Total	Total Incl O&P
0010	**WATER SUPPLY, HDPE**									
0011	Butt fusion joints, SDR 21 40' lengths not including excavation or backfill									
0100	4" diameter	B-22A	400	.100	L.F.	4.13	4.19	1.66	9.98	12.80
0200	6" diameter		380	.105		8.20	4.41	1.74	14.35	17.70
0300	8" diameter		320	.125		10.65	5.25	2.07	17.97	22
0400	10" diameter		300	.133		20	5.60	2.21	27.81	33
0500	12" diameter		260	.154		22.50	6.45	2.55	31.50	37
0600	14" diameter	B-22B	220	.182		38.50	7.60	5.05	51.15	60
0700	16" diameter		180	.222		44	9.30	6.15	59.45	69
0800	18" diameter		140	.286		49.50	11.95	7.90	69.35	81

33 11 Water Utility Distribution Piping

33 11 13 – Public Water Utility Distribution Piping

33 11 13.35 Water Supply, HDPE

		Crew	Daily Output	Labor-Hours	Unit	Material	2014 Bare Costs Labor	Equipment	Total	Total Incl O&P
0900	24" diameter	B-22B	100	.400	L.F.	70.50	16.75	11.10	98.35	116
1000	Fittings									
1100	Elbows, 90 degrees									
1200	4" diameter	B-22A	32	1.250	Ea.	23.50	52.50	20.50	96.50	129
1300	6" diameter		28	1.429		55	60	23.50	138.50	179
1400	8" diameter		24	1.667		148	70	27.50	245.50	300
1500	10" diameter		18	2.222		550	93	37	680	790
1600	12" diameter		12	3.333		550	140	55	745	880
1700	14" diameter	B-22B	9	4.444		760	186	123	1,069	1,250
1800	16" diameter		6	6.667		870	279	185	1,334	1,600
1900	18" diameter		4	10		985	420	277	1,682	2,025
2000	24" diameter		3	13.333		1,325	560	370	2,255	2,725
2100	Tees									
2200	4" diameter	B-22A	30	1.333	Ea.	28	56	22	106	142
2300	6" diameter		26	1.538		68.50	64.50	25.50	158.50	202
2400	8" diameter		22	1.818		175	76	30	281	345
2500	10" diameter		15	2.667		231	112	44	387	475
2600	12" diameter		10	4		865	168	66	1,099	1,275
2700	14" diameter	B-22B	8	5		1,025	210	139	1,374	1,600
2800	16" diameter		6	6.667		1,200	279	185	1,664	1,950
2900	18" diameter		4	10		1,400	420	277	2,097	2,500
3000	24" diameter		2	20		1,875	840	555	3,270	3,975
4100	Caps									
4110	4" diameter	B-22A	34	1.176	Ea.	17.65	49.50	19.45	86.60	117
4120	6" diameter		30	1.333		41	56	22	119	156
4130	8" diameter		26	1.538		71	64.50	25.50	161	205
4150	10" diameter		20	2		310	84	33	427	510
4160	12" diameter		14	2.857		320	120	47.50	487.50	585

33 11 13.45 Water Supply, Copper Pipe

		Crew	Daily Output	Labor-Hours	Unit	Material	2014 Bare Costs Labor	Equipment	Total	Total Incl O&P
0010	**WATER SUPPLY, COPPER PIPE**									
0020	Not including excavation or backfill									
2000	Tubing, type K, 20' joints, 3/4" diameter	Q-1	400	.040	L.F.	6.75	2.07		8.82	10.60
2200	1" diameter		320	.050		9	2.59		11.59	13.80
3000	1-1/2" diameter		265	.060		14.35	3.13		17.48	20.50
3020	2" diameter		230	.070		22	3.60		25.60	30
3040	2-1/2" diameter		146	.110		32.50	5.70		38.20	44.50
3060	3" diameter		134	.119		45	6.20		51.20	59.50
4012	4" diameter		95	.168		75	8.70		83.70	95
4016	6" diameter	Q-2	80	.300		161	16.10		177.10	202
5000	Tubing, type L									
5108	2" diameter	Q-1	230	.070	L.F.	17.65	3.60		21.25	25
6010	3" diameter		134	.119		36	6.20		42.20	49
6012	4" diameter		95	.168		59.50	8.70		68.20	78.50
6016	6" diameter	Q-2	80	.300		142	16.10		158.10	181
7020	Fittings, brass, corporation stops, 3/4" diameter	1 Plum	19	.421	Ea.	45	24		69	86
7040	1" diameter		16	.500		45.50	29		74.50	93.50
7060	1-1/2" diameter		13	.615		143	35.50		178.50	211
7080	2" diameter		11	.727		211	42		253	295
7100	Curb stops, 3/4" diameter		19	.421		48	24		72	89.50
7120	1" diameter		16	.500		77	29		106	128
7140	1-1/2" diameter		13	.615		220	35.50		255.50	296
7160	2" diameter		11	.727		330	42		372	430

33 11 Water Utility Distribution Piping

33 11 13 – Public Water Utility Distribution Piping

33 11 13.45 Water Supply, Copper Pipe

		Crew	Daily Output	Labor-Hours	Unit	Material	2014 Bare Costs Labor	Equipment	Total	Total Incl O&P
7165	Fittings, brass, corporation stops, no lead, 3/4" diameter	1 Plum	19	.421	Ea.	76	24		100	120
7166	1" diameter		16	.500		100	29		129	154
7167	1-1/2" diameter		13	.615		220	35.50		255.50	296
7168	2" diameter		11	.727		330	42		372	430
7170	Curb stops, no lead, 3/4" diameter		19	.421		103	24		127	151
7171	1" diameter		16	.500		147	29		176	206
7172	1-1/2" diameter		13	.615		275	35.50		310.50	360
7173	2" diameter		11	.727		290	42		332	385

33 12 Water Utility Distribution Equipment

33 12 13 – Water Service Connections

33 12 13.15 Tapping, Crosses and Sleeves

		Crew	Daily Output	Labor-Hours	Unit	Material	Labor	Equipment	Total	Total Incl O&P
0010	**TAPPING, CROSSES AND SLEEVES**									
4000	Drill and tap pressurized main (labor only)									
4100	6" main, 1" to 2" service	Q-1	3	5.333	Ea.		276		276	415
4150	8" main, 1" to 2" service	"	2.75	5.818	"		300		300	455
4500	Tap and insert gate valve									
4600	8" main, 4" branch	B-21	3.20	8.750	Ea.	370	44		414	620
4650	6" branch		2.70	10.370		440	52		492	730
4700	10" Main, 4" branch		2.70	10.370		440	52		492	730
4750	6" branch		2.35	11.915		505	59.50		564.50	840
4800	12" main, 6" branch		2.35	11.915		505	59.50		564.50	840
4850	8" branch		2.35	11.915		505	59.50		564.50	840
7020	Crosses, 4" x 4"		37	.757		1,550	32	3.78	1,585.78	1,750
7030	6" x 4"		25	1.120		1,325	47.50	5.60	1,378.10	1,525
7040	6" x 6"		25	1.120		1,325	47.50	5.60	1,378.10	1,525
7060	8" x 6"		21	1.333		1,600	56.50	6.65	1,663.15	1,875
7080	8" x 8"		21	1.333		1,675	56.50	6.65	1,738.15	1,950
7100	10" x 6"		21	1.333		2,625	56.50	6.65	2,688.15	3,000
7120	10" x 10"		21	1.333		2,825	56.50	6.65	2,888.15	3,200
7140	12" x 6"		18	1.556		2,625	65.50	7.80	2,698.30	3,000
7160	12" x 12"		18	1.556		3,675	65.50	7.80	3,748.30	4,125
7180	14" x 6"		16	1.750		7,150	74	8.75	7,232.75	7,975
7200	14" x 14"		16	1.750		7,550	74	8.75	7,632.75	8,450
7220	16" x 6"		14	2		7,550	84.50	10	7,644.50	8,475
7240	16" x 10"		14	2		7,750	84.50	10	7,844.50	8,700
7260	16" x 16"		14	2		8,150	84.50	10	8,244.50	9,125
7280	18" x 6"		10	2.800		11,500	118	14	11,632	12,800
7300	18" x 12"		10	2.800		11,500	118	14	11,632	12,900
7320	18" x 18"		10	2.800		10,500	118	14	10,632	11,800
7340	20" x 6"		8	3.500		9,250	148	17.50	9,415.50	10,400
7360	20" x 12"		8	3.500		9,925	148	17.50	10,090.50	11,100
7380	20" x 20"		8	3.500		14,500	148	17.50	14,665.50	16,200
7400	24" x 6"		6	4.667		12,000	197	23.50	12,220.50	13,500
7420	24" x 12"		6	4.667		12,100	197	23.50	12,320.50	13,700
7440	24" x 18"		6	4.667		18,100	197	23.50	18,320.50	20,200
7460	24" x 24"		6	4.667		18,800	197	23.50	19,020.50	21,000
7600	Cut-in sleeves with rubber gaskets, 4"		18	1.556		435	65.50	7.80	508.30	585
7620	6"		12	2.333		480	98.50	11.65	590.15	690
7640	8"		10	2.800		640	118	14	772	900
7660	10"		10	2.800		895	118	14	1,027	1,175

33 12 Water Utility Distribution Equipment

33 12 13 – Water Service Connections

33 12 13.15 Tapping, Crosses and Sleeves		Crew	Daily Output	Labor-Hours	Unit	Material	2014 Bare Costs Labor	Equipment	Total	Total Incl O&P
7680	12"	B-21	9	3.111	Ea.	1,200	131	15.55	1,346.55	1,550
7800	Cut-in valves with rubber gaskets, 4"		18	1.556		475	65.50	7.80	548.30	630
7820	6"		12	2.333		645	98.50	11.65	755.15	875
7840	8"		10	2.800		1,000	118	14	1,132	1,300
7860	10"		10	2.800		1,050	118	14	1,182	1,350
7880	12"		9	3.111		1,250	131	15.55	1,396.55	1,600
7900	Tapping Valve 4 inch, MJ, ductile iron		18	1.556		570	65.50	7.80	643.30	740
7920	6 inch, MJ, ductile iron		12	2.333		790	98.50	11.65	900.15	1,025
8000	Sleeves with rubber gaskets, 4" x 4"		37	.757		870	32	3.78	905.78	1,025
8010	6" x 4"		25	1.120		865	47.50	5.60	918.10	1,025
8020	6" x 6"		25	1.120		1,025	47.50	5.60	1,078.10	1,200
8030	8" x 4"		21	1.333		880	56.50	6.65	943.15	1,075
8040	8" x 6"		21	1.333		1,025	56.50	6.65	1,088.15	1,225
8060	8" x 8"		21	1.333		1,350	56.50	6.65	1,413.15	1,575
8070	10" x 4"		21	1.333		970	56.50	6.65	1,033.15	1,175
8080	10" x 6"		21	1.333		1,050	56.50	6.65	1,113.15	1,250
8090	10" x 8"		21	1.333		1,525	56.50	6.65	1,588.15	1,775
8100	10" x 10"		21	1.333		2,225	56.50	6.65	2,288.15	2,550
8110	12" x 4"		18	1.556		1,075	65.50	7.80	1,148.30	1,300
8120	12" x 6"		18	1.556		1,675	65.50	7.80	1,748.30	1,925
8130	12" x 8"		18	1.556		1,350	65.50	7.80	1,423.30	1,600
8135	12" x 10"		18	1.556		2,250	65.50	7.80	2,323.30	2,575
8140	12" x 12"		18	1.556		2,900	65.50	7.80	2,973.30	3,275
8160	14" x 6"		16	1.750		1,775	74	8.75	1,857.75	2,075
8180	14" x 14"		16	1.750		2,400	74	8.75	2,482.75	2,775
8200	16" x 6"		14	2		2,075	84.50	10	2,169.50	2,425
8220	16" x 10"		14	2		3,325	84.50	10	3,419.50	3,825
8240	16" x 16"		14	2		9,300	84.50	10	9,394.50	10,300
8260	18" x 6"		10	2.800		2,125	118	14	2,257	2,525
8280	18" x 12"		10	2.800		3,525	118	14	3,657	4,075
8300	18" x 18"		10	2.800		12,000	118	14	12,132	13,400
8320	20" x 6"		8	3.500		2,475	148	17.50	2,640.50	2,950
8340	20" x 12"		8	3.500		4,175	148	17.50	4,340.50	4,850
8360	20" x 20"		8	3.500		12,000	148	17.50	12,165.50	13,400
8380	24" x 6"		6	4.667		2,600	197	23.50	2,820.50	3,175
8400	24" x 12"		6	4.667		4,325	197	23.50	4,545.50	5,100
8420	24" x 18"		6	4.667		10,700	197	23.50	10,920.50	12,100
8440	24" x 24"		6	4.667		15,600	197	23.50	15,820.50	17,500
8800	Hydrant valve box, 6' long	B-20	20	1.200		259	49		308	360
8820	8' long		18	1.333		335	54.50		389.50	450
8830	Valve box w/lid 4' deep		14	1.714		116	70		186	236
8840	Valve box and large base w/lid		14	1.714		325	70		395	470

33 12 16 – Water Utility Distribution Valves

33 12 16.10 Valves		Crew	Daily Output	Labor-Hours	Unit	Material	Labor	Equipment	Total	Total Incl O&P
0010	**VALVES**, water distribution									
0011	See Sections 22 05 23.20 and 22 05 23.60									
3000	Butterfly valves with boxes, cast iron, mech. jt.									
3100	4" diameter	B-6	6	4	Ea.	830	160	61	1,051	1,225
3140	6" diameter		6	4		965	160	61	1,186	1,375
3180	8" diameter		6	4		1,150	160	61	1,371	1,575
3300	10" diameter		6	4		1,625	160	61	1,846	2,100
3340	12" diameter		6	4		1,675	160	61	1,896	2,175

33 12 Water Utility Distribution Equipment

33 12 16 – Water Utility Distribution Valves

33 12 16.10 Valves

		Crew	Daily Output	Labor-Hours	Unit	Material	2014 Bare Costs Labor	Equipment	Total	Total Incl O&P
3400	14" diameter	B-6	4	6	Ea.	4,025	241	91.50	4,357.50	4,900
3440	16" diameter		4	6		3,550	241	91.50	3,882.50	4,375
3460	18" diameter		4	6		6,150	241	91.50	6,482.50	7,225
3480	20" diameter		4	6		5,375	241	91.50	5,707.50	6,375
3500	24" diameter		4	6		7,500	241	91.50	7,832.50	8,725
3600	With lever operator									
3610	4" diameter	B-6	6	4	Ea.	715	160	61	936	1,100
3614	6" diameter		6	4		850	160	61	1,071	1,250
3616	8" diameter		6	4		1,025	160	61	1,246	1,450
3618	10" diameter		6	4		1,500	160	61	1,721	1,975
3620	12" diameter		6	4		1,550	160	61	1,771	2,050
3622	14" diameter		4	6		3,700	241	91.50	4,032.50	4,550
3624	16" diameter		4	6		3,225	241	91.50	3,557.50	4,025
3626	18" diameter		4	6		5,825	241	91.50	6,157.50	6,875
3628	20" diameter		4	6		5,050	241	91.50	5,382.50	6,025
3630	24" diameter		4	6		7,175	241	91.50	7,507.50	8,350
3700	Check valves, flanged									
3710	4" diameter	B-6	6	4	Ea.	940	160	61	1,161	1,350
3714	6" diameter		6	4		1,375	160	61	1,596	1,825
3716	8" diameter		6	4		2,300	160	61	2,521	2,850
3718	10" diameter		6	4		3,300	160	61	3,521	3,950
3720	12" diameter		6	4		3,825	160	61	4,046	4,550
3722	14" diameter		4	6		5,250	241	91.50	5,582.50	6,250
3724	16" diameter		4	6		6,600	241	91.50	6,932.50	7,725
3726	18" diameter		4	6		9,775	241	91.50	10,107.50	11,300
3728	20" diameter		4	6		11,400	241	91.50	11,732.50	13,000
3730	24" diameter		4	6		16,300	241	91.50	16,632.50	18,400
3800	Gate valves, C.I., 250 PSI, mechanical joint, w/boxes									
3810	4" diameter	B-6	6	4	Ea.	395	160	61	616	750
3814	6" diameter		6	4		520	160	61	741	890
3816	8" diameter		6	4		920	160	61	1,141	1,325
3818	10" diameter		6	4		1,125	160	61	1,346	1,550
3820	12" diameter		6	4		1,475	160	61	1,696	1,950
3822	14" diameter		4	6		4,700	241	91.50	5,032.50	5,650
3824	16" diameter		4	6		5,075	241	91.50	5,407.50	6,075
3826	18" diameter		4	6		8,275	241	91.50	8,607.50	9,575
3828	20" diameter		4	6		11,700	241	91.50	12,032.50	13,300
3830	24" diameter		4	6		14,600	241	91.50	14,932.50	16,600
3831	30" diameter		4	6		38,400	241	91.50	38,732.50	42,700
3832	36" diameter		4	6		56,000	241	91.50	56,332.50	62,000
3880	Sleeve, for tapping mains, 8" x 4", add					880			880	970
3884	10" x 6", add					1,050			1,050	1,150
3888	12" x 6", add					1,675			1,675	1,825
3892	12" x 8", add					1,350			1,350	1,500

33 12 16.20 Valves

		Crew	Daily Output	Labor-Hours	Unit	Material	Labor	Equipment	Total	Total Incl O&P
0010	**VALVES**									
0011	Special trim or use									
9000	Valves, gate valve, N.R.S. PIV with post, 4" diameter	B-6	6	4	Ea.	1,400	160	61	1,621	1,850
9020	6" diameter		6	4		1,525	160	61	1,746	2,000
9040	8" diameter		6	4		1,925	160	61	2,146	2,425
9060	10" diameter		6	4		2,100	160	61	2,321	2,650
9080	12" diameter		6	4		2,475	160	61	2,696	3,050

33 12 Water Utility Distribution Equipment

33 12 16 – Water Utility Distribution Valves

33 12 16.20 Valves		Crew	Daily Output	Labor-Hours	Unit	Material	2014 Bare Costs Labor	Equipment	Total	Total Incl O&P
9100	14" diameter	B-6	6	4	Ea.	5,475	160	61	5,696	6,350
9120	OS&Y, 4" diameter		6	4		730	160	61	951	1,125
9140	6" diameter		6	4		990	160	61	1,211	1,425
9160	8" diameter		6	4		1,550	160	61	1,771	2,025
9180	10" diameter		6	4		2,475	160	61	2,696	3,025
9200	12" diameter		6	4		3,725	160	61	3,946	4,425
9220	14" diameter		4	6		5,075	241	91.50	5,407.50	6,075
9400	Check valves, rubber disc, 2-1/2" diameter		6	4		710	160	61	931	1,100
9420	3" diameter		6	4		725	160	61	946	1,125
9440	4" diameter		6	4		940	160	61	1,161	1,350
9480	6" diameter		6	4		1,375	160	61	1,596	1,825
9500	8" diameter		6	4		2,300	160	61	2,521	2,850
9520	10" diameter		6	4		3,300	160	61	3,521	3,950
9540	12" diameter		6	4		3,825	160	61	4,046	4,550
9542	14" diameter		4	6		5,250	241	91.50	5,582.50	6,250
9700	Detector check valves, reducing, 4" diameter		6	4		1,325	160	61	1,546	1,775
9720	6" diameter		6	4		2,025	160	61	2,246	2,550
9740	8" diameter		6	4		2,975	160	61	3,196	3,575
9760	10" diameter		6	4		5,250	160	61	5,471	6,100
9800	Galvanized, 4" diameter		6	4		1,550	160	61	1,771	2,025
9820	6" diameter		6	4		2,075	160	61	2,296	2,600
9840	8" diameter		6	4		3,250	160	61	3,471	3,900
9860	10" diameter		6	4		5,000	160	61	5,221	5,800

33 12 19 – Water Utility Distribution Fire Hydrants

33 12 19.10 Fire Hydrants

			Crew	Daily Output	Labor-Hours	Unit	Material	Labor	Equipment	Total	Total Incl O&P
0010	**FIRE HYDRANTS**	G3010-410									
0020	Mechanical joints unless otherwise noted										
1000	Fire hydrants, two way; excavation and backfill not incl.										
1100	4-1/2" valve size, depth 2'-0"		B-21	10	2.800	Ea.	1,650	118	14	1,782	2,025
1200	4'-6"			9	3.111		1,975	131	15.55	2,121.55	2,400
1260	6'-0"			7	4		2,100	169	20	2,289	2,575
1340	8'-0"			6	4.667		2,225	197	23.50	2,445.50	2,775
1420	10'-0"			5	5.600		2,375	236	28	2,639	3,025
2000	5-1/4" valve size, depth 2'-0"			10	2.800		1,775	118	14	1,907	2,175
2080	4'-0"			9	3.111		1,900	131	15.55	2,046.55	2,300
2160	6'-0"			7	4		2,050	169	20	2,239	2,550
2240	8'-0"			6	4.667		2,175	197	23.50	2,395.50	2,700
2320	10'-0"			5	5.600		2,500	236	28	2,764	3,150
2350	For threeway valves, add						7%				
2400	Lower barrel extensions with stems, 1'-0"		B-20	14	1.714		355	70		425	500
2440	2'-0"			13	1.846		430	75.50		505.50	590
2480	3'-0"			12	2		770	81.50		851.50	970
2520	4'-0"			10	2.400		815	98		913	1,050
5000	Indicator post										
5020	Adjustable, valve size 4" to 14", 4' bury		B-21	10	2.800	Ea.	800	118	14	932	1,075
5060	8' bury			7	4		1,100	169	20	1,289	1,500
5080	10' bury			6	4.667		1,150	197	23.50	1,370.50	1,600
5100	12' bury			5	5.600		1,400	236	28	1,664	1,950
5120	14' bury			4	7		1,525	295	35	1,855	2,175
5500	Non-adjustable, valve size 4" to 14", 3' bury			10	2.800		800	118	14	932	1,075
5520	3'-6" bury			10	2.800		800	118	14	932	1,075
5540	4' bury			9	3.111		800	131	15.55	946.55	1,100

33 16 Water Utility Storage Tanks

33 16 13 – Aboveground Water Utility Storage Tanks

33 16 13.16 Prestressed Conc. Water Storage Tanks

		Crew	Daily Output	Labor-Hours	Unit	Material	2014 Bare Costs Labor	Equipment	Total	Total Incl O&P
0010	**PRESTRESSED CONC. WATER STORAGE TANKS**									
0020	Not including fdn., pipe or pumps, 250,000 gallons				Ea.				299,000	329,500

33 21 Water Supply Wells

33 21 13 – Public Water Supply Wells

33 21 13.10 Wells and Accessories

		Crew	Daily Output	Labor-Hours	Unit	Material	Labor	Equipment	Total	Total Incl O&P
0010	**WELLS & ACCESSORIES**									
0011	Domestic									
0100	Drilled, 4" to 6" diameter	B-23	120	.333	L.F.		12.35	24	36.35	45.50
0200	8" diameter	"	95.20	.420	"		15.55	30.50	46.05	57.50
0400	Gravel pack well, 40' deep, incl. gravel & casing, complete									
0500	24" diameter casing x 18" diameter screen	B-23	.13	307	Total	38,800	11,400	22,200	72,400	84,500
0600	36" diameter casing x 18" diameter screen		.12	333	"	40,000	12,400	24,000	76,400	89,500
0800	Observation wells, 1-1/4" riser pipe		163	.245	V.L.F.	20.50	9.10	17.70	47.30	56
0900	For flush Buffalo roadway box, add	1 Skwk	16.60	.482	Ea.	51.50	23		74.50	92
1200	Test well, 2-1/2" diameter, up to 50' deep (15 to 50 GPM)	B-23	1.51	26.490	"	810	980	1,900	3,690	4,525
1300	Over 50' deep, add	"	121.80	.328	L.F.	21.50	12.15	23.50	57.15	68.50
1500	Pumps, installed in wells to 100' deep, 4" submersible									
1510	1/2 H.P.	Q-1	3.22	4.969	Ea.	450	257		707	885
1520	3/4 H.P.		2.66	6.015		590	310		900	1,125
1600	1 H.P.		2.29	6.987		680	360		1,040	1,300
1700	1-1/2 H.P.	Q-22	1.60	10		735	520	415	1,670	2,050
1800	2 H.P.		1.33	12.030		890	625	500	2,015	2,475
1900	3 H.P.		1.14	14.035		1,375	725	580	2,680	3,250
2000	5 H.P.		1.14	14.035		2,300	725	580	3,605	4,275
3000	Pump, 6" submersible, 25' to 150' deep, 25 H.P., 249 to 297 GPM		.89	17.978		7,175	930	745	8,850	10,100
3100	25' to 500' deep, 30 H.P., 100 to 300 GPM		.73	21.918		8,600	1,125	910	10,635	12,200
8000	Steel well casing	B-23A	3020	.008	Lb.	1.22	.33	.90	2.45	2.83
8110	Well screen assembly, stainless steel, 2" diameter		273	.088	L.F.	81	3.64	9.90	94.54	106
8120	3" diameter		253	.095		141	3.93	10.70	155.63	174
8130	4" diameter		200	.120		177	4.97	13.55	195.52	217
8140	5" diameter		168	.143		187	5.90	16.10	209	233
8150	6" diameter		126	.190		212	7.90	21.50	241.40	269
8160	8" diameter		98.50	.244		288	10.10	27.50	325.60	360
8170	10" diameter		73	.329		360	13.60	37	410.60	455
8180	12" diameter		62.50	.384		410	15.90	43.50	469.40	520
8190	14" diameter		54.30	.442		460	18.30	50	528.30	590
8200	16" diameter		48.30	.497		510	20.50	56	586.50	655
8210	18" diameter		39.20	.612		635	25.50	69	729.50	815
8220	20" diameter		31.20	.769		710	32	87	829	930
8230	24" diameter		23.80	1.008		855	42	114	1,011	1,125
8240	26" diameter		21	1.143		930	47.50	129	1,106.50	1,250
8250	PVC well casing, 2" diameter		280	.086		1.66	3.55	9.65	14.86	17.95
8252	3" diameter		260	.092		3.40	3.82	10.40	17.62	21
8254	4" diameter		205	.117		4.89	4.85	13.20	22.94	27.50
8256	6" diameter		130	.185		8.50	7.65	21	37.15	44
8258	8" diameter		100	.240		12.95	9.95	27	49.90	59.50
8300	Slotted PVC, 1-1/4" diameter		521	.046		2.41	1.91	5.20	9.52	11.25
8310	1-1/2" diameter		488	.049		2.95	2.04	5.55	10.54	12.45
8320	2" diameter		273	.088		4.03	3.64	9.90	17.57	21
8330	3" diameter		253	.095		5.85	3.93	10.70	20.48	24

33 21 Water Supply Wells

33 21 13 – Public Water Supply Wells

33 21 13.10 Wells and Accessories

		Crew	Daily Output	Labor-Hours	Unit	Material	2014 Bare Costs Labor	Equipment	Total	Total Incl O&P
8340	4" diameter	B-23A	200	.120	L.F.	6.75	4.97	13.55	25.27	30
8350	5" diameter		168	.143		14.05	5.90	16.10	36.05	42.50
8360	6" diameter		126	.190		16.70	7.90	21.50	46.10	54
8370	8" diameter		98.50	.244		26	10.10	27.50	63.60	74
8400	Artificial gravel pack, 2" screen, 6" casing	B-23B	174	.138		3.61	5.70	17.50	26.81	32
8405	8" casing		111	.216		4.63	8.95	27.50	41.08	49
8410	10" casing		74.50	.322		6.60	13.35	41	60.95	73
8415	12" casing		60	.400		9.25	16.55	51	76.80	91.50
8420	14" casing		50.20	.478		10.80	19.80	60.50	91.10	109
8425	16" casing		40.70	.590		13.60	24.50	75	113.10	135
8430	18" casing		36	.667		16.35	27.50	84.50	128.35	154
8435	20" casing		29.50	.814		19.25	33.50	103	155.75	187
8440	24" casing		25.70	.934		22	38.50	119	179.50	215
8445	26" casing		24.60	.976		24.50	40.50	124	189	225
8450	30" casing		20	1.200		28	49.50	152	229.50	275
8455	36" casing		16.40	1.463		31.50	60.50	186	278	335
8500	Develop well		8	3	Hr.	243	124	380	747	875
8550	Pump test well		8	3		81	124	380	585	700
8560	Standby well	B-23A	8	3		76.50	124	340	540.50	645
8570	Standby, drill rig		8	3			124	340	464	560
8580	Surface seal well, concrete filled		1	24	Ea.	835	995	2,700	4,530	5,425
8590	Well test pump, install & remove	B-23	1	40			1,475	2,875	4,350	5,450
8600	Well sterilization, chlorine	2 Clab	1	16		120	585		705	1,025
8610	Well water pressure switch	1 Clab	12	.667		49.50	24.50		74	92
8630	Well, water pressure switch with manual reset	"	12	.667		35	24.50		59.50	76
9950	See Section 31 23 19.40 for wellpoints									
9960	See Section 31 23 19.30 for drainage wells									

33 21 13.20 Water Supply Wells, Pumps

		Crew	Daily Output	Labor-Hours	Unit	Material	Labor	Equipment	Total	Total Incl O&P
0010	**WATER SUPPLY WELLS, PUMPS**									
0011	With pressure control									
1000	Deep well, jet, 42 gal. galvanized tank									
1040	3/4 HP	1 Plum	.80	10	Ea.	1,100	575		1,675	2,100
3000	Shallow well, jet, 30 gal. galvanized tank									
3040	1/2 HP	1 Plum	2	4	Ea.	895	230		1,125	1,325

33 31 Sanitary Utility Sewerage Piping

33 31 13 – Public Sanitary Utility Sewerage Piping

33 31 13.10 Sewage Collection, Valves

		Crew	Daily Output	Labor-Hours	Unit	Material	Labor	Equipment	Total	Total Incl O&P
0010	**SEWAGE COLLECTION, VALVES**									
1000	Backwater sewer line valve									
1010	Offset type, bronze swing check assy and bronze cover									
1020	B&S connections									
1040	2" size	2 Skwk	14	1.143	Ea.	555	54		609	695
1050	3" size		12.50	1.280		860	60.50		920.50	1,050
1060	4" size		11	1.455		1,075	69		1,144	1,275
1070	6" size	3 Skwk	15	1.600		1,525	75.50		1,600.50	1,800
1080	8" size	"	7.50	3.200		2,150	151		2,301	2,600
1110	Backwater drainage control									
1120	Offset type, w/bronze manually operated shear gate									
1130	B&S connections									
1160	4" size	2 Skwk	11	1.455	Ea.	1,900	69		1,969	2,175

33 31 Sanitary Utility Sewerage Piping

33 31 13 – Public Sanitary Utility Sewerage Piping

33 31 13.10 Sewage Collection, Valves		Crew	Daily Output	Labor-Hours	Unit	Material	2014 Bare Costs Labor	Equipment	Total	Total Incl O&P
1170	6" size	3 Skwk	15	1.600	Ea.	2,400	75.50		2,475.50	2,750

33 31 13.15 Sewage Collection, Concrete Pipe

0010	**SEWAGE COLLECTION, CONCRETE PIPE**									
0020	See Section 33 41 13.60 for sewage/drainage collection, concrete pipe									

33 31 13.25 Sewage Collection, Polyvinyl Chloride Pipe

0010	**SEWAGE COLLECTION, POLYVINYL CHLORIDE PIPE**									
0020	Not including excavation or backfill									
2000	20' lengths, SDR 35, B&S, 4" diameter	B-20	375	.064	L.F.	1.46	2.62		4.08	5.65
2040	6" diameter	"	350	.069		3.29	2.80		6.09	7.95
2120	10" diameter	B-21	330	.085		11.40	3.58	.42	15.40	18.50
2160	12" diameter		320	.088		12.75	3.69	.44	16.88	20
2200	15" diameter		240	.117		13	4.92	.58	18.50	22.50
4000	Piping, DWV PVC, no exc./bkfill., 10' L, Sch 40, 4" diameter	B-20	375	.064		4.64	2.62		7.26	9.15
4010	6" diameter		350	.069		10.20	2.80		13	15.55
4020	8" diameter		335	.072		19.70	2.93		22.63	26

33 36 Utility Septic Tanks

33 36 13 – Utility Septic Tank and Effluent Wet Wells

33 36 13.13 Concrete Utility Septic Tank

0010	**CONCRETE UTILITY SEPTIC TANK**	G3020-300								
0011	Not including excavation or piping									
0015	Septic tanks, precast, 1,000 gallon	B-21	8	3.500	Ea.	1,075	148	17.50	1,240.50	1,450
0060	1,500 gallon		7	4		1,500	169	20	1,689	1,925
0100	2,000 gallon		5	5.600		1,975	236	28	2,239	2,550
0200	5,000 gallon	B-13	3.50	16		9,475	640	213	10,328	11,600
0300	15,000 gallon, 4 piece	B-13B	1.70	32.941		21,700	1,325	670	23,695	26,700
0400	25,000 gallon, 4 piece		1.10	50.909		42,200	2,025	1,025	45,250	50,500
0500	40,000 gallon, 4 piece		.80	70		54,000	2,800	1,425	58,225	65,500
0520	50,000 gallon, 5 piece	B-13C	.60	93.333		62,000	3,725	2,850	68,575	77,500
0640	75,000 gallon, cast in place	C-14C	.25	448		75,500	19,700	130	95,330	113,500
0660	100,000 gallon	"	.15	746		93,500	32,800	217	126,517	153,500
1150	Leaching field chambers, 13' x 3'-7" x 1'-4", standard	B-13	16	3.500		480	140	46.50	666.50	790
1200	Heavy duty, 8' x 4' x 1'-6"		14	4		284	159	53.50	496.50	615
1300	13' x 3'-9" x 1'-6"		12	4.667		1,125	186	62	1,373	1,600
1350	20' x 4' x 1'-6"		5	11.200		1,175	445	149	1,769	2,125
1400	Leaching pit, precast concrete, 3' diameter, 3' deep	B-21	8	3.500		710	148	17.50	875.50	1,025
1500	6' diameter, 3' section		4.70	5.957		885	251	30	1,166	1,400
2000	Velocity reducing pit, precast conc., 6' diameter, 3' deep		4.70	5.957		1,600	251	30	1,881	2,175

33 36 13.19 Polyethylene Utility Septic Tank

0010	**POLYETHYLENE UTILITY SEPTIC TANK**									
0015	High density polyethylene, 1,000 gallon	B-21	8	3.500	Ea.	1,300	148	17.50	1,465.50	1,675
0020	1,250 gallon		8	3.500		1,400	148	17.50	1,565.50	1,800
0025	1,500 gallon		7	4		1,650	169	20	1,839	2,100

33 36 19 – Utility Septic Tank Effluent Filter

33 36 19.13 Utility Septic Tank Effluent Tube Filter

0010	**UTILITY SEPTIC TANK EFFLUENT TUBE FILTER**									
3000	Effluent filter, 4" diameter	1 Skwk	8	1	Ea.	55.50	47.50		103	135
3020	6" diameter	"	7	1.143	"	230	54		284	335

33 36 Utility Septic Tanks

33 36 33 – Utility Septic Tank Drainage Field

33 36 33.13 Utility Septic Tank Tile Drainage Field

		Crew	Daily Output	Labor-Hours	Unit	Material	2014 Bare Costs Labor	Equipment	Total	Total Incl O&P
0010	**UTILITY SEPTIC TANK TILE DRAINAGE FIELD**									
0015	Distribution box, concrete, 5 outlets	2 Clab	20	.800	Ea.	78.50	29.50		108	131
0020	7 outlets		16	1		78.50	36.50		115	143
0025	9 outlets		8	2		475	73.50		548.50	640
0115	Distribution boxes, HDPE, 5 outlets		20	.800		64.50	29.50		94	116
0120	8 outlets		10	1.600		68.50	58.50		127	166
0240	Distribution boxes, Outlet Flow Leveler	1 Clab	50	.160		2.22	5.85		8.07	11.50
0300	Precast concrete, galley, 4' x 4' x 4'	B-21	16	1.750		241	74	8.75	323.75	390

33 36 50 – Drainage Field Systems

33 36 50.10 Drainage Field Excavation and Fill

		Crew	Daily Output	Labor-Hours	Unit	Material	Labor	Equipment	Total	Total Incl O&P
0010	**DRAINAGE FIELD EXCAVATION AND FILL**									
2200	Septic tank & drainage field excavation with 3/4 c.y backhoe	B-12F	145	.110	C.Y.		4.79	4.57	9.36	12.30
2400	4' trench for disposal field, 3/4 C.Y. backhoe	"	335	.048	L.F.		2.08	1.98	4.06	5.35
2600	Gravel fill, run of bank	B-6	150	.160	C.Y.	19.90	6.40	2.44	28.74	34.50
2800	Crushed stone, 3/4"	"	150	.160	"	34.50	6.40	2.44	43.34	50

33 41 Storm Utility Drainage Piping

33 41 13 – Public Storm Utility Drainage Piping

33 41 13.40 Piping, Storm Drainage, Corrugated Metal

		Crew	Daily Output	Labor-Hours	Unit	Material	Labor	Equipment	Total	Total Incl O&P
0010	**PIPING, STORM DRAINAGE, CORRUGATED METAL**									
0020	Not including excavation or backfill									
2000	Corrugated metal pipe, galvanized									
2020	Bituminous coated with paved invert, 20' lengths									
2040	8" diameter, 16 ga.	B-14	330	.145	L.F.	8.70	5.65	1.11	15.46	19.40
2060	10" diameter, 16 ga.		260	.185		9.05	7.15	1.41	17.61	22.50
2080	12" diameter, 16 ga.		210	.229		11.10	8.85	1.74	21.69	27.50
2100	15" diameter, 16 ga.		200	.240		15.25	9.30	1.83	26.38	33
2120	18" diameter, 16 ga.		190	.253		16.75	9.80	1.92	28.47	35.50
2140	24" diameter, 14 ga.		160	.300		21.50	11.60	2.28	35.38	44
2160	30" diameter, 14 ga.	B-13	120	.467		28	18.60	6.20	52.80	66
2180	36" diameter, 12 ga.		120	.467		35.50	18.60	6.20	60.30	74.50
2200	48" diameter, 12 ga.		100	.560		53	22.50	7.45	82.95	100
2220	60" diameter, 10 ga.	B-13B	75	.747		80	30	15.15	125.15	150
2240	72" diameter, 8 ga.	"	45	1.244		95.50	49.50	25.50	170.50	209
2250	End sections, 8" diameter, 16 ga.	B-14	20	2.400	Ea.	43.50	93	18.25	154.75	211
2255	10" diameter, 16 ga.		20	2.400		60	93	18.25	171.25	229
2260	12" diameter, 16 ga.		18	2.667		120	103	20.50	243.50	315
2265	15" diameter, 16 ga.		18	2.667		205	103	20.50	328.50	410
2270	18" diameter, 16 ga.		16	3		240	116	23	379	470
2275	24" diameter, 16 ga.	B-13	16	3.500		280	140	46.50	466.50	575
2280	30" diameter, 16 ga.		14	4		490	159	53.50	702.50	840
2285	36" diameter, 14 ga.		14	4		730	159	53.50	942.50	1,100
2290	48" diameter, 14 ga.		10	5.600		1,850	223	74.50	2,147.50	2,450
2292	60" diameter, 14 ga.		6	9.333		1,950	370	124	2,444	2,850
2294	72" diameter, 14 ga.	B-13B	5	11.200		3,125	445	227	3,797	4,375
2300	Bends or elbows, 8" diameter	B-14	28	1.714		117	66.50	13.05	196.55	245
2320	10" diameter		25	1.920		146	74.50	14.60	235.10	290
2340	12" diameter, 16 ga.		23	2.087		169	81	15.90	265.90	325
2342	18" diameter, 16 ga.		20	2.400		244	93	18.25	355.25	430
2344	24" diameter, 14 ga.		16	3		335	116	23	474	575

33 41 Storm Utility Drainage Piping

33 41 13 – Public Storm Utility Drainage Piping

33 41 13.40 Piping, Storm Drainage, Corrugated Metal		Crew	Daily Output	Labor-Hours	Unit	Material	2014 Bare Costs Labor	Equipment	Total	Total Incl O&P
2346	30" diameter, 14 ga.	B-14	15	3.200	Ea.	430	124	24.50	578.50	690
2348	36" diameter, 14 ga.	B-13	15	3.733		580	149	49.50	778.50	920
2350	48" diameter, 12 ga.	"	12	4.667		805	186	62	1,053	1,250
2352	60" diameter, 10 ga.	B-13B	10	5.600		1,075	223	114	1,412	1,650
2354	72" diameter, 10 ga.	"	6	9.333		1,325	370	189	1,884	2,250
2360	Wyes or tees, 8" diameter	B-14	25	1.920		164	74.50	14.60	253.10	310
2380	10" diameter		21	2.286		205	88.50	17.40	310.90	380
2400	12" diameter, 16 ga.		19	2.526		244	98	19.25	361.25	440
2410	18" diameter, 16 ga.		16	3		325	116	23	464	560
2412	24" diameter, 14 ga.		16	3		485	116	23	624	740
2414	30" diameter, 14 ga.	B-13	12	4.667		645	186	62	893	1,075
2416	36" diameter, 14 ga.		11	5.091		795	203	68	1,066	1,250
2418	48" diameter, 12 ga.		10	5.600		1,175	223	74.50	1,472.50	1,700
2420	60" diameter, 10 ga.	B-13B	8	7		1,700	279	142	2,121	2,450
2422	72" diameter, 10 ga.	"	5	11.200		2,050	445	227	2,722	3,175
2500	Galvanized, uncoated, 20' lengths									
2520	8" diameter, 16 ga.	B-14	355	.135	L.F.	7.85	5.25	1.03	14.13	17.85
2540	10" diameter, 16 ga.		280	.171		9	6.65	1.30	16.95	21.50
2560	12" diameter, 16 ga.		220	.218		10	8.45	1.66	20.11	26
2580	15" diameter, 16 ga.		220	.218		12.50	8.45	1.66	22.61	28.50
2600	18" diameter, 16 ga.		205	.234		15.10	9.05	1.78	25.93	32.50
2620	24" diameter, 14 ga.		175	.274		19	10.60	2.09	31.69	39.50
2640	30" diameter, 14 ga.	B-13	130	.431		25	17.15	5.75	47.90	60.50
2660	36" diameter, 12 ga.		130	.431		32	17.15	5.75	54.90	68
2680	48" diameter, 12 ga.		110	.509		47.50	20.50	6.80	74.80	91
2690	60" diameter, 10 ga.	B-13B	78	.718		72	28.50	14.55	115.05	139
2695	72" diameter, 10 ga.	"	60	.933		86	37	18.95	141.95	173
2711	Bends or elbows, 12" diameter, 16 ga.	B-14	30	1.600	Ea.	146	62	12.20	220.20	269
2712	15" diameter, 16 ga.		25.04	1.917		180	74	14.60	268.60	330
2714	18" diameter, 16 ga.		20	2.400		202	93	18.25	313.25	385
2716	24" diameter, 14 ga.		16	3		293	116	23	432	525
2718	30" diameter, 14 ga.		15	3.200		340	124	24.50	488.50	595
2720	36" diameter, 14 ga.	B-13	15	3.733		515	149	49.50	713.50	850
2722	48" diameter, 12 ga.		12	4.667		680	186	62	928	1,100
2724	60" diameter, 10 ga.		10	5.600		1,050	223	74.50	1,347.50	1,575
2726	72" diameter, 10 ga.		6	9.333		1,350	370	124	1,844	2,175
2728	Wyes or tees, 12" diameter, 16 ga.	B-14	22.48	2.135		193	82.50	16.25	291.75	360
2730	18" diameter, 16 ga.		15	3.200		282	124	24.50	430.50	530
2732	24" diameter, 14 ga.		15	3.200		450	124	24.50	598.50	715
2734	30" diameter, 14 ga.		14	3.429		580	133	26	739	870
2736	36" diameter, 14 ga.	B-13	14	4		755	159	53.50	967.50	1,125
2738	48" diameter, 12 ga.		12	4.667		1,075	186	62	1,323	1,550
2740	60" diameter, 10 ga.		10	5.600		1,575	223	74.50	1,872.50	2,150
2742	72" diameter, 10 ga.		6	9.333		1,850	370	124	2,344	2,750
2780	End sections, 8" diameter	B-14	35	1.371		73	53	10.45	136.45	173
2785	10" diameter		35	1.371		77	53	10.45	140.45	178
2790	12" diameter		35	1.371		114	53	10.45	177.45	218
2800	18" diameter		30	1.600		115	62	12.20	189.20	236
2810	24" diameter	B-13	25	2.240		215	89.50	30	334.50	405
2820	30" diameter		25	2.240		330	89.50	30	449.50	530
2825	36" diameter		20	2.800		480	112	37.50	629.50	740
2830	48" diameter		10	5.600		950	223	74.50	1,247.50	1,475
2835	60" diameter	B-13B	5	11.200		1,650	445	227	2,322	2,725

33 41 Storm Utility Drainage Piping

33 41 13 – Public Storm Utility Drainage Piping

33 41 13.40 Piping, Storm Drainage, Corrugated Metal		Crew	Daily Output	Labor-Hours	Unit	Material	2014 Bare Costs Labor	Equipment	Total	Total Incl O&P
2840	72" diameter	B-13B	4	14	Ea.	1,975	560	284	2,819	3,325
2850	Couplings, 12" diameter					10.60			10.60	11.65
2855	18" diameter					15.15			15.15	16.65
2860	24" diameter					20			20	22
2865	30" diameter					25			25	27.50
2870	36" diameter					32			32	35
2875	48" diameter					47.50			47.50	52.50
2880	60" diameter					60			60	66
2885	72" diameter					72.50			72.50	80

33 41 13.60 Sewage/Drainage Collection, Concrete Pipe

		Crew	Daily Output	Labor-Hours	Unit	Material	Labor	Equipment	Total	Total Incl O&P
0010	**SEWAGE/DRAINAGE COLLECTION, CONCRETE PIPE**									
0020	Not including excavation or backfill									
1000	Non-reinforced pipe, extra strength, B&S or T&G joints									
1010	6" diameter	B-14	265.04	.181	L.F.	5.60	7	1.38	13.98	18.45
1020	8" diameter		224	.214		6.15	8.30	1.63	16.08	21.50
1030	10" diameter		216	.222		6.85	8.60	1.69	17.14	22.50
1040	12" diameter		200	.240		8.15	9.30	1.83	19.28	25.50
1050	15" diameter		180	.267		11.25	10.35	2.03	23.63	30.50
1060	18" diameter		144	.333		13.20	12.90	2.54	28.64	37
1070	21" diameter		112	.429		16.30	16.60	3.26	36.16	47
1080	24" diameter		100	.480		22	18.60	3.65	44.25	56.50
2000	Reinforced culvert, class 3, no gaskets									
2010	12" diameter	B-14	150	.320	L.F.	10	12.40	2.44	24.84	32.50
2020	15" diameter		150	.320		13	12.40	2.44	27.84	36
2030	18" diameter		132	.364		16.75	14.10	2.77	33.62	43
2035	21" diameter		120	.400		21	15.50	3.04	39.54	50.50
2040	24" diameter		100	.480		25	18.60	3.65	47.25	60
2045	27" diameter	B-13	92	.609		36	24.50	8.10	68.60	85.50
2050	30" diameter		88	.636		40.50	25.50	8.50	74.50	93
2060	36" diameter		72	.778		55	31	10.35	96.35	119
2070	42" diameter	B-13B	72	.778		74	31	15.80	120.80	146
2080	48" diameter		64	.875		88	35	17.75	140.75	170
2090	60" diameter		48	1.167		135	46.50	23.50	205	247
2100	72" diameter		40	1.400		205	56	28.50	289.50	345
2120	84" diameter		32	1.750		250	70	35.50	355.50	420
2140	96" diameter		24	2.333		300	93	47.50	440.50	525
2200	With gaskets, class 3, 12" diameter	B-21	168	.167		11	7.05	.83	18.88	24
2220	15" diameter		160	.175		14.30	7.40	.88	22.58	28
2230	18" diameter		152	.184		18.45	7.80	.92	27.17	33.50
2240	24" diameter		136	.206		31.50	8.70	1.03	41.23	49
2260	30" diameter	B-13	88	.636		48	25.50	8.50	82	101
2270	36" diameter	"	72	.778		64.50	31	10.35	105.85	129
2290	48" diameter	B-13B	64	.875		100	35	17.75	152.75	183
2310	72" diameter	"	40	1.400		224	56	28.50	308.50	365
2330	Flared ends, 12" diameter	B-21	31	.903	Ea.	225	38	4.52	267.52	310
2340	15" diameter		25	1.120		266	47.50	5.60	319.10	370
2400	18" diameter		20	1.400		305	59	7	371	435
2420	24" diameter		14	2		370	84.50	10	464.50	545
2440	36" diameter	B-13	10	5.600		790	223	74.50	1,087.50	1,300
3080	Radius pipe, add to pipe prices, 12" to 60" diameter				L.F.	50%				
3090	Over 60" diameter, add				"	20%				
3500	Reinforced elliptical, 8' lengths, C507 class 3									

33 41 Storm Utility Drainage Piping

33 41 13 – Public Storm Utility Drainage Piping

	33 41 13.60 Sewage/Drainage Collection, Concrete Pipe	Crew	Daily Output	Labor-Hours	Unit	Material	2014 Bare Costs Labor	Equipment	Total	Total Incl O&P
3520	14" x 23" inside, round equivalent 18" diameter	B-21	82	.341	L.F.	40.50	14.40	1.71	56.61	68.50
3530	24" x 38" inside, round equivalent 30" diameter	B-13	58	.966		61.50	38.50	12.85	112.85	141
3540	29" x 45" inside, round equivalent 36" diameter		52	1.077		78	43	14.35	135.35	167
3550	38" x 60" inside, round equivalent 48" diameter		38	1.474		133	58.50	19.65	211.15	258
3560	48" x 76" inside, round equivalent 60" diameter		26	2.154		185	86	28.50	299.50	370
3570	58" x 91" inside, round equivalent 72" diameter	▼	22	2.545	▼	270	101	34	405	490
3780	Concrete slotted pipe, class 4 mortar joint									
3800	12" diameter	B-21	168	.167	L.F.	27	7.05	.83	34.88	41.50
3840	18" diameter	"	152	.184	"	31.50	7.80	.92	40.22	47.50
3900	Concrete slotted pipe, Class 4 O-ring joint									
3940	12" diameter	B-21	168	.167	L.F.	27	7.05	.83	34.88	41.50
3960	18" diameter	"	152	.184	"	31.50	7.80	.92	40.22	47.50

33 42 Culverts

33 42 16 – Concrete Culverts

33 42 16.15 Oval Arch Culverts

		Crew	Daily Output	Labor-Hours	Unit	Material	Labor	Equipment	Total	Total Incl O&P
0010	**OVAL ARCH CULVERTS**									
3000	Corrugated galvanized or aluminum, coated & paved									
3020	17" x 13", 16 ga., 15" equivalent	B-14	200	.240	L.F.	13.25	9.30	1.83	24.38	31
3040	21" x 15", 16 ga., 18" equivalent		150	.320		16.05	12.40	2.44	30.89	39.50
3060	28" x 20", 14 ga., 24" equivalent		125	.384		24	14.85	2.92	41.77	52
3080	35" x 24", 14 ga., 30" equivalent	▼	100	.480		29.50	18.60	3.65	51.75	65
3100	42" x 29", 12 ga., 36" equivalent	B-13	100	.560		35.50	22.50	7.45	65.45	81
3120	49" x 33", 12 ga., 42" equivalent		90	.622		41	25	8.30	74.30	92.50
3140	57" x 38", 12 ga., 48" equivalent	▼	75	.747	▼	57	30	9.95	96.95	119
3160	Steel, plain oval arch culverts, plain									
3180	17" x 13", 16 ga., 15" equivalent	B-14	225	.213	L.F.	11.95	8.25	1.62	21.82	27.50
3200	21" x 15", 16 ga., 18" equivalent		175	.274		14.45	10.60	2.09	27.14	34.50
3220	28" x 20", 14 ga., 24" equivalent		150	.320		21.50	12.40	2.44	36.34	45
3240	35" x 24", 14 ga., 30" equivalent	B-13	108	.519		26.50	20.50	6.90	53.90	68
3260	42" x 29", 12 ga., 36" equivalent		108	.519		32	20.50	6.90	59.40	74.50
3280	49" x 33", 12 ga., 42" equivalent		92	.609		37	24.50	8.10	69.60	87
3300	57" x 38", 12 ga., 48" equivalent	▼	75	.747	▼	51.50	30	9.95	91.45	113
3320	End sections, 17" x 13"		22	2.545	Ea.	144	101	34	279	355
3340	42" x 29"	▼	17	3.294	"	395	131	44	570	685
3360	Multi-plate arch, steel	B-20	1690	.014	Lb.	1.30	.58		1.88	2.33

33 44 Storm Utility Water Drains

33 44 13 – Utility Area Drains

33 44 13.13 Catchbasins

		Crew	Daily Output	Labor-Hours	Unit	Material	Labor	Equipment	Total	Total Incl O&P
0010	**CATCHBASINS**									
0011	Not including footing & excavation									
1600	Frames & grates, C.I., 24" square, 500 lb.	B-6	7.80	3.077	Ea.	340	123	47	510	615
1700	26" D shape, 600 lb.		7	3.429		505	138	52	695	825
1800	Light traffic, 18" diameter, 100 lb.		10	2.400		122	96.50	36.50	255	320
1900	24" diameter, 300 lb.		8.70	2.759		196	111	42	349	430
2000	36" diameter, 900 lb.		5.80	4.138		570	166	63	799	950
2100	Heavy traffic, 24" diameter, 400 lb.		7.80	3.077		244	123	47	414	510
2200	36" diameter, 1150 lb.	▼	3	8	▼	795	320	122	1,237	1,500

33 44 Storm Utility Water Drains

33 44 13 – Utility Area Drains

33 44 13.13 Catchbasins

		Crew	Daily Output	Labor-Hours	Unit	Material	2014 Bare Costs Labor	Equipment	Total	Total Incl O&P
2300	Mass. State standard, 26" diameter, 475 lb.	B-6	7	3.429	Ea.	265	138	52	455	560
2400	30" diameter, 620 lb.		7	3.429		345	138	52	535	650
2500	Watertight, 24" diameter, 350 lb.		7.80	3.077		320	123	47	490	590
2600	26" diameter, 500 lb.		7	3.429		425	138	52	615	735
2700	32" diameter, 575 lb.	↓	6	4	↓	850	160	61	1,071	1,250
2800	3 piece cover & frame, 10" deep,									
2900	1200 lb., for heavy equipment	B-6	3	8	Ea.	1,050	320	122	1,492	1,775
3000	Raised for paving 1-1/4" to 2" high									
3100	4 piece expansion ring									
3200	20" to 26" diameter	1 Clab	3	2.667	Ea.	157	97.50		254.50	325
3300	30" to 36" diameter	"	3	2.667	"	216	97.50		313.50	390
3320	Frames and covers, existing, raised for paving, 2", including									
3340	row of brick, concrete collar, up to 12" wide frame	B-6	18	1.333	Ea.	44.50	53.50	20.50	118.50	154
3360	20" to 26" wide frame		11	2.182		66.50	87.50	33	187	244
3380	30" to 36" wide frame	↓	9	2.667		82.50	107	40.50	230	300
3400	Inverts, single channel brick	D-1	3	5.333		96	220		316	440
3500	Concrete		5	3.200		103	132		235	315
3600	Triple channel, brick		2	8		147	330		477	665
3700	Concrete	↓	3	5.333	↓	139	220		359	485

33 46 Subdrainage

33 46 16 – Subdrainage Piping

33 46 16.25 Piping, Subdrainage, Corrugated Metal

			Crew	Daily Output	Labor-Hours	Unit	Material	Labor	Equipment	Total	Total Incl O&P
0010	**PIPING, SUBDRAINAGE, CORRUGATED METAL**	G1030-805									
0021	Not including excavation and backfill										
2010	Aluminum, perforated										
2020	6" diameter, 18 ga.		B-20	380	.063	L.F.	6.55	2.58		9.13	11.20
2200	8" diameter, 16 ga.		"	370	.065		8.60	2.65		11.25	13.55
2220	10" diameter, 16 ga.		B-21	360	.078		10.75	3.28	.39	14.42	17.30
2240	12" diameter, 16 ga.			285	.098		12	4.15	.49	16.64	20
2260	18" diameter, 16 ga.		↓	205	.137	↓	18	5.75	.68	24.43	29.50
3000	Uncoated galvanized, perforated										
3020	6" diameter, 18 ga.		B-20	380	.063	L.F.	6.05	2.58		8.63	10.65
3200	8" diameter, 16 ga.		"	370	.065		8.35	2.65		11	13.25
3220	10" diameter, 16 ga.		B-21	360	.078		8.85	3.28	.39	12.52	15.20
3240	12" diameter, 16 ga.			285	.098		9.85	4.15	.49	14.49	17.75
3260	18" diameter, 16 ga.		↓	205	.137		15.05	5.75	.68	21.48	26
4000	Steel, perforated, asphalt coated										
4020	6" diameter 18 ga.		B-20	380	.063	L.F.	6.55	2.58		9.13	11.20
4030	8" diameter 18 ga.		"	370	.065		8.60	2.65		11.25	13.55
4040	10" diameter 16 ga.		B-21	360	.078		10.10	3.28	.39	13.77	16.60
4050	12" diameter 16 ga.			285	.098		11.10	4.15	.49	15.74	19.15
4060	18" diameter 16 ga.		↓	205	.137		17.15	5.75	.68	23.58	28.50

33 46 16.30 Piping, Subdrainage, Plastic

		Crew	Daily Output	Labor-Hours	Unit	Material	Labor	Equipment	Total	Total Incl O&P
0010	**PIPING, SUBDRAINAGE, PLASTIC**									
0020	Not including excavation and backfill									
2100	Perforated PVC, 4" diameter	B-14	314	.153	L.F.	1.46	5.90	1.16	8.52	12
2110	6" diameter		300	.160		3.29	6.20	1.22	10.71	14.45
2120	8" diameter		290	.166		6.30	6.40	1.26	13.96	18.20
2130	10" diameter		280	.171		9.15	6.65	1.30	17.10	21.50
2140	12" diameter	↓	270	.178	↓	12.75	6.90	1.35	21	26

33 49 Storm Drainage Structures

33 49 13 – Storm Drainage Manholes, Frames, and Covers

33 49 13.10 Storm Drainage Manholes, Frames and Covers

		Crew	Daily Output	Labor-Hours	Unit	Material	2014 Bare Costs Labor	Equipment	Total	Total Incl O&P
0010	**STORM DRAINAGE MANHOLES, FRAMES & COVERS**									
0020	Excludes footing, excavation, backfill (See line items for frame & cover)									
0050	Brick, 4' inside diameter, 4' deep	D-1	1	16	Ea.	510	660		1,170	1,550
0100	6' deep		.70	22.857		725	945		1,670	2,250
0150	8' deep		.50	32	↓	935	1,325		2,260	3,050
0200	For depths over 8', add		4	4	V.L.F.	83.50	165		248.50	345
0400	Concrete blocks (radial), 4' I.D., 4' deep		1.50	10.667	Ea.	370	440		810	1,075
0500	6' deep		1	16		495	660		1,155	1,550
0600	8' deep		.70	22.857		620	945		1,565	2,125
0700	For depths over 8', add	↓	5.50	2.909	V.L.F.	65	120		185	255
0800	Concrete, cast in place, 4' x 4', 8" thick, 4' deep	C-14H	2	24	Ea.	505	1,075	16.55	1,596.55	2,250
0900	6' deep		1.50	32		725	1,450	22	2,197	3,050
1000	8' deep		1	48	↓	1,050	2,175	33	3,258	4,500
1100	For depths over 8', add		8	6	V.L.F.	118	271	4.14	393.14	550
1110	Precast, 4' I.D., 4' deep	B-22	4.10	7.317	Ea.	775	315	51	1,141	1,375
1120	6' deep		3	10		970	425	70	1,465	1,800
1130	8' deep		2	15	↓	1,100	640	105	1,845	2,325
1140	For depths over 8', add		16	1.875	V.L.F.	125	80	13.15	218.15	275
1150	5' I.D., 4' deep	B-6	3	8	Ea.	1,625	320	122	2,067	2,400
1160	6' deep		2	12		1,675	480	183	2,338	2,775
1170	8' deep		1.50	16	↓	2,350	640	244	3,234	3,850
1180	For depths over 8', add		12	2	V.L.F.	277	80	30.50	387.50	460
1190	6' I.D., 4' deep		2	12	Ea.	2,125	480	183	2,788	3,275
1200	6' deep		1.50	16		2,550	640	244	3,434	4,050
1210	8' deep		1	24	↓	3,150	965	365	4,480	5,325
1220	For depths over 8', add		8	3	V.L.F.	375	120	45.50	540.50	645
1250	Slab tops, precast, 8" thick									
1300	4' diameter manhole	B-6	8	3	Ea.	245	120	45.50	410.50	505
1400	5' diameter manhole		7.50	3.200		405	128	48.50	581.50	695
1500	6' diameter manhole	↓	7	3.429		625	138	52	815	955
3800	Steps, heavyweight cast iron, 7" x 9"	1 Bric	40	.200		19.05	9.10		28.15	35
3900	8" x 9"		40	.200		23	9.10		32.10	39
3928	12" x 10-1/2"		40	.200		26.50	9.10		35.60	43.50
4000	Standard sizes, galvanized steel		40	.200		21.50	9.10		30.60	37.50
4100	Aluminum		40	.200		24	9.10		33.10	40
4150	Polyethylene	↓	40	.200	↓	24.50	9.10		33.60	41

33 51 Natural-Gas Distribution

33 51 13 – Natural-Gas Piping

33 51 13.10 Piping, Gas Service and Distribution, P.E.

		Crew	Daily Output	Labor-Hours	Unit	Material	Labor	Equipment	Total	Total Incl O&P
0010	**PIPING, GAS SERVICE AND DISTRIBUTION, POLYETHYLENE**									
0020	Not including excavation or backfill									
1000	60 psi coils, compression coupling @ 100', 1/2" diameter, SDR 11	B-20A	608	.053	L.F.	.61	2.35		2.96	4.26
1010	1" diameter, SDR 11		544	.059		1.41	2.63		4.04	5.55
1040	1-1/4" diameter, SDR 11		544	.059		1.76	2.63		4.39	5.95
1100	2" diameter, SDR 11		488	.066		2.85	2.93		5.78	7.60
1160	3" diameter, SDR 11	↓	408	.078		5.60	3.51		9.11	11.50
1500	60 PSI 40' joints with coupling, 3" diameter, SDR 11	B-21A	408	.098		5.55	4.49	1.18	11.22	14.25
1540	4" diameter, SDR 11		352	.114		12.15	5.20	1.37	18.72	23
1600	6" diameter, SDR 11		328	.122		33.50	5.60	1.47	40.57	47
1640	8" diameter, SDR 11	↓	272	.147	↓	51.50	6.75	1.77	60.02	68.50

33 52 Liquid Fuel Distribution

33 52 16 – Gasoline Distribution

33 52 16.13 Gasoline Piping		Crew	Daily Output	Labor-Hours	Unit	Material	2014 Bare Costs Labor	Equipment	Total	Total Incl O&P
0010	**GASOLINE PIPING**									
0020	Primary containment pipe, fiberglass-reinforced									
0030	Plastic pipe 15' & 30' lengths									
0040	2" diameter	Q-6	425	.056	L.F.	5.95	3.08		9.03	11.20
0050	3" diameter		400	.060		10.40	3.28		13.68	16.35
0060	4" diameter	↓	375	.064	↓	13.70	3.49		17.19	20.50
0100	Fittings									
0110	Elbows, 90° & 45°, bell-ends, 2"	Q-6	24	1	Ea.	43.50	54.50		98	130
0120	3" diameter		22	1.091		55	59.50		114.50	151
0130	4" diameter		20	1.200		70	65.50		135.50	176
0200	Tees, bell ends, 2"		21	1.143		60.50	62.50		123	161
0210	3" diameter		18	1.333		64	73		137	181
0230	Flanges bell ends, 2"		24	1		33.50	54.50		88	120
0240	3" diameter		22	1.091		39	59.50		98.50	133
0250	4" diameter		20	1.200		45	65.50		110.50	149
0260	Sleeve couplings, 2"		21	1.143		12.40	62.50		74.90	108
0270	3" diameter		18	1.333		17.80	73		90.80	130
0280	4" diameter		15	1.600		23	87.50		110.50	158
0290	Threaded adapters 2"		21	1.143		18	62.50		80.50	114
0300	3" diameter		18	1.333		34	73		107	148
0310	4" diameter		15	1.600		38	87.50		125.50	174
0320	Reducers, 2"		27	.889		27.50	48.50		76	104
0330	3" diameter		22	1.091		27.50	59.50		87	120
0340	4" diameter	↓	20	1.200	↓	37	65.50		102.50	140
1010	Gas station product line for secondary containment (double wall)									
1100	Fiberglass reinforced plastic pipe 25' lengths									
1120	Pipe, plain end, 3" diameter	Q-6	375	.064	L.F.	26.50	3.49		29.99	35
1130	4" diameter		350	.069		32	3.74		35.74	41
1140	5" diameter		325	.074		35.50	4.03		39.53	45
1150	6" diameter	↓	300	.080	↓	39	4.37		43.37	49.50
1200	Fittings									
1230	Elbows, 90° & 45°, 3" diameter	Q-6	18	1.333	Ea.	136	73		209	260
1240	4" diameter		16	1.500		167	82		249	310
1250	5" diameter		14	1.714		184	93.50		277.50	345
1260	6" diameter		12	2		204	109		313	390
1270	Tees, 3" diameter		15	1.600		166	87.50		253.50	315
1280	4" diameter		12	2		202	109		311	385
1290	5" diameter		9	2.667		315	146		461	570
1300	6" diameter		6	4		375	218		593	745
1310	Couplings, 3" diameter		18	1.333		55	73		128	171
1320	4" diameter		16	1.500		119	82		201	255
1330	5" diameter		14	1.714		214	93.50		307.50	375
1340	6" diameter		12	2		315	109		424	515
1350	Cross-over nipples, 3" diameter		18	1.333		10.60	73		83.60	122
1360	4" diameter		16	1.500		12.75	82		94.75	138
1370	5" diameter		14	1.714		15.90	93.50		109.40	159
1380	6" diameter		12	2		19.10	109		128.10	186
1400	Telescoping, reducers, concentric 4" x 3"		18	1.333		47.50	73		120.50	163
1410	5" x 4"		17	1.412		95.50	77		172.50	221
1420	6" x 5"	↓	16	1.500	↓	234	82		316	380

33 61 Hydronic Energy Distribution

33 61 13 – Underground Hydronic Energy Distribution

33 61 13.20 Pipe Conduit, Prefabricated/Preinsulated		Crew	Daily Output	Labor-Hours	Unit	Material	2014 Bare Costs Labor	Equipment	Total	Total Incl O&P
0010	**PIPE CONDUIT, PREFABRICATED/PREINSULATED**	R221113-70								
0020	Does not include trenching, fittings or crane.									
0300	For cathodic protection, add 12 to 14%									
0310	of total built-up price (casing plus service pipe)									
0580	Polyurethane insulated system, 250°F. max. temp.									
0620	Black steel service pipe, standard wt., 1/2" insulation									
0660	3/4" diam. pipe size	Q-17	54	.296	L.F.	57	15.60	1.03	73.63	87.50
0670	1" diam. pipe size		50	.320		63	16.85	1.11	80.96	95.50
0680	1-1/4" diam. pipe size		47	.340		70	17.90	1.18	89.08	105
0690	1-1/2" diam. pipe size		45	.356		76	18.70	1.23	95.93	113
0700	2" diam. pipe size		42	.381		79	20	1.32	100.32	118
0710	2-1/2" diam. pipe size		34	.471		80	25	1.63	106.63	127
0720	3" diam. pipe size		28	.571		93	30	1.98	124.98	150
0730	4" diam. pipe size		22	.727		117	38.50	2.52	158.02	190
0740	5" diam. pipe size		18	.889		149	47	3.08	199.08	238
0750	6" diam. pipe size	Q-18	23	1.043		173	57	2.40	232.40	280
0760	8" diam. pipe size		19	1.263		254	69	2.91	325.91	385
0770	10" diam. pipe size		16	1.500		320	82	3.45	405.45	485
0780	12" diam. pipe size		13	1.846		400	101	4.25	505.25	595
0790	14" diam. pipe size		11	2.182		445	119	5	569	675
0800	16" diam. pipe size		10	2.400		510	131	5.50	646.50	770
0810	18" diam. pipe size		8	3		590	164	6.90	760.90	905
0820	20" diam. pipe size		7	3.429		655	187	7.90	849.90	1,000
0830	24" diam. pipe size		6	4		805	218	9.20	1,032.20	1,225
0900	For 1" thick insulation, add					10%				
0940	For 1-1/2" thick insulation, add					13%				
0980	For 2" thick insulation, add					20%				
1500	Gland seal for system, 3/4" diam. pipe size	Q-17	32	.500	Ea.	775	26.50	1.73	803.23	890
1510	1" diam. pipe size		32	.500		775	26.50	1.73	803.23	890
1540	1-1/4" diam. pipe size		30	.533		825	28	1.85	854.85	955
1550	1-1/2" diam. pipe size		30	.533		825	28	1.85	854.85	955
1560	2" diam. pipe size		28	.571		980	30	1.98	1,011.98	1,125
1570	2-1/2" diam. pipe size		26	.615		1,050	32.50	2.13	1,084.63	1,225
1580	3" diam. pipe size		24	.667		1,125	35	2.31	1,162.31	1,300
1590	4" diam. pipe size		22	.727		1,350	38.50	2.52	1,391.02	1,525
1600	5" diam. pipe size		19	.842		1,650	44.50	2.91	1,697.41	1,900
1610	6" diam. pipe size	Q-18	26	.923		1,750	50.50	2.12	1,802.62	2,000
1620	8" diam. pipe size		25	.960		2,050	52.50	2.21	2,104.71	2,325
1630	10" diam. pipe size		23	1.043		2,375	57	2.40	2,434.40	2,725
1640	12" diam. pipe size		21	1.143		2,625	62.50	2.63	2,690.13	3,000
1650	14" diam. pipe size		19	1.263		2,950	69	2.91	3,021.91	3,350
1660	16" diam. pipe size		18	1.333		3,500	73	3.07	3,576.07	3,975
1670	18" diam. pipe size		16	1.500		3,700	82	3.45	3,785.45	4,200
1680	20" diam. pipe size		14	1.714		4,200	93.50	3.94	4,297.44	4,775
1690	24" diam. pipe size		12	2		4,675	109	4.60	4,788.60	5,325
2000	Elbow, 45° for system									
2020	3/4" diam. pipe size	Q-17	14	1.143	Ea.	490	60	3.95	553.95	630
2040	1" diam. pipe size		13	1.231		500	65	4.26	569.26	655
2050	1-1/4" diam. pipe size		11	1.455		570	76.50	5.05	651.55	745
2060	1-1/2" diam. pipe size		9	1.778		595	93.50	6.15	694.65	805
2070	2" diam. pipe size		6	2.667		620	140	9.25	769.25	905
2080	2-1/2" diam. pipe size		4	4		670	211	13.85	894.85	1,075
2090	3" diam. pipe size		3.50	4.571		775	241	15.80	1,031.80	1,225

33 61 Hydronic Energy Distribution

33 61 13 – Underground Hydronic Energy Distribution

33 61 13.20 Pipe Conduit, Prefabricated/Preinsulated

		Crew	Daily Output	Labor-Hours	Unit	Material	2014 Bare Costs Labor	Equipment	Total	Total Incl O&P
2100	4" diam. pipe size	Q-17	3	5.333	Ea.	905	281	18.45	1,204.45	1,450
2110	5" diam. pipe size	↓	2.80	5.714		1,175	300	19.75	1,494.75	1,750
2120	6" diam. pipe size	Q-18	4	6		1,325	330	13.80	1,668.80	1,950
2130	8" diam. pipe size		3	8		1,925	435	18.40	2,378.40	2,775
2140	10" diam. pipe size		2.40	10		2,350	545	23	2,918	3,425
2150	12" diam. pipe size		2	12		3,075	655	27.50	3,757.50	4,425
2160	14" diam. pipe size		1.80	13.333		3,825	730	30.50	4,585.50	5,350
2170	16" diam. pipe size		1.60	15		4,550	820	34.50	5,404.50	6,300
2180	18" diam. pipe size		1.30	18.462		5,725	1,000	42.50	6,767.50	7,875
2190	20" diam. pipe size		1	24		7,200	1,300	55	8,555	9,950
2200	24" diam. pipe size	↓	.70	34.286		9,050	1,875	79	11,004	12,900
2260	For elbow, 90°, add					25%				
2300	For tee, straight, add					85%	30%			
2340	For tee, reducing, add					170%	30%			
2380	For weldolet, straight, add				↓	50%				
2400	Polyurethane insulation, 1"									
2410	FRP carrier and casing									
2420	4"	Q-5	18	.889	L.F.	54.50	47		101.50	131
2422	6"	Q-6	23	1.043		94.50	57		151.50	190
2424	8"		19	1.263		154	69		223	273
2426	10"		16	1.500		207	82		289	350
2428	12"	↓	13	1.846	↓	270	101		371	450
2430	FRP carrier and PVC casing									
2440	4"	Q-5	18	.889	L.F.	21.50	47		68.50	94
2444	8"	Q-6	19	1.263		45	69		114	154
2446	10"		16	1.500		64.50	82		146.50	195
2448	12"	↓	13	1.846		82.50	101		183.50	243
2450	PVC carrier and casing									
2460	4"	Q-1	36	.444	L.F.	10.45	23		33.45	46.50
2462	6"	"	29	.552		14.60	28.50		43.10	59
2464	8"	Q-2	36	.667		20.50	36		56.50	76.50
2466	10"		32	.750		28	40.50		68.50	91.50
2468	12"	↓	31	.774	↓	32.50	41.50		74	98.50

33 63 Steam Energy Distribution

33 63 13 – Underground Steam and Condensate Distribution Piping

33 63 13.10 Calcium Silicate Insulated System

		Crew	Daily Output	Labor-Hours	Unit	Material	Labor	Equipment	Total	Total Incl O&P
0010	**CALCIUM SILICATE INSULATED SYSTEM**									
0011	High temp. (1200 degrees F)									
2840	Steel casing with protective exterior coating									
2850	6-5/8" diameter	Q-18	52	.462	L.F.	106	25	1.06	132.06	155
2860	8-5/8" diameter		50	.480		115	26	1.10	142.10	168
2870	10-3/4" diameter		47	.511		135	28	1.17	164.17	191
2880	12-3/4" diameter		44	.545		147	30	1.25	178.25	208
2890	14" diameter		41	.585		166	32	1.35	199.35	232
2900	16" diameter		39	.615		178	33.50	1.42	212.92	248
2910	18" diameter		36	.667		196	36.50	1.53	234.03	273
2920	20" diameter		34	.706		221	38.50	1.62	261.12	305
2930	22" diameter		32	.750		305	41	1.73	347.73	400
2940	24" diameter		29	.828		350	45	1.90	396.90	455
2950	26" diameter	↓	26	.923	↓	400	50.50	2.12	452.62	520

33 63 Steam Energy Distribution

33 63 13 – Underground Steam and Condensate Distribution Piping

33 63 13.10 Calcium Silicate Insulated System

		Crew	Daily Output	Labor-Hours	Unit	Material	2014 Bare Costs Labor	Equipment	Total	Total Incl O&P
2960	28" diameter	Q-18	23	1.043	L.F.	495	57	2.40	554.40	635
2970	30" diameter		21	1.143		530	62.50	2.63	595.13	675
2980	32" diameter		19	1.263		595	69	2.91	666.91	760
2990	34" diameter		18	1.333		600	73	3.07	676.07	775
3000	36" diameter		16	1.500		645	82	3.45	730.45	840
3040	For multi-pipe casings, add					10%				
3060	For oversize casings, add					2%				
3400	Steel casing gland seal, single pipe									
3420	6-5/8" diameter	Q-18	25	.960	Ea.	1,475	52.50	2.21	1,529.71	1,700
3440	8-5/8" diameter		23	1.043		1,725	57	2.40	1,784.40	2,000
3450	10-3/4" diameter		21	1.143		1,925	62.50	2.63	1,990.13	2,225
3460	12-3/4" diameter		19	1.263		2,275	69	2.91	2,346.91	2,600
3470	14" diameter		17	1.412		2,525	77	3.25	2,605.25	2,900
3480	16" diameter		16	1.500		2,950	82	3.45	3,035.45	3,375
3490	18" diameter		15	1.600		3,275	87.50	3.68	3,366.18	3,725
3500	20" diameter		13	1.846		3,625	101	4.25	3,730.25	4,125
3510	22" diameter		12	2		4,075	109	4.60	4,188.60	4,650
3520	24" diameter		11	2.182		4,575	119	5	4,699	5,200
3530	26" diameter		10	2.400		5,225	131	5.50	5,361.50	5,925
3540	28" diameter		9.50	2.526		5,900	138	5.80	6,043.80	6,700
3550	30" diameter		9	2.667		6,000	146	6.15	6,152.15	6,825
3560	32" diameter		8.50	2.824		6,750	154	6.50	6,910.50	7,675
3570	34" diameter		8	3		7,350	164	6.90	7,520.90	8,350
3580	36" diameter		7	3.429		7,825	187	7.90	8,019.90	8,900
3620	For multi-pipe casings, add					5%				
4000	Steel casing anchors, single pipe									
4020	6-5/8" diameter	Q-18	8	3	Ea.	1,325	164	6.90	1,495.90	1,700
4040	8-5/8" diameter		7.50	3.200		1,375	175	7.35	1,557.35	1,800
4050	10-3/4" diameter		7	3.429		1,800	187	7.90	1,994.90	2,300
4060	12-3/4" diameter		6.50	3.692		1,925	202	8.50	2,135.50	2,450
4070	14" diameter		6	4		2,275	218	9.20	2,502.20	2,850
4080	16" diameter		5.50	4.364		2,625	238	10.05	2,873.05	3,275
4090	18" diameter		5	4.800		2,975	262	11.05	3,248.05	3,675
4100	20" diameter		4.50	5.333		3,275	291	12.25	3,578.25	4,050
4110	22" diameter		4	6		3,650	330	13.80	3,993.80	4,525
4120	24" diameter		3.50	6.857		3,975	375	15.75	4,365.75	4,950
4130	26" diameter		3	8		4,500	435	18.40	4,953.40	5,625
4140	28" diameter		2.50	9.600		4,950	525	22	5,497	6,250
4150	30" diameter		2	12		5,250	655	27.50	5,932.50	6,800
4160	32" diameter		1.50	16		6,275	875	37	7,187	8,300
4170	34" diameter		1	24		7,050	1,300	55	8,405	9,775
4180	36" diameter		1	24		7,675	1,300	55	9,030	10,500
4220	For multi-pipe, add					5%	20%			
4800	Steel casing elbow									
4820	6-5/8" diameter	Q-18	15	1.600	Ea.	1,800	87.50	3.68	1,891.18	2,125
4830	8-5/8" diameter		15	1.600		1,925	87.50	3.68	2,016.18	2,250
4850	10-3/4" diameter		14	1.714		2,325	93.50	3.94	2,422.44	2,700
4860	12-3/4" diameter		13	1.846		2,750	101	4.25	2,855.25	3,175
4870	14" diameter		12	2		2,950	109	4.60	3,063.60	3,425
4880	16" diameter		11	2.182		3,225	119	5	3,349	3,725
4890	18" diameter		10	2.400		3,675	131	5.50	3,811.50	4,250
4900	20" diameter		9	2.667		3,925	146	6.15	4,077.15	4,550
4910	22" diameter		8	3		4,225	164	6.90	4,395.90	4,900

33 63 Steam Energy Distribution

33 63 13 – Underground Steam and Condensate Distribution Piping

33 63 13.10 Calcium Silicate Insulated System		Crew	Daily Output	Labor-Hours	Unit	Material	2014 Bare Costs Labor	Equipment	Total	Total Incl O&P
4920	24" diameter	Q-18	7	3.429	Ea.	4,650	187	7.90	4,844.90	5,425
4930	26" diameter		6	4		5,100	218	9.20	5,327.20	5,950
4940	28" diameter		5	4.800		5,500	262	11.05	5,773.05	6,450
4950	30" diameter		4	6		5,525	330	13.80	5,868.80	6,575
4960	32" diameter		3	8		6,275	435	18.40	6,728.40	7,600
4970	34" diameter		2	12		6,900	655	27.50	7,582.50	8,625
4980	36" diameter	↓	2	12	↓	7,350	655	27.50	8,032.50	9,125
5500	Black steel service pipe, std. wt., 1" thick insulation									
5510	3/4" diameter pipe size	Q-17	54	.296	L.F.	48	15.60	1.03	64.63	77
5540	1" diameter pipe size		50	.320		49.50	16.85	1.11	67.46	81
5550	1-1/4" diameter pipe size		47	.340		56	17.90	1.18	75.08	90
5560	1-1/2" diameter pipe size		45	.356		61	18.70	1.23	80.93	97
5570	2" diameter pipe size		42	.381		67.50	20	1.32	88.82	106
5580	2-1/2" diameter pipe size		34	.471		71	25	1.63	97.63	118
5590	3" diameter pipe size		28	.571		81	30	1.98	112.98	137
5600	4" diameter pipe size		22	.727		104	38.50	2.52	145.02	175
5610	5" diameter pipe size	↓	18	.889		140	47	3.08	190.08	228
5620	6" diameter pipe size	Q-18	23	1.043	↓	153	57	2.40	212.40	257
6000	Black steel service pipe, std. wt., 1-1/2" thick insul.									
6010	3/4" diameter pipe size	Q-17	54	.296	L.F.	48.50	15.60	1.03	65.13	78
6040	1" diameter pipe size		50	.320		54	16.85	1.11	71.96	86
6050	1-1/4" diameter pipe size		47	.340		61	17.90	1.18	80.08	95.50
6060	1-1/2" diameter pipe size		45	.356		66.50	18.70	1.23	86.43	103
6070	2" diameter pipe size		42	.381		72	20	1.32	93.32	111
6080	2-1/2" diameter pipe size		34	.471		77	25	1.63	103.63	124
6090	3" diameter pipe size		28	.571		86	30	1.98	117.98	143
6100	4" diameter pipe size		22	.727		110	38.50	2.52	151.02	182
6110	5" diameter pipe size	↓	18	.889		140	47	3.08	190.08	228
6120	6" diameter pipe size	Q-18	23	1.043		160	57	2.40	219.40	265
6130	8" diameter pipe size		19	1.263		230	69	2.91	301.91	360
6140	10" diameter pipe size		16	1.500		288	82	3.45	373.45	445
6150	12" diameter pipe size	↓	13	1.846		345	101	4.25	450.25	535
6190	For 2" thick insulation, add					15%				
6220	For 2-1/2" thick insulation, add					25%				
6260	For 3" thick insulation, add				↓	30%				
6800	Black steel service pipe, ex. hvy. wt., 1" thick insul.									
6820	3/4" diameter pipe size	Q-17	50	.320	L.F.	50.50	16.85	1.11	68.46	82
6840	1" diameter pipe size		47	.340		53.50	17.90	1.18	72.58	87.50
6850	1-1/4" diameter pipe size		44	.364		62	19.15	1.26	82.41	98.50
6860	1-1/2" diameter pipe size		42	.381		65.50	20	1.32	86.82	104
6870	2" diameter pipe size		40	.400		69.50	21	1.38	91.88	110
6880	2-1/2" diameter pipe size		31	.516		86	27	1.79	114.79	138
6890	3" diameter pipe size		27	.593		97	31	2.05	130.05	156
6900	4" diameter pipe size		21	.762		128	40	2.64	170.64	203
6910	5" diameter pipe size	↓	17	.941		179	49.50	3.26	231.76	275
6920	6" diameter pipe size	Q-18	22	1.091	↓	181	59.50	2.51	243.01	292
7400	Black steel service pipe, ex. hvy. wt., 1-1/2" thick insul.									
7420	3/4" diameter pipe size	Q-17	50	.320	L.F.	42	16.85	1.11	59.96	72.50
7440	1" diameter pipe size		47	.340		48	17.90	1.18	67.08	81
7450	1-1/4" diameter pipe size		44	.364		55.50	19.15	1.26	75.91	91.50
7460	1-1/2" diameter pipe size		42	.381		61	20	1.32	82.32	99
7470	2" diameter pipe size		40	.400		67.50	21	1.38	89.88	108
7480	2-1/2" diameter pipe size	↓	31	.516		73	27	1.79	101.79	123

33 63 Steam Energy Distribution

33 63 13 − Underground Steam and Condensate Distribution Piping

33 63 13.10 Calcium Silicate Insulated System		Crew	Daily Output	Labor-Hours	Unit	Material	2014 Bare Costs Labor	Equipment	Total	Total Incl O&P
7490	3" diameter pipe size	Q-17	27	.593	L.F.	85.50	31	2.05	118.55	143
7500	4" diameter pipe size		21	.762		112	40	2.64	154.64	186
7510	5" diameter pipe size	↓	17	.941		154	49.50	3.26	206.76	249
7520	6" diameter pipe size	Q-18	22	1.091		170	59.50	2.51	232.01	280
7530	8" diameter pipe size		18	1.333		255	73	3.07	331.07	395
7540	10" diameter pipe size		15	1.600		305	87.50	3.68	396.18	470
7550	12" diameter pipe size	↓	13	1.846		370	101	4.25	475.25	565
7590	For 2" thick insulation, add					13%				
7640	For 2-1/2" thick insulation, add					18%				
7680	For 3" thick insulation, add				↓	24%				

33 63 13.20 Combined Steam Pipe With Condensate Return

		Crew	Daily Output	Labor-Hours	Unit	Material	Labor	Equipment	Total	Total Incl O&P
0010	**COMBINED STEAM PIPE WITH CONDENSATE RETURN**									
9010	8" and 4" in 24" case	Q-18	100	.240	L.F.	540	13.10	.55	553.65	610
9020	6" and 3" in 20" case		104	.231		365	12.60	.53	378.13	425
9030	3" and 1-1/2" in 16" case		110	.218		275	11.90	.50	287.40	320
9040	2" and 1-1/4" in 12-3/4" case		114	.211		239	11.50	.48	250.98	281
9050	1-1/2" and 1-1/4" in 10-3/4" case	↓	116	.207	↓	237	11.30	.48	248.78	279
9100	Steam pipe only (no return)									
9110	6" in 18" case	Q-18	108	.222	L.F.	300	12.15	.51	312.66	350
9120	4" in 14" case		112	.214		239	11.70	.49	251.19	281
9130	3" in 14" case		114	.211		221	11.50	.48	232.98	261
9140	2-1/2" in 12-3/4" case		118	.203		206	11.10	.47	217.57	244
9150	2" in 12-3/4" case	↓	122	.197	↓	197	10.75	.45	208.20	234

Estimating Tips

Products such as conveyors, material handling cranes and hoists, as well as other items specified in this division, require trained installers. The general contractor may not have any choice as to who will perform the installation or when it will be performed. Long lead times are often required for these products, making early decisions in purchasing and scheduling necessary.

The installation of this type of equipment may require the embedment of mounting hardware during construction of floors, structural walls, or interior walls/partitions. Electrical connections will require coordination with the electrical contractor.

Reference Numbers

Reference numbers are shown in shaded boxes at the beginning of some major classifications. These numbers refer to related items in the Reference Section. The reference information may be an estimating procedure, an alternate pricing method, or technical information.

Note: Not all subdivisions listed here necessarily appear in this publication.

No part of this publication may be reproduced, stored in a retrieval system, or transmitted in any form or by any means without prior written permission of RSMeans, a division of Reed Construction Data.

41 22 Cranes and Hoists

41 22 23 – Hoists

41 22 23.10 Material Handling		Crew	Daily Output	Labor-Hours	Unit	Material	2014 Bare Costs Labor	Equipment	Total	Total Incl O&P
0010	**MATERIAL HANDLING**, cranes, hoists and lifts									
1500	Cranes, portable hydraulic, floor type, 2,000 lb. capacity				Ea.	3,500			3,500	3,875
1600	4,000 lb. capacity					4,025			4,025	4,425
1800	Movable gantry type, 12' to 15' range, 2,000 lb. capacity					2,050			2,050	2,250
1900	6,000 lb. capacity					4,825			4,825	5,300
2100	Hoists, electric overhead, chain, hook hung, 15' lift, 1 ton cap.					2,875			2,875	3,175
2200	3 ton capacity					3,375			3,375	3,725
2500	5 ton capacity					7,800			7,800	8,575
2600	For hand-pushed trolley, add					15%				
2700	For geared trolley, add					30%				
2800	For motor trolley, add					75%				
3000	For lifts over 15', 1 ton, add				L.F.	25			25	27.50
3100	5 ton, add				"	52			52	57
3300	Lifts, scissor type, portable, electric, 36" high, 2,000 lb.				Ea.	3,475			3,475	3,825
3400	48" high, 4,000 lb.				"	4,175			4,175	4,600

Division 44 - Pollution Control Equipment

Estimating Tips

This section involves equipment and construction costs for air noise and odor pollution control systems.

These systems may be interrelated and care must be taken that the complete systems are estimated. For example, air pollution equipment may include dust and air-entrained particles that have to be collected. The vacuum systems could be noisy, requiring silencers to reduce noise pollution, and the collected solids have to be disposed of to prevent solid pollution.

Reference Numbers

Reference numbers are shown in shaded boxes at the beginning of some major classifications. These numbers refer to related items in the Reference Section. The reference information may be an estimating procedure, an alternate pricing method, or technical information.

Note: Not all subdivisions listed here necessarily appear in this publication.

44 11 Particulate Control Equipment

44 11 16 – Fugitive Dust Barrier Systems

44 11 16.10 Dust Collection Systems		Crew	Daily Output	Labor-Hours	Unit	Material	2014 Bare Costs Labor	Equipment	Total	Total Incl O&P
0010	**DUST COLLECTION SYSTEMS** Commercial/Industrial									
0120	Central vacuum units									
0130	Includes stand, filters and motorized shaker									
0200	500 CFM, 10" inlet, 2 HP	Q-20	2.40	8.333	Ea.	4,175	415		4,590	5,225
0220	1000 CFM, 10" inlet, 3 HP		2.20	9.091		4,400	455		4,855	5,525
0240	1500 CFM, 10" inlet, 5 HP		2	10		4,625	500		5,125	5,850
0260	3000 CFM, 13" inlet, 10 HP		1.50	13.333		12,900	665		13,565	15,200
0280	5000 CFM, 16" inlet, 2 @ 10 HP	↓	1	20	↓	13,700	1,000		14,700	16,600
1000	Vacuum tubing, galvanized									
1100	2-1/8" OD, 16 ga.	Q-9	440	.036	L.F.	2.95	1.79		4.74	6
1110	2-1/2" OD, 16 ga.		420	.038		3.24	1.88		5.12	6.40
1120	3" OD, 16 ga.		400	.040		4.13	1.97		6.10	7.55
1130	3-1/2" OD, 16 ga.		380	.042		5.85	2.07		7.92	9.60
1140	4" OD, 16 ga.		360	.044		6	2.19		8.19	9.95
1150	5" OD, 14 ga.		320	.050		10.80	2.46		13.26	15.60
1160	6" OD, 14 ga.		280	.057		12.10	2.81		14.91	17.60
1170	8" OD, 14 ga.		200	.080		18.30	3.94		22.24	26
1180	10" OD, 12 ga.		160	.100		35.50	4.92		40.42	46.50
1190	12" OD, 12 ga.		120	.133		47	6.55		53.55	61.50
1200	14" OD, 12 ga.	↓	80	.200	↓	53.50	9.85		63.35	73.50
1940	Hose, flexible wire reinforced rubber									
1956	3" diam.	Q-9	400	.040	L.F.	8.95	1.97		10.92	12.85
1960	4" diam.		360	.044		10.35	2.19		12.54	14.75
1970	5" diam.		320	.050		13.50	2.46		15.96	18.60
1980	6" diam.	↓	280	.057	↓	14.10	2.81		16.91	19.80
2000	90° Elbow, slip fit									
2110	2-1/8" diam.	Q-9	70	.229	Ea.	10.85	11.25		22.10	29
2120	2-1/2" diam.		65	.246		15.20	12.10		27.30	35
2130	3" diam.		60	.267		20.50	13.15		33.65	42.50
2140	3-1/2" diam.		55	.291		25.50	14.30		39.80	50
2150	4" diam.		50	.320		31.50	15.75		47.25	58.50
2160	5" diam.		45	.356		60	17.50		77.50	92.50
2170	6" diam.		40	.400		82.50	19.70		102.20	121
2180	8" diam.	↓	30	.533		157	26.50		183.50	212
2400	45° Elbow, slip fit									
2410	2-1/8" diam.	Q-9	70	.229	Ea.	9.50	11.25		20.75	27.50
2420	2-1/2" diam.		65	.246		14	12.10		26.10	34
2430	3" diam.		60	.267		16.90	13.15		30.05	38.50
2440	3-1/2" diam.		55	.291		21.50	14.30		35.80	45.50
2450	4" diam.		50	.320		27.50	15.75		43.25	54.50
2460	5" diam.		45	.356		46.50	17.50		64	77.50
2470	6" diam.		40	.400		63	19.70		82.70	99
2480	8" diam.	↓	35	.457	↓	123	22.50		145.50	171
2800	90° TY, slip fit thru 6" diam.									
2810	2-1/8" diam.	Q-9	42	.381	Ea.	17.75	18.75		36.50	48
2820	2-1/2" diam.		39	.410		26.50	20		46.50	60
2830	3" diam.		36	.444		37	22		59	74
2840	3-1/2" diam.		33	.485		39	24		63	79
2850	4" diam.		30	.533		41.50	26.50		68	86
2860	5" diam.		27	.593		70.50	29		99.50	122
2870	6" diam.	↓	24	.667		101	33		134	161
2880	8" diam., butt end					155			155	170
2890	10" diam., butt end				↓	475			475	525

44 11 Particulate Control Equipment

44 11 16 – Fugitive Dust Barrier Systems

44 11 16.10 Dust Collection Systems		Crew	Daily Output	Labor-Hours	Unit	Material	2014 Bare Costs Labor	Equipment	Total	Total Incl O&P
2900	12" diam., butt end				Ea.	475			475	525
2910	14" diam., butt end					855			855	940
2920	6" x 4" diam., butt end					106			106	116
2930	8" x 4" diam., butt end					200			200	220
2940	10" x 4" diam., butt end					263			263	289
2950	12" x 4" diam., butt end					300			300	330
3100	90° Elbow, butt end segmented									
3110	8" diam., butt end, segmented				Ea.	157			157	172
3120	10" diam., butt end, segmented					390			390	430
3130	12" diam., butt end, segmented					495			495	545
3140	14" diam., butt end, segmented					620			620	680
3200	45° Elbow, butt end segmented									
3210	8" diam., butt end, segmented				Ea.	123			123	136
3220	10" diam., butt end, segmented					282			282	310
3230	12" diam., butt end, segmented					288			288	315
3240	14" diam., butt end, segmented					585			585	645
3400	All butt end fittings require one coupling per joint.									
3410	Labor for fitting included with couplings.									
3460	Compression coupling, galvanized, neoprene gasket									
3470	2-1/8" diam.	Q-9	44	.364	Ea.	10.85	17.90		28.75	39.50
3480	2-1/2" diam.		44	.364		10.85	17.90		28.75	39.50
3490	3" diam.		38	.421		15.60	20.50		36.10	48.50
3500	3-1/2" diam.		35	.457		17.45	22.50		39.95	53.50
3510	4" diam.		33	.485		18.95	24		42.95	57.50
3520	5" diam.		29	.552		21.50	27		48.50	65.50
3530	6" diam.		26	.615		25.50	30.50		56	74.50
3540	8" diam.		22	.727		46	36		82	105
3550	10" diam.		20	.800		67.50	39.50		107	135
3560	12" diam.		18	.889		84	44		128	160
3570	14" diam.		16	1		119	49		168	206
3800	Air gate valves, galvanized									
3810	2-1/8" diam.	Q-9	30	.533	Ea.	111	26.50		137.50	162
3820	2-1/2" diam.		28	.571		119	28		147	174
3830	3" diam.		26	.615		127	30.50		157.50	187
3840	4" diam.		23	.696		147	34.50		181.50	215
3850	6" diam.		18	.889		205	44		249	292

Division Notes

	CREW	DAILY OUTPUT	LABOR-HOURS	UNIT	BARE COSTS MAT.	LABOR	EQUIP.	TOTAL	TOTAL INCL O&P

Estimating Tips

This division contains information about water and wastewater equipment and systems, which was formerly located in Division 44.

The main areas of focus are total wastewater treatment plants and components of wastewater treatment plants. In addition, there are assemblies such as sewage treatment lagoons that can be found in some publications under G30 Site Mechanical Utilities.

Also included in this section are oil/water separators for wastewater treatment.

Reference Numbers

Reference numbers are shown in shaded boxes at the beginning of some major classifications. These numbers refer to related items in the Reference Section. The reference information may be an estimating procedure, an alternate pricing method, or technical information.

Note: Not all subdivisions listed here necessarily appear in this publication.

46 07 Packaged Water and Wastewater Treatment Equipment

46 07 53 – Packaged Wastewater Treatment Equipment

46 07 53.10 Biological Packaged Water Treatment Plants

		Crew	Daily Output	Labor-Hours	Unit	Material	2014 Bare Costs Labor	Equipment	Total	Total Incl O&P
0010	**BIOLOGICAL PACKAGED WATER TREATMENT PLANTS**									
0011	Not including fencing or external piping									
0020	Steel packaged, blown air aeration plants									
0100	1,000 GPD				Gal.				55	60.50
0200	5,000 GPD								22	24
0300	15,000 GPD								22	24
0400	30,000 GPD								15.40	16.95
0500	50,000 GPD								11	12.10
0600	100,000 GPD								9.90	10.90
0700	200,000 GPD								8.80	9.70
0800	500,000 GPD				▼				7.70	8.45
1000	Concrete, extended aeration, primary and secondary treatment									
1010	10,000 GPD				Gal.				22	24
1100	30,000 GPD								15.40	16.95
1200	50,000 GPD								11	12.10
1400	100,000 GPD								9.90	10.90
1500	500,000 GPD				▼				7.70	8.45
1700	Municipal wastewater treatment facility									
1720	1.0 MGD				Gal.				11	12.10
1740	1.5 MGD								10.60	11.65
1760	2.0 MGD								10	11
1780	3.0 MGD								7.80	8.60
1800	5.0 MGD				▼				5.80	6.70
2000	Holding tank system, not incl. excavation or backfill									
2010	Recirculating chemical water closet	2 Plum	4	4	Ea.	520	230		750	920
2100	For voltage converter, add	"	16	1		295	57.50		352.50	410
2200	For high level alarm, add	1 Plum	7.80	1.026	▼	114	59		173	215

46 07 53.20 Wastewater Treatment System

		Crew	Daily Output	Labor-Hours	Unit	Material	Labor	Equipment	Total	Total Incl O&P
0010	**WASTEWATER TREATMENT SYSTEM**									
0020	Fiberglass, 1,000 gallon	B-21	1.29	21.705	Ea.	4,075	915	109	5,099	6,025
0100	1,500 gallon	"	1.03	27.184	"	8,150	1,150	136	9,436	10,900

46 23 Grit Removal And Handling Equipment

46 23 23 – Vortex Grit Removal Equipment

46 23 23.10 Rainwater Filters

		Crew	Daily Output	Labor-Hours	Unit	Material	Labor	Equipment	Total	Total Incl O&P
0010	**RAINWATER FILTERS**									
0100	42 gal./min	B-21	3.50	8	Ea.	42,300	340	40	42,680	47,200
0200	65 gal./min		3.50	8		42,400	340	40	42,780	47,200
0300	208 gal./min	▼	3.50	8	▼	42,400	340	40	42,780	47,300

46 25 Oil and Grease Separation and Removal Equipment

46 25 13 – Coalescing Oil-Water Separators

46 25 13.20 Oil/Water Separators		Crew	Daily Output	Labor-Hours	Unit	Material	2014 Bare Costs Labor	Equipment	Total	Total Incl O&P
0010	**OIL/WATER SEPARATORS**									
0020	Underground, tank only									
0030	Excludes excavation, backfill, & piping									
0100	200 GPM	B-21	3.50	8	Ea.	42,200	340	40	42,580	47,000
0110	400 GPM		3.25	8.615		57,500	365	43	57,908	64,000
0120	600 GPM		2.75	10.182		65,500	430	51	65,981	72,500
0130	800 GPM		2.50	11.200		72,500	475	56	73,031	80,500
0140	1000 GPM		2	14		86,500	590	70	87,160	96,000
0150	1200 GPM		1.50	18.667		92,500	790	93.50	93,383.50	103,500
0160	1500 GPM	▼	1	28	▼	113,500	1,175	140	114,815	127,000

Division Notes

Assemblies Section

Table of Contents

Table No.	Page
D SERVICES	

D2010 Plumbing Fixtures
D2010 110 Water Closet Systems	438
D2010 120 Water Closets, Group	439
D2010 210 Urinal Systems	440
D2010 220 Urinal Systems, Battery Mount	441
D2010 310 Lavatory Systems	442
D2010 320 Lavatory Systems, Battery Mount	443
D2010 410 Kitchen Sink Systems	444
D2010 420 Laundry Sink Systems	445
D2010 430 Laboratory Sink Systems	446
D2010 440 Service Sink Systems	447
D2010 510 Bathtub Systems	448
D2010 610 Group Wash Fountain Systems	449
D2010 710 Shower Systems	450
D2010 810 Drinking Fountain Systems	451
D2010 820 Water Cooler Systems	452
D2010 920 Two Fixture Bathroom, Two Wall Plumbing	453
D2010 922 Two Fixture Bathroom, One Wall Plumbing	453
D2010 924 Three Fixture Bathroom, One Wall Plumbing	454
D2010 926 Three Fixture Bathroom, Two Wall Plumbing	454
D2010 928 Four Fixture Bathroom, Two Wall Plumbing	455
D2010 930 Four Fixture Bathroom, Three Wall Plumbing	455
D2010 932 Five Fixture Bathroom, Two Wall Plumbing	456
D2010 934 Five Fixture Bathroom, Three Wall Plumbing	456
D2010 936 Five Fixture Bathroom, One Wall Plumbing	456
D2010 951 Sample Estimate	457

D2020 Domestic Water Distribution
D2020 210 Electric Water Heaters - Residential Systems	458
D2020 220 Gas Fired Water Heaters - Residential Systems	459
D2020 230 Oil Fired Water Heaters - Residential Systems	460
D2020 240 Electric Water Heaters - Commercial Systems	461
D2020 250 Gas Fired Water Heaters - Commercial Systems	462
D2020 260 Oil Fired Water Heaters - Commercial Systems	463
D2020 265 Solar, Closed Loop, Add-On Hot Water Systems	465
D2020 270 Solar, Drainback, Hot Water Systems	467
D2020 275 Solar, Draindown, Hot Water Systems	469
D2020 280 Solar Recirculation, Domestic Hot Water Systems	471
D2020 285 Thermosyphon, Hot Water	473
D2020 290 Solar Hot Water, Air To Water Heat Exchange	475
D2020 295 Solar, Closed Loop, Hot Water Systems	477

D2040 Rain Water Drainage
D2040 210 Roof Drain Systems	478

D3010 Energy Supply
D3010 510 Apartment Building Heating - Fin Tube Radiation	480
D3010 650 Solar, Closed Loop, Space/Hot Water Systems	483
D3010 660 Solar Swimming Pool Heater Systems	485
D3010 675 Air To Water Heat Exchange	487

D3020 Heat Generating Systems
D3020 102 Small Heating Systems, Hydronic, Electric Boilers	488

D3090 Other HVAC Systems/Equip
D3090 320 Garage Exhaust Systems	489

D4010 Sprinklers
D4010 310 Dry Pipe Sprinkler Systems	490
D4010 350 Preaction Sprinkler Systems	493
D4010 370 Deluge Sprinkler Systems	496
D4010 390 Firecycle Sprinkler Systems	499
D4010 410 Wet Pipe Sprinkler Systems	502
D4010 412 Wet Sprinkler Seismic Components	505
D4010 413 Seismic Wet Pipe Flexible Feed Sprinkler Systems	508
D4010 414 Seismic Wet Pipe Flexible Feed Sprinkler Systems	510

D4020 Standpipes
D4020 310 Wet Standpipe Risers, Class I	512
Wet Standpipe Risers, Class II	512
Wet Standpipe Risers, Class III	513
D4020 330 Dry Standpipe Risers, Class I	514
Dry Standpipe Risers, Class II	514
Dry Standpipe Risers, Class III	515
D4020 410 Fire Hose Equipment	516

D4090 Other Fire Protection Systems
D4090 910 Fire Suppression Unit Components	517
D4090 920 FM200 Systems	517

G BUILDING SITEWORK

G1030 Site Earthwork
G1030 805 Trenching Common Earth	520
G1030 806 Trenching Loam & Sandy Clay	522
G1030 807 Trenching Sand & Gravel	524
G1030 815 Pipe Bedding	526

G2040 Site Development
G2040 920 Swimming Pools	528

G2050 Landscaping
G2050 710 Site Irrigation	529
G2050 720 Site Irrigation	532

G3010 Water Supply
G3010 120 Water Service	533
G3010 410 Fire Hydrants	535

G3020 Sanitary Sewer
G3020 302 Septic Systems	537
G3020 710 Aerated Sewage Lagoon	539

G3030 Storm Sewer
G3030 210 Manholes & Catch Basins	541

How RSMeans Assemblies Data Works

Assemblies estimating provides a fast and reasonably accurate way to develop construction costs. An assembly is the grouping of individual work items, with appropriate quantities, to provide a cost for a major construction component in a convenient unit of measure.

An assemblies estimate is often used during early stages of design development to compare the cost impact of various design alternatives on total building cost.

Assemblies estimates are also used as an efficient tool to verify construction estimates.

Assemblies estimates do not require a completed design or detailed drawings. Instead, they are based on the general size of the structure and other known parameters of the project. The degree of accuracy of an assemblies estimate is generally within +/- 15%.

RSMeans assemblies are identified by a **unique 12-character identifier**. The assemblies are numbered using UNIFORMAT II, ASTM Standard E1557. The first 6 characters represent this system to Level 4. The last 6 characters represent further breakdown by RSMeans in order to arrange items in understandable groups of similar tasks. Line numbers are consistent across all RSMeans publications, so a line number in any RSMeans assemblies data set will always refer to the same work.

RSMeans assemblies **Narrative descriptions** are shown in a hierarchical structure to make them readable. In order to read a complete description, read up through the indents to the top of the section. Include everything that is above and to the left that is not contradicted by information below.

Information is available in the Reference Section to assist the estimator with estimating procedures, alternate pricing methods, and additional technical information.

The **Reference Box** indicates the exact location of this information in the Reference Section at the back of the book. The "R" stands for "reference," and the remaining characters are the RSMeans line numbers.

For supplemental customizable square foot estimating forms, visit: http://www.reedconstructiondata.com/rsmeans/assemblies

Most assemblies consist of three major elements: a graphic, the system components, and the cost data itself. The **Graphic** is a visual representation showing the typical appearance of the assembly in question, frequently accompanied by additional explanatory technical information describing the class of items. The **System Components** is a listing of the individual tasks that make up the assembly, including the quantity and unit of measure for each item, along with the cost of material and installation. The **Assemblies Data** below lists prices for other similar systems with dimensional and/or size variations.

All RSMeans assemblies costs represent the cost for the installing contractor. An allowance for profit has been added to all material, labor, and equipment rental costs. A markup for labor burdens, including workers' compensation, fixed overhead, and business overhead, is included with installation costs.

The data published in RSMeans print books represents a "national average" cost. This data should be modified to the project location using the **City Cost Indexes** or **Location Factors** tables found in the Reference Section in the back of the book.

All RSMeans assemblies data includes a typical **Unit of Measure** used for estimating that item. For instance, while the unit of measure for the graphic shown here is Each, some A/C systems are estimated by the square foot (S.F.). The estimator needs to take special care that the unit in the data matches the unit in the takeoff. Abbreviations can be found in the Reference Section.

System components are listed separately to detail what is included in the development of the total system price.

D40 Fire Protection

D4020 Standpipes

System Components	QUANTITY	UNIT	COST PER FLOOR MAT.	INST.	TOTAL
SYSTEM D4020 310 0560 WET STANDPIPE RISER, CLASS I, STEEL, BLACK, SCH. 40 PIPE, 10' HEIGHT 4" DIAMETER PIPE, ONE FLOOR					
Pipe, steel, black, schedule 40, threaded, 4" diam.	20.000	L.F.	600	700	1,300
Pipe, Tee, malleable iron, black, 150 lb. threaded, 4" pipe size	2.000	Ea.	584	630	1,214
Pipe, 90° elbow, malleable iron, black, 150 lb threaded 4" pipe size	1.000	Ea.	185	209	394
Pipe, nipple, steel, black, schedule 40, 2-1/2" pipe size x 3" long	2.000	Ea.	24.50	156	180.50
Fire valve, gate, 300 lb., brass w/handwheel, 2-1/2" pipe size	1.000	Ea.	198	95.50	293.50
Fire valve, pressure restricting, adj, rgh brs, 2-1/2" pipe size	1.000	Ea.	352	191	543
Valve, swing check, w/ball drip, CI w/brs. ftngs., 4" pipe size	1.000	Ea.	365	400	765
Standpipe conn wall dble. flush brs. w/plugs & chains 2-1/2"x2-1/2"x4"	1.000	Ea.	545	241	786
Valve, swing check, bronze, 125 lb, regrinding disc, 2-1/2" pipe size	1.000	Ea.	835	83.50	918.50
Roof manifold, fire, w/valves & caps, horiz/vert brs 2-1/2"x2-1/2"x4"	1.000	Ea.	176	251	427
Fire, hydrolator, vent & drain, 2-1/2" pipe size	1.000	Ea.	74.50	56	130.50
Valve, gate, iron body 125 lb., OS&Y, threaded, 4" pipe size	1.000	Ea.	805	415	1,220
TOTAL			4,744	3,428	8,172

D4020 310	Wet Standpipe Risers, Class I		COST PER FLOOR		
			MAT.	INST.	TOTAL
0550	Wet standpipe risers, Class I, steel, black sch. 40, 10' height				
0560	4" diameter pipe, one floor		4,750	3,425	8,175
0580	Additional floors		1,075	1,075	2,150
0600	6" diameter pipe, one floor		8,075	6,025	14,100
0620	Additional floors	R211226 -10	1,950	1,675	3,625
0640	8" diameter pipe, one floor		12,500	7,250	19,750
0660	Additional floors	R211226 -20	2,950	2,050	5,000
0680					

How RSMeans Assemblies Data Works (Continued)

Sample Estimate

This sample demonstrates the elements of an estimate, including a tally of the RSMeans data lines. Published assemblies costs include all markups for labor burden and profit for the installing contractor. This estimate adds a summary of the markups applied by a general contractor on the installing contractors' work. These figures represent the total cost to the owner. The RSMeans location factor is added at the bottom of the estimate to adjust the cost of the work to a specific location

Work performed: The body of the estimate shows the RSMeans data selected, including line numbers, a brief description of each item, its takeoff quantity and unit, and the total installed cost, including the installing contractor's overhead and profit.

Location Factor: RSMeans published data is based on national average costs. If necessary, adjust the total cost of the project using a location factor from the "Location Factor" table or the "City Cost Indexes" table, found in the Reference Section. Use location factors if the work is general, covering the work of multiple trades. If the work is by a single trade (e.g., masonry) use the more specific data found in the City Cost Indexes.

To adjust costs by location factors, multiply the base cost by the factor and divide by 100.

Contingency: A factor for contingency may be added to any estimate to represent the cost of unknowns that may occur between the time that the estimate is performed and the time the project is constructed. The amount of the allowance will depend on the stage of design at which the estimate is done, and the contractor's assessment of the risk invloved.

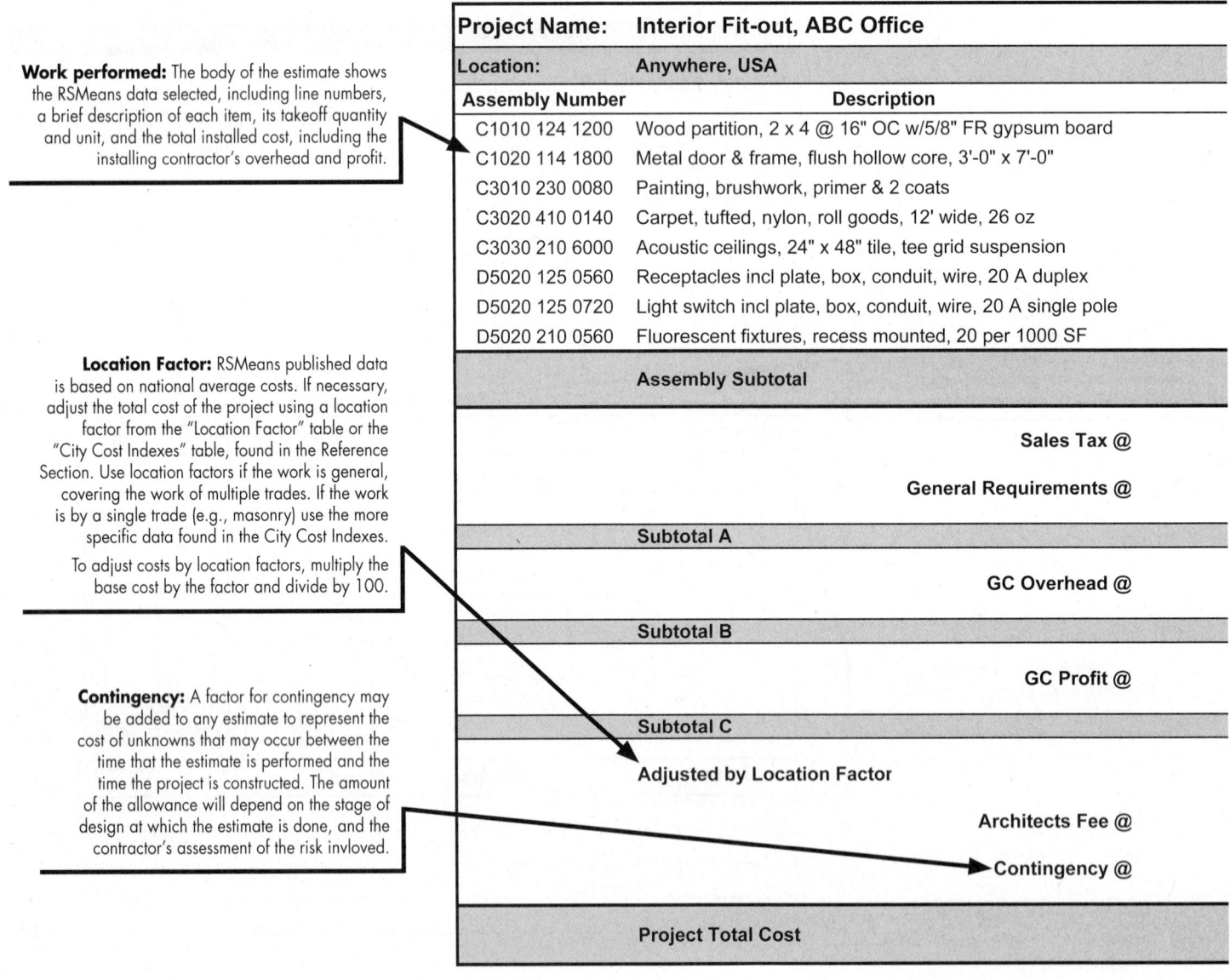

Project Name:	Interior Fit-out, ABC Office
Location:	Anywhere, USA
Assembly Number	Description
C1010 124 1200	Wood partition, 2 x 4 @ 16" OC w/5/8" FR gypsum board
C1020 114 1800	Metal door & frame, flush hollow core, 3'-0" x 7'-0"
C3010 230 0080	Painting, brushwork, primer & 2 coats
C3020 410 0140	Carpet, tufted, nylon, roll goods, 12' wide, 26 oz
C3030 210 6000	Acoustic ceilings, 24" x 48" tile, tee grid suspension
D5020 125 0560	Receptacles incl plate, box, conduit, wire, 20 A duplex
D5020 125 0720	Light switch incl plate, box, conduit, wire, 20 A single pole
D5020 210 0560	Fluorescent fixtures, recess mounted, 20 per 1000 SF

Assembly Subtotal

Sales Tax @

General Requirements @

Subtotal A

GC Overhead @

Subtotal B

GC Profit @

Subtotal C

Adjusted by Location Factor

Architects Fee @

Contingency @

Project Total Cost

This estimate is based on an interactive spreadsheet. A copy of this spreadsheet is located on the RSMeans website at
http://www.reedconstructiondata.com/rsmeans/extras/546011.
You are free to download it and adjust it to your methodology.

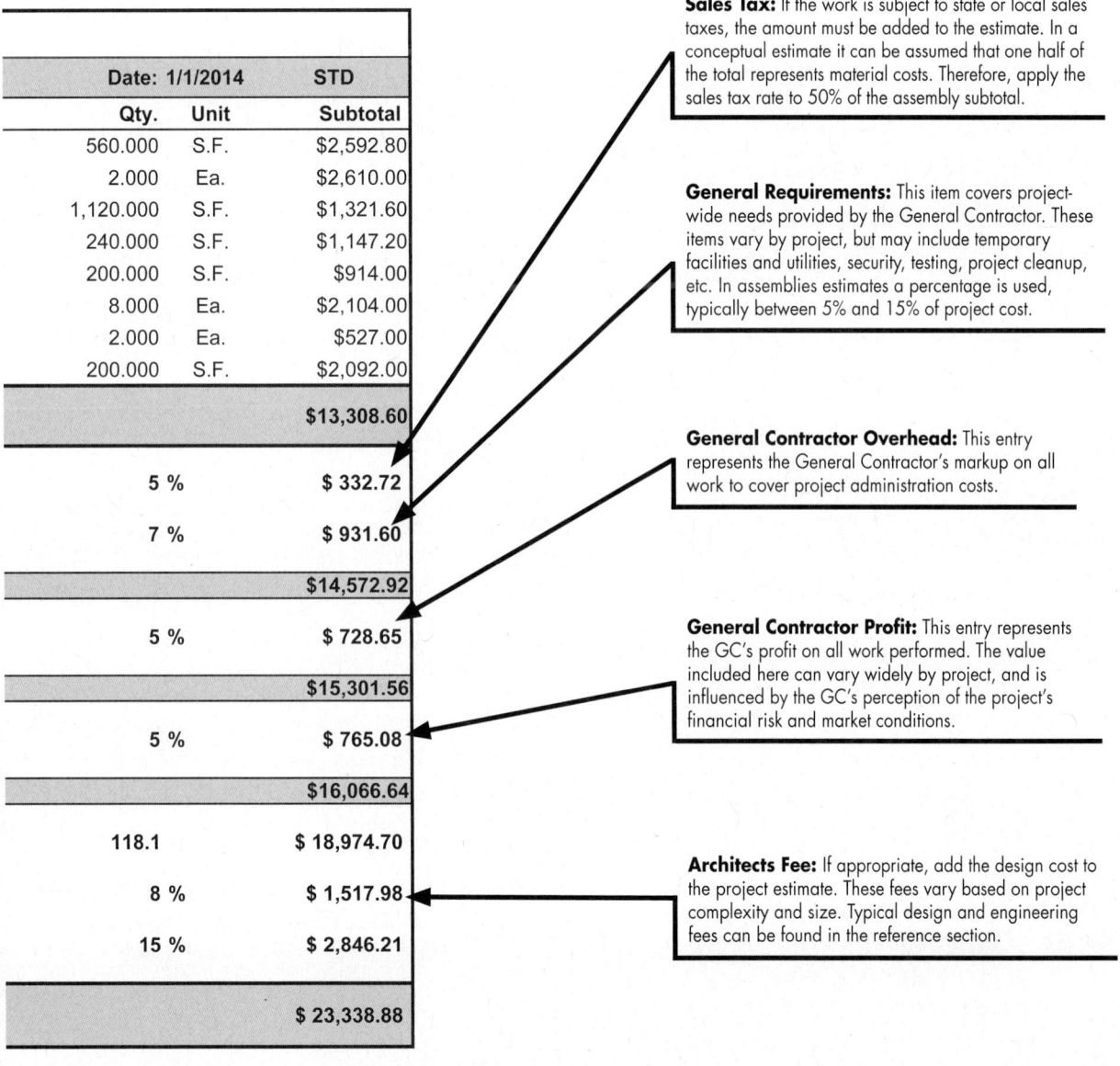

D Services

D20 Plumbing

D2010 Plumbing Fixtures

Systems are complete with trim seat and rough-in (supply, waste and vent) for connection to supply branches and waste mains.

One Piece Wall Hung

Supply

Waste/Vent

Floor Mount

System Components	QUANTITY	UNIT	MAT.	INST.	TOTAL
SYSTEM D2010 110 1880					
WATER CLOSET, VITREOUS CHINA					
TANK TYPE, WALL HUNG, TWO PIECE					
Water closet, tank type vit china wall hung 2 pc. w/seat supply & stop	1.000	Ea.	690	236	926
Pipe Steel galvanized, schedule 40, threaded, 2" diam.	4.000	L.F.	53.40	78.20	131.60
Pipe, CI soil, no hub, cplg 10' OC, hanger 5' OC, 4" diam.	2.000	L.F.	37.30	43	80.30
Pipe, coupling, standard coupling, CI soil, no hub, 4" diam.	2.000	Ea.	44	76	120
Copper tubing type L solder joint, hangar 10' O.C., 1/2" diam.	6.000	L.F.	22.26	51.60	73.86
Wrought copper 90° elbow for solder joints 1/2" diam.	2.000	Ea.	6.78	70	76.78
Wrought copper Tee for solder joints 1/2" diam.	1.000	Ea.	5.80	53.50	59.30
Supports/carrier, water closet, siphon jet, horiz, single, 4" waste	1.000	Ea.	720	130	850
TOTAL			1,579.54	738.30	2,317.84

D2010 110	Water Closet Systems		MAT.	INST.	TOTAL
1800	Water closet, vitreous china				
1840	Tank type, wall hung				
1880	Close coupled two piece	R224000 -30	1,575	740	2,315
1920	Floor mount, one piece		1,550	785	2,335
1960	One piece low profile	R224000 -40	1,600	785	2,385
2000	Two piece close coupled		660	785	1,445
2040	Bowl only with flush valve				
2080	Wall hung		1,925	835	2,760
2120	Floor mount		885	800	1,685
2160	Floor mount, ADA compliant with 18" high bowl		880	820	1,700

D20 Plumbing

D2010 Plumbing Fixtures

Systems are complete with trim, seat, flush valve and rough-in (supply, waste and vent) for connection to supply branches and waste mains.

Side by Side

Back to Back

Supply | Waste/Vent | Supply | Waste/Vent

System Components	QUANTITY	UNIT	COST EACH MAT.	INST.	TOTAL
SYSTEM D2010 120 1760					
WATER CLOSETS, BATTERY MOUNT, WALL HUNG, SIDE BY SIDE, FIRST CLOSET					
Water closet, bowl only w/flush valve, seat, wall hung	1.000	Ea.	830	216	1,046
Pipe, CI soil, no hub, cplg 10' OC, hanger 5' OC, 4" diam.	3.000	L.F.	55.95	64.50	120.45
Coupling, standard, CI, soil, no hub, 4" diam.	2.000	Ea.	44	76	120
Copper tubing, type L, solder joints, hangers 10' OC, 1" diam.	6.000	L.F.	51.90	61.20	113.10
Copper tubing, type DWV, solder joints, hangers 10'OC, 2" diam.	6.000	L.F.	112.50	94.80	207.30
Wrought copper 90° elbow for solder joints 1" diam.	1.000	Ea.	18.70	43.50	62.20
Wrought copper Tee for solder joints 1" diam.	1.000	Ea.	43	69.50	112.50
Support/carrier, siphon jet, horiz, adjustable single, 4" pipe	1.000	Ea.	720	130	850
Valve, gate, bronze, 125 lb, NRS, soldered 1" diam.	1.000	Ea.	82.50	36.50	119
Wrought copper, DWV, 90° elbow, 2" diam.	1.000	Ea.	64	69.50	133.50
TOTAL			2,022.55	861.50	2,884.05

D2010 120	Water Closets, Group		COST EACH MAT.	INST.	TOTAL
1760	Water closets, battery mount, wall hung, side by side, first closet	R224000-30	2,025	860	2,885
1800	Each additional water closet, add		1,925	815	2,740
3000	Back to back, first pair of closets	R224000-40	3,525	1,150	4,675
3100	Each additional pair of closets, back to back		3,475	1,125	4,600

D20 Plumbing

D2010 Plumbing Fixtures

Systems are complete with trim, flush valve and rough-in (supply, waste and vent) for connection to supply branches and waste mains.

Stall Type

Supply   **Waste/Vent**

Wall Hung

System Components	QUANTITY	UNIT	MAT.	INST.	TOTAL
SYSTEM D2010 210 2000					
URINAL, VITREOUS CHINA, WALL HUNG					
Urinal, wall hung, vitreous china, incl. hanger	1.000	Ea.	350	415	765
Pipe, steel, galvanized, schedule 40, threaded, 1-1/2" diam.	5.000	L.F.	49.75	78.25	128
Copper tubing type DWV, solder joint, hangers 10' OC, 2" diam.	3.000	L.F.	56.25	47.40	103.65
Combination Y & 1/8 bend for CI soil pipe, no hub, 3" diam.	1.000	Ea.	20		20
Pipe, CI, no hub, cplg. 10' OC, hanger 5' OC, 3" diam.	4.000	L.F.	58.60	78.20	136.80
Pipe coupling standard, CI soil, no hub, 3" diam.	3.000	Ea.	37.60	66	103.60
Copper tubing type L, solder joint, hanger 10' OC 3/4" diam.	5.000	L.F.	28.25	45.75	74
Wrought copper 90° elbow for solder joints 3/4" diam.	1.000	Ea.	7.60	36.50	44.10
Wrought copper Tee for solder joints, 3/4" diam.	1.000	Ea.	13.95	58	71.95
TOTAL			622	825.10	1,447.10

D2010 210	Urinal Systems		MAT.	INST.	TOTAL
2000	Urinal, vitreous china, wall hung	R224000 -30	620	825	1,445
2040	Stall type		1,425	990	2,415

D20 Plumbing

D2010 Plumbing Fixtures

Systems are complete with trim, flush valve and rough-in (supply, waste and vent) for connection to supply branches and waste mains.

Side by Side · Back to Back

Waste/Vent · Supply · Supply · Waste/Vent

System Components	QUANTITY	UNIT	MAT.	INST.	TOTAL
SYSTEM D2010 220 1760					
URINALS, BATTERY MOUNT, WALL HUNG, SIDE BY SIDE, FIRST URINAL					
Urinal, wall hung, vitreous china, with hanger & trim	1.000	Ea.	350	415	765
No hub cast iron soil pipe, 3" diameter	4.000	L.F.	58.60	78.20	136.80
No hub cast iron sanitary tee, 3" diameter	1.000	Ea.	20		20
No hub coupling, 3" diameter	2.000	Ea.	37.60	66	103.60
Copper tubing, type L, 3/4" diameter	5.000	L.F.	28.25	45.75	74
Copper tubing, type DWV, 2" diameter	2.000	L.F.	37.50	31.60	69.10
Copper 90° elbow, 3/4" diameter	2.000	Ea.	15.20	73	88.20
Copper 90° elbow, type DWV, 2" diameter	2.000	Ea.	64	69.50	133.50
Galvanized steel pipe, 1-1/2" diameter	5.000	L.F.	49.75	78.25	128
Cast iron drainage elbow, 90°, 1-1/2" diameter	1.000	Ea.	13.95	58	71.95
TOTAL			674.85	915.30	1,590.15

D2010 220	Urinal Systems, Battery Mount		MAT.	INST.	TOTAL
1760	Urinals, battery mount, side by side, first urinal		675	915	1,590
1800	Each additional urinal, add		700	890	1,590
2000	Back to back, first pair of urinals	R224000	1,275	1,500	2,775
2100	Each additional pair of urinals, back to back	-30	975	1,175	2,150

441

D20 Plumbing

D2010 Plumbing Fixtures

Systems are complete with trim and rough-in (supply, waste and vent) to connect to supply branches and waste mains.

Vanity Top

Supply   Waste/Vent

Wall Hung

System Components	QUANTITY	UNIT	MAT.	INST.	TOTAL
SYSTEM D2010 310 1560					
LAVATORY W/TRIM, VANITY TOP, P.E. ON C.I., 20" X 18"					
Lavatory w/trim, PE on CI, white, vanity top, 20" x 18" oval	1.000	Ea.	360	196	556
Pipe, steel, galvanized, schedule 40, threaded, 1-1/4" diam.	4.000	L.F.	34	56.20	90.20
Copper tubing type DWV, solder joint, hanger 10' OC 1-1/4" diam.	4.000	L.F.	43.60	46.40	90
Wrought copper DWV, Tee, sanitary, 1-1/4" diam.	1.000	Ea.	65	77	142
P trap w/cleanout, 20 ga., 1-1/4" diam.	1.000	Ea.	232	38.50	270.50
Copper tubing type L, solder joint, hanger 10' OC 1/2" diam.	10.000	L.F.	37.10	86	123.10
Wrought copper 90° elbow for solder joints 1/2" diam.	2.000	Ea.	6.78	70	76.78
Wrought copper Tee for solder joints, 1/2" diam.	2.000	Ea.	11.60	107	118.60
Stop, chrome, angle supply, 1/2" diam.	2.000	Ea.	28.40	63	91.40
TOTAL			818.48	740.10	1,558.58

D2010 310	Lavatory Systems	MAT.	INST.	TOTAL
1560	Lavatory w/trim, vanity top, PE on CI 20" x 18", Vanity top by others.	820	740	1,560
1600	19" x 16" oval	650	740	1,390
1640	18" round	930	740	1,670
1680	Cultured marble, 19" x 17"	650	740	1,390
1720	25" x 19"	685	740	1,425
1760	Stainless, self-rimming, 25" x 22"	865	740	1,605
1800	17" x 22" R224000-30	850	740	1,590
1840	Steel enameled, 20" x 17"	675	760	1,435
1880	19" round	645	760	1,405
1920	Vitreous china, 20" x 16"	745	775	1,520
1960	19" x 16"	750	775	1,525
2000	22" x 13"	750	775	1,525
2040	Wall hung, PE on CI, 18" x 15"	960	815	1,775
2080	19" x 17"	1,050	815	1,865
2120	20" x 18"	825	815	1,640
2160	Vitreous china, 18" x 15"	775	840	1,615
2200	19" x 17"	730	840	1,570
2240	24" x 20"	820	840	1,660
2300	20" x 27", handicap	1,500	905	2,405

442

D20 Plumbing

D2010 Plumbing Fixtures

Systems are complete with trim, flush valve and rough-in (supply, waste and vent) for connection to supply branches and waste mains.

Waste/Vent

Supply
(Two supply systems required)

Supply
(Two supply systems required)

Waste/Vent

System Components	QUANTITY	UNIT	MAT.	INST.	TOTAL
SYSTEM D2010 320 1760					
LAVATORIES, BATTERY MOUNT, WALL HUNG, SIDE BY SIDE, FIRST LAVATORY					
Lavatory w/trim wall hung PE on CI 20" x 18"	1.000	Ea.	305	156	461
Stop, chrome, angle supply, 3/8" diameter	2.000	Ea.	28.40	58	86.40
Concealed arm support	1.000	Ea.	510	116	626
P trap w/cleanout, 20 ga. C.P., 1-1/4" diameter	1.000	Ea.	34	38.50	72.50
Copper tubing, type L, 1/2" diameter	10.000	L.F.	37.10	86	123.10
Copper tubing, type DWV, 1-1/4" diameter	4.000	L.F.	43.60	46.40	90
Copper 90° elbow, 1/2" diameter	2.000	Ea.	6.78	70	76.78
Copper tee, 1/2" diameter	2.000	Ea.	11.60	107	118.60
DWV copper sanitary tee, 1-1/4" diameter	2.000	Ea.	130	154	284
Galvanized steel pipe, 1-1/4" diameter	4.000	L.F.	34	56.20	90.20
Black cast iron 90° elbow, 1-1/4" diameter	1.000	Ea.	11	57	68
TOTAL			1,151.48	945.10	2,096.58

D2010 320	Lavatory Systems, Battery Mount	MAT.	INST.	TOTAL
1760	Lavatories, battery mount, side by side, first lavatory	1,150	945	2,095
1800	Each additional lavatory, add	1,025	655	1,680
2000	Back to back, first pair of lavatories	1,825	1,500	3,325
2100	Each additional pair of lavatories, back to back	1,750	1,250	3,000

D20 Plumbing

D2010 Plumbing Fixtures

Systems are complete with trim and rough-in (supply, waste and vent) to connect to supply branches and waste mains.

Countertop Single Bowl

Supply

Waste/Vent

Countertop Double Bowl

System Components	QUANTITY	UNIT	COST EACH MAT.	COST EACH INST.	COST EACH TOTAL
SYSTEM D2010 410 1720					
KITCHEN SINK W/TRIM, COUNTERTOP, P.E. ON C.I., 24" X 21", SINGLE BOWL					
Kitchen sink, counter top, PE on CI, 1 bowl, 24" x 21" OD	1.000	Ea.	300	224	524
Pipe, steel, galvanized, schedule 40, threaded, 1-1/4" diam.	4.000	L.F.	34	56.20	90.20
Copper tubing, type DWV, solder, hangers 10' OC 1-1/2" diam.	6.000	L.F.	82.20	77.10	159.30
Wrought copper, DWV, Tee, sanitary, 1-1/2" diam.	1.000	Ea.	80.50	87	167.50
P trap, standard, copper, 1-1/2" diam.	1.000	Ea.	225	41	266
Copper tubing, type L, solder joints, hangers 10' OC 1/2" diam.	10.000	L.F.	37.10	86	123.10
Wrought copper 90° elbow for solder joints 1/2" diam.	2.000	Ea.	6.78	70	76.78
Wrought copper Tee for solder joints, 1/2" diam.	2.000	Ea.	11.60	107	118.60
Stop, angle supply, chrome, 1/2" CTS	2.000	Ea.	28.40	63	91.40
TOTAL			805.58	811.30	1,616.88

D2010 410	Kitchen Sink Systems	MAT.	INST.	TOTAL
1720	Kitchen sink w/trim, countertop, PE on CI, 24"x21", single bowl	805	810	1,615
1760	30" x 21" single bowl	1,125	810	1,935
1800	32" x 21" double bowl	910	875	1,785
1880	Stainless steel, 19" x 18" single bowl	1,150	810	1,960
1920	25" x 22" single bowl	1,225	810	2,035
1960	33" x 22" double bowl	1,575	875	2,450
2000	43" x 22" double bowl	1,775	885	2,660
2040	44" x 22" triple bowl	2,025	925	2,950
2080	44" x 24" corner double bowl	1,450	885	2,335
2120	Steel, enameled, 24" x 21" single bowl	1,050	810	1,860
2160	32" x 21" double bowl	1,075	875	1,950
2240	Raised deck, PE on CI, 32" x 21", dual level, double bowl	995	1,100	2,095
2280	42" x 21" dual level, triple bowl	1,850	1,200	3,050

D20 Plumbing

D2010 Plumbing Fixtures

Systems are complete with trim and rough-in (supply, waste and vent) to connect to supply branches and waste mains.

Single Compartment Sink   Supply   Waste/Vent   Double Compartment Sink

System Components	QUANTITY	UNIT	MAT.	INST.	TOTAL
SYSTEM D2010 420 1760					
LAUNDRY SINK W/TRIM, PE ON CI, BLACK IRON FRAME					
24" X 20" OD, SINGLE COMPARTMENT					
Laundry sink PE on CI w/trim & frame, 24" x 21" OD, 1 compartment	1.000	Ea.	635	209	844
Pipe, steel, galvanized, schedule 40, threaded, 1-1/4" diam	4.000	L.F.	34	56.20	90.20
Copper tubing, type DWV, solder joint, hanger 10' OC 1-1/2"diam	6.000	L.F.	82.20	77.10	159.30
Wrought copper, DWV, Tee, sanitary, 1-1/2" diam	1.000	Ea.	80.50	87	167.50
P trap, standard, copper, 1-1/2" diam	1.000	Ea.	225	41	266
Copper tubing type L, solder joints, hangers 10' OC, 1/2" diam	10.000	L.F.	37.10	86	123.10
Wrought copper 90° elbow for solder joints 1/2" diam	2.000	Ea.	6.78	70	76.78
Wrought copper Tee for solder joints, 1/2" diam	2.000	Ea.	11.60	107	118.60
Stop, angle supply, 1/2" diam	2.000	Ea.	28.40	63	91.40
TOTAL			1,140.58	796.30	1,936.88

D2010 420	Laundry Sink Systems		MAT.	INST.	TOTAL
1740	Laundry sink w/trim, PE on CI, black iron frame				
1760	24" x 20", single compartment		1,150	795	1,945
1800	24" x 23" single compartment	R224000	1,175	795	1,970
1840	48" x 21" double compartment	-30	1,450	865	2,315
1920	Molded stone, on wall, 22" x 21" single compartment		655	795	1,450
1960	45"x 21" double compartment		900	865	1,765
2040	Plastic, on wall or legs, 18" x 23" single compartment		655	780	1,435
2080	20" x 24" single compartment		675	780	1,455
2120	36" x 23" double compartment		735	840	1,575
2160	40" x 24" double compartment		840	840	1,680

D20 Plumbing

D2010 Plumbing Fixtures

Corrosion resistant laboratory sink systems are complete with trim and rough-in (supply, waste and vent) to connect to supply branches and waste mains.

Laboratory Sink

Supply   Waste/Vent

Polypropylene Cup Sink

System Components	QUANTITY	UNIT	MAT.	INST.	TOTAL
SYSTEM D2010 430 1600					
LABORATORY SINK W/TRIM, STAINLESS STEEL, SINGLE BOWL					
DOUBLE DRAINBOARD, 54" X 24" O.D.					
Sink w/trim, stainless steel, 1 bowl, 2 drainboards 54" x 24" OD	1.000	Ea.	1,875	415	2,290
Pipe, polypropylene, schedule 40, acid resistant 1-1/2" diam.	10.000	L.F.	87	184	271
Tee, sanitary, polypropylene, acid resistant, 1-1/2" diam.	1.000	Ea.	18.80	69.50	88.30
P trap, polypropylene, acid resistant, 1-1/2" diam.	1.000	Ea.	60	41	101
Copper tubing type L, solder joint, hanger 10' O.C. 1/2" diam.	10.000	L.F.	37.10	86	123.10
Wrought copper 90° elbow for solder joints 1/2" diam.	2.000	Ea.	6.78	70	76.78
Wrought copper Tee for solder joints, 1/2" diam.	2.000	Ea.	11.60	107	118.60
Stop, angle supply, chrome, 1/2" diam.	2.000	Ea.	28.40	63	91.40
TOTAL			2,124.68	1,035.50	3,160.18

D2010 430	Laboratory Sink Systems	MAT.	INST.	TOTAL
1580	Laboratory sink w/trim, stainless steel, single bowl,			
1590	Stainless steel, single bowl,			
1600	Double drainboard, 54" x 24" O.D.	2,125	1,025	3,150
1640	Single drainboard, 47" x 24"O.D.	1,550	1,025	2,575
1670	Stainless steel, double bowl,			
1680	70" x 24" O.D.	2,325	1,025	3,350
1750	Polyethylene, single bowl,			
1760	Flanged, 14-1/2" x 14-1/2" O.D.	555	935	1,490
1800	18-1/2" x 18-1/2" O.D.	655	935	1,590
1840	23-1/2" x 20-1/2" O.D.	700	935	1,635
1920	Polypropylene, cup sink, oval, 7" x 4" O.D.	375	825	1,200
1960	10" x 4-1/2" O.D.	400	825	1,225

D20 Plumbing

D2010 Plumbing Fixtures

Wall Hung

Supply

Waste/Vent

Corner, Floor

Corrosion resistant laboratory sink systems are complete with trim and rough-in (supply, waste and vent) to connect to supply branches and waste mains.

System Components	QUANTITY	UNIT	MAT.	INST.	TOTAL
SYSTEM D2010 440 4260					
SERVICE SINK, PE ON CI, CORNER FLOOR, 28"X28", W/RIM GUARD & TRIM					
Service sink, corner floor, PE on CI, 28" x 28", w/rim guard & trim	1.000	Ea.	1,100	284	1,384
Copper tubing type DWV, solder joint, hanger 10'OC 3" diam.	6.000	L.F.	231	129	360
Copper tubing type DWV, solder joint, hanger 10'OC 2" diam	4.000	L.F.	75	63.20	138.20
Wrought copper DWV, Tee, sanitary, 3" diam.	1.000	Ea.	410	179	589
P trap with cleanout & slip joint, copper 3" diam	1.000	Ea.	835	63	898
Copper tubing, type L, solder joints, hangers 10' OC, 1/2" diam	10.000	L.F.	37.10	86	123.10
Wrought copper 90° elbow for solder joints 1/2" diam	2.000	Ea.	6.78	70	76.78
Wrought copper Tee for solder joints, 1/2" diam	2.000	Ea.	11.60	107	118.60
Stop, angle supply, chrome, 1/2" diam	2.000	Ea.	28.40	63	91.40
TOTAL			2,734.88	1,044.20	3,779.08

D2010 440	Service Sink Systems		MAT.	INST.	TOTAL
4260	Service sink w/trim, PE on CI, corner floor, 28" x 28", w/rim guard	R224000 -30	2,725	1,050	3,775
4300	Wall hung w/rim guard, 22" x 18"		2,900	1,225	4,125
4340	24" x 20"		2,975	1,225	4,200
4380	Vitreous china, wall hung 22" x 20"		2,725	1,225	3,950

D20 Plumbing

D2010 Plumbing Fixtures

Systems are complete with trim and rough-in (supply, waste and vent) to connect to supply branches and waste mains.

Recessed Bathtub

Supply

Waste/Vent   Corner Bathtub

System Components	QUANTITY	UNIT	MAT.	INST.	TOTAL
SYSTEM D2010 510 2000					
BATHTUB, RECESSED, PORCELAIN ENAMEL ON CAST IRON,, 48" x 42"					
Bath tub, porcelain enamel on cast iron, w/fittings, 48" x 42"	1.000	Ea.	2,900	315	3,215
Pipe, steel, galvanized, schedule 40, threaded, 1-1/4" diam.	4.000	L.F.	34	56.20	90.20
Pipe, CI no hub soil w/couplings 10' OC, hangers 5' OC, 4" diam.	3.000	L.F.	55.95	64.50	120.45
Combination Y and 1/8 bend for C.I. soil pipe, no hub, 4" pipe size	1.000	Ea.	52		52
Drum trap, 3" x 5", copper, 1-1/2" diam.	1.000	Ea.	315	43.50	358.50
Copper tubing type L, solder joints, hangers 10' OC 1/2" diam.	10.000	L.F.	37.10	86	123.10
Wrought copper 90° elbow, solder joints, 1/2" diam.	2.000	Ea.	6.78	70	76.78
Wrought copper Tee, solder joints, 1/2" diam.	2.000	Ea.	11.60	107	118.60
Stop, angle supply, 1/2" diameter	2.000	Ea.	28.40	63	91.40
Copper tubing type DWV, solder joints, hanger 10' OC 1-1/2" diam.	3.000	L.F.	41.10	38.55	79.65
Pipe coupling, standard, C.I. soil no hub, 4" pipe size	2.000	Ea.	44	76	120
TOTAL			3,525.93	919.75	4,445.68

D2010 510	Bathtub Systems		MAT.	INST.	TOTAL
2000	Bathtub, recessed, P.E. on Cl., 48" x 42"	R224000 -30	3,525	920	4,445
2040	72" x 36"		3,625	1,025	4,650
2080	Mat bottom, 5' long		1,850	890	2,740
2120	5'-6" long		2,575	920	3,495
2160	Corner, 48" x 42"		3,525	890	4,415
2200	Formed steel, enameled, 4'-6" long		1,175	820	1,995

D20 Plumbing

D2010 Plumbing Fixtures

Circular Fountain

Supply

Waste/Vent

Semi-Circular Fountain

Systems are complete with trim, flush valve and rough-in (supply, waste and vent) for connection to supply branches and waste mains.

System Components	QUANTITY	UNIT	COST EACH MAT.	COST EACH INST.	COST EACH TOTAL
SYSTEM D2010 610 1760					
GROUP WASH FOUNTAIN, PRECAST TERRAZZO					
CIRCULAR, 36" DIAMETER					
Wash fountain, group, precast terrazzo, foot control 36" diam.	1.000	Ea.	6,375	650	7,025
Copper tubing type DWV, solder joint, hanger 10' OC, 2" diam.	10.000	L.F.	187.50	158	345.50
P trap, standard, copper, 2" diam.	1.000	Ea.	345	46.50	391.50
Wrought copper, Tee, sanitary, 2" diam.	1.000	Ea.	94	99.50	193.50
Copper tubing type L, solder joint, hanger 10' OC 1/2" diam.	20.000	L.F.	74.20	172	246.20
Wrought copper 90° elbow for solder joints 1/2" diam.	3.000	Ea.	10.17	105	115.17
Wrought copper Tee for solder joints, 1/2" diam.	2.000	Ea.	11.60	107	118.60
TOTAL			7,097.47	1,338	8,435.47

D2010 610	Group Wash Fountain Systems	MAT.	INST.	TOTAL
1740	Group wash fountain, precast terrazzo			
1760	Circular, 36" diameter	7,100	1,350	8,450
1800	54" diameter	10,600	1,475	12,075
1840	Semi-circular, 36" diameter	5,975	1,350	7,325
1880	54" diameter	8,175	1,475	9,650
1960	Stainless steel, circular, 36" diameter	6,050	1,250	7,300
2000	54" diameter	8,475	1,375	9,850
2040	Semi-circular, 36" diameter	6,000	1,250	7,250
2080	54" diameter	6,675	1,375	8,050
2160	Thermoplastic, circular, 36" diameter	4,975	1,025	6,000
2200	54" diameter	5,950	1,175	7,125
2240	Semi-circular, 36" diameter	4,975	1,025	6,000
2280	54" diameter	5,825	1,175	7,000

R224000-30

D20 Plumbing

D2010 Plumbing Fixtures

Systems are complete with trim and rough-in (supply, waste and vent) for connection to supply branches and waste mains.

Three Wall

Supply   Waste/Vent

Corner Angle

System Components	QUANTITY	UNIT	MAT.	INST.	TOTAL
SYSTEM D2010 710 1560					
SHOWER, STALL, BAKED ENAMEL, MOLDED STONE RECEPTOR, 30" SQUARE					
Shower stall, enameled steel, molded stone receptor, 30" square	1.000	Ea.	1,275	241	1,516
Copper tubing type DWV, solder joints, hangers 10' OC, 2" diam.	6.000	L.F.	82.20	77.10	159.30
Wrought copper DWV, Tee, sanitary, 2" diam.	1.000	Ea.	80.50	87	167.50
Trap, standard, copper, 2" diam.	1.000	Ea.	225	41	266
Copper tubing type L, solder joint, hanger 10' OC 1/2" diam.	16.000	L.F.	59.36	137.60	196.96
Wrought copper 90° elbow for solder joints 1/2" diam.	3.000	Ea.	10.17	105	115.17
Wrought copper Tee for solder joints, 1/2" diam.	2.000	Ea.	11.60	107	118.60
Stop and waste, straightway, bronze, solder joint 1/2" diam.	2.000	Ea.	27.10	58	85.10
TOTAL			1,770.93	853.70	2,624.63

D2010 710	Shower Systems	MAT.	INST.	TOTAL
1560	Shower, stall, baked enamel, molded stone receptor, 30" square	1,775	855	2,630
1600	32" square	1,775	865	2,640
1640	Terrazzo receptor, 32" square	1,975	865	2,840
1680	36" square	2,125	875	3,000
1720	36" corner angle	2,400	875	3,275
1800	Fiberglass one piece, three walls, 32" square	1,025	840	1,865
1840	36" square	1,050	840	1,890
1880	Polypropylene, molded stone receptor, 30" square	1,200	1,250	2,450
1920	32" square	1,200	1,250	2,450
1960	Built-in head, arm, bypass, stops and handles	117	320	437
2050	Shower, stainless steel panels, handicap			
2100	w/fixed and handheld head, control valves, grab bar, and seat	5,650	3,825	9,475
2500	Shower, group with six heads, thermostatic mix valves & balancing valve	5,450	935	6,385
2520	Five heads	4,525	850	5,375

R224000-30

D20 Plumbing

D2010 Plumbing Fixtures

Wall Mounted, No Back

Supply   Waste/Vent

Wall Mounted, Low Back

Systems are complete with trim and rough-in (supply, waste and vent) to connect to supply branches and waste mains.

System Components	QUANTITY	UNIT	MAT.	INST.	TOTAL
SYSTEM D2010 810 1800					
DRINKING FOUNTAIN, ONE BUBBLER, WALL MOUNTED					
NON RECESSED, BRONZE, NO BACK					
Drinking fountain, wall mount, bronze, 1 bubbler	1.000	Ea.	1,100	174	1,274
Copper tubing, type L, solder joint, hanger 10' OC 3/8" diam.	5.000	L.F.	17.25	41.50	58.75
Stop, supply, straight, chrome, 3/8" diam.	1.000	Ea.	13.55	29	42.55
Wrought copper 90° elbow for solder joints 3/8" diam.	1.000	Ea.	10.15	31.50	41.65
Wrought copper Tee for solder joints, 3/8" diam.	1.000	Ea.	17.15	49.50	66.65
Copper tubing, type DWV, solder joint, hanger 10' OC 1-1/4" diam.	4.000	L.F.	43.60	46.40	90
P trap, standard, copper drainage, 1-1/4" diam.	1.000	Ea.	232	38.50	270.50
Wrought copper, DWV, Tee, sanitary, 1-1/4" diam.	1.000	Ea.	65	77	142
TOTAL			1,498.70	487.40	1,986.10

D2010 810	Drinking Fountain Systems		MAT.	INST.	TOTAL
1740	Drinking fountain, one bubbler, wall mounted				
1760	Non recessed				
1800	Bronze, no back	R224000-30	1,500	485	1,985
1840	Cast iron, enameled, low back		1,400	485	1,885
1880	Fiberglass, 12" back		2,450	485	2,935
1920	Stainless steel, no back		1,550	485	2,035
1960	Semi-recessed, poly marble		1,450	485	1,935
2040	Stainless steel		1,650	485	2,135
2080	Vitreous china		1,375	485	1,860
2120	Full recessed, poly marble		2,150	485	2,635
2200	Stainless steel		2,075	485	2,560
2240	Floor mounted, pedestal type, aluminum		2,875	665	3,540
2320	Bronze		2,525	665	3,190
2360	Stainless steel		2,325	665	2,990

D20 Plumbing

D2010 Plumbing Fixtures

Systems are complete with trim and rough-in (supply, waste and vent) for connection to supply branches and waste mains.

Wall Hung | Supply | Waste/Vent | Floor Mounted

System Components	QUANTITY	UNIT	MAT.	INST.	TOTAL
SYSTEM D2010 820 1840					
WATER COOLER, ELECTRIC, SELF CONTAINED, WALL HUNG, 8.2 G.P.H.					
Water cooler, wall mounted, 8.2 GPH	1.000	Ea.	795	315	1,110
Copper tubing type DWV, solder joint, hanger 10' OC 1-1/4" diam.	4.000	L.F.	43.60	46.40	90
Wrought copper DWV, Tee, sanitary 1-1/4" diam.	1.000	Ea.	65	77	142
P trap, copper drainage, 1-1/4" diam.	1.000	Ea.	232	38.50	270.50
Copper tubing type L, solder joint, hanger 10' OC 3/8" diam.	5.000	L.F.	17.25	41.50	58.75
Wrought copper 90° elbow for solder joints 3/8" diam.	1.000	Ea.	10.15	31.50	41.65
Wrought copper Tee for solder joints, 3/8" diam.	1.000	Ea.	17.15	49.50	66.65
Stop and waste, straightway, bronze, solder, 3/8" diam.	1.000	Ea.	13.55	29	42.55
TOTAL			1,193.70	628.40	1,822.10

D2010 820	Water Cooler Systems		MAT.	INST.	TOTAL
1840	Water cooler, electric, wall hung, 8.2 G.P.H.	R224000 -30	1,200	630	1,830
1880	Dual height, 14.3 G.P.H.		1,475	645	2,120
1920	Wheelchair type, 7.5 G.P.H.		1,400	630	2,030
1960	Semi recessed, 8.1 G.P.H.		1,175	630	1,805
2000	Full recessed, 8 G.P.H.	R224000 -50	2,525	675	3,200
2040	Floor mounted, 14.3 G.P.H.		1,325	545	1,870
2080	Dual height, 14.3 G.P.H.		1,675	665	2,340
2120	Refrigerated compartment type, 1.5 G.P.H.		1,950	545	2,495

D20 Plumbing

D2010 Plumbing Fixtures

Two Fixture Bathroom Systems consisting of a lavatory, water closet, and rough-in service piping.
- Prices for plumbing and fixtures only.

*Common wall is with an adjacent bathroom.

System Components	QUANTITY	UNIT	MAT.	INST.	TOTAL
SYSTEM D2010 920 1180					
BATHROOM, LAVATORY & WATER CLOSET, 2 WALL PLUMBING, STAND ALONE					
Two piece, closed coupled	1.000	Ea.	256	236	492
Water closet, rough-in waste & vent	1.000	Set	365	410	775
Lavatory w/ftngs., wall hung, white, PE on CI, 20" x 18"	1.000	Ea.	305	156	461
Lavatory, rough-in waste & vent	1.000	Set	520	755	1,275
Copper tubing type L, solder joint, hanger 10' OC 1/2" diam.	10.000	L.F.	37.10	86	123.10
Pipe, steel, galvanized, schedule 40, threaded, 2" diam.	12.000	L.F.	160.20	234.60	394.80
Pipe, CI soil, no hub, coupling 10' OC, hanger 5' OC, 4" diam.	7.000	L.F.	129.85	161	290.85
TOTAL			1,773.15	2,038.60	3,811.75

D2010 920	Two Fixture Bathroom, Two Wall Plumbing	MAT.	INST.	TOTAL
1180	Bathroom, lavatory & water closet, 2 wall plumbing, stand alone	1,775	2,050	3,825
1200	Share common plumbing wall*	1,600	1,750	3,350

D2010 922	Two Fixture Bathroom, One Wall Plumbing		MAT.	INST.	TOTAL
2220	Bathroom, lavatory & water closet, one wall plumbing, stand alone	R224000-30	1,650	1,825	3,475
2240	Share common plumbing wall*		1,450	1,550	3,000
2260		R224000-40			
2280					

D20 Plumbing

D2010 Plumbing Fixtures

Three Fixture Bathroom Systems consisting of a lavatory, water closet, bathtub or shower and rough-in service piping.
- Prices for plumbing and fixtures only.

*Common wall is with an adjacent bathroom.

System Components	QUANTITY	UNIT	MAT.	INST.	TOTAL
SYSTEM D2010 924 1170					
BATHROOM, LAVATORY, WATER CLOSET & BATHTUB					
ONE WALL PLUMBING, STAND ALONE					
Wtr closet, rough-in, supply, waste and vent	1.000	Set	365	410	775
Wtr closet, 2 pc close cpld vit china flr mntd w/seat supply & stop	1.000	Ea.	256	236	492
Lavatory w/ftngs, wall hung, white, PE on CI, 20" x 18"	1.000	Ea.	305	156	461
Lavatory, rough-in waste & vent	1.000	Set	520	755	1,275
Bathtub, white PE on CI, w/ftgs, mat bottom, recessed, 5' long	1.000	Ea.	1,225	284	1,509
Baths, rough-in waste and vent	1.000	Set	481.50	544.50	1,026
TOTAL			3,152.50	2,385.50	5,538

D2010 924	Three Fixture Bathroom, One Wall Plumbing		MAT.	INST.	TOTAL
1150	Bathroom, three fixture, one wall plumbing	R224000 -30			
1160	Lavatory, water closet & bathtub				
1170	Stand alone	R224000 -40	3,150	2,375	5,525
1180	Share common plumbing wall *		2,700	1,725	4,425

D2010 926	Three Fixture Bathroom, Two Wall Plumbing	MAT.	INST.	TOTAL
2130	Bathroom, three fixture, two wall plumbing			
2140	Lavatory, water closet & bathtub			
2160	Stand alone	3,175	2,425	5,600
2180	Long plumbing wall common *	2,850	1,925	4,775
3610	Lavatory, bathtub & water closet			
3620	Stand alone	3,425	2,750	6,175
3640	Long plumbing wall common *	3,225	2,475	5,700
4660	Water closet, corner bathtub & lavatory			
4680	Stand alone	4,875	2,450	7,325
4700	Long plumbing wall common *	4,350	1,850	6,200
6100	Water closet, stall shower & lavatory			
6120	Stand alone	3,575	2,750	6,325
6140	Long plumbing wall common *	3,400	2,525	5,925
7060	Lavatory, corner stall shower & water closet			
7080	Stand alone	3,925	2,425	6,350
7100	Short plumbing wall common *	3,225	1,625	4,850

D20 Plumbing

D2010 Plumbing Fixtures

Four Fixture Bathroom Systems consisting of a lavatory, water closet, bathtub, shower and rough-in service piping.

- Prices for plumbing and fixtures only.

*Common wall is with an adjacent bathroom.

System Components	QUANTITY	UNIT	COST EACH MAT.	COST EACH INST.	COST EACH TOTAL
SYSTEM D2010 928 1160					
BATHROOM, BATHTUB, WATER CLOSET, STALL SHOWER & LAVATORY					
TWO WALL PLUMBING, STAND ALONE					
Wtr closet, 2 pc close cpld vit china flr mntd w/seat supply & stop	1.000	Ea.	256	236	492
Water closet, rough-in waste & vent	1.000	Set	365	410	775
Lavatory w/ftngs, wall hung, white PE on CI, 20" x 18"	1.000	Ea.	305	156	461
Lavatory, rough-in waste & vent	1.000	Set	52	75.50	127.50
Bathtub, white PE on CI, w/ftgs, mat bottom, recessed, 5' long	1.000	Ea.	1,225	284	1,509
Baths, rough-in waste and vent	1.000	Set	535	605	1,140
Shower stall, bkd enam, molded stone receptor, door & trim 32" sq.	1.000	Ea.	1,275	250	1,525
Shower stall, rough-in supply, waste & vent	1.000	Set	590	610	1,200
TOTAL			4,603	2,626.50	7,229.50

D2010 928		Four Fixture Bathroom, Two Wall Plumbing		MAT.	INST.	TOTAL
1140	Bathroom, four fixture, two wall plumbing		R224000 -30			
1150		Bathtub, water closet, stall shower & lavatory				
1160		Stand alone	R224000 -40	4,600	2,625	7,225
1180		Long plumbing wall common *		4,075	2,025	6,100
2260		Bathtub, lavatory, corner stall shower & water closet				
2280		Stand alone		5,225	2,650	7,875
2320		Long plumbing wall common *		4,700	2,025	6,725
3620		Bathtub, stall shower, lavatory & water closet				
3640		Stand alone		5,075	3,300	8,375
3660		Long plumbing wall (opp. door) common *		4,525	2,700	7,225

D2010 930		Four Fixture Bathroom, Three Wall Plumbing	MAT.	INST.	TOTAL
4680	Bathroom, four fixture, three wall plumbing				
4700		Bathtub, stall shower, lavatory & water closet			
4720		Stand alone	5,900	3,650	9,550
4760		Long plumbing wall (opposite door) common *	5,725	3,350	9,075

D20 Plumbing

D2010 Plumbing Fixtures

Five Fixture Bathroom Systems consisting of two lavatories, a water closet, bathtub, shower and rough-in service piping.

- Prices for plumbing and fixtures only.

*Common wall is with an adjacent bathroom.

System Components	QUANTITY	UNIT	MAT.	INST.	TOTAL
SYSTEM D2010 932 1360					
BATHROOM, BATHTUB, WATER CLOSET, STALL SHOWER & TWO LAVATORIES					
TWO WALL PLUMBING, STAND ALONE					
Wtr closet, 2 pc close cpld vit china flr mntd incl seat, supply & stop	1.000	Ea.	256	236	492
Water closet, rough-in waste & vent	1.000	Set	365	410	775
Lavatory w/ftngs, wall hung, white PE on CI, 20" x 18"	2.000	Ea.	610	312	922
Lavatory, rough-in waste & vent	2.000	Set	1,040	1,510	2,550
Bathtub, white PE on CI, w/ftgs, mat bottom, recessed, 5' long	1.000	Ea.	1,225	284	1,509
Baths, rough-in waste and vent	1.000	Set	535	605	1,140
Shower stall, bkd enam molded stone receptor, door & ftng, 32" sq.	1.000	Ea.	1,275	250	1,525
Shower stall, rough-in supply, waste & vent	1.000	Set	590	610	1,200
TOTAL			5,896	4,217	10,113

D2010 932	Five Fixture Bathroom, Two Wall Plumbing	MAT.	INST.	TOTAL
1320	Bathroom, five fixture, two wall plumbing			
1340	Bathtub, water closet, stall shower & two lavatories R224000 -30			
1360	Stand alone	5,900	4,225	10,125
1400	One short plumbing wall common *	5,350	3,600	8,950
1500	Bathtub, two lavatories, corner stall shower & water closet			
1520	Stand alone	6,525	4,225	10,750
1540	Long plumbing wall common*	5,875	3,275	9,150

D2010 934	Five Fixture Bathroom, Three Wall Plumbing	MAT.	INST.	TOTAL
2360	Bathroom, five fixture, three wall plumbing			
2380	Water closet, bathtub, two lavatories & stall shower			
2400	Stand alone	6,525	4,225	10,750
2440	One short plumbing wall common *	5,975	3,625	9,600

D2010 936	Five Fixture Bathroom, One Wall Plumbing	MAT.	INST.	TOTAL
4080	Bathroom, five fixture, one wall plumbing			
4100	Bathtub, two lavatories, corner stall shower & water closet			
4120	Stand alone	6,200	3,775	9,975
4160	Share common wall *	5,225	2,450	7,675

D20 Plumbing

D2010 Plumbing Fixtures

Example of Plumbing Cost Calculations: The bathroom system includes the individual fixtures such as bathtub, lavatory, shower and water closet. These fixtures are listed below as separate items merely as a checklist.

D2010 951 — Plumbing Systems 20 Unit, 2 Story Apartment Building

	FIXTURE	SYSTEM	LINE	QUANTITY	UNIT	COST EACH MAT.	COST EACH INST.	TOTAL
0440	Bathroom	D2010 926	3640	20	Ea.	64,500	49,700	114,200
0480	Bathtub							
0520	Booster pump[1]	not req'd.						
0560	Drinking fountain							
0600	Garbage disposal[1]	not incl.						
0660								
0680	Grease interceptor							
0720	Water heater	D2020 250	2140	1	Ea.	14,100	3,400	17,500
0760	Kitchen sink	D2010 410	1960	20	Ea.	31,700	17,500	49,200
0800	Laundry sink	D2010 420	1840	4	Ea.	5,800	3,450	9,250
0840	Lavatory							
0900								
0920	Roof drain, 1 floor	D2040 210	4200	2	Ea.	1,850	1,800	3,650
0960	Roof drain, add'l floor	D2040 210	4240	20	L.F.	370	460	830
1000	Service sink	D2010 440	4300	1	Ea.	2,900	1,225	4,125
1040	Sewage ejector[1]	not req'd.						
1080	Shower							
1100								
1160	Sump pump							
1200	Urinal							
1240	Water closet							
1320								
1360	SUB TOTAL					121,500	77,500	199,000
1481	Water controls	R221113-40		10%[2]		12,100	7,750	19,850
1521	Pipe & fittings[3]	R221113-40		30%[2]		36,400	23,300	59,700
1560	Other							
1601	Quality/complexity	R221113-40		15%[2]		18,200	11,600	29,800
1680								
1720	TOTAL					188,500	120,500	309,000
1741								

[1]**Note:** Cost for items such as booster pumps, backflow preventers, sewage ejectors, water meters, etc., may be obtained from the unit price section in the front of this book.

Water controls, pipe and fittings, and the Quality/Complexity factors come from Table R221113-40.

[2]Percentage of subtotal.

[3]Long, easily discernable runs of pipe would be more accurately priced from Unit Price Section 22 11 13. If this is done, reduce the miscellaneous percentage in proportion.

D20 Plumbing

D2020 Domestic Water Distribution

Installation includes piping and fittings within 10' of heater. Electric water heaters do not require venting.

1 Kilowatt hour will raise:			
Gallons of Water	Degrees F	Gallons of Water	Degrees F
4.1	100°	6.8	60°
4.5	90°	8.2	50°
5.1	80°	10.0	40°
5.9	70°		

System Components

System Components	QUANTITY	UNIT	MAT.	INST.	TOTAL
SYSTEM D2020 210 1780					
ELECTRIC WATER HEATER, RESIDENTIAL, 100°F RISE					
10 GALLON TANK, 7 GPH					
Water heater, residential electric, glass lined tank, 10 gal.	1.000	Ea.	360	300	660
Copper tubing, type L, solder joint, hanger 10' OC 1/2" diam.	30.000	L.F.	111.30	258	369.30
Wrought copper 90° elbow for solder joints 1/2" diam.	4.000	Ea.	13.56	140	153.56
Wrought copper Tee for solder joints, 1/2" diam.	2.000	Ea.	11.60	107	118.60
Union, wrought copper, 1/2" diam.	2.000	Ea.	83	73	156
Valve, gate, bronze, 125 lb, NRS, soldered 1/2" diam.	2.000	Ea.	103	58	161
Relief valve, bronze, press & temp, self-close, 3/4" IPS	1.000	Ea.	278	25	303
Wrought copper adapter, CTS to MPT 3/4" IPS	1.000	Ea.	11.90	41	52.90
Copper tubing, type L, solder joints, 3/4" diam.	1.000	L.F.	5.65	9.15	14.80
Wrought copper 90° elbow for solder joints 3/4" diam.	1.000	Ea.	7.60	36.50	44.10
TOTAL			985.61	1,047.65	2,033.26

D2020 210	Electric Water Heaters - Residential Systems		MAT.	INST.	TOTAL
1760	Electric water heater, residential, 100°F rise				
1780	10 gallon tank, 7 GPH	R224000-10	985	1,050	2,035
1820	20 gallon tank, 7 GPH	R224000-20	1,200	1,125	2,325
1860	30 gallon tank, 7 GPH		1,525	1,175	2,700
1900	40 gallon tank, 8 GPH	R224000-40	2,175	1,300	3,475
1940	52 gallon tank, 10 GPH		2,300	1,300	3,600
1980	66 gallon tank, 13 GPH		3,625	1,475	5,100
2020	80 gallon tank, 16 GPH		3,800	1,550	5,350
2060	120 gallon tank, 23 GPH		5,200	1,825	7,025

D20 Plumbing

D2020 Domestic Water Distribution

Installation includes piping and fittings within 10' of heater. Gas heaters require vent piping (not included with these units).

System Components	QUANTITY	UNIT	COST EACH MAT.	COST EACH INST.	TOTAL
SYSTEM D2020 220 2260					
GAS FIRED WATER HEATER, RESIDENTIAL, 100°F RISE					
30 GALLON TANK, 32 GPH					
Water heater, residential, gas, glass lined tank, 30 gallon	1.000	Ea.	985	350	1,335
Copper tubing, type L, solder joint, hanger 10' OC, 3/4" diam	33.000	L.F.	186.45	301.95	488.40
Wrought copper 90° elbow for solder joints, 3/4" diam	5.000	Ea.	38	182.50	220.50
Wrought copper Tee for solder joints, 3/4" diam	2.000	Ea.	27.90	116	143.90
Wrought copper union for soldered joints, 3/4" diam.	2.000	Ea.	105	77	182
Valve bronze, 125 lb., NRS, soldered 3/4" diam	2.000	Ea.	116	70	186
Relief valve, press & temp, bronze, self-close, 3/4" diam	1.000	Ea.	278	25	303
Wrought copper, adapter, CTS to MPT 3/4" IPS	1.000	Ea.	11.90	41	52.90
Pipe steel black, schedule 40, threaded, 1/2" diam	10.000	L.F.	30	110.50	140.50
Pipe, 90° elbow, malleable iron black, 150 lb., threaded, 1/2" diam	2.000	Ea.	6.74	93	99.74
Pipe, union with brass seat, malleable iron black, 1/2" diam	1.000	Ea.	14.90	49.50	64.40
Valve, gas stop w/o check, brass, 1/2" IPS	1.000	Ea.	14.50	29	43.50
TOTAL			1,814.39	1,445.45	3,259.84

D2020 220	Gas Fired Water Heaters - Residential Systems		MAT.	INST.	TOTAL
2200	Gas fired water heater, residential, 100°F rise				
2260	30 gallon tank, 32 GPH	R224000-10	1,825	1,450	3,275
2300	40 gallon tank, 32 GPH		2,150	1,625	3,775
2340	50 gallon tank, 63 GPH	R224000-20	2,200	1,625	3,825
2380	75 gallon tank, 63 GPH		3,075	1,800	4,875
2420	100 gallon tank, 63 GPH	R224000-40	3,250	1,900	5,150

D20 Plumbing

D2020 Domestic Water Distribution

Installation includes piping and fittings within 10' of heater. Oil fired heaters require vent piping (not included in these prices).

System Components	QUANTITY	UNIT	MAT.	INST.	TOTAL
SYSTEM D2020 230 2220					
OIL FIRED WATER HEATER, RESIDENTIAL, 100°F RISE					
30 GALLON TANK, 103 GPH					
Water heater, residential, oil glass lined tank, 30 Gal	1.000	Ea.	1,275	350	1,625
Copper tubing, type L, solder joint, hanger 10' O.C. 3/4" diam.	33.000	L.F.	186.45	301.95	488.40
Wrought copper 90° elbow for solder joints 3/4" diam.	5.000	Ea.	38	182.50	220.50
Wrought copper Tee for solder joints, 3/4" diam.	2.000	Ea.	27.90	116	143.90
Wrought copper union for soldered joints, 3/4" diam.	2.000	Ea.	105	77	182
Valve, gate, bronze, 125 lb, NRS, soldered 3/4" diam.	2.000	Ea.	116	70	186
Relief valve, bronze, press & temp, self-close, 3/4" IPS	1.000	Ea.	278	25	303
Wrought copper adapter, CTS to MPT, 3/4" IPS	1.000	Ea.	11.90	41	52.90
Copper tubing, type L, solder joint, hanger 10' OC 3/8" diam.	10.000	L.F.	34.50	83	117.50
Wrought copper 90° elbow for solder joints 3/8" diam.	2.000	Ea.	20.30	63	83.30
Valve, globe, fusible, 3/8" diam.	1.000	Ea.	14.35	29.50	43.85
TOTAL			2,107.40	1,338.95	3,446.35

D2020 230	Oil Fired Water Heaters - Residential Systems	MAT.	INST.	TOTAL
2200	Oil fired water heater, residential, 100°F rise			
2220	30 gallon tank, 103 GPH	2,100	1,350	3,450
2260	50 gallon tank, 145 GPH	2,675	1,500	4,175
2300	70 gallon tank, 164 GPH R224000-20	3,750	1,700	5,450
2340	85 gallon tank, 181 GPH	11,200	1,725	12,925

D20 Plumbing

D2020 Domestic Water Distribution

Systems below include piping and fittings within 10' of heater. Electric water heaters do not require venting.

System Components	QUANTITY	UNIT	COST EACH MAT.	COST EACH INST.	TOTAL
SYSTEM D2020 240 1820					
ELECTRIC WATER HEATER, COMMERCIAL, 100°F RISE					
50 GALLON TANK, 9 KW, 37 GPH					
Water heater, commercial, electric, 50 Gal, 9 KW, 37 GPH	1.000	Ea.	4,600	385	4,985
Copper tubing, type L, solder joint, hanger 10' OC, 3/4" diam	34.000	L.F.	192.10	311.10	503.20
Wrought copper 90° elbow for solder joints 3/4" diam	5.000	Ea.	38	182.50	220.50
Wrought copper Tee for solder joints, 3/4" diam	2.000	Ea.	27.90	116	143.90
Wrought copper union for soldered joints, 3/4" diam.	2.000	Ea.	105	77	182
Valve, gate, bronze, 125 lb, NRS, soldered 3/4" diam	2.000	Ea.	116	70	186
Relief valve, bronze, press & temp, self-close, 3/4" IPS	1.000	Ea.	278	25	303
Wrought copper adapter, copper tubing to male, 3/4" IPS	1.000	Ea.	11.90	41	52.90
TOTAL			5,368.90	1,207.60	6,576.50

D2020 240	Electric Water Heaters - Commercial Systems		MAT.	INST.	TOTAL
1800	Electric water heater, commercial, 100°F rise				
1820	50 gallon tank, 9 KW 37 GPH		5,375	1,200	6,575
1860	80 gal, 12 KW 49 GPH		7,150	1,500	8,650
1900	36 KW 147 GPH	R224000 -10	9,400	1,625	11,025
1940	120 gal, 36 KW 147 GPH		10,300	1,750	12,050
1980	150 gal, 120 KW 490 GPH	R224000 -20	29,900	1,875	31,775
2020	200 gal, 120 KW 490 GPH		31,400	1,925	33,325
2060	250 gal, 150 KW 615 GPH	R224000 -40	38,700	2,250	40,950
2100	300 gal, 180 KW 738 GPH		44,300	2,350	46,650
2140	350 gal, 30 KW 123 GPH		29,000	2,550	31,550
2180	180 KW 738 GPH		39,700	2,550	42,250
2220	500 gal, 30 KW 123 GPH		37,400	3,000	40,400
2260	240 KW 984 GPH		53,500	3,000	56,500
2300	700 gal, 30 KW 123 GPH		31,100	3,425	34,525
2340	300 KW 1230 GPH		45,000	3,425	48,425
2380	1000 gal, 60 KW 245 GPH		39,200	4,750	43,950
2420	480 KW 1970 GPH		60,500	4,750	65,250
2460	1500 gal, 60 KW 245 GPH		79,500	5,875	85,375
2500	480 KW 1970 GPH		108,000	5,875	113,875

D20 Plumbing

D2020 Domestic Water Distribution

Units may be installed in multiples for increased capacity.

Included below is the heater with self-energizing gas controls, safety pilots, insulated jacket, hi-limit aquastat and pressure relief valve.

Installation includes piping and fittings within 10' of heater. Gas heaters require vent piping (not included in these prices).

System Components	QUANTITY	UNIT	MAT.	INST.	TOTAL
SYSTEM D2020 250 1780					
GAS FIRED WATER HEATER, COMMERCIAL, 100°F RISE					
75.5 MBH INPUT, 63 GPH					
Water heater, commercial, gas, 75.5 MBH, 63 GPH	1.000	Ea.	3,850	495	4,345
Copper tubing, type L, solder joint, hanger 10' OC, 1-1/4" diam	30.000	L.F.	370.50	360	730.50
Wrought copper 90° elbow for solder joints 1-1/4" diam	4.000	Ea.	112	186	298
Wrought copper tee for solder joints, 1-1/4" diam	2.000	Ea.	118	154	272
Wrought copper union for soldered joints, 1-1/4" diam	2.000	Ea.	312	99	411
Valve, gate, bronze, 125 lb, NRS, soldered 1-1/4" diam	2.000	Ea.	252	93	345
Relief valve, bronze, press & temp, self-close, 3/4" IPS	1.000	Ea.	278	25	303
Copper tubing, type L, solder joints, 3/4" diam	8.000	L.F.	45.20	73.20	118.40
Wrought copper 90° elbow for solder joints 3/4" diam	1.000	Ea.	7.60	36.50	44.10
Wrought copper, adapter, CTS to MPT, 3/4" IPS	1.000	Ea.	11.90	41	52.90
Pipe steel black, schedule 40, threaded, 3/4" diam	10.000	L.F.	34.90	114	148.90
Pipe, 90° elbow, malleable iron black, 150 lb threaded, 3/4" diam	2.000	Ea.	8.14	99	107.14
Pipe, union with brass seat, malleable iron black, 3/4" diam	1.000	Ea.	17.10	53.50	70.60
Valve, gas stop w/o check, brass, 3/4" IPS	1.000	Ea.	17.05	31.50	48.55
TOTAL			5,434.39	1,860.70	7,295.09

D2020 250	Gas Fired Water Heaters - Commercial Systems		MAT.	INST.	TOTAL
1760	Gas fired water heater, commercial, 100°F rise				
1780	75.5 MBH input, 63 GPH		5,425	1,850	7,275
1860	100 MBH input, 91 GPH		7,600	1,950	9,550
1980	155 MBH input, 150 GPH	R224000 -10	10,600	2,250	12,850
2060	200 MBH input, 192 GPH		12,000	2,725	14,725
2140	300 MBH input, 278 GPH	R224000 -20	14,100	3,400	17,500
2180	390 MBH input, 374 GPH		16,700	3,425	20,125
2220	500 MBH input, 480 GPH	R224000 -40	21,900	3,700	25,600
2260	600 MBH input, 576 GPH		25,000	3,975	28,975

D20 Plumbing

D2020 Domestic Water Distribution

Units may be installed in multiples for increased capacity.

Included below is the heater, wired-in flame retention burners, cadmium cell primary controls, hi-limit controls, ASME pressure relief valves, draft controls, and insulated jacket.

Oil fired water heater systems include piping and fittings within 10' of heater. Oil fired heaters require vent piping (not included in these systems).

System Components	QUANTITY	UNIT	COST EACH MAT.	COST EACH INST.	TOTAL
SYSTEM D2020 260 1820 OIL FIRED WATER HEATER, COMMERCIAL, 100°F RISE 140 GAL., 140 MBH INPUT, 134 GPH					
Water heater, commercial, oil, 140 gal., 140 MBH input, 134 GPH	1.000	Ea.	21,900	590	22,490
Copper tubing, type L, solder joint, hanger 10' OC, 3/4" diam.	34.000	L.F.	192.10	311.10	503.20
Wrought copper 90° elbow for solder joints 3/4" diam.	5.000	Ea.	38	182.50	220.50
Wrought copper Tee for solder joints, 3/4" diam.	2.000	Ea.	27.90	116	143.90
Wrought copper union for soldered joints, 3/4" diam.	2.000	Ea.	105	77	182
Valve, bronze, 125 lb, NRS, soldered 3/4" diam.	2.000	Ea.	116	70	186
Relief valve, bronze, press & temp, self-close, 3/4" IPS	1.000	Ea.	278	25	303
Wrought copper adapter, copper tubing to male, 3/4" IPS	1.000	Ea.	11.90	41	52.90
Copper tubing, type L, solder joint, hanger 10' OC, 3/8" diam.	10.000	L.F.	34.50	83	117.50
Wrought copper 90° elbow for solder joints 3/8" diam.	2.000	Ea.	20.30	63	83.30
Valve, globe, fusible, 3/8" IPS	1.000	Ea.	14.35	29.50	43.85
TOTAL			22,738.05	1,588.10	24,326.15

D2020 260	Oil Fired Water Heaters - Commercial Systems		MAT.	INST.	TOTAL
1800	Oil fired water heater, commercial, 100°F rise				
1820	140 gal., 140 MBH input, 134 GPH		22,700	1,600	24,300
1900	140 gal., 255 MBH input, 247 GPH		24,900	1,975	26,875
1940	140 gal., 270 MBH input, 259 GPH	R224000-10	30,400	2,275	32,675
1980	140 gal., 400 MBH input, 384 GPH		31,900	2,625	34,525
2060	140 gal., 720 MBH input, 691 GPH	R224000-20	33,600	2,725	36,325
2100	221 gal., 300 MBH input, 288 GPH		44,700	3,000	47,700
2140	221 gal., 600 MBH input, 576 GPH	R224000-40	49,500	3,025	52,525
2180	221 gal., 800 MBH input, 768 GPH		50,500	3,150	53,650
2220	201 gal., 1000 MBH input, 960 GPH		51,500	3,175	54,675
2260	201 gal., 1250 MBH input, 1200 GPH		52,500	3,250	55,750
2300	201 gal., 1500 MBH input, 1441 GPH		56,500	3,325	59,825
2340	411 gal., 600 MBH input, 576 GPH		57,000	3,425	60,425
2380	411 gal., 800 MBH input, 768 GPH		57,500	3,500	61,000
2420	411 gal., 1000 MBH input, 960 GPH		63,000	4,025	67,025
2460	411 gal., 1250 MBH input, 1200 GPH		64,500	4,125	68,625
2500	397 gal., 1500 MBH input, 1441 GPH		65,000	4,275	69,275

D20 Plumbing

D2020 Domestic Water Distribution

In this closed-loop indirect collection system, fluid with a low freezing temperature, such as propylene glycol, transports heat from the collectors to water storage. The transfer fluid is contained in a closed-loop consisting of collectors, supply and return piping, and a remote heat exchanger. The heat exchanger transfers heat energy from the fluid in the collector loop to potable water circulated in a storage loop. A typical two-or-three panel system contains 5 to 6 gallons of heat transfer fluid.

When the collectors become approximately 20°F warmer than the storage temperature, a controller activates the circulator on the collector and storage loops. The circulators will move the fluid and potable water through the heat exchanger until heat collection no longer occurs. At that point, the system shuts down. Since the heat transfer medium is a fluid with a very low freezing temperature, there is no need for it to be drained from the system between periods of collection.

D20 Plumbing

D2020 Domestic Water Distribution

System Components		QUANTITY	UNIT	COST EACH		
				MAT.	INST.	TOTAL
SYSTEM D2020 265 2760						
SOLAR, CLOSED LOOP, ADD-ON HOT WATER SYS., EXTERNAL HEAT EXCHANGER						
3/4" TUBING, TWO 3'X7' BLACK CHROME COLLECTORS						
A,B,G,L,K,M	Heat exchanger fluid-fluid pkg incl 2 circulators, expansion tank, Check valve, relief valve, controller, hi temp cutoff, & 2 sensors	1.000	Ea.	845	500	1,345
C	Thermometer, 2" dial	3.000	Ea.	72	130.50	202.50
D, T	Fill & drain valve, brass, 3/4" connection	1.000	Ea.	10.70	29	39.70
E	Air vent, manual, 1/8" fitting	2.000	Ea.	56	43	99
F	Air purger	1.000	Ea.	51.50	58	109.50
H	Strainer, Y type, bronze body, 3/4" IPS	1.000	Ea.	37.50	37	74.50
I	Valve, gate, bronze, NRS, soldered 3/4" diam	6.000	Ea.	348	210	558
J	Neoprene vent flashing	2.000	Ea.	28.20	70	98.20
N-1, N	Relief valve temp & press, 150 psi 210°F self-closing 3/4" IPS	1.000	Ea.	23	23	46
O	Pipe covering, urethane, ultraviolet cover, 1" wall 3/4" diam	20.000	L.F.	60.20	125	185.20
P	Pipe covering, fiberglass, all service jacket, 1" wall, 3/4" diam	50.000	L.F.	53.50	249	302.50
Q	Collector panel solar energy blk chrome on copper, 1/8" temp glass 3'x7'	2.000	Ea.	2,200	264	2,464
	Roof clamps for solar energy collector panels	2.000	Set	6.16	35.80	41.96
R	Valve, swing check, bronze, regrinding disc, 3/4" diam	2.000	Ea.	208	70	278
S	Pressure gauge, 60 psi, 2" dial	1.000	Ea.	26.50	21.50	48
U	Valve, water tempering, bronze, sweat connections, 3/4" diam	1.000	Ea.	139	35	174
W-2, V	Tank water storage w/heating element, drain, relief valve, existing	1.000	Ea.			
	Copper tubing type L, solder joint, hanger 10' OC 3/4" diam	20.000	L.F.	113	183	296
	Copper tubing, type M, solder joint, hanger 10' OC 3/4" diam	70.000	L.F.	306.60	623	929.60
	Sensor wire, #22-2 conductor multistranded	.500	C.L.F.	7.83	32	39.83
	Solar energy heat transfer fluid, propylene glycol anti-freeze	6.000	Gal.	102	150	252
	Wrought copper fittings & solder, 3/4" diam	76.000	Ea.	577.60	2,774	3,351.60
	TOTAL			5,272.29	5,662.80	10,935.09

D2020 265	Solar, Closed Loop, Add-On Hot Water Systems		COST EACH		
			MAT.	INST.	TOTAL
2550	Solar, closed loop, add-on hot water system, external heat exchanger				
2570	3/8" tubing, 3 ea. 4' x 4'-4" vacuum tube collectors		5,975	5,300	11,275
2580	1/2" tubing, 4 ea. 4 x 4'-4" vacuum tube collectors		6,425	5,750	12,175
2600	2 ea. 3'x7' black chrome collectors	R235616 -60	4,675	5,425	10,100
2620	3 ea. 3'x7' black chrome collectors		5,775	5,575	11,350
2640	2 ea. 3'x7' flat black collectors		4,075	5,450	9,525
2660	3 ea. 3'x7' flat black collectors		4,875	5,600	10,475
2700	3/4" tubing, 3 ea. 3'x7' black chrome collectors		6,375	5,825	12,200
2720	3 ea. 3'x7' flat black absorber plate collectors		5,475	5,825	11,300
2740	2 ea. 4'x9' flat black w/plastic glazing collectors		5,075	5,850	10,925
2760	2 ea. 3'x7' black chrome collectors		5,275	5,675	10,950
2780	1" tubing, 4 ea 2'x9' plastic absorber & glazing collectors		6,900	6,600	13,500
2800	4 ea. 3'x7' black chrome absorber collectors		8,825	6,625	15,450
2820	4 ea. 3'x7' flat black absorber collectors		7,625	6,650	14,275

D20 Plumbing

D2020 Domestic Water Distribution

In the drainback indirect-collection system, the heat transfer fluid is distilled water contained in a loop consisting of collectors, supply and return piping, and an unpressurized holding tank. A large heat exchanger containing incoming potable water is immersed in the holding tank. When a controller activates solar collection, the distilled water is pumped through the collectors and heated and pumped back down to the holding tank. When the temperature differential between the water in the collectors and water in storage is such that collection no longer occurs, the pump turns off and gravity causes the distilled water in the collector loop to drain back to the holding tank. All the loop piping is pitched so that the water can drain out of the collectors and piping and not freeze there. As hot water is needed in the home, incoming water first flows through the holding tank with the immersed heat exchanger and is warmed and then flows through a conventional heater for any supplemental heating that is necessary.

D20 Plumbing

D2020 Domestic Water Distribution

System Components		QUANTITY	UNIT	COST EACH MAT.	COST EACH INST.	TOTAL
SYSTEM D2020 270 2760						
SOLAR, DRAINBACK, ADD ON, HOT WATER, IMMERSED HEAT EXCHANGER						
3/4" TUBING, THREE EA 3'X7' BLACK CHROME COLLECTOR						
A, B	Differential controller 2 sensors, thermostat, solar energy system	1.000	Ea.	625	58	683
C	Thermometer 2" dial	3.000	Ea.	72	130.50	202.50
D, T	Fill & drain valve, brass, 3/4" connection	1.000	Ea.	10.70	29	39.70
E-1	Automatic air vent 1/8" fitting	1.000	Ea.	18.95	21.50	40.45
H	Strainer, Y type, bronze body, 3/4" IPS	1.000	Ea.	37.50	37	74.50
I	Valve, gate, bronze, NRS, soldered 3/4" diam	2.000	Ea.	116	70	186
J	Neoprene vent flashing	2.000	Ea.	28.20	70	98.20
L	Circulator, solar heated liquid, 1/20 HP	1.000	Ea.	177	104	281
N	Relief valve temp. & press. 150 psi 210°F self-closing 3/4" IPS	1.000	Ea.	23	23	46
O	Pipe covering, urethane, ultraviolet cover, 1" wall, 3/4" diam	20.000	L.F.	60.20	125	185.20
P	Pipe covering, fiberglass, all service jacket, 1" wall, 3/4" diam	50.000	L.F.	53.50	249	302.50
Q	Collector panel solar energy blk chrome on copper, 1/8" temp glas 3'x7'	3.000	Ea.	3,300	396	3,696
	Roof clamps for solar energy collector panels	3.000	Set	9.24	53.70	62.94
R	Valve, swing check, bronze, regrinding disc, 3/4" diam	1.000	Ea.	104	35	139
U	Valve, water tempering, bronze sweat connections, 3/4" diam	1.000	Ea.	139	35	174
V	Tank, water storage w/heating element, drain, relief valve, existing	1.000	Ea.			
W	Tank, water storage immersed heat exchr elec 2"x1/2# insul 120 gal	1.000	Ea.	2,025	495	2,520
X	Valve, globe, bronze, rising stem, 3/4" diam, soldered	3.000	Ea.	444	105	549
Y	Flow control valve	1.000	Ea.	132	31.50	163.50
Z	Valve, ball, bronze, solder 3/4" diam, solar loop flow control	1.000	Ea.	24.50	35	59.50
	Copper tubing, type L, solder joint, hanger 10' OC 3/4" diam	20.000	L.F.	113	183	296
	Copper tubing, type M, solder joint, hanger 10' OC 3/4" diam	70.000	L.F.	306.60	623	929.60
	Sensor wire, #22-2 conductor, multistranded	.500	C.L.F.	7.83	32	39.83
	Wrought copper fittings & solder, 3/4" diam	76.000	Ea.	577.60	2,774	3,351.60
	TOTAL			8,404.82	5,715.20	14,120.02

D2020 270	Solar, Drainback, Hot Water Systems		COST EACH MAT.	COST EACH INST.	TOTAL
2550	Solar, drainback, hot water, immersed heat exchanger				
2560	3/8" tubing, 3 ea. 4' x 4'-4" vacuum tube collectors		7,675	5,150	12,825
2580	1/2" tubing, 4 ea 4' x 4'-4" vacuum tube collectors, 80 gal tank	R235616	8,125	5,600	13,725
2600	120 gal tank	-60	8,375	5,675	14,050
2640	2 ea. 3'x7' blk chrome collectors, 80 gal tank		6,375	5,275	11,650
2660	3 ea. 3'x7' blk chrome collectors, 120 gal tank		7,750	5,500	13,250
2700	2 ea. 3'x7' flat blk collectors, 120 gal tank		6,025	5,350	11,375
2720	3 ea. 3'x7' flat blk collectors, 120 gal tank		6,850	5,500	12,350
2760	3/4" tubing, 3 ea 3'x7' black chrome collectors, 120 gal tank		8,400	5,725	14,125
2780	3 ea. 3'x7' flat black absorber collectors, 120 gal tank		7,500	5,725	13,225
2800	2 ea. 4'x9' flat blk w/plastic glazing collectors 120 gal tank		7,100	5,750	12,850
2840	1" tubing, 4 ea. 2'x9' plastic absorber & glazing collectors, 120 gal tank		9,050	6,500	15,550
2860	4 ea. 3'x7' black chrome absorber collectors, 120 gal tank		11,000	6,525	17,525
2880	4 ea. 3'x7' flat black absorber collectors, 120 gal tank		9,775	6,550	16,325

D20 Plumbing

D2020 Domestic Water Distribution

In the draindown direct-collection system, incoming domestic water is heated in the collectors. When the controller activates solar collection, domestic water is first heated as it flows through the collectors and is then pumped to storage. When conditions are no longer suitable for heat collection, the pump shuts off and the water in the loop drains down and out of the system by means of solenoid valves and properly pitched piping.

D20 Plumbing

D2020 Domestic Water Distribution

System Components

		QUANTITY	UNIT	COST EACH MAT.	COST EACH INST.	TOTAL
	SYSTEM D2020 275 2760					
	SOLAR, DRAINDOWN, HOT WATER, DIRECT COLLECTION					
	3/4" TUBING, THREE 3'X7' BLACK CHROME COLLECTORS					
A, B	Differential controller, 2 sensors, thermostat, solar energy system	1.000	Ea.	625	58	683
A-1	Solenoid valve, solar heating loop, brass, 3/4" diam, 24 volts	3.000	Ea.	390	231	621
B-1	Solar energy sensor, freeze prevention	1.000	Ea.	26	21.50	47.50
C	Thermometer, 2" dial	3.000	Ea.	72	130.50	202.50
E-1	Vacuum relief valve, 3/4" diam	1.000	Ea.	32.50	21.50	54
F-1	Air vent, automatic, 1/8" fitting	1.000	Ea.	18.95	21.50	40.45
H	Strainer, Y type, bronze body, 3/4" IPS	1.000	Ea.	37.50	37	74.50
I	Valve, gate, bronze, NRS, soldered, 3/4" diam	2.000	Ea.	116	70	186
J	Vent flashing neoprene	2.000	Ea.	28.20	70	98.20
K	Circulator, solar heated liquid, 1/25 HP	1.000	Ea.	300	89.50	389.50
N	Relief valve temp & press 150 psi 210°F self-closing 3/4" IPS	1.000	Ea.	23	23	46
O	Pipe covering, urethane, ultraviolet cover, 1" wall, 3/4" diam	20.000	L.F.	60.20	125	185.20
P	Pipe covering, fiberglass, all service jacket, 1" wall, 3/4" diam	50.000	L.F.	53.50	249	302.50
	Roof clamps for solar energy collector panels	3.000	Set	9.24	53.70	62.94
Q	Collector panel solar energy blk chrome on copper, 1/8" temp glass 3'x7'	3.000	Ea.	3,300	396	3,696
R	Valve, swing check, bronze, regrinding disc, 3/4" diam, soldered	2.000	Ea.	208	70	278
T	Drain valve, brass, 3/4" connection	2.000	Ea.	21.40	58	79.40
U	Valve, water tempering, bronze, sweat connections, 3/4" diam	1.000	Ea.	139	35	174
W-2, W	Tank, water storage elec elem 2"x1/2# insul 120 gal	1.000	Ea.	2,025	495	2,520
X	Valve, globe, bronze, rising stem, 3/4" diam, soldered	1.000	Ea.	148	35	183
	Copper tubing, type L, solder joints, hangers 10' OC 3/4" diam	20.000	L.F.	113	183	296
	Copper tubing, type M, solder joints, hangers 10' OC 3/4" diam	70.000	L.F.	306.60	623	929.60
	Sensor wire, #22-2 conductor, multistranded	.500	C.L.F.	7.83	32	39.83
	Wrought copper fittings & solder, 3/4" diam	76.000	Ea.	577.60	2,774	3,351.60
	TOTAL			8,638.52	5,902.20	14,540.72

D2020 275	Solar, Draindown, Hot Water Systems		MAT.	INST.	TOTAL
2550	Solar, draindown, hot water				
2560	3/8" tubing, 3 ea. 4' x 4'-4" vacuum tube collectors, 80 gal tank		7,425	5,650	13,075
2580	1/2" tubing, 4 ea. 4' x 4'-4" vacuum tube collectors, 80 gal tank	R235616	8,425	5,800	14,225
2600	120 gal tank	-60	8,675	5,875	14,550
2640	2 ea. 3'x7' black chrome collectors, 80 gal tank		6,675	5,475	12,150
2660	3 ea. 3'x7' black chrome collectors, 120 gal tank		8,025	5,675	13,700
2700	2 ea. 3'x7' flat black collectors, 120 gal tank		6,325	5,550	11,875
2720	3 ea. 3'x7' flat black collectors, 120 gal tank		7,050	5,675	12,725
2760	3/4" tubing, 3 ea. 3'x7' black chrome collectors, 120 gal tank		8,650	5,900	14,550
2780	3 ea. 3'x7' flat collectors, 120 gal tank		7,750	5,925	13,675
2800	2 ea. 4'x9' flat black & plastic glazing collectors, 120 gal tank		7,325	5,950	13,275
2840	1" tubing, 4 ea. 2'x9' plastic absorber & glazing collectors, 120 gal tank		12,800	7,825	20,625
2860	4 ea. 3'x7' black chrome absorber collectors, 120 gal tank		12,400	7,425	19,825
2880	4 ea. 3'x7' flat black absorber collectors, 120 gal tank		10,100	6,725	16,825

D20 Plumbing

D2020 Domestic Water Distribution

In the recirculation system, (a direct-collection system), incoming domestic water is heated in the collectors. When the controller activates solar collection, domestic water is heated as it flows through the collectors and then it flows back to storage. When conditions are not suitable for heat collection, the pump shuts off and the flow of the water stops. In this type of system, water remains in the collector loop at all times. A "frost sensor" at the collector activates the circulation of warm water from storage through the collectors when protection from freezing is required.

D20 Plumbing

D2020 Domestic Water Distribution

System Components

		QUANTITY	UNIT	COST EACH MAT.	COST EACH INST.	TOTAL
	SYSTEM D2020 280 2820					
	SOLAR, RECIRCULATION, HOT WATER					
	3/4" TUBING, TWO 3'X7' BLACK CHROME COLLECTORS					
A, B	Differential controller 2 sensors, thermostat, for solar energy system	1.000	Ea.	625	58	683
A-1	Solenoid valve, solar heating loop, brass, 3/4" IPS, 24 volts	2.000	Ea.	260	154	414
B-1	Solar energy sensor freeze prevention	1.000	Ea.	26	21.50	47.50
C	Thermometer, 2" dial	3.000	Ea.	72	130.50	202.50
D	Drain valve, brass, 3/4" connection	1.000	Ea.	10.70	29	39.70
F-1	Air vent, automatic, 1/8" fitting	1.000	Ea.	18.95	21.50	40.45
H	Strainer, Y type, bronze body, 3/4" IPS	1.000	Ea.	37.50	37	74.50
I	Valve, gate, bronze, 125 lb, soldered 3/4" diam	3.000	Ea.	174	105	279
J	Vent flashing, neoprene	2.000	Ea.	28.20	70	98.20
L	Circulator, solar heated liquid, 1/20 HP	1.000	Ea.	177	104	281
N	Relief valve, temp & press 150 psi 210° F self-closing 3/4" IPS	2.000	Ea.	46	46	92
O	Pipe covering, urethane, ultraviolet cover, 1" wall, 3/4" diam	20.000	L.F.	60.20	125	185.20
P	Pipe covering, fiberglass, all service jacket, 1" wall 3/4" diam	50.000	L.F.	53.50	249	302.50
Q	Collector panel solar energy blk chrome on copper, 1/8" temp glass 3'x7'	2.000	Ea.	2,200	264	2,464
	Roof clamps for solar energy collector panels	2.000	Set	6.16	35.80	41.96
R	Valve, swing check, bronze, 125 lb, regrinding disc, soldered 3/4" diam	2.000	Ea.	208	70	278
U	Valve, water tempering, bronze, sweat connections, 3/4" diam	1.000	Ea.	139	35	174
V	Tank, water storage, w/heating element, drain, relief valve, existing	1.000	Ea.			
X	Valve, globe, bronze, 125 lb, 3/4" diam	2.000	Ea.	296	70	366
	Copper tubing, type L, solder joints, hangers 10' OC 3/4" diam	20.000	L.F.	113	183	296
	Copper tubing, type M, solder joints, hangers 10' OC 3/4" diam	70.000	L.F.	306.60	623	929.60
	Wrought copper fittings & solder, 3/4" diam	76.000	Ea.	577.60	2,774	3,351.60
	Sensor wire, #22-2 conductor, multistranded	.500	C.L.F.	7.83	32	39.83
	TOTAL			**5,443.24**	**5,237.30**	**10,680.54**

D2020 280	Solar Recirculation, Domestic Hot Water Systems		MAT.	INST.	TOTAL
2550	Solar recirculation, hot water				
2560	3/8" tubing, 3 ea. 4' x 4'-4" vacuum tube collectors		6,075	4,875	10,950
2580	1/2" tubing, 4 ea. 4' x 4'-4" vacuum tube collectors		6,550	5,325	11,875
2640	2 ea. 3'x7' black chrome collectors	R235616-60	4,800	5,000	9,800
2660	3 ea. 3'x7' black chrome collectors		5,900	5,150	11,050
2700	2 ea. 3'x7' flat black collectors		4,200	5,025	9,225
2720	3 ea. 3'x7' flat black collectors		5,000	5,175	10,175
2760	3/4" tubing, 3 ea. 3'x7' black chrome collectors		6,550	5,375	11,925
2780	3 ea. 3'x7' flat black absorber plate collectors		5,650	5,400	11,050
2800	2 ea. 4'x9' flat black w/plastic glazing collectors		5,250	5,425	10,675
2820	2 ea. 3'x7' black chrome collectors		5,450	5,225	10,675
2840	1" tubing, 4 ea. 2'x9' black plastic absorber & glazing collectors		7,325	6,175	13,500
2860	4 ea. 3'x7' black chrome absorber collectors		9,250	6,200	15,450
2880	4 ea. 3'x7' flat black absorber collectors		8,050	6,225	14,275

D20 Plumbing

D2020 Domestic Water Distribution

The thermosyphon domestic hot water system, a direct collection system, operates under city water pressure and does not require pumps for system operation. An insulated water storage tank is located above the collectors. As the sun heats the collectors, warm water in them rises by means of natural convection; the colder water in the storage tank flows into the collectors by means of gravity. As long as the sun is shining the water continues to flow through the collectors and to become warmer.

To prevent freezing, the system must be drained or the collectors covered with an insulated lid when the temperature drops below 32°F.

D20 Plumbing

D2020 Domestic Water Distribution

System Components	QUANTITY	UNIT	COST EACH MAT.	COST EACH INST.	COST EACH TOTAL
SYSTEM D2020 285 0960 **SOLAR, THERMOSYPHON, WATER HEATER** 3/4" TUBING, TWO 3'X7' BLACK CHROME COLLECTORS					
D-1 Framing lumber, fir, 2" x 6" x 8', tank cradle	.008	M.B.F.	5	7.24	12.24
F-1 Framing lumber, fir, 2" x 4" x 24', sleepers	.016	M.B.F.	10	22	32
I Valve, gate, bronze, 125 lb, soldered 1/2" diam	2.000	Ea.	103	58	161
J Vent flashing, neoprene	4.000	Ea.	56.40	140	196.40
O Pipe covering, urethane, ultraviolet cover, 1" wall, 1/2"diam	40.000	L.F.	92.40	244	336.40
P Pipe covering fiberglass all service jacket 1" wall 1/2" diam	160.000	L.F.	156.80	764.80	921.60
Q Collector panel solar, blk chrome on copper, 3/16" temp glass 3'-6" x 7.5'	2.000	Ea.	1,600	278	1,878
Y Flow control valve, globe, bronze, 125#, soldered, 1/2" diam	1.000	Ea.	111	29	140
U Valve, water tempering, bronze, sweat connections, 1/2" diam	1.000	Ea.	120	29	149
W Tank, water storage, solar energy system, 80 Gal, 2" x 1/2 lb insul	1.000	Ea.	1,775	435	2,210
Copper tubing type L, solder joints, hangers 10' OC 1/2" diam	150.000	L.F.	556.50	1,290	1,846.50
Copper tubing type M, solder joints, hangers 10' OC 1/2" diam	50.000	L.F.	145.50	415	560.50
Sensor wire, #22-2 gauge multistranded	.500	C.L.F.	7.83	32	39.83
Wrought copper fittings & solder, 1/2" diam	75.000	Ea.	254.25	2,625	2,879.25
TOTAL			4,993.68	6,369.04	11,362.72

D2020 285	Thermosyphon, Hot Water		COST EACH MAT.	COST EACH INST.	COST EACH TOTAL
0960 0970	Solar, thermosyphon, hot water, two collector system	R235616 -60	5,000	6,375	11,375

473

D20 Plumbing

D2020 Domestic Water Distribution

This domestic hot water pre-heat system includes heat exchanger with a circulating pump, blower, air-to-water, coil and controls, mounted in the upper collector manifold. Heat from the hot air coming out of the collectors is transferred through the heat exchanger. For each degree of DHW preheating gained, one degree less heating is needed from the fuel fired water heater. The system is simple, inexpensive to operate and can provide a substantial portion of DHW requirements for modest additional cost.

D20 Plumbing

D2020 Domestic Water Distribution

System Components		QUANTITY	UNIT	COST EACH MAT.	COST EACH INST.	TOTAL
	SYSTEM D2020 290 2560 SOLAR, HOT WATER, AIR TO WATER HEAT EXCHANGE THREE COLLECTORS, OPTICAL BLACK ON ALUM., 7.5' x 3.5', 80 GAL TANK					
A, B	Differential controller, 2 sensors, thermostat, solar energy system	1.000	Ea.	625	58	683
C	Thermometer, 2" dial	2.000	Ea.	48	87	135
C-1	Heat exchanger, air to fluid, up flow 70 MBH	1.000	Ea.	395	360	755
E	Air vent, manual, for solar energy system 1/8" fitting	1.000	Ea.	28	21.50	49.50
F	Air purger	1.000	Ea.	51.50	58	109.50
G	Expansion tank	1.000	Ea.	75	21.50	96.50
H	Strainer, Y type, bronze body, 1/2" IPS	1.000	Ea.	30.50	35.50	66
I	Valve, gate, bronze, 125 lb, NRS, soldered 1/2" diam	5.000	Ea.	257.50	145	402.50
N	Relief valve, temp & pressure solar 150 psi 210°F self-closing	1.000	Ea.	23	23	46
P	Pipe covering, fiberglass, all service jacket, 1" wall, 1/2" diam	60.000	L.F.	58.80	286.80	345.60
Q	Collector panel solar energy, air, black on alum plate, 7.5' x 3.5'	3.000	Ea.	2,400	417	2,817
B-1, R-1	Shutter damper for solar heater circulator	1.000	Ea.	61.50	100	161.50
B-1, R-1	Shutter motor for solar heater circulator	1.000	Ea.	153	75	228
R	Backflow preventer, 1/2" pipe size	2.000	Ea.	166	87	253
T	Drain valve, brass, 3/4" connection	1.000	Ea.	10.70	29	39.70
U	Valve, water tempering, bronze, sweat connections, 1/2" diam	1.000	Ea.	120	29	149
W	Tank, water storage, solar energy system, 80 Gal, 2" x 2 lb insul	1.000	Ea.	1,775	435	2,210
V	Tank, water storage, w/heating element, drain, relief valve, existing	1.000	System			
X	Valve, globe, bronze, 125 lb, rising stem, 1/2" diam	1.000	Ea.	111	29	140
	Copper tubing type M, solder joints, hangers 10' OC, 1/2" diam	50.000	L.F.	145.50	415	560.50
	Copper tubing type L, solder joints, hangers 10' OC 1/2" diam	10.000	L.F.	37.10	86	123.10
	Wrought copper fittings & solder, 1/2" diam	10.000	Ea.	33.90	350	383.90
	Sensor wire, #22-2 conductor multistranded	.500	C.L.F.	7.83	32	39.83
Q-1, Q-2	Ductwork, fiberglass, aluminized jacket, 1-1/2" thick, 8" diam	32.000	S.F.	180.80	214.40	395.20
Q-3	Manifold for flush mount solar energy collector panels, air	6.000	L.F.	888	45	933
	TOTAL			7,682.63	3,439.70	11,122.33

D2020 290	Solar Hot Water, Air To Water Heat Exchange		COST EACH MAT.	INST.	TOTAL
2550	Solar hot water, air to water heat exchange				
2560	Three collectors, optical black on aluminum, 7.5' x 3.5', 80 Gal tank		7,675	3,450	11,125
2580	Four collectors, optical black on aluminum, 7.5'x3.5', 80 Gal tank	R235616 -60	8,825	3,650	12,475
2600	Four collectors, optical black on aluminum, 7.5' x 3.5', 120 Gal tank		9,075	3,700	12,775

D20 Plumbing

D2020 Domestic Water Distribution

In this closed-loop indirect collection system, fluid with a low freezing temperature, such as propylene glycol, transports heat from the collectors to water storage. The transfer fluid is contained in a closed-loop consisting of collectors, supply and return piping, and a heat exchanger immersed in the storage tank. A typical two-or-three panel system contains 5 to 6 gallons of heat transfer fluid.

When the collectors become approximately 20°F warmer than the storage temperature, a controller activates the circulator. The circulator moves the fluid continuously through the collectors until the temperature difference between the collectors and storage is such that heat collection no longer occurs; at that point, the circulator shuts off. Since the heat transfer fluid has a very low freezing temperature, there is no need for it to be drained from the collectors between periods of collection.

D20 Plumbing

D2020 Domestic Water Distribution

System Components	QUANTITY	UNIT	MAT.	INST.	TOTAL
SYSTEM D2020 295 2760					
SOLAR, CLOSED LOOP, HOT WATER SYSTEM, IMMERSED HEAT EXCHANGER					
3/4" TUBING, THREE 3' X 7' BLACK CHROME COLLECTORS					
A, B Differential controller, 2 sensors, thermostat, solar energy system	1.000	Ea.	625	58	683
C Thermometer 2" dial	3.000	Ea.	72	130.50	202.50
D, T Fill & drain valves, brass, 3/4" connection	3.000	Ea.	32.10	87	119.10
E Air vent, manual, 1/8" fitting	1.000	Ea.	28	21.50	49.50
F Air purger	1.000	Ea.	51.50	58	109.50
G Expansion tank	1.000	Ea.	75	21.50	96.50
I Valve, gate, bronze, NRS, soldered 3/4" diam	3.000	Ea.	174	105	279
J Neoprene vent flashing	2.000	Ea.	28.20	70	98.20
K Circulator, solar heated liquid, 1/25 HP	1.000	Ea.	300	89.50	389.50
N-1, N Relief valve, temp & press 150 psi 210°F self-closing 3/4" IPS	2.000	Ea.	46	46	92
O Pipe covering, urethane ultraviolet cover, 1" wall, 3/4" diam	20.000	L.F.	60.20	125	185.20
P Pipe covering, fiberglass, all service jacket, 1" wall, 3/4" diam	50.000	L.F.	53.50	249	302.50
Roof clamps for solar energy collector panel	3.000	Set	9.24	53.70	62.94
Q Collector panel solar blk chrome on copper, 1/8" temp glass, 3'x7'	3.000	Ea.	3,300	396	3,696
R-1 Valve, swing check, bronze, regrinding disc, 3/4" diam, soldered	1.000	Ea.	104	35	139
S Pressure gauge, 60 psi, 2-1/2" dial	1.000	Ea.	26.50	21.50	48
U Valve, water tempering, bronze, sweat connections, 3/4" diam	1.000	Ea.	139	35	174
W-2, W Tank, water storage immersed heat exchr elec elem 2"x2# insul 120 Gal	1.000	Ea.	2,025	495	2,520
X Valve, globe, bronze, rising stem, 3/4" diam, soldered	1.000	Ea.	148	35	183
Copper tubing type L, solder joint, hanger 10' OC 3/4" diam	20.000	L.F.	113	183	296
Copper tubing, type M, solder joint, hanger 10' OC 3/4" diam	70.000	L.F.	306.60	623	929.60
Sensor wire, #22-2 conductor multistranded	.500	C.L.F.	7.83	32	39.83
Solar energy heat transfer fluid, propylene glycol, anti-freeze	6.000	Gal.	102	150	252
Wrought copper fittings & solder, 3/4" diam	76.000	Ea.	577.60	2,774	3,351.60
TOTAL			**8,404.27**	**5,894.20**	**14,298.47**

D2020 295	Solar, Closed Loop, Hot Water Systems		MAT.	INST.	TOTAL
2550	Solar, closed loop, hot water system, immersed heat exchanger				
2560	3/8" tubing, 3 ea. 4' x 4'-4" vacuum tube collectors, 80 gal. tank		7,750	5,350	13,100
2580	1/2" tubing, 4 ea. 4' x 4'-4" vacuum tube collectors, 80 gal. tank		8,225	5,800	14,025
2600	120 gal. tank		8,475	5,850	14,325
2640	2 ea. 3'x7' black chrome collectors, 80 gal. tank	R235616 -60	6,525	5,525	12,050
2660	120 gal. tank		6,725	5,525	12,250
2700	2 ea. 3'x7' flat black collectors, 120 gal. tank		6,125	5,550	11,675
2720	3 ea. 3'x7' flat black collectors, 120 gal. tank		6,925	5,700	12,625
2760	3/4" tubing, 3 ea. 3'x7' black chrome collectors, 120 gal. tank		8,400	5,900	14,300
2780	3 ea. 3'x7' flat black collectors, 120 gal. tank		7,500	5,925	13,425
2800	2 ea. 4'x9' flat black w/plastic glazing collectors 120 gal. tank		7,100	5,925	13,025
2840	1" tubing, 4 ea. 2'x9' plastic absorber & glazing collectors 120 gal. tank		8,825	6,650	15,475
2860	4 ea. 3'x7' black chrome collectors, 120 gal. tank		10,800	6,675	17,475
2880	4 ea. 3'x7' flat black absorber collectors, 120 gal. tank		9,575	6,700	16,275

D20 Plumbing

D2040 Rain Water Drainage

Design Assumptions: Vertical conductor size is based on a maximum rate of rainfall of 4" per hour. To convert roof area to other rates multiply "Max. S.F. Roof Area" shown by four and divide the result by desired local rate. The answer is the local roof area that may be handled by the indicated pipe diameter.

Basic cost is for roof drain, 10' of vertical leader and 10' of horizontal, plus connection to the main.

Pipe Dia.	Max. S.F. Roof Area	Gallons per Min.
2"	544	23
3"	1610	67
4"	3460	144
5"	6280	261
6"	10,200	424
8"	22,000	913

System Components	QUANTITY	UNIT	MAT.	INST.	TOTAL
SYSTEM D2040 210 1880					
ROOF DRAIN, DWV PVC PIPE, 2" DIAM., 10' HIGH					
Drain, roof, main, PVC, dome type 2" pipe size	1.000	Ea.	144	89.50	233.50
Clamp, roof drain, underdeck	1.000	Ea.	30	52	82
Pipe, Tee, PVC DWV, schedule 40, 2" pipe size	1.000	Ea.	10.10	62.50	72.60
Pipe, PVC, DWV, schedule 40, 2" diam.	20.000	L.F.	74.60	420	494.60
Pipe, elbow, PVC schedule 40, 2" diam.	2.000	Ea.	6.36	69	75.36
TOTAL			265.06	693	958.06

D2040 210	Roof Drain Systems	MAT.	INST.	TOTAL
1880	Roof drain, DWV PVC, 2" diam., piping, 10' high	265	695	960
1920	For each additional foot add	3.73	21	24.73
1960	3" diam., 10' high	355	810	1,165
2000	For each additional foot add	6.65	23.50	30.15
2040	4" diam., 10' high	445	915	1,360
2080	For each additional foot add	9.15	26	35.15
2120	5" diam., 10' high	1,400	1,050	2,450
2160	For each additional foot add	31	29	60
2200	6" diam., 10' high	1,350	1,175	2,525
2240	For each additional foot add	18.95	32	50.95
2280	8" diam., 10' high	3,025	1,975	5,000
2320	For each additional foot add	46.50	40.50	87
3940	C.I., soil, single hub, service wt., 2" diam. piping, 10' high	565	760	1,325
3980	For each additional foot add	10.25	19.85	30.10
4120	3" diam., 10' high	790	830	1,620
4160	For each additional foot add	14.25	21	35.25
4200	4" diam., 10' high	925	905	1,830
4240	For each additional foot add	18.55	23	41.55
4280	5" diam., 10' high	1,300	990	2,290
4320	For each additional foot add	25.50	25.50	51
4360	6" diam., 10' high	1,650	1,050	2,700
4400	For each additional foot add	32	26.50	58.50
4440	8" diam., 10' high	3,200	2,175	5,375
4480	For each additional foot add	48.50	45	93.50
6040	Steel galv. sch 40 threaded, 2" diam. piping, 10' high	720	740	1,460
6080	For each additional foot add	13.35	19.55	32.90
6120	3" diam., 10' high	1,450	1,050	2,500
6160	For each additional foot add	28.50	29	57.50

D20 Plumbing

D2040 Rain Water Drainage

D2040 210	Roof Drain Systems	COST EACH		
		MAT.	INST.	TOTAL
6200	4" diam., 10' high	2,075	1,375	3,450
6240	For each additional foot add	41.50	35	76.50
6280	5" diam., 10' high	2,400	1,125	3,525
6320	For each additional foot add	64.50	34	98.50
6360	6" diam, 10' high	2,825	1,475	4,300
6400	For each additional foot add	70	46.50	116.50
6440	8" diam., 10' high	4,475	2,125	6,600
6480	For each additional foot add	82	52.50	134.50

D30 HVAC

D3010 Energy Supply

Basis for Heat Loss Estimate, Apartment Type Structures:

1. Masonry walls and flat roof are insulated. U factor is assumed at .08.
2. Window glass area taken as BOCA minimum, 1/10th of floor area. Double insulating glass with 1/4" air space, U = .65.
3. Infiltration = 0.3 C.F. per hour per S.F. of net wall.
4. Concrete floor loss is 2 BTUH per S.F.
5. Temperature difference taken as 70°F.
6. Ventilating or makeup air has not been included and must be added if desired. Air shafts are not used.

System Components	QUANTITY	UNIT	MAT.	INST.	TOTAL
SYSTEM D3010 510 1760					
HEATING SYSTEM, FIN TUBE RADIATION, FORCED HOT WATER					
1,000 S.F. AREA, 10,000 C.F. VOLUME					
Boiler, oil fired, CI, burner, ctrls/insul/breech/pipe/ftng/valves, 109 MBH	1.000	Ea.	4,200	3,937.50	8,137.50
Circulating pump, CI flange connection, 1/12 HP	1.000	Ea.	390	209	599
Expansion tank, painted steel, ASME 18 Gal capacity	1.000	Ea.	785	91	876
Storage tank, steel, above ground, 275 Gal capacity w/supports	1.000	Ea.	540	254	794
Copper tubing type L, solder joint, hanger 10' OC, 3/4" diam	100.000	L.F.	565	915	1,480
Radiation, 3/4" copper tube w/alum fin baseboard pkg, 7" high	30.000	L.F.	243	660	903
Pipe covering, calcium silicate w/cover, 1' wall, 3/4' diam	100.000	L.F.	363	675	1,038
TOTAL			7,086	6,741.50	13,827.50
COST PER S.F.			7.09	6.74	13.83

D3010 510	Apartment Building Heating - Fin Tube Radiation	MAT.	INST.	TOTAL
1740	Heating systems, fin tube radiation, forced hot water			
1760	1,000 S.F. area, 10,000 C.F. volume	7.08	6.75	13.83
1800	10,000 S.F. area, 100,000 C.F. volume	3.33	4.06	7.39
1840	20,000 S.F. area, 200,000 C.F. volume	3.64	4.55	8.19
1880	30,000 S.F. area, 300,000 C.F. volume	3.52	4.40	7.92

D30 HVAC

D3010 Energy Supply

Many styles of active solar energy systems exist. Those shown on the following page represent the majority of systems now being installed in different regions of the country.

The five active domestic hot water (DHW) systems which follow are typically specified as two or three panel systems with additional variations being the type of glazing and size of the storage tanks. Various combinations have been costed for the user's evaluation and comparison. The basic specifications from which the following systems were developed satisfy the construction detail requirements specified in the HUD Intermediate Minimum Property Standards (IMPS). If these standards are not complied with, the renewable energy system's costs could be significantly lower than shown.

To develop the system's specifications and costs it was necessary to make a number of assumptions about the systems. Certain systems are more appropriate to one climatic region than another or the systems may require modifications to be usable in particular locations. Specific instances in which the systems are not appropriate throughout the country as specified, or in which modification will be needed include the following:

- The freeze protection mechanisms provided in the DHW systems vary greatly. In harsh climates, where freeze is a major concern, a closed-loop indirect collection system may be more appropriate than a direct collection system.
- The thermosyphon water heater system described cannot be used when temperatures drop below 32°F.
- In warm climates it may be necessary to modify the systems installed to prevent overheating.

For each renewable resource (solar) system a schematic diagram and descriptive summary of the system is provided along with a list of all the components priced as part of the system. The costs were developed based on these specifications.

Considerations affecting costs which may increase or decrease beyond the estimates presented here include the following:

- Special structural qualities (allowance for earthquake, future expansion, high winds, and unusual spans or shapes);
- Isolated building site or rough terrain that would affect the transportation of personnel, material, or equipment;
- Unusual climatic conditions during the construction process;
- Substitution of other materials or system components for those used in the system specifications.

D30 HVAC

D3010 Energy Supply

In this closed-loop indirect collection system, fluid with a low freezing temperature, propylene glycol, transports heat from the collectors to water storage. The transfer fluid is contained in a closed loop consisting of collectors, supply and return piping, and a heat exchanger immersed in the storage tank.

When the collectors become approximately 20°F warmer than the storage temperature, the controller activates the circulator. The circulator moves the fluid continuously until the temperature difference between fluid in the collectors and storage is such that the collection will no longer occur and then the circulator turns off. Since the heat transfer fluid has a very low freezing temperature, there is no need for it to be drained from the collectors between periods of collection.

D30 HVAC

D3010 Energy Supply

System Components	QUANTITY	UNIT	MAT.	INST.	TOTAL
SYSTEM D3010 650 2750					
SOLAR, CLOSED LOOP, SPACE/HOT WATER					
1" TUBING, TEN 3'X7' BLK CHROME ON COPPER ABSORBER COLLECTORS					
A, B — Differential controller 2 sensors, thermostat, solar energy system	2.000	Ea.	1,250	116	1,366
C — Thermometer, 2" dial	10.000	Ea.	240	435	675
C-1 — Heat exchanger, solar energy system, fluid to air, up flow, 80 MBH	1.000	Ea.	525	415	940
D, T — Fill & drain valves, brass, 3/4" connection	5.000	Ea.	53.50	145	198.50
D-1 — Fan center	1.000	Ea.	128	142	270
E — Air vent, manual, 1/8" fitting	1.000	Ea.	28	21.50	49.50
E-2 — Thermostat, 2 stage for sensing room temperature	1.000	Ea.	197	88.50	285.50
F — Air purger	2.000	Ea.	103	116	219
F-1 — Controller, liquid temperature, solar energy system	1.000	Ea.	118	139	257
G — Expansion tank	2.000	Ea.	150	43	193
I — Valve, gate, bronze, 125 lb, soldered, 1" diam	5.000	Ea.	412.50	182.50	595
J — Vent flashing, neoprene	2.000	Ea.	28.20	70	98.20
K — Circulator, solar heated liquid, 1/25 HP	2.000	Ea.	600	179	779
L — Circulator, solar heated liquid, 1/20 HP	1.000	Ea.	177	104	281
N — Relief valve, temp & pressure 150 psi 210°F self-closing	4.000	Ea.	92	92	184
N-1 — Relief valve, pressure poppet, bronze, 30 psi, 3/4" IPS	3.000	Ea.	241.50	75	316.50
O — Pipe covering, urethane, ultraviolet cover, 1" wall, 1" diam	50.000	L.F.	150.50	312.50	463
P — Pipe covering, fiberglass, all service jacket, 1" wall, 1" diam	60.000	L.F.	64.20	298.80	363
Q — Collector panel solar energy blk chrome on copper 1/8" temp glass 3'x7'	10.000	Ea.	11,000	1,320	12,320
Roof clamps for solar energy collector panel	10.000	Set	30.80	179	209.80
R — Valve, swing check, bronze, 125 lb, regrinding disc, 3/4" & 1" diam	4.000	Ea.	600	146	746
S — Pressure gage, 0-60 psi, for solar energy system	2.000	Ea.	53	43	96
U — Valve, water tempering, bronze, sweat connections, 3/4" diam	1.000	Ea.	139	35	174
W-2, W-1, W — Tank, water storage immersed heat xchr elec elem 2"x1/2# ins 120 gal	4.000	Ea.	8,100	1,980	10,080
X — Valve, globe, bronze, 125 lb, rising stem, 1" diam	3.000	Ea.	699	109.50	808.50
Y — Valve, flow control	1.000	Ea.	132	31.50	163.50
Copper tubing, type M, solder joint, hanger 10' OC 1" diam	110.000	L.F.	786.50	1,094.50	1,881
Copper tubing, type L, solder joint, hanger 10' OC 3/4" diam	20.000	L.F.	113	183	296
Wrought copper fittings & solder, 3/4" & 1" diam	121.000	Ea.	2,262.70	5,263.50	7,526.20
Sensor, wire, #22-2 conductor, multistranded	.700	C.L.F.	10.96	44.80	55.76
Ductwork, galvanized steel, for heat exchanger	8.000	Lb.	5.84	63.60	69.44
Solar energy heat transfer fluid propylene glycol, anti-freeze	25.000	Gal.	425	625	1,050
TOTAL			**28,916.20**	**14,093.20**	**43,009.40**

D3010 650	Solar, Closed Loop, Space/Hot Water Systems		MAT.	INST.	TOTAL
2540	Solar, closed loop, space/hot water				
2550	1/2" tubing, 12 ea. 4'x4'4" vacuum tube collectors		26,600	13,200	39,800
2600	3/4" tubing, 12 ea. 4'x4'4" vacuum tube collectors		28,200	13,500	41,700
2650	10 ea. 3' x 7' black chrome absorber collectors	R235616 -60	27,400	13,100	40,500
2700	10 ea. 3' x 7' flat black absorber collectors		24,400	13,200	37,600
2750	1" tubing, 10 ea. 3' x 7' black chrome absorber collectors		28,900	14,100	43,000
2800	10 ea. 3' x 7' flat black absorber collectors		26,000	14,200	40,200
2850	6 ea. 4' x 9' flat black w/plastic glazing collectors		24,000	14,100	38,100
2900	12 ea. 2' x 9' plastic absorber and glazing collectors		25,400	14,300	39,700

D30 HVAC

D3010 Energy Supply

This draindown pool system uses a differential thermostat similar to those used in solar domestic hot water and space heating applications. To heat the pool, the pool water passes through the conventional pump-filter loop and then flows through the collectors. When collection is not possible, or when the pool temperature is reached, all water drains from the solar loop back to the pool through the existing piping. The modes are controlled by solenoid valves or other automatic valves in conjunction with a vacuum breaker relief valve, which facilitates draindown.

D30 HVAC

D3010 Energy Supply

System Components	QUANTITY	UNIT	COST EACH MAT.	COST EACH INST.	TOTAL
SYSTEM D3010 660 2640					
SOLAR SWIMMING POOL HEATER, ROOF MOUNTED COLLECTORS					
TEN 4' X 10' FULLY WETTED UNGLAZED PLASTIC ABSORBERS					
A Differential thermostat/controller, 110V, adj pool pump system	1.000	Ea.	330	350	680
A-1 Solenoid valve, PVC, normally 1 open 1 closed (included)	2.000	Ea.			
B Sensor, thermistor type (included)	2.000	Ea.			
E-1 Valve, vacuum relief	1.000	Ea.	32.50	21.50	54
Q Collector panel, solar energy, plastic, liquid full wetted, 4' x 10'	10.000	Ea.	2,720	2,500	5,220
R Valve, ball check, PVC, socket, 1-1/2" diam	1.000	Ea.	114	35	149
Z Valve, ball, PVC, socket, 1-1/2" diam	3.000	Ea.	213	105	318
Pipe, PVC, sch 40, 1-1/2" diam	80.000	L.F.	420	1,544	1,964
Pipe fittings, PVC sch 40, socket joint, 1-1/2" diam	10.000	Ea.	20.40	350	370.40
Sensor wire, #22-2 conductor, multistranded	.500	C.L.F.	7.83	32	39.83
Roof clamps for solar energy collector panels	10.000	Set	30.80	179	209.80
Roof strap, teflon for solar energy collector panels	26.000	L.F.	676	88.14	764.14
TOTAL			4,564.53	5,204.64	9,769.17

D3010 660	Solar Swimming Pool Heater Systems		MAT.	INST.	TOTAL
2530	Solar swimming pool heater systems, roof mounted collectors				
2540	10 ea. 3'x7' black chrome absorber, 1/8" temp. glass		12,800	4,025	16,825
2560	10 ea. 4'x8' black chrome absorber, 3/16" temp. glass	R235616 -60	14,300	4,800	19,100
2580	10 ea. 3'8"x6' flat black absorber, 3/16" temp. glass		9,850	4,100	13,950
2600	10 ea. 4'x9' flat black absorber, plastic glazing		11,800	4,975	16,775
2620	10 ea. 2'x9' rubber absorber, plastic glazing		14,200	5,375	19,575
2640	10 ea. 4'x10' fully wetted unglazed plastic absorber		4,575	5,200	9,775
2660	Ground mounted collectors				
2680	10 ea. 3'x7' black chrome absorber, 1/8" temp. glass		12,900	4,475	17,375
2700	10 ea. 4'x8' black chrome absorber, 3/16" temp. glass		14,400	5,250	19,650
2720	10 ea. 3'8"x6' flat blk absorber, 3/16" temp. glass		9,900	4,550	14,450
2740	10 ea. 4'x9' flat blk absorber, plastic glazing		11,900	5,450	17,350
2760	10 ea. 2'x9' rubber absorber, plastic glazing		14,200	5,675	19,875
2780	10 ea. 4'x10' fully wetted unglazed plastic absorber		4,625	5,675	10,300

D30 HVAC

D3010 Energy Supply

The complete Solar Air Heating System provides maximum savings of conventional fuel with both space heating and year-round domestic hot water heating. It allows for the home air conditioning to operate simultaneously and independently from the solar domestic water heating in summer. The system's modes of operation are:

Mode 1: The building is heated directly from the collectors with air circulated by the Solar Air Mover.

Mode 2: When heat is not needed in the building, the dampers change within the air mover to circulate the air from the collectors to the rock storage bin.

Mode 3: When heat is not available from the collector array and is available in rock storage, the air mover draws heated air from rock storage and directs it into the building. When heat is not available from the collectors or the rock storage bin, the auxiliary heating unit will provide heat for the building. The size of the collector array is typically 25% the size of the main floor area.

D30 HVAC

D3010 Energy Supply

System Components		QUANTITY	UNIT	MAT.	INST.	TOTAL
	SYSTEM D3010 675 1210					
	SOLAR, SPACE/HOT WATER, AIR TO WATER HEAT EXCHANGE					
A, B	Differential controller 2 sensors thermos., solar energy sys liquid loop	1.000	Ea.	625	58	683
A-1, B	Differential controller 2 sensors 6 station solar energy sys air loop	1.000	Ea.	259	232	491
B-1	Solar energy sensor, freeze prevention	1.000	Ea.	26	21.50	47.50
C	Thermometer for solar energy system, 2" dial	2.000	Ea.	48	87	135
C-1	Heat exchanger, solar energy system, air to fluid, up flow, 70 MBH	1.000	Ea.	395	360	755
D	Drain valve, brass, 3/4" connection	2.000	Ea.	21.40	58	79.40
E	Air vent, manual, for solar energy system 1/8" fitting	1.000	Ea.	28	21.50	49.50
E-1	Thermostat, 2 stage for sensing room temperature	1.000	Ea.	197	88.50	285.50
F	Air purger	1.000	Ea.	51.50	58	109.50
G	Expansion tank, for solar energy system	1.000	Ea.	75	21.50	96.50
I	Valve, gate, bronze, 125 lb, NRS, soldered 3/4" diam	2.000	Ea.	116	70	186
K	Circulator, solar heated liquid, 1/25 HP	1.000	Ea.	300	89.50	389.50
N	Relief valve temp & press 150 psi 210°F self-closing, 3/4" IPS	1.000	Ea.	23	23	46
N-1	Relief valve, pressure, poppet, bronze, 30 psi, 3/4" IPS	1.000	Ea.	80.50	25	105.50
P	Pipe covering, fiberglass, all service jacket, 1" wall, 3/4" diam	60.000	L.F.	64.20	298.80	363
Q-3	Manifold for flush mount solar energy collector panels	20.000	L.F.	2,960	150	3,110
Q	Collector panel solar energy, air, black on alum. plate, 7.5' x 3.5'	10.000	Ea.	8,000	1,390	9,390
R	Valve, swing check, bronze, 125 lb, regrinding disc, 3/4" diam	2.000	Ea.	208	70	278
S	Pressure gage, 2" dial, for solar energy system	1.000	Ea.	26.50	21.50	48
U	Valve, water tempering, bronze, sweat connections, 3/4" diam	1.000	Ea.	139	35	174
W-2, W	Tank, water storage, solar, elec element 2"x1/2# insul, 80 Gal	1.000	Ea.	1,775	435	2,210
X	Valve, globe, bronze, 125 lb, soldered, 3/4" diam	1.000	Ea.	148	35	183
	Copper tubing type L, solder joints, hangers 10' OC 3/4" diam	10.000	L.F.	56.50	91.50	148
	Copper tubing type M, solder joints, hangers 10' OC 3/4" diam	60.000	L.F.	262.80	534	796.80
	Wrought copper fittings & solder, 3/4" diam	26.000	Ea.	197.60	949	1,146.60
	Sensor wire, #22-2 conductor multistranded	1.200	C.L.F.	18.78	76.80	95.58
Q-1	Duct work, rigid fiberglass, rectangular	400.000	S.F.	340	2,140	2,480
	Duct work, spiral preformed, steel, PVC coated both sides, 12" x 10"	8.000	Ea.	208	400	608
R-2	Shutter/damper for solar heater circulator	9.000	Ea.	553.50	900	1,453.50
K-1	Shutter motor for solar heater circulator blower	9.000	Ea.	1,377	675	2,052
K-1	Fan, solar energy heated air circulator, space & DHW system	1.000	Ea.	1,700	2,400	4,100
Z-1	Tank, solar energy air storage, 6'-3"H 7'x7' = 306 CF/2000 Gal	1.000	Ea.	15,700	1,550	17,250
Z-2	Crushed stone 1-1/2"	11.000	C.Y.	341	86.46	427.46
C-2	Thermometer, remote probe, 2" dial	1.000	Ea.	36.50	87	123.50
R-1	Solenoid valve	1.000	Ea.	130	77	207
V, R-4, R-3	Furnace, supply diffusers, return grilles, existing	1.000	Ea.	15,700	1,550	17,250
	TOTAL			52,187.78	15,165.56	67,353.34

D3010 675	Air To Water Heat Exchange		MAT.	INST.	TOTAL
1210 1220	Solar, air to water heat exchange, for space/hot water heating	R235616 -60	52,000	15,200	67,200

487

D30 HVAC

D3020 Heat Generating Systems

Boiler   Baseboard Radiation

Small Electric Boiler System Considerations:
1. Terminal units are fin tube baseboard radiation rated at 720 BTU/hr with 200° water temperature or 820 BTU/hr steam.
2. Primary use being for residential or smaller supplementary areas, the floor levels are based on 7-1/2' ceiling heights.
3. All distribution piping is copper for boilers through 205 MBH. All piping for larger systems is steel pipe.

System Components

System Components	QUANTITY	UNIT	MAT.	INST.	TOTAL
SYSTEM D3020 102 1120 **SMALL HEATING SYSTEM, HYDRONIC, ELECTRIC BOILER** **1,480 S.F., 61 MBH, STEAM, 1 FLOOR**					
Boiler, electric steam, std cntrls, trim, ftngs and valves, 18 KW, 61.4 MBH	1.000	Ea.	5,032.50	1,760	6,792.50
Copper tubing type L, solder joint, hanger 10'OC, 1-1/4" diam	160.000	L.F.	1,976	1,920	3,896
Radiation, 3/4" copper tube w/alum fin baseboard pkg 7" high	60.000	L.F.	486	1,320	1,806
Rough in baseboard panel or fin tube with valves & traps	10.000	Set	3,550	6,650	10,200
Pipe covering, calcium silicate w/cover, 1" wall 1-1/4" diam	160.000	L.F.	580.80	1,112	1,692.80
Low water cut-off, quick hookup, in gage glass tappings	1.000	Ea.	260	44	304
TOTAL			11,885.30	12,806	24,691.30
COST PER S.F.			8.03	8.65	16.68

D3020 102 Small Heating Systems, Hydronic, Electric Boilers

		COST PER S.F. MAT.	INST.	TOTAL
1100	Small heating systems, hydronic, electric boilers			
1120	Steam, 1 floor, 1480 S.F., 61 M.B.H.	8.04	8.65	16.69
1160	3,000 S.F., 123 M.B.H.	6.30	7.60	13.90
1200	5,000 S.F., 205 M.B.H.	5.55	7	12.55
1240	2 floors, 12,400 S.F., 512 M.B.H.	4.47	6.95	11.42
1280	3 floors, 24,800 S.F., 1023 M.B.H.	4.96	6.90	11.86
1360	Hot water, 1 floor, 1,000 S.F., 41 M.B.H.	12.55	4.82	17.37
1400	2,500 S.F., 103 M.B.H.	9.55	8.65	18.20
1440	2 floors, 4,850 S.F., 205 M.B.H.	9.30	10.35	19.65
1480	3 floors, 9,700 S.F., 410 M.B.H.	9.90	10.70	20.60

D30 HVAC

D3090 Other HVAC Systems/Equip

Cast Iron Garage Exhaust System

Dual Exhaust System

System Components	QUANTITY	UNIT	COST EACH MAT.	INST.	TOTAL
SYSTEM D3090 320 1040					
GARAGE, EXHAUST, SINGLE 3" EXHAUST OUTLET, CARS & LIGHT TRUCKS					
A Outlet top assy, for engine exhaust system with adapters and ftngs, 3" diam	1.000	Ea.	265	42	307
F Bullnose (guide) for engine exhaust system, 3" diam	1.000	Ea.	30.50		30.50
G Galvanized flexible tubing for engine exhaust system, 3" diam	8.000	L.F.	102		102
H Adapter for metal tubing end of engine exhaust system, 3" tail pipe	1.000	Ea.	54.50		54.50
J Pipe, sewer, cast iron, push-on joint, 8" diam.	18.000	L.F.	981	765	1,746
Excavating utility trench, chain trencher, 8" wide, 24" deep	18.000	L.F.		33.30	33.30
Backfill utility trench by hand, incl. compaction, 8" wide 24" deep	18.000	L.F.		46.26	46.26
Stand for blower, concrete over polystyrene core, 6" high	1.000	Ea.	5.70	21	26.70
AC&V duct spiral reducer 10"x8"	1.000	Ea.	22.50	37.50	60
AC&V duct spiral reducer 12"x10"	1.000	Ea.	26	50	76
AC&V duct spiral preformed 45° elbow, 8" diam	1.000	Ea.	8.90	43	51.90
AC&V utility fan, belt drive, 3 phase, 2000 CFM, 1 HP	2.000	Ea.	2,600	660	3,260
Safety switch, heavy duty fused, 240V, 3 pole, 30 amp	1.000	Ea.	153	200	353
TOTAL			4,249.10	1,898.06	6,147.16

D3090 320	Garage Exhaust Systems	COST PER BAY MAT.	INST.	TOTAL
1040	Garage, single 3" exhaust outlet, cars & light trucks, one bay	4,250	1,900	6,150
1060	Additional bays up to seven bays	995	510	1,505
1500	4" outlet, trucks, one bay	4,275	1,900	6,175
1520	Additional bays up to six bays	1,025	510	1,535
1600	5" outlet, diesel trucks, one bay	4,600	1,900	6,500
1650	Additional single bays up to six	1,450	605	2,055
1700	Two adjoining bays	4,600	1,900	6,500
2000	Dual exhaust, 3" outlets, pair of adjoining bays	5,325	2,375	7,700
2100	Additional pairs of adjoining bays	1,625	605	2,230

D40 Fire Protection

D4010 Sprinklers

Dry Pipe System: A system employing automatic sprinklers attached to a piping system containing air under pressure, the release of which from the opening of sprinklers permits the water pressure to open a valve known as a "dry pipe valve". The water then flows into the piping system and out the opened sprinklers.

All areas are assumed to be open.

System Components	QUANTITY	UNIT	MAT.	INST.	TOTAL
SYSTEM D4010 310 0580					
DRY PIPE SPRINKLER, STEEL, BLACK, SCH. 40 PIPE					
LIGHT HAZARD, ONE FLOOR, 2000 S.F.					
Valve, gate, iron body 125 lb., OS&Y, flanged, 4" pipe size	1.000	Ea.	603.75	311.25	915
Valve, swing check, bronze, 125 lb, regrinding disc, 2-1/2" pipe size	1.000	Ea.	626.25	62.63	688.88
Valve, angle, bronze, 150 lb., rising stem, threaded, 2" pipe size	1.000	Ea.	588.75	47.25	636
*Alarm valve, 2-1/2" pipe size	1.000	Ea.	1,275	300	1,575
Alarm, water motor, complete with gong	1.000	Ea.	322.50	125.25	447.75
Fire alarm horn, electric	1.000	Ea.	49.88	71.63	121.51
Valve swing check w/balldrip CI with brass trim, 4" pipe size	1.000	Ea.	273.75	300	573.75
Pipe, steel, black, schedule 40, 4" diam.	10.000	L.F.	176.25	267.30	443.55
Dry pipe valve, trim & gauges, 4" pipe size	1.000	Ea.	2,250	900	3,150
Pipe, steel, black, schedule 40, threaded, cplg & hngr 10'OC 2-1/2" diam.	20.000	L.F.	231.75	375	606.75
Pipe, steel, black, schedule 40, threaded, cplg & hngr 10'OC 2" diam.	12.500	L.F.	92.34	183.28	275.62
Pipe, steel, black, schedule 40, threaded, cplg & hngr 10'OC 1-1/4" diam.	37.500	L.F.	177.19	395.16	572.35
Pipe, steel, black, schedule 40, threaded, cplg & hngr 10'OC 1" diam.	112.000	L.F.	428.40	1,100.40	1,528.80
Pipe Tee, malleable iron black, 150 lb. threaded, 4" pipe size	2.000	Ea.	438	472.50	910.50
Pipe Tee, malleable iron black, 150 lb. threaded, 2-1/2" pipe size	2.000	Ea.	123	208.50	331.50
Pipe Tee, malleable iron black, 150 lb. threaded, 2" pipe size	1.000	Ea.	28.50	85.50	114
Pipe Tee, malleable iron black, 150 lb. threaded, 1-1/4" pipe size	5.000	Ea.	67.31	335.63	402.94
Pipe Tee, malleable iron black, 150 lb. threaded, 1" pipe size	4.000	Ea.	33.15	261	294.15
Pipe 90° elbow malleable iron black, 150 lb. threaded, 1" pipe size	6.000	Ea.	31.73	240.75	272.48
Sprinkler head dry 1/2" orifice 1" NPT, 3" to 4-3/4" length	12.000	Ea.	1,536	576	2,112
Air compressor, 200 Gal sprinkler system capacity, 1/3 HP	1.000	Ea.	607.50	386.25	993.75
*Standpipe connection, wall, flush, brs. w/plug & chain 2-1/2"x2-1/2"	1.000	Ea.	125.25	180.75	306
Valve gate bronze, 300 psi, NRS, class 150, threaded, 1" pipe size	1.000	Ea.	87	27.38	114.38
TOTAL			10,173.25	7,213.41	17,386.66
COST PER S.F.			5.09	3.61	8.70

*Not included in systems under 2000 S.F.

D4010 310	Dry Pipe Sprinkler Systems		MAT.	INST.	TOTAL
0520	Dry pipe sprinkler systems, steel, black, sch. 40 pipe				
0530	Light hazard, one floor, 500 S.F.		9.30	6.10	15.40
0560	1000 S.F.	R211313-10	5.35	3.58	8.93
0580	2000 S.F.		5.10	3.60	8.70
0600	5000 S.F.	R211313-20	2.68	2.44	5.12
0620	10,000 S.F.		1.92	2.04	3.96

D40 Fire Protection

D4010 Sprinklers

D4010 310	Dry Pipe Sprinkler Systems		MAT.	INST.	TOTAL
0640	50,000 S.F.	R211313 -30	1.48	1.82	3.30
0660	Each additional floor, 500 S.F.		2.05	3	5.05
0680	1000 S.F.		1.79	2.46	4.25
0700	2000 S.F.		1.73	2.29	4.02
0720	5000 S.F.		1.41	1.94	3.35
0740	10,000 S.F.		1.33	1.81	3.14
0760	50,000 S.F.		1.20	1.62	2.82
1000	Ordinary hazard, one floor, 500 S.F.		9.50	6.15	15.65
1020	1000 S.F.		5.50	3.61	9.11
1040	2000 S.F.		5.20	3.75	8.95
1060	5000 S.F.		3.11	2.63	5.74
1080	10,000 S.F.		2.47	2.70	5.17
1100	50,000 S.F.		2.23	2.52	4.75
1140	Each additional floor, 500 S.F.		2.24	3.07	5.31
1160	1000 S.F.		2.07	2.75	4.82
1180	2000 S.F.		2.04	2.50	4.54
1200	5000 S.F.		1.92	2.16	4.08
1220	10,000 S.F.		1.71	2.14	3.85
1240	50,000 S.F.		1.64	1.83	3.47
1500	Extra hazard, one floor, 500 S.F.		13.15	7.65	20.80
1520	1000 S.F.		8.05	5.55	13.60
1540	2000 S.F.		5.80	4.85	10.65
1560	5000 S.F.		3.60	3.69	7.29
1580	10,000 S.F.		3.76	3.50	7.26
1600	50,000 S.F.		4.10	3.38	7.48
1660	Each additional floor, 500 S.F.		3.12	3.80	6.92
1680	1000 S.F.		3.07	3.62	6.69
1700	2000 S.F.		2.82	3.63	6.45
1720	5000 S.F.		2.42	3.18	5.60
1740	10,000 S.F.		2.87	2.87	5.74
1760	50,000 S.F.		2.92	2.76	5.68
2020	Grooved steel, black, sch. 40 pipe, light hazard, one floor, 2000 S.F.		4.87	3.08	7.95
2060	10,000 S.F.		1.87	1.75	3.62
2100	Each additional floor, 2000 S.F.		1.72	1.83	3.55
2150	10,000 S.F.		1.28	1.52	2.80
2200	Ordinary hazard, one floor, 2000 S.F.		5.10	3.28	8.38
2250	10,000 S.F.		2.30	2.28	4.58
2300	Each additional floor, 2000 S.F.		1.97	2.03	4
2350	10,000 S.F.		1.71	2.05	3.76
2400	Extra hazard, one floor, 2000 S.F.		5.75	4.13	9.88
2450	10,000 S.F.		3.28	2.91	6.19
2500	Each additional floor, 2000 S.F.		2.78	2.98	5.76
2550	10,000 S.F.		2.53	2.53	5.06
3050	Grooved steel, black, sch. 10 pipe, light hazard, one floor, 2000 S.F.		4.77	3.07	7.84
3100	10,000 S.F.		1.80	1.73	3.53
3150	Each additional floor, 2000 S.F.		1.62	1.82	3.44
3200	10,000 S.F.		1.21	1.50	2.71
3250	Ordinary hazard, one floor, 2000 S.F.		5.05	3.27	8.32
3300	10,000 S.F.		2.20	2.25	4.45
3350	Each additional floor, 2000 S.F.		1.88	2.02	3.90
3400	10,000 S.F.		1.61	2.02	3.63
3450	Extra hazard, one floor, 2000 S.F.		5.70	4.12	9.82
3500	10,000 S.F.		3.07	2.86	5.93
3550	Each additional floor, 2000 S.F.		2.72	2.97	5.69
3600	10,000 S.F.		2.42	2.49	4.91
4050	Copper tubing, type M, light hazard, one floor, 2000 S.F.		6.15	3.07	9.22
4100	10,000 S.F.		2.89	1.75	4.64
4150	Each additional floor, 2000 S.F.		3	1.86	4.86

D40 Fire Protection

D4010 Sprinklers

D4010 310	Dry Pipe Sprinkler Systems	MAT.	INST.	TOTAL
4200	10,000 S.F.	2.30	1.53	3.83
4250	Ordinary hazard, one floor, 2000 S.F.	6.55	3.41	9.96
4300	10,000 S.F.	3.54	2.08	5.62
4350	Each additional floor, 2000 S.F.	3.86	2.12	5.98
4400	10,000 S.F.	2.79	1.81	4.60
4450	Extra hazard, one floor, 2000 S.F.	7.80	4.15	11.95
4500	10,000 S.F.	7.90	3.18	11.08
4550	Each additional floor, 2000 S.F.	4.81	3	7.81
4600	10,000 S.F.	6.15	2.77	8.92
5050	Copper tubing, type M, T-drill system, light hazard, one floor			
5060	2000 S.F.	5.95	2.86	8.81
5100	10,000 S.F.	2.47	1.46	3.93
5150	Each additional floor, 2000 S.F.	2.83	1.65	4.48
5200	10,000 S.F.	1.88	1.24	3.12
5250	Ordinary hazard, one floor, 2000 S.F.	5.90	2.92	8.82
5300	10,000 S.F.	3.18	1.86	5.04
5350	Each additional floor, 2000 S.F.	2.74	1.67	4.41
5400	10,000 S.F.	2.48	1.54	4.02
5450	Extra hazard, one floor, 2000 S.F.	6.65	3.44	10.09
5500	10,000 S.F.	5.45	2.34	7.79
5550	Each additional floor, 2000 S.F.	3.70	2.29	5.99
5600	10,000 S.F.	3.70	1.93	5.63

D40 Fire Protection

D4010 Sprinklers

Pre-Action System: A system employing automatic sprinklers attached to a piping system containing air that may or may not be under pressure, with a supplemental heat responsive system of generally more sensitive characteristics than the automatic sprinklers themselves, installed in the same areas as the sprinklers. Actuation of the heat responsive system, as from a fire, opens a valve which permits water to flow into the sprinkler piping system and to be discharged from those sprinklers which were opened by heat from the fire.

All areas are assumed to be open.

System Components	QUANTITY	UNIT	COST EACH MAT.	COST EACH INST.	TOTAL
SYSTEM D4010 350 0580					
PREACTION SPRINKLER SYSTEM, STEEL BLACK SCH. 40 PIPE					
LIGHT HAZARD, 1 FLOOR, 2000 S.F.					
Valve, gate, iron body 125 lb., OS&Y, flanged, 4" pipe size	1.000	Ea.	603.75	311.25	915
*Valve, swing check w/ball drip CI with brass trim 4" pipe size	1.000	Ea.	273.75	300	573.75
Valve, swing check, bronze, 125 lb, regrinding disc, 2-1/2" pipe size	1.000	Ea.	626.25	62.63	688.88
Valve, angle, bronze, 150 lb., rising stem, threaded, 2" pipe size	1.000	Ea.	588.75	47.25	636
*Alarm valve, 2-1/2" pipe size	1.000	Ea.	1,275	300	1,575
Alarm, water motor, complete with gong	1.000	Ea.	322.50	125.25	447.75
Fire alarm horn, electric	1.000	Ea.	49.88	71.63	121.51
Thermostatic release for release line	2.000	Ea.	1,050	50.25	1,100.25
Pipe, steel, black, schedule 40, 4" diam.	10.000	L.F.	176.25	267.30	443.55
Dry pipe valve, trim & gauges, 4" pipe size	1.000	Ea.	2,250	900	3,150
Pipe, steel, black, schedule 40, threaded, cplg. & hngr. 10'OC 2-1/2" diam.	20.000	L.F.	231.75	375	606.75
Pipe steel black, schedule 40, threaded, cplg. & hngr. 10'OC 2" diam.	12.500	L.F.	92.34	183.28	275.62
Pipe, steel, black, schedule 40, threaded, cplg. & hngr. 10'OC 1-1/4" diam.	37.500	L.F.	177.19	395.16	572.35
Pipe, steel, black, schedule 40, threaded, cplg. & hngr. 10'OC 1" diam.	112.000	L.F.	428.40	1,100.40	1,528.80
Pipe, Tee, malleable iron, black, 150 lb. threaded, 4" diam.	2.000	Ea.	438	472.50	910.50
Pipe, Tee, malleable iron, black, 150 lb. threaded, 2-1/2" pipe size	2.000	Ea.	123	208.50	331.50
Pipe, Tee, malleable iron, black, 150 lb. threaded, 2" pipe size	1.000	Ea.	28.50	85.50	114
Pipe, Tee, malleable iron, black, 150 lb. threaded, 1-1/4" pipe size	5.000	Ea.	67.31	335.63	402.94
Pipe, Tee, malleable iron, black, 150 lb. threaded, 1" pipe size	4.000	Ea.	33.15	261	294.15
Pipe, 90° elbow, malleable iron, blk., 150 lb. threaded, 1" pipe size	6.000	Ea.	31.73	240.75	272.48
Sprinkler head, std. spray, brass 135°-286°F 1/2" NPT, 3/8" orifice	12.000	Ea.	182.40	504	686.40
Air compressor auto complete 200 Gal sprinkler sys. cap., 1/3 HP	1.000	Ea.	607.50	386.25	993.75
*Standpipe conn.,wall, flush, brass w/plug & chain 2-1/2" x 2-1/2"	1.000	Ea.	125.25	180.75	306
Valve, gate, bronze, 300 psi, NRS, class 150, threaded, 1" pipe size	1.000	Ea.	87	27.38	114.38
TOTAL			9,869.65	7,191.66	17,061.31
COST PER S.F.			4.93	3.60	8.53

*Not included in systems under 2000 S.F.

D4010 350	Preaction Sprinkler Systems		COST PER S.F. MAT.	INST.	TOTAL
0520	Preaction sprinkler systems, steel, black, sch. 40 pipe				
0530	Light hazard, one floor, 500 S.F.		9.20	4.87	14.07
0560	1000 S.F.	R211313 -10	5.40	3.67	9.07
0580	2000 S.F.		4.94	3.59	8.53

493

D40 Fire Protection

D4010 Sprinklers

D4010 350	Preaction Sprinkler Systems		MAT.	INST.	TOTAL
0600	5000 S.F.	R211313-20	2.56	2.43	4.99
0620	10,000 S.F.		1.81	2.04	3.85
0640	50,000 S.F.	R211313-30	1.38	1.81	3.19
0660	Each additional floor, 500 S.F.		2.29	2.67	4.96
0680	1000 S.F.		1.81	2.46	4.27
0700	2000 S.F.		1.74	2.29	4.03
0720	5000 S.F.		1.29	1.93	3.22
0740	10,000 S.F.		1.22	1.81	3.03
0760	50,000 S.F.		1.18	1.66	2.84
1000	Ordinary hazard, one floor, 500 S.F.		9.50	5.30	14.80
1020	1000 S.F.		5.35	3.60	8.95
1040	2000 S.F.		5.30	3.75	9.05
1060	5000 S.F.		2.85	2.61	5.46
1080	10,000 S.F.		2.16	2.68	4.84
1100	50,000 S.F.		1.92	2.51	4.43
1140	Each additional floor, 500 S.F.		2.61	3.08	5.69
1160	1000 S.F.		1.79	2.48	4.27
1180	2000 S.F.		1.62	2.47	4.09
1200	5000 S.F.		1.78	2.32	4.10
1220	10,000 S.F.		1.57	2.45	4.02
1240	50,000 S.F.		1.48	2.17	3.65
1500	Extra hazard, one floor, 500 S.F.		12.80	6.75	19.55
1520	1000 S.F.		7.40	5.10	12.50
1540	2000 S.F.		5.30	4.83	10.13
1560	5000 S.F.		3.36	3.98	7.34
1580	10,000 S.F.		3.30	3.88	7.18
1600	50,000 S.F.		3.57	3.77	7.34
1660	Each additional floor, 500 S.F.		3.16	3.80	6.96
1680	1000 S.F.		2.59	3.60	6.19
1700	2000 S.F.		2.33	3.61	5.94
1720	5000 S.F.		1.95	3.19	5.14
1740	10,000 S.F.		2.25	2.90	5.15
1760	50,000 S.F.		2.19	2.73	4.92
2020	Grooved steel, black, sch. 40 pipe, light hazard, one floor, 2000 S.F.		4.88	3.08	7.96
2060	10,000 S.F.		1.76	1.75	3.51
2100	Each additional floor of 2000 S.F.		1.73	1.83	3.56
2150	10,000 S.F.		1.17	1.52	2.69
2200	Ordinary hazard, one floor, 2000 S.F.		4.96	3.27	8.23
2250	10,000 S.F.		1.99	2.26	4.25
2300	Each additional floor, 2000 S.F.		1.81	2.02	3.83
2350	10,000 S.F.		1.40	2.03	3.43
2400	Extra hazard, one floor, 2000 S.F.		5.25	4.11	9.36
2450	10,000 S.F.		2.66	2.88	5.54
2500	Each additional floor, 2000 S.F.		2.29	2.96	5.25
2550	10,000 S.F.		1.88	2.49	4.37
3050	Grooved steel, black, sch. 10 pipe light hazard, one floor, 2000 S.F.		4.78	3.07	7.85
3100	10,000 S.F.		1.69	1.73	3.42
3150	Each additional floor, 2000 S.F.		1.63	1.82	3.45
3200	10,000 S.F.		1.10	1.50	2.60
3250	Ordinary hazard, one floor, 2000 S.F.		4.80	3.06	7.86
3300	10,000 S.F.		1.63	2.22	3.85
3350	Each additional floor, 2000 S.F.		1.72	2.01	3.73
3400	10,000 S.F.		1.30	2	3.30
3450	Extra hazard, one floor, 2000 S.F.		5.20	4.10	9.30
3500	10,000 S.F.		2.42	2.82	5.24
3550	Each additional floor, 2000 S.F.		2.23	2.95	5.18
3600	10,000 S.F.		1.77	2.45	4.22
4050	Copper tubing, type M, light hazard, one floor, 2000 S.F.		6.15	3.07	9.22

D40 Fire Protection

D4010 Sprinklers

D4010 350	Preaction Sprinkler Systems	MAT.	INST.	TOTAL
4100	10,000 S.F.	2.78	1.75	4.53
4150	Each additional floor, 2000 S.F.	3.02	1.86	4.88
4200	10,000 S.F.	1.93	1.52	3.45
4250	Ordinary hazard, one floor, 2000 S.F.	6.40	3.40	9.80
4300	10,000 S.F.	3.23	2.06	5.29
4350	Each additional floor, 2000 S.F.	3.05	1.88	4.93
4400	10,000 S.F.	2.28	1.65	3.93
4450	Extra hazard, one floor, 2000 S.F.	7.30	4.13	11.43
4500	10,000 S.F.	7.20	3.13	10.33
4550	Each additional floor, 2000 S.F.	4.32	2.98	7.30
4600	10,000 S.F.	5.50	2.73	8.23
5050	Copper tubing, type M, T-drill system, light hazard, one floor			
5060	2000 S.F.	5.95	2.86	8.81
5100	10,000 S.F.	2.36	1.46	3.82
5150	Each additional floor, 2000 S.F.	2.84	1.65	4.49
5200	10,000 S.F.	1.77	1.24	3.01
5250	Ordinary hazard, one floor, 2000 S.F.	5.75	2.91	8.66
5300	10,000 S.F.	2.87	1.84	4.71
5350	Each additional floor, 2000 S.F.	2.60	1.67	4.27
5400	10,000 S.F.	2.28	1.61	3.89
5450	Extra hazard, one floor, 2000 S.F.	6.20	3.42	9.62
5500	10,000 S.F.	4.76	2.29	7.05
5550	Each additional floor, 2000 S.F.	3.21	2.27	5.48
5600	10,000 S.F.	3.05	1.89	4.94

D40 Fire Protection

D4010 Sprinklers

Deluge System: A system employing open sprinklers attached to a piping system connected to a water supply through a valve which is opened by the operation of a heat responsive system installed in the same areas as the sprinklers. When this valve opens, water flows into the piping system and discharges from all sprinklers attached thereto.

All areas are assumed to be open.

System Components	QUANTITY	UNIT	MAT.	INST.	TOTAL
SYSTEM D4010 370 0580					
DELUGE SPRINKLER SYSTEM, STEEL BLACK SCH. 40 PIPE					
LIGHT HAZARD, 1 FLOOR, 2000 S.F.					
Valve, gate, iron body 125 lb., OS&Y, flanged, 4" pipe size	1.000	Ea.	603.75	311.25	915
Valve, swing check w/ball drip, CI w/brass ftngs., 4" pipe size	1.000	Ea.	273.75	300	573.75
Valve, swing check, bronze, 125 lb, regrinding disc, 2-1/2" pipe size	1.000	Ea.	626.25	62.63	688.88
Valve, angle, bronze, 150 lb., rising stem, threaded, 2" pipe size	1.000	Ea.	588.75	47.25	636.
*Alarm valve, 2-1/2" pipe size	1.000	Ea.	1,275	300	1,575
Alarm, water motor, complete with gong	1.000	Ea.	322.50	125.25	447.75
Fire alarm horn, electric	1.000	Ea.	49.88	71.63	121.51
Thermostatic release for release line	2.000	Ea.	1,050	50.25	1,100.25
Pipe, steel, black, schedule 40, 4" diam.	10.000	L.F.	176.25	267.30	443.55
Deluge valve trim, pressure relief, emergency release, gauge, 4" pipe size	1.000	Ea.	3,693.75	900	4,593.75
Deluge system, monitoring panel w/deluge valve & trim	1.000	Ea.	8,475	27.75	8,502.75
Pipe, steel, black, schedule 40, threaded, cplg & hngr 10' OC 2-1/2" diam.	20.000	L.F.	231.75	375	606.75
Pipe, steel, black, schedule 40, threaded, cplg & hngr 10' OC 2" diam.	12.500	L.F.	92.34	183.28	275.62
Pipe, steel, black, schedule 40, threaded, cplg & hngr 10' OC 1-1/4" diam.	37.500	L.F.	177.19	395.16	572.35
Pipe, steel, black, schedule 40, threaded, cplg & hngr 10' OC 1" diam.	112.000	L.F.	428.40	1,100.40	1,528.80
Pipe, Tee, malleable iron, black, 150 lb. threaded, 4" pipe size	2.000	Ea.	438	472.50	910.50
Pipe, Tee, malleable iron, black, 150 lb. threaded, 2-1/2" pipe size	2.000	Ea.	123	208.50	331.50
Pipe, Tee, malleable iron, black, 150 lb. threaded, 2" pipe size	1.000	Ea.	28.50	85.50	114
Pipe, Tee, malleable iron, black, 150 lb. threaded, 1-1/4" pipe size	5.000	Ea.	67.31	335.63	402.94
Pipe, Tee, malleable iron, black, 150 lb. threaded, 1" pipe size	4.000	Ea.	33.15	261	294.15
Pipe, 90° elbow, malleable iron, black, 150 lb. threaded 1" pipe size	6.000	Ea.	31.73	240.75	272.48
Sprinkler head, std spray, brass 135°-286°F 1/2" NPT, 3/8" orifice	9.720	Ea.	182.40	504	686.40
Air compressor, auto, complete, 200 Gal sprinkler sys. cap., 1/3 HP	1.000	Ea.	607.50	386.25	993.75
*Standpipe connection, wall, flush w/plug & chain 2-1/2" x 2-1/2"	1.000	Ea.	125.25	180.75	306
Valve, gate, bronze, 300 psi, NRS, class 150, threaded, 1" pipe size	1.000	Ea.	87	27.38	114.38
TOTAL			19,788.40	7,219.41	27,007.81
COST PER S.F.			9.89	3.61	13.50

*Not included in systems under 2000 S.F.

D4010 370	Deluge Sprinkler Systems		MAT.	INST.	TOTAL
0520	Deluge sprinkler systems, steel, black, sch. 40 pipe				
0530	Light hazard, one floor, 500 S.F.		28	4.93	32.93
0560	1000 S.F.	R211313 -10	14.65	3.53	18.18
0580	2000 S.F.		9.90	3.60	13.50

D40 Fire Protection

D4010 Sprinklers

D4010 370	Deluge Sprinkler Systems		MAT.	INST.	TOTAL
0600	5000 S.F.	R211313 -20	4.54	2.44	6.98
0620	10,000 S.F.		2.80	2.04	4.84
0640	50,000 S.F.	R211313 -30	1.58	1.81	3.39
0660	Each additional floor, 500 S.F.		2.29	2.67	4.96
0680	1000 S.F.		1.81	2.46	4.27
0700	2000 S.F.		1.74	2.29	4.03
0720	5000 S.F.		1.29	1.93	3.22
0740	10,000 S.F.		1.22	1.81	3.03
0760	50,000 S.F.		1.18	1.66	2.84
1000	Ordinary hazard, one floor, 500 S.F.		29	5.65	34.65
1020	1000 S.F.		14.65	3.63	18.28
1040	2000 S.F.		10.25	3.76	14.01
1060	5000 S.F.		4.83	2.62	7.45
1080	10,000 S.F.		3.15	2.68	5.83
1100	50,000 S.F.		2.17	2.55	4.72
1140	Each additional floor, 500 S.F.		2.61	3.08	5.69
1160	1000 S.F.		1.79	2.48	4.27
1180	2000 S.F.		1.62	2.47	4.09
1200	5000 S.F.		1.66	2.14	3.80
1220	10,000 S.F.		1.53	2.17	3.70
1240	50,000 S.F.		1.40	1.98	3.38
1500	Extra hazard, one floor, 500 S.F.		31.50	6.85	38.35
1520	1000 S.F.		17.10	5.30	22.40
1540	2000 S.F.		10.30	4.84	15.14
1560	5000 S.F.		5.05	3.67	8.72
1580	10,000 S.F.		4.16	3.54	7.70
1600	50,000 S.F.		4.24	3.46	7.70
1660	Each additional floor, 500 S.F.		3.16	3.80	6.96
1680	1000 S.F.		2.59	3.60	6.19
1700	2000 S.F.		2.33	3.61	5.94
1720	5000 S.F.		1.95	3.19	5.14
1740	10,000 S.F.		2.30	3.01	5.31
1760	50,000 S.F.		2.31	2.91	5.22
2000	Grooved steel, black, sch. 40 pipe, light hazard, one floor				
2020	2000 S.F.		9.85	3.09	12.94
2060	10,000 S.F.		2.77	1.77	4.54
2100	Each additional floor, 2,000 S.F.		1.73	1.83	3.56
2150	10,000 S.F.		1.17	1.52	2.69
2200	Ordinary hazard, one floor, 2000 S.F.		4.96	3.27	8.23
2250	10,000 S.F.		2.98	2.26	5.24
2300	Each additional floor, 2000 S.F.		1.81	2.02	3.83
2350	10,000 S.F.		1.40	2.03	3.43
2400	Extra hazard, one floor, 2000 S.F.		10.25	4.12	14.37
2450	10,000 S.F.		3.68	2.89	6.57
2500	Each additional floor, 2000 S.F.		2.29	2.96	5.25
2550	10,000 S.F.		1.88	2.49	4.37
3000	Grooved steel, black, sch. 10 pipe, light hazard, one floor				
3050	2000 S.F.		9.15	2.94	12.09
3100	10,000 S.F.		2.68	1.73	4.41
3150	Each additional floor, 2000 S.F.		1.63	1.82	3.45
3200	10,000 S.F.		1.10	1.50	2.60
3250	Ordinary hazard, one floor, 2000 S.F.		9.85	3.27	13.12
3300	10,000 S.F.		2.62	2.22	4.84
3350	Each additional floor, 2000 S.F.		1.72	2.01	3.73
3400	10,000 S.F.		1.30	2	3.30
3450	Extra hazard, one floor, 2000 S.F.		10.15	4.11	14.26
3500	10,000 S.F.		3.41	2.82	6.23
3550	Each additional floor, 2000 S.F.		2.23	2.95	5.18

D40 Fire Protection

D4010 Sprinklers

D4010 370 Deluge Sprinkler Systems

		COST PER S.F.		
		MAT.	INST.	TOTAL
3600	10,000 S.F.	1.77	2.45	4.22
4000	Copper tubing, type M, light hazard, one floor			
4050	2000 S.F.	11.10	3.08	14.18
4100	10,000 S.F.	3.77	1.75	5.52
4150	Each additional floor, 2000 S.F.	3.01	1.85	4.86
4200	10,000 S.F.	1.93	1.52	3.45
4250	Ordinary hazard, one floor, 2000 S.F.	11.35	3.41	14.76
4300	10,000 S.F.	4.22	2.06	6.28
4350	Each additional floor, 2000 S.F.	3.05	1.88	4.93
4400	10,000 S.F.	2.28	1.65	3.93
4450	Extra hazard, one floor, 2000 S.F.	12.25	4.14	16.39
4500	10,000 S.F.	8.25	3.15	11.40
4550	Each additional floor, 2000 S.F.	4.32	2.98	7.30
4600	10,000 S.F.	5.50	2.73	8.23
5000	Copper tubing, type M, T-drill system, light hazard, one floor			
5050	2000 S.F.	10.95	2.87	13.82
5100	10,000 S.F.	3.35	1.46	4.81
5150	Each additional floor, 2000 S.F.	2.44	1.68	4.12
5200	10,000 S.F.	1.77	1.24	3.01
5250	Ordinary hazard, one floor, 2000 S.F.	10.70	2.92	13.62
5300	10,000 S.F.	3.86	1.84	5.70
5350	Each additional floor, 2000 S.F.	2.58	1.66	4.24
5400	10,000 S.F.	2.28	1.61	3.89
5450	Extra hazard, one floor, 2000 S.F.	11.15	3.43	14.58
5500	10,000 S.F.	5.75	2.29	8.04
5550	Each additional floor, 2000 S.F.	3.21	2.27	5.48
5600	10,000 S.F.	3.05	1.89	4.94

D40 Fire Protection

D4010 Sprinklers

Firecycle is a fixed fire protection sprinkler system utilizing water as its extinguishing agent. It is a time delayed, recycling, preaction type which automatically shuts the water off when heat is reduced below the detector operating temperature and turns the water back on when that temperature is exceeded.

The system senses a fire condition through a closed circuit electrical detector system which controls water flow to the fire automatically. Batteries supply up to 90 hour emergency power supply for system operation. The piping system is dry (until water is required) and is monitored with pressurized air. Should any leak in the system piping occur, an alarm will sound, but water will not enter the system until heat is sensed by a Firecycle detector.

All areas are assumed to be open.

System Components	QUANTITY	UNIT	COST EACH MAT.	INST.	TOTAL
SYSTEM D4010 390 0580					
FIRECYCLE SPRINKLER SYSTEM, STEEL BLACK SCH. 40 PIPE					
LIGHT HAZARD, ONE FLOOR, 2000 S.F.					
Valve, gate, iron body 125 lb., OS&Y, flanged, 4" pipe size	1.000	Ea.	603.75	311.25	915
Valve, angle, bronze, 150 lb., rising stem, threaded, 2" pipe size	1.000	Ea.	588.75	47.25	636
Valve, swing check, bronze, 125 lb, regrinding disc, 2-1/2" pipe size	1.000	Ea.	626.25	62.63	688.88
*Alarm valve, 2-1/2" pipe size	1.000	Ea.	1,275	300	1,575
Alarm, water motor, complete with gong	1.000	Ea.	322.50	125.25	447.75
Pipe, steel, black, schedule 40, 4" diam.	10.000	L.F.	176.25	267.30	443.55
Fire alarm, horn, electric	1.000	Ea.	49.88	71.63	121.51
Pipe, steel, black, schedule 40, threaded, cplg & hngr 10' OC 2-1/2" diam.	20.000	L.F.	231.75	375	606.75
Pipe, steel, black, schedule 40, threaded, cplg & hngr 10' OC 2" diam.	12.500	L.F.	92.34	183.28	275.62
Pipe, steel, black, schedule 40, threaded, cplg & hngr 10' OC 1-1/4" diam.	37.500	L.F.	177.19	395.16	572.35
Pipe, steel, black, schedule 40, threaded, cplg & hngr 10' OC 1" diam.	112.000	L.F.	428.40	1,100.40	1,528.80
Pipe, Tee, malleable iron, black, 150 lb. threaded, 4" pipe size	2.000	Ea.	438	472.50	910.50
Pipe, Tee, malleable iron, black, 150 lb. threaded, 2-1/2" pipe size	2.000	Ea.	123	208.50	331.50
Pipe, Tee, malleable iron, black, 150 lb. threaded, 2" pipe size	1.000	Ea.	28.50	85.50	114
Pipe, Tee, malleable iron, black, 150 lb. threaded, 1-1/4" pipe size	5.000	Ea.	67.31	335.63	402.94
Pipe, Tee, malleable iron, black, 150 lb. threaded, 1" pipe size	4.000	Ea.	33.15	261	294.15
Pipe, 90° elbow, malleable iron, black, 150 lb. threaded, 1" pipe size	6.000	Ea.	31.73	240.75	272.48
Sprinkler head std spray, brass 135°-286°F 1/2" NPT, 3/8" orifice	12.000	Ea.	182.40	504	686.40
Firecycle controls, incls panel, battery, solenoid valves, press switches	1.000	Ea.	16,275	1,912.50	18,187.50
Detector, firecycle system	2.000	Ea.	1,125	63	1,188
Firecycle pkg, swing check & flow control valves w/trim 4" pipe size	1.000	Ea.	4,443.75	900	5,343.75
Air compressor, auto, complete, 200 Gal sprinkler sys. cap., 1/3 HP	1.000	Ea.	607.50	386.25	993.75
*Standpipe connection, wall, flush, brass w/plug & chain 2-1/2"x2-1/2"	1.000	Ea.	125.25	180.75	306
Valve, gate, bronze 300 psi, NRS, class 150, threaded, 1" diam.	1.000	Ea.	87	27.38	114.38
TOTAL			28,139.65	8,816.91	36,956.56
COST PER S.F.			14.07	4.41	18.48

*Not included in systems under 2000 S.F.

D4010 390	Firecycle Sprinkler Systems	COST PER S.F. MAT.	INST.	TOTAL
0520	Firecycle sprinkler systems, steel black sch. 40 pipe			
0530	Light hazard, one floor, 500 S.F.	46	9.50	55.50

D40 Fire Protection

D4010 Sprinklers

D4010 390	Firecycle Sprinkler Systems		MAT.	INST.	TOTAL
0560	1000 S.F.	R211313-10	24	6	30
0580	2000 S.F.		14.05	4.40	18.45
0600	5000 S.F.	R211313-20	6.25	2.76	9.01
0620	10,000 S.F.		3.70	2.21	5.91
0640	50,000 S.F.	R211313-30	1.81	1.84	3.65
0660	Each additional floor of 500 S.F.		2.36	2.68	5.04
0680	1000 S.F.		1.84	2.46	4.30
0700	2000 S.F.		1.50	2.28	3.78
0720	5000 S.F.		1.32	1.94	3.26
0740	10,000 S.F.		1.30	1.82	3.12
0760	50,000 S.F.		1.23	1.65	2.88
1000	Ordinary hazard, one floor, 500 S.F.		46.50	9.90	56.40
1020	1000 S.F.		23.50	5.90	29.40
1040	2000 S.F.		14.15	4.55	18.70
1060	5000 S.F.		6.50	2.94	9.44
1080	10,000 S.F.		4.05	2.85	6.90
1100	50,000 S.F.		2.59	2.84	5.43
1140	Each additional floor, 500 S.F.		2.68	3.09	5.77
1160	1000 S.F.		1.82	2.48	4.30
1180	2000 S.F.		1.84	2.27	4.11
1200	5000 S.F.		1.69	2.15	3.84
1220	10,000 S.F.		1.48	2.13	3.61
1240	50,000 S.F.		1.43	1.88	3.31
1500	Extra hazard, one floor, 500 S.F.		49.50	11.40	60.90
1520	1000 S.F.		25.50	7.40	32.90
1540	2000 S.F.		14.45	5.65	20.10
1560	5000 S.F.		6.75	3.99	10.74
1580	10,000 S.F.		5.15	4.02	9.17
1600	50,000 S.F.		4.37	4.45	8.82
1660	Each additional floor, 500 S.F.		3.23	3.81	7.04
1680	1000 S.F.		2.62	3.60	6.22
1700	2000 S.F.		2.37	3.61	5.98
1720	5000 S.F.		1.98	3.20	5.18
1740	10,000 S.F.		2.33	2.91	5.24
1760	50,000 S.F.		2.32	2.80	5.12
2020	Grooved steel, black, sch. 40 pipe, light hazard, one floor				
2030	2000 S.F.		14	3.89	17.89
2060	10,000 S.F.		3.94	2.65	6.59
2100	Each additional floor, 2000 S.F.		1.77	1.83	3.60
2150	10,000 S.F.		1.25	1.53	2.78
2200	Ordinary hazard, one floor, 2000 S.F.		14.10	4.08	18.18
2250	10,000 S.F.		4.16	2.56	6.72
2300	Each additional floor, 2000 S.F.		1.85	2.02	3.87
2350	10,000 S.F.		1.48	2.04	3.52
2400	Extra hazard, one floor, 2000 S.F.		14.40	4.92	19.32
2450	10,000 S.F.		4.50	3.02	7.52
2500	Each additional floor, 2000 S.F.		2.33	2.96	5.29
2550	10,000 S.F.		1.96	2.50	4.46
3050	Grooved steel, black, sch. 10 pipe light hazard, one floor,				
3060	2000 S.F.		13.90	3.88	17.78
3100	10,000 S.F.		3.58	1.90	5.48
3150	Each additional floor, 2000 S.F.		1.67	1.82	3.49
3200	10,000 S.F.		1.18	1.51	2.69
3250	Ordinary hazard, one floor, 2000 S.F.		14	4.07	18.07
3300	10,000 S.F.		3.78	2.40	6.18
3350	Each additional floor, 2000 S.F.		1.76	2.01	3.77
3400	10,000 S.F.		1.38	2.01	3.39
3450	Extra hazard, one floor, 2000 S.F.		14.35	4.91	19.26

D40 Fire Protection

D4010 Sprinklers

D4010 390	Firecycle Sprinkler Systems	COST PER S.F.		
		MAT.	INST.	TOTAL
3500	10,000 S.F.	4.29	2.97	7.26
3550	Each additional floor, 2000 S.F.	2.27	2.95	5.22
3600	10,000 S.F.	1.85	2.46	4.31
4060	Copper tubing, type M, light hazard, one floor, 2000 S.F.	15.30	3.88	19.18
4100	10,000 S.F.	4.67	1.92	6.59
4150	Each additional floor, 2000 S.F.	3.05	1.86	4.91
4200	10,000 S.F.	2.27	1.54	3.81
4250	Ordinary hazard, one floor, 2000 S.F.	15.55	4.21	19.76
4300	10,000 S.F.	5.10	2.23	7.33
4350	Each additional floor, 2000 S.F.	3.09	1.88	4.97
4400	10,000 S.F.	2.33	1.63	3.96
4450	Extra hazard, one floor, 2000 S.F.	16.45	4.94	21.39
4500	10,000 S.F.	9.15	3.33	12.48
4550	Each additional floor, 2000 S.F.	4.36	2.98	7.34
4600	10,000 S.F.	5.55	2.74	8.29
5060	Copper tubing, type M, T-drill system, light hazard, one floor 2000 S.F.	15.10	3.67	18.77
5100	10,000 S.F.	4.25	1.63	5.88
5150	Each additional floor, 2000 S.F.	3.06	1.75	4.81
5200	10,000 S.F.	1.85	1.25	3.10
5250	Ordinary hazard, one floor, 2000 S.F.	14.85	3.72	18.57
5300	10,000 S.F.	4.76	2.01	6.77
5350	Each additional floor, 2000 S.F.	2.62	1.66	4.28
5400	10,000 S.F.	2.36	1.62	3.98
5450	Extra hazard, one floor, 2000 S.F.	15.30	4.23	19.53
5500	10,000 S.F.	6.65	2.44	9.09
5550	Each additional floor, 2000 S.F.	3.25	2.27	5.52
5600	10,000 S.F.	3.13	1.90	5.03

D40 Fire Protection
D4010 Sprinklers

Wet Pipe System. A system employing automatic sprinklers attached to a piping system containing water and connected to a water supply so that water discharges immediately from sprinklers opened by heat from a fire.

All areas are assumed to be open.

System Components	QUANTITY	UNIT	MAT.	INST.	TOTAL
SYSTEM D4010 410 0580					
WET PIPE SPRINKLER, STEEL, BLACK, SCH. 40 PIPE					
LIGHT HAZARD, ONE FLOOR, 2000 S.F.					
Valve, gate, iron body, 125 lb., OS&Y, flanged, 4" diam.	1.000	Ea.	603.75	311.25	915
Valve, swing check, bronze, 125 lb, regrinding disc, 2-1/2" pipe size	1.000	Ea.	626.25	62.63	688.88
Valve, angle, bronze, 150 lb., rising stem, threaded, 2" diam.	1.000	Ea.	588.75	47.25	636
*Alarm valve, 2-1/2" pipe size	1.000	Ea.	1,275	300	1,575
Alarm, water motor, complete with gong	1.000	Ea.	322.50	125.25	447.75
Valve, swing check, w/balldrip CI with brass trim 4" pipe size	1.000	Ea.	273.75	300	573.75
Pipe, steel, black, schedule 40, 4" diam.	10.000	L.F.	176.25	267.30	443.55
*Flow control valve, trim & gauges, 4" pipe size	1.000	Set	4,500	682.50	5,182.50
Fire alarm horn, electric	1.000	Ea.	49.88	71.63	121.51
Pipe, steel, black, schedule 40, threaded, cplg & hngr 10' OC, 2-1/2" diam.	20.000	L.F.	231.75	375	606.75
Pipe, steel, black, schedule 40, threaded, cplg & hngr 10' OC, 2" diam.	12.500	L.F.	92.34	183.28	275.62
Pipe, steel, black, schedule 40, threaded, cplg & hngr 10' OC, 1-1/4" diam.	37.500	L.F.	177.19	395.16	572.35
Pipe, steel, black, schedule 40, threaded cplg & hngr 10' OC, 1" diam.	112.000	L.F.	428.40	1,100.40	1,528.80
Pipe Tee, malleable iron black, 150 lb. threaded, 4" pipe size	2.000	Ea.	438	472.50	910.50
Pipe Tee, malleable iron black, 150 lb. threaded, 2-1/2" pipe size	2.000	Ea.	123	208.50	331.50
Pipe Tee, malleable iron black, 150 lb. threaded, 2" pipe size	1.000	Ea.	28.50	85.50	114
Pipe Tee, malleable iron black, 150 lb. threaded, 1-1/4" pipe size	5.000	Ea.	67.31	335.63	402.94
Pipe Tee, malleable iron black, 150 lb. threaded, 1" pipe size	4.000	Ea.	33.15	261	294.15
Pipe 90° elbow, malleable iron black, 150 lb. threaded, 1" pipe size	6.000	Ea.	31.73	240.75	272.48
Sprinkler head, standard spray, brass 135°-286°F 1/2" NPT, 3/8" orifice	12.000	Ea.	182.40	504	686.40
Valve, gate, bronze, NRS, class 150, threaded, 1" pipe size	1.000	Ea.	87	27.38	114.38
*Standpipe connection, wall, single, flush w/plug & chain 2-1/2"x2-1/2"	1.000	Ea.	125.25	180.75	306
TOTAL			10,462.15	6,537.66	16,999.81
COST PER S.F.			5.23	3.27	8.50

*Not included in systems under 2000 S.F.

D4010 410	Wet Pipe Sprinkler Systems		MAT.	INST.	TOTAL
0520	Wet pipe sprinkler systems, steel, black, sch. 40 pipe				
0530	Light hazard, one floor, 500 S.F.		2.80	3.14	5.94
0560	1000 S.F.	R211313 -10	5.85	3.25	9.10
0580	2000 S.F.		5.25	3.26	8.51
0600	5000 S.F.	R211313 -20	2.47	2.29	4.76
0620	10,000 S.F.		1.60	1.96	3.56

D40 Fire Protection

D4010 Sprinklers

D4010 410	Wet Pipe Sprinkler Systems		MAT.	INST.	TOTAL
0640	50,000 S.F.	R211313 -30	1.08	1.77	2.85
0660	Each additional floor, 500 S.F.		1.26	2.68	3.94
0680	1000 S.F.	R211313 -40	1.30	2.49	3.79
0700	2000 S.F.		1.22	2.26	3.48
0720	5000 S.F.		.87	1.91	2.78
0740	10,000 S.F.		.85	1.79	2.64
0760	50,000 S.F.		.70	1.40	2.10
1000	Ordinary hazard, one floor, 500 S.F.		3.08	3.35	6.43
1020	1000 S.F.		5.80	3.18	8.98
1040	2000 S.F.		5.35	3.41	8.76
1060	5000 S.F.		2.76	2.47	5.23
1080	10,000 S.F.		1.95	2.60	4.55
1100	50,000 S.F.		1.58	2.44	4.02
1140	Each additional floor, 500 S.F.		1.56	3.03	4.59
1160	1000 S.F.		1.26	2.45	3.71
1180	2000 S.F.		1.36	2.46	3.82
1200	5000 S.F.		1.38	2.34	3.72
1220	10,000 S.F.		1.20	2.43	3.63
1240	50,000 S.F.		1.14	2.15	3.29
1500	Extra hazard, one floor, 500 S.F.		11.20	5.20	16.40
1520	1000 S.F.		7	4.53	11.53
1540	2000 S.F.		5.65	4.64	10.29
1560	5000 S.F.		3.47	4.07	7.54
1580	10,000 S.F.		3.03	3.84	6.87
1600	50,000 S.F.		3.39	3.71	7.10
1660	Each additional floor, 500 S.F.		2.11	3.75	5.86
1680	1000 S.F.		2.06	3.57	5.63
1700	2000 S.F.		1.81	3.58	5.39
1720	5000 S.F.		1.53	3.17	4.70
1740	10,000 S.F.		1.88	2.88	4.76
1760	50,000 S.F.		1.87	2.76	4.63
2020	Grooved steel, black sch. 40 pipe, light hazard, one floor, 2000 S.F.		5.20	2.75	7.95
2060	10,000 S.F.		2	1.74	3.74
2100	Each additional floor, 2000 S.F.		1.21	1.80	3.01
2150	10,000 S.F.		.80	1.50	2.30
2200	Ordinary hazard, one floor, 2000 S.F.		5.25	2.94	8.19
2250	10,000 S.F.		1.78	2.18	3.96
2300	Each additional floor, 2000 S.F.		1.29	1.99	3.28
2350	10,000 S.F.		1.03	2.01	3.04
2400	Extra hazard, one floor, 2000 S.F.		5.55	3.78	9.33
2450	10,000 S.F.		2.42	2.79	5.21
2500	Each additional floor, 2000 S.F.		1.77	2.93	4.70
2550	10,000 S.F.		1.51	2.47	3.98
3050	Grooved steel, black sch. 10 pipe, light hazard, one floor, 2000 S.F.		5.10	2.74	7.84
3100	10,000 S.F.		1.48	1.65	3.13
3150	Each additional floor, 2000 S.F.		1.11	1.79	2.90
3200	10,000 S.F.		.73	1.48	2.21
3250	Ordinary hazard, one floor, 2000 S.F.		5.20	2.93	8.13
3300	10,000 S.F.		1.68	2.15	3.83
3350	Each additional floor, 2000 S.F.		1.20	1.98	3.18
3400	10,000 S.F.		.93	1.98	2.91
3450	Extra hazard, one floor, 2000 S.F.		5.50	3.77	9.27
3500	10,000 S.F.		2.21	2.74	4.95
3550	Each additional floor, 2000 S.F.		1.71	2.92	4.63
3600	10,000 S.F.		1.40	2.43	3.83
4050	Copper tubing, type M, light hazard, one floor, 2000 S.F.		6.45	2.74	9.19
4100	10,000 S.F.		2.57	1.67	4.24
4150	Each additional floor, 2000 S.F.		2.49	1.83	4.32

D40 Fire Protection

D4010 Sprinklers

D4010 410	Wet Pipe Sprinkler Systems	MAT.	INST.	TOTAL
4200	10,000 S.F.	1.82	1.51	3.33
4250	Ordinary hazard, one floor, 2000 S.F.	6.70	3.07	9.77
4300	10,000 S.F.	3.02	1.98	5
4350	Each additional floor, 2000 S.F.	2.82	2.01	4.83
4400	10,000 S.F.	2.11	1.77	3.88
4450	Extra hazard, one floor, 2000 S.F.	7.60	3.80	11.40
4500	10,000 S.F.	7	3.05	10.05
4550	Each additional floor, 2000 S.F.	3.80	2.95	6.75
4600	10,000 S.F.	5.10	2.71	7.81
5050	Copper tubing, type M, T-drill system, light hazard, one floor			
5060	2000 S.F.	6.30	2.53	8.83
5100	10,000 S.F.	2.15	1.38	3.53
5150	Each additional floor, 2000 S.F.	2.32	1.62	3.94
5200	10,000 S.F.	1.40	1.22	2.62
5250	Ordinary hazard, one floor, 2000 S.F.	6.05	2.58	8.63
5300	10,000 S.F.	2.66	1.76	4.42
5350	Each additional floor, 2000 S.F.	2.06	1.63	3.69
5400	10,000 S.F.	1.91	1.59	3.50
5450	Extra hazard, one floor, 2000 S.F.	6.50	3.09	9.59
5500	10,000 S.F.	4.55	2.21	6.76
5550	Each additional floor, 2000 S.F.	2.79	2.28	5.07
5600	10,000 S.F.	2.68	1.87	4.55

D40 Fire Protection

D4010 Sprinklers

Plan

Longitudinal Movement

Elevation

Horizontal Views

System Components	QUANTITY	UNIT	COST PER EACH MAT.	INST.	TOTAL
SYSTEM D4010 412 1000					
WET SPRINKLER SYSTEM, SCHEDULE 10 BLACK STEEL GROOVED PIPE					
SEPARATION ASSEMBLY, 2 INCH DIAMETER					
Hanger support for 2 inch pipe	2.000	Ea.	45	76	121
Pipe roller support yoke	2.000	Ea.	22.20	17.90	40.10
Spool pieces 2 inch	3.000	Ea.	20.25	48.45	68.70
2 inch grooved elbows	6.000	Ea.	198	168	366
Coupling grooved joint 2 inch pipe	10.000	Ea.	325	139	464
Roll grooved joint labor only	8.000	Ea.		86.40	86.40
TOTAL			610.45	535.75	1,146.20

D4010 412	Wet Sprinkler Seismic Components	MAT.	INST.	TOTAL
1000	Wet sprinkler sys, Sch. 10, blk steel grooved, separation assembly, 2 inch	610	535	1,145
1100	2-1/2 inch	670	595	1,265
1200	3 inch	900	685	1,585
1300	4 inch	1,125	845	1,970
1400	6 inch	2,300	1,225	3,525
1500	8 inch	4,125	1,425	5,550
2000	Wet sprinkler sys, Sch. 10, blk steel grooved, flexible coupling, 2-1/2 inch	475	297	772
2100	3 inch	495	350	845

D40 Fire Protection

D4010 Sprinklers

D4010 412	Wet Sprinkler Seismic Components	MAT.	INST.	TOTAL
		COST PER EACH		
2200	4 inch	660	465	1,125
2300	6 inch	945	650	1,595
2400	8 inch	1,375	815	2,190
2990	Allowance for seismic movement by providing annular space in concrete wall			
2995	If additional cores are performed for the same size and at the same time reduce			
2996	price of additional cores by 50 %. If different size cores reduce by 30 %.			
3000	Wet pipe sprinkler systems, 2 inch black steel, annular space 6 inch wall	17.50	620	637.50
3010	8 inch wall	17.75	625	642.75
3020	10 inch wall	17.95	630	647.95
3030	12 inch wall	18.20	635	653.20
3100	Wet pipe sprinkler systems, 2-1/2/3 inch black steel, annular space 6" wall	17.50	635	652.50
3110	8 inch wall	17.75	640	657.75
3120	10 inch wall	17.95	650	667.95
3130	12 inch wall	18.20	660	678.20
3200	Wet pipe sprinkler systems, 4 inch black steel, annular space 6" wall	17.50	710	727.50
3210	8 inch wall	17.75	715	732.75
3220	10 inch wall	17.95	720	737.95
3230	12 inch wall	18.20	725	743.20
3300	Wet pipe sprinkler systems, 6 inch black steel, annular space 6" wall	17.50	735	752.50
3310	8 inch wall	17.75	745	762.75
3320	10 inch wall	17.95	750	767.95
3330	12 inch wall	18.20	755	773.20
3400	Wet pipe sprinkler systems, 8 inch black steel, annular space 6" wall	17.50	765	782.50
3410	8 inch wall	17.75	770	787.75
3420	10 inch wall	17.95	775	792.95
3430	12 inch wall	18.20	780	798.20
3996	Lateral, longitudinal and 4-way seismic strut braces			
4000	Wet pipe sprinkler systems, 2 inch pipe double lateral strut brace	164	291	455
4010	2-1/2 inch	168	291	459
4020	3 inch	171	291	462
4030	4 inch	196	295	491
4040	6 inch	227	296	523
4050	8 inch	248	296	544
4100	Wet pipe sprinkler systems, 2 inch pipe longitudinal strut brace	164	291	455
4110	2-1/2 inch	168	291	459
4120	3 inch	171	291	462
4130	4 inch	196	295	491
4140	6 inch	227	296	523
4150	8 inch	248	296	544
4200	Wet pipe sprinkler systems, 2 inch pipe single lateral strut brace	112	192	304
4210	2-1/2 inch	116	192	308
4220	3 inch	119	192	311
4230	4 inch	141	195	336
4240	6 inch	172	196	368
4250	8 inch	193	196	389
4300	Wet pipe sprinkler systems, 2 inch pipe longitudinal wire brace	90	200	290
4310	2-1/2 inch	93.50	201	294.50
4320	3 inch	96.50	201	297.50
4330	4 inch	122	207	329
4340	6 inch	152	205	357
4350	8 inch	174	206	380
4400	Wet pipe sprinkler systems, 2 inch pipe 4-way wire brace	119	310	429
4410	2-1/2 inch	123	310	433
4420	3 inch	126	310	436
4430	4 inch	159	320	479
4440	6 inch	188	315	503
4450	8 inch	210	315	525
4500	Wet pipe sprinkler systems, 2 inch pipe 4-way strut brace	268	490	758

D40 Fire Protection

D4010 Sprinklers

D4010 412	Wet Sprinkler Seismic Components	COST PER EACH		
		MAT.	INST.	TOTAL
4510	2-1/2 inch	272	490	762
4520	3 inch	275	490	765
4530	4 inch	305	495	800
4540	6 inch	335	495	830
4550	8 inch	360	495	855

D40 Fire Protection

D4010 Sprinklers

Wet Pipe System. A system employing automatic sprinklers attached to a piping system containing water and connected to a water supply so that water discharges immediately from sprinklers opened by heat from a fire.

All areas are assumed to be open.

System Components	QUANTITY	UNIT	COST PER S.F. MAT.	COST PER S.F. INST.	COST PER S.F. TOTAL
SYSTEM D4010 413 0580					
SEISMIC WET PIPE SPRINKLER, STEEL, BLACK, SCH. 40 PIPE					
LIGHT HAZARD, ONE FLOOR, 3' FLEXIBLE FEED, 2000 S.F.					
Valve, gate, iron body, 125 lb., OS&Y, flanged, 4" diam.	1.000	Ea.	.23	.12	.35
Valve, swing check, bronze, 125 lb, regrinding disc, 2-1/2" pipe size	1.000	Ea.	.23	.02	.25
Valve, angle, bronze, 150 lb., rising stem, threaded, 2" diam.	1.000	Ea.	.22	.02	.24
*Alarm valve, 2-1/2" pipe size	1.000	Ea.	.48	.11	.59
Alarm, water motor, complete with gong	1.000	Ea.	.12	.05	.17
Valve, swing check, w/balldrip CI with brass trim 4" pipe size	1.000	Ea.	.10	.11	.21
Pipe, steel, black, schedule 40, 4" diam.	10.000	L.F.	.07	.10	.17
*Flow control valve, trim & gauges, 4" pipe size	1.000	Set	1.69	.26	1.95
Fire alarm horn, electric	1.000	Ea.	.02	.03	.05
Pipe, steel, black, schedule 40, threaded, cplg & hngr 10' OC, 2-1/2" diam.	20.000	L.F.	.09	.14	.23
Pipe, steel, black, schedule 40, threaded, cplg & hngr 10' OC, 2" diam.	12.500	L.F.	.03	.07	.10
Pipe, steel, black, schedule 40, threaded, cplg & hngr 10' OC, 1-1/4" diam.	37.500	L.F.	.07	.15	.22
Pipe Tee, malleable iron black, 150 lb. threaded, 4" pipe size	2.000	Ea.	.16	.18	.34
Pipe Tee, malleable iron black, 150 lb. threaded, 2-1/2" pipe size	2.000	Ea.	.05	.08	.13
Pipe Tee, malleable iron black, 150 lb. threaded, 2" pipe size	1.000	Ea.	.01	.03	.04
Pipe Tee, malleable iron black, 150 lb. threaded, 1-1/4" pipe size	5.000	Ea.	.03	.13	.16
Pipe Tee, malleable iron black, 150 lb. threaded, 1" pipe size	4.000	Ea.	.01	.10	.11
Pipe 90° elbow, malleable iron black, 150 lb. threaded, 1" pipe size	6.000	Ea.	.01	.09	.10
Sprinkler head, standard spray, brass 135°-286°F 1/2" NPT, 3/8" orifice	12.000	Ea.	.07	.19	.26
Valve, gate, bronze, NRS, class 150, threaded, 1" pipe size	1.000	Ea.	.03	.01	.04
*Standpipe connection, wall, single, flush w/plug & chain 2-1/2"x2-1/2"	1.000	Ea.	.05	.07	.12
Flexible connector 36 inches long	12.000	Ea.	.35	.10	.45
TOTAL			4.12	2.16	6.28

*Not included in systems under 2000 S.F.

D4010 413	Seismic Wet Pipe Flexible Feed Sprinkler Systems	COST PER S.F. MAT.	COST PER S.F. INST.	COST PER S.F. TOTAL
0520	Seismic wet pipe 3' flexible feed sprinkler systems, steel, black, sch. 40 pipe			
0525	Note: See D4010-412 for strut braces, flexible joints and separation assemblies			
0530	Light hazard, one floor, 3' flexible feed, 500 S.F.	2.87	2.53	5.40
0560	1000 S.F.	6.05	2.86	8.91
0580	2000 S.F.	4.12	2.16	6.28
0600	5000 S.F.	2.69	2.01	4.70
0620	10,000 S.F.	1.84	1.83	3.67
0640	50,000 S.F.	1.26	1.54	2.80

D40 Fire Protection

D4010 Sprinklers

D4010 413	Seismic Wet Pipe Flexible Feed Sprinkler Systems	MAT.	INST.	TOTAL
		\multicolumn{3}{c	}{COST PER S.F.}
0660	Each additional floor, 500 S.F.	1.44	2.10	3.54
0680	1000 S.F.	1.58	2.14	3.72
0700	2000 S.F.	1.36	1.81	3.17
0720	5000 S.F.	1.09	1.63	2.72
0740	10,000 S.F.	1.09	1.66	2.75
0760	50,000 S.F.	.93	1.31	2.24
1000	Ordinary hazard, one floor, 500 S.F.	3.26	2.77	6.03
1020	1000 S.F.	6.05	2.76	8.81
1040	2000 S.F.	5.65	3.15	8.80
1060	5000 S.F.	3.10	2.29	5.39
1080	10,000 S.F.	2.32	2.48	4.80
1100	50,000 S.F.	1.89	2.20	4.09
1140	Each additional floor, 500 S.F.	1.74	2.45	4.19
1160	1000 S.F.	1.51	2.03	3.54
1180	2000 S.F.	1.67	2.20	3.87
1200	5000 S.F.	1.72	2.16	3.88
1220	10,000 S.F.	1.57	2.31	3.88
1240	50,000 S.F.	1.45	1.91	3.36
1500	Extra hazard, one floor, 500 S.F.	11.45	4.29	15.74
1520	1000 S.F.	7.25	3.66	10.91
1540	2000 S.F.	6.10	4.23	10.33
1560	5000 S.F.	3.93	3.81	7.74
1580	10,000 S.F.	3.56	3.64	7.20
1600	50,000 S.F.	3.93	3.48	7.41
1660	Each additional floor, 500 S.F.	2.35	2.85	5.20
1680	1000 S.F.	2.31	2.70	5.01
1700	2000 S.F.	2.26	3.17	5.43
1720	5000 S.F.	1.99	2.91	4.90
1740	10,000 S.F.	2.41	2.68	5.09
1760	50,000 S.F.	2.41	2.53	4.94
2020	Grooved steel, black sch. 40 pipe, light hazard, one floor, 2000 S.F.	5.35	2.39	7.74
2060	10,000 S.F.	2.01	1.58	3.59
2100	Each additional floor, 2000 S.F.	1.36	1.44	2.80
2150	10,000 S.F.	.81	1.34	2.15
2200	Ordinary hazard, one floor, 2000 S.F.	5.60	2.74	8.34
2250	10,000 S.F.	2.16	2.10	4.26
2300	Each additional floor, 2000 S.F.	1.60	1.79	3.39
2350	10,000 S.F.	1.41	1.93	3.34
2400	Extra hazard, one floor, 2000 S.F.	6	3.46	9.46
2450	10,000 S.F.	2.96	2.65	5.61
2500	Each additional floor, 2000 S.F.	2.22	2.61	4.83
2550	10,000 S.F.	2.05	2.33	4.38
3050	Grooved steel black sch. 10 pipe, light hazard, one floor, 2000 S.F.	5.25	2.38	7.63
3100	10,000 S.F.	1.73	1.55	3.28
3150	Each additional floor, 2000 S.F.	1.26	1.43	2.69
3200	10,000 S.F.	.74	1.32	2.06
3250	Ordinary hazard, one floor, 2000 S.F.	5.50	2.73	8.23
3300	10,000 S.F.	2.06	2.07	4.13
3350	Each additional floor, 2000 S.F.	1.51	1.78	3.29
3400	10,000 S.F.	1.31	1.90	3.21
3450	Extra hazard, one floor, 2000 S.F.	5.95	3.45	9.40
3500	10,000 S.F.	2.75	2.60	5.35
3550	Each additional floor, 2000 S.F.	2.16	2.60	4.76
3600	10,000 S.F.	1.94	2.29	4.23

D40 Fire Protection

D4010 Sprinklers

Wet Pipe System. A system employing automatic sprinklers attached to a piping system containing water and connected to a water supply so that water discharges immediately from sprinklers opened by heat from a fire.

All areas are assumed to be open.

System Components	QUANTITY	UNIT	MAT.	INST.	TOTAL
SYSTEM D4010 414 0580					
SEISMIC WET PIPE SPRINKLER, STEEL, BLACK, SCH. 40 PIPE					
LIGHT HAZARD, ONE FLOOR, 4' FLEXIBLE FEED, 2000 S.F.					
Valve, gate, iron body, 125 lb., OS&Y, flanged, 4" diam.	1.000	Ea.	.23	.12	.35
Valve, swing check, bronze, 125 lb, regrinding disc, 2-1/2" pipe size	1.000	Ea.	.23	.02	.25
Valve, angle, bronze, 150 lb., rising stem, threaded, 2" diam.	1.000	Ea.	.22	.02	.24
*Alarm valve, 2-1/2" pipe size	1.000	Ea.	.48	.11	.59
Alarm, water motor, complete with gong	1.000	Ea.	.12	.05	.17
Valve, swing check, w/balldrip CI with brass trim 4" pipe size	1.000	Ea.	.10	.11	.21
Pipe, steel, black, schedule 40, 4" diam.	10.000	L.F.	.07	.10	.17
*Flow control valve, trim & gauges, 4" pipe size	1.000	Set	1.69	.26	1.95
Fire alarm horn, electric	1.000	Ea.	.02	.03	.05
Pipe, steel, black, schedule 40, threaded, cplg & hngr 10' OC, 2-1/2" diam.	20.000	L.F.	.09	.14	.23
Pipe, steel, black, schedule 40, threaded, cplg & hngr 10' OC, 2" diam.	12.500	L.F.	.03	.07	.10
Pipe, steel, black, schedule 40, threaded, cplg & hngr 10' OC, 1-1/4" diam.	37.500	L.F.	.07	.15	.22
Pipe Tee, malleable iron black, 150 lb. threaded, 4" pipe size	2.000	Ea.	.16	.18	.34
Pipe Tee, malleable iron black, 150 lb. threaded, 2-1/2" pipe size	2.000	Ea.	.05	.08	.13
Pipe Tee, malleable iron black, 150 lb. threaded, 2" pipe size	1.000	Ea.	.01	.03	.04
Pipe Tee, malleable iron black, 150 lb. threaded, 1-1/4" pipe size	5.000	Ea.	.03	.13	.16
Pipe Tee, malleable iron black, 150 lb. threaded, 1" pipe size	4.000	Ea.	.01	.10	.11
Pipe 90° elbow, malleable iron black, 150 lb. threaded, 1" pipe size	6.000	Ea.	.01	.09	.10
Sprinkler head, standard spray, brass 135°-286°F 1/2" NPT, 3/8" orifice	12.000	Ea.	.07	.19	.26
Valve, gate, bronze, NRS, class 150, threaded, 1" pipe size	1.000	Ea.	.03	.01	.04
*Standpipe connection, wall, single, flush w/plug & chain 2-1/2"x2-1/2"	1.000	Ea.	.05	.07	.12
Flexible connector 48 inches long	12.000	Ea.	.40	.14	.54
TOTAL			4.17	2.20	6.37

*Not included in systems under 2000 S.F.

D4010 414	Seismic Wet Pipe Flexible Feed Sprinkler Systems	MAT.	INST.	TOTAL
0520	Seismic wet pipe 4' flexible feed sprinkler systems, steel, black, sch. 40 pipe			
0525	Note: See D4010-412 for strut braces, flexible joints and separation assemblies			
0530	Light hazard, one floor, 4' flexible feed, 500 S.F.	2.92	2.57	5.49
0560	1000 S.F.	6.10	2.90	9
0580	2000 S.F.	4.17	2.20	6.37
0600	5000 S.F.	2.75	2.05	4.80
0620	10,000 S.F.	1.88	1.87	3.75
0640	50,000 S.F.	1.31	1.57	2.88

D40 Fire Protection

D4010 Sprinklers

D4010 414	Seismic Wet Pipe Flexible Feed Sprinkler Systems	MAT.	INST.	TOTAL
		COST PER S.F.		
0660	Each additional floor, 500 S.F.	1.51	2.15	3.66
0680	1000 S.F.	1.64	2.18	3.82
0700	2000 S.F.	1.41	1.85	3.26
0720	5000 S.F.	1.15	1.67	2.82
0740	10,000 S.F.	1.13	1.70	2.83
0760	50,000 S.F.	.98	1.34	2.32
1000	Ordinary hazard, one floor, 500 S.F.	3.33	2.82	6.15
1020	1000 S.F.	6.15	2.80	8.95
1040	2000 S.F.	5.70	3.20	8.90
1060	5000 S.F.	3.17	2.34	5.51
1080	10,000 S.F.	2.38	2.52	4.90
1100	50,000 S.F.	1.96	2.25	4.21
1140	Each additional floor, 500 S.F.	1.81	2.50	4.31
1160	1000 S.F.	1.57	2.07	3.64
1180	2000 S.F.	1.74	2.25	3.99
1200	5000 S.F.	1.79	2.21	4
1220	10,000 S.F.	1.63	2.35	3.98
1240	50,000 S.F.	1.52	1.96	3.48
1500	Extra hazard, one floor, 500 S.F.	11.55	4.36	15.91
1520	1000 S.F.	7.35	3.73	11.08
1540	2000 S.F.	6.20	4.30	10.50
1560	5000 S.F.	4.02	3.87	7.89
1580	10,000 S.F.	3.67	3.72	7.39
1600	50,000 S.F.	4.04	3.56	7.60
1660	Each additional floor, 500 S.F.	2.45	2.92	5.37
1680	1000 S.F.	2.41	2.77	5.18
1700	2000 S.F.	2.36	3.24	5.60
1720	5000 S.F.	2.08	2.97	5.05
1740	10,000 S.F.	2.52	2.76	5.28
1760	50,000 S.F.	2.52	2.61	5.13
2020	Grooved steel, black sch. 40 pipe, light hazard, one floor, 2000 S.F.	5.40	2.43	7.83
2060	10,000 S.F.	2.03	1.59	3.62
2100	Each additional floor, 2000 S.F.	1.41	1.48	2.89
2150	10,000 S.F.	.83	1.35	2.18
2200	Ordinary hazard, one floor, 2000 S.F.	5.65	2.79	8.44
2250	10,000 S.F.	2.22	2.14	4.36
2300	Each additional floor, 2000 S.F.	1.67	1.84	3.51
2350	10,000 S.F.	1.47	1.97	3.44
2400	Extra hazard, one floor, 2000 S.F.	6.10	3.53	9.63
2450	10,000 S.F.	3.07	2.73	5.80
2500	Each additional floor, 2000 S.F.	2.32	2.68	5
2550	10,000 S.F.	2.16	2.41	4.57
3050	Grooved steel black sch. 10 pipe, light hazard, one floor, 2000 S.F.	5.30	2.42	7.72
3100	10,000 S.F.	1.77	1.59	3.36
3150	Each additional floor, 2000 S.F.	1.31	1.47	2.78
3200	10,000 S.F.	.76	1.33	2.09
3250	Ordinary hazard, one floor, 2000 S.F.	5.55	2.78	8.33
3300	10,000 S.F.	2.12	2.11	4.23
3350	Each additional floor, 2000 S.F.	1.58	1.83	3.41
3400	10,000 S.F.	1.37	1.94	3.31
3450	Extra hazard, one floor, 2000 S.F.	6.05	3.52	9.57
3500	10,000 S.F.	2.86	2.68	5.54
3550	Each additional floor, 2000 S.F.	2.26	2.67	4.93
3600	10,000 S.F.	2.05	2.37	4.42

D40 Fire Protection

D4020 Standpipes

System Components	QUANTITY	UNIT	COST PER FLOOR MAT.	COST PER FLOOR INST.	COST PER FLOOR TOTAL
SYSTEM D4020 310 0560 **WET STANDPIPE RISER, CLASS I, STEEL, BLACK, SCH. 40 PIPE, 10' HEIGHT 4" DIAMETER PIPE, ONE FLOOR**					
Pipe, steel, black, schedule 40, threaded, 4" diam.	20.000	L.F.	600	700	1,300
Pipe, Tee, malleable iron, black, 150 lb. threaded, 4" pipe size	2.000	Ea.	584	630	1,214
Pipe, 90° elbow, malleable iron, black, 150 lb threaded 4" pipe size	1.000	Ea.	185	209	394
Pipe, nipple, steel, black, schedule 40, 2-1/2" pipe size x 3" long	2.000	Ea.	24.50	156	180.50
Fire valve, gate, 300 lb., brass w/handwheel, 2-1/2" pipe size	1.000	Ea.	198	95.50	293.50
Fire valve, pressure restricting, adj, rgh brs, 2-1/2" pipe size	1.000	Ea.	352	191	543
Valve, swing check, w/ball drip, CI w/brs. ftngs., 4" pipe size	1.000	Ea.	365	400	765
Standpipe conn wall dble. flush brs. w/plugs & chains 2-1/2"x2-1/2"x4"	1.000	Ea.	545	241	786
Valve, swing check, bronze, 125 lb, regrinding disc, 2-1/2" pipe size	1.000	Ea.	835	83.50	918.50
Roof manifold, fire, w/valves & caps, horiz/vert brs 2-1/2"x2-1/2"x4"	1.000	Ea.	176	251	427
Fire, hydrolator, vent & drain, 2-1/2" pipe size	1.000	Ea.	74.50	56	130.50
Valve, gate, iron body 125 lb., OS&Y, threaded, 4" pipe size	1.000	Ea.	805	415	1,220
TOTAL			4,744	3,428	8,172

D4020 310	Wet Standpipe Risers, Class I		MAT.	INST.	TOTAL
0550	Wet standpipe risers, Class I, steel, black sch. 40, 10' height				
0560	4" diameter pipe, one floor		4,750	3,425	8,175
0580	Additional floors		1,075	1,075	2,150
0600	6" diameter pipe, one floor		8,075	6,025	14,100
0620	Additional floors	R211226 -10	1,950	1,675	3,625
0640	8" diameter pipe, one floor		12,500	7,250	19,750
0660	Additional floors	R211226 -20	2,950	2,050	5,000
0680					

D4020 310	Wet Standpipe Risers, Class II	MAT.	INST.	TOTAL
1030	Wet standpipe risers, Class II, steel, black sch. 40, 10' height			
1040	2" diameter pipe, one floor	1,875	1,250	3,125
1060	Additional floors	550	480	1,030
1080	2-1/2" diameter pipe, one floor	2,750	1,800	4,550
1100	Additional floors	635	555	1,190
1120				

D40 Fire Protection

D4020 Standpipes

D4020 310	Wet Standpipe Risers, Class III	COST PER FLOOR		
		MAT.	INST.	TOTAL
1530	Wet standpipe risers, Class III, steel, black sch. 40, 10' height			
1540	4" diameter pipe, one floor	4,850	3,425	8,275
1560	Additional floors	915	895	1,810
1580	6" diameter pipe, one floor	8,175	6,025	14,200
1600	Additional floors	2,000	1,675	3,675
1620	8" diameter pipe, one floor	12,600	7,250	19,850
1640	Additional floors	3,000	2,050	5,050

D40 Fire Protection

D4020 Standpipes

- Roof connections with hose gate valves (for combustible roof)
- Hose connections on each floor (size based on class of service)
- Siamese inlet connections (for fire department use)
- Roof
- Check Valve

System Components	QUANTITY	UNIT	COST PER FLOOR MAT.	COST PER FLOOR INST.	COST PER FLOOR TOTAL
SYSTEM D4020 330 0540					
DRY STANDPIPE RISER, CLASS I, PIPE, STEEL, BLACK, SCH 40, 10' HEIGHT					
4" DIAMETER PIPE, ONE FLOOR					
Pipe, steel, black, schedule 40, threaded, 4" diam.	20.000	L.F.	600	700	1,300
Pipe, Tee, malleable iron, black, 150 lb. threaded, 4" pipe size	2.000	Ea.	584	630	1,214
Pipe, 90° elbow, malleable iron, black, 150 lb threaded 4" pipe size	1.000	Ea.	185	209	394
Pipe, nipple, steel, black, schedule 40, 2-1/2" pipe size x 3" long	2.000	Ea.	24.50	156	180.50
Fire valve gate NRS 300 lb., brass w/handwheel, 2-1/2" pipe size	1.000	Ea.	198	95.50	293.50
Fire valve, pressure restricting, adj, rgh brs, 2-1/2" pipe size	1.000	Ea.	176	95.50	271.50
Standpipe conn wall dble. flush brs. w/plugs & chains 2-1/2"x2-1/2"x4"	1.000	Ea.	545	241	786
Valve swing check w/ball drip Cl w/brs. ftngs., 4"pipe size	1.000	Ea.	365	400	765
Roof manifold, fire, w/valves & caps, horiz/vert brs 2-1/2"x2-1/2"x4"	1.000	Ea.	176	251	427
TOTAL			2,853.50	2,778	5,631.50

D4020 330	Dry Standpipe Risers, Class I		COST PER FLOOR MAT.	COST PER FLOOR INST.	COST PER FLOOR TOTAL
0530	Dry standpipe riser, Class I, steel, black sch. 40, 10' height				
0540	4" diameter pipe, one floor		2,850	2,775	5,625
0560	Additional floors		990	1,000	1,990
0580	6" diameter pipe, one floor		5,775	4,775	10,550
0600	Additional floors	R211226 -10	1,875	1,625	3,500
0620	8" diameter pipe, one floor		9,225	5,775	15,000
0640	Additional floors	R211226 -20	2,875	1,975	4,850
0660					

D4020 330	Dry Standpipe Risers, Class II	COST PER FLOOR MAT.	COST PER FLOOR INST.	COST PER FLOOR TOTAL
1030	Dry standpipe risers, Class II, steel, black sch. 40, 10' height			
1040	2" diameter pipe, one floor	1,525	1,275	2,800
1060	Additional floor	475	420	895
1080	2-1/2" diameter pipe, one floor	2,175	1,500	3,675
1100	Additional floors	560	500	1,060
1120				

D40 Fire Protection

D4020 Standpipes

D4020 330	Dry Standpipe Risers, Class III	COST PER FLOOR			
		MAT.	INST.	TOTAL
1530	Dry standpipe risers, Class III, steel, black sch. 40, 10' height			
1540	4" diameter pipe, one floor	2,900	2,725	5,625
1560	Additional floors	850	915	1,765
1580	6" diameter pipe, one floor	5,850	4,775	10,625
1600	Additional floors	1,925	1,625	3,550
1620	8" diameter pipe, one floor	9,275	5,775	15,050
1640	Additional floor	2,925	1,975	4,900

D40 Fire Protection

D4020 Standpipes

D4020 410	Fire Hose Equipment	MAT.	INST.	TOTAL
0100	Adapters, reducing, 1 piece, FxM, hexagon, cast brass, 2-1/2" x 1-1/2"	58		58
0200	Pin lug, 1-1/2" x 1"	50		50
0250	3" x 2-1/2"	131		131
0300	For polished chrome, add 75% mat.			
0400	Cabinets, D.S. glass in door, recessed, steel box, not equipped			
0500	Single extinguisher, steel door & frame	115	151	266
0550	Stainless steel door & frame	215	151	366
0600	Valve, 2-1/2" angle, steel door & frame	142	100	242
0650	Aluminum door & frame	172	100	272
0700	Stainless steel door & frame	231	100	331
0750	Hose rack assy, 2-1/2" x 1-1/2" valve & 100' hose, steel door & frame	265	201	466
0800	Aluminum door & frame	390	201	591
0850	Stainless steel door & frame	520	201	721
0900	Hose rack assy & extinguisher, 2-1/2"x1-1/2" valve & hose, steel door & frame	330	241	571
0950	Aluminum	500	241	741
1000	Stainless steel	605	241	846
1550	Compressor, air, dry pipe system, automatic, 200 gal., 3/4 H.P.	810	515	1,325
1600	520 gal., 1 H.P.	835	515	1,350
1650	Alarm, electric pressure switch (circuit closer)	92.50	26	118.50
2500	Couplings, hose, rocker lug, cast brass, 1-1/2"	56		56
2550	2-1/2"	73		73
3000	Escutcheon plate, for angle valves, polished brass, 1-1/2"	16.70		16.70
3050	2-1/2"	27		27
3500	Fire pump, electric, w/controller, fittings, relief valve			
3550	4" pump, 30 H.P., 500 G.P.M.	17,100	3,750	20,850
3600	5" pump, 40 H.P., 1000 G.P.M.	18,800	4,250	23,050
3650	5" pump, 100 H.P., 1000 G.P.M.	25,900	4,725	30,625
3700	For jockey pump system, add	2,775	605	3,380
5000	Hose, per linear foot, synthetic jacket, lined,			
5100	300 lb. test, 1-1/2" diameter	3.40	.46	3.86
5150	2-1/2" diameter	5.85	.55	6.40
5200	500 lb. test, 1-1/2" diameter	3.50	.46	3.96
5250	2-1/2" diameter	6.20	.55	6.75
5500	Nozzle, plain stream, polished brass, 1-1/2" x 10"	50		50
5550	2-1/2" x 15" x 13/16" or 1-1/2"	105		105
5600	Heavy duty combination adjustable fog and straight stream w/handle 1-1/2"	465		465
5650	2-1/2" direct connection	575		575
6000	Rack, for 1-1/2" diameter hose 100 ft. long, steel	64	60.50	124.50
6050	Brass	109	60.50	169.50
6500	Reel, steel, for 50 ft. long 1-1/2" diameter hose	137	86	223
6550	For 75 ft. long 2-1/2" diameter hose	222	86	308
7050	Siamese, w/plugs & chains, polished brass, sidewalk, 4" x 2-1/2" x 2-1/2"	665	480	1,145
7100	6" x 2-1/2" x 2-1/2"	810	605	1,415
7200	Wall type, flush, 4" x 2-1/2" x 2-1/2"	545	241	786
7250	6" x 2-1/2" x 2-1/2"	755	262	1,017
7300	Projecting, 4" x 2-1/2" x 2-1/2"	495	241	736
7350	6" x 2-1/2" x 2-1/2"	840	262	1,102
7400	For chrome plate, add 15% mat.			
8000	Valves, angle, wheel handle, 300 Lb., rough brass, 1-1/2"	94.50	56	150.50
8050	2-1/2"	183	95.50	278.50
8100	Combination pressure restricting, 1-1/2"	89.50	56	145.50
8150	2-1/2"	183	95.50	278.50
8200	Pressure restricting, adjustable, satin brass, 1-1/2"	125	56	181
8250	2-1/2"	176	95.50	271.50
8300	Hydrolator, vent and drain, rough brass, 1-1/2"	74.50	56	130.50
8350	2-1/2"	74.50	56	130.50
8400	Cabinet assy, incls. adapter, rack, hose, and nozzle	820	365	1,185

D40 Fire Protection

D4090 Other Fire Protection Systems

General: Automatic fire protection (suppression) systems other than water sprinklers may be desired for special environments, high risk areas, isolated locations or unusual hazards. Some typical applications would include:

Paint dip tanks
Securities vaults
Electronic data processing
Tape and data storage
Transformer rooms
Spray booths
Petroleum storage
High rack storage

Piping and wiring costs are dependent on the individual application and must be added to the component costs shown below.

All areas are assumed to be open.

D4090 910	Fire Suppression Unit Components	COST EACH		
		MAT.	INST.	TOTAL
0020	Detectors with brackets			
0040	Fixed temperature heat detector	55.50	80	135.50
0060	Rate of temperature rise detector	55.50	80	135.50
0080	Ion detector (smoke) detector	120	103	223
0200	Extinguisher agent			
0240	200 lb FM200, container	8,300	315	8,615
0280	75 lb carbon dioxide cylinder	1,350	209	1,559
0320	Dispersion nozzle			
0340	FM200 1-1/2" dispersion nozzle	71.50	49.50	121
0380	Carbon dioxide 3" x 5" dispersion nozzle	71.50	38.50	110
0420	Control station			
0440	Single zone control station with batteries	1,850	640	2,490
0470	Multizone (4) control station with batteries	3,500	1,275	4,775
0490				
0500	Electric mechanical release	179	335	514
0520				
0550	Manual pull station	64.50	116	180.50
0570				
0640	Battery standby power 10" x 10" x 17"	450	160	610
0700				
0740	Bell signalling device	85.50	80	165.50

D4090 920	FM200 Systems	COST PER C.F.		
		MAT.	INST.	TOTAL
0820	Average FM200 system, minimum			1.88
0840	Maximum			3.74

G Building Sitework

No part of this publication may be reproduced, stored in a retrieval system, or transmitted in any form or by any means without prior written permission of RSMeans, a division of Reed Construction Data.

G10 Site Preparation

G1030 Site Earthwork

Trenching Systems are shown on a cost per linear foot basis. The systems include: excavation; backfill and removal of spoil; and compaction for various depths and trench bottom widths. The backfill has been reduced to accommodate a pipe of suitable diameter and bedding.

The slope for trench sides varies from none to 1:1.

The Expanded System Listing shows Trenching Systems that range from 2' to 12' in width. Depths range from 2' to 25'.

System Components	QUANTITY	UNIT	EQUIP.	LABOR	TOTAL
SYSTEM G1030 805 1310					
TRENCHING COMMON EARTH, NO SLOPE, 2' WIDE, 2' DP, 3/8 C.Y. BUCKET					
Excavation, trench, hyd. backhoe, track mtd., 3/8 C.Y. bucket	.148	B.C.Y.	.40	1.03	1.43
Backfill and load spoil, from stockpile	.153	L.C.Y.	.12	.31	.43
Compaction by vibrating plate, 6" lifts, 4 passes	.118	E.C.Y.	.03	.38	.41
Remove excess spoil, 8 C.Y. dump truck, 2 mile roundtrip	.040	L.C.Y.	.15	.15	.30
TOTAL			.70	1.87	2.57

G1030 805	Trenching Common Earth	EQUIP.	LABOR	TOTAL
1310	Trenching, common earth, no slope, 2' wide, 2' deep, 3/8 C.Y. bucket	.70	1.87	2.57
1320	3' deep, 3/8 C.Y. bucket	.99	2.82	3.81
1330	4' deep, 3/8 C.Y. bucket	1.29	3.78	5.07
1340	6' deep, 3/8 C.Y. bucket	1.68	4.92	6.60
1350	8' deep, 1/2 C.Y. bucket	2.22	6.55	8.77
1360	10' deep, 1 C.Y. bucket	3.48	7.75	11.23
1400	4' wide, 2' deep, 3/8 C.Y. bucket	1.61	3.70	5.31
1410	3' deep, 3/8 C.Y. bucket	2.21	5.60	7.81
1420	4' deep, 1/2 C.Y. bucket	2.57	6.30	8.87
1430	6' deep, 1/2 C.Y. bucket	4.15	10.15	14.30
1440	8' deep, 1/2 C.Y. bucket	6.65	13.05	19.70
1450	10' deep, 1 C.Y. bucket	8.05	16.20	24.25
1460	12' deep, 1 C.Y. bucket	10.35	20.50	30.85
1470	15' deep, 1-1/2 C.Y. bucket	9.40	18.50	27.90
1480	18' deep, 2-1/2 C.Y. bucket	13.05	26	39.05
1520	6' wide, 6' deep, 5/8 C.Y. bucket w/trench box	8.60	14.95	23.55
1530	8' deep, 3/4 C.Y. bucket	11.40	19.75	31.15
1540	10' deep, 1 C.Y. bucket	11.55	20.50	32.05
1550	12' deep, 1-1/2 C.Y. bucket	12.40	22	34.40
1560	16' deep, 2-1/2 C.Y. bucket	17.10	27.50	44.60
1570	20' deep, 3-1/2 C.Y. bucket	22	33	55
1580	24' deep, 3-1/2 C.Y. bucket	26	39.50	65.50
1640	8' wide, 12' deep, 1-1/2 C.Y. bucket w/trench box	17.40	28	45.40
1650	15' deep, 1-1/2 C.Y. bucket	22.50	36.50	59
1660	18' deep, 2-1/2 C.Y. bucket	25	36.50	61.50
1680	24' deep, 3-1/2 C.Y. bucket	35.50	51	86.50
1730	10' wide, 20' deep, 3-1/2 C.Y. bucket w/trench box	28.50	48.50	77
1740	24' deep, 3-1/2 C.Y. bucket	42.50	58.50	101
1780	12' wide, 20' deep, 3-1/2 C.Y. bucket w/trench box	45	61.50	106.50
1790	25' deep, bucket	56	78.50	134.50
1800	1/2 to 1 slope, 2' wide, 2' deep, 3/8 C.Y. bucket	.99	2.82	3.81
1810	3' deep, 3/8 C.Y. bucket	1.67	4.97	6.64
1820	4' deep, 3/8 C.Y. bucket	2.49	7.60	10.09
1840	6' deep, 3/8 C.Y. bucket	4.02	12.35	16.37

G10 Site Preparation

G1030 Site Earthwork

G1030 805	Trenching Common Earth	EQUIP.	LABOR	TOTAL
1860	8' deep, 1/2 C.Y. bucket	6.40	19.75	26.15
1880	10' deep, 1 C.Y. bucket	11.95	27.50	39.45
2300	4' wide, 2' deep, 3/8 C.Y. bucket	1.91	4.65	6.56
2310	3' deep, 3/8 C.Y. bucket	2.88	7.75	10.63
2320	4' deep, 1/2 C.Y. bucket	3.63	9.65	13.28
2340	6' deep, 1/2 C.Y. bucket	6.95	18	24.95
2360	8' deep, 1/2 C.Y. bucket	12.85	26.50	39.35
2380	10' deep, 1 C.Y. bucket	17.80	37	54.80
2400	12' deep, 1 C.Y. bucket	24.50	50.50	75
2430	15' deep, 1-1/2 C.Y. bucket	26.50	54	80.50
2460	18' deep, 2-1/2 C.Y. bucket	46.50	84.50	131
2840	6' wide, 6' deep, 5/8 C.Y. bucket w/trench box	12.70	22.50	35.20
2860	8' deep, 3/4 C.Y. bucket	18.35	33.50	51.85
2880	10' deep, 1 C.Y. bucket	18.60	34	52.60
2900	12' deep, 1-1/2 C.Y. bucket	23.50	44.50	68
2940	16' deep, 2-1/2 C.Y. bucket	39	65.50	104.50
2980	20' deep, 3-1/2 C.Y. bucket	55	88.50	143.50
3020	24' deep, 3-1/2 C.Y. bucket	77.50	121	198.50
3100	8' wide, 12' deep, 1-1/2 C.Y. bucket w/trench box	29	51	80
3120	15' deep, 1-1/2 C.Y. bucket	42.50	74	116.50
3140	18' deep, 2-1/2 C.Y. bucket	54	88	142
3180	24' deep, 3-1/2 C.Y. bucket	87	133	220
3270	10' wide, 20' deep, 3-1/2 C.Y. bucket w/trench box	56	102	158
3280	24' deep, 3-1/2 C.Y. bucket	96	145	241
3370	12' wide, 20' deep, 3-1/2 C.Y. bucket w/trench box	81	118	199
3380	25' deep, 3-1/2 C.Y. bucket	112	169	281
3500	1 to 1 slope, 2' wide, 2' deep, 3/8 C.Y. bucket	1.29	3.78	5.07
3520	3' deep, 3/8 C.Y. bucket	3.72	8.50	12.22
3540	4' deep, 3/8 C.Y. bucket	3.68	11.40	15.08
3560	6' deep, 1/2 C.Y. bucket	4.02	12.35	16.37
3580	8' deep, 1/2 C.Y. bucket	7.95	24.50	32.45
3600	10' deep, 1 C.Y. bucket	20.50	47	67.50
3800	4' wide, 2' deep, 3/8 C.Y. bucket	2.21	5.60	7.81
3820	3' deep, 3/8 C.Y. bucket	3.55	9.90	13.45
3840	4' deep, 1/2 C.Y. bucket	4.67	12.90	17.57
3860	6' deep, 1/2 C.Y. bucket	9.75	26	35.75
3880	8' deep, 1/2 C.Y. bucket	19.10	40	59.10
3900	10' deep, 1 C.Y. bucket	27.50	58	85.50
3920	12' deep, 1 C.Y. bucket	40.50	84	124.50
3940	15' deep, 1-1/2 C.Y. bucket	43.50	89.50	133
3960	18' deep, 2-1/2 C.Y. bucket	64	116	180
4030	6' wide, 6' deep, 5/8 C.Y. bucket w/trench box	16.60	30	46.60
4040	8' deep, 3/4 C.Y. bucket	24	41.50	65.50
4050	10' deep, 1 C.Y. bucket	27	50	77
4060	12' deep, 1-1/2 C.Y. bucket	36	68	104
4070	16' deep, 2-1/2 C.Y. bucket	61	104	165
4080	20' deep, 3-1/2 C.Y. bucket	88.50	144	232.50
4090	24' deep, 3-1/2 C.Y. bucket	129	203	332
4500	8' wide, 12' deep, 1-1/2 C.Y. bucket w/trench box	41	74	115
4550	15' deep, 1-1/2 C.Y. bucket	62	111	173
4600	18' deep, 2-1/2 C.Y. bucket	81.50	136	217.50
4650	24' deep, 3-1/2 C.Y. bucket	139	214	353
4800	10' wide, 20' deep, 3-1/2 C.Y. bucket w/trench box	83	155	238
4850	24' deep, 3-1/2 C.Y. bucket	148	226	374
4950	12' wide, 20' deep, 3-1/2 C.Y. bucket w/trench box	117	175	292
4980	25' deep, 3-1/2 C.Y. bucket	168	256	424

G10 Site Preparation

G1030 Site Earthwork

Trenching Systems are shown on a cost per linear foot basis. The systems include: excavation; backfill and removal of spoil; and compaction for various depths and trench bottom widths. The backfill has been reduced to accommodate a pipe of suitable diameter and bedding.

The slope for trench sides varies from none to 1:1.

The Expanded System Listing shows Trenching Systems that range from 2' to 12' in width. Depths range from 2' to 25'.

System Components	QUANTITY	UNIT	COST PER L.F. EQUIP.	LABOR	TOTAL
SYSTEM G1030 806 1310					
TRENCHING LOAM & SANDY CLAY, NO SLOPE, 2' WIDE, 2' DP, 3/8 C.Y. BUCKET					
Excavation, trench, hyd. backhoe, track mtd., 3/8 C.Y. bucket	.148	B.C.Y.	.37	.95	1.32
Backfill and load spoil, from stockpile	.165	L.C.Y.	.13	.34	.47
Compaction by vibrating plate 18" wide, 6" lifts, 4 passes	.118	E.C.Y.	.03	.38	.41
Remove excess spoil, 8 C.Y. dump truck, 2 mile roundtrip	.042	L.C.Y.	.16	.16	.32
TOTAL			.70	1.87	2.57

G1030 806	Trenching Loam & Sandy Clay	EQUIP.	LABOR	TOTAL
1310	Trenching, loam & sandy clay, no slope, 2' wide, 2' deep, 3/8 C.Y. bucket	.69	1.83	2.52
1320	3' deep, 3/8 C.Y. bucket	1.08	3.03	4.11
1330	4' deep, 3/8 C.Y. bucket	1.26	3.69	4.95
1340	6' deep, 3/8 C.Y. bucket	1.80	4.40	6.20
1350	8' deep, 1/2 C.Y. bucket	2.39	5.80	8.20
1360	10' deep, 1 C.Y. bucket	2.69	6.25	8.95
1400	4' wide, 2' deep, 3/8 C.Y. bucket	1.60	3.63	5.25
1410	3' deep, 3/8 C.Y. bucket	2.18	5.50	7.70
1420	4' deep, 1/2 C.Y. bucket	2.56	6.20	8.75
1430	6' deep, 1/2 C.Y. bucket	4.42	9.05	13.45
1440	8' deep, 1/2 C.Y. bucket	6.50	12.85	19.35
1450	10' deep, 1 C.Y. bucket	6.45	13.20	19.65
1460	12' deep, 1 C.Y. bucket	8.10	16.40	24.50
1470	15' deep, 1-1/2 C.Y. bucket	9.80	19.10	29
1480	18' deep, 2-1/2 C.Y. bucket	11.60	21	32.50
1520	6' wide, 6' deep, 5/8 C.Y. bucket w/trench box	8.35	14.75	23
1530	8' deep, 3/4 C.Y. bucket	11	19.35	30.50
1540	10' deep, 1 C.Y. bucket	10.60	19.70	30.50
1550	12' deep, 1-1/2 C.Y. bucket	12.10	22	34
1560	16' deep, 2-1/2 C.Y. bucket	16.70	28	44.50
1570	20' deep, 3-1/2 C.Y. bucket	20	33	53
1580	24' deep, 3-1/2 C.Y. bucket	25.50	40	65.50
1640	8' wide, 12' deep, 1-1/4 C.Y. bucket w/trench box	17.15	28	45
1650	15' deep, 1-1/2 C.Y. bucket	21	35.50	56.50
1660	18' deep, 2-1/2 C.Y. bucket	25.50	40	65.50
1680	24' deep, 3-1/2 C.Y. bucket	34.50	52	86.50
1730	10' wide, 20' deep, 3-1/2 C.Y. bucket w/trench box	35	52	87
1740	24' deep, 3-1/2 C.Y. bucket	43.50	64	108
1780	12' wide, 20' deep, 3-1/2 C.Y. bucket w/trench box	42	61.50	104
1790	25' deep, 3-1/2 C.Y. bucket	54.50	79.50	134
1800	1/2:1 slope, 2' wide, 2' deep, 3/8 C.Y. bucket	.98	2.76	3.74
1810	3' deep, 3/8 C.Y. bucket	1.63	4.86	6.50
1820	4' deep, 3/8 C.Y. bucket	2.43	7.40	9.85
1840	6' deep, 3/8 C.Y. bucket	4.32	11	15.30

G10 Site Preparation

G1030 Site Earthwork

G1030 806	Trenching Loam & Sandy Clay	COST PER L.F. EQUIP.	LABOR	TOTAL
1860	8' deep, 1/2 C.Y. bucket	6.85	17.65	24.50
1880	10' deep, 1 C.Y. bucket	9.15	22	31
2300	4' wide, 2' deep, 3/8 C.Y. bucket	1.90	4.56	6.45
2310	3' deep, 3/8 C.Y. bucket	2.84	7.60	10.45
2320	4' deep, 1/2 C.Y. bucket	3.59	9.45	13.05
2340	6' deep, 1/2 C.Y. bucket	7.40	16.15	23.50
2360	8' deep, 1/2 C.Y. bucket	12.50	26	38.50
2380	10' deep, 1 C.Y. bucket	14.20	30.50	44.50
2400	12' deep, 1 C.Y. bucket	23.50	50	73.50
2430	15' deep, 1-1/2 C.Y. bucket	27.50	56	83.50
2460	18' deep, 2-1/2 C.Y. bucket	46	85.50	132
2840	6' wide, 6' deep, 5/8 C.Y. bucket w/trench box	12.10	22.50	34.50
2860	8' deep, 3/4 C.Y. bucket	17.70	33	50.50
2880	10' deep, 1 C.Y. bucket	18.95	37	56
2900	12' deep, 1-1/2 C.Y. bucket	23.50	45	68.50
2940	16' deep, 2-1/2 C.Y. bucket	38	66.50	105
2980	20' deep, 3-1/2 C.Y. bucket	53	89.50	143
3020	24' deep, 3-1/2 C.Y. bucket	75	123	198
3100	8' wide, 12' deep, 1-1/2 C.Y. bucket w/trench box	28.50	51	79.50
3120	15' deep, 1-1/2 C.Y. bucket	39	72	111
3140	18' deep, 2-1/2 C.Y. bucket	52.50	89	142
3180	24' deep, 3-1/2 C.Y. bucket	84.50	135	220
3270	10' wide, 20' deep, 3-1/2 C.Y. bucket w/trench box	68	109	177
3280	24' deep, 3-1/2 C.Y. bucket	93.50	147	241
3320	12' wide, 20' deep, 3-1/2 C.Y. bucket w/trench box	75	118	193
3380	25' deep, 3-1/2 C.Y. bucket w/trench box	102	159	261
3500	1:1 slope, 2' wide, 2' deep, 3/8 C.Y. bucket	1.26	3.69	4.95
3520	3' deep, 3/8 C.Y. bucket	2.29	6.95	9.25
3540	4' deep, 3/8 C.Y. bucket	3.57	11.10	14.65
3560	6' deep, 1/2 C.Y. bucket	4.32	11	15.30
3580	8' deep, 1/2 C.Y. bucket	11.35	29.50	41
3600	10' deep, 1 C.Y. bucket	15.65	38	53.50
3800	4' wide, 2' deep, 3/8 C.Y. bucket	2.18	5.50	7.70
3820	3' deep, 1/2 C.Y. bucket	3.49	9.70	13.20
3840	4' deep, 1/2 C.Y. bucket	4.61	12.65	17.25
3860	6' deep, 1/2 C.Y. bucket	10.35	23	33.50
3880	8' deep, 1/2 C.Y. bucket	18.55	39.50	58
3900	10' deep, 1 C.Y. bucket	22	47.50	69.50
3920	12' deep, 1 C.Y. bucket	31.50	67	98.50
3940	15' deep, 1-1/2 C.Y. bucket	45.50	92.50	138
3960	18' deep, 2-1/2 C.Y. bucket	63	118	181
4030	6' wide, 6' deep, 5/8 C.Y. bucket w/trench box	15.85	30	46
4040	8' deep, 3/4 C.Y. bucket	24.50	46.50	71
4050	10' deep, 1 C.Y. bucket	27.50	54.50	82
4060	12' deep, 1-1/2 C.Y. bucket	35	68.50	104
4070	16' deep, 2-1/2 C.Y. bucket	59.50	105	165
4080	20' deep, 3-1/2 C.Y. bucket	86	146	232
4090	24' deep, 3-1/2 C.Y. bucket	125	206	330
4500	8' wide, 12' deep, 1-1/4 C.Y. bucket w/trench box	40	74	114
4550	15' deep, 1-1/2 C.Y. bucket	57	108	165
4600	18' deep, 2-1/2 C.Y. bucket	79.50	138	218
4650	24' deep, 3-1/2 C.Y. bucket	134	217	350
4800	10' wide, 20' deep, 3-1/2 C.Y. bucket w/trench box	101	166	267
4850	24' deep, 3-1/2 C.Y. bucket	143	230	375
4950	12' wide, 20' deep, 3-1/2 C.Y. bucket w/trench box	108	175	283
4980	25' deep, 3-1/2 C.Y. bucket	163	259	420

G10 Site Preparation

G1030 Site Earthwork

Trenching Systems are shown on a cost per linear foot basis. The systems include: excavation; backfill and removal of spoil; and compaction for various depths and trench bottom widths. The backfill has been reduced to accommodate a pipe of suitable diameter and bedding.

The slope for trench sides varies from none to 1:1.

The Expanded System Listing shows Trenching Systems that range from 2' to 12' in width. Depths range from 2' to 25'.

System Components	QUANTITY	UNIT	COST PER L.F. EQUIP.	LABOR	TOTAL
SYSTEM G1030 807 1310					
TRENCHING SAND & GRAVEL, NO SLOPE, 2' WIDE, 2' DEEP, 3/8 C.Y. BUCKET					
Excavation, trench, hyd. backhoe, track mtd., 3/8 C.Y. bucket	.148	B.C.Y.	.36	.94	1.30
Backfill and load spoil, from stockpile	.140	L.C.Y.	.11	.29	.40
Compaction by vibrating plate 18" wide, 6" lifts, 4 passes	.118	E.C.Y.	.03	.38	.41
Remove excess spoil, 8 C.Y. dump truck, 2 mile roundtrip	.035	L.C.Y.	.14	.14	.28
TOTAL			.64	1.75	2.39

G1030 807	Trenching Sand & Gravel	EQUIP.	LABOR	TOTAL
1310	Trenching, sand & gravel, no slope, 2' wide, 2' deep, 3/8 C.Y. bucket	.64	1.75	2.39
1320	3' deep, 3/8 C.Y. bucket	1.02	2.93	3.95
1330	4' deep, 3/8 C.Y. bucket	1.18	3.52	4.70
1340	6' deep, 3/8 C.Y. bucket	1.72	4.19	5.90
1350	8' deep, 1/2 C.Y. bucket	2.26	5.55	7.80
1360	10' deep, 1 C.Y. bucket	2.51	5.90	8.40
1400	4' wide, 2' deep, 3/8 C.Y. bucket	1.46	3.42	4.88
1410	3' deep, 3/8 C.Y. bucket	2	5.20	7.20
1420	4' deep, 1/2 C.Y. bucket	2.35	5.85	8.20
1430	6' deep, 1/2 C.Y. bucket	4.18	8.65	12.85
1440	8' deep, 1/2 C.Y. bucket	6.15	12.20	18.35
1450	10' deep, 1 C.Y. bucket	6.10	12.45	18.55
1460	12' deep, 1 C.Y. bucket	7.65	15.55	23
1470	15' deep, 1-1/2 C.Y. bucket	9.25	18	27.50
1480	18' deep, 2-1/2 C.Y. bucket	11	19.70	30.50
1520	6' wide, 6' deep, 5/8 C.Y. bucket w/trench box	7.85	13.95	22
1530	8' deep, 3/4 C.Y. bucket	10.40	18.40	29
1540	10' deep, 1 C.Y. bucket	10	18.60	28.50
1550	12' deep, 1-1/2 C.Y. bucket	11.40	21	32.50
1560	16' deep, 2 C.Y. bucket	16	26	42
1570	20' deep, 3-1/2 C.Y. bucket	19.15	31	50
1580	24' deep, 3-1/2 C.Y. bucket	24	37.50	61.50
1640	8' wide, 12' deep, 1-1/2 C.Y. bucket w/trench box	16.10	26.50	42.50
1650	15' deep, 1-1/2 C.Y. bucket	19.55	33.50	53
1660	18' deep, 2-1/2 C.Y. bucket	24	37.50	61.50
1680	24' deep, 3-1/2 C.Y. bucket	32.50	48.50	81
1730	10' wide, 20' deep, 3-1/2 C.Y. bucket w/trench box	32.50	48.50	81
1740	24' deep, 3-1/2 C.Y. bucket	41	60	101
1780	12' wide, 20' deep, 3-1/2 C.Y. bucket w/trench box	39.50	57.50	97
1790	25' deep, 3-1/2 C.Y. bucket	51.50	74.50	126
1800	1/2:1 slope, 2' wide, 2' deep, 3/8 C.Y. bucket	.91	2.63	3.54
1810	3' deep, 3/8 C.Y. bucket	1.53	4.64	6.15
1820	4' deep, 3/8 C.Y. bucket	2.28	7.10	9.40
1840	6' deep, 3/8 C.Y. bucket	4.11	10.55	14.65

G10 Site Preparation

G1030 Site Earthwork

G1030 807	Trenching Sand & Gravel	COST PER L.F. EQUIP.	LABOR	TOTAL
1860	8' deep, 1/2 C.Y. bucket	6.55	16.85	23.50
1880	10' deep, 1 C.Y. bucket	8.60	21	29.50
2300	4' wide, 2' deep, 3/8 C.Y. bucket	1.73	4.31	6.05
2310	3' deep, 3/8 C.Y. bucket	2.61	7.20	9.80
2320	4' deep, 1/2 C.Y. bucket	3.31	8.95	12.25
2340	6' deep, 1/2 C.Y. bucket	7.05	15.45	22.50
2360	8' deep, 1/2 C.Y. bucket	11.85	24.50	36.50
2380	10' deep, 1 C.Y. bucket	13.40	28.50	42
2400	12' deep, 1 C.Y. bucket	22.50	47.50	70
2430	15' deep, 1-1/2 C.Y. bucket	26	52.50	78.50
2460	18' deep, 2-1/2 C.Y. bucket	43.50	80.50	124
2840	6' wide, 6' deep, 5/8 C.Y. bucket w/trench box	11.45	21	32.50
2860	8' deep, 3/4 C.Y. bucket	16.80	31	48
2880	10' deep, 1 C.Y. bucket	17.95	35	53
2900	12' deep, 1-1/2 C.Y. bucket	22.50	42.50	65
2940	16' deep, 2 C.Y. bucket	36.50	62.50	99
2980	20' deep, 3-1/2 C.Y. bucket	50.50	84	135
3020	24' deep, 3-1/2 C.Y. bucket	71.50	115	187
3100	8' wide, 12' deep, 1-1/4 C.Y. bucket w/trench box	27	48.50	75.50
3120	15' deep, 1-1/2 C.Y. bucket	36.50	67.50	104
3140	18' deep, 2-1/2 C.Y. bucket	50	83.50	134
3180	24' deep, 3-1/2 C.Y. bucket	80	126	206
3270	10' wide, 20' deep, 3-1/2 C.Y. bucket w/trench box	64	102	166
3280	24' deep, 3-1/2 C.Y. bucket	88.50	138	227
3370	12' wide, 20' deep, 3-1/2 C.Y. bucket w/trench box	71.50	113	185
3380	25' deep, 3-1/2 C.Y. bucket	103	159	262
3500	1:1 slope, 2' wide, 2' deep, 3/8 C.Y. bucket	2.05	4.29	6.35
3520	3' deep, 3/8 C.Y. bucket	2.14	6.65	8.80
3540	4' deep, 3/8 C.Y. bucket	3.36	10.65	14
3560	6' deep, 3/8 C.Y. bucket	4.11	10.55	14.65
3580	8' deep, 1/2 C.Y. bucket	10.80	28	39
3600	10' deep, 1 C.Y. bucket	14.65	35.50	50
3800	4' wide, 2' deep, 3/8 C.Y. bucket	2	5.20	7.20
3820	3' deep, 3/8 C.Y. bucket	3.23	9.20	12.45
3840	4' deep, 1/2 C.Y. bucket	4.26	12	16.25
3860	6' deep, 1/2 C.Y. bucket	9.90	22	32
3880	8' deep, 1/2 C.Y. bucket	17.60	37.50	55
3900	10' deep, 1 C.Y. bucket	21	45	66
3920	12' deep, 1 C.Y. bucket	30	63.50	93.50
3940	15' deep, 1-1/2 C.Y. bucket	43	87.50	131
3960	18' deep, 2-1/2 C.Y. bucket	59.50	111	171
4030	6' wide, 6' deep, 5/8 C.Y. bucket w/trench box	15.05	28.50	43.50
4040	8' deep, 3/4 C.Y. bucket	23	44	67
4050	10' deep, 1 C.Y. bucket	26	51.50	77.50
4060	12' deep, 1-1/2 C.Y. bucket	33	64.50	97.50
4070	16' deep, 2 C.Y. bucket	57	99	156
4080	20' deep, 3-1/2 C.Y. bucket	81.50	137	219
4090	24' deep, 3-1/2 C.Y. bucket	119	193	310
4500	8' wide, 12' deep, 1-1/2 C.Y. bucket w/trench box	38	70	108
4550	15' deep, 1-1/2 C.Y. bucket	53.50	102	156
4600	18' deep, 2-1/2 C.Y. bucket	76	129	205
4650	24' deep, 3-1/2 C.Y. bucket	127	204	330
4800	10' wide, 20' deep, 3-1/2 C.Y. bucket w/trench box	95.50	155	251
4850	24' deep, 3-1/2 C.Y. bucket	136	216	350
4950	12' wide, 20' deep, 3-1/2 C.Y. bucket w/trench box	102	164	266
4980	25' deep, 3-1/2 C.Y. bucket	154	243	395

G10 Site Preparation

G1030 Site Earthwork

The Pipe Bedding System is shown for various pipe diameters. Compacted bank sand is used for pipe bedding and to fill 12" over the pipe. No backfill is included. Various side slopes are shown to accommodate different soil conditions. Pipe sizes vary from 6" to 84" diameter.

System Components	QUANTITY	UNIT	COST PER L.F. MAT.	COST PER L.F. INST.	COST PER L.F. TOTAL
SYSTEM G1030 815 1440					
PIPE BEDDING, SIDE SLOPE 0 TO 1, 1' WIDE, PIPE SIZE 6" DIAMETER					
Borrow, bank sand, 2 mile haul, machine spread	.086	C.Y.	1.65	.67	2.32
Compaction, vibrating plate	.086	C.Y.		.21	.21
TOTAL			1.65	.88	2.53

G1030 815	Pipe Bedding	MAT.	INST.	TOTAL
1440	Pipe bedding, side slope 0 to 1, 1' wide, pipe size 6" diameter	1.65	.88	2.53
1460	2' wide, pipe size 8" diameter	3.57	1.91	5.48
1480	Pipe size 10" diameter	3.65	1.96	5.61
1500	Pipe size 12" diameter	3.72	2	5.72
1520	3' wide, pipe size 14" diameter	6.05	3.23	9.28
1540	Pipe size 15" diameter	6.10	3.25	9.35
1560	Pipe size 16" diameter	6.15	3.29	9.44
1580	Pipe size 18" diameter	6.25	3.36	9.61
1600	4' wide, pipe size 20" diameter	8.90	4.77	13.67
1620	Pipe size 21" diameter	9	4.81	13.81
1640	Pipe size 24" diameter	9.20	4.93	14.13
1660	Pipe size 30" diameter	9.35	5	14.35
1680	6' wide, pipe size 32" diameter	16	8.60	24.60
1700	Pipe size 36" diameter	16.40	8.75	25.15
1720	7' wide, pipe size 48" diameter	26	13.95	39.95
1740	8' wide, pipe size 60" diameter	31.50	16.95	48.45
1760	10' wide, pipe size 72" diameter	44	23.50	67.50
1780	12' wide, pipe size 84" diameter	58.50	31	89.50
2140	Side slope 1/2 to 1, 1' wide, pipe size 6" diameter	3.08	1.64	4.72
2160	2' wide, pipe size 8" diameter	5.25	2.79	8.04
2180	Pipe size 10" diameter	5.60	3.01	8.61
2200	Pipe size 12" diameter	5.95	3.18	9.13
2220	3' wide, pipe size 14" diameter	8.60	4.60	13.20
2240	Pipe size 15" diameter	8.80	4.69	13.49
2260	Pipe size 16" diameter	9	4.82	13.82
2280	Pipe size 18" diameter	9.50	5.05	14.55
2300	4' wide, pipe size 20" diameter	12.50	6.70	19.20
2320	Pipe size 21" diameter	12.80	6.85	19.65
2340	Pipe size 24" diameter	13.60	7.25	20.85
2360	Pipe size 30" diameter	15.10	8.05	23.15
2380	6' wide, pipe size 32" diameter	22	11.90	33.90
2400	Pipe size 36" diameter	23.50	12.65	36.15
2420	7' wide, pipe size 48" diameter	37	19.75	56.75
2440	8' wide, pipe size 60" diameter	47	25	72
2460	10' wide, pipe size 72" diameter	64	34.50	98.50
2480	12' wide, pipe size 84" diameter	84	45	129
2620	Side slope 1 to 1, 1' wide, pipe size 6" diameter	4.51	2.41	6.92
2640	2' wide, pipe size 8" diameter	6.95	3.71	10.66

G10 Site Preparation

G1030 Site Earthwork

G1030 815	Pipe Bedding	MAT.	INST.	TOTAL
2660	Pipe size 10" diameter	7.55	4.03	11.58
2680	Pipe size 12" diameter	8.20	4.40	12.60
2700	3' wide, pipe size 14" diameter	11.15	5.95	17.10
2720	Pipe size 15" diameter	11.50	6.15	17.65
2740	Pipe size 16" diameter	11.85	6.35	18.20
2760	Pipe size 18" diameter	12.70	6.80	19.50
2780	4' wide, pipe size 20" diameter	16.10	8.60	24.70
2800	Pipe size 21" diameter	16.55	8.85	25.40
2820	Pipe size 24" diameter	17.95	9.60	27.55
2840	Pipe size 30" diameter	20.50	11.10	31.60
2860	6' wide, pipe size 32" diameter	28.50	15.25	43.75
2880	Pipe size 36" diameter	31	16.50	47.50
2900	7' wide, pipe size 48" diameter	47.50	25.50	73
2920	8' wide, pipe size 60" diameter	62	33	95
2940	10' wide, pipe size 72" diameter	84.50	45	129.50
2960	12' wide, pipe size 84" diameter	110	59	169

COST PER L.F.

G20 Site Improvements

G2040 Site Development

- Chain Link Fence
- Concrete Sidewalk
- Pool Equipment (Building not included in price)
- 12' x 30' Swimming Pool
- Fence Gate

The Swimming Pool System is a complete package. Everything from excavation to deck hardware is included in system costs. Below are three basic types of pool systems: residential, motel, and municipal. Systems elements include: excavation, pool materials, installation, deck hardware, pumps and filters, sidewalk, and fencing.

The Expanded System Listing shows three basic types of pools with a variety of finishes and basic materials. These systems are either vinyl lined with metal sides; gunite shell with a cement plaster finish or tile finish; or concrete sided with vinyl lining. Pool sizes listed here vary from 12' x 30' to 60' x 82.5'. All costs are on a per unit basis.

System Components	QUANTITY	UNIT	MAT.	INST.	TOTAL
SYSTEM G2040 920 1000					
SWIMMING POOL, RESIDENTIAL, CONC. SIDES, VINYL LINED, 12' X 30'					
Swimming pool, residential, in ground including equipment	360.000	S.F.	8,460	5,158.80	13,618.80
4" thick reinforced concrete sidewalk, broom finish, no base	400.000	S.F.	728	1,028	1,756
Chain link fence, residential, 3' high	124.000	L.F.	272.80	410.44	683.24
Fence gate, chain link	1.000	Ea.	89	138.50	227.50
TOTAL			9,549.80	6,735.74	16,285.54

G2040 920	Swimming Pools	MAT.	INST.	TOTAL
1000	Swimming pool, residential class, concrete sides, vinyl lined, 12' x 30'	9,550	6,725	16,275
1100	16' x 32'	10,800	7,625	18,425
1200	20' x 40'	14,000	9,750	23,750
1500	Tile finish, 12' x 30'	21,400	28,400	49,800
1600	16' x 32'	21,800	29,800	51,600
1700	20' x 40'	32,600	41,900	74,500
2000	Metal sides, vinyl lined, 12' x 30'	8,125	4,425	12,550
2100	16' x 32'	9,225	4,975	14,200
2200	20' x 40'	11,900	6,300	18,200
3000	Gunite shell, cement plaster finish, 12' x 30'	16,700	12,200	28,900
3100	16' x 32'	21,200	16,000	37,200
3200	20' x 40'	29,400	15,800	45,200
4000	Motel class, concrete sides, vinyl lined, 20' x 40'	20,200	13,600	33,800
4100	28' x 60'	28,400	18,900	47,300
4500	Tile finish, 20' x 40'	38,400	48,500	86,900
4600	28' x 60'	61,000	78,500	139,500
5000	Metal sides, vinyl lined, 20' x 40'	21,000	9,975	30,975
5100	28' x 60'	28,200	13,500	41,700
6000	Gunite shell, cement plaster finish, 20' x 40'	55,000	39,200	94,200
6100	28' x 60'	82,000	58,500	140,500
7000	Municipal class, gunite shell, cement plaster finish, 42' x 75'	221,500	139,500	361,000
7100	60' x 82.5'	284,500	179,000	463,500
7500	Concrete walls, tile finish, 42' x 75'	248,500	185,000	433,500
7600	60' x 82.5'	324,000	245,500	569,500
7700	Tile finish and concrete gutter, 42' x 75'	272,500	185,000	457,500
7800	60' x 82.5'	353,000	245,500	598,500
7900	Tile finish and stainless gutter, 42' x 75'	319,000	185,000	504,000
8000	60' x 82.5'	472,000	283,500	755,500

G20 Site Improvements

G2050 Landscaping

There are three basic types of Site Irrigation Systems: pop-up, riser mounted and quick coupling. Sprinkler heads are spray, impact or gear driven. Each system includes: the hardware for spraying the water; the pipe and fittings needed to deliver the water; and all other accessory equipment such as valves, couplings, nipples, and nozzles. Excavation heads and backfill costs are also included in the system.
The Expanded System Listing shows a wide variety of Site Irrigation Systems.

System Components	QUANTITY	UNIT	COST PER S.F. MAT.	COST PER S.F. INST.	COST PER S.F. TOTAL
SYSTEM G2050 710 1000					
SITE IRRIGATION, POP UP SPRAY, PLASTIC, 10' RADIUS, 1000 S.F., PVC PIPE					
Excavation, chain trencher	54.000	L.F.		43.74	43.74
Pipe, fittings & nipples, PVC Schedule 40, 1" diameter	5.000	Ea.	4.25	110	114.25
Fittings, bends or elbows, 4" diameter	5.000	Ea.	54.25	245	299.25
Couplings PVC plastic, high pressure, 1" diameter	5.000	Ea.	4.25	140	144.25
Valves, bronze, globe, 125 lb. rising stem, threaded, 1" diameter	1.000	Ea.	233	36.50	269.50
Head & nozzle, pop-up spray, PVC plastic	5.000	Ea.	17.90	97.75	115.65
Backfill by hand with compaction	54.000	L.F.		35.10	35.10
Total cost per 1,000 S.F.			313.65	708.09	1,021.74
Total cost per S.F.			.31	.71	1.02

G2050 710	Site Irrigation	MAT.	INST.	TOTAL
1000	Site irrigation, pop up spray, 10' radius, 1000 S.F., PVC pipe	.32	.71	1.03
1100	Polyethylene pipe	.34	.63	.97
1200	14' radius, 8000 S.F., PVC pipe	.11	.31	.42
1300	Polyethylene pipe	.11	.29	.40
1400	18' square, 1000 S.F. PVC pipe	.30	.49	.79
1500	Polyethylene pipe	.31	.42	.73
1600	24' square, 8000 S.F., PVC pipe	.09	.19	.28
1700	Polyethylene pipe	.09	.18	.27
1800	4' x 30' strip, 200 S.F., PVC pipe	1.46	2	3.46
1900	Polyethylene pipe	1.48	1.85	3.33
2000	6' x 40' strip, 800 S.F., PVC pipe	.39	.67	1.06
2200	Economy brass, 11' radius, 1000 S.F., PVC pipe	.37	.69	1.06
2300	Polyethylene pipe	.38	.63	1.01
2400	14' radius, 8000 S.F., PVC pipe	.13	.31	.44
2500	Polyethylene pipe	.13	.29	.42
2600	3' x 28' strip, 200 S.F., PVC pipe	1.60	2	3.60
2700	Polyethylene pipe	1.62	1.85	3.47
2800	7' x 36' strip, 1000 S.F., PVC pipe	.34	.53	.87
2900	Polyethylene pipe	.34	.49	.83
3000	Hd brass, 11' radius, 1000 S.F., PVC pipe	.42	.69	1.11
3100	Polyethylene pipe	.43	.63	1.06
3200	14' radius, 8000 S.F., PVC pipe	.15	.31	.46
3300	Polyethylene pipe	.15	.29	.44
3400	Riser mounted spray, plastic,10'radius, 1000S.F., PVC pipe	.33	.69	1.02
3500	Polyethylene pipe	.34	.63	.97
3600	12' radius, 5000 S.F., PVC pipe	.15	.40	.55

G20 Site Improvements

G2050 Landscaping

G2050 710	Site Irrigation	COST PER S.F.		
		MAT.	INST.	TOTAL
3700	Polyethylene pipe	.15	.37	.52
3800	19' square, 2000 S.F., PVC pipe	.16	.21	.37
3900	Polyethylene pipe	.16	.19	.35
4000	24' square, 8000 S.F., PVC pipe	.09	.19	.28
4100	Polyethylene pipe	.09	.18	.27
4200	5' x 32' strip, 300 S.F., PVC pipe	.99	1.35	2.34
4300	Polyethylene pipe	1	1.25	2.25
4400	6' x 40' strip, 800 S.F., PVC pipe	.39	.67	1.06
4500	Polyethylene pipe	.40	.61	1.01
4600	Brass, 11' radius, 1000 S.F., PVC pipe	.37	.69	1.06
4700	Polyethylene pipe	.38	.63	1.01
4800	14' radius, 8000 S.F., PVC pipe	.13	.31	.44
4900	Polythylene pipe	.13	.29	.42
5000	Pop up gear drive stream type, plastic, 30' radius, 10,000 S.F., PVC pipe	.25	.59	.84
5020	Polyethylene pipe	.13	.17	.30
5040	40,000 S.F., PVC pipe	.20	.60	.80
5060	Polyethylene pipe	.09	.17	.26
5080	Riser mounted gear drive stream type, plas., 30'rad., 10,000 S.F., PVC pipe	.24	.59	.83
5100	Polyethylene pipe	.12	.17	.29
5120	40,000 S.F., PVC pipe	.20	.60	.80
5140	Polyethylene pipe	.09	.17	.26
5200	Q.C. valve thread type w/impact head, brass, 75'rad, 20,000 S.F., PVC pipe	.11	.22	.33
5250	Polyethylene pipe	.06	.05	.11
5300	100,000 S.F., PVC pipe	.13	.36	.49
5350	Polyethylene pipe	.06	.08	.14
5400	Q.C. valve lug type w/impact head, brass, 75'rad, 20,000 S.F., PVC pipe	.11	.22	.33
5450	Polyethylene pipe	.06	.05	.11
5500	100,000 S.F. PVC pipe	.13	.36	.49
5550	Polyethylene pipe	.06	.08	.14
6000	Site irrigation, pop up impact type, plastic, 40'rad., 10000 S.F., PVC pipe	.16	.37	.53
6100	Polyethylene pipe	.08	.10	.18
6200	45,000 S.F., PVC pipe	.13	.42	.55
6300	Polyethylene pipe	.05	.11	.16
6400	High medium volume brass, 40' radius, 10000 S.F., PVC pipe	.17	.37	.54
6500	Polyethylene pipe	.09	.10	.19
6600	45,000 S.F. PVC pipe	.14	.42	.56
6700	Polyethylene pipe	.06	.11	.17
6800	High volume brass, 60' radius, 25,000 S.F., PVC pipe	.12	.30	.42
6900	Polyethylene pipe	.05	.08	.13
7000	100,000 S.F., PVC pipe	.11	.29	.40
7100	Polyethylene pipe	.06	.07	.13
7200	Riser mounted part/full impact type, plas., 40'rad., 10,000 S.F., PVC pipe	.15	.37	.52
7300	Polyethylene pipe	.07	.20	.27
7400	45,000 S.F., PVC pipe	.12	.42	.54
7500	Polyethylene	.04	.11	.15
7600	Low medium volume brass, 40' radius, 10,000 S.F., PVC pipe	.16	.37	.53
7700	Polyethylene pipe	.08	.10	.18
7800	45,000 S.F., PVC pipe	.13	.42	.55
7900	Polyethylene pipe	.05	.11	.16
8000	Medium volume brass, 50' radius, 30,000 S.F., PVC pipe	.12	.31	.43
8100	Polyethylene pipe	.05	.08	.13
8200	70,000 S.F., PVC pipe	.17	.51	.68
8300	Polyethylene pipe	.08	.12	.20
8400	Riser mounted full only impact type, plas., 40'rad., 10000 S.F., PVC pipe	.15	.37	.52
8500	Polyethylene pipe	.07	.10	.17
8600	45,000 S.F., PVC pipe	.12	.42	.54
8700	Polyethylene pipe	.04	.11	.15
8800	Low medium volume brass, 40' radius, 10,000 S.F., PVC pipe	.16	.37	.53

G20 Site Improvements

G2050 Landscaping

G2050 710	Site Irrigation	COST PER S.F.		
		MAT.	INST.	TOTAL
8900	Polyethylene pipe	.08	.10	.18
9000	45,000 S.F., PVC pipe	.13	.42	.55
9100	Polyethylene pipe	.05	.11	.16
9200	High volume brass, 80' radius, 75,000 S.F., PVC pipe	.08	.22	.30
9300	Polyethylene pipe	.01	.02	.03
9400	150,000 S.F., PVC pipe	.08	.19	.27
9500	Polyethylene pipe	.01	.02	.03
9600	Very high volume brass, 100' radius, 100,000 S.F., PVC pipe	.07	.16	.23
9700	Polyethylene pipe	.01	.02	.03
9800	200,000 S.F., PVC pipe	.07	.16	.23
9900	Polyethylene pipe	.01	.02	.03

G20 Site Improvements

G2050 Landscaping

Drip irrigation systems can be for either lawns or individual plants. Shown below are drip irrigation systems for lawns. These drip type systems are most cost effective where there are water restrictions on use or the cost of water is very high. The supply uses PVC piping with the laterals polyethylene drip lines.

System Components	QUANTITY	UNIT	COST PER S.F. MAT.	COST PER S.F. INST.	COST PER S.F. TOTAL
SYSTEM G2050 720 1000					
SITE IRRIGATION, DRIP SYSTEM, 800 S.F. (20' X 40')					
Excavate supply and header lines 4" x 18" deep	50.000	L.F.		43	43
Excavate drip lines 4" x 4" deep	5.000	L.F.		728	728
Install supply manifold (stop valve, strainer, PRV, adapters)	1.000	Ea.	18.60	147	165.60
Install 3/4" PVC supply line and risers to 4" drip level	75.000	L.F.	29.25	146.25	175.50
Install 3/4" PVC fittings for supply line	33.000	Ea.	61.71	483.45	545.16
Install PVC to drip line adapters	14.000	Ea.	15.12	205.10	220.22
Install 1/4 inch drip lines with hose clamps	560.000	L.F.	123.20	274.40	397.60
Backfill trenches	3.180	C.Y.		130.38	130.38
Cleanup area when completed	1.000	Ea.		113	113
Total cost per 800 S.F.			247.88	2,270.58	2,518.46
Total cost per S.F.			.31	2.84	3.15

G2050 720	Site Irrigation	MAT.	INST.	TOTAL
1000	Site irrigation, drip lawn watering system, 800 S.F. (20' x 40' area)	.31	2.84	3.15
1100	1000 S.F. (20' x 50' area)	.28	2.55	2.83
1200	1600 S.F. (40' x 40' area), 2 zone	.32	2.72	3.04
1300	2000 S.F. (40' x 50' area), 2 zone	.30	2.45	2.75
1400	2400 S.F. (60' x 40' area), 3 zone	.32	2.68	3
1500	3000 S.F. (60' x 50' area), 3 zone	.30	2.41	2.71
1600	3200 S.F. (2 x 40' x 40'), 4 zone	.32	2.67	2.99
1700	4000 S.F. (2 x 40' x 50'), 4 zone	.29	2.42	2.71
1800	4800 S.F. (2 x 60' x 40'), 6 zone with control	.33	2.65	2.98
1900	6000 S.F. (2 x 60' x 50'), 6 zone with control	.30	2.37	2.67
2000	6400 S.F. (4 x 40' x 40'), 8 zone with control	.34	2.69	3.03
2100	8000 S.F. (4 x 40' x 50'), 8 zone with control	.29	2.44	2.73
2200	3200 S.F. (2 x 40' x 40'), 4 zone with control	.34	2.74	3.08
2300	4000 S.F. (2 x 40' x 50'), 4 zone with control	.30	2.47	2.77
2400	2400 S.F. (60' x 40'), 3 zone with control	.34	2.76	3.10
2500	3000 S.F. (60' x 50'), 3 zone with control	.32	2.48	2.80
2600	4800 S.F. (2 x 60' x 40'), 6 zone, manual	.31	2.61	2.92
2700	6000 S.F. (2 x 60' x 50'), 6 zone, manual	.29	2.35	2.64
2800	1000 S.F. (10' x 100' area)	.18	1.59	1.77
2900	2000 S.F. (2 x 10' x 100' area)	.20	1.53	1.73

G30 Site Mechanical Utilities

G3010 Water Supply

The Water Service Systems are for copper service taps from 1" to 2" diameter into pressurized mains from 6" to 8" diameter. Costs are given for two offsets with depths varying from 2' to 10'. Included in system components are excavation and backfill and required curb stops with boxes.

System Components	QUANTITY	UNIT	COST EACH MAT.	COST EACH INST.	COST EACH TOTAL
SYSTEM G3010 120 1000					
WATER SERVICE, 6" MAIN, 1" COPPER SERVICE, 10' OFFSET, 2' DEEP					
Trench excavation, 1/2 C.Y. backhoe, 1 laborer	2.220	C.Y.		16.52	16.52
Drill & tap pressurized main, 6", 1" to 2" service	1.000	Ea.		415	415
Corporation stop, 1" diameter	1.000	Ea.	50	43.50	93.50
Saddles, cast iron, 3/4" & 1", add	1.000	Ea.	99		99
Copper tubing, type K, 1" diameter	11.000	L.F.	108.90	43.01	151.91
Curb stop, brass, 1" diameter	1.000	Ea.	84.50	43.50	128
Curb box, cast iron, 1" diameter	1.000	Ea.	57	58	115
Backfill by hand, no compaction, heavy soil	2.220	C.Y.		5.62	5.62
Compaction, vibrating plate	2.220	C.Y.		17.71	17.71
TOTAL			399.40	642.86	1,042.26

G3010 120	Water Service	MAT.	INST.	TOTAL
1000	Water service, 6" main, 1" copper service, 10' offset, 2' deep	400	645	1,045
1040	4' deep	400	710	1,110
1060	6' deep	400	800	1,200
1080	8' deep	400	925	1,325
1100	10' deep	400	985	1,385
1200	20' offset, 2' deep	500	725	1,225
1240	4' deep	500	855	1,355
1260	6' deep	500	1,050	1,550
1280	8' deep	500	1,275	1,775
1300	10' deep	500	1,550	2,050
2000	1-1/2" copper service, 10' offset, 2' deep	785	700	1,485
2040	4' deep	785	770	1,555
2060	6' deep	785	860	1,645
2080	8' deep	785	980	1,765
2100	10' deep	785	1,125	1,910
2200	20' offset, 2' deep	940	790	1,730
2240	4' deep	940	925	1,865
2260	6' deep	940	1,125	2,065
2280	8' deep	940	1,350	2,290
2300	10' deep	940	1,625	2,565
3000	6" main, 2" copper service, 10' offset, 2' deep	1,075	730	1,805
3040	4' deep	1,075	795	1,870
3060	6' deep	1,075	885	1,960
3080	8' deep	1,075	1,000	2,075
3100	10' deep	1,075	1,150	2,225
3200	20' offset, 2' deep	1,325	825	2,150
3240	4' deep	1,325	955	2,280
3260	6' deep	1,325	1,150	2,475

533

G30 Site Mechanical Utilities

G3010 Water Supply

G3010 120	Water Service	MAT.	INST.	TOTAL
3280	8' deep	1,325	1,375	2,700
3300	10' deep	1,325	1,650	2,975
4000	8" main, 1" copper service, 10' offset, 2' deep	400	685	1,085
4040	4' deep	400	750	1,150
4060	6' deep	400	840	1,240
4080	8' deep	400	965	1,365
4100	10' deep	400	1,100	1,500
4200	20' offset, 2' deep	500	765	1,265
4240	4' deep	500	895	1,395
4260	6' deep	500	1,075	1,575
4280	8' deep	500	1,325	1,825
4300	10' deep	500	1,600	2,100
5000	1-1/2" copper service, 10' offset, 2' deep	785	740	1,525
5040	4' deep	785	810	1,595
5060	6' deep	785	900	1,685
5080	8' deep	785	1,025	1,810
5100	10' deep	785	1,150	1,935
5200	20' offset, 2' deep	940	830	1,770
5240	4' deep	940	965	1,905
5260	6' deep	940	1,150	2,090
5280	8' deep	940	1,400	2,340
5300	10' deep	940	1,675	2,615
6000	2" copper service, 10' offset, 2' deep	1,075	770	1,845
6040	4' deep	1,075	835	1,910
6060	6' deep	1,075	925	2,000
6080	8' deep	1,075	1,050	2,125
6100	10' deep	1,075	1,175	2,250
6200	20' offset, 2' deep	1,325	865	2,190
6240	4' deep	1,325	995	2,320
6260	6' deep	1,325	1,200	2,525
6280	8' deep	1,325	1,425	2,750
6300	10' deep	1,325	1,700	3,025

G30 Site Mechanical Utilities

G3010 Water Supply

The Fire Hydrant Systems include: four different hydrants with three different lengths of offsets, at several depths of trenching. Excavation and backfill is included with each system as well as thrust blocks as necessary and pipe bedding. Finally spreading of excess material and fine grading complete the components.

System Components	QUANTITY	UNIT	MAT.	INST.	TOTAL
SYSTEM G3010 410 1500					
HYDRANT, 4-1/2" VALVE SIZE, TWO WAY, 10' OFFSET, 2' DEEP					
Excavation, trench, 1 C.Y. hydraulic backhoe	3.556	C.Y.		28.30	28.30
Sleeve, 12" x 6"	1.000	Ea.	1,825	109.55	1,934.55
Ductile iron pipe, 6" diameter	8.000	L.F.	156.40	166.08	322.48
Gate valve, 6"	1.000	Ea.	1,675	313	1,988
Hydrant valve box, 6' long	1.000	Ea.	285	76	361
4-1/2" valve size, depth 2'-0"	1.000	Ea.	1,825	197.40	2,022.40
Thrust blocks, at valve, shoe and sleeve	1.600	C.Y.	289.60	435.65	725.25
Borrow, crushed stone, 3/8"	4.089	C.Y.	114	31.44	145.44
Backfill and compact, dozer, air tamped	4.089	C.Y.		69.35	69.35
Spread excess excavated material	.204	C.Y.		8.36	8.36
Fine grade, hand	400.000	S.F.		262.50	262.50
TOTAL			6,170	1,697.63	7,867.63

G3010 410	Fire Hydrants	MAT.	INST.	TOTAL
1000	Hydrant, 4-1/2" valve size, two way, 0' offset, 2' deep	6,025	1,175	7,200
1100	4' deep	6,350	1,200	7,550
1200	6' deep	6,500	1,250	7,750
1300	8' deep	6,725	1,300	8,025
1400	10' deep	6,900	1,375	8,275
1500	10' offset, 2' deep	6,175	1,700	7,875
1600	4' deep	6,500	2,000	8,500
1700	6' deep	6,650	2,525	9,175
1800	8' deep	6,875	3,325	10,200
1900	10' deep	7,050	4,975	12,025
2000	20' offset, 2' deep	6,375	2,250	8,625
2100	4' deep	6,700	2,725	9,425
2200	6' deep	6,850	3,525	10,375
2300	8' deep	7,075	4,675	11,750
2400	10' deep	7,250	6,225	13,475
2500	Hydrant, 4-1/2" valve size, three way, 0' offset, 2' deep	6,150	1,200	7,350
2600	4' deep	6,500	1,200	7,700
2700	6' deep	6,650	1,275	7,925
2800	8' deep	6,900	1,325	8,225
2900	10' deep	7,075	1,400	8,475
3000	10' offset, 2' deep	6,300	1,725	8,025
3100	4' deep	6,650	2,000	8,650
3200	6' deep	6,800	2,550	9,350
3300	8' deep	7,050	3,350	10,400

G30 Site Mechanical Utilities

G3010 Water Supply

G3010 410	Fire Hydrants	COST PER EACH		
		MAT.	INST.	TOTAL
3400	10' deep	7,225	5,025	12,250
3500	20' offset, 2' deep	6,500	2,275	8,775
3600	4' deep	6,850	2,750	9,600
3700	6' deep	7,000	3,550	10,550
3800	8' deep	7,250	4,700	11,950
3900	10' deep	7,425	6,250	13,675
5000	5-1/4" valve size, two way, 0' offset, 2' deep	6,175	1,175	7,350
5100	4' deep	6,275	1,200	7,475
5200	6' deep	6,475	1,250	7,725
5300	8' deep	6,650	1,300	7,950
5400	10' deep	7,025	1,375	8,400
5500	10' offset, 2' deep	6,325	1,700	8,025
5600	4' deep	6,425	2,000	8,425
5700	6' deep	6,625	2,525	9,150
5800	8' deep	6,800	3,325	10,125
5900	10' deep	7,175	4,975	12,150
6000	20' offset, 2' deep	6,525	2,250	8,775
6100	4' deep	6,625	2,725	9,350
6200	6' deep	6,825	3,525	10,350
6300	8' deep	7,000	4,675	11,675
6400	10' deep	7,375	6,225	13,600
6500	Three way, 0' offset, 2' deep	6,300	1,200	7,500
6600	4' deep	6,400	1,200	7,600
6700	6' deep	6,625	1,275	7,900
6800	8' deep	6,800	1,325	8,125
6900	10' deep	7,200	1,400	8,600
7000	10' offset, 2' deep	6,450	1,725	8,175
7100	4' deep	6,575	2,000	8,575
7200	6' deep	6,775	2,550	9,325
7300	8' deep	6,975	3,350	10,325
7400	10' deep	7,375	5,025	12,400
7500	20' offset, 2' deep	6,650	2,275	8,925
7600	4' deep	6,750	2,475	9,225
7700	6' deep	6,975	3,550	10,525

G30 Site Mechanical Utilities

G3020 Sanitary Sewer

The Septic System includes: a septic tank; leaching field; concrete distribution boxes; plus excavation and gravel backfill.

The Expanded System Listing shows systems with tanks ranging from 1000 gallons to 2000 gallons. Tanks are either concrete or fiberglass. Cost is on a complete unit basis.

System Components	QUANTITY	UNIT	MAT.	INST.	TOTAL
SYSTEM G3020 302 0300					
1000 GALLON TANK, LEACHING FIELD, 600 S.F., 50' FROM BLDG.					
Mobilization	2.000	Ea.		544	544
Septic tank, precast, 1000 gallon	1.000	Ea.	1,200	247.25	1,447.25
Effluent filter	1.000	Ea.	61	73.50	134.50
Distribution box, precast, 5 outlets	1.000	Ea.	86	56.50	142.50
Flow leveler	3.000	Ea.	7.32	27.15	34.47
Sewer pipe, PVC, SDR 35, 4" diameter	40.000	L.F.	257.60	646.40	904
Tee	1.000	Ea.	14.85	73.50	88.35
Elbow	2.000	Ea.	21.70	98	119.70
Viewport cap	1.000	Ea.	8.45	24.50	32.95
Filter fabric	67.000	S.Y.	136.68	25.46	162.14
Detectable marking tape	1.600	C.L.F.	10	4.83	14.83
Excavation	160.000	C.Y.		1,968	1,968
Backfill	133.000	L.C.Y.		380.38	380.38
Spoil	55.000	L.C.Y.		350.35	350.35
Compaction	113.000	E.C.Y.		927.73	927.73
Stone fill, 3/4" to 1-1/2"	39.000	C.Y.	1,462.50	488.67	1,951.17
TOTAL			**3,266.10**	**5,936.22**	**9,202.32**

G3020 302	Septic Systems	MAT.	INST.	TOTAL
0300	1000 gal. concrete septic tank, 600 S.F. leaching field, 50 feet from bldg.	3,275	5,950	9,225
0301	200 feet from building	3,500	7,575	11,075
0305	750 S.F. leaching field, 50 feet from building	3,700	6,550	10,250
0306	200 feet from building	3,950	8,225	12,175
0400	1250 gal. concrete septic tank, 600 S.F. leaching field, 50 feet from bldg.	3,300	6,200	9,500
0401	200 feet from building	3,550	7,875	11,425
0405	750 S.F. leaching field, 50 feet from building	3,725	6,850	10,575
0406	200 feet from building	3,975	8,525	12,500
0410	1000 S.F. leaching field, 50 feet from building	4,475	7,975	12,450
0411	200 feet from building	4,725	9,650	14,375
0500	1500 gal. concrete septic tank, 600 S.F. leaching field, 50 feet from bldg.	3,725	6,475	10,200
0501	200 feet from building	3,975	8,125	12,100
0505	750 S.F. leaching field, 50 feet from building	4,150	7,100	11,250
0506	200 feet from building	4,400	9,625	14,025
0510	1000 S.F. leaching field, 50 feet from building	4,900	8,275	13,175
0511	200 feet from building	5,150	9,925	15,075
0515	1100 S.F. leaching field, 50 feet from building	5,175	8,675	13,850
0516	200 feet from building	5,425	10,500	15,925
0600	2000 gal concrete tank, 1200 S.F. leaching field, 50 feet from building	6,000	9,225	15,225
0601	200 feet from building	6,250	10,900	17,150

G30 Site Mechanical Utilities

G3020 Sanitary Sewer

G3020 302 Septic Systems

		MAT.	INST.	TOTAL
0605	1500 S.F. leaching field, 50 feet from building	6,875	10,600	17,475
0606	200 feet from building	7,125	12,300	19,425
0710	2500 gal concrete tank, 1500 S.F. leaching field, 50 feet from building	7,450	11,300	18,750
0711	200 feet from building	7,700	13,000	20,700
0715	1600 S.F. leaching field, 50 feet from building	7,775	11,700	19,475
0716	200 feet from building	8,025	13,400	21,425
0720	1800 S.F. leaching field, 50 feet from building	8,350	12,700	21,050
0721	200 feet from building	8,600	14,400	23,000
1300	1000 gal. 2 compartment tank, 600 S.F. leaching field, 50 feet from bldg.	3,400	5,950	9,350
1301	200 feet from building	3,625	7,575	11,200
1305	750 S.F. leaching field, 50 feet from building	3,825	6,550	10,375
1306	200 feet from building	4,075	8,225	12,300
1400	1250 gal. 2 compartment tank, 600 S.F. leaching field, 50 feet from bldg.	3,775	6,200	9,975
1401	200 feet from building	4,025	7,875	11,900
1405	750 S.F. leaching field, 50 feet from building	4,200	6,850	11,050
1406	200 feet from building	4,450	8,525	12,975
1410	1000 S.F. leaching field, 50 feet from building	4,950	7,975	12,925
1411	200 feet from building	5,200	9,650	14,850
1500	1500 gal. 2 compartment tank, 600 S.F. leaching field, 50 feet from bldg.	4,150	6,475	10,625
1501	200 feet from building	4,400	8,125	12,525
1505	750 S.F. leaching field, 50 feet from building	4,575	7,100	11,675
1506	200 feet from building	4,825	9,625	14,450
1510	1000 S.F. leaching field, 50 feet from building	5,325	8,275	13,600
1511	200 feet from building	5,575	9,925	15,500
1515	1100 S.F. leaching field, 50 feet from building	5,600	8,675	14,275
1516	200 feet from building	5,850	10,500	16,350
1600	2000 gal 2 compartment tank, 1200 S.F. leaching field, 50 feet from bldg.	6,475	9,225	15,700
1601	200 feet from building	6,725	10,900	17,625
1605	1500 S.F. leaching field, 50 feet from building	7,350	10,600	17,950
1606	200 feet from building	7,600	12,300	19,900
1710	2500 gal 2 compartment tank, 1500 S.F. leaching field, 50 feet from bldg.	8,025	11,300	19,325
1711	200 feet from building	8,275	13,000	21,275
1715	1600 S.F. leaching field, 50 feet from building	8,350	11,700	20,050
1716	200 feet from building	8,600	13,400	22,000
1720	1800 S.F. leaching field, 50 feet from building	8,925	12,700	21,625
1721	200 feet from building	9,175	14,400	23,575

G30 Site Mechanical Utilities

G3020 Sanitary Sewer

Diagram showing cross-section of aerated sewage lagoon with Grass, Lagoon Liner, Aerator, and Gravel labeled.

System Components	QUANTITY	UNIT	MAT.	INST.	TOTAL
SYSTEM G3020 710 1000					
AERATED DOMESTIC SEWAGE LAGOON, COMMON EARTH, 500,000 GPD					
Supervision of lagoon construction	45.000	Day		26,325	26,325
Quality Control of Earthwork per day	20.000	Day		9,000	9,000
Anchoring aerators per cell, 6 anchors per cell	2.000	Ea.	8,050	7,510	15,560
Excavate lagoon	7778.000	B.C.Y.		18,744.98	18,744.98
Finish grade slopes for seeding	25.000	M.S.F.		712.50	712.50
Spread & grade top of lagoon banks for compaction	454.000	M.S.F.		27,694	27,694
Compact lagoon banks	7127.000	E.C.Y.		5,345.25	5,345.25
Install pond liner	71.000	M.S.F.	45,440	78,100	123,540
Provide water for compaction and dust control	1556.000	E.C.Y.	2,053.92	3,205.36	5,259.28
Dispose of excess material on-site	846.000	L.C.Y.		2,994.84	2,994.84
Seed non-lagoon side of banks	25.000	M.S.F.	712.50	712.50	1,425
Place gravel on top of banks	1269.000	S.Y.	7,677.45	1,827.36	9,504.81
Provide 6' high fence around lagoon	1076.000	L.F.	22,596	7,456.68	30,052.68
Provide 12' wide fence gate	1.000	Opng.	475	515.50	990.50
Surface aerator, 30 HP, 50 lbs/HR, 900 RPM	4.000	Ea.	146,400	12,720	159,120
Temporary Fencing	1269.000	L.F.	4,124.25	2,867.94	6,992.19
TOTAL			**237,529.12**	**205,731.91**	**443,261.03**

G3020 710	Aerated Sewage Lagoon	MAT.	INST.	TOTAL
1000	500,000 GPD Domestic Sewage Aerated Lagoon, common earth, excluding power	237,500	205,500	443,000
1100	600,000 GPD	319,000	238,500	557,500
1200	700,000 GPD	325,000	257,000	582,000
1300	800,000 GPD	405,000	286,000	691,000
1400	900,000 GPD	410,000	306,000	716,000
1500	1 MGD	459,500	386,000	845,500
2000	500,000 GPD sandy clay & loam	237,000	203,500	440,500
2100	600,000 GPD	319,000	236,000	555,000
2200	700,000 GPD	325,000	252,000	577,000
2300	800,000 GPD	405,000	282,500	687,500
2400	900,000 GPD	410,000	302,500	712,500
2500	1 MGD	564,000	478,000	1,042,000
3000	500,000 GPD sand & gravel	237,000	201,500	438,500
3100	600,000 GPD	319,000	234,500	553,500
3200	700,000 GPD	325,000	251,500	576,500
3300	800,000 GPD	411,000	280,500	691,500
3400	900,000 GPD	410,000	297,500	707,500
3500	1 MGD	460,000	381,000	841,000
5000	500,000 GPD Lagoon aerators power supply	8,500	11,400	19,900
5100	600,000 GPD	10,200	14,000	24,200

G30 Site Mechanical Utilities

G3020 Sanitary Sewer

G3020 710	Aerated Sewage Lagoon	COST EACH		
		MAT.	INST.	TOTAL
5200	700,000 GPD	10,200	14,000	24,200
5300	800,000 GPD	17,000	22,800	39,800
5400	900,000 GPD	17,000	22,800	39,800
5500	1 MGD	17,000	22,800	39,800
5600	Lagoons, aerated without liner			
6000	500,000 GPD Domestic Sewage Aerated Lagoon, common earth, no power/liner	191,500	121,500	313,000
6100	600,000 GPD	269,000	146,500	415,500
6200	700,000 GPD	270,500	155,500	426,000
6300	800,000 GPD	345,500	177,500	523,000
6400	900,000 GPD	345,500	188,000	533,500
6500	1 MGD	368,500	218,000	586,500
7000	500,000 GPD sandy clay & loam	191,500	119,000	310,500
7100	600,000 GPD	269,000	144,500	413,500
7200	700,000 GPD	270,500	152,500	423,000
7300	800,000 GPD	345,500	174,500	520,000
7400	900,000 GPD	355,000	187,500	542,500
7500	1 MGD	368,500	214,500	583,000

G30 Site Mechanical Utilities

G3030 Storm Sewer

Manhole | Catch Basin

The Manhole and Catch Basin System includes: excavation with a backhoe; a formed concrete footing; frame and cover; cast iron steps and compacted backfill.

The Expanded System Listing shows manholes that have a 4', 5' and 6' inside diameter riser. Depths range from 4' to 14'. Construction material shown is either concrete, concrete block, precast concrete, or brick.

System Components	QUANTITY	UNIT	COST PER EACH MAT.	COST PER EACH INST.	COST PER EACH TOTAL
SYSTEM G3030 210 1920					
MANHOLE/CATCH BASIN, BRICK, 4' I.D. RISER, 4' DEEP					
Excavation, hydraulic backhoe, 3/8 C.Y. bucket	14.815	B.C.Y.		117.93	117.93
Trim sides and bottom of excavation	64.000	S.F.		60.80	60.80
Forms in place, manhole base, 4 uses	20.000	SFCA	14.20	104	118.20
Reinforcing in place footings, #4 to #7	.019	Ton	20.90	22.80	43.70
Concrete, 3000 psi	.925	C.Y.	100.83		100.83
Place and vibrate concrete, footing, direct chute	.925	C.Y.		48.40	48.40
Catch basin or MH, brick, 4' ID, 4' deep	1.000	Ea.	560	1,000	1,560
Catch basin or MH steps; heavy galvanized cast iron	1.000	Ea.	21	13.90	34.90
Catch basin or MH frame and cover	1.000	Ea.	268	240.50	508.50
Fill, granular	12.954	L.C.Y.	297.94		297.94
Backfill, spread with wheeled front end loader	12.954	L.C.Y.		32.78	32.78
Backfill compaction, 12" lifts, air tamp	12.954	E.C.Y.		115.29	115.29
TOTAL			1,282.87	1,756.40	3,039.27

G3030 210	Manholes & Catch Basins	MAT.	INST.	TOTAL
1920	Manhole/catch basin, brick, 4' I.D. riser, 4' deep	1,275	1,750	3,025
1940	6' deep	1,825	2,450	4,275
1960	8' deep	2,425	3,375	5,800
1980	10' deep	2,875	4,150	7,025
3000	12' deep	3,575	4,500	8,075
3020	14' deep	4,425	6,300	10,725
3200	Block, 4' I.D. riser, 4' deep	1,125	1,425	2,550
3220	6' deep	1,575	2,000	3,575
3240	8' deep	2,075	2,800	4,875
3260	10' deep	2,475	3,425	5,900
3280	12' deep	3,150	4,350	7,500
3300	14' deep	3,950	5,325	9,275
4620	Concrete, cast-in-place, 4' I.D. riser, 4' deep	1,275	2,450	3,725
4640	6' deep	1,825	3,250	5,075
4660	8' deep	2,550	4,700	7,250
4680	10' deep	3,075	5,825	8,900
4700	12' deep	3,875	7,225	11,100
4720	14' deep	4,775	8,650	13,425
5820	Concrete, precast, 4' I.D. riser, 4' deep	1,575	1,300	2,875
5840	6' deep	2,100	1,750	3,850

G30 Site Mechanical Utilities

G3030 Storm Sewer

G3030 210	Manholes & Catch Basins	COST PER EACH		
		MAT.	INST.	TOTAL
5860	8' deep	2,625	2,425	5,050
5880	10' deep	3,150	3,000	6,150
5900	12' deep	3,975	3,700	7,675
5920	14' deep	4,900	4,700	9,600
6000	5' I.D. riser, 4' deep	2,575	1,475	4,050
6020	6' deep	2,925	2,075	5,000
6040	8' deep	4,100	2,750	6,850
6060	10' deep	5,225	3,475	8,700
6080	12' deep	6,450	4,450	10,900
6100	14' deep	7,800	5,400	13,200
6200	6' I.D. riser, 4' deep	3,250	1,950	5,200
6220	6' deep	4,100	2,575	6,675
6240	8' deep	5,225	3,600	8,825
6260	10' deep	6,625	4,550	11,175
6280	12' deep	8,125	5,725	13,850
6300	14' deep	9,775	6,950	16,725

Reference Section

All the reference information is in one section, making it easy to find what you need to know ... and easy to use the book on a daily basis. This section is visually identified by a vertical gray bar on the page edges.

In this Reference Section, we've included Equipment Rental Costs, a listing of rental and operating costs; Crew Listings, a full listing of all crews and equipment, and their costs; Historical Cost Indexes for cost comparisons over time; City Cost Indexes and Location Factors for adjusting costs to the region you are in; Reference Tables, where you will find explanations, estimating information and procedures, or technical data; Change Orders, information on pricing changes to contract documents; Square Foot Costs that allow you to make a rough estimate for the overall cost of a project; and an explanation of all the Abbreviations in the book.

Table of Contents

Construction Equipment Rental Costs	545
Crew Listings	557
Historical Cost Indexes	593
City Cost Indexes	594
Location Factors	637
Reference Tables	643
R01 General Requirements	643
R02 Existing Conditions	658
R03 Concrete	659
R05 Metals	660
R13 Special Construction	660
R21 Fire Suppression	660

Reference Tables (cont.)

R22 Plumbing	664
R23 Heating, Ventilating & Air Conditioning	674
R31 Earthwork	678
R33 Utilities	679
Change Orders	680
Square Foot Costs	685
Abbreviations	695

Equipment Rental Costs

Estimating Tips

- This section contains the average costs to rent and operate hundreds of pieces of construction equipment. This is useful information when estimating the time and material requirements of any particular operation in order to establish a unit or total cost. Equipment costs include not only rental, but also operating costs for equipment under normal use.

Rental Costs

- Equipment rental rates are obtained from industry sources throughout North America–contractors, suppliers, dealers, manufacturers, and distributors.
- Rental rates vary throughout the country, with larger cities generally having lower rates. Lease plans for new equipment are available for periods in excess of six months, with a percentage of payments applying toward purchase.
- Monthly rental rates vary from 2% to 5% of the purchase price of the equipment depending on the anticipated life of the equipment and its wearing parts.
- Weekly rental rates are about 1/3 the monthly rates, and daily rental rates are about 1/3 the weekly rate.
- Rental rates can also be treated as reimbursement costs for contractor-owned equipment. Owned equipment costs include depreciation, loan payments, interest, taxes, insurance, storage, and major repairs.

Operating Costs

- The operating costs include parts and labor for routine servicing, such as repair and replacement of pumps, filters and worn lines. Normal operating expendables, such as fuel, lubricants, tires and electricity (where applicable), are also included.
- Extraordinary operating expendables with highly variable wear patterns, such as diamond bits and blades, are excluded. These costs can be found as material costs in the Unit Price section.
- The hourly operating costs listed do not include the operator's wages.

Equipment Cost/Day

- Any power equipment required by a crew is shown in the Crew Listings with a daily cost.
- The daily cost of equipment needed by a crew is based on dividing the weekly rental rate by 5 (number of working days in the week), and then adding the hourly operating cost times 8 (the number of hours in a day). This "Equipment Cost/Day" is shown in the far right column of the Equipment Rental pages.
- If equipment is needed for only one or two days, it is best to develop your own cost by including components for daily rent and hourly operating cost. This is important when the listed Crew for a task does not contain the equipment needed, such as a crane for lifting mechanical heating/cooling equipment up onto a roof.
- If the quantity of work is less than the crew's Daily Output shown for a Unit Price line item that includes a bare unit equipment cost, it is recommended to estimate one day's rental cost and operating cost for equipment shown in the Crew Listing for that line item.

Mobilization/Demobilization

- The cost to move construction equipment from an equipment yard or rental company to the jobsite and back again is not included in equipment rental costs listed in the Reference section, nor in the bare equipment cost of any Unit Price line item, nor in any equipment costs shown in the Crew listings.
- Mobilization (to the site) and demobilization (from the site) costs can be found in the Unit Price section.
- If a piece of equipment is already at the jobsite, it is not appropriate to utilize mobil./demob. costs again in an estimate.

No part of this publication may be reproduced, stored in a retrieval system, or transmitted in any form or by any means without prior written permission of RSMeans, a division of Reed Construction Data.

01 54 | Construction Aids

01 54 33 | Equipment Rental

				UNIT	HOURLY OPER. COST	RENT PER DAY	RENT PER WEEK	RENT PER MONTH	EQUIPMENT COST/DAY	
10	0010	**CONCRETE EQUIPMENT RENTAL** without operators	R015433 -10							10
	0200	Bucket, concrete lightweight, 1/2 C.Y.		Ea.	.80	23	69	207	20.20	
	0300	1 C.Y.			.85	27	81	243	23	
	0400	1-1/2 C.Y.			1.10	36.50	110	330	30.80	
	0500	2 C.Y.			1.20	43.50	130	390	35.60	
	0580	8 C.Y.			5.95	252	755	2,275	198.60	
	0600	Cart, concrete, self-propelled, operator walking, 10 C.F.			3.40	55	165	495	60.20	
	0700	Operator riding, 18 C.F.			5.65	93.50	280	840	101.20	
	0800	Conveyer for concrete, portable, gas, 16" wide, 26' long			13.50	122	365	1,100	181	
	0900	46' long			13.90	147	440	1,325	199.20	
	1000	56' long			14.05	155	465	1,400	205.40	
	1100	Core drill, electric, 2-1/2 H.P., 1" to 8" bit diameter			1.50	57	171	515	46.20	
	1150	11 H.P., 8" to 18" cores			5.90	115	345	1,025	116.20	
	1200	Finisher, concrete floor, gas, riding trowel, 96" wide			14.05	142	425	1,275	197.40	
	1300	Gas, walk-behind, 3 blade, 36" trowel			2.35	20	60	180	30.80	
	1400	4 blade, 48" trowel			4.70	26.50	80	240	53.60	
	1500	Float, hand-operated (Bull float) 48" wide			.08	13.35	40	120	8.65	
	1570	Curb builder, 14 H.P., gas, single screw			14.65	250	750	2,250	267.20	
	1590	Double screw			15.30	287	860	2,575	294.40	
	1600	Floor grinder, concrete and terrazzo, electric, 22" path			2.19	134	402	1,200	97.90	
	1700	Edger, concrete, electric, 7" path			1.02	51.50	155	465	39.15	
	1750	Vacuum pick-up system for floor grinders, wet/dry			1.47	81.50	245	735	60.75	
	1800	Mixer, powered, mortar and concrete, gas, 6 C.F., 18 H.P.			9.20	117	350	1,050	143.60	
	1900	10 C.F., 25 H.P.			11.60	140	420	1,250	176.80	
	2000	16 C.F.			11.95	162	485	1,450	192.60	
	2100	Concrete, stationary, tilt drum, 2 C.Y.			6.80	228	685	2,050	191.40	
	2120	Pump, concrete, truck mounted 4" line 80' boom			24.55	855	2,570	7,700	710.40	
	2140	5" line, 110' boom			31.75	1,125	3,400	10,200	934	
	2160	Mud jack, 50 C.F. per hr.			7.35	127	380	1,150	134.80	
	2180	225 C.F. per hr.			9.65	145	435	1,300	164.20	
	2190	Shotcrete pump rig, 12 C.Y./hr.			15.10	220	660	1,975	252.80	
	2200	35 C.Y./hr.			17.80	237	710	2,125	284.40	
	2600	Saw, concrete, manual, gas, 18 H.P.			7.25	43.50	130	390	84	
	2650	Self-propelled, gas, 30 H.P.			14.40	100	300	900	175.20	
	2675	V-groove crack chaser, manual, gas, 6 H.P.			2.55	17.35	52	156	30.80	
	2700	Vibrators, concrete, electric, 60 cycle, 2 H.P.			.43	8.65	26	78	8.65	
	2800	3 H.P.			.61	11.65	35	105	11.90	
	2900	Gas engine, 5 H.P.			2.20	16	48	144	27.20	
	3000	8 H.P.			3.00	15	45	135	33	
	3050	Vibrating screed, gas engine, 8 H.P.			2.78	68.50	205	615	63.25	
	3120	Concrete transit mixer, 6 x 4, 250 H.P., 8 C.Y., rear discharge			62.50	565	1,690	5,075	838	
	3200	Front discharge			73.60	690	2,065	6,200	1,002	
	3300	6 x 6, 285 H.P., 12 C.Y., rear discharge			72.65	650	1,950	5,850	971.20	
	3400	Front discharge			75.95	695	2,085	6,250	1,025	
20	0010	**EARTHWORK EQUIPMENT RENTAL** without operators	R015433 -10							20
	0040	Aggregate spreader, push type 8' to 12' wide		Ea.	3.25	25.50	76	228	41.20	
	0045	Tailgate type, 8' wide			3.10	32.50	98	294	44.40	
	0055	Earth auger, truck-mounted, for fence & sign posts, utility poles			19.75	445	1,335	4,000	425	
	0060	For borings and monitoring wells			47.30	680	2,040	6,125	786.40	
	0070	Portable, trailer mounted			3.15	32.50	98	294	44.80	
	0075	Truck-mounted, for caissons, water wells			100.90	2,900	8,720	26,200	2,551	
	0080	Horizontal boring machine, 12" to 36" diameter, 45 H.P.			24.60	193	580	1,750	312.80	
	0090	12" to 48" diameter, 65 H.P.			34.85	335	1,005	3,025	479.80	
	0095	Auger, for fence posts, gas engine, hand held			.55	6	18	54	8	
	0100	Excavator, diesel hydraulic, crawler mounted, 1/2 C.Y. cap.			25.00	415	1,240	3,725	448	
	0120	5/8 C.Y. capacity			33.05	550	1,650	4,950	594.40	
	0140	3/4 C.Y. capacity			36.00	625	1,870	5,600	662	
	0150	1 C.Y. capacity			50.60	685	2,050	6,150	814.80	

01 54 | Construction Aids

01 54 33 | Equipment Rental

		UNIT	HOURLY OPER. COST	RENT PER DAY	RENT PER WEEK	RENT PER MONTH	EQUIPMENT COST/DAY
0200	1-1/2 C.Y. capacity	Ea.	60.80	910	2,725	8,175	1,031
0300	2 C.Y. capacity		68.50	1,050	3,135	9,400	1,175
0320	2-1/2 C.Y. capacity		104.75	1,325	3,995	12,000	1,637
0325	3-1/2 C.Y. capacity		147.35	2,100	6,315	18,900	2,442
0330	4-1/2 C.Y. capacity		176.65	2,725	8,150	24,500	3,043
0335	6 C.Y. capacity		224.40	2,875	8,640	25,900	3,523
0340	7 C.Y. capacity		226.60	2,975	8,915	26,700	3,596
0342	Excavator attachments, bucket thumbs		3.20	245	735	2,200	172.60
0345	Grapples		2.75	193	580	1,750	138
0347	Hydraulic hammer for boom mounting, 5000 ft lb.		14.30	425	1,280	3,850	370.40
0349	11,000 ft lb.		23.20	750	2,250	6,750	635.60
0350	Gradall type, truck mounted, 3 ton @ 15' radius, 5/8 C.Y.		43.45	890	2,670	8,000	881.60
0370	1 C.Y. capacity		47.65	1,050	3,120	9,350	1,005
0400	Backhoe-loader, 40 to 45 H.P., 5/8 C.Y. capacity		14.95	233	700	2,100	259.60
0450	45 H.P. to 60 H.P., 3/4 C.Y. capacity		23.90	290	870	2,600	365.20
0460	80 H.P., 1-1/4 C.Y. capacity		26.20	315	940	2,825	397.60
0470	112 H.P., 1-1/2 C.Y. capacity		41.65	645	1,935	5,800	720.20
0482	Backhoe-loader attachment, compactor, 20,000 lb.		5.75	142	425	1,275	131
0485	Hydraulic hammer, 750 ft lb.		3.25	96.50	290	870	84
0486	Hydraulic hammer, 1200 ft lb.		6.30	217	650	1,950	180.40
0500	Brush chipper, gas engine, 6" cutter head, 35 H.P.		11.95	107	320	960	159.60
0550	Diesel engine, 12" cutter head, 130 H.P.		28.20	285	855	2,575	396.60
0600	15" cutter head, 165 H.P.		33.90	330	985	2,950	468.20
0750	Bucket, clamshell, general purpose, 3/8 C.Y.		1.25	38.50	115	345	33
0800	1/2 C.Y.		1.35	45	135	405	37.80
0850	3/4 C.Y.		1.50	53.50	160	480	44
0900	1 C.Y.		1.55	58.50	175	525	47.40
0950	1-1/2 C.Y.		2.55	80	240	720	68.40
1000	2 C.Y.		2.65	88.50	265	795	74.20
1010	Bucket, dragline, medium duty, 1/2 C.Y.		.75	23.50	70	210	20
1020	3/4 C.Y.		.75	24.50	74	222	20.80
1030	1 C.Y.		.80	26	78	234	22
1040	1-1/2 C.Y.		1.20	40	120	360	33.60
1050	2 C.Y.		1.25	45	135	405	37
1070	3 C.Y.		1.95	61.50	185	555	52.60
1200	Compactor, manually guided 2-drum vibratory smooth roller, 7.5 H.P.		7.35	207	620	1,850	182.80
1250	Rammer/tamper, gas, 8"		2.70	46.50	140	420	49.60
1260	15"		3.00	53.50	160	480	56
1300	Vibratory plate, gas, 18" plate, 3000 lb. blow		2.65	24.50	73	219	35.80
1350	21" plate, 5000 lb. blow		3.30	32.50	98	294	46
1370	Curb builder/extruder, 14 H.P., gas, single screw		14.65	250	750	2,250	267.20
1390	Double screw		15.30	287	860	2,575	294.40
1500	Disc harrow attachment, for tractor		.43	72.50	217	650	46.85
1810	Feller buncher, shearing & accumulating trees, 100 H.P.		42.50	650	1,945	5,825	729
1860	Grader, self-propelled, 25,000 lb.		39.55	595	1,780	5,350	672.40
1910	30,000 lb.		43.25	600	1,805	5,425	707
1920	40,000 lb.		64.65	1,000	3,015	9,050	1,120
1930	55,000 lb.		84.65	1,600	4,800	14,400	1,637
1950	Hammer, pavement breaker, self-propelled, diesel, 1000 to 1250 lb.		30.25	355	1,060	3,175	454
2000	1300 to 1500 lb.		45.38	705	2,120	6,350	787.05
2050	Pile driving hammer, steam or air, 4150 ft lb. @ 225 bpm		10.35	475	1,425	4,275	367.80
2100	8750 ft lb. @ 145 bpm		12.35	660	1,980	5,950	494.80
2150	15,000 ft lb. @ 60 bpm		13.95	795	2,380	7,150	587.60
2200	24,450 ft lb. @ 111 bpm		14.95	880	2,640	7,925	647.60
2250	Leads, 60' high for pile driving hammers up to 20,000 ft lb.		3.30	80.50	242	725	74.80
2300	90' high for hammers over 20,000 ft lb.		4.95	141	424	1,275	124.40
2350	Diesel type hammer, 22,400 ft lb.		26.93	535	1,610	4,825	537.45
2400	41,300 ft lb.		36.30	585	1,760	5,275	642.40

01 54 | Construction Aids

01 54 33 | Equipment Rental

			UNIT	HOURLY OPER. COST	RENT PER DAY	RENT PER WEEK	RENT PER MONTH	EQUIPMENT COST/DAY
20	2450	141,000 ft lb.	Ea.	55.55	1,050	3,160	9,475	1,076
	2500	Vib. elec. hammer/extractor, 200 kW diesel generator, 34 H.P.		56.25	635	1,910	5,725	832
	2550	80 H.P.		103.45	930	2,785	8,350	1,385
	2600	150 H.P.		196.65	1,775	5,335	16,000	2,640
	2800	Log chipper, up to 22" diameter, 600 H.P.		64.95	630	1,895	5,675	898.60
	2850	Logger, for skidding & stacking logs, 150 H.P.		53.45	820	2,460	7,375	919.60
	2860	Mulcher, diesel powered, trailer mounted		24.70	213	640	1,925	325.60
	2900	Rake, spring tooth, with tractor		18.31	335	1,007	3,025	347.90
	3000	Roller, vibratory, tandem, smooth drum, 20 H.P.		8.75	148	445	1,325	159
	3050	35 H.P.		11.25	253	760	2,275	242
	3100	Towed type vibratory compactor, smooth drum, 50 H.P.		26.25	335	1,010	3,025	412
	3150	Sheepsfoot, 50 H.P.		27.60	370	1,115	3,350	443.80
	3170	Landfill compactor, 220 H.P.		91.85	1,500	4,485	13,500	1,632
	3200	Pneumatic tire roller, 80 H.P.		17.00	355	1,065	3,200	349
	3250	120 H.P.		25.45	605	1,815	5,450	566.60
	3300	Sheepsfoot vibratory roller, 240 H.P.		69.65	1,100	3,305	9,925	1,218
	3320	340 H.P.		101.85	1,600	4,830	14,500	1,781
	3350	Smooth drum vibratory roller, 75 H.P.		25.15	595	1,790	5,375	559.20
	3400	125 H.P.		32.85	715	2,150	6,450	692.80
	3410	Rotary mower, brush, 60", with tractor		22.65	305	915	2,750	364.20
	3420	Rototiller, walk-behind, gas, 5 H.P.		2.23	66	198	595	57.45
	3422	8 H.P.		3.30	91.50	275	825	81.40
	3440	Scrapers, towed type, 7 C.Y. capacity		5.70	110	330	990	111.60
	3450	10 C.Y. capacity		6.45	150	450	1,350	141.60
	3500	15 C.Y. capacity		6.85	173	520	1,550	158.80
	3525	Self-propelled, single engine, 14 C.Y. capacity		113.60	1,525	4,600	13,800	1,829
	3550	Dual engine, 21 C.Y. capacity		174.50	2,000	6,015	18,000	2,599
	3600	31 C.Y. capacity		232.10	2,850	8,545	25,600	3,566
	3640	44 C.Y. capacity		287.55	3,650	10,960	32,900	4,492
	3650	Elevating type, single engine, 11 C.Y. capacity		71.20	1,025	3,050	9,150	1,180
	3700	22 C.Y. capacity		138.15	2,225	6,690	20,100	2,443
	3710	Screening plant 110 H.P. w/5' x 10' screen		31.63	435	1,300	3,900	513.05
	3720	5' x 16' screen		34.85	540	1,615	4,850	601.80
	3850	Shovel, crawler-mounted, front-loading, 7 C.Y. capacity		246.75	2,900	8,700	26,100	3,714
	3855	12 C.Y. capacity		354.50	3,950	11,840	35,500	5,204
	3860	Shovel/backhoe bucket, 1/2 C.Y.		2.45	66.50	200	600	59.60
	3870	3/4 C.Y.		2.50	73.50	220	660	64
	3880	1 C.Y.		2.60	83.50	250	750	70.80
	3890	1-1/2 C.Y.		2.75	96.50	290	870	80
	3910	3 C.Y.		3.10	130	390	1,175	102.80
	3950	Stump chipper, 18" deep, 30 H.P.		7.06	176	528	1,575	162.10
	4110	Dozer, crawler, torque converter, diesel 80 H.P.		29.75	405	1,215	3,650	481
	4150	105 H.P.		36.10	505	1,515	4,550	591.80
	4200	140 H.P.		52.10	780	2,335	7,000	883.80
	4260	200 H.P.		77.35	1,175	3,530	10,600	1,325
	4310	300 H.P.		100.55	1,700	5,065	15,200	1,817
	4360	410 H.P.		135.65	2,200	6,620	19,900	2,409
	4370	500 H.P.		174.30	2,875	8,610	25,800	3,116
	4380	700 H.P.		261.35	4,000	11,970	35,900	4,485
	4400	Loader, crawler, torque conv., diesel, 1-1/2 C.Y., 80 H.P.		31.35	455	1,360	4,075	522.80
	4450	1-1/2 to 1-3/4 C.Y., 95 H.P.		34.15	585	1,760	5,275	625.20
	4510	1-3/4 to 2-1/4 C.Y., 130 H.P.		54.60	870	2,615	7,850	959.80
	4530	2-1/2 to 3-1/4 C.Y., 190 H.P.		67.00	1,100	3,275	9,825	1,191
	4560	3-1/2 to 5 C.Y., 275 H.P.		86.10	1,400	4,215	12,600	1,532
	4610	Front end loader, 4WD, articulated frame, diesel 1 to 1-1/4 C.Y., 70 H.P.		19.90	227	680	2,050	295.20
	4620	1-1/2 to 1-3/4 C.Y., 95 H.P.		25.30	285	855	2,575	373.40
	4650	1-3/4 to 2 C.Y., 130 H.P.		29.45	365	1,100	3,300	455.60
	4710	2-1/2 to 3-1/2 C.Y., 145 H.P.		33.35	425	1,280	3,850	522.80

01 54 | Construction Aids

01 54 33 | Equipment Rental

		Unit	Hourly Oper. Cost	Rent Per Day	Rent Per Week	Rent Per Month	Equipment Cost/Day
4730	3 to 4-1/2 C.Y., 185 H.P.	Ea.	42.85	550	1,655	4,975	673.80
4760	5-1/4 to 5-3/4 C.Y., 270 H.P.		67.80	845	2,535	7,600	1,049
4810	7 to 9 C.Y., 475 H.P.		116.25	1,775	5,295	15,900	1,989
4870	9 - 11 C.Y., 620 H.P.		155.75	2,425	7,285	21,900	2,703
4880	Skid steer loader, wheeled, 10 C.F., 30 H.P. gas		10.90	145	435	1,300	174.20
4890	1 C.Y., 78 H.P., diesel		20.35	243	730	2,200	308.80
4892	Skid-steer attachment, auger		.50	83.50	251	755	54.20
4893	Backhoe		.66	111	332	995	71.70
4894	Broom		.74	124	372	1,125	80.30
4895	Forks		.23	38.50	115	345	24.85
4896	Grapple		.56	92.50	278	835	60.10
4897	Concrete hammer		1.06	177	531	1,600	114.70
4898	Tree spade		1.03	172	515	1,550	111.25
4899	Trencher		.67	112	335	1,000	72.35
4900	Trencher, chain, boom type, gas, operator walking, 12 H.P.		5.40	46.50	140	420	71.20
4910	Operator riding, 40 H.P.		20.55	292	875	2,625	339.40
5000	Wheel type, diesel, 4' deep, 12" wide		87.50	845	2,530	7,600	1,206
5100	6' deep, 20" wide		93.25	1,925	5,760	17,300	1,898
5150	Chain type, diesel, 5' deep, 8" wide		38.85	560	1,685	5,050	647.80
5200	Diesel, 8' deep, 16" wide		156.45	3,550	10,620	31,900	3,376
5202	Rock trencher, wheel type, 6" wide x 18" deep		21.50	330	985	2,950	369
5206	Chain type, 18" wide x 7' deep		112.60	2,825	8,445	25,300	2,590
5210	Tree spade, self-propelled		14.00	267	800	2,400	272
5250	Truck, dump, 2-axle, 12 ton, 8 C.Y. payload, 220 H.P.		35.15	227	680	2,050	417.20
5300	Three axle dump, 16 ton, 12 C.Y. payload, 400 H.P.		62.40	320	965	2,900	692.20
5310	Four axle dump, 25 ton, 18 C.Y. payload, 450 H.P.		73.20	470	1,410	4,225	867.60
5350	Dump trailer only, rear dump, 16-1/2 C.Y.		5.40	138	415	1,250	126.20
5400	20 C.Y.		5.85	157	470	1,400	140.80
5450	Flatbed, single axle, 1-1/2 ton rating		28.65	68.50	205	615	270.20
5500	3 ton rating		34.25	98.50	295	885	333
5550	Off highway rear dump, 25 ton capacity		74.55	1,200	3,590	10,800	1,314
5600	35 ton capacity		83.10	1,350	4,025	12,100	1,470
5610	50 ton capacity		103.40	1,600	4,800	14,400	1,787
5620	65 ton capacity		106.30	1,575	4,735	14,200	1,797
5630	100 ton capacity		154.35	2,775	8,315	24,900	2,898
6000	Vibratory plow, 25 H.P., walking		9.25	60	180	540	110
0010	**GENERAL EQUIPMENT RENTAL** without operators	R015433-10					
0150	Aerial lift, scissor type, to 15' high, 1000 lb. cap., electric	Ea.	3.10	51.50	155	465	55.80
0160	To 25' high, 2000 lb. capacity		3.55	66.50	200	600	68.40
0170	Telescoping boom to 40' high, 500 lb. capacity, diesel		17.00	315	945	2,825	325
0180	To 45' high, 500 lb. capacity		18.85	355	1,070	3,200	364.80
0190	To 60' high, 600 lb. capacity		21.25	475	1,430	4,300	456
0195	Air compressor, portable, 6.5 CFM, electric		.56	12.65	38	114	12.10
0196	Gasoline		.79	19	57	171	17.70
0200	Towed type, gas engine, 60 CFM		13.55	48.50	145	435	137.40
0300	160 CFM		15.80	51.50	155	465	157.40
0400	Diesel engine, rotary screw, 250 CFM		17.05	110	330	990	202.40
0500	365 CFM		23.10	133	400	1,200	264.80
0550	450 CFM		29.35	167	500	1,500	334.80
0600	600 CFM		51.70	232	695	2,075	552.60
0700	750 CFM		51.90	240	720	2,150	559.20
0800	For silenced models, small sizes, add to rent		3%	5%	5%	5%	
0900	Large sizes, add to rent		5%	7%	7%	7%	
0930	Air tools, breaker, pavement, 60 lb.	Ea.	.50	9.65	29	87	9.80
0940	80 lb.		.50	10.35	31	93	10.20
0950	Drills, hand (jackhammer) 65 lb.		.60	17.35	52	156	15.20
0960	Track or wagon, swing boom, 4" drifter		63.45	885	2,660	7,975	1,040
0970	5" drifter		75.50	1,075	3,255	9,775	1,255

01 54 | Construction Aids

01 54 33 | Equipment Rental

			UNIT	HOURLY OPER. COST	RENT PER DAY	RENT PER WEEK	RENT PER MONTH	EQUIPMENT COST/DAY	
40	0975	Track mounted quarry drill, 6" diameter drill	Ea.	129.75	1,600	4,830	14,500	2,004	40
	0980	Dust control per drill		.97	24	72	216	22.15	
	0990	Hammer, chipping, 12 lb.		.55	26	78	234	20	
	1000	Hose, air with couplings, 50' long, 3/4" diameter		.03	5	15	45	3.25	
	1100	1" diameter		.04	6.35	19	57	4.10	
	1200	1-1/2" diameter		.05	9	27	81	5.80	
	1300	2" diameter		.07	12	36	108	7.75	
	1400	2-1/2" diameter		.11	19	57	171	12.30	
	1410	3" diameter		.14	23	69	207	14.90	
	1450	Drill, steel, 7/8" x 2'		.05	8.65	26	78	5.60	
	1460	7/8" x 6'		.05	9	27	81	5.80	
	1520	Moil points		.02	3.33	10	30	2.15	
	1525	Pneumatic nailer w/accessories		.47	31	93	279	22.35	
	1530	Sheeting driver for 60 lb. breaker		.03	5.65	17	51	3.65	
	1540	For 90 lb. breaker		.11	7.35	22	66	5.30	
	1550	Spade, 25 lb.		.45	7	21	63	7.80	
	1560	Tamper, single, 35 lb.		.55	36.50	109	325	26.20	
	1570	Triple, 140 lb.		.82	54.50	164	490	39.35	
	1580	Wrenches, impact, air powered, up to 3/4" bolt		.40	12.65	38	114	10.80	
	1590	Up to 1-1/4" bolt		.50	23.50	70	210	18	
	1600	Barricades, barrels, reflectorized, 1 to 99 barrels		.03	4.60	13.80	41.50	3	
	1610	100 to 200 barrels		.02	3.53	10.60	32	2.30	
	1620	Barrels with flashers, 1 to 99 barrels		.03	5.25	15.80	47.50	3.40	
	1630	100 to 200 barrels		.03	4.20	12.60	38	2.75	
	1640	Barrels with steady burn type C lights		.04	7	21	63	4.50	
	1650	Illuminated board, trailer mounted, with generator		3.55	130	390	1,175	106.40	
	1670	Portable barricade, stock, with flashers, 1 to 6 units		.03	5.25	15.80	47.50	3.40	
	1680	25 to 50 units		.03	4.90	14.70	44	3.20	
	1685	Butt fusion machine, wheeled, 1.5 HP electric, 2" - 8" diameter pipe		2.62	167	500	1,500	120.95	
	1690	Tracked, 20 HP diesel, 4"-12" diameter pipe		11.10	490	1,465	4,400	381.80	
	1695	83 HP diesel, 8" - 24" diameter pipe		30.30	975	2,930	8,800	828.40	
	1700	Carts, brick, hand powered, 1000 lb. capacity		.44	72.50	218	655	47.10	
	1800	Gas engine, 1500 lb., 7-1/2' lift		4.10	115	345	1,025	101.80	
	1822	Dehumidifier, medium, 6 lb./hr., 150 CFM		.99	62	186	560	45.10	
	1824	Large, 18 lb./hr., 600 CFM		2.01	126	378	1,125	91.70	
	1830	Distributor, asphalt, trailer mounted, 2000 gal., 38 H.P. diesel		10.15	330	985	2,950	278.20	
	1840	3000 gal., 38 H.P. diesel		11.65	360	1,075	3,225	308.20	
	1850	Drill, rotary hammer, electric		1.00	26.50	80	240	24	
	1860	Carbide bit, 1-1/2" diameter, add to electric rotary hammer		.02	3.61	10.84	32.50	2.35	
	1865	Rotary, crawler, 250 H.P.		150.45	2,100	6,285	18,900	2,461	
	1870	Emulsion sprayer, 65 gal., 5 H.P. gas engine		2.90	98	294	880	82	
	1880	200 gal., 5 H.P. engine		7.85	163	490	1,475	160.80	
	1900	Floor auto-scrubbing machine, walk-behind, 28" path		4.52	293	880	2,650	212.15	
	1930	Floodlight, mercury vapor, or quartz, on tripod, 1000 watt		.41	19.35	58	174	14.90	
	1940	2000 watt		.78	39	117	350	29.65	
	1950	Floodlights, trailer mounted with generator, 1 - 300 watt light		3.65	71.50	215	645	72.20	
	1960	2 - 1000 watt lights		4.90	98.50	295	885	98.20	
	2000	4 - 300 watt lights		4.60	93.50	280	840	92.80	
	2005	Foam spray rig, incl. box trailer, compressor, generator, proportioner		36.53	490	1,465	4,400	585.25	
	2020	Forklift, straight mast, 12' lift, 5000 lb., 2 wheel drive, gas		25.35	203	610	1,825	324.80	
	2040	21' lift, 5000 lb., 4 wheel drive, diesel		21.15	242	725	2,175	314.20	
	2050	For rough terrain, 42' lift, 35' reach, 9000 lb., 110 H.P.		29.95	485	1,460	4,375	531.60	
	2060	For plant, 4 ton capacity, 80 H.P., 2 wheel drive, gas		15.50	93.50	280	840	180	
	2080	10 ton capacity, 120 H.P., 2 wheel drive, diesel		24.30	162	485	1,450	291.40	
	2100	Generator, electric, gas engine, 1.5 kW to 3 kW		3.95	11.35	34	102	38.40	
	2200	5 kW		5.10	15.35	46	138	50	
	2300	10 kW		9.65	36.50	110	330	99.20	
	2400	25 kW		11.00	86.50	260	780	140	

01 54 | Construction Aids

01 54 33 | Equipment Rental

			UNIT	HOURLY OPER. COST	RENT PER DAY	RENT PER WEEK	RENT PER MONTH	EQUIPMENT COST/DAY	
40	2500	Diesel engine, 20 kW	Ea.	12.55	68.50	205	615	141.40	40
	2600	50 kW		24.35	103	310	930	256.80	
	2700	100 kW		45.15	130	390	1,175	439.20	
	2800	250 kW		88.85	237	710	2,125	852.80	
	2850	Hammer, hydraulic, for mounting on boom, to 500 ft lb.		2.55	75	225	675	65.40	
	2860	1000 ft lb.		4.35	127	380	1,150	110.80	
	2900	Heaters, space, oil or electric, 50 MBH		1.93	7.65	23	69	20.05	
	3000	100 MBH		3.61	10.65	32	96	35.30	
	3100	300 MBH		10.58	38.50	115	345	107.65	
	3150	500 MBH		17.40	45	135	405	166.20	
	3200	Hose, water, suction with coupling, 20' long, 2" diameter		.02	3	9	27	1.95	
	3210	3" diameter		.03	4.33	13	39	2.85	
	3220	4" diameter		.03	5	15	45	3.25	
	3230	6" diameter		.11	17.65	53	159	11.50	
	3240	8" diameter		.20	33.50	100	300	21.60	
	3250	Discharge hose with coupling, 50' long, 2" diameter		.01	1.33	4	12	.90	
	3260	3" diameter		.01	2.33	7	21	1.50	
	3270	4" diameter		.02	3.67	11	33	2.35	
	3280	6" diameter		.06	9.35	28	84	6.10	
	3290	8" diameter		.20	33.50	100	300	21.60	
	3295	Insulation blower		.77	6	18	54	9.75	
	3300	Ladders, extension type, 16' to 36' long		.14	22.50	68	204	14.70	
	3400	40' to 60' long		.19	31	93	279	20.10	
	3405	Lance for cutting concrete		2.52	89	267	800	73.55	
	3407	Lawn mower, rotary, 22", 5 H.P.		2.10	59	177	530	52.20	
	3408	48" self propelled		2.92	75	225	675	68.35	
	3410	Level, electronic, automatic, with tripod and leveling rod		1.49	99	297	890	71.30	
	3430	Laser type, for pipe and sewer line and grade		.73	48.50	145	435	34.85	
	3440	Rotating beam for interior control		.97	64	192	575	46.15	
	3460	Builder's optical transit, with tripod and rod		.10	16.65	50	150	10.80	
	3500	Light towers, towable, with diesel generator, 2000 watt		4.60	93.50	280	840	92.80	
	3600	4000 watt		4.90	98.50	295	885	98.20	
	3700	Mixer, powered, plaster and mortar, 6 C.F., 7 H.P.		2.90	19	57	171	34.60	
	3800	10 C.F., 9 H.P.		3.05	30.50	91	273	42.60	
	3850	Nailer, pneumatic		.47	31	93	279	22.35	
	3900	Paint sprayers complete, 8 CFM		.94	62.50	187	560	44.90	
	4000	17 CFM		1.57	104	313	940	75.15	
	4020	Pavers, bituminous, rubber tires, 8' wide, 50 H.P., diesel		31.85	500	1,500	4,500	554.80	
	4030	10' wide, 150 H.P.		107.95	1,825	5,460	16,400	1,956	
	4050	Crawler, 8' wide, 100 H.P., diesel		90.45	1,825	5,500	16,500	1,824	
	4060	10' wide, 150 H.P.		115.10	2,250	6,715	20,100	2,264	
	4070	Concrete paver, 12' to 24' wide, 250 H.P.		102.05	1,525	4,600	13,800	1,736	
	4080	Placer-spreader-trimmer, 24' wide, 300 H.P.		147.05	2,475	7,405	22,200	2,657	
	4100	Pump, centrifugal gas pump, 1-1/2" diam., 65 GPM		3.95	50	150	450	61.60	
	4200	2" diameter, 130 GPM		5.30	55	165	495	75.40	
	4300	3" diameter, 250 GPM		5.60	56.50	170	510	78.80	
	4400	6" diameter, 1500 GPM		29.35	177	530	1,600	340.80	
	4500	Submersible electric pump, 1-1/4" diameter, 55 GPM		.38	16.65	50	150	13.05	
	4600	1-1/2" diameter, 83 GPM		.42	19	57	171	14.75	
	4700	2" diameter, 120 GPM		1.43	23.50	71	213	25.65	
	4800	3" diameter, 300 GPM		2.55	41.50	125	375	45.40	
	4900	4" diameter, 560 GPM		10.65	160	480	1,450	181.20	
	5000	6" diameter, 1590 GPM		15.75	215	645	1,925	255	
	5100	Diaphragm pump, gas, single, 1-1/2" diameter		1.18	50.50	152	455	39.85	
	5200	2" diameter		4.25	63.50	190	570	72	
	5300	3" diameter		4.25	63.50	190	570	72	
	5400	Double, 4" diameter		6.35	108	325	975	115.80	
	5450	Pressure washer 5 GPM, 3000 psi		4.80	51.50	155	465	69.40	

551

01 54 | Construction Aids

01 54 33 | Equipment Rental

			UNIT	HOURLY OPER. COST	RENT PER DAY	RENT PER WEEK	RENT PER MONTH	EQUIPMENT COST/DAY	
40	5460	7 GPM, 3000 psi	Ea.	6.25	60	180	540	86	40
	5500	Trash pump, self-priming, gas, 2" diameter		4.50	21.50	64	192	48.80	
	5600	Diesel, 4" diameter		9.30	90	270	810	128.40	
	5650	Diesel, 6" diameter		25.55	150	450	1,350	294.40	
	5655	Grout Pump		26.80	260	780	2,350	370.40	
	5700	Salamanders, L.P. gas fired, 100,000 Btu		3.81	13.65	41	123	38.70	
	5705	50,000 Btu		2.39	10.35	31	93	25.30	
	5720	Sandblaster, portable, open top, 3 C.F. capacity		.55	26.50	80	240	20.40	
	5730	6 C.F. capacity		.95	40	120	360	31.60	
	5740	Accessories for above		.13	21.50	64	192	13.85	
	5750	Sander, floor		.71	17	51	153	15.90	
	5760	Edger		.57	18.35	55	165	15.55	
	5800	Saw, chain, gas engine, 18" long		2.25	22	66	198	31.20	
	5900	Hydraulic powered, 36" long		.75	65	195	585	45	
	5950	60" long		.75	66.50	200	600	46	
	6000	Masonry, table mounted, 14" diameter, 5 H.P.		1.31	56.50	170	510	44.50	
	6050	Portable cut-off, 8 H.P.		2.40	33.50	100	300	39.20	
	6100	Circular, hand held, electric, 7-1/4" diameter		.22	4.67	14	42	4.55	
	6200	12" diameter		.22	8	24	72	6.55	
	6250	Wall saw, w/hydraulic power, 10 H.P.		10.45	60	180	540	119.60	
	6275	Shot blaster, walk-behind, 20" wide		4.80	293	880	2,650	214.40	
	6280	Sidewalk broom, walk-behind		2.32	70	210	630	60.55	
	6300	Steam cleaner, 100 gallons per hour		3.70	76.50	230	690	75.60	
	6310	200 gallons per hour		5.25	95	285	855	99	
	6340	Tar Kettle/Pot, 400 gallons		10.50	76.50	230	690	130	
	6350	Torch, cutting, acetylene-oxygen, 150' hose, excludes gases		.30	15	45	135	11.40	
	6360	Hourly operating cost includes tips and gas		19.00				152	
	6410	Toilet, portable chemical		.12	20.50	62	186	13.35	
	6420	Recycle flush type		.15	24.50	74	222	16	
	6430	Toilet, fresh water flush, garden hose,		.18	29.50	89	267	19.25	
	6440	Hoisted, non-flush, for high rise		.15	24.50	73	219	15.80	
	6465	Tractor, farm with attachment		21.90	277	830	2,500	341.20	
	6500	Trailers, platform, flush deck, 2 axle, 25 ton capacity		5.45	117	350	1,050	113.60	
	6600	40 ton capacity		7.00	160	480	1,450	152	
	6700	3 axle, 50 ton capacity		7.55	177	530	1,600	166.40	
	6800	75 ton capacity		9.35	232	695	2,075	213.80	
	6810	Trailer mounted cable reel for high voltage line work		5.35	255	764	2,300	195.60	
	6820	Trailer mounted cable tensioning rig		10.64	505	1,520	4,550	389.10	
	6830	Cable pulling rig		71.33	2,825	8,510	25,500	2,273	
	6900	Water tank trailer, engine driven discharge, 5000 gallons		6.95	143	430	1,300	141.60	
	6925	10,000 gallons		9.50	198	595	1,775	195	
	6950	Water truck, off highway, 6000 gallons		89.80	775	2,330	7,000	1,184	
	7010	Tram car for high voltage line work, powered, 2 conductor		6.27	138	415	1,250	133.15	
	7020	Transit (builder's level) with tripod		.10	16.65	50	150	10.80	
	7030	Trench box, 3000 lb., 6' x 8'		.56	93.50	280	840	60.50	
	7040	7200 lb., 6' x 20'		.78	130	389	1,175	84.05	
	7050	8000 lb., 8' x 16'		1.08	180	540	1,625	116.65	
	7060	9500 lb., 8' x 20'		1.21	201	603	1,800	130.30	
	7065	11,000 lb., 8' x 24'		1.27	211	633	1,900	136.75	
	7070	12,000 lb., 10' x 20'		1.36	227	680	2,050	146.90	
	7100	Truck, pickup, 3/4 ton, 2 wheel drive		15.25	58.50	175	525	157	
	7200	4 wheel drive		15.55	73.50	220	660	168.40	
	7250	Crew carrier, 9 passenger		21.60	86.50	260	780	224.80	
	7290	Flat bed truck, 20,000 lb. GVW		22.30	125	375	1,125	253.40	
	7300	Tractor, 4 x 2, 220 H.P.		31.05	197	590	1,775	366.40	
	7410	330 H.P.		46.05	272	815	2,450	531.40	
	7500	6 x 4, 380 H.P.		52.85	315	950	2,850	612.80	
	7600	450 H.P.		64.05	385	1,150	3,450	742.40	

01 54 | Construction Aids

01 54 33 | Equipment Rental

		Description	UNIT	HOURLY OPER. COST	RENT PER DAY	RENT PER WEEK	RENT PER MONTH	EQUIPMENT COST/DAY	
40	7610	Tractor, with A frame, boom and winch, 225 H.P.	Ea.	33.95	272	815	2,450	434.60	40
	7620	Vacuum truck, hazardous material, 2500 gallons		12.00	292	875	2,625	271	
	7625	5,000 gallons		14.27	410	1,230	3,700	360.15	
	7650	Vacuum, HEPA, 16 gallon, wet/dry		.83	18	54	162	17.45	
	7655	55 gallon, wet/dry		.77	27	81	243	22.35	
	7660	Water tank, portable		.16	26.50	80	240	17.30	
	7690	Sewer/catch basin vacuum, 14 C.Y., 1500 gallons		18.61	615	1,850	5,550	518.90	
	7700	Welder, electric, 200 amp		3.64	16.35	49	147	38.90	
	7800	300 amp		5.41	20	60	180	55.30	
	7900	Gas engine, 200 amp		13.90	23.50	70	210	125.20	
	8000	300 amp		15.90	24.50	74	222	142	
	8100	Wheelbarrow, any size		.08	13	39	117	8.45	
	8200	Wrecking ball, 4000 lb.		2.30	68.50	205	615	59.40	
50	0010	**HIGHWAY EQUIPMENT RENTAL** without operators	R015433-10						50
	0050	Asphalt batch plant, portable drum mixer, 100 ton/hr.	Ea.	79.15	1,450	4,320	13,000	1,497	
	0060	200 ton/hr.		90.32	1,525	4,590	13,800	1,641	
	0070	300 ton/hr.		107.39	1,800	5,390	16,200	1,937	
	0100	Backhoe attachment, long stick, up to 185 H.P., 10.5' long		.35	23	69	207	16.60	
	0140	Up to 250 H.P., 12' long		.37	24.50	74	222	17.75	
	0180	Over 250 H.P., 15' long		.52	34.50	103	310	24.75	
	0200	Special dipper arm, up to 100 H.P., 32' long		1.06	70.50	211	635	50.70	
	0240	Over 100 H.P., 33' long		1.32	87.50	263	790	63.15	
	0280	Catch basin/sewer cleaning truck, 3 ton, 9 C.Y., 1000 gal.		46.25	390	1,170	3,500	604	
	0300	Concrete batch plant, portable, electric, 200 C.Y./hr.		22.08	510	1,525	4,575	481.65	
	0520	Grader/dozer attachment, ripper/scarifier, rear mounted, up to 135 H.P.		3.45	65	195	585	66.60	
	0540	Up to 180 H.P.		4.05	81.50	245	735	81.40	
	0580	Up to 250 H.P.		4.45	93.50	280	840	91.60	
	0700	Pvmt. removal bucket, for hyd. excavator, up to 90 H.P.		1.90	51.50	155	465	46.20	
	0740	Up to 200 H.P.		2.10	73.50	220	660	60.80	
	0780	Over 200 H.P.		2.25	85	255	765	69	
	0900	Aggregate spreader, self-propelled, 187 H.P.		53.75	685	2,055	6,175	841	
	1000	Chemical spreader, 3 C.Y.		3.40	43.50	130	390	53.20	
	1900	Hammermill, traveling, 250 H.P.		77.73	2,050	6,120	18,400	1,846	
	2000	Horizontal borer, 3" diameter, 13 H.P. gas driven		6.25	55	165	495	83	
	2150	Horizontal directional drill, 20,000 lb. thrust, 78 H.P. diesel		31.55	675	2,030	6,100	658.40	
	2160	30,000 lb. thrust, 115 H.P.		39.40	1,025	3,105	9,325	936.20	
	2170	50,000 lb. thrust, 170 H.P.		56.45	1,325	3,970	11,900	1,246	
	2190	Mud trailer for HDD, 1500 gallons, 175 H.P., gas		35.35	153	460	1,375	374.80	
	2200	Hydromulcher, diesel, 3000 gallon, for truck mounting		24.05	247	740	2,225	340.40	
	2300	Gas, 600 gallon		9.20	98.50	295	885	132.60	
	2400	Joint & crack cleaner, walk behind, 25 H.P.		4.35	50	150	450	64.80	
	2500	Filler, trailer mounted, 400 gallons, 20 H.P.		9.60	212	635	1,900	203.80	
	3000	Paint striper, self-propelled, 40 gallon, 22 H.P.		7.60	157	470	1,400	154.80	
	3100	120 gallon, 120 H.P.		24.80	400	1,195	3,575	437.40	
	3200	Post drivers, 6" I-Beam frame, for truck mounting		19.65	380	1,145	3,425	386.20	
	3400	Road sweeper, self-propelled, 8' wide, 90 H.P.		40.30	570	1,715	5,150	665.40	
	3450	Road sweeper, vacuum assisted, 4 C.Y., 220 gallons		77.55	625	1,870	5,600	994.40	
	4000	Road mixer, self-propelled, 130 H.P.		47.80	775	2,320	6,950	846.40	
	4100	310 H.P.		84.40	2,100	6,335	19,000	1,942	
	4220	Cold mix paver, incl. pug mill and bitumen tank, 165 H.P.		98.80	2,325	6,965	20,900	2,183	
	4240	Pavement brush, towed		3.20	93.50	280	840	81.60	
	4250	Paver, asphalt, wheel or crawler, 130 H.P., diesel		98.45	2,300	6,905	20,700	2,169	
	4300	Paver, road widener, gas 1' to 6', 67 H.P.		49.40	905	2,715	8,150	938.20	
	4400	Diesel, 2' to 14', 88 H.P.		63.70	1,050	3,170	9,500	1,144	
	4600	Slipform pavers, curb and gutter, 2 track, 75 H.P.		52.00	805	2,410	7,225	898	
	4700	4 track, 165 H.P.		41.65	730	2,185	6,550	770.20	
	4800	Median barrier, 215 H.P.		54.00	890	2,670	8,000	966	

01 54 | Construction Aids

01 54 33 | Equipment Rental

		Description	UNIT	HOURLY OPER. COST	RENT PER DAY	RENT PER WEEK	RENT PER MONTH	EQUIPMENT COST/DAY	
50	4901	Trailer, low bed, 75 ton capacity	Ea.	10.10	232	695	2,075	219.80	50
	5000	Road planer, walk behind, 10" cutting width, 10 H.P.		4.00	33.50	100	300	52	
	5100	Self-propelled, 12" cutting width, 64 H.P.		11.45	113	340	1,025	159.60	
	5120	Traffic line remover, metal ball blaster, truck mounted, 115 H.P.		48.75	750	2,245	6,725	839	
	5140	Grinder, truck mounted, 115 H.P.		55.45	810	2,435	7,300	930.60	
	5160	Walk-behind, 11 H.P.		4.55	53.50	160	480	68.40	
	5200	Pavement profiler, 4' to 6' wide, 450 H.P.		258.10	3,350	10,045	30,100	4,074	
	5300	8' to 10' wide, 750 H.P.		405.10	4,400	13,180	39,500	5,877	
	5400	Roadway plate, steel, 1" x 8' x 20'		.08	13	39	117	8.45	
	5600	Stabilizer, self-propelled, 150 H.P.		50.10	610	1,835	5,500	767.80	
	5700	310 H.P.		100.25	1,700	5,125	15,400	1,827	
	5800	Striper, truck mounted, 120 gallon paint, 460 H.P.		71.05	485	1,460	4,375	860.40	
	5900	Thermal paint heating kettle, 115 gallons		5.95	25.50	77	231	63	
	6000	Tar kettle, 330 gallon, trailer mounted		8.70	58.50	175	525	104.60	
	7000	Tunnel locomotive, diesel, 8 to 12 ton		32.40	585	1,760	5,275	611.20	
	7005	Electric, 10 ton		26.55	670	2,010	6,025	614.40	
	7010	Muck cars, 1/2 C.Y. capacity		2.05	24.50	74	222	31.20	
	7020	1 C.Y. capacity		2.30	32.50	98	294	38	
	7030	2 C.Y. capacity		2.45	36.50	110	330	41.60	
	7040	Side dump, 2 C.Y. capacity		2.65	45	135	405	48.20	
	7050	3 C.Y. capacity		3.55	51.50	155	465	59.40	
	7060	5 C.Y. capacity		5.10	65	195	585	79.80	
	7100	Ventilating blower for tunnel, 7-1/2 H.P.		2.05	51.50	155	465	47.40	
	7110	10 H.P.		2.22	53.50	160	480	49.75	
	7120	20 H.P.		3.43	69.50	208	625	69.05	
	7140	40 H.P.		5.59	98.50	295	885	103.70	
	7160	60 H.P.		8.54	152	455	1,375	159.30	
	7175	75 H.P.		11.25	207	620	1,850	214	
	7180	200 H.P.		22.56	305	910	2,725	362.50	
	7800	Windrow loader, elevating		55.95	1,325	3,960	11,900	1,240	
60	0010	**LIFTING AND HOISTING EQUIPMENT RENTAL** without operators	R015433-10						60
	0120	Aerial lift truck, 2 person, to 80'		Ea.	26.75	715	2,145	6,425	643
	0140	Boom work platform, 40' snorkel	R015433-15		16.10	277	830	2,500	294.80
	0150	Crane, flatbed mounted, 3 ton capacity			16.55	193	580	1,750	248.40
	0200	Crane, climbing, 106' jib, 6000 lb. capacity, 410 fpm	R312316-45		37.76	1,625	4,860	14,600	1,274
	0300	101' jib, 10,250 lb. capacity, 270 fpm			44.26	2,050	6,160	18,500	1,586
	0500	Tower, static, 130' high, 106' jib, 6200 lb. capacity at 400 fpm			41.56	1,875	5,620	16,900	1,456
	0600	Crawler mounted, lattice boom, 1/2 C.Y., 15 tons at 12' radius			37.52	640	1,920	5,750	684.15
	0700	3/4 C.Y., 20 tons at 12' radius			50.03	800	2,400	7,200	880.25
	0800	1 C.Y., 25 tons at 12' radius			66.70	1,075	3,195	9,575	1,173
	0900	1-1/2 C.Y., 40 tons at 12' radius			66.70	1,075	3,225	9,675	1,179
	1000	2 C.Y., 50 tons at 12' radius			70.70	1,250	3,765	11,300	1,319
	1100	3 C.Y., 75 tons at 12' radius			75.65	1,475	4,435	13,300	1,492
	1200	100 ton capacity, 60' boom			85.65	1,700	5,090	15,300	1,703
	1300	165 ton capacity, 60' boom			109.15	2,000	5,980	17,900	2,069
	1400	200 ton capacity, 70' boom			132.10	2,500	7,480	22,400	2,553
	1500	350 ton capacity, 80' boom			184.60	3,725	11,210	33,600	3,719
	1600	Truck mounted, lattice boom, 6 x 4, 20 tons at 10' radius			37.31	1,100	3,310	9,925	960.50
	1700	25 tons at 10' radius			40.35	1,200	3,610	10,800	1,045
	1800	8 x 4, 30 tons at 10' radius			43.85	1,275	3,840	11,500	1,119
	1900	40 tons at 12' radius			46.92	1,325	4,010	12,000	1,177
	2000	60 tons at 15' radius			53.24	1,425	4,240	12,700	1,274
	2050	82 tons at 15' radius			60.06	1,525	4,540	13,600	1,388
	2100	90 tons at 15' radius			67.57	1,650	4,940	14,800	1,529
	2200	115 tons at 15' radius			76.34	1,850	5,520	16,600	1,715
	2300	150 tons at 18' radius			84.40	1,950	5,815	17,400	1,838
	2350	165 tons at 18' radius			90.25	2,050	6,160	18,500	1,954
	2400	Truck mounted, hydraulic, 12 ton capacity			43.00	530	1,595	4,775	663

01 54 | Construction Aids

01 54 33 | Equipment Rental

			UNIT	HOURLY OPER. COST	RENT PER DAY	RENT PER WEEK	RENT PER MONTH	EQUIPMENT COST/DAY	
60	2500	25 ton capacity	Ea.	45.35	640	1,915	5,750	745.80	60
	2550	33 ton capacity		45.90	655	1,960	5,875	759.20	
	2560	40 ton capacity		59.35	765	2,290	6,875	932.80	
	2600	55 ton capacity		77.50	860	2,585	7,750	1,137	
	2700	80 ton capacity		101.35	1,400	4,205	12,600	1,652	
	2720	100 ton capacity		94.95	1,450	4,325	13,000	1,625	
	2740	120 ton capacity		109.95	1,550	4,665	14,000	1,813	
	2760	150 ton capacity		128.55	2,050	6,135	18,400	2,255	
	2800	Self-propelled, 4 x 4, with telescoping boom, 5 ton		17.75	230	690	2,075	280	
	2900	12-1/2 ton capacity		32.65	365	1,100	3,300	481.20	
	3000	15 ton capacity		33.35	385	1,160	3,475	498.80	
	3050	20 ton capacity		36.30	450	1,350	4,050	560.40	
	3100	25 ton capacity		37.90	505	1,520	4,550	607.20	
	3150	40 ton capacity		46.50	570	1,705	5,125	713	
	3200	Derricks, guy, 20 ton capacity, 60' boom, 75' mast		27.18	395	1,190	3,575	455.45	
	3300	100' boom, 115' mast		42.63	685	2,050	6,150	751.05	
	3400	Stiffleg, 20 ton capacity, 70' boom, 37' mast		29.70	515	1,550	4,650	547.60	
	3500	100' boom, 47' mast		45.64	825	2,480	7,450	861.10	
	3550	Helicopter, small, lift to 1250 lb. maximum, w/pilot		100.87	3,200	9,610	28,800	2,729	
	3600	Hoists, chain type, overhead, manual, 3/4 ton		.10	.33	1	3	1	
	3900	10 ton		.75	6	18	54	9.60	
	4000	Hoist and tower, 5000 lb. cap., portable electric, 40' high		4.77	228	685	2,050	175.15	
	4100	For each added 10' section, add		.11	18	54	162	11.70	
	4200	Hoist and single tubular tower, 5000 lb. electric, 100' high		6.47	320	957	2,875	243.15	
	4300	For each added 6'-6" section, add		.19	31	93	279	20.10	
	4400	Hoist and double tubular tower, 5000 lb., 100' high		6.95	350	1,054	3,150	266.40	
	4500	For each added 6'-6" section, add		.21	34.50	103	310	22.30	
	4550	Hoist and tower, mast type, 6000 lb., 100' high		7.48	365	1,093	3,275	278.45	
	4570	For each added 10' section, add		.13	21.50	64	192	13.85	
	4600	Hoist and tower, personnel, electric, 2000 lb., 100' @ 125 fpm		15.90	970	2,910	8,725	709.20	
	4700	3000 lb., 100' @ 200 fpm		18.13	1,100	3,290	9,875	803.05	
	4800	3000 lb., 150' @ 300 fpm		20.13	1,225	3,690	11,100	899.05	
	4900	4000 lb., 100' @ 300 fpm		20.82	1,250	3,760	11,300	918.55	
	5000	6000 lb., 100' @ 275 fpm		22.44	1,325	3,950	11,900	969.50	
	5100	For added heights up to 500', add	L.F.	.01	1.67	5	15	1.10	
	5200	Jacks, hydraulic, 20 ton	Ea.	.05	2	6	18	1.60	
	5500	100 ton		.40	11.65	35	105	10.20	
	6100	Jacks, hydraulic, climbing w/50' jackrods, control console, 30 ton cap.		1.97	131	394	1,175	94.55	
	6150	For each added 10' jackrod section, add		.05	3.33	10	30	2.40	
	6300	50 ton capacity		3.17	211	633	1,900	151.95	
	6350	For each added 10' jackrod section, add		.06	4	12	36	2.90	
	6500	125 ton capacity		8.30	555	1,660	4,975	398.40	
	6550	For each added 10' jackrod section, add		.57	37.50	113	340	27.15	
	6600	Cable jack, 10 ton capacity with 200' cable		1.65	110	329	985	79	
	6650	For each added 50' of cable, add		.20	13	39	117	9.40	
70	0010	**WELLPOINT EQUIPMENT RENTAL** without operators	R015433-10						70
	0020	Based on 2 months rental							
	0100	Combination jetting & wellpoint pump, 60 H.P. diesel	Ea.	18.14	325	976	2,925	340.30	
	0200	High pressure gas jet pump, 200 H.P., 300 psi	"	43.54	278	834	2,500	515.10	
	0300	Discharge pipe, 8" diameter	L.F.	.01	.53	1.59	4.77	.40	
	0350	12" diameter		.01	.78	2.33	7	.55	
	0400	Header pipe, flows up to 150 GPM, 4" diameter		.01	.48	1.44	4.32	.35	
	0500	400 GPM, 6" diameter		.01	.56	1.69	5.05	.40	
	0600	800 GPM, 8" diameter		.01	.78	2.33	7	.55	
	0700	1500 GPM, 10" diameter		.01	.82	2.46	7.40	.55	
	0800	2500 GPM, 12" diameter		.02	1.55	4.65	13.95	1.10	
	0900	4500 GPM, 16" diameter		.03	1.98	5.95	17.85	1.45	

01 54 | Construction Aids

01 54 33 | Equipment Rental

			UNIT	HOURLY OPER. COST	RENT PER DAY	RENT PER WEEK	RENT PER MONTH	EQUIPMENT COST/DAY	
70	0950	For quick coupling aluminum and plastic pipe, add	L.F.	.03	2.05	6.16	18.50	1.45	70
	1100	Wellpoint, 25' long, with fittings & riser pipe, 1-1/2" or 2" diameter		.06	4.10	12.29	37	2.95	
	1200	Wellpoint pump, diesel powered, 4" suction, 20 H.P.		7.71	188	563	1,700	174.30	
	1300	6" suction, 30 H.P.		10.54	233	698	2,100	223.90	
	1400	8" suction, 40 H.P.		14.24	320	957	2,875	305.30	
	1500	10" suction, 75 H.P.		21.97	375	1,119	3,350	399.55	
	1600	12" suction, 100 H.P.		31.24	590	1,770	5,300	603.90	
	1700	12" suction, 175 H.P.		46.78	655	1,970	5,900	768.25	
80	0010	**MARINE EQUIPMENT RENTAL** without operators	Ea.						80
	0200	Barge, 400 Ton, 30' wide x 90' long		17.50	1,075	3,205	9,625	781	
	0240	800 Ton, 45' wide x 90' long		21.20	1,300	3,900	11,700	949.60	
	2000	Tugboat, diesel, 100 H.P.		38.95	217	650	1,950	441.60	
	2040	250 H.P.		81.55	395	1,185	3,550	889.40	
	2080	380 H.P.		160.00	1,175	3,545	10,600	1,989	
	3000	Small work boat, gas, 16-foot, 50 H.P.		19.15	60	180	540	189.20	
	4000	Large, diesel, 48-foot, 200 H.P.		90.80	1,250	3,745	11,200	1,475	

Reference: R015433-10

Crews

Crew No.	Bare Costs Hr.	Bare Costs Daily	Incl. Subs O&P Hr.	Incl. Subs O&P Daily	Cost Per Labor-Hour Bare Costs	Cost Per Labor-Hour Incl. O&P
Crew A-1					Bare Costs	Incl. O&P
1 Building Laborer	$36.65	$293.20	$56.55	$452.40	$36.65	$56.55
1 Concrete Saw, Gas Manual		84.00		92.40	10.50	11.55
8 L.H., Daily Totals		$377.20		$544.80	$47.15	$68.10
Crew A-1A	Hr.	Daily	Hr.	Daily	Bare Costs	Incl. O&P
1 Skilled Worker	$47.30	$378.40	$73.25	$586.00	$47.30	$73.25
1 Shot Blaster, 20"		214.40		235.84	26.80	29.48
8 L.H., Daily Totals		$592.80		$821.84	$74.10	$102.73
Crew A-1B	Hr.	Daily	Hr.	Daily	Bare Costs	Incl. O&P
1 Building Laborer	$36.65	$293.20	$56.55	$452.40	$36.65	$56.55
1 Concrete Saw		175.20		192.72	21.90	24.09
8 L.H., Daily Totals		$468.40		$645.12	$58.55	$80.64
Crew A-1C	Hr.	Daily	Hr.	Daily	Bare Costs	Incl. O&P
1 Building Laborer	$36.65	$293.20	$56.55	$452.40	$36.65	$56.55
1 Chain Saw, Gas, 18"		31.20		34.32	3.90	4.29
8 L.H., Daily Totals		$324.40		$486.72	$40.55	$60.84
Crew A-1D	Hr.	Daily	Hr.	Daily	Bare Costs	Incl. O&P
1 Building Laborer	$36.65	$293.20	$56.55	$452.40	$36.65	$56.55
1 Vibrating Plate, Gas, 18"		35.80		39.38	4.47	4.92
8 L.H., Daily Totals		$329.00		$491.78	$41.13	$61.47
Crew A-1E	Hr.	Daily	Hr.	Daily	Bare Costs	Incl. O&P
1 Building Laborer	$36.65	$293.20	$56.55	$452.40	$36.65	$56.55
1 Vibrating Plate, Gas, 21"		46.00		50.60	5.75	6.33
8 L.H., Daily Totals		$339.20		$503.00	$42.40	$62.88
Crew A-1F	Hr.	Daily	Hr.	Daily	Bare Costs	Incl. O&P
1 Building Laborer	$36.65	$293.20	$56.55	$452.40	$36.65	$56.55
1 Rammer/Tamper, Gas, 8"		49.60		54.56	6.20	6.82
8 L.H., Daily Totals		$342.80		$506.96	$42.85	$63.37
Crew A-1G	Hr.	Daily	Hr.	Daily	Bare Costs	Incl. O&P
1 Building Laborer	$36.65	$293.20	$56.55	$452.40	$36.65	$56.55
1 Rammer/Tamper, Gas, 15"		56.00		61.60	7.00	7.70
8 L.H., Daily Totals		$349.20		$514.00	$43.65	$64.25
Crew A-1H	Hr.	Daily	Hr.	Daily	Bare Costs	Incl. O&P
1 Building Laborer	$36.65	$293.20	$56.55	$452.40	$36.65	$56.55
1 Exterior Steam Cleaner		75.60		83.16	9.45	10.40
8 L.H., Daily Totals		$368.80		$535.56	$46.10	$66.94
Crew A-1J	Hr.	Daily	Hr.	Daily	Bare Costs	Incl. O&P
1 Building Laborer	$36.65	$293.20	$56.55	$452.40	$36.65	$56.55
1 Cultivator, Walk-Behind, 5 H.P.		57.45		63.20	7.18	7.90
8 L.H., Daily Totals		$350.65		$515.60	$43.83	$64.45
Crew A-1K	Hr.	Daily	Hr.	Daily	Bare Costs	Incl. O&P
1 Building Laborer	$36.65	$293.20	$56.55	$452.40	$36.65	$56.55
1 Cultivator, Walk-Behind, 8 H.P.		81.40		89.54	10.18	11.19
8 L.H., Daily Totals		$374.60		$541.94	$46.83	$67.74
Crew A-1M	Hr.	Daily	Hr.	Daily	Bare Costs	Incl. O&P
1 Building Laborer	$36.65	$293.20	$56.55	$452.40	$36.65	$56.55
1 Snow Blower, Walk-Behind		60.55		66.61	7.57	8.33
8 L.H., Daily Totals		$353.75		$519.01	$44.22	$64.88

Crew No.	Bare Costs Hr.	Bare Costs Daily	Incl. Subs O&P Hr.	Incl. Subs O&P Daily	Cost Per Labor-Hour Bare Costs	Cost Per Labor-Hour Incl. O&P
Crew A-2	Hr.	Daily	Hr.	Daily	Bare Costs	Incl. O&P
2 Laborers	$36.65	$586.40	$56.55	$904.80	$36.60	$56.27
1 Truck Driver (light)	36.50	292.00	55.70	445.60		
1 Flatbed Truck, Gas, 1.5 Ton		270.20		297.22	11.26	12.38
24 L.H., Daily Totals		$1148.60		$1647.62	$47.86	$68.65
Crew A-2A	Hr.	Daily	Hr.	Daily	Bare Costs	Incl. O&P
2 Laborers	$36.65	$586.40	$56.55	$904.80	$36.60	$56.27
1 Truck Driver (light)	36.50	292.00	55.70	445.60		
1 Flatbed Truck, Gas, 1.5 Ton		270.20		297.22		
1 Concrete Saw		175.20		192.72	18.56	20.41
24 L.H., Daily Totals		$1323.80		$1840.34	$55.16	$76.68
Crew A-2B	Hr.	Daily	Hr.	Daily	Bare Costs	Incl. O&P
1 Truck Driver (light)	$36.50	$292.00	$55.70	$445.60	$36.50	$55.70
1 Flatbed Truck, Gas, 1.5 Ton		270.20		297.22	33.77	37.15
8 L.H., Daily Totals		$562.20		$742.82	$70.28	$92.85
Crew A-3A	Hr.	Daily	Hr.	Daily	Bare Costs	Incl. O&P
1 Truck Driver (light)	$36.50	$292.00	$55.70	$445.60	$36.50	$55.70
1 Pickup Truck, 4 x 4, 3/4 Ton		168.40		185.24	21.05	23.16
8 L.H., Daily Totals		$460.40		$630.84	$57.55	$78.86
Crew A-3B	Hr.	Daily	Hr.	Daily	Bare Costs	Incl. O&P
1 Equip. Oper. (medium)	$48.90	$391.20	$74.15	$593.20	$43.23	$65.72
1 Truck Driver (heavy)	37.55	300.40	57.30	458.40		
1 Dump Truck, 12 C.Y., 400 H.P.		692.20		761.42		
1 F.E. Loader, W.M., 2.5 C.Y.		522.80		575.08	75.94	83.53
16 L.H., Daily Totals		$1906.60		$2388.10	$119.16	$149.26
Crew A-3C	Hr.	Daily	Hr.	Daily	Bare Costs	Incl. O&P
1 Equip. Oper. (light)	$47.05	$376.40	$71.35	$570.80	$47.05	$71.35
1 Loader, Skid Steer, 78 H.P.		308.80		339.68	38.60	42.46
8 L.H., Daily Totals		$685.20		$910.48	$85.65	$113.81
Crew A-3D	Hr.	Daily	Hr.	Daily	Bare Costs	Incl. O&P
1 Truck Driver (light)	$36.50	$292.00	$55.70	$445.60	$36.50	$55.70
1 Pickup Truck, 4 x 4, 3/4 Ton		168.40		185.24		
1 Flatbed Trailer, 25 Ton		113.60		124.96	35.25	38.77
8 L.H., Daily Totals		$574.00		$755.80	$71.75	$94.47
Crew A-3E	Hr.	Daily	Hr.	Daily	Bare Costs	Incl. O&P
1 Equip. Oper. (crane)	$50.25	$402.00	$76.20	$609.60	$43.90	$66.75
1 Truck Driver (heavy)	37.55	300.40	57.30	458.40		
1 Pickup Truck, 4 x 4, 3/4 Ton		168.40		185.24	10.53	11.58
16 L.H., Daily Totals		$870.80		$1253.24	$54.42	$78.33
Crew A-3F	Hr.	Daily	Hr.	Daily	Bare Costs	Incl. O&P
1 Equip. Oper. (crane)	$50.25	$402.00	$76.20	$609.60	$43.90	$66.75
1 Truck Driver (heavy)	37.55	300.40	57.30	458.40		
1 Pickup Truck, 4 x 4, 3/4 Ton		168.40		185.24		
1 Truck Tractor, 6x4, 380 H.P.		612.80		674.08		
1 Lowbed Trailer, 75 Ton		219.80		241.78	62.56	68.82
16 L.H., Daily Totals		$1703.40		$2169.10	$106.46	$135.57

Crews

Crew No.	Bare Costs Hr.	Daily	Incl. Subs O&P Hr.	Daily	Cost Per Labor-Hour Bare Costs	Incl. O&P
Crew A-3G	Hr.	Daily	Hr.	Daily	Bare Costs	Incl. O&P
1 Equip. Oper. (crane)	$50.25	$402.00	$76.20	$609.60	$43.90	$66.75
1 Truck Driver (heavy)	37.55	300.40	57.30	458.40		
1 Pickup Truck, 4 x 4, 3/4 Ton		168.40		185.24		
1 Truck Tractor, 6x4, 450 H.P.		742.40		816.64		
1 Lowbed Trailer, 75 Ton		219.80		241.78	70.66	77.73
16 L.H., Daily Totals		$1833.00		$2311.66	$114.56	$144.48
Crew A-3H	Hr.	Daily	Hr.	Daily	Bare Costs	Incl. O&P
1 Equip. Oper. (crane)	$50.25	$402.00	$76.20	$609.60	$50.25	$76.20
1 Hyd. Crane, 12 Ton (Daily)		874.00		961.40	109.25	120.18
8 L.H., Daily Totals		$1276.00		$1571.00	$159.50	$196.38
Crew A-3I	Hr.	Daily	Hr.	Daily	Bare Costs	Incl. O&P
1 Equip. Oper. (crane)	$50.25	$402.00	$76.20	$609.60	$50.25	$76.20
1 Hyd. Crane, 25 Ton (Daily)		1003.00		1103.30	125.38	137.91
8 L.H., Daily Totals		$1405.00		$1712.90	$175.63	$214.11
Crew A-3J	Hr.	Daily	Hr.	Daily	Bare Costs	Incl. O&P
1 Equip. Oper. (crane)	$50.25	$402.00	$76.20	$609.60	$50.25	$76.20
1 Hyd. Crane, 40 Ton (Daily)		1240.00		1364.00	155.00	170.50
8 L.H., Daily Totals		$1642.00		$1973.60	$205.25	$246.70
Crew A-3K	Hr.	Daily	Hr.	Daily	Bare Costs	Incl. O&P
1 Equip. Oper. (crane)	$50.25	$402.00	$76.20	$609.60	$46.90	$71.10
1 Equip. Oper. (oiler)	43.55	348.40	66.00	528.00		
1 Hyd. Crane, 55 Ton (Daily)		1480.00		1628.00		
1 P/U Truck, 3/4 Ton (Daily)		182.00		200.20	103.88	114.26
16 L.H., Daily Totals		$2412.40		$2965.80	$150.78	$185.36
Crew A-3L	Hr.	Daily	Hr.	Daily	Bare Costs	Incl. O&P
1 Equip. Oper. (crane)	$50.25	$402.00	$76.20	$609.60	$46.90	$71.10
1 Equip. Oper. (oiler)	43.55	348.40	66.00	528.00		
1 Hyd. Crane, 80 Ton (Daily)		2211.00		2432.10		
1 P/U Truck, 3/4 Ton (Daily)		182.00		200.20	149.56	164.52
16 L.H., Daily Totals		$3143.40		$3769.90	$196.46	$235.62
Crew A-3M	Hr.	Daily	Hr.	Daily	Bare Costs	Incl. O&P
1 Equip. Oper. (crane)	$50.25	$402.00	$76.20	$609.60	$46.90	$71.10
1 Equip. Oper. (oiler)	43.55	348.40	66.00	528.00		
1 Hyd. Crane, 100 Ton (Daily)		2200.00		2420.00		
1 P/U Truck, 3/4 Ton (Daily)		182.00		200.20	148.88	163.76
16 L.H., Daily Totals		$3132.40		$3757.80	$195.78	$234.86
Crew A-3N	Hr.	Daily	Hr.	Daily	Bare Costs	Incl. O&P
1 Equip. Oper. (crane)	$50.25	$402.00	$76.20	$609.60	$50.25	$76.20
1 Tower Cane (monthly)		1100.00		1210.00	137.50	151.25
8 L.H., Daily Totals		$1502.00		$1819.60	$187.75	$227.45
Crew A-3P	Hr.	Daily	Hr.	Daily	Bare Costs	Incl. O&P
1 Equip. Oper. (light)	$47.05	$376.40	$71.35	$570.80	$47.05	$71.35
1 A.T. Forklift, 42' lift		531.60		584.76	66.45	73.09
8 L.H., Daily Totals		$908.00		$1155.56	$113.50	$144.44
Crew A-4	Hr.	Daily	Hr.	Daily	Bare Costs	Incl. O&P
2 Carpenters	$45.85	$733.60	$70.75	$1132.00	$43.75	$67.00
1 Painter, Ordinary	39.55	316.40	59.50	476.00		
24 L.H., Daily Totals		$1050.00		$1608.00	$43.75	$67.00

Crew No.	Bare Costs Hr.	Daily	Incl. Subs O&P Hr.	Daily	Cost Per Labor-Hour Bare Costs	Incl. O&P
Crew A-5	Hr.	Daily	Hr.	Daily	Bare Costs	Incl. O&P
2 Laborers	$36.65	$586.40	$56.55	$904.80	$36.63	$56.46
.25 Truck Driver (light)	36.50	73.00	55.70	111.40		
.25 Flatbed Truck, Gas, 1.5 Ton		67.55		74.31	3.75	4.13
18 L.H., Daily Totals		$726.95		$1090.51	$40.39	$60.58
Crew A-6	Hr.	Daily	Hr.	Daily	Bare Costs	Incl. O&P
1 Instrument Man	$47.30	$378.40	$73.25	$586.00	$45.67	$70.13
1 Rodman/Chainman	44.05	352.40	67.00	536.00		
1 Level, Electronic		71.30		78.43	4.46	4.90
16 L.H., Daily Totals		$802.10		$1200.43	$50.13	$75.03
Crew A-7	Hr.	Daily	Hr.	Daily	Bare Costs	Incl. O&P
1 Chief of Party	$59.90	$479.20	$92.15	$737.20	$50.42	$77.47
1 Instrument Man	47.30	378.40	73.25	586.00		
1 Rodman/Chainman	44.05	352.40	67.00	536.00		
1 Level, Electronic		71.30		78.43	2.97	3.27
24 L.H., Daily Totals		$1281.30		$1937.63	$53.39	$80.73
Crew A-8	Hr.	Daily	Hr.	Daily	Bare Costs	Incl. O&P
1 Chief of Party	$59.90	$479.20	$92.15	$737.20	$48.83	$74.85
1 Instrument Man	47.30	378.40	73.25	586.00		
2 Rodmen/Chainmen	44.05	704.80	67.00	1072.00		
1 Level, Electronic		71.30		78.43	2.23	2.45
32 L.H., Daily Totals		$1633.70		$2473.63	$51.05	$77.30
Crew A-9	Hr.	Daily	Hr.	Daily	Bare Costs	Incl. O&P
1 Asbestos Foreman	$51.65	$413.20	$80.35	$642.80	$51.21	$79.69
7 Asbestos Workers	51.15	2864.40	79.60	4457.60		
64 L.H., Daily Totals		$3277.60		$5100.40	$51.21	$79.69
Crew A-10A	Hr.	Daily	Hr.	Daily	Bare Costs	Incl. O&P
1 Asbestos Foreman	$51.65	$413.20	$80.35	$642.80	$51.32	$79.85
2 Asbestos Workers	51.15	818.40	79.60	1273.60		
24 L.H., Daily Totals		$1231.60		$1916.40	$51.32	$79.85
Crew A-10B	Hr.	Daily	Hr.	Daily	Bare Costs	Incl. O&P
1 Asbestos Foreman	$51.65	$413.20	$80.35	$642.80	$51.27	$79.79
3 Asbestos Workers	51.15	1227.60	79.60	1910.40		
32 L.H., Daily Totals		$1640.80		$2553.20	$51.27	$79.79
Crew A-10C	Hr.	Daily	Hr.	Daily	Bare Costs	Incl. O&P
3 Asbestos Workers	$51.15	$1227.60	$79.60	$1910.40	$51.15	$79.60
1 Flatbed Truck, Gas, 1.5 Ton		270.20		297.22	11.26	12.38
24 L.H., Daily Totals		$1497.80		$2207.62	$62.41	$91.98
Crew A-10D	Hr.	Daily	Hr.	Daily	Bare Costs	Incl. O&P
2 Asbestos Workers	$51.15	$818.40	$79.60	$1273.60	$49.02	$75.35
1 Equip. Oper. (crane)	50.25	402.00	76.20	609.60		
1 Equip. Oper. (oiler)	43.55	348.40	66.00	528.00		
1 Hydraulic Crane, 33 Ton		759.20		835.12	23.73	26.10
32 L.H., Daily Totals		$2328.00		$3246.32	$72.75	$101.45
Crew A-11	Hr.	Daily	Hr.	Daily	Bare Costs	Incl. O&P
1 Asbestos Foreman	$51.65	$413.20	$80.35	$642.80	$51.21	$79.69
7 Asbestos Workers	51.15	2864.40	79.60	4457.60		
2 Chip. Hammers, 12 Lb., Elec.		40.00		44.00	.63	.69
64 L.H., Daily Totals		$3317.60		$5144.40	$51.84	$80.38

Crews

Crew No.		Bare Costs		Incl. Subs O&P		Cost Per Labor-Hour	
Crew A-12	Hr.	Daily	Hr.	Daily	Bare Costs	Incl. O&P
1 Asbestos Foreman	$51.65	$413.20	$80.35	$642.80	$51.21	$79.69
7 Asbestos Workers	51.15	2864.40	79.60	4457.60		
1 Trk-Mtd Vac, 14 CY, 1500 Gal.		518.90		570.79		
1 Flatbed Truck, 20,000 GVW		253.40		278.74	12.07	13.27
64 L.H., Daily Totals		$4049.90		$5949.93	$63.28	$92.97

Crew No.		Bare Costs		Incl. Subs O&P		Cost Per Labor-Hour	
Crew A-13	Hr.	Daily	Hr.	Daily	Bare Costs	Incl. O&P
1 Equip. Oper. (light)	$47.05	$376.40	$71.35	$570.80	$47.05	$71.35
1 Trk-Mtd Vac, 14 CY, 1500 Gal.		518.90		570.79		
1 Flatbed Truck, 20,000 GVW		253.40		278.74	96.54	106.19
8 L.H., Daily Totals		$1148.70		$1420.33	$143.59	$177.54

Crew B-1	Hr.	Daily	Hr.	Daily	Bare Costs	Incl. O&P
1 Labor Foreman (outside)	$38.65	$309.20	$59.65	$477.20	$37.32	$57.58
2 Laborers	36.65	586.40	56.55	904.80		
24 L.H., Daily Totals		$895.60		$1382.00	$37.32	$57.58

Crew B-1A	Hr.	Daily	Hr.	Daily	Bare Costs	Incl. O&P
1 Labor Foreman (outside)	$38.65	$309.20	$59.65	$477.20	$37.32	$57.58
2 Laborers	36.65	586.40	56.55	904.80		
2 Cutting Torches		22.80		25.08		
2 Sets of Gases		304.00		334.40	13.62	14.98
24 L.H., Daily Totals		$1222.40		$1741.48	$50.93	$72.56

Crew B-1B	Hr.	Daily	Hr.	Daily	Bare Costs	Incl. O&P
1 Labor Foreman (outside)	$38.65	$309.20	$59.65	$477.20	$40.55	$62.24
2 Laborers	36.65	586.40	56.55	904.80		
1 Equip. Oper. (crane)	50.25	402.00	76.20	609.60		
2 Cutting Torches		22.80		25.08		
2 Sets of Gases		304.00		334.40		
1 Hyd. Crane, 12 Ton		663.00		729.30	30.93	34.02
32 L.H., Daily Totals		$2287.40		$3080.38	$71.48	$96.26

Crew B-1C	Hr.	Daily	Hr.	Daily	Bare Costs	Incl. O&P
1 Labor Foreman (outside)	$38.65	$309.20	$59.65	$477.20	$37.32	$57.58
2 Laborers	36.65	586.40	56.55	904.80		
1 Aerial Lift Truck, 60' Boom		456.00		501.60	19.00	20.90
24 L.H., Daily Totals		$1351.60		$1883.60	$56.32	$78.48

Crew B-1D	Hr.	Daily	Hr.	Daily	Bare Costs	Incl. O&P
2 Laborers	$36.65	$586.40	$56.55	$904.80	$36.65	$56.55
1 Small Work Boat, Gas, 50 H.P.		189.20		208.12		
1 Pressure Washer, 7 GPM		86.00		94.60	17.20	18.92
16 L.H., Daily Totals		$861.60		$1207.52	$53.85	$75.47

Crew B-1E	Hr.	Daily	Hr.	Daily	Bare Costs	Incl. O&P
1 Labor Foreman (outside)	$38.65	$309.20	$59.65	$477.20	$37.15	$57.33
3 Laborers	36.65	879.60	56.55	1357.20		
1 Work Boat, Diesel, 200 H.P.		1475.00		1622.50		
2 Pressure Washer, 7 GPM		172.00		189.20	51.47	56.62
32 L.H., Daily Totals		$2835.80		$3646.10	$88.62	$113.94

Crew B-1F	Hr.	Daily	Hr.	Daily	Bare Costs	Incl. O&P
2 Skilled Workers	$47.30	$756.80	$73.25	$1172.00	$43.75	$67.68
1 Laborer	36.65	293.20	56.55	452.40		
1 Small Work Boat, Gas, 50 H.P.		189.20		208.12		
1 Pressure Washer, 7 GPM		86.00		94.60	11.47	12.61
24 L.H., Daily Totals		$1325.20		$1927.12	$55.22	$80.30

Crew B-1G	Hr.	Daily	Hr.	Daily	Bare Costs	Incl. O&P
2 Laborers	$36.65	$586.40	$56.55	$904.80	$36.65	$56.55
1 Small Work Boat, Gas, 50 H.P.		189.20		208.12	11.82	13.01
16 L.H., Daily Totals		$775.60		$1112.92	$48.48	$69.56

Crew B-1H	Hr.	Daily	Hr.	Daily	Bare Costs	Incl. O&P
2 Skilled Workers	$47.30	$756.80	$73.25	$1172.00	$43.75	$67.68
1 Laborer	36.65	293.20	56.55	452.40		
1 Small Work Boat, Gas, 50 H.P.		189.20		208.12	7.88	8.67
24 L.H., Daily Totals		$1239.20		$1832.52	$51.63	$76.36

Crew B-1J	Hr.	Daily	Hr.	Daily	Bare Costs	Incl. O&P
1 Labor Foreman (inside)	$37.15	$297.20	$57.30	$458.40	$36.90	$56.92
1 Laborer	36.65	293.20	56.55	452.40		
16 L.H., Daily Totals		$590.40		$910.80	$36.90	$56.92

Crew B-1K	Hr.	Daily	Hr.	Daily	Bare Costs	Incl. O&P
1 Carpenter Foreman (inside)	$46.35	$370.80	$71.50	$572.00	$46.10	$71.13
1 Carpenter	45.85	366.80	70.75	566.00		
16 L.H., Daily Totals		$737.60		$1138.00	$46.10	$71.13

Crew B-2	Hr.	Daily	Hr.	Daily	Bare Costs	Incl. O&P
1 Labor Foreman (outside)	$38.65	$309.20	$59.65	$477.20	$37.05	$57.17
4 Laborers	36.65	1172.80	56.55	1809.60		
40 L.H., Daily Totals		$1482.00		$2286.80	$37.05	$57.17

Crew B-2A	Hr.	Daily	Hr.	Daily	Bare Costs	Incl. O&P
1 Labor Foreman (outside)	$38.65	$309.20	$59.65	$477.20	$37.32	$57.58
2 Laborers	36.65	586.40	56.55	904.80		
1 Aerial Lift Truck, 60' Boom		456.00		501.60	19.00	20.90
24 L.H., Daily Totals		$1351.60		$1883.60	$56.32	$78.48

Crew B-3	Hr.	Daily	Hr.	Daily	Bare Costs	Incl. O&P
1 Labor Foreman (outside)	$38.65	$309.20	$59.65	$477.20	$39.33	$60.25
2 Laborers	36.65	586.40	56.55	904.80		
1 Equip. Oper. (medium)	48.90	391.20	74.15	593.20		
2 Truck Drivers (heavy)	37.55	600.80	57.30	916.80		
1 Crawler Loader, 3 C.Y.		1191.00		1310.10		
2 Dump Trucks, 12 C.Y., 400 H.P.		1384.40		1522.84	53.65	59.02
48 L.H., Daily Totals		$4463.00		$5724.94	$92.98	$119.27

Crew B-3A	Hr.	Daily	Hr.	Daily	Bare Costs	Incl. O&P
4 Laborers	$36.65	$1172.80	$56.55	$1809.60	$39.10	$60.07
1 Equip. Oper. (medium)	48.90	391.20	74.15	593.20		
1 Hyd. Excavator, 1.5 C.Y.		1031.00		1134.10	25.77	28.35
40 L.H., Daily Totals		$2595.00		$3536.90	$64.88	$88.42

Crew B-3B	Hr.	Daily	Hr.	Daily	Bare Costs	Incl. O&P
2 Laborers	$36.65	$586.40	$56.55	$904.80	$39.94	$61.14
1 Equip. Oper. (medium)	48.90	391.20	74.15	593.20		
1 Truck Driver (heavy)	37.55	300.40	57.30	458.40		
1 Backhoe Loader, 80 H.P.		397.60		437.36		
1 Dump Truck, 12 C.Y., 400 H.P.		692.20		761.42	34.06	37.46
32 L.H., Daily Totals		$2367.80		$3155.18	$73.99	$98.60

Crew B-3C	Hr.	Daily	Hr.	Daily	Bare Costs	Incl. O&P
3 Laborers	$36.65	$879.60	$56.55	$1357.20	$39.71	$60.95
1 Equip. Oper. (medium)	48.90	391.20	74.15	593.20		
1 Crawler Loader, 4 C.Y.		1532.00		1685.20	47.88	52.66
32 L.H., Daily Totals		$2802.80		$3635.60	$87.59	$113.61

Crews

Crew No.	Bare Costs		Incl. Subs O&P		Cost Per Labor-Hour	

Crew B-4	Hr.	Daily	Hr.	Daily	Bare Costs	Incl. O&P
1 Labor Foreman (outside)	$38.65	$309.20	$59.65	$477.20	$37.13	$57.19
4 Laborers	36.65	1172.80	56.55	1809.60		
1 Truck Driver (heavy)	37.55	300.40	57.30	458.40		
1 Truck Tractor, 220 H.P.		366.40		403.04		
1 Flatbed Trailer, 40 Ton		152.00		167.20	10.80	11.88
48 L.H., Daily Totals		$2300.80		$3315.44	$47.93	$69.07

Crew B-5	Hr.	Daily	Hr.	Daily	Bare Costs	Incl. O&P
1 Labor Foreman (outside)	$38.65	$309.20	$59.65	$477.20	$40.44	$62.02
4 Laborers	36.65	1172.80	56.55	1809.60		
2 Equip. Oper. (medium)	48.90	782.40	74.15	1186.40		
1 Air Compressor, 250 cfm		202.40		222.64		
2 Breakers, Pavement, 60 lb.		19.60		21.56		
2 -50' Air Hoses, 1.5"		11.60		12.76		
1 Crawler Loader, 3 C.Y.		1191.00		1310.10	25.44	27.98
56 L.H., Daily Totals		$3689.00		$5040.26	$65.88	$90.00

Crew B-5A	Hr.	Daily	Hr.	Daily	Bare Costs	Incl. O&P
1 Labor Foreman (outside)	$38.65	$309.20	$59.65	$477.20	$39.88	$61.10
6 Laborers	36.65	1759.20	56.55	2714.40		
2 Equip. Oper. (medium)	48.90	782.40	74.15	1186.40		
1 Equip. Oper. (light)	47.05	376.40	71.35	570.80		
2 Truck Drivers (heavy)	37.55	600.80	57.30	916.80		
1 Air Compressor, 365 cfm		264.80		291.28		
2 Breakers, Pavement, 60 lb.		19.60		21.56		
8 -50' Air Hoses, 1"		32.80		36.08		
2 Dump Trucks, 8 C.Y., 220 H.P.		834.40		917.84	12.00	13.20
96 L.H., Daily Totals		$4979.60		$7132.36	$51.87	$74.30

Crew B-5B	Hr.	Daily	Hr.	Daily	Bare Costs	Incl. O&P
1 Powderman	$47.30	$378.40	$73.25	$586.00	$42.96	$65.58
2 Equip. Oper. (medium)	48.90	782.40	74.15	1186.40		
3 Truck Drivers (heavy)	37.55	901.20	57.30	1375.20		
1 F.E. Loader, W.M.,2.5 C.Y.		522.80		575.08		
3 Dump Trucks, 12 C.Y., 400 H.P.		2076.60		2284.26		
1 Air Compressor, 365 CFM		264.80		291.28	59.67	65.64
48 L.H., Daily Totals		$4926.20		$6298.22	$102.63	$131.21

Crew B-5C	Hr.	Daily	Hr.	Daily	Bare Costs	Incl. O&P
3 Laborers	$36.65	$879.60	$56.55	$1357.20	$40.97	$62.58
1 Equip. Oper. (medium)	48.90	391.20	74.15	593.20		
2 Truck Drivers (heavy)	37.55	600.80	57.30	916.80		
1 Equip. Oper. (crane)	50.25	402.00	76.20	609.60		
1 Equip. Oper. (oiler)	43.55	348.40	66.00	528.00		
2 Dump Trucks, 12 C.Y., 400 H.P.		1384.40		1522.84		
1 Crawler Loader, 4 C.Y.		1532.00		1685.20		
1 S.P. Crane, 4x4, 25 Ton		607.20		667.92	55.06	60.56
64 L.H., Daily Totals		$6145.60		$7880.76	$96.03	$123.14

Crew B-5D	Hr.	Daily	Hr.	Daily	Bare Costs	Incl. O&P
1 Labor Foreman (outside)	$38.65	$309.20	$59.65	$477.20	$40.08	$61.43
4 Laborers	36.65	1172.80	56.55	1809.60		
2 Equip. Oper. (medium)	48.90	782.40	74.15	1186.40		
1 Truck Driver (heavy)	37.55	300.40	57.30	458.40		
1 Air Compressor, 250 cfm		202.40		222.64		
2 Breakers, Pavement, 60 lb.		19.60		21.56		
2 -50' Air Hoses, 1.5"		11.60		12.76		
1 Crawler Loader, 3 C.Y.		1191.00		1310.10		
1 Dump Truck, 12 C.Y., 400 H.P.		692.20		761.42	33.08	36.38
64 L.H., Daily Totals		$4681.60		$6260.08	$73.15	$97.81

Crew B-6	Hr.	Daily	Hr.	Daily	Bare Costs	Incl. O&P
2 Laborers	$36.65	$586.40	$56.55	$904.80	$40.12	$61.48
1 Equip. Oper. (light)	47.05	376.40	71.35	570.80		
1 Backhoe Loader, 48 H.P.		365.20		401.72	15.22	16.74
24 L.H., Daily Totals		$1328.00		$1877.32	$55.33	$78.22

Crew B-6A	Hr.	Daily	Hr.	Daily	Bare Costs	Incl. O&P
.5 Labor Foreman (outside)	$38.65	$154.60	$59.65	$238.60	$41.95	$64.21
1 Laborer	36.65	293.20	56.55	452.40		
1 Equip. Oper. (medium)	48.90	391.20	74.15	593.20		
1 Vacuum Truck, 5000 Gal.		360.15		396.17	18.01	19.81
20 L.H., Daily Totals		$1199.15		$1680.37	$59.96	$84.02

Crew B-6B	Hr.	Daily	Hr.	Daily	Bare Costs	Incl. O&P
2 Labor Foremen (outside)	$38.65	$618.40	$59.65	$954.40	$37.32	$57.58
4 Laborers	36.65	1172.80	56.55	1809.60		
1 S.P. Crane, 4x4, 5 Ton		280.00		308.00		
1 Flatbed Truck, Gas, 1.5 Ton		270.20		297.22		
1 Butt Fusion Mach., 4"-12" diam.		381.80		419.98	19.42	21.36
48 L.H., Daily Totals		$2723.20		$3789.20	$56.73	$78.94

Crew B-6C	Hr.	Daily	Hr.	Daily	Bare Costs	Incl. O&P
2 Labor Foremen (outside)	$38.65	$618.40	$59.65	$954.40	$37.32	$57.58
4 Laborers	36.65	1172.80	56.55	1809.60		
1 S.P. Crane, 4x4, 12 Ton		481.20		529.32		
1 Flatbed Truck, Gas, 3 Ton		333.00		366.30		
1 Butt Fusion Mach., 8"-24" diam.		828.40		911.24	34.22	37.64
48 L.H., Daily Totals		$3433.80		$4570.86	$71.54	$95.23

Crew B-7	Hr.	Daily	Hr.	Daily	Bare Costs	Incl. O&P
1 Labor Foreman (outside)	$38.65	$309.20	$59.65	$477.20	$39.02	$60.00
4 Laborers	36.65	1172.80	56.55	1809.60		
1 Equip. Oper. (medium)	48.90	391.20	74.15	593.20		
1 Brush Chipper, 12", 130 H.P.		396.60		436.26		
1 Crawler Loader, 3 C.Y.		1191.00		1310.10		
2 Chain Saws, Gas, 36" Long		90.00		99.00	34.95	38.45
48 L.H., Daily Totals		$3550.80		$4725.36	$73.97	$98.44

Crew B-7A	Hr.	Daily	Hr.	Daily	Bare Costs	Incl. O&P
2 Laborers	$36.65	$586.40	$56.55	$904.80	$40.12	$61.48
1 Equip. Oper. (light)	47.05	376.40	71.35	570.80		
1 Rake w/Tractor		347.90		382.69		
2 Chain Saw, Gas, 18"		62.40		68.64	17.10	18.81
24 L.H., Daily Totals		$1373.10		$1926.93	$57.21	$80.29

Crew B-7B	Hr.	Daily	Hr.	Daily	Bare Costs	Incl. O&P
1 Labor Foreman (outside)	$38.65	$309.20	$59.65	$477.20	$38.81	$59.61
4 Laborers	36.65	1172.80	56.55	1809.60		
1 Equip. Oper. (medium)	48.90	391.20	74.15	593.20		
1 Truck Driver (heavy)	37.55	300.40	57.30	458.40		
1 Brush Chipper, 12", 130 H.P.		396.60		436.26		
1 Crawler Loader, 3 C.Y.		1191.00		1310.10		
2 Chain Saws, Gas, 36" Long		90.00		99.00		
1 Dump Truck, 8 C.Y., 220 H.P.		417.20		458.92	37.41	41.15
56 L.H., Daily Totals		$4268.40		$5642.68	$76.22	$100.76

Crews

Crew No.	Bare Costs		Incl. Subs O&P		Cost Per Labor-Hour	
Crew B-7C	Hr.	Daily	Hr.	Daily	Bare Costs	Incl. O&P
1 Labor Foreman (outside)	$38.65	$309.20	$59.65	$477.20	$38.81	$59.61
4 Laborers	36.65	1172.80	56.55	1809.60		
1 Equip. Oper. (medium)	48.90	391.20	74.15	593.20		
1 Truck Driver (heavy)	37.55	300.40	57.30	458.40		
1 Brush Chipper, 12", 130 H.P.		396.60		436.26		
1 Crawler Loader, 3 C.Y.		1191.00		1310.10		
2 Chain Saws, Gas, 36" Long		90.00		99.00		
1 Dump Truck, 12 C.Y., 400 H.P.		692.20		761.42	42.32	46.55
56 L.H., Daily Totals		$4543.40		$5945.18	$81.13	$106.16
Crew B-8	Hr.	Daily	Hr.	Daily	Bare Costs	Incl. O&P
1 Labor Foreman (outside)	$38.65	$309.20	$59.65	$477.20	$41.05	$62.71
2 Laborers	36.65	586.40	56.55	904.80		
2 Equip. Oper. (medium)	48.90	782.40	74.15	1186.40		
1 Equip. Oper. (oiler)	43.55	348.40	66.00	528.00		
2 Truck Drivers (heavy)	37.55	600.80	57.30	916.80		
1 Hyd. Crane, 25 Ton		745.80		820.38		
1 Crawler Loader, 3 C.Y.		1191.00		1310.10		
2 Dump Trucks, 12 C.Y., 400 H.P.		1384.40		1522.84	51.89	57.08
64 L.H., Daily Totals		$5948.40		$7666.52	$92.94	$119.79
Crew B-9	Hr.	Daily	Hr.	Daily	Bare Costs	Incl. O&P
1 Labor Foreman (outside)	$38.65	$309.20	$59.65	$477.20	$37.05	$57.17
4 Laborers	36.65	1172.80	56.55	1809.60		
1 Air Compressor, 250 cfm		202.40		222.64		
2 Breakers, Pavement, 60 lb.		19.60		21.56		
2 -50' Air Hoses, 1.5"		11.60		12.76	5.84	6.42
40 L.H., Daily Totals		$1715.60		$2543.76	$42.89	$63.59
Crew B-9A	Hr.	Daily	Hr.	Daily	Bare Costs	Incl. O&P
2 Laborers	$36.65	$586.40	$56.55	$904.80	$36.95	$56.80
1 Truck Driver (heavy)	37.55	300.40	57.30	458.40		
1 Water Tank Trailer, 5000 Gal.		141.60		155.76		
1 Truck Tractor, 220 H.P.		366.40		403.04		
2 -50' Discharge Hoses, 3"		3.00		3.30	21.29	23.42
24 L.H., Daily Totals		$1397.80		$1925.30	$58.24	$80.22
Crew B-9B	Hr.	Daily	Hr.	Daily	Bare Costs	Incl. O&P
2 Laborers	$36.65	$586.40	$56.55	$904.80	$36.95	$56.80
1 Truck Driver (heavy)	37.55	300.40	57.30	458.40		
2 -50' Discharge Hoses, 3"		3.00		3.30		
1 Water Tank Trailer, 5000 Gal.		141.60		155.76		
1 Truck Tractor, 220 H.P.		366.40		403.04		
1 Pressure Washer		69.40		76.34	24.18	26.60
24 L.H., Daily Totals		$1467.20		$2001.64	$61.13	$83.40
Crew B-9D	Hr.	Daily	Hr.	Daily	Bare Costs	Incl. O&P
1 Labor Foreman (outside)	$38.65	$309.20	$59.65	$477.20	$37.05	$57.17
4 Common Laborers	36.65	1172.80	56.55	1809.60		
1 Air Compressor, 250 cfm		202.40		222.64		
2 -50' Air Hoses, 1.5"		11.60		12.76		
2 Air Powered Tampers		52.40		57.64	6.66	7.33
40 L.H., Daily Totals		$1748.40		$2579.84	$43.71	$64.50
Crew B-10	Hr.	Daily	Hr.	Daily	Bare Costs	Incl. O&P
1 Equip. Oper. (medium)	$48.90	$391.20	$74.15	$593.20	$44.82	$68.28
.5 Laborer	36.65	146.60	56.55	226.20		
12 L.H., Daily Totals		$537.80		$819.40	$44.82	$68.28

Crew No.	Bare Costs		Incl. Subs O&P		Cost Per Labor-Hour	
Crew B-10A	Hr.	Daily	Hr.	Daily	Bare Costs	Incl. O&P
1 Equip. Oper. (medium)	$48.90	$391.20	$74.15	$593.20	$44.82	$68.28
.5 Laborer	36.65	146.60	56.55	226.20		
1 Roller, 2-Drum, W.B., 7.5 H.P.		182.80		201.08	15.23	16.76
12 L.H., Daily Totals		$720.60		$1020.48	$60.05	$85.04
Crew B-10B	Hr.	Daily	Hr.	Daily	Bare Costs	Incl. O&P
1 Equip. Oper. (medium)	$48.90	$391.20	$74.15	$593.20	$44.82	$68.28
.5 Laborer	36.65	146.60	56.55	226.20		
1 Dozer, 200 H.P.		1325.00		1457.50	110.42	121.46
12 L.H., Daily Totals		$1862.80		$2276.90	$155.23	$189.74
Crew B-10C	Hr.	Daily	Hr.	Daily	Bare Costs	Incl. O&P
1 Equip. Oper. (medium)	$48.90	$391.20	$74.15	$593.20	$44.82	$68.28
.5 Laborer	36.65	146.60	56.55	226.20		
1 Dozer, 200 H.P.		1325.00		1457.50		
1 Vibratory Roller, Towed, 23 Ton		412.00		453.20	144.75	159.22
12 L.H., Daily Totals		$2274.80		$2730.10	$189.57	$227.51
Crew B-10D	Hr.	Daily	Hr.	Daily	Bare Costs	Incl. O&P
1 Equip. Oper. (medium)	$48.90	$391.20	$74.15	$593.20	$44.82	$68.28
.5 Laborer	36.65	146.60	56.55	226.20		
1 Dozer, 200 H.P.		1325.00		1457.50		
1 Sheepsft. Roller, Towed		443.80		488.18	147.40	162.14
12 L.H., Daily Totals		$2306.60		$2765.08	$192.22	$230.42
Crew B-10E	Hr.	Daily	Hr.	Daily	Bare Costs	Incl. O&P
1 Equip. Oper. (medium)	$48.90	$391.20	$74.15	$593.20	$44.82	$68.28
.5 Laborer	36.65	146.60	56.55	226.20		
1 Tandem Roller, 5 Ton		159.00		174.90	13.25	14.57
12 L.H., Daily Totals		$696.80		$994.30	$58.07	$82.86
Crew B-10F	Hr.	Daily	Hr.	Daily	Bare Costs	Incl. O&P
1 Equip. Oper. (medium)	$48.90	$391.20	$74.15	$593.20	$44.82	$68.28
.5 Laborer	36.65	146.60	56.55	226.20		
1 Tandem Roller, 10 Ton		242.00		266.20	20.17	22.18
12 L.H., Daily Totals		$779.80		$1085.60	$64.98	$90.47
Crew B-10G	Hr.	Daily	Hr.	Daily	Bare Costs	Incl. O&P
1 Equip. Oper. (medium)	$48.90	$391.20	$74.15	$593.20	$44.82	$68.28
.5 Laborer	36.65	146.60	56.55	226.20		
1 Sheepsfoot Roller, 240 H.P.		1218.00		1339.80	101.50	111.65
12 L.H., Daily Totals		$1755.80		$2159.20	$146.32	$179.93
Crew B-10H	Hr.	Daily	Hr.	Daily	Bare Costs	Incl. O&P
1 Equip. Oper. (medium)	$48.90	$391.20	$74.15	$593.20	$44.82	$68.28
.5 Laborer	36.65	146.60	56.55	226.20		
1 Diaphragm Water Pump, 2"		72.00		79.20		
1 -20' Suction Hose, 2"		1.95		2.15		
2 -50' Discharge Hoses, 2"		1.80		1.98	6.31	6.94
12 L.H., Daily Totals		$613.55		$902.73	$51.13	$75.23
Crew B-10I	Hr.	Daily	Hr.	Daily	Bare Costs	Incl. O&P
1 Equip. Oper. (medium)	$48.90	$391.20	$74.15	$593.20	$44.82	$68.28
.5 Laborer	36.65	146.60	56.55	226.20		
1 Diaphragm Water Pump, 4"		115.80		127.38		
1 -20' Suction Hose, 4"		3.25		3.58		
2 -50' Discharge Hoses, 4"		4.70		5.17	10.31	11.34
12 L.H., Daily Totals		$661.55		$955.52	$55.13	$79.63

Crews

Crew No.	Bare Costs		Incl. Subs O&P		Cost Per Labor-Hour	
Crew B-10J	Hr.	Daily	Hr.	Daily	Bare Costs	Incl. O&P
1 Equip. Oper. (medium)	$48.90	$391.20	$74.15	$593.20	$44.82	$68.28
.5 Laborer	36.65	146.60	56.55	226.20		
1 Centrifugal Water Pump, 3"		78.80		86.68		
1 -20' Suction Hose, 3"		2.85		3.13		
2 -50' Discharge Hoses, 3"		3.00		3.30	7.05	7.76
12 L.H., Daily Totals		$622.45		$912.51	$51.87	$76.04

Crew B-10K	Hr.	Daily	Hr.	Daily	Bare Costs	Incl. O&P
1 Equip. Oper. (medium)	$48.90	$391.20	$74.15	$593.20	$44.82	$68.28
.5 Laborer	36.65	146.60	56.55	226.20		
1 Centr. Water Pump, 6"		340.80		374.88		
1 -20' Suction Hose, 6"		11.50		12.65		
2 -50' Discharge Hoses, 6"		12.20		13.42	30.38	33.41
12 L.H., Daily Totals		$902.30		$1220.35	$75.19	$101.70

Crew B-10L	Hr.	Daily	Hr.	Daily	Bare Costs	Incl. O&P
1 Equip. Oper. (medium)	$48.90	$391.20	$74.15	$593.20	$44.82	$68.28
.5 Laborer	36.65	146.60	56.55	226.20		
1 Dozer, 80 H.P.		481.00		529.10	40.08	44.09
12 L.H., Daily Totals		$1018.80		$1348.50	$84.90	$112.38

Crew B-10M	Hr.	Daily	Hr.	Daily	Bare Costs	Incl. O&P
1 Equip. Oper. (medium)	$48.90	$391.20	$74.15	$593.20	$44.82	$68.28
.5 Laborer	36.65	146.60	56.55	226.20		
1 Dozer, 300 H.P.		1817.00		1998.70	151.42	166.56
12 L.H., Daily Totals		$2354.80		$2818.10	$196.23	$234.84

Crew B-10N	Hr.	Daily	Hr.	Daily	Bare Costs	Incl. O&P
1 Equip. Oper. (medium)	$48.90	$391.20	$74.15	$593.20	$44.82	$68.28
.5 Laborer	36.65	146.60	56.55	226.20		
1 F.E. Loader, T.M., 1.5 C.Y.		522.80		575.08	43.57	47.92
12 L.H., Daily Totals		$1060.60		$1394.48	$88.38	$116.21

Crew B-10O	Hr.	Daily	Hr.	Daily	Bare Costs	Incl. O&P
1 Equip. Oper. (medium)	$48.90	$391.20	$74.15	$593.20	$44.82	$68.28
.5 Laborer	36.65	146.60	56.55	226.20		
1 F.E. Loader, T.M., 2.25 C.Y.		959.80		1055.78	79.98	87.98
12 L.H., Daily Totals		$1497.60		$1875.18	$124.80	$156.26

Crew B-10P	Hr.	Daily	Hr.	Daily	Bare Costs	Incl. O&P
1 Equip. Oper. (medium)	$48.90	$391.20	$74.15	$593.20	$44.82	$68.28
.5 Laborer	36.65	146.60	56.55	226.20		
1 Crawler Loader, 3 C.Y.		1191.00		1310.10	99.25	109.18
12 L.H., Daily Totals		$1728.80		$2129.50	$144.07	$177.46

Crew B-10Q	Hr.	Daily	Hr.	Daily	Bare Costs	Incl. O&P
1 Equip. Oper. (medium)	$48.90	$391.20	$74.15	$593.20	$44.82	$68.28
.5 Laborer	36.65	146.60	56.55	226.20		
1 Crawler Loader, 4 C.Y.		1532.00		1685.20	127.67	140.43
12 L.H., Daily Totals		$2069.80		$2504.60	$172.48	$208.72

Crew B-10R	Hr.	Daily	Hr.	Daily	Bare Costs	Incl. O&P
1 Equip. Oper. (medium)	$48.90	$391.20	$74.15	$593.20	$44.82	$68.28
.5 Laborer	36.65	146.60	56.55	226.20		
1 F.E. Loader, W.M., 1 C.Y.		295.20		324.72	24.60	27.06
12 L.H., Daily Totals		$833.00		$1144.12	$69.42	$95.34

Crew B-10S	Hr.	Daily	Hr.	Daily	Bare Costs	Incl. O&P
1 Equip. Oper. (medium)	$48.90	$391.20	$74.15	$593.20	$44.82	$68.28
.5 Laborer	36.65	146.60	56.55	226.20		
1 F.E. Loader, W.M., 1.5 C.Y.		373.40		410.74	31.12	34.23
12 L.H., Daily Totals		$911.20		$1230.14	$75.93	$102.51

Crew B-10T	Hr.	Daily	Hr.	Daily	Bare Costs	Incl. O&P
1 Equip. Oper. (medium)	$48.90	$391.20	$74.15	$593.20	$44.82	$68.28
.5 Laborer	36.65	146.60	56.55	226.20		
1 F.E. Loader, W.M., 2.5 C.Y.		522.80		575.08	43.57	47.92
12 L.H., Daily Totals		$1060.60		$1394.48	$88.38	$116.21

Crew B-10U	Hr.	Daily	Hr.	Daily	Bare Costs	Incl. O&P
1 Equip. Oper. (medium)	$48.90	$391.20	$74.15	$593.20	$44.82	$68.28
.5 Laborer	36.65	146.60	56.55	226.20		
1 F.E. Loader, W.M., 5.5 C.Y.		1049.00		1153.90	87.42	96.16
12 L.H., Daily Totals		$1586.80		$1973.30	$132.23	$164.44

Crew B-10V	Hr.	Daily	Hr.	Daily	Bare Costs	Incl. O&P
1 Equip. Oper. (medium)	$48.90	$391.20	$74.15	$593.20	$44.82	$68.28
.5 Laborer	36.65	146.60	56.55	226.20		
1 Dozer, 700 H.P.		4485.00		4933.50	373.75	411.13
12 L.H., Daily Totals		$5022.80		$5752.90	$418.57	$479.41

Crew B-10W	Hr.	Daily	Hr.	Daily	Bare Costs	Incl. O&P
1 Equip. Oper. (medium)	$48.90	$391.20	$74.15	$593.20	$44.82	$68.28
.5 Laborer	36.65	146.60	56.55	226.20		
1 Dozer, 105 H.P.		591.80		650.98	49.32	54.25
12 L.H., Daily Totals		$1129.60		$1470.38	$94.13	$122.53

Crew B-10X	Hr.	Daily	Hr.	Daily	Bare Costs	Incl. O&P
1 Equip. Oper. (medium)	$48.90	$391.20	$74.15	$593.20	$44.82	$68.28
.5 Laborer	36.65	146.60	56.55	226.20		
1 Dozer, 410 H.P.		2409.00		2649.90	200.75	220.82
12 L.H., Daily Totals		$2946.80		$3469.30	$245.57	$289.11

Crew B-10Y	Hr.	Daily	Hr.	Daily	Bare Costs	Incl. O&P
1 Equip. Oper. (medium)	$48.90	$391.20	$74.15	$593.20	$44.82	$68.28
.5 Laborer	36.65	146.60	56.55	226.20		
1 Vibr. Roller, Towed, 12 Ton		559.20		615.12	46.60	51.26
12 L.H., Daily Totals		$1097.00		$1434.52	$91.42	$119.54

Crew B-11A	Hr.	Daily	Hr.	Daily	Bare Costs	Incl. O&P
1 Equipment Oper. (med.)	$48.90	$391.20	$74.15	$593.20	$42.77	$65.35
1 Laborer	36.65	293.20	56.55	452.40		
1 Dozer, 200 H.P.		1325.00		1457.50	82.81	91.09
16 L.H., Daily Totals		$2009.40		$2503.10	$125.59	$156.44

Crew B-11B	Hr.	Daily	Hr.	Daily	Bare Costs	Incl. O&P
1 Equipment Oper. (light)	$47.05	$376.40	$71.35	$570.80	$41.85	$63.95
1 Laborer	36.65	293.20	56.55	452.40		
1 Air Powered Tamper		26.20		28.82		
1 Air Compressor, 365 cfm		264.80		291.28		
2 -50' Air Hoses, 1.5"		11.60		12.76	18.91	20.80
16 L.H., Daily Totals		$972.20		$1356.06	$60.76	$84.75

Crew B-11C	Hr.	Daily	Hr.	Daily	Bare Costs	Incl. O&P
1 Equipment Oper. (med.)	$48.90	$391.20	$74.15	$593.20	$42.77	$65.35
1 Laborer	36.65	293.20	56.55	452.40		
1 Backhoe Loader, 48 H.P.		365.20		401.72	22.82	25.11
16 L.H., Daily Totals		$1049.60		$1447.32	$65.60	$90.46

Crews

Crew No.	Bare Costs		Incl. Subs O&P		Cost Per Labor-Hour	
Crew B-11J	Hr.	Daily	Hr.	Daily	Bare Costs	Incl. O&P
1 Equipment Oper. (med.)	$48.90	$391.20	$74.15	$593.20	$42.77	$65.35
1 Laborer	36.65	293.20	56.55	452.40		
1 Grader, 30,000 Lbs.		707.00		777.70		
1 Ripper, Beam & 1 Shank		81.40		89.54	49.27	54.20
16 L.H., Daily Totals		$1472.80		$1912.84	$92.05	$119.55
Crew B-11K	Hr.	Daily	Hr.	Daily	Bare Costs	Incl. O&P
1 Equipment Oper. (med.)	$48.90	$391.20	$74.15	$593.20	$42.77	$65.35
1 Laborer	36.65	293.20	56.55	452.40		
1 Trencher, Chain Type, 8' D		3376.00		3713.60	211.00	232.10
16 L.H., Daily Totals		$4060.40		$4759.20	$253.78	$297.45
Crew B-11L	Hr.	Daily	Hr.	Daily	Bare Costs	Incl. O&P
1 Equipment Oper. (med.)	$48.90	$391.20	$74.15	$593.20	$42.77	$65.35
1 Laborer	36.65	293.20	56.55	452.40		
1 Grader, 30,000 Lbs.		707.00		777.70	44.19	48.61
16 L.H., Daily Totals		$1391.40		$1823.30	$86.96	$113.96
Crew B-11M	Hr.	Daily	Hr.	Daily	Bare Costs	Incl. O&P
1 Equipment Oper. (med.)	$48.90	$391.20	$74.15	$593.20	$42.77	$65.35
1 Laborer	36.65	293.20	56.55	452.40		
1 Backhoe Loader, 80 H.P.		397.60		437.36	24.85	27.34
16 L.H., Daily Totals		$1082.00		$1482.96	$67.63	$92.69
Crew B-11N	Hr.	Daily	Hr.	Daily	Bare Costs	Incl. O&P
1 Labor Foreman (outside)	$38.65	$309.20	$59.65	$477.20	$40.19	$61.31
2 Equipment Operators (med.)	48.90	782.40	74.15	1186.40		
6 Truck Drivers (heavy)	37.55	1802.40	57.30	2750.40		
1 F.E. Loader, W.M., 5.5 C.Y.		1049.00		1153.90		
1 Dozer, 410 H.P.		2409.00		2649.90		
6 Dump Trucks, Off Hwy., 50 Ton		10722.00		11794.20	196.94	216.64
72 L.H., Daily Totals		$17074.00		$20012.00	$237.14	$277.94
Crew B-11Q	Hr.	Daily	Hr.	Daily	Bare Costs	Incl. O&P
1 Equipment Operator (med.)	$48.90	$391.20	$74.15	$593.20	$44.82	$68.28
.5 Laborer	36.65	146.60	56.55	226.20		
1 Dozer, 140 H.P.		883.80		972.18	73.65	81.02
12 L.H., Daily Totals		$1421.60		$1791.58	$118.47	$149.30
Crew B-11R	Hr.	Daily	Hr.	Daily	Bare Costs	Incl. O&P
1 Equipment Operator (med.)	$48.90	$391.20	$74.15	$593.20	$44.82	$68.28
.5 Laborer	36.65	146.60	56.55	226.20		
1 Dozer, 200 H.P.		1325.00		1457.50	110.42	121.46
12 L.H., Daily Totals		$1862.80		$2276.90	$155.23	$189.74
Crew B-11S	Hr.	Daily	Hr.	Daily	Bare Costs	Incl. O&P
1 Equipment Operator (med.)	$48.90	$391.20	$74.15	$593.20	$44.82	$68.28
.5 Laborer	36.65	146.60	56.55	226.20		
1 Dozer, 300 H.P.		1817.00		1998.70		
1 Ripper, Beam & 1 Shank		81.40		89.54	158.20	174.02
12 L.H., Daily Totals		$2436.20		$2907.64	$203.02	$242.30
Crew B-11T	Hr.	Daily	Hr.	Daily	Bare Costs	Incl. O&P
1 Equipment Operator (med.)	$48.90	$391.20	$74.15	$593.20	$44.82	$68.28
.5 Laborer	36.65	146.60	56.55	226.20		
1 Dozer, 410 H.P.		2409.00		2649.90		
1 Ripper, Beam & 2 Shanks		91.60		100.76	208.38	229.22
12 L.H., Daily Totals		$3038.40		$3570.06	$253.20	$297.51

Crew No.	Bare Costs		Incl. Subs O&P		Cost Per Labor-Hour	
Crew B-11U	Hr.	Daily	Hr.	Daily	Bare Costs	Incl. O&P
1 Equipment Operator (med.)	$48.90	$391.20	$74.15	$593.20	$44.82	$68.28
.5 Laborer	36.65	146.60	56.55	226.20		
1 Dozer, 520 H.P.		3116.00		3427.60	259.67	285.63
12 L.H., Daily Totals		$3653.80		$4247.00	$304.48	$353.92
Crew B-11V	Hr.	Daily	Hr.	Daily	Bare Costs	Incl. O&P
3 Laborers	$36.65	$879.60	$56.55	$1357.20	$36.65	$56.55
1 Roller, 2-Drum, W.B., 7.5 H.P.		182.80		201.08	7.62	8.38
24 L.H., Daily Totals		$1062.40		$1558.28	$44.27	$64.93
Crew B-11W	Hr.	Daily	Hr.	Daily	Bare Costs	Incl. O&P
1 Equipment Operator (med.)	$48.90	$391.20	$74.15	$593.20	$38.42	$58.64
1 Common Laborer	36.65	293.20	56.55	452.40		
10 Truck Drivers (heavy)	37.55	3004.00	57.30	4584.00		
1 Dozer, 200 H.P.		1325.00		1457.50		
1 Vibratory Roller, Towed, 23 Ton		412.00		453.20		
10 Dump Trucks, 8 C.Y., 220 H.P.		4172.00		4589.20	61.55	67.71
96 L.H., Daily Totals		$9597.40		$12129.50	$99.97	$126.35
Crew B-11Y	Hr.	Daily	Hr.	Daily	Bare Costs	Incl. O&P
1 Labor Foreman (outside)	$38.65	$309.20	$59.65	$477.20	$40.96	$62.76
5 Common Laborers	36.65	1466.00	56.55	2262.00		
3 Equipment Operators (med.)	48.90	1173.60	74.15	1779.60		
1 Dozer, 80 H.P.		481.00		529.10		
2 Roller, 2-Drum, W.B., 7.5 H.P.		365.60		402.16		
4 Vibrating Plate, Gas, 21"		184.00		202.40	14.31	15.75
72 L.H., Daily Totals		$3979.40		$5652.46	$55.27	$78.51
Crew B-12A	Hr.	Daily	Hr.	Daily	Bare Costs	Incl. O&P
1 Equip. Oper. (crane)	$50.25	$402.00	$76.20	$609.60	$43.45	$66.38
1 Laborer	36.65	293.20	56.55	452.40		
1 Hyd. Excavator, 1 C.Y.		814.80		896.28	50.92	56.02
16 L.H., Daily Totals		$1510.00		$1958.28	$94.38	$122.39
Crew B-12B	Hr.	Daily	Hr.	Daily	Bare Costs	Incl. O&P
1 Equip. Oper. (crane)	$50.25	$402.00	$76.20	$609.60	$43.45	$66.38
1 Laborer	36.65	293.20	56.55	452.40		
1 Hyd. Excavator, 1.5 C.Y.		1031.00		1134.10	64.44	70.88
16 L.H., Daily Totals		$1726.20		$2196.10	$107.89	$137.26
Crew B-12C	Hr.	Daily	Hr.	Daily	Bare Costs	Incl. O&P
1 Equip. Oper. (crane)	$50.25	$402.00	$76.20	$609.60	$43.45	$66.38
1 Laborer	36.65	293.20	56.55	452.40		
1 Hyd. Excavator, 2 C.Y.		1175.00		1292.50	73.44	80.78
16 L.H., Daily Totals		$1870.20		$2354.50	$116.89	$147.16
Crew B-12D	Hr.	Daily	Hr.	Daily	Bare Costs	Incl. O&P
1 Equip. Oper. (crane)	$50.25	$402.00	$76.20	$609.60	$43.45	$66.38
1 Laborer	36.65	293.20	56.55	452.40		
1 Hyd. Excavator, 3.5 C.Y.		2442.00		2686.20	152.63	167.89
16 L.H., Daily Totals		$3137.20		$3748.20	$196.07	$234.26
Crew B-12E	Hr.	Daily	Hr.	Daily	Bare Costs	Incl. O&P
1 Equip. Oper. (crane)	$50.25	$402.00	$76.20	$609.60	$43.45	$66.38
1 Laborer	36.65	293.20	56.55	452.40		
1 Hyd. Excavator, .5 C.Y.		448.00		492.80	28.00	30.80
16 L.H., Daily Totals		$1143.20		$1554.80	$71.45	$97.17

Crews

Crew No.	Bare Costs		Incl. Subs O&P		Cost Per Labor-Hour	
Crew B-12F	Hr.	Daily	Hr.	Daily	Bare Costs	Incl. O&P
1 Equip. Oper. (crane)	$50.25	$402.00	$76.20	$609.60	$43.45	$66.38
1 Laborer	36.65	293.20	56.55	452.40		
1 Hyd. Excavator, .75 C.Y.		662.00		728.20	41.38	45.51
16 L.H., Daily Totals		$1357.20		$1790.20	$84.83	$111.89

Crew B-12G	Hr.	Daily	Hr.	Daily	Bare Costs	Incl. O&P
1 Equip. Oper. (crane)	$50.25	$402.00	$76.20	$609.60	$43.45	$66.38
1 Laborer	36.65	293.20	56.55	452.40		
1 Crawler Crane, 15 Ton		684.15		752.57		
1 Clamshell Bucket, .5 C.Y.		37.80		41.58	45.12	49.63
16 L.H., Daily Totals		$1417.15		$1856.15	$88.57	$116.01

Crew B-12H	Hr.	Daily	Hr.	Daily	Bare Costs	Incl. O&P
1 Equip. Oper. (crane)	$50.25	$402.00	$76.20	$609.60	$43.45	$66.38
1 Laborer	36.65	293.20	56.55	452.40		
1 Crawler Crane, 25 Ton		1173.00		1290.30		
1 Clamshell Bucket, 1 C.Y.		47.40		52.14	76.28	83.90
16 L.H., Daily Totals		$1915.60		$2404.44	$119.72	$150.28

Crew B-12I	Hr.	Daily	Hr.	Daily	Bare Costs	Incl. O&P
1 Equip. Oper. (crane)	$50.25	$402.00	$76.20	$609.60	$43.45	$66.38
1 Laborer	36.65	293.20	56.55	452.40		
1 Crawler Crane, 20 Ton		880.25		968.27		
1 Dragline Bucket, .75 C.Y.		20.80		22.88	56.32	61.95
16 L.H., Daily Totals		$1596.25		$2053.16	$99.77	$128.32

Crew B-12J	Hr.	Daily	Hr.	Daily	Bare Costs	Incl. O&P
1 Equip. Oper. (crane)	$50.25	$402.00	$76.20	$609.60	$43.45	$66.38
1 Laborer	36.65	293.20	56.55	452.40		
1 Gradall, 5/8 C.Y.		881.60		969.76	55.10	60.61
16 L.H., Daily Totals		$1576.80		$2031.76	$98.55	$126.99

Crew B-12K	Hr.	Daily	Hr.	Daily	Bare Costs	Incl. O&P
1 Equip. Oper. (crane)	$50.25	$402.00	$76.20	$609.60	$43.45	$66.38
1 Laborer	36.65	293.20	56.55	452.40		
1 Gradall, 3 Ton, 1 C.Y.		1005.00		1105.50	62.81	69.09
16 L.H., Daily Totals		$1700.20		$2167.50	$106.26	$135.47

Crew B-12L	Hr.	Daily	Hr.	Daily	Bare Costs	Incl. O&P
1 Equip. Oper. (crane)	$50.25	$402.00	$76.20	$609.60	$43.45	$66.38
1 Laborer	36.65	293.20	56.55	452.40		
1 Crawler Crane, 15 Ton		684.15		752.57		
1 F.E. Attachment, .5 C.Y.		59.60		65.56	46.48	51.13
16 L.H., Daily Totals		$1438.95		$1880.13	$89.93	$117.51

Crew B-12M	Hr.	Daily	Hr.	Daily	Bare Costs	Incl. O&P
1 Equip. Oper. (crane)	$50.25	$402.00	$76.20	$609.60	$43.45	$66.38
1 Laborer	36.65	293.20	56.55	452.40		
1 Crawler Crane, 20 Ton		880.25		968.27		
1 F.E. Attachment, .75 C.Y.		64.00		70.40	59.02	64.92
16 L.H., Daily Totals		$1639.45		$2100.68	$102.47	$131.29

Crew B-12N	Hr.	Daily	Hr.	Daily	Bare Costs	Incl. O&P
1 Equip. Oper. (crane)	$50.25	$402.00	$76.20	$609.60	$43.45	$66.38
1 Laborer	36.65	293.20	56.55	452.40		
1 Crawler Crane, 25 Ton		1173.00		1290.30		
1 F.E. Attachment, 1 C.Y.		70.80		77.88	77.74	85.51
16 L.H., Daily Totals		$1939.00		$2430.18	$121.19	$151.89

Crew B-12O	Hr.	Daily	Hr.	Daily	Bare Costs	Incl. O&P
1 Equip. Oper. (crane)	$50.25	$402.00	$76.20	$609.60	$43.45	$66.38
1 Laborer	36.65	293.20	56.55	452.40		
1 Crawler Crane, 40 Ton		1179.00		1296.90		
1 F.E. Attachment, 1.5 C.Y.		80.00		88.00	78.69	86.56
16 L.H., Daily Totals		$1954.20		$2446.90	$122.14	$152.93

Crew B-12P	Hr.	Daily	Hr.	Daily	Bare Costs	Incl. O&P
1 Equip. Oper. (crane)	$50.25	$402.00	$76.20	$609.60	$43.45	$66.38
1 Laborer	36.65	293.20	56.55	452.40		
1 Crawler Crane, 40 Ton		1179.00		1296.90		
1 Dragline Bucket, 1.5 C.Y.		33.60		36.96	75.79	83.37
16 L.H., Daily Totals		$1907.80		$2395.86	$119.24	$149.74

Crew B-12Q	Hr.	Daily	Hr.	Daily	Bare Costs	Incl. O&P
1 Equip. Oper. (crane)	$50.25	$402.00	$76.20	$609.60	$43.45	$66.38
1 Laborer	36.65	293.20	56.55	452.40		
1 Hyd. Excavator, 5/8 C.Y.		594.40		653.84	37.15	40.87
16 L.H., Daily Totals		$1289.60		$1715.84	$80.60	$107.24

Crew B-12S	Hr.	Daily	Hr.	Daily	Bare Costs	Incl. O&P
1 Equip. Oper. (crane)	$50.25	$402.00	$76.20	$609.60	$43.45	$66.38
1 Laborer	36.65	293.20	56.55	452.40		
1 Hyd. Excavator, 2.5 C.Y.		1637.00		1800.70	102.31	112.54
16 L.H., Daily Totals		$2332.20		$2862.70	$145.76	$178.92

Crew B-12T	Hr.	Daily	Hr.	Daily	Bare Costs	Incl. O&P
1 Equip. Oper. (crane)	$50.25	$402.00	$76.20	$609.60	$43.45	$66.38
1 Laborer	36.65	293.20	56.55	452.40		
1 Crawler Crane, 75 Ton		1492.00		1641.20		
1 F.E. Attachment, 3 C.Y.		102.80		113.08	99.67	109.64
16 L.H., Daily Totals		$2290.00		$2816.28	$143.13	$176.02

Crew B-12V	Hr.	Daily	Hr.	Daily	Bare Costs	Incl. O&P
1 Equip. Oper. (crane)	$50.25	$402.00	$76.20	$609.60	$43.45	$66.38
1 Laborer	36.65	293.20	56.55	452.40		
1 Crawler Crane, 75 Ton		1492.00		1641.20		
1 Dragline Bucket, 3 C.Y.		52.60		57.86	96.54	106.19
16 L.H., Daily Totals		$2239.80		$2761.06	$139.99	$172.57

Crew B-12Y	Hr.	Daily	Hr.	Daily	Bare Costs	Incl. O&P
1 Equip. Oper. (crane)	$50.25	$402.00	$76.20	$609.60	$41.18	$63.10
2 Laborers	36.65	586.40	56.55	904.80		
1 Hyd. Excavator, 3.5 C.Y.		2442.00		2686.20	101.75	111.93
24 L.H., Daily Totals		$3430.40		$4200.60	$142.93	$175.03

Crew B-12Z	Hr.	Daily	Hr.	Daily	Bare Costs	Incl. O&P
1 Equip. Oper. (crane)	$50.25	$402.00	$76.20	$609.60	$41.18	$63.10
2 Laborers	36.65	586.40	56.55	904.80		
1 Hyd. Excavator, 2.5 C.Y.		1637.00		1800.70	68.21	75.03
24 L.H., Daily Totals		$2625.40		$3315.10	$109.39	$138.13

Crew B-13	Hr.	Daily	Hr.	Daily	Bare Costs	Incl. O&P
1 Labor Foreman (outside)	$38.65	$309.20	$59.65	$477.20	$39.86	$61.15
4 Laborers	36.65	1172.80	56.55	1809.60		
1 Equip. Oper. (crane)	50.25	402.00	76.20	609.60		
1 Equip. Oper. (oiler)	43.55	348.40	66.00	528.00		
1 Hyd. Crane, 25 Ton		745.80		820.38	13.32	14.65
56 L.H., Daily Totals		$2978.20		$4244.78	$53.18	$75.80

Crews

Crew No.	Bare Costs		Incl. Subs O&P		Cost Per Labor-Hour	
Crew B-13A	Hr.	Daily	Hr.	Daily	Bare Costs	Incl. O&P
1 Labor Foreman (outside)	$38.65	$309.20	$59.65	$477.20	$40.69	$62.24
2 Laborers	36.65	586.40	56.55	904.80		
2 Equipment Operators (med.)	48.90	782.40	74.15	1186.40		
2 Truck Drivers (heavy)	37.55	600.80	57.30	916.80		
1 Crawler Crane, 75 Ton		1492.00		1641.20		
1 Crawler Loader, 4 C.Y.		1532.00		1685.20		
2 Dump Trucks, 8 C.Y., 220 H.P.		834.40		917.84	68.90	75.79
56 L.H., Daily Totals		$6137.20		$7729.44	$109.59	$138.03

Crew B-13B	Hr.	Daily	Hr.	Daily	Bare Costs	Incl. O&P
1 Labor Foreman (outside)	$38.65	$309.20	$59.65	$477.20	$39.86	$61.15
4 Laborers	36.65	1172.80	56.55	1809.60		
1 Equip. Oper. (crane)	50.25	402.00	76.20	609.60		
1 Equip. Oper. (oiler)	43.55	348.40	66.00	528.00		
1 Hyd. Crane, 55 Ton		1137.00		1250.70	20.30	22.33
56 L.H., Daily Totals		$3369.40		$4675.10	$60.17	$83.48

Crew B-13C	Hr.	Daily	Hr.	Daily	Bare Costs	Incl. O&P
1 Labor Foreman (outside)	$38.65	$309.20	$59.65	$477.20	$39.86	$61.15
4 Laborers	36.65	1172.80	56.55	1809.60		
1 Equip. Oper. (crane)	50.25	402.00	76.20	609.60		
1 Equip. Oper. (oiler)	43.55	348.40	66.00	528.00		
1 Crawler Crane, 100 Ton		1703.00		1873.30	30.41	33.45
56 L.H., Daily Totals		$3935.40		$5297.70	$70.28	$94.60

Crew B-13D	Hr.	Daily	Hr.	Daily	Bare Costs	Incl. O&P
1 Laborer	$36.65	$293.20	$56.55	$452.40	$43.45	$66.38
1 Equip. Oper. (crane)	50.25	402.00	76.20	609.60		
1 Hyd. Excavator, 1 C.Y.		814.80		896.28		
1 Trench Box		84.05		92.45	56.18	61.80
16 L.H., Daily Totals		$1594.05		$2050.74	$99.63	$128.17

Crew B-13E	Hr.	Daily	Hr.	Daily	Bare Costs	Incl. O&P
1 Laborer	$36.65	$293.20	$56.55	$452.40	$43.45	$66.38
1 Equip. Oper. (crane)	50.25	402.00	76.20	609.60		
1 Hyd. Excavator, 1.5 C.Y.		1031.00		1134.10		
1 Trench Box		84.05		92.45	69.69	76.66
16 L.H., Daily Totals		$1810.25		$2288.55	$113.14	$143.03

Crew B-13F	Hr.	Daily	Hr.	Daily	Bare Costs	Incl. O&P
1 Laborer	$36.65	$293.20	$56.55	$452.40	$43.45	$66.38
1 Equip. Oper. (crane)	50.25	402.00	76.20	609.60		
1 Hyd. Excavator, 3.5 C.Y.		2442.00		2686.20		
1 Trench Box		84.05		92.45	157.88	173.67
16 L.H., Daily Totals		$3221.25		$3840.66	$201.33	$240.04

Crew B-13G	Hr.	Daily	Hr.	Daily	Bare Costs	Incl. O&P
1 Laborer	$36.65	$293.20	$56.55	$452.40	$43.45	$66.38
1 Equip. Oper. (crane)	50.25	402.00	76.20	609.60		
1 Hyd. Excavator, .75 C.Y.		662.00		728.20		
1 Trench Box		84.05		92.45	46.63	51.29
16 L.H., Daily Totals		$1441.25		$1882.66	$90.08	$117.67

Crew B-13H	Hr.	Daily	Hr.	Daily	Bare Costs	Incl. O&P
1 Laborer	$36.65	$293.20	$56.55	$452.40	$43.45	$66.38
1 Equip. Oper. (crane)	50.25	402.00	76.20	609.60		
1 Gradall, 5/8 C.Y.		881.60		969.76		
1 Trench Box		84.05		92.45	60.35	66.39
16 L.H., Daily Totals		$1660.85		$2124.22	$103.80	$132.76

Crew B-13I	Hr.	Daily	Hr.	Daily	Bare Costs	Incl. O&P
1 Laborer	$36.65	$293.20	$56.55	$452.40	$43.45	$66.38
1 Equip. Oper. (crane)	50.25	402.00	76.20	609.60		
1 Gradall, 3 Ton, 1 C.Y.		1005.00		1105.50		
1 Trench Box		84.05		92.45	68.07	74.87
16 L.H., Daily Totals		$1784.25		$2259.95	$111.52	$141.25

Crew B-13J	Hr.	Daily	Hr.	Daily	Bare Costs	Incl. O&P
1 Laborer	$36.65	$293.20	$56.55	$452.40	$43.45	$66.38
1 Equip. Oper. (crane)	50.25	402.00	76.20	609.60		
1 Hyd. Excavator, 2.5 C.Y.		1637.00		1800.70		
1 Trench Box		84.05		92.45	107.57	118.32
16 L.H., Daily Totals		$2416.25		$2955.16	$151.02	$184.70

Crew B-14	Hr.	Daily	Hr.	Daily	Bare Costs	Incl. O&P
1 Labor Foreman (outside)	$38.65	$309.20	$59.65	$477.20	$38.72	$59.53
4 Laborers	36.65	1172.80	56.55	1809.60		
1 Equip. Oper. (light)	47.05	376.40	71.35	570.80		
1 Backhoe Loader, 48 H.P.		365.20		401.72	7.61	8.37
48 L.H., Daily Totals		$2223.60		$3259.32	$46.33	$67.90

Crew B-14A	Hr.	Daily	Hr.	Daily	Bare Costs	Incl. O&P
1 Equip. Oper. (crane)	$50.25	$402.00	$76.20	$609.60	$45.72	$69.65
.5 Laborer	36.65	146.60	56.55	226.20		
1 Hyd. Excavator, 4.5 C.Y.		3043.00		3347.30	253.58	278.94
12 L.H., Daily Totals		$3591.60		$4183.10	$299.30	$348.59

Crew B-14B	Hr.	Daily	Hr.	Daily	Bare Costs	Incl. O&P
1 Equip. Oper. (crane)	$50.25	$402.00	$76.20	$609.60	$45.72	$69.65
.5 Laborer	36.65	146.60	56.55	226.20		
1 Hyd. Excavator, 6 C.Y.		3523.00		3875.30	293.58	322.94
12 L.H., Daily Totals		$4071.60		$4711.10	$339.30	$392.59

Crew B-14C	Hr.	Daily	Hr.	Daily	Bare Costs	Incl. O&P
1 Equip. Oper. (crane)	$50.25	$402.00	$76.20	$609.60	$45.72	$69.65
.5 Laborer	36.65	146.60	56.55	226.20		
1 Hyd. Excavator, 7 C.Y.		3596.00		3955.60	299.67	329.63
12 L.H., Daily Totals		$4144.60		$4791.40	$345.38	$399.28

Crew B-14F	Hr.	Daily	Hr.	Daily	Bare Costs	Incl. O&P
1 Equip. Oper. (crane)	$50.25	$402.00	$76.20	$609.60	$45.72	$69.65
.5 Laborer	36.65	146.60	56.55	226.20		
1 Hyd. Shovel, 7 C.Y.		3714.00		4085.40	309.50	340.45
12 L.H., Daily Totals		$4262.60		$4921.20	$355.22	$410.10

Crew B-14G	Hr.	Daily	Hr.	Daily	Bare Costs	Incl. O&P
1 Equip. Oper. (crane)	$50.25	$402.00	$76.20	$609.60	$45.72	$69.65
.5 Laborer	36.65	146.60	56.55	226.20		
1 Hyd. Shovel, 12 C.Y.		5204.00		5724.40	433.67	477.03
12 L.H., Daily Totals		$5752.60		$6560.20	$479.38	$546.68

Crew B-14J	Hr.	Daily	Hr.	Daily	Bare Costs	Incl. O&P
1 Equip. Oper. (medium)	$48.90	$391.20	$74.15	$593.20	$44.82	$68.28
.5 Laborer	36.65	146.60	56.55	226.20		
1 F.E. Loader, 8 C.Y.		1989.00		2187.90	165.75	182.32
12 L.H., Daily Totals		$2526.80		$3007.30	$210.57	$250.61

Crews

Crew No.		Bare Costs		Incl. Subs O&P		Cost Per Labor-Hour	
Crew B-14K	Hr.	Daily	Hr.	Daily	Bare Costs	Incl. O&P
1 Equip. Oper. (medium)	$48.90	$391.20	$74.15	$593.20	$44.82	$68.28
.5 Laborer	36.65	146.60	56.55	226.20		
1 F.E. Loader, 10 C.Y.		2703.00		2973.30	225.25	247.78
12 L.H., Daily Totals		$3240.80		$3792.70	$270.07	$316.06

Crew B-15	Hr.	Daily	Hr.	Daily	Bare Costs	Incl. O&P
1 Equipment Oper. (med.)	$48.90	$391.20	$74.15	$593.20	$40.66	$62.01
.5 Laborer	36.65	146.60	56.55	226.20		
2 Truck Drivers (heavy)	37.55	600.80	57.30	916.80		
2 Dump Trucks, 12 C.Y., 400 H.P.		1384.40		1522.84		
1 Dozer, 200 H.P.		1325.00		1457.50	96.76	106.44
28 L.H., Daily Totals		$3848.00		$4716.54	$137.43	$168.45

Crew B-16	Hr.	Daily	Hr.	Daily	Bare Costs	Incl. O&P
1 Labor Foreman (outside)	$38.65	$309.20	$59.65	$477.20	$37.38	$57.51
2 Laborers	36.65	586.40	56.55	904.80		
1 Truck Driver (heavy)	37.55	300.40	57.30	458.40		
1 Dump Truck, 12 C.Y., 400 H.P.		692.20		761.42	21.63	23.79
32 L.H., Daily Totals		$1888.20		$2601.82	$59.01	$81.31

Crew B-17	Hr.	Daily	Hr.	Daily	Bare Costs	Incl. O&P
2 Laborers	$36.65	$586.40	$56.55	$904.80	$39.48	$60.44
1 Equip. Oper. (light)	47.05	376.40	71.35	570.80		
1 Truck Driver (heavy)	37.55	300.40	57.30	458.40		
1 Backhoe Loader, 48 H.P.		365.20		401.72		
1 Dump Truck, 8 C.Y., 220 H.P.		417.20		458.92	24.45	26.90
32 L.H., Daily Totals		$2045.60		$2794.64	$63.92	$87.33

Crew B-17A	Hr.	Daily	Hr.	Daily	Bare Costs	Incl. O&P
2 Labor Foremen (outside)	$38.65	$618.40	$59.65	$954.40	$39.38	$60.82
6 Laborers	36.65	1759.20	56.55	2714.40		
1 Skilled Worker Foreman (out)	49.30	394.40	76.35	610.80		
1 Skilled Worker	47.30	378.40	73.25	586.00		
80 L.H., Daily Totals		$3150.40		$4865.60	$39.38	$60.82

Crew B-17B	Hr.	Daily	Hr.	Daily	Bare Costs	Incl. O&P
2 Laborers	$36.65	$586.40	$56.55	$904.80	$39.48	$60.44
1 Equip. Oper. (light)	47.05	376.40	71.35	570.80		
1 Truck Driver (heavy)	37.55	300.40	57.30	458.40		
1 Backhoe Loader, 48 H.P.		365.20		401.72		
1 Dump Truck, 12 C.Y., 400 H.P.		692.20		761.42	33.04	36.35
32 L.H., Daily Totals		$2320.60		$3097.14	$72.52	$96.79

Crew B-18	Hr.	Daily	Hr.	Daily	Bare Costs	Incl. O&P
1 Labor Foreman (outside)	$38.65	$309.20	$59.65	$477.20	$37.32	$57.58
2 Laborers	36.65	586.40	56.55	904.80		
1 Vibrating Plate, Gas, 21"		46.00		50.60	1.92	2.11
24 L.H., Daily Totals		$941.60		$1432.60	$39.23	$59.69

Crew B-19	Hr.	Daily	Hr.	Daily	Bare Costs	Incl. O&P
1 Pile Driver Foreman (outside)	$46.40	$371.20	$73.60	$588.80	$46.01	$71.70
4 Pile Drivers	44.40	1420.80	70.40	2252.80		
2 Equip. Oper. (crane)	50.25	804.00	76.20	1219.20		
1 Equip. Oper. (oiler)	43.55	348.40	66.00	528.00		
1 Crawler Crane, 40 Ton		1179.00		1296.90		
1 Lead, 90' High		124.40		136.84		
1 Hammer, Diesel, 22k ft-lb		537.45		591.20	28.76	31.64
64 L.H., Daily Totals		$4785.25		$6613.73	$74.77	$103.34

Crew No.		Bare Costs		Incl. Subs O&P		Cost Per Labor-Hour	
Crew B-19A	Hr.	Daily	Hr.	Daily	Bare Costs	Incl. O&P
1 Pile Driver Foreman (outside)	$46.40	$371.20	$73.60	$588.80	$46.01	$71.70
4 Pile Drivers	44.40	1420.80	70.40	2252.80		
2 Equip. Oper. (crane)	50.25	804.00	76.20	1219.20		
1 Equip. Oper. (oiler)	43.55	348.40	66.00	528.00		
1 Crawler Crane, 75 Ton		1492.00		1641.20		
1 Lead, 90' high		124.40		136.84		
1 Hammer, Diesel, 41k ft-lb		642.40		706.64	35.29	38.82
64 L.H., Daily Totals		$5203.20		$7073.48	$81.30	$110.52

Crew B-19B	Hr.	Daily	Hr.	Daily	Bare Costs	Incl. O&P
1 Pile Driver Foreman (outside)	$46.40	$371.20	$73.60	$588.80	$46.01	$71.70
4 Pile Drivers	44.40	1420.80	70.40	2252.80		
2 Equip. Oper. (crane)	50.25	804.00	76.20	1219.20		
1 Equip. Oper. (oiler)	43.55	348.40	66.00	528.00		
1 Crawler Crane, 40 Ton		1179.00		1296.90		
1 Lead, 90' High		124.40		136.84		
1 Hammer, Diesel, 22k ft-lb		537.45		591.20		
1 Barge, 400 Ton		781.00		859.10	40.97	45.06
64 L.H., Daily Totals		$5566.25		$7472.84	$86.97	$116.76

Crew B-19C	Hr.	Daily	Hr.	Daily	Bare Costs	Incl. O&P
1 Pile Driver Foreman (outside)	$46.40	$371.20	$73.60	$588.80	$46.01	$71.70
4 Pile Drivers	44.40	1420.80	70.40	2252.80		
2 Equip. Oper. (crane)	50.25	804.00	76.20	1219.20		
1 Equip. Oper. (oiler)	43.55	348.40	66.00	528.00		
1 Crawler Crane, 75 Ton		1492.00		1641.20		
1 Lead, 90' High		124.40		136.84		
1 Hammer, Diesel, 41k ft-lb		642.40		706.64		
1 Barge, 400 Ton		781.00		859.10	47.50	52.25
64 L.H., Daily Totals		$5984.20		$7932.58	$93.50	$123.95

Crew B-20	Hr.	Daily	Hr.	Daily	Bare Costs	Incl. O&P
1 Labor Foreman (outside)	$38.65	$309.20	$59.65	$477.20	$40.87	$63.15
1 Skilled Worker	47.30	378.40	73.25	586.00		
1 Laborer	36.65	293.20	56.55	452.40		
24 L.H., Daily Totals		$980.80		$1515.60	$40.87	$63.15

Crew B-20A	Hr.	Daily	Hr.	Daily	Bare Costs	Incl. O&P
1 Labor Foreman (outside)	$38.65	$309.20	$59.65	$477.20	$44.73	$68.16
1 Laborer	36.65	293.20	56.55	452.40		
1 Plumber	57.55	460.40	86.90	695.20		
1 Plumber Apprentice	46.05	368.40	69.55	556.40		
32 L.H., Daily Totals		$1431.20		$2181.20	$44.73	$68.16

Crew B-21	Hr.	Daily	Hr.	Daily	Bare Costs	Incl. O&P
1 Labor Foreman (outside)	$38.65	$309.20	$59.65	$477.20	$42.21	$65.01
1 Skilled Worker	47.30	378.40	73.25	586.00		
1 Laborer	36.65	293.20	56.55	452.40		
.5 Equip. Oper. (crane)	50.25	201.00	76.20	304.80		
.5 S.P. Crane, 4x4, 5 Ton		140.00		154.00	5.00	5.50
28 L.H., Daily Totals		$1321.80		$1974.40	$47.21	$70.51

Crew B-21A	Hr.	Daily	Hr.	Daily	Bare Costs	Incl. O&P
1 Labor Foreman (outside)	$38.65	$309.20	$59.65	$477.20	$45.83	$69.77
1 Laborer	36.65	293.20	56.55	452.40		
1 Plumber	57.55	460.40	86.90	695.20		
1 Plumber Apprentice	46.05	368.40	69.55	556.40		
1 Equip. Oper. (crane)	50.25	402.00	76.20	609.60		
1 S.P. Crane, 4x4, 12 Ton		481.20		529.32	12.03	13.23
40 L.H., Daily Totals		$2314.40		$3320.12	$57.86	$83.00

Crews

Crew No.	Bare Costs Hr.	Bare Costs Daily	Incl. Subs O&P Hr.	Incl. Subs O&P Daily	Cost Per Labor-Hour Bare Costs	Cost Per Labor-Hour Incl. O&P
Crew B-21B						
1 Labor Foreman (outside)	$38.65	$309.20	$59.65	$477.20	$39.77	$61.10
3 Laborers	36.65	879.60	56.55	1357.20		
1 Equip. Oper. (crane)	50.25	402.00	76.20	609.60		
1 Hyd. Crane, 12 Ton		663.00		729.30	16.57	18.23
40 L.H., Daily Totals		$2253.80		$3173.30	$56.34	$79.33
Crew B-21C						
1 Labor Foreman (outside)	$38.65	$309.20	$59.65	$477.20	$39.86	$61.15
4 Laborers	36.65	1172.80	56.55	1809.60		
1 Equip. Oper. (crane)	50.25	402.00	76.20	609.60		
1 Equip. Oper. (oiler)	43.55	348.40	66.00	528.00		
2 Cutting Torches		22.80		25.08		
2 Sets of Gases		304.00		334.40		
1 Lattice Boom Crane, 90 Ton		1529.00		1681.90	33.14	36.45
56 L.H., Daily Totals		$4088.20		$5465.78	$73.00	$97.60
Crew B-22						
1 Labor Foreman (outside)	$38.65	$309.20	$59.65	$477.20	$42.74	$65.76
1 Skilled Worker	47.30	378.40	73.25	586.00		
1 Laborer	36.65	293.20	56.55	452.40		
.75 Equip. Oper. (crane)	50.25	301.50	76.20	457.20		
.75 S.P. Crane, 4x4, 5 Ton		210.00		231.00	7.00	7.70
30 L.H., Daily Totals		$1492.30		$2203.80	$49.74	$73.46
Crew B-22A						
1 Labor Foreman (outside)	$38.65	$309.20	$59.65	$477.20	$41.90	$64.44
1 Skilled Worker	47.30	378.40	73.25	586.00		
2 Laborers	36.65	586.40	56.55	904.80		
1 Equipment Operator, Crane	50.25	402.00	76.20	609.60		
1 S.P. Crane, 4x4, 5 Ton		280.00		308.00		
1 Butt Fusion Mach., 4"-12" diam.		381.80		419.98	16.55	18.20
40 L.H., Daily Totals		$2337.80		$3305.58	$58.45	$82.64
Crew B-22B						
1 Labor Foreman (outside)	$38.65	$309.20	$59.65	$477.20	$41.90	$64.44
1 Skilled Worker	47.30	378.40	73.25	586.00		
2 Laborers	36.65	586.40	56.55	904.80		
1 Equip. Oper. (crane)	50.25	402.00	76.20	609.60		
1 S.P. Crane, 4x4, 5 Ton		280.00		308.00		
1 Butt Fusion Mach., 8"-24" diam.		828.40		911.24	27.71	30.48
40 L.H., Daily Totals		$2784.40		$3796.84	$69.61	$94.92
Crew B-22C						
1 Skilled Worker	$47.30	$378.40	$73.25	$586.00	$41.98	$64.90
1 Laborer	36.65	293.20	56.55	452.40		
1 Butt Fusion Mach., 2"-8" diam.		120.95		133.04	7.56	8.32
16 L.H., Daily Totals		$792.55		$1171.44	$49.53	$73.22
Crew B-23						
1 Labor Foreman (outside)	$38.65	$309.20	$59.65	$477.20	$37.05	$57.17
4 Laborers	36.65	1172.80	56.55	1809.60		
1 Drill Rig, Truck-Mounted		2551.00		2806.10		
1 Flatbed Truck, Gas, 3 Ton		333.00		366.30	72.10	79.31
40 L.H., Daily Totals		$4366.00		$5459.20	$109.15	$136.48

Crew No.	Bare Costs Hr.	Bare Costs Daily	Incl. Subs O&P Hr.	Incl. Subs O&P Daily	Cost Per Labor-Hour Bare Costs	Cost Per Labor-Hour Incl. O&P
Crew B-23A						
1 Labor Foreman (outside)	$38.65	$309.20	$59.65	$477.20	$41.40	$63.45
1 Laborer	36.65	293.20	56.55	452.40		
1 Equip. Oper. (medium)	48.90	391.20	74.15	593.20		
1 Drill Rig, Truck-Mounted		2551.00		2806.10		
1 Pickup Truck, 3/4 Ton		157.00		172.70	112.83	124.12
24 L.H., Daily Totals		$3701.60		$4501.60	$154.23	$187.57
Crew B-23B						
1 Labor Foreman (outside)	$38.65	$309.20	$59.65	$477.20	$41.40	$63.45
1 Laborer	36.65	293.20	56.55	452.40		
1 Equip. Oper. (medium)	48.90	391.20	74.15	593.20		
1 Drill Rig, Truck-Mounted		2551.00		2806.10		
1 Pickup Truck, 3/4 Ton		157.00		172.70		
1 Centr. Water Pump, 6"		340.80		374.88	127.03	139.74
24 L.H., Daily Totals		$4042.40		$4876.48	$168.43	$203.19
Crew B-24						
1 Cement Finisher	$44.05	$352.40	$65.10	$520.80	$42.18	$64.13
1 Laborer	36.65	293.20	56.55	452.40		
1 Carpenter	45.85	366.80	70.75	566.00		
24 L.H., Daily Totals		$1012.40		$1539.20	$42.18	$64.13
Crew B-25						
1 Labor Foreman (outside)	$38.65	$309.20	$59.65	$477.20	$40.17	$61.63
7 Laborers	36.65	2052.40	56.55	3166.80		
3 Equip. Oper. (medium)	48.90	1173.60	74.15	1779.60		
1 Asphalt Paver, 130 H.P.		2169.00		2385.90		
1 Tandem Roller, 10 Ton		242.00		266.20		
1 Roller, Pneum. Whl., 12 Ton		349.00		383.90	31.36	34.50
88 L.H., Daily Totals		$6295.20		$8459.60	$71.54	$96.13
Crew B-25B						
1 Labor Foreman (outside)	$38.65	$309.20	$59.65	$477.20	$40.90	$62.67
7 Laborers	36.65	2052.40	56.55	3166.80		
4 Equip. Oper. (medium)	48.90	1564.80	74.15	2372.80		
1 Asphalt Paver, 130 H.P.		2169.00		2385.90		
2 Tandem Rollers, 10 Ton		484.00		532.40		
1 Roller, Pneum. Whl., 12 Ton		349.00		383.90	31.27	34.40
96 L.H., Daily Totals		$6928.40		$9319.00	$72.17	$97.07
Crew B-25C						
1 Labor Foreman (outside)	$38.65	$309.20	$59.65	$477.20	$41.07	$62.93
3 Laborers	36.65	879.60	56.55	1357.20		
2 Equip. Oper. (medium)	48.90	782.40	74.15	1186.40		
1 Asphalt Paver, 130 H.P.		2169.00		2385.90		
1 Tandem Roller, 10 Ton		242.00		266.20	50.23	55.25
48 L.H., Daily Totals		$4382.20		$5672.90	$91.30	$118.19
Crew B-25D						
1 Labor Foreman (outside)	$38.65	$309.20	$59.65	$477.20	$41.15	$63.05
3 Laborers	36.65	879.60	56.55	1357.20		
2.125 Equip. Oper. (medium)	48.90	831.30	74.15	1260.55		
.125 Truck Driver (heavy)	37.55	37.55	57.30	57.30		
.125 Truck Tractor, 6x4, 380 H.P.		76.60		84.26		
.125 Dist. Tanker, 3000 Gallon		38.52		42.38		
1 Asphalt Paver, 130 H.P.		2169.00		2385.90		
1 Tandem Roller, 10 Ton		242.00		266.20	50.52	55.57
50 L.H., Daily Totals		$4583.77		$5930.99	$91.68	$118.62

Crews

Crew No.		Bare Costs		Incl. Subs O&P		Cost Per Labor-Hour	
Crew B-25E	Hr.	Daily	Hr.	Daily	Bare Costs	Incl. O&P
1 Labor Foreman (outside)	$38.65	$309.20	$59.65	$477.20	$41.23	$63.15
3 Laborers	36.65	879.60	56.55	1357.20		
2.250 Equip. Oper. (medium)	48.90	880.20	74.15	1334.70		
.25 Truck Driver (heavy)	37.55	75.10	57.30	114.60		
.25 Truck Tractor, 6x4, 380 H.P.		153.20		168.52		
.25 Dist. Tanker, 3000 Gallon		77.05		84.75		
1 Asphalt Paver, 130 H.P.		2169.00		2385.90		
1 Tandem Roller, 10 Ton		242.00		266.20	50.79	55.87
52 L.H., Daily Totals		$4785.35		$6189.07	$92.03	$119.02
Crew B-26	Hr.	Daily	Hr.	Daily	Bare Costs	Incl. O&P
1 Labor Foreman (outside)	$38.65	$309.20	$59.65	$477.20	$41.00	$62.90
6 Laborers	36.65	1759.20	56.55	2714.40		
2 Equip. Oper. (medium)	48.90	782.40	74.15	1186.40		
1 Rodman (reinf.)	50.65	405.20	79.55	636.40		
1 Cement Finisher	44.05	352.40	65.10	520.80		
1 Grader, 30,000 Lbs.		707.00		777.70		
1 Paving Mach. & Equip.		2657.00		2922.70	38.23	42.05
88 L.H., Daily Totals		$6972.40		$9235.60	$79.23	$104.95
Crew B-26A	Hr.	Daily	Hr.	Daily	Bare Costs	Incl. O&P
1 Labor Foreman (outside)	$38.65	$309.20	$59.65	$477.20	$41.00	$62.90
6 Laborers	36.65	1759.20	56.55	2714.40		
2 Equip. Oper. (medium)	48.90	782.40	74.15	1186.40		
1 Rodman (reinf.)	50.65	405.20	79.55	636.40		
1 Cement Finisher	44.05	352.40	65.10	520.80		
1 Grader, 30,000 Lbs.		707.00		777.70		
1 Paving Mach. & Equip.		2657.00		2922.70		
1 Concrete Saw		175.20		192.72	40.22	44.24
88 L.H., Daily Totals		$7147.60		$9428.32	$81.22	$107.14
Crew B-26B	Hr.	Daily	Hr.	Daily	Bare Costs	Incl. O&P
1 Labor Foreman (outside)	$38.65	$309.20	$59.65	$477.20	$41.66	$63.84
6 Laborers	36.65	1759.20	56.55	2714.40		
3 Equip. Oper. (medium)	48.90	1173.60	74.15	1779.60		
1 Rodman (reinf.)	50.65	405.20	79.55	636.40		
1 Cement Finisher	44.05	352.40	65.10	520.80		
1 Grader, 30,000 Lbs.		707.00		777.70		
1 Paving Mach. & Equip.		2657.00		2922.70		
1 Concrete Pump, 110' Boom		934.00		1027.40	44.77	49.25
96 L.H., Daily Totals		$8297.60		$10856.20	$86.43	$113.09
Crew B-26C	Hr.	Daily	Hr.	Daily	Bare Costs	Incl. O&P
1 Labor Foreman (outside)	$38.65	$309.20	$59.65	$477.20	$40.22	$61.77
6 Laborers	36.65	1759.20	56.55	2714.40		
1 Equip. Oper. (medium)	48.90	391.20	74.15	593.20		
1 Rodman (reinf.)	50.65	405.20	79.55	636.40		
1 Cement Finisher	44.05	352.40	65.10	520.80		
1 Paving Mach. & Equip.		2657.00		2922.70		
1 Concrete Saw		175.20		192.72	35.40	38.94
80 L.H., Daily Totals		$6049.40		$8057.42	$75.62	$100.72
Crew B-27	Hr.	Daily	Hr.	Daily	Bare Costs	Incl. O&P
1 Labor Foreman (outside)	$38.65	$309.20	$59.65	$477.20	$37.15	$57.33
3 Laborers	36.65	879.60	56.55	1357.20		
1 Berm Machine		294.40		323.84	9.20	10.12
32 L.H., Daily Totals		$1483.20		$2158.24	$46.35	$67.44

Crew No.		Bare Costs		Incl. Subs O&P		Cost Per Labor-Hour	
Crew B-28	Hr.	Daily	Hr.	Daily	Bare Costs	Incl. O&P
2 Carpenters	$45.85	$733.60	$70.75	$1132.00	$42.78	$66.02
1 Laborer	36.65	293.20	56.55	452.40		
24 L.H., Daily Totals		$1026.80		$1584.40	$42.78	$66.02
Crew B-29	Hr.	Daily	Hr.	Daily	Bare Costs	Incl. O&P
1 Labor Foreman (outside)	$38.65	$309.20	$59.65	$477.20	$39.86	$61.15
4 Laborers	36.65	1172.80	56.55	1809.60		
1 Equip. Oper. (crane)	50.25	402.00	76.20	609.60		
1 Equip. Oper. (oiler)	43.55	348.40	66.00	528.00		
1 Gradall, 5/8 C.Y.		881.60		969.76	15.74	17.32
56 L.H., Daily Totals		$3114.00		$4394.16	$55.61	$78.47
Crew B-30	Hr.	Daily	Hr.	Daily	Bare Costs	Incl. O&P
1 Equip. Oper. (medium)	$48.90	$391.20	$74.15	$593.20	$41.33	$62.92
2 Truck Drivers (heavy)	37.55	600.80	57.30	916.80		
1 Hyd. Excavator, 1.5 C.Y.		1031.00		1134.10		
2 Dump Trucks, 12 C.Y., 400 H.P.		1384.40		1522.84	100.64	110.71
24 L.H., Daily Totals		$3407.40		$4166.94	$141.97	$173.62
Crew B-31	Hr.	Daily	Hr.	Daily	Bare Costs	Incl. O&P
1 Labor Foreman (outside)	$38.65	$309.20	$59.65	$477.20	$38.89	$60.01
3 Laborers	36.65	879.60	56.55	1357.20		
1 Carpenter	45.85	366.80	70.75	566.00		
1 Air Compressor, 250 cfm		202.40		222.64		
1 Sheeting Driver		5.30		5.83		
2 -50' Air Hoses, 1.5"		11.60		12.76	5.48	6.03
40 L.H., Daily Totals		$1774.90		$2641.63	$44.37	$66.04
Crew B-32	Hr.	Daily	Hr.	Daily	Bare Costs	Incl. O&P
1 Laborer	$36.65	$293.20	$56.55	$452.40	$45.84	$69.75
3 Equip. Oper. (medium)	48.90	1173.60	74.15	1779.60		
1 Grader, 30,000 Lbs.		707.00		777.70		
1 Tandem Roller, 10 Ton		242.00		266.20		
1 Dozer, 200 H.P.		1325.00		1457.50	71.06	78.17
32 L.H., Daily Totals		$3740.80		$4733.40	$116.90	$147.92
Crew B-32A	Hr.	Daily	Hr.	Daily	Bare Costs	Incl. O&P
1 Laborer	$36.65	$293.20	$56.55	$452.40	$44.82	$68.28
2 Equip. Oper. (medium)	48.90	782.40	74.15	1186.40		
1 Grader, 30,000 Lbs.		707.00		777.70		
1 Roller, Vibratory, 25 Ton		692.80		762.08	58.33	64.16
24 L.H., Daily Totals		$2475.40		$3178.58	$103.14	$132.44
Crew B-32B	Hr.	Daily	Hr.	Daily	Bare Costs	Incl. O&P
1 Laborer	$36.65	$293.20	$56.55	$452.40	$44.82	$68.28
2 Equip. Oper. (medium)	48.90	782.40	74.15	1186.40		
1 Dozer, 200 H.P.		1325.00		1457.50		
1 Roller, Vibratory, 25 Ton		692.80		762.08	84.08	92.48
24 L.H., Daily Totals		$3093.40		$3858.38	$128.89	$160.77
Crew B-32C	Hr.	Daily	Hr.	Daily	Bare Costs	Incl. O&P
1 Labor Foreman (outside)	$38.65	$309.20	$59.65	$477.20	$43.11	$65.87
2 Laborers	36.65	586.40	56.55	904.80		
3 Equip. Oper. (medium)	48.90	1173.60	74.15	1779.60		
1 Grader, 30,000 Lbs.		707.00		777.70		
1 Tandem Roller, 10 Ton		242.00		266.20		
1 Dozer, 200 H.P.		1325.00		1457.50	47.38	52.11
48 L.H., Daily Totals		$4343.20		$5663.00	$90.48	$117.98

Crews

Crew No.	Bare Costs Hr.	Bare Costs Daily	Incl. Subs O&P Hr.	Incl. Subs O&P Daily	Cost Per Labor-Hour Bare Costs	Cost Per Labor-Hour Incl. O&P
Crew B-33A						
1 Equip. Oper. (medium)	$48.90	$391.20	$74.15	$593.20	$45.40	$69.12
.5 Laborer	36.65	146.60	56.55	226.20		
.25 Equip. Oper. (medium)	48.90	97.80	74.15	148.30		
1 Scraper, Towed, 7 C.Y.		111.60		122.76		
1.25 Dozers, 300 H.P.		2271.25		2498.38	170.20	187.22
14 L.H., Daily Totals		$3018.45		$3588.84	$215.60	$256.35
Crew B-33B						
1 Equip. Oper. (medium)	$48.90	$391.20	$74.15	$593.20	$45.40	$69.12
.5 Laborer	36.65	146.60	56.55	226.20		
.25 Equip. Oper. (medium)	48.90	97.80	74.15	148.30		
1 Scraper, Towed, 10 C.Y.		141.60		155.76		
1.25 Dozers, 300 H.P.		2271.25		2498.38	172.35	189.58
14 L.H., Daily Totals		$3048.45		$3621.84	$217.75	$258.70
Crew B-33C						
1 Equip. Oper. (medium)	$48.90	$391.20	$74.15	$593.20	$45.40	$69.12
.5 Laborer	36.65	146.60	56.55	226.20		
.25 Equip. Oper. (medium)	48.90	97.80	74.15	148.30		
1 Scraper, Towed, 15 C.Y.		158.80		174.68		
1.25 Dozers, 300 H.P.		2271.25		2498.38	173.57	190.93
14 L.H., Daily Totals		$3065.65		$3640.76	$218.97	$260.05
Crew B-33D						
1 Equip. Oper. (medium)	$48.90	$391.20	$74.15	$593.20	$45.40	$69.12
.5 Laborer	36.65	146.60	56.55	226.20		
.25 Equip. Oper. (medium)	48.90	97.80	74.15	148.30		
1 S.P. Scraper, 14 C.Y.		1829.00		2011.90		
.25 Dozer, 300 H.P.		454.25		499.68	163.09	179.40
14 L.H., Daily Totals		$2918.85		$3479.28	$208.49	$248.52
Crew B-33E						
1 Equip. Oper. (medium)	$48.90	$391.20	$74.15	$593.20	$45.40	$69.12
.5 Laborer	36.65	146.60	56.55	226.20		
.25 Equip. Oper. (medium)	48.90	97.80	74.15	148.30		
1 S.P. Scraper, 21 C.Y.		2599.00		2858.90		
.25 Dozer, 300 H.P.		454.25		499.68	218.09	239.90
14 L.H., Daily Totals		$3688.85		$4326.27	$263.49	$309.02
Crew B-33F						
1 Equip. Oper. (medium)	$48.90	$391.20	$74.15	$593.20	$45.40	$69.12
.5 Laborer	36.65	146.60	56.55	226.20		
.25 Equip. Oper. (medium)	48.90	97.80	74.15	148.30		
1 Elev. Scraper, 11 C.Y.		1180.00		1298.00		
.25 Dozer, 300 H.P.		454.25		499.68	116.73	128.41
14 L.H., Daily Totals		$2269.85		$2765.38	$162.13	$197.53
Crew B-33G						
1 Equip. Oper. (medium)	$48.90	$391.20	$74.15	$593.20	$45.40	$69.12
.5 Laborer	36.65	146.60	56.55	226.20		
.25 Equip. Oper. (medium)	48.90	97.80	74.15	148.30		
1 Elev. Scraper, 22 C.Y.		2443.00		2687.30		
.25 Dozer, 300 H.P.		454.25		499.68	206.95	227.64
14 L.H., Daily Totals		$3532.85		$4154.68	$252.35	$296.76

Crew No.	Bare Costs Hr.	Bare Costs Daily	Incl. Subs O&P Hr.	Incl. Subs O&P Daily	Cost Per Labor-Hour Bare Costs	Cost Per Labor-Hour Incl. O&P
Crew B-33H						
.5 Laborer	$36.65	$146.60	$56.55	$226.20	$45.40	$69.12
1 Equipment Operator (med.)	48.90	391.20	74.15	593.20		
.25 Equipment Operator (med.)	48.90	97.80	74.15	148.30		
1 S.P. Scraper, 44 C.Y.		4492.00		4941.20		
.25 Dozer, 410 H.P.		602.25		662.48	363.88	400.26
14 L.H., Daily Totals		$5729.85		$6571.38	$409.27	$469.38
Crew B-33J						
1 Equipment Operator (med.)	$48.90	$391.20	$74.15	$593.20	$48.90	$74.15
1 S.P. Scraper, 14 C.Y.		1829.00		2011.90	228.63	251.49
8 L.H., Daily Totals		$2220.20		$2605.10	$277.52	$325.64
Crew B-33K						
1 Equipment Operator (med.)	$48.90	$391.20	$74.15	$593.20	$45.40	$69.12
.25 Equipment Operator (med.)	48.90	97.80	74.15	148.30		
.5 Laborer	36.65	146.60	56.55	226.20		
1 S.P. Scraper, 31 C.Y.		3566.00		3922.60		
.25 Dozer, 410 H.P.		602.25		662.48	297.73	327.51
14 L.H., Daily Totals		$4803.85		$5552.77	$343.13	$396.63
Crew B-34A						
1 Truck Driver (heavy)	$37.55	$300.40	$57.30	$458.40	$37.55	$57.30
1 Dump Truck, 8 C.Y., 220 H.P.		417.20		458.92	52.15	57.37
8 L.H., Daily Totals		$717.60		$917.32	$89.70	$114.67
Crew B-34B						
1 Truck Driver (heavy)	$37.55	$300.40	$57.30	$458.40	$37.55	$57.30
1 Dump Truck, 12 C.Y., 400 H.P.		692.20		761.42	86.53	95.18
8 L.H., Daily Totals		$992.60		$1219.82	$124.08	$152.48
Crew B-34C						
1 Truck Driver (heavy)	$37.55	$300.40	$57.30	$458.40	$37.55	$57.30
1 Truck Tractor, 6x4, 380 H.P.		612.80		674.08		
1 Dump Trailer, 16.5 C.Y.		126.20		138.82	92.38	101.61
8 L.H., Daily Totals		$1039.40		$1271.30	$129.93	$158.91
Crew B-34D						
1 Truck Driver (heavy)	$37.55	$300.40	$57.30	$458.40	$37.55	$57.30
1 Truck Tractor, 6x4, 380 H.P.		612.80		674.08		
1 Dump Trailer, 20 C.Y.		140.80		154.88	94.20	103.62
8 L.H., Daily Totals		$1054.00		$1287.36	$131.75	$160.92
Crew B-34E						
1 Truck Driver (heavy)	$37.55	$300.40	$57.30	$458.40	$37.55	$57.30
1 Dump Truck, Off Hwy., 25 Ton		1314.00		1445.40	164.25	180.68
8 L.H., Daily Totals		$1614.40		$1903.80	$201.80	$237.97
Crew B-34F						
1 Truck Driver (heavy)	$37.55	$300.40	$57.30	$458.40	$37.55	$57.30
1 Dump Truck, Off Hwy., 35 Ton		1470.00		1617.00	183.75	202.13
8 L.H., Daily Totals		$1770.40		$2075.40	$221.30	$259.43
Crew B-34G						
1 Truck Driver (heavy)	$37.55	$300.40	$57.30	$458.40	$37.55	$57.30
1 Dump Truck, Off Hwy., 50 Ton		1787.00		1965.70	223.38	245.71
8 L.H., Daily Totals		$2087.40		$2424.10	$260.93	$303.01

Crews

Crew No.		Bare Costs		Incl. Subs O&P		Cost Per Labor-Hour	
Crew B-34H	Hr.	Daily	Hr.	Daily	Bare Costs	Incl. O&P	
1 Truck Driver (heavy)	$37.55	$300.40	$57.30	$458.40	$37.55	$57.30
1 Dump Truck, Off Hwy., 65 Ton		1797.00		1976.70	224.63	247.09
8 L.H., Daily Totals		$2097.40		$2435.10	$262.18	$304.39

Crew B-34I	Hr.	Daily	Hr.	Daily	Bare Costs	Incl. O&P
1 Truck Driver (heavy)	$37.55	$300.40	$57.30	$458.40	$37.55	$57.30
1 Dump Truck, 18 C.Y., 450 H.P.		867.60		954.36	108.45	119.30
8 L.H., Daily Totals		$1168.00		$1412.76	$146.00	$176.60

Crew B-34J	Hr.	Daily	Hr.	Daily	Bare Costs	Incl. O&P
1 Truck Driver (heavy)	$37.55	$300.40	$57.30	$458.40	$37.55	$57.30
1 Dump Truck, Off Hwy., 100 Ton		2898.00		3187.80	362.25	398.48
8 L.H., Daily Totals		$3198.40		$3646.20	$399.80	$455.77

Crew B-34K	Hr.	Daily	Hr.	Daily	Bare Costs	Incl. O&P
1 Truck Driver (heavy)	$37.55	$300.40	$57.30	$458.40	$37.55	$57.30
1 Truck Tractor, 6x4, 450 H.P.		742.40		816.64		
1 Lowbed Trailer, 75 Ton		219.80		241.78	120.28	132.30
8 L.H., Daily Totals		$1262.60		$1516.82	$157.82	$189.60

Crew B-34L	Hr.	Daily	Hr.	Daily	Bare Costs	Incl. O&P
1 Equip. Oper. (light)	$47.05	$376.40	$71.35	$570.80	$47.05	$71.35
1 Flatbed Truck, Gas, 1.5 Ton		270.20		297.22	33.77	37.15
8 L.H., Daily Totals		$646.60		$868.02	$80.83	$108.50

Crew B-34M	Hr.	Daily	Hr.	Daily	Bare Costs	Incl. O&P
1 Equip. Oper. (light)	$47.05	$376.40	$71.35	$570.80	$47.05	$71.35
1 Flatbed Truck, Gas, 3 Ton		333.00		366.30	41.63	45.79
8 L.H., Daily Totals		$709.40		$937.10	$88.67	$117.14

Crew B-34N	Hr.	Daily	Hr.	Daily	Bare Costs	Incl. O&P
1 Truck Driver (heavy)	$37.55	$300.40	$57.30	$458.40	$37.55	$57.30
1 Dump Truck, 8 C.Y., 220 H.P.		417.20		458.92		
1 Flatbed Trailer, 40 Ton		152.00		167.20	71.15	78.27
8 L.H., Daily Totals		$869.60		$1084.52	$108.70	$135.57

Crew B-34P	Hr.	Daily	Hr.	Daily	Bare Costs	Incl. O&P
1 Pipe Fitter	$58.50	$468.00	$88.35	$706.80	$47.97	$72.73
1 Truck Driver (light)	36.50	292.00	55.70	445.60		
1 Equip. Oper. (medium)	48.90	391.20	74.15	593.20		
1 Flatbed Truck, Gas, 3 Ton		333.00		366.30		
1 Backhoe Loader, 48 H.P.		365.20		401.72	29.09	32.00
24 L.H., Daily Totals		$1849.40		$2513.62	$77.06	$104.73

Crew B-34Q	Hr.	Daily	Hr.	Daily	Bare Costs	Incl. O&P
1 Pipe Fitter	$58.50	$468.00	$88.35	$706.80	$48.42	$73.42
1 Truck Driver (light)	36.50	292.00	55.70	445.60		
1 Equip. Oper. (crane)	50.25	402.00	76.20	609.60		
1 Flatbed Trailer, 25 Ton		113.60		124.96		
1 Dump Truck, 8 C.Y., 220 H.P.		417.20		458.92		
1 Hyd. Crane, 25 Ton		745.80		820.38	53.19	58.51
24 L.H., Daily Totals		$2438.60		$3166.26	$101.61	$131.93

Crew B-34R	Hr.	Daily	Hr.	Daily	Bare Costs	Incl. O&P
1 Pipe Fitter	$58.50	$468.00	$88.35	$706.80	$48.42	$73.42
1 Truck Driver (light)	36.50	292.00	55.70	445.60		
1 Equip. Oper. (crane)	50.25	402.00	76.20	609.60		
1 Flatbed Trailer, 25 Ton		113.60		124.96		
1 Dump Truck, 8 C.Y., 220 H.P.		417.20		458.92		
1 Hyd. Crane, 25 Ton		745.80		820.38		
1 Hyd. Excavator, 1 C.Y.		814.80		896.28	87.14	95.86
24 L.H., Daily Totals		$3253.40		$4062.54	$135.56	$169.27

Crew B-34S	Hr.	Daily	Hr.	Daily	Bare Costs	Incl. O&P
2 Pipe Fitters	$58.50	$936.00	$88.35	$1413.60	$51.20	$77.55
1 Truck Driver (heavy)	37.55	300.40	57.30	458.40		
1 Equip. Oper. (crane)	50.25	402.00	76.20	609.60		
1 Flatbed Trailer, 40 Ton		152.00		167.20		
1 Truck Tractor, 6x4, 380 H.P.		612.80		674.08		
1 Hyd. Crane, 80 Ton		1652.00		1817.20		
1 Hyd. Excavator, 2 C.Y.		1175.00		1292.50	112.24	123.47
32 L.H., Daily Totals		$5230.20		$6432.58	$163.44	$201.02

Crew B-34T	Hr.	Daily	Hr.	Daily	Bare Costs	Incl. O&P
2 Pipe Fitters	$58.50	$936.00	$88.35	$1413.60	$51.20	$77.55
1 Truck Driver (heavy)	37.55	300.40	57.30	458.40		
1 Equip. Oper. (crane)	50.25	402.00	76.20	609.60		
1 Flatbed Trailer, 40 Ton		152.00		167.20		
1 Truck Tractor, 6x4, 380 H.P.		612.80		674.08		
1 Hyd. Crane, 80 Ton		1652.00		1817.20	75.53	83.08
32 L.H., Daily Totals		$4055.20		$5140.08	$126.72	$160.63

Crew B-35	Hr.	Daily	Hr.	Daily	Bare Costs	Incl. O&P
1 Labor Foreman (outside)	$38.65	$309.20	$59.65	$477.20	$45.66	$69.76
1 Skilled Worker	47.30	378.40	73.25	586.00		
1 Welder (plumber)	57.55	460.40	86.90	695.20		
1 Laborer	36.65	293.20	56.55	452.40		
1 Equip. Oper. (crane)	50.25	402.00	76.20	609.60		
1 Equip. Oper. (oiler)	43.55	348.40	66.00	528.00		
1 Welder, Electric, 300 amp		55.30		60.83		
1 Hyd. Excavator, .75 C.Y.		662.00		728.20	14.94	16.44
48 L.H., Daily Totals		$2908.90		$4137.43	$60.60	$86.20

Crew B-35A	Hr.	Daily	Hr.	Daily	Bare Costs	Incl. O&P
1 Labor Foreman (outside)	$38.65	$309.20	$59.65	$477.20	$44.37	$67.87
2 Laborers	36.65	586.40	56.55	904.80		
1 Skilled Worker	47.30	378.40	73.25	586.00		
1 Welder (plumber)	57.55	460.40	86.90	695.20		
1 Equip. Oper. (crane)	50.25	402.00	76.20	609.60		
1 Equip. Oper. (oiler)	43.55	348.40	66.00	528.00		
1 Welder, Gas Engine, 300 amp		142.00		156.20		
1 Crawler Crane, 75 Ton		1492.00		1641.20	29.18	32.10
56 L.H., Daily Totals		$4118.80		$5598.20	$73.55	$99.97

Crew B-36	Hr.	Daily	Hr.	Daily	Bare Costs	Incl. O&P
1 Labor Foreman (outside)	$38.65	$309.20	$59.65	$477.20	$41.95	$64.21
2 Laborers	36.65	586.40	56.55	904.80		
2 Equip. Oper. (medium)	48.90	782.40	74.15	1186.40		
1 Dozer, 200 H.P.		1325.00		1457.50		
1 Aggregate Spreader		41.20		45.32		
1 Tandem Roller, 10 Ton		242.00		266.20	40.20	44.23
40 L.H., Daily Totals		$3286.20		$4337.42	$82.16	$108.44

Crews

Crew No.		Bare Costs		Incl. Subs O&P		Cost Per Labor-Hour	
Crew B-36A	Hr.	Daily	Hr.	Daily	Bare Costs	Incl. O&P
1 Labor Foreman (outside)	$38.65	$309.20	$59.65	$477.20	$43.94	$67.05
2 Laborers	36.65	586.40	56.55	904.80		
4 Equip. Oper. (medium)	48.90	1564.80	74.15	2372.80		
1 Dozer, 200 H.P.		1325.00		1457.50		
1 Aggregate Spreader		41.20		45.32		
1 Tandem Roller, 10 Ton		242.00		266.20		
1 Roller, Pneum. Whl., 12 Ton		349.00		383.90	34.95	38.45
56 L.H., Daily Totals		$4417.60		$5907.72	$78.89	$105.50

Crew B-36B	Hr.	Daily	Hr.	Daily	Bare Costs	Incl. O&P
1 Labor Foreman (outside)	$38.65	$309.20	$59.65	$477.20	$43.14	$65.83
2 Laborers	36.65	586.40	56.55	904.80		
4 Equip. Oper. (medium)	48.90	1564.80	74.15	2372.80		
1 Truck Driver (heavy)	37.55	300.40	57.30	458.40		
1 Grader, 30,000 Lbs.		707.00		777.70		
1 F.E. Loader, Crl, 1.5 C.Y.		625.20		687.72		
1 Dozer, 300 H.P.		1817.00		1998.70		
1 Roller, Vibratory, 25 Ton		692.80		762.08		
1 Truck Tractor, 6x4, 450 H.P.		742.40		816.64		
1 Water Tank Trailer, 5000 Gal.		141.60		155.76	73.84	81.23
64 L.H., Daily Totals		$7486.80		$9411.80	$116.98	$147.06

Crew B-36C	Hr.	Daily	Hr.	Daily	Bare Costs	Incl. O&P
1 Labor Foreman (outside)	$38.65	$309.20	$59.65	$477.20	$44.58	$67.88
3 Equip. Oper. (medium)	48.90	1173.60	74.15	1779.60		
1 Truck Driver (heavy)	37.55	300.40	57.30	458.40		
1 Grader, 30,000 Lbs.		707.00		777.70		
1 Dozer, 300 H.P.		1817.00		1998.70		
1 Roller, Vibratory, 25 Ton		692.80		762.08		
1 Truck Tractor, 6x4, 450 H.P.		742.40		816.64		
1 Water Tank Trailer, 5000 Gal.		141.60		155.76	102.52	112.77
40 L.H., Daily Totals		$5884.00		$7226.08	$147.10	$180.65

Crew B-36D	Hr.	Daily	Hr.	Daily	Bare Costs	Incl. O&P
1 Labor Foreman (outside)	$38.65	$309.20	$59.65	$477.20	$46.34	$70.53
3 Equip. Oper. (medium)	48.90	1173.60	74.15	1779.60		
1 Grader, 30,000 Lbs.		707.00		777.70		
1 Dozer, 300 H.P.		1817.00		1998.70		
1 Roller, Vibratory, 25 Ton		692.80		762.08	100.53	110.58
32 L.H., Daily Totals		$4699.60		$5795.28	$146.86	$181.10

Crew B-37	Hr.	Daily	Hr.	Daily	Bare Costs	Incl. O&P
1 Labor Foreman (outside)	$38.65	$309.20	$59.65	$477.20	$38.72	$59.53
4 Laborers	36.65	1172.80	56.55	1809.60		
1 Equip. Oper. (light)	47.05	376.40	71.35	570.80		
1 Tandem Roller, 5 Ton		159.00		174.90	3.31	3.64
48 L.H., Daily Totals		$2017.40		$3032.50	$42.03	$63.18

Crew B-37A	Hr.	Daily	Hr.	Daily	Bare Costs	Incl. O&P
2 Laborers	$36.65	$586.40	$56.55	$904.80	$36.60	$56.27
1 Truck Driver (light)	36.50	292.00	55.70	445.60		
1 Flatbed Truck, Gas, 1.5 Ton		270.20		297.22		
1 Tar Kettle, T.M.		104.60		115.06	15.62	17.18
24 L.H., Daily Totals		$1253.20		$1762.68	$52.22	$73.44

Crew B-37B	Hr.	Daily	Hr.	Daily	Bare Costs	Incl. O&P
3 Laborers	$36.65	$879.60	$56.55	$1357.20	$36.61	$56.34
1 Truck Driver (light)	36.50	292.00	55.70	445.60		
1 Flatbed Truck, Gas, 1.5 Ton		270.20		297.22		
1 Tar Kettle, T.M.		104.60		115.06	11.71	12.88
32 L.H., Daily Totals		$1546.40		$2215.08	$48.33	$69.22

Crew B-37C	Hr.	Daily	Hr.	Daily	Bare Costs	Incl. O&P
2 Laborers	$36.65	$586.40	$56.55	$904.80	$36.58	$56.13
2 Truck Drivers (light)	36.50	584.00	55.70	891.20		
2 Flatbed Trucks, Gas, 1.5 Ton		540.40		594.44		
1 Tar Kettle, T.M.		104.60		115.06	20.16	22.17
32 L.H., Daily Totals		$1815.40		$2505.50	$56.73	$78.30

Crew B-37D	Hr.	Daily	Hr.	Daily	Bare Costs	Incl. O&P
1 Laborer	$36.65	$293.20	$56.55	$452.40	$36.58	$56.13
1 Truck Driver (light)	36.50	292.00	55.70	445.60		
1 Pickup Truck, 3/4 Ton		157.00		172.70	9.81	10.79
16 L.H., Daily Totals		$742.20		$1070.70	$46.39	$66.92

Crew B-37E	Hr.	Daily	Hr.	Daily	Bare Costs	Incl. O&P
3 Laborers	$36.65	$879.60	$56.55	$1357.20	$39.84	$60.94
1 Equip. Oper. (light)	47.05	376.40	71.35	570.80		
1 Equip. Oper. (medium)	48.90	391.20	74.15	593.20		
2 Truck Drivers (light)	36.50	584.00	55.70	891.20		
4 Barrels w/ Flasher		13.60		14.96		
1 Concrete Saw		175.20		192.72		
1 Rotary Hammer Drill		24.00		26.40		
1 Hammer Drill Bit		2.35		2.59		
1 Loader, Skid Steer, 30 H.P.		174.20		191.62		
1 Conc. Hammer Attach.		114.70		126.17		
1 Vibrating Plate, Gas, 18"		35.80		39.38		
2 Flatbed Trucks, Gas, 1.5 Ton		540.40		594.44	19.29	21.22
56 L.H., Daily Totals		$3311.45		$4600.68	$59.13	$82.15

Crew B-37F	Hr.	Daily	Hr.	Daily	Bare Costs	Incl. O&P
3 Laborers	$36.65	$879.60	$56.55	$1357.20	$36.61	$56.34
1 Truck Driver (light)	36.50	292.00	55.70	445.60		
4 Barrels w/ Flasher		13.60		14.96		
1 Concrete Mixer, 10 C.F.		176.80		194.48		
1 Air Compressor, 60 cfm		137.40		151.14		
1 -50' Air Hose, 3/4"		3.25		3.58		
1 Spade (Chipper)		7.80		8.58		
1 Flatbed Truck, Gas, 1.5 Ton		270.20		297.22	19.03	20.94
32 L.H., Daily Totals		$1780.65		$2472.76	$55.65	$77.27

Crew B-37G	Hr.	Daily	Hr.	Daily	Bare Costs	Incl. O&P
1 Labor Foreman (outside)	$38.65	$309.20	$59.65	$477.20	$38.72	$59.53
4 Laborers	36.65	1172.80	56.55	1809.60		
1 Equip. Oper. (light)	47.05	376.40	71.35	570.80		
1 Berm Machine		294.40		323.84		
1 Tandem Roller, 5 Ton		159.00		174.90	9.45	10.39
48 L.H., Daily Totals		$2311.80		$3356.34	$48.16	$69.92

Crew B-37H	Hr.	Daily	Hr.	Daily	Bare Costs	Incl. O&P
1 Labor Foreman (outside)	$38.65	$309.20	$59.65	$477.20	$38.72	$59.53
4 Laborers	36.65	1172.80	56.55	1809.60		
1 Equip. Oper. (light)	47.05	376.40	71.35	570.80		
1 Tandem Roller, 5 Ton		159.00		174.90		
1 Flatbed Trucks, Gas, 1.5 Ton		270.20		297.22		
1 Tar Kettle, T.M.		104.60		115.06	11.12	12.23
48 L.H., Daily Totals		$2392.20		$3444.78	$49.84	$71.77

Crews

Crew No.		Bare Costs		Incl. Subs O&P		Cost Per Labor-Hour	
Crew B-37I	Hr.	Daily	Hr.	Daily	Bare Costs	Incl. O&P
3 Laborers	$36.65	$879.60	$56.55	$1357.20	$39.84	$60.94
1 Equip. Oper. (light)	47.05	376.40	71.35	570.80		
1 Equip. Oper. (medium)	48.90	391.20	74.15	593.20		
2 Truck Drivers (light)	36.50	584.00	55.70	891.20		
4 Barrels w/ Flasher		13.60		14.96		
1 Concrete Saw		175.20		192.72		
1 Rotary Hammer Drill		24.00		26.40		
1 Hammer Drill Bit		2.35		2.59		
1 Air Compressor, 60 cfm		137.40		151.14		
1 -50' Air Hose, 3/4"		3.25		3.58		
1 Spade (Chipper)		7.80		8.58		
1 Loader, Skid Steer, 30 H.P.		174.20		191.62		
1 Conc. Hammer Attach.		114.70		126.17		
1 Concrete Mixer, 10 C.F.		176.80		194.48		
1 Vibrating Plate, Gas, 18"		35.80		39.38		
2 Flatbed Trucks, Gas, 1.5 Ton		540.40		594.44	25.10	27.61
56 L.H., Daily Totals		$3636.70		$4958.45	$64.94	$88.54

Crew No.		Bare Costs		Incl. Subs O&P		Cost Per Labor-Hour	
Crew B-37J	Hr.	Daily	Hr.	Daily	Bare Costs	Incl. O&P
1 Labor Foreman (outside)	$38.65	$309.20	$59.65	$477.20	$38.72	$59.53
4 Laborers	36.65	1172.80	56.55	1809.60		
1 Equip. Oper. (light)	47.05	376.40	71.35	570.80		
1 Air Compressor, 60 cfm		137.40		151.14		
1 -50' Air Hose, 3/4"		3.25		3.58		
2 Concrete Mixer, 10 C.F.		353.60		388.96		
2 Flatbed Trucks, Gas, 1.5 Ton		540.40		594.44		
1 Shot Blaster, 20"		214.40		235.84	26.02	28.62
48 L.H., Daily Totals		$3107.45		$4231.56	$64.74	$88.16

Crew B-37K	Hr.	Daily	Hr.	Daily	Bare Costs	Incl. O&P
1 Labor Foreman (outside)	$38.65	$309.20	$59.65	$477.20	$38.72	$59.53
4 Laborers	36.65	1172.80	56.55	1809.60		
1 Equip. Oper. (light)	47.05	376.40	71.35	570.80		
1 Air Compressor, 60 cfm		137.40		151.14		
1 -50' Air Hose, 3/4"		3.25		3.58		
2 Flatbed Trucks, Gas, 1.5 Ton		540.40		594.44		
1 Shot Blaster, 20"		214.40		235.84	18.66	20.52
48 L.H., Daily Totals		$2753.85		$3842.59	$57.37	$80.05

Crew B-38	Hr.	Daily	Hr.	Daily	Bare Costs	Incl. O&P
1 Labor Foreman (outside)	$38.65	$309.20	$59.65	$477.20	$41.58	$63.65
2 Laborers	36.65	586.40	56.55	904.80		
1 Equip. Oper. (light)	47.05	376.40	71.35	570.80		
1 Equip. Oper. (medium)	48.90	391.20	74.15	593.20		
1 Backhoe Loader, 48 H.P.		365.20		401.72		
1 Hyd. Hammer, (1200 lb.)		180.40		198.44		
1 F.E. Loader, W.M., 4 C.Y.		673.80		741.18		
1 Pvmt. Rem. Bucket		60.80		66.88	32.01	35.21
40 L.H., Daily Totals		$2943.40		$3954.22	$73.58	$98.86

Crew B-39	Hr.	Daily	Hr.	Daily	Bare Costs	Incl. O&P
1 Labor Foreman (outside)	$38.65	$309.20	$59.65	$477.20	$38.72	$59.53
4 Laborers	36.65	1172.80	56.55	1809.60		
1 Equip. Oper. (light)	47.05	376.40	71.35	570.80		
1 Air Compressor, 250 cfm		202.40		222.64		
2 Breakers, Pavement, 60 lb.		19.60		21.56		
2 -50' Air Hoses, 1.5"		11.60		12.76	4.87	5.35
48 L.H., Daily Totals		$2092.00		$3114.56	$43.58	$64.89

Crew B-40	Hr.	Daily	Hr.	Daily	Bare Costs	Incl. O&P
1 Pile Driver Foreman (outside)	$46.40	$371.20	$73.60	$588.80	$46.01	$71.70
4 Pile Drivers	44.40	1420.80	70.40	2252.80		
2 Equip. Oper. (crane)	50.25	804.00	76.20	1219.20		
1 Equip. Oper. (oiler)	43.55	348.40	66.00	528.00		
1 Crawler Crane, 40 Ton		1179.00		1296.90		
1 Vibratory Hammer & Gen.		2640.00		2904.00	59.67	65.64
64 L.H., Daily Totals		$6763.40		$8789.70	$105.68	$137.34

Crew B-40B	Hr.	Daily	Hr.	Daily	Bare Costs	Incl. O&P
1 Labor Foreman (outside)	$38.65	$309.20	$59.65	$477.20	$40.40	$61.92
3 Laborers	36.65	879.60	56.55	1357.20		
1 Equip. Oper. (crane)	50.25	402.00	76.20	609.60		
1 Equip. Oper. (oiler)	43.55	348.40	66.00	528.00		
1 Lattice Boom Crane, 40 Ton		1177.00		1294.70	24.52	26.97
48 L.H., Daily Totals		$3116.20		$4266.70	$64.92	$88.89

Crew B-41	Hr.	Daily	Hr.	Daily	Bare Costs	Incl. O&P
1 Labor Foreman (outside)	$38.65	$309.20	$59.65	$477.20	$37.95	$58.44
4 Laborers	36.65	1172.80	56.55	1809.60		
.25 Equip. Oper. (crane)	50.25	100.50	76.20	152.40		
.25 Equip. Oper. (oiler)	43.55	87.10	66.00	132.00		
.25 Crawler Crane, 40 Ton		294.75		324.23	6.70	7.37
44 L.H., Daily Totals		$1964.35		$2895.43	$44.64	$65.81

Crew B-42	Hr.	Daily	Hr.	Daily	Bare Costs	Incl. O&P
1 Labor Foreman (outside)	$38.65	$309.20	$59.65	$477.20	$41.27	$64.78
4 Laborers	36.65	1172.80	56.55	1809.60		
1 Equip. Oper. (crane)	50.25	402.00	76.20	609.60		
1 Equip. Oper. (oiler)	43.55	348.40	66.00	528.00		
1 Welder	51.10	408.80	90.20	721.60		
1 Hyd. Crane, 25 Ton		745.80		820.38		
1 Welder, Gas Engine, 300 amp		142.00		156.20		
1 Horz. Boring Csg. Mch.		479.80		527.78	21.37	23.51
64 L.H., Daily Totals		$4008.80		$5650.36	$62.64	$88.29

Crew B-43	Hr.	Daily	Hr.	Daily	Bare Costs	Incl. O&P
1 Labor Foreman (outside)	$38.65	$309.20	$59.65	$477.20	$40.40	$61.92
3 Laborers	36.65	879.60	56.55	1357.20		
1 Equip. Oper. (crane)	50.25	402.00	76.20	609.60		
1 Equip. Oper. (oiler)	43.55	348.40	66.00	528.00		
1 Drill Rig, Truck-Mounted		2551.00		2806.10	53.15	58.46
48 L.H., Daily Totals		$4490.20		$5778.10	$93.55	$120.38

Crew B-44	Hr.	Daily	Hr.	Daily	Bare Costs	Incl. O&P
1 Pile Driver Foreman (outside)	$46.40	$371.20	$73.60	$588.80	$45.14	$70.52
4 Pile Drivers	44.40	1420.80	70.40	2252.80		
2 Equip. Oper. (crane)	50.25	804.00	76.20	1219.20		
1 Laborer	36.65	293.20	56.55	452.40		
1 Crawler Crane, 40 Ton		1179.00		1296.90		
1 Lead, 60' High		74.80		82.28		
1 Hammer, Diesel, 15K ft.-lbs.		587.60		646.36	28.77	31.65
64 L.H., Daily Totals		$4730.60		$6538.74	$73.92	$102.17

Crew B-45	Hr.	Daily	Hr.	Daily	Bare Costs	Incl. O&P
1 Equip. Oper. (medium)	$48.90	$391.20	$74.15	$593.20	$43.23	$65.72
1 Truck Driver (heavy)	37.55	300.40	57.30	458.40		
1 Dist. Tanker, 3000 Gallon		308.20		339.02		
1 Truck Tractor, 6x4, 380 H.P.		612.80		674.08	57.56	63.32
16 L.H., Daily Totals		$1612.60		$2064.70	$100.79	$129.04

Crews

Crew No.		Bare Costs		Incl. Subs O&P		Cost Per Labor-Hour	
Crew B-46	Hr.	Daily	Hr.	Daily	Bare Costs	Incl. O&P
1 Pile Driver Foreman (outside)	$46.40	$371.20	$73.60	$588.80	$40.86	$64.01
2 Pile Drivers	44.40	710.40	70.40	1126.40		
3 Laborers	36.65	879.60	56.55	1357.20		
1 Chain Saw, Gas, 36" Long		45.00		49.50	.94	1.03
48 L.H., Daily Totals		$2006.20		$3121.90	$41.80	$65.04
Crew B-47	Hr.	Daily	Hr.	Daily	Bare Costs	Incl. O&P
1 Blast Foreman (outside)	$38.65	$309.20	$59.65	$477.20	$40.78	$62.52
1 Driller	36.65	293.20	56.55	452.40		
1 Equip. Oper. (light)	47.05	376.40	71.35	570.80		
1 Air Track Drill, 4"		1040.00		1144.00		
1 Air Compressor, 600 cfm		552.60		607.86		
2 -50' Air Hoses, 3"		29.80		32.78	67.60	74.36
24 L.H., Daily Totals		$2601.20		$3285.04	$108.38	$136.88
Crew B-47A	Hr.	Daily	Hr.	Daily	Bare Costs	Incl. O&P
1 Drilling Foreman (outside)	$38.65	$309.20	$59.65	$477.20	$44.15	$67.28
1 Equip. Oper. (heavy)	50.25	402.00	76.20	609.60		
1 Equip. Oper. (oiler)	43.55	348.40	66.00	528.00		
1 Air Track Drill, 5"		1255.00		1380.50	52.29	57.52
24 L.H., Daily Totals		$2314.60		$2995.30	$96.44	$124.80
Crew B-47C	Hr.	Daily	Hr.	Daily	Bare Costs	Incl. O&P
1 Laborer	$36.65	$293.20	$56.55	$452.40	$41.85	$63.95
1 Equip. Oper. (light)	47.05	376.40	71.35	570.80		
1 Air Compressor, 750 cfm		559.20		615.12		
2 -50' Air Hoses, 3"		29.80		32.78		
1 Air Track Drill, 4"		1040.00		1144.00	101.81	111.99
16 L.H., Daily Totals		$2298.60		$2815.10	$143.66	$175.94
Crew B-47E	Hr.	Daily	Hr.	Daily	Bare Costs	Incl. O&P
1 Labor Foreman (outside)	$38.65	$309.20	$59.65	$477.20	$37.15	$57.33
3 Laborers	36.65	879.60	56.55	1357.20		
1 Flatbed Truck, Gas, 3 Ton		333.00		366.30	10.41	11.45
32 L.H., Daily Totals		$1521.80		$2200.70	$47.56	$68.77
Crew B-47G	Hr.	Daily	Hr.	Daily	Bare Costs	Incl. O&P
1 Labor Foreman (outside)	$38.65	$309.20	$59.65	$477.20	$39.75	$61.02
2 Laborers	36.65	586.40	56.55	904.80		
1 Equip. Oper. (light)	47.05	376.40	71.35	570.80		
1 Air Track Drill, 4"		1040.00		1144.00		
1 Air Compressor, 600 cfm		552.60		607.86		
2 -50' Air Hoses, 3"		29.80		32.78		
1 Gunite Pump Rig		370.40		407.44	62.27	68.50
32 L.H., Daily Totals		$3264.80		$4144.88	$102.03	$129.53
Crew B-47H	Hr.	Daily	Hr.	Daily	Bare Costs	Incl. O&P
1 Skilled Worker Foreman (out)	$49.30	$394.40	$76.35	$610.80	$47.80	$74.03
3 Skilled Workers	47.30	1135.20	73.25	1758.00		
1 Flatbed Truck, Gas, 3 Ton		333.00		366.30	10.41	11.45
32 L.H., Daily Totals		$1862.60		$2735.10	$58.21	$85.47

Crew No.		Bare Costs		Incl. Subs O&P		Cost Per Labor-Hour	
Crew B-48	Hr.	Daily	Hr.	Daily	Bare Costs	Incl. O&P
1 Labor Foreman (outside)	$38.65	$309.20	$59.65	$477.20	$41.35	$63.26
3 Laborers	36.65	879.60	56.55	1357.20		
1 Equip. Oper. (crane)	50.25	402.00	76.20	609.60		
1 Equip. Oper. (oiler)	43.55	348.40	66.00	528.00		
1 Equip. Oper. (light)	47.05	376.40	71.35	570.80		
1 Centr. Water Pump, 6"		340.80		374.88		
1 -20' Suction Hose, 6"		11.50		12.65		
1 -50' Discharge Hose, 6"		6.10		6.71		
1 Drill Rig, Truck-Mounted		2551.00		2806.10	51.95	57.15
56 L.H., Daily Totals		$5225.00		$6743.14	$93.30	$120.41
Crew B-49	Hr.	Daily	Hr.	Daily	Bare Costs	Incl. O&P
1 Labor Foreman (outside)	$38.65	$309.20	$59.65	$477.20	$42.91	$65.99
3 Laborers	36.65	879.60	56.55	1357.20		
2 Equip. Oper. (crane)	50.25	804.00	76.20	1219.20		
2 Equip. Oper. (oilers)	43.55	696.80	66.00	1056.00		
1 Equip. Oper. (light)	47.05	376.40	71.35	570.80		
2 Pile Drivers	44.40	710.40	70.40	1126.40		
1 Hyd. Crane, 25 Ton		745.80		820.38		
1 Centr. Water Pump, 6"		340.80		374.88		
1 -20' Suction Hose, 6"		11.50		12.65		
1 -50' Discharge Hose, 6"		6.10		6.71		
1 Drill Rig, Truck-Mounted		2551.00		2806.10	41.54	45.69
88 L.H., Daily Totals		$7431.60		$9827.52	$84.45	$111.68
Crew B-50	Hr.	Daily	Hr.	Daily	Bare Costs	Incl. O&P
2 Pile Driver Foremen (outside)	$46.40	$742.40	$73.60	$1177.60	$43.80	$68.40
6 Pile Drivers	44.40	2131.20	70.40	3379.20		
2 Equip. Oper. (crane)	50.25	804.00	76.20	1219.20		
1 Equip. Oper. (oiler)	43.55	348.40	66.00	528.00		
3 Laborers	36.65	879.60	56.55	1357.20		
1 Crawler Crane, 40 Ton		1179.00		1296.90		
1 Lead, 60' High		74.80		82.28		
1 Hammer, Diesel, 15K ft.-lbs.		587.60		646.36		
1 Air Compressor, 600 cfm		552.60		607.86		
2 -50' Air Hoses, 3"		29.80		32.78		
1 Chain Saw, Gas, 36" Long		45.00		49.50	22.04	24.25
112 L.H., Daily Totals		$7374.40		$10376.88	$65.84	$92.65
Crew B-51	Hr.	Daily	Hr.	Daily	Bare Costs	Incl. O&P
1 Labor Foreman (outside)	$38.65	$309.20	$59.65	$477.20	$36.96	$56.92
4 Laborers	36.65	1172.80	56.55	1809.60		
1 Truck Driver (light)	36.50	292.00	55.70	445.60		
1 Flatbed Truck, Gas, 1.5 Ton		270.20		297.22	5.63	6.19
48 L.H., Daily Totals		$2044.20		$3029.62	$42.59	$63.12
Crew B-52	Hr.	Daily	Hr.	Daily	Bare Costs	Incl. O&P
1 Carpenter Foreman (outside)	$47.85	$382.80	$73.85	$590.80	$42.50	$65.17
1 Carpenter	45.85	366.80	70.75	566.00		
3 Laborers	36.65	879.60	56.55	1357.20		
1 Cement Finisher	44.05	352.40	65.10	520.80		
.5 Rodman (reinf.)	50.65	202.60	79.55	318.20		
.5 Equip. Oper. (medium)	48.90	195.60	74.15	296.60		
.5 Crawler Loader, 3 C.Y.		595.50		655.05	10.63	11.70
56 L.H., Daily Totals		$2975.30		$4304.65	$53.13	$76.87
Crew B-53	Hr.	Daily	Hr.	Daily	Bare Costs	Incl. O&P
1 Equip. Oper. (light)	$47.05	$376.40	$71.35	$570.80	$47.05	$71.35
1 Trencher, Chain, 12 H.P.		71.20		78.32	8.90	9.79
8 L.H., Daily Totals		$447.60		$649.12	$55.95	$81.14

573

Crews

Crew No.	Bare Costs Hr.	Bare Costs Daily	Incl. Subs O&P Hr.	Incl. Subs O&P Daily	Cost Per Labor-Hour Bare Costs	Cost Per Labor-Hour Incl. O&P
Crew B-54	Hr.	Daily	Hr.	Daily	Bare Costs	Incl. O&P
1 Equip. Oper. (light)	$47.05	$376.40	$71.35	$570.80	$47.05	$71.35
1 Trencher, Chain, 40 H.P.		339.40		373.34	42.42	46.67
8 L.H., Daily Totals		$715.80		$944.14	$89.47	$118.02
Crew B-54A	Hr.	Daily	Hr.	Daily	Bare Costs	Incl. O&P
.17 Labor Foreman (outside)	$38.65	$52.56	$59.65	$81.12	$47.41	$72.04
1 Equipment Operator (med.)	48.90	391.20	74.15	593.20		
1 Wheel Trencher, 67 H.P.		1206.00		1326.60	128.85	141.73
9.36 L.H., Daily Totals		$1649.76		$2000.92	$176.26	$213.77
Crew B-54B	Hr.	Daily	Hr.	Daily	Bare Costs	Incl. O&P
.25 Labor Foreman (outside)	$38.65	$77.30	$59.65	$119.30	$46.85	$71.25
1 Equipment Operator (med.)	48.90	391.20	74.15	593.20		
1 Wheel Trencher, 150 H.P.		1898.00		2087.80	189.80	208.78
10 L.H., Daily Totals		$2366.50		$2800.30	$236.65	$280.03
Crew B-54C	Hr.	Daily	Hr.	Daily	Bare Costs	Incl. O&P
1 Laborer	$36.65	$293.20	$56.55	$452.40	$42.77	$65.35
1 Equipment Operator (med.)	48.90	391.20	74.15	593.20		
1 Wheel Trencher, 67 H.P.		1206.00		1326.60	75.38	82.91
16 L.H., Daily Totals		$1890.40		$2372.20	$118.15	$148.26
Crew B-54D	Hr.	Daily	Hr.	Daily	Bare Costs	Incl. O&P
1 Laborer	$36.65	$293.20	$56.55	$452.40	$42.77	$65.35
1 Equipment Operator (med.)	48.90	391.20	74.15	593.20		
1 Rock Trencher, 6" Width		369.00		405.90	23.06	25.37
16 L.H., Daily Totals		$1053.40		$1451.50	$65.84	$90.72
Crew B-54E	Hr.	Daily	Hr.	Daily	Bare Costs	Incl. O&P
1 Laborer	$36.65	$293.20	$56.55	$452.40	$42.77	$65.35
1 Equipment Operator (med.)	48.90	391.20	74.15	593.20		
1 Rock Trencher, 18" Width		2590.00		2849.00	161.88	178.06
16 L.H., Daily Totals		$3274.40		$3894.60	$204.65	$243.41
Crew B-55	Hr.	Daily	Hr.	Daily	Bare Costs	Incl. O&P
2 Laborers	$36.65	$586.40	$56.55	$904.80	$36.60	$56.27
1 Truck Driver (light)	36.50	292.00	55.70	445.60		
1 Truck-Mounted Earth Auger		786.40		865.04		
1 Flatbed Truck, Gas, 3 Ton		333.00		366.30	46.64	51.31
24 L.H., Daily Totals		$1997.80		$2581.74	$83.24	$107.57
Crew B-56	Hr.	Daily	Hr.	Daily	Bare Costs	Incl. O&P
1 Laborer	$36.65	$293.20	$56.55	$452.40	$41.85	$63.95
1 Equip. Oper. (light)	47.05	376.40	71.35	570.80		
1 Air Track Drill, 4"		1040.00		1144.00		
1 Air Compressor, 600 cfm		552.60		607.86		
1 -50' Air Hose, 3"		14.90		16.39	100.47	110.52
16 L.H., Daily Totals		$2277.10		$2791.45	$142.32	$174.47

Crew No.	Bare Costs Hr.	Bare Costs Daily	Incl. Subs O&P Hr.	Incl. Subs O&P Daily	Cost Per Labor-Hour Bare Costs	Cost Per Labor-Hour Incl. O&P
Crew B-57	Hr.	Daily	Hr.	Daily	Bare Costs	Incl. O&P
1 Labor Foreman (outside)	$38.65	$309.20	$59.65	$477.20	$42.13	$64.38
2 Laborers	36.65	586.40	56.55	904.80		
1 Equip. Oper. (crane)	50.25	402.00	76.20	609.60		
1 Equip. Oper. (light)	47.05	376.40	71.35	570.80		
1 Equip. Oper. (oiler)	43.55	348.40	66.00	528.00		
1 Crawler Crane, 25 Ton		1173.00		1290.30		
1 Clamshell Bucket, 1 C.Y.		47.40		52.14		
1 Centr. Water Pump, 6"		340.80		374.88		
1 -20' Suction Hose, 6"		11.50		12.65		
20 -50' Discharge Hoses, 6"		122.00		134.20	35.31	38.84
48 L.H., Daily Totals		$3717.10		$4954.57	$77.44	$103.22
Crew B-58	Hr.	Daily	Hr.	Daily	Bare Costs	Incl. O&P
2 Laborers	$36.65	$586.40	$56.55	$904.80	$40.12	$61.48
1 Equip. Oper. (light)	47.05	376.40	71.35	570.80		
1 Backhoe Loader, 48 H.P.		365.20		401.72		
1 Small Helicopter, w/ Pilot		2729.00		3001.90	128.93	141.82
24 L.H., Daily Totals		$4057.00		$4879.22	$169.04	$203.30
Crew B-59	Hr.	Daily	Hr.	Daily	Bare Costs	Incl. O&P
1 Truck Driver (heavy)	$37.55	$300.40	$57.30	$458.40	$37.55	$57.30
1 Truck Tractor, 220 H.P.		366.40		403.04		
1 Water Tank Trailer, 5000 Gal.		141.60		155.76	63.50	69.85
8 L.H., Daily Totals		$808.40		$1017.20	$101.05	$127.15
Crew B-59A	Hr.	Daily	Hr.	Daily	Bare Costs	Incl. O&P
2 Laborers	$36.65	$586.40	$56.55	$904.80	$36.95	$56.80
1 Truck Driver (heavy)	37.55	300.40	57.30	458.40		
1 Water Tank Trailer, 5000 Gal.		141.60		155.76		
1 Truck Tractor, 220 H.P.		366.40		403.04	21.17	23.28
24 L.H., Daily Totals		$1394.80		$1922.00	$58.12	$80.08
Crew B-60	Hr.	Daily	Hr.	Daily	Bare Costs	Incl. O&P
1 Labor Foreman (outside)	$38.65	$309.20	$59.65	$477.20	$42.84	$65.38
2 Laborers	36.65	586.40	56.55	904.80		
1 Equip. Oper. (crane)	50.25	402.00	76.20	609.60		
2 Equip. Oper. (light)	47.05	752.80	71.35	1141.60		
1 Equip. Oper. (oiler)	43.55	348.40	66.00	528.00		
1 Crawler Crane, 40 Ton		1179.00		1296.90		
1 Lead, 60' High		74.80		82.28		
1 Hammer, Diesel, 15K ft.-lbs.		587.60		646.36		
1 Backhoe Loader, 48 H.P.		365.20		401.72	39.40	43.34
56 L.H., Daily Totals		$4605.40		$6088.46	$82.24	$108.72
Crew B-61	Hr.	Daily	Hr.	Daily	Bare Costs	Incl. O&P
1 Labor Foreman (outside)	$38.65	$309.20	$59.65	$477.20	$39.13	$60.13
3 Laborers	36.65	879.60	56.55	1357.20		
1 Equip. Oper. (light)	47.05	376.40	71.35	570.80		
1 Cement Mixer, 2 C.Y.		191.40		210.54		
1 Air Compressor, 160 cfm		157.40		173.14	8.72	9.59
40 L.H., Daily Totals		$1914.00		$2788.88	$47.85	$69.72
Crew B-62	Hr.	Daily	Hr.	Daily	Bare Costs	Incl. O&P
2 Laborers	$36.65	$586.40	$56.55	$904.80	$40.12	$61.48
1 Equip. Oper. (light)	47.05	376.40	71.35	570.80		
1 Loader, Skid Steer, 30 H.P.		174.20		191.62	7.26	7.98
24 L.H., Daily Totals		$1137.00		$1667.22	$47.38	$69.47

Crews

Crew No.	Bare Costs Hr.	Bare Costs Daily	Incl. Subs O&P Hr.	Incl. Subs O&P Daily	Cost Per Labor-Hour Bare Costs	Cost Per Labor-Hour Incl. O&P
Crew B-63	Hr.	Daily	Hr.	Daily	Bare Costs	Incl. O&P
4 Laborers	$36.65	$1172.80	$56.55	$1809.60	$38.73	$59.51
1 Equip. Oper. (light)	47.05	376.40	71.35	570.80		
1 Loader, Skid Steer, 30 H.P.		174.20		191.62	4.36	4.79
40 L.H., Daily Totals		$1723.40		$2572.02	$43.09	$64.30
Crew B-63B	Hr.	Daily	Hr.	Daily	Bare Costs	Incl. O&P
1 Labor Foreman (inside)	$37.15	$297.20	$57.30	$458.40	$39.38	$60.44
2 Laborers	36.65	586.40	56.55	904.80		
1 Equip. Oper. (light)	47.05	376.40	71.35	570.80		
1 Loader, Skid Steer, 78 H.P.		308.80		339.68	9.65	10.62
32 L.H., Daily Totals		$1568.80		$2273.68	$49.02	$71.05
Crew B-64	Hr.	Daily	Hr.	Daily	Bare Costs	Incl. O&P
1 Laborer	$36.65	$293.20	$56.55	$452.40	$36.58	$56.13
1 Truck Driver (light)	36.50	292.00	55.70	445.60		
1 Power Mulcher (small)		159.60		175.56		
1 Flatbed Truck, Gas, 1.5 Ton		270.20		297.22	26.86	29.55
16 L.H., Daily Totals		$1015.00		$1370.78	$63.44	$85.67
Crew B-65	Hr.	Daily	Hr.	Daily	Bare Costs	Incl. O&P
1 Laborer	$36.65	$293.20	$56.55	$452.40	$36.58	$56.13
1 Truck Driver (light)	36.50	292.00	55.70	445.60		
1 Power Mulcher (Large)		325.60		358.16		
1 Flatbed Truck, Gas, 1.5 Ton		270.20		297.22	37.24	40.96
16 L.H., Daily Totals		$1181.00		$1553.38	$73.81	$97.09
Crew B-66	Hr.	Daily	Hr.	Daily	Bare Costs	Incl. O&P
1 Equip. Oper. (light)	$47.05	$376.40	$71.35	$570.80	$47.05	$71.35
1 Loader-Backhoe, 40 H.P.		259.60		285.56	32.45	35.70
8 L.H., Daily Totals		$636.00		$856.36	$79.50	$107.05
Crew B-67	Hr.	Daily	Hr.	Daily	Bare Costs	Incl. O&P
1 Millwright	$48.10	$384.80	$70.85	$566.80	$47.58	$71.10
1 Equip. Oper. (light)	47.05	376.40	71.35	570.80		
1 Forklift, R/T, 4,000 Lb.		314.20		345.62	19.64	21.60
16 L.H., Daily Totals		$1075.40		$1483.22	$67.21	$92.70
Crew B-67B	Hr.	Daily	Hr.	Daily	Bare Costs	Incl. O&P
1 Millwright Foreman (inside)	$48.60	$388.80	$71.60	$572.80	$48.35	$71.22
1 Millwright	48.10	384.80	70.85	566.80		
16 L.H., Daily Totals		$773.60		$1139.60	$48.35	$71.22
Crew B-68	Hr.	Daily	Hr.	Daily	Bare Costs	Incl. O&P
2 Millwrights	$48.10	$769.60	$70.85	$1133.60	$47.75	$71.02
1 Equip. Oper. (light)	47.05	376.40	71.35	570.80		
1 Forklift, R/T, 4,000 Lb.		314.20		345.62	13.09	14.40
24 L.H., Daily Totals		$1460.20		$2050.02	$60.84	$85.42
Crew B-68A	Hr.	Daily	Hr.	Daily	Bare Costs	Incl. O&P
1 Millwright Foreman (inside)	$48.60	$388.80	$71.60	$572.80	$48.27	$71.10
2 Millwrights	48.10	769.60	70.85	1133.60		
1 Forklift, 8,000 Lb.		180.00		198.00	7.50	8.25
24 L.H., Daily Totals		$1338.40		$1904.40	$55.77	$79.35

Crew No.	Bare Costs Hr.	Bare Costs Daily	Incl. Subs O&P Hr.	Incl. Subs O&P Daily	Cost Per Labor-Hour Bare Costs	Cost Per Labor-Hour Incl. O&P
Crew B-68B	Hr.	Daily	Hr.	Daily	Bare Costs	Incl. O&P
1 Millwright Foreman (inside)	$48.60	$388.80	$71.60	$572.80	$52.37	$78.11
2 Millwrights	48.10	769.60	70.85	1133.60		
2 Electricians	53.35	853.60	79.85	1277.60		
2 Plumbers	57.55	920.80	86.90	1390.40		
1 Forklift, 5,000 Lb.		324.80		357.28	5.80	6.38
56 L.H., Daily Totals		$3257.60		$4731.68	$58.17	$84.49
Crew B-68C	Hr.	Daily	Hr.	Daily	Bare Costs	Incl. O&P
1 Millwright Foreman (inside)	$48.60	$388.80	$71.60	$572.80	$51.90	$77.30
1 Millwright	48.10	384.80	70.85	566.80		
1 Electrician	53.35	426.80	79.85	638.80		
1 Plumber	57.55	460.40	86.90	695.20		
1 Forklift, 5,000 Lb.		324.80		357.28	10.15	11.16
32 L.H., Daily Totals		$1985.60		$2830.88	$62.05	$88.47
Crew B-68D	Hr.	Daily	Hr.	Daily	Bare Costs	Incl. O&P
1 Labor Foreman (inside)	$37.15	$297.20	$57.30	$458.40	$40.28	$61.73
1 Laborer	36.65	293.20	56.55	452.40		
1 Equip. Oper. (light)	47.05	376.40	71.35	570.80		
1 Forklift, 5,000 Lb.		324.80		357.28	13.53	14.89
24 L.H., Daily Totals		$1291.60		$1838.88	$53.82	$76.62
Crew B-68E	Hr.	Daily	Hr.	Daily	Bare Costs	Incl. O&P
1 Struc. Steel Foreman (inside)	$51.60	$412.80	$91.05	$728.40	$51.20	$90.37
3 Struc. Steel Workers	51.10	1226.40	90.20	2164.80		
1 Welder	51.10	408.80	90.20	721.60		
1 Forklift, 8,000 Lb.		180.00		198.00	4.50	4.95
40 L.H., Daily Totals		$2228.00		$3812.80	$55.70	$95.32
Crew B-68F	Hr.	Daily	Hr.	Daily	Bare Costs	Incl. O&P
1 Skilled Worker Foreman (out)	$49.30	$394.40	$76.35	$610.80	$47.97	$74.28
2 Skilled Workers	47.30	756.80	73.25	1172.00		
1 Forklift, 5,000 Lb.		324.80		357.28	13.53	14.89
24 L.H., Daily Totals		$1476.00		$2140.08	$61.50	$89.17
Crew B-69	Hr.	Daily	Hr.	Daily	Bare Costs	Incl. O&P
1 Labor Foreman (outside)	$38.65	$309.20	$59.65	$477.20	$40.40	$61.92
3 Laborers	36.65	879.60	56.55	1357.20		
1 Equip. Oper. (crane)	50.25	402.00	76.20	609.60		
1 Equip. Oper. (oiler)	43.55	348.40	66.00	528.00		
1 Hyd. Crane, 80 Ton		1652.00		1817.20	34.42	37.86
48 L.H., Daily Totals		$3591.20		$4789.20	$74.82	$99.78
Crew B-69A	Hr.	Daily	Hr.	Daily	Bare Costs	Incl. O&P
1 Labor Foreman (outside)	$38.65	$309.20	$59.65	$477.20	$40.26	$61.42
3 Laborers	36.65	879.60	56.55	1357.20		
1 Equip. Oper. (medium)	48.90	391.20	74.15	593.20		
1 Concrete Finisher	44.05	352.40	65.10	520.80		
1 Curb/Gutter Paver, 2-Track		898.00		987.80	18.71	20.58
48 L.H., Daily Totals		$2830.40		$3936.20	$58.97	$82.00
Crew B-69B	Hr.	Daily	Hr.	Daily	Bare Costs	Incl. O&P
1 Labor Foreman (outside)	$38.65	$309.20	$59.65	$477.20	$40.26	$61.42
3 Laborers	36.65	879.60	56.55	1357.20		
1 Equip. Oper. (medium)	48.90	391.20	74.15	593.20		
1 Cement Finisher	44.05	352.40	65.10	520.80		
1 Curb/Gutter Paver, 4-Track		770.20		847.22	16.05	17.65
48 L.H., Daily Totals		$2702.60		$3795.62	$56.30	$79.08

575

Crews

Crew No.	Bare Costs		Incl. Subs O&P		Cost Per Labor-Hour	
Crew B-70	Hr.	Daily	Hr.	Daily	Bare Costs	Incl. O&P
1 Labor Foreman (outside)	$38.65	$309.20	$59.65	$477.20	$42.19	$64.54
3 Laborers	36.65	879.60	56.55	1357.20		
3 Equip. Oper. (medium)	48.90	1173.60	74.15	1779.60		
1 Grader, 30,000 Lbs.		707.00		777.70		
1 Ripper, Beam & 1 Shank		81.40		89.54		
1 Road Sweeper, S.P., 8' wide		665.40		731.94		
1 F.E. Loader, W.M., 1.5 C.Y.		373.40		410.74	32.63	35.89
56 L.H., Daily Totals		$4189.60		$5623.92	$74.81	$100.43

Crew No.	Bare Costs		Incl. Subs O&P		Cost Per Labor-Hour	
Crew B-70A	Hr.	Daily	Hr.	Daily	Bare Costs	Incl. O&P
1 Laborer	$36.65	$293.20	$56.55	$452.40	$46.45	$70.63
4 Equip. Oper. (medium)	48.90	1564.80	74.15	2372.80		
1 Grader, 40,000 Lbs.		1120.00		1232.00		
1 F.E. Loader, W.M., 2.5 C.Y.		522.80		575.08		
1 Dozer, 80 H.P.		481.00		529.10		
1 Roller, Pneum. Whl., 12 Ton		349.00		383.90	61.82	68.00
40 L.H., Daily Totals		$4330.80		$5545.28	$108.27	$138.63

Crew No.	Bare Costs		Incl. Subs O&P		Cost Per Labor-Hour	
Crew B-71	Hr.	Daily	Hr.	Daily	Bare Costs	Incl. O&P
1 Labor Foreman (outside)	$38.65	$309.20	$59.65	$477.20	$42.19	$64.54
3 Laborers	36.65	879.60	56.55	1357.20		
3 Equip. Oper. (medium)	48.90	1173.60	74.15	1779.60		
1 Pvmt. Profiler, 750 H.P.		5877.00		6464.70		
1 Road Sweeper, S.P., 8' wide		665.40		731.94		
1 F.E. Loader, W.M., 1.5 C.Y.		373.40		410.74	123.50	135.85
56 L.H., Daily Totals		$9278.20		$11221.38	$165.68	$200.38

Crew No.	Bare Costs		Incl. Subs O&P		Cost Per Labor-Hour	
Crew B-72	Hr.	Daily	Hr.	Daily	Bare Costs	Incl. O&P
1 Labor Foreman (outside)	$38.65	$309.20	$59.65	$477.20	$43.02	$65.74
3 Laborers	36.65	879.60	56.55	1357.20		
4 Equip. Oper. (medium)	48.90	1564.80	74.15	2372.80		
1 Pvmt. Profiler, 750 H.P.		5877.00		6464.70		
1 Hammermill, 250 H.P.		1846.00		2030.60		
1 Windrow Loader		1240.00		1364.00		
1 Mix Paver 165 H.P.		2183.00		2401.30		
1 Roller, Pneum. Whl., 12 Ton		349.00		383.90	179.61	197.57
64 L.H., Daily Totals		$14248.60		$16851.70	$222.63	$263.31

Crew No.	Bare Costs		Incl. Subs O&P		Cost Per Labor-Hour	
Crew B-73	Hr.	Daily	Hr.	Daily	Bare Costs	Incl. O&P
1 Labor Foreman (outside)	$38.65	$309.20	$59.65	$477.20	$44.56	$67.94
2 Laborers	36.65	586.40	56.55	904.80		
5 Equip. Oper. (medium)	48.90	1956.00	74.15	2966.00		
1 Road Mixer, 310 H.P.		1942.00		2136.20		
1 Tandem Roller, 10 Ton		242.00		266.20		
1 Hammermill, 250 H.P.		1846.00		2030.60		
1 Grader, 30,000 Lbs.		707.00		777.70		
.5 F.E. Loader, W.M., 1.5 C.Y.		186.70		205.37		
.5 Truck Tractor, 220 H.P.		183.20		201.52		
.5 Water Tank Trailer, 5000 Gal.		70.80		77.88	80.90	88.99
64 L.H., Daily Totals		$8029.30		$10043.47	$125.46	$156.93

Crew No.	Bare Costs		Incl. Subs O&P		Cost Per Labor-Hour	
Crew B-74	Hr.	Daily	Hr.	Daily	Bare Costs	Incl. O&P
1 Labor Foreman (outside)	$38.65	$309.20	$59.65	$477.20	$43.25	$65.92
1 Laborer	36.65	293.20	56.55	452.40		
4 Equip. Oper. (medium)	48.90	1564.80	74.15	2372.80		
2 Truck Drivers (heavy)	37.55	600.80	57.30	916.80		
1 Grader, 30,000 Lbs.		707.00		777.70		
1 Ripper, Beam & 1 Shank		81.40		89.54		
2 Stabilizers, 310 H.P.		3654.00		4019.40		
1 Flatbed Truck, Gas, 3 Ton		333.00		366.30		
1 Chem. Spreader, Towed		53.20		58.52		
1 Roller, Vibratory, 25 Ton		692.80		762.08		
1 Water Tank Trailer, 5000 Gal.		141.60		155.76		
1 Truck Tractor, 220 H.P.		366.40		403.04	94.21	103.63
64 L.H., Daily Totals		$8797.40		$10851.54	$137.46	$169.56

Crew No.	Bare Costs		Incl. Subs O&P		Cost Per Labor-Hour	
Crew B-75	Hr.	Daily	Hr.	Daily	Bare Costs	Incl. O&P
1 Labor Foreman (outside)	$38.65	$309.20	$59.65	$477.20	$44.06	$67.16
1 Laborer	36.65	293.20	56.55	452.40		
4 Equip. Oper. (medium)	48.90	1564.80	74.15	2372.80		
1 Truck Driver (heavy)	37.55	300.40	57.30	458.40		
1 Grader, 30,000 Lbs.		707.00		777.70		
1 Ripper, Beam & 1 Shank		81.40		89.54		
2 Stabilizers, 310 H.P.		3654.00		4019.40		
1 Dist. Tanker, 3000 Gallon		308.20		339.02		
1 Truck Tractor, 6x4, 380 H.P.		612.80		674.08		
1 Roller, Vibratory, 25 Ton		692.80		762.08	108.15	118.96
56 L.H., Daily Totals		$8523.80		$10422.62	$152.21	$186.12

Crew No.	Bare Costs		Incl. Subs O&P		Cost Per Labor-Hour	
Crew B-76	Hr.	Daily	Hr.	Daily	Bare Costs	Incl. O&P
1 Dock Builder Foreman (outside)	$46.40	$371.20	$73.60	$588.80	$45.83	$71.56
5 Dock Builders	44.40	1776.00	70.40	2816.00		
2 Equip. Oper. (crane)	50.25	804.00	76.20	1219.20		
1 Equip. Oper. (oiler)	43.55	348.40	66.00	528.00		
1 Crawler Crane, 50 Ton		1319.00		1450.90		
1 Barge, 400 Ton		781.00		859.10		
1 Hammer, Diesel, 15K ft.-lbs.		587.60		646.36		
1 Lead, 60' High		74.80		82.28		
1 Air Compressor, 600 cfm		552.60		607.86		
2 -50' Air Hoses, 3"		29.80		32.78	46.46	51.10
72 L.H., Daily Totals		$6644.40		$8831.28	$92.28	$122.66

Crew No.	Bare Costs		Incl. Subs O&P		Cost Per Labor-Hour	
Crew B-76A	Hr.	Daily	Hr.	Daily	Bare Costs	Incl. O&P
1 Labor Foreman (outside)	$38.65	$309.20	$59.65	$477.20	$39.46	$60.58
5 Laborers	36.65	1466.00	56.55	2262.00		
1 Equip. Oper. (crane)	50.25	402.00	76.20	609.60		
1 Equip. Oper. (oiler)	43.55	348.40	66.00	528.00		
1 Crawler Crane, 50 Ton		1319.00		1450.90		
1 Barge, 400 Ton		781.00		859.10	32.81	36.09
64 L.H., Daily Totals		$4625.60		$6186.80	$72.28	$96.67

Crew No.	Bare Costs		Incl. Subs O&P		Cost Per Labor-Hour	
Crew B-77	Hr.	Daily	Hr.	Daily	Bare Costs	Incl. O&P
1 Labor Foreman (outside)	$38.65	$309.20	$59.65	$477.20	$37.02	$57.00
3 Laborers	36.65	879.60	56.55	1357.20		
1 Truck Driver (light)	36.50	292.00	55.70	445.60		
1 Crack Cleaner, 25 H.P.		64.80		71.28		
1 Crack Filler, Trailer Mtd.		203.80		224.18		
1 Flatbed Truck, Gas, 3 Ton		333.00		366.30	15.04	16.54
40 L.H., Daily Totals		$2082.40		$2941.76	$52.06	$73.54

Crews

Crew No.	Bare Costs Hr.	Daily	Incl. Subs O&P Hr.	Daily	Cost Per Labor-Hour Bare Costs	Incl. O&P
Crew B-78					Bare Costs	Incl. O&P
1 Labor Foreman (outside)	$38.65	$309.20	$59.65	$477.20	$36.96	$56.92
4 Laborers	36.65	1172.80	56.55	1809.60		
1 Truck Driver (light)	36.50	292.00	55.70	445.60		
1 Paint Striper, S.P., 40 Gallon		154.80		170.28		
1 Flatbed Truck, Gas, 3 Ton		333.00		366.30		
1 Pickup Truck, 3/4 Ton		157.00		172.70	13.43	14.78
48 L.H., Daily Totals		$2418.80		$3441.68	$50.39	$71.70
Crew B-78A	Hr.	Daily	Hr.	Daily	Bare Costs	Incl. O&P
1 Equip. Oper. (light)	$47.05	$376.40	$71.35	$570.80	$47.05	$71.35
1 Line Rem. (Metal Balls) 115 H.P.		839.00		922.90	104.88	115.36
8 L.H., Daily Totals		$1215.40		$1493.70	$151.93	$186.71
Crew B-78B	Hr.	Daily	Hr.	Daily	Bare Costs	Incl. O&P
2 Laborers	$36.65	$586.40	$56.55	$904.80	$37.81	$58.19
.25 Equip. Oper. (light)	47.05	94.10	71.35	142.70		
1 Pickup Truck, 3/4 Ton		157.00		172.70		
1 Line Rem.,11 H.P.,Walk Behind		68.40		75.24		
.25 Road Sweeper, S.P., 8' wide		166.35		182.99	21.76	23.94
18 L.H., Daily Totals		$1072.25		$1478.43	$59.57	$82.13
Crew B-78C	Hr.	Daily	Hr.	Daily	Bare Costs	Incl. O&P
1 Labor Foreman (outside)	$38.65	$309.20	$59.65	$477.20	$36.96	$56.92
4 Laborers	36.65	1172.80	56.55	1809.60		
1 Truck Driver (light)	36.50	292.00	55.70	445.60		
1 Paint Striper, T.M., 120 Gal.		860.40		946.44		
1 Flatbed Truck, Gas, 3 Ton		333.00		366.30		
1 Pickup Truck, 3/4 Ton		157.00		172.70	28.13	30.95
48 L.H., Daily Totals		$3124.40		$4217.84	$65.09	$87.87
Crew B-78D	Hr.	Daily	Hr.	Daily	Bare Costs	Incl. O&P
2 Labor Foremen (outside)	$38.65	$618.40	$59.65	$954.40	$37.03	$57.09
7 Laborers	36.65	2052.40	56.55	3166.80		
1 Truck Driver (light)	36.50	292.00	55.70	445.60		
1 Paint Striper, T.M., 120 Gal.		860.40		946.44		
1 Flatbed Truck, Gas, 3 Ton		333.00		366.30		
3 Pickup Trucks, 3/4 Ton		471.00		518.10		
1 Air Compressor, 60 cfm		137.40		151.14		
1 -50' Air Hose, 3/4"		3.25		3.58		
1 Breakers, Pavement, 60 lb.		9.80		10.78	22.69	24.95
80 L.H., Daily Totals		$4777.65		$6563.14	$59.72	$82.04
Crew B-78E	Hr.	Daily	Hr.	Daily	Bare Costs	Incl. O&P
2 Labor Foremen (outside)	$38.65	$618.40	$59.65	$954.40	$36.97	$57.00
9 Laborers	36.65	2638.80	56.55	4071.60		
1 Truck Driver (light)	36.50	292.00	55.70	445.60		
1 Paint Striper, T.M., 120 Gal.		860.40		946.44		
1 Flatbed Truck, Gas, 3 Ton		333.00		366.30		
4 Pickup Trucks, 3/4 Ton		628.00		690.80		
2 Air Compressor, 60 cfm		274.80		302.28		
2 -50' Air Hose, 3/4"		6.50		7.15		
2 Breakers, Pavement, 60 lb.		19.60		21.56	22.11	24.32
96 L.H., Daily Totals		$5671.50		$7806.13	$59.08	$81.31

Crew No.	Bare Costs Hr.	Daily	Incl. Subs O&P Hr.	Daily	Cost Per Labor-Hour Bare Costs	Incl. O&P
Crew B-78F	Hr.	Daily	Hr.	Daily	Bare Costs	Incl. O&P
2 Labor Foremen (outside)	$38.65	$618.40	$59.65	$954.40	$36.92	$56.93
11 Laborers	36.65	3225.20	56.55	4976.40		
1 Truck Driver (light)	36.50	292.00	55.70	445.60		
1 Paint Striper, T.M., 120 Gal.		860.40		946.44		
1 Flatbed Truck, Gas, 3 Ton		333.00		366.30		
7 Pickup Trucks, 3/4 Ton		1099.00		1208.90		
3 Air Compressor, 60 cfm		412.20		453.42		
3 -50' Air Hose, 3/4"		9.75		10.73		
3 Breakers, Pavement, 60 lb.		29.40		32.34	24.50	26.95
112 L.H., Daily Totals		$6879.35		$9394.52	$61.42	$83.88
Crew B-79	Hr.	Daily	Hr.	Daily	Bare Costs	Incl. O&P
1 Labor Foreman (outside)	$38.65	$309.20	$59.65	$477.20	$37.02	$57.00
3 Laborers	36.65	879.60	56.55	1357.20		
1 Truck Driver (light)	36.50	292.00	55.70	445.60		
1 Paint Striper, T.M., 120 Gal.		860.40		946.44		
1 Heating Kettle, 115 Gallon		63.00		69.30		
1 Flatbed Truck, Gas, 3 Ton		333.00		366.30		
2 Pickup Trucks, 3/4 Ton		314.00		345.40	39.26	43.19
40 L.H., Daily Totals		$3051.20		$4007.44	$76.28	$100.19
Crew B-79A	Hr.	Daily	Hr.	Daily	Bare Costs	Incl. O&P
1.5 Equip. Oper. (light)	$47.05	$564.60	$71.35	$856.20	$47.05	$71.35
.5 Line Remov. (Grinder) 115 H.P.		465.30		511.83		
1 Line Rem. (Metal Balls) 115 H.P.		839.00		922.90	108.69	119.56
12 L.H., Daily Totals		$1868.90		$2290.93	$155.74	$190.91
Crew B-79B	Hr.	Daily	Hr.	Daily	Bare Costs	Incl. O&P
1 Laborer	$36.65	$293.20	$56.55	$452.40	$36.65	$56.55
1 Set of Gases		152.00		167.20	19.00	20.90
8 L.H., Daily Totals		$445.20		$619.60	$55.65	$77.45
Crew B-79C	Hr.	Daily	Hr.	Daily	Bare Costs	Incl. O&P
1 Labor Foreman (outside)	$38.65	$309.20	$59.65	$477.20	$36.91	$56.87
5 Laborers	36.65	1466.00	56.55	2262.00		
1 Truck Driver (light)	36.50	292.00	55.70	445.60		
1 Paint Striper, T.M., 120 Gal.		860.40		946.44		
1 Heating Kettle, 115 Gallon		63.00		69.30		
1 Flatbed Truck, Gas, 3 Ton		333.00		366.30		
3 Pickup Trucks, 3/4 Ton		471.00		518.10		
1 Air Compressor, 60 cfm		137.40		151.14		
1 -50' Air Hose, 3/4"		3.25		3.58		
1 Breakers, Pavement, 60 lb.		9.80		10.78	33.53	36.89
56 L.H., Daily Totals		$3945.05		$5250.44	$70.45	$93.76
Crew B-79D	Hr.	Daily	Hr.	Daily	Bare Costs	Incl. O&P
2 Labor Foremen (outside)	$38.65	$618.40	$59.65	$954.40	$37.13	$57.22
5 Laborers	36.65	1466.00	56.55	2262.00		
1 Truck Driver (light)	36.50	292.00	55.70	445.60		
1 Paint Striper, T.M., 120 Gal.		860.40		946.44		
1 Heating Kettle, 115 Gallon		63.00		69.30		
1 Flatbed Truck, Gas, 3 Ton		333.00		366.30		
4 Pickup Trucks, 3/4 Ton		628.00		690.80		
1 Air Compressor, 60 cfm		137.40		151.14		
1 -50' Air Hose, 3/4"		3.25		3.58		
1 Breakers, Pavement, 60 lb.		9.80		10.78	31.79	34.97
64 L.H., Daily Totals		$4411.25		$5900.34	$68.93	$92.19

Crews

Crew No.	Bare Costs	Incl. Subs O&P	Cost Per Labor-Hour

Crew B-79E	Hr.	Daily	Hr.	Daily	Bare Costs	Incl. O&P
2 Labor Foremen (outside)	$38.65	$618.40	$59.65	$954.40	$37.03	$57.09
7 Laborers	36.65	2052.40	56.55	3166.80		
1 Truck Driver (light)	36.50	292.00	55.70	445.60		
1 Paint Striper, T.M., 120 Gal.		860.40		946.44		
1 Heating Kettle, 115 Gallon		63.00		69.30		
1 Flatbed Truck, Gas, 3 Ton		333.00		366.30		
5 Pickup Trucks, 3/4 Ton		785.00		863.50		
2 Air Compressors, 60 cfm		274.80		302.28		
2 -50' Air Hoses, 3/4"		6.50		7.15		
2 Breakers, Pavement, 60 lb.		19.60		21.56	29.28	32.21
80 L.H., Daily Totals		$5305.10		$7143.33	$66.31	$89.29

Crew B-80	Hr.	Daily	Hr.	Daily	Bare Costs	Incl. O&P
1 Labor Foreman (outside)	$38.65	$309.20	$59.65	$477.20	$39.71	$60.81
1 Laborer	36.65	293.20	56.55	452.40		
1 Truck Driver (light)	36.50	292.00	55.70	445.60		
1 Equip. Oper. (light)	47.05	376.40	71.35	570.80		
1 Flatbed Truck, Gas, 3 Ton		333.00		366.30		
1 Earth Auger, Truck-Mtd.		425.00		467.50	23.69	26.06
32 L.H., Daily Totals		$2028.80		$2779.80	$63.40	$86.87

Crew B-80A	Hr.	Daily	Hr.	Daily	Bare Costs	Incl. O&P
3 Laborers	$36.65	$879.60	$56.55	$1357.20	$36.65	$56.55
1 Flatbed Truck, Gas, 3 Ton		333.00		366.30	13.88	15.26
24 L.H., Daily Totals		$1212.60		$1723.50	$50.52	$71.81

Crew B-80B	Hr.	Daily	Hr.	Daily	Bare Costs	Incl. O&P
3 Laborers	$36.65	$879.60	$56.55	$1357.20	$39.25	$60.25
1 Equip. Oper. (light)	47.05	376.40	71.35	570.80		
1 Crane, Flatbed Mounted, 3 Ton		248.40		273.24	7.76	8.54
32 L.H., Daily Totals		$1504.40		$2201.24	$47.01	$68.79

Crew B-80C	Hr.	Daily	Hr.	Daily	Bare Costs	Incl. O&P
2 Laborers	$36.65	$586.40	$56.55	$904.80	$36.60	$56.27
1 Truck Driver (light)	36.50	292.00	55.70	445.60		
1 Flatbed Truck, Gas, 1.5 Ton		270.20		297.22		
1 Manual Fence Post Auger, Gas		8.00		8.80	11.59	12.75
24 L.H., Daily Totals		$1156.60		$1656.42	$48.19	$69.02

Crew B-81	Hr.	Daily	Hr.	Daily	Bare Costs	Incl. O&P
1 Laborer	$36.65	$293.20	$56.55	$452.40	$41.03	$62.67
1 Equip. Oper. (medium)	48.90	391.20	74.15	593.20		
1 Truck Driver (heavy)	37.55	300.40	57.30	458.40		
1 Hydromulcher, T.M., 3000 Gal.		340.40		374.44		
1 Truck Tractor, 220 H.P.		366.40		403.04	29.45	32.40
24 L.H., Daily Totals		$1691.60		$2281.48	$70.48	$95.06

Crew B-81A	Hr.	Daily	Hr.	Daily	Bare Costs	Incl. O&P
1 Laborer	$36.65	$293.20	$56.55	$452.40	$36.58	$56.13
1 Truck Driver (light)	36.50	292.00	55.70	445.60		
1 Hydromulcher, T.M., 600 Gal.		132.60		145.86		
1 Flatbed Truck, Gas, 3 Ton		333.00		366.30	29.10	32.01
16 L.H., Daily Totals		$1050.80		$1410.16	$65.67	$88.14

Crew B-82	Hr.	Daily	Hr.	Daily	Bare Costs	Incl. O&P
1 Laborer	$36.65	$293.20	$56.55	$452.40	$41.85	$63.95
1 Equip. Oper. (light)	47.05	376.40	71.35	570.80		
1 Horiz. Borer, 6 H.P.		83.00		91.30	5.19	5.71
16 L.H., Daily Totals		$752.60		$1114.50	$47.04	$69.66

Crew B-82A	Hr.	Daily	Hr.	Daily	Bare Costs	Incl. O&P
1 Laborer	$36.65	$293.20	$56.55	$452.40	$41.85	$63.95
1 Equip. Oper. (light)	47.05	376.40	71.35	570.80		
1 Flatbed Truck, Gas, 3 Ton		333.00		366.30		
1 Flatbed Trailer, 25 Ton		113.60		124.96		
1 Horiz. Dir. Drill, 20k lb. Thrust		658.40		724.24	69.06	75.97
16 L.H., Daily Totals		$1774.60		$2238.70	$110.91	$139.92

Crew B-82B	Hr.	Daily	Hr.	Daily	Bare Costs	Incl. O&P
2 Laborers	$36.65	$586.40	$56.55	$904.80	$40.12	$61.48
1 Equip. Oper. (light)	47.05	376.40	71.35	570.80		
1 Flatbed Truck, Gas, 3 Ton		333.00		366.30		
1 Flatbed Trailer, 25 Ton		113.60		124.96		
1 Horiz. Dir. Drill, 30k lb. Thrust		936.20		1029.82	57.62	63.38
24 L.H., Daily Totals		$2345.60		$2996.68	$97.73	$124.86

Crew B-82C	Hr.	Daily	Hr.	Daily	Bare Costs	Incl. O&P
2 Laborers	$36.65	$586.40	$56.55	$904.80	$40.12	$61.48
1 Equip. Oper. (light)	47.05	376.40	71.35	570.80		
1 Flatbed Truck, Gas, 3 Ton		333.00		366.30		
1 Flatbed Trailer, 25 Ton		113.60		124.96		
1 Horiz. Dir. Drill, 50k lb. Thrust		1246.00		1370.60	70.53	77.58
24 L.H., Daily Totals		$2655.40		$3337.46	$110.64	$139.06

Crew B-82D	Hr.	Daily	Hr.	Daily	Bare Costs	Incl. O&P
1 Equip. Oper. (light)	$47.05	$376.40	$71.35	$570.80	$47.05	$71.35
1 Mud Trailer for HDD, 1500 Gal.		374.80		412.28	46.85	51.53
8 L.H., Daily Totals		$751.20		$983.08	$93.90	$122.89

Crew B-83	Hr.	Daily	Hr.	Daily	Bare Costs	Incl. O&P
1 Tugboat Captain	$48.90	$391.20	$74.15	$593.20	$42.77	$65.35
1 Tugboat Hand	36.65	293.20	56.55	452.40		
1 Tugboat, 250 H.P.		889.40		978.34	55.59	61.15
16 L.H., Daily Totals		$1573.80		$2023.94	$98.36	$126.50

Crew B-84	Hr.	Daily	Hr.	Daily	Bare Costs	Incl. O&P
1 Equip. Oper. (medium)	$48.90	$391.20	$74.15	$593.20	$48.90	$74.15
1 Rotary Mower/Tractor		364.20		400.62	45.52	50.08
8 L.H., Daily Totals		$755.40		$993.82	$94.42	$124.23

Crew B-85	Hr.	Daily	Hr.	Daily	Bare Costs	Incl. O&P
3 Laborers	$36.65	$879.60	$56.55	$1357.20	$39.28	$60.22
1 Equip. Oper. (medium)	48.90	391.20	74.15	593.20		
1 Truck Driver (heavy)	37.55	300.40	57.30	458.40		
1 Aerial Lift Truck, 80'		643.00		707.30		
1 Brush Chipper, 12", 130 H.P.		396.60		436.26		
1 Pruning Saw, Rotary		6.55		7.21	26.15	28.77
40 L.H., Daily Totals		$2617.35		$3559.57	$65.43	$88.99

Crew B-86	Hr.	Daily	Hr.	Daily	Bare Costs	Incl. O&P
1 Equip. Oper. (medium)	$48.90	$391.20	$74.15	$593.20	$48.90	$74.15
1 Stump Chipper, S.P.		162.10		178.31	20.26	22.29
8 L.H., Daily Totals		$553.30		$771.51	$69.16	$96.44

Crew B-86A	Hr.	Daily	Hr.	Daily	Bare Costs	Incl. O&P
1 Equip. Oper. (medium)	$48.90	$391.20	$74.15	$593.20	$48.90	$74.15
1 Grader, 30,000 Lbs.		707.00		777.70	88.38	97.21
8 L.H., Daily Totals		$1098.20		$1370.90	$137.28	$171.36

Crews

Crew No.	Bare Costs Hr.	Bare Costs Daily	Incl. Subs O&P Hr.	Incl. Subs O&P Daily	Cost Per Labor-Hour Bare Costs	Cost Per Labor-Hour Incl. O&P
Crew B-86B	Hr.	Daily	Hr.	Daily	Bare Costs	Incl. O&P
1 Equip. Oper. (medium)	$48.90	$391.20	$74.15	$593.20	$48.90	$74.15
1 Dozer, 200 H.P.		1325.00		1457.50	165.63	182.19
8 L.H., Daily Totals		$1716.20		$2050.70	$214.53	$256.34
Crew B-87	Hr.	Daily	Hr.	Daily	Bare Costs	Incl. O&P
1 Laborer	$36.65	$293.20	$56.55	$452.40	$46.45	$70.63
4 Equip. Oper. (medium)	48.90	1564.80	74.15	2372.80		
2 Feller Bunchers, 100 H.P.		1458.00		1603.80		
1 Log Chipper, 22" Tree		898.60		988.46		
1 Dozer, 105 H.P.		591.80		650.98		
1 Chain Saw, Gas, 36" Long		45.00		49.50	74.83	82.32
40 L.H., Daily Totals		$4851.40		$6117.94	$121.29	$152.95
Crew B-88	Hr.	Daily	Hr.	Daily	Bare Costs	Incl. O&P
1 Laborer	$36.65	$293.20	$56.55	$452.40	$47.15	$71.64
6 Equip. Oper. (medium)	48.90	2347.20	74.15	3559.20		
2 Feller Bunchers, 100 H.P.		1458.00		1603.80		
1 Log Chipper, 22" Tree		898.60		988.46		
2 Log Skidders, 50 H.P.		1839.20		2023.12		
1 Dozer, 105 H.P.		591.80		650.98		
1 Chain Saw, Gas, 36" Long		45.00		49.50	86.30	94.93
56 L.H., Daily Totals		$7473.00		$9327.46	$133.45	$166.56
Crew B-89	Hr.	Daily	Hr.	Daily	Bare Costs	Incl. O&P
1 Equip. Oper. (light)	$47.05	$376.40	$71.35	$570.80	$41.77	$63.52
1 Truck Driver (light)	36.50	292.00	55.70	445.60		
1 Flatbed Truck, Gas, 3 Ton		333.00		366.30		
1 Concrete Saw		175.20		192.72		
1 Water Tank, 65 Gal.		17.30		19.03	32.84	36.13
16 L.H., Daily Totals		$1193.90		$1594.45	$74.62	$99.65
Crew B-89A	Hr.	Daily	Hr.	Daily	Bare Costs	Incl. O&P
1 Skilled Worker	$47.30	$378.40	$73.25	$586.00	$41.98	$64.90
1 Laborer	36.65	293.20	56.55	452.40		
1 Core Drill (Large)		116.20		127.82	7.26	7.99
16 L.H., Daily Totals		$787.80		$1166.22	$49.24	$72.89
Crew B-89B	Hr.	Daily	Hr.	Daily	Bare Costs	Incl. O&P
1 Equip. Oper. (light)	$47.05	$376.40	$71.35	$570.80	$41.77	$63.52
1 Truck Driver (light)	36.50	292.00	55.70	445.60		
1 Wall Saw, Hydraulic, 10 H.P.		119.60		131.56		
1 Generator, Diesel, 100 kW		439.20		483.12		
1 Water Tank, 65 Gal.		17.30		19.03		
1 Flatbed Truck, Gas, 3 Ton		333.00		366.30	56.82	62.50
16 L.H., Daily Totals		$1577.50		$2016.41	$98.59	$126.03
Crew B-90	Hr.	Daily	Hr.	Daily	Bare Costs	Incl. O&P
1 Labor Foreman (outside)	$38.65	$309.20	$59.65	$477.20	$39.73	$60.83
3 Laborers	36.65	879.60	56.55	1357.20		
2 Equip. Oper. (light)	47.05	752.80	71.35	1141.60		
2 Truck Drivers (heavy)	37.55	600.80	57.30	916.80		
1 Road Mixer, 310 H.P.		1942.00		2136.20		
1 Dist. Truck, 2000 Gal.		278.20		306.02	34.69	38.16
64 L.H., Daily Totals		$4762.60		$6335.02	$74.42	$98.98

Crew No.	Bare Costs Hr.	Bare Costs Daily	Incl. Subs O&P Hr.	Incl. Subs O&P Daily	Cost Per Labor-Hour Bare Costs	Cost Per Labor-Hour Incl. O&P
Crew B-90A	Hr.	Daily	Hr.	Daily	Bare Costs	Incl. O&P
1 Labor Foreman (outside)	$38.65	$309.20	$59.65	$477.20	$43.94	$67.05
2 Laborers	36.65	586.40	56.55	904.80		
4 Equip. Oper. (medium)	48.90	1564.80	74.15	2372.80		
2 Graders, 30,000 Lbs.		1414.00		1555.40		
1 Tandem Roller, 10 Ton		242.00		266.20		
1 Roller, Pneum. Whl., 12 Ton		349.00		383.90	35.80	39.38
56 L.H., Daily Totals		$4465.40		$5960.30	$79.74	$106.43
Crew B-90B	Hr.	Daily	Hr.	Daily	Bare Costs	Incl. O&P
1 Labor Foreman (outside)	$38.65	$309.20	$59.65	$477.20	$43.11	$65.87
2 Laborers	36.65	586.40	56.55	904.80		
3 Equip. Oper. (medium)	48.90	1173.60	74.15	1779.60		
1 Roller, Pneum. Whl., 12 Ton		349.00		383.90		
1 Road Mixer, 310 H.P.		1942.00		2136.20	47.73	52.50
48 L.H., Daily Totals		$4360.20		$5681.70	$90.84	$118.37
Crew B-90C	Hr.	Daily	Hr.	Daily	Bare Costs	Incl. O&P
1 Labor Foreman (outside)	$38.65	$309.20	$59.65	$477.20	$40.42	$61.84
4 Laborers	36.65	1172.80	56.55	1809.60		
3 Equip. Oper. (medium)	48.90	1173.60	74.15	1779.60		
3 Truck Drivers (heavy)	37.55	901.20	57.30	1375.20		
3 Road Mixers, 310 H.P.		5826.00		6408.60	66.20	72.83
88 L.H., Daily Totals		$9382.80		$11850.20	$106.62	$134.66
Crew B-90D	Hr.	Daily	Hr.	Daily	Bare Costs	Incl. O&P
1 Labor Foreman (outside)	$38.65	$309.20	$59.65	$477.20	$39.84	$61.02
6 Laborers	36.65	1759.20	56.55	2714.40		
3 Equip. Oper. (medium)	48.90	1173.60	74.15	1779.60		
3 Truck Drivers (heavy)	37.55	901.20	57.30	1375.20		
3 Road Mixers, 310 H.P.		5826.00		6408.60	56.02	61.62
104 L.H., Daily Totals		$9969.20		$12755.00	$95.86	$122.64
Crew B-90E	Hr.	Daily	Hr.	Daily	Bare Costs	Incl. O&P
1 Labor Foreman (outside)	$38.65	$309.20	$59.65	$477.20	$41.06	$62.84
4 Laborers	36.65	1172.80	56.55	1809.60		
3 Equip. Oper. (medium)	48.90	1173.60	74.15	1779.60		
1 Truck Driver (heavy)	37.55	300.40	57.30	458.40		
1 Road Mixers, 310 H.P.		1942.00		2136.20	26.97	29.67
72 L.H., Daily Totals		$4898.00		$6661.00	$68.03	$92.51
Crew B-91	Hr.	Daily	Hr.	Daily	Bare Costs	Incl. O&P
1 Labor Foreman (outside)	$38.65	$309.20	$59.65	$477.20	$43.14	$65.83
2 Laborers	36.65	586.40	56.55	904.80		
4 Equip. Oper. (medium)	48.90	1564.80	74.15	2372.80		
1 Truck Driver (heavy)	37.55	300.40	57.30	458.40		
1 Dist. Tanker, 3000 Gallon		308.20		339.02		
1 Truck Tractor, 6x4, 380 H.P.		612.80		674.08		
1 Aggreg. Spreader, S.P.		841.00		925.10		
1 Roller, Pneum. Whl., 12 Ton		349.00		383.90		
1 Tandem Roller, 10 Ton		242.00		266.20	36.77	40.44
64 L.H., Daily Totals		$5113.80		$6801.50	$79.90	$106.27
Crew B-91B	Hr.	Daily	Hr.	Daily	Bare Costs	Incl. O&P
1 Laborer	$36.65	$293.20	$56.55	$452.40	$42.77	$65.35
1 Equipment Oper. (med.)	48.90	391.20	74.15	593.20		
1 Road Sweeper, Vac. Assist.		994.40		1093.84	62.15	68.36
16 L.H., Daily Totals		$1678.80		$2139.44	$104.93	$133.72

Crews

Crew No.		Bare Costs		Incl. Subs O&P		Cost Per Labor-Hour	
Crew B-91C	Hr.	Daily	Hr.	Daily	Bare Costs	Incl. O&P
1 Laborer	$36.65	$293.20	$56.55	$452.40	$36.58	$56.13
1 Truck Driver (light)	36.50	292.00	55.70	445.60		
1 Catch Basin Cleaning Truck		604.00		664.40	37.75	41.52
16 L.H., Daily Totals		$1189.20		$1562.40	$74.33	$97.65

Crew B-91D	Hr.	Daily	Hr.	Daily	Bare Costs	Incl. O&P
1 Labor Foreman (outside)	$38.65	$309.20	$59.65	$477.20	$41.65	$63.67
5 Laborers	36.65	1466.00	56.55	2262.00		
5 Equip. Oper. (medium)	48.90	1956.00	74.15	2966.00		
2 Truck Drivers (heavy)	37.55	600.80	57.30	916.80		
1 Aggreg. Spreader, S.P.		841.00		925.10		
2 Truck Tractor, 6x4, 380 H.P.		1225.60		1348.16		
2 Dist. Tanker, 3000 Gallon		616.40		678.04		
2 Pavement Brush, Towed		163.20		179.52		
2 Roller, Pneum. Whl., 12 Ton		698.00		767.80	34.08	37.49
104 L.H., Daily Totals		$7876.20		$10520.62	$75.73	$101.16

Crew B-92	Hr.	Daily	Hr.	Daily	Bare Costs	Incl. O&P
1 Labor Foreman (outside)	$38.65	$309.20	$59.65	$477.20	$37.15	$57.33
3 Laborers	36.65	879.60	56.55	1357.20		
1 Crack Cleaner, 25 H.P.		64.80		71.28		
1 Air Compressor, 60 cfm		137.40		151.14		
1 Tar Kettle, T.M.		104.60		115.06		
1 Flatbed Truck, Gas, 3 Ton		333.00		366.30	19.99	21.99
32 L.H., Daily Totals		$1828.60		$2538.18	$57.14	$79.32

Crew B-93	Hr.	Daily	Hr.	Daily	Bare Costs	Incl. O&P
1 Equip. Oper. (medium)	$48.90	$391.20	$74.15	$593.20	$48.90	$74.15
1 Feller Buncher, 100 H.P.		729.00		801.90	91.13	100.24
8 L.H., Daily Totals		$1120.20		$1395.10	$140.03	$174.39

Crew B-94A	Hr.	Daily	Hr.	Daily	Bare Costs	Incl. O&P
1 Laborer	$36.65	$293.20	$56.55	$452.40	$36.65	$56.55
1 Diaphragm Water Pump, 2"		72.00		79.20		
1 -20' Suction Hose, 2"		1.95		2.15		
2 -50' Discharge Hoses, 2"		1.80		1.98	9.47	10.42
8 L.H., Daily Totals		$368.95		$535.73	$46.12	$66.97

Crew B-94B	Hr.	Daily	Hr.	Daily	Bare Costs	Incl. O&P
1 Laborer	$36.65	$293.20	$56.55	$452.40	$36.65	$56.55
1 Diaphragm Water Pump, 4"		115.80		127.38		
1 -20' Suction Hose, 4"		3.25		3.58		
2 -50' Discharge Hoses, 4"		4.70		5.17	15.47	17.02
8 L.H., Daily Totals		$416.95		$588.52	$52.12	$73.57

Crew B-94C	Hr.	Daily	Hr.	Daily	Bare Costs	Incl. O&P
1 Laborer	$36.65	$293.20	$56.55	$452.40	$36.65	$56.55
1 Centrifugal Water Pump, 3"		78.80		86.68		
1 -20' Suction Hose, 3"		2.85		3.13		
2 -50' Discharge Hoses, 3"		3.00		3.30	10.58	11.64
8 L.H., Daily Totals		$377.85		$545.51	$47.23	$68.19

Crew B-94D	Hr.	Daily	Hr.	Daily	Bare Costs	Incl. O&P
1 Laborer	$36.65	$293.20	$56.55	$452.40	$36.65	$56.55
1 Centr. Water Pump, 6"		340.80		374.88		
1 -20' Suction Hose, 6"		11.50		12.65		
2 -50' Discharge Hoses, 6"		12.20		13.42	45.56	50.12
8 L.H., Daily Totals		$657.70		$853.35	$82.21	$106.67

Crew C-1	Hr.	Daily	Hr.	Daily	Bare Costs	Incl. O&P
3 Carpenters	$45.85	$1100.40	$70.75	$1698.00	$43.55	$67.20
1 Laborer	36.65	293.20	56.55	452.40		
32 L.H., Daily Totals		$1393.60		$2150.40	$43.55	$67.20

Crew C-2	Hr.	Daily	Hr.	Daily	Bare Costs	Incl. O&P
1 Carpenter Foreman (outside)	$47.85	$382.80	$73.85	$590.80	$44.65	$68.90
4 Carpenters	45.85	1467.20	70.75	2264.00		
1 Laborer	36.65	293.20	56.55	452.40		
48 L.H., Daily Totals		$2143.20		$3307.20	$44.65	$68.90

Crew C-2A	Hr.	Daily	Hr.	Daily	Bare Costs	Incl. O&P
1 Carpenter Foreman (outside)	$47.85	$382.80	$73.85	$590.80	$44.35	$67.96
3 Carpenters	45.85	1100.40	70.75	1698.00		
1 Cement Finisher	44.05	352.40	65.10	520.80		
1 Laborer	36.65	293.20	56.55	452.40		
48 L.H., Daily Totals		$2128.80		$3262.00	$44.35	$67.96

Crew C-3	Hr.	Daily	Hr.	Daily	Bare Costs	Incl. O&P
1 Rodman Foreman (outside)	$52.65	$421.20	$82.70	$661.60	$46.95	$73.17
4 Rodmen (reinf.)	50.65	1620.80	79.55	2545.60		
1 Equip. Oper. (light)	47.05	376.40	71.35	570.80		
2 Laborers	36.65	586.40	56.55	904.80		
3 Stressing Equipment		30.60		33.66		
.5 Grouting Equipment		82.10		90.31	1.76	1.94
64 L.H., Daily Totals		$3117.50		$4806.77	$48.71	$75.11

Crew C-4	Hr.	Daily	Hr.	Daily	Bare Costs	Incl. O&P
1 Rodman Foreman (outside)	$52.65	$421.20	$82.70	$661.60	$51.15	$80.34
3 Rodmen (reinf.)	50.65	1215.60	79.55	1909.20		
3 Stressing Equipment		30.60		33.66	.96	1.05
32 L.H., Daily Totals		$1667.40		$2604.46	$52.11	$81.39

Crew C-4A	Hr.	Daily	Hr.	Daily	Bare Costs	Incl. O&P
2 Rodmen (reinf.)	$50.65	$810.40	$79.55	$1272.80	$50.65	$79.55
4 Stressing Equipment		40.80		44.88	2.55	2.81
16 L.H., Daily Totals		$851.20		$1317.68	$53.20	$82.36

Crew C-5	Hr.	Daily	Hr.	Daily	Bare Costs	Incl. O&P
1 Rodman Foreman (outside)	$52.65	$421.20	$82.70	$661.60	$49.86	$77.59
4 Rodmen (reinf.)	50.65	1620.80	79.55	2545.60		
1 Equip. Oper. (crane)	50.25	402.00	76.20	609.60		
1 Equip. Oper. (oiler)	43.55	348.40	66.00	528.00		
1 Hyd. Crane, 25 Ton		745.80		820.38	13.32	14.65
56 L.H., Daily Totals		$3538.20		$5165.18	$63.18	$92.24

Crew C-6	Hr.	Daily	Hr.	Daily	Bare Costs	Incl. O&P
1 Labor Foreman (outside)	$38.65	$309.20	$59.65	$477.20	$38.22	$58.49
4 Laborers	36.65	1172.80	56.55	1809.60		
1 Cement Finisher	44.05	352.40	65.10	520.80		
2 Gas Engine Vibrators		66.00		72.60	1.38	1.51
48 L.H., Daily Totals		$1900.40		$2880.20	$39.59	$60.00

Crews

Crew No.		Bare Costs		Incl. Subs O&P		Cost Per Labor-Hour	
Crew C-7	Hr.	Daily	Hr.	Daily	Bare Costs	Incl. O&P
1 Labor Foreman (outside)	$38.65	$309.20	$59.65	$477.20	$39.82	$60.85
5 Laborers	36.65	1466.00	56.55	2262.00		
1 Cement Finisher	44.05	352.40	65.10	520.80		
1 Equip. Oper. (medium)	48.90	391.20	74.15	593.20		
1 Equip. Oper. (oiler)	43.55	348.40	66.00	528.00		
2 Gas Engine Vibrators		66.00		72.60		
1 Concrete Bucket, 1 C.Y.		23.00		25.30		
1 Hyd. Crane, 55 Ton		1137.00		1250.70	17.03	18.73
72 L.H., Daily Totals		$4093.20		$5729.80	$56.85	$79.58

Crew C-7A	Hr.	Daily	Hr.	Daily	Bare Costs	Incl. O&P
1 Labor Foreman (outside)	$38.65	$309.20	$59.65	$477.20	$37.13	$57.13
5 Laborers	36.65	1466.00	56.55	2262.00		
2 Truck Drivers (heavy)	37.55	600.80	57.30	916.80		
2 Conc. Transit Mixers		2050.00		2255.00	32.03	35.23
64 L.H., Daily Totals		$4426.00		$5911.00	$69.16	$92.36

Crew C-7B	Hr.	Daily	Hr.	Daily	Bare Costs	Incl. O&P
1 Labor Foreman (outside)	$38.65	$309.20	$59.65	$477.20	$39.46	$60.58
5 Laborers	36.65	1466.00	56.55	2262.00		
1 Equipment Operator, Crane	50.25	402.00	76.20	609.60		
1 Equipment Oiler	43.55	348.40	66.00	528.00		
1 Conc. Bucket, 2 C.Y.		35.60		39.16		
1 Lattice Boom Crane, 165 Ton		1954.00		2149.40	31.09	34.20
64 L.H., Daily Totals		$4515.20		$6065.36	$70.55	$94.77

Crew C-7C	Hr.	Daily	Hr.	Daily	Bare Costs	Incl. O&P
1 Labor Foreman (outside)	$38.65	$309.20	$59.65	$477.20	$39.96	$61.34
5 Laborers	36.65	1466.00	56.55	2262.00		
2 Equipment Operators (med.)	48.90	782.40	74.15	1186.40		
2 F.E. Loaders, W.M., 4 C.Y.		1347.60		1482.36	21.06	23.16
64 L.H., Daily Totals		$3905.20		$5407.96	$61.02	$84.50

Crew C-7D	Hr.	Daily	Hr.	Daily	Bare Costs	Incl. O&P
1 Labor Foreman (outside)	$38.65	$309.20	$59.65	$477.20	$38.69	$59.51
5 Laborers	36.65	1466.00	56.55	2262.00		
1 Equip. Oper. (medium)	48.90	391.20	74.15	593.20		
1 Concrete Conveyer		205.40		225.94	3.67	4.03
56 L.H., Daily Totals		$2371.80		$3558.34	$42.35	$63.54

Crew C-8	Hr.	Daily	Hr.	Daily	Bare Costs	Incl. O&P
1 Labor Foreman (outside)	$38.65	$309.20	$59.65	$477.20	$40.80	$61.95
3 Laborers	36.65	879.60	56.55	1357.20		
2 Cement Finishers	44.05	704.80	65.10	1041.60		
1 Equip. Oper. (medium)	48.90	391.20	74.15	593.20		
1 Concrete Pump (Small)		710.40		781.44	12.69	13.95
56 L.H., Daily Totals		$2995.20		$4250.64	$53.49	$75.90

Crew C-8A	Hr.	Daily	Hr.	Daily	Bare Costs	Incl. O&P
1 Labor Foreman (outside)	$38.65	$309.20	$59.65	$477.20	$39.45	$59.92
3 Laborers	36.65	879.60	56.55	1357.20		
2 Cement Finishers	44.05	704.80	65.10	1041.60		
48 L.H., Daily Totals		$1893.60		$2876.00	$39.45	$59.92

Crew C-8B	Hr.	Daily	Hr.	Daily	Bare Costs	Incl. O&P
1 Labor Foreman (outside)	$38.65	$309.20	$59.65	$477.20	$39.50	$60.69
3 Laborers	36.65	879.60	56.55	1357.20		
1 Equip. Oper. (medium)	48.90	391.20	74.15	593.20		
1 Vibrating Power Screed		63.25		69.58		
1 Roller, Vibratory, 25 Ton		692.80		762.08		
1 Dozer, 200 H.P.		1325.00		1457.50	52.03	57.23
40 L.H., Daily Totals		$3661.05		$4716.76	$91.53	$117.92

Crew C-8C	Hr.	Daily	Hr.	Daily	Bare Costs	Incl. O&P
1 Labor Foreman (outside)	$38.65	$309.20	$59.65	$477.20	$40.26	$61.42
3 Laborers	36.65	879.60	56.55	1357.20		
1 Cement Finisher	44.05	352.40	65.10	520.80		
1 Equip. Oper. (medium)	48.90	391.20	74.15	593.20		
1 Shotcrete Rig, 12 C.Y./hr		252.80		278.08		
1 Air Compressor, 160 cfm		157.40		173.14		
4 -50' Air Hoses, 1"		16.40		18.04		
4 -50' Air Hoses, 2"		31.00		34.10	9.53	10.49
48 L.H., Daily Totals		$2390.00		$3451.76	$49.79	$71.91

Crew C-8D	Hr.	Daily	Hr.	Daily	Bare Costs	Incl. O&P
1 Labor Foreman (outside)	$38.65	$309.20	$59.65	$477.20	$41.60	$63.16
1 Laborer	36.65	293.20	56.55	452.40		
1 Cement Finisher	44.05	352.40	65.10	520.80		
1 Equipment Oper. (light)	47.05	376.40	71.35	570.80		
1 Air Compressor, 250 cfm		202.40		222.64		
2 -50' Air Hoses, 1"		8.20		9.02	6.58	7.24
32 L.H., Daily Totals		$1541.80		$2252.86	$48.18	$70.40

Crew C-8E	Hr.	Daily	Hr.	Daily	Bare Costs	Incl. O&P
1 Labor Foreman (outside)	$38.65	$309.20	$59.65	$477.20	$39.95	$60.96
3 Laborers	36.65	879.60	56.55	1357.20		
1 Cement Finisher	44.05	352.40	65.10	520.80		
1 Equipment Oper. (light)	47.05	376.40	71.35	570.80		
1 Shotcrete Rig, 35 C.Y./hr		284.40		312.84		
1 Air Compressor, 250 cfm		202.40		222.64		
4 -50' Air Hoses, 1"		16.40		18.04		
4 -50' Air Hoses, 2"		31.00		34.10	11.13	12.24
48 L.H., Daily Totals		$2451.80		$3513.62	$51.08	$73.20

Crew C-10	Hr.	Daily	Hr.	Daily	Bare Costs	Incl. O&P
1 Laborer	$36.65	$293.20	$56.55	$452.40	$41.58	$62.25
2 Cement Finishers	44.05	704.80	65.10	1041.60		
24 L.H., Daily Totals		$998.00		$1494.00	$41.58	$62.25

Crew C-10B	Hr.	Daily	Hr.	Daily	Bare Costs	Incl. O&P
3 Laborers	$36.65	$879.60	$56.55	$1357.20	$39.61	$59.97
2 Cement Finishers	44.05	704.80	65.10	1041.60		
1 Concrete Mixer, 10 C.F.		176.80		194.48		
2 Trowels, 48" Walk-Behind		107.20		117.92	7.10	7.81
40 L.H., Daily Totals		$1868.40		$2711.20	$46.71	$67.78

Crew C-10C	Hr.	Daily	Hr.	Daily	Bare Costs	Incl. O&P
1 Laborer	$36.65	$293.20	$56.55	$452.40	$41.58	$62.25
2 Cement Finishers	44.05	704.80	65.10	1041.60		
1 Trowel, 48" Walk-Behind		53.60		58.96	2.23	2.46
24 L.H., Daily Totals		$1051.60		$1552.96	$43.82	$64.71

Crews

Crew No.		Bare Costs		Incl. Subs O&P		Cost Per Labor-Hour	
Crew C-10D	Hr.	Daily	Hr.	Daily	Bare Costs	Incl. O&P
1 Laborer	$36.65	$293.20	$56.55	$452.40	$41.58	$62.25
2 Cement Finishers	44.05	704.80	65.10	1041.60		
1 Vibrating Power Screed		63.25		69.58		
1 Trowel, 48" Walk-Behind		53.60		58.96	4.87	5.36
24 L.H., Daily Totals		$1114.85		$1622.54	$46.45	$67.61
Crew C-10E	Hr.	Daily	Hr.	Daily	Bare Costs	Incl. O&P
1 Laborer	$36.65	$293.20	$56.55	$452.40	$41.58	$62.25
2 Cement Finishers	44.05	704.80	65.10	1041.60		
1 Vibrating Power Screed		63.25		69.58		
1 Cement Trowel, 96" Ride-On		197.40		217.14	10.86	11.95
24 L.H., Daily Totals		$1258.65		$1780.71	$52.44	$74.20
Crew C-10F	Hr.	Daily	Hr.	Daily	Bare Costs	Incl. O&P
1 Laborer	$36.65	$293.20	$56.55	$452.40	$41.58	$62.25
2 Cement Finishers	44.05	704.80	65.10	1041.60		
1 Aerial Lift Truck, 60' Boom		456.00		501.60	19.00	20.90
24 L.H., Daily Totals		$1454.00		$1995.60	$60.58	$83.15
Crew C-11	Hr.	Daily	Hr.	Daily	Bare Costs	Incl. O&P
1 Struc. Steel Foreman (outside)	$53.10	$424.80	$93.70	$749.60	$50.39	$86.34
6 Struc. Steel Workers	51.10	2452.80	90.20	4329.60		
1 Equip. Oper. (crane)	50.25	402.00	76.20	609.60		
1 Equip. Oper. (oiler)	43.55	348.40	66.00	528.00		
1 Lattice Boom Crane, 150 Ton		1838.00		2021.80	25.53	28.08
72 L.H., Daily Totals		$5466.00		$8238.60	$75.92	$114.43
Crew C-12	Hr.	Daily	Hr.	Daily	Bare Costs	Incl. O&P
1 Carpenter Foreman (outside)	$47.85	$382.80	$73.85	$590.80	$45.38	$69.81
3 Carpenters	45.85	1100.40	70.75	1698.00		
1 Laborer	36.65	293.20	56.55	452.40		
1 Equip. Oper. (crane)	50.25	402.00	76.20	609.60		
1 Hyd. Crane, 12 Ton		663.00		729.30	13.81	15.19
48 L.H., Daily Totals		$2841.40		$4080.10	$59.20	$85.00
Crew C-13	Hr.	Daily	Hr.	Daily	Bare Costs	Incl. O&P
1 Struc. Steel Worker	$51.10	$408.80	$90.20	$721.60	$49.35	$83.72
1 Welder	51.10	408.80	90.20	721.60		
1 Carpenter	45.85	366.80	70.75	566.00		
1 Welder, Gas Engine, 300 amp		142.00		156.20	5.92	6.51
24 L.H., Daily Totals		$1326.40		$2165.40	$55.27	$90.22
Crew C-14	Hr.	Daily	Hr.	Daily	Bare Costs	Incl. O&P
1 Carpenter Foreman (outside)	$47.85	$382.80	$73.85	$590.80	$44.90	$69.13
5 Carpenters	45.85	1834.00	70.75	2830.00		
4 Laborers	36.65	1172.80	56.55	1809.60		
4 Rodmen (reinf.)	50.65	1620.80	79.55	2545.60		
2 Cement Finishers	44.05	704.80	65.10	1041.60		
1 Equip. Oper. (crane)	50.25	402.00	76.20	609.60		
1 Equip. Oper. (oiler)	43.55	348.40	66.00	528.00		
1 Hyd. Crane, 80 Ton		1652.00		1817.20	11.47	12.62
144 L.H., Daily Totals		$8117.60		$11772.40	$56.37	$81.75

Crew No.		Bare Costs		Incl. Subs O&P		Cost Per Labor-Hour	
Crew C-14A	Hr.	Daily	Hr.	Daily	Bare Costs	Incl. O&P
1 Carpenter Foreman (outside)	$47.85	$382.80	$73.85	$590.80	$46.01	$71.06
16 Carpenters	45.85	5868.80	70.75	9056.00		
4 Rodmen (reinf.)	50.65	1620.80	79.55	2545.60		
2 Laborers	36.65	586.40	56.55	904.80		
1 Cement Finisher	44.05	352.40	65.10	520.80		
1 Equip. Oper. (medium)	48.90	391.20	74.15	593.20		
1 Gas Engine Vibrator		33.00		36.30		
1 Concrete Pump (Small)		710.40		781.44	3.72	4.09
200 L.H., Daily Totals		$9945.80		$15028.94	$49.73	$75.14
Crew C-14B	Hr.	Daily	Hr.	Daily	Bare Costs	Incl. O&P
1 Carpenter Foreman (outside)	$47.85	$382.80	$73.85	$590.80	$45.94	$70.83
16 Carpenters	45.85	5868.80	70.75	9056.00		
4 Rodmen (reinf.)	50.65	1620.80	79.55	2545.60		
2 Laborers	36.65	586.40	56.55	904.80		
2 Cement Finishers	44.05	704.80	65.10	1041.60		
1 Equip. Oper. (medium)	48.90	391.20	74.15	593.20		
1 Gas Engine Vibrator		33.00		36.30		
1 Concrete Pump (Small)		710.40		781.44	3.57	3.93
208 L.H., Daily Totals		$10298.20		$15549.74	$49.51	$74.76
Crew C-14C	Hr.	Daily	Hr.	Daily	Bare Costs	Incl. O&P
1 Carpenter Foreman (outside)	$47.85	$382.80	$73.85	$590.80	$43.92	$67.77
6 Carpenters	45.85	2200.80	70.75	3396.00		
2 Rodmen (reinf.)	50.65	810.40	79.55	1272.80		
4 Laborers	36.65	1172.80	56.55	1809.60		
1 Cement Finisher	44.05	352.40	65.10	520.80		
1 Gas Engine Vibrator		33.00		36.30	.29	.32
112 L.H., Daily Totals		$4952.20		$7626.30	$44.22	$68.09
Crew C-14D	Hr.	Daily	Hr.	Daily	Bare Costs	Incl. O&P
1 Carpenter Foreman (outside)	$47.85	$382.80	$73.85	$590.80	$45.63	$70.35
18 Carpenters	45.85	6602.40	70.75	10188.00		
2 Rodmen (reinf.)	50.65	810.40	79.55	1272.80		
2 Laborers	36.65	586.40	56.55	904.80		
1 Cement Finisher	44.05	352.40	65.10	520.80		
1 Equip. Oper. (medium)	48.90	391.20	74.15	593.20		
1 Gas Engine Vibrator		33.00		36.30		
1 Concrete Pump (Small)		710.40		781.44	3.72	4.09
200 L.H., Daily Totals		$9869.00		$14888.14	$49.34	$74.44
Crew C-14E	Hr.	Daily	Hr.	Daily	Bare Costs	Incl. O&P
1 Carpenter Foreman (outside)	$47.85	$382.80	$73.85	$590.80	$45.10	$69.85
2 Carpenters	45.85	733.60	70.75	1132.00		
4 Rodmen (reinf.)	50.65	1620.80	79.55	2545.60		
3 Laborers	36.65	879.60	56.55	1357.20		
1 Cement Finisher	44.05	352.40	65.10	520.80		
1 Gas Engine Vibrator		33.00		36.30	.38	.41
88 L.H., Daily Totals		$4002.20		$6182.70	$45.48	$70.26
Crew C-14F	Hr.	Daily	Hr.	Daily	Bare Costs	Incl. O&P
1 Labor Foreman (outside)	$38.65	$309.20	$59.65	$477.20	$41.81	$62.59
2 Laborers	36.65	586.40	56.55	904.80		
6 Cement Finishers	44.05	2114.40	65.10	3124.80		
1 Gas Engine Vibrator		33.00		36.30	.46	.50
72 L.H., Daily Totals		$3043.00		$4543.10	$42.26	$63.10

Crews

Crew No.	Bare Costs		Incl. Subs O&P		Cost Per Labor-Hour	
Crew C-14G	Hr.	Daily	Hr.	Daily	Bare Costs	Incl. O&P
1 Labor Foreman (outside)	$38.65	$309.20	$59.65	$477.20	$41.16	$61.88
2 Laborers	36.65	586.40	56.55	904.80		
4 Cement Finishers	44.05	1409.60	65.10	2083.20		
1 Gas Engine Vibrator		33.00		36.30	.59	.65
56 L.H., Daily Totals		$2338.20		$3501.50	$41.75	$62.53
Crew C-14H	Hr.	Daily	Hr.	Daily	Bare Costs	Incl. O&P
1 Carpenter Foreman (outside)	$47.85	$382.80	$73.85	$590.80	$45.15	$69.42
2 Carpenters	45.85	733.60	70.75	1132.00		
1 Rodman (reinf.)	50.65	405.20	79.55	636.40		
1 Laborer	36.65	293.20	56.55	452.40		
1 Cement Finisher	44.05	352.40	65.10	520.80		
1 Gas Engine Vibrator		33.00		36.30	.69	.76
48 L.H., Daily Totals		$2200.20		$3368.70	$45.84	$70.18
Crew C-14L	Hr.	Daily	Hr.	Daily	Bare Costs	Incl. O&P
1 Carpenter Foreman (outside)	$47.85	$382.80	$73.85	$590.80	$42.80	$65.80
6 Carpenters	45.85	2200.80	70.75	3396.00		
4 Laborers	36.65	1172.80	56.55	1809.60		
1 Cement Finisher	44.05	352.40	65.10	520.80		
1 Gas Engine Vibrator		33.00		36.30	.34	.38
96 L.H., Daily Totals		$4141.80		$6353.50	$43.14	$66.18
Crew C-14M	Hr.	Daily	Hr.	Daily	Bare Costs	Incl. O&P
1 Carpenter Foreman (outside)	$47.85	$382.80	$73.85	$590.80	$44.56	$68.41
2 Carpenters	45.85	733.60	70.75	1132.00		
1 Rodman (reinf.)	50.65	405.20	79.55	636.40		
2 Laborers	36.65	586.40	56.55	904.80		
1 Cement Finisher	44.05	352.40	65.10	520.80		
1 Equip. Oper. (medium)	48.90	391.20	74.15	593.20		
1 Gas Engine Vibrator		33.00		36.30		
1 Concrete Pump (Small)		710.40		781.44	11.62	12.78
64 L.H., Daily Totals		$3595.00		$5195.74	$56.17	$81.18
Crew C-15	Hr.	Daily	Hr.	Daily	Bare Costs	Incl. O&P
1 Carpenter Foreman (outside)	$47.85	$382.80	$73.85	$590.80	$43.14	$66.08
2 Carpenters	45.85	733.60	70.75	1132.00		
3 Laborers	36.65	879.60	56.55	1357.20		
2 Cement Finishers	44.05	704.80	65.10	1041.60		
1 Rodman (reinf.)	50.65	405.20	79.55	636.40		
72 L.H., Daily Totals		$3106.00		$4758.00	$43.14	$66.08
Crew C-16	Hr.	Daily	Hr.	Daily	Bare Costs	Incl. O&P
1 Labor Foreman (outside)	$38.65	$309.20	$59.65	$477.20	$40.80	$61.95
3 Laborers	36.65	879.60	56.55	1357.20		
2 Cement Finishers	44.05	704.80	65.10	1041.60		
1 Equip. Oper. (medium)	48.90	391.20	74.15	593.20		
1 Gunite Pump Rig		370.40		407.44		
2 -50' Air Hoses, 3/4"		6.50		7.15		
2 -50' Air Hoses, 2"		15.50		17.05	7.01	7.71
56 L.H., Daily Totals		$2677.20		$3900.84	$47.81	$69.66
Crew C-16A	Hr.	Daily	Hr.	Daily	Bare Costs	Incl. O&P
1 Laborer	$36.65	$293.20	$56.55	$452.40	$43.41	$65.22
2 Cement Finishers	44.05	704.80	65.10	1041.60		
1 Equip. Oper. (medium)	48.90	391.20	74.15	593.20		
1 Gunite Pump Rig		370.40		407.44		
2 -50' Air Hoses, 3/4"		6.50		7.15		
2 -50' Air Hoses, 2"		15.50		17.05		
1 Aerial Lift Truck, 60' Boom		456.00		501.60	26.51	29.16
32 L.H., Daily Totals		$2237.60		$3020.44	$69.92	$94.39

Crew No.	Bare Costs		Incl. Subs O&P		Cost Per Labor-Hour	
Crew C-17	Hr.	Daily	Hr.	Daily	Bare Costs	Incl. O&P
2 Skilled Worker Foremen (out)	$49.30	$788.80	$76.35	$1221.60	$47.70	$73.87
8 Skilled Workers	47.30	3027.20	73.25	4688.00		
80 L.H., Daily Totals		$3816.00		$5909.60	$47.70	$73.87
Crew C-17A	Hr.	Daily	Hr.	Daily	Bare Costs	Incl. O&P
2 Skilled Worker Foremen (out)	$49.30	$788.80	$76.35	$1221.60	$47.73	$73.90
8 Skilled Workers	47.30	3027.20	73.25	4688.00		
.125 Equip. Oper. (crane)	50.25	50.25	76.20	76.20		
.125 Hyd. Crane, 80 Ton		206.50		227.15	2.55	2.80
81 L.H., Daily Totals		$4072.75		$6212.95	$50.28	$76.70
Crew C-17B	Hr.	Daily	Hr.	Daily	Bare Costs	Incl. O&P
2 Skilled Worker Foremen (out)	$49.30	$788.80	$76.35	$1221.60	$47.76	$73.93
8 Skilled Workers	47.30	3027.20	73.25	4688.00		
.25 Equip. Oper. (crane)	50.25	100.50	76.20	152.40		
.25 Hyd. Crane, 80 Ton		413.00		454.30		
.25 Trowel, 48" Walk-Behind		13.40		14.74	5.20	5.72
82 L.H., Daily Totals		$4342.90		$6531.04	$52.96	$79.65
Crew C-17C	Hr.	Daily	Hr.	Daily	Bare Costs	Incl. O&P
2 Skilled Worker Foremen (out)	$49.30	$788.80	$76.35	$1221.60	$47.79	$73.95
8 Skilled Workers	47.30	3027.20	73.25	4688.00		
.375 Equip. Oper. (crane)	50.25	150.75	76.20	228.60		
.375 Hyd. Crane, 80 Ton		619.50		681.45	7.46	8.21
83 L.H., Daily Totals		$4586.25		$6819.65	$55.26	$82.16
Crew C-17D	Hr.	Daily	Hr.	Daily	Bare Costs	Incl. O&P
2 Skilled Worker Foremen (out)	$49.30	$788.80	$76.35	$1221.60	$47.82	$73.98
8 Skilled Workers	47.30	3027.20	73.25	4688.00		
.5 Equip. Oper. (crane)	50.25	201.00	76.20	304.80		
.5 Hyd. Crane, 80 Ton		826.00		908.60	9.83	10.82
84 L.H., Daily Totals		$4843.00		$7123.00	$57.65	$84.80
Crew C-17E	Hr.	Daily	Hr.	Daily	Bare Costs	Incl. O&P
2 Skilled Worker Foremen (out)	$49.30	$788.80	$76.35	$1221.60	$47.70	$73.87
8 Skilled Workers	47.30	3027.20	73.25	4688.00		
1 Hyd. Jack with Rods		94.55		104.01	1.18	1.30
80 L.H., Daily Totals		$3910.55		$6013.60	$48.88	$75.17
Crew C-18	Hr.	Daily	Hr.	Daily	Bare Costs	Incl. O&P
.125 Labor Foreman (outside)	$38.65	$38.65	$59.65	$59.65	$36.87	$56.89
1 Laborer	36.65	293.20	56.55	452.40		
1 Concrete Cart, 10 C.F.		60.20		66.22	6.69	7.36
9 L.H., Daily Totals		$392.05		$578.27	$43.56	$64.25
Crew C-19	Hr.	Daily	Hr.	Daily	Bare Costs	Incl. O&P
.125 Labor Foreman (outside)	$38.65	$38.65	$59.65	$59.65	$36.87	$56.89
1 Laborer	36.65	293.20	56.55	452.40		
1 Concrete Cart, 18 C.F.		101.20		111.32	11.24	12.37
9 L.H., Daily Totals		$433.05		$623.37	$48.12	$69.26
Crew C-20	Hr.	Daily	Hr.	Daily	Bare Costs	Incl. O&P
1 Labor Foreman (outside)	$38.65	$309.20	$59.65	$477.20	$39.36	$60.21
5 Laborers	36.65	1466.00	56.55	2262.00		
1 Cement Finisher	44.05	352.40	65.10	520.80		
1 Equip. Oper. (medium)	48.90	391.20	74.15	593.20		
2 Gas Engine Vibrators		66.00		72.60		
1 Concrete Pump (Small)		710.40		781.44	12.13	13.34
64 L.H., Daily Totals		$3295.20		$4707.24	$51.49	$73.55

Crews

Crew No.	Bare Costs Hr.	Bare Costs Daily	Incl. Subs O&P Hr.	Incl. Subs O&P Daily	Cost Per Labor-Hour Bare Costs	Cost Per Labor-Hour Incl. O&P
Crew C-21	Hr.	Daily	Hr.	Daily	Bare Costs	Incl. O&P
1 Labor Foreman (outside)	$38.65	$309.20	$59.65	$477.20	$39.36	$60.21
5 Laborers	36.65	1466.00	56.55	2262.00		
1 Cement Finisher	44.05	352.40	65.10	520.80		
1 Equip. Oper. (medium)	48.90	391.20	74.15	593.20		
2 Gas Engine Vibrators		66.00		72.60		
1 Concrete Conveyer		205.40		225.94	4.24	4.66
64 L.H., Daily Totals		$2790.20		$4151.74	$43.60	$64.87
Crew C-22	Hr.	Daily	Hr.	Daily	Bare Costs	Incl. O&P
1 Rodman Foreman (outside)	$52.65	$421.20	$82.70	$661.60	$50.85	$79.75
4 Rodmen (reinf.)	50.65	1620.80	79.55	2545.60		
.125 Equip. Oper. (crane)	50.25	50.25	76.20	76.20		
.125 Equip. Oper. (oiler)	43.55	43.55	66.00	66.00		
.125 Hyd. Crane, 25 Ton		93.22		102.55	2.22	2.44
42 L.H., Daily Totals		$2229.03		$3451.95	$53.07	$82.19
Crew C-23	Hr.	Daily	Hr.	Daily	Bare Costs	Incl. O&P
2 Skilled Worker Foremen (out)	$49.30	$788.80	$76.35	$1221.60	$47.62	$73.44
6 Skilled Workers	47.30	2270.40	73.25	3516.00		
1 Equip. Oper. (crane)	50.25	402.00	76.20	609.60		
1 Equip. Oper. (oiler)	43.55	348.40	66.00	528.00		
1 Lattice Boom Crane, 90 Ton		1529.00		1681.90	19.11	21.02
80 L.H., Daily Totals		$5338.60		$7557.10	$66.73	$94.46
Crew C-23A	Hr.	Daily	Hr.	Daily	Bare Costs	Incl. O&P
1 Labor Foreman (outside)	$38.65	$309.20	$59.65	$477.20	$41.15	$62.99
2 Laborers	36.65	586.40	56.55	904.80		
1 Equip. Oper. (crane)	50.25	402.00	76.20	609.60		
1 Equip. Oper. (oiler)	43.55	348.40	66.00	528.00		
1 Crawler Crane, 100 Ton		1703.00		1873.30		
3 Conc. Buckets, 8 C.Y.		595.80		655.38	57.47	63.22
40 L.H., Daily Totals		$3944.80		$5048.28	$98.62	$126.21
Crew C-24	Hr.	Daily	Hr.	Daily	Bare Costs	Incl. O&P
2 Skilled Worker Foremen (out)	$49.30	$788.80	$76.35	$1221.60	$47.62	$73.44
6 Skilled Workers	47.30	2270.40	73.25	3516.00		
1 Equip. Oper. (crane)	50.25	402.00	76.20	609.60		
1 Equip. Oper. (oiler)	43.55	348.40	66.00	528.00		
1 Lattice Boom Crane, 150 Ton		1838.00		2021.80	22.98	25.27
80 L.H., Daily Totals		$5647.60		$7897.00	$70.59	$98.71
Crew C-25	Hr.	Daily	Hr.	Daily	Bare Costs	Incl. O&P
2 Rodmen (reinf.)	$50.65	$810.40	$79.55	$1272.80	$39.90	$64.58
2 Rodmen Helpers	29.15	466.40	49.60	793.60		
32 L.H., Daily Totals		$1276.80		$2066.40	$39.90	$64.58
Crew C-27	Hr.	Daily	Hr.	Daily	Bare Costs	Incl. O&P
2 Cement Finishers	$44.05	$704.80	$65.10	$1041.60	$44.05	$65.10
1 Concrete Saw		175.20		192.72	10.95	12.05
16 L.H., Daily Totals		$880.00		$1234.32	$55.00	$77.14
Crew C-28	Hr.	Daily	Hr.	Daily	Bare Costs	Incl. O&P
1 Cement Finisher	$44.05	$352.40	$65.10	$520.80	$44.05	$65.10
1 Portable Air Compressor, Gas		17.70		19.47	2.21	2.43
8 L.H., Daily Totals		$370.10		$540.27	$46.26	$67.53
Crew C-29	Hr.	Daily	Hr.	Daily	Bare Costs	Incl. O&P
1 Laborer	$36.65	$293.20	$56.55	$452.40	$36.65	$56.55
1 Pressure Washer		69.40		76.34	8.68	9.54
8 L.H., Daily Totals		$362.60		$528.74	$45.33	$66.09

Crew No.	Bare Costs Hr.	Bare Costs Daily	Incl. Subs O&P Hr.	Incl. Subs O&P Daily	Cost Per Labor-Hour Bare Costs	Cost Per Labor-Hour Incl. O&P
Crew C-30	Hr.	Daily	Hr.	Daily	Bare Costs	Incl. O&P
1 Laborer	$36.65	$293.20	$56.55	$452.40	$36.65	$56.55
1 Concrete Mixer, 10 C.F.		176.80		194.48	22.10	24.31
8 L.H., Daily Totals		$470.00		$646.88	$58.75	$80.86
Crew C-31	Hr.	Daily	Hr.	Daily	Bare Costs	Incl. O&P
1 Cement Finisher	$44.05	$352.40	$65.10	$520.80	$44.05	$65.10
1 Grout Pump		370.40		407.44	46.30	50.93
8 L.H., Daily Totals		$722.80		$928.24	$90.35	$116.03
Crew C-32	Hr.	Daily	Hr.	Daily	Bare Costs	Incl. O&P
1 Cement Finisher	$44.05	$352.40	$65.10	$520.80	$40.35	$60.83
1 Laborer	36.65	293.20	56.55	452.40		
1 Crack Chaser Saw, Gas, 6 H.P.		30.80		33.88		
1 Vacuum Pick-Up System		60.75		66.83	5.72	6.29
16 L.H., Daily Totals		$737.15		$1073.91	$46.07	$67.12
Crew D-1	Hr.	Daily	Hr.	Daily	Bare Costs	Incl. O&P
1 Bricklayer	$45.60	$364.80	$69.50	$556.00	$41.30	$62.95
1 Bricklayer Helper	37.00	296.00	56.40	451.20		
16 L.H., Daily Totals		$660.80		$1007.20	$41.30	$62.95
Crew D-2	Hr.	Daily	Hr.	Daily	Bare Costs	Incl. O&P
3 Bricklayers	$45.60	$1094.40	$69.50	$1668.00	$42.50	$64.85
2 Bricklayer Helpers	37.00	592.00	56.40	902.40		
.5 Carpenter	45.85	183.40	70.75	283.00		
44 L.H., Daily Totals		$1869.80		$2853.40	$42.50	$64.85
Crew D-3	Hr.	Daily	Hr.	Daily	Bare Costs	Incl. O&P
3 Bricklayers	$45.60	$1094.40	$69.50	$1668.00	$42.34	$64.57
2 Bricklayer Helpers	37.00	592.00	56.40	902.40		
.25 Carpenter	45.85	91.70	70.75	141.50		
42 L.H., Daily Totals		$1778.10		$2711.90	$42.34	$64.57
Crew D-4	Hr.	Daily	Hr.	Daily	Bare Costs	Incl. O&P
1 Bricklayer	$45.60	$364.80	$69.50	$556.00	$41.66	$63.41
2 Bricklayer Helpers	37.00	592.00	56.40	902.40		
1 Equip. Oper. (light)	47.05	376.40	71.35	570.80		
1 Grout Pump, 50 C.F./hr.		134.80		148.28	4.21	4.63
32 L.H., Daily Totals		$1468.00		$2177.48	$45.88	$68.05
Crew D-5	Hr.	Daily	Hr.	Daily	Bare Costs	Incl. O&P
1 Bricklayer	$45.60	$364.80	$69.50	$556.00	$45.60	$69.50
8 L.H., Daily Totals		$364.80		$556.00	$45.60	$69.50
Crew D-6	Hr.	Daily	Hr.	Daily	Bare Costs	Incl. O&P
3 Bricklayers	$45.60	$1094.40	$69.50	$1668.00	$41.48	$63.26
3 Bricklayer Helpers	37.00	888.00	56.40	1353.60		
.25 Carpenter	45.85	91.70	70.75	141.50		
50 L.H., Daily Totals		$2074.10		$3163.10	$41.48	$63.26
Crew D-7	Hr.	Daily	Hr.	Daily	Bare Costs	Incl. O&P
1 Tile Layer	$41.95	$335.60	$61.85	$494.80	$37.60	$55.42
1 Tile Layer Helper	33.25	266.00	49.00	392.00		
16 L.H., Daily Totals		$601.60		$886.80	$37.60	$55.42
Crew D-8	Hr.	Daily	Hr.	Daily	Bare Costs	Incl. O&P
3 Bricklayers	$45.60	$1094.40	$69.50	$1668.00	$42.16	$64.26
2 Bricklayer Helpers	37.00	592.00	56.40	902.40		
40 L.H., Daily Totals		$1686.40		$2570.40	$42.16	$64.26

Crews

Crew No.		Bare Costs		Incl. Subs O&P		Cost Per Labor-Hour	
Crew D-9	Hr.	Daily	Hr.	Daily	Bare Costs	Incl. O&P
3 Bricklayers	$45.60	$1094.40	$69.50	$1668.00	$41.30	$62.95
3 Bricklayer Helpers	37.00	888.00	56.40	1353.60		
48 L.H., Daily Totals		$1982.40		$3021.60	$41.30	$62.95
Crew D-10	Hr.	Daily	Hr.	Daily	Bare Costs	Incl. O&P
1 Bricklayer Foreman (outside)	$47.60	$380.80	$72.55	$580.40	$45.11	$68.66
1 Bricklayer	45.60	364.80	69.50	556.00		
1 Bricklayer Helper	37.00	296.00	56.40	451.20		
1 Equip. Oper. (crane)	50.25	402.00	76.20	609.60		
1 S.P. Crane, 4x4, 12 Ton		481.20		529.32	15.04	16.54
32 L.H., Daily Totals		$1924.80		$2726.52	$60.15	$85.20
Crew D-11	Hr.	Daily	Hr.	Daily	Bare Costs	Incl. O&P
1 Bricklayer Foreman (outside)	$47.60	$380.80	$72.55	$580.40	$43.40	$66.15
1 Bricklayer	45.60	364.80	69.50	556.00		
1 Bricklayer Helper	37.00	296.00	56.40	451.20		
24 L.H., Daily Totals		$1041.60		$1587.60	$43.40	$66.15
Crew D-12	Hr.	Daily	Hr.	Daily	Bare Costs	Incl. O&P
1 Bricklayer Foreman (outside)	$47.60	$380.80	$72.55	$580.40	$41.80	$63.71
1 Bricklayer	45.60	364.80	69.50	556.00		
2 Bricklayer Helpers	37.00	592.00	56.40	902.40		
32 L.H., Daily Totals		$1337.60		$2038.80	$41.80	$63.71
Crew D-13	Hr.	Daily	Hr.	Daily	Bare Costs	Incl. O&P
1 Bricklayer Foreman (outside)	$47.60	$380.80	$72.55	$580.40	$43.88	$66.97
1 Bricklayer	45.60	364.80	69.50	556.00		
2 Bricklayer Helpers	37.00	592.00	56.40	902.40		
1 Carpenter	45.85	366.80	70.75	566.00		
1 Equip. Oper. (crane)	50.25	402.00	76.20	609.60		
1 S.P. Crane, 4x4, 12 Ton		481.20		529.32	10.03	11.03
48 L.H., Daily Totals		$2587.60		$3743.72	$53.91	$77.99
Crew E-1	Hr.	Daily	Hr.	Daily	Bare Costs	Incl. O&P
1 Welder Foreman (outside)	$53.10	$424.80	$93.70	$749.60	$50.42	$85.08
1 Welder	51.10	408.80	90.20	721.60		
1 Equip. Oper. (light)	47.05	376.40	71.35	570.80		
1 Welder, Gas Engine, 300 amp		142.00		156.20	5.92	6.51
24 L.H., Daily Totals		$1352.00		$2198.20	$56.33	$91.59
Crew E-2	Hr.	Daily	Hr.	Daily	Bare Costs	Incl. O&P
1 Struc. Steel Foreman (outside)	$53.10	$424.80	$93.70	$749.60	$50.19	$85.24
4 Struc. Steel Workers	51.10	1635.20	90.20	2886.40		
1 Equip. Oper. (crane)	50.25	402.00	76.20	609.60		
1 Equip. Oper. (oiler)	43.55	348.40	66.00	528.00		
1 Lattice Boom Crane, 90 Ton		1529.00		1681.90	27.30	30.03
56 L.H., Daily Totals		$4339.40		$6455.50	$77.49	$115.28
Crew E-3	Hr.	Daily	Hr.	Daily	Bare Costs	Incl. O&P
1 Struc. Steel Foreman (outside)	$53.10	$424.80	$93.70	$749.60	$51.77	$91.37
1 Struc. Steel Worker	51.10	408.80	90.20	721.60		
1 Welder	51.10	408.80	90.20	721.60		
1 Welder, Gas Engine, 300 amp		142.00		156.20	5.92	6.51
24 L.H., Daily Totals		$1384.40		$2349.00	$57.68	$97.88

Crew No.		Bare Costs		Incl. Subs O&P		Cost Per Labor-Hour	
Crew E-3A	Hr.	Daily	Hr.	Daily	Bare Costs	Incl. O&P
1 Struc. Steel Foreman (outside)	$53.10	$424.80	$93.70	$749.60	$51.77	$91.37
1 Struc. Steel Worker	51.10	408.80	90.20	721.60		
1 Welder	51.10	408.80	90.20	721.60		
1 Welder, Gas Engine, 300 amp		142.00		156.20		
1 Aerial Lift Truck, 40' Boom		325.00		357.50	19.46	21.40
24 L.H., Daily Totals		$1709.40		$2706.50	$71.22	$112.77
Crew E-4	Hr.	Daily	Hr.	Daily	Bare Costs	Incl. O&P
1 Struc. Steel Foreman (outside)	$53.10	$424.80	$93.70	$749.60	$51.60	$91.08
3 Struc. Steel Workers	51.10	1226.40	90.20	2164.80		
1 Welder, Gas Engine, 300 amp		142.00		156.20	4.44	4.88
32 L.H., Daily Totals		$1793.20		$3070.60	$56.04	$95.96
Crew E-5	Hr.	Daily	Hr.	Daily	Bare Costs	Incl. O&P
2 Struc. Steel Foremen (outside)	$53.10	$849.60	$93.70	$1499.20	$50.66	$87.08
5 Struc. Steel Workers	51.10	2044.00	90.20	3608.00		
1 Equip. Oper. (crane)	50.25	402.00	76.20	609.60		
1 Welder	51.10	408.80	90.20	721.60		
1 Equip. Oper. (oiler)	43.55	348.40	66.00	528.00		
1 Lattice Boom Crane, 90 Ton		1529.00		1681.90		
1 Welder, Gas Engine, 300 amp		142.00		156.20	20.89	22.98
80 L.H., Daily Totals		$5723.80		$8804.50	$71.55	$110.06
Crew E-6	Hr.	Daily	Hr.	Daily	Bare Costs	Incl. O&P
3 Struc. Steel Foremen (outside)	$53.10	$1274.40	$93.70	$2248.80	$50.70	$87.29
9 Struc. Steel Workers	51.10	3679.20	90.20	6494.40		
1 Equip. Oper. (crane)	50.25	402.00	76.20	609.60		
1 Welder	51.10	408.80	90.20	721.60		
1 Equip. Oper. (oiler)	43.55	348.40	66.00	528.00		
1 Equip. Oper. (light)	47.05	376.40	71.35	570.80		
1 Lattice Boom Crane, 90 Ton		1529.00		1681.90		
1 Welder, Gas Engine, 300 amp		142.00		156.20		
1 Air Compressor, 160 cfm		157.40		173.14		
2 Impact Wrenches		36.00		39.60	14.57	16.02
128 L.H., Daily Totals		$8353.60		$13224.04	$65.26	$103.31
Crew E-7	Hr.	Daily	Hr.	Daily	Bare Costs	Incl. O&P
1 Struc. Steel Foreman (outside)	$53.10	$424.80	$93.70	$749.60	$50.66	$87.08
4 Struc. Steel Workers	51.10	1635.20	90.20	2886.40		
1 Equip. Oper. (crane)	50.25	402.00	76.20	609.60		
1 Equip. Oper. (oiler)	43.55	348.40	66.00	528.00		
1 Welder Foreman (outside)	53.10	424.80	93.70	749.60		
2 Welders	51.10	817.60	90.20	1443.20		
1 Lattice Boom Crane, 90 Ton		1529.00		1681.90		
2 Welder, Gas Engine, 300 amp		284.00		312.40	22.66	24.93
80 L.H., Daily Totals		$5865.80		$8960.70	$73.32	$112.01
Crew E-8	Hr.	Daily	Hr.	Daily	Bare Costs	Incl. O&P
1 Struc. Steel Foreman (outside)	$53.10	$424.80	$93.70	$749.60	$50.45	$86.35
4 Struc. Steel Workers	51.10	1635.20	90.20	2886.40		
1 Welder Foreman (outside)	53.10	424.80	93.70	749.60		
4 Welders	51.10	1635.20	90.20	2886.40		
1 Equip. Oper. (crane)	50.25	402.00	76.20	609.60		
1 Equip. Oper. (oiler)	43.55	348.40	66.00	528.00		
1 Equip. Oper. (light)	47.05	376.40	71.35	570.80		
1 Lattice Boom Crane, 90 Ton		1529.00		1681.90		
4 Welder, Gas Engine, 300 amp		568.00		624.80	20.16	22.18
104 L.H., Daily Totals		$7343.80		$11287.10	$70.61	$108.53

Crews

Crew No.		Bare Costs		Incl. Subs O&P		Cost Per Labor-Hour	
Crew E-9	Hr.	Daily	Hr.	Daily	Bare Costs	Incl. O&P
2 Struc. Steel Foremen (outside)	$53.10	$849.60	$93.70	$1499.20	$50.70	$87.29
5 Struc. Steel Workers	51.10	2044.00	90.20	3608.00		
1 Welder Foreman (outside)	53.10	424.80	93.70	749.60		
5 Welders	51.10	2044.00	90.20	3608.00		
1 Equip. Oper. (crane)	50.25	402.00	76.20	609.60		
1 Equip. Oper. (oiler)	43.55	348.40	66.00	528.00		
1 Equip. Oper. (light)	47.05	376.40	71.35	570.80		
1 Lattice Boom Crane, 90 Ton		1529.00		1681.90		
5 Welder, Gas Engine, 300 amp		710.00		781.00	17.49	19.24
128 L.H., Daily Totals		$8728.20		$13636.10	$68.19	$106.53

Crew E-10	Hr.	Daily	Hr.	Daily	Bare Costs	Incl. O&P
1 Welder Foreman (outside)	$53.10	$424.80	$93.70	$749.60	$52.10	$91.95
1 Welder	51.10	408.80	90.20	721.60		
1 Welder, Gas Engine, 300 amp		142.00		156.20		
1 Flatbed Truck, Gas, 3 Ton		333.00		366.30	29.69	32.66
16 L.H., Daily Totals		$1308.60		$1993.70	$81.79	$124.61

Crew E-11	Hr.	Daily	Hr.	Daily	Bare Costs	Incl. O&P
2 Painters, Struc. Steel	$40.50	$648.00	$73.05	$1168.80	$41.17	$68.50
1 Building Laborer	36.65	293.20	56.55	452.40		
1 Equip. Oper. (light)	47.05	376.40	71.35	570.80		
1 Air Compressor, 250 cfm		202.40		222.64		
1 Sandblaster, Portable, 3 C.F.		20.40		22.44		
1 Set Sand Blasting Accessories		13.85		15.23	7.40	8.13
32 L.H., Daily Totals		$1554.25		$2452.32	$48.57	$76.63

Crew E-11A	Hr.	Daily	Hr.	Daily	Bare Costs	Incl. O&P
2 Painters, Struc. Steel	$40.50	$648.00	$73.05	$1168.80	$41.17	$68.50
1 Building Laborer	36.65	293.20	56.55	452.40		
1 Equip. Oper. (light)	47.05	376.40	71.35	570.80		
1 Air Compressor, 250 cfm		202.40		222.64		
1 Sandblaster, Portable, 3 C.F.		20.40		22.44		
1 Set Sand Blasting Accessories		13.85		15.23		
1 Aerial Lift Truck, 60' Boom		456.00		501.60	21.65	23.81
32 L.H., Daily Totals		$2010.25		$2953.92	$62.82	$92.31

Crew E-11B	Hr.	Daily	Hr.	Daily	Bare Costs	Incl. O&P
2 Painters, Struc. Steel	$40.50	$648.00	$73.05	$1168.80	$39.22	$67.55
1 Building Laborer	36.65	293.20	56.55	452.40		
2 Paint Sprayer, 8 C.F.M.		89.80		98.78		
1 Aerial Lift Truck, 60' Boom		456.00		501.60	22.74	25.02
24 L.H., Daily Totals		$1487.00		$2221.58	$61.96	$92.57

Crew E-12	Hr.	Daily	Hr.	Daily	Bare Costs	Incl. O&P
1 Welder Foreman (outside)	$53.10	$424.80	$93.70	$749.60	$50.08	$82.53
1 Equip. Oper. (light)	47.05	376.40	71.35	570.80		
1 Welder, Gas Engine, 300 amp		142.00		156.20	8.88	9.76
16 L.H., Daily Totals		$943.20		$1476.60	$58.95	$92.29

Crew E-13	Hr.	Daily	Hr.	Daily	Bare Costs	Incl. O&P
1 Welder Foreman (outside)	$53.10	$424.80	$93.70	$749.60	$51.08	$86.25
.5 Equip. Oper. (light)	47.05	188.20	71.35	285.40		
1 Welder, Gas Engine, 300 amp		142.00		156.20	11.83	13.02
12 L.H., Daily Totals		$755.00		$1191.20	$62.92	$99.27

Crew E-14	Hr.	Daily	Hr.	Daily	Bare Costs	Incl. O&P
1 Welder Foreman (outside)	$53.10	$424.80	$93.70	$749.60	$53.10	$93.70
1 Welder, Gas Engine, 300 amp		142.00		156.20	17.75	19.52
8 L.H., Daily Totals		$566.80		$905.80	$70.85	$113.22

Crew No.		Bare Costs		Incl. Subs O&P		Cost Per Labor-Hour	
Crew E-16	Hr.	Daily	Hr.	Daily	Bare Costs	Incl. O&P
1 Welder Foreman (outside)	$53.10	$424.80	$93.70	$749.60	$52.10	$91.95
1 Welder	51.10	408.80	90.20	721.60		
1 Welder, Gas Engine, 300 amp		142.00		156.20	8.88	9.76
16 L.H., Daily Totals		$975.60		$1627.40	$60.98	$101.71

Crew E-17	Hr.	Daily	Hr.	Daily	Bare Costs	Incl. O&P
1 Struc. Steel Foreman (outside)	$53.10	$424.80	$93.70	$749.60	$52.10	$91.95
1 Structural Steel Worker	51.10	408.80	90.20	721.60		
16 L.H., Daily Totals		$833.60		$1471.20	$52.10	$91.95

Crew E-18	Hr.	Daily	Hr.	Daily	Bare Costs	Incl. O&P
1 Struc. Steel Foreman (outside)	$53.10	$424.80	$93.70	$749.60	$51.06	$87.69
3 Structural Steel Workers	51.10	1226.40	90.20	2164.80		
1 Equipment Operator (med.)	48.90	391.20	74.15	593.20		
1 Lattice Boom Crane, 20 Ton		960.50		1056.55	24.01	26.41
40 L.H., Daily Totals		$3002.90		$4564.15	$75.07	$114.10

Crew E-19	Hr.	Daily	Hr.	Daily	Bare Costs	Incl. O&P
1 Struc. Steel Foreman (outside)	$53.10	$424.80	$93.70	$749.60	$50.42	$85.08
1 Structural Steel Worker	51.10	408.80	90.20	721.60		
1 Equip. Oper. (light)	47.05	376.40	71.35	570.80		
1 Lattice Boom Crane, 20 Ton		960.50		1056.55	40.02	44.02
24 L.H., Daily Totals		$2170.50		$3098.55	$90.44	$129.11

Crew E-20	Hr.	Daily	Hr.	Daily	Bare Costs	Incl. O&P
1 Struc. Steel Foreman (outside)	$53.10	$424.80	$93.70	$749.60	$50.30	$85.86
5 Structural Steel Workers	51.10	2044.00	90.20	3608.00		
1 Equip. Oper. (crane)	50.25	402.00	76.20	609.60		
1 Equip. Oper. (oiler)	43.55	348.40	66.00	528.00		
1 Lattice Boom Crane, 40 Ton		1177.00		1294.70	18.39	20.23
64 L.H., Daily Totals		$4396.20		$6789.90	$68.69	$106.09

Crew E-22	Hr.	Daily	Hr.	Daily	Bare Costs	Incl. O&P
1 Skilled Worker Foreman (out)	$49.30	$394.40	$76.35	$610.80	$47.97	$74.28
2 Skilled Workers	47.30	756.80	73.25	1172.00		
24 L.H., Daily Totals		$1151.20		$1782.80	$47.97	$74.28

Crew E-24	Hr.	Daily	Hr.	Daily	Bare Costs	Incl. O&P
3 Structural Steel Workers	$51.10	$1226.40	$90.20	$2164.80	$50.55	$86.19
1 Equipment Operator (med.)	48.90	391.20	74.15	593.20		
1 Hyd. Crane, 25 Ton		745.80		820.38	23.31	25.64
32 L.H., Daily Totals		$2363.40		$3578.38	$73.86	$111.82

Crew E-25	Hr.	Daily	Hr.	Daily	Bare Costs	Incl. O&P
1 Welder Foreman (outside)	$53.10	$424.80	$93.70	$749.60	$53.10	$93.70
1 Cutting Torch		11.40		12.54	1.43	1.57
8 L.H., Daily Totals		$436.20		$762.14	$54.52	$95.27

Crew F-3	Hr.	Daily	Hr.	Daily	Bare Costs	Incl. O&P
4 Carpenters	$45.85	$1467.20	$70.75	$2264.00	$46.73	$71.84
1 Equip. Oper. (crane)	50.25	402.00	76.20	609.60		
1 Hyd. Crane, 12 Ton		663.00		729.30	16.57	18.23
40 L.H., Daily Totals		$2532.20		$3602.90	$63.31	$90.07

Crew F-4	Hr.	Daily	Hr.	Daily	Bare Costs	Incl. O&P
4 Carpenters	$45.85	$1467.20	$70.75	$2264.00	$46.20	$70.87
1 Equip. Oper. (crane)	50.25	402.00	76.20	609.60		
1 Equip. Oper. (oiler)	43.55	348.40	66.00	528.00		
1 Hyd. Crane, 55 Ton		1137.00		1250.70	23.69	26.06
48 L.H., Daily Totals		$3354.60		$4652.30	$69.89	$96.92

Crews

Crew No.	Bare Costs		Incl. Subs O&P		Cost Per Labor-Hour	
Crew F-5	Hr.	Daily	Hr.	Daily	Bare Costs	Incl. O&P
1 Carpenter Foreman (outside)	$47.85	$382.80	$73.85	$590.80	$46.35	$71.53
3 Carpenters	45.85	1100.40	70.75	1698.00		
32 L.H., Daily Totals		$1483.20		$2288.80	$46.35	$71.53
Crew F-6	Hr.	Daily	Hr.	Daily	Bare Costs	Incl. O&P
2 Carpenters	$45.85	$733.60	$70.75	$1132.00	$43.05	$66.16
2 Building Laborers	36.65	586.40	56.55	904.80		
1 Equip. Oper. (crane)	50.25	402.00	76.20	609.60		
1 Hyd. Crane, 12 Ton		663.00		729.30	16.57	18.23
40 L.H., Daily Totals		$2385.00		$3375.70	$59.63	$84.39
Crew F-7	Hr.	Daily	Hr.	Daily	Bare Costs	Incl. O&P
2 Carpenters	$45.85	$733.60	$70.75	$1132.00	$41.25	$63.65
2 Building Laborers	36.65	586.40	56.55	904.80		
32 L.H., Daily Totals		$1320.00		$2036.80	$41.25	$63.65
Crew G-1	Hr.	Daily	Hr.	Daily	Bare Costs	Incl. O&P
1 Roofer Foreman (outside)	$41.15	$329.20	$70.05	$560.40	$36.58	$62.26
4 Roofers Composition	39.15	1252.80	66.65	2132.80		
2 Roofer Helpers	29.15	466.40	49.60	793.60		
1 Application Equipment		199.20		219.12		
1 Tar Kettle/Pot		130.00		143.00		
1 Crew Truck		224.80		247.28	9.89	10.88
56 L.H., Daily Totals		$2602.40		$4096.20	$46.47	$73.15
Crew G-2	Hr.	Daily	Hr.	Daily	Bare Costs	Incl. O&P
1 Plasterer	$41.95	$335.60	$63.15	$505.20	$38.60	$58.57
1 Plasterer Helper	37.20	297.60	56.00	448.00		
1 Building Laborer	36.65	293.20	56.55	452.40		
1 Grout Pump, 50 C.F./hr.		134.80		148.28	5.62	6.18
24 L.H., Daily Totals		$1061.20		$1553.88	$44.22	$64.75
Crew G-2A	Hr.	Daily	Hr.	Daily	Bare Costs	Incl. O&P
1 Roofer Composition	$39.15	$313.20	$66.65	$533.20	$34.98	$57.60
1 Roofer Helper	29.15	233.20	49.60	396.80		
1 Building Laborer	36.65	293.20	56.55	452.40		
1 Foam Spray Rig, Trailer-Mtd.		585.25		643.77		
1 Pickup Truck, 3/4 Ton		157.00		172.70	30.93	34.02
24 L.H., Daily Totals		$1581.85		$2198.88	$65.91	$91.62
Crew G-3	Hr.	Daily	Hr.	Daily	Bare Costs	Incl. O&P
2 Sheet Metal Workers	$54.70	$875.20	$83.55	$1336.80	$45.67	$70.05
2 Building Laborers	36.65	586.40	56.55	904.80		
32 L.H., Daily Totals		$1461.60		$2241.60	$45.67	$70.05
Crew G-4	Hr.	Daily	Hr.	Daily	Bare Costs	Incl. O&P
1 Labor Foreman (outside)	$38.65	$309.20	$59.65	$477.20	$37.32	$57.58
2 Building Laborers	36.65	586.40	56.55	904.80		
1 Flatbed Truck, Gas, 1.5 Ton		270.20		297.22		
1 Air Compressor, 160 cfm		157.40		173.14	17.82	19.60
24 L.H., Daily Totals		$1323.20		$1852.36	$55.13	$77.18
Crew G-5	Hr.	Daily	Hr.	Daily	Bare Costs	Incl. O&P
1 Roofer Foreman (outside)	$41.15	$329.20	$70.05	$560.40	$35.55	$60.51
2 Roofers Composition	39.15	626.40	66.65	1066.40		
2 Roofer Helpers	29.15	466.40	49.60	793.60		
1 Application Equipment		199.20		219.12	4.98	5.48
40 L.H., Daily Totals		$1621.20		$2639.52	$40.53	$65.99

Crew No.	Bare Costs		Incl. Subs O&P		Cost Per Labor-Hour	
Crew G-6A	Hr.	Daily	Hr.	Daily	Bare Costs	Incl. O&P
2 Roofers Composition	$39.15	$626.40	$66.65	$1066.40	$39.15	$66.65
1 Small Compressor, Electric		12.10		13.31		
2 Pneumatic Nailers		44.70		49.17	3.55	3.90
16 L.H., Daily Totals		$683.20		$1128.88	$42.70	$70.56
Crew G-7	Hr.	Daily	Hr.	Daily	Bare Costs	Incl. O&P
1 Carpenter	$45.85	$366.80	$70.75	$566.00	$45.85	$70.75
1 Small Compressor, Electric		12.10		13.31		
1 Pneumatic Nailer		22.35		24.59	4.31	4.74
8 L.H., Daily Totals		$401.25		$603.89	$50.16	$75.49
Crew H-1	Hr.	Daily	Hr.	Daily	Bare Costs	Incl. O&P
2 Glaziers	$44.05	$704.80	$67.00	$1072.00	$47.58	$78.60
2 Struc. Steel Workers	51.10	817.60	90.20	1443.20		
32 L.H., Daily Totals		$1522.40		$2515.20	$47.58	$78.60
Crew H-2	Hr.	Daily	Hr.	Daily	Bare Costs	Incl. O&P
2 Glaziers	$44.05	$704.80	$67.00	$1072.00	$41.58	$63.52
1 Building Laborer	36.65	293.20	56.55	452.40		
24 L.H., Daily Totals		$998.00		$1524.40	$41.58	$63.52
Crew H-3	Hr.	Daily	Hr.	Daily	Bare Costs	Incl. O&P
1 Glazier	$44.05	$352.40	$67.00	$536.00	$39.35	$60.35
1 Helper	34.65	277.20	53.70	429.60		
16 L.H., Daily Totals		$629.60		$965.60	$39.35	$60.35
Crew H-4	Hr.	Daily	Hr.	Daily	Bare Costs	Incl. O&P
1 Carpenter	$45.85	$366.80	$70.75	$566.00	$42.87	$65.75
1 Carpenter Helper	34.65	277.20	53.70	429.60		
.5 Electrician	53.35	213.40	79.85	319.40		
20 L.H., Daily Totals		$857.40		$1315.00	$42.87	$65.75
Crew J-1	Hr.	Daily	Hr.	Daily	Bare Costs	Incl. O&P
3 Plasterers	$41.95	$1006.80	$63.15	$1515.60	$40.05	$60.29
2 Plasterer Helpers	37.20	595.20	56.00	896.00		
1 Mixing Machine, 6 C.F.		143.60		157.96	3.59	3.95
40 L.H., Daily Totals		$1745.60		$2569.56	$43.64	$64.24
Crew J-2	Hr.	Daily	Hr.	Daily	Bare Costs	Incl. O&P
3 Plasterers	$41.95	$1006.80	$63.15	$1515.60	$40.14	$60.19
2 Plasterer Helpers	37.20	595.20	56.00	896.00		
1 Lather	40.60	324.80	59.70	477.60		
1 Mixing Machine, 6 C.F.		143.60		157.96	2.99	3.29
48 L.H., Daily Totals		$2070.40		$3047.16	$43.13	$63.48
Crew J-3	Hr.	Daily	Hr.	Daily	Bare Costs	Incl. O&P
1 Terrazzo Worker	$41.65	$333.20	$61.40	$491.20	$38.23	$56.35
1 Terrazzo Helper	34.80	278.40	51.30	410.40		
1 Floor Grinder, 22" Path		97.90		107.69		
1 Terrazzo Mixer		192.60		211.86	18.16	19.97
16 L.H., Daily Totals		$902.10		$1221.15	$56.38	$76.32
Crew J-4	Hr.	Daily	Hr.	Daily	Bare Costs	Incl. O&P
2 Cement Finishers	$44.05	$704.80	$65.10	$1041.60	$41.58	$62.25
1 Laborer	36.65	293.20	56.55	452.40		
1 Floor Grinder, 22" Path		97.90		107.69		
1 Floor Edger, 7" Path		39.15		43.06		
1 Vacuum Pick-Up System		60.75		66.83	8.24	9.07
24 L.H., Daily Totals		$1195.80		$1711.58	$49.83	$71.32

Crews

Crew No.		Bare Costs		Incl. Subs O&P		Cost Per Labor-Hour	
Crew J-4A	Hr.	Daily	Hr.	Daily	Bare Costs	Incl. O&P
2 Cement Finishers	$44.05	$704.80	$65.10	$1041.60	$40.35	$60.83
2 Laborers	36.65	586.40	56.55	904.80		
1 Floor Grinder, 22" Path		97.90		107.69		
1 Floor Edger, 7" Path		39.15		43.06		
1 Vacuum Pick-Up System		60.75		66.83		
1 Floor Auto Scrubber		212.15		233.37	12.81	14.09
32 L.H., Daily Totals		$1701.15		$2397.34	$53.16	$74.92
Crew J-4B	Hr.	Daily	Hr.	Daily	Bare Costs	Incl. O&P
1 Laborer	$36.65	$293.20	$56.55	$452.40	$36.65	$56.55
1 Floor Auto Scrubber		212.15		233.37	26.52	29.17
8 L.H., Daily Totals		$505.35		$685.76	$63.17	$85.72
Crew J-6	Hr.	Daily	Hr.	Daily	Bare Costs	Incl. O&P
2 Painters	$39.55	$632.80	$59.50	$952.00	$40.70	$61.73
1 Building Laborer	36.65	293.20	56.55	452.40		
1 Equip. Oper. (light)	47.05	376.40	71.35	570.80		
1 Air Compressor, 250 cfm		202.40		222.64		
1 Sandblaster, Portable, 3 C.F.		20.40		22.44		
1 Set Sand Blasting Accessories		13.85		15.23	7.40	8.13
32 L.H., Daily Totals		$1539.05		$2235.51	$48.10	$69.86
Crew J-7	Hr.	Daily	Hr.	Daily	Bare Costs	Incl. O&P
2 Painters	$39.55	$632.80	$59.50	$952.00	$39.55	$59.50
1 Floor Belt Sander		15.90		17.49		
1 Floor Sanding Edger		15.55		17.11	1.97	2.16
16 L.H., Daily Totals		$664.25		$986.60	$41.52	$61.66
Crew K-1	Hr.	Daily	Hr.	Daily	Bare Costs	Incl. O&P
1 Carpenter	$45.85	$366.80	$70.75	$566.00	$41.17	$63.23
1 Truck Driver (light)	36.50	292.00	55.70	445.60		
1 Flatbed Truck, Gas, 3 Ton		333.00		366.30	20.81	22.89
16 L.H., Daily Totals		$991.80		$1377.90	$61.99	$86.12
Crew K-2	Hr.	Daily	Hr.	Daily	Bare Costs	Incl. O&P
1 Struc. Steel Foreman (outside)	$53.10	$424.80	$93.70	$749.60	$46.90	$79.87
1 Struc. Steel Worker	51.10	408.80	90.20	721.60		
1 Truck Driver (light)	36.50	292.00	55.70	445.60		
1 Flatbed Truck, Gas, 3 Ton		333.00		366.30	13.88	15.26
24 L.H., Daily Totals		$1458.60		$2283.10	$60.77	$95.13
Crew L-1	Hr.	Daily	Hr.	Daily	Bare Costs	Incl. O&P
1 Electrician	$53.35	$426.80	$79.85	$638.80	$55.45	$83.38
1 Plumber	57.55	460.40	86.90	695.20		
16 L.H., Daily Totals		$887.20		$1334.00	$55.45	$83.38
Crew L-2	Hr.	Daily	Hr.	Daily	Bare Costs	Incl. O&P
1 Carpenter	$45.85	$366.80	$70.75	$566.00	$40.25	$62.23
1 Carpenter Helper	34.65	277.20	53.70	429.60		
16 L.H., Daily Totals		$644.00		$995.60	$40.25	$62.23
Crew L-3	Hr.	Daily	Hr.	Daily	Bare Costs	Incl. O&P
1 Carpenter	$45.85	$366.80	$70.75	$566.00	$49.94	$76.22
.5 Electrician	53.35	213.40	79.85	319.40		
.5 Sheet Metal Worker	54.70	218.80	83.55	334.20		
16 L.H., Daily Totals		$799.00		$1219.60	$49.94	$76.22

Crew No.		Bare Costs		Incl. Subs O&P		Cost Per Labor-Hour	
Crew L-3A	Hr.	Daily	Hr.	Daily	Bare Costs	Incl. O&P
1 Carpenter Foreman (outside)	$47.85	$382.80	$73.85	$590.80	$50.13	$77.08
.5 Sheet Metal Worker	54.70	218.80	83.55	334.20		
12 L.H., Daily Totals		$601.60		$925.00	$50.13	$77.08
Crew L-4	Hr.	Daily	Hr.	Daily	Bare Costs	Incl. O&P
2 Skilled Workers	$47.30	$756.80	$73.25	$1172.00	$43.08	$66.73
1 Helper	34.65	277.20	53.70	429.60		
24 L.H., Daily Totals		$1034.00		$1601.60	$43.08	$66.73
Crew L-5	Hr.	Daily	Hr.	Daily	Bare Costs	Incl. O&P
1 Struc. Steel Foreman (outside)	$53.10	$424.80	$93.70	$749.60	$51.26	$88.70
5 Struc. Steel Workers	51.10	2044.00	90.20	3608.00		
1 Equip. Oper. (crane)	50.25	402.00	76.20	609.60		
1 Hyd. Crane, 25 Ton		745.80		820.38	13.32	14.65
56 L.H., Daily Totals		$3616.60		$5787.58	$64.58	$103.35
Crew L-5A	Hr.	Daily	Hr.	Daily	Bare Costs	Incl. O&P
1 Struc. Steel Foreman (outside)	$53.10	$424.80	$93.70	$749.60	$51.39	$87.58
2 Structural Steel Workers	51.10	817.60	90.20	1443.20		
1 Equip. Oper. (crane)	50.25	402.00	76.20	609.60		
1 S.P. Crane, 4x4, 25 Ton		607.20		667.92	18.98	20.87
32 L.H., Daily Totals		$2251.60		$3470.32	$70.36	$108.45
Crew L-5B	Hr.	Daily	Hr.	Daily	Bare Costs	Incl. O&P
1 Struc. Steel Foreman (outside)	$53.10	$424.80	$93.70	$749.60	$52.53	$83.63
2 Structural Steel Workers	51.10	817.60	90.20	1443.20		
2 Electricians	53.35	853.60	79.85	1277.60		
2 Steamfitters/Pipefitters	58.50	936.00	88.35	1413.60		
1 Equip. Oper. (crane)	50.25	402.00	76.20	609.60		
1 Equip. Oper. (oiler)	43.55	348.40	66.00	528.00		
1 Hyd. Crane, 80 Ton		1652.00		1817.20	22.94	25.24
72 L.H., Daily Totals		$5434.40		$7838.80	$75.48	$108.87
Crew L-6	Hr.	Daily	Hr.	Daily	Bare Costs	Incl. O&P
1 Plumber	$57.55	$460.40	$86.90	$695.20	$56.15	$84.55
.5 Electrician	53.35	213.40	79.85	319.40		
12 L.H., Daily Totals		$673.80		$1014.60	$56.15	$84.55
Crew L-7	Hr.	Daily	Hr.	Daily	Bare Costs	Incl. O&P
2 Carpenters	$45.85	$733.60	$70.75	$1132.00	$44.29	$67.99
1 Building Laborer	36.65	293.20	56.55	452.40		
.5 Electrician	53.35	213.40	79.85	319.40		
28 L.H., Daily Totals		$1240.20		$1903.80	$44.29	$67.99
Crew L-8	Hr.	Daily	Hr.	Daily	Bare Costs	Incl. O&P
2 Carpenters	$45.85	$733.60	$70.75	$1132.00	$48.19	$73.98
.5 Plumber	57.55	230.20	86.90	347.60		
20 L.H., Daily Totals		$963.80		$1479.60	$48.19	$73.98
Crew L-9	Hr.	Daily	Hr.	Daily	Bare Costs	Incl. O&P
1 Labor Foreman (inside)	$37.15	$297.20	$57.30	$458.40	$41.83	$66.78
2 Building Laborers	36.65	586.40	56.55	904.80		
1 Struc. Steel Worker	51.10	408.80	90.20	721.60		
.5 Electrician	53.35	213.40	79.85	319.40		
36 L.H., Daily Totals		$1505.80		$2404.20	$41.83	$66.78

Crews

Crew No.		Bare Costs		Incl. Subs O&P		Cost Per Labor-Hour	
Crew L-10	Hr.	Daily	Hr.	Daily	Bare Costs	Incl. O&P
1 Struc. Steel Foreman (outside)	$53.10	$424.80	$93.70	$749.60	$51.48	$86.70
1 Structural Steel Worker	51.10	408.80	90.20	721.60		
1 Equip. Oper. (crane)	50.25	402.00	76.20	609.60		
1 Hyd. Crane, 12 Ton		663.00		729.30	27.63	30.39
24 L.H., Daily Totals		$1898.60		$2810.10	$79.11	$117.09
Crew L-11	Hr.	Daily	Hr.	Daily	Bare Costs	Incl. O&P
2 Wreckers	$36.65	$586.40	$61.60	$985.60	$42.65	$67.69
1 Equip. Oper. (crane)	50.25	402.00	76.20	609.60		
1 Equip. Oper. (light)	47.05	376.40	71.35	570.80		
1 Hyd. Excavator, 2.5 C.Y.		1637.00		1800.70		
1 Loader, Skid Steer, 78 H.P.		308.80		339.68	60.81	66.89
32 L.H., Daily Totals		$3310.60		$4306.38	$103.46	$134.57
Crew M-1	Hr.	Daily	Hr.	Daily	Bare Costs	Incl. O&P
3 Elevator Constructors	$74.15	$1779.60	$110.95	$2662.80	$70.44	$105.39
1 Elevator Apprentice	59.30	474.40	88.70	709.60		
5 Hand Tools		48.00		52.80	1.50	1.65
32 L.H., Daily Totals		$2302.00		$3425.20	$71.94	$107.04
Crew M-3	Hr.	Daily	Hr.	Daily	Bare Costs	Incl. O&P
1 Electrician Foreman (outside)	$55.35	$442.80	$82.85	$662.80	$55.92	$84.14
1 Common Laborer	36.65	293.20	56.55	452.40		
.25 Equipment Operator (med.)	48.90	97.80	74.15	148.30		
1 Elevator Constructor	74.15	593.20	110.95	887.60		
1 Elevator Apprentice	59.30	474.40	88.70	709.60		
.25 S.P. Crane, 4x4, 20 Ton		140.10		154.11	4.12	4.53
34 L.H., Daily Totals		$2041.50		$3014.81	$60.04	$88.67
Crew M-4	Hr.	Daily	Hr.	Daily	Bare Costs	Incl. O&P
1 Electrician Foreman (outside)	$55.35	$442.80	$82.85	$662.80	$55.31	$83.24
1 Common Laborer	36.65	293.20	56.55	452.40		
.25 Equipment Operator, Crane	50.25	100.50	76.20	152.40		
.25 Equip. Oper. (oiler)	43.55	87.10	66.00	132.00		
1 Elevator Constructor	74.15	593.20	110.95	887.60		
1 Elevator Apprentice	59.30	474.40	88.70	709.60		
.25 S.P. Crane, 4x4, 40 Ton		178.25		196.07	4.95	5.45
36 L.H., Daily Totals		$2169.45		$3192.88	$60.26	$88.69
Crew Q-1	Hr.	Daily	Hr.	Daily	Bare Costs	Incl. O&P
1 Plumber	$57.55	$460.40	$86.90	$695.20	$51.80	$78.22
1 Plumber Apprentice	46.05	368.40	69.55	556.40		
16 L.H., Daily Totals		$828.80		$1251.60	$51.80	$78.22
Crew Q-1A	Hr.	Daily	Hr.	Daily	Bare Costs	Incl. O&P
.25 Plumber Foreman (outside)	$59.55	$119.10	$89.90	$179.80	$57.95	$87.50
1 Plumber	57.55	460.40	86.90	695.20		
10 L.H., Daily Totals		$579.50		$875.00	$57.95	$87.50
Crew Q-1C	Hr.	Daily	Hr.	Daily	Bare Costs	Incl. O&P
1 Plumber	$57.55	$460.40	$86.90	$695.20	$50.83	$76.87
1 Plumber Apprentice	46.05	368.40	69.55	556.40		
1 Equip. Oper. (medium)	48.90	391.20	74.15	593.20		
1 Trencher, Chain Type, 8' D		3376.00		3713.60	140.67	154.73
24 L.H., Daily Totals		$4596.00		$5558.40	$191.50	$231.60
Crew Q-2	Hr.	Daily	Hr.	Daily	Bare Costs	Incl. O&P
2 Plumbers	$57.55	$920.80	$86.90	$1390.40	$53.72	$81.12
1 Plumber Apprentice	46.05	368.40	69.55	556.40		
24 L.H., Daily Totals		$1289.20		$1946.80	$53.72	$81.12

Crew No.		Bare Costs		Incl. Subs O&P		Cost Per Labor-Hour	
Crew Q-3	Hr.	Daily	Hr.	Daily	Bare Costs	Incl. O&P
1 Plumber Foreman (inside)	$58.05	$464.40	$87.65	$701.20	$54.80	$82.75
2 Plumbers	57.55	920.80	86.90	1390.40		
1 Plumber Apprentice	46.05	368.40	69.55	556.40		
32 L.H., Daily Totals		$1753.60		$2648.00	$54.80	$82.75
Crew Q-4	Hr.	Daily	Hr.	Daily	Bare Costs	Incl. O&P
1 Plumber Foreman (inside)	$58.05	$464.40	$87.65	$701.20	$54.80	$82.75
1 Plumber	57.55	460.40	86.90	695.20		
1 Welder (plumber)	57.55	460.40	86.90	695.20		
1 Plumber Apprentice	46.05	368.40	69.55	556.40		
1 Welder, Electric, 300 amp		55.30		60.83	1.73	1.90
32 L.H., Daily Totals		$1808.90		$2708.83	$56.53	$84.65
Crew Q-5	Hr.	Daily	Hr.	Daily	Bare Costs	Incl. O&P
1 Steamfitter	$58.50	$468.00	$88.35	$706.80	$52.65	$79.50
1 Steamfitter Apprentice	46.80	374.40	70.65	565.20		
16 L.H., Daily Totals		$842.40		$1272.00	$52.65	$79.50
Crew Q-6	Hr.	Daily	Hr.	Daily	Bare Costs	Incl. O&P
2 Steamfitters	$58.50	$936.00	$88.35	$1413.60	$54.60	$82.45
1 Steamfitter Apprentice	46.80	374.40	70.65	565.20		
24 L.H., Daily Totals		$1310.40		$1978.80	$54.60	$82.45
Crew Q-7	Hr.	Daily	Hr.	Daily	Bare Costs	Incl. O&P
1 Steamfitter Foreman (inside)	$59.00	$472.00	$89.10	$712.80	$55.70	$84.11
2 Steamfitters	58.50	936.00	88.35	1413.60		
1 Steamfitter Apprentice	46.80	374.40	70.65	565.20		
32 L.H., Daily Totals		$1782.40		$2691.60	$55.70	$84.11
Crew Q-8	Hr.	Daily	Hr.	Daily	Bare Costs	Incl. O&P
1 Steamfitter Foreman (inside)	$59.00	$472.00	$89.10	$712.80	$55.70	$84.11
1 Steamfitter	58.50	468.00	88.35	706.80		
1 Welder (steamfitter)	58.50	468.00	88.35	706.80		
1 Steamfitter Apprentice	46.80	374.40	70.65	565.20		
1 Welder, Electric, 300 amp		55.30		60.83	1.73	1.90
32 L.H., Daily Totals		$1837.70		$2752.43	$57.43	$86.01
Crew Q-9	Hr.	Daily	Hr.	Daily	Bare Costs	Incl. O&P
1 Sheet Metal Worker	$54.70	$437.60	$83.55	$668.40	$49.23	$75.17
1 Sheet Metal Apprentice	43.75	350.00	66.80	534.40		
16 L.H., Daily Totals		$787.60		$1202.80	$49.23	$75.17
Crew Q-10	Hr.	Daily	Hr.	Daily	Bare Costs	Incl. O&P
2 Sheet Metal Workers	$54.70	$875.20	$83.55	$1336.80	$51.05	$77.97
1 Sheet Metal Apprentice	43.75	350.00	66.80	534.40		
24 L.H., Daily Totals		$1225.20		$1871.20	$51.05	$77.97
Crew Q-11	Hr.	Daily	Hr.	Daily	Bare Costs	Incl. O&P
1 Sheet Metal Foreman (inside)	$55.20	$441.60	$84.30	$674.40	$52.09	$79.55
2 Sheet Metal Workers	54.70	875.20	83.55	1336.80		
1 Sheet Metal Apprentice	43.75	350.00	66.80	534.40		
32 L.H., Daily Totals		$1666.80		$2545.60	$52.09	$79.55
Crew Q-12	Hr.	Daily	Hr.	Daily	Bare Costs	Incl. O&P
1 Sprinkler Installer	$55.40	$443.20	$83.70	$669.60	$49.85	$75.33
1 Sprinkler Apprentice	44.30	354.40	66.95	535.60		
16 L.H., Daily Totals		$797.60		$1205.20	$49.85	$75.33

Crews

Crew No.	Bare Costs		Incl. Subs O&P		Cost Per Labor-Hour	
Crew Q-13	Hr.	Daily	Hr.	Daily	Bare Costs	Incl. O&P
1 Sprinkler Foreman (inside)	$55.90	$447.20	$84.45	$675.60	$52.75	$79.70
2 Sprinkler Installers	55.40	886.40	83.70	1339.20		
1 Sprinkler Apprentice	44.30	354.40	66.95	535.60		
32 L.H., Daily Totals		$1688.00		$2550.40	$52.75	$79.70
Crew Q-14	Hr.	Daily	Hr.	Daily	Bare Costs	Incl. O&P
1 Asbestos Worker	$51.15	$409.20	$79.60	$636.80	$46.02	$71.63
1 Asbestos Apprentice	40.90	327.20	63.65	509.20		
16 L.H., Daily Totals		$736.40		$1146.00	$46.02	$71.63
Crew Q-15	Hr.	Daily	Hr.	Daily	Bare Costs	Incl. O&P
1 Plumber	$57.55	$460.40	$86.90	$695.20	$51.80	$78.22
1 Plumber Apprentice	46.05	368.40	69.55	556.40		
1 Welder, Electric, 300 amp		55.30		60.83	3.46	3.80
16 L.H., Daily Totals		$884.10		$1312.43	$55.26	$82.03
Crew Q-16	Hr.	Daily	Hr.	Daily	Bare Costs	Incl. O&P
2 Plumbers	$57.55	$920.80	$86.90	$1390.40	$53.72	$81.12
1 Plumber Apprentice	46.05	368.40	69.55	556.40		
1 Welder, Electric, 300 amp		55.30		60.83	2.30	2.53
24 L.H., Daily Totals		$1344.50		$2007.63	$56.02	$83.65
Crew Q-17	Hr.	Daily	Hr.	Daily	Bare Costs	Incl. O&P
1 Steamfitter	$58.50	$468.00	$88.35	$706.80	$52.65	$79.50
1 Steamfitter Apprentice	46.80	374.40	70.65	565.20		
1 Welder, Electric, 300 amp		55.30		60.83	3.46	3.80
16 L.H., Daily Totals		$897.70		$1332.83	$56.11	$83.30
Crew Q-17A	Hr.	Daily	Hr.	Daily	Bare Costs	Incl. O&P
1 Steamfitter	$58.50	$468.00	$88.35	$706.80	$51.85	$78.40
1 Steamfitter Apprentice	46.80	374.40	70.65	565.20		
1 Equip. Oper. (crane)	50.25	402.00	76.20	609.60		
1 Hyd. Crane, 12 Ton		663.00		729.30		
1 Welder, Electric, 300 amp		55.30		60.83	29.93	32.92
24 L.H., Daily Totals		$1962.70		$2671.73	$81.78	$111.32
Crew Q-18	Hr.	Daily	Hr.	Daily	Bare Costs	Incl. O&P
2 Steamfitters	$58.50	$936.00	$88.35	$1413.60	$54.60	$82.45
1 Steamfitter Apprentice	46.80	374.40	70.65	565.20		
1 Welder, Electric, 300 amp		55.30		60.83	2.30	2.53
24 L.H., Daily Totals		$1365.70		$2039.63	$56.90	$84.98
Crew Q-19	Hr.	Daily	Hr.	Daily	Bare Costs	Incl. O&P
1 Steamfitter	$58.50	$468.00	$88.35	$706.80	$52.88	$79.62
1 Steamfitter Apprentice	46.80	374.40	70.65	565.20		
1 Electrician	53.35	426.80	79.85	638.80		
24 L.H., Daily Totals		$1269.20		$1910.80	$52.88	$79.62
Crew Q-20	Hr.	Daily	Hr.	Daily	Bare Costs	Incl. O&P
1 Sheet Metal Worker	$54.70	$437.60	$83.55	$668.40	$50.05	$76.11
1 Sheet Metal Apprentice	43.75	350.00	66.80	534.40		
.5 Electrician	53.35	213.40	79.85	319.40		
20 L.H., Daily Totals		$1001.00		$1522.20	$50.05	$76.11
Crew Q-21	Hr.	Daily	Hr.	Daily	Bare Costs	Incl. O&P
2 Steamfitters	$58.50	$936.00	$88.35	$1413.60	$54.29	$81.80
1 Steamfitter Apprentice	46.80	374.40	70.65	565.20		
1 Electrician	53.35	426.80	79.85	638.80		
32 L.H., Daily Totals		$1737.20		$2617.60	$54.29	$81.80

Crew No.	Bare Costs		Incl. Subs O&P		Cost Per Labor-Hour	
Crew Q-22	Hr.	Daily	Hr.	Daily	Bare Costs	Incl. O&P
1 Plumber	$57.55	$460.40	$86.90	$695.20	$51.80	$78.22
1 Plumber Apprentice	46.05	368.40	69.55	556.40		
1 Hyd. Crane, 12 Ton		663.00		729.30	41.44	45.58
16 L.H., Daily Totals		$1491.80		$1980.90	$93.24	$123.81
Crew Q-22A	Hr.	Daily	Hr.	Daily	Bare Costs	Incl. O&P
1 Plumber	$57.55	$460.40	$86.90	$695.20	$47.63	$72.30
1 Plumber Apprentice	46.05	368.40	69.55	556.40		
1 Laborer	36.65	293.20	56.55	452.40		
1 Equip. Oper. (crane)	50.25	402.00	76.20	609.60		
1 Hyd. Crane, 12 Ton		663.00		729.30	20.72	22.79
32 L.H., Daily Totals		$2187.00		$3042.90	$68.34	$95.09
Crew Q-23	Hr.	Daily	Hr.	Daily	Bare Costs	Incl. O&P
1 Plumber Foreman (outside)	$59.55	$476.40	$89.90	$719.20	$55.33	$83.65
1 Plumber	57.55	460.40	86.90	695.20		
1 Equip. Oper. (medium)	48.90	391.20	74.15	593.20		
1 Lattice Boom Crane, 20 Ton		960.50		1056.55	40.02	44.02
24 L.H., Daily Totals		$2288.50		$3064.15	$95.35	$127.67
Crew R-1	Hr.	Daily	Hr.	Daily	Bare Costs	Incl. O&P
1 Electrician Foreman	$53.85	$430.80	$80.60	$644.80	$47.20	$71.26
3 Electricians	53.35	1280.40	79.85	1916.40		
2 Helpers	34.65	554.40	53.70	859.20		
48 L.H., Daily Totals		$2265.60		$3420.40	$47.20	$71.26
Crew R-1A	Hr.	Daily	Hr.	Daily	Bare Costs	Incl. O&P
1 Electrician	$53.35	$426.80	$79.85	$638.80	$44.00	$66.78
1 Helper	34.65	277.20	53.70	429.60		
16 L.H., Daily Totals		$704.00		$1068.40	$44.00	$66.78
Crew R-2	Hr.	Daily	Hr.	Daily	Bare Costs	Incl. O&P
1 Electrician Foreman	$53.85	$430.80	$80.60	$644.80	$47.64	$71.96
3 Electricians	53.35	1280.40	79.85	1916.40		
2 Helpers	34.65	554.40	53.70	859.20		
1 Equip. Oper. (crane)	50.25	402.00	76.20	609.60		
1 S.P. Crane, 4x4, 5 Ton		280.00		308.00	5.00	5.50
56 L.H., Daily Totals		$2947.60		$4338.00	$52.64	$77.46
Crew R-3	Hr.	Daily	Hr.	Daily	Bare Costs	Incl. O&P
1 Electrician Foreman	$53.85	$430.80	$80.60	$644.80	$52.93	$79.42
1 Electrician	53.35	426.80	79.85	638.80		
.5 Equip. Oper. (crane)	50.25	201.00	76.20	304.80		
.5 S.P. Crane, 4x4, 5 Ton		140.00		154.00	7.00	7.70
20 L.H., Daily Totals		$1198.60		$1742.40	$59.93	$87.12
Crew R-4	Hr.	Daily	Hr.	Daily	Bare Costs	Incl. O&P
1 Struc. Steel Foreman (outside)	$53.10	$424.80	$93.70	$749.60	$51.95	$88.83
3 Struc. Steel Workers	51.10	1226.40	90.20	2164.80		
1 Electrician	53.35	426.80	79.85	638.80		
1 Welder, Gas Engine, 300 amp		142.00		156.20	3.55	3.90
40 L.H., Daily Totals		$2220.00		$3709.40	$55.50	$92.73

Crews

Crew No.		Bare Costs		Incl. Subs O&P		Cost Per Labor-Hour	
Crew R-5	Hr.	Daily	Hr.	Daily	Bare Costs	Incl. O&P	
1 Electrician Foreman	$53.85	$430.80	$80.60	$644.80	$46.60	$70.41
4 Electrician Linemen	53.35	1707.20	79.85	2555.20		
2 Electrician Operators	53.35	853.60	79.85	1277.60		
4 Electrician Groundmen	34.65	1108.80	53.70	1718.40		
1 Crew Truck		224.80		247.28		
1 Flatbed Truck, 20,000 GVW		253.40		278.74		
1 Pickup Truck, 3/4 Ton		157.00		172.70		
.2 Hyd. Crane, 55 Ton		227.40		250.14		
.2 Hyd. Crane, 12 Ton		132.60		145.86		
.2 Earth Auger, Truck-Mtd.		85.00		93.50		
1 Tractor w/Winch		434.60		478.06	17.21	18.93
88 L.H., Daily Totals		$5615.20		$7862.28	$63.81	$89.34

Crew R-6	Hr.	Daily	Hr.	Daily	Bare Costs	Incl. O&P
1 Electrician Foreman	$53.85	$430.80	$80.60	$644.80	$46.60	$70.41
4 Electrician Linemen	53.35	1707.20	79.85	2555.20		
2 Electrician Operators	53.35	853.60	79.85	1277.60		
4 Electrician Groundmen	34.65	1108.80	53.70	1718.40		
1 Crew Truck		224.80		247.28		
1 Flatbed Truck, 20,000 GVW		253.40		278.74		
1 Pickup Truck, 3/4 Ton		157.00		172.70		
.2 Hyd. Crane, 55 Ton		227.40		250.14		
.2 Hyd. Crane, 12 Ton		132.60		145.86		
.2 Earth Auger, Truck-Mtd.		85.00		93.50		
1 Tractor w/Winch		434.60		478.06		
3 Cable Trailers		586.80		645.48		
.5 Tensioning Rig		194.55		214.01		
.5 Cable Pulling Rig		1136.50		1250.15	39.01	42.91
88 L.H., Daily Totals		$7533.05		$9971.92	$85.60	$113.32

Crew R-7	Hr.	Daily	Hr.	Daily	Bare Costs	Incl. O&P
1 Electrician Foreman	$53.85	$430.80	$80.60	$644.80	$37.85	$58.18
5 Electrician Groundmen	34.65	1386.00	53.70	2148.00		
1 Crew Truck		224.80		247.28	4.68	5.15
48 L.H., Daily Totals		$2041.60		$3040.08	$42.53	$63.34

Crew R-8	Hr.	Daily	Hr.	Daily	Bare Costs	Incl. O&P
1 Electrician Foreman	$53.85	$430.80	$80.60	$644.80	$47.20	$71.26
3 Electrician Linemen	53.35	1280.40	79.85	1916.40		
2 Electrician Groundmen	34.65	554.40	53.70	859.20		
1 Pickup Truck, 3/4 Ton		157.00		172.70		
1 Crew Truck		224.80		247.28	7.95	8.75
48 L.H., Daily Totals		$2647.40		$3840.38	$55.15	$80.01

Crew R-9	Hr.	Daily	Hr.	Daily	Bare Costs	Incl. O&P
1 Electrician Foreman	$53.85	$430.80	$80.60	$644.80	$44.06	$66.87
1 Electrician Lineman	53.35	426.80	79.85	638.80		
2 Electrician Operators	53.35	853.60	79.85	1277.60		
4 Electrician Groundmen	34.65	1108.80	53.70	1718.40		
1 Pickup Truck, 3/4 Ton		157.00		172.70		
1 Crew Truck		224.80		247.28	5.97	6.56
64 L.H., Daily Totals		$3201.80		$4699.58	$50.03	$73.43

Crew R-10	Hr.	Daily	Hr.	Daily	Bare Costs	Incl. O&P
1 Electrician Foreman	$53.85	$430.80	$80.60	$644.80	$50.32	$75.62
4 Electrician Linemen	53.35	1707.20	79.85	2555.20		
1 Electrician Groundman	34.65	277.20	53.70	429.60		
1 Crew Truck		224.80		247.28		
3 Tram Cars		399.45		439.39	13.01	14.31
48 L.H., Daily Totals		$3039.45		$4316.27	$63.32	$89.92

Crew R-11	Hr.	Daily	Hr.	Daily	Bare Costs	Incl. O&P
1 Electrician Foreman	$53.85	$430.80	$80.60	$644.80	$50.59	$76.11
4 Electricians	53.35	1707.20	79.85	2555.20		
1 Equip. Oper. (crane)	50.25	402.00	76.20	609.60		
1 Common Laborer	36.65	293.20	56.55	452.40		
1 Crew Truck		224.80		247.28		
1 Hyd. Crane, 12 Ton		663.00		729.30	15.85	17.44
56 L.H., Daily Totals		$3721.00		$5238.58	$66.45	$93.55

Crew R-12	Hr.	Daily	Hr.	Daily	Bare Costs	Incl. O&P
1 Carpenter Foreman (inside)	$46.35	$370.80	$71.50	$572.00	$43.30	$67.73
4 Carpenters	45.85	1467.20	70.75	2264.00		
4 Common Laborers	36.65	1172.80	56.55	1809.60		
1 Equip. Oper. (medium)	48.90	391.20	74.15	593.20		
1 Steel Worker	51.10	408.80	90.20	721.60		
1 Dozer, 200 H.P.		1325.00		1457.50		
1 Pickup Truck, 3/4 Ton		157.00		172.70	16.84	18.52
88 L.H., Daily Totals		$5292.80		$7590.60	$60.15	$86.26

Crew R-13	Hr.	Daily	Hr.	Daily	Bare Costs	Incl. O&P
1 Electrician Foreman	$53.85	$430.80	$80.60	$644.80	$51.43	$77.18
3 Electricians	53.35	1280.40	79.85	1916.40		
.25 Equip. Oper. (crane)	50.25	100.50	76.20	152.40		
1 Equipment Oiler	43.55	348.40	66.00	528.00		
.25 Hydraulic Crane, 33 Ton		189.80		208.78	4.52	4.97
42 L.H., Daily Totals		$2349.90		$3450.38	$55.95	$82.15

Crew R-15	Hr.	Daily	Hr.	Daily	Bare Costs	Incl. O&P
1 Electrician Foreman	$53.85	$430.80	$80.60	$644.80	$52.38	$78.56
4 Electricians	53.35	1707.20	79.85	2555.20		
1 Equipment Oper. (light)	47.05	376.40	71.35	570.80		
1 Aerial Lift Truck, 40' Boom		325.00		357.50	6.77	7.45
48 L.H., Daily Totals		$2839.40		$4128.30	$59.15	$86.01

Crew R-18	Hr.	Daily	Hr.	Daily	Bare Costs	Incl. O&P
.25 Electrician Foreman	$53.85	$107.70	$80.60	$161.20	$41.88	$63.82
1 Electrician	53.35	426.80	79.85	638.80		
2 Helpers	34.65	554.40	53.70	859.20		
26 L.H., Daily Totals		$1088.90		$1659.20	$41.88	$63.82

Crew R-19	Hr.	Daily	Hr.	Daily	Bare Costs	Incl. O&P
.5 Electrician Foreman	$53.85	$215.40	$80.60	$322.40	$53.45	$80.00
2 Electricians	53.35	853.60	79.85	1277.60		
20 L.H., Daily Totals		$1069.00		$1600.00	$53.45	$80.00

Crew R-21	Hr.	Daily	Hr.	Daily	Bare Costs	Incl. O&P
1 Electrician Foreman	$53.85	$430.80	$80.60	$644.80	$53.36	$79.89
3 Electricians	53.35	1280.40	79.85	1916.40		
.1 Equip. Oper. (medium)	48.90	39.12	74.15	59.32		
.1 S.P. Crane, 4x4, 25 Ton		60.72		66.79	1.85	2.04
32.8 L.H., Daily Totals		$1811.04		$2687.31	$55.21	$81.93

Crew R-22	Hr.	Daily	Hr.	Daily	Bare Costs	Incl. O&P
.66 Electrician Foreman	$53.85	$284.33	$80.60	$425.57	$45.40	$68.73
2 Electricians	53.35	853.60	79.85	1277.60		
2 Helpers	34.65	554.40	53.70	859.20		
37.28 L.H., Daily Totals		$1692.33		$2562.37	$45.40	$68.73

Crews

Crew No.	Bare Costs		Incl. Subs O & P		Cost Per Labor-Hour	
Crew R-30	Hr.	Daily	Hr.	Daily	Bare Costs	Incl. O&P
.25 Electrician Foreman (outside)	$55.35	$110.70	$82.85	$165.70	$43.23	$65.74
1 Electrician	53.35	426.80	79.85	638.80		
2 Laborers, (Semi-Skilled)	36.65	586.40	56.55	904.80		
26 L.H., Daily Totals		$1123.90		$1709.30	$43.23	$65.74
Crew R-31	Hr.	Daily	Hr.	Daily	Bare Costs	Incl. O&P
1 Electrician	$53.35	$426.80	$79.85	$638.80	$53.35	$79.85
1 Core Drill, Electric, 2.5 H.P.		46.20		50.82	5.78	6.35
8 L.H., Daily Totals		$473.00		$689.62	$59.13	$86.20
Crew W-41E	Hr.	Daily	Hr.	Daily	Bare Costs	Incl. O&P
.5 Plumber Foreman (outside)	$59.55	$238.20	$89.90	$359.60	$49.59	$75.36
1 Plumber	57.55	460.40	86.90	695.20		
1 Laborer	36.65	293.20	56.55	452.40		
20 L.H., Daily Totals		$991.80		$1507.20	$49.59	$75.36

Historical Cost Indexes

The table below lists both the RSMeans® historical cost index based on Jan. 1, 1993 = 100 as well as the computed value of an index based on Jan. 1, 2014 costs. Since the Jan. 1, 2014 figure is estimated, space is left to write in the actual index figures as they become available through either the quarterly *RSMeans Construction Cost Indexes* or as printed in the *Engineering News-Record*. To compute the actual index based on Jan. 1, 2014 = 100, divide the historical cost index for a particular year by the actual Jan. 1, 2014 construction cost index. Space has been left to advance the index figures as the year progresses.

Year	Historical Cost Index Jan. 1, 1993 = 100 Est.	Actual	Current Index Based on Jan. 1, 2014 = 100 Est.	Actual	Year	Historical Cost Index Jan. 1, 1993 = 100 Actual	Current Index Based on Jan. 1, 2014 = 100 Est.	Actual	Year	Historical Cost Index Jan. 1, 1993 = 100 Actual	Current Index Based on Jan. 1, 2014 = 100 Est.	Actual
Oct 2014*					July 1999	117.6	58.0		July 1981	70.0	34.5	
July 2014*					1998	115.1	56.8		1980	62.9	31.0	
April 2014*					1997	112.8	55.6		1979	57.8	28.5	
Jan 2014*	202.7		100.0	100.0	1996	110.2	54.4		1978	53.5	26.4	
July 2013		201.2	99.3		1995	107.6	53.1		1977	49.5	24.4	
2012		194.6	96.0		1994	104.4	51.5		1976	46.9	23.1	
2011		191.2	94.3		1993	101.7	50.2		1975	44.8	22.1	
2010		183.5	90.5		1992	99.4	49.1		1974	41.4	20.4	
2009		180.1	88.9		1991	96.8	47.8		1973	37.7	18.6	
2008		180.4	89.0		1990	94.3	46.5		1972	34.8	17.2	
2007		169.4	83.6		1989	92.1	45.5		1971	32.1	15.8	
2006		162.0	79.9		1988	89.9	44.3		1970	28.7	14.2	
2005		151.6	74.8		1987	87.7	43.3		1969	26.9	13.3	
2004		143.7	70.9		1986	84.2	41.6		1968	24.9	12.3	
2003		132.0	65.1		1985	82.6	40.8		1967	23.5	11.6	
2002		128.7	63.5		1984	82.0	40.4		1966	22.7	11.2	
2001		125.1	61.7		1983	80.2	39.5		1965	21.7	10.7	
2000		120.9	59.6		1982	76.1	37.6		1964	21.2	10.5	

Adjustments to Costs

The "Historical Cost Index" can be used to convert national average building costs at a particular time to the approximate building costs for some other time.

Example:

Estimate and compare construction costs for different years in the same city.

To estimate the national average construction cost of a building in 1970, knowing that it cost $900,000 in 2014:

INDEX in 1970 = 28.7
INDEX in 2014 = 202.7

Note: The city cost indexes for Canada can be used to convert U.S. national averages to local costs in Canadian dollars.

Time Adjustment Using the Historical Cost Indexes:

$$\frac{\text{Index for Year A}}{\text{Index for Year B}} \times \text{Cost in Year B} = \text{Cost in Year A}$$

$$\frac{\text{INDEX 1970}}{\text{INDEX 2014}} \times \text{Cost 2014} = \text{Cost 1970}$$

$$\frac{28.7}{202.7} \times \$900{,}000 = .142 \times \$900{,}000 = \$128{,}000$$

The construction cost of the building in 1970 is $128,000.

*Historical Cost Index updates and other resources are provided on the following website.
http://www.reedconstructiondata.com/rsmeans/chgnotice/456321

City Cost Indexes

How to Use the City Cost Indexes

What you should know before you begin

RSMeans City Cost Indexes (CCI) are an extremely useful tool to use when you want to compare costs from city to city and region to region.

This publication contains average construction cost indexes for 731 U.S. and Canadian cities covering over 930 three-digit zip code locations, as listed directly under each city.

Keep in mind that a City Cost Index number is a percentage ratio of a specific city's cost to the national average cost of the same item at a stated time period.

In other words, these index figures represent relative construction factors (or, if you prefer, multipliers) for Material and Installation costs, as well as the weighted average for Total In Place costs for each CSI MasterFormat division. Installation costs include both labor and equipment rental costs. When estimating equipment rental rates only, for a specific location, use 01 54 33 EQUIPMENT RENTAL COSTS in the Reference Section at the back of the book.

The 30 City Average Index is the average of 30 major U.S. cities and serves as a National Average.

Index figures for both material and installation are based on the 30 major city average of 100 and represent the cost relationship as of July 1, 2013. The index for each division is computed from representative material and labor quantities for that division. The weighted average for each city is a weighted total of the components listed above it, but does not include relative productivity between trades or cities.

As changes occur in local material prices, labor rates, and equipment rental rates (including fuel costs), the impact of these changes should be accurately measured by the change in the City Cost Index for each particular city (as compared to the 30 City Average).

Therefore, if you know (or have estimated) building costs in one city today, you can easily convert those costs to expected building costs in another city.

In addition, by using the Historical Cost Index, you can easily convert National Average building costs at a particular time to the approximate building costs for some other time. The City Cost Indexes can then be applied to calculate the costs for a particular city.

Quick Calculations

Location Adjustment Using the City Cost Indexes:

$$\frac{\text{Index for City A}}{\text{Index for City B}} \times \text{Cost in City B} = \text{Cost in City A}$$

Time Adjustment for the National Average Using the Historical Cost Index:

$$\frac{\text{Index for Year A}}{\text{Index for Year B}} \times \text{Cost in Year B} = \text{Cost in Year A}$$

Adjustment from the National Average:

$$\frac{\text{Index for City A}}{100} \times \text{National Average Cost} = \text{Cost in City A}$$

Since each of the other RSMeans publications contains many different items, any *one* item multiplied by the particular city index may give incorrect results. However, the larger the number of items compiled, the closer the results should be to actual costs for that particular city.

The City Cost Indexes for Canadian cities are calculated using Canadian material and equipment prices and labor rates, in Canadian dollars. Therefore, indexes for Canadian cities can be used to convert U.S. National Average prices to local costs in Canadian dollars.

How to use this section

1. Compare costs from city to city.

In using the RSMeans Indexes, remember that an index number is not a fixed number but a ratio: It's a percentage ratio of a building component's cost at any stated time to the National Average cost of that same component at the same time period. Put in the form of an equation:

$$\frac{\text{Specific City Cost}}{\text{National Average Cost}} \times 100 = \text{City Index Number}$$

Therefore, when making cost comparisons between cities, do not subtract one city's index number from the index number of another city and read the result as a percentage difference. Instead, divide one city's index number by that of the other city. The resulting number may then be used as a multiplier to calculate cost differences from city to city.

The formula used to find cost differences between cities for the purpose of comparison is as follows:

$$\frac{\text{City A Index}}{\text{City B Index}} \times \text{City B Cost (Known)} = \text{City A Cost (Unknown)}$$

In addition, you can use RSMeans CCI to calculate and compare costs division by division between cities using the same basic formula. (Just be sure that you're comparing similar divisions.)

2. Compare a specific city's construction costs with the National Average.

When you're studying construction location feasibility, it's advisable to compare a prospective project's cost index with an index of the National Average cost.

For example, divide the weighted average index of construction costs of a specific city by that of the 30 City Average, which = 100.

$$\frac{\text{City Index}}{100} = \% \text{ of National Average}$$

As a result, you get a ratio that indicates the relative cost of construction in that city in comparison with the National Average.

3. Convert U.S. National Average to actual costs in Canadian City.

$$\frac{\text{Index for Canadian City}}{100} \times \text{National Average Cost} = \text{Cost in Canadian City in \$ CAN}$$

4. **Adjust construction cost data based on a National Average.**
When you use a source of construction cost data which is based on a National Average (such as RSMeans cost data publications), it is necessary to adjust those costs to a specific location.

$$\frac{\text{City Index}}{100} \times \text{``Book'' Cost Based on National Average Costs} = \text{City Cost (Unknown)}$$

5. **When applying the City Cost Indexes to demolition projects, use the appropriate division installation index.** For example, for removal of existing doors and windows, use Division 8 (Openings) index.

What you might like to know about how we developed the Indexes

The information presented in the CCI is organized according to the Construction Specifications Institute (CSI) MasterFormat 2012 classification system.

To create a reliable index, RSMeans researched the building type most often constructed in the United States and Canada. Because it was concluded that no one type of building completely represented the building construction industry, nine different types of buildings were combined to create a composite model.

The exact material, labor, and equipment quantities are based on detailed analyses of these nine building types, and then each quantity is weighted in proportion to expected usage. These various material items, labor hours, and equipment rental rates are thus combined to form a composite building representing as closely as possible the actual usage of materials, labor, and equipment used in the North American building construction industry.

The following structures were chosen to make up that composite model:

1. Factory, 1 story
2. Office, 2-4 story
3. Store, Retail
4. Town Hall, 2-3 story
5. High School, 2-3 story
6. Hospital, 4-8 story
7. Garage, Parking
8. Apartment, 1-3 story
9. Hotel/Motel, 2-3 story

For the purposes of ensuring the timeliness of the data, the components of the index for the composite model have been streamlined. They currently consist of:

- specific quantities of 66 commonly used construction materials;
- specific labor-hours for 21 building construction trades; and
- specific days of equipment rental for 6 types of construction equipment (normally used to install the 66 material items by the 21 trades.) Fuel costs and routine maintenance costs are included in the equipment cost.

A sophisticated computer program handles the updating of all costs for each city on a quarterly basis. Material and equipment price quotations are gathered quarterly from 731 cities in the United States and Canada. These prices and the latest negotiated labor wage rates for 21 different building trades are used to compile the quarterly update of the City Cost Index.

The 30 major U.S. cities used to calculate the National Average are:

Atlanta, GA	Memphis, TN
Baltimore, MD	Milwaukee, WI
Boston, MA	Minneapolis, MN
Buffalo, NY	Nashville, TN
Chicago, IL	New Orleans, LA
Cincinnati, OH	New York, NY
Cleveland, OH	Philadelphia, PA
Columbus, OH	Phoenix, AZ
Dallas, TX	Pittsburgh, PA
Denver, CO	St. Louis, MO
Detroit, MI	San Antonio, TX
Houston, TX	San Diego, CA
Indianapolis, IN	San Francisco, CA
Kansas City, MO	Seattle, WA
Los Angeles, CA	Washington, DC

What the CCI does not indicate

The weighted average for each city is a total of the divisional components weighted to reflect typical usage, but it does not include the productivity variations between trades or cities.

In addition, the CCI does not take into consideration factors such as the following:

- managerial efficiency
- competitive conditions
- automation
- restrictive union practices
- unique local requirements
- regional variations due to specific building codes

City Cost Indexes

		UNITED STATES 30 CITY AVERAGE			ALABAMA ANNISTON 362			ALABAMA BIRMINGHAM 350 - 352			ALABAMA BUTLER 369			ALABAMA DECATUR 356			ALABAMA DOTHAN 363		
DIVISION		MAT.	INST.	TOTAL	MAT.	INST.	TOTAL	MAT.	INST.	TOTAL	MAT.	INST.	TOTAL	MAT.	INST.	TOTAL	MAT.	INST.	TOTAL
015433	CONTRACTOR EQUIPMENT		100.0	100.0		101.6	101.6		101.7	101.7		98.8	98.8		101.6	101.6		98.8	98.8
0241, 31 - 34	SITE & INFRASTRUCTURE, DEMOLITION	100.0	100.0	100.0	87.7	94.3	92.3	88.8	94.9	93.1	101.0	88.3	92.1	81.3	92.8	89.3	98.7	88.3	91.4
0310	Concrete Forming & Accessories	100.0	100.0	100.0	92.7	48.2	54.3	93.3	70.2	73.4	89.0	43.6	49.8	94.7	46.4	53.0	97.9	43.0	50.5
0320	Concrete Reinforcing	100.0	100.0	100.0	87.0	87.4	87.2	92.9	87.8	90.4	92.0	47.4	69.7	87.0	79.5	83.3	92.0	46.8	69.4
0330	Cast-in-Place Concrete	100.0	100.0	100.0	102.0	50.5	80.6	99.7	74.7	89.3	99.4	55.7	81.3	93.7	65.1	81.9	99.4	47.0	77.7
03	CONCRETE	100.0	100.0	100.0	99.8	57.9	79.1	95.1	76.1	85.7	100.6	50.4	75.8	91.4	60.7	76.3	100.0	47.0	73.9
04	MASONRY	100.0	100.0	100.0	90.5	67.9	76.5	89.9	76.2	81.4	95.3	49.4	66.9	88.4	52.1	66.0	96.6	40.4	61.8
05	METALS	100.0	100.0	100.0	96.9	92.5	95.5	97.0	94.7	96.3	95.8	78.1	90.4	99.0	87.7	95.5	95.9	76.9	90.0
06	WOOD, PLASTICS & COMPOSITES	100.0	100.0	100.0	92.7	43.7	65.1	96.3	68.3	80.5	87.3	43.4	62.5	98.0	43.9	67.5	99.6	43.4	67.9
07	THERMAL & MOISTURE PROTECTION	100.0	100.0	100.0	100.1	53.8	81.3	95.9	81.4	90.0	100.1	57.6	82.9	93.1	59.3	79.4	100.1	49.9	79.7
08	OPENINGS	100.0	100.0	100.0	93.9	50.9	83.8	95.8	72.3	90.3	94.0	45.2	82.4	98.8	51.6	87.6	94.0	45.2	82.5
0920	Plaster & Gypsum Board	100.0	100.0	100.0	87.5	42.4	56.8	97.0	67.8	77.1	84.9	42.0	55.7	98.1	42.6	60.4	95.2	42.0	59.1
0950, 0980	Ceilings & Acoustic Treatment	100.0	100.0	100.0	87.8	42.4	57.6	98.4	67.8	78.0	87.8	42.0	57.3	95.7	42.6	60.4	87.8	42.0	57.3
0960	Flooring	100.0	100.0	100.0	104.1	41.0	85.2	109.0	76.2	99.2	109.2	57.0	93.6	106.0	51.5	89.7	114.9	28.0	88.9
0970, 0990	Wall Finishes & Painting/Coating	100.0	100.0	100.0	91.4	32.1	55.8	93.5	56.8	71.4	91.4	54.5	69.2	89.3	68.2	76.7	91.4	54.5	69.2
09	FINISHES	100.0	100.0	100.0	92.7	44.0	65.6	99.4	69.3	82.6	95.9	46.7	68.5	95.9	48.8	69.7	98.7	41.0	66.6
COVERS	DIVS. 10 - 14, 25, 28, 41, 43, 44, 46	100.0	100.0	100.0	100.0	70.1	94.0	100.0	87.4	97.5	100.0	42.5	88.5	100.0	44.4	88.9	100.0	42.6	88.5
21, 22, 23	FIRE SUPPRESSION, PLUMBING & HVAC	100.0	100.0	100.0	99.9	56.6	82.6	100.0	62.8	85.1	97.3	34.0	71.9	100.0	41.7	76.6	97.3	33.9	71.9
26, 27, 3370	ELECTRICAL, COMMUNICATIONS & UTIL.	100.0	100.0	100.0	90.7	58.7	74.0	99.7	62.8	80.4	92.6	41.2	65.7	94.5	66.5	79.9	91.5	58.8	74.4
MF2010	WEIGHTED AVERAGE	100.0	100.0	100.0	96.5	63.0	81.9	97.5	74.1	87.3	96.8	50.5	76.6	96.6	59.1	80.3	96.9	50.5	76.7

		ALABAMA EVERGREEN 364			ALABAMA GADSDEN 359			ALABAMA HUNTSVILLE 357 - 358			ALABAMA JASPER 355			ALABAMA MOBILE 365 - 366			ALABAMA MONTGOMERY 360 - 361		
DIVISION		MAT.	INST.	TOTAL	MAT.	INST.	TOTAL	MAT.	INST.	TOTAL	MAT.	INST.	TOTAL	MAT.	INST.	TOTAL	MAT.	INST.	TOTAL
015433	CONTRACTOR EQUIPMENT		98.8	98.8		101.6	101.6		101.6	101.6		101.6	101.6		98.8	98.8		98.8	98.8
0241, 31 - 34	SITE & INFRASTRUCTURE, DEMOLITION	101.5	88.6	92.4	86.7	93.8	91.7	81.1	93.8	90.0	86.3	93.6	91.4	94.0	89.3	90.7	94.0	89.2	90.6
0310	Concrete Forming & Accessories	85.6	45.2	50.8	87.0	43.3	49.3	94.7	71.1	74.3	92.2	35.3	43.0	97.1	55.0	60.7	96.9	45.0	52.1
0320	Concrete Reinforcing	92.0	47.4	69.7	92.2	87.7	90.0	87.0	80.5	83.8	87.0	86.5	86.8	89.8	79.4	84.6	94.9	87.1	91.0
0330	Cast-in-Place Concrete	99.5	58.3	82.4	93.8	66.7	82.5	91.3	69.6	82.3	103.9	44.8	79.3	104.3	67.4	88.9	105.4	57.4	85.5
03	CONCRETE	100.9	52.0	76.8	95.7	61.4	78.8	90.3	73.3	81.9	98.9	50.1	74.8	97.0	65.3	81.4	98.3	58.9	78.8
04	MASONRY	95.4	54.1	69.8	86.7	62.3	71.6	89.7	67.7	76.1	84.2	42.0	58.1	93.8	55.1	69.9	90.4	50.5	65.7
05	METALS	95.9	78.1	90.4	96.9	94.0	96.0	99.0	90.5	96.4	96.9	91.1	95.1	97.9	90.6	95.6	97.0	92.8	95.7
06	WOOD, PLASTICS & COMPOSITES	83.6	43.4	60.9	87.8	39.6	60.6	98.0	74.3	84.6	94.8	32.3	59.6	98.2	54.1	73.3	97.7	43.4	67.0
07	THERMAL & MOISTURE PROTECTION	100.1	57.9	83.0	93.1	72.4	84.7	93.0	78.5	87.1	93.2	46.1	74.1	99.8	71.3	88.2	100.2	67.8	87.0
08	OPENINGS	94.0	45.2	82.4	95.4	49.8	84.6	98.8	68.2	91.5	95.3	48.7	84.3	97.2	58.6	88.1	98.3	54.0	87.8
0920	Plaster & Gypsum Board	84.2	42.0	55.5	89.8	38.2	54.7	98.1	73.9	81.7	94.2	30.7	51.0	92.1	53.1	65.5	91.5	42.0	57.8
0950, 0980	Ceilings & Acoustic Treatment	87.8	42.0	57.3	92.1	38.2	56.2	97.6	73.9	81.9	92.1	30.7	51.2	93.3	53.1	66.5	95.9	42.0	60.0
0960	Flooring	107.0	57.0	92.1	102.0	76.2	94.3	106.0	76.2	97.1	104.3	41.0	85.3	114.6	59.5	98.1	110.8	59.5	95.5
0970, 0990	Wall Finishes & Painting/Coating	91.4	54.5	69.2	89.3	62.0	72.9	89.3	59.7	71.5	89.3	39.4	59.3	95.0	55.4	71.2	91.2	54.5	69.1
09	FINISHES	95.3	47.9	68.9	93.3	49.4	68.8	96.4	71.1	82.3	94.5	35.2	61.5	99.0	54.7	74.3	99.0	47.1	70.1
COVERS	DIVS. 10 - 14, 25, 28, 41, 43, 44, 46	100.0	44.1	88.8	100.0	79.5	95.9	100.0	85.3	97.0	100.0	40.8	88.1	100.0	83.9	96.8	100.0	79.4	95.9
21, 22, 23	FIRE SUPPRESSION, PLUMBING & HVAC	97.3	36.3	72.9	102.0	36.9	75.9	100.0	52.9	81.1	102.0	53.8	82.6	100.0	61.5	84.5	100.1	34.8	73.9
26, 27, 3370	ELECTRICAL, COMMUNICATIONS & UTIL.	90.1	41.2	64.6	94.5	62.8	78.0	95.7	66.5	80.4	94.1	62.8	77.7	93.4	58.5	75.2	93.9	62.3	77.4
MF2010	WEIGHTED AVERAGE	96.5	51.9	77.1	96.7	60.8	81.1	96.7	70.8	85.4	97.0	56.7	79.4	97.8	65.9	83.9	97.9	58.1	80.5

		ALABAMA PHENIX CITY 368			ALABAMA SELMA 367			ALABAMA TUSCALOOSA 354			ALASKA ANCHORAGE 995 - 996			ALASKA FAIRBANKS 997			ALASKA JUNEAU 998		
DIVISION		MAT.	INST.	TOTAL	MAT.	INST.	TOTAL	MAT.	INST.	TOTAL	MAT.	INST.	TOTAL	MAT.	INST.	TOTAL	MAT.	INST.	TOTAL
015433	CONTRACTOR EQUIPMENT		98.8	98.8		98.8	98.8		101.6	101.6		113.5	113.5		113.5	113.5		113.5	113.5
0241, 31 - 34	SITE & INFRASTRUCTURE, DEMOLITION	105.2	89.3	94.0	98.6	89.1	92.0	81.6	94.0	90.3	125.5	128.8	127.8	116.4	128.8	125.1	132.0	128.8	129.8
0310	Concrete Forming & Accessories	92.6	39.9	47.1	90.3	43.5	49.9	94.6	48.8	55.1	127.8	119.0	120.2	133.0	118.7	120.7	129.2	118.9	120.3
0320	Concrete Reinforcing	91.9	63.3	77.7	92.0	86.2	89.1	87.0	87.7	87.3	148.0	111.5	129.8	142.1	111.5	126.8	122.7	111.5	117.1
0330	Cast-in-Place Concrete	99.5	57.9	82.2	99.4	47.2	77.8	95.1	68.5	84.1	131.0	117.4	125.3	129.2	117.7	124.4	136.1	117.4	128.3
03	CONCRETE	103.6	52.4	78.3	99.4	54.5	77.3	92.1	64.5	78.5	143.0	116.3	129.8	126.5	116.3	121.4	139.9	116.2	128.2
04	MASONRY	95.3	41.8	62.2	98.7	39.3	62.0	88.6	65.5	74.3	180.9	122.2	144.6	182.2	122.2	145.1	167.6	122.2	139.5
05	METALS	95.8	83.5	92.0	95.8	90.3	94.1	98.2	94.1	96.9	114.6	102.5	110.9	118.8	102.6	113.8	117.6	102.4	112.9
06	WOOD, PLASTICS & COMPOSITES	92.5	36.3	60.8	89.3	43.4	63.4	98.0	45.4	68.3	117.1	117.9	117.6	125.0	117.4	120.7	119.8	117.9	118.7
07	THERMAL & MOISTURE PROTECTION	100.4	64.9	86.0	100.0	53.0	80.9	93.1	74.4	85.5	169.9	115.5	147.8	169.2	116.6	147.9	163.4	115.5	143.9
08	OPENINGS	93.9	44.6	82.3	93.9	54.0	84.5	98.8	59.8	89.6	119.7	115.2	118.7	116.2	115.2	116.0	116.6	115.2	116.3
0920	Plaster & Gypsum Board	89.0	34.7	52.1	86.8	42.0	56.3	98.1	44.2	61.4	126.3	118.3	120.9	148.9	117.8	127.7	131.2	118.3	122.4
0950, 0980	Ceilings & Acoustic Treatment	87.8	34.7	52.5	87.8	42.0	57.3	97.6	44.2	62.1	126.5	118.3	121.0	121.7	117.8	119.1	123.5	118.3	120.0
0960	Flooring	111.3	59.5	95.8	109.8	29.0	85.6	106.0	76.2	97.1	150.1	128.3	143.6	140.7	128.3	137.0	142.1	128.3	138.0
0970, 0990	Wall Finishes & Painting/Coating	91.4	54.5	69.2	91.4	54.5	69.2	89.3	46.9	63.8	146.5	113.3	126.5	143.9	121.9	130.7	143.3	113.3	125.2
09	FINISHES	97.5	43.2	67.2	96.0	41.2	65.5	96.3	52.0	71.6	143.1	120.5	130.5	139.4	121.2	129.2	139.6	120.5	129.0
COVERS	DIVS. 10 - 14, 25, 28, 41, 43, 44, 46	100.0	79.0	95.8	100.0	42.5	88.5	100.0	81.3	96.2	100.0	112.5	102.5	100.0	112.5	102.5	100.0	112.5	102.5
21, 22, 23	FIRE SUPPRESSION, PLUMBING & HVAC	97.3	35.1	72.4	97.3	34.3	72.1	100.0	34.6	73.8	100.2	104.3	101.9	100.0	106.7	102.8	100.1	100.6	100.3
26, 27, 3370	ELECTRICAL, COMMUNICATIONS & UTIL.	92.1	71.2	81.2	91.2	41.2	65.1	95.0	62.8	78.2	120.5	115.6	117.9	134.7	115.6	124.7	125.5	115.6	120.3
MF2010	WEIGHTED AVERAGE	97.3	55.7	79.2	96.7	50.8	76.7	96.7	62.0	81.6	121.3	114.3	118.3	120.7	115.0	118.2	120.6	113.5	117.5

596

City Cost Indexes

DIVISION		ALASKA KETCHIKAN 999			ARIZONA CHAMBERS 865			ARIZONA FLAGSTAFF 860			ARIZONA GLOBE 855			ARIZONA KINGMAN 864			ARIZONA MESA/TEMPE 852		
		MAT.	INST.	TOTAL	MAT.	INST.	TOTAL	MAT.	INST.	TOTAL	MAT.	INST.	TOTAL	MAT.	INST.	TOTAL	MAT.	INST.	TOTAL
015433	CONTRACTOR EQUIPMENT		113.5	113.5		92.6	92.6		92.6	92.6		94.2	94.2		92.6	92.6		94.2	94.2
0241, 31 - 34	SITE & INFRASTRUCTURE, DEMOLITION	170.9	128.8	141.3	67.2	96.6	87.9	84.8	96.8	93.3	100.7	98.4	99.1	67.2	96.8	88.0	91.4	98.6	96.4
0310	Concrete Forming & Accessories	124.5	118.9	119.7	98.2	59.2	64.5	103.7	66.4	71.5	96.3	59.3	64.3	96.5	66.4	70.5	99.3	69.6	73.6
0320	Concrete Reinforcing	111.6	111.5	111.5	100.0	83.5	91.7	99.8	83.5	91.7	110.1	83.5	96.8	100.1	83.5	91.8	110.8	83.5	97.2
0330	Cast-in-Place Concrete	261.0	117.4	201.4	90.3	73.5	83.3	90.4	73.8	83.5	93.8	73.6	85.4	90.0	73.8	83.3	94.6	73.9	86.0
03	CONCRETE	214.6	116.2	166.0	96.5	68.9	82.8	116.1	72.2	94.4	109.8	68.9	89.7	96.1	72.2	84.3	101.7	73.7	87.8
04	MASONRY	191.0	122.2	148.4	90.6	62.7	73.3	90.7	62.8	73.4	110.3	62.7	80.9	90.6	62.8	73.4	110.6	62.8	81.0
05	METALS	119.0	102.4	113.9	95.1	75.0	88.9	95.6	75.7	89.5	92.4	75.5	87.2	95.7	75.8	89.6	92.7	76.4	87.7
06	WOOD, PLASTICS & COMPOSITES	115.7	117.9	117.0	96.1	56.3	73.6	102.2	65.7	81.6	90.0	56.4	71.0	91.6	65.7	77.0	93.4	69.9	80.2
07	THERMAL & MOISTURE PROTECTION	172.3	115.5	149.2	95.1	65.1	82.9	96.7	68.2	85.1	99.3	64.6	85.2	95.0	65.7	83.1	98.6	66.6	85.6
08	OPENINGS	118.1	115.2	117.4	108.1	65.7	98.1	108.2	70.8	99.4	102.2	65.8	93.6	108.3	70.8	99.4	102.3	73.2	95.4
0920	Plaster & Gypsum Board	137.4	118.3	124.4	85.6	55.0	64.8	89.1	64.7	72.5	94.2	55.0	67.5	79.1	64.7	69.3	96.4	69.0	77.7
0950, 0980	Ceilings & Acoustic Treatment	114.7	118.3	117.1	107.8	55.0	72.7	108.7	64.7	79.4	96.4	55.0	68.8	108.7	64.7	79.4	96.4	69.0	78.1
0960	Flooring	140.7	128.3	137.0	92.1	40.3	76.6	94.1	40.4	78.1	100.0	40.3	82.1	91.1	54.7	80.2	101.1	50.4	86.0
0970, 0990	Wall Finishes & Painting/Coating	143.9	113.3	125.5	97.8	56.3	72.9	97.8	56.3	72.9	108.9	56.3	77.3	97.8	56.3	72.9	108.9	56.3	77.3
09	FINISHES	140.7	120.5	129.5	95.7	54.0	72.4	98.8	59.6	76.9	97.6	54.1	73.4	94.7	62.0	76.5	97.1	64.0	78.7
COVERS	DIVS. 10 - 14, 25, 28, 41, 43, 44, 46	100.0	112.5	102.5	100.0	82.3	96.5	100.0	83.3	96.7	100.0	82.5	96.5	100.0	83.3	96.7	100.0	84.0	96.8
21, 22, 23	FIRE SUPPRESSION, PLUMBING & HVAC	98.4	100.6	99.3	97.0	78.9	89.7	100.2	80.0	92.1	95.2	78.9	88.7	97.0	80.0	90.2	100.1	79.0	91.6
26, 27, 3370	ELECTRICAL, COMMUNICATIONS & UTIL.	134.6	115.6	124.7	102.2	72.6	86.7	101.1	65.8	82.7	98.7	65.8	81.6	102.2	65.8	83.2	95.3	65.8	79.9
MF2010	WEIGHTED AVERAGE	132.0	113.5	124.0	97.3	71.9	86.3	101.0	72.9	88.8	99.0	71.2	86.9	97.3	73.2	86.8	98.8	73.8	87.9

DIVISION		ARIZONA PHOENIX 850,853			ARIZONA PRESCOTT 863			ARIZONA SHOW LOW 859			ARIZONA TUCSON 856 - 857			ARKANSAS BATESVILLE 725			ARKANSAS CAMDEN 717		
		MAT.	INST.	TOTAL	MAT.	INST.	TOTAL	MAT.	INST.	TOTAL	MAT.	INST.	TOTAL	MAT.	INST.	TOTAL	MAT.	INST.	TOTAL
015433	CONTRACTOR EQUIPMENT		94.7	94.7		92.6	92.6		94.2	94.2		94.2	94.2		89.0	89.0		89.0	89.0
0241, 31 - 34	SITE & INFRASTRUCTURE, DEMOLITION	91.9	98.8	96.7	73.5	96.6	89.7	102.8	98.4	99.7	87.1	98.6	95.1	73.4	84.9	81.5	73.3	84.5	81.2
0310	Concrete Forming & Accessories	100.2	69.7	73.9	99.8	53.5	59.8	103.2	69.3	74.0	99.5	69.5	73.6	85.2	45.9	51.3	83.9	31.4	38.6
0320	Concrete Reinforcing	109.0	83.7	96.3	99.8	83.5	91.6	110.8	83.5	97.2	91.8	83.5	87.6	86.7	65.8	76.3	88.2	65.7	76.9
0330	Cast-in-Place Concrete	94.7	74.0	86.1	90.3	73.5	83.3	93.9	73.6	85.5	97.4	73.9	87.6	81.1	46.1	66.6	81.2	39.0	63.7
03	CONCRETE	101.3	73.8	87.7	101.7	66.3	84.2	112.1	73.4	93.0	99.7	73.6	86.8	85.2	50.5	68.0	86.9	41.5	64.5
04	MASONRY	98.0	64.6	77.3	90.7	62.7	73.4	110.4	62.7	80.9	95.9	62.8	75.4	98.7	41.5	63.3	111.6	31.9	62.3
05	METALS	94.2	77.2	88.9	95.6	74.7	89.2	92.2	75.7	87.1	93.5	76.3	88.2	98.0	66.9	88.4	97.3	66.5	87.8
06	WOOD, PLASTICS & COMPOSITES	94.4	69.9	80.6	97.5	48.5	69.9	98.0	69.9	82.1	93.7	69.9	80.3	86.3	46.5	63.9	89.8	30.4	56.3
07	THERMAL & MOISTURE PROTECTION	98.2	68.7	86.2	95.5	64.2	82.8	99.5	66.5	86.1	99.6	65.8	85.9	100.2	44.8	77.7	96.1	36.4	71.9
08	OPENINGS	104.2	73.2	96.9	108.2	61.5	97.2	101.3	73.2	94.7	98.3	73.2	92.4	94.6	46.0	83.1	93.5	40.6	81.0
0920	Plaster & Gypsum Board	98.5	69.0	78.4	85.7	47.0	59.4	98.6	69.0	78.4	101.9	69.0	79.5	81.0	45.4	56.7	80.2	28.7	45.2
0950, 0980	Ceilings & Acoustic Treatment	103.5	69.0	80.5	106.9	47.0	67.1	96.4	69.0	78.1	97.3	69.0	78.4	90.6	45.4	60.5	86.2	28.7	47.9
0960	Flooring	101.4	52.7	86.8	92.9	40.3	77.1	102.5	45.9	85.5	91.9	45.0	77.9	104.6	60.1	91.3	111.2	40.3	90.0
0970, 0990	Wall Finishes & Painting/Coating	108.9	62.0	80.7	97.8	56.3	72.9	108.9	56.3	77.3	110.1	56.3	77.8	99.8	42.0	65.1	99.5	50.8	70.2
09	FINISHES	99.1	65.0	80.1	96.2	49.4	70.1	99.2	63.0	79.0	95.4	62.9	77.3	89.4	47.8	66.4	92.1	34.2	59.8
COVERS	DIVS. 10 - 14, 25, 28, 41, 43, 44, 46	100.0	84.0	96.8	100.0	81.4	96.3	100.0	84.0	96.8	100.0	84.0	96.8	100.0	40.6	88.1	100.0	36.5	87.3
21, 22, 23	FIRE SUPPRESSION, PLUMBING & HVAC	100.0	80.0	92.0	100.2	78.9	91.7	95.2	79.0	88.7	100.1	79.0	91.6	95.3	52.1	77.9	95.3	53.5	78.5
26, 27, 3370	ELECTRICAL, COMMUNICATIONS & UTIL.	102.2	68.3	84.5	100.7	65.8	82.5	95.7	65.8	80.1	97.6	65.8	81.0	95.0	64.0	78.8	91.5	62.0	76.1
MF2010	WEIGHTED AVERAGE	99.4	74.8	88.7	98.9	69.7	86.2	99.1	73.5	88.0	97.6	73.6	87.1	94.2	55.1	77.2	94.5	50.4	75.3

DIVISION		ARKANSAS FAYETTEVILLE 727			ARKANSAS FORT SMITH 729			ARKANSAS HARRISON 726			ARKANSAS HOT SPRINGS 719			ARKANSAS JONESBORO 724			ARKANSAS LITTLE ROCK 720 - 722		
		MAT.	INST.	TOTAL	MAT.	INST.	TOTAL	MAT.	INST.	TOTAL	MAT.	INST.	TOTAL	MAT.	INST.	TOTAL	MAT.	INST.	TOTAL
015433	CONTRACTOR EQUIPMENT		89.0	89.0		89.0	89.0		89.0	89.0		89.0	89.0		107.5	107.5		89.0	89.0
0241, 31 - 34	SITE & INFRASTRUCTURE, DEMOLITION	72.8	86.4	82.4	77.7	86.3	83.8	77.9	84.9	82.8	76.4	85.9	83.0	96.9	100.3	99.3	87.9	86.3	86.8
0310	Concrete Forming & Accessories	80.6	39.4	45.1	100.2	57.7	63.5	89.7	45.8	51.8	81.2	36.5	42.6	88.7	49.2	54.6	95.3	60.4	65.2
0320	Concrete Reinforcing	86.7	71.6	79.2	87.7	72.3	80.0	86.3	65.7	76.0	86.5	65.8	76.2	83.9	64.8	74.4	92.7	66.5	79.6
0330	Cast-in-Place Concrete	81.1	50.2	68.3	92.8	70.4	83.5	90.0	44.2	71.0	83.0	39.5	64.9	88.3	55.9	74.9	91.5	70.5	82.8
03	CONCRETE	84.9	50.1	67.7	93.1	65.3	79.4	92.6	49.8	71.4	90.7	44.0	67.6	89.8	55.9	73.1	94.7	65.4	80.2
04	MASONRY	89.1	42.1	60.0	94.1	50.0	66.8	99.0	39.7	62.3	83.9	34.1	53.1	90.5	42.6	60.9	93.9	50.0	66.7
05	METALS	98.0	69.3	89.1	100.3	71.1	91.3	99.1	66.6	89.1	97.3	66.9	87.9	94.6	78.8	89.7	100.5	69.2	90.8
06	WOOD, PLASTICS & COMPOSITES	81.9	36.3	56.2	103.9	60.4	79.4	92.3	46.5	66.5	86.8	36.5	58.4	90.1	49.4	67.1	96.3	63.9	78.1
07	THERMAL & MOISTURE PROTECTION	101.0	46.2	78.8	101.4	54.5	82.4	100.5	43.5	77.4	96.3	38.6	72.9	105.2	50.0	82.8	97.7	54.9	80.3
08	OPENINGS	94.6	45.8	83.0	95.4	57.1	86.4	95.4	46.7	83.9	93.5	42.4	81.4	97.4	53.5	87.0	96.2	58.9	87.4
0920	Plaster & Gypsum Board	78.7	34.9	48.9	86.6	59.7	68.3	85.5	45.4	58.2	78.8	35.1	49.0	95.4	48.0	63.1	92.1	63.3	72.5
0950, 0980	Ceilings & Acoustic Treatment	90.6	34.9	53.5	92.4	59.7	70.6	92.4	45.4	61.1	86.2	35.1	52.2	93.8	48.0	63.3	95.8	63.3	74.2
0960	Flooring	102.0	60.1	89.5	111.0	61.9	96.3	106.9	60.1	92.9	110.0	60.1	95.1	74.5	54.4	68.5	109.1	61.9	95.0
0970, 0990	Wall Finishes & Painting/Coating	99.8	30.0	57.9	99.8	55.0	72.8	99.8	42.0	65.1	99.5	50.8	70.2	88.3	49.4	64.9	100.9	56.4	74.2
09	FINISHES	88.9	40.9	62.2	93.0	58.4	73.8	91.8	47.8	67.3	91.8	41.3	63.7	86.5	49.6	66.0	95.9	60.7	76.3
COVERS	DIVS. 10 - 14, 25, 28, 41, 43, 44, 46	100.0	50.0	90.0	100.0	76.8	95.4	100.0	50.5	90.1	100.0	37.5	87.5	100.0	46.3	89.3	100.0	77.2	95.4
21, 22, 23	FIRE SUPPRESSION, PLUMBING & HVAC	95.3	50.6	77.4	100.2	50.1	80.1	95.3	49.8	77.0	95.3	49.1	76.8	100.4	52.9	81.4	100.0	56.4	82.5
26, 27, 3370	ELECTRICAL, COMMUNICATIONS & UTIL.	88.8	52.4	69.8	92.3	66.1	78.6	93.5	38.0	64.5	93.5	69.4	81.0	99.0	64.0	80.7	99.3	69.5	83.7
MF2010	WEIGHTED AVERAGE	93.0	52.9	75.6	96.8	61.7	81.5	95.5	51.1	76.1	93.8	52.3	75.8	96.3	59.4	80.2	98.0	63.8	83.1

597

City Cost Indexes

		ARKANSAS												CALIFORNIA					
	DIVISION	PINE BLUFF			RUSSELLVILLE			TEXARKANA			WEST MEMPHIS			ALHAMBRA			ANAHEIM		
		716			728			718			723			917 - 918			928		
		MAT.	INST.	TOTAL	MAT.	INST.	TOTAL	MAT.	INST.	TOTAL	MAT.	INST.	TOTAL	MAT.	INST.	TOTAL	MAT.	INST.	TOTAL
015433	CONTRACTOR EQUIPMENT		89.0	89.0		89.0	89.0		89.7	89.7		107.5	107.5		98.9	98.9		100.0	100.0
0241, 31 - 34	SITE & INFRASTRUCTURE, DEMOLITION	78.5	86.3	84.0	74.5	84.9	81.8	90.2	86.9	87.9	103.2	100.3	101.2	96.4	109.7	105.7	96.5	107.7	104.4
0310	Concrete Forming & Accessories	80.9	60.3	63.2	85.9	56.0	60.1	87.6	41.0	47.4	94.4	49.5	55.6	115.1	116.9	116.6	105.5	123.1	120.7
0320	Concrete Reinforcing	88.1	66.5	77.3	87.3	65.6	76.5	87.7	65.8	76.8	83.9	64.9	74.4	107.4	114.9	111.1	93.2	114.8	103.9
0330	Cast-in-Place Concrete	83.0	70.4	77.8	85.0	46.0	68.8	90.5	44.0	71.2	92.6	56.0	77.4	94.7	119.1	104.8	91.3	121.7	104.0
03	CONCRETE	91.5	65.4	78.6	88.4	54.9	71.9	89.1	47.6	68.6	97.3	56.0	76.9	104.0	116.4	110.1	103.1	120.0	111.4
04	MASONRY	120.2	50.0	76.8	93.3	38.0	59.1	98.1	33.7	58.2	76.8	42.6	55.7	122.3	118.6	120.0	77.4	116.2	101.4
05	METALS	98.0	69.1	89.1	98.0	66.5	88.3	90.8	67.0	83.5	93.6	79.1	89.2	84.0	99.9	88.9	103.1	100.2	102.2
06	WOOD, PLASTICS & COMPOSITES	86.3	63.9	73.7	87.7	60.4	72.3	95.2	43.0	65.7	96.3	49.4	69.9	96.6	113.1	105.9	98.5	121.4	111.4
07	THERMAL & MOISTURE PROTECTION	96.4	54.9	79.6	101.2	44.7	78.3	97.1	45.3	76.1	105.6	50.0	83.1	95.2	117.2	104.2	97.9	121.3	107.4
08	OPENINGS	93.5	58.9	85.3	94.6	56.2	85.5	98.3	46.7	86.1	97.4	53.5	87.0	89.8	115.5	95.9	100.3	120.2	105.0
0920	Plaster & Gypsum Board	78.4	63.3	68.1	81.0	59.7	66.5	82.2	41.7	54.6	98.0	48.0	64.0	102.1	113.7	110.0	100.6	122.0	115.2
0950, 0980	Ceilings & Acoustic Treatment	86.2	63.3	71.0	90.6	59.7	70.0	89.7	41.7	57.8	91.8	48.0	62.7	102.5	113.7	109.9	106.7	122.0	116.9
0960	Flooring	109.7	61.9	95.4	104.3	60.1	91.1	111.9	51.9	94.0	76.7	54.4	70.0	93.0	109.2	97.8	105.1	114.1	107.8
0970, 0990	Wall Finishes & Painting/Coating	99.5	56.4	73.6	99.8	35.7	61.3	99.5	31.0	58.3	88.3	50.8	65.8	101.7	108.4	105.7	100.4	106.2	103.9
09	FINISHES	91.8	60.7	74.5	90.0	55.3	70.7	94.3	41.3	64.7	87.7	49.8	66.6	98.0	114.0	106.9	102.0	119.6	111.8
COVERS	DIVS. 10 - 14, 25, 28, 41, 43, 44, 46	100.0	77.2	95.4	100.0	42.1	88.4	100.0	33.5	86.7	100.0	46.4	89.3	100.0	110.1	102.0	100.0	111.3	102.3
21, 22, 23	FIRE SUPPRESSION, PLUMBING & HVAC	100.2	52.4	81.1	95.3	51.9	77.9	100.2	54.6	81.9	95.5	66.5	83.9	95.1	113.7	102.6	100.0	118.0	107.2
26, 27, 3370	ELECTRICAL, COMMUNICATIONS & UTIL.	91.6	69.5	80.1	92.2	45.7	68.0	93.4	36.8	63.9	100.6	65.8	82.4	120.2	119.9	120.0	91.4	105.9	99.0
MF2010	WEIGHTED AVERAGE	96.9	62.9	82.1	94.1	54.4	76.8	95.7	49.9	75.7	95.6	62.7	81.2	98.2	113.9	105.0	98.9	114.1	105.6

		CALIFORNIA																	
	DIVISION	BAKERSFIELD			BERKELEY			EUREKA			FRESNO			INGLEWOOD			LONG BEACH		
		932 - 933			947			955			936 - 938			903 - 905			906 - 908		
		MAT.	INST.	TOTAL	MAT.	INST.	TOTAL	MAT.	INST.	TOTAL	MAT.	INST.	TOTAL	MAT.	INST.	TOTAL	MAT.	INST.	TOTAL
015433	CONTRACTOR EQUIPMENT		97.9	97.9		98.5	98.5		97.6	97.6		97.9	97.9		95.7	95.7		95.7	95.7
0241, 31 - 34	SITE & INFRASTRUCTURE, DEMOLITION	97.5	105.3	103.0	119.7	106.0	110.1	108.5	101.6	103.7	99.4	104.8	103.2	92.6	102.2	99.3	99.8	102.2	101.5
0310	Concrete Forming & Accessories	104.6	122.7	120.2	118.3	148.0	143.9	114.9	130.6	128.4	103.9	131.7	127.9	113.1	116.9	116.4	108.0	116.9	115.7
0320	Concrete Reinforcing	108.4	114.7	111.5	92.5	116.2	104.3	101.9	115.4	108.6	84.7	115.4	100.0	108.6	114.9	111.7	107.6	114.9	111.2
0330	Cast-in-Place Concrete	92.1	120.7	104.0	127.8	122.0	125.4	99.1	116.3	106.2	97.5	117.3	105.7	87.6	120.7	101.1	99.7	120.2	108.2
03	CONCRETE	101.1	119.4	110.1	110.4	131.2	120.6	114.4	121.4	117.8	101.5	122.4	111.8	95.7	116.8	106.1	105.0	116.8	110.8
04	MASONRY	93.2	115.7	107.1	119.2	131.9	127.0	99.4	123.4	114.3	96.3	121.3	111.7	75.8	118.7	102.4	84.5	118.7	105.7
05	METALS	105.8	99.4	103.8	106.5	100.3	104.6	102.9	97.8	101.3	105.9	100.4	104.2	93.8	100.6	95.9	93.7	100.6	95.8
06	WOOD, PLASTICS & COMPOSITES	93.8	121.5	109.4	110.4	154.0	135.0	112.3	135.9	125.6	105.3	135.9	122.5	106.0	113.2	110.1	99.5	113.2	107.2
07	THERMAL & MOISTURE PROTECTION	104.0	111.9	107.2	103.2	135.9	116.5	101.9	116.0	107.6	94.0	113.6	101.9	100.9	118.5	108.1	101.1	118.5	108.2
08	OPENINGS	96.5	116.2	101.2	96.8	140.5	107.1	99.8	112.7	102.8	96.7	124.0	103.1	91.0	115.5	96.8	91.0	115.5	96.8
0920	Plaster & Gypsum Board	98.9	122.0	114.7	102.8	155.2	138.4	106.0	136.9	127.0	97.5	136.9	124.3	103.9	113.7	110.5	99.8	113.7	109.2
0950, 0980	Ceilings & Acoustic Treatment	109.6	122.0	117.9	104.1	155.2	138.1	112.1	136.9	128.6	106.0	136.9	126.5	101.7	113.7	109.7	101.7	113.7	109.7
0960	Flooring	104.9	109.2	106.2	108.7	126.4	114.0	109.3	118.7	112.1	110.4	135.9	118.0	98.9	109.2	102.0	96.5	109.2	100.3
0970, 0990	Wall Finishes & Painting/Coating	100.8	104.2	102.8	102.7	142.1	126.4	102.0	47.2	69.1	115.0	108.6	111.1	102.1	108.4	105.9	102.1	108.4	105.9
09	FINISHES	102.9	118.7	111.7	103.9	145.5	127.1	107.9	122.2	115.9	104.3	131.9	119.7	101.4	114.1	108.4	100.6	114.1	108.1
COVERS	DIVS. 10 - 14, 25, 28, 41, 43, 44, 46	100.0	111.3	102.3	100.0	125.1	105.0	100.0	121.4	104.3	100.0	121.4	104.3	100.0	110.2	102.0	100.0	110.2	102.0
21, 22, 23	FIRE SUPPRESSION, PLUMBING & HVAC	100.1	108.3	103.4	95.2	144.7	115.0	95.1	99.1	96.7	100.2	109.6	104.0	94.7	113.7	102.4	94.7	113.7	102.4
26, 27, 3370	ELECTRICAL, COMMUNICATIONS & UTIL.	102.0	102.3	102.2	109.2	142.6	126.6	98.8	112.6	106.0	91.7	102.9	97.6	105.5	119.9	113.0	105.2	119.9	112.9
MF2010	WEIGHTED AVERAGE	100.8	110.6	105.1	103.4	133.1	116.4	101.6	111.6	106.0	100.0	114.5	106.3	95.7	113.4	103.4	97.1	113.4	104.2

		CALIFORNIA																	
	DIVISION	LOS ANGELES			MARYSVILLE			MODESTO			MOJAVE			OAKLAND			OXNARD		
		900 - 902			959			953			935			946			930		
		MAT.	INST.	TOTAL	MAT.	INST.	TOTAL	MAT.	INST.	TOTAL	MAT.	INST.	TOTAL	MAT.	INST.	TOTAL	MAT.	INST.	TOTAL
015433	CONTRACTOR EQUIPMENT		99.2	99.2		97.6	97.6		97.6	97.6		97.9	97.9		98.5	98.5		96.8	96.8
0241, 31 - 34	SITE & INFRASTRUCTURE, DEMOLITION	100.3	105.7	104.1	104.8	104.5	104.6	99.6	104.5	103.0	93.1	105.3	101.7	126.2	106.0	112.0	99.7	103.4	102.3
0310	Concrete Forming & Accessories	109.9	123.2	121.4	105.0	131.8	128.2	101.3	131.8	127.7	116.3	113.5	113.9	107.2	148.0	142.4	107.7	123.1	121.0
0320	Concrete Reinforcing	109.4	115.0	112.2	101.9	115.3	108.6	105.5	115.5	110.5	102.0	114.6	108.3	94.6	116.2	105.4	100.1	114.8	107.4
0330	Cast-in-Place Concrete	94.6	120.9	105.5	110.7	117.6	113.6	99.1	117.6	106.8	84.4	120.7	99.4	121.2	122.0	121.5	98.2	121.0	107.6
03	CONCRETE	100.9	119.9	110.3	115.1	122.6	118.8	105.9	122.6	114.1	93.9	115.4	104.5	109.9	131.2	120.4	102.1	119.8	110.8
04	MASONRY	90.0	119.9	108.5	100.3	118.2	111.4	99.3	122.3	113.5	94.7	115.7	107.7	126.6	131.9	129.9	97.5	116.5	109.3
05	METALS	100.6	102.0	101.0	102.4	101.1	102.0	99.6	101.4	100.2	102.6	99.3	101.6	100.8	100.3	100.7	100.7	100.0	100.5
06	WOOD, PLASTICS & COMPOSITES	105.7	121.4	114.5	99.5	135.9	120.0	95.3	135.9	118.2	105.0	109.2	107.4	98.4	154.0	129.8	99.6	121.5	111.9
07	THERMAL & MOISTURE PROTECTION	100.3	121.1	108.7	101.4	119.3	108.6	101.0	118.5	108.1	100.1	110.2	104.2	101.2	135.9	115.3	103.3	119.4	109.8
08	OPENINGS	97.8	120.2	103.1	99.1	124.7	105.2	98.0	124.7	104.3	92.1	109.5	96.2	96.8	140.5	107.2	94.5	120.2	100.6
0920	Plaster & Gypsum Board	103.5	122.0	116.1	98.3	136.9	124.5	101.0	136.9	125.4	108.9	109.4	109.2	97.4	155.2	136.7	101.8	122.0	115.6
0950, 0980	Ceilings & Acoustic Treatment	111.6	122.0	118.5	111.2	136.9	128.3	106.7	136.9	126.8	107.4	109.4	108.7	107.0	155.2	139.1	109.9	122.0	118.0
0960	Flooring	96.8	114.1	101.9	105.2	111.3	107.0	105.6	113.5	107.9	110.4	109.2	110.0	104.4	126.4	111.0	102.9	114.1	106.2
0970, 0990	Wall Finishes & Painting/Coating	101.1	108.4	105.5	102.0	118.3	111.8	102.0	118.3	111.8	100.5	104.2	102.7	102.7	142.1	126.4	100.5	121.2	100.9
09	FINISHES	102.9	119.8	112.3	105.0	128.9	118.3	104.0	129.2	118.1	104.6	111.3	108.3	103.1	145.5	126.7	102.3	119.1	111.6
COVERS	DIVS. 10 - 14, 25, 28, 41, 43, 44, 46	100.0	111.3	102.3	100.0	121.4	104.3	100.0	121.4	104.3	100.0	110.0	102.0	100.0	125.1	105.0	100.0	111.5	102.3
21, 22, 23	FIRE SUPPRESSION, PLUMBING & HVAC	100.0	118.0	107.2	95.1	109.8	101.0	100.0	108.8	103.5	95.2	108.3	100.4	100.1	144.7	118.0	100.1	118.0	107.3
26, 27, 3370	ELECTRICAL, COMMUNICATIONS & UTIL.	103.6	120.6	112.5	95.4	109.9	103.0	97.9	106.5	102.4	90.5	102.3	96.6	108.3	142.6	126.2	96.2	109.9	103.3
MF2010	WEIGHTED AVERAGE	100.1	116.5	107.3	100.8	115.1	107.0	100.4	114.9	106.7	96.8	108.6	101.9	103.9	133.1	116.6	99.6	114.2	105.9

City Cost Indexes

CALIFORNIA

DIVISION		PALM SPRINGS 922			PALO ALTO 943			PASADENA 910 - 912			REDDING 960			RICHMOND 948			RIVERSIDE 925		
		MAT.	INST.	TOTAL	MAT.	INST.	TOTAL	MAT.	INST.	TOTAL	MAT.	INST.	TOTAL	MAT.	INST.	TOTAL	MAT.	INST.	TOTAL
015433	CONTRACTOR EQUIPMENT		98.8	98.8		98.5	98.5		98.9	98.9		97.6	97.6		98.5	98.5		98.8	98.8
0241, 31 - 34	SITE & INFRASTRUCTURE, DEMOLITION	88.2	105.7	100.5	115.6	106.0	108.8	93.4	109.7	104.8	109.5	104.5	106.0	125.4	106.0	111.8	95.0	105.7	102.5
0310	Concrete Forming & Accessories	102.1	116.9	114.9	105.4	144.5	139.2	104.9	116.9	115.2	105.0	129.6	126.2	120.9	147.8	144.2	105.8	123.1	120.7
0320	Concrete Reinforcing	107.1	114.7	110.9	92.5	115.9	104.2	108.3	114.9	111.6	104.7	115.3	110.0	92.5	116.0	104.2	104.1	114.7	109.4
0330	Cast-in-Place Concrete	87.1	121.6	101.4	108.1	122.0	113.8	89.9	119.1	102.0	110.9	117.6	113.7	124.3	122.0	123.4	94.7	121.7	105.9
03	CONCRETE	97.7	117.2	107.3	99.3	129.6	114.2	99.4	116.4	107.8	115.2	121.5	118.3	112.0	131.0	121.4	103.8	120.0	111.8
04	MASONRY	74.9	115.9	100.3	101.7	127.8	117.8	106.8	118.6	114.1	110.3	118.2	115.2	119.0	127.8	124.4	75.9	115.9	100.7
05	METALS	103.8	99.9	102.6	98.3	99.8	98.8	84.1	99.9	88.9	100.5	101.1	100.7	98.4	99.8	98.8	103.4	100.1	102.4
06	WOOD, PLASTICS & COMPOSITES	93.2	113.3	104.5	95.8	149.6	126.1	83.3	113.1	100.1	97.8	132.8	117.5	114.2	154.0	136.7	98.5	121.4	111.4
07	THERMAL & MOISTURE PROTECTION	97.6	119.0	106.3	100.8	133.4	114.0	95.0	117.2	104.0	108.6	118.9	112.8	101.4	134.2	114.7	98.1	121.2	107.5
08	OPENINGS	96.6	115.5	101.1	96.8	137.5	106.5	89.8	115.5	95.9	102.0	123.0	107.0	96.9	139.9	107.1	99.1	120.2	104.1
0920	Plaster & Gypsum Board	95.5	113.7	107.9	95.6	150.6	133.0	95.0	113.7	107.7	104.5	133.7	124.4	104.9	155.2	139.1	99.9	122.0	115.0
0950, 0980	Ceilings & Acoustic Treatment	104.1	113.7	110.5	105.0	150.6	135.4	102.5	113.7	109.9	131.8	133.7	133.1	105.0	155.2	138.4	111.0	122.0	118.4
0960	Flooring	107.7	109.2	108.1	103.4	126.4	110.3	88.9	109.2	95.0	105.4	111.3	107.2	110.3	126.4	115.1	109.1	114.1	110.6
0970, 0990	Wall Finishes & Painting/Coating	98.7	113.7	107.7	102.7	142.1	126.4	101.7	108.4	105.7	104.1	118.3	112.6	102.7	142.1	126.4	98.7	106.2	103.2
09	FINISHES	100.7	114.7	108.5	101.4	142.9	124.5	95.7	114.0	105.9	110.9	127.1	119.9	105.4	145.5	127.7	103.5	119.6	112.5
COVERS	DIVS. 10 - 14, 25, 28, 41, 43, 44, 46	100.0	110.4	102.1	100.0	124.7	104.9	100.0	110.1	102.0	100.0	121.0	104.2	100.0	125.1	105.0	100.0	111.3	102.3
21, 22, 23	FIRE SUPPRESSION, PLUMBING & HVAC	95.1	113.6	102.5	95.2	143.3	114.5	95.1	113.7	102.6	100.1	109.8	104.0	95.2	144.7	115.0	100.0	118.0	107.2
26, 27, 3370	ELECTRICAL, COMMUNICATIONS & UTIL.	95.0	106.0	100.8	108.1	153.8	132.0	116.8	119.9	118.4	97.8	109.9	104.1	108.8	132.4	121.1	91.6	106.0	99.1
MF2010	WEIGHTED AVERAGE	96.7	111.6	103.2	99.5	133.1	114.2	96.3	113.9	104.0	103.6	114.6	108.4	102.5	131.2	115.0	98.9	114.0	105.5

CALIFORNIA

DIVISION		SACRAMENTO 942,956 - 958			SALINAS 939			SAN BERNARDINO 923 - 924			SAN DIEGO 919 - 921			SAN FRANCISCO 940 - 941			SAN JOSE 951		
		MAT.	INST.	TOTAL	MAT.	INST.	TOTAL	MAT.	INST.	TOTAL	MAT.	INST.	TOTAL	MAT.	INST.	TOTAL	MAT.	INST.	TOTAL
015433	CONTRACTOR EQUIPMENT		98.0	98.0		97.9	97.9		98.8	98.8		99.0	99.0		108.8	108.8		98.5	98.5
0241, 31 - 34	SITE & INFRASTRUCTURE, DEMOLITION	101.0	111.5	108.3	112.3	105.1	107.2	75.1	105.7	96.6	99.7	103.3	102.2	128.6	111.8	116.8	132.0	99.7	109.3
0310	Concrete Forming & Accessories	105.6	132.4	128.7	110.9	132.8	129.8	109.5	116.9	115.9	103.8	112.0	110.9	106.9	149.0	143.2	107.4	147.9	142.3
0320	Concrete Reinforcing	87.6	115.9	101.6	100.6	115.9	108.3	104.1	114.7	109.4	104.9	114.7	109.8	107.8	116.6	112.2	92.8	116.2	104.5
0330	Cast-in-Place Concrete	102.5	118.4	109.1	97.0	117.9	105.7	65.5	121.6	88.8	98.5	108.7	102.7	124.3	123.5	124.0	115.1	121.6	117.8
03	CONCRETE	99.8	122.7	111.1	112.4	123.2	117.7	77.7	117.2	97.2	104.6	110.7	107.6	113.5	132.7	123.0	111.8	131.4	121.5
04	MASONRY	103.4	122.3	115.1	94.4	127.4	114.8	82.4	115.9	103.1	97.1	114.2	107.7	127.2	136.2	132.8	126.8	131.9	130.0
05	METALS	96.4	95.7	96.2	105.3	102.5	104.4	103.3	99.9	102.3	99.6	101.3	100.1	107.1	110.1	108.0	97.8	107.1	100.6
06	WOOD, PLASTICS & COMPOSITES	92.3	136.1	117.0	104.4	136.1	122.3	102.3	113.3	108.5	97.0	108.8	103.6	98.4	154.2	129.9	107.5	153.8	133.6
07	THERMAL & MOISTURE PROTECTION	108.5	121.1	113.6	100.6	125.9	110.9	96.8	119.0	105.8	99.7	106.6	102.5	102.9	139.0	117.6	97.4	138.4	114.1
08	OPENINGS	110.7	124.8	114.1	95.6	130.7	103.9	96.6	115.5	101.1	96.8	111.4	100.3	101.8	140.6	110.7	89.5	140.3	101.5
0920	Plaster & Gypsum Board	92.5	136.9	122.7	102.9	137.1	126.2	101.7	113.7	109.9	102.3	108.9	106.8	99.8	155.2	137.5	96.2	155.2	136.3
0950, 0980	Ceilings & Acoustic Treatment	105.0	136.9	126.2	107.4	137.1	127.2	106.7	113.7	111.4	106.9	108.9	109.1	114.8	155.2	141.7	104.8	155.2	138.3
0960	Flooring	103.3	119.7	108.2	105.1	123.3	110.6	111.0	109.2	110.4	96.2	114.1	101.5	104.4	126.4	111.0	98.7	126.4	107.0
0970, 0990	Wall Finishes & Painting/Coating	100.3	118.3	111.1	101.6	142.1	126.0	98.7	106.2	103.2	98.3	106.4	103.2	102.7	152.6	132.7	102.3	142.1	126.2
09	FINISHES	99.8	130.4	116.9	104.2	133.7	120.6	101.8	113.9	108.6	101.0	111.3	106.7	105.2	146.8	128.4	102.7	145.3	126.4
COVERS	DIVS. 10 - 14, 25, 28, 41, 43, 44, 46	100.0	121.9	104.4	100.0	121.5	104.3	100.0	110.2	102.1	100.0	109.1	101.8	100.0	125.6	105.1	100.0	124.7	105.0
21, 22, 23	FIRE SUPPRESSION, PLUMBING & HVAC	100.0	121.9	108.7	95.2	115.3	103.2	95.1	113.8	102.6	100.0	116.3	106.5	100.0	171.6	128.7	100.0	143.3	117.3
26, 27, 3370	ELECTRICAL, COMMUNICATIONS & UTIL.	103.4	111.3	107.5	91.7	122.8	107.9	95.0	103.7	99.6	102.2	98.9	100.5	108.3	161.4	136.1	101.0	153.8	128.6
MF2010	WEIGHTED AVERAGE	101.2	118.6	108.8	100.2	120.3	108.9	94.6	111.2	101.9	100.2	109.1	104.1	106.1	143.8	122.5	102.2	134.6	116.3

CALIFORNIA

DIVISION		SAN LUIS OBISPO 934			SAN MATEO 944			SAN RAFAEL 949			SANTA ANA 926 - 927			SANTA BARBARA 931			SANTA CRUZ 950		
		MAT.	INST.	TOTAL	MAT.	INST.	TOTAL	MAT.	INST.	TOTAL	MAT.	INST.	TOTAL	MAT.	INST.	TOTAL	MAT.	INST.	TOTAL
015433	CONTRACTOR EQUIPMENT		97.9	97.9		98.5	98.5		98.5	98.5		98.8	98.8		97.9	97.9		98.5	98.5
0241, 31 - 34	SITE & INFRASTRUCTURE, DEMOLITION	104.7	105.3	105.1	122.8	106.0	111.0	113.7	111.6	112.2	86.7	105.7	100.0	99.6	105.3	103.6	131.6	99.5	109.0
0310	Concrete Forming & Accessories	118.0	116.9	117.0	111.2	147.9	142.9	115.2	147.8	143.4	109.7	116.9	115.9	108.6	123.1	121.1	107.5	133.0	129.5
0320	Concrete Reinforcing	102.0	114.8	108.4	92.5	116.1	104.3	93.1	116.1	104.6	107.6	114.7	111.2	100.1	114.8	107.4	115.0	115.9	115.5
0330	Cast-in-Place Concrete	104.2	120.8	111.1	120.4	122.0	121.0	140.1	121.2	132.2	83.6	121.6	99.4	97.8	120.9	107.4	114.3	119.8	116.6
03	CONCRETE	110.1	116.9	113.4	108.4	131.1	119.6	129.7	130.6	130.2	95.2	117.2	106.1	102.0	119.7	110.7	115.0	124.1	119.5
04	MASONRY	96.1	117.4	109.3	118.7	130.8	126.2	98.0	130.8	118.3	72.6	116.2	99.6	94.9	117.4	108.8	130.4	127.5	128.6
05	METALS	103.2	99.7	102.1	98.2	100.1	98.8	99.6	97.8	99.1	103.4	99.9	102.3	101.1	100.0	100.8	105.1	105.8	105.3
06	WOOD, PLASTICS & COMPOSITES	107.5	113.4	110.8	103.5	154.0	132.0	100.4	153.8	130.5	103.9	113.3	109.2	99.6	121.5	111.9	107.5	136.2	123.7
07	THERMAL & MOISTURE PROTECTION	100.8	117.5	107.6	101.2	136.2	115.4	105.2	132.9	116.5	97.9	119.1	106.5	100.2	119.0	107.9	97.4	128.5	110.0
08	OPENINGS	93.9	111.8	98.1	96.8	139.9	107.0	108.2	139.8	115.7	96.0	115.5	100.6	95.2	120.2	101.1	90.7	130.8	100.2
0920	Plaster & Gypsum Board	110.4	113.7	112.6	100.8	155.2	137.8	102.5	155.2	138.4	102.4	113.7	110.1	101.8	122.0	115.6	103.7	137.1	126.4
0950, 0980	Ceilings & Acoustic Treatment	107.4	113.7	111.6	105.0	155.2	138.4	113.0	155.2	141.1	106.7	113.7	111.4	109.9	122.0	118.0	108.5	137.1	127.5
0960	Flooring	111.3	109.2	110.7	106.0	126.4	112.1	115.1	121.3	117.0	111.5	109.2	110.8	104.4	109.2	105.8	102.8	123.3	108.9
0970, 0990	Wall Finishes & Painting/Coating	100.5	101.1	100.9	102.7	142.1	126.4	99.0	142.1	124.9	98.7	106.2	103.2	100.5	101.2	100.9	102.5	142.1	126.3
09	FINISHES	106.1	112.6	109.7	103.4	145.5	126.8	105.8	144.4	127.3	103.3	113.9	109.2	102.9	118.2	111.4	105.7	133.8	121.3
COVERS	DIVS. 10 - 14, 25, 28, 41, 43, 44, 46	100.0	120.4	104.1	100.0	125.2	105.0	100.0	124.5	104.9	100.0	110.4	102.1	100.0	111.5	102.3	100.0	121.8	104.4
21, 22, 23	FIRE SUPPRESSION, PLUMBING & HVAC	95.2	113.8	102.6	95.2	142.8	114.3	95.1	164.7	123.0	95.1	113.8	102.6	100.1	118.0	107.3	100.0	115.3	106.1
26, 27, 3370	ELECTRICAL, COMMUNICATIONS & UTIL.	90.5	106.0	98.6	108.1	143.6	126.7	105.1	116.4	111.0	95.0	105.9	100.7	89.4	111.8	101.1	99.9	122.8	111.9
MF2010	WEIGHTED AVERAGE	99.3	111.6	104.7	101.7	132.7	115.2	104.2	133.6	117.0	96.5	111.6	103.1	98.9	114.6	105.7	104.1	120.4	111.2

599

City Cost Indexes

		CALIFORNIA															COLORADO		
DIVISION		SANTA ROSA			STOCKTON			SUSANVILLE			VALLEJO			VAN NUYS			ALAMOSA		
		954			952			961			945			913 - 916			811		
		MAT.	INST.	TOTAL	MAT.	INST.	TOTAL	MAT.	INST.	TOTAL	MAT.	INST.	TOTAL	MAT.	INST.	TOTAL	MAT.	INST.	TOTAL
015433	CONTRACTOR EQUIPMENT		98.2	98.2		97.6	97.6		97.6	97.6		98.5	98.5		98.9	98.9		93.8	93.8
0241, 31 - 34	SITE & INFRASTRUCTURE, DEMOLITION	100.1	104.5	103.2	99.4	104.5	103.0	115.9	104.5	107.9	100.5	111.4	108.2	109.9	109.7	109.7	132.1	89.0	101.9
0310	Concrete Forming & Accessories	103.1	146.8	140.8	105.3	131.8	128.2	106.2	129.6	126.4	105.6	146.5	140.9	110.9	116.9	116.1	105.2	69.0	73.9
0320	Concrete Reinforcing	102.7	116.4	109.6	105.5	115.5	110.5	104.7	115.3	110.0	94.3	116.3	105.3	108.3	114.9	111.6	108.0	78.4	93.2
0330	Cast-in-Place Concrete	108.7	119.2	113.1	96.6	117.6	105.3	100.9	117.6	107.8	111.6	119.6	114.9	94.8	119.1	104.9	100.6	79.3	91.8
03	CONCRETE	114.9	130.0	122.4	104.9	122.6	113.6	117.6	121.5	119.5	105.2	129.5	117.2	113.8	116.4	115.1	115.5	74.6	95.3
04	MASONRY	98.6	133.7	120.3	99.2	122.3	113.5	108.3	118.2	114.4	74.6	132.4	110.4	122.3	118.6	120.0	123.2	72.7	92.0
05	METALS	103.7	104.3	103.9	99.8	101.4	100.3	99.6	101.1	100.1	99.6	97.6	99.0	83.2	99.9	88.3	97.2	81.1	92.2
06	WOOD, PLASTICS & COMPOSITES	94.6	153.6	127.9	100.8	135.9	120.6	99.5	132.8	118.3	89.8	153.8	125.9	91.6	113.1	103.8	93.8	69.0	79.8
07	THERMAL & MOISTURE PROTECTION	98.3	134.5	113.0	101.1	118.5	108.2	109.2	118.9	113.2	103.0	134.0	115.6	95.8	117.2	104.5	105.1	78.4	94.2
08	OPENINGS	97.4	140.2	107.6	98.0	124.7	104.3	101.9	123.0	106.9	110.0	140.3	117.2	89.7	115.5	95.8	98.4	75.3	92.9
0920	Plaster & Gypsum Board	97.2	155.2	136.7	101.0	136.9	125.4	105.4	133.7	124.7	97.3	155.2	136.7	99.9	113.7	109.3	75.8	67.9	70.4
0950, 0980	Ceilings & Acoustic Treatment	106.7	155.2	139.0	113.7	136.9	129.1	123.9	133.7	130.4	115.0	155.2	141.8	99.8	113.7	109.1	104.2	67.9	80.1
0960	Flooring	108.1	117.0	110.8	105.6	113.5	107.9	105.9	111.3	107.5	111.2	126.4	115.7	91.1	109.2	96.5	112.4	55.8	95.5
0970, 0990	Wall Finishes & Painting/Coating	98.7	142.1	124.8	102.0	118.3	111.8	104.1	118.3	112.6	99.9	142.1	125.3	101.7	108.4	105.7	113.0	24.1	59.6
09	FINISHES	102.6	143.0	125.1	105.5	129.2	118.7	110.4	127.1	119.7	102.8	144.7	126.1	97.7	114.0	106.8	104.9	61.2	80.5
COVERS	DIVS. 10 - 14, 25, 28, 41, 43, 44, 46	100.0	123.2	104.7	100.0	121.4	104.3	100.0	121.1	104.2	100.0	123.7	104.7	100.0	110.1	102.0	100.0	89.1	97.8
21, 22, 23	FIRE SUPPRESSION, PLUMBING & HVAC	95.1	163.4	122.5	100.0	108.8	103.5	95.2	109.8	101.0	100.0	128.8	111.6	95.1	113.7	102.6	95.2	72.6	86.1
26, 27, 3370	ELECTRICAL, COMMUNICATIONS & UTIL.	95.4	116.6	106.4	97.9	110.9	104.7	98.1	109.9	104.3	100.6	125.1	113.4	116.8	119.9	118.4	96.8	74.2	85.0
MF2010	WEIGHTED AVERAGE	100.3	133.4	114.7	100.5	115.5	107.0	102.6	114.6	107.8	100.6	127.2	112.2	99.1	113.9	105.5	102.0	74.6	90.1

		COLORADO																	
DIVISION		BOULDER			COLORADO SPRINGS			DENVER			DURANGO			FORT COLLINS			FORT MORGAN		
		803			808 - 809			800 - 802			813			805			807		
		MAT.	INST.	TOTAL	MAT.	INST.	TOTAL	MAT.	INST.	TOTAL	MAT.	INST.	TOTAL	MAT.	INST.	TOTAL	MAT.	INST.	TOTAL
015433	CONTRACTOR EQUIPMENT		94.9	94.9		93.4	93.4		97.7	97.7		93.8	93.8		94.9	94.9		94.9	94.9
0241, 31 - 34	SITE & INFRASTRUCTURE, DEMOLITION	92.4	95.1	94.3	94.5	93.1	93.5	93.7	101.2	99.0	125.8	89.0	100.0	104.3	94.7	97.5	95.3	94.6	94.8
0310	Concrete Forming & Accessories	102.0	81.7	84.4	92.9	77.4	79.5	98.5	77.7	80.5	111.4	69.2	74.9	99.3	76.4	79.5	102.5	76.6	80.2
0320	Concrete Reinforcing	102.0	78.5	90.3	101.2	82.7	92.0	101.2	82.7	92.0	108.0	78.4	93.2	102.1	78.6	90.3	102.2	78.4	90.3
0330	Cast-in-Place Concrete	102.4	79.8	93.0	105.2	82.9	95.9	99.6	82.6	92.5	115.7	79.4	100.6	115.6	78.6	100.2	100.5	78.6	91.4
03	CONCRETE	104.7	80.6	92.8	108.1	80.6	94.6	102.4	80.6	91.7	117.8	74.8	96.5	115.3	77.8	96.8	103.2	77.8	90.7
04	MASONRY	93.1	73.2	80.8	94.1	73.0	81.0	96.1	73.3	82.0	110.9	72.4	87.1	110.4	76.9	89.7	107.5	72.9	86.1
05	METALS	97.7	84.4	93.6	100.8	86.5	96.4	103.4	86.7	98.2	97.2	81.2	92.2	98.9	81.3	93.5	97.5	81.1	92.4
06	WOOD, PLASTICS & COMPOSITES	93.3	84.9	88.6	83.8	79.1	81.2	91.0	79.0	84.3	102.6	69.0	83.7	90.8	79.0	84.1	93.3	79.0	85.2
07	THERMAL & MOISTURE PROTECTION	104.4	80.5	94.7	105.3	80.3	95.1	103.7	74.6	91.9	105.0	78.3	94.2	104.8	72.2	91.6	104.4	79.9	94.4
08	OPENINGS	101.5	83.9	97.3	105.8	82.0	100.2	106.5	81.9	100.7	105.3	75.3	98.2	101.4	80.7	96.5	101.4	80.7	96.5
0920	Plaster & Gypsum Board	104.0	84.6	90.8	86.1	78.6	81.0	99.6	78.6	85.3	88.1	67.9	74.3	98.0	78.6	84.8	104.0	78.6	86.7
0950, 0980	Ceilings & Acoustic Treatment	98.0	84.6	89.1	105.1	78.6	87.4	108.2	78.6	88.5	104.2	67.9	80.1	98.0	78.6	85.0	98.0	78.6	85.0
0960	Flooring	103.7	85.3	98.2	95.8	69.5	87.9	100.5	86.3	96.3	116.8	55.8	98.5	100.4	55.8	87.1	104.1	55.8	89.7
0970, 0990	Wall Finishes & Painting/Coating	100.8	68.0	81.1	100.5	41.4	65.0	100.8	77.3	86.7	113.0	24.1	59.6	100.8	41.1	64.9	100.8	54.7	73.1
09	FINISHES	102.6	81.2	90.7	99.4	72.0	84.2	103.1	79.0	89.7	106.9	61.2	81.5	101.4	68.9	83.3	102.6	70.4	84.7
COVERS	DIVS. 10 - 14, 25, 28, 41, 43, 44, 46	100.0	90.0	98.0	100.0	89.7	97.9	100.0	89.4	97.9	100.0	89.1	97.8	100.0	89.4	97.9	100.0	89.4	97.9
21, 22, 23	FIRE SUPPRESSION, PLUMBING & HVAC	95.2	77.6	88.1	100.2	72.7	89.2	100.0	79.9	91.9	95.2	84.9	91.0	100.1	77.5	91.0	95.2	77.5	88.1
26, 27, 3370	ELECTRICAL, COMMUNICATIONS & UTIL.	96.9	85.2	90.8	100.5	83.0	91.4	102.1	85.2	93.3	96.2	71.1	83.1	96.9	85.1	90.7	97.2	85.1	90.9
MF2010	WEIGHTED AVERAGE	98.6	82.0	91.3	101.3	79.3	91.7	101.6	82.6	93.3	102.4	76.9	91.2	102.1	79.6	92.3	99.1	79.6	90.6

		COLORADO																	
DIVISION		GLENWOOD SPRINGS			GOLDEN			GRAND JUNCTION			GREELEY			MONTROSE			PUEBLO		
		816			804			815			806			814			810		
		MAT.	INST.	TOTAL	MAT.	INST.	TOTAL	MAT.	INST.	TOTAL	MAT.	INST.	TOTAL	MAT.	INST.	TOTAL	MAT.	INST.	TOTAL
015433	CONTRACTOR EQUIPMENT		96.5	96.5		94.9	94.9		96.5	96.5		94.9	94.9		95.1	95.1		93.8	93.8
0241, 31 - 34	SITE & INFRASTRUCTURE, DEMOLITION	140.8	95.9	109.3	104.4	94.9	97.7	125.5	95.5	104.5	91.7	94.1	93.4	134.6	92.1	104.8	117.7	91.5	99.3
0310	Concrete Forming & Accessories	102.3	76.7	80.2	95.2	76.5	79.1	110.1	76.2	80.8	97.5	80.2	82.5	101.8	76.4	79.9	107.3	77.8	81.8
0320	Concrete Reinforcing	106.9	78.4	92.7	102.2	78.4	90.3	107.3	78.3	92.8	102.0	77.0	89.5	106.8	78.3	92.6	103.3	82.7	93.1
0330	Cast-in-Place Concrete	100.6	78.8	91.5	100.5	78.6	91.4	111.4	78.1	97.5	96.6	58.5	80.8	100.6	78.7	91.5	99.9	83.7	93.2
03	CONCRETE	120.6	77.9	99.6	113.8	77.8	96.0	114.1	77.4	96.0	99.9	72.2	86.3	111.6	77.7	94.9	104.3	81.1	92.8
04	MASONRY	97.5	72.9	82.3	110.3	72.9	87.2	129.6	72.7	94.4	104.6	49.7	70.6	103.9	72.7	84.6	94.1	73.2	81.2
05	METALS	96.9	81.8	92.2	97.7	81.2	92.6	98.4	80.5	92.9	98.9	78.5	92.6	96.0	80.8	91.3	100.0	87.4	96.1
06	WOOD, PLASTICS & COMPOSITES	89.1	79.1	83.4	86.2	79.0	82.1	100.4	79.1	88.4	88.4	84.9	86.4	90.2	79.2	84.0	96.3	79.4	86.8
07	THERMAL & MOISTURE PROTECTION	105.0	79.5	94.6	105.3	74.2	92.7	104.1	68.8	89.8	104.3	65.0	88.3	105.1	79.5	94.7	103.6	80.0	94.0
08	OPENINGS	104.3	80.8	98.7	101.4	80.7	96.5	105.0	80.8	99.3	101.4	83.9	97.3	105.5	80.8	99.7	100.1	82.1	95.9
0920	Plaster & Gypsum Board	111.5	78.6	89.1	95.8	78.6	84.1	123.4	78.6	92.9	96.5	84.6	88.4	75.0	78.6	77.4	80.2	78.6	79.1
0950, 0980	Ceilings & Acoustic Treatment	103.4	78.6	86.9	98.0	78.6	85.0	103.4	78.6	86.9	98.0	84.6	89.1	104.2	78.6	87.2	113.1	78.6	90.1
0960	Flooring	111.7	51.4	93.7	98.3	55.8	85.6	116.2	55.8	98.1	99.5	55.8	86.4	114.6	46.2	94.2	113.4	86.3	105.3
0970, 0990	Wall Finishes & Painting/Coating	112.9	54.7	77.9	100.8	68.0	81.1	112.9	68.0	85.9	100.8	25.6	55.6	113.0	24.1	59.6	113.0	38.4	67.1
09	FINISHES	109.9	69.8	87.5	101.3	72.6	85.3	110.8	72.0	89.2	100.1	70.7	83.7	105.4	65.3	83.1	105.8	74.9	88.6
COVERS	DIVS. 10 - 14, 25, 28, 41, 43, 44, 46	100.0	89.5	97.9	100.0	89.4	97.9	100.0	89.5	97.9	100.0	90.0	98.0	100.0	89.8	98.0	100.0	90.3	98.1
21, 22, 23	FIRE SUPPRESSION, PLUMBING & HVAC	95.2	84.8	91.0	95.2	77.0	87.9	100.0	84.3	93.7	100.1	77.5	91.0	95.2	84.8	91.0	100.0	72.8	89.1
26, 27, 3370	ELECTRICAL, COMMUNICATIONS & UTIL.	93.5	71.1	81.8	97.2	85.1	90.9	95.9	55.1	74.6	96.9	85.1	90.7	95.9	57.7	76.0	96.9	74.3	85.1
MF2010	WEIGHTED AVERAGE	102.2	79.3	92.2	100.6	79.6	91.4	104.4	76.7	92.4	99.7	76.1	89.4	101.2	76.5	90.5	100.9	78.6	91.2

600

City Cost Indexes

DIVISION		COLORADO SALIDA 812			CONNECTICUT																
					BRIDGEPORT 066			BRISTOL 060			HARTFORD 061			MERIDEN 064			NEW BRITAIN 060		
		MAT.	INST.	TOTAL	MAT.	INST.	TOTAL	MAT.	INST.	TOTAL	MAT.	INST.	TOTAL	MAT.	INST.	TOTAL	MAT.	INST.	TOTAL
015433	CONTRACTOR EQUIPMENT		95.1	95.1		100.6	100.6		100.6	100.6		100.6	100.6		101.1	101.1		100.6	100.6
0241, 31 - 34	SITE & INFRASTRUCTURE, DEMOLITION	125.5	92.4	102.3	111.7	104.0	106.3	110.9	104.0	106.1	106.0	104.0	104.6	108.9	104.7	105.9	111.1	104.0	106.1
0310	Concrete Forming & Accessories	110.0	76.4	81.0	98.2	123.8	120.3	98.2	123.7	120.3	95.1	123.7	119.8	97.9	123.7	120.2	98.6	123.7	120.3
0320	Concrete Reinforcing	106.6	78.3	92.5	110.0	129.3	119.7	110.0	129.3	119.7	109.5	129.3	119.4	110.0	129.3	119.7	110.0	129.3	119.7
0330	Cast-in-Place Concrete	115.2	78.7	100.1	103.7	127.5	113.6	97.2	127.4	109.8	99.0	127.4	110.8	93.5	127.4	107.6	98.8	127.4	110.7
03	CONCRETE	112.8	77.7	95.5	105.3	125.8	115.4	102.2	125.7	113.8	102.8	125.8	114.1	100.5	125.7	112.9	103.0	125.7	114.2
04	MASONRY	131.2	72.7	95.0	98.1	134.8	120.8	90.8	134.8	118.0	94.7	134.8	119.5	90.4	134.8	117.9	92.1	134.8	118.6
05	METALS	95.7	80.8	91.1	102.0	125.1	109.1	102.0	125.1	109.1	107.8	125.1	113.1	99.5	125.1	107.4	98.6	125.1	106.8
06	WOOD, PLASTICS & COMPOSITES	97.4	79.2	87.2	99.1	122.9	112.5	99.1	122.9	112.5	92.6	122.9	109.7	99.1	122.9	112.5	99.1	122.9	112.5
07	THERMAL & MOISTURE PROTECTION	104.1	79.5	94.1	100.3	129.2	112.0	100.4	126.9	111.2	104.8	126.9	113.8	100.4	126.9	111.2	100.4	126.9	111.2
08	OPENINGS	98.5	80.8	94.3	107.0	132.2	113.0	107.0	132.2	113.0	106.1	132.2	112.2	109.4	132.2	114.8	107.0	132.2	113.0
0920	Plaster & Gypsum Board	75.4	78.6	77.5	99.9	123.0	115.6	99.9	123.0	115.6	95.4	123.0	114.2	101.5	123.0	116.1	99.9	123.0	115.6
0950, 0980	Ceilings & Acoustic Treatment	104.2	78.6	87.2	97.9	123.0	114.6	97.9	123.0	114.6	101.3	123.0	115.7	102.1	123.0	116.0	97.9	123.0	114.6
0960	Flooring	119.1	46.2	97.3	99.3	133.6	109.6	99.3	133.6	109.6	97.8	133.6	108.5	99.3	133.6	109.6	99.3	133.6	109.6
0970, 0990	Wall Finishes & Painting/Coating	113.0	24.1	59.6	96.4	122.3	112.0	96.4	122.3	112.0	97.3	122.3	112.3	96.4	122.3	112.0	96.4	122.3	112.0
09	FINISHES	105.6	65.3	83.2	100.8	125.1	114.3	100.8	125.1	114.3	99.7	125.1	113.8	101.9	125.1	114.8	100.8	125.1	114.3
COVERS	DIVS. 10 - 14, 25, 28, 41, 43, 44, 46	100.0	89.9	98.0	100.0	110.4	102.1	100.0	110.5	102.1	100.0	110.5	102.1	100.0	110.5	102.1	100.0	110.5	102.1
21, 22, 23	FIRE SUPPRESSION, PLUMBING & HVAC	95.2	72.6	86.1	100.0	116.0	106.4	100.0	116.0	106.4	100.0	116.0	106.4	95.1	116.0	103.5	100.0	116.0	106.4
26, 27, 3370	ELECTRICAL, COMMUNICATIONS & UTIL.	96.1	74.2	84.7	99.2	108.7	104.2	99.2	111.3	105.5	98.6	112.3	105.7	99.2	111.3	105.5	99.3	111.3	105.6
MF2010	WEIGHTED AVERAGE	101.7	76.2	90.6	101.8	120.2	109.8	101.1	120.4	109.5	102.0	120.6	110.1	99.6	120.5	108.7	100.8	120.4	109.3

DIVISION		CONNECTICUT																	
		NEW HAVEN 065			NEW LONDON 063			NORWALK 068			STAMFORD 069			WATERBURY 067			WILLIMANTIC 062		
		MAT.	INST.	TOTAL	MAT.	INST.	TOTAL	MAT.	INST.	TOTAL	MAT.	INST.	TOTAL	MAT.	INST.	TOTAL	MAT.	INST.	TOTAL
015433	CONTRACTOR EQUIPMENT		101.1	101.1		101.1	101.1		100.6	100.6		100.6	100.6		100.6	100.6		100.6	100.6
0241, 31 - 34	SITE & INFRASTRUCTURE, DEMOLITION	110.9	104.7	106.5	103.1	104.7	104.2	111.5	104.0	106.2	112.2	104.0	106.4	111.4	104.0	106.2	111.6	104.0	106.3
0310	Concrete Forming & Accessories	98.0	123.7	120.2	97.9	123.7	120.2	98.2	124.2	120.6	98.2	124.2	120.7	98.2	123.7	120.2	98.2	123.7	120.2
0320	Concrete Reinforcing	110.0	129.3	119.7	86.3	129.3	107.7	110.0	129.3	119.7	110.0	129.3	119.7	110.0	129.3	119.7	110.0	129.3	119.7
0330	Cast-in-Place Concrete	100.5	127.4	111.7	85.7	127.4	103.0	102.0	128.9	113.2	103.7	128.9	114.2	103.7	127.4	113.6	96.9	127.4	109.6
03	CONCRETE	117.2	125.7	121.4	90.2	125.7	107.7	104.5	126.5	115.3	105.3	126.5	115.8	105.3	125.7	115.4	102.1	125.7	113.7
04	MASONRY	91.0	134.8	118.1	89.4	134.8	117.5	90.6	134.8	118.0	91.3	134.8	118.2	91.3	134.8	118.2	90.6	134.8	118.0
05	METALS	98.8	125.1	106.9	98.5	125.1	106.7	102.0	125.5	109.3	102.0	125.6	109.3	102.0	125.0	109.1	101.8	125.0	108.9
06	WOOD, PLASTICS & COMPOSITES	99.1	122.9	112.5	99.1	122.9	112.5	99.1	122.9	112.5	99.1	122.9	112.5	99.1	122.9	112.5	99.1	122.9	112.5
07	THERMAL & MOISTURE PROTECTION	100.5	126.7	111.1	100.4	126.9	111.1	100.4	129.4	112.2	100.4	129.4	112.2	100.4	126.7	111.1	100.6	125.7	110.8
08	OPENINGS	107.0	132.2	113.0	109.9	132.2	115.2	107.0	132.2	113.0	107.0	132.2	113.0	107.0	132.2	113.0	110.0	132.2	115.2
0920	Plaster & Gypsum Board	99.9	123.0	115.6	99.9	123.0	115.6	99.9	123.0	115.6	99.9	123.0	115.6	99.9	123.0	115.6	99.9	123.0	115.6
0950, 0980	Ceilings & Acoustic Treatment	97.9	123.0	114.6	95.9	123.0	113.9	97.9	123.0	114.6	97.9	123.0	114.6	97.9	123.0	114.6	95.9	123.0	113.9
0960	Flooring	99.3	133.6	109.6	99.3	133.6	109.6	99.3	133.6	109.6	99.3	133.6	109.6	99.3	133.6	109.6	99.3	135.7	110.2
0970, 0990	Wall Finishes & Painting/Coating	96.4	122.3	112.0	96.4	122.3	112.0	96.4	122.3	112.0	96.4	122.3	112.0	96.4	122.3	112.0	96.4	122.3	112.0
09	FINISHES	100.8	125.1	114.4	99.8	125.1	113.9	100.8	125.1	114.4	100.9	125.1	114.4	100.7	125.1	114.3	100.5	125.4	114.4
COVERS	DIVS. 10 - 14, 25, 28, 41, 43, 44, 46	100.0	110.5	102.1	100.0	110.5	102.1	100.0	110.6	102.1	100.0	110.7	102.1	100.0	110.4	102.1	100.0	110.5	102.1
21, 22, 23	FIRE SUPPRESSION, PLUMBING & HVAC	100.0	116.0	106.4	95.1	116.0	103.5	100.0	116.0	106.4	100.0	116.1	106.4	100.0	116.0	106.4	100.0	116.0	106.4
26, 27, 3370	ELECTRICAL, COMMUNICATIONS & UTIL.	99.2	111.3	105.5	95.9	111.3	103.9	99.2	156.4	129.1	99.2	163.8	132.9	98.7	108.7	103.9	99.2	112.2	106.0
MF2010	WEIGHTED AVERAGE	102.2	120.5	110.2	97.7	120.5	107.6	101.4	126.9	112.5	101.5	127.9	113.0	101.4	120.1	109.6	101.4	120.6	109.7

DIVISION		D.C. WASHINGTON 200 - 205			DELAWARE									FLORIDA					
					DOVER 199			NEWARK 197			WILMINGTON 198			DAYTONA BEACH 321			FORT LAUDERDALE 333		
		MAT.	INST.	TOTAL	MAT.	INST.	TOTAL	MAT.	INST.	TOTAL	MAT.	INST.	TOTAL	MAT.	INST.	TOTAL	MAT.	INST.	TOTAL
015433	CONTRACTOR EQUIPMENT		104.4	104.4		117.0	117.0		117.0	117.0		117.2	117.2		98.8	98.8		91.7	91.7
0241, 31 - 34	SITE & INFRASTRUCTURE, DEMOLITION	106.8	94.0	97.8	99.7	112.3	108.6	99.1	112.3	108.4	92.4	112.6	106.6	107.0	90.4	95.4	93.6	78.3	82.9
0310	Concrete Forming & Accessories	99.5	79.2	81.9	97.1	101.5	100.9	97.2	101.5	100.9	98.3	101.5	101.1	97.3	69.7	73.5	95.5	70.1	73.6
0320	Concrete Reinforcing	109.0	91.3	100.2	98.5	105.2	101.9	93.8	105.2	99.5	96.4	105.2	100.8	92.8	79.0	85.9	89.8	74.4	82.1
0330	Cast-in-Place Concrete	121.4	88.1	107.5	99.1	100.6	99.7	84.4	100.6	91.1	88.9	100.6	93.8	86.6	72.0	80.5	90.9	78.4	85.7
03	CONCRETE	109.6	85.8	97.9	99.9	102.8	101.3	92.1	102.8	97.4	94.8	102.8	98.7	88.2	73.4	80.9	90.7	75.0	82.9
04	MASONRY	95.9	80.0	86.1	103.3	97.2	99.5	99.9	97.2	98.2	103.6	97.2	99.6	92.3	66.8	76.5	89.9	69.4	77.3
05	METALS	100.1	107.3	102.3	103.1	117.0	107.4	103.7	117.0	107.8	102.9	117.0	107.3	96.4	92.8	95.3	95.5	91.5	94.2
06	WOOD, PLASTICS & COMPOSITES	103.5	77.2	88.7	95.0	101.1	98.4	97.6	101.1	99.6	98.2	101.1	99.8	99.7	70.8	83.4	96.0	68.7	80.6
07	THERMAL & MOISTURE PROTECTION	102.4	86.0	95.8	101.9	110.2	105.3	104.2	110.2	106.7	102.2	110.2	105.5	99.6	74.9	89.6	96.9	82.3	91.0
08	OPENINGS	100.3	87.6	97.3	94.3	110.5	98.1	93.4	110.5	97.4	96.6	110.5	99.9	98.9	69.1	91.8	96.9	67.1	89.9
0920	Plaster & Gypsum Board	105.8	76.5	85.9	105.9	101.0	102.6	103.4	101.0	101.8	104.8	101.0	102.2	95.1	70.3	78.3	98.5	68.2	77.9
0950, 0980	Ceilings & Acoustic Treatment	107.8	76.5	87.0	104.2	101.0	102.0	95.0	101.0	99.0	102.2	101.0	101.4	90.7	70.3	77.2	91.3	68.2	75.9
0960	Flooring	104.7	95.3	101.9	99.1	109.0	102.1	95.5	109.0	99.6	101.4	109.0	103.7	115.2	74.9	103.2	115.4	70.6	102.0
0970, 0990	Wall Finishes & Painting/Coating	104.6	84.0	92.2	97.3	105.3	102.1	99.5	105.3	103.0	97.3	105.3	102.1	99.4	75.7	85.2	96.5	71.7	81.6
09	FINISHES	101.6	81.7	90.5	101.4	102.8	102.2	97.4	102.8	100.4	102.0	102.8	102.4	100.8	70.9	84.1	99.4	69.6	82.8
COVERS	DIVS. 10 - 14, 25, 28, 41, 43, 44, 46	100.0	98.4	99.7	100.0	89.7	97.9	100.0	89.7	97.9	100.0	89.7	97.9	100.0	83.5	96.7	100.0	86.5	97.3
21, 22, 23	FIRE SUPPRESSION, PLUMBING & HVAC	100.1	92.9	97.2	100.0	115.8	106.3	100.2	115.8	106.5	100.2	115.8	106.4	100.0	77.8	91.1	100.0	68.0	87.2
26, 27, 3370	ELECTRICAL, COMMUNICATIONS & UTIL.	102.7	102.6	102.6	95.2	109.0	102.4	97.7	109.0	103.6	96.6	109.0	103.1	94.4	56.2	74.5	92.6	72.7	82.2
MF2010	WEIGHTED AVERAGE	101.6	91.6	97.2	99.7	108.1	103.3	98.8	108.1	102.8	99.4	108.1	103.2	97.4	74.4	87.3	96.4	73.9	86.6

601

City Cost Indexes

		FLORIDA																	
	DIVISION	FORT MYERS			GAINESVILLE			JACKSONVILLE			LAKELAND			MELBOURNE			MIAMI		
		339,341			326,344			320,322			338			329			330 - 332,340		
		MAT.	INST.	TOTAL	MAT.	INST.	TOTAL	MAT.	INST.	TOTAL	MAT.	INST.	TOTAL	MAT.	INST.	TOTAL	MAT.	INST.	TOTAL
015433	CONTRACTOR EQUIPMENT		98.8	98.8		98.8	98.8		98.8	98.8		98.8	98.8		98.8	98.8		91.7	91.7
0241, 31 - 34	SITE & INFRASTRUCTURE, DEMOLITION	104.3	89.5	93.9	115.4	89.5	97.2	107.0	89.8	94.9	106.2	89.9	94.7	113.9	89.7	96.9	96.7	78.1	83.6
0310	Concrete Forming & Accessories	91.1	76.6	78.6	92.3	55.5	60.5	97.1	55.9	61.5	87.4	77.1	78.5	93.5	71.4	74.4	100.7	70.5	74.6
0320	Concrete Reinforcing	90.8	95.7	93.3	98.5	66.9	82.7	92.8	67.0	79.9	93.1	96.6	94.8	93.9	79.0	86.5	95.9	74.5	85.2
0330	Cast-in-Place Concrete	94.8	69.4	84.3	99.4	63.6	84.5	87.4	69.5	80.0	97.0	70.8	86.1	104.3	74.3	91.9	91.7	79.4	86.6
03	CONCRETE	91.3	78.6	85.1	98.7	62.0	80.6	88.6	64.2	76.6	93.0	79.5	86.3	99.0	75.0	87.1	92.4	75.5	84.0
04	MASONRY	84.1	62.9	71.0	106.7	62.3	79.2	92.3	62.3	73.7	100.3	77.1	85.9	91.2	71.0	78.7	90.3	73.2	79.7
05	METALS	97.6	98.6	97.9	95.3	86.6	92.7	95.0	87.0	92.5	97.5	99.5	98.1	104.4	93.0	100.9	95.9	90.3	94.1
06	WOOD, PLASTICS & COMPOSITES	92.5	79.2	85.0	93.2	53.0	70.5	99.7	53.0	73.4	87.5	79.2	82.8	94.9	70.8	81.3	103.4	68.7	83.8
07	THERMAL & MOISTURE PROTECTION	96.8	81.3	90.5	99.9	62.5	84.7	99.8	63.1	84.9	96.7	86.0	92.4	100.0	77.3	90.8	98.1	76.5	89.4
08	OPENINGS	97.9	76.2	92.8	97.1	54.0	86.9	98.9	57.2	89.0	97.9	76.9	92.9	98.1	72.4	92.0	99.1	67.0	91.5
0920	Plaster & Gypsum Board	93.9	79.0	83.8	90.8	51.9	64.4	95.1	51.9	65.8	90.5	79.0	82.7	90.8	70.3	76.9	96.2	68.2	77.1
0950, 0980	Ceilings & Acoustic Treatment	85.8	79.0	81.3	86.1	51.9	63.4	90.7	51.9	64.9	85.8	79.0	81.3	88.1	70.3	76.3	95.9	68.2	77.4
0960	Flooring	112.1	56.1	95.4	112.6	43.9	92.1	115.2	65.7	100.4	109.8	57.6	94.2	112.8	74.9	101.5	118.5	74.3	105.3
0970, 0990	Wall Finishes & Painting/Coating	101.6	62.1	77.8	99.4	62.1	77.0	99.4	66.3	79.5	101.6	62.1	77.8	99.4	93.9	96.1	93.5	71.7	80.4
09	FINISHES	98.9	71.5	83.6	99.4	52.9	73.5	100.9	57.7	76.8	97.9	71.8	83.3	99.6	73.9	85.3	101.9	70.8	84.5
COVERS	DIVS. 10 - 14, 25, 28, 41, 43, 44, 46	100.0	74.6	94.9	100.0	81.9	96.4	100.0	81.4	96.3	100.0	74.6	94.9	100.0	84.9	97.0	100.0	87.1	97.4
21, 22, 23	FIRE SUPPRESSION, PLUMBING & HVAC	97.4	65.7	84.7	99.8	65.5	85.5	100.0	65.5	86.2	97.4	82.5	91.5	100.0	79.9	92.0	100.0	67.9	87.1
26, 27, 3370	ELECTRICAL, COMMUNICATIONS & UTIL.	94.7	64.3	78.9	94.7	73.5	83.6	94.1	63.7	78.2	92.8	63.1	77.3	95.7	69.5	82.0	96.6	75.5	85.6
MF2010	WEIGHTED AVERAGE	96.4	74.1	86.7	98.7	68.0	85.3	97.2	67.7	84.3	97.1	79.3	89.4	99.9	77.9	90.3	97.7	74.6	87.6

		FLORIDA																	
	DIVISION	ORLANDO			PANAMA CITY			PENSACOLA			SARASOTA			ST. PETERSBURG			TALLAHASSEE		
		327 - 328,347			324			325			342			337			323		
		MAT.	INST.	TOTAL	MAT.	INST.	TOTAL	MAT.	INST.	TOTAL	MAT.	INST.	TOTAL	MAT.	INST.	TOTAL	MAT.	INST.	TOTAL
015433	CONTRACTOR EQUIPMENT		98.8	98.8		98.8	98.8		98.8	98.8		98.8	98.8		98.8	98.8		98.8	98.8
0241, 31 - 34	SITE & INFRASTRUCTURE, DEMOLITION	108.9	89.6	95.4	119.4	88.6	97.8	118.9	89.1	98.0	111.3	89.6	96.1	107.8	89.3	94.8	105.5	88.9	93.8
0310	Concrete Forming & Accessories	99.8	73.7	77.3	96.4	44.9	52.0	94.4	53.0	58.6	96.2	76.7	79.4	95.0	52.2	58.1	100.4	45.1	52.7
0320	Concrete Reinforcing	95.9	76.2	86.1	97.0	74.9	85.9	99.4	75.4	87.4	92.8	96.5	94.6	93.1	88.8	91.0	102.5	66.8	84.6
0330	Cast-in-Place Concrete	107.4	71.8	92.6	91.9	57.7	77.7	113.4	67.0	94.2	104.0	70.6	90.1	98.0	65.5	84.5	93.5	57.3	78.5
03	CONCRETE	99.0	74.6	86.9	96.8	56.7	77.0	106.3	63.6	85.2	97.2	79.2	88.3	94.6	65.2	80.1	94.3	55.2	75.0
04	MASONRY	97.1	66.8	78.4	97.0	48.1	66.7	114.9	55.4	78.1	92.6	77.1	83.0	138.8	49.9	83.8	93.8	54.8	69.6
05	METALS	95.2	91.3	94.0	96.2	88.2	93.7	97.3	89.6	94.9	99.8	99.1	99.6	98.4	94.6	97.2	93.2	86.2	91.1
06	WOOD, PLASTICS & COMPOSITES	99.3	76.9	86.7	98.5	42.7	67.0	96.1	52.7	71.6	99.4	79.2	88.0	97.4	51.3	71.4	98.3	42.3	66.7
07	THERMAL & MOISTURE PROTECTION	98.9	75.5	89.4	100.1	57.5	82.8	100.0	63.4	85.1	98.4	86.0	93.4	96.9	58.7	81.4	105.5	72.7	92.2
08	OPENINGS	100.0	70.9	93.1	96.8	47.6	85.1	96.7	58.3	87.6	99.6	76.0	94.0	97.9	62.4	89.5	102.0	48.5	89.3
0920	Plaster & Gypsum Board	99.8	76.7	84.0	93.7	41.4	58.1	94.8	51.7	65.5	95.7	79.0	84.3	96.5	50.2	65.0	105.5	40.9	61.6
0950, 0980	Ceilings & Acoustic Treatment	99.6	76.7	84.3	88.1	41.4	57.0	88.1	51.7	63.9	91.3	79.0	83.1	87.8	50.2	62.7	99.5	40.9	60.5
0960	Flooring	112.1	74.9	101.0	114.8	43.8	93.5	110.4	64.1	96.6	117.1	58.9	99.7	114.2	56.2	96.8	112.1	63.1	97.4
0970, 0990	Wall Finishes & Painting/Coating	96.2	71.7	81.5	99.4	66.3	79.5	99.4	66.3	79.5	102.2	62.1	78.1	101.6	62.1	77.8	96.4	66.3	78.3
09	FINISHES	103.1	74.1	86.9	101.3	45.8	70.4	99.8	55.7	75.3	102.6	72.0	85.6	100.4	52.9	74.0	103.5	49.3	73.3
COVERS	DIVS. 10 - 14, 25, 28, 41, 43, 44, 46	100.0	84.2	96.8	100.0	47.4	89.5	100.0	47.5	89.5	100.0	74.6	94.9	100.0	57.3	91.4	100.0	67.6	93.5
21, 22, 23	FIRE SUPPRESSION, PLUMBING & HVAC	100.0	57.8	83.1	100.0	53.4	81.3	100.0	53.6	81.4	99.9	64.1	85.5	100.0	59.7	83.9	100.0	39.8	75.9
26, 27, 3370	ELECTRICAL, COMMUNICATIONS & UTIL.	98.9	61.4	79.3	93.1	60.7	76.2	97.0	56.8	76.0	95.7	63.0	78.6	92.8	63.0	77.3	99.2	60.8	79.1
MF2010	WEIGHTED AVERAGE	99.3	71.4	87.1	98.5	59.1	81.3	100.8	62.4	84.1	99.3	75.3	88.8	100.2	64.7	84.7	98.7	58.0	81.0

		FLORIDA				GEORGIA												
	DIVISION	TAMPA			WEST PALM BEACH			ALBANY			ATHENS			ATLANTA			AUGUSTA		
		335 - 336,346			334,349			317,398			306			300 - 303,399			308 - 309		
		MAT.	INST.	TOTAL	MAT.	INST.	TOTAL	MAT.	INST.	TOTAL	MAT.	INST.	TOTAL	MAT.	INST.	TOTAL	MAT.	INST.	TOTAL
015433	CONTRACTOR EQUIPMENT		98.8	98.8		91.7	91.7		92.5	92.5		93.8	93.8		94.3	94.3		93.8	93.8
0241, 31 - 34	SITE & INFRASTRUCTURE, DEMOLITION	108.3	89.8	95.3	90.5	78.4	82.0	96.0	79.8	84.6	93.9	93.2	93.4	90.8	94.7	93.5	87.9	92.9	91.4
0310	Concrete Forming & Accessories	98.0	77.3	80.1	99.2	69.7	73.7	90.9	44.5	50.9	90.6	47.0	52.9	94.1	75.0	77.6	91.6	67.0	70.4
0320	Concrete Reinforcing	89.8	96.6	93.2	92.3	74.2	83.3	89.6	81.4	85.5	98.6	80.4	89.5	97.9	82.4	90.2	99.0	74.1	86.6
0330	Cast-in-Place Concrete	95.8	70.9	85.5	86.4	75.5	81.9	92.4	55.3	77.0	107.4	56.5	86.3	107.4	71.1	92.3	101.4	50.6	80.3
03	CONCRETE	93.2	79.6	86.5	87.8	73.7	80.9	91.0	56.7	74.1	103.8	57.3	80.8	101.3	75.1	88.4	96.8	63.0	80.1
04	MASONRY	90.9	77.1	82.4	89.5	67.7	76.0	93.8	49.9	66.6	77.7	54.7	63.5	90.9	66.9	76.1	91.2	44.0	62.0
05	METALS	97.4	99.7	98.1	94.6	90.9	93.5	96.6	86.4	93.5	94.7	74.6	85.8	91.6	77.9	87.4	90.5	72.7	85.0
06	WOOD, PLASTICS & COMPOSITES	101.7	79.2	89.0	101.3	68.7	82.9	93.0	38.9	62.5	90.3	41.9	63.0	94.4	77.5	84.9	91.7	72.6	80.9
07	THERMAL & MOISTURE PROTECTION	97.1	86.0	92.6	96.7	71.4	86.4	96.5	61.4	82.3	93.6	53.8	77.5	93.5	73.0	85.2	93.3	58.1	79.0
08	OPENINGS	99.1	79.2	94.4	96.2	67.1	89.3	98.8	45.0	86.0	90.4	46.9	80.1	95.8	73.4	90.5	90.4	63.7	84.1
0920	Plaster & Gypsum Board	99.2	79.0	85.5	103.2	68.2	79.4	100.1	37.4	57.5	96.9	40.4	58.4	99.1	77.2	84.2	98.0	72.1	80.4
0950, 0980	Ceilings & Acoustic Treatment	91.3	79.0	83.1	85.8	68.2	74.1	93.5	37.4	56.2	94.1	40.4	58.4	94.1	77.2	82.8	95.1	72.1	79.8
0960	Flooring	115.4	57.6	98.1	117.4	67.7	102.5	116.2	48.7	96.0	94.5	54.9	82.7	95.8	66.4	87.0	94.8	47.4	80.6
0970, 0990	Wall Finishes & Painting/Coating	101.6	62.1	77.8	96.5	71.7	81.6	97.6	53.9	71.3	96.1	46.6	66.6	96.1	85.9	89.9	96.1	46.9	66.6
09	FINISHES	102.0	71.8	85.2	99.1	69.0	82.4	101.6	44.2	69.6	93.5	46.8	67.5	93.8	74.8	83.2	93.3	61.9	75.8
COVERS	DIVS. 10 - 14, 25, 28, 41, 43, 44, 46	100.0	84.9	97.0	100.0	86.5	97.3	100.0	80.3	96.1	100.0	79.4	95.9	100.0	85.7	97.1	100.0	80.6	96.1
21, 22, 23	FIRE SUPPRESSION, PLUMBING & HVAC	100.0	82.6	93.0	97.4	64.0	84.0	100.0	69.5	87.7	95.1	71.1	85.5	100.0	71.9	88.7	100.0	62.6	85.0
26, 27, 3370	ELECTRICAL, COMMUNICATIONS & UTIL.	92.6	63.1	77.2	93.8	72.7	82.8	96.1	60.1	77.3	99.5	71.2	84.7	98.8	72.7	85.2	100.1	63.2	80.9
MF2010	WEIGHTED AVERAGE	97.9	79.8	90.0	95.3	72.3	85.3	97.5	62.5	82.3	94.7	65.1	81.8	96.8	75.3	87.5	95.6	64.8	82.2

602

City Cost Indexes

		GEORGIA																	
		COLUMBUS			DALTON			GAINESVILLE			MACON			SAVANNAH			STATESBORO		
	DIVISION	318 - 319			307			305			310 - 312			313 - 314			304		
		MAT.	INST.	TOTAL	MAT.	INST.	TOTAL	MAT.	INST.	TOTAL	MAT.	INST.	TOTAL	MAT.	INST.	TOTAL	MAT.	INST.	TOTAL
015433	CONTRACTOR EQUIPMENT		92.5	92.5		105.5	105.5		93.8	93.8		102.8	102.8		93.5	93.5		93.2	93.2
0241, 31 - 34	SITE & INFRASTRUCTURE, DEMOLITION	95.9	79.9	84.6	94.2	97.5	96.5	93.7	93.2	93.3	96.6	94.8	95.3	97.6	81.2	86.1	95.0	77.6	82.7
0310	Concrete Forming & Accessories	90.8	55.6	60.4	83.8	47.8	52.7	93.8	43.9	50.7	90.3	53.1	58.2	92.5	51.0	56.7	78.7	52.6	56.2
0320	Concrete Reinforcing	89.8	81.8	85.8	98.1	76.3	87.2	98.4	80.3	89.4	91.0	81.5	86.2	96.8	74.2	85.5	97.7	43.1	70.4
0330	Cast-in-Place Concrete	92.1	54.8	76.6	104.3	51.0	82.2	112.8	53.8	88.3	90.9	66.7	80.8	100.2	55.8	81.8	107.1	60.7	87.9
03	CONCRETE	90.8	61.5	76.4	102.9	55.8	79.7	105.7	54.9	80.6	90.4	64.4	77.6	95.9	58.4	77.4	103.2	55.3	79.5
04	MASONRY	93.7	56.2	70.5	78.8	37.3	53.2	86.3	55.1	67.0	105.6	45.3	68.3	90.3	51.4	66.2	80.2	41.3	56.1
05	METALS	96.1	87.5	93.5	91.6	84.3	89.3	90.0	73.8	85.0	92.2	87.1	90.6	93.3	83.8	90.3	95.0	74.0	88.5
06	WOOD, PLASTICS & COMPOSITES	93.0	53.3	70.6	74.1	48.5	59.7	94.2	38.8	62.9	101.2	52.4	73.7	105.9	47.0	72.7	67.6	55.1	60.6
07	THERMAL & MOISTURE PROTECTION	96.5	64.6	83.5	95.6	52.5	78.1	93.6	55.8	78.2	94.9	63.2	82.0	95.6	57.4	80.1	94.2	52.1	77.1
08	OPENINGS	98.8	60.2	89.6	91.2	50.7	81.6	90.4	40.9	78.7	97.6	53.7	87.2	103.0	49.6	90.4	92.2	41.6	80.2
0920	Plaster & Gypsum Board	100.1	52.3	67.6	84.0	47.3	59.0	99.1	37.2	57.0	105.5	51.3	68.6	98.1	45.8	62.5	84.1	54.1	63.7
0950, 0980	Ceilings & Acoustic Treatment	93.5	52.3	66.1	106.9	47.3	67.2	94.1	37.2	56.2	88.4	51.3	63.7	101.5	45.8	64.4	102.2	54.1	70.2
0960	Flooring	116.2	51.3	96.8	94.7	47.8	80.7	95.7	47.4	81.3	91.7	48.7	78.8	113.7	47.2	93.8	112.7	45.7	92.7
0970, 0990	Wall Finishes & Painting/Coating	97.6	67.9	79.7	87.3	61.1	71.5	96.1	46.9	66.6	99.8	53.9	72.2	96.0	59.5	74.1	94.0	40.1	61.6
09	FINISHES	101.5	54.7	75.4	101.7	48.1	71.9	94.0	42.6	65.3	89.7	51.0	68.2	101.7	49.9	72.9	104.7	50.1	74.3
COVERS	DIVS. 10 - 14, 25, 28, 41, 43, 44, 46	100.0	81.9	96.4	100.0	23.4	84.6	100.0	40.5	88.1	100.0	80.3	96.1	100.0	80.0	96.0	100.0	44.7	88.9
21, 22, 23	FIRE SUPPRESSION, PLUMBING & HVAC	100.0	64.7	85.8	95.2	58.2	80.4	95.1	71.0	85.5	100.0	68.2	87.2	100.0	62.9	85.2	95.7	58.0	80.5
26, 27, 3370	ELECTRICAL, COMMUNICATIONS & UTIL.	96.2	71.3	83.2	109.8	69.3	88.7	99.5	71.2	84.7	95.2	62.7	78.2	101.0	58.7	78.9	99.6	58.7	78.3
MF2010	WEIGHTED AVERAGE	97.4	66.6	84.0	96.4	60.1	80.6	95.2	62.8	81.1	96.2	65.9	83.0	98.4	62.0	82.6	96.5	56.9	79.2

		GEORGIA							HAWAII						IDAHO		
		VALDOSTA			WAYCROSS			HILO			HONOLULU			STATES & POSS., GUAM			BOISE		
	DIVISION	316			315			967			968			969			836 - 837		
		MAT.	INST.	TOTAL	MAT.	INST.	TOTAL	MAT.	INST.	TOTAL	MAT.	INST.	TOTAL	MAT.	INST.	TOTAL	MAT.	INST.	TOTAL
015433	CONTRACTOR EQUIPMENT		92.5	92.5		92.5	92.5		98.6	98.6		98.6	98.6		162.1	162.1		97.9	97.9
0241, 31 - 34	SITE & INFRASTRUCTURE, DEMOLITION	104.6	79.9	87.3	101.5	78.6	85.4	126.9	106.4	112.5	136.4	106.4	115.3	165.7	103.6	122.1	82.4	97.1	92.7
0310	Concrete Forming & Accessories	81.4	45.0	50.0	83.1	66.3	68.6	108.1	135.3	131.5	119.2	135.3	133.1	110.3	64.7	70.9	100.7	79.2	82.2
0320	Concrete Reinforcing	91.7	78.8	85.3	91.7	75.0	83.3	109.0	121.6	115.3	116.9	121.6	119.2	188.7	31.1	110.0	101.9	81.2	91.6
0330	Cast-in-Place Concrete	90.5	56.5	76.4	102.2	49.0	80.1	186.3	126.3	161.4	153.8	126.3	142.4	161.5	106.6	138.7	91.1	89.4	90.4
03	CONCRETE	95.4	56.9	76.4	98.7	62.8	81.0	148.2	128.4	138.4	141.5	128.4	135.0	149.6	73.5	112.0	100.7	83.2	92.1
04	MASONRY	99.3	51.0	69.4	100.0	39.9	62.8	127.9	128.5	128.3	128.1	128.5	128.4	182.7	44.3	97.1	118.9	82.4	96.3
05	METALS	95.7	85.4	92.5	94.8	79.3	90.0	107.6	107.9	107.7	120.8	107.9	116.8	138.8	76.3	119.5	101.1	81.2	95.0
06	WOOD, PLASTICS & COMPOSITES	81.0	38.5	57.0	82.8	74.1	77.9	101.3	137.9	122.0	119.2	137.9	129.7	113.8	68.6	88.3	88.9	77.7	82.6
07	THERMAL & MOISTURE PROTECTION	96.7	63.3	83.1	96.5	51.2	78.1	108.0	124.8	114.8	123.3	124.8	123.9	124.5	67.7	101.5	95.0	81.2	89.4
08	OPENINGS	94.7	43.5	82.6	94.7	59.2	86.3	99.7	132.7	107.5	108.8	132.7	114.4	104.3	55.2	92.6	99.1	72.5	92.8
0920	Plaster & Gypsum Board	92.0	37.1	54.6	92.4	73.7	79.7	107.7	138.9	128.9	143.9	138.9	140.5	213.3	56.8	106.8	84.9	76.9	79.5
0950, 0980	Ceilings & Acoustic Treatment	90.8	37.1	55.1	88.9	73.7	78.8	121.3	138.9	133.0	132.4	138.9	136.7	240.6	56.8	118.3	107.2	76.9	87.0
0960	Flooring	110.1	48.7	91.7	111.2	30.4	87.1	125.5	138.2	129.3	144.2	138.2	142.4	147.9	47.2	117.8	95.5	85.2	92.4
0970, 0990	Wall Finishes & Painting/Coating	97.6	53.3	70.9	97.6	46.9	67.1	99.1	145.5	127.0	107.4	145.5	130.3	104.1	36.7	63.6	102.9	40.3	65.3
09	FINISHES	99.0	44.5	68.7	98.6	58.7	76.4	115.6	138.1	128.1	130.3	138.1	134.6	192.9	60.0	118.8	97.8	76.5	85.9
COVERS	DIVS. 10 - 14, 25, 28, 41, 43, 44, 46	100.0	79.1	95.8	100.0	52.9	90.6	100.0	116.4	103.3	100.0	116.4	103.3	100.0	76.3	95.3	100.0	89.8	98.0
21, 22, 23	FIRE SUPPRESSION, PLUMBING & HVAC	100.0	71.4	88.5	96.8	58.3	81.4	100.1	109.4	103.8	100.2	109.4	103.8	102.1	38.1	76.4	100.0	72.5	89.0
26, 27, 3370	ELECTRICAL, COMMUNICATIONS & UTIL.	94.2	57.6	75.1	98.7	58.7	77.8	105.2	124.0	115.0	106.5	124.0	115.7	151.3	42.2	94.3	96.6	74.6	85.1
MF2010	WEIGHTED AVERAGE	97.5	62.6	82.3	97.3	60.7	81.3	110.4	121.0	115.0	114.9	121.0	117.6	131.2	58.9	99.7	99.8	79.4	90.9

		IDAHO															ILLINOIS		
		COEUR D'ALENE			IDAHO FALLS			LEWISTON			POCATELLO			TWIN FALLS			BLOOMINGTON		
	DIVISION	838			834			835			832			833			617		
		MAT.	INST.	TOTAL	MAT.	INST.	TOTAL	MAT.	INST.	TOTAL	MAT.	INST.	TOTAL	MAT.	INST.	TOTAL	MAT.	INST.	TOTAL
015433	CONTRACTOR EQUIPMENT		92.5	92.5		97.9	97.9		92.5	92.5		97.9	97.9		97.9	97.9		101.4	101.4
0241, 31 - 34	SITE & INFRASTRUCTURE, DEMOLITION	79.9	92.3	88.6	80.3	96.9	91.9	86.3	93.1	91.1	83.1	97.1	92.9	89.4	98.0	95.5	96.3	97.8	97.3
0310	Concrete Forming & Accessories	113.8	82.2	86.5	94.9	79.0	81.1	118.9	83.1	88.0	100.9	78.9	81.9	101.8	56.0	62.3	85.5	119.2	114.6
0320	Concrete Reinforcing	109.5	96.9	103.2	103.9	81.0	92.4	109.5	97.1	103.3	102.3	81.2	91.8	104.2	81.0	92.6	95.8	111.2	103.5
0330	Cast-in-Place Concrete	98.5	86.4	93.5	86.8	75.0	81.9	102.3	86.8	95.9	93.7	89.2	91.8	96.1	64.4	82.9	98.7	113.8	104.9
03	CONCRETE	107.5	86.3	97.1	92.9	78.0	85.6	111.1	86.9	99.2	99.9	83.0	91.5	107.5	64.2	86.1	97.2	115.8	106.4
04	MASONRY	120.4	85.3	98.7	114.1	82.4	94.5	120.8	85.3	98.8	116.6	82.4	95.4	119.0	82.4	96.4	113.4	118.3	116.4
05	METALS	95.2	87.1	92.7	108.9	79.6	99.9	94.6	87.0	92.6	109.0	80.9	100.3	109.0	79.5	99.9	97.3	113.3	102.2
06	WOOD, PLASTICS & COMPOSITES	92.7	81.8	86.5	83.0	79.4	80.9	97.9	81.8	88.8	88.9	77.7	82.6	89.8	48.3	66.4	88.1	118.1	105.0
07	THERMAL & MOISTURE PROTECTION	147.5	81.5	120.7	94.6	71.7	85.3	147.7	81.5	120.8	95.0	73.9	86.4	95.7	75.2	87.4	100.3	113.3	105.6
08	OPENINGS	117.8	73.5	107.4	102.8	68.6	94.7	117.8	76.1	107.9	99.8	67.5	92.1	102.8	49.6	90.2	93.1	106.6	96.3
0920	Plaster & Gypsum Board	152.1	81.2	103.9	75.3	78.7	77.6	153.2	81.2	104.3	77.6	76.9	77.1	78.5	46.7	56.8	90.3	118.6	109.5
0950, 0980	Ceilings & Acoustic Treatment	138.9	81.2	100.5	105.1	78.7	87.5	138.9	81.2	100.5	113.1	76.9	89.0	107.8	46.7	67.1	94.3	118.6	110.5
0960	Flooring	137.5	46.7	110.3	95.7	43.9	80.2	140.2	92.5	126.0	98.7	85.2	94.6	99.7	43.9	83.0	98.0	120.9	104.8
0970, 0990	Wall Finishes & Painting/Coating	122.3	66.3	88.7	102.9	41.2	65.8	122.3	66.3	88.7	102.8	42.0	66.2	102.9	38.1	63.9	95.3	127.8	114.8
09	FINISHES	166.2	74.0	114.8	95.5	69.5	81.0	167.3	83.0	120.4	98.9	76.7	86.5	98.9	50.8	72.1	97.4	121.1	110.6
COVERS	DIVS. 10 - 14, 25, 28, 41, 43, 44, 46	100.0	89.7	97.9	100.0	48.9	89.8	100.0	89.7	97.9	100.0	89.8	98.0	100.0	45.4	89.1	100.0	103.9	100.8
21, 22, 23	FIRE SUPPRESSION, PLUMBING & HVAC	99.5	84.3	93.4	101.0	72.3	89.5	100.7	87.2	95.3	100.0	72.4	88.9	100.0	70.8	88.3	95.0	107.8	100.1
26, 27, 3370	ELECTRICAL, COMMUNICATIONS & UTIL.	88.7	79.4	83.9	88.7	70.9	79.4	86.7	80.8	83.6	94.1	70.9	82.0	90.2	61.6	75.2	93.5	96.8	95.3
MF2010	WEIGHTED AVERAGE	107.6	83.3	97.0	99.6	75.5	89.1	108.3	85.5	98.4	100.8	78.4	91.1	101.8	68.5	87.4	96.9	109.9	102.6

603

City Cost Indexes

ILLINOIS

DIVISION		CARBONDALE 629 MAT.	INST.	TOTAL	CENTRALIA 628 MAT.	INST.	TOTAL	CHAMPAIGN 618-619 MAT.	INST.	TOTAL	CHICAGO 606-608 MAT.	INST.	TOTAL	DECATUR 625 MAT.	INST.	TOTAL	EAST ST. LOUIS 620-622 MAT.	INST.	TOTAL
015433	CONTRACTOR EQUIPMENT		107.7	107.7		107.7	107.7		102.1	102.1		93.9	93.9		102.1	102.1		107.7	107.7
0241, 31-34	SITE & INFRASTRUCTURE, DEMOLITION	100.2	98.6	99.0	100.5	99.2	99.6	105.2	98.4	100.4	105.3	95.4	98.4	92.2	98.6	96.7	102.6	99.2	100.2
0310	Concrete Forming & Accessories	95.5	107.9	106.2	97.4	114.8	112.4	91.6	117.1	113.7	96.6	159.8	151.1	97.7	116.9	114.2	92.8	115.0	112.0
0320	Concrete Reinforcing	89.8	100.4	95.1	89.8	113.3	101.5	95.8	105.1	100.5	95.5	163.2	129.3	87.9	105.9	96.9	89.7	113.4	101.5
0330	Cast-in-Place Concrete	91.8	101.1	95.7	92.3	118.5	103.2	114.4	109.2	112.2	103.4	149.6	122.6	100.2	111.9	105.1	93.8	118.5	104.1
03	CONCRETE	87.0	104.8	95.8	87.5	116.3	101.7	109.6	112.1	110.8	99.2	155.5	127.0	97.5	113.1	105.2	88.4	116.5	102.3
04	MASONRY	79.4	107.7	96.9	79.5	117.4	103.0	136.2	118.4	125.2	101.8	158.0	136.6	75.9	115.5	100.4	79.7	117.4	103.0
05	METALS	94.3	113.7	100.3	94.4	120.8	102.5	97.3	108.7	100.8	95.0	135.9	107.6	97.9	109.6	101.5	95.4	121.0	103.3
06	WOOD, PLASTICS & COMPOSITES	106.0	106.1	106.1	108.7	112.3	110.7	95.0	116.5	107.1	98.2	159.6	132.8	106.2	116.5	112.0	103.2	112.3	108.3
07	THERMAL & MOISTURE PROTECTION	98.0	101.6	99.5	98.0	112.3	103.8	100.9	114.5	106.5	99.2	147.5	118.8	104.0	109.6	106.3	98.0	111.9	103.6
08	OPENINGS	88.3	110.5	93.6	88.4	117.3	95.2	93.7	112.1	98.1	101.3	162.6	115.8	99.3	112.4	102.4	88.4	117.3	95.3
0920	Plaster & Gypsum Board	95.7	106.2	102.9	96.8	112.6	107.6	92.5	116.9	109.1	89.7	161.3	138.4	98.4	116.9	111.0	94.6	112.6	106.9
0950, 0980	Ceilings & Acoustic Treatment	96.2	106.2	102.9	96.2	112.6	107.1	94.3	116.9	109.4	101.5	161.3	141.3	102.4	116.9	112.1	96.2	112.6	107.1
0960	Flooring	121.3	122.4	121.6	122.3	117.9	120.9	100.7	122.4	107.1	93.1	150.7	110.3	108.6	118.8	111.7	120.2	117.9	119.5
0970, 0990	Wall Finishes & Painting/Coating	104.5	99.5	101.5	104.5	105.2	104.9	95.3	111.1	104.8	93.8	156.1	131.2	96.1	108.4	103.5	104.5	105.2	104.9
09	FINISHES	102.4	108.9	106.0	102.8	114.4	109.3	99.3	118.0	109.7	98.7	158.7	132.1	102.9	117.3	110.9	102.0	114.5	109.0
COVERS	DIVS. 10-14, 25, 28, 41, 43, 44, 46	100.0	104.1	100.8	100.0	106.5	101.3	100.0	103.1	100.6	100.0	125.0	105.0	100.0	103.0	100.6	100.0	106.5	101.3
21, 22, 23	FIRE SUPPRESSION, PLUMBING & HVAC	95.0	104.3	98.7	95.0	94.4	94.8	95.0	105.3	99.1	99.9	136.1	114.4	100.0	98.9	99.5	99.9	98.4	99.3
26, 27, 3370	ELECTRICAL, COMMUNICATIONS & UTIL.	94.0	106.7	100.6	95.5	104.3	100.1	96.8	94.5	95.6	96.1	137.1	117.5	98.4	91.3	94.7	95.0	104.3	99.9
MF2010	WEIGHTED AVERAGE	93.9	106.2	99.2	94.1	108.4	100.4	100.2	108.0	103.6	98.9	141.9	117.6	98.2	105.9	101.6	95.5	109.3	101.5

ILLINOIS

DIVISION		EFFINGHAM 624 MAT.	INST.	TOTAL	GALESBURG 614 MAT.	INST.	TOTAL	JOLIET 604 MAT.	INST.	TOTAL	KANKAKEE 609 MAT.	INST.	TOTAL	LA SALLE 613 MAT.	INST.	TOTAL	NORTH SUBURBAN 600-603 MAT.	INST.	TOTAL
015433	CONTRACTOR EQUIPMENT		102.1	102.1		101.4	101.4		92.0	92.0		92.0	92.0		101.4	101.4		92.0	92.0
0241, 31-34	SITE & INFRASTRUCTURE, DEMOLITION	97.5	98.2	98.0	98.7	97.7	98.0	105.2	94.7	97.8	98.9	93.9	95.4	98.1	98.4	98.3	104.4	93.7	96.8
0310	Concrete Forming & Accessories	102.2	115.5	113.7	91.5	119.0	115.3	98.4	163.3	154.4	92.4	141.3	134.6	105.1	125.1	122.4	97.8	153.4	145.8
0320	Concrete Reinforcing	90.7	101.0	95.9	95.4	111.1	103.2	95.5	155.7	125.6	96.3	146.5	121.4	95.5	143.7	119.6	95.5	157.3	126.4
0330	Cast-in-Place Concrete	99.9	107.3	103.0	101.6	107.9	104.2	103.3	149.4	122.4	96.4	132.5	111.4	101.5	123.6	110.7	103.4	143.5	120.0
03	CONCRETE	98.4	109.9	104.1	100.1	113.7	106.8	99.3	155.5	127.1	93.7	138.2	115.6	100.9	127.9	114.3	99.3	149.4	124.0
04	MASONRY	83.8	107.6	98.5	113.5	119.7	117.4	105.0	151.1	133.5	101.3	139.8	125.1	113.5	125.6	121.0	101.8	146.6	129.5
05	METALS	95.1	106.1	98.5	97.3	112.8	102.1	92.9	130.6	104.6	92.9	124.7	102.7	97.4	130.3	107.5	94.1	130.1	105.2
06	WOOD, PLASTICS & COMPOSITES	108.8	116.5	113.1	94.8	118.2	108.0	99.7	164.8	136.4	92.8	140.7	119.8	110.0	123.5	117.6	98.2	154.1	129.7
07	THERMAL & MOISTURE PROTECTION	103.6	107.6	105.2	100.5	109.9	104.3	99.1	144.5	117.5	98.4	135.6	113.5	100.6	119.4	108.2	99.5	140.8	116.3
08	OPENINGS	93.8	111.2	97.9	93.1	112.2	97.6	99.0	163.5	114.2	91.9	147.0	105.0	93.1	128.3	101.5	99.0	157.1	112.8
0920	Plaster & Gypsum Board	98.3	116.9	111.0	92.5	118.7	110.3	86.7	166.8	141.2	84.1	142.0	123.5	99.2	124.2	116.2	89.7	155.8	134.6
0950, 0980	Ceilings & Acoustic Treatment	96.2	116.9	110.0	94.3	118.7	110.5	101.5	166.8	144.9	101.5	142.0	128.4	94.3	124.2	114.2	101.5	155.8	137.6
0960	Flooring	109.7	122.4	113.5	100.6	120.9	106.7	92.7	142.9	107.7	90.1	134.3	103.3	106.5	124.8	112.0	93.1	142.9	108.0
0970, 0990	Wall Finishes & Painting/Coating	96.1	104.2	101.0	95.3	96.0	95.7	92.0	151.4	127.7	92.0	127.8	113.5	95.3	127.8	114.8	93.8	151.4	128.4
09	FINISHES	101.9	116.8	110.2	98.6	117.8	109.3	98.1	160.0	132.6	96.6	137.1	119.2	101.2	124.7	114.3	98.7	152.5	128.7
COVERS	DIVS. 10-14, 25, 28, 41, 43, 44, 46	100.0	71.7	94.3	100.0	103.8	100.8	100.0	125.4	105.1	100.0	117.6	103.5	100.0	104.4	100.9	100.0	120.4	104.1
21, 22, 23	FIRE SUPPRESSION, PLUMBING & HVAC	95.1	103.5	98.4	95.0	105.7	99.3	100.0	132.7	113.1	95.1	126.6	107.7	95.0	124.8	107.0	99.9	130.9	112.3
26, 27, 3370	ELECTRICAL, COMMUNICATIONS & UTIL.	96.0	106.6	101.6	94.5	87.3	90.7	95.4	141.2	119.3	90.1	140.4	116.4	91.6	140.4	117.1	95.2	131.5	114.2
MF2010	WEIGHTED AVERAGE	96.3	106.3	100.6	97.6	107.6	102.0	98.4	140.7	116.8	94.8	130.8	110.5	97.7	125.1	109.6	98.5	135.9	114.8

ILLINOIS

DIVISION		PEORIA 615-616 MAT.	INST.	TOTAL	QUINCY 623 MAT.	INST.	TOTAL	ROCK ISLAND 612 MAT.	INST.	TOTAL	ROCKFORD 610-611 MAT.	INST.	TOTAL	SOUTH SUBURBAN 605 MAT.	INST.	TOTAL	SPRINGFIELD 626-627 MAT.	INST.	TOTAL
015433	CONTRACTOR EQUIPMENT		101.4	101.4		102.1	102.1		101.4	101.4		101.4	101.4		92.0	92.0		102.1	102.1
0241, 31-34	SITE & INFRASTRUCTURE, DEMOLITION	99.2	97.8	98.2	96.3	98.1	97.6	96.8	96.8	96.8	98.5	98.8	98.7	104.4	93.7	96.8	99.2	98.6	98.8
0310	Concrete Forming & Accessories	94.5	119.9	116.4	100.0	113.8	111.9	93.2	101.3	100.2	98.9	130.1	125.8	97.8	153.4	145.8	98.1	117.3	114.6
0320	Concrete Reinforcing	92.9	111.2	102.0	90.4	105.6	98.0	95.4	104.2	99.8	87.8	138.2	113.0	95.5	157.3	126.4	92.8	106.1	99.4
0330	Cast-in-Place Concrete	98.5	116.0	105.8	100.0	104.2	101.8	99.4	98.4	99.0	100.9	126.4	111.5	103.4	143.5	120.0	95.2	108.5	100.7
03	CONCRETE	97.3	116.9	106.9	98.0	109.0	103.4	98.1	101.2	99.6	97.8	130.1	113.7	99.3	149.4	124.0	95.6	112.1	103.7
04	MASONRY	113.5	120.0	117.5	104.2	104.6	104.4	113.4	96.2	102.7	88.8	134.3	117.0	101.8	146.6	129.5	86.1	119.0	106.5
05	METALS	100.0	113.6	104.2	95.1	108.4	99.2	97.3	108.3	100.7	100.0	127.8	108.6	94.1	130.1	105.2	95.5	109.7	99.9
06	WOOD, PLASTICS & COMPOSITES	102.5	118.2	111.4	106.1	116.5	112.0	96.4	100.6	98.8	102.5	127.8	116.8	98.2	154.1	129.7	103.0	116.5	110.6
07	THERMAL & MOISTURE PROTECTION	101.2	113.8	106.3	103.6	104.1	103.8	100.4	98.5	99.7	103.5	128.4	113.6	99.5	140.8	116.3	105.2	112.8	108.3
08	OPENINGS	99.2	116.7	103.4	94.6	112.4	98.8	93.1	101.0	95.0	99.2	132.5	107.1	99.0	157.1	112.8	101.3	112.4	103.9
0920	Plaster & Gypsum Board	96.3	118.7	111.5	96.8	116.9	110.5	92.5	100.5	97.9	96.3	128.5	118.2	89.7	155.8	134.6	97.1	116.9	110.6
0950, 0980	Ceilings & Acoustic Treatment	100.5	118.7	112.6	96.2	116.9	110.0	94.3	100.5	98.4	100.5	128.5	119.2	101.5	155.8	137.6	106.6	116.9	113.5
0960	Flooring	103.8	120.9	108.9	108.6	110.1	109.1	101.7	107.5	103.4	103.8	124.8	110.1	93.1	142.9	108.0	113.1	109.8	112.1
0970, 0990	Wall Finishes & Painting/Coating	95.3	127.8	114.8	96.1	105.1	101.5	95.3	96.0	95.7	95.3	135.3	119.3	93.8	151.4	128.4	94.6	105.1	100.9
09	FINISHES	101.4	121.4	112.6	101.3	113.6	108.2	98.8	101.9	100.6	101.4	129.8	117.2	98.7	152.5	128.7	106.1	115.4	111.3
COVERS	DIVS. 10-14, 25, 28, 41, 43, 44, 46	100.0	104.3	100.9	100.0	72.7	94.5	100.0	97.2	99.4	100.0	111.6	102.3	100.0	120.4	104.1	100.0	103.3	100.7
21, 22, 23	FIRE SUPPRESSION, PLUMBING & HVAC	99.9	105.8	102.3	95.1	100.8	97.4	95.0	99.8	96.9	100.0	114.9	106.0	99.9	130.9	112.3	99.9	103.4	101.3
26, 27, 3370	ELECTRICAL, COMMUNICATIONS & UTIL.	95.5	97.4	96.5	93.2	81.3	87.0	86.5	95.7	91.3	95.8	127.0	112.1	95.2	131.5	114.2	101.0	92.8	96.8
MF2010	WEIGHTED AVERAGE	99.9	110.4	104.5	96.9	101.6	99.0	96.6	99.8	98.0	98.9	123.5	109.6	98.5	135.9	114.8	99.0	107.2	102.5

City Cost Indexes

DIVISION		ANDERSON 460			BLOOMINGTON 474			COLUMBUS 472			EVANSVILLE 476-477			FORT WAYNE 467-468			GARY 463-464		
		MAT.	INST.	TOTAL	MAT.	INST.	TOTAL	MAT.	INST.	TOTAL	MAT.	INST.	TOTAL	MAT.	INST.	TOTAL	MAT.	INST.	TOTAL
015433	CONTRACTOR EQUIPMENT		96.7	96.7		87.2	87.2		87.2	87.2		116.1	116.1		96.7	96.7		96.7	96.7
0241, 31 - 34	SITE & INFRASTRUCTURE, DEMOLITION	90.5	96.0	94.4	82.7	94.8	91.2	79.4	94.6	90.1	87.8	123.5	112.9	91.5	96.0	94.6	91.0	99.2	96.8
0310	Concrete Forming & Accessories	98.8	80.8	83.3	101.9	81.1	83.9	95.9	78.9	81.2	95.4	81.9	83.7	97.1	75.4	78.3	98.9	115.2	113.0
0320	Concrete Reinforcing	98.0	83.7	90.9	87.1	83.4	85.2	87.5	83.4	85.4	95.4	76.7	86.1	98.0	76.8	87.4	98.0	110.1	104.0
0330	Cast-in-Place Concrete	104.8	78.1	93.7	100.8	78.0	91.3	100.4	72.5	88.8	96.3	87.7	92.8	111.3	83.7	99.9	109.5	112.4	110.7
03	CONCRETE	96.3	80.9	88.7	102.3	80.3	91.4	101.6	77.3	89.6	102.7	83.1	93.0	99.3	79.1	89.3	98.5	113.1	105.7
04	MASONRY	84.8	79.7	81.7	90.5	76.4	81.8	90.4	76.2	81.6	86.4	83.3	84.5	88.2	77.3	81.4	86.1	112.8	102.7
05	METALS	92.9	89.8	92.0	95.9	78.7	90.6	95.9	78.0	90.4	89.2	84.4	87.8	92.9	86.4	90.9	92.9	108.0	97.6
06	WOOD, PLASTICS & COMPOSITES	108.2	80.8	92.8	118.0	81.1	97.2	112.5	78.3	93.2	97.4	80.8	88.0	107.9	74.8	89.2	105.7	115.0	110.9
07	THERMAL & MOISTURE PROTECTION	105.4	77.1	93.9	95.3	79.7	88.9	94.9	79.4	88.6	99.3	84.7	93.4	105.2	70.0	94.6	104.0	109.0	106.0
08	OPENINGS	95.0	81.9	91.9	105.0	82.0	99.6	100.8	80.5	96.0	98.3	79.8	93.9	95.0	74.7	90.2	95.0	119.4	100.8
0920	Plaster & Gypsum Board	94.3	80.6	85.0	98.1	81.2	86.6	94.7	78.2	83.6	93.1	79.6	83.9	93.5	74.4	80.5	87.6	115.8	106.8
0950, 0980	Ceilings & Acoustic Treatment	89.6	80.6	83.7	83.9	81.2	82.1	83.9	78.3	80.2	88.8	79.6	82.7	89.6	74.4	79.5	89.6	115.8	107.1
0960	Flooring	99.9	86.6	96.0	105.8	76.4	97.0	101.4	76.4	93.9	100.8	82.2	95.2	99.9	80.2	94.0	99.9	123.6	107.0
0970, 0990	Wall Finishes & Painting/Coating	97.3	70.9	81.4	89.6	83.0	85.7	89.6	83.0	85.7	94.7	86.7	89.9	97.3	74.4	83.5	97.3	126.6	114.9
09	FINISHES	95.0	80.9	87.1	96.2	80.4	87.4	94.5	78.7	85.7	95.2	82.4	88.1	94.8	76.1	84.4	94.0	118.3	107.5
COVERS	DIVS. 10 - 14, 25, 28, 41, 43, 44, 46	100.0	91.5	98.3	100.0	89.1	97.8	100.0	88.7	97.7	100.0	94.1	98.8	100.0	89.5	97.9	100.0	105.3	101.1
21, 22, 23	FIRE SUPPRESSION, PLUMBING & HVAC	100.0	75.8	90.3	99.8	79.0	91.4	94.9	79.0	88.5	100.0	82.2	92.0	100.0	72.4	88.9	100.0	105.9	102.3
26, 27, 3370	ELECTRICAL, COMMUNICATIONS & UTIL.	85.9	88.4	87.2	98.2	83.5	90.5	97.3	85.2	91.0	94.5	86.3	90.2	86.6	78.4	82.3	97.7	110.2	104.2
MF2010	WEIGHTED AVERAGE	95.5	83.1	90.1	98.7	81.5	91.2	96.6	80.9	89.7	96.5	86.6	92.1	96.0	79.2	88.7	96.8	110.1	102.6

DIVISION		INDIANAPOLIS 461-462			KOKOMO 469			LAFAYETTE 479			LAWRENCEBURG 470			MUNCIE 473			NEW ALBANY 471		
		MAT.	INST.	TOTAL	MAT.	INST.	TOTAL	MAT.	INST.	TOTAL	MAT.	INST.	TOTAL	MAT.	INST.	TOTAL	MAT.	INST.	TOTAL
015433	CONTRACTOR EQUIPMENT		93.3	93.3		96.7	96.7		87.2	87.2		105.0	105.0		95.7	95.7		95.1	95.1
0241, 31 - 34	SITE & INFRASTRUCTURE, DEMOLITION	90.3	99.3	96.6	87.1	95.8	93.2	80.2	94.5	90.3	78.0	111.0	101.2	82.5	95.0	91.3	75.3	97.4	90.8
0310	Concrete Forming & Accessories	98.0	85.1	86.9	102.4	77.8	81.2	93.5	82.8	84.3	92.4	77.2	79.3	93.5	80.5	82.3	90.2	75.1	77.2
0320	Concrete Reinforcing	100.0	85.4	92.7	88.7	83.7	86.2	87.1	83.6	85.4	86.3	76.6	81.5	96.3	83.6	90.0	87.7	79.6	83.6
0330	Cast-in-Place Concrete	103.3	84.7	95.6	103.7	81.0	94.3	100.9	80.7	92.5	94.4	76.6	87.0	105.9	77.7	94.2	97.4	75.7	88.4
03	CONCRETE	99.5	84.8	92.3	93.1	80.5	86.9	101.8	81.9	92.0	95.0	77.4	86.3	101.1	80.6	91.0	100.2	76.5	88.5
04	MASONRY	90.3	81.3	84.7	84.4	80.0	81.7	95.5	79.4	85.8	75.6	76.0	75.9	92.4	79.8	84.6	82.3	69.5	74.4
05	METALS	93.6	81.7	89.9	89.5	89.3	89.4	94.3	78.6	89.5	91.1	85.2	89.3	97.7	89.9	95.3	93.0	82.7	89.8
06	WOOD, PLASTICS & COMPOSITES	105.3	85.3	94.0	111.6	76.6	91.9	109.5	83.2	94.7	95.8	77.0	85.2	111.4	80.6	94.0	97.9	76.0	85.5
07	THERMAL & MOISTURE PROTECTION	100.9	82.6	93.5	105.0	77.1	93.7	94.9	81.5	89.5	99.2	78.3	90.7	97.9	78.0	89.8	86.8	71.4	80.6
08	OPENINGS	103.1	84.8	98.8	90.0	79.6	87.5	99.2	83.1	95.4	100.4	76.8	94.8	98.5	81.7	94.5	97.9	78.2	93.2
0920	Plaster & Gypsum Board	88.5	85.1	86.2	99.8	76.3	83.8	91.8	83.3	86.0	72.6	76.8	75.5	93.1	80.6	84.6	91.1	75.6	80.6
0950, 0980	Ceilings & Acoustic Treatment	91.3	85.1	87.2	89.6	76.3	80.8	79.4	83.3	82.0	93.2	76.8	82.3	84.8	80.6	82.0	88.8	75.6	80.0
0960	Flooring	101.7	86.6	97.2	103.6	89.9	99.5	100.4	88.1	96.8	74.8	86.6	78.3	100.7	86.6	96.5	98.3	64.0	88.1
0970, 0990	Wall Finishes & Painting/Coating	99.5	83.0	89.6	97.3	73.3	82.9	89.6	95.1	92.9	90.6	73.2	80.1	89.6	70.9	78.4	94.7	84.5	88.6
09	FINISHES	95.4	85.4	89.8	96.6	79.3	87.0	92.9	85.2	88.6	85.8	78.9	82.0	94.0	80.7	86.6	94.5	74.3	83.2
COVERS	DIVS. 10 - 14, 25, 28, 41, 43, 44, 46	100.0	93.0	98.6	100.0	89.1	97.8	100.0	91.3	98.3	100.0	44.9	89.0	100.0	90.9	98.2	100.0	44.3	88.8
21, 22, 23	FIRE SUPPRESSION, PLUMBING & HVAC	99.9	80.1	92.0	95.1	79.2	88.7	94.9	80.2	89.0	95.6	75.7	87.6	99.8	75.8	90.2	95.1	76.9	87.8
26, 27, 3370	ELECTRICAL, COMMUNICATIONS & UTIL.	100.7	88.3	94.2	90.4	78.8	84.4	96.8	82.6	89.4	92.3	75.3	83.4	90.2	80.3	85.0	92.8	77.6	84.9
MF2010	WEIGHTED AVERAGE	98.3	85.1	92.6	93.3	82.0	88.4	96.3	82.9	90.5	93.3	79.3	87.2	97.3	81.8	90.6	94.4	77.0	86.8

DIVISION		SOUTH BEND 465-466			TERRE HAUTE 478			WASHINGTON 475			BURLINGTON 526			CARROLL 514			CEDAR RAPIDS 522-524		
		MAT.	INST.	TOTAL	MAT.	INST.	TOTAL	MAT.	INST.	TOTAL	MAT.	INST.	TOTAL	MAT.	INST.	TOTAL	MAT.	INST.	TOTAL
015433	CONTRACTOR EQUIPMENT		105.0	105.0		116.1	116.1		116.1	116.1		100.2	100.2		100.2	100.2		96.8	96.8
0241, 31 - 34	SITE & INFRASTRUCTURE, DEMOLITION	92.9	95.9	95.0	89.7	124.2	113.9	89.2	121.5	111.9	96.5	97.3	97.1	85.8	97.3	93.9	98.5	96.0	96.8
0310	Concrete Forming & Accessories	102.6	79.8	82.9	96.3	78.1	80.6	97.1	80.6	82.9	97.3	77.3	80.1	85.2	51.6	56.2	103.2	81.0	84.1
0320	Concrete Reinforcing	101.1	79.1	90.1	95.4	83.8	89.6	88.2	50.3	69.3	94.3	85.4	89.9	95.1	81.1	88.1	95.1	86.8	90.9
0330	Cast-in-Place Concrete	96.3	80.7	89.8	93.3	87.0	90.7	101.5	86.6	95.3	109.6	56.3	87.5	109.6	61.0	89.4	109.9	81.3	98.0
03	CONCRETE	90.8	81.3	86.1	105.6	82.6	94.2	111.1	77.1	94.3	101.7	72.5	87.3	100.5	61.7	81.3	101.8	82.8	92.4
04	MASONRY	90.7	79.1	83.5	93.6	80.5	85.5	86.4	82.3	83.9	97.2	66.0	77.9	99.0	73.5	83.2	102.8	77.5	87.1
05	METALS	93.0	99.8	95.1	89.9	88.1	89.4	84.7	69.6	80.0	87.0	94.9	89.4	87.0	90.4	88.1	89.3	93.8	90.7
06	WOOD, PLASTICS & COMPOSITES	105.0	79.2	90.4	99.5	76.4	86.5	100.0	80.7	89.1	95.3	78.4	85.7	81.8	45.9	61.6	101.8	80.5	89.8
07	THERMAL & MOISTURE PROTECTION	97.5	81.9	91.2	99.4	81.9	92.3	99.5	84.4	93.4	103.8	75.6	92.4	104.1	70.5	90.5	104.7	79.9	94.6
08	OPENINGS	91.0	78.9	88.1	98.9	79.4	94.3	95.6	69.2	89.3	92.4	72.1	87.6	96.6	54.7	86.7	97.6	81.7	93.8
0920	Plaster & Gypsum Board	87.0	78.9	81.5	93.1	75.0	80.8	93.0	79.5	83.8	100.7	77.7	85.1	95.9	44.3	60.8	105.3	80.1	88.2
0950, 0980	Ceilings & Acoustic Treatment	93.4	78.9	83.8	88.8	75.0	79.6	82.5	79.5	80.5	110.6	77.7	88.7	110.6	44.3	66.5	114.1	80.1	91.5
0960	Flooring	97.4	89.9	95.1	100.8	86.6	96.6	101.7	77.8	94.6	111.2	38.5	89.5	105.6	33.8	84.2	127.9	79.7	113.5
0970, 0990	Wall Finishes & Painting/Coating	87.6	87.4	87.5	94.7	83.7	88.1	94.7	85.2	89.0	99.0	80.2	87.7	99.0	79.9	87.5	100.4	73.7	84.3
09	FINISHES	95.3	82.4	88.1	95.2	79.9	86.6	94.3	81.4	87.1	107.2	70.4	86.7	103.3	49.4	73.3	113.2	80.0	94.7
COVERS	DIVS. 10 - 14, 25, 28, 41, 43, 44, 46	100.0	90.7	98.1	100.0	93.2	98.6	100.0	94.0	98.8	100.0	89.2	97.8	100.0	66.9	93.4	100.0	90.2	98.0
21, 22, 23	FIRE SUPPRESSION, PLUMBING & HVAC	99.9	77.4	90.9	100.0	80.4	92.1	95.1	79.3	88.7	95.3	76.9	87.9	95.3	74.2	86.8	100.0	80.7	92.4
26, 27, 3370	ELECTRICAL, COMMUNICATIONS & UTIL.	97.6	86.6	91.9	92.6	83.6	87.9	93.1	86.3	89.5	100.7	69.3	84.3	101.4	81.3	90.7	98.2	81.3	89.3
MF2010	WEIGHTED AVERAGE	95.7	84.2	90.7	97.2	85.8	92.2	95.0	83.3	89.9	96.6	76.9	88.0	96.4	72.2	85.9	99.4	83.4	92.4

605

City Cost Indexes

			IOWA

		COUNCIL BLUFFS			CRESTON			DAVENPORT			DECORAH			DES MOINES			DUBUQUE		
DIVISION		515			508			527 - 528			521			500 - 503,509			520		
		MAT.	INST.	TOTAL	MAT.	INST.	TOTAL	MAT.	INST.	TOTAL	MAT.	INST.	TOTAL	MAT.	INST.	TOTAL	MAT.	INST.	TOTAL
015433	CONTRACTOR EQUIPMENT		96.3	96.3		100.2	100.2		100.2	100.2		100.2	100.2		101.8	101.8		95.7	95.7
0241, 31 - 34	SITE & INFRASTRUCTURE, DEMOLITION	103.6	92.3	95.7	86.7	96.4	93.5	96.8	99.3	98.6	95.1	96.3	95.9	95.5	99.7	98.5	96.4	93.3	94.2
0310	Concrete Forming & Accessories	84.6	70.7	72.6	84.5	67.2	69.6	102.5	91.2	92.7	94.9	46.5	53.1	103.2	78.9	82.2	85.9	73.8	75.4
0320	Concrete Reinforcing	97.0	79.3	88.2	93.3	81.2	87.3	95.1	100.0	97.5	94.3	79.0	86.7	102.7	81.4	92.0	93.7	86.6	90.2
0330	Cast-in-Place Concrete	114.3	75.1	98.0	102.1	65.4	86.9	105.9	91.7	100.0	106.7	59.3	87.0	89.8	88.0	89.1	107.6	100.4	104.6
03	CONCRETE	104.1	74.7	89.6	96.9	70.2	83.7	99.9	93.5	96.7	99.6	58.4	79.3	95.0	83.2	89.2	98.6	86.2	92.4
04	MASONRY	104.5	75.9	86.8	96.9	81.4	87.3	100.0	85.5	91.1	117.7	72.7	89.8	90.7	80.4	84.3	103.7	73.7	85.1
05	METALS	94.3	90.2	93.1	89.4	90.6	89.8	89.3	103.9	93.8	87.1	87.7	87.3	95.2	92.9	94.5	88.0	93.5	89.7
06	WOOD, PLASTICS & COMPOSITES	80.6	69.7	74.4	79.1	63.1	70.1	101.9	90.3	95.3	92.3	37.9	61.6	98.8	77.4	86.8	82.2	71.7	76.3
07	THERMAL & MOISTURE PROTECTION	104.2	68.4	89.6	106.2	78.8	95.1	104.2	87.4	97.4	104.0	54.7	84.0	106.6	79.5	92.0	104.4	71.3	91.0
08	OPENINGS	96.6	75.5	91.6	106.6	64.4	96.6	97.6	91.4	96.1	95.2	49.8	84.5	101.9	82.8	97.4	96.6	79.3	92.5
0920	Plaster & Gypsum Board	95.9	69.0	77.6	94.1	62.0	72.2	105.3	90.0	94.9	99.6	36.0	56.4	91.9	76.8	81.6	95.9	71.1	79.0
0950, 0980	Ceilings & Acoustic Treatment	110.6	69.0	82.9	111.5	62.0	78.5	114.1	90.0	98.1	110.6	36.0	61.0	115.0	76.8	89.6	110.6	71.1	84.3
0960	Flooring	104.3	82.1	97.6	104.2	33.8	83.2	113.5	98.7	109.1	111.1	48.0	92.2	109.4	87.6	102.9	118.8	77.6	106.5
0970, 0990	Wall Finishes & Painting/Coating	95.3	66.9	78.2	95.8	79.9	86.3	99.0	97.6	98.1	99.0	33.6	59.7	91.2	79.9	84.4	99.4	65.1	78.8
09	FINISHES	104.1	72.0	86.2	102.4	61.5	79.6	109.1	93.2	100.2	106.9	43.6	71.6	105.5	80.4	91.5	108.4	72.9	88.6
COVERS	DIVS. 10 - 14, 25, 28, 41, 43, 44, 46	100.0	88.9	97.8	100.0	71.4	94.3	100.0	93.1	98.6	100.0	85.2	97.0	100.0	90.3	98.1	100.0	88.6	97.7
21, 22, 23	FIRE SUPPRESSION, PLUMBING & HVAC	100.2	75.8	90.4	95.3	80.3	89.2	100.2	93.2	97.4	95.3	74.5	86.9	100.0	79.8	91.9	100.2	75.4	90.2
26, 27, 3370	ELECTRICAL, COMMUNICATIONS & UTIL.	103.7	81.6	92.1	94.8	81.3	87.8	95.5	91.2	93.2	98.1	45.4	70.6	106.9	81.3	93.6	102.2	78.4	89.8
MF2010	WEIGHTED AVERAGE	100.2	78.8	90.9	96.7	77.9	88.5	98.4	93.4	96.2	97.4	65.6	83.5	99.4	83.9	92.7	98.5	80.4	90.6

			IOWA

		FORT DODGE			MASON CITY			OTTUMWA			SHENANDOAH			SIBLEY			SIOUX CITY		
DIVISION		505			504			525			516			512			510 - 511		
		MAT.	INST.	TOTAL	MAT.	INST.	TOTAL	MAT.	INST.	TOTAL	MAT.	INST.	TOTAL	MAT.	INST.	TOTAL	MAT.	INST.	TOTAL
015433	CONTRACTOR EQUIPMENT		100.2	100.2		100.2	100.2		95.7	95.7		96.3	96.3		100.2	100.2		100.2	100.2
0241, 31 - 34	SITE & INFRASTRUCTURE, DEMOLITION	94.6	95.0	94.9	94.6	96.2	95.8	96.8	91.4	93.0	102.0	91.4	94.6	106.0	95.0	98.3	107.6	95.8	99.3
0310	Concrete Forming & Accessories	85.1	46.0	51.3	89.6	46.4	52.3	92.9	71.9	74.8	86.2	57.8	61.7	86.6	38.7	45.3	103.2	67.9	72.7
0320	Concrete Reinforcing	93.3	67.4	80.4	93.2	80.6	86.9	94.3	85.6	90.0	97.0	68.6	82.8	97.0	67.3	82.2	95.1	78.2	86.7
0330	Cast-in-Place Concrete	95.8	45.0	74.7	95.8	58.7	80.4	110.3	54.8	87.3	110.5	61.1	90.0	108.3	45.7	82.3	109.0	56.7	87.3
03	CONCRETE	92.6	51.1	72.1	92.9	58.5	75.9	101.2	69.4	85.5	101.5	62.1	82.1	100.5	48.1	74.6	101.2	66.9	84.3
04	MASONRY	95.9	38.5	60.4	108.7	69.4	84.4	100.3	57.1	73.6	104.0	75.9	86.6	122.1	38.8	70.6	97.2	55.6	71.4
05	METALS	89.5	81.9	87.2	89.5	88.8	89.3	86.9	92.0	88.5	93.3	84.3	90.5	87.2	81.4	85.4	89.3	88.2	89.0
06	WOOD, PLASTICS & COMPOSITES	79.6	46.2	60.8	83.9	37.9	58.0	89.4	78.2	83.1	82.2	54.0	66.3	83.1	36.7	56.9	101.8	68.2	82.9
07	THERMAL & MOISTURE PROTECTION	105.6	58.5	86.5	105.2	64.6	88.7	104.6	65.2	88.6	103.5	66.6	88.5	103.8	49.2	81.6	104.3	64.9	88.3
08	OPENINGS	100.8	50.1	88.8	93.3	50.2	83.1	96.6	76.2	91.8	88.1	55.5	80.4	93.4	44.9	81.9	97.6	66.9	90.3
0920	Plaster & Gypsum Board	94.1	44.6	60.4	94.1	36.0	54.6	97.0	77.7	83.9	95.9	52.8	66.6	95.9	34.7	54.3	105.3	67.2	79.4
0950, 0980	Ceilings & Acoustic Treatment	111.5	44.6	67.0	111.5	36.0	61.3	110.6	77.7	88.7	110.6	52.8	72.1	110.6	34.7	60.1	114.1	67.2	82.9
0960	Flooring	105.8	48.0	88.5	107.9	48.0	90.0	121.8	51.1	100.6	104.9	34.9	84.0	106.4	34.6	85.0	113.6	54.7	96.0
0970, 0990	Wall Finishes & Painting/Coating	95.8	63.1	76.1	95.8	30.7	56.7	99.4	79.9	87.7	95.3	66.9	78.2	99.0	63.1	77.4	99.0	63.1	77.4
09	FINISHES	104.4	47.1	72.5	105.0	42.9	70.4	109.5	69.2	87.0	104.2	53.4	75.9	106.7	38.9	68.9	110.7	64.8	85.1
COVERS	DIVS. 10 - 14, 25, 28, 41, 43, 44, 46	100.0	79.8	96.0	100.0	84.8	97.0	100.0	84.2	96.8	100.0	67.9	93.6	100.0	78.9	95.8	100.0	86.9	97.4
21, 22, 23	FIRE SUPPRESSION, PLUMBING & HVAC	95.3	65.8	83.4	95.3	71.9	85.9	95.3	71.3	85.7	95.3	75.1	87.2	95.3	69.5	84.9	100.2	76.3	90.6
26, 27, 3370	ELECTRICAL, COMMUNICATIONS & UTIL.	101.9	42.9	71.1	100.9	57.0	77.9	100.5	67.8	83.4	98.1	80.6	89.0	98.1	42.4	69.1	98.2	74.7	85.9
MF2010	WEIGHTED AVERAGE	96.6	58.8	80.2	96.5	66.6	83.5	97.3	73.0	86.7	97.1	72.1	86.2	97.7	57.5	80.2	99.1	73.6	88.0

			IOWA		KANSAS

		SPENCER			WATERLOO			BELLEVILLE			COLBY			DODGE CITY			EMPORIA		
DIVISION		513			506 - 507			669			677			678			668		
		MAT.	INST.	TOTAL	MAT.	INST.	TOTAL	MAT.	INST.	TOTAL	MAT.	INST.	TOTAL	MAT.	INST.	TOTAL	MAT.	INST.	TOTAL
015433	CONTRACTOR EQUIPMENT		100.2	100.2		100.2	100.2		103.1	103.1		103.1	103.1		103.1	103.1		101.4	101.4
0241, 31 - 34	SITE & INFRASTRUCTURE, DEMOLITION	106.1	95.0	98.3	99.5	96.1	97.1	111.0	94.7	99.5	109.4	94.8	99.1	111.9	94.5	99.7	102.8	92.2	95.3
0310	Concrete Forming & Accessories	92.8	38.7	46.1	101.6	48.1	55.4	95.0	55.3	60.7	102.2	59.6	65.4	95.3	59.5	64.4	86.5	66.7	69.4
0320	Concrete Reinforcing	97.0	67.3	82.2	94.0	86.5	90.2	99.6	56.1	77.9	101.2	56.1	78.7	98.8	56.1	77.5	98.2	56.7	77.5
0330	Cast-in-Place Concrete	108.3	45.7	82.3	102.6	55.0	82.9	118.9	56.9	93.2	114.3	57.0	90.6	116.4	56.7	91.6	114.9	54.8	90.0
03	CONCRETE	100.9	48.1	74.9	98.4	59.1	79.0	117.7	57.4	88.0	114.9	59.3	87.5	116.2	59.1	88.0	110.0	61.8	86.2
04	MASONRY	122.1	38.8	70.6	96.6	73.0	82.0	94.9	59.8	73.2	96.9	59.8	73.9	104.4	60.0	76.9	100.5	68.4	80.7
05	METALS	87.1	81.4	85.4	91.8	92.8	92.1	97.8	78.4	91.9	98.2	78.7	92.2	99.7	77.8	92.9	97.6	79.9	92.1
06	WOOD, PLASTICS & COMPOSITES	89.3	36.7	59.6	98.6	37.9	64.4	96.4	53.3	72.1	103.9	59.2	78.7	95.3	59.2	74.9	87.7	68.2	76.7
07	THERMAL & MOISTURE PROTECTION	104.6	49.2	82.1	105.4	71.4	91.6	98.4	62.3	83.7	101.8	62.9	86.1	101.9	62.3	85.8	97.0	77.4	89.1
08	OPENINGS	104.3	44.9	90.2	94.2	53.5	84.5	96.4	49.4	85.3	99.5	52.6	88.4	99.5	52.6	88.4	94.4	57.5	85.6
0920	Plaster & Gypsum Board	97.0	34.7	54.7	103.9	36.0	57.7	98.0	51.8	66.6	100.7	57.9	71.6	94.8	57.9	69.7	95.1	67.1	76.0
0950, 0980	Ceilings & Acoustic Treatment	110.6	34.7	60.1	115.0	36.0	62.5	91.6	51.8	65.1	94.3	57.9	70.1	94.3	57.9	70.1	91.6	67.1	75.3
0960	Flooring	109.1	34.6	86.8	113.3	76.9	102.4	106.4	39.9	86.5	105.7	39.9	86.0	102.2	39.9	83.6	101.8	37.1	82.4
0970, 0990	Wall Finishes & Painting/Coating	99.0	63.1	77.4	95.8	79.4	86.0	97.4	39.8	62.8	97.4	39.8	62.8	97.4	39.8	62.8	97.4	39.8	62.8
09	FINISHES	107.6	38.9	69.3	108.8	54.4	78.5	101.3	50.4	72.9	101.5	53.9	75.0	99.8	53.9	74.2	98.6	58.6	76.3
COVERS	DIVS. 10 - 14, 25, 28, 41, 43, 44, 46	100.0	78.9	95.8	100.0	85.3	97.1	100.0	41.8	88.3	100.0	42.5	88.5	100.0	42.5	88.5	100.0	41.5	88.3
21, 22, 23	FIRE SUPPRESSION, PLUMBING & HVAC	95.3	69.5	84.9	100.2	80.0	92.1	95.1	73.1	86.3	95.1	68.7	84.5	100.0	68.7	87.4	95.1	73.4	86.4
26, 27, 3370	ELECTRICAL, COMMUNICATIONS & UTIL.	99.9	42.4	69.9	97.0	57.0	76.1	93.3	66.0	79.1	97.9	66.0	81.3	94.7	73.4	83.6	90.8	73.1	81.5
MF2010	WEIGHTED AVERAGE	99.2	57.5	81.0	98.3	70.9	86.4	99.4	65.6	84.7	100.2	65.6	85.1	101.7	66.5	86.4	97.8	70.0	85.7

City Cost Indexes

		KANSAS																	
DIVISION		FORT SCOTT			HAYS			HUTCHINSON			INDEPENDENCE			KANSAS CITY			LIBERAL		
		667			676			675			673			660 - 662			679		
		MAT.	INST.	TOTAL	MAT.	INST.	TOTAL	MAT.	INST.	TOTAL	MAT.	INST.	TOTAL	MAT.	INST.	TOTAL	MAT.	INST.	TOTAL
015433	CONTRACTOR EQUIPMENT		102.3	102.3		103.1	103.1		103.1	103.1		103.1	103.1		99.7	99.7		103.1	103.1
0241, 31 - 34	SITE & INFRASTRUCTURE, DEMOLITION	99.4	93.0	94.9	115.5	94.8	100.9	93.4	94.6	94.3	113.1	94.6	100.1	93.5	92.9	93.1	115.2	94.6	100.7
0310	Concrete Forming & Accessories	102.6	78.6	81.9	99.7	59.6	65.1	89.9	59.5	63.6	110.8	67.8	73.7	99.3	96.3	96.7	95.8	59.5	64.5
0320	Concrete Reinforcing	97.5	96.3	96.9	98.8	56.1	77.5	98.8	56.1	77.5	98.2	63.0	80.6	94.6	97.8	96.2	100.2	56.1	78.2
0330	Cast-in-Place Concrete	106.6	55.7	85.5	90.2	57.0	76.5	83.6	54.2	71.4	116.8	54.5	91.0	91.4	99.3	94.7	90.3	54.2	75.3
03	CONCRETE	105.2	74.6	90.1	106.7	59.3	83.3	89.1	58.3	73.9	117.4	63.3	90.7	97.3	97.9	97.6	108.9	58.3	83.9
04	MASONRY	101.0	61.4	76.5	104.3	59.8	76.8	96.4	59.6	73.6	94.0	65.7	76.5	103.6	99.5	101.1	102.4	59.6	75.9
05	METALS	97.5	94.4	96.5	97.8	78.7	91.9	97.6	77.7	91.5	97.5	81.3	92.5	104.9	100.6	103.5	98.1	77.7	91.8
06	WOOD, PLASTICS & COMPOSITES	106.5	85.4	94.6	100.6	59.2	77.2	89.9	59.2	72.6	114.7	69.6	89.3	102.4	96.7	99.1	95.9	59.2	75.2
07	THERMAL & MOISTURE PROTECTION	97.8	82.8	91.7	102.1	62.9	86.2	100.9	62.2	85.2	101.9	82.4	94.0	97.5	99.8	98.4	102.3	62.2	86.0
08	OPENINGS	94.4	80.1	91.0	99.4	52.6	88.3	99.4	52.6	88.3	97.2	59.8	88.4	95.6	90.5	94.4	99.5	52.6	88.4
0920	Plaster & Gypsum Board	100.6	84.9	89.9	98.1	57.9	70.7	93.6	57.9	69.3	107.8	68.6	81.1	93.5	96.5	95.5	95.5	57.9	69.9
0950, 0980	Ceilings & Acoustic Treatment	91.6	84.9	87.2	94.3	57.9	70.1	94.3	57.9	70.1	94.3	68.6	77.2	91.6	96.5	94.9	94.3	57.9	70.1
0960	Flooring	117.6	40.1	94.4	104.7	39.9	85.3	99.5	39.9	81.7	109.9	39.9	89.0	93.5	100.5	95.6	102.4	39.9	83.7
0970, 0990	Wall Finishes & Painting/Coating	99.2	39.8	63.5	97.4	39.8	62.8	97.4	39.8	62.8	97.4	39.8	62.8	105.6	69.2	83.7	97.4	39.8	62.8
09	FINISHES	104.0	68.4	84.2	101.4	53.9	74.9	97.3	53.9	73.1	103.8	60.0	79.4	98.5	93.1	95.4	100.7	53.9	74.6
COVERS	DIVS. 10 - 14, 25, 28, 41, 43, 44, 46	100.0	47.4	89.5	100.0	42.5	88.5	100.0	42.5	88.5	100.0	43.7	88.7	100.0	63.5	92.7	100.0	42.5	88.5
21, 22, 23	FIRE SUPPRESSION, PLUMBING & HVAC	95.1	69.3	84.7	95.1	68.7	84.5	95.1	68.7	84.5	95.1	71.1	85.5	99.9	96.8	98.6	95.1	68.7	84.5
26, 27, 3370	ELECTRICAL, COMMUNICATIONS & UTIL.	90.1	73.1	81.2	96.8	66.0	80.7	91.7	66.0	78.3	94.0	73.1	83.0	95.5	97.3	96.4	94.7	73.4	83.6
MF2010	WEIGHTED AVERAGE	97.7	74.3	87.5	99.6	65.6	84.8	95.8	65.4	82.5	99.9	70.3	87.0	99.4	95.7	97.7	99.5	66.4	85.1

		KANSAS										KENTUCKY							
DIVISION		SALINA			TOPEKA			WICHITA			ASHLAND			BOWLING GREEN			CAMPTON		
		674			664 - 666			670 - 672			411 - 412			421 - 422			413 - 414		
		MAT.	INST.	TOTAL	MAT.	INST.	TOTAL	MAT.	INST.	TOTAL	MAT.	INST.	TOTAL	MAT.	INST.	TOTAL	MAT.	INST.	TOTAL
015433	CONTRACTOR EQUIPMENT		103.1	103.1		101.4	101.4		103.1	103.1		98.0	98.0		95.1	95.1		101.5	101.5
0241, 31 - 34	SITE & INFRASTRUCTURE, DEMOLITION	102.2	94.8	97.0	97.9	91.2	93.2	98.7	93.3	94.9	108.9	87.0	93.5	75.6	97.7	91.1	83.8	98.8	94.3
0310	Concrete Forming & Accessories	91.8	55.7	60.6	97.2	41.8	49.4	97.8	51.7	58.0	89.0	105.7	103.4	86.5	84.0	84.4	90.4	84.6	85.4
0320	Concrete Reinforcing	98.2	62.6	80.4	97.7	97.2	97.5	97.5	75.8	86.7	89.1	106.1	97.6	86.4	89.5	88.0	87.3	105.0	96.1
0330	Cast-in-Place Concrete	101.0	54.7	81.8	96.1	46.5	75.6	94.7	52.7	77.3	88.7	102.9	94.6	88.0	97.5	91.9	98.1	75.2	88.6
03	CONCRETE	103.0	58.0	80.8	98.3	55.4	77.1	97.6	57.8	78.0	96.1	105.2	100.6	94.4	89.7	92.1	98.2	85.2	91.8
04	MASONRY	118.4	59.8	82.1	97.6	56.9	72.4	94.9	51.1	67.8	92.2	105.1	100.2	94.0	81.6	86.3	91.6	70.2	78.4
05	METALS	99.5	81.5	93.9	101.6	96.4	100.0	101.6	83.7	96.1	92.1	109.8	97.6	93.7	86.9	91.6	93.0	93.7	93.2
06	WOOD, PLASTICS & COMPOSITES	91.5	53.3	70.0	95.4	38.1	63.1	95.3	52.5	71.2	79.0	105.4	93.9	92.3	83.9	87.5	90.8	91.1	90.9
07	THERMAL & MOISTURE PROTECTION	101.4	62.6	85.7	100.6	69.1	87.8	100.0	57.6	82.8	91.3	99.3	94.5	86.7	82.1	84.8	99.6	69.4	87.3
08	OPENINGS	99.4	49.7	87.6	98.6	55.3	88.4	101.6	56.6	91.0	97.0	99.3	97.5	97.9	79.4	93.5	99.3	89.8	97.1
0920	Plaster & Gypsum Board	93.6	51.8	65.2	97.6	36.1	55.8	92.4	51.0	64.2	61.2	105.6	91.4	86.4	83.7	84.6	86.4	90.2	89.0
0950, 0980	Ceilings & Acoustic Treatment	94.3	51.8	66.0	97.6	36.1	56.7	99.4	51.0	67.2	83.8	105.6	98.3	88.8	83.7	85.4	88.8	90.2	89.7
0960	Flooring	100.7	63.1	89.5	106.9	75.0	97.3	108.3	64.1	95.1	80.1	104.2	87.3	96.5	87.6	93.9	98.6	42.0	81.7
0970, 0990	Wall Finishes & Painting/Coating	97.4	39.8	62.8	99.2	53.7	71.9	96.3	49.0	67.9	96.9	103.4	100.8	94.7	74.3	82.4	94.7	66.8	77.9
09	FINISHES	98.5	54.8	74.2	103.1	47.0	71.8	102.7	53.5	75.3	82.5	105.7	95.4	93.1	84.0	88.0	93.9	75.7	83.8
COVERS	DIVS. 10 - 14, 25, 28, 41, 43, 44, 46	100.0	54.2	90.8	100.0	48.4	89.7	100.0	51.1	90.2	100.0	96.3	99.3	100.0	59.8	91.9	100.0	57.3	91.4
21, 22, 23	FIRE SUPPRESSION, PLUMBING & HVAC	100.0	68.7	87.5	100.0	69.0	87.5	99.8	64.3	85.6	95.0	92.9	94.1	100.0	84.3	93.7	95.1	81.5	89.7
26, 27, 3370	ELECTRICAL, COMMUNICATIONS & UTIL.	94.5	73.3	83.4	94.2	73.1	83.2	99.0	73.3	85.6	90.7	99.6	95.4	93.3	82.1	87.4	90.6	67.7	78.6
MF2010	WEIGHTED AVERAGE	100.4	67.0	85.9	99.4	66.5	85.1	99.9	65.1	84.8	94.0	100.1	96.6	95.5	84.8	90.8	95.1	80.2	88.6

		KENTUCKY																	
DIVISION		CORBIN			COVINGTON			ELIZABETHTOWN			FRANKFORT			HAZARD			HENDERSON		
		407 - 409			410			427			406			417 - 418			424		
		MAT.	INST.	TOTAL	MAT.	INST.	TOTAL	MAT.	INST.	TOTAL	MAT.	INST.	TOTAL	MAT.	INST.	TOTAL	MAT.	INST.	TOTAL
015433	CONTRACTOR EQUIPMENT		101.5	101.5		105.0	105.0		95.1	95.1		101.5	101.5		101.5	101.5		116.1	116.1
0241, 31 - 34	SITE & INFRASTRUCTURE, DEMOLITION	83.0	98.5	93.9	79.4	112.3	102.5	70.4	97.8	89.6	87.1	99.7	95.9	81.8	100.1	94.6	78.4	124.1	110.5
0310	Concrete Forming & Accessories	88.9	72.3	74.6	86.0	92.9	92.0	81.6	82.4	82.3	99.5	73.4	77.0	87.1	85.2	85.6	93.4	83.5	84.8
0320	Concrete Reinforcing	87.0	72.1	79.5	85.9	94.0	90.0	86.9	93.8	90.3	99.8	93.7	96.8	87.7	105.0	96.3	86.6	87.6	87.1
0330	Cast-in-Place Concrete	92.2	59.9	78.8	93.9	99.6	96.3	79.6	74.6	77.5	87.2	70.6	80.3	94.4	84.2	90.2	77.8	92.5	83.9
03	CONCRETE	92.6	68.4	80.7	96.5	95.5	96.0	86.6	82.0	84.3	93.0	76.6	84.9	95.0	88.7	91.9	91.6	87.4	89.6
04	MASONRY	89.8	63.0	73.2	105.4	103.3	104.1	78.5	76.8	77.4	87.2	80.8	83.3	89.3	72.8	79.1	97.8	94.5	95.8
05	METALS	93.4	79.8	89.2	91.0	91.7	91.2	92.8	88.8	91.6	95.8	89.5	93.9	93.0	93.7	93.2	84.4	88.6	85.7
06	WOOD, PLASTICS & COMPOSITES	88.5	77.7	82.4	88.7	86.0	87.2	87.4	83.9	85.4	105.5	70.3	85.6	87.7	91.1	89.6	95.1	80.8	87.1
07	THERMAL & MOISTURE PROTECTION	98.8	65.6	85.3	99.4	95.4	97.8	86.4	78.8	83.3	98.8	77.1	90.0	99.5	71.3	88.1	99.0	93.6	96.8
08	OPENINGS	99.3	67.2	91.7	101.3	85.7	97.6	97.9	85.3	94.9	102.1	78.2	96.4	99.7	80.3	95.1	96.0	81.8	92.7
0920	Plaster & Gypsum Board	90.6	76.4	80.9	69.4	86.1	80.8	85.6	83.7	84.3	100.9	68.8	79.0	85.6	90.2	88.7	89.7	79.7	82.9
0950, 0980	Ceilings & Acoustic Treatment	86.1	76.4	79.6	92.3	86.1	88.2	88.8	83.7	85.4	96.8	68.8	78.1	88.8	90.2	89.7	82.5	79.7	80.6
0960	Flooring	100.6	42.0	83.1	72.7	86.2	76.7	94.1	87.6	92.2	108.0	55.4	92.3	97.1	42.7	80.9	100.1	87.6	96.4
0970, 0990	Wall Finishes & Painting/Coating	94.2	50.5	67.9	90.6	84.1	86.7	94.7	77.8	84.6	93.7	81.5	86.5	94.7	66.8	77.9	94.7	98.1	96.7
09	FINISHES	93.9	65.0	77.8	84.8	90.5	87.9	92.0	82.5	86.7	101.6	70.3	84.2	93.2	76.5	83.9	92.6	85.7	88.8
COVERS	DIVS. 10 - 14, 25, 28, 41, 43, 44, 46	100.0	49.7	89.9	100.0	102.8	100.6	100.0	86.6	97.3	100.0	67.2	93.4	100.0	58.1	91.6	100.0	67.1	93.4
21, 22, 23	FIRE SUPPRESSION, PLUMBING & HVAC	95.1	72.9	86.2	95.6	90.6	93.6	95.4	81.9	90.0	100.1	84.1	93.7	95.1	82.4	90.0	95.4	79.3	88.9
26, 27, 3370	ELECTRICAL, COMMUNICATIONS & UTIL.	90.5	82.1	86.1	94.6	79.6	86.8	90.4	82.1	86.1	99.0	89.2	93.9	90.6	59.7	74.5	92.7	82.2	87.2
MF2010	WEIGHTED AVERAGE	94.4	73.3	85.2	95.2	93.2	94.3	92.2	83.7	88.5	97.9	82.4	91.1	94.6	79.9	88.2	93.0	87.9	90.8

607

City Cost Indexes

KENTUCKY

	DIVISION	LEXINGTON 403-405 MAT.	INST.	TOTAL	LOUISVILLE 400-402 MAT.	INST.	TOTAL	OWENSBORO 423 MAT.	INST.	TOTAL	PADUCAH 420 MAT.	INST.	TOTAL	PIKEVILLE 415-416 MAT.	INST.	TOTAL	SOMERSET 425-426 MAT.	INST.	TOTAL
015433	CONTRACTOR EQUIPMENT		101.5	101.5		95.1	95.1		116.1	116.1		116.1	116.1		98.0	98.0		101.5	101.5
0241, 31 - 34	SITE & INFRASTRUCTURE, DEMOLITION	85.0	100.9	96.2	75.7	97.8	91.3	87.9	124.1	113.3	80.9	123.6	110.9	119.6	86.0	96.0	75.1	99.3	92.1
0310	Concrete Forming & Accessories	101.5	74.3	78.0	100.6	83.0	85.4	91.8	83.4	84.5	89.9	81.8	82.9	97.7	90.9	91.9	88.0	77.6	79.0
0320	Concrete Reinforcing	95.5	93.5	94.5	97.1	93.8	95.4	86.6	90.2	88.4	87.1	89.0	88.0	89.6	105.9	97.7	86.9	93.7	90.3
0330	Cast-in-Place Concrete	94.3	93.9	94.2	95.8	75.4	87.4	90.7	96.1	92.9	82.9	88.3	85.1	97.5	98.7	98.0	77.8	100.3	87.1
03	CONCRETE	95.8	84.9	90.5	96.7	82.5	89.7	103.4	89.1	96.4	96.0	85.5	90.8	109.6	97.1	103.5	81.8	88.6	85.2
04	MASONRY	88.4	73.4	79.1	88.9	78.2	82.3	90.5	92.6	91.8	93.5	90.0	91.3	90.0	98.5	95.2	84.7	78.1	80.6
05	METALS	95.8	89.9	93.9	97.6	88.9	94.9	85.9	90.2	87.2	83.0	88.4	84.6	92.0	108.5	97.1	92.9	89.3	91.8
06	WOOD, PLASTICS & COMPOSITES	105.1	70.3	85.5	103.1	83.9	92.3	93.0	80.8	86.1	90.6	80.9	85.2	88.6	90.4	89.6	88.3	77.7	82.3
07	THERMAL & MOISTURE PROTECTION	99.1	85.7	93.6	97.9	79.5	90.4	99.3	93.9	97.1	99.0	80.9	91.7	91.9	81.5	87.7	99.0	71.8	87.9
08	OPENINGS	99.5	75.3	93.8	99.0	85.3	95.8	96.0	83.9	93.1	95.2	79.4	91.5	97.7	88.4	95.5	98.7	79.6	94.2
0920	Plaster & Gypsum Board	100.6	68.8	79.0	93.2	83.7	86.7	88.2	79.7	82.4	87.1	79.8	82.1	64.5	90.2	82.0	85.6	76.4	79.4
0950, 0980	Ceilings & Acoustic Treatment	90.5	68.8	76.1	94.3	83.7	87.2	82.5	79.7	80.6	82.5	79.8	80.7	83.8	90.2	88.0	88.8	76.4	80.5
0960	Flooring	105.0	71.6	95.0	102.9	87.6	98.3	99.5	87.6	96.0	98.5	60.4	87.1	83.8	104.2	89.9	97.4	42.0	80.8
0970, 0990	Wall Finishes & Painting/Coating	94.2	82.8	87.3	93.9	77.8	84.2	94.7	98.1	96.7	94.7	81.4	86.7	96.9	83.0	88.5	94.7	77.8	84.6
09	FINISHES	97.5	73.6	84.2	97.9	83.5	89.9	92.9	85.2	88.6	92.1	77.8	84.1	84.9	93.0	89.4	92.6	70.9	80.5
COVERS	DIVS. 10 - 14, 25, 28, 41, 43, 44, 46	100.0	88.4	97.7	100.0	87.1	97.4	100.0	100.9	100.2	100.0	63.4	92.7	100.0	57.8	91.5	100.0	60.0	92.0
21, 22, 23	FIRE SUPPRESSION, PLUMBING & HVAC	100.0	80.6	92.2	100.0	82.6	93.0	100.0	78.4	91.3	95.4	82.9	90.4	95.0	89.0	92.6	95.4	79.1	88.8
26, 27, 3370	ELECTRICAL, COMMUNICATIONS & UTIL.	93.4	82.1	87.5	100.0	82.1	90.6	92.7	82.2	87.3	95.1	81.8	88.2	93.7	75.1	84.0	91.0	82.1	86.4
MF2010	WEIGHTED AVERAGE	97.0	82.4	90.7	97.8	84.2	91.9	95.6	88.9	92.7	93.2	86.2	90.2	96.2	90.0	93.5	92.7	81.5	87.8

LOUISIANA

	DIVISION	ALEXANDRIA 713-714 MAT.	INST.	TOTAL	BATON ROUGE 707-708 MAT.	INST.	TOTAL	HAMMOND 704 MAT.	INST.	TOTAL	LAFAYETTE 705 MAT.	INST.	TOTAL	LAKE CHARLES 706 MAT.	INST.	TOTAL	MONROE 712 MAT.	INST.	TOTAL
015433	CONTRACTOR EQUIPMENT		89.7	89.7		89.3	89.3		89.9	89.9		89.9	89.9		89.3	89.3		89.7	89.7
0241, 31 - 34	SITE & INFRASTRUCTURE, DEMOLITION	96.3	86.7	89.5	96.0	86.0	89.0	99.1	86.9	90.5	100.1	87.0	90.9	100.8	85.9	90.3	96.3	85.9	89.0
0310	Concrete Forming & Accessories	83.0	44.2	49.5	94.1	61.7	66.1	78.1	46.9	51.2	95.2	54.9	60.4	96.1	57.5	62.8	82.5	43.9	49.2
0320	Concrete Reinforcing	89.4	56.0	72.7	95.1	59.0	77.1	93.3	59.4	76.4	94.6	59.0	76.8	94.6	59.2	76.9	88.4	55.6	72.0
0330	Cast-in-Place Concrete	94.6	50.3	76.2	91.8	58.7	78.0	93.9	43.9	73.2	93.5	47.6	74.4	98.4	67.4	85.5	94.6	56.7	78.8
03	CONCRETE	94.7	49.7	72.5	95.9	60.7	78.5	93.3	49.1	71.5	94.4	53.8	74.4	96.7	61.8	79.5	94.5	51.5	73.3
04	MASONRY	116.0	54.2	77.8	97.6	54.7	71.0	99.0	51.8	69.8	99.0	53.0	70.5	98.3	59.5	74.3	110.9	48.4	72.2
05	METALS	89.5	71.1	83.9	106.0	70.1	94.9	97.6	69.5	88.9	96.7	70.0	88.5	96.7	70.5	88.7	89.5	68.5	83.0
06	WOOD, PLASTICS & COMPOSITES	89.8	42.7	63.2	98.5	65.3	79.8	84.8	48.1	64.1	106.3	57.0	78.5	104.4	59.1	78.8	89.1	43.2	63.2
07	THERMAL & MOISTURE PROTECTION	97.5	54.1	79.9	95.8	62.4	82.2	96.9	57.0	80.7	97.4	60.8	82.5	96.5	59.5	81.5	97.5	54.7	80.1
08	OPENINGS	99.8	45.8	87.0	98.6	60.0	89.5	92.7	52.5	83.2	96.3	53.9	86.3	96.3	55.7	86.7	99.7	48.8	87.7
0920	Plaster & Gypsum Board	79.0	41.4	53.4	99.4	64.6	75.7	100.1	46.8	63.8	109.6	56.0	73.2	109.6	58.1	74.6	78.6	41.9	53.6
0950, 0980	Ceilings & Acoustic Treatment	87.1	41.4	56.7	103.1	64.6	77.4	98.7	46.8	64.2	97.0	56.0	69.7	98.0	58.1	71.5	87.1	41.9	57.0
0960	Flooring	110.0	66.9	97.1	106.0	66.9	94.3	98.6	66.9	89.1	106.8	66.9	94.9	106.8	74.3	97.1	109.5	58.1	94.2
0970, 0990	Wall Finishes & Painting/Coating	99.5	65.8	79.2	96.1	47.3	66.8	103.4	49.3	70.9	103.4	59.2	76.8	103.4	51.0	71.9	99.5	49.1	69.2
09	FINISHES	93.2	50.0	69.1	100.6	61.6	78.9	97.8	50.9	71.7	101.1	57.5	76.8	101.3	59.9	78.2	93.0	46.5	67.1
COVERS	DIVS. 10 - 14, 25, 28, 41, 43, 44, 46	100.0	51.4	90.3	100.0	80.9	96.2	100.0	46.6	89.3	100.0	79.4	95.9	100.0	80.4	96.1	100.0	47.0	89.4
21, 22, 23	FIRE SUPPRESSION, PLUMBING & HVAC	100.2	56.5	82.7	100.0	59.5	83.8	95.2	43.7	74.5	100.1	60.7	84.3	100.1	61.7	84.7	100.2	54.1	81.7
26, 27, 3370	ELECTRICAL, COMMUNICATIONS & UTIL.	90.5	58.2	73.6	103.0	60.8	81.0	98.2	57.3	76.8	99.5	66.6	82.3	98.9	66.6	82.0	92.2	59.6	75.2
MF2010	WEIGHTED AVERAGE	96.8	57.8	79.8	100.4	63.6	84.4	96.3	54.9	78.3	98.5	62.7	82.9	98.6	65.0	84.0	96.7	56.4	79.2

LOUISIANA / MAINE

	DIVISION	NEW ORLEANS 700-701 MAT.	INST.	TOTAL	SHREVEPORT 710-711 MAT.	INST.	TOTAL	THIBODAUX 703 MAT.	INST.	TOTAL	AUGUSTA 043 MAT.	INST.	TOTAL	BANGOR 044 MAT.	INST.	TOTAL	BATH 045 MAT.	INST.	TOTAL
015433	CONTRACTOR EQUIPMENT		90.2	90.2		89.7	89.7		89.9	89.9		100.6	100.6		100.6	100.6		100.6	100.6
0241, 31 - 34	SITE & INFRASTRUCTURE, DEMOLITION	105.2	88.4	93.4	100.8	86.0	90.4	101.4	87.0	91.3	89.2	100.8	97.3	91.0	101.0	98.0	88.9	100.8	97.2
0310	Concrete Forming & Accessories	97.7	66.5	70.8	98.8	44.7	52.2	90.1	64.3	67.8	99.2	97.0	97.3	93.6	98.8	98.1	89.7	97.2	96.2
0320	Concrete Reinforcing	96.7	60.1	78.4	92.6	56.1	74.4	93.3	59.2	76.3	96.4	110.1	103.2	90.1	111.5	100.8	89.2	111.2	100.2
0330	Cast-in-Place Concrete	99.7	70.1	87.4	99.4	53.4	80.3	100.9	51.3	80.3	97.9	60.3	82.3	80.2	113.0	93.8	80.2	60.3	72.0
03	CONCRETE	97.8	67.1	82.6	99.6	50.9	75.6	98.4	59.3	79.1	104.6	86.2	95.5	94.4	105.3	99.8	94.5	86.4	90.5
04	MASONRY	101.8	60.7	76.4	102.0	51.0	70.4	124.8	49.3	78.1	96.8	64.8	77.0	106.0	103.9	104.7	112.3	85.8	95.9
05	METALS	109.5	73.0	98.3	93.3	69.2	85.9	97.6	69.3	88.9	97.8	88.0	94.8	88.7	90.4	89.2	87.2	89.6	87.9
06	WOOD, PLASTICS & COMPOSITES	99.7	67.9	81.8	101.7	43.7	69.0	92.9	69.7	79.8	98.6	107.7	103.7	95.1	98.7	97.1	89.7	107.7	99.9
07	THERMAL & MOISTURE PROTECTION	96.1	69.0	85.1	97.1	52.2	78.9	96.7	54.6	79.6	104.3	66.8	89.1	100.2	86.0	94.5	100.2	73.2	89.2
08	OPENINGS	96.2	66.4	89.2	98.8	46.4	86.4	97.1	64.2	89.3	104.2	90.0	100.8	102.5	85.0	98.4	102.5	90.0	99.5
0920	Plaster & Gypsum Board	108.2	67.2	80.3	90.0	42.5	57.7	102.7	69.1	79.8	100.8	107.4	105.2	98.8	98.0	98.2	95.1	107.4	103.3
0950, 0980	Ceilings & Acoustic Treatment	98.7	67.2	77.8	91.3	42.5	58.8	98.7	69.1	79.0	101.5	107.4	105.4	91.6	98.0	95.9	89.6	107.4	101.4
0960	Flooring	105.3	66.9	93.8	113.4	62.4	98.1	104.3	44.9	86.5	103.1	56.3	89.1	98.9	113.0	103.1	97.3	56.3	85.0
0970, 0990	Wall Finishes & Painting/Coating	108.8	65.5	82.8	95.5	45.5	65.4	104.7	50.6	72.2	101.0	64.7	79.2	95.3	45.1	65.1	95.3	45.1	65.1
09	FINISHES	102.4	66.6	82.4	98.2	47.4	69.9	100.1	60.1	77.8	101.4	87.7	93.8	98.6	96.3	97.3	97.1	85.6	90.7
COVERS	DIVS. 10 - 14, 25, 28, 41, 43, 44, 46	100.0	83.2	96.6	100.0	77.3	95.5	100.0	81.3	96.3	100.0	103.3	100.7	100.0	110.0	102.0	100.0	103.3	100.7
21, 22, 23	FIRE SUPPRESSION, PLUMBING & HVAC	100.0	65.8	86.3	100.0	58.9	83.5	95.2	61.7	81.7	99.9	65.6	86.2	100.1	75.9	90.4	95.2	65.7	83.4
26, 27, 3370	ELECTRICAL, COMMUNICATIONS & UTIL.	103.2	72.9	87.4	97.7	67.2	81.8	96.7	72.8	84.2	99.3	80.0	89.2	98.5	77.2	87.4	96.4	80.0	87.9
MF2010	WEIGHTED AVERAGE	101.5	69.8	87.7	98.4	59.6	81.5	98.7	64.9	84.0	100.3	80.8	91.8	97.6	90.8	94.7	96.1	83.0	90.4

City Cost Indexes

		MAINE																	
	DIVISION	HOULTON			KITTERY			LEWISTON			MACHIAS			PORTLAND			ROCKLAND		
		047			039			042			046			040 - 041			048		
		MAT.	INST.	TOTAL	MAT.	INST.	TOTAL	MAT.	INST.	TOTAL	MAT.	INST.	TOTAL	MAT.	INST.	TOTAL	MAT.	INST.	TOTAL
015433	CONTRACTOR EQUIPMENT		100.6	100.6		100.6	100.6		100.6	100.6		100.6	100.6		100.6	100.6		100.6	100.6
0241, 31 - 34	SITE & INFRASTRUCTURE, DEMOLITION	90.7	101.8	98.5	85.6	100.8	96.3	88.2	101.0	97.2	90.1	101.8	98.3	86.9	101.0	96.8	86.3	101.8	97.2
0310	Concrete Forming & Accessories	97.1	103.6	102.7	90.4	97.7	96.7	98.9	98.9	98.9	94.4	103.5	102.3	95.8	98.9	98.4	95.4	103.5	102.4
0320	Concrete Reinforcing	90.1	109.9	100.0	84.5	111.3	97.8	110.8	111.5	111.1	90.1	109.9	100.0	107.5	111.5	109.5	90.1	109.9	100.0
0330	Cast-in-Place Concrete	80.3	72.6	77.1	81.4	61.2	73.0	81.9	113.0	94.8	80.2	72.0	76.8	95.5	113.0	102.7	81.9	72.0	77.8
03	CONCRETE	95.5	93.3	94.4	90.1	87.0	88.6	95.1	105.3	100.1	94.9	93.1	94.0	100.8	105.3	103.0	92.2	93.1	92.6
04	MASONRY	91.0	76.3	81.9	101.0	86.8	92.2	91.7	103.9	99.2	91.0	76.3	81.9	101.3	103.9	102.9	85.9	76.3	80.0
05	METALS	87.4	87.9	87.5	87.1	89.8	87.9	91.7	90.4	91.3	87.4	87.9	87.6	99.4	90.4	96.6	87.3	87.9	87.5
06	WOOD, PLASTICS & COMPOSITES	99.1	107.7	104.0	88.7	107.7	99.4	101.2	98.7	99.8	96.1	107.7	102.6	96.1	98.7	97.5	96.9	107.7	103.0
07	THERMAL & MOISTURE PROTECTION	100.3	74.0	89.6	103.2	72.5	90.7	100.1	86.0	94.4	100.3	73.0	89.2	101.4	86.0	95.2	100.0	73.0	89.1
08	OPENINGS	102.6	88.2	99.2	102.7	90.0	99.6	105.6	86.0	100.7	102.6	88.2	99.2	105.7	85.0	100.8	102.5	88.2	99.1
0920	Plaster & Gypsum Board	101.0	107.4	105.3	94.7	107.4	103.3	104.5	98.0	100.1	99.5	107.4	104.8	101.0	98.0	99.0	99.5	107.4	104.8
0950, 0980	Ceilings & Acoustic Treatment	89.6	107.4	101.4	90.6	107.4	101.8	101.4	98.0	99.1	89.6	107.4	101.4	104.4	98.0	100.1	89.6	107.4	101.4
0960	Flooring	100.1	52.7	85.9	97.5	58.7	85.9	101.6	113.0	105.0	99.3	52.7	85.4	101.9	113.0	105.2	99.6	52.7	85.6
0970, 0990	Wall Finishes & Painting/Coating	95.3	140.1	122.2	95.4	39.3	61.7	95.3	45.1	65.1	95.3	140.1	122.2	93.4	45.1	64.4	95.3	140.1	122.2
09	FINISHES	98.8	99.7	99.3	96.2	85.6	90.3	101.8	96.3	98.7	98.3	99.7	99.1	102.5	96.3	99.0	98.0	99.7	98.9
COVERS	DIVS. 10 - 14, 25, 28, 41, 43, 44, 46	100.0	108.9	101.8	100.0	103.6	100.7	100.0	110.0	102.0	100.0	108.9	101.8	100.0	110.0	102.0	100.0	108.9	101.8
21, 22, 23	FIRE SUPPRESSION, PLUMBING & HVAC	95.2	74.1	86.8	95.2	77.3	88.0	100.1	75.9	90.4	95.2	74.1	86.8	100.0	75.9	90.3	95.2	74.1	86.8
26, 27, 3370	ELECTRICAL, COMMUNICATIONS & UTIL.	100.6	80.0	89.9	98.0	80.0	88.6	100.6	80.1	89.9	100.6	80.0	89.9	103.3	80.1	91.2	100.5	80.0	89.8
MF2010	WEIGHTED AVERAGE	95.9	86.6	91.8	95.1	85.7	91.0	98.3	91.2	95.2	95.7	86.5	91.7	100.8	91.2	96.6	95.0	86.5	91.3

		MAINE			MARYLAND													
	DIVISION	WATERVILLE			ANNAPOLIS			BALTIMORE			COLLEGE PARK			CUMBERLAND			EASTON		
		049			214			210 - 212			207 - 208			215			216		
		MAT.	INST.	TOTAL	MAT.	INST.	TOTAL	MAT.	INST.	TOTAL	MAT.	INST.	TOTAL	MAT.	INST.	TOTAL	MAT.	INST.	TOTAL
015433	CONTRACTOR EQUIPMENT		100.6	100.6		99.3	99.3		102.8	102.8		104.3	104.3		99.3	99.3		99.3	99.3
0241, 31 - 34	SITE & INFRASTRUCTURE, DEMOLITION	90.8	100.8	97.8	98.2	90.9	93.1	99.8	94.8	96.3	103.7	94.1	96.9	91.4	90.9	91.1	98.2	88.0	91.0
0310	Concrete Forming & Accessories	89.2	97.0	96.0	98.1	70.3	74.1	100.3	72.6	76.4	86.2	71.2	73.2	91.6	80.8	82.3	89.5	71.7	74.2
0320	Concrete Reinforcing	90.1	110.1	100.1	104.0	82.1	93.0	104.5	82.1	93.3	109.0	80.5	94.7	87.7	74.6	81.2	87.0	80.6	83.8
0330	Cast-in-Place Concrete	80.3	60.3	72.0	110.8	77.1	96.8	111.3	77.9	97.4	122.5	78.0	104.0	93.6	85.7	90.3	104.0	49.2	81.2
03	CONCRETE	96.0	86.2	91.1	103.7	75.9	90.0	104.2	77.2	90.9	104.9	76.6	93.2	89.4	82.2	85.9	97.0	66.6	82.0
04	MASONRY	100.7	64.8	78.5	96.1	72.9	81.7	98.4	72.9	82.6	108.4	69.4	84.3	94.6	84.4	88.3	108.1	43.7	68.2
05	METALS	87.3	88.0	87.5	104.7	93.8	101.3	101.5	94.3	99.3	87.0	97.5	90.2	99.2	91.2	96.7	99.4	88.1	95.9
06	WOOD, PLASTICS & COMPOSITES	89.2	107.7	99.6	97.6	70.6	82.4	97.3	73.7	84.0	87.6	70.8	78.1	86.3	79.8	82.6	84.1	79.1	81.3
07	THERMAL & MOISTURE PROTECTION	100.3	66.8	86.7	101.1	79.6	92.4	99.1	80.2	91.4	102.6	78.5	92.9	99.5	80.4	91.7	99.6	60.4	83.7
08	OPENINGS	102.6	90.0	99.6	99.3	78.1	94.3	97.7	79.8	93.5	94.7	75.2	90.1	96.0	78.7	91.9	94.4	72.9	89.3
0920	Plaster & Gypsum Board	95.1	107.4	103.4	100.5	70.0	79.7	103.4	73.0	82.8	97.1	70.0	78.6	100.7	79.5	86.3	100.3	78.8	85.7
0950, 0980	Ceilings & Acoustic Treatment	89.6	107.4	101.4	100.6	70.0	80.2	98.1	73.0	81.4	98.1	70.0	79.4	100.1	79.5	86.4	100.1	78.8	85.9
0960	Flooring	97.0	56.3	84.8	100.7	78.6	94.1	100.0	78.6	93.6	97.9	80.5	92.7	96.2	92.5	95.1	95.4	52.6	82.6
0970, 0990	Wall Finishes & Painting/Coating	95.3	64.7	76.9	91.7	80.9	85.2	94.6	80.9	86.4	104.6	77.5	88.3	94.4	75.0	82.7	94.4	77.5	84.3
09	FINISHES	97.2	87.7	91.9	97.2	72.1	83.2	99.9	73.9	85.4	96.3	72.8	83.2	97.6	82.3	89.1	97.8	70.5	82.6
COVERS	DIVS. 10 - 14, 25, 28, 41, 43, 44, 46	100.0	103.3	100.7	100.0	85.6	97.1	100.0	86.3	97.3	100.0	83.8	96.8	100.0	90.2	98.0	100.0	77.2	95.4
21, 22, 23	FIRE SUPPRESSION, PLUMBING & HVAC	95.2	65.6	83.3	100.0	81.7	92.6	100.0	81.7	92.7	95.3	85.0	91.2	95.1	73.0	86.2	95.1	68.6	84.5
26, 27, 3370	ELECTRICAL, COMMUNICATIONS & UTIL.	100.6	80.0	89.8	96.7	91.7	94.1	101.4	91.7	96.3	102.2	99.8	100.9	96.6	81.6	88.8	96.1	64.4	79.5
MF2010	WEIGHTED AVERAGE	96.2	80.8	89.5	100.3	81.8	92.3	100.5	82.8	92.8	97.6	83.9	91.7	96.0	81.9	89.9	97.4	69.3	85.2

		MARYLAND															MASSACHUSETTS		
	DIVISION	ELKTON			HAGERSTOWN			SALISBURY			SILVER SPRING			WALDORF			BOSTON		
		219			217			218			209			206			020 - 022, 024		
		MAT.	INST.	TOTAL	MAT.	INST.	TOTAL	MAT.	INST.	TOTAL	MAT.	INST.	TOTAL	MAT.	INST.	TOTAL	MAT.	INST.	TOTAL
015433	CONTRACTOR EQUIPMENT		99.3	99.3		99.3	99.3		99.3	99.3		97.2	97.2		97.2	97.2		106.3	106.3
0241, 31 - 34	SITE & INFRASTRUCTURE, DEMOLITION	85.8	88.7	87.8	89.8	91.6	91.0	98.1	88.0	91.0	91.3	87.5	88.6	98.0	87.1	90.3	100.8	108.2	106.0
0310	Concrete Forming & Accessories	95.5	82.6	84.4	90.6	76.3	78.3	103.5	52.9	59.8	94.8	71.4	74.6	102.4	69.6	74.1	100.9	142.0	136.4
0320	Concrete Reinforcing	87.0	101.6	94.3	87.7	74.6	81.2	87.0	65.6	76.3	107.6	80.4	94.0	108.3	80.4	94.4	108.7	157.1	132.8
0330	Cast-in-Place Concrete	84.2	76.5	81.0	89.2	85.7	87.8	104.0	47.9	80.7	125.4	79.8	106.5	140.5	76.9	114.1	104.6	152.2	124.3
03	CONCRETE	82.6	84.7	83.7	86.0	80.2	83.1	97.9	55.0	76.7	108.0	77.1	92.7	118.7	75.3	97.3	108.3	147.2	127.5
04	MASONRY	94.1	59.9	73.0	100.3	84.4	90.4	107.7	48.1	70.9	107.3	72.2	85.9	92.7	67.6	77.2	108.6	162.1	141.7
05	METALS	99.5	99.2	99.4	99.3	91.3	96.8	99.5	82.4	94.2	91.2	93.9	92.1	91.2	93.8	92.0	100.8	132.1	110.5
06	WOOD, PLASTICS & COMPOSITES	91.3	88.8	89.9	85.4	73.5	78.7	101.5	56.3	76.0	94.5	70.1	80.8	102.3	70.1	84.1	100.3	141.7	123.6
07	THERMAL & MOISTURE PROTECTION	99.2	75.1	89.4	99.2	78.6	90.8	99.8	65.2	85.7	105.6	84.4	97.0	106.1	82.2	96.4	100.4	151.9	123.5
08	OPENINGS	94.4	81.4	91.3	94.4	74.5	89.6	94.6	62.8	87.1	86.8	74.7	84.0	87.5	74.7	84.5	99.2	144.4	109.9
0920	Plaster & Gypsum Board	103.6	88.8	93.6	100.3	73.0	81.8	110.0	55.3	72.8	102.9	70.0	80.5	106.2	70.0	81.6	104.2	142.4	130.2
0950, 0980	Ceilings & Acoustic Treatment	100.1	88.8	92.6	101.0	73.0	82.4	100.1	55.3	70.3	107.8	70.0	82.6	107.8	70.0	82.6	101.8	142.4	128.8
0960	Flooring	97.6	59.5	86.2	95.8	92.5	94.8	101.0	65.9	90.5	104.0	80.5	97.0	107.5	80.5	99.4	98.6	183.1	123.9
0970, 0990	Wall Finishes & Painting/Coating	94.4	77.5	84.3	94.4	77.5	84.3	94.4	77.5	84.3	111.6	77.5	91.1	111.6	77.5	91.1	98.6	150.8	130.0
09	FINISHES	98.0	78.4	87.1	97.5	78.9	87.1	100.7	57.3	76.5	96.8	72.3	83.2	98.5	71.7	83.6	102.1	151.0	129.3
COVERS	DIVS. 10 - 14, 25, 28, 41, 43, 44, 46	100.0	61.4	92.3	100.0	89.5	97.9	100.0	39.7	87.9	100.0	83.1	96.6	100.0	81.4	96.3	100.0	118.6	103.7
21, 22, 23	FIRE SUPPRESSION, PLUMBING & HVAC	95.1	79.3	88.8	100.0	81.4	92.5	95.1	61.2	81.5	95.3	86.3	91.7	95.3	83.7	90.6	100.1	127.0	110.9
26, 27, 3370	ELECTRICAL, COMMUNICATIONS & UTIL.	97.9	91.7	94.6	96.4	81.6	88.7	94.8	68.7	81.2	99.4	99.8	99.6	96.6	99.8	98.3	100.4	138.6	120.6
MF2010	WEIGHTED AVERAGE	95.1	81.9	89.4	96.9	82.8	90.7	97.8	63.3	82.8	96.9	83.8	91.2	97.5	82.2	90.9	101.8	138.1	117.6

609

City Cost Indexes

MASSACHUSETTS

DIVISION		BROCKTON 023			BUZZARDS BAY 025			FALL RIVER 027			FITCHBURG 014			FRAMINGHAM 017			GREENFIELD 013		
		MAT.	INST.	TOTAL	MAT.	INST.	TOTAL	MAT.	INST.	TOTAL	MAT.	INST.	TOTAL	MAT.	INST.	TOTAL	MAT.	INST.	TOTAL
015433	CONTRACTOR EQUIPMENT		102.4	102.4		102.4	102.4		103.3	103.3		100.6	100.6		101.8	101.8		100.6	100.6
0241, 31 - 34	SITE & INFRASTRUCTURE, DEMOLITION	95.7	104.8	102.1	85.8	104.8	99.1	94.8	104.9	101.9	85.8	104.7	99.1	82.5	104.5	97.9	89.6	103.2	99.2
0310	Concrete Forming & Accessories	102.6	131.5	127.5	100.2	131.4	127.1	102.6	131.7	127.7	94.8	126.0	121.7	102.0	131.2	127.2	93.2	112.1	109.5
0320	Concrete Reinforcing	106.8	157.0	131.9	85.6	165.2	125.4	106.8	165.2	136.0	85.4	156.0	120.6	85.3	156.7	121.0	88.6	126.0	107.3
0330	Cast-in-Place Concrete	98.6	152.4	120.9	82.0	152.4	111.2	95.4	152.9	119.3	85.7	148.3	111.7	85.8	140.8	108.6	88.2	129.2	105.2
03	CONCRETE	104.8	142.5	123.4	88.7	144.0	115.9	103.3	144.3	123.5	86.8	138.3	112.2	89.7	138.4	113.7	90.6	119.9	105.1
04	MASONRY	102.4	159.6	137.8	95.0	159.6	134.9	103.3	159.5	138.1	93.0	155.2	131.5	98.6	155.3	133.7	96.9	133.2	119.4
05	METALS	99.4	129.7	108.8	94.4	132.9	106.3	99.4	133.3	109.9	95.6	126.2	105.0	95.6	129.0	105.9	97.8	109.2	101.3
06	WOOD, PLASTICS & COMPOSITES	102.0	129.4	117.5	98.9	129.4	116.1	102.0	129.6	117.6	95.7	122.7	111.0	102.2	129.1	117.4	93.5	109.8	102.7
07	THERMAL & MOISTURE PROTECTION	101.7	148.3	120.6	101.0	146.7	119.5	101.6	145.9	119.6	100.8	140.1	116.8	100.9	145.2	118.9	100.9	121.2	109.1
08	OPENINGS	98.1	137.7	107.4	94.0	135.4	103.8	98.1	135.5	106.9	104.5	133.9	111.5	94.7	137.6	104.8	104.5	114.7	106.9
0920	Plaster & Gypsum Board	95.4	129.7	118.7	91.1	129.7	117.3	95.4	129.7	118.7	97.3	122.8	114.6	99.9	129.7	120.1	98.0	109.4	105.8
0950, 0980	Ceilings & Acoustic Treatment	100.9	129.7	120.1	90.0	129.7	116.4	100.9	129.7	120.1	90.6	122.8	112.0	90.6	129.7	116.6	99.5	109.4	106.1
0960	Flooring	100.4	183.1	125.1	98.9	183.1	124.1	99.8	183.1	124.7	98.2	183.1	123.6	99.6	183.1	124.5	97.5	149.9	113.2
0970, 0990	Wall Finishes & Painting/Coating	96.0	135.1	119.5	96.0	135.1	119.5	96.0	135.1	119.5	95.1	135.1	119.2	96.1	135.1	119.6	95.1	108.5	103.1
09	FINISHES	100.0	141.4	123.0	96.0	141.4	121.3	99.9	141.6	123.1	96.1	137.4	119.1	96.7	141.2	121.5	98.3	118.7	109.7
COVERS	DIVS. 10 - 14, 25, 28, 41, 43, 44, 46	100.0	116.4	103.3	100.0	116.4	103.3	100.0	117.0	103.4	100.0	102.5	100.5	100.0	115.9	103.2	100.0	105.1	101.0
21, 22, 23	FIRE SUPPRESSION, PLUMBING & HVAC	100.1	111.2	104.5	95.2	111.2	101.6	100.1	111.3	104.6	95.8	113.6	103.0	95.8	120.4	105.7	95.8	100.8	97.8
26, 27, 3370	ELECTRICAL, COMMUNICATIONS & UTIL.	100.7	100.7	100.7	97.6	100.7	99.2	100.6	100.7	100.6	100.4	104.2	102.4	96.7	129.2	113.7	100.4	92.8	96.4
MF2010	WEIGHTED AVERAGE	100.4	126.3	111.7	94.9	126.7	108.8	100.2	126.8	111.8	96.3	124.6	108.6	95.5	131.0	111.0	97.5	110.2	103.0

MASSACHUSETTS

DIVISION		HYANNIS 026			LAWRENCE 019			LOWELL 018			NEW BEDFORD 027			PITTSFIELD 012			SPRINGFIELD 010 - 011		
		MAT.	INST.	TOTAL	MAT.	INST.	TOTAL	MAT.	INST.	TOTAL	MAT.	INST.	TOTAL	MAT.	INST.	TOTAL	MAT.	INST.	TOTAL
015433	CONTRACTOR EQUIPMENT		102.4	102.4		102.4	102.4		100.6	100.6		103.3	103.3		100.6	100.6		100.6	100.6
0241, 31 - 34	SITE & INFRASTRUCTURE, DEMOLITION	91.9	104.8	101.0	94.5	104.8	101.7	93.7	104.7	101.5	93.4	104.9	101.5	94.7	103.1	100.6	94.2	103.2	100.5
0310	Concrete Forming & Accessories	94.6	131.4	126.3	103.2	131.7	127.8	100.0	131.5	127.2	102.6	131.7	127.7	99.9	111.5	109.9	100.2	111.7	110.1
0320	Concrete Reinforcing	85.6	165.2	125.4	105.6	146.6	126.0	106.5	146.3	126.4	106.8	165.2	136.0	87.9	126.0	106.9	106.5	126.0	116.2
0330	Cast-in-Place Concrete	90.0	152.4	115.9	99.2	149.8	120.2	90.2	149.7	114.9	84.1	152.9	112.6	98.4	128.3	110.8	93.9	128.5	108.3
03	CONCRETE	95.0	144.0	119.2	104.9	139.8	122.1	96.3	139.5	117.6	97.9	144.3	120.8	97.2	119.3	108.1	98.0	119.4	108.6
04	MASONRY	101.4	159.6	137.7	103.3	159.6	138.1	92.4	155.2	131.3	101.3	159.5	137.3	93.0	131.5	116.8	92.7	131.9	117.0
05	METALS	95.9	132.9	107.3	98.2	125.8	106.7	98.1	123.0	105.8	99.4	133.3	109.9	98.0	109.2	101.4	100.6	109.2	103.2
06	WOOD, PLASTICS & COMPOSITES	91.9	129.4	113.1	102.7	129.4	117.7	101.9	129.4	117.4	102.0	129.6	117.6	101.9	109.8	106.4	101.9	109.8	106.4
07	THERMAL & MOISTURE PROTECTION	101.2	146.7	119.7	101.3	147.9	120.2	101.2	145.8	119.3	101.6	145.9	119.6	101.2	120.4	109.0	101.2	120.6	109.1
08	OPENINGS	94.6	135.4	104.3	98.4	134.9	107.1	105.7	134.9	112.6	98.1	135.5	106.9	105.7	114.7	107.8	105.7	114.7	107.8
0920	Plaster & Gypsum Board	86.7	129.7	115.9	102.8	129.7	121.1	102.8	129.7	121.1	95.4	129.7	118.7	102.8	109.4	107.3	102.8	109.4	107.3
0950, 0980	Ceilings & Acoustic Treatment	92.9	129.7	117.4	101.5	129.7	120.2	101.5	129.7	120.2	100.9	129.7	120.1	100.4	109.4	106.8	101.5	109.4	106.8
0960	Flooring	96.8	183.1	122.6	100.2	183.1	125.0	100.2	183.1	125.0	99.8	183.1	124.7	100.4	149.9	115.2	99.9	149.9	114.8
0970, 0990	Wall Finishes & Painting/Coating	96.0	135.1	119.5	95.2	135.1	119.2	95.1	135.1	119.2	96.0	135.1	119.5	95.1	108.5	103.1	96.5	108.5	103.7
09	FINISHES	95.8	141.4	121.2	100.4	141.4	123.2	100.3	141.4	123.2	99.7	141.6	123.0	100.4	118.3	110.4	100.3	118.4	110.4
COVERS	DIVS. 10 - 14, 25, 28, 41, 43, 44, 46	100.0	116.4	103.3	100.0	116.5	103.3	100.0	116.5	103.3	100.0	117.0	103.4	100.0	104.5	100.9	100.0	104.6	100.9
21, 22, 23	FIRE SUPPRESSION, PLUMBING & HVAC	100.1	111.2	104.5	100.0	120.4	108.2	100.0	118.0	107.3	100.1	111.3	104.6	100.0	100.0	100.0	100.0	100.2	100.1
26, 27, 3370	ELECTRICAL, COMMUNICATIONS & UTIL.	98.1	100.7	99.4	99.4	129.2	115.0	99.9	129.2	115.2	101.5	100.7	101.1	99.9	92.8	96.2	99.9	92.8	96.2
MF2010	WEIGHTED AVERAGE	97.6	126.7	110.3	100.1	131.3	113.7	99.4	130.1	112.8	99.6	126.8	111.4	99.6	109.6	103.9	100.0	109.8	104.3

MASSACHUSETTS / MICHIGAN

DIVISION		WORCESTER 015 - 016			ANN ARBOR 481			BATTLE CREEK 490			BAY CITY 487			DEARBORN 481			DETROIT 482		
		MAT.	INST.	TOTAL	MAT.	INST.	TOTAL	MAT.	INST.	TOTAL	MAT.	INST.	TOTAL	MAT.	INST.	TOTAL	MAT.	INST.	TOTAL
015433	CONTRACTOR EQUIPMENT		100.6	100.6		110.7	110.7		102.6	102.6		110.7	110.7		110.7	110.7		98.7	98.7
0241, 31 - 34	SITE & INFRASTRUCTURE, DEMOLITION	94.1	104.7	101.5	78.7	97.2	91.6	89.4	87.9	88.4	70.2	96.4	88.6	78.4	97.4	91.7	92.0	99.0	96.9
0310	Concrete Forming & Accessories	100.5	126.1	122.6	97.8	111.4	109.6	97.5	87.1	88.5	97.9	89.0	90.2	97.7	116.2	113.6	99.5	116.2	113.9
0320	Concrete Reinforcing	106.5	155.2	130.8	95.7	123.7	109.7	92.2	96.0	94.1	95.7	122.8	109.3	95.7	123.8	109.8	97.2	123.8	110.5
0330	Cast-in-Place Concrete	93.3	148.3	116.2	86.2	109.5	95.9	95.1	100.0	97.1	82.5	92.0	86.5	84.3	112.5	96.0	91.4	112.5	100.1
03	CONCRETE	97.8	138.2	117.7	92.7	113.2	102.8	96.0	92.6	94.3	91.0	97.0	94.0	91.8	116.4	103.9	95.5	115.2	105.2
04	MASONRY	92.2	155.2	131.2	97.2	108.2	104.0	99.1	86.5	91.3	96.8	87.3	90.9	97.1	113.6	107.3	95.5	113.6	106.7
05	METALS	100.6	126.0	108.4	93.7	118.2	101.2	95.8	88.6	93.6	94.3	115.0	100.7	93.7	118.5	101.4	96.3	101.2	97.8
06	WOOD, PLASTICS & COMPOSITES	102.4	122.7	113.9	94.9	112.4	104.8	100.3	86.3	92.4	94.9	88.7	91.4	94.9	116.9	107.3	97.2	116.9	108.3
07	THERMAL & MOISTURE PROTECTION	101.2	140.1	117.0	100.0	108.5	103.4	96.7	85.5	92.1	97.8	94.2	96.3	98.5	113.5	104.6	97.3	113.5	103.9
08	OPENINGS	105.7	133.7	112.3	97.7	110.0	100.6	94.2	82.7	91.5	97.7	94.8	97.0	97.7	112.4	101.1	100.0	112.1	102.8
0920	Plaster & Gypsum Board	102.8	122.8	116.4	106.2	111.8	110.0	93.0	82.0	85.6	106.2	87.3	93.4	106.2	116.4	113.2	109.0	116.4	114.0
0950, 0980	Ceilings & Acoustic Treatment	101.5	122.8	115.7	93.1	111.8	105.5	96.5	82.0	86.9	94.1	87.3	89.6	93.1	116.4	108.6	96.8	116.4	109.8
0960	Flooring	100.2	183.1	125.0	96.1	117.6	102.5	103.7	93.6	100.7	96.1	81.9	91.8	95.7	116.0	101.8	97.1	116.0	102.7
0970, 0990	Wall Finishes & Painting/Coating	95.1	135.1	119.2	86.5	100.8	95.1	97.2	81.8	87.9	86.5	83.7	84.9	86.5	99.0	94.0	88.5	99.0	94.8
09	FINISHES	100.3	137.4	121.0	94.3	111.7	104.0	97.7	87.5	92.0	94.1	86.5	89.9	94.2	114.8	105.7	96.8	114.8	106.8
COVERS	DIVS. 10 - 14, 25, 28, 41, 43, 44, 46	100.0	108.6	101.7	100.0	106.1	101.2	100.0	98.2	99.6	100.0	92.6	98.5	100.0	107.5	101.5	100.0	107.7	101.5
21, 22, 23	FIRE SUPPRESSION, PLUMBING & HVAC	100.0	113.1	105.3	100.0	99.9	100.0	100.0	87.7	95.1	100.0	84.6	93.8	100.0	108.8	103.5	100.0	103.6	101.4
26, 27, 3370	ELECTRICAL, COMMUNICATIONS & UTIL.	99.9	104.2	102.1	96.0	110.0	103.4	90.3	80.4	85.1	95.0	91.1	92.9	96.0	105.2	100.8	97.2	105.1	101.3
MF2010	WEIGHTED AVERAGE	100.0	124.6	110.7	96.4	107.9	101.4	96.8	87.4	92.7	95.8	92.5	94.4	96.2	110.9	102.6	97.9	109.3	102.8

City Cost Indexes

MICHIGAN

DIVISION		FLINT 484-485 MAT.	INST.	TOTAL	GAYLORD 497 MAT.	INST.	TOTAL	GRAND RAPIDS 493,495 MAT.	INST.	TOTAL	IRON MOUNTAIN 498-499 MAT.	INST.	TOTAL	JACKSON 492 MAT.	INST.	TOTAL	KALAMAZOO 491 MAT.	INST.	TOTAL
015433	CONTRACTOR EQUIPMENT		110.7	110.7		105.1	105.1		102.6	102.6		94.2	94.2		105.1	105.1		102.6	102.6
0241, 31-34	SITE & INFRASTRUCTURE, DEMOLITION	68.2	96.5	88.1	83.4	85.7	85.0	89.5	88.0	88.4	92.1	94.1	93.5	105.8	88.0	93.3	89.7	87.9	88.4
0310	Concrete Forming & Accessories	100.6	91.4	92.7	95.4	77.7	80.1	97.0	84.7	86.4	87.5	86.1	86.3	92.4	88.3	88.9	97.5	86.8	88.2
0320	Concrete Reinforcing	95.7	123.2	109.5	85.9	109.3	97.6	98.6	95.9	97.2	85.7	95.4	90.6	83.5	123.1	103.3	92.2	93.6	92.9
0330	Cast-in-Place Concrete	86.8	94.7	90.1	94.9	87.8	92.0	100.1	97.6	99.1	112.3	87.2	101.9	94.8	93.8	94.4	97.0	99.8	98.2
03	CONCRETE	93.2	99.1	96.1	92.4	88.2	90.3	99.3	90.7	95.1	101.8	88.4	95.2	87.2	97.3	92.2	99.4	92.0	95.8
04	MASONRY	97.3	96.8	97.0	109.6	80.5	91.6	96.1	86.6	90.2	95.6	87.6	90.7	89.7	92.8	91.7	97.8	86.5	90.8
05	METALS	93.7	115.7	100.5	97.2	111.7	101.6	93.0	88.2	91.5	96.6	91.1	94.9	97.3	113.6	102.4	95.8	87.5	93.3
06	WOOD, PLASTICS & COMPOSITES	98.2	89.9	93.5	92.8	76.9	83.8	101.5	83.1	91.1	87.5	86.1	86.7	91.5	86.6	88.7	100.3	86.3	92.4
07	THERMAL & MOISTURE PROTECTION	97.8	97.9	97.9	95.1	76.2	87.4	99.5	77.4	90.5	98.6	83.1	92.3	94.5	94.5	94.5	96.7	85.5	92.1
08	OPENINGS	97.7	95.5	97.1	94.8	76.0	90.3	100.6	85.6	97.0	101.3	77.3	95.5	93.8	93.9	93.8	94.2	82.1	91.3
0920	Plaster & Gypsum Board	107.7	88.6	94.7	92.8	74.9	80.6	94.9	78.8	83.9	53.7	86.1	75.7	91.0	84.9	86.9	93.0	82.0	85.6
0950, 0980	Ceilings & Acoustic Treatment	93.1	88.6	90.1	95.7	74.9	81.9	105.1	78.8	87.6	94.6	86.1	88.9	95.7	84.9	88.5	96.5	82.0	86.9
0960	Flooring	96.1	94.9	95.7	95.7	93.9	95.1	103.6	84.3	97.8	119.6	95.6	112.4	94.4	84.3	91.4	103.7	77.1	95.8
0970, 0990	Wall Finishes & Painting/Coating	86.5	87.3	87.0	93.1	83.7	87.5	98.9	80.0	87.6	111.4	71.8	87.6	93.1	98.8	96.5	97.2	81.8	87.9
09	FINISHES	93.6	90.7	92.0	97.5	79.8	87.7	101.7	83.8	91.7	99.7	86.4	92.3	98.6	88.0	92.7	97.7	84.2	90.2
COVERS	DIVS. 10-14, 25, 28, 41, 43, 44, 46	100.0	93.9	98.8	100.0	90.7	98.1	100.0	98.2	99.6	100.0	89.5	97.9	100.0	93.7	98.7	100.0	98.2	99.6
21, 22, 23	FIRE SUPPRESSION, PLUMBING & HVAC	100.0	90.6	96.2	95.5	80.1	89.3	100.0	83.8	93.5	95.4	86.0	91.6	95.5	88.8	92.8	100.0	81.6	92.6
26, 27, 3370	ELECTRICAL, COMMUNICATIONS & UTIL.	96.0	98.3	97.2	88.4	78.9	83.4	95.2	72.3	83.3	94.5	86.3	90.2	92.2	110.0	101.5	90.2	83.7	86.8
MF2010	WEIGHTED AVERAGE	96.1	96.8	96.4	95.5	84.4	90.7	98.1	84.5	92.2	97.5	87.4	93.1	95.0	95.9	95.4	97.1	85.9	92.2

MICHIGAN / MINNESOTA

DIVISION		LANSING 488-489 MAT.	INST.	TOTAL	MUSKEGON 494 MAT.	INST.	TOTAL	ROYAL OAK 480,483 MAT.	INST.	TOTAL	SAGINAW 486 MAT.	INST.	TOTAL	TRAVERSE CITY 496 MAT.	INST.	TOTAL	BEMIDJI 566 MAT.	INST.	TOTAL
015433	CONTRACTOR EQUIPMENT		110.7	110.7		102.6	102.6		96.2	96.2		110.7	110.7		94.2	94.2		99.0	99.0
0241, 31-34	SITE & INFRASTRUCTURE, DEMOLITION	90.5	96.6	94.8	87.5	87.9	87.8	83.0	96.8	92.7	71.2	96.4	88.9	78.4	93.6	89.0	94.0	98.3	97.0
0310	Concrete Forming & Accessories	98.0	89.2	90.4	97.9	85.6	87.3	93.6	112.7	110.1	97.8	88.8	90.0	87.5	75.7	77.3	88.7	91.4	91.0
0320	Concrete Reinforcing	99.0	123.0	111.0	92.8	96.1	94.5	86.8	123.8	105.2	95.7	122.8	109.3	87.0	92.9	90.0	97.0	108.0	102.5
0330	Cast-in-Place Concrete	100.5	92.9	97.3	94.8	95.8	95.2	75.5	108.3	89.1	85.2	91.9	88.0	87.6	80.0	84.5	104.1	104.2	104.1
03	CONCRETE	100.0	97.4	98.7	94.3	90.6	92.5	79.9	112.2	95.8	92.2	96.9	94.5	83.9	80.8	82.4	96.9	100.0	98.4
04	MASONRY	95.3	93.3	94.1	96.3	89.4	92.0	92.0	112.1	104.5	98.7	87.3	91.6	93.7	82.3	86.7	97.2	104.8	101.9
05	METALS	91.4	115.1	98.7	93.6	88.5	92.0	96.8	99.4	97.6	93.8	114.8	100.3	96.5	89.7	94.4	90.2	121.0	99.7
06	WOOD, PLASTICS & COMPOSITES	96.4	88.0	91.6	96.5	85.2	90.1	90.4	113.1	103.2	91.2	88.7	89.8	87.5	75.2	80.6	71.5	87.5	80.6
07	THERMAL & MOISTURE PROTECTION	98.6	90.6	95.3	95.8	80.4	89.5	96.5	112.3	102.9	98.6	94.2	96.9	97.8	72.8	87.6	105.6	97.1	102.1
08	OPENINGS	98.7	94.4	97.7	93.4	86.5	91.8	97.6	110.4	100.6	95.6	94.8	95.4	101.3	71.4	94.3	97.5	109.2	100.2
0920	Plaster & Gypsum Board	92.2	86.6	88.4	74.3	80.9	78.8	103.2	112.5	109.5	106.2	87.3	93.4	53.7	74.9	68.1	100.6	87.4	91.6
0950, 0980	Ceilings & Acoustic Treatment	96.8	86.6	90.0	97.4	80.9	86.4	92.3	112.5	105.8	93.1	87.3	89.3	94.6	74.9	81.5	137.1	87.4	104.0
0960	Flooring	102.8	84.3	97.2	102.5	84.3	97.1	93.7	116.0	100.4	96.1	81.9	91.8	119.6	93.9	111.9	107.8	123.0	112.3
0970, 0990	Wall Finishes & Painting/Coating	96.3	81.5	87.4	95.6	80.0	86.3	87.9	98.8	94.5	86.5	83.7	84.9	111.4	45.6	71.9	95.3	97.9	96.9
09	FINISHES	98.2	87.0	91.9	94.3	84.0	88.6	93.3	112.2	103.8	93.9	86.5	89.8	98.6	74.3	85.1	111.3	97.1	103.4
COVERS	DIVS. 10-14, 25, 28, 41, 43, 44, 46	100.0	93.3	98.7	100.0	98.0	99.6	100.0	104.6	100.9	100.0	92.6	98.5	100.0	86.8	97.4	100.0	99.3	99.9
21, 22, 23	FIRE SUPPRESSION, PLUMBING & HVAC	99.9	88.9	95.5	99.9	83.1	93.1	95.4	108.0	100.5	100.0	84.2	93.7	95.4	79.4	89.0	95.5	84.8	91.2
26, 27, 3370	ELECTRICAL, COMMUNICATIONS & UTIL.	99.2	91.4	95.1	90.7	77.2	83.6	98.3	104.5	101.5	93.5	91.1	92.2	89.9	78.9	84.2	104.4	105.2	104.8
MF2010	WEIGHTED AVERAGE	97.7	94.1	96.1	95.6	85.5	91.2	94.2	107.5	100.0	95.7	92.4	94.2	94.6	81.0	88.6	97.7	99.5	98.5

MINNESOTA

DIVISION		BRAINERD 564 MAT.	INST.	TOTAL	DETROIT LAKES 565 MAT.	INST.	TOTAL	DULUTH 556-558 MAT.	INST.	TOTAL	MANKATO 560 MAT.	INST.	TOTAL	MINNEAPOLIS 553-555 MAT.	INST.	TOTAL	ROCHESTER 559 MAT.	INST.	TOTAL
015433	CONTRACTOR EQUIPMENT		101.6	101.6		99.0	99.0		100.4	100.4		101.6	101.6		103.8	103.8		100.4	100.4
0241, 31-34	SITE & INFRASTRUCTURE, DEMOLITION	94.2	103.0	100.4	92.2	98.6	96.7	96.3	100.4	99.2	91.1	103.7	99.9	97.9	105.6	103.3	97.0	100.0	99.1
0310	Concrete Forming & Accessories	90.6	93.6	93.2	85.7	93.5	92.5	100.4	112.2	110.6	98.4	101.1	100.7	101.2	127.4	123.8	102.8	106.6	106.1
0320	Concrete Reinforcing	95.9	108.2	102.0	97.0	108.0	102.5	102.6	108.5	105.5	95.8	118.2	107.0	97.2	118.5	107.8	94.8	118.2	106.5
0330	Cast-in-Place Concrete	113.1	108.2	111.1	101.1	107.5	103.7	107.4	107.0	107.2	104.2	109.0	106.2	110.5	117.3	113.3	109.0	100.9	105.6
03	CONCRETE	100.8	102.4	101.6	94.5	102.1	98.2	101.3	110.3	105.7	96.5	107.9	102.1	103.4	122.6	112.9	100.7	107.7	104.1
04	MASONRY	119.6	113.5	115.8	119.9	110.8	114.2	104.6	117.0	112.3	108.6	110.5	109.8	114.7	125.3	121.3	107.0	112.5	110.4
05	METALS	91.3	121.0	100.5	90.2	120.9	99.7	93.4	123.3	102.6	91.2	126.7	102.1	98.6	130.0	108.3	95.2	128.4	105.4
06	WOOD, PLASTICS & COMPOSITES	89.5	87.3	88.3	68.5	87.5	79.3	103.6	111.9	108.3	99.0	98.7	98.8	109.0	126.0	118.6	109.0	105.2	106.9
07	THERMAL & MOISTURE PROTECTION	104.7	109.9	106.8	105.4	110.1	107.3	100.1	116.9	106.9	105.0	100.6	103.2	102.1	128.6	112.9	103.5	102.2	102.9
08	OPENINGS	84.8	109.0	90.5	97.4	109.2	100.2	98.3	118.2	103.0	89.1	117.7	95.9	94.9	133.8	104.1	93.5	122.5	100.4
0920	Plaster & Gypsum Board	84.6	87.4	86.5	99.5	87.4	91.3	91.1	112.6	105.8	89.5	99.1	96.0	99.3	127.0	118.2	99.5	105.8	103.8
0950, 0980	Ceilings & Acoustic Treatment	66.7	87.4	80.5	137.1	87.4	104.0	98.7	112.6	108.0	66.7	99.1	88.3	98.5	127.0	117.5	95.0	105.8	102.2
0960	Flooring	106.7	123.0	111.6	106.6	123.0	111.5	107.1	124.3	112.3	108.3	123.0	112.7	98.3	123.0	105.7	103.2	123.0	109.1
0970, 0990	Wall Finishes & Painting/Coating	89.6	97.9	94.6	95.3	97.9	96.9	91.0	111.3	103.2	101.0	105.3	103.6	96.5	128.1	115.5	91.3	105.3	99.7
09	FINISHES	93.1	98.5	96.1	110.7	98.6	104.0	101.4	114.7	108.8	94.8	105.3	100.7	100.5	127.2	115.4	99.4	109.5	105.0
COVERS	DIVS. 10-14, 25, 28, 41, 43, 44, 46	100.0	100.9	100.2	100.0	101.2	100.2	100.0	100.1	100.0	100.0	101.1	100.2	100.0	109.1	101.8	100.0	102.6	100.5
21, 22, 23	FIRE SUPPRESSION, PLUMBING & HVAC	94.8	87.8	92.0	95.5	87.6	92.3	99.9	97.2	98.8	94.8	88.5	92.3	99.9	116.9	106.7	100.0	96.0	98.4
26, 27, 3370	ELECTRICAL, COMMUNICATIONS & UTIL.	101.9	104.9	103.4	104.1	69.1	85.8	98.4	104.9	101.8	108.8	89.2	98.5	101.8	109.0	105.6	98.2	89.2	93.5
MF2010	WEIGHTED AVERAGE	96.3	102.3	98.9	98.3	96.7	97.6	99.0	108.6	103.2	96.6	102.4	99.1	100.6	119.9	109.0	98.8	104.9	101.5

611

City Cost Indexes

		MINNESOTA																	MISSISSIPPI		
	DIVISION	SAINT PAUL			ST. CLOUD			THIEF RIVER FALLS			WILLMAR			WINDOM			BILOXI		
		550 - 551			563			567			562			561			395		
		MAT.	INST.	TOTAL	MAT.	INST.	TOTAL	MAT.	INST.	TOTAL	MAT.	INST.	TOTAL	MAT.	INST.	TOTAL	MAT.	INST.	TOTAL
015433	CONTRACTOR EQUIPMENT		100.4	100.4		101.6	101.6		99.0	99.0		101.6	101.6		101.6	101.6		99.3	99.3
0241, 31 - 34	SITE & INFRASTRUCTURE, DEMOLITION	99.1	101.1	100.5	89.7	105.1	100.5	92.9	98.2	96.7	89.1	102.8	98.7	83.3	101.7	96.3	98.8	89.6	92.3
0310	Concrete Forming & Accessories	102.0	124.9	121.8	87.7	120.4	116.0	89.4	90.8	90.6	87.5	93.1	92.3	92.0	86.8	87.5	95.6	54.1	59.8
0320	Concrete Reinforcing	98.5	118.5	108.5	95.9	118.4	107.1	97.4	107.9	102.7	95.6	118.0	106.8	95.6	117.1	106.4	95.8	58.4	77.1
0330	Cast-in-Place Concrete	120.2	116.7	118.7	99.9	115.8	106.5	103.2	89.5	97.5	101.4	86.9	95.4	87.7	67.8	79.4	108.7	54.8	86.3
03	CONCRETE	110.6	121.3	115.9	92.3	118.9	105.4	95.6	94.6	95.1	92.2	96.7	94.4	83.2	87.2	85.2	100.0	56.8	78.7
04	MASONRY	115.6	125.3	121.6	105.3	122.1	115.7	97.2	104.8	101.9	109.7	113.5	112.0	119.4	91.2	102.0	90.3	45.1	62.4
05	METALS	94.1	129.8	105.1	92.0	128.0	103.1	90.3	120.3	99.6	91.1	125.8	101.8	91.0	123.5	101.0	91.9	82.4	89.0
06	WOOD, PLASTICS & COMPOSITES	102.8	122.7	114.0	86.8	117.6	104.2	72.5	87.5	81.0	86.6	87.5	87.1	90.8	84.8	87.4	102.3	57.4	77.0
07	THERMAL & MOISTURE PROTECTION	101.8	128.1	112.5	104.8	120.1	111.0	106.5	100.0	103.9	104.6	110.8	107.1	104.6	91.0	99.1	97.0	55.4	80.1
08	OPENINGS	97.1	132.0	105.3	89.1	129.2	98.6	97.5	109.2	100.2	90.1	93.5	88.3	90.2	92.0	90.6	97.2	58.6	88.0
0920	Plaster & Gypsum Board	87.3	123.8	112.1	84.6	118.6	107.7	100.2	87.4	91.5	84.6	87.6	86.7	84.6	84.8	84.7	104.4	56.5	71.8
0950, 0980	Ceilings & Acoustic Treatment	98.5	123.8	115.4	66.7	118.6	101.2	137.1	87.4	104.0	66.7	87.6	80.6	66.7	84.8	78.7	95.9	56.5	69.7
0960	Flooring	99.5	123.0	106.5	103.3	123.0	109.2	107.5	123.0	112.1	105.1	123.0	110.4	107.3	123.0	112.0	114.4	56.1	97.0
0970, 0990	Wall Finishes & Painting/Coating	99.9	127.1	116.3	101.0	128.1	117.3	95.3	97.9	96.9	95.3	97.9	96.9	95.3	105.3	101.3	94.7	44.4	64.5
09	FINISHES	99.5	125.1	113.8	92.6	121.9	108.9	111.1	97.1	103.3	92.7	96.2	94.6	92.8	93.2	93.0	100.8	53.6	74.5
COVERS	DIVS. 10 - 14, 25, 28, 41, 43, 44, 46	100.0	108.5	101.7	100.0	107.2	101.4	100.0	99.2	99.8	100.0	100.8	100.2	100.0	95.6	99.1	100.0	56.9	91.4
21, 22, 23	FIRE SUPPRESSION, PLUMBING & HVAC	99.9	114.9	105.9	99.7	116.3	106.3	95.5	84.5	91.1	94.8	108.4	100.2	94.8	85.3	91.0	100.0	50.1	80.0
26, 27, 3370	ELECTRICAL, COMMUNICATIONS & UTIL.	95.2	111.4	103.7	101.9	111.4	106.9	101.4	69.1	84.6	101.9	87.2	94.2	108.8	89.1	98.5	98.5	57.6	77.1
MF2010	WEIGHTED AVERAGE	100.2	118.8	108.3	96.3	117.8	105.7	97.3	93.8	95.7	94.9	103.0	98.4	95.3	93.3	94.4	97.7	59.1	80.9

		MISSISSIPPI																
	DIVISION	CLARKSDALE			COLUMBUS			GREENVILLE			GREENWOOD			JACKSON			LAUREL		
		386			397			387			389			390 - 392			394		
		MAT.	INST.	TOTAL	MAT.	INST.	TOTAL	MAT.	INST.	TOTAL	MAT.	INST.	TOTAL	MAT.	INST.	TOTAL	MAT.	INST.	TOTAL
015433	CONTRACTOR EQUIPMENT		99.3	99.3		99.3	99.3		99.3	99.3		99.3	99.3		99.3	99.3		99.3	99.3
0241, 31 - 34	SITE & INFRASTRUCTURE, DEMOLITION	95.6	89.4	91.2	97.1	89.6	91.9	101.3	89.6	93.1	98.3	89.1	91.8	95.9	89.6	91.5	102.9	89.5	93.5
0310	Concrete Forming & Accessories	85.3	50.4	55.2	84.4	52.4	56.8	82.0	65.2	67.5	94.3	50.4	56.4	94.6	58.9	63.8	84.4	54.2	58.3
0320	Concrete Reinforcing	100.0	42.9	71.5	102.7	43.9	73.3	100.6	62.2	81.4	100.0	54.2	77.1	101.2	57.5	79.4	103.3	40.3	71.9
0330	Cast-in-Place Concrete	103.0	54.4	82.8	110.8	59.5	89.5	106.0	56.5	85.5	110.7	54.1	87.2	98.9	57.8	81.8	108.4	56.1	86.7
03	CONCRETE	95.9	52.1	74.3	101.6	54.9	78.6	101.1	63.0	82.3	102.1	53.9	78.3	96.1	59.7	78.2	104.0	53.9	79.3
04	MASONRY	93.5	45.8	64.0	114.7	52.5	76.2	138.0	50.3	83.8	94.1	45.6	64.1	95.7	50.3	67.6	110.9	49.0	72.6
05	METALS	90.5	74.9	85.7	88.9	75.8	84.9	91.5	84.1	89.3	90.5	76.4	86.2	98.5	82.3	93.5	89.0	74.8	84.6
06	WOOD, PLASTICS & COMPOSITES	85.6	52.4	66.9	87.2	53.4	68.1	82.2	69.7	75.1	98.7	52.4	72.6	99.6	61.0	77.8	88.3	55.9	70.0
07	THERMAL & MOISTURE PROTECTION	95.9	48.0	76.5	97.0	52.3	78.8	96.2	61.0	81.9	96.3	47.9	76.6	97.1	60.3	82.1	97.1	54.0	79.6
08	OPENINGS	96.8	50.6	85.8	96.5	50.7	85.7	96.8	62.9	88.8	96.8	51.9	86.2	101.0	56.8	90.5	93.5	51.5	83.6
0920	Plaster & Gypsum Board	87.7	51.4	63.0	92.7	52.4	65.3	87.3	69.2	75.0	98.9	51.4	66.6	97.4	60.2	72.1	92.7	54.9	67.0
0950, 0980	Ceilings & Acoustic Treatment	87.8	51.4	63.6	90.3	52.4	65.1	89.7	69.2	76.1	87.8	51.4	63.6	98.4	60.2	73.0	90.3	54.9	66.8
0960	Flooring	111.6	53.8	94.3	108.2	60.3	93.9	109.9	53.8	93.1	117.1	53.8	98.2	111.9	56.1	95.2	106.8	53.8	91.0
0970, 0990	Wall Finishes & Painting/Coating	97.4	47.1	67.2	94.7	54.1	70.3	97.4	67.3	79.3	97.4	47.1	67.2	91.7	67.3	77.0	94.7	62.2	75.2
09	FINISHES	96.5	51.0	71.1	96.3	54.4	73.0	97.0	64.3	78.8	99.9	51.0	72.7	100.6	59.6	77.8	96.4	55.5	73.6
COVERS	DIVS. 10 - 14, 25, 28, 41, 43, 44, 46	100.0	57.2	91.4	100.0	58.4	91.7	100.0	60.0	92.0	100.0	57.2	91.4	100.0	59.1	91.8	100.0	38.5	87.7
21, 22, 23	FIRE SUPPRESSION, PLUMBING & HVAC	97.9	42.2	75.6	97.3	38.3	73.7	100.0	48.6	79.4	97.9	40.2	74.7	100.0	58.3	83.3	97.4	46.8	77.1
26, 27, 3370	ELECTRICAL, COMMUNICATIONS & UTIL.	95.6	47.5	70.5	95.7	50.3	72.0	95.6	64.2	79.2	95.6	44.5	68.9	100.1	64.2	81.3	97.3	58.3	77.0
MF2010	WEIGHTED AVERAGE	95.8	53.8	77.5	97.1	55.1	78.8	99.4	63.1	83.6	97.0	53.3	78.0	99.1	63.5	83.6	97.2	57.1	79.7

		MISSISSIPPI									MISSOURI								
	DIVISION	MCCOMB			MERIDIAN			TUPELO			BOWLING GREEN			CAPE GIRARDEAU			CHILLICOTHE		
		396			393			388			633			637			646		
		MAT.	INST.	TOTAL	MAT.	INST.	TOTAL	MAT.	INST.	TOTAL	MAT.	INST.	TOTAL	MAT.	INST.	TOTAL	MAT.	INST.	TOTAL
015433	CONTRACTOR EQUIPMENT		99.3	99.3		99.3	99.3		99.3	99.3		106.6	106.6		106.6	106.6		102.1	102.1
0241, 31 - 34	SITE & INFRASTRUCTURE, DEMOLITION	90.3	89.2	89.6	94.3	89.9	91.2	93.1	89.3	90.4	90.1	95.1	93.6	91.7	95.0	94.1	105.0	94.6	97.7
0310	Concrete Forming & Accessories	84.4	52.2	56.6	81.7	64.8	67.1	82.5	52.5	56.6	94.2	88.8	89.5	87.3	75.1	76.8	89.9	95.2	94.5
0320	Concrete Reinforcing	103.9	42.3	73.1	102.7	57.5	80.1	97.9	54.2	76.1	103.2	100.8	102.0	104.5	88.8	96.7	97.4	110.0	103.7
0330	Cast-in-Place Concrete	96.2	53.9	78.7	102.9	60.1	85.2	103.0	55.8	83.4	92.6	85.4	89.6	91.6	87.3	89.8	97.9	87.8	93.7
03	CONCRETE	91.6	52.5	72.3	95.8	63.2	79.7	95.4	55.4	75.7	95.9	91.0	93.5	95.0	83.4	89.3	102.1	95.9	99.1
04	MASONRY	116.2	45.1	72.2	90.0	54.5	68.0	127.9	48.7	78.9	122.1	97.9	107.1	118.5	79.0	94.1	103.6	100.6	101.7
05	METALS	89.1	73.1	84.2	90.0	82.5	87.7	90.4	76.6	86.2	96.1	113.8	101.6	97.2	107.3	100.3	96.9	110.5	101.1
06	WOOD, PLASTICS & COMPOSITES	87.2	55.3	69.2	84.6	66.7	74.5	82.8	53.4	66.2	95.3	88.0	91.2	88.1	71.5	78.7	96.0	95.4	95.7
07	THERMAL & MOISTURE PROTECTION	96.6	47.2	76.6	96.8	62.6	82.9	95.9	53.8	78.8	103.2	98.8	101.4	103.1	83.5	95.1	97.8	97.0	97.4
08	OPENINGS	96.6	52.2	86.1	96.5	65.5	89.2	96.8	51.8	86.1	97.0	98.5	97.3	97.0	75.5	91.9	96.0	98.1	96.5
0920	Plaster & Gypsum Board	92.7	54.3	66.6	92.7	66.1	74.6	87.3	52.4	63.6	97.9	87.6	90.9	96.4	70.6	78.8	97.4	95.1	95.8
0950, 0980	Ceilings & Acoustic Treatment	90.3	54.3	66.4	92.3	66.1	74.8	87.8	52.4	64.2	97.1	87.6	90.8	97.1	70.6	79.5	92.2	95.1	94.1
0960	Flooring	108.2	53.8	91.9	106.7	56.1	91.6	110.1	53.8	93.3	98.4	98.5	98.4	95.2	85.6	92.3	96.0	104.4	98.5
0970, 0990	Wall Finishes & Painting/Coating	94.7	43.2	63.8	94.7	77.8	84.5	97.4	60.0	74.9	102.3	102.8	102.6	102.3	74.8	85.8	96.2	91.3	93.3
09	FINISHES	95.7	52.1	71.4	96.0	65.2	78.8	96.0	53.8	72.5	100.8	91.2	95.4	99.6	75.4	86.1	99.9	96.4	97.9
COVERS	DIVS. 10 - 14, 25, 28, 41, 43, 44, 46	100.0	60.7	92.1	100.0	61.1	92.2	100.0	58.4	91.7	100.0	90.2	98.0	100.0	92.3	98.5	100.0	90.9	98.2
21, 22, 23	FIRE SUPPRESSION, PLUMBING & HVAC	97.3	35.3	72.5	100.0	60.4	84.1	98.1	44.2	76.5	95.0	97.7	96.1	99.9	96.6	98.6	95.2	96.3	96.8
26, 27, 3370	ELECTRICAL, COMMUNICATIONS & UTIL.	94.1	50.6	71.4	97.3	58.2	76.9	95.4	50.4	71.9	98.0	80.8	89.1	98.0	103.9	101.1	93.7	80.5	86.8
MF2010	WEIGHTED AVERAGE	95.7	52.9	77.1	96.2	65.3	82.7	97.3	56.1	79.4	98.1	94.7	96.6	99.1	90.6	95.4	97.7	96.2	97.1

City Cost Indexes

		\multicolumn{18}{c	}{MISSOURI}
\multicolumn{2}{	c	}{DIVISION}	\multicolumn{3}{c	}{COLUMBIA 652}	\multicolumn{3}{c	}{FLAT RIVER 636}	\multicolumn{3}{c	}{HANNIBAL 634}	\multicolumn{3}{c	}{HARRISONVILLE 647}	\multicolumn{3}{c	}{JEFFERSON CITY 650 - 651}	\multicolumn{3}{c	}{JOPLIN 648}
		MAT.	INST.	TOTAL	MAT.	INST.	TOTAL	MAT.	INST.	TOTAL	MAT.	INST.	TOTAL	MAT.	INST.	TOTAL	MAT.	INST.	TOTAL
---	---	---	---	---	---	---	---	---	---	---	---	---	---	---	---	---	---	---	---
015433	CONTRACTOR EQUIPMENT		107.7	107.7		106.6	106.6		106.6	106.6		102.1	102.1		107.7	107.7		105.8	105.8
0241, 31 - 34	SITE & INFRASTRUCTURE, DEMOLITION	100.0	96.6	97.6	92.8	95.2	94.5	88.0	95.1	93.0	96.1	95.6	95.8	99.2	96.6	97.4	105.6	99.4	101.3
0310	Concrete Forming & Accessories	85.5	80.0	80.7	100.1	86.2	88.1	92.4	79.3	81.1	87.3	100.9	99.1	95.6	80.0	82.1	101.6	73.7	77.5
0320	Concrete Reinforcing	98.1	116.3	107.2	104.5	108.0	106.3	102.7	100.8	101.7	97.0	110.1	103.5	100.4	113.1	106.7	100.6	101.7	101.1
0330	Cast-in-Place Concrete	89.8	87.1	88.7	95.6	94.2	95.0	87.6	85.2	86.6	100.3	106.1	102.7	95.3	87.1	91.9	106.0	79.2	94.9
03	CONCRETE	86.8	90.6	88.7	98.8	94.2	96.6	92.2	86.8	89.5	98.5	104.8	101.6	91.9	90.0	91.0	101.7	81.7	91.8
04	MASONRY	133.2	88.7	105.7	119.0	80.5	95.2	113.5	97.9	103.8	98.3	105.3	102.6	97.0	88.7	91.9	96.6	86.8	90.5
05	METALS	97.1	119.7	104.1	96.0	115.8	102.1	96.1	113.5	101.5	97.3	111.5	101.7	96.6	118.4	103.3	100.0	102.2	100.7
06	WOOD, PLASTICS & COMPOSITES	95.4	76.3	84.6	103.2	84.5	92.7	93.3	75.5	83.2	92.5	99.8	96.6	102.0	76.3	87.5	107.5	71.8	87.4
07	THERMAL & MOISTURE PROTECTION	95.5	88.1	92.5	103.4	93.8	99.5	103.0	96.9	100.5	97.1	105.8	100.6	100.8	88.1	95.7	97.1	79.6	90.0
08	OPENINGS	94.7	93.3	94.4	96.9	98.8	97.4	97.0	84.3	94.0	96.1	102.9	97.7	94.2	92.5	93.8	97.2	78.2	92.7
0920	Plaster & Gypsum Board	85.3	75.5	78.7	103.8	84.1	90.4	96.7	74.7	81.8	92.5	99.6	97.3	89.2	75.5	79.9	103.5	70.7	81.2
0950, 0980	Ceilings & Acoustic Treatment	97.9	75.5	83.0	97.1	84.1	88.4	97.1	74.7	82.2	92.2	99.6	97.1	102.1	75.5	84.4	93.0	70.7	78.2
0960	Flooring	109.7	97.5	106.1	101.3	85.6	96.6	97.7	98.5	98.0	91.7	104.4	95.5	116.0	97.5	110.5	121.0	76.3	107.6
0970, 0990	Wall Finishes & Painting/Coating	102.4	78.8	88.2	102.3	80.0	88.9	102.3	102.8	102.6	101.1	105.3	103.6	99.6	78.8	87.1	95.8	76.5	84.2
09	FINISHES	97.5	81.6	88.6	102.5	84.1	92.3	100.2	83.8	91.1	97.5	101.7	99.9	101.7	81.6	90.5	106.1	74.5	88.4
COVERS	DIVS. 10 - 14, 25, 28, 41, 43, 44, 46	100.0	95.6	99.1	100.0	94.3	98.8	100.0	88.8	97.8	100.0	93.1	98.6	100.0	95.6	99.1	100.0	89.6	97.9
21, 22, 23	FIRE SUPPRESSION, PLUMBING & HVAC	99.9	97.8	99.0	95.0	97.6	96.1	95.0	97.6	96.1	95.1	100.5	97.2	99.9	98.7	99.4	100.1	72.7	89.1
26, 27, 3370	ELECTRICAL, COMMUNICATIONS & UTIL.	98.4	86.6	92.2	102.4	104.0	103.2	96.8	82.9	89.6	100.4	99.7	100.1	104.3	86.6	95.1	91.5	69.7	80.1
MF2010	WEIGHTED AVERAGE	98.6	93.5	96.4	99.0	95.8	97.6	97.1	92.6	95.1	97.3	102.2	99.5	98.4	93.4	96.2	99.5	81.0	91.5

		\multicolumn{18}{c	}{MISSOURI}
\multicolumn{2}{	c	}{DIVISION}	\multicolumn{3}{c	}{KANSAS CITY 640 - 641}	\multicolumn{3}{c	}{KIRKSVILLE 635}	\multicolumn{3}{c	}{POPLAR BLUFF 639}	\multicolumn{3}{c	}{ROLLA 654 - 655}	\multicolumn{3}{c	}{SEDALIA 653}	\multicolumn{3}{c	}{SIKESTON 638}
		MAT.	INST.	TOTAL	MAT.	INST.	TOTAL	MAT.	INST.	TOTAL	MAT.	INST.	TOTAL	MAT.	INST.	TOTAL	MAT.	INST.	TOTAL
---	---	---	---	---	---	---	---	---	---	---	---	---	---	---	---	---	---	---	---
015433	CONTRACTOR EQUIPMENT		103.4	103.4		99.0	99.0		101.3	101.3		107.7	107.7		99.7	99.7		101.3	101.3
0241, 31 - 34	SITE & INFRASTRUCTURE, DEMOLITION	98.2	98.0	98.0	90.9	91.2	91.1	78.4	95.0	90.1	99.0	96.6	97.3	94.9	92.5	93.2	81.7	95.1	91.1
0310	Concrete Forming & Accessories	100.3	107.8	106.8	85.2	79.3	80.1	85.3	75.0	76.4	92.7	94.0	93.8	90.1	79.3	80.8	86.4	75.0	76.5
0320	Concrete Reinforcing	95.5	117.1	106.3	103.4	101.9	102.6	106.5	97.6	102.0	98.6	97.7	98.2	97.1	109.4	103.2	105.9	101.1	103.6
0330	Cast-in-Place Concrete	98.6	107.8	102.4	95.5	85.1	91.2	73.5	87.8	79.5	91.8	96.4	93.7	95.9	84.5	91.2	78.6	87.8	82.4
03	CONCRETE	97.8	109.7	103.7	110.2	86.4	98.5	86.0	84.6	85.3	88.6	96.5	92.5	101.1	87.6	94.4	89.6	85.3	87.5
04	MASONRY	100.4	109.1	105.8	125.6	88.7	102.8	117.1	79.1	93.6	109.4	89.4	97.0	115.2	86.6	97.5	116.8	79.1	93.5
05	METALS	107.7	114.9	109.9	95.7	105.2	98.6	96.3	102.9	98.3	96.5	111.9	101.3	95.1	109.3	99.5	96.6	104.4	99.0
06	WOOD, PLASTICS & COMPOSITES	106.7	107.5	107.1	81.1	76.3	78.4	80.3	71.5	75.3	103.0	94.9	98.5	95.4	76.3	84.6	81.9	71.5	76.0
07	THERMAL & MOISTURE PROTECTION	97.2	108.6	101.8	109.8	89.1	101.4	108.3	85.3	99.0	95.6	95.5	95.6	101.8	93.7	98.5	108.5	83.9	98.5
08	OPENINGS	103.5	109.8	105.0	102.0	85.6	98.1	103.0	77.9	97.1	94.7	90.4	93.7	99.0	87.3	96.2	103.0	78.9	97.3
0920	Plaster & Gypsum Board	100.2	107.5	105.1	91.8	75.5	80.7	92.3	70.6	77.5	87.2	94.7	92.3	80.4	75.5	77.1	94.1	70.6	78.1
0950, 0980	Ceilings & Acoustic Treatment	99.3	107.5	104.7	95.3	75.5	82.1	97.1	70.6	79.5	97.9	94.7	95.8	97.9	75.5	83.0	97.1	70.6	79.5
0960	Flooring	96.6	109.3	100.4	76.5	97.5	82.8	90.2	83.5	88.2	113.0	97.5	108.3	86.4	97.5	89.7	90.6	83.5	88.5
0970, 0990	Wall Finishes & Painting/Coating	101.1	112.6	108.0	97.6	80.4	87.2	97.1	74.8	83.7	102.4	102.8	102.7	102.4	102.4	102.4	97.1	74.8	83.7
09	FINISHES	101.4	108.5	105.4	100.0	81.7	89.8	99.6	75.1	85.9	98.8	95.5	96.9	95.7	83.7	89.0	100.3	75.1	86.2
COVERS	DIVS. 10 - 14, 25, 28, 41, 43, 44, 46	100.0	101.3	100.3	100.0	89.0	97.8	100.0	92.4	98.5	100.0	91.4	98.3	100.0	91.5	98.3	100.0	92.4	98.5
21, 22, 23	FIRE SUPPRESSION, PLUMBING & HVAC	100.0	106.7	102.7	95.1	97.1	95.9	95.1	96.0	95.4	95.0	98.0	96.2	94.9	93.8	94.5	95.1	96.1	95.5
26, 27, 3370	ELECTRICAL, COMMUNICATIONS & UTIL.	102.0	100.1	101.0	97.0	82.9	89.6	97.2	103.9	100.7	96.6	86.6	91.4	98.2	99.7	99.0	96.3	103.9	100.3
MF2010	WEIGHTED AVERAGE	101.6	106.7	103.8	100.3	90.0	95.8	97.0	90.3	94.1	96.3	95.6	96.0	98.1	92.6	95.7	97.5	90.6	94.5

		\multicolumn{9}{c	}{MISSOURI}	\multicolumn{9}{c	}{MONTANA}
\multicolumn{2}{	c	}{DIVISION}	\multicolumn{3}{c	}{SPRINGFIELD 656 - 658}	\multicolumn{3}{c	}{ST. JOSEPH 644 - 645}	\multicolumn{3}{c	}{ST. LOUIS 630 - 631}	\multicolumn{3}{c	}{BILLINGS 590 - 591}	\multicolumn{3}{c	}{BUTTE 597}	\multicolumn{3}{c	}{GREAT FALLS 594}
		MAT.	INST.	TOTAL	MAT.	INST.	TOTAL	MAT.	INST.	TOTAL	MAT.	INST.	TOTAL	MAT.	INST.	TOTAL	MAT.	INST.	TOTAL
---	---	---	---	---	---	---	---	---	---	---	---	---	---	---	---	---	---	---	---
015433	CONTRACTOR EQUIPMENT		102.3	102.3		102.1	102.1		107.6	107.6		99.3	99.3		99.0	99.0		99.0	99.0
0241, 31 - 34	SITE & INFRASTRUCTURE, DEMOLITION	97.2	94.4	95.2	99.8	93.7	95.5	91.9	98.4	96.5	95.9	97.1	96.8	103.3	96.9	98.8	107.1	96.9	100.0
0310	Concrete Forming & Accessories	98.2	74.7	77.9	100.2	91.1	92.3	100.2	103.6	103.1	101.3	71.6	75.7	88.4	73.3	75.4	101.1	71.6	75.6
0320	Concrete Reinforcing	94.3	116.0	105.1	94.4	109.8	102.1	95.5	114.0	104.7	93.9	82.8	88.3	101.9	82.9	92.4	94.0	82.8	88.4
0330	Cast-in-Place Concrete	97.5	78.8	89.7	98.5	103.6	100.6	91.6	105.0	97.2	120.2	68.9	98.9	132.3	70.3	106.6	139.8	65.0	108.7
03	CONCRETE	97.7	84.6	91.2	97.5	99.5	98.5	94.5	106.8	100.5	106.9	73.6	90.5	110.6	74.9	93.0	115.9	72.2	94.3
04	MASONRY	89.5	87.2	88.1	100.1	96.7	98.0	99.7	112.5	107.6	117.9	75.0	91.4	113.3	77.9	91.4	117.5	76.7	92.3
05	METALS	101.2	107.5	103.1	103.7	110.7	105.9	101.9	120.5	107.7	107.5	91.0	102.4	100.6	91.1	97.7	103.8	91.1	99.9
06	WOOD, PLASTICS & COMPOSITES	103.4	72.8	86.1	107.5	89.7	97.4	103.1	100.8	101.8	99.5	70.0	82.8	86.6	72.5	78.6	100.7	70.0	83.4
07	THERMAL & MOISTURE PROTECTION	99.9	81.1	92.2	97.7	97.4	97.5	103.2	107.6	105.0	108.1	71.2	93.1	108.0	72.5	93.6	108.5	69.8	92.8
08	OPENINGS	101.6	81.2	96.8	101.7	98.3	100.9	97.2	110.5	100.4	93.9	68.7	88.0	92.3	71.3	87.3	95.0	68.1	88.6
0920	Plaster & Gypsum Board	88.3	71.9	77.1	104.8	89.1	94.1	104.7	100.8	102.1	99.9	69.3	79.1	100.3	71.9	81.0	109.7	69.3	82.2
0950, 0980	Ceilings & Acoustic Treatment	97.9	71.9	80.6	98.4	89.1	92.2	102.4	100.8	101.3	99.8	69.3	79.5	107.0	71.9	83.6	108.8	69.3	82.5
0960	Flooring	110.5	76.3	100.3	99.8	104.4	101.2	101.1	99.4	100.6	106.0	71.5	95.7	103.9	75.5	95.4	111.0	75.5	100.4
0970, 0990	Wall Finishes & Painting/Coating	96.2	76.5	84.4	96.2	88.6	91.6	102.3	106.5	104.8	99.0	88.5	92.7	97.2	52.3	70.2	97.2	88.5	92.0
09	FINISHES	101.0	75.1	86.6	102.5	92.7	97.0	103.7	102.5	103.0	102.7	73.0	86.2	103.4	71.2	85.5	107.4	73.7	88.6
COVERS	DIVS. 10 - 14, 25, 28, 41, 43, 44, 46	100.0	93.0	98.6	100.0	96.9	99.4	100.0	102.3	100.5	100.0	96.1	99.2	100.0	96.4	99.3	100.0	96.1	99.2
21, 22, 23	FIRE SUPPRESSION, PLUMBING & HVAC	99.9	73.0	89.1	100.1	89.5	95.8	100.0	107.6	103.0	100.1	74.9	90.0	100.2	70.7	88.4	100.2	72.2	89.0
26, 27, 3370	ELECTRICAL, COMMUNICATIONS & UTIL.	102.9	70.3	85.9	100.5	80.5	90.1	100.5	107.1	103.9	96.8	73.9	84.8	104.4	72.2	87.6	96.9	72.2	84.0
MF2010	WEIGHTED AVERAGE	99.9	82.1	92.1	100.7	93.9	97.8	99.6	107.6	103.1	102.2	77.9	91.6	102.0	77.2	91.2	103.4	77.1	92.0

613

City Cost Indexes

MONTANA

DIVISION		HAVRE 595			HELENA 596			KALISPELL 599			MILES CITY 593			MISSOULA 598			WOLF POINT 592		
		MAT.	INST.	TOTAL	MAT.	INST.	TOTAL	MAT.	INST.	TOTAL	MAT.	INST.	TOTAL	MAT.	INST.	TOTAL	MAT.	INST.	TOTAL
015433	CONTRACTOR EQUIPMENT		99.0	99.0		99.0	99.0		99.0	99.0		99.0	99.0		99.0	99.0		99.0	99.0
0241, 31 - 34	SITE & INFRASTRUCTURE, DEMOLITION	111.0	96.7	100.9	100.1	96.6	97.7	93.4	96.8	95.8	99.9	96.3	97.4	85.8	96.6	93.4	118.0	96.4	102.9
0310	Concrete Forming & Accessories	81.6	69.1	70.8	104.4	66.7	71.9	91.6	71.5	74.3	99.6	67.6	72.0	91.6	71.5	74.2	92.2	67.8	71.2
0320	Concrete Reinforcing	102.7	82.8	92.8	106.8	82.8	94.8	104.6	83.4	94.0	102.3	82.8	92.6	103.6	83.4	93.5	103.7	82.8	93.3
0330	Cast-in-Place Concrete	142.5	61.8	109.0	106.9	64.3	89.2	114.9	64.0	93.8	125.8	61.6	99.1	97.5	66.2	84.5	140.8	62.0	108.1
03	CONCRETE	119.1	70.0	94.9	104.5	69.8	87.3	100.0	72.0	86.2	107.4	69.3	88.6	88.0	72.7	80.4	123.0	69.6	96.6
04	MASONRY	114.2	71.1	87.5	110.7	77.1	89.9	112.2	78.3	91.2	119.4	66.5	86.7	135.6	77.9	99.9	120.5	67.2	87.5
05	METALS	96.8	90.9	95.0	102.7	90.0	98.7	96.6	91.4	95.0	95.9	91.1	94.4	97.2	91.3	95.4	96.0	91.1	94.5
06	WOOD, PLASTICS & COMPOSITES	78.4	70.0	73.7	103.7	63.8	81.2	89.9	70.0	78.7	97.8	70.0	82.1	89.9	70.0	78.7	89.5	70.0	78.5
07	THERMAL & MOISTURE PROTECTION	108.3	69.4	92.5	105.4	72.2	91.9	107.7	74.3	94.1	108.0	66.8	91.3	107.4	80.6	96.5	108.7	67.1	91.8
08	OPENINGS	92.4	68.1	86.6	96.2	66.3	89.2	92.3	68.3	86.6	91.8	68.0	86.2	92.3	68.0	86.6	91.9	67.8	86.2
0920	Plaster & Gypsum Board	96.2	69.3	77.9	97.9	62.9	74.1	100.3	69.3	79.2	109.0	69.3	82.0	100.3	69.3	79.2	104.2	69.3	80.4
0950, 0980	Ceilings & Acoustic Treatment	107.0	69.3	81.9	109.7	62.9	78.5	107.0	69.3	81.9	106.1	69.3	81.6	107.0	69.3	81.9	106.1	69.3	81.6
0960	Flooring	101.3	75.5	93.6	106.1	60.2	92.4	105.7	75.5	96.7	110.7	71.5	99.0	105.7	75.5	96.7	107.1	71.5	96.5
0970, 0990	Wall Finishes & Painting/Coating	97.2	49.7	68.7	93.5	48.2	66.3	97.2	48.2	67.8	97.2	48.2	67.8	97.2	63.3	76.8	97.2	49.7	68.7
09	FINISHES	102.8	67.9	83.4	104.7	63.0	81.5	103.3	69.4	84.4	106.2	65.9	83.8	102.7	71.0	85.1	105.9	66.2	83.8
COVERS	DIVS. 10 - 14, 25, 28, 41, 43, 44, 46	100.0	72.6	94.5	100.0	95.6	99.1	100.0	96.2	99.2	100.0	92.6	98.5	100.0	96.1	99.2	100.0	92.9	98.6
21, 22, 23	FIRE SUPPRESSION, PLUMBING & HVAC	95.3	69.3	84.9	100.1	70.9	88.4	95.3	70.3	85.3	95.3	69.7	85.0	100.0	69.9	88.0	95.3	70.0	85.2
26, 27, 3370	ELECTRICAL, COMMUNICATIONS & UTIL.	96.9	72.7	84.3	103.8	72.7	87.6	101.2	71.0	85.4	96.9	77.4	86.7	102.2	69.3	85.0	96.9	77.4	86.7
MF2010	WEIGHTED AVERAGE	100.6	74.2	89.1	102.0	75.1	90.2	98.4	76.3	88.8	99.5	74.7	88.7	99.4	76.4	89.4	101.7	74.9	90.0

NEBRASKA

DIVISION		ALLIANCE 693			COLUMBUS 686			GRAND ISLAND 688			HASTINGS 689			LINCOLN 683 - 685			MCCOOK 690		
		MAT.	INST.	TOTAL	MAT.	INST.	TOTAL	MAT.	INST.	TOTAL	MAT.	INST.	TOTAL	MAT.	INST.	TOTAL	MAT.	INST.	TOTAL
015433	CONTRACTOR EQUIPMENT		97.5	97.5		101.4	101.4		101.4	101.4		101.4	101.4		101.4	101.4		101.4	101.4
0241, 31 - 34	SITE & INFRASTRUCTURE, DEMOLITION	98.8	98.4	98.6	99.1	92.0	94.1	103.8	92.8	96.1	102.6	92.0	95.2	93.6	92.8	93.0	102.1	91.9	95.0
0310	Concrete Forming & Accessories	88.3	57.7	61.9	95.9	68.1	71.9	95.4	74.0	76.9	98.5	73.6	77.0	97.5	68.9	72.8	93.9	57.3	62.3
0320	Concrete Reinforcing	110.1	85.0	97.6	101.7	76.9	89.3	101.1	76.8	89.0	101.1	77.5	89.3	99.2	76.3	87.7	103.1	76.5	89.8
0330	Cast-in-Place Concrete	109.4	61.3	89.4	114.9	61.3	92.6	121.6	68.6	99.6	121.6	64.6	97.9	99.8	71.0	87.9	117.6	60.6	94.0
03	CONCRETE	120.2	64.5	92.7	106.9	68.2	87.8	111.7	73.4	92.8	111.9	71.9	92.1	98.0	71.9	85.1	109.5	63.1	86.6
04	MASONRY	109.7	74.9	88.2	112.6	79.0	91.8	105.7	77.4	88.2	114.2	89.5	98.9	95.9	68.4	78.9	105.5	74.7	86.5
05	METALS	101.2	79.1	94.4	91.9	85.2	89.8	93.6	87.0	91.5	94.4	86.5	91.9	95.6	86.2	92.7	96.0	84.2	92.4
06	WOOD, PLASTICS & COMPOSITES	88.3	53.6	68.6	99.7	67.2	81.4	99.0	73.9	84.8	102.7	73.9	86.4	98.3	67.2	80.7	97.1	53.4	72.5
07	THERMAL & MOISTURE PROTECTION	106.4	69.5	91.4	104.4	70.7	90.7	104.5	74.2	92.2	104.5	85.4	96.8	100.4	70.1	88.1	100.8	67.4	87.2
08	OPENINGS	92.8	59.4	84.9	94.1	66.7	87.6	94.2	70.9	88.7	94.2	70.3	88.5	103.2	63.5	93.8	93.2	57.5	84.7
0920	Plaster & Gypsum Board	81.0	51.9	61.2	88.7	66.1	73.3	88.0	73.0	77.8	89.9	73.0	78.4	90.1	66.1	73.8	92.5	51.9	64.9
0950, 0980	Ceilings & Acoustic Treatment	98.5	51.9	67.5	93.3	66.1	75.2	93.3	73.0	79.8	93.3	73.0	79.8	99.3	66.1	77.2	94.3	51.9	66.1
0960	Flooring	104.5	83.6	98.2	102.5	102.2	102.4	102.3	109.2	104.4	103.4	102.2	103.1	107.6	85.1	100.9	101.6	83.6	96.2
0970, 0990	Wall Finishes & Painting/Coating	174.1	56.0	103.1	91.9	65.1	75.8	91.9	69.1	78.2	91.9	65.1	75.8	100.0	79.6	87.7	95.3	48.8	67.3
09	FINISHES	102.8	60.8	79.4	98.5	72.9	84.3	98.7	78.8	87.6	99.2	76.9	86.8	102.3	72.7	85.8	99.6	59.9	77.5
COVERS	DIVS. 10 - 14, 25, 28, 41, 43, 44, 46	100.0	67.0	93.4	100.0	88.3	97.7	100.0	89.3	97.9	100.0	89.1	97.8	100.0	88.6	97.7	100.0	66.3	93.2
21, 22, 23	FIRE SUPPRESSION, PLUMBING & HVAC	95.3	72.0	85.9	95.1	77.1	87.9	100.0	78.5	91.4	95.1	77.2	87.9	99.9	78.5	91.3	95.0	77.1	87.8
26, 27, 3370	ELECTRICAL, COMMUNICATIONS & UTIL.	91.7	71.0	80.9	92.0	82.4	87.0	90.7	65.9	77.8	90.1	85.6	87.7	103.2	65.9	83.7	93.9	71.0	81.9
MF2010	WEIGHTED AVERAGE	100.4	71.6	87.9	97.5	77.9	88.9	99.1	77.8	89.9	98.5	81.2	90.9	99.5	75.3	89.0	98.1	72.1	86.8

NEBRASKA / NEVADA

DIVISION		NORFOLK 687			NORTH PLATTE 691			OMAHA 680 - 681			VALENTINE 692			CARSON CITY 897			ELKO 898		
		MAT.	INST.	TOTAL	MAT.	INST.	TOTAL	MAT.	INST.	TOTAL	MAT.	INST.	TOTAL	MAT.	INST.	TOTAL	MAT.	INST.	TOTAL
015433	CONTRACTOR EQUIPMENT		92.7	92.7		101.4	101.4		92.7	92.7		95.6	95.6		97.9	97.9		97.9	97.9
0241, 31 - 34	SITE & INFRASTRUCTURE, DEMOLITION	82.3	91.1	88.5	103.4	92.0	95.4	86.4	91.8	90.2	86.4	95.9	93.1	78.7	99.6	93.4	61.9	98.4	87.5
0310	Concrete Forming & Accessories	82.4	72.7	74.0	96.3	73.1	76.3	93.5	72.9	75.7	84.5	57.1	60.8	105.3	94.6	96.1	111.2	83.2	87.0
0320	Concrete Reinforcing	101.8	66.6	84.3	102.5	77.4	90.0	104.1	76.9	90.5	103.1	66.3	84.8	107.5	118.0	112.7	106.9	113.9	110.4
0330	Cast-in-Place Concrete	116.1	63.8	94.4	117.6	65.9	96.2	92.6	77.1	86.2	104.3	58.6	85.3	104.1	86.2	96.7	98.7	80.9	91.3
03	CONCRETE	105.7	68.7	87.5	109.6	72.1	91.1	94.6	75.4	85.1	107.0	60.0	83.8	104.1	96.0	100.1	98.6	88.3	93.5
04	MASONRY	118.7	78.7	93.9	93.2	86.1	88.8	98.9	79.9	87.1	105.3	74.8	86.4	116.6	89.3	99.7	114.9	73.1	89.0
05	METALS	95.2	74.1	88.7	95.1	86.1	92.3	95.4	79.5	90.5	107.3	74.1	97.1	90.1	104.0	94.4	92.8	99.5	94.8
06	WOOD, PLASTICS & COMPOSITES	82.8	73.4	77.5	99.1	73.9	84.9	91.2	72.1	80.5	82.8	52.9	65.9	91.6	96.6	94.4	99.4	83.7	90.6
07	THERMAL & MOISTURE PROTECTION	104.4	74.9	92.4	100.8	84.1	94.0	100.0	79.8	91.8	101.6	69.1	88.4	102.7	91.2	98.0	99.0	76.6	89.9
08	OPENINGS	95.9	67.7	89.2	92.5	70.3	87.3	101.4	71.9	94.4	95.0	56.5	85.9	99.6	109.4	101.9	100.6	87.3	97.4
0920	Plaster & Gypsum Board	88.7	73.0	78.0	92.5	73.0	79.3	99.5	71.7	80.6	94.7	51.9	65.6	91.3	96.4	94.8	93.3	83.1	86.4
0950, 0980	Ceilings & Acoustic Treatment	106.9	73.0	84.4	94.3	73.0	80.1	112.8	71.7	85.5	113.1	51.9	72.4	101.8	96.4	98.2	102.3	83.1	89.5
0960	Flooring	131.7	102.2	122.9	102.4	102.2	102.3	129.9	83.4	116.0	133.3	81.3	117.7	100.8	109.9	103.6	103.7	52.1	88.3
0970, 0990	Wall Finishes & Painting/Coating	164.9	65.1	104.9	95.3	65.1	77.1	148.8	68.2	100.4	171.4	67.5	108.9	102.7	82.5	90.6	104.2	97.9	100.4
09	FINISHES	120.0	76.9	96.0	99.9	76.9	87.1	119.6	74.0	94.2	123.7	61.1	88.8	100.4	98.0	98.1	99.4	78.9	88.0
COVERS	DIVS. 10 - 14, 25, 28, 41, 43, 44, 46	100.0	87.9	97.6	100.0	68.6	93.7	100.0	88.1	97.6	100.0	64.3	92.8	100.0	92.0	98.4	100.0	64.5	92.9
21, 22, 23	FIRE SUPPRESSION, PLUMBING & HVAC	94.9	76.9	87.7	99.9	77.1	90.8	99.8	78.0	91.1	94.7	76.9	87.6	100.0	81.4	92.6	97.7	81.2	91.1
26, 27, 3370	ELECTRICAL, COMMUNICATIONS & UTIL.	90.9	82.4	86.5	92.0	78.9	85.2	97.9	82.4	89.8	89.0	88.6	88.8	100.4	97.4	98.8	98.1	97.3	97.7
MF2010	WEIGHTED AVERAGE	99.2	77.5	89.8	98.5	79.6	90.2	99.6	79.3	90.7	100.6	73.6	88.9	99.1	93.8	96.8	97.6	86.1	92.6

614

City Cost Indexes

		NEVADA									NEW HAMPSHIRE								
DIVISION		ELY			LAS VEGAS			RENO			CHARLESTON			CLAREMONT			CONCORD		
		893			889 - 891			894 - 895			036			037			032 - 033		
		MAT.	INST.	TOTAL	MAT.	INST.	TOTAL	MAT.	INST.	TOTAL	MAT.	INST.	TOTAL	MAT.	INST.	TOTAL	MAT.	INST.	TOTAL
015433	CONTRACTOR EQUIPMENT		97.9	97.9		97.9	97.9		97.9	97.9		100.6	100.6		100.6	100.6		100.6	100.6
0241, 31 - 34	SITE & INFRASTRUCTURE, DEMOLITION	67.0	99.9	90.1	69.7	101.6	92.1	67.0	99.6	89.9	87.4	99.2	95.7	81.2	99.2	93.9	94.5	101.4	99.3
0310	Concrete Forming & Accessories	104.2	105.9	105.6	105.4	111.0	110.3	100.5	94.5	95.3	88.1	82.9	83.6	93.8	82.9	84.4	95.9	94.3	94.5
0320	Concrete Reinforcing	105.7	115.2	110.4	97.5	121.8	109.6	99.9	121.4	110.6	84.5	95.2	89.8	84.5	95.2	89.8	92.9	95.8	94.4
0330	Cast-in-Place Concrete	105.9	106.7	106.2	102.6	107.5	104.6	112.0	86.2	101.3	97.4	72.1	86.9	89.5	72.1	82.3	111.5	92.2	103.5
03	CONCRETE	106.6	107.6	107.1	101.8	111.4	106.6	106.3	96.6	101.5	99.5	81.6	90.6	91.5	81.6	86.6	106.0	93.7	100.0
04	MASONRY	119.0	104.2	109.9	108.1	102.8	104.8	114.2	89.3	98.8	87.8	82.3	84.4	87.3	82.3	84.2	97.6	102.1	100.4
05	METALS	92.7	104.8	96.4	99.8	109.3	102.7	94.2	105.3	97.6	95.2	90.6	93.8	95.2	90.6	93.8	101.2	94.0	99.0
06	WOOD, PLASTICS & COMPOSITES	90.3	106.3	99.3	89.2	110.4	101.2	85.2	96.6	91.6	86.9	92.3	90.0	93.3	92.3	92.7	94.8	93.3	93.9
07	THERMAL & MOISTURE PROTECTION	99.6	95.9	98.1	112.5	102.0	108.2	99.1	91.2	95.9	102.8	72.0	90.3	102.7	72.0	90.2	104.9	91.4	99.4
08	OPENINGS	100.5	99.6	100.3	99.6	117.9	104.0	98.6	104.2	99.9	102.6	78.0	96.8	103.8	78.0	97.7	104.1	88.4	100.4
0920	Plaster & Gypsum Board	90.0	106.4	101.1	87.8	110.7	103.4	82.0	96.4	91.8	94.7	91.4	92.5	95.8	91.4	92.8	98.4	92.4	94.4
0950, 0980	Ceilings & Acoustic Treatment	102.3	106.4	105.0	111.1	110.7	110.8	107.6	96.4	100.1	90.6	91.4	91.2	90.6	91.4	91.2	95.9	92.4	93.6
0960	Flooring	101.5	58.2	88.6	93.3	109.9	98.3	98.8	109.9	102.1	96.3	32.7	77.3	98.2	32.7	78.6	97.1	112.2	101.6
0970, 0990	Wall Finishes & Painting/Coating	104.2	123.6	115.9	107.2	123.6	117.0	104.2	82.5	91.2	95.4	46.7	66.2	95.4	46.9	66.3	95.1	97.5	96.5
09	FINISHES	98.9	99.4	99.2	98.4	112.2	106.1	98.0	96.2	97.0	94.9	71.3	81.8	95.1	71.3	81.9	95.9	98.0	97.0
COVERS	DIVS. 10 - 14, 25, 28, 41, 43, 44, 46	100.0	65.0	93.0	100.0	106.3	101.3	100.0	92.0	98.4	100.0	90.0	98.0	100.0	90.0	98.0	100.0	104.8	101.0
21, 22, 23	FIRE SUPPRESSION, PLUMBING & HVAC	97.7	106.7	101.3	100.2	105.1	102.1	100.0	81.5	92.6	95.2	40.4	73.2	95.2	40.5	73.3	99.9	86.1	94.4
26, 27, 3370	ELECTRICAL, COMMUNICATIONS & UTIL.	98.4	109.7	104.3	102.7	120.6	112.0	98.8	97.4	98.0	99.6	54.1	75.8	99.6	54.1	75.8	99.6	80.4	89.6
MF2010	WEIGHTED AVERAGE	98.7	103.4	100.8	100.2	109.4	104.2	99.0	93.8	96.7	96.9	70.0	85.2	96.0	70.1	84.7	100.8	92.3	97.1

		NEW HAMPSHIRE															NEW JERSEY		
DIVISION		KEENE			LITTLETON			MANCHESTER			NASHUA			PORTSMOUTH			ATLANTIC CITY		
		034			035			031			030			038			082,084		
		MAT.	INST.	TOTAL	MAT.	INST.	TOTAL	MAT.	INST.	TOTAL	MAT.	INST.	TOTAL	MAT.	INST.	TOTAL	MAT.	INST.	TOTAL
015433	CONTRACTOR EQUIPMENT		100.6	100.6		100.6	100.6		100.6	100.6		100.6	100.6		100.6	100.6		98.8	98.8
0241, 31 - 34	SITE & INFRASTRUCTURE, DEMOLITION	95.2	99.5	98.2	81.3	99.5	94.1	93.6	101.4	99.1	96.5	101.4	100.0	90.1	101.5	98.1	98.0	104.4	102.5
0310	Concrete Forming & Accessories	92.6	84.9	85.9	103.4	84.9	87.4	98.4	94.8	95.3	100.3	94.8	95.5	89.4	95.1	94.3	110.1	128.4	125.9
0320	Concrete Reinforcing	84.5	95.3	89.9	85.2	95.3	90.3	106.7	95.8	101.3	105.5	95.8	100.7	84.5	95.9	90.2	79.1	121.6	100.3
0330	Cast-in-Place Concrete	97.9	74.4	88.1	87.9	74.4	82.3	111.9	115.3	113.3	92.6	115.3	102.0	87.9	115.4	99.3	84.2	132.2	104.1
03	CONCRETE	99.2	83.2	91.3	91.0	83.3	87.2	108.6	101.9	105.3	99.5	101.9	100.6	91.0	102.1	96.4	92.9	127.3	109.9
04	MASONRY	91.7	85.9	88.1	96.9	85.9	90.1	94.9	102.1	99.4	91.4	102.1	98.0	87.7	102.1	96.6	99.2	125.4	115.4
05	METALS	95.8	91.2	94.4	95.9	91.2	94.4	103.8	94.5	100.9	100.8	94.5	98.9	97.3	95.1	96.7	95.6	106.5	99.0
06	WOOD, PLASTICS & COMPOSITES	91.6	92.3	92.0	102.9	92.3	96.9	95.1	93.3	94.1	101.4	93.3	96.8	88.2	93.3	91.1	115.6	130.2	123.9
07	THERMAL & MOISTURE PROTECTION	103.2	75.0	91.8	102.8	73.5	90.9	106.9	94.9	102.0	103.6	94.9	100.1	103.3	115.6	108.3	104.6	123.2	112.2
08	OPENINGS	101.1	82.9	96.8	104.8	78.7	98.6	106.9	88.4	102.6	105.8	88.4	101.6	106.5	85.1	101.4	102.3	125.4	107.7
0920	Plaster & Gypsum Board	95.1	91.4	92.6	108.4	91.4	96.9	98.7	92.4	94.4	103.8	92.4	96.1	94.7	92.4	93.2	107.3	130.5	123.1
0950, 0980	Ceilings & Acoustic Treatment	90.6	91.4	91.2	90.6	91.4	91.2	98.6	92.4	94.5	101.4	92.4	95.4	91.6	92.4	92.2	88.9	130.5	116.6
0960	Flooring	97.9	53.6	84.7	106.2	32.7	84.2	98.2	112.2	102.4	100.9	112.2	104.2	96.5	112.2	101.2	102.2	150.6	116.6
0970, 0990	Wall Finishes & Painting/Coating	95.4	46.7	66.2	95.4	61.2	74.9	98.2	97.5	97.8	95.4	97.5	96.7	95.4	97.5	96.7	93.0	117.3	107.6
09	FINISHES	96.7	76.2	85.3	99.0	73.8	85.0	98.5	98.0	98.2	101.0	98.0	99.3	95.9	98.0	97.1	99.4	132.5	117.8
COVERS	DIVS. 10 - 14, 25, 28, 41, 43, 44, 46	100.0	91.2	98.2	100.0	99.2	99.8	100.0	104.8	101.0	100.0	104.8	101.0	100.0	104.8	101.0	100.0	110.5	102.1
21, 22, 23	FIRE SUPPRESSION, PLUMBING & HVAC	95.2	44.2	74.7	95.2	65.2	83.2	99.9	86.2	94.4	100.1	86.2	94.5	100.1	86.2	94.5	99.6	119.6	107.6
26, 27, 3370	ELECTRICAL, COMMUNICATIONS & UTIL.	99.6	64.7	81.4	100.8	57.3	78.1	101.6	80.4	90.6	102.2	80.4	90.8	100.1	80.4	89.8	92.7	141.4	118.1
MF2010	WEIGHTED AVERAGE	97.4	74.0	87.2	97.1	77.2	88.4	102.1	93.6	98.4	100.6	93.6	97.6	98.2	94.2	96.4	98.1	123.5	109.2

		NEW JERSEY																	
DIVISION		CAMDEN			DOVER			ELIZABETH			HACKENSACK			JERSEY CITY			LONG BRANCH		
		081			078			072			076			073			077		
		MAT.	INST.	TOTAL	MAT.	INST.	TOTAL	MAT.	INST.	TOTAL	MAT.	INST.	TOTAL	MAT.	INST.	TOTAL	MAT.	INST.	TOTAL
015433	CONTRACTOR EQUIPMENT		98.8	98.8		100.6	100.6		100.6	100.6		100.6	100.6		98.8	98.8		98.4	98.4
0241, 31 - 34	SITE & INFRASTRUCTURE, DEMOLITION	99.1	104.7	103.0	108.2	105.4	106.3	112.7	105.4	107.6	108.9	105.5	106.5	98.0	105.4	103.2	102.6	105.3	104.5
0310	Concrete Forming & Accessories	101.7	128.4	124.7	97.4	129.0	124.7	109.3	129.0	126.3	97.4	128.9	124.6	101.1	129.0	125.2	101.7	128.7	125.0
0320	Concrete Reinforcing	103.8	121.8	112.8	80.0	128.6	104.3	80.0	128.6	104.3	80.0	128.6	104.3	103.8	128.6	116.2	80.0	128.6	104.3
0330	Cast-in-Place Concrete	81.6	132.1	102.6	102.2	127.3	112.6	87.8	127.3	104.2	99.9	127.3	111.3	79.8	127.4	99.5	88.7	132.1	106.7
03	CONCRETE	94.1	127.3	110.4	99.7	127.4	113.3	95.3	127.4	111.1	97.8	127.3	112.4	93.2	127.2	110.0	97.1	128.7	112.7
04	MASONRY	89.2	125.4	111.8	87.3	125.9	111.2	101.2	125.9	116.5	90.8	125.9	112.5	80.8	125.9	108.7	94.4	125.4	113.6
05	METALS	100.7	106.6	102.5	95.6	113.6	101.2	97.1	113.6	102.2	95.7	113.4	101.2	100.8	111.2	104.0	95.7	111.1	100.5
06	WOOD, PLASTICS & COMPOSITES	104.2	130.2	118.9	101.5	130.2	117.7	117.1	130.2	124.5	101.5	130.2	117.7	102.3	130.2	118.0	103.4	130.1	118.5
07	THERMAL & MOISTURE PROTECTION	104.5	122.7	111.9	101.5	130.6	113.4	101.7	130.6	113.5	101.3	123.3	110.3	101.1	130.6	113.1	101.2	123.4	110.3
08	OPENINGS	104.6	125.4	109.5	108.8	127.9	113.4	106.9	127.9	111.9	106.2	127.9	111.4	104.5	127.9	110.0	100.7	127.9	107.1
0920	Plaster & Gypsum Board	102.8	130.5	121.6	99.9	130.5	120.7	107.7	130.5	123.2	99.9	130.5	120.7	103.2	130.5	121.8	101.7	130.5	121.3
0950, 0980	Ceilings & Acoustic Treatment	99.7	130.5	120.2	88.6	130.5	116.5	90.6	130.5	117.2	88.6	130.5	116.5	99.5	130.5	120.1	88.6	130.5	116.5
0960	Flooring	98.7	150.6	114.2	96.8	172.3	119.4	101.2	172.3	122.5	96.8	172.3	119.4	97.7	172.3	120.0	97.9	172.3	120.1
0970, 0990	Wall Finishes & Painting/Coating	93.0	117.3	107.6	94.6	119.3	109.4	94.6	119.3	109.4	94.6	119.3	109.4	94.7	119.3	109.5	94.7	117.3	108.3
09	FINISHES	100.0	132.5	118.1	97.1	135.5	118.5	100.4	135.5	119.9	96.9	135.5	118.4	99.8	135.5	119.7	97.9	136.0	119.2
COVERS	DIVS. 10 - 14, 25, 28, 41, 43, 44, 46	100.0	110.5	102.1	100.0	113.6	102.7	100.0	113.6	102.7	100.0	113.6	102.7	100.0	113.6	102.7	100.0	110.4	102.1
21, 22, 23	FIRE SUPPRESSION, PLUMBING & HVAC	100.0	119.7	107.9	99.6	123.4	109.1	100.0	122.8	109.1	99.6	123.3	109.1	100.0	123.4	109.3	99.6	122.6	108.8
26, 27, 3370	ELECTRICAL, COMMUNICATIONS & UTIL.	97.9	141.4	120.6	90.9	137.6	115.3	91.5	137.6	115.6	90.9	139.7	116.4	95.5	139.7	118.6	90.6	131.5	111.9
MF2010	WEIGHTED AVERAGE	99.4	123.6	109.9	98.7	125.4	110.3	99.5	125.3	110.7	98.4	125.5	110.2	98.5	125.5	110.2	97.8	124.1	109.2

615

City Cost Indexes

		NEW JERSEY																	
		NEW BRUNSWICK			NEWARK			PATERSON			POINT PLEASANT			SUMMIT			TRENTON		
DIVISION		088 - 089			070 - 071			074 - 075			087			079			085 - 086		
		MAT.	INST.	TOTAL	MAT.	INST.	TOTAL	MAT.	INST.	TOTAL	MAT.	INST.	TOTAL	MAT.	INST.	TOTAL	MAT.	INST.	TOTAL
---	---	---	---	---	---	---	---	---	---	---	---	---	---	---	---	---	---	---	---
015433	CONTRACTOR EQUIPMENT		98.4	98.4		100.6	100.6		100.6	100.6		98.4	98.4		100.6	100.6		98.4	98.4
0241, 31 - 34	SITE & INFRASTRUCTURE, DEMOLITION	111.8	105.3	107.3	114.4	105.4	108.1	110.9	105.5	107.1	113.5	105.3	107.7	110.1	105.4	106.8	96.6	105.3	102.7
0310	Concrete Forming & Accessories	104.6	128.9	125.6	97.2	129.0	124.6	99.4	128.9	124.8	99.3	120.3	117.4	100.2	129.0	125.0	99.9	128.6	124.6
0320	Concrete Reinforcing	80.0	128.6	104.3	103.3	128.6	116.0	103.8	128.6	116.2	80.0	128.5	104.2	80.0	128.6	104.3	103.3	115.7	109.5
0330	Cast-in-Place Concrete	103.9	132.3	115.7	109.6	127.3	117.0	101.6	127.3	112.3	103.9	128.4	114.1	84.9	127.3	102.5	99.5	132.0	113.0
03	CONCRETE	109.8	128.9	119.3	106.9	127.4	117.0	103.4	127.3	115.2	109.5	123.7	116.5	92.3	127.4	109.6	102.3	126.3	114.1
04	MASONRY	96.6	125.9	114.7	93.4	125.9	113.5	87.8	125.9	111.4	86.4	124.6	110.0	89.3	125.9	112.0	92.1	125.4	112.7
05	METALS	95.7	111.2	100.4	102.8	113.6	106.1	95.7	113.5	101.2	95.7	110.7	100.3	95.6	113.6	101.1	100.4	105.7	102.0
06	WOOD, PLASTICS & COMPOSITES	108.9	130.1	120.9	99.0	130.2	116.6	104.1	130.2	118.8	101.6	119.1	111.5	105.3	130.2	119.3	101.4	130.1	117.6
07	THERMAL & MOISTURE PROTECTION	104.8	129.9	115.0	102.0	130.6	113.6	101.8	123.3	110.4	104.8	122.1	111.8	101.9	130.6	113.6	104.2	123.1	111.9
08	OPENINGS	97.2	127.9	104.5	107.5	127.9	112.3	111.6	127.9	115.4	99.1	122.2	104.6	113.5	127.9	117.0	106.7	124.4	110.9
0920	Plaster & Gypsum Board	103.9	130.5	122.0	99.9	130.5	120.7	103.2	130.5	121.8	99.4	119.1	112.8	101.7	130.5	121.3	99.9	130.5	120.7
0950, 0980	Ceilings & Acoustic Treatment	88.9	130.5	116.6	102.4	130.5	121.1	99.5	130.5	120.1	88.9	119.1	109.0	88.6	130.5	116.5	101.4	130.5	120.8
0960	Flooring	100.0	172.3	121.6	98.5	172.3	120.6	97.7	172.3	120.0	97.8	150.6	113.6	98.0	172.3	120.2	99.3	172.3	121.1
0970, 0990	Wall Finishes & Painting/Coating	93.0	119.3	108.8	96.0	119.3	110.0	94.6	119.3	109.4	93.0	117.3	107.6	94.6	119.3	109.4	98.3	117.3	109.7
09	FINISHES	99.7	135.4	119.6	98.6	135.5	119.1	100.0	135.5	119.8	98.5	125.9	113.7	98.0	135.5	118.9	100.1	136.0	120.1
COVERS	DIVS. 10 - 14, 25, 28, 41, 43, 44, 46	100.0	113.4	102.7	100.0	113.6	102.7	100.0	113.6	102.7	100.0	104.2	100.8	100.0	113.6	102.7	100.0	110.4	102.1
21, 22, 23	FIRE SUPPRESSION, PLUMBING & HVAC	99.6	122.9	108.9	100.0	122.8	109.1	100.0	123.4	109.4	99.6	122.5	108.7	99.6	122.8	108.9	100.0	122.5	109.0
26, 27, 3370	ELECTRICAL, COMMUNICATIONS & UTIL.	93.4	137.6	116.5	99.2	139.7	120.3	95.5	137.6	117.5	92.7	131.5	112.9	91.5	137.6	115.6	101.4	138.7	120.8
MF2010	WEIGHTED AVERAGE	99.7	125.2	110.8	101.9	125.6	112.2	100.2	125.2	111.1	99.2	121.4	108.8	98.7	125.3	110.3	100.8	124.0	110.9

		NEW JERSEY			NEW MEXICO													
		VINELAND			ALBUQUERQUE			CARRIZOZO			CLOVIS			FARMINGTON			GALLUP		
DIVISION		080,083			870 - 872			883			881			874			873		
		MAT.	INST.	TOTAL	MAT.	INST.	TOTAL	MAT.	INST.	TOTAL	MAT.	INST.	TOTAL	MAT.	INST.	TOTAL	MAT.	INST.	TOTAL
---	---	---	---	---	---	---	---	---	---	---	---	---	---	---	---	---	---	---	---
015433	CONTRACTOR EQUIPMENT		98.8	98.8		109.6	109.6		109.6	109.6		109.6	109.6		109.6	109.6		109.6	109.6
0241, 31 - 34	SITE & INFRASTRUCTURE, DEMOLITION	102.7	104.7	104.1	82.2	104.1	97.6	102.9	104.1	103.7	91.2	104.1	100.2	88.4	104.1	99.4	96.0	104.1	101.7
0310	Concrete Forming & Accessories	96.8	128.4	124.1	101.4	65.0	70.0	99.2	65.0	69.7	99.1	64.9	69.6	101.4	65.0	70.0	101.5	65.0	70.0
0320	Concrete Reinforcing	79.1	118.5	98.7	100.5	71.1	85.8	109.9	71.1	90.5	111.2	71.1	91.2	109.9	71.1	90.5	105.2	71.1	88.2
0330	Cast-in-Place Concrete	90.7	132.2	107.9	96.9	71.2	86.2	95.4	71.2	85.3	95.3	71.1	85.3	97.8	71.2	86.8	92.1	71.2	83.4
03	CONCRETE	97.6	126.7	111.9	101.9	69.4	85.9	119.1	69.4	94.6	107.6	69.3	88.7	105.7	69.4	87.8	112.5	69.4	91.2
04	MASONRY	87.9	125.4	111.1	99.5	60.6	75.4	101.6	60.6	76.2	101.7	60.6	76.2	106.3	60.6	78.0	95.4	60.6	73.8
05	METALS	95.6	106.0	98.8	103.8	87.9	98.9	98.4	87.9	95.2	98.1	87.8	94.9	101.4	87.9	97.2	100.6	87.9	96.6
06	WOOD, PLASTICS & COMPOSITES	98.5	130.2	116.4	92.8	65.5	77.4	88.9	65.5	75.7	88.9	65.5	75.7	92.9	65.5	77.4	92.9	65.5	77.4
07	THERMAL & MOISTURE PROTECTION	104.4	123.2	112.1	99.4	71.8	88.2	100.7	71.8	89.0	99.6	71.8	88.3	99.6	71.8	88.3	100.5	71.8	88.8
08	OPENINGS	98.6	125.0	104.9	101.4	68.5	93.6	98.6	68.5	91.5	98.8	68.5	91.6	104.0	68.5	95.6	104.1	68.5	95.7
0920	Plaster & Gypsum Board	98.0	130.5	120.1	88.3	64.1	71.8	75.4	64.1	67.7	75.4	64.1	67.7	81.6	64.1	69.7	81.6	64.1	69.7
0950, 0980	Ceilings & Acoustic Treatment	88.9	130.5	116.6	106.3	64.1	78.2	104.2	64.1	77.5	104.2	64.1	77.5	103.1	64.1	77.2	103.1	64.1	77.2
0960	Flooring	97.1	150.6	113.1	100.2	67.2	90.3	100.2	67.2	90.4	100.2	67.2	90.4	101.7	67.2	91.4	101.7	67.2	91.4
0970, 0990	Wall Finishes & Painting/Coating	93.0	117.3	107.6	109.1	68.1	84.4	102.9	68.1	82.0	102.9	68.1	82.0	102.9	68.1	82.0	102.9	68.1	82.0
09	FINISHES	97.2	132.5	116.8	98.3	65.5	80.0	99.0	65.5	80.3	97.6	65.5	79.7	96.9	65.5	79.4	98.1	65.5	79.9
COVERS	DIVS. 10 - 14, 25, 28, 41, 43, 44, 46	100.0	110.5	102.1	100.0	76.3	95.2	100.0	76.3	95.2	100.0	76.3	95.2	100.0	76.3	95.2	100.0	76.3	95.2
21, 22, 23	FIRE SUPPRESSION, PLUMBING & HVAC	99.6	119.6	107.6	100.2	70.7	88.4	97.2	70.7	86.6	97.2	70.5	86.5	100.1	70.7	88.3	97.1	70.7	86.6
26, 27, 3370	ELECTRICAL, COMMUNICATIONS & UTIL.	92.7	141.4	118.1	88.5	72.9	80.4	90.6	72.9	81.4	88.1	72.9	80.2	86.8	72.9	79.6	86.0	72.9	79.2
MF2010	WEIGHTED AVERAGE	97.5	123.4	108.8	99.2	73.6	88.0	100.1	73.6	88.5	98.1	73.5	87.4	99.7	73.6	88.3	99.3	73.6	88.1

		NEW MEXICO																	
		LAS CRUCES			LAS VEGAS			ROSWELL			SANTA FE			SOCORRO			TRUTH/CONSEQUENCES		
DIVISION		880			877			882			875			878			879		
		MAT.	INST.	TOTAL	MAT.	INST.	TOTAL	MAT.	INST.	TOTAL	MAT.	INST.	TOTAL	MAT.	INST.	TOTAL	MAT.	INST.	TOTAL
---	---	---	---	---	---	---	---	---	---	---	---	---	---	---	---	---	---	---	---
015433	CONTRACTOR EQUIPMENT		86.0	86.0		109.6	109.6		109.6	109.6		109.6	109.6		109.6	109.6		86.1	86.1
0241, 31 - 34	SITE & INFRASTRUCTURE, DEMOLITION	91.6	83.8	86.1	87.8	104.1	99.2	93.3	104.1	100.9	92.9	104.1	100.7	84.3	104.1	98.2	101.9	83.8	89.2
0310	Concrete Forming & Accessories	95.6	63.8	68.2	101.4	65.0	70.0	99.2	65.0	69.7	100.1	65.0	69.8	101.5	65.0	70.0	98.7	63.8	68.6
0320	Concrete Reinforcing	107.3	71.0	89.1	106.9	71.1	89.0	111.2	71.1	91.2	106.0	71.1	88.6	109.1	71.1	90.1	102.8	71.0	86.9
0330	Cast-in-Place Concrete	89.9	63.7	79.0	95.2	71.2	85.2	95.3	71.2	85.3	103.2	71.2	89.9	93.2	71.2	84.1	102.1	63.7	86.1
03	CONCRETE	86.1	65.9	76.1	103.1	69.4	86.4	108.3	69.4	89.1	105.6	69.4	87.7	102.0	69.4	85.9	95.1	65.9	80.7
04	MASONRY	97.5	60.2	74.4	95.6	60.6	73.9	112.2	60.6	80.3	99.5	60.6	75.4	95.5	60.6	73.9	93.1	60.2	72.8
05	METALS	97.1	81.3	92.2	100.3	87.9	96.4	99.3	87.9	95.8	97.7	87.9	94.7	100.6	87.9	96.6	100.1	81.4	94.3
06	WOOD, PLASTICS & COMPOSITES	78.8	64.4	70.7	92.9	65.5	77.4	88.9	65.5	75.7	94.6	65.5	78.2	92.9	65.5	77.4	84.6	64.4	73.2
07	THERMAL & MOISTURE PROTECTION	86.2	67.2	78.5	99.2	71.8	88.1	99.7	71.8	88.4	101.7	71.8	89.5	99.1	71.8	88.0	86.9	67.2	78.9
08	OPENINGS	91.5	67.9	85.9	100.3	68.5	92.8	98.6	68.5	91.5	102.4	68.5	94.4	100.1	68.5	92.7	93.3	67.9	87.3
0920	Plaster & Gypsum Board	74.2	64.1	67.3	81.6	64.1	69.7	75.4	64.1	67.7	91.6	64.1	72.9	81.6	64.1	69.7	83.4	64.1	70.3
0950, 0980	Ceilings & Acoustic Treatment	91.4	64.1	73.2	103.1	64.1	77.2	104.2	64.1	77.5	103.6	64.1	77.3	103.1	64.1	77.2	90.4	64.1	72.9
0960	Flooring	129.3	67.2	110.7	101.7	67.2	91.4	100.2	67.2	90.4	108.8	67.2	96.4	101.7	67.2	91.4	132.2	67.2	112.7
0970, 0990	Wall Finishes & Painting/Coating	90.1	68.1	76.8	102.9	68.1	82.0	102.9	68.1	82.0	107.7	68.1	83.9	102.9	68.1	82.0	93.7	68.1	78.3
09	FINISHES	107.1	64.6	83.4	96.8	65.5	79.3	97.7	65.5	79.8	101.9	65.5	81.6	96.7	65.5	79.3	109.0	64.6	84.3
COVERS	DIVS. 10 - 14, 25, 28, 41, 43, 44, 46	100.0	73.6	94.7	100.0	76.3	95.2	100.0	76.3	95.2	100.0	76.3	95.2	100.0	76.3	95.2	100.0	73.6	94.7
21, 22, 23	FIRE SUPPRESSION, PLUMBING & HVAC	100.4	70.5	88.4	97.1	70.7	86.6	100.0	70.7	88.3	100.1	70.7	88.3	97.1	70.7	86.6	97.0	70.5	86.4
26, 27, 3370	ELECTRICAL, COMMUNICATIONS & UTIL.	89.9	72.9	81.0	88.5	72.9	80.4	89.6	72.9	80.9	100.8	72.9	86.3	86.5	72.9	79.4	90.1	72.9	81.1
MF2010	WEIGHTED AVERAGE	95.8	70.4	84.7	97.7	73.6	87.2	99.8	73.6	88.4	100.5	73.6	88.8	97.3	73.6	87.0	97.0	70.4	85.4

City Cost Indexes

DIVISION		NEW MEXICO TUCUMCARI 884 MAT.	INST.	TOTAL	NEW YORK ALBANY 120-122 MAT.	INST.	TOTAL	BINGHAMTON 137-139 MAT.	INST.	TOTAL	BRONX 104 MAT.	INST.	TOTAL	BROOKLYN 112 MAT.	INST.	TOTAL	BUFFALO 140-142 MAT.	INST.	TOTAL
015433	CONTRACTOR EQUIPMENT		109.6	109.6		113.5	113.5		114.4	114.4		112.8	112.8		114.1	114.1		97.0	97.0
0241, 31 - 34	SITE & INFRASTRUCTURE, DEMOLITION	90.8	104.1	100.1	75.0	107.1	97.5	92.8	94.3	93.9	110.9	123.1	119.5	117.0	128.4	125.0	98.8	98.3	98.4
0310	Concrete Forming & Accessories	99.1	64.9	69.6	100.7	96.1	96.7	102.2	83.3	85.9	103.3	171.7	162.3	108.6	171.2	162.6	97.4	115.5	113.1
0320	Concrete Reinforcing	109.0	71.1	90.1	105.6	106.3	106.0	93.5	96.4	94.9	96.7	184.6	140.6	94.9	184.6	139.7	102.1	103.0	102.5
0330	Cast-in-Place Concrete	95.3	71.1	85.3	88.1	100.9	93.4	100.9	127.9	112.1	94.6	173.8	127.5	103.0	172.0	131.7	105.5	119.2	111.2
03	CONCRETE	106.7	69.3	88.3	101.0	100.3	100.6	96.2	102.7	99.4	96.5	173.2	134.4	108.3	172.0	139.7	104.6	113.7	109.1
04	MASONRY	112.9	60.6	80.5	90.9	100.3	96.7	105.8	126.6	118.7	94.3	176.4	145.1	116.4	176.4	153.5	107.3	119.6	114.9
05	METALS	98.1	87.8	94.9	101.3	109.5	103.9	96.0	118.2	102.9	100.8	149.3	115.8	104.4	140.3	115.4	99.5	95.6	98.3
06	WOOD, PLASTICS & COMPOSITES	88.9	65.5	75.7	97.7	93.6	95.4	107.3	79.3	91.5	100.1	171.4	140.4	109.3	171.1	144.1	98.7	115.2	108.0
07	THERMAL & MOISTURE PROTECTION	99.5	71.8	88.3	99.4	96.7	98.3	106.9	96.3	102.6	108.5	163.2	130.7	108.9	162.7	130.8	101.5	110.4	105.1
08	OPENINGS	98.5	68.5	91.4	99.9	92.0	98.0	90.7	81.6	88.5	87.0	172.2	107.1	88.0	166.5	106.6	91.8	103.3	94.6
0920	Plaster & Gypsum Board	75.4	64.1	67.7	105.0	93.1	96.9	111.6	78.2	88.9	105.4	173.3	151.6	106.9	173.3	152.1	103.2	115.4	111.5
0950, 0980	Ceilings & Acoustic Treatment	104.2	64.1	77.5	95.9	93.1	94.0	94.6	78.2	83.7	82.8	173.3	143.0	90.2	173.3	145.5	101.6	115.4	110.8
0960	Flooring	100.2	67.2	90.4	97.0	105.2	99.4	106.2	93.1	102.3	96.6	185.6	123.2	113.2	185.6	134.8	101.0	120.1	106.7
0970, 0990	Wall Finishes & Painting/Coating	102.9	68.1	82.0	96.8	89.4	92.4	93.3	95.3	94.5	97.9	148.0	128.0	124.2	148.0	138.5	102.4	112.5	108.5
09	FINISHES	97.6	65.5	79.7	97.8	96.7	97.2	96.7	84.9	90.1	95.3	172.0	138.1	111.3	171.8	145.0	101.0	116.9	109.9
COVERS	DIVS. 10 - 14, 25, 28, 41, 43, 44, 46	100.0	76.3	95.2	100.0	96.0	99.2	100.0	93.5	98.7	100.0	132.7	106.6	100.0	132.0	106.4	100.0	106.1	101.2
21, 22, 23	FIRE SUPPRESSION, PLUMBING & HVAC	97.2	70.5	86.5	100.0	98.8	99.5	100.4	87.3	95.1	100.2	164.7	126.1	99.8	164.4	125.7	100.0	97.1	98.8
26, 27, 3370	ELECTRICAL, COMMUNICATIONS & UTIL.	90.6	72.9	81.4	96.0	99.3	97.7	98.9	107.9	103.6	96.1	175.0	137.3	98.7	175.0	138.5	97.5	101.9	99.8
MF2010	WEIGHTED AVERAGE	98.8	73.5	87.8	98.6	100.2	99.3	98.1	99.4	98.6	98.0	163.9	126.7	102.6	162.9	128.9	99.7	105.8	102.4

DIVISION		NEW YORK ELMIRA 148-149 MAT.	INST.	TOTAL	FAR ROCKAWAY 116 MAT.	INST.	TOTAL	FLUSHING 113 MAT.	INST.	TOTAL	GLENS FALLS 128 MAT.	INST.	TOTAL	HICKSVILLE 115,117,118 MAT.	INST.	TOTAL	JAMAICA 114 MAT.	INST.	TOTAL
015433	CONTRACTOR EQUIPMENT		116.0	116.0		114.1	114.1		114.1	114.1		113.5	113.5		114.1	114.1		114.1	114.1
0241, 31 - 34	SITE & INFRASTRUCTURE, DEMOLITION	93.3	94.3	94.0	120.2	128.4	125.9	120.2	128.4	126.0	65.2	106.7	94.3	110.1	127.3	122.2	114.5	128.4	124.3
0310	Concrete Forming & Accessories	83.9	86.3	86.0	94.9	171.2	160.8	98.7	171.2	161.3	87.1	86.9	86.9	91.5	144.5	137.3	98.7	171.2	161.3
0320	Concrete Reinforcing	100.5	90.6	95.6	94.9	184.6	139.7	96.5	184.6	140.5	97.0	94.1	95.5	94.9	184.4	139.6	94.9	184.6	139.7
0330	Cast-in-Place Concrete	94.7	96.4	95.4	111.4	172.0	136.6	111.4	172.0	136.6	81.7	96.9	88.0	94.7	165.6	124.1	103.0	172.0	131.7
03	CONCRETE	95.0	92.3	93.7	114.3	172.0	142.8	114.9	172.0	143.1	89.9	92.6	91.3	100.1	157.8	128.6	107.6	172.0	139.4
04	MASONRY	101.4	93.7	96.6	119.7	176.4	154.8	113.9	176.4	152.6	90.1	93.2	92.0	110.5	166.0	144.9	118.1	176.4	154.2
05	METALS	95.4	114.7	101.3	104.4	140.3	115.5	104.4	140.3	115.5	94.8	105.1	98.0	105.9	138.2	115.8	104.4	140.3	115.5
06	WOOD, PLASTICS & COMPOSITES	86.3	84.9	85.5	92.2	171.1	136.7	97.1	171.1	138.8	86.1	84.9	85.4	88.7	141.5	118.5	97.1	171.1	138.8
07	THERMAL & MOISTURE PROTECTION	102.2	84.1	94.8	108.8	162.7	130.7	108.9	162.7	130.7	93.0	92.8	92.9	108.5	154.3	127.1	108.7	162.7	130.6
08	OPENINGS	92.4	83.8	90.3	86.7	166.5	105.6	86.7	166.5	105.6	91.3	84.5	89.7	86.7	150.3	101.8	86.7	166.5	105.6
0920	Plaster & Gypsum Board	100.2	84.2	89.3	95.1	173.3	148.3	97.8	173.3	149.1	95.2	84.2	87.7	94.8	142.7	127.4	97.8	173.3	149.1
0950, 0980	Ceilings & Acoustic Treatment	93.4	84.2	87.2	78.7	173.3	141.6	78.7	173.3	141.6	85.9	84.2	84.8	77.7	142.7	121.0	78.7	173.3	141.6
0960	Flooring	92.9	93.1	93.0	108.5	185.6	131.6	110.0	185.6	132.6	86.1	105.2	91.8	107.6	145.3	118.9	110.0	185.6	132.6
0970, 0990	Wall Finishes & Painting/Coating	101.7	87.6	93.2	124.2	148.0	138.5	124.2	148.0	138.5	94.1	89.3	91.2	124.2	148.0	138.5	124.2	148.0	138.5
09	FINISHES	96.1	87.4	91.3	106.4	171.8	142.8	107.1	171.8	143.1	89.1	89.8	89.5	104.9	143.7	126.5	106.6	171.8	142.9
COVERS	DIVS. 10 - 14, 25, 28, 41, 43, 44, 46	100.0	93.4	98.7	100.0	132.0	106.4	100.0	132.0	106.4	100.0	92.7	98.5	100.0	125.1	105.0	100.0	132.0	106.4
21, 22, 23	FIRE SUPPRESSION, PLUMBING & HVAC	95.3	87.8	92.3	94.9	164.4	122.8	94.9	164.4	122.8	95.4	92.0	94.0	99.8	151.8	120.6	94.9	164.4	122.8
26, 27, 3370	ELECTRICAL, COMMUNICATIONS & UTIL.	95.5	96.9	96.2	106.1	175.0	142.1	106.1	175.0	142.1	91.1	99.3	95.4	98.0	135.8	117.7	96.9	175.0	137.7
MF2010	WEIGHTED AVERAGE	95.9	93.1	94.7	102.4	162.9	128.7	102.3	162.9	128.7	92.6	95.0	93.6	100.7	146.6	120.7	100.6	162.9	127.7

DIVISION		NEW YORK JAMESTOWN 147 MAT.	INST.	TOTAL	KINGSTON 124 MAT.	INST.	TOTAL	LONG ISLAND CITY 111 MAT.	INST.	TOTAL	MONTICELLO 127 MAT.	INST.	TOTAL	MOUNT VERNON 105 MAT.	INST.	TOTAL	NEW ROCHELLE 108 MAT.	INST.	TOTAL
015433	CONTRACTOR EQUIPMENT		94.0	94.0		114.1	114.1		114.1	114.1		114.1	114.1		112.8	112.8		112.8	112.8
0241, 31 - 34	SITE & INFRASTRUCTURE, DEMOLITION	94.5	93.8	94.0	122.1	124.3	123.7	118.1	128.4	125.3	117.5	124.2	122.2	117.9	118.8	118.6	117.1	118.8	118.3
0310	Concrete Forming & Accessories	83.9	81.2	81.6	88.7	106.0	103.6	103.1	171.2	161.9	95.7	105.0	103.7	93.5	128.5	123.7	108.6	128.5	125.8
0320	Concrete Reinforcing	100.7	98.6	99.6	97.4	140.6	119.0	94.9	184.6	139.7	96.6	140.4	118.5	95.2	183.2	139.2	95.3	183.2	139.2
0330	Cast-in-Place Concrete	98.3	91.3	95.4	110.4	131.3	119.1	106.3	172.0	133.6	103.4	126.8	113.1	105.7	128.9	115.3	105.7	128.9	115.3
03	CONCRETE	98.0	88.0	93.0	112.5	120.8	116.6	110.7	172.0	140.9	107.2	118.8	112.9	105.9	137.8	121.6	105.4	137.8	121.4
04	MASONRY	109.1	91.1	98.0	104.0	136.8	124.3	113.3	176.4	152.3	98.0	131.0	118.4	100.0	126.1	116.2	100.0	126.1	116.2
05	METALS	92.7	90.9	92.2	103.3	116.5	107.4	104.4	140.3	115.4	103.3	116.1	107.2	100.6	134.6	111.1	100.9	134.6	111.3
06	WOOD, PLASTICS & COMPOSITES	84.8	78.6	81.3	87.3	98.6	93.7	103.3	171.1	141.5	95.2	98.6	97.1	89.3	131.7	113.2	107.6	131.7	121.2
07	THERMAL & MOISTURE PROTECTION	101.6	84.0	94.5	114.3	136.3	123.3	108.8	162.7	130.7	114.0	134.0	122.1	109.4	135.8	120.1	109.4	135.8	120.7
08	OPENINGS	92.2	82.5	89.9	92.8	117.5	98.7	86.7	166.5	105.6	88.5	117.5	95.4	87.0	145.0	100.7	87.0	145.0	100.7
0920	Plaster & Gypsum Board	93.2	77.7	82.6	97.2	98.6	98.1	102.6	173.3	150.7	99.5	98.6	98.9	100.1	132.3	122.0	112.8	132.3	126.1
0950, 0980	Ceilings & Acoustic Treatment	90.7	77.7	82.0	74.3	98.6	90.5	78.7	173.3	141.6	74.3	98.6	90.5	81.1	132.3	115.2	81.1	132.3	115.2
0960	Flooring	95.9	89.7	94.1	103.1	73.8	94.4	111.4	185.6	133.6	105.4	73.8	95.9	89.0	180.7	116.4	95.2	180.7	120.8
0970, 0990	Wall Finishes & Painting/Coating	103.0	91.3	96.0	125.9	113.8	118.6	124.2	148.0	138.5	125.9	113.8	118.6	96.1	148.0	127.3	96.1	148.0	127.3
09	FINISHES	95.6	83.0	88.6	103.7	98.4	100.7	107.9	171.8	143.5	104.2	97.8	100.7	92.8	139.7	118.9	96.0	139.7	120.3
COVERS	DIVS. 10 - 14, 25, 28, 41, 43, 44, 46	100.0	94.8	99.0	100.0	109.7	101.9	100.0	132.0	106.4	100.0	108.9	101.8	100.0	117.8	103.6	100.0	117.8	103.6
21, 22, 23	FIRE SUPPRESSION, PLUMBING & HVAC	95.2	84.6	91.0	95.3	121.1	105.6	99.8	164.4	125.7	95.3	117.9	104.4	95.4	118.6	104.7	95.4	118.6	104.7
26, 27, 3370	ELECTRICAL, COMMUNICATIONS & UTIL.	94.3	90.4	92.3	92.7	113.7	103.6	97.5	175.0	137.9	92.7	113.7	103.6	94.4	158.9	128.1	94.4	158.9	128.1
MF2010	WEIGHTED AVERAGE	96.0	87.8	92.4	100.7	118.3	108.4	102.2	162.9	128.6	99.4	116.5	106.8	97.9	133.3	113.3	98.2	133.3	113.5

617

City Cost Indexes

			NEW YORK
	DIVISION		NEW YORK 100 - 102		NIAGARA FALLS 143		PLATTSBURGH 129		POUGHKEEPSIE 125 - 126		QUEENS 110		RIVERHEAD 119
		MAT.	INST.	TOTAL	MAT.	INST.	TOTAL	MAT.	INST.	TOTAL	MAT.	INST.	TOTAL	MAT.	INST.	TOTAL	MAT.	INST.	TOTAL
---	---	---	---	---	---	---	---	---	---	---	---	---	---	---	---	---	---	---	---
015433	CONTRACTOR EQUIPMENT		113.2	113.2		94.0	94.0		99.2	99.2		114.1	114.1		114.1	114.1		114.1	114.1
0241, 31 - 34	SITE & INFRASTRUCTURE, DEMOLITION	119.6	123.8	122.5	96.6	95.4	95.8	89.9	102.7	98.9	118.3	124.0	122.3	113.4	128.4	123.9	111.3	127.3	122.5
0310	Concrete Forming & Accessories	107.7	179.3	169.5	83.9	113.2	109.2	91.9	83.8	84.9	88.7	148.2	140.0	91.7	171.2	160.3	96.0	144.5	137.9
0320	Concrete Reinforcing	102.4	184.7	143.5	99.3	99.3	99.3	101.3	93.6	97.5	97.4	140.9	119.1	96.5	184.6	140.5	96.7	184.4	140.5
0330	Cast-in-Place Concrete	106.4	179.2	136.7	101.6	118.7	108.7	100.1	92.7	97.0	106.9	127.3	115.4	97.9	172.0	128.7	96.3	165.6	125.0
03	CONCRETE	106.7	178.5	142.1	100.2	111.7	105.9	104.1	88.4	96.3	109.6	138.2	123.7	103.3	172.0	137.2	100.8	157.8	128.9
04	MASONRY	104.1	178.4	150.0	117.3	118.8	118.2	85.9	87.3	86.8	97.8	128.1	116.6	107.8	176.4	150.2	115.8	166.0	146.9
05	METALS	114.1	149.5	125.0	95.4	92.3	94.5	99.2	85.5	95.0	103.3	118.5	108.0	104.4	140.3	115.4	106.3	138.2	116.1
06	WOOD, PLASTICS & COMPOSITES	104.6	180.6	147.5	84.7	110.9	99.5	94.1	81.9	87.2	87.3	158.2	127.3	88.8	171.1	135.3	94.0	141.5	120.8
07	THERMAL & MOISTURE PROTECTION	108.5	165.3	131.5	101.7	108.0	104.3	108.4	89.8	100.9	114.3	138.8	124.2	108.5	162.7	130.5	109.2	154.3	127.5
08	OPENINGS	92.6	177.5	112.7	92.2	100.2	94.1	98.8	80.1	94.4	92.8	149.1	106.2	86.7	166.5	105.6	86.7	150.3	101.8
0920	Plaster & Gypsum Board	112.2	182.7	160.2	93.2	111.0	105.3	119.4	80.6	93.0	97.2	160.0	139.9	94.8	173.3	148.2	96.1	142.7	127.8
0950, 0980	Ceilings & Acoustic Treatment	101.5	182.7	155.5	90.7	111.0	104.2	100.8	80.6	87.4	74.3	160.0	131.3	78.7	173.3	141.6	78.5	142.7	121.3
0960	Flooring	97.6	185.6	123.9	95.9	113.8	101.3	109.9	105.2	108.5	103.1	148.0	116.6	107.6	185.6	130.9	108.5	145.3	119.5
0970, 0990	Wall Finishes & Painting/Coating	97.9	156.4	133.1	103.0	109.7	107.0	123.3	76.4	95.1	125.9	113.8	118.6	124.2	148.0	138.5	124.2	148.0	138.5
09	FINISHES	101.2	178.9	144.5	95.7	113.6	105.7	109.9	86.4	92.9	103.5	146.3	127.3	105.3	171.8	142.3	105.5	143.7	126.8
COVERS	DIVS. 10 - 14, 25, 28, 41, 43, 44, 46	100.0	137.1	107.4	100.0	104.7	100.9	100.0	48.9	89.8	100.0	113.3	102.7	100.0	132.0	106.4	100.0	125.1	105.0
21, 22, 23	FIRE SUPPRESSION, PLUMBING & HVAC	100.1	165.7	126.4	95.2	97.1	95.9	95.3	92.0	94.0	95.3	117.9	104.4	99.8	164.4	125.7	100.0	151.8	120.7
26, 27, 3370	ELECTRICAL, COMMUNICATIONS & UTIL.	104.2	175.0	141.1	92.9	99.3	96.3	89.2	90.4	89.8	92.7	113.7	103.6	98.0	175.0	138.2	99.7	135.8	118.5
MF2010	WEIGHTED AVERAGE	103.8	166.4	131.1	96.9	103.9	100.0	97.3	88.4	93.4	100.0	127.5	112.0	100.8	162.9	127.8	101.4	146.6	121.1

			NEW YORK
	DIVISION		ROCHESTER 144 - 146		SCHENECTADY 123		STATEN ISLAND 103		SUFFERN 109		SYRACUSE 130 - 132		UTICA 133 - 135
		MAT.	INST.	TOTAL	MAT.	INST.	TOTAL	MAT.	INST.	TOTAL	MAT.	INST.	TOTAL	MAT.	INST.	TOTAL	MAT.	INST.	TOTAL
---	---	---	---	---	---	---	---	---	---	---	---	---	---	---	---	---	---	---	---
015433	CONTRACTOR EQUIPMENT		116.4	116.4		113.5	113.5		112.8	112.8		112.8	112.8		113.5	113.5		113.5	113.5
0241, 31 - 34	SITE & INFRASTRUCTURE, DEMOLITION	77.8	109.6	100.2	74.9	107.1	97.5	122.7	123.1	123.0	113.8	119.1	117.5	91.8	106.6	102.2	70.8	104.5	94.4
0310	Concrete Forming & Accessories	104.8	92.4	94.1	103.1	96.1	97.1	93.0	171.7	160.9	102.1	131.2	127.2	101.1	91.2	92.5	102.4	82.2	85.0
0320	Concrete Reinforcing	101.9	90.7	96.3	96.0	106.3	101.2	96.7	184.6	140.6	95.3	140.9	118.1	94.4	94.8	94.6	94.4	91.9	93.2
0330	Cast-in-Place Concrete	97.6	98.9	98.1	98.6	100.9	99.5	105.7	173.8	134.0	102.5	134.8	115.9	93.8	105.4	98.6	85.8	98.3	91.0
03	CONCRETE	107.3	95.2	101.3	104.5	100.3	102.4	107.6	173.2	140.0	102.4	133.3	117.6	99.5	97.5	98.6	97.8	90.6	94.2
04	MASONRY	100.2	98.6	99.2	88.1	100.3	95.7	106.7	176.4	149.8	99.4	138.2	123.4	98.1	106.3	103.2	90.0	97.7	94.8
05	METALS	101.4	104.5	102.3	98.8	109.5	102.1	98.7	149.3	114.3	98.7	118.1	104.7	99.5	105.0	101.2	97.5	103.8	99.5
06	WOOD, PLASTICS & COMPOSITES	103.8	90.6	96.4	106.0	93.6	99.0	87.8	171.4	135.0	99.7	131.7	117.7	103.7	87.6	94.6	103.7	79.4	90.0
07	THERMAL & MOISTURE PROTECTION	100.5	98.0	99.5	94.2	96.7	95.3	108.8	163.2	130.9	109.3	140.9	122.1	102.5	97.9	100.6	90.9	94.4	92.3
08	OPENINGS	96.7	86.9	94.4	96.9	92.0	95.7	87.0	172.2	107.1	87.0	135.5	98.5	92.5	80.4	89.6	95.2	79.2	91.4
0920	Plaster & Gypsum Board	111.1	90.3	96.9	106.6	93.1	97.4	100.2	173.3	149.9	104.6	132.3	123.5	101.9	87.0	91.7	101.9	78.6	86.0
0950, 0980	Ceilings & Acoustic Treatment	99.4	90.3	93.3	92.0	93.1	92.8	82.8	173.3	143.0	81.1	132.3	115.2	94.6	87.0	89.5	94.6	78.6	83.9
0960	Flooring	92.3	104.9	96.1	92.6	105.2	96.3	92.9	185.6	120.6	92.0	145.3	107.9	93.6	94.9	94.0	90.9	95.0	92.1
0970, 0990	Wall Finishes & Painting/Coating	102.9	97.4	99.6	94.1	89.4	91.3	97.9	148.0	128.0	96.1	121.7	111.5	98.3	97.0	97.5	90.8	97.0	94.5
09	FINISHES	103.1	95.2	98.7	94.5	96.7	95.8	94.7	172.0	137.8	93.8	130.2	114.1	96.0	92.1	93.8	94.1	85.4	89.2
COVERS	DIVS. 10 - 14, 25, 28, 41, 43, 44, 46	100.0	97.0	99.4	100.0	96.0	99.2	100.0	132.7	106.6	100.0	121.8	104.4	100.0	96.3	99.3	100.0	91.9	98.4
21, 22, 23	FIRE SUPPRESSION, PLUMBING & HVAC	100.0	87.9	95.1	100.2	98.8	99.6	100.2	164.7	126.1	95.4	119.6	105.1	100.2	91.2	96.6	100.2	88.8	95.6
26, 27, 3370	ELECTRICAL, COMMUNICATIONS & UTIL.	99.2	94.3	96.7	95.5	99.3	97.5	96.1	175.0	137.3	102.3	121.5	112.3	99.1	97.7	98.4	97.3	97.7	97.5
MF2010	WEIGHTED AVERAGE	100.3	95.6	98.3	97.8	100.2	98.8	99.7	163.9	127.6	98.0	126.2	110.3	98.5	97.0	97.9	96.7	93.2	95.1

			NEW YORK		NORTH CAROLINA
	DIVISION		WATERTOWN 136		WHITE PLAINS 106		YONKERS 107		ASHEVILLE 287 - 288		CHARLOTTE 281 - 282		DURHAM 277
		MAT.	INST.	TOTAL	MAT.	INST.	TOTAL	MAT.	INST.	TOTAL	MAT.	INST.	TOTAL	MAT.	INST.	TOTAL	MAT.	INST.	TOTAL
---	---	---	---	---	---	---	---	---	---	---	---	---	---	---	---	---	---	---	---
015433	CONTRACTOR EQUIPMENT		113.5	113.5		112.8	112.8		112.8	112.8		95.6	95.6		95.6	95.6		100.8	100.8
0241, 31 - 34	SITE & INFRASTRUCTURE, DEMOLITION	78.1	106.1	97.8	110.6	118.8	116.4	119.1	118.8	118.9	102.4	76.6	84.3	106.0	76.7	85.4	102.6	85.3	90.4
0310	Concrete Forming & Accessories	87.5	87.4	87.4	107.1	128.5	125.6	107.4	128.5	125.7	95.4	42.1	49.4	98.6	43.7	51.2	100.0	45.1	52.7
0320	Concrete Reinforcing	95.0	94.8	94.9	95.3	183.2	139.2	99.1	183.2	141.1	94.4	63.0	78.7	100.3	58.3	79.3	94.8	57.6	76.2
0330	Cast-in-Place Concrete	99.8	91.8	96.5	93.8	128.9	108.4	105.0	128.9	114.9	110.7	51.9	86.3	115.3	50.4	88.4	104.1	47.6	80.7
03	CONCRETE	110.2	91.2	100.8	96.1	137.8	116.7	105.6	137.9	121.5	102.3	51.2	77.1	104.9	50.5	78.1	98.8	50.1	74.8
04	MASONRY	91.0	100.2	96.7	98.9	126.1	115.7	103.6	126.1	117.5	81.7	44.5	58.7	88.9	52.1	66.1	85.5	38.6	56.4
05	METALS	97.6	104.8	99.8	100.3	134.6	110.9	109.9	134.6	117.5	94.2	82.7	90.7	95.0	81.3	90.8	108.6	80.2	99.9
06	WOOD, PLASTICS & COMPOSITES	85.0	86.3	85.8	105.6	131.7	120.3	105.4	131.7	120.3	96.4	41.2	65.2	101.6	42.8	68.4	99.5	45.7	69.1
07	THERMAL & MOISTURE PROTECTION	91.2	94.9	92.7	109.2	135.8	120.0	109.5	135.8	120.2	103.8	44.5	79.7	98.5	47.5	77.8	106.6	46.6	82.2
08	OPENINGS	95.2	83.2	92.4	87.0	145.0	100.7	90.0	145.3	103.1	94.7	44.6	82.9	99.9	45.5	87.1	97.8	47.7	85.9
0920	Plaster & Gypsum Board	92.2	85.7	87.7	107.6	132.3	124.4	111.8	132.3	125.8	97.5	39.2	57.8	97.8	40.9	59.1	108.8	43.8	64.6
0950, 0980	Ceilings & Acoustic Treatment	94.6	85.7	88.6	81.1	132.3	115.2	99.7	132.3	121.4	91.6	39.2	56.7	96.0	40.9	59.3	95.1	43.8	61.0
0960	Flooring	84.8	95.0	87.9	93.8	180.7	119.8	93.4	180.7	119.5	101.5	43.5	84.1	101.1	44.2	84.1	106.2	43.5	87.4
0970, 0990	Wall Finishes & Painting/Coating	90.8	91.4	91.2	96.1	148.0	127.3	96.1	148.0	127.3	107.1	41.1	67.4	107.3	50.4	73.1	103.4	38.2	64.2
09	FINISHES	91.7	88.8	90.1	94.3	139.7	119.5	99.4	139.7	121.8	96.8	41.5	66.0	97.0	43.8	67.4	100.2	43.8	68.8
COVERS	DIVS. 10 - 14, 25, 28, 41, 43, 44, 46	100.0	93.8	98.8	100.0	117.8	103.6	100.0	120.5	104.1	100.0	78.2	95.6	100.0	78.6	95.7	100.0	73.8	94.8
21, 22, 23	FIRE SUPPRESSION, PLUMBING & HVAC	100.2	81.5	92.7	100.4	118.6	107.7	100.4	118.6	107.7	100.4	53.5	81.6	100.0	55.3	82.1	100.3	53.1	81.4
26, 27, 3370	ELECTRICAL, COMMUNICATIONS & UTIL.	99.1	90.5	94.6	94.4	158.9	128.1	102.4	158.9	131.9	96.8	57.0	76.0	95.7	59.9	77.1	94.4	45.6	68.9
MF2010	WEIGHTED AVERAGE	98.1	91.9	95.4	98.0	133.3	113.4	102.5	133.4	116.0	97.6	55.9	79.4	98.7	57.6	80.8	100.2	54.4	80.2

City Cost Indexes

DIVISION		NORTH CAROLINA																	
		ELIZABETH CITY 279			FAYETTEVILLE 283			GASTONIA 280			GREENSBORO 270,272 - 274			HICKORY 286			KINSTON 285		
		MAT.	INST.	TOTAL	MAT.	INST.	TOTAL	MAT.	INST.	TOTAL	MAT.	INST.	TOTAL	MAT.	INST.	TOTAL	MAT.	INST.	TOTAL
015433	CONTRACTOR EQUIPMENT		105.0	105.0		100.8	100.8		95.6	95.6		100.8	100.8		100.8	100.8		100.8	100.8
0241, 31 - 34	SITE & INFRASTRUCTURE, DEMOLITION	107.1	87.0	93.0	101.6	85.2	90.1	102.2	76.6	84.3	102.5	85.3	90.4	101.3	85.1	90.0	100.4	85.2	89.7
0310	Concrete Forming & Accessories	85.5	43.3	49.1	95.0	60.4	65.1	101.9	39.5	48.0	99.7	45.4	52.8	91.6	37.9	45.2	87.9	43.0	49.2
0320	Concrete Reinforcing	92.8	47.3	70.1	98.0	57.7	77.9	94.8	58.3	76.6	93.7	57.8	75.8	94.4	57.9	76.1	93.9	47.2	70.6
0330	Cast-in-Place Concrete	104.3	47.4	80.7	116.1	48.9	88.2	108.2	53.4	85.5	103.4	48.2	80.5	110.7	48.7	84.9	106.9	45.4	81.3
03	CONCRETE	98.8	47.3	73.4	104.4	57.3	81.2	100.9	49.7	75.6	98.3	50.4	74.7	102.0	47.3	75.0	98.9	46.4	73.0
04	MASONRY	97.5	48.8	67.4	84.7	39.6	56.8	85.8	51.7	64.7	82.9	42.1	57.6	72.2	44.4	55.0	77.5	49.0	59.8
05	METALS	96.1	76.9	90.2	113.2	80.4	103.1	94.9	81.2	90.7	101.9	80.4	95.3	94.3	79.9	89.9	93.2	76.0	87.9
06	WOOD, PLASTICS & COMPOSITES	82.0	44.5	60.9	95.9	66.6	79.4	105.5	37.4	67.1	99.2	45.9	69.1	90.9	36.0	59.9	87.7	43.9	63.0
07	THERMAL & MOISTURE PROTECTION	105.9	44.1	80.8	103.7	46.6	80.5	104.0	46.0	80.5	106.3	43.7	80.9	104.1	42.5	79.1	103.9	43.7	79.5
08	OPENINGS	95.0	39.3	81.8	94.8	59.1	86.4	98.4	42.0	85.1	97.8	47.8	86.0	94.8	37.6	81.2	94.9	44.2	82.9
0920	Plaster & Gypsum Board	100.1	42.0	60.6	102.8	65.4	77.3	105.1	35.3	57.6	110.5	44.0	65.3	97.5	33.9	54.2	97.6	42.0	59.8
0950, 0980	Ceilings & Acoustic Treatment	95.1	42.0	59.7	92.5	65.4	74.5	95.2	35.3	55.3	95.1	44.0	61.1	91.6	33.9	53.2	95.2	42.0	59.8
0960	Flooring	98.3	23.6	76.0	101.7	43.5	84.3	104.8	43.5	86.5	106.2	40.4	86.5	101.4	33.5	81.1	98.7	23.0	76.1
0970, 0990	Wall Finishes & Painting/Coating	103.4	42.9	67.0	107.1	33.9	63.1	107.1	41.1	67.4	103.4	32.3	60.7	107.1	41.1	67.4	107.1	40.1	66.8
09	FINISHES	97.1	39.5	65.0	97.7	55.5	74.2	99.5	39.4	66.0	100.5	42.8	68.3	96.9	36.2	63.1	96.8	38.8	64.5
COVERS	DIVS. 10 - 14, 25, 28, 41, 43, 44, 46	100.0	72.2	94.4	100.0	76.0	95.2	100.0	77.9	95.6	100.0	77.8	95.6	100.0	77.6	95.5	100.0	73.4	94.7
21, 22, 23	FIRE SUPPRESSION, PLUMBING & HVAC	95.3	52.2	78.0	100.3	52.8	81.3	100.4	53.5	81.6	100.2	53.3	81.4	95.5	53.3	78.6	95.5	52.3	78.2
26, 27, 3370	ELECTRICAL, COMMUNICATIONS & UTIL.	94.1	35.3	63.4	95.6	51.9	72.8	96.2	58.7	76.6	93.4	57.0	74.4	94.2	58.7	75.7	94.0	47.2	69.6
MF2010	WEIGHTED AVERAGE	96.9	52.2	77.4	100.9	58.5	82.5	98.4	56.1	80.0	98.8	56.3	80.3	95.6	54.9	77.9	95.3	53.7	77.2

DIVISION		NORTH CAROLINA																NORTH DAKOTA		
		MURPHY 289			RALEIGH 275 - 276			ROCKY MOUNT 278			WILMINGTON 284			WINSTON-SALEM 271			BISMARCK 585		
		MAT.	INST.	TOTAL	MAT.	INST.	TOTAL	MAT.	INST.	TOTAL	MAT.	INST.	TOTAL	MAT.	INST.	TOTAL	MAT.	INST.	TOTAL
015433	CONTRACTOR EQUIPMENT		95.6	95.6		100.8	100.8		100.8	100.8		95.6	95.6		100.8	100.8		99.0	99.0
0241, 31 - 34	SITE & INFRASTRUCTURE, DEMOLITION	103.6	76.5	84.5	103.7	85.3	90.8	105.0	85.3	91.2	103.6	76.8	84.8	102.8	85.3	90.5	100.7	97.6	98.5
0310	Concrete Forming & Accessories	102.6	40.2	48.7	98.7	47.6	54.6	92.1	45.2	51.6	96.8	49.9	56.3	101.6	49.0	56.2	103.6	41.3	49.8
0320	Concrete Reinforcing	93.9	46.9	70.4	100.0	57.8	78.9	92.8	57.6	75.2	95.1	57.6	76.4	93.7	57.6	75.7	105.6	83.2	94.4
0330	Cast-in-Place Concrete	114.5	45.0	85.7	109.3	52.0	85.5	102.0	48.3	79.7	110.3	50.2	85.3	105.9	50.6	83.0	109.5	48.4	84.1
03	CONCRETE	105.3	45.0	75.5	102.0	52.8	77.7	99.5	50.3	75.2	102.2	53.1	77.9	99.6	52.8	76.5	103.6	53.0	78.6
04	MASONRY	74.5	41.9	54.4	88.1	42.9	60.2	76.2	40.2	53.9	72.4	42.4	53.9	83.0	39.3	56.0	110.9	57.4	77.8
05	METALS	92.1	75.3	87.0	94.8	80.4	90.4	95.3	79.6	90.5	93.9	80.2	89.7	99.4	80.2	93.5	97.7	85.4	93.9
06	WOOD, PLASTICS & COMPOSITES	106.2	40.5	69.2	97.9	48.6	70.1	89.5	45.7	64.8	98.8	50.1	71.3	99.2	50.4	71.7	87.3	35.6	58.1
07	THERMAL & MOISTURE PROTECTION	104.0	41.6	78.6	100.6	47.9	79.2	106.3	41.6	80.1	103.8	47.9	81.1	106.3	44.3	81.2	109.3	50.8	85.6
08	OPENINGS	94.7	39.8	81.7	97.0	48.5	85.5	94.3	42.9	82.1	94.9	51.7	84.6	97.8	50.3	86.5	100.9	47.9	88.4
0920	Plaster & Gypsum Board	104.2	38.5	59.5	102.5	46.8	64.6	102.1	43.8	62.5	100.3	48.3	65.0	110.5	48.7	68.5	94.3	33.8	53.1
0950, 0980	Ceilings & Acoustic Treatment	91.6	38.5	56.3	95.9	46.8	63.2	92.4	43.8	60.1	92.5	48.3	63.1	95.1	48.7	64.2	136.9	33.8	68.3
0960	Flooring	105.2	24.5	81.1	103.4	43.5	85.4	101.8	22.2	78.0	102.3	45.3	85.2	106.2	43.5	87.4	109.8	73.4	98.9
0970, 0990	Wall Finishes & Painting/Coating	107.1	40.2	66.9	102.1	37.7	63.4	103.4	39.3	64.9	107.1	39.5	66.5	103.4	37.5	63.8	94.1	30.0	55.6
09	FINISHES	99.0	36.8	64.3	99.7	45.7	69.6	97.9	40.2	65.7	97.6	47.8	69.9	100.5	46.7	70.5	111.7	43.8	73.9
COVERS	DIVS. 10 - 14, 25, 28, 41, 43, 44, 46	100.0	77.5	95.5	100.0	74.5	94.9	100.0	74.4	94.9	100.0	75.7	95.1	100.0	79.3	95.9	100.0	84.8	97.0
21, 22, 23	FIRE SUPPRESSION, PLUMBING & HVAC	95.5	52.0	78.0	100.0	53.6	81.4	95.3	53.9	78.7	100.4	55.1	82.2	100.2	53.5	81.5	100.2	61.3	84.6
26, 27, 3370	ELECTRICAL, COMMUNICATIONS & UTIL.	97.8	30.0	62.4	96.5	41.3	67.7	96.1	40.6	67.1	97.0	51.9	73.4	93.4	57.0	74.4	102.2	73.3	87.1
MF2010	WEIGHTED AVERAGE	96.4	49.2	75.8	98.2	55.1	79.4	96.0	53.2	77.4	97.2	56.6	79.5	98.5	57.1	80.5	102.0	64.0	85.5

DIVISION		NORTH DAKOTA																	
		DEVILS LAKE 583			DICKINSON 586			FARGO 580 - 581			GRAND FORKS 582			JAMESTOWN 584			MINOT 587		
		MAT.	INST.	TOTAL	MAT.	INST.	TOTAL	MAT.	INST.	TOTAL	MAT.	INST.	TOTAL	MAT.	INST.	TOTAL	MAT.	INST.	TOTAL
015433	CONTRACTOR EQUIPMENT		99.0	99.0		99.0	99.0		99.0	99.0		99.0	99.0		99.0	99.0		99.0	99.0
0241, 31 - 34	SITE & INFRASTRUCTURE, DEMOLITION	103.3	95.2	97.6	111.5	93.7	99.0	98.4	97.7	97.9	107.3	93.7	97.8	102.3	93.7	96.3	104.8	97.6	99.8
0310	Concrete Forming & Accessories	102.6	36.0	45.1	91.7	34.9	42.7	102.5	42.2	50.5	95.6	34.8	43.1	93.2	34.6	42.6	91.4	66.0	69.5
0320	Concrete Reinforcing	101.9	83.5	92.7	102.8	41.0	71.9	94.5	83.3	88.9	100.3	83.3	91.8	102.5	48.3	75.4	103.8	83.6	93.7
0330	Cast-in-Place Concrete	122.9	47.0	91.4	111.2	42.0	83.8	120.9	50.3	79.9	111.2	45.1	83.8	121.4	45.0	89.7	111.2	47.5	84.7
03	CONCRETE	110.8	50.2	80.9	110.1	41.0	76.0	104.7	54.0	79.7	107.3	48.7	78.4	109.3	42.2	76.2	105.8	63.7	85.0
04	MASONRY	111.7	64.4	82.4	113.4	59.2	79.8	97.9	57.4	72.9	106.2	63.8	80.0	123.8	33.4	67.8	105.6	64.2	80.0
05	METALS	95.1	84.6	91.9	95.1	64.1	85.5	99.0	86.0	95.0	95.1	79.3	90.2	95.1	65.1	85.8	95.4	86.3	92.6
06	WOOD, PLASTICS & COMPOSITES	88.7	32.8	57.1	76.6	32.8	51.9	86.8	36.1	58.2	80.8	32.8	53.7	78.5	32.8	52.7	76.4	68.7	72.0
07	THERMAL & MOISTURE PROTECTION	106.7	49.4	83.4	107.1	47.3	82.8	108.7	51.8	85.6	106.9	49.4	83.5	106.6	41.5	80.1	106.6	56.0	86.1
08	OPENINGS	96.6	40.8	83.4	96.5	30.9	81.0	98.8	48.2	86.8	96.6	40.8	83.4	96.6	33.0	81.5	96.7	66.0	89.4
0920	Plaster & Gypsum Board	115.2	30.9	57.9	105.2	30.9	54.7	91.5	34.3	52.6	106.7	30.9	55.1	106.3	30.9	55.0	105.2	67.9	79.8
0950, 0980	Ceilings & Acoustic Treatment	142.1	30.9	68.1	142.1	30.9	68.1	138.7	34.3	69.2	142.1	30.9	68.1	142.1	30.9	68.1	142.1	67.9	92.7
0960	Flooring	117.0	36.5	92.9	110.5	36.5	88.4	109.8	73.4	98.9	112.4	36.5	89.7	111.2	36.5	88.9	110.3	86.4	103.1
0970, 0990	Wall Finishes & Painting/Coating	94.9	23.0	51.7	94.9	23.0	51.7	91.0	72.2	79.7	94.9	28.9	55.2	94.9	23.0	51.7	94.9	25.2	53.0
09	FINISHES	117.6	32.9	70.4	115.3	32.9	69.4	112.3	48.8	76.9	115.5	33.6	69.9	114.7	32.9	69.1	114.5	65.4	87.1
COVERS	DIVS. 10 - 14, 25, 28, 41, 43, 44, 46	100.0	33.3	86.6	100.0	33.5	86.7	100.0	84.9	97.0	100.0	33.4	86.7	100.0	82.8	96.6	100.0	88.5	97.7
21, 22, 23	FIRE SUPPRESSION, PLUMBING & HVAC	95.5	74.6	87.1	95.5	67.1	84.1	100.1	77.6	91.1	100.4	33.5	73.8	95.5	35.8	71.5	100.4	61.2	84.7
26, 27, 3370	ELECTRICAL, COMMUNICATIONS & UTIL.	99.3	36.7	66.6	109.1	74.2	90.9	104.8	67.7	85.5	103.3	54.0	78.0	99.3	36.6	66.6	106.9	77.5	91.5
MF2010	WEIGHTED AVERAGE	101.0	58.6	82.5	101.9	57.9	82.7	101.7	67.6	86.9	101.8	51.6	79.9	101.0	45.1	76.7	101.8	71.0	88.4

619

City Cost Indexes

		NORTH DAKOTA			OHIO													
		WILLISTON			AKRON			ATHENS			CANTON			CHILLICOTHE			CINCINNATI		
	DIVISION	588			442 - 443			457			446 - 447			456			451 - 452		
		MAT.	INST.	TOTAL	MAT.	INST.	TOTAL	MAT.	INST.	TOTAL	MAT.	INST.	TOTAL	MAT.	INST.	TOTAL	MAT.	INST.	TOTAL
015433	CONTRACTOR EQUIPMENT		99.0	99.0		95.3	95.3		90.7	90.7		95.3	95.3		100.3	100.3		100.0	100.0
0241, 31 - 34	SITE & INFRASTRUCTURE, DEMOLITION	105.3	93.7	97.2	96.8	102.4	100.7	111.6	92.6	98.3	96.9	102.0	100.4	96.5	104.0	101.8	93.6	103.5	100.5
0310	Concrete Forming & Accessories	97.3	34.9	43.5	100.2	92.2	93.3	94.1	85.8	86.9	100.2	82.3	84.7	96.2	92.1	92.6	98.1	80.6	83.0
0320	Concrete Reinforcing	104.8	41.0	72.9	93.9	93.3	93.6	90.1	86.6	88.4	93.9	76.5	85.2	87.0	81.0	84.0	92.2	81.7	87.0
0330	Cast-in-Place Concrete	111.2	45.2	83.8	92.4	97.4	94.5	111.9	92.2	103.7	93.3	95.2	94.1	101.6	96.1	99.3	93.6	93.6	93.6
03	CONCRETE	107.3	41.0	74.6	92.9	93.5	93.2	106.4	87.6	97.2	93.4	85.3	89.4	100.8	91.3	96.1	95.3	85.5	90.4
04	MASONRY	100.0	59.2	74.7	92.1	94.1	93.3	82.7	89.5	87.0	92.8	84.6	87.7	90.1	96.7	94.2	89.8	85.5	87.1
05	METALS	95.2	64.2	85.7	94.8	82.2	90.9	101.6	78.6	94.5	94.8	75.1	88.7	93.7	84.9	91.0	95.9	84.6	92.4
06	WOOD, PLASTICS & COMPOSITES	82.2	32.8	54.3	99.1	91.4	94.7	83.4	86.8	85.3	99.5	80.8	88.9	96.0	90.6	92.9	98.5	78.5	87.2
07	THERMAL & MOISTURE PROTECTION	106.8	47.3	82.7	111.2	95.8	104.9	99.7	91.9	96.6	112.3	91.6	103.9	101.2	94.3	98.4	99.5	89.9	95.6
08	OPENINGS	96.7	30.9	81.1	115.2	92.5	109.8	99.9	83.6	96.1	108.5	77.2	101.1	91.9	84.0	90.0	99.9	78.9	94.9
0920	Plaster & Gypsum Board	106.7	30.9	55.1	102.0	90.8	94.4	93.9	86.1	88.6	103.1	79.9	87.3	96.6	90.5	92.5	98.3	78.1	84.6
0950, 0980	Ceilings & Acoustic Treatment	142.1	30.9	68.1	98.7	90.8	93.5	102.0	86.1	91.5	98.7	79.9	86.2	96.3	90.5	92.5	97.2	78.1	84.5
0960	Flooring	113.2	36.5	90.3	98.6	92.5	96.7	124.5	100.8	117.4	98.7	82.0	93.7	101.5	100.8	101.3	102.5	90.0	98.7
0970, 0990	Wall Finishes & Painting/Coating	94.9	32.2	57.2	97.1	103.9	101.2	104.9	93.1	97.8	97.1	82.8	88.5	102.2	89.8	94.7	102.2	84.8	91.7
09	FINISHES	115.6	34.0	70.1	100.1	93.3	96.4	102.9	90.1	95.8	100.4	81.8	90.0	100.4	93.7	96.6	100.8	82.2	90.4
COVERS	DIVS. 10 - 14, 25, 28, 41, 43, 44, 46	100.0	33.5	86.7	100.0	95.3	99.1	100.0	53.3	90.6	100.0	93.1	98.6	100.0	93.1	98.6	100.0	91.4	98.3
21, 22, 23	FIRE SUPPRESSION, PLUMBING & HVAC	95.5	67.1	84.1	100.0	94.3	97.7	95.2	52.4	78.0	100.0	82.5	93.0	95.6	94.0	94.9	100.0	81.3	92.5
26, 27, 3370	ELECTRICAL, COMMUNICATIONS & UTIL.	103.7	74.2	88.3	99.3	92.8	95.9	97.8	98.9	98.3	98.5	87.8	92.9	97.3	85.4	91.1	96.1	79.7	87.6
MF2010	WEIGHTED AVERAGE	100.3	58.0	81.9	99.8	93.4	97.0	99.1	80.6	91.1	99.2	85.0	93.0	96.4	92.2	94.5	97.8	84.8	92.1

		OHIO																
		CLEVELAND			COLUMBUS			DAYTON			HAMILTON			LIMA			LORAIN		
	DIVISION	441			430 - 432			453 - 454			450			458			440		
		MAT.	INST.	TOTAL	MAT.	INST.	TOTAL	MAT.	INST.	TOTAL	MAT.	INST.	TOTAL	MAT.	INST.	TOTAL	MAT.	INST.	TOTAL
015433	CONTRACTOR EQUIPMENT		95.6	95.6		94.9	94.9		95.3	95.3		100.3	100.3		93.1	93.1		95.3	95.3
0241, 31 - 34	SITE & INFRASTRUCTURE, DEMOLITION	96.7	102.9	101.1	95.2	99.3	98.1	92.4	103.4	100.1	92.3	103.6	100.2	105.2	92.5	96.3	96.3	101.3	99.8
0310	Concrete Forming & Accessories	100.3	97.9	98.2	99.1	85.9	87.7	98.1	79.0	81.6	98.1	80.1	82.5	94.0	84.3	85.6	100.3	85.3	87.3
0320	Concrete Reinforcing	94.3	93.7	94.0	101.6	81.7	91.7	92.2	80.6	86.4	92.2	81.7	87.0	90.1	79.1	84.6	93.9	93.6	93.7
0330	Cast-in-Place Concrete	90.7	105.5	96.8	90.6	95.9	92.8	87.0	86.0	86.6	93.3	92.8	93.1	102.8	97.2	100.5	88.0	103.5	94.4
03	CONCRETE	92.2	99.0	95.5	95.8	88.3	92.1	92.2	81.5	86.9	95.1	85.0	90.1	99.3	87.6	93.5	90.8	92.6	91.7
04	MASONRY	96.6	105.1	101.9	97.9	94.8	96.0	89.3	81.7	84.6	89.6	84.0	86.1	112.9	81.7	93.6	88.8	98.5	94.8
05	METALS	96.3	85.2	92.9	95.9	80.5	91.2	95.2	77.8	89.8	95.2	85.1	92.1	101.6	80.4	95.1	95.4	83.3	91.6
06	WOOD, PLASTICS & COMPOSITES	98.2	94.8	96.2	97.5	83.7	89.7	99.7	77.3	87.1	98.5	78.5	87.2	83.3	84.0	83.7	99.1	82.0	89.4
07	THERMAL & MOISTURE PROTECTION	109.8	108.1	109.1	101.8	95.7	99.1	104.5	86.4	97.2	101.3	88.7	96.2	99.3	92.5	96.5	112.2	101.1	107.7
08	OPENINGS	104.5	94.3	102.1	103.0	80.4	97.7	99.3	78.0	94.2	96.9	78.9	92.6	99.9	78.3	94.8	108.5	87.4	103.5
0920	Plaster & Gypsum Board	101.2	94.3	96.5	97.6	83.2	87.8	98.3	76.8	83.7	98.3	78.1	84.6	93.9	83.2	86.6	102.0	81.2	87.8
0950, 0980	Ceilings & Acoustic Treatment	97.0	94.3	95.2	100.4	83.2	89.0	98.2	76.8	84.0	97.2	78.1	84.5	101.0	83.2	89.2	98.7	81.2	87.1
0960	Flooring	98.4	105.1	100.4	94.7	87.9	92.6	105.2	80.7	97.9	102.5	90.0	98.7	123.6	88.3	113.1	98.7	105.1	100.6
0970, 0990	Wall Finishes & Painting/Coating	97.1	105.2	102.0	100.9	93.3	96.3	102.2	85.2	92.0	102.2	84.8	91.7	104.9	83.8	92.2	97.1	105.2	102.0
09	FINISHES	99.7	99.6	99.6	97.5	86.6	91.4	101.7	79.5	89.3	100.7	81.9	90.2	101.9	84.7	92.3	100.2	90.4	94.7
COVERS	DIVS. 10 - 14, 25, 28, 41, 43, 44, 46	100.0	102.1	100.4	100.0	93.8	98.8	100.0	90.9	98.2	100.0	90.9	98.2	100.0	94.1	98.8	100.0	98.3	99.7
21, 22, 23	FIRE SUPPRESSION, PLUMBING & HVAC	100.0	99.8	99.9	100.0	92.3	96.9	100.8	81.7	93.1	100.5	80.5	92.5	95.2	84.7	91.0	100.0	85.2	94.0
26, 27, 3370	ELECTRICAL, COMMUNICATIONS & UTIL.	98.8	106.2	102.6	97.1	88.1	92.4	94.7	82.7	88.4	95.1	81.0	87.8	98.1	78.4	87.8	98.7	86.7	92.4
MF2010	WEIGHTED AVERAGE	98.9	100.0	99.4	98.6	89.7	94.7	97.5	83.2	91.3	97.4	84.5	91.8	99.6	84.4	93.0	98.8	90.4	95.2

		OHIO																
		MANSFIELD			MARION			SPRINGFIELD			STEUBENVILLE			TOLEDO			YOUNGSTOWN		
	DIVISION	448 - 449			433			455			439			434 - 436			444 - 445		
		MAT.	INST.	TOTAL	MAT.	INST.	TOTAL	MAT.	INST.	TOTAL	MAT.	INST.	TOTAL	MAT.	INST.	TOTAL	MAT.	INST.	TOTAL
015433	CONTRACTOR EQUIPMENT		95.3	95.3		94.5	94.5		95.3	95.3		99.3	99.3		97.3	97.3		95.3	95.3
0241, 31 - 34	SITE & INFRASTRUCTURE, DEMOLITION	92.7	102.2	99.3	91.1	97.8	95.8	92.7	102.1	99.3	134.7	107.7	115.8	94.4	99.8	98.2	96.7	103.0	101.1
0310	Concrete Forming & Accessories	90.3	83.5	84.4	95.7	81.3	83.3	98.1	83.3	85.4	97.6	87.9	89.2	99.0	92.5	93.4	100.2	85.9	87.9
0320	Concrete Reinforcing	85.4	75.6	80.5	93.7	81.5	87.6	92.2	78.9	85.6	91.1	85.9	88.5	101.6	87.2	94.4	93.9	86.1	90.0
0330	Cast-in-Place Concrete	85.6	94.5	89.3	82.8	92.9	87.0	89.4	87.6	87.9	89.9	94.7	91.9	90.6	100.2	94.6	91.5	97.1	93.8
03	CONCRETE	85.6	85.4	85.5	88.0	85.1	86.5	93.3	83.1	88.3	91.7	89.3	90.5	95.8	93.9	94.9	92.5	89.3	90.9
04	MASONRY	91.2	93.7	92.7	100.0	92.4	95.3	89.5	81.7	84.6	86.5	93.8	91.0	106.6	96.8	100.6	92.4	91.1	91.6
05	METALS	95.6	75.5	89.4	95.0	78.1	89.8	95.1	77.0	89.5	91.6	79.9	87.9	95.7	86.9	93.0	94.8	79.3	90.0
06	WOOD, PLASTICS & COMPOSITES	86.8	82.0	84.1	93.5	81.7	86.9	101.1	83.4	91.1	89.3	86.1	87.5	97.5	91.5	94.1	99.1	84.1	90.6
07	THERMAL & MOISTURE PROTECTION	111.6	93.6	104.3	101.4	82.0	93.5	104.4	87.0	97.4	113.5	96.6	106.6	103.4	100.9	102.4	112.4	92.8	104.4
08	OPENINGS	108.8	76.7	101.2	97.1	77.6	92.5	97.4	78.6	92.9	97.5	82.9	94.0	100.4	88.8	97.6	108.5	84.2	102.7
0920	Plaster & Gypsum Board	95.2	81.2	85.7	95.4	81.2	85.7	98.3	83.2	88.0	95.5	85.1	88.4	97.6	91.2	93.2	102.0	83.3	89.3
0950, 0980	Ceilings & Acoustic Treatment	99.6	81.2	87.4	100.4	81.2	87.7	98.2	83.2	88.2	97.3	85.1	89.2	100.4	91.2	94.3	98.7	83.3	88.5
0960	Flooring	94.6	104.7	97.6	93.5	71.0	86.8	105.2	80.7	97.9	122.7	99.8	115.9	93.9	101.5	96.2	98.7	92.1	96.7
0970, 0990	Wall Finishes & Painting/Coating	97.1	88.5	91.9	100.9	48.5	69.4	102.2	85.2	92.0	114.2	102.6	107.2	100.9	102.1	101.6	97.1	92.6	94.4
09	FINISHES	98.0	87.7	92.3	96.4	76.1	85.1	101.7	83.1	91.4	114.3	91.2	101.4	97.2	94.8	95.9	100.3	87.3	93.0
COVERS	DIVS. 10 - 14, 25, 28, 41, 43, 44, 46	100.0	93.2	98.6	100.0	54.9	91.0	100.0	91.6	98.3	100.0	95.2	99.0	100.0	96.3	99.3	100.0	94.0	98.8
21, 22, 23	FIRE SUPPRESSION, PLUMBING & HVAC	95.1	83.4	90.4	95.2	89.2	93.0	100.8	81.7	93.1	95.5	91.9	94.0	100.1	99.2	99.7	100.0	88.4	95.3
26, 27, 3370	ELECTRICAL, COMMUNICATIONS & UTIL.	96.2	77.7	86.5	91.2	77.7	84.1	94.7	85.6	90.0	86.4	112.7	100.1	97.2	101.6	99.5	98.7	82.9	90.5
MF2010	WEIGHTED AVERAGE	96.6	85.6	91.8	95.0	83.7	90.1	97.5	84.2	91.7	96.8	94.5	95.8	98.7	96.4	97.7	99.1	88.3	94.4

620

City Cost Indexes

DIVISION		OHIO ZANESVILLE 437-438			OKLAHOMA ARDMORE 734			CLINTON 736			DURANT 747			ENID 737			GUYMON 739		
		MAT.	INST.	TOTAL	MAT.	INST.	TOTAL	MAT.	INST.	TOTAL	MAT.	INST.	TOTAL	MAT.	INST.	TOTAL	MAT.	INST.	TOTAL
015433	CONTRACTOR EQUIPMENT		94.5	94.5		82.5	82.5		81.6	81.6		81.6	81.6		81.6	81.6		81.6	81.6
0241, 31 - 34	SITE & INFRASTRUCTURE, DEMOLITION	93.9	99.5	97.9	94.7	91.4	92.4	96.1	90.0	91.8	92.5	89.7	90.6	97.6	90.0	92.2	99.9	89.7	92.7
0310	Concrete Forming & Accessories	92.8	82.6	84.0	94.4	41.3	48.6	92.9	47.7	53.9	85.7	45.1	50.6	96.6	34.8	43.3	100.3	45.9	53.3
0320	Concrete Reinforcing	93.1	85.3	89.2	87.6	80.2	83.9	88.1	80.2	84.2	92.3	80.6	86.4	87.5	80.2	83.9	88.1	80.2	84.1
0330	Cast-in-Place Concrete	87.2	93.3	89.8	99.3	44.2	76.4	95.9	45.9	75.2	92.1	43.5	71.9	95.9	47.2	75.7	96.0	43.9	74.3
03	CONCRETE	91.5	86.5	89.0	97.5	50.0	74.1	97.1	53.5	75.6	92.9	51.5	72.5	97.7	48.2	73.3	100.7	51.9	76.7
04	MASONRY	96.8	83.4	88.5	93.4	58.1	71.5	118.8	58.1	81.2	92.4	63.4	74.5	100.6	58.1	74.3	95.9	54.6	70.3
05	METALS	96.4	80.3	91.4	93.3	70.6	86.3	93.4	70.6	86.4	91.4	71.3	85.2	94.8	70.5	87.3	93.9	70.3	86.6
06	WOOD, PLASTICS & COMPOSITES	89.6	81.7	85.2	100.7	38.7	65.7	99.7	47.3	70.1	89.5	44.0	63.8	103.6	30.0	62.0	107.5	47.0	73.4
07	THERMAL & MOISTURE PROTECTION	101.5	91.1	97.3	106.1	61.0	87.8	106.1	62.0	88.2	99.9	62.3	84.6	106.2	60.3	87.6	106.5	58.4	87.0
08	OPENINGS	97.1	80.2	93.1	93.9	49.6	83.4	93.9	54.0	84.4	93.9	52.7	84.1	93.9	44.4	82.1	94.0	49.8	83.5
0920	Plaster & Gypsum Board	92.8	81.2	84.9	84.3	37.5	52.4	84.0	46.3	58.3	79.1	42.9	54.5	85.1	28.4	46.5	85.7	46.0	58.7
0950, 0980	Ceilings & Acoustic Treatment	100.4	81.2	87.6	86.2	37.5	53.8	86.2	46.3	59.7	85.4	42.9	57.1	86.2	28.4	47.8	88.9	46.0	60.4
0960	Flooring	92.0	86.9	90.5	110.5	43.9	90.6	109.4	41.9	89.2	106.9	63.1	93.8	111.2	41.9	90.5	112.9	24.8	86.5
0970, 0990	Wall Finishes & Painting/Coating	100.9	76.0	85.9	96.5	52.4	70.0	96.5	52.4	70.0	100.1	52.4	71.4	96.5	52.4	70.0	96.5	33.8	58.8
09	FINISHES	95.9	82.2	88.3	93.8	41.0	64.4	93.6	45.8	67.0	91.8	47.7	67.2	94.3	35.4	61.5	95.8	39.7	64.6
COVERS	DIVS. 10 - 14, 25, 28, 41, 43, 44, 46	100.0	90.0	98.0	100.0	78.0	95.6	100.0	79.0	95.8	100.0	78.3	95.7	100.0	77.1	95.4	100.0	77.8	95.5
21, 22, 23	FIRE SUPPRESSION, PLUMBING & HVAC	95.2	89.7	93.0	95.4	66.3	83.7	95.4	66.3	83.7	95.4	65.8	83.5	100.3	66.3	86.6	95.4	64.2	82.9
26, 27, 3370	ELECTRICAL, COMMUNICATIONS & UTIL.	91.5	78.2	84.6	89.7	70.4	79.6	90.7	70.4	80.1	94.7	70.4	82.0	90.7	70.4	80.1	92.3	63.0	77.0
MF2010	WEIGHTED AVERAGE	95.5	85.6	91.2	95.2	62.3	80.9	96.5	63.6	82.2	94.3	64.0	81.1	97.2	60.9	81.4	96.3	60.5	80.7

DIVISION		OKLAHOMA LAWTON 735			MCALESTER 745			MIAMI 743			MUSKOGEE 744			OKLAHOMA CITY 730-731			PONCA CITY 746		
		MAT.	INST.	TOTAL	MAT.	INST.	TOTAL	MAT.	INST.	TOTAL	MAT.	INST.	TOTAL	MAT.	INST.	TOTAL	MAT.	INST.	TOTAL
015433	CONTRACTOR EQUIPMENT		82.5	82.5		81.6	81.6		89.7	89.7		89.7	89.7		82.7	82.7		81.6	81.6
0241, 31 - 34	SITE & INFRASTRUCTURE, DEMOLITION	94.3	91.4	92.2	86.3	89.9	88.9	87.8	87.2	87.4	88.0	87.0	87.3	93.7	91.8	92.3	93.0	89.9	90.8
0310	Concrete Forming & Accessories	100.2	45.3	52.8	83.6	44.6	49.9	96.8	67.7	71.7	101.3	34.1	43.3	98.2	55.5	61.3	92.2	47.3	53.4
0320	Concrete Reinforcing	87.7	80.2	84.0	91.9	80.2	86.1	90.5	80.3	85.4	91.3	79.7	85.5	92.8	80.2	86.5	91.3	80.3	85.8
0330	Cast-in-Place Concrete	92.8	47.3	73.9	80.8	46.0	66.3	84.6	48.3	69.6	85.7	46.6	69.4	94.5	47.7	75.1	94.6	45.5	74.2
03	CONCRETE	93.8	52.9	73.6	83.4	52.1	68.0	88.2	64.0	76.2	90.0	48.3	69.4	97.3	57.5	77.7	95.0	53.1	74.3
04	MASONRY	95.9	58.1	72.5	110.7	57.1	77.5	95.2	57.2	71.7	113.0	46.0	71.5	98.7	57.4	73.1	87.8	57.1	68.8
05	METALS	98.5	70.6	89.9	91.3	70.5	84.9	91.3	82.8	88.6	92.7	81.0	89.1	92.0	70.7	85.4	91.3	70.8	84.9
06	WOOD, PLASTICS & COMPOSITES	106.4	44.1	71.3	86.9	44.0	62.7	102.4	74.6	86.7	107.0	31.4	64.3	100.3	58.2	76.5	97.6	47.0	69.1
07	THERMAL & MOISTURE PROTECTION	106.1	61.7	88.1	99.6	61.0	83.9	99.9	64.9	85.7	100.0	49.5	79.5	98.8	62.8	84.2	100.1	66.7	86.5
08	OPENINGS	95.4	52.5	85.1	93.9	52.5	84.1	93.9	69.3	88.1	93.9	42.8	81.8	94.8	59.7	86.5	93.9	53.8	84.4
0920	Plaster & Gypsum Board	87.1	43.0	57.1	77.9	42.9	54.1	85.0	74.3	77.7	87.2	29.9	48.1	92.9	57.5	68.8	83.5	46.0	58.0
0950, 0980	Ceilings & Acoustic Treatment	94.2	43.0	60.1	85.4	42.9	57.1	85.4	74.3	78.0	94.3	29.7	51.3	97.5	57.5	70.9	85.4	46.0	59.2
0960	Flooring	113.1	41.9	91.8	105.8	41.9	86.7	112.8	63.1	97.9	115.3	40.5	92.9	109.5	41.9	89.3	109.9	41.9	89.6
0970, 0990	Wall Finishes & Painting/Coating	96.5	52.4	70.0	100.1	38.6	63.1	100.1	79.1	87.5	100.1	34.9	60.9	97.9	52.4	70.6	100.1	52.4	71.4
09	FINISHES	96.4	43.8	67.1	90.8	42.0	63.6	93.7	68.8	79.8	96.7	33.7	61.6	98.6	52.0	72.6	93.4	45.3	66.6
COVERS	DIVS. 10 - 14, 25, 28, 41, 43, 44, 46	100.0	78.6	95.7	100.0	78.3	95.6	100.0	82.1	96.4	100.0	76.5	95.3	100.0	80.0	96.0	100.0	78.6	95.7
21, 22, 23	FIRE SUPPRESSION, PLUMBING & HVAC	100.3	66.3	86.7	95.4	61.5	81.8	95.4	61.8	81.9	100.3	60.5	84.3	100.0	66.0	86.4	95.4	61.7	81.9
26, 27, 3370	ELECTRICAL, COMMUNICATIONS & UTIL.	92.3	70.4	80.9	93.1	66.1	79.0	94.6	66.1	79.7	92.7	53.7	72.3	97.3	70.4	83.3	92.7	70.0	80.8
MF2010	WEIGHTED AVERAGE	97.5	63.3	82.6	93.7	61.1	79.5	94.0	68.2	82.8	96.6	56.3	79.0	97.3	65.4	83.4	94.3	62.5	80.5

DIVISION		OKLAHOMA POTEAU 749			SHAWNEE 748			TULSA 740-741			WOODWARD 738			OREGON BEND 977			EUGENE 974		
		MAT.	INST.	TOTAL	MAT.	INST.	TOTAL	MAT.	INST.	TOTAL	MAT.	INST.	TOTAL	MAT.	INST.	TOTAL	MAT.	INST.	TOTAL
015433	CONTRACTOR EQUIPMENT		89.0	89.0		81.6	81.6		89.7	89.7		81.6	81.6		97.9	97.9		97.9	97.9
0241, 31 - 34	SITE & INFRASTRUCTURE, DEMOLITION	74.3	85.9	82.4	96.0	89.9	91.7	94.2	87.4	89.4	96.3	90.0	91.9	106.8	102.8	103.9	96.9	102.7	101.0
0310	Concrete Forming & Accessories	90.0	41.9	48.5	85.5	44.7	50.3	101.4	41.1	49.4	93.1	47.8	54.0	111.2	100.6	102.0	107.6	100.4	101.4
0320	Concrete Reinforcing	92.4	80.3	86.3	91.3	80.2	85.8	91.6	80.2	85.9	87.5	80.3	83.9	95.7	101.5	98.6	99.7	101.5	100.6
0330	Cast-in-Place Concrete	84.6	45.7	68.5	97.6	43.7	75.2	93.3	47.0	74.1	95.9	46.1	75.2	94.2	104.1	98.3	91.2	104.0	96.5
03	CONCRETE	90.6	51.6	71.3	96.6	51.4	74.3	95.3	51.7	73.8	97.4	53.6	75.8	104.8	101.6	103.2	96.3	101.5	98.9
04	MASONRY	95.5	57.2	71.8	111.9	57.1	78.0	96.1	61.0	74.4	89.5	58.1	70.1	98.7	104.5	102.3	96.0	104.5	101.3
05	METALS	91.3	82.4	88.5	91.2	70.4	84.8	95.9	82.0	91.6	93.5	71.0	86.5	90.7	96.6	92.5	91.4	96.4	93.0
06	WOOD, PLASTICS & COMPOSITES	94.3	40.0	63.7	89.3	44.0	63.7	106.2	37.5	67.4	99.8	47.0	70.0	103.3	100.3	101.6	98.8	100.3	99.6
07	THERMAL & MOISTURE PROTECTION	100.0	61.0	84.2	100.0	60.3	83.9	100.0	62.6	84.8	106.2	67.1	90.3	104.6	96.3	101.2	103.9	93.4	99.6
08	OPENINGS	93.9	50.5	83.6	93.9	52.5	84.1	95.4	48.1	84.2	93.9	53.8	84.4	93.8	103.8	96.2	94.2	103.8	96.5
0920	Plaster & Gypsum Board	82.0	38.7	52.5	79.1	42.9	54.5	87.2	36.0	52.4	84.6	46.0	58.4	103.7	100.2	101.3	101.1	100.2	100.5
0950, 0980	Ceilings & Acoustic Treatment	85.4	38.7	54.3	85.4	42.9	57.1	94.3	36.0	55.5	88.9	46.0	60.4	92.3	100.2	97.5	93.3	100.2	97.9
0960	Flooring	109.3	63.1	95.4	106.9	33.3	84.9	114.0	43.6	92.9	109.4	43.9	89.8	107.6	105.3	106.9	106.1	105.3	105.9
0970, 0990	Wall Finishes & Painting/Coating	100.1	52.4	71.4	100.1	34.9	60.9	100.1	41.4	64.8	96.5	52.4	70.0	104.6	72.3	85.2	104.6	72.3	85.2
09	FINISHES	91.5	45.3	65.8	92.1	40.1	63.1	96.5	39.7	64.9	94.4	45.9	67.4	103.3	98.1	100.4	101.7	98.1	99.7
COVERS	DIVS. 10 - 14, 25, 28, 41, 43, 44, 46	100.0	78.1	95.6	100.0	78.3	95.7	100.0	79.3	95.8	100.0	79.0	95.8	100.0	101.9	100.4	100.0	101.9	100.4
21, 22, 23	FIRE SUPPRESSION, PLUMBING & HVAC	95.4	61.7	81.9	95.4	65.8	83.5	100.3	63.5	85.5	95.4	66.3	83.7	95.1	102.6	98.1	100.0	102.6	101.0
26, 27, 3370	ELECTRICAL, COMMUNICATIONS & UTIL.	92.8	66.1	78.9	94.8	70.4	82.1	94.7	66.1	79.8	92.2	70.4	80.8	95.8	96.3	96.1	94.5	96.3	95.4
MF2010	WEIGHTED AVERAGE	93.5	62.1	79.8	95.7	62.2	81.1	97.4	62.2	82.1	95.3	63.8	81.6	97.3	100.5	98.7	97.1	100.4	98.5

City Cost Indexes

OREGON

DIVISION		KLAMATH FALLS 976			MEDFORD 975			PENDLETON 978			PORTLAND 970 - 972			SALEM 973			VALE 979		
		MAT.	INST.	TOTAL	MAT.	INST.	TOTAL	MAT.	INST.	TOTAL	MAT.	INST.	TOTAL	MAT.	INST.	TOTAL	MAT.	INST.	TOTAL
015433	CONTRACTOR EQUIPMENT		97.9	97.9		97.9	97.9		95.4	95.4		97.9	97.9		97.9	97.9		95.4	95.4
0241, 31 - 34	SITE & INFRASTRUCTURE, DEMOLITION	110.8	102.7	105.1	104.5	102.7	103.3	103.2	96.4	98.4	99.4	102.8	101.7	92.9	102.8	99.8	91.4	96.3	94.9
0310	Concrete Forming & Accessories	104.1	100.3	100.9	103.2	100.3	100.7	104.4	100.7	101.2	108.8	100.6	101.7	107.1	100.6	101.5	110.9	99.7	101.2
0320	Concrete Reinforcing	95.7	101.5	98.6	97.3	101.5	99.4	95.1	101.5	98.3	100.5	101.5	101.0	106.1	101.5	103.8	92.7	101.4	97.0
0330	Cast-in-Place Concrete	94.2	104.0	98.3	94.2	104.0	98.3	95.0	105.4	99.3	93.7	104.1	98.0	88.4	104.1	94.9	74.9	105.0	87.4
03	CONCRETE	107.6	101.5	104.6	102.2	101.5	101.8	88.4	102.1	95.2	97.7	101.6	99.7	96.0	101.6	98.8	74.7	101.5	88.0
04	MASONRY	110.8	104.5	106.9	93.1	104.5	100.2	101.7	104.6	103.5	96.5	104.5	101.5	103.3	104.5	104.1	100.2	104.6	102.9
05	METALS	90.7	96.3	92.5	91.0	96.3	92.6	97.1	97.1	97.1	92.4	96.6	93.7	97.8	96.6	97.4	97.0	96.0	96.7
06	WOOD, PLASTICS & COMPOSITES	93.8	100.3	97.5	92.6	100.3	97.0	95.9	100.4	98.4	92.8	100.3	100.1	96.7	100.3	98.7	104.8	100.4	102.3
07	THERMAL & MOISTURE PROTECTION	104.7	94.8	100.7	104.4	94.8	100.5	97.4	95.4	96.6	103.8	98.1	101.5	101.7	96.3	99.5	96.9	94.8	96.1
08	OPENINGS	93.9	103.8	96.2	96.3	103.8	98.1	89.5	103.9	92.9	92.2	103.8	95.0	95.9	103.8	97.8	89.4	95.8	90.9
0920	Plaster & Gypsum Board	98.2	100.2	99.6	97.5	100.2	99.3	85.2	100.2	95.4	100.9	100.2	100.4	97.3	100.2	99.3	91.9	100.2	97.5
0950, 0980	Ceilings & Acoustic Treatment	100.3	100.2	100.2	107.3	100.2	102.6	64.6	100.2	88.3	95.3	100.2	98.5	100.7	100.2	100.4	64.6	100.2	88.3
0960	Flooring	105.0	105.3	105.1	104.4	105.3	104.7	72.5	105.3	82.3	103.6	105.3	104.1	104.3	105.3	104.6	74.2	105.3	83.5
0970, 0990	Wall Finishes & Painting/Coating	104.6	59.4	77.4	104.6	59.4	77.4	96.1	66.9	78.5	104.3	66.9	81.8	102.8	72.3	84.5	96.1	72.3	81.8
09	FINISHES	104.0	96.7	99.9	104.4	96.7	100.1	71.8	97.6	86.2	101.4	97.5	99.2	100.5	98.1	99.2	72.3	98.2	86.7
COVERS	DIVS. 10 - 14, 25, 28, 41, 43, 44, 46	100.0	101.8	100.4	100.0	101.8	100.4	100.0	92.2	98.4	100.0	101.9	100.4	100.0	101.9	100.4	100.0	102.1	100.4
21, 22, 23	FIRE SUPPRESSION, PLUMBING & HVAC	95.1	102.6	98.1	100.0	102.6	101.0	96.9	115.3	104.3	100.0	102.6	101.0	100.0	102.6	101.1	96.9	71.5	86.7
26, 27, 3370	ELECTRICAL, COMMUNICATIONS & UTIL.	94.6	81.3	87.6	97.9	81.3	89.2	87.3	96.5	92.1	94.7	103.2	99.1	102.1	96.3	99.1	87.3	73.9	80.3
MF2010	WEIGHTED AVERAGE	98.2	98.2	98.2	98.4	98.2	98.3	93.1	102.5	97.2	97.2	101.4	99.1	99.0	100.5	99.7	91.3	89.8	90.7

PENNSYLVANIA

DIVISION		ALLENTOWN 181			ALTOONA 166			BEDFORD 155			BRADFORD 167			BUTLER 160			CHAMBERSBURG 172		
		MAT.	INST.	TOTAL	MAT.	INST.	TOTAL	MAT.	INST.	TOTAL	MAT.	INST.	TOTAL	MAT.	INST.	TOTAL	MAT.	INST.	TOTAL
015433	CONTRACTOR EQUIPMENT		113.5	113.5		113.5	113.5		110.5	110.5		113.5	113.5		113.5	113.5		112.7	112.7
0241, 31 - 34	SITE & INFRASTRUCTURE, DEMOLITION	90.7	105.4	101.0	94.2	105.5	102.1	103.7	102.2	102.7	89.5	104.8	100.3	85.7	106.9	100.6	85.1	102.7	97.4
0310	Concrete Forming & Accessories	100.5	114.2	112.3	85.7	80.4	81.1	85.2	82.7	83.0	87.9	82.5	83.2	87.2	95.7	94.5	89.5	81.8	82.9
0320	Concrete Reinforcing	94.4	111.5	103.0	91.5	106.0	98.7	89.7	83.5	86.6	93.5	106.4	99.9	92.2	112.5	102.3	88.7	106.3	97.5
0330	Cast-in-Place Concrete	85.0	101.5	91.8	94.7	87.5	91.7	109.6	71.2	93.6	90.4	92.9	91.5	83.7	98.1	89.6	94.9	71.0	85.0
03	CONCRETE	94.3	110.0	102.1	89.5	89.1	89.3	103.4	80.2	91.9	96.0	92.0	94.0	81.7	100.8	91.1	103.3	84.1	93.8
04	MASONRY	94.2	102.2	99.2	96.9	65.6	77.5	114.6	88.8	98.5	94.0	88.5	90.6	98.9	98.5	98.7	94.7	88.9	91.1
05	METALS	99.9	123.1	107.0	93.7	118.3	101.3	97.5	105.6	100.0	97.8	118.2	104.1	93.4	123.4	102.7	99.5	116.5	104.7
06	WOOD, PLASTICS & COMPOSITES	103.0	117.3	111.1	81.2	82.7	82.0	83.7	82.6	83.1	87.6	80.3	83.5	82.6	94.8	89.5	92.5	82.7	87.0
07	THERMAL & MOISTURE PROTECTION	102.5	120.4	109.7	101.8	92.0	97.8	104.0	88.6	97.8	102.4	92.3	98.3	101.5	100.5	101.1	102.8	74.0	91.1
08	OPENINGS	92.5	116.3	98.1	86.4	90.1	87.3	92.1	83.6	90.1	92.6	92.3	92.5	86.4	105.7	91.0	89.7	84.8	88.6
0920	Plaster & Gypsum Board	99.7	117.6	111.8	91.1	81.9	84.8	91.6	81.9	85.0	91.9	79.4	83.4	91.1	94.4	93.3	99.4	81.9	87.5
0950, 0980	Ceilings & Acoustic Treatment	85.7	117.6	106.9	90.1	81.9	84.6	95.7	81.9	86.5	88.4	79.4	82.4	91.1	94.4	93.3	90.3	81.9	84.7
0960	Flooring	93.6	97.6	94.8	87.4	99.3	91.0	96.9	89.4	94.7	88.5	105.8	93.6	88.2	82.5	86.5	93.7	47.0	79.7
0970, 0990	Wall Finishes & Painting/Coating	98.3	72.1	82.5	94.6	108.9	103.2	104.7	83.3	91.9	98.3	92.6	94.9	94.6	108.9	103.2	101.4	83.3	90.7
09	FINISHES	93.8	107.1	101.2	92.2	85.9	88.7	98.3	82.6	89.6	91.9	85.7	88.5	92.1	93.5	92.9	93.9	75.4	83.6
COVERS	DIVS. 10 - 14, 25, 28, 41, 43, 44, 46	100.0	104.5	100.9	100.0	95.3	99.1	100.0	98.6	99.7	100.0	99.3	99.9	100.0	102.3	100.5	100.0	95.2	99.0
21, 22, 23	FIRE SUPPRESSION, PLUMBING & HVAC	100.2	110.6	104.4	99.8	84.3	93.6	95.0	89.1	92.7	95.4	90.9	93.6	95.0	93.9	94.5	95.4	89.0	92.8
26, 27, 3370	ELECTRICAL, COMMUNICATIONS & UTIL.	98.2	98.9	98.6	88.3	111.1	100.2	93.7	111.1	102.8	91.7	111.1	101.8	88.9	110.1	100.0	89.4	88.3	88.9
MF2010	WEIGHTED AVERAGE	97.6	108.8	102.4	94.3	92.7	93.6	98.0	92.6	95.7	95.3	96.8	96.0	92.1	102.2	96.5	96.0	89.7	93.2

PENNSYLVANIA

DIVISION		DOYLESTOWN 189			DUBOIS 158			ERIE 164 - 165			GREENSBURG 156			HARRISBURG 170 - 171			HAZLETON 182		
		MAT.	INST.	TOTAL	MAT.	INST.	TOTAL	MAT.	INST.	TOTAL	MAT.	INST.	TOTAL	MAT.	INST.	TOTAL	MAT.	INST.	TOTAL
015433	CONTRACTOR EQUIPMENT		94.6	94.6		110.5	110.5		113.5	113.5		110.5	110.5		112.7	112.7		113.5	113.5
0241, 31 - 34	SITE & INFRASTRUCTURE, DEMOLITION	103.5	91.4	95.1	108.7	102.5	104.4	91.1	105.9	101.5	99.8	104.7	103.2	85.2	104.5	98.7	84.1	105.7	99.3
0310	Concrete Forming & Accessories	84.7	129.1	123.0	84.7	83.9	84.0	99.9	88.3	89.9	91.5	95.6	95.1	99.8	89.4	90.8	82.3	90.2	89.1
0320	Concrete Reinforcing	91.3	135.0	113.1	89.1	106.7	97.9	93.5	106.5	100.0	89.1	112.5	100.8	98.1	106.7	102.4	91.7	110.5	101.0
0330	Cast-in-Place Concrete	80.3	87.6	83.4	105.7	93.5	100.6	93.1	82.6	88.7	101.7	97.5	100.0	97.1	96.8	97.0	80.3	92.7	85.5
03	CONCRETE	90.0	115.6	102.6	104.8	92.8	98.9	88.6	90.9	89.7	98.4	95.0	99.4	99.2	96.5	97.9	87.0	96.1	91.5
04	MASONRY	97.2	127.9	116.2	114.6	89.6	99.2	86.7	92.8	90.5	125.3	98.5	108.7	95.0	90.6	92.3	106.2	96.5	100.2
05	METALS	97.4	125.6	106.1	97.5	117.6	103.7	93.9	117.0	101.0	97.5	121.9	105.0	106.1	120.3	110.5	99.6	121.4	106.3
06	WOOD, PLASTICS & COMPOSITES	83.1	131.5	110.4	82.6	82.4	82.5	99.2	86.7	92.1	90.5	94.7	92.9	100.1	88.7	93.7	81.6	88.6	85.6
07	THERMAL & MOISTURE PROTECTION	100.6	132.8	113.7	104.3	95.5	100.7	102.3	93.4	98.7	103.9	100.1	102.4	106.2	112.2	108.6	102.0	107.6	104.2
08	OPENINGS	94.7	140.0	105.4	92.1	93.5	92.4	86.6	93.0	88.1	92.1	105.7	95.3	95.8	95.4	95.7	93.0	94.5	93.4
0920	Plaster & Gypsum Board	89.6	132.2	118.6	90.5	81.7	84.5	99.7	86.0	90.4	92.9	94.4	93.9	103.4	88.1	93.0	90.0	88.0	88.7
0950, 0980	Ceilings & Acoustic Treatment	84.9	132.2	116.4	95.7	81.7	86.4	85.7	86.0	85.9	94.8	94.4	94.5	99.4	88.1	91.9	86.7	88.0	87.6
0960	Flooring	78.6	136.8	96.0	96.7	105.8	99.4	92.3	93.5	92.7	99.8	69.3	90.7	98.7	93.4	97.1	86.2	96.3	89.2
0970, 0990	Wall Finishes & Painting/Coating	97.8	70.5	81.4	104.7	108.9	107.2	103.3	92.6	96.9	104.7	108.9	107.2	102.3	91.5	95.8	98.3	105.6	102.7
09	FINISHES	85.1	123.5	106.5	98.6	89.3	93.4	94.3	89.2	91.5	98.7	91.9	94.9	98.8	89.9	93.8	90.2	91.0	90.6
COVERS	DIVS. 10 - 14, 25, 28, 41, 43, 44, 46	100.0	71.9	94.4	100.0	98.6	99.7	100.0	100.9	100.2	100.0	102.1	100.4	100.0	96.4	99.3	100.0	99.1	99.8
21, 22, 23	FIRE SUPPRESSION, PLUMBING & HVAC	95.0	125.4	107.2	95.0	89.5	92.8	99.8	93.1	97.1	95.0	92.8	94.1	100.0	90.7	96.3	95.4	98.8	96.7
26, 27, 3370	ELECTRICAL, COMMUNICATIONS & UTIL.	91.1	126.9	109.8	94.3	111.1	103.1	90.1	87.4	88.7	94.3	111.1	103.1	96.8	88.4	92.4	92.6	91.5	92.0
MF2010	WEIGHTED AVERAGE	94.6	120.8	106.0	98.4	97.1	97.8	94.1	95.0	94.5	98.0	101.6	99.5	99.6	95.9	98.0	95.0	98.8	96.7

City Cost Indexes

PENNSYLVANIA

DIVISION		INDIANA 157			JOHNSTOWN 159			KITTANNING 162			LANCASTER 175-176			LEHIGH VALLEY 180			MONTROSE 188		
		MAT.	INST.	TOTAL	MAT.	INST.	TOTAL	MAT.	INST.	TOTAL	MAT.	INST.	TOTAL	MAT.	INST.	TOTAL	MAT.	INST.	TOTAL
015433	CONTRACTOR EQUIPMENT		110.5	110.5		110.5	110.5		113.5	113.5		112.7	112.7		113.5	113.5		113.5	113.5
0241, 31 - 34	SITE & INFRASTRUCTURE, DEMOLITION	97.9	103.3	101.7	104.2	103.9	104.0	88.2	106.8	101.3	77.7	104.4	96.5	87.9	105.1	99.9	86.6	103.4	98.4
0310	Concrete Forming & Accessories	85.8	86.3	86.2	84.7	84.2	84.3	87.2	95.8	94.6	91.6	88.9	89.3	94.2	113.6	110.9	83.3	88.5	87.8
0320	Concrete Reinforcing	88.4	112.6	100.5	89.7	112.2	100.9	92.2	112.6	102.4	88.4	106.6	97.5	91.7	111.4	101.5	96.0	108.4	102.2
0330	Cast-in-Place Concrete	99.7	97.3	98.7	110.6	92.9	103.2	86.9	98.0	91.5	80.8	97.7	87.8	86.9	100.9	92.7	85.2	90.9	87.6
03	CONCRETE	95.9	96.3	96.1	104.2	93.7	99.0	84.1	100.8	92.4	90.9	96.5	93.7	93.3	109.4	101.2	92.0	94.1	93.1
04	MASONRY	110.3	100.2	104.1	111.1	88.0	96.8	101.7	100.2	100.8	99.9	90.8	94.3	94.2	101.7	98.8	94.1	96.0	95.3
05	METALS	97.6	121.5	105.0	97.5	120.1	104.5	93.5	123.4	102.7	99.5	119.7	105.7	99.5	121.7	106.4	97.9	115.3	103.2
06	WOOD, PLASTICS & COMPOSITES	84.5	82.4	83.3	82.6	82.6	82.6	82.6	94.8	89.5	95.2	88.7	91.5	94.4	117.3	107.3	82.4	88.3	85.8
07	THERMAL & MOISTURE PROTECTION	103.8	97.9	101.4	104.0	93.7	99.8	101.6	100.9	101.3	102.4	98.5	100.8	102.3	103.6	102.8	102.0	92.4	98.1
08	OPENINGS	92.1	95.2	92.8	92.1	91.8	92.0	86.4	102.0	90.1	89.7	99.3	92.0	93.0	116.3	98.5	89.8	93.8	90.8
0920	Plaster & Gypsum Board	92.0	81.7	85.0	90.3	81.9	84.6	91.1	94.4	93.3	101.3	88.1	92.3	93.0	117.6	109.7	90.4	87.7	88.6
0950, 0980	Ceilings & Acoustic Treatment	95.7	81.7	86.4	94.8	81.9	86.2	91.1	94.4	93.3	90.3	88.1	88.8	86.7	117.6	107.2	88.4	87.7	88.0
0960	Flooring	97.4	105.8	99.9	96.7	100.0	97.7	88.2	105.8	93.5	94.5	93.4	94.2	90.9	97.6	92.9	86.7	58.9	78.4
0970, 0990	Wall Finishes & Painting/Coating	104.7	108.9	107.2	104.7	108.9	107.2	94.6	108.9	103.2	101.9	58.4	75.7	98.3	67.9	80.0	98.3	105.6	102.7
09	FINISHES	98.0	90.2	93.7	97.9	88.7	92.8	92.3	97.5	95.2	93.9	86.2	89.6	92.2	105.5	99.6	91.0	84.8	87.5
COVERS	DIVS. 10 - 14, 25, 28, 41, 43, 44, 46	100.0	100.7	100.1	100.0	98.8	99.8	100.0	102.3	100.5	100.0	96.3	99.3	100.0	104.1	100.8	100.0	99.1	99.8
21, 22, 23	FIRE SUPPRESSION, PLUMBING & HVAC	95.0	92.7	94.1	95.0	89.8	92.9	95.0	96.1	95.4	95.4	90.8	93.5	95.4	110.3	101.4	95.4	98.0	96.4
26, 27, 3370	ELECTRICAL, COMMUNICATIONS & UTIL.	94.3	111.1	103.1	94.3	111.1	103.1	88.3	111.1	100.2	90.9	45.1	67.0	92.6	133.5	113.9	91.7	97.2	94.5
MF2010	WEIGHTED AVERAGE	96.9	100.1	98.3	97.9	97.3	97.7	92.5	103.4	97.3	94.8	89.3	92.4	95.4	112.5	102.9	94.4	97.1	95.6

PENNSYLVANIA

DIVISION		NEW CASTLE 161			NORRISTOWN 194			OIL CITY 163			PHILADELPHIA 190-191			PITTSBURGH 150-152			POTTSVILLE 179		
		MAT.	INST.	TOTAL	MAT.	INST.	TOTAL	MAT.	INST.	TOTAL	MAT.	INST.	TOTAL	MAT.	INST.	TOTAL	MAT.	INST.	TOTAL
015433	CONTRACTOR EQUIPMENT		113.5	113.5		98.0	98.0		113.5	113.5		97.5	97.5		111.8	111.8		112.7	112.7
0241, 31 - 34	SITE & INFRASTRUCTURE, DEMOLITION	86.1	106.9	100.7	92.4	99.8	97.6	84.7	105.0	99.0	99.9	99.2	99.4	103.4	106.7	105.7	80.4	104.6	97.4
0310	Concrete Forming & Accessories	87.2	97.4	96.0	83.5	130.8	124.3	87.2	83.5	84.0	98.7	139.9	134.3	99.0	98.7	98.7	83.2	90.0	89.1
0320	Concrete Reinforcing	91.0	93.3	92.2	92.5	142.2	117.3	92.2	93.1	92.6	100.5	142.2	121.3	90.1	112.9	101.5	87.7	103.8	95.7
0330	Cast-in-Place Concrete	84.4	96.9	89.6	79.7	127.5	99.6	82.0	96.5	88.1	94.8	131.7	110.1	105.7	97.7	102.4	86.0	97.9	91.0
03	CONCRETE	82.0	97.6	89.7	86.9	131.4	108.9	80.5	91.2	85.7	98.2	136.9	117.3	101.7	102.0	101.9	94.6	96.6	95.6
04	MASONRY	98.1	98.5	98.4	106.5	125.7	118.4	98.2	95.2	96.3	95.5	130.2	117.0	103.3	103.3	103.3	94.5	92.9	93.5
05	METALS	93.5	114.0	99.8	99.4	130.9	109.1	93.5	112.1	99.2	100.7	131.2	110.1	98.9	122.9	106.3	99.7	118.9	105.6
06	WOOD, PLASTICS & COMPOSITES	82.6	98.1	91.4	81.5	131.4	109.7	82.6	80.3	81.3	97.2	141.5	122.2	100.5	98.0	99.1	84.9	88.6	87.0
07	THERMAL & MOISTURE PROTECTION	101.5	97.0	99.7	103.9	132.6	115.6	101.4	94.9	98.8	102.7	135.7	116.1	104.1	102.0	103.3	102.5	107.1	104.3
08	OPENINGS	86.4	98.0	89.2	88.6	142.0	101.2	86.4	80.7	85.1	94.4	147.5	106.9	95.5	107.5	98.3	89.8	92.5	90.4
0920	Plaster & Gypsum Board	91.1	97.8	95.7	88.2	132.2	118.1	91.1	79.4	83.2	97.2	142.6	128.1	99.8	97.8	98.5	95.7	88.0	90.5
0950, 0980	Ceilings & Acoustic Treatment	91.1	97.8	95.6	90.5	132.2	118.3	91.1	79.4	83.3	91.6	142.6	125.6	95.7	97.8	97.1	90.3	88.0	88.8
0960	Flooring	88.2	58.5	79.4	91.6	139.5	105.9	88.2	105.8	93.5	98.5	139.5	110.7	102.8	108.0	104.3	91.1	93.4	91.8
0970, 0990	Wall Finishes & Painting/Coating	94.6	108.9	103.2	100.8	147.4	128.8	94.6	108.9	103.2	105.9	154.0	134.8	104.7	121.2	114.6	101.9	105.6	104.1
09	FINISHES	92.1	91.2	91.6	92.9	133.9	115.7	92.0	88.3	89.9	98.2	141.6	122.4	100.9	102.1	101.6	92.4	91.9	92.1
COVERS	DIVS. 10 - 14, 25, 28, 41, 43, 44, 46	100.0	102.6	100.5	100.0	111.7	102.3	100.0	100.4	100.1	100.0	120.5	104.1	100.0	102.5	100.5	100.0	96.7	99.3
21, 22, 23	FIRE SUPPRESSION, PLUMBING & HVAC	95.0	93.5	94.4	95.2	125.8	107.4	95.0	92.6	94.0	100.0	130.5	112.2	99.9	101.2	100.4	95.4	98.2	96.5
26, 27, 3370	ELECTRICAL, COMMUNICATIONS & UTIL.	88.9	88.7	88.8	92.0	148.7	121.6	90.8	110.1	100.9	98.6	148.7	124.7	96.7	111.1	104.2	89.0	91.1	90.1
MF2010	WEIGHTED AVERAGE	92.1	97.2	94.4	94.9	129.5	110.0	92.1	97.0	94.3	98.9	133.4	113.9	99.7	105.8	102.3	94.7	98.0	96.1

PENNSYLVANIA

DIVISION		READING 195-196			SCRANTON 184-185			STATE COLLEGE 168			STROUDSBURG 183			SUNBURY 178			UNIONTOWN 154		
		MAT.	INST.	TOTAL	MAT.	INST.	TOTAL	MAT.	INST.	TOTAL	MAT.	INST.	TOTAL	MAT.	INST.	TOTAL	MAT.	INST.	TOTAL
015433	CONTRACTOR EQUIPMENT		117.0	117.0		113.5	113.5		112.7	112.7		113.5	113.5		113.5	113.5		110.5	110.5
0241, 31 - 34	SITE & INFRASTRUCTURE, DEMOLITION	96.9	111.5	107.1	91.1	105.9	101.5	81.6	104.2	97.5	85.6	103.5	98.2	90.9	105.6	101.2	98.5	104.7	102.8
0310	Concrete Forming & Accessories	99.0	90.5	91.7	100.7	90.1	91.5	85.4	80.1	80.9	88.7	89.5	89.4	95.8	88.9	89.9	79.0	95.7	93.4
0320	Concrete Reinforcing	93.8	103.9	98.8	94.4	108.7	101.6	92.7	106.7	99.7	94.7	115.2	104.9	90.1	106.7	98.4	89.1	112.5	100.8
0330	Cast-in-Place Concrete	71.5	98.2	82.6	88.7	92.9	90.5	85.6	66.6	77.7	83.6	73.4	79.4	94.0	96.5	95.0	99.7	97.5	98.8
03	CONCRETE	86.2	96.9	91.5	96.1	95.8	96.0	96.3	81.9	89.2	90.8	89.8	90.3	98.2	96.2	97.2	95.5	100.6	98.0
04	MASONRY	96.3	94.0	94.9	94.6	96.5	95.8	99.5	79.2	87.0	91.8	101.3	97.7	93.9	90.4	91.7	127.7	98.5	109.6
05	METALS	99.7	119.2	105.7	101.9	121.2	107.9	97.6	119.0	104.2	99.6	117.8	105.2	99.4	120.0	105.8	97.3	122.0	104.9
06	WOOD, PLASTICS & COMPOSITES	100.3	88.5	93.7	103.0	88.2	94.7	89.7	82.7	85.7	88.6	88.3	88.4	93.2	88.6	90.6	76.4	94.7	86.7
07	THERMAL & MOISTURE PROTECTION	104.3	113.6	108.1	102.4	96.3	99.9	101.6	92.1	97.7	102.1	85.8	95.5	103.5	105.8	104.4	103.7	100.2	102.3
08	OPENINGS	93.1	98.3	94.3	92.5	93.7	92.8	89.6	90.1	89.7	93.0	89.1	92.1	89.9	93.4	90.7	92.0	105.7	95.3
0920	Plaster & Gypsum Board	100.4	88.0	92.0	101.9	87.6	92.2	93.1	81.9	85.5	91.4	87.7	88.9	95.2	88.0	90.3	88.4	94.4	92.5
0950, 0980	Ceilings & Acoustic Treatment	81.5	88.0	85.8	94.6	87.6	89.9	85.8	81.9	83.2	84.9	87.7	86.8	86.8	88.0	87.6	94.8	94.4	94.5
0960	Flooring	95.5	93.4	94.9	93.6	108.9	98.2	91.4	99.3	93.8	89.0	52.3	78.0	91.8	89.3	91.0	94.4	105.8	97.8
0970, 0990	Wall Finishes & Painting/Coating	99.5	105.6	103.1	98.3	105.6	102.7	98.3	108.9	104.7	98.3	66.0	78.9	101.9	105.6	104.1	104.7	108.9	107.2
09	FINISHES	94.0	91.5	92.6	96.0	94.6	95.2	91.2	85.9	88.3	90.9	80.3	85.0	93.1	90.6	91.7	96.5	97.3	96.9
COVERS	DIVS. 10 - 14, 25, 28, 41, 43, 44, 46	100.0	99.0	99.8	100.0	99.1	99.8	100.0	91.8	98.4	100.0	63.5	92.7	100.0	95.9	99.2	100.0	102.1	100.4
21, 22, 23	FIRE SUPPRESSION, PLUMBING & HVAC	100.2	107.3	103.0	100.2	98.8	99.6	95.4	84.6	91.1	95.4	100.5	97.4	95.4	90.4	93.4	95.0	92.8	94.1
26, 27, 3370	ELECTRICAL, COMMUNICATIONS & UTIL.	98.7	91.1	94.7	98.2	97.2	97.7	90.8	111.1	101.4	92.6	142.5	118.6	89.3	93.0	91.2	91.3	111.1	101.6
MF2010	WEIGHTED AVERAGE	97.1	101.1	98.8	98.3	99.7	98.9	94.9	93.0	94.1	94.9	102.0	98.0	95.5	96.3	95.8	97.1	102.2	99.4

623

City Cost Indexes

		PENNSYLVANIA																	
	DIVISION	WASHINGTON			WELLSBORO			WESTCHESTER			WILKES-BARRE			WILLIAMSPORT			YORK		
		153			169			193			186 - 187			177			173 - 174		
		MAT.	INST.	TOTAL	MAT.	INST.	TOTAL	MAT.	INST.	TOTAL	MAT.	INST.	TOTAL	MAT.	INST.	TOTAL	MAT.	INST.	TOTAL
015433	CONTRACTOR EQUIPMENT		110.5	110.5		113.5	113.5		98.0	98.0		113.5	113.5		113.5	113.5		112.7	112.7
0241, 31 - 34	SITE & INFRASTRUCTURE, DEMOLITION	98.6	104.7	102.9	93.1	103.3	100.3	98.0	97.2	97.4	83.7	105.7	99.2	82.7	104.2	97.8	81.1	104.5	97.5
0310	Concrete Forming & Accessories	85.9	95.8	94.5	87.3	87.1	87.1	90.1	129.3	123.9	91.3	90.4	90.5	92.3	61.5	65.7	86.6	89.5	89.1
0320	Concrete Reinforcing	89.1	112.6	100.9	92.7	108.3	100.5	91.7	109.7	100.7	93.5	110.5	102.0	89.4	66.6	78.0	90.1	106.7	98.4
0330	Cast-in-Place Concrete	99.7	97.6	98.8	89.7	88.8	89.3	88.2	126.7	104.2	80.3	92.6	85.4	79.6	74.3	77.4	86.4	97.9	91.2
03	CONCRETE	96.0	100.6	98.3	98.5	92.8	95.7	94.3	124.4	109.1	87.9	96.2	92.0	86.0	68.8	77.5	95.8	96.9	96.4
04	MASONRY	109.8	100.2	103.9	100.2	90.4	94.2	101.2	125.7	116.3	106.6	96.1	100.1	86.4	94.2	91.2	94.9	90.8	92.4
05	METALS	97.2	122.3	105.0	97.7	115.4	103.1	99.4	114.9	104.2	97.8	121.8	105.2	99.5	100.2	99.7	101.1	120.3	107.0
06	WOOD, PLASTICS & COMPOSITES	84.6	94.7	90.3	87.1	88.6	87.9	88.8	131.4	112.8	91.0	88.6	89.7	89.6	52.2	68.5	88.7	88.7	88.7
07	THERMAL & MOISTURE PROTECTION	103.8	100.6	102.5	102.6	90.3	97.6	104.3	131.7	115.4	102.0	107.4	104.2	103.0	103.3	103.1	102.5	112.2	106.5
08	OPENINGS	92.0	105.7	95.3	92.5	94.0	92.9	88.6	125.7	97.4	89.8	100.4	92.3	89.9	54.8	81.6	89.8	95.4	91.1
0920	Plaster & Gypsum Board	91.8	94.4	93.5	91.3	88.0	89.1	90.0	132.2	118.7	92.3	88.0	89.4	95.7	50.5	64.9	96.8	88.1	90.9
0950, 0980	Ceilings & Acoustic Treatment	94.8	94.4	94.5	85.8	88.0	87.3	90.5	132.2	118.3	88.4	88.0	88.2	90.3	50.5	63.8	89.3	88.1	88.5
0960	Flooring	97.5	105.8	100.0	88.2	51.7	77.3	94.4	139.5	107.9	89.8	96.3	91.7	90.8	50.6	78.8	92.3	93.4	92.6
0970, 0990	Wall Finishes & Painting/Coating	104.7	106.8	106.0	98.3	105.6	102.7	100.8	147.4	128.8	98.3	105.6	102.7	101.9	105.6	104.1	101.9	91.5	95.6
09	FINISHES	97.8	97.3	97.6	91.6	82.6	86.6	94.4	131.9	115.3	91.9	92.3	92.1	92.9	63.1	76.3	92.7	89.9	91.1
COVERS	DIVS. 10 - 14, 25, 28, 41, 43, 44, 46	100.0	102.1	100.4	100.0	95.8	99.2	100.0	115.5	103.1	100.0	99.0	99.8	100.0	93.2	98.6	100.0	96.5	99.3
21, 22, 23	FIRE SUPPRESSION, PLUMBING & HVAC	95.0	96.0	95.4	95.4	89.6	93.0	95.2	125.5	107.3	95.4	98.7	96.7	95.4	91.4	93.8	100.2	90.8	96.4
26, 27, 3370	ELECTRICAL, COMMUNICATIONS & UTIL.	93.7	111.1	102.8	91.7	83.3	87.3	91.9	117.6	105.3	92.6	91.5	92.0	89.8	47.7	67.8	90.9	88.4	89.6
MF2010	WEIGHTED AVERAGE	96.7	103.2	99.5	95.9	92.3	94.3	95.8	121.7	107.1	94.7	99.2	96.7	93.6	79.3	87.4	96.5	96.0	96.3

		PUERTO RICO			RHODE ISLAND						SOUTH CAROLINA								
	DIVISION	SAN JUAN			NEWPORT			PROVIDENCE			AIKEN			BEAUFORT			CHARLESTON		
		009			028			029			298			299			294		
		MAT.	INST.	TOTAL	MAT.	INST.	TOTAL	MAT.	INST.	TOTAL	MAT.	INST.	TOTAL	MAT.	INST.	TOTAL	MAT.	INST.	TOTAL
015433	CONTRACTOR EQUIPMENT		90.2	90.2		102.1	102.1		102.1	102.1		100.5	100.5		100.5	100.5		100.5	100.5
0241, 31 - 34	SITE & INFRASTRUCTURE, DEMOLITION	132.3	90.5	103.0	90.6	104.2	100.2	92.0	104.2	100.6	118.8	87.1	96.5	113.9	85.3	93.8	97.9	86.0	89.5
0310	Concrete Forming & Accessories	93.7	18.1	28.4	102.6	120.6	118.1	103.9	120.6	118.3	98.6	68.9	73.0	97.5	38.8	46.8	96.5	64.4	68.8
0320	Concrete Reinforcing	191.6	13.1	102.5	106.8	153.9	130.3	103.4	153.9	128.6	94.8	67.3	81.0	93.9	26.6	60.3	93.7	61.2	77.5
0330	Cast-in-Place Concrete	103.7	31.2	73.6	80.3	124.1	98.5	96.9	124.1	108.2	80.9	70.8	76.7	80.9	47.8	67.2	94.9	50.2	76.3
03	CONCRETE	110.5	22.7	67.2	96.1	127.4	111.6	103.6	127.4	115.3	101.3	70.4	86.0	98.6	41.7	70.5	94.1	60.2	77.3
04	MASONRY	83.6	17.2	42.5	96.4	128.9	116.5	100.6	128.9	118.1	73.9	62.5	66.8	87.4	33.8	54.3	88.6	41.6	59.6
05	METALS	117.2	34.9	91.8	99.4	126.8	107.8	105.5	126.8	112.8	92.9	85.1	90.5	92.9	69.3	85.6	94.8	81.6	90.7
06	WOOD, PLASTICS & COMPOSITES	97.0	17.9	52.4	101.9	119.8	112.0	104.2	119.8	113.0	99.2	70.1	82.8	97.6	38.7	64.3	96.4	70.0	81.5
07	THERMAL & MOISTURE PROTECTION	129.7	21.3	85.7	101.5	120.4	109.2	102.3	120.4	109.6	105.3	67.7	90.0	105.0	39.6	78.5	104.2	48.2	81.5
08	OPENINGS	150.1	15.1	118.2	98.1	128.0	105.1	100.9	128.0	107.3	92.5	66.3	86.3	92.6	37.5	79.5	96.0	64.8	88.6
0920	Plaster & Gypsum Board	154.5	15.3	59.8	94.7	119.8	111.8	97.0	119.8	112.5	104.6	69.1	80.4	108.1	36.6	59.4	109.8	68.9	82.0
0950, 0980	Ceilings & Acoustic Treatment	230.8	15.3	87.4	95.3	119.8	111.6	95.2	119.8	111.6	96.0	69.1	78.1	99.5	36.6	57.7	99.5	68.9	79.1
0960	Flooring	225.0	13.4	161.7	99.8	140.3	111.9	100.4	140.3	112.4	104.8	69.7	94.3	106.1	52.1	90.0	105.8	59.3	91.9
0970, 0990	Wall Finishes & Painting/Coating	203.3	17.2	91.4	96.0	124.2	112.9	94.4	124.2	112.3	105.7	71.9	85.4	105.7	35.4	63.5	105.7	68.3	83.2
09	FINISHES	214.4	17.4	104.6	98.6	124.7	113.2	98.0	124.7	112.9	102.4	69.5	84.0	103.5	40.7	68.5	101.6	64.9	81.2
COVERS	DIVS. 10 - 14, 25, 28, 41, 43, 44, 46	100.0	16.5	83.3	100.0	105.8	101.2	100.0	105.8	101.2	100.0	73.6	94.7	100.0	72.1	94.4	100.0	70.7	94.1
21, 22, 23	FIRE SUPPRESSION, PLUMBING & HVAC	102.8	14.2	67.3	100.1	108.4	103.4	99.9	108.4	103.3	95.5	64.5	83.0	95.5	37.3	72.1	100.4	54.9	82.1
26, 27, 3370	ELECTRICAL, COMMUNICATIONS & UTIL.	126.9	13.5	67.7	101.5	100.4	100.9	102.5	100.4	101.4	96.8	67.4	81.5	100.7	34.4	66.1	99.0	90.9	94.8
MF2010	WEIGHTED AVERAGE	122.1	24.6	79.7	99.0	116.6	106.6	101.5	116.6	108.1	96.4	70.4	85.1	97.1	45.7	74.7	97.7	66.4	84.1

		SOUTH CAROLINA																SOUTH DAKOTA		
	DIVISION	COLUMBIA			FLORENCE			GREENVILLE			ROCK HILL			SPARTANBURG			ABERDEEN		
		290 - 292			295			296			297			293			574		
		MAT.	INST.	TOTAL	MAT.	INST.	TOTAL	MAT.	INST.	TOTAL	MAT.	INST.	TOTAL	MAT.	INST.	TOTAL	MAT.	INST.	TOTAL
015433	CONTRACTOR EQUIPMENT		100.5	100.5		100.5	100.5		100.5	100.5		100.5	100.5		100.5	100.5		99.0	99.0
0241, 31 - 34	SITE & INFRASTRUCTURE, DEMOLITION	98.0	85.7	89.5	107.9	86.0	92.5	102.9	85.6	90.7	100.4	84.6	89.3	102.7	85.6	90.7	94.3	94.5	94.4
0310	Concrete Forming & Accessories	95.4	46.1	52.9	84.6	46.3	51.5	96.0	46.1	52.9	94.5	39.6	47.1	99.0	46.2	53.5	98.9	37.5	45.9
0320	Concrete Reinforcing	96.8	61.0	78.9	93.3	61.2	77.3	93.2	45.9	69.6	94.1	45.6	69.8	93.2	58.9	76.1	100.5	39.3	69.9
0330	Cast-in-Place Concrete	97.9	50.7	78.3	80.9	50.0	68.1	80.9	49.9	68.0	80.9	44.5	65.8	80.9	50.0	68.1	110.5	44.5	83.1
03	CONCRETE	95.9	52.2	74.3	93.3	52.0	72.9	92.2	49.1	70.9	90.3	44.2	67.5	92.4	51.6	72.3	103.7	41.7	73.1
04	MASONRY	85.5	38.3	56.3	74.0	41.6	54.0	72.0	41.6	53.2	94.1	35.2	57.7	74.0	41.6	54.0	104.8	55.8	74.5
05	METALS	92.1	80.8	88.6	93.7	81.2	89.8	93.7	75.7	88.1	92.9	72.4	86.6	93.7	80.4	89.6	98.4	65.2	88.2
06	WOOD, PLASTICS & COMPOSITES	100.4	45.8	69.6	82.1	45.8	61.6	96.0	45.8	67.7	94.3	39.8	63.5	100.3	45.8	69.6	96.3	36.8	62.8
07	THERMAL & MOISTURE PROTECTION	100.0	44.2	77.4	104.5	45.7	80.6	104.4	45.7	80.6	104.3	40.5	78.4	104.5	45.7	80.6	104.3	48.8	81.8
08	OPENINGS	97.5	51.7	86.7	92.6	51.7	82.9	92.5	48.1	82.0	92.6	40.3	80.2	92.5	51.1	82.7	93.9	37.6	80.6
0920	Plaster & Gypsum Board	102.7	44.0	62.8	97.4	44.0	61.0	103.1	44.0	62.9	102.4	37.7	58.4	106.1	44.0	63.8	106.1	35.1	57.8
0950, 0980	Ceilings & Acoustic Treatment	101.2	44.0	63.1	96.9	44.0	61.7	96.0	44.0	61.4	96.0	37.7	57.2	96.0	44.0	61.4	108.2	35.1	59.6
0960	Flooring	99.7	44.4	83.1	98.4	44.4	82.3	103.7	58.2	90.1	102.9	45.3	85.7	105.0	58.2	91.0	111.1	51.2	93.2
0970, 0990	Wall Finishes & Painting/Coating	102.4	68.3	81.9	105.7	68.3	83.2	105.7	68.3	83.2	105.7	40.0	66.2	105.7	68.3	83.2	95.6	37.9	60.9
09	FINISHES	99.7	47.7	70.7	98.5	48.1	70.4	100.2	50.4	72.5	99.6	40.2	66.5	101.0	50.4	72.8	106.0	39.7	69.1
COVERS	DIVS. 10 - 14, 25, 28, 41, 43, 44, 46	100.0	67.9	93.6	100.0	67.9	93.6	100.0	67.9	93.6	100.0	66.2	93.2	100.0	67.9	93.6	100.0	41.7	88.3
21, 22, 23	FIRE SUPPRESSION, PLUMBING & HVAC	100.0	53.2	81.2	100.4	53.2	81.5	100.4	53.1	81.4	95.5	45.2	75.3	100.4	53.2	81.5	100.2	40.6	76.3
26, 27, 3370	ELECTRICAL, COMMUNICATIONS & UTIL.	99.0	61.1	79.3	96.8	61.1	78.2	99.0	58.0	77.6	99.0	57.9	77.6	99.0	58.0	77.6	99.2	52.4	74.7
MF2010	WEIGHTED AVERAGE	97.2	57.3	79.8	96.2	57.8	79.4	96.2	56.5	78.9	95.6	51.2	76.3	96.4	57.4	79.5	100.1	50.6	78.6

City Cost Indexes

SOUTH DAKOTA

DIVISION		MITCHELL 573			MOBRIDGE 576			PIERRE 575			RAPID CITY 577			SIOUX FALLS 570 - 571			WATERTOWN 572		
		MAT.	INST.	TOTAL	MAT.	INST.	TOTAL	MAT.	INST.	TOTAL	MAT.	INST.	TOTAL	MAT.	INST.	TOTAL	MAT.	INST.	TOTAL
015433	CONTRACTOR EQUIPMENT		99.0	99.0		99.0	99.0		99.0	99.0		99.0	99.0		100.0	100.0		99.0	99.0
0241, 31 - 34	SITE & INFRASTRUCTURE, DEMOLITION	91.2	94.5	93.5	91.2	94.5	93.5	93.0	94.6	94.1	92.8	94.6	94.1	92.4	96.2	95.1	91.1	94.5	93.5
0310	Concrete Forming & Accessories	98.1	37.8	46.1	88.8	37.7	44.7	101.3	39.3	47.8	106.7	37.5	47.0	104.1	41.4	50.0	85.4	37.5	44.1
0320	Concrete Reinforcing	99.9	47.8	73.9	102.5	39.3	70.9	104.3	71.5	87.9	94.0	71.6	82.8	101.7	71.5	86.6	97.1	39.4	68.3
0330	Cast-in-Place Concrete	107.4	43.3	80.8	107.4	44.5	81.3	96.8	43.2	74.6	106.5	43.8	80.5	102.8	44.0	78.4	107.4	47.0	82.3
03	CONCRETE	101.3	42.9	72.5	101.1	41.8	71.8	97.3	48.1	73.0	100.6	47.6	74.4	99.2	49.3	74.6	100.0	42.6	71.7
04	MASONRY	93.9	54.2	69.4	102.5	55.8	73.6	103.2	54.4	73.0	102.6	57.4	74.6	97.5	54.5	70.9	127.0	58.5	84.6
05	METALS	97.4	65.4	87.6	97.5	65.2	87.5	99.7	80.1	93.6	100.2	80.7	94.2	101.4	80.3	94.9	97.4	65.3	87.6
06	WOOD, PLASTICS & COMPOSITES	95.2	37.4	62.6	83.9	37.1	57.5	98.2	37.6	64.0	100.1	34.1	62.9	96.2	40.2	64.6	80.3	36.8	55.8
07	THERMAL & MOISTURE PROTECTION	104.1	46.8	80.9	104.2	48.8	81.7	105.1	47.1	81.5	104.7	49.0	82.1	109.3	49.4	85.0	104.0	49.5	81.9
08	OPENINGS	92.7	38.2	79.8	95.1	37.2	81.4	101.6	47.3	88.7	97.5	45.4	85.2	98.9	48.7	87.0	92.7	37.3	79.6
0920	Plaster & Gypsum Board	104.5	35.7	57.7	97.2	35.4	55.1	97.2	35.9	55.5	105.6	32.3	55.8	97.3	38.6	57.4	95.2	35.1	54.3
0950, 0980	Ceilings & Acoustic Treatment	104.6	35.7	58.7	108.2	35.4	59.7	106.4	35.9	59.5	110.8	32.3	58.6	107.3	38.6	61.6	104.6	35.1	58.4
0960	Flooring	110.7	51.2	93.0	106.4	51.2	89.9	110.5	36.5	88.4	110.4	80.1	101.4	105.9	76.3	97.1	105.1	51.2	89.0
0970, 0990	Wall Finishes & Painting/Coating	95.6	41.5	63.1	95.6	42.6	63.7	99.1	45.5	66.9	95.6	45.5	65.5	93.9	45.5	64.8	95.6	37.9	60.9
09	FINISHES	104.8	40.4	68.9	103.3	40.3	68.2	105.5	38.2	68.0	106.3	44.9	72.1	103.8	47.5	72.4	101.9	39.7	67.2
COVERS	DIVS. 10 - 14, 25, 28, 41, 43, 44, 46	100.0	38.7	87.7	100.0	41.7	88.3	100.0	77.9	95.6	100.0	77.8	95.6	100.0	78.3	95.7	100.0	41.7	88.3
21, 22, 23	FIRE SUPPRESSION, PLUMBING & HVAC	95.3	40.3	73.2	95.3	40.8	73.4	100.0	64.6	85.8	100.2	65.1	86.1	100.1	39.4	75.8	95.3	40.8	73.4
26, 27, 3370	ELECTRICAL, COMMUNICATIONS & UTIL.	97.5	45.0	70.1	99.2	45.9	71.4	104.6	52.4	77.4	95.7	52.5	73.1	97.4	76.9	86.7	96.7	45.9	70.2
MF2010	WEIGHTED AVERAGE	97.5	49.5	76.6	98.1	49.8	77.1	100.8	59.2	82.7	100.0	60.3	82.7	100.0	58.8	82.1	98.5	50.2	77.5

TENNESSEE

DIVISION		CHATTANOOGA 373 - 374			COLUMBIA 384			COOKEVILLE 385			JACKSON 383			JOHNSON CITY 376			KNOXVILLE 377 - 379		
		MAT.	INST.	TOTAL	MAT.	INST.	TOTAL	MAT.	INST.	TOTAL	MAT.	INST.	TOTAL	MAT.	INST.	TOTAL	MAT.	INST.	TOTAL
015433	CONTRACTOR EQUIPMENT		104.8	104.8		98.8	98.8		98.8	98.8		104.9	104.9		98.7	98.7		98.7	98.7
0241, 31 - 34	SITE & INFRASTRUCTURE, DEMOLITION	103.4	99.6	100.7	86.7	88.1	87.7	92.0	87.6	88.9	95.1	97.9	97.1	110.7	88.4	95.1	89.5	88.9	89.1
0310	Concrete Forming & Accessories	97.9	56.6	62.3	82.3	63.5	66.1	82.5	36.3	42.6	89.0	46.0	51.9	84.6	39.4	45.6	96.4	42.4	49.8
0320	Concrete Reinforcing	89.9	63.2	76.6	87.7	63.2	75.4	87.7	63.2	75.4	87.6	64.0	75.8	90.5	62.1	76.3	89.9	62.2	76.1
0330	Cast-in-Place Concrete	100.0	60.4	83.5	91.8	52.4	75.5	104.0	44.3	79.2	101.6	50.1	80.2	80.4	61.0	72.4	93.9	48.8	75.2
03	CONCRETE	93.8	60.8	77.5	93.6	61.2	77.6	103.2	46.3	75.1	94.7	52.7	74.0	100.2	53.2	77.0	91.2	50.3	71.1
04	MASONRY	108.3	49.7	72.1	118.0	55.5	79.3	112.3	44.5	70.4	118.2	48.0	74.8	123.3	45.6	75.2	85.2	53.0	65.3
05	METALS	97.7	87.4	94.5	91.5	87.6	90.3	91.6	87.5	90.3	93.8	87.6	91.9	94.9	86.4	92.3	98.3	86.7	94.7
06	WOOD, PLASTICS & COMPOSITES	107.1	58.6	79.8	73.0	65.9	69.0	73.2	34.4	51.3	88.9	46.4	65.0	79.1	37.8	55.8	93.4	37.8	62.0
07	THERMAL & MOISTURE PROTECTION	100.1	57.6	82.8	93.7	60.2	80.1	94.1	46.5	74.8	96.0	55.0	79.3	95.2	53.5	78.3	93.3	58.2	79.0
08	OPENINGS	101.7	57.1	91.2	92.5	59.3	84.6	92.5	43.2	80.8	100.1	52.9	88.9	98.0	46.2	85.7	95.3	46.2	83.7
0920	Plaster & Gypsum Board	82.0	57.8	65.6	85.1	65.3	71.6	85.1	32.8	49.5	86.2	45.2	58.5	99.3	36.3	56.5	106.9	36.3	58.9
0950, 0980	Ceilings & Acoustic Treatment	97.1	57.8	71.0	81.5	65.3	70.7	81.5	32.8	49.1	90.3	45.2	60.3	93.3	36.3	55.4	94.2	36.3	55.7
0960	Flooring	98.2	57.7	86.1	94.5	20.9	72.5	94.5	59.7	84.1	91.6	41.5	76.6	97.2	41.4	80.5	101.3	56.8	88.0
0970, 0990	Wall Finishes & Painting/Coating	97.0	53.9	71.1	87.7	34.1	55.5	87.7	37.3	57.4	89.2	50.6	66.0	94.3	43.0	63.5	94.3	49.6	67.4
09	FINISHES	94.5	56.6	73.4	92.3	53.2	70.5	92.8	39.3	63.0	91.9	44.7	65.6	99.1	38.9	65.6	92.1	44.6	65.6
COVERS	DIVS. 10 - 14, 25, 28, 41, 43, 44, 46	100.0	42.4	88.5	100.0	49.2	89.8	100.0	42.1	88.4	100.0	44.2	88.8	100.0	44.3	88.8	100.0	46.8	89.3
21, 22, 23	FIRE SUPPRESSION, PLUMBING & HVAC	100.2	61.9	84.9	97.4	79.8	90.3	97.4	74.2	88.1	100.1	69.5	87.8	100.0	60.4	84.1	100.0	65.7	86.2
26, 27, 3370	ELECTRICAL, COMMUNICATIONS & UTIL.	101.9	69.8	85.1	92.6	58.8	75.0	94.4	63.1	78.0	99.5	59.9	78.8	91.9	47.4	68.7	97.8	57.3	76.6
MF2010	WEIGHTED AVERAGE	99.5	65.5	84.7	95.3	67.5	83.2	96.4	60.4	80.7	98.3	62.5	82.7	99.2	56.7	80.7	96.2	60.5	80.6

TENNESSEE / TEXAS

DIVISION		MCKENZIE 382			MEMPHIS 375,380 - 381			NASHVILLE 370 - 372			ABILENE 795 - 796			AMARILLO 790 - 791			AUSTIN 786 - 787		
		MAT.	INST.	TOTAL	MAT.	INST.	TOTAL	MAT.	INST.	TOTAL	MAT.	INST.	TOTAL	MAT.	INST.	TOTAL	MAT.	INST.	TOTAL
015433	CONTRACTOR EQUIPMENT		98.8	98.8		103.2	103.2		105.2	105.2		89.7	89.7		89.7	89.7		89.0	89.0
0241, 31 - 34	SITE & INFRASTRUCTURE, DEMOLITION	91.7	87.7	88.9	93.0	94.6	94.1	97.7	100.3	99.6	94.0	86.7	88.8	92.7	87.7	89.2	95.8	86.8	89.5
0310	Concrete Forming & Accessories	90.1	39.0	46.0	95.4	64.7	68.9	99.3	65.7	70.3	100.3	38.9	47.3	101.2	52.5	59.2	96.9	54.3	60.2
0320	Concrete Reinforcing	87.8	63.9	75.9	95.5	64.5	80.0	97.2	63.6	80.4	89.4	51.7	70.6	93.2	51.7	72.5	91.1	49.4	70.2
0330	Cast-in-Place Concrete	101.8	58.3	83.7	92.4	61.9	79.8	89.4	66.0	79.7	100.1	47.4	78.2	92.5	53.9	76.5	93.3	61.7	80.2
03	CONCRETE	101.9	52.4	77.5	91.6	65.2	78.6	90.6	66.9	78.9	98.3	45.5	72.2	97.5	53.7	75.9	92.7	56.7	74.9
04	MASONRY	116.5	47.0	73.5	100.1	63.2	77.3	91.0	59.3	71.4	98.7	56.9	72.9	100.5	57.7	74.0	99.5	45.8	66.3
05	METALS	91.6	87.8	90.4	99.0	89.3	96.0	99.5	88.7	96.2	100.6	68.5	90.7	96.1	68.9	87.7	97.3	64.8	87.3
06	WOOD, PLASTICS & COMPOSITES	82.1	37.1	56.7	96.4	66.6	79.6	106.2	66.5	83.8	104.2	37.3	66.4	98.2	54.7	73.7	93.4	58.1	73.5
07	THERMAL & MOISTURE PROTECTION	94.1	49.5	76.0	95.5	64.0	82.7	97.1	61.7	82.7	101.2	49.6	80.2	101.8	53.5	82.2	101.1	49.6	80.2
08	OPENINGS	92.5	45.1	81.3	101.3	64.6	92.6	101.5	64.7	92.8	92.8	41.0	80.5	94.3	52.8	84.5	101.9	55.1	90.8
0920	Plaster & Gypsum Board	88.0	35.6	52.4	92.8	65.9	74.5	95.0	65.9	75.2	86.2	35.8	51.9	92.4	53.8	66.2	88.6	57.3	67.3
0950, 0980	Ceilings & Acoustic Treatment	81.5	35.6	51.0	95.4	65.9	75.8	92.1	65.9	74.6	93.3	35.8	55.1	102.3	53.8	70.0	99.4	57.3	71.4
0960	Flooring	97.2	40.1	80.1	98.9	36.7	80.3	97.4	65.4	87.9	113.4	77.9	102.8	108.3	77.9	99.2	105.5	40.2	86.0
0970, 0990	Wall Finishes & Painting/Coating	87.7	50.6	65.4	91.9	56.0	70.3	95.3	73.6	82.2	98.8	50.2	69.6	98.9	49.5	69.2	102.8	40.8	65.6
09	FINISHES	93.9	39.1	63.3	96.5	58.4	75.3	94.8	66.4	79.0	96.2	46.4	68.5	99.3	56.9	75.6	100.1	50.3	72.4
COVERS	DIVS. 10 - 14, 25, 28, 41, 43, 44, 46	100.0	28.2	85.6	100.0	82.1	96.4	100.0	82.6	96.5	100.0	76.1	95.2	100.0	67.7	93.5	100.0	78.3	95.7
21, 22, 23	FIRE SUPPRESSION, PLUMBING & HVAC	97.4	69.8	86.3	100.0	74.9	90.0	100.0	83.2	93.3	100.3	46.0	78.5	100.1	50.8	80.3	100.1	56.1	82.5
26, 27, 3370	ELECTRICAL, COMMUNICATIONS & UTIL.	94.1	64.4	78.6	100.9	64.9	82.1	97.6	63.1	79.6	94.9	48.7	70.7	98.7	64.6	80.9	97.0	62.1	78.8
MF2010	WEIGHTED AVERAGE	96.6	60.4	80.9	98.5	71.3	86.7	97.9	74.1	87.5	98.3	53.6	78.8	98.2	60.0	81.6	98.6	59.1	81.4

625

City Cost Indexes

TEXAS

DIVISION		BEAUMONT 776-777			BROWNWOOD 768			BRYAN 778			CHILDRESS 792			CORPUS CHRISTI 783-784			DALLAS 752-753		
		MAT.	INST.	TOTAL	MAT.	INST.	TOTAL	MAT.	INST.	TOTAL	MAT.	INST.	TOTAL	MAT.	INST.	TOTAL	MAT.	INST.	TOTAL
015433	CONTRACTOR EQUIPMENT		91.2	91.2		89.7	89.7		91.2	91.2		89.7	89.7		96.2	96.2		99.1	99.1
0241, 31-34	SITE & INFRASTRUCTURE, DEMOLITION	89.0	87.6	88.0	100.5	86.1	90.4	80.7	90.3	87.5	103.1	87.0	91.8	138.3	82.8	99.3	107.7	88.2	94.0
0310	Concrete Forming & Accessories	106.7	48.4	56.4	97.7	33.0	41.8	84.5	45.5	50.8	98.7	61.9	66.9	98.8	39.8	47.9	97.5	64.1	68.7
0320	Concrete Reinforcing	89.0	41.5	65.3	91.7	50.7	71.3	91.3	49.5	70.4	89.5	51.0	70.3	84.6	47.2	65.9	99.1	52.8	76.0
0330	Cast-in-Place Concrete	86.8	51.7	72.2	105.3	42.7	79.3	69.6	65.7	68.0	102.7	49.4	80.6	109.1	46.9	83.3	96.4	54.8	79.1
03	CONCRETE	96.2	49.3	73.1	105.2	40.8	73.4	80.7	54.5	67.8	107.8	56.2	82.3	99.0	45.5	72.6	99.6	60.1	80.1
04	MASONRY	99.8	57.3	73.5	130.0	48.3	79.5	137.2	56.3	87.2	102.9	49.8	70.0	88.6	51.4	65.6	101.9	56.5	73.8
05	METALS	95.2	64.6	85.8	97.3	62.8	86.6	94.8	70.5	87.3	98.1	65.7	88.1	92.7	75.0	87.2	100.6	79.6	94.1
06	WOOD, PLASTICS & COMPOSITES	114.0	48.4	77.0	101.5	29.2	60.7	79.0	39.6	56.8	103.5	67.5	83.2	116.8	38.2	72.4	102.3	67.6	82.7
07	THERMAL & MOISTURE PROTECTION	103.5	55.5	84.0	100.1	44.1	77.4	95.4	56.2	79.5	101.7	48.8	80.3	102.3	48.4	80.4	92.1	62.5	80.1
08	OPENINGS	94.4	45.2	82.8	91.6	37.0	78.7	97.4	45.9	85.2	90.0	56.6	82.1	106.7	39.5	90.8	103.9	58.4	93.2
0920	Plaster & Gypsum Board	97.4	47.3	63.4	87.8	27.5	46.8	84.7	38.3	53.1	85.8	66.9	73.0	93.9	36.7	54.9	96.7	66.9	76.4
0950, 0980	Ceilings & Acoustic Treatment	95.5	47.3	63.5	91.7	27.5	49.0	89.0	38.3	55.3	91.5	66.9	75.2	91.5	36.7	55.0	103.4	66.9	79.1
0960	Flooring	109.1	74.3	98.7	109.2	45.5	90.1	84.2	62.9	77.9	111.6	41.6	90.7	115.4	64.2	100.1	107.8	48.6	90.1
0970, 0990	Wall Finishes & Painting/Coating	96.1	48.6	67.6	96.3	29.8	56.3	92.6	62.5	74.5	98.8	38.6	62.7	115.0	57.7	80.6	105.5	51.5	73.1
09	FINISHES	90.2	52.7	69.4	95.3	34.0	61.2	79.5	48.8	62.4	96.4	56.8	74.4	103.7	45.6	71.3	100.4	60.4	78.1
COVERS	DIVS. 10-14, 25, 28, 41, 43, 44, 46	100.0	79.4	95.9	100.0	37.3	87.4	100.0	79.8	96.0	100.0	70.0	94.0	100.0	78.3	95.7	100.0	82.0	96.4
21, 22, 23	FIRE SUPPRESSION, PLUMBING & HVAC	100.2	59.7	83.9	95.3	49.1	76.8	95.3	65.9	83.5	95.4	52.3	78.1	100.2	40.0	76.0	100.0	62.1	84.8
26, 27, 3370	ELECTRICAL, COMMUNICATIONS & UTIL.	92.1	70.8	81.0	93.2	43.7	67.4	90.2	67.7	78.4	94.9	43.4	68.0	91.8	61.4	76.0	95.7	65.2	79.7
MF2010	WEIGHTED AVERAGE	96.7	61.2	81.2	98.5	48.4	76.7	94.1	63.0	80.5	97.9	56.9	80.0	99.7	53.7	79.6	100.1	65.8	85.2

TEXAS

DIVISION		DEL RIO 788			DENTON 762			EASTLAND 764			EL PASO 798-799,885			FORT WORTH 760-761			GALVESTON 775		
		MAT.	INST.	TOTAL	MAT.	INST.	TOTAL	MAT.	INST.	TOTAL	MAT.	INST.	TOTAL	MAT.	INST.	TOTAL	MAT.	INST.	TOTAL
015433	CONTRACTOR EQUIPMENT		89.0	89.0		95.7	95.7		89.7	89.7		89.8	89.8		89.7	89.7		100.1	100.1
0241, 31-34	SITE & INFRASTRUCTURE, DEMOLITION	119.1	85.4	95.4	100.5	79.3	85.6	103.1	86.0	91.1	95.1	86.1	88.8	97.4	87.1	90.1	108.0	88.1	94.1
0310	Concrete Forming & Accessories	95.3	36.5	44.6	104.6	32.3	42.2	98.5	35.8	44.4	97.2	43.8	51.1	96.9	63.5	68.1	94.2	65.6	69.5
0320	Concrete Reinforcing	85.2	46.7	66.0	93.0	51.0	72.1	92.0	50.7	71.3	97.5	45.8	71.7	97.3	52.7	75.0	90.8	62.8	76.8
0330	Cast-in-Place Concrete	117.4	42.8	86.4	81.5	44.4	66.1	111.3	43.3	83.1	96.6	41.3	73.7	98.6	50.7	78.7	92.3	66.9	81.8
03	CONCRETE	119.8	41.7	81.3	82.8	42.0	62.7	110.0	42.4	76.6	99.9	44.3	72.5	100.7	57.7	79.5	97.2	67.1	82.3
04	MASONRY	103.6	50.0	70.4	138.8	47.1	82.1	99.9	48.2	67.9	96.5	46.9	65.8	98.4	56.4	72.4	95.9	56.4	71.5
05	METALS	92.5	60.8	82.7	96.9	75.5	90.3	97.1	63.2	86.6	95.3	62.6	85.2	98.8	68.5	89.5	96.3	88.8	94.0
06	WOOD, PLASTICS & COMPOSITES	97.0	34.6	61.8	112.9	31.9	67.2	108.3	32.8	65.7	99.9	45.8	69.4	99.5	67.5	81.4	96.4	66.4	79.4
07	THERMAL & MOISTURE PROTECTION	99.2	43.6	76.6	98.3	44.1	76.3	100.5	44.2	77.6	98.9	54.5	80.9	98.4	52.6	79.8	94.5	65.3	82.6
08	OPENINGS	99.3	33.9	83.9	107.8	37.5	91.1	72.9	39.0	64.9	91.2	43.9	80.0	98.7	58.3	89.2	101.8	64.0	92.9
0920	Plaster & Gypsum Board	89.7	33.0	51.2	92.3	30.2	50.0	87.8	31.2	49.3	93.7	44.6	60.3	92.0	66.9	74.9	92.0	65.7	74.1
0950, 0980	Ceilings & Acoustic Treatment	87.8	33.0	51.4	96.2	30.2	52.2	91.7	31.2	51.4	95.9	44.6	61.7	102.3	66.9	78.8	91.6	65.7	74.4
0960	Flooring	97.3	34.7	78.6	102.5	43.5	84.8	141.2	45.5	112.6	114.0	63.5	98.9	137.1	41.4	108.5	98.1	62.9	87.6
0970, 0990	Wall Finishes & Painting/Coating	101.3	38.4	63.5	106.2	29.6	60.2	97.8	30.5	57.4	93.2	34.9	58.2	95.2	51.1	68.7	104.3	47.0	69.9
09	FINISHES	95.8	36.0	62.5	92.0	33.5	59.4	104.8	36.3	66.6	99.2	46.5	69.9	107.4	58.8	80.3	88.8	63.0	74.4
COVERS	DIVS. 10-14, 25, 28, 41, 43, 44, 46	100.0	37.1	87.4	100.0	36.0	87.2	100.0	38.4	87.7	100.0	77.3	95.4	100.0	81.7	96.3	100.0	84.3	96.9
21, 22, 23	FIRE SUPPRESSION, PLUMBING & HVAC	95.2	46.9	75.8	95.3	37.9	72.3	95.3	49.4	76.9	100.0	34.2	73.6	100.0	57.7	83.1	95.2	66.0	83.5
26, 27, 3370	ELECTRICAL, COMMUNICATIONS & UTIL.	94.0	30.1	60.6	95.6	43.8	68.6	93.1	43.8	67.3	92.7	54.9	73.0	94.5	62.4	77.8	91.7	69.5	80.1
MF2010	WEIGHTED AVERAGE	99.4	46.2	76.3	98.2	46.6	75.7	96.5	49.1	75.9	97.2	50.6	76.9	99.6	62.5	83.5	96.3	69.7	84.7

TEXAS

DIVISION		GIDDINGS 789			GREENVILLE 754			HOUSTON 770-772			HUNTSVILLE 773			LAREDO 780			LONGVIEW 756		
		MAT.	INST.	TOTAL	MAT.	INST.	TOTAL	MAT.	INST.	TOTAL	MAT.	INST.	TOTAL	MAT.	INST.	TOTAL	MAT.	INST.	TOTAL
015433	CONTRACTOR EQUIPMENT		89.0	89.0		96.7	96.7		100.0	100.0		91.2	91.2		89.0	89.0		91.5	91.5
0241, 31-34	SITE & INFRASTRUCTURE, DEMOLITION	105.0	85.4	91.3	101.0	83.5	88.7	106.7	87.9	93.5	95.8	87.9	90.3	99.1	86.8	90.5	97.3	90.3	92.4
0310	Concrete Forming & Accessories	93.0	55.5	60.7	89.1	27.3	35.7	96.3	65.8	69.9	91.8	55.4	60.4	95.3	40.0	47.5	85.5	28.1	35.9
0320	Concrete Reinforcing	85.8	46.8	66.3	99.6	24.9	62.3	90.6	63.3	77.0	91.5	50.5	71.0	85.2	47.6	66.4	98.4	22.7	60.6
0330	Cast-in-Place Concrete	99.5	42.9	76.0	87.8	37.7	67.0	89.5	66.9	80.1	95.6	48.2	75.9	84.1	59.3	73.8	102.1	35.0	74.2
03	CONCRETE	96.5	50.2	73.7	92.2	32.6	62.8	94.9	67.2	81.3	103.8	52.8	78.7	91.4	49.1	70.6	108.4	30.8	70.1
04	MASONRY	111.8	50.0	73.5	159.0	46.1	89.1	95.8	66.6	77.7	135.8	51.6	83.7	97.3	58.6	73.3	154.6	42.0	85.0
05	METALS	92.0	61.8	82.6	97.9	63.6	87.3	99.2	89.7	96.3	94.7	66.3	85.9	95.0	64.0	85.5	91.2	50.5	78.7
06	WOOD, PLASTICS & COMPOSITES	96.0	59.4	75.4	92.3	26.3	55.1	98.8	66.4	80.5	88.3	58.2	71.3	97.0	38.1	63.7	86.2	28.4	53.6
07	THERMAL & MOISTURE PROTECTION	99.9	47.7	78.7	92.2	39.5	70.8	94.1	68.1	83.6	96.3	48.2	76.8	98.2	51.9	79.4	93.5	34.9	69.7
08	OPENINGS	98.5	49.0	86.7	102.1	26.0	84.1	104.4	64.0	94.8	97.4	48.5	85.8	99.2	39.4	85.0	91.7	26.0	76.1
0920	Plaster & Gypsum Board	88.6	58.7	68.3	90.4	24.5	45.5	94.5	65.7	74.9	89.5	57.4	67.7	90.8	36.7	54.0	88.7	26.8	46.6
0950, 0980	Ceilings & Acoustic Treatment	87.8	58.7	68.4	99.0	24.5	49.4	97.0	65.7	76.2	89.0	57.4	67.9	92.2	36.7	55.2	95.4	26.8	49.7
0960	Flooring	97.8	34.7	79.0	103.7	47.8	87.0	99.1	62.9	88.3	87.4	34.7	71.7	97.1	64.2	87.3	108.1	35.1	86.2
0970, 0990	Wall Finishes & Painting/Coating	101.3	30.5	58.8	105.5	25.1	57.2	104.3	62.5	79.2	92.6	31.3	55.8	101.3	54.4	73.1	95.1	29.3	55.6
09	FINISHES	94.7	50.2	69.9	96.9	30.4	59.8	96.0	64.7	78.5	81.9	49.9	64.1	95.2	45.1	67.3	101.0	29.4	61.1
COVERS	DIVS. 10-14, 25, 28, 41, 43, 44, 46	100.0	42.9	88.6	100.0	34.4	86.9	100.0	84.3	96.9	100.0	43.3	88.7	100.0	76.3	95.2	100.0	21.9	84.4
21, 22, 23	FIRE SUPPRESSION, PLUMBING & HVAC	95.3	62.5	82.1	95.2	43.5	74.4	100.0	66.1	86.4	95.3	63.3	82.5	100.1	38.8	75.5	95.1	27.5	68.0
26, 27, 3370	ELECTRICAL, COMMUNICATIONS & UTIL.	90.8	43.4	66.1	92.3	27.1	58.3	93.7	67.8	80.2	90.2	43.4	65.8	94.2	59.9	76.3	92.7	53.2	72.0
MF2010	WEIGHTED AVERAGE	96.4	55.6	78.6	99.4	42.2	74.5	98.7	70.9	86.6	97.2	56.9	79.6	97.1	53.7	78.2	99.0	40.5	73.5

626

City Cost Indexes

TEXAS

DIVISION		LUBBOCK 793-794			LUFKIN 759			MCALLEN 785			MCKINNEY 750			MIDLAND 797			ODESSA 797		
		MAT.	INST.	TOTAL	MAT.	INST.	TOTAL	MAT.	INST.	TOTAL	MAT.	INST.	TOTAL	MAT.	INST.	TOTAL	MAT.	INST.	TOTAL
015433	CONTRACTOR EQUIPMENT		98.2	98.2		91.5	91.5		96.4	96.4		96.7	96.7		98.2	98.2		89.7	89.7
0241, 31-34	SITE & INFRASTRUCTURE, DEMOLITION	116.1	85.0	94.3	92.5	91.8	92.0	142.5	82.9	100.6	97.4	83.6	87.7	118.7	85.0	95.1	94.2	87.0	89.1
0310	Concrete Forming & Accessories	99.1	51.0	57.6	88.6	56.8	61.2	99.9	35.8	44.6	88.2	36.3	43.4	103.0	51.0	58.2	100.2	50.9	57.7
0320	Concrete Reinforcing	90.5	51.7	71.1	100.2	62.4	81.3	84.8	47.1	66.0	99.6	51.1	75.4	91.4	50.9	71.2	89.4	50.9	70.2
0330	Cast-in-Place Concrete	100.3	48.4	78.8	91.3	44.7	72.0	118.5	45.3	88.1	82.4	45.2	66.9	106.6	48.6	82.5	100.1	47.5	78.3
03	CONCRETE	96.9	52.0	74.7	100.8	54.3	77.8	106.6	43.1	75.3	87.6	44.0	66.0	101.6	51.9	77.1	98.3	50.7	74.8
04	MASONRY	98.1	55.7	71.8	118.7	52.3	77.6	103.2	51.4	71.1	171.0	48.4	95.1	114.8	42.0	69.7	98.7	42.0	63.6
05	METALS	104.1	80.1	96.7	98.0	69.1	89.1	92.4	74.5	86.9	97.8	74.4	90.6	102.4	79.0	95.2	99.9	67.1	89.8
06	WOOD, PLASTICS & COMPOSITES	104.1	54.9	76.3	94.3	59.4	74.6	115.4	33.6	69.2	91.0	36.4	60.2	109.1	54.9	78.5	104.2	54.7	76.3
07	THERMAL & MOISTURE PROTECTION	90.8	52.7	75.3	93.3	48.0	74.9	102.6	45.0	79.2	92.0	45.9	73.3	91.0	47.4	73.3	101.2	46.6	79.0
08	OPENINGS	103.6	49.4	90.8	69.6	53.4	65.8	103.4	36.8	87.6	102.0	39.8	87.3	102.5	49.1	89.9	92.8	49.0	82.4
0920	Plaster & Gypsum Board	86.7	53.8	64.3	87.7	58.7	68.0	94.9	31.9	52.0	89.6	34.8	52.4	88.3	53.8	64.8	86.2	53.8	64.2
0950, 0980	Ceilings & Acoustic Treatment	95.1	53.8	67.6	88.3	58.7	68.6	92.2	31.9	52.1	99.0	34.8	56.3	92.4	53.8	66.7	93.3	53.8	67.0
0960	Flooring	107.0	39.3	86.7	143.4	38.2	112.0	114.9	65.0	100.0	103.3	35.6	83.1	108.2	48.6	90.4	113.4	48.6	94.0
0970, 0990	Wall Finishes & Painting/Coating	110.2	49.5	73.7	95.1	37.8	60.6	115.0	28.9	63.2	105.5	41.2	66.9	110.2	49.5	73.7	98.8	49.5	69.2
09	FINISHES	98.8	48.8	70.9	109.1	52.0	77.3	104.3	39.8	68.4	96.4	36.7	63.1	99.2	50.7	72.1	96.2	50.6	70.8
COVERS	DIVS. 10-14, 25, 28, 41, 43, 44, 46	100.0	77.8	95.6	100.0	70.8	94.2	100.0	76.0	95.2	100.0	39.7	87.9	100.0	78.6	95.7	100.0	78.3	95.6
21, 22, 23	FIRE SUPPRESSION, PLUMBING & HVAC	99.8	45.2	77.9	95.1	63.5	82.4	95.3	30.8	69.4	95.2	41.4	73.6	94.9	45.3	75.0	100.3	45.3	78.2
26, 27, 3370	ELECTRICAL, COMMUNICATIONS & UTIL.	93.6	49.6	70.6	94.0	43.4	67.6	91.6	34.2	61.6	92.4	45.2	67.8	93.5	40.4	65.8	95.0	40.3	66.4
MF2010	WEIGHTED AVERAGE	100.0	56.3	81.0	95.9	59.1	79.9	99.7	46.6	76.6	99.3	48.9	77.3	99.9	53.6	79.7	98.2	52.5	78.3

TEXAS

DIVISION		PALESTINE 758			SAN ANGELO 769			SAN ANTONIO 781-782			TEMPLE 765			TEXARKANA 755			TYLER 757		
		MAT.	INST.	TOTAL	MAT.	INST.	TOTAL	MAT.	INST.	TOTAL	MAT.	INST.	TOTAL	MAT.	INST.	TOTAL	MAT.	INST.	TOTAL
015433	CONTRACTOR EQUIPMENT		91.5	91.5		89.7	89.7		91.5	91.5		89.7	89.7		91.5	91.5		91.5	91.5
0241, 31-34	SITE & INFRASTRUCTURE, DEMOLITION	98.6	91.1	93.3	97.0	88.1	90.8	101.8	91.4	94.5	86.0	85.9	85.9	87.0	91.1	89.9	96.6	91.5	93.1
0310	Concrete Forming & Accessories	80.1	55.8	59.1	98.0	36.5	44.9	94.5	54.8	60.3	101.3	39.7	48.1	95.4	34.0	42.4	90.3	33.1	40.9
0320	Concrete Reinforcing	97.7	50.7	74.2	91.6	50.9	71.3	91.8	49.1	70.5	91.8	47.4	69.6	97.6	52.4	75.0	98.4	53.0	75.7
0330	Cast-in-Place Concrete	83.5	42.6	66.5	99.3	46.6	77.4	84.5	65.4	76.6	81.4	48.4	67.7	84.1	43.4	67.2	100.2	42.5	76.3
03	CONCRETE	102.9	50.9	77.2	100.4	43.9	72.5	92.6	58.3	75.7	86.0	45.2	65.9	93.6	42.0	68.1	108.0	41.4	75.1
04	MASONRY	113.5	48.5	73.3	126.6	41.6	74.0	98.9	66.5	78.8	137.9	48.4	82.5	175.4	44.7	94.5	164.8	54.4	96.5
05	METALS	97.7	62.7	86.9	97.4	66.5	87.9	96.1	68.8	87.6	97.1	62.8	86.5	91.1	66.9	83.6	97.6	68.0	88.5
06	WOOD, PLASTICS & COMPOSITES	84.7	59.4	70.4	101.8	36.1	64.7	96.7	52.8	71.9	111.0	39.4	70.6	98.1	32.3	61.0	96.0	30.5	59.0
07	THERMAL & MOISTURE PROTECTION	93.7	46.4	74.5	100.0	42.1	76.5	95.9	64.3	83.1	99.6	45.0	77.4	93.2	47.2	74.5	93.6	53.2	77.2
08	OPENINGS	69.5	50.7	65.1	91.7	38.8	79.2	99.9	51.7	88.5	69.6	41.8	63.0	91.7	37.9	78.9	69.5	36.9	61.8
0920	Plaster & Gypsum Board	84.4	58.7	66.9	87.8	34.6	51.6	88.8	51.8	63.6	87.8	38.0	53.9	93.2	30.7	50.7	87.7	28.9	47.7
0950, 0980	Ceilings & Acoustic Treatment	88.3	58.7	68.6	91.7	34.6	53.7	91.2	51.8	65.0	91.7	38.0	56.0	95.4	30.7	52.4	88.3	28.9	48.7
0960	Flooring	135.3	58.9	112.5	109.2	34.7	86.9	97.6	64.3	87.6	143.4	41.7	113.0	115.6	59.4	98.7	145.3	37.0	112.9
0970, 0990	Wall Finishes & Painting/Coating	95.1	30.5	56.3	96.3	49.5	68.2	98.5	54.4	72.0	97.8	38.5	62.2	95.1	39.4	61.6	95.1	39.4	61.6
09	FINISHES	106.9	54.4	77.7	95.0	36.4	62.4	94.8	55.4	72.8	104.1	39.6	68.2	103.1	38.1	66.9	109.9	32.6	66.9
COVERS	DIVS. 10-14, 25, 28, 41, 43, 44, 46	100.0	40.7	88.1	100.0	75.0	95.0	100.0	80.6	96.1	100.0	36.8	87.3	100.0	74.9	95.0	100.0	74.7	94.9
21, 22, 23	FIRE SUPPRESSION, PLUMBING & HVAC	95.1	58.5	80.4	95.3	44.6	75.0	100.0	64.5	85.8	95.3	48.2	76.4	95.1	32.9	70.2	95.1	57.7	80.1
26, 27, 3370	ELECTRICAL, COMMUNICATIONS & UTIL.	90.1	43.4	65.7	97.2	45.8	70.4	94.3	60.0	76.4	94.3	64.9	79.0	93.8	61.1	76.8	92.7	55.0	73.0
MF2010	WEIGHTED AVERAGE	95.4	55.8	78.2	98.2	49.5	77.0	97.5	64.5	83.1	95.0	52.7	76.6	98.4	49.7	77.2	98.9	54.6	79.6

TEXAS / UTAH

DIVISION		VICTORIA 779			WACO 766-767			WAXAHACHIE 751			WHARTON 774			WICHITA FALLS 763			LOGAN 843		
		MAT.	INST.	TOTAL	MAT.	INST.	TOTAL	MAT.	INST.	TOTAL	MAT.	INST.	TOTAL	MAT.	INST.	TOTAL	MAT.	INST.	TOTAL
015433	CONTRACTOR EQUIPMENT		99.0	99.0		89.7	89.7		96.7	96.7		100.1	100.1		89.7	89.7		97.2	97.2
0241, 31-34	SITE & INFRASTRUCTURE, DEMOLITION	112.7	82.3	91.4	94.2	87.0	89.2	99.0	84.4	88.8	118.4	85.5	95.3	94.9	86.7	89.1	93.4	96.2	95.4
0310	Concrete Forming & Accessories	94.0	35.1	43.2	99.9	39.6	47.9	88.2	59.5	63.5	88.7	36.7	43.9	99.9	37.2	45.8	104.8	58.1	64.5
0320	Concrete Reinforcing	87.2	30.3	58.8	91.5	49.4	70.5	99.6	51.1	75.4	90.7	50.2	70.5	91.5	51.7	71.6	102.3	80.2	91.3
0330	Cast-in-Place Concrete	103.6	41.0	77.7	88.1	65.8	78.8	86.9	41.7	68.1	106.5	51.2	83.5	94.1	44.7	73.6	86.9	72.8	81.0
03	CONCRETE	104.5	38.5	71.9	93.0	51.7	72.6	91.3	53.2	72.5	108.8	46.4	78.0	95.8	43.8	70.1	107.9	67.9	88.2
04	MASONRY	112.4	34.1	64.0	98.3	53.7	70.7	159.7	48.5	90.9	97.0	44.6	64.6	98.8	56.9	72.9	103.9	59.7	76.5
05	METALS	94.9	70.3	87.3	99.4	67.0	89.4	97.9	77.0	91.4	96.3	78.5	90.8	99.4	68.5	89.8	101.0	78.2	94.0
06	WOOD, PLASTICS & COMPOSITES	99.9	35.9	63.8	109.2	35.1	67.4	91.0	67.5	77.8	89.4	32.6	57.3	109.2	35.0	67.3	81.5	56.2	67.2
07	THERMAL & MOISTURE PROTECTION	98.4	39.5	74.5	100.5	46.9	78.8	92.1	46.3	73.5	94.8	49.3	76.3	100.5	50.5	80.2	97.9	61.8	83.3
08	OPENINGS	101.7	35.2	85.9	81.5	42.5	72.2	102.1	56.8	91.3	101.8	37.4	86.6	81.5	40.3	71.7	94.3	57.1	85.5
0920	Plaster & Gypsum Board	89.3	34.3	51.9	88.3	33.6	51.1	90.1	66.9	74.3	86.8	30.9	48.8	88.3	33.4	51.0	75.6	54.8	61.4
0950, 0980	Ceilings & Acoustic Treatment	92.5	34.3	53.8	93.5	33.6	53.6	100.7	66.9	78.2	91.6	30.9	51.2	93.5	33.4	53.5	105.1	54.8	71.6
0960	Flooring	97.8	38.2	80.0	142.3	45.1	113.3	103.3	38.2	83.8	96.1	45.5	80.9	143.5	77.9	123.9	100.2	48.6	84.8
0970, 0990	Wall Finishes & Painting/Coating	104.5	37.8	64.4	97.8	49.8	69.0	105.5	37.8	64.8	104.3	31.3	60.5	100.5	49.5	69.9	102.9	62.0	78.3
09	FINISHES	87.0	36.0	58.6	104.7	40.2	68.8	97.1	54.8	73.5	88.3	36.7	59.5	105.3	45.0	71.7	98.5	56.0	74.8
COVERS	DIVS. 10-14, 25, 28, 41, 43, 44, 46	100.0	36.6	87.3	100.0	78.2	95.6	100.0	40.6	88.1	100.0	38.7	87.7	100.0	65.2	93.0	100.0	54.1	90.8
21, 22, 23	FIRE SUPPRESSION, PLUMBING & HVAC	95.3	31.2	69.6	100.2	54.2	81.8	95.2	40.5	73.2	95.2	63.4	82.5	100.2	48.4	79.5	100.0	69.0	87.6
26, 27, 3370	ELECTRICAL, COMMUNICATIONS & UTIL.	96.5	39.7	66.9	97.4	65.1	80.5	92.4	65.2	78.2	95.3	43.8	68.4	99.1	61.0	79.2	95.4	74.3	84.4
MF2010	WEIGHTED AVERAGE	98.2	42.7	74.1	97.2	57.3	79.8	99.3	56.4	80.6	98.1	53.8	78.8	97.8	55.0	79.2	99.7	68.8	86.2

627

City Cost Indexes

		UTAH													VERMONT					
DIVISION		OGDEN			PRICE			PROVO			SALT LAKE CITY			BELLOWS FALLS			BENNINGTON		
		842,844			845			846 - 847			840 - 841			051			052		
		MAT.	INST.	TOTAL	MAT.	INST.	TOTAL	MAT.	INST.	TOTAL	MAT.	INST.	TOTAL	MAT.	INST.	TOTAL	MAT.	INST.	TOTAL
015433	CONTRACTOR EQUIPMENT		97.2	97.2		96.3	96.3		96.3	96.3		97.2	97.2		100.6	100.6		100.6	100.6
0241, 31 - 34	SITE & INFRASTRUCTURE, DEMOLITION	82.2	96.2	92.1	90.8	94.6	93.5	89.8	94.6	93.2	81.8	96.2	91.9	82.6	100.1	94.9	82.0	100.1	94.7
0310	Concrete Forming & Accessories	104.8	58.1	64.5	107.1	49.4	57.3	106.3	58.0	64.6	107.3	58.0	64.7	100.2	87.5	89.3	97.8	109.1	107.6
0320	Concrete Reinforcing	102.0	80.2	91.1	109.8	80.1	95.0	110.7	80.2	95.5	104.3	80.2	92.3	83.0	87.3	85.2	83.0	87.3	85.2
0330	Cast-in-Place Concrete	88.2	72.8	81.8	87.0	60.0	75.8	87.0	72.7	81.1	96.4	72.7	86.6	89.5	113.1	99.3	89.5	113.1	99.3
03	CONCRETE	97.5	67.9	82.9	109.3	59.6	84.8	107.8	67.8	88.1	116.9	67.8	92.6	93.5	96.2	94.8	93.3	105.8	99.5
04	MASONRY	97.8	59.7	74.2	109.1	59.7	78.5	109.3	59.7	78.6	111.0	59.7	79.3	94.6	96.9	96.0	102.1	96.9	98.9
05	METALS	101.5	78.2	94.3	98.3	77.6	91.9	99.3	78.0	92.7	106.1	78.0	97.4	95.6	89.3	93.6	95.5	89.2	93.6
06	WOOD, PLASTICS & COMPOSITES	81.5	56.2	67.2	84.4	45.3	62.3	82.9	56.2	67.8	83.2	56.2	68.0	102.2	85.5	92.8	99.2	115.1	108.1
07	THERMAL & MOISTURE PROTECTION	96.9	61.8	82.6	99.5	59.4	83.3	99.5	61.8	84.2	103.8	61.8	86.8	98.0	84.8	92.7	98.0	83.1	92.0
08	OPENINGS	94.3	57.1	85.5	98.3	50.1	86.9	98.3	57.1	88.5	96.1	57.1	86.9	103.5	88.1	99.9	103.5	104.2	103.7
0920	Plaster & Gypsum Board	75.6	54.8	61.4	77.9	43.5	54.5	76.0	54.8	61.6	83.0	54.8	63.8	99.8	84.4	89.3	97.9	114.9	109.5
0950, 0980	Ceilings & Acoustic Treatment	105.1	54.8	71.6	105.1	43.5	64.2	105.1	54.8	71.6	99.2	54.8	69.6	92.3	84.4	87.1	92.3	114.9	107.3
0960	Flooring	98.0	48.6	83.3	101.2	36.1	81.8	109.9	48.6	85.3	102.2	48.6	86.2	101.6	106.5	103.1	100.8	106.5	102.5
0970, 0990	Wall Finishes & Painting/Coating	102.9	62.0	78.3	102.9	38.3	64.1	102.9	49.0	70.5	106.5	49.0	71.9	95.7	106.8	102.3	95.7	106.8	102.3
09	FINISHES	96.5	56.0	74.0	99.5	44.5	68.9	99.1	54.6	74.3	98.2	54.6	73.9	97.2	92.6	94.6	96.7	110.1	104.2
COVERS	DIVS. 10 - 14, 25, 28, 41, 43, 44, 46	100.0	54.1	90.8	100.0	47.0	89.4	100.0	54.0	90.8	100.0	54.0	90.8	100.0	100.3	100.1	100.0	103.5	100.7
21, 22, 23	FIRE SUPPRESSION, PLUMBING & HVAC	100.0	69.0	87.6	97.5	68.6	85.9	100.0	69.0	87.6	100.1	69.0	87.6	95.2	98.5	96.5	95.2	98.4	96.5
26, 27, 3370	ELECTRICAL, COMMUNICATIONS & UTIL.	95.7	74.3	84.5	101.2	74.3	87.1	96.2	61.2	77.9	98.6	61.2	79.1	98.6	91.5	94.9	98.6	62.6	79.8
MF2010	WEIGHTED AVERAGE	97.9	68.8	85.2	100.1	65.3	84.9	100.1	66.7	85.6	102.2	66.8	86.8	96.6	94.7	95.8	96.9	95.3	96.2

		VERMONT																
DIVISION		BRATTLEBORO			BURLINGTON			GUILDHALL			MONTPELIER			RUTLAND			ST. JOHNSBURY		
		053			054			059			056			057			058		
		MAT.	INST.	TOTAL	MAT.	INST.	TOTAL	MAT.	INST.	TOTAL	MAT.	INST.	TOTAL	MAT.	INST.	TOTAL	MAT.	INST.	TOTAL
015433	CONTRACTOR EQUIPMENT		100.6	100.6		100.6	100.6		100.6	100.6		100.6	100.6		100.6	100.6		100.6	100.6
0241, 31 - 34	SITE & INFRASTRUCTURE, DEMOLITION	83.4	100.1	95.1	86.2	100.0	95.9	81.7	98.7	93.6	86.6	99.8	95.8	85.9	99.8	95.6	81.7	98.7	93.6
0310	Concrete Forming & Accessories	100.5	87.4	89.2	98.0	89.6	90.8	98.1	80.1	82.6	101.8	86.1	88.2	100.8	89.6	91.1	96.5	80.1	82.4
0320	Concrete Reinforcing	82.2	87.3	84.7	105.7	87.2	96.4	83.8	87.2	85.5	89.0	87.2	88.1	103.7	87.2	95.5	82.2	87.2	84.7
0330	Cast-in-Place Concrete	92.3	113.1	100.9	102.6	112.0	106.5	86.8	103.2	93.6	105.0	111.9	107.9	87.7	111.9	97.8	86.8	103.2	93.6
03	CONCRETE	95.5	96.2	95.9	104.0	96.7	100.4	91.0	89.5	90.3	102.8	95.2	99.0	96.9	96.7	96.8	90.7	89.5	90.1
04	MASONRY	101.1	96.9	98.5	104.8	95.4	99.0	101.3	79.6	87.8	100.1	95.4	97.2	85.8	95.4	91.7	124.9	79.6	96.9
05	METALS	95.5	89.2	93.6	104.0	88.4	99.2	95.6	88.3	93.3	100.2	88.3	96.5	101.0	88.4	97.1	95.6	88.3	93.3
06	WOOD, PLASTICS & COMPOSITES	102.6	85.5	92.9	98.3	90.3	93.8	98.5	85.5	91.2	96.4	85.5	90.3	102.7	90.3	95.7	93.6	85.5	89.0
07	THERMAL & MOISTURE PROTECTION	98.1	80.0	90.8	104.3	82.5	95.4	97.9	72.6	87.6	105.0	82.0	95.6	98.2	82.5	91.8	97.8	72.6	87.6
08	OPENINGS	103.5	88.1	99.9	107.5	86.4	102.5	103.5	83.8	98.8	103.8	83.8	99.1	106.6	86.4	101.8	103.5	83.8	98.8
0920	Plaster & Gypsum Board	99.8	84.4	89.3	100.5	89.3	92.9	106.1	84.4	91.3	105.4	84.4	91.1	100.5	89.3	92.9	108.0	84.4	91.9
0950, 0980	Ceilings & Acoustic Treatment	92.3	84.4	87.1	96.0	89.3	91.6	92.3	84.4	87.1	102.5	84.4	90.5	97.0	89.3	91.9	92.3	84.4	87.1
0960	Flooring	101.7	106.5	103.2	103.1	106.5	104.1	104.6	106.5	105.2	104.0	106.5	104.8	101.6	106.5	103.1	107.7	106.5	107.3
0970, 0990	Wall Finishes & Painting/Coating	95.7	106.8	102.3	100.4	85.3	91.4	95.7	85.3	89.4	100.5	85.3	91.4	95.7	85.3	89.4	95.7	85.3	89.4
09	FINISHES	97.3	92.6	94.7	99.1	92.7	95.5	98.8	85.9	91.6	101.0	89.8	94.8	98.4	92.7	95.2	99.9	85.9	92.1
COVERS	DIVS. 10 - 14, 25, 28, 41, 43, 44, 46	100.0	100.3	100.1	100.0	100.2	100.1	100.0	94.3	98.9	100.0	99.7	99.9	100.0	100.2	100.0	100.0	94.3	98.9
21, 22, 23	FIRE SUPPRESSION, PLUMBING & HVAC	95.2	98.4	96.5	99.9	73.3	89.3	95.2	65.3	83.2	95.0	73.3	86.3	100.1	73.3	89.3	95.2	65.3	83.2
26, 27, 3370	ELECTRICAL, COMMUNICATIONS & UTIL.	98.6	91.5	94.9	98.0	62.6	79.5	98.6	62.6	79.8	98.0	62.6	79.5	98.6	62.6	79.8	98.6	62.6	79.8
MF2010	WEIGHTED AVERAGE	97.2	94.5	96.0	101.5	85.1	94.4	96.8	79.2	89.1	99.2	84.3	92.7	99.2	85.1	93.0	97.9	79.2	89.8

		VERMONT			VIRGINIA														
DIVISION		WHITE RIVER JCT.			ALEXANDRIA			ARLINGTON			BRISTOL			CHARLOTTESVILLE			CULPEPER		
		050			223			222			242			229			227		
		MAT.	INST.	TOTAL	MAT.	INST.	TOTAL	MAT.	INST.	TOTAL	MAT.	INST.	TOTAL	MAT.	INST.	TOTAL	MAT.	INST.	TOTAL
015433	CONTRACTOR EQUIPMENT		100.6	100.6		102.1	102.1		100.8	100.8		100.8	100.8		105.1	105.1		100.8	100.8
0241, 31 - 34	SITE & INFRASTRUCTURE, DEMOLITION	85.7	98.8	94.9	113.5	89.3	96.5	123.7	87.3	98.2	107.8	85.7	92.3	112.6	87.7	95.1	111.2	87.2	94.3
0310	Concrete Forming & Accessories	95.0	80.7	82.7	93.3	73.6	76.3	92.5	73.0	75.7	88.4	43.4	49.6	86.8	49.6	54.7	83.9	72.3	73.9
0320	Concrete Reinforcing	83.0	87.2	85.1	85.0	87.0	86.0	95.9	87.0	91.5	95.9	67.3	81.6	95.3	73.0	84.2	95.9	86.9	91.4
0330	Cast-in-Place Concrete	92.3	104.1	97.2	107.5	82.2	97.0	104.6	81.4	95.0	104.1	49.1	81.3	108.3	56.1	86.6	107.1	68.8	91.2
03	CONCRETE	97.3	90.1	93.7	101.2	80.1	90.8	105.6	79.6	92.8	102.1	51.7	77.2	102.8	57.9	80.7	100.1	74.9	87.6
04	MASONRY	112.6	81.2	93.1	86.0	72.9	77.8	99.1	71.3	81.9	89.1	52.3	66.3	112.3	54.2	76.3	100.9	71.2	82.5
05	METALS	95.6	88.3	93.3	102.5	97.0	100.8	101.1	97.4	99.9	99.9	85.8	95.6	100.2	90.6	97.2	100.3	95.4	98.8
06	WOOD, PLASTICS & COMPOSITES	95.9	85.5	90.0	98.1	71.9	83.3	94.4	71.9	81.7	86.6	40.9	60.8	85.1	46.6	63.4	83.7	71.9	77.1
07	THERMAL & MOISTURE PROTECTION	98.2	73.3	88.1	103.2	82.4	94.7	105.1	81.2	95.4	104.7	61.3	87.1	104.3	69.0	90.0	104.5	78.7	94.0
08	OPENINGS	103.5	83.8	98.8	96.4	75.6	91.5	94.6	75.6	90.1	97.4	46.8	85.5	95.7	52.7	85.5	96.0	75.6	91.2
0920	Plaster & Gypsum Board	96.5	84.4	88.3	108.4	70.9	82.9	104.4	70.9	81.6	100.1	38.9	58.4	100.1	44.1	62.0	100.3	70.9	80.3
0950, 0980	Ceilings & Acoustic Treatment	92.3	84.4	87.1	97.7	70.9	79.9	95.1	70.9	79.0	94.2	38.9	57.4	94.2	44.1	60.8	95.1	70.9	79.0
0960	Flooring	99.7	106.5	101.8	106.7	84.5	100.0	105.2	82.7	98.5	101.9	67.7	91.7	100.6	67.7	90.7	100.6	82.7	95.2
0970, 0990	Wall Finishes & Painting/Coating	95.7	85.3	89.4	120.5	79.9	96.1	120.5	79.9	96.1	106.3	55.5	75.7	106.3	77.7	89.1	120.5	77.7	94.8
09	FINISHES	96.5	86.3	90.8	102.9	75.5	87.6	102.6	74.7	87.0	99.2	48.2	70.8	98.8	54.6	74.2	99.6	74.3	85.5
COVERS	DIVS. 10 - 14, 25, 28, 41, 43, 44, 46	100.0	94.9	99.0	100.0	88.5	97.7	100.0	86.0	97.2	100.0	69.5	93.9	100.0	78.0	95.6	100.0	86.0	97.2
21, 22, 23	FIRE SUPPRESSION, PLUMBING & HVAC	95.2	66.1	83.5	100.3	87.7	95.2	100.3	86.8	94.9	95.4	51.6	77.8	95.4	68.7	84.7	95.4	73.2	86.5
26, 27, 3370	ELECTRICAL, COMMUNICATIONS & UTIL.	98.6	62.6	79.8	95.8	99.2	97.6	93.4	99.2	96.5	95.5	36.2	64.6	95.4	70.8	82.6	98.0	99.2	98.6
MF2010	WEIGHTED AVERAGE	97.9	79.7	90.0	99.8	85.4	93.5	100.6	84.7	93.6	98.1	55.6	79.6	99.2	67.3	85.3	98.6	80.8	90.9

628

City Cost Indexes

		VIRGINIA																
	DIVISION	FAIRFAX			FARMVILLE			FREDERICKSBURG			GRUNDY			HARRISONBURG			LYNCHBURG		
		220 - 221			239			224 - 225			246			228			245		
		MAT.	INST.	TOTAL	MAT.	INST.	TOTAL	MAT.	INST.	TOTAL	MAT.	INST.	TOTAL	MAT.	INST.	TOTAL	MAT.	INST.	TOTAL
015433	CONTRACTOR EQUIPMENT		100.8	100.8		105.1	105.1		100.8	100.8		100.8	100.8		100.8	100.8		100.8	100.8
0241, 31 - 34	SITE & INFRASTRUCTURE, DEMOLITION	122.4	87.3	97.8	107.2	87.2	93.2	110.8	87.2	94.2	105.6	84.4	90.7	119.4	85.9	95.9	106.5	85.9	92.1
0310	Concrete Forming & Accessories	86.8	72.8	74.7	100.4	44.6	52.3	86.8	70.8	73.0	91.5	37.2	44.6	82.9	43.5	48.9	88.4	61.6	65.3
0320	Concrete Reinforcing	95.9	87.0	91.5	94.7	51.5	73.1	96.7	86.9	91.8	94.6	52.0	73.3	95.9	66.9	81.5	95.3	67.3	81.3
0330	Cast-in-Place Concrete	104.6	81.3	94.9	106.8	54.1	85.0	106.2	64.7	89.0	104.1	48.1	80.9	104.6	58.8	85.6	104.1	58.4	85.2
03	CONCRETE	105.3	79.4	92.5	101.4	50.9	76.5	99.8	72.8	86.5	100.9	45.3	73.5	103.2	54.9	79.3	100.8	63.0	82.1
04	MASONRY	99.0	71.3	81.8	95.6	45.6	64.7	99.9	71.2	82.1	90.9	48.1	64.4	97.4	51.2	68.8	104.9	54.1	73.5
05	METALS	100.4	97.0	99.3	100.0	78.2	93.3	100.3	96.2	99.0	99.9	71.0	91.0	100.2	85.1	95.5	100.1	86.3	95.9
06	WOOD, PLASTICS & COMPOSITES	86.6	71.9	78.3	101.9	45.0	69.8	86.6	69.6	77.0	89.6	35.8	59.3	82.7	39.8	58.5	86.6	64.0	73.9
07	THERMAL & MOISTURE PROTECTION	105.0	73.9	92.4	105.3	50.2	83.0	104.5	79.1	94.2	104.7	46.1	80.9	104.9	64.5	88.5	104.5	65.1	88.5
08	OPENINGS	94.6	75.6	90.1	94.9	44.0	82.9	95.7	74.0	90.6	97.4	35.2	82.7	96.0	50.6	85.3	96.0	59.4	87.3
0920	Plaster & Gypsum Board	100.3	70.9	80.3	109.3	42.5	63.8	100.3	68.5	78.7	100.1	33.7	54.9	100.1	37.8	57.7	100.1	62.7	74.7
0950, 0980	Ceilings & Acoustic Treatment	95.1	70.9	79.0	92.5	42.5	59.2	95.1	68.5	77.4	94.2	33.7	53.9	94.2	37.8	56.7	94.2	62.7	73.3
0960	Flooring	102.2	82.7	96.4	108.6	61.4	94.5	102.2	82.7	96.4	103.2	34.3	82.6	100.3	75.3	92.9	101.9	67.7	91.7
0970, 0990	Wall Finishes & Painting/Coating	120.5	79.9	96.1	107.7	39.6	66.7	120.5	77.7	94.8	106.3	35.9	64.0	120.5	55.5	81.4	106.3	55.5	75.7
09	FINISHES	101.2	74.7	86.4	101.8	47.0	71.3	100.1	72.9	85.0	99.3	35.9	64.0	100.0	48.9	71.5	99.0	61.5	78.1
COVERS	DIVS. 10 - 14, 25, 28, 41, 43, 44, 46	100.0	86.0	97.2	100.0	46.8	89.3	100.0	81.9	96.4	100.0	44.8	88.9	100.0	69.3	93.8	100.0	72.7	94.5
21, 22, 23	FIRE SUPPRESSION, PLUMBING & HVAC	95.4	86.8	91.9	95.3	44.3	74.9	95.4	86.5	91.8	95.4	63.5	82.6	95.4	67.1	84.1	95.4	68.4	84.6
26, 27, 3370	ELECTRICAL, COMMUNICATIONS & UTIL.	96.6	99.2	98.0	90.4	49.4	69.0	93.6	99.2	96.6	95.5	45.7	69.5	95.7	98.7	97.3	96.6	53.9	74.3
MF2010	WEIGHTED AVERAGE	99.3	84.4	92.8	97.8	53.3	78.5	98.2	83.0	91.6	98.0	53.4	78.6	98.8	68.1	85.5	98.6	66.1	84.5

		VIRGINIA																
	DIVISION	NEWPORT NEWS			NORFOLK			PETERSBURG			PORTSMOUTH			PULASKI			RICHMOND		
		236			233 - 235			238			237			243			230 - 232		
		MAT.	INST.	TOTAL	MAT.	INST.	TOTAL	MAT.	INST.	TOTAL	MAT.	INST.	TOTAL	MAT.	INST.	TOTAL	MAT.	INST.	TOTAL
015433	CONTRACTOR EQUIPMENT		105.1	105.1		105.7	105.7		105.1	105.1		105.0	105.0		100.8	100.8		105.0	105.0
0241, 31 - 34	SITE & INFRASTRUCTURE, DEMOLITION	106.2	88.4	93.7	105.7	89.4	94.3	109.7	88.6	94.9	104.7	88.2	93.1	104.9	85.1	91.0	103.5	88.6	93.0
0310	Concrete Forming & Accessories	99.7	65.4	70.1	99.5	65.6	70.2	93.0	58.3	63.1	88.9	54.2	58.9	91.5	39.5	46.6	100.7	58.3	64.1
0320	Concrete Reinforcing	94.4	72.8	83.6	99.5	72.8	86.2	94.1	73.2	83.6	94.1	72.8	83.4	94.6	66.7	80.7	100.7	73.2	87.0
0330	Cast-in-Place Concrete	103.8	66.8	88.5	111.9	66.9	93.2	110.3	54.2	87.0	102.8	66.7	87.8	104.1	48.9	81.2	98.8	54.2	80.3
03	CONCRETE	98.6	68.6	83.8	103.3	68.7	86.2	103.6	61.2	82.7	97.3	63.5	80.7	100.9	49.6	75.6	97.3	61.2	79.5
04	MASONRY	90.6	54.9	68.5	99.1	54.9	71.7	103.3	53.6	72.5	96.0	54.9	70.6	85.9	49.1	63.1	92.1	53.5	68.2
05	METALS	102.2	90.4	98.6	102.0	90.5	98.4	100.1	91.1	97.3	101.2	89.7	97.7	100.0	83.8	95.0	106.2	91.1	101.5
06	WOOD, PLASTICS & COMPOSITES	100.6	67.7	82.0	97.0	67.7	80.5	91.4	59.8	73.6	87.6	52.6	67.8	89.6	37.6	60.3	101.6	59.8	78.0
07	THERMAL & MOISTURE PROTECTION	105.3	69.7	90.8	101.9	69.7	88.8	105.3	69.6	90.8	105.3	67.0	89.7	104.7	51.0	82.9	103.7	69.6	89.8
08	OPENINGS	95.3	63.3	87.7	94.2	63.3	86.9	94.6	59.9	86.4	95.4	55.0	85.8	97.5	43.2	84.6	99.0	59.9	89.8
0920	Plaster & Gypsum Board	110.2	65.8	80.0	106.3	65.8	78.8	101.7	57.7	71.8	102.0	50.2	66.8	100.1	35.5	56.1	104.7	57.7	72.8
0950, 0980	Ceilings & Acoustic Treatment	96.1	65.8	75.9	97.7	65.8	76.5	93.4	57.7	69.7	96.1	50.2	65.6	94.2	35.5	55.1	100.4	57.7	72.0
0960	Flooring	108.6	67.7	96.4	103.5	67.7	92.8	104.7	72.5	95.1	101.7	67.7	91.5	103.2	67.7	92.6	109.4	72.5	98.4
0970, 0990	Wall Finishes & Painting/Coating	107.7	42.9	68.8	105.8	77.7	88.9	107.7	77.7	89.7	107.7	77.7	89.7	106.3	55.5	75.7	107.0	77.7	89.4
09	FINISHES	102.5	63.5	80.8	100.7	67.4	82.2	100.1	62.5	79.1	99.5	57.9	76.3	99.3	44.7	68.9	104.4	62.5	81.1
COVERS	DIVS. 10 - 14, 25, 28, 41, 43, 44, 46	100.0	75.4	95.1	100.0	75.4	95.1	100.0	78.2	95.6	100.0	73.7	94.7	100.0	45.0	89.0	100.0	78.2	95.6
21, 22, 23	FIRE SUPPRESSION, PLUMBING & HVAC	100.2	64.1	85.7	100.0	64.7	85.8	95.3	66.9	83.9	100.2	64.7	86.0	95.4	63.4	82.5	100.0	66.9	86.7
26, 27, 3370	ELECTRICAL, COMMUNICATIONS & UTIL.	93.1	60.3	76.0	96.9	59.7	77.4	93.3	70.8	81.5	91.4	59.7	74.9	95.5	55.7	74.7	96.2	70.8	83.0
MF2010	WEIGHTED AVERAGE	99.2	68.2	85.7	100.0	68.8	86.4	98.6	68.9	85.6	98.6	66.1	84.5	97.7	58.3	80.6	100.4	68.9	86.7

		VIRGINIA									WASHINGTON								
	DIVISION	ROANOKE			STAUNTON			WINCHESTER			CLARKSTON			EVERETT			OLYMPIA		
		240 - 241			244			226			994			982			985		
		MAT.	INST.	TOTAL	MAT.	INST.	TOTAL	MAT.	INST.	TOTAL	MAT.	INST.	TOTAL	MAT.	INST.	TOTAL	MAT.	INST.	TOTAL
015433	CONTRACTOR EQUIPMENT		100.8	100.8		105.1	105.1		100.8	100.8		90.4	90.4		102.1	102.1		102.1	102.1
0241, 31 - 34	SITE & INFRASTRUCTURE, DEMOLITION	105.2	85.9	91.7	109.1	87.3	93.8	118.1	87.2	96.4	97.1	90.0	92.1	89.4	110.3	104.1	90.2	110.4	104.4
0310	Concrete Forming & Accessories	97.9	61.8	66.8	91.1	51.1	56.6	85.2	70.8	72.8	114.8	68.8	75.1	114.9	99.6	101.7	103.1	99.8	100.2
0320	Concrete Reinforcing	95.6	67.4	81.5	95.3	50.5	72.9	95.3	86.3	90.8	105.7	87.4	96.6	110.2	100.2	105.2	116.1	100.2	108.2
0330	Cast-in-Place Concrete	118.3	58.5	93.5	108.3	52.0	84.9	104.6	64.7	88.1	97.0	83.2	91.3	99.8	103.8	101.5	89.2	105.4	96.0
03	CONCRETE	105.5	63.1	84.6	102.3	53.0	78.0	102.6	72.7	87.9	110.7	77.4	94.3	98.7	100.7	99.7	96.9	101.3	99.1
04	MASONRY	92.1	54.1	68.6	100.6	52.3	70.7	95.1	60.0	73.3	97.2	80.6	87.0	111.9	99.0	103.9	103.1	99.9	101.1
05	METALS	102.2	86.5	97.4	100.2	81.7	94.5	100.3	94.8	98.6	88.1	81.2	86.0	104.3	91.7	100.4	104.2	91.6	100.3
06	WOOD, PLASTICS & COMPOSITES	99.0	64.0	79.3	89.6	51.0	67.8	85.1	69.6	76.4	103.3	65.2	81.8	112.1	99.3	104.9	94.6	99.3	97.2
07	THERMAL & MOISTURE PROTECTION	104.3	67.2	89.3	104.3	53.0	83.5	105.0	76.2	93.3	142.7	76.9	116.0	100.5	100.5	100.5	101.6	97.8	100.1
08	OPENINGS	96.4	59.4	87.6	96.0	49.2	84.9	97.5	74.7	92.1	106.7	69.7	98.0	101.9	99.8	101.4	103.5	99.8	102.6
0920	Plaster & Gypsum Board	108.4	62.7	77.3	100.1	48.6	65.1	100.3	68.5	78.7	130.6	64.0	85.3	110.7	99.2	102.9	103.1	99.2	100.5
0950, 0980	Ceilings & Acoustic Treatment	97.7	62.7	74.5	94.2	48.6	63.9	95.1	68.5	77.4	97.9	64.0	75.3	99.9	99.2	99.5	106.3	99.2	101.6
0960	Flooring	106.7	67.7	95.0	102.7	40.2	84.0	101.6	82.7	96.0	98.3	48.0	83.2	117.9	93.3	110.6	109.2	93.4	104.5
0970, 0990	Wall Finishes & Painting/Coating	106.3	55.5	75.7	106.3	35.3	63.6	120.5	89.1	101.6	106.4	59.8	78.4	104.8	90.1	96.0	110.3	90.1	98.1
09	FINISHES	101.8	62.4	79.8	99.2	47.6	70.4	100.5	74.2	85.9	114.8	62.6	85.7	109.0	97.1	102.4	104.6	97.4	100.6
COVERS	DIVS. 10 - 14, 25, 28, 41, 43, 44, 46	100.0	72.7	94.5	100.0	72.6	94.5	100.0	85.7	97.1	100.0	77.7	95.5	100.0	99.6	99.9	100.0	100.1	100.0
21, 22, 23	FIRE SUPPRESSION, PLUMBING & HVAC	100.3	65.8	86.5	95.4	60.1	81.2	95.4	86.8	91.9	95.4	84.7	91.1	100.1	100.3	100.2	100.0	100.8	100.3
26, 27, 3370	ELECTRICAL, COMMUNICATIONS & UTIL.	95.5	55.7	74.7	94.3	74.6	84.0	94.1	99.2	96.8	94.9	94.4	94.6	103.0	98.5	100.7	101.0	99.7	100.3
MF2010	WEIGHTED AVERAGE	100.2	66.0	85.3	98.4	62.7	82.9	98.7	82.1	91.5	100.6	80.7	92.0	102.1	99.6	101.0	101.0	100.0	100.6

629

City Cost Indexes

WASHINGTON

DIVISION		RICHLAND 993 MAT.	INST.	TOTAL	SEATTLE 980-981,987 MAT.	INST.	TOTAL	SPOKANE 990-992 MAT.	INST.	TOTAL	TACOMA 983-984 MAT.	INST.	TOTAL	VANCOUVER 986 MAT.	INST.	TOTAL	WENATCHEE 988 MAT.	INST.	TOTAL
015433	CONTRACTOR EQUIPMENT		90.4	90.4		102.0	102.0		90.4	90.4		102.1	102.1		97.1	97.1		102.1	102.1
0241, 31-34	SITE & INFRASTRUCTURE, DEMOLITION	99.6	90.8	93.4	94.5	108.6	104.4	98.7	90.8	93.2	92.4	110.4	105.0	102.3	98.1	99.3	101.4	109.2	106.9
0310	Concrete Forming & Accessories	114.9	78.4	83.4	106.9	100.3	101.2	119.8	78.0	83.7	105.5	99.8	100.6	106.2	92.1	94.0	107.2	77.5	81.6
0320	Concrete Reinforcing	101.4	87.6	94.5	112.2	100.3	106.3	102.1	87.6	94.9	109.0	100.2	104.6	109.9	100.0	104.9	109.8	89.3	99.6
0330	Cast-in-Place Concrete	97.2	85.5	92.4	103.6	105.6	104.5	101.0	85.3	94.5	102.6	105.4	103.8	114.6	99.5	108.3	104.8	77.4	93.4
03	CONCRETE	110.3	82.6	96.6	102.0	101.7	101.8	112.6	82.3	97.6	100.5	101.3	100.9	110.5	95.9	103.3	108.4	79.8	94.3
04	MASONRY	99.3	83.2	89.3	109.2	99.9	103.4	99.9	83.8	89.9	106.7	99.9	102.5	107.3	96.7	100.7	109.1	91.5	98.2
05	METALS	88.5	84.1	87.1	104.5	93.4	101.1	90.8	83.7	88.6	106.0	91.6	101.6	103.4	92.1	99.9	103.5	86.1	98.1
06	WOOD, PLASTICS & COMPOSITES	103.4	75.7	87.8	102.2	99.3	100.6	112.1	75.7	91.6	101.3	99.3	100.2	93.1	91.9	92.5	102.9	75.9	87.6
07	THERMAL & MOISTURE PROTECTION	144.1	79.7	117.9	97.6	100.9	99.0	140.6	79.1	115.6	100.2	98.7	99.6	100.6	92.5	97.3	100.8	82.2	93.3
08	OPENINGS	109.2	71.2	100.2	105.2	99.8	103.9	109.7	71.7	100.7	102.7	99.8	102.0	98.5	94.4	97.5	102.2	71.5	94.9
0920	Plaster & Gypsum Board	130.6	74.8	92.6	106.9	99.2	101.7	121.5	74.8	89.7	111.5	99.2	103.1	109.4	91.9	97.5	114.0	75.1	87.5
0950, 0980	Ceilings & Acoustic Treatment	104.8	74.8	84.8	101.8	99.2	100.1	99.8	74.8	83.2	103.5	99.2	100.6	98.4	91.9	94.1	95.3	75.1	81.8
0960	Flooring	98.7	81.6	93.5	110.9	101.0	108.0	92.8	86.9	94.8	111.4	93.4	106.0	118.1	104.0	113.9	114.1	63.8	99.1
0970, 0990	Wall Finishes & Painting/Coating	106.4	66.2	82.2	104.2	90.1	95.7	106.6	72.4	86.0	104.8	90.1	96.0	107.2	68.4	83.9	104.8	66.2	81.6
09	FINISHES	116.6	76.7	94.4	108.1	98.9	102.9	114.3	78.5	94.3	108.2	97.4	102.2	105.6	91.9	98.0	108.5	72.9	88.7
COVERS	DIVS. 10-14, 25, 28, 41, 43, 44, 46	100.0	87.2	97.4	100.0	100.1	100.0	100.0	87.1	97.4	100.0	100.1	100.0	100.0	62.0	92.4	100.0	78.5	95.7
21, 22, 23	FIRE SUPPRESSION, PLUMBING & HVAC	100.3	109.7	104.1	100.0	117.1	106.8	100.2	85.9	94.5	100.1	100.8	100.4	100.2	88.6	95.5	95.2	87.9	92.3
26, 27, 3370	ELECTRICAL, COMMUNICATIONS & UTIL.	92.1	94.4	93.3	102.4	107.0	104.8	90.5	79.1	84.5	102.8	99.7	101.2	107.5	101.0	104.1	103.7	98.5	101.0
MF2010	WEIGHTED AVERAGE	102.2	89.6	96.7	102.5	104.8	103.5	102.5	82.7	93.8	102.3	100.0	101.3	103.1	93.2	98.8	102.0	87.1	95.5

WASHINGTON / WEST VIRGINIA

DIVISION		YAKIMA 989 MAT.	INST.	TOTAL	BECKLEY 258-259 MAT.	INST.	TOTAL	BLUEFIELD 247-248 MAT.	INST.	TOTAL	BUCKHANNON 262 MAT.	INST.	TOTAL	CHARLESTON 250-253 MAT.	INST.	TOTAL	CLARKSBURG 263-264 MAT.	INST.	TOTAL
015433	CONTRACTOR EQUIPMENT		102.1	102.1		100.8	100.8		100.8	100.8		100.8	100.8		100.8	100.8		100.8	100.8
0241, 31-34	SITE & INFRASTRUCTURE, DEMOLITION	94.7	109.8	105.3	97.9	89.2	91.8	99.6	89.2	92.3	105.8	89.2	94.1	99.0	90.0	92.7	106.3	89.2	94.3
0310	Concrete Forming & Accessories	106.0	95.3	96.8	86.3	92.0	91.2	88.3	91.9	91.4	87.7	92.2	91.6	99.6	93.4	94.2	85.3	92.1	91.1
0320	Concrete Reinforcing	109.5	88.8	99.2	92.3	89.4	90.9	94.2	84.9	89.6	94.8	85.0	89.9	100.6	89.5	95.1	94.8	84.4	89.6
0330	Cast-in-Place Concrete	109.6	84.4	99.2	102.0	102.1	102.0	101.9	102.1	102.0	101.6	100.3	101.0	100.6	102.7	101.4	111.3	98.4	105.9
03	CONCRETE	105.1	90.0	97.7	96.5	95.5	96.0	96.9	94.7	95.8	99.7	94.2	97.0	98.1	96.4	97.2	103.7	93.4	98.6
04	MASONRY	100.4	81.6	88.8	90.8	93.4	92.4	87.5	93.4	91.1	97.4	92.3	94.2	89.2	95.3	93.0	100.7	92.3	95.5
05	METALS	104.2	85.7	98.5	101.2	98.2	100.3	100.2	96.5	99.1	100.4	97.3	99.4	99.8	98.9	99.5	100.4	96.9	99.3
06	WOOD, PLASTICS & COMPOSITES	101.7	99.3	100.4	85.1	92.8	89.4	88.6	92.8	91.0	87.9	92.8	90.6	101.1	92.8	96.4	84.4	92.8	89.1
07	THERMAL & MOISTURE PROTECTION	100.3	81.6	92.7	107.1	89.4	99.9	107.0	89.4	98.4	104.7	90.0	98.7	103.6	90.2	98.2	104.6	88.9	98.3
08	OPENINGS	102.2	81.0	97.2	96.1	85.0	93.5	97.9	83.9	94.6	97.9	83.9	94.6	98.1	85.0	95.0	97.9	85.1	94.9
0920	Plaster & Gypsum Board	111.1	99.2	103.0	93.3	92.4	92.7	99.3	92.4	94.6	99.7	92.4	94.7	97.5	92.4	94.0	97.5	92.4	94.0
0950, 0980	Ceilings & Acoustic Treatment	97.3	99.2	98.6	93.4	92.4	92.7	92.4	92.4	92.4	94.2	92.4	93.0	102.2	92.4	95.7	94.2	92.4	93.0
0960	Flooring	112.4	59.1	96.4	100.5	114.1	104.6	99.8	114.1	104.1	99.5	114.1	103.9	104.4	114.1	107.3	98.6	114.1	103.2
0970, 0990	Wall Finishes & Painting/Coating	104.8	72.4	85.3	107.1	93.5	98.9	106.3	93.5	98.6	106.3	93.8	98.8	106.6	93.5	98.7	106.3	93.8	98.8
09	FINISHES	107.2	86.5	95.7	97.8	96.9	97.3	97.5	96.9	97.2	98.4	96.7	97.4	102.4	97.4	99.6	97.7	96.7	97.1
COVERS	DIVS. 10-14, 25, 28, 41, 43, 44, 46	100.0	97.3	99.5	100.0	56.6	91.3	100.0	56.6	91.3	100.0	96.7	99.3	100.0	97.7	99.5	100.0	96.7	99.3
21, 22, 23	FIRE SUPPRESSION, PLUMBING & HVAC	100.1	109.1	103.7	95.5	91.5	93.9	95.4	91.5	93.8	95.4	92.3	94.2	100.1	92.9	97.2	95.4	92.2	94.1
26, 27, 3370	ELECTRICAL, COMMUNICATIONS & UTIL.	105.9	94.4	99.9	91.2	91.9	91.6	94.4	91.9	93.1	95.8	96.8	96.3	96.3	91.9	94.0	95.8	96.8	96.3
MF2010	WEIGHTED AVERAGE	102.5	94.4	99.0	96.9	92.0	94.8	97.1	91.7	94.7	98.2	93.6	96.2	99.0	94.1	96.9	98.8	93.5	96.5

WEST VIRGINIA

DIVISION		GASSAWAY 266 MAT.	INST.	TOTAL	HUNTINGTON 255-257 MAT.	INST.	TOTAL	LEWISBURG 249 MAT.	INST.	TOTAL	MARTINSBURG 254 MAT.	INST.	TOTAL	MORGANTOWN 265 MAT.	INST.	TOTAL	PARKERSBURG 261 MAT.	INST.	TOTAL
015433	CONTRACTOR EQUIPMENT		100.8	100.8		100.8	100.8		100.8	100.8		100.8	100.8		100.8	100.8		100.8	100.8
0241, 31-34	SITE & INFRASTRUCTURE, DEMOLITION	103.2	89.2	93.3	102.4	90.4	94.0	115.3	89.2	97.0	101.4	89.7	93.2	100.6	89.8	93.0	108.8	89.9	95.5
0310	Concrete Forming & Accessories	87.2	92.2	91.5	98.2	96.0	96.3	85.6	91.8	91.0	86.3	82.4	83.0	85.6	91.8	91.0	89.8	90.0	90.0
0320	Concrete Reinforcing	94.8	85.0	89.9	93.7	90.6	92.2	94.8	84.9	89.9	92.3	83.0	87.6	94.8	84.3	89.6	94.2	84.8	89.5
0330	Cast-in-Place Concrete	106.3	101.0	104.1	111.1	103.5	108.0	101.9	102.1	102.0	106.8	96.0	102.3	101.5	100.1	101.0	103.8	97.1	101.0
03	CONCRETE	100.2	94.4	97.3	101.8	98.0	99.9	105.8	94.6	100.3	99.9	88.0	94.1	96.6	93.8	95.3	101.6	92.1	96.9
04	MASONRY	101.5	92.3	95.8	89.8	99.9	96.0	90.3	93.4	92.2	91.7	85.6	87.9	117.9	92.3	102.0	78.3	90.7	86.0
05	METALS	100.3	97.2	99.4	103.6	99.5	102.3	100.3	96.5	99.1	101.6	94.9	99.5	100.4	96.7	99.3	101.0	96.8	99.7
06	WOOD, PLASTICS & COMPOSITES	87.0	92.8	90.2	97.2	95.4	96.2	84.7	92.8	89.3	85.1	81.3	82.9	84.7	92.8	89.3	88.0	88.8	88.4
07	THERMAL & MOISTURE PROTECTION	104.4	90.0	98.6	107.3	91.4	100.8	105.2	89.4	98.8	107.4	78.2	95.5	104.5	88.9	98.2	104.4	89.6	98.4
08	OPENINGS	96.1	83.9	93.2	95.4	86.6	93.4	97.9	83.9	94.6	98.1	73.9	92.4	99.1	85.1	95.8	96.9	81.8	93.3
0920	Plaster & Gypsum Board	99.0	92.4	94.5	102.1	95.1	97.4	97.5	92.4	94.0	93.9	80.5	84.8	97.5	92.4	94.0	100.1	88.3	92.0
0950, 0980	Ceilings & Acoustic Treatment	94.2	92.4	93.0	96.1	95.1	95.4	94.2	92.4	93.0	96.1	80.5	85.7	94.2	92.4	93.0	94.2	88.3	90.3
0960	Flooring	99.3	114.1	103.8	107.9	120.3	111.6	98.7	114.1	103.3	100.5	114.1	104.6	98.7	114.1	103.3	102.4	114.1	105.9
0970, 0990	Wall Finishes & Painting/Coating	106.3	93.5	98.6	107.1	93.5	98.9	106.3	77.0	88.7	107.1	42.2	68.1	106.3	93.8	98.8	106.3	93.5	98.6
09	FINISHES	97.9	96.6	97.2	101.7	100.5	101.0	98.8	95.1	96.7	98.7	82.3	89.6	97.3	96.7	96.9	99.3	94.8	96.8
COVERS	DIVS. 10-14, 25, 28, 41, 43, 44, 46	100.0	96.7	99.3	100.0	98.6	99.7	100.0	56.6	91.3	100.0	95.4	99.1	100.0	65.4	93.1	100.0	97.1	99.4
21, 22, 23	FIRE SUPPRESSION, PLUMBING & HVAC	95.4	91.4	93.8	100.4	89.3	95.9	95.4	91.5	93.8	95.5	86.7	91.9	95.4	92.2	94.1	100.2	88.1	95.4
26, 27, 3370	ELECTRICAL, COMMUNICATIONS & UTIL.	95.8	91.9	93.8	94.9	96.1	95.5	91.8	91.9	91.8	96.9	82.8	89.6	96.0	96.8	96.4	96.0	92.8	94.3
MF2010	WEIGHTED AVERAGE	98.2	92.8	95.8	99.8	95.2	97.8	98.4	91.4	95.4	98.3	86.1	93.0	98.8	92.6	96.1	98.9	91.4	95.6

630

City Cost Indexes

DIVISION		WEST VIRGINIA									WISCONSIN								
		PETERSBURG			ROMNEY			WHEELING			BELOIT			EAU CLAIRE			GREEN BAY		
		268			267			260			535			547			541 - 543		
		MAT.	INST.	TOTAL	MAT.	INST.	TOTAL	MAT.	INST.	TOTAL	MAT.	INST.	TOTAL	MAT.	INST.	TOTAL	MAT.	INST.	TOTAL
015433	CONTRACTOR EQUIPMENT		100.8	100.8		100.8	100.8		100.8	100.8		102.6	102.6		101.2	101.2		99.0	99.0
0241, 31 - 34	SITE & INFRASTRUCTURE, DEMOLITION	99.9	89.8	92.8	102.6	89.8	93.6	109.5	89.9	95.7	95.8	108.6	104.8	94.6	103.6	100.9	98.2	99.6	99.2
0310	Concrete Forming & Accessories	88.7	91.6	91.2	84.8	91.6	90.7	91.5	92.6	92.4	100.1	102.1	101.9	99.7	101.3	101.1	107.8	100.5	101.5
0320	Concrete Reinforcing	94.2	84.9	89.6	94.8	76.3	85.6	93.6	84.3	89.0	96.6	117.1	106.9	93.7	100.9	97.3	91.9	93.2	92.6
0330	Cast-in-Place Concrete	101.5	96.3	99.4	106.3	96.3	102.1	103.8	100.9	102.6	103.7	101.9	102.9	102.1	99.5	101.0	105.6	100.9	103.7
03	CONCRETE	96.8	92.5	94.7	100.0	91.0	95.5	101.6	94.4	98.1	99.9	104.9	102.3	98.8	100.7	99.7	101.9	99.5	100.7
04	MASONRY	92.9	92.3	92.5	90.7	92.3	91.7	100.2	91.7	94.9	98.5	104.8	102.4	89.8	102.7	97.8	120.3	101.9	108.9
05	METALS	100.5	96.6	99.3	100.5	93.7	98.4	101.2	96.9	99.9	94.2	108.4	98.6	93.1	102.7	96.0	95.4	99.1	96.5
06	WOOD, PLASTICS & COMPOSITES	88.9	92.8	91.1	83.8	92.8	88.9	89.6	92.8	91.4	98.2	101.6	100.1	105.2	101.6	103.2	110.0	101.6	105.3
07	THERMAL & MOISTURE PROTECTION	104.5	84.5	96.4	104.6	81.2	95.1	104.7	89.5	98.5	103.4	97.9	101.2	103.2	88.1	97.1	105.0	87.1	97.7
08	OPENINGS	99.1	83.9	95.5	99.0	82.0	95.0	97.6	85.1	94.7	103.6	110.2	105.1	101.0	98.1	100.3	97.4	97.8	97.5
0920	Plaster & Gypsum Board	99.7	92.4	94.7	97.1	92.4	93.9	100.1	92.4	94.8	92.8	101.9	99.0	101.4	101.9	101.8	96.6	101.9	100.2
0950, 0980	Ceilings & Acoustic Treatment	94.2	92.4	93.0	94.2	92.4	93.0	94.2	92.4	93.0	92.5	101.9	98.8	101.7	101.9	101.9	92.7	101.9	98.9
0960	Flooring	100.3	114.1	104.5	98.5	114.1	103.2	103.2	114.1	106.4	101.9	120.3	107.4	94.2	116.2	100.8	113.3	116.2	114.2
0970, 0990	Wall Finishes & Painting/Coating	106.3	93.8	98.8	106.3	93.8	98.8	106.3	93.8	98.8	91.8	99.3	96.3	88.0	88.3	88.2	98.1	85.0	90.2
09	FINISHES	98.1	96.7	97.3	97.4	96.7	97.0	99.5	96.9	98.1	99.7	105.1	102.7	99.2	103.2	101.4	103.3	100.8	102.9
COVERS	DIVS. 10 - 14, 25, 28, 41, 43, 44, 46	100.0	65.3	93.1	100.0	96.7	99.3	100.0	97.1	99.4	100.0	98.2	99.6	100.0	98.5	99.7	100.0	98.1	99.6
21, 22, 23	FIRE SUPPRESSION, PLUMBING & HVAC	95.4	91.2	93.7	95.4	90.8	93.6	100.3	92.8	97.3	100.0	100.0	100.0	100.1	87.5	95.1	100.4	84.3	93.9
26, 27, 3370	ELECTRICAL, COMMUNICATIONS & UTIL.	99.2	82.9	90.7	98.5	82.9	90.3	93.1	96.8	95.0	102.8	87.3	94.7	103.9	86.9	95.0	98.2	82.9	90.2
MF2010	WEIGHTED AVERAGE	98.0	90.1	94.6	98.2	90.4	94.8	99.8	93.8	97.2	99.6	101.9	100.6	98.7	96.4	97.7	100.5	94.1	97.7

DIVISION		WISCONSIN																	
		KENOSHA			LA CROSSE			LANCASTER			MADISON			MILWAUKEE			NEW RICHMOND		
		531			546			538			537			530,532			540		
		MAT.	INST.	TOTAL	MAT.	INST.	TOTAL	MAT.	INST.	TOTAL	MAT.	INST.	TOTAL	MAT.	INST.	TOTAL	MAT.	INST.	TOTAL
015433	CONTRACTOR EQUIPMENT		100.5	100.5		101.2	101.2		102.6	102.6		102.6	102.6		90.8	90.8		101.6	101.6
0241, 31 - 34	SITE & INFRASTRUCTURE, DEMOLITION	101.8	105.5	104.4	88.5	103.6	99.1	95.0	108.5	104.4	91.3	108.6	103.4	98.0	98.7	98.5	91.7	104.0	100.3
0310	Concrete Forming & Accessories	106.8	102.5	103.1	86.7	100.8	98.9	99.5	99.7	99.7	103.7	101.7	102.0	102.2	120.1	117.6	95.2	102.8	101.8
0320	Concrete Reinforcing	96.4	98.6	97.5	93.4	92.9	93.1	97.9	92.8	95.4	102.3	93.1	97.7	100.5	98.9	99.7	91.0	100.7	95.9
0330	Cast-in-Place Concrete	112.8	106.0	109.9	91.8	97.0	94.0	103.0	101.0	102.2	94.3	101.8	97.4	102.3	110.6	105.7	106.3	83.6	96.9
03	CONCRETE	104.6	103.0	103.8	90.1	98.1	94.1	99.6	98.9	99.2	97.1	100.1	98.6	100.3	111.9	106.0	96.6	95.9	96.3
04	MASONRY	95.9	115.5	108.1	89.0	102.7	97.5	98.6	104.9	102.5	99.4	104.8	102.8	122.2	120.5	113.5	115.4	102.8	107.6
05	METALS	95.0	101.6	97.0	93.0	98.9	94.8	91.8	96.1	93.1	93.5	98.1	94.9	98.1	94.9	97.1	93.2	101.5	95.8
06	WOOD, PLASTICS & COMPOSITES	101.8	98.7	100.1	90.2	101.6	96.7	97.5	101.6	99.8	97.0	101.6	99.6	101.1	121.3	112.5	94.0	104.9	100.2
07	THERMAL & MOISTURE PROTECTION	103.4	108.4	105.4	102.7	87.2	96.4	103.1	83.5	95.2	106.5	94.0	101.4	99.9	113.0	105.2	104.9	93.8	100.4
08	OPENINGS	97.6	102.8	98.9	101.0	86.7	97.6	99.2	87.5	96.4	105.1	101.6	104.3	106.2	115.1	108.3	86.9	99.9	90.0
0920	Plaster & Gypsum Board	83.2	99.0	93.9	95.9	101.9	100.0	91.5	101.9	98.6	94.2	101.9	99.5	99.9	122.1	115.0	85.5	105.5	99.1
0950, 0980	Ceilings & Acoustic Treatment	92.5	99.0	96.8	100.8	101.9	101.6	87.2	101.9	97.0	102.4	101.9	102.1	96.1	122.1	113.4	65.8	105.5	92.2
0960	Flooring	120.9	112.4	118.4	88.3	116.2	96.7	101.6	112.6	104.9	103.5	112.6	106.2	101.6	116.1	105.9	106.5	116.2	109.4
0970, 0990	Wall Finishes & Painting/Coating	101.5	115.2	109.7	88.0	75.5	80.5	91.8	67.8	77.4	89.8	99.3	95.5	93.2	125.6	112.7	101.0	88.3	93.4
09	FINISHES	104.6	105.5	105.1	96.1	101.8	99.3	98.3	98.4	98.3	101.0	103.9	102.6	102.5	121.0	112.8	93.7	103.3	99.0
COVERS	DIVS. 10 - 14, 25, 28, 41, 43, 44, 46	100.0	101.6	100.3	100.0	98.5	99.7	100.0	50.6	90.1	100.0	98.2	99.6	100.0	104.8	101.0	100.0	98.4	99.7
21, 22, 23	FIRE SUPPRESSION, PLUMBING & HVAC	100.2	96.9	98.9	100.1	87.4	95.0	95.1	87.2	92.0	100.0	96.3	98.5	100.0	102.5	101.0	94.8	86.8	91.6
26, 27, 3370	ELECTRICAL, COMMUNICATIONS & UTIL.	103.4	99.6	101.4	104.2	87.1	95.3	102.6	87.1	94.5	103.6	94.3	98.8	102.9	101.9	102.4	102.0	87.1	94.3
MF2010	WEIGHTED AVERAGE	100.2	102.9	101.4	97.2	95.0	96.3	97.4	93.5	95.7	99.5	99.8	99.6	100.9	107.9	104.0	96.2	95.9	96.1

DIVISION		WISCONSIN																	
		OSHKOSH			PORTAGE			RACINE			RHINELANDER			SUPERIOR			WAUSAU		
		549			539			534			545			548			544		
		MAT.	INST.	TOTAL	MAT.	INST.	TOTAL	MAT.	INST.	TOTAL	MAT.	INST.	TOTAL	MAT.	INST.	TOTAL	MAT.	INST.	TOTAL
015433	CONTRACTOR EQUIPMENT		99.0	99.0		102.6	102.6		102.6	102.6		99.0	99.0		101.6	101.6		99.0	99.0
0241, 31 - 34	SITE & INFRASTRUCTURE, DEMOLITION	90.0	99.6	96.7	85.8	108.4	101.6	95.6	109.3	105.2	102.1	99.6	100.3	88.6	104.0	99.4	85.9	99.6	95.5
0310	Concrete Forming & Accessories	91.0	99.9	98.7	92.1	100.7	99.5	100.4	102.5	102.2	88.4	99.9	98.3	93.4	94.0	94.0	90.3	100.5	99.1
0320	Concrete Reinforcing	92.0	93.1	92.6	98.0	93.0	95.5	96.6	98.6	97.6	92.2	92.2	92.2	91.0	92.7	91.8	92.2	92.9	92.5
0330	Cast-in-Place Concrete	97.9	100.8	99.1	88.5	103.0	94.5	101.7	105.7	103.3	111.0	101.0	106.9	100.1	100.0	100.1	91.3	100.9	95.3
03	CONCRETE	91.7	99.6	95.3	87.5	100.1	93.7	98.9	102.9	100.9	103.6	99.0	101.3	91.2	96.1	93.6	86.8	99.4	93.0
04	MASONRY	102.5	102.2	102.3	97.5	104.8	102.1	98.6	115.5	109.0	118.7	102.7	108.8	114.7	106.3	109.5	102.1	101.9	102.0
05	METALS	93.4	98.4	94.9	92.5	97.1	93.9	95.9	101.5	97.6	93.3	97.8	94.7	94.1	99.4	95.7	93.2	98.9	94.9
06	WOOD, PLASTICS & COMPOSITES	89.8	101.6	96.5	88.4	101.6	95.8	98.4	98.7	98.6	87.4	101.6	95.4	92.4	92.2	92.3	89.3	101.6	96.3
07	THERMAL & MOISTURE PROTECTION	104.2	87.2	97.3	102.6	93.8	99.0	103.4	107.9	105.2	105.0	85.2	96.9	104.5	94.3	100.4	104.1	85.8	96.6
08	OPENINGS	93.4	97.8	94.5	99.3	94.2	98.1	103.6	102.8	103.4	93.5	88.8	92.4	86.4	92.5	87.8	93.6	97.7	94.6
0920	Plaster & Gypsum Board	83.9	101.9	96.2	85.4	101.9	96.7	92.8	99.0	97.0	83.9	101.9	96.2	85.4	92.4	90.2	83.9	101.9	96.2
0950, 0980	Ceilings & Acoustic Treatment	92.7	101.9	98.9	89.9	101.9	97.9	92.5	99.0	96.8	92.7	101.9	98.9	66.7	92.4	83.8	92.7	101.9	98.9
0960	Flooring	105.1	116.2	108.4	98.2	120.3	104.8	101.9	112.4	105.0	104.3	116.2	107.9	107.8	124.7	112.9	104.9	116.2	108.3
0970, 0990	Wall Finishes & Painting/Coating	95.3	85.0	89.1	91.8	67.8	77.4	91.8	115.3	105.9	95.3	61.7	75.1	89.6	103.7	98.1	95.3	85.0	89.1
09	FINISHES	98.4	100.9	99.8	96.4	100.5	98.7	99.7	105.5	103.0	99.3	99.1	99.2	93.1	100.7	97.3	98.1	102.6	100.6
COVERS	DIVS. 10 - 14, 25, 28, 41, 43, 44, 46	100.0	66.1	93.2	100.0	64.8	92.9	100.0	101.6	100.3	100.0	56.4	91.3	100.0	96.7	99.3	100.0	98.1	99.6
21, 22, 23	FIRE SUPPRESSION, PLUMBING & HVAC	95.5	83.4	90.6	95.1	96.1	95.5	100.0	96.9	98.8	95.5	87.0	92.1	94.8	90.4	93.0	95.5	86.3	91.8
26, 27, 3370	ELECTRICAL, COMMUNICATIONS & UTIL.	102.6	81.0	91.3	106.8	94.3	100.3	102.6	95.2	98.7	101.9	82.6	91.8	107.2	99.3	103.1	103.8	82.6	92.7
MF2010	WEIGHTED AVERAGE	96.3	92.4	94.6	96.1	97.9	96.9	99.7	102.6	101.0	98.7	92.4	95.9	96.1	97.6	96.7	95.7	94.4	95.2

631

City Cost Indexes

DIVISION		WYOMING																	
		CASPER 826			CHEYENNE 820			NEWCASTLE 827			RAWLINS 823			RIVERTON 825			ROCK SPRINGS 829 - 831		
		MAT.	INST.	TOTAL	MAT.	INST.	TOTAL	MAT.	INST.	TOTAL	MAT.	INST.	TOTAL	MAT.	INST.	TOTAL	MAT.	INST.	TOTAL
015433	CONTRACTOR EQUIPMENT		97.9	97.9		97.9	97.9		97.9	97.9		97.9	97.9		97.9	97.9		97.9	97.9
0241, 31 - 34	SITE & INFRASTRUCTURE, DEMOLITION	97.2	95.2	95.8	90.1	95.2	93.7	82.0	94.8	91.0	95.8	94.8	95.1	89.3	94.8	93.2	85.8	94.7	92.1
0310	Concrete Forming & Accessories	104.4	43.5	51.9	106.6	54.7	61.8	97.1	63.5	68.1	101.3	63.5	68.6	96.0	63.0	67.5	103.5	62.9	68.4
0320	Concrete Reinforcing	108.6	70.5	89.6	100.8	71.3	86.1	108.4	71.7	90.1	108.1	71.7	89.9	109.1	71.4	90.2	109.1	71.4	90.3
0330	Cast-in-Place Concrete	98.6	72.6	87.8	91.1	72.7	83.5	92.0	62.1	79.6	92.1	62.1	79.6	92.0	52.6	75.7	92.0	52.3	75.5
03	CONCRETE	103.0	59.4	81.5	100.1	64.6	82.6	100.6	64.8	82.9	114.3	64.8	89.9	108.9	61.3	85.4	101.1	61.2	81.4
04	MASONRY	100.0	34.5	59.5	95.8	47.6	65.9	92.6	55.7	69.7	92.6	55.7	69.7	92.6	49.9	66.1	147.5	48.7	86.4
05	METALS	98.1	71.0	89.7	101.5	72.4	92.5	96.6	72.4	89.2	96.7	72.4	89.2	96.8	71.6	89.0	97.5	71.9	89.6
06	WOOD, PLASTICS & COMPOSITES	96.3	41.0	65.1	95.6	55.6	73.0	85.6	68.4	75.9	89.6	68.4	77.6	84.5	68.4	75.5	93.6	68.4	79.4
07	THERMAL & MOISTURE PROTECTION	105.2	49.4	82.5	99.7	54.4	81.3	101.2	55.9	82.8	102.7	55.9	83.7	102.1	59.7	84.9	101.3	59.3	84.2
08	OPENINGS	102.1	48.9	89.5	102.3	56.9	91.6	106.2	60.8	95.5	105.9	60.8	95.2	106.1	60.8	95.3	106.5	60.8	95.7
0920	Plaster & Gypsum Board	93.1	39.1	56.3	84.6	54.2	63.9	80.3	67.4	71.5	80.7	67.4	71.6	80.3	67.4	71.5	90.0	67.4	74.6
0950, 0980	Ceilings & Acoustic Treatment	108.9	39.1	62.4	100.3	54.2	69.6	102.3	67.4	79.0	102.3	67.4	79.0	102.3	67.4	79.0	102.3	67.4	79.0
0960	Flooring	109.6	33.6	86.8	107.4	54.0	91.5	101.0	52.6	86.5	103.8	52.6	88.5	100.6	52.6	86.2	105.9	52.6	89.9
0970, 0990	Wall Finishes & Painting/Coating	100.0	44.0	66.3	106.9	44.0	69.1	102.6	56.6	74.9	102.6	56.6	74.9	102.6	56.6	74.9	102.6	56.6	74.9
09	FINISHES	106.4	40.1	69.5	103.2	52.8	75.1	97.9	60.6	77.1	100.2	60.6	78.2	98.7	60.6	77.5	100.6	60.3	78.2
COVERS	DIVS. 10 - 14, 25, 28, 41, 43, 44, 46	100.0	88.9	97.8	100.0	90.6	98.1	100.0	91.5	98.3	100.0	91.5	98.3	100.0	58.0	91.6	100.0	82.7	96.5
21, 22, 23	FIRE SUPPRESSION, PLUMBING & HVAC	100.0	65.4	86.1	100.0	65.5	86.2	97.5	64.8	84.3	97.5	64.8	84.3	97.5	64.7	84.3	100.0	64.2	85.6
26, 27, 3370	ELECTRICAL, COMMUNICATIONS & UTIL.	96.0	65.6	80.1	99.1	70.2	84.0	96.8	70.6	83.2	96.8	70.6	83.2	96.8	64.4	79.9	95.0	64.4	79.0
MF2010	WEIGHTED AVERAGE	100.4	60.7	83.1	100.1	65.7	85.2	98.2	67.8	84.9	100.2	67.8	86.1	99.3	64.9	84.3	101.8	65.4	85.9

DIVISION		WYOMING												CANADA					
		SHERIDAN 828			WHEATLAND 822			WORLAND 824			YELLOWSTONE NAT'L PA 821			BARRIE, ONTARIO			BATHURST, NEW BRUNSWICK		
		MAT.	INST.	TOTAL	MAT.	INST.	TOTAL	MAT.	INST.	TOTAL	MAT.	INST.	TOTAL	MAT.	INST.	TOTAL	MAT.	INST.	TOTAL
015433	CONTRACTOR EQUIPMENT		97.9	97.9		97.9	97.9		97.9	97.9		97.9	97.9		101.6	101.6		101.9	101.9
0241, 31 - 34	SITE & INFRASTRUCTURE, DEMOLITION	89.8	95.2	93.6	86.4	94.8	92.3	83.9	94.7	91.5	84.0	94.9	91.7	115.3	101.0	105.3	98.4	97.3	97.6
0310	Concrete Forming & Accessories	104.3	51.6	58.8	99.1	51.1	57.7	99.2	62.7	67.7	99.2	64.0	68.8	125.1	91.5	96.1	103.7	65.1	70.4
0320	Concrete Reinforcing	109.1	71.8	90.4	108.4	71.7	90.1	109.1	71.1	90.1	111.0	71.1	91.1	174.2	86.5	130.4	139.2	58.3	98.8
0330	Cast-in-Place Concrete	95.1	72.7	85.8	96.0	61.8	81.8	92.0	52.2	75.5	92.0	54.2	76.3	166.0	90.4	134.6	136.1	62.8	105.7
03	CONCRETE	108.8	63.3	86.3	105.4	59.2	82.6	100.8	61.0	81.2	101.1	62.3	81.9	151.0	90.3	121.1	130.9	63.7	97.7
04	MASONRY	92.8	57.2	70.8	92.9	55.2	69.6	94.4	44.9	63.1	92.6	48.5	65.3	165.2	99.8	124.7	165.6	66.1	104.0
05	METALS	100.3	72.9	91.9	96.6	72.2	89.1	96.8	71.2	88.9	97.4	71.2	89.3	108.2	91.4	103.0	104.4	72.6	94.6
06	WOOD, PLASTICS & COMPOSITES	97.0	51.6	71.4	87.5	52.1	67.5	87.5	68.4	76.8	87.5	68.4	76.8	120.6	90.1	103.4	97.8	65.4	79.5
07	THERMAL & MOISTURE PROTECTION	102.4	56.5	83.7	101.5	55.0	82.6	101.3	56.0	82.9	100.8	57.5	83.2	112.4	91.0	103.7	104.6	63.6	87.9
08	OPENINGS	106.8	52.5	93.9	104.8	54.7	93.0	106.4	60.8	95.6	99.8	60.8	90.6	92.5	88.6	91.5	89.2	57.2	81.6
0920	Plaster & Gypsum Board	105.6	50.0	67.8	80.3	50.5	60.0	80.3	67.4	71.5	80.5	67.4	71.6	151.2	89.8	109.4	151.8	64.2	92.2
0950, 0980	Ceilings & Acoustic Treatment	105.2	50.0	68.5	102.3	50.5	67.8	102.3	67.4	79.0	103.1	67.4	79.3	95.2	89.8	91.6	100.9	64.2	76.5
0960	Flooring	105.0	52.6	89.3	102.5	51.7	87.3	102.5	52.6	87.6	102.5	52.6	87.6	130.7	100.7	121.7	117.8	48.0	96.9
0970, 0990	Wall Finishes & Painting/Coating	105.1	44.0	68.4	102.6	35.2	62.1	102.6	56.6	74.9	102.6	56.6	74.9	111.9	91.8	99.8	108.4	52.7	74.9
09	FINISHES	106.2	49.7	74.7	98.6	48.3	70.6	98.3	60.5	77.2	98.5	61.4	77.8	117.3	93.5	104.0	113.4	60.9	84.2
COVERS	DIVS. 10 - 14, 25, 28, 41, 43, 44, 46	100.0	90.1	98.0	100.0	82.0	96.4	100.0	57.8	91.5	100.0	59.0	91.8	139.2	76.2	126.6	131.1	66.9	118.3
21, 22, 23	FIRE SUPPRESSION, PLUMBING & HVAC	97.5	65.5	84.6	97.5	64.5	84.2	97.5	64.4	84.2	97.5	66.2	84.9	102.2	101.5	101.9	102.5	71.3	90.0
26, 27, 3370	ELECTRICAL, COMMUNICATIONS & UTIL.	99.5	64.5	81.2	96.8	70.2	82.9	96.8	64.4	79.9	95.7	64.4	79.4	116.5	92.6	104.0	114.2	63.3	87.6
MF2010	WEIGHTED AVERAGE	100.9	65.2	85.4	98.7	64.6	83.9	98.3	64.1	83.4	97.7	65.2	83.5	117.3	94.9	107.5	112.1	68.6	93.1

DIVISION		CANADA																	
		BRANDON, MANITOBA			BRANTFORD, ONTARIO			BRIDGEWATER, NOVA SCOTIA			CALGARY, ALBERTA			CAP-DE-LA-MADELEINE, QUEBEC			CHARLESBOURG, QUEBEC		
		MAT.	INST.	TOTAL	MAT.	INST.	TOTAL	MAT.	INST.	TOTAL	MAT.	INST.	TOTAL	MAT.	INST.	TOTAL	MAT.	INST.	TOTAL
015433	CONTRACTOR EQUIPMENT		103.9	103.9		101.6	101.6		101.5	101.5		106.4	106.4		102.5	102.5		102.5	102.5
0241, 31 - 34	SITE & INFRASTRUCTURE, DEMOLITION	117.9	99.9	105.3	114.9	101.3	105.4	99.9	98.9	99.2	124.6	105.7	111.3	95.9	100.6	99.2	95.9	100.6	99.2
0310	Concrete Forming & Accessories	126.8	74.1	81.3	124.9	99.1	102.6	97.1	73.2	76.5	127.7	96.8	101.0	130.3	90.0	95.5	130.3	90.0	95.5
0320	Concrete Reinforcing	189.5	54.9	122.3	167.9	85.2	126.6	142.9	48.2	95.6	130.9	74.1	102.5	142.9	81.4	112.2	142.9	81.4	112.2
0330	Cast-in-Place Concrete	146.1	77.7	117.7	162.0	112.0	141.3	166.5	73.2	127.8	201.5	106.2	162.0	130.8	99.3	117.7	130.8	99.3	117.7
03	CONCRETE	138.9	72.4	106.1	147.1	100.9	124.3	144.7	69.3	107.5	167.1	95.9	132.0	128.7	91.7	110.5	128.7	91.7	110.5
04	MASONRY	186.5	68.8	113.7	167.9	104.4	128.6	163.2	73.2	107.5	203.4	90.7	133.7	163.7	88.6	117.2	163.7	88.6	117.2
05	METALS	117.8	77.8	105.5	105.5	92.2	101.4	104.6	75.3	95.6	130.4	89.2	117.7	103.8	88.8	99.2	103.8	88.8	99.2
06	WOOD, PLASTICS & COMPOSITES	124.4	75.1	96.6	119.8	97.9	107.4	89.2	72.6	79.9	107.8	96.5	101.4	131.6	89.9	108.1	131.6	89.9	108.1
07	THERMAL & MOISTURE PROTECTION	109.1	73.3	94.6	112.6	96.6	106.1	107.5	71.4	92.9	121.9	94.1	110.6	106.2	91.8	100.3	106.2	91.8	100.3
08	OPENINGS	97.0	66.3	89.7	93.6	94.9	93.9	86.6	66.4	81.8	94.2	86.1	92.3	94.7	83.0	91.9	94.7	83.0	91.9
0920	Plaster & Gypsum Board	129.5	73.9	91.7	142.7	97.8	112.1	141.8	71.7	94.1	140.9	95.9	110.3	172.3	89.4	115.9	172.3	89.4	115.9
0950, 0980	Ceilings & Acoustic Treatment	102.5	73.9	83.5	97.4	97.8	97.6	97.4	71.7	80.3	145.9	95.9	112.6	97.4	89.4	92.1	97.4	89.4	92.1
0960	Flooring	136.3	71.5	116.9	130.5	100.7	121.6	113.1	68.5	99.8	133.3	94.3	121.7	130.5	101.4	121.8	130.5	101.4	121.8
0970, 0990	Wall Finishes & Painting/Coating	113.1	60.0	81.2	109.0	100.8	104.1	109.0	65.5	82.9	114.2	112.4	113.1	109.0	93.7	99.8	109.0	93.7	99.8
09	FINISHES	118.8	72.5	93.0	115.6	100.1	107.0	109.9	71.8	88.7	129.1	98.5	112.0	118.9	92.8	104.4	118.9	92.8	104.4
COVERS	DIVS. 10 - 14, 25, 28, 41, 43, 44, 46	131.1	69.1	118.7	131.1	78.4	120.6	131.1	68.2	118.5	131.1	98.4	124.6	131.1	87.3	122.4	131.1	87.3	122.4
21, 22, 23	FIRE SUPPRESSION, PLUMBING & HVAC	102.5	85.9	95.9	102.5	104.5	103.3	102.5	86.3	96.0	101.3	92.2	97.6	102.9	93.0	98.9	102.9	93.0	98.9
26, 27, 3370	ELECTRICAL, COMMUNICATIONS & UTIL.	120.0	71.3	94.6	112.9	92.1	102.0	118.7	65.9	91.1	114.2	100.8	107.2	112.6	74.3	92.6	112.6	74.3	92.6
MF2010	WEIGHTED AVERAGE	118.7	77.3	100.7	115.5	98.8	108.2	113.4	76.1	97.2	124.7	95.4	111.9	112.8	89.4	102.6	112.8	89.4	102.6

632

City Cost Indexes

CANADA

DIVISION		CHARLOTTETOWN, PRINCE EDWARD ISLAND			CHICOUTIMI, QUEBEC			CORNER BROOK, NEWFOUNDLAND			CORNWALL, ONTARIO			DALHOUSIE, NEW BRUNSWICK			DARTMOUTH, NOVA SCOTIA		
		MAT.	INST.	TOTAL	MAT.	INST.	TOTAL	MAT.	INST.	TOTAL	MAT.	INST.	TOTAL	MAT.	INST.	TOTAL	MAT.	INST.	TOTAL
015433	CONTRACTOR EQUIPMENT		101.5	101.5		102.5	102.5		103.1	103.1		101.6	101.6		101.9	101.9		101.5	101.5
0241, 31 - 34	SITE & INFRASTRUCTURE, DEMOLITION	115.8	96.4	102.2	96.1	99.7	98.6	120.7	97.9	104.7	113.0	100.8	104.5	97.6	97.3	97.4	111.0	98.9	102.5
0310	Concrete Forming & Accessories	105.6	59.5	65.8	131.4	92.8	98.1	106.9	63.0	69.0	122.8	91.6	95.9	103.2	65.3	70.5	98.9	73.2	76.8
0320	Concrete Reinforcing	146.3	48.3	97.3	105.0	96.4	100.7	173.8	50.1	112.0	167.9	84.9	126.5	144.1	58.4	101.3	181.7	48.2	115.7
0330	Cast-in-Place Concrete	164.6	62.0	122.0	128.2	99.2	116.1	170.1	71.7	129.2	145.7	102.2	127.7	137.6	62.8	106.6	164.2	73.2	126.4
03	CONCRETE	153.1	59.1	106.7	120.1	95.6	108.0	167.6	64.5	116.8	139.3	94.1	117.0	135.2	63.8	100.0	152.9	69.3	111.6
04	MASONRY	182.6	63.7	109.0	167.4	94.1	122.0	183.2	65.3	110.2	166.7	95.7	122.7	158.4	66.1	101.3	196.5	73.2	120.2
05	METALS	127.0	68.6	108.9	104.2	92.3	100.6	118.3	74.0	104.6	105.5	90.8	101.0	101.7	72.8	92.8	118.3	75.3	105.0
06	WOOD, PLASTICS & COMPOSITES	88.5	59.1	71.9	132.6	93.2	110.4	104.3	62.1	80.5	118.6	91.0	103.0	97.7	65.4	79.5	94.6	72.6	82.2
07	THERMAL & MOISTURE PROTECTION	119.5	61.7	96.1	105.0	95.9	101.3	112.7	63.2	92.6	112.4	91.1	103.8	110.8	63.6	91.6	110.4	71.4	94.6
08	OPENINGS	93.4	51.1	83.4	93.8	80.5	90.6	102.9	58.5	92.4	94.7	88.2	93.1	90.3	57.2	82.4	88.3	66.4	83.1
0920	Plaster & Gypsum Board	144.0	57.7	85.3	176.1	92.8	119.4	160.5	60.8	92.7	208.0	90.7	128.2	148.2	64.2	91.1	156.0	71.7	98.6
0950, 0980	Ceilings & Acoustic Treatment	121.3	57.7	79.0	100.1	92.8	95.2	103.4	60.8	75.0	100.9	90.7	94.1	100.2	64.2	76.3	110.5	71.7	84.7
0960	Flooring	116.1	64.0	100.5	132.3	101.4	123.0	122.1	58.3	103.0	130.5	99.2	121.1	116.3	73.4	103.5	118.0	68.5	103.2
0970, 0990	Wall Finishes & Painting/Coating	112.9	43.9	71.4	108.4	105.5	106.7	113.0	63.1	83.0	109.0	93.9	99.9	109.9	52.7	75.5	113.0	65.5	84.4
09	FINISHES	118.9	59.1	85.6	120.3	96.5	107.0	119.5	62.2	87.6	125.3	93.6	107.6	112.7	65.9	86.6	118.0	71.8	92.3
COVERS	DIVS. 10 - 14, 25, 28, 41, 43, 44, 46	131.1	66.4	118.2	131.1	88.0	122.5	131.1	67.2	118.3	131.1	75.8	120.1	131.1	66.9	118.3	131.1	68.2	118.5
21, 22, 23	FIRE SUPPRESSION, PLUMBING & HVAC	102.7	65.2	87.7	102.5	93.7	99.0	102.5	73.4	90.8	102.9	102.3	102.6	102.5	71.3	90.0	102.5	86.3	96.0
26, 27, 3370	ELECTRICAL, COMMUNICATIONS & UTIL.	111.9	53.6	81.5	111.0	86.5	98.2	117.4	59.7	87.2	113.9	93.1	103.0	115.3	59.8	86.3	122.2	65.9	92.8
MF2010	WEIGHTED AVERAGE	120.5	64.0	95.9	111.9	93.1	103.7	122.2	69.0	99.0	115.5	95.1	106.7	112.1	68.8	93.2	119.7	76.1	100.7

CANADA

DIVISION		EDMONTON, ALBERTA			FORT MCMURRAY, ALBERTA			FREDERICTON, NEW BRUNSWICK			GATINEAU, QUEBEC			GRANBY, QUEBEC			HALIFAX, NOVA SCOTIA		
		MAT.	INST.	TOTAL	MAT.	INST.	TOTAL	MAT.	INST.	TOTAL	MAT.	INST.	TOTAL	MAT.	INST.	TOTAL	MAT.	INST.	TOTAL
015433	CONTRACTOR EQUIPMENT		106.4	106.4		103.7	103.7		101.9	101.9		102.5	102.5		102.5	102.5		101.5	101.5
0241, 31 - 34	SITE & INFRASTRUCTURE, DEMOLITION	146.0	105.7	117.7	120.4	101.9	107.4	100.8	97.3	98.3	95.7	100.5	99.1	96.2	100.5	99.2	105.2	98.8	100.7
0310	Concrete Forming & Accessories	127.2	96.8	101.0	123.0	94.3	98.2	122.7	65.6	73.5	130.3	89.8	95.4	130.3	89.8	95.3	104.9	78.3	81.9
0320	Concrete Reinforcing	140.3	74.1	107.2	155.3	74.1	114.7	136.1	58.5	97.4	151.1	81.3	116.3	151.1	81.3	116.3	147.6	62.8	105.3
0330	Cast-in-Place Concrete	201.0	106.2	161.7	215.2	104.8	169.4	134.8	62.9	104.9	129.0	99.3	116.7	133.3	99.2	119.2	165.6	81.0	130.5
03	CONCRETE	168.4	95.9	132.6	170.0	94.3	132.7	133.1	64.0	99.0	129.2	91.7	110.7	131.2	91.6	111.7	146.0	76.9	111.9
04	MASONRY	193.2	90.7	129.8	207.6	89.6	134.6	192.4	67.6	115.2	163.6	88.6	117.2	163.9	88.6	117.3	193.7	83.4	125.5
05	METALS	130.4	89.2	117.7	130.8	89.0	117.9	123.0	73.3	107.7	103.8	88.7	99.2	103.8	88.6	99.1	130.3	81.6	115.3
06	WOOD, PLASTICS & COMPOSITES	108.6	96.5	101.8	113.7	93.8	102.5	112.9	65.4	86.1	131.6	89.9	108.1	131.6	89.9	108.1	88.7	77.4	82.3
07	THERMAL & MOISTURE PROTECTION	127.2	94.1	113.8	120.1	92.6	108.9	117.3	64.6	95.9	106.2	91.8	100.3	106.2	90.2	99.7	118.6	77.9	102.1
08	OPENINGS	94.4	86.1	92.4	94.7	84.6	92.3	89.3	56.1	81.5	94.7	78.2	90.8	94.7	78.2	90.8	86.5	72.0	83.1
0920	Plaster & Gypsum Board	145.5	95.9	111.8	140.9	93.3	108.5	153.7	64.2	92.8	140.5	89.4	105.8	143.0	89.4	106.6	140.8	76.6	97.1
0950, 0980	Ceilings & Acoustic Treatment	150.5	95.9	114.2	107.2	93.3	97.9	113.3	64.2	80.6	97.4	89.4	92.1	97.4	89.4	92.1	116.1	76.6	89.8
0960	Flooring	135.0	94.3	122.8	130.5	94.3	119.7	128.5	76.7	113.0	130.5	101.4	121.8	130.5	101.4	121.8	118.2	90.0	109.8
0970, 0990	Wall Finishes & Painting/Coating	114.6	112.4	113.3	109.1	97.5	102.2	112.6	67.2	85.3	109.0	93.7	99.8	109.0	93.7	99.8	113.9	82.0	94.7
09	FINISHES	133.3	98.5	113.9	118.9	95.0	105.6	121.4	68.1	91.7	114.5	92.8	102.4	114.9	92.8	102.6	117.2	81.2	97.2
COVERS	DIVS. 10 - 14, 25, 28, 41, 43, 44, 46	131.1	98.4	124.6	131.1	97.5	124.4	131.1	66.9	118.3	131.1	87.3	122.4	131.1	87.3	122.4	131.1	69.6	118.8
21, 22, 23	FIRE SUPPRESSION, PLUMBING & HVAC	101.2	92.2	97.6	102.9	101.0	102.2	102.6	81.1	94.0	102.9	93.0	98.9	102.5	93.0	98.7	100.7	77.7	91.5
26, 27, 3370	ELECTRICAL, COMMUNICATIONS & UTIL.	110.3	100.8	105.4	107.1	87.5	96.8	117.2	78.9	97.2	112.6	74.3	92.6	113.2	74.3	92.9	120.1	80.5	99.4
MF2010	WEIGHTED AVERAGE	125.0	95.4	112.1	124.1	94.2	111.1	118.1	74.0	98.9	112.5	89.2	102.3	112.8	89.1	102.4	119.9	80.6	102.8

CANADA

DIVISION		HAMILTON, ONTARIO			HULL, QUEBEC			JOLIETTE, QUEBEC			KAMLOOPS, BRITISH COLUMBIA			KINGSTON, ONTARIO			KITCHENER, ONTARIO		
		MAT.	INST.	TOTAL	MAT.	INST.	TOTAL	MAT.	INST.	TOTAL	MAT.	INST.	TOTAL	MAT.	INST.	TOTAL	MAT.	INST.	TOTAL
015433	CONTRACTOR EQUIPMENT		108.3	108.3		102.5	102.5		102.5	102.5		105.6	105.6		103.9	103.9		103.8	103.8
0241, 31 - 34	SITE & INFRASTRUCTURE, DEMOLITION	117.0	112.7	114.0	95.7	100.5	99.1	96.3	100.6	99.3	118.4	104.0	108.3	113.0	104.5	107.1	101.2	105.1	104.0
0310	Concrete Forming & Accessories	127.9	94.2	98.8	130.3	89.8	95.4	130.3	90.0	95.5	123.5	92.8	97.0	122.9	91.7	96.0	118.0	86.6	90.9
0320	Concrete Reinforcing	153.0	93.1	123.1	151.1	81.3	116.3	142.9	81.4	112.2	112.0	79.2	95.6	167.9	84.9	126.5	103.3	93.0	98.2
0330	Cast-in-Place Concrete	151.7	100.2	130.3	129.0	99.3	116.7	134.4	99.3	119.8	116.3	103.1	110.8	145.7	102.2	127.6	140.7	90.1	119.7
03	CONCRETE	140.5	96.0	118.6	129.2	91.7	110.7	130.4	91.7	111.3	136.9	93.9	115.7	141.2	94.1	118.0	118.8	89.2	104.2
04	MASONRY	193.0	99.7	135.3	163.6	88.6	117.2	164.0	88.6	117.3	170.7	96.6	124.9	173.6	95.7	125.4	159.0	95.3	119.6
05	METALS	121.5	92.7	112.6	103.8	88.7	99.2	103.8	88.8	99.2	106.2	88.7	100.8	107.2	90.7	102.1	113.9	92.5	107.3
06	WOOD, PLASTICS & COMPOSITES	105.6	93.7	98.9	131.6	89.9	108.1	131.6	89.9	108.1	101.3	91.3	95.7	118.6	91.1	103.1	109.7	85.3	96.0
07	THERMAL & MOISTURE PROTECTION	124.3	95.0	112.4	106.2	91.8	100.3	106.2	91.8	100.3	121.2	88.6	108.0	112.4	92.2	104.2	112.1	91.6	103.8
08	OPENINGS	91.9	92.1	91.9	94.7	78.2	90.8	94.7	83.0	91.9	91.0	87.7	90.2	94.7	87.9	93.1	85.7	85.8	85.7
0920	Plaster & Gypsum Board	172.8	93.5	118.9	140.5	89.4	105.8	172.3	89.4	115.9	124.3	90.6	101.4	211.3	90.8	129.4	141.4	84.9	102.9
0950, 0980	Ceilings & Acoustic Treatment	130.0	93.5	105.7	97.4	89.4	92.1	97.4	89.4	92.1	97.4	90.6	92.9	114.3	90.8	98.7	105.3	84.9	91.7
0960	Flooring	132.6	104.1	124.1	130.5	101.4	121.8	130.5	101.4	121.8	129.8	56.6	107.9	130.5	99.2	121.1	124.4	104.1	118.3
0970, 0990	Wall Finishes & Painting/Coating	114.2	102.5	107.2	109.0	93.7	99.8	109.0	93.7	99.8	109.0	85.8	95.0	109.0	86.8	95.6	107.9	92.5	98.6
09	FINISHES	129.3	97.1	111.3	114.5	92.8	102.4	118.9	92.8	104.4	115.0	86.0	98.9	128.7	92.9	108.7	114.3	90.1	100.8
COVERS	DIVS. 10 - 14, 25, 28, 41, 43, 44, 46	131.1	100.4	125.0	131.1	87.3	122.4	131.1	87.3	122.4	131.1	97.5	124.5	131.1	75.8	120.1	131.1	98.3	124.6
21, 22, 23	FIRE SUPPRESSION, PLUMBING & HVAC	101.4	85.5	95.1	102.5	93.0	98.7	102.5	93.0	98.7	102.5	96.5	100.1	102.9	102.5	102.7	100.5	87.6	95.3
26, 27, 3370	ELECTRICAL, COMMUNICATIONS & UTIL.	111.6	96.0	103.5	114.5	74.3	93.5	113.2	74.3	92.9	116.7	83.9	99.6	113.9	91.7	102.3	114.4	93.6	103.6
MF2010	WEIGHTED AVERAGE	119.3	95.3	108.8	112.6	89.2	102.4	113.0	89.4	102.7	114.9	92.4	105.1	116.6	95.2	107.3	111.6	92.0	103.1

633

City Cost Indexes

CANADA

DIVISION		LAVAL, QUEBEC			LETHBRIDGE, ALBERTA			LLOYDMINSTER, ALBERTA			LONDON, ONTARIO			MEDICINE HAT, ALBERTA			MONCTON, NEW BRUNSWICK		
		MAT.	INST.	TOTAL	MAT.	INST.	TOTAL	MAT.	INST.	TOTAL	MAT.	INST.	TOTAL	MAT.	INST.	TOTAL	MAT.	INST.	TOTAL
015433	CONTRACTOR EQUIPMENT		102.5	102.5		103.7	103.7		103.7	103.7		103.9	103.9		103.7	103.7		101.9	101.9
0241, 31 - 34	SITE & INFRASTRUCTURE, DEMOLITION	96.1	100.5	99.2	113.0	102.5	105.6	112.9	101.9	105.2	116.2	105.3	108.5	111.6	101.9	104.8	97.8	97.5	97.6
0310	Concrete Forming & Accessories	130.5	89.8	95.4	124.2	94.4	98.5	122.4	84.4	89.6	126.1	88.0	93.2	124.1	84.3	89.8	103.7	65.9	71.0
0320	Concrete Reinforcing	151.1	81.3	116.3	155.3	74.1	114.8	155.3	74.0	114.7	130.9	93.0	112.0	155.3	74.0	114.7	139.2	61.5	100.4
0330	Cast-in-Place Concrete	133.3	99.3	119.2	161.4	104.8	137.9	149.7	101.0	129.5	161.6	98.1	135.2	149.7	101.0	129.5	131.1	79.8	109.8
03	CONCRETE	131.2	91.7	111.7	144.7	94.4	119.9	139.1	88.6	114.2	141.5	92.5	117.3	139.2	88.6	114.2	128.5	70.8	100.1
04	MASONRY	163.8	88.6	117.3	181.6	89.6	124.7	163.2	82.9	113.5	196.0	97.0	134.7	163.2	82.9	113.5	165.2	66.2	103.9
05	METALS	103.9	88.7	99.2	124.2	89.0	113.4	106.4	88.9	101.0	122.9	92.0	113.4	106.4	88.8	100.9	104.4	81.5	97.4
06	WOOD, PLASTICS & COMPOSITES	131.8	89.9	108.2	117.3	93.8	104.1	113.7	83.8	96.9	108.5	86.4	96.0	117.3	83.8	98.4	97.8	65.4	79.5
07	THERMAL & MOISTURE PROTECTION	106.8	91.8	100.7	117.4	92.6	107.3	114.3	88.0	103.7	125.2	92.8	112.0	120.9	88.0	107.6	108.9	65.8	91.4
08	OPENINGS	94.7	78.2	90.8	94.7	84.6	92.3	94.7	79.2	91.0	91.1	87.0	90.1	94.7	79.2	91.0	89.2	62.4	82.8
0920	Plaster & Gypsum Board	143.3	89.4	106.7	131.4	93.3	105.5	126.2	83.0	96.8	173.0	86.0	113.8	129.2	83.0	97.7	151.8	64.2	92.2
0950, 0980	Ceilings & Acoustic Treatment	97.4	89.4	92.1	106.3	93.3	97.6	97.4	83.0	87.8	130.1	86.0	100.7	97.4	83.0	87.8	100.9	64.2	76.5
0960	Flooring	130.5	101.4	121.8	130.5	94.3	119.7	130.5	94.3	119.7	132.6	104.1	124.1	130.5	94.3	119.7	117.8	73.4	104.5
0970, 0990	Wall Finishes & Painting/Coating	109.0	93.7	99.8	109.0	106.5	107.5	109.1	83.0	93.4	112.6	99.3	104.6	109.0	83.0	93.4	108.4	52.7	74.9
09	FINISHES	114.9	92.8	102.6	116.5	95.9	105.0	113.9	85.7	98.2	129.0	91.9	108.3	114.2	85.7	98.3	113.4	65.9	86.9
COVERS	DIVS. 10 - 14, 25, 28, 41, 43, 44, 46	131.1	87.3	122.4	131.1	97.5	124.4	131.1	94.1	123.7	131.1	99.0	124.7	131.1	94.1	123.7	131.1	66.9	118.3
21, 22, 23	FIRE SUPPRESSION, PLUMBING & HVAC	100.5	93.0	97.5	102.7	97.6	100.7	102.9	97.6	100.8	101.5	82.8	94.0	102.5	94.2	99.2	102.5	71.6	90.1
26, 27, 3370	ELECTRICAL, COMMUNICATIONS & UTIL.	114.4	74.3	93.5	108.4	87.5	97.5	106.1	87.5	96.4	109.7	93.6	101.3	106.1	87.5	96.4	118.8	63.3	89.8
MF2010	WEIGHTED AVERAGE	112.4	89.2	102.3	118.7	93.7	107.8	113.9	90.3	103.6	119.5	91.9	107.5	114.0	89.6	103.4	112.3	71.4	94.5

CANADA

DIVISION		MONTREAL, QUEBEC			MOOSE JAW, SASKATCHEWAN			NEW GLASGOW, NOVA SCOTIA			NEWCASTLE, NEW BRUNSWICK			NORTH BAY, ONTARIO			OSHAWA, ONTARIO		
		MAT.	INST.	TOTAL	MAT.	INST.	TOTAL	MAT.	INST.	TOTAL	MAT.	INST.	TOTAL	MAT.	INST.	TOTAL	MAT.	INST.	TOTAL
015433	CONTRACTOR EQUIPMENT		104.0	104.0		100.2	100.2		101.5	101.5		101.9	101.9		101.6	101.6		103.8	103.8
0241, 31 - 34	SITE & INFRASTRUCTURE, DEMOLITION	107.7	100.0	102.3	112.7	96.5	101.3	104.4	98.9	100.5	98.4	97.3	97.6	119.5	100.4	106.1	112.9	104.5	107.0
0310	Concrete Forming & Accessories	131.4	93.1	98.4	107.3	61.9	68.1	98.8	73.2	76.8	103.7	65.3	70.6	127.4	89.0	94.3	123.8	91.1	95.6
0320	Concrete Reinforcing	135.8	96.4	116.1	109.6	63.8	86.7	173.8	48.2	111.1	139.2	58.4	98.8	205.4	84.4	145.0	163.7	87.1	125.4
0330	Cast-in-Place Concrete	174.5	100.5	143.8	145.3	72.1	114.9	164.2	73.2	126.4	136.1	62.9	105.7	158.1	88.4	129.1	162.7	89.2	132.2
03	CONCRETE	148.7	96.2	122.8	122.3	66.5	94.8	151.6	69.3	111.0	130.9	63.9	97.8	154.2	88.1	121.6	145.4	89.9	118.0
04	MASONRY	172.0	94.1	123.8	161.7	65.5	102.2	182.8	73.2	115.0	165.6	66.1	104.0	188.0	91.5	128.3	162.1	96.0	121.2
05	METALS	124.5	92.8	114.7	102.7	74.9	94.1	116.0	75.3	103.4	104.4	72.8	94.7	117.0	90.4	108.8	105.5	92.6	101.5
06	WOOD, PLASTICS & COMPOSITES	120.2	93.5	105.1	99.0	60.4	77.2	94.6	72.6	82.2	97.8	65.4	79.5	127.3	89.6	106.0	117.2	90.1	101.9
07	THERMAL & MOISTURE PROTECTION	118.4	96.5	109.5	105.6	65.5	89.3	110.4	71.4	94.6	108.9	63.6	90.5	115.7	87.4	104.2	113.1	89.2	103.4
08	OPENINGS	92.8	82.4	90.3	90.1	58.1	82.5	88.3	66.4	83.1	89.2	57.2	81.6	95.9	85.7	93.5	92.2	89.9	91.7
0920	Plaster & Gypsum Board	144.9	92.8	109.5	121.3	59.1	79.0	154.0	71.7	98.0	151.8	64.2	92.2	151.5	89.2	109.1	146.6	89.8	108.0
0950, 0980	Ceilings & Acoustic Treatment	115.8	92.8	100.5	97.4	59.1	71.9	102.5	71.7	82.0	100.9	64.2	76.5	102.5	89.2	93.6	100.8	89.8	93.5
0960	Flooring	138.7	101.4	127.5	120.8	64.4	104.0	118.0	68.5	103.2	117.8	73.4	104.5	136.3	99.2	125.2	127.7	106.8	121.5
0970, 0990	Wall Finishes & Painting/Coating	112.9	105.7	108.6	109.0	68.2	84.5	113.0	65.5	84.4	108.4	52.7	74.9	113.0	93.1	101.1	107.9	106.6	107.1
09	FINISHES	123.1	96.7	108.4	110.3	62.7	83.8	116.0	71.8	91.4	113.4	65.9	86.9	121.6	91.8	105.0	115.7	95.7	104.5
COVERS	DIVS. 10 - 14, 25, 28, 41, 43, 44, 46	131.1	88.6	122.6	131.1	66.1	118.1	131.1	68.2	118.5	131.1	66.9	118.3	131.1	74.5	119.8	131.1	99.7	124.8
21, 22, 23	FIRE SUPPRESSION, PLUMBING & HVAC	101.3	93.8	98.3	102.9	77.8	92.8	102.5	86.3	96.0	102.5	71.3	90.0	102.5	100.3	101.6	100.5	102.9	101.5
26, 27, 3370	ELECTRICAL, COMMUNICATIONS & UTIL.	113.8	86.5	99.6	115.4	64.3	88.7	117.7	65.9	90.7	113.4	63.3	87.3	117.7	93.0	104.8	115.6	92.6	103.6
MF2010	WEIGHTED AVERAGE	119.2	93.4	107.9	111.2	70.8	93.6	117.7	76.1	99.6	112.1	69.3	93.5	120.5	92.9	108.5	114.6	96.1	106.5

CANADA

DIVISION		OTTAWA, ONTARIO			OWEN SOUND, ONTARIO			PETERBOROUGH, ONTARIO			PORTAGE LA PRAIRIE, MANITOBA			PRINCE ALBERT, SASKATCHEWAN			PRINCE GEORGE, BRITISH COLUMBIA		
		MAT.	INST.	TOTAL	MAT.	INST.	TOTAL	MAT.	INST.	TOTAL	MAT.	INST.	TOTAL	MAT.	INST.	TOTAL	MAT.	INST.	TOTAL
015433	CONTRACTOR EQUIPMENT		103.8	103.8		101.6	101.6		101.6	101.6		103.9	103.9		100.2	100.2		105.6	105.6
0241, 31 - 34	SITE & INFRASTRUCTURE, DEMOLITION	110.8	105.0	106.7	115.3	100.9	105.2	114.9	100.8	105.0	124.3	99.9	104.0	107.8	96.7	100.0	121.8	104.0	109.3
0310	Concrete Forming & Accessories	126.4	90.9	95.8	125.1	87.5	92.8	124.9	90.1	94.9	124.3	73.6	80.6	107.3	61.7	67.9	113.0	87.5	91.0
0320	Concrete Reinforcing	144.0	93.0	118.5	174.2	86.5	130.4	167.9	84.9	126.5	155.3	54.9	105.2	114.6	63.8	89.2	112.0	79.2	95.6
0330	Cast-in-Place Concrete	161.6	99.0	135.6	166.0	84.3	132.1	162.0	90.1	132.2	149.7	77.1	119.6	131.6	72.0	106.9	145.8	103.1	128.0
03	CONCRETE	143.6	94.1	119.2	151.0	86.4	119.2	147.1	89.3	118.6	132.7	72.0	102.7	116.7	66.3	91.9	150.0	91.5	121.2
04	MASONRY	175.1	97.0	126.8	165.2	97.3	123.2	167.9	98.4	124.9	166.3	67.7	105.3	160.8	65.5	101.8	172.9	96.6	125.7
05	METALS	123.6	91.9	113.8	108.2	91.2	103.0	105.5	90.9	101.0	106.4	77.7	97.5	102.8	74.7	94.1	106.2	88.8	100.8
06	WOOD, PLASTICS & COMPOSITES	108.8	90.5	98.5	120.6	86.2	101.2	119.8	88.2	102.0	117.1	75.1	93.4	99.0	60.4	77.2	101.3	83.9	91.5
07	THERMAL & MOISTURE PROTECTION	127.5	92.7	113.4	112.4	88.2	102.6	112.6	93.1	104.7	106.1	72.8	92.6	105.5	64.4	88.8	115.3	87.9	104.1
08	OPENINGS	93.8	89.2	92.7	92.5	85.0	90.7	93.6	85.7	92.2	94.7	66.3	88.0	89.8	58.1	81.6	91.0	83.7	89.3
0920	Plaster & Gypsum Board	192.8	90.2	123.0	151.2	85.7	106.7	142.7	87.8	105.4	123.6	73.9	89.8	121.3	59.1	79.0	124.3	83.0	96.2
0950, 0980	Ceilings & Acoustic Treatment	134.7	90.2	105.1	95.2	85.7	88.9	97.4	87.8	91.0	97.4	73.9	81.8	97.4	59.1	71.9	97.4	83.0	87.8
0960	Flooring	126.9	99.2	118.6	130.7	100.7	121.7	130.5	99.2	121.1	130.5	71.5	112.8	120.8	64.4	104.0	126.1	77.6	111.6
0970, 0990	Wall Finishes & Painting/Coating	112.4	94.1	101.4	111.9	91.8	99.8	109.0	95.5	100.9	109.1	60.0	79.6	109.0	58.2	78.5	109.0	85.8	95.0
09	FINISHES	130.7	92.9	109.7	117.3	90.6	102.4	115.6	92.5	102.7	113.4	72.3	90.5	110.3	61.6	83.2	113.9	85.1	97.9
COVERS	DIVS. 10 - 14, 25, 28, 41, 43, 44, 46	131.1	97.1	124.3	139.2	75.0	126.4	131.1	76.0	120.1	131.1	68.7	118.6	131.1	66.1	118.1	131.1	97.1	124.3
21, 22, 23	FIRE SUPPRESSION, PLUMBING & HVAC	101.5	83.2	94.2	102.2	100.2	101.4	102.5	103.9	103.1	102.5	85.4	95.6	102.9	70.1	89.7	102.5	96.5	100.1
26, 27, 3370	ELECTRICAL, COMMUNICATIONS & UTIL.	109.0	94.4	101.4	117.9	91.7	104.2	112.9	92.6	102.3	115.1	61.5	87.1	115.4	64.3	88.7	113.0	83.9	97.8
MF2010	WEIGHTED AVERAGE	119.2	92.5	107.5	117.4	93.0	106.8	115.5	94.9	106.5	113.9	75.6	97.2	110.3	69.0	92.3	115.9	91.7	105.3

City Cost Indexes

CANADA

DIVISION		QUEBEC, QUEBEC			RED DEER, ALBERTA			REGINA, SASKATCHEWAN			RIMOUSKI, QUEBEC			ROUYN-NORANDA, QUEBEC			SAINT HYACINTHE, QUEBEC		
		MAT.	INST.	TOTAL	MAT.	INST.	TOTAL	MAT.	INST.	TOTAL	MAT.	INST.	TOTAL	MAT.	INST.	TOTAL	MAT.	INST.	TOTAL
015433	CONTRACTOR EQUIPMENT		104.4	104.4		103.7	103.7		100.2	100.2		102.5	102.5		102.5	102.5		102.5	102.5
0241, 31 - 34	SITE & INFRASTRUCTURE, DEMOLITION	103.7	100.0	101.1	111.6	101.9	104.8	123.5	98.4	105.9	96.1	99.7	98.6	95.7	100.5	99.1	96.1	100.5	99.2
0310	Concrete Forming & Accessories	131.8	93.4	98.6	139.3	84.3	91.9	114.1	90.4	93.7	130.3	92.8	98.0	130.3	89.8	95.4	130.3	89.8	95.4
0320	Concrete Reinforcing	136.6	96.4	116.5	155.3	74.0	114.7	132.1	83.4	107.8	107.7	96.4	102.1	151.1	81.3	116.3	151.1	81.3	116.3
0330	Cast-in-Place Concrete	142.8	100.9	125.4	149.7	101.0	129.5	173.6	94.9	140.9	135.7	99.2	120.5	129.0	99.3	116.7	133.3	99.3	119.2
03	CONCRETE	133.9	96.4	115.4	140.3	88.6	114.8	150.3	90.6	120.9	125.3	95.6	110.7	129.2	91.7	110.7	131.2	91.7	111.7
04	MASONRY	172.0	94.1	123.8	163.2	82.9	113.5	184.7	94.4	128.8	163.4	94.1	120.5	163.6	88.6	117.2	163.8	88.6	117.3
05	METALS	124.4	93.0	114.7	106.4	88.8	100.9	125.6	85.7	113.3	103.4	92.3	100.0	103.8	88.7	99.2	103.8	88.7	99.2
06	WOOD, PLASTICS & COMPOSITES	119.9	93.5	105.0	117.3	83.8	98.4	98.3	91.3	94.4	131.6	93.2	110.0	131.6	89.9	108.1	131.6	89.9	108.1
07	THERMAL & MOISTURE PROTECTION	118.9	96.6	109.9	130.7	88.0	113.4	117.9	85.6	104.8	106.2	95.9	102.0	106.2	91.8	100.3	106.6	91.8	100.6
08	OPENINGS	94.0	90.3	93.1	94.7	79.2	91.0	90.8	80.2	88.3	94.2	80.5	91.0	94.7	78.2	90.8	94.7	78.2	90.8
0920	Plaster & Gypsum Board	164.8	92.8	115.8	129.2	83.0	97.7	147.8	91.0	109.2	172.1	92.8	118.1	140.3	89.4	105.7	142.7	89.4	106.5
0950, 0980	Ceilings & Acoustic Treatment	122.8	92.8	102.8	97.4	83.0	87.8	123.9	91.0	102.0	96.5	92.8	94.0	96.5	89.4	91.8	96.5	89.4	91.8
0960	Flooring	134.3	101.4	124.5	133.2	94.3	121.6	132.5	66.7	112.8	130.5	101.4	121.8	130.5	101.4	121.8	130.5	101.4	121.8
0970, 0990	Wall Finishes & Painting/Coating	115.4	105.7	109.5	109.0	83.0	93.4	112.9	90.9	99.7	109.0	105.7	107.0	109.0	93.7	99.8	109.0	93.7	99.8
09	FINISHES	126.3	96.7	109.8	115.0	85.7	98.7	125.5	86.8	103.9	118.7	96.5	106.3	114.3	92.8	102.3	114.6	92.8	102.5
COVERS	DIVS. 10 - 14, 25, 28, 41, 43, 44, 46	131.1	88.7	122.6	131.1	94.1	123.7	131.1	73.5	119.6	131.1	88.0	122.5	131.1	87.3	122.4	131.1	87.3	122.4
21, 22, 23	FIRE SUPPRESSION, PLUMBING & HVAC	101.2	93.8	98.3	102.5	94.2	99.2	100.7	84.2	94.1	102.5	93.7	99.0	102.5	93.0	98.7	98.9	93.0	96.5
26, 27, 3370	ELECTRICAL, COMMUNICATIONS & UTIL.	110.9	86.5	98.2	106.1	87.5	96.4	115.4	91.4	102.9	113.2	86.5	99.3	113.2	74.3	92.9	114.0	74.3	93.3
MF2010	WEIGHTED AVERAGE	117.5	93.8	107.2	114.5	89.6	103.6	120.3	88.4	106.4	112.3	93.1	103.9	112.5	89.2	102.3	111.9	89.2	102.0

CANADA

DIVISION		SAINT JOHN, NEW BRUNSWICK			SARNIA, ONTARIO			SASKATOON, SASKATCHEWAN			SAULT STE MARIE, ONTARIO			SHERBROOKE, QUEBEC			SOREL, QUEBEC		
		MAT.	INST.	TOTAL	MAT.	INST.	TOTAL	MAT.	INST.	TOTAL	MAT.	INST.	TOTAL	MAT.	INST.	TOTAL	MAT.	INST.	TOTAL
015433	CONTRACTOR EQUIPMENT		101.9	101.9		101.6	101.6		100.2	100.2		101.6	101.6		102.5	102.5		102.5	102.5
0241, 31 - 34	SITE & INFRASTRUCTURE, DEMOLITION	98.5	98.5	98.5	113.4	100.9	104.6	108.7	98.4	101.5	103.7	100.4	101.4	96.2	100.5	99.2	96.3	100.6	99.3
0310	Concrete Forming & Accessories	123.2	66.3	74.1	123.7	97.4	101.0	107.4	90.4	92.7	113.0	87.0	90.6	130.3	89.8	95.4	130.3	90.0	95.5
0320	Concrete Reinforcing	139.2	61.9	100.6	119.4	86.3	102.9	115.5	83.4	99.5	107.8	84.6	96.2	151.1	81.3	116.3	142.9	81.4	112.2
0330	Cast-in-Place Concrete	134.1	80.7	111.9	149.5	103.7	130.5	141.4	94.9	122.1	134.3	88.5	115.3	133.3	99.3	119.2	134.4	99.3	119.8
03	CONCRETE	131.4	71.4	101.8	133.3	97.4	115.6	122.5	90.6	106.7	117.3	87.3	102.5	131.2	91.7	111.7	130.4	91.7	111.3
04	MASONRY	185.8	72.2	115.5	179.4	101.2	131.0	167.2	94.4	122.2	164.3	94.5	121.1	163.9	88.6	117.3	164.0	88.6	117.3
05	METALS	104.3	82.6	97.6	105.5	91.3	101.1	101.9	85.7	96.9	104.7	90.8	100.4	103.8	88.7	99.2	103.8	88.8	99.2
06	WOOD, PLASTICS & COMPOSITES	121.4	64.5	89.3	118.9	96.8	106.4	98.0	91.3	94.3	106.0	86.5	95.0	131.6	89.9	108.1	131.6	89.9	108.1
07	THERMAL & MOISTURE PROTECTION	109.2	69.6	93.1	112.6	96.7	106.2	106.8	85.6	98.2	111.5	89.4	102.5	106.2	91.8	100.3	106.2	91.8	100.3
08	OPENINGS	89.1	60.8	82.4	95.7	91.5	94.7	90.1	80.2	87.7	87.2	85.3	86.8	94.7	78.2	90.8	94.7	83.0	91.9
0920	Plaster & Gypsum Board	169.4	63.3	97.2	164.7	96.7	118.4	129.7	91.0	103.4	131.7	86.0	100.6	142.7	89.4	106.5	172.1	89.4	115.9
0950, 0980	Ceilings & Acoustic Treatment	107.2	63.3	78.0	102.7	96.7	98.7	119.7	91.0	100.6	97.4	86.0	89.8	96.5	89.4	91.8	96.5	89.4	91.8
0960	Flooring	128.4	73.4	112.0	130.5	108.5	123.9	120.7	66.7	104.6	123.6	102.7	117.3	130.5	101.4	121.8	130.5	101.4	121.8
0970, 0990	Wall Finishes & Painting/Coating	108.4	78.8	90.6	109.0	108.0	108.4	109.9	90.9	98.5	109.0	100.0	103.6	109.0	93.7	99.8	109.0	93.7	99.8
09	FINISHES	120.2	68.5	91.4	119.8	101.1	109.4	116.5	86.8	99.9	111.4	91.3	100.2	114.6	92.8	102.5	118.7	92.8	104.3
COVERS	DIVS. 10 - 14, 25, 28, 41, 43, 44, 46	131.1	67.3	118.3	131.1	77.5	120.4	131.1	73.5	119.6	131.1	98.4	124.6	131.1	87.3	122.4	131.1	87.3	122.4
21, 22, 23	FIRE SUPPRESSION, PLUMBING & HVAC	102.5	81.0	93.9	102.5	110.2	105.6	100.6	84.2	94.0	102.5	96.8	100.2	102.9	93.0	98.9	102.5	93.0	98.7
26, 27, 3370	ELECTRICAL, COMMUNICATIONS & UTIL.	121.9	90.3	105.4	115.7	94.9	104.8	117.1	91.4	103.7	114.4	93.0	103.2	113.2	74.3	92.9	113.2	74.3	92.9
MF2010	WEIGHTED AVERAGE	114.6	78.3	98.8	115.3	99.5	108.4	111.3	88.4	101.3	110.7	93.1	103.0	112.8	89.2	102.5	113.0	89.4	102.7

CANADA

DIVISION		ST CATHARINES, ONTARIO			ST JEROME, QUEBEC			ST JOHNS, NEWFOUNDLAND			SUDBURY, ONTARIO			SUMMERSIDE, PRINCE EDWARD ISLAND			SYDNEY, NOVA SCOTIA		
		MAT.	INST.	TOTAL	MAT.	INST.	TOTAL	MAT.	INST.	TOTAL	MAT.	INST.	TOTAL	MAT.	INST.	TOTAL	MAT.	INST.	TOTAL
015433	CONTRACTOR EQUIPMENT		101.6	101.6		102.5	102.5		103.1	103.1		101.6	101.6		101.5	101.5		101.5	101.5
0241, 31 - 34	SITE & INFRASTRUCTURE, DEMOLITION	101.8	101.8	101.8	95.7	100.5	99.1	116.4	99.9	104.8	101.9	101.3	101.5	112.2	96.4	101.1	100.1	98.9	99.2
0310	Concrete Forming & Accessories	115.9	93.9	96.9	130.3	89.8	95.4	113.8	82.4	86.7	111.9	87.4	90.7	99.1	59.6	65.0	98.8	73.2	76.8
0320	Concrete Reinforcing	104.3	93.0	98.7	151.1	81.3	116.3	161.9	74.4	118.2	105.1	93.0	99.0	171.5	48.3	110.0	173.8	48.2	111.1
0330	Cast-in-Place Concrete	134.4	100.1	120.1	129.0	99.3	116.7	166.6	87.6	133.8	135.5	96.2	119.2	155.8	62.0	116.9	126.6	73.2	104.5
03	CONCRETE	115.8	95.8	105.9	129.2	91.7	110.7	159.3	83.0	121.6	116.2	91.6	104.0	156.5	59.1	108.5	133.8	69.3	102.0
04	MASONRY	158.6	99.7	122.1	163.6	88.6	117.2	187.1	89.8	126.9	158.6	94.9	119.2	181.9	63.7	108.8	180.0	73.2	113.9
05	METALS	104.6	92.4	100.8	103.8	88.7	99.2	126.2	83.4	113.0	104.6	91.9	100.7	116.0	68.6	101.4	116.0	75.3	103.4
06	WOOD, PLASTICS & COMPOSITES	107.4	93.6	99.6	131.6	89.9	108.1	101.4	80.9	89.8	103.5	86.5	93.9	94.9	59.1	74.7	94.6	72.6	82.2
07	THERMAL & MOISTURE PROTECTION	112.1	95.9	105.5	106.2	91.8	100.3	118.6	84.9	105.0	111.5	90.3	102.9	109.9	62.5	90.7	110.4	71.4	94.6
08	OPENINGS	85.2	90.7	86.5	94.7	78.2	90.8	99.3	73.4	93.2	86.0	85.3	85.8	99.1	51.1	87.8	88.3	66.4	83.1
0920	Plaster & Gypsum Board	129.0	93.4	104.8	140.3	89.4	105.7	154.5	80.1	103.9	133.6	86.0	101.2	154.4	57.7	88.6	154.0	71.7	98.0
0950, 0980	Ceilings & Acoustic Treatment	100.8	93.4	95.9	96.5	89.4	91.8	116.8	80.1	92.4	95.5	86.0	89.2	102.5	57.7	72.7	102.5	71.7	82.0
0960	Flooring	123.0	100.7	116.3	130.5	101.4	121.8	122.3	60.3	103.8	120.9	102.7	115.4	118.1	64.0	101.9	118.0	68.5	103.2
0970, 0990	Wall Finishes & Painting/Coating	107.9	102.5	104.7	109.0	93.7	99.8	112.8	75.8	90.6	107.9	93.5	99.3	113.0	43.9	71.4	113.0	65.5	84.4
09	FINISHES	111.2	96.4	103.0	114.3	92.8	102.3	121.1	78.0	97.1	110.1	90.7	99.3	117.0	59.1	84.7	116.0	71.8	91.4
COVERS	DIVS. 10 - 14, 25, 28, 41, 43, 44, 46	131.1	76.3	120.1	131.1	87.3	122.4	131.1	73.2	119.5	131.1	98.5	124.6	131.1	66.4	118.2	131.1	68.2	118.5
21, 22, 23	FIRE SUPPRESSION, PLUMBING & HVAC	100.5	84.4	94.1	102.5	93.0	98.7	100.8	73.9	90.0	102.5	85.6	95.8	102.5	65.2	87.6	102.5	86.3	96.0
26, 27, 3370	ELECTRICAL, COMMUNICATIONS & UTIL.	116.0	94.5	104.7	113.8	74.3	93.2	115.9	72.4	93.2	113.1	94.8	103.6	116.7	53.6	83.7	117.7	65.9	90.7
MF2010	WEIGHTED AVERAGE	109.7	93.0	102.4	112.5	89.2	102.3	122.0	80.4	103.9	109.9	91.7	101.9	119.9	64.1	95.4	115.6	76.1	98.4

635

City Cost Indexes

CANADA

DIVISION		THUNDER BAY, ONTARIO			TIMMINS, ONTARIO			TORONTO, ONTARIO			TROIS RIVIERES, QUEBEC			TRURO, NOVA SCOTIA			VANCOUVER, BRITISH COLUMBIA		
		MAT.	INST.	TOTAL	MAT.	INST.	TOTAL	MAT.	INST.	TOTAL	MAT.	INST.	TOTAL	MAT.	INST.	TOTAL	MAT.	INST.	TOTAL
015433	CONTRACTOR EQUIPMENT		101.6	101.6		101.6	101.6		103.9	103.9		102.5	102.5		101.5	101.5		112.1	112.1
0241, 31 - 34	SITE & INFRASTRUCTURE, DEMOLITION	106.8	101.7	103.2	114.9	100.4	104.7	137.0	105.8	115.1	100.9	100.6	100.7	100.1	98.9	99.2	120.3	108.2	111.8
0310	Concrete Forming & Accessories	123.7	90.8	95.3	124.9	89.0	93.9	128.5	99.5	103.4	133.2	90.0	95.9	97.1	73.3	76.5	122.0	91.6	95.8
0320	Concrete Reinforcing	93.2	91.9	92.5	167.9	84.4	126.2	151.3	93.2	122.3	173.8	81.4	127.6	142.9	48.2	95.6	142.2	81.7	112.0
0330	Cast-in-Place Concrete	147.9	98.5	127.4	162.0	88.4	131.5	150.7	108.2	133.1	131.1	99.3	117.9	168.3	73.3	128.8	157.7	99.2	133.4
03	CONCRETE	124.2	93.7	109.2	147.1	88.1	118.0	139.8	101.1	120.7	136.8	91.7	114.6	145.5	69.3	107.9	148.1	92.7	120.8
04	MASONRY	159.3	99.0	122.0	167.9	91.5	120.6	194.2	104.1	138.4	183.8	88.6	124.9	163.4	73.2	107.6	182.3	93.0	127.0
05	METALS	104.4	91.3	100.4	105.5	90.4	100.9	125.1	93.1	115.2	115.1	88.8	107.0	104.6	75.3	95.6	141.4	91.4	126.0
06	WOOD, PLASTICS & COMPOSITES	117.2	89.5	101.6	119.8	89.6	102.7	111.5	98.7	104.3	139.8	89.9	111.7	89.2	72.6	79.9	104.1	91.7	97.1
07	THERMAL & MOISTURE PROTECTION	112.4	93.0	104.5	112.6	87.4	102.4	124.3	99.6	114.2	109.3	91.8	102.2	107.5	71.4	92.9	130.0	88.5	113.2
08	OPENINGS	84.3	87.8	85.1	93.6	85.7	91.7	91.8	96.8	93.0	97.0	83.0	93.7	86.6	66.4	81.8	94.5	87.9	93.0
0920	Plaster & Gypsum Board	162.4	89.2	112.6	142.7	89.2	106.3	155.3	98.6	116.7	186.1	89.4	120.3	141.8	71.7	94.1	140.0	90.9	106.6
0950, 0980	Ceilings & Acoustic Treatment	95.5	89.2	91.3	97.4	89.2	91.9	125.4	98.6	107.6	101.6	89.4	93.5	97.4	71.7	80.3	150.0	90.9	110.7
0960	Flooring	127.7	107.8	121.7	130.5	99.2	121.1	132.8	110.4	126.1	136.3	101.4	125.8	113.1	68.5	99.8	129.2	97.7	119.8
0970, 0990	Wall Finishes & Painting/Coating	107.9	92.1	98.4	109.0	93.1	99.5	114.8	106.6	109.9	113.0	93.7	101.4	109.0	65.5	82.9	114.0	97.7	104.2
09	FINISHES	116.4	94.2	104.0	115.6	91.8	102.4	127.3	102.5	113.5	125.3	92.8	107.2	109.9	71.8	88.7	129.1	93.6	109.3
COVERS	DIVS. 10 - 14, 25, 28, 41, 43, 44, 46	131.1	76.0	120.1	131.1	74.5	119.8	131.1	102.1	125.3	131.1	87.3	122.4	131.1	68.2	118.5	131.1	96.9	124.3
21, 22, 23	FIRE SUPPRESSION, PLUMBING & HVAC	100.5	84.3	94.0	102.5	100.3	101.6	101.2	94.6	98.6	102.5	93.0	98.7	102.5	86.3	96.0	101.3	82.3	93.7
26, 27, 3370	ELECTRICAL, COMMUNICATIONS & UTIL.	114.4	93.0	103.2	114.4	93.0	103.2	111.5	96.6	103.7	118.0	74.3	95.2	112.9	65.9	88.4	111.3	81.3	95.6
MF2010	WEIGHTED AVERAGE	110.9	91.7	102.6	115.6	92.9	105.7	120.2	99.0	111.0	117.9	89.4	105.5	113.0	76.1	96.9	123.3	90.0	108.8

CANADA

DIVISION		VICTORIA, BRITISH COLUMBIA			WHITEHORSE, YUKON			WINDSOR, ONTARIO			WINNIPEG, MANITOBA			YARMOUTH, NOVA SCOTIA			YELLOWKNIFE, NWT		
		MAT.	INST.	TOTAL	MAT.	INST.	TOTAL	MAT.	INST.	TOTAL	MAT.	INST.	TOTAL	MAT.	INST.	TOTAL	MAT.	INST.	TOTAL
015433	CONTRACTOR EQUIPMENT		108.7	108.7		101.7	101.7		101.6	101.6		105.9	105.9		101.5	101.5		101.6	101.6
0241, 31 - 34	SITE & INFRASTRUCTURE, DEMOLITION	121.5	107.7	111.8	123.0	97.4	105.0	97.3	101.6	100.3	119.5	101.3	106.7	104.2	98.9	100.5	138.7	101.7	112.8
0310	Concrete Forming & Accessories	112.9	91.0	94.0	117.7	60.9	68.7	123.7	90.0	94.6	131.5	68.1	76.8	98.8	73.2	76.8	116.1	82.0	86.7
0320	Concrete Reinforcing	113.0	81.6	97.3	153.9	61.9	107.9	102.2	93.0	97.6	141.0	57.4	99.3	173.8	48.2	111.1	154.5	64.7	109.6
0330	Cast-in-Place Concrete	146.4	98.6	126.6	191.7	71.3	141.7	137.7	100.3	122.1	181.5	73.8	136.8	162.4	73.2	125.4	231.5	89.4	172.5
03	CONCRETE	152.5	91.9	122.6	158.2	65.4	112.4	117.6	94.1	106.0	153.4	68.7	111.6	150.8	69.3	110.6	176.9	81.7	129.9
04	MASONRY	170.1	93.0	122.4	205.6	63.9	117.9	158.7	97.8	121.0	179.6	69.3	111.4	182.6	73.2	114.9	209.1	76.2	126.8
05	METALS	104.3	87.4	99.1	129.3	74.6	112.5	104.5	91.8	100.6	143.3	74.6	122.1	116.0	75.3	103.4	129.3	79.8	114.0
06	WOOD, PLASTICS & COMPOSITES	101.6	91.6	96.0	100.8	59.6	77.5	117.2	89.1	101.4	111.1	69.0	87.3	94.6	72.6	82.2	100.3	83.7	90.9
07	THERMAL & MOISTURE PROTECTION	115.8	88.3	104.7	125.8	63.4	100.5	112.2	92.2	104.0	118.7	70.7	99.2	110.4	71.4	94.6	122.1	80.0	105.0
08	OPENINGS	91.8	84.1	90.0	95.4	56.8	86.3	84.1	88.1	85.0	93.1	61.6	85.6	88.3	66.4	83.1	94.3	70.5	88.7
0920	Plaster & Gypsum Board	121.7	90.9	100.7	146.1	58.2	86.3	147.7	88.8	107.6	139.7	67.5	90.6	154.0	71.7	98.0	176.0	83.1	112.8
0950, 0980	Ceilings & Acoustic Treatment	99.3	90.9	93.7	127.7	58.2	81.4	95.5	88.8	91.0	143.6	67.5	93.0	102.5	71.7	82.0	133.5	83.1	99.9
0960	Flooring	125.9	77.6	111.5	140.3	62.5	117.0	127.7	105.0	120.9	131.3	74.0	114.1	118.0	68.5	103.2	134.7	94.2	122.6
0970, 0990	Wall Finishes & Painting/Coating	109.9	97.7	102.6	124.2	57.2	83.9	107.9	95.4	100.4	112.9	55.4	78.4	113.0	65.5	84.4	121.6	82.0	97.8
09	FINISHES	114.1	90.1	100.7	130.9	60.6	91.7	114.0	93.5	102.6	127.8	68.3	94.6	116.0	71.8	91.4	134.9	83.7	106.4
COVERS	DIVS. 10 - 14, 25, 28, 41, 43, 44, 46	131.1	72.9	119.5	131.1	65.4	118.0	131.1	75.4	120.0	131.1	67.9	118.5	131.1	68.2	118.5	131.1	68.7	118.6
21, 22, 23	FIRE SUPPRESSION, PLUMBING & HVAC	102.5	82.2	94.4	102.9	76.4	92.3	100.5	83.9	93.9	101.2	65.7	87.0	102.5	86.3	96.0	103.0	96.0	100.2
26, 27, 3370	ELECTRICAL, COMMUNICATIONS & UTIL.	113.8	81.3	96.8	134.3	63.7	97.5	119.3	94.5	106.4	112.9	66.9	88.9	117.7	65.9	90.7	132.1	87.2	108.7
MF2010	WEIGHTED AVERAGE	115.9	88.1	103.8	126.3	69.8	101.7	110.2	91.7	102.2	123.6	70.8	100.6	117.6	76.1	99.6	128.8	85.8	110.1

Location Factors

Costs shown in RSMeans cost data publications are based on national averages for materials and installation. To adjust these costs to a specific location, simply multiply the base cost by the factor and divide by 100 for that city. The data is arranged alphabetically by state and postal zip code numbers. For a city not listed, use the factor for a nearby city with similar economic characteristics.

STATE/ZIP	CITY	MAT.	INST.	TOTAL
ALABAMA				
350-352	Birmingham	97.5	74.1	87.3
354	Tuscaloosa	96.7	62.0	81.6
355	Jasper	97.0	56.7	79.4
356	Decatur	96.6	59.1	80.3
357-358	Huntsville	96.7	70.8	85.4
359	Gadsden	96.7	60.8	81.1
360-361	Montgomery	97.9	58.1	80.5
362	Anniston	96.5	63.0	81.9
363	Dothan	96.9	50.5	76.7
364	Evergreen	96.5	51.9	77.1
365-366	Mobile	97.8	65.9	83.9
367	Selma	96.7	50.8	76.7
368	Phenix City	97.3	55.7	79.2
369	Butler	96.8	50.5	76.6
ALASKA				
995-996	Anchorage	121.3	114.3	118.3
997	Fairbanks	120.7	115.0	118.2
998	Juneau	120.6	113.5	117.5
999	Ketchikan	132.0	113.5	124.0
ARIZONA				
850,853	Phoenix	99.4	74.8	88.7
851,852	Mesa/Tempe	98.8	73.8	87.9
855	Globe	99.0	71.2	86.9
856-857	Tucson	97.6	73.6	87.1
859	Show Low	99.1	73.5	88.0
860	Flagstaff	101.0	72.9	88.8
863	Prescott	98.9	69.7	86.2
864	Kingman	97.3	73.2	86.8
865	Chambers	97.3	71.9	86.3
ARKANSAS				
716	Pine Bluff	96.9	62.9	82.1
717	Camden	94.5	50.4	75.3
718	Texarkana	95.7	49.9	75.7
719	Hot Springs	93.8	52.3	75.8
720-722	Little Rock	98.0	63.8	83.1
723	West Memphis	95.6	62.7	81.2
724	Jonesboro	96.3	59.4	80.2
725	Batesville	94.2	55.1	77.2
726	Harrison	95.5	51.1	76.1
727	Fayetteville	93.0	52.9	75.6
728	Russellville	94.1	54.4	76.8
729	Fort Smith	96.8	61.7	81.5
CALIFORNIA				
900-902	Los Angeles	100.1	116.5	107.3
903-905	Inglewood	95.7	113.4	103.4
906-908	Long Beach	97.1	113.4	104.2
910-912	Pasadena	96.3	113.9	104.0
913-916	Van Nuys	99.1	113.9	105.5
917-918	Alhambra	98.2	113.9	105.0
919-921	San Diego	100.2	109.1	104.1
922	Palm Springs	96.7	111.6	103.2
923-924	San Bernardino	94.6	111.2	101.9
925	Riverside	98.9	114.0	105.5
926-927	Santa Ana	96.5	111.6	103.1
928	Anaheim	98.9	114.1	105.6
930	Oxnard	99.6	114.2	105.9
931	Santa Barbara	98.9	114.6	105.7
932-933	Bakersfield	100.8	110.6	105.1
934	San Luis Obispo	99.3	111.6	104.7
935	Mojave	96.8	108.6	101.9
936-938	Fresno	100.0	114.5	106.3
939	Salinas	100.2	120.3	108.9
940-941	San Francisco	106.1	143.8	122.5
942,956-958	Sacramento	101.2	118.6	108.8
943	Palo Alto	99.5	133.1	114.2
944	San Mateo	101.7	132.7	115.2
945	Vallejo	100.6	127.2	112.2
946	Oakland	103.9	133.1	116.6
947	Berkeley	103.4	133.1	116.4
948	Richmond	102.5	131.2	115.0
949	San Rafael	104.2	133.6	117.0
950	Santa Cruz	104.1	120.4	111.2

STATE/ZIP	CITY	MAT.	INST.	TOTAL
CALIFORNIA (CONT'D)				
951	San Jose	102.2	134.6	116.3
952	Stockton	100.5	115.5	107.0
953	Modesto	100.4	114.9	106.7
954	Santa Rosa	100.3	133.4	114.7
955	Eureka	101.6	111.6	106.0
959	Marysville	100.8	115.1	107.0
960	Redding	103.6	114.6	108.4
961	Susanville	102.6	114.6	107.8
COLORADO				
800-802	Denver	101.6	82.6	93.3
803	Boulder	98.6	82.0	91.3
804	Golden	100.6	79.6	91.4
805	Fort Collins	102.1	79.6	92.3
806	Greeley	99.7	76.1	89.4
807	Fort Morgan	99.1	79.6	90.6
808-809	Colorado Springs	101.3	79.3	91.7
810	Pueblo	100.9	78.6	91.2
811	Alamosa	102.0	74.6	90.1
812	Salida	101.7	76.2	90.6
813	Durango	102.4	76.8	91.2
814	Montrose	101.2	76.5	90.5
815	Grand Junction	104.4	76.7	92.4
816	Glenwood Springs	102.2	79.3	92.2
CONNECTICUT				
060	New Britain	100.8	120.4	109.3
061	Hartford	102.0	120.6	110.1
062	Willimantic	101.4	120.6	109.7
063	New London	97.7	120.5	107.6
064	Meriden	99.6	120.5	108.7
065	New Haven	102.2	120.5	110.2
066	Bridgeport	101.8	120.2	109.8
067	Waterbury	101.4	120.1	109.6
068	Norwalk	101.4	126.9	112.5
069	Stamford	101.5	127.9	113.0
D.C.				
200-205	Washington	101.6	91.6	97.2
DELAWARE				
197	Newark	98.8	108.1	102.8
198	Wilmington	99.4	108.1	103.2
199	Dover	99.7	108.1	103.3
FLORIDA				
320,322	Jacksonville	97.2	67.7	84.3
321	Daytona Beach	97.4	74.4	87.3
323	Tallahassee	98.7	58.0	81.0
324	Panama City	98.5	59.1	81.3
325	Pensacola	100.8	62.4	84.1
326,344	Gainesville	98.7	68.0	85.3
327-328,347	Orlando	99.3	71.4	87.1
329	Melbourne	99.9	77.9	90.3
330-332,340	Miami	97.7	74.6	87.6
333	Fort Lauderdale	96.4	73.9	86.6
334,349	West Palm Beach	95.3	72.3	85.3
335-336,346	Tampa	97.9	79.8	90.0
337	St. Petersburg	100.2	64.7	84.7
338	Lakeland	97.1	79.3	89.4
339,341	Fort Myers	96.4	74.1	86.7
342	Sarasota	99.3	75.3	88.8
GEORGIA				
300-303,399	Atlanta	96.8	75.3	87.5
304	Statesboro	96.5	56.9	79.2
305	Gainesville	95.2	62.8	81.1
306	Athens	94.7	65.1	81.8
307	Dalton	96.4	60.1	80.6
308-309	Augusta	95.6	64.8	82.2
310-312	Macon	96.2	65.9	83.0
313-314	Savannah	98.4	62.0	82.6
315	Waycross	97.3	60.7	81.3
316	Valdosta	97.5	62.6	82.3
317,398	Albany	97.5	62.5	82.3
318-319	Columbus	97.4	66.6	84.0

637

Location Factors

STATE/ZIP	CITY	MAT.	INST.	TOTAL
HAWAII				
967	Hilo	110.4	121.0	115.0
968	Honolulu	114.9	121.0	117.6
STATES & POSS.				
969	Guam	131.2	58.9	99.7
IDAHO				
832	Pocatello	100.8	78.4	91.1
833	Twin Falls	101.9	68.5	87.4
834	Idaho Falls	99.6	75.5	89.1
835	Lewiston	108.3	85.5	98.4
836-837	Boise	99.8	79.4	90.9
838	Coeur d'Alene	107.6	83.3	97.0
ILLINOIS				
600-603	North Suburban	98.5	135.9	114.8
604	Joliet	98.4	140.7	116.8
605	South Suburban	98.5	135.9	114.8
606-608	Chicago	98.9	141.9	117.6
609	Kankakee	94.8	130.8	110.5
610-611	Rockford	98.9	123.5	109.6
612	Rock Island	96.6	99.8	98.0
613	La Salle	97.7	125.1	109.6
614	Galesburg	97.6	107.6	102.0
615-616	Peoria	99.9	110.4	104.5
617	Bloomington	96.9	109.9	102.6
618-619	Champaign	100.2	108.0	103.6
620-622	East St. Louis	95.5	109.3	101.5
623	Quincy	96.9	101.6	99.0
624	Effingham	96.3	106.3	100.6
625	Decatur	98.2	105.9	101.6
626-627	Springfield	99.0	107.2	102.5
628	Centralia	94.1	108.4	100.4
629	Carbondale	93.9	106.2	99.2
INDIANA				
460	Anderson	95.5	83.1	90.1
461-462	Indianapolis	98.3	85.1	92.6
463-464	Gary	96.8	110.1	102.6
465-466	South Bend	95.7	84.2	90.7
467-468	Fort Wayne	96.0	79.2	88.7
469	Kokomo	93.3	82.0	88.4
470	Lawrenceburg	93.3	79.3	87.2
471	New Albany	94.4	77.0	86.8
472	Columbus	96.6	80.9	89.7
473	Muncie	97.3	81.8	90.6
474	Bloomington	98.7	81.5	91.2
475	Washington	95.0	83.3	89.9
476-477	Evansville	96.5	86.6	92.1
478	Terre Haute	97.2	85.8	92.2
479	Lafayette	96.3	82.9	90.5
IOWA				
500-503,509	Des Moines	99.4	83.9	92.7
504	Mason City	96.5	66.6	83.5
505	Fort Dodge	96.6	58.8	80.2
506-507	Waterloo	98.3	70.9	86.4
508	Creston	96.7	77.9	88.5
510-511	Sioux City	99.1	73.6	88.0
512	Sibley	97.7	57.5	80.2
513	Spencer	99.2	57.5	81.0
514	Carroll	96.4	72.2	85.9
515	Council Bluffs	100.2	78.8	90.9
516	Shenandoah	97.1	72.1	86.2
520	Dubuque	98.5	80.4	90.6
521	Decorah	97.4	65.6	83.5
522-524	Cedar Rapids	99.4	83.4	92.4
525	Ottumwa	97.3	73.0	86.7
526	Burlington	96.6	76.9	88.0
527-528	Davenport	98.4	93.4	96.2
KANSAS				
660-662	Kansas City	99.4	95.7	97.7
664-666	Topeka	99.4	66.5	85.1
667	Fort Scott	97.7	74.3	87.5
668	Emporia	97.8	70.0	85.7
669	Belleville	99.4	65.6	84.7
670-672	Wichita	99.9	65.1	84.8
673	Independence	99.9	70.3	87.0
674	Salina	100.4	67.0	85.9
675	Hutchinson	95.8	65.4	82.5
676	Hays	99.6	65.6	84.8
677	Colby	100.2	65.6	85.1

STATE/ZIP	CITY	MAT.	INST.	TOTAL
KANSAS (CONT'D)				
678	Dodge City	101.7	66.5	86.4
679	Liberal	99.5	66.4	85.1
KENTUCKY				
400-402	Louisville	97.8	84.2	91.9
403-405	Lexington	97.0	82.4	90.7
406	Frankfort	97.9	82.4	91.1
407-409	Corbin	94.4	73.3	85.2
410	Covington	95.2	93.2	94.3
411-412	Ashland	94.0	100.1	96.6
413-414	Campton	95.1	80.2	88.6
415-416	Pikeville	96.2	90.0	93.5
417-418	Hazard	94.6	79.9	88.2
420	Paducah	93.2	86.2	90.2
421-422	Bowling Green	95.5	84.8	90.8
423	Owensboro	95.6	88.9	92.7
424	Henderson	93.0	87.9	90.8
425-426	Somerset	92.7	81.5	87.8
427	Elizabethtown	92.2	83.7	88.5
LOUISIANA				
700-701	New Orleans	101.5	69.8	87.7
703	Thibodaux	98.7	64.9	84.0
704	Hammond	96.3	54.9	78.3
705	Lafayette	98.5	62.7	82.9
706	Lake Charles	98.6	65.0	84.0
707-708	Baton Rouge	100.4	63.6	84.4
710-711	Shreveport	98.4	59.6	81.5
712	Monroe	96.7	56.4	79.2
713-714	Alexandria	96.8	57.8	79.8
MAINE				
039	Kittery	95.1	85.7	91.0
040-041	Portland	100.8	91.2	96.6
042	Lewiston	98.3	91.2	95.2
043	Augusta	100.3	80.8	91.8
044	Bangor	97.6	90.8	94.7
045	Bath	96.1	83.0	90.4
046	Machias	95.7	86.5	91.7
047	Houlton	95.9	86.6	91.8
048	Rockland	95.0	86.5	91.3
049	Waterville	96.2	80.8	89.5
MARYLAND				
206	Waldorf	97.5	82.2	90.9
207-208	College Park	97.6	83.9	91.7
209	Silver Spring	96.9	83.8	91.2
210-212	Baltimore	100.5	82.8	92.8
214	Annapolis	100.3	81.8	92.3
215	Cumberland	96.0	81.9	89.9
216	Easton	97.4	69.3	85.2
217	Hagerstown	96.9	82.8	90.7
218	Salisbury	97.8	63.3	82.8
219	Elkton	95.1	81.9	89.4
MASSACHUSETTS				
010-011	Springfield	100.0	109.8	104.3
012	Pittsfield	99.6	109.6	103.9
013	Greenfield	97.5	110.2	103.0
014	Fitchburg	96.3	124.6	108.6
015-016	Worcester	100.0	124.6	110.7
017	Framingham	95.5	131.0	111.0
018	Lowell	99.4	130.1	112.8
019	Lawrence	100.1	131.3	113.7
020-022, 024	Boston	101.8	138.1	117.6
023	Brockton	100.4	126.3	111.7
025	Buzzards Bay	94.9	126.7	108.8
026	Hyannis	97.6	126.7	110.3
027	New Bedford	99.6	126.8	111.4
MICHIGAN				
480,483	Royal Oak	94.2	107.5	100.0
481	Ann Arbor	96.4	107.9	101.4
482	Detroit	97.9	109.3	102.8
484-485	Flint	96.1	96.8	96.4
486	Saginaw	95.7	92.4	94.2
487	Bay City	95.8	92.5	94.4
488-489	Lansing	97.7	94.1	96.1
490	Battle Creek	96.8	87.4	92.7
491	Kalamazoo	97.1	85.9	92.2
492	Jackson	95.0	95.9	95.4
493,495	Grand Rapids	98.1	84.5	92.2
494	Muskegon	95.6	85.5	91.2

Location Factors

STATE/ZIP	CITY	MAT.	INST.	TOTAL
MICHIGAN (CONT'D)				
496	Traverse City	94.6	81.0	88.6
497	Gaylord	95.5	84.4	90.7
498-499	Iron mountain	97.5	87.4	93.1
MINNESOTA				
550-551	Saint Paul	100.2	118.8	108.3
553-555	Minneapolis	100.6	119.9	109.0
556-558	Duluth	99.0	108.6	103.2
559	Rochester	98.8	104.9	101.5
560	Mankato	96.6	102.4	99.1
561	Windom	95.3	93.3	94.4
562	Willmar	94.9	103.0	98.4
563	St. Cloud	96.3	117.8	105.7
564	Brainerd	96.3	102.3	98.9
565	Detroit Lakes	98.3	96.7	97.6
566	Bemidji	97.7	99.5	98.5
567	Thief River Falls	97.3	93.8	95.7
MISSISSIPPI				
386	Clarksdale	95.8	53.8	77.5
387	Greenville	99.4	63.1	83.6
388	Tupelo	97.3	56.1	79.4
389	Greenwood	97.0	53.3	78.0
390-392	Jackson	99.1	63.5	83.6
393	Meridian	96.2	65.3	82.7
394	Laurel	97.2	57.1	79.7
395	Biloxi	97.7	59.1	80.9
396	Mccomb	95.7	52.9	77.1
397	Columbus	97.1	55.1	78.8
MISSOURI				
630-631	St. Louis	99.6	107.6	103.1
633	Bowling Green	98.1	94.7	96.6
634	Hannibal	97.1	92.6	95.1
635	Kirksville	100.3	90.0	95.8
636	Flat River	99.0	95.8	97.6
637	Cape Girardeau	99.1	90.6	95.4
638	Sikeston	97.5	90.6	94.5
639	Poplar Bluff	97.0	90.3	94.1
640-641	Kansas City	101.6	106.7	103.8
644-645	St. Joseph	100.7	93.9	97.8
646	Chillicothe	97.7	96.2	97.1
647	Harrisonville	97.3	102.2	99.5
648	Joplin	99.5	81.0	91.5
650-651	Jefferson City	98.4	93.4	96.2
652	Columbia	98.6	93.5	96.4
653	Sedalia	98.1	92.6	95.7
654-655	Rolla	96.3	95.6	96.0
656-658	Springfield	99.9	82.1	92.1
MONTANA				
590-591	Billings	102.2	77.9	91.6
592	Wolf Point	101.7	74.9	90.0
593	Miles City	99.5	74.7	88.7
594	Great Falls	103.4	77.1	92.0
595	Havre	100.6	74.2	89.1
596	Helena	102.0	75.1	90.2
597	Butte	102.0	77.2	91.2
598	Missoula	99.4	76.4	89.4
599	Kalispell	98.4	76.3	88.8
NEBRASKA				
680-681	Omaha	99.6	79.3	90.7
683-685	Lincoln	99.5	75.3	89.0
686	Columbus	97.5	77.9	88.9
687	Norfolk	99.2	77.5	89.8
688	Grand Island	99.1	77.8	89.9
689	Hastings	98.5	81.2	90.9
690	Mccook	98.1	72.1	86.8
691	North Platte	98.5	79.6	90.2
692	Valentine	100.6	73.6	88.9
693	Alliance	100.4	71.6	87.9
NEVADA				
889-891	Las Vegas	100.2	109.4	104.2
893	Ely	98.7	103.4	100.8
894-895	Reno	99.0	93.8	96.7
897	Carson City	99.1	93.8	96.8
898	Elko	97.6	86.1	92.6
NEW HAMPSHIRE				
030	Nashua	100.6	93.6	97.6
031	Manchester	102.1	93.6	98.4

STATE/ZIP	CITY	MAT.	INST.	TOTAL
NEW HAMPSHIRE (CONT'D)				
032-033	Concord	100.8	92.3	97.1
034	Keene	97.4	74.0	87.2
035	Littleton	97.1	77.2	88.4
036	Charleston	96.9	70.1	85.2
037	Claremont	96.0	70.1	84.7
038	Portsmouth	98.2	94.2	96.4
NEW JERSEY				
070-071	Newark	101.9	125.6	112.2
072	Elizabeth	99.5	125.3	110.7
073	Jersey City	98.5	125.5	110.2
074-075	Paterson	100.2	125.2	111.1
076	Hackensack	98.4	125.5	110.2
077	Long Branch	97.8	124.1	109.2
078	Dover	98.7	125.4	110.3
079	Summit	98.7	125.3	110.3
080,083	Vineland	97.5	123.4	108.8
081	Camden	99.4	123.6	109.9
082,084	Atlantic City	98.1	123.5	109.2
085-086	Trenton	100.8	124.0	110.9
087	Point Pleasant	99.2	121.4	108.8
088-089	New Brunswick	99.7	125.2	110.8
NEW MEXICO				
870-872	Albuquerque	99.2	73.6	88.0
873	Gallup	99.3	73.6	88.1
874	Farmington	99.7	73.6	88.3
875	Santa Fe	100.5	73.6	88.8
877	Las Vegas	97.7	73.6	87.2
878	Socorro	97.3	73.6	87.0
879	Truth/Consequences	97.0	70.4	85.4
880	Las Cruces	95.8	70.4	84.7
881	Clovis	98.1	73.5	87.4
882	Roswell	99.8	73.6	88.4
883	Carrizozo	100.1	73.6	88.5
884	Tucumcari	98.8	73.5	87.8
NEW YORK				
100-102	New York	103.8	166.4	131.1
103	Staten Island	99.7	163.9	127.6
104	Bronx	98.0	163.9	126.7
105	Mount Vernon	97.9	133.3	113.3
106	White Plains	98.0	133.3	113.4
107	Yonkers	102.5	133.4	116.0
108	New Rochelle	98.2	133.3	113.5
109	Suffern	98.0	126.2	110.3
110	Queens	100.8	162.9	127.8
111	Long Island City	102.2	162.9	128.6
112	Brooklyn	102.6	162.9	128.9
113	Flushing	102.3	162.9	128.7
114	Jamaica	100.6	162.9	127.7
115,117,118	Hicksville	100.7	146.6	120.7
116	Far Rockaway	102.4	162.9	128.7
119	Riverhead	101.4	146.6	121.1
120-122	Albany	98.6	100.2	99.3
123	Schenectady	97.8	100.2	98.8
124	Kingston	100.7	118.3	108.4
125-126	Poughkeepsie	100.0	127.5	112.0
127	Monticello	99.4	116.5	106.8
128	Glens Falls	92.6	95.0	93.6
129	Plattsburgh	97.3	88.4	93.4
130-132	Syracuse	98.5	97.0	97.9
133-135	Utica	96.7	93.2	95.1
136	Watertown	98.1	91.9	95.4
137-139	Binghamton	98.1	99.4	98.6
140-142	Buffalo	99.7	105.8	102.4
143	Niagara Falls	96.9	103.9	100.0
144-146	Rochester	100.3	95.6	98.3
147	Jamestown	96.0	87.8	92.4
148-149	Elmira	95.9	93.1	94.7
NORTH CAROLINA				
270,272-274	Greensboro	98.8	56.3	80.3
271	Winston-Salem	98.5	57.1	80.5
275-276	Raleigh	98.2	55.1	79.4
277	Durham	100.2	54.4	80.2
278	Rocky Mount	96.0	53.2	77.4
279	Elizabeth City	96.9	52.2	77.4
280	Gastonia	98.4	56.1	80.0
281-282	Charlotte	98.7	57.6	80.8
283	Fayetteville	100.9	58.5	82.5
284	Wilmington	97.2	56.6	79.5
285	Kinston	95.3	53.7	77.2

Location Factors

STATE/ZIP	CITY	MAT.	INST.	TOTAL
NORTH CAROLINA (CONT'D)				
286	Hickory	95.6	54.9	77.9
287-288	Asheville	97.6	55.9	79.4
289	Murphy	96.4	49.2	75.8
NORTH DAKOTA				
580-581	Fargo	101.7	67.6	86.9
582	Grand Forks	101.8	51.6	79.9
583	Devils Lake	101.0	58.6	82.5
584	Jamestown	101.0	45.1	76.7
585	Bismarck	102.0	64.0	85.5
586	Dickinson	101.9	57.9	82.7
587	Minot	101.8	71.0	88.4
588	Williston	100.3	58.0	81.9
OHIO				
430-432	Columbus	98.6	89.7	94.7
433	Marion	95.0	83.7	90.1
434-436	Toledo	98.7	96.4	97.7
437-438	Zanesville	95.5	85.6	91.2
439	Steubenville	96.8	94.5	95.8
440	Lorain	98.8	90.4	95.2
441	Cleveland	98.9	100.0	99.4
442-443	Akron	99.8	93.4	97.0
444-445	Youngstown	99.1	88.3	94.4
446-447	Canton	99.2	85.0	93.0
448-449	Mansfield	96.6	85.6	91.8
450	Hamilton	97.4	84.5	91.8
451-452	Cincinnati	97.8	84.8	92.1
453-454	Dayton	97.5	83.2	91.3
455	Springfield	97.5	84.2	91.7
456	Chillicothe	96.4	92.2	94.5
457	Athens	99.1	80.6	91.1
458	Lima	99.6	84.4	93.0
OKLAHOMA				
730-731	Oklahoma City	97.3	65.4	83.4
734	Ardmore	95.2	62.3	80.9
735	Lawton	97.5	63.3	82.6
736	Clinton	96.5	63.6	82.2
737	Enid	97.2	60.9	81.4
738	Woodward	95.3	63.8	81.6
739	Guymon	96.3	60.5	80.7
740-741	Tulsa	97.4	62.2	82.1
743	Miami	94.0	68.3	82.8
744	Muskogee	96.6	56.3	79.0
745	Mcalester	93.7	61.1	79.5
746	Ponca City	94.3	62.5	80.5
747	Durant	94.3	64.0	81.1
748	Shawnee	95.7	62.2	81.1
749	Poteau	93.5	62.1	79.8
OREGON				
970-972	Portland	97.2	101.4	99.1
973	Salem	99.0	100.5	99.7
974	Eugene	97.1	100.4	98.5
975	Medford	98.4	98.2	98.3
976	Klamath Falls	98.2	98.2	98.2
977	Bend	97.3	100.5	98.7
978	Pendleton	93.1	102.5	97.2
979	Vale	91.3	89.8	90.7
PENNSYLVANIA				
150-152	Pittsburgh	99.7	105.8	102.3
153	Washington	96.7	103.2	99.5
154	Uniontown	97.1	102.2	99.4
155	Bedford	98.0	92.6	95.7
156	Greensburg	98.0	101.6	99.5
157	Indiana	96.9	100.1	98.3
158	Dubois	98.4	97.1	97.8
159	Johnstown	97.9	97.3	97.7
160	Butler	92.1	102.2	96.5
161	New Castle	92.1	97.2	94.4
162	Kittanning	92.5	103.4	97.3
163	Oil City	92.1	97.0	94.3
164-165	Erie	94.1	95.0	94.5
166	Altoona	94.3	92.7	93.6
167	Bradford	95.3	96.8	96.0
168	State College	94.9	93.0	94.1
169	Wellsboro	95.9	92.3	94.3
170-171	Harrisburg	99.6	95.9	98.0
172	Chambersburg	96.0	89.7	93.2
173-174	York	96.5	96.0	96.3
175-176	Lancaster	94.8	89.3	92.4

STATE/ZIP	CITY	MAT.	INST.	TOTAL
PENNSYLVANIA (CONT'D)				
177	Williamsport	93.6	79.3	87.4
178	Sunbury	95.5	96.3	95.8
179	Pottsville	94.7	98.0	96.1
180	Lehigh Valley	95.4	112.5	102.9
181	Allentown	97.6	108.8	102.4
182	Hazleton	95.0	98.8	96.7
183	Stroudsburg	94.9	102.0	98.0
184-185	Scranton	98.3	99.7	98.9
186-187	Wilkes-Barre	94.7	99.2	96.7
188	Montrose	94.4	97.1	95.6
189	Doylestown	94.6	120.8	106.0
190-191	Philadelphia	98.9	133.4	113.9
193	Westchester	95.8	121.7	107.1
194	Norristown	94.9	129.5	110.0
195-196	Reading	97.1	101.1	98.8
PUERTO RICO				
009	San Juan	122.1	24.6	79.7
RHODE ISLAND				
028	Newport	99.0	116.6	106.6
029	Providence	101.5	116.6	108.1
SOUTH CAROLINA				
290-292	Columbia	97.2	57.3	79.8
293	Spartanburg	96.4	57.4	79.5
294	Charleston	97.7	66.4	84.1
295	Florence	96.2	57.8	79.4
296	Greenville	96.2	56.5	78.9
297	Rock Hill	95.6	51.2	76.3
298	Aiken	96.4	70.4	85.1
299	Beaufort	97.1	45.7	74.7
SOUTH DAKOTA				
570-571	Sioux Falls	100.0	58.8	82.1
572	Watertown	98.5	50.2	77.5
573	Mitchell	97.5	49.5	76.6
574	Aberdeen	100.1	50.6	78.6
575	Pierre	100.8	59.2	82.7
576	Mobridge	98.1	49.8	77.1
577	Rapid City	100.0	60.3	82.7
TENNESSEE				
370-372	Nashville	97.9	74.1	87.5
373-374	Chattanooga	99.5	65.5	84.7
375,380-381	Memphis	98.5	71.3	86.7
376	Johnson City	99.2	56.7	80.7
377-379	Knoxville	96.2	60.5	80.6
382	Mckenzie	96.6	60.4	80.9
383	Jackson	98.3	62.5	82.7
384	Columbia	95.3	67.5	83.2
385	Cookeville	96.4	60.4	80.7
TEXAS				
750	Mckinney	99.3	48.9	77.3
751	Waxahackie	99.3	56.4	80.6
752-753	Dallas	100.1	65.8	85.2
754	Greenville	99.4	42.2	74.5
755	Texarkana	98.4	49.7	77.2
756	Longview	99.0	40.5	73.5
757	Tyler	98.9	54.6	79.6
758	Palestine	95.4	55.8	78.2
759	Lufkin	95.9	59.1	79.9
760-761	Fort Worth	99.6	62.5	83.5
762	Denton	98.2	46.6	75.7
763	Wichita Falls	97.8	55.0	79.2
764	Eastland	96.5	49.1	75.9
765	Temple	95.0	52.7	76.6
766-767	Waco	97.2	57.3	79.8
768	Brownwood	98.5	48.4	76.7
769	San Angelo	98.2	49.5	77.0
770-772	Houston	98.7	70.9	86.6
773	Huntsville	97.2	56.9	79.6
774	Wharton	98.1	53.8	78.8
775	Galveston	96.3	69.7	84.7
776-777	Beaumont	96.7	61.2	81.2
778	Bryan	94.1	63.0	80.5
779	Victoria	98.2	42.7	74.1
780	Laredo	97.1	53.7	78.2
781-782	San Antonio	97.5	64.5	83.1
783-784	Corpus Christi	99.7	53.7	79.6
785	Mc Allen	99.7	46.6	76.6
786-787	Austin	98.6	59.1	81.4

Location Factors

STATE/ZIP	CITY	MAT.	INST.	TOTAL
TEXAS (CONT'D)				
788	Del Rio	99.4	46.2	76.3
789	Giddings	96.4	55.6	78.6
790-791	Amarillo	98.2	60.0	81.6
792	Childress	97.9	56.9	80.0
793-794	Lubbock	100.0	56.3	81.0
795-796	Abilene	98.3	53.6	78.8
797	Midland	99.9	53.6	79.7
798-799,885	El Paso	97.2	50.6	76.9
UTAH				
840-841	Salt Lake City	102.2	66.8	86.8
842,844	Ogden	97.9	68.8	85.2
843	Logan	99.7	68.8	86.2
845	Price	100.1	65.3	84.9
846-847	Provo	100.1	66.7	85.6
VERMONT				
050	White River Jct.	97.9	79.7	90.0
051	Bellows Falls	96.6	94.7	95.8
052	Bennington	96.9	95.3	96.2
053	Brattleboro	97.2	94.5	96.0
054	Burlington	101.5	85.1	94.4
056	Montpelier	99.2	84.3	92.7
057	Rutland	99.2	85.1	93.0
058	St. Johnsbury	97.9	79.2	89.8
059	Guildhall	96.8	79.2	89.1
VIRGINIA				
220-221	Fairfax	99.3	84.4	92.8
222	Arlington	100.6	84.7	93.6
223	Alexandria	99.8	85.4	93.5
224-225	Fredericksburg	98.2	83.0	91.6
226	Winchester	98.7	82.1	91.5
227	Culpeper	98.6	80.8	90.9
228	Harrisonburg	98.8	68.1	85.5
229	Charlottesville	99.2	67.3	85.3
230-232	Richmond	100.4	68.9	86.7
233-235	Norfolk	100.0	68.8	86.4
236	Newport News	99.2	68.2	85.7
237	Portsmouth	98.6	66.1	84.5
238	Petersburg	98.6	68.9	85.6
239	Farmville	97.8	53.3	78.5
240-241	Roanoke	100.2	66.0	85.3
242	Bristol	98.1	55.6	79.6
243	Pulaski	97.7	58.3	80.6
244	Staunton	98.4	62.7	82.9
245	Lynchburg	98.6	66.1	84.5
246	Grundy	98.0	53.4	78.6
WASHINGTON				
980-981,987	Seattle	102.5	104.8	103.5
982	Everett	102.1	99.6	101.0
983-984	Tacoma	102.3	100.0	101.3
985	Olympia	101.0	100.0	100.6
986	Vancouver	103.1	93.2	98.8
988	Wenatchee	102.0	87.1	95.5
989	Yakima	102.5	94.4	99.0
990-992	Spokane	102.5	82.7	93.8
993	Richland	102.2	89.6	96.7
994	Clarkston	100.6	80.7	92.0
WEST VIRGINIA				
247-248	Bluefield	97.1	91.7	94.7
249	Lewisburg	98.4	91.4	95.4
250-253	Charleston	99.0	94.1	96.9
254	Martinsburg	98.3	86.1	93.0
255-257	Huntington	99.8	95.2	97.8
258-259	Beckley	96.9	92.0	94.8
260	Wheeling	99.8	93.8	97.2
261	Parkersburg	98.9	91.4	95.6
262	Buckhannon	98.2	93.6	96.2
263-264	Clarksburg	98.8	93.5	96.5
265	Morgantown	98.8	92.6	96.1
266	Gassaway	98.2	92.8	95.8
267	Romney	98.2	90.4	94.8
268	Petersburg	98.0	90.1	94.6
WISCONSIN				
530,532	Milwaukee	100.9	107.9	104.0
531	Kenosha	100.2	102.9	101.4
534	Racine	99.7	102.6	101.0
535	Beloit	99.6	101.9	100.6
537	Madison	99.5	99.8	99.6

STATE/ZIP	CITY	MAT.	INST.	TOTAL
WISCONSIN (CONT'D)				
538	Lancaster	97.4	93.5	95.7
539	Portage	96.1	97.9	96.9
540	New Richmond	96.2	95.9	96.1
541-543	Green Bay	100.5	94.1	97.7
544	Wausau	95.7	94.4	95.2
545	Rhinelander	98.7	92.4	95.9
546	La Crosse	97.2	95.0	96.3
547	Eau Claire	98.7	96.4	97.7
548	Superior	96.1	97.6	96.7
549	Oshkosh	96.3	92.4	94.6
WYOMING				
820	Cheyenne	100.1	65.7	85.2
821	Yellowstone Nat'l Park	97.7	65.2	83.5
822	Wheatland	98.7	64.6	83.9
823	Rawlins	100.2	67.8	86.1
824	Worland	98.3	64.1	83.4
825	Riverton	99.3	64.9	84.3
826	Casper	100.4	60.7	83.1
827	Newcastle	98.2	67.8	84.9
828	Sheridan	100.9	65.2	85.4
829-831	Rock Springs	101.8	65.4	85.9
CANADIAN FACTORS (reflect Canadian currency)				
ALBERTA				
	Calgary	124.7	95.4	111.9
	Edmonton	125.0	95.4	112.1
	Fort McMurray	124.1	94.2	111.1
	Lethbridge	118.7	93.7	107.8
	Lloydminster	113.9	90.3	103.6
	Medicine Hat	114.0	89.6	103.4
	Red Deer	114.5	89.6	103.6
BRITISH COLUMBIA				
	Kamloops	114.9	92.4	105.1
	Prince George	115.9	91.7	105.3
	Vancouver	123.3	90.0	108.8
	Victoria	115.9	88.1	103.8
MANITOBA				
	Brandon	118.7	77.3	100.7
	Portage la Prairie	113.9	75.6	97.2
	Winnipeg	123.6	70.8	100.6
NEW BRUNSWICK				
	Bathurst	112.1	68.6	93.1
	Dalhousie	112.1	68.8	93.2
	Fredericton	118.1	74.0	98.9
	Moncton	112.3	71.4	94.5
	Newcastle	112.1	69.3	93.5
	St. John	114.6	78.3	98.8
NEWFOUNDLAND				
	Corner Brook	122.2	69.0	99.0
	St Johns	122.0	80.4	103.9
NORTHWEST TERRITORIES				
	Yellowknife	128.8	85.8	110.1
NOVA SCOTIA				
	Bridgewater	113.4	76.1	97.2
	Dartmouth	119.7	76.1	100.7
	Halifax	119.9	80.6	102.8
	New Glasgow	117.7	76.1	99.6
	Sydney	115.6	76.1	98.4
	Truro	113.0	76.1	96.9
	Yarmouth	117.6	76.1	99.6
ONTARIO				
	Barrie	117.3	94.9	107.5
	Brantford	115.5	98.8	108.2
	Cornwall	115.5	95.1	106.7
	Hamilton	119.3	95.3	108.8
	Kingston	116.6	95.2	107.3
	Kitchener	111.6	92.0	103.1
	London	119.5	91.9	107.5
	North Bay	120.5	92.9	108.5
	Oshawa	114.6	96.1	106.5
	Ottawa	119.2	92.5	107.5
	Owen Sound	117.4	93.0	106.8
	Peterborough	115.5	94.9	106.5
	Sarnia	115.3	99.5	108.4

Location Factors

STATE/ZIP	CITY	MAT.	INST.	TOTAL
ONTARIO (CONT'D)				
	Sault Ste Marie	110.7	93.1	103.0
	St. Catharines	109.7	93.0	102.4
	Sudbury	109.9	91.7	101.9
	Thunder Bay	110.9	91.7	102.6
	Timmins	115.6	92.9	105.7
	Toronto	120.2	99.0	111.0
	Windsor	110.2	91.7	102.2
PRINCE EDWARD ISLAND				
	Charlottetown	120.5	64.0	95.9
	Summerside	119.5	64.1	95.4
QUEBEC				
	Cap-de-la-Madeleine	112.8	89.4	102.6
	Charlesbourg	112.8	89.4	102.6
	Chicoutimi	111.9	93.1	103.7
	Gatineau	112.5	89.2	102.3
	Granby	112.8	89.1	102.4
	Hull	112.6	89.2	102.4
	Joliette	113.0	89.4	102.7
	Laval	112.4	89.2	102.3
	Montreal	119.2	93.4	107.9
	Quebec	117.5	93.8	107.2
	Rimouski	112.3	93.1	103.9
	Rouyn-Noranda	112.5	89.2	102.3
	Saint Hyacinthe	111.9	89.2	102.0
	Sherbrooke	112.8	89.2	102.5
	Sorel	113.0	89.4	102.7
	St Jerome	112.5	89.2	102.3
	Trois Rivieres	117.9	89.4	105.5
SASKATCHEWAN				
	Moose Jaw	111.2	70.8	93.6
	Prince Albert	110.3	69.0	92.3
	Regina	120.3	88.4	106.4
	Saskatoon	111.3	88.4	101.3
YUKON				
	Whitehorse	126.3	69.8	101.7

General Requirements — R0111 Summary of Work

R011105-05 Tips for Accurate Estimating

1. Use pre-printed or columnar forms for orderly sequence of dimensions and locations and for recording telephone quotations.
2. Use only the front side of each paper or form except for certain pre-printed summary forms.
3. Be consistent in listing dimensions: For example, length x width x height. This helps in rechecking to ensure that, the total length of partitions is appropriate for the building area.
4. Use printed (rather than measured) dimensions where given.
5. Add up multiple printed dimensions for a single entry where possible.
6. Measure all other dimensions carefully.
7. Use each set of dimensions to calculate multiple related quantities.
8. Convert foot and inch measurements to decimal feet when listing. Memorize decimal equivalents to .01 parts of a foot (1/8" equals approximately .01").
9. Do not "round off" quantities until the final summary.
10. Mark drawings with different colors as items are taken off.
11. Keep similar items together, different items separate.
12. Identify location and drawing numbers to aid in future checking for completeness.
13. Measure or list everything on the drawings or mentioned in the specifications.
14. It may be necessary to list items not called for to make the job complete.
15. Be alert for: Notes on plans such as N.T.S. (not to scale); changes in scale throughout the drawings; reduced size drawings; discrepancies between the specifications and the drawings.
16. Develop a consistent pattern of performing an estimate. For example:
    a. Start the quantity takeoff at the lower floor and move to the next higher floor.
    b. Proceed from the main section of the building to the wings.
    c. Proceed from south to north or vice versa, clockwise or counterclockwise.
    d. Take off floor plan quantities first, elevations next, then detail drawings.
17. List all gross dimensions that can be either used again for different quantities, or used as a rough check of other quantities for verification (exterior perimeter, gross floor area, individual floor areas, etc.).
18. Utilize design symmetry or repetition (repetitive floors, repetitive wings, symmetrical design around a center line, similar room layouts, etc.). Note: Extreme caution is needed here so as not to omit or duplicate an area.
19. Do not convert units until the final total is obtained. For instance, when estimating concrete work, keep all units to the nearest cubic foot, then summarize and convert to cubic yards.
20. When figuring alternatives, it is best to total all items involved in the basic system, then total all items involved in the alternates. Therefore you work with positive numbers in all cases. When adds and deducts are used, it is often confusing whether to add or subtract a portion of an item; especially on a complicated or involved alternate.

R011105-10 Unit Gross Area Requirements

The figures in the table below indicate typical ranges in square feet as a function of the "occupant" unit. This table is best used in the preliminary design stages to help determine the probable size requirement for the total project.

Building Type	Unit	Gross Area in S.F. 1/4	Median	3/4
Apartments	Unit	660	860	1,100
Auditorium & Play Theaters	Seat	18	25	38
Bowling Alleys	Lane		940	
Churches & Synagogues	Seat	20	28	39
Dormitories	Bed	200	230	275
Fraternity & Sorority Houses	Bed	220	315	370
Garages, Parking	Car	325	355	385
Hospitals	Bed	685	850	1,075
Hotels	Rental Unit	475	600	710
Housing for the elderly	Unit	515	635	755
Housing, Public	Unit	700	875	1,030
Ice Skating Rinks	Total	27,000	30,000	36,000
Motels	Rental Unit	360	465	620
Nursing Homes	Bed	290	350	450
Restaurants	Seat	23	29	39
Schools, Elementary	Pupil	65	77	90
Junior High & Middle		85	110	129
Senior High		102	130	145
Vocational		110	135	195
Shooting Ranges	Point		450	
Theaters & Movies	Seat		15	

General Requirements — R0111 Summary of Work

R011105-20 Floor Area Ratios

Table below lists commonly used gross to net area and net to gross area ratios expressed in % for various building types.

Building Type	Gross to Net Ratio	Net to Gross Ratio	Building Type	Gross to Net Ratio	Net to Gross Ratio
Apartment	156	64	School Buildings (campus type)		
Bank	140	72	Administrative	150	67
Church	142	70	Auditorium	142	70
Courthouse	162	61	Biology	161	62
Department Store	123	81	Chemistry	170	59
Garage	118	85	Classroom	152	66
Hospital	183	55	Dining Hall	138	72
Hotel	158	63	Dormitory	154	65
Laboratory	171	58	Engineering	164	61
Library	132	76	Fraternity	160	63
Office	135	75	Gymnasium	142	70
Restaurant	141	70	Science	167	60
Warehouse	108	93	Service	120	83
			Student Union	172	59

The gross area of a building is the total floor area based on outside dimensions.

The net area of a building is the usable floor area for the function intended and excludes such items as stairways, corridors and mechanical rooms. In the case of a commercial building, it might be considered as the "leasable area."

R011105-30 Occupancy Determinations

		S.F. Required per Person		
	Description	BOCA	SBC	UBC
Assembly Areas	Fixed Seats	**	6	7
	Movable Seats		15	15
	Concentrated	7		
	Unconcentrated	15		
	Standing Space	3		
Educational	Unclassified			
	Classrooms	20	40	20
	Shop Areas	50	100	50
Institutional	Unclassified		125	
	In-Patient Areas	240		
	Sleeping Areas	120		
Mercantile	Basement	30	30	20
	Ground Floor	30	30	30
	Upper Floors	60	60	50
Office		100	100	100

BOCA=Building Officials & Code Administrators
SBC=Southern Building Code
UBC=Uniform Building Code

** The occupancy load for assembly area with fixed seats shall be determined by the number of fixed seats installed.

General Requirements R0111 Summary of Work

R011105-40 Weather Data and Design Conditions

City	Latitude (1) 0	Latitude (1) 1'	Winter Temperatures (1) Med. of Annual Extremes	Winter Temperatures (1) 99%	Winter Temperatures (1) 97½%	Winter Degree Days (2)	Summer (Design Dry Bulb) Temperatures and Relative Humidity 1%	2½%	5%
UNITED STATES									
Albuquerque, NM	35	0	5.1	12	16	4,400	96/61	94/61	92/61
Atlanta, GA	33	4	11.9	17	22	3,000	94/74	92/74	90/73
Baltimore, MD	39	2	7	14	17	4,600	94/75	91/75	89/74
Birmingham, AL	33	3	13	17	21	2,600	96/74	94/75	92/74
Bismarck, ND	46	5	-32	-23	-19	8,800	95/68	91/68	88/67
Boise, ID	43	3	1	3	10	5,800	96/65	94/64	91/64
Boston, MA	42	2	-1	6	9	5,600	91/73	88/71	85/70
Burlington, VT	44	3	-17	-12	-7	8,200	88/72	85/70	82/69
Charleston, WV	38	2	3	7	11	4,400	92/74	90/73	87/72
Charlotte, NC	35	1	13	18	22	3,200	95/74	93/74	91/74
Casper, WY	42	5	-21	-11	-5	7,400	92/58	90/57	87/57
Chicago, IL	41	5	-8	-3	2	6,600	94/75	91/74	88/73
Cincinnati, OH	39	1	0	1	6	4,400	92/73	90/72	88/72
Cleveland, OH	41	2	-3	1	5	6,400	91/73	88/72	86/71
Columbia, SC	34	0	16	20	24	2,400	97/76	95/75	93/75
Dallas, TX	32	5	14	18	22	2,400	102/75	100/75	97/75
Denver, CO	39	5	-10	-5	1	6,200	93/59	91/59	89/59
Des Moines, IA	41	3	-14	-10	-5	6,600	94/75	91/74	88/73
Detroit, MI	42	2	-3	3	6	6,200	91/73	88/72	86/71
Great Falls, MT	47	3	-25	-21	-15	7,800	91/60	88/60	85/59
Hartford, CT	41	5	-4	3	7	6,200	91/74	88/73	85/72
Houston, TX	29	5	24	28	33	1,400	97/77	95/77	93/77
Indianapolis, IN	39	4	-7	-2	2	5,600	92/74	90/74	87/73
Jackson, MS	32	2	16	21	25	2,200	97/76	95/76	93/76
Kansas City, MO	39	1	-4	2	6	4,800	99/75	96/74	93/74
Las Vegas, NV	36	1	18	25	28	2,800	108/66	106/65	104/65
Lexington, KY	38	0	-1	3	8	4,600	93/73	91/73	88/72
Little Rock, AR	34	4	11	15	20	3,200	99/76	96/77	94/77
Los Angeles, CA	34	0	36	41	43	2,000	93/70	89/70	86/69
Memphis, TN	35	0	10	13	18	3,200	98/77	95/76	93/76
Miami, FL	25	5	39	44	47	200	91/77	90/77	89/77
Milwaukee, WI	43	0	-11	-8	-4	7,600	90/74	87/73	84/71
Minneapolis, MN	44	5	-22	-16	-12	8,400	92/75	89/73	86/71
New Orleans, LA	30	0	28	29	33	1,400	93/78	92/77	90/77
New York, NY	40	5	6	11	15	5,000	92/74	89/73	87/72
Norfolk, VA	36	5	15	20	22	3,400	93/77	91/76	89/76
Oklahoma City, OK	35	2	4	9	13	3,200	100/74	97/74	95/73
Omaha, NE	41	2	-13	-8	-3	6,600	94/76	91/75	88/74
Philadelphia, PA	39	5	6	10	14	4,400	93/75	90/74	87/72
Phoenix, AZ	33	3	27	31	34	1,800	109/71	107/71	105/71
Pittsburgh, PA	40	3	-1	3	7	6,000	91/72	88/71	86/70
Portland, ME	43	4	-10	-6	-1	7,600	87/72	84/71	81/69
Portland, OR	45	4	18	17	23	4,600	89/68	85/67	81/65
Portsmouth, NH	43	1	-8	-2	2	7,200	89/73	85/71	83/70
Providence, RI	41	4	-1	5	9	6,000	89/73	86/72	83/70
Rochester, NY	43	1	-5	1	5	6,800	91/73	88/71	85/70
Salt Lake City, UT	40	5	0	3	8	6,000	97/62	95/62	92/61
San Francisco, CA	37	5	36	38	40	3,000	74/63	71/62	69/61
Seattle, WA	47	4	22	22	27	5,200	85/68	82/66	78/65
Sioux Falls, SD	43	4	-21	-15	-11	7,800	94/73	91/72	88/71
St. Louis, MO	38	4	-3	3	8	5,000	98/75	94/75	91/75
Tampa, FL	28	0	32	36	40	680	92/77	91/77	90/76
Trenton, NJ	40	1	4	11	14	5,000	91/75	88/74	85/73
Washington, DC	38	5	7	14	17	4,200	93/75	91/74	89/74
Wichita, KS	37	4	-3	3	7	4,600	101/72	98/73	96/73
Wilmington, DE	39	4	5	10	14	5,000	92/74	89/74	87/73
ALASKA									
Anchorage	61	1	-29	-23	-18	10,800	71/59	68/58	66/56
Fairbanks	64	5	-59	-51	-47	14,280	82/62	78/60	75/59
CANADA									
Edmonton, Alta.	53	3	-30	-29	-25	11,000	85/66	82/65	79/63
Halifax, N.S.	44	4	-4	1	5	8,000	79/66	76/65	74/64
Montreal, Que.	45	3	-20	-16	-10	9,000	88/73	85/72	83/71
Saskatoon, Sask.	52	1	-35	-35	-31	11,000	89/68	86/66	83/65
St. John's, N.F.	47	4	1	3	7	8,600	77/66	75/65	73/64
Saint John, N.B.	45	2	-15	-12	-8	8,200	80/67	77/65	75/64
Toronto, Ont.	43	4	-10	-5	-1	7,000	90/73	87/72	85/71
Vancouver, B.C.	49	1	13	15	19	6,000	79/67	77/66	74/65
Winnipeg, Man.	49	5	-31	-30	-27	10,800	89/73	86/71	84/70

(1) Handbook of Fundamentals, ASHRAE, Inc., NY 1989
(2) Local Climatological Annual Survey, USDC Env. Science Services Administration, Asheville, NC

General Requirements   R0111 Summary of Work

R011105-50 Metric Conversion Factors

Description: This table is primarily for converting customary U.S. units in the left hand column to SI metric units in the right hand column. In addition, conversion factors for some commonly encountered Canadian and non-SI metric units are included.

If You Know		Multiply By		To Find
Length	Inches	x	25.4[a]	=	Millimeters
	Feet	x	0.3048[a]	=	Meters
	Yards	x	0.9144[a]	=	Meters
	Miles (statute)	x	1.609	=	Kilometers
Area	Square inches	x	645.2	=	Square millimeters
	Square feet	x	0.0929	=	Square meters
	Square yards	x	0.8361	=	Square meters
Volume	Cubic inches	x	16,387	=	Cubic millimeters
(Capacity)	Cubic feet	x	0.02832	=	Cubic meters
	Cubic yards	x	0.7646	=	Cubic meters
	Gallons (U.S. liquids)[b]	x	0.003785	=	Cubic meters[c]
	Gallons (Canadian liquid)[b]	x	0.004546	=	Cubic meters[c]
	Ounces (U.S. liquid)[b]	x	29.57	=	Milliliters[c,d]
	Quarts (U.S. liquid)[b]	x	0.9464	=	Liters[c,d]
	Gallons (U.S. liquid)[b]	x	3.785	=	Liters[c,d]
Force	Kilograms force[d]	x	9.807	=	Newtons
	Pounds force	x	4.448	=	Newtons
	Pounds force	x	0.4536	=	Kilograms force[d]
	Kips	x	4448	=	Newtons
	Kips	x	453.6	=	Kilograms force[d]
Pressure,	Kilograms force per square centimeter[d]	x	0.09807	=	Megapascals
Stress,	Pounds force per square inch (psi)	x	0.006895	=	Megapascals
Strength	Kips per square inch	x	6.895	=	Megapascals
(Force per unit area)	Pounds force per square inch (psi)	x	0.07031	=	Kilograms force per square centimeter[d]
	Pounds force per square foot	x	47.88	=	Pascals
	Pounds force per square foot	x	4.882	=	Kilograms force per square meter[d]
Bending	Inch-pounds force	x	0.01152	=	Meter-kilograms force[d]
Moment	Inch-pounds force	x	0.1130	=	Newton-meters
Or Torque	Foot-pounds force	x	0.1383	=	Meter-kilograms force[d]
	Foot-pounds force	x	1.356	=	Newton-meters
	Meter-kilograms force[d]	x	9.807	=	Newton-meters
Mass	Ounces (avoirdupois)	x	28.35	=	Grams
	Pounds (avoirdupois)	x	0.4536	=	Kilograms
	Tons (metric)	x	1000	=	Kilograms
	Tons, short (2000 pounds)	x	907.2	=	Kilograms
	Tons, short (2000 pounds)	x	0.9072	=	Megagrams[e]
Mass per	Pounds mass per cubic foot	x	16.02	=	Kilograms per cubic meter
Unit	Pounds mass per cubic yard	x	0.5933	=	Kilograms per cubic meter
Volume	Pounds mass per gallon (U.S. liquid)[b]	x	119.8	=	Kilograms per cubic meter
	Pounds mass per gallon (Canadian liquid)[b]	x	99.78	=	Kilograms per cubic meter
Temperature	Degrees Fahrenheit	(F-32)/1.8		=	Degrees Celsius
	Degrees Fahrenheit	(F+459.67)/1.8		=	Degrees Kelvin
	Degrees Celsius	C+273.15		=	Degrees Kelvin

[a] The factor given is exact
[b] One U.S. gallon = 0.8327 Canadian gallon
[c] 1 liter = 1000 milliliters = 1000 cubic centimeters
  1 cubic decimeter = 0.001 cubic meter
[d] Metric but not SI unit
[e] Called "tonne" in England and "metric ton" in other metric countries

General Requirements — R0111 Summary of Work

R011105-60 Weights and Measures

Measures of Length
1 Mile = 1760 Yards = 5280 Feet
1 Yard = 3 Feet = 36 inches
1 Foot = 12 Inches
1 Mil = 0.001 Inch
1 Fathom = 2 Yards = 6 Feet
1 Rod = 5.5 Yards = 16.5 Feet
1 Hand = 4 Inches
1 Span = 9 Inches
1 Micro-inch = One Millionth Inch or 0.000001 Inch
1 Micron = One Millionth Meter + 0.00003937 Inch

Surveyor's Measure
1 Mile = 8 Furlongs = 80 Chains
1 Furlong = 10 Chains = 220 Yards
1 Chain = 4 Rods = 22 Yards = 66 Feet = 100 Links
1 Link = 7.92 Inches

Square Measure
1 Square Mile = 640 Acres = 6400 Square Chains
1 Acre = 10 Square Chains = 4840 Square Yards = 43,560 Sq. Ft.
1 Square Chain = 16 Square Rods = 484 Square Yards = 4356 Sq. Ft.
1 Square Rod = 30.25 Square Yards = 272.25 Square Feet = 625 Square Lines
1 Square Yard = 9 Square Feet
1 Square Foot = 144 Square Inches
An Acre equals a Square 208.7 Feet per Side

Cubic Measure
1 Cubic Yard = 27 Cubic Feet
1 Cubic Foot = 1728 Cubic Inches
1 Cord of Wood = 4 x 4 x 8 Feet = 128 Cubic Feet
1 Perch of Masonry = 16½ x 1½ x 1 Foot = 24.75 Cubic Feet

Avoirdupois or Commercial Weight
1 Gross or Long Ton = 2240 Pounds
1 Net or Short ton = 2000 Pounds
1 Pound = 16 Ounces = 7000 Grains
1 Ounce = 16 Drachms = 437.5 Grains
1 Stone = 14 Pounds

Shipping Measure
For Measuring Internal Capacity of a Vessel:
  1 Register Ton = 100 Cubic Feet

For Measurement of Cargo:
  Approximately 40 Cubic Feet of Merchandise is considered a Shipping Ton, unless that bulk would weigh more than 2000 Pounds, in which case Freight Charge may be based upon weight.

40 Cubic Feet = 32.143 U.S. Bushels = 31.16 Imp. Bushels

Liquid Measure
1 Imperial Gallon = 1.2009 U.S. Gallon = 277.42 Cu. In.
1 Cubic Foot = 7.48 U.S. Gallons

General Requirements — R0111 Summary of Work

R011110-30 Engineering Fees

Typical **Structural Engineering Fees** based on type of construction and total project size. These fees are included in Architectural Fees.

Type of Construction	Total Project Size (in thousands of dollars)			
	$500	$500-$1,000	$1,000-$5,000	Over $5000
Industrial buildings, factories & warehouses	Technical payroll times 2.0 to 2.5	1.60%	1.25%	1.00%
Hotels, apartments, offices, dormitories, hospitals, public buildings, food stores		2.00%	1.70%	1.20%
Museums, banks, churches and cathedrals		2.00%	1.75%	1.25%
Thin shells, prestressed concrete, earthquake resistive		2.00%	1.75%	1.50%
Parking ramps, auditoriums, stadiums, convention halls, hangars & boiler houses		2.50%	2.00%	1.75%
Special buildings, major alterations, underpinning & future expansion		Add to above 0.5%	Add to above 0.5%	Add to above 0.5%

For complex reinforced concrete or unusually complicated structures, add 20% to 50%.

Typical **Mechanical and Electrical Engineering Fees** are based on the size of the subcontract. The fee structure for both are shown below. These fees are included in Architectural Fees.

Type of Construction	Subcontract Size							
	$25,000	$50,000	$100,000	$225,000	$350,000	$500,000	$750,000	$1,000,000
Simple structures	6.4%	5.7%	4.8%	4.5%	4.4%	4.3%	4.2%	4.1%
Intermediate structures	8.0	7.3	6.5	5.6	5.1	5.0	4.9	4.8
Complex structures	10.1	9.0	9.0	8.0	7.5	7.5	7.0	7.0

For renovations, add 15% to 25% to applicable fee.

General Requirements — R0121 Allowances

R012153-10 Repair and Remodeling

Cost figures are based on new construction utilizing the most cost-effective combination of labor, equipment and material with the work scheduled in proper sequence to allow the various trades to accomplish their work in an efficient manner.

The costs for repair and remodeling work must be modified due to the following factors that may be present in any given repair and remodeling project.

1. Equipment usage curtailment due to the physical limitations of the project, with only hand-operated equipment being used.
2. Increased requirement for shoring and bracing to hold up the building while structural changes are being made and to allow for temporary storage of construction materials on above-grade floors.
3. Material handling becomes more costly due to having to move within the confines of an enclosed building. For multi-story construction, low capacity elevators and stairwells may be the only access to the upper floors.
4. Large amount of cutting and patching and attempting to match the existing construction is required. It is often more economical to remove entire walls rather than create many new door and window openings. This sort of trade-off has to be carefully analyzed.
5. Cost of protection of completed work is increased since the usual sequence of construction usually cannot be accomplished.
6. Economies of scale usually associated with new construction may not be present. If small quantities of components must be custom fabricated due to job requirements, unit costs will naturally increase. Also, if only small work areas are available at a given time, job scheduling between trades becomes difficult and subcontractor quotations may reflect the excessive start-up and shut-down phases of the job.
7. Work may have to be done on other than normal shifts and may have to be done around an existing production facility which has to stay in production during the course of the repair and remodeling.
8. Dust and noise protection of adjoining non-construction areas can involve substantial special protection and alter usual construction methods.
9. Job may be delayed due to unexpected conditions discovered during demolition or removal. These delays ultimately increase construction costs.
10. Piping and ductwork runs may not be as simple as for new construction. Wiring may have to be snaked through walls and floors.
11. Matching "existing construction" may be impossible because materials may no longer be manufactured. Substitutions may be expensive.
12. Weather protection of existing structure requires additional temporary structures to protect building at openings.
13. On small projects, because of local conditions, it may be necessary to pay a tradesman for a minimum of four hours for a task that is completed in one hour.

All of the above areas can contribute to increased costs for a repair and remodeling project. Each of the above factors should be considered in the planning, bidding and construction stage in order to minimize the increased costs associated with repair and remodeling jobs.

General Requirements — R0129 Payment Procedures

R012909-80 Sales Tax by State

State sales tax on materials is tabulated below (5 states have no sales tax). Many states allow local jurisdictions, such as a county or city, to levy additional sales tax.

Some projects may be sales tax exempt, particularly those constructed with public funds.

State	Tax (%)	State	Tax (%)	State	Tax (%)	State	Tax (%)
Alabama	4	Illinois	6.25	Montana	0	Rhode Island	7
Alaska	0	Indiana	7	Nebraska	5.5	South Carolina	6
Arizona	5.6	Iowa	6	Nevada	6.85	South Dakota	4
Arkansas	6	Kansas	6.3	New Hampshire	0	Tennessee	7
California	7.25	Kentucky	6	New Jersey	7	Texas	6.25
Colorado	2.9	Louisiana	4	New Mexico	5.13	Utah	5.95
Connecticut	6.35	Maine	5	New York	4	Vermont	6
Delaware	0	Maryland	6	North Carolina	4.75	Virginia	5
District of Columbia	6	Massachusetts	6.25	North Dakota	5	Washington	6.5
Florida	6	Michigan	6	Ohio	5.5	West Virginia	6
Georgia	4	Minnesota	6.88	Oklahoma	4.5	Wisconsin	5
Hawaii	4	Mississippi	7	Oregon	0	Wyoming	4
Idaho	6	Missouri	4.23	Pennsylvania	6	Average	5.06%

Sales Tax by Province (Canada)

GST - a value-added tax, which the government imposes on most goods and services provided in or imported into Canada. PST - a retail sales tax, which three of the provinces impose on the price of most goods and some services. QST - a value-added tax, similar to the federal GST, which Quebec imposes. HST - Five provinces have combined their retail sales tax with the federal GST into one harmonized tax.

Province	PST (%)	QST (%)	GST (%)	HST (%)
Alberta	0	0	5	0
British Columbia	7	0	5	0
Manitoba	7	0	5	0
New Brunswick	0	0	0	13
Newfoundland	0	0	0	13
Northwest Territories	0	0	5	0
Nova Scotia	0	0	0	15
Ontario	0	0	0	13
Prince Edward Island	0	0	5	14
Quebec	0	9.975	5	0
Saskatchewan	5	0	5	0
Yukon	0	0	5	0

General Requirements — R0129 Payment Procedures

R012909-85 Unemployment Taxes and Social Security Taxes

State unemployment tax rates vary not only from state to state, but also with the experience rating of the contractor. The federal unemployment tax rate is 6.0% of the first $7,000 of wages. This is reduced by a credit of up to 5.4% for timely payment to the state. The minimum federal unemployment tax is 0.6% after all credits.

Social security (FICA) for 2014 is estimated at time of publication to be 7.65% of wages up to $115,800.

R012909-90 Overtime

One way to improve the completion date of a project or eliminate negative float from a schedule is to compress activity duration times. This can be achieved by increasing the crew size or working overtime with the proposed crew.

To determine the costs of working overtime to compress activity duration times, consider the following examples. Below is an overtime efficiency and cost chart based on a five, six, or seven day week with an eight through twelve hour day. Payroll percentage increases for time and one half and double time are shown for the various working days.

Days per Week	Hours per Day	1st Week	2nd Week	3rd Week	4th Week	Average 4 Weeks	@ 1-1/2 Times	@ 2 Times
	8	100%	100%	100%	100%	100 %	100 %	100 %
	9	100	100	95	90	96.25	105.6	111.1
5	10	100	95	90	85	91.25	110.0	120.0
	11	95	90	75	65	81.25	113.6	127.3
	12	90	85	70	60	76.25	116.7	133.3
	8	100	100	95	90	96.25	108.3	116.7
	9	100	95	90	85	92.50	113.0	125.9
6	10	95	90	85	80	87.50	116.7	133.3
	11	95	85	70	65	78.75	119.7	139.4
	12	90	80	65	60	73.75	122.2	144.4
	8	100	95	85	75	88.75	114.3	128.6
	9	95	90	80	70	83.75	118.3	136.5
7	10	90	85	75	65	78.75	121.4	142.9
	11	85	80	65	60	72.50	124.0	148.1
	12	85	75	60	55	68.75	126.2	152.4

General Requirements — R0131 Project Management & Coordination

R013113-40 Builder's Risk Insurance

Builder's Risk Insurance is insurance on a building during construction. Premiums are paid by the owner or the contractor. Blasting, collapse and underground insurance would raise total insurance costs above those listed. Floater policy for materials delivered to the job runs $.75 to $1.25 per $100 value. Contractor equipment insurance runs $.50 to $1.50 per $100 value. Insurance for miscellaneous tools to $1,500 value runs from $3.00 to $7.50 per $100 value.

Tabulated below are New England Builder's Risk insurance rates in dollars per $100 value for $1,000 deductible. For $25,000 deductible, rates can be reduced 13% to 34%. On contracts over $1,000,000, rates may be lower than those tabulated. Policies are written annually for the total completed value in place. For "all risk" insurance (excluding flood, earthquake and certain other perils) add $.025 to total rates below.

Coverage	Frame Construction (Class 1) Range	Frame Construction (Class 1) Average	Brick Construction (Class 4) Range	Brick Construction (Class 4) Average	Fire Resistive (Class 6) Range	Fire Resistive (Class 6) Average
Fire Insurance	$.350 to $.850	$.600	$.158 to $.189	$.174	$.052 to $.080	$.070
Extended Coverage	.115 to .200	.158	.080 to .105	.101	.081 to .105	.100
Vandalism	.012 to .016	.014	.008 to .011	.011	.008 to .011	.010
Total Annual Rate	$.477 to $1.066	$.772	$.246 to $.305	$.286	$.141 to $.196	$.180

General Requirements — R0131 Project Management & Coordination

R013113-50 General Contractor's Overhead

There are two distinct types of overhead on a construction project: Project Overhead and Main Office Overhead. Project Overhead includes those costs at a construction site not directly associated with the installation of construction materials. Examples of Project Overhead costs include the following:

1. Superintendent
2. Construction office and storage trailers
3. Temporary sanitary facilities
4. Temporary utilities
5. Security fencing
6. Photographs
7. Clean up
8. Performance and payment bonds

The above Project Overhead items are also referred to as General Requirements and therefore are estimated in Division 1. Division 1 is the first division listed in the CSI MasterFormat but it is usually the last division estimated. The sum of the costs in Divisions 1 through 49 is referred to as the sum of the direct costs.

All construction projects also include indirect costs. The primary components of indirect costs are the contractor's Main Office Overhead and profit. The amount of the Main Office Overhead expense varies depending on the the following:

1. Owner's compensation
2. Project managers and estimator's wages
3. Clerical support wages
4. Office rent and utilities
5. Corporate legal and accounting costs
6. Advertising
7. Automobile expenses
8. Association dues
9. Travel and entertainment expenses

These costs are usually calculated as a percentage of annual sales volume. This percentage can range from 35% for a small contractor doing less than $500,000 to 5% for a large contractor with sales in excess of $100 million.

General Requirements R0131 Project Management & Coordination

R013113-60 Workers' Compensation Insurance Rates by Trade

The table below tabulates the national averages for workers' compensation insurance rates by trade and type of building. The average "Insurance Rate" is multiplied by the "% of Building Cost" for each trade. This produces the "Workers' Compensation" cost by % of total labor cost, to be added for each trade by building type to determine the weighted average workers' compensation rate for the building types analyzed.

Trade	Insurance Rate (% Labor Cost) Range	Average	% of Building Cost Office Bldgs.	Schools & Apts.	Mfg.	Workers' Compensation Office Bldgs.	Schools & Apts.	Mfg.
Excavation, Grading, etc.	3.9 % to 18.7%	9.7%	4.8%	4.9%	4.5%	0.47%	0.48%	0.44%
Piles & Foundations	5.8 to 37	14.7	7.1	5.2	8.7	1.04	0.76	1.28
Concrete	4.6 to 38.6	13.1	5.0	14.8	3.7	0.66	1.94	0.48
Masonry	5.1 to 35	13.5	6.9	7.5	1.9	0.93	1.01	0.26
Structural Steel	5.8 to 108.2	34.6	10.7	3.9	17.6	3.70	1.35	6.09
Miscellaneous & Ornamental Metals	5.3 to 31.2	12.0	2.8	4.0	3.6	0.34	0.48	0.43
Carpentry & Millwork	5.5 to 34.9	15.4	3.7	4.0	0.5	0.57	0.62	0.08
Metal or Composition Siding	5.8 to 68.3	18.2	2.3	0.3	4.3	0.42	0.05	0.78
Roofing	5.8 to 93.3	31.3	2.3	2.6	3.1	0.72	0.81	0.97
Doors & Hardware	4.2 to 34.8	11.1	0.9	1.4	0.4	0.10	0.16	0.04
Sash & Glazing	5.3 to 36.8	13.1	3.5	4.0	1.0	0.46	0.52	0.13
Lath & Plaster	3.7 to 35	11.6	3.3	6.9	0.8	0.38	0.80	0.09
Tile, Marble & Floors	3.3 to 27.1	8.5	2.6	3.0	0.5	0.22	0.26	0.04
Acoustical Ceilings	3.4 to 36.7	8.1	2.4	0.2	0.3	0.19	0.02	0.02
Painting	5 to 33	11.5	1.5	1.6	1.6	0.17	0.18	0.18
Interior Partitions	5.5 to 34.9	15.4	3.9	4.3	4.4	0.60	0.66	0.68
Miscellaneous Items	2.5 to 150	14.4	5.2	3.7	9.7	0.75	0.53	1.40
Elevators	1.6 to 14.4	5.7	2.1	1.1	2.2	0.12	0.06	0.13
Sprinklers	3.2 to 18.3	7.2	0.5	—	2.0	0.04	—	0.14
Plumbing	2.6 to 17.5	7.1	4.9	7.2	5.2	0.35	0.51	0.37
Heat., Vent., Air Conditioning	4 to 18.7	8.8	13.5	11.0	12.9	1.19	0.97	1.14
Electrical	2.8 to 13.2	5.8	10.1	8.4	11.1	0.59	0.49	0.64
Total	1.6 % to 150%	—	100.0%	100.0%	100.0%	14.01%	12.66%	15.81%
				Overall Weighted Average	14.16%			

Workers' Compensation Insurance Rates by States

The table below lists the weighted average workers' compensation base rate for each state with a factor comparing this with the national average of 14%.

State	Weighted Average	Factor	State	Weighted Average	Factor	State	Weighted Average	Factor
Alabama	20.3%	145	Kentucky	14.9%	106	North Dakota	9.6%	69
Alaska	14.3	102	Louisiana	19.3	138	Ohio	8.6	61
Arizona	13.1	94	Maine	12.7	91	Oklahoma	12.9	92
Arkansas	8.6	61	Maryland	14.6	104	Oregon	12.3	88
California	26.8	191	Massachusetts	12.0	86	Pennsylvania	18.1	129
Colorado	7.5	54	Michigan	16.9	121	Rhode Island	12.6	90
Connecticut	24.6	176	Minnesota	23.5	168	South Carolina	19.6	140
Delaware	12.4	89	Mississippi	13.8	99	South Dakota	15.5	111
District of Columbia	11.5	82	Missouri	15.4	110	Tennessee	12.5	89
Florida	11.1	79	Montana	8.9	64	Texas	9.5	68
Georgia	29.2	209	Nebraska	18.3	131	Utah	9.1	65
Hawaii	8.4	60	Nevada	9.0	64	Vermont	13.6	97
Idaho	10.7	76	New Hampshire	20.8	149	Virginia	9.9	71
Illinois	28.1	201	New Jersey	15.0	107	Washington	10.9	78
Indiana	5.7	41	New Mexico	15.0	107	West Virginia	8.3	59
Iowa	15.1	108	New York	17.9	128	Wisconsin	13.6	97
Kansas	8.7	62	North Carolina	17.4	124	Wyoming	6.3	45
			Weighted Average for U.S. is	14.2% of payroll = 100%				

Rates in the following table are the base or manual costs per $100 of payroll for workers' compensation in each state. Rates are usually applied to straight time wages only and not to premium time wages and bonuses.

The weighted average skilled worker rate for 35 trades is 14%. For bidding purposes, apply the full value of workers' compensation directly to total labor costs, or if labor is 38%, materials 42% and overhead and profit 20% of total cost, carry 38/80 × 14% =6.6% of cost (before overhead and profit) into overhead. Rates vary not only from state to state but also with the experience rating of the contractor.

Rates are the most current available at the time of publication.

General Requirements — R0131 Project Management & Coordination

R013113-60 Workers' Compensation Insurance Rates by Trade and State (cont.)

State	Carpentry — 3 stories or less 5651	Carpentry — interior cab. work 5437	Carpentry — general 5403	Concrete Work — NOC 5213	Concrete Work — flat (flr., sdwk.) 5221	Electrical Wiring — inside 5190	Excavation — earth NOC 6217	Excavation — rock 6217	Glaziers 5462	Insulation Work 5479	Lathing 5443	Masonry 5022	Painting & Decorating 5474	Pile Driving 6003	Plastering 5480	Plumbing 5183
AL	34.03	17.01	24.59	13.79	10.37	8.87	16.42	16.42	19.75	19.84	9.24	23.79	15.91	18.57	15.88	8.73
AK	17.14	9.44	11.58	10.41	9.44	5.63	8.23	8.23	32.03	14.36	11.73	10.86	10.79	21.68	11.31	6.20
AZ	21.40	9.27	20.28	10.31	6.93	6.08	7.07	7.07	10.80	14.99	5.47	13.01	12.78	10.80	8.64	6.84
AR	12.80	4.77	9.18	6.63	5.16	4.03	6.82	6.82	8.14	7.67	3.99	8.30	7.08	8.10	13.00	4.55
CA	34.87	34.83	34.86	16.55	16.55	13.24	18.65	18.65	24.16	16.48	17.83	25.13	23.59	23.18	35.04	17.48
CO	8.95	4.93	6.07	6.68	4.75	2.92	5.49	5.49	6.00	7.27	3.95	8.29	6.56	6.51	5.69	4.16
CT	30.95	18.48	31.07	26.49	14.43	7.89	16.02	16.02	20.24	20.60	9.92	31.53	18.60	20.17	18.75	11.28
DE	12.46	12.46	9.93	10.07	8.80	4.53	7.70	7.70	10.90	9.93	6.60	11.53	13.39	15.03	10.90	6.60
DC	8.92	7.25	6.94	12.65	7.38	4.69	8.02	8.02	9.61	6.63	9.83	9.32	5.82	17.36	8.00	7.80
FL	11.32	8.62	11.92	12.51	6.20	6.17	7.30	7.30	10.56	10.45	5.53	10.92	10.23	17.84	9.68	5.63
GA	68.34	20.32	28.40	21.36	15.09	12.05	17.32	17.32	19.39	18.80	11.94	35.11	33.04	26.44	19.65	12.76
HI	8.51	6.92	12.69	7.05	4.02	4.47	5.40	5.40	8.22	5.20	6.26	7.01	6.19	7.32	8.33	5.31
ID	14.63	6.84	12.43	11.48	5.28	4.43	7.25	7.25	8.81	9.06	4.46	10.55	9.38	7.65	8.43	5.10
IL	33.34	21.75	28.57	38.59	16.83	10.03	14.53	14.53	25.55	21.41	14.32	28.94	18.57	24.53	32.84	12.51
IN	8.36	4.21	6.98	4.56	4.15	2.83	3.92	3.92	5.29	6.15	3.85	5.14	4.97	7.23	3.71	2.56
IA	15.46	11.35	19.21	14.85	11.16	5.20	11.01	11.01	15.76	9.89	6.48	14.30	10.13	11.38	21.73	7.80
KS	14.92	7.39	9.74	8.07	5.70	4.46	5.55	5.55	7.54	8.17	4.09	8.38	7.75	7.56	6.20	5.23
KY	18.75	13.00	20.00	8.50	6.50	6.00	14.00	14.00	22.12	11.15	7.60	7.50	9.00	16.03	12.93	5.00
LA	31.05	16.86	25.02	13.35	12.37	7.85	14.11	14.11	19.19	15.06	7.99	17.31	17.64	18.85	14.84	7.10
ME	12.24	8.87	16.18	16.43	8.70	4.74	12.34	12.34	14.82	12.79	5.09	10.88	12.44	9.00	9.99	5.90
MD	11.40	15.89	15.00	19.22	6.00	5.23	8.11	8.11	13.77	12.42	6.15	10.11	6.40	15.19	9.17	6.65
MA	8.68	5.23	9.61	18.85	6.24	2.84	4.35	4.35	9.58	7.78	5.27	10.55	5.09	12.92	4.68	3.50
MI	18.41	10.45	18.41	18.27	12.58	5.82	13.28	13.28	11.67	10.92	10.45	18.82	13.49	37.00	10.88	7.18
MN	20.27	20.82	34.13	11.58	13.15	5.83	11.00	11.00	36.85	18.37	11.55	15.59	17.55	27.58	11.58	8.09
MS	22.76	9.99	13.34	9.60	6.67	6.18	9.49	9.49	14.94	10.15	5.94	13.12	11.01	13.14	10.43	8.17
MO	19.66	11.10	12.81	12.38	10.34	6.15	9.96	9.96	9.47	11.55	7.44	15.13	11.42	13.38	12.15	9.00
MT	9.67	6.59	10.80	6.22	5.67	3.71	7.41	7.41	7.50	8.67	4.15	7.62	8.22	7.95	6.80	4.79
NE	21.80	13.15	21.48	18.23	11.20	8.03	17.38	17.38	12.05	21.53	8.35	18.73	13.00	13.93	13.68	8.73
NV	10.41	6.82	12.39	10.83	5.26	4.33	8.75	8.75	6.55	7.55	3.41	6.63	7.10	8.43	6.21	6.76
NH	24.24	11.06	19.63	28.40	16.47	5.63	12.61	12.61	14.12	12.63	7.30	16.19	20.93	18.53	11.20	9.49
NJ	18.69	13.18	18.69	15.51	11.69	5.13	10.83	10.83	11.21	12.63	15.13	18.71	11.28	15.79	15.13	6.96
NM	25.01	9.56	18.92	11.73	9.43	6.05	7.22	7.22	10.95	10.56	6.11	15.74	12.91	12.66	14.20	6.70
NY	12.54	10.74	19.18	23.71	17.15	7.85	12.16	12.16	18.72	10.88	12.86	21.24	14.99	20.13	10.02	10.83
NC	30.12	9.54	15.76	16.11	9.31	10.56	15.41	15.41	14.28	13.03	8.11	13.05	14.08	16.61	14.12	10.73
ND	9.54	4.31	9.54	5.43	5.43	3.75	4.13	4.13	10.50	10.50	9.93	6.52	6.00	10.59	9.93	6.11
OH	7.73	4.31	5.52	4.67	4.92	3.43	4.79	4.79	8.83	11.38	36.65	6.93	8.15	6.99	7.80	3.52
OK	18.82	10.32	12.49	13.50	7.66	6.82	9.48	9.48	11.79	11.31	5.13	14.63	10.66	11.24	14.01	6.82
OR	22.30	8.52	11.92	10.51	10.42	5.35	9.08	9.08	12.64	10.88	6.09	14.37	11.81	11.44	10.81	6.35
PA	19.22	19.22	15.60	19.54	13.58	7.39	11.11	11.11	14.34	15.60	14.34	16.04	17.13	19.44	14.34	9.19
RI	9.84	11.55	13.75	12.56	9.68	4.58	8.16	8.16	10.61	12.48	5.17	10.07	8.82	19.30	8.69	6.75
SC	29.32	13.80	17.10	12.84	8.21	9.82	12.01	12.01	12.92	14.13	9.57	15.39	16.10	15.67	14.47	9.40
SD	20.86	17.96	20.35	16.10	10.49	5.08	9.75	9.75	10.84	14.71	6.95	14.65	11.22	12.63	10.35	10.02
TN	22.71	10.14	11.54	10.75	7.08	5.49	10.05	10.05	11.07	10.34	6.60	11.69	8.27	12.08	11.04	5.17
TX	10.12	7.33	10.12	8.08	5.99	5.98	7.60	7.60	8.25	9.27	4.15	8.69	7.28	10.26	7.28	5.27
UT	14.72	6.59	7.50	7.34	6.57	3.56	6.79	6.79	7.45	7.15	6.41	8.96	7.73	6.72	5.56	4.17
VT	15.69	10.21	13.77	13.60	11.57	4.37	11.83	11.83	12.40	15.34	5.79	11.92	8.78	10.43	9.35	7.38
VA	12.38	8.14	8.89	10.51	5.76	4.50	7.40	7.40	8.02	7.27	5.45	7.98	8.93	8.31	6.69	5.07
WA	9.18	9.18	9.18	8.55	8.55	3.18	6.68	6.68	13.99	6.79	9.18	13.26	12.26	21.68	10.43	4.33
WV	14.85	5.54	8.34	7.68	4.51	3.53	6.46	6.46	6.74	6.29	4.33	8.39	6.03	8.09	6.54	3.52
WI	12.52	13.14	17.06	11.98	7.59	5.08	7.84	7.84	13.88	11.22	4.00	13.69	12.43	28.24	10.16	5.95
WY	5.82	5.82	5.82	5.82	5.82	5.82	5.82	5.82	5.82	5.82	5.82	5.82	5.82	5.82	5.82	5.82
AVG.	18.19	11.07	15.38	13.15	8.92	5.83	9.69	9.69	13.15	11.67	8.12	13.49	11.54	14.65	11.55	7.08

General Requirements — R0131 Project Management & Coordination

R013113-60 Workers' Compensation Insurance Rates by Trade and State (cont.)

State	Roofing 5551	Sheet Metal Work (HVAC) 5538	Steel Erection — door & sash 5102	Steel Erection — inter., ornam. 5102	Steel Erection — structure 5040	Steel Erection — NOC 5057	Tile Work — (interior ceramic) 5348	Waterproofing 9014	Wrecking 5701	Boiler-making 3726	Mill-wrights 3724	Structural Painters 5037	Elevator Constructor 5160	Truck Drivers 7219	Sprinkler Installer 5188
AL	54.71	10.34	12.52	12.52	40.75	21.56	12.35	9.27	40.75	11.93	12.83	55.64	10.09	23.31	8.40
AK	25.49	8.05	7.82	7.82	29.28	26.40	5.33	5.77	29.28	8.56	8.24	31.37	4.06	11.75	5.13
AZ	29.22	10.31	13.61	13.61	26.91	12.89	4.50	6.21	26.91	9.07	7.41	59.76	5.09	8.01	5.48
AR	16.03	4.74	8.34	8.34	15.40	11.28	6.69	2.58	15.40	4.94	5.76	25.69	2.73	10.96	5.72
CA	71.13	18.78	18.41	18.41	35.28	19.69	14.59	23.59	19.69	8.98	13.58	35.28	5.07	27.13	18.32
CO	15.60	4.16	5.36	5.36	23.64	10.59	4.10	3.23	18.40	4.37	3.87	22.88	2.78	7.22	3.85
CT	58.74	12.93	21.24	21.24	65.13	29.83	15.63	6.43	65.13	19.44	14.12	66.07	11.07	22.67	10.49
DE	26.71	5.55	13.16	13.16	23.27	13.16	8.00	11.53	23.27	6.17	6.19	23.27	6.19	11.13	6.60
DC	17.05	6.45	23.33	23.33	26.02	7.91	6.08	4.38	26.02	15.31	7.62	35.16	9.65	14.34	4.00
FL	18.17	7.69	9.91	9.91	23.04	12.59	5.63	5.26	23.04	9.07	6.05	39.44	3.54	9.24	6.41
GA	93.33	16.61	22.17	22.17	60.89	25.88	13.50	11.64	60.89	19.54	14.79	72.60	7.53	19.39	15.17
HI	17.24	4.93	7.60	7.60	21.77	6.88	6.52	5.40	21.77	4.82	7.71	24.83	2.28	10.54	3.49
ID	24.73	7.46	12.31	12.31	25.42	10.44	6.15	4.84	25.42	5.30	5.61	29.98	3.26	9.46	4.65
IL	47.55	15.21	31.16	31.16	84.68	19.36	27.09	7.88	84.68	23.23	15.10	101.23	14.42	19.73	16.65
IN	11.57	4.32	5.32	5.32	11.59	4.80	3.26	2.77	11.59	3.85	3.70	13.20	1.58	5.75	3.20
IA	28.01	8.53	9.15	9.15	52.79	14.30	9.38	6.15	25.61	14.19	8.59	87.69	6.14	13.60	7.75
KS	16.26	5.64	6.71	6.71	18.97	11.93	5.23	4.63	11.96	6.01	7.60	26.96	3.22	7.88	5.30
KY	34.00	12.00	11.72	11.72	36.00	11.02	16.73	3.50	36.00	12.15	8.35	150.00	9.56	10.00	10.32
LA	43.53	14.84	20.14	20.14	44.26	12.79	9.99	7.91	44.20	15.48	10.94	55.57	6.11	23.25	11.22
ME	22.72	6.77	8.92	8.92	37.18	12.21	5.76	4.64	37.18	8.08	6.58	54.29	4.39	8.15	6.70
MD	29.05	11.90	11.92	11.92	59.02	17.92	6.49	5.84	32.91	14.58	9.95	43.60	4.46	16.02	11.88
MA	30.99	5.72	6.89	6.89	54.08	33.00	5.81	2.48	23.75	15.42	5.98	23.51	4.21	8.28	4.13
MI	41.18	9.83	12.28	12.28	37.00	11.27	11.73	6.46	37.00	9.41	8.67	11.27	5.13	9.09	8.21
MN	67.71	17.30	9.82	9.82	108.16	7.70	14.69	7.95	7.70	10.92	12.24	36.60	6.41	11.39	7.62
MS	31.61	7.46	15.08	15.08	38.45	7.53	8.45	5.85	38.45	7.77	9.97	38.74	4.78	52.03	7.53
MO	36.73	8.64	12.33	12.33	47.55	24.29	10.41	5.92	47.55	14.45	9.66	60.85	6.20	12.44	8.43
MT	29.42	4.80	6.98	6.98	18.72	5.89	5.47	4.64	5.89	5.03	5.91	28.63	3.01	6.06	3.67
NE	35.08	12.88	14.80	14.80	59.03	14.13	8.68	6.73	67.83	16.33	11.35	66.18	7.13	15.98	10.30
NV	16.59	7.34	7.06	7.06	19.32	12.20	4.42	4.23	12.20	4.79	6.15	28.67	4.30	10.30	5.47
NH	36.82	8.93	14.28	14.28	95.03	28.57	12.35	7.17	95.03	8.91	9.04	54.18	6.29	36.82	10.68
NJ	40.10	7.22	12.87	12.87	17.25	13.36	9.55	6.63	16.86	4.18	9.20	23.37	7.00	15.75	5.89
NM	35.07	9.63	11.79	11.79	45.52	16.76	7.73	7.53	45.52	8.00	8.00	42.38	8.41	14.60	7.75
NY	35.96	13.56	19.25	19.25	35.26	19.08	10.35	7.49	18.10	21.41	10.22	42.40	12.20	14.37	6.45
NC	35.39	11.87	14.43	14.43	45.05	22.54	7.62	6.10	45.05	18.41	9.71	78.32	8.90	17.84	10.60
ND	17.63	6.11	10.59	10.59	10.90	10.59	10.50	17.63	8.18	3.33	10.50	10.59	3.33	8.79	6.11
OH	14.89	4.00	6.58	6.58	11.33	5.84	4.85	4.58	5.84	2.76	4.52	24.79	2.13	10.02	3.68
OK	22.44	9.62	10.98	10.98	29.21	15.15	6.69	6.41	32.94	9.25	6.21	52.38	3.88	11.26	6.40
OR	27.22	6.85	7.08	7.08	22.04	13.44	9.10	6.73	22.04	9.10	9.28	11.81	3.70	14.41	6.02
PA	38.35	8.92	19.10	19.10	30.01	19.10	11.69	16.04	30.01	7.95	8.85	30.01	8.85	16.69	9.19
RI	23.59	8.48	7.37	7.37	35.11	24.46	5.33	4.98	22.15	11.33	6.32	39.60	4.85	10.13	5.85
SC	46.45	10.88	15.24	15.24	31.33	24.73	7.98	5.42	31.33	11.24	10.59	66.07	8.64	15.07	10.12
SD	34.39	7.29	14.74	14.74	33.52	23.05	6.19	5.30	33.52	22.67	11.38	66.74	6.69	10.62	8.14
TN	29.78	8.13	13.42	13.42	16.07	18.88	7.46	4.09	16.07	8.10	7.46	40.34	4.92	10.23	6.10
TX	16.22	11.69	7.60	7.60	22.99	9.48	4.68	5.57	7.13	4.30	5.95	7.34	3.34	12.75	5.27
UT	23.93	5.78	5.71	5.71	24.61	11.81	5.75	4.31	20.14	5.91	4.77	21.36	3.71	7.05	4.16
VT	27.01	7.30	12.01	12.01	34.39	18.99	7.01	7.90	34.39	8.67	7.28	48.29	6.07	13.40	6.27
VA	25.97	5.47	7.46	7.46	29.39	10.80	5.76	3.30	29.39	4.92	7.38	27.90	5.39	9.03	5.30
WA	21.14	4.47	8.09	8.09	8.09	8.09	7.86	21.14	8.09	2.77	6.15	12.26	2.46	8.52	4.70
WV	19.62	5.62	7.13	7.13	19.39	9.55	4.86	3.11	20.70	5.62	6.64	35.86	4.23	8.93	3.76
WI	28.04	9.43	12.08	12.08	20.72	16.12	13.78	5.07	20.72	9.68	8.34	24.08	4.11	10.15	4.64
WY	5.82	5.82	5.82	5.82	5.82	5.82	5.82	5.82	5.82	5.82	5.82	5.82	5.82	5.82	5.82
AVG.	31.29	8.79	11.95	11.95	34.64	15.21	8.54	6.86	29.21	9.87	8.39	41.47	5.67	13.67	7.24

General Requirements — R0131 Project Management & Coordination

R013113-60 Workers' Compensation (cont.) (Canada in Canadian dollars)

Province		Alberta	British Columbia	Manitoba	Ontario	New Brunswick	Newfndld. & Labrador	Northwest Territories	Nova Scotia	Prince Edward Island	Quebec	Saskatchewan	Yukon
Carpentry—3 stories or less	Rate	5.71	3.27	3.98	4.55	3.17	7.81	4.99	6.85	6.42	12.03	3.43	6.63
	Code	42143	721028	40102	723	238130	4226	4-41	4226	403	80110	B1226	202
Carpentry—interior cab. work	Rate	1.89	2.80	3.98	4.55	3.17	4.07	4.99	5.87	3.94	12.03	3.43	6.63
	Code	42133	721021	40102	723	238350	4279	4-41	4274	402	80110	B1127	202
CARPENTRY—general	Rate	5.71	3.27	3.98	4.55	3.17	4.07	4.99	6.85	6.42	12.03	3.43	6.63
	Code	42143	721028	40102	723	238130	4299	4-41	4226	403	80110	B1202	202
CONCRETE WORK—NOC	Rate	3.88	3.57	6.50	18.31	3.17	7.81	4.99	5.75	3.94	1.92	2.99	6.63
	Code	42104	721010	40110	748	238110	4224	4-41	4224	402	80100	B1314	203
CONCRETE WORK—flat (flr. sidewalk)	Rate	3.88	3.57	6.50	18.31	3.17	7.81	4.99	5.75	3.94	11.92	2.99	6.63
	Code	42104	721010	40110	748	238110	4224	4-41	4224	402	80100	B1314	203
ELECTRICAL Wiring—inside	Rate	1.56	1.18	1.65	3.69	1.28	2.29	2.44	1.97	3.94	5.04	1.83	3.89
	Code	42124	721019	40203	704	238210	4261	4-46	4261	402	80170	B1105	206
EXCAVATION—earth NOC	Rate	1.86	3.79	4.00	5.29	3.17	2.93	3.88	3.89	3.94	6.40	2.19	3.89
	Code	40604	721031	40706	711	238190	4214	4-43	4214	402	80030	R1106	207
EXCAVATION—rock	Rate	1.86	3.79	4.00	5.29	3.17	2.93	3.88	3.89	3.94	6.40	2.19	3.89
	Code	40604	721031	40706	711	238190	4214	4-43	4214	402	80030	R1106	207
GLAZIERS	Rate	2.23	3.79	3.98	10.25	3.17	5.10	4.99	5.22	3.94	15.02	2.99	3.89
	Code	42121	715020	40109	751	238150	4233	4-41	4233	402	80150	B1304	212
INSULATION WORK	Rate	1.62	9.87	3.98	10.25	3.17	5.10	4.99	5.22	3.94	12.03	3.43	6.63
	Code	42184	721029	40102	751	238310	4234	4-41	4234	402	80110	B1207	202
LATHING	Rate	4.20	7.84	3.98	4.55	3.17	4.07	4.99	5.87	3.94	12.03	2.99	6.63
	Code	42135	721042	40102	723	238390	4279	4-41	4271	402	80110	B1316	202
MASONRY	Rate	3.88	3.30	3.98	12.70	3.17	5.10	4.99	5.22	3.94	11.92	2.99	6.63
	Code	42102	721037	40102	741	238140	4231	4-41	4231	402	80100	B1318	202
PAINTING & DECORATING	Rate	3.16	3.62	3.11	7.51	3.17	4.07	4.99	5.87	3.94	12.03	3.43	6.63
	Code	42111	721041	40105	719	238320	4275	4-41	4275	402	80110	B1201	202
PILE DRIVING	Rate	3.88	3.42	4.00	7.03	3.17	2.93	3.88	5.75	6.42	6.40	2.99	6.63
	Code	42159	722004	40706	732	238190	4129	4-43	4221	403	80030	B1310	202
PLASTERING	Rate	4.20	7.84	4.45	7.51	3.17	4.07	4.99	5.87	3.94	12.03	3.43	6.63
	Code	42135	721042	40108	719	238390	4271	4-41	4271	402	80110	B1221	202
PLUMBING	Rate	1.56	2.55	2.78	4.16	1.28	2.47	2.44	2.81	3.94	5.89	1.83	3.89
	Code	42122	721043	40204	707	238220	4241	4-46	4241	401	80160	B1101	214
ROOFING	Rate	4.71	5.42	7.22	14.80	3.17	7.81	4.99	9.43	3.94	18.52	2.99	6.63
	Code	42118	721036	40403	728	238160	4236	4-41	4236	402	80130	B1320	202
SHEET METAL WORK (HVAC)	Rate	1.56	2.55	7.22	4.16	1.28	2.47	2.44	2.81	3.94	5.89	1.83	3.89
	Code	42117	721043	40402	707	238220	4244	4-46	4244	401	80160	B1107	208
STEEL ERECTION—door & sash	Rate	1.62	10.62	6.82	18.31	3.17	7.81	4.99	5.22	3.94	16.70	2.99	6.63
	Code	42106	722005	40502	748	238120	4227	-37	4227	402	80080	B1322	202
STEEL ERECTION—inter., ornam.	Rate	1.62	10.62	6.82	18.31	3.17	7.81	4.99	5.22	3.94	16.70	2.99	6.63
	Code	42106	722005	40502	748	238120	4227	4-41	4227	402	80080	B1322	202
STEEL ERECTION—structure	Rate	1.62	10.62	6.82	18.31	3.17	7.81	4.99	5.22	3.94	16.70	2.99	6.63
	Code	42106	722005	40502	748	238120	4227	4-41	4227	402	80080	B1322	202
STEEL ERECTION—NOC	Rate	1.62	10.62	6.82	18.31	3.17	7.81	4.99	5.22	3.94	16.70	2.99	6.63
	Code	42106	722005	40502	748	238120	4227	4-41	4227	402	80080	B1322	202
TILE WORK—inter. (ceramic)	Rate	3.02	3.65	1.86	7.51	3.17	4.07	4.99	5.87	3.94	12.03	2.99	6.63
	Code	42113	721054	40103	719	238340	4276	4-41	4276	402	80110	B1301	202
WATERPROOFING	Rate	3.16	3.62	3.98	4.55	3.17	4.07	4.99	5.22	3.94	18.52	3.13	6.63
	Code	42139	721016	40102	723	238190	4299	4-41	4239	402	80130	B1217	202
WRECKING	Rate	1.86	5.64	6.90	18.31	3.17	2.93	3.88	3.89	3.94	12.03	2.19	6.63
	Code	40604	721005	40106	748	238190	4211	4-43	4211	402	80110	R1125	202

General Requirements — R0131 Project Management & Coordination

R013113-80 Performance Bond

This table shows the cost of a Performance Bond for a construction job scheduled to be completed in 12 months. Add 1% of the premium cost per month for jobs requiring more than 12 months to complete. The rates are "standard" rates offered to contractors that the bonding company considers financially sound and capable of doing the work. Preferred rates are offered by some bonding companies based upon financial strength of the contractor. Actual rates vary from contractor to contractor and from bonding company to bonding company. Contractors should prequalify through a bonding agency before submitting a bid on a contract that requires a bond.

Contract Amount	Building Construction Class B Projects	Highways & Bridges Class A New Construction	Class A-1 Highway Resurfacing
First $100,000 bid	$25.00 per M	$15.00 per M	$9.40 per M
Next 400,000 bid	$2,500 plus $15.00 per M	$1,500 plus $10.00 per M	$940 plus $7.20 per M
Next 2,000,000 bid	8,500 plus 10.00 per M	5,500 plus 7.00 per M	3,820 plus 5.00 per M
Next 2,500,000 bid	28,500 plus 7.50 per M	19,500 plus 5.50 per M	15,820 plus 4.50 per M
Next 2,500,000 bid	47,250 plus 7.00 per M	33,250 plus 5.00 per M	28,320 plus 4.50 per M
Over 7,500,000 bid	64,750 plus 6.00 per M	45,750 plus 4.50 per M	39,570 plus 4.00 per M

General Requirements — R0154 Construction Aids

R015423-10 Steel Tubular Scaffolding

On new construction, tubular scaffolding is efficient up to 60' high or five stories. Above this it is usually better to use a hung scaffolding if construction permits. Swing scaffolding operations may interfere with tenants. In this case, the tubular is more practical at all heights.

In repairing or cleaning the front of an existing building the cost of tubular scaffolding per S.F. of building front increases as the height increases above the first tier. The first tier cost is relatively high due to leveling and alignment.

The minimum efficient crew for erecting and dismantling is three workers. They can set up and remove 18 frame sections per day up to 5 stories high. For 6 to 12 stories high, a crew of four is most efficient. Use two or more on top and two on the bottom for handing up or hoisting. They can also set up and remove 18 frame sections per day. At 7' horizontal spacing, this will run about 800 S.F. per day of erecting and dismantling. Time for placing and removing planks must be added to the above. A crew of three can place and remove 72 planks per day up to 5 stories. For over 5 stories, a crew of four can place and remove 80 planks per day.

The table below shows the number of pieces required to erect tubular steel scaffolding for 1000 S.F. of building frontage. This area is made up of a scaffolding system that is 12 frames (11 bays) long by 2 frames high.

For jobs under twenty-five frames, add 50% to rental cost. Rental rates will be lower for jobs over three months duration. Large quantities for long periods can reduce rental rates by 20%.

Description of Component	Number of Pieces for 1000 S.F. of Building Front	Unit
5' Wide Standard Frame, 6'-4" High	24	Ea.
Leveling Jack & Plate	24	
Cross Brace	44	
Side Arm Bracket, 21"	12	
Guardrail Post	12	
Guardrail, 7' section	22	
Stairway Section	2	
Stairway Starter Bar	1	
Stairway Inside Handrail	2	
Stairway Outside Handrail	2	
Walk-Thru Frame Guardrail	2	

Scaffolding is often used as falsework over 15' high during construction of cast-in-place concrete beams and slabs. Two foot wide scaffolding is generally used for heavy beam construction. The span between frames depends upon the load to be carried with a maximum span of 5'.

Heavy duty shoring frames with a capacity of 10,000#/leg can be spaced up to 10' O.C. depending upon form support design and loading.

Scaffolding used as horizontal shoring requires less than half the material required with conventional shoring.

On new construction, erection is done by carpenters.

Rolling towers supporting horizontal shores can reduce labor and speed the job. For maintenance work, catwalks with spans up to 70' can be supported by the rolling towers.

General Requirements — R0154 Construction Aids

R015433-10 Contractor Equipment

Rental Rates shown elsewhere in the book pertain to late model high quality machines in excellent working condition, rented from equipment dealers. Rental rates from contractors may be substantially lower than the rental rates from equipment dealers depending upon economic conditions; for older, less productive machines, reduce rates by a maximum of 15%. Any overtime must be added to the base rates. For shift work, rates are lower. Usual rule of thumb is 150% of one shift rate for two shifts; 200% for three shifts.

For periods of less than one week, operated equipment is usually more economical to rent than renting bare equipment and hiring an operator.

Costs to move equipment to a job site (mobilization) or from a job site (demobilization) are not included in rental rates, nor in any Equipment costs on any Unit Price line items or crew listings. These costs can be found elsewhere. If a piece of equipment is already at a job site, it is not appropriate to utilize mob/demob costs in an estimate again.

Rental rates vary throughout the country with larger cities generally having lower rates. Lease plans for new equipment are available for periods in excess of six months with a percentage of payments applying toward purchase.

Rental rates can also be treated as reimbursement costs for contractor-owned equipment. Owned equipment costs include depreciation, loan payments, interest, taxes, insurance, storage, and major repairs.

Monthly rental rates vary from 2% to 5% of the cost of the equipment depending on the anticipated life of the equipment and its wearing parts. Weekly rates are about 1/3 the monthly rates and daily rental rates about 1/3 the weekly rate.

The hourly operating costs for each piece of equipment include costs to the user such as fuel, oil, lubrication, normal expendables for the equipment, and a percentage of mechanic's wages chargeable to maintenance. The hourly operating costs listed do not include the operator's wages.

The daily cost for equipment used in the standard crews is figured by dividing the weekly rate by five, then adding eight times the hourly operating cost to give the total daily equipment cost, not including the operator. This figure is in the right hand column of the Equipment listings under Equipment Cost/Day.

Pile Driving rates shown for pile hammer and extractor do not include leads, crane, boiler or compressor. Vibratory pile driving requires an added field specialist during set-up and pile driving operation for the electric model. The hydraulic model requires a field specialist for set-up only. Up to 125 reuses of sheet piling are possible using vibratory drivers. For normal conditions, crane capacity for hammer type and size are as follows.

Crane Capacity	Hammer Type and Size		
	Air or Steam	Diesel	Vibratory
25 ton	to 8,750 ft.-lb.		70 H.P.
40 ton	15,000 ft.-lb.	to 32,000 ft.-lb.	170 H.P.
60 ton	25,000 ft.-lb.		300 H.P.
100 ton		112,000 ft.-lb.	

Cranes should be specified for the job by size, building and site characteristics, availability, performance characteristics, and duration of time required.

Backhoes & Shovels rent for about the same as equivalent size cranes but maintenance and operating expense is higher. Crane operators rate must be adjusted for high boom heights. Average adjustments: for 150' boom add 2% per hour; over 185', add 4% per hour; over 210', add 6% per hour; over 250', add 8% per hour and over 295', add 12% per hour.

Tower Cranes of the climbing or static type have jibs from 50' to 200' and capacities at maximum reach range from 4,000 to 14,000 pounds. Lifting capacities increase up to maximum load as the hook radius decreases.

Typical rental rates, based on purchase price are about 2% to 3% per month.

Erection and dismantling runs between 500 and 2000 labor hours. Climbing operation takes 10 labor hours per 20' climb. Crane dead time is about 5 hours per 40' climb. If crane is bolted to side of the building add cost of ties and extra mast sections. Climbing cranes have from 80' to 180' of mast while static cranes have 80' to 800' of mast.

Truck Cranes can be converted to tower cranes by using tower attachments. Mast heights over 400' have been used.

A single 100' high material **Hoist and Tower** can be erected and dismantled in about 400 labor hours; a double 100' high hoist and tower in about 600 labor hours. Erection times for additional heights are 3 and 4 labor hours per vertical foot respectively up to 150', and 4 to 5 labor hours per vertical foot over 150' high. A 40' high portable Buck hoist takes about 160 labor hours to erect and dismantle. Additional heights take 2 labor hours per vertical foot to 80' and 3 labor hours per vertical foot for the next 100'. Most material hoists do not meet local code requirements for carrying personnel.

A 150' high **Personnel Hoist** requires about 500 to 800 labor hours to erect and dismantle. Budget erection time at 5 labor hours per vertical foot for all trades. Local code requirements or labor scarcity requiring overtime can add up to 50% to any of the above erection costs.

Earthmoving Equipment: The selection of earthmoving equipment depends upon the type and quantity of material, moisture content, haul distance, haul road, time available, and equipment available. Short haul cut and fill operations may require dozers only, while another operation may require excavators, a fleet of trucks, and spreading and compaction equipment. Stockpiled material and granular material are easily excavated with front end loaders. Scrapers are most economically used with hauls between 300' and 1-1/2 miles if adequate haul roads can be maintained. Shovels are often used for blasted rock and any material where a vertical face of 8' or more can be excavated. Special conditions may dictate the use of draglines, clamshells, or backhoes. Spreading and compaction equipment must be matched to the soil characteristics, the compaction required and the rate the fill is being supplied.

Existing Conditions — R0241 Demolition

R024119-10 Demolition Defined

Whole Building Demolition - Demolition of the whole building with no concern for any particular building element, component, or material type being demolished. This type of demolition is accomplished with large pieces of construction equipment that break up the structure, load it into trucks and haul it to a disposal site, but disposal or dump fees are not included. Demolition of below-grade foundation elements, such as footings, foundation walls, grade beams, slabs on grade, etc., is not included. Certain mechanical equipment containing flammable liquids or ozone-depleting refrigerants, electric lighting elements, communication equipment components, and other building elements may contain hazardous waste, and must be removed, either selectively or carefully, as hazardous waste before the building can be demolished.

Foundation Demolition - Demolition of below-grade foundation footings, foundation walls, grade beams, and slabs on grade. This type of demolition is accomplished by hand or pneumatic hand tools, and does not include saw cutting, or handling, loading, hauling, or disposal of the debris.

Gutting - Removal of building interior finishes and electrical/mechanical systems down to the load-bearing and sub-floor elements of the rough building frame, with no concern for any particular building element, component, or material type being demolished. This type of demolition is accomplished by hand or pneumatic hand tools, and includes loading into trucks, but not hauling, disposal or dump fees, scaffolding, or shoring. Certain mechanical equipment containing flammable liquids or ozone-depleting refrigerants, electric lighting elements, communication equipment components, and other building elements may contain hazardous waste, and must be removed, either selectively or carefully, as hazardous waste, before the building is gutted.

Selective Demolition - Demolition of a selected building element, component, or finish, with some concern for surrounding or adjacent elements, components, or finishes (see the first Subdivision (s) at the beginning of appropriate Divisions). This type of demolition is accomplished by hand or pneumatic hand tools, and does not include handling, loading, storing, hauling, or disposal of the debris, scaffolding, or shoring. "Gutting" methods may be used in order to save time, but damage that is caused to surrounding or adjacent elements, components, or finishes may have to be repaired at a later time.

Careful Removal - Removal of a piece of service equipment, building element or component, or material type, with great concern for both the removed item and surrounding or adjacent elements, components or finishes. The purpose of careful removal may be to protect the removed item for later re-use, preserve a higher salvage value of the removed item, or replace an item while taking care to protect surrounding or adjacent elements, components, connections, or finishes from cosmetic and/or structural damage. An approximation of the time required to perform this type of removal is 1/3 to 1/2 the time it would take to install a new item of like kind (see Reference Number R220105-10). This type of removal is accomplished by hand or pneumatic hand tools, and does not include loading, hauling, or storing the removed item, scaffolding, shoring, or lifting equipment.

Cutout Demolition - Demolition of a small quantity of floor, wall, roof, or other assembly, with concern for the appearance and structural integrity of the surrounding materials. This type of demolition is accomplished by hand or pneumatic hand tools, and does not include saw cutting, handling, loading, hauling, or disposal of debris, scaffolding, or shoring.

Rubbish Handling - Work activities that involve handling, loading or hauling of debris. Generally, the cost of rubbish handling must be added to the cost of all types of demolition, with the exception of whole building demolition.

Minor Site Demolition - Demolition of site elements outside the footprint of a building. This type of demolition is accomplished by hand or pneumatic hand tools, or with larger pieces of construction equipment, and may include loading a removed item onto a truck (check the Crew for equipment used). It does not include saw cutting, hauling or disposal of debris, and, sometimes, handling or loading.

R024119-20 Dumpsters

Dumpster rental costs on construction sites are presented in two ways.

The cost per week rental includes the delivery of the dumpster; its pulling or emptying once per week, and its final removal. The assumption is made that the dumpster contractor could choose to empty a dumpster by simply bringing in an empty unit and removing the full one. These costs also include the disposal of the materials in the dumpster.

The Alternate Pricing can be used when actual planned conditions are not approximated by the weekly numbers. For example, these lines can be used when a dumpster is needed for 4 weeks and will need to be emptied 2 or 3 times per week. Conversely the Alternate Pricing lines can be used when a dumpster will be rented for several weeks or months but needs to be emptied only a few times over this period.

Existing Conditions — R0265 Underground Storage Tank Removal

R026510-20 Underground Storage Tank Removal

Underground Storage Tank Removal can be divided into two categories: Non-Leaking and Leaking. Prior to removing an underground storage tank, tests should be made, with the proper authorities present, to determine whether a tank has been leaking or the surrounding soil has been contaminated.

To safely remove Liquid Underground Storage Tanks:
1. Excavate to the top of the tank.
2. Disconnect all piping.
3. Open all tank vents and access ports.
4. Remove all liquids and/or sludge.
5. Purge the tank with an inert gas.
6. Provide access to the inside of the tank and clean out the interior using proper personal protective equipment (PPE).
7. Excavate soil surrounding the tank using proper PPE for on-site personnel.
8. Pull and properly dispose of the tank.
9. Clean up the site of all contaminated material.
10. Install new tanks or close the excavation.

Existing Conditions   R0282 Asbestos Remediation

R028213-20   Asbestos Removal Process

Asbestos removal is accomplished by a specialty contractor who understands the federal and state regulations regarding the handling and disposal of the material. The process of asbestos removal is divided into many individual steps. An accurate estimate can be calculated only after all the steps have been priced.

The steps are generally as follows:

1. Obtain an asbestos abatement plan from an industrial hygienist.
2. Monitor the air quality in and around the removal area and along the path of travel between the removal area and transport area. This establishes the background contamination.
3. Construct a two part decontamination chamber at entrance to removal area.
4. Install a HEPA filter to create a negative pressure in the removal area.
5. Install wall, floor and ceiling protection as required by the plan, usually 2 layers of fireproof 6 mil polyethylene.
6. Industrial hygienist visually inspects work area to verify compliance with plan.
7. Provide temporary supports for conduit and piping affected by the removal process.
8. Proceed with asbestos removal and bagging process. Monitor air quality as described in Step #2. Discontinue operations when contaminate levels exceed applicable standards.
9. Document the legal disposal of materials in accordance with EPA standards.
10. Thoroughly clean removal area including all ledges, crevices and surfaces.
11. Post abatement inspection by industrial hygienist to verify plan compliance.
12. Provide a certificate from a licensed industrial hygienist attesting that contaminate levels are within acceptable standards before returning area to regular use.

Concrete   R0330 Cast-In-Place Concrete

R033053-60   Maximum Depth of Frost Penetration in Inches

Metals — R0505 Common Work Results for Metals

R050521-20 Welded Structural Steel

Usual weight reductions with welded design run 10% to 20% compared with bolted or riveted connections. This amounts to about the same total cost compared with bolted structures since field welding is more expensive than bolts. For normal spans of 18' to 24' figure 6 to 7 connections per ton.

Trusses — For welded trusses add 4% to weight of main members for connections. Up to 15% less steel can be expected in a welded truss compared to one that is shop bolted. Cost of erection is the same whether shop bolted or welded.

General — Typical electrodes for structural steel welding are E6010, E6011, E60T and E70T. Typical buildings vary between 2# to 8# of weld rod per ton of steel. Buildings utilizing continuous design require about three times as much welding as conventional welded structures. In estimating field erection by welding, it is best to use the average linear feet of weld per ton to arrive at the welding cost per ton. The type, size and position of the weld will have a direct bearing on the cost per linear foot. A typical field welder will deposit 1.8# to 2# of weld rod per hour manually. Using semiautomatic methods can increase production by as much as 50% to 75%.

Special Construction — R1311 Swimming Pools

R131113-20 Swimming Pools

Pool prices given per square foot of surface area include pool structure, filter and chlorination equipment, pumps, related piping, ladders/steps, maintenance kit, skimmer and vacuum system. Decks and electrical service to equipment are not included.

Residential in-ground pool construction can be divided into two categories: vinyl lined and gunite. Vinyl lined pool walls are constructed of different materials including wood, concrete, plastic or metal. The bottom is often graded with sand over which the vinyl liner is installed. Vermiculite or soil cement bottoms may be substituted for an added cost.

Gunite pool construction is used both in residential and municipal installations. These structures are steel reinforced for strength and finished with a white cement limestone plaster.

Municipal pools will have a higher cost because plumbing codes require more expensive materials, chlorination equipment and higher filtration rates.

Municipal pools greater than 1,800 S.F. require gutter systems to control waves. This gutter may be formed into the concrete wall. Often a vinyl/stainless steel gutter or gutter/wall system is specified, which will raise the pool cost.

Competition pools usually require tile bottoms and sides with contrasting lane striping, which will also raise the pool cost.

Fire Suppression — R2112 Fire-Suppression Standpipes

R211226-10 Standpipe Systems

The basis for standpipe system design is National Fire Protection Association NFPA 14, however, the authority having jurisdiction should be consulted for special conditions, local requirements and approval.

Standpipe systems, properly designed and maintained, are an effective and valuable time saving aid for extinquishing fires, especially in the upper stories of tall buildings, the interior of large commercial or industrial malls, or other areas where construction features or access make the laying of temporary hose lines time consuming and/or hazardous. Standpipes are frequently installed with automatic sprinkler systems for maximum protection.

There are three general classes of service for standpipe systems:
Class I for use by fire departments and personnel with special training for heavy streams (2-1/2" hose connections).
Class II for use by building occupants until the arrival of the fire department (1-1/2" hose connector with hose).
Class III for use by either fire departments and trained personnel or by the building occupants (both 2-1/2" and 1-1/2" hose connections or one 2-1/2" hose valve with an easily removable 2-1/2" by 1-1/2" adapter).

Standpipe systems are also classified by the way water is supplied to the system. The four basic types are:
Type 1: Wet standpipe system having supply valve open and water pressure maintained at all times.
Type 2: Standpipe system so arranged through the use of approved devices as to admit water to the system automatically by opening a hose valve.
Type 3: Standpipe system arranged to admit water to the system through manual operation of approved remote control devices located at each hose station.
Type 4: Dry standpipe having no permanent water supply.

Fire Suppression — R2112 Fire-Suppression Standpipes

R211226-20 NFPA 14 Basic Standpipe Design

Class	Design-Use	Pipe Size Minimums	Water Supply Minimums
Class I	2 1/2" hose connection on each floor. All areas within 150' of an exit in every exit stairway. Fire Department Trained Personnel	Height to 100', 4" dia. Heights above 100', 6" dia. (275' max. except with pressure regulators 400' max.)	For each standpipe riser 500 GPM flow. For common supply pipe allow 500 GPM for first standpipe plus 250 GPM for each additional standpipe (2500 GPM max. total). 30 min. duration 65 PSI at 500 GPM
Class II	1 1/2" hose connection with hose on each floor. All areas within 130' of hose connection measured along path of hose travel. Occupant personnel	Height to 50', 2" dia. Height above 50', 2 1/2" dia.	For each standpipe riser 100 GPM flow. For multiple riser common supply pipe 100 GPM. 300 min. duration, 65 PSI at 100 GPM
Class III	Both of above. Class I valved connections will meet Class III with additional 2 1/2" by 1 1/2" adapter and 1 1/2" hose.	Same as Class I	Same as Class I

*Note: Where 2 or more standpipes are installed in the same building or section of building they shall be interconnected at the bottom.

Combined Systems

Combined systems are systems where the risers supply both automatic sprinklers and 2-1/2" hose connection outlets for fire department use. In such a system the sprinkler spacing pattern shall be in accordance with NFPA 13 while the risers and supply piping will be sized in accordance with NFPA 14. When the building is completely sprinklered the risers may be sized by hydraulic calculation. The minimum size riser for buildings not completely sprinklered is 6".

The minimum water supply of a completely sprinklered, light hazard, high-rise occupancy building will be 500 GPM while the supply required for other types of completely sprinklered high-rise buildings is 1000 GPM.

General System Requirements

1. Approved valves will be provided at the riser for controlling branch lines to hose outlets.
2. A hose valve will be provided at each outlet for attachment of hose.
3. Where pressure at any standpipe outlet exceeds 100 PSI a pressure reducer must be installed to limit the pressure to 100 PSI. Note that the pressure head due to gravity in 100' of riser is 43.4 PSI. This must be overcome by city pressure, fire pumps, or gravity tanks to provide adequate pressure at the top of the riser.
4. Each hose valve on a wet system having linen hose shall have an automatic drip connection to prevent valve leakage from entering the hose.
5. Each riser will have a valve to isolate it from the rest of the system.
6. One or more fire department connections as an auxiliary supply shall be provided for each Class I or Class III standpipe system. In buildings having two or more zones, a connection will be provided for each zone.
7. There will be no shutoff valve in the fire department connection, but a check valve will be located in the line before it joins the system.
8. All hose connections street side will be identified on a cast plate or fitting as to purpose.

Fire Suppression

R2113 Fire-Suppression Sprinkler Systems

R211313-10 Sprinkler Systems (Automatic)

Sprinkler systems may be classified by type as follows:

1. **Wet Pipe System.** A system employing automatic sprinklers attached to a piping system containing water and connected to a water supply so that water discharges immediately from sprinklers opened by a fire.
2. **Dry Pipe System.** A system employing automatic sprinklers attached to a piping system containing air under pressure, the release of which as from the opening of sprinklers permits the water pressure to open a valve known as a "dry pipe valve". The water then flows into the piping system and out the opened sprinklers.
3. **Pre-Action System.** A system employing automatic sprinklers attached to a piping system containing air that may or may not be under pressure, with a supplemental heat responsive system of generally more sensitive characteristics than the automatic sprinklers themselves, installed in the same areas as the sprinklers; actuation of the heat responsive system, as from a fire, opens a valve which permits water to flow into the sprinkler piping system and to be discharged from any sprinklers which may be open.
4. **Deluge System.** A system employing open sprinklers attached to a piping system connected to a water supply through a valve which is opened by the operation of a heat responsive system installed in the same areas as the sprinklers. When this valve opens, water flows into the piping system and discharges from all sprinklers attached thereto.
5. **Combined Dry Pipe and Pre-Action Sprinkler System.** A system employing automatic sprinklers attached to a piping system containing air under pressure with a supplemental heat responsive system of generally more sensitive characteristics than the automatic sprinklers themselves, installed in the same areas as the sprinklers; operation of the heat responsive system, as from a fire, actuates tripping devices which open dry pipe valves simultaneously and without loss of air pressure in the system. Operation of the heat responsive system also opens approved air exhaust valves at the end of the feed main which facilitates the filling of the system with water which usually precedes the opening of sprinklers. The heat responsive system also serves as an automatic fire alarm system.
6. **Limited Water Supply System.** A system employing automatic sprinklers and conforming to these standards but supplied by a pressure tank of limited capacity.
7. **Chemical Systems.** Systems using halon, carbon dioxide, dry chemical or high expansion foam as selected for special requirements. Agent may extinguish flames by chemically inhibiting flame propagation, suffocate flames by excluding oxygen, interrupting chemical action of oxygen uniting with fuel or sealing and cooling the combustion center.
8. **Firecycle System.** Firecycle is a fixed fire protection sprinkler system utilizing water as its extinguishing agent. It is a time delayed, recycling, preaction type which automatically shuts the water off when heat is reduced below the detector operating temperature and turns the water back on when that temperature is exceeded. The system senses a fire condition through a closed circuit electrical detector system which controls water flow to the fire automatically. Batteries supply up to 90 hour emergency power supply for system operation. The piping system is dry (until water is required) and is monitored with pressurized air. Should any leak in the system piping occur, an alarm will sound, but water will not enter the system until heat is sensed by a firecycle detector.

Area coverage sprinkler systems may be laid out and fed from the supply in any one of several patterns as shown below. It is desirable, if possible, to utilize a central feed and achieve a shorter flow path from the riser to the furthest sprinkler. This permits use of the smallest sizes of pipe possible with resulting savings.

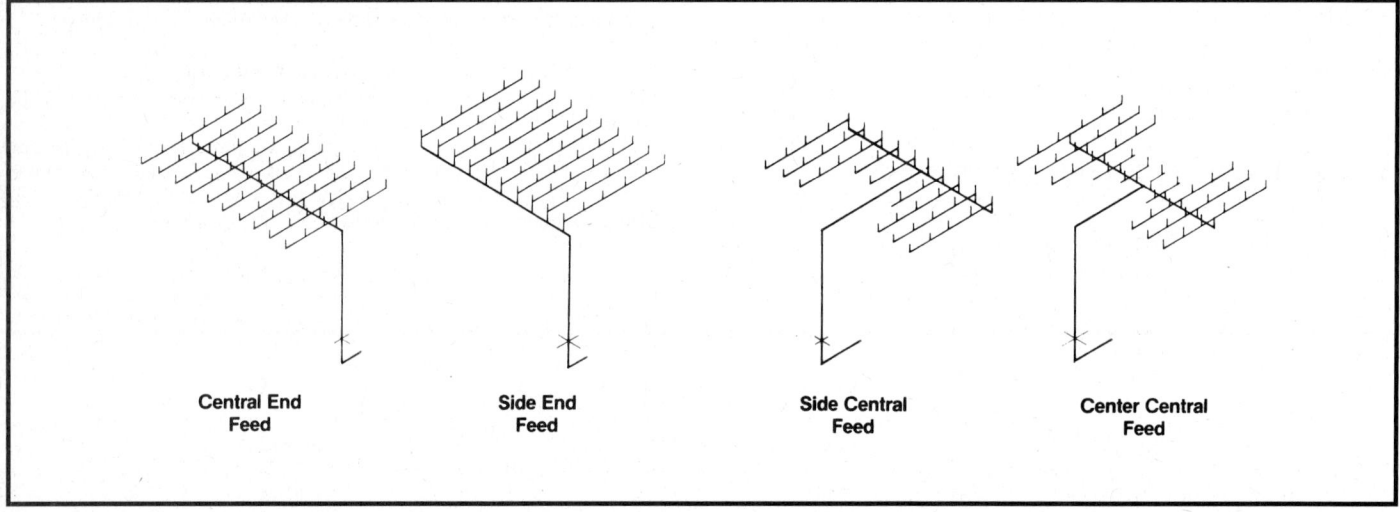

Central End Feed Side End Feed Side Central Feed Center Central Feed

Fire Suppression — R2113 Fire-Suppression Sprinkler Systems

R211313-20 System Classification

System Classification
Rules for installation of sprinkler systems vary depending on the classification of occupancy falling into one of three categories as follows:

Light Hazard Occupancy
The protection area allotted per sprinkler should not exceed 225 S.F., with the maximum distance between lines and sprinklers on lines being 15'. The sprinklers do not need to be staggered. Branch lines should not exceed eight sprinklers on either side of a cross main. Each large area requiring more than 100 sprinklers and without a sub-dividing partition should be supplied by feed mains or risers sized for ordinary hazard occupancy.
Maximum system area = 52,000 S.F.

Included in this group are:
- Churches
- Clubs
- Educational
- Hospitals
- Institutional
- Libraries (except large stack rooms)
- Museums
- Nursing Homes
- Offices
- Residential
- Restaurants
- Theaters and Auditoriums (except stages and prosceniums)
- Unused Attics

Ordinary Hazard Occupancy
The protection area allotted per sprinkler shall not exceed 130 S.F. of noncombustible ceiling and 130 S.F. of combustible ceiling. The maximum allowable distance between sprinkler lines and sprinklers on line is 15'. Sprinklers shall be staggered if the distance between heads exceeds 12'. Branch lines should not exceed eight sprinklers on either side of a cross main.
Maximum system area = 52,000 S.F.

Included in this group are:

Group 1
- Automotive Parking and Showrooms
- Bakeries
- Beverage manufacturing
- Canneries
- Dairy Products Manufacturing/Processing
- Electronic Plans
- Glass and Glass Products Manufacturing
- Laundries
- Restaurant Service Areas

Group 2
- Cereal Mills
- Chemical Plants—Ordinary
- Confectionery Products
- Distilleries
- Dry Cleaners
- Feed Mills
- Horse Stables
- Leather Goods Manufacturing
- Libraries—Large Stack Room Areas
- Machine Shops
- Metal Working
- Mercantile
- Paper and Pulp Mills
- Paper Process Plants
- Piers and Wharves
- Post Offices
- Printing and Publishing
- Repair Garages
- Stages
- Textile Manufacturing
- Tire Manufacturing
- Tobacco Products Manufacturing
- Wood Machining
- Wood Product Assembly

Extra Hazard Occupancy
The protection area allotted per sprinkler shall not exceed 100 S.F. of noncombustible ceiling and 100 S.F. of combustible ceiling. The maximum allowable distance between lines and between sprinklers on lines is 12'. Sprinklers on alternate lines shall be staggered if the distance between sprinklers on lines exceeds 8'. Branch lines should not exceed six sprinklers on either side of a cross main.
Maximum system area:
- Design by pipe schedule = 25,000 S.F.
- Design by hydraulic calculation = 40,000 S.F.

Included in this group are:

Group 1
- Aircraft hangars
- Combustible Hydraulic Fluid Use Area
- Die Casting
- Metal Extruding
- Plywood/Particle Board Manufacturing
- Printing (inks with flash points < 100 degrees F
- Rubber Reclaiming, Compounding, Drying, Milling, Vulcanizing
- Saw Mills
- Textile Picking, Opening, Blending, Garnetting, Carding, Combing of Cotton, Synthetics, Wood Shoddy, or Burlap
- Upholstering with Plastic Foams

Group 2
- Asphalt Saturating
- Flammable Liquids Spraying
- Flow Coating
- Manufactured/Modular Home Building Assemblies (where finished enclosure is present and has combustible interiors)
- Open Oil Quenching
- Plastics Processing
- Solvent Cleaning
- Varnish and Paint Dipping

Fire Suppression — R2113 Fire-Suppression Sprinkler Systems

R211313-30 Sprinkler Quantities for Various Sizes and Types of Pipe

Sprinkler Quantities: The table below lists the usual maximum number of sprinkler heads for each size of copper and steel pipe for both wet and dry systems. These quantities may be adjusted to meet individual structural needs or local code requirements. Maximum area on any one floor for one system is: light hazard and ordinary hazard 52,000 S.F., extra hazardous 25,000 S.F.

Pipe Size Diameter	Light Hazard Occupancy — Steel Pipe	Light Hazard Occupancy — Copper Pipe	Ordinary Hazard Occupancy — Steel Pipe	Ordinary Hazard Occupancy — Copper Pipe	Extra Hazard Occupancy — Steel Pipe	Extra Hazard Occupancy — Copper Pipe
1"	2 sprinklers	2 sprinklers	2 sprinklers	2 sprinklers	1 sprinklers	1 sprinklers
1-1/4"	3	3	3	3	2	2
1-1/2"	5	5	5	5	5	5
2"	10	12	10	12	8	8
2-1/2"	30	40	20	25	15	20
3"	60	65	40	45	27	30
3-1/2"	100	115	65	75	40	45
4"			100	115	55	65
5"			160	180	90	100
6"			275	300	150	170

Dry Pipe Systems: A dry pipe system should be installed where a wet pipe system is impractical, as in rooms or buildings which cannot be properly heated.

The use of an approved dry pipe system is more desirable than shutting off the water supply during cold weather.

Not more than 750 gallons of system capacity should be controlled by one dry pipe valve. Where two or more dry pipe valves are used, systems should preferably be divided horizontally.

R211313-40 Adjustment for Sprinkler/Standpipe Installations

Quality/Complexity Multiplier (For all installations)

Economy installation, add	0 to 5%
Good quality, medium complexity, add	5 to 15%
Above average quality and complexity, add	15 to 25%

Plumbing — R2201 Operation & Maintenance of Plumbing

R220102-20 Labor Adjustment Factors

Labor Adjustment Factors are provided for Divisions 21, 22, and 23 to assist the mechanical estimator account for the various complexities and special conditions of any particular project. While a single percentage has been entered on each line of Division 22 01 02.20, it should be understood that these are just suggested midpoints of ranges of values commonly used by mechanical estimators. They may be increased or decreased depending on the severity of the special conditions.

The group for "existing occupied buildings", has been the subject of requests for explanation. Actually there are two stages to this group: buildings that are existing and "finished" but unoccupied, and those that also are occupied. Buildings that are "finished" may result in higher labor costs due to the workers having to be more careful not to damage finished walls, ceilings, floors, etc. and may necessitate special protective coverings and barriers. Also corridor bends and doorways may not accommodate long pieces of pipe or larger pieces of equipment. Work above an already hung ceiling can be very time consuming. The addition of occupants may force the work to be done on premium time (nights and/or weekends), eliminate the possible use of some preferred tools such as pneumatic drivers, powder charged drivers etc. The estimator should evaluate the access to the work area and just how the work is going to be accomplished to arrive at an increase in labor costs over "normal" new construction productivity.

R220105-10 Demolition (Selective vs. Removal for Replacement)

Demolition can be divided into two basic categories.

One type of demolition involves the removal of material with no concern for its replacement. The labor-hours to estimate this work are found under "Selective Demolition" in the Fire Protection, Plumbing and HVAC Divisions. It is selective in that individual items or all the material installed as a system or trade grouping such as plumbing or heating systems are removed. This may be accomplished by the easiest way possible, such as sawing, torch cutting, or sledge hammer as well as simple unbolting.

The second type of demolition is the removal of some item for repair or replacement. This removal may involve careful draining, opening of unions, disconnecting and tagging of electrical connections, capping of pipes/ducts to prevent entry of debris or leakage of the material contained as well as transport of the item away from its in-place location to a truck/dumpster. An approximation of the time required to accomplish this type of demolition is to use half of the time indicated as necessary to install a new unit. For example; installation of a new pump might be listed as requiring 6 labor-hours so if we had to estimate the removal of the old pump we would allow an additional 3 hours for a total of 9 hours. That is, the complete replacement of a defective pump with a new pump would be estimated to take 9 labor-hours.

Plumbing — R2205 Common Work Results for Plumbing

R220523-80 Valve Materials

VALVE MATERIALS

Bronze:
Bronze is one of the oldest materials used to make valves. It is most commonly used in hot and cold water systems and other non-corrosive services. It is often used as a seating surface in larger iron body valves to ensure tight closure.

Carbon Steel:
Carbon steel is a high strength material. Therefore, valves made from this metal are used in higher pressure services, such as steam lines up to 600 psi at 850°F. Many steel valves are available with butt-weld ends for economy and are generally used in high pressure steam service as well as other higher pressure non-corrosive services.

Forged Steel:
Valves from tough carbon steel are used in service up to 2000 psi and temperatures up to 1000°F in Gate, Globe and Check valves.

Iron:
Valves are normally used in medium to large pipe lines to control non-corrosive fluid and gases, where pressures do not exceed 250 psi at 450° or 500 psi cold water, oil or gas.

Stainless Steel:
Developed steel alloys can be used in over 90% corrosive services.

Plastic PVC:
This is used in a great variety of valves generally in high corrosive service with lower temperatures and pressures.

VALVE SERVICE PRESSURES

Pressure ratings on valves provide an indication of the safe operating pressure for a valve at some elevated temperature. This temperature is dependent upon the materials used and the fabrication of the valve. When specific data is not available, a good "rule-of-thumb" to follow is the temperature of saturated steam on the primary rating indicated on the valve body. Example: The valve has the number 150S printed on the side indicating 150 psi and hence, a maximum operating temperature of 367°F (temperature of saturated steam and 150 psi).

DEFINITIONS

1. "WOG" – Water, oil, gas (cold working pressures).
2. "SWP" – Steam working pressure.
3. 100% area (full port) – means the area through the valve is equal to or greater than the area of standard pipe.
4. "Standard Opening" – means that the area through the valve is less than the area of standard pipe and therefore these valves should be used only where restriction of flow is unimportant.
5. "Round Port" – means the valve has a full round opening through the plug and body, of the same size and area as standard pipe.
6. "Rectangular Port" – valves have rectangular shaped ports through the plug body. The area of the port is either equal to 100% of the area of standard pipe, or restricted (standard opening). In either case it is clearly marked.
7. "ANSI" – American National Standards Institute.

R220523-90 Valve Selection Considerations

INTRODUCTION: In any piping application, valve performance is critical. Valves should be selected to give the best performance at the lowest cost.

The following is a list of performance characteristics generally expected of valves.

1. Stopping flow or starting it.
2. Throttling flow (Modulation).
3. Flow direction changing.
4. Checking backflow (Permitting flow in only one direction).
5. Relieving or regulating pressure.

In order to properly select the right valve, some facts must be determined.

A. What liquid or gas will flow through the valve?
B. Does the fluid contain suspended particles?
C. Does the fluid remain in liquid form at all times?
D. Which metals does fluid corrode?
E. What are the pressure and temperature limits? (As temperature and pressure rise, so will the price of the valve.)
F. Is there constant line pressure?
G. Is the valve merely an on-off valve?
H. Will checking of backflow be required?
I. Will the valve operate frequently or infrequently?

Valves are classified by design type into such classifications as Gate, Globe, Angle, Check, Ball, Butterfly and Plug. They are also classified by end connection, stem, pressure restrictions and material such as bronze, cast iron, etc. Each valve has a specific use. A quality valve used correctly will provide a lifetime of trouble-free service, but a high quality valve installed in the wrong service may require frequent attention.

Plumbing — R2205 Common Work Results for Plumbing

STEM TYPES

(OS & Y)—Rising Stem-Outside Screw and Yoke

Offers a visual indication of whether the valve is open or closed. Recommended where high temperatures, corrosives, and solids in the line might cause damage to inside-valve stem threads. The stem threads are engaged by the yoke bushing so the stem rises through the hand wheel as it is turned.

(R.S.)—Rising Stem-Inside Screw

Adequate clearance for operation must be provided because both the hand wheel and the stem rise.
The valve wedge position is indicated by the position of the stem and hand wheel.

(N.R.S.)—Non-Rising Stem-Inside Screw

A minimum clearance is required for operating this type of valve. Excessive wear or damage to stem threads inside the valve may be caused by heat, corrosion, and solids. Because the hand wheel and stem do not rise, wedge position cannot be visually determined.

VALVE TYPES

Gate Valves

Provide full flow, minute pressure drop, minimum turbulence and minimum fluid trapped in the line.
They are normally used where operation is infrequent.

Globe Valves

Globe valves are designed for throttling and/or frequent operation with positive shut-off. Particular attention must be paid to the several types of seating materials available to avoid unnecessary wear. The seats must be compatible with the fluid in service and may be composition or metal. The configuration of the Globe valve opening causes turbulence which results in increased resistance. Most bronze Globe valves are rising stem-inside screw, but they are also available on O.S. & Y.

Angle Valves

The fundamental difference between the Angle valve and the Globe valve is the fluid flow through the Angle valve. It makes a 90° turn and offers less resistance to flow than the Globe valve while replacing an elbow. An Angle valve thus reduces the number of joints and installation time.

Plumbing — R2205 Common Work Results for Plumbing

Check Valves

Check valves are designed to prevent backflow by automatically seating when the direction of fluid is reversed.

Swing Check valves are generally installed with Gate-valves, as they provide comparable full flow. Usually recommended for lines where flow velocities are low and should not be used on lines with pulsating flow. Recommended for horizontal installation, or in vertical lines only where flow is upward.

Lift Check Valves

These are commonly used with Globe and Angle valves since they have similar diaphragm seating arrangements and are recommended for preventing backflow of steam, air, gas and water, and on vapor lines with high flow velocities. For horizontal lines, horizontal lift checks should be used and vertical lift checks for vertical lines.

Ball Valves

Ball valves are light and easily installed, yet because of modern elastomeric seats, provide tight closure. Flow is controlled by rotating up to 90° a drilled ball which fits tightly against resilient seals. This ball seats with flow in either direction, and valve handle indicates the degree of opening. Recommended for frequent operation readily adaptable to automation, ideal for installation where space is limited.

Butterfly Valves

Butterfly valves provide bubble-tight closure with excellent throttling characteristics. They can be used for full-open, closed and for throttling applications.

The Butterfly valve consists of a disc within the valve body which is controlled by a shaft. In its closed position, the valve disc seals against a resilient seat. The disc position throughout the full 90° rotation is visually indicated by the position of the operator.

A Butterfly valve is only a fraction of the weight of a Gate valve and requires no gaskets between flanges in most cases. Recommended for frequent operation and adaptable to automation where space is limited.

Wafer and Lug type bodies when installed between two pipe flanges, can be easily removed from the line. The pressure of the bolted flanges holds the valve in place.
Locating lugs makes installation easier.

Plug Valves

Lubricated plug valves, because of the wide range of service to which they are adapted, may be classified as all purpose valves. They can be safely used at all pressure and vacuums, and at all temperatures up to the limits of available lubricants. They are the most satisfactory valves for the handling of gritty suspensions and many other destructive, erosive, corrosive and chemical solutions.

Plumbing

R2211 Facility Water Distribution

R221113-40 Plumbing Approximations for Quick Estimating

Water Control
Water Meter; Backflow Preventer, ... 10 to 15% of Fixtures
Shock Absorbers; Vacuum Breakers;
Mixer.

Pipe And Fittings .... 30 to 60% of Fixtures

> **Note:** Lower percentage for compact buildings or larger buildings with plumbing in one area.
> Larger percentage for large buildings with plumbing spread out.
> In extreme cases pipe may be more than 100% of fixtures.
> Percentages **do not** include special purpose or process piping.

Plumbing Labor
1 & 2 Story Residential .... Rough-in Labor = 80% of Materials
Apartment Buildings .... Rough-in Labor = 90 to 100% of Materials
Labor for handling and placing fixtures is approximately 25 to 30% of fixtures

Quality/Complexity Multiplier (for all installations)
Economy installation, add. .... 0 to 5%
Good quality, medium complexity, add .... 5 to 15%
Above average quality and complexity, add .... 15 to 25%

R221113-50 Pipe Material Considerations

1. Malleable fittings should be used for gas service.
2. Malleable fittings are used where there are stresses/strains due to expansion and vibration.
3. Cast fittings may be broken as an aid to disassembling of heating lines frozen by long use, temperature and minerals.
4. Cast iron pipe is extensively used for underground and submerged service.
5. Type M (light wall) copper tubing is available in hard temper only and is used for nonpressure and less severe applications than K and L.
6. Type L (medium wall) copper tubing, available hard or soft for interior service.
7. Type K (heavy wall) copper tubing, available in hard or soft temper for use where conditions are severe. For underground and interior service.
8. Hard drawn tubing requires fewer hangers or supports but should not be bent. Silver brazed fittings are recommended, however soft solder is normally used.
9. Type DMV (very light wall) copper tubing designed for drainage, waste and vent plus other non-critical pressure services.

Domestic/Imported Pipe and Fittings Cost

The prices shown in this publication for steel/cast iron pipe and steel, cast iron, malleable iron fittings are based on domestic production sold at the normal trade discounts. The above listed items of foreign manufacture may be available at prices of 1/3 to 1/2 those shown. Some imported items after minor machining or finishing operations are being sold as domestic to further complicate the system.

Caution: Most pipe prices in this book also include a coupling and pipe hangers which for the larger sizes can add significantly to the per foot cost and should be taken into account when comparing "book cost" with quoted supplier's cost.

Plumbing — R2211 Facility Water Distribution

R221113-70 Piping to 10' High

When taking off pipe, it is important to identify the different material types and joining procedures, as well as distances between supports and components required for proper support.

During the takeoff, measure through all fittings. Do not subtract the lengths of the fittings, valves, or strainers, etc. This added length plus the final rounding of the totals will compensate for nipples and waste.

When rounding off totals always increase the actual amount to correspond with manufacturer's shipping lengths.

A. Both red brass and yellow brass pipe are normally furnished in 12' lengths, plain end. The Unit Price section includes in the linear foot costs two field threads and one coupling per 10' length. A carbon steel clevis type hanger assembly every 10' is also prorated into the linear foot costs, including both material and labor.

B. Cast iron soil pipe is furnished in either 5' or 10' lengths. For pricing purposes, the Unit Price section features 10' lengths with a joint and a carbon steel clevis hanger assembly every 5' prorated into the per foot costs of both material and labor.

Three methods of joining are considered: lead and oakum poured joints or push-on gasket type joints for the bell and spigot pipe and a joint clamp for the no-hub soil pipe. The labor and material costs for each of these individual joining procedures are also prorated into the linear costs per foot.

C. Copper tubing covers types K, L, M, and DWV which are furnished in 20' lengths. Means pricing data is based on a tubing cut each length and a coupling and two soft soldered joints every 10'. A carbon steel, clevis type hanger assembly every 10' is also prorated into the per foot costs. The prices for refrigeration tubing are for materials only. Labor for full lengths may be based on the type L labor but short cut measures in tight areas can increase the installation labor-hours from 20 to 40%.

D. Corrosion-resistant piping does not lend itself to one particular standard of hanging or support assembly due to its diversity of application and placement. The several varieties of corrosion-resistant piping do not include any material or labor costs for hanger assemblies (See the Unit Price section for appropriate selection).

E. Glass pipe is furnished in standard lengths either 5' or 10' long, beaded on one end. Special orders for diverse lengths beaded on both ends are also available. For pricing purposes, R.S. Means features 10' lengths with a coupling and a carbon steel band hanger assembly every 10' prorated into the per foot linear costs.

Glass pipe is also available with conical ends and standard lengths ranging from 6" through 3' in 6" increments, then up to 10' in 12" increments. Special lengths can be customized for particular installation requirements.

For pricing purposes, Means has based the labor and material pricing on 10' lengths. Included in these costs per linear foot are the prorated costs for a flanged assembly every 10' consisting of two flanges, a gasket, two insertable seals, and the required number of bolts and nuts. A carbon steel band hanger assembly based on 10' center lines has also been prorated into the costs per foot for labor and materials.

F. Plastic pipe of several compositions and joining methods are considered. Fiberglass reinforced pipe (FRP) is priced, based on 10' lengths (20' lengths are also available), with coupling and epoxy joints every 10'. FRP is furnished in both "General Service" and "High Strength." A carbon steel clevis hanger assembly, 3 for every 10', is built into the prorated labor and material costs on a per foot basis.

The PVC and CPVC pipe schedules 40, 80 and 120, plus SDR ratings are all based on 20' lengths with a coupling installed every 10', as well as a carbon steel clevis hanger assembly every 3'. The PVC and ABS type DWV piping is based on 10' lengths with solvent weld couplings every 10', and with carbon steel clevis hanger assemblies, 3 for every 10'. The rest of the plastic piping in this section is based on flexible 100' coils and does not include any coupling or supports.

This section ends with PVC drain and sewer piping based on 10' lengths with bell and spigot ends and O-ring type, push-on joints.

G. Stainless steel piping includes both weld end and threaded piping, both in the type 304 and 316 specification and in the following schedules, 5, 10, 40, 80, and 160. Although this piping is usually furnished in 20' lengths, this cost grouping has a joint (either heli-arc butt-welded or threads and coupling) every 10'. A carbon steel clevis type hanger assembly is also included at 10' intervals and prorated into the linear foot costs.

H. Carbon steel pipe includes both black and galvanized. This section encompasses schedules 40 (standard) and 80 (extra heavy).

Several common methods of joining steel pipe — such as thread and coupled, butt welded, and flanged (150 lb. weld neck flanges) are also included.

For estimating purposes, it is assumed that the piping is purchased in 20' lengths and that a compatible joint is made up every 10'. These joints are prorated into the labor and material costs per linear foot. The following hanger and support assemblies every 10' are also included: carbon steel clevis for the T & C pipe, and single rod roll type for both the welded and flanged piping. All of these hangers are oversized to accommodate pipe insulation 3/4" thick through 5" pipe size and 1-1/2" thick from 6" through 12" pipe size.

I. Grooved joint steel pipe is priced both black and galvanized, in schedules 10, 40, and 80, furnished in 20' lengths. This section describes two joining methods: cut groove and roll groove. The schedule 10 piping is roll-grooved, while the heavier schedules are cut-grooved. The labor and material costs are prorated into per linear foot prices, including a coupled joint every 10', as well as a carbon steel clevis hanger assembly.

Notes:

The pipe hanger assemblies mentioned in the preceding paragraphs include the described hanger; appropriately sized steel, box-type insert and nut; plus 18" of threaded hanger rod.

C clamps are used when the pipe is to be supported from steel shapes rather than anchored in the slab. C clamps are slightly less costly than inserts. However, to save time in estimating, it is advisable to use the given line number cost, rather than substituting a C clamp for the insert.

Add to piping labor for elevated installation:

10' to 14.5' high	10%	30' to 34.5' high	40%
15' to 19.5' high	20%	35' to 39.5' high	50%
20' to 24.5' high	25%	40' and higher	55%
25' to 29.5' high	35%		

When using the percentage adds for elevated piping installations as shown above, bear in mind that the given heights are for the pipe supports, even though the insert, anchor, or clamp may be several feet higher than the pipe itself.

An allowance has been included in the piping installation time for testing and minor tightening of leaking joints, fittings, stuffing boxes, packing glands, etc. For extraordinary test requirements such as x-rays, prolonged pressure or demonstration tests, a percentage of the piping labor, based on the estimator's experience, must be added to the labor total. A testing service specializing in weld x-rays should be consulted for pricing if it is an estimate requirement. Equipment installation time includes start-up with associated adjustments.

Plumbing — R2213 Facility Sanitary Sewerage

R221316-10 Drainage Requirements

Drainage lines must have a slope to maintain flow for proper operation. This slope should not be less than 1/4" per foot for 3" diameter or smaller pipe and not less than 1/8" per foot for 4" diameter or larger pipe. The capacity of building drainage systems is calculated on a basis of "drainage fixture units" (d.f.u.) as per the following chart.

Type of Fixture	d.f.u. Value	Type of Fixture	d.f.u. Value
Automatic clothes washer (2" standpipe)	3	Service sink (trap standard)	3
Bathroom group (water closet, lavatory and		Service sink (P trap)	2
bathtub or shower) tank type closet	6	Urinal, pedestal, syphon jet blowout	6
Bathtub (with or without overhead shower)	2	Urinal, wall hung	4
Clinic sink	6	Urinal, stall washout	4
Combination sink & tray with food disposal	4	Wash sink (cir. or mult.) per faucet set	2
Dental unit or cuspidor	1	Water closet, tank operated	4
Dental lavatory	1	Water closet, valve operated	6
Drinking fountain	1/2	Fixtures not listed above	
Dishwasher, domestic	2	Trap size 1-1/4" or smaller	1
Floor drains with 2" waste	3	Trap size 1-1/2"	2
Kitchen sink, domestic with one 1-1/2" trap	2	Trap size 2"	3
Kitchen sink, domestic with food disposal	2	Trap size 2-1/2"	4
Lavatory with 1-1/4" waste	1	Trap size 3"	5
Laundry tray (1 or 2 compartment)	2	Trap size 4"	6
Shower stall, domestic	2		

For continuous or nearly continuous flow into the sytem from a pump, air conditioning equipment or other item, allow 2 fixture units for each gallon per minute of flow.

When the "drainage fixture units" (d.f.u.) for each horizontal branch or vertical stack is computed from the table above, the appropriate pipe size for each branch or stack is determined from the table below.

R221316-20 Allowable Fixture Units (d.f.u.) for Branches and Stacks

Pipe Diam.	Horiz. Branch (not incl. drains)	Stack Size for 3 Stories or 3 Levels	Stack size for Over 3 levels	Maximum for 1 Story building Stack
1-1/2"	3	4	8	2
2"	6	10	24	6
2-1/2"	12	20	42	9
3"	20*	48*	72*	20*
4"	160	240	500	90
5"	360	540	1100	200
6"	620	960	1900	350
8"	1400	2200	3600	600
10"	2500	3800	5600	1000
12"	3900	6000	8400	1500
15"	7000			

*Not more than two water closets or bathroom groups within each branch interval nor more than six water closets or bathroom groups on the stack.

Stacks sized for the total may be reduced as load decreases at each story to a minimum diameter of 1/2 the maximum diameter.

Plumbing — R2240 Plumbing Fixtures

R224000-10 Hot Water Consumption Rates

Type of Building	Size Factor	Maximum Hourly Demand	Average Day Demand
Apartment Dwellings	No. of Apartments: Up to 20 21 to 50 51 to 75 76 to 100 101 to 200 201 up	12.0 Gal. per apt. 10.0 Gal. per apt. 8.5 Gal. per apt. 7.0 Gal. per apt. 6.0 Gal. per apt. 5.0 Gal. per apt.	42.0 Gal. per apt. 40.0 Gal. per apt. 38.0 Gal. per apt. 37.0 Gal. per apt. 36.0 Gal. per apt. 35.0 Gal. per apt.
Dormitories	Men Women	3.8 Gal. per man 5.0 Gal. per woman	13.1 Gal. per man 12.3 Gal. per woman
Hospitals	Per bed	23.0 Gal. per patient	90.0 Gal. per patient
Hotels	Single room with bath Double room with bath	17.0 Gal. per unit 27.0 Gal. per unit	50.0 Gal. per unit 80.0 Gal. per unit
Motels	No. of units: Up to 20 21 to 100 101 Up	6.0 Gal. per unit 5.0 Gal. per unit 4.0 Gal. per unit	20.0 Gal. per unit 14.0 Gal. per unit 10.0 Gal. per unit
Nursing Homes		4.5 Gal. per bed	18.4 Gal. per bed
Office buildings		0.4 Gal. per person	1.0 Gal. per person
Restaurants	Full meal type Drive-in snack type	1.5 Gal./max. meals/hr. 0.7 Gal./max. meals/hr.	2.4 Gal. per meal 0.7 Gal. per meal
Schools	Elementary Secondary & High	0.6 Gal. per student 1.0 Gal. per student	0.6 Gal. per student 1.8 Gal. per student

For evaluation purposes, recovery rate and storage capacity are inversely proportional. Water heaters should be sized so that the maximum hourly demand anticipated can be met in addition to allowance for the heat loss from the pipes and storage tank.

R224000-20 Fixture Demands in Gallons Per Fixture Per Hour

Table below is based on 140° F final temperature except for dishwashers in public places (*) where 180° F water is mandatory.

Fixture	Apartment House	Club	Gym	Hospital	Hotel	Indust. Plant	Office	Private Home	School
Bathtubs	20	20	30	20	20			20	
Dishwashers, automatic	15	50-150*		50-150*	50-200*	20-100*		15	20-100*
Kitchen sink	10	20		20	30	20	20	10	20
Laundry, stationary tubs	20	28		28	28			20	
Laundry, automatic wash	75	75		100	150			75	
Private lavatory	2	2	2	2	2	2	2	2	2
Public lavatory	4	6	8	6	8	12	6		15
Showers	30	150	225	75	75	225	30	30	225
Service sink	20	20		20	30	20	20	15	20
Demand factor	0.30	0.30	0.40	0.25	0.25	0.40	0.30	0.30	0.40
Storage capacity factor	1.25	0.90	1.00	0.60	0.80	1.00	2.00	0.70	1.00

To obtain the probable maximum demand multiply the total demands for the fixtures (gal./fixture/hour) by the demand factor. The heater should have a heating capacity in gallons per hour equal to this maximum. The storage tank should have a capacity in gallons equal to the probable maximum demand multiplied by the storage capacity factor.

Plumbing

R2240 Plumbing Fixtures

R224000-30 Minimum Plumbing Fixture Requirements

Minimum Plumbing Fixture Requirements

Type of Building or Occupancy (2)	Water Closets (14) (Fixtures per Person) Male	Water Closets (14) (Fixtures per Person) Female	Urinals (5,10) (Fixtures per Person) Male	Urinals (5,10) (Fixtures per Person) Female	Lavatories (Fixtures per Person) Male	Lavatories (Fixtures per Person) Female	Bathtubs or Showers (Fixtures per Person)	Drinking Fountains (Fixtures per Person) (3, 13)
Assembly Places- Theatres, Auditoriums, Convention Halls, etc.-for permanent employee use	1: 1 - 15 2: 16 - 35 3: 36 - 55	1: 1 - 15 2: 16 - 35 3: 36 - 55	0: 1 - 9 1: 10 - 50		1 per 40	1 per 40		
	Over 55, add 1 fixture for each additional 40 persons		Add one fixture for each additional 50 males					
Assembly Places- Theatres, Auditoriums, Convention Halls, etc. - for public use	1: 1 - 100 2: 101 - 200 3: 201 - 400	3: 1 - 50 4: 51 - 100 8: 101 - 200 11: 201 - 400	1: 1 - 100 2: 101 - 200 3: 201 - 400 4: 401 - 600		1: 1 - 200 2: 201 - 400 3: 401 - 750	1: 1 - 200 2: 201 - 400 3: 401 - 750		1: 1 - 150 2: 151 - 400 3: 401 - 750
	Over 400, add 1 fixture for each additional 500 males and 1 for each additional 125 females		Over 600, add 1 fixture for each additional 300 males		Over 750, add 1 fixture for each additional 500 persons			Over 750, add one fixture for each additional 500 persons
Dormitories (9) School or Labor	1 per 10	1 per 8	1 per 25		1 per 12	1 per 12	1 per 8	1 per 150 (12)
	Add 1 fixture for each additional 25 males (over 10) and 1 for each additional 20 females (over 8)		Over 150, add 1 fixture for each additional 50 males		Over 12 add 1 fixture for each additional 20 males and 1 for each 15 additional females		For females add 1 bathtub per 30. Over 150, add 1 per 20	
Dormitories- for Staff Use	1: 1 - 15 2: 16 - 35 3: 36 - 55	1: 1 - 15 2: 16 - 35 4: 36 - 55	1 per 50		1 per 40	1 per 40	1 per 8	
	Over 55, add 1 fixture for each additional 40 persons							
Dwellings: Single Dwelling	1 per dwelling				1 per dwelling		1 per dwelling	
Multiple Dwelling or Apartment House	1 per dwelling or apartment unit				1 per dwelling or apartment unit		1 per dwelling or apartment unit	
Hospital Waiting rooms	1 per room				1 per room			1 per 150 (12)
Hospitals- for employee use	1: 1 - 15 2: 16 - 35 3: 36 - 55	1: 1 - 15 3: 16 - 35 4: 36 - 55	0: 1 - 9 1: 10 - 50		1 per 40	1 per 40		
	Over 55, add 1 fixture for each additional 40 persons		Add 1 fixture for each additional 50 males					
Hospitals: Individual Room Ward Room	1 per room 1 per 8 patients				1 per room 1 per 10 patients		1 per room 1 per 20 patients	1 per 150 (12)
Industrial (6) Warehouses Workshops, Foundries and similar establishments- for employee use	1: 1 -10 2: 11 - 25 3: 26 - 50 4: 51 - 75 5: 76 - 100	1: 1 -10 2: 11 - 25 3: 26 - 50 4: 51 - 75 5: 76 - 100			Up to 100, per 10 persons Over 100, 1 per 15 persons (7, 8)		1 shower for each 15 persons exposed to excessive heat or to skin contamination with poisonous, infectious or irritating material	1 per 150 (12)
	Over 100, add 1 fixture for each additional 30 persons							
Institutional - Other than Hospitals or Penal Institutions (on each occupied floor)	1 per 25	1 per 20	0: 1 - 9 1: 10 - 50 Add 1 fixture for each additional 50 males		1 per 10	1 per 10	1 per 8	1 per 150 (12)
Institutional - Other than Hospitals or Penal Institutions (on each occupied floor)- for employee use	1: 1 - 15 2: 16 - 35 3: 36 - 55	1: 1 - 15 3: 16 - 35 4: 36 - 55	0: 1 - 9 1: 10 - 50		1 per 40	1 per 40	1 per 8	1 per 150 (12)
	Over 55, add 1 fixture for each additional 40 persons		Add 1 fixture for each additional 50 males					
Office or Public Buildings	1: 1 - 100 2: 101 - 200 3: 201 - 400	3: 1 - 50 4: 51 - 100 8: 101 - 200 11: 201 - 400	1: 1 - 100 2: 101 - 200 3: 201 - 400 4: 401 - 600		1: 1 - 200 2: 201 - 400 3: 401 - 750	1: 1 - 200 2: 201 - 400 3: 401 - 750		1 per 150 (12)
	Over 400, add 1 fixture for each additional 500 males and 1 for each additional 150 females		Over 600, add 1 fixture for each additional 300 males		Over 750, add 1 fixture for each additional 500 persons			
Office or Public Buildings - for employee use	1: 1 - 15 2: 16 - 35 3: 36 - 55	1: 1 - 15 3: 16 - 35 4: 36 - 55	0: 1 - 9 1: 10 - 50		1 per 40	1 per 40		
	Over 55, add 1 fixture for each additional 40 persons		Add 1 fixture for each additional 50 males					

Plumbing

R2240 Plumbing Fixtures

R224000-30 Minimum Plumbing Fixture Requirements (cont.)

Minimum Plumbing Fixture Requirements

Type of Building or Occupancy	Water Closets (14) (Fixtures per Person) Male	Water Closets (14) (Fixtures per Person) Female	Urinals (5, 10) (Fixtures per Person) Male	Urinals (5, 10) (Fixtures per Person) Female	Lavatories (Fixtures per Person) Male	Lavatories (Fixtures per Person) Female	Bathtubs or Showers (Fixtures per Person)	Drinking Fountains (Fixtures per Person) (3, 13)
Penal Institutions - for employee use	1: 1 - 15 2: 16 - 35 3: 36 - 55 Over 55, add 1 fixture for each additional 40 persons	1: 1 - 15 3: 16 - 35 4: 36 - 55	0: 1 - 9 1: 10 - 50 Add 1 fixture for each additional 50 males		1 per 40	1 per 40		1 per 150 (12)
Penal Institutions - for prison use Cell	1 per cell	1 per cell			1 per cell	1 per cell		1 per cellblock floor
Exercise room	1 per exercise room	1 per exercise room	1 per exercise room		1 per exercise room	1 per exercise room		1 per exercise room
Restaurants, Pubs and Lounges (11)	1: 1 - 50 2: 51 - 150 3: 151 - 300 Over 300, add 1 fixture for each additional 200 persons	1: 1 - 50 2: 51 - 150 4: 151 - 300	1: 1 - 150 Over 150, add 1 fixture for each additional 150 males		1: 1 - 150 2: 151 - 200 3: 201 - 400 Over 400, add 1 fixture for each additional 400 persons	1: 1 - 150 2: 151 - 200 3: 201 - 400		
Schools - for staff use All Schools	1: 1 - 15 2: 16 - 35 3: 36 - 55 Over 55, add 1 fixture for each additional 40 persons	1: 1 - 15 3: 16 - 35 4: 36 - 55	1 per 50		1 per 40	1 per 40		
Schools - for student use: Nursery	1: 1 - 20 2: 21 - 50 Over 50, add 1 fixture for each additional 50 persons	1: 1 - 20 2: 21 - 50			1: 1 - 25 2: 26 - 50 Over 50, add 1 fixture for each additional 50 persons	1: 1 - 25 2: 26 - 50		1 per 150 (12)
Elementary	1 per 30	1 per 25	1 per 75		1 per 35	1 per 35		1 per 150 (12)
Secondary	1 per 40	1 per 30	1 per 35		1 per 40	1 per 40		1 per 150 (12)
Others (Colleges, Universities, Adult Centers, etc.	1 per 40	1 per 30	1 per 35		1 per 40	1 per 40		1 per 150 (12)
Worship Places Educational and Activities Unit	1 per 150	1 per 75	1 per 150		1 per 2 water closets			1 per 150 (12)
Worship Places Principal Assembly Place	1 per 150	1 per 75	1 per 150		1 per 2 water closets			1 per 150 (12)

Notes:
1. The figures shown are based upon one (1) fixture being the minimum required for the number of persons indicated or any fraction thereof.
2. Building categories not shown on this table shall be considered separately by the Administrative Authority.
3. Drinking fountains shall not be installed in toilet rooms.
4. Laundry trays. One (1) laundry tray or one (1) automatic washer standpipe for each dwelling unit or one (1) laundry trays or one (1) automatic washer standpipes, or combination thereof, for each twelve (12) apartments. Kitchen sinks, one (1) for each dwelling or apartment unit.
5. For each urinal added in excess of the minimum required, one water closet may be deducted. The number of water closets shall not be reduced to less than two-thirds (2/3) of the minimum requirement.
6. As required by ANSI Z4.1-1968, Sanitation in Places of Employment.
7. Where there is exposure to skin contamination with poisonous, infectious, or irritating materials, provide one (1) lavatory for each five (5) persons.
8. Twenty-four (24) lineal inches of wash sink or eighteen (18) inches of a circular basin, when provided with water outlets for such space shall be considered equivalent to one (1) lavatory.
9. Laundry trays, one (1) for each fifty (50) persons. Service sinks, one (1) for each hundred (100) persons.
10. General. In applying this schedule of facilities, consideration shall be given to the accessibility of the fixtures. Conformity purely on a numerical basis may not result in an installation suited to the need of the individual establishment. For example, schools should be provided with toilet facilities on each floor having classrooms.
    a. Surrounding materials, wall and floor space to a point two (2) feet in front of urinal lip and four (4) feet above the floor, and at least two (2) feet to each side of the urinal shall be lined with non-absorbent materials.
    b. Trough urinals shall be prohibited.
11. A restaurant is defined as a business which sells food to be consumed on the premises.
    a. The number of occupants for a drive-in restaurant shall be considered as equal to the number of parking stalls.
    b. Employee toilet facilities shall not to be included in the above restaurant requirements. Hand washing facilities shall be available in the kitchen for employees.
12. Where food is consumed indoors, water stations may be substituted for drinking fountains. Offices, or public buildings for use by more than six (6) persons shall have one (1) drinking fountain for the first one hundred fifty (150) persons and one additional fountain for each three hundred (300) persons thereafter.
13. There shall be a minimum of one (1) drinking fountain per occupied floor in schools, theaters, auditoriums, dormitories, offices of public building.
14. The total number of water closets for females shall be at least equal to the total number of water closets and urinals required for males.

Plumbing

R2240 Plumbing Fixtures

R224000-40 Plumbing Fixture Installation Time

Item	Rough-In	Set	Total Hours	Item	Rough-In	Set	Total Hours
Bathtub	5	5	10	Shower head only	2	1	3
Bathtub and shower, cast iron	6	6	12	Shower drain	3	1	4
Fire hose reel and cabinet	4	2	6	Shower stall, slate		15	15
Floor drain to 4 inch diameter	3	1	4	Slop sink	5	3	8
Grease trap, single, cast iron	5	3	8	Test 6 fixtures			14
Kitchen gas range		4	4	Urinal, wall	6	2	8
Kitchen sink, single	4	4	8	Urinal, pedestal or floor	6	4	10
Kitchen sink, double	6	6	12	Water closet and tank	4	3	7
Laundry tubs	4	2	6	Water closet and tank, wall hung	5	3	8
Lavatory wall hung	5	3	8	Water heater, 45 gals. gas, automatic	5	2	7
Lavatory pedestal	5	3	8	Water heaters, 65 gals. gas, automatic	5	2	7
Shower and stall	6	4	10	Water heaters, electric, plumbing only	4	2	6

Fixture prices in front of book are based on the cost per fixture set in place. The rough-in cost, which must be added for each fixture, includes carrier, if required, some supply, waste and vent pipe connecting fittings and stops. The lengths of rough-in pipe are nominal runs which would connect to the larger runs and stacks. The supply runs and DWV runs and stacks must be accounted for in separate entries. In the eastern half of the United States it is common for the plumber to carry these to a point 5' outside the building.

R224000-50 Water Cooler Application

Type of Service	Requirement
Office, School or Hospital	12 persons per gallon per hour
Office, Lobby or Department Store	4 or 5 gallons per hour per fountain
Light manufacturing	7 persons per gallon per hour
Heavy manufacturing	5 persons per gallon per hour
Hot heavy manufacturing	4 persons per gallon per hour
Hotel	.08 gallons per hour per room
Theatre	1 gallon per hour per 100 seats

Heating, Ventilating & A.C.

R2305 Common Work Results for HVAC

R230500-10 Subcontractors

On the unit cost pages of the R.S. Means Cost Data books, the last column is entitled "Total Incl. O&P". This is normally the cost of the installing contractor. In the HVAC Division, this is the cost of the mechanical contractor. If the particular work being estimated is to be performed by a sub to the mechanical contractor, the mechanical's profit and handling charge (usually 10%) is added to the total of the last column.

Heating, Ventilating & A.C. R2356 Solar Energy Heating Equipment

R235616-60 Solar Heating (Space and Hot Water)

Collectors should face as close to due South as possible, however, variations of up to 20 degrees on either side of true South are acceptable. Local climate and collector type may influence the choice between east or west deviations. Obviously they should be located so they are not shaded from the sun's rays. Incline collectors at a slope of latitude minus 5 degrees for domestic hot water and latitude plus 15 degrees for space heating.

Flat plate collectors consist of a number of components as follows: Insulation to reduce heat loss through the bottom and sides of the collector. The enclosure which contains all the components in this assembly is usually weatherproof and prevents dust, wind and water from coming in contact with the absorber plate. The cover plate usually consists of one or more layers of a variety of glass or plastic and reduces the reradiation by creating an air space which traps the heat between the cover and the absorber plates.

The absorber plate must have a good thermal bond with the fluid passages. The absorber plate is usually metallic and treated with a surface coating which improves absorptivity. Black or dark paints or selective coatings are used for this purpose, and the design of this passage and plate combination helps determine a solar system's effectiveness.

Heat transfer fluid passage tubes are attached above and below or integral with an absorber plate for the purpose of transferring thermal energy from the absorber plate to a heat transfer medium. The heat exchanger is a device for transferring thermal energy from one fluid to another.

Piping and storage tanks should be well insulated to minimize heat losses.

Size domestic water heating storage tanks to hold 20 gallons of water per user, minimum, plus 10 gallons per dishwasher or washing machine. For domestic water heating an optimum collector size is approximately 3/4 square foot of area per gallon of water storage. For space heating of residences and small commercial applications the collector is commonly sized between 30% and 50% of the internal floor area. For space heating of large commercial applications, collector areas less than 30% of the internal floor area can still provide significant heat reductions.

A supplementary heat source is recommended for Northern states for December through February.

The solar energy transmission per square foot of collector surface varies greatly with the material used. Initial cost, heat transmittance and useful life are obviously interrelated.

Heating, Ventilating & A.C. R2383 Radiant Heating Units

Fig. R238313-11

R238313-10 Heat Trace Systems

Before you can determine the cost of a HEAT TRACE installation, the method of attachment must be established. There are (4) common methods:

1. Cable is simply attached to the pipe with polyester tape every 12'.
2. Cable is attached with a continuous cover of 2" wide aluminum tape.
3. Cable is attached with factory extruded heat transfer cement and covered with metallic raceway with clips every 10'.
4. Cable is attached between layers of pipe insulation using either clips or polyester tape.

Example: Components for method 3 must include:
A. Heat trace cable by voltage and watts per linear foot.
B. Heat transfer cement, 1 gallon per 60 linear feet of cover.
C. Metallic raceway by size and type.
D. Raceway clips by size of pipe.

When taking off linear foot lengths of cable add the following for each valve in the system. (E)

In all of the above methods each component of the system must be priced individually.

SCREWED OR WELDED VALVE:	FLANGED VALVE:	BUTTERFLY VALVES:
1/2" = 6"	1/2" = 1' -0"	1/2" = 0'
3/4" = 9"	3/4" = 1' -6"	3/4" = 0'
1" = 1' -0"	1" = 2' -0"	1" = 1' -0"
1-1/2" = 1' -6"	1-1/2" = 2' -6"	1-1/2" = 1' -6"
2" = 2'	2" = 2' -6"	2" = 2' -0"
2-1/2" = 2' -6"	2-1/2" = 3' -0"	2-1/2" = 2' -6"
3" = 2' -6"	3" = 3' -6"	3" = 2' -6"
4" = 4' -0"	4" = 4' -0"	4" = 3' -0"
6" = 7' -0"	6" = 8' -0"	6" = 3' -6"
8" = 9' -6"	8" = 11' -0"	8" = 4' -0"
10" = 12' -6"	10" = 14' -0"	10" = 4' -0"
12" = 15' -0"	12" = 16' -6"	12" = 5' -0"
14" = 18' -0"	14" = 19' -6"	14" = 5' -6"
16" = 21' -6"	16" = 23' -0"	16" = 6' -0"
18" = 25' -6"	18" = 27' -0"	18" = 6' -6"
20" = 28' -6"	20" = 30' -0"	20" = 7' -0"
24" = 34' -0"	24" = 36' -0"	24" = 8' -0"
30" = 40' -0"	30" = 42' -0"	30" = 10' -0"

Heating, Ventilating & A.C. R2383 Radiant Heating Units

R238313-10 Heat Trace Systems (cont.)

Add the following quantities of heat transfer cement to linear foot totals for each valve:

Nominal Valve Size	Gallons of Cement per Valve
1/2"	0.14
3/4"	0.21
1"	0.29
1-1/2"	0.36
2"	0.43
2-1/2"	0.70
3"	0.71
4"	1.00
6"	1.43
8"	1.48
10"	1.50
12"	1.60
14"	1.75
16"	2.00
18"	2.25
20"	2.50
24"	3.00
30"	3.75

The following must be added to the list of components to accurately price HEAT TRACE systems:

1. Expediter fitting and clamp fasteners (F)
2. Junction box and nipple connected to expediter fitting (G)
3. Field installed terminal blocks within junction box
4. Ground lugs
5. Piping from power source to expediter fitting
6. Controls
7. Thermostats
8. Branch wiring
9. Cable splices
10. End of cable terminations
11. Branch piping fittings and boxes

Deduct the following percentages from labor if cable lengths in the same area exceed:

150' to 250' 10%   351' to 500' 20%
251' to 350' 15%   Over 500' 25%

Add the following percentages to labor for elevated installations:

15' to 20' high 10%   31' to 35' high 40%
21' to 25' high 20%   36' to 40' high 50%
26' to 30' high 30%   Over 40' high 60%

R238313-20 Spiral-Wrapped Heat Trace Cable (Pitch Table)

In order to increase the amount of heat, occasionally heat trace cable is wrapped in a spiral fashion around a pipe; increasing the number of feet of heater cable per linear foot of pipe.

Engineers first determine the heat loss per foot of pipe (based on the insulating material, its thickness, and the temperature differential across it). A ratio is then calculated by the formula:

$$\text{Feet of Heat Trace per Foot of Pipe} = \frac{\text{Watts/Foot of Heat Loss}}{\text{Watts/Foot of the Cable}}$$

The linear distance between wraps (pitch) is then taken from a chart or table. Generally, the pitch is listed on a drawing leaving the estimator to calculate the total length of heat tape required. An approximation may be taken from this table.

Feet of Heat Trace Per Foot of Pipe

Pitch In Inches	1	1¼	1½	2	2½	3	4	6	8	10	12	14	16	18	20	24
3.5	1.80															
4	1.65															
5	1.46	1.60	1.80													
6	1.34	1.45	1.55	1.75												
7	1.25	1.35	1.43	1.57	1.75											
8	1.20	1.28	1.34	1.45	1.60	1.80										
9	1.16	1.23	1.28	1.37	1.51	1.68										
10	1.13	1.19	1.24	1.32	1.44	1.57	1.82									
15	1.06	1.08	1.10	1.15	1.21	1.29	1.42	1.78								
20	1.04	1.05	1.06	1.08	1.13	1.17	1.25	1.49	1.73							
25		1.04	1.04	1.06	1.08	1.11	1.17	1.33	1.51	1.72						
30				1.04	1.05	1.07	1.12	1.24	1.37	1.54	1.70	1.80				
35						1.06	1.09	1.17	1.28	1.42	1.54	1.64	1.78			
40							1.07	1.14	1.22	1.33	1.44	1.52	1.64	1.75		
50							1.05	1.09	1.15	1.22	1.29	1.35	1.44	1.53	1.64	1.83
60								1.06	1.11	1.16	1.21	1.25	1.31	1.39	1.46	1.62
70								1.05	1.08	1.12	1.17	1.19	1.24	1.30	1.35	1.47
80									1.06	1.09	1.13	1.15	1.19	1.24	1.28	1.38
90									1.04	1.06	1.10	1.13	1.16	1.19	1.23	1.32
100										1.05	1.08	1.10	1.13	1.15	1.19	1.23

Note: Common practice would normally limit the lower end of the table to 5% of additional heat and above 80% an engineer would likely opt for two (2) parallel cables.

EARTHWORK

R3123 Excavation & Fill

R312319-90 Wellpoints

A single stage wellpoint system is usually limited to dewatering an average 15' depth below normal ground water level. Multi-stage systems are employed for greater depth with the pumping equipment installed only at the lowest header level. Ejectors with unlimited lift capacity can be economical when two or more stages of wellpoints can be replaced or when horizontal clearance is restricted, such as in deep trenches or tunneling projects, and where low water flows are expected. Wellpoints are usually spaced on 2-1/2' to 10' centers along a header pipe. Wellpoint spacing, header size, and pump size are all determined by the expected flow as dictated by soil conditions.

In almost all soils encountered in wellpoint dewatering, the wellpoints may be jetted into place. Cemented soils and stiff clays may require sand wicks about 12" in diameter around each wellpoint to increase efficiency and eliminate weeping into the excavation. These sand wicks require 1/2 to 3 C.Y. of washed filter sand and are installed by using a 12" diameter steel casing and hole puncher jetted into the ground 2' deeper than the wellpoint. Rock may require predrilled holes.

Labor required for the complete installation and removal of a single stage wellpoint system is in the range of 3/4 to 2 labor-hours per linear foot of header, depending upon jetting conditions, wellpoint spacing, etc.

Continuous pumping is necessary except in some free draining soil where temporary flooding is permissible (as in trenches which are backfilled after each day's work). Good practice requires provision of a stand-by pump during the continuous pumping operation.

Systems for continuous trenching below the water table should be installed three to four times the length of expected daily progress to ensure uninterrupted digging, and header pipe size should not be changed during the job.

For pervious free draining soils, deep wells in place of wellpoints may be economical because of lower installation and maintenance costs. Daily production ranges between two to three wells per day, for 25' to 40' depths, to one well per day for depths over 50'.

Detailed analysis and estimating for any dewatering problem is available at no cost from wellpoint manufacturers. Major firms will quote "sufficient equipment" quotes or their affiliates offer lump sum proposals to cover complete dewatering responsibility.

Description for 200' System with 8" Header		Quantities
Equipment & Material	Wellpoints 25' long, 2" diameter @ 5' O.C.	40 Each
	Header pipe, 8" diameter	200 L.F.
	Discharge pipe, 8" diameter	100 L.F.
	8" valves	3 Each
	Combination jetting & wellpoint pump (standby)	1 Each
	Wellpoint pump, 8" diameter	1 Each
	Transportation to and from site	1 Day
	Fuel for 30 days x 60 gal./day	1800 Gallons
	Lubricants for 30 days x 16 lbs./day	480 Lbs.
	Sand for points	40 C.Y.
Labor	Technician to supervise installation	1 Week
	Labor for installation and removal of system	300 Labor-hours
	4 Operators straight time 40 hrs./wk. for 4.33 wks.	693 Hrs.
	4 Operators overtime 2 hrs./wk. for 4.33 wks.	35 Hrs.

Earthwork — R3141 Shoring

R314116-40 Wood Sheet Piling

Wood sheet piling may be used for depths to 20' where there is no ground water. If moderate ground water is encountered Tongue & Groove sheeting will help to keep it out. When considerable ground water is present, steel sheeting must be used.

For estimating purposes on trench excavation, sizes are as follows:

Depth	Sheeting	Wales	Braces	B.F. per S.F.
To 8'	3 x 12's	6 x 8's, 2 line	6 x 8's, @ 10'	4.0 @ 8'
8' x 12'	3 x 12's	10 x 10's, 2 line	10 x 10's, @ 9'	5.0 average
12' to 20'	3 x 12's	12 x 12's, 3 line	12 x 12's, @ 8'	7.0 average

Sheeting to be toed in at least 2' depending upon soil conditions. A five person crew with an air compressor and sheeting driver can drive and brace 440 SF/day at 8' deep, 360 SF/day at 12' deep, and 320 SF/day at 16' deep.

For normal soils, piling can be pulled in 1/3 the time to install. Pulling difficulty increases with the time in the ground. Production can be increased by high pressure jetting.

R314116-45 Steel Sheet Piling

Limiting weights are 22 to 38#/S.F. of wall surface with 27#/S.F. average for usual types and sizes. (Weights of piles themselves are from 30.7#/L.F. to 57#/L.F. but they are 15" to 21" wide.) Lightweight sections 12" to 28" wide from 3 ga. to 12 ga. thick are also available for shallow excavations. Piles may be driven two at a time with an impact or vibratory hammer (use vibratory to pull) hung from a crane without leads. A reasonable estimate of the life of steel sheet piling is 10 uses with up to 125 uses possible if a vibratory hammer is used. Used piling costs from 50% to 80% of new piling depending on location and market conditions. Sheet piling and H piles can be rented for about 30% of the delivered mill price for the first month and 5% per month thereafter. Allow 1 labor-hour per pile for cleaning and trimming after driving. These costs increase with depth and hydrostatic head. Vibratory drivers are faster in wet granular soils and are excellent for pile extraction. Pulling difficulty increases with the time in the ground and may cost more than driving. It is often economical to abandon the sheet piling, especially if it can be used as the outer wall form. Allow about 1/3 additional length or more for toeing into ground. Add bracing, waler and strut costs. Waler costs can equal the cost per ton of sheeting.

Utilities — R3311 Water Utility Distribution Piping

R331113-80 Piping Designations

There are several systems currently in use to describe pipe and fittings. The following paragraphs will help to identify and clarify classifications of piping systems used for water distribution.

Piping may be classified by schedule. Piping schedules include 5S, 10S, 10, 20, 30, Standard, 40, 60, Extra Strong, 80, 100, 120, 140, 160 and Double Extra Strong. These schedules are dependent upon the pipe wall thickness. The wall thickness of a particular schedule may vary with pipe size.

Ductile iron pipe for water distribution is classified by Pressure Classes such as Class 150, 200, 250, 300 and 350. These classes are actually the rated water working pressure of the pipe in pounds per square inch (psi). The pipe in these pressure classes is designed to withstand the rated water working pressure plus a surge allowance of 100 psi.

The American Water Works Association (AWWA) provides standards for various types of **plastic pipe**. C-900 is the specification for polyvinyl chloride (PVC) piping used for water distribution in sizes ranging from 4" through 12". C-901 is the specification for polyethylene (PE) pressure pipe, tubing and fittings used for water distribution in sizes ranging from 1/2" through 3". C-905 is the specification for PVC piping sizes 14" and greater.

PVC pressure-rated pipe is identified using the standard dimensional ratio (SDR) method. This method is defined by the American Society for Testing and Materials (ASTM) Standard D 2241. This pipe is available in SDR numbers 64, 41, 32.5, 26, 21, 17, and 13.5. Pipe with an SDR of 64 will have the thinnest wall while pipe with an SDR of 13.5 will have the thickest wall. When the pressure rating (PR) of a pipe is given in psi, it is based on a line supplying water at 73 degrees F.

The National Sanitation Foundation (NSF) seal of approval is applied to products that can be used with potable water. These products have been tested to ANSI/NSF Standard 14.

Valves and strainers are classified by American National Standards Institute (ANSI) Classes. These Classes are 125, 150, 200, 250, 300, 400, 600, 900, 1500 and 2500. Within each class there is an operating pressure range dependent upon temperature. Design parameters should be compared to the appropriate material dependent, pressure-temperature rating chart for accurate valve selection.

Change Orders

Change Order Considerations
A change order is a written document, usually prepared by the design professional, and signed by the owner, the architect/engineer, and the contractor. A change order states the agreement of the parties to: an addition, deletion, or revision in the work; an adjustment in the contract sum, if any; or an adjustment in the contract time, if any. Change orders, or "extras" in the construction process occur after execution of the construction contract and impact architects/engineers, contractors, and owners.

Change orders that are properly recognized and managed can ensure orderly, professional, and profitable progress for all who are involved in the project. There are many causes for change orders and change order requests. In all cases, change orders or change order requests should be addressed promptly and in a precise and prescribed manner. The following paragraphs include information regarding change order pricing and procedures.

The Causes of Change Orders
Reasons for issuing change orders include:
- Unforeseen field conditions that require a change in the work
- Correction of design discrepancies, errors, or omissions in the contract documents
- Owner-requested changes, either by design criteria, scope of work, or project objectives
- Completion date changes for reasons unrelated to the construction process
- Changes in building code interpretations, or other public authority requirements that require a change in the work
- Changes in availability of existing or new materials and products

Procedures
Properly written contract documents must include the correct change order procedures for all parties—owners, design professionals and contractors—to follow in order to avoid costly delays and litigation.

Being "in the right" is not always a sufficient or acceptable defense. The contract provisions requiring notification and documentation must be adhered to within a defined or reasonable time frame.

The appropriate method of handling change orders is by a written proposal and acceptance by all parties involved. Prior to starting work on a project, all parties should identify their authorized agents who may sign and accept change orders, as well as any limits placed on their authority.

Time may be a critical factor when the need for a change arises. For such cases, the contractor might be directed to proceed on a "time and materials" basis, rather than wait for all paperwork to be processed—a delay that could impede progress. In this situation, the contractor must still follow the prescribed change order procedures including, but not limited to, notification and documentation.

All forms used for change orders should be dated and signed by the proper authority. Lack of documentation can be very costly, especially if legal judgments are to be made, and if certain field personnel are no longer available. For time and material change orders, the contractor should keep accurate daily records of all labor and material allocated to the change.

Owners or awarding authorities who do considerable and continual building construction (such as the federal government) realize the inevitability of change orders for numerous reasons, both predictable and unpredictable. As a result, the federal government, the American Institute of Architects (AIA), the Engineers Joint Contract Documents Committee (EJCDC) and other contractor, legal, and technical organizations have developed standards and procedures to be followed by all parties to achieve contract continuance and timely completion, while being financially fair to all concerned.

In addition to the change order standards put forth by industry associations, there are also many books available on the subject.

Pricing Change Orders
When pricing change orders, regardless of their cause, the most significant factor is when the change occurs. The need for a change may be perceived in the field or requested by the architect/engineer *before* any of the actual installation has begun, or may evolve or appear *during* construction when the item of work in question is partially installed. In the latter cases, the original sequence of construction is disrupted, along with all contiguous and supporting systems. Change orders cause the greatest impact when they occur *after* the installation has been completed and must be uncovered, or even replaced. Post-completion changes may be caused by necessary design changes, product failure, or changes in the owner's requirements that are not discovered until the building or the systems begin to function.

Specified procedures of notification and record keeping must be adhered to and enforced regardless of the stage of construction: *before, during,* or *after* installation. Some bidding documents anticipate change orders by requiring that unit prices including overhead and profit percentages—for additional as well as deductible changes—be listed. Generally these unit prices do not fully take into account the ripple effect, or impact on other trades, and should be used for general guidance only.

When pricing change orders, it is important to classify the time frame in which the change occurs. There are two basic time frames for change orders: *pre-installation change orders*, which occur before the start of construction, and *post-installation change orders*, which involve reworking after the original installation. Change orders that occur between these stages may be priced according to the extent of work completed using a combination of techniques developed for pricing *pre-* and *post-installation* changes.

The following factors are the basis for a check list to use when preparing a change order estimate.

Factors To Consider When Pricing Change Orders

As an estimator begins to prepare a change order, the following questions should be reviewed to determine their impact on the final price.

General

- *Is the change order work* pre-installation *or* post-installation?

  Change order work costs vary according to how much of the installation has been completed. Once workers have the project scoped in their minds, even though they have not started, it can be difficult to refocus.

Consequently they may spend more than the normal amount of time understanding the change. Also, modifications to work in place, such as trimming or refitting, usually take more time than was initially estimated. The greater the amount of work in place, the more reluctant workers are to change it. Psychologically they may resent the change and as a result the rework takes longer than normal. Post-installation change order estimates must include demolition of existing work as required to accomplish the change. If the work is performed at a later time, additional obstacles, such as building finishes, may be present which must be protected. Regardless of whether the change occurs pre-installation or post-installation, attempt to isolate the identifiable factors and price them separately. For example, add shipping costs that may be required pre-installation or any demolition required post-installation. Then analyze the potential impact on productivity of psychological and/or learning curve factors and adjust the output rates accordingly. One approach is to break down the typical workday into segments and quantify the impact on each segment. The following chart may be useful as a guide:

Activities (Productivity) Expressed as Percentages of a Workday				
Task	Means Mechanical Cost Data (for New Construction)	Pre-Installation Change Orders	Post-Installation Change Orders
1. Study plans	3%	6%	6%
2. Material procurement	3%	3%	3%
3. Receiving and storing	3%	3%	3%
4. Mobilization	5%	5%	5%
5. Site movement	5%	5%	8%
6. Layout and marking	8%	10%	12%
7. Actual installation	64%	59%	54%
8. Clean-up	3%	3%	3%
9. Breaks—non-productive	6%	6%	6%
Total	100%	100%	100%

Change Order Installation Efficiency

The labor-hours expressed (for new construction) are based on average installation time, using an efficiency level of approximately 60%–65%. For change order situations, adjustments to this efficiency level should reflect the daily labor-hour allocation for that particular occurrence.

If any of the specific percentages expressed in the above chart do not apply to a particular project situation, then those percentage points should be reallocated to the appropriate task(s). Example: Using data for new construction, assume there is no new material being utilized. The percentages for Tasks 2 and 3 would therefore be reallocated to other tasks. If the time required for Tasks 2 and 3 can now be applied to installation, we can add the time allocated for *Material Procurement* and *Receiving and Storing* to the *Actual Installation* time for new construction, thereby increasing the Actual Installation percentage.

This chart shows that, due to reduced productivity, labor costs will be higher than those for new construction by 5%–15% for pre-installation change orders and by 15%–25% for post-installation change orders. Each job and change order is unique and must be examined individually. Many factors, covered elsewhere in this section, can each have a significant impact on productivity and change order costs. All such factors should be considered in every case.

- *Will the change substantially delay the original completion date?*

  A significant change in the project may cause the original completion date to be extended. The extended schedule may subject the contractor to new wage rates dictated by relevant labor contracts. Project supervision and other project overhead must also be extended beyond the original completion date. The schedule extension may also put installation into a new weather season. For example, underground piping scheduled for October installation was delayed until January. As a result, frost penetrated the trench area, thereby changing the degree of difficulty of the task. Changes and delays may have a ripple effect throughout the project. This effect must be analyzed and negotiated with the owner.

- *What is the net effect of a deduct change order?*

  In most cases, change orders resulting in a deduction or credit reflect only bare costs. The contractor may retain the overhead and profit based on the original bid.

Materials

- *Will you have to pay more or less for the new material, required by the change order, than you paid for the original purchase?*

  The same material prices or discounts will usually apply to materials purchased for change orders as new construction. In some instances, however, the contractor may forfeit the advantages of competitive pricing for change orders. Consider the following example:

  A contractor purchased over $20,000 worth of fan coil units for an installation, and obtained the maximum discount. Some time later it was determined the project required an additional matching unit. The contractor has to purchase this unit from the original supplier to ensure a match. The supplier at this time may not discount the unit because of the small quantity, and the fact that he is no longer in a competitive situation. The impact of quantity on purchase can add between 0% and 25% to material prices and/or subcontractor quotes.

- *If materials have been ordered or delivered to the job site, will they be subject to a cancellation charge or restocking fee?*

  Check with the supplier to determine if ordered materials are subject to a cancellation charge. Delivered materials not used as result of a change order may be subject to a restocking fee if returned to the supplier. Common restocking charges run between 20% and 40%. Also, delivery charges to return the goods to the supplier must be added.

Labor

- *How efficient is the existing crew at the actual installation?*

  Is the same crew that performed the initial work going to do the change order? Possibly the change consists of the installation of a unit identical to one already installed; therefore, the change should take less time. Be sure to consider this potential productivity increase and modify the productivity rates accordingly.

- *If the crew size is increased, what impact will that have on supervision requirements?*

  Under most bargaining agreements or management practices, there is a point at which a working foreman is replaced by a nonworking foreman. This replacement increases project overhead by adding a nonproductive worker. If additional workers are added to accelerate the project or to perform changes while maintaining the schedule, be sure to add additional supervision time if warranted. Calculate the hours involved and the additional cost directly if possible.

- *What are the other impacts of increased crew size?*

  The larger the crew, the greater the potential for productivity to decrease. Some of the factors that cause this productivity loss are: overcrowding (producing restrictive conditions in the working space), and possibly a shortage of any special tools and equipment required. Such factors affect not only the crew working on the elements directly involved in the change order, but other crews whose movement may also be hampered.

  As the crew increases, check its basic composition for changes by the addition or deletion of apprentices or nonworking foreman, and quantify the potential effects of equipment shortages or other logistical factors.

- *As new crews, unfamiliar with the project, are brought onto the site, how long will it take them to become oriented to the project requirements?*

  The orientation time for a new crew to become 100% effective varies with the site and type of project. Orientation is easiest at a new construction site, and most difficult at existing, very restrictive renovation sites. The type of work also affects orientation time. When all elements of the work are exposed, such as concrete or masonry work, orientation is decreased. When the work is concealed or less visible, such as existing electrical systems, orientation takes longer. Usually orientation can be accomplished in one day or less. Costs for added orientation should be itemized and added to the total estimated cost.

- *How much actual production can be gained by working overtime?*

  Short term overtime can be used effectively to accomplish more work in a day. However, as overtime is scheduled to run beyond several weeks, studies have shown marked decreases in output. The following chart shows the effect of long term overtime on worker efficiency. If the anticipated change requires extended overtime to keep the job on schedule, these factors can be used as a guide to predict the impact on time and cost. Add project overhead, particularly supervision, that may also be incurred.

| Days per Week | Hours per Day | Production Efficiency ||||| Payroll Cost Factors ||
		1 Week	2 Weeks	3 Weeks	4 Weeks	Average 4 Weeks	@ 1-1/2 Times	@ 2 Times
5	8	100%	100%	100%	100%	100%	100%	100%
	9	100	100	95	90	96.25	105.6	111.1
	10	100	95	90	85	91.25	110.0	120.0
	11	95	90	75	65	81.25	113.6	127.3
	12	90	85	70	60	76.25	116.7	133.3
6	8	100	100	95	90	96.25	108.3	116.7
	9	100	95	90	85	92.50	113.0	125.9
	10	95	90	85	80	87.50	116.7	133.3
	11	95	85	70	65	78.75	119.7	139.4
	12	90	80	65	60	73.75	122.2	144.4
7	8	100	95	85	75	88.75	114.3	128.6
	9	95	90	80	70	83.75	118.3	136.5
	10	90	85	75	65	78.75	121.4	142.9
	11	85	80	65	60	72.50	124.0	148.1
	12	85	75	60	55	68.75	126.2	152.4

Effects of Overtime

Caution: Under many labor agreements, Sundays and holidays are paid at a higher premium than the normal overtime rate.

The use of long-term overtime is counterproductive on almost any construction job; that is, the longer the period of overtime, the lower the actual production rate. Numerous studies have been conducted, and while they have resulted in slightly different numbers, all reach the same conclusion. The figure above tabulates the effects of overtime work on efficiency.

As illustrated, there can be a difference between the *actual* payroll cost per hour and the *effective* cost per hour for overtime work. This is due to the reduced production efficiency with the increase in weekly hours beyond 40. This difference between actual and effective cost results from overtime work over a prolonged period. Short-term overtime work does not result in as great a reduction in efficiency and, in such cases, effective cost may not vary significantly from the actual payroll cost. As the total hours per week are increased on a regular basis, more time is lost due to fatigue, lowered morale, and an increased accident rate.

As an example, assume a project where workers are working 6 days a week, 10 hours per day. From the figure above (based on productivity studies), the average effective productive hours over a 4-week period are:

$$0.875 \times 60 = 52.5$$

Depending upon the locale and day of week, overtime hours may be paid at time and a half or double time. For time and a half, the overall (average) *actual* payroll cost (including regular and overtime hours) is determined as follows:

$$\frac{40 \text{ reg. hrs.} + (20 \text{ overtime hrs.} \times 1.5)}{60 \text{ hrs.}} = 1.167$$

Based on 60 hours, the payroll cost per hour will be 116.7% of the normal rate at 40 hours per week. However, because the effective production (efficiency) for 60 hours is reduced to the equivalent of 52.5 hours, the effective cost of overtime is calculated as follows:

For time and a half:

$$\frac{40 \text{ reg. hrs.} + (20 \text{ overtime hrs.} \times 1.5)}{52.5 \text{ hrs.}} = 1.33$$

Installed cost will be 133% of the normal rate (for labor).

Thus, when figuring overtime, the actual cost per unit of work will be higher than the apparent overtime payroll dollar increase, due to the reduced productivity of the longer workweek. These efficiency calculations are true only for those cost factors determined by hours worked. Costs that are applied weekly or monthly, such as equipment rentals, will not be similarly affected.

Equipment

- *What equipment is required to complete the change order?*

Change orders may require extending the rental period of equipment already on the job site, or the addition of special equipment brought in to accomplish the change work. In either case, the additional rental charges and operator labor charges must be added.

Summary

The preceding considerations and others you deem appropriate should be analyzed and applied to a change order estimate. The impact of each should be quantified and listed on the estimate to form an audit trail.

Change orders that are properly identified, documented, and managed help to ensure the orderly, professional and profitable progress of the work. They also minimize potential claims or disputes at the end of the project.

Square Foot Costs

Estimating Tips

- The cost figures in this Square Foot Cost section were derived from approximately 11,000 projects contained in the RSMeans database of completed construction projects. They include the contractor's overhead and profit, but do not generally include architectural fees or land costs. The figures have been adjusted to January of the current year. New projects are added to our files each year, and outdated projects are discarded. For this reason, certain costs may not show a uniform annual progression. In no case are all subdivisions of a project listed.

- These projects were located throughout the U.S. and reflect a tremendous variation in square foot (S.F.) and cubic foot (C.F.) costs. This is due to differences, not only in labor and material costs, but also in individual owners' requirements. For instance, a bank in a large city would have different features than one in a rural area. This is true of all the different types of buildings analyzed. Therefore, caution should be exercised when using these square foot costs. For example, for courthouses, costs in the database are local courthouse costs and will not apply to the larger, more elaborate federal courthouses. As a general rule, the projects in the 1/4 column do not include any site work or equipment, while the projects in the 3/4 column may include both equipment and site work. The median figures do not generally include site work.

- None of the figures "go with" any others. All individual cost items were computed and tabulated separately. Thus, the sum of the median figures for plumbing, HVAC and electrical will not normally total up to the total mechanical and electrical costs arrived at by separate analysis and tabulation of the projects.

- Each building was analyzed as to total and component costs and percentages. The figures were arranged in ascending order with the results tabulated as shown. The 1/4 column shows that 25% of the projects had lower costs and 75% had higher. The 3/4 column shows that 75% of the projects had lower costs and 25% had higher. The median column shows that 50% of the projects had lower costs and 50% had higher.

- There are two times when square foot costs are useful. The first is in the conceptual stage when no details are available. Then square foot costs make a useful starting point. The second is after the bids are in and the costs can be worked back into their appropriate categories for information purposes. As soon as details become available in the project design, the square foot approach should be discontinued and the project priced as to its particular components. When more precision is required, or for estimating the replacement cost of specific buildings, the current edition of *RSMeans Square Foot Costs* should be used.

- In using the figures in this section, it is recommended that the median column be used for preliminary figures if no additional information is available. The median figures, when multiplied by the total city construction cost index figures (see city cost indexes) and then multiplied by the project size modifier at the end of this section, should present a fairly accurate base figure, which would then have to be adjusted in view of the estimator's experience, local economic conditions, code requirements, and the owner's particular requirements. There is no need to factor the percentage figures, as these should remain constant from city to city. All tabulations mentioning air conditioning had at least partial air conditioning.

- The editors of this book would greatly appreciate receiving cost figures on one or more of your recent projects, which would then be included in the averages for next year. All cost figures received will be kept confidential, except that they will be averaged with other similar projects to arrive at square foot cost figures for next year's book. See the last page of the book for details and the discount available for submitting one or more of your projects.

50 17 | Square Foot Costs

50 17 00 | S.F. Costs

			UNIT	UNIT COSTS 1/4	UNIT COSTS MEDIAN	UNIT COSTS 3/4	% OF TOTAL 1/4	% OF TOTAL MEDIAN	% OF TOTAL 3/4	
01	0010	APARTMENTS Low Rise (1 to 3 story)	S.F.	73	92.50	123				01
	0020	Total project cost	C.F.	6.55	8.70	10.75				
	0100	Site work	S.F.	5.35	8.55	15	6.05%	10.55%	13.95%	
	0500	Masonry		1.44	3.55	5.80	1.54%	3.92%	6.50%	
	1500	Finishes		7.75	10.65	13.15	9.05%	10.75%	12.85%	
	1800	Equipment		2.40	3.63	5.40	2.71%	3.99%	5.95%	
	2720	Plumbing		5.70	7.30	9.30	6.65%	8.95%	10.05%	
	2770	Heating, ventilating, air conditioning		3.63	4.47	6.55	4.20%	5.60%	7.60%	
	2900	Electrical		4.25	5.65	7.65	5.20%	6.65%	8.35%	
	3100	Total: Mechanical & Electrical		15.10	19.60	24	16.05%	18.20%	23%	
	9000	Per apartment unit, total cost	Apt.	68,000	104,000	153,500				
	9500	Total: Mechanical & Electrical	"	12,900	20,300	26,500				
02	0010	APARTMENTS Mid Rise (4 to 7 story)	S.F.	97	117	144				02
	0020	Total project costs	C.F.	7.55	10.45	14.30				
	0100	Site work	S.F.	3.88	7.85	15.15	5.25%	6.70%	9.20%	
	0500	Masonry		4.05	8.90	12.15	5.10%	7.25%	10.50%	
	1500	Finishes		12.70	17	20	10.70%	13.50%	17.70%	
	1800	Equipment		2.71	4.22	5.55	2.54%	3.47%	4.31%	
	2500	Conveying equipment		2.20	2.70	3.26	2.05%	2.27%	2.69%	
	2720	Plumbing		5.70	9.10	9.65	5.70%	7.20%	8.95%	
	2900	Electrical		6.40	8.70	10.55	6.35%	7.20%	8.95%	
	3100	Total: Mechanical & Electrical		20.50	25.50	31	18.25%	21%	23%	
	9000	Per apartment unit, total cost	Apt.	109,500	129,500	214,000				
	9500	Total: Mechanical & Electrical	"	20,700	24,000	25,100				
03	0010	APARTMENTS High Rise (8 to 24 story)	S.F.	110	127	152				03
	0020	Total project costs	C.F.	10.70	12.45	14.65				
	0100	Site work	S.F.	3.99	6.45	9	2.58%	4.84%	6.15%	
	0500	Masonry		6.35	11.55	14.40	4.74%	9.65%	11.05%	
	1500	Finishes		12.20	15.25	18	9.75%	11.80%	13.70%	
	1800	Equipment		3.54	4.35	5.75	2.78%	3.49%	4.35%	
	2500	Conveying equipment		2.50	3.79	5.15	2.23%	2.78%	3.37%	
	2720	Plumbing		7.05	9.55	11.70	6.80%	7.20%	10.45%	
	2900	Electrical		7.55	9.55	12.90	6.45%	7.65%	8.80%	
	3100	Total: Mechanical & Electrical		22.50	29	35	17.95%	22.50%	24.50%	
	9000	Per apartment unit, total cost	Apt.	114,000	126,000	174,500				
	9500	Total: Mechanical & Electrical	"	22,700	28,200	29,900				
04	0010	AUDITORIUMS	S.F.	115	158	230				04
	0020	Total project costs	C.F.	7.15	9.95	14.30				
	2720	Plumbing	S.F.	6.80	10	11.95	5.85%	7.20%	8.70%	
	2900	Electrical		9.20	13.30	22	6.85%	9.50%	11.40%	
	3100	Total: Mechanical & Electrical		61	81.50	99	24.50%	27.50%	31%	
05	0010	AUTOMOTIVE SALES	S.F.	84.50	115	143				05
	0020	Total project costs	C.F.	5.55	6.70	8.65				
	2720	Plumbing	S.F.	3.86	6.70	7.30	2.89%	6.05%	6.50%	
	2770	Heating, ventilating, air conditioning		5.95	9.10	9.80	4.61%	10%	10.35%	
	2900	Electrical		6.80	10.70	15.70	7.25%	8.80%	12.15%	
	3100	Total: Mechanical & Electrical		19	30.50	36.50	17.30%	20.50%	22%	
06	0010	BANKS	S.F.	166	207	263				06
	0020	Total project costs	C.F.	11.85	16.10	21				
	0100	Site work	S.F.	19	30	42.50	7.90%	12.95%	17%	
	0500	Masonry		7.55	16.15	29	3.36%	6.90%	10.05%	
	1500	Finishes		14.85	22.50	28	5.85%	8.65%	11.70%	
	1800	Equipment		6.25	13.75	28	1.34%	5.55%	10.50%	
	2720	Plumbing		5.20	7.45	10.85	2.82%	3.90%	4.93%	
	2770	Heating, ventilating, air conditioning		9.85	13.20	17.60	4.86%	7.15%	8.50%	
	2900	Electrical		15.75	21	27.50	8.25%	10.20%	12.20%	
	3100	Total: Mechanical & Electrical		38	50.50	61	16.55%	19.45%	24%	
	3500	See also division 11020 & 11030 (MF2004 11 16 00 & 11 17 00)								

50 17 | Square Foot Costs

50 17 00 | S.F. Costs

			UNIT	UNIT COSTS 1/4	MEDIAN	3/4	% OF TOTAL 1/4	MEDIAN	3/4
13	0010	**CHURCHES**	S.F.	112	143	185			
	0020	Total project costs	C.F.	6.90	8.75	11.50			
	1800	Equipment	S.F.	1.21	3.17	6.80	.93%	2.11%	4.31%
	2720	Plumbing		4.35	6.10	8.95	3.51%	4.96%	6.25%
	2770	Heating, ventilating, air conditioning		10.15	13.25	18.75	7.50%	10%	12%
	2900	Electrical		9.55	12.95	17.65	7.30%	8.80%	10.90%
	3100	Total: Mechanical & Electrical		29.50	39	51.50	18.30%	22%	24.50%
	3500	See also division 11040 (MF2004 11 91 00)							
15	0010	**CLUBS, COUNTRY**	S.F.	120	144	182			
	0020	Total project costs	C.F.	9.65	11.75	16.25			
	2720	Plumbing	S.F.	7.25	10.75	24.50	5.60%	7.90%	10%
	2900	Electrical		9.40	12.90	16.80	7%	8.95%	11%
	3100	Total: Mechanical & Electrical		50	62.50	66	19%	26.50%	29.50%
17	0010	**CLUBS, SOCIAL Fraternal**	S.F.	101	138	184			
	0020	Total project costs	C.F.	5.95	9.05	10.80			
	2720	Plumbing	S.F.	6	7.50	11.30	5.60%	6.90%	8.55%
	2770	Heating, ventilating, air conditioning		7.05	10.50	13.45	8.20%	9.25%	14.40%
	2900	Electrical		6.85	11.80	13.55	5.95%	9.30%	10.55%
	3100	Total: Mechanical & Electrical		37	40	51	21%	23%	23.50%
18	0010	**CLUBS, Y.M.C.A.**	S.F.	132	164	247			
	0020	Total project costs	C.F.	5.55	9.25	13.80			
	2720	Plumbing	S.F.	7.60	15.10	16.95	5.65%	7.60%	10.85%
	2900	Electrical		9.85	12.75	22	6.45%	8%	9.20%
	3100	Total: Mechanical & Electrical		40	45.50	74.50	20.50%	21.50%	28.50%
19	0010	**COLLEGES Classrooms & Administration**	S.F.	116	164	212			
	0020	Total project costs	C.F.	6.95	12.35	19.85			
	0500	Masonry	S.F.	9.15	16.85	20.50	5.10%	8.05%	10.50%
	2720	Plumbing		5.20	12.65	23.50	5.10%	6.60%	8.95%
	2900	Electrical		9.85	15.55	21.50	7.70%	9.85%	12%
	3100	Total: Mechanical & Electrical		41.50	57	76	23%	28%	31.50%
21	0010	**COLLEGES Science, Engineering, Laboratories**	S.F.	225	264	305			
	0020	Total project costs	C.F.	12.95	18.85	21.50			
	1800	Equipment	S.F.	6.10	28.50	31	2%	6.45%	12.65%
	2900	Electrical		18.60	26.50	40.50	7.10%	9.40%	12.10%
	3100	Total: Mechanical & Electrical		69	82	127	28.50%	31.50%	41%
	3500	See also division 11600 (MF2004 11 53 00)							
23	0010	**COLLEGES Student Unions**	S.F.	144	195	236			
	0020	Total project costs	C.F.	8	10.50	13			
	3100	Total: Mechanical & Electrical	S.F.	54	58.50	69	23.50%	26%	29%
25	0010	**COMMUNITY CENTERS**	S.F.	120	147	199			
	0020	Total project costs	C.F.	7.75	11.05	14.35			
	1800	Equipment	S.F.	2.40	4.70	7.75	1.47%	3.01%	5.30%
	2720	Plumbing		5.60	9.85	13.45	4.85%	7%	8.95%
	2770	Heating, ventilating, air conditioning		8.45	13.10	18.80	6.80%	10.35%	12.90%
	2900	Electrical		10.05	13.25	19.50	7.15%	8.90%	10.40%
	3100	Total: Mechanical & Electrical		33	42	60	18.80%	23%	30%
28	0010	**COURT HOUSES**	S.F.	171	202	278			
	0020	Total project costs	C.F.	13.15	15.70	19.80			
	2720	Plumbing	S.F.	8.15	11.35	12.90	5.95%	7.45%	8.20%
	2900	Electrical		18.20	20.50	30	9.05%	10.65%	12.15%
	3100	Total: Mechanical & Electrical		51	66.50	73	22.50%	26.50%	30%
30	0010	**DEPARTMENT STORES**	S.F.	63.50	85.50	108			
	0020	Total project costs	C.F.	3.39	4.40	5.95			
	2720	Plumbing	S.F.	1.97	2.49	3.78	1.82%	4.21%	5.90%
	2770	Heating, ventilating, air conditioning		4.79	8.90	13.35	8.20%	9.10%	14.80%

50 17 | Square Foot Costs

50 17 00 | S.F. Costs

			UNIT	UNIT COSTS 1/4	MEDIAN	3/4	% OF TOTAL 1/4	MEDIAN	3/4	
30	2900	Electrical	S.F.	7.25	9.95	11.75	9.05%	12.15%	14.95%	30
	3100	Total: Mechanical & Electrical		12.80	16.35	28.50	13.20%	21.50%	50%	
31	0010	**DORMITORIES Low Rise (1 to 3 story)**	S.F.	120	166	208				31
	0020	Total project costs	C.F.	6.80	10.95	16.45				
	2720	Plumbing	S.F.	7.20	9.65	12.20	8.05%	9%	9.65%	
	2770	Heating, ventilating, air conditioning		7.65	9.15	12.20	4.61%	8.05%	10%	
	2900	Electrical		7.95	12.10	16.55	6.40%	8.65%	9.50%	
	3100	Total: Mechanical & Electrical		41.50	44.50	69	22%	25%	27%	
	9000	Per bed, total cost	Bed	51,000	56,500	121,000				
32	0010	**DORMITORIES Mid Rise (4 to 8 story)**	S.F.	147	192	237				32
	0020	Total project costs	C.F.	16.25	17.85	21.50				
	2900	Electrical	S.F.	15.65	17.80	24	8.20%	10.20%	11.95%	
	3100	Total: Mechanical & Electrical	"	44	87.50	88.50	25%	30.50%	35.50%	
	9000	Per bed, total cost	Bed	21,000	47,900	277,000				
34	0010	**FACTORIES**	S.F.	55.50	83	128				34
	0020	Total project costs	C.F.	3.58	5.35	8.85				
	0100	Site work	S.F.	6.35	11.60	18.40	6.95%	11.45%	17.95%	
	2720	Plumbing		3.01	5.60	9.25	3.73%	6.05%	8.10%	
	2770	Heating, ventilating, air conditioning		5.85	8.40	11.35	5.25%	8.45%	11.35%	
	2900	Electrical		6.95	10.95	16.75	8.10%	10.50%	14.20%	
	3100	Total: Mechanical & Electrical		19.80	30.50	40	21%	28.50%	35.50%	
36	0010	**FIRE STATIONS**	S.F.	111	153	209				36
	0020	Total project costs	C.F.	6.45	8.90	11.85				
	0500	Masonry	S.F.	15.10	29	37.50	7.50%	10.95%	15.55%	
	1140	Roofing		3.61	9.75	11.10	1.90%	4.94%	5.05%	
	1580	Painting		2.80	4.18	4.28	1.37%	1.57%	2.07%	
	1800	Equipment		1.38	2.64	4.87	.62%	1.63%	3.42%	
	2720	Plumbing		6.20	9.85	13.95	5.85%	7.35%	9.45%	
	2770	Heating, ventilating, air conditioning		6.15	9.95	15.35	5.15%	7.40%	9.40%	
	2900	Electrical		7.95	14.10	19	6.90%	8.60%	10.60%	
	3100	Total: Mechanical & Electrical		40.50	52	59	18.40%	23%	27%	
37	0010	**FRATERNITY HOUSES & Sorority Houses**	S.F.	111	142	195				37
	0020	Total project costs	C.F.	11	11.50	13.85				
	2720	Plumbing	S.F.	8.35	9.60	17.55	6.80%	8%	10.85%	
	2900	Electrical		7.30	15.80	19.35	6.60%	9.90%	10.65%	
38	0010	**FUNERAL HOMES**	S.F.	117	159	289				38
	0020	Total project costs	C.F.	11.90	13.25	25.50				
	2900	Electrical	S.F.	5.15	9.45	10.35	3.58%	4.44%	5.95%	
39	0010	**GARAGES, COMMERCIAL (Service)**	S.F.	66	102	141				39
	0020	Total project costs	C.F.	4.35	6.40	9.35				
	1800	Equipment	S.F.	3.73	8.40	13.05	2.21%	4.62%	6.80%	
	2720	Plumbing		4.58	7.05	12.85	5.45%	7.85%	10.65%	
	2730	Heating & ventilating		4.51	7.80	9.95	5.25%	6.85%	8.20%	
	2900	Electrical		6.30	9.60	13.80	7.15%	9.25%	10.85%	
	3100	Total: Mechanical & Electrical		11.85	26.50	39	12.35%	17.40%	26%	
40	0010	**GARAGES, MUNICIPAL (Repair)**	S.F.	97	129	182				40
	0020	Total project costs	C.F.	6.05	7.70	13.25				
	0500	Masonry	S.F.	5.10	17.85	27.50	4.03%	9.15%	12.50%	
	2720	Plumbing		4.36	8.35	15.75	3.59%	6.70%	7.95%	
	2730	Heating & ventilating		7.45	10.80	21	6.15%	7.45%	13.50%	
	2900	Electrical		7.15	11.30	16.40	6.65%	8.15%	11.15%	
	3100	Total: Mechanical & Electrical		33	46	66.50	21.50%	23%	28.50%	
41	0010	**GARAGES, PARKING**	S.F.	38.50	55	95				41
	0020	Total project costs	C.F.	3.55	4.82	7				

50 17 | Square Foot Costs

50 17 00 | S.F. Costs

				UNIT COSTS			% OF TOTAL		
			UNIT	1/4	MEDIAN	3/4	1/4	MEDIAN	3/4
41	2720	Plumbing	S.F.	.65	1.66	2.56	1.72%	2.70%	3.85%
	2900	Electrical		2.08	2.74	4	4.52%	5.40%	6.35%
	3100	Total: Mechanical & Electrical	↓	4.03	5.90	7.35	7%	8.90%	11.05%
	3200								
	9000	Per car, total cost	Car	16,000	20,000	25,500			
43	0010	**GYMNASIUMS**	S.F.	105	140	193			
	0020	Total project costs	C.F.	5.25	7.10	8.75			
	1800	Equipment	S.F.	2.50	4.68	8.45	1.76%	3.26%	6.70%
	2720	Plumbing		5.75	7.75	10.20	4.65%	6.40%	7.75%
	2770	Heating, ventilating, air conditioning		6.25	10.90	22	5.15%	9.05%	11.10%
	2900	Electrical		8.35	11	14.65	6.75%	8.50%	10.30%
	3100	Total: Mechanical & Electrical	↓	28.50	40	48.50	19.75%	23.50%	29%
	3500	See also division 11480 (MF2004 11 67 00)							
46	0010	**HOSPITALS**	S.F.	202	253	350			
	0020	Total project costs	C.F.	15.30	19.05	27			
	1800	Equipment	S.F.	5.10	9.85	17	.80%	2.53%	4.80%
	2720	Plumbing		17.40	24.50	31.50	7.60%	9.10%	10.85%
	2770	Heating, ventilating, air conditioning		25.50	32.50	45.50	7.80%	12.95%	16.65%
	2900	Electrical		22.50	30	44.50	10%	11.75%	14.10%
	3100	Total: Mechanical & Electrical	↓	65	90.50	134	28%	33.50%	37%
	9000	Per bed or person, total cost	Bed	233,500	322,000	371,000			
	9900	See also division 11700 (MF2004 11 71 00)							
48	0010	**HOUSING For the Elderly**	S.F.	99	126	154			
	0020	Total project costs	C.F.	7.10	9.85	12.55			
	0100	Site work	S.F.	6.90	10.75	15.75	5.05%	7.90%	12.10%
	0500	Masonry		1.87	11.30	16.50	1.30%	6.05%	11%
	1800	Equipment		2.40	3.30	5.25	1.88%	3.23%	4.43%
	2510	Conveying systems		2.24	3.24	4.40	1.78%	2.20%	2.81%
	2720	Plumbing		7.40	9.40	11.85	8.15%	9.55%	10.50%
	2730	Heating, ventilating, air conditioning		3.78	5.35	8	3.30%	5.60%	7.25%
	2900	Electrical		7.40	10.05	12.85	7.30%	8.50%	10.25%
	3100	Total: Mechanical & Electrical	↓	25.50	31.50	40.50	18.10%	22.50%	29%
	9000	Per rental unit, total cost	Unit	92,500	108,000	120,000			
	9500	Total: Mechanical & Electrical	"	20,600	23,600	27,600			
50	0010	**HOUSING Public (Low Rise)**	S.F.	83.50	116	151			
	0020	Total project costs	C.F.	6.60	9.25	11.50			
	0100	Site work	S.F.	10.60	15.30	25	8.35%	11.75%	16.50%
	1800	Equipment		2.27	3.70	5.65	2.26%	3.03%	4.24%
	2720	Plumbing		6	7.95	10.05	7.15%	9.05%	11.60%
	2730	Heating, ventilating, air conditioning		3.02	5.85	6.40	4.26%	6.05%	6.45%
	2900	Electrical		5.05	7.50	10.45	5.10%	6.55%	8.25%
	3100	Total: Mechanical & Electrical	↓	24	31	34.50	14.50%	17.55%	26.50%
	9000	Per apartment, total cost	Apt.	91,500	104,000	131,000			
	9500	Total: Mechanical & Electrical	"	19,500	24,100	26,700			
51	0010	**ICE SKATING RINKS**	S.F.	71.50	167	183			
	0020	Total project costs	C.F.	5.25	5.35	6.20			
	2720	Plumbing	S.F.	2.66	5	5.10	3.12%	3.23%	5.65%
	2900	Electrical	↓	7.65	11.70	12.40	6.30%	10.15%	15.05%
52	0010	**JAILS**	S.F.	185	280	360			
	0020	Total project costs	C.F.	19.60	27.50	32			
	1800	Equipment	S.F.	8.45	25	42.50	2.80%	6.95%	9.80%
	2720	Plumbing		21	28	37	7%	8.90%	13.35%
	2770	Heating, ventilating, air conditioning		19.60	26	50.50	7.50%	9.45%	17.75%
	2900	Electrical		23	31	39	9.40%	11.70%	15.25%
	3100	Total: Mechanical & Electrical	↓	61	108	128	28%	30%	34%
53	0010	**LIBRARIES**	S.F.	140	186	243			
	0020	Total project costs	C.F.	9.40	11.75	15			

689

50 17 | Square Foot Costs

50 17 00	S.F. Costs	UNIT	UNIT COSTS 1/4	MEDIAN	3/4	% OF TOTAL 1/4	MEDIAN	3/4
0500	Masonry	S.F.	9.30	19	31.50	5.60%	7.15%	10.95%
1800	Equipment		1.88	5.05	7.60	.28%	1.39%	4.07%
2720	Plumbing		5.05	7.20	10	3.38%	4.60%	5.70%
2770	Heating, ventilating, air conditioning		11.20	19	23	7.80%	10.95%	12.80%
2900	Electrical		14.10	18.65	25	8.40%	10.55%	12%
3100	Total: Mechanical & Electrical		46	56	65.50	21%	24%	28%
0010	**LIVING, ASSISTED**	S.F.	127	151	176			
0020	Total project costs	C.F.	10.70	12.50	14.20			
0500	Masonry	S.F.	3.68	4.44	5.45	2.36%	3.16%	3.86%
1800	Equipment		2.81	3.35	4.30	2.12%	2.45%	2.87%
2720	Plumbing		10.65	14.25	14.75	6.05%	8.15%	10.60%
2770	Heating, ventilating, air conditioning		12.65	13.20	14.45	7.95%	9.35%	9.70%
2900	Electrical		12.45	14	16.15	9%	9.95%	10.65%
3100	Total: Mechanical & Electrical		34	40.50	47	24%	28.50%	31.50%
0010	**MEDICAL CLINICS**	S.F.	129	159	203			
0020	Total project costs	C.F.	9.45	12.25	16.30			
1800	Equipment	S.F.	1.65	7.35	11.40	1.05%	2.94%	6.35%
2720	Plumbing		8.60	12.10	16.15	6.15%	8.40%	10.10%
2770	Heating, ventilating, air conditioning		10.25	13.40	19.70	6.65%	8.85%	11.35%
2900	Electrical		11.10	15.75	20.50	8.10%	10%	12.20%
3100	Total: Mechanical & Electrical		35.50	48	66	22%	27%	33.50%
3500	See also division 11700 (MF2004 11 71 00)							
0010	**MEDICAL OFFICES**	S.F.	121	151	185			
0020	Total project costs	C.F.	9.05	12.30	16.60			
1800	Equipment	S.F.	4.01	7.95	11.15	.66%	4.87%	6.50%
2720	Plumbing		6.70	10.35	13.95	5.60%	6.80%	8.50%
2770	Heating, ventilating, air conditioning		8.10	11.70	15.45	6.10%	8%	9.70%
2900	Electrical		9.70	14.20	19.75	7.50%	9.80%	11.70%
3100	Total: Mechanical & Electrical		26.50	38.50	55	19.30%	23%	28.50%
0010	**MOTELS**	S.F.	76.50	111	145			
0020	Total project costs	C.F.	6.80	9.15	14.95			
2720	Plumbing	S.F.	7.75	9.90	11.80	9.45%	10.60%	12.55%
2770	Heating, ventilating, air conditioning		4.74	7.05	12.65	5.60%	5.60%	10%
2900	Electrical		7.25	9.15	11.40	7.45%	9.05%	10.45%
3100	Total: Mechanical & Electrical		21.50	31	53	18.50%	24%	25.50%
5000								
9000	Per rental unit, total cost	Unit	39,000	74,000	80,000			
9500	Total: Mechanical & Electrical	"	7,600	11,500	13,400			
0010	**NURSING HOMES**	S.F.	120	155	193			
0020	Total project costs	C.F.	9.45	11.80	16.15			
1800	Equipment	S.F.	3.28	5	8.40	2%	3.62%	4.99%
2720	Plumbing		10.30	15.60	18.80	8.75%	10.10%	12.70%
2770	Heating, ventilating, air conditioning		10.85	16.45	22	9.70%	11.45%	11.80%
2900	Electrical		11.90	14.95	20.50	9.40%	10.60%	12.50%
3100	Total: Mechanical & Electrical		28.50	39.50	66.50	26%	28%	30%
9000	Per bed or person, total cost	Bed	53,500	67,000	86,000			
0010	**OFFICES Low Rise (1 to 4 story)**	S.F.	101	132	172			
0020	Total project costs	C.F.	7.25	9.95	13.10			
0100	Site work	S.F.	8.15	14.20	21	5.95%	9.70%	13.55%
0500	Masonry		3.92	7.90	14.40	2.61%	5.40%	8.45%
1800	Equipment		.91	2.11	5.75	.57%	1.50%	3.42%
2720	Plumbing		3.62	5.60	8.20	3.66%	4.50%	6.10%
2770	Heating, ventilating, air conditioning		8	11.15	16.35	7.20%	10.30%	11.70%
2900	Electrical		8.30	11.85	16.85	7.45%	9.65%	11.40%
3100	Total: Mechanical & Electrical		23	31.50	48	18.20%	22%	27%
0010	**OFFICES Mid Rise (5 to 10 story)**	S.F.	107	130	176			
0020	Total project costs	C.F.	7.60	9.70	13.75			

50 17 | Square Foot Costs

50 17 00 | S.F. Costs

			UNIT	UNIT COSTS 1/4	UNIT COSTS MEDIAN	UNIT COSTS 3/4	% OF TOTAL 1/4	% OF TOTAL MEDIAN	% OF TOTAL 3/4	
62	2720	Plumbing	S.F.	3.24	5	7.20	2.83%	3.74%	4.50%	62
	2770	Heating, ventilating, air conditioning		8.15	11.65	18.60	7.65%	9.40%	11%	
	2900	Electrical		7.95	10.20	15.40	6.35%	7.80%	10%	
	3100	Total: Mechanical & Electrical		20.50	26.50	50.50	18.95%	21%	27.50%	
63	0010	**OFFICES High Rise (11 to 20 story)**	S.F.	132	166	204				63
	0020	Total project costs	C.F.	9.20	11.50	16.50				
	2900	Electrical	S.F.	8	9.75	14.50	5.80%	7.85%	10.50%	
	3100	Total: Mechanical & Electrical		26	34.50	58.50	16.90%	23.50%	34%	
64	0010	**POLICE STATIONS**	S.F.	158	208	265				64
	0020	Total project costs	C.F.	12.60	15.40	21				
	0500	Masonry	S.F.	15.80	26	33	7.80%	9.10%	11.35%	
	1800	Equipment		2.27	11.05	17.50	.98%	3.35%	6.70%	
	2720	Plumbing		8.85	17.60	22	5.65%	6.90%	10.75%	
	2770	Heating, ventilating, air conditioning		12.95	18.35	26	5.85%	10.55%	11.70%	
	2900	Electrical		17.25	25	32.50	9.80%	11.70%	14.50%	
	3100	Total: Mechanical & Electrical		65.50	69	92	28.50%	31.50%	32.50%	
65	0010	**POST OFFICES**	S.F.	125	154	196				65
	0020	Total project costs	C.F.	7.50	9.50	10.80				
	2720	Plumbing	S.F.	5.05	6.95	8.80	4.24%	5.30%	5.60%	
	2770	Heating, ventilating, air conditioning		7.85	10.85	12.05	6.65%	7.15%	9.35%	
	2900	Electrical		10.30	14.50	17.20	7.25%	9%	11%	
	3100	Total: Mechanical & Electrical		29.50	39	44	16.25%	18.80%	22%	
66	0010	**POWER PLANTS**	S.F.	865	1,150	2,100				66
	0020	Total project costs	C.F.	24	52	111				
	2900	Electrical	S.F.	61	130	193	9.30%	12.75%	21.50%	
	8100	Total: Mechanical & Electrical		152	495	1,100	32.50%	32.50%	52.50%	
67	0010	**RELIGIOUS EDUCATION**	S.F.	101	131	161				67
	0020	Total project costs	C.F.	5.60	8	10				
	2720	Plumbing	S.F.	4.20	5.95	8.40	4.40%	5.30%	7.10%	
	2770	Heating, ventilating, air conditioning		10.60	12	16.95	10.05%	11.45%	12.35%	
	2900	Electrical		7.40	11.25	14.85	7.60%	9.05%	10.35%	
	3100	Total: Mechanical & Electrical		32.50	42.50	52	22%	23%	27%	
69	0010	**RESEARCH Laboratories & Facilities**	S.F.	154	217	315				69
	0020	Total project costs	C.F.	11.20	22	26.50				
	1800	Equipment	S.F.	6.70	13.15	32	.94%	4.58%	9.10%	
	2720	Plumbing		10.95	19.30	31	6.15%	8.30%	10.80%	
	2770	Heating, ventilating, air conditioning		13.55	45.50	54	7.25%	16.50%	17.50%	
	2900	Electrical		18.10	29	47.50	9.55%	11.15%	14.60%	
	3100	Total: Mechanical & Electrical		57.50	102	144	29.50%	37%	41%	
70	0010	**RESTAURANTS**	S.F.	146	188	245				70
	0020	Total project costs	C.F.	12.25	16.10	21				
	1800	Equipment	S.F.	8.65	23	35	6.10%	13%	15.65%	
	2720	Plumbing		11.55	14	18.40	6.10%	8.15%	9%	
	2770	Heating, ventilating, air conditioning		14.70	20.50	24.50	9.20%	12%	12.40%	
	2900	Electrical		15.40	19	24.50	8.35%	10.55%	11.55%	
	3100	Total: Mechanical & Electrical		47.50	51	66	21%	25%	29.50%	
	9000	Per seat unit, total cost	Seat	5,350	7,125	8,450				
	9500	Total: Mechanical & Electrical	"	1,350	1,775	2,100				
72	0010	**RETAIL STORES**	S.F.	68	91.50	121				72
	0020	Total project costs	C.F.	4.62	6.60	9.15				
	2720	Plumbing	S.F.	2.47	4.12	7.05	3.26%	4.60%	6.80%	
	2770	Heating, ventilating, air conditioning		5.35	7.30	10.95	6.75%	8.75%	10.15%	
	2900	Electrical		6.15	8.40	12.10	7.25%	9.90%	11.60%	
	3100	Total: Mechanical & Electrical		16.30	21	28	17.05%	21%	23.50%	
74	0010	**SCHOOLS Elementary**	S.F.	111	138	169				74
	0020	Total project costs	C.F.	7.25	9.30	12.05				
	0500	Masonry	S.F.	7.85	16.95	25	4.89%	10.50%	14%	
	1800	Equipment		2.58	4.93	9.30	1.83%	3.13%	4.61%	

691

50 17 | Square Foot Costs

50 17 00 | S.F. Costs

			UNIT	UNIT COSTS 1/4	MEDIAN	3/4	% OF TOTAL 1/4	MEDIAN	3/4	
74	2720	Plumbing	S.F.	6.40	9.05	12.05	5.70%	7.15%	9.35%	74
	2730	Heating, ventilating, air conditioning		9.60	15.25	21.50	8.15%	10.80%	14.90%	
	2900	Electrical		10.50	14	17.85	8.45%	10.05%	11.85%	
	3100	Total: Mechanical & Electrical	▼	38	46.50	59	25%	27.50%	30%	
	9000	Per pupil, total cost	Ea.	12,800	19,000	42,500				
	9500	Total: Mechanical & Electrical	"	3,600	4,550	11,500				
76	0010	**SCHOOLS Junior High & Middle**	S.F.	115	142	173				76
	0020	Total project costs	C.F.	7.30	9.40	10.55				
	0500	Masonry	S.F.	12.30	18.40	22.50	7.55%	11.10%	14.30%	
	1800	Equipment		3.14	5.90	8.65	1.80%	3.03%	4.80%	
	2720	Plumbing		6.70	8.25	10.20	5.30%	6.80%	7.25%	
	2770	Heating, ventilating, air conditioning		13.35	16.20	28.50	8.90%	11.55%	14.20%	
	2900	Electrical		11.70	14.45	18.25	8.05%	9.55%	10.60%	
	3100	Total: Mechanical & Electrical	▼	36.50	47	58.50	23%	25.50%	29.50%	
	9000	Per pupil, total cost	Ea.	14,500	19,100	25,600				
78	0010	**SCHOOLS Senior High**	S.F.	120	147	184				78
	0020	Total project costs	C.F.	7.20	10.10	17.05				
	1800	Equipment	S.F.	3.17	7.40	10.25	1.88%	2.67%	4.30%	
	2720	Plumbing		6.70	10.05	18.35	5.60%	6.90%	8.30%	
	2770	Heating, ventilating, air conditioning		11.50	15.65	24	8.95%	11.60%	15%	
	2900	Electrical		12.05	16.05	23.50	8.70%	10.35%	12.50%	
	3100	Total: Mechanical & Electrical	▼	39.50	47	76	24%	26.50%	28.50%	
	9000	Per pupil, total cost	Ea.	11,200	22,900	28,600				
80	0010	**SCHOOLS Vocational**	S.F.	96.50	140	173				80
	0020	Total project costs	C.F.	6	8.60	11.90				
	0500	Masonry	S.F.	4.24	14	21.50	3.20%	4.61%	10.95%	
	1800	Equipment		3.02	7.50	10.35	1.24%	3.10%	4.26%	
	2720	Plumbing		6.15	9.20	13.55	5.40%	6.90%	8.55%	
	2770	Heating, ventilating, air conditioning		8.65	16.05	27	8.60%	11.90%	14.65%	
	2900	Electrical		10.05	13.75	18.80	8.45%	11%	13.20%	
	3100	Total: Mechanical & Electrical	▼	38	66.50	85	27.50%	29.50%	31%	
	9000	Per pupil, total cost	Ea.	13,400	36,000	53,500				
83	0010	**SPORTS ARENAS**	S.F.	84.50	113	174				83
	0020	Total project costs	C.F.	4.58	8.20	10.60				
	2720	Plumbing	S.F.	4.22	7.40	15.65	4.35%	6.35%	9.40%	
	2770	Heating, ventilating, air conditioning		10.55	12.45	17.30	8.80%	10.20%	13.55%	
	2900	Electrical	▼	7.80	11.90	15.40	7.65%	9.75%	11.90%	
85	0010	**SUPERMARKETS**	S.F.	78	90	106				85
	0020	Total project costs	C.F.	4.34	5.25	7.95				
	2720	Plumbing	S.F.	4.35	5.50	6.40	5.40%	6%	7.45%	
	2770	Heating, ventilating, air conditioning		6.40	8.50	10.35	8.60%	8.65%	9.60%	
	2900	Electrical		9.15	11.15	13.25	10.40%	12.45%	13.60%	
	3100	Total: Mechanical & Electrical	▼	25	26.50	35.50	23.50%	27.50%	28.50%	
86	0010	**SWIMMING POOLS**	S.F.	126	295	450				86
	0020	Total project costs	C.F.	10.10	12.60	13.75				
	2720	Plumbing	S.F.	11.65	13.35	17.90	4.80%	9.70%	20.50%	
	2900	Electrical		9.60	15.35	34	6.05%	6.95%	7.60%	
	3100	Total: Mechanical & Electrical	▼	60	79	104	11.15%	14.10%	23.50%	
87	0010	**TELEPHONE EXCHANGES**	S.F.	157	238	310				87
	0020	Total project costs	C.F.	10.40	16.75	23				
	2720	Plumbing	S.F.	7.05	10.90	16	4.52%	5.80%	6.90%	
	2770	Heating, ventilating, air conditioning		16.40	33	41	11.80%	16.05%	18.40%	
	2900	Electrical		17.05	27	48	10.90%	14%	17.85%	
	3100	Total: Mechanical & Electrical	▼	50.50	95.50	135	29.50%	33.50%	44.50%	
91	0010	**THEATERS**	S.F.	99.50	135	199				91
	0020	Total project costs	C.F.	4.87	7.20	10.55				

50 17 | Square Foot Costs

		50 17 00	S.F. Costs	UNIT	UNIT COSTS 1/4	UNIT COSTS MEDIAN	UNIT COSTS 3/4	% OF TOTAL 1/4	% OF TOTAL MEDIAN	% OF TOTAL 3/4	
91	2720		Plumbing	S.F.	3.28	3.80	15.55	2.92%	4.70%	6.80%	91
	2770		Heating, ventilating, air conditioning		10.25	12.40	15.35	8%	12.25%	13.40%	
	2900		Electrical		9.20	12.45	25.50	8.05%	9.35%	12.25%	
	3100		Total: Mechanical & Electrical		23.50	32	37.50	21.50%	25.50%	27.50%	
94	0010	**TOWN HALLS City Halls & Municipal Buildings**		S.F.	110	149	195				94
	0020		Total project costs	C.F.	7.95	12.40	18.10				
	2720		Plumbing	S.F.	4.91	9.20	16.10	4.31%	5.95%	7.95%	
	2770		Heating, ventilating, air conditioning		8.85	17.60	26	7.05%	9.05%	13.45%	
	2900		Electrical		11.15	15.40	21	8.05%	9.45%	11.65%	
	3100		Total: Mechanical & Electrical		38.50	49	75	22%	26.50%	31%	
97	0010	**WAREHOUSES & Storage Buildings**		S.F.	44	63	93				97
	0020		Total project costs	C.F.	2.30	3.60	5.95				
	0100		Site work	S.F.	4.53	9	13.55	6.05%	12.95%	19.55%	
	0500		Masonry		2.49	6.25	13.45	3.60%	7.35%	12%	
	1800		Equipment		.67	1.52	8.55	.72%	1.69%	5.55%	
	2720		Plumbing		1.46	2.62	4.91	2.90%	4.80%	6.55%	
	2730		Heating, ventilating, air conditioning		1.67	4.71	6.30	2.41%	5%	8.90%	
	2900		Electrical		2.60	4.89	8	5.15%	7.20%	10.05%	
	3100		Total: Mechanical & Electrical		7.25	11.10	22	13.30%	18.90%	26%	
99	0010	**WAREHOUSE & OFFICES Combination**		S.F.	54	72	99				99
	0020		Total project costs	C.F.	2.76	4.01	5.90				
	1800		Equipment	S.F.	.93	1.81	2.69	.42%	1.19%	2.40%	
	2720		Plumbing		2.08	3.63	5.40	3.74%	4.76%	6.30%	
	2770		Heating, ventilating, air conditioning		3.29	5.15	7.20	5%	5.65%	10.05%	
	2900		Electrical		3.68	5.45	8.50	5.85%	8%	10%	
	3100		Total: Mechanical & Electrical		10.25	15.55	24.50	14.55%	19.95%	24.50%	

Square Foot Project Size Modifier

One factor that affects the S.F. cost of a particular building is the size. In general, for buildings built to the same specifications in the same locality, the larger building will have the lower S.F. cost. This is due mainly to the decreasing contribution of the exterior walls, plus the economy of scale usually achievable in larger buildings. The area conversion scale shown below will give a factor to convert costs for the typical size building to an adjusted cost for the particular project.

The square foot base size table lists the median costs, most typical project size in our accumulated data, and the range in size of the projects.

The size factor for your project is determined by dividing your project area in S.F. by the typical project size for the particular building type. With this factor, enter the area conversion scale at the appropriate size factor and determine the appropriate cost multiplier for your building size.

Example: Determine the cost per S.F. for a 100,000 S.F. Mid-rise apartment building.

$$\frac{\text{Proposed building area} = 100,000 \text{ S.F.}}{\text{Typical size from below} = 50,000 \text{ S.F.}} = 2.00$$

Enter area conversion scale at 2.0, intersect curve, read horizontally the appropriate cost multiplier of .94. Size adjusted cost becomes .94 × $117.00 = $110.00 based on national average costs.

Note: For size factors less than .50, the cost multiplier is 1.1
For size factors greater than 3.5, the cost multiplier is .90

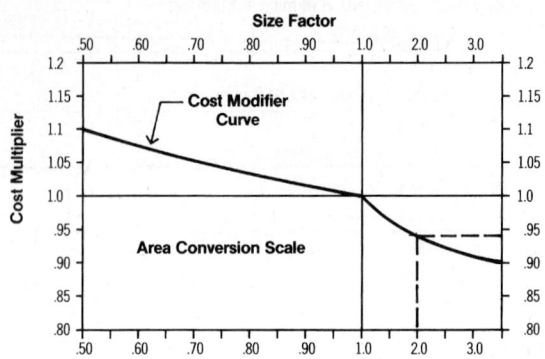

Square Foot Base Size

Building Type	Median Cost per S.F.	Typical Size Gross S.F.	Typical Range Gross S.F.	Building Type	Median Cost per S.F.	Typical Size Gross S.F.	Typical Range Gross S.F.
Apartments, Low Rise	$ 92.50	21,000	9,700 - 37,200	Jails	$ 280.00	40,000	5,500 - 145,000
Apartments, Mid Rise	117.00	50,000	32,000 - 100,000	Libraries	186.00	12,000	7,000 - 31,000
Apartments, High Rise	127.00	145,000	95,000 - 600,000	Living, Assisted	151.00	32,300	23,500 - 50,300
Auditoriums	158.00	25,000	7,600 - 39,000	Medical Clinics	159.00	7,200	4,200 - 15,700
Auto Sales	115.00	20,000	10,800 - 28,600	Medical Offices	151.00	6,000	4,000 - 15,000
Banks	207.00	4,200	2,500 - 7,500	Motels	111.00	40,000	15,800 - 120,000
Churches	143.00	17,000	2,000 - 42,000	Nursing Homes	155.00	23,000	15,000 - 37,000
Clubs, Country	144.00	6,500	4,500 - 15,000	Offices, Low Rise	132.00	20,000	5,000 - 80,000
Clubs, Social	138.00	10,000	6,000 - 13,500	Offices, Mid Rise	130.00	120,000	20,000 - 300,000
Clubs, YMCA	164.00	28,300	12,800 - 39,400	Offices, High Rise	166.00	260,000	120,000 - 800,000
Colleges (Class)	164.00	50,000	15,000 - 150,000	Police Stations	208.00	10,500	4,000 - 19,000
Colleges (Science Lab)	264.00	45,600	16,600 - 80,000	Post Offices	154.00	12,400	6,800 - 30,000
College (Student Union)	195.00	33,400	16,000 - 85,000	Power Plants	1150.00	7,500	1,000 - 20,000
Community Center	147.00	9,400	5,300 - 16,700	Religious Education	131.00	9,000	6,000 - 12,000
Court Houses	202.00	32,400	17,800 - 106,000	Research	217.00	19,000	6,300 - 45,000
Dept. Stores	85.50	90,000	44,000 - 122,000	Restaurants	188.00	4,400	2,800 - 6,000
Dormitories, Low Rise	166.00	25,000	10,000 - 95,000	Retail Stores	91.50	7,200	4,000 - 17,600
Dormitories, Mid Rise	192.00	85,000	20,000 - 200,000	Schools, Elementary	138.00	41,000	24,500 - 55,000
Factories	83.00	26,400	12,900 - 50,000	Schools, Jr. High	142.00	92,000	52,000 - 119,000
Fire Stations	153.00	5,800	4,000 - 8,700	Schools, Sr. High	147.00	101,000	50,500 - 175,000
Fraternity Houses	142.00	12,500	8,200 - 14,800	Schools, Vocational	140.00	37,000	20,500 - 82,000
Funeral Homes	159.00	10,000	4,000 - 20,000	Sports Arenas	113.00	15,000	5,000 - 40,000
Garages, Commercial	102.00	9,300	5,000 - 13,600	Supermarkets	90.00	44,000	12,000 - 60,000
Garages, Municipal	129.00	8,300	4,500 - 12,600	Swimming Pools	295.00	20,000	10,000 - 32,000
Garages, Parking	55.00	163,000	76,400 - 225,300	Telephone Exchange	238.00	4,500	1,200 - 10,600
Gymnasiums	140.00	19,200	11,600 - 41,000	Theaters	135.00	10,500	8,800 - 17,500
Hospitals	253.00	55,000	27,200 - 125,000	Town Halls	149.00	10,800	4,800 - 23,400
House (Elderly)	126.00	37,000	21,000 - 66,000	Warehouses	63.00	25,000	8,000 - 72,000
Housing (Public)	116.00	36,000	14,400 - 74,400	Warehouse & Office	72.00	25,000	8,000 - 72,000
Ice Rinks	167.00	29,000	27,200 - 33,600				

Abbreviations

A	Area Square Feet; Ampere	Brk., brk	Brick	Csc	Cosecant
AAFES	Army and Air Force Exchange Service	brkt	Bracket	C.S.F.	Hundred Square Feet
ABS	Acrylonitrile Butadiene Stryrene; Asbestos Bonded Steel	Brng.	Bearing	CSI	Construction Specifications Institute
A.C., AC	Alternating Current; Air-Conditioning; Asbestos Cement; Plywood Grade A & C	Brs.	Brass	CT	Current Transformer
		Brz.	Bronze	CTS	Copper Tube Size
		Bsn.	Basin	Cu	Copper, Cubic
		Btr.	Better	Cu. Ft.	Cubic Foot
ACI	American Concrete Institute	Btu	British Thermal Unit	cw	Continuous Wave
ACR	Air Conditioning Refrigeration	BTUH	BTU per Hour	C.W.	Cool White; Cold Water
ADA	Americans with Disabilities Act	Bu.	bushels	Cwt.	100 Pounds
AD	Plywood, Grade A & D	BUR	Built-up Roofing	C.W.X.	Cool White Deluxe
Addit.	Additional	BX	Interlocked Armored Cable	C.Y.	Cubic Yard (27 cubic feet)
Adj.	Adjustable	°C	degree centegrade	C.Y./Hr.	Cubic Yard per Hour
af	Audio-frequency	c	Conductivity, Copper Sweat	Cyl.	Cylinder
AFUE	Annual Fuel Utilization Efficiency	C	Hundred; Centigrade	d	Penny (nail size)
AGA	American Gas Association	C/C	Center to Center, Cedar on Cedar	D	Deep; Depth; Discharge
Agg.	Aggregate	C-C	Center to Center	Dis., Disch.	Discharge
A.H., Ah	Ampere Hours	Cab	Cabinet	Db	Decibel
A hr.	Ampere-hour	Cair.	Air Tool Laborer	Dbl.	Double
A.H.U., AHU	Air Handling Unit	Cal.	caliper	DC	Direct Current
A.I.A.	American Institute of Architects	Calc	Calculated	DDC	Direct Digital Control
AIC	Ampere Interrupting Capacity	Cap.	Capacity	Demob.	Demobilization
Allow.	Allowance	Carp.	Carpenter	d.f.t.	Dry Film Thickness
alt., alt	Alternate	C.B.	Circuit Breaker	d.f.u.	Drainage Fixture Units
Alum.	Aluminum	C.C.A.	Chromate Copper Arsenate	D.H.	Double Hung
a.m.	Ante Meridiem	C.C.F.	Hundred Cubic Feet	DHW	Domestic Hot Water
Amp.	Ampere	cd	Candela	DI	Ductile Iron
Anod.	Anodized	cd/sf	Candela per Square Foot	Diag.	Diagonal
ANSI	American National Standards Institute	CD	Grade of Plywood Face & Back	Diam., Dia	Diameter
		CDX	Plywood, Grade C & D, exterior glue	Distrib.	Distribution
APA	American Plywood Association	Cefi.	Cement Finisher	Div.	Division
Approx.	Approximate	Cem.	Cement	Dk.	Deck
Apt.	Apartment	CF	Hundred Feet	D.L.	Dead Load; Diesel
Asb.	Asbestos	C.F.	Cubic Feet	DLH	Deep Long Span Bar Joist
A.S.B.C.	American Standard Building Code	CFM	Cubic Feet per Minute	dlx	Deluxe
Asbe.	Asbestos Worker	CFRP	Carbon Fiber Reinforced Plastic	Do.	Ditto
ASCE.	American Society of Civil Engineers	c.g.	Center of Gravity	DOP	Dioctyl Phthalate Penetration Test (Air Filters)
A.S.H.R.A.E.	American Society of Heating, Refrig. & AC Engineers	CHW	Chilled Water; Commercial Hot Water	Dp., dp	Depth
		C.I., CI	Cast Iron	D.P.S.T.	Double Pole, Single Throw
ASME	American Society of Mechanical Engineers	C.I.P., CIP	Cast in Place	Dr.	Drive
		Circ.	Circuit	DR	Dimension Ratio
ASTM	American Society for Testing and Materials	C.L.	Carload Lot	Drink.	Drinking
		CL	Chain Link	D.S.	Double Strength
Attchmt.	Attachment	Clab.	Common Laborer	D.S.A.	Double Strength A Grade
Avg., Ave.	Average	Clam	Common maintenance laborer	D.S.B.	Double Strength B Grade
AWG	American Wire Gauge	C.L.F.	Hundred Linear Feet	Dty.	Duty
AWWA	American Water Works Assoc.	CLF	Current Limiting Fuse	DWV	Drain Waste Vent
Bbl.	Barrel	CLP	Cross Linked Polyethylene	DX	Deluxe White, Direct Expansion
B&B, BB	Grade B and Better; Balled & Burlapped	cm	Centimeter	dyn	Dyne
		CMP	Corr. Metal Pipe	e	Eccentricity
B&S	Bell and Spigot	CMU	Concrete Masonry Unit	E	Equipment Only; East; emissivity
B.&W.	Black and White	CN	Change Notice	Ea.	Each
b.c.c.	Body-centered Cubic	Col.	Column	EB	Encased Burial
B.C.Y.	Bank Cubic Yards	CO₂	Carbon Dioxide	Econ.	Economy
BE	Bevel End	Comb.	Combination	E.C.Y	Embankment Cubic Yards
B.F.	Board Feet	comm.	Commercial, Communication	EDP	Electronic Data Processing
Bg. cem.	Bag of Cement	Compr.	Compressor	EIFS	Exterior Insulation Finish System
BHP	Boiler Horsepower; Brake Horsepower	Conc.	Concrete	E.D.R.	Equiv. Direct Radiation
		Cont., cont	Continuous; Continued, Container	Eq.	Equation
B.I.	Black Iron	Corr.	Corrugated	EL	elevation
bidir.	bidirectional	Cos	Cosine	Elec.	Electrician; Electrical
Bit., Bitum.	Bituminous	Cot	Cotangent	Elev.	Elevator; Elevating
Bit., Conc.	Bituminous Concrete	Cov.	Cover	EMT	Electrical Metallic Conduit; Thin Wall Conduit
Bk.	Backed	C/P	Cedar on Paneling		
Bkrs.	Breakers	CPA	Control Point Adjustment	Eng.	Engine, Engineered
Bldg., bldg	Building	Cplg.	Coupling	EPDM	Ethylene Propylene Diene Monomer
Blk.	Block	CPM	Critical Path Method		
Bm.	Beam	CPVC	Chlorinated Polyvinyl Chloride	EPS	Expanded Polystyrene
Boil.	Boilermaker	C.Pr.	Hundred Pair	Eqhv.	Equip. Oper., Heavy
bpm	Blows per Minute	CRC	Cold Rolled Channel	Eqlt.	Equip. Oper., Light
BR	Bedroom	Creos.	Creosote	Eqmd.	Equip. Oper., Medium
Brg.	Bearing	Crpt.	Carpet & Linoleum Layer	Eqmm.	Equip. Oper., Master Mechanic
Brhe.	Bricklayer Helper	CRT	Cathode-ray Tube	Eqol.	Equip. Oper., Oilers
Bric.	Bricklayer	CS	Carbon Steel, Constant Shear Bar Joist	Equip.	Equipment
				ERW	Electric Resistance Welded

Abbreviations

E.S.	Energy Saver	H	High Henry	Lath.	Lather
Est.	Estimated	HC	High Capacity	Lav.	Lavatory
esu	Electrostatic Units	H.D., HD	Heavy Duty; High Density	lb.; #	Pound
E.W.	Each Way	H.D.O.	High Density Overlaid	L.B., LB	Load Bearing; L Conduit Body
EWT	Entering Water Temperature	HDPE	High density polyethelene plastic	L. & E.	Labor & Equipment
Excav.	Excavation	Hdr.	Header	lb./hr.	Pounds per Hour
excl	Excluding	Hdwe.	Hardware	lb./L.F.	Pounds per Linear Foot
Exp., exp	Expansion, Exposure	H.I.D., HID	High Intensity Discharge	lbf/sq.in.	Pound-force per Square Inch
Ext., ext	Exterior; Extension	Help.	Helper Average	L.C.L.	Less than Carload Lot
Extru.	Extrusion	HEPA	High Efficiency Particulate Air Filter	L.C.Y.	Loose Cubic Yard
f.	Fiber stress			Ld.	Load
F	Fahrenheit; Female; Fill	Hg	Mercury	LE	Lead Equivalent
Fab., fab	Fabricated; fabric	HIC	High Interrupting Capacity	LED	Light Emitting Diode
FBGS	Fiberglass	HM	Hollow Metal	L.F.	Linear Foot
F.C.	Footcandles	HMWPE	high molecular weight polyethylene	L.F. Nose	Linear Foot of Stair Nosing
f.c.c.	Face-centered Cubic			L.F. Rsr	Linear Foot of Stair Riser
f'c.	Compressive Stress in Concrete; Extreme Compressive Stress	HO	High Output	Lg.	Long; Length; Large
		Horiz.	Horizontal	L & H	Light and Heat
F.E.	Front End	H.P., HP	Horsepower; High Pressure	LH	Long Span Bar Joist
FEP	Fluorinated Ethylene Propylene (Teflon)	H.P.F.	High Power Factor	L.H.	Labor Hours
		Hr.	Hour	L.L., LL	Live Load
F.G.	Flat Grain	Hrs./Day	Hours per Day	L.L.D.	Lamp Lumen Depreciation
F.H.A.	Federal Housing Administration	HSC	High Short Circuit	lm	Lumen
Fig.	Figure	Ht.	Height	lm/sf	Lumen per Square Foot
Fin.	Finished	Htg.	Heating	lm/W	Lumen per Watt
FIPS	Female Iron Pipe Size	Htrs.	Heaters	LOA	Length Over All
Fixt.	Fixture	HVAC	Heating, Ventilation & Air-Conditioning	log	Logarithm
FJP	Finger jointed and primed			L-O-L	Lateralolet
Fl. Oz.	Fluid Ounces	Hvy.	Heavy	long.	longitude
Flr.	Floor	HW	Hot Water	L.P., LP	Liquefied Petroleum; Low Pressure
FM	Frequency Modulation; Factory Mutual	Hyd.; Hydr.	Hydraulic	L.P.F.	Low Power Factor
		Hz	Hertz (cycles)	LR	Long Radius
Fmg.	Framing	I.	Moment of Inertia	L.S.	Lump Sum
FM/UL	Factory Mutual/Underwriters Labs	IBC	International Building Code	Lt.	Light
Fdn.	Foundation	I.C.	Interrupting Capacity	Lt. Ga.	Light Gauge
FNPT	Female National Pipe Thread	ID	Inside Diameter	L.T.L.	Less than Truckload Lot
Fori.	Foreman, Inside	I.D.	Inside Dimension; Identification	Lt. Wt.	Lightweight
Foro.	Foreman, Outside	I.F.	Inside Frosted	L.V.	Low Voltage
Fount.	Fountain	I.M.C.	Intermediate Metal Conduit	M	Thousand; Material; Male; Light Wall Copper Tubing
fpm	Feet per Minute	In.	Inch		
FPT	Female Pipe Thread	Incan.	Incandescent	M²CA	Meters Squared Contact Area
Fr	Frame	Incl.	Included; Including	m/hr.; M.H.	Man-hour
F.R.	Fire Rating	Int.	Interior	mA	Milliampere
FRK	Foil Reinforced Kraft	Inst.	Installation	Mach.	Machine
FSK	Foil/scrim/kraft	Insul., insul	Insulation/Insulated	Mag. Str.	Magnetic Starter
FRP	Fiberglass Reinforced Plastic	I.P.	Iron Pipe	Maint.	Maintenance
FS	Forged Steel	I.P.S., IPS	Iron Pipe Size	Marb.	Marble Setter
FSC	Cast Body; Cast Switch Box	IPT	Iron Pipe Threaded	Mat; Mat'l.	Material
Ft., ft	Foot; Feet	I.W.	Indirect Waste	Max.	Maximum
Ftng.	Fitting	J	Joule	MBF	Thousand Board Feet
Ftg.	Footing	J.I.C.	Joint Industrial Council	MBH	Thousand BTU's per hr.
Ft lb.	Foot Pound	K	Thousand; Thousand Pounds; Heavy Wall Copper Tubing, Kelvin	MC	Metal Clad Cable
Furn.	Furniture			MCC	Motor Control Center
FVNR	Full Voltage Non-Reversing	K.A.H.	Thousand Amp. Hours	M.C.F.	Thousand Cubic Feet
FVR	Full Voltage Reversing	kcmil	Thousand Circular Mils	MCFM	Thousand Cubic Feet per Minute
FXM	Female by Male	KD	Knock Down	M.C.M.	Thousand Circular Mils
Fy.	Minimum Yield Stress of Steel	K.D.A.T.	Kiln Dried After Treatment	MCP	Motor Circuit Protector
g	Gram	kg	Kilogram	MD	Medium Duty
G	Gauss	kG	Kilogauss	MDF	Medium-density fibreboard
Ga.	Gauge	kgf	Kilogram Force	M.D.O.	Medium Density Overlaid
Gal., gal.	Gallon	kHz	Kilohertz	Med.	Medium
gpm, GPM	Gallon per Minute	Kip	1000 Pounds	MF	Thousand Feet
Galv., galv	Galvanized	KJ	Kiljoule	M.F.B.M.	Thousand Feet Board Measure
GC/MS	Gas Chromatograph/Mass Spectrometer	K.L.	Effective Length Factor	Mfg.	Manufacturing
		K.L.F.	Kips per Linear Foot	Mfrs.	Manufacturers
Gen.	General	Km	Kilometer	mg	Milligram
GFI	Ground Fault Interrupter	KO	Knock Out	MGD	Million Gallons per Day
GFRC	Glass Fiber Reinforced Concrete	K.S.F.	Kips per Square Foot	MGPH	Thousand Gallons per Hour
Glaz.	Glazier	K.S.I.	Kips per Square Inch	MH, M.H.	Manhole; Metal Halide; Man-Hour
GPD	Gallons per Day	kV	Kilovolt	MHz	Megahertz
gpf	Gallon per flush	kVA	Kilovolt Ampere	Mi.	Mile
GPH	Gallons per Hour	kVAR	Kilovar (Reactance)	MI	Malleable Iron; Mineral Insulated
GPM	Gallons per Minute	KW	Kilowatt	MIPS	Male Iron Pipe Size
GR	Grade	KWh	Kilowatt-hour	mj	Mechanical Joint
Gran.	Granular	L	Labor Only; Length; Long; Medium Wall Copper Tubing	m	Meter
Grnd.	Ground			mm	Millimeter
GVW	Gross Vehicle Weight	Lab.	Labor	Mill.	Millwright
GWB	Gypsum wall board	lat	Latitude	Min., min.	Minimum, minute

Abbreviations

Misc.	Miscellaneous	PDCA	Painting and Decorating Contractors of America	SC	Screw Cover
ml	Milliliter, Mainline			SCFM	Standard Cubic Feet per Minute
M.L.F.	Thousand Linear Feet	P.E., PE	Professional Engineer; Porcelain Enamel; Polyethylene; Plain End	Scaf.	Scaffold
Mo.	Month			Sch., Sched.	Schedule
Mobil.	Mobilization			S.C.R.	Modular Brick
Mog.	Mogul Base	P.E.C.I.	Porcelain Enamel on Cast Iron	S.D.	Sound Deadening
MPH	Miles per Hour	Perf.	Perforated	SDR	Standard Dimension Ratio
MPT	Male Pipe Thread	PEX	Cross linked polyethylene	S.E.	Surfaced Edge
MRGWB	Moisture Resistant Gypsum Wallboard	Ph.	Phase	Sel.	Select
		P.I.	Pressure Injected	SER, SEU	Service Entrance Cable
MRT	Mile Round Trip	Pile.	Pile Driver	S.F.	Square Foot
ms	Millisecond	Pkg.	Package	S.F.C.A.	Square Foot Contact Area
M.S.F.	Thousand Square Feet	Pl.	Plate	S.F. Flr.	Square Foot of Floor
Mstz.	Mosaic & Terrazzo Worker	Plah.	Plasterer Helper	S.F.G.	Square Foot of Ground
M.S.Y.	Thousand Square Yards	Plas.	Plasterer	S.F. Hor.	Square Foot Horizontal
Mtd., mtd., mtd	Mounted	plf	Pounds Per Linear Foot	SFR	Square Feet of Radiation
Mthe.	Mosaic & Terrazzo Helper	Pluh.	Plumbers Helper	S.F. Shlf.	Square Foot of Shelf
Mtng.	Mounting	Plum.	Plumber	S4S	Surface 4 Sides
Mult.	Multi; Multiply	Ply.	Plywood	Shee.	Sheet Metal Worker
M.V.A.	Million Volt Amperes	p.m.	Post Meridiem	Sin.	Sine
M.V.A.R.	Million Volt Amperes Reactance	Pntd.	Painted	Skwk.	Skilled Worker
MV	Megavolt	Pord.	Painter, Ordinary	SL	Saran Lined
MW	Megawatt	pp	Pages	S.L.	Slimline
MXM	Male by Male	PP, PPL	Polypropylene	Sldr.	Solder
MYD	Thousand Yards	P.P.M.	Parts per Million	SLH	Super Long Span Bar Joist
N	Natural; North	Pr.	Pair	S.N.	Solid Neutral
nA	Nanoampere	P.E.S.B.	Pre-engineered Steel Building	SO	Stranded with oil resistant inside insulation
NA	Not Available; Not Applicable	Prefab.	Prefabricated		
N.B.C.	National Building Code	Prefin.	Prefinished	S-O-L	Socketolet
NC	Normally Closed	Prop.	Propelled	sp	Standpipe
NEMA	National Electrical Manufacturers Assoc.	PSF, psf	Pounds per Square Foot	S.P.	Static Pressure; Single Pole; Self-Propelled
		PSI, psi	Pounds per Square Inch		
NEHB	Bolted Circuit Breaker to 600V.	PSIG	Pounds per Square Inch Gauge	Spri.	Sprinkler Installer
NFPA	National Fire Protection Association	PSP	Plastic Sewer Pipe	spwg	Static Pressure Water Gauge
NLB	Non-Load-Bearing	Pspr.	Painter, Spray	S.P.D.T.	Single Pole, Double Throw
NM	Non-Metallic Cable	Psst.	Painter, Structural Steel	SPF	Spruce Pine Fir; Sprayed Polyurethane Foam
nm	Nanometer	P.T.	Potential Transformer		
No.	Number	P. & T.	Pressure & Temperature	S.P.S.T.	Single Pole, Single Throw
NO	Normally Open	Ptd.	Painted	SPT	Standard Pipe Thread
N.O.C.	Not Otherwise Classified	Ptns.	Partitions	Sq.	Square; 100 Square Feet
Nose.	Nosing	Pu	Ultimate Load	Sq. Hd.	Square Head
NPT	National Pipe Thread	PVC	Polyvinyl Chloride	Sq. In.	Square Inch
NQOD	Combination Plug-on/Bolt on Circuit Breaker to 240V.	Pvmt.	Pavement	S.S.	Single Strength; Stainless Steel
		PRV	Pressure Relief Valve	S.S.B.	Single Strength B Grade
N.R.C., NRC	Noise Reduction Coefficient/ Nuclear Regulator Commission	Pwr.	Power	sst, ss	Stainless Steel
		Q	Quantity Heat Flow	Sswk.	Structural Steel Worker
N.R.S.	Non Rising Stem	Qt.	Quart	Sswl.	Structural Steel Welder
ns	Nanosecond	Quan., Qty.	Quantity	St.; Stl.	Steel
nW	Nanowatt	Q.C.	Quick Coupling	STC	Sound Transmission Coefficient
OB	Opposing Blade	r	Radius of Gyration	Std.	Standard
OC	On Center	R	Resistance	Stg.	Staging
OD	Outside Diameter	R.C.P.	Reinforced Concrete Pipe	STK	Select Tight Knot
O.D.	Outside Dimension	Rect.	Rectangle	STP	Standard Temperature & Pressure
ODS	Overhead Distribution System	recpt.	receptacle	Stpi.	Steamfitter, Pipefitter
O.G.	Ogee	Reg.	Regular	Str.	Strength; Starter; Straight
O.H.	Overhead	Reinf.	Reinforced	Strd.	Stranded
O&P	Overhead and Profit	Req'd.	Required	Struct.	Structural
Oper.	Operator	Res.	Resistant	Sty.	Story
Opng.	Opening	Resi.	Residential	Subj.	Subject
Orna.	Ornamental	RF	Radio Frequency	Subs.	Subcontractors
OSB	Oriented Strand Board	RFID	Radio-frequency identification	Surf.	Surface
OS&Y	Outside Screw and Yoke	Rgh.	Rough	Sw.	Switch
OSHA	Occupational Safety and Health Act	RGS	Rigid Galvanized Steel	Swbd.	Switchboard
		RHW	Rubber, Heat & Water Resistant; Residential Hot Water	S.Y.	Square Yard
Ovhd.	Overhead			Syn.	Synthetic
OWG	Oil, Water or Gas	rms	Root Mean Square	S.Y.P.	Southern Yellow Pine
Oz.	Ounce	Rnd.	Round	Sys.	System
P.	Pole; Applied Load; Projection	Rodm.	Rodman	t.	Thickness
p.	Page	Rofc.	Roofer, Composition	T	Temperature; Ton
Pape.	Paperhanger	Rofp.	Roofer, Precast	Tan	Tangent
P.A.P.R.	Powered Air Purifying Respirator	Rohe.	Roofer Helpers (Composition)	T.C.	Terra Cotta
PAR	Parabolic Reflector	Rots.	Roofer, Tile & Slate	T & C	Threaded and Coupled
P.B., PB	Push Button	R.O.W.	Right of Way	T.D.	Temperature Difference
Pc., Pcs.	Piece, Pieces	RPM	Revolutions per Minute	Tdd	Telecommunications Device for the Deaf
P.C.	Portland Cement; Power Connector	R.S.	Rapid Start		
P.C.F.	Pounds per Cubic Foot	Rsr	Riser	T.E.M.	Transmission Electron Microscopy
PCM	Phase Contrast Microscopy	RT	Round Trip	temp	Temperature, Tempered, Temporary
		S.	Suction; Single Entrance; South	TFFN	Nylon Jacketed Wire
		SBS	Styrene Butadiere Styrene		

Abbreviations

TFE	Tetrafluoroethylene (Teflon)	U.L., UL	Underwriters Laboratory	w/	With
T. & G.	Tongue & Groove; Tar & Gravel	Uld.	unloading	W.C., WC	Water Column; Water Closet
Th., Thk.	Thick	Unfin.	Unfinished	W.F.	Wide Flange
Thn.	Thin	UPS	Uninterruptible Power Supply	W.G.	Water Gauge
Thrded	Threaded	URD	Underground Residential Distribution	Wldg.	Welding
Tilf.	Tile Layer, Floor			W. Mile	Wire Mile
Tilh.	Tile Layer, Helper	US	United States	W-O-L	Weldolet
THHN	Nylon Jacketed Wire	USGBC	U.S. Green Building Council	W.R.	Water Resistant
THW.	Insulated Strand Wire	USP	United States Primed	Wrck.	Wrecker
THWN	Nylon Jacketed Wire	UTMCD	Uniform Traffic Manual For Control Devices	W.S.P.	Water, Steam, Petroleum
T.L., TL	Truckload			WT., Wt.	Weight
T.M.	Track Mounted	UTP	Unshielded Twisted Pair	WWF	Welded Wire Fabric
Tot.	Total	V	Volt	XFER	Transfer
T-O-L	Threadolet	VA	Volt Amperes	XFMR	Transformer
tmpd	Tempered	V.C.T.	Vinyl Composition Tile	XHD	Extra Heavy Duty
TPO	Thermoplastic Polyolefin	VAV	Variable Air Volume	XHHW,	Cross-Linked Polyethylene Wire
T.S.	Trigger Start	VC	Veneer Core	XLPE	Insulation
Tr.	Trade	VDC	Volts Direct Current	XLP	Cross-linked Polyethylene
Transf.	Transformer	Vent.	Ventilation	Xport	Transport
Trhv.	Truck Driver, Heavy	Vert.	Vertical	Y	Wye
Trlr	Trailer	V.F.	Vinyl Faced	yd	Yard
Trlt.	Truck Driver, Light	V.G.	Vertical Grain	yr	Year
TTY	Teletypewriter	VHF	Very High Frequency	Δ	Delta
TV	Television	VHO	Very High Output	%	Percent
T.W.	Thermoplastic Water Resistant Wire	Vib.	Vibrating	~	Approximately
		VLF	Vertical Linear Foot	Ø	Phase; diameter
UCI	Uniform Construction Index	VOC	Volitile Organic Compound	@	At
UF	Underground Feeder	Vol.	Volume	#	Pound; Number
UGND	Underground Feeder	VRP	Vinyl Reinforced Polyester	<	Less Than
UHF	Ultra High Frequency	W	Wire; Watt; Wide; West	>	Greater Than
U.I.	United Inch			Z	zone

Index

A

Abandon catch basin	22
Abatement asbestos	26
equipment asbestos	25
ABC extinguisher portable	67
extinguisher wheeled	67
Aboveground storage tank	
concrete	319
ABS DWV pipe	210
Absorber shock	244
vibration	121
A/C packaged terminal	344
removal fan coil	307
removal roof top	308
removal thru-the-wall	308
removal window	308
Accelerator sprinkler system	94
Access road and parking area	15
Accessories bathroom	65, 66
compressor	273
fuel oil	316
polypropylene	226
roof	51
toilet	65
Acid resistant drain	267
resistant pipe	300
Acoustical sealant	55
ACR tubing copper	147
Adapter fire hose swivel	89
polybutylene	225
Adjustment factors	8
factors labor	100
for sprinkler/standpipe installations	664
Aerated sewage lagoon	539, 540
Aeration sewage	428
Aerator	296
Aerial lift	549
lift truck	554
survey	22
Aggregate spreader	546, 553
Air automatic vent	320
cock	321
compressor	273, 549
compressor dental	76
compressor package	273
compressor portable	549
compressor removal	101
compressor sprinkler system	94
conditioner cooling & heating	344
conditioner direct expansion	346
conditioner general	306
conditioner maintenance	17
conditioner packaged terminal	344
conditioner portable	345
conditioner receptacle	359
conditioner removal	306
conditioner thru-wall	344
conditioner window	345
conditioner wiring	361
conditioning ventilating	345
control	320
curtain removal	306
deaerators	320
eliminator	320
filter removal	306
filtration	25
hose	550
line couplers	273
purging scoop	320
quality	73
sampling	25
scoop	320
separators	320
spade	550
unit removal make up	308
vent automatic	320
vent convector or baseboard	347
Airless sprayer	25
Air-source heat pumps	345
Alarm fire	93, 366
panels and devices	366
panels and devices fire	366
residential	360
sprinkler	366
standpipe	93, 366
system medical	297
valve sprinkler	86
vent	316
water motor	93
All fuel chimney	335, 338
service jacket insulation	125, 130
Alteration fee	8
Aluminum flashing	50, 265
insulation jacketing	138
pipe	410, 413
rivet	44
salvage	102
weld rod	45
Anchor bolt	33
chemical	38, 42
epoxy	38, 42
expansion	42
hollow wall	42
lead screw	43
machinery	35
metal nailing	42
nailing	42
nylon nailing	42
pipe conduit casing	418
plastic screw	43
screw	42
toggle bolt	42
wedge	43
Angle valve bronze	103
valve plastic	108
Anti-freeze inhibited	329
Anti-siphon device	244
Appliance	72
plumbing	274, 276, 278, 295, 296
residential	72, 360
Arch culvert oval	412
Architectural fee	8
Area cleanup	27
Asbestos abatement	26
abatement equipment	25
area decontamination	27
demolition	27
disposal	28
encapsulation	28
remediation plan/method	25
removal	26
removal process	659
Ash receiver	66
Asphalt distributor	550
paver	551
plant portable	553
Asphaltic binder	386
concrete	386
concrete paving	386
pavement	386
wearing course	386
Aspirator dental	76
Atomizer water	25
Attachment ripper	553
Attachments skidsteer	549
Auger earth	546
Auto park drain	266
Automatic fire-suppression	95
flush	289
sprinkler system	388
washing machine	73
Automotive equipment	70
exhaust system	489
lift	84
spray painting booth	70
Autopsy equipment	76
Auto-scrub machine	550
Axe fire	91

B

B and S pipe gasket	258
Backer rod	54
Backfill	373
trench	371
Backflow preventer	243
preventer removal	101
preventer solar	343
Backhoe	547
bucket	548
extension	553
trenching	520, 522, 524
Backwater drainage control	407
sewer valve	407
valve	270
Bag disposable	25
glove	26
Balance valve circuit setter	320
Balancing hydronic	327
Ball check valve	109
check valve plastic	109
valve	109
valve bronze	103, 104
valve grooved-joint	191
valve plastic	108
valve slow close	91
wrecking	553
Baptistry	76
Bar front	82
grab	65
towel	65
tub	65
Barber chair	70
equipment	70
Barge construction	556
Barometric damper	332
Barrels flasher	550
reflectorized	550
Barricade	16
flasher	550
tape	16
Barriers and enclosures	15
dust	16
separation	26, 28
slipform paver	553
Base cabinet	78
sink	78
stabilizer	554
vanity	78
Baseboard heat	347
heat electric	349
removal hydronic	307
Baseplate scaffold	13
shoring	13
Basket type strainer	324
Bath hospital	292
paraffin	75
perineal	292
redwood	80
spa	80
steam	81
whirlpool	75, 80
Bathroom	282
accessories	65, 66
exhaust fan	329
faucet	284
five fixture	456
fixture	279, 280, 282, 283
four fixture	455
heater & fan	329
soaking	282
system	454-456
three fixture	454
two fixture	453
Bathtub	448
bar	65
enclosure	64
removal	101
residential	282
Beam attachment welded	111
clamp	111
flange clip	120
Bedding pipe	373
Bedpan cleanser	292
Bell & spigot pipe	249
Bellow expansion joint	322
Bibb hose	320
Bidet	279
removal	101
rough-in	279
Binder asphaltic	386
Bituminous coating	392
paver	551
Blaster shot	552
Blind flange	174
structural bolt	45
Block manhole	414
manhole/catch basin	541
Bloodbank refrigeration	74
Blowdown removal boiler	306
Blower insulation	551
Board insulation	314
mineral wool	314
Boat work	556
Boiler	339
blowdown removal	306
condensing	339
cut off low water	315
demolition	306
drain	320
electric	339
electric steam	339
gas-fired	339
gas pulse	339
general	100
hot water	339, 340
insulation	314
oil-fired	340
removal	306
steam	339, 340
steam electric	488
Bolt & gasket set	173
anchor	33
blind structural	45
Bond performance	11, 656
Boom truck	554
Booster system water pressure	248
Booth painting	70
Borer horizontal	553
Boring horizontal	396
machine horizontal	546
service	396
Borosilicate pipe	298
Borrow fill	373
loading and/or spreading	373
Bottled water cooler	296
Boundary and survey marker	22
Bowl toilet	279
Box buffalo	406
distribution	409

699

Index

flow leveler distribution	409
storage	12
trench	552
utility	395
Brace cross	46
Bracing shoring	13
Bracket pipe	110
scaffold	13
Braided bronze hose	238
steel hose	239
Brass pipe	144
pipe fitting	144
salvage	102
valve	103
Break glass station	366
Breaker pavement	549
vacuum	244
Breeching insulation	314
Brick cart	550
catch basin	414
manhole	414
Bridge sidewalk	14
Bronze angle valve	103
ball valve	104
body strainer	325
butterfly valve	104
globe valve	105
swing check valve	104
valve	103
Broom finish concrete	38
sidewalk	552
Brush chipper	547
cutter	547, 548
Bubbler	294
drinking	295
Bucket backhoe	548
clamshell	547
concrete	546
dragline	547
excavator	553
pavement	553
Buffalo box	406
Buggy concrete	546
Builder's risk insurance	650
Building permit	11
portable	12
sprinkler	94
temporary	12
Built-in range	72
shower	283
Bulldozer	548
Bullfloat concrete	546
Buncher feller	547
Bunk house trailer	12
Burner gas conversion	341
gun type	341
oil	341
residential	341
Butt fusion machine	550
Butterfly valve	308, 403
valve bronze	104
valve grooved-joint	192
Butyl caulking	55
rubber	54

C

Cabinet base	78
convector heater	350
fire equipment	66
hardboard	78
heater electric	351
hose rack	66
hotel	66
kitchen	78
medicine	66
shower	63
unit heater	351
valve	67
Cable hanger	120
jack	555
pulling	552
tensioning	552
trailer	552
Calcium silicate insulated system	417
silicate insulation	314, 417
silicate pipe covering	127
Cap HDPE piping	401
vent	265, 316
Car tram	552
Carbon dioxide extinguisher	95
Carriers and supports removal	101
fixture support	289
pipe	112, 116
Carrier/support fixture	289
Cart brick	550
concrete	546
mounted extinguisher	67
Cartridge style water filter	274
Car wash water heater	277
Casework custom	78
Casing anchor pipe conduit	418
Cast-in-place concrete	36, 37
iron cleanout	123
iron fitting	172, 250
iron manhole cover	412
iron pipe	249
iron pipe fitting	161, 250
iron pipe gasket	258
iron radiator	347
iron trap	263
iron vent caps	265
Caster scaffold	13
Cast iron weld rod	45
Catch basin	412
basin and manhole	542
basin brick	414
basin cleaner	553
basin precast	414
basin removal	22
basin vacuum	553
basin/manhole	541
Cathodic protection	363
Catwalk scaffold	14
Caulking	55
lead	249
oakum	249
polyurethane	55
sealant	54
Ceiling fan	329
heater	73
Cell equipment	70
prison	82
Cellular glass insulation	126
Cement liner	395
Central station AHU removal	306
vacuum	71
vacuum unit	424
Centrifugal in-line motor	247
pump	246, 551
Chain fall hoist	555
hoist	422
link fence	16
saw	552
trencher	371, 549
Chair barber	70
dental	76
hydraulic	76
pipe support	116
Chamber decontamination	26, 28
Channel steel	114
Charges disposal	28
powder	44
Check lift valve	107
swing valve	104, 310
valve	86, 312, 404, 405
valve ball	109
valve grooved-joint	191
valve silent	106, 310
valve wafer	106, 311
Checkered plate	46
plate cover	46
plate platform	46
Chemical anchor	38, 42
dilution tank	303
dry extinguisher	67
spreader	553
storage tank	303
toilet	552
water closet	428
Chemical-waste dilution tank	303
Chilled water pipe underground	416
Chiller remote water	296
removal water	308
Chimney all fuel	335, 338
all fuel vent	338
positive pressure	335, 338
prefabricated metal	335
vent	333
vent fitting	338
vent oval	335
Chipper brush	547
log	548
stump	548
Chipping hammer	550
Chloride accelerator concrete	37
Chlorination system	296
Church equipment	76
CI acid resistant drain	267
roof drain system	478
Circuit setter	320
Circular saw	552
Circulating pump	327
Clamp beam	111
pipe	111, 392
pipe repair	393
riser	112
valve	108
Clamshell bucket	547
Classroom sink	287
sink rough-in	287
Clean control joint	33
ductwork	18
tank	24
Cleaner catch basin	553
crack	553
steam	552
Cleaning & disposal equipment	74
duct	18
internal pipe	394
up	17
Cleanout and drain removal	101
cast iron	123
floor	122
pipe	122
polypropylene	124
PVC	124
tee	123
wall type	123
Clean up area	27
Clevis pipe hanger	113
Climbing crane	554
hydraulic jack	555
Clinic service sink	292
Clock timer	360
Closet water	279, 287
Clothes dryer commercial	71
Coat hook	65
Coating bituminous	392
epoxy	392
polyethylene	392
Cock air	321
drains and specialties	320
gas	103
gauge	321
pet	321
Coin laundry water heater	278
CO_2 extinguisher	67
fire extinguishing system	95
Cold mix paver	553
Collection concrete pipe sewage	408
sewage/drainage	411
system dust	424
Collector solar energy system	341, 342
Combination device	358
Comfort station	82
Command dog	17
Commercial dishwasher	74
gas water heater	277
lavatories	287
oil-fired water heater	278
sinks	287
toilet accessories	65
water heater	276, 278, 461-463
water heater electric	276
Commissioning	17
Compact fill	373
Compactor earth	547
landfill	548
Companion flange	174
Compartment shower	63
toilet	62
Compensation workers'	10
Component control	315
Compressed air equipment automotive	70
Compression joint fitting	152
seal	53
tank	323
Compressor accessories	273
air	273, 549
air package	273
reciprocating	273
reciprocating hermetic	273
removal air	101
removal refrigerant	308
Computer room unit removal	306
Concrete asphaltic	386
broom finish	38
bucket	546
buggy	546
bullfloat	546
cart	546
cast-in-place	36, 37
catch basin	414
chloride accelerator	37
conveyer	546
core drilling	39
direct chute	37
distribution box	409
drill	354
drilling	39
equipment rental	546
fiber reinforcing	37
float finish	38
floor edger	546
floor grinder	546
hand trowel finish	38
hole cutting	355
hole drilling	354
hot weather	37
impact drilling	39

Index

insert hanger	112
lance	551
machine trowel finish	38
manhole	414
mixer	546
non-chloride accelerator	37
paver	551
pipe	411
pipe removal	22
placement footing	37
placing	37
pump	546
ready mix	37
reinforcing	35
retarder	37
saw	546
septic tank	408
short load	37
spreader	551
tank aboveground storage	319
trowel	546
truck	546
truck holding	37
utility vault	395
vibrator	546
water reducer	37
water storage tanks	406
winter	37
Condensate meter rmvl. steam	308
pump pneumatic	329
removal pump	328
Condenser pad	352
removal	306
Condensing boiler	339
unit removal	306
Conditioner general air	306
maintenance air	17
portable air	345
Conditioning ventilating air	345
Conduit & fitting flexible	356
fitting pipe	416
pipe prefabricated	419
prefabricated pipe	416, 417
preinsulated pipe	416
Connection fire-department	88
motor	356
standpipe	88
Connector flexible	238
gas	238
water copper tubing	238
Construction barge	556
cost index	9
management fee	8
photography	11
sump hole	372
temporary	15
time	8
Contained pipe	233
Contaminated soil	24
Contingencies	8
Contractor overhead	17
Control & motor starter	362
air	320
component	315
draft	332
flow check	323
hand/off/automatic	363
joint	32
joint clean	33
joint sawn	32
joint sealant	33
package firecycle	93
pressure sprinkler	94
station	363
station stop/start	363
valve heating	316, 323

water level	315
Control component	316
Convector cabinet heater	351
hydronic heating	347
Cooking equipment	74
range	72
Cooler beverage	73
removal evaporative	307
water	295
Cooling tower removal	306
Copper ACR tubing	147
drum trap	264
DWV tubing	147
fitting	148
flashing	50, 265
grooved-joint	156
pipe	145, 401
rivet	44
salvage	102
tube	145, 401
tube fittings	148
tubing oxygen class	146
Core drill	546
drilling concrete	39
Cornice drain	270
Corporation stop	401
Corrosion resistance	392
resistant fitting	300
resistant lab equipment	303
resistant pipe	298, 299
resistant pipe fitting	300
resistant trap	264
Corrosion-resistant sink	287
Corrugated metal pipe	409, 413
pipe	413
steel pipe	413
Cost mark-up	10
Countertop sink	282
Coupler air line	273
Coupling fire hose	90
mechanical	184
plastic	222
polypropylene	226
Course wearing	386
Cover checkered plate	46
manhole	412
stair tread	16
Covering calcium silicate pipe	127
protection saddle	117
CPVC pipe	211
valve	108
Crack cleaner	553
filler	553
Crane	422
climbing	554
crawler	554
crew daily	12
hydraulic	422, 554
material handling	422
tower	12, 554
truck mounted	554
Crawler crane	554
drill rotary	550
shovel	548
Crew daily crane	12
forklift	12
Crosses	402
brace	46
Crowbar fire	91
Cubicle shower	283
toilet	63
Culvert end	411
reinforced	411
Cup sink	288, 446
Curb builder	546
extruder	547

prefabricated	352
slipform paver	553
stop	401
Curtain removal air	306
rod	65
Custom casework	78
Cut in sleeve	402
in valve	403
off low water boiler	315
pipe groove	192
Cutter brush	547, 548
Cutting and drilling	354
torch	552

D

Daily crane crew	12
Damper barometric	332
Darkroom equipment	72
sink	72
Deaerators	320
air	320
Deck drain	266
roof	48
Decontamination asbestos area	27
chamber	26, 28
enclosure	26-28
equipment	25
Decorator device	357
switch	357
Deep seal trap	263
Dehumidifier	550
removal	306
Dehydrator package sprinkler	94
Delivery charge	15
Deluge sprinkler	496, 498
sprinkler monitoring panel	95
valve assembly sprinkler	95
Demo hydronic convector & radiator	307
Demolish remove pavement and curb	22
Demolition	101, 658
asbestos	27
boiler	306
furnace	307
hammer	547
HVAC	306
mold contaminated area	29
pavement	22
plumbing	101
site	22
torch cutting	23
Dental chair	76
equipment	76
metal interceptor	268
office equipment	76
Derrick crane guyed	555
crane stiffleg	555
Detection intrusion	366
leak	367
probe leak	367
system	366
tape	396
Detector check valve	405
fire & heat	366
meter water supply	242
smoke	366
temperature rise	366
Detention equipment	70
Detour sign	16
Device anti-siphon	244
combination	358
decorator	357
GFI	359

receptacle	358
residential	357
Dewater	372, 373
pumping	372
Dewatering	372, 678
equipment	555
Diaphragm pump	551
valve	108
Dielectric union	321
Diesel hammer	547
tugboat	556
Diffuser pump suction	326
suction	326
Dilution tank chemical	303
tank chemical-waste	303
Dimmer switch	357
Direct chute concrete	37
Directional drill horizontal	553
Disc harrow	547
Discharge hose	551
Dishwasher commercial	74
residential	72
water heater	277
Dispenser hot water	275
napkin	65
soap	65
toilet tissue	65
towel	65
Dispersion nozzle	95
Disposable bag	25
Disposal asbestos	28
charges	28
field	408
garbage	72
Distiller water	75
Distribution box	409
box concrete	409
box flow leveler	409
box HDPE	409
pipe water	396, 398
Distributor asphalt	550
Ditching	372
Diving stand	80
Dog command	17
Dome drain	271
Domestic hot water temp maint.	120
hot water	458-460
water booster system	248
Door bell residential	359
fire	366
shower	63, 64
Double wall pipe	229, 415
Dozer	548
Draft control	332
damper vent	332
Dragline bucket	547
Drain	263, 270
acid resistant	267
boiler	320
CI acid resistant	267
deck	266
dome	271
expansion joint roof	271
facility trench	271
floor	266
main	271
pipe	372
PVC acid resistant	267
roof	270
sanitary	266
scupper	271
sediment bucket	266
shower	263
terrace	271
trench	271
Drainage control backwater	407

701

Index

field 409
fitting 261
fixture units 670
pipe 298, 299, 408, 409, 411
trap 263, 264
Drill concrete 354
  core 546
  main 402
  quarry 550
  steel 550
  tap main 402
  track 549
Drilling concrete 39
  concrete impact 39
  horizontal 396
  plaster 58
  rig 546
  steel 43
Drinking bubbler 295
  fountain 294, 295
  fountain deck rough-in .. 295
  fountain floor rough-in . 295
  fountain handicap 294
  fountain support 289
  fountain wall rough-in . 294
Drip irrigation 387
  irrigation subsurface 387
  pan elbow 262
Driver post 553
  sheeting 550
Drum trap copper 264
Dry cleaning equipment 70
  pipe sprinkler head 94
  standpipe riser 514
Dryer accessory washer 285
  commercial clothes 71
  hand 65
  industrial 70
  receptacle 359
  residential 73
  vent 73, 286
Dry-pipe sprinkler system ... 94
Dual flush flushometer . 285, 289
  flush valve toilet 289
Duck tarpaulin 15
Duct cleaning 18
  insulation 313
Ductile iron fitting 397
  iron fitting mech joint . 397
  iron grooved-joint 187
  iron pipe 396
Ductless removal split 308
Ductwork clean 18
  removal 306
Dump charges 23
  truck 549
Dumpsters 23, 658
Dumptruck off highway 549
Duplex style strainer 325
Dust barrier 16
  collection systems 424
DWV pipe ABS 210
  PVC pipe 210, 408
  tubing copper 147

E

Earth auger 546
  compactor 547
  scraper 548
  vibrator 370
Earthquake shutoff gas 341
  valve gas 341
Earthwork equipment rental . 546
Ecclesiastical equipment ... 76

Economizer shower head water . 283
Edger concrete floor 546
Efficiency flush valve high . 289
Effluent filter septic system . 408
Ejector pump 269
Elbow drip pan 262
  HDPE piping 401
  pipe 187
Electric baseboard heater . 349
  boiler 339, 488
  boiler small hydronic .. 488
  cabinet heater 351
  commercial water heater . 276
  fire pump 516
  generator 550
  heater 73
  heater removal 307
  heating 349
  hoist 422
  hot air heater 350
  infrared heater 346
  pool heater 340
  unit heater 350
  water cooler 295
  water heater 276
  water heater commercial . 461
  water heater residential . 275, 458
  water heater tankless, gas . 275
Electrical fee 8
  installation drilling ... 354
  knockout 355
  laboratory 75
Electricity temporary 11
Electrofusion welded PP pipe . 301
Elevated installation add . 100
  pipe add 100
Elevator fee 8
Eliminator air 320
Emergency equipment
  laboratory 75
  eyewash 293
  shower 292
Employer liability 10
Emulsion sprayer 550
Encapsulation asbestos 28
  pipe 28
Enclosure bathtub 64
  decontamination 26-28
  shower 64
End culvert 411
Energy circulator air solar . 341
  distribution pipe steam . 417
  system air purger solar . 343
  system air vent solar .. 343
  system balancing valve solar . 343
  system control valve solar . 343
  system controller solar . 342
  system expansion tank solar . 343
  system gauge pressure solar . 343
  system heat exchanger solar . 342
  system storage tank solar . 342
  system vacuum relief solar . 343
Engine exhaust elimination . 330
Engineering fees 8, 648
Entrainment eliminator 269
Entrance screen 62
Epoxy anchor 38, 42
  coating 392
  fiberglass wound pipe . 299
  resin fitting 214
  welded wire 36
Equipment 70, 546
  automotive 70
  barber 70
  cell 70
  cleaning & disposal 74

darkroom 72
dental 76
dental office 76
detention 70
dry cleaning 70
ecclesiastical 76
fire 516
fire hose 89, 516
fire-extinguishing 95
foundation formwork 32
hospital 75
insulation 314
insurance 10
laboratory emergency 75
laundry 70, 71
lubrication 70
medical 75, 76
medical sterilizing 75
mobilization 14
pad 36
refrigerated storage 73
rental 546, 657
rental concrete 546
rental earthwork 546
rental general 549
rental highway 553
rental lifting 554
rental marine 556
rental wellpoint 555
roof support 119
safety eye/face 293
swimming pool 80
Escutcheon 242
  plate 92, 93
  sprinkler 93
Estimate electrical heating . 349
  plumbing 457
Estimating 643
Ethylene glycol 329
Evaporative cooler removal . 307
Evaporator removal 307
Excavating trench 370
  utility trench 371
Excavation 370
  hand 370, 372
  septic tank 409
  tractor 370
  trench 370, 371, 520-522, 524
Excavator bucket 553
  hydraulic 546
Exchanger heat 344
  removal heat 307
Exhaust elimination engine . 330
  elimination garage 330
  hood 72
  system 330
  systems removal 307
Expansion anchor 42
  joint 322
  joint bellows 322
  joint grooved-joint ... 191
  joint roof drain 271
  shield 42
  tank 322
  tank steel 322
Extension backhoe 553
  ladder 551
Exterior sprinkler 388
Extinguisher cart mounted . 67
  chemical dry 67
  CO₂ 67
  fire 67, 95
  FM200 fire 96
  installation 67
  portable ABC 67
  pressurized fire 67

standard 67
wheeled ABC 67
Extra work 10
Extruder curb 547
Eye wash fountain 293
  wash portable 293
Eye/face wash safety equipment . 293
Eyewash emergency 293
  safety equipment 293

F

Fabric-backed flashing 50
Face wash fountain 292
Facility trench drain 271
Factor 8
Fan bathroom exhaust 329
  ceiling 329
  paddle 361
  removal 307
  residential 361
  ventilation 361
  wiring 361
Fan coil A/C removal 307
Faucet & fitting 284, 289, 292
  bathroom 284
  gooseneck 292
  laundry 284
  lavatory 284
  medical 292
F&C plates 242
Fee architectural 8
  engineering 8
Feller buncher 547
Fence chain link 16
  plywood 16
  temporary 16
  wire 16
Fiber reinforcing concrete . 37
Fiberglass insulation 313
  panel 15
  pipe covering 129
  rainwater storage tank . 272
  tank 318
  trench drain 271
  underground storage tank . 248
Field disposal 408
  drainage 409
  office 12
  personnel 10
Fill 373
  borrow 373
  by borrow & utility bedding . 373
  gravel 373
Filler crack 553
  joint 54
Fillet welding 44
Film equipment 72
Filter iron removal 274
  oil 316
  rainwater 428
  removal air 306
  swimming pool 296
  water 274
Filtration air 25
  equipment 80
Finishing floor concrete .. 38
Fin tube radiation apartment
building 480
Fire & heat detectors 366
  alarm 93, 366
  alarm panels and devices . 366
  axe 91
  call pullbox 366
  crowbar 91

Index

door 366	pipe conduit 416	drain 266	Furnace demolition 307
equipment cabinet 66	plastic 214, 219	drain removal 101	Fused pipe 400
extinguisher 67, 95	plastic pipe 86	heating radiant 347	Fusible oil valve 316
extinguisher portable 66, 67	polybutylene pipe 224	heating removal radiant 307	Fusion machine butt 550
extinguishing ... 490-495, 497-504,	polypropylene 226, 301	receptor 267	welded pipe fittings 301
508, 510, 512, 514-517	PVC 216, 399	sander 552	
extinguishing system 86, 89,	PVC pipe 226	scupper 271	**G**
90, 95, 97	refrigeration 154	Flow check control 323	
horn 366	removal piping 101	fitting hydronic 324	Galley septic 409
hose 90	stainless steel 200	meter 327	Galvanized fitting 262
hose adapter 89	stainless steel pipe 200	Flue chimney metal 333	flashing 265
hose coupling 90	steel 161, 172	prefab metal 333	pipe fitting 262
hose equipment 89	steel carbon pipe 175	shutter damper removal 307	welded wire 36
hose gate valve 92	steel pipe 175	Fluid heat transfer 342	Gantry crane 422
hose nipple 90	tube 194	Flush automatic 289	Garage exhaust elimination 330
hose nozzle 90	vent chimney 333, 338	flushometer dual 285, 289	exhaust system 489
hose rack 91	weld 179	HET 289	ventilation 330
hose reel 91	weld joint pipe 179	valve 289	Garbage disposal 72
hose storage cabinet 66	welded steel pipe 183	FM200 fire extinguisher 96	Gas cocks 103
hose storage house 91	Fixture bathroom ... 279, 280, 282, 283	fire suppression 517	connector 238
hose valve 91	carrier/support 289	system 517	conversion burner 341
hose wye 92	institution 293	Foam core DWV ABS pipe 210	earthquake valve 341
hydrant 405, 535	plumbing 267, 268, 275, 279,	core DWV PVC pipe 210	fired boiler 339
hydrant building 88	281, 288, 294, 327, 407	pipe covering 132	fired infrared heater 346
panel 366	prison 293	rubber pipe covering 133	fired water heater commercial . 462
protection 66	removal 101	spray rig 550	fired water heater residential .. 459
protection system	removal prison 102	Foam-water system components .. 95	meter 316
classification 663	residential 360	Folding door shower 64	meter residential 316
pump 96, 97	support handicap 289	Food service equipment 74	pipe 414
retardant pipe 300	Flange blind 174	Foot valve 109	piping medical 297
signal bell 366	clip beam 120	valve oil 316	pressure regulators 240
sprinkler head 94	companion 174	Footing concrete placement 37	pulse combustion 339
suppression 517	gasket 173	formwork continuous 32	regulator 241
tool 91	grooved-joint 190	formwork spread 32	safety valve 341
valve hydrant 92	PVC 224	reinforcing 35	station tank 318
Firecycle sprinkler 500	slip on escutcheon 242	spread 36	stop valve 103
system control package 93	stainless steel 205	Forklift 550	vent 333
valve package 95	steel 181, 190	crew 12	vent fitting 333
Fire-department connection 88	steel weld-on 181	Formwork continuous footing ... 32	water heater commercial 277
Fire-extinguishing equipment ... 95	threaded 174	equipment foundation 32	water heater instantaneous ... 277
Firestop wood 48	Flanged fitting 173	slab edge 32	water heater residential 277
Firestopping 52	pipe 158	slab haunch 37	water heater tankless 277
Fire-suppression automatic 95	Flare joint fitting 154	slab on grade 32	Gasket & bolt set 174
hose reel 91	Flasher barrels 550	slab thickened edge 37	B and S pipe 258
pipe fitting 86	barricade 550	slab trench 32	cast iron pipe 258
plastic pipe 86	Flashing 50	slab turndown 37	flange 173
Fitting adapter pipe 150	aluminum 50, 265	sleeve 32	joint 53
cast iron 172, 250	copper 50, 265	spread footing 32	joint pipe 249
cast iron pipe 161, 250	fabric-backed 50	Foundation mat 36	joint pipe fitting 254
compression joint 152	galvanized 265	mat concrete placement 38	neoprene 53
copper 148	laminated sheet 50	Fountain drinking 294, 295	toilet 285
copper pipe 148	lead 266	eye wash 293	Gasoline piping 415
corrosion resistant 300	masonry 50	face wash 292	Gate box 403
drainage 261	mastic-backed 50	lighting 297	valve 309, 312, 404
drainage special pipe 261	neoprene 266	nozzle 297	valve grooved-joint 192
ductile iron 397	paperbacked 50	pump 296	valve soldered 104
DWV pipe 219	plastic sheet 51	wash 288	Gauge cock 321
epoxy resin 214	PVC 51	water pump 296	fuel oil 316
flanged 173	sheet metal 50	Frame scaffold 13	pressure 315
flanged cast iron pipe 172	stainless 50	Framing lightweight 46	pressure/vacuum 315
flare joint 154	vent 265	lightweight angle 46	vacuum 315
forged steel pipe 179	vent chimney 334, 338	lightweight channel 46	Gear driven sprinkler 389
galvanized 262	Flatbed truck 549, 552	miscellaneous wood 48	General air conditioners 306
gas vent 333	truck crane 554	pipe support 46	boiler 100
glass 298	Flexible conduit & fitting 356	slotted channel 46	contractor's overhead 651
grooved-joint 157, 185	connector 238	wood 48	equipment rental 549
grooved-joint pipe 187	metal hose 238	Freezer 73	fill 373
grooved unit 156, 188	sprinkler connector 93	Frost penetration 659	maintenance 17
HDPE 230	Float finish concrete 38	Fuel oil gauge 316	Generator electric 550
insert pipe 225	valve 323	oil pump 317	steam 75
malleable iron 165	Floater equipment 10	oil pump & motor set 317	GFI receptacle 359
malleable iron pipe 166, 183	Floodlight trailer 550	oil specialties 316	Gland seal pipe conduit 416
mechanical joint pipe 183	tripod 550	oil valve 316	Glass door shower 64
monoflow 324	Floor cleaning 17	tank 318	filler 295
pipe 162, 172, 261,	cleanout 122	Fume elimination welding 332	fitting 298
398, 399, 416, 418	concrete finishing 38	hood removal weld 307	

703

Index

insulation cellular 126
lined water heater 73
mirror 65
P trap 264
pipe 101, 298
pipe fitting 298
pipe removal 101
process supply pipe 298
shower stall 64
Glassware sterilizer 74
 washer 74
Globe valve 309, 312
 valve bronze 105
Glove bag 26
Glycol ethylene 329
 propylene 329
Gooseneck faucet 292
Grab bar 65
Gradall 547
Grader motorized 547
Grading 379
Gravel fill 373
 pack well 406
Grease interceptor 268
 trap 268
Grille painting 58
Grinder concrete floor 546
 pump/sewage ejector removal . 101
 system pump 269
Grit removal and handling
 equipment 428
Groove cut labor steel 192
 roll labor steel 193
Grooved-joint check valve ... 191
 joint copper 156
 joint ductile iron 187
 joint fitting 157, 185
 joint flange 190
 joint pipe 185
 joint plug valve 192
 joint reducer 157
 joint strainer 191
 joint suction diffuser 191
 reducer 188
 unit fitting 156, 188
Ground box hydrant 245
 hydrant removal 101
 post hydrant 246
Group shower 288
 wash fountain 288, 449
Grout pump 552
Guard service 17
 snow 51
 sprinkler head 94
Guardrail scaffold 13
 temporary 16
Guide pipe 112
Gun type burner 341
Guyed derrick crane 555

H

Hair interceptor 268
Hammer chipping 550
 demolition 547
 diesel 547
 drill rotary 550
 hydraulic 547, 551
 pile 547
 vibratory 548
Hammermill 553
Hand dryer 65
 excavation 370, 372
 hole 395
 trowel finish concrete 38

Handicap drinking fountain ... 294
 fixture support 289
 lavatory 292
 shower 283, 292
 tub shower 282
 water cooler 295
Handling material 422
Hand/off/automatic control ... 363
Hanger & support pipe 110
 assembly pipe 115
 cable 120
 clevis 113
 pipe 110
 plastic pipe 119
 rod coupling 115
 rod pipe 115
 rod socket 118
 roll 116
 split ring 114
 trapeze 116
 U-bolt pipe 118
 U-hook 119
 vibration absorbing 121
Hanger/support removal pipe ... 101
Hanging lintel 46
Hardboard cabinet 78
Harrow disc 547
Hat and coat strip 65
Hauling 374
 cycle 374
Haunch slab 37
Hazardous waste cleanup 24
 waste disposal 24
HDPE distribution box 409
 fittings 230
 pipe 229
 pipe dual wall 233
 pipe liner 395
 piping 400
 piping caps 401
 piping elbows 401
 piping tees 401
Head sprinkler 93, 389
Header pipe wellpoint 555
Heat baseboard 347
 electric baseboard 349
 exchanger 344
 exchanger plate-type 344
 exchanger removal 307
 exchanger shell-type 344
 pipes 344
 pump 345
 pump air-source 345
 pump removal 307
 pump residential 361
 recovery package removal ... 307
 temporary 11
 therapy 75
 trace system 349
 transfer fluid 342
 transfer package removal ... 307
Heater & fan bathroom 329
 cabinet convector 350
 ceiling mount unit 351
 electric 73
 electric hot air 350
 gas fired infrared 346
 gas residential water 277
 gas water 277
 infrared 346
 infrared quartz 350
 quartz 350
 removal electric 307
 removal space 308
 sauna 81
 space 551

swimming pool 340
swimming pool solar 485
terminal 346
tubular infrared 346
vertical discharge 350
wall mounted 351
water 73, 277, 361
water residential electric ... 275
Heating 125, 129, 316, 338,
     340, 342, 344, 347
 & ventilating unit removal ... 307
 control valve 323
 electric 349
 estimate electrical 349
 fin tube residential 480
 hot water solar 467
 hydronic 339, 340, 347
 industrial 339
 insulation 128, 314
 kettle 554
 panel radiant 350
 solar energy 475, 485
Heavy duty shoring 13
Helicopter 555
HET flush 289
High efficiency flush valve ... 289
 efficiency urinal 289
 impact pipe 208
Highway equipment rental 553
Historical Cost Index 9
Hoist 422
 chain 422
 chain fall 555
 electric 422
 overhead 422
 personnel 555
 tower 555
Holding tank 428
Hole cutting electrical 355
 drilling electrical 354
Hollow wall anchor 42
Hood and ventilation equipment 332
 exhaust 72
 range 72
Hook coat 65
 robe 65
 type pipe outlet 185
Horizontal borer 553
 boring 396
 boring machine 546
 directional drill 553
 drilling 396
Horn fire 366
Hose & hydrant house 91
 adapter fire 89
 air 550
 bibb 320
 bibb sillcock 285
 braided bronze 238
 braided stainless steel ... 239
 braided steel 239
 discharge 551
 equipment 89
 fire 90
 metal flexible 238
 nozzle 90
 rack 91
 rack cabinet 66
 reel fire 91
 reel fire-suppression 91
 suction 551
 valve cabinet 67
 water 551
Hospital bath 292
 equipment 75
 sink 292

water closet 291
whirlpool 292
Hot air heater electric 350
 tub 80
 water boiler 339, 340
 water dispenser 275
 water dispenser removal 101
 water heating 339, 347
 water temp maint domestic ... 120
 water-water exchange ... 344
 weather concrete 37
Hotel cabinet 66
Hot water commercial 461-463
 heating forced 480
 solar energy 465, 467, 469,
    471, 473, 477, 483
 system 458, 460-463
House fire hose storage 91
 hose & hydrant 91
Hubbard tank 75
Humidifier 73
 removal 307
HVAC demolition 306
 equipment insulation 314
 piping specialties .. 239, 240, 269,
    315, 322, 324, 326
Hydrant building fire 88
 fire 405
 ground box 245
 ground post 246
 removal 22
 removal ground 101
 removal wall 101
 screw type valve 92
 tool 91
 wall 245
 water 245
Hydraulic chair 76
 crane 422, 554
 excavator 546
 hammer 547, 551
 jack 555
 jack climbing 555
 lift 84
 pressure test 313
Hydrocumulator 248
Hydromulcher 553
Hydronic balancing 327
 baseboard removal 307
 convector & radiator demo. ... 307
 energy underground 417
 flow fitting 324
 heating 339, 340, 347
 heating control valve ... 323
 heating convector 347
 unit heater removal 307

I

Ice machine 74
Icemaker 72, 74
Impact sprinkler 389
 wrench 550
Index construction cost 9
Indicator thermoflo 327
 valve post 405
Indoor water storage tank ... 249
Induced draft fan removal 307
Industrial dryer 70
 heating 339
 safety fixture 292
 safety fixture removal 101
Inert gas 24
Infrared heater 346
 heater electric 346

704

Index

heater gas fired 346
heater, tubular 346
quartz heater 350
Insert fitting PVC 226
 hanger concrete 112
Inspection internal pipe 394
Installation add elevated 100
 extinguisher 67
Instantaneous gas water heaters .. 277
 water heater 275
Institution fixture 293
Insulated protectors ADA 124
Insulation 129
 all service jacket 125, 130
 blower 551
 board 314
 boiler 314
 breeching 314
 calcium silicate 314
 duct 313
 equipment 314
 fiberglass 313
 heating 128, 314
 HVAC equipment 314
 jacket stainless-steel 142
 jacketing 134
 jacketing aluminum 138
 jacketing protective 134
 pipe 124
 piping 124
 polyethylene 132, 313
 polyolefin 133
 removal 26
 vapor proof 126
 vaporproof 133
 water heater 124
Insurance 10, 650
 builder risk 10
 equipment 10
 public liability 10
Interceptor 268
 drain oil 267
 grease 268
 hair 268
 lint 268
 metal recovery 268
 oil 268
 sand 268
Internal cleaning pipe 394
Interval timer 357
Intrusion detection 366
 systems 366
Invert manhole 413
Iron alloy mechanical joint
 pipe 299
 body valve 308
 removal filter 274
Irrigation drip 387
 system sprinkler 388

J

J hook clamp w/nail 120
Jack cable 555
 hydraulic 555
Jacketing insulation 134
Jackhammer 549
Jacking 396
Jet water system 407
Job condition 9
Jockey pump fire 96
Joint control 32
 expansion 322
 filler 54
 gaskets 53

grooved-joint expansion 191
 push-on 396
 restraint 397, 400
 sealer 53, 54
 Tyton 396

K

Kettle 74
 heating 554
 tar 552, 554
Kitchen cabinet 78
 equipment 74
 equipment fee 8
 sink 282, 444
 sink faucet 284
 sink residential 282
 unit commercial 71
Knee action mixing valve 292
Knockout electrical 355

L

Label pipe & valve 122
Labor adjustment factors 100
 index 9
 modifiers 100
 pipe groove roll 193
 pipe threading st. steel 199
 pipe threading steel 160
 pipe weld stainless steel 199
 pipe weld steel 160
Laboratory safety equipment 75
 sink 74, 75, 287, 446
Ladder extension 551
 rolling 14
 towel 66
Lag screw 44
 screw shield 42
Laminated sheet flashing 50
Lance concrete 551
Landfill compactor 548
Landscape fee 8
Laser level 551
Latex caulking 54
Laundry equipment 70, 71
 faucet 284
 sink 283, 445
 water heater 278
 water heater coin 278
Lavatory 442
 commercial 287
 faucet 284
 handicap 292
 pedestal type 281
 prison/institution 293
 removal 101
 residential 280, 287
 sink 280
 support 289
 system battery mount 443
 vanity top 280
 wall hung 281
Lawn mower 551
 sprinkler 389, 529
Leaching field 537
 field chamber 408
 pit 408
Lead caulking 249
 flashing 50, 266
 salvage 102
 screw anchor 43
Leader line wye 92
Leads pile 547

Leak detection 367
 detection probe 367
 detection tank 367
Level laser 551
Leveling jack shoring 13
Liability employer 10
 insurance 651
Lift aerial 549
 automotive 84
 hydraulic 84
 scissor 422, 549
 truck aerial 554
Lifting equipment rental 554
Light dental 76
 stand 25
 temporary 11
 tower 551
 underwater 80
Lighting fountain 297
 outlet 359
 residential 359
 temporary 11
Lightning suppressor 357
Lightweight angle framing 46
 channel framing 46
 framing 46
Line remover traffic 554
Lined valve 107
Liner cement 395
 pipe 395
Link seals pipe 241
Lint interceptor 268
Lintel hanging 46
Load center residential 356
Loader skidsteer 549
 tractor 548
 vacuum 25
 wheeled 548
 windrow 554
Locking receptacle 359
Locomotive tunnel 554
Log chipper 548
 skidder 548
Lowbed trailer 554
Lubricated plug valve 311
Lubrication equipment 70

M

Machine auto-scrub 550
 screw 44
 trowel finish concrete 38
 welding 553
Machinery anchor 35
Magnetic motor starter 362
Main drain 271
Maintenance cable
 temperature 120
 general 17
 mechanical 17
Make up air unit removal 308
Malleable iron fitting 165
 iron pipe fitting 165
Management fee construction 8
Manhole 414
 brick 414
 concrete 414
 cover 412
 frame and cover 412
 invert 413
 raise 413
 removal 22
 step 414
Manifold roof fire valve 88
Marble screen 63

shower stall 64
 sink 442
Marine equipment rental 556
Marker boundary and survey 22
 pipe 122
Mark-up cost 10
Mason scaffold 13
Masonry flashing 50
 manhole 414
 saw 552
Mastic-backed flashing 50
Mat concrete placement
 foundation 36, 38
Material handling 422
 index 9
Mechanical 664
 coupling 184
 equipment demolition 307
 fee 8
 maintenance 17
 roof supports 119
Medical alarm systems 297
 clean copper tubing 146
 equipment 75, 76
 faucet 292
 gas piping 297
 gas specialties removal 101
 nitrogen or oxygen 297
 sterilizer 75
 sterilizing equipment 75
 vacuum pump 297
 vacuum system 297
Medicine cabinet 66
Metal flexible hose 238
 flue chimney 333
 interceptor dental 268
 nailing anchor 42
 pipe 409
 pipe removal 101
 recovery interceptor 268
 support assemblies 58
 toilet partition 62
 valve removal 102
Meter 316
 flow 327
 gas 316
 Venturi flow 327
 water supply 242
 water supply detector 242
 water supply domestic 242
Metric conversion factors 646
Microtunneling 395
Mineral wool board 314
 wool pipe covering 124
Minor site demolition 22
Mirror 65
 glass 65
Miscellaneous painting 58
 pump 247
Mixer concrete 546
 mortar 546, 551
 plaster 551
 road 553
Mixing shower valve 283
 valve 240, 285
 valve shower 283
Mobilization equipment 14
 or demobilization 14
Modification to cost 9
Modifiers labor 100
Module tub-shower 282
Moil point 550
Mold abatement work area 28
 contaminated area demolition .. 29
Monoflow fitting 324
 tee fitting 324

705

Index

Monument survey	22
Mop holder strip	65
sink	288
Mortar mixer	546, 551
Mortuary equipment	76
Motor centrifugal in-line	247
connections	356
starter	362
starter & control	362
starter enclosed & heated	362
starter magnetic	362
starter w/circuit protector	362
starter w/fused switch	362
Motorized grader	547
zone valve	316
Mount vibration absorbing	121
Mounting board plywood	48
Mower lawn	551
Muck car tunnel	554
Mud pump	546
trailer	553
Mulcher power	548
Mylar tarpaulin	15

N

Nailer pneumatic	550, 551
wood	48
Nailing anchor	42
Napkin dispenser	65
Needle valve	110
Neoprene flashing	266
gasket	53
Nipple fire hose	90
pipe	165
Nitrogen or oxygen medical	297
No hub pipe	250
hub pipe fitting	259
Non-chloride accelerator concrete	37
Nondestructive pipe testing	313
Nonfusible oil valve	316
Nozzle dispersion	95
fire hose	90
fog	90
fountain	297
playpipe	90
Nut threaded	114
Nylon nailing anchor	42

O

Oakum caulking	249
Observation well	406
Occupancy determinations	644
Off highway dumptruck	549
Office & storage space	12
field	12
overhead	10
trailer	12
Oil burner	341
filter	316
fired boiler	340
fired infrared heater	346
fired water heater commercial	463
fired water heater residential	460
interceptor	268
interceptor drain	267
shutoff valve	316
specialties fuel	316
valve fuel	316
Oil-fired boiler	340

water heater	278
water heater commercial	278
water heater residential	278
Oil/water separator	429
Omitted work	10
Opening roof frame	46
Operating room equipment	76
OSHA testing	27
Outlet lighting	359
Oval arch culverts	412
chimney vent	335
Oven	72, 74
Overhaul	23
Overhead & profit	10, 11
contractor	17, 651
hoist	422
office	10
Overtime	9, 10, 650
Oxygen lance cutting	23

P

P trap	264
trap glass	264
trap running	263
Packaged terminal A/C removal	308
terminal air conditioner	344
water heater system	277
Packaging waste	28
Pad condenser	352
equipment	36
prefabricated	352
vibration absorbing	121
Paddle fan	361
Paint & coatings	58
interior miscellaneous	58
sprayer	551
striper	553, 554
Painting booth	70
grille	58
miscellaneous	58
pipe	58
swimming pool	80
Panel and device alarm	366
fiberglass	15
fire	366
radiant heat	350
Paperbacked flashing	50
Paperholder	65
Paraffin bath	75
Parking lots paving	386
Partition shower	63
toilet	62, 63
toilet stone	63
Patient care equipment	75
Pavement asphaltic	386
breaker	549
bucket	553
demolition	22
planer	554
profiler	554
replacement	386
widener	553
Paver asphalt	551
bituminous	551
cold mix	553
concrete	551
roof	51
shoulder	553
Paving asphaltic concrete	386
parking lots	386
Pedestal type lavatory	281
Perforated aluminum pipe	413
pipe	413
PVC pipe	413

Performance bond	11, 656
Perineal bath	292
Permit building	11
Personal respirator	25
Personnel field	10
hoist	555
Pet cock	321
PEX pipe	213
tubing	212, 347
tubing fittings	348
Photography	11
construction	11
time lapse	11
Pickup truck	552
Pick-up vacuum	546
Pigtail steam syphon	321
Pilaster toilet partition	63
Pile driving	657
hammer	547
leads	547
Piling sheet	679
Pin powder	44
Pipe & fittings	103, 106, 109, 111, 123, 145, 147, 148, 157, 164, 167, 175, 186, 195, 201, 208, 219, 242, 243, 245, 251, 264, 266, 270, 284, 290, 298, 300, 310-312
& valve labels	122
acid resistant	300
add elevated	100
aluminum	410, 413
and fittings	668
bedding	373, 526, 527
bedding trench	372
bracket	110
brass	144
bursting	395
carrier	112, 116
cast iron	249
clamp	111, 392
clamp plastic	119
cleaning internal	394
cleanout	122
concrete	411
conduit gland seal	416
conduit prefabricated	416, 417
conduit preinsulated	416
conduit system	416, 419
contained	233
copper	145, 401
corrosion resistant	299
corrugated	413
corrugated metal	409, 413
covering	124
covering calcium silicate	127
covering fiberglass	129
covering foam	132
covering mineral wool	124
CPVC	211
double wall	229, 415
drain	372
drainage	298, 299, 408, 409, 411
dual wall HDPE	233
ductile iron	396
DWV PVC	210, 408
elbow	187
encapsulation	28
epoxy fiberglass wound	299
fire retardant	300
fire-suppression plastic	86
fitting	162, 172, 261, 398, 399, 416, 418
fitting adapter	150
fitting brass	144
fitting cast iron	161, 250
fitting copper	148

fitting corrosion resistant	300
fitting drainage special	261
fitting DWV	219
fitting fire-suppression	86
fitting flanged cast iron	172
fitting forged steel	179
fitting galvanized	262
fitting gasket joint	254
fitting glass	298
fitting grooved-joint	187
fitting insert	225
fitting malleable iron	166, 183
fitting mechanical joint	183
fitting no hub	258
fitting pitched	261
fitting plastic	86, 214, 222, 223, 398
fitting polybutylene	224
fitting polypropylene	226
fitting PVC	226
fitting soil	250
fitting stainless steel	200
fitting steel	161, 175
fitting steel carbon	175
fitting weld joint	179
fitting weld steel	175
fitting welded steel	183
flanged	158
foam core DWV ABS	210
foam core DWV PVC	210
freezing	393
fused	400
gas	414
glass	101, 298
groove cut	192
groove roll labor	193
grooved-joint	185
guide	112
hanger	110
hanger & support	110
hanger assembly	115
hanger clevis	113
hanger plastic	119
hanger rod	115
hanger strap	118
hanger U-bolt	118
hanger/support removal	101
HDPE	229
heat	344
high impact	208
inspection internal	394
insulation	124
insulation removal	26
internal cleaning	394
iron alloy mechanical joint	299
liner	395
liner HDPE	395
lining	395
link seals	241
marker	122
metal	409
nipple	165
nipple steel	165
no hub	250
painting	58
perforated aluminum	413
PEX	213
plastic	86, 208, 211
plastic FRP	208
polyethylene	398, 414
polypropylene	213, 300
prefabricated	416, 417
preinsulated	416, 417
process pressure	298
proxylene	300
PVC	208, 398, 408

Index

PVC pressure	209	pipe clamp	119	gunite shell	528	winter	15
reinforced concrete	411	pipe fire-suppression	86	heater electric	340	worker	25
relay	372	pipe fitting	214, 222, 223, 398	heaters swimming	340	Protective insulation jacketing	134
removal	22	pipe hanger	119	swimming	80	Protector ADA insulated	124
removal glass	101	pipe polyethylene	229	Pop up sprinkler	389	Proxylene pipe	300
removal metal	101	pipe removal	102	Porcelain enamel sink	442	P&T relief valve	106
removal plastic	102	pipe weld	232	Portable air compressor	549	PT/AC unit	344
repair	392	screw anchor	43	asphalt plant	553	Pulse boiler gas	339
repair clamp	393	sheet flashing	51	building	12	combustion gas	339
residential PVC	212	toilet partition	63	eye wash	293	Pump	327, 406
ring hanger	113	trap	264	fire extinguisher	66, 67	and motor set	317
rodding	394	trench drains	271	Positive pressure chimney	335, 338	centrifugal	246, 551
roll single	116	valve	108	Post driver	553	circulating	327
roof support	119	valve removal	103	indicator valve	405	concrete	546
seal	241	vent caps	265	Potable water softener	274	condensate removal	328
sewage	408, 409, 411	Plastic-laminate toilet		water treatment	274	contractor	372
sewage collection PVC	408	compartment	62	Potable-water storage	249	diaphragm	551
shock absorber	244	Plate checkered	46	storage tank	248	fire	96, 97
single hub	249	escutcheon	92, 93	Powder actuated tool	44	fire jockey	96
sleeve	241	F&C	242	charge	44	fountain	296
sleeve plastic	32	roadway	554	pin	44	fuel oil	317
sleeve sheet metal	32	Plate-type heat exchanger	344	Power mulcher	548	general utility	246
sleeve steel	32	Platform checkered plate	46	temporary	11	grinder system	269
soil	249	trailer	552	trowel	546	grout	552
stainless steel	195	Playpipe fire hose	91	Precast catch basin	414	heat	345
steam	420	Plow vibrator	549	receptor	64	in-line centrifugal	327
steel	157, 410	Plug valve	311	septic tank	408	miscellaneous	247
subdrainage	413	valve grooved-joint	192	Prefab metal flue	333	mud	546
subdrainage plastic	413	Plumbing	243	Prefabricated comfort station	82	operator	373
support	111	appliance	274, 276, 278, 295, 296	metal chimney	335	pneumatic ejector	329
support chair	116	approximations for quick		pad	352	pressure booster	248
support framing	46	estimating	668	pipe	416, 417	removal	102
support saddle	116	demolition	101	Preinsulated pipe	416, 417	removal heat	307
tee	187	fixture	267, 268, 275,	Pressure booster package		rotary	247
testing	313		279, 281, 288, 294, 327,	water	248	sewage ejector	269
testing nondestructive	313		407, 444, 451, 674	booster pump	248	shallow well	407
threading st steel labor	199	fixture removal	101	booster system water	248	shotcrete	546
threading steel labor	160	fixture requirements	672, 673	gauge	315	submersible	272, 406, 551
water	396	laboratory	75	pipe	209	sump	73, 272
water distribution	401	system estimate	457	pipe process	298	trash	552
weld joint	158, 195	Plywood fence	16	pipe PVC	209	turbine	247
weld stainless steel labor	199	mounting board	48	powered condensate pump	329	vacuum	297
weld steel labor	160	sheathing roof & walls	48	reducing valve steam	240	water	327, 406, 551
wrapping	392	sidewalk	16	reducing valve water	106	water supply well	407
X-ray	313	Pneumatic condensate pump	329	regulator	240	wellpoint	555
Pipebursting	395	ejector pumps	329	regulator gas	240	Pumping	372
Piping designations	679	nailer	550, 551	regulator oil	241	dewater	372
Piping excavation	520, 522, 524	steam condensate pump	329	relief valve	106	Push-on joint	396
fitting removal	101	Pocket pitch	51	restricting valve	92	Putlog scaffold	14
gas service polyethylene	414	Point moil	550	switch well water	407	PVC acid resistant drain	267
gasoline	415	of use water heater	275	test	313	cleanout	124
HDPE	400	Polybutylene adapter	225	test hydraulic	313	fitting	216, 399
insulation	124	Polyethylene coating	392	valve relief	106	flange	224
specialties HVAC	239, 240, 269,	insulation	132, 313	washer	551	flashing	51
	315, 322, 324, 326	pipe	398, 414, 529, 532	Pressure/vacuum gauge	315	insert fitting	226
storm drainage	409	plastic pipe	229	Pressurized fire extinguisher	67	pipe	208, 398, 408, 529, 532
to 10′ high	669	septic tank	408	Preventer backflow	243	pipe perforated	413
Pit leaching	408	tarpaulin	15	removal backflow	101	pipe residential	212
sump	372	Polymer trench drain	271	Primer trap	265	roof drain system	478
Pitch pocket	51	Polyolefin insulation	133	Prison cell	82	trap	264
Pitched pipe fitting	261	Polypropylene accessories	226	fixture	293	union	224
Placing concrete	37	cleanout	124	fixture removal	102	valve	108
Plan remediation	25	couplings	226	toilet	70	well casing	406
Planer pavement	554	pipe	213, 300	work inside	9		
Plank scaffolding	13	pipe fittings	226, 301	Prison/institution lavatory	293		
Plant screening	548	shower	283	shower	294	**Q**	
Plant-mix asphalt paving	386	sink	287	sink	294		
Plaster drilling	58	tools	228	Process supply pipe glass	298	Quarry drill	550
mixer	551	trap	264	Product piping	415	Quartz heater	350
Plastic angle valve	108	tubing	213	Profiler pavement	554	Quick coupling sprinkler	389
ball check valve	109	valve	109	Project overhead	11		
ball valve	108	Polysulfide caulking	55	sign	17		
coupling	222	Polyurethane caulking	55	Promenade drain	266	**R**	
fitting	214, 219	Polyvinyl tarpaulin	15	Property line survey	22		
FRP pipe	208	Pool concrete side/vinyl lined	528	Propylene glycol	329	Raceway	354-356
pipe	86, 208, 211	filtration swimming	296	Protection fire	66	Rack hose	91

707

Index

Radiant floor heating 347
   floor heating removal 307
   heating panel 350
Radiator air vent 320
   cast iron 347
   supply valve 323
   valve 323
Rainwater filter 428
   storage tank fiberglass 272
Raise manhole 413
   manhole frame 413
Rake tractor 548
Rammer/tamper 547
Ramp temporary 15
Range cooking 72
   hood 72
   receptacle 359
Ready mix concrete 37, 648
Receiver ash 66
Receptacle air conditioner 359
   device 358
   dryer 359
   GFI 359
   locking 359
   range 359
   telephone 359
   television 359
   waste 66
   weatherproof 359
Receptor & interceptor removal . 102
   floor 267
   precast 64
   shower 64, 267, 283, 288
   terrazzo 64
Reciprocating compressor 273
   hermetic compressor 273
Recirculating chemical toilet 428
Recovery package removal heat .. 307
Reducer grooved 188
   grooved-joint 157
Redwood bath 80
   tub 80
Reflectorized barrels 550
Refrigerant compressor removal . 308
   removal 308
Refrigerated storage equipment .. 73
Refrigeration bloodbank 74
   fitting 154
   tubing 147
Refrigerator compartment cooler . 296
Regulator gas 241
   oil pressure 241
   pressure 240
Reinforced concrete pipe 411
   culvert 411
Reinforcing concrete 35
   footing 35
Relay pipe 372
Release emergency sprinkler 93
Relief valve P&T 106
   valve self-closing 106
   valve temperature 106
   vent removal 308
Relining sewer 395
Remediation plan 25
   plan/method asbestos 25
Removal air conditioner 306
   asbestos 26
   bathtub 101
   bidet 101
   boiler 306
   catch basin 22
   concrete pipe 22
   condenser 306
   dehumidifier 306
   ductwork 306

evaporator 307
fan 307
fixture 101
heating & ventilating unit ... 307
humidifier 307
hydrant 22
insulation 26
lavatory 101
pipe 22
pipe insulation 26
plumbing fixtures 101
pump 102
refrigerant 308
shower 101
sidewalk 22
sink 101
solar heating system 102
sprinkler system 102
steel pipe 22
tank 24
urinal 101
ventilator 308
wash fountain 101
water closet 101
water fountain 101
water heater 102
water softener 103
whirlpool/hot tub 101
Rental equipment 546
Repair pipe 392
Replacement pavement 386
Residential alarm 360
   appliance 72, 360
   applications 356
   bathtubs 282
   burner 341
   device 357
   dishwasher 72
   door bell 359
   dryer 73
   fan 361
   fin tube heating 480
   fixture 360
   gas meter 316
   gas water heater 277
   heat pump 361
   kitchen sinks 282
   lavatories 280, 287
   lighting 359
   load center 356
   oil-fired water heater .. 278, 460
   pool 528
   PVC pipe 212
   roof jack 330
   service 356
   sink 282
   smoke detector 360
   sprinkler 94
   switch 357
   transition 330
   wall cap 329
   wash bowl 280
   washer 72
   water heater 73, 278, 361, 459
   water heater electric ... 276, 458
   wiring 356, 360
Resistance corrosion 392
Resistant trap corrosion 264
Respirator 25
   personal 25
Retarder concrete 37
Rigid joint sealant 55
Ring hanger pipe 113
Ripper attachment 553
Riser clamp 112
   mounted sprinkler 389

pipe wellpoint 556
standpipe 512, 515
Rivet 44
   aluminum 44
   copper 44
   stainless 45
   steel 45
   tool 45
Road mixer 553
   sweeper 553
   temporary 15
Roadway plate 554
Robe hook 65
Rock trencher 549
Rod backer 54
   coupling hanger 115
   curtain 65
   pipe hanger 115
   socket 118
   threaded 115
   tie 46
   weld 45
Roll hanger 116
Roller sheepsfoot 548
   tandem 548
   vibratory 548
Rolling ladder 14
   tower scaffold 14
Roof accessories 51
   drain 270
   drain removal 101
   drain system 478, 479
   fire valve manifold 88
   flashing vent chimney 338
   frame opening 46
   jack residential 330
   paver 51
   paver and support 51
   sheathing 48
   support equipment 119
   support mechanical 119
   support pipe 119
Rooftop A/C removal 308
Room unit removal computer 306
Rotary crawler drill 550
   hammer drill 550
   pump 247
Rototiller 548
Rough-in bidet 279
   classroom sink 287
   drinking fountain deck 295
   drinking fountain floor 295
   drinking fountain wall 294
   sink corrosion resistant 288
   sink countertop 282
   sink cup 288
   sink raised deck 282
   sink service floor 288
   sink service wall 288
   tub 283
Rubber pipe insulation 132
Rubbish handling 23
Running trap 263

S

S trap 263
Saddle covering protection 117
   pipe support 116
Safety equipment laboratory ... 75
   eye/face equipment 293
   fixture industrial 292
   fixture removal industrial .. 101
   shower 292
   valve gas 341

Salamander 552
Sales tax 9, 649
Sampling air 25
Sand backfill 526
   interceptor 268
Sandblasting equipment 552
Sander floor 552
Sanitary drain 266
   sewer 539
   tee 259
Sauna 81
Saw chain 552
   circular 552
   concrete 546
   masonry 552
Sawn control joint 32
Scaffold baseplate 13
   bracket 13
   caster 13
   catwalk 14
   frame 13
   guardrail 13
   mason 13
   putlog 14
   rolling tower 14
   specialties 14
   stairway 13
   wood plank 13
Scaffolding 656
   plank 13
   tubular 13
Scissor lift 422, 549
Scoop air 320
   air purging 320
Scraper earth 548
Screed, gas engine,
   8HP vibrating 546
Screen entrance 62
   urinal 63
Screening plant 548
Screw anchor 42
   lag 44
   machine 44
Scrub station 76
Scupper drain 271
   floor 271
Seal compression 53
   pipe 241
Sealant acoustical 55
   caulking 54
   control joint 33
   rigid joint 55
   tape 55
Sealer joint 53, 54
Seat toilet 280
   water closet 280
Secondary treatment plant 428
Security shower 294
Sediment bucket drain 266
   strainer trap 265
   strainer Y valve 110
Seismic fire extinguishing 505
   wet pipe system 508, 509
Self-closing relief valve 106
Self-contained single pkg.
   A/C removal 308
Semi-steel valve 311
Sentry dog 17
Separation barrier 26, 28
Separator 269
   air 320
   oil/water 429
Septic galley 409
   system 408, 537
   system effluent filter 408
   tank 408, 537

Index

tank concrete	408
tank polyethylene	408
tank precast	408
trench	537
Service boring	396
residential	356
sink	288, 447
sink clinic	292
sink faucet	289
station equipment	70
tap	533
Setter circuit	320
Sewage aeration	428
collection concrete pipe	408
collection PVC pipe	408
collection valves	407
ejector pump	269
municipal waste water	428
pipe	408, 409, 411
treatment plants	428
Sewage/drainage collection	411
Sewer relining	395
valve backwater	407
Sheathing	48
roof	48
roof & walls plywood	48
wall	48
Sheepsfoot roller	548
Sheet metal flashing	50
piling	679
Sheeting driver	550
wood	372, 679
Shelf bathroom	65
Shell-type heat exchanger	344
Shield expansion	42
lag screw	42
Shift work	9
Shock absorber	244
absorber pipe	244
Shoring baseplate	13
bracing	13
heavy duty	13
leveling jack	13
slab	14
Short load concrete	37
Shot blaster	552
Shotcrete pump	546
Shoulder paver	553
Shovel crawler	548
Shower arm	283
built-in	283
by-pass valve	284
cabinet	63
compartments	63
cubicle	283
door	63, 64
drain	263
emergency	292
enclosure	64
glass door	64
group	288
handicap	283, 292
partition	63
polypropylene	283
prison/institution	294
receptor	64, 267, 283, 288
removal	101
safety	292
security	294
spray	285
stall	283
surround	64
system	450
Shower head water economizer	283
Shower/tub control set	285
Shower/tub valve spout set	285
Shutoff gas earthquake	341
Shutoff valve oil	316
Shutter damper removal flue	307
Sidewalk bridge	14
broom	552
removal	22
temporary	16
Sign	17
detour	16
project	17
Signal bell fire	366
Silent check valve	106, 310
Silicone	54
caulking	55
Sillcock hose bibb	285
Simplex style strainer	324
Single hub pipe	249
pipe roll	116
pkg. A/C rmvl. self-contained	308
Sink barber	70
base	78
classroom	287
commercial	287
corrosion resistant rough-in	288
corrosion-resistant	287
countertop	282
countertop rough-in	282
cup	288
cup rough-in	288
darkroom	72
faucet kitchen	284
hospital	292
kitchen	282
laboratory	74, 75, 287
laundry	283
lavatories	280
mop	288
polypropylene	287
prison/institution	294
raised deck rough-in	282
removal	101
residential	282
service	288
service floor rough-in	288
service wall rough-in	288
slop	288
stainless steel	287
support	290
waste treatment	267
Site demolition	22
improvement	387
irrigation	529, 532
utility	526
Sitework manhole	541
Skidder log	548
Skidsteer attachments	549
loader	549
Slab edge formwork	32
haunch	37
haunch formwork	37
on grade	37
on grade formwork	32
shoring	14
thickened edge	37
thickened edge formwork	37
trench formwork	32
turndown formwork	37
Sleeve	241, 403
and tap	397
cut in	402
formwork	32
pipe	241
plastic pipe	32
sheet metal pipe	32
steel pipe	32
wall	241
Sliding door shower	64
mirror	66
Slip on escutcheon flange	242
Slipform paver barrier	553
paver curb	553
Slop sink	288
Slotted channel framing	46
pipe	412
Slow close ball valve	92
Small tool	15
Smoke detector	366
vent chimney	333
Snow guard	51
Snubber valve	321
Soaking bathroom	282
tub	282
Soap dispenser	65
holder	66
Socket hanger rod	118
Socket-O-Let	181
Softener water	274
Soil decontamination	24
pipe	249
Solar air to water heat exchange	487
backflow preventer	343
closed loop add-on hot water	465
closed loop hot water system	477
closed loop space/hot water	483
drain back hot water	466
draindown hot water	469
energy	342
energy circulator air	341
energy system	465, 471, 473, 477, 483
energy system air purger	343
energy system air vent	343
energy system balancing valve	343
energy system collector	341, 342
energy system control valve	343
energy system controller	342
energy system expansion tank	343
energy system gauge pressure	343
energy system heat exchanger	342
energy system storage tank	342
energy system thermometer	343
energy system vacuum relief	343
heating	675
heating system	341
heating system removal	102
hot water	475
recirculation hot water	471
swimming pool heater	485
system solenoid valve	343
Solar energy pool heater	485
water heater	475
Solder	145
Source heat pump water	345
Spa bath	80
Space heater	551
heater removal	308
office & storage	12
Spade air	550
tree	549
Specialties	65
piping HVAC	238, 269, 322, 324, 326
scaffold	14
Split ductless removal	308
ring hanger	114
Spotter	379
Spray painting booth automotive	70
rig foam	550
shower	285
substrate	27
Sprayer airless	25
emulsion	550
paint	551
Spread footing	36
footing formwork	32
Spreader aggregate	546, 553
chemical	553
concrete	551
Sprinkler alarm	366
ball valve	92
cabinet	93
connector flexible	93
control pressure	94
dehydrator package	94
deluge system	497, 498
dry pipe	490
dry pipe system	490, 492
escutcheon	93
firecycle system	500, 501
flow control valve	86
head	93, 389
head guard	94
head wrench	94
irrigation system	388
lawn	389
line tester	88
monitoring panel deluge	95
preaction system	493
quantities	664
release emergency	93
residential	94
system	93, 388, 662
system accelerator	94
system automatic	388
system dry-pipe	94
system removal	102
trim valve	92
underground	387
valve	86, 94
wet pipe system	502-504, 508, 510
Stabilizer base	554
Stainless flashing	50
rivet	45
screen	62
steel fitting	200
steel flange	205
steel pipe	195
steel shelf	65
steel sink	287, 442, 444
steel tubing	193, 194
steel valve	312
weld rod	45
Stainless-steel insulation jacket	142
Stair temporary protection	16
Stairway scaffold	13
Stall shower	283
toilet	62, 63
type urinal	286
urinal	286
Standard extinguisher	67
Standpipe alarm	93, 366
connection	88
equipment	516
systems	660
Starter board & switch	363
motor	362
Station AHU removal central	306
control	363
hospital	75
Steam bath	81
bath residential	81
boiler	339, 340
boiler electric	339
cleaner	552
condensate meter removal	308

709

Index

condensate pump
  pneumatic 329
  energy distribution pipe 417
  jacketed kettle 74
  pipe 420
  pipe underground 416
  pressure reducing valve 240
  radiator valve 323
  syphon pigtail 321
Steel channel 114
  drill 550
  drilling 43
  expansion tank 322
  fitting 161, 172
  flange 181, 190
  pipe 157, 410
  pipe corrugated 413
  pipe fitting 161, 175
  pipe nipple 165
  pipe removal 22
  rivet 45
  roof drain system 478, 479
  salvage 102
  sink 442, 444
  tank 322
  tank above ground 319
  underground storage tank 317
  weld rod 45
  well casing 406
  well screen 406
Step manhole 414
Sterilizer barber 70
  dental 76
  glassware 74
  medical 75
Stiffleg derrick crane 555
Stone toilet compartment 63
Stop corporation 401
  valve 285
  water supply 285
Stop/start control station 363
Storage box 12
  potable-water 249
  tank 318
  tank chemical 303
  tank concrete aboveground ... 319
  tank fiberglass rainwater 272
  tank fiberglass underground .. 248
  tank potable-water 248
  tank steel underground 317
  tank underground 318
Storm drainage manhole frames . 414
  drainage piping 409
  drainage system 479
Strainer basket type 324
  bronze body 325
  duplex style 325
  grooved-joint 191
  simplex style 324
  Y type bronze body 325
  Y type iron body 326
Strap pipe hanger 118
Strapless outlet 185
Strap-on T outlet 189, 190
Strip footing 36
Striper paint 553, 554
Structural fee 8
  welding 44
Stump chipper 548
Subcontractor O&P 11
Subcontractors 674
Subdrainage pipe 413
  plastic pipe 413
  system 413
Submersible pump 272, 406, 551
  sump pump 272

Subsurface drip irrigation 387
Suction diffuser 326
  diffuser grooved-joint 191
  diffuser pump 326
  hose 551
Sump hole construction 372
  pit 372
  pump 73, 272
  pump submersible 272
Supply copper pipe water 401
  ductile iron pipe water 396
  HDPE water 400
  polyethylene pipe water 398
Support carrier fixture 289
  drinking fountain 289
  framing pipe 46
  lavatory 289
  pipe 111
  sink 290
  urinal 290
  water closet 290
  water cooler 291
Suppressor lightning 357
Surfactant 27
Surround shower 64
  tub 64
Survey aerial 22
  monument 22
  property line 22
  topographic 22
Sweeper road 553
Swimming pool 80
  pool equipment 80
  pool filter 296
  pool filtration 296
  pool heaters 340
  pool painting 80
  pool residential 528
  pools 660
Swing check valve 104
  check valve bronze 104
Switch decorator 357
  dimmer 357
  residential 357
  tamper 94
  well water pressure 407
Switchboard electric 363
Swivel adapter fire hose 89
System boiler steam 488
  component foam-water 95
  deluge sprinkler 497, 498
  drinking fountain 451
  dry pipe sprinkler 490-492
  exhaust 330
  fire-extinguishing 86, 89,
                          90, 95, 97
  fire sprinkler ... 490-504, 508, 510
  firecycle sprinkler 499-501
  FM200 fire suppression 517
  grinder pump 269
  heat trace 349
  pipe conduit 416, 419
  plumbing estimate 457
  preaction sprinkler 493-495
  removal exhaust 307
  riser dry standpipe 515
  septic 408
  sitework catch basin 541
  sitework swimming pool 528
  sitework trenching .. 520, 522, 524
  solar energy 471, 485
  solenoid valve solar 343
  sprinkler 86, 93-95
  standpipe riser dry 514, 515
  standpipe riser wet 512
  storm drainage 478

  subdrainage 413
  wet pipe sprinkler 502-504

T

T outlet strap-on 189, 190
Tag valve 122
Tamper 370, 550
  switch 94
Tandem roller 548
Tank above ground steel 319
  aboveground storage conc. ... 319
  clean 24
  compression 323
  darkroom 72
  disposal 24
  expansion 322
  fiberglass 318
  fiberglass rainwater storage .. 272
  fiberglass underground
    storage 248
  holding 428
  horizontal above ground 319
  Hubbard 75
  leak detection 367
  potable-water storage 248
  removal 24
  septic 408
  steel 322
  steel underground storage ... 317
  storage 318
  water 318, 406, 553
  water heater storage 249
  water htr. liquid container ... 102
  water storage solar 343
Tankless electric water heaters .. 275
  gas water heaters 277
Tank/wtr. htr. liquid
container removal 102
Tap and sleeve 397
Tape barricade 16
  detection 396
  sealant 55
  underground 396
Tapping crosses and sleeves 402
  main 402
  valves 403
Tar kettle 552, 554
Tarpaulin 15
  duck 15
  Mylar 15
  polyethylene 15
  polyvinyl 15
Tax 9
  sales 9
  social security 10
  unemployment 10
Taxes 650
Tee cleanout 123
  fitting monoflow 324
  HDPE piping 401
  pipe 187
Telephone receptacle 359
Television receptacle 359
Temperature maintenance
cable 120
  relief valve 106
  rise detector 366
Tempering valve 107
  valve water 107, 240
Temporary barricade 16
  building 12
  construction 11, 15
  electricity 11
  facility 16

  fence 16
  guardrail 16
  heat 11
  light 11
  lighting 11
  power 11
  ramp 15
  road 15
  toilet 552
  utilities 11
Terminal A/C packaged 344
  A/C removal packaged 308
  heater 346
Terne coated flashing 50
Terrace drain 271
Terrazzo receptor 64
Test pressure 313
  well 406
Tester sprinkler line 88
Testing OSHA 27
  pipe 313
Thermoflo indicator 327
Thermometer 315
  solar energy system 343
  well 185
Thermostat contactor
combination 350
  integral 349
  wire 361
Thermosyphon hot water 473
Thickened edge slab 37
Threaded flange 174
  nut 114
  rod 115
Threading labor st. steel 199
  labor steel 160
Thread-O-Let 181
Thru-the-wall A/C removal ... 308
Thru-wall air conditioner 344
Tie rod 46
Time lapse photography 11
Timer clock 360
  interval 357
Toggle bolt anchor 42
Toilet 428
  accessories commercial 65
  bowl 279
  chemical 552
  compartment plastic-laminate .. 62
  compartment stone 63
  dual flush valve 289
  gasket 285
  partition 62, 63
  prison 70
  seat 280
  stall 62, 63
  stone partition 63
  temporary 552
  tissue dispenser 65
Tool fire 91
  hydrant 91
  polypropylene 228
  powder actuated 44
  rivet 45
  small 15
Topographical surveys 22
Torch cutting 552
  cutting demolition 23
Towel bar 65
  dispenser 65
Tower crane 12, 554, 657
  hoist 555
  light 551
  removal cooling 306
Track drill 549
Tractor loader 548

Index

rake 548
truck 552
Traffic line remover 554
Trailer bunk house 12
  floodlight 550
  lowbed 554
  mud 553
  office 12
  platform 552
  truck 552
  water 552
Tram car 552
Transfer package removal heat ... 307
Transition residential 330
Transparent pipe 298
Trap cast iron 263
  deep seal 263
  drainage 263, 264
  grease 268
  plastic 264
  polypropylene 264
  primer 265
  PVC 264
  sediment strainer 265
Trapeze hanger 116
Trash pump 552
Treatment plant secondary 428
  plant sewage 428
  potable-water 274
  salts water 274
Tree spade 549
Trench backfill 371, 526, 527
  box 552
  disposal field 409
  drain 271
  drain fiberglass 271
  drain plastic 271
  drain polymer 271
  excavating 370
  excavation 370, 371
  utility 371
Trencher chain 371, 549
  rock 549
  wheel 549
Trenching 372, 521
  common earth 520, 521
  loam & sandy clay 522, 523
  sand & gravel 524, 525
Tripod floodlight 550
Trowel concrete 546
  power 546
Truck boom 554
  concrete 546
  crane flatbed 554
  dump 549
  flatbed 549, 552
  holding concrete 37
  mounted crane 554
  pickup 552
  tractor 552
  trailer 552
  vacuum 553
  winch 553
Tub bar 65
  hot 80
  redwood 80
  rough-in 283
  shower handicap 282
  soaking 282
  surround 64
Tube copper 145, 401
  fitting 194
  fittings copper 148
Tubing copper 145
  fitting PEX 348
  oxygen class copper 146

PEX 212, 347
  polypropylene 213
  refrigeration 147
  stainless steel 193, 194
Tub-shower module 282
Tubular scaffolding 13
Tugboat diesel 556
Tumbler holder 66
Tunnel locomotive 554
  muck car 554
  ventilator 554
Turbine pump 247
  water meter 242
Turnbuckle 115
Turndown slab 37
TV inspection sewer pipeline 394
Tyton joint 396

U

U-bolt pipe hanger 118
U-hook hanger 119
Underdrain 414
Underground chilled water pipe .. 416
  hydronic energy 417
  piping 526
  sprinkler 387
  steam pipe 416
  storage tank 318
  storage tank fiberglass 248
  storage tank removal 658
  storage tank steel 317
  tape 396
Underwater light 80
Unemployment tax 10
Union dielectric 321
  PVC 224
Unit commercial kitchen 71
  gross area requirements 643
  heater cabinet 351
  heater electric 350
  heater removal hydronic 307
  removal condensing 306
  removal valance 307
Urinal 286
  high efficiency 289
  removal 101
  screen 63
  stall 286
  stall type 286
  support 290
  system 440
  system battery mount 441
  wall hung 286
  waterless 286
  watersaving 286
Utensil washer medical 75
  washer-sanitizer 74
Utility accessories 396
  box 395
  excavation 520, 522, 524
  structure 541
  temporary 11
  trench 371
  trench excavating 371

V

Vacuum breaker 244
  catch basin 553
  central 71
  cleaning 71
  gauge 315
  loader 25

pick-up 546
pump 297
pump medical 297
system medical 297
truck 553
unit central 424
wet/dry 553
Valance unit removal 307
Valve 103, 108, 266, 308,
                   310, 312, 343, 404
  assembly dry pipe sprinkler 95
  assembly sprinkler deluge 95
  backwater 270
  balancing & shut off 323
  ball 109
  ball check 109
  box 403
  brass 103
  bronze 103
  bronze butterfly 104
  butterfly 308, 403
  cabinet 67
  cap fire 92
  check 86, 312, 404, 405
  check lift 107
  check swing 310
  circuit setter balance 320
  clamp 108
  control solar energy system ... 343
  CPVC 108
  cut-in 403
  diaphragm 108
  fire hose 91
  float 323
  flush 289
  foot 109
  foot oil 316
  fusible oil 316
  gas stop 103
  gate 309, 312, 404
  globe 309, 312
  grooved-joint butterfly 192
  grooved-joint gate 192
  heating control 323
  hot water radiator 323
  hydrant fire 92
  hydrant screw type 92
  iron body 308
  lined 107
  lubricated plug 311
  materials 665-667
  mixing 240, 285
  mixing shower 283
  motorized zone 316
  needle 110
  nonfusible oil 316
  oil shutoff 316
  package firecycle 95
  plastic 108
  plug 311
  polypropylene 109
  PVC 108
  radiator 323
  radiator supply 323
  relief pressure 106
  removal metal 102
  removal plastic 103
  selection considerations 665
  semi-steel 311
  sewage collection 407
  shower by-pass 284
  shower mixing 283
  slow close ball 92
  snubber 321
  soldered gate 104
  spout set shower-tub 285

sprinkler 86, 94
sprinkler alarm 86
sprinkler ball 92
sprinkler flow control 86
sprinkler trim 92
stainless steel 312
steam radiator 323
stop 285
swing check 104
tag 122
tapping 403
tempering 107
washing machine 285
water pressure 106
water supply 285
water tempering 240
water utility distribution 403
Y sediment strainer 110
Y-check 110
Vanity base 78
  top lavatory 280, 442
Vapor proof insulation 126
Vaporproof insulation 133
Vent air automatic 320
  alarm 316
  automatic air 320
  cap 265, 316
  cap cast iron 265
  cap plastic 265
  chimney 333
  chimney all fuel 338
  chimney fitting 333, 338
  chimney flashing 334, 338
  chimney prefab. removal 308
  convector or baseboard air ... 347
  draft damper 332
  dryer 73, 286
  flashing 265
  flashing and caps removal 103
  gas 333
  metal chimney 333
  radiator air 320
  removal relief 308
Ventilation equip. and hood 332
  fan 361
  garage 330
Ventilator removal 308
  tunnel 554
Venturi flow meter 327
Vertical discharge heater 350
Vibrating screed, gas
  engine, 8HP 546
Vibration absorber 121
  absorbing hanger 121
  absorbing mount 121
  absorbing pad 121
Vibrator concrete 546
  earth 370
  plow 549
Vibratory hammer 548
  roller 548
Vitreous china lavatory 442
Vitreous china service sink 447

W

Wafer check valve 106, 311
Wall cap residential 329
  heater 73
  hung lavatory 281
  hung urinal 286
  hydrant 245
  hydrant removal 101
  sheathing 48
  sleeve 241

711

Index

type cleanout	123	
urn ash receiver	66	
Wash bowl residential	280	
fountain	288	
fountain group	288	
fountain removal	101	
safety equipment eye/face	293	
Washer	114	
and extractor	71	
commercial	71	
darkroom	72	
dryer accessories	285	
pressure	551	
residential	72	
Washer-sanitizer utensil	74	
Washing machine automatic	72	
machine valve	285	
Waste cleanup hazardous	24	
disposal hazardous	24	
packaging	28	
receptacle	66	
treatment sink	267	
Wastewater treatment system	428	
Watchdog	17	
Watchman service	17	
Water atomizer	25	
booster system domestic	248	
bubbler	451	
chiller remote	296	
chiller removal	308	
closet	279, 287	
closet chemical	428	
closet healthcare	291	
closet hospital	291	
closet removal	101	
closet seat	280	
closet support	290, 291	
cooler	295, 452, 674	
cooler bottled	296	
cooler electric	295	
cooler handicap	295	
cooler support	291	
copper tubing connector	238	
dispenser hot	275	
distiller	75	
distribution pipe	396, 398, 401	
filter	274	
filter commercial removal	103	
filter dirt and rust	274	
flow indicator	86	
fountain removal	101	
hammer arrester	244	
heater	73, 277, 361	
heater car wash	277	
heater commercial	276, 278	
heater dishwasher	277	
heater electric	276	
heater gas	277	
heater gas residential	277	
heater instantaneous	275	
heater insulation	124	
heater laundry	278	
heater oil-fired	278	
heater point of use	275	
heater removal	102	
heater residential	73, 278, 361	
heater residential electric	275	
heater storage tank	249	
heater system packaged	277	
heater under the sink	275	
heater wrap kit	124	
heating hot	339, 347	
hose	551	
hydrant	245	
level control	315	
meter turbine	242	
motor alarm	93	
pipe	396	
pressure booster package	248	
pressure booster system	248	
pressure reducing valve	106	
pressure relief valve	106	
pressure switch well	407	
pressure valve	106	
pump	327, 406, 551	
pump fire	97	
pump fountain	296	
pumping	372	
purification	274	
reducer concrete	37	
service	533	
softener	274	
softener potable	274	
softener removal	103	
source heat pump	345	
storage solar tank	343	
storage tank indoor	249	
supply copper pipe	401	
supply detector meter	242	
supply domestic meter	242	
supply ductile iron pipe	396	
supply HDPE	400	
supply meter	242	
supply polyethylene pipe	398	
supply PVC pipe	398	
supply stop	285	
supply valve	285	
supply well pump	407	
tank	318, 406, 553	
tempering valve	107, 240	
trailer	552	
treatment	428	
treatment potable	274	
treatment salts	274	
utility distribution valves	403	
well	406	
Water closet	439	
battery mount	439	
system	438	
Water heater electric	458, 461	
gas-fired	459, 462	
oil-fired	460, 463	
solar energy	467, 469, 471, 473, 477, 483	
Waterless urinal	286	
Watersaving urinal	286	
Water-water exchange hot	344	
Wearing course	386	
Weather data and design conditions	645	
Weatherproof receptacle	359	
Wedge anchor	43	
Weights and measures	647	
Weld fitting	179	
fume hood removal	307	
joint pipe	158, 195	
plastic pipe	232	
rod	45	
rod aluminum	45	
rod cast-iron	45	
rod stainless	45	
rod steel	45	
X-ray	313	
Welded beam attachment	111	
structural steel	660	
wire epoxy	36	
wire fabric	36	
wire galvanized	36	
Welding fillet	44	
fume elimination	332	
labor stainless-steel	199	
labor steel	160	
machine	553	
structural	44	
Weld-O-Let	181	
Weld-on flange steel	181	
Well	372	
& accessory	406	
casing PVC	406	
casing steel	406	
gravel pack	406	
pump	406	
pump shallow	407	
screen steel	406	
thermometer	185	
water	406	
water pressure switch	407	
Wellpoint	373	
discharge pipe	555	
equipment rental	555	
header pipe	555	
pump	555	
riser pipe	556	
Wellpoints	678	
Wet sprinkler seismic component	505-507	
sprinkler system	505	
standpipe riser	512	
Wet/dry vacuum	553	
Wheel trencher	549	
Wheelbarrow	553	
Wheeled loader	548	
Whirlpool bath	75, 80	
hospital	292	
Whirlpool/hot tub removal	101	
Widener pavement	553	
Winch truck	553	
Window A/C removal	308	
air conditioner	345	
Windrow loader	554	
Winter concrete	37	
protection	15	
Wire fence	16	
thermostat	361	
Wiring air conditioner	361	
fan	361	
residential	356, 360	
Wood firestop	48	
framing	48	
framing miscellaneous	48	
nailer	48	
plank scaffold	13	
sheet piling	679	
sheeting	372, 679	
Work boat	556	
extra	10	
inside prison	9	
Worker protection	25	
Workers' compensation	10, 652-655	
Wrapping pipe	392	
Wrecking ball	553	
Wrench impact	550	
sprinkler head	94	
Wye fire hose	92	

X

X-ray pipe	313
weld	313

Y

Y sediment strainer valve	110
type bronze body strainer	325
type iron body strainer	326
Y check valve	110

Notes

Notes

Notes

Notes

Notes

Notes

Division Notes

	CREW	DAILY OUTPUT	LABOR-HOURS	UNIT	MAT.	LABOR	EQUIP.	TOTAL	TOTAL INCL O&P

Division Notes

Division Notes

Division Notes

		CREW	DAILY OUTPUT	LABOR-HOURS	UNIT	BARE COSTS MAT.	LABOR	EQUIP.	TOTAL	TOTAL INCL O&P

Division Notes

	CREW	DAILY OUTPUT	LABOR-HOURS	UNIT	MAT.	LABOR	EQUIP.	TOTAL	TOTAL INCL O&P

Division Notes

		CREW	DAILY OUTPUT	LABOR-HOURS	UNIT	BARE COSTS MAT.	LABOR	EQUIP.	TOTAL	TOTAL INCL O&P

Reed Construction Data/RSMeans— a tradition of excellence in construction cost information and services since 1942

Table of Contents
Annual Cost Guides
Seminars
New Titles
Reference Books
Order Form

For more information visit the RSMeans website at **www.rsmeans.com**

Unit prices according to the latest MasterFormat!

Book Selection Guide The following table provides definitive information on the content of each cost data publication. The number of lines of data provided in each unit price or assemblies division, as well as the number of reference tables and crews, is listed for each book. The presence of other elements such as equipment rental costs, historical cost indexes, city cost indexes, square foot models, or cross-referenced indexes is also indicated. You can use the table to help select the RSMeans book that has the quantity and type of information you most need in your work.

Unit Cost Divisions	Building Construction	Mechanical	Electrical	Commercial Renovation	Square Foot	Site Work Landsc.	Green Building	Interior	Concrete Masonry	Open Shop	Heavy Construction	Light Commercial	Facilities Construction	Plumbing	Residential
1	582	395	411	508		521	196	319	463	581	511	244	1054	411	180
2	789	279	85	741		1005	180	399	222	788	746	481	1233	289	257
3	1689	347	231	1087		1477	995	329	2040	1688	1685	479	1787	317	385
4	963	21	0	932		729	179	620	1159	931	621	521	1176	0	435
5	1873	158	155	1068		811	1779	1073	716	1873	1034	954	1874	204	723
6	2404	18	18	2062		110	560	1523	278	2400	126	2085	2076	22	2612
7	1531	215	128	1569		584	747	516	509	1528	26	1282	1633	227	1018
8	2107	81	45	2530		268	1140	1768	112	2093	0	2109	2842	0	1511
9	1929	72	26	1726		313	449	2014	387	1870	15	1644	2179	54	1430
10	1006	17	10	623		215	27	818	158	1006	29	511	1098	235	224
11	1026	208	165	489		126	54	878	28	1010	0	230	1046	169	110
12	557	0	2	327		248	151	1669	12	545	19	364	1718	23	317
13	729	117	114	246		347	128	253	66	728	253	82	756	71	80
14	277	36	0	224		0	0	261	0	277	0	12	297	16	6
21	90	0	16	37		0	0	250	0	87	0	68	427	432	220
22	1159	7329	156	1170		1578	1099	849	20	1123	1687	861	7171	9082	712
23	1182	6901	586	886		160	898	741	38	1106	111	759	5013	1830	434
26	1379	499	10134	1039		816	661	1142	55	1351	562	1305	10083	444	621
27	72	0	298	34		13	0	60	0	72	39	52	280	0	4
28	97	58	137	72		0	21	79	0	100	0	41	152	44	26
31	1498	735	610	803		3206	289	7	1208	1443	3285	600	1560	660	611
32	782	54	8	859		4396	352	362	286	753	1824	407	1660	166	455
33	509	1020	542	209		2036	41	0	231	508	2024	120	1585	1211	118
34	109	0	47	4		192	0	0	31	64	195	0	130	0	0
35	18	0	0	0		327	0	0	0	18	442	0	83	0	0
41	60	0	0	32		0	0	22	0	60	30	0	67	14	0
44	75	83	0	0		0	0	0	0	0	0	0	75	75	0
46	23	16	0	0		274	151	0	0	23	264	0	33	33	0
48	12	0	27	0		0	27	0	0	12	19	12	12	0	12
Totals	**24527**	**18659**	**13951**	**19277**		**19752**	**10124**	**15952**	**8019**	**24038**	**15547**	**15223**	**49100**	**16029**	**12501**

Assem Div	Building Construction	Mechanical	Electrical	Commercial Renovation	Square Foot	Site Work Landscape	Assemblies	Green Building	Interior	Concrete Masonry	Heavy Construction	Light Commercial	Facilities Construction	Plumbing	Asm Div	Residential
A		15	0	188	150	577	598	0	0	536	571	154	24	0	1	368
B		0	0	848	2498	0	5658	56	329	1975	368	2089	174	0	2	211
C		0	0	647	872	0	1251	0	1576	146	0	762	249	0	3	588
D		1067	943	712	1856	72	2532	330	822	0	0	1342	1106	1081	4	851
E		0	0	86	260	0	300	0	5	0	0	257	5	0	5	392
F		0	0	0	114	0	114	0	0	0	0	114	3	0	6	357
G		522	442	318	262	3740	663	0	0	534	1558	199	288	758	7	307
															8	760
															9	80
															10	0
															11	0
															12	0
Totals		**1604**	**1385**	**2799**	**6012**	**4389**	**11116**	**386**	**2732**	**3191**	**2497**	**4917**	**1849**	**1839**		**3914**

Reference Section	Building Construction Costs	Mechanical	Electrical	Commercial Renovation	Square Foot	Site Work Landscape	Assem.	Green Building	Interior	Concrete Masonry	Open Shop	Heavy Construction	Light Commercial	Facilities Construction	Plumbing	Resi.
Reference Tables	yes	yes	yes	yes	no	yes	yes	yes	yes	yes	yes	yes	yes	yes	yes	yes
Models					111			25					50			28
Crews	561	561	561	540		561		558	561	561	538	561	538	540	561	538
Equipment Rental Costs	yes	yes	yes	yes		yes		yes	yes	yes	yes	yes	yes	yes	yes	yes
Historical Cost Indexes	yes	yes	yes	yes		yes		yes	yes	yes	yes	yes	yes	yes	yes	no
City Cost Indexes	yes	yes	yes	yes		yes		yes	yes	yes	yes	yes	yes	yes	yes	yes

Annual Cost Guides

Cost guides also available in CD and at rsmeansonline.com

For more information visit the RSMeans website at www.rsmeans.com

Unit prices according to the latest MasterFormat!

RSMeans Building Construction Cost Data 2014

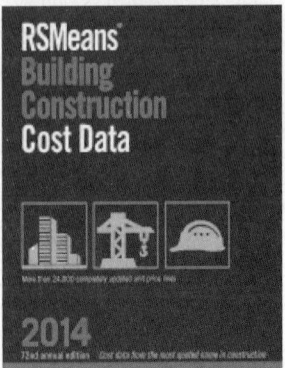

Offers you unchallenged unit price reliability in an easy-to-use format. Whether used for verifying complete, finished estimates or for periodic checks, it supplies more cost facts better and faster than any comparable source. More than 24,000 unit prices have been updated for 2014. The City Cost Indexes and Location Factors cover more than 930 areas, for indexing to any project location in North America. Order and get *RSMeans Quarterly Update Service* FREE.

$194.95 | Available Sept. 2013 | Catalog no. 60014

RSMeans Green Building Cost Data 2014

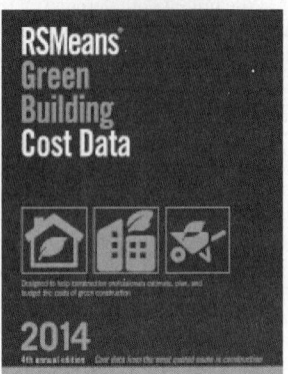

Estimate, plan, and budget the costs of green building for both new commercial construction and renovation work with this fourth edition of *RSMeans Green Building Cost Data*. More than 10,000 unit costs for a wide array of green building products plus assemblies costs. Easily identified cross references to LEED and Green Globes building rating systems criteria.

$160.95 | Available Nov. 2013 | Catalog no. 60554

RSMeans Mechanical Cost Data 2014

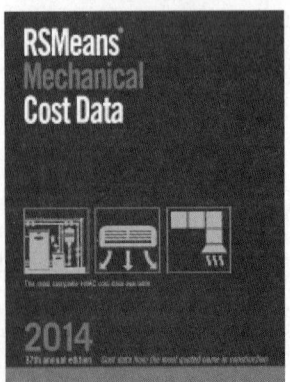

Total unit and systems price guidance for mechanical construction . . . materials, parts, fittings, and complete labor cost information. Includes prices for piping, heating, air conditioning, ventilation, and all related construction.

Plus new 2014 unit costs for:

- Thousands of installed HVAC/controls assemblies components
- "On-site" Location Factors for more than 930 cities and towns in the U.S. and Canada
- Crews, labor, and equipment

$191.95 | Available Oct. 2013 | Catalog no. 60024

RSMeans Facilities Construction Cost Data 2014

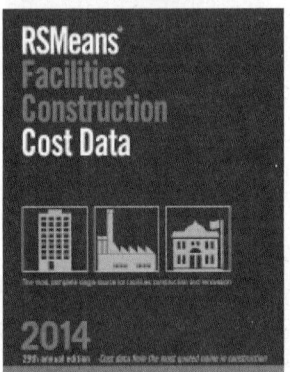

For the maintenance and construction of commercial, industrial, municipal, and institutional properties. Costs are shown for new and remodeling construction and are broken down into materials, labor, equipment, and overhead and profit. Special emphasis is given to sections on mechanical, electrical, furnishings, site work, building maintenance, finish work, and demolition.

More than 49,000 unit costs, plus assemblies costs and a comprehensive Reference Section are included.

$496.95 | Available Nov. 2013 | Catalog no. 60204

RSMeans Square Foot Costs 2014
Accurate and Easy-to-Use

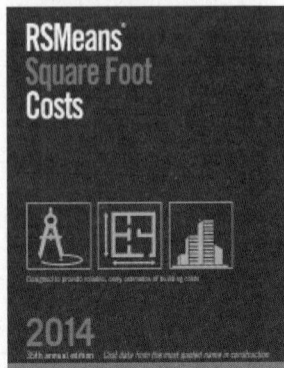

- **Updated price information** based on nationwide figures from suppliers, estimators, labor experts, and contractors
- New green building models
- Realistic graphics, offering true-to-life illustrations of building projects
- Extensive information on using square foot cost data, including sample estimates and alternate pricing methods

$205.95 | Available Oct. 2013 | Catalog no. 60054

RSMeans Commercial Renovation Cost Data 2014
Commercial/Multifamily Residential

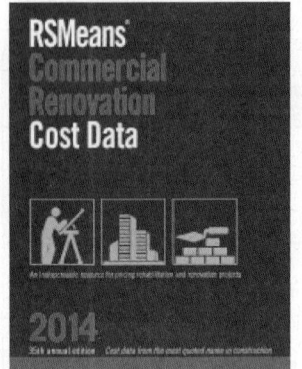

Use this valuable tool to estimate commercial and multifamily residential renovation and remodeling.

Includes: New costs for hundreds of unique methods, materials, and conditions that only come up in repair and remodeling, PLUS:

- Unit costs for more than 19,000 construction components
- Installed costs for more than 2,000 assemblies
- More than 930 "on-site" localization factors for the U.S. and Canada

$156.95 | Available Oct. 2013 | Catalog no. 60044

Annual Cost Guides

Cost guides also available in CD and at rsmeansonline.com

For more information visit the RSMeans website at **www.rsmeans.com**

Unit prices according to the latest MasterFormat!

RSMeans Electrical Cost Data 2014

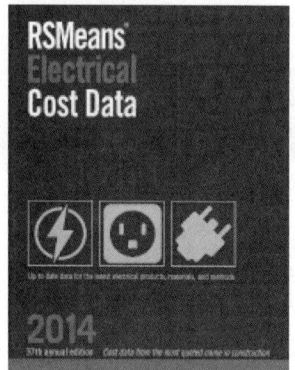

Pricing information for every part of electrical cost planning. More than 13,000 unit and systems costs with design tables; clear specifications and drawings; engineering guides; illustrated estimating procedures; complete labor-hour and materials costs for better scheduling and procurement; and the latest electrical products and construction methods.

- A variety of special electrical systems, including cathodic protection
- Costs for maintenance, demolition, HVAC/mechanical, specialties, equipment, and more

$196.95 | Available Oct. 2013 | Catalog no. 60034

RSMeans Electrical Change Order Cost Data 2014

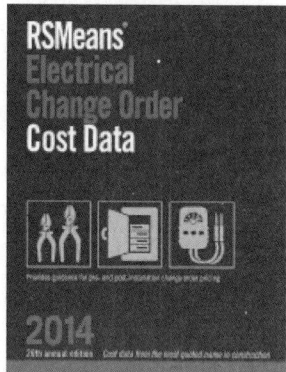

RSMeans Electrical Change Order Cost Data provides you with electrical unit prices exclusively for pricing change orders—based on the recent, direct experience of contractors and suppliers. Analyze and check your own change order estimates against the experience others have had doing the same work. It also covers productivity analysis and change order cost justifications. With useful information for calculating the effects of change orders and dealing with their administration.

$187.95 | Available Dec. 2013 | Catalog no. 60234

RSMeans Assemblies Cost Data 2014

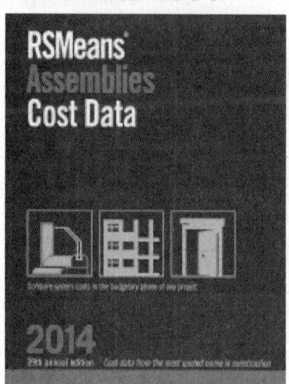

RSMeans Assemblies Cost Data takes the guesswork out of preliminary or conceptual estimates. Now you don't have to try to calculate the assembled cost by working up individual component costs. We've done all the work for you.

Presents detailed illustrations, descriptions, specifications, and costs for every conceivable building assembly—over 350 types in all—arranged in the easy-to-use UNIFORMAT II system. Each illustrated "assembled" cost includes a complete grouping of materials and associated installation costs, including the installing contractor's overhead and profit.

$315.95 | Available Sept. 2013 | Catalog no. 60064

RSMeans Open Shop Building Construction Cost Data 2014

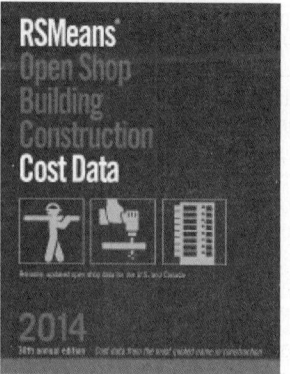

The latest costs for accurate budgeting and estimating of new commercial and residential construction . . . renovation work . . . change orders . . . cost engineering.

RSMeans Open Shop "BCCD" will assist you to:

- Develop benchmark prices for change orders
- Plug gaps in preliminary estimates and budgets
- Estimate complex projects
- Substantiate invoices on contracts
- Price ADA-related renovations

$165.95 | Available Dec. 2013 | Catalog no. 60154

RSMeans Residential Cost Data 2014

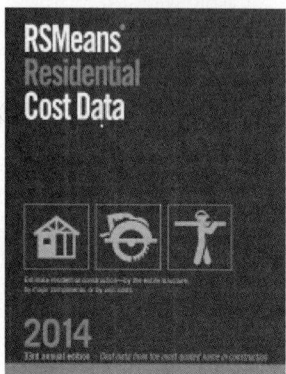

Contains square foot costs for 28 basic home models with the look of today, plus hundreds of custom additions and modifications you can quote right off the page. Includes costs for more than 700 residential systems. Complete with blank estimating forms, sample estimates, and step-by-step instructions.

Contains line items for cultured stone and brick, PVC trim lumber, and TPO roofing.

$139.95 | Available Oct. 2013 | Catalog no. 60174

RSMeans Site Work & Landscape Cost Data 2014

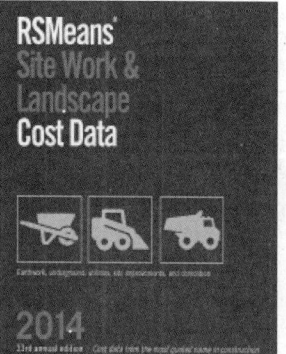

Includes unit and assemblies costs for earthwork, sewerage, piped utilities, site improvements, drainage, paving, trees and shrubs, street openings/repairs, underground tanks, and more. Contains 78 types of assemblies costs for accurate conceptual estimates.

Includes:

- Estimating for infrastructure improvements
- Environmentally-oriented construction
- ADA-mandated handicapped access
- Hazardous waste line items

$187.95 | Available Dec. 2013 | Catalog no. 60284

Annual Cost Guides

Cost guides also available in CD and at rsmeansonline.com

For more information visit the RSMeans website at www.rsmeans.com

Unit prices according to the latest MasterFormat!

RSMeans Facilities Maintenance & Repair Cost Data 2014

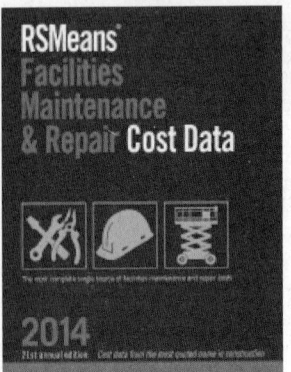

RSMeans Facilities Maintenance & Repair Cost Data gives you a complete system to manage and plan your facility repair and maintenance costs and budget efficiently. Guidelines for auditing a facility and developing an annual maintenance plan. Budgeting is included, along with reference tables on cost and management, and information on frequency and productivity of maintenance operations.

The only nationally recognized source of maintenance and repair costs. Developed in cooperation with the Civil Engineering Research Laboratory (CERL) of the Army Corps of Engineers.

$434.95 | Available Nov. 2013 | Catalog no. 60304

RSMeans Concrete & Masonry Cost Data 2014

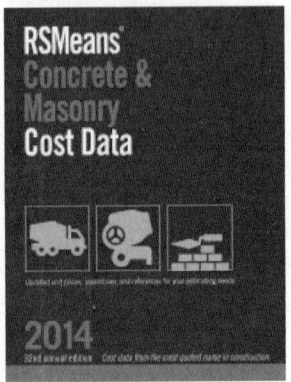

Provides you with cost facts for virtually all concrete/masonry estimating needs, from complicated formwork to various sizes and face finishes of brick and block—all in great detail. The comprehensive Unit Price Section contains more than 8,000 selected entries. Also contains an Assemblies [Cost] Section, and a detailed Reference Section that supplements the cost data.

$177.95 | Available Dec. 2013 | Catalog no. 60114

RSMeans Construction Cost Indexes 2014

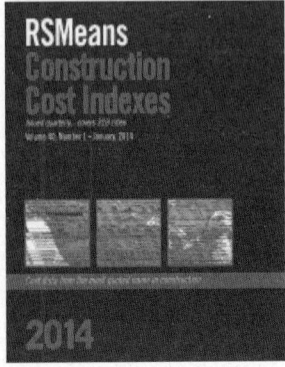

What materials and labor costs will change unexpectedly this year? By how much?

- Breakdowns for 318 major cities
- National averages for 30 key cities
- Expanded five major city indexes
- Historical construction cost indexes

$362.00 per year (subscription) | Catalog no. 50144
$93.50 individual quarters | Catalog no. 60144 A,B,C,D

RSMeans Light Commercial Cost Data 2014

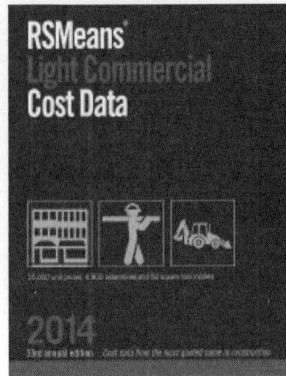

Specifically addresses the light commercial market, which is a specialized niche in the construction industry. Aids you, the owner/designer/contractor, in preparing all types of estimates—from budgets to detailed bids. Includes new advances in methods and materials.

Assemblies Section allows you to evaluate alternatives in the early stages of design/planning.

More than 15,000 unit costs ensure that you have the prices you need . . . when you need them.

$144.95 | Available Nov. 2013 | Catalog no. 60184

RSMeans Labor Rates for the Construction Industry 2014

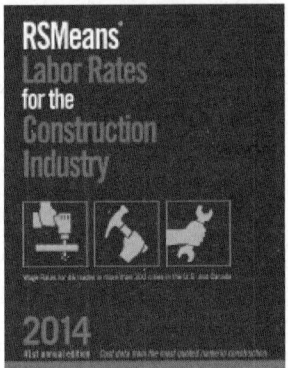

Complete information for estimating labor costs, making comparisons, and negotiating wage rates by trade for more than 300 U.S. and Canadian cities. With 46 construction trades listed by local union number in each city, and historical wage rates included for comparison. Each city chart lists the county and is alphabetically arranged with handy visual flip tabs for quick reference.

$424.95 | Available Dec. 2013 | Catalog no. 60124

RSMeans Interior Cost Data 2014

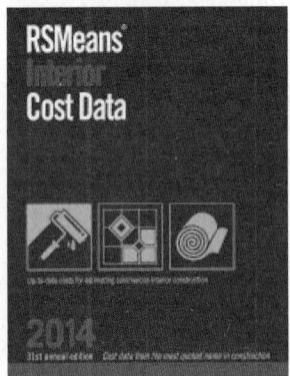

Provides you with prices and guidance needed to make accurate interior work estimates. Contains costs on materials, equipment, hardware, custom installations, furnishings, and labor costs . . . for new and remodel commercial and industrial interior construction, including updated information on office furnishings, and reference information.

$196.95 | Available Nov. 2013 | Catalog no. 60094

Annual Cost Guides

Cost guides also available in CD and at rsmeansonline.com

For more information visit the RSMeans website at www.rsmeans.com

Unit prices according to the latest MasterFormat!

RSMeans Heavy Construction Cost Data 2014

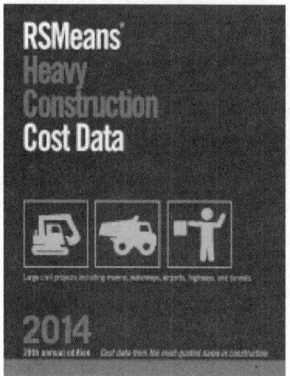

A comprehensive guide to heavy construction costs. Includes costs for highly specialized projects such as tunnels, dams, highways, airports, and waterways. Information on labor rates, equipment, and materials costs is included. Features unit price costs, systems costs, and numerous reference tables for costs and design.

$192.95 | Available Dec. 2013 | Catalog no. 60164

RSMeans Plumbing Cost Data 2014

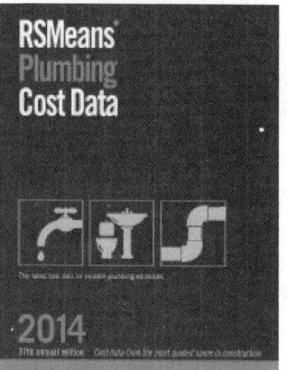

Comprehensive unit prices and assemblies for plumbing, irrigation systems, commercial and residential fire protection, point-of-use water heaters, and the latest approved materials. This publication and its companion, *RSMeans Mechanical Cost Data*, provide full-range cost estimating coverage for all the mechanical trades.

Now contains updated costs for potable water and radiant heat systems and high efficiency fixtures.

$194.95 | Available Oct. 2013 | Catalog no. 60214

2014 RSMeans Seminar Schedule

Note: call for exact dates and details.

Location	Dates	Location	Dates
Seattle, WA	January and August	Jacksonville FL	September
Dallas/Ft. Worth, TX	January	Dallas, TX	September
Austin, TX	February	Charleston, SC	October
Anchorage, AK	March and September	Houston, TX	October
Las Vegas, NV	March and October	Atlanta, GA	November
Washington, DC	April and September	Baltimore, MD	November
Phoenix, AZ	April	Orlando, FL	November
Kansas City, MO	April	San Diego, CA	December
Toronto	May	San Antonio, TX	December
Denver, CO	May	Raleigh, NC	December
San Francisco, CA	June		
Bethesda, MD	June		
Columbus, GA	June		
El Segundo, CA	August		

1-800-334-3509, Press 1

Professional Development

For more information visit the RSMeans website at **www.rsmeans.com**

eLearning Training Sessions

Learn how to use *RSMeans Online*™ or *RSMeans CostWorks*® CD from the convenience of your home or office. Our eLearning training sessions let you join a training conference call and share the instructors' desktops, so you can view the presentation and step-by-step instructions on your own computer screen. The live webinars are held from 9 a.m. to 4 p.m. eastern standard time, with a one-hour break for lunch.

For these sessions, you must have a computer with high speed Internet access and a compatible Web browser. Learn more at www.rsmeansonline.com or call for a schedule: 1-800-334-3509 and press 1. Webinars are generally held on selected Wednesdays each month.

RSMeans Online™
$299 per person

RSMeans CostWorks® CD
$299 per person

Professional Development

For more information visit the RSMeans website at **www.rsmeans.com**

RSMeans Online™ Training

Construction estimating is vital to the decision-making process at each state of every project. RSMeansOnline works the way you do. It's systematic, flexible and intuitive. In this one day class you will see how you can estimate any phase of any project faster and better.

Some of what you'll learn:
- Customizing RSMeansOnline
- Making the most of RSMeans "Circle Reference" numbers
- How to integrate your cost data
- Generate reports, exporting estimates to MS Excel, sharing, collaborating and more

Also available: RSMeans Online™ training webinar

Facilities Construction Estimating

In this *two-day* course, professionals working in facilities management can get help with their daily challenges to establish budgets for all phases of a project.

Some of what you'll learn:
- Determining the full scope of a project
- Identifying the scope of risks and opportunities
- Creative solutions to estimating issues
- Organizing estimates for presentation and discussion
- Special techniques for repair/remodel and maintenance projects
- Negotiating project change orders

Who should attend: facility managers, engineers, contractors, facility tradespeople, planners, and project managers.

Scheduling with Microsoft Project

Two days of hands-on training gives you a look at the basics to putting it all together with MS Project to create a complete project schedule. Learn to work better and faster, manage changes and updates, enhance tracking, generate actionable reports, and boost control of your time and budget.

Some of what you'll learn:
- Defining task/activity relationships: link tasks, establish predecessor/successor relationships and create logs
- Baseline scheduling: the essential what/when
- Using MS Project to determine the critical path
- Integrate RSMeans data with MS Project
- The dollar-loaded schedule: enhance the reliability of this "must-know" technique for scheduling public projects.

Who should attend: contractors' project-management teams, architects and engineers, project owners and their representatives, and those interested in improving their project planning and scheduling and management skills.

Maintenance & Repair Estimating for Facilities

This *two-day* course teaches attendees how to plan, budget, and estimate the cost of ongoing and preventive maintenance and repair for existing buildings and grounds.

Some of what you'll learn:
- The most financially favorable maintenance, repair, and replacement scheduling and estimating
- Auditing and value engineering facilities
- Preventive planning and facilities upgrading
- Determining both in-house and contract-out service costs
- Annual, asset-protecting M&R plan

Who should attend: facility managers, maintenance supervisors, buildings and grounds superintendents, plant managers, planners, estimators, and others involved in facilities planning and budgeting.

Practical Project Management for Construction Professionals

In this *two-day* course, acquire the essential knowledge and develop the skills to effectively and efficiently execute the day-to-day responsibilities of the construction project manager.

Covers:
- General conditions of the construction contract
- Contract modifications: change orders and construction change directives
- Negotiations with subcontractors and vendors
- Effective writing: notification and communications
- Dispute resolution: claims and liens

Who should attend: architects, engineers, owner's representatives, project managers.

Mechanical & Electrical Estimating

This *two-day* course teaches attendees how to prepare more accurate and complete mechanical/electrical estimates, avoiding the pitfalls of omission and double-counting, while understanding the composition and rationale within the RSMeans mechanical/electrical database.

Some of what you'll learn:
- The unique way mechanical and electrical systems are interrelated
- M&E estimates—conceptual, planning, budgeting, and bidding stages
- Order of magnitude, square foot, assemblies, and unit price estimating
- Comparative cost analysis of equipment and design alternatives

Who should attend: architects, engineers, facilities managers, mechanical and electrical contractors, and others who need a highly reliable method for developing, understanding, and evaluating mechanical and electrical contracts.

Professional Development

For more information visit the RSMeans website at www.rsmeans.com

Unit Price Estimating

This interactive *two-day* seminar teaches attendees how to interpret project information and process it into final, detailed estimates with the greatest accuracy level.

The most important credential an estimator can take to the job is the ability to visualize construction and estimate accurately.

Some of what you'll learn:
- Interpreting the design in terms of cost
- The most detailed, time-tested methodology for accurate pricing
- Key cost drivers—material, labor, equipment, staging, and subcontracts
- Understanding direct and indirect costs for accurate job cost accounting and change order management

Who should attend: corporate and government estimators and purchasers, architects, engineers, and others who need to produce accurate project estimates.

RSMeans CostWorks® CD Training

This one-day course helps users become more familiar with the functionality of *RSMeans CostWork*s program. Each menu, icon, screen, and function found in the program is explained in depth. Time is devoted to hands-on estimating exercises.

Some of what you'll learn:
- Searching the database using all navigation methods
- Exporting RSMeans data to your preferred spreadsheet format
- Viewing crews, assembly components, and much more
- Automatically regionalizing the database

This training session requires you to bring a laptop computer to class.

When you register for this course you will receive an outline for your laptop requirements.

Also offering web training for CostWorks CD!

Facilities Est. Using RSMeans CostWorks® CD

Combines hands-on skill building with best estimating practices and real-life problems. Brings you up-to-date with key concepts, and provides tips, pointers, and guidelines to save time and avoid cost oversights and errors.

Some of what you'll learn:
- Estimating process concepts
- Customizing and adapting RSMeans cost data
- Establishing scope of work to account for all known variables
- Budget estimating: when, why, and how
- Site visits: what to look for—what you can't afford to overlook
- How to estimate repair and remodeling variables

This training session requires you to bring a laptop computer to class.

Who should attend: facility managers, architects, engineers, contractors, facility tradespeople, planners, project managers and anyone involved with JOC, SABRE, or IDIQ.

Conceptual Estimating Using RSMeans CostWorks® CD

This *two day* class uses the leading industry data and a powerful software package to develop highly accurate conceptual estimates for your construction projects. All attendees must bring a laptop computer loaded with the current year *Square Foot Model Costs* and the *Assemblies Cost Data* CostWorks titles.

Some of what you'll learn:
- Introduction to conceptual estimating
- Types of conceptual estimates
- Helpful hints
- Order of magnitude estimating
- Square foot estimating
- Assemblies estimating

Who should attend: architects, engineers, contractors, construction estimators, owner's representatives, and anyone looking for an electronic method for performing square foot estimating.

Assessing Scope of Work for Facility Construction Estimating

This *two-day* practical training program addresses the vital importance of understanding the SCOPE of projects in order to produce accurate cost estimates for facility repair and remodeling.

Some of what you'll learn:
- Discussions of site visits, plans/specs, record drawings of facilities, and site-specific lists
- Review of CSI divisions, including means, methods, materials, and the challenges of scoping each topic
- Exercises in SCOPE identification and SCOPE writing for accurate estimating of projects
- Hands-on exercises that require SCOPE, take-off, and pricing

Who should attend: corporate and government estimators, planners, facility managers, and others who need to produce accurate project estimates.

Unit Price Est. Using RSMeans CostWorks® CD

Step-by-step instruction and practice problems to identify and track key cost drivers—material, labor, equipment, staging, and subcontractors—for each specific task. Learn the most detailed, time-tested methodology for accurately "pricing" these variables, their impact on each other and on total cost.

Some of what you'll learn:
- Unit price cost estimating
- Order of magnitude, square foot, and assemblies estimating
- Quantity takeoff
- Direct and indirect construction costs
- Development of contractor's bill rates
- How to use *RSMeans Building Construction Cost Data*

This training session requires you to bring a laptop computer to class.

Who should attend: architects, engineers, corporate and government estimators, facility managers, and government procurement staff.

Professional Development

For more information visit the RSMeans website at **www.rsmeans.com**

Registration Information

Register early . . . Save up to $100! Register 30 days before the start date of a seminar and save $100 off your total fee. *Note: This discount can be applied only once per order. It cannot be applied to team discount registrations or any other special offer.*

How to register Register by phone today! The RSMeans toll-free number for making reservations is **1-800-334-3509, Press 1.**

Two day seminar registration fee - $935. One-day *RSMeans CostWorks*® training registration fee - $375. To register by mail, complete the registration form and return, with your full fee, to: RSMeans Seminars, 700 Longwater Drive, Norwell, MA 02061.

Government pricing All federal government employees save off the regular seminar price. Other promotional discounts cannot be combined with the government discount.

Team discount program for two to four seminar registrations. Call for pricing: 1-800-334-3509, Press 1

Multiple course discounts When signing up for two or more courses, call for pricing.

Refund policy Cancellations will be accepted up to ten business days prior to the seminar start. There are no refunds for cancellations received later than ten working days prior to the first day of the seminar. A $150 processing fee will be applied for all cancellations. Written notice of cancellation is required. Substitutions can be made at any time before the session starts. **No-shows are subject to the full seminar fee.**

AACE approved courses Many seminars described and offered here have been approved for 14 hours (1.4 recertification credits) of credit by the AACE International Certification Board toward meeting the continuing education requirements for recertification as a Certified Cost Engineer/Certified Cost Consultant.

AIA Continuing Education We are registered with the AIA Continuing Education System (AIA/CES) and are committed to developing quality learning activities in accordance with the CES criteria. Many seminars meet the AIA/CES criteria for Quality Level 2. AIA members may receive (14) learning units (LUs) for each two-day RSMeans course.

Daily course schedule The first day of each seminar session begins at 8:30 a.m. and ends at 4:30 p.m. The second day begins at 8:00 a.m. and ends at 4:00 p.m. Participants are urged to bring a hand-held calculator, since many actual problems will be worked out in each session.

Continental breakfast Your registration includes the cost of a continental breakfast, and a morning and afternoon refreshment break. These informal segments allow you to discuss topics of mutual interest with other seminar attendees. (You are free to make your own lunch and dinner arrangements.)

Hotel/transportation arrangements RSMeans arranges to hold a block of rooms at most host hotels. To take advantage of special group rates when making your reservation, be sure to mention that you are attending the RSMeans seminar. You are, of course, free to stay at the lodging place of your choice. **(Hotel reservations and transportation arrangements should be made directly by seminar attendees.)**

Important Class sizes are limited, so please register as soon as possible.

Note: Pricing subject to change.

Registration Form

ADDS-1000

Call 1-800-334-3509, Press 1 to register or FAX this form 1-800-632-6732. Visit our website: www.rsmeans.com

Please register the following people for the RSMeans construction seminars as shown here. We understand that we must make our own hotel reservations if overnight stays are necessary.

☐ Full payment of $_____ enclosed.

☐ Bill me.

Please print name of registrant(s).
(To appear on certificate of completion)

_____

_____

_____

P.O. #: _____
GOVERNMENT AGENCIES MUST SUPPLY PURCHASE ORDER NUMBER OR TRAINING FORM.

Firm name_____

Address_____

City/State/Zip_____

Telephone no._____ Fax no._____

E-mail address_____

Charge registration(s) to: ☐ MasterCard ☐ VISA ☐ American Express

Account no._____ Exp. date_____

Cardholder's signature_____

Seminar name_____

Seminar City_____

Please mail check to: RSMeans Seminars, 700 Longwater Drive, Norwell, MA 02061 USA

New Titles - Reference Books

For more information visit the RSMeans website at www.rsmeans.com

Solar Energy: Technologies & Project Delivery for Buildings

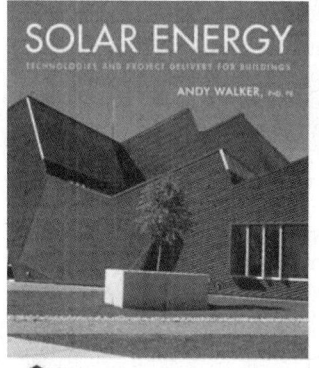

by Andy Walker

An authoritative reference on the design of solar energy systems in building projects, with applications, operating principles, and simple tools for the construction, engineering, and design professionals, the book simplifies the solar design and engineering process and provides sample documentation for the complete design of a solar energy system for buildings.

$85.00 | 320 pages, hardover | Catalog no. 67365

Project Scheduling & Management for Construction, 4th Edition

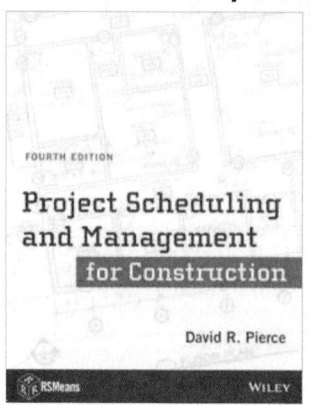

by David R. Pierce

First published in 1988 by RSMeans, the new edition of *Project Scheduling and Management for Construction* has been substantially revised for both professionals and students enrolled in construction management and civil engineering programs. While retaining its emphasis on developing practical, professional-level scheduling skills, the new edition is a relatable, real world case study.

$95.00 | 272 pages, softcover | Catalog no. 67367

Estimating Building Costs, 2nd Edition
For the Residential & Light Commercial Construction Professional

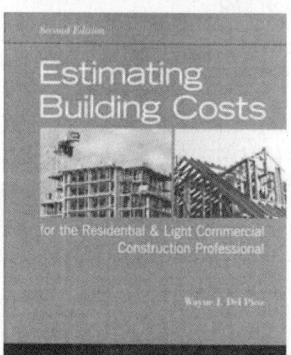

by Wayne J. Del Pico

This book guides readers through the entire estimating process, explaining in detail how to put together a reliable estimate that can be used not only for budgeting, but also for developing a schedule, managing a project, dealing with contingencies, and ultimately making a profit.

Completely revised and updated to reflect the CSI MasterFormat 2010 system, this practical guide describes estimating techniques for each building system and how to apply them according to the latest industry standards.

$75.00 | 398 pages, softcover | Catalog no. 67343A

Project Control: Integrating Cost & Schedule in Construction

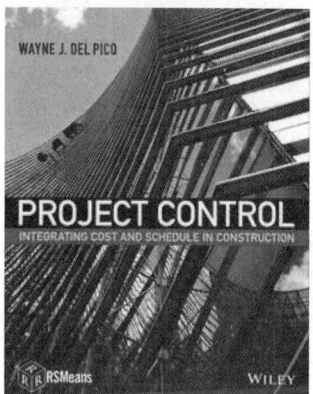

by Wayne J. Del Pico

Written by a seasoned professional in the field, *Project Control: Integrating Cost and Schedule in Construction* fills a void in the area of project control as applied in the construction industry today. It demonstrates how productivity models for an individual project are created, monitored, and controlled, and how corrective actions are implemented as deviations from the baseline occur.

$95.00 | 240 pages, softcover | Catalog no. 67366

How to Estimate with Means Data & CostWorks, 4th Edition

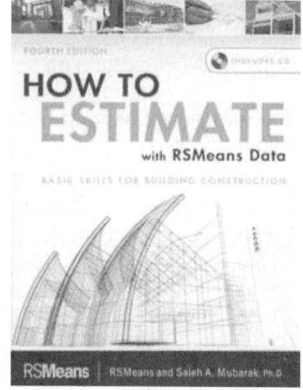

by RSMeans and Saleh A. Mubarak, Ph.D.

This step-by-step guide takes you through all the major construction items with extensive coverage of site work, concrete and masonry, wood and metal framing, doors and windows, and other divisions. The only construction cost estimating handbook that uses the most popular source of construction data, RSMeans, this indispensible guide features access to the instructional version of CostWorks in electronic form, enabling you to practice techniques to solve real-world estimating problems.

$75.00 | 292 pages, softcover | Includes CostWorks CD
Catalog no. 67324C

RSMeans Cost Data, Student Edition

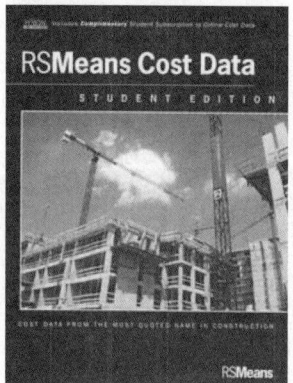

This book provides a thorough introduction to cost estimating in a self-contained print and online package. With clear explanations and a hands-on, example-driven approach, it is the ideal reference for students and new professionals.

Features include:

- Commercial and residential construction cost data in print and online formats
- Complete how-to guidance on the essentials of cost estimating
- A supplemental website with plans, problem sets, and a full sample estimate

$99.00 | 512 pages, softcover | Catalog no. 67363

Reference Books

For more information visit the RSMeans website at **www.rsmeans.com**

Value Engineering: Practical Applications
For Design, Construction, Maintenance & Operations

by Alphonse Dell'Isola, PE

A tool for immediate application—for engineers, architects, facility managers, owners, and contractors. Includes making the case for VE—the management briefing; integrating VE into planning, budgeting, and design; conducting life cycle costing; using VE methodology in design review and consultant selection; case studies; VE workbook; and a life cycle costing program on disk.

$85.00 | Over 450 pages, illustrated, softcover | Catalog no. 67319A

Risk Management for Design & Construction

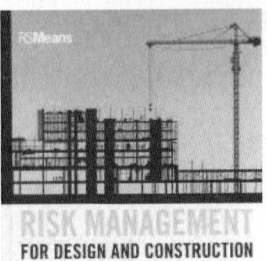

Introduces risk as a central pillar of project management and shows how a project manager can prepare to deal with uncertainty. Experts in the field use clear, straightforward terminology to demystify the concepts of project uncertainty and risk. Includes:

- Integrated cost and schedule risk analysis
- An introduction to a ready-to-use system of analyzing a project's risks and tools to proactively manage risks
- A methodology that was developed and used by the Washington State Department of Transportation
- Case studies and examples on the proper application of principles
- Information about combining value analysis with risk analysis

$125.00 | Over 250 pages, softcover | Catalog no. 67359

How Your House Works, 2nd Edition

by Charlie Wing

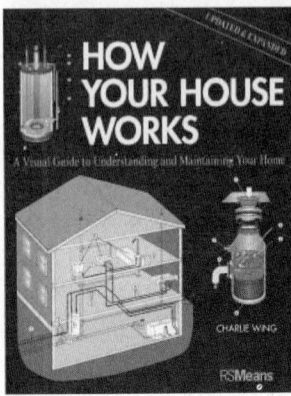

Knowledge of your home's systems helps you control repair and construction costs and makes sure the correct elements are being installed or replaced. This book uncovers the mysteries behind just about every major appliance and building element in your house. See-through, cross-section drawings in full color show you exactly how these things should be put together and how they function, including what to check if they don't work. It just might save you having to call in a professional.

$22.95 | 192 pages, softcover | Catalog no. 67351A

Complete Book of Framing, 2nd Edition

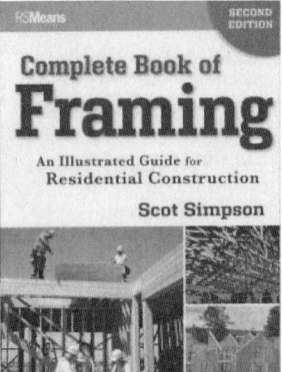

by Scot Simpson

This updated, easy-to-learn guide to rough carpentry and framing is written by an expert with more than thirty years of framing experience. Starting with the basics, this book begins with types of lumber, nails, and what tools are needed, followed by detailed, fully illustrated steps for framing each building element. Framer-Friendly Tips throughout the book show how to get a task done right—and more easily.

$29.95 | 352 pages, softcover | Catalog no. 67353A

Green Home Improvement

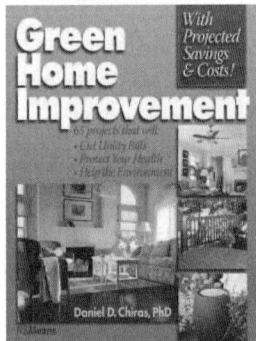

by Daniel D. Chiras, PhD

With energy costs rising and environmental awareness increasing, people are looking to make their homes greener. This book, with 65 projects and actual costs and projected savings help homeowners prioritize their green improvements.

Projects range from simple water savers that cost only a few dollars, to bigger-ticket items such as HVAC systems. With color photos and cost estimates, each project compares options and describes the work involved, the benefits, and the savings.

$34.95 | 320 pages, illustrated, softcover | Catalog no. 67355

Life Cycle Costing for Facilities

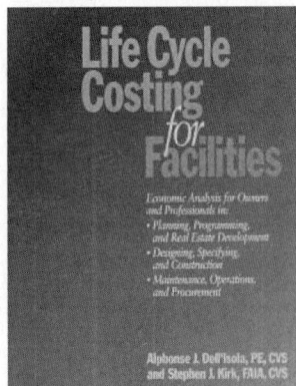

by Alphonse Dell'Isola and Dr. Steven Kirk

Guidance for achieving higher quality design and construction projects at lower costs! Cost-cutting efforts often sacrifice quality to yield the cheapest product. Life cycle costing enables building designers and owners to achieve both. The authors of this book show how LCC can work for a variety of projects — from roads to HVAC upgrades to different types of buildings.

$99.95 | 396 pages, hardcover | Catalog no. 67341

Reference Books

For more information visit the RSMeans website at www.rsmeans.com

Builder's Essentials: Plan Reading & Material Takeoff
For Residential and Light Commercial Construction

by Wayne J. DelPico

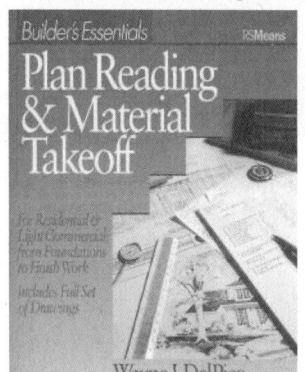

A valuable tool for understanding plans and specs, and accurately calculating material quantities. Step-by-step instructions and takeoff procedures based on a full set of working drawings.

$45.00 | Over 420 pages, softcover | Catalog no. 67307

Concrete Repair & Maintenance Illustrated

by Peter Emmons

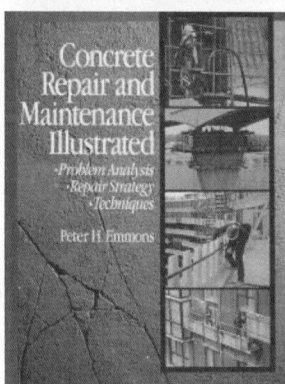

Hundreds of illustrations show users how to analyze, repair, clean, and maintain concrete structures for optimal performance and cost effectiveness. From parking garages to roads and bridges to structural concrete, this comprehensive book describes the causes, effects, and remedies for concrete wear and failure. Invaluable for planning jobs, selecting materials, and training employees, this book is a must-have for concrete specialists, general contractors, facility managers, civil and structural engineers, and architects.

$75.00 | 300 pages, illustrated, softcover | Catalog no. 67146

Electrical Estimating Methods, 3rd Edition

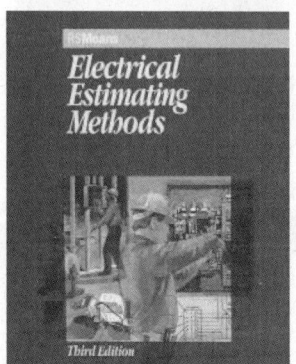

Expanded edition includes sample estimates and cost information in keeping with the latest version of the CSI MasterFormat and UNIFORMAT II. Complete coverage of fiber optic and uninterruptible power supply electrical systems, broken down by components, and explained in detail. Includes a new chapter on computerized estimating methods. A practical companion to *RSMeans Electrical Cost Data.*

$75.00 | Over 325 pages, hardcover | Catalog no. 67230B

Means Unit Price Estimating Methods, 4th Edition

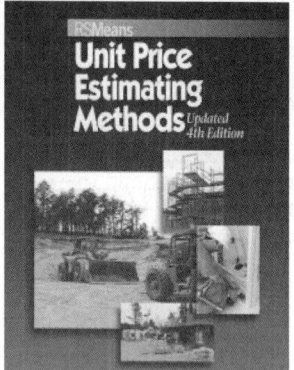

Includes cost data and estimating examples, updated to reflect changes to the CSI numbering system and new features of RSMeans cost data. It describes the most productive, universally accepted ways to estimate, and uses checklists and forms to illustrate shortcuts and timesavers. A model estimate demonstrates procedures. A new chapter explores computer estimating alternatives.

$75.00 | Over 350 pages, illustrated, softcover | Catalog no. 67303B

Square Foot & UNIFORMAT Assemblies Estimating Methods, 3rd Edition

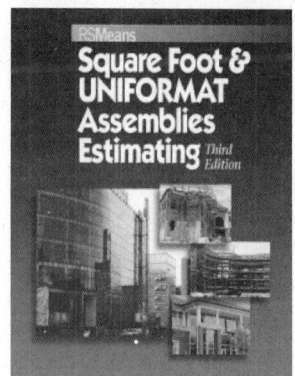

Develop realistic square foot and assemblies costs for budgeting and construction funding. The new edition features updated guidance on square foot and assemblies estimating using UNIFORMAT II. An essential reference for anyone who performs conceptual estimates.

$75.00 | Over 300 pages, illustrated, softcover | Catalog no. 67145B

Residential & Light Commercial Construction Standards, 3rd Edition

Updated by RSMeans and contributing authors

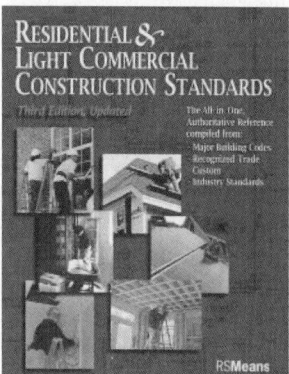

This book provides authoritative requirements and recommendations compiled from leading professional associations, industry publications, and building code organizations. This all-in-one reference helps establish a standard for workmanship, quickly resolve disputes, and avoid defect claims. Updated third edition includes new coverage of green building, seismic, hurricane, and mold-resistant construction.

$65.00 | Over 550 pages, illustrated, softcover | Catalog no. 67322B

Reference Books

For more information visit the RSMeans website at **www.rsmeans.com**

Construction Business Management

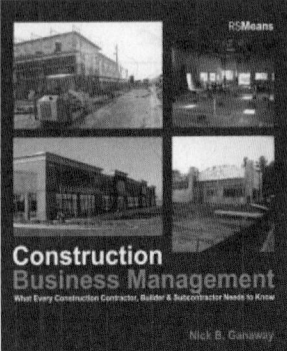

by Nick Ganaway

Only 43% of construction firms stay in business after four years. Make sure your company thrives with valuable guidance from a pro with 25 years of success as a commercial contractor. Find out what it takes to build all aspects of a business that is profitable, enjoyable, and enduring. With a bonus chapter on retail construction.

$55.00 | 200 pages, softcover | Catalog no. 67352

The Practice of Cost Segregation Analysis

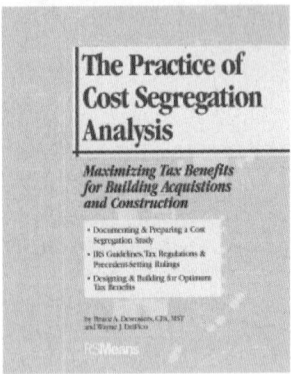

by Bruce A. Desrosiers and Wayne J. DelPico

This expert guide walks you through the practice of cost segregation analysis, which enables property owners to defer taxes and benefit from "accelerated cost recovery" through depreciation deductions on assets that are properly identified and classified.

With a glossary of terms, sample cost segregation estimates for various building types, key information resources, and updates via a dedicated website, this book is a critical resource for anyone involved in cost segregation analysis.

$105.00 | Over 225 pages | Catalog no. 67345

Green Building: Project Planning & Cost Estimating, 3rd Edition

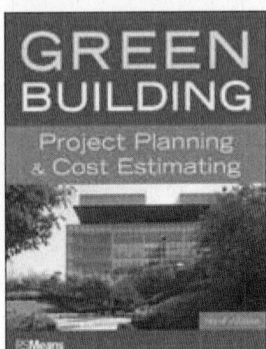

Since the widely read first edition of this book, green building has gone from a growing trend to a major force in design and construction.

This new edition has been updated with the latest in green building technologies, design concepts, standards, and costs. Full-color with all new case studies—plus a new chapter on commercial real estate.

$110.00 | Over 450 pages, softcover | Catalog no. 67338B

Project Scheduling & Management for Construction, 3rd Edition

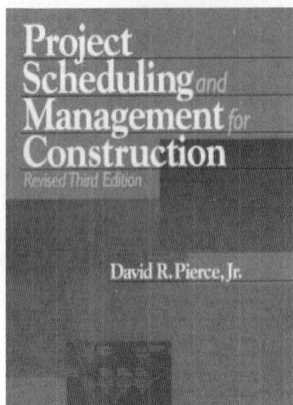

by David R. Pierce, Jr.

A comprehensive yet easy-to-follow guide to construction project scheduling and control—from vital project management principles through the latest scheduling, tracking, and controlling techniques. The author is a leading authority on scheduling, with years of field and teaching experience at leading academic institutions. Spend a few hours with this book and come away with a solid understanding of this essential management topic.

$64.95 | Over 300 pages, illustrated, hardcover | Catalog no. 67247B

Preventive Maintenance for Multi-Family Housing

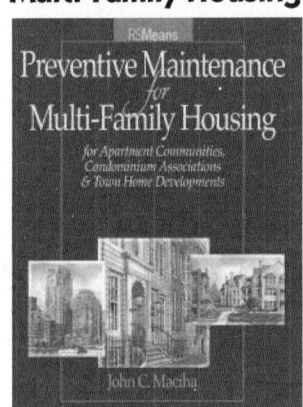

by John C. Maciha

Prepared by one of the nation's leading experts on multi-family housing.

This complete PM system for apartment and condominium communities features expert guidance, checklists for buildings and grounds maintenance tasks and their frequencies, a reusable wall chart to track maintenance, and a dedicated website featuring customizable electronic forms. A must-have for anyone involved with multi-family housing maintenance and upkeep.

$95.00 | 225 pages | Catalog no. 67346

Job Order Contracting: Expediting Construction Project Delivery

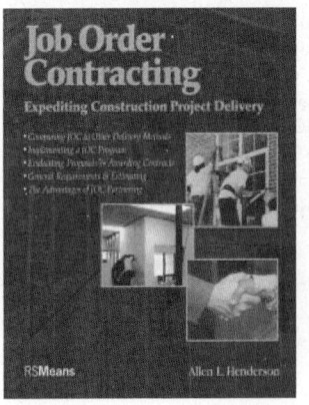

by Allen L. Henderson

Expert guidance to help you implement JOC—fast becoming the preferred project delivery method for repair and renovation, minor new construction, and maintenance projects in the public sector and in many states and municipalities. The author, a leading JOC expert and practitioner, shows how to:

- Establish a JOC program
- Evaluate proposals & award contracts
- Handle general requirements and estimating
- Partner for maximum benefits

$89.95 | 192 pages, illustrated, hardcover | Catalog no. 67348

Reference Books

For more information visit the RSMeans website at www.rsmeans.com

Construction Supervision

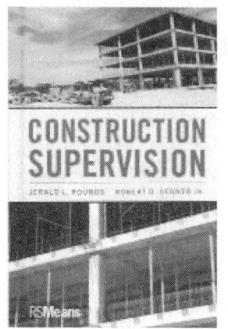

This brand new title inspires supervisory excellence with proven tactics and techniques applied by thousands of construction supervisors over the past decade. Recognizing the unique and critical role the supervisor plays in project success, the book's leadership guidelines carve out a practical blueprint for motivating work performance and increasing productivity through effective communication.

Features:

- A unique focus on field supervision and crew management
- Coverage of supervision from the foreman to the superintendent level
- An overview of technical skills whose mastery will build confidence and success for the supervisor
- A detailed view of "soft" management and communication skills

$95.00 | Over 450 pages, softcover | Catalog no. 67358

The Homeowner's Guide to Mold

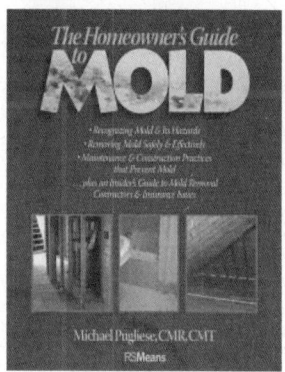

By Michael Pugliese

Expert guidance to protect your health and your home.

Mold, whether caused by leaks, humidity or flooding, is a real health and financial issue—for homeowners and contractors.

This full-color book explains:

- Construction and maintenance practices to prevent mold
- How to inspect for and remove mold
- Mold remediation procedures and costs
- What to do after a flood
- How to deal with insurance companies

$21.95 | 144 pages, softcover | Catalog no. 67344

Landscape Estimating Methods, 5th Edition

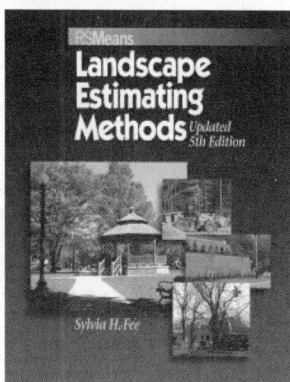

Answers questions about preparing competitive landscape construction estimates, with up-to-date cost estimates and the new MasterFormat classification system. Expanded and revised to address the latest materials and methods, including new coverage on approaches to green building. Includes:

- Step-by-step explanation of the estimating process
- Sample forms and worksheets that save time and prevent errors

$75.00 | Over 350 pages, softcover | Catalog no. 67295C

Building & Renovating Schools

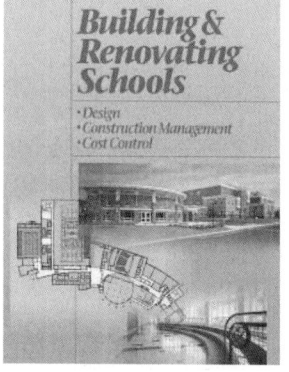

This all-inclusive guide covers every step of the school construction process—from initial planning, needs assessment, and design, right through moving into the new facility. A must-have resource for anyone concerned with new school construction or renovation. With square foot cost models for elementary, middle, and high school facilities, and real-life case studies of recently completed school projects.

The contributors to this book—architects, construction project managers, contractors, and estimators who specialize in school construction—provide start-to-finish, expert guidance on the process.

$99.95 | Over 425 pages, hardcover | Catalog no. 67342

Universal Design Ideas for Style, Comfort & Safety

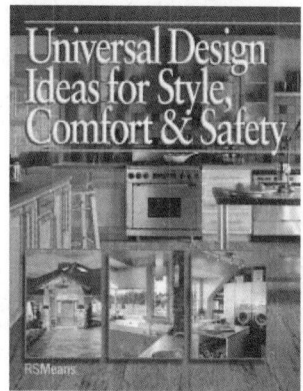

by RSMeans and Lexicon Consulting, Inc.

Incorporating universal design when building or remodeling helps people of any age and physical ability more fully and safely enjoy their living spaces. This book shows how universal design can be artfully blended into the most attractive homes. It discusses specialized products like adjustable countertops and chair lifts, as well as simple ways to enhance a home's safety and comfort. With color photos and expert guidance, every area of the home is covered. Includes budget estimates that give an idea how much projects will cost.

$21.95 | 160 pages, illustrated, softcover | Catalog no. 67354

Plumbing Estimating Methods, 3rd Edition

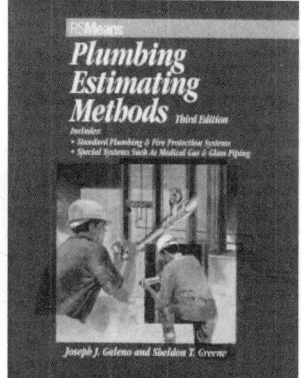

by Joseph J. Galeno and Sheldon T. Greene

Updated and revised! This practical guide walks you through a plumbing estimate, from basic materials and installation methods through change order analysis. *Plumbing Estimating Methods* covers residential, commercial, industrial, and medical systems, and features sample takeoff and estimate forms and detailed illustrations of systems and components.

$65.00 | 330+ pages, softcover | Catalog no. 67283B

Reference Books

For more information visit the RSMeans website at **www.rsmeans.com**

Understanding & Negotiating Construction Contracts

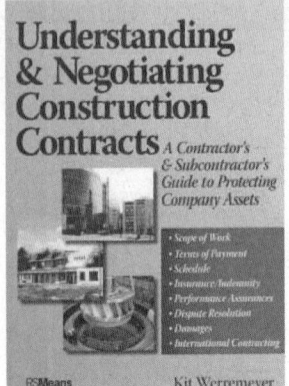

by Kit Werremeyer

Take advantage of the author's 30 years' experience in small-to-large (including international) construction projects. Learn how to identify, understand, and evaluate high risk terms and conditions typically found in all construction contracts—then negotiate to lower or eliminate the risk, improve terms of payment, and reduce exposure to claims and disputes. The author avoids "legalese" and gives real-life examples from actual projects.

$80.00 | 300 pages, softcover | Catalog no. 67350

The Gypsum Construction Handbook

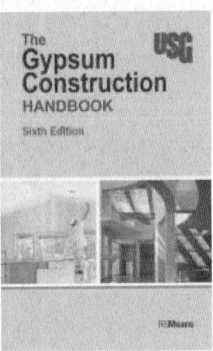

by USG

An invaluable reference of construction procedures for gypsum drywall, cement board, veneer plaster, and conventional plaster. This new edition includes the newest product developments, installation methods, fire- and sound-rated construction information, illustrated framing-to-finish application instructions, estimating and planning information, and more. Great for architects and engineers; contractors, builders, and dealers; apprentices and training programs; building inspectors and code officials; and anyone interested in gypsum construction. Features information on tools and safety practices, a glossary of construction terms, and a list of important agencies and associations. Also available in Spanish.

$40.00 | Over 575 pages, illustrated, softcover | Catalog no. 67357A

Mechanical Estimating Methods, 4th Edition

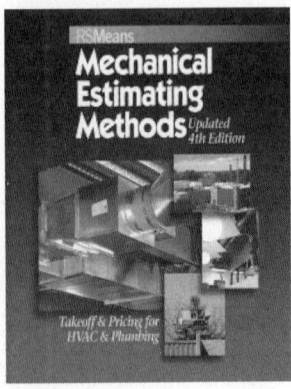

Completely updated, this guide assists you in making a review of plans, specs, and bid packages, with suggestions for takeoff procedures, listings, substitutions, and pre-bid scheduling for all components of HVAC. Includes suggestions for budgeting labor and equipment usage. Compares materials and construction methods to allow you to select the best option.

$70.00 per copy | Over 350 pages, illustrated, softcover | Catalog no. 67294B

Means Illustrated Construction Dictionary

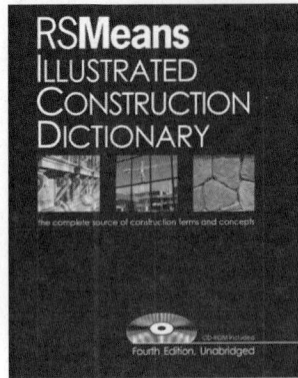

Unabridged 4th Edition, with CD-ROM

Long regarded as the industry's finest, *Means Illustrated Construction Dictionary* is now even better. With nearly 20,000 terms and more than 1,400 illustrations and photos, it is the clear choice for the most comprehensive and current information. The companion CD-ROM that comes with this new edition adds extra features such as: larger graphics and expanded definitions.

$99.95 | Over 790 pages, illust., hardcover | Catalog no. 67292B

RSMeans Estimating Handbook, 3rd Edition

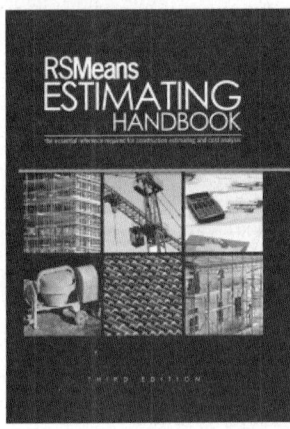

Widely used in the industry for tasks ranging from routine estimates to special cost analysis projects, this handbook has been completely updated and reorganized with new and expanded technical information.

RSMeans Estimating Handbook will help construction professionals:

- Evaluate architectural plans and specifications
- Prepare accurate quantity takeoffs
- Compare design alternatives and costs
- Perform value engineering
- Double-check estimates and quotes
- Estimate change orders

$110.00 | Over 900 pages, hardcover | Catalog No. 67276B

2014 Order Form

ORDER TOLL FREE 1-800-334-3509
OR FAX 1-800-632-6732

For more information visit the RSMeans website at www.rsmeans.com

Qty.	Book no.	COST ESTIMATING BOOKS	Unit Price	Total
	60064	Assemblies Cost Data 2014	$315.95	
	60014	Building Construction Cost Data 2014	194.95	
	60044	Commercial Renovaion Cost Data 2014	156.95	
	60114	Concrete & Masonry Cost Data 2014	177.95	
	50144	Construction Cost Indexes 2014 (subscription)	362.00	
	60144A	Construction Cost Index–January 2014	93.50	
	60144B	Construction Cost Index–April 2014	93.50	
	60144C	Construction Cost Index–July 2014	93.50	
	60144D	Construction Cost Index–October 2014	93.50	
	60344	Contr. Pricing Guide: Resid. R & R Costs 2014	39.95	
	60234	Electrical Change Order Cost Data 2014	187.95	
	60034	Electrical Cost Data 2014	196.95	
	60204	Facilities Construction Cost Data 2014	496.95	
	60304	Facilities Maintenance & Repair Cost Data 2014	434.95	
	60554	Green Building Cost Data 2014	160.95	
	60164	Heavy Construction Cost Data 2014	192.95	
	60094	Interior Cost Data 2014	196.95	
	60124	Labor Rates for the Const. Industry 2014	424.95	
	60184	Light Commercial Cost Data 2014	144.95	
	60024	Mechanical Cost Data 2014	191.95	
	60154	Open Shop Building Const. Cost Data 2014	165.95	
	60214	Plumbing Cost Data 2014	194.95	
	60174	Residential Cost Data 2014	139.95	
	60284	Site Work & Landscape Cost Data 2014	187.95	
	60054	Square Foot Costs 2014	205.95	
	62013	Yardsticks for Costing (2013)	185.95	
	62014	Yardsticks for Costing (2014)	189.95	
		REFERENCE BOOKS		
	67329	Bldrs Essentials: Best Bus. Practices for Bldrs	40.00	
	67307	Bldrs Essentials: Plan Reading & Takeoff	35.95	
	67342	Building & Renovating Schools	99.95	
	67261A	The Building Prof. Guide to Contract Documents	75.00	
	67339	Building Security: Strategies & Costs	95.00	
	67353A	Complete Book of Framing, 2nd Ed.	29.95	
	67146	Concrete Repair & Maintenance Illustrated	75.00	
	67352	Construction Business Management	55.00	
	67364	Construction Law	110.00	
	67358	Construction Supervision	95.00	
	67363	Cost Data Student Edition	99.00	
	67314	Cost Planning & Est. for Facil. Maint.	95.00	
	67230B	Electrical Estimating Methods, 3rd Ed.	75.00	
	67343A	Est. Bldg. Costs for Resi. & Lt. Comm., 2nd Ed.	75.00	
	67276B	Estimating Handbook, 3rd Ed.	110.00	
	67318	Facilities Operations & Engineering Reference	115.00	
	67301	Facilities Planning & Relocation	89.95	
	67338B	Green Building: Proj. Planning & Cost Est., 3rd Ed.	110.00	
	67355	Green Home Improvement	34.95	
	67357A	The Gypsum Construction Handbook	40.00	
	67148	Heavy Construction Handbook	110.00	
	67308E	Home Improvement Costs–Int. Projects, 9th Ed.	24.95	
	67309E	Home Improvement Costs–Ext. Projects, 9th Ed.	24.95	

Qty.	Book no.	REFERENCE BOOKS (Cont.)	Unit Price	Total
	67344	Homeowner's Guide to Mold	21.95	
	67324C	How to Est.w/Means Data & CostWorks, 4th Ed.	75.00	
	67351A	How Your House Works, 2nd Ed.	22.95	
	67282A	Illustrated Const. Dictionary, Condensed, 2nd Ed.	65.00	
	67292B	Illustrated Const. Dictionary, w/CD-ROM, 4th Ed.	99.95	
	67362	Illustrated Const. Dictionary, Student Ed.	55.00	
	67348	Job Order Contracting	89.95	
	67295C	Landscape Estimating Methods, 5th Ed.	75.00	
	67341	Life Cycle Costing for Facilities	99.95	
	67294B	Mechanical Estimating Methods, 4th Ed.	70.00	
	67283B	Plumbing Estimating Methods, 3rd Ed.	65.00	
	67345	Practice of Cost Segregation Analysis	105.00	
	67366	Project Control: Integrating Cost & Sched.	95.00	
	67346	Preventive Maint. for Multi-Family Housing	95.00	
	67367	Project Scheduling & Manag. for Constr., 4th Ed.	95.00	
	67322B	Resi. & Light Commercial Const. Stds., 3rd Ed.	65.00	
	67359	Risk Management for Design & Construction	125.00	
	67365	Solar Energy: Technology & Project Delivery	85.00	
	67145B	Sq. Ft. & Assem. Estimating Methods, 3rd Ed.	75.00	
	67321	Total Productive Facilities Management	85.00	
	67350	Understanding and Negotiating Const. Contracts	80.00	
	67303B	Unit Price Estimating Methods, 4th Ed.	75.00	
	67354	Universal Design	21.95	
	67319A	Value Engineering: Practical Applications	85.00	

MA residents add 6.25% state sales tax
Shipping & Handling**
Total (U.S. Funds)*

Prices are subject to change and are for U.S. delivery only. *Canadian customers may call for current prices. **Shipping & handling charges: Add 7% of total order for check and credit card payments. Add 9% of total order for invoiced orders.

Send order to: ADDV-1000

Name (please print) _____

Company _____

☐ Company
☐ Home Address _____
_____

City/State/Zip _____

Phone # _____ P.O. # _____

(Must accompany all orders being billed)

Mail to: RSMeans, 700 Longwater Drive, Norwell, MA 02061

RSMeans Project Cost Report

By filling out this report, your project data will contribute to the database that supports the RSMeans® Project Cost Square Foot Data. When you fill out this form, RSMeans will provide a $50 discount off one of the RSMeans products advertised in the preceding pages. Please complete the form including all items where you have cost data, and all the items marked (☑).

$50.00 Discount per product for each report you submit

Project Description (NEW construction only, please)

☑ Building Use (Office School . . .) _____

☑ Address (City, State) _____
_____

☑ Total Building Area (SF) _____

☑ Ground Floor (SF) _____

☑ Frame (Wood, Steel . . .) _____

☑ Exterior Wall (Brick, Tilt-up . . .) _____

☑ Basement: (check one) ☐ Full ☐ Partial ☐ None

☑ Number of Stories _____

☑ Floor-to-Floor Height _____

☑ Volume (C.F.) _____

% Air Conditioned _____ Tons _____

Total Project Cost $ _____

Owner _____

Architect _____

General Contractor _____

☑ Bid Date _____

☑ Typical Bay Size _____

☑ Occupant Capacity _____

☑ Labor Force: _____ % Union _____ % Non-Union

☑ Project Description (Circle one number in each line.)

    1. Economy   2. Average   3. Custom   4. Luxury

    1. Square   2. Rectangular   3. Irregular   4. Very Irregular

Comments _____
_____

A	☑	General Conditions	$
B	☑	Site Work	$
C	☑	Concrete	$
D	☑	Masonry	$
E	☑	Metals	$
F	☑	Wood & Plastics	$
G	☑	Thermal & Moisture Protection	$
GR		Roofing & Flashing	$
H	☑	Doors and Windows	$
J	☑	Finishes	$
JP		Painting & Wall Covering	$

K	☑	Specialties	$
L	☑	Equipment	$
M	☑	Furnishings	$
N	☑	Special Construction	$
P	☑	Conveying Systems	$
Q	☑	Mechanical	$
QP		Plumbing	$
QB		HVAC	$
R	☑	Electrical	$
S	☑	Mech./Elec. Combined	$

Please specify the RSMeans product you wish to receive.
Complete the address information.

Product Name _____

Product Number _____

Your Name _____

Title _____

Company _____
   ☐ Company
   ☐ Home   Street Address _____

City, State, Zip _____

Email Address _____

Return by mail or fax 888-492-6770.

Method of Payment:

Credit Card # _____

Expiration Date _____

Check _____

Purchase Order _____

Phone Number _____
(if you would prefer a sales call)

RSMeans
a division of Reed Construction Data
Square Foot Costs Department
700 Longwater Drive
Norwell, MA 02061
www.RSMeans.com